Table of Atomic Masses*

Element	Symbol	Atomic Number	Atomic Mass	Element	Symbol	Atomic Number	Atomic Mass	Element	Symbol	Atomic Number	Atomic Mass
Actinium	Ac	89	(227)†	Hafnium	Hf	72	178.5	Promethium	Pm	61	(145)
Aluminum	Al	13	26.98	Helium	He	2	4.003	Protactinium	Pa	91	(231)
Americium	Am	95	(243)	Holmium	Ho	67	164.9	Radium	Ra	88	226
Antimony	Sb	51	121.8	Hydrogen	H	1	1.008	Radon	Rn	86	(222)
Argon	Ar	18	39.95	Indium	In	49	114.8	Rhenium	Re	75	186.2
Arsenic	As	33	74.92	Iodine	I	53	126.9	Rhodium	Rh	45	102.9
Astatine	At	85	(210)	Iridium	Ir	77	192.2	Rubidium	Rb	37	85.47
Barium	Ba	56	137.3	Iron	Fe	26	55.85	Ruthenium	Ru	44	101.1
Berkelium	Bk	97	(247)	Krypton	Kr	36	83.80	Samarium	Sm	62	150.4
Beryllium	Be	4	9.012	Lanthanum	La	57	138.9	Scandium	Sc	21	44.96
Bismuth	Bi	83	209.0	Lawrencium	Lr	103	(260)	Selenium	Se	34	78.96
Boron	B	5	10.81	Lead	Pb	82	207.2	Silicon	Si	14	28.09
Bromine	Br	35	79.90	Lithium	Li	3	6.941	Silver	Ag	47	107.9
Cadmium	Cd	48	112.4	Lutetium	Lu	71	175.0	Sodium	Na	11	22.99
Calcium	Ca	20	40.08	Magnesium	Mg	12	24.31	Strontium	Sr	38	87.62
Californium	Cf	98	(251)	Manganese	Mn	25	54.94	Sulfur	S	16	32.06
Carbon	C	6	12.01	Mendelevium	Md	101	(258)	Tantalum	Ta	73	180.9
Cerium	Ce	58	140.1	Mercury	Hg	80	200.6	Technetium	Tc	43	(98)
Cesium	Cs	55	132.9	Molybdenum	Mo	42	95.94	Tellurium	Te	52	127.6
Chlorine	Cl	17	35.45	Neodymium	Nd	60	144.2	Terbium	Tb	65	158.9
Chromium	Cr	24	52.00	Neon	Ne	10	20.18	Thallium	Tl	81	204.4
Cobalt	Co	27	58.93	Neptunium	Np	93	(237)	Thorium	Th	90	232.0
Copper	Cu	29	63.55	Nickel	Ni	28	58.69	Thulium	Tm	69	168.9
Curium	Cm	96	(247)	Niobium	Nb	41	92.91	Tin	Sn	50	118.7
Dysprosium	Dy	66	162.5	Nitrogen	N	7	14.01	Titanium	Ti	22	47.88
Einsteinium	Es	99	(252)	Nobelium	No	102	(259)	Tungsten	W	74	183.9
Erbium	Er	68	167.3	Osmium	Os	76	190.2	Uranium	U	92	238.0
Europium	Eu	63	152.0	Oxygen	O	8	16.00	Vanadium	V	23	50.94
Fermium	Fm	100	(257)	Palladium	Pd	46	106.4	Xenon	Xe	54	131.3
Fluorine	F	9	19.00	Phosphorus	P	15	30.97	Ytterbium	Yb	70	173.0
Francium	Fr	87	(223)	Platinum	Pt	78	195.1	Yttrium	Y	39	88.91
Gadolinium	Gd	64	157.3	Plutonium	Pu	94	(244)	Zinc	Zn	30	65.38
Gallium	Ga	31	69.72	Polonium	Po	84	(209)	Zirconium	Zr	40	91.22
Germanium	Ge	32	72.59	Potassium	K	19	39.10				
Gold	Au	79	197.0	Praseodymium	Pr	59	140.9				

*The values given here are to four significant figures.
†A value given in parentheses denotes the mass of the longest-lived isotope.

CHEMISTRY

SECOND EDITION

CHEMISTRY

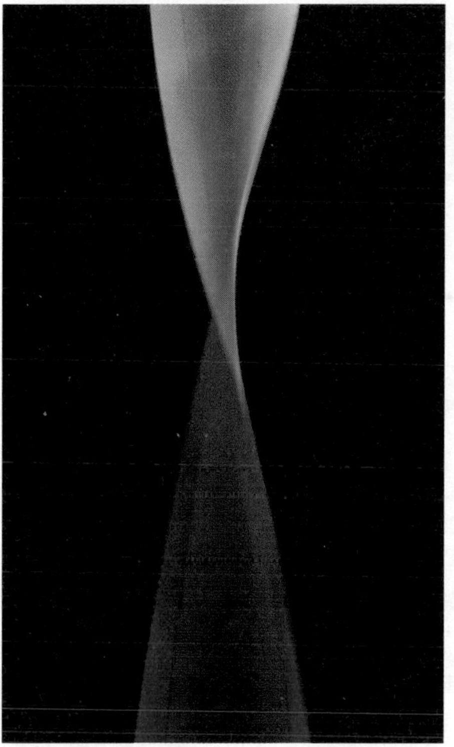

Steven S. Zumdahl

UNIVERSITY OF ILLINOIS

D. C. HEATH AND COMPANY Lexington, Massachusetts Toronto

To Mary Le Quesne
My courageous and talented editor.
Her memory inspires us to work harder.

Acquisitions Editor: Mary Le Quesne
Production Editor: Jill E. Hobbs
Designer: Henry Rachlin
Production Coordinator: Mike O'Dea
Photo Researcher: Martha L. Shethar
Text Permissions Editor: Margaret Roll

Cover: Ted Hansen/Slide Graphics of New England

Published simultaneously in Canada.

Printed in the United States of America.

International Standard Book Number: 0-669-16708-8

Library of Congress Catalog Card Number: 88-81734

10 9 8 7 6 5 4

TO THE PROFESSOR

BECAUSE of the enthusiastic response to the first edition of this text, the second edition of *Chemistry* continues in the same spirit: a presentation of the concepts of chemistry in a clear, interesting, and student-friendly manner. Much fine-tuning, as well as substantial rewriting in some areas, went into this revision, and I have incorporated many constructive suggestions from instructors who used the previous edition.

Since visual material is especially important for learning general chemistry, which typically presents a pictorial view of chemical concepts, the use of color has been markedly increased in this edition. Special illustrations and color photographs integrate descriptive chemistry with chemical principles, and many new *Chemical Impact* features emphasize practical applications of newly learned concepts.

Important Features of the Second Edition

- With a strong **problem-solving orientation,** this text talks to the student about how to approach and solve chemical problems. I have made a strong pitch to the student for using a thoughtful and logical approach rather than simply memorizing procedures. In particular, an innovative method is given for dealing with acid-base equilibria, the material the typical student finds most difficult and frustrating. The key to this approach involves first deciding what species are present in solution, then thinking about the chemical properties of these species. This method provides a general framework for approaching all types of solution equilibria.

  The text contains 280 sample exercises, with many more examples given in the discussions leading to sample exercises or used to illustrate general strategies. When a specific strategy is presented, it is summarized, and the sample exercise that follows it reinforces the step-by-step attack on the problem.

- I have presented a thorough **treatment of reactions** that occur in solution, including acid-base reactions. This material appears in Chapter 4, directly after the chapter on chemical stoichiometry, to emphasize the connection between solution reactions and chemical reactions in general. The early presentation of this material provides an opportunity to cover some interesting descriptive chemistry and also supports the lab, which typically involves a great deal of aqueous chemistry. Chapter 4 also includes oxidation-reduction reactions, because a large number of interesting and important chemical reactions involve redox processes. However, coverage of oxidation-reduction is optional at this point and depends on the needs of a specific course.

- **Descriptive chemistry** and chemical principles are thoroughly integrated in this text. Chemical models may appear sterile and confusing without the observations that stimulated their invention. On the other hand, facts without organizing principles may seem overwhelming. A combination of observations and models can make chemistry both interesting and understandable. In addition, in those chapters that deal with the chemistry of the elements systematically, I have made a continuous effort to show how properties and models correlate. Descriptive chemistry is presented in a variety of ways—as applications of the principles in separate sections, in sample exercises and exercise sets, in photographs, and in *Chemical Impact* features.

- Throughout the book a strong **emphasis on models** prevails. Coverage includes how they are constructed, how they are tested, and what we learn when they inevitably fail. Models are developed naturally, with pertinent observations always presented first to show why a particular model was invented. In addition, I have tried to present chemistry as a human activity carried out by real people, many of whom are described in the text or in *Chemical Impacts*.

- Everyday-life **applications** of chemistry that should be of interest to the students taking general chemistry appear throughout the text. For example, the *Chemical Impact* "Refurbishing the Lady" illustrates the chemistry behind the restoration of the Statue of Liberty, while features on hot and cold packs give examples of exo- and endothermic reactions.
- A unique feature of this text is the chapter on **industrial chemistry** (Chapter 24). This chapter describes several chemical industries ranging from the manufacture of polymers to winemaking. Students can see the actual application of chemical principles in settings where factors such as safety, environmental impact, and economics must be considered in addition to the traditional aspects of chemistry. Whether employed as a whole or used as individual sections at relevant points in the course, the chapter offers an unusual perspective on chemical concepts in action. The *Instructor's Guide* offers suggestions for creative usage of this material.
- Judging from the favorable comments of instructors and students who have used the first edition, the text seemed to work very well in a variety of courses. I was especially pleased that **readability** was cited as a key strength when students were asked to assess the text. Thus, although the text has been fine-tuned in many areas, I have endeavored to build on the basic descriptions, strategies, analogies, and explanations that were successful in the first edition.

Changes in the Second Edition

- Approximately 45% more exercises have been added. Most of the new exercises are straightforward, drill-type problems to help students test their grasp of the fundamental concepts. A large proportion of the exercises are grouped by topic, and, although they emphasize fundamental principles, the context is often a real-life application of chemistry. This approach makes the problem more realistic and provides a means for introducing more descriptive chemistry. The problems in the additional exercises section are not grouped by topic and tend to be more complicated than the earlier exercises. This helps the student gain experience in recognizing various applications of the concepts covered in the chapter and in synthesizing several concepts.
- By request of instructors and students, I have increased the number of problem-solving strategies within the text. Strategies are followed by sample exercises with step-by-step solutions to reinforce the logical approach.
- Chapter 4 has been extensively rewritten from the first edition to simplify and clarify this important material as much as possible. In particular, great pains have been taken to explain how to predict the products of reactions. Problem-solving strategies and sample exercises have been increased, while a more detailed discussion of qualitative analysis moves to later chapters. A careful reading of this chapter will show that its length is due mainly to careful explanations of the concepts related to reactions in solution.
- The introductory sections on kinetics in Chapter 12 present the rate laws in a simpler fashion. These revisions emphasize that there are two different types of rate laws.
- Following suggestions from reviewers, material on buffers in Chapter 15 has been extensively revised. New flow charts visually clarify what is happening when changes occur in these systems.
- Material on the concept of formal charges has been added to Chapter 8.
- Data tables are now collected at the back of the text for easy use by students.
- Sixteen new *Chemical Impact* features bring the total number to 44. They discuss scientists and real world applications of chemical concepts.

Flexibility of Topic Order

The order of topics in the text was chosen because it is preferred by the majority of instructors. However, I consciously constructed the book so that many other orders are possible. For example, at the University of Illinois, for a two-semester sequence we use the chapters in the order 1–6, 13–15, 7–9, 21, 20, 12, 10, 11, 16, 17, and 22. Sections of Chapters 18, 19, 23, and 24 are used throughout the two semesters as appropriate. This order, chosen because of the way the laboratory is organized, is not necessarily recommended, but it illustrates the flexibility of order built into the text.

Some specific points about topic order:

- About half of chemistry courses present kinetics before equilibria, while the other half present equilibria first. This text is written to accommodate either order.
- The introductory aspects of thermodynamics are presented relatively early (in Chapter 6) because of the importance of energy in various chemical processes and models, but the more subtle thermodynamic concepts are left until later (Chapter 16). These two chapters may be used together if desired.
- To make the book more flexible, the derivation of the ideal gas law from the kinetic-molecular theory and quantitative analysis using spectroscopy are presented in the appendices. While mainstream general chemistry courses typically do not cover this material, some courses may find it appropriate. By using the optional material in the appendices and by assigning the more difficult end-of-chapter exercises (from the additional exercises section), an instructor will find the level of the text appropriate for many majors courses or for other courses requiring a more extensive coverage of these topics.
- Because some courses cover bonding using only a Lewis-structure approach, orbitals are not presented in the introductory chapter on bonding (Chapter 8). In Chapter 9 both hybridization and the molecular orbital model are covered, but either or both of these topics may be omitted if desired.
- Chapter 4 can be tailored to fit the specific course involved. Used in its entirety where it stands in the book, it provides interesting examples of descriptive chemistry and supports the laboratory program. Material in this chapter can also be skipped entirely or covered at some later point whenever appropriate. For example, the sections on oxidation and reduction can be taught with electrochemistry. Although many instructors prefer early introduction of this concept, these sections can be omitted without complication since the next few chapters do not depend on this material.

Supplements

An extensive supplements package has been designed to make this book more useful to both student and instructor. These supplements include:

- **New to this Edition:** *Study Guide,* by Paul B. Kelter of the University of Wisconsin at Oshkosh. Written to be a self-study aid for students, this guide includes alternate strategies for solving various types of problems, supplemental explanations for the most difficult material, and self-tests. There are approximately 400 worked examples and 700 practice problems (with answers) designed to give students mastery and confidence.
- *Solutions Guide,* by Kenneth C. Brooks and Steven S. Zumdahl, both of the University of Illinois (Urbana), provides detailed solutions for two-thirds of the end-of-chapter exercises (designated by colored question numbers or letters) using the strategies emphasized in the text. To ensure the accuracy of the solutions, this supplement and the *Complete Solutions Guide* were checked independently by several instructors.
- *Complete Solutions Guide,* by Kenneth C. Brooks and Steven S. Zumdahl, presents detailed solutions for **all** of the end-of-chapter exercises in the text for the convenience of faculty and staff involved in instruction and for instructors who wish their students to have solutions for all exercises. Departmental approval is required for the sale of the *Complete Solutions Guide* to students.
- *Instructor's Guide,* by Kenneth C. Brooks, includes suggestions for alternative orders of topics, amplification of strategies used in various chapters, suggested sources of additional information, and a section on notes for teaching assistants.
- **New to this Edition:** *Solving Equilibrium Problems with Applications to Qualitative Analysis,* by Steven S. Zumdahl. Revised from a text successfully used by thousands of students over the last ten years, this book offers thorough, step-by-step procedures for solving problems related to equilibria taking place both in the gas phase and in solution. Containing hundreds of sample exercises, test exercises with complete solutions and end-of-chapter exercises with answers, the text utilizes the same problem-solving methods found in *Chemistry* and is an excellent source of additional drill-type problems. The last chapter presents an exploratory qualitative analysis experiment with explanations based on the principles of aqueous equilibria.

- *Experimental Chemistry,* Second Edition, by James F. Hall of the University of Lowell, provides an extensively revised laboratory program compatible with the text. The 43 experiments present a wide variety of chemistry, and many experiments offer choices of procedures. Safety is strongly emphasized throughout the program.
- *Instructor's Resource Guide* for *Experimental Chemistry,* Second Edition, by James F. Hall, contains tips including hints on running experiments, approximate times for each experiment, and answers to all pre-lab and post-lab questions posed in the laboratory guide.
- *Test Item File,* available to adopters, offers a printed version of more than 2000 exam questions referenced to the appropriate text section. Questions are available in both multiple-choice and open-ended formats.
- *Computerized Testing* presents the *Test Item File* questions in a new computerized testing program by ESA Test. Instructors can produce chapter tests, midterms, and final exams easily and with excellent graphics capability. The instructor can also edit existing questions or add new ones as desired, or preview questions on screen and add them to the test with a single keystroke. The testing program is available for Apple II, Macintosh, and IBM computers.
- *Heath Chemical Lecture Demonstrations,* a set of videotapes prepared by Paul B. Kelter, gives the instructor a choice of 50 chemistry lecture demonstrations. Designed to be both informative and motivational, the demonstrations depict the practical applications of chemical principles and consistently emphasize laboratory safety.
- *University of Illinois Film Center general chemistry videotapes,* which may also be purchased by adopters, reinforce the major concepts found in the text.
- *Transparencies,* a set of 90 transparencies (many in color), are available to adopters of the second edition of the text.

Acknowledgments

The success of this text is due to the efforts of many people. It is truly a privilege to work with Mary Le Quesne, Senior Science Editor, who has made immeasurable contributions to this project both in terms of content and spirit. Mary's knowledge, energy, enthusiasm, and unfailing good humor must be seen to be believed. I also want to thank Jill Hobbs, Production Editor, who always remained calm and who furnished an amazing level of organization. Jill's efforts improved the book in many different ways.

I owe much to Ken Brooks, who collaborated in writing the end-of-chapter exercises and who has contributed greatly to the project through his enthusiastic attitude and unerring suggestions concerning content and approach. In addition, Glenn Vogel of Ithaca College read through the exercise sets and provided many helpful suggestions. My thanks also to my colleagues in the general chemistry program at the University of Illinois, particularly Roxy Wilson, Barbara Whitmarsh and Tom Hummel, who have made many constructive suggestions. I am especially grateful to my wife Eunice who is truly a partner in this project, providing constant moral support and spending long hours discussing, typing, checking, and proofreading. I also greatly appreciate the continuing understanding and support of Whitney and Leslie.

Many other people at D. C. Heath supplied invaluable assistance during this project: Jim Hamann, Senior Product Manager, whose energy and thorough understanding of the marketplace have contributed greatly to the success of this book; Martha Shethar, Photo Editor, who showed real creativity in securing the photos for the second edition; and Henry Rachlin, Designer, who developed a beautiful and functional design.

Finally, I want to extend special thanks to the following people who reviewed all or part of the manuscript at various stages:

Edmund W. Benson, Central Michigan University
Michael J. Carlo, Angelo State University
John Gelder, Oklahoma State University
James F. Hall, University of Lowell
Thomas Hummel, University of Illinois
Wilbert Hutton, Iowa State University of Science and Technology
Paul Kelter, University of Wisconsin–Oshkosh
Stanley Marcus, Cornell University
Thomas Richardson, North Georgia College
Maurice E. Schwartz, University of Notre Dame
Joseph A. Stanko, University of South Florida
Francis Timmers, University of Wisconsin–Oshkosh
Richard Treptow, Chicago State University
Glenn Vogel, Ithaca College
Barbara Whitmarsh, University of Illinois
Roxy Wilson, University of Illinois
Orville Ziebarth, Mankato State University

In addition, many users and other instructors submitted useful information. These include Paul G. Abajian, Johnson State College; Ed Acheson, Millikin University; Marvin J. Albanik, Essex Community College; John F. Albrecht, University of Wisconsin Center–Richmond; J. F. Allen, Clemson University; C. P. Anderson, University of Connecticut–Avery Point; James K. Archer, Northeast Texas Community College; Margaret M. Augustin, Rogers State College; Ron Backus, American River College; Dennis Baker, Olivet College; Robert C. Benz, Lake Erie College; Robert D. Bingham, Kansas Technical Institute; Rita G. Blatt, Pennsylvania State University; Robert Boggess, Radford University; E. Bollenbach, Northwestern Connecticut Community College; Allan P. Bonomy, Daytona Beach Community College; Ralph R. Booth, Davis and Elkins College; Lawrence A. Bottomley, Georgia Institute of Technology; Milton L. Bradley, Delta State University; W. Scott Briggs, Western Washington University; Joe Brundage, Cuesta College; Guy Buccino, Waterbury State Technical College; John D. Bugay, Kilgore College; Arthur L. Bycer, Valley Forge Military Junior College; B. Edward Cain, Rochester Institute of Technology; Stephen C. Carlson, Lansing Community College; John J. Casazza, St. Thomas Aquinas College; Mary C. Cavallaro, Salem State College; James C. Chang, University of Northern Illinois; Allan Childs, Northwest Community College; Betty Jo Chitester, Villa Maria College; Richard P. Ciula, California State University–Fresno; Wallace Coker, Gadsden State Junior College; Warren A. Colson, Framingham State College; Alan D. Cooper, Worcester State College; J. Ray Cozort, Northeast Mississippi Junior College; Joseph Crockett, Bridgewater College; Thomas E. Crumm, Indiana University of Pennsylvania; J. M. Cubina, New York Institute of Technology; Clyde Davis, Humboldt State University; Dennis D. Davis, New Mexico State University; Phillip H. Davis, University of Tennessee–Martin; Edwin L. DeYoung, The Loop College; George Eastland, Saginaw Valley State University; George Eng, University of the District of Columbia; Donald R. Evers, Iowa Central Community College; Terry Eyrich, Merced College.

Paul Farnham, College of the Redwoods; Larry Ferren, Olivet Nazarene University; Walter Flanders, SUNY–Morrisville; Aline B. Frappier, Community College of Rhode Island; William M. Frase, University of Cincinnati; Clara Wright Frazier, Louisburg College; Mark Freilich, Memphis State University; Helmuth Fuchs, SUNY–Farmingdale; Martin Fuller, Austin College; Alan Gabrielli, Southern College of Technology; Barbara A. Gage, Prince George's Community College; Susan Gannaway, Eastern College; Roy Garvey, North Dakota State University; Carole R. Gatz, Portland State University; Rudy Gerlach, Muskingum College; Marcia L. Gillette, Indiana University at Kokomo; Robert V. Glynn; David E. Goldberg, Brooklyn College; George W. Goth, Skyline College; Thomas A. Gover, Gustavus Adolphus College; Lawrence Gries, Queen's College of CUNY; Thomas J. Greenbow, Southeastern Massachusetts University; Cecil N. Hammonds, Penn Valley College; Michael D. Hampton, University of Central Florida; Steven E. Hannum, George Fox College; Gordon Harrach, Western Nebraska Community College; David W. Herlocker, Western Maryland College; Christine K. F. Hermann, Radford University; Stan Herowkski, Niagara County College; Fred Hilgeman, Southwestern University; Donald P. Hoster, Community College of Baltimore; Norman W. Hunter, Western Kentucky University; Peter E. Hurlimann, Citrus College; Albert W. Jache, Marquette University; Patricia Jackson, Lorain County Community College; C. Jankowski, Northeastern University; Thomas V. Jeffries, Campbellsville College; Clark Jenkins, North Arkansas Community College; Albert J. Judge, Jr., Lakeland Community College; Catherine A. Keenan, Chaffey College; Leslie A. Kinsland, University of Southwestern Louisiana; Soter G. Kokalis, Wm. Rainey Harper Community College; George Kraus, Charles County Community College; R. M. Kren, University of Michigan–Flint; Bette A. Kreuz, University of Michigan–Dearborn.

Joseph Lechner, Mt. Vernon Nazarene College; Cindy H. Lillie, Cedar Crest College; E. C. Lindblad, Dana College; Mary K. Linde, Clark Technical College; Diana Malone, Clark College; Ann Manner, Greater New Haven State Technical College; Kevin H. Mayo, Temple University; Glenda B. Michaels, Western State College; Everett J. Miller, SUNY–Cobleskill; Lewis Milner, North Central Technical College; Carl Minnier, Essex Community College; Qui-Chee Mir, Yakima Valley College; C. Jonathan Mitschele, St. Joseph's College; L. O. Morgan, University of Texas–Austin; Moher M. Nasr, Linderwood College; Thomas J. Ouellete, Washburn University; Keith K. Parker, Western Montana College; Earl F. Pearson, Western Kentucky University; James Petrich, San Antonio College; Gerry Prody, Western Washington University; Earl Pye, California State Polytechnical University, Pomona; Muriel H. Ramsden, Union County College; J. Rawlings, Auburn University at Montgomery; Paul E. Reinbold, Southern Nazarene University; James W. Rhoades, Crowder College; Steve Richter, Marycrest College; Lyman Rickard, High Point College; Bobby J. Roberson, Brewer State Junior College; Robert Rudd, Northwestern Michigan College; Michael J. Saliby, University of New Haven; Stanley F. Sarner, Delaware Technical & Community College; Arlyne Sarquis, Miami University, Middletown Campus; David W. Schroder, Lincoln College; Anthony Scioly, Siena Heights College; Raymond B. Scott, Mary Washington College; A. B. Sears, Darton College; C. Duane Sell, W. R. Harper College; David B. Shaw, Madison Area Technical College; Conrad F. Shiba, Centre College; Angel M. Sierra, Foothill Community College; Narendra Singh, Methodist College; Ram P. Singhal, Wichita State University; Trudie Jo Slapar, Vincennes University; David Stanislawski, Raymond Walters College; Mabel-Ruth Stephanic, Oklahoma State University; Robert P. Stewart, Jr., Miami University; Troy J. Stewart, Alcom State University; Darrell R. Strait, Southwest Baptist University; W. D. Timberlake, Los Angeles Harbor College; Charles A. Trapp, University of Louisville; A. R. Trujillo, College of Eastern Utah; John L. Tyvoll, Salisbury State University; Richard Uthe, General College, University of Minnesota; Janet B. VanDoren, University of Akron; Anna M. Wartman, Augustana College; L. Westmoreland, Central State University; G. K. Wittenberg, Glendale Community College; Sidney H. Young, University of South Alabama; Noel Zaugg, Ricks College.

TO THE STUDENT

THE MAJOR PURPOSE of this book, of course, is to help you learn chemistry. However, this main thrust is closely linked to two other goals: to show how important and how interesting the subject is; and to show how to think like a chemist. To solve complicated problems the chemist uses logic, trial and error, intuition, and, above all, patience. A chemist is used to being wrong. The important thing is to learn from a mistake, recheck assumptions, and try again. A chemist thrives on puzzles that seem to defy solutions.

Many of you using this text do not plan to be practicing chemists. However, the non-chemist can benefit from the chemist's attitude. Problem-solving is important in all professions and in all walks of life. The techniques you will learn from this book will serve you well in any career you choose. Thus, I believe that the study of chemistry has much to offer the non-major, including an understanding of many fascinating and important phenomena and a chance to hone problem-solving skills.

This book attempts to present chemistry in a manner that is sensible to the novice. Chemistry is not the result of an inspired vision. It is the product of countless observations and many attempts, using logic and trial and error, to account for these observations. In this book the concepts are developed in a natural way: the observations come first and then models are constructed to explain the observed behavior.

Models are a central theme in this book. The uses and limitations of models are emphasized and science is treated as a human activity, subject to all the normal human foibles. Mistakes are discussed as well as successes.

A central theme of this book is a thoughtful, systematic approach to problem-solving. Learning encompasses much more than simply memorizing facts. Truly educated people use their factual knowledge as a starting point—a base for creative approaches to solving problems.

As you have probably heard, learning chemistry can be frustrating. It is easy to lose sight of how really important and interesting chemistry is. You *can* learn chemistry and even enjoy the process, but you must understand that finesse works much better than brute force. Chemistry can be difficult not so much because the concepts are hard, but because it deals with complicated systems. Before a problem can be solved, a lot of facts must be sifted to find the pertinent ones. There is no alternative to thinking things through.

Read through the material in the text carefully. For most concepts, illustrations or photos will help you visualize what is going on. Often a given type of problem is "walked through" in the text before the corresponding sample exercises appear. Strategies for solving problems are given throughout the text.

Thoroughly examine the sample exercises and the problem-solving strategies, which are enclosed by blue lines. The strategies summarize the approach taken in the text; the sample exercises follow the strategies step-by-step. Schematics in Chapter 15 also illustrate the logical pathways to solving aqueous equilibrium problems.

Throughout the text, I have used margin notes to highlight key points, to comment on an application of the text material, or to reference material in other parts of the book. The boxed features called *Chemical Impacts* discuss especially interesting applications of chemistry to the everyday world.

Each chapter has a summary and key terms list for review, and the glossary gives a quick reference for definitions.

Learning chemistry requires working the end-of-chapter exercises assigned by your professor. Answers to exercises denoted by blue question numbers or letters are in the back of the book, while complete solutions to those exercises are in the *Solutions Guide*.

The *Study Guide* contains extra practice problems and many examples. The supplement *Solving Equilibrium Problems with Applications to Qualitative Analysis* reinforces in great detail the text's step-by-step approach to solving equilibrium problems and contains many worked examples and self-quiz questions to help you assess your level of proficiency.

It is very important to use the exercises to your best

advantage. Your main goal should not be to simply get the correct answer but to *understand the process* for getting the answer. Memorizing the solutions for specific problems is not a very good way to prepare for an exam. There are too many pigeonholes required to cover every possible problem type. Look within the problem for the solution. Use the concepts you have learned along with a systematic, logical approach to find the solution. Learn to trust yourself to think it out. You will make mistakes, but the important thing is to learn from these errors. The only way to gain confidence is to do lots of practice problems and use these to diagnose your weaknesses.

Be patient, thoughtful, and work hard to understand rather than simply memorize. I wish you an interesting and satisfying year.

ABOUT THE AUTHOR

STEVEN S. ZUMDAHL received his B.S. degree in Chemistry from Wheaton College (Illinois) in 1964 and his Ph.D. in Chemistry from the University of Illinois (Urbana) in 1968.

In 20 years of teaching he has been a faculty member at the University of Colorado (Boulder), Parkland College (Illinois), and the University of Illinois (Urbana). Currently he is Professor of Chemistry and Director of General Chemistry at the University of Illinois.

Professor Zumdahl is known at the University of Illinois for his rapport with students and for his outstanding teaching ability. During his tenure at the University, he has received the University of Illinois Award for Excellence in Teaching, the Liberal Arts and Sciences College Award for Distinguished Teaching, and the School of Chemical Sciences Teaching Award (three times).

Dr. Z., as he is known to his students, is an avid cyclist who rides his graphite/Kevlar composite racing bicycle about 10,000 miles a year. His favorite cycling routes include the mountain passes around Breckenridge, Colorado. He also collects and restores classic automobiles, including a 1963 Corvette which has been his daily transportation for more than 20 years.

CONTENTS

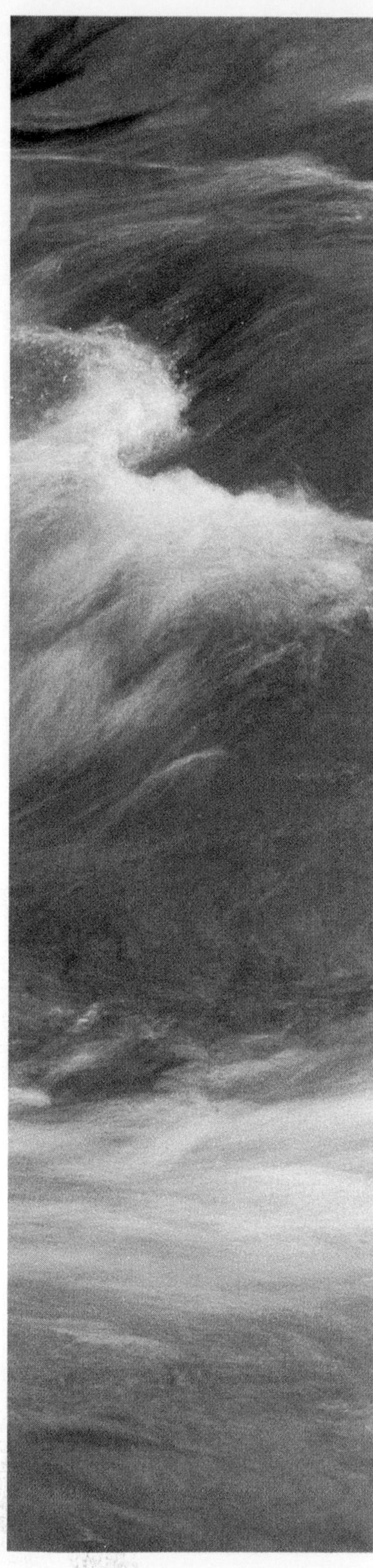

CHAPTER 3 **Stoichiometry** 73

CHAPTER 4 **Types of Chemical Reactions and Solution Stoichiometry** 119

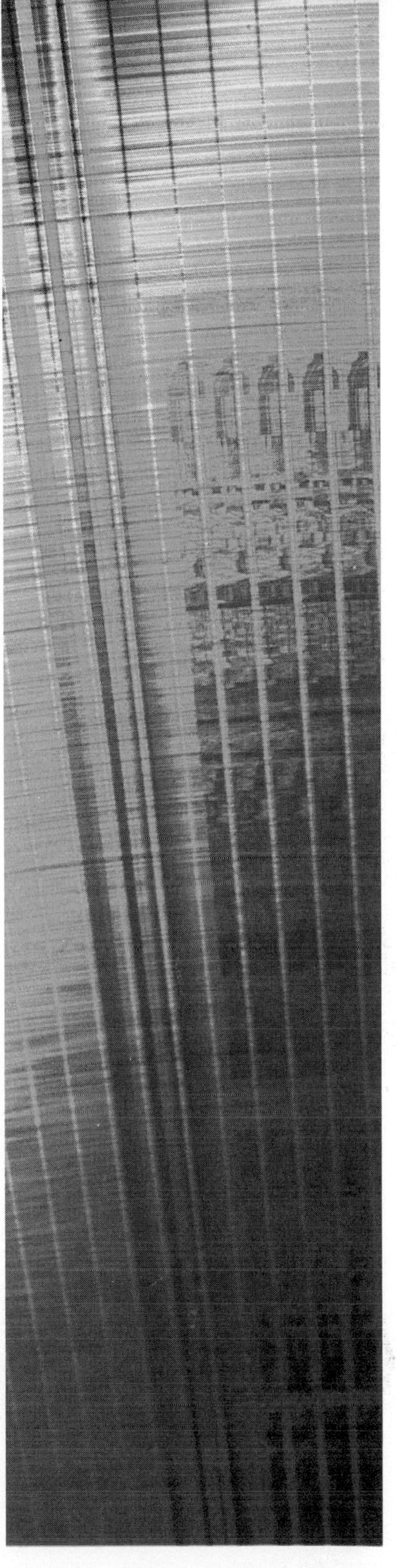

CHAPTER 5 Gases 175

CHAPTER 6 Thermochemistry 221

CHAPTER 7 Atomic Structure and Periodicity 263

CHAPTER 9 Covalent Bonding: Orbitals 383

CHAPTER 10 Liquids and Solids 417

CHAPTER 11 Properties of Solutions 473

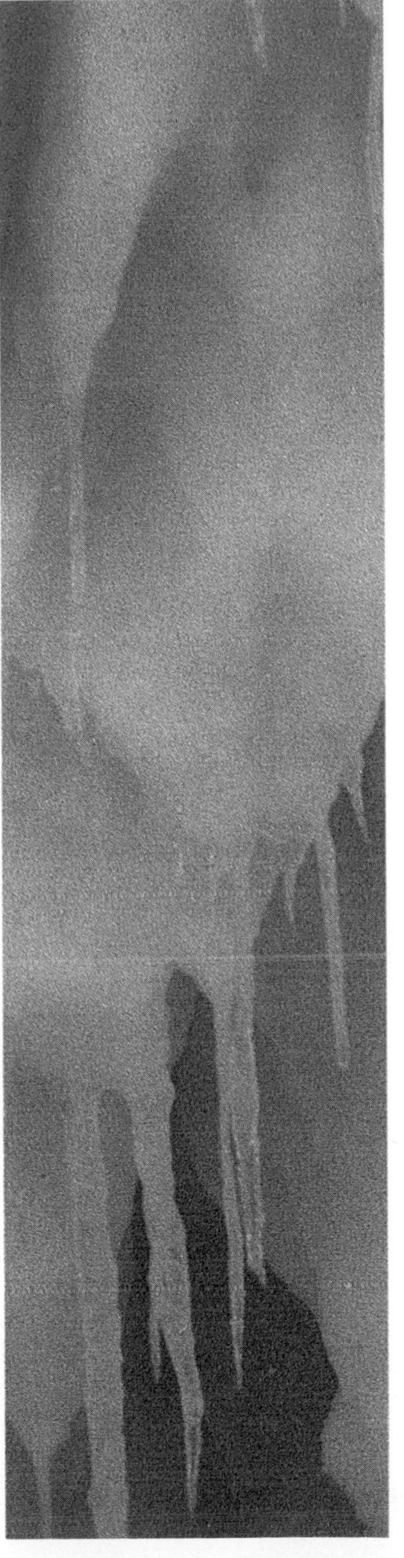

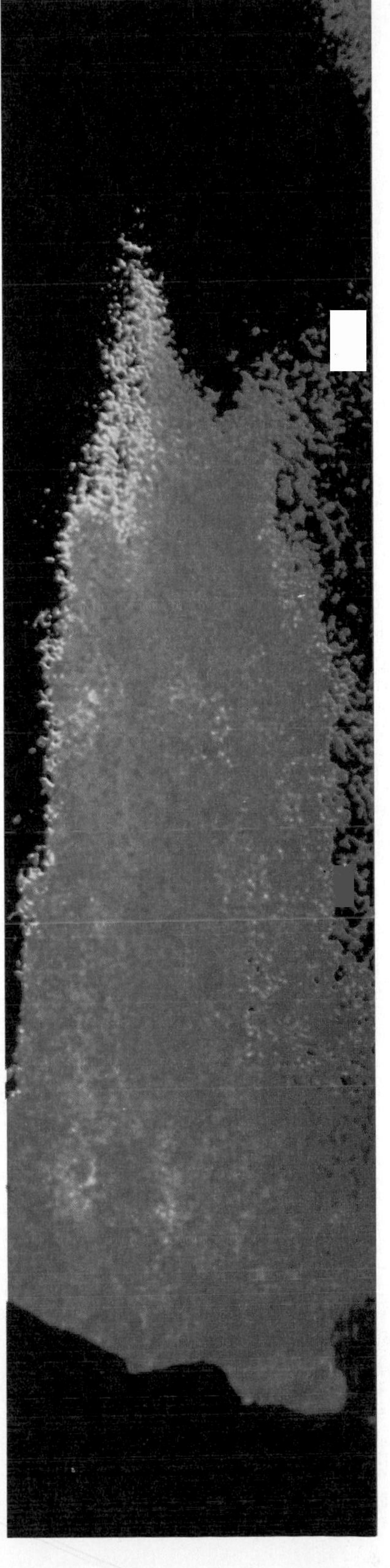

CHAPTER 19 The Representative Elements: Groups 5A Through 8A 851

CHAPTER 20 Transition Metals and Coordination Chemistry 893

CHAPTER 21 The Nucleus: A Chemist's View 939

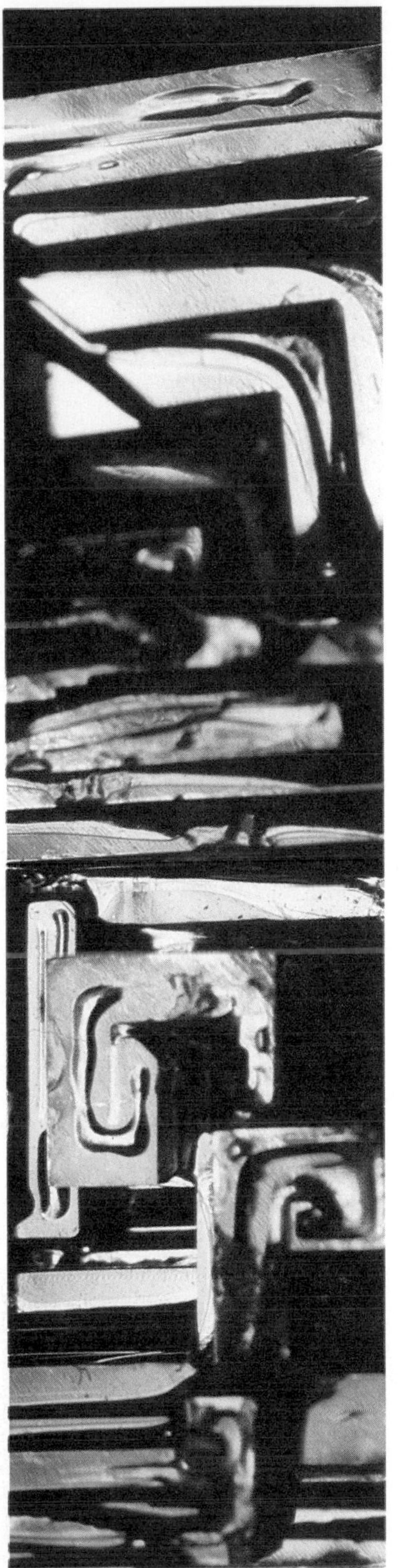

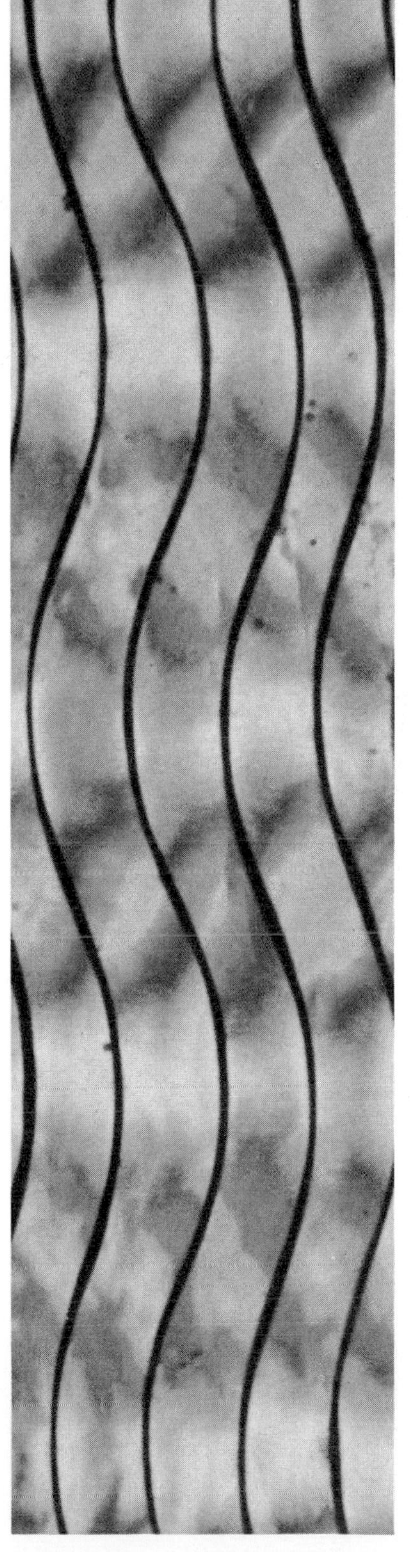

CHEMISTRY

CHAPTER 1

Chemical Foundations

Chemistry deals with all the materials of the universe and the changes that these materials undergo. Practitioners of this broad discipline are involved in activities as diverse as looking for molecules in space, exploring the fundamental particles of matter, making new materials, and trying to find out how organisms, such as humans, work.

Chemists deal with many different phenomena and do research to try to solve a wide variety of problems. These are a few of the questions that chemists are currently considering:

Can DNA be modified to cure diseases such as sickle-cell anemia and diabetes?

What causes acid rain, and how can it be prevented?

How can we save the millions of books containing acidic paper that are now crumbling on library shelves everywhere?

Can highly selective pesticides be synthesized?

Can our large supplies of coal be economically turned into natural gas?

Why do certain substances cause cancer and others seem to protect us from cancer?

Do the freons used as refrigerants pose a danger to the protective ozone layer in the atmosphere?

How does lithium affect mental health?

For melting ice on roads can we replace salt, which damages cars and roads, with a substance made from corn?

Can batteries be designed that would make electric cars feasible?

How can bacteria be modified so that they produce useful chemicals, such as insulin?

Can fluffy cellulose, which has no calories, replace 50% of the flour in baked goods?

CONTENTS

< Rose Bengal dye emits a brilliant red light when it is excited by an energy source. Rose Bengal was first synthesized around 1880 as a dye for wool.

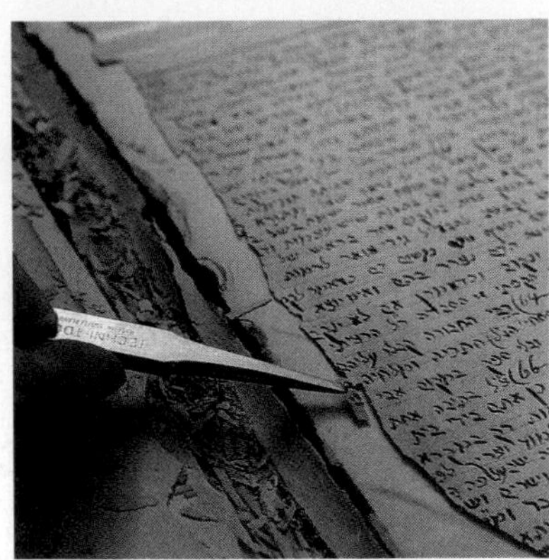

Acidic paper literally crumbles with age.

Figure 1.1 on the following page shows four of the diverse areas in which chemists participate.

Chemical knowledge is expanding at a phenomenal rate. At present nearly 400,000 research reports in the various areas of chemistry are published each year. This amounts to more than a thousand articles per day. Clearly it is not humanly possible to master the entirety of chemical knowledge; however, it is important to understand at least the basics of chemistry. Chemistry is at the heart of the processes that sustain our life and allow us to learn, work, keep warm, and travel. In fact, chemistry is vital to most of the activities of our lives. Understanding the chemical processes that maintain life and control the events in the world around us is essential to our continued well-being.

This textbook introduces you to the types of problems chemists deal with and the methods and models used to solve them. A major goal of this book is to acquaint you with the fundamental assumptions that chemists make in describing the behavior of matter and how these principles are used to solve real-life problems. Another goal is to help you become a better problem solver. As you will see, chemical systems tend to be complex, but critical thought, intelligent approximations, and creative use of fundamental knowledge can make them comprehensible. You cannot simply rely on memorization. You have to *think* to solve chemistry problems. This makes chemistry challenging and sometimes frustrating but also gives you a chance to improve your problem-solving skills, which will come in handy in other parts of your life.

This text stresses thoughtful problem solving rather than rote memorization.

Thus this textbook will

1. help you understand the fundamental models of chemistry,
2. show you how chemistry is involved in the real world, and
3. help you develop your skills as a problem solver.

In this chapter we will develop a general idea of science as a discipline and will demonstrate how to handle one of the most important tools of science, the measurement.

1.1 The Scientific Method

Purpose

- To identify the principal operations and limitations of the scientific method.

Ideally, a science such as chemistry grows and progresses systematically by a process called the **scientific method.** This approach in its simplest form consists of several distinct operations:

1. *Making observations*. The observations may be *qualitative* (the sky is blue; mercury is a liquid at room temperature) or *quantitative* (the pressure of the gas is 1 atmosphere; the temperature of the water is 54°C). A quantitative observa-

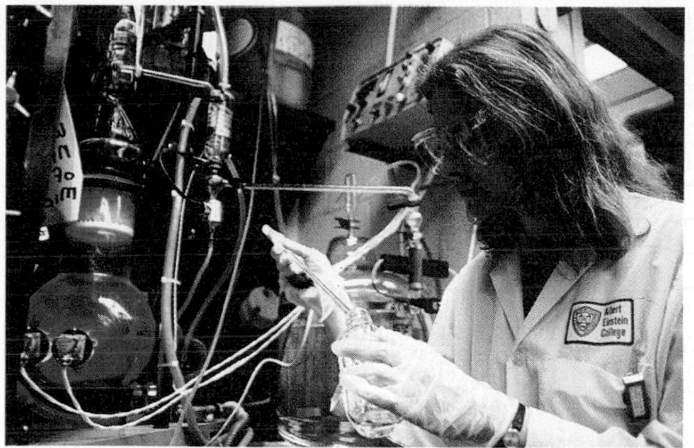

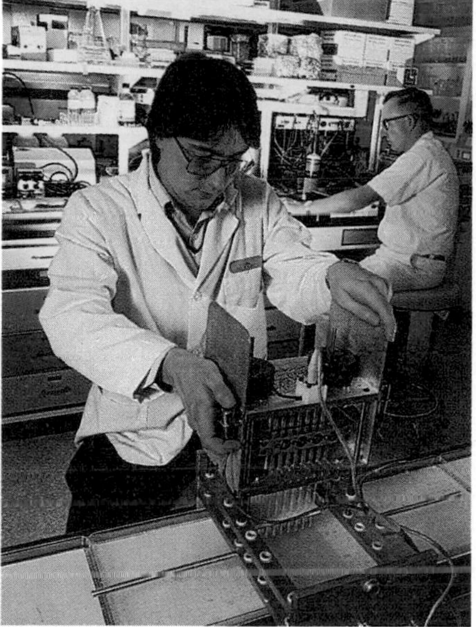

Figure 1.1

Chemists at work (clockwise from upper left): A Ph.D. researcher at work in a molecular biology lab. Blending perfumes. Lasers are used in many types of chemical studies. Researchers operate a multi-sample liquid transfer device in a lab devoted to finding anti-arthritis drugs.

tion is called a **measurement.** We will discuss measurements in more detail later.

2. *Looking for patterns in the observations*. This process often results in the formulation of a **natural law.** A natural law is a statement that expresses generally observed behavior. For example, studies of innumerable chemical reactions have shown that the substances present after a reaction have the same total mass as that of the substances present before the reaction took place. These observations can be generalized as a natural law called the law of conservation of mass.

   A natural law is often stated in terms of a mathematical formula. For example, Robert Boyle (Fig. 1.2) observed that under certain conditions the volume of a gas is inversely proportional to its pressure. This can be stated symbolically as

$$V \propto \frac{1}{P}$$

   where V represents the gas volume, P represents the pressure of the gas, and the symbol $\propto$ means "proportional to."

Figure 1.2

Robert Boyle (1627–1691) was born in Ireland. He became especially interested in experiments involving air and developed an air pump with which he produced evacuated cylinders. He used these cylinders to show that a feather and a lump of lead fall at the same rate in the absence of air resistance and that sound cannot be produced in a vacuum. His most famous experiments involved careful measurements of the volume of a gas as a function of pressure. In his book *The Sceptical Chemist,* Boyle urged that the ancient view of elements as mystical substances should be abandoned and that an element should instead be defined as anything that cannot be broken down into simpler substances. This conception was an important step in the development of modern chemistry.

A second formula that states the same law is

$$V = k\left(\frac{1}{P}\right)$$

where k is called the proportionality constant.

3. *Formulating theories*. A **theory** (often called a **model**) consists of a set of assumptions put forth to explain the observed behavior of matter. At first the set of assumptions is often called a **hypothesis.** If the tentative hypothesis survives the tests of many experiments, we gain confidence in its value and call it a theory or a model. It is important to distinguish between observations and theories. An observation is a *fact* that endures. The accuracy of a measurement may be increased by the development of a more sensitive measuring device, but the measurement should always remain fundamentally the same. A theory is an *interpretation*—a speculation—as to why nature behaves in a particular way. Theories inevitably change as more facts become known. An example of this process is given in the Chemical Impact feature in this chapter.
4. *Designing experiments to test the theories*. Ideally, science is self-correcting, continuously testing its models. It is important to remember that models are human inventions. They are an attempt to explain observed natural behavior in terms of ideas from human experience. Models by their very nature are imperfect. We must continue to do experiments and refine our models in light of new observations if we hope to approach a more correct understanding of natural phenomena.

The scientific method is represented in its simplest form in Fig. 1.3.

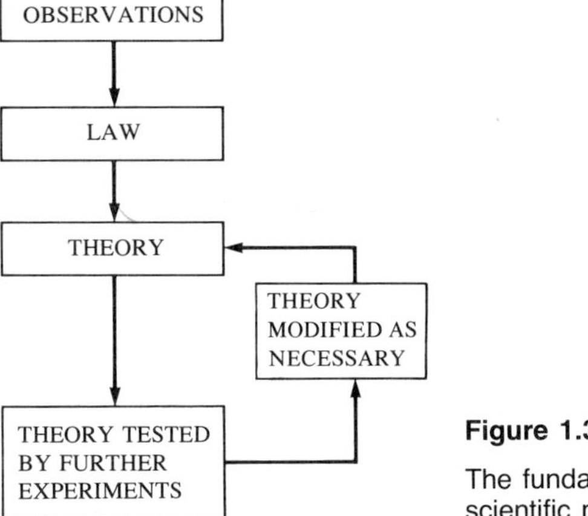

Figure 1.3

The fundamental steps of the scientific method.

The preceding discussion describes the ideal scientific method. However, it is important to understand that science does not always progress smoothly and efficiently. Scientists are human: they have prejudices; they misinterpret data; they become emotionally attached to their theories and thus nonobjective; and they play politics. Science is affected by profit motives, budgets, fads, wars, and religious beliefs. Galileo, for example, was forced to recant his astronomical observations in the face of strong religious resistance. Lavoisier, the father of modern chemistry,

Chemical Impact

Observations, Theories, and the Planets

Humans have always been fascinated by the heavens, by the behavior of the sun by day and the stars by night. Although more accurately measured now because of precise instruments, the basic *observations* of these events have remained the same over the past 4000 years. However, our *interpretation* of the events has changed dramatically. For example, about 2000 B.C. the Egyptians postulated that the sun was a boat inhabited by the god Ra, who daily sailed across the sky.

Over the years patterns in the changes in the heavens were recognized and, through marvelous devices such as Stonehenge in England, were connected to the seasons of the year. People also noted that seven objects seemed to move against the background of "fixed stars." These objects, actually the sun, the moon, and the planets Mercury, Venus, Mars, Jupiter, and Saturn, were called the "wanderers." The planets appeared to move from west to east, except Mars, which seemed to slow down and even move backwards for a few weeks.

One of the first explanations for these observations came from Eudoxus, born in 400 B.C. He imagined the earth as fixed, with the planets attached to a nested set of transparent spheres that moved at different rates around the earth. The stars were attached to the outermost sphere. This model, although clever, still did not account for the strange behavior of Mars. Five hundred years later, Ptolemy, a Greek scholar, worked out a plan more complex than that of Eudoxus, in which the planets were attached to the edges of spheres that "rolled around" the spheres of Eudoxus (see figure). This model accounted for the behavior of all of the planets, including the apparent reversals in the motion of Mars.

Because of human prejudice that the earth should be the center of the universe, Ptolemy's model was assumed to be correct for more than a thousand years, and its wide acceptance actually inhibited the advancement of astronomy. Finally in 1543 a Polish cleric, Nicholas Copernicus, postulated that the earth was only one of the planets, all of which revolved around the sun. This "demotion" of the earth's status produced violent opposition to the new model, and in fact Copernicus's writings were "corrected" by religious officials before scholars were allowed to use them.

The Copernican theory persisted and was finally given a solid mathematical base by Johannes Kepler. Kepler postulated elliptical rather than circular orbits for the planets in order to account more completely for their observed motions. Kepler's hypotheses were in turn further refined 36 years after his death by Isaac Newton, who recognized that the concept of gravitation could account for the positions and motions of the planets. However, even the brilliant models of Newton were found incomplete by Albert Einstein, who showed that Newton's mechanics was a special case of a much more general model called quantum mechanics.

Thus the same basic observations were made for several thousand years, but the explanations—the models—have changed remarkably from the Egyptians' boat of Ra to Einstein's relativity.

The lesson here is that our models will inevitably change, and we should expect them to do so. They can help us make scientific progress, or they can inhibit progress if we become too attached to them. Although the fundamental facts of chemistry will remain the same, the models in a chemistry text written a century from now will certainly be quite different from the ones presented here.

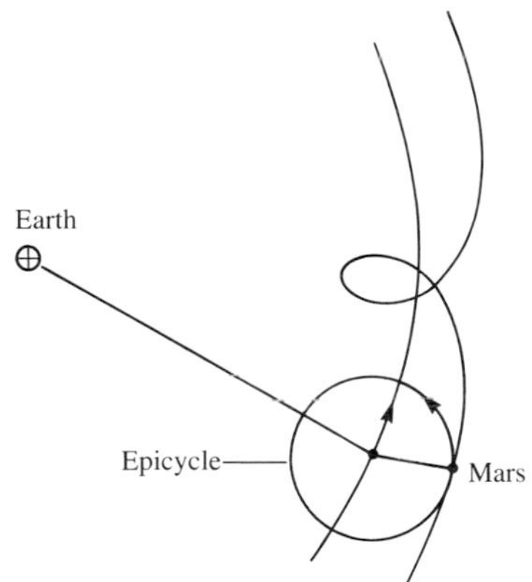

Ptolemy's earth-centered system.

was beheaded because of his political affiliations. Great progress in the chemistry of nitrogen fertilizers resulted from the desire to produce explosives to fight wars. The progress of science is often affected more by the frailties of humans and their institutions than by the limitations of scientific measuring devices. The scientific method is only as effective as the humans using it. It does not automatically lead to progress.

1.2 Units of Measurement

Purpose

- To describe the SI system of units and prefixes.

Making observations is fundamental to all science. A quantitative observation, or **measurement,** always consists of two parts: a *number* and a scale (called a *unit*). Both parts must be present for the measurement to be meaningful.

In this textbook we will use measurements of mass, length, time, temperature, electric current, and the amount of a substance, among others. Scientists recognized long ago that standard systems of units had to be adopted if measurements were to be useful. If every scientist had his or her own set of units, complete chaos would result. Unfortunately, different standards were adopted in different parts of the world. The two most widely used systems are the *English system* used in the United States and the *metric system* used by most of the rest of the industrialized world. This duality obviously causes a good deal of trouble; for example, parts as simple as bolts are not interchangeable between machines built using the different systems. As a result, the United States has begun to adopt the metric system.

Most scientists in all countries have for many years used the metric system. In 1960 an international agreement set up a system of units called the *International System* (*le Système International* in French), or the **SI system.** This system is based on the metric system and units derived from the metric system. The fundamental SI units are listed in Table 1.1. We will discuss how to manipulate these units later in this chapter.

Because the fundamental units are not always convenient (expressing the mass of a pin in kilograms is awkward), the SI system employs prefixes to change the size of the unit. These are listed in Table 1.2.

The Fundamental SI Units

Physical quantity	Name of unit	Abbreviation
Mass	kilogram	kg
Length	meter	m
Time	second	s
Temperature	Kelvin	K
Electric current	ampere	A
Amount of substance	mole	mol
Luminous intensity	candela	cd

Table 1.1

The Prefixes Used in the SI System. Those most commonly encountered are shown in color.

Prefix	Symbol	Meaning	Exponential notation*
exa	E	1,000,000,000,000,000,000	10^{18}
peta	P	1,000,000,000,000,000	10^{15}
tera	T	1,000,000,000,000	10^{12}
giga	G	1,000,000,000	10^{9}
mega	M	1,000,000	10^{6}
kilo	k	1,000	10^{3}
hecto	h	100	10^{2}
deka	da	10	10^{1}
—	—	1	10^{0}
deci	d	0.1	10^{-1}
centi	c	0.01	10^{-2}
milli	m	0.001	10^{-3}
micro	μ	0.000001	10^{-6}
nano	n	0.000000001	10^{-9}
pico	p	0.000000000001	10^{-12}
femto	f	0.000000000000001	10^{-15}
atto	a	0.000000000000000001	10^{-18}

*See Appendix 1.1 if you need a review of exponential notation.

Table 1.2

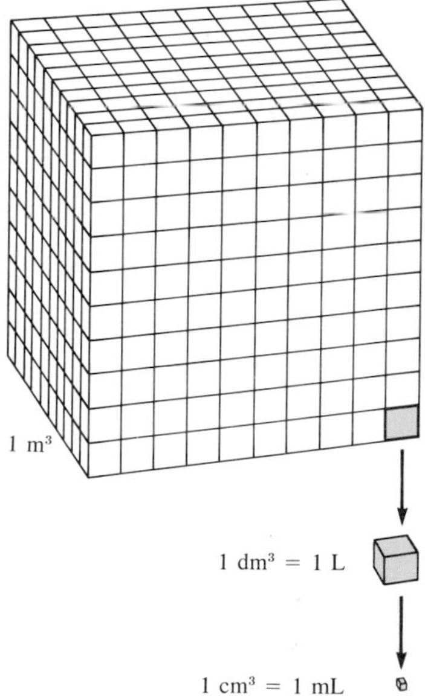

Figure 1.4

The largest cube has sides 1 m in length and a volume of 1 m^3. The middle-sized cube has sides 1 dm in length and a volume of 1 dm^3, or 1 L. The smallest cube has sides 1 cm in length and a volume of 1 cm^3, or 1 mL.

One physical quantity that is very important in chemistry is *volume*, which is not a fundamental SI unit but is derived from length. A cube that measures 1 meter on each edge is represented in Fig. 1.4. This cube has a volume of $(1\ m)^3 = 1\ m^3$ or, recognizing that there are 10 decimeters in a meter, the volume is $(10\ dm)^3 = 1000\ dm^3$. A cubic decimeter, dm^3, is commonly called a liter (L), which is a unit of volume slightly larger than a quart. As shown in Fig. 1.4, there are 1000 liters in a cube with a volume of 1 cubic meter. Similarly since 1 decimeter equals 10 centimeters, the liter $(1\ dm)^3$ can be divided into 1000 cubes each with a volume of 1 cubic centimeter. Thus 1 liter contains 1000 cubic centimeters, or 1000 milliliters.

Chemical laboratory work frequently requires measurement of the volumes of solutions or of pure liquids. Several devices for the accurate determination of liquid volume are shown in Fig. 1.5.

An important point concerning measurements is the relationship between mass and weight. Although these terms are often used interchangeably by chemists, they are not the same. The mass of an object represents the quantity of matter in that object (see Table 1.3). More precisely, ***mass** measures the resistance of an object to a change in its state of motion*. Mass is measured by the force necessary to give an object a given acceleration. On earth we use the force that gravity exerts on an object to measure its mass. We call this force the object's **weight.** Since weight is the response of mass to gravity, it varies with the strength of the gravitational field. Your body mass is the same on the earth or on the moon. However, your weight would be much less on the moon than on earth because of the moon's smaller gravitational field.

Some Examples of Commonly Used Units

Length	A dime is 1 mm thick.
	A quarter is 2.5 cm in diameter.
	The average height of an adult man is 1.8 m.
Mass	A nickel has a mass of about 5 g.
	A 120-lb person has a mass of about 55 kg.
Volume	A 12-oz can of soda has a volume of about 360 mL.

Table 1.3

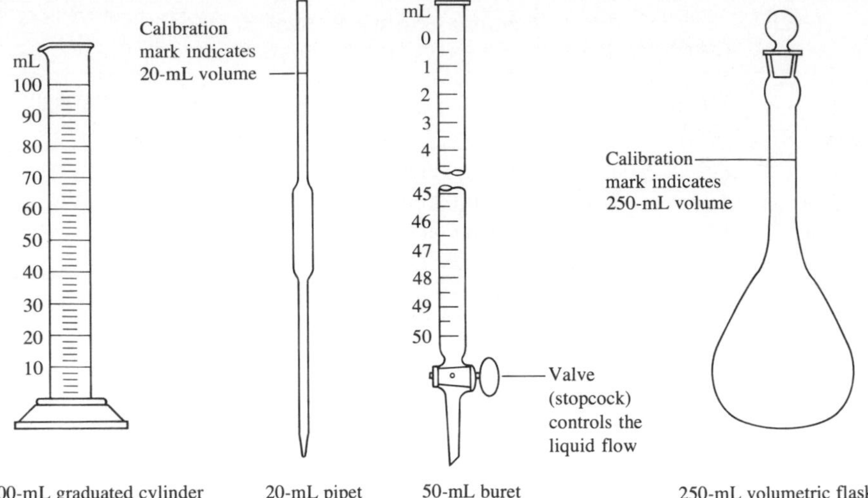

Figure 1.5

Common types of laboratory equipment used to measure liquid volume.

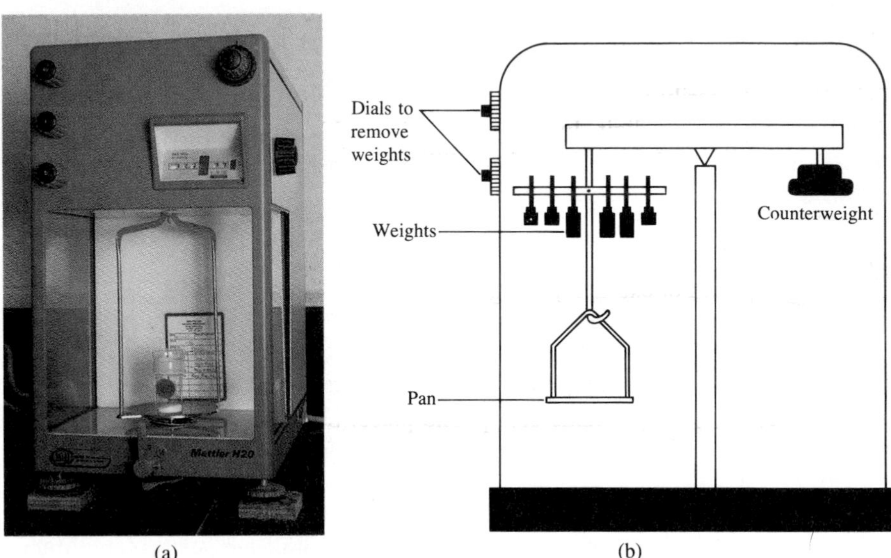

Figure 1.6

(a) Actual appearance of a single-pan analytical balance.
(b) Schematic view showing its principle of operation. When an object is placed on the balance pan, "weights" (masses) are removed from holders above the pan to restore the balance. The mass of the object equals the mass of the removed weights.

When we weigh something on a chemical balance (see Fig. 1.6), we are really comparing the mass of that object to a standard mass. Thus the terms weight and mass have come to be used interchangeably.

1.3 Uncertainty in Measurement

Purpose

- To identify causes of uncertainty in measurement.
- To show how significant figures are used.
- To compare precision and accuracy in measurement.

The number associated with a measurement is obtained using some measuring device. For example, consider the measurement of the volume of a liquid in a buret, as shown in Fig. 1.7, where the scale is greatly magnified. The volume is about 22.15 milliliters. Note that we must estimate the last number by interpolating between the 0.1-milliliter marks. Since the last number is estimated, its value may be different if another person makes the same measurement. If five different people read the same volume, the results might be as follows:

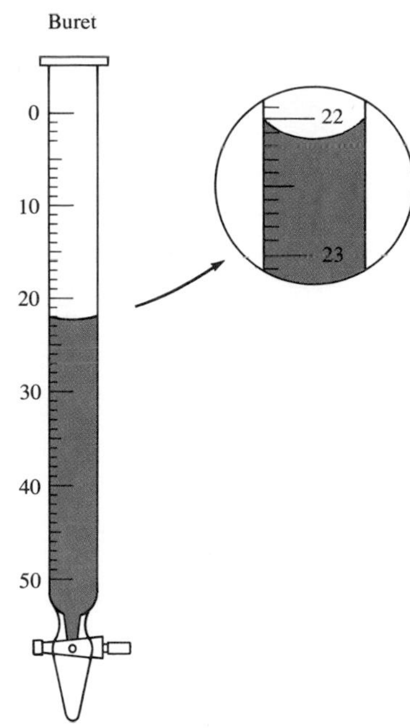

Figure 1.7

Measurement of volume using a buret. The volume is read at the bottom of the liquid curve (called the meniscus).

Person	Result of measurement
1	22.15 mL
2	22.14 mL
3	22.16 mL
4	22.17 mL
5	22.16 mL

Note from these results that the first three numbers (22.1) remain the same regardless of who makes the measurement; these are called certain digits. However, the digit to the right of the 1 must be estimated and thus varies; it is called an uncertain digit. We customarily report a measurement by recording all of the certain digits plus the *first* uncertain digit. In our example it would not make any sense to try to record the volume to thousandths of a milliliter because the value for hundredths of a milliliter must be estimated when using the buret.

It is very important to realize that a *measurement always has some degree of* ***uncertainty***. The uncertainty of a measurement depends on the precision of the measuring device. For example, using a bathroom scale, you might estimate that the mass of a grapefruit is approximately 1.5 pounds. Weighing the same grapefruit on a highly precise balance might produce a result of 1.476 pounds. In the first case, the uncertainty occurs in the tenths of a pound place; in the second case, the uncertainty occurs in the thousandths of a pound place. Suppose we weigh two similar grapefruit on the two devices and obtain the following results:

A measurement always has some degree of uncertainty.

	Bathroom scale	Balance
Grapefruit 1	1.5 lb	1.476 lb
Grapefruit 2	1.5 lb	1.518 lb

Do the two grapefruits have the same mass? The answer depends on which set of results you consider. Thus a conclusion based on a series of measurements depends on the certainty of those measurements. For this reason, it is important to indicate the uncertainty in any measurement. This is done by always recording the certain digits and the first uncertain digit (the estimated number). These numbers are called the **significant figures** of a measurement.

The convention of significant figures automatically indicates the uncertainty in a measurement. The uncertainty in the last number (the estimated number) is usually assumed to be ± 1 unless otherwise indicated. For example, the measurement 1.86 kilograms really means 1.86 ± 0.01 kilograms.

Uncertainty in measurement is discussed in more detail in Appendix 1.5.

Sample Exercise 1.1

In analyzing a sample of polluted water, a chemist measured out a 25.00-mL water sample with a pipet (Fig. 1.5). At another point in the analysis, the chemist used a graduated cylinder (Fig. 1.5) to measure 25 mL of a solution. What is the difference between 25.00 mL and 25 mL?

Solution

Even though the two volume measurements appear to be equal, they really convey different information. The quantity 25 mL means that the volume is between 24 mL and 26 mL, whereas the quantity 25.00 mL means that the volume is between 24.99 mL and 25.01 mL. The pipet measures volume with much greater precision than does the graduated cylinder.

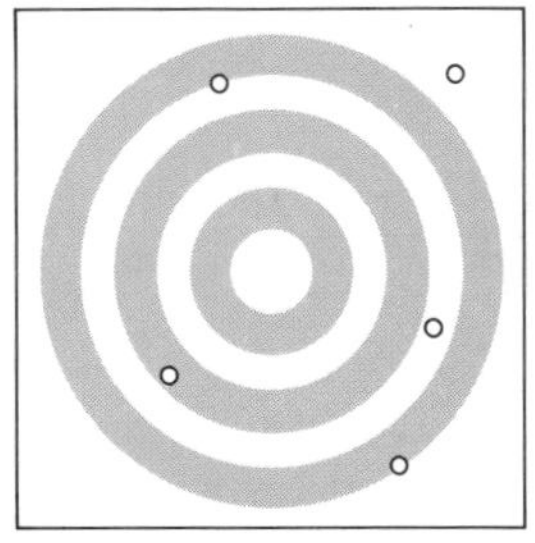

(a)

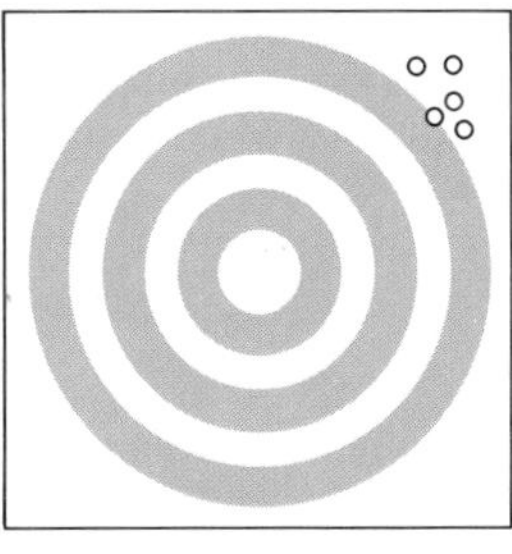

(b)

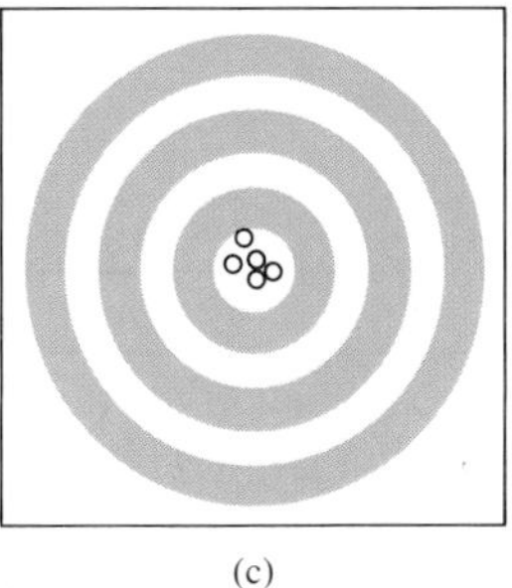

(c)

Figure 1.8

Shooting targets show the difference between *precise* and *accurate.* (a) Neither accurate nor precise (large random errors). (b) Precise but not accurate (small random errors, large systematic error). (c) Bull's-eye! Both precise and accurate (small random errors, no systematic error).

Precision and Accuracy

Two terms often used to describe uncertainty in measurements are precision and accuracy. Although these words are frequently used interchangeably in everyday life, they have different meanings in the scientific context. **Accuracy** refers to the agreement of a particular value with the true value. **Precision** refers to the degree of agreement among several measurements of the same quantity. Precision reflects the *reproducibility* of a given type of measurement. The difference between these terms is illustrated by the results of three different target practices shown in Fig. 1.8.

Two different types of errors are also introduced in Fig. 1.8. A **random error** (also called an indeterminate error) means that a measurement has an equal probability of being high or low. This type of error occurs in estimating the value of the last digit of a measurement. The second type of error is called **systematic error** (or determinate error). This type of error occurs in the same direction each time; it is either always high or always low. Figure 1.8(a) indicates large random errors (poor technique). Figure 1.8(b) indicates small random errors but a large systematic error, and Figure 1.8(c) indicates small random errors and no systematic error.

In quantitative work, precision is often used as an indication of accuracy; we assume that the *average* of a series of precise measurements (which should "average out" the random errors because of their equal probability of being high or low) is accurate, or close to the "true" value. However, this assumption is only valid if systematic errors are absent. Suppose we weigh a piece of brass five times on a very precise balance and obtain the following results:

Weighing	Result
1	2.486 g
2	2.487 g
3	2.485 g
4	2.484 g
5	2.488 g

Normally, we would assume that the true mass of the piece of brass is very close to 2.486 grams, which is the average of the five results. However, if the balance has a defect causing it to give a result that is consistently 1.000 gram too high (a systematic error of +1.000 gram), then 2.486 grams would be seriously in error. The point here is that high precision among several measurements is an indication of accuracy *only* if you can be sure that systematic errors are absent.

Precision is an indication of accuracy only if there are no systematic errors.

Sample Exercise 1.2

To check the accuracy of a graduated cylinder, a student filled the cylinder to the 25-mL mark using water delivered from a buret (Fig. 1.5), and then read the volume delivered. Following are the results of five trials:

Trial	Volume shown by graduated cylinder	Volume shown by the buret
1	25 mL	26.54 mL
2	25 mL	26.51 mL
3	25 mL	26.60 mL
4	25 mL	26.49 mL
5	25 mL	26.57 mL
Average	25 mL	26.54 mL

Is the graduated cylinder accurate?

Solution

The results of the trials show very good precision (for a graduated cylinder). The student has good technique. However, note that the average value measured by the buret is significantly different from 25 mL. Thus this graduated cylinder is not very accurate. It produces a systematic error (in this case, the indicated result is low for each measurement).

1.4 Significant Figures and Calculations

Purpose

- To show how to determine the number of significant figures in a calculated result.

Calculating the final result for an experiment usually involves adding, subtracting, multiplying, or dividing the results of various types of measurements. Thus it is very important that the uncertainty in the final result is known correctly. To ensure that this is the case, we use rules that involve counting the significant figures in each number and determining the correct number of significant figures in the final result.

Rules for Counting Significant Figures

1. *Nonzero integers*. Nonzero integers always count as significant figures.
2. *Zeros*. There are three classes of zeros:

Leading zeros are never significant figures.

   a. *Leading zeros* are zeros that *precede* all of the nonzero digits. They do not count as significant figures. In the number 0.0025 the three zeros simply indicate the position of the decimal point. This number has only two significant figures.

Captive zeros are always significant figures.

   b. *Captive zeros* are zeros *between* nonzero digits. They always count as significant figures. The number 1.008 has four significant figures.

Trailing zeros are sometimes significant figures.

   c. *Trailing zeros* are zeros at the *right end* of the number. They are significant only if the number contains a decimal point. The number 100 has only one significant figure, whereas the number 1.00×10^2 has three significant figures. The number one hundred written as 100. also has three significant figures.

Exact numbers never limit the number of significant figures in a calculation.

3. *Exact numbers*. Many times calculations involve numbers that were not obtained using measuring devices but were determined by counting: 10 experiments, 3 apples, 8 molecules. Such numbers are called *exact numbers*. They can be assumed to have an infinite number of significant figures. Other examples of exact numbers are the 2 in $2\pi r$ (the circumference of a circle) and the 4 and the 3 in $\frac{4}{3}\pi r^3$ (the volume of a sphere). Exact numbers can also arise from definitions. For example, one inch is defined as exactly 2.54 centimeters. Thus in the statement 1 in = 2.54 cm, neither the 2.54 nor the 1 limits the number of significant figures when used in a calculation.

Exponential notation is reviewed in Appendix 1.1.

Note that the number 1.00×10^2 above is written in **exponential notation.** This type of notation has at least two advantages: the number of significant figures can be easily indicated, and fewer zeros are needed to write a very large or very small number. For example, the number 0.000060 is much more conveniently represented as 6.0×10^{-5}. (The number has two significant figures.)

The following rules apply to the determination of the number of significant figures in the result of a calculation.

Rules for Significant Figures in Mathematical Operations*

1. *For multiplication or division* the number of significant figures in the result is the same as the number in the least precise measurement used in the calculation. For example, consider this calculation:

$$4.56 \times 1.4 = 6.38 \xrightarrow{\text{Corrected}} 6.4$$

Limiting term has two significant figures (1.4)

Two significant figures (6.4)

The correct product has only two significant figures, since 1.4 has two significant figures.

---

*Although these rules work well for most cases, they can give misleading results in certain cases. For a discussion of this see: L. M. Schwartz, Propagation of Significant Figures, *J. Chem. Ed.* **62** (1985): 693.

2. *For addition or subtraction* the result has the same number of decimal places as the least precise measurement used in the calculation. For example, consider the following sum:

$$\begin{array}{r} 12.11 \\ 18.0 \\ 1.013 \\ \hline 31.123 \end{array} \xrightarrow{\text{Corrected}} 31.1$$

18.0 ←limiting term has one decimal place

31.1 ← One decimal place

The correct result is 31.1, since 18.0 has only one decimal place.

Note that for multiplication and division significant figures are counted. For addition and subtraction the decimal places are counted.

In most calculations you will need to round off numbers to obtain the correct number of significant figures. The following rules should be applied for rounding off.

Rules for Rounding Off

1. In a series of calculations, carry the extra digits through to the final result, *then* round off.*
2. If the digit to be removed
   a. is less than 5, the preceding digit stays the same. For example, 1.33 rounds to 1.3.
   b. is equal to or greater than 5, the preceding digit is increased by 1. For example, 1.36 rounds to 1.4.

Rule 2 is consistent with the operation of electronic calculators.

Sample Exercise 1.3

Give the number of significant figures for each of the following results.

a. A student's extraction procedure on tea yields 0.0105 g of caffeine.

b. A chemist records a weight of 0.050080 g in an analysis.

c. In an experiment a span of time is determined to be 8.050×10^{-3} s.

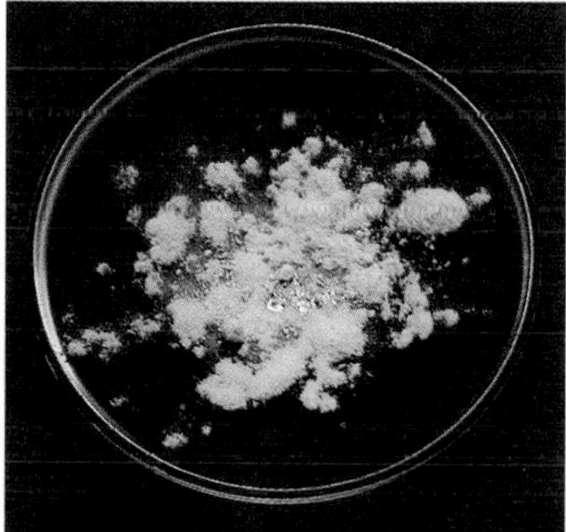

Caffeine, which occurs in such items as tea, coffee, and cola nuts, is a respiratory and cardiac stimulant.

Solution

a. The number contains three significant figures. The zeros to the left of the 1 are leading zeros and are not significant, but the remaining zero (a captive zero) is significant.

b. The number contains five significant figures. The leading zeros (to the left of the 5) are not significant. The captive zeros between the 5 and the 8 are significant, and the trailing zero to the right of the 8 is significant because the number contains a decimal point.

c. This number has four significant figures. Both zeros are significant.

---

*This practice will not usually be followed in the Sample Exercises in this text because we want to show the correct number of significant figures in each step of the problems. However, in the answers to the end-of-chapter exercises, only the final answer is rounded off.

Sample Exercise 1.4

Carry out the following mathematical operations and give each result with the correct number of significant figures.

a. $1.05 \times 10^{-3} \div 6.135$

b. $21 - 13.8$

c. As part of a lab assignment to determine the value of the gas constant (R), a student measured the pressure (P), volume (V), and temperature (T) for an amount of gas, where

$$R = \frac{PV}{T}$$

The following values were obtained: $P = 2.560$, $T = 275.15$, and $V = 8.8$. (Gases will be discussed in detail in Chapter 5; we will not be concerned at this time about the units for these quantities.) Calculate R to the correct number of significant figures.

Solution

a. The result is 1.71×10^{-4}, which has three significant figures because the term with the least precision (1.05×10^{-3}) has three significant figures.

b. The result is 7 with no decimal place because the number with the least number of decimal places (21) has none.

c.

$$R = \frac{PV}{T} = \frac{(2.560)(8.8)}{275.15} = 8.2 \times 10^{-2}$$

The final result has two significant figures because 8.8 (the least precise measurement) has two significant figures.

There is a useful lesson to be learned from the last part of Sample Exercise 1.4. The student measured the pressure and temperature to greater precision than the volume. A more precise value of R could have been obtained if a more precise measurement of V had been made. As it is, the efforts expended to measure P and T very precisely were wasted. Remember that a series of measurements to obtain some final result should all be done to about the same precision.

1.5 Dimensional Analysis

Purpose

- To show how to convert units between the English and metric systems.

It is often necessary to convert results from one system of units to another. The best way to do this is by a method called the *unit factor method,* or more commonly **dimensional analysis.** To illustrate the use of this method, we will consider several unit conversions. Some equivalents in the English and metric systems are listed in Table 1.4. A more complete list of conversion factors given to more significant figures appears inside the back cover of the text.

Consider a pin measuring 2.85 centimeters in length. What is its length in inches? To solve this problem we must use the equivalence statement

$$2.54 \text{ cm} = 1 \text{ in}$$

If we divide both sides of this equation by 2.54 cm, we get

$$\frac{2.54 \text{ cm}}{2.54 \text{ cm}} = 1 = \frac{1 \text{ in}}{2.54 \text{ cm}}$$

Note that the expression 1 in/2.54 cm equals 1. This expression is called a **unit factor.** Since 1 inch and 2.54 centimeters are exactly equivalent, multiplying any expression by this unit factor will not change its value.

The pin has a length of 2.85 centimeters. Multiplying this length by the unit factor gives

$$2.85 \cancel{\text{cm}} \times \frac{1 \text{ in}}{2.54 \cancel{\text{cm}}} = \frac{2.85}{2.54} \text{ in} = 1.12 \text{ in}$$

Note that the centimeter units cancel to give inches for the result. This is exactly what we wanted to accomplish. Note also that the result has three significant figures as required by the number 2.85. Recall that the 1 and 2.54 in the conversion factor are exact numbers by definition.

English-Metric Equivalents	
Length	
	1 m = 1.094 yd
	2.54 cm = 1 in
Mass	
	1 kg = 2.205 lb
	453.6 g = 1 lb
Volume	
	1 L = 1.06 qt
	1 ft^3 = 28.32 L

Table 1.4

Sample Exercise 1.5

A pencil is 7.00 in long. What is its length in centimeters?

Solution

In this case we want to convert from inches to centimeters. Therefore, we must use the reciprocal of the unit factor used above to do the opposite conversion:

$$7.00 \cancel{\text{in}} \times \frac{2.54 \text{ cm}}{1 \cancel{\text{in}}} = (7.00)(2.54) \text{ cm} = 17.8 \text{ cm}$$

Here the inch units cancel, leaving centimeters as requested.

Note that two unit factors can be derived from each equivalence statement. For example, from the equivalence statement 2.54 cm = 1 in, the two unit factors are:

$$\frac{2.54 \text{ cm}}{1 \text{ in}} \quad \text{and} \quad \frac{1 \text{ in}}{2.54 \text{ cm}}$$

How do you choose which one to use in a given situation? Simply look at the *direction* of the required change. To change from inches to centimeters, the inches must cancel. Thus the factor 2.54 cm/1 in is used. To change from centimeters to inches, centimeters must cancel, and the factor 1 in/2.54 cm is appropriate.

Consider the direction of the required change to select the correct unit factor.

Converting from One Unit to Another

- To convert from one unit to another, use the equivalence statement that relates the two units.
- Derive the appropriate unit factor by looking at the direction of the required change (to cancel the unwanted units).
- Multiply the quantity to be converted by the unit factor to give the quantity with the desired units.

Sample Exercise 1.6

You want to order a bicycle with a 25.5-in frame, but the sizes in the catalog are given only in centimeters. What size should you order?

Solution

You need to go from inches to centimeters, so 2.54 cm/1 in is appropriate:

$$25.5\ \cancel{\text{in}} \times \frac{2.54\ \text{cm}}{1\ \cancel{\text{in}}} = 64.8\ \text{cm}$$

To ensure that the conversion procedure is clear, a multistep problem is considered in Sample Exercise 1.7.

Sample Exercise 1.7

A student has entered a 10.0-km run. How long is the run in miles?

Solution

This conversion can be accomplished in several different ways. Since we have the equivalence statement 1 m = 1.094 yd, we will proceed by a path that uses this fact. Before we start any calculations, let us consider our strategy. We have kilometers, which we want to change to miles. We can do this by the following route:

$$\text{kilometers} \longrightarrow \text{meters} \longrightarrow \text{yards} \longrightarrow \text{miles}$$

To proceed in this way, we need the following equivalence statements:

$$1\ \text{km} = 1000\ \text{m}$$
$$1\ \text{m} = 1.094\ \text{yd}$$
$$1760\ \text{yd} = 1\ \text{mi}$$

To make sure the process is clear, we will proceed step by step:

Kilometers to Meters:

$$10.0\ \cancel{\text{km}} \times \frac{1000\ \text{m}}{1\ \cancel{\text{km}}} = 1.00 \times 10^4\ \text{m}$$

Sample Exercise 1.7, continued

Meters to Yards:

$$1.00 \times 10^4 \cancel{m} \times \frac{1.094 \text{ yd}}{1 \cancel{m}} = 1.094 \times 10^4 \text{ yd}$$

Note that we should have only three significant figures in the result. Since this is an intermediate result, however, we will carry the extra figures. Remember, round off only the final result.

Yards to Miles:

$$1.094 \times 10^4 \cancel{yd} \times \frac{1 \text{ mi}}{1760 \cancel{yd}} = 6.216 \text{ mi}$$

Note in this case that 1 mi equals exactly 1760 yd *by designation*. Thus 1760 is an exact number of four significant figures.

Since the distance was originally given as 10.0 km, the result can have only three significant figures and should be rounded to 6.22 mi. Thus

$$10.0 \text{ km} = 6.22 \text{ mi}$$

Alternatively, we can combine the steps:

$$10.0 \cancel{km} \times \frac{1000 \cancel{m}}{1 \cancel{km}} \times \frac{1.094 \cancel{yd}}{1 \cancel{m}} \times \frac{1 \text{ mi}}{1760 \cancel{yd}} = 6.22 \text{ mi}$$

In using dimensional analysis your verification that everything has been done correctly is that you end up with the correct units. *In doing chemistry problems, you should always include the units for the quantities used.* Always check to see that the units cancel to give the correct units for the final result. This provides a very valuable check, especially for complicated problems.

Study the procedures for unit conversions in the following Sample Exercises.

Sample Exercise 1.8

The speed limit on highways in the United States is 55 mi/h. What number would be posted in kilometers per hour?

Solution

$$\frac{55 \cancel{mi}}{\text{h}} \times \frac{1760 \cancel{yd}}{1 \cancel{mi}} \times \frac{1 \cancel{m}}{1.094 \cancel{yd}} \times \frac{1 \text{ km}}{1000 \cancel{m}} = 88 \text{ km/h}$$

Note that all units cancel except the desired kilometers per hour.

Sample Exercise 1.9

A Japanese car is advertised as having a gas mileage of 15 km/L. Convert this rating to miles per gallon.

Sample Exercise 1.9, continued

Solution

$$\frac{15\ \cancel{\text{km}}}{\cancel{\text{L}}} \times \frac{1000\ \cancel{\text{m}}}{1\ \cancel{\text{km}}} \times \frac{1.094\ \cancel{\text{yd}}}{1\ \cancel{\text{m}}} \times \frac{1\ \text{mi}}{1760\ \cancel{\text{yd}}} \times \frac{1\ \cancel{\text{L}}}{1.06\ \cancel{\text{qt}}} \times \frac{4\ \cancel{\text{qt}}}{1\ \text{gal}} = 35\ \text{mi/gal}$$

1.6 Temperature

Purpose

- To demonstrate conversions among the Fahrenheit, Celsius, and Kelvin temperature scales.

Three systems for measuring temperature are widely used: the Celsius scale, the Kelvin scale, and the Fahrenheit scale. The first two temperature systems are used in the physical sciences, and the third is used in most of the engineering sciences. Our purpose here is to define the three temperature scales and show how conversions from one scale to another can be performed.

The three temperature scales are defined and compared in Fig. 1.9. Note that the size of the temperature unit (the degree) is the same for the Kelvin and Celsius scales. The fundamental difference between these two temperature scales is in their zero points. Conversion between these two scales simply requires an adjustment for the different zero points.

$T_K = T_C + 273.15$

$T_C = T_K - 273.15$

Temperature (Kelvin) = temperature (Celsius) + 273.15

or Temperature (Celsius) = temperature (Kelvin) − 273.15

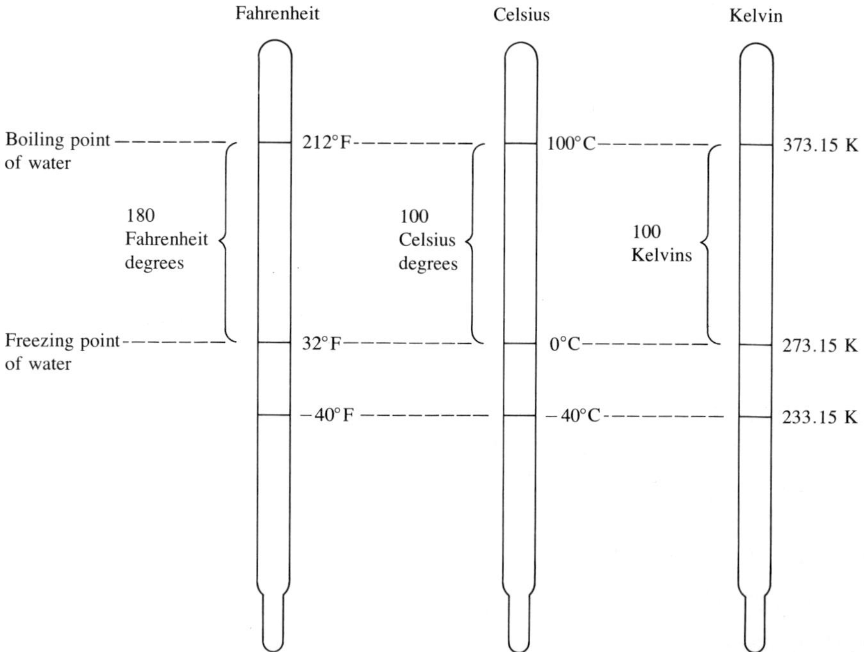

Figure 1.9

The three major temperature scales.

For example, to convert 300.00 K to the Celsius scale, we do the following calculation:

$$300.00 - 273.15 = 26.85°C$$

Note that, in expressing temperature in Celsius units, the designation °C is used. The degree symbol is not used when writing temperature in terms of the Kelvin scale. The unit of temperature on this scale is called a Kelvin and is symbolized by K.

To convert between the Fahrenheit and Celsius scales is somewhat more complicated because both the degree sizes and the zero points are different. Thus we need to consider two adjustments: one for degree size, and one for the zero point. First, we must account for the difference in degree size. This can be done by reconsidering Fig. 1.9. Notice that since 212°F = 100°C and 32°F = 0°C,

$$212 - 32 = 180 \text{ Fahrenheit degrees} = 100 - 0 = 100 \text{ Celsius degrees}$$

Thus 180° on the Fahrenheit scale are equivalent to 100° on the Celsius scale, and the unit factor is

$$\frac{180°F}{100°C} \quad \text{or} \quad \frac{9°F}{5°C}$$

or the reciprocal, depending on the direction in which we need to go.

Next, we must consider the different zero points. Since 32°F = 0°C, we obtain the corresponding Celsius temperature by first subtracting 32 from the Fahrenheit temperature to account for the different zero points. Then the unit factor is applied to adjust for the difference in the degree size. This process is summarized by the equation

$$(T_F - 32°F)\frac{5°C}{9°F} = T_C \qquad (1.1)$$

where T_F and T_C represent a given temperature on the Fahrenheit and Celsius scales, respectively. In the opposite conversion, we first correct for degree size and then correct for the different zero point. This process can be summarized in the following general equation:

$$T_F = T_C \times \frac{9°F}{5°C} + 32°F \qquad (1.2)$$

Equations (1.1) and (1.2) are really the same equation in different forms. See if you can obtain Equation (1.2) by starting with Equation (1.1) and rearranging.

At this point it is worthwhile to weigh the two alternatives for learning to do temperature conversions: you can simply memorize the equations, or you can take the time to learn the differences between the temperature scales and to understand the processes involved in converting from one scale to another. The latter approach may take a little more effort, but the understanding you gain will stick with you much longer than the memorized formulas. This choice will also apply to many of the other chemical concepts. Try to think things through!

Understand the process in converting from one temperature scale to another; do not simply memorize the equations.

Sample Exercise 1.10

Normal body temperature is 98.6°F. Convert this temperature to the Celsius and Kelvin scales.

Sample Exercise 1.10, continued

Solution

Rather than simply using the formulas to solve this problem, we will proceed by thinking it through. The situation is diagrammed in Fig. 1.10. First, we want to convert 98.6°F to the Celsius scale. The number of Fahrenheit degrees between 32.0°F and 98.6°F is 66.6°F. We must convert this difference to Celsius degrees:

$$66.6°F \times \frac{5°C}{9°F} = 37.0°C$$

Thus 98.6°F corresponds to 37.0°C.

Now we can convert to the Kelvin scale:

$$T_K = T_C + 273.15 = 37.0 + 273.15 = 310.2 \text{ K}$$

Note that the final answer has only one decimal place (37.0 is limiting).

Sample Exercise 1.11

One interesting feature of the Celsius and Fahrenheit scales is that −40°C and −40°F represent the same temperature, as shown in Fig. 1.9. Verify that this is true.

Solution

The difference between 32°F and −40°F is 72°F. The difference between 0°C and −40°C is 40°C. The ratio of these is

$$\frac{72°F}{40°C} = \frac{8 \times 9°F}{8 \times 5°C} = \frac{9°F}{5°C}$$

as required. Thus −40°C is equivalent to −40°F.

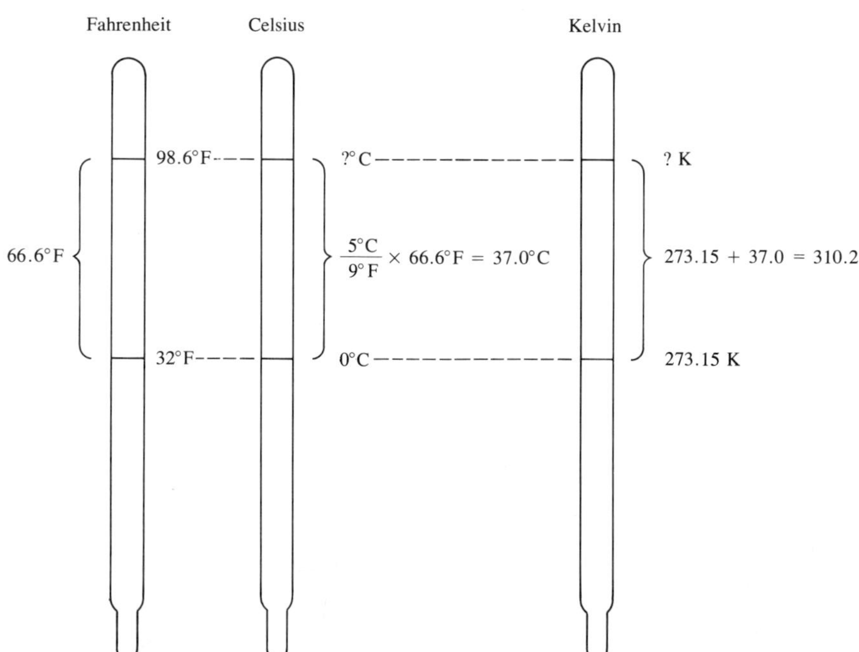

Figure 1.10

Normal body temperature on the Fahrenheit, Celsius, and Kelvin scales.

Since, as shown in Sample Exercise 1.11, −40° on both the Fahrenheit and Celsius scales represents the same temperature, this point can be used as a reference point (like 0°C and 32°F) for a relationship between the two scales:

$$\frac{\text{Number of Fahrenheit degrees}}{\text{Number of Celsius degrees}} = \frac{T_F - (-40)}{T_C - (-40)} = \frac{9°F}{5°C}$$

or

$$\frac{T_F + 40}{T_C + 40} = \frac{9°F}{5°C} \tag{1.3}$$

where T_F and T_C represent the same temperature (but not the same number). This equation can be used to convert Fahrenheit temperatures to Celsius and vice versa and may be easier to remember than Equations (1.1) and (1.2).

Sample Exercise 1.12

Liquid nitrogen, which is often used as a coolant for low-temperature experiments, has a boiling point of 77 K. What is this temperature on the Fahrenheit scale?

Solution

We will first convert 77 K to the Celsius scale:

$$T_C = T_K - 273.15 = 77 - 273.15 = -196°C$$

To convert to the Fahrenheit scale we will use Equation (1.3):

$$\frac{T_F + 40}{T_C + 40} = \frac{9°F}{5°C}$$

$$\frac{T_F + 40}{-196°C + 40} = \frac{9°F}{5°C}$$

$$\frac{T_F + 40}{-156°C} = \frac{9°F}{5°C}$$

$$T_F + 40 = \frac{9°F}{5°C}(-156°C) = -281°F$$

$$T_F = -281°F - 40 = -321°F$$

Liquid nitrogen is so cold that water condenses out of the surrounding air, forming a cloud as the nitrogen is poured.

1.7 Density

Purpose

- To illustrate calculations involving density.

A property of matter that is often used by chemists as an "identification tag" for a substance is **density,** the mass per unit volume:

$$\text{Density} = \frac{\text{mass}}{\text{volume}}$$

The density of a liquid can be easily determined by weighing an accurately known volume of the liquid. This procedure is illustrated in Sample Exercise 1.13.

Sample Exercise 1.13

A chemist, trying to identify the main component in a commercial record cleaner, finds that 25.00 cm^3 of the substance has a mass of 19.625 g at 20°C. The following are the names and densities of the compounds that might be the main component of the cleaner:

Compound	Density in g/cm^3 at 20°C
Chloroform	1.492
Diethyl ether	0.714
Ethanol	0.789
Isopropyl alcohol	0.785
Toluene	0.867

Which of these compounds is the most likely to be the main component of the record cleaner?

Solution

To identify the unknown substance, we must determine its density. This can be done by using the definition of density.

$$\text{Density} = \frac{\text{mass}}{\text{volume}} = \frac{19.625\ \text{g}}{25.00\ \text{cm}^3} = 0.7850\ \text{g/cm}^3$$

This density corresponds exactly to that of isopropyl alcohol, which is therefore the most likely main component of the record cleaner. However, note that the density of ethanol is also very close. In order to be sure that the compound is isopropyl alcohol, we should run several more density experiments. (In the modern laboratory many other types of tests could be done to distinguish between these two liquids.)

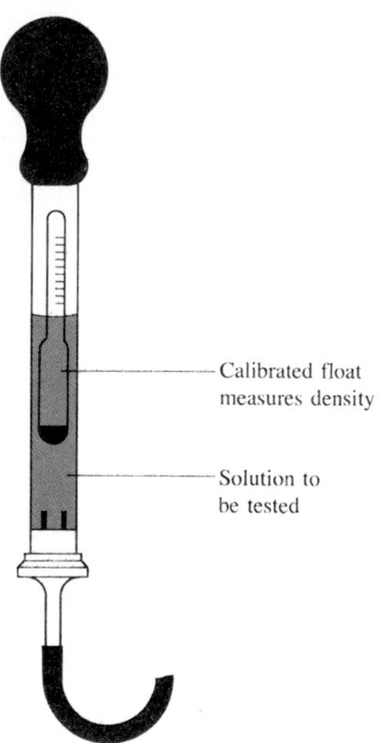

Figure 1.11

A hydrometer of the type used to test the density of the sulfuric acid solution in a car's battery and of the antifreeze in a car's cooling system. The float is calibrated and sinks to a depth that is related to the density of the solution being tested.

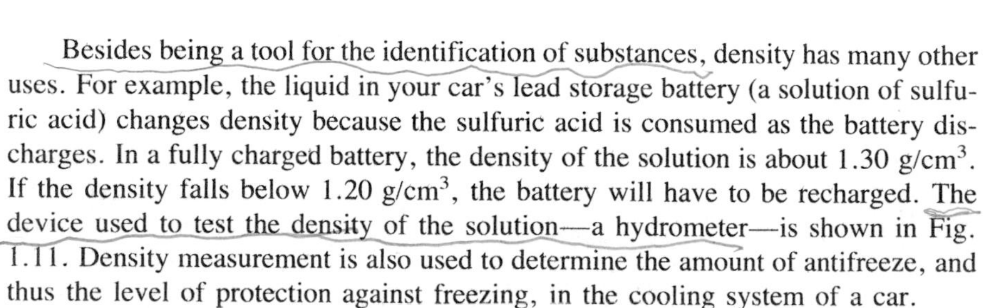

Besides being a tool for the identification of substances, density has many other uses. For example, the liquid in your car's lead storage battery (a solution of sulfuric acid) changes density because the sulfuric acid is consumed as the battery discharges. In a fully charged battery, the density of the solution is about 1.30 g/cm^3. If the density falls below 1.20 g/cm^3, the battery will have to be recharged. The device used to test the density of the solution—a hydrometer—is shown in Fig. 1.11. Density measurement is also used to determine the amount of antifreeze, and thus the level of protection against freezing, in the cooling system of a car.

The densities of various common substances are given in Table 1.5.

Densities of Various Common Substances at 20°C		
Substance	Physical state	Density (g/cm^3)
Oxygen	gas	0.00133*
Hydrogen	gas	0.000084*
Ethanol	liquid	0.789
Benzene	liquid	0.880
Water	liquid	1.000
Magnesium	solid	1.74
Salt (sodium chloride)	solid	2.16
Aluminum	solid	2.70
Iron	solid	7.87
Copper	solid	8.96
Silver	solid	10.5
Lead	solid	11.34
Mercury	liquid	13.6
Gold	solid	19.32

*At 1 atmosphere pressure

Table 1.5

Because gold has a higher density than sand, it settles to the bottom of a prospector's pan.

1.8 Classification of Matter

Purpose

- To show how matter can be classified into subgroups.

Before we can hope to understand the changes we see going on around us—the growth of plants, the rusting of steel, the aging of people, rain becoming more acidic—we must find out how matter is organized. **Matter,** the material of the universe, is clearly complex and has many levels of organization. In this section we introduce the basic ideas about the structure of matter and its behavior.

We will start by considering the definitions of the fundamental properties of matter. Matter exists in three **states:** solid, liquid, and gas. A *solid* is rigid; it has a fixed volume and shape. A *liquid* has a definite volume but no specific shape; it assumes the shape of its container. A *gas* has no fixed volume or shape; it takes on the shape and volume of its container. In contrast to liquids and solids, which are only slightly compressible, gases are highly compressible; it is relatively easy to decrease the volume of a gas. Table 1.5 gives the state of some common substances at 20°C.

Most of the matter around us consists of **mixtures** of pure substances. Wood, gasoline, wine, soil, and air are all mixtures. The main characteristic of a mixture is that it has *variable composition.* For example, wood is a mixture of many substances, the proportions of which vary depending on the type of wood and where it grows. Mixtures can be classified as **homogeneous** (the same throughout) or **heterogeneous** (containing regions with differing properties).

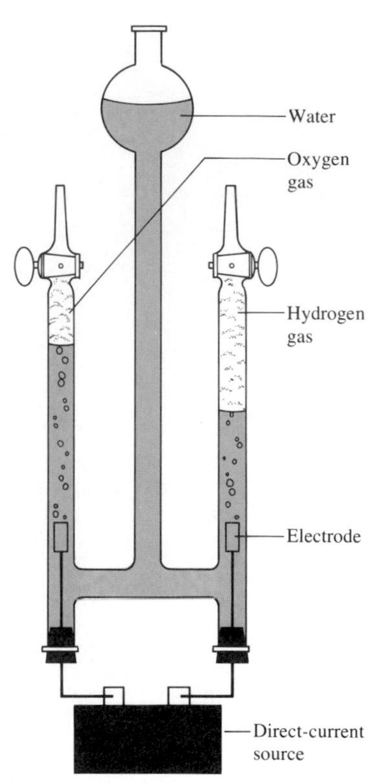

Figure 1.12

Electrolysis is an example of a chemical process. It will be discussed in detail in Chapter 17.

A homogeneous mixture is called a **solution.** Air is a solution consisting of a mixture of gases. Wine is a complex liquid solution. Brass is a solid solution of copper and zinc. Sand in water, dust suspended in the air, and iced tea with ice cubes are all examples of heterogeneous mixtures. Heterogeneous mixtures can usually be separated into two or more homogeneous mixtures or pure substances (for example, the ice cubes can be separated from the tea).

Mixtures can be separated into pure substances by physical methods. A **pure substance** is one with constant composition. Water is a good illustration of these ideas. As we will discuss in detail later, pure water is composed solely of H_2O molecules, but the water found in nature (ground water or the water in a lake or ocean) is really a mixture. Sea water, for example, contains large amounts of dissolved minerals. Boiling sea water produces steam, which can be condensed to pure water, leaving the minerals behind as solids. The dissolved minerals in sea water can also be separated out by freezing the mixture, since pure water freezes out. The processes of boiling and freezing cause **physical changes:** when water freezes or boils, it changes its state but remains water; it is still composed of H_2O molecules. A physical change is a change in the form of a substance, not in its chemical composition. A physical change can be used to separate a mixture into pure compounds, but it will not break compounds into elements.

It should be noted that absolute purity is an ideal. Because water, for example, inevitably comes into contact with other materials when it is synthesized or separated from a mixture, it is never absolutely pure. With great care, however, substances can be obtained in very nearly pure form.

Pure substances contain compounds or elements. A **compound** is a substance with *constant composition* that can be broken down into elements by chemical processes. An example of a chemical process is the electrolysis of water (see Fig. 1.12), in which an electric current is passed through water to break it down into the elements hydrogen and oxygen. This process produces a **chemical change** because the water molecules have been broken down. The water is gone, and in its place we have the elements hydrogen and oxygen. **Elements** cannot be decomposed into simpler substances by chemical or physical means.

We have seen that the matter around us has various levels of organization. The most fundamental substances we have discussed so far are elements. As we will see in later chapters, elements also have structure: they are composed of atoms, which in turn are composed of nuclei and electrons. Even the nucleus has structure: it is composed of protons and neutrons. And even these can be broken down further,

The element mercury (left) combines with the element iodine (center) to form the compound mercuric iodide (right). This is an example of a chemical change.

into elementary particles called quarks. However, we need not concern ourselves with such details at this point. Figure 1.13 summarizes our discussion of the organization of matter.

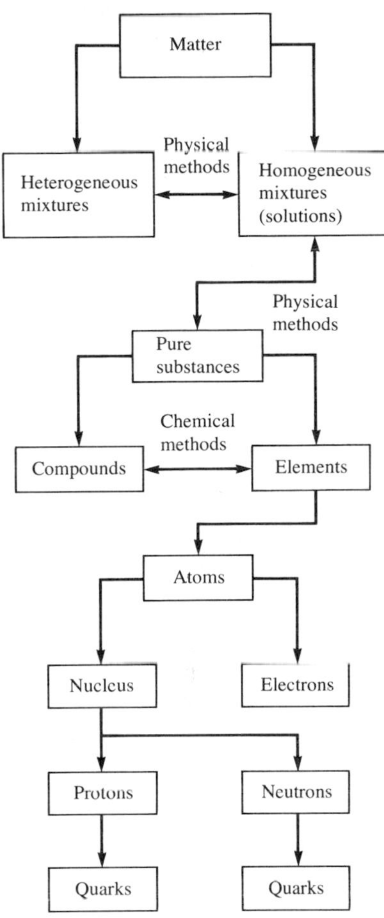

Figure 1.13

The organization of matter.

1.9 Separation of Mixtures

Purpose

- To introduce some physical methods for separating mixtures.

One of the fundamental operations of chemistry is the identification of substances. For example, several years ago, a toxic chemical commonly called dioxin, which is a by-product of the manufacture of certain industrial chemicals, was mistakenly mixed with the waste oil used to keep the dust down on the roads of several rural communities. Because of dioxin's toxicity, public safety required that the exact dioxin content of the soil in these communities be determined. Chemists were asked to find a way to separate the tiny amount of dioxin (approximately one microgram per gram of soil) from a very complicated soil sample.

This example is really quite typical of the problems faced by analytical chemists. The separation of mixtures is a crucial part of any chemical analysis. Separations are also important in many other situations. One example is the refining of petroleum; the thick, sticky liquid pumped from the earth must be separated into various components to be useful. Another example is the production of pure water from the mixture sea water.

One of the most important methods for separating the components of a liquid mixture is **distillation,** a process that depends on differences in the ease of vaporization (how readily liquids become gases) of the components. In simple distillation the liquid mixture is heated in a device such as that shown in Fig. 1.14 on page 26. The most volatile component vaporizes at the lowest temperature, and the vapor passes through a cooled tube (a condenser), where it condenses back into its liquid state.

The term *volatile* refers to the ease with which a liquid can be changed to its vapor.

The simple, one-stage distillation apparatus shown in Fig. 1.14 works very well when only one component of the mixture is volatile. For example, a mixture of water and sand is easily separated by boiling off the water. Water containing dissolved minerals behaves in much the same way. As the water is boiled off, the minerals remain behind as nonvolatile solids. Simple distillation of sea water using the sun as the heat source is an excellent way to desalinate (remove the minerals from) sea water.

However, when a liquid mixture contains several volatile components, simple, one-step distillation does not give a pure substance in the receiving flask. Some vapor from *all* of the components will inevitably be condensed. Nevertheless, the liquid in the receiving flask will be richer in the most volatile component than the original liquid. By redistilling several times, we can eventually approach a pure liquid of the most volatile mixture component. In fact, these repeated distillations can be accomplished in one process using a fractional distillation apparatus, as shown in Fig. 1.15 on page 26. This device is different from the simple distillation apparatus in that a column has been placed between the boiling flask and the condenser. This column contains glass beads or some other material that provides a

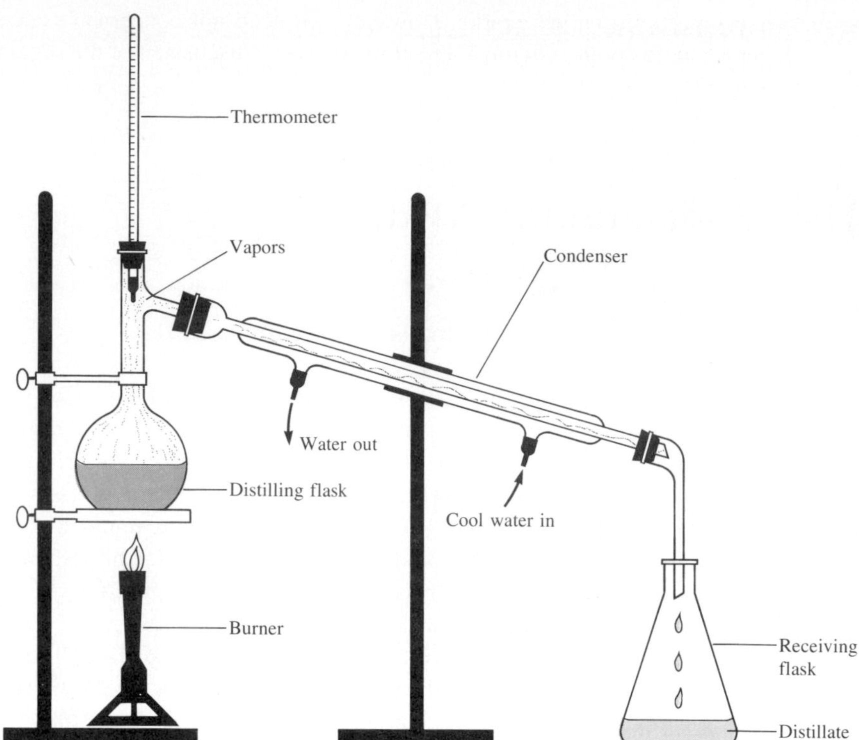

Figure 1.14

Simple laboratory distillation apparatus. Cool water circulates through the outer portion of the condenser, causing vapors from the distilling flask to condense into a liquid. The nonvolatile component of the mixture remains in the distilling flask.

large surface area. As the vapor proceeds up the column, it condenses and revaporizes many times; each time is a simple distillation. By the time the vapor reaches the top of the column, it consists essentially of molecules of the most volatile component and can be condensed to give the pure liquid.

Fractional distillation is an important technique in the laboratory and in the chemical plant. For example, fractional distillation carried out in a distillation tower such as is illustrated in Fig. 1.16 is used to separate the fractions of petroleum. The components are taken off the tower at various points, with the most volatile leaving nearest the top.

Another method of separation is simple **filtration,** which is used when a mixture consists of a solid and a liquid. The mixture is poured onto a mesh, such as filter paper, which passes the liquid and leaves the solid behind (see Fig. 1.17).

A third method of separation is called **chromatography.** Chromatography is the general name applied to a series of methods that employ a system with two *phases* (states) of matter: a mobile phase and a stationary phase. The stationary phase is a solid, and the mobile phase is either a liquid or a gas. The separation process occurs because the components of the mixture have different affinities for the two phases and thus move through the system at different rates. A component with a high affinity for the mobile phase moves relatively quickly through the chromatographic system, whereas one with a high affinity for the solid phase moves more slowly.

For example, consider the separation of components A and B, which are both dissolved in a liquid (called the solvent). A porous solid (the stationary phase) is packed into a column, and the solution is poured on top, as shown in Fig. 1.18. Pure solvent is then used to wash the substances through the column at different rates.

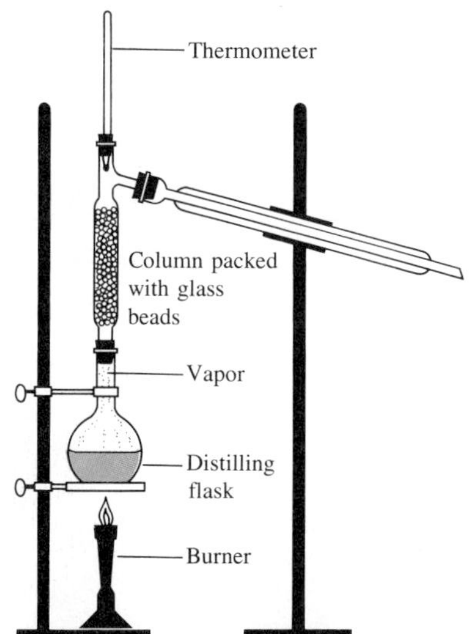

Figure 1.15

A fractional distillation apparatus. The distilling flask contains more than one volatile component.

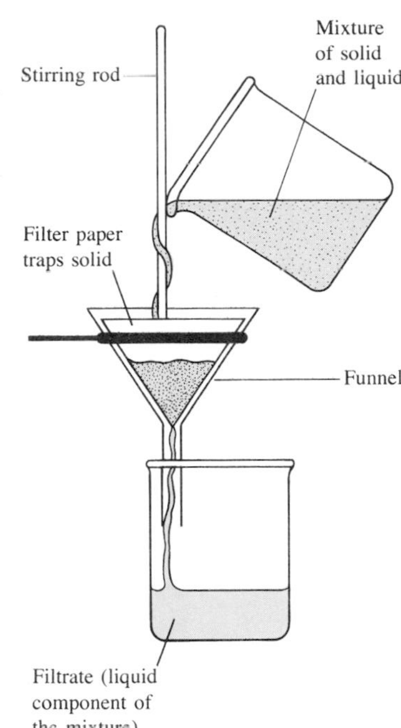

Figure 1.16 (far left)

Petroleum distillation tower. The petroleum industry is discussed in Chapter 24.

Figure 1.17 (left)

Filtration separates a liquid from a solid. The liquid passes through the filter paper, and the solid particles are trapped. Filter paper is available in a wide range of pore sizes; some filter paper has pores small enough to remove bacteria.

For example, if A is strongly attracted to the surface of the solid and if B dissolves readily in the solvent, B will move through the column much faster than A and can be collected first at the bottom of the column. Eventually A will come through and can be collected. In this way A and B can be separated. This is an example of **liquid-column chromatography.**

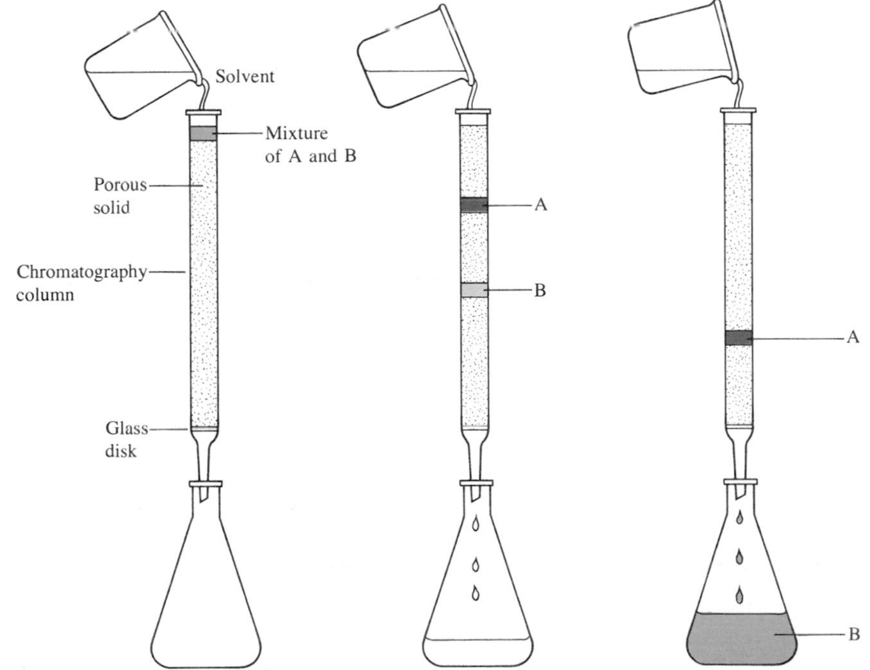

Figure 1.18

(a) A chromatography column separates a mixture of A and B, seen here at the top of the column. (b) Since B has a higher affinity than A does for the mobile phase, the band of B moves more quickly down the column. (c) Finally, B is washed off the column into the receiving flask, while A remains on the column.

Gaseous mixtures can be separated using a very similar type of column. Once again the stationary phase is a solid, but the mobile phase is a gas, called the *carrier gas,* that contains the gaseous substances to be separated. The different vapors interact to various degrees with the stationary phase. Those that interact to the least extent with the solid exit the column first (see Fig. 1.19). This method of separation is called **gas chromatography.** It can be used, for example, in analyzing the exhaust from an automobile to test for environmentally harmful gases. The effects of auto pollution are discussed in Chapter 5.

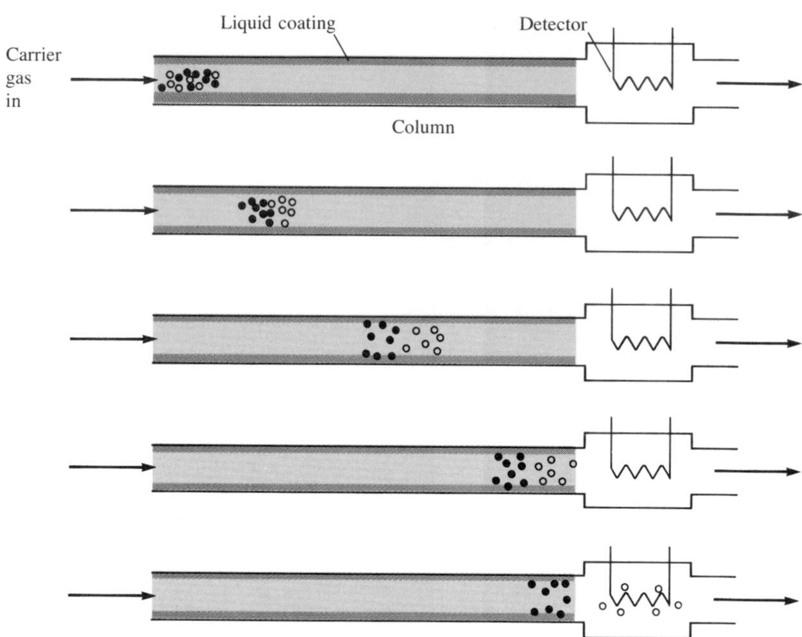

Figure 1.19

Separation of a gaseous mixture is accomplished by using an unreactive carrier gas (mobile phase) to drive the mixture through the column, a narrow tube coated with a nonvolatile liquid (stationary phase).

Because the components of the mixture interact to different extents with the stationary phase, they move along the column at different rates, and separation occurs. As they exit the column, the components pass through a detector that signals their presence.

Another type of chromatography, **paper chromatography,** employs a strip of porous paper, such as filter paper, as the stationary phase. A drop of the mixture to be separated is placed on the paper, which is then dipped into a liquid (the mobile phase) that travels up the paper as though it were a wick (see Fig. 1.20). This method of separating a mixture is often used by biochemists, who study the chemistry of living systems.

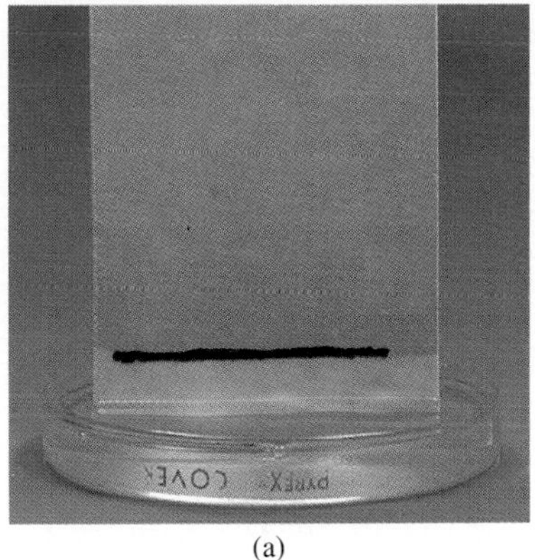

(a)

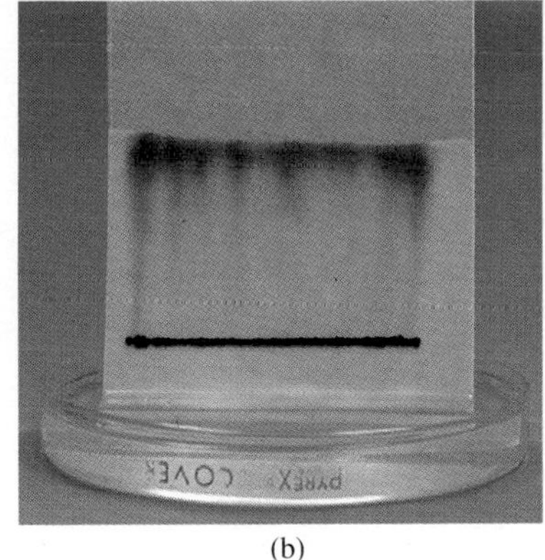

(b)

(c)

Figure 1.20

Paper chromatography of ink. (a) A line of the mixture to be separated is placed at one end of a sheet of porous paper. (b) The paper acts as a wick to draw up the liquid. (c) The component with the weakest attraction for the paper travels faster than those that cling to the paper.

FOR REVIEW

Summary

Ideally, the progress of any science occurs by using the scientific method, which first makes observations of nature, then looks for patterns in the observations that often lead to the formulation of natural laws. A theory or model is then constructed to provide an interpretation of the behavior of nature. The theory is tested by further experiments and modified as necessary.

Quantitative observations are called measurements and always consist of a number and a scale or unit. Measurement always involves some uncertainty. Drawing conclusions from the results of measurements requires that the extent of the uncertainty be known. This information is conveyed by using the correct number of significant figures. The preferred system of units among scientists is the SI system. Converting results from one set of units to another is accomplished with unit factors (dimensional analysis).

Matter can exist in three states—solid, liquid, and gas—and has many levels of organization. Mixtures of compounds can be separated by methods involving only physical changes, such as distillation, filtration, and various forms of chromatography. On the other hand, compounds can be broken down to elements only through chemical changes.

Key Terms

Section 1.1
scientific method
measurement
natural law
theory
model
hypothesis

Section 1.2
SI system
mass
weight

Section 1.3
uncertainty
significant figures
accuracy
precision
random error
systematic error

Section 1.4
exponential notation

Section 1.5
dimensional analysis
unit factor

Section 1.7
density

Section 1.8
matter
states (of matter)
homogeneous mixture
heterogeneous mixture
solution
pure substance
physical change
compound
chemical change
element

Section 1.9
distillation
filtration
chromatography
liquid-column chromatography
gas chromatography
paper chromatography

Exercises

A blue exercise number indicates that the answer to that exercise appears at the back of this book and a solution appears in the Solutions Guide.

Scientific Method

1. Define and explain the differences between the following words:
   a. law and theory
   b. theory and experiment
   c. qualitative and quantitative
   d. hypothesis and theory

2. Is the scientific method suitable for solving problems only in the sciences? Explain.

3. Confronted with the box shown in the diagram, you wish to discover something about its internal workings. You have no tools and cannot open the box. You pull on rope B, and it moves rather freely. When you pull on rope A, rope C appears to be pulled slightly into the box. When you pull on rope C, rope A almost disappears into the box.*

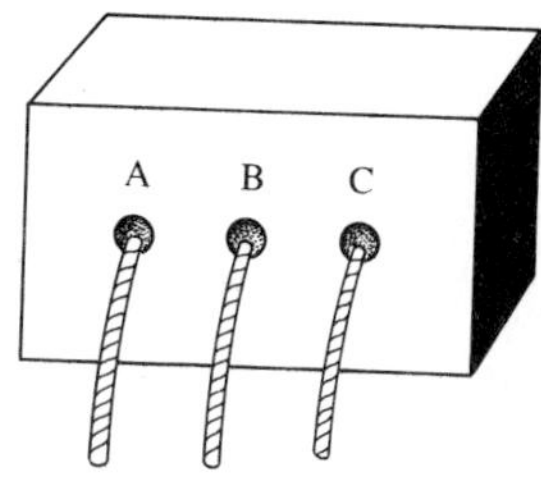

   a. Based on these observations, construct a model for the interior mechanism of the box.
   b. What further experiments could you do to refine your model?

4. Which of the following statements (hypotheses) could be tested by quantitative measurement?
   a. Ty Cobb was a better hitter than Pete Rose.
   b. Ivory soap is $99\frac{44}{100}\%$ pure.
   c. Rolaids consumes 47 times its weight in excess stomach acid.

5. Paracelsas, a sixteenth century alchemist and healer, adopted as his slogan: "The patients are your textbook, the sickbed is your study." Is this view consistent with using the scientific method?

Uncertainty, Precision, Accuracy, and Significant Figures

6. Which of the following are exact numbers?
   a. The elevation of Denver, Colorado, the Mile High City, is *5280* ft.
   b. There are *12* eggs in a dozen.
   c. One yard is equal to *0.9144* m.
   d. The announced attendance at a football game was *52,806*.
   e. In 1983, *1,759* Ph.D.s in chemistry were awarded in the United States.
   f. The budget deficit of the U.S. Government in fiscal year 1986 was *$200 billion*.

*From Yoder, Suydam, and Snavely, *Chemistry* (New York: Harcourt Brace Jovanovich, 1975), pp. 9–11.

7. What is the difference between precision and accuracy?

8. A professor weighed 15 pennies on a balance and recorded the following masses:

3.112 g	3.109 g	3.059 g
2.467 g	3.079 g	2.518 g
3.129 g	2.545 g	3.050 g
3.053 g	3.054 g	3.072 g
3.081 g	3.131 g	3.064 g

Curious about the results, he looked at the dates on each penny. Two of the light pennies were minted in 1983 and one in 1982. The dates on the 12 heavier pennies ranged from 1970 to 1982. Two of the 12 heavier pennies were minted in 1982.

a. Do you think the Bureau of the Mint changed the way they made pennies? Why or why not?
b. The professor calculated the average mass of the 12 heavy pennies. He expressed this average as 3.0828 ± 0.0482 g. What is wrong with the numbers in this result, and how should the value be expressed?

9. The chemist in Sample Exercise 1.13 did some further experiments. She found that the pipet used to measure the volume of the record cleaner is accurate to ± 0.03 cm^3. The mass measurement is accurate to ± 0.002 g. Are these measurements sufficiently precise for the chemist to distinguish between isopropyl alcohol and ethanol?

10. How many significant figures are in each of the following?

a. 0.0012	d. 106	g. 2006
b. 437,000	e. 125,904,000	h. 3050
c. 900.0	f. 1.0012	i. 0.001060

11. Express each of the numbers in Exercise 10 in exponential form.

12. Using exponential notation express the number 582,000,000 to

a. one significant figure
b. two significant figures
c. three significant figures
d. five significant figures
e. seven significant figures

13. Perform the indicated mathematical operations and express each result to the correct number of significant figures.

a. 3.894×2.16
b. $2.96 + 8.1 + 5.0214$
c. $485 \div 9.231$
d. 2.46×2
e. $9.146 - 9.137$
f. $2.17 + 4.32 + 401.278 + 21.826$

14. Perform the following mathematical operations and express each result to the correct number of significant figures.

a. $4.184 \times 100.62 \times (25.27 - 24.16)$
b. $\dfrac{8.925 - 8.904}{8.925} \times 100$
(This type of calculation is done many times in calculating a percentage error. Assume this example is such a calculation; thus 100 can be considered to be an exact number.)
c. $(9.04 - 8.23 + 21.954 + 81.0) \div 3.1416$
d. $\dfrac{9.2 \times 100.65}{8.321 + 4.026}$
e. $0.1654 + 2.07 - 2.114$
f. $8.27(4.987 - 4.962)$
g. $\dfrac{9.5 + 4.1 + 2.8 + 3.175}{4}$
(Assume this operation is taking the average of four numbers. Thus 4 in the denominator is exact.)
h. $\dfrac{9.025 - 9.024}{9.025} \times 100$ (100 is exact)

15. Perform the following mathematical operations and express the result to the correct number of significant figures.

a. $6.022 \times 10^{23} \times 1.05 \times 10^{2}$
b. $\dfrac{6.6262 \times 10^{-34} \times 2.998 \times 10^{8}}{2.54 \times 10^{-9}}$
c. $1.285 \times 10^{-2} + 1.24 \times 10^{-3} + 1.879 \times 10^{-1}$
d. $1.285 \times 10^{-2} - 1.24 \times 10^{-3}$
e. $\dfrac{(1.00866 - 1.00728)}{6.02205 \times 10^{23}}$
f. $\dfrac{9.875 \times 10^{2} - 9.795 \times 10^{2}}{9.875 \times 10^{2}} \times 100$ (100 is exact)
g. $\dfrac{9.42 \times 10^{2} + 8.234 \times 10^{2} + 1.625 \times 10^{3}}{3}$ (3 is exact)

16. Assume the following numbers are known to ± 1 in the last digit. Perform the following divisions and estimate the maximum uncertainty in each result. That is, determine the largest and smallest number that could occur for each case. (See Appendix 1.5.)

a. $103 \div 101$
b. $101 \div 99$
c. $99 \div 101$

Considering your calculated error limits, to how many significant figures should each result be expressed? Do any of these constitute an exception to the rule for division listed in Section 1.4? Rationalize any discrepancies.

Many times errors are expressed in terms of percentage. The percent error is the absolute value of the difference of the true value and the experimental value, divided by the true value, and multiplied by 100.

$$\text{percent error} = \frac{|\text{true value} - \text{experimental value}|}{\text{true value}} \times 100$$

17. Calculate the percent error for the following measurements:

a. The density of an aluminum block determined in an experiment was 2.64 g/cm^3. (True value 2.70 g/cm^3)

b. The experimental determination of iron in iron ore was 16.48%. (True value 16.12%)
c. A balance measured the mass of a 1.000 g standard as 0.9981 g.

18. A rule of thumb in designing experiments is to avoid using a result that is the small difference between two large measured quantities. In terms of uncertainties in measurement, why is this good advice?

19. Complete the following conversions:
a. 1 km = ________ m = ________ mm = ________ pm
b. 1 g = ________ kg = ________ mg = ________ ng
c. 1 mL = ________ L = ________ dm^3 = ________ cm^3
d. 1 mg = ________ kg = ________ g = ________ μg
= ________ ng = ________ pg = ________ fg
e. 1 s = ________ ms = ________ ns

20. a. How many kilograms are in one teragram?
b. How many nanometers are in 6.50×10^2 terameters?
c. How many kilograms are in 25 femtograms?
d. How many liters are in 8.0 cubic decimeters?
e. How many microliters are in one milliliter?
f. How many picograms are in one microgram?

21. Many atomic dimensions are expressed in angstroms. ($1\ \text{Å} = 1 \times 10^{-8}$ cm). What is the angstrom equal to in the SI units nm and pm?

22. Perform the following unit conversions:
a. In the 1987–88 NBA season the long and the short of it ranged from 7 ft 6 in tall Manute Bol to 5 ft 2 in tall Tyrone Bogues. What are these basketball players' heights in centimeters and meters? (These heights are $\pm\frac{1}{2}$ in. Use this to calculate the uncertainty of the heights in cm. Use this result to determine the number of significant figures to use in your answers.)
b. The circumference of the earth is 25,000 mi at the equator. What is the circumference in kilometers? in meters?
c. A rectangular solid measures 1.0 m by 5.6 cm by 2.1 dm. Express its volume in cubic meters, liters, cubic inches, and cubic feet.

23. According to *The Sporting News,* the fastest recorded fastball was thrown by Nolan Ryan and had a velocity of 100.8 mi/h.
a. Calculate the velocity in meters/second.
b. How long would it take for this pitch to travel the 60 ft 6 in from the mound to home plate?

24. In Shakespeare's *Richard III,* the First Murderer says:

"Take that, and that! [*Stabs Clarence*]
If that is not enough, I'll drown you in the malmsey butt within!"

Given that 1 butt = 126 gal, in how many liters of malmsey (a foul brew similar to mead) was the unfortunate Clarence about to be drowned?

25. Science fiction often uses nautical analogies to describe space travel. If the starship U.S.S. Enterprise is traveling at warp factor 1.71, what is its speed in knots? (Warp 1.71 = 5.00 times the speed of light; speed of light = 3.00×10^8 m/s; 1 knot = 2000 yd/h, exactly.)

26. Convert each of the following expressions to appropriate SI units:
a. an 18-gal gas tank
b. an automobile engine with a displacement of $4.00 \times 10^2\ \text{in}^3$
c. a $\frac{1}{4}$-in (0.25) diameter tube
d. a 14,110-ft elevation above sea level, the height of Pikes Peak in Colorado
e. a 198-lb man
f. a speed of 45 mi/h

27. Chemists often give the composition of a mixture as mass per unit volume. For example, if 1.5 g of table sugar is dissolved in 5.0 mL of water, we say there is 1.5 g/5.0 mL = 0.30 g/mL of sugar in the solution. Express this concentration in each of the following units.
a. mg/mL
b. kg/m^3
c. μg/mL
d. ng/mL
e. μg/μL

28. Use the following exact conversion factors to do the stated conversions:

6 ft = 1 fathom
100 fathoms = 1 cable length
10 cable lengths = 1 nautical mile
3 nautical miles = 1 league

a. 928 miles = ________ leagues = ________ km
b. 1.0 cable length = ________ km = ________ cm
c. The *Green Monster,* as the left field wall in Boston's Fenway Park is called, is 315 ft from home plate down the left field line and 27 ft high. What are these distances in meters? How far from home plate is the Green Monster in cable lengths? How high is it in fathoms?

29. Use the following exact conversion factors to perform the stated calculations.

$5\frac{1}{2}$ yards = 1 rod
40 rods = 1 furlong
8 furlongs = 1 mile

a. The Kentucky Derby race is 1.25 miles. How long is the race in rods, furlongs, meters, and kilometers?
b. A marathon race is 26 miles, 385 yards. What is this distance in rods, furlongs, meters, and kilometers?

30. Although the preferred SI unit of area is the m^2, land is often measured in the metric system in hectares (ha). One hectare is

equal to 10,000 m^2. In the English system of measure, land is often measured in acres (1 acre = 160 rod^2). Use these exact conversions and those given in Exercise 29 to calculate:

a. 1 ha = ________ km^2
b. the area of a 5.5 acre plot of land in hectares, m^2, and km^2
c. A lot with dimensions 120 ft by 75 ft is to be sold for \$6500. What is the price per acre? What is the price per hectare?
d. A plot of land is 27 ft wide by 1.4 furlongs long. What is the area in square feet, square rods, acres, m^2, and hectares?

31. Precious metals and gems are measured in troy weights in the English system:

24 grains = 1 pennyweight (exact)
20 pennyweight = 1 troy ounce (exact)
12 troy ounces = 1 troy pound (exact)
1 grain = 0.0648 grams
1 carat = 0.200 grams

a. The most common English unit of mass is the pound avoirdupois (the pound discussed in Chapter 1). What is one troy pound in kilograms and in pounds?
b. What is the mass of a troy ounce of gold in grams and carats?
c. The density of gold is 19.3 g/cm^3. What is the volume of a troy pound of gold?

32. Apothecaries (druggists) use the following set of measures in the English system:

20 grains ap = 1 scruple (exact)
3 scruples = 1 dram ap (exact)
8 dram ap = 1 oz ap (exact)
1 dram ap = 3.888 g

a. Is an apothocary grain the same as a troy grain? (See Exercise 31.)
b. 1 oz ap = ________ oz troy
c. An aspirin tablet contains 5.00×10^2 mg of active ingredient. How many gr ap of active ingredient does it contain? How many scruples?
d. What is the mass of 1 scruple in grams?

33. The long ton is equal to 2240 lb exactly. The metric tonne is equal to 1000 kg exactly. What is the mass of a long ton in kilograms and in metric tonnes?

34. The speed of light is 3.00×10^8 m/s. How far will a beam of light travel in 1.00 ns?

35. On the distant planet of Majipoor, the smallest monetary unit is the weight. The monetary scale is:

100 weights = 1 crown
20 crowns = 1 royal

a. A horse costs 85.0 crowns. How much does the horse cost in royals and weights? How many horses can be purchased for 50 royals? How much of the 50 royals would be left after purchasing the horses?
b. Haigus hides cost 13 crowns per bundle. A bundle contains 25 haigus hides. What is the cost of a single haigus hide in weights? What is the cost of 288 bundles of haigus hides in royals?

Temperature

36. Ethylene glycol is the main component in automobile antifreeze. To monitor the temperature of an auto cooling system, you intend to use a meter that reads from 0 to 100. You devise a new temperature scale based on the approximate melting and boiling points of a typical antifreeze solution (−45°C and 115°C). You wish these points to correspond to 0°A and 100°A, respectively.

a. Derive an expression for converting between °A and °C.
b. Derive an expression for converting between °F and °A.
c. At what temperature would your thermometer and a Celsius thermometer give the same numerical reading?
d. Your thermometer reads 86°A. What is the temperature in °C and °F?
e. What is a temperature of 45°C in °A?

37. A person has a temperature of 102.5°F. What is this temperature on the Celsius scale? on the Kelvin scale?

38. Many chemical quantities are specified as being measured at 25°C. What is this temperature on the Fahrenheit scale?

39. Helium boils at about 4 K. What is this temperature in °F?

40. A thermometer gives a reading of 20.6°C ± 0.1°C. What is the temperature in °F? What is the uncertainty?

41. A thermometer gives a reading of 96.1°F ±0.2°F. What is the temperature in °C? What is the uncertainty?

Density

42. The density of aluminum is 2.70 g/cm^3. Express this value in units of kg/m^3 and lb/ft^3.

43. The diameter of a hydrogen nucleus is 1.0×10^{-3} pm and its mass is 1.67×10^{-24} g. What is the density of the nucleus in grams per cubic centimeter?

44. Hydrogen gas (H_2) at 0°C and 1 atmosphere pressure has a density of 0.045 g/L. What is the density in g/cm^3? kg/m^3?

45. On October 21, 1982, the Bureau of the Mint changed the composition of pennies (see Exercise 8). Instead of an alloy of 95% Cu and 5% Zn by mass, a core of 99.2% Zn and 0.8% Cu with a thin shell of copper was adopted. The overall

composition of the new penny was 97.6% Zn and 2.4% Cu by mass. Does this account for the difference in mass among the pennies in Exercise 8, assuming that each type of penny is the same size? (Density of Cu = 8.96 g/cm^3; density of Zn = 7.14 g/cm^3.)

46. Diamonds are measured in carats, and 1 carat = 0.200 g. The density of diamond is 3.51 g/cm^3. What is the volume of a 5.0-carat diamond?

47. In each of the following pairs, which has the greater mass? (See Table 1.5.)
a. 1.0 kg of feathers or 1.0 kg of lead
b. 1.0 mL of mercury or 1.0 mL of water
c. 19.32 mL of water or 1.00 mL of gold
d. 0.1 mL of copper or 1.0 L of benzene

48. In each of the following pairs, which has the greater volume?
a. 1.0 kg of feathers or 1.0 kg lead
b. 100 g of gold or 100 g of water
c. 1.0 L of copper or 1.0 L of mercury

49. The mass and volume of an object were measured and found to be 16.50 g ±0.02 g and 15.5 cm^3 ±0.2 cm^3, respectively. Calculate the density with appropriate error limits. (See Appendix 1.5.)

50. The density of an irregularly shaped object was determined as follows. The mass of the object was found to be 28.90 g ±0.03 g. A graduated cylinder was partially filled with water. The reading of the level of the water was 6.4 cm^3 ±0.1 cm^3. The object was dropped in the cylinder and the level of the water rose to 9.8 cm^3 ±0.1 cm^3. What is the density of the object with appropriate error limits? (See Appendix 1.5.)

51. A rectangular block has dimensions* 29 mm × 35 mm × 100. mm. The mass of the block is 615.0 g ±0.2 g. What are the volume and density of the block? What is the uncertainty in the volume and density? Assume the dimensions of the block are known to ±1 mm.

Classification and Separation of Matter

52. What are some of the differences between a solid, a liquid, and a gas?

53. What is the difference between homogeneous and heterogeneous matter? Classify each of the following as homogeneous or heterogeneous:
a. soil
b. the atmosphere
c. a carbonated soft drink
d. gasoline
e. gold
f. a solution of ethanol and water.

54. Classify each of the following as a mixture or a pure substance:

a. water	d. iron	g. wine
b. blood	e. brass	h. leather
c. the oceans	f. uranium	i. table salt (NaCl)

55. Of the pure substances in Exercise 54, which are elements, and which are compounds?

56. Distinguish between physical changes and chemical changes.

57. Why is the separation of mixtures into pure or relatively pure substances so important when performing a chemical analysis?

Additional Exercises

58. The sun is estimated to have a mass of 2 × 10^{36} kg. Assuming it to be a sphere of average radius 6.96 × 10^5 km, calculate the average density of the sun in units of
a. kg/km^3 b. kg/m^3 c. g/cm^3

59. In the opening scenes of the movie *Raiders of the Lost Ark,* Indiana Jones tries to remove a gold idol from a booby-trapped pedestal. He replaces the idol with a bag of sand of approximately equal volume. (Density of gold = 19.32 g/cm^3; density of sand ≈ 3 g/cm^3.)
a. Did he have a reasonable chance of not activating the mass-sensitive booby trap?
b. In a later scene he and an unscrupulous guide play catch with the idol. Assume that the volume of the idol is about 1.0 L. If it were solid gold, what mass would the idol have? Is playing catch with it plausible?

60. An automobile has a gas mileage of 36.1 mi/gal. How many kilometers can it travel on 1.0 L of gasoline?

61. According to the *Official Rules of Baseball,* a baseball must have a circumference not more than 9.25 in or less than 9.00 in and a mass not more than 5.25 oz or less than 5.00 oz. What range of densities can a baseball be expected to have? How would you express this range as a single number with an accompanying uncertainty limit?

62. What is wrong with the following statement?
''The results of the experiment did not agree with the theory. Something must be wrong with the experiment.''

63. If the price of gasoline at a service station is $1.09/gal, what is the cost per liter?

64. Why is it incorrect to say that the results of a measurement were accurate but not precise?

*Recall from Section 1.4 that a decimal point in a number makes the trailing zeros significant. Thus 100. has 3 significant digits.

65. The world record for the hundred yard dash is 9.1 s. What is the average velocity in units of ft/s, mi/h, m/s, and km/h? At this speed how long would it take to run 1.00×10^2 m?

66. If the amount of mercury in a polluted lake is 0.4 μg Hg/mL, what is the total mass in kg of mercury in the lake? (The lake has a surface area of 100 mi^2 and an average depth of 20 ft.)

67. What is 0 K in °F?

68. A 16-oz container of baking soda cost 65¢. What is the cost of the baking soda per gram and per kilogram?

69. Two spherical objects have the same mass. One floats on water, the other sinks. Which object has the greater diameter? Explain your answer.

70. The density of osmium (the densest metal) is 22.57 g/cm^3. What is the mass of a block of osmium with dimensions 5.00 cm $\times$ 4.00 cm $\times$ 2.50 cm? What volume would be occupied by 1.00 kg of osmium?

71. One metal object is a cube with edges of 3.00 cm and a mass of 140.4 g. A second metal object is a sphere with radius 1.42 cm and a mass of 61.6 g. Are these objects made of the same or different metals? Assume the calculated densities are accurate to $\pm$1.00%.

CHAPTER 2

Atoms, Molecules, and Ions

Where does one start to learn chemistry? Clearly we must learn some essential vocabulary and something about the origins of the science before we can proceed very far. Thus, whereas Chapter 1 provided background on the fundamental ideas and procedures of science in general, Chapter 2 provides the specific chemical background necessary for understanding the material in the next few chapters of this text. The coverage of these topics will be necessarily brief at this point. However we will develop these ideas more fully as it becomes appropriate.

One major goal of this chapter is to present the systems for naming common chemical compounds. These systems will provide you with the vocabulary necessary to understand this book and to pursue your laboratory studies.

Because chemistry is concerned first and foremost with chemical changes, we want to proceed as quickly as possible to a study of chemical reactions. Therefore, chemical reactions will be covered in detail in Chapters 3 and 4, chapters that should help you understand the work you will be doing in the laboratory. However, before we can discuss reactions, we must consider some fundamental ideas about atoms and how they combine.

CONTENTS

< A polarized light micrograph (×32) of crystallized sulfur. Vast underground deposits of elemental sulfur are found in Texas and Louisiana.

The Alchemists, an oil on slate painting by Jan van der Straet in 1570.

2.1 The Early History of Chemistry

Purpose

- To give a brief account of early chemical discoveries.

Chemistry has been important since ancient times. The processing of natural ores to produce metals for ornaments and weapons and the use of embalming fluids are two examples of chemical phenomena that were utilized prior to 1000 B.C.

The Greeks were the first to try to explain why chemical changes occur. By about 400 B.C. they had proposed that all matter was composed of four fundamental substances: fire, earth, water, and air. The Greeks also considered the question of whether matter is continuous, and thus infinitely divisible into smaller pieces, or composed of small indivisible particles. One supporter of the latter position was Democritus, who used the term *atomos* (which later became *atoms*) to describe these ultimate particles. However, because the Greeks had no experiments to test their ideas, no definitive conclusion about the divisibility of matter was reached.

The next 2000 years of chemical history were dominated by a pseudoscience called alchemy. Alchemists were often mystics and fakes who were obsessed with the idea of turning cheap metals into gold. However, this period also saw important discoveries: elements such as mercury, sulfur, and antimony were discovered, and alchemists learned how to prepare the mineral acids.

The foundations for modern chemistry were laid in the sixteenth century with the development of systematic metallurgy (extraction of metals from ores) by a German, Georg Bauer, and the medicinal application of minerals by a Swiss alchemist called Paracelsus.

The first person to perform truly quantitative physical experiments was Robert Boyle (1627–1691), who carefully measured the relationship between the pressure and volume of gases. When Boyle published his book, *The Sceptical Chemist,* in 1661, the quantitative sciences of physics and chemistry were born. In addition to his results on the quantitative behavior of gases, Boyle's other major contribution to chemistry consisted of his ideas about the chemical elements. Boyle held no preconceived notion about the number of elements. In his view a substance was an element unless it could be broken down into two or more simpler substances. As Boyle's experimental definition of an element became generally accepted, the list of known elements began to grow, and the Greek system of four elements finally died. Although Boyle was an excellent scientist, he was not always right. For example, he clung to the alchemist's views that metals were not true elements and that a way would eventually be found to change one metal to another.

The phenomenon of combustion evoked intense interest in the seventeenth and eighteenth centuries. The German chemist Georg Stahl (1660–1734) suggested that a substance he called phlogiston flowed out of the burning material. Stahl postulated that a substance burning in a closed container eventually stopped burning, because the air in the container became saturated with phlogiston. Oxygen gas, discovered by Joseph Priestley (1733–1804), an English clergyman and scientist (Fig. 2.1), was found to support vigorous combustion and was thus supposed to be low in phlogiston. In fact, oxygen was originally called "dephlogisticated air."

Figure 2.1

Joseph Priestley was born in England on March 13, 1733, and showed a great talent for science and languages from an early age. Priestley performed many important scientific experiments, among them the discovery that the gas produced by the fermentation of grain (later identified as carbon dioxide) could be dissolved in water to produce the pleasant drink called *seltzer*. Also, as a result of meeting Benjamin Franklin in London in 1766, Priestley became interested in electricity and was the first to observe that graphite was an electrical conductor. However, Priestley's greatest discovery occurred in 1774 when he isolated oxygen by heating mercuric oxide.

Because of his nonconformist political views (he supported both the American and French revolutions), he was forced to leave England (a mob burned his house in Birmingham in 1791). He spent his last decade peacefully in the United States, and died in 1804.

2.2 Fundamental Chemical Laws

Purpose

- To describe and illustrate the laws of conservation of mass, definite proportion, and multiple proportions.

By the late eighteenth century, combustion had been studied extensively; the gases carbon dioxide, nitrogen, hydrogen, and oxygen had been discovered; and the list of elements continued to grow. However, it was Antoine Lavoisier (1743–1794), a French chemist (Fig. 2.2), who finally explained the true nature of combustion, thus clearing the way for the tremendous progress that was made near the end of the eighteenth century. Lavoisier, like Boyle, regarded measurement as the

Figure 2.2

Antoine Lavoisier was born in Paris on August 26, 1743. Although Lavoisier's father wanted his son to follow him into the legal profession, young Lavoisier was fascinated by science. From the beginning of his scientific career, Lavoisier recognized the importance of accurate measurements. His careful weighings showed that mass was conserved in chemical reactions and that combustion involves reaction with oxygen. Also, he wrote the first modern chemistry textbook. It is not surprising that Lavoisier is often called the father of modern chemistry.

To help support his scientific work, Lavoisier invested in a private tax-collecting firm and married the daughter of one of the company executives. His connection to the tax collectors proved fatal, for radical French revolutionaries demanded his execution, which occurred on the guillotine on May 8, 1794.

essential operation of chemistry. His experiments, in which he carefully weighed the reactants and products of various reactions, suggested that *mass was neither created nor destroyed*. Lavoisier's discovery of this **law of conservation of mass** was the basis for the developments in chemistry in the nineteenth century.

Lavoisier's quantitative experiments showed that combustion involved oxygen (which Lavoisier named), not phlogiston. He also discovered that life was supported by a process that also involved oxygen and was similar in many ways to combustion. In 1789 Lavoisier published the first modern chemistry textbook, *Elementary Treatise on Chemistry*, in which he presented a unified picture of the chemical knowledge assembled up to that time. Unfortunately, in the same year the text was published, the French Revolution broke out. Lavoisier, who had been associated with collecting taxes for the government, was executed on the guillotine as an enemy of the people in 1794.

After 1800 chemistry was dominated by scientists who, following Lavoisier's lead, performed careful weighing experiments to study the course of chemical reactions and to determine the composition of various chemical compounds. One of these chemists, a Frenchman, Joseph Proust (1754–1826), showed that *a given compound always contains exactly the same proportion of elements by weight*. For example, Proust found that the substance copper carbonate is always 5.3 parts copper to 4 parts oxygen to 1 part carbon (by mass). The principle of the constant composition of compounds, originally called Proust's law, is now known as the **law of definite proportion.**

Figure 2.3

John Dalton (1766–1844), an Englishman, began teaching at a Quaker school when he was 12. His fascination with science included an intense interest in meteorology (he kept careful daily weather records for 46 years), which led to an interest in the gases of the air and their ultimate components, atoms. Dalton is best known for his atomic theory, in which he postulated that the fundamental differences among atoms are their masses. He was the first to prepare a table of relative atomic weights.

Dalton was a humble man with several apparent handicaps: he was poor; he was not articulate; he was not a skilled experimentalist; and he was color-blind, a terrible problem for a chemist. In spite of these disadvantages, he helped to revolutionize the science of chemistry.

Proust's discovery stimulated John Dalton (1766–1844), an English schoolteacher (Fig. 2.3), to think about atoms. Dalton reasoned that if elements were composed of tiny individual particles, a given compound should always contain the same combination of these atoms. This concept explained why the same relative masses of elements were always found in a given compound.

But Dalton discovered another principle that convinced him even more of the existence of atoms. He noted, for example, that carbon and oxygen formed two different compounds that contained different relative amounts of carbon and oxygen as shown by the following data:

	Weight of oxygen that combines with 1 g of carbon
Compound I	1.33 g
Compound II	2.66 g

Dalton noted that compound II contained twice as much oxygen per gram of carbon as compound I, a fact that could be easily explained in terms of atoms. Compound I might be CO, and compound II might be CO_2.* This principle, which was found to apply to compounds of other elements as well, became known as the **law of multiple proportions:** *when two elements form a series of compounds, the ratios of the masses of the second element that combine with 1 gram of the first element can always be reduced to small whole numbers*.

*Subscripts are used to show the numbers of atoms present. The number 1 is understood and thus is not written. The symbols for the elements and the writing of chemical formulas will be illustrated further in Sections 2.6 and 2.7.

To make sure the significance of this observation is clear, in Sample Exercise 2.1 we will consider data for a series of compounds consisting of nitrogen and oxygen.

Sample Exercise 2.1

The following data were collected for several compounds of nitrogen and oxygen:

	Mass of nitrogen that combines with 1 g of oxygen
Compound I	1.750 g
Compound II	0.8750 g
Compound III	0.4375 g

Show how these data illustrate the law of multiple proportions.

Solution

For the law of multiple proportions to hold, the ratios of the masses of nitrogen combining with 1 gram of oxygen in each pair of compounds should be small whole numbers. We therefore compute the ratios as follows:

$$\frac{\text{I}}{\text{II}} = \frac{1.750}{0.875} = \frac{2}{1}$$

$$\frac{\text{II}}{\text{III}} = \frac{0.875}{0.4375} = \frac{2}{1}$$

$$\frac{\text{I}}{\text{III}} = \frac{1.750}{0.4375} = \frac{4}{1}$$

These results support the law of multiple proportions.

The significance of the data in Sample Exercise 2.1 is that compound I contains twice as much nitrogen (N) per gram of oxygen (O) as does compound II and that compound II contains twice as much nitrogen per gram of oxygen as does compound III. In terms of the numbers of atoms combining, these data can be explained by any of the following formulas.

Compound I	N_2O		NO		N_4O_2
Compound II	NO	or	NO_2	or	N_2O_2
Compound III	NO_2		NO_4		N_2O_4

In fact an infinite number of other possibilities exists. Dalton could not deduce absolute formulas from the available data on relative masses. However, the data on the composition of compounds in terms of the relative masses of the elements supported his hypothesis that each element consisted of a certain type of atom and that compounds were formed from specific combinations of atoms.

2.3 Dalton's Atomic Theory

Purpose

- To describe Dalton's theory of atoms and show the significance of Gay-Lussac's experiments.

These statements are a modern paraphrase of Dalton's ideas.

In 1808 Dalton published *A New System of Chemical Philosophy,* in which he presented his theory of atoms:

1. *Each element is made up of tiny particles called atoms.*
2. *The atoms of a given element are identical; the atoms of different elements are different in some fundamental way or ways.*
3. *Chemical compounds are formed when atoms combine with each other. A given compound always has the same relative numbers and types of atoms.*
4. *Chemical reactions involve reorganization of the atoms—changes in the way they are bound together. The atoms themselves are not changed in a chemical reaction.*

Joseph Louis Gay-Lussac, a French physicist and chemist, was remarkably versatile. Although he is now primarily known for his studies on the combining of volumes of gases, Gay-Lussac was instrumental in the studies of many of the other properties of gases. Some of Gay-Lussac's motivation to learn about gases arose from his passion for ballooning. In fact, he made ascents to heights of over four miles to collect air samples, setting altitude records that stood for about 50 years. Gay-Lussac also was the co-discoverer of boron and the developer of a process for manufacturing sulfuric acid. As chief assayer of the French mint, Gay-Lussac developed many techniques for chemical analysis and invented many types of glassware now used routinely in labs. Gay-Lussac spent his last twenty years as a lawmaker in the French government.

It is instructive to consider Dalton's reasoning on the relative masses of the atoms of the various elements. In Dalton's time water was known to be composed of the elements hydrogen and oxygen, with 8 grams of oxygen present for every 1 gram of hydrogen. If the formula for water were OH, an oxygen atom would have to have 8 times the mass of a hydrogen atom. However, if the formula for water were H_2O (two atoms of hydrogen for every oxygen atom), this would mean that each atom of oxygen is 16 times as heavy as *each* atom of hydrogen (since the ratio of the mass of one oxygen to that of *two* hydrogens is 8 to 1). Because the formula for water was not then known, Dalton could not specify the relative masses of oxygen and hydrogen unambiguously. To solve the problem, Dalton made a fundamental assumption: he decided that nature would be as simple as possible. This assumption led him to conclude that the formula for water should be OH. He thus assigned hydrogen a mass of 1 and oxygen a mass of 8.

Using similar reasoning for other compounds, Dalton prepared the first table of **atomic masses** (called **atomic weights** by chemists, since mass is often determined by comparison to a standard mass—a process called *weighing*). Many of the masses were later proved to be wrong because of Dalton's incorrect assumptions about the formulas of certain compounds, but the construction of a table of masses was an important step forward.

Although not recognized as such for many years, the keys to determining absolute formulas for compounds were provided in the experimental work of the French chemist Joseph Gay-Lussac (1778–1850) and by the hypothesis of an Italian chemist named Amadeo Avogadro (1776–1856). In 1809 Gay-Lussac performed experiments in which he measured (under the same conditions of temperature and pressure) the volumes of gases that reacted with each other. For example, Gay-Lussac found that 2 volumes of hydrogen react with 1 volume of oxygen to form 2 volumes of gaseous water and that 1 volume of hydrogen reacts with 1 volume of chlorine to form 2 volumes of hydrogen chloride. These results are represented schematically in Fig. 2.4.

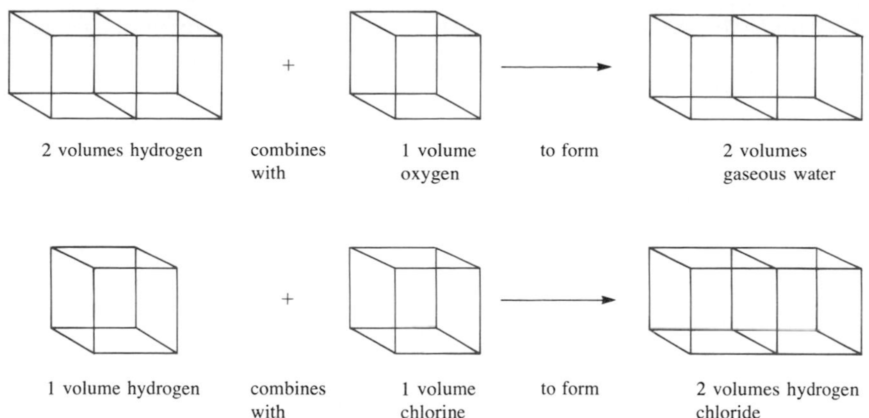

Figure 2.4

A representation of some of Gay-Lussac's experimental results on combining gas volumes.

In 1811 Avogadro interpreted these results by proposing that, *at the same temperature and pressure, equal volumes of different gases contain the same number of particles*. This assumption (called **Avogadro's hypothesis**) makes sense if the distances between the particles in a gas are very great compared to the sizes of the particles. Under these conditions the volume of a gas is determined by the number of molecules present, not by the size of the individual particles.

If Avogadro's hypothesis is correct, Gay-Lussac's result,

2 volumes of hydrogen react with 1 volume of oxygen
$\rightarrow$ 2 volumes of water vapor

can be expressed as follows:

2 molecules* of hydrogen react with 1 molecule of oxygen
$\rightarrow$ 2 molecules of water

These observations can best be explained by assuming that gaseous hydrogen, oxygen, and chlorine are all composed of diatomic (two-atom) molecules: H_2, O_2, and Cl_2, respectively. Gay-Lussac's results can then be represented as shown in Fig. 2.5. (Note that this reasoning suggests that the formula for water is H_2O, not OH as Dalton believed.)

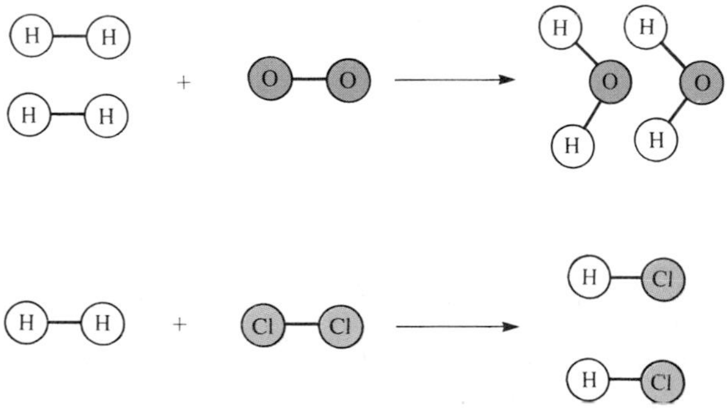

Figure 2.5

A representation of combining gases at the molecular level. The circles represent atoms in the molecules.

*A molecule is a collection of atoms (see Section 2.6).

Figure 2.6

J. J. Thomson (1856–1940) was an English physicist at Cambridge University. He received the Nobel Prize in physics in 1906.

Unfortunately, Avogadro's interpretations were not accepted by most chemists and a half-century of confusion followed, in which many different assumptions were made about formulas and atomic masses.

During the nineteenth century painstaking measurements were made of the masses of various elements that combined to form compounds. From these experiments a list of relative atomic masses could be determined. One of the chemists involved in adding to this list was a Swede named Jöns Jakob Berzelius (1779–1848), who discovered the elements cerium, selenium, silicon, and thorium and developed the modern symbols for the elements used in writing the formulas of compounds.

2.4 Early Experiments to Characterize the Atom

Purpose

- To summarize the experiments that characterized the structure of the atom.

Dalton's atomic theory caused chemistry to become more systematic and more sensible. The concept of atoms was clearly a good idea, and scientists became very interested in the structure of the atom.

The Electron

The first important experiments that led to an understanding of the composition of the atom were done by the English physicist J. J. Thomson (Fig. 2.6), who studied electrical discharges in partially evacuated tubes called *cathode-ray tubes* (Fig. 2.7) during the period from 1898 to 1903. Thomson found that, when high voltage was applied to the tube, a "ray" he called a **cathode ray** (because it emanated from the negative electrode, or cathode) was produced. Because this ray was produced at the negative electrode and was repelled by the negative pole of an applied electric field (see Fig. 2.8), Thomson postulated that the ray was a stream of negatively charged particles, which he called **electrons.** By experiments in which he measured the deflection of the beam of electrons in a magnetic field, Thomson determined the *charge-to-mass ratio* of an electron:

$$\frac{e}{m} = -1.76 \times 10^8 \text{ C/g}$$

where e represents the charge on the electron in coulombs and m represents the electron mass in grams.

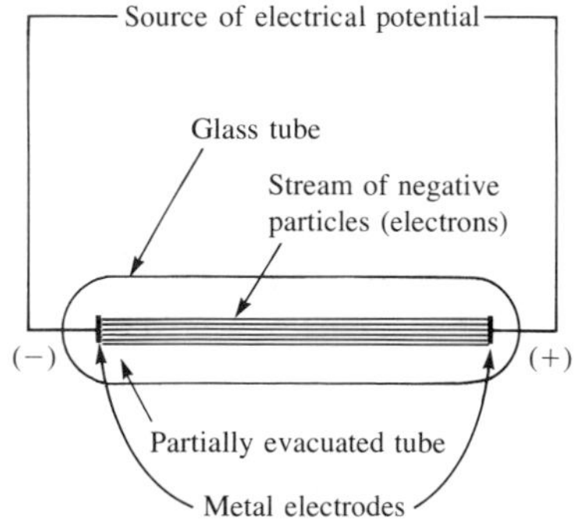

Figure 2.7

Schematic of a cathode-ray tube. The fast-moving electrons excite the gas in the tube, causing a glow between the electrodes.

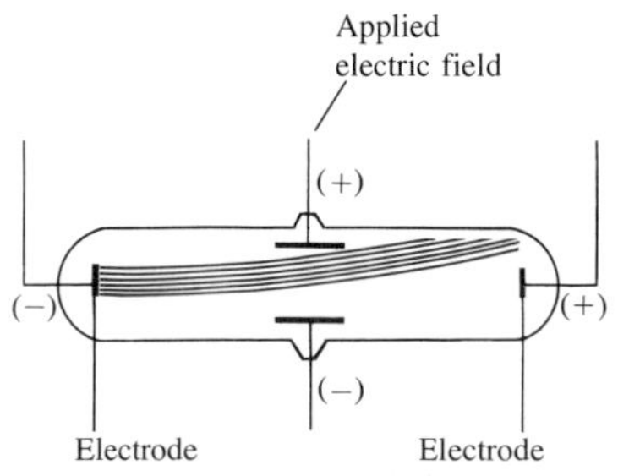

Figure 2.8

Deflection of cathode rays by an applied electric field.

One of Thomson's primary goals in his cathode-ray tube experiments was to gain an understanding of the structure of the atom. He reasoned that, since electrons could be produced from electrodes made of various types of metals, *all* atoms must contain electrons. Since atoms were known to be electrically neutral, Thomson further assumed that atoms also must contain some positive charge. Thomson postulated that an atom consisted of a diffuse cloud of positive charge with the negative electrons embedded randomly in it. This model, shown in Fig. 2.9, is often called the *plum pudding model* because the electrons are like raisins dispersed in a pudding (the positive charge cloud), as in plum pudding, a favorite English dessert.*

In 1909 Robert Millikan (1868–1953), working at the University of Chicago, performed very clever experiments involving charged oil drops. These experiments allowed him to determine the magnitude of the electron charge (see Fig. 2.10). With this value and the charge-to-mass ratio determined by Thomson, Millikan was able to calculate the mass of the electron as 9.11×10^{-28} gram.

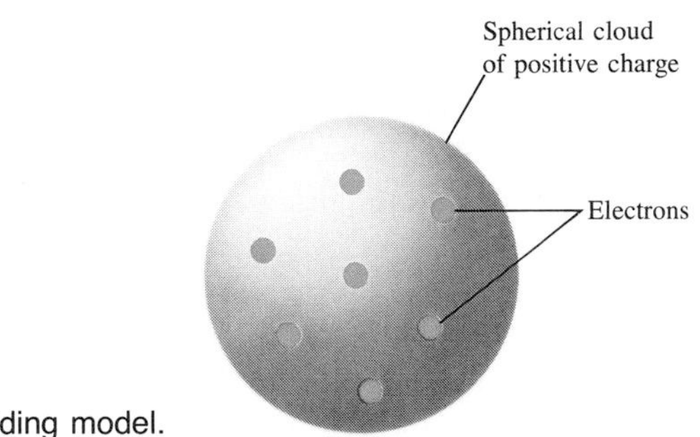

Figure 2.9

Thomson's plum pudding model.

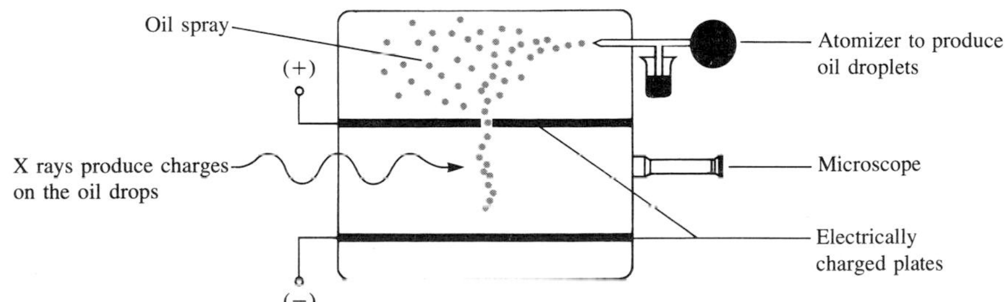

Figure 2.10

A schematic representation of the apparatus Millikan used to determine the charge on the electron. The fall of charged oil droplets due to gravity can be halted by adjusting the voltage across the two plates. The voltage and the mass of an oil drop can then be used to calculate the charge on the oil drop. Millikan's experiments showed that the charge on an oil drop is always a whole-number multiple of the electron charge.

Radioactivity

In the late nineteenth century, it was discovered that certain elements produce high-energy radiation. For example, in 1896 the French scientist Henri Becquerel found that a piece of a mineral containing uranium could produce its image on a photographic plate in the absence of light. He attributed this phenomenon to a spontaneous emission of radiation by the uranium, which he called **radioactivity.** Studies in the early twentieth century demonstrated three types of radioactive emission: gamma (γ) rays, beta (β) particles, and alpha (α) particles. A γ ray is high-energy "light"; a β particle is a high-speed electron; and an α particle has a 2+ charge, that is, a charge twice that of the electron and with the opposite sign. The mass of an α particle is 7300 times that of the electron. More modes of radioactivity are now

*Although J. J. Thomson is generally given credit for this model, the idea was apparently first suggested by the English mathematician and physicist William Thomson (better known as Lord Kelvin and not related to J. J. Thomson).

Chemical Impact

Berzelius, Selenium, and Silicon

Jöns Jakob Berzelius (see figure) was probably the best experimental chemist of his generation and, given the crudeness of his laboratory equipment, maybe the best of all time. Unlike Lavoisier, who could afford to buy the best laboratory equipment available, Berzelius worked with minimal equipment in very plain surroundings. One of Berzelius's students described the Swedish chemist's workplace: "The laboratory consisted of two ordinary rooms with the very simplest arrangements; there were neither furnaces nor hoods, neither water system nor gas. Against the walls stood some closets with the chemicals, in the middle the mercury trough and the blast lamp table. Beside this was the sink consisting of a stone water holder with a stopcock and a pot standing under it. [Next door in the kitchen] stood a small heating furnace."

In these simple facilities Berzelius performed more than 2000 experiments over a ten-year period to determine accurate atomic masses for the 50 elements then known. His success can be seen from the data in the table at right. These remarkably accurate values attest to his experimental skills and patience.

Besides his table of atomic weights, Berzelius made many other major contributions to chemistry. The most important of these was the invention of a simple set of symbols for the elements along with a system for writing the formulas of compounds to replace the awkward symbolic representations of the alchemists (see table on page 47). Although some chemists, including Dalton, objected to the new system, it was gradually adopted and forms the basis of the system we use today.

In addition to these accomplishments, Berzelius also discovered the elements cerium, thorium, selenium, and silicon. Of these elements, selenium and silicon are particularly important in today's world. Berzelius discovered selenium in 1817 in connection with his studies of sulfuric acid. For years selenium's toxicity has been known, but only recently have we become aware that it may have a positive effect on human health. Studies have shown that trace amounts of selenium in the diet may protect people from heart disease and cancer. One study based on data from 27 countries showed an inverse relationship between the cancer death rate and the selenium content of soil in a particular region (low cancer death rate in areas with high selenium content). Another research paper reported an inverse relationship between the selenium content of the

Jöns Jakob Berzelius (1779–1848).

Comparison of Several of Berzelius's Atomic Weights with the Modern Values

Element	Atomic weight	
	Berzelius's value	Current value
Chlorine	35.41	35.45
Copper	63.00	63.55
Hydrogen	1.00	1.01
Lead	207.12	207.2
Nitrogen	14.05	14.01
Oxygen	16.00	16.00
Potassium	39.19	39.10
Silver	108.12	107.87
Sulfur	32.18	32.06

Chemical Impact

The Alchemists' Symbols for Some Common Elements and Compounds

Substance	Alchemists' symbol
Silver	☽
Lead	♄
Tin	♃
Platinum	☽☉
Sulfuric acid	+⊕
Alcohol	🜅
Sea salt	🜔

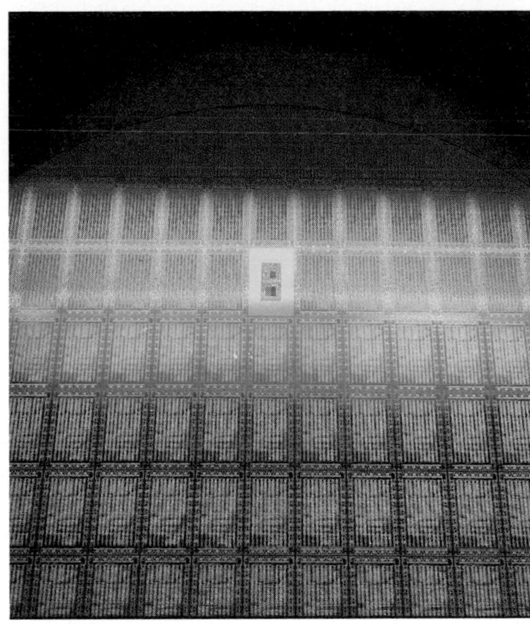

A silicon chip.

blood and the incidence of breast cancer in women. Selenium is also found in the heart muscle and may play an important role in proper heart function. Because of these and other studies, selenium's reputation has improved, and many scientists are now studying its function in the human body.

Silicon is the second most abundant element in the earth's crust, exceeded only by oxygen. As we will see in Chapter 10, compounds involving silicon bonded to oxygen make up most of the earth's sand, rock, and soil. Berzelius prepared silicon in its pure form in 1824 by heating silicon tetrafluoride (SiF_4) with potassium metal. Today, silicon forms the basis for the modern microelectronics industry centered near San Francisco in a place that has come to be known as "Silicon Valley." The technology of the silicon chip (see figure) with its printed circuits has transformed computers from room-sized monsters with thousands of unreliable vacuum tubes to desktop and notebook-sized units with trouble free "solid-state" circuitry*.

*For further reading see Bernard Jaffe, *Crucibles: The Story of Chemistry* (Premiere Book, 1957).

known, and we will discuss them in Chapter 21. Here we will consider only α particles because they were used in some crucial early experiments.

The Nuclear Atom

In 1911 Ernest Rutherford (Fig. 2.11), who performed many of the pioneering experiments to explore radioactivity, carried out an experiment to test Thomson's plum pudding model. The experiment involved directing α particles at a thin sheet

Figure 2.11

Ernest Rutherford (1871–1937) was born on a farm in New Zealand. In 1895 he placed second in a scholarship competition to attend Cambridge University, but was awarded the scholarship when the winner decided to stay home and get married. As a scientist in England, Rutherford did much of the early work on characterizing radioactivity. He named the α and β particles and the γ ray and coined the term *half-life* to describe an important attribute of radioactive elements. His experiments on the behavior of α particles striking thin metal foils led him to postulate the nuclear atom. He also invented the name *proton* for the nucleus of the hydrogen atom. He received the Nobel Prize in chemistry in 1908.

of metal foil, as illustrated in Fig. 2.12. Rutherford reasoned that, if Thomson's model were accurate, the massive α particles should crash through the thin foil like cannonballs through gauze, as shown in Fig. 2.13(a). He expected the α particles to travel through the foil with, at the most, very minor deflections in their paths. The results of the experiment were very different from those Rutherford anticipated. Although most of the α particles passed straight through, many of the particles were deflected at large angles, as shown in Fig. 2.13(b), and some were reflected, never hitting the detector. This outcome was a great surprise to Rutherford. (He wrote that this result was comparable to shooting a howitzer at a piece of paper and having the shell reflected back.)

Rutherford knew from these results that the plum pudding model for the atom could not be correct. The large deflections of the α particles could only be caused by a center of concentrated positive charge, as illustrated in Fig. 2.13(b). Most of the α particles pass directly through the foil because the atom is mostly open space. The deflected α particles are those that had a "close encounter" with the positive center

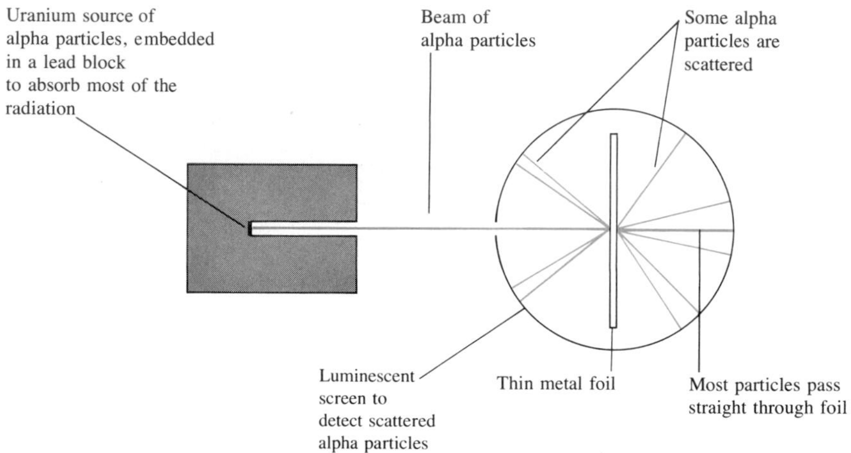

Figure 2.12

Rutherford's experiment on α-particle bombardment of metal foil.

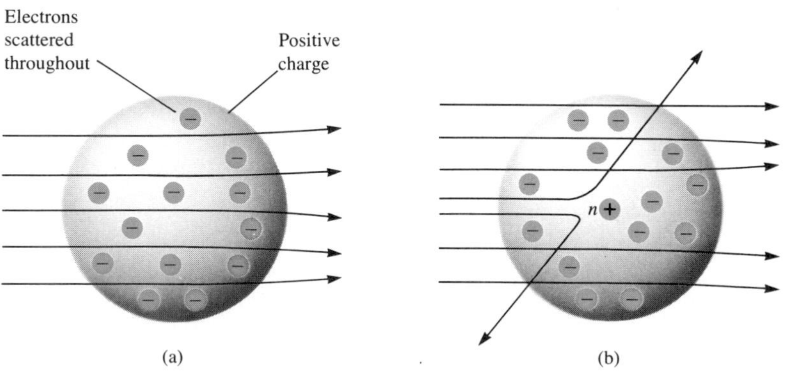

Figure 2.13

(a) The expected results of the metal foil experiment if Thomson's model were correct. (b) Actual results.

of the atom, and the few reflected α particles are those that made a ''direct hit'' on the much more massive positive center.

In Rutherford's mind these results could only be explained in terms of a **nuclear atom**—an atom with a dense center of positive charge (the **nucleus**) with electrons moving around at a distance that is large relative to the nuclear radius.

2.5 The Modern View of Atomic Structure: An Introduction

Purpose

- To describe features of subatomic particles.
- To explain the use of the symbol $^{A}_{Z}X$ to describe a given atom.

In the years since Thomson and Rutherford, a great deal has been learned about atomic structure. Because much of this material will be covered in detail in later chapters, only an introduction will be given here. The simplest view of the atom is that it consists of a tiny nucleus (with a diameter of about 10^{-13} cm) and electrons that move about the nucleus at an average distance of about 10^{-8} cm away from it (see Fig. 2.14).

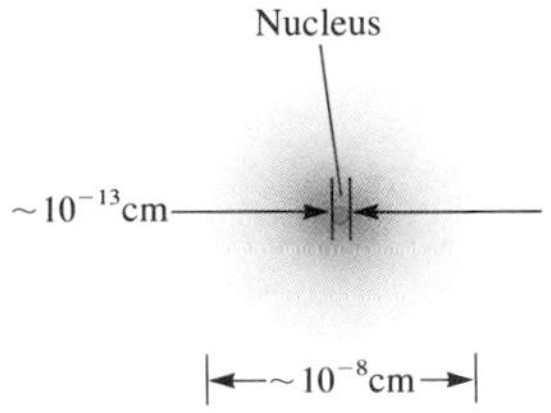

As we will see later, the chemistry of an atom mainly results from its electrons. For this reason chemists can be satisfied with a relatively crude nuclear model. The nucleus is assumed to contain **protons,** which have a positive charge equal in magnitude to the electron's negative charge, and **neutrons,** which have virtually the same mass as a proton but no charge. The masses and charges of the electron, proton, and neutron are shown in Table 2.1.

Two striking things about the nucleus are its small size compared to the overall size of the atom and its extremely high density. The tiny nucleus accounts for almost all of the atom's mass. Its great density is dramatically demonstrated by the fact that a piece of nuclear material about the size of a pea would have a mass of 250 million tons!

An important question to consider at this point is ''*If all atoms are composed of these same components, why do different atoms have different chemical properties?*'' The answer to this question lies in the number and the arrangement of the

Figure 2.14

A nuclear atom viewed in cross section.

The Mass and Charge of the Electron, Proton, and Neutron

Particle	Mass	Charge*
Electron	9.11×10^{-28} g	1−
Proton	1.67×10^{-24} g	1+
Neutron	1.67×10^{-24} g	none

*The magnitude of the charge of the electron and the proton is 1.60×10^{-19} C.

Table 2.1

electrons. The electrons comprise most of the atomic volume and thus are the parts that "intermingle" when atoms combine to form molecules. Therefore, the number of electrons possessed by a given atom greatly affects its ability to interact with other atoms. As a result, the atoms of different elements, which have different numbers of protons and electrons, show different chemical behavior.

The *chemistry* of an atom arises from its electrons.

A sodium atom has 11 protons in its nucleus. Since atoms have no net charge, the number of electrons must equal the number of protons. Therefore, a sodium atom has 11 electrons moving around its nucleus. It is *always* true that a sodium atom has 11 protons and 11 electrons. However, each sodium atom also has neutrons in its nucleus, and different types of sodium atoms exist that have different numbers of neutrons. For example, consider the sodium atoms represented in Fig. 2.15. These two atoms are **isotopes,** or *atoms with the same number of protons but different numbers of neutrons*. Note that the symbol for one particular type of sodium atom is written

Mass number → $^{A}_{Z}X$ ← Element
Atomic number ↗

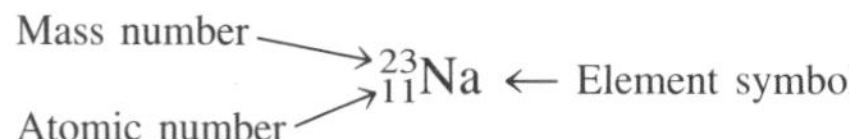

where the **atomic number** Z (number of protons) is written as a subscript and the **mass number** A (the total number of protons and neutrons) is written as a superscript. (The particular atom represented here is called "sodium twenty-three." It has 11 electrons, 11 protons, and 12 neutrons.) Because the chemistry of an atom is due to its electrons, isotopes show almost identical chemical properties. In nature elements are usually found as a mixture of isotopes.

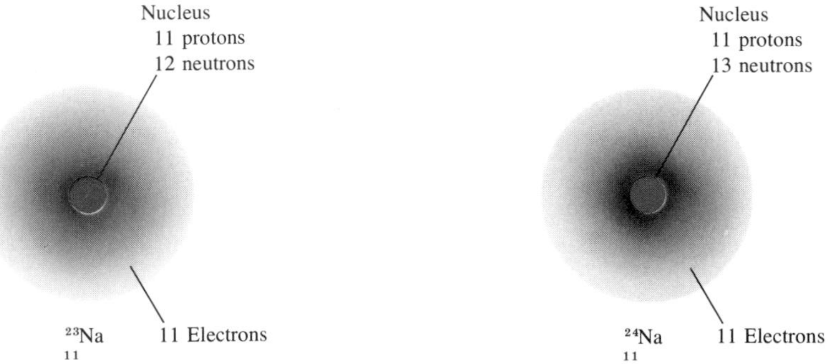

Figure 2.15

Two isotopes of sodium. Both have eleven protons and electrons, but they differ in the number of neutrons in their nuclei.

Sample Exercise 2.2

Write the symbol for the atom that has an atomic number of 9 and a mass number of 19. How many electrons and how many neutrons does this atom have?

Solution

The atomic number 9 means the atom has 9 protons. This element is called fluorine, symbolized by F. The atom is represented as

$$^{19}_{9}F$$

and is called "fluorine nineteen." Since the atom has 9 protons, it must also have 9 electrons to achieve electrical neutrality. The mass number gives the total number of protons and neutrons, which means that this atom has 10 neutrons.

2.6 Molecules and Ions

Purpose

- To introduce basic ideas of bonding in molecules.
- To show various ways of representing molecules.

From a chemist's viewpoint the most interesting characteristic of an atom is its ability to combine with other atoms to form compounds. It was John Dalton who first recognized that chemical compounds were collections of atoms, but he could not determine the structure of atoms or their means for binding to each other. During the twentieth century we have learned that atoms have electrons and that these electrons participate in bonding one atom to another. We will discuss bonding thoroughly in Chapters 8 and 9; here we will introduce some simple bonding ideas that will be useful in the next few chapters.

The forces that hold atoms together in compounds are called **chemical bonds.** One way that atoms can form bonds is by *sharing electrons*. These bonds are called **covalent bonds,** and the resulting collection of atoms is called a **molecule.** Molecules can be represented in several different ways. The simplest method is the **chemical formula,** in which the symbols for the elements are used to indicate the types of atoms present and subscripts are used to indicate the relative numbers of atoms. For example, the formula for carbon dioxide is CO_2, meaning that each molecule contains 1 atom of carbon and 2 atoms of oxygen.

Examples of molecules that contain covalent bonds are hydrogen (H_2), water (H_2O), oxygen (O_2), ammonia (NH_3), and methane (CH_4). More information about a molecule is given by its **structural formula,** in which the individual bonds are shown (indicated by lines). Structural formulas may or may not indicate the actual shape of the molecule. For example, water might be represented as

H—O—H or H/O\H

The structure on the right shows the actual shape of the water molecule. Scientists know from experimental evidence that the molecule looks like this. (We will study the shapes of molecules further in Chapter 8.) Other examples of structural formulas are

Ammonia: H—N(—H)—H or N with H, H, H

Methane: H—C(—H)(—H)—H or C with H, H, H, H

In the actual structures on the right the central atom and the solid lines are understood to be in the plane of the paper. Atoms connected to the central atom by dashed lines are behind the plane of the paper, and atoms connected to the central atom by wedges are in front of the plane of the page.

In a compound composed of molecules, the individual molecules move around as independent units. For example, a sample of methane gas is represented in Fig. 2.16 using **space-filling models.** These models show the relative sizes of the atoms as well as their relative orientation in the molecule. Figure 2.17 shows examples.

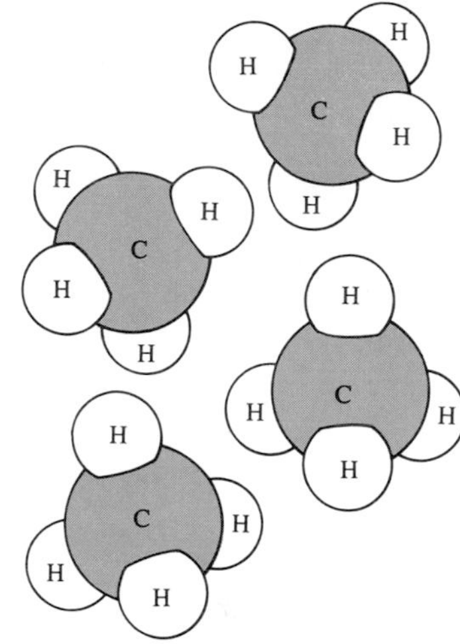

Figure 2.16

Space-filling model of methane gas. This type of model shows both the relative sizes of the atoms in the molecule and their spatial relationships.

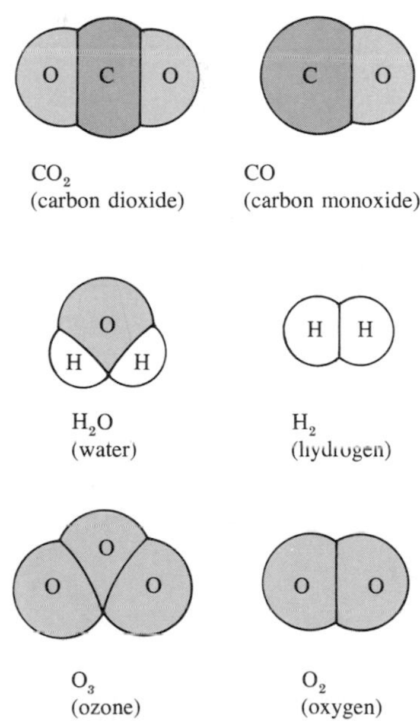

Figure 2.17

Space-filling models of various molecules.

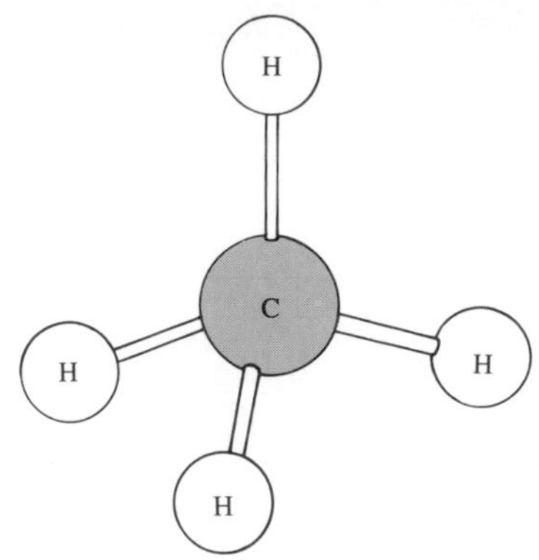

Figure 2.18

Ball-and-stick model of methane.

Ball-and-stick models are also used to represent molecules. The ball-and-stick structure of methane is shown in Fig. 2.18.

A second type of chemical bond results from attractions among ions. An **ion** is an atom or group of atoms that has a net positive or negative charge. The best known ionic compound is common table salt, or sodium chloride, which forms when neutral chlorine and sodium react.

(upper left) Sodium metal is soft enough to be cut easily with a knife. (upper right) Chlorine is a green gas. (lower left) Sodium and chlorine react vigorously to form sodium chloride (lower right).

To see how ions are formed, consider what happens when an electron is transferred from sodium to chlorine (the neutrons in the nuclei will be ignored):

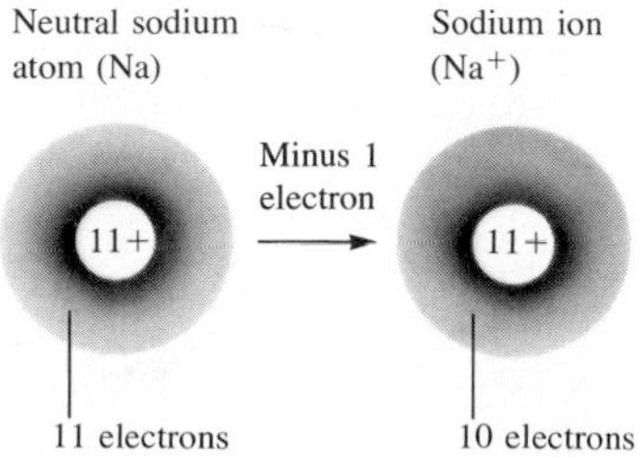

With one electron stripped off, the sodium, with its 11 protons and only 10 electrons, now has a net 1+ charge—it has become a *positive ion*. A positive ion is called a **cation.** The sodium ion is written as Na^+, and the process can be represented in shorthand form as

$$Na \rightarrow Na^+ + e^-$$

If an electron is added to chlorine,

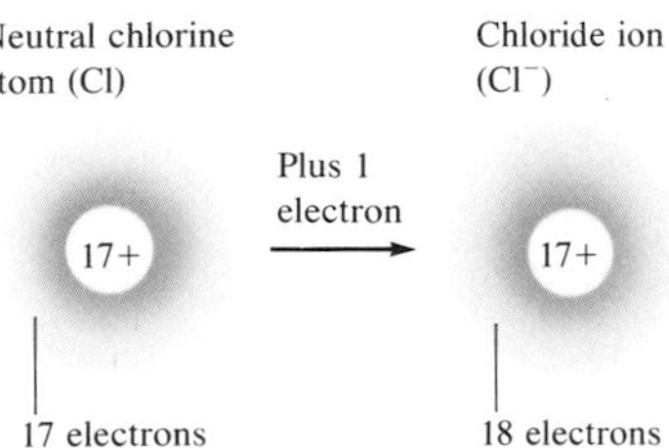

the 18 electrons produce a net 1− charge; the chlorine has become an *ion with a negative charge*—an **anion.** The chloride ion is written as Cl^-, and the process is represented as

$$Cl + e^- \rightarrow Cl^-$$

Because anions and cations have opposite charges, they attract each other. This *force of attraction between oppositely charged ions* is called an **ionic bond.** Sodium metal and chlorine gas (a green gas composed of Cl_2 molecules) react to form solid sodium chloride, which contains many Na^+ and Cl^- ions packed together as shown in Fig. 2.19(a). The solid forms the beautiful colorless cubic crystals shown in Fig. 2.19(b).

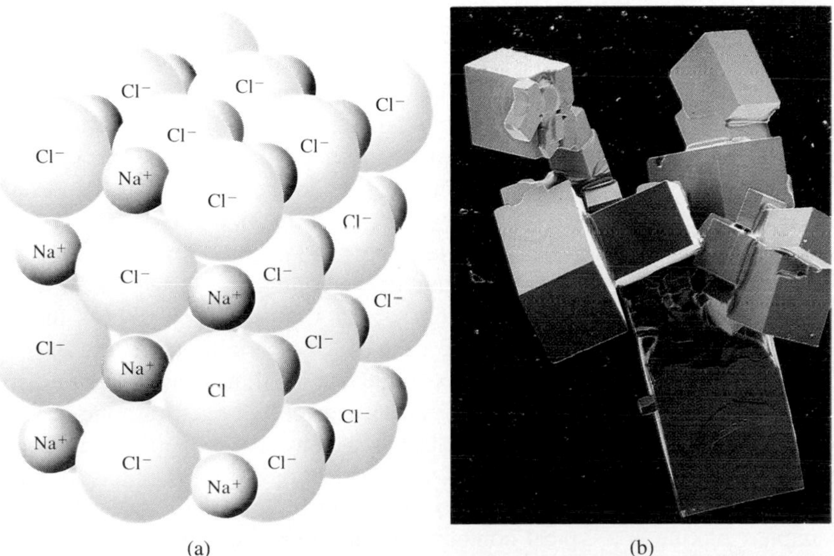

Figure 2.19

(a) The arrangement of sodium ions (Na^+) and chloride ions (Cl^-) in the ionic compound sodium chloride.
(b) Sodium chloride forms cubic crystals.

A solid consisting of oppositely charged ions is called an *ionic solid*, or a *salt*. Ionic solids can consist of simple ions, as in sodium chloride, or of **polyatomic** (many-atom) **ions,** as in ammonium nitrate (NH_4NO_3), which contains ammonium cations (NH_4^+) and nitrate anions (NO_3^-). The ball-and-stick models of these ions are shown in Fig. 2.20.

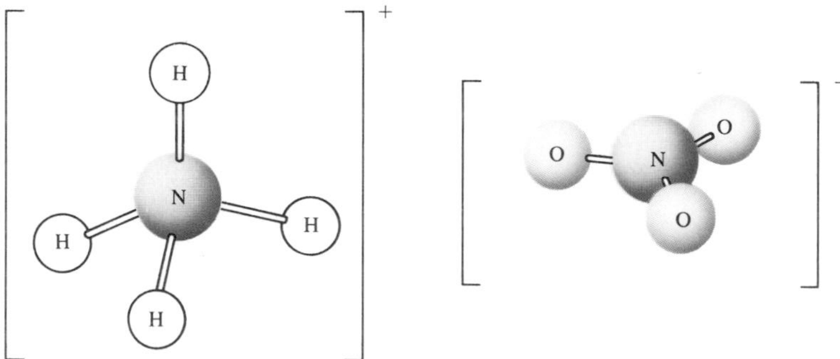

Figure 2.20

Ball-and-stick models of the ammonium ion (NH_4^+) and the nitrate ion (NO_3^-).

2.7 An Introduction to the Periodic Table

Purpose

- To introduce various features of the periodic table.

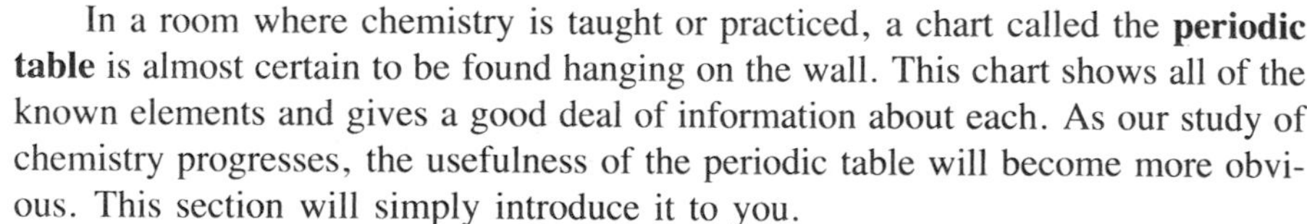

In a room where chemistry is taught or practiced, a chart called the **periodic table** is almost certain to be found hanging on the wall. This chart shows all of the known elements and gives a good deal of information about each. As our study of chemistry progresses, the usefulness of the periodic table will become more obvious. This section will simply introduce it to you.

Gold is one of only a few metals that occur naturally as a free element, sometimes in spectacular form. The largest gold nugget ever mined in Colorado (found near Breckenridge) weighs 13 pounds.

A simple version of the periodic table is shown in Fig. 2.21. The letters given in the boxes are the symbols for the elements. The number shown above each symbol is the atomic number (number of protons) for that element. For example, carbon (C) has atomic number 6, and lead (Pb) has atomic number 82. Most of the elements are **metals.** Metals have characteristic physical properties such as efficient conduction of heat and electricity, malleability (they can be hammered into thin sheets), ductility (they can be pulled into wires), and (often) a lustrous appearance. Chemically, metals tend to *lose* electrons to form positive ions. For example, copper is a typical metal. It is lustrous (although it tarnishes readily); it is an excellent conductor of electricity (it is widely used in electrical wires); and it is readily formed into various shapes, such as pipes for water systems. Copper is also found in many salts, such as the beautifully blue copper sulfate (see page 58 for a photograph), in which copper is present as Cu^{2+} ions. Copper is a member of the transition metals—the metals shown in the center of the periodic table.

Metals tend to form positive ions; nonmetals tend to form negative ions.

The relatively few **nonmetals** appear in the upper right-hand corner of the table (to the right of the heavy line in Fig. 2.21), except hydrogen, a nonmetal that is grouped with the metals. The nonmetals lack the physical properties that characterize the metals. Chemically, they tend to *gain* electrons to form anions in reactions with metals. Nonmetals often bond to each other by forming covalent bonds. For example, chlorine is a typical nonmetal. Under normal conditions it exists as Cl_2

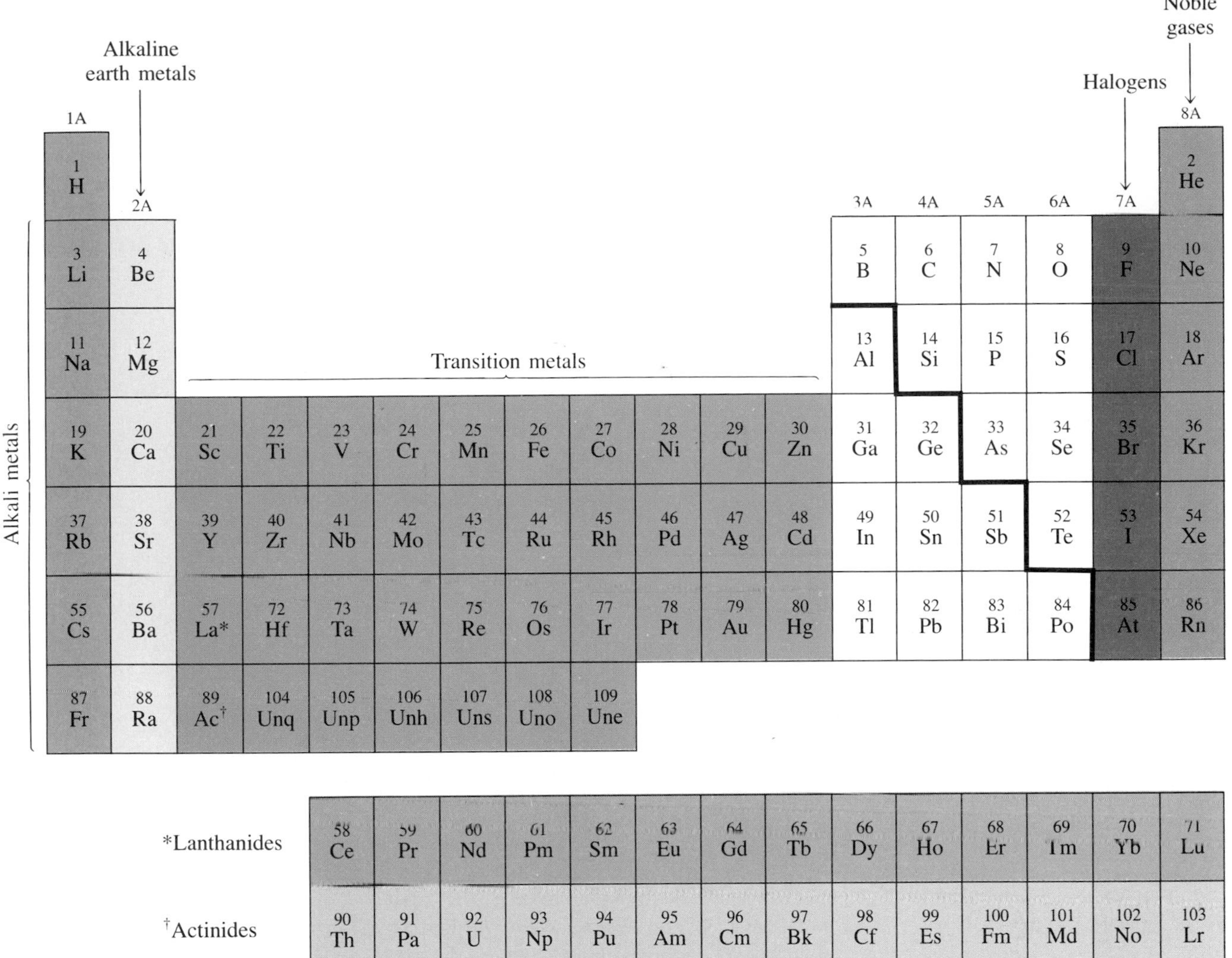

Figure 2.21

The periodic table. Although rutherfordium and hahnium have been proposed as names for elements 104 and 105, an attempt is being made to name the elements after 103 systematically, using a letter abbreviation for each number. In this system the symbol for element 104 is Unq, and for 105 is Unp. The systematic symbols are shown for elements 106–109.

molecules; it reacts with metals to form salts containing Cl^- ions (NaCl, for example); and it forms covalent bonds with nonmetals (for example, hydrogen chloride gas, or HCl).

The periodic table is arranged so that elements in the same vertical columns (called **groups** or **families**) have *similar chemical properties*. For example, all of the **alkali metals,** members of Group 1A—lithium (Li), sodium (Na), potassium (K), rubidium (Rb), cesium (Cs), and francium (Fr)—are very active elements that readily form ions with a 1+ charge when they react with nonmetals. The members of Group 2A—beryllium (Be), magnesium (Mg), calcium (Ca), strontium (Sr), barium (Ba), and radium (Ra)—are called the **alkaline earth metals.** They all form

Elements in the same vertical column in the periodic table form a group (or family) and have similar properties.

ions with a 2+ charge when they react with nonmetals. The **halogens,** the members of Group 7A—fluorine (F), chlorine (Cl), bromine (Br), iodine (I), and astatine (At)—all form diatomic molecules. Fluorine, chlorine, bromine, and iodine all react with metals to form salts containing ions with a 1− charge (F^-, Cl^-, Br^-, and I^-). The members of Group 8A—helium (He), neon (Ne), argon (Ar), krypton (Kr), xenon (Xe), and radon (Rn)—are known as the **noble gases.** They all exist under normal conditions as monatomic (single-atom) gases and have little chemical reactivity.

The horizontal rows of elements in the periodic table are called **periods.** Horizontal row one is called the first period (it contains H and He); row two is called the second period (elements Li through Ne); and so on.

Three members of the halogen family: iodine (purple), bromine (reddish-brown), and chlorine (green).

Another format of the periodic table will be discussed in Section 7.11.

We will learn much more about the periodic table as we continue with our study of chemistry. Meanwhile, when an element is introduced in this text, you should always note its position on the periodic table.

2.8 Naming Simple Compounds

Purpose

- To demonstrate how to name compounds given their formulas, and to write formulas given their names.

When chemistry was an infant science, there was no system for naming compounds. Names such as sugar of lead, blue vitrol, quicklime, Epsom salts, milk of magnesia, gypsum, and laughing gas were coined by early chemists. Such names are called *common names*. As chemistry grew, it became clear that using common names for compounds would lead to unacceptable chaos. More than four million chemical compounds are currently known. Memorizing common names for these compounds would be an impossible task.

The solution, of course, is to adopt a *system* for naming compounds in which the name tells something about the composition of the compound. After learning the system, a chemist given a formula should be able to name the compound, or given a name should be able to construct the compound's formula. In this section we will specify the most important rules for naming compounds other than organic compounds (those based on chains of carbon atoms).

The systematic naming of organic compounds will be discussed in Chapter 22.

We will begin with the systems for naming inorganic **binary compounds**—compounds composed of two elements.

Binary Ionic Compounds (Type I)

Binary ionic compounds contain a positive ion (cation) always written first in the formula and a negative ion (anion). To name these compounds the following rules apply:

1. The cation is always named first and the anion second.
2. A monatomic (meaning from one atom) cation takes its name from the name of the element. For example, Na^+ is called sodium in the names of compounds containing this ion.
3. A monatomic anion is named by taking the first part of the element name and adding *-ide*. Thus the Cl^- ion is called chloride.

A monatomic cation has the same name as its parent element.

Some common monatomic cations and anions and their names are given in Table 2.2.

Common Monatomic Cations and Anions

Cation	Name	Anion	Name
H^+	hydrogen	H^-	hydride
Li^+	lithium	F^-	fluoride
Na^+	sodium	Cl^-	chloride
K^+	potassium	Br^-	bromide
Cs^+	cesium	I^-	iodide
Be^{2+}	beryllium	O^{2-}	oxide
Mg^{2+}	magnesium	S^{2-}	sulfide
Ca^{2+}	calcium	N^{3-}	nitride
Ba^{2+}	barium	P^{3-}	phosphide
Al^{3+}	aluminum		
Ag^+	silver		

Table 2.2

The rules for naming binary ionic compounds are illustrated by the following examples:

Compound	*Ions Present*	*Name*
NaCl	Na^+, Cl^-	sodium chloride
KI	K^+, I^-	potassium iodide
CaS	Ca^{2+}, S^{2-}	calcium sulfide
Li_3N	Li^+, N^{3-}	lithium nitride
CsBr	Cs^+, Br^-	cesium bromide
MgO	Mg^{2+}, O^{2-}	magnesium oxide

In formulas of ionic compounds, simple ions are represented by the element symbol: Cl means Cl^-, Na means Na^+, etc.

Sample Exercise 2.3

Name each binary compound.

a. CsF **b.** $AlCl_3$ **c.** LiH

Solution

a. CsF is cesium fluoride.
b. $AlCl_3$ is aluminum chloride.
c. LiH is lithium hydride.

Notice that, in each case, the cation is named first, then the anion is named.

Type II binary ionic compounds contain a metal that can form more than one type of cation.

Binary Ionic Compounds (Type II)

In the ionic compounds considered above (Type I), the metal present forms only a single type of cation. That is, sodium forms only Na^+, calcium forms only Ca^{2+}, and so on. However, as we will see in more detail later in the text, there are many metals that can form more than one type of positive ion and thus more than one type of ionic compound with a given anion. For example, the compound $FeCl_2$ contains Fe^{2+} ions, and the compound $FeCl_3$ contains Fe^{3+} ions. In a case such as this, the *charge on the metal ion must be specified*. The systematic names for these two iron compounds are iron(II) chloride and iron(III) chloride, respectively, where the *Roman numeral indicates the charge* of the cation.

Another system for naming these ionic compounds that is seen in the older literature was used for metals that form only two ions. *The ion with the higher charge has a name ending in -ic, and the one with the lower charge has a name ending in -ous*. In this system, for example, Fe^{3+} is called the ferric ion, and Fe^{2+} is called the ferrous ion. The names for $FeCl_3$ and $FeCl_2$ are then ferric chloride and ferrous chloride, respectively.

Table 2.3 gives both names for many Type II common cations. The system that uses Roman numerals will be used exclusively in this text.

(top) A dish of copper(II) sulfate. (bottom) A close-up photo of copper(II) sulfate crystals.

Table 2.3 Common Type II Cations

Ion	Systematic name	Alternate name
Fe^{3+}	iron(III)	ferric
Fe^{2+}	iron(II)	ferrous
Cu^{2+}	copper(II)	cupric
Cu^+	copper(I)	cuprous
Co^{3+}	cobalt(III)	cobaltic
Co^{2+}	cobalt(II)	cobaltous
Sn^{4+}	tin(IV)	stannic
Sn^{2+}	tin(II)	stannous
Pb^{4+}	lead(IV)	plumbic
Pb^{2+}	lead(II)	plumbous
Hg^{2+}	mercury(II)	mercuric
Hg_2^{2+}*	mercury(I)	mercurous

*Note that mercury(I) ions always occur bound together to form Hg_2^{2+}.

Sample Exercise 2.4

Give the systematic name of each of the following compounds.

a. CuCl **b.** HgO **c.** Fe_2O_3 **d.** MnO_2 **e.** $SnCl_4$

Solution

All of these compounds include a metal that can form more than one type of cation; thus we must first determine the charge on each cation. This can be done by recognizing that a compound must be electrically neutral; that is, the positive and negative charge must exactly balance.

A compound must be electrically neutral.

a. In CuCl, for example, since the anion is Cl^-, the cation must be Cu^+. The name is copper(I) chloride, where the Roman numeral I indicates the 1+ charge on Cu^+.

b. In HgO, since the anion is oxide, O^{2-}, the mercury cation must be Hg^{2+} to give a net charge of zero as required. Thus the name is mercury(II) oxide.

c. In Fe_2O_3, the three O^{2-} ions carry a total charge of 6−, and the two iron cations must carry a total charge of 6+. Thus each iron ion is Fe^{3+}, and the name is iron(III) oxide.

d. In the compound MnO_2, the cation has a 4+ charge, and the name is manganese(IV) oxide.

e. In $SnCl_4$, the cation also has a 4+ charge, so the name is tin(IV) chloride.

Note that the use of a Roman numeral in a systematic name is required only in cases where more than one ionic compound forms between a given pair of elements. This most commonly occurs for compounds containing transition metals, which often form more than one cation. *Elements that form only one cation do not need to be identified by a Roman numeral.* Common metals that do not require Roman numerals are the Group 1A elements, which form only 1+ ions; the Group 2A elements, which form only 2+ ions; and aluminum, which forms only Al^{3+}.

A compound containing a transition metal must have a Roman numeral in its name.

As shown in Sample Exercise 2.4, when a metal ion is present that forms more than one type of cation, the charge on the metal ion must be determined by balancing the positive and negative charges of the compound. To do this you must be able to recognize the common cations and anions and know their charges (see Tables 2.2 and 2.4).

Sample Exercise 2.5

Give the systematic name of each of the following compounds.

a. $CoBr_2$ **b.** $CaCl_2$ **c.** Al_2O_3 **d.** $CrCl_3$

Solution

Compound	*Name*	*Comment*
a. $CoBr_2$	cobalt(II) bromide	Cobalt is a transition metal; the compound name must have a Roman numeral. The two Br^- ions must be balanced by a Co^{2+} cation.

Sample Exercise 2.5, continued

b. $CaCl_2$	calcium chloride	Calcium, an alkaline earth metal, forms only the Ca^{2+} ion. A Roman numeral is not necessary.
c. Al_2O_3	aluminum oxide	Aluminum forms only Al^{3+}. A Roman numeral is not necessary.
d. $CrCl_3$	chromium(III) chloride	Chromium is a transition metal. The compound name must have a Roman numeral. $CrCl_3$ contains Cr^{3+}.

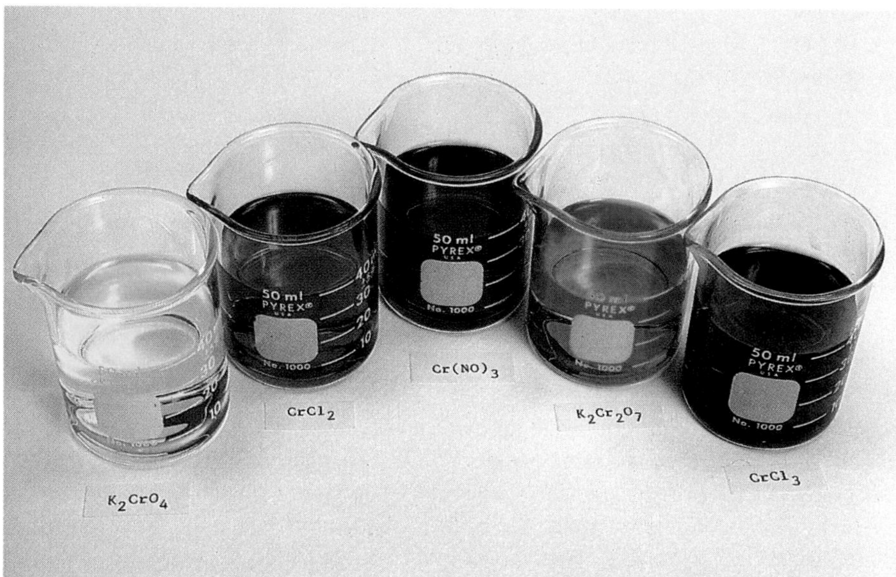

Various chromium compounds dissolved in water.

Ionic Compounds with Polyatomic Ions

Polyatomic ion formulas must be memorized.

We have not yet considered ionic compounds that contain polyatomic ions. For example, the compound ammonium nitrate, NH_4NO_3, contains the polyatomic ions NH_4^+ and NO_3^-. Polyatomic ions are assigned special names that you *must memorize* in order to name the compounds containing them. The most important polyatomic ions and their names are listed in Table 2.4.

Note in Table 2.4 that several series of anions contain an atom of a given element and different numbers of oxygen atoms. These anions are called **oxyanions.** When there are two members in such a series, the name of the one with the smaller number of oxygen atoms ends in *-ite* and the name of the one with the larger number ends in *-ate,* for example, sulfite (SO_3^{2-}) and sulfate (SO_4^{2-}). When more than two oxyanions make up a series, *hypo-* (less than) and *per-* (more than) are used as prefixes to name the members of the series with the fewest and the most oxygen atoms, respectively. The best example involves the oxyanions containing chlorine as shown in Table 2.4.

Common Polyatomic Ions

Ion	Name	Ion	Name
NH_4^+	ammonium	CO_3^{2-}	carbonate
NO_2^-	nitrite	HCO_3^-	hydrogen carbonate (bicarbonate is a widely used common name)
NO_3^-	nitrate		
SO_3^{2-}	sulfite		
SO_4^{2-}	sulfate	ClO^-	hypochlorite
HSO_4^-	hydrogen sulfate (bisulfate is a widely used common name)	ClO_2^-	chlorite
		ClO_3^-	chlorate
		ClO_4^-	perchlorate
OH^-	hydroxide	$C_2H_3O_2^-$	acetate
CN^-	cyanide	MnO_4^-	permanganate
PO_4^{3-}	phosphate	$Cr_2O_7^{2-}$	dichromate
HPO_4^{2-}	hydrogen phosphate	CrO_4^{2-}	chromate
$H_2PO_4^-$	dihydrogen phosphate	O_2^{2-}	peroxide

Table 2.4

Sample Exercise 2.6

Give the systematic name of each of the following compounds.

a. Na_2SO_4 **b.** KH_2PO_4 **c.** $Fe(NO_3)_3$ **d.** $Mn(OH)_2$
e. Na_2SO_3 **f.** Na_2CO_3 **g.** $NaHCO_3$ **h.** $CsClO_4$
i. $NaOCl$ **j.** Na_2SeO_4 **k.** $KBrO_3$

Solution

Compound	*Name*	*Comment*
a. Na_2SO_4	sodium sulfate	
b. KH_2PO_4	potassium dihydrogen phosphate	
c. $Fe(NO_3)_3$	iron(III) nitrate	Transition metal—name must contain a Roman numeral. Fe^{3+} ion balances three NO_3^- ions.
d. $Mn(OH)_2$	manganese(II) hydroxide	Transition metal—name must contain a Roman numeral. Mn^{2+} ion balances two OH^- ions.
e. Na_2SO_3	sodium sulfite	
f. Na_2CO_3	sodium carbonate	
g. $NaHCO_3$	sodium hydrogen carbonate	Often called sodium bicarbonate.
h. $CsClO_4$	cesium perchlorate	
i. $NaOCl$	sodium hypochlorite	
j. Na_2SeO_4	sodium selenate	Atoms in the same group, like sulfur and selenium, often form similar ions that are named similarly. Thus SeO_4^{2-} is selenate, like SO_4^{2-} (sulfate).
k. $KBrO_3$	potassium bromate	As above, BrO_3^- is bromate, like ClO_3^- (chlorate).

Binary Covalent Compounds

In binary covalent compounds the element names follow the same rules as for binary ionic compounds.

Binary covalent compounds are formed between *two nonmetals*. Although these compounds do not contain ions, they are named very similarly to binary ionic compounds.

In naming binary covalent compounds the following rules apply:

1. The first element in the formula is named first using the full element name.
2. The second element is named as if it were an anion.
3. Prefixes are used to denote the numbers of atoms present. These prefixes are given in Table 2.5.
4. The prefix *mono-* is never used for naming the first element. For example, CO is called carbon monoxide, *not* monocarbon monoxide.

Prefixes Used to Indicate Number in Chemical Names

Prefix	Number indicated
mono-	1
di-	2
tri-	3
tetra-	4
penta-	5
hexa-	6
hepta-	7
octa-	8

Table 2.5

To see how these rules apply we will now consider the names of the several covalent compounds formed by nitrogen and oxygen:

Compound	*Systematic Name*	*Common Name*
N_2O	dinitrogen monoxide	nitrous oxide
NO	nitrogen monoxide	nitric oxide
NO_2	nitrogen dioxide	
N_2O_3	dinitrogen trioxide	
N_2O_4	dinitrogen tetroxide	
N_2O_5	dinitrogen pentoxide	

Notice from the above examples that to avoid awkward pronunciations, the final *o* or *a* of the prefix is often dropped when the element begins with a vowel. For example, N_2O_4 is called dinitrogen tetroxide, *not* dinitrogen tetr*a*oxide, and CO is called carbon monoxide, *not* carbon mon*o*oxide.

Water and ammonia are always called by their common names.

Some compounds are always referred to by their common names. The two best examples are water and ammonia. The systematic names for H_2O and NH_3 are never used.

Sample Exercise 2.7

Name each of the following compounds.

a. PCl_5 **b.** PCl_3 **c.** SF_6 **d.** SO_3 **e.** SO_2 **f.** CO_2

Solution

Compound	*Name*
a. PCl_5	phosphorus pentachloride
b. PCl_3	phosphorus trichloride
c. SF_6	sulfur hexafluoride
d. SO_3	sulfur trioxide
e. SO_2	sulfur dioxide
f. CO_2	carbon dioxide

The rules for naming simple ionic and covalent compounds are summarized in Table 2.6. Notice that prefixes to indicate the number of atoms are used only in binary covalent compounds (those containing two nonmetals).

Sample Exercise 2.8 contains compounds from the various classes to give you practice in deciding which set of rules to use.

Binary Ionic Compounds (Cation and Anion Are Monatomic)

	Composition	Cation name	Anion name
Type I	A metal and a nonmetal. Metal forms only one type of cation.	Full metal name	Root of nonmetal name plus *-ide*
Type II	A metal and a nonmetal. Metal forms more than one type of cation.	Full metal name followed by Roman numeral indicating cation's charge	Root of nonmetal name plus *-ide*

Ionic Compounds with Polyatomic Ions
Polyatomic ions have special names. Any monatomic ions present are named by the rules for binary ionic compounds.

Binary Covalent Compounds (contain two nonmetals)
The first element is named using the full element name. The second element is named as if it were a monatomic anion. Prefixes are used to indicate the numbers of atoms of each element present.

Table 2.6

Sample Exercise 2.8

Give the systematic name for each of the following compounds.

a. P_4O_{10} **b.** Nb_2O_5 **c.** Li_2O_2 **d.** $Ti(NO_3)_4$

Solution

Compound	*Name*	*Comment*
a. P_4O_{10}	tetraphosphorus decoxide	Binary covalent compound (two nonmetals) so prefixes are used. The *a* in deca- is dropped here.
b. Nb_2O_5	niobium(V) oxide	Contains Nb^{5+} and O^{2-} ions. Niobium is a transition metal and requires a Roman numeral.
c. Li_2O_2	lithium peroxide	Contains the Li^+ and O_2^{2-} (peroxide) ions.
d. $Ti(NO_3)_4$	titanium(IV) nitrate	Contains the Ti^{4+} and NO_3 ions. Titanium is a transition metal and requires a Roman numeral.

Formulas from Names

So far we have started with the chemical formula of a compound and decided on its systematic name. The reverse process is also important. For example, given the name calcium hydroxide, we can write the formula as $Ca(OH)_2$ since we know that calcium only forms Ca^{2+} ions and that, since hydroxide is OH^-, two of these anions will be required to give a neutral compound. Similarly the name iron(II) oxide implies the formula FeO, since the Roman numeral II indicates the presence of Fe^{2+} and since the oxide ion is O^{2-}.

Sample Exercise 2.9

Given the following systematic names, write the formula for each compound.

a. ammonium sulfate **b.** vanadium(V) fluoride **c.** dioxygen difluoride
d. rubidium peroxide **e.** gallium oxide

Solution

Name	*Chemical Formula*	*Comment*
a. ammonium sulfate	$(NH_4)_2SO_4$	Two ammonium ions (NH_4^+) are required for each sulfate ion (SO_4^{2-}) to achieve charge balance.
b. vanadium(V) fluoride	VF_5	The compound contains V^{5+} ions and requires five F^- ions for charge balance.
c. dioxygen difluoride	O_2F_2	The prefix di- indicates two of each atom.
d. rubidium peroxide	Rb_2O_2	Since rubidium is in group 1A, it forms only 1+ ions. Thus two Rb^+ ions are needed to balance the 2− charge on the peroxide ion (O_2^{2-}).
e. gallium oxide	Ga_2O_3	Since gallium is in group 3A, like aluminum, it forms 3+ ions. Two Ga^{3+} ions are required to balance the charge on three O^{2-} ions.

Acids

Acids can be recognized by the hydrogen that appears first in the formula.

When dissolved in water, certain molecules produce a solution containing free H^+ ions (protons). These substances, acids, will be discussed in detail in Chapters 4, 14, and 15. Here we will simply present the rules for naming acids.

An acid can be viewed as a molecule with one or more H^+ ions attached to an anion. The rules for naming acids depend on whether or not the anion contains oxygen. If the *anion does not contain oxygen,* the acid is named with the prefix *hydro-* and the suffix *-ic*. For example, when gaseous HCl is dissolved in water, it forms hydrochloric acid. Similarly, HCN and H_2S dissolved in water are called hydrocyanic and hydrosulfuric acids, respectively.

When the *anion contains oxygen,* the acid name is formed from the root name of the anion with a suffix of *-ic* or *-ous*. If the anion name ends in *-ate,* the *-ate* is replaced by *-ic* (or sometimes *-ric*). For example, H_2SO_4 contains the sulfate anion (SO_4^{2-}) and is called sulfuric acid; H_3PO_4 contains the phosphate anion (PO_4^{3-}) and is called phosphoric acid; and $HC_2H_3O_2$ contains the acetate ion ($C_2H_3O_2^-$) and is called acetic acid. If the anion has an *-ite* ending, the *-ite* is replaced by *-ous*. For example, H_2SO_3, which contains sulfite (SO_3^{2-}), is named sulfurous acid; and HNO_2, which contains nitrite (NO_2^-), is named nitrous acid. The application of these rules can be seen in the names of the acids of the oxyanions of chlorine:

Acid	*Anion*	*Name*
$HClO_4$	perchlor*ate*	perchlor*ic* acid
$HClO_3$	chlor*ate*	chlor*ic* acid
$HClO_2$	chlor*ite*	chlor*ous* acid
$HClO$	hypochlor*ite*	hypochlor*ous* acid

The names of the most important acids are given in Tables 2.7 and 2.8.

Names of Acids That Do Not Contain Oxygen

Acid	Name
HF	hydrofluoric acid
HCl	hydrochloric acid
HBr	hydrobromic acid
HI	hydroiodic acid
HCN	hydrocyanic acid
H_2S	hydrosulfuric acid

Table 2.7

Names of Some Oxygen-Containing Acids

Acid	Name
HNO_3	nitric acid
HNO_2	nitrous acid
H_2SO_4	sulfuric acid
H_2SO_3	sulfurous acid
H_3PO_4	phosphoric acid
$HC_2H_3O_2$	acetic acid

Table 2.8

FOR REVIEW

Summary

Three fundamental laws formed the basis for early chemistry: the law of conservation of mass (matter can neither be created nor destroyed), the law of definite proportion (a given compound always contains exactly the same proportion of elements by mass), and the law of multiple proportions (if two elements *A* and *B* form a series of compounds, the ratios of the masses of *A* that combine with 1 gram of *B* can always be represented by small whole numbers). Dalton accounted for these laws in his atomic theory. He postulated that all elements are composed of atoms; that all atoms of a given element are identical; that chemical compounds are formed when atoms combine; and that the atoms themselves are not changed in a chemical reaction, but are just reorganized.

Atoms consist of a dense nucleus containing protons and neutrons, surrounded by electrons that occupy a large volume relative to the size of the nucleus. Electrons have a relatively small mass (1/1840 of the proton mass) and a negative charge. Protons have a positive charge equal in magnitude (but opposite in sign) to that on the electron. A neutron has virtually the same mass as a proton but no charge.

Isotopes are atoms with the same number of protons (thus constituting the same element) but different numbers of neutrons. That is, isotopes have the same atomic number but different mass numbers (total numbers of neutrons and protons).

Atoms combine to form molecules by sharing electrons to form covalent bonds. A molecule can be described by a chemical formula showing the numbers and types of atoms involved, a structural formula (showing which atoms are joined to each other), or by ball-and-stick or space-filling models that show the exact positions of the atoms in space. When an atom loses one or more electrons, it forms a positive ion called a cation. If an atom gains electrons, it becomes a negatively charged anion. The interaction of oppositely charged ions to form an ionic compound is called ionic bonding.

The periodic table arranges the elements in order of increasing atomic number, and elements having similar chemical properties fall into vertical columns, or groups. Most of the elements are metals, which tend to form cations in ionic compounds with nonmetals, which are elements that tend to form anions.

Compounds are named using systems of rules. These rules vary depending on the type of compound being named. For binary ionic compounds (those that contain a metal and a nonmetal), the metal is named first, followed by a name derived from the root name of the nonmetal. In a compound (Type I) containing a metal that always forms the same cation (sodium always forms Na^+, magnesium always forms Mg^{2+}, and so on) the name of the metal is sufficient. In a compound (Type II) that contains a metal that can form more than one cation (iron can form Fe^{2+} and Fe^{3+} and so on) the metal name is followed by a Roman numeral that indicates the cation charge.

Polyatomic ions have special names that must be memorized. In compounds containing polyatomic ions, the cation name is given first, followed by the anion name. For binary covalent compounds (those that contain two nonmetals) the first element is named using the entire element name and the second element is named as if it were an anion. Prefixes are used to indicate the numbers of atoms present.

Suggested Reading

1. Isaac Asimov, *A Short Story of Chemistry* (Anchor Books, Doubleday and Co., Inc., 1965).
2. Robert P. Crease and Charles C. Mann, *The Second Creation,* Chapters 1 and 2 (Macmillan Publishing Co., 1986).
3. Enid S. Lipeles, "The Chemical Contributions of Amadeo Avogadro" *J. Chem. Ed.* **60** (1983): 127.
4. Sir William Roscoe, *John Dalton and the Rise of Modern Chemistry,* 1896.

Key Terms

Section 2.2
law of conservation of mass
law of definite proportion
law of multiple proportions

Section 2.3
atomic masses
atomic weights
Avogadro's hypothesis

Section 2.4
cathode ray
electron
radioactivity
nuclear atom
nucleus

Section 2.5
proton
neutron
isotopes
atomic number
mass number

Section 2.6
chemical bond
covalent bond
molecule
chemical formula
structural formula
space-filling model
ball-and-stick model
ion
cation
anion
ionic bond
polyatomic ion

Section 2.7
periodic table
metal
nonmetal
group (family)
period
alkali metals
alkaline earth metals
halogens
noble gases

Section 2.8
binary compounds
binary ionic compounds
oxyanions
binary covalent compounds
acid

Exercises

A blue exercise number indicates that the answer to that exercise appears at the back of this book and a solution appears in the Solutions Guide.

Development of the Atomic Theory

1. Several compounds containing only sulfur (S) and fluorine (F) are known. Three of them have the following compositions:

   i. 1.188 g of F for every 1.000 g of S
   ii. 2.375 g of F for every 1.000 g of S
   iii. 3.563 g of F for every 1.000 g of S

   How do these data illustrate the law of multiple proportions?

2. A reaction of 1 L of chlorine gas (Cl_2) with 3 L of fluorine gas (F_2) yields 2 L of a gaseous product. All gas volumes are at the same temperature and pressure. What is the formula of the gaseous product?

3. When mixtures of gaseous H_2 and gaseous Cl_2 react, a product forms that has the same properties regardless of the relative amounts of H_2 and Cl_2 used.

   a. How is this result interpreted in terms of the law of definite proportion?
   b. When a volume of H_2 reacts with an equal volume of Cl_2 at the same temperature and pressure, what volume of product having the formula HCl is formed?

4. Early tables of atomic weights were generated by measuring the mass of a substance that reacts with 1 g of oxygen. Given the following data and taking the atomic weight of hydrogen as 1.00, generate a table of relative atomic weights for these elements.

Element	Mass that combines with 1.00 g oxygen	Assumed formula
Hydrogen	0.1260 g	HO
Sodium	2.8750 g	NaO
Magnesium	1.5000 g	MgO

   How do your values compare with those in the periodic table? How do you account for any differences?

5. The vitamin niacin (nicotinic acid, $C_6H_5NO_2$) can be isolated from a variety of natural sources; for example, liver, yeast, milk, and whole grain. It can also be synthesized from commercially available materials. Which source of nicotinic acid, from a nutritional view, is best for use in a multivitamin tablet? Why?

6. How does Dalton's atomic theory account for each of the following?
   a. the law of conservation of mass
   b. the law of definite proportion
   c. the law of multiple proportions

7. What refinements had to be made in Dalton's atomic theory to account for Gay-Lussac's results on the combining volumes of gases?

8. One of the best indications of a "good" theory is that it raises more questions for further experimentation than it originally answered. Does this apply to Dalton's atomic theory? If so, in what ways?

9. Dalton assumed that all atoms of the same element were identical in all of their properties. How have we had to modify this?

The Nature of the Atom

10. What evidence led to the conclusion that cathode rays had a negative charge?

11. Is there a difference between a cathode ray and a β particle?

12. From the information in this chapter on the mass of the proton, the mass of the electron, and the sizes of the nucleus and the atom, calculate the densities of a hydrogen nucleus and a hydrogen atom.

13. A chemist in a galaxy far, far away performed the Millikan oil drop experiment and got the following results for the charge on various drops. What is the charge of the electron in zirkombs?

2.56×10^{-12} zirkombs	7.68×10^{-12} zirkombs
3.84×10^{-12} zirkombs	5.12×10^{-12} zirkombs

14. What discoveries were made by J. J. Thomson, Henri Becquerel, and Lord Rutherford? How did Dalton's model of the atom have to be modified to account for these discoveries?

Elements and the Periodic Table

15. What is the distinction between atomic number and mass number? Between mass number and atomic weight?

16. Distinguish between the terms *family* and *period* in the periodic table. For which of the above terms is the term *group* also used?

17. What are the symbols of the following elements: gold, silver, mercury, potassium, iron, antimony, tungsten?

18. What are the symbols of the following nonmetals: fluorine, chlorine, bromine, sulfur, oxygen, phosphorus?

19. What are the symbols of the following metals: sodium, beryllium, manganese, chromium, uranium?

20. Give the name of the metal corresponding to each symbol: Sn, Pt, Co, Ni, Mg, Ba, K.

21. Give the names of the nonmetals with the symbols: As, I, Xe, He, C, Si.

22. List the noble gas elements. Which of the noble gases has only radioactive isotopes? (This situation is indicated on most periodic tables by parentheses around the mass of the element. See inside front cover.)

23. Which lanthanide element and which transition element have only radioactive isotopes? (See Exercise 22.)

24. How many elements are there in
   a. the second period of the periodic table?
   b. the third period?
   c. the fourth period?
   d. the iron group?
   e. the oxygen family?
   f. the nickel group?

25. The elements in one of the groups in the periodic table are often called the *coinage metals*. Identify the elements in this group based on your own experience.

26. Give the number of protons and neutrons in the nucleus of each of the following atoms:

a. $^{238}_{94}Pu$	c. $^{52}_{24}Cr$	e. $^{60}_{27}Co$
b. $^{65}_{29}Cu$	d. $^{4}_{2}He$	f. $^{54}_{24}Cr$

27. Using the periodic table, give the number of protons and neutrons in the nucleus of each of the following atoms:

a. ^{15}N	c. ^{207}Pb	e. ^{107}Ag
b. ^{3}H	d. ^{151}Eu	f. ^{109}Ag

28. Identify each of the following elements:

a. $^{31}_{15}X$	c. $^{39}_{19}X$
b. $^{127}_{53}X$	d. $^{173}_{70}X$

29. How many protons, neutrons, and electrons are in each of the following atoms or ions?

a. $^{24}_{12}Mg$	d. $^{59}_{27}Co^{3+}$	g. $^{79}_{34}Se^{2-}$
b. $^{24}_{12}Mg^{2+}$	e. $^{59}_{27}Co$	h. $^{63}_{28}Ni$
c. $^{59}_{27}Co^{2+}$	f. $^{79}_{34}Se$	i. $^{59}_{28}Ni^{2+}$

30. An ion contains 50 protons, 68 neutrons, and 48 electrons. What is its symbol?

31. An atom has 9 protons and 10 neutrons in the nucleus. What is its symbol?

32. What is the symbol for an ion with 63 protons, 60 electrons, and 88 neutrons?

33. What is the symbol of an ion with 16 protons, 18 neutrons, and 18 electrons?

34. Complete the following table:

Symbol	Number of protons in nucleus	Number of neutrons in nucleus	Number of electrons	Net charge
	33	42		3+
$^{128}_{52}Te^{2-}$			54	
	16	16	16	
	81	123		1+
$^{195}_{78}Pt$				

35. Complete the following table:

Symbol	Number of protons in nucleus	Number of neutrons in nucleus	Number of electrons	Net charge
$^{238}_{92}U$				
	20	20		2+
	23	28	20	
$^{89}_{39}Y$				
	35	44	36	
	15	16		3−

36. Classify the following elements as metals or nonmetals:

Mg	Si	Rn
Ti	Ge	Eu
Au	B	Am
Bi	At	Br

37. The distinction between metals and nonmetals is really not a clear one. Some elements, called *metalloids,* are intermediate in their properties. Which elements in Exercise 36 would you reclassify as metalloids? What other elements in the periodic table would you expect to be metalloids?

38. Which of the following sets of elements are all in the same group in the periodic table?
a. Fe, Ru, Os
b. Rh, Pd, Ag
c. Sn, As, S
d. Se, Te, Po
e. N, P, O
f. C, Si, Ge
g. Rb, Sn
h. Mg, Ca

39. Would you expect each of the following atoms to gain or lose electrons when forming ions? What is the most likely ion each will form?
a. Na
b. Sr
c. Ba
d. I
e. Al
f. S
g. B
h. Cs
i. Se

40. Consider the elements of the carbon family: C, Si, Ge, Sn, and Pb. What is the trend in metallic character as one goes down a group in the periodic table?

41. What is the trend in metallic character going from left to right across a period in the periodic table?

Nomenclature

42. Name the following compounds:
a. CrO_3
b. Cr_2O_3
c. Al_2O_3
d. SeO_2
e. SeO_3
f. NI_3
g. PCl_3
h. SF_2

43. Name the following compounds:
a. NiO
b. Fe_2O_3
c. CeO_2
d. Ce_2O_3
e. Ag_2S
f. MnO_2
g. NaH
h. H_2S

44. Name the following compounds:
a. NaCl
b. $MgCl_2$
c. RbBr
d. CsF
e. AlI_3
f. HI
g. NO
h. NF_3
i. N_2F_4
j. N_2Cl_2
k. SiF_4
l. H_2Se

45. Name the following compounds:
a. $KClO_4$
b. $Ca_3(PO_4)_2$
c. $Al_2(SO_4)_3$
d. $Pb(NO_3)_2$
e. $BaSO_3$
f. $NaNO_2$
g. $KMnO_4$
h. $K_2Cr_2O_7$
i. $AuCl_3$
j. HIO_2
k. TiO_2
l. NiS

46. Name the following compounds:

a. HNO_3
b. HNO_2
c. H_3PO_4
d. H_3PO_3
e. $NaHSO_4$
f. $Ca(HSO_3)_2$
g. $NaBrO_3$
h. $Fe(IO_4)_3$
i. $Ru(NO_3)_3$
j. V_2O_5
k. $PtCl_4$
l. $FePO_4$

47. Write formulas for the following compounds:

a. cesium bromide
b. barium sulfate
c. ammonium chloride
d. chlorine monoxide
e. silicon tetrachloride
f. chlorine trifluoride
g. beryllium oxide
h. magnesium fluoride

48. Write formulas for the following compounds:

a. ammonium hydrogen phosphate
b. mercury(I) sulfide
c. silicon dioxide
d. sodium sulfite
e. aluminum hydrogen sulfate
f. nitrogen trichloride
g. hydrobromic acid
h. bromous acid
i. perbromic acid
j. potassium hydrogen sulfide
k. calcium iodide
l. cesium perchlorate

49. Write formulas for the following compounds:

a. sulfur difluoride
b. sulfur hexafluoride
c. sodium dihydrogen phosphate
d. lithium nitride
e. chromium(III) carbonate
f. tin(II) fluoride
g. ammonium acetate
h. ammonium hydrogen sulfate
i. cobalt(III) nitrate
j. mercury(I) chloride
k. potassium chlorate
l. sodium hydride

50. Write formulas for the following compounds:

a. sodium hydroxide
b. aluminum hydroxide
c. hydrogen cyanide
d. sodium peroxide
e. copper(II) acetate
f. carbon tetrafluoride
g. lead(II) oxide
h. lead(IV) oxide
i. lead dioxide (common name)
j. copper(I) bromide
k. germanium dioxide
l. gallium arsenide

51. The formulas and common names for several substances are given below. What are the systematic names for these substances?

a. sugar of lead	$Pb(C_2H_3O_2)_2$
b. blue vitrol	$CuSO_4$
c. quicklime	CaO
d. Epsom salts	$MgSO_4$
e. milk of magnesia	$Mg(OH)_2$
f. gypsum	$CaSO_4$
g. laughing gas	N_2O

Additional Exercises

52. Insulin is a complex protein molecule produced by the pancreas in all vertebrates. It is a hormone that regulates carbohydrate metabolism. Inability to produce insulin results in diabetes mellitus. Diabetes is treated by prescribing injections of insulin. Given the law of definite proportion, would you expect any differences in chemical activity between human insulin extracted from pancreatic tissue and human insulin produced by genetically engineered bacteria? Why or why not?

53. Technetium (Tc) was the first synthetically produced element. Technetium comes from the Greek word for artificial. It was first produced by Perries and Segré in 1937 in Berkeley, California, by bombarding a molybdenum plate with ^2H nuclei. Elemental technetium is produced from ammonium pertechnetate. How many protons and neutrons are in the nuclei of ^{98}Tc and ^{99}Tc? What is the formula of ammonium pertechnetate?

54. The early alchemists used to do an experiment in which water was boiled for several days in a sealed glass container. Eventually, some solid residue would begin to appear in the bottom of the flask. This result was interpreted to mean that some of the water in the flask had been converted into earth. When Lavoisier repeated this experiment, he found that the water weighed the same before and after heating, and the weight of the flask plus the solid residue equaled the original weight of the flask. Were the alchemists correct? Explain what really happened. (This experiment is described in the article by A. F. Scott, in *Scientific American,* January 1984.)

55. Elements in the same family often form oxyanions of the same general formula. The anions are named in a similar fashion. What are the names of the oxyanions of selenium and tellurium: SeO_4^{2-}, SeO_3^{2-}, TeO_4^{2-}, TeO_3^{2-}?

56. By analogy to phosphorus compounds, name the following: Na_3AsO_4, H_3AsO_4, $Mg_3(SbO_4)_2$.

57. The mass of beryllium that combines with 1.000 g of oxygen to form beryllium oxide is 0.5633 g. When atomic weights were first being measured, it was thought that the formula of beryllium oxide was Be_2O_3. What would be the atomic mass of beryllium if this were the case?

58. Predict the formula and name of a binary compound formed from the following elements:

a. Ba and O
b. Li and H
c. In and F
d. As and S
e. B and O
f. O and F
g. Al and H
h. In and P

59. Hydrazine, ammonia, and hydrogen azide all contain only nitrogen and hydrogen. The mass of hydrogen that combines with 1.00 g of nitrogen for each compound is 1.44×10^{-1} g, 2.16×10^{-1} g, and 2.40×10^{-2} g, respectively. Show how these data illustrate the law of multiple proportions.

60. Identify each of the following elements:
a. A noble gas with 54 protons in the nucleus.
b. A member of the same family as oxygen. The anion with a 2− charge contains 36 electrons.
c. A member of the alkaline earth metal family. The 2+ ion contains 18 electrons.
d. A transition metal with 42 protons in the nucleus.
e. A radioactive element with 94 protons and 94 electrons in the neutral atom.

61. Chlorine has two natural isotopes: $^{37}_{17}Cl$ and $^{35}_{17}Cl$. Hydrogen reacts with chlorine to form the compound HCl. Would a given amount of hydrogen react with different masses of the two chlorine isotopes? Does this conflict with the law of definite proportion? Why or why not?

CHAPTER 3

Stoichiometry

Chemical reactions have a profound effect on our lives. There are many examples: food is converted to energy in the human body; nitrogen and hydrogen are combined to form ammonia, which is used as a fertilizer; fuels and plastics are produced from petroleum; the starch in plants is synthesized from carbon dioxide and water using energy from sunlight; human insulin is produced in laboratories by bacteria; cancer is induced in humans by substances from our environment; and so on, in a seemingly endless list. The central activity of chemistry is to understand chemical changes such as these, and the study of reactions occupies a central place in this book. We will examine why reactions occur, how fast they occur, and the specific pathways they follow.

In this chapter we will consider the quantities of materials consumed and produced in chemical reactions. This area of study is called **chemical stoichiometry** (pronounced stoy·kē·om′·etry). To understand chemical stoichiometry you must first understand the system of atomic masses and the composition of chemical compounds.

CONTENTS

< The volcano reaction. Orange ammonium dichromate decomposes to green chromium(III) oxide, releasing so much energy that the product is red-hot.

3.1 Atomic Masses

Purpose

- To describe the modern atomic mass scale and explain how atomic masses are determined experimentally.

As we saw in Chapter 2, the first quantitative information about atomic masses came from the work of Dalton, Gay-Lussac, Lavoisier, Avogadro, and Berzelius. By observing the proportions in which elements combine to form various compounds, nineteenth century chemists calculated relative atomic masses. The modern system of atomic masses, instituted in 1961, is based on ^{12}C ("carbon twelve") as the standard. In this system, *^{12}C is assigned a mass of exactly 12 atomic mass units* (amu), and the masses of all other atoms are given relative to this standard.

The most accurate method currently available for comparing the masses of atoms involves the use of the **mass spectrometer.** In this instrument, diagrammed in Fig. 3.1, atoms or molecules are vaporized and passed into a beam of high-speed electrons. The high-speed electrons knock electrons off the atoms or molecules being analyzed and change them to positive ions. An applied electric field then accelerates these ions into a magnetic field. Because an accelerating ion produces its own magnetic field, an interaction with the applied magnetic field occurs, which tends to change the path of the ion. The amount of path deflection for each ion depends on its mass—the most massive ions are deflected the smallest amount—which causes the ions to separate, as shown in Fig. 3.1. A comparison of the positions where the ions hit the detector plate gives very accurate values of their relative masses. For example, when ^{12}C and ^{13}C are analyzed in a mass spectrometer, the ratio of their masses is found to be

$$\frac{\text{Mass } ^{13}\text{C}}{\text{Mass } ^{12}\text{C}} = 1.0836129$$

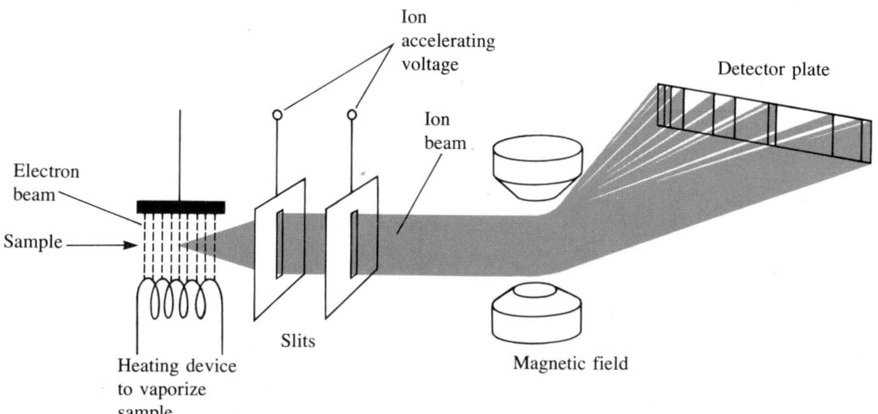

Figure 3.1

Schematic diagram of a mass spectrometer.

Since the atomic mass unit is defined such that the mass of ^{12}C is *exactly* 12 atomic mass units, then on this same scale,

$$\text{Mass of } ^{13}\text{C} = (1.0836129)(\underset{\substack{\uparrow \\ \text{Exact number,} \\ \text{by definition}}}{12} \text{ amu}) = 13.003355 \text{ amu}$$

The masses of other atoms can be determined in a similar fashion.

The mass for each element is given in the table inside the front cover of this book. This value, even though it is actually a mass, is (for historical reasons) called the **atomic weight** for each element.

Look at the value of the atomic weight of carbon given in this table. You probably expect to see 12 since we said the system of atomic masses is based on ^{12}C. However, the number given for carbon is not 12 but 12.01. Why? The reason for this apparent discrepancy is that the carbon found on earth (natural carbon) is a mixture of the isotopes ^{12}C, ^{13}C, and ^{14}C. All three isotopes have six protons, but they have six, seven, and eight neutrons, respectively. Because natural carbon is a mixture of isotopes, the atomic weight we use for carbon is an *average value* based on its isotopic composition.

Most elements occur in nature as mixtures of isotopes; thus atomic weights are usually average values.

The average atomic mass for carbon is computed as follows. It is known that natural carbon is composed of 98.89% ^{12}C atoms and 1.11% ^{13}C atoms. The amount of ^{14}C is negligibly small at this level of precision. Using the masses of ^{12}C (exactly 12 amu) and ^{13}C (13.003355 amu), we can calculate the average atomic mass for natural carbon as follows:

$$98.89\% \text{ of } 12 \text{ amu} + 1.11\% \text{ of } 13.0034 \text{ amu}$$
$$= (0.9889)(12 \text{ amu}) + (0.0111)(13.0034 \text{ amu}) = 12.01 \text{ amu}$$

This is called the atomic weight of carbon.

Even though natural carbon does not contain a single atom with mass 12.01, for stoichiometric purposes, we consider carbon to be composed of only one type of atom with a mass of 12.01. We do this so that we can count atoms of natural carbon by weighing a sample of carbon. Since it is very important that you understand this, we'll illustrate with the nonchemical example of jelly beans. It is much easier to weigh out 3000 grams of jelly beans (with an average mass of 3 grams per jelly bean) than to count out 1000 of them. Note that none of the jelly beans has to have a mass of 3 grams for this method to work; only the *average* mass must be 3 grams. We extend this same principle to counting atoms. For natural carbon with an average mass of 12.01 atomic mass units, to obtain 1000 atoms would require weighing out 12,010 atomic mass units of natural carbon (a mixture of ^{12}C and ^{13}C).

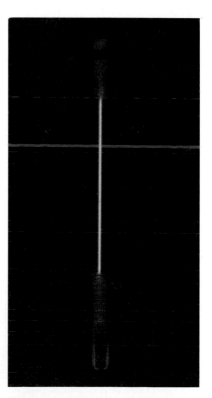

Neon gas glowing in a discharge tube.

As in the case of carbon, the mass for each element given in the table inside the front cover of the book is an average value based on the isotopic composition of the naturally occurring element. For instance, the mass listed for hydrogen (1.008) is the average mass for natural hydrogen, which is a mixture of ^{1}H and ^{2}H (deuterium). *No* atom of hydrogen actually has the mass 1.008.

In addition to being used for determining accurate mass values for individual atoms, the mass spectrometer is also used to determine the isotopic composition of a natural element. For example, when a sample of natural neon is injected into a mass spectrometer, the results shown in Fig. 3.2 are obtained. The areas of the

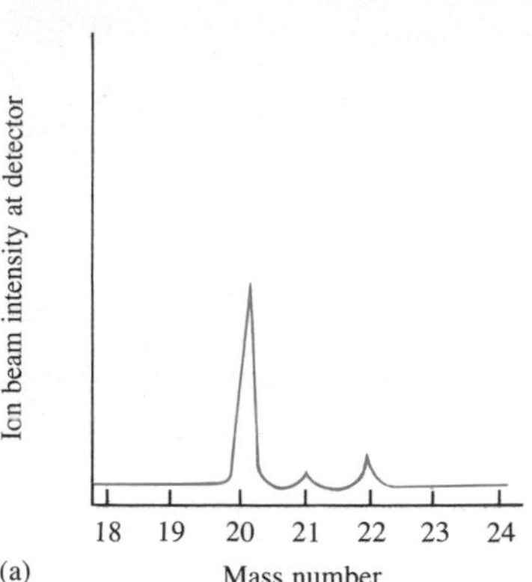

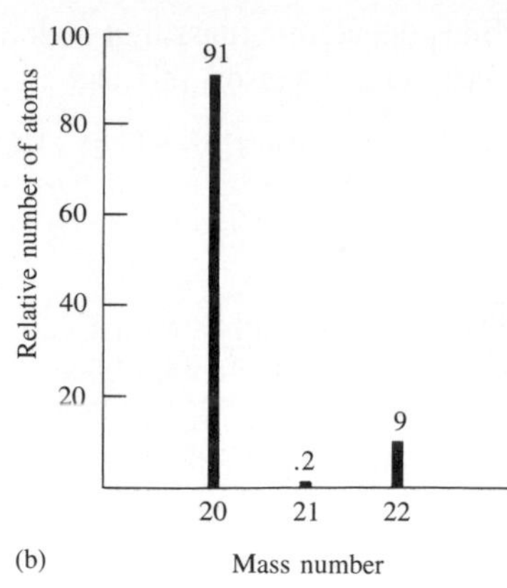

Figure 3.2

The relative intensities of the signals recorded when natural neon is injected into a mass spectrometer, represented in terms of (a) "peaks" and (b) a bar graph. The relative areas of the peaks are 0.9092 (^{20}Ne), 0.00257 (^{21}Ne), and 0.0882 (^{22}Ne); natural neon is therefore 90.92% ^{20}Ne, 0.257% ^{21}Ne, and 8.82% ^{22}Ne.

"peaks" or the heights of the bars indicate the relative numbers of $^{20}_{10}Ne$, $^{21}_{10}Ne$, and $^{22}_{10}Ne$ atoms.

Sample Exercise 3.1

When a sample of natural copper is vaporized and injected into a mass spectrometer, the results shown in Fig. 3.3 are obtained. Use these data to compute the average mass of natural copper. (The mass values for ^{63}Cu and ^{65}Cu are 62.93 amu and 64.93 amu, respectively.)

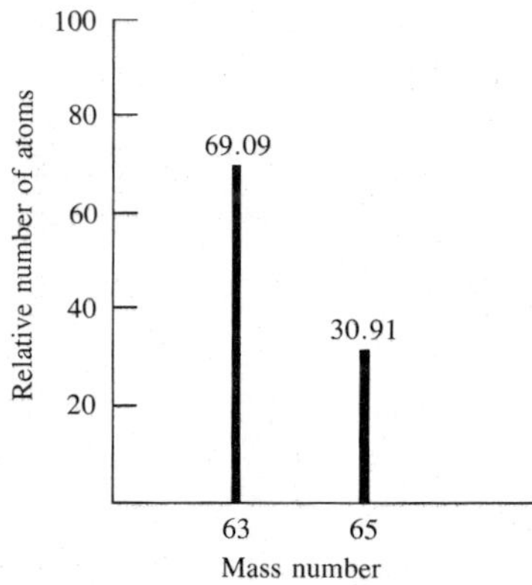

Figure 3.3

Mass spectrum of natural copper.

Solution

As shown by the graph, of every 100 atoms of natural copper, on the average 69.09 are ^{63}Cu and 30.91 are ^{65}Cu. Thus the average mass of 100 atoms of natural copper is

$$(69.09 \cancel{\text{atoms}})\left(62.93 \frac{\text{amu}}{\cancel{\text{atom}}}\right) + (30.91 \cancel{\text{atoms}})\left(64.93 \frac{\text{amu}}{\cancel{\text{atom}}}\right) = 6355 \text{ amu}$$

The average mass per atom is

$$\frac{6355 \text{ amu}}{100 \text{ atoms}} = 63.55 \text{ amu/atom}$$

This mass value is used in doing calculations involving the chemistry of copper and is the value given in the table inside the front cover of this book.

3.2 The Mole

Purpose

- To explain the importance of the mole concept.
- To show how to convert among moles, mass, and number of particles for a given sample.

Because samples of matter typically contain so many atoms, a unit of measure called the mole has been established to use in counting atoms. The **mole** (abbreviated mol) is defined as *the number equal to the number of carbon atoms in exactly 12 grams of pure* ^{12}C. Techniques such as mass spectrometry, which count atoms very precisely, have been used to determine this number as 6.02214×10^{23} (6.022×10^{23} will be sufficient for our purposes). This number is called **Avogadro's number** to honor his contributions to chemistry. *One mole of something consists of* 6.022×10^{23} *units of that substance*. Just as a dozen eggs is 12 eggs, a mole of eggs is 6.022×10^{23} eggs.

The SI definition of the mole is the amount of a substance that contains as many entities as there are in exactly 12 g of carbon-12.

Avogadro's number is 6.022×10^{23}. One mole of anything is 6.022×10^{23} units of that substance.

The magnitude of the number 6.022×10^{23} is very difficult to imagine. To give you some idea, 1 mole of seconds represents a span of time 4 million times as long as the earth has already existed, and 1 mole of marbles is enough to cover the entire earth to a depth of 50 miles! However, since atoms are so tiny, a mole of atoms or molecules is a perfectly manageable quantity to use in a reaction (see Fig. 3.4).

How do we use the mole in chemical calculations? Recall that Avogadro's number is defined as the number of atoms in exactly 12 grams of ^{12}C. This means that 12 grams of ^{12}C contain 6.022×10^{23} atoms. It also means that a 12.01-gram sample of natural carbon contains 6.022×10^{23} atoms (a mixture of ^{12}C, ^{13}C, and ^{14}C atoms, with an average mass of 12.01). Since the ratio of the masses of the samples (12 g/12.01 g) is the same as the ratio of the masses of the individual components (12 amu/12.01 amu), the two samples contain the *same number* of components.

To be sure this point is clear, think of oranges with an average mass of 0.5 pound each and grapefruit with an average mass of 1.0 pound each. Any two sacks for which the sack of grapefruit weighs twice as much as the sack of oranges will contain the same number of pieces of fruit. The same idea extends to atoms. Compare natural carbon (average mass of 12.01) and natural helium (average mass of 4.003). A sample of 12.01 grams of natural carbon contains the same number of

Figure 3.4

Samples containing one mole each of iron, sulfur, mercury, and iodine.

atoms as 4.003 grams of natural helium. Both samples contain 1 mole of atoms (6.022×10^{23}). Table 3.1 gives more examples that illustrate this basic idea.

Comparison of 1-Mole Samples of Various Elements		
Element	Number of atoms present	Mass of sample (g)
Aluminum	6.022×10^{23}	26.98
Gold	6.022×10^{23}	196.97
Iron	6.022×10^{23}	55.85
Sulfur	6.022×10^{23}	32.06
Boron	6.022×10^{23}	10.81
Xenon	6.022×10^{23}	131.30

Table 3.1

The mass of 1 mole of an element is equal to its atomic mass in grams.

Thus the mole is defined such that a sample of a natural element with a mass equal to the element's atomic mass expressed in grams contains 1 mole of atoms. This definition also fixes the relationship between the atomic mass unit and the gram. Since 6.022×10^{23} atoms of carbon (each with a mass of 12 amu) have a mass of 12 g this means that

$$(6.022 \times 10^{23} \cancel{\text{atoms}})\left(\frac{12 \text{ amu}}{\cancel{\text{atom}}}\right) = 12 \text{ g}$$

and

$$6.022 \times 10^{23} \text{ amu} = \underset{\substack{\uparrow \\ \text{Exact} \\ \text{number}}}{1 \text{ g}}$$

This relationship can be used to derive the unit factor needed to convert between atomic mass units and grams.

Sample Exercise 3.2

Americium is an element that does not occur naturally. It can be made in very small amounts in a device called a particle accelerator. Compute the mass in grams of a sample of americium containing six atoms.

Solution

From the table inside the front cover of the book, we note that one americium atom has a mass of 243 amu. Thus the mass of six atoms is

$$6 \cancel{\text{atoms}} \times 243 \frac{\text{amu}}{\cancel{\text{atom}}} = 1.46 \times 10^3 \text{ amu}$$

Using the relationship

$$6.022 \times 10^{23} \text{ amu} = 1 \text{ g}$$

we write the conversion factor for converting atomic mass units to grams:

$$\frac{1 \text{ g}}{6.022 \times 10^{23} \text{ amu}}$$

Sample Exercise 3.2, continued

The mass of six americium atoms in grams is

$$1.46 \times 10^3 \cancel{\text{amu}} \times \frac{1 \text{ g}}{6.022 \times 10^{23} \cancel{\text{amu}}} = 2.42 \times 10^{-21} \text{ g}$$

To do chemical calculations you must understand what the mole means and how to determine the number of moles in a given mass of a substance. These procedures are illustrated in Sample Exercises 3.3 and 3.4.

Sample Exercise 3.3

Aluminum (Al) is a metal with a high strength-to-weight ratio and a high resistance to corrosion; thus it is often used for structural purposes. Compute both the number of moles of atoms and the number of atoms in a 10.0-g sample of aluminum.

Aluminum alloys are used for many high-quality bicycle components, such as this chain wheel.

Solution

The mass of 1 mol (6.022×10^{23} atoms) of aluminum is 26.98 g. The sample we are considering has a mass of 10.0 g. Since the mass is less than 26.98 g, this sample contains less than 1 mol of aluminum atoms. We can calculate the number of moles of aluminum atoms in 10.0 g as follows:

$$10.0 \cancel{\text{g Al}} \times \frac{1 \text{ mol Al}}{26.98 \cancel{\text{g Al}}} = 0.371 \text{ mol Al}$$

The number of atoms in 10.0 g (0.371 mol) of aluminum is

$$0.371 \cancel{\text{mol Al}} \times \frac{6.022 \times 10^{23} \text{ atoms}}{1 \cancel{\text{mol Al}}} = 2.23 \times 10^{23} \text{ atoms}$$

Sample Exercise 3.4

A silicon chip used in an integrated circuit of a microcomputer has a mass of 5.68 mg. How many silicon (Si) atoms are present in this chip?

Solution

The strategy for doing this problem is to convert from milligrams of silicon to grams of silicon, then to moles of silicon, and finally to atoms of silicon:

$$5.68 \cancel{\text{mg Si}} \times \frac{1 \text{ g Si}}{1000 \cancel{\text{mg Si}}} = 5.68 \times 10^{-3} \text{ g Si}$$

$$5.68 \times 10^{-3} \cancel{\text{g Si}} \times \frac{1 \text{ mol Si}}{28.08 \cancel{\text{g Si}}} = 2.02 \times 10^{-4} \text{ mol Si}$$

$$2.02 \times 10^{-4} \cancel{\text{mol Si}} \times \frac{6.022 \times 10^{23} \text{ atoms}}{1 \cancel{\text{mol Si}}} = 1.22 \times 10^{20} \text{ atoms}$$

It always makes sense to think about orders of magnitude as you do a calculation. In Sample Exercise 3.4, 5.68 mg of silicon is clearly much less than 1 mole of silicon (which has a mass of 28.08 grams), so the final answer of 1.22×10^{20} atoms (compared to 6.022×10^{23} atoms) is at least in the right direction. Paying careful attention to units and making this type of check can help you detect an inverted conversion factor or a number that was incorrectly entered into your calculator.

Always check to see if your answer is sensible.

Sample Exercise 3.5

Cobalt (Co) is a metal that is added to steel to improve its resistance to corrosion. Calculate both the number of moles in a sample of cobalt containing 5.00×10^{20} atoms and the mass of the sample.

Solution

Note that the sample of 5.00×10^{20} atoms of cobalt is less than 1 mol (6.022×10^{23} atoms) of cobalt. What fraction of a mole it represents can be determined as follows:

$$5.00 \times 10^{20} \cancel{\text{ atoms Co}} \times \frac{1 \text{ mol Co}}{6.022 \times 10^{23} \cancel{\text{ atoms Co}}} = 8.30 \times 10^{-4} \text{ mol Co}$$

Since the mass of 1 mol of cobalt atoms is 58.93 g, the mass of 5.00×10^{20} atoms can be determined as follows:

$$8.30 \times 10^{-4} \cancel{\text{ mol Co}} \times \frac{58.93 \text{ g Co}}{1 \cancel{\text{ mol Co}}} = 4.89 \times 10^{-2} \text{ g Co}$$

3.3 Molecular Weight/Molar Mass

Purpose

- To show how to calculate molecular weights.
- To show how to convert among molecular weight, moles, and number of particles in a given sample.

A chemical compound is, ultimately, a collection of atoms. For example, methane (the major component of natural gas) consists of molecules that each contain one carbon and four hydrogen atoms (CH_4). How can we calculate the mass of 1 mole of methane; that is, what is the mass of 6.022×10^{23} CH_4 molecules? Since each CH_4 molecule contains one carbon atom and four hydrogen atoms, 1 mole of CH_4 molecules consists of 1 mole of carbon atoms and 4 moles of hydrogen atoms. The mass of 1 mole of methane can be found by summing the masses of carbon and hydrogen present:

The atomic weight for carbon to five significant digits is 12.011.

$$\begin{aligned} \text{Mass of 1 mol of C} &= 12.011 \text{ g} \\ \text{Mass of 4 mol of H} &= \underline{4 \times 1.008 \text{ g}} \\ \text{Mass of 1 mol of } CH_4 &= 16.043 \text{ g} \end{aligned}$$

Since the number 16.043 represents the mass of one mole of methane molecules, it makes sense to call it the *molar mass* for methane. However, traditionally the term *molecular weight* has been used to describe the mass of one mole of a substance. Thus the terms **molar mass** and **molecular weight** mean exactly the same thing: *the mass in grams of one mole of a compound*. The molecular weight of a known substance is obtained by summing the masses of the component atoms as we did for methane.

A substance's molecular weight or molar mass is the mass in grams of 1 mole of the substance.

Some substances exist as a collection of ions rather than as separate molecules. An example is ordinary table salt, sodium chloride (NaCl), which is composed of an array of Na^+ and Cl^- ions. There are no NaCl molecules present. However, in this book, for convenience, we will apply the term *molecular weight* to both ionic and molecular substances. Thus we will refer to 58.44 (22.99 + 35.45) as the molecular weight for NaCl. In some books the term *formula weight* is used instead of molecular weight for ionic compounds.

Sample Exercise 3.6

Juglone, a dye known for centuries, is produced from the husks of black walnuts. It is also a natural herbicide (weed killer) that kills off competitive plants around the black walnut tree but does not affect grass and other noncompetitive plants. The formula for juglone is $C_{10}H_6O_3$.

a. Calculate the molecular weight (molar mass) of juglone.

b. A sample of 1.56×10^{-2} g of pure juglone was extracted from black walnut husks. How many moles of juglone does this sample represent?

Solution

a. The molecular weight (molar mass) is obtained by summing the masses of the component atoms. In 1 mol of juglone there are 10 mol of carbon atoms, 6 mol of hydrogen atoms, and 3 mol of oxygen atoms:

$$\begin{aligned} 10\text{ C:} \quad 10 \times 12.011 \text{ g} &= 120.11 \text{ g} \\ 6\text{ H:} \quad 6 \times 1.008 \text{ g} &= 6.048 \text{ g} \\ 3\text{ O:} \quad 3 \times 15.999 \text{ g} &= \underline{47.997 \text{ g}} \\ \text{Mass of 1 mol of } C_{10}H_6O_3 &= 174.16 \text{ g} \end{aligned}$$

The atomic weight for oxygen to five significant digits is 15.999.

The mass of 1 mol of juglone is 174.16 g, which is the molecular weight.

b. The mass of 1 mol of this compound is 174.16 g; thus 1.56×10^{-2} g is much less than a mole. The exact fraction of a mole can be determined as follows:

$$1.56 \times 10^{-2} \cancel{\text{g juglone}} \times \frac{1 \text{ mol juglone}}{174.16 \cancel{\text{g juglone}}} = 8.96 \times 10^{-5} \text{ mol juglone}$$

Sample Exercise 3.7

Calcium carbonate ($CaCO_3$), also called calcite, is the principal mineral found in limestone, marble, chalk, pearls, and the shells of marine animals such as clams.

a. Calculate the molecular weight (molar mass) of calcium carbonate.

Close-up of abalone shell.

Sample Exercise 3.7, continued

b. A certain sample of calcium carbonate contains 4.86 mol. What is the mass in grams of this sample? What is the mass of the CO_3^{2-} ions present?

Solution

a. Calcium carbonate is an ionic compound composed of Ca^{2+} and CO_3^{2-} ions. In 1 mol of calcium carbonate, there are 1 mol of Ca^{2+} ions and 1 mol of CO_3^{2-} ions. The molecular weight is calculated by summing the masses of the components:

$$\begin{aligned} &1\ Ca^{2+}: && 1 \times 40.08\ \text{g} = 40.08\ \text{g} \\ &1\ CO_3^{2-}: && \\ &1\ C: && 1 \times 12.011\ \text{g} = 12.011\ \text{g} \\ &3\ O: && 3 \times 15.999\ \text{g} = \underline{47.997\ \text{g}} \\ &\text{Mass of 1 mol of } CaCO_3 && = 100.09\ \text{g} \end{aligned}$$

Thus the mass of 1 mol of $CaCO_3$ (1 mol Ca^{2+} plus 1 mol CO_3^{2-}) is 100.09 g. This is the molecular weight.

b. The mass of 1 mol of $CaCO_3$ is 100.09 g. The sample contains nearly 5 mol, or close to 500 g. The exact amount is determined as follows:

$$4.86\ \cancel{\text{mol } CaCO_3} \times \frac{100.09\ \text{g } CaCO_3}{1\ \cancel{\text{mol } CaCO_3}} = 486\ \text{g } CaCO_3$$

To find the mass of carbonate ions (CO_3^{2-}) present in this sample, we must realize that 4.86 mol of $CaCO_3$ contains 4.86 mol of Ca^{2+} ions and 4.86 mol of CO_3^{2-} ions. The mass of 1 mol of CO_3^{2-} ions is

$$\begin{aligned} 1\ C:\ 1 \times 12.011 &= 12.011\ \text{g} \\ 3\ O:\ 3 \times 15.999 &= \underline{47.997\ \text{g}} \\ \text{Mass of 1 mol of } CO_3^{2-} &= 60.008\ \text{g} \end{aligned}$$

Thus the mass of 4.86 mol of CO_3^{2-} ions is

$$4.86\ \cancel{\text{mol } CO_3^{2-}} \times \frac{60.008\ \text{g } CO_3^{2-}}{1\ \cancel{\text{mol } CO_3^{2-}}} = 292\ \text{g } CO_3^{2-}$$

Sample Exercise 3.8

Isopentyl acetate ($C_7H_{14}O_2$), the compound responsible for the scent of bananas, can be produced commercially. Interestingly, bees release about 1 μg (1×10^{-6} g) of this compound when they sting, in order to attract other bees to join the attack. How many molecules of isopentyl acetate are released in a typical bee sting? How many atoms of carbon are present?

Solution

Since we are given a mass of isopentyl acetate and want the number of molecules, we must first compute the molecular weight:

Sample Exercise 3.8, continued

$$7 \cancel{\text{mol}}\ \text{C} \times 12.011 \frac{\text{g}}{\cancel{\text{mol}}} = 84.077 \text{ g C}$$

$$14 \cancel{\text{mol}}\ \text{H} \times 1.0079 \frac{\text{g}}{\cancel{\text{mol}}} = 14.111 \text{ g H}$$

The atomic weight for hydrogen to five significant digits is 1.0079.

$$2 \cancel{\text{mol}}\ \text{O} \times 15.999 \frac{\text{g}}{\cancel{\text{mol}}} = \underline{31.998 \text{ g O}}$$

$$130.186 \text{ g}$$

This means that 1 mol of isopentyl acetate (6.022×10^{23} molecules) has a mass of 130.186 g.

To find the number of molecules released in a sting, we must first determine the number of moles of isopentyl acetate in 1×10^{-6} g:

$$1 \times 10^{-6} \cancel{\text{g } C_7H_{14}O_2} \times \frac{1 \text{ mol } C_7H_{14}O_2}{130.186 \cancel{\text{g } C_7H_{14}O_2}} = 8 \times 10^{-9} \text{ mol } C_7H_{14}O_2$$

Since 1 mol is 6.022×10^{23} units, we can determine the number of molecules:

$$8 \times 10^{-9} \cancel{\text{mol } C_7H_{14}O_2} \times \frac{6.022 \times 10^{23} \text{ molecules}}{1 \cancel{\text{mol } C_7H_{14}O_2}} = 5 \times 10^{15} \text{ molecules}$$

To determine the number of carbon atoms present, we must multiply the number of molecules by 7 since each molecule of isopentyl acetate contains seven carbon atoms:

$$5 \times 10^{15} \cancel{\text{molecules}} \times \frac{7 \text{ carbon atoms}}{\cancel{\text{molecule}}} = 4 \times 10^{16} \text{ carbon atoms}$$

Note: In keeping with our practice of always showing the correct number of significant figures, we have rounded off after each step. However, if extra digits are carried throughout this problem, the final answer rounds to 3×10^{16}.

3.4 Percent Composition of Compounds

Purpose

- To demonstrate the calculation of the mass percent of a given element in a compound.

So far we have discussed the composition of a compound in terms of the numbers of its constituent atoms. It is often useful to know a compound's composition in terms of the masses of its elements. We can obtain this information from the formula of the compound by comparing the mass of each element present in 1 mole of the compound to the total mass of 1 mole of the compound.

For example, for ethanol, which has the formula C_2H_5OH, the mass of each element present and the molecular weight are obtained as follows:

$$\text{Mass of C} = 2\,\cancel{\text{mol}} \times 12.011\frac{\text{g}}{\cancel{\text{mol}}} = 24.022\text{ g}$$

$$\text{Mass of H} = 6\,\cancel{\text{mol}} \times 1.008\frac{\text{g}}{\cancel{\text{mol}}} = 6.048\text{ g}$$

$$\text{Mass of O} = 1\,\cancel{\text{mol}} \times 15.999\frac{\text{g}}{\cancel{\text{mol}}} = \underline{15.999\text{ g}}$$

$$\text{Mass of 1 mol of } C_2H_5OH = 46.069\text{ g}$$

The *mass percent* (often called the weight percent) of carbon in ethanol can be computed by comparing the mass of carbon in 1 mole of ethanol to the total mass of 1 mole of ethanol and multiplying the result by 100:

$$\text{Mass percent of C} = \frac{\text{mass of C in 1 mol } C_2H_5OH}{\text{mass of 1 mol } C_2H_5OH} \times 100$$

$$= \frac{24.022\text{ g}}{46.069\text{ g}} \times 100 = 52.144\%$$

The mass percents of hydrogen and oxygen in ethanol are obtained in a similar manner:

$$\text{Mass percent of H} = \frac{\text{mass of H in 1 mol } C_2H_5OH}{\text{mass of 1 mol } C_2H_5OH} \times 100$$

$$= \frac{6.048\text{ g}}{46.069\text{ g}} \times 100 = 13.13\%$$

$$\text{Mass percent of O} = \frac{\text{mass of O in 1 mol } C_2H_5OH}{\text{mass of 1 mol } C_2H_5OH} \times 100$$

$$= \frac{15.999\text{ g}}{46.069\text{ g}} \times 100 = 34.728\%$$

Notice that the percentages add up to 100% if rounded to two decimal places; this is the check of the calculations.

Sample Exercise 3.9

Carvone is a substance that occurs in two forms having different arrangements of the atoms but the same molecular formula ($C_{10}H_{14}O$) and weight. One type of carvone gives caraway seeds their characteristic smell, and the other type is responsible for the smell of spearmint oil. Compute the mass percent of each element in carvone.

Solution

The masses of the elements in 1 mol of carvone are:

Sample Exercise 3.9, continued

$$\text{Mass of C in 1 mol} = 10\,\cancel{\text{mol}} \times 12.011\,\frac{\text{g}}{\cancel{\text{mol}}} = 120.11 \text{ g}$$

$$\text{Mass of H in 1 mol} = 14\,\cancel{\text{mol}} \times 1.008\,\frac{\text{g}}{\cancel{\text{mol}}} = 14.11 \text{ g}$$

$$\text{Mass of O in 1 mol} = 1\,\cancel{\text{mol}} \times 15.999\,\frac{\text{g}}{\cancel{\text{mol}}} = \underline{15.999 \text{ g}}$$

$$\text{Mass of 1 mol of } C_{10}H_{14}O = 150.22 \text{ g}$$

Next we find the fraction of the total mass contributed by each element and convert it to a percentage:

$$\text{Mass percent of C} = \frac{120.11 \text{ g C}}{150.22 \text{ g } C_{10}H_{14}O} \times 100 = 79.956\%$$

$$\text{Mass percent of H} = \frac{14.11 \text{ g H}}{150.22 \text{ g } C_{10}H_{14}O} \times 100 = 9.393\%$$

$$\text{Mass percent of O} = \frac{15.999 \text{ g O}}{150.22 \text{ g } C_{10}H_{14}O} \times 100 = 10.650\%$$

Check: Sum the individual mass percent values—they should total to 100% within round-off errors. In this case, the percentages add up to 99.999%.

Sample Exercise 3.10

Penicillin, the first of a now large number of antibiotics (antibacterial agents), was discovered accidentally by the Scottish bacteriologist Alexander Fleming in 1928, but he was never able to isolate it as a pure compound. This and similar antibiotics have saved millions of lives that might have been lost to infections. Penicillin F has the formula $C_{14}H_{20}N_2SO_4$. Compute the mass percent of each element.

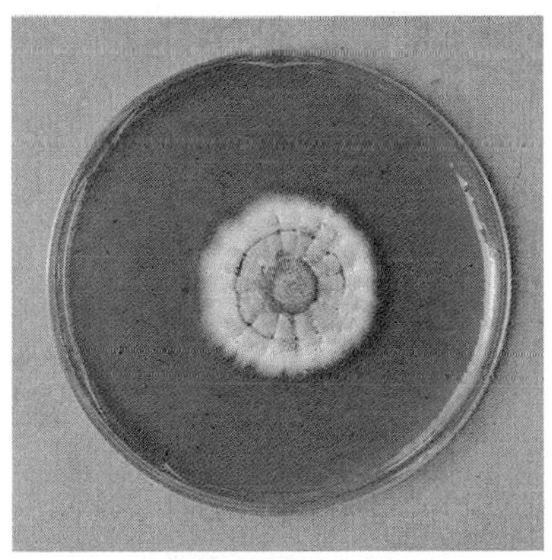

Penicillin is isolated from a mold that can be grown in large quantities in fermentation tanks.

Solution

The molecular weight of penicillin F is computed as follows:

$$\text{C:} \quad 14\,\cancel{\text{mol}} \times 12.011\,\frac{\text{g}}{\cancel{\text{mol}}} = 168.15 \text{ g}$$

$$\text{H:} \quad 20\,\cancel{\text{mol}} \times 1.008\,\frac{\text{g}}{\cancel{\text{mol}}} = 20.16 \text{ g}$$

$$\text{N:} \quad 2\,\cancel{\text{mol}} \times 14.007\,\frac{\text{g}}{\cancel{\text{mol}}} = 28.014 \text{ g}$$

$$\text{S:} \quad 1\,\cancel{\text{mol}} \times 32.06\,\frac{\text{g}}{\cancel{\text{mol}}} = 32.06 \text{ g}$$

$$\text{O:} \quad 4\,\cancel{\text{mol}} \times 15.999\,\frac{\text{g}}{\cancel{\text{mol}}} = \underline{63.996 \text{ g}}$$

$$\text{Mass of 1 mol of } C_{14}H_{20}N_2SO_4 = 312.38 \text{ g}$$

Sample Exercise 3.10, continued

$$\text{Mass percent of C} = \frac{168.15 \text{ g C}}{312.38 \text{ g } C_{14}H_{20}N_2SO_4} \times 100 = 53.829\%$$

$$\text{Mass percent of H} = \frac{20.16 \text{ g H}}{312.38 \text{ g } C_{14}H_{20}N_2SO_4} \times 100 = 6.454\%$$

$$\text{Mass percent of N} = \frac{28.014 \text{ g N}}{312.38 \text{ g } C_{14}H_{20}N_2SO_4} \times 100 = 8.968\%$$

$$\text{Mass percent of S} = \frac{32.06 \text{ g S}}{312.38 \text{ g } C_{14}H_{20}N_2SO_4} \times 100 = 10.26\%$$

$$\text{Mass percent of O} = \frac{63.996 \text{ g O}}{312.38 \text{ g } C_{14}H_{20}N_2SO_4} \times 100 = 20.487\%$$

Check: The percentages add up to 100.00%.

3.5 Determining the Formula of a Compound

Purpose

- To demonstrate the calculation of the empirical formula of a compound.
- To show how to obtain the molecular formula, given the empirical formula and the molecular weight.

When a new compound is prepared, one of the first items of interest is the formula of the compound. This is most often determined by taking a weighed sample of the compound and either decomposing it into its component elements or reacting it with oxygen to produce substances such as CO_2, H_2O and N_2, which are then collected and weighed. A device for doing this type of analysis is shown in Fig. 3.5. The results of such analyses provide the mass of each type of element in the compound, which can be used to determine the mass percent of each element.

We will see how information of this type can be used to compute the formula of a compound. Suppose a substance has been prepared that is composed of carbon, hydrogen, and nitrogen. When 0.1156 gram of this compound is reacted with oxygen, 0.1638 gram of carbon dioxide (CO_2) and 0.1676 gram of water (H_2O) are collected. Assuming that all of the carbon in the compound is converted to CO_2, we

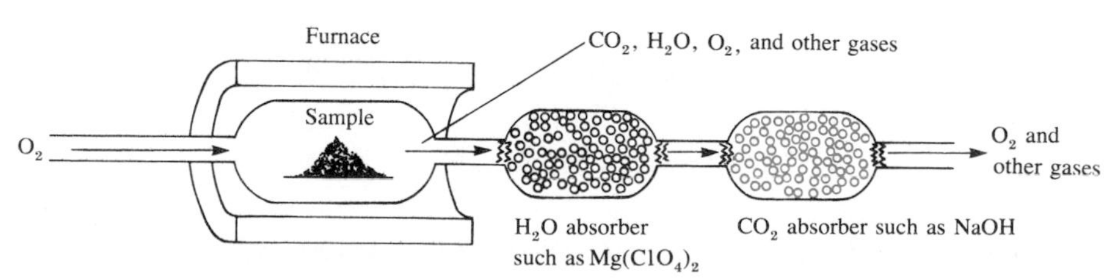

Figure 3.5

A schematic diagram of the combustion device used to analyze substances for carbon and hydrogen. The sample is burned in the presence of excess oxygen, which converts all of its carbon to carbon dioxide and all of its hydrogen to water. These are collected by absorption using appropriate materials, and their amounts are determined by measuring the increase in weights of the absorbents.

can determine the mass of carbon originally present in the 0.1156-gram sample. To do this, we must use the fraction (by mass) of carbon in CO_2. The molecular weight of CO_2 is

$$\text{C:}\quad 1\ \cancel{\text{mol}} \times 12.011 \frac{\text{g}}{\cancel{\text{mol}}} = 12.011\ \text{g}$$

$$\text{O:}\quad 2\ \cancel{\text{mol}} \times 15.999 \frac{\text{g}}{\cancel{\text{mol}}} = \underline{31.998\ \text{g}}$$

$$\text{Molecular weight of } CO_2 = 44.009\ \text{g}$$

The fraction of carbon present by mass is

$$\frac{\text{Mass of C}}{\text{total mass of } CO_2} = \frac{12.011\ \text{g C}}{44.009\ \text{g } CO_2}$$

This factor can now be used to determine the mass of carbon in 0.1638 gram of CO_2:

$$0.1638\ \cancel{\text{g } CO_2} \times \frac{12.011\ \text{g C}}{44.009\ \cancel{\text{g } CO_2}} = 0.04470\ \text{g C}$$

Remember that this carbon originally came from the 0.1156-gram sample of unknown compound. Thus the mass percent of carbon in this compound is

$$\frac{0.04470\ \text{g C}}{0.1156\ \text{g compound}} \times 100 = 38.67\%\ \text{C}$$

The same procedure can be used to find the mass percent of hydrogen in the unknown compound. We assume that all of the hydrogen present in the original 0.1156 gram of compound was converted to H_2O. The molecular weight of H_2O is 18.015 grams, and the fraction of hydrogen by mass in H_2O is

$$\frac{\text{Mass of H}}{\text{mass of } H_2O} = \frac{2.016\ \text{g H}}{18.015\ \text{g } H_2O}$$

Therefore, the mass of hydrogen in 0.1676 gram of H_2O is

$$0.1676\ \cancel{\text{g } H_2O} \times \frac{2.016\ \text{g H}}{18.015\ \cancel{\text{g } H_2O}} = 0.01876\ \text{g H}$$

And the mass percent of hydrogen in the compound is

$$\frac{0.01876\ \text{g H}}{0.1156\ \text{g compound}} \times 100 = 16.23\%\ \text{H}$$

The unknown compound contains only carbon, hydrogen, and nitrogen. So far we have determined that it is 38.67% carbon and 16.23% hydrogen. The remainder must be nitrogen:

$$100.00\% - (\underset{\%\text{ C}}{\underset{\uparrow}{38.67\%}} + \underset{\%\text{ H}}{\underset{\uparrow}{16.23\%}}) = 45.10\%\ \text{N}$$

We have determined that the compound contains 38.67% carbon, 16.23% hydrogen, and 45.10% nitrogen. Next we use these data to obtain the formula.

Quartz crystals.

Since the formula of a compound indicates the *numbers* of atoms in the compound, we must convert the masses of the elements to numbers of atoms. The easiest way to do this is to work with 100.00 grams of the compound. In the present case, 38.67% carbon by mass means 38.67 grams of carbon per 100.00 grams of compound; 16.23% hydrogen means 16.23 grams of hydrogen per 100.00 grams of compound; and so on. To determine the formula, we must calculate the number of carbon atoms in 38.67 grams of carbon, the number of hydrogen atoms in 16.23 grams of hydrogen, and the number of nitrogen atoms in 45.10 grams of nitrogen. We can do this as follows:

$$38.67\ \cancel{\text{g C}} \times \frac{1\ \text{mol C}}{12.011\ \cancel{\text{g C}}} = 3.220\ \text{mol C}$$

$$16.23\ \cancel{\text{g H}} \times \frac{1\ \text{mol H}}{1.008\ \cancel{\text{g H}}} = 16.10\ \text{mol H}$$

$$45.10\ \cancel{\text{g N}} \times \frac{1\ \text{mol N}}{14.007\ \cancel{\text{g N}}} = 3.220\ \text{mol N}$$

Thus 100.00 grams of this compound contains 3.220 moles of carbon atoms, 16.10 moles of hydrogen atoms, and 3.220 moles of nitrogen atoms.

We can find the smallest *whole number ratio* of atoms in this compound by dividing each of the mole values above by the smallest of the three:

$$\text{C:}\quad \frac{3.220}{3.220} = 1$$

$$\text{H:}\quad \frac{16.10}{3.220} = 5$$

$$\text{N:}\quad \frac{3.220}{3.220} = 1$$

Thus the formula of this compound can be written CH_5N. This formula is called the **empirical formula.** It represents the *simplest whole number ratio of the various types of atoms in a compound.*

If this compound is molecular, then the formula might well be CH_5N. It might also be $C_2H_{10}N_2$, or $C_3H_{15}N_3$, and so on, that is, some multiple of the simplest whole number ratio. Each of these alternatives also has the correct relative numbers of atoms. Any molecule that can be represented as $(CH_5N)_x$, where x is an integer, has the empirical formula CH_5N. To be able to specify the exact formula of the molecule involved, the **molecular formula,** we must know the molecular weight.

Molecular formula = (empirical formula)$_x$, where x is an integer.

Suppose we know that this compound with empirical formula CH_5N has a molecular weight of 31.06. How do we determine which of the possible choices represents the molecular formula? Since the molecular formula is always a whole-number multiple of the empirical formula, we must first find the empirical formula weight for CH_5N:

$$\begin{aligned} \text{1 C:}&\quad 1 \times 12.011\ \text{g} = 12.011\ \text{g} \\ \text{5 H:}&\quad 5 \times 1.008\ \text{g} = 5.040\ \text{g} \\ \text{1 N:}&\quad 1 \times 14.007\ \text{g} = \underline{14.007\ \text{g}} \\ \text{Formula weight of } CH_5N &= 31.058\ \text{g} \end{aligned}$$

This is the same as the known molecular weight of the compound. Thus in this case the empirical formula and the molecular formula are the same; this substance consists of molecules with the formula CH_5N. It is quite common for the empirical and molecular formulas to be different; some examples where this is the case are shown in Fig. 3.6.

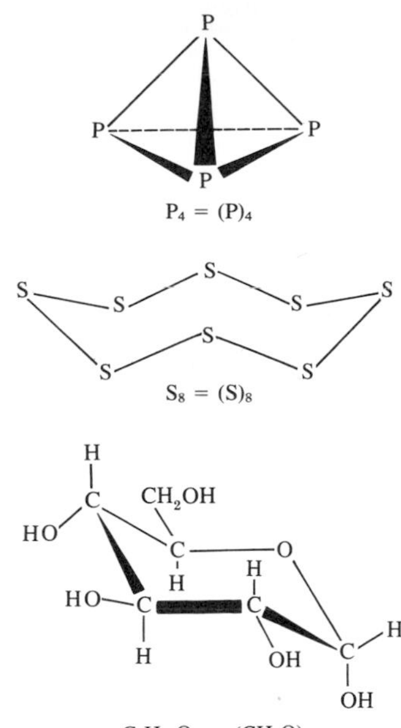

Figure 3.6

Examples of substances whose empirical and molecular formulas differ. Notice that molecular formula = (empirical formula)$_x$ where x is an integer.

Sample Exercise 3.11

Determine the empirical and molecular formulas for a compound that gives the following analysis (in mass percents):

71.65% Cl   24.27% C   4.07% H

The molecular weight is known to be 98.96 g/mol.

Solution

First, we convert the mass percents to masses in grams. In 100.00 g of the compound, there are 71.65 g of chlorine, 24.27 g of carbon, and 4.07 g of hydrogen. We use these masses to compute the moles of atoms present:

$$71.65 \cancel{\text{g Cl}} \times \frac{1 \text{ mol Cl}}{35.453 \cancel{\text{g Cl}}} = 2.021 \text{ mol Cl}$$

$$24.27 \cancel{\text{g C}} \times \frac{1 \text{ mol C}}{12.011 \cancel{\text{g C}}} = 2.021 \text{ mol C}$$

$$4.07 \cancel{\text{g H}} \times \frac{1 \text{ mol H}}{1.008 \cancel{\text{g H}}} = 4.04 \text{ mol H}$$

Dividing each mole value by 2.02 (the smallest number of moles present), we obtain the empirical formula $ClCH_2$.

To determine the molecular formula, we must compare the empirical formula weight to the molecular weight. The empirical formula weight is 49.48 (confirm this). The molecular weight is known to be 98.96.

$$\frac{\text{Molecular weight}}{\text{empirical formula weight}} = \frac{98.96}{49.48} = 2$$

$$\text{Molecular formula} = (ClCH_2)_2 = Cl_2C_2H_4$$

This substance is composed of molecules with the formula $Cl_2C_2H_4$.

Notice that the method we employ here allows us to determine the molecular formula of a compound but not its structural formula. The compound $Cl_2C_2H_4$ is called dichloroethane. There are two forms of this compound, shown in Fig. 3.7. The form in (b) is used as an additive to gasoline to help prevent knocking in engines.

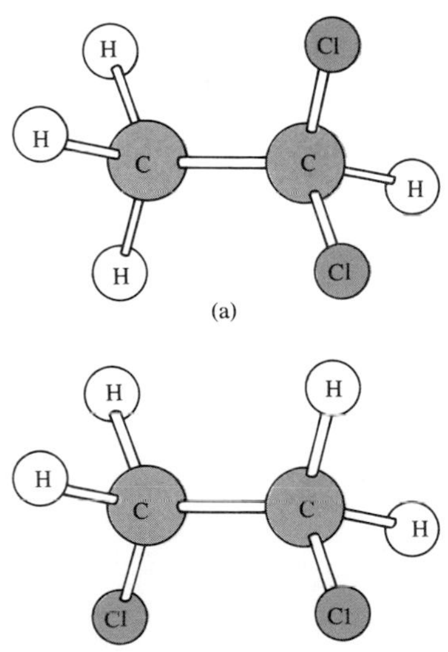

Figure 3.7

The two forms of dichloroethane.

Sample Exercise 3.12

A white powder is analyzed and found to contain 43.64% phosphorus and 56.36% oxygen by mass. The compound has a molecular weight of 283.88. What are the compound's empirical and molecular formulas?

Sample Exercise 3.12, continued

Solution

In 100.00 g of this compound, there are 43.64 g of phosphorus and 56.36 g of oxygen. In terms of moles, in 100.00 g of compound we have

$$43.64 \not{g}\not{P} \times \frac{1 \text{ mol P}}{30.97 \not{g}\not{P}} = 1.409 \text{ mol P}$$

$$56.36 \not{g}\not{O} \times \frac{1 \text{ mol O}}{15.999 \not{g}\not{O}} = 3.523 \text{ mol O}$$

Dividing both mole values by the smaller one gives

$$\frac{1.409}{1.409} = 1 \text{ P} \quad \text{and} \quad \frac{3.523}{1.409} = 2.5 \text{ O}$$

This yields the formula $PO_{2.5}$. Since compounds must contain whole numbers of atoms, the empirical formula should contain only whole numbers. To obtain the simplest set of whole numbers, we multiply both numbers by 2 to give the empirical formula P_2O_5.

To obtain the molecular formula, we must compare the empirical formula weight to the molecular weight. The empirical formula weight for P_2O_5 is 141.94.

$$\frac{\text{Molecular weight}}{\text{empirical formula weight}} = \frac{283.88}{141.94} = 2$$

The molecular formula is $(P_2O_5)_2$, or P_4O_{10}.

The structural formula of this interesting compound is given in Fig. 3.8.

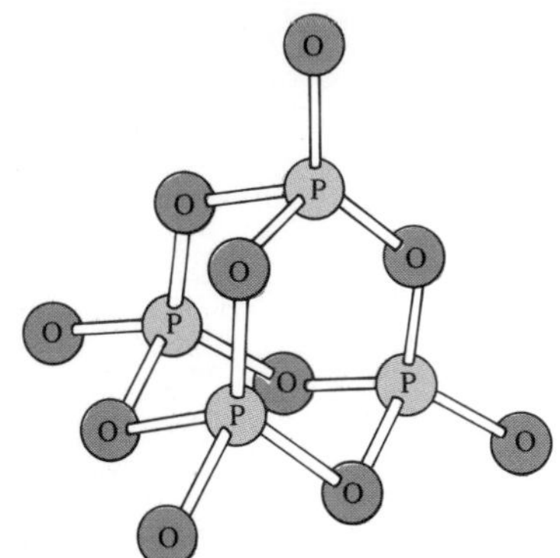

Figure 3.8

The structural formula of P_4O_{10}. Note that the oxygen atoms act as "bridges" between the phosphorus atoms. This compound has a great affinity for water and is often used as a desiccant, or drying agent.

In Sample Exercises 3.11 and 3.12 we found the molecular formula by comparing the empirical formula weight to the molecular weight. There is an alternate way to obtain the molecular formula. We know the molecular weight (molar mass) of the compound is 98.96. Thus one mole of the compound weighs 98.96 grams. Since we also know the mass percentages of each element, we can compute the mass of each element present in a mole of compound.

$$\text{Chlorine:} \quad \frac{71.65 \text{ g Cl}}{100.0 \not{g} \text{ compound}} \times \frac{98.96 \not{g}}{\text{mol}} = \frac{70.90 \text{ g Cl}}{\text{mol compound}}$$

$$\text{Carbon:} \quad \frac{24.27 \text{ g C}}{100.0 \not{g} \text{ compound}} \times \frac{98.96 \not{g}}{\text{mol}} = \frac{24.02 \text{ g C}}{\text{mol compound}}$$

$$\text{Hydrogen:} \quad \frac{4.07 \text{ g H}}{100.0 \not{g} \text{ compound}} \times \frac{98.96 \not{g}}{\text{mol}} = \frac{4.03 \text{ g H}}{\text{mol compound}}$$

Now we can compute moles of atoms present per mole of compound:

$$\text{Chlorine:} \quad \frac{70.90 \not{g}\,\not{Cl}}{\text{mol compound}} \times \frac{1 \text{ mol Cl}}{35.453 \not{g}\,\not{Cl}} = \frac{2.000 \text{ mol Cl}}{\text{mol compound}}$$

Carbon: $\frac{24.02\ \cancel{g\ C}}{\text{mol compound}} \times \frac{1\ \text{mol C}}{12.011\ \cancel{g\ C}} = \frac{2.000\ \text{mol C}}{\text{mol compound}}$

Hydrogen: $\frac{4.03\ \cancel{g\ H}}{\text{mol compound}} \times \frac{1\ \text{mol H}}{1.008\ \cancel{g\ H}} = \frac{4.00\ \text{mol H}}{\text{mol compound}}$

Thus one mole of the compound contains 2 mol Cl atoms, 2 mol C atoms, and 4 mol H atoms, and the molecular formula is $Cl_2C_2H_4$ as obtained in Sample Exercise 3.11.

Sample Exercise 3.13

Caffeine, a stimulant found in coffee, tea, and chocolate, contains 49.48% carbon, 5.15% hydrogen, 28.87% nitrogen, and 16.49% oxygen, by mass and has a molecular weight of 194.2. Determine the molecular formula of caffeine.

Solution

We will first determine the mass of each element in one mole (194.2 g) of caffeine:

$$\frac{49.48\ \text{g C}}{100.0\ \cancel{g}\ \text{caffeine}} \times \frac{194.2\ \cancel{g}}{\text{mol}} = \frac{96.09\ \text{g C}}{\text{mol caffeine}}$$

$$\frac{5.15\ \text{g H}}{100.0\ \cancel{g}\ \text{caffeine}} \times \frac{194.2\ \cancel{g}}{\text{mol}} = \frac{10.0\ \text{g H}}{\text{mol caffeine}}$$

$$\frac{28.87\ \text{g N}}{100.0\ \cancel{g}\ \text{caffeine}} \times \frac{194.2\ \cancel{g}}{\text{mol}} = \frac{56.07\ \text{g N}}{\text{mol caffeine}}$$

$$\frac{16.49\ \text{g O}}{100.0\ \cancel{g}\ \text{caffeine}} \times \frac{194.2\ \cancel{g}}{\text{mol}} = \frac{32.02\ \text{g O}}{\text{mol caffeine}}$$

Now we will convert to moles:

Carbon: $\frac{96.09\ \cancel{g\ C}}{\text{mol caffeine}} \times \frac{1\ \text{mol C}}{12.011\ \cancel{g\ C}} = \frac{8.000\ \text{mol C}}{\text{mol caffeine}}$

Hydrogen: $\frac{10.0\ \cancel{g\ H}}{\text{mol caffeine}} \times \frac{1\ \text{mol H}}{1.008\ \cancel{g\ H}} = \frac{9.92\ \text{mol H}}{\text{mol caffeine}}$

Nitrogen: $\frac{56.07\ \cancel{g\ N}}{\text{mol caffeine}} \times \frac{1\ \text{mol N}}{14.01\ \cancel{g\ N}} = \frac{4.002\ \text{mol N}}{\text{mol caffeine}}$

Oxygen: $\frac{32.02\ \cancel{g\ O}}{\text{mol caffeine}} \times \frac{1\ \text{mol O}}{16.00\ \cancel{g\ O}} = \frac{2.001\ \text{mol O}}{\text{mol caffeine}}$

Rounding the numbers to integers gives the molecular formula for caffeine: $C_8H_{10}N_4O_2$.

The methods for obtaining empirical and molecular formulas are summarized on the following page.

Empirical Formula Determination

- Since mass percentage gives the number of grams of a particular element per 100 grams of compound, base the calculation on 100 grams of compound. Each percent will then represent the mass in grams of that element.
- Determine the number of moles of each element present in 100 grams of compound using the atomic weights of the elements present.
- Divide each value of the number of moles by the smallest of the values. If each resulting number is a whole number (after appropriate rounding), these numbers represent the subscripts of the elements in the empirical formula.
- If the numbers obtained in the previous step are not whole numbers, multiply each number by an integer so that the results are all whole numbers.

Numbers very close to whole numbers, such as 9.92 and 1.08, should be rounded to whole numbers. Numbers such as 2.25, 4.33, and 2.72 should not be rounded to whole numbers.

Molecular Formula Determination

Method One

- Obtain the empirical formula.
- Compute the empirical formula weight.
- Calculate the ratio:

$$\frac{\text{molecular weight}}{\text{empirical formula weight}}$$

- The integer from the previous step represents the number of empirical formula units in one molecule. When the empirical formula subscripts are multiplied by this integer, the molecular formula results.

Method Two

- Using the mass percentages and the molecular weight, determine the mass of each element present in one mole of compound.
- Determine the number of moles of each element present in one mole of compound.
- The integers from the previous step represent the subscripts in the molecular formula.

Carbon burning in oxygen.

3.6 Chemical Equations

Purpose

- To identify the characteristics of a chemical reaction and the information given by a chemical equation.

Chemical Reactions

A chemical change involves a reorganization of the atoms in one or more substances. For example, when the methane (CH_4) in natural gas combines with oxygen (O_2) in the air and burns, carbon dioxide (CO_2) and water (H_2O) are formed. This process is represented by a **chemical equation** with the **reactants** (here methane and oxygen) on the left side of an arrow and the **products** (carbon dioxide and water) on the right side:

$$\underbrace{CH_4 + O_2}_{\text{Reactants}} \rightarrow \underbrace{CO_2 + H_2O}_{\text{Products}}$$

Notice that the atoms have been reorganized. *Bonds have been broken, and new ones formed.* It is important to recognize that *in a chemical reaction atoms are neither created nor destroyed. All atoms present in the reactants must be accounted for among the products.* In other words, there must be the same number of each type of atom on the product side and on the reactant side of the arrow. Making sure that this rule is obeyed is called **balancing a chemical equation** for a reaction.

The equation (shown above) for the reaction between CH_4 and O_2 is not balanced. As we will see in the next section, the equation can be balanced to produce

$$CH_4 + 2O_2 \rightarrow CO_2 + 2H_2O$$

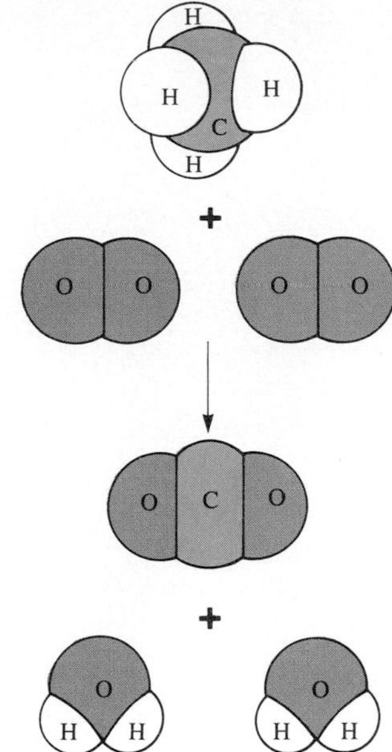

Figure 3.9

The reaction between methane and oxygen to give water and carbon dioxide. Note that no atoms have been gained or lost in the reaction. The reaction simply reorganizes the atoms.

This reaction is shown more graphically in Fig. 3.9. We can check that the equation is balanced by comparing the number of each type of atom on both sides:

$$CH_4 + 2O_2 \longrightarrow CO_2 + 2H_2O$$

CH_4: 1 C, 4 H; $2O_2$: 4 O; CO_2: 1 C, 2 O; $2H_2O$: 4 H, 2 O

To summarize our check, we have

Reactants	Products
1 C	1 C
4 H	4 H
4 O	4 O

The Meaning of a Chemical Equation

The chemical equation for a reaction gives two important types of information: the nature of the reactants and products, and the relative numbers of each.

The nature of the reactants and products must be determined by experiment. For example, the products of a reaction might be separated by one of the physical methods discussed in Chapter 1 and then identified. Besides specifying the compounds involved in the reaction, the equation often gives the *physical states* of the reactants and products:

State	Symbol
Solid	(*s*)
Liquid	(*l*)
Gas	(*g*)
Dissolved in water (in aqueous solution)	(*aq*)

Hydrochloric acid reacts with solid sodium hydrogen carbonate to produce gaseous carbon dioxide.

For example, when hydrochloric acid in aqueous solution is added to solid sodium hydrogen carbonate, the products carbon dioxide gas, liquid water, and sodium chloride (which dissolves in the water) are formed:

$$HCl(aq) + NaHCO_3(s) \longrightarrow CO_2(g) + H_2O(l) + NaCl(aq)$$

The relative numbers of reactants and products in a reaction are given by the *coefficients* in the balanced equation. (The coefficients can be determined since we know that the same number of each type of atom must occur on both sides of the equation.) For example, the balanced equation

$$CH_4(g) + 2O_2(g) \rightarrow CO_2(g) + 2H_2O(g)$$

can be interpreted in several equivalent ways, as shown in Table 3.2. Note that the total mass is 80 grams for both reactants and products. We should expect this, since chemical reactions involve only a rearrangement of atoms. Atoms, and therefore mass, are conserved in a chemical reaction.

Information Conveyed by the Balanced Equation for the Combustion of Methane

Reactants	$\longrightarrow$	Products
$CH_4(g) + 2O_2(g)$	$\longrightarrow$	$CO_2(g) + 2H_2O(g)$
1 molecule CH_4 + 2 molecules O_2	$\longrightarrow$	1 molecule CO_2 + 2 molecules H_2O
1 mol CH_4 molecules + 2 mol O_2 molecules	$\longrightarrow$	1 mol CO_2 molecules + 2 mol H_2O molecules
6.022×10^{23} CH_4 molecules + $2(6.022 \times 10^{23})$ O_2 molecules	$\longrightarrow$	6.022×10^{23} CO_2 molecules + $2(6.022 \times 10^{23})$ H_2O molecules
16 g CH_4 + 2(32 g) O_2	$\longrightarrow$	44 g CO_2 + 2(18 g) H_2O
80 g reactants	$\longrightarrow$	80 g products

Table 3.2

From this discussion you can see that a balanced chemical equation gives you a great deal of information.

3.7 Balancing Chemical Equations

Purpose

- To show how to write a balanced equation to describe a chemical reaction.

An unbalanced chemical equation is of limited use. Whenever you see an equation, you should ask yourself whether or not it is balanced. The principle that lies at the heart of the balancing process is that atoms are conserved in a chemical reaction. The same number of each type of atom must be found among the reactants and products. It is also important to recognize that the identity of the reactants and products of a reaction have been determined by experimental observation. For example, when liquid ethanol is burned in the presence of sufficient oxygen gas, the products will always be carbon dioxide and water. When the equation for this reaction is balanced, the *identities* of the reactants and products must not be changed. *The formulas of the compounds must never be changed in balancing a chemical equation.* That is, the subscripts in a formula cannot be changed, nor can atoms be added or subtracted from a formula.

Most chemical equations can be balanced by inspection, that is, by trial and error. It is always best to start with the most complicated molecules (those containing the greatest number of atoms). For example, consider the reaction of ethanol with oxygen, given by the unbalanced equation

In balancing equations, start with the most complicated molecule.

$$C_2H_5OH(l) + O_2(g) \rightarrow CO_2(g) + H_2O(g)$$

The most complicated molecule here is C_2H_5OH. We will begin by balancing the products that contain the atoms in C_2H_5OH. Since C_2H_5OH contains two carbon atoms, we place a 2 before the CO_2 to balance the carbon atoms:

$$\underset{\text{2 C atoms}}{C_2H_5OH(l)} + O_2(g) \rightarrow \underset{\text{2 C atoms}}{2CO_2(g)} + H_2O(g)$$

Since C_2H_5OH contains six hydrogen atoms, the hydrogen atoms can be balanced by placing a 3 before the H_2O:

$$\underset{(5+1)\text{ H}}{C_2H_5OH(l)} + O_2(g) \rightarrow 2CO_2(g) + \underset{(3\times 2)\text{ H}}{3H_2O(g)}$$

Last, we balance the oxygen atoms. Note that the right side of the above equation contains seven oxygen atoms, while the left side has only three. We can correct this by putting a 3 before the O_2 to produce the balanced equation:

$$\underset{1\text{ O}}{C_2H_5OH(l)} + \underset{6\text{ O}}{3O_2(g)} \rightarrow \underset{(2\times 2)\text{ O}}{2CO_2(g)} + \underset{3\text{ O}}{3H_2O(g)}$$

Now we check:

$$C_2H_5OH(l) + 3O_2(g) \rightarrow 2CO_2(g) + 3H_2O(g)$$

2 C atoms	2 C atoms
6 H atoms	6 H atoms
7 O atoms	7 O atoms

The equation is balanced.

Writing and Balancing the Equation for a Chemical Reaction

- STEP 1
  Determine what reaction is occurring. What are the reactants, the products, and the states involved?
- STEP 2
  Write the *unbalanced* equation that summarizes the information from Step 1.

- STEP 3
  Balance the equation by inspection, starting with the most complicated molecule(s). Determine what coefficients are necessary so that the same number of each type of atom appears on both reactant and product sides. Do not change the identities (formulas) of any of the reactants or products.

Sample Exercise 3.14

Chromate and dichromate compounds are carcinogens (cancer inducing agents) and should be handled very carefully.

Chromium compounds show a variety of bright colors. When solid ammonium dichromate, $(NH_4)_2Cr_2O_7$, a vivid orange compound, is ignited, a spectacular reaction occurs, as shown in the two photographs. Although the reaction is somewhat more complex, let's assume here that the products are solid chromium(III) oxide, nitrogen gas (consists of N_2 molecules), and water vapor. Balance the equation for this reaction.

Solution

STEP 1

From the description given, the reactant is solid ammonium dichromate, $(NH_4)_2Cr_2O_7(s)$, and the products are nitrogen gas, $N_2(g)$, water vapor, $H_2O(g)$, and solid chromium(III) oxide, $Cr_2O_3(s)$. The formula for chromium(III) oxide can be determined by recognizing that the Roman numeral III means that Cr^{3+} ions are present. For a neutral compound the formula must then be Cr_2O_3 since each oxide ion is O^{2-}.

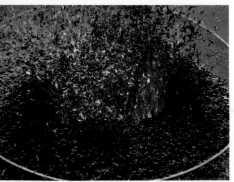

Decomposition of ammonium dichromate.

STEP 2

The unbalanced equation is

$$(NH_4)_2Cr_2O_7(s) \longrightarrow Cr_2O_3(s) + N_2(g) + H_2O(g)$$

STEP 3

Note that nitrogen and chromium are balanced (two nitrogen atoms and two chromium atoms on each side), but hydrogen and oxygen are not. A coefficient of 4 for H_2O balances the hydrogen atoms:

$$\underset{(4 \times 2)\ H}{(NH_4)_2Cr_2O_7(s)} \rightarrow Cr_2O_3(s) + N_2(g) + \underset{(4 \times 2)\ H}{4H_2O(g)}$$

Note that in balancing the hydrogen we have also balanced the oxygen, since there are seven oxygen atoms in the reactants and in the products.

Check:

$$\underset{\text{Reactant atoms}}{2\ N,\ 8\ H,\ 2\ Cr,\ 7\ O} \rightarrow \underset{\text{Product atoms}}{2\ N,\ 8\ H,\ 2\ Cr,\ 7\ O}$$

The equation is balanced.

Sample Exercise 3.15

Propane (C_3H_8), a liquid at 25°C under high pressure, is often used as a fuel in rural areas where there is no natural gas pipeline. When liquid propane is released from

Sample Exercise 3.15, continued

its storage tank, the pressure on it is lower and it changes to a gas. Propane gas reacts with oxygen gas (a combustion reaction) in a furnace to produce gaseous carbon dioxide and water vapor. Balance the equation for this reaction.

Propane burns readily in the presence of oxygen giving this characteristic blue flame.

Solution

The reactants are propane and oxygen; the products are carbon dioxide and water. All are in the gaseous state. Thus the unbalanced equation for the reaction is

$$C_3H_8(g) + O_2(g) \rightarrow CO_2(g) + H_2O(g)$$

We start with C_3H_8 since it is the most complicated molecule. Since C_3H_8 contains three carbon atoms per molecule, a coefficient of 3 is needed for CO_2:

$$C_3H_8(g) + O_2(g) \rightarrow 3CO_2(g) + H_2O(g)$$

Also, each C_3H_8 molecule contains eight hydrogen atoms, so a coefficient of 4 is required for H_2O:

$$C_3H_8(g) + O_2(g) \rightarrow 3CO_2(g) + 4H_2O(g)$$

The final element to be balanced is oxygen. Note that the left side of the equation now has two oxygen atoms, and the right side has ten. We can balance the oxygen using a coefficient of 5 for O_2:

$$C_3H_8(g) + 5O_2(g) \rightarrow 3CO_2(g) + 4H_2O(g)$$

Check:

$$\underset{\text{Reactant atoms}}{3\text{ C, }8\text{ H, }10\text{ O}} \longrightarrow \underset{\text{Product atoms}}{3\text{ C, }8\text{ H, }10\text{ O}}$$

Sample Exercise 3.16

In the presence of a red-hot platinum catalyst at 1000°C, ammonia gas, $NH_3(g)$, reacts with oxygen gas to form gaseous nitric oxide, $NO(g)$, and water vapor. This reaction is the first step in the commercial production of nitric acid by the Ostwald process. Balance the equation for this reaction.

The Ostwald process is described in Section 19.2.

Solution

The unbalanced equation for the reaction is

$$NH_3(g) + O_2(g) \rightarrow NO(g) + H_2O(g)$$

To balance the equation, let's begin with hydrogen. A coefficient of 2 for NH_3 and a coefficient of 3 for H_2O give six atoms of hydrogen on both sides:

$$2NH_3(g) + O_2(g) \rightarrow NO(g) + 3H_2O(g)$$

The nitrogen can be balanced with a coefficient of 2 for NO:

$$2NH_3(g) + O_2(g) \rightarrow 2NO(g) + 3H_2O(g)$$

Finally, note that there are two atoms of oxygen on the left and five on the right. The oxygen can be balanced with a coefficient of $\frac{5}{2}$ for O_2:

$$2NH_3(g) + \frac{5}{2}O_2(g) \rightarrow 2NO(g) + 3H_2O(g)$$

Chemical Impact

Sulfuric Acid: The Most Important Chemical

More sulfuric acid (H_2SO_4) is produced in the world than any other chemical. About 40 million tons (4×10^{10} kilograms) is manufactured annually in the United States (see Fig. 3.10). Sulfuric acid is used in the production of fertilizers, explosives, petroleum products, detergents, dyes, insecticides, drugs, plastics, steel, storage batteries, and many other materials. The largest amount of sulfuric acid is used in the production of phosphate fertilizers. In this process calcium phosphate, $Ca_3(PO_4)_2$, in phosphate rock, which cannot be used by plants because of its insolubility in ground water, is converted to forms that will dissolve in water, thus making the phosphate available to plants. This reaction can be represented as

$$Ca_3(PO_4)_2(s) + 3H_2SO_4(aq) \rightarrow 3CaSO_4(s) + 2H_3PO_4(aq)$$

The mixture of $CaSO_4$ and H_3PO_4 (phosphoric acid) is dried, pulverized, and spread on fields, where the phosphate is dissolved by rainfall.

Sulfuric acid is produced by a sequence of three simple reactions.

1. The combustion of sulfur to form sulfur dioxide:

$$S(s) + O_2(g) \rightarrow SO_2(g)$$

2. The conversion of sulfur dioxide to sulfur trioxide:

$$2SO_2(g) + O_2(g) \rightarrow 2SO_3(g)$$

3. The combination of sulfur trioxide with water:

$$SO_3(g) + H_2O(l) \rightarrow H_2SO_4(aq)$$

Over 90% of sulfuric acid is produced commercially by the *contact process*. This name was coined because the sulfur dioxide and oxygen molecules react in *contact* with the surface of solid vanadium(V) oxide (V_2O_5).

Because gaseous sulfur trioxide reacts so violently with water, it is absorbed during the production process by a sulfuric acid solution rather than by pure water. The sulfur trioxide is added to a flowing solution of sulfuric acid to which water is constantly added to keep the concentration at 98% sulfuric acid by mass. This is the substance sold as *concentrated sulfuric acid.*

One remarkable property of sulfuric acid is its great affinity for water. For example, when it is mixed with sugar, it dehydrates the sugar (takes the water out) and forms a column of black carbon (see Fig. 19.22). Because of this high affinity for water, sulfuric acid is often used as a drying agent in the production of explosives, dyes, detergents, and various anhydrous (water-free) materials.

The violence with which sulfuric acid and water combine makes the dilution of concentrated sulfuric acid potentially hazardous. The addition of water to the concentrated acid produces a vigorous reaction, which often causes acid droplets to spew in all directions. For obvious reasons, this must be avoided. *Always add the acid to water when diluting* so that any accidental splattering will involve dilute acid rather than concentrated acid.

Figure 3.10
A sulfuric acid plant.

Sample Exercise 3.16, continued

However, the convention is to have whole number coefficients. We simply multiply the entire equation by 2.

$$4NH_3(g) + 5O_2(g) \rightarrow 4NO(g) + 6H_2O(g)$$

Check: There are 4 N, 12 H, and 10 O on both sides, so the equation is balanced.

3.8 Stoichiometric Calculations: Amounts of Reactants and Products

Purpose

- To show how to calculate masses of reactants and products using the chemical equation.

As we have seen in previous sections of this chapter, the coefficients in chemical equations represent *numbers* of molecules, not masses of molecules. However, in the laboratory or chemical plant, when a reaction is to be run, the amounts of substances needed cannot be determined by counting molecules directly. Counting is always done by weighing. In this section we will see how chemical equations can be used to deal with *masses* of reacting chemicals.

To develop the principles involved in dealing with the stoichiometry of reactions, we will again consider the reaction of propane with oxygen to produce carbon dioxide and water (see Sample Exercise 3.15). What mass of oxygen will react with 96.1 grams of propane? The first thing that we must do in calculations involving chemical reactions is *write the balanced chemical equation* for the reaction. In this case the balanced equation is:

Before doing any calculations involving a chemical reaction, be sure the equation for the reaction is balanced.

$$C_3H_8(g) + 5O_2(g) \rightarrow 3CO_2(g) + 4H_2O(g)$$

This equation means that 1 mole of C_3H_8 will react with 5 moles of O_2 to produce 3 moles of CO_2 and 4 moles of H_2O. To use this equation to find the masses of reactants and products, we must be able to convert between masses and moles of substances. Thus we must first ask: *How many moles of propane are present in 96.1 grams of propane?* The molecular weight of propane to three significant figures is 44.1 ($3 \times 12.011 + 8 \times 1.008$). The moles of propane can be calculated as follows:

$$96.1 \cancel{\text{g } C_3H_8} \times \frac{1 \text{ mol } C_3H_8}{44.1 \cancel{\text{g } C_3H_8}} = 2.18 \text{ mol } C_3H_8$$

Next we must take into account the fact that each mole of propane reacts with 5 moles of oxygen. The best way to do this is to use the balanced equation to construct a **mole ratio.** In this case we want to convert from moles of propane to moles of oxygen. From the balanced equation we see that 5 moles of O_2 are required for each mole of C_3H_8, so the appropriate ratio is

$$\frac{5 \text{ mol } O_2}{1 \text{ mol } C_3H_8}$$

Multiplying the number of moles of C_3H_8 by this factor gives the number of moles of O_2 required:

$$2.18 \cancel{\text{mol } C_3H_8} \times \frac{5 \text{ mol } O_2}{1 \cancel{\text{mol } C_3H_8}} = 10.9 \text{ mol } O_2$$

Notice that the mole ratio is set up so that the moles of C_3H_8 cancel out, and the units that result are moles of O_2.

Since the original question asked for the mass of oxygen needed to react with 96.1 grams of propane, the 10.9 moles of O_2 must be converted to *grams*. Since the molecular weight of O_2 is 32.0,

$$10.9 \cancel{\text{mol } O_2} \times \frac{32.0 \text{ g } O_2}{1 \cancel{\text{mol } O_2}} = 349 \text{ g } O_2$$

Therefore, 349 grams of oxygen is required to burn 96.1 grams of propane.

This example can be extended by asking: What mass of carbon dioxide is produced when 96.1 grams of propane is combusted with oxygen? In this case we must convert between moles of propane and moles of carbon dioxide. This can be done by looking at the balanced equation, which shows that 3 moles of CO_2 are produced for each mole of C_3H_8 reacted. The mole ratio needed to convert from moles of propane to moles of carbon dioxide is

$$\frac{3 \text{ mol } CO_2}{1 \text{ mol } C_3H_8}$$

The conversion is

$$2.18 \cancel{\text{mol } C_3H_8} \times \frac{3 \text{ mol } CO_2}{1 \cancel{\text{mol } C_3H_8}} = 6.54 \text{ mol } CO_2$$

Then, using the molecular weight of CO_2 (44.0 grams), we calculate the mass of CO_2 produced:

$$6.54 \cancel{\text{mol } CO_2} \times \frac{44.0 \text{ g } CO_2}{1 \cancel{\text{mol } CO_2}} = 288 \text{ g } CO_2$$

Let's review the sequence of steps to find the mass of carbon dioxide produced from 96.1 grams of propane.

$$96.1 \text{ g } C_3H_8 \xrightarrow{\frac{1 \text{ mol } C_3H_8}{44.1 \text{ g } C_3H_8}} 2.18 \text{ mol } C_3H_8 \xrightarrow{\frac{3 \text{ mol } CO_2}{1 \text{ mol } C_3H_8}} 6.54 \text{ mol } CO_2$$

$$\xrightarrow{\frac{44.0 \text{ g } CO_2}{1 \text{ mol } CO_2}} 288 \text{ g } CO_2$$

The Steps for Calculating Masses of Reactants and Products in Chemical Reactions

- STEP 1
  Balance the equation for the reaction.
- STEP 2
  Convert known masses of the substances to moles.
- STEP 3
  Use the balanced equation to set up the appropriate mole ratios.
- STEP 4
  Use the mole ratios to calculate the number of moles of the desired reactant or product.
- STEP 5
  Convert from moles back to grams if required by the problem.

Sample Exercise 3.17

Solid lithium hydroxide is used in space vehicles to remove exhaled carbon dioxide from the living environment to form solid lithium carbonate and liquid water. What mass of gaseous carbon dioxide can be absorbed by 1.00 kg of lithium hydroxide?

Solution

STEP 1

Using the description of the reaction, we can write the unbalanced equation:

$$\mathrm{LiOH}(s) + \mathrm{CO_2}(g) \rightarrow \mathrm{Li_2CO_3}(s) + \mathrm{H_2O}(l)$$

The balanced equation is

$$2\mathrm{LiOH}(s) + \mathrm{CO_2}(g) \rightarrow \mathrm{Li_2CO_3}(s) + \mathrm{H_2O}(l)$$

STEP 2

We convert the given mass of LiOH to moles, using the molecular weight of LiOH (6.941 + 15.999 + 1.008 = 23.95):

$$1.00\ \cancel{\mathrm{kg\ LiOH}} \times \frac{1000\ \cancel{\mathrm{g\ LiOH}}}{1\ \cancel{\mathrm{kg\ LiOH}}} \times \frac{1\ \mathrm{mol\ LiOH}}{23.95\ \cancel{\mathrm{g\ LiOH}}} = 41.8\ \mathrm{mol\ LiOH}$$

STEP 3

The appropriate mole ratio is

$$\frac{1\ \mathrm{mol\ CO_2}}{2\ \mathrm{mol\ LiOH}}$$

STEP 4

We calculate the moles of CO_2 needed to react with the given mass of LiOH, using this mole ratio:

$$41.8\ \cancel{\mathrm{mol\ LiOH}} \times \frac{1\ \mathrm{mol\ CO_2}}{2\ \cancel{\mathrm{mol\ LiOH}}} = 20.9\ \mathrm{mol\ CO_2}$$

In a space vehicle, such as the space shuttle Discovery, exhaled carbon dioxide must be removed from the air.

Sample Exercise 3.17, continued

STEP 5

We calculate the mass of CO_2, using its molecular weight (44.0):

$$20.9 \cancel{\text{mol } CO_2} \times \frac{44.0 \text{ g } CO_2}{1 \cancel{\text{mol } CO_2}} = 9.20 \times 10^2 \text{ g } CO_2$$

Thus 920 g of $CO_2(g)$ will be absorbed by 1.00 kg of $LiOH(s)$.

Sample Exercise 3.18

Baking soda ($NaHCO_3$) is often used as an antacid. It neutralizes excess hydrochloric acid secreted by the stomach.

$$NaHCO_3(s) + HCl(aq) \rightarrow NaCl(aq) + H_2O(l) + CO_2(aq)$$

Milk of magnesia, which is an aqueous suspension of magnesium hydroxide, is also used as an antacid.

$$Mg(OH)_2(s) + 2HCl(aq) \rightarrow 2H_2O(l) + MgCl_2(aq)$$

Which is the more effective antacid per gram, $NaHCO_3$ or $Mg(OH)_2$?

Solution

To answer the question, we need to determine the amount of HCl neutralized per gram of $NaHCO_3$ and per gram of $Mg(OH)_2$.

Using the molecular weight of $NaHCO_3$ (84.01 g/mol), we can determine the moles of $NaHCO_3$ in 1.00 g of $NaHCO_3$:

$$1.00 \cancel{\text{g } NaHCO_3} \times \frac{1 \text{ mol } NaHCO_3}{84.01 \cancel{\text{g } NaHCO_3}} = 1.19 \times 10^{-2} \text{ mol } NaHCO_3$$

Next we determine the moles of HCl, using the mole ratio 1 mol HCl/1 mol $NaHCO_3$:

$$1.19 \times 10^{-2} \cancel{\text{mol } NaHCO_3} \times \frac{1 \text{ mol HCl}}{1 \cancel{\text{mol } NaHCO_3}} = 1.19 \times 10^{-2} \text{ mol HCl}$$

Thus 1.00 g of $NaHCO_3$ will neutralize 1.19×10^{-2} mol HCl.

Using the molecular weight of $Mg(OH)_2$ (58.32), we determine the moles of $Mg(OH)_2$ in 1.00 g:

$$1.00 \cancel{\text{g } Mg(OH)_2} \times \frac{1 \text{ mol } Mg(OH)_2}{58.32 \cancel{\text{g } Mg(OH)_2}} = 1.71 \times 10^{-2} \text{ mol } Mg(OH)_2$$

To determine the moles of HCl that will react with this amount of $Mg(OH)_2$, we use the mole ratio 2 mol HCl/1 mol $Mg(OH)_2$:

$$1.71 \times 10^{-2} \cancel{\text{mol } Mg(OH)_2} \times \frac{2 \text{ mol HCl}}{1 \cancel{\text{mol } Mg(OH)_2}} = 3.42 \times 10^{-2} \text{ mol HCl}$$

Thus 1.00 g of $Mg(OH)_2$ will neutralize 3.42×10^{-2} mol HCl, and it is a better antacid per gram than $NaHCO_3$.

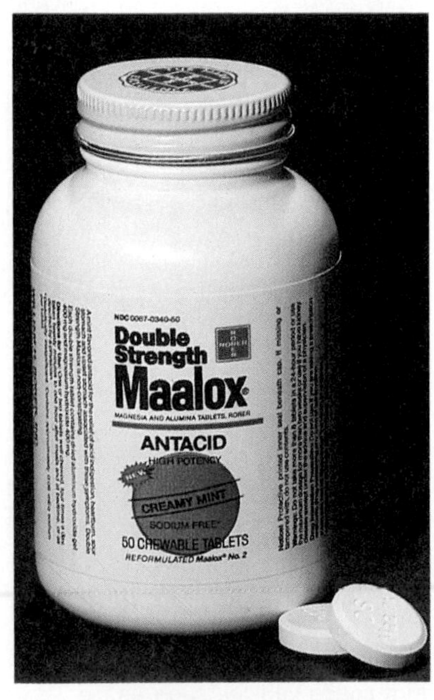

The antacid Maalox contains a mixture of magnesium hydroxide and aluminum hydroxide.

3.9 Calculations Involving a Limiting Reactant

Purpose

- To show how to recognize the limiting reactant.
- To demonstrate the use of the limiting reactant in stoichiometric calculations.

When chemicals are mixed together to undergo a reaction, they are often mixed in **stoichiometric quantities,** that is, in exactly the correct amounts so that all reactants "run out" (are used up) at the same time. To clarify this concept, let's consider the production of hydrogen for use in the manufacture of ammonia by the **Haber process.** Ammonia, a very important fertilizer itself and a starting material for other fertilizers, is made by combining nitrogen from the air with hydrogen according to the equation:

The details of the Haber process are discussed in Section 19.2.

$$N_2(g) + 3H_2(g) \rightarrow 2NH_3(g)$$

The hydrogen for this process is produced from the reaction of methane with water:

$$CH_4(g) + H_2O(g) \rightarrow 3H_2(g) + CO(g)$$

What mass of water is required to react *exactly* with 2.50×10^3 kilograms of methane? That is, how much water will just use up all of the 2.50×10^3 kilograms of methane, leaving no methane or water remaining? Using the principles developed in the last section, we can calculate that if 2.50×10^3 kilograms of methane is mixed with 2.81×10^3 kilograms of water, both reactants will "run out" at the same time. The reactants have been mixed in stoichiometric quantities.

If, on the other hand, 2.50×10^3 kilograms of methane is mixed with 3.00×10^3 kilograms of water, the methane will be consumed before the water runs out. The water will be in *excess*. In this case, the quantity of products formed will be determined by the quantity of methane present. Once the methane is consumed, no more products can be formed, even though some water still remains. In this situation, because the amount of methane *limits* the amount of products that can be formed, it is called the **limiting reactant,** or **limiting reagent.** In any stoichiometry problem, it is essential to determine which reactant is the limiting one in order to calculate correctly the amounts of products that will be formed.

Suppose 25.0 kilograms of nitrogen and 5.00 kilograms of hydrogen are mixed and reacted to form ammonia. How do we calculate the mass of ammonia produced when this reaction is run to completion?

To solve this problem, we must use the balanced equation

$$N_2(g) + 3H_2(g) \rightarrow 2NH_3(g)$$

to determine whether nitrogen or hydrogen is a limiting reactant and then to determine the amount of ammonia that is formed. We first calculate the moles of reactants present:

$$25.0\ \cancel{\text{kg N}_2} \times \frac{1000\ \cancel{\text{g N}_2}}{1\ \cancel{\text{kg N}_2}} \times \frac{1\ \text{mol N}_2}{28.0\ \cancel{\text{g N}_2}} = 8.93 \times 10^2\ \text{mol N}_2$$

$$5.00\ \cancel{\text{kg H}_2} \times \frac{1000\ \cancel{\text{g H}_2}}{1\ \cancel{\text{kg H}_2}} \times \frac{1\ \text{mol H}_2}{2.016\ \cancel{\text{g H}_2}} = 2.48 \times 10^3\ \text{mol H}_2$$

Since 1 mole of N_2 reacts with 3 moles of H_2, the number of moles of H_2 that will react exactly with 8.93×10^2 moles of N_2 is

$$8.93 \times 10^2 \cancel{\text{mol N}_2} \times \frac{3 \text{ mol H}_2}{1 \cancel{\text{mol N}_2}} = 2.68 \times 10^3 \text{ mol H}_2$$

Thus 8.93×10^2 moles of N_2 require 2.68×10^3 moles of H_2 to react completely. However, in this case, only 2.48×10^3 moles of H_2 are present. This means that the hydrogen will be consumed before the nitrogen. Thus hydrogen is the *limiting reactant* in this particular situation, and we must use the amount of hydrogen to compute the quantity of ammonia formed:

Always check to see which reactant is limiting.

$$2.48 \times 10^3 \cancel{\text{mol H}_2} \times \frac{2 \text{ mol NH}_3}{3 \cancel{\text{mol H}_2}} = 1.65 \times 10^3 \text{ mol NH}_3$$

Converting moles to kilograms gives

$$1.65 \times 10^3 \cancel{\text{mol NH}_3} \times \frac{17.0 \text{ g NH}_3}{1 \cancel{\text{mol NH}_3}} = 2.80 \times 10^4 \text{ g NH}_3 = 28.0 \text{ kg NH}_3$$

Note that to determine the limiting reactant, we could have started instead with the given amount of hydrogen and calculated the moles of nitrogen required:

$$2.48 \times 10^3 \cancel{\text{mol H}_2} \times \frac{1 \text{ mol N}_2}{3 \cancel{\text{mol H}_2}} = 8.27 \times 10^2 \text{ mol N}_2$$

Thus 2.48×10^3 moles of H_2 require 8.27×10^2 moles of N_2. Since 8.93×10^2 moles of N_2 are actually present, the nitrogen is in excess. The hydrogen will run out first, and thus again we find that hydrogen limits the amount of ammonia formed.

A related, but simpler way to determine which reactant is limiting is to compare the mole ratio of the substances required by the balanced equation to the mole ratio of reactants actually present. For example, in this case the mole ratio of H_2 to N_2 required by the balanced equation is

$$\frac{3 \text{ mol H}_2}{1 \text{ mol N}_2}$$

That is,

$$\frac{\text{mol H}_2}{\text{mol N}_2}(\text{required}) = \frac{3}{1} = 3$$

In this experiment, we have 2.48×10^3 mol H_2 and 8.93×10^2 mol N_2. Thus the ratio

$$\frac{\text{mol H}_2}{\text{mol N}_2}(\text{actual}) = \frac{2.48 \times 10^3}{8.93 \times 10^2} = 2.78$$

Since 2.78 is less than 3, the actual mole ratio of H_2 to N_2 is too small, and H_2 must be limiting. If the actual H_2 to N_2 mole ratio had been greater than 3, then the H_2 would have been in excess and the N_2 would be limiting.

Sample Exercise 3.19

Nitrogen gas can be prepared by passing gaseous ammonia over solid copper(II) oxide at high temperatures. The other products of the reaction are solid copper and water vapor. If 18.1 g of NH_3 is reacted with 90.4 g of CuO, which is the limiting reactant? How many grams of N_2 will be formed?

Solution

From the description of the problem, we obtain the balanced equation:

$$2NH_3(g) + 3CuO(s) \rightarrow N_2(g) + 3Cu(s) + 3H_2O(g)$$

Next we must compute the moles of NH_3 (mol. wt. = 17.0) and of CuO (mol. wt. = 79.5):

$$18.1\ \cancel{\text{g NH}_3} \times \frac{1\ \text{mol NH}_3}{17.0\ \cancel{\text{g NH}_3}} = 1.06\ \text{mol NH}_3$$

$$90.4\ \cancel{\text{g CuO}} \times \frac{1\ \text{mol CuO}}{79.5\ \cancel{\text{g CuO}}} = 1.14\ \text{mol CuO}$$

To determine the limiting reactant, we use the mole ratio for CuO and NH_3:

$$1.06\ \cancel{\text{mol NH}_3} \times \frac{3\ \text{mol CuO}}{2\ \cancel{\text{mol NH}_3}} = 1.59\ \text{mol CuO}$$

Thus 1.59 mol of CuO is required to react with 1.06 mol of NH_3. Since only 1.14 mol of CuO is actually present, the amount of CuO is limiting; CuO will run out before NH_3 does. We can verify this conclusion by comparing the mole ratio of CuO and NH_3 required by the balanced equation

$$\frac{\text{mol CuO}}{\text{mol NH}_3}(\text{required}) = \frac{3}{2} = 1.5$$

to the mole ratio actually present

$$\frac{\text{mol CuO}}{\text{mol NH}_3}(\text{actual}) = \frac{1.14}{1.06} = 1.08$$

Since the actual ratio is too small (smaller than 1.5), CuO is the limiting reactant.

Since CuO is the limiting reactant, it is the amount of CuO that must be used in calculating the amount of N_2 formed. From the balanced equation the mole ratio between CuO and N_2 is

$$\frac{1\ \text{mol N}_2}{3\ \text{mol CuO}}$$

$$1.14\ \cancel{\text{mol CuO}} \times \frac{1\ \text{mol N}_2}{3\ \cancel{\text{mol CuO}}} = 0.380\ \text{mol N}_2$$

Using the molecular weight of N_2 (28.0), we can calculate the mass of N_2 produced:

$$0.380\ \cancel{\text{mol N}_2} \times \frac{28.0\ \text{g N}_2}{1\ \cancel{\text{mol N}_2}} = 10.6\ \text{g N}_2$$

The amount of a given product formed when the limiting reactant is completely consumed is called the **theoretical yield** of that product. In Sample Exercise 3.19, 10.6 grams of nitrogen is the theoretical yield. This is the *maximum amount* of nitrogen that can be produced from the quantities of reactants used. Actually, the amount of product predicted by the theoretical yield is seldom obtained because of side reactions (other reactions that involve one or more of the reactants or products) and other complications. The *actual yield* of product is often given as a percentage of the theoretical yield. This is called the **percent yield:**

Percent yield is important as an indicator of the efficiency of a particular laboratory or industrial reaction.

$$\frac{\text{Actual yield}}{\text{theoretical yield}} \times 100 = \text{percent yield}$$

For example, if the reaction considered in Sample Exercise 3.19 actually gave 6.63 grams of nitrogen instead of the predicted 10.6 grams, the percent yield of nitrogen would be

$$\frac{6.63 \cancel{\text{g N}_2}}{10.6 \cancel{\text{g N}_2}} \times 100 = 62.5\%$$

Sample Exercise 3.20

Methanol (CH_3OH), also called methyl alcohol, is the simplest alcohol. It is used as a fuel in race cars and is a potential replacement for gasoline. Methanol can be manufactured by combination of gaseous carbon monoxide and hydrogen. Suppose 68.5 kg of $CO(g)$ is reacted with 8.60 kg of $H_2(g)$. Calculate the theoretical yield of methanol. If 3.57×10^4 g of CH_3OH is actually produced, what is the percent yield of methanol?

Solution

First, we must find out which reactant is limiting. The balanced equation is

$$2H_2(g) + CO(g) \rightarrow CH_3OH(l)$$

Next, we must calculate the moles of reactants:

Methanol is used as a fuel in Indianapolis-type racing cars.

$$68.5 \cancel{\text{kg CO}} \times \frac{1000 \cancel{\text{g CO}}}{1 \cancel{\text{kg CO}}} \times \frac{1 \text{ mol CO}}{28.01 \cancel{\text{g CO}}} = 2.45 \times 10^3 \text{ mol CO}$$

$$8.60 \cancel{\text{kg H}_2} \times \frac{1000 \cancel{\text{g H}_2}}{1 \cancel{\text{kg H}_2}} \times \frac{1 \text{ mol H}_2}{2.016 \cancel{\text{g H}_2}} = 4.27 \times 10^3 \text{ mol H}_2$$

To determine which reactant is limiting, we compare the mole ratio of H_2 and CO required by the balanced equation

$$\frac{\text{mol H}_2}{\text{mol CO}}(\text{required}) = \frac{2}{1} = 2$$

to the actual mole ratio

$$\frac{\text{mol H}_2}{\text{mol CO}}(\text{actual}) = \frac{4.27 \times 10^3}{2.45 \times 10^3} = 1.74$$

Since the actual mole ratio of H_2 to CO is smaller than the required ratio, *H_2 is limiting*. We therefore must use the amount of H_2 and the mole ratio between H_2 and

Sample Exercise 3.20, continued

CH_3OH to determine the maximum amount of methanol that can be produced:

$$4.27 \times 10^3 \cancel{\text{mol } H_2} \times \frac{1 \text{ mol } CH_3OH}{2 \cancel{\text{mol } H_2}} = 2.14 \times 10^3 \text{ mol } CH_3OH$$

Using the molecular weight of CH_3OH (32.04 g), we can calculate the theoretical yield in grams:

$$2.14 \times 10^3 \cancel{\text{mol } CH_3OH} \times \frac{32.04 \text{ g } CH_3OH}{1 \cancel{\text{mol } CH_3OH}} = 6.86 \times 10^4 \text{ g } CH_3OH$$

Thus, from the amounts of reactants given, the maximum amount of CH_3OH that can be formed is 6.86×10^4 g. This is the *theoretical yield.*

The percent yield is then

$$\frac{\text{Actual yield (grams)}}{\text{theoretical yield (grams)}} \times 100 = \frac{3.57 \times 10^4 \cancel{\text{g } CH_3OH}}{6.86 \times 10^4 \cancel{\text{g } CH_3OH}} \times 100 = 52.0\%$$

Steps for Solving a Stoichiometry Problem Involving Masses of Reactants and Products

- STEP 1
  Write and balance the equation for the reaction.
- STEP 2
  Convert known masses of substances to moles.
- STEP 3
  By comparing the mole ratio of reactants required by the balanced equation to the mole ratio of reactants actually present, determine which reactant is limiting.
- STEP 4
  Using the amount of the limiting reactant and the appropriate mole ratios, compute the number of moles of the desired product.
- STEP 5
  Convert from moles to grams, using the molecular weight.

Sample Exercise 3.21

Potassium chromate, a bright yellow solid, is produced by the reaction of solid chromite ore ($FeCr_2O_4$) with solid potassium carbonate and gaseous oxygen at high temperatures. The other products of the reaction are solid iron(III) oxide and gaseous carbon dioxide. In a particular experiment, 169 kg of chromite ore, 298 kg of potassium carbonate, and 75.0 kg of oxygen were sealed into a reaction vessel and reacted at a high temperature. The amount of potassium chromate obtained was 194 kg. Calculate the percent yield of potassium chromate.

Chromite ore shown with metallic chromium.

Sample Exercise 3.21, continued

Solution

STEP 1

The unbalanced equation, which can be written from the above description of the reaction, is

$$FeCr_2O_4(s) + K_2CO_3(s) + O_2(g) \rightarrow K_2CrO_4(s) + Fe_2O_3(s) + CO_2(g)$$

The balanced equation is

$$4FeCr_2O_4(s) + 8K_2CO_3(s) + 7O_2(g) \rightarrow 8K_2CrO_4(s) + 2Fe_2O_3(s) + 8CO_2(g)$$

STEP 2

The numbers of moles of the various reactants are:

$$169 \cancel{\text{kg FeCr}_2\text{O}_4} \times \frac{1000 \cancel{\text{g FeCr}_2\text{O}_4}}{1 \cancel{\text{kg FeCr}_2\text{O}_4}} \times \frac{1 \text{ mol FeCr}_2\text{O}_4}{223.84 \cancel{\text{g FeCr}_2\text{O}_4}} = 7.55 \times 10^2 \text{ mol FeCr}_2\text{O}_4$$

$$298 \cancel{\text{kg K}_2\text{CO}_3} \times \frac{1000 \cancel{\text{g K}_2\text{CO}_3}}{1 \cancel{\text{kg K}_2\text{CO}_3}} \times \frac{1 \text{ mol K}_2\text{CO}_3}{138.21 \cancel{\text{g K}_2\text{CO}_3}} = 2.16 \times 10^3 \text{ mol K}_2\text{CO}_3$$

$$75.0 \cancel{\text{kg O}_2} \times \frac{1000 \cancel{\text{g O}_2}}{1 \cancel{\text{kg O}_2}} \times \frac{1 \text{ mol O}_2}{32.00 \cancel{\text{g O}_2}} = 2.34 \times 10^3 \text{ mol O}_2$$

STEP 3

Now we must determine which of the three reactants is limiting. To do this, we will compare the mole ratios of reactants required by the balanced equation to the actual mole ratios. For the reactants K_2CO_3 and $FeCr_2O_4$ the required mole ratio is

$$\frac{\text{mol K}_2\text{CO}_3}{\text{mol FeCr}_2\text{O}_4}(\text{required}) = \frac{8}{4} = 2$$

while the actual mole ratio is

$$\frac{\text{mol K}_2\text{CO}_3}{\text{mol FeCr}_2\text{O}_4}(\text{actual}) = \frac{2.16 \times 10^3}{7.55 \times 10^2} = 2.86$$

Since the actual mole ratio is greater than that required, the K_2CO_3 is in excess compared to $FeCr_2O_4$. Thus either $FeCr_2O_4$ or O_2 must be limiting. To determine which of these will limit the amounts of products, we compare the required mole ratio

$$\frac{\text{mol O}_2}{\text{mol FeCr}_2\text{O}_4}(\text{required}) = \frac{7}{4} = 1.75$$

to the actual mole ratio

$$\frac{\text{mol O}_2}{\text{mol FeCr}_2\text{O}_4}(\text{actual}) = \frac{2.34 \times 10^3}{7.55 \times 10^2} = 3.10$$

Sample Exercise 3.21, continued

Since the actual mole ratio of O_2 to $FeCr_2O_4$ exceeds what is required, the O_2 is in excess of the $FeCr_2O_4$. Thus $FeCr_2O_4$ is the limiting reagent. Since this is the first case we have considered that involves three reactants, we will verify this conclusion by calculating the amounts (in moles) of K_2CO_3 and O_2 required by the amount of $FeCr_2O_4$ present:

$$7.55 \times 10^2 \cancel{\text{mol FeCr}_2\text{O}_4} \times \frac{8 \text{ mol K}_2\text{CO}_3}{4 \cancel{\text{mol FeCr}_2\text{O}_4}} = 1.51 \times 10^3 \text{ mol K}_2\text{CO}_3 \text{ required}$$

↑ Compare to 2.16×10^3 mol K_2CO_3 present

$$7.55 \times 10^2 \cancel{\text{mol FeCr}_2\text{O}_4} \times \frac{7 \text{ mol O}_2}{4 \cancel{\text{mol FeCr}_2\text{O}_4}} = 1.32 \times 10^3 \text{ mol O}_2 \text{ required}$$

↑ Compare to 2.34×10^3 mol O_2 present

STEP 4

Thus more K_2CO_3 and O_2 are present than required. These reactants are in excess, so $FeCr_2O_4$ is the limiting reactant. We must use the amount of $FeCr_2O_4$ to calculate the maximum amount of K_2CrO_4 that can be formed:

$$7.55 \times 10^2 \cancel{\text{mol FeCr}_2\text{O}_4} \times \frac{8 \text{ mol K}_2\text{CrO}_4}{4 \cancel{\text{mol FeCr}_2\text{O}_4}} = 1.51 \times 10^3 \text{ mol K}_2\text{CrO}_4$$

STEP 5

Using the molecular weight of K_2CrO_4, we can determine the mass.

$$1.51 \times 10^3 \cancel{\text{mol K}_2\text{CrO}_4} \times \frac{194.19 \text{ g K}_2\text{CrO}_4}{1 \cancel{\text{mol K}_2\text{CrO}_4}} = 2.93 \times 10^5 \text{ g K}_2\text{CrO}_4$$

This represents the theoretical yield of K_2CrO_4. The actual yield was 194 kg, or 1.94×10^5 g. Thus the percent yield is

$$\frac{1.94 \times 10^5 \cancel{\text{g K}_2\text{CrO}_4}}{2.93 \times 10^5 \cancel{\text{g K}_2\text{CrO}_4}} \times 100 = 66.2\%$$

FOR REVIEW

Summary

Stoichiometry deals with the quantities of materials consumed and produced in chemical reactions. It is, in effect, chemical arithmetic. Atomic weights (masses) are based on the average value of the masses of the naturally occurring isotopes. All atomic weights are based on a mass scale that assigns exactly 12 atomic mass units to a ^{12}C atom.

A mole is a unit of measure equal to the number of carbon atoms in exactly 12 grams of pure ^{12}C. This number has been determined experimentally to be 6.02214×10^{23}, which is called Avogadro's number. One mole of any substance contains Avogadro's number of units. One mole of an element has a mass equal to the element's atomic weight in grams.

The molecular weight or molar mass of a compound is the mass in grams of one mole of the compound and is computed by summing the atomic weights of its constituent atoms.

Percent composition represents the mass percent of each element in a compound:

$$\text{Mass percent} = \frac{\text{mass of element in 1 mol of substance}}{\text{mass of 1 mol of substance}} \times 100$$

The empirical formula is the simplest whole number ratio of the various types of atoms in a compound and can be derived from the percent composition of the compound. The molecular formula is the exact formula of a molecule of a substance and is always an integer multiple of the empirical formula.

In a chemical reaction atoms are neither created nor destroyed; they are merely reorganized. All atoms present in the reactants must be accounted for among the products. A chemical equation represents the chemical reaction showing reactants on the left side of an arrow and products on the right. A balanced equation gives the relative numbers of reactant and product molecules.

Amounts of reactants consumed and products formed can be calculated from the balanced equation for a reaction by using the mole ratios relating the reactants and products. A limiting reactant (reagent) is the one that is consumed first, and thus it determines how much product will be formed.

The theoretical yield of a product is the maximum amount that can be produced with a given amount of the limiting reactant. The actual yield is always less than this value and is usually represented as the percent yield, a percentage of the theoretical yield.

Key Terms

chemical stoichiometry

Section 3.1

mass spectrometer
atomic weight

Section 3.2

mole
Avogadro's number

Section 3.3

molar mass
molecular weight

Section 3.5

empirical formula
molecular formula

Section 3.6

chemical equation
reactants
products
balancing a chemical equation

Section 3.8

mole ratio

Section 3.9

stoichiometric quantities
Haber process
limiting reactant
(limiting reagent)
theoretical yield
percent yield

Exercises

A blue exercise number indicates that the answer to that exercise appears at the back of this book and a solution appears in the Solutions Guide.

Atomic Masses and the Mass Spectrometer

1. The atomic mass of boron (B) is given in the periodic table as 10.81, yet no single atom of boron has a mass of 10.81 amu. Explain.

2. The element magnesium (Mg) has three stable isotopes with the following masses and abundances:

Isotope	Mass (amu)	Abundance
^{24}Mg	23.9850	78.99%
^{25}Mg	24.9858	10.00%
^{26}Mg	25.9826	11.01%

Calculate the average atomic mass (the atomic weight) of magnesium from these data.

3. The element europium exists in nature as two isotopes: ^{151}Eu has a mass of 150.9196 amu, and ^{153}Eu has a mass of 152.9209 amu. The average atomic mass of europium is 151.96 amu. Calculate the relative abundance of the two europium isotopes.

4. The element silver (Ag) has two naturally occurring isotopes: ^{107}Ag with a mass of 106.905 amu, and ^{109}Ag. Silver consists of 51.82% ^{107}Ag and has an average atomic mass of 107.868 amu. Calculate the mass of ^{109}Ag.

5. Assume that element ^{117}(Uus) is synthesized and that it has the following stable isotopes.

^{284}Uus (283.4 amu)	34.60%
^{285}Uus (284.7 amu)	21.20%
^{288}Uus (287.8 amu)	44.20%

What is the value of the atomic weight that would be listed on the periodic table?

6. The listed atomic mass of chromium is 52.00 amu. Does this mean that chromium can only have a single isotope with mass 52.00 amu? Why or why not?

7. An element consists of 90.51% of an isotope with a mass of 19.992 amu, 0.27% of an isotope with a mass of 20.994 amu, and 9.22% of an isotope with a mass of 21.990 amu. Calculate the average atomic mass and identify the element.

8. The mass spectrum of bromine (Br_2) consists of three peaks with the following relative sizes:

Mass (amu)	Relative size
157.84	0.2534
159.84	0.5000
161.84	0.2466

How do you interpret these data?

9. Naturally occurring tellurium (Te) has the following isotopic abundances:

Isotope	Abundance	Mass (amu)
^{120}Te	0.09%	119.90
^{122}Te	2.46%	121.90
^{123}Te	0.87%	122.90
^{124}Te	4.61%	123.90
^{125}Te	6.99%	124.90
^{126}Te	18.71%	125.90
^{128}Te	31.79%	127.90
^{130}Te	34.48%	129.91

Draw the mass spectrum of H_2Te, assuming that the only hydrogen isotope present is ^1H (mass, 1.008).

Moles and Molecular Weights

10. Give the number of moles of each element present in 1.0 mol of each of the following substances:
    a. NH_3
    b. N_2H_4
    c. $(NH_4)_2Cr_2O_7$
    d. $CoCl_2 \cdot 6H_2O$
    e. $Ca_3(PO_4)_2$
    f. Na_2HPO_4

11. How many atoms of each element are present in 1.0 mol of each of the substances in Exercise 10?

12. How many grams of each element are present in 1.0 mol of each of the substances in Exercise 10?

13. How many moles of each element are present in 1.0 g of each of the substances in Exercise 10?

14. How many atoms of each element are present in 1.0 g of each of the substances in Exercise 10?

15. Determine the mass in grams of:
    a. 3.00×10^{20} N_2 molecules
    b. 3.00×10^{-3} mol of N_2
    c. 1.5×10^2 mol of N_2
    d. a single N_2 molecule
    e. 2.00×10^{-15} mol of N_2
    f. 18.0 picomoles of N_2
    g. 5.0 nanomoles of N_2

16. Aluminum metal is produced by passing an electric current through a solution of aluminum oxide (Al_2O_3) dissolved in molten cryolite (Na_3AlF_6). Calculate the molecular weights of Al_2O_3 and Na_3AlF_6.

17. Ascorbic acid, or vitamin C ($C_6H_8O_6$), is an essential vitamin. It cannot be stored by the body and must be present in the diet. What is the molecular weight of ascorbic acid? Vitamin C tablets are often taken as a dietary supplement. If a typical tablet contains 500.0 mg of vitamin C, how many moles and how many molecules of vitamin C does it contain?

18. How many moles are represented by each of these samples?
    a. 100 molecules (exactly) of H_2O
    b. 100.0 g of H_2O
    c. 500 atoms (exactly) of Fe
    d. 500.0 g of Fe
    e. 150 molecules (exactly) of O_2
    f. 150.0 g of Fe_2O_3
    g. 10.0 mg of NO_2
    h. 1.0 femtomoles of NO_2
    i. 1.5×10^{16} molecules of BF_3
    j. 2.6 mg of BF_3

19. Aspartame is an artificial sweetener that is 160 times sweeter than sucrose (table sugar) when dissolved in water. It is marketed as Nutra-Sweet. The molecular formula of aspartame is $C_{14}H_{18}N_2O_5$.
    a. Calculate the molecular weight of aspartame.
    b. How many moles of molecules are in 10.0 g of aspartame?
    c. What is the mass in grams of 1.56 mol of aspartame?
    d. How many molecules are in 5.0 mg of aspartame?
    e. How many atoms of nitrogen are in 1.2 g of aspartame?
    f. What is the mass in grams of 1.0×10^9 molecules of aspartame?
    g. What is the mass in grams of one molecule of aspartame?

20. Dimethylnitrosamine, $(CH_3)_2N_2O$, is a carcinogenic (cancer-causing) substance that may be formed in foods, beverages, or gastric juices from the reaction of nitrite ion (used as a food preservative) with other substances.
    a. What is the molecular weight of dimethylnitrosamine?
    b. How many moles of $(CH_3)_2N_2O$ molecules are in 250 mg of dimethylnitrosamine?
    c. What is the mass of 0.050 mol of dimethylnitrosamine?
    d. How many atoms of hydrogen are in 1.0 mol of dimethylnitrosamine?
    e. What is the mass of 1.0×10^6 molecules of dimethylnitrosamine?
    f. What is the mass in grams of one molecule of dimethylnitrosamine?

21. The molecular formula of acetylsalicylic acid (aspirin), one of the most commonly used pain relievers, is $C_9H_8O_4$.
    a. Calculate the molecular weight of aspirin.
    b. A typical aspirin tablet contains 500 mg of $C_9H_8O_4$. How many moles of $C_9H_8O_4$ molecules and how many molecules of acetylsalicylic acid are in a 500-mg tablet?

22. Chloral hydrate ($C_2H_3Cl_3O_2$) is a drug that is used as a sedative and hypnotic. It is the compound used to make "Mickey Finns" in detective stories.
    a. Calculate the molecular weight of chloral hydrate.
    b. How many moles of $C_2H_3Cl_3O_2$ molecules are in 500.0 g of chloral hydrate?
    c. What is the mass in grams of 2.0×10^{-2} mol of chloral hydrate?
    d. How many chlorine atoms are in 5.0 g of chloral hydrate?
    e. What mass of chloral hydrate would contain 1.0 g of Cl?
    f. What is the mass of exactly 500 molecules of chloral hydrate?

23. Calculate the molecular weight of each of the following substances:
    a. $CuCl_2 \cdot 6H_2O$ (The $\cdot 6H_2O$ indicates the presence of six water molecules.)
    b. $NaBrO_3$
    c. $(C_2F_4)_{500}$ (the polymer Teflon), which is made of the unit:

$$-\overset{\displaystyle F}{\underset{\displaystyle F}{\overset{|}{\underset{|}{C}}}}-\overset{\displaystyle F}{\underset{\displaystyle F}{\overset{|}{\underset{|}{C}}}}-$$

    d. H_3PO_4
    e. $CaCO_3$

24. How many atoms of carbon are present in 1.0 g of each of the following?
    a. CH_4O
    b. CH_3CO_2H
    c. Na_2CO_3
    d. $C_6H_{12}O_6$
    e. $C_{12}H_{22}O_{11}$
    f. $CHCl_3$

25. In the spring of 1984, concern arose over the presence of ethylene dibromide, or EDB, in grains and cereals. EDB has the molecular formula $C_2H_4Br_2$ and until 1984 was commonly used as a plant fumigant. The federal limit for EDB in finished cereal products is 30.0 parts per billion (ppb), where 1.0 ppb $= 10^{-9}$ g of EDB for every 1.0 g of sample. How many molecules of EDB are in 1.0 lb of flour if 30.0 ppb of EDB are present?

Percent Composition

26. Calculate the percent composition by mass of the substances in Exercises 1 and 59 in Chapter 2.

27. In 1987 the first substance to act as a superconductor at a temperature above that of liquid nitrogen (77 K) was discovered. The approximate formula of this substance is $YBa_2Cu_3O_7$. Calculate the percent composition by mass of this material.

28. Arrange the following substances in order of the increasing percent of phosphorus.
    a. Na_3PO_4
    b. PH_3
    c. P_4O_{10}
    d. $(NPCl_2)_3$

29. There are several important compounds that contain only nitrogen and oxygen. Calculate the mass percent of nitrogen in each of the following:

a. NO, a gas formed by the reaction of N_2 and O_2 in internal combustion engines
b. NO_2, a brown gas mainly responsible for the brownish color of photochemical smog
c. N_2O_4, a colorless liquid used as a fuel in space shuttles
d. N_2O, a colorless gas used as an anesthetic by dentists (known as laughing gas)

30. Photocells utilize a semiconducting material that will produce an electric current or a change in resistance on exposure to light. Compounds of cadmium (Cd) with an element from Group 6A of the periodic table are used in many common photocells. Calculate the mass percent of Cd in CdS, CdSe, and CdTe.

31. Vitamin B_{12}, cyanocobalamin, is essential for human nutrition. It is concentrated in animal tissue but not in higher plants. Although nutritional requirements for the vitamin are quite low, people who abstain completely from animal products may develop a deficiency anemia. Cyanocobalamin is the form used in vitamin supplements. It contains 4.34% cobalt by mass. Calculate the molecular weight (molar mass) of cyanocobalamin assuming there is one atom of cobalt in every molecule of cyanocobalamin.

32. Hemoglobin is the protein that transports oxygen in mammals. Hemoglobin is 0.342% Fe by mass, and each hemoglobin molecule contains four iron atoms. Calculate the molecular weight (molar mass) of hemoglobin.

33. The first binary compounds containing a noble gas element were prepared at the Argonne National Laboratories near Chicago in 1964. The first compound prepared was XeF_4. Two other xenon fluorides, XeF_2 and XeF_6, are also known. Calculate the mass percent of fluorine in each of these three compounds.

34. Calculate the percent composition by mass of the following compounds that are important starting materials for synthetic polymers?
a. $C_3H_4O_2$ (acrylic acid, from which acrylic plastics are made)
b. $C_4H_6O_2$ (methyl acrylate, from which plexiglass is made)
c. C_3H_3N (acrylonitrile, from which orlon is made)

35. Calculate the percent composition by mass of ethylene dibromide, aluminum oxide, and cryolite, compounds mentioned in Exercises 16 and 25.

Empirical and Molecular Formulas

36. What is the difference between the empirical and molecular formulas of a compound? Can they ever be the same? Why or why not?

37. Determine the molecular formulas to which the following empirical formulas and molecular weights pertain:
a. SNH, 188.32
b. $NPCl_2$, 347.66
c. CoC_4O_4, 341.94
d. SN, 184.27

38. Express the composition of each compound as the mass percents of its elements.
a. formaldehyde, CH_2O
b. glucose, $C_6H_{12}O_6$
c. acetic acid, $HC_2H_3O_2$

39. Considering your answer to Exercise 38, which type of formula, empirical or molecular, can be obtained from elemental analysis that gives percent composition?

40. What is the empirical formula for each of the following compounds (where the common names are given)?
a. vitamin C, $C_6H_8O_6$
b. benzene, C_6H_6
c. acetylene, C_2H_2
d. phosphorus(V) oxide, P_4O_{10}
e. glucose, $C_6H_{12}O_6$
f. acetic acid, $HC_2H_3O_2$

41. A compound that contains only carbon, hydrogen, and oxygen is 48.38% C and 8.12% H by mass. What is the empirical formula of this substance?

42. A compound contains only carbon, hydrogen, nitrogen, and oxygen. Combustion of 0.157 g of the compound produced 0.213 g of CO_2 and 0.0310 g of H_2O. In another experiment, it is found that 0.103 g of the compound produces 0.0230 g of NH_3. What is the empirical formula of the compound? *Hint:* Combustion involves reacting with excess O_2. Assume that all of the carbon ends up in CO_2 and all of the hydrogen ends up in H_2O. Also assume that all of the nitrogen ends up in the NH_3 in the second experiment.

43. A confiscated white substance, suspected of being cocaine, was purified by a forensic chemist and subjected to elemental analysis. Combustion of a 50.86-mg sample yielded 150.0 mg of CO_2 and 46.05 mg of H_2O. Analysis for nitrogen showed that the compound contained 9.39% N by mass. The formula of cocaine is $C_{17}H_{21}NO_4$. Can the forensic chemist conclude that the suspected compound is cocaine?

44. Calculate the empirical formula for each of the substances in Exercises 1 and 59 in Chapter 2.

45. The active ingredient in photographic fixer solution contains sodium, sulfur, and oxygen. Analysis of a sample shows that the sample contains 0.979 g Na, 1.365 g S, and 1.021 g O. What is the empirical formula of this substance?

46. A sample of urea contains 1.121 g N, 0.161 g H, 0.480 g C, and 0.640 g O. What is the empirical formula of urea?

47. A compound that contains only nitrogen and oxygen is 30.4% N by mass; the molecular weight (molar mass) of the compound is 92. What is the empirical formula of the compound? What is the molecular formula of the compound?

48. One of the most commonly used white pigments in paint is a compound of titanium and oxygen that contains 59.9% Ti by mass. Calculate the empirical formula of the compound.

49. A compound contains only carbon, hydrogen, and oxygen. Combustion of 10.68 mg of the compound yields 16.01 mg of CO_2 and 4.37 mg of H_2O. The molecular weight (molar mass) of the compound is 176.1. What are the empirical and molecular formulas of the compound?

50. A compound containing only sulfur and nitrogen is 69.6% S by mass; the molecular weight (molar mass) is 184. What are the empirical and molecular formulas of the compound?

51. Cumene is a hydrocarbon that is used in the production of acetone and phenol in the chemical industry. Combustion of 47.6 mg of cumene produces 156.8 mg of CO_2 and 42.8 mg of water. The molecular weight (molar mass) is between 115 and 125. Determine the empirical and molecular formulas.

Balancing Chemical Equations

52. Silicon is produced for the chemical and electronics industries by the following reactions. Give the balanced equation for each reaction.

a. $SiO_2(s) + C(s) \xrightarrow[\text{arc furnace}]{\text{Electric}} Si(s) + CO(g)$

b. Silicon tetrachloride is reacted with very pure magnesium, producing silicon and magnesium chloride.

c. $Na_2SiF_6(s) + Na(s) \rightarrow Si(s) + NaF(s)$

53. Balance the following reactions that occur during steel production in a blast furnace:

a. $Fe_2O_3 + CO \rightarrow Fe_3O_4 + CO_2$

b. $Fe_3O_4 + CO \rightarrow FeO + CO_2$

54. Give the balanced equation for each of the following chemical reactions:

a. $HNO_3(l) + P_4O_{10}(s) \rightarrow (HPO_3)_3(l) + N_2O_5(g)$

b. $I_4O_9(s) \rightarrow I_2O_6(s) + I_2(s) + O_2(g)$

c. Solid iron(III) sulfide reacts with gaseous hydrogen chloride to form solid iron(III) chloride and hydrogen sulfide gas.

d. Carbon disulfide liquid reacts with ammonia gas to produce hydrogen sulfide gas and solid ammonium thiocyanate, (NH_4SCN).

55. Balance the following equations representing combustion reactions:

a. $C_{12}H_{22}O_{11}(s) + O_2(g) \rightarrow CO_2(g) + H_2O(g)$

b. $C_6H_6(l) + O_2(g) \rightarrow CO_2(g) + H_2O(g)$

c. $Fe(s) + O_2(g) \rightarrow Fe_2O_3(s)$

d. $C_4H_{10}(g) + O_2(g) \rightarrow CO_2(g) + H_2O(g)$

e. $FeO(s) + O_2(g) \rightarrow Fe_2O_3(s)$

56. Balance the following equations:

a. $Ca(OH)_2(aq) + H_3PO_4(aq) \rightarrow H_2O(l) + Ca_3(PO_4)_2(s)$

b. $Al(OH)_3(s) + HCl(aq) \rightarrow AlCl_3(aq) + H_2O(l)$

c. $AgNO_3(aq) + H_2SO_4(aq) \rightarrow Ag_2SO_4(s) + HNO_3(aq)$

57. Phosphorus occurs naturally in the form of fluorapatite, $CaF_2 \cdot 3Ca_3(PO_4)_2$, the dot indicating 1 part CaF_2 to 3 parts $Ca_3(PO_4)_2$. This mineral is reacted with an aqueous solution of sulfuric acid in the preparation of a fertilizer. The products are phosphoric acid, hydrogen fluoride, and gypsum, $CaSO_4 \cdot 2H_2O$. Write and balance the chemical equation describing this process.

58. Balance the following equations:

a. $Cr(s) + S_8(s) \rightarrow Cr_2S_3(s)$

b. $NaHCO_3(s) \xrightarrow{\text{Heat}} Na_2CO_3(s) + CO_2(g) + H_2O(g)$

c. $KClO_3(s) \xrightarrow{\text{Heat}} KCl(s) + O_2(g)$

d. $Eu(s) + HF(g) \rightarrow EuF_3(s) + H_2(g)$

59. Write a balanced chemical equation that describes each of the following:

a. Pure indium metal reacts with oxygen gas to form solid indium(III) oxide.

b. The fermentation of fruit juice to produce wine involves the conversion of glucose ($C_6H_{12}O_6$) to ethanol (C_2H_6O) and carbon dioxide (CO_2).

c. Potassium metal reacts with water to produce aqueous potassium hydroxide and hydrogen gas.

d. Barium oxide reacts with nitric acid to produce barium nitrate and water.

60. Balance the following equations:

a. $Cu(s) + AgNO_3(aq) \rightarrow Ag(s) + Cu(NO_3)_2(aq)$

b. $Zn(s) + HCl(aq) \rightarrow ZnCl_2(aq) + H_2(g)$

c. $Au_2S_3(s) + H_2(g) \rightarrow Au(s) + H_2S(g)$

d. $Ca(s) + H_2O(l) \rightarrow Ca(OH)_2(aq) + H_2(g)$

61. Lead hydrogen arsenate, an inorganic insecticide still used against the potato beetle, is usually produced using the following reaction:

$$Pb(NO_3)_2(aq) + H_3AsO_4(aq) \rightarrow PbHAsO_4(s) + HNO_3(aq)$$

Balance this equation.

62. The electrolysis of concentrated brine solutions is an important source of NaOH, H_2, and Cl_2 for the chemical industry. The reaction is

$$NaCl(aq) + H_2O(l) \xrightarrow{\text{Electricity}} Cl_2(g) + H_2(g) + NaOH(aq)$$

Balance this equation.

Reaction Stoichiometry

63. Calculate the masses of Cr_2O_3, N_2, and H_2O produced from 10.8 g of $(NH_4)_2Cr_2O_7$ in an ammonium dichromate volcano reaction as described in Sample Exercise 3.14.

64. Over the years the thermite reaction has been used for welding railroad rails, in incendiary bombs, and to ignite solid fueled rocket motors. The reaction is:

$$Fe_2O_3(s) + 2Al(s) \rightarrow 2Fe(l) + Al_2O_3(s)$$

What masses of iron(III) oxide and aluminum must be used to produce 15.0 g of iron? What mass of aluminum oxide would be produced?

65. The reusable booster rockets of the U.S. space shuttle employ a mixture of aluminum and ammonium perchlorate for fuel. A possible equation for this reaction is

$$3Al(s) + 3NH_4ClO_4(s) \rightarrow Al_2O_3(s) + AlCl_3(s) + 3NO(g) + 6H_2O(g)$$

What mass of NH_4ClO_4 should be used in the fuel mixture for every kilogram of Al?

66. Nitric acid is produced commercially by the Ostwald process. The three steps of the Ostwald process are shown in the following equations:

$$4NH_3(g) + 5O_2(g) \rightarrow 4NO(g) + 6H_2O(g)$$

$$2NO(g) + O_2(g) \rightarrow 2NO_2(g)$$

$$3NO_2(g) + H_2O(l) \rightarrow 2HNO_3(aq) + NO(g)$$

What mass of NH_3 must be used to produce 1.0×10^6 kg of HNO_3 by the Ostwald process, assuming 100% yield in each reaction?

67. One of the major commercial uses of sulfuric acid is in the production of phosphoric acid and calcium sulfate. The phosphoric acid is used for fertilizer. The reaction is

$$Ca_3(PO_4)_2(s) + 3H_2SO_4(aq) \rightarrow 3CaSO_4(s) + 2H_3PO_4(aq)$$

What mass of concentrated sulfuric acid (98% H_2SO_4 by mass) must be used to react completely with 1.0×10^2 g of $Ca_3(PO_4)_2$?

68. One of relatively few reactions that takes place directly between two solids at room temperature is

$$Ba(OH)_2 \cdot 8H_2O(s) + NH_4SCN(s) \rightarrow Ba(SCN)_2(s) + H_2O(l) + NH_3(g)$$

The $\cdot 8H_2O$ in $Ba(OH)_2 \cdot 8H_2O$ indicates the presence of eight water molecules. This compound is called barium hydroxide octahydrate.

a. Balance the equation.
b. What mass of ammonium thiocyanate (NH_4SCN) must be used if it is to react completely with 6.5 g of barium hydroxide octahydrate?
c. What masses of the three products would be produced in the reaction of part b?

69. Elixirs such as Alka-Seltzer use the reaction of sodium bicarbonate with citric acid in aqueous solution to produce a fizz:

$$3NaHCO_3(aq) + C_6H_8O_7(aq) \rightarrow 3CO_2(g) + 3H_2O(l) + Na_3C_6H_5O_7(aq)$$

a. What mass of $C_6H_8O_7$ should be used for every 1.0×10^2 mg of $NaHCO_3$?
b. What mass of $CO_2(g)$ would be produced from such a mixture?

70. Aspirin ($C_9H_8O_4$) is synthesized by reacting salicylic acid ($C_7H_6O_3$) with acetic anhydride ($C_4H_6O_3$). The balanced reaction is

$$2C_7H_6O_3 + C_4H_6O_3 \rightarrow 2C_9H_8O_4 + H_2O$$

a. What mass of acetic anhydride will react completely with 1.00×10^2 g of salicylic acid?
b. What mass of aspirin will be produced in this reaction?

71. a. Write the balanced equation for the combustion of isooctane (C_8H_{18}) to produce water vapor and carbon dioxide gas.
b. Assuming gasoline is 100% isooctane, with a density of 0.692 g/mL, what mass of carbon dioxide is produced by the combustion of 1.2×10^{10} gal of gasoline (the approximate annual consumption of gasoline in the United States)?

Limiting Reactants and Percent Yield

72. Consider the reaction:

$$Mg(s) + I_2(s) \rightarrow MgI_2(s)$$

Identify the limiting reagent in each of the reaction mixtures below:

a. 100 atoms of Mg and 100 molecules of I_2
b. 150 atoms of Mg and 100 molecules of I_2
c. 200 atoms of Mg and 300 molecules of I_2
d. 0.16 mol Mg and 0.25 mol I_2
e. 0.14 mol Mg and 0.14 mol I_2
f. 0.12 mol Mg and 0.08 mol I_2
g. 6.078 g Mg and 63.455 g I_2
h. 1.00 g Mg and 2.00 g I_2
i. 1.00 g Mg and 20.00 g I_2

73. Consider the reaction

$$2H_2(g) + O_2(g) \rightarrow 2H_2O(g)$$

Identify the limiting reagent in each of the reaction mixtures given below:

a. 50 molecules H_2 and 25 molecules O_2
b. 100 molecules H_2 and 40 molecules O_2
c. 100 molecules H_2 and 100 molecules O_2
d. 0.50 mol H_2 and 0.75 mol O_2
e. 0.80 mol H_2 and 0.75 mol O_2
f. 1.0 g H_2 and 0.25 mol O_2
g. 5.00 g H_2 and 56.00 g O_2

74. Consider the reaction

$$4Al(s) + 3O_2(g) \rightarrow 2Al_2O_3(s)$$

Identify the limiting reagent in each of the following reaction mixtures:

a. 1.0 mol Al and 1.0 mol O_2
b. 2.0 mol Al and 4.0 mol O_2
c. 0.50 mol Al and 0.75 mol O_2
d. 64.75 g Al and 115.21 g O_2
e. 75.89 g Al and 112.25 g O_2
f. 51.28 g Al and 118.22 g O_2

75. When copper is heated with an excess of sulfur, copper(I) sulfide is the product. When 1.50 g of copper was heated with excess sulfur, 1.76 g of copper(I) sulfide was isolated. What is the theoretical yield? What was the actual percent yield?

76. Aluminum burns in bromine, producing aluminum bromide:

$$2Al(s) + 3Br_2(l) \rightarrow 2AlBr_3(s)$$

When 6.0 g of aluminum was reacted with an excess of bromine, 50.3 g of aluminum bromide was isolated. Calculate the theoretical and percent yield of this reaction.

77. When a mixture of silver metal and sulfur is heated, silver sulfide is formed:

$$16Ag(s) + S_8(s) \xrightarrow{\text{Heat}} 8Ag_2S(s)$$

a. What mass of Ag_2S is produced from a mixture of 2.0 g of Ag and 2.0 g of S?
b. What mass of which reactant is left unreacted?

78. Ammonia is produced from the reaction of nitrogen and hydrogen according to the following balanced chemical equation:

$$N_2(g) + 3H_2(g) \rightarrow 2NH_3(g)$$

a. What mass of ammonia is produced from a mixture of 1.00×10^3 g of N_2 and 5.00×10^2 g of H_2?
b. What mass of which starting material remains unreacted?

79. Hydrogen cyanide is produced industrially from the reaction of gaseous ammonia, oxygen, and methane.

$$2NH_3(g) + 3O_2(g) + 2CH_4(g) \rightarrow 2HCN(g) + 6H_2O(g)$$

If 5.00×10^3 kg each of NH_3, O_2, and CH_4 are reacted, what mass of HCN and of H_2O would be produced, assuming 100% yield?

80. The production capacity for acrylonitrile (C_3H_3N) in the United States is over 2 billion pounds per year. Acrylonitrile, the building block for polyacrylonitrile fibers and a variety of plastics, is produced from gaseous propylene, ammonia, and oxygen.

$$2C_3H_6(g) + 2NH_3(g) + 3O_2(g) \rightarrow 2C_3H_3N(g) + 6H_2O(g)$$

a. What mass of acrylonitrile can be produced from a mixture of 5.00×10^2 g of propylene, 5.00×10^2 g of ammonia, and 1.00×10^3 g of oxygen?
b. What mass of water is produced, and what masses of which starting materials are left in excess?

81. Anthraquinone ($C_{14}H_8O_2$), an important intermediate in the dye industry, is produced from the reaction of benzene (C_6H_6) with phthalic anhydride ($C_8H_4O_3$), followed by dehydration with sulfuric acid. The overall reaction is

$$C_8H_4O_3(s) + C_6H_6(l) \rightarrow C_{14}H_8O_2(s) + H_2O(l)$$

a. What mass of benzene will react completely with 2.00×10^3 g of phthalic anhydride?
b. What mass of anthroquinone and of water will be produced assuming 100% yield?
c. If 1.96×10^3 g of anthroquinone is actually obtained, what is the percent yield of the reaction?

82. A student prepared aspirin in a laboratory experiment using the reaction in Exercise 70. The student reacted 1.50 g salicylic acid with 2.00 g of acetic anhydride. The yield was 1.50 g of aspirin. Calculate the theoretical yield and the percent yield for this experiment.

83. Hexamethylenediamine, $C_6H_{16}N_2$, is one of the starting materials for the production of nylon. It can be prepared from adipic acid, $C_6H_{10}O_4$, by the following overall reaction:

$$C_6H_{10}O_4(l) + 2NH_3(g) + 4H_2(g) \rightarrow C_6H_{16}N_2(l) + 4H_2O(l)$$

a. What mass of hexamethylenediamine can be produced from 1.00×10^3 g of adipic acid?
b. What is the percent yield if 765 g of hexamethylenediamine is made from 1.00×10^3 g of adipic acid?

Additional Exercises

84. Only one isotope of this element occurs in nature. One atom of this isotope has a mass of 9.123×10^{-23} g. Identify the element and give its atomic weight.

85. Automobile safety glass is made by laminating a sheet of polyvinylbutyral between two thin sheets of glass. Polyvinylbutyral has the general formula (the fragment shown is called a monomer unit):

$$\left[-CHCH_2CHCH_2- \right]_n \quad \text{(the two CH carbons each bonded to O; the two O atoms joined through } CH(CH_2CH_2CH_3))$$

If the average molecular weight of a sample of polyvinylbutyral is 100,000, what is the average number of monomer units present in the molecule?

86. The compound cisplatin, $Pt(NH_3)_2Cl_2$, has been extensively studied as a potential antitumor agent [see the *Journal of Chemical Education:* **54** (1977): 739].
   a. Calculate the elemental percent composition by mass of cisplatin.
   b. Cisplatin is synthesized as follows:

$$K_2PtCl_4(aq) + 2NH_3(aq) \rightarrow Pt(NH_3)_2Cl_2(s) + 2KCl(aq)$$

What mass of cisplatin can be made from 65 g of K_2PtCl_4 and sufficient NH_3? What mass of KCl is also produced?

87. Indium oxide contains 4.784 g of indium for every gram of oxygen. In 1869, when Mendeleev first presented his version of the periodic table, he proposed that the formula of indium oxide was In_2O_3. Before that time, it was thought that the formula was InO. What values for the atomic mass of indium are obtained using these two formulas?

88. A compound composed of only antimony and oxygen is 83.53% Sb by mass. The molecular weight is between 550 and 600. Calculate the empirical and molecular formulas of the compound.

89. When aluminum metal is heated with an element from Group 6A of the periodic table, an ionic compound forms. When the experiment is performed with an unknown Group 6A element, the product is 18.56% Al by mass. What is the formula of the compound?

90. Commercial bronze, an alloy of Zn and Cu, reacts with hydrochloric acid as follows:

$$Zn(s) + 2HCl(aq) \rightarrow ZnCl_2(aq) + H_2(g)$$

(Cu does not react with HCl.) When 0.5065 g of a certain bronze alloy is reacted with excess HCl, 0.0985 g of $ZnCl_2$ is eventually isolated.
   a. What is the composition of the bronze by mass?
   b. How could this result be checked without changing the above procedure?

91. A salt contains only barium and one of the halide ions. A 0.158-g sample of the salt was dissolved in water, and an excess of sulfuric acid was added to form barium sulfate ($BaSO_4$), which was filtered, dried, and weighed. Its mass was found to be 0.124 g. What is the formula of the barium halide?

92. Terephthalic acid is an important chemical used in the manufacture of polyesters and plasticizers. It contains only C, H, and O. Combustion of 19.81 mg of terephthalic acid produced 41.98 mg CO_2 and 6.45 mg of H_2O. The molecular weight (molar mass) of terephthalic acid is 166. Calculate the empirical and molecular formulas for terephthalic acid.

93. Bromine (Br_2) is produced by reacting sea water with chlorine (Cl_2).

$$2Br^-(aq) + Cl_2(aq) \rightarrow Br_2(aq) + 2Cl^-(aq)$$

What mass of Cl_2 is needed to produce 1.0 kg of Br_2?

94. Calcium oxide, or lime, is produced by the thermal decomposition of limestone.

$$CaCO_3(s) \xrightarrow{\text{Heat}} CaO(s) + CO_2(g)$$

What mass of CaO can be produced from 2.00×10^3 kg of limestone?

95. Consider the following unbalanced equation:

$$Ca_3(PO_4)_2(s) + H_2SO_4(aq) \rightarrow CaSO_4(s) + H_3PO_4(aq)$$

What masses of calcium sulfate and phosphoric acid can be produced from the reaction of 1.0 kg of calcium phosphate with 1.0 kg of concentrated sulfuric acid (98% H_2SO_4 by mass)?

96. Impure nickel can be purified by first forming the compound $Ni(CO)_4$, which is then decomposed by heating to yield very pure nickel. Metallic nickel reacts directly with gaseous carbon monoxide as follows:

$$Ni(s) + 4CO(g) \rightarrow Ni(CO)_4(g)$$

Other metals present do not react. If 94.2 g of a metal mixture produces 98.4 g of $Ni(CO)_4$, what is the mass percent of nickel in the original sample?

97. Mercury and bromine will react with each other to produce mercury(II) bromide:

$$Hg(l) + Br_2(l) \rightarrow HgBr_2(s)$$

   a. What mass of $HgBr_2$ is produced from the reaction of 10.0 g Hg and 9.00 g Br_2? What mass of which reagent is left unreacted?
   b. What mass of $HgBr_2$ is produced from the reaction of 5.00 mL of mercury (density = 13.5 g/mL) and 5.00 mL of bromine (density = 3.2 g/mL)?

Types of Chemical Reactions and Solution Stoichiometry

Much of the chemistry that affects each of us occurs among substances dissolved in water. For example, virtually all of the chemistry that makes life possible occurs in an aqueous environment. Also, various tests for illnesses involve aqueous reactions. Modern medical practice depends heavily on analyses of blood and other body fluids. In addition to the common tests for sugar, cholesterol, and iron, analyses for specific chemical markers allow detection of many diseases before more obvious symptoms occur.

Aqueous chemistry is also important in our environment. In recent years contamination of the ground water by substances such as chloroform and nitrates has been widely publicized. Also, pesticides have been discovered in mothers' milk.

Thus, in order to understand the chemistry that occurs in such diverse places as the human body, the ground waters, the oceans, the local water treatment plant, your hair as you shampoo it, and many more, we must understand how substances dissolved in water react with each other.

However, before we can understand solution reactions, we need to discuss the nature of solutions in which water is the dissolving medium, or *solvent*. These are called **aqueous solutions.** In this chapter we will study the nature of materials after they are dissolved in water, and various types of reactions that occur among these substances. You will see that the procedures developed in Chapter 3 to deal with chemical reactions work very well for reactions that take place in aqueous solutions. To understand the types of reactions that occur in aqueous solutions, we must first explore the types of species present. This requires an understanding of the nature of water.

< Water is one of the most important compounds on earth because it can dissolve so many substances. Even crystal clear mountain stream water contains some dissolved salts.

CONTENTS

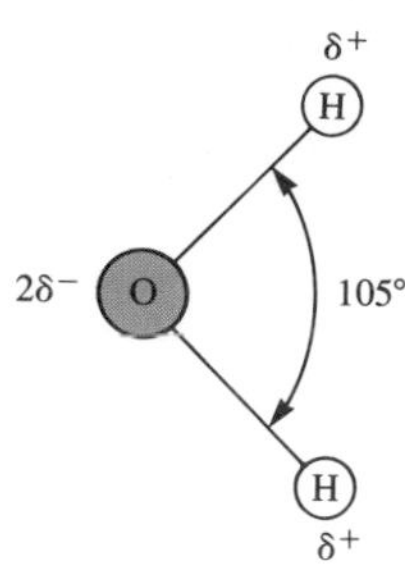

Figure 4.1

The water molecule is polar.

4.1 Water, the Common Solvent

Purpose

- To show why the polar nature of water makes it an effective solvent.

Water is one of the most important substances on earth. It is, of course, crucial for sustaining the reactions that keep us alive, but it also affects our lives in many indirect ways. Water helps moderate the earth's temperature; it cools automobile engines, nuclear power plants, and many industrial processes; it provides a means of transportation on the earth's surface and a medium for the growth of a myriad of creatures we use as food; and much more.

One of the most valuable functions of water involves its ability to dissolve many different substances. For example, salt disappears when you sprinkle it into the water used to cook vegetables, as does sugar when you add it to your iced tea. In each case the ''disappearing'' substance is definitely still present—you can taste it. What happens when a solid dissolves? To understand this process we need to consider the nature of water. Liquid water consists of a collection of H_2O molecules. An individual H_2O molecule is ''bent'' or V-shaped, with an H—O—H angle of about 105°:

H 105° H
O

The O—H bonds in the water molecule are covalent bonds formed by electron sharing between the oxygen and hydrogen atoms. However, the electrons of the bond are not shared equally between these atoms. For reasons we will discuss in later chapters, oxygen has a greater attraction for electrons than does hydrogen. If the electrons were shared equally between the two atoms, both would be electrically neutral because, on the average, the number of electrons around each would equal the number of protons in that nucleus. However, because the oxygen atom has a greater attraction for electrons, the shared electrons tend to spend more time close to the oxygen than to either of the hydrogens. Thus, the oxygen atom gains a slight excess of negative charge, and the hydrogen atoms are slightly positive. This is shown in Fig. 4.1, where δ (delta) indicates a *partial* charge *(less than one unit of charge)*. Because of this unequal charge distribution, water is said to be a **polar molecule.** It is this polarity that gives water its great ability to dissolve compounds.

A schematic of an ionic solid dissolving in water is shown in Fig. 4.2. Note that the ''positive ends'' of the water molecules are attracted to the negatively charged anions while the ''negative ends'' are attracted to the positively charged cations. This process is called **hydration.** The hydration of its ions tends to cause a salt to ''fall apart'' in the water, to dissolve. The strong forces present among the positive and negative ions of the solid are replaced by strong water-ion interactions.

It is very important to remember that when ionic substances (salts) dissolve in water, they break up into the *individual* cations and anions. For instance, when ammonium nitrate (NH_4NO_3) dissolves in water, the resulting solution contains

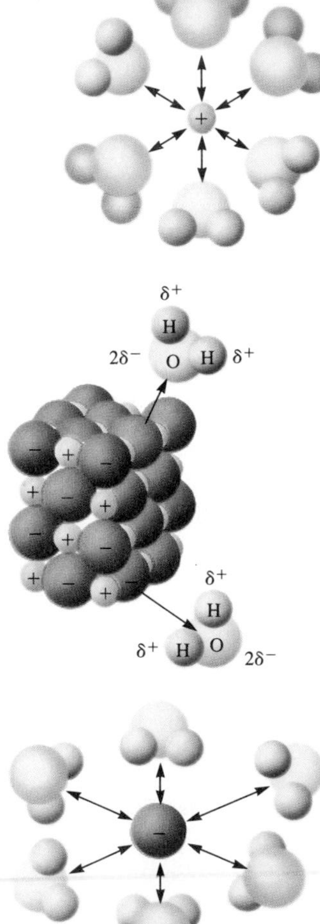

Figure 4.2

Polar water molecules interact with the positive and negative ions of a salt.

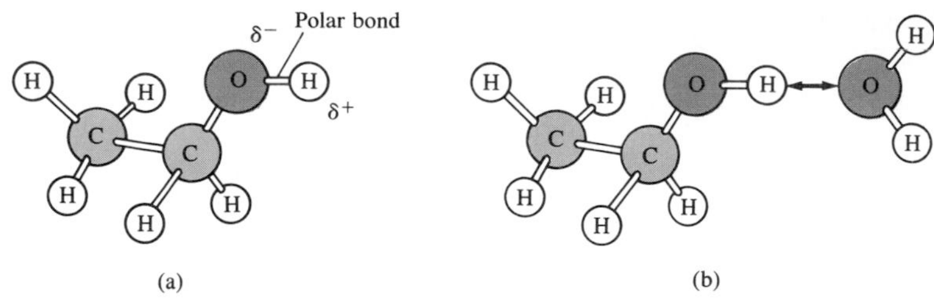

Figure 4.3

(a) The ethanol molecule contains a polar O—H bond similar to those in the water molecule. (b) The polar water molecule interacts strongly with the polar O—H bond in ethanol. This is a case of "like dissolving like."

NH_4^+ and NO_3^- ions floating around independently. This process can be represented as

$$NH_4NO_3(s) \xrightarrow{H_2O(l)} NH_4^+(aq) + NO_3^-(aq)$$

where (aq) designates that the ions are hydrated by unspecified numbers of water molecules.

Water also dissolves many nonionic substances. Ethanol (C_2H_5OH), for example, is very soluble in water. Wine, beer, and mixed drinks are aqueous solutions of alcohol and other substances. Why is ethanol so soluble in water? The answer lies in the structure of the alcohol molecule, which is shown in Fig. 4.3(a). The molecule contains a polar O—H bond like those in water, which makes it very compatible with water. The interaction of water with ethanol is represented in Fig. 4.3(b).

Many substances do not dissolve in water. Pure water will not, for example, dissolve animal fat because fat molecules are nonpolar and do not interact effectively with polar water molecules. In general, polar and ionic substances are soluble in water. "Like dissolves like" is a useful rule for predicting solubility. We will explore the basis for this generalization when we discuss the details of solution formation in Chapter 11.

4.2 The Nature of Aqueous Solutions: Strong and Weak Electrolytes

Purpose

- To characterize strong electrolytes, weak electrolytes, and nonelectrolytes.

As we discussed in Chapter 2, a solution is a homogeneous mixture. It is the same throughout (the first sip of a cup of coffee is the same as the last), but its composition can be varied by changing the amount of dissolved substances (one can make weak or strong coffee). In this section we will consider what happens when a substance, the **solute,** is dissolved in liquid water, the **solvent.**

One useful property for characterizing a solution is its **electrical conductivity,** its ability to conduct an electric current. This characteristic can be checked conveniently using an apparatus like the one shown in Fig. 4.4 on page 122. If the solution in the container conducts electricity, the bulb lights. Some solutions conduct current very efficiently, and the bulb shines very brightly; these solutions contain **strong electrolytes.** Other solutions conduct only a small current, and the bulb glows dimly; these solutions contain **weak electrolytes.** Some solutions permit no current to flow, and the bulb remains unlit; these solutions contain **nonelectrolytes.**

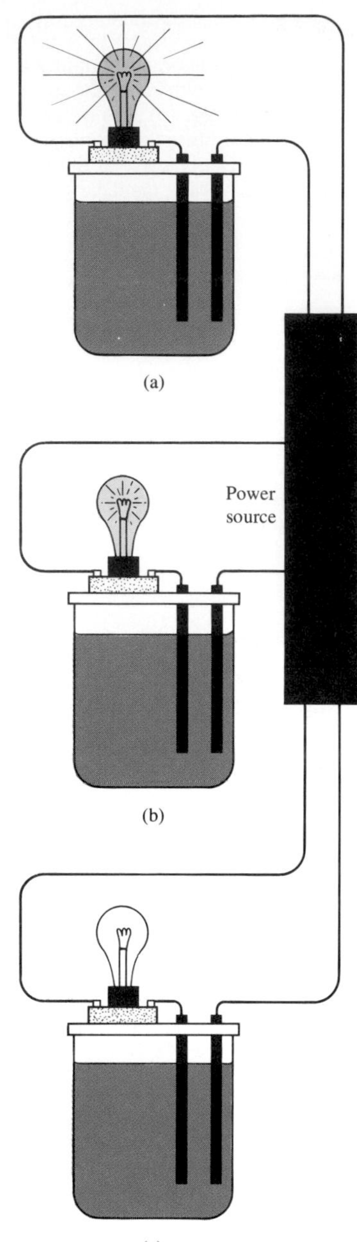

Figure 4.4

Electrical conductivity of aqueous solutions. To complete the circuit and allow current to flow, there must be charge carriers (ions) in the solution. (a) Strong electrolyte in solution. (b) Weak electrolyte in solution. (c) Nonelectrolyte in solution.

The Arrhenius definition of an acid: a substance that produces H^+ ions in solution.

The basis for the conductivity properties of solutions was first correctly identified by Svante Arrhenius, then a Swedish graduate student in physics, who carried out research on the nature of solutions at the University of Uppsala in the early 1880s. Arrhenius came to believe that the conductivity of solutions arose from the presence of ions, an idea that was at first scorned by the majority of the scientific establishment. However, in the late 1890s, when atoms were found to contain charged particles, the ionic theory suddenly made sense and became widely accepted. The story of Arrhenius and his struggle to get the scientific establishment to accept his theory is described in the Chemical Impact feature.

As Arrhenius postulated, the extent to which a solution can conduct an electrical current depends directly on the number of ions present. Some materials, such as sodium chloride, readily produce ions in aqueous solution and are thus strong electrolytes. Other substances, such as ammonia, produce relatively few ions when dissolved in water and are weak electrolytes. A third class of materials, such as sugar, form virtually no ions when dissolved in water and are nonelectrolytes.

Strong Electrolytes

We will consider several classes of strong electrolytes: (1) soluble salts, (2) strong acids, and (3) strong bases.

As shown in Fig. 4.2, a salt consists of an array of cations and anions, which separate and become hydrated when the salt dissolves. **Solubility** is usually measured in terms of the mass (grams) of solute that dissolve per given volume of solvent *or* in terms of the number of moles of solute that dissolve in a given volume of solution. Some salts, such as NaCl, KCl, and NH_4Cl, are very soluble in water. For example, approximately 357 grams of NaCl will dissolve in a liter of water at 25°C. On the other hand, many salts are only very slightly soluble in water; for example, silver chloride (AgCl) dissolves in water only to a slight extent (approximately 2×10^{-3} gram per liter at 25°C). We will consider only soluble salts at this point.

One of Arrhenius's most important discoveries concerned the nature of **acids.** Acidic behavior was first associated with the sour taste of citrus fruits. In fact, the word *acid* comes directly from the Latin word *acidus,* meaning sour. The *mineral acids* sulfuric acid (H_2SO_4) and nitric acid (HNO_3), so named because they were originally obtained by the treatment of minerals, were discovered around 1300.

Acids were known for hundreds of years before the time of Arrhenius, but no one had recognized their essential nature. In his studies of solutions, Arrhenius found that when the substances HCl, HNO_3, and H_2SO_4 were dissolved in water, they behaved as strong electrolytes. He postulated that this was the result of ionization reactions in water, for example:

$$\mathrm{HCl} \xrightarrow{\mathrm{H_2O}} \mathrm{H^+}(aq) + \mathrm{Cl^-}(aq)$$

$$\mathrm{HNO_3} \xrightarrow{\mathrm{H_2O}} \mathrm{H^+}(aq) + \mathrm{NO_3^-}(aq)$$

$$\mathrm{H_2SO_4} \xrightarrow{\mathrm{H_2O}} \mathrm{H^+}(aq) + \mathrm{HSO_4^-}(aq)$$

Thus Arrhenius proposed that an *acid is a substance that produces H^+ ions (protons) when it is dissolved in water.*

Studies of conductivity show that when HCl, HNO_3, and H_2SO_4 are placed in water *virtually every molecule* dissociates to give ions. These substances are strong

Chemical Impact

Arrhenius, A Man With Solutions

Science is a human endeavor—subject to human frailties and governed by personalities, politics, and prejudices. One of the best illustrations of the often bumpy path of the advancement of scientific knowledge is the story of Swedish chemist Svante Arrhenius.

When Arrhenius began his doctorate at the University of Uppsala around 1880, he chose to study the passage of electricity through solutions. This was a problem that had baffled scientists for a century. The first experiments had been done in the 1770s by Cavendish, who compared the conductivity of salt solutions with that of rain water, using his own physiological reaction to the electric shocks he received! Arrhenius had an array of instruments to measure electric current, but the process of carefully weighing, measuring, and recording data from a multitude of experiments was a tedious one.

After his long series of experiments were performed, Arrhenius quit his laboratory bench and returned to his country home to try to formulate a model that could account for his data. He wrote, "I got the idea in the night of the 17th of May in the year 1883, and I could not sleep that night until I had worked through the whole problem." His idea was that ions were responsible for conducting electricity through a solution.

Back at Uppsala, Arrhenius took his doctoral dissertation containing the new theory to his advisor, Professor Cleve, an eminent chemist and the discoverer of the elements holmium and thulium. Cleve's uninterested response was what Arrhenius had expected. It was in keeping with Cleve's resistance to new ideas—he had not even accepted Mendeleev's periodic table, introduced ten years earlier.

Svante August Arrhenius

It is a longstanding custom that before a doctoral degree is granted the dissertation must be defended before a panel of professors. Although this procedure is still followed at most universities today, the problems are usually worked out in private with the evaluating professors before the actual defense. However, when Arrhenius did it, the dissertation defense was an open debate, which could be rancorous and humiliating. Knowing that it would be unwise to antagonize his professors, Arrhenius downplayed his convictions about his new theory as he defended his dissertation. His diplomacy paid off: he was awarded his degree, albeit reluctantly, as the professors still did not believe his model and considered him to be a marginal scientist, at best.

Such a setback could have ended his scientific career, but Arrhenius was a crusader; he was determined to see his theory triumph. He promptly embarked on a political campaign, enlisting the aid of several prominent scientists, to get his theory accepted.

Ultimately, the ionic theory triumphed. Arrhenius's fame spread, and honors were heaped on him, culminating in the Nobel Prize in chemistry in 1903. Not one to rest on his laurels, Arrhenius turned to new fields, including astronomy; he formulated a new theory that the solar system may have come into being through the collision of stars. His exceptional versatility led him to study the use of serums to fight disease, energy resources and conservation, and the origin of life.

Additional insight on Arrhenius and his scientific career can be obtained from his address on receiving the Willard Gibbs Award. *See Journal of American Chemical Society* **36** (1912): 353.

Acids and bases are discussed further in Chapters 14 and 15.

electrolytes and are thus called **strong acids.** All three are very important chemicals, and much more will be said about them as we proceed. However, at this point the following facts are important:

> Sulfuric acid, nitric acid, and hydrochloric acid are aqueous solutions and should be written in chemical equations as $H_2SO_4(aq)$, $HNO_3(aq)$, and $HCl(aq)$, respectively, although they often appear without the (aq) symbol.
>
> A strong acid is one that completely dissociates into its ions. This means that if 100 molecules of HCl are dissolved in water, 100 H^+ ions and 100 Cl^- ions are produced. Virtually no HCl molecules exist in aqueous solution.
>
> Sulfuric acid is a special case. The formula H_2SO_4 indicates that this acid can produce two H^+ ions per molecule when dissolved in water. However, only the first H^+ ion is completely dissociated. The second H^+ ion can be pulled off under certain conditions, which we will discuss later. Thus a solution of H_2SO_4 dissolved in water contains mostly H^+ ions and HSO_4^- ions.

Perchloric acid, $HClO_4(aq)$, is another strong acid.

Strong electrolytes dissociate completely in aqueous solution.

Another important class of strong electrolytes are the **strong bases,** soluble compounds containing the *hydroxide ion* (OH^-) that completely dissociate when dissolved in water. Solutions containing bases have a bitter taste and a slippery feel. The most commonly used basic solutions are produced when solid sodium hydroxide (NaOH) or potassium hydroxide (KOH) is dissolved in water to produce ions as follows:

$$NaOH(s) \xrightarrow{H_2O} Na^+(aq) + OH^-(aq)$$

$$KOH(s) \xrightarrow{H_2O} K^+(aq) + OH^-(aq)$$

Weak Electrolytes

Weak electrolytes dissociate only to a small extent in aqueous solution.

Weak electrolytes are substances that produce relatively few ions when dissolved in water. The most common weak electrolytes are weak acids and weak bases.

The main acidic component of vinegar is acetic acid ($HC_2H_3O_2$). The formula is written so as to indicate that acetic acid has two chemically distinct types of hydrogen atoms. Formulas for acids are often written with the acidic hydrogen atom or atoms (any that will produce H^+ ions in solution) listed first. If any nonacidic hydrogens are present, they are written later in the formula. Thus the formula $HC_2H_3O_2$ indicates one acidic and three nonacidic hydrogen atoms. The dissociation reaction for acetic acid in water can be written as follows:

$$HC_2H_3O_2(aq) \xrightarrow{H_2O} H^+(aq) + C_2H_3O_2^-(aq)$$

Acetic acid is very different from the strong acids in that only about 1% of its molecules dissociate in aqueous solution. This means that when 100 molecules of $HC_2H_3O_2$ are dissolved in water, approximately 99 molecules of $HC_2H_3O_2$ remain intact and only one H^+ ion and one $C_2H_3O_2^-$ ion are produced.

Because acetic acid is a weak electrolyte, it is called a **weak acid.** Any acid, such as acetic acid, that *dissociates only to a slight extent in aqueous solution is called a weak acid.* In Chapter 14 we will explore the subject of weak acids in detail.

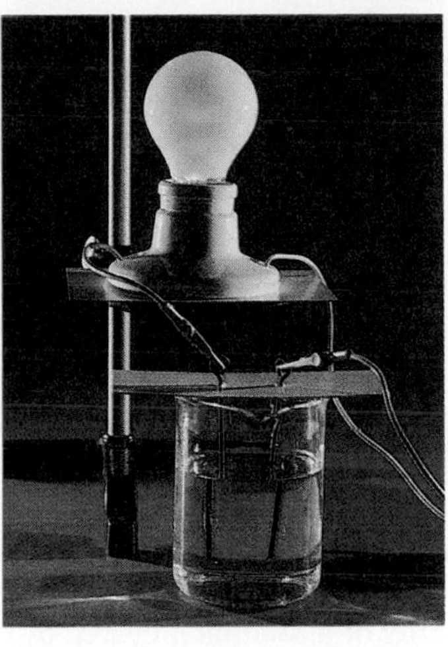

(left) A hydrochloric acid solution freely conducts electricity. The bulb glows brightly. (right) An acetic acid solution is a weak conductor of electricity. The bulb glows dimly.

The most common **weak base** is ammonia (NH_3). When ammonia is dissolved in water, it reacts as follows:

$$NH_3(aq) + H_2O(l) \rightarrow NH_4^+(aq) + OH^-(aq)$$

The solution is *basic* since OH^- ions are produced. Ammonia is called a **weak base** because *the resulting solution is a weak electrolyte*—very few ions are formed. In fact, for every 100 molecules of NH_3 that are dissolved, only one NH_4^+ ion and one OH^- ion are produced; 99 molecules of NH_3 remain unreacted.

Nonelectrolytes

Nonelectrolytes are substances that dissolve in water but do not produce any ions. An example of a nonelectrolyte is ethanol (see Fig. 4.3 for the structural formula). When ethanol dissolves, entire C_2H_5OH molecules are dispersed in the water. Since the molecules do not break up into ions, the resulting solution does not conduct an electric current. Another common nonelectrolyte is table sugar (sucrose, $C_{12}H_{22}O_{11}$), which is very soluble in water, but which produces no ions when it dissolves.

4.3 The Composition of Solutions

Purpose

- To define molarity and demonstrate calculations on the compositions of solutions.

Chemical reactions often take place when two solutions are mixed. To perform stoichiometric calculations in such cases we must know two things: (1) the *nature of the reaction,* which depends on the exact forms the chemicals take when dissolved; and (2) the *amounts of chemicals* present in the solutions, that is, the composition of each solution.

The composition of a solution can be described in many different ways, as we will see in Chapter 11. At this point we will consider only the most commonly used expression of concentration, **molarity** (M), which is defined as *moles of solute per volume of solution in liters:*

$$M = \text{molarity} = \frac{\text{moles of solute}}{\text{liters of solution}}$$

A solution that is 1.0 molar (written as 1.0 M) contains 1.0 mole of solute per liter of solution.

Sample Exercise 4.1

Calculate the molarity of a solution prepared by dissolving 11.5 g of solid NaOH in enough water to make 1.50 L of solution.

Solution

To find the molarity of the solution, we first compute the number of moles of solute using the molecular weight of NaOH (40.0):

$$11.5\ \cancel{\text{g NaOH}} \times \frac{1\ \text{mol NaOH}}{40.0\ \cancel{\text{g NaOH}}} = 0.288\ \text{mol NaOH}$$

Then we divide by the volume of the solution in liters:

$$\text{Molarity} = \frac{\text{mol solute}}{\text{L solution}} = \frac{0.288\ \text{mol NaOH}}{1.50\ \text{L solution}} = 0.192\ M\ \text{NaOH}$$

Sample Exercise 4.2

Calculate the molarity of a solution prepared by bubbling 1.56 g of gaseous HCl into enough water to make 26.8 mL of solution.

Solution

First we calculate the number of moles of HCl (mol. wt. = 36.5):

$$1.56\ \cancel{\text{g HCl}} \times \frac{1\ \text{mol HCl}}{36.5\ \cancel{\text{g HCl}}} = 4.27 \times 10^{-2}\ \text{mol HCl}$$

Next we must change the volume of the solution to liters:

$$26.8\ \cancel{\text{mL}} \times \frac{1\ \text{L}}{1000\ \cancel{\text{mL}}} = 2.68 \times 10^{-2}\ \text{L}$$

Finally we divide the moles of solute by the liters of solution:

$$\text{Molarity} = \frac{4.27 \times 10^{-2}\ \text{mol HCl}}{2.68 \times 10^{-2}\ \text{L solution}} = 1.59\ M\ \text{HCl}$$

It is important to realize that the description of a solution's composition may not accurately reflect the true chemical nature of the solution. Solution concentration is always given in terms of the form of the solute *before* it dissolves. For example, when a solution is described as being 1.0 M NaCl, this means that the solution was prepared by dissolving 1.0 mole of solid NaCl in enough water to make 1.0 liter of solution; it does not mean that the solution contains 1.0 mole of NaCl units. Actually the solution contains 1.0 mole of Na^+ ions and 1.0 mole of Cl^- ions.

Sample Exercise 4.3

Give the concentrations of all of the ions in each of the following solutions:

a. 0.50 M $Co(NO_3)_2$ **b.** 1 M $Fe(ClO_4)_3$

Solution

a. When solid $Co(NO_3)_2$ dissolves, the cobalt(II) cation and the nitrate anions separate:

$$Co(NO_3)_2(s) \xrightarrow{H_2O} Co^{2+}(aq) + 2NO_3^-(aq)$$

For each mole of $Co(NO_3)_2$ that is dissolved, the solution contains 1 mol of Co^{2+} ions and 2 mol of NO_3^- ions. Thus a solution that is 0.50 M $Co(NO_3)_2$ contains 0.50 M Co^{2+} and (2 × 0.50) M NO_3^-, or 1.0 M NO_3^-.

b. When solid $Fe(ClO_4)_3$ dissolves, the iron(III) cation and the perchlorate anions separate:

$$Fe(ClO_4)_3(s) \xrightarrow{H_2O} Fe^{3+}(aq) + 3ClO_4^-(aq)$$

Thus a solution that is 1 M $Fe(ClO_4)_3$ contains 1 M Fe^{3+} ions and 3 M ClO_4^- ions.

An aqueous solution of $Co(NO_3)_2$.

Often we need to determine the number of moles of solute present in a given volume of a solution of known molarity. The procedure for doing this is easily derived from the definition of molarity. If we multiply the molarity of a solution by the volume (in liters) of a particular sample of the solution, we get the moles of solute present in that sample:

$$\text{Liters of solution} \times \text{molarity} = \cancel{\text{liters of solution}} \times \frac{\text{moles of solute}}{\cancel{\text{liters of solution}}}$$

$$= \text{moles of solute}$$

$$M = \frac{\text{moles of solute}}{\text{liters of solution}}$$

This procedure is demonstrated in Sample Exercises 4.4 through 4.7.

Sample Exercise 4.4

How many moles of Ag^+ ions are present in 25 mL of a 0.75 M $AgNO_3$ solution?

Sample Exercise 4.4, continued

Solution

To solve this problem we must first recognize that a 0.75 M $AgNO_3$ contains 0.75 M Ag^+ ions and 0.75 M NO_3^- ions. To find the moles of Ag^+ in 25 mL of a 0.75 M Ag^+ solution, we must express the volume in liters:

$$25 \cancel{\text{mL}} \times \frac{1 \text{ L}}{1000 \cancel{\text{mL}}} = 2.5 \times 10^{-2} \text{ L}$$

and multiply it by the molarity:

$$2.5 \times 10^{-2} \cancel{\text{L solution}} \times \frac{0.75 \text{ mol Ag}^+}{\text{L solution}} = 1.9 \times 10^{-2} \text{ mol Ag}^+$$

Sample Exercise 4.5

Calculate the number of moles of Cl^- ions in 1.75 L of 1.0×10^{-3} M $AlCl_3$.

Solution

When solid $AlCl_3$ dissolves, it produces ions as follows:

$$\text{AlCl}_3(s) \xrightarrow{\text{H}_2\text{O}} \text{Al}^{3+}(aq) + 3\text{Cl}^-(aq)$$

Thus a 1.0×10^{-3} M $AlCl_3$ solution contains 1.0×10^{-3} M Al^{3+} ions and 3.0×10^{-3} M Cl^- ions.

To calculate the moles of Cl^- ions in 1.75 L of the 1.0×10^{-3} M $AlCl_3$ solution, we must multiply the volume times the molarity:

$$1.75 \text{ L solution} \times 3.0 \times 10^{-3}\ M \text{ Cl}^- = 1.75 \cancel{\text{L solution}} \times \frac{3.0 \times 10^{-3} \text{ mol Cl}^-}{\cancel{\text{L solution}}}$$

$$= 5.2 \times 10^{-3} \text{ mol Cl}^-$$

Sample Exercise 4.6

Formalin is an aqueous solution of formaldehyde (HCHO). At high concentrations it is used as a preservative for biological specimens. How many grams of formaldehyde are in 2.5 L of 12.3 M formalin?

Solution

We must first determine the number of moles of formaldehyde in 2.5 L of 12.3 M formalin. Remember that volume of solution (in liters) times molarity yields moles of solute. In this case the volume of solution is 2.5 L, and the molarity is 12.3 mol of HCHO per liter of solution.

$$2.5 \cancel{\text{L solution}} \times \frac{12.3 \text{ mol HCHO}}{\cancel{\text{L solution}}} = 31 \text{ mol HCHO}$$

Sample Exercise 4.6, continued

Next, using the molecular weight of HCHO (30.0), we convert 31 mol of HCHO to grams:

$$31\ \cancel{\text{mol HCHO}} \times \frac{30.0\ \text{g HCHO}}{1\ \cancel{\text{mol HCHO}}} = 9.3 \times 10^2\ \text{g HCHO}$$

Thus 2.5 L of 12.3 *M* formalin contains 9.3×10^2 g of formaldehyde.

Sample Exercise 4.7

Typical blood serum is about 0.14 *M* NaCl. What volume of blood contains 1.0 mg of NaCl?

Solution

We must first determine the number of moles represented by 1.0 mg of NaCl (mol. wt. = 58.45):

$$1.0\ \cancel{\text{mg NaCl}} \times \frac{1\ \cancel{\text{g NaCl}}}{1000\ \cancel{\text{mg NaCl}}} \times \frac{1\ \text{mol NaCl}}{58.45\ \cancel{\text{g NaCl}}} = 1.7 \times 10^{-5}\ \text{mol NaCl}$$

Next we must determine what volume of 0.14 *M* NaCl solution contains 1.7×10^{-5} mol of NaCl. There is some volume, call it V, that when multiplied by the molarity of this solution will yield 1.7×10^{-5} mol of NaCl, that is:

$$V \times \frac{0.14\ \text{mol NaCl}}{\text{L solution}} = 1.7 \times 10^{-5}\ \text{mol NaCl}$$

We want to solve for the volume:

$$V = \frac{1.7 \times 10^{-5}\ \cancel{\text{mol NaCl}}}{\dfrac{0.14\ \cancel{\text{mol NaCl}}}{\text{L solution}}} = 1.2 \times 10^{-4}\ \text{L solution}$$

Thus 0.12 mL of blood contains 1.7×10^{-5} mol of NaCl, or 1.0 mg of NaCl.

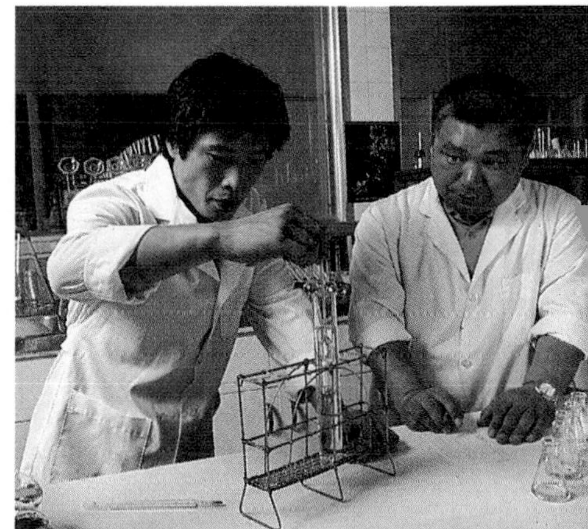

Winemaking has become a science as well as an art.

A **standard solution** is a solution *whose concentration is accurately known.* Standard solutions, often used in chemical analysis, can be prepared as shown in Fig. 4.5 on the following page and in Sample Exercise 4.8.

Sample Exercise 4.8

To analyze the alcohol content of a certain wine, a chemist needs 1.00 L of an aqueous 0.200 *M* $K_2Cr_2O_7$ (potassium dichromate) solution. How much solid $K_2Cr_2O_7$ must be weighed out to make this solution?

Sample Exercise 4.8, continued

Solution

We must first determine the moles of $K_2Cr_2O_7$ required:

$$1.00 \text{ L } \cancel{\text{solution}} \times \frac{0.200 \text{ mol } K_2Cr_2O_7}{\text{L } \cancel{\text{solution}}} = 0.200 \text{ mol } K_2Cr_2O_7$$

This amount can be converted to grams using the molecular weight of $K_2Cr_2O_7$ (294.2):

$$0.200 \cancel{\text{mol } K_2Cr_2O_7} \times \frac{294.2 \text{ g } K_2Cr_2O_7}{\cancel{\text{mol } K_2Cr_2O_7}} = 58.8 \text{ g } K_2Cr_2O_7$$

Thus to make 1.00 L of 0.200 M $K_2Cr_2O_7$, the chemist must weigh out 58.8 g of $K_2Cr_2O_7$, put it in a 1.00 L volumetric flask, and add water to the mark on the flask.

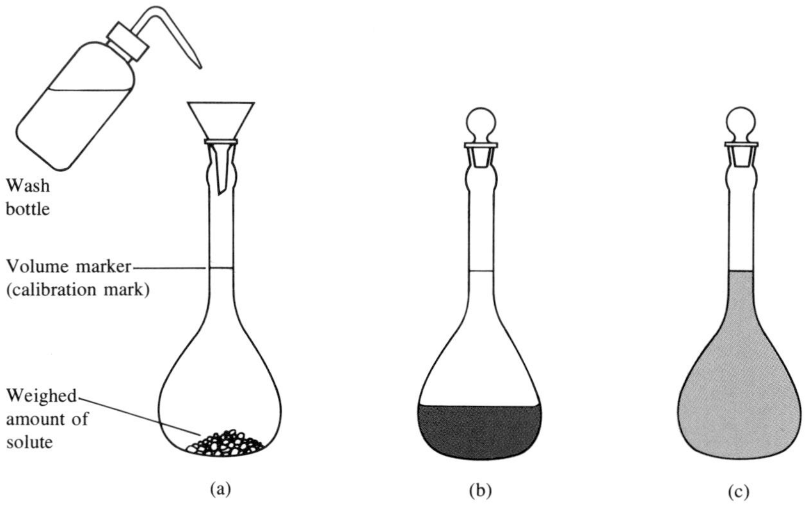

Figure 4.5

Steps involved in the preparation of a standard solution. (a) A weighed amount of a substance (the solute) is put into the volumetric flask, and a small quantity of water is added. (b) The solid is dissolved in the water by gently swirling the flask (*with the stopper in place*). (c) More water is added, until the level of the solution just reaches the mark etched on the neck of the flask.

Dilution

Dilution with water doesn't alter the number of moles of solute present.

To save time and space in the laboratory, routinely used solutions are often purchased or prepared in concentrated form (called *stock solutions*). Water is then added to achieve the molarity desired for a particular solution. This process is called **dilution.** For example, the common acids are purchased as concentrated solutions and diluted as needed. A typical dilution calculation involves determining how much water must be added to an amount of stock solution to achieve a solution of the desired concentration. The key to doing these calculations is that, since only water is added in the dilution, all of the solute in the final dilute solution must come from the concentrated stock solution, that is:

$$\text{Moles of solute after dilution} = \text{moles of solute before dilution}$$

For example, suppose we need to prepare 500. milliliters of 1.00 M acetic acid ($HC_2H_3O_2$) from a 17.5 M stock solution of acetic acid. What volume of the stock solution is required? The first step is to determine the number of moles of acetic acid in the final solution by multiplying the volume by the molarity:

$$500.\ \cancel{\text{mL solution}} \times \frac{1\ \cancel{\text{L solution}}}{1000\ \cancel{\text{mL solution}}} \times \frac{1.00\ \text{mol } HC_2H_3O_2}{\cancel{\text{L solution}}} = 0.500\ \text{mol } HC_2H_3O_2$$

Thus we need to use a volume of 17.5 M acetic acid that contains 0.500 mole of $HC_2H_3O_2$, that is:

$$V \times \frac{17.5\ \text{mol } HC_2H_3O_2}{\text{L solution}} = 0.500\ \text{mol } HC_2H_3O_2$$

Solving for V gives

$$V = \frac{0.500\ \cancel{\text{mol } HC_2H_3O_2}}{\dfrac{17.5\ \cancel{\text{mol } HC_2H_3O_2}}{\text{L solution}}} = 0.0286\ \text{L or } 28.6\ \text{mL solution}$$

Thus, to make 500 milliliters of a 1.00 M acetic acid solution, we can take 28.6 milliliters of 17.5 M acetic acid and dilute it to a total volume of 500 milliliters.

A dilution procedure typically involves two types of glassware: a pipet and a volumetric flask. A pipet is a device for accurately measuring and transferring a given volume of solution. There are two common types of pipets: *volumetric pipets* and *measuring pipets,* as shown in Fig. 4.6. To measure out volumes such as 28.6 mL, a volume for which a volumetric pipet is not available, we would use a measuring pipet to deliver 28.6 mL of 17.5 M acetic acid into a 500-mL volumetric flask and then add water to the mark, as shown in Fig. 4.7.

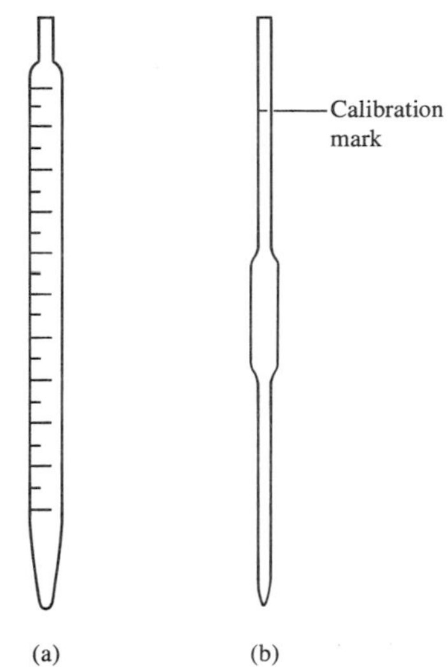

Figure 4.6

(a) A measuring pipet is graduated and can be used to measure various volumes of liquid accurately. (b) A volumetric pipet is designed to measure *one* volume accurately. When filled to the mark, it delivers the volume indicated on the pipet (in this case, 20 mL).

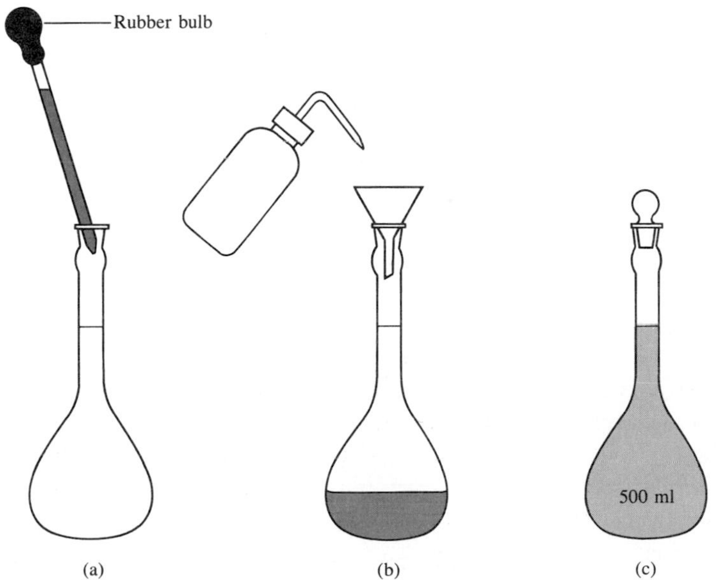

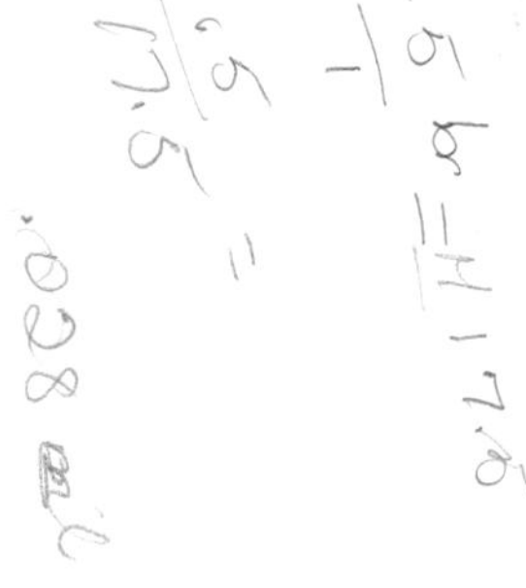

Figure 4.7

(a) A measuring pipet is used to transfer 28.6 mL of 17.5 M acetic acid solution to a volumetric flask. (b) Water is added to the flask to the calibration mark. (c) The resulting solution is 1 M acetic acid.

Sample Exercise 4.9

What volume of 16 *M* sulfuric acid must be used to prepare 1.5 L of a 0.10 *M* H_2SO_4 solution?

Solution

We must first determine the moles of H_2SO_4 in 1.5 L of 0.10 *M* H_2SO_4:

$$1.5\ \cancel{\text{L solution}} \times \frac{0.10\ \text{mol H}_2\text{SO}_4}{\cancel{\text{L solution}}} = 0.15\ \text{mol H}_2\text{SO}_4$$

Next we must find the volume of 16 *M* H_2SO_4 that contains 0.15 mol of H_2SO_4:

$$V \times \frac{16\ \text{mol H}_2\text{SO}_4}{\text{L solution}} = 0.15\ \text{mol H}_2\text{SO}_4$$

Solving for *V* gives

$$V = \frac{0.15\ \cancel{\text{mol H}_2\text{SO}_4}}{\dfrac{16\ \cancel{\text{mol H}_2\text{SO}_4}}{1\ \text{L solution}}} = 9.4 \times 10^{-3}\ \text{L or 9.4 mL solution}$$

Thus, to make 1.5 L of 0.10 *M* H_2SO_4 using 16 *M* H_2SO_4, we must take 9.4 mL of the concentrated acid and dilute it with water to 1.5 L. The correct way to do this is to add the 9.4 mL of acid to about 1 L of water and then dilute to 1.5 L by adding more water.

4.4 Types of Chemical Reactions

Purpose

- To introduce the types of solution reactions.

Although we have considered many reactions so far in this text, we have examined only a tiny fraction of the millions of possible chemical reactions. In order to make sense of all these reactions we need some system for grouping reactions into classes. Although there are many different ways to do this, we will use the system most commonly used by practicing chemists. They divide reactions into the following groups:

Types of Solution Reactions

- Precipitation reactions
- Acid/base reactions
- Oxidation/reduction reactions

Virtually all reactions can be put into one of these classes. We will define and illustrate each type in the following sections.

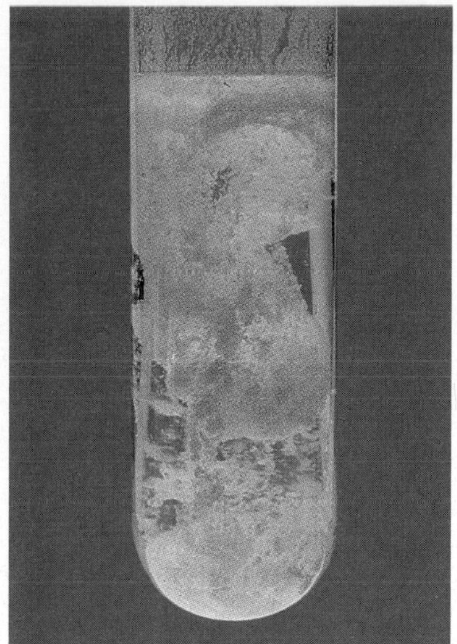

Figure 4.8

When yellow aqueous potassium chromate is added to a colorless barium nitrate solution, brownish-yellow barium chromate precipitates.

4.5 Precipitation Reactions

Purpose

- To show how to predict whether a solid will form in a solution reaction.

When two solutions are mixed, an insoluble substance sometimes forms; that is, a solid forms and separates from the solution. Such a reaction is called a **precipitation reaction** and the solid that forms is called a **precipitate.** For example, a precipitation reaction occurs when an aqueous solution of potassium chromate ($K_2CrO_4(aq)$), which is yellow, is added to a colorless aqueous solution containing barium nitrate ($Ba(NO_3)_2(aq)$). As shown in Fig. 4.8, when these solutions are mixed, a brownish-yellow solid forms. What is the equation that describes this chemical change? To write the equation we must know the identities of the reactants and products. The reactants have already been described: $K_2CrO_4(aq)$ and $Ba(NO_3)_2(aq)$. Is there some way we can predict the identities of the products? In particular, what is the brownish-yellow solid?

The best way to predict the identity of this solid is to think carefully about what products are possible. To do this we need to know what species are present in the mixed solution after the reactant solutions are mixed. First, let's think about the nature of each reactant solution. The designation $Ba(NO_3)_2(aq)$ means that barium nitrate (a white solid) has been dissolved in water. Notice that barium nitrate contains the Ba^{2+} and NO_3^- ions. *Remember:* **In virtually every case, when a solid containing ions dissolves in water, the ions separate** and move around independently. That is, $Ba(NO_3)_2(aq)$ does not contain $Ba(NO_3)_2$ units; it contains separated Ba^{2+} and NO_3^- ions.

When ionic compounds dissolve in water, the *resulting solution contains the separated ions.*

Similarly, since solid potassium chromate contains the K^+ and CrO_4^{2-} ions, an aqueous solution of potassium chromate (which is prepared by dissolving solid K_2CrO_4 in water) contains these separated ions.

We can represent the mixing of $K_2CrO_4(aq)$ and $Ba(NO_3)_2(aq)$ in two ways. First, we can write:

$$K_2CrO_4(aq) + Ba(NO_3)_2(aq) \rightarrow \text{products}$$

However, a much more accurate representation is:

$$\underbrace{2K^+(aq) + CrO_4^{2-}(aq)}_{\text{The ions in } K_2CrO_4(aq)} + \underbrace{Ba^{2+}(aq) + 2NO_3^-(aq)}_{\text{The ions in } Ba(NO_3)_2(aq)} \rightarrow \text{products}$$

Thus the mixed solution contains the ions:

$$K^+, CrO_4^{2-}, Ba^{2+}, NO_3^-$$

How can some or all of these ions combine to form the brownish-yellow solid observed when the original solutions are mixed? This is not an easy question to answer. In fact, predicting the products of a chemical reaction is one of the hardest things a beginning chemistry student is asked to do. Even an experienced chemist, when confronted with a new reaction, is often not sure what will happen. The chemist tries to think of the various possibilities, then considers the likelihood of each possibility, and then makes a prediction (an educated guess). Only after identifying each product *experimentally* is the chemist sure what reaction has taken place. However, it is very useful to be able to make an educated guess because it provides a place to start. It tells us what kinds of products we are most likely to find.

We already know some things that will help us predict the products.

1. We know that when ions form a solid compound, the compound must have a zero net charge. Thus the products of this reaction must contain *both anions and cations*. For example, K^+ and Ba^{2+} could not combine to form the solid, nor could $CrO_4{}^{2-}$ and $NO_3{}^-$.
2. Most ionic materials contain only two types of ions: one type of cation, and one type of anion (for example, NaCl, KOH, Na_2SO_4, K_2CrO_4, $Co(NO_3)_2$, NH_4Cl, Na_2CO_3).

The possible combinations of a given cation and a given anion from the list of ions K^+, $CrO_4{}^{2-}$, Ba^{2+}, $NO_3{}^-$ are:

$$K_2CrO_4,\ KNO_3,\ BaCrO_4,\ Ba(NO_3)_2$$

Which of these possibilities is most likely to represent the brownish-yellow solid? We know it's not K_2CrO_4 or $Ba(NO_3)_2$. They are the reactants. They were present (dissolved) in the separate solutions that were mixed. The only real possibilities for the solid that formed are:

$$KNO_3 \quad \text{and} \quad BaCrO_4$$

To decide which of these most likely represents the brownish-yellow solid, we need more facts. An experienced chemist knows that the K^+ ion and the $NO_3{}^-$ ion are both colorless. Thus if the solid is KNO_3, it should be white not brownish-yellow. On the other hand the $CrO_4{}^{2-}$ ion is yellow (note in Fig. 4.8 that $K_2CrO_4(aq)$ is yellow). Thus the brownish-yellow solid is almost certainly $BaCrO_4$. Further tests show that this is the case. So far we have determined that one product of the reaction between $K_2CrO_4(aq)$ and $Ba(NO_3)_2(aq)$ is $BaCrO_4(s)$, but what happened to the K^+ and $NO_3{}^-$ ions? The answer is that these ions are left dissolved in the solution. That is, KNO_3 does not form a solid when the K^+ and $NO_3{}^-$ ions are present in this much water. In other words, if we took the white solid ($KNO_3(s)$) and put it in the same quantity of water as is present in the mixed solution, it would dissolve. Thus when we mix $K_2CrO_4(aq)$ and $Ba(NO_3)_2(aq)$, $BaCrO_4(s)$ forms but KNO_3 is left behind in solution (we write it as $KNO_3(aq)$). Thus the equation for this precipitation reaction is:

$$K_2CrO_4(aq) + Ba(NO_3)_2(aq) \rightarrow BaCrO_4(s) + KNO_3(aq)$$

If we filtered off the solid $BaCrO_4$ and then evaporated the water, the white solid KNO_3 would be obtained.

Now let's consider another example. When an aqueous solution of silver nitrate is added to an aqueous solution of potassium chloride, a white precipitate forms. We can represent what we know so far as:

$$AgNO_3(aq) + KCl(aq) \rightarrow \text{unknown white solid}$$

Remembering that when ionic substances dissolve in water, the ions separate, we can write

$$\underbrace{Ag^+, NO_3^-}_{\text{In silver nitrate solution}} + \underbrace{K^+, Cl^-}_{\text{In potassium chloride solution}} \rightarrow \underbrace{Ag^+, NO_3^-, K^+, Cl^-}_{\text{Combined solution, before reaction}} \rightarrow \text{white solid}$$

Since we know the white solid must contain both positive and negative ions, the possible compounds that can be assembled from this collection of ions are:

$$AgNO_3,\ KCl,\ AgCl,\ KNO_3$$

Since $AgNO_3$ and KCl are the substances dissolved in the reactant solutions, we know that they do not represent the white solid product. The only real possibilities are:

$$AgCl \quad \text{and} \quad KNO_3$$

From the example considered above we know that KNO_3 is quite soluble in water. Thus, solid KNO_3 will not form when the reactant solutions are mixed. The product must be AgCl(*s*) (which can be proved by experiment to be true). The equation for the reaction now can be written:

$$AgNO_3(aq) + KCl(aq) \rightarrow AgCl(s) + KNO_3(aq)$$

Notice that to do these two examples, we had to know both concepts (solids always have a zero net charge) and facts (KNO_3 is very soluble in water, the CrO_4^{2-} is yellow, and so on).

Doing chemistry requires both understanding ideas and remembering facts.

Predicting the identity of the solid product in a precipitation reaction requires knowledge of the solubilities of common ionic substances. To help you predict the products of precipitation reactions, some simple solubility rules are given in Table 4.1. You should memorize these rules.

Simple Rules for Solubility of Salts in Water
1. Most nitrate (NO_3^-) salts are soluble.
2. Most salts of Na^+, K^+, and NH_4^+ are soluble.
3. Most chloride salts are soluble. Notable exceptions are AgCl, $PbCl_2$, and Hg_2Cl_2.
4. Most sulfate salts are soluble. Notable exceptions are $BaSO_4$, $PbSO_4$, and $CaSO_4$.
5. Most hydroxide salts are only slightly soluble. The important soluble hydroxides are NaOH, KOH, and $Ca(OH)_2$ (marginally soluble).
6. Most sulfide (S^{2-}), carbonate (CO_3^{2-}), and phosphate (PO_4^{3-}) salts are only slightly soluble.

Table 4.1

The phrase "slightly soluble" used in the solubility rules in Table 4.1 means that the tiny amount of solid that dissolves is not noticeable. The solid appears to be insoluble to the naked eye. Thus the terms insoluble and slightly soluble are often used interchangeably.

Note that the information in Table 4.1 allows us to identify the white solid formed when solutions of $AgNO_3$ and KCl are mixed as AgCl, since Rules 1 and 2 indicate that KNO_3 is soluble and Rule 3 states that AgCl is (virtually) insoluble. Fig. 4.9 shows the results of mixing silver nitrate and potassium chloride solutions.

Figure 4.9

Precipitation of silver chloride by mixing solutions of silver nitrate and potassium chloride. The K^+ and NO_3^- ions remain in solution.

When solutions containing ionic substances are mixed, it will be helpful in determining the products if you think in terms of *ion interchange*. For example, in the above discussion we considered the results of mixing $AgNO_3(aq)$ and $KCl(aq)$. In determining the products, we took the cation from one reactant and combined it with the anion of the other reactant:

$$Ag^+ + NO_3^- + K^+ + Cl^- \rightarrow$$

Possible solid products

The solubility rules in Table 4.1 allow us to predict whether either product is a solid.

Focus on the ions in solution before any reaction occurs.

The key to dealing with the chemistry of aqueous solutions is first to *focus on the actual components of the solution before reaction* and then figure out how those components will react with each other. Sample Exercise 4.10 shows how to do this for three different reactions.

Sample Exercise 4.10

Using the solubility rules in Table 4.1, predict what will happen when the following pairs of solutions are mixed.

a. $KNO_3(aq)$ and $BaCl_2(aq)$

b. $Na_2SO_4(aq)$ and $Pb(NO_3)_2(aq)$

c. $KOH(aq)$ and $Fe(NO_3)_3(aq)$

Solution

a. $KNO_3(aq)$ stands for an aqueous solution obtained by dissolving solid KNO_3 in water to form a solution containing the hydrated ions $K^+(aq)$ and $NO_3^-(aq)$. Likewise $BaCl_2(aq)$ is a solution formed by dissolving solid $BaCl_2$ in water to produce $Ba^{2+}(aq)$ and $Cl^-(aq)$. When these two solutions are mixed, the resulting solution will contain the ions K^+, NO_3^-, Ba^{2+}, and Cl^-. All will be hydrated, but (aq) is omitted for simplicity. To look for possible solid products combine the cation from one reactant with the anion from the other:

$$K^+ + NO_3^- + Ba^{2+} + Cl^- \rightarrow$$

Possible solid products

Note from Table 4.1 that the rules predict that both KCl and $Ba(NO_3)_2$ are soluble in water. Thus no precipitate will form when $KNO_3(aq)$ and $BaCl_2(aq)$ are mixed. All of the ions will remain dissolved in the solution. No reaction occurs.

Lead sulfate is a white solid.

b. Using the same procedures as in part a, we find that the ions present in the combined solution before any reaction occurs are Na^+, SO_4^{2-}, Pb^{2+}, and NO_3^-. The possible salts that could form as solids are:

$$Na^+ + SO_4^{2-} + Pb^{2+} + NO_3^- \rightarrow$$

Sample Exercise 4.10, continued

The compound $NaNO_3$ is soluble but $PbSO_4$ is insoluble (see Rule 4 in Table 4.1). When these solutions are mixed, $PbSO_4$ will precipitate from the solution. The balanced equation is:

$$Na_2SO_4(aq) + Pb(NO_3)_2(aq) \rightarrow PbSO_4(s) + 2NaNO_3(aq)$$

c. The combined solution (before any reaction occurs) contains the ions K^+, OH^-, Fe^{3+}, and NO_3^-. The salts that might precipitate are KNO_3 and $Fe(OH)_3$. The solubility rules in Table 4.1 indicate that both K^+ and NO_3^- salts are soluble. However, $Fe(OH)_3$ is only slightly soluble (Rule 5) and hence will precipitate. The balanced equation is:

$$3KOH(aq) + Fe(NO_3)_3(aq) \rightarrow Fe(OH)_3(s) + 3KNO_3(aq)$$

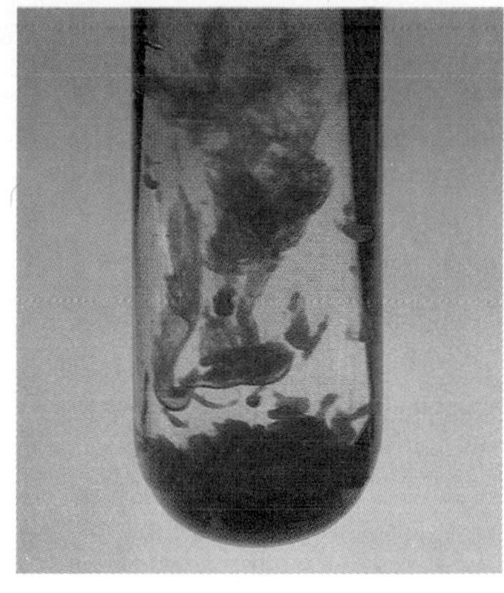

Solid $Fe(OH)_3$ forms when aqueous KOH and $Fe(NO_3)_3$ are mixed.

4.6 Describing Reactions in Solution

Purpose

- To describe reactions in solution by molecular, complete ionic, and net ionic equations.

In this section we will consider the types of equations used to represent reactions in solution. For example, when we mix aqueous potassium chromate with aqueous barium nitrate, a reaction occurs to form a precipitate ($BaCrO_4$) and dissolved potassium nitrate. So far we have written the **molecular equation** for this reaction:

$$K_2CrO_4(aq) + Ba(NO_3)_2(aq) \rightarrow BaCrO_4(s) + KNO_3(aq)$$

Although this equation shows the reactants and products of the reaction, it does not give a very clear picture of what actually occurs in solution. As we have seen, aqueous solutions of potassium chromate, barium nitrate, and potassium nitrate contain the individual ions, not molecules as is implied by the molecular equation. Thus the **complete ionic equation**

$$2K^+(aq) + CrO_4^{2-}(aq) + Ba^{2+}(aq) + 2NO_3^-(aq) \rightarrow BaCrO_4(s) + 2K^+(aq) + 2NO_3^-(aq)$$

better represents the actual forms of the reactants and products in solution. *In a complete ionic equation, all substances that are strong electrolytes are represented as ions.*

A strong electrolyte is a substance that completely breaks apart into ions when dissolved in water.

The complete ionic equation reveals that only some of the ions participate in the reaction. The K^+ and NO_3^- ions are present in solution both before and after the reaction. Ions such as these that do not participate directly in a reaction in solution are called **spectator ions.** The ions that participate in this reaction are the Ba^{2+} and CrO_4^{2-} ions, which combine to form solid $BaCrO_4$:

$$Ba^{2+}(aq) + CrO_4^{2-}(aq) \rightarrow BaCrO_4(s)$$

Net ionic equations include only those components that undergo changes in the reaction.

This equation, called the **net ionic equation,** includes only those solution components directly involved in the reaction. Chemists usually write the net ionic equation for a reaction in solution because it gives the actual forms of the reactants and products and only includes the species that undergo a change.

Three Types of Equations Are Used to Describe Reactions in Solution:

- The *molecular equation* gives the overall reaction stoichiometry but not necessarily the actual forms of the reactants and products in solution.
- The *complete ionic equation* represents as ions all reactants and products that are strong electrolytes.
- The *net ionic equation* includes only those solution components undergoing a change. Spectator ions are not included.

Sample Exercise 4.11

For each of the following reactions write the molecular equation, the complete ionic equation, and the net ionic equation.

a. Aqueous potassium chloride is added to aqueous silver nitrate to form a silver chloride precipitate plus aqueous potassium nitrate.

b. Aqueous potassium hydroxide is mixed with aqueous iron(III) nitrate to form a precipitate of iron(III) hydroxide and aqueous potassium nitrate.

Solution

a. *Molecular*

$$KCl(aq) + AgNO_3(aq) \rightarrow AgCl(s) + KNO_3(aq)$$

Complete Ionic (Remember, any ionic compound dissolved in water will be present as the separated ions.)

$$\underbrace{K^+(aq)}_{\uparrow\ \text{Spectator ion}} + Cl^-(aq) + Ag^+(aq) + \underbrace{NO_3^-(aq)}_{\uparrow\ \text{Spectator ion}} \rightarrow$$

$$\underbrace{AgCl(s)}_{\uparrow\ \text{Solid, not written as separate ions}} + \underbrace{K^+(aq)}_{\uparrow\ \text{Spectator ion}} + \underbrace{NO_3^-(aq)}_{\uparrow\ \text{Spectator ion}}$$

Canceling the spectator ions

$$\cancel{K^+}(aq) + Cl^-(aq) + Ag^+(aq) + \cancel{NO_3^-}(aq) \rightarrow AgCl(s) + \cancel{K^+}(aq) + \cancel{NO_3^-}(aq)$$

gives the following net ionic equation:

Net Ionic

$$Cl^-(aq) + Ag^+(aq) \rightarrow AgCl(s)$$

Sample Exercise 4.11, continued

b. *Molecular*

$$3KOH(aq) + Fe(NO_3)_3(aq) \rightarrow Fe(OH)_3(s) + 3KNO_3(aq)$$

Complete Ionic

$$3K^+(aq) + 3OH^-(aq) + Fe^{3+}(aq) + 3NO_3^-(aq) \rightarrow Fe(OH)_3(s) + 3K^+(aq) + 3NO_3^-(aq)$$

Net Ionic

$$3OH^-(aq) + Fe^{3+}(aq) \rightarrow Fe(OH)_3(s)$$

4.7 Selective Precipitation

Purpose

- To describe how selective precipitation can be used to separate cations from a mixture in solution.

We can use the fact that salts have different solubilities to separate mixtures of ions. For example, suppose we have an aqueous solution containing the cations Ag^+, Ba^{2+}, and Fe^{3+}, and the anion NO_3^-. We want to separate the cations by precipitating them one at a time, a process called **selective precipitation.**

How can the separation of these cations be accomplished? We can perform some preliminary tests and observe the reactivity of each cation toward the anions Cl^-, SO_4^{2-}, and OH^-. For example, to test the reactivity of Ag^+ toward Cl^-, we can mix the $AgNO_3$ solution with aqueous KCl or NaCl. As we have seen, this produces a precipitate. When we carry out tests of this type using all of the possible combinations, we obtain the results in Table 4.2.

Testing the Reactivity of the Cations Ag^+, Ba^{2+}, and Fe^{3+} with the Anions Cl^-, SO_4^{2-}, and OH^-

Cation	Test solution (anion) NaCl(*aq*) (Cl^-)	$Na_2SO_4(aq)$ (SO_4^{2-})	NaOH(*aq*) (OH^-)
Ag^+	White precipitate (AgCl)	No reaction	White precipitate that turns brown ($AgOH \rightarrow Ag_2O$; White, Brown)
Ba^{2+}	No reaction	White precipitate ($BaSO_4$)	No reaction
Fe^{3+}	Yellow color but no solid	No reaction	Reddish-brown precipitate [$Fe(OH)_3$]

Table 4.2

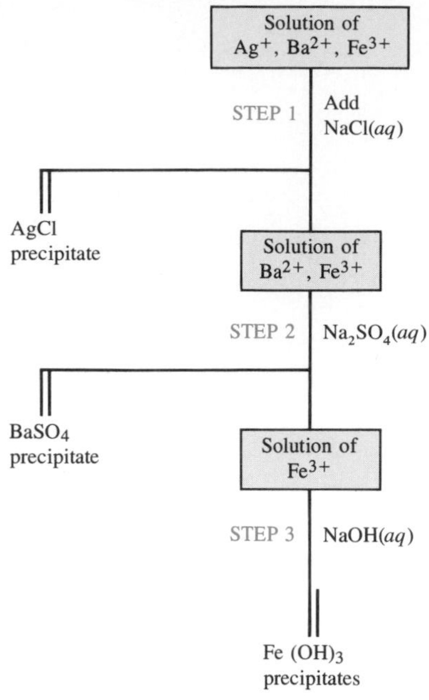

Figure 4.10

Selective precipitation of Ag^+, Ba^{2+}, and Fe^{3+} ions. In this schematic representation a double line means that a solid forms, and a single line designates a solution.

After studying these results, we might proceed to separate the cations as follows:

STEP 1

Add an aqueous solution of NaCl to the solution containing the Ag^+, Ba^{2+}, and Fe^{3+} ions. Solid AgCl will form and can be removed, leaving Ba^{2+} and Fe^{3+} ions in solution.

STEP 2

Add an aqueous solution of Na_2SO_4 to the solution containing the Ba^{2+} and Fe^{3+} ions. Solid $BaSO_4$ will form and can be removed, leaving only Fe^{3+} ions in solution.

STEP 3

Add an aqueous solution of NaOH to the solution containing the Fe^{3+} ions. Solid $Fe(OH)_3$ will form and can be removed.

Steps 1–3 are represented schematically in Fig. 4.10.

Note that adding the anions in this order precipitates the cations one at a time and thus separates them. The process whereby mixtures of ions are separated and identified is called **qualitative analysis.** In this example, the qualitative analysis was carried out by selective precipitation, but it can also be accomplished using other separation techniques such as chromatography (see Section 1.9).

4.8 Stoichiometry of Precipitation Reactions

Purpose

- To demonstrate stoichiometric calculations involving precipitation reactions.

In Chapter 3 we covered the principles of chemical stoichiometry: the procedures for calculating quantities of reactants and products involved in a chemical reaction. Recall that in performing these calculations, we first convert all quantities to moles and then use the coefficients of the balanced equation. In cases where reactants are mixed we must determine which reactant is limiting, since the reactant that is consumed first will limit the amounts of products formed. *These same principles apply to reactions that take place in solutions.* However, there are two points about solution reactions that need special emphasis. The first is that it is sometimes difficult to tell immediately what reaction will occur when two solutions are mixed. Usually we must do some thinking about the various possibilities and then decide what happens. The first step in this process *always* should be to write down the species that are actually present in the solution, as we did in Section 4.5.

The second special point about solution reactions is that to obtain the moles of reactants we must use the volume of a particular solution and its molarity. This procedure was covered in Section 4.3.

We will introduce stoichiometric calculations for reactions in solution in Sample Exercise 4.12.

Sample Exercise 4.12

Calculate the mass of solid NaCl that must be added to 1.50 L of a 0.100 *M* $AgNO_3$ solution to precipitate all of the Ag^+ ions in the form of AgCl.

Solution

When added to the $AgNO_3$ solution (which contains Ag^+ and NO_3^- ions), the solid NaCl dissolves to yield Na^+ and Cl^- ions. Thus the mixed solution contains the ions:

$$Ag^+, NO_3^-, Na^+, \text{ and } Cl^-$$

Since from Table 4.1 $NaNO_3$ is soluble and AgCl is insoluble, solid AgCl forms according to the following net ionic reaction:

$$Ag^+(aq) + Cl^-(aq) \rightarrow AgCl(s)$$

In this case we must add enough Cl^- ions to react with all of the Ag^+ ions present. Thus we must calculate the moles of Ag^+ ions present in 1.50 L of a 0.100 *M* $AgNO_3$ solution (remember that a 0.100 *M* $AgNO_3$ solution contains 0.100 *M* Ag^+ ions and 0.100 *M* NO_3^- ions):

$$1.50 \cancel{L} \times \frac{0.100 \text{ mol } Ag^+}{\cancel{L}} = 0.150 \text{ mol } Ag^+$$

Since Ag^+ and Cl^- react in a 1:1 ratio, 0.150 mol of Cl^- ions and thus 0.150 mol of NaCl is required. We calculate the mass of NaCl required as follows:

$$0.150 \cancel{\text{mol NaCl}} \times \frac{58.4 \text{ g NaCl}}{\cancel{\text{mol NaCl}}} = 8.76 \text{ g NaCl}$$

Notice from Sample Exercise 4.12 that the procedures for doing stoichiometric calculations for solution reactions are very similar to those for other types of reactions. It is useful to think in terms of the following steps for reactions in solution.

Stoichiometry Steps for Reactions in Solution

- Identify the species present in the combined solution and determine what reaction occurs.
- Write the balanced equation for the reaction.
- Calculate the moles of reactants.
- Determine which reactant is limiting.
- Calculate the moles of product or products, as required.
- Convert to grams or other units, as required.

Sample Exercise 4.13

When aqueous solutions of Na_2SO_4 and $Pb(NO_3)_2$ are mixed, $PbSO_4$ precipitates. Calculate the mass of $PbSO_4$ formed when 1.25 L of 0.0500 *M* $Pb(NO_3)_2$ and 2.00 L of 0.0250 *M* Na_2SO_4 are mixed.

Solution

STEP 1

When the aqueous solutions of Na_2SO_4 (containing Na^+ and SO_4^{2-} ions) and $Pb(NO_3)_2$ (containing Pb^{2+} and NO_3^- ions) are mixed, the mixed solution contains the ions: Na^+, SO_4^{2-}, Pb^{2+}, and NO_3^-. Since $NaNO_3$ is soluble and $PbSO_4$ is insoluble (see Table 4.1), solid $PbSO_4$ will form.

STEP 2

The net ionic equation is:

$$Pb^{2+}(aq) + SO_4^{2-}(aq) \rightarrow PbSO_4(s)$$

STEP 3

Since 0.0500 *M* $Pb(NO_3)_2$ contains 0.0500 *M* Pb^{2+} ions, we can calculate the moles of Pb^{2+} ions in 1.25 L of this solution as follows:

$$1.25\ \cancel{\text{L}} \times \frac{0.0500 \text{ mol } Pb^{2+}}{\cancel{\text{L}}} = 0.0625 \text{ mol } Pb^{2+}$$

The 0.0250 *M* Na_2SO_4 solution contains 0.0250 *M* SO_4^{2-} ions, and the number of moles of SO_4^{2-} ions in 2.00 L of this solution is

$$2.00\ \cancel{\text{L}} \times \frac{0.0250 \text{ mol } SO_4^{2-}}{\cancel{\text{L}}} = 0.0500 \text{ mol } SO_4^{2-}$$

STEP 4

Because Pb^{2+} and SO_4^{2-} react in a 1:1 ratio, the amount of SO_4^{2-} will be limiting.

STEP 5

Since the Pb^{2+} ions are present in excess, only 0.0500 mol of solid $PbSO_4$ will be formed.

STEP 6

The mass of $PbSO_4$ formed can be calculated using the molecular weight of $PbSO_4$ (303.3):

$$0.0500\ \cancel{\text{mol } PbSO_4} \times \frac{303.3 \text{ g } PbSO_4}{1\ \cancel{\text{mol } PbSO_4}} = 15.2 \text{ g } PbSO_4$$

One method for determining the amount of a given substance present in a solution is to form a precipitate that includes the substance. The precipitate is then filtered, dried, and weighed. This process, called **gravimetric analysis,** is illustrated in Sample Exercise 4.14.

Sample Exercise 4.14

Phosphorite, also called phosphate rock, is a mineral containing PO_4^{3-} and OH^- anions and Ca^{2+} cations. It is treated with sulfuric acid in the manufacture of phosphate fertilizers (see Chapter 3). A chemist finds the calcium content in an impure sample of phosphate rock by weighing out a 0.4367-g sample, dissolving it in water, and precipitating the Ca^{2+} ions as the insoluble hydrated salt* $CaC_2O_4 \cdot H_2O$ ($C_2O_4^{2-}$ is called the oxalate ion). After being filtered and dried (at a temperature of about 100°C so that the extraneous water is driven off but not the water of hydration), the $CaC_2O_4 \cdot H_2O$ precipitate weighed 0.2920 g. Calculate the mass percent of calcium in the sample of phosphate rock.

Solution

This is a straightforward stoichiometry problem. The gravimetric procedure can be summarized as follows:

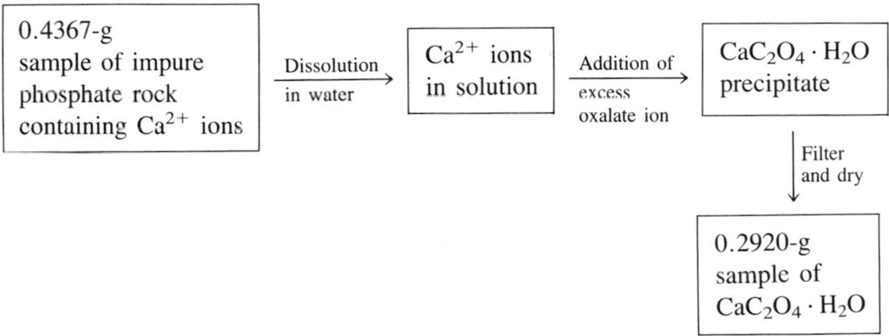

In this analysis excess $C_2O_4^{2-}$ ions are added to ensure that all Ca^{2+} ions are precipitated. Thus the number of moles of Ca^{2+} ions in the original sample determines the number of moles of $CaC_2O_4 \cdot H_2O$ formed. Using the molecular weight of $CaC_2O_4 \cdot H_2O$ (146.12), we can calculate the moles of $CaC_2O_4 \cdot H_2O$:

$$0.2920\ \cancel{\text{g CaC}_2\text{O}_4 \cdot \text{H}_2\text{O}} \times \frac{1\ \text{mol CaC}_2\text{O}_4 \cdot \text{H}_2\text{O}}{146.12\ \cancel{\text{g CaC}_2\text{O}_4 \cdot \text{H}_2\text{O}}} = 1.998 \times 10^{-3}\ \text{mol CaC}_2\text{O}_4 \cdot \text{H}_2\text{O}$$

Thus the original sample of impure phosphate rock contained 1.998×10^{-3} mol of Ca^{2+} ions, which we convert to grams:

$$1.998 \times 10^{-3}\ \cancel{\text{mol Ca}^{2+}} \times \frac{40.08\ \text{g Ca}^{2+}}{1\ \cancel{\text{mol Ca}^{2+}}} = 8.009 \times 10^{-2}\ \text{g Ca}^{2+}$$

The mass percent of calcium in the original sample is then

$$\frac{8.009 \times 10^{-2}\ \cancel{\text{g}}}{0.4367\ \cancel{\text{g}}} \times 100 = 18.34\%$$

---

*Hydrated salts contain one or more H_2O molecules per formula unit in addition to the cations and anions. A dot is used in the formula of these salts.

4.9 Acid-Base Reactions

Purpose

- To show how to do calculations involved in acid-base volumetric analysis.

Earlier in this chapter we considered Arrhenius's concept of acids and bases: an acid is a substance that produces H^+ ions when dissolved in water, and a base is a substance that produces OH^- ions. Although these ideas are fundamentally correct, it is convenient to have a more general definition of a base, which covers substances that do not produce OH^- ions. Such a definition was provided by Brönsted and Lowry, who defined acids and bases as follows:

The Brönsted-Lowry concept of acids and bases will be discussed in detail in Chapter 14.

An acid is a proton donor.

A base is a proton acceptor.

How do we recognize acid-base reactions? One of the most difficult tasks for someone inexperienced in chemistry is to predict what reaction might occur when two solutions are mixed. With precipitation reactions, we found that the best way to deal with this problem is to focus on the species actually present in the mixed solution. This also applies to acid-base reactions. For example, when an aqueous solution of hydrogen chloride (HCl) is mixed with an aqueous solution of sodium hydroxide (NaOH), the combined solution contains the ions H^+, Cl^-, Na^+ and OH^-, since HCl is a strong acid and NaOH is a strong base. How can we predict what reaction occurs, if any? First, will NaCl precipitate? From Table 4.1 we know that NaCl is soluble in water and thus will not precipitate. The Na^+ and Cl^- ions are spectator ions. On the other hand, because water is a nonelectrolyte, large quantities of H^+ and OH^- ions cannot coexist in solution. They will react to form H_2O molecules:

$$H^+(aq) + OH^-(aq) \rightarrow H_2O(l)$$

This is the net ionic equation for the reaction that occurs when aqueous solutions of HCl and NaOH are mixed.

Next, consider mixing an aqueous solution of acetic acid ($HC_2H_3O_2$) with an aqueous solution of potassium hydroxide (KOH). In our earlier discussion of conductivity, we said that an aqueous solution of acetic acid is a weak electrolyte. This tells us that acetic acid does not dissociate into ions to any great extent. In fact, in aqueous solution 99% of the $HC_2H_3O_2$ molecules remain undissociated. However, when solid KOH is dissolved in water, it dissociates completely to produce K^+ and OH^- ions. So in the solution formed by mixing aqueous solutions of $HC_2H_3O_2$ and KOH, *before any reaction occurs* the principal species are H_2O, $HC_2H_3O_2$, K^+, and OH^-. What reaction will occur? A possible precipitation reaction involves K^+ and OH^-, but we know that KOH is soluble. Another possibility is a reaction involving the hydroxide ion and a proton donor. Is there a source of protons in the solution? The answer is yes—the $HC_2H_3O_2$ molecules. The OH^- ion has such a strong affinity for protons that it can strip them from the $HC_2H_3O_2$ molecules. Thus the net ionic equation for the reaction is

$$OH^-(aq) + HC_2H_3O_2(aq) \rightarrow H_2O(l) + C_2H_3O_2^-(aq)$$

This reaction illustrates a very important general principle: *the hydroxide ion is such a strong base that for purposes of stoichiometry it is assumed to react completely with any weak acid dissolved in water*. Of course, OH^- ions also react completely with the H^+ ions in the solutions of strong acids.

We will now deal with the stoichiometry of acid-base reactions in aqueous solutions. The procedure is fundamentally the same as that used previously.

Performing Calculations for Acid-Base Reactions

- List the species present in the combined solution *before reaction*, and decide what reaction will occur.
- Write the balanced net ionic equation for this reaction.
- Change the given quantities of reactants to moles. For reactions in solution, use the volumes of the original solutions and their molarities.
- Determine the limiting reactant where appropriate.
- Calculate the moles of the required reactant or product.
- Convert to grams or a volume of solution, as required by the problem.

An acid-base reaction is often called a **neutralization reaction.** When just enough base is added to react exactly with the acid in a solution, we say the acid has been *neutralized*.

Sample Exercise 4.15

What volume of a 0.100 *M* HCl solution is needed to neutralize 25.0 mL of a 0.350 *M* NaOH solution?

Solution

STEP 1

The species present in the mixed solutions before any reaction occurs are

$$\underbrace{H^+, Cl^-,}_{\text{From HCl}(aq)} \qquad \underbrace{Na^+, OH^-,}_{\text{From NaOH}(aq)} \qquad \text{and } H_2O$$

What reaction will occur? The two possibilities are

$$Na^+(aq) + Cl^-(aq) \rightarrow NaCl(s)$$
$$H^+(aq) + OH^-(aq) \rightarrow H_2O(l)$$

Since we know that NaCl is soluble, the first reaction does not take place (Na^+ and Cl^- are spectator ions). However, as we have seen before, the reaction of the H^+ and OH^- ions to form H_2O does occur.

STEP 2

The balanced net ionic equation for the reaction is

$$H^+(aq) + OH^-(aq) \rightarrow H_2O(l)$$

Sample Exercise 4.15, continued

STEP 3

Next we calculate the number of moles of OH^- ions in the 25.0-mL sample of 0.350 *M* NaOH:

$$25.0\ \cancel{\text{mL NaOH}} \times \frac{1\ \cancel{\text{L}}}{1000\ \cancel{\text{mL}}} \times \frac{0.350\ \text{mol OH}^-}{\cancel{\text{L NaOH}}} = 8.75 \times 10^{-3}\ \text{mol OH}^-$$

STEP 4

This problem requires the addition of just enough H^+ ions to react exactly with the OH^- ions present. Thus we need not be concerned with determining a limiting reactant.

STEP 5

Since H^+ and OH^- ions react in a 1:1 ratio, 8.75×10^{-3} mol of H^+ ions is required to neutralize the OH^- ions present.

STEP 6

The volume (*V*) of 0.100 *M* HCl required to furnish this amount of H^+ ions can be calculated as follows:

$$V \times \frac{0.100\ \text{mol H}^+}{\text{L}} = 8.75 \times 10^{-3}\ \text{mol H}^+$$

Solving for *V* gives

$$V = \frac{8.75 \times 10^{-3}\ \cancel{\text{mol H}^+}}{\dfrac{0.100\ \cancel{\text{mol H}^+}}{\text{L}}} = 8.75 \times 10^{-2}\ \text{L}$$

Thus 8.75×10^{-2} L (87.5 mL) of 0.100 *M* HCl is required to neutralize 25.0 mL of 0.350 *M* NaOH.

Sample Exercise 4.16

In a certain experiment 28.0 mL of 0.250 *M* HNO_3 and 53.0 mL of 0.320 *M* KOH are mixed. Calculate the amount of water formed in the resulting reaction. What is the concentration of H^+ or OH^- ions in excess after the reaction goes to completion?

Solution

The ions available for reaction are

$$\underbrace{H^+,\ NO_3^-,}_{\text{From HNO}_3\text{ solution}} \quad \underbrace{K^+,\ OH^-,}_{\text{From KOH solution}} \quad \text{and } H_2O$$

Since KNO_3 is soluble, K^+ and NO_3^- are spectator ions, so the net ionic equation is

$$H^+(aq) + OH^-(aq) \rightarrow H_2O(l)$$

Sample Exercise 4.16, continued

We next compute the amounts of H^+ and OH^- ions present.

$$28.0\ \cancel{\text{mL HNO}_3} \times \frac{1\ \cancel{\text{L}}}{1000\ \cancel{\text{mL}}} \times \frac{0.250\ \text{mol H}^+}{\cancel{\text{L HNO}_3}} = 7.00 \times 10^{-3}\ \text{mol H}^+$$

$$53.0\ \cancel{\text{mL KOH}} \times \frac{1\ \cancel{\text{L}}}{1000\ \cancel{\text{mL}}} \times \frac{0.320\ \text{mol OH}^-}{\cancel{\text{L KOH}}} = 1.70 \times 10^{-2}\ \text{mol OH}^-$$

Since H^+ and OH^- react in a 1:1 ratio, the limiting reactant is H^+. This means that 7.00×10^{-3} mol of H^+ ions will react with 7.00×10^{-3} mol of OH^- ions to form 7.00×10^{-3} mol of H_2O.

The amount of OH^- ions in excess is obtained from the following difference:

$$\text{Original amount} - \text{amount consumed} = \text{amount in excess}$$

$$1.70 \times 10^{-2}\ \text{mol OH}^- - 7.00 \times 10^{-3}\ \text{mol OH}^- = 1.00 \times 10^{-2}\ \text{mol OH}^-$$

The volume of the combined solution is the sum of the individual volumes:

$$\text{Original volume of HNO}_3 + \text{original volume of KOH} = \text{total volume}$$

$$28.0\ \text{mL} + 53.0\ \text{mL} = 81.0\ \text{mL} = 8.10 \times 10^{-2}\ \text{L}$$

Thus the molarity of OH^- ions in excess is

$$M = \frac{\text{mol OH}^-}{\text{L solution}} = \frac{1.00 \times 10^{-2}\ \text{mol OH}^-}{8.10 \times 10^{-2}\ \text{L}} = 0.123\ M\ \text{OH}^-$$

Acid-Base Titrations

Acid-base **titrations** are an example of **volumetric analysis,** a technique in which one solution is used to analyze another. The solution used to carry out the analysis is called the **titrant** and is delivered from a device called a *buret* (see Figure 1.5 in Chapter 1), which measures the volume accurately. The point in the titration at which enough titrant has been added to react exactly with the substance being determined is called the **equivalence point** or the **stoichiometric point.** This point is often marked by the change in color of a chemical called an **indicator.** The titration procedure is illustrated in Fig. 4.11 on the following page.

Titrations will be discussed further in Chapter 15.

The following requirements must be met for a titration to be successful:

The concentration of the titrant must be known. Such a titrant is called a *standard solution.*

The exact reaction between titrant and substance being analyzed must be known.

The stoichiometric (equivalence) point must be known. An indicator that changes color at, or very near, the stoichiometric point is often used.

The point at which the indicator changes color is called the **end point.** The goal is to choose an indicator whose end point coincides with the stoichiometric

Ideally, the end point and stoichiometric point should coincide.

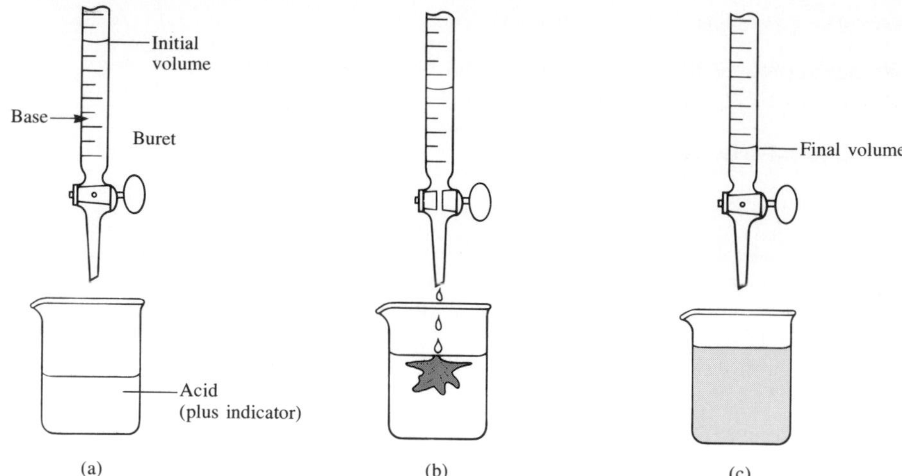

Figure 4.11

The titration of an acid with a base. (a) The titrant (the base) is in the buret, and the beaker contains the acid solution along with a small amount of indicator. (b) As base is added drop by drop to the acid solution in the beaker during the titration, the indicator changes color as each drop hits the acid solution; but the color disappears on mixing. (c) The stoichiometric (equivalence) point is marked by a permanent indicator color change. The volume of base added is the difference between the final and initial buret readings.

point. An indicator very commonly used for acid-base titrations is *phenolphthalein,* which is colorless in acid and turns pink at the end point when an acid is titrated with a base.

The volume of titrant required to reach the stoichiometric point must be known as accurately as possible.

We will deal with acid-base titrations only briefly here, but will return to the topic of titrations and indicators in more detail in Chapter 15. When a substance being analyzed contains an acid, the amount of acid present is usually determined by titration with a standard solution containing hydroxide ions, illustrated in Sample Exercise 4.18. In Sample Exercise 4.17 we will show how to determine the concentration of a sodium hydroxide solution.

Sample Exercise 4.17

A student carries out an experiment to standardize (determine the exact concentration of) a sodium hydroxide solution. To do this the student weighs out a 1.3009-g sample of potassium hydrogen phthalate (KHP, mol. wt. = 204.22), a compound with the formula $KHC_8H_4O_4$ which has one acidic hydrogen. The student dissolves the KHP in distilled water and titrates it with the sodium hydroxide solution. The difference between the final and initial buret readings indicates that 41.20 mL of the sodium hydroxide solution is required just to react with the 1.3009 g of KHP. Calculate the concentration of the sodium hydroxide solution.

Solution

KHP is a potassium salt that dissolves in water to give the K^+ and $HC_8H_4O_4^-$ ions. The net ionic reaction between aqueous sodium hydroxide (contains Na^+ and OH^- ions) and KHP is

$$HC_8H_4O_4^-(aq) + OH^-(aq) \rightarrow H_2O(l) + C_8H_4O_4^{2-}(aq)$$

Sample Exercise 4.17, continued

Since the reaction is 1:1, we know that 41.20 mL of the sodium hydroxide solution contains exactly the same number of moles of OH^- as there are moles of $HC_8H_4O_4^-$ in 1.3009 g of $KHC_8H_4O_4$.

We calculate the moles of $KHC_8H_4O_4$ in the usual way:

$$1.3009 \cancel{\text{g KHC}_8\text{H}_4\text{O}_4} \times \frac{1 \text{ mol KHC}_8\text{H}_4\text{O}_4}{204.22 \cancel{\text{g KHC}_8\text{H}_4\text{O}_4}} = 6.3701 \times 10^{-3} \text{ mol KHC}_8\text{H}_4\text{O}_4$$

This means that 6.3701×10^{-3} mol OH^- must be added to react with the 6.3701×10^{-3} mol $HC_8H_4O_4^-$. Thus 41.20 mL (4.120×10^{-2} L) of the sodium hydroxide solution contains 6.3701×10^{-3} mol OH^- (and Na^+). The concentration of the sodium hydroxide solution is

$$\text{molarity of NaOH} = \frac{\text{mol NaOH}}{\text{L solution}} = \frac{6.3701 \times 10^{-3} \text{ mol NaOH}}{4.120 \times 10^{-2} \text{ L}}$$
$$= 0.1546\ M$$

This standard sodium hydroxide solution can now be used in other experiments (see Sample Exercise 4.18).

Sample Exercise 4.18

An environmental chemist analyzed the effluent (the released waste material) from an industrial process known to produce the compounds carbon tetrachloride (CCl_4) and benzoic acid ($HC_7H_5O_2$), a weak acid that has one acidic hydrogen atom per molecule. A sample of this effluent weighing 0.3518 g was placed in water and shaken vigorously to dissolve the benzoic acid. The resulting aqueous solution required 10.59 mL of 0.1546 *M* NaOH for neutralization. Calculate the mass percent of $HC_7H_5O_2$ in the original sample.

Solution

In this case the sample was a mixture containing CCl_4 and $HC_7H_5O_2$, and it was titrated with OH^- ions. Clearly CCl_4 is not an acid (it contains no hydrogen atoms); so we can assume it does not react with OH^- ions. However, $HC_7H_5O_2$ is an acid and donates one H^+ ion per molecule to react with an OH^- ion as follows:

$$HC_7H_5O_2(aq) + OH^-(aq) \rightarrow H_2O(l) + C_7H_5O_2^-(aq)$$

Although $HC_7H_5O_2$ is a weak acid, the OH^- ion is such a strong base that we can assume that each OH^- ion added will react with a $HC_7H_5O_2$ molecule until all of the benzoic acid is consumed.

We must first determine the number of moles of OH^- ions required to react with all of the $HC_7H_5O_2$:

$$10.59 \cancel{\text{mL NaOH}} \times \frac{1 \cancel{\text{L}}}{1000 \cancel{\text{mL}}} \times \frac{0.1546 \text{ mol OH}^-}{\cancel{\text{L NaOH}}} = 1.637 \times 10^{-3} \text{ mol OH}^-$$

Sample Exercise 4.18, continued

This number is also the number of moles of $HC_7H_5O_2$ present. The number of grams of the acid is calculated using its molecular weight (122.125):

$$1.637 \times 10^{-3} \cancel{\text{mol } HC_7H_5O_2} \times \frac{122.125 \text{ g } HC_7H_5O_2}{1 \cancel{\text{mol } HC_7H_5O_2}} = 0.1999 \text{ g } HC_7H_5O_2$$

The mass percent of $HC_7H_5O_2$ in the original sample is

$$\frac{0.1999 \cancel{g}}{0.3518 \cancel{g}} \times 100 = 56.82\%$$

The first step in the analysis of a complex solution is to write down the components and focus on the chemistry of each one.

Chemical systems often seem difficult to deal with simply because there are many components. Solving a problem involving a solution where several components are present is simplified if you *think* about the *chemistry* involved. *The key to success is to write down all the components in the solution and focus on the chemistry of each one.* We have been emphasizing this approach in dealing with the reactions between ions in solution. Make it a habit to write down the components of solutions.

4.10 Oxidation-Reduction Reactions

Purpose

- To characterize oxidation-reduction reactions.
- To describe how to assign oxidation states.
- To identify oxidizing and reducing agents.

Oxidation of copper metal by nitric acid. The copper atoms lose two electrons to form Cu^{2+} ions, which give a blue color in water.

As we have seen, many important substances are ionic. Sodium chloride, for example, can be formed by the reaction of elemental sodium and chlorine:

$$2Na(s) + Cl_2(g) \rightarrow 2NaCl(s)$$

In this reaction an electron is transferred from a sodium atom to a chlorine atom, yielding a Na^+ ion and a Cl^- ion. *Reactions like this one, in which one or more electrons are transferred, are called* **oxidation-reduction reactions,** or **redox reactions.**

Many important chemical reactions involve oxidation and reduction. In fact, most reactions used for energy production are redox reactions. In humans the oxidation of sugars, fats, and proteins provides the energy necessary for life. Combustion reactions, which provide most of the energy to power our civilization, also involve oxidation and reduction. An example is the reaction of methane with oxygen:

$$CH_4(g) + 2O_2(g) \rightarrow CO_2(g) + 2H_2O(g) + \text{energy}$$

Even though none of the reactants or products in this reaction is ionic, the reaction is still assumed to involve a transfer of electrons from carbon to oxygen. To explain this, we must introduce the concept of oxidation states.

Oxidation States

The concept of **oxidation states** provides a way to keep track of electrons in oxidation-reduction reactions. Oxidation states are defined by a set of rules, most of which describe how to divide up the shared electrons in compounds containing covalent bonds. However, before we discuss these rules we need to discuss the distribution of electrons in a bond.

Recall from the discussion of the water molecule in Section 4.1 that oxygen has a greater attraction for electrons than does hydrogen, causing the O—H bonds in the water molecule to be polar. This phenomenon occurs in other bonds as well, and we will discuss the topic of polarity in detail in Chapter 8. For now we will be satisfied with some general guidelines to help us keep track of electrons in oxidation-reduction reactions. The nonmetals with the highest attraction for shared electrons are in the upper right-hand corner of the periodic table. They are fluorine, oxygen, nitrogen, and chlorine. The relative ability of these atoms to attract shared electrons is

$$F > O > N \approx Cl$$

↗ Greatest attraction for electrons ↖ Least attraction for electrons

That is, fluorine attracts shared electrons to the greatest extent, followed by oxygen, then nitrogen and chlorine.

The rules for assigning oxidation states are given in Table 4.3. Application of these rules allows the assignment of oxidation states in most compounds. The principles are illustrated by Sample Exercise 4.19.

Table 4.3

Rules for Assigning Oxidation States
1. The oxidation state of an atom in an element is 0. For example, the oxidation state of each atom in the substances Na(*s*), $O_2(g)$, $O_3(g)$, and Hg(*l*) is 0.
2. The oxidation state of a monatomic ion is the same as its charge. For example, the oxidation state of the Na^+ ion is +1.
3. Oxygen is assigned an oxidation state of −2 in its covalent compounds, such as CO, CO_2, SO_2, and SO_3. The exception to this rule occurs in peroxides (compounds containing the O_2^{2-} group), where each oxygen is assigned an oxidation state of −1. The best-known example of a peroxide is hydrogen peroxide (H_2O_2).
4. In its covalent compounds with nonmetals, hydrogen is assigned an oxidation state of +1. For example, in the compounds HCl, NH_3, H_2O, and CH_4, hydrogen is assigned an oxidation state of +1.
5. In binary compounds the element with the greater attraction for the electrons in the bond is assigned a negative oxidation state equal to its charge in its ionic compounds. For example, fluorine is always assigned an oxidation state of −1. That is, for purposes of counting electrons, fluorine is assumed to be F^-. Nitrogen is usually assigned −3. For example, in NH_3, nitrogen is assigned an oxidation state of −3; in H_2S, sulfur is assigned an oxidation state of −2; in HI, iodine is assigned an oxidation state of −1; and so on.
6. The sum of the oxidation states must be zero for an electrically neutral compound and must be equal to the overall charge for an ionic species. For example, the sum of the oxidation states for the hydrogen and oxygen atoms in water is 0; the sum of the oxidation states for the carbon and oxygen atoms in CO_3^{2-} is −2; and the sum of oxidation states for the nitrogen and hydrogen atoms in NH_4^+ is +1.

Sample Exercise 4.19

Assign oxidation states to all atoms in the following:

a. CO_2 **b.** SF_6 **c.** NO_3^-

Solution

Actual charges are given as $n-$ or $n+$. Oxidation states (*not actual charges*) are given as $-n$ or $+n$. For example, for NO_3^-, the overall charge is 1−; the oxidation state for N is +5 and for O is −2.

a. The rule that takes precedence here is that oxygen is assigned an oxidation state of −2. The oxidation state for carbon can be determined by recognizing that since CO_2 has no charge, the sum of the oxidation states for oxygen and carbon must be 0. Since each oxygen is −2 and there are two oxygen atoms, the carbon atom must be assigned an oxidation state of +4:

$$\underset{+4 \nearrow \quad \nwarrow -2 \text{ for each oxygen}}{CO_2}$$

Check: $+4 + 2(-2) = 0$

b. Since fluorine has the greater attraction for the shared electrons, we assign its oxidation state first. Since its charge in ionic compounds is 1−, we assign −1 as the oxidation state of each fluorine atom. The sulfur must then be assigned an oxidation state of +6 to balance the total of −6 from the fluorine atoms:

$$\underset{+6 \nearrow \quad \nwarrow -1 \text{ for each fluorine}}{SF_6}$$

Check: $+6 + 6(-1) = 0$

c. Since oxygen has a greater attraction than does nitrogen for the shared electrons, we assign its oxidation state of −2 first. Because the sum of the oxidation states of the three oxygens is −6 and the net charge on the NO_3^- ion is 1−, the nitrogen must have an oxidation state of +5:

$$\underset{+5 \nearrow \quad \nwarrow -2 \text{ for each oxygen}}{NO_3^-}$$

Check: $+5 + 3(-2) = -1$

Next let us consider the oxidation states of the atoms in Fe_3O_4, which is the main component in magnetite, an iron ore that accounts for the reddish color of many types of rocks and soils. We assign each oxygen atom its usual oxidation state of −2. The three iron atoms must yield a total of +8 to balance the total of −8 from the four oxygens. This means that each iron atom has an oxidation state of $+\frac{8}{3}$. A noninteger value for oxidation state may seem strange since charge is expressed in whole numbers. However, although they are rare, noninteger oxidation states do occur because of the rather arbitrary way that electrons are divided up by the rules in Table 4.3. For Fe_3O_4, for example, the rules assume that all of the iron atoms are equal, when in fact this compound can best be viewed as containing four O^{2-} ions, two Fe^{3+} ions, and one Fe^{2+} ion per formula unit. (Note that the "average" charge on iron works out to be $\frac{8}{3}+$ which is equal to the oxidation state we determined above.) Noninteger oxidation states should not intimidate you. They are used in the same way as integer oxidation states—for keeping track of electrons.

Magnetite is a magnetic ore containing Fe_3O_4.

Chemical Impact

Aging: Does It Involve Oxidation?

Although aging is supposed to bring wisdom, almost no one wants to get old. Along with wisdom, aging brings wrinkles, loss of physical strength, and greater susceptibility to disease.

Why do we age? No one knows for certain, but many scientists think that oxidation plays a major role. The oxygen molecule and other oxidizing agents in the body apparently can extract single electrons from the large molecules that make up cell walls, thus making them very reactive. Subsequently these activated molecules can link up, changing the properties of the cell wall. At some point enough of these reactions have occurred that the body's immune system comes to view the changed cell as an "enemy" and destroys it. This is particularly detrimental to the organism when the cells involved are irreplaceable. Nerve cells, for example, fall into this category.

The body has defenses against oxidation, such as vitamin E, a well-known antioxidant. Studies have shown that red blood cells age much faster than normal when they are deficient in vitamin E. Based on studies such as these, some have suggested large doses of vitamin E as a preventative against aging, but there is no evidence that this potentially dangerous practice has any impact on aging.

Oxidation is only one possible cause for aging. Research continues on many fronts to try to discover why we get "older" as time passes.

Suggested Reading

1. Durk Pearson and Sandy Shaw, *Life Extension*. New York: Warner Books, Inc., 1983.
2. Morton Rothstein, "Biochemical Studies of Aging," *Chemical and Engineering News*, August 11, 1986, p. 26.

The Characteristics of Oxidation-Reduction Reactions

Oxidation-reduction reactions are characterized by a transfer of electrons. In some cases, the transfer occurs in a literal sense to form ions, such as in this reaction:

$$2Na(s) + Cl_2(g) \rightarrow 2NaCl(s)$$

However, sometimes the transfer occurs in a more formal sense, such as in the combustion of methane (the oxidation state for each atom is given):

$$CH_4(g) + 2O_2(g) \rightarrow CO_2(g) + 2H_2O(g)$$

Oxidation state: C in CH_4: -4; H: $+1$ (each H); O in O_2: 0; C in CO_2: $+4$; O: -2 (each O); H in H_2O: $+1$ (each H); O: -2

Note that the oxidation state for oxygen in O_2 is 0 because it is in elemental form. In this reaction there are no ionic compounds, but we can still describe the process in terms of a transfer of electrons. Note that carbon undergoes a change in oxidation state from -4 in CH_4 to $+4$ in CO_2. Such a change can be accounted for by a loss of eight electrons (the symbol e^- stands for an electron):

$$\underset{-4}{CH_4} \rightarrow \underset{+4}{CO_2} + 8e^-$$

On the other hand, each oxygen changes from an oxidation state of 0 in O_2 to -2 in

H_2O and CO_2, signifying a gain of two electrons per atom. Since four oxygen atoms are involved, this is a gain of eight electrons:

$$\underset{0}{2O_2} + 8e^- \rightarrow CO_2 + 2H_2O \qquad 4(-2) = -8$$

No change occurs in the oxidation state of hydrogen, and it is not formally involved in the electron transfer process.

Oxidation is an increase in oxidation state. Reduction is a decrease in oxidation state.

With this background, we can now define some important terms. **Oxidation** is an *increase* in oxidation state (a loss of electrons). **Reduction** is a *decrease* in oxidation state (a gain of electrons). Thus in the reaction

$$\underset{0}{2Na(s)} + \underset{0}{Cl_2(g)} \rightarrow 2NaCl(s) \quad (+1, -1)$$

A helpful mnemonic device is OIL RIG (*O*xidation *I*nvolves *L*oss; *R*eduction *I*nvolves *G*ain).

sodium is oxidized and chlorine is reduced. In addition, Cl_2 is called the **oxidizing agent (electron acceptor)** and Na is called the **reducing agent (electron donor).**

Concerning the reaction

$$CH_4(g) + 2O_2(g) \rightarrow CO_2(g) + 2H_2O(g)$$
$$CH_4: -4, +1 \quad O_2: 0 \quad CO_2: +4, -2 \quad H_2O: +1, -2$$

we can say the following:

An oxidizing agent is reduced and a reducing agent is oxidized in a redox reaction.

Carbon is oxidized because there is an increase in its oxidation state (carbon has formally lost electrons).

Oxygen is reduced because there has been a decrease in its oxidation state (oxygen has formally gained electrons).

CH_4 is the reducing agent.

O_2 is the oxidizing agent.

Note that when the oxidizing or reducing agent is named, the *whole compound* is specified, not just the element that undergoes the change in oxidation state.

Sample Exercise 4.20

When powdered aluminum metal is mixed with pulverized iodine crystals and a drop of water is added, the resulting reaction produces a great deal of energy. The mixture bursts into flames, and a purple smoke of I_2 vapor is produced from the excess iodine. The equation for the reaction is

$$2Al(s) + 3I_2(s) \rightarrow 2AlI_3(s)$$

For this reaction, identify the atoms that are oxidized and reduced, and specify the oxidizing and reducing agents.

2. This means that three oxygen atoms are needed to balance the increase in the oxidation state of the single carbon atom. We can write this relationship as follows:

$$\underset{\text{1 carbon atom}}{\underset{\uparrow}{CH_3OH(l)}} + \underset{\text{3 oxygen atoms}}{\underset{\uparrow}{\tfrac{3}{2}O_2(g)}} \rightarrow \text{products}$$

The rest of the equation can be balanced by inspection:

$$CH_3OH(l) + \tfrac{3}{2}O_2(g) \rightarrow CO_2(g) + 2H_2O(g)$$

We write it in conventional format (multiply through by 2):

$$2CH_3OH(l) + 3O_2(g) \rightarrow 2CO_2(g) + 4H_2O(g)$$

In using the oxidation states method to balance an oxidation-reduction equation, we find the coefficients for the reactants that will make the total increase in oxidation state balance the total decrease. The remainder of the equation is then balanced by inspection.

The procedures for balancing an oxidation-reduction reaction by the oxidation states method are summarized below.

Balancing an Oxidation-Reduction Reaction by the Oxidation States Method

- Assign the oxidation states of all atoms.
- Decide which element is oxidized and determine the increase in oxidation state.
- Decide which element is reduced and determine the decrease in oxidation state.
- Choose coefficients for the species containing the atom oxidized and the atom reduced such that the total increase in oxidation state equals the total decrease in oxidation state.
- Balance the remainder of the equation by inspection.

Sample Exercise 4.22

Because metals are so reactive, very few are found in nature in pure form. Metallurgy involves reducing the metal ions in ores to the elemental form. In Sample Exercise 4.21, we considered the treatment of the lead-containing ore galena. The production of manganese from the ore pyrolusite, which contains MnO_2, uses aluminum as the reducing agent. Using oxidation states, balance the equation for this process.

$$MnO_2(s) + Al(s) \rightarrow Mn(s) + Al_2O_3(s)$$

Solution

STEP 1

First we assign oxidation states:

$$MnO_2(s) + Al(s) \rightarrow Mn(s) + Al_2O_3(s)$$

$MnO_2(s)$: Mn	$MnO_2(s)$: O	$Al(s)$	$Mn(s)$	$Al_2O_3(s)$: Al	$Al_2O_3(s)$: O
+4	−2 (each O)	0	0	+3 (each Al)	−2 (each O)

Sample Exercise 4.22, continued

STEPS 2 AND 3

Each Mn atom undergoes a decrease in oxidation state of 4 (from +4 to 0), while each Al atom undergoes an increase of 3 (from 0 to +3).

STEP 4

Thus we need three Mn atoms for every four Al atoms in order to balance the increase and decrease in oxidation states:

$$\text{Increase} = 4(3) = \text{decrease} = 3(4)$$

$$3MnO_2(s) + 4Al(s) \rightarrow \text{products}$$

STEP 5

We balance the rest of the equation by inspection:

$$3MnO_2(s) + 4Al(s) \rightarrow 3Mn(s) + 2Al_2O_3(s)$$

The Half-Reaction Method

For oxidation-reduction reactions that occur in aqueous solution, it is often useful to separate the reaction into two **half-reactions:** one involving oxidation and the other involving reduction. For example, consider the unbalanced equation for the oxidation-reduction reaction between cerium(IV) ion and tin(II) ion:

$$Ce^{4+}(aq) + Sn^{2+}(aq) \rightarrow Ce^{3+}(aq) + Sn^{4+}(aq)$$

This reaction can be separated into a half-reaction involving the substance being *reduced,*

$$Ce^{4+}(aq) \rightarrow Ce^{3+}(aq)$$

and one involving the substance being *oxidized,*

$$Sn^{2+}(aq) \rightarrow Sn^{4+}(aq)$$

The general procedure is to balance the equations for the half-reactions separately and then to add them to obtain the overall balanced equation. The half-reaction method for balancing oxidation-reduction equations differs slightly depending on whether the reaction takes place in acidic or basic solution.

The Half-Reaction Method for Balancing Equations for Oxidation-Reduction Reactions Occurring in Acidic Solution

- Write the equations for the oxidation and reduction half-reactions.
- For each half-reaction:
  a. Balance all of the elements except hydrogen and oxygen.
  b. Balance oxygen using H_2O.
  c. Balance hydrogen using H^+.
  d. Balance the charge using electrons.

- If necessary, multiply one or both balanced half-reactions by an integer to equalize the number of electrons transferred in the two half-reactions.
- Add the half-reactions, and cancel identical species.
- Check that the elements and charges balance.

We will illustrate this method by balancing the equation for the reaction between permanganate and iron(II) ions in acidic solution:

$$MnO_4^-(aq) + Fe^{2+}(aq) \xrightarrow{\text{Acid}} Fe^{3+}(aq) + Mn^{2+}(aq)$$

This reaction is used to analyze iron ore for its iron content.

Step 1: Identify and write equations for the half-reactions. The oxidation states for the half-reaction involving the permanganate ion show that manganese is reduced:

$$\underset{+7\ -2\,(\text{each O})}{MnO_4^-} \rightarrow \underset{+2}{Mn^{2+}}$$

This is the *reduction half-reaction.* The other half-reaction involves the oxidation of iron(II) to iron(III) ion and is the *oxidation half-reaction:*

$$\underset{+2}{Fe^{2+}} \rightarrow \underset{+3}{Fe^{3+}}$$

Step 2: Balance each half-reaction. For the reduction reaction, we have

$$MnO_4^-(aq) \rightarrow Mn^{2+}(aq)$$

a. The manganese is balanced.
b. We balance oxygen by adding $4H_2O$ to the right side of the equation:

$$MnO_4^-(aq) \rightarrow Mn^{2+}(aq) + 4H_2O(l)$$

c. Next we balance hydrogen by adding $8H^+$ to the left side:

$$8H^+(aq) + MnO_4^-(aq) \rightarrow Mn^{2+}(aq) + 4H_2O(l)$$

d. All of the elements have been balanced, but we need to balance the charge using electrons. At this point we have the following charges for reactants and products in the reduction half-reaction:

$$\underbrace{8H^+(aq) + MnO_4^-(aq)}_{\substack{8+\ +\ 1-\\ 7+}} \rightarrow \underbrace{Mn^{2+}(aq) + 4H_2O(l)}_{\substack{2+\ +\ 0\\ 2+}}$$

We can equalize the charges by adding five electrons to the left side:

$$\underbrace{5e^- + 8H^+(aq) + MnO_4^-(aq)}_{2+} \rightarrow \underbrace{Mn^{2+}(aq) + 4H_2O(l)}_{2+}$$

Both the *elements* and the *charges* are now balanced, so this represents the balanced reduction half-reaction. The fact that five electrons appear on the reactant side of the equation makes sense since five electrons are required to reduce MnO_4^- (Mn has an oxidation state of +7) to Mn^{2+} (Mn has an oxidation state of +2).

For the oxidation reaction,

$$Fe^{2+}(aq) \rightarrow Fe^{3+}(aq)$$

the elements are balanced, and we must simply balance the charge.

$$\underbrace{Fe^{2+}(aq)}_{2+} \rightarrow \underbrace{Fe^{3+}(aq)}_{3+}$$

One electron is needed on the right side to give a net 2+ charge on both sides:

$$\underbrace{Fe^{2+}(aq)}_{2+} \rightarrow \underbrace{Fe^{3+}(aq) + e^-}_{2+}$$

The number of electrons gained in the reduction half-reaction must equal the number of electrons lost in the oxidation half-reaction.

Step 3: Equalize the electron transfer in the two half-reactions. Since the reduction half-reaction involves a transfer of five electrons and the oxidation half-reaction involves a transfer of only one electron, the oxidation half-reaction must be multiplied by 5:

$$5Fe^{2+}(aq) \rightarrow 5Fe^{3+}(aq) + 5e^-$$

Step 4: Add the half-reactions. The half-reactions are added to give

$$5e^- + 5Fe^{2+}(aq) + MnO_4^-(aq) + 8H^+(aq)$$
$$\rightarrow 5Fe^{3+}(aq) + Mn^{2+}(aq) + 4H_2O(l) + 5e^-$$

Note that the electrons cancel out (as they must) to give the final balanced equation:

$$5Fe^{2+}(aq) + MnO_4^-(aq) + 8H^+(aq) \rightarrow 5Fe^{3+}(aq) + Mn^{2+}(aq) + 4H_2O(l)$$

Step 5: Check that elements and charges balance.

Elements balance: 5 Fe, 1 Mn, 4 O, 8 H $\rightarrow$ 5 Fe, 1 Mn, 4 O, 8 H

Charges balance: $5(2+) + (1-) + 8(1+) = 17+ \rightarrow 5(3+) + (2+) + 0 = 17+$

The equation is balanced.

Sample Exercise 4.23

Potassium dichromate ($K_2Cr_2O_7$) is a bright orange compound that can be reduced to a green solution of Cr^{3+} ions. This color change can be used to test a person's breath for alcohol (C_2H_5OH) content.* The reactants and products of the oxidation-reduction reaction are

$$H^+(aq) + Cr_2O_7^{2-}(aq) + C_2H_5OH(l) \rightarrow Cr^{3+}(aq) + CO_2(g) + H_2O(l)$$

Balance this equation using the half-reaction method.

*For an experiment based on the breathalyzer *see* William C. Timmer, An Experiment in Forensic Chemistry, *J. Chem. Ed.* **63** (1986): 897.

Sample Exercise 4.23, continued

Solution

STEP 1

The reduction half-reaction is

$$Cr_2O_7^{2-}(aq) \rightarrow Cr^{3+}(aq)$$

Chromium is reduced from an oxidation state of +6 in $Cr_2O_7^{2-}$ to one of +3 in Cr^{3+}.

The oxidation half-reaction is

$$C_2H_5OH(l) \rightarrow CO_2(g)$$

Carbon is oxidized from an oxidation state of −2 in C_2H_5OH to +4 in CO_2.

STEP 2

Balancing all elements except hydrogen and oxygen in the first half-reaction, we have

$$Cr_2O_7^{2-}(aq) \rightarrow 2Cr^{3+}(aq)$$

Balancing oxygen using H_2O, we have

$$Cr_2O_7^{2-}(aq) \rightarrow 2Cr^{3+}(aq) + 7H_2O(l)$$

Balancing hydrogen using H^+, we have

$$14H^+(aq) + Cr_2O_7^{2-}(aq) \rightarrow 2Cr^{3+}(aq) + 7H_2O(l)$$

Balancing the charge using electrons, we have

$$6e^- + 14H^+(aq) + Cr_2O_7^{2-}(aq) \rightarrow 2Cr^{3+}(aq) + 7H_2O(l)$$

Next, we turn to the oxidation half-reaction

$$C_2H_5OH(l) \rightarrow CO_2(g)$$

Balancing carbon, we have

$$C_2H_5OH(l) \rightarrow 2CO_2(g)$$

Balancing oxygen using H_2O, we have

$$C_2H_5OH(l) + 3H_2O(l) \rightarrow 2CO_2(g)$$

Balancing hydrogen using H^+, we have

$$C_2H_5OH(l) + 3H_2O(l) \rightarrow 2CO_2(g) + 12H^+(aq)$$

Balancing the charge using electrons, we have

$$C_2H_5OH(l) + 3H_2O(l) \rightarrow 2CO_2(g) + 12H^+(aq) + 12e^-$$

STEP 3

In the reduction half-reaction there are 6 electrons on the left-hand side, and there are 12 electrons on the right-hand side of the oxidation half-reaction.

Sample Exercise 4.23, continued

Thus we multiply the reduction half-reaction by 2 to give

$$12e^- + 28H^+(aq) + 2Cr_2O_7^{2-}(aq) \rightarrow 4Cr^{3+}(aq) + 14H_2O(l)$$

STEP 4

Adding the half-reactions, and canceling identical species, we have

Reduction half-reaction:

$$12e^- + 28H^+(aq) + 2Cr_2O_7^{2-}(aq) \rightarrow 4Cr^{3+}(aq) + 14H_2O(l)$$

Oxidation half-reaction:

$$C_2H_5OH(l) + 3H_2O(l) \rightarrow 2CO_2(g) + 12H^+(aq) + 12e^-$$

$$16H^+(aq) + 2Cr_2O_7^{2-}(aq) + C_2H_5OH(l) \rightarrow 4Cr^{3+} + 11H_2O(l) + 2CO_2(g)$$

STEP 5

Check that elements and charges balance.

Elements balance: 22 H, 4 Cr, 15 O, 2 C $\rightarrow$ 22 H, 4 Cr, 15 O, 2 C

Charges balance: $+16 + 2(-2) + 0 = +12 \rightarrow 4(+3) + 0 + 0 = +12$

Oxidation-reduction reactions can occur in basic as well as in acidic solutions. The half-reaction method for balancing equations is slightly different in such cases.

The Half-Reaction Method for Balancing Equations for Oxidation-Reduction Reactions Occurring in Basic Solution

- Use the half-reaction method as specified for acidic solutions to obtain the final balanced equation *as if H^+ ions were present.*
- To both sides of the equation obtained above, add a number of OH^- ions that is equal to the number of H^+ ions. (We want to eliminate H^+ by forming H_2O.)
- Form H_2O on the side containing both H^+ and OH^- ions, and eliminate the number of H_2O molecules that appear on both sides of the equation.
- Check that elements and charges balance.

We will illustrate how the rules are applied in Sample Exercise 4.24.

Sample Exercise 4.24

Silver is sometimes found in nature as large nuggets; more often it is found mixed with other metals and their ores. Cyanide ion is often used to extract the silver by the following reaction that occurs in basic solution:

$$Ag(s) + CN^-(aq) + O_2(g) \xrightarrow{\text{Basic}} Ag(CN)_2^-(aq)$$

Sample Exercise 4.24, continued

Balance this equation using the half-reaction method.

Solution

STEP 1

Balance the equation as if H^+ ions were present. Balance the oxidation half-reaction:

$$CN^-(aq) + Ag(s) \rightarrow Ag(CN)_2{}^-(aq)$$

Balance carbon and nitrogen:

$$2CN^-(aq) + Ag(s) \rightarrow Ag(CN)_2{}^-(aq)$$

Balance the charge:

$$2CN^-(aq) + Ag(s) \rightarrow Ag(CN)_2{}^-(aq) + e^-$$

Balance the reduction half-reaction:

$$O_2(g) \rightarrow$$

Balance oxygen:

$$O_2(g) \rightarrow 2H_2O(l)$$

Balance hydrogen:

$$O_2(g) + 4H^+(aq) \rightarrow 2H_2O(l)$$

Balance the charge:

$$4e^- + O_2(g) + 4H^+(aq) \rightarrow 2H_2O(l)$$

Multiply the balanced oxidation half-reaction by 4:

$$8CN^-(aq) + 4Ag(s) \rightarrow 4Ag(CN)_2{}^-(aq) + 4e^-$$

Add the half-reactions, and cancel identical species:

Oxidation half-reaction:

$$8CN^-(aq) + 4Ag(s) \rightarrow 4Ag(CN)_2{}^-(aq) + 4e^-$$

Reduction half-reaction:

$$4e^- + O_2(g) + 4H^+(aq) \rightarrow 2H_2O(l)$$

$$8CN^-(aq) + 4Ag(s) + O_2(g) + 4H^+(aq) \rightarrow 4Ag(CN)_2{}^-(aq) + 2H_2O(l)$$

STEP 2

Add OH^- ions to both sides of balanced equation. We need to add $4OH^-$ to each side:

$$8CN^-(aq) + 4Ag(s) + O_2(g) + \underbrace{4H^+(aq) + 4OH^-(aq)}_{4H_2O(l)}$$

$$\rightarrow 4Ag(CN)_2{}^-(aq) + 2H_2O(l) + 4OH^-(aq)$$

Sample Exercise 4.24, continued

STEP 3

Eliminate as many H_2O molecules as possible.

$$8CN^-(aq) + 4Ag(s) + O_2(g) + 2H_2O(l) \rightarrow 4Ag(CN)_2^-(aq) + 4OH^-(aq)$$

STEP 4

Check that elements and charges balance.

Elements balance: 8 C, 8 N, 4 Ag, 4 O, 4 H → 8 C, 8 N, 4 Ag, 4 O, 4 H

Charges balance: $8(1-) + 0 + 0 + 0 = 8- \rightarrow 4(1-) + 4(1-) = 8-$

4.12 Simple Oxidation-Reduction Titrations

Purpose

- To learn to do the calculations associated with oxidation-reduction titrations.

Solid potassium permanganate.

Oxidation-reduction reactions are commonly used as the basis for volumetric analytical procedures. For example, a reducing substance can be titrated with a solution of a strong oxidizing agent, or vice versa. Three of the most frequently used oxidizing agents are aqueous solutions of *potassium permanganate* ($KMnO_4$), *potassium dichromate* ($K_2Cr_2O_7$), and *cerium hydrogen sulfate* [$Ce(HSO_4)_4$].

Sample Exercise 4.25

Cerium(IV) ion is a strong oxidizing agent that accepts one electron to produce cerium(III) ion:

$$Ce^{4+}(aq) + e^- \rightarrow Ce^{3+}(aq)$$

A solution containing an unknown concentration of Sn^{2+} ions was titrated with a solution containing Ce^{4+} ions, which oxidize the Sn^{2+} ions to Sn^{4+} ions. In one titration, 1.00 L of the unknown solution required 46.45 mL of a 0.1050 *M* Ce^{4+} solution to reach the stoichiometric point. Calculate the concentration of Sn^{2+} ions in the unknown solution.

Solution

The unbalanced equation for the titration reaction is

$$Ce^{4+}(aq) + Sn^{2+}(aq) \rightarrow Ce^{3+}(aq) + Sn^{4+}(aq)$$

The balanced equation is

$$2Ce^{4+}(aq) + Sn^{2+}(aq) \rightarrow 2Ce^{3+}(aq) + Sn^{4+}(aq)$$

Sample Exercise 4.25, continued

We can obtain the number of moles of Ce^{4+} ions from the volume and molarity of the Ce^{4+} solution used as the titrant:

$$46.45\ \cancel{\text{mL}} \times \frac{1\ \cancel{\text{L}}}{1000\ \cancel{\text{mL}}} \times \frac{0.1050\ \text{mol Ce}^{4+}}{\cancel{\text{L}}} = 4.877 \times 10^{-3}\ \text{mol Ce}^{4+}$$

The number of moles of Sn^{2+} ions can be obtained by applying the appropriate mole ratio from the balanced equation:

$$4.877 \times 10^{-3}\ \cancel{\text{mol Ce}^{4+}} \times \frac{1\ \text{mol Sn}^{2+}}{2\ \cancel{\text{mol Ce}^{4+}}} = 2.439 \times 10^{-3}\ \text{mol Sn}^{2+}$$

This value represents the quantity of Sn^{2+} ions in 1.00 L of solution. Thus the concentration of Sn^{2+} in the unknown solution is

$$\text{Molarity} = \frac{\text{mol Sn}^{2+}}{\text{L solution}} = \frac{2.439 \times 10^{-3}\ \text{mol Sn}^{2+}}{1.00\ \text{L}} = 2.44 \times 10^{-3}\ M$$

Another strong oxidizing agent, the permanganate ion (MnO_4^-), can undergo several different reactions. The one that occurs in acid solution is the most commonly used:

$$MnO_4^-(aq) + 8H^+(aq) + 5e^- \rightarrow Mn^{2+}(aq) + 4H_2O(l)$$

Permanganate has the advantage of being its own indicator—the MnO_4^- ion is intensely purple, and the Mn^{2+} ion is almost colorless. A typical titration using permanganate is illustrated in Fig. 4.12. As long as some reducing agent remains in the solution being titrated, the solution remains colorless (assuming all other species present are colorless), since the purple MnO_4^- ion being added is converted to the essentially colorless Mn^{2+} ion. However, when all the reducing agent has been consumed, the next drop of permanganate titrant will turn the solution being titrated light purple (pink). Thus the end point (where the color change indicates the titration should stop) occurs approximately one drop beyond the stoichiometric point (the actual point at which all of the reducing agent has been consumed).

Sample Exercise 4.26 describes a typical volumetric analysis using permanganate.

Figure 4.12

Permanganate being introduced into a flask of reducing agent.

Sample Exercise 4.26

Iron ores contain iron oxide minerals, which often contain a mixture of Fe^{2+} and Fe^{3+} ions. Such an ore can be analyzed for its iron content by dissolving it in acidic solution, reducing all the iron to Fe^{2+} ions, and then titrating with a standard solution of potassium permanganate. In the resulting solution, MnO_4^- is reduced to Mn^{2+}, and Fe^{2+} is oxidized to Fe^{3+}. A sample of iron ore weighing 0.3500 g was dissolved in acidic solution, and all of the iron was reduced to Fe^{2+}. Then the solution was titrated with a $1.621 \times 10^{-2}\ M$ $KMnO_4$ solution. The titration required 41.56 mL of the permanganate solution to reach the light purple end point. Determine the mass percent of iron in the iron ore.

Sample Exercise 4.26, continued

Solution

First we write the equation for the reaction:

$$H^+(aq) + MnO_4^-(aq) + Fe^{2+}(aq) \rightarrow Fe^{3+}(aq) + Mn^{2+}(aq) + H_2O(l)$$

Using the half-reaction method, we balance the equation:

$$8H^+(aq) + MnO_4^-(aq) + 5Fe^{2+}(aq) \rightarrow 5Fe^{3+}(aq) + Mn^{2+}(aq) + 4H_2O(l)$$

The number of moles of MnO_4^- ion required in the titration is found from the volume and concentration of permanganate solution used:

$$41.56\,\cancel{\text{mL}} \times \frac{1\,\cancel{\text{L}}}{1000\,\cancel{\text{mL}}} \times \frac{1.621 \times 10^{-2}\text{ mol MnO}_4^-}{\cancel{\text{L}}} = 6.737 \times 10^{-4}\text{ mol MnO}_4^-$$

The balanced equation shows that five times as much Fe^{2+} as MnO_4^- is required:

$$6.737 \times 10^{-4}\,\cancel{\text{mol MnO}_4^-} \times \frac{5\text{ mol Fe}^{2+}}{1\,\cancel{\text{mol MnO}_4^-}} = 3.368 \times 10^{-3}\text{ mol Fe}^{2+}$$

This means that the 0.3500-g sample of iron ore contained 3.368×10^{-3} mol of iron. The mass of iron present is

$$3.368 \times 10^{-3}\,\cancel{\text{mol Fe}} \times \frac{55.85\text{ g Fe}}{1\,\cancel{\text{mol Fe}}} = 0.1881\text{ g Fe}$$

The mass percent of iron in the iron ore is

$$\frac{0.1881\,\cancel{\text{g}}}{0.3500\,\cancel{\text{g}}} \times 100 = 53.74\%$$

FOR REVIEW

Summary

Many important chemical reactions occur in aqueous solutions. Water is a good solvent for ionic solids because the polar water molecules interact with the ions. This assists in the dissolving process. One characteristic property of a solution is its electrical conductivity. Solutions of strong electrolytes conduct current easily since there is 100% dissociation to ions. Weak electrolytes dissociate only to a small extent and their solutions have poor conductivity. Nonelectrolytes produce no ions when they dissolve and conduct no current.

The Arrhenius definition of an acid is a substance that produces H^+ ions (protons) in solution; a base is a substance that produces OH^- ions. The Brönsted-Lowry concept defines an acid as a proton donor and a base as a proton acceptor. A strong acid dissociates completely in water, and a weak acid dissociates only very slightly.

One means of describing solution concentration is molarity:

$$\text{Molarity } (M) = \frac{\text{mol of solute}}{\text{L of solution}}$$

Since molarity relates the volume of the solution and the moles of solute, it allows the principles of stoichiometry to be applied to reactions in solution. A standard solution is one where the molarity is accurately known. When a solution is diluted, only solvent is added, which means that

$$\text{Moles of solute after dilution} = \text{moles of solute before dilution}$$

Three types of equations describe reactions in solution: molecular equations, which show reactants, products, and stoichiometry on a molecular basis; complete ionic equations, in which all reactants and products that are strong electrolytes are represented as ions; and net ionic equations, in which only those solution components undergoing a change are represented. Spectator ions (those ions remaining unchanged in a reaction) are not included in a net ionic equation.

General rules concerning solubility help to predict the outcomes of precipitation reactions. One common technique of qualitative analysis is the separating and identifying of a particular cation in a mixture of ions in solution by selective precipitation. Gravimetric analysis is a technique in which the substance to be determined is separated from solution as a precipitate and weighed. Volumetric analysis involves titration of a solution containing the substance to be determined. One type of volumetric analysis utilizes neutralization (acid-base) reactions; the unknown concentration of an acid is determined by titration with a base of known concentration (or vice versa). The point in the titration where just enough titrant has been added to react exactly with the substance being determined is called the stoichiometric (equivalence) point.

Oxidation-reduction (redox) reactions involve a transfer of electrons. One method for keeping track of electrons is to use oxidation states for atoms in compounds. A series of rules is used to assign these oxidation states. Oxidation is an increase in oxidation state (a loss of electrons). Reduction is a decrease in oxidation state (a gain of electrons). An oxidizing agent accepts electrons, and a reducing agent donates electrons.

Oxidation-reduction equations can be balanced using either the oxidation states method or the half-reaction method that splits a reaction into two parts (the oxidation half-reaction and the reduction half-reaction). A balanced equation can be used to do the stoichiometric calculations for an oxidation-reduction titration.

Key Terms

aqueous solution

Section 4.1

polar molecule
hydration

Section 4.2

solute
solvent
electrical conductivity
strong electrolyte
weak electrolyte
nonelectrolyte
solubility
acid
strong acid
strong base
weak acid
weak base

Section 4.3

molarity
standard solution
dilution

Section 4.5

precipitation reaction
precipitate

Section 4.6

molecular equation
complete ionic equation
spectator ions
net ionic equation

Section 4.7

selective precipitation
qualitative analysis

Section 4.8

gravimetric analysis

Section 4.9

neutralization reaction
titration
volumetric analysis
titrant
stoichiometric (equivalence) point
indicator
end point

Section 4.10

oxidation-reduction (redox) reaction
oxidation state
oxidation
reduction
oxidizing agent (electron acceptor)
reducing agent (electron donor)

Section 4.11

half-reactions

Exercises

A blue exercise number indicates that the answer to that exercise appears at the back of this book and a solution appears in the Solutions Guide.

Aqueous Solutions: Strong and Weak Electrolytes

1. Show how each of the following strong electrolytes "breaks up" when it dissolves in water.
   a. NaBr
   b. $MgCl_2$
   c. $Al(NO_3)_3$
   d. $(NH_4)_2SO_4$
   e. HI
   f. $FeSO_4$
   g. $KMnO_4$
   h. $HClO_4$
   i. $NH_4C_2H_3O_2$ (ammonium acetate)

2. Distinguish between the terms *slightly soluble* and *weak electrolyte*.

3. Classify each of the following as a strong or a weak base:
   a. $Ca(OH)_2$
   b. KOH
   c. TlOH (water-soluble)
   d. NH_3

4. How would you determine experimentally whether a substance is a strong or a weak electrolyte?

5. Calcium chloride is a strong electrolyte and is used to "salt" streets in the winter to melt ice and snow. Write a reaction to show how this substance breaks apart when it dissolves in water.

6. Commercial cold packs and hot packs are available for treating athletic injuries. Both types contain a pouch of water and a dry chemical. When the pack is struck, the pouch of water breaks, dissolving the chemical, and the solution becomes either hot or cold. Many hot packs use magnesium sulfate, and many cold packs use ammonium nitrate. Write reactions to show how these strong electrolytes break apart in water.

Solution Concentration: Molarity

7. Describe how you would prepare 1.0 L of each of the following solutions:
   a. 0.10 *M* NaCl from solid NaCl
   b. 0.10 *M* NaCl from a 2.5 *M* stock solution
   c. 0.20 *M* $NaIO_3$ from solid $NaIO_3$
   d. 0.010 *M* $NaIO_3$ from a 0.20 *M* stock solution
   e. 0.050 *M* potassium hydrogen phthalate ($KHC_8H_4O_4$; abbreviated KHP) from solid potassium hydrogen phthalate
   f. 0.040 *M* KHP from a 0.50 *M* stock solution

8. How would you prepare 1.00 L of a 0.50 *M* solution of each of the following?
   a. H_2SO_4 from "concentrated" (18 *M*) sulfuric acid
   b. HCl from "concentrated" (12 *M*) reagent
   c. $NiCl_2$ from the salt $NiCl_2 \cdot 6H_2O$
   d. HNO_3 from "concentrated" (16 *M*) reagent
   e. Sodium carbonate from the pure solid

9. Calculate the molarity of each of these solutions:
   a. 5.623 g of $NaHCO_3$ is dissolved in enough water to make 250.0 mL of solution.
   b. 184.6 mg of $K_2Cr_2O_7$ is dissolved in enough water to make 500.0 mL of solution.
   c. 0.1025 g of copper metal is dissolved in 35 mL of concentrated HNO_3 to form Cu^{2+} ions and then water is added to make a total volume of 200.0 mL. (Calculate the molarity of Cu^{2+}.)

10. Calculate the concentration of all ions present in each of the following solutions of strong electrolytes:
    a. 0.15 *M* $CaCl_2$
    b. 0.26 *M* $Al(NO_3)_3$
    c. 0.25 *M* $K_2Cr_2O_7$
    d. 2.0×10^{-3} *M* $Al_2(SO_4)_3$

11. Calculate the concentration of all ions present in each of the following solutions of strong electrolytes:
    a. 0.100 g of $MgCl_2$ in 100.0 mL of solution
    b. 55.1 mg of NH_4Br in 500.0 mL of solution
    c. 5.47 g of Na_2S in 1.00 L of solution
    d. 0.208 g of $AlCl_3$ in 250.0 mL of solution

12. Disodium aurothiomalate ($Na_2C_4H_3O_4SAu$) has the trade name Myocrisin and is used in the treatment of rheumatoid arthritis. Patients receive weekly intramuscular injections of 50.0 mg of Myocrisin in 0.500 mL of solution. During treatment, serum levels of gold are as high as 300.0 μg of gold per 100.0 mL of serum. Calculate the above two concentrations of Myocrisin in units of molarity.

13. A solution is prepared by dissolving 10.8 g of ammonium sulfate in enough water to make 100.0 mL of stock solution. A 10.00 mL sample of this stock solution is added to 50.00 mL of water. Calculate the concentration of ammonium ions and sulfate ions in the final solution.

14. A solution of ethanol (C_2H_5OH) in water is prepared by dissolving 10.0 mL of ethanol (density: 0.79 g/cm^3) in enough water to make 250.0 mL of solution. What is the molarity of the ethanol in this solution?

15. The units of parts per million (ppm) and parts per billion (ppb) are commonly used by environmental chemists. In general, 1 ppm means 1 part of solution for every 10^6 parts of solution. (Both solute and solution are measured using the same units.) Mathematically, by mass

$$\text{ppm} = \frac{\mu\text{g solute}}{\text{g solution}} = \frac{\text{mg solute}}{\text{kg solution}}$$

In the case of very dilute aqueous solutions, a concentration of 1.0 ppm is equal to 1.0 μg of solute per 1.0 mL which equals 1.0 g of solution. Parts per billion is defined in a similar fashion. Calculate the molarity of each of the following aqueous solutions.
    a. 5.0 ppb Hg in H_2O
    b. 1.0 ppb $CHCl_3$ in H_2O

c. 10.0 ppm As in H_2O
d. 0.10 ppm DDT ($C_{14}H_9Cl_5$) in H_2O

16. a. 150 mg of Na_2CO_3 is dissolved in H_2O to give 1.0 L of solution. What is the concentration of Na^+ in parts per million?
b. 2.5 mg of dioctyl phthalate ($C_{24}H_{38}O_4$), a plasticizer, is dissolved in H_2O to give 500.0 mL of solution. What is the concentration of dioctyl phthalate in parts per billion? (See Exercise 15 for definitions.)

17. A standard is prepared for the analysis of fluoxymesterone ($C_{20}H_{29}FO_3$), an anabolic steroid. A stock solution is prepared by dissolving 10.0 mg of fluoxymesterone in a total volume of 500.0 mL. A 100.0-μL aliquot (portion) of this solution is diluted to a final volume of 100.0 mL. Calculate the concentration of the final solution in parts per billion and in terms of molarity. (See Exercise 15 for definitions.)

18. The World Health Organization lists the maximum desirable concentrations of Mg^{2+} and Ca^{2+} in drinking water as 30. and 75 mg/L, respectively. The maximum permissible concentrations are 1250 and 200. mg/L for the same ions. Calculate these concentrations in mol/L. Recall that a decimal point makes trailing zeros count as significant digits.

19. Low concentrations of fluoride ion in drinking water furnish excellent protection against dental caries. There are no apparent deleterious effects, even over many years, provided the concentration of fluoride ion is at or below 1 ppm (1 mg/L). However, at 2–3 ppm a brown mottling of teeth can occur, and harmful toxic effects are possible at 50. ppm. Calculate these concentrations in mol/L. (See Exercise 15 for definitions.)

20. A solution is prepared by dissolving 0.5842 g of oxalic acid ($H_2C_2O_4$) in enough water to make 100.0 mL of solution. A 10.00-mL aliquot (portion) of this solution is then diluted to a final volume of 250.0 mL. What is the molarity of the final oxalic acid solution?

21. In the spectroscopic analysis of many substances, a series of standard solutions of known concentration are measured in order to generate a calibration curve. How would you prepare standard solutions containing 10.0, 25.0, 50.0, 75.0, and 100. ppm of copper from a commercially produced 1000.0-ppm solution? Assume each solution has a final volume of 100.0 mL.

22. A stock solution containing Mn^{2+} ions is prepared by dissolving 1.584 g of pure manganese metal in nitric acid and diluting to a final volume of 1.00 L. The following solutions are prepared by dilution:

For solution A, 50.00 mL of stock solution is diluted to 1000.0 mL.

For solution B, 10.00 mL of A is diluted to 250.0 mL.

For solution C, 10.00 mL of B is diluted to 500.0 mL.

Calculate the concentrations of the stock solution and solutions A, B, and C.

23. You have a stock solution that is 0.200 *M* HCl. Which of the following procedures would give a final solution that is most accurately 1.00×10^{-4} *M* HCl? Why?

A 10.00-mL portion (aliquot) of the stock solution is diluted to 200.0 mL using a volumetric pipet and flask, followed by two 10:1 dilutions using volumetric pipet and flask. Assume all volume measurements are accurate to $\pm$.01 mL.

A 0.050-mL aliquot of the stock solution is diluted to a final volume of 100.0 mL. Assume the 0.050 mL aliquot can be measured to ± 1 μL.

Precipitation Reactions

24. Predict the products and write the balanced molecular equation, complete ionic equation, and the net ionic equation for each of the following:
a. $BaCl_2(aq) + Na_2SO_4(aq) \rightarrow$
b. $Pb(NO_3)_2(aq) + HCl(aq) \rightarrow$
c. $AgNO_3(aq) + Na_3PO_4(aq) \rightarrow$
d. $NaOH(aq) + Fe(NO_3)_3(aq) \rightarrow$
e. $FeCl_3(aq) + Na_2S(aq) \rightarrow$
f. $NiSO_4(aq) + KOH(aq) \rightarrow$

25. Write net ionic equations for each of the following:
a. $AgNO_3(aq) + KI(aq) \rightarrow$
b. $CuSO_4(aq) + Na_2S(aq) \rightarrow$
c. $CoCl_2(aq) + NaOH(aq) \rightarrow$
d. $NiCl_2(aq) + HNO_3(aq) \rightarrow$

26. Write net ionic equations for the reaction, if any, that occurs when aqueous solutions of the following are mixed:
a. ammonium sulfate and barium nitrate
b. lead nitrate and sodium chloride
c. sodium phosphate and potassium nitrate
d. sodium bromide and hydrochloric acid
e. copper(II) chloride and sodium hydroxide

27. How would you separate the following ions in aqueous solution by selective precipitation?
a. Ag^+, Ba^{2+}, Cr^{3+}
b. Ag^+, Pb^{2+}, Cu^{2+}
c. Hg_2^{2+}, Ni^{2+}

28. A lake may be polluted with Pb^{2+} ions. How might you qualitatively test for Pb^{2+}?

29. A sample may contain any or all of the following ions: Hg_2^{2+}, Ba^{2+}, and Mn^{2+}. No precipitate formed when an aqueous solution of NaCl or Na_2SO_4 was added to the sample solution. A precipitate formed when the sample solution was made basic with NaOH. Which ion or ions are present in the sample solution?

30. What mass of solid AgBr is produced when 100.0 mL of 0.150 *M* $AgNO_3$ is added to 20.0 mL of 1.00 *M* NaBr?

31. What volume of 0.100 *M* NaOH is required to precipitate all of the nickel(II) ion from 150.0 mL of 0.250 *M* $Ni(NO_3)_2$?

32. Aluminum can be determined gravimetrically by reaction with a solution of 8-hydroxyquinoline (C_9H_7NO). The net ionic equation is:

$$Al^{3+}(aq) + 3C_9H_7NO(aq) \rightarrow Al(C_9H_6NO)_3(s) + 3H^+(aq)$$

A mass of 0.1248 g of $Al(C_9H_6NO)_3$ was obtained by precipitating all of the Al^{3+} from a solution prepared by dissolving 1.8571 g of a mineral. What is the mass percent of aluminum in the mineral?

33. What mass of barium sulfate is produced when 100.0 mL of a 0.100 *M* solution of barium chloride is mixed with 100.0 mL of a 0.100 *M* solution of iron(III) sulfate?

34. During the developing process, silver bromide is removed from photographic film by the fixer. The major component of the fixer is sodium thiosulfate. The net ionic equation for the reaction is

$$AgBr(s) + 2S_2O_3^{2-}(aq) \rightarrow Ag(S_2O_3)_2^{3-}(aq) + Br^-(aq)$$

What mass of AgBr can be dissolved by 1.00 L of 0.200 *M* $Na_2S_2O_3$?

35. A 100.0-mL aliquot of 0.200 *M* aqueous potassium hydroxide is mixed with 100.0 mL of 0.200 *M* aqueous magnesium nitrate.
   a. Write a balanced chemical equation for any reaction that occurs.
   b. What precipitate forms?
   c. What mass of that precipitate is produced?

36. The thallium (present as Tl_2SO_4) in a 9.486 g pesticide sample was precipitated as thallium(I) iodide. Calculate the percentage of Tl_2SO_4 in the sample if 0.1824 g of TlI was recovered.

37. Saccharin ($C_7H_5NO_3S$) is sometimes dispensed in tablet form. Ten tablets with a total mass of 0.5894 g were dissolved in water. They were oxidized to convert all of the sulfur to sulfate ion, which was precipitated by adding an excess of barium chloride solution. The mass of $BaSO_4$ obtained was 0.5032 g. What is the average mass of saccharin per tablet? What is the average mass percent of saccharin in the tablets?

38. Chlorisondiamine ($C_{14}H_{18}Cl_6N_2$) is a drug used in the treatment of hypertension. A 1.28 g sample of a medication containing the drug was treated to destroy the organic material and release all of the chlorine as chloride ion. When the filtered solution containing chloride ion was treated with an excess of silver nitrate, 0.104 g of silver chloride was recovered. Calculate the mass percent of chlorisondiamine in the medication, assuming the drug is the only source of chloride.

Acid-Base Reactions

39. Complete and balance each acid-base reaction:
   a. $H_3PO_4(aq) + NaOH(aq) \rightarrow$
      Contains three acidic hydrogens
   b. $HClO_4(aq) + Mg(OH)_2(s) \rightarrow$
   c. $Al(OH)_3(s) + H_2SO_4(aq) \rightarrow$
   d. $HCN(aq) + NaOH(aq) \rightarrow$
   e. $H_2Se(aq) + Ba(OH)_2(s) \rightarrow$
      Contains two acidic hydrogens
   f. $H_2SO_3(aq) + NaOH(aq) \rightarrow$
      Contains two acidic hydrogens
   g. $H_2C_2O_4(aq) + NaOH(aq) \rightarrow$
      Contains two acidic hydrogens

40. Write balanced equations (all three types) for the reactions that occur when the following aqueous solutions are mixed:
   a. ammonia and nitric acid
   b. barium hydroxide and hydrochloric acid
   c. perchloric acid ($HClO_4(aq)$) and solid iron(III) hydroxide
   d. solid silver hydroxide and hydrobromic acid

41. A solution contains an unknown concentration of citric acid, $H_3C_6H_5O_7$, which has three acidic hydrogens. It takes 41.28 mL of 0.250 *M* sodium hydroxide to titrate (react completely with) 10.00 mL of the citric acid solution. The reaction is

$$H_3C_6H_5O_7(aq) + 3NaOH(aq) \rightarrow 3H_2O(l) + Na_3C_6H_5O_7(aq)$$

What is the molarity of the original citric acid solution?

42. Some of the substances commonly used in stomach antacids are MgO, $Mg(OH)_2$, and $Al(OH)_3$.
   a. Write a balanced equation for the neutralization of hydrochloric acid by each of these substances.
   b. Which of these substances will neutralize the greatest amount of 0.1 *M* HCl per gram?

43. Carminic acid, a naturally occurring red pigment extracted from the cochineal insect, contains only carbon, hydrogen, and oxygen. It was commonly used as a dye in the first half of the nineteenth century. It is 53.66% C and 4.09% H by mass. A titration required 18.02 mL of 0.0406 *M* NaOH to neutralize 0.3602 g of carminic acid. Assuming that there is only one acidic hydrogen per molecule, what is the molecular formula of carminic acid?

44. What volume of each of the following acids will react completely with 50.00 mL of 0.200 *M* NaOH?
   a. 0.100 *M* HCl
   b. 0.100 *M* H_2SO_3 (2 acidic hydrogens)
   c. 0.200 *M* H_3PO_4 (3 acidic hydrogens)
   d. 0.150 *M* HNO_3
   e. 0.200 *M* $HC_2H_3O_2$ (1 acidic hydrogen)
   f. 0.300 *M* H_2SO_4 (2 acidic hydrogens)

45. Sodium hydroxide solution is standardized by titrating a pure sample of potassium hydrogen phthalate (KHP), an acid with one acidic hydrogen and a molecular weight of 204.22. It

takes 20.46 mL of a sodium hydroxide solution to titrate a 0.1082-g sample of KHP. What is the molarity of the sodium hydroxide?

46. A 0.500-L sample of H_2SO_4 solution was analyzed by taking a 100.0-mL aliquot and adding 50.0 mL of 0.213 *M* NaOH. After the reaction occurred, an excess of OH^- ions remained in the solution. The excess base required 13.21 mL of 0.103 *M* HCl for neutralization. Calculate the molarity of the original sample of H_2SO_4.

47. Calcium metal will react with water as follows:

$$Ca(s) + 2H_2O(l) \rightarrow Ca(OH)_2(aq) + H_2(g)$$

What is the molarity of hydroxide ions in the solution formed when 4.25 g of calcium metal is dissolved in enough water to make a final volume of 225 mL?

48. A 10.00-mL sample of sulfuric acid from an automobile battery requires 35.08 mL of 2.12 *M* sodium hydroxide solution for complete neutralization. What is the molarity of the sulfuric acid? The reaction is

$$H_2SO_4(aq) + 2NaOH(aq) \rightarrow 2H_2O(l) + Na_2SO_4(aq)$$

49. A 10.00-mL sample of vinegar, an aqueous solution of acetic acid ($HC_2H_3O_2$), is titrated with 0.5062 *M* NaOH, and 16.58 mL is required to reach the end point.
   a. What is the molarity of the acetic acid?
   b. If the density of the vinegar is 1.006 g/cm^3, what is the mass percent of acetic acid in the vinegar?

50. A 25.00-mL sample of hydrochloric acid solution takes 24.16 mL of 0.106 *M* sodium hydroxide for complete neutralization. What is the concentration of the hydrochloric acid solution?

51. A solution is prepared by dissolving 15 g of NaOH in 150 mL of 0.25 *M* nitric acid. Will the final solution be acidic, basic, or neutral? Calculate the concentrations of all of the ions present in the solution after the reaction has occurred.

52. A 50.00-mL sample of an ammonia solution is analyzed by titration with HCl. The reaction is

$$NH_3(aq) + H^+(aq) \rightarrow NH_4^+(aq)$$

It took 39.47 mL of 0.0984 *M* HCl to titrate (react completely with) the ammonia. What is the concentration of the original ammonia solution?

53. Hydrochloric acid (75 mL of 0.25 *M*) is added to 225 mL of 0.055 *M* $Ba(OH)_2$ solution. What is the concentration of the excess H^+ or OH^- left in this solution?

Oxidation-Reduction Reactions

54. Assign oxidation states to all atoms in each compound.
   a. $KMnO_4$
   b. NiO_2
   c. $K_4Fe(CN)_6$ (Fe only)
   d. $(NH_4)_2HPO_4$
   e. P_4O_6
   f. Fe_3O_4
   g. $XeOF_4$
   h. SF_4
   i. CO
   j. $Na_2C_2O_4$

55. Assign an oxidation state to chlorine in each of the following anions: OCl^-, ClO_2^-, ClO_3^-, and ClO_4^-.

56. Assign an oxidation state to nitrogen in each of the following:
   a. Li_3N
   b. NH_3
   c. N_2H_4
   d. NO
   e. N_2O
   f. NO_2
   g. NO_2^-
   h. NO_3^-
   i. N_2

57. Assign oxidation states to all atoms in each of the following:
   a. UO_2^{2+}
   b. As_2O_3
   c. $NaBiO_3$
   d. As_4
   e. $HAsO_2$
   f. $Mg_2P_2O_7$
   g. $Na_2S_2O_3$
   h. Hg_2Cl_2
   i. $Ca(NO_3)_2$

58. Tell which of the following are oxidation-reduction reactions. For those that are, identify the oxidizing agent, the reducing agent, the substance being oxidized, and the substance being reduced.
   a. $CH_4(g) + 2O_2(g) \rightarrow CO_2(g) + 2H_2O(g)$
   b. $Zn(s) + 2HCl(aq) \rightarrow ZnCl_2(aq) + H_2(g)$
   c. $Cr_2O_7^{2-}(aq) + 2OH^-(aq) \rightarrow 2CrO_4^{2-}(aq) + H_2O(l)$
   d. $O_3(g) + NO(g) \rightarrow O_2(g) + NO_2(g)$
   e. $2H_2O_2(l) \rightarrow 2H_2O(l) + O_2(g)$
   f. $2CuCl(aq) \rightarrow CuCl_2(aq) + Cu(s)$
   g. $HCl(g) + NH_3(g) \rightarrow NH_4Cl(s)$
   h. $SiCl_4(l) + 2H_2O(l) \rightarrow 4HCl(aq) + SiO_2(s)$
   i. $SiCl_4(l) + 2Mg(s) \rightarrow 2MgCl_2(s) + Si(s)$

59. Balance each of the following oxidation-reduction reactions using the oxidation states method:
   a. $C_2H_6(g) + O_2(g) \rightarrow CO_2(g) + H_2O(g)$
   b. $Mg(s) + HCl(aq) \rightarrow Mg^{2+}(aq) + 2Cl^-(aq) + H_2(g)$
   c. $Cu(s) + Ag^+(aq) \rightarrow Cu^{2+}(aq) + Ag(s)$
   d. $Cu(s) + HNO_3(aq) \rightarrow Cu(NO_3)_2(aq) + NO(g)$
   e. $Zn(s) + H_2SO_4(aq) \rightarrow ZnSO_4(aq) + H_2(g)$

60. Balance the following oxidation-reduction reactions, which occur in acidic solution, using the half-reaction method.
   a. $Cu(s) + HNO_3(aq) \rightarrow Cu^{2+}(aq) + NO(g)$
   b. $Cr_2O_7^{2-}(aq) + Cl^-(aq) \rightarrow Cr^{3+}(aq) + Cl_2(g)$
   c. $Pb(s) + PbO_2(s) + H_2SO_4(aq) \rightarrow PbSO_4(s)$
   d. $Mn^{2+}(aq) + NaBiO_3(s) \rightarrow Bi^{3+}(aq) + MnO_4^-(aq)$
   e. $H_3AsO_4(aq) + Zn(s) \rightarrow AsH_3(g) + Zn^{2+}(aq)$
   f. $As_2O_3(s) + NO_3^-(aq) \rightarrow H_3AsO_4(aq) + NO(g)$
   g. $Br^-(aq) + MnO_4^-(aq) \rightarrow Br_2(l) + Mn^{2+}(aq)$
   h. $CH_3OH(aq) + Cr_2O_7^{2-}(aq) \rightarrow CH_2O(aq) + Cr^{3+}(aq)$

61. Balance the following oxidation-reduction reactions, which occur in basic solution, using the half-reaction method.
   a. $Al(s) + MnO_4^-(aq) \rightarrow MnO_2(s) + Al(OH)_4^-(aq)$
   b. $Cl_2(g) \rightarrow Cl^-(aq) + ClO^-(aq)$
   c. $NO_2^-(aq) + Al(s) \rightarrow NH_3(g) + AlO_2^-(aq)$
   d. $MnO_4^-(aq) + S^{2-}(aq) \rightarrow MnS(s) + S(s)$
   e. $CN^-(aq) + MnO_4^-(aq) \rightarrow CNO^-(aq) + MnO_2(s)$

62. Balance the following equations by the half-reaction method:

a. $Fe(s) + HCl(aq) \rightarrow HFeCl_4(aq) + H_2(g)$

b. $IO_3^-(aq) + I^-(aq) \xrightarrow{\text{(acid)}} I_3^-(aq)$

c. $Cr(NCS)_6^{4-}(aq) + Ce^{4+}(aq) \xrightarrow{\text{(acid)}}$
$Cr^{3+}(aq) + Ce^{3+}(aq) + NO_3^-(aq) + CO_2(g) + SO_4^{2-}(aq)$

d. $CrI_3(s) + Cl_2(g) \xrightarrow{\text{(base)}}$
$CrO_4^-(aq) + IO_4^-(aq) + Cl^-(aq)$

e. $Fe(CN)_6^{4-}(aq) + Ce^{4+}(aq) \xrightarrow{\text{(base)}}$
$Ce(OH)_3(s) + Fe(OH)_3(s) + CO_3^{2-}(aq) + NO_3^-(aq)$

f. $Fe(OH)_2(s) + H_2O_2(aq) \xrightarrow{\text{(base)}} Fe(OH)_3(s)$

63. The Ostwald process for the commercial production of nitric acid involves the following three steps:

$$4NH_3(g) + 5O_2(g) \rightarrow 4NO(g) + 6H_2O(g)$$
$$2NO(g) + O_2(g) \rightarrow 2NO_2(g)$$
$$3NO_2(g) + H_2O(l) \rightarrow 2HNO_3(aq) + NO(g)$$

a. Which reactions in the Ostwald process are oxidation-reduction reactions?

b. Identify the oxidizing agent and the reducing agent in those that are.

c. How much nitric oxide, NO, can be produced from a mixture of

5.0×10^6 g of ammonia and 5.0×10^7 g of O_2?

64. Chlorine gas was first prepared in 1774 by C. W. Scheele by oxidizing hydrochloric acid with manganese(IV) oxide. The reaction is

$$NaCl(aq) + H_2SO_4(aq) + MnO_2(s) \rightarrow Na_2SO_4(aq) + MnCl_2(aq) + H_2O(l) + Cl_2(g)$$

Balance this reaction by the oxidation number method.

65. One of the classical methods for the determination of the manganese content in steel is to convert all the manganese to the deeply colored permanganate ion and then to measure the absorption of light. The steel is dissolved in nitric acid, producing the manganese(II) ion and nitrogen dioxide gas. This solution is then reacted with an acidic solution containing periodate ion; the products are the permanganate and iodate ions. Write balanced chemical equations for both of these steps.

66. Gold metal will not dissolve in either concentrated nitric acid or concentrated hydrochloric acid. It will dissolve, however, in *aqua regia,* a mixture of the two concentrated acids. The products of the reaction are the $AuCl_4^-$ ion and gaseous NO. Write a balanced equation for the dissolution of gold in aqua regia.

67. A solution of permanganate is standardized by titration with oxalic acid ($H_2C_2O_4$). It required 28.97 mL of the permanganate solution to react completely with 0.1058 g of oxalic acid. The unbalanced equation for the reaction is

$$MnO_4^-(aq) + H_2C_2O_4(aq) \xrightarrow{\text{Acid}} Mn^{2+}(aq) + CO_2(g)$$

What is the molarity of the permanganate solution?

68. A 50.00-mL sample of solution containing Fe^{2+} ions is titrated with a 0.0216 M $KMnO_4$ solution. It required 20.62 mL of the $KMnO_4$ solution to oxidize all of the Fe^{2+} ions to Fe^{3+} ions by the reaction:

$$MnO_4^-(aq) + Fe^{2+}(aq) \xrightarrow{\text{Acid}} Mn^{2+}(aq) + Fe^{3+}(aq) \quad \text{(Unbalanced)}$$

a. What was the concentration of Fe^{2+} ions in the sample solution?

b. What volume of 0.0150 M $K_2Cr_2O_7$ solution would it take to do the same titration? The reaction is

$$Cr_2O_7^{2-}(aq) + Fe^{2+}(aq) \xrightarrow{\text{Acid}} Cr^{3+}(aq) + Fe^{3+}(aq) \quad \text{(Unbalanced)}$$

69. The iron content of iron ore can be determined by titration with standard $KMnO_4$ solution. The iron ore is dissolved in HCl, and all of the iron is reduced to Fe^{2+} ions. This solution is then titrated with $KMnO_4$ solution, producing Fe^{3+} and Mn^{2+} ions in acidic solution. If it required 41.95 mL of 0.0205 M $KMnO_4$ to titrate a solution made from 0.6128 g of iron ore, what is the mass percent of iron in the iron ore?

70. The legal definition of intoxication in some states is a blood alcohol (C_2H_5OH) level of 0.1% by mass or higher. It required 35.48 mL of 0.05182 M $K_2Cr_2O_7$ to titrate a 50.02-g sample of blood.

a. Assuming only C_2H_5OH reacts with $K_2Cr_2O_7$, was the person from whom this blood was taken legally intoxicated? (See Sample Exercise 4.23.)

b. What would be your answer to part a if instead it required 48.02 mL of the same $K_2Cr_2O_7$ dichromate solution to titrate 48.91 g of blood?

71. Stibnite (Sb_2S_3) is the most important ore containing antimony. A 0.506 g sample of ore was chemically treated to produce antimony(III) ions in solution. The antimony(III) was oxidized to antimony(V) by adding 25.00 mL of 0.0233 M $KMnO_4$ solution. The excess $KMnO_4$ was titrated with 0.0843 M Fe^{2+}; 2.58 mL was required, producing $Fe^{3+}(aq)$ and $Mn^{2+}(aq)$. All reactions were carried out in acidic solutions. Calculate the mass percent of Sb_2S_3 in the sample.

72. What mass of CO_2 is produced from the reaction of 0.500 g of $Na_2C_2O_4$ with 50.00 mL of 0.0200 M $KMnO_4$ solution in the presence of acid? (See Exercise 67.)

73. A piece of copper metal (1.50 g) is placed in 250 mL of a 0.20 M $AgNO_3$ solution. Will all of the copper dissolve in this solution? The net ionic equation is

$$Cu(s) + 2Ag^+(aq) \rightarrow 2Ag(s) + Cu^{2+}(aq)$$

Additional Exercises

74. In most of its ionic compounds, cobalt is either Co(II) or Co(III). One such compound, containing chloride ion and waters of hydration, was analyzed, and the following results were obtained. A 0.256-g sample of the compound was dissolved in water, and excess silver nitrate was added. The silver chloride was filtered, dried, and weighed, and it had a mass of 0.308 g. A second sample of 0.416 g of the compound was dissolved in water, and an excess of sodium hydroxide was added. The hydroxide salt was filtered and heated in a flame, forming cobalt(III) oxide. The mass of cobalt(III) oxide formed was 0.145 g.
   a. Write balanced net ionic equations for the three reactions described.
   b. What is the percent composition, by mass, of the compound?
   c. Assuming the compound contains one cobalt atom per formula unit, what is the molecular formula?

75. One of the harmful effects of acid rain is the deterioration of structures or statues made of marble or limestone, both essentially calcium carbonate. The reaction of calcium carbonate with sulfuric acid yields carbon dioxide, water, and calcium sulfate. Since calcium sulfate is soluble in water, part of the object is washed away by the rain. Write a balanced chemical equation for the reaction of sulfuric acid with calcium carbonate.

76. Triiodide ions are generated in solution by the following reaction in acidic solution:

$$IO_3^-(aq) + I^-(aq) \rightarrow I_3^-(aq)$$

Triiodide ion is determined by titration with a sodium thiosulfate ($Na_2S_2O_3$) solution. The products are iodide ion and tetrathionate ion ($S_4O_6^{2-}$).
   a. Balance the equation for the reaction of IO_3^- with I^- ions.
   b. Write and balance the equation for the reaction of $S_2O_3^{2-}$ with I_3^- in acidic solution.
   c. A 25.00-mL sample of a 0.0100 *M* solution of KIO_3 is reacted with an excess of KI. It requires 32.04 mL of $Na_2S_2O_3$ solution to titrate the I_3^- ions present. What is the molarity of the $Na_2S_2O_3$ solution?
   d. How would you prepare 500.0 mL of the KIO_3 solution in part c, using pure dry KIO_3?

77. The zinc in a 1.200-g sample of foot powder was precipitated as $ZnNH_4PO_4$. Strong heating of the precipitate yielded 0.4089 g of $Zn_2P_2O_7$. Calculate the mass percent of zinc in the sample of foot powder.

78. Lead chromate ($PbCrO_4$) is used as a pigment in paints and is commonly called chrome yellow. What mass of chrome yellow is produced when 100.0 mL of 0.4100 *M* sodium chromate is mixed with 100.0 mL of 0.3200 *M* lead(II) nitrate?

79. Rust stains can be removed by washing a surface with a dilute solution of oxalic acid ($H_2C_2O_4$). The reaction is

$$Fe_2O_3(s) + 6H_2C_2O_4(aq) \rightarrow 2Fe(C_2O_4)_3^{3-}(aq) + 3H_2O(l) + 6H^+(aq)$$

   a. Is this an oxidation-reduction reaction?
   b. What mass of rust can be removed by 1.0 L of a 0.14 *M* solution of oxalic acid?

80. A mixture contains only NaCl and $Fe(NO_3)_3$. A 0.456-g sample of the mixture is dissolved in water, and an excess of NaOH is added, producing a precipitate of $Fe(OH)_3$. The precipitate is filtered, dried, and weighed. Its mass is 0.107 g. Calculate:
   a. the mass of iron in the sample
   b. the mass of $Fe(NO_3)_3$ in the sample
   c. the mass percent of $Fe(NO_3)_3$ in the sample

81. A sample is a mixture of KCl and KBr. When 0.1024 g of the sample is dissolved in water and reacted with excess silver nitrate, 0.1889 g of solid is obtained. What is the composition by mass percent of the mixture?

82. What mass of $Fe(OH)_3$ would be produced by reacting 75.0 mL of 0.105 *M* $Fe(NO_3)_3$ with 125 mL of 0.150 *M* NaOH?

83. You are given a compound that is either iron(II) sulfate or iron(III) sulfate. How could you determine the identity of the compound using a dilute solution of potassium permanganate?

84. You are given a solid that is a mixture of Na_2SO_4 and K_2SO_4. A 0.205-g sample of the mixture is dissolved in water. An excess of an aqueous solution of $BaCl_2$ is added. The $BaSO_4$ that is formed is filtered, dried, and weighed. Its mass is 0.298 g. What mass of SO_4^{2-} ion is in the sample? What is the mass percent of SO_4^{2-} ion in the sample? What are the percent compositions by mass of Na_2SO_4 and K_2SO_4 in the sample?

85. A solution is prepared by dissolving 0.150 ±0.003 g of NaCl in a volumetric flask. The volume of the flask is 100.0 ±0.5 mL. What is the molarity of NaCl in the solution? What is the range of values for the molarity of NaCl? Express this as a value ± the uncertainty. (See Appendix 1.5.)

86. It took 25.06 ±0.05 mL of a sodium hydroxide solution to titrate a 0.4016 g sample of KHP (see Exercise 45). Calculate the concentration and uncertainty in the concentration of the sodium hydroxide solution. (See Appendix 1.5.)

87. You wish to prepare one liter of a 0.02 *M* potassium iodate solution. You require that the final concentration be within 1% of 0.02 *M* and that the concentration must be known accurately to the fourth decimal place. How would you prepare this solution? Specify the glassware you would use, the accuracy needed for the balance, and the ranges of acceptable masses of KIO_3 that can be used.

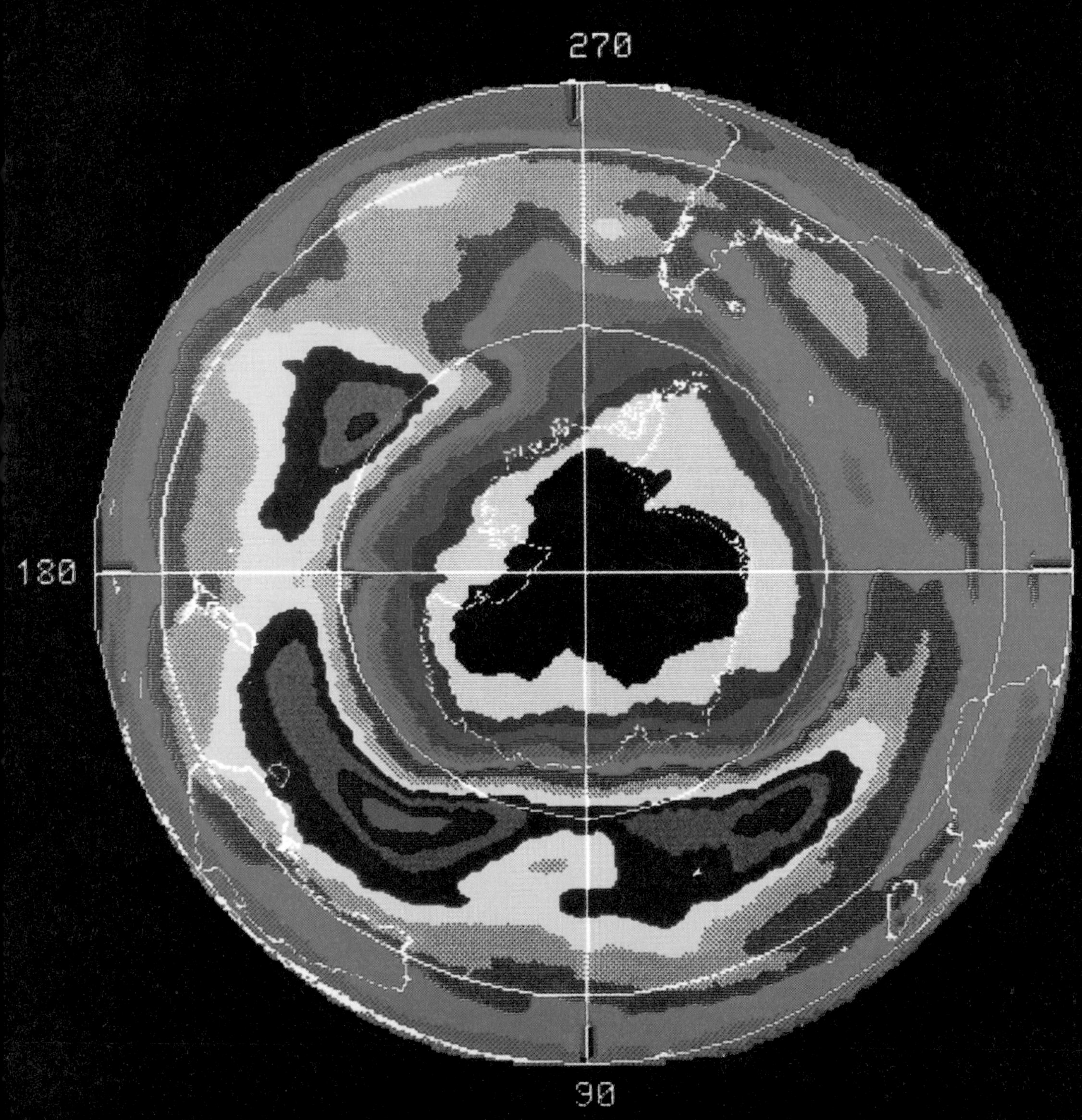
270
180
90

CHAPTER 5

Gases

Matter exists in three distinct physical states: gas, liquid, and solid. Of these, the gaseous state is the easiest to describe both experimentally and theoretically. In particular, the study of gases provides an excellent example of the scientific method in action. It illustrates how observations lead to natural laws, which in turn can be accounted for by models. Then, as more accurate measurements become available, the models are modified.

In addition to providing a good illustration of the scientific method, gases are important in their own right. For example, gases are often produced in chemical reactions and thus must be dealt with in stoichiometric calculations. Also, the earth's atmosphere is a mixture of gases, mainly elemental nitrogen and oxygen; it both supports life and acts as a waste receptacle for the exhaust gases that accompany many industrial processes.

Therefore, it is important to understand the behavior of gases. We will pursue this goal in this chapter considering the properties of gases, the laws and models that describe the behavior of gases, and finally the reactions that occur among the gases in the atmosphere.

CONTENTS

< A pictorial representation of the ozone concentrations over Antarctica, from data taken by the Nimbus-7 satellite.

5.1 Early Experiments

Purpose

- To describe how barometers and manometers operate.
- To define and demonstrate the use of the various units of pressure.

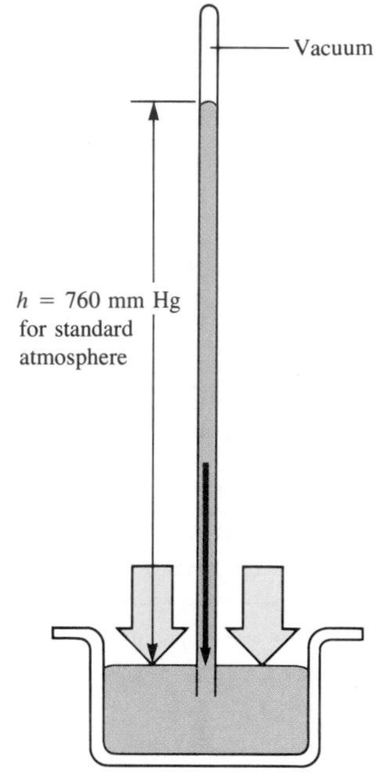

Figure 5.1

A torricellian barometer. The tube, completely filled with mercury, is inverted in a dish of mercury. Mercury flows out of the tube until the pressure of the column of mercury (shown by black arrow) "standing on the surface" of the mercury in the dish is equal to the pressure of the air (shown by yellow arrows) on the rest of the surface of the mercury in the dish.

Even though the Greeks considered "air" to be one of the four fundamental elements and various alchemists obtained "airs," or "vapors," in their experiments, careful study of these elusive substances proved difficult. The first person to attempt a scientific study of the "vapors" produced in chemical reactions was a Flemish physician named Jan Baptista Van Helmont (1577–1644). Thinking that air and similar substances must be akin to the "chaos" from which, according to Greek myth, the universe was created, Van Helmont described these substances using the Flemish word for *chaos,* which was *gas.*

Van Helmont extensively studied a gas he obtained from burning wood, which he called "gas sylvestre" and which we now know as carbon dioxide, and noted that this substance was similar in many ways but not identical to air. By the end of his life, the importance of gases, especially air, was becoming more apparent. In 1643 an Italian physicist named Evangelista Torricelli (1608–1647), who had been a student of Galileo, performed experiments that showed that *the air in the atmosphere exerts pressure.* (In fact, as we will see, all gases exert pressure.) Torricelli designed the first **barometer** by filling a tube that was closed at one end with mercury and then inverting it in a dish of mercury (see Fig. 5.1). He observed that a column of mercury approximately 760 millimeters long always remained in the tube, as a result of the pressure of the atmosphere.

A few years later Otto von Guericke, a German physicist, invented an air pump, often called a vacuum pump, which he used in a famous demonstration for the King of Prussia in 1654. Guericke placed two hemispheres together and pumped the air out of the resulting sphere through a valve, which was subsequently closed. He then dramatically showed that teams of horses could not pull the hemispheres apart. However, after secretly opening the valve to let air in, Guericke was able to separate the hemispheres easily by hand. The King of Prussia was so impressed by Guericke's cleverness that he awarded him a lifetime pension.

Units of Pressure

Because instruments used for measuring pressure, such as the **manometer** (see Fig. 5.2), so often utilize columns of mercury because of its high density, the most commonly used units for pressure are based on the height of the mercury column (in millimeters) the gas pressure can support. The unit **mm Hg** (millimeters of mercury) is often called the **torr** in honor of Torricelli. A related unit for pressure is the **standard atmosphere:**

$$1 \text{ standard atmosphere} = 1 \text{ atm} = 760 \text{ mm Hg} = 760 \text{ torr}$$

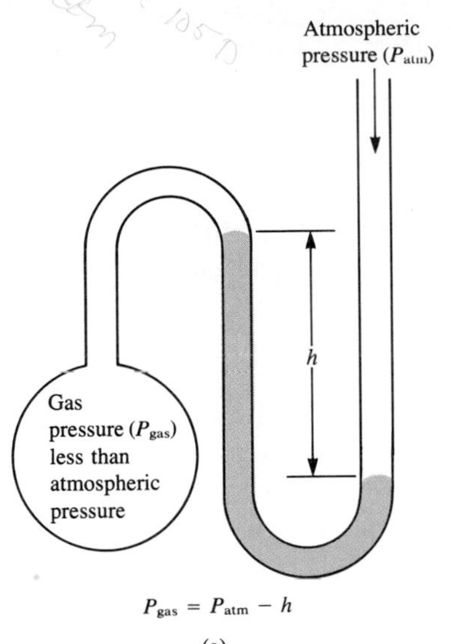

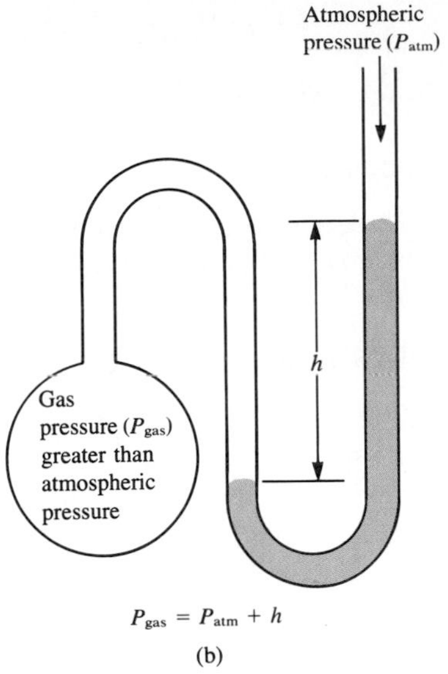

Figure 5.2

A simple manometer, a device for measuring the pressure of a gas in a container. The pressure of the gas is given by h (the difference in mercury levels) in units of torr (equivalent to mm Hg). (a) Gas pressure = atmospheric pressure − h. (b) Gas pressure = atmospheric pressure + h.

However, since pressure is defined as force per unit area,

$$\text{Pressure} = \frac{\text{force}}{\text{area}}$$

the fundamental units of pressure involve units of force divided by units of area. In the SI system, the unit of force is the newton (N) and the unit of area is meters squared (m^2). (For a review of the SI system, see Chapter 1.) Thus the unit of pressure in the SI system is newtons per meter squared (N/m^2), and is called the **pascal (Pa)** (see Fig. 5.3). In terms of pascals the standard atmosphere is

$$1 \text{ standard atmosphere} = 101{,}325 \text{ Pa}$$

Thus 1 atmosphere is about 10^5 pascals. Since the pascal is so small, and since it is not commonly used in the United States, we will use it sparingly in this book. However, converting from torrs or atmospheres to pascals is straightforward, as shown in Sample Exercise 5.1.

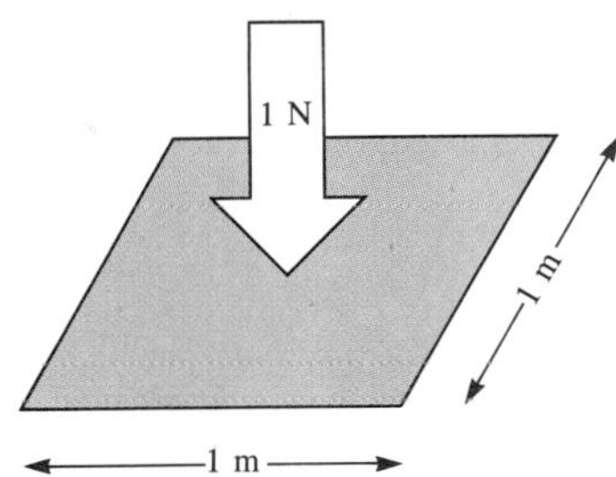

Figure 5.3

The SI system defines a force of 1 newton (N) acting on an area of 1 square meter to be a pascal.

Sample Exercise 5.1

The pressure of a gas is measured as 49 torr. Represent this pressure in both atmospheres and pascals.

1 atm = 760 mm Hg
= 760 torr
= 101,325 Pa
= 29.92 in Hg
= 14.7 lb/in²

Solution

$$49 \text{ torr} \times \frac{1 \text{ atm}}{760 \text{ torr}} = 6.4 \times 10^{-2} \text{ atm}$$

$$6.4 \times 10^{-2} \text{ atm} \times \frac{101{,}325 \text{ Pa}}{1 \text{ atm}} = 6.5 \times 10^{3} \text{ Pa}$$

5.2 The Gas Laws of Boyle, Charles, and Avogadro

Purpose

- To describe certain laws that relate the volume, pressure, and temperature of a gas and to do calculations involving these laws.

Boyle's Law

The first quantitative experiments on gases were performed by an Irish chemist, Robert Boyle (1627–1691). Using a J-shaped tube closed at one end (Fig. 5.4), which he reportedly set up in the multistory entryway of his house, Boyle studied the relationship between the pressure of the trapped gas and its volume. Representative values from Boyle's experiments are given in Table 5.1. Boyle recognized from these data that the product of the pressure and volume for the trapped air sample was constant within the accuracies of his measurements (note the third column in Table 5.1). This behavior can be represented by the equation

$$PV = k$$

which is called **Boyle's law** and where k is a constant at a specific temperature for a given sample of air.

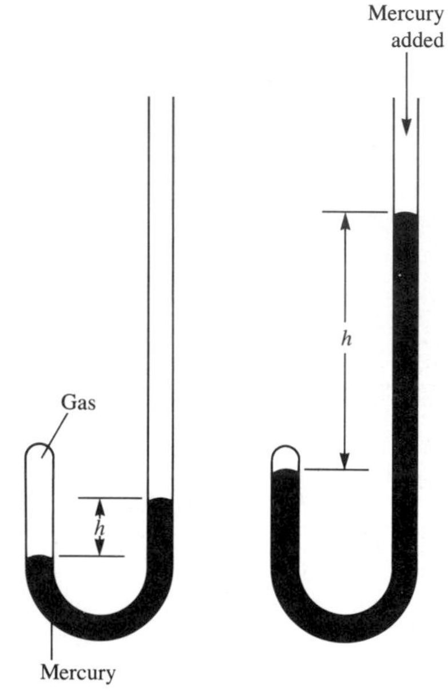

Figure 5.4

A J-tube similar to the one used by Boyle.

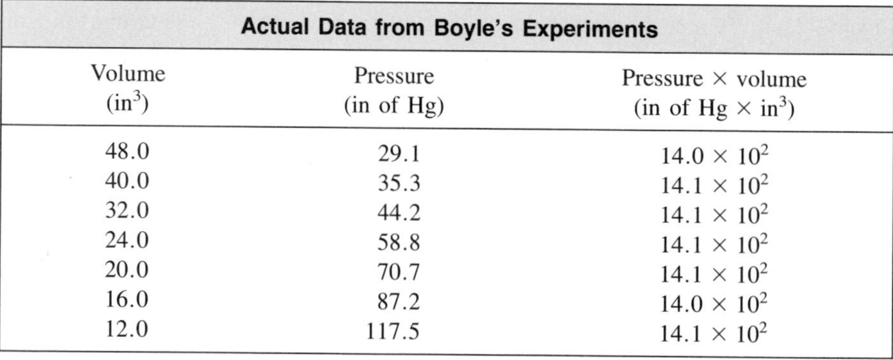

Actual Data from Boyle's Experiments		
Volume (in^3)	Pressure (in of Hg)	Pressure × volume (in of Hg × in^3)
48.0	29.1	14.0×10^2
40.0	35.3	14.1×10^2
32.0	44.2	14.1×10^2
24.0	58.8	14.1×10^2
20.0	70.7	14.1×10^2
16.0	87.2	14.0×10^2
12.0	117.5	14.1×10^2

Table 5.1

Boyle's law: $V \propto 1/P$ at constant temperature.

It is convenient to represent the data in Table 5.1 using two different plots. Figure 5.5(a) shows a plot of P versus V, which forms a curve called a hyperbola. Notice that as the pressure drops by half, the volume doubles. Thus there is an *inverse relationship* between pressure and volume. The second type of plot can be obtained by rearranging Boyle's law to give

$$V = \frac{k}{P}$$

which is the equation for a straight line of the type

$$y = mx + b$$

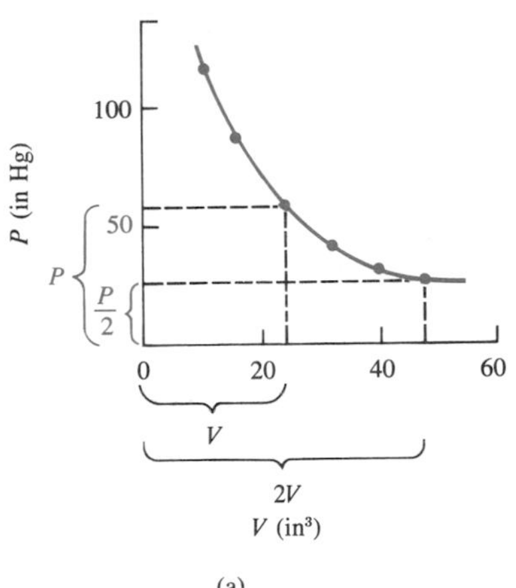

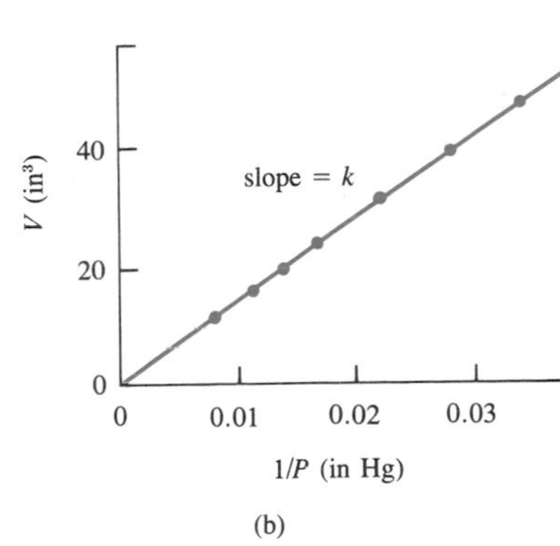

Figure 5.5

Plotting Boyle's data from Table 5.1. (a) A plot of P versus V shows that the volume doubles as the pressure is halved. (b) A plot of V versus $1/P$ gives a straight line. The slope of this line equals the value of the constant k.

where m represents the slope and b the intercept of the straight line. In this case, $y = V$, $x = 1/P$, $m = k$, and $b = 0$. Thus a plot of V versus $1/P$ using Boyle's data gives a straight line with an intercept of zero, as shown in Fig. 5.5(b).

Graphing is reviewed in Appendix 1.3.

Actually Boyle's law only approximately describes the relationship between pressure and volume for a gas. Highly accurate measurements on various gases at a given temperature have shown that the product PV is not quite constant but changes with pressure. Results for several gases are shown in Fig. 5.6. Note the small changes that occur in the product PV as the pressure is varied. Such changes become very significant at pressures much higher than normal atmospheric pressure. We will discuss these deviations and the reasons for them in detail in Section 5.8. *A gas that obeys Boyle's law is called an* **ideal gas.** We will describe the characteristics of an ideal gas more completely in Section 5.3.

One common use of Boyle's law is to predict the new volume of a gas when the pressure is changed (at constant temperature), or vice versa.

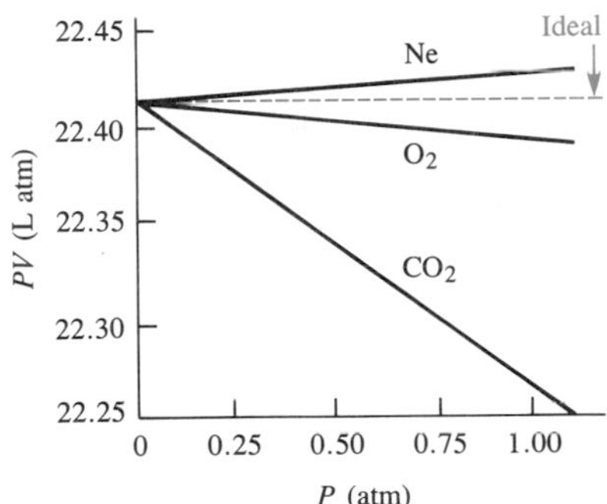

Figure 5.6

A plot of PV versus P for several gases. An ideal gas is expected to have a constant value of PV, as shown by the dotted line. Carbon dioxide shows the largest change in PV, and this change is actually quite small: PV changes from about 22.39 L atm at 0.25 atm to 22.26 L atm at 1.00 atm. Thus Boyle's law is a good approximation at these relatively low pressures.

Sample Exercise 5.2

Freon-12 (the common name for the compound CCl_2F_2) is widely used in refrigeration systems. Consider a 1.53-L sample of gaseous CCl_2F_2 at a pressure of 5.6×10^3 Pa. If the pressure is changed to 1.5×10^4 Pa at a constant temperature, what will be the new volume of the gas?

Solution

We can solve this problem using Boyle's law,

$$PV = k$$

which can also be written as

$$P_1V_1 = k = P_2V_2 \quad \text{or} \quad P_1V_1 = P_2V_2$$

where the subscripts 1 and 2 represent two states (conditions) of the gas (both at the same temperature). In this case

$$P_1 = 5.6 \times 10^3 \text{ Pa} \qquad P_2 = 1.5 \times 10^4 \text{ Pa}$$
$$V_1 = 1.53 \text{ L} \qquad V_2 = ?$$

Sample Exercise 5.2, continued

We can solve the preceding equation for V_2:

$$V_2 = \frac{P_1V_1}{P_2} = \frac{5.6 \times 10^3 \cancel{\text{Pa}} \times 1.53 \text{ L}}{1.5 \times 10^4 \cancel{\text{Pa}}} = 0.57 \text{ L}$$

The new volume will be 0.57 L.

Always check that your answer makes physical (common!) sense.

The fact that the volume decreases in Sample Exercise 5.2 makes sense because the pressure was increased. *To eliminate errors, make it a habit to check whether an answer to a problem makes physical sense.*

We mentioned above that Boyle's law is only approximately true for real gases. To determine the significance of the deviations, studies of the effect of changing pressure on the volume of a gas are often done, as shown in Sample Exercise 5.3.

Sample Exercise 5.3

To study how closely gaseous ammonia obeys Boyle's Law, several volume measurements were made at various pressures, using 1.0 mol of NH_3 gas at a temperature of 0°C. Using the results listed below, calculate the Boyle's law constant for NH_3 at the various pressures.

Experiment	Pressure (atm)	Volume (L)
1	0.13	172.1
2	0.25	89.28
3	0.30	74.35
4	0.50	44.49
5	0.75	29.55
6	1.00	22.08

Solution

To determine how closely NH_3 gas follows Boyle's law under these conditions, we calculate the value of k (in L atm) for each set of values:

Experiment	1	2	3	4	5	6
$k = PV$	22.37	22.32	22.31	22.24	22.16	22.08

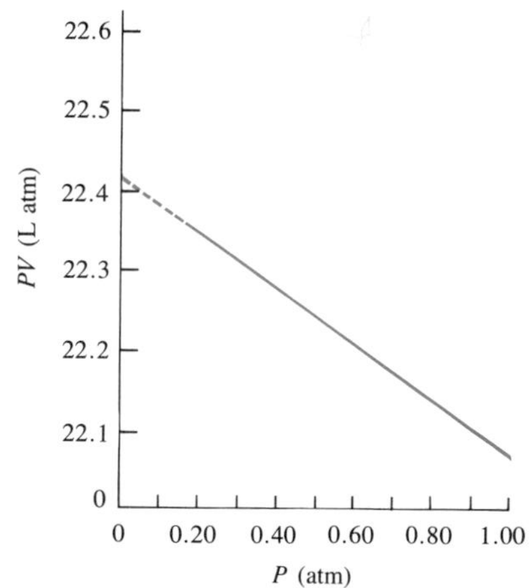

Figure 5.7

A plot of *PV* versus *P* for 1 mol of ammonia. The dashed line shows the extrapolation of the data to zero pressure to give the "ideal" value of *PV* of 22.41 L atm.

Although the deviations from true Boyle's law behavior are quite small at these low pressures, note that the value of k changes regularly in one direction as the pressure is increased. Thus to calculate the "ideal" value of k for NH_3, we can plot PV versus P, as shown in Fig. 5.7, and extrapolate (extend the line beyond the experimental points) back to zero pressure, where for reasons we will discuss later, a gas behaves most ideally. The value of k obtained by this extrapolation is 22.41 L atm. Notice that this is the same value obtained from similar plots for the gases CO_2, O_2, and Ne at 0°C, as shown in Fig. 5.6.

Charles's Law

In the century following Boyle's findings, scientists continued to study the properties of gases. One of these scientists was a French physicist, Jacques Charles (1746–1823), who was the first person to fill a balloon with hydrogen gas and who made the first solo balloon flight. Charles found in 1787 that the volume of a gas at constant pressure increases *linearly* with the temperature of the gas. That is, a plot of the volume of a gas (at constant pressure) versus its temperature (°C) gives a straight line. This behavior is shown for several gases in Fig. 5.8. A very interesting feature of these plots is that the volumes of all the gases extrapolate to zero at the same temperature, −273.2°C.

On the Kelvin temperature scale this point is defined as 0 K, which leads to the following relationship between the Kelvin and Celsius scales:

$$K = 0°C + 273$$

When the volumes of the gases shown in Fig. 5.8 are plotted versus temperature on the Kelvin scale, the plots in Fig. 5.9 result. In this case, the volume of each gas is *directly proportional to temperature* and extrapolates to zero when the temperature is 0 K. This behavior is represented by the equation known as **Charles's law,**

Charles's law: $V \propto T$ (expressed in K) at constant pressure.

$$V = bT$$

where T is in Kelvins and b is a proportionality constant.

Before we illustrate the uses of Charles's law, let us consider the importance of 0 K. At temperatures below this point, the extrapolated volumes would become negative. Since a gas cannot have a negative volume, this suggests that 0 K has a special significance. In fact, 0 K is called **absolute zero,** and there is much evidence to suggest that this temperature cannot be attained. Temperatures of approximately 0.0001 K have been produced in laboratories, but 0 K has never been reached.

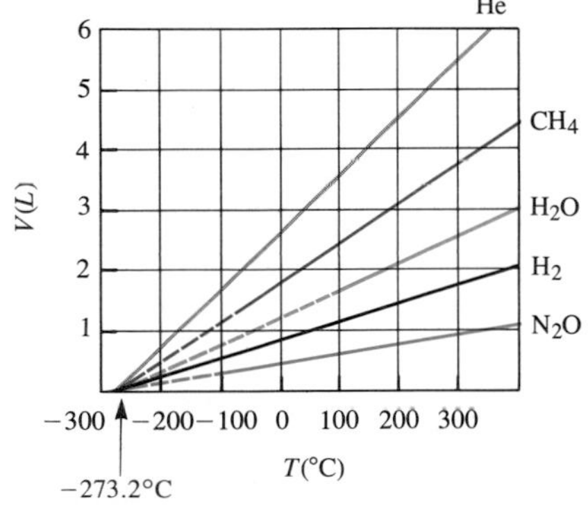

Figure 5.8

Plots of V versus T (°C) for several gases. The solid lines represent experimental measurements on gases. The dashed lines represent extrapolation of the data into regions where these gases would become liquids or solids.

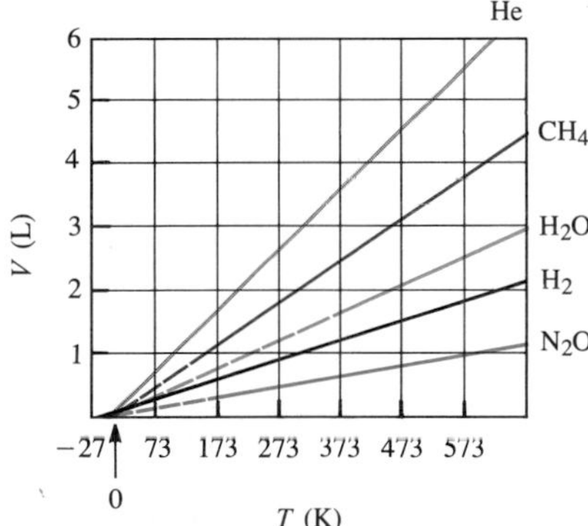

Figure 5.9

Plots of V versus T as in Fig. 5.8, except here the Kelvin scale is used for temperature.

Sample Exercise 5.4

A sample of gas at 15°C and 1 atm has a volume of 2.58 L. What volume will this gas occupy at 38°C and 1 atm?

Solution

Charles's law, which describes the dependence of the volume of a gas on temperature at constant pressure, can be used to solve this problem. Charles's law in the form $V = bT$ can be rearranged to

$$\frac{V}{T} = b$$

An equivalent statement is

$$\frac{V_1}{T_1} = b = \frac{V_2}{T_2}$$

where the subscripts 1 and 2 represent two states for a given sample of gas at constant pressure. In this case, we are given the following (note that the temperature values *must* be changed to the Kelvin scale):

$$T_1 = 15°\text{C} + 273 = 288\ \text{K} \qquad T_2 = 38°\text{C} + 273 = 311\ \text{K}$$
$$V_1 = 2.58\ \text{L} \qquad V_2 = ?$$

Solving for V_2 gives

$$V_2 = \left(\frac{T_2}{T_1}\right)V_1 = \left(\frac{311\ \cancel{\text{K}}}{288\ \cancel{\text{K}}}\right)2.58\ \text{L} = 2.79\ \text{L}$$

Check: The new volume is greater than the initial volume, which makes physical sense because the gas will expand as it is heated.

Avogadro's Law

In Chapter 2 we told how in 1811 the Italian chemist Avogadro postulated that equal volumes of gases at the same temperature and pressure contain the same number of "particles." This is called **Avogadro's law,** which is illustrated by Fig. 5.10 and can be stated mathematically as

$$V = an$$

where V is the volume of the gas, n is the number of moles, and a is a proportionality constant. This equation states that *for a gas at constant temperature and pressure the volume is directly proportional to the number of moles of gas*. This relationship has been verified by experiment and is obeyed closely by gases at low pressures.

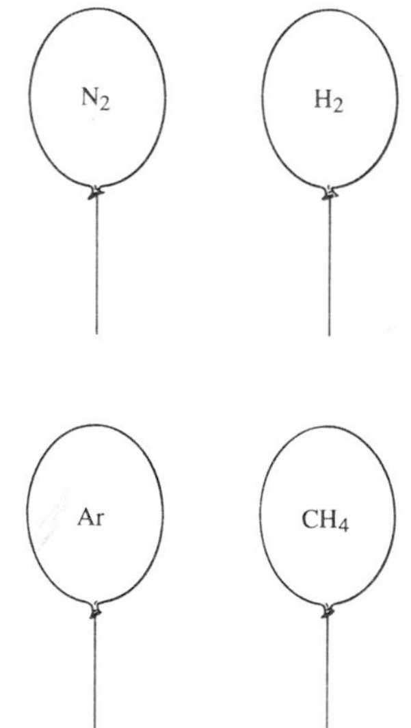

Figure 5.10

These balloons each hold 1.0 L of gas at 25°C and 1 atm. Each balloon contains 0.041 mol of gas, or 2.5 × 10^{22} molecules.

Sample Exercise 5.5

Suppose we have a 12.2-L sample containing 0.50 mol of oxygen gas (O_2) at a pressure of 1 atm and a temperature of 25°C. If all of this O_2 is converted to ozone (O_3) at the same temperature and pressure, what will be the volume of the ozone?

Sample Exercise 5.5, continued

Solution

The balanced equation for the reaction is

$$3O_2(g) \rightarrow 2O_3(g)$$

To calculate the moles of O_3 produced, we must use the appropriate mole ratio:

$$0.50\ \cancel{\text{mol } O_2} \times \frac{2 \text{ mol } O_3}{3\ \cancel{\text{mol } O_2}} = 0.33 \text{ mol } O_3$$

Avogadro's law states that $V = an$, which can be rearranged to give

$$\frac{V}{n} = a$$

Since a is a constant, an alternative representation is

$$\frac{V_1}{n_1} = a = \frac{V_2}{n_2}$$

where V_1 is the volume of n_1 moles of O_2 gas and V_2 is the volume of n_2 moles of O_3 gas. In this case we have

$$n_1 = 0.50 \text{ mol} \qquad n_2 = 0.33 \text{ mol}$$
$$V_1 = 12.2 \text{ L} \qquad V_2 = ?$$

Solving for V_2 gives

$$V_2 = \left(\frac{n_2}{n_1}\right)V_1 = \left(\frac{0.33\ \cancel{\text{mol}}}{0.50\ \cancel{\text{mol}}}\right) 12.2 \text{ L} = 8.1 \text{ L}$$

Check: Note that the volume decreases, as it should, since fewer moles of gas will be present after O_2 is converted to O_3.

5.3 The Ideal Gas Law

Purpose

- To define the ideal gas law.
- To show how to do calculations involving the ideal gas law.

We have considered three laws that describe the behavior of gases as revealed by experimental observations:

Boyle's law:	$V = \frac{k}{P}$	(at constant T and n)
Charles's law:	$V = bT$	(at constant P and n)
Avogadro's law:	$V = an$	(at constant T and P)

These relationships show how the volume of a gas depends on pressure, temperature, and number of moles of gas present and can be combined as follows:

$$V = R\left(\frac{Tn}{P}\right)$$

$R = 0.08206 \frac{\text{L atm}}{\text{K mol}}$

where R is the combined proportionality constant called the **universal gas constant.** When the pressure is expressed in atmospheres and the volume in liters, R has the value 0.08206 L atm/K mol. The above equation can be rearranged to the more familiar form of the **ideal gas law:**

$$PV = nRT$$

The ideal gas law is an *equation of state* for a gas, where the state of the gas is its condition at a given time. A particular *state* of a gas is described by its pressure, volume, temperature, and number of moles. Knowledge of any three of these properties is enough to completely define the state of a gas, since the fourth property can then be determined from the equation for the ideal gas law.

The ideal gas law applies best at pressures smaller than 1 atm.

It is important to recognize that the ideal gas law is an empirical equation—it is based on experimental measurements of the properties of gases. A gas that obeys this equation is said to behave *ideally*. That is, this equation defines the behavior of any ideal gas. Most gases obey this equation closely at pressures below 1 atmosphere. Unless you are given information to the contrary, you should assume ideal gas behavior when working problems involving gases in this text.

The ideal gas law can be used to solve a variety of problems. Sample Exercise 5.6 demonstrates one type, where you arc asked to find one property characterizing the state of a gas, given the other three.

The reaction of zinc with hydrochloric acid to produce bubbles of hydrogen gas.

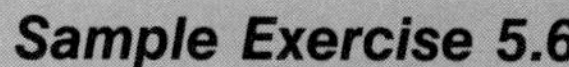

Sample Exercise 5.6

A sample of hydrogen gas (H_2) has a volume of 8.56 L at a temperature of 0°C and a pressure of 1.5 atm. Calculate the moles of H_2 present in this gas sample.

Solution

Solving the ideal gas law for n gives

$$n = \frac{PV}{RT}$$

In this case $P = 1.5$ atm, $V = 8.56$ L, $T = 0°\text{C} + 273 = 273$ K, and $R =$ 0.08206 L atm/K mol. Thus

$$n = \frac{(1.5\ \cancel{\text{atm}})(8.56\ \cancel{\text{L}})}{\left(0.08206\ \frac{\cancel{\text{L}}\ \cancel{\text{atm}}}{\cancel{\text{K}}\ \text{mol}}\right)(273\ \cancel{\text{K}})} = 0.57 \text{ mol}$$

The ideal gas law is also used to calculate the changes that will occur when the conditions of the gas are changed.

Sample Exercise 5.7

Suppose we have a sample of ammonia gas with a volume of 3.5 L at a pressure of 1.68 atm. The gas is compressed to a volume of 1.35 L at a constant temperature. Use the ideal gas law to calculate the final pressure.

Solution

The basic assumption we make when using the ideal gas law to describe a change in state for a gas is that the equation applies equally well to both the initial and the final states. In dealing with a change in state, we always *place the variables that change on one side of the equals sign and the constants on the other*. In this case the pressure and volume change, and the temperature and the number of moles remain constant (as does R, by definition). Thus we write the ideal gas law as

$$\underbrace{PV}_{\text{Change}} = \underbrace{nRT}_{\text{Remain constant}}$$

Since n and T remain the same in this case, we can write $P_1V_1 = nRT$ and $P_2V_2 = nRT$. Combining these gives

$$P_1V_1 = nRT = P_2V_2 \qquad \text{or} \qquad P_1V_1 = P_2V_2$$

We are given $P_1 = 1.68$ atm, $V_1 = 3.5$ L, $V_2 = 1.35$ L. Solving for P_2 thus gives

$$P_2 = \left(\frac{V_1}{V_2}\right)P_1 = \left(\frac{3.5\ \cancel{\text{L}}}{1.35\ \cancel{\text{L}}}\right) 1.68 \text{ atm} = 4.4 \text{ atm}$$

Check: Does this answer make sense? The volume was decreased (at constant temperature), which means that the pressure should increase, as the result of the calculation indicates. Note that the calculated final pressure is 4.4 atm, which is higher than 1 atm, where most gases can be assumed to behave ideally. Because of this we might find that if we *measured* the pressure of this gas sample, the observed pressure would differ slightly from 4.4 atm.

Sample Exercise 5.8

A sample of methane gas that has a volume of 3.8 L at 5°C is heated to 86°C at constant pressure. Calculate its new volume.

Solution

To solve this problem we take the ideal gas law and segregate the changing variables and the constants on opposite sides of the equation. In this case volume and temperature change, and number of moles and pressure (and of course R) remain constant. Thus $PV = nRT$ becomes

$$\frac{V}{T} = \frac{nR}{P}$$

which leads to

$$\frac{V_1}{T_1} = \frac{nR}{P} \quad \text{and} \quad \frac{V_2}{T_2} = \frac{nR}{P}$$

Sample Exercise 5.8, continued

Combining these gives

$$\frac{V_1}{T_1} = \frac{nR}{P} = \frac{V_2}{T_2} \quad \text{or} \quad \frac{V_1}{T_1} = \frac{V_2}{T_2}$$

We are given

$$T_1 = 5°\text{C} + 273 = 278 \text{ K} \qquad T_2 = 86°\text{C} + 273 = 359 \text{ K}$$
$$V_1 = 3.8 \text{ L} \qquad V_2 = ?$$

Thus

$$V_2 = \frac{T_2V_1}{T_1} = \frac{(359 \cancel{\text{K}})(3.8 \text{ L})}{278 \cancel{\text{K}}} = 4.9 \text{ L}$$

Check: Is the answer sensible? In this case the temperature was increased (at constant pressure) so the volume should increase. Thus the answer makes sense.

The problem in Sample Exercise 5.8 could be described as a "Charles's law problem," and the problem in Sample Exercise 5.7 might be said to be a "Boyle's law problem." In both cases, however, we started with the ideal gas law. The real advantage of using the ideal gas law is that you need to remember only *one* equation to do virtually any problem dealing with gases.

Sample Exercise 5.9

A sample of diborane gas (B_2H_6), a substance that bursts into flames when exposed to air, has a pressure of 345 torr at a temperature of −15°C and a volume of 3.48 L. If conditions are changed so that the temperature is 36°C and the pressure is 468 torr, what will be the volume of the sample?

Solution

Since for this sample pressure, temperature, and volume all change while the number of moles remains constant, we use the ideal gas law in the form

$$\frac{PV}{T} = nR$$

which leads to

$$\frac{P_1V_1}{T_1} = nR = \frac{P_2V_2}{T_2} \quad \text{or} \quad \frac{P_1V_1}{T_1} = \frac{P_2V_2}{T_2}$$

Then

$$V_2 = \frac{T_2P_1V_1}{T_1P_2}$$

We have

$$P_1 = 345 \text{ torr} \qquad P_2 = 468 \text{ torr}$$
$$T_1 = -15°\text{C} + 273 = 258 \text{ K} \qquad T_2 = 36°\text{C} + 273 = 309 \text{ K}$$
$$V_1 = 3.48 \text{ L} \qquad V_2 = ?$$

Sample Exercise 5.9, continued

Thus

$$V_2 = \frac{(309\ \cancel{K})(345\ \cancel{torr})(3.48\ L)}{(258\ \cancel{K})(468\ \cancel{torr})} = 3.07\ L$$

Since the equation used in Sample Exercise 5.9 involved a *ratio* of pressures, it was unnecessary to convert pressures to units of atmospheres. The units of torrs cancel out. (You will obtain the same answer by inserting $P_1 = \frac{345}{760}$ and $P_2 = \frac{468}{760}$ into the equation.) However, temperature *must always* be converted to the Kelvin scale since this conversion involves *addition* of 273 and the conversion factor does not cancel out. Be careful.

Always convert the temperature to the Kelvin scale when applying the ideal gas law.

One of the many other types of problems dealing with gases that can be solved using the ideal gas law is illustrated in Sample Exercise 5.10.

Sample Exercise 5.10

A sample containing 0.35 mol of argon gas at a temperature of 13°C and a pressure of 568 torr is heated to 56°C and a pressure of 897 torr. Calculate the change in volume that occurs.

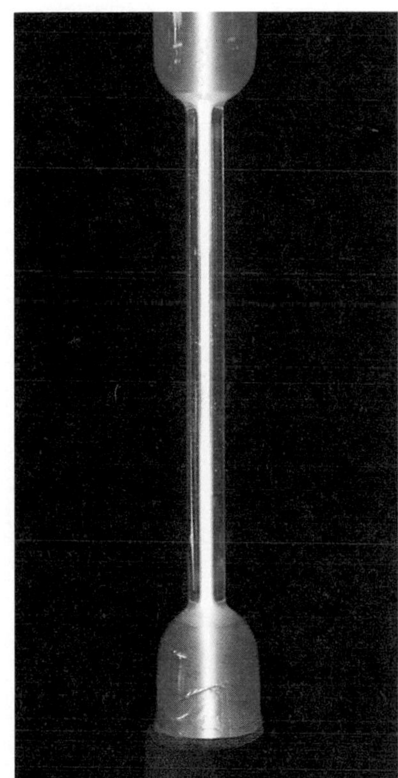

Argon glowing in a discharge tube.

Solution

We use the ideal gas law to find the volume for each set of conditions:

State 1	State 2
$n_1 = 0.35$ mol	$n_2 = 0.35$ mol
$P_1 = 568\ \cancel{torr} \times \frac{1\ atm}{760\ \cancel{torr}} = 0.747$ atm	$P_2 = 897\ \cancel{torr} \times \frac{1\ atm}{760\ \cancel{torr}} = 1.18$ atm
$T_1 = 13°C + 273 = 286$ K	$T_2 = 56°C + 273 = 329$ K

Solving the ideal gas law for volume gives

$$V_1 = \frac{n_1RT_1}{P_1} = \frac{(0.35\ \cancel{mol})(0.08206\ L\ \cancel{atm}/\cancel{K}\ \cancel{mol})(286\ \cancel{K})}{(0.747\ \cancel{atm})} = 11\ L$$

and

$$V_2 = \frac{n_2RT_2}{P_2} = \frac{(0.35\ \cancel{mol})(0.08206\ L\ \cancel{atm}/\cancel{K}\ \cancel{mol})(329\ \cancel{K})}{(1.18\ \cancel{atm})} = 8.0\ L$$

Thus, in going from state 1 to state 2, the volume changes from 11 L to 8.0 L. The change in volume, ΔV (Δ is the Greek letter delta), is then

$$\Delta V = V_2 - V_1 = 8.0\ L - 11\ L = -3\ L$$

The *change* in volume is negative here, since the volume decreases. Note that for this problem (unlike Sample Exercise 5.9) the pressures must be converted from torrs to atmospheres as required by the atmosphere part of the units for R, since each volume was found separately and the conversion factor does not cancel.

5.4 Gas Stoichiometry

Purpose

- To define the molar volume for an ideal gas.
- To define STP.
- To show how to do stoichiometric calculations for reactions involving gases.
- To show how to calculate molecular weight from gas density.

Suppose we have 1 mole of an ideal gas at 0°C (273.2 K) and 1 atmosphere. Then, from the ideal gas law, the volume of the gas is given by

$$V = \frac{nRT}{P} = \frac{(1.000\ \cancel{\text{mol}})(0.08206\ \text{L}\ \cancel{\text{atm}}/\cancel{\text{K}}\ \cancel{\text{mol}})(273.2\ \cancel{\text{K}})}{1.000\ \cancel{\text{atm}}} = 22.42\ \text{L}$$

This volume of 22.42 liters is called the **molar volume** of an ideal gas. The measured molar volumes of several gases are listed in Table 5.2. Note that the molar volumes of some of the gases are very close to the ideal value, while others deviate significantly. Later in this chapter we will discuss some of the reasons for the deviations.

Molar Volumes for Various Gases at 0°C and 1 atm.

Gas	Molar volume (L)
Oxygen (O_2)	22.397
Nitrogen (N_2)	22.402
Hydrogen (H_2)	22.433
Helium (He)	22.434
Argon (Ar)	22.397
Carbon dioxide (CO_2)	22.260
Ammonia (NH_3)	22.079

Table 5.2

STP: 0°C and 1 atm

The conditions 0°C and 1 atmosphere are called **standard temperature and pressure** (abbreviated **STP**). Properties of gases are often given under these conditions. For example, the molar volume of an ideal gas is 22.42 liters only at STP (see Fig. 5.11).

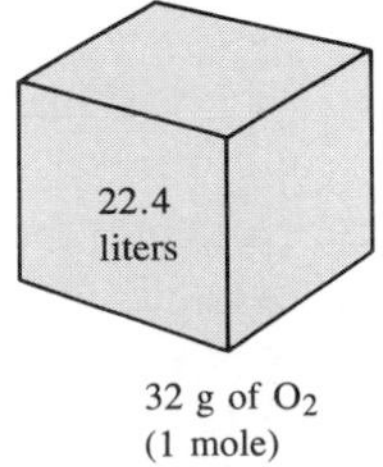

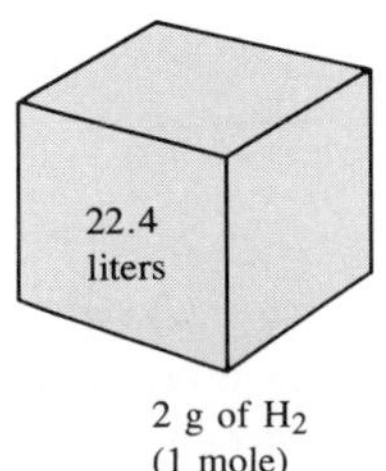

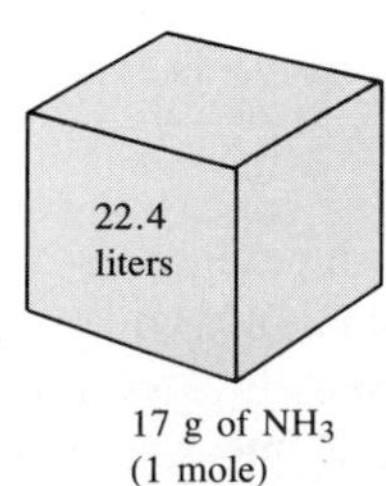

Figure 5.11

A mole of any gas occupies a volume of approximately 22.4 L at STP.

Sample Exercise 5.11

A sample of nitrogen gas has a volume of 1.75 L at STP. How many moles of N_2 are present?

Solution

We could solve this problem using the ideal gas equation, but we can take a shortcut by using the molar volume of an ideal gas at STP. Since 1 mol of an ideal gas at STP has a volume of 22.42 L, 1.75 L of N_2 at STP will contain less than 1 mol. We can find how many moles using the ratio of 1.75 L to 22.42 L:

$$1.75 \cancel{\text{L N}_2} \times \frac{1 \text{ mol N}_2}{22.42 \cancel{\text{L N}_2}} = 7.81 \times 10^{-2} \text{ mol N}_2$$

Many chemical reactions involve gases. By assuming ideal behavior for these gases, we can carry out stoichiometric calculations if the pressure, volume, and temperature of the gases are known.

Sample Exercise 5.12

Quicklime (CaO) is produced by the thermal decomposition of calcium carbonate ($CaCO_3$). Calculate the volume of CO_2 at STP produced from the decomposition of 152 g of $CaCO_3$ according to the reaction

$$CaCO_3(s) \rightarrow CaO(s) + CO_2(g)$$

Solution

We use the same strategy we used in the stoichiometry problems earlier in the book. That is, we compute the number of moles of $CaCO_3$ consumed and the number of moles of CO_2 produced. The moles of CO_2 can then be converted to volume using the molar volume of an ideal gas.

Using the molecular weight of $CaCO_3$ (100.1), we can calculate the number of moles of $CaCO_3$:

$$152 \cancel{\text{g CaCO}_3} \times \frac{1 \text{ mol CaCO}_3}{100.1 \cancel{\text{g CaCO}_3}} = 1.52 \text{ mol CaCO}_3$$

Since each mole of $CaCO_3$ produces a mole of CO_2, 1.52 mole of CO_2 will be formed. We can compute the volume of CO_2 at STP by using the molar volume:

$$1.52 \cancel{\text{mol CO}_2} \times \frac{22.42 \text{ L CO}_2}{1 \cancel{\text{mol CO}_2}} = 34.1 \text{ L CO}_2$$

Thus the decomposition of 152 g of $CaCO_3$ will produce 34.1 L of CO_2 at STP.

When lime (CaO) is heated to incandescence with a hot flame, it gives an intense, bright light. Such lights were used by 1800s' theaters to spotlight actors, originating the phrase "in the limelight."

Note that in Sample Exercise 5.12 the final step involved calculation of the volume of gas from the number of moles. Since the conditions were specified as STP, we were able to use the molar volume of a gas at STP. If the conditions of a problem are different from STP, the ideal gas law must be used to compute the volume.

Remember that the molar volume of an ideal gas is 22.42 L *only* at STP.

Sample Exercise 5.13

A sample of methane gas having a volume of 2.80 L at 25°C and 1.65 atm was mixed with a sample of oxygen gas having a volume of 35.0 L at 31°C and 1.25 atm. The mixture was ignited to form carbon dioxide and water. Calculate the volume of CO_2 formed at a pressure of 2.50 atm and a temperature of 125°C.

Solution

From the description of the reaction, the unbalanced equation is

$$CH_4(g) + O_2(g) \rightarrow CO_2(g) + H_2O(g)$$

This can be balanced to give

$$CH_4(g) + 2O_2(g) \rightarrow CO_2(g) + 2H_2O(g)$$

Next we must find the limiting reactant, which requires calculating the numbers of moles of the reactants. Thus we convert the given volumes of methane and oxygen to moles using the ideal gas law:

$$n_{CH_4} = \frac{PV}{RT} = \frac{(1.65\ \text{atm})(2.80\ \text{L})}{(0.08206\ \text{L atm/K mol})(298\ \text{K})} = 0.189\ \text{mol}$$

$$n_{O_2} = \frac{PV}{RT} = \frac{(1.25\ \text{atm})(35.0\ \text{L})}{(0.08206\ \text{L atm/K mol})(304\ \text{K})} = 1.75\ \text{mol}$$

In the balanced equation for the combustion reaction, 1 mol of CH_4 requires 2 mol of O_2. Thus the moles of O_2 required by 0.189 mol of CH_4 can be calculated as follows:

$$0.189\ \text{mol CH}_4 \times \frac{2\ \text{mol O}_2}{1\ \text{mol CH}_4} = 0.378\ \text{mol O}_2$$

Since 1.75 mol of O_2 are available, O_2 is in excess. The limiting reactant is CH_4. The number of moles of CH_4 available must be used to calculate the number of moles of CO_2 produced:

$$0.189\ \text{mol CH}_4 \times \frac{1\ \text{mol CO}_2}{1\ \text{mol CH}_4} = 0.189\ \text{mol CO}_2$$

Since the conditions stated are not STP, we must use the ideal gas law to calculate the volume:

$$V = \frac{nRT}{P}$$

In this case $n = 0.189$ mol, $T = 125°C + 273 = 398$ K, $P = 2.50$ atm, and $R = 0.08206$ L atm/K mol. Thus

$$V = \frac{(0.189\ \text{mol})(0.08206\ \text{L atm/K mol})(398\ \text{K})}{2.50\ \text{atm}} = 2.47\ \text{L}$$

This represents the volume of CO_2 produced under these conditions.

Molecular Weight of a Gas

One very important use of the ideal gas law is in the calculation of the molecular weight of a gas from its measured density. To see the relationship between gas density and molecular weight consider that the number of moles of gas, n, can be expressed as

$$n = \frac{\text{grams of gas}}{\text{mol. wt.}} = \frac{\text{mass}}{\text{mol. wt.}} = \frac{m}{\text{mol. wt.}}$$

where mol. wt. is the molecular weight of the gas (grams per mole). Substitution into the ideal gas equation gives:

$$P = \frac{nRT}{V} = \frac{(m/\text{mol. wt.})RT}{V} = \frac{m(RT)}{V(\text{mol. wt.})}$$

But m/V is the gas density, d, in units of grams per liter. Thus

$$\text{density} = \frac{\text{mass}}{\text{volume}}$$

$$P = \frac{dRT}{\text{mol. wt.}}$$

or

$$\text{mol. wt.} = \frac{dRT}{P} \qquad (5.1)$$

Thus, if the density of a gas at a given temperature and pressure is known, its molecular weight can be calculated.

Sample Exercise 5.14

The density of a gas was measured at 1.50 atm and 27°C and found to be 1.95 g/L. Calculate the molecular weight of the gas.

Solution

Using Equation (5.1), we calculate the molecular weight as follows:

$$\text{mol. wt.} = \frac{dRT}{P} = \frac{\left(1.95\,\frac{\text{g}}{\cancel{\text{L}}}\right)\left(0.08206\,\frac{\cancel{\text{L}}\,\cancel{\text{atm}}}{\cancel{\text{K}}\,\text{mol}}\right)(300\,\cancel{\text{K}})}{1.50\,\cancel{\text{atm}}} = 32.0\text{ g/mol}$$

Check: These are the units expected for molecular weight.

You could memorize the equation involving gas density and molecular weight, but it is better simply to remember the ideal gas equation, the definition of density, and the relationship between number of moles and molecular weight. You can then derive this equation when you need it. This approach proves you understand the concepts and means one less equation to memorize.

5.5 Dalton's Law of Partial Pressures

Purpose

- To state the relationship between partial pressures and total pressure and between partial pressure and mole fraction.

Among the experiments that led John Dalton to propose the atomic theory were his studies of mixtures of gases. In 1803 Dalton summarized his observations in this statement: *For a mixture of gases in a container, the total pressure exerted is the sum of the pressures that each gas would exert if it were alone.* This statement, known as **Dalton's law of partial pressures,** can be expressed as follows

$$P_{\text{TOTAL}} = P_1 + P_2 + P_3 + \cdots$$

where the subscripts refer to the individual gases (gas 1, gas 2, etc.). The pressures P_1, P_2, P_3, and so on are called **partial pressures;** that is, each one is the pressure that gas would exert if it were alone in the container.

Assuming that each gas behaves ideally, the partial pressure of each gas can be calculated from the ideal gas law:

$$P_1 = \frac{n_1RT}{V}, \qquad P_2 = \frac{n_2RT}{V}, \qquad P_3 = \frac{n_3RT}{V}, \qquad \cdots$$

The total pressure of the mixture, P_{TOTAL}, can be represented as

$$\begin{aligned} P_{\text{TOTAL}} = P_1 + P_2 + P_3 + \cdots &= \frac{n_1RT}{V} + \frac{n_2RT}{V} + \frac{n_3RT}{V} + \cdots \\ &= (n_1 + n_2 + n_3 + \cdots)\left(\frac{RT}{V}\right) \\ &= n_{\text{TOTAL}}\left(\frac{RT}{V}\right) \end{aligned}$$

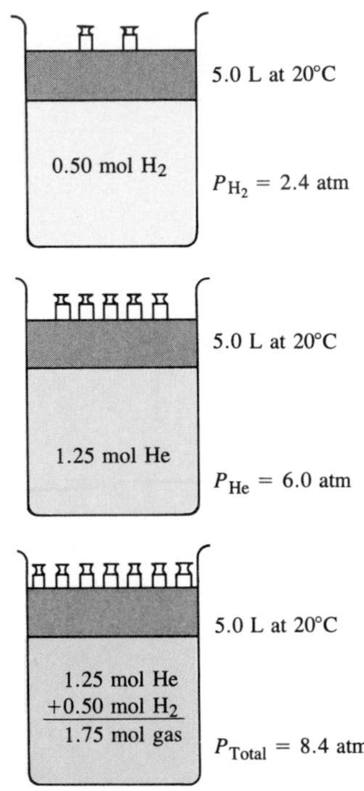

Figure 5.12

The partial pressure of each gas in a mixture of gases in a container depends on the number of moles of that gas. The total pressure is the sum of the partial pressures and depends on the total moles of gas particles present, no matter what they are.

where n_{TOTAL} is the sum of the numbers of moles of the various gases. Thus, for a mixture of ideal gases, it is the *total number of moles of particles* that is important, not the identity or composition of the individual gas particles. This idea is illustrated in Fig. 5.12.

This important experimental result indicates some fundamental characteristics of an ideal gas. The fact that the pressure exerted by an ideal gas is not affected by the identity, and thus the structure of the gas particles, tells us two things about ideal gases: (1) the volume of the individual gas particle must not be important; and (2) the forces among the particles must not be important. If these factors were important, the pressure exerted by the gas would depend on the nature of the individual particles. These observations will strongly influence the model that we will construct to explain ideal gas behavior.

Sample Exercise 5.15

Scuba diver.

Mixtures of helium and oxygen are used in scuba diving tanks to help prevent "the bends." For a particular dive 46 L of O_2 at 25°C and 1.0 atm was pumped along with 12 L of He at 25°C and 1.0 atm into a tank with a volume of 5.0 L. Calculate the partial pressure of each gas and the total pressure in the tank at 25°C.

Solution

The first step is to calculate the number of moles of each gas using the ideal gas law in the form:

$$n = \frac{PV}{RT}$$

$$n_{O_2} = \frac{(1.0\ \cancel{\text{atm}})(46\ \cancel{\text{L}})}{(0.08206\ \cancel{\text{L}}\ \cancel{\text{atm}}/\cancel{\text{K}}\ \text{mol})(298\ \cancel{\text{K}})} = 1.9\ \text{mol}$$

$$n_{He} = \frac{(1.0\ \cancel{\text{atm}})(12\ \cancel{\text{L}})}{(0.08206\ \cancel{\text{L}}\ \cancel{\text{atm}}/\cancel{\text{K}}\ \text{mol})(298\ \cancel{\text{K}})} = 0.49\ \text{mol}$$

The tank containing the mixture has a volume of 5.0 L, and the temperature is 25°C. We can use these data and the ideal gas law to calculate the partial pressure of each gas:

$$P = \frac{nRT}{V}$$

$$P_{O_2} = \frac{(1.9\ \cancel{\text{mol}})(0.08206\ \cancel{\text{L}}\ \text{atm}/\cancel{\text{K}}\ \cancel{\text{mol}})(298\ \cancel{\text{K}})}{5.0\ \cancel{\text{L}}} = 9.3\ \text{atm}$$

$$P_{He} = \frac{(0.49\ \cancel{\text{mol}})(0.08206\ \cancel{\text{L}}\ \text{atm}/\cancel{\text{K}}\ \cancel{\text{mol}})(298\ \cancel{\text{K}})}{5.0\ \cancel{\text{L}}} = 2.4\ \text{atm}$$

The total pressure is the sum of the partial pressures:

$$P_{\text{TOTAL}} = P_{O_2} + P_{He} = 9.3\ \text{atm} + 2.4\ \text{atm} = 11.7\ \text{atm}$$

At this point we need to define the **mole fraction:** *the ratio of the number of moles of a given component in a mixture to the total number of moles in the mixture.* The Greek letter chi (χ) is used to symbolize the mole fraction. For example, for a given component in a mixture, the mole fraction (χ_1) is

$$\chi_1 = \frac{n_1}{n_{\text{TOTAL}}} = \frac{n_1}{n_1 + n_2 + n_3 + \cdots}$$

From the ideal gas equation we know that the number of moles of a gas is directly proportional to the pressure of the gas since

$$n = P\left(\frac{V}{RT}\right)$$

(top) When water is placed in a can and boiled, water vapor displaces much of the air in the can. Since the can is open to the atmosphere, $P_{atm} = P_{air} + P_{H_2O}$. After boiling for a considerable time, $P_{H_2O} \gg P_{air}$. (bottom) When the can is sealed and allowed to cool, the water vapor in the can condenses, decreasing P_{H_2O}. As the pressure inside the can decreases, the atmospheric pressure crumples the can.

That is, for each component in the mixture,

$$n_1 = P_1\left(\frac{V}{RT}\right), \qquad n_2 = P_2\left(\frac{V}{RT}\right), \qquad \cdots$$

Therefore, we can represent the mole fraction in terms of pressures:

$$\chi_1 = \frac{n_1}{n_{\text{TOTAL}}} = \frac{\overbrace{P_1(V/RT)}^{n_1}}{\underbrace{P_1(V/RT)}_{n_1} + \underbrace{P_2(V/RT)}_{n_2} + \underbrace{P_3(V/RT)}_{n_3} + \cdots}$$

$$= \frac{(V/RT)P_1}{(V/RT)(P_1 + P_2 + P_3 + \cdots)}$$

$$= \frac{P_1}{P_1 + P_2 + P_3 + \cdots} = \frac{P_1}{P_{\text{TOTAL}}}$$

Similarly,

$$\chi_2 = \frac{n_2}{n_{\text{TOTAL}}} = \frac{P_2}{P_{\text{TOTAL}}}$$

and so on. Thus the mole fraction of a particular component in a mixture of ideal gases is directly related to its partial pressure.

Sample Exercise 5.16

The partial pressure of oxygen in air was observed to be 156 torr when the atmospheric pressure was 743 torr. Calculate the mole fraction of O_2 present.

Solution

The mole fraction of O_2 can be calculated from the equation

$$\chi_{O_2} = \frac{P_{O_2}}{P_{\text{TOTAL}}} = \frac{156\ \cancel{\text{torr}}}{743\ \cancel{\text{torr}}} = 0.210$$

Note that mole fraction has no units.

The expression for the mole fraction,

$$\chi_1 = \frac{P_1}{P_{\text{TOTAL}}}$$

can be rearranged:

$$P_1 = \chi_1 \times P_{\text{TOTAL}}$$

That is, *the partial pressure of a particular component of a gaseous mixture is the mole fraction of that component times the total pressure.*

Sample Exercise 5.17

The mole fraction of nitrogen in the air is 0.7808. Calculate the partial pressure of N_2 in air when the atmospheric pressure is 760. torr.

Solution

The partial pressure of N_2 can be calculated as follows:

$$P_{N_2} = \chi_{N_2} \times P_{TOTAL} = 0.7808 \times 760.\ \text{torr} = 593\ \text{torr}$$

A mixture of gases occurs whenever a gas is collected by displacement of water. For example, Fig. 5.13 shows the collection of oxygen gas produced by the decomposition of solid potassium chlorate. In this situation the gas in the bottle will be a mixture of water vapor and the gas being collected. Water vapor is present because molecules of water escape from the surface of the liquid and collect in the space above the liquid. Molecules of water also return to the liquid. When the rate of escape equals the rate of return, the number of water molecules in the vapor state remains constant, and thus the pressure of water vapor remains constant. This pressure, which depends on temperature, is called the vapor pressure of water.

Vapor pressure will be discussed in detail in Chapter 10. A table of water vapor pressure values is given in Section 10.8.

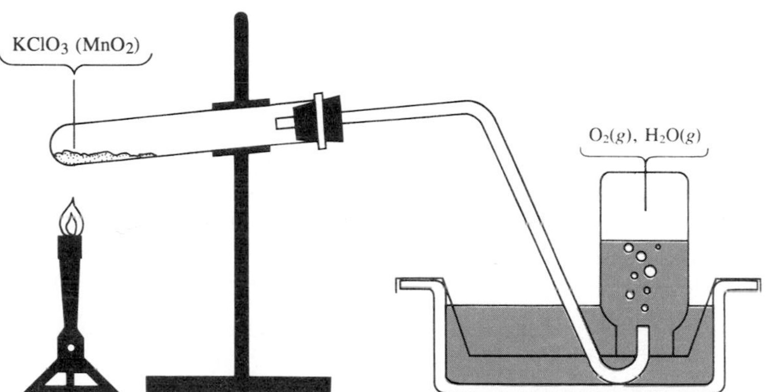

Figure 5.13

The production of oxygen by thermal decomposition of $KClO_3$. The MnO_2 is mixed with the $KClO_3$ to make the reaction faster.

Sample Exercise 5.18

A sample of solid potassium chlorate ($KClO_3$) was heated in a test tube (see Fig. 5.13) and decomposed according to the following reaction:

$$2KClO_3(s) \rightarrow 2KCl(s) + 3O_2(g)$$

The oxygen produced was collected by displacement of water at 22°C at a total pressure of 754 torr. The volume of the gas collected was 0.65 L, and the vapor pressure of water at 22°C is 21 torr. Calculate the partial pressure of O_2 in the gas collected and the mass of $KClO_3$ in the sample that was decomposed.

Solution

First we find the partial pressure of O_2 from Dalton's law of partial pressures:

$$P_{TOTAL} = P_{O_2} + P_{H_2O} = P_{O_2} + 21\ \text{torr} = 754\ \text{torr}$$

Chemical Impact

Scuba Diving

Oxygen is essential to our existence, but surprisingly, it can be harmful under certain circumstances. At sea level the partial pressure of this life-sustaining gas is 0.21 atmosphere,* and in each normal breath we take in about 0.02 mole of O_2 molecules. Our bodies operate very effectively under these conditions, but the situation changes if we subject ourselves to greater pressures, for example, by diving in deep water. A scuba diver at a depth of 100 feet experiences approximately an extra 3 atmospheres of pressure, and at 300 feet the pressure is about 10 atmospheres. This increased pressure affects the ear canals and squeezes the lungs, but a more serious effect involves the increased partial pressure of oxygen in air breathed at these pressures. Elevated concentrations of oxygen are actually quite harmful for reasons that are not well understood. The symptoms of oxygen poisoning include confusion, impaired vision and hearing, and nausea. Clearly, it is possible to get too much of a good thing.

At a depth of 300 feet under the ocean's surface, the partial pressure of oxygen is about 2 atmospheres (0.21×10 atmospheres) for a diver inhaling compressed air. This is much too high, and the oxygen must be diluted by another gas. Although it might seem to be an ideal candidate, nitrogen cannot be used because at elevated pressures a large quantity of nitrogen dissolves in the blood, producing nitrogen narcosis, often called "rapture of the deep," which has an effect not unlike that from imbibing too many martinis. The increased solubility of nitrogen in blood at high pressures is also responsible for an agonizing condition called "the bends," which results when a diver makes too rapid an ascent. The diver's joints become locked in a bent position, hence the name. Just as a bottle of soda fizzes when the pressure inside the bottle is released by opening the cap, the excess dissolved nitrogen in the blood stream forms bubbles that can stop blood flow and impair the nervous system.

Helium is the diluting agent most often used in scuba tanks. It is inert, and its solubility in blood is much lower than that of oxygen or nitrogen. However, it has the effect of raising the pitch of the human voice, producing a Donald Duck effect. This occurs because the pitch of the voice depends on the density of the gas surrounding the vocal cords: lower density yields a higher pitch. Because the mass of a He atom is much smaller than the masses of N_2 and O_2, the density of helium gas is much lower than that of air.

*This result is obtained from Dalton's law. Since the mole fraction of O_2 in air is 0.21, the partial pressure of O_2 is 0.21×1 atm = 0.21 atm.

Sample Exercise 5.18, continued

Thus

$$P_{O_2} = 754 \text{ torr} - 21 \text{ torr} = 733 \text{ torr}$$

Now we use the ideal gas law to find the number of moles of O_2:

$$n_{O_2} = \frac{P_{O_2}V}{RT}$$

In this case

$$P_{O_2} = 733 \text{ torr} = \frac{733 \cancel{\text{torr}}}{760 \cancel{\text{torr}}/\text{atm}} = 0.964 \text{ atm}$$

$$V = 0.650 \text{ L}$$
$$T = 22°\text{C} + 273 = 295 \text{ K}$$
$$R = 0.08206 \text{ L atm/K mol}$$

Sample Exercise 5.18, continued

Thus

$$n_{O_2} = \frac{(0.964 \text{ atm})(0.650 \text{ L})}{(0.08206 \text{ L atm/K mol})(295 \text{ K})} = 2.59 \times 10^{-2} \text{ mol}$$

Next we want to calculate the moles of $KClO_3$ needed to produce this quantity of O_2. From the balanced equation for the decomposition of $KClO_3$, we have a mole ratio of 2 mol $KClO_3$/3 mol O_2. The moles of $KClO_3$ can be calculated as follows:

$$2.59 \times 10^{-2} \text{ mol } O_2 \times \frac{2 \text{ mol } KClO_3}{3 \text{ mol } O_2} = 1.73 \times 10^{-2} \text{ mol } KClO_3$$

Using the molecular weight of $KClO_3$ (122.6), we calculate the grams of $KClO_3$:

$$1.73 \times 10^{-2} \text{ mol } KClO_3 \times \frac{122.6 \text{ g } KClO_3}{1 \text{ mol } KClO_3} = 2.12 \text{ g } KClO_3$$

Thus the original sample contained 2.12 g of $KClO_3$.

5.6 The Kinetic Molecular Theory of Gases

Purpose

- To present the basic postulates of the kinetic molecular theory.
- To define temperature.
- To show how to calculate and use root mean square velocity.

We have so far considered the behavior of gases from an experimental point of view. Based on observations from different types of experiments, we know that at pressures of less than 1 atmosphere most gases obey the ideal gas law. Now we want to construct a model to explain this behavior.

Before we do this let's briefly review the scientific method. Recall that a law is a way of generalizing behavior that has been observed in many experiments. Laws are very useful, since they allow us to predict the behavior of similar systems. For example, if a chemist prepares a new gaseous compound, a measurement of the gas density at known pressure and temperature can provide a reliable value for the compound's molecular weight.

However, although laws summarize observed behavior, they do not tell us *why* nature behaves in the observed fashion. This is the central question for scientists. To try to answer this question, we construct theories (build models). The models in chemistry consist of speculations about what the individual atoms or molecules (microscopic particles) might be doing to cause the observed behavior of the macroscopic systems (collections of very large numbers of atoms and molecules).

A model is considered successful if it explains the known behavior in question and predicts correctly the results of future experiments. It is important to understand that a model can never be proved absolutely true. In fact, *any model is an approximation* by its very nature and is doomed to fail at some point. Models range from the simple to the extraordinarily complex. We use simple models to predict approximate behavior and more complicated models to account very precisely for observed quantitative behavior. In this text, we will stress simple models that provide a picture of what might be happening and that fit the most important experimental results.

An example of this type of model is the **kinetic molecular theory,** a simple model that attempts to explain the properties of an ideal gas. This model is based on speculations about the behavior of the individual gas particles (atoms or molecules). The postulates of the kinetic molecular theory can be stated as follows:

> Gases consist of particles, which have the following properties:
>
> The particles are so small compared to the distances between them that *the volume of the individual particles can be assumed to be negligible* (zero).
>
> *The particles are in constant motion. The collisions of the particles with the walls of the container are the cause of the pressure exerted by the gas.*
>
> *The particles are assumed to exert no forces on each other;* they are assumed neither to attract nor to repel each other.
>
> *The average kinetic energy of a collection of gas particles is assumed to be directly proportional to the Kelvin temperature of the gas.*

Of course, real gases do not conform to these assumptions, but we will see now that these postulates do indeed explain *ideal* gas behavior.

The true test of a model is how well its predictions fit the experimental observations. The postulates of the kinetic molecular model picture an ideal gas as consisting of particles having no volume and no attraction for each other, and it is assumed that the gas produces pressure on its container by collisions with the walls. To test the validity of this model we need to consider the question: When we apply the principles of physics to a collection of these gas particles, can we derive an expression for pressure that agrees with the ideal gas law? Yes, we can. As shown in detail in Appendix 2, we can apply the definitions of velocity, momentum, force, and pressure to the collection of particles in an ideal gas and *derive* the following expression for pressure:

$$P = \frac{2}{3}\left[\frac{nN_A(\frac{1}{2}m\overline{u^2})}{V}\right]$$

where P is the pressure of the gas, n is the number of moles of gas, N_A is Avogadro's number, m is the mass of each particle, $\overline{u^2}$ is the average of the square of the velocities of the particles, and V is the volume of the container.

Kinetic energy (KE) given by the equation $KE = \frac{1}{2}mu^2$ is the energy due to the motion of a particle. We will discuss this further in Section 6.1.

The quantity $\frac{1}{2}\,m\overline{u^2}$ represents the average kinetic energy of a gas particle. If the average kinetic energy of an individual particle is multiplied by N_A, the number of particles in a mole, we get the average kinetic energy for a mole of gas particles:

$$(\text{KE})_{\text{avg}} = N_A(\tfrac{1}{2}m\overline{u^2})$$

Using this definition, we can rewrite the expression for pressure as

$$P = \frac{2}{3}\left(\frac{n\,(\text{KE})_{\text{avg}}}{V}\right) \quad \text{or} \quad \frac{PV}{n} = \frac{2}{3}\,(\text{KE})_{\text{avg}}$$

(a)

(b)

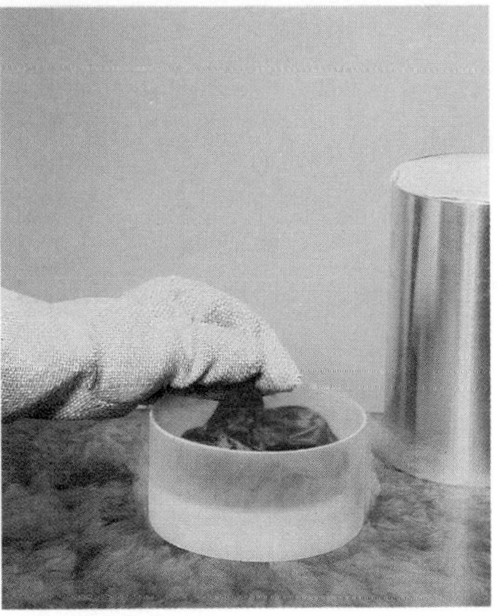

(c)

(a) A balloon filled with air at room temperature. (b) The balloon is dipped into liquid nitrogen at 77 K. (c) The balloon collapses as the molecules inside slow down due to the decreased temperature. Slower molecules produce a lower pressure.

The fourth postulate of the kinetic molecular theory is that the average kinetic energy of the particles in the gas sample is directly proportional to the temperature in Kelvins. Thus since $(\text{KE})_{\text{avg}} \propto T$ we can write

$$\frac{PV}{n} = \frac{2}{3}(\text{KE})_{\text{avg}} \propto T \quad \text{or} \quad \frac{PV}{n} \propto T$$

Note that this expression has been *derived* from the assumptions of the kinetic molecular theory. How does it compare to the ideal gas law—the equation obtained from experiment? Compare the ideal gas law,

$$\frac{PV}{n} = RT \qquad \text{From experiment}$$

with the result from the kinetic molecular theory,

$$\frac{PV}{n} \propto T \qquad \text{From theory}$$

These expressions have exactly the same form if R, the universal gas constant, is considered the proportionality constant in the second case.

The agreement between the results from the ideal gas law and the kinetic molecular theory gives us confidence in the validity of the model. The characteristics we have assumed for ideal gas particles must agree, at least under certain conditions, with their actual behavior.

The Meaning of Temperature

As we have seen, the assumption of the kinetic molecular theory that the Kelvin temperature is a measure of the average kinetic energy of the gas particles is supported by the agreement of the model with the ideal gas law. In fact, the exact relationship between temperature and average kinetic energy can be obtained by combining the equations

$$\frac{PV}{n} = RT = \frac{2}{3}(\text{KE})_{\text{avg}}$$

which yields the expression

$$(\text{KE})_{\text{avg}} = \frac{3}{2}RT$$

This is a very important relationship. It summarizes the meaning of the Kelvin temperature of a gas: the Kelvin temperature is an index of the random motions of the particles of a gas, with higher temperature meaning greater motion. (As we will see in Chapter 10, temperature is an index of the random motions in solids and liquids as well as in gases.)

Root Mean Square Velocity

In the equation from the kinetic molecular theory, the average velocity of the gas particles is a special kind of average. The symbol $\overline{u^2}$ means the average of the *squares* of the particle velocities. The square root of $\overline{u^2}$ is called the **root mean square velocity** and is symbolized by u_{rms}:

$$u_{\text{rms}} = \sqrt{\overline{u^2}}$$

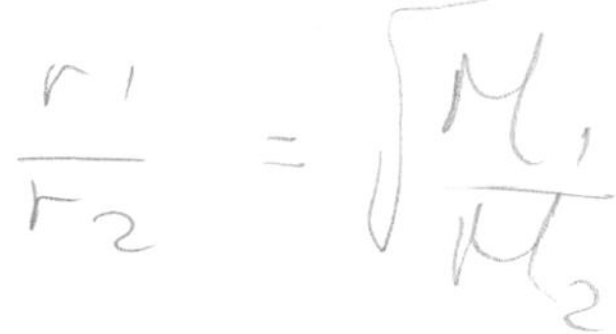

We can obtain an expression for u_{rms} from the equations

$$(\text{KE})_{\text{avg}} = N_A(\tfrac{1}{2}m\overline{u^2}) \quad \text{and} \quad (\text{KE})_{\text{avg}} = \frac{3}{2}RT$$

Combination of these equations gives

$$N_A(\tfrac{1}{2}m\overline{u^2}) = \frac{3}{2}RT \quad \text{or} \quad \overline{u^2} = \frac{3RT}{N_A m}$$

Taking the square root of both sides of the last equation produces

$$\sqrt{\overline{u^2}} = u_{\text{rms}} = \sqrt{\frac{3RT}{N_A m}}$$

In this expression m represents the mass in kilograms of a single gas particle. When N_A, the number of particles in a mole, is multiplied by m, the product is the mass of a *mole* of gas particles in *kilograms*. We will call this quantity M. Substituting M for $N_A m$ in the equation for u_{rms}, we obtain

$$u_{\text{rms}} = \sqrt{\frac{3RT}{M}}$$

$R = 0.08206 \dfrac{\text{L atm}}{\text{K mol}}$

$R = 8.3145 \dfrac{\text{J}}{\text{K mol}}$

Before we can use this equation, we need to consider the units for R. So far we have used 0.08206 L atm/K mol as the value of R. But to obtain the desired units (meters per second) for u_{rms}, R must be expressed in different units. As we will see in more detail in Chapter 6, the energy unit most often used in the SI system is the joule (J). A **joule** is defined as a kilogram meter squared per second squared (kg m^2/s^2). When R is converted to include the unit of joules, it has the value 8.3145 J/K mol. When R in these units is used in the expression $\sqrt{3RT/M}$, u_{rms} is in the units of meters per second as desired.

Sample Exercise 5.19

Calculate the root mean square velocity for the atoms in a sample of helium gas at 25°C.

Solution

The formula for root mean square velocity is

$$u_{rms} = \sqrt{\frac{3RT}{M}}$$

In this case $T = 25°C + 273 = 298$ K, $R = 8.3145$ J/K mol, and M is the mass of a mole of helium in kg:

$$M = 4.00 \frac{\cancel{g}}{\text{mol}} \times \frac{1 \text{ kg}}{1000 \cancel{g}} = 4.00 \times 10^{-3} \text{ kg/mol}$$

Thus

$$u_{rms} = \sqrt{\frac{3\left(8.3145 \frac{\text{J}}{\cancel{\text{K}} \cancel{\text{mol}}}\right)(298 \cancel{\text{K}})}{4.00 \times 10^{-3} \frac{\text{kg}}{\cancel{\text{mol}}}}} = \sqrt{1.86 \times 10^6 \frac{\text{J}}{\text{kg}}}$$

Since the units of J are kg m^2/s^2, this expression becomes

$$\sqrt{1.86 \times 10^6 \frac{\cancel{\text{kg}} \text{ m}^2}{\cancel{\text{kg}} \text{ s}^2}} = 1.36 \times 10^3 \text{ m/s}$$

Figure 5.14

Path of one particle in a gas. Any given particle will continuously change its course as a result of collisions with other particles, as well as with the walls of the container.

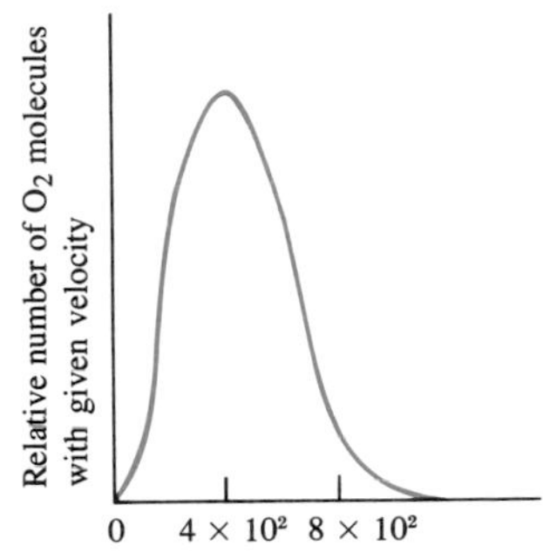

Figure 5.15

A plot of the relative number of O_2 molecules that have a given velocity at STP.

So far we have said nothing about the range of velocities actually found in a gas sample. In a real gas there are large numbers of collisions between particles. For example, as we will see in the next section, when an odorous gas such as ammonia is released in a room, it takes some time for the odor to permeate the air. This delay results from collisions between the NH_3 molecules and O_2 and N_2 molecules in the air, which slow the mixing process.

If the path of a particular gas particle could be monitored, it would look very erratic, something like that shown in Fig. 5.14. The average distance a particle travels between collisions in a particular gas sample is called the *mean free path*. It is typically a very small distance (1×10^{-7} m for O_2 at STP). One effect of the many collisions among gas particles is to produce a large range of velocities as the particles collide and exchange kinetic energy. Although u_{rms} for oxygen gas at STP is approximately 500 meters per second, the majority of O_2 molecules do not have this velocity. The actual distribution of molecular velocities for oxygen gas at STP is shown in Fig. 5.15. This figure shows the relative number of gas molecules having each particular velocity.

We are also interested in the effect of *temperature* on the velocity distribution in a gas. Figure 5.16 shows the velocity distribution for nitrogen gas at three temperatures. Note that as the temperature is increased, the curve peak, which reflects the average velocity, moves toward higher values and the range of velocities becomes much larger.

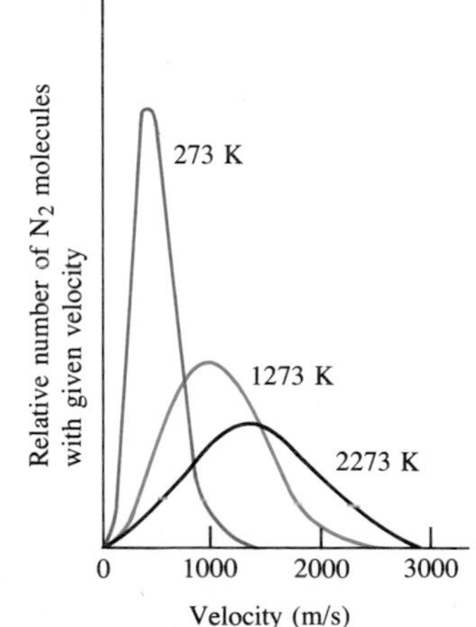

Figure 5.16

A plot of the relative number of N_2 molecules that have a given velocity at three temperatures. Note that as the temperature increases, both the average velocity (reflected by the curve's peak) and the spread of velocities increase.

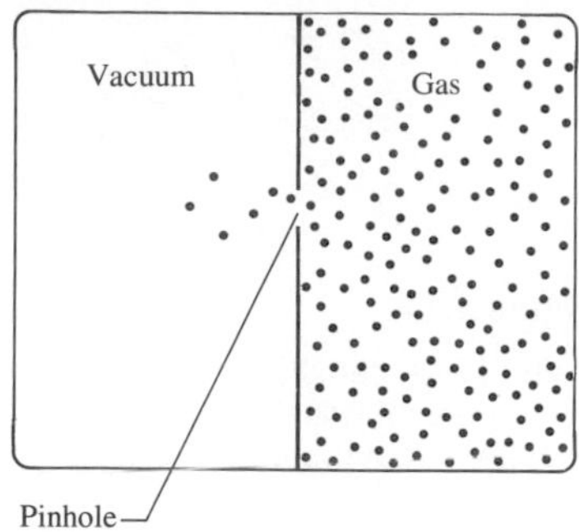

Figure 5.17

The effusion of a gas into an evacuated chamber. The rate of effusion (the rate at which the gas is transferred across the barrier through the pin hole) is inversely proportional to the square root of the molecular weight of the gas.

5.7 Effusion and Diffusion

Purpose

- To describe effusion and diffusion.
- To show the relationship between effusion and molecular weight.

We have seen that the postulates of the kinetic molecular theory combined with the appropriate physical principles produce an equation that successfully fits the experimentally observed behavior of an ideal gas. Two phenomena involving gases provide further tests of this model.

Diffusion is the term used to describe the mixing of gases. When a small amount of pungent-smelling ammonia is released at the front of a classroom, it takes some time before everyone in the room can smell it, because time is required for the ammonia to mix with the air. The rate of diffusion is the rate of the mixing of gases. **Effusion** is the term used to describe the passage of a gas through a tiny orifice into an evacuated chamber, as shown in Fig. 5.17. The rate of effusion measures the speed at which the gas is transferred into the chamber.

Effusion

In Graham's law the units for molecular weight can be g/mol or kg/mol, since the units cancel in the ratio $\sqrt{M_2}/\sqrt{M_1}$.

Thomas Graham (1805–1869), a Scottish chemist, found experimentally that the rate of effusion of a gas is inversely proportional to the square root of the mass of its particles. Stated in another way, the relative rates of effusion of two gases at the same temperature are given by the inverse ratio of the square roots of the masses of the gas particles:

$$\frac{\text{Rate of effusion for gas 1}}{\text{Rate of effusion for gas 2}} = \frac{\sqrt{M_2}}{\sqrt{M_1}}$$

where M_1 and M_2 represent the atomic or molecular weights of the gases as appropriate. This is called **Graham's law of effusion.**

Figure 5.18

An aerial view of the gas diffusion uranium-enriching plant at Oak Ridge, Tennessee.

Sample Exercise 5.20

Calculate the ratio of the effusion rates of hydrogen gas (H_2) and uranium hexafluoride (UF_6), a gas used in the enrichment process to produce fuel for nuclear reactors (see Fig. 5.18).

Solution

First, we need to compute the molecular weights: mol. wt. of H_2 = 2.016, and mol. wt. of UF_6 = 352.02. From Graham's law,

$$\frac{\text{Rate of effusion for } H_2}{\text{Rate of effusion for } UF_6} = \frac{\sqrt{M_{UF_6}}}{\sqrt{M_{H_2}}} = \sqrt{\frac{352.02}{2.016}} = 13.2$$

The effusion rate of the very light H_2 molecules is about 13 times that of the very heavy UF_6 molecules.

Does the kinetic molecular model for gases correctly predict the relative effusion rates of gases summarized by Graham's law? To answer this question we must recognize that the effusion rate for a gas depends directly on the average velocity of its particles. The faster the gas particles are moving, the more likely they are to pass through the effusion orifice. This reasoning leads to the following *prediction* for two gases at the same temperature (T):

$$\frac{\text{Effusion rate for gas 1}}{\text{Effusion rate for gas 2}} = \frac{u_{\text{rms}} \text{ for gas 1}}{u_{\text{rms}} \text{ for gas 2}} = \frac{\sqrt{\dfrac{3RT}{M_1}}}{\sqrt{\dfrac{3RT}{M_2}}} = \frac{\sqrt{M_2}}{\sqrt{M_1}}$$

This is Graham's law, and thus the kinetic molecular model fits the experimental results for the effusion of gases.

Diffusion

Diffusion is frequently illustrated by the lecture demonstration represented in Fig. 5.19, in which two cotton plugs soaked in ammonia and hydrochloric acid are simultaneously placed at the ends of a long tube. A white ring of ammonium chloride (NH_4Cl) forms where the NH_3 and HCl molecules meet several minutes later:

$$NH_3(g) + HCl(g) \rightarrow \underset{\text{White solid}}{NH_4Cl(s)}$$

When $HCl(g)$ and $NH_3(g)$ meet in the tube, a white ring of $NH_4Cl(s)$ forms.

The distances traveled by the NH_3 and HCl molecules are measured, and their ratio is determined. Careful technique produces a value of 1.5 for this ratio, which is exactly the result predicted by the kinetic molecular model. Note that the relative distances traveled by the two types of molecules are directly related to the relative velocities:

$$\frac{\text{Distance traveled by } NH_3}{\text{Distance traveled by HCl}} = \frac{u_{\text{rms}} \text{ for } NH_3}{u_{\text{rms}} \text{ for HCl}} = \sqrt{\frac{M_{HCl}}{M_{NH_3}}} = \sqrt{\frac{36.5}{17}} = 1.5$$

The diffusion of the gases through the tube is surprisingly slow in light of the fact that the velocities of HCl and NH_3 molecules at 25°C are about 450 and 660 meters per second, respectively. Why does it take so long for the NH_3 and HCl molecules to meet? The answer is that the tube contains air and the NH_3 and HCl molecules undergo many collisions with O_2 and N_2 molecules as they travel through the tube. Thus, diffusion is quite complicated to describe theoretically. However, since *relative* diffusion rates of gases depend on the relative velocities of the gas particles, this phenomenon can be treated using the simple kinetic molecular model for gases, as shown by the preceding example.

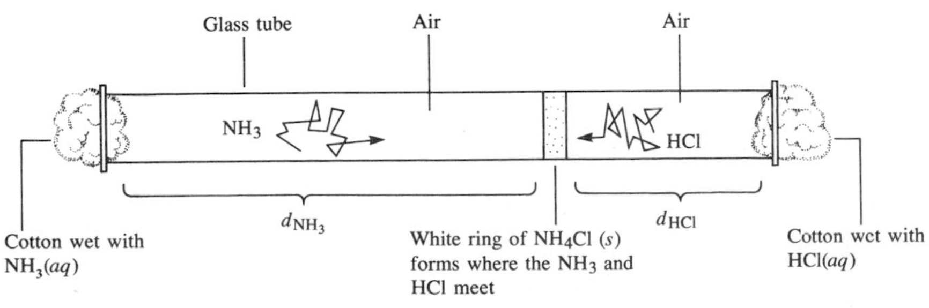

Figure 5.19

A demonstration of the relative diffusion rates of NH_3 and HCl molecules through air. Two cotton plugs, one dipped in $HCl(aq)$ and one dipped in $NH_3(aq)$, are simultaneously inserted into the ends of the tube. Gaseous NH_3 and HCl vaporizing from the cotton plugs diffuse toward each other and, where they meet, react to form $NH_4Cl(s)$.

5.8 Real Gases

Purpose

- To describe how real gases deviate from ideal behavior.
- To show how van der Waals's equation allows for real conditions.

An ideal gas is a hypothetical concept. No gas *exactly* follows the ideal gas law, although many gases come very close at low pressures. Thus ideal gas behavior can best be thought of as the behavior *approached by* ***real gases*** under certain conditions.

We have seen that a very simple model, the kinetic molecular theory, by making some rather drastic assumptions (no interparticle interactions and zero volume for the gas particles), successfully explains ideal behavior. However, it is important to examine real gas behavior to see how it differs from that predicted by the ideal gas law and to determine what modifications are needed in the kinetic molecular theory to explain the observed behavior. Since a model is an approximation and will inevitably fail, we must be ready to learn from such failures. In fact, we often learn more about nature from the failures of our models than from their successes.

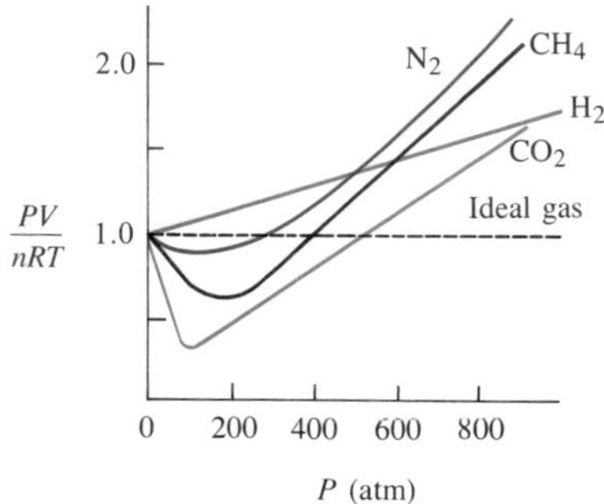

Figure 5.20

Plots of PV/nRT versus P for several gases (200 K). Note the significant deviations from ideal behavior ($PV/nRT = 1$). The behavior is close to ideal only at low pressures (less than 1 atm).

We must first examine the experimentally observed behavior of real gases. This is most often done by measuring pressure, volume, temperature, and number of moles for a gas and observing how the quantity PV/nRT depends on pressure. Plots of PV/nRT versus P are shown for several gases in Fig. 5.20. For an ideal gas PV/nRT equals 1 under all conditions, but notice that for real gases, PV/nRT approaches 1 only at very low pressures. To illustrate the effect of temperature, PV/nRT is plotted versus P for nitrogen gas at several temperatures in Fig. 5.21. Note that the behavior of the gas appears to become more nearly ideal as the temperature is increased. The most important conclusion to be drawn from these figures is that a real gas typically exhibits behavior that is closest to ideal behavior at *low pressures* and *high temperatures*.

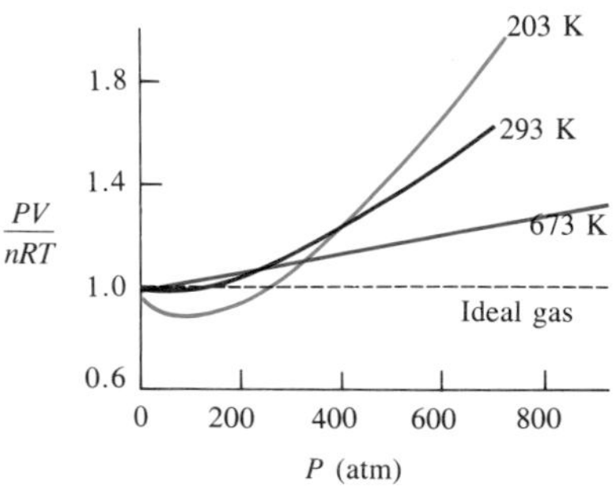

Figure 5.21

Plots of PV/nRT versus P for nitrogen gas at three temperatures. Note that, although nonideal behavior is evident in each case, the deviations are smaller at the higher temperatures.

How can we modify the assumptions of the kinetic molecular theory to fit the behavior of real gases? An equation for real gases was developed in 1873 by Johannes van der Waals, a physics professor at the University of Amsterdam who in 1910 received a Nobel Prize for his work. To follow his analyses, we start with the ideal gas law,

$$P = \frac{nRT}{V}$$

Remember that this equation describes the behavior of a hypothetical gas consisting of volumeless entities that do not interact with each other. In contrast, a real gas consists of atoms or molecules that have finite volumes. This means that the volume available to a given particle in a real gas is less than the volume of the container because the gas particles themselves take up some of the space. To account for this, van der Waals represented the actual volume as the volume of the container, V, minus a correction factor for the volume of the molecules, nb, where n is the number of moles of gas and b is an empirical constant (one determined by fitting the equation to the experimental results). Thus, the volume *actually available* to a given gas molecule is given by the difference $V - nb$.

This modification of the ideal gas equation leads to the equation

$$P' = \frac{nRT}{(V - nb)}$$

P′ is corrected for the finite volume of the particles. The attractive forces have not yet been taken into account.

The volume of the gas particles has now been taken into account.

The next step is to allow for the attractions that occur among the particles in a real gas. The effect of these attractions is to make the observed pressure P_{obs} smaller than it would be if the gas particles did not interact:

$$P_{obs} = (P' - \text{correction factor}) = \left(\frac{nRT}{V - nb} - \text{correction factor}\right)$$

This effect can be understood using the following model. When gas particles come close together, attractive forces occur, which cause the particles to hit the wall with slightly less force than they would in the absence of these interactions (see Fig. 5.22).

The attractive forces among molecules will be discussed in Chapter 10.

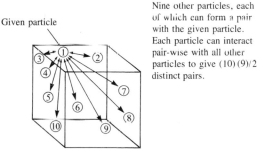

The size of the correction factor will depend on the concentration of gas molecules defined in terms of moles of gas particles per liter (n/V). The higher the concentration, the more likely it will be that a pair of gas particles will be close enough to attract each other. For large numbers of particles, the number of interacting *pairs* of particles depends on the square of the number of particles and thus on the square of the concentration, or $(n/V)^2$. This can be justified as follows: In a gas sample containing N particles, there are $N - 1$ partners available for each particle, as shown in Fig. 5.23. Since the 1· · ·2 pair is the same as the 2· · ·1 pair, this analysis counts each pair twice. Thus for N particles there are $N(N - 1)/2$ pairs. If N is a

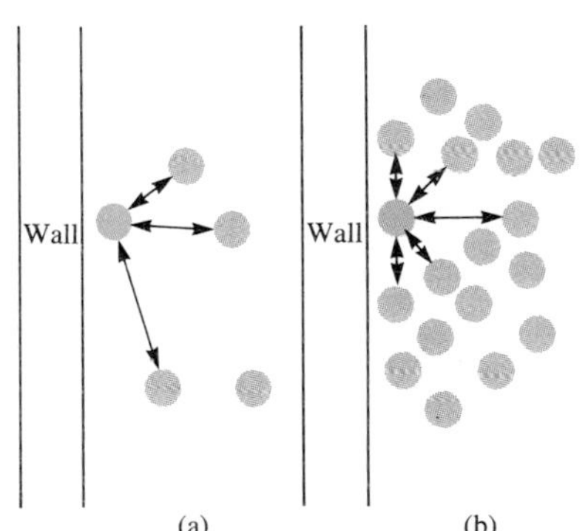

Figure 5.22

(a) Gas at low concentration—relatively few interactions between particles. The indicated gas particle exerts a pressure on the wall close to that predicted for an ideal gas.
(b) Gas at high concentration—many more interactions between particles. The indicated gas particle exerts a much lower pressure on the wall than would be expected in the absence of interactions.

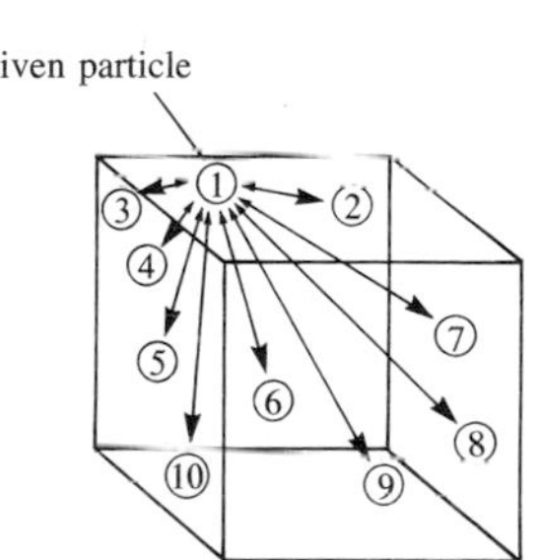

Given particle
Nine other particles, each of which can form a pair with the given particle. Each particle can interact pair-wise with all other particles to give (10)(9)/2 distinct pairs.
Gas sample with 10 particles

Figure 5.23

Illustration of pair-wise interactions among gas particles. In a sample with 10 particles, each particle has 9 possible partners, to give 10(9)/2 = 45 distinct pairs. The factor of $\frac{1}{2}$ arises because when particle 1 is the particle of interest we count the ①· ·② pair, and when particle ② is the particle of interest we count the ②· ·① pair. However ①· ·② and ②· ·① are the same pair that we thus have counted twice. Therefore we must divide by 2 to get the number of pairs.

We have now corrected for both the finite volume and the attractive forces of the particles.

very large number, $N - 1$ approximately equals N, giving $N^2/2$ possible pairs. Thus the correction to the ideal pressure for the attractions of the particles has the form

$$P_{obs} = P' - a\left(\frac{n}{V}\right)^2$$

where a is a proportionality constant (which includes the factor of $\frac{1}{2}$ from $N^2/2$). The value of a for a given real gas can be determined from observing the actual behavior of that gas. Inserting the corrections for both the volume of the particles and the attractions of the particles gives the equation

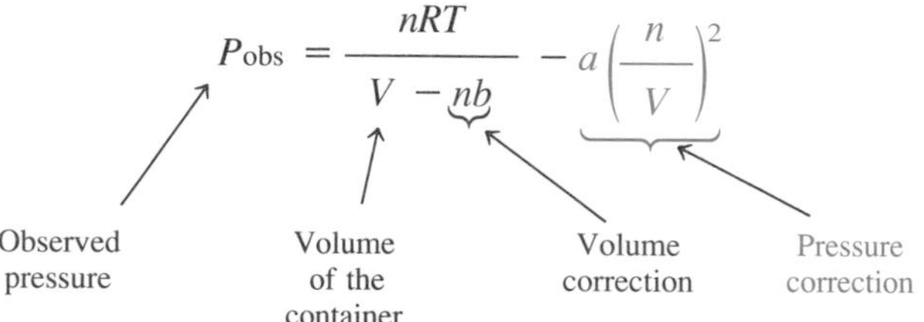

This equation can be rearranged to give the **van der Waals equation:**

P_{obs} is usually called just P.

$$\underbrace{\left[P_{obs} + a\left(\frac{n}{V}\right)^2\right]}_{\substack{\text{Corrected pressure} \\ P_{ideal}}}\underbrace{(V - nb)}_{\substack{\text{Corrected volume} \\ V_{ideal}}} = nRT$$

The values of the weighting factors, a and b, are determined for a given gas by fitting experimental behavior. That is, a and b are varied until the best fit of the observed pressure is obtained under various conditions. The values of a and b for various gases are given in Table 5.3.

Values of the van der Waals Constants for Some Common Gases

Gas	$a\left(\frac{\text{atm L}^2}{\text{mol}^2}\right)$	$b\left(\frac{\text{L}}{\text{mol}}\right)$
He	0.034	0.0237
Ne	0.211	0.0171
Ar	1.35	0.0322
Kr	2.32	0.0398
Xe	4.19	0.0511
H_2	0.244	0.0266
N_2	1.39	0.0391
O_2	1.36	0.0318
Cl_2	6.49	0.0562
CO_2	3.59	0.0427
CH_4	2.25	0.0428
NH_3	4.17	0.0371
H_2O	5.46	0.0305

Table 5.3

Experimental studies indicate that the changes van der Waals made in the basic assumptions of the kinetic molecular theory corrected the major flaws in the model. First, consider the effects of volume. For a gas at low pressure (large volume), the volume of the container is very large compared to the volumes of the gas particles. That is, the volume available to the gas is essentially equal to the volume of the container, and the gas behaves ideally. On the other hand, for a gas at high pressure (small volume), the volume of the particles becomes significant, so that the volume available to the gas is significantly less than the container volume. This is illustrated in Fig. 5.24. It is observed that the volume correction constant b generally increases with the size of the gas molecule. This gives further support to these arguments.

The fact that a real gas tends to behave more ideally at high temperatures can also be explained in terms of the van der Waals model. At high temperatures the particles are moving so rapidly that the effects of interparticle interactions are not very important.

The corrections made by van der Waals to the kinetic molecular theory make physical sense, and this fact makes us confident that we understand the fundamentals of gas behavior at the particle level. This is significant since so much important chemistry takes place in the gas phase. In fact, the mixture of gases called the atmosphere is vital to our existence. In the next section we consider some of the important reactions that occur in the atmosphere.

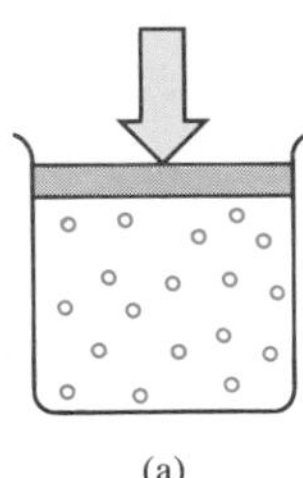

(a)

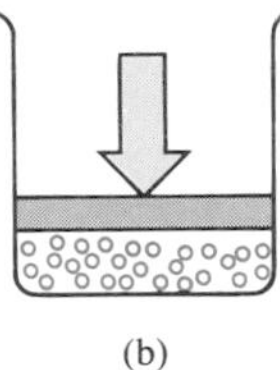

(b)

Figure 5.24

The volume taken up by the gas particles themselves is less important at (a) large container volumes (low pressure) than at (b) small container volumes (high pressure).

5.9 Chemistry in the Atmosphere

Purpose

- To characterize the composition of the atmosphere.
- To describe some of the chemistry of air pollution.

The most important gases to us are those in the **atmosphere** that surrounds the earth's surface. The principal components are N_2 and O_2, but the atmosphere contains many other important gases, such as H_2O and CO_2. The average composition of the earth's atmosphere near sea level, with the water vapor removed, is shown in Table 5.4 on the following page. Because of gravitational effects, the composition of the earth's atmosphere is not constant: heavier molecules tend to be near the earth's surface, and light molecules tend to migrate to higher altitudes and eventually to escape into space. The atmosphere is a highly complex and dynamic system, but for convenience we divide it into several layers based on the way the temperature changes with altitude. (The lowest layer, called the troposphere, is shown in Fig. 5.25 on the following page.) Note that, in contrast to the complex temperature profile of the atmosphere, the pressure decreases in a regular way with increasing altitude.

The chemistry occurring in the higher levels of the atmosphere is mostly determined by the effects of high-energy radiation and particles from the sun and other sources in space. In fact, the upper atmosphere serves as an important shield to prevent this high-energy radiation from reaching the earth, where it would damage the relatively fragile molecules sustaining life. In particular, the ozone in the upper atmosphere helps prevent high-energy ultraviolet radiation from penetrating to the earth. Intensive research is in progress to determine the natural factors that control the ozone concentration and how it is affected by chemicals released into the atmosphere.

The Aurora Borealis over Alaska.

The chemistry occurring in the *troposphere,* the layer of atmosphere closest to the earth's surface, is strongly influenced by human activities. Millions of tons of

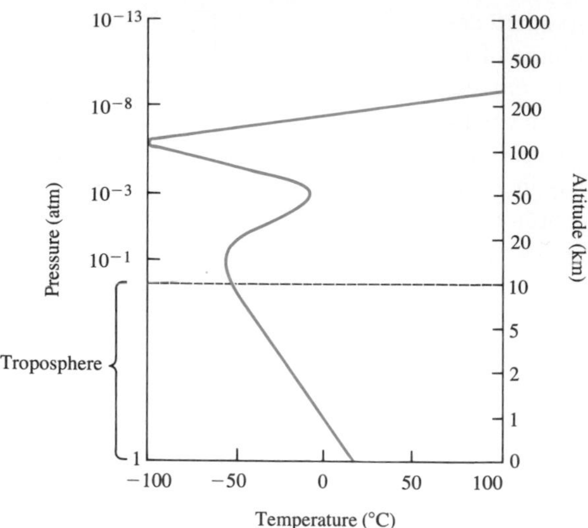

Figure 5.25

The variation of temperature and pressure with altitude. Note that the pressure steadily decreases with altitude, but the temperature increases and decreases.

Component	Mole fraction
N_2	0.78084
O_2	0.20948
Ar	0.00934
CO_2	0.000345
Ne	0.00001818
He	0.00000524
CH_4	0.00000168
Kr	0.00000114
H_2	0.0000005
NO	0.0000005
Xe	0.000000087

Atmospheric Composition near Sea Level (Dry Air)*

*As found in the atmosphere, air contains various amounts of water vapor depending on conditions.

Table 5.4

gases and particulates are released into the troposphere by our highly industrial civilization. Actually, it is amazing that the atmosphere can absorb so much material with relatively small permanent changes.

Significant changes, however, have occurred. Severe **air pollution** is found around many large cities, and it is probable that long-range changes in the planet's weather are taking place. We will discuss some of the long-range effects of pollution in Chapter 6. In this section we will deal with the short-term, localized effects of pollution.

The two main sources of pollution are transportation and production of electricity. The combustion of petroleum in vehicles produces CO, CO_2, NO, and NO_2, along with unburned molecules from petroleum. When this mixture is trapped close to the ground in stagnant air, reactions occur that produce chemicals irritating and harmful to living systems.

The complex chemistry of polluted air appears to center around *nitric oxide* (NO). At the high temperatures in the gasoline and diesel engines of cars and trucks, N_2 and O_2 react to form a small quantity of NO which is emitted into the air with the exhaust gases (see Fig. 5.26). This NO is oxidized in air to NO_2 which in turn absorbs energy from sunlight and breaks up into nitric oxide and free oxygen atoms:

$$NO_2(g) \xrightarrow[\text{energy}]{\text{Radiant}} NO(g) + O(g)$$

Oxygen atoms are very reactive and can combine with O_2 to form *ozone:*

$$O(g) + O_2(g) \rightarrow O_3(g)$$

Ozone is also very reactive. It can react with the unburned hydrocarbons in the polluted air to produce chemicals that cause the eyes to water and burn and are harmful to the respiratory system.

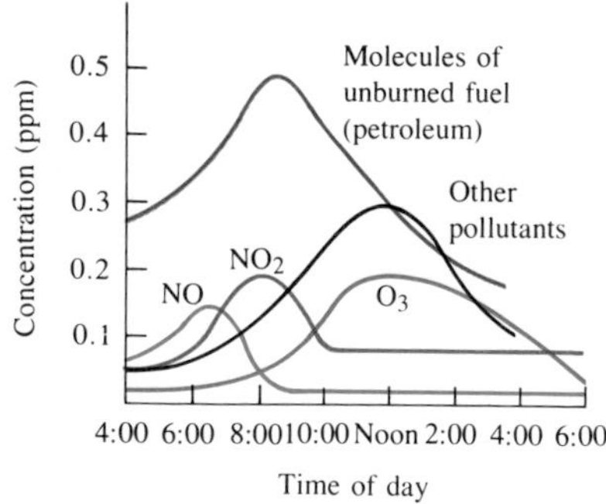

Figure 5.26

Concentration (in molecules per million molecules of "air") of some smog components versus time of day. After P. A. Leighton, "Photochemistry of Air Pollution" in *Physical Chemistry: A Series of Monographs,* ed. Eric Hutchinson and P. Van Rysselberghe, Vol. IX, New York: Academic Press, 1961.

The end-product of this whole process is often referred to as **photochemical smog,** so called because light is required to initiate some of the reactions. The production of photochemical smog can be more clearly understood by examining as a group the reactions discussed above:

$$NO_2(g) \rightarrow NO(g) + O(g)$$
$$O(g) + O_2(g) \rightarrow O_3(g)$$
$$NO(g) + \tfrac{1}{2}O_2(g) \rightarrow NO_2(g)$$
$$\text{Net reaction:} \quad \tfrac{3}{2}O_2(g) \rightarrow O_3(g)$$

Although represented here as O_2, the actual oxidant is an organic peroxide such as CH_3COO, formed by reaction of O_2 with organic pollutants.

Note that the NO_2 molecules assist in the formation of ozone without being themselves used up. The ozone then produces other pollutants.

We can observe this process by analyzing polluted air at various times during a day (see Fig. 5.26). As people drive to work between 6 and 8 a.m., the amounts of NO, NO_2, and unburned molecules from petroleum increase. Later as the decomposition of NO_2 occurs, the concentration of ozone and other pollutants builds up. Current efforts to combat the formation of photochemical smog are focused on cutting down the amounts of molecules from unburned fuel in automobile exhaust and designing engines that produce less nitric oxide (see Fig. 5.27).

Figure 5.27

Our various modes of transportation produce large amounts of nitrogen oxides, which facilitate the formation of photochemical smog.

The other major source of pollution results from burning coal to produce electricity. Much of the coal found in the Midwest contains significant quantities of sulfur, which, when burned, produces sulfur dioxide:

$$S \text{ (In coal)} + O_2(g) \rightarrow SO_2(g)$$

A further oxidation reaction occurs when sulfur dioxide is changed to sulfur trioxide in the air:*

$$2SO_2(g) + O_2(g) \rightarrow 2SO_3(g)$$

*This reaction is very slow unless solid particles are present. See Chapter 12 for a discussion.

Figure 5.28

Trees in North Carolina damaged by acid rain.

The production of sulfur trioxide is significant because it can combine with droplets of water in the air to form sulfuric acid:

$$SO_3(g) + H_2O(l) \rightarrow H_2SO_4(aq)$$

Sulfuric acid is very corrosive to both living things and building materials. Another result of this type of pollution is **acid rain** (see Fig. 5.28). In many parts of the northeastern United States and southeastern Canada, the acid rain has caused some freshwater lakes to become too acidic for fish to live.

The problem of sulfur dioxide pollution is made more complicated by the energy crisis. As petroleum supplies dwindle and the price increases, our dependence on coal will grow. As supplies of low-sulfur coal are used up, high-sulfur coal will be utilized. One way to use high-sulfur coal without further harming the air quality is to remove the sulfur dioxide from the exhaust gas by means of a system called a *scrubber* before it is emitted from the power plant stack. A common method of *scrubbing* is to blow powdered limestone ($CaCO_3$) into the combustion chamber, where it is decomposed to lime and carbon dioxide:

$$CaCO_3(s) \rightarrow CaO(s) + CO_2(g)$$

The lime then combines with the sulfur dioxide to form calcium sulfite:

$$CaO(s) + SO_2(g) \rightarrow CaSO_3(s)$$

To remove the calcium sulfite and any remaining unreacted sulfur dioxide, an aqueous suspension of lime is injected into the combustion chamber and the stack to produce a *slurry* (a thick suspension) as shown in Fig. 5.29.

Unfortunately, there are many problems associated with scrubbing. The systems are complicated and expensive and consume a great deal of energy. The large quantities of calcium sulfite produced in the process present a disposal problem. With a typical scrubber approximately 1 ton of calcium sulfite per year is produced per person served by the power plant. Since no use has yet been found for this calcium sulfite, it is usually buried in a landfill. As a result of these difficulties, air pollution by sulfur dioxide continues to be a major problem, one that is expensive in terms of damage to the environment and human health as well as in monetary terms.

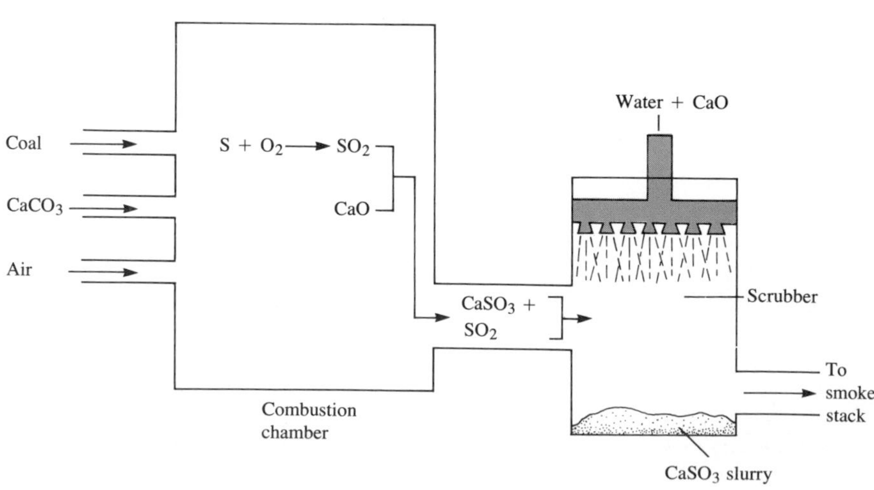

Figure 5.29

A schematic diagram of the process for scrubbing sulfur dioxide from stack gases in power plants.

Chemical Impact

Acid Rain: A Growing Problem

Rainwater, even in pristine, wilderness areas is slightly acidic, because some of the carbon dioxide present in the atmosphere dissolves in the raindrops to produce H^+ ions by the following reaction:

$$H_2O(l) + CO_2(g) \rightarrow H^+(aq) + HCO_3^-(aq)$$

This process produces only very small concentrations of H^+ ion in the rainwater. However, gases such as NO_2 and SO_2, which are by-products of energy use, can produce significantly higher H^+ concentrations. Nitrogen dioxide reacts with water to give a mixture of nitrous acid and nitric acid

$$2NO_2(g) + H_2O(l) \rightarrow HNO_2(aq) + HNO_3(aq)$$

Sulfur dioxide is oxidized to sulfur trioxide, which then reacts with water to form sulfuric acid

$$2SO_2(g) + O_2(g) \rightarrow 2SO_3(g)$$

$$SO_3(g) + H_2O(l) \rightarrow H_2SO_4(aq)$$

The damage caused by the acid formed in polluted air is a growing worldwide problem. Lakes are dying in Norway, the forests are sick in Germany, and buildings and statues are deteriorating all over the world.

For example, the Field Museum in Chicago contains more white Georgia marble than any other structure in the world. But nearly seventy years of exposure to the elements has taken such a toll on it that the building is now undergoing a multi-million-dollar renovation to replace the damaged marble with freshly quarried material.

What is the chemistry of the deterioration of marble by sulfuric acid? Marble is produced by geological processes at high temperatures and pressures from limestone, a sedimentary rock formed by slow deposition of calcium carbonate from the shells of marine organisms. Limestone and marble are chemically identical ($CaCO_3$) but differ in physical properties because limestone is composed of smaller particles of calcium carbonate and is thus more porous and more workable. Although both limestone and marble are used for buildings, marble can be polished to a higher sheen and is often preferred for decorative purposes.

Both marble and limestone react with sulfuric acid to form calcium sulfate. The process can be represented most simply as

$$CaCO_3(s) + H_2SO_4(aq) \rightarrow Ca^{2+}(aq) + SO_4^{2-}(aq) + H_2O(l) + CO_2(g)$$

In this equation the calcium sulfate is represented by separate hydrated ions because calcium sulfate is quite water soluble and dissolves in rainwater. Thus in areas bathed by rainwater the marble slowly dissolves away.

In areas of the building protected from the rain, the calcium sulfate can form the mineral gypsum, $CaSO_4 \cdot 2H_2O$. The $\cdot 2H_2O$ in the formula of gypsum indicates the presence of two water molecules (called waters of hydration) for each $CaSO_4$ formula unit in the solid. The smooth surface

Sculpture on the Field Museum of Chicago marred by acid rain and soot.

of the marble is thus replaced by a thin layer of gypsum, a more porous material that binds soot and dust.

What can be done to protect limestone and marble structures from this kind of damage? Of course one approach is to lower sulfur dioxide emissions from power plants (Fig. 5.29). In addition, scientists are experimenting with coatings to protect marble from the acidic atmosphere. However, a coating can do more harm than good unless it "breathes." If moisture trapped beneath the coating freezes, the expanding ice can fracture the marble. Needless to say, it is difficult to find a coating that will allow water to pass but not acid, but the search continues.

Suggested Reading

A. Elena Charola, "Acid Rain Effects on Stone Monuments," *J. Chem. Ed.* **64** (1987), p. 436.

FOR REVIEW

Summary

Of the three states of matter, the gaseous state is the most easily understood experimentally and theoretically. A gas can be described quantitatively by specifying four properties: volume, pressure, temperature, and the amount of gas present.

Atmospheric pressure is measured with a barometer, and in the laboratory gas pressure is usually measured with a manometer. Both devices employ columns filled with mercury, and the most commonly used units of pressure are based on the height of such a column in millimeters: 1 mm Hg equals 1 torr; 1 standard atmosphere equals 760 torr. In the SI system, the unit of pressure is the pascal (1 N/m^2).

Early experiments on the behavior of gases dealt with the relationship among the four characteristic properties. Boyle's law states that the volume of a given amount of a gas is inversely proportional to its pressure (at constant temperature). Charles's law states that, for a given amount of gas at a constant pressure, the volume is directly proportional to the temperature in Kelvins. At a temperature of −273.2°C (0 K), the volume of a gas extrapolates to zero, and this temperature is called absolute zero. Avogadro's law states that equal volumes of gases at the same temperature and pressure contain the same number of particles.

These three laws can be combined into the ideal gas law:

$$PV = nRT$$

where R is called the universal gas constant (0.08206 L atm/K mol). This equation allows calculation of any one of the characteristic properties, given the other three. The ideal gas law describes the limiting behavior approached by real gases at low pressures and high temperatures.

The molar volume (22.42 liters) is the volume of an ideal gas at STP (standard temperature and pressure) or 0°C and 1 atmosphere.

The pressure of a gaseous mixture is described by Dalton's law of partial pressures, which states that the total pressure of a mixture of gases in a container is the sum of the pressures that each gas would exert if it were alone (its partial pressure), that is,

$$P_{\text{TOTAL}} = P_1 + P_2 + P_3 + \cdots$$

The mole fraction of a given component in a mixture of gases can be calculated from the ratio of the partial pressure of that component to the total gas pressure.

The kinetic molecular theory of gases is a theoretical model constructed to account for ideal gas behavior. This model assumes that the volume of the gas particles is zero, that there are no interactions between particles, and that the particles are in constant motion, colliding with the container walls to produce pressure. In addition, this model shows that the average kinetic energy of the gas particles is directly proportional to the Kelvin temperature of the gas.

The root mean square velocity for the particles in a gas can be calculated from the temperature and the molecular weight of the gas. Within any gas sample there is a range of velocities, since the gas particles collide and exchange kinetic energy. An increase in temperature increases the average velocity of the gas particles. Relative rates of diffusion, the mixing of gases, and effusion, the passage of a gas through a small orifice into an empty chamber, are inversely proportional to the square roots of the atomic or molecular weights of the gases in question.

Real gases behave ideally only at high temperatures and low pressures. The van der Waals equation is a modification of the ideal gas equation that takes into consideration that real gas particles take up volume and interact with each other.

Key Terms

Section 5.1
barometer
manometer
mm Hg
torr
standard atmosphere
pascal

Section 5.2
Boyle's law
ideal gas
Charles's law
absolute zero
Avogadro's law

Section 5.3
universal gas constant
ideal gas law

Section 5.4
molar volume
standard temperature and pressure (STP)

Section 5.5
Dalton's law of partial pressures
partial pressure
mole fraction

Section 5.6
kinetic molecular theory
root mean square velocity
joule

Section 5.7
diffusion
effusion
Graham's law of effusion

Section 5.8
real gas
van der Waals equation

Section 5.9
atmosphere
air pollution
photochemical smog
acid rain

Exercises

A blue exercise number indicates that the answer to that exercise appears at the back of this book and a solution appears in the Solutions Guide.

Pressure

1. Atmospheric pressures announced during weather forecasts are typically given in inches of mercury. Calculate the value of 1 standard atmosphere in inches of mercury.

2. A gauge on a compressed gas cylinder reads 2200 psi (pounds per square inch; 1 atm = 14.7 psi). Express this pressure in each of the following units:
   a. standard atmospheres
   b. megapascals (MPa)
   c. torr

3. A sealed-tube manometer as shown below can be used to measure pressures below atmospheric pressure. The tube above the mercury is evacuated. When there is a vacuum in the flask, the mercury levels in both arms of the U-tube are equal. If a gaseous sample is introduced into the flask, the mercury levels are different. The difference, h, is a measure of the pressure of the gas inside the flask. If h is equal to 6.5 cm, calculate the pressure in the flask in torr, pascals, and atmospheres.

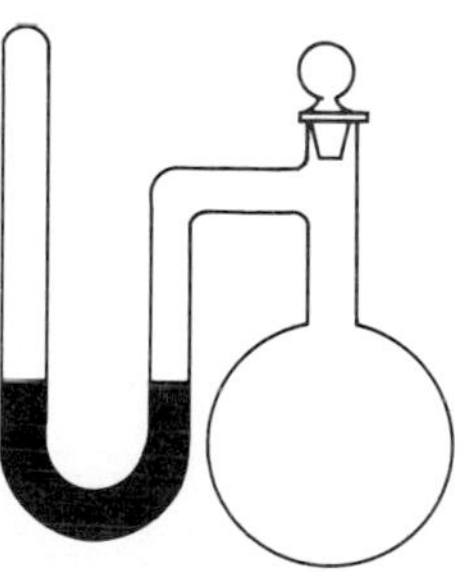

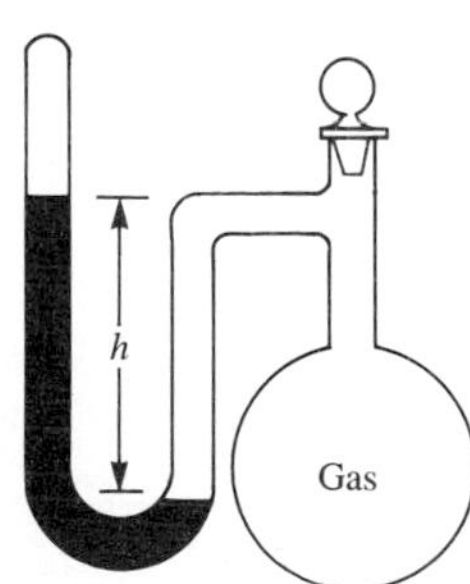

4. Freon-12 (CF_2Cl_2) is commonly used as the refrigerant in central home air conditioners. The system is initially charged to a pressure of 70 psi. Express this pressure in each of the following units (remember that 1 atm = 14.7 psi):
   a. mm Hg
   b. atm
   c. Pa
   d. kPa
   e. MPa

5. A diagram for an open-tube manometer is shown below.

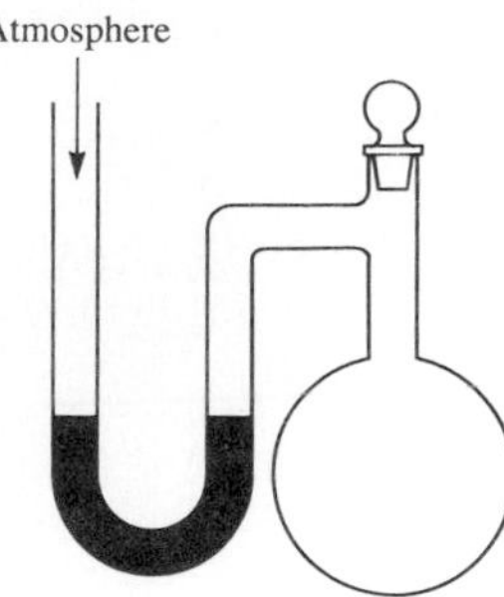

If the flask is open to the atmosphere, the mercury levels are equal. For each of the following situations where a gas is contained in the flask, calculate the pressure in the flask in torr, atmospheres, and pascals.

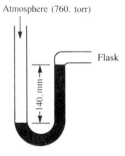

a. Atmosphere (760. torr)

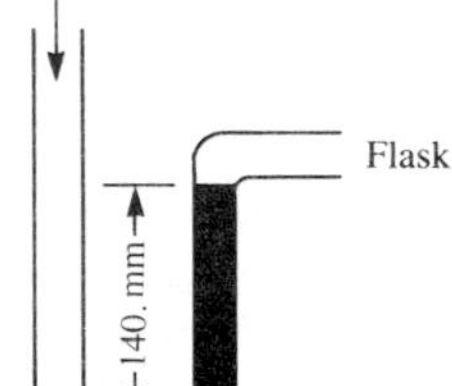

b. Atmosphere (760. torr)

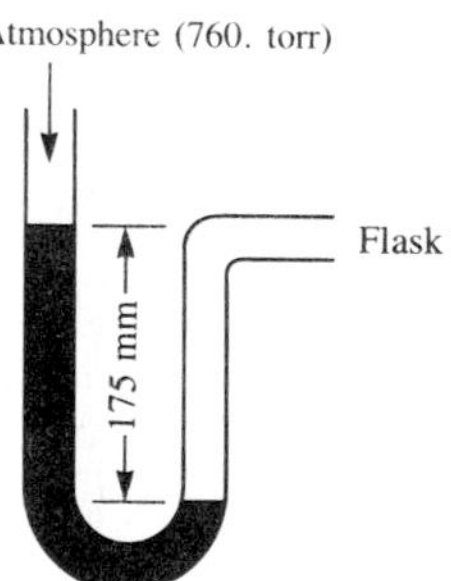

c. Calculate the pressures in the flask in parts a and b (in torr) if the atmospheric pressure is 635 torr.

6. The gravitational force exerted by an object is given by

$$F = mg$$

where F is the force in newtons, m is the mass in kilograms, and g is the acceleration due to gravity, or 9.81 m/s^2. Calculate the force exerted per unit of area by a column of mercury (density = 13.59 g/cm^3) 76.0 cm high. How high would a column of water (density = 1.00 g/cm^3) have to be to exert the same force?

7. If the open-tube manometer in Exercise 5 contains a nonvolatile silicone oil (density = 1.30 g/cm^3) instead of mercury, what are the pressures in the flask as shown in parts (a) and (b) in torr, atmospheres, and pascals?

8. What advantage would there be in using a less dense fluid than mercury in a manometer used to measure relatively small differences in pressure? (See Exercise 7.)

Gas Laws

9. Calculate the value of the universal gas constant, R, in each of the following units:
   a. $\dfrac{\text{L Pa}}{\text{K mol}}$
   b. $\dfrac{\text{m}^3\text{ Pa}}{\text{K mol}}$
   c. What other SI unit can replace m^3 Pa in part b?

10. Draw a qualitative graph to show how the first property varies with the second in each of the following (assume one mole of an ideal gas):
    a. PV versus V with constant T
    b. P versus T with constant V
    c. T versus V with constant P
    d. P versus V with constant T
    e. P versus $1/V$ with constant T
    f. PV/T versus P

11. Consider the flask diagrammed below. What are the final partial pressures of H_2 and N_2 after the stopcock between the two flasks is opened? (Assume the final volume is 1.5 L.) What is the total pressure (in torr)?

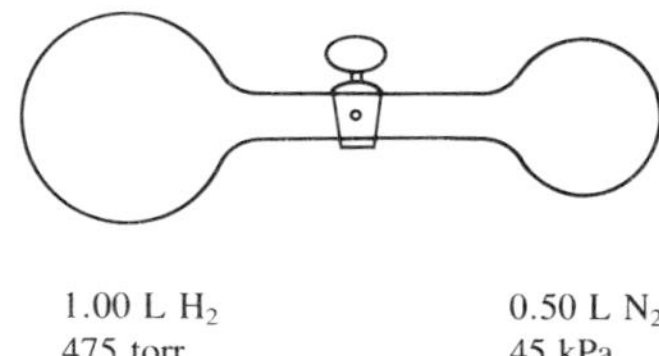

12. A steel cylinder contains 150 mol of argon gas at a temperature of 25°C and a pressure of 7.5 MPa. After some of the argon has been used, the pressure is 2.0 MPa at a temperature of 19°C. What mass of argon remains in the cylinder?

13. A 5.0-L flask contains 0.6 g of O_2 at a temperature of 22°C. What is the pressure (in atm) inside the flask?

14. A balloon is filled with helium gas at 25°C. The balloon expands until the pressure is equal to the barometric pressure of 720. torr. The balloon rises to an altitude of 6000 ft where the pressure is 605 torr and the temperature is 15°C. What is the change in the volume of the balloon as it ascends to 6000 ft?

15. A 2.00-L flask contains helium gas at a pressure of 685 torr and a temperature of 0°C. What would be the pressure in the flask if each of the following occurred?
a. The temperature is increased to 150.°C.
b. The temperature is lowered to 77 K.

16. A piece of solid carbon dioxide, with a mass of 7.8 g, is placed in a 4.0-L otherwise empty container at 27°C. What is the pressure in the container after all of the carbon dioxide vaporizes? If 7.8 g of solid carbon dioxide were placed in the same container but it already contained air at 740 torr, what would be the partial pressure of carbon dioxide and the total pressure in the container after the carbon dioxide vaporizes?

17. A balloon is filled to a volume of 7.00×10^2 mL at a temperature of 20.0°C. The balloon is then cooled to a temperature of 1.00×10^2 K. What is the final volume?

18. An ideal gas is in a cylinder with a volume of 5.0×10^2 mL at a temperature of 30.0°C and a pressure of 710 torr. The gas is compressed to a volume of 25 mL, and the temperature is raised to 820°C. What is the new pressure?

19. The steel bomb of a bomb calorimeter, which has a volume of 75 mL, is charged with oxygen to a pressure of 145 atm at 22°C. Calculate the moles of oxygen in the bomb.

20. Complete the following table for an ideal gas:

	P (atm)	$V(L)$	$n(mol)$	T
a.	5.00	—	2.00	155°C
b.	0.300	2.00	—	155 K
c.	4.47	25.0	2.01	—
d.	—	2.25	10.5	75°C

21. Complete the following table for an ideal gas:

	P	V	n	T
a.	875 torr	275 mL	0.0105 mol	—
b.	—	0.100 L	0.200 mol	38°C
c.	2.50 atm	—	3.00 mol	565°C
d.	688 torr	986 mL	—	565 K

22. A container is filled with an ideal gas to a pressure of 40.0 atm at 0°C.
a. What will be the pressure in the container if it is heated to 45°C?
b. At what temperature would the pressure be 1.50×10^2 atm?
c. At what temperature would the pressure be 25.0 atm?

23. A 2.50-L container is filled with 175 g of argon.
a. If the pressure is 10.0 atm, what is the temperature?
b. If the temperature is 225 K, what is the pressure?

24. Calculate the volume occupied by 1.00 mol of an ideal gas at STP.

25. A compressed gas cylinder, at 13.7 MPa and 23°C, is in a room where a fire raises the temperature to 450°C. What is the new pressure in the cylinder?

26. A bicycle tire is filled with air to a pressure of 610 kPa, at a temperature of 19°C. Riding the bike on asphalt on a hot day increases the temperature of the tire to 58°C. The volume of the tire increases by 4.0%. What is the new pressure in the bicycle tire?

27. A hot-air balloon is filled with air to a volume of 4.00×10^3 m^3 at 745 torr and 21°C. The air in the balloon is then heated to 62°C, and the balloon expands to a volume of 4.20×10^3 m^3. What is the ratio of the number of moles of air in the heated balloon to the original number of moles of air in the balloon? *Hint:* openings in the balloon allow air to flow in and out. Thus the pressure in the balloon is always the same as that of the atmosphere.

28. A sample of oxygen gas is collected over water at 25°C and a total pressure of 641 torr. The volume of gas collected is 500.0 mL. What mass of oxygen is collected? (At 25°C the vapor pressure of water is 23.8 torr.)

29. At 0°C, a 1.0-L flask contains 5.0×10^{-2} mol of N_2, 1.5×10^2 mg of O_2, and NO at a concentration of 5.0×10^{18} molecules/cm^3. What is the partial pressure of each gas, and what is the total pressure in the flask?

30. At STP, 1.0 L of Br_2 reacts completely with 3.0 L of F_2, producing 2.0 L of a product. What is the formula of the product? (All substances are gases.)

31. A sample of nitrogen gas was collected over water at 20°C and a total pressure of 1.00 atm. A total volume of 2.50×10^2 mL was collected. What mass of nitrogen was collected? (At 20.°C the vapor pressure of water is 17.5 torr.)

32. A 0.586 g sample of helium is collected over water at 25°C and 1.0 atm total pressure. What volume is occupied by the wet helium gas? (At 25°C the vapor pressure of water is 23.8 torr.)

33. A mixture of 1.00 g of H_2 and 1.00 g of He exerts a pressure of 0.480 atm. What is the partial pressure of each gas present in the mixture?

34. The partial pressure of $CH_4(g)$ is 0.175 atm and $O_2(g)$ is 0.250 atm in a mixture of the two gases.
a. What is the mole fraction of each gas in the mixture?
b. If the mixture occupies a volume of 10.5 L at 65°C, calculate the total number of moles of gas in the mixture.
c. Calculate the number of grams of each gas in the mixture.

Gas Density, Molecular Weight, and Reaction Stoichiometry

35. Silicon tetrachloride, $SiCl_4$, and trichlorosilane, $SiHCl_3$, are both starting materials for the production of electronics grade silicon. Calculate the densities of pure $SiCl_4$ and pure $SiHCl_3$ vapor at 85°C and 758 torr.

36. Calculate the density of ammonia gas at 27°C and 635 torr.

37. A compound contains only nitrogen and hydrogen and is 87.4% nitrogen by mass. A gaseous sample of the compound has a density of 0.977 g/L at 710. torr and 100°C. What is the molecular formula of the compound?

38. A compound has the empirical formula CHCl. A 256-mL flask, at 373 K and 750 torr, contains 0.80 g of the gaseous compound. Give the molecular formula.

39. One of the chemical controversies of the 19th century concerned the element beryllium (Be). Berzelius originally claimed that beryllium was a trivalent element (forming Be^{3+} ions) and that it gave an oxide with the formula Be_2O_3. This resulted in a calculated atomic weight of 13.5 for beryllium. In formulating his periodic table, Mendeleev proposed that beryllium was divalent (forming Be^{2+} ions) and that it gave an oxide with the formula BeO. This assumption gives an atomic weight of 9.0. In 1894 A. Combes (*Comptes Rendes* 1894, p. 1221) reacted beryllium with the anion $C_5H_7O_2^-$ and measured the density of the gaseous product. Combes's data for two different experiments are as follows:

	I	II
Mass	0.2022 g	0.2224 g
Volume	22.6 cm^3	26.0 cm^3
Temperature	13°C	17°C
Pressure	765.2 mm Hg	764.6 mm Hg

If beryllium is a divalent metal, the molecular formula of the product will be $Be(C_5H_7O_2)_2$; if it is trivalent, the formula will be $Be(C_5H_7O_2)_3$. Show how Combes's data help to confirm that beryllium is a divalent metal.

40. Discrepancies in the experimental values of the molecular weight of nitrogen provided some of the first evidence for the existence of the noble gases. If pure nitrogen is collected from the decomposition of ammonium nitrite,

$$NH_4NO_2(s) \xrightarrow{\text{Heat}} N_2(g) + 2H_2O(g)$$

its measured molecular weight is 28.01. If O_2, CO_2, and H_2O are removed from air, the remaining gas has an average molecular weight of 28.15. Assuming this discrepancy is solely a result of contamination with argon (atomic wt. = 39.95), calculate the ratio of moles of Ar to moles of N_2 in air.

41. Oxygen gas can be produced in small quantities in the laboratory from the thermal decomposition of potassium chlorate:

$$2KClO_3(s) \xrightarrow{\text{Heat}} 2KCl(s) + 3O_2(g)$$

If 3.7 g of $KClO_3$ is heated, what volume of gas will be collected over water at 27°C and 735 torr? (At 27°C the vapor pressure of water is 26.7 torr.)

42. In 1983 production of ammonia in the United States totaled 27.37×10^9 lb (*Chemical & Engineering News,* May 7, 1984). Assuming 100% yield, calculate the volumes of N_2 and H_2 measured at STP required to produce this much ammonia by the reaction:

$$N_2(g) + 3H_2(g) \rightarrow 2NH_3(g)$$

43. Urea (H_2NCONH_2) is used extensively as a nitrogen source in fertilizers. It is produced commercially from the reaction of ammonia and carbon dioxide.

$$2NH_3(g) + CO_2(g) \xrightarrow[\text{Pressure}]{\text{Heat}} H_2NCONH_2(s) + H_2O(g)$$

Ammonia is added to an evacuated 5.0-L flask, to a pressure of 9.0 atm, at a temperature of 23°C. Carbon dioxide is added to the ammonia, to give a total pressure of 14.0 atm. The flask is then heated. What mass of urea is produced in this reaction, assuming 100% yield?

44. Calculate the volume of O_2, at STP, required for the complete combustion of 125 g of octane (C_8H_{18}) to CO_2 and H_2O.

45. The method used by Joseph Priestley to obtain oxygen made use of the thermal decomposition of mercuric oxide:

$$2HgO(s) \xrightarrow{\text{Heat}} 2Hg(l) + O_2(g)$$

What volume of oxygen gas, measured at 30°C and 725 torr, can be produced from the complete decomposition of 4.1 g of mercuric oxide?

46. Hydrogen cyanide is commercially prepared by the reaction of methane, $CH_4(g)$, ammonia, $NH_3(g)$, and oxygen, $O_2(g)$, at high temperature. The other product is gaseous water.

a. Write a chemical equation for the reaction.
b. What volume of HCN(g) can be obtained from 20.0 L $CH_4(g)$, 20.0 L $NH_3(g)$, and 20.0 L $O_2(g)$? The volumes of all gases are measured at the same temperature and pressure.

47. Xenon and fluorine will react to form binary compounds when a mixture of the two gases is heated to 400°C in a nickel reaction vessel. A 100.0-mL nickel container is filled with xenon and fluorine, to partial pressures of 1.24 atm and 10.10 atm, respectively, and a temperature of 25°C. The reaction vessel is heated to 400°C and then cooled to a temperature at which F_2 is a gas and the xenon fluorides are nonvolatile solids. The remaining F_2 gas is transferred to another 100.0-

mL nickel container where the pressure of F_2 at 25°C is 7.62 atm. Assuming all of the xenon has reacted, what is the formula of the product?

48. The nitrogen content of organic compounds can be determined by the Dumas method. The compound in question is first reacted by passage over hot CuO(*s*):

$$\text{Compound} \xrightarrow[\text{CuO}(s)]{\text{Hot}} N_2(g) + CO_2(g) + H_2O(g)$$

The product gas is then passed through a concentrated solution of KOH to remove the CO_2. After passage through the KOH solution the gas contains N_2 and is saturated with water vapor. In a given experiment a 0.253-g sample of a compound produced 31.8 mL of N_2 saturated with water vapor at 25°C and 726 torr. What is the mass percent nitrogen in the compound? (The vapor pressure of water at 25°C is 23.8 torr.)

49. An organic compound contains C, H, N, and O. Combustion of 0.1023 g of the compound yielded 0.2766 g of CO_2 and 0.0991 g of H_2O. A sample of 0.4831 g of the compound was analyzed for nitrogen by the Dumas method. At STP, 27.6 mL of dry N_2 was obtained. In a third experiment, the density of the compound as a gas was found to be 4.02 g/L at 127°C and 256 torr. What are the empirical formula and the molecular formula of the compound?

50. Consider the first step in the industrial production of nitric acid:

$$4NH_3(g) + 5O_2(g) \rightarrow 4NO(g) + 6H_2O(g)$$

a. What volume of NO, measured at 1.0 atm and 1000°C, can be produced from 10.0 L of NH_3 and excess O_2 measured at the same temperature and pressure?
b. What volume of O_2 measured at STP is consumed in reacting with 10.0 kg of NH_3?
c. What mass of NO is produced from the reaction of 5.00×10^2 L of NH_3, measured at 250.°C and 3.00 atm, with excess O_2?
d. What mass of H_2O is produced from the reaction of 65 L of NH_3 and 75 L of O_2, both measured at STP?
e. How many moles of NO are produced from the mixture in part d?

Kinetic Molecular Theory and Real Gases

51. Using the postulates of the kinetic molecular theory, give a molecular interpretation of Boyle's law, Charles's law, and Dalton's law of partial pressures.

52. Calculate the average kinetic energies of CH_4 and N_2 molecules, at 273 K and 546 K.

53. Calculate the root mean square velocities of CH_4 and N_2 molecules, at 273 K and 546 K.

54. Do all of the molecules in a one-mole sample of $CH_4(g)$ have the same energy at 273 K?

55. Consider three identical flasks filled with different gases.

Flask A: CO at 760 torr and 0°C
Flask B: N_2 at 250 torr and 0°C
Flask C: H_2 at 100 torr and 0°C

a. In which flask will the molecules have the greatest average kinetic energy?
b. In which flask will the molecules have the greatest root mean square velocity?
c. Which flask will have the greatest number of collisions per second with the walls of the container?

56. One way of separating oxygen isotopes is by gaseous diffusion of carbon monoxide. Calculate the relative rates of diffusion of $^{12}C^{16}O$, $^{12}C^{17}O$, and $^{12}C^{18}O$. Name some advantages and disadvantages of separating oxygen isotopes by gaseous diffusion of carbon dioxide instead of carbon monoxide.

57. Consider a 1.0-L container of neon gas at STP. Will the average kinetic energy, root mean square velocity, frequency of collisions of gas molecules with each other, frequency of collisions of gas molecules with the walls of the container, and energy of impact of gas molecules with the container increase, decrease, or remain the same under each of the following conditions:

a. The temperature is increased to 100°C.
b. The temperature is decreased to 50°C.
c. The volume is decreased to 0.5 L.
d. The number of moles of neon is doubled.

58. The diffusion rate of an unknown gas is measured and found to be 31.50 mL/min. Under identical experimental conditions, the diffusion rate of O_2 is found to be 30.50 mL/min. If the choices are CH_4, CO, NO, CO_2, and NO_2, what is the identity of the unknown gas?

59. The rate of diffusion of a particular gas was measured to be 24.0 mL/min. Under the same conditions the rate of diffusion of pure methane, CH_4, gas is 47.8 mL/min. What is the molecular weight of the unknown gas?

60. It took 4.5 minutes for 1.0 L of helium to effuse through a porous barrier. How long will it take for 1.0 L of Cl_2 gas to effuse under identical conditions?

61. Calculate the pressure exerted by 0.5000 mol of N_2 in a 1.000-L container at 25.0°C

a. using the ideal gas law
b. using the van der Waals equation
c. Compare the results.

62. Calculate the pressure exerted by 0.5000 mol of N_2 in a 10.00 L container at 25.0°C

a. using the ideal gas law
b. using the van der Waals equation
c. Compare the results.
d. Compare the results to those in the previous problem.

63. Use values of the van der Waals constants for NH_3 (Table 5.3) and show how much of its deviation from ideal behavior can be accounted for by the van der Waals equation. Use the data for pressure and volume given in Sample Exercise 5.3.

64. We state that the ideal gas law tends to hold best at low pressures and high temperatures. Show how the van der Waals equation simplifies to the ideal gas law under these conditions.

Atmosphere Chemistry

65. Use the data in Table 5.4 to calculate the partial pressure of NO in dry air assuming the total pressure is 1.0 atm. Assuming a temperature of 0°C, calculate the number of molecules of NO per cubic centimeter.

66. Atmospheric scientists often use mixing ratios to express the concentrations of trace compounds in air. Mixing ratios are often expressed as ppmv (parts per million volume):

$$\text{ppmv of } X = \frac{\text{vol. of } X \text{ at STP}}{\text{total vol. of air at STP}} \times 10^6$$

In November 1983, the concentration of carbon monoxide in the air in downtown Denver, Colorado, reached 3.0×10^2 ppmv. The atmospheric pressure at that time was 628 torr and the temperature was 0°C.
a. What was the partial pressure of CO?
b. What was the concentration of CO in molecules per cubic meter?
c. What was the concentration of CO in molecules per cubic centimeter?

67. A 10.0-L sample of air was collected at an altitude of 1.00×10^2 km. What would be the volume of the sample at STP? (See Fig. 5.25.)

68. Write reactions to show how the nitric and sulfuric acids in acid rain react with marble and limestone. Both marble and limestone are primarily calcium carbonate.

69. In a particular urban area the NO_2 concentration reached a maximum value of 0.2 ppm by volume. (See Exercise 66.) Assume this concentration exists to an altitude of 5.0×10^3 ft over an area of 5.0 mi by 10.0 mi. Assuming 1.0 atm pressure and 0°C temperature, calculate the total mass of NO_2 in the atmosphere over this city.

70. Trace organic compounds in the atmosphere are first concentrated and then measured by gas chromatography. In the concentration step, several liters of air are pumped through a tube containing a porous substance that traps organic compounds. The tube is then connected to a gas chromatograph and heated to release the trapped compounds. The organic compounds are separated in the column and the amounts are measured. In an analysis for benzene and toluene in air, a 2.0-L sample of air, at 748 torr and 23°C, was passed through the trap. The gas chromatography analysis showed that this air sample contained 101 ng of benzene (C_6H_6) and 274 ng of toluene (C_7H_8). Calculate the mixing ratio (see Exercise 66) and number of molecules per cubic centimeter for both benzene and toluene.

Additional Exercises

71. Rationalize the following observations:
a. Aerosol cans will explode if heated.
b. You can drink through a soda straw.
c. A thin-walled can will collapse when the air inside is removed by a vacuum pump.

72. Why do manufacturers produce different types of tennis balls for high and low altitudes?

73. In the "Méthode Champenoise," grape juice is fermented in a wine bottle to produce sparkling wine. The reaction is

$$C_6H_{12}O_6(aq) \rightarrow 2C_2H_5OH(aq) + 2CO_2(g)$$

Fermentation of 750. mL of grape juice (density = 1.0 g/cm^3) is allowed to take place in a bottle with a total volume of 825 mL until 12% by volume is ethanol (C_2H_5OH). Assuming that the CO_2 is insoluble in H_2O (actually, a wrong assumption), what would be the pressure of CO_2 inside the wine bottle at 25°C? (The density of ethanol is 0.79 g/cm^3.)

74. A compressed gas cylinder contains 5.00×10^2 g of argon gas. The pressure inside the cylinder is 2050. psi at a temperature of 18°C. How much gas remains in the cylinder if the pressure is 980. psi at a temperature of 26°C?

75. The total mass that can be lifted by a balloon is given by the difference between the mass of air displaced by the balloon and the mass of the gas inside the balloon. Consider a hot-air balloon that approximates a sphere 5.0 m in diameter and contains air heated to 65°C. The surrounding air temperature is 21°C. The pressure in the balloon is equal to the atmospheric pressure, which is 745 torr.
a. What total mass can the balloon lift? Assume the average molecular weight of air is 29.0. (*Hint:* heated air is less dense than cool air.)
b. If the balloon is filled with enough helium at 21°C and 745 torr to achieve the same volume as in part a, what total mass can the balloon lift?
c. What mass could the hot-air balloon in part a lift if it were on the ground in Denver, Colorado, where a typical atmospheric pressure is 630. torr?

76. An important process for the production of acrylonitrile (C_3H_3N) (1983 U.S. production was 2.15×10^9 lb) is given by the following reaction:

$$2C_3H_6(g) + 2NH_3(g) + 3O_2(g) \rightarrow 2C_3H_3N(g) + 6H_2O(g)$$

A 150.-L reactor is charged to the following partial pressures at 25°C.

$$P_{C_3H_6} = 0.50 \text{ MPa}$$
$$P_{NH_3} = 0.80 \text{ MPa}$$
$$P_{O_2} = 1.50 \text{ MPa}$$

What mass of acrylonitrile can be produced from this mixture (MPa = 10^6 Pa)?

77. The oxides of Group 2A metals (symbolized by M here) react with carbon dioxide according to the following reaction:

$$MO(s) + CO_2(g) \rightarrow MCO_3(s)$$

A 2.85-g sample containing only MgO and CuO is placed in a 3.0-L container. The container is filled with CO_2 to a pressure of 740 torr, at 20.°C. After the reaction has gone to completion, the pressure inside the flask is 390 torr, also at 20.°C. What is the mass percent of MgO in the mixture? Assume only the MgO reacts with CO_2.

78. Nitrous oxide (N_2O) can be produced by the thermal decomposition of ammonium nitrate:

$$NH_4NO_3(s) \xrightarrow{\text{Heat}} N_2O(g) + 2H_2O(l)$$

What volume of N_2O, collected over water, at a total pressure of 94 kPa and 22°C, can be produced from the thermal decomposition of 2.6 g of NH_4NO_3? (The vapor pressure of water at 22°C is 21 torr.)

79. Formaldehyde (CH_2O) is sometimes released from foamed insulation currently being used in homes. The federal standard for the allowable amount of CH_2O in air is 1.0 ppbv (parts per billion volume; similar to ppmv as defined in Exercise 66). How many molecules per cubic centimeter is this at STP? If the concentration of formaldehyde in a room is 1.0 ppbv, what total mass of formaldehyde is present at STP if the room measures 18 ft × 24 ft × 8 ft?

80. Acetylene gas, $C_2H_2(g)$, can be produced by reacting solid calcium carbide, CaC_2, with water. The products are acetylene and calcium hydroxide. What volume of wet acetylene is collected at 25°C and 715 torr when 5.20 g of calcium carbide is reacted with an excess of water? (At 25°C the vapor pressure of water is 23.8 torr.)

81. Consider the three flasks in the diagram below. Assuming the connecting tubes have negligible volume, what is the partial pressure of each gas and the total pressure when all of the stopcocks are opened?

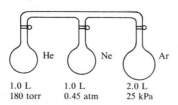

He	Ne	Ar
1.0 L	1.0 L	2.0 L
180 torr	0.45 atm	25 kPa

82. A 1.00-L container is evacuated until the pressure is 1.00×10^{-6} torr at 22°C. How many molecules of gas remain in the container? How many molecules are there per cubic centimeter?

83. A compound contains only C, H, and N. It is 58.51% C and 7.37% H by mass. Helium effuses through a porous frit 3.20 times as fast as the compound does. Determine the empirical and molecular formulas of this compound.

84. A compound contains only C, H, and N. A chemist analyzes it by doing the following experiments.

i. Complete combustion of 35.0 mg of the compound produced 33.5 mg of CO_2 and 41.1 mg of H_2O.

ii. A 65.2-mg sample of the compound was analyzed for nitrogen by the Dumas method, giving 35.6 mL of N_2 at 740. torr and 25°C.

iii. The effusion rate of the compound as a gas was measured and found to be 24.6 mL/min. The effusion rate of argon gas, under identical conditions, is 26.4 mL/min.

What is the formula of the compound?

85. A 15.0-L tank is filled with H_2 to a pressure of 2.00×10^2 atm. How many balloons (each 2.00 L) can be inflated to a pressure of 1.00 atm from the tank? Assume that there is no temperature change and that the tank cannot be emptied below 1.00 atm pressure.

86. Consider the following diagram:

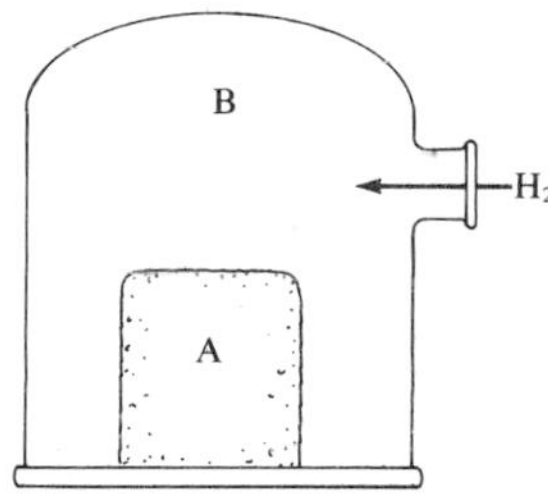

A porous container (A), filled with air at STP, is contained in a large enclosed container (B) which is flushed with $H_2(g)$. What will happen to the pressure inside container (A)? Explain your answer.

87. Without looking at tables of values, which of the following gases would you expect to have the largest value of the van der Waals constant *b*: H_2, N_2, CH_4, C_2H_6, or C_3H_8?

88. From the values in Table 5.3 for the van der Waals constant *a* for the gases H_2, CO_2, N_2, and CH_4, predict which of these gas molecules show the strongest intermolecular attractions.

89. Derive a linear relationship between gas density and temperature and use it to estimate the value of absolute zero temperature (in °C to the nearest 0.1°C) from an air sample whose density is 1.293 g/L at 0.0°C and 0.946 g/L at 100.0°C. Assume air obeys the ideal gas law and that the pressure is held constant.

CHAPTER 6

Thermochemistry

Energy is the essence of our very existence as individuals and as a society. The food that we eat furnishes the energy that allows us to live, work, and play, as the coal and oil consumed by manufacturing and transportation systems power our modern industrialized civilization.

Huge quantities of carbon-based fossil fuels have been available for the taking. This abundance of fuels has led to a world society with a voracious appetite for energy, one that consumes millions of barrels of petroleum every day. We are now dangerously dependent on the dwindling supplies of oil, and this dependence is an important source of tension among nations in today's world. In an incredibly short time, we have moved from a period of ample and cheap supplies of petroleum to one of high prices and uncertain supplies. If our present standard of living is to be maintained, we must find alternatives to petroleum. To do this we need to know the relationship between chemistry and energy, which we explore in this chapter.

There are additional problems with fossil fuels. The waste products from burning fossil fuels significantly affect our environment. For example, when a carbon-based fuel is burned, the carbon reacts with oxygen to form carbon dioxide, which is released into the atmosphere. Although much of this carbon dioxide is consumed in various natural processes such as photosynthesis and the formation of carbonate minerals, the amount of carbon dioxide in the atmosphere is steadily increasing. This increase is significant because atmospheric carbon dioxide absorbs heat radiated from the earth's surface and radiates it back toward the earth. This is an important mechanism for controlling the earth's temperature. Many scientists fear that an increase in the concentration of carbon dioxide will warm the earth, causing significant changes in climate. In addition, impurities in the fuels react with components of the air to produce air pollution. We discussed some aspects of this problem in Chapter 5.

In this chapter we will cover the fundamental concepts of energy and take a brief look at the practical aspects of energy supply and pollution. Additional theoretical aspects of energy will be presented in Chapter 16.

CONTENTS

< Solar panels at the Solar One facility, located in the desert near Barstow, California. This solar-thermal operation uses the sun's energy to produce steam to power an electricity-producing turbine.

6.1 The Nature of Energy

Purpose

- To describe the energy flow between a system and its surroundings.
- To discuss the first law of thermodynamics.
- To show how to calculate the work that results from changing the volume of a gas at constant pressure.

Although the concept of energy is quite familiar, energy itself is rather difficult to define concisely. We will define **energy** as the *capacity to do work or to produce heat*. In this chapter we will concentrate on the heat transfer that accompanies chemical processes.

The total energy content of the universe is constant.

One of the most important characteristics of energy is that it is conserved. The **law of conservation of energy** states that *energy can be converted from one form to another but can be neither created nor destroyed*. That is, the energy of the universe is constant. Energy can be classified as either potential energy or kinetic energy. **Potential energy** is energy due to position or composition. For example, water behind a dam has potential energy that can be converted to work when the water flows down through turbines, thereby creating electricity. Attractive or repulsive forces also lead to potential energy. The energy released when gasoline is burned results from differences in attractive forces between nuclei and electrons in the reactants and products. The **kinetic energy** of an object is due to the motion of the object and depends on the mass of the object (m) and its velocity (v): $KE = \frac{1}{2}mv^2$.

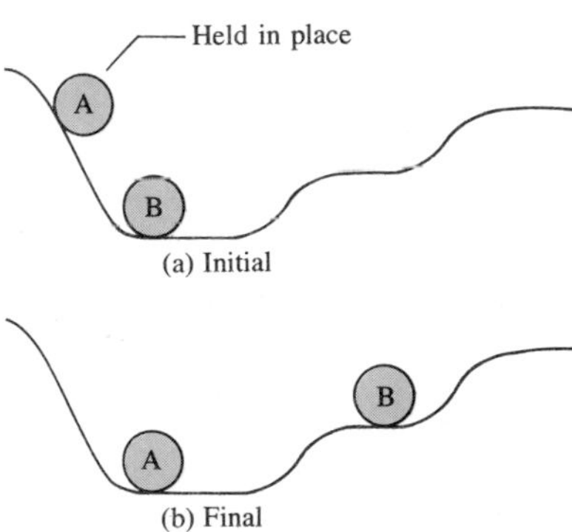

Figure 6.1

(a) In the initial positions, ball A has a higher potential energy than ball B. (b) After A has rolled down the hill, the potential energy lost by A has been converted to random motions of the components of the hill (frictional heating) and to the increase in the potential energy of B.

Energy can be converted from one form to another. For example, consider the two balls in Fig. 6.1(a). Ball A, because of its higher position, has more potential energy than ball B. When A is released, it moves down the hill, strikes B, and eventually the arrangement shown in Fig. 6.1(b) is achieved. What has happened in going from the initial to the final arrangement? The potential energy of A has decreased, but since energy is conserved, all the energy lost by A must be accounted for. How is this energy distributed?

Initially, the potential energy of A is changed to kinetic energy as the ball rolls down the hill. Part of this kinetic energy has been transferred to B, raising B to a higher final position. Thus B has increased potential energy. However, since the final position of B is lower than the original position of A, some of the energy is still unaccounted for. Both balls in their final positions are at rest, so the missing energy cannot be due to their motions. What has happened to the remaining energy?

The answer lies in the interaction between the hill's surface and the ball. As A rolls down the hill, some of its kinetic energy is transferred to the surface of the hill as heat. This transfer of energy is called *frictional heating*. The temperature of the hill is increased very slightly as the ball rolls down.

Heat involves a *transfer* of energy.

Heat and temperature are decidedly different. As we saw in Chapter 5, temperature is a property that reflects the random motions of the particles in a particular substance. **Heat,** on the other hand, involves the *transfer* of energy between two objects due to a temperature difference. Heat is not a substance contained by an object, although we often talk of heat as if this were true.

Note that in going from the initial to the final arrangements in Fig. 6.1, ball B gains potential energy because work was done by ball A on B. **Work** is defined as force acting over a distance. Work is required to raise B from its original position to its final one. Part of the original energy stored as potential energy in A has been transferred through work to B, thereby increasing B's potential energy. Thus, there are two ways to transfer energy: through work and through heat.

In rolling to the bottom of the hill shown in Fig. 6.1, ball A will always lose the same amount of potential energy. However, the way that this energy transfer is divided between work and heat depends on the specific conditions—the **pathway.** For example, the surface of the hill might be so rough that the energy of A is expended completely through frictional heating; A is moving so slowly when it hits B that it cannot move B to the next level. In this case no work is done. Regardless of the condition of the hill's surface, the *total energy* transferred will be constant. However, the amounts of heat and work will differ. Energy change is independent of the pathway; however, work and heat are both dependent on the pathway. *A property that is independent of the pathway* is called a **state function.** Energy is a state function. Work and heat are not state functions.

Energy is a state function; work and heat are not.

Chemical Energy

The ideas we have just illustrated using mechanical examples also apply to chemical systems. The combustion of methane, for example, is used to heat many homes in the United States:

$$CH_4(g) + 2O_2(g) \rightarrow CO_2(g) + 2H_2O(g) + \text{energy (heat)}$$

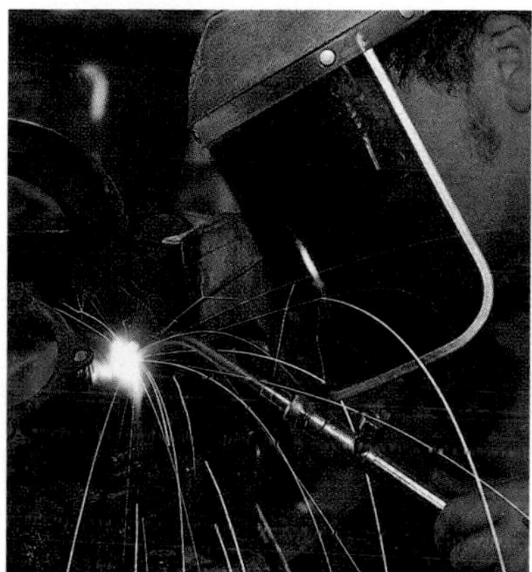

Oxy-acetylene welding. The reaction of acetylene, C_2H_2, with oxygen furnishes the energy for welding steel.

To discuss this reaction we divide the universe into two parts: the system and the surroundings. The **system** is the part of the universe on which we wish to focus attention; the **surroundings** include everything else in the universe. In this case we define the system as the reactants and products of the reaction. The surroundings consist of the reaction container (a furnace, for example), the room, and anything else other than the reactants and products.

When a reaction results in the evolution of heat, it is said to be **exothermic** (*exo-* is a prefix meaning "out of"); that is, energy flows *out of the system*. For example, in the combustion of methane, energy flows out of the system as heat. Reactions that absorb energy from the surroundings are said to be **endothermic.** When the heat flow is *into a system,* the process is endothermic. For example, the formation of nitric oxide from nitrogen and oxygen is endothermic:

$$N_2(g) + O_2(g) + \text{energy (heat)} \rightarrow 2NO(g)$$

Where does the energy, released as heat, come from in an exothermic reaction? The answer lies in the difference in potential energy between the products and the reactants. Which has lower potential energy, the reactants or the products? We know that total energy is conserved and that energy flows from the system into the surroundings in an exothermic reaction. This means that *the energy gained by the surroundings must be equal to the energy lost by the system*. The energy content of the system for methane combustion decreases, which means that 1 mole of CO_2 and 2 moles of H_2O molecules (the products) possess less potential energy than do 1 mole of CH_4 and 2 moles of O_2 molecules (the reactants). The heat flow into the surroundings results from a lowering of the potential energy of the reaction system. This always holds true. *An exothermic reaction means that potential energy stored in chemical bonds is being converted to thermal energy (random kinetic energy) via heat.*

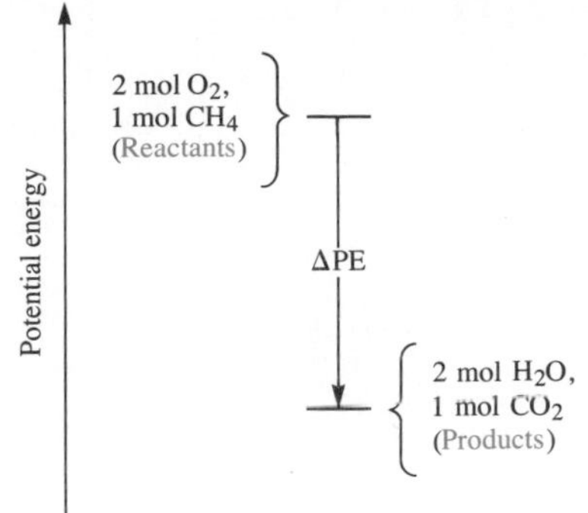

Figure 6.2

The combustion of methane releases the quantity of energy Δ(PE) to the surroundings via heat flow. This is an exothermic process.

The energy situation in the combustion of methane is shown in Fig. 6.2, where Δ(PE) (Δ is the Greek letter delta) represents the *change* in potential energy stored in the bonds of the products as compared to the bonds of the reactants. This is the quantity of energy that is transferred to the surroundings through heat.

For an endothermic reaction the situation is reversed, as shown in Fig. 6.3. Energy that flows into the system as heat is used to increase the potential energy of the system. In this case the products have higher potential energy than the reactants.

The study of energy and its interconversions is called **thermodynamics.** The law of conservation of energy is often called the **first law of thermodynamics** and stated as follows: *the energy of the universe is constant*.

The **internal energy** (E) of a system is defined by the equation:

$$\Delta E = q + w$$

where ΔE represents the change in the system's internal energy, q represents heat, and w represents work. Thus internal energy is a property of a system that can be changed by a flow of work, heat, or both.

Potential energy
2 mol NO
(Products)
ΔPE
1 mol N_2,
1 mol O_2
(Reactants)

Figure 6.3

The energy diagram for the reaction of nitrogen and oxygen to form nitric oxide. This is an endothermic process.

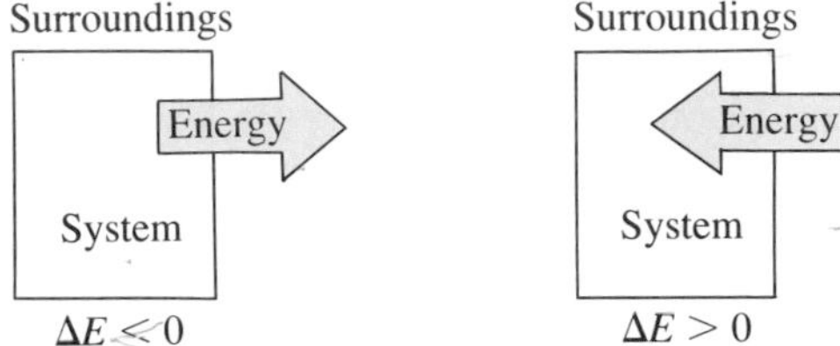

Thermodynamic quantities always consist of two parts: a *number,* giving the magnitude of the change; and a *sign,* indicating the direction of the flow. *The sign reflects the system's point of view*. For example, if a quantity of energy flows *into* the system via heat (an endothermic process), q will be equal to $+x$, where the *positive* sign indicates the *system's energy is increasing*. On the other hand, when energy flows *out of* the system via heat (an exothermic process), q will be equal to $-x$, where the *negative* sign indicates the *system's energy is decreasing*.

The same conventions also apply in this text to the flow of work. If the system does work on the surroundings (energy flows out of the system), w is negative. If the surroundings do work on the system (energy flows into the system), w is positive.

We define work from the system's point of view so it is consistent for all thermodynamic quantities. That is, using this convention, the signs of both q and w reflect what happens to the system; thus we use $\Delta E = q + w$.

The convention in this text is to take the system's point of view: $q = -x$ means an exothermic process, and $q = +x$ means an endothermic one.

In this text we *always* take the system's point of view. Energy entering the system is positive—the energy of the system is increased. This convention is not followed in every area of science. For example, engineers are in the business of designing machines to do work, that is, to make the system (the machine) transfer energy to its surroundings through work. Because of this, engineers define work from the surroundings' point of view. In their convention, work that flows out of the system is treated as positive because the energy of the surroundings has increased. The first law of thermodynamics is then written $\Delta E = q - w'$, where w' signifies the work from the surroundings' point of view.

Sample Exercise 6.1

Calculate ΔE for a system undergoing an endothermic process in which 15.6 kJ of heat flows and where 1.4 kJ of work is done on the system.

Solution

We use the equation

$$\Delta E = q + w$$

where $q = +15.6$ kJ since the process is endothermic, and $w = +1.4$ kJ since work is done on the system. Thus

$$\Delta E = 15.6 \text{ kJ} + 1.4 \text{ kJ} = 17.0 \text{ kJ}$$

The system has gained 17.0 kJ of energy.

The joule (J) is the fundamental SI unit for energy:

$$\text{J} = \frac{\text{kg m}^2}{\text{s}^2}$$

One kilojoule (kJ) = 10^3 J.

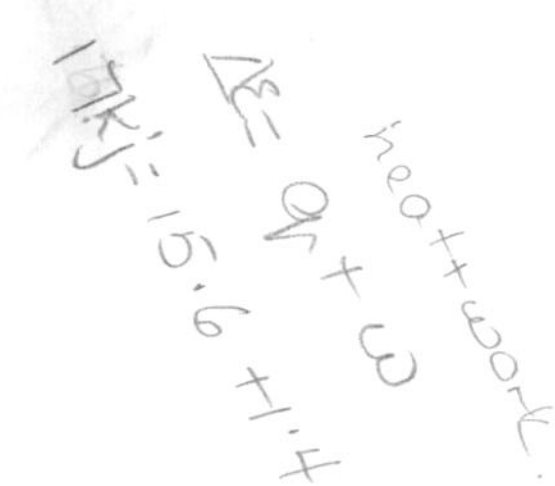

A common type of work associated with chemical processes is work done by a gas (through *expansion*) or work done to a gas (through *compression*). For example, in an automobile engine the heat from the combustion of the gasoline expands the gases in the cylinder to push back the piston, and this motion is then translated into the motion of the car.

Suppose we have a gas confined to a cylindrical container with a movable piston as shown in Fig. 6.4, where F is the force acting on a piston of area A. Since pressure is defined as force per unit area, the pressure of the gas is

$$P = \frac{F}{A}$$

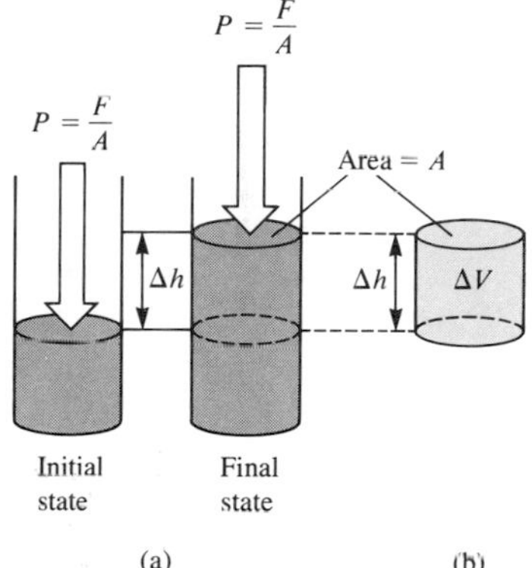

Figure 6.4

(a) The piston, moving a distance Δh against a pressure P, does work on the surroundings. (b) Since the volume of a cylinder is the area of the base times its height, the change in volume of the gas is given by $\Delta h \times A = \Delta V$.

Work is defined as force applied over a distance, so if the piston moves a distance Δh, as shown in Fig. 6.4, then the work done is

$$\text{Work} = \text{force} \times \text{distance} = F \times \Delta h$$

Since $P = F/A$ or $F = P \times A$, then

$$\text{Work} = F \times \Delta h = P \times A \times \Delta h$$

Since the volume of a cylinder equals the area of the piston times the height of the cylinder (Fig. 6.4), the change in volume ΔV resulting from the piston moving a distance Δh is

$$\Delta V = \text{final volume} - \text{initial volume} = A \times \Delta h$$

Substituting $\Delta V = A \times \Delta h$ into the expression for work gives

$$\text{Work} = P \times A \times \Delta h = P\Delta V$$

This gives us the *magnitude* (size) of the work required to expand a gas ΔV against a pressure P.

What about the sign of the work? The gas (the system) is expanding, pushing back the piston against pressure. Thus the system is doing work on the surroundings, so from the system's point of view the sign of the work should be negative.

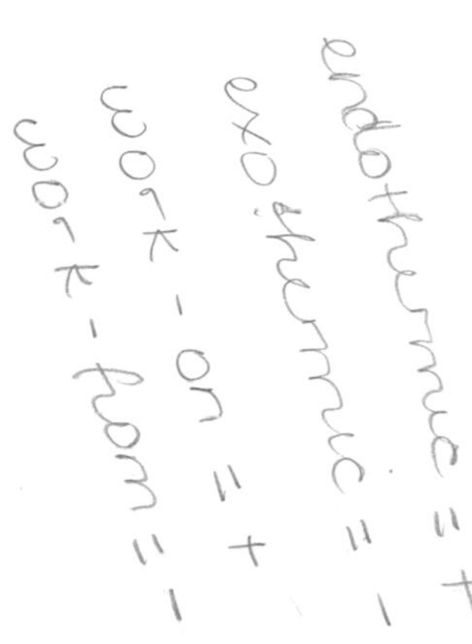

w and $P\Delta V$ have opposite signs since when the gas expands (ΔV is positive), work flows into the surroundings (w is negative).

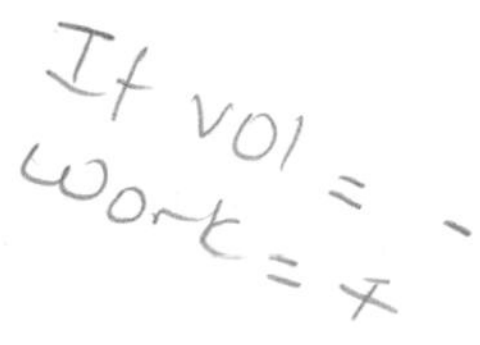

For an *expanding* gas, ΔV is a positive quantity because the volume increases. Thus ΔV and w must have opposite signs, and we have the equation

$$w = -P\Delta V$$

Note that for a gas expanding against an external pressure P, w is a negative quantity as required since work flows out of the system. When a gas is *compressed*, ΔV is a negative quantity (the volume decreases), which makes w a positive quantity (work flows into the system).

Sample Exercise 6.2

Calculate the work associated with the expansion of a gas from 46 L to 64 L at a constant external pressure of 15 atm.

Solution

For a gas at constant pressure,

$$w = -P\Delta V$$

In this case $P = 15$ atm and $\Delta V = 64 - 46 = 18$ L. Hence

$$w = -15 \text{ atm} \times 18 \text{ L} = -270 \text{ L atm}$$

Note that since the gas expands, it does work on its surroundings. Energy flows out of the gas so w is a negative quantity.

For an ideal gas, work occurs only when its volume changes. Thus, if a gas is heated at constant volume, the pressure increases but no work occurs.

Sample Exercise 6.3

A balloon is being inflated to its full extent by heating the air inside it. In the final stages of this process, the volume of the balloon changes from 4.00×10^6 L to 4.50×10^6 L by addition of 1.3×10^8 J of energy as heat. Assuming the balloon expands against a constant pressure of 1.0 atm, calculate ΔE for the process. (To convert between L atm and J use 1 L atm = 101.3 J.)

Solution

To calculate ΔE we use the equation

$$\Delta E = q + w$$

Since the problem states that 1.3×10^8 J of energy is *added* as heat,

$$q = +1.3 \times 10^8 \text{ J}$$

The work done can be calculated from the expression

$$w = -P\Delta V$$

In this case $P = 1.0$ atm and

$$\begin{aligned}\Delta V &= V_{\text{final}} - V_{\text{initial}} \\ &= 4.50 \times 10^6 \text{ L} - 4.00 \times 10^6 \text{ L} = 0.50 \times 10^6 \text{ L} = 5.0 \times 10^5 \text{ L}\end{aligned}$$

Thus

$$w = -1.0 \text{ atm} \times 5.0 \times 10^5 \text{ L} = -5.0 \times 10^5 \text{ L atm}$$

A propane burner is used to heat the air in a hot air balloon.

Sample Exercise 6.3, continued

Note that the negative sign for w makes sense, since the gas is expanding and doing work on the surroundings.

To calculate ΔE we must sum q and w. However, q is given in units of J and w is given in units of L atm. We must change the work to units of joules:

$$w = -5.0 \times 10^5 \ \cancel{\text{L atm}} \times \frac{101.3 \text{ J}}{\cancel{\text{L atm}}} = -5.1 \times 10^7 \text{ J}$$

Then

$$\Delta E = q + w = (+1.3 \times 10^8 \text{ J}) + (-5.1 \times 10^7 \text{ J}) = 8 \times 10^7 \text{ J}$$

Since more energy is added through heating than the gas expends doing work, there is a net increase in the internal energy of the gas in the balloon. Hence ΔE is positive.

6.2 Enthalpy and Calorimetry

Purpose

- To define enthalpy and demonstrate calculations of the change in enthalpy in a chemical reaction.
- To show how a change in enthalpy is measured by calorimetry.

Enthalpy

So far we have discussed the internal energy of a system. A less familiar property of a system is its **enthalpy** (H), which is defined as

$$H = E + PV$$

where E is the internal energy of the system, P is the pressure of the system, and V is the volume of the system.

Since internal energy, pressure, and volume are all state functions, enthalpy is also a state function. But what exactly is enthalpy? To help answer this question, consider a process carried out at constant pressure and where the only work allowed is pressure-volume work ($w = -P\Delta V$). Under these conditions, the expression

Enthalpy is a state function. A change in enthalpy does not depend on the pathway between two states.

Recall from the previous section that w and $P\Delta V$ have opposite signs:

$$w = -P\Delta V$$

$$\Delta E = q + w$$

becomes

$$\Delta E = q - P\Delta V$$

or

$$q = \Delta E + P\Delta V$$

We will now relate q to a change in enthalpy. Since the definition of enthalpy is $H = E + PV$, we can say

$$\text{Change in } H = (\text{change in } E) + (\text{change in } PV)$$

or

$$\Delta H = \Delta E + \Delta(PV)$$

The change in enthalpy of a system has no easily interpreted meaning except at constant pressure, where ΔH = heat.

Since P is constant, the change in PV is due only to a change in volume, and

$$\Delta(PV) = P\Delta V$$

Thus

$$\Delta H = \Delta E + P\Delta V$$

This expression is identical to the one we obtained for q:

$$q = \Delta E + P\Delta V$$

Thus, for a process carried out at constant pressure and where the only work allowed is that from a volume change, we have

$$\Delta H = q$$

$\Delta H = q$ *only* at constant pressure.

At constant pressure, the change in enthalpy (ΔH) *of the system is equal to the energy flow as heat*. This means that, for a reaction studied at constant pressure, the flow of heat is a measure of the change in enthalpy for the system. This is why the terms *heat of reaction* and *change in enthalpy* are used interchangeably for reactions studied at constant pressure.

For a chemical reaction the enthalpy change is given by the equation

$$\Delta H = H_{\text{products}} - H_{\text{reactants}}$$

At constant pressure exothermic means ΔH is negative; endothermic means ΔH is positive.

In a case in which the products of a reaction have a greater enthalpy than the reactants, ΔH will be positive, heat will be absorbed by the system and the reaction is endothermic. On the other hand, if the enthalpy of the products is less than that of the reactants, ΔH will be negative. In this case the overall decrease in enthalpy is achieved by the generation of heat, and the reaction is exothermic.

Sample Exercise 6.4

When 1 mol of methane (CH_4) is burned at constant pressure, 890 kJ of energy is released as heat. Calculate ΔH for a process in which a 5.8-g sample of methane is burned at constant pressure.

Solution

At constant pressure, 890 kJ of energy per mole of CH_4 is produced as heat:

$$\Delta H = -890 \text{ kJ/mol } CH_4$$

Note that the minus sign indicates an exothermic process. In this case a 5.8-g sample of CH_4 (mol. wt. = 16.0) is burned. Since this amount is smaller than 1 mol, less than 890 kJ will be released as heat. The actual value can be calculated as follows:

$$5.8 \cancel{\text{g CH}_4} \times \frac{1 \text{ mol CH}_4}{16.0 \cancel{\text{g CH}_4}} = 0.36 \text{ mol CH}_4$$

and

$$0.36 \cancel{\text{mol CH}_4} \times \frac{-890 \text{ kJ}}{\cancel{\text{mol CH}_4}} = -320 \text{ kJ}$$

Thus, when a 5.8-g sample of CH_4 is burned at constant pressure,

$$\Delta H = \text{heat flow} = -320 \text{ kJ}$$

Calorimetry

We can determine the heat associated with a chemical reaction experimentally using a device called a **calorimeter. Calorimetry,** the science of measuring heat, is based on observing the temperature change when a body absorbs or discharges energy as heat. Substances differ in their responses to being heated. One substance might require a great deal of heat to raise its temperature by one degree, while another will exhibit the same temperature change after absorbing relatively little heat. The **heat capacity** (*C*) of a substance is a measure of this property and is expressed as

$$C = \frac{\text{heat absorbed}}{\text{increase in temperature}}$$

When an element or a compound is heated, the energy required will depend on the amount of the substance present (for example, it takes twice as much energy to raise the temperature of two grams of water by one degree as it takes to raise the temperature of one gram of water by one degree). Thus in giving the heat capacity of a substance the amount of substance must be specified. If the heat capacity is given *per gram* of the substance, it is called the **specific heat capacity** and its units are J/°C g. If the heat capacity is given *per mole* of the substance, it is called the **molar heat capacity** which has the units J/°C mol. The specific heat capacities of some common substances are given in Table 6.1.

Specific heat capacity: the energy required to raise the temperature of one gram of a substance by one degree Celsius.

Molar heat capacity: the energy required to raise the temperature of one mole of a substance by one degree Celsius.

The Specific Heat Capacities of Some Common Substances

Substance	Specific heat capacity (J/°C g)
$H_2O(l)$	4.18
$H_2O(s)$	2.03
$Al(s)$	0.89
$Fe(s)$	0.45
$Hg(l)$	0.14
$C(s)$	0.71

Table 6.1

Although the calorimeters used for highly accurate work are precision instruments, a very simple calorimeter can be used to examine the fundamentals of calorimetry. All we need are two nested Styrofoam cups with a cover through which a stirrer and thermometer can be inserted, as shown in Fig. 6.5. This is called a *coffee-cup calorimeter*. The outer cup is used to provide extra insulation. The inner cup holds the solution in which the reaction occurs.

Measurement of heat using a simple calorimeter such as that shown in Fig. 6.5 is an example of **constant pressure calorimetry,** since the pressure (atmospheric pressure) remains constant during the process. Constant pressure calorimetry is used in determining the changes in enthalpy (heats of reactions) occurring in solution. Under these conditions, the change in enthalpy equals the heat, as we have seen.

For example, suppose we mix 50.0 mL of 1.0 *M* HCl at 25.0°C with 50.0 mL of 1.0 *M* NaOH also at 25°C in a calorimeter. After the reactants are mixed by stirring, the temperature is observed to increase to 31.9°C. As we saw in Section 4.9, the net ionic equation for this reaction is

$$H^+(aq) + OH^-(aq) \rightarrow H_2O(l)$$

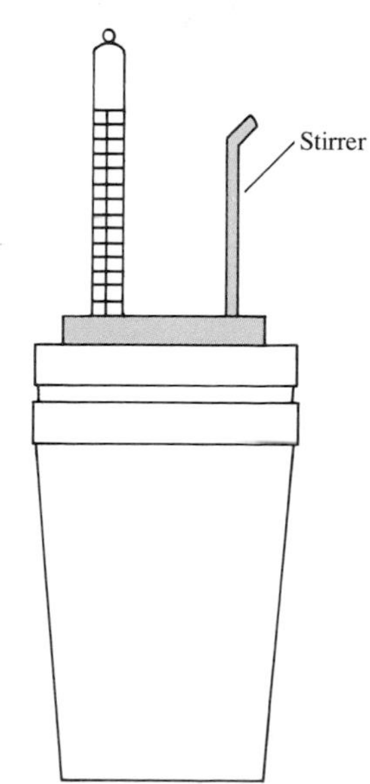

Figure 6.5

A coffee-cup calorimeter made of two Styrofoam cups.

When these reactants (each originally at the same temperature) are mixed, the temperature of the mixed solution is observed to increase. This means that the

If two reactants at the same temperature are mixed and the resulting solution gets warmer, this means the reaction taking place is exothermic. An endothermic reaction cools the solution.

chemical reaction must be releasing energy as heat, which increases the random motions of the solution components, which in turn increases the temperature. The quantity of energy released can be determined from the temperature increase, the mass of solution, and the specific heat capacity of the solution. For an approximate result, we will assume that the calorimeter does not absorb or leak any heat and that the solution can be treated as if it were pure water with a density of 1.0 g/mL.

We also need to know the heat required to raise the temperature of a given amount of water by 1°C. Table 6.1 lists the specific heat capacity of water as 4.18 J/°C g. This means that 4.18 J of energy is required to raise the temperature of 1 gram of water by 1°C.

Based on these assumptions and definitions, we can calculate the heat (change in enthalpy) for the neutralization reaction:

An infra-red thermogram shows heat leakage from a house.

Energy released by the reaction
= energy absorbed by the solution
= specific heat capacity × mass of solution × increase in temperature

where the increase in temperature = 31.9°C − 25.0°C = 6.9°C, and the mass of solution = 100.0 mL × 1.0 g/mL = 1.0×10^2 g. Thus

$$\text{Energy released} = \left(4.18 \frac{\text{J}}{\cancel{^\circ\text{C}}\,\cancel{\text{g}}}\right)(1.0 \times 10^2\ \cancel{\text{g}})(6.9\cancel{^\circ\text{C}})$$

$$= 2.9 \times 10^3\ \text{J}$$

How much energy would have been released if twice these amounts of solutions had been mixed? The answer is that twice as much energy would have been produced. The heat of a reaction is an *extensive property;* it depends directly on the amount of substance, in this case on the amounts of reactants. In contrast, an *intensive property* is not related to the amount of a substance. For example, temperature is an intensive property.

Enthalpies of reaction are often expressed in terms of moles of reacting substances. The number of moles of H^+ ions consumed in the above experiment is

$$50.0\ \cancel{\text{mL}} \times \frac{1\ \cancel{\text{L}}}{1000\ \cancel{\text{mL}}} \times \frac{1.0\ \text{mol}}{\cancel{\text{L}}}\text{H}^+ = 5.0 \times 10^{-2}\ \text{mol H}^+$$

Thus 2.9×10^3 J heat was released when 5.0×10^{-2} moles of H^+ ions reacted, or $\frac{2.9 \times 10^3\ \text{J}}{5.0 \times 10^{-2}\ \text{mol H}^+} = 5.8 \times 10^4$ J of heat released per 1.0 mol of H^+ ions neutralized. Thus the *magnitude* of the enthalpy of reaction per mole for

Notice that in this example we mentally keep track of the direction of the energy flow and assign the correct sign at the end of the calculation.

$$\text{H}^+(aq) + \text{OH}^-(aq) \rightarrow \text{H}_2\text{O}(l)$$

at constant pressure is 58 kJ/mol. Since heat is *evolved,* $\Delta H = -58$ kJ/mol.

Sample Exercise 6.5

When 1.00 L of 1.00 *M* $Ba(NO_3)_2$ solution at 25.0°C is added to 1.00 L of 1.00 *M* Na_2SO_4 solution at 25°C in a calorimeter, the white solid $BaSO_4$ forms and the temperature of the mixture increases to 28.1°C. Assuming that the calorimeter absorbs only a negligible quantity of heat, that the specific heat capacity of the solution is 4.18 J/°C g, and that the density of the final solution is 1.0 g/mL, calculate the enthalpy change per mole of $BaSO_4$ formed.

Sample Exercise 6.5, continued

Solution

The ions present before any reaction are Ba^{2+}, NO_3^-, Na^+, and SO_4^{2-}. The Na^+ and NO_3^- ions are spectator ions, since $NaNO_3$ is very soluble in water and will not precipitate under these conditions. The net ionic equation of the reaction is therefore

$$Ba^{2+}(aq) + SO_4^{2-}(aq) \rightarrow BaSO_4(s)$$

Since the temperature increases, the formation of the solid $BaSO_4$ must be exothermic; ΔH will be negative.

Heat evolved by reaction
= heat absorbed by solution
= specific heat capacity × mass of solution × increase in temperature

Since 1.00 L of each solution is used, the total solution volume is 2.00 L and

$$\text{Mass of solution} = 2.00\,\cancel{\text{L}} \times \frac{1000\,\cancel{\text{mL}}}{1\,\cancel{\text{L}}} \times \frac{1.0\text{ g}}{\cancel{\text{mL}}} = 2.0 \times 10^3\text{ g}$$

$$\text{Temperature increase} = 28.1°\text{C} - 25.0°\text{C} = 3.1°\text{C}$$

$$\text{Heat evolved} = (4.18\text{ J}/\cancel{°\text{C}}\,\cancel{\text{g}})(2.0 \times 10^3\,\cancel{\text{g}})(3.1\cancel{°\text{C}}) = 2.6 \times 10^4\text{ J}$$

Thus $$q = \Delta H = -2.6 \times 10^4\text{ J}$$

Since 1.0 L of 1.0 *M* $Ba(NO_3)_2$ contains 1 mol of Ba^{2+} ions, and 1.0 L of 1.0 *M* Na_2SO_4 contains 1.0 mol of SO_4^{2-} ions, 1.0 mol of solid $BaSO_4$ is formed in this experiment. Thus the enthalpy change per mole of $BaSO_4$ formed is

$$\Delta H = -2.6 \times 10^4\text{ J/mol} = -26\text{ kJ/mol}$$

Calorimetry experiments can also be performed at **constant volume.** For example, when a photographic flashbulb flashes, the bulb becomes very hot, since the reaction of the zirconium or magnesium wire with the oxygen inside the bulb is exothermic. The reaction occurs inside the flashbulb, which is rigid and does not change volume. Under these conditions no work is done, since the volume must change for pressure-volume work to be performed. To study the energy changes in reactions under conditions of constant volume, a **bomb calorimeter** (Fig. 6.6) is used. Weighed reactants are placed inside a rigid steel container (the "bomb") and ignited. The energy change is determined by measuring the increase in the temperature of the water and other calorimeter parts. For a constant volume process, the change in volume (ΔV) is equal to zero, and so work (which is $-P\Delta V$) is also equal to zero. Furthermore,

$$\Delta E = q + w = q$$

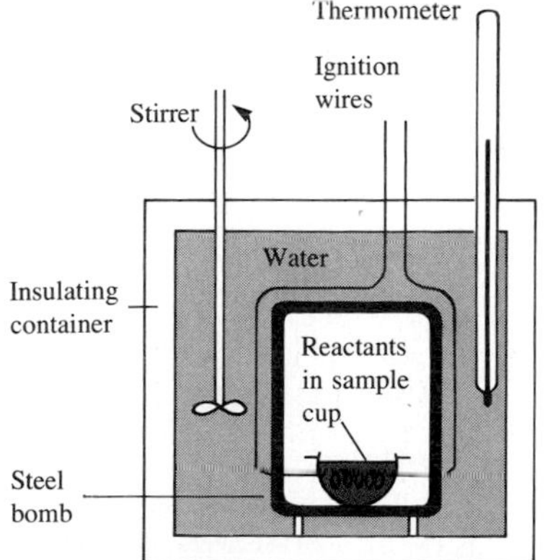

Figure 6.6

A bomb calorimeter. The reaction is carried out inside a rigid steel "bomb," and the heat evolved is absorbed by the surrounding water and other calorimeter parts. The quantity of energy produced by the reaction can be calculated from the temperature increase.

Suppose we wish to measure the energy of combustion of octane (C_8H_{18}), a component of gasoline. A 0.5269-g sample of octane is placed in a bomb calorimeter known to have a heat capacity of 11.3 kJ/°C. This means that 11.3 kJ of energy is required to raise the temperature of the water and other parts of the calorimeter by 1°C. The octane is ignited in the presence of excess oxygen, and the temperature increase of the calorimeter is 2.25°C. The amount of energy released is calculated as follows.

Energy released by the reaction

$= $ temperature increase $\times$ energy required to change the temperature by 1°C

$= \Delta T \times$ heat capacity of calorimeter

$= 2.25\cancel{°C} \times 11.3\ kJ/\cancel{°C} = 25.4\ kJ$

This means that 25.4 kJ of energy was released by the combustion of 0.5269 g of octane. The number of moles of octane is

$$0.5269\ \cancel{g\ octane} \times \frac{1\ mol\ octane}{114.2\ \cancel{g\ octane}} = 4.614 \times 10^{-3}\ mol\ octane$$

Since 25.4 kJ of energy was released for 4.614×10^{-3} mol of octane, the energy released per mole is

$$\frac{25.4\ kJ}{4.614 \times 10^{-3}\ mol} = 5.50 \times 10^{3}\ kJ/mol$$

Since the reaction is exothermic, ΔE is negative:

$$\Delta E_{combustion} = -5.50 \times 10^{3}\ kJ/mol$$

Note that since no work is done in this case, ΔE is equal to the heat.

$$\Delta E = q + w = q \qquad \text{since } w = 0$$

Thus $q = -5.50 \times 10^{3}$ kJ/mol.

Sample Exercise 6.6

Hydrogen's potential as a fuel is discussed in Section 6.6.

It has been suggested that hydrogen gas obtained by the decomposition of water might be a substitute for natural gas (principally methane). To compare the energies of combustion of these fuels, the following experiment was carried out using a bomb calorimeter with a heat capacity of 11.3 kJ/°C. When a 1.50-g sample of methane gas was burned with excess oxygen in the calorimeter, the temperature increased by 7.3°C. When a 1.15-g sample of hydrogen gas was burned with excess oxygen, the temperature increase was 14.3°C. Calculate the energy of combustion per gram for hydrogen and methane.

Solution

We calculate the energy of combustion for methane using the heat capacity of the calorimeter (11.3 kJ/°C) and the observed temperature increase of 7.3°C:

$$\text{Energy } \textit{evolved} \text{ in the combustion of 1.5 g of } CH_4 = (11.3\ kJ/°C)(7.3°C) = 83\ kJ$$

$$\text{Energy } \textit{evolved} \text{ in the combustion of 1 g of } CH_4 = \frac{83\ kJ}{1.5\ g} = 55\ kJ/g$$

The direction of energy flow is indicated by words in this example. Using signs:

$\Delta E_{combustion} = -55$ kJ/g

for methane and

$\Delta E_{combustion} = -141$ kJ/g

for hydrogen.

Similarly, for hydrogen:

$$\text{Energy } \textit{evolved} \text{ in the combustion of 1.15 g of } H_2 = (11.3\ kJ/°C)(14.3°C) = 162\ kJ$$

$$\text{Energy } \textit{released} \text{ in the combustion of 1 g of } H_2 = \frac{162\ kJ}{1.15\ g} = 141\ kJ/g$$

Chemical Impact

Firewalking: Magic or Science?

For millennia people have been amazed at the ability of Eastern mystics to walk across beds of glowing coals without any apparent discomfort. Even in the United States, thousands of people have performed feats of firewalking as part of motivational seminars. How is this possible? Do firewalkers have supernatural powers?

Actually, there are good scientific explanations, based on the concepts covered in this chapter, of why firewalking is possible. The first important factor concerns the heat capacity of feet. Because human tissue is mainly composed of water, it has a relatively large specific heat capacity. This means that a large amount of energy must be transferred from the coals to significantly change the temperature of the feet. During the brief contact between feet and coals, there is relatively little time for energy flow so the feet do not reach a high enough temperature to cause damage.

Secondly, although the surface of the coals has a very high temperature, the red hot layer is very thin. Therefore, the quantity of energy available to heat the feet is smaller than might be expected. This factor points up the difference between temperature and heat. Temperature reflects the *intensity* of the random kinetic energy in a given sample of matter. The amount of energy available for heat flow, on the other hand, depends on the quantity of matter at a given temperature—ten grams of matter at a given temperature contains ten times as much thermal energy as one gram of the same matter. This is why the tiny spark from a sparkler does not hurt when it hits your hand. The spark has a very high temperature but has so little mass that no significant energy transfer occurs to your hand. This same argument applies to the very thin hot layer on the coals.

A third factor that aids firewalkers is the so-called Leidenfrost effect—the phenomenon that allows water droplets to skate around on a hot griddle for a surprisingly long time. The part of the droplet in contact with the hot surface vaporizes first, providing a gaseous layer that both allows the droplet to skate around and acts as a barrier through which energy does not readily flow to the rest of the droplet. The perspiration on the feet of the presumably tense firewalker would have the same effect. In addition, because firewalking is often done at night, with moist grass surrounding the bed of coals, the firewalker's feet are probably damp before the walk, providing ample moisture for the Leidenfrost effect to occur.

Thus, although firewalking is an impressive feat, there are several sound scientific reasons why anyone should be able to do it with the proper training and a properly prepared bed of coals.

Firewalking ritual in Fiji.

Sample Exercise 6.6, continued

The energy released in the combustion of 1 g of hydrogen is approximately 2.5 times that of 1 g of methane, indicating that hydrogen gas is a potentially useful fuel.

6.3 Hess's Law

Purpose

- To discuss the characteristics of enthalpy changes.
- To show how to calculate ΔH for a chemical reaction.

ΔH is not dependent on the reaction pathway.

Since enthalpy is a state function, the change in enthalpy in going from some initial state to some final state is independent of the pathway. This means that *in going from a particular set of reactants to a particular set of products, the change in enthalpy is the same whether the reaction takes place in one step or in a series of steps*. This principle is known as **Hess's law** and can be illustrated by examining the oxidation of nitrogen to produce nitrogen dioxide. The overall reaction can be written in one step where the enthalpy change is represented by ΔH_1.

$$N_2(g) + 2O_2(g) \rightarrow 2NO_2(g) \qquad \Delta H_1 = 68 \text{ kJ}$$

This reaction can also be carried out in two distinct steps, with enthalpy changes designated by ΔH_2 and ΔH_3:

$$\begin{array}{lll} & N_2(g) + O_2(g) \rightarrow 2NO(g) & \Delta H_2 = 180 \text{ kJ} \\ & 2NO(g) + O_2(g) \rightarrow 2NO_2(g) & \Delta H_3 = -112 \text{ kJ} \\ \hline \text{Net reaction:} & N_2(g) + 2O_2(g) \rightarrow 2NO_2(g) & \Delta H_2 + \Delta H_3 = 68 \text{ kJ} \end{array}$$

Note that the sum of the two steps gives the net, or overall, reaction and that

$$\Delta H_1 = \Delta H_2 + \Delta H_3 = 68 \text{ kJ}$$

The principle of Hess's law is shown schematically in Fig. 6.7.

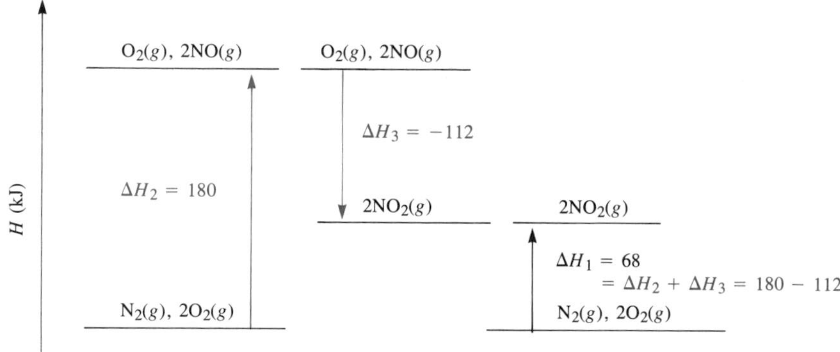

Figure 6.7

The principle of Hess's law. The same change in enthalpy occurs when nitrogen and oxygen react to form nitrogen dioxide, regardless of whether the reaction occurs in one or two steps.

Characteristics of Enthalpy Changes

In order to use Hess's law to compute enthalpy changes for reactions, it is important to understand two characteristics of ΔH for a reaction:

Reversing a reaction changes the sign of ΔH.

1. If a reaction is reversed, the sign of ΔH is also reversed.
2. The magnitude of ΔH is directly proportional to the quantities of reactants and products in a reaction. If the coefficients in a balanced reaction are multiplied by an integer, the value of ΔH is multiplied by the same integer.

Both of these rules follow in a straightforward way from the properties of enthalpy changes. The first rule can be explained by recalling that the *sign* of ΔH indicates the *direction* of the heat flow at constant pressure. If the direction of the reaction is reversed, the direction of the heat flow will also be reversed. To see this, consider the preparation of xenon tetrafluoride, which was the first binary compound made from a noble gas:

$$Xe(g) + 2F_2(g) \rightarrow XeF_4(s) \qquad \Delta H = -251 \text{ kJ}$$

This reaction is exothermic, and 251 kJ of energy flows into the surroundings as heat. On the other hand, if the colorless XeF_4 crystals are decomposed into the elements, according to the equation

$$XeF_4(s) \rightarrow Xe(g) + 2F_2(g)$$

the opposite energy flow will occur because 251 kJ of energy will have to be added to the system to produce this endothermic reaction. Thus, for this reaction $\Delta H = +251$ kJ.

The second rule comes from the fact that ΔH is an extensive property, depending on the amount of substances reacting. For example, since 251 kJ of energy is evolved for the reaction

$$Xe(g) + 2F_2(g) \rightarrow XeF_4(s)$$

then for a preparation involving twice the quantities of reactants and products, or

$$2Xe(g) + 4F_2(g) \rightarrow 2XeF_4(s)$$

twice as much heat would be evolved:

$$\Delta H = 2(-251 \text{ kJ}) = -502 \text{ kJ}$$

Crystals of xenon tetrafluoride, the first reported binary compound containing a noble gas element.

Sample Exercise 6.7

Carbon occurs in two forms: *graphite*, the soft, black, slippery material used in ''lead'' pencils and as a lubricant for locks; and *diamond*, the brilliant, hard gemstone. Using the enthalpies of combustion for graphite (−394 kJ/mol) and diamond (−396 kJ/mol), calculate ΔH for the conversion of graphite to diamond:

$$C_{\text{graphite}}(s) \rightarrow C_{\text{diamond}}(s)$$

Solution

The combustion reactions are

$$C_{\text{graphite}}(s) + O_2(g) \rightarrow CO_2(g) \qquad \Delta H = -394 \text{ kJ}$$
$$C_{\text{diamond}}(s) + O_2(g) \rightarrow CO_2(g) \qquad \Delta H = -396 \text{ kJ}$$

Note that if we reverse the second reaction (which means we must change the sign of ΔH) and sum the two reactions, we obtain the desired reaction:

$$\begin{array}{ll} C_{\text{graphite}}(s) + O_2(g) \rightarrow CO_2(g) & \Delta H = -394 \text{ kJ} \\ CO_2(g) \rightarrow C_{\text{diamond}}(s) + O_2(g) & \Delta H = -(-396 \text{ kJ}) \\ \hline C_{\text{graphite}}(s) \rightarrow C_{\text{diamond}}(s) & \Delta H = 2 \text{ kJ} \end{array}$$

Thus 2 kJ of energy is required to change 1 mol of graphite to diamond. This process is endothermic.

Sample Exercise 6.8

Diborane (B_2H_6) is a highly reactive boron hydride, which was once considered as a possible rocket fuel for the U.S. space program. Calculate ΔH for the synthesis of diborane from its elements, according to the equation

$$2B(s) + 3H_2(g) \rightarrow B_2H_6(g)$$

using the following data:

Reaction	ΔH
(a) $2B(s) + \frac{3}{2}O_2(g) \rightarrow B_2O_3(s)$	-1273 kJ
(b) $B_2H_6(g) + 3O_2(g) \rightarrow B_2O_3(s) + 3H_2O(g)$	-2035 kJ
(c) $H_2(g) + \frac{1}{2}O_2(g) \rightarrow H_2O(l)$	-286 kJ
(d) $H_2O(l) \rightarrow H_2O(g)$	44 kJ

Solution

To obtain ΔH for the required reaction, we must somehow combine equations (a), (b), (c), and (d) to produce that reaction, and add the corresponding ΔH values. This can best be done by focusing on the reactants and products of the required reaction. The reactants are $B(s)$ and $H_2(g)$, and the product is $B_2H_6(g)$. How can we obtain the correct equation? Reaction (a) has $B(s)$ as a reactant, which is needed in the required equation. Thus reaction (a) will be used as it is. Reaction (b) has $B_2H_6(g)$ as a reactant, but this substance is needed as a product. Thus reaction (b) must be reversed, and the sign of ΔH changed accordingly. Up to this point we have

$$\begin{array}{lll}
\text{(a)} & 2B(s) + \frac{3}{2}O_2(g) \rightarrow B_2O_3(s) & \Delta H = -1273 \text{ kJ} \\
-\text{(b)} & B_2O_3(s) + 3H_2O(g) \rightarrow B_2H_6(g) + 3O_2(g) & \Delta H = -(-2035 \text{ kJ}) \\
\hline
\text{Sum:} & B_2O_3(s) + 2B(s) + \frac{3}{2}O_2(g) + 3H_2O(g) \rightarrow B_2O_3(s) + B_2H_6(g) + 3O_2(g) & \Delta H = 762 \text{ kJ}
\end{array}$$

Deleting the species that occur on both sides gives:

$$2B(s) + 3H_2O(g) \rightarrow B_2H_6(g) + \frac{3}{2}O_2(g) \qquad \Delta H = 762 \text{ kJ}$$

We are closer to the required reaction, but we still need to remove $H_2O(g)$ and $O_2(g)$, and introduce $H_2(g)$ as a reactant. We can do this using reactions (c) and (d). If we multiply reaction (c) and its ΔH value by 3 and add the result to the above equation, we have

$$\begin{array}{lll}
 & 2B(s) + 3H_2O(g) \rightarrow B_2H_6(g) + \frac{3}{2}O_2(g) & \Delta H = 762 \text{ kJ} \\
3 \times \text{(c)} & 3[H_2(g) + \frac{1}{2}O_2(g) \rightarrow H_2O(l)] & \Delta H = 3(-286 \text{ kJ}) \\
\hline
\text{Sum:} & 2B(s) + 3H_2(g) + \frac{3}{2}O_2(g) + 3H_2O(g) \rightarrow B_2H_6(g) + \frac{3}{2}O_2(g) + 3H_2O(l) & \Delta H = -96 \text{ kJ}
\end{array}$$

We can cancel the $\frac{3}{2}O_2(g)$ on both sides, but cannot cancel the H_2O because it is gaseous on one side and liquid on the other. This can be solved by adding reaction (d), multiplied by 3:

$$\begin{array}{lll}
 & 2B(s) + 3H_2(g) + 3H_2O(g) \rightarrow B_2H_6(g) + 3H_2O(l) & \Delta H = -96 \text{ kJ} \\
3 \times \text{(d)} & 3[H_2O(l) \rightarrow H_2O(g)] & \Delta H = 3(44 \text{ kJ}) \\
\hline
 & 2B(s) + 3H_2(g) + 3H_2O(g) + 3H_2O(l) \rightarrow B_2H_6(g) + 3H_2O(l) + 3H_2O(g) & \Delta H = +36 \text{ kJ}
\end{array}$$

Sample Exercise 6.8, continued

This gives the reaction required by the problem:

$$2B(s) + 3H_2(g) \rightarrow B_2H_6(g) \qquad \Delta H = +36 \text{ kJ}$$

Thus ΔH for the synthesis of 1 mol of diborane is +36 kJ.

Hints for Using Hess's Law

Calculations involving Hess's law typically require that several reactions be manipulated and combined to finally give the reaction of interest. In doing this you should work *backwards* from the required reaction, using the reactants and products to decide how to manipulate the other given reactions at your disposal. Reverse any reactions as needed to give the required reactants and products, and multiply reactions to give the correct numbers of reactants and products. This is a process that involves some trial and error but can be very systematic if you always use the final reaction to guide what you do.

6.4 Standard Enthalpies of Formation

Purpose

- To define standard states.
- To show how to use standard enthalpies of formation to calculate ΔH for a reaction.

For a reaction under conditions of constant pressure, we can obtain the enthalpy change using a calorimeter. However, this can be a very difficult process. In fact, in some cases it is impossible, since some reactions do not lend themselves to such study. An example is the conversion of solid carbon from its graphite form to its diamond form:

$$C_{\text{graphite}}(s) \rightarrow C_{\text{diamond}}(s)$$

The value of ΔH for this process cannot be obtained by direct measurement in a calorimeter, because it is much too slow under normal conditions. However, as we saw in Sample Exercise 6.7, ΔH for this process can be calculated from heats of combustion. This is only one example of how useful it is to be able to *calculate* ΔH values for chemical reactions. We will next show how to do this using standard enthalpies of formation.

The **standard enthalpy of formation** (ΔH_f°) of a compound is defined as the *change in enthalpy that accompanies the formation of 1 mole of a compound from its elements with all substances in their standard states at 25°C.*

A *superscript zero* on a thermodynamic function, for example ΔH°, indicates that the corresponding process has been carried out under standard conditions. The

standard state for a substance is a reference state that must be defined because thermodynamic functions often depend on the concentrations (or pressures) of the substances involved. Thus to properly compare the thermodynamic properties of two substances, we must use a common reference state. Also for most thermodynamic properties, we can only measure *changes* in that property. For example, we have no method for determining absolute values of enthalpy. We can only measure enthalpy changes (ΔH values) by performing heat flow experiments.

Standard state is *not* the same as standard temperature and pressure (STP) for a gas.

The Following Are the Conventional Definitions of Standard States:

- For a gas, the standard state is a pressure of exactly 1 atmosphere.
- For a substance present in a solution, the standard state is a concentration of exactly 1 *M*.
- For a pure substance in a condensed state (liquid or solid), the standard state is the pure liquid or solid.
- For an element, the standard state is the form in which the element exists under conditions of 1 atmosphere and 25°C.

The standard state for oxygen is $O_2(g)$ at a pressure of 1 atmosphere; the standard state for sodium is Na(*s*); the standard state for mercury is Hg(*l*), and so on.

Several important characteristics of the definition of the enthalpy of formation will become clearer if we again consider the formation of nitrogen dioxide from elements in their standard states:

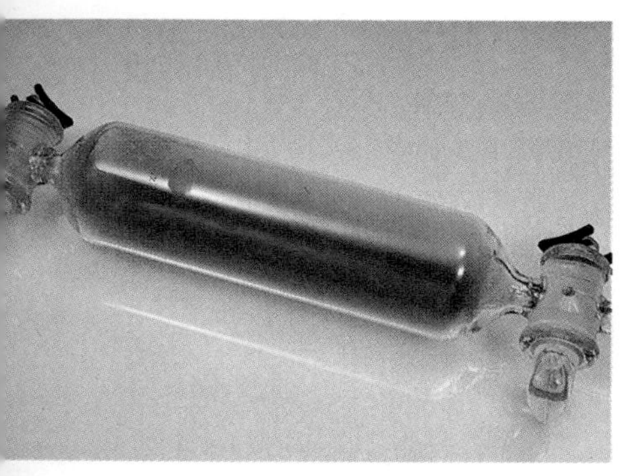
Brown nitrogen dioxide gas.

$$\tfrac{1}{2}N_2(g) + O_2(g) \rightarrow NO_2(g) \qquad \Delta H_f^\circ = 34 \text{ kJ/mol}$$

Note that the reaction is written so that both elements are in their standard states, and 1 mole of product is formed. Enthalpies of formation are *always* given per mole of product. The product also must be in its standard state.

The formation reaction for methanol is written as

$$C(s) + 2H_2(g) + \tfrac{1}{2}O_2(g) \rightarrow CH_3OH(l) \qquad \Delta H_f^\circ = -239 \text{ kJ/mol}$$

The standard state of carbon is graphite, the standard states for oxygen and hydrogen are the diatomic gases, and the standard state for methanol is the liquid.

The ΔH_f° values for some common substances are shown in Table 6.2. More values are found in Appendix 4. The importance of tabulated ΔH_f° values is that

Standard Enthalpies of Formation For Several Compounds at 25°C

Compound	ΔH_f° (kJ/mol)
$NH_3(g)$	−46
$NO_2(g)$	34
$H_2O(l)$	−286
$Al_2O_3(s)$	−1676
$Fe_2O_3(s)$	−826
$CO_2(g)$	−394
$CH_3OH(l)$	−239
$C_8H_{18}(l)$	−269

Table 6.2

enthalpies for many reactions can be calculated using these numbers. To see how this is done, let's calculate the standard enthalpy change for the combustion of methane:

$$CH_4(g) + 2O_2(g) \rightarrow CO_2(g) + 2H_2O(l)$$

Enthalpy is a state function, so we can invoke Hess's law and choose *any* convenient pathway from reactants to products and then sum the enthalpy changes. A convenient pathway, shown in Fig. 6.8, involves taking the reactants apart to the respective elements in their standard states in reactions (a) and (b), and then forming the products from these elements in reactions (c) and (d). This general pathway will work for any reaction, since atoms are conserved in a chemical reaction.

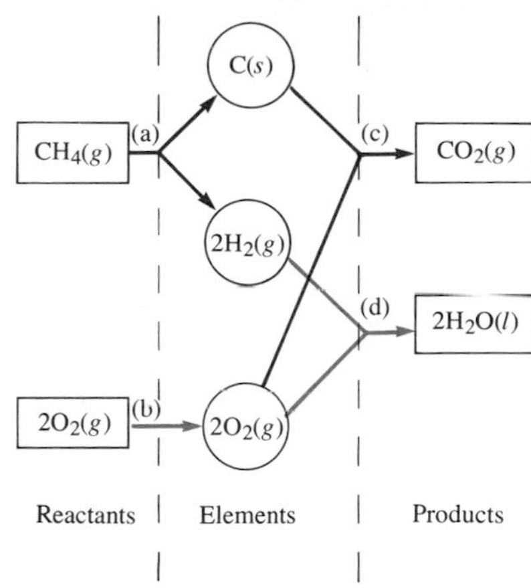

Figure 6.8

In this pathway for the combustion of methane, the reactants are first taken apart in reactions (a) and (b) to form the constituent elements in their standard states, which are then used to assemble the products in reactions (c) and (d).

Note from Fig. 6.8 that reaction (a), where methane is taken apart into its elements,

$$CH_4(g) \rightarrow C(s) + 2H_2(g)$$

is just the reverse of the formation reaction for methane:

$$C(s) + 2H_2(g) \rightarrow CH_4(g) \qquad \Delta H_f^\circ = -75 \text{ kJ/mol}$$

Since reversing a reaction means changing the sign of ΔH, but keeping the magnitude the same, ΔH for reaction (a) is $-\Delta H_f^\circ$ or 75 kJ. Thus $\Delta H^\circ_{(a)} = 75$ kJ.

Next we consider reaction (b). Here oxygen is already an element in its standard state, so no change is needed. Thus $\Delta H^\circ_{(b)} = 0$.

The next steps, reactions (c) and (d), use the elements formed in reactions (a) and (b) to form the products. Note that reaction (c) is simply the formation reaction for carbon dioxide:

$$C(s) + O_2(g) \rightarrow CO_2(g) \qquad \Delta H_f^\circ = -394 \text{ kJ/mol}$$

and

$$\Delta H^\circ_{(c)} = \Delta H_f^\circ \text{ for } CO_2(g) = -394 \text{ kJ}$$

Reaction (d) is the formation reaction for water:

$$H_2(g) + \tfrac{1}{2}O_2(g) \rightarrow H_2O(l) \qquad \Delta H_f^\circ = -286 \text{ kJ/mol}$$

However, since 2 moles of water are required in the balanced equation, we must form 2 moles of water from the elements:

$$2H_2(g) + O_2(g) \rightarrow 2H_2O(l)$$

Thus

$$\Delta H^\circ_{(d)} = 2 \times \Delta H_f^\circ \text{ for } H_2O(l) = 2(-286 \text{ kJ}) = -572 \text{ kJ}$$

We have now completed the pathway from the reactants to the products. The change in enthalpy for the reaction is the sum of the ΔH values (including their signs) for the steps:

$$\begin{aligned}\Delta H^\circ_{\text{reaction}} &= \Delta H^\circ_{(a)} + \Delta H^\circ_{(b)} + \Delta H^\circ_{(c)} + \Delta H^\circ_{(d)} \\ &= -\Delta H_f^\circ \text{ for } CH_4(g) + 0 + \Delta H_f^\circ \text{ for } CO_2(g) + 2 \times \Delta H_f^\circ \text{ for } H_2O(l) \\ &= -(-75 \text{ kJ}) + 0 + (-394 \text{ kJ}) + (-572 \text{ kJ}) \\ &= -891 \text{ kJ}\end{aligned}$$

Let's examine carefully the pathway we used in this example. First, the reactants were broken down into the elements. This involved reversing the formation reactions and thus switching the signs of the enthalpies of formation. The products

Subtraction means to reverse the sign and add.

were then constructed from these elements. This involved formation reactions and thus enthalpies of formation. We can summarize this entire process as follows: *The enthalpy change for a given reaction can be calculated by subtracting the enthalpies of formation of the reactants from the enthalpies of formation of the products.* Remember to multiply the enthalpies of formation by integers as required by the balanced equation. This statement can be represented symbolically as follows:

$$\Delta H^\circ_{\text{reaction}} = \Sigma \Delta H_f^\circ(\text{products}) - \Sigma \Delta H_f^\circ(\text{reactants}) \quad (6.1)$$

where the symbol Σ (sigma) means "to take the sum of the terms."

Elements are not included in the calculation since elements require no change in form. We have in effect *defined* the enthalpy of formation of an element in its standard state as zero, since we have chosen this as our reference point for calculating enthalpy changes in reactions.

Keep in Mind the Following Key Concepts When Doing Enthalpy Calculations:

- When a reaction is reversed, the magnitude of ΔH remains the same, but its sign changes.
- When the balanced equation for a reaction is multiplied by an integer, the value of ΔH for that reaction must be multiplied by the same integer.
- The change in enthalpy for a given reaction can be calculated from the enthalpies of formation of the reactants and products:

$$\Delta H^\circ_{\text{reaction}} = \Sigma \Delta H_f^\circ(\text{products}) - \Sigma \Delta H_f^\circ(\text{reactants})$$

- Elements in their standard states are not included in the $\Delta H_{\text{reaction}}$ calculations. That is, ΔH_f° for an element in its standard state is zero.

Elements in their standard states are not included in enthalpy calculations using ΔH_f° values.

Sample Exercise 6.9

Using the standard enthalpies of formation listed in Table 6.2, calculate the standard enthalpy change for the overall reaction that occurs when ammonia is burned in air to form nitrogen dioxide and water. This is the first step in the manufacture of nitric acid.

$$4NH_3(g) + 7O_2(g) \rightarrow 4NO_2(g) + 6H_2O(l)$$

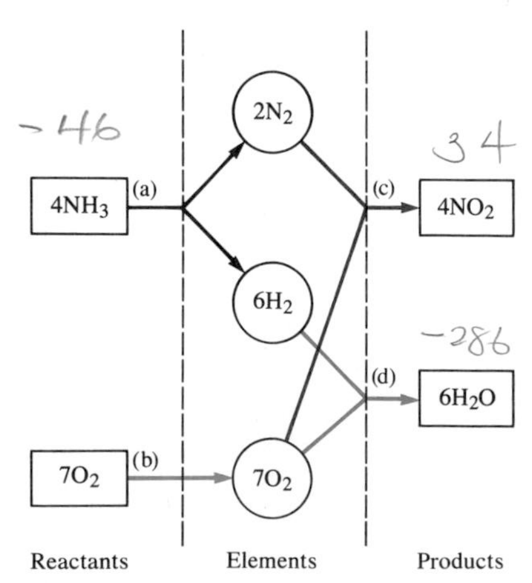

Figure 6.9

A pathway for the combustion of ammonia.

Solution

Let's use the pathway in which the reactants are broken down into elements in their standard states, which are then used to form the products (see Fig. 6.9).

STEP 1

Decomposition of $NH_3(g)$ into elements [reaction (a) in Fig. 6.9]. The first step is to decompose 4 mol of NH_3 into N_2 and H_2:

$$4NH_3(g) \rightarrow 2N_2(g) + 6H_2(g)$$

The above reaction is 4 times the *reverse* of the formation reaction for NH_3:

$$\tfrac{1}{2}N_2(g) + \tfrac{3}{2}H_2(g) \rightarrow NH_3(g) \qquad \Delta H_f^\circ = -46 \text{ kJ/mol}$$

Sample Exercise 6.9, continued

Thus

$$\Delta H°_{(a)} = 4 \text{ mol}[-(-46 \text{ kJ/mol})] = 184 \text{ kJ}$$

STEP 2

Elemental oxygen [reaction (b) in Fig. 6.9]. Since $O_2(g)$ is an element in its standard state, $\Delta H°_{(b)} = 0$.

We now have the elements $N_2(g)$, $H_2(g)$, and $O_2(g)$, which can be combined to form the products of the overall reaction.

STEP 3

Synthesis of $NO_2(g)$ from elements [reaction (c) in Fig. 6.9]. The overall reaction equation has 4 mol of NO_2. Thus the required reaction is 4 times the formation reaction for NO_2:

$$4 \times [\tfrac{1}{2}N_2(g) + O_2(g) \rightarrow NO_2(g)]$$

and

$$\Delta H°_{(c)} = 4 \times \Delta H_f° \text{ for } NO_2(g)$$

From Table 6.2, $\Delta H_f°$ for $NO_2(g) = 34$ kJ/mol and

$$\Delta H°_{(c)} = 4 \text{ mol} \times 34 \text{ kJ/mol} = 136 \text{ kJ}$$

STEP 4

Synthesis of $H_2O(l)$ from elements [reaction (d) in Fig. 6.9]. Since the overall reaction equation has 6 mol of $H_2O(l)$, the required reaction is 6 times the formation reaction for $H_2O(l)$:

$$6 \times [H_2(g) + \tfrac{1}{2}O_2(g) \rightarrow H_2O(l)]$$

and

$$\Delta H°_{(d)} = 6 \times \Delta H_f° \text{ for } H_2O(l)$$

From Table 6.2 $\Delta H_f°$ for $H_2O(l) = -286$ kJ/mol and

$$\Delta H°_{(d)} = 6 \text{ mol } (-286 \text{ kJ/mol}) = -1716 \text{ kJ}$$

To summarize, we have done the following:

$$\begin{matrix} 4NH_3(g) \xrightarrow{\Delta H°_{(a)}} \\ 7O_2(g) \xrightarrow{\Delta H°_{(b)} = 0} \end{matrix} \left\{ \begin{matrix} 2N_2(g) + 6H_2(g) \\ 7O_2(g) \end{matrix} \right\} \begin{matrix} \xrightarrow{\Delta H°_{(c)}} 4NO_2(g) \\ \xrightarrow{\Delta H°_{(d)}} 6H_2O(l) \end{matrix}$$

Elements in their standard states

We add the $\Delta H°$ values for the steps to get $\Delta H°$ for the overall reaction:

$$\begin{aligned} \Delta H°_{\text{reaction}} &= \Delta H°_{(a)} + \Delta H°_{(b)} + \Delta H°_{(c)} + \Delta H°_{(d)} \\ &= 4 \times -\Delta H_f°(NH_3(g)) + 0 + 4 \times \Delta H_f°(NO_2(g)) \\ &\quad + 6 \times \Delta H_f°(H_2O(l)) \\ &= 4 \times \Delta H_f°(NO_2(g)) + 6 \times \Delta H_f°(H_2O(l)) - 4 \times \Delta H_f°(NH_3(g)) \\ &= \Delta H_f°(\text{products}) - \Delta H_f°(\text{reactants}) \end{aligned}$$

Sample Exercise 6.9, continued

Remember elemental reactants and products do not need to be included, since ΔH_f° for an element in its standard state is zero. Note that we have again obtained Equation (6.1). The final solution is

$$\Delta H^\circ_{\text{reaction}} = 6 \times (-286\text{ kJ}) + 4 \times (34\text{ kJ}) - 4 \times (-46\text{ kJ})$$
$$= -1396\text{ kJ}$$

Now that we have shown the basis for Equation (6.1), we will make direct use of it to calculate ΔH for reactions in succeeding exercises.

Sample Exercise 6.10

Using enthalpies of formation, calculate the standard change in enthalpy for the thermite reaction,

$$2\text{Al}(s) + \text{Fe}_2\text{O}_3(s) \rightarrow \text{Al}_2\text{O}_3(s) + 2\text{Fe}(s)$$

This reaction occurs when a mixture of powdered aluminum and iron(III) oxide are ignited with a magnesium fuse.

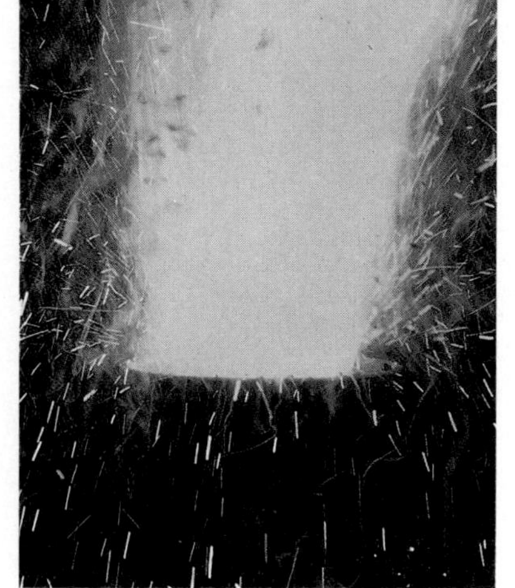

The thermite reaction is one of the most energetic chemical reactions known.

Solution

We use Equation (6.1):

$$\Delta H^\circ = \Sigma\Delta H_f^\circ(\text{products}) - \Sigma\Delta H_f^\circ(\text{reactants})$$

where

$$\Delta H_f^\circ \text{ for Fe}_2\text{O}_3(s) = -826\text{ kJ/mol}$$
$$\Delta H_f^\circ \text{ for Al}_2\text{O}_3(s) = -1676\text{ kJ/mol}$$
$$\Delta H_f^\circ \text{ for Al}(s) = \Delta H_f^\circ \text{ for Fe}(s) = 0$$

Thus

$$\Delta H^\circ_{\text{reaction}} = \Delta H_f^\circ \text{ for Al}_2\text{O}_3(s) - \Delta H_f^\circ \text{ for Fe}_2\text{O}_3(s)$$
$$= -1676\text{ kJ} - (-826\text{ kJ}) = -850\text{ kJ}$$

This reaction is so highly exothermic that the iron produced is initially molten. This process is often used as a lecture demonstration and also has been used in welding massive steel objects such as ships' propellers.

Sample Exercise 6.11

Methanol (CH_3OH) is often used as a fuel in high-performance engines. Using the data in Table 6.2, compare the standard enthalpy of combustion per gram of methanol to that per gram of gasoline. Gasoline is actually a mixture of compounds, but assume for this problem that gasoline is pure liquid octane (C_8H_{18}).

Solution

The combustion reaction for methanol is

$$2\text{CH}_3\text{OH}(l) + 3\text{O}_2(g) \rightarrow 2\text{CO}_2(g) + 4\text{H}_2\text{O}(l)$$

Sample Exercise 6.11, continued

Using the standard enthalpies of formation from Table 6.2 and Equation (6.1), we have

$$\begin{aligned}\Delta H^\circ_{\text{reaction}} &= 2 \times \Delta H_f^\circ \text{ for } CO_2(g) + 4 \times \Delta H_f^\circ \text{ for } H_2O(l) \\ &\quad - 2 \times \Delta H_f^\circ \text{ for } CH_3OH(l) \\ &= 2 \times (-394 \text{ kJ}) + 4 \times (-286 \text{ kJ}) - 2 \times (-239 \text{ kJ}) \\ &= -1454 \text{ kJ}\end{aligned}$$

Thus 1454 kJ of heat is evolved when 2 mol of methanol burn. The molecular weight of methanol is 32.0. This means that 1454 kJ of energy is produced when 64.0 g of methanol burns. The enthalpy of combustion per gram of methanol is

$$\frac{-1454 \text{ kJ}}{64.0 \text{ g}} = -22.7 \text{ kJ/g}$$

The combustion reaction for octane is

$$2C_8H_{18}(l) + 25O_2(g) \longrightarrow 16CO_2(g) + 18H_2O(l)$$

Using the standard enthalpies of information from Table 6.2 and Equation (6.1), we have

$$\begin{aligned}\Delta H^\circ_{\text{reaction}} &= 16 \times \Delta H_f^\circ \text{ for } CO_2(g) + 18 \times \Delta H_f^\circ \text{ for } H_2O(l) \\ &\quad - 2 \times \Delta H_f^\circ \text{ for } C_8H_{18}(l) \\ &= 16 \times (-394 \text{ kJ}) + 18 \times (-286 \text{ kJ}) - 2 \times (-269 \text{ kJ}) \\ &= -1.09 \times 10^4 \text{ kJ}\end{aligned}$$

This is the amount of heat evolved when 2 mol of octane burn. Since the molecular weight of octane is 114.2, the enthalpy of combustion per gram of octane is

$$\frac{-1.09 \times 10^4 \text{ kJ}}{2(114.2 \text{ g})} = -47.8 \text{ kJ/g}$$

The enthalpy of combustion per gram of octane is approximately twice that per gram of methanol. On this basis, gasoline appears to be superior to methanol for use in a racing car, where weight considerations are usually very important. Why then is methanol in fact used in racing cars? The answer is that methanol burns much more smoothly than gasoline in high-performance engines, and this advantage more than compensates for its weight disadvantage.

6.5 Present Sources of Energy

Purpose

- To discuss fossil fuels and the effects of their use on climate.

Woody plants, coal, petroleum, and natural gas provide a vast resource of energy that originally came from the sun. By the process of photosynthesis, plants store energy that can be claimed by burning the plants themselves or the decay

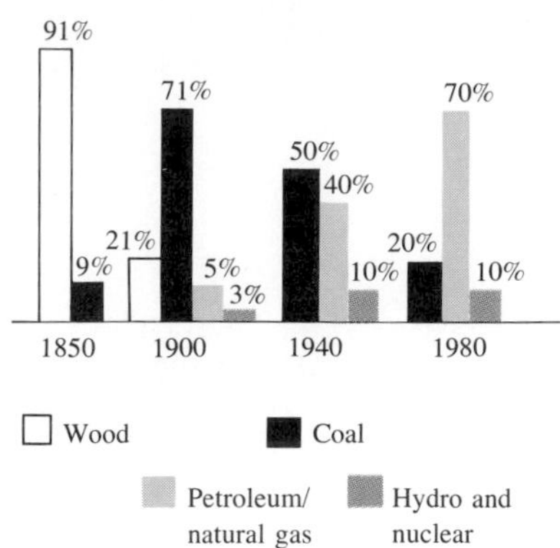

Figure 6.10

Energy sources used in the United States.

products that have been converted to **fossil fuels.** Although the United States currently depends heavily on petroleum for energy, this dependency is a relatively recent phenomenon, as shown in Fig. 6.10. In this section we discuss some sources of energy and their effects on the environment.

Petroleum and Natural Gas

Although how they were produced is not completely understood, petroleum and natural gas were most likely formed from the remains of marine organisms that lived approximately 500 million years ago. **Petroleum** is a thick, dark liquid composed mostly of compounds called *hydrocarbons* that contain carbon and hydrogen. (Carbon is unique among elements in the extent to which it can bond to itself to form chains of various lengths.) Table 6.3 gives the formulas and names for several common hydrocarbons. **Natural gas,** usually associated with petroleum deposits, consists mostly of methane but also contains significant amounts of ethane, propane, and butane.

Names and Formulas for Some Common Hydrocarbons	
Formula	Name
CH_4	Methane
C_2H_6	Ethane
C_3H_8	Propane
C_4H_{10}	Butane
C_5H_{12}	Pentane
C_6H_{14}	Hexane
C_7H_{16}	Heptane
C_8H_{18}	Octane

Table 6.3

Uses of the Various Petroleum Fractions	
Petroleum fraction in terms of numbers of carbon atoms	Major uses
C_5–C_{10}	Gasoline
C_{10}–C_{18}	Kerosene Jet fuel
C_{15}–C_{25}	Diesel fuel Heating oil Lubricating oil
$>C_{25}$	Asphalt

Table 6.4

The composition of petroleum varies somewhat, but it is mostly hydrocarbons having chains of from 5 to more than 25 carbons. To be used efficiently, the petroleum must be separated into fractions by boiling. The lighter molecules (having the lowest boiling points) can be boiled off, leaving the heavier ones behind (see Figure 1.16, Chapter 1). The uses of various petroleum fractions are shown in Table 6.4.

The petroleum era began when the demand for lamp oil during the Industrial Revolution outstripped the traditional sources: animal fats and whale oil. In response to this increased demand, Edwin Drake drilled the first oil well in 1859 at Titusville, Pennsylvania. The petroleum from this well was refined to produce *kerosene* (fraction C_{10}–C_{18}), which served as an excellent lamp oil. *Gasoline* (fraction C_5–C_{10}) had limited use and was often discarded. However, the development of the electric light decreased the need for kerosene. Also, the advent of the "horseless carriage" with its gasoline-powered engine signaled the birth of the gasoline age.

As gasoline became more important, new ways were sought to increase the yield of gasoline obtained from each barrel of petroleum. William Burton invented a process at Standard Oil of Indiana called *pyrolytic (high-temperature) cracking*. In this process the heavier molecules of the kerosene fraction are heated to about 700°C, causing them to break (crack) into the smaller molecules of hydrocarbons in the gasoline fraction. As cars became larger, more efficient internal combustion engines were designed. Because of the uneven burning of the gasoline then available, these engines "knocked," producing unwanted noise and even engine damage. Intensive research to find additives that would promote smoother burning produced tetraethyl lead, $(C_2H_5)_4Pb$, a very effective "antiknock" agent.

The addition of tetraethyl lead to gasoline became a common practice, and by 1960 gasoline contained as much as 3 grams of lead per gallon. As we have discovered so often in recent years, technological advances can produce environmental problems. To prevent air pollution from automobile exhaust, catalytic converters have been added to car exhaust systems. The effectiveness of these converters, however, is destroyed by lead. The use of leaded gasoline has also greatly increased the amount of lead in the environment, where it can be ingested by animals and humans. For these reasons the use of lead in gasoline is being phased out. This has required extensive (and expensive) modifications of engines and of the gasoline refining process.

Coal

Coal was formed from the remains of plants that were buried and subjected to pressure and heat over long periods of time. Plant materials have a high content of cellulose, a complex molecule whose empirical formula is CH_2O but whose molecular weight is around 500,000. After the plants and trees that flourished on the earth at various times and places died and were buried, chemical changes gradually lowered the oxygen and hydrogen content of the cellulose molecules. Coal "matures" through four stages: lignite, subbituminous, bituminous, and anthracite. Each stage has a higher carbon to oxygen and carbon to hydrogen ratio; that is, the relative carbon content gradually increases. Typical elemental compositions of the various coals are given in Table 6.5. The energy available from the combustion of a given mass of coal increases as the carbon content increases. Anthracite is the most valuable coal, and lignite the least.

Coal has variable composition depending on both its age and location.

Elemental Composition of Various Types of Coal

	Mass percent of each element				
Type of coal	C	H	O	N	S
Lignite	71	4	23	1	1
Subbituminous	77	5	16	1	1
Bituminous	80	6	8	1	5
Anthracite	92	3	3	1	1

Table 6.5

Coal train in Illinois.

Coal is an important and plentiful fuel in the United States, currently furnishing approximately 20% of our energy. As the supply of petroleum dwindles, the share of the energy supply from coal is expected to increase to approximately 30% by the year 2000. However, coal is expensive and dangerous to mine underground, and the strip mining of fertile farmland in the Midwest or of scenic land in the West causes obvious problems. In addition, the burning of coal, especially high-sulfur coal, yields air pollutants such as sulfur dioxide, which in turn can lead to acid rain, as we learned in Chapter 5. However, even if coal were pure carbon, the carbon dioxide produced when it was burned would still have significant effects on the earth's climate.

Effects of Carbon Dioxide on Climate

The earth receives a tremendous quantity of radiant energy from the sun, about 30% of which is reflected into space by the earth's atmosphere. The remaining energy passes through the atmosphere to the earth's surface. Some of this energy is absorbed by plants to drive photosynthesis and some by the oceans to evaporate water, but most of it is absorbed by soil, rock, and water to increase the temperature of the earth's surface. This energy is in turn radiated from the heated surface mainly as *infrared radiation,* often called heat radiation.

The electromagnetic spectrum including visible and infrared radiation is discussed in Chapter 7.

The atmosphere, like window glass, is transparent to visible light, but does not allow all of the infrared radiation to pass through. Molecules in the atmosphere, principally H_2O and CO_2, strongly absorb infrared radiation and radiate it back

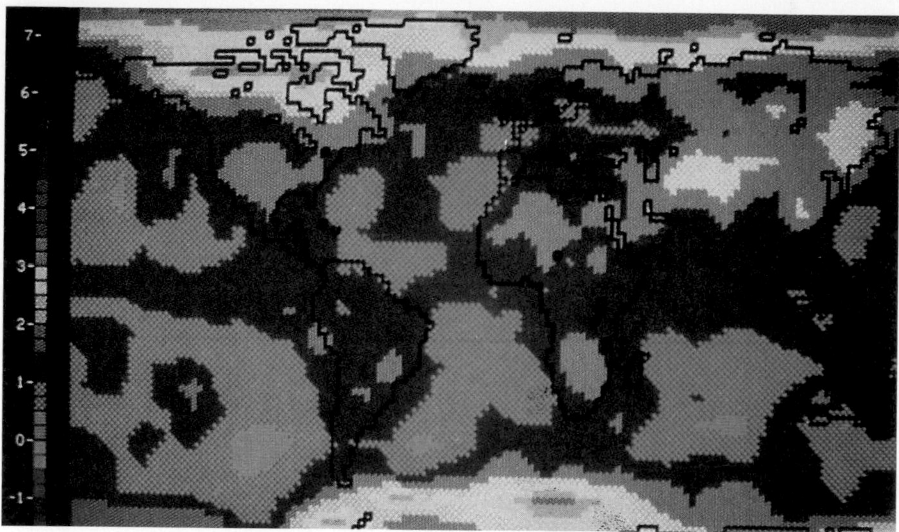

Changes in annual mean temperature between 1958 and 2010–19.

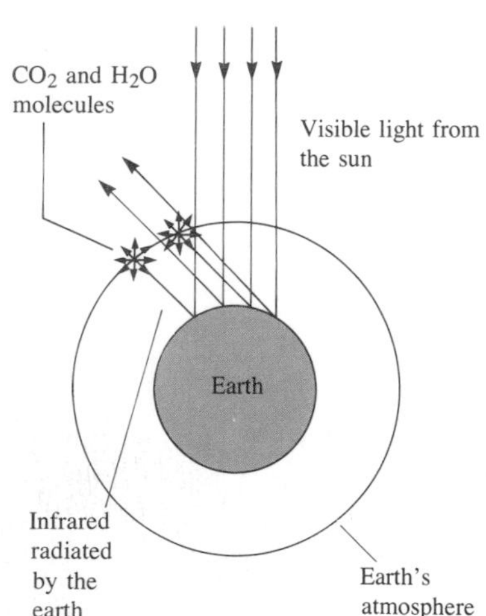

Figure 6.11

The earth's atmosphere is transparent to visible light from the sun. This visible light strikes the earth, and part of it is changed to infrared radiation. This infrared radiation from the earth's surface is strongly absorbed by CO_2, H_2O, and other molecules present in smaller amounts (for example, CH_4 and N_2O) in the atmosphere. In effect, the atmosphere traps some of the energy, acting like the glass in a greenhouse and keeping the earth warmer than it would otherwise be.

The average tmperature of the earth's surface is 288 K. It would be ~255 K without the "greenhouse gases."

toward the earth, as shown in Fig. 6.11. A net amount of thermal energy is thus retained by the earth's atmosphere. This causes the earth to be much warmer than it would be without its atmosphere. In a way the atmosphere acts like the glass of a greenhouse, which is transparent to visible light but absorbs infrared radiation, thus raising the temperature inside the building. This **greenhouse effect** is seen even more spectacularly on Venus, where the dense atmosphere is mainly responsible for the high surface temperature of that planet.

Thus the temperature of the earth's surface is controlled to a significant extent by the carbon dioxide and water content of the atmosphere. The effect of atmospheric moisture (humidity) is apparent in the Midwest. In summer, when the humidity is high, the heat of the sun is retained well into the night, giving very high nighttime temperatures. On the other hand, in winter, the coldest temperatures always occur on clear nights, when the low humidity allows efficient radiation of energy back into space.

The atmosphere's water content is controlled by the water cycle (evaporation and precipitation), and the average remains constant over the years. However, as fossil fuels have been used more extensively, the carbon dioxide concentration has increased by about 16% from 1880 to 1980. Projections indicate that the carbon dioxide content of the atmosphere may be double what it was in 1880 by the twenty-first century. This *could* increase the earth's average temperature by as much as 3°C, causing dramatic changes in climate and greatly affecting the growth of food crops.

How well can we predict long-term effects? Because weather has been studied for a period of time that is miniscule compared to the age of the earth, the factors that control the earth's climate in the long range are not clearly understood. For example, we do not understand what causes the earth's periodic ice ages. So indeed it is difficult to estimate the impact of the increasing carbon dioxide levels.

In fact, the variation in the earth's average temperature over the last century is somewhat confusing. In the northern latitudes during the last century, the average temperature rose by 0.8°C over a period of 60 years, then cooled by 0.5°C during the next 25 years, and finally warmed by 0.2°C in the last 15 years. Such fluctuations do not match the steady increase in carbon dioxide. However, in southern latitudes and near the equator, during the last century, the average temperature showed a steady rise totaling 0.4°C. This figure is in reasonable agreement with the predicted effect of the increasing carbon dioxide concentration over that period. Although the problem is complex, it appears that we really should be concerned about levels of carbon dioxide in the atmosphere as we consider major new sources of energy.

6.6 New Energy Sources

Purpose

- To discuss energy alternatives.
- To compare the available energy of various fuels.

As we search for the energy sources of the future, we need to consider economic, climatic, and supply factors. There are several potential energy sources: the sun (solar), nuclear processes (fission and fusion), biomass (plants), and synthetic fuels. Direct use of the sun's radiant energy to heat our homes and run our factories and transportation systems seems a sensible long-term goal. But what do we do now? Conservation of fossil fuels is one obvious step, but substitutes for fossil fuels must also be found. We will discuss some alternative sources of energy here. Nuclear power will be considered in Chapter 21.

Photovoltaic cells for the conversion of radiant energy to electrical energy.

Coal Conversion

One alternative energy source involves using a traditional fuel—coal—in new ways. Since transportation costs for solid coal are high, more energy-efficient fuels are being developed from coal. One possibility is to produce a gaseous fuel. Substances like coal that contain large molecules have high boiling points and tend to be solids or thick liquids. To convert coal from a solid to a gas therefore requires reducing the size of the molecules; the coal structure must be broken down in a process called *coal gasification*. This is done by treating the coal with oxygen and steam at high temperatures to break many of the carbon-carbon bonds. These bonds are replaced by carbon-hydrogen and carbon-oxygen bonds as the coal fragments react with the water and oxygen. The process is represented in Fig. 6.12 on page 248. The desired product is a mixture of carbon monoxide and hydrogen called *synthetic gas,* or **syngas,** and methane (CH_4) gas. Since all the components of this product can react with oxygen to release heat in a combustion reaction, this gas is a useful fuel.

One of the most important considerations in designing an industrial process is the efficient use of energy. In coal gasification some of the reactions are exothermic:

An industrial process must be energy-efficient.

$$C(s) + 2H_2(g) \rightarrow CH_4(g) \qquad \Delta H^\circ = -75 \text{ kJ}$$
$$C(s) + \tfrac{1}{2}O_2(g) \rightarrow CO(g) \qquad \Delta H^\circ = -111 \text{ kJ}$$
$$C(s) + O_2(g) \rightarrow CO_2(g) \qquad \Delta H^\circ = -394 \text{ kJ}$$

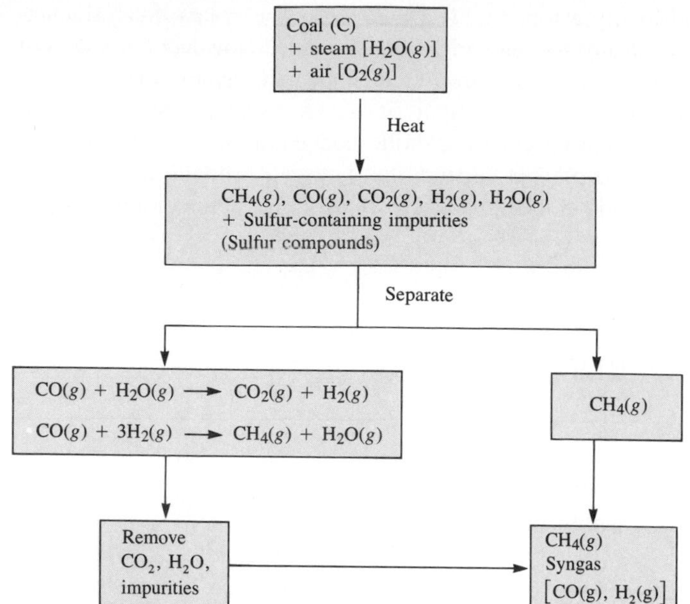

Figure 6.12

Coal gasification. Reaction of coal with a mixture of steam and air breaks down the large hydrocarbon molecules in the coal to smaller gaseous molecules, which can be used as fuels.

Other gasification reactions are endothermic, for example:

$$C(s) + H_2O(g) \rightarrow H_2(g) + CO(g) \qquad \Delta H° = 131 \text{ kJ}$$

If such conditions as the rate of feed of coal, air, and steam are carefully controlled, the correct temperature can be maintained in the process without using any external energy source. That is, an energy balance is maintained.

As we said, syngas can be used directly as a fuel, but it is also important as a raw material to produce other fuels. For example, syngas can be directly converted to methanol:

$$CO(g) + 2H_2(g) \rightarrow CH_3OH(l)$$

Methanol is used in the production of synthetic fibers and plastics and can also be used as a fuel. In addition, it can be converted directly to gasoline. Approximately half of South Africa's gasoline supply comes from methanol produced from syngas.

In addition to coal gasification, the formation of *coal slurries* is another new use of coal. A slurry is a suspension of fine particles in a liquid, and coal must be pulverized and mixed with water to form a slurry. The slurry can be handled, stored, and burned in ways similar to those used for *residual oil,* a heavy fuel oil from petroleum accounting for 13% of U.S. petroleum imports. One hope is that coal slurries might replace solid coal and residual oil as fuels for electricity-generating power plants. However, the water needed for slurries might place an unacceptable burden on water resources, especially in the western states.

Hydrogen as a Fuel

If you have ever seen a lecture demonstration where hydrogen-oxygen mixtures were ignited or if you have seen a newsfilm or an old photo of the Hindenburg disaster, you have witnessed a demonstration of hydrogen's potential as a fuel. The combustion reaction is

$$H_2(g) + \tfrac{1}{2}O_2(g) \rightarrow H_2O(l) \qquad \Delta H° = -286 \text{ kJ}$$

As we saw in Sample Exercise 6.6, the heat of combustion of $H_2(g)$ per gram is approximately 2.5 times that of natural gas. In addition, hydrogen has a real advantage over fossil fuels in that the only product of hydrogen combustion is water; fossil fuels also produce carbon dioxide. But even though it appears that hydrogen is a very logical choice for a major future fuel, there are three main problems: the cost of production, storage, and transport.

First let's look at the production problem. Although hydrogen is very abundant on earth, virtually none of it exists as the free gas. Currently, the main source of hydrogen gas is from the treatment of natural gas with steam:

$$CH_4(g) + H_2O(g) \rightarrow 3H_2(g) + CO(g)$$

We can calculate ΔH for this reaction using Equation (6.1):

$$\begin{aligned}\Delta H^\circ &= \Sigma\Delta H_f^\circ(\text{products}) - \Sigma\Delta H_f^\circ(\text{reactants})\\ &= \Delta H_f^\circ \text{ for } CO(g) - \Delta H_f^\circ \text{ for } CH_4(g) - \Delta H_f^\circ \text{ for } H_2O(g)\\ &= -111\text{ kJ} - (-75\text{ kJ}) - (-242\text{ kJ}) = 206\text{ kJ}\end{aligned}$$

Note that this reaction is highly endothermic; treating methane with steam is not an efficient way to obtain hydrogen for fuel. It would be much more economical to burn the methane directly.

A virtually inexhaustible supply of hydrogen exists in the waters of the world's oceans. However, the reaction

$$H_2O(l) \rightarrow H_2(g) + \tfrac{1}{2}O_2(g)$$

requires 286 kJ of energy per mole of liquid water, and under current circumstances large-scale production of hydrogen from water is not economically feasible. However, several methods for such production are currently being studied: electrolysis of water, thermal decomposition of water, and biological decomposition of water.

Electrolysis of water involves passing an electrical current through it, as shown in Fig. 1.12 in Chapter 1. The present cost of electricity makes the hydrogen produced by electrolysis too expensive to be competitive as a fuel. However, if in the future we develop more efficient sources of electricity, this situation could change.

Electrolysis will be discussed in Chapter 17.

Thermal decomposition is another method for producing hydrogen from water. This involves heating the water to several thousand degrees where it spontaneously decomposes into hydrogen and oxygen. However, attaining temperatures in this range would be very expensive even if a practical heat source and a suitable reaction container were available.

In the thermochemical decomposition of water, chemical reactions, as well as heat, are used to "split" water into its components. One such system involves the following reactions (the temperature required for each is given in parentheses):

$$\begin{array}{rl} 2HI \rightarrow I_2 + H_2 & (425^\circ C)\\ 2H_2O + SO_2 + I_2 \rightarrow H_2SO_4 + 2HI & (90^\circ C)\\ H_2SO_4 \rightarrow SO_2 + H_2O + \tfrac{1}{2}O_2 & (825^\circ C)\\ \hline \text{Net reaction: } H_2O \rightarrow H_2 + \tfrac{1}{2}O_2 & \end{array}$$

Note that the HI is not consumed in the net reaction. Note also that the maximum temperature required is 825°C, a temperature that is feasible if a nuclear reactor is used as a heat source. A current research goal is to find a system for which the required temperatures are low enough that sunlight can be used as the energy source.

Green plants, such as these wildflowers at the base of Torrey's Peak in Colorado, store energy from the sun.

But what about the systems on earth that biologically decompose water without the aid of electricity or high temperatures? In the process of photosynthesis, green plants absorb carbon dioxide and water and use them along with energy from the sun to produce the substances needed for growth. Scientists have studied photosynthesis for years, hoping to get answers to humanity's food and energy shortages. At present much of this research involves attempts to modify the photosynthetic process so that plants will release hydrogen gas from water instead of using the hydrogen to produce complex compounds. Small-scale experiments have shown that under certain conditions plants do produce hydrogen gas, but the yields are far from being commercially useful. Thus economical production of hydrogen gas remains unrealized.

The storage and transportation of hydrogen present two problems. First, on metal surfaces the H_2 molecule decomposes to atoms. Since the atoms are so small, they can migrate into the metal, causing structural changes that make it brittle. This might lead to a pipeline failure if hydrogen were pumped under high pressure.

A second problem is the relatively small amount of energy that is available *per unit volume* of hydrogen. Although the energy available per gram of hydrogen is significantly greater than that per gram of methane, the energy available per given volume of hydrogen is about one-third that available from the same volume of methane. This is demonstrated in Sample Exercise 6.12.

Sample Exercise 6.12

Compare the energy available from the combustion of a given volume of methane and hydrogen at the same temperature and pressure.

Solution

In Sample Exercise 6.6, we calculated the heat released for the combustion of methane and hydrogen: 55 kJ/g CH_4 and 141 kJ/g H_2. We also know from our study of gases that 1 mol of $H_2(g)$ has the same volume as 1 mol of $CH_4(g)$ at the same temperature and pressure (assuming ideal behavior). Thus, for molar volumes of both gases under the same conditions of temperature and pressure:

$$\frac{\text{Enthalpy of combustion of 1 molar volume of } H_2(g)}{\text{Enthalpy of combustion of 1 molar volume of } CH_4(g)}$$

$$= \frac{\text{enthalpy of combustion per mole of } H_2}{\text{enthalpy of combustion per mole of } CH_4}$$

$$= \frac{(-141 \cancel{\text{kJ}}/\cancel{\text{g}})(2.02 \cancel{\text{g } H_2}/\cancel{\text{mol } H_2})}{(-55 \cancel{\text{kJ}}/\cancel{\text{g}})(16.04 \cancel{\text{g } CH_4}/\cancel{\text{mol } CH_4})}$$

$$= \frac{-285}{-882} \approx \frac{1}{3}$$

Thus about three times the volume of hydrogen is needed to furnish the same energy as a given volume of methane.

Could hydrogen be considered as a potential fuel for automobiles? This is an intriguing question. The internal combustion engines in automobiles can be easily adapted to burn hydrogen. However, the primary difficulty is the storage of enough hydrogen to give an automobile a reasonable range. This is illustrated by Sample Exercise 6.13.

Sample Exercise 6.13

Assuming that the combustion of hydrogen gas provides three times as much energy per gram as gasoline, calculate the volume of liquid H_2 (density = 0.0710 g/mL) required to furnish the energy contained in 80.0 L (about 20 gal) of gasoline (density = 0.740 g/mL). Calculate also the volume that this hydrogen would occupy as a gas at 1.00 atm and 25°C.

Solution

The mass of 80.0 L of gasoline is

$$80.0\ \text{L} \times \frac{1000\ \text{mL}}{1\ \text{L}} \times \frac{0.740\ \text{g}}{\text{mL}} = 59{,}200\ \text{g}$$

Since H_2 furnishes three times as much energy per gram as gasoline, only a third as much liquid hydrogen is needed to furnish the same energy:

$$\text{Mass of } H_2(l) \text{ needed} = \frac{59{,}200\ \text{g}}{3} = 19{,}700\ \text{g}$$

Since density = mass/volume, then volume = mass/density, and the volume of $H_2(l)$ needed is

$$V = \frac{19{,}700\ \text{g}}{0.0710\ \text{g/mL}} = 2.77 \times 10^5\ \text{mL} = 277\ \text{L}$$

Thus 277 L of liquid H_2 is needed to furnish the same energy of combustion as 80.0 L of gasoline.

To calculate the volume that this hydrogen would occupy as a gas at 1.00 atm and 25°C, we use the ideal gas law:

$$PV = nRT$$

In this case $P = 1.00$ atm, $T = 273 + 25°\text{C} = 298$ K, and $R = 0.08206$ L atm/K mol. Also,

$$n = 19{,}700\ \text{g}\ H_2 \times \frac{1\ \text{mol}\ H_2}{2.02\ \text{g}\ H_2} = 9.75 \times 10^3\ \text{mol}\ H_2$$

Thus

$$V = \frac{nRT}{P} = \frac{(9.75 \times 10^3\ \text{mol})(0.08206\ \text{L atm/K mol})(298\ \text{K})}{1.00\ \text{atm}} = 2.38 \times 10^5\ \text{L} = 238{,}000\ \text{L}$$

At 1 atm and 25°C, the hydrogen gas needed to replace 20 gal of gasoline occupies a volume of 238,000 L.

Chemical Impact

Heat Packs

A skier is trapped by a sudden snowstorm. After building a snow cave for protection, she realizes her hands and feet are freezing; she is in danger of frostbite. Then she remembers the four small packs in her pocket. She removes the plastic cover from each one to reveal a small paper packet. She places one packet in each boot and one in each mitten. Soon her hands and feet are toasty warm.

These ''magic'' packets of energy contain a mixture of powdered iron, activated carbon, sodium chloride, cellulose (sawdust), and zeolite, all moistened by a little water. The paper cover is permeable to air.

The exothermic reaction that produces the heat is a very common one—the rusting of iron. The overall reaction can be represented as

$$4Fe(s) + 3O_2(g) \rightarrow 2Fe_2O_3(s) \qquad \Delta H° = -1652 \text{ kJ}$$

although in reality it is somewhat more complicated. When the plastic envelope is removed, O_2 molecules penetrate the paper and the reaction begins.

The oxidation of iron by oxygen occurs naturally. Any steel surface exposed to the atmosphere inevitably rusts. However, this process is quite slow—much too slow to be useful in hot packs. However, if the iron is ground into a fine powder, the resulting increase in surface area causes the reaction with oxygen to be fast enough to warm hands and feet. The packet can produce heat for up to six hours.

A popular brand of portable hot pack.

You can see from Sample Exercise 6.13 that an automobile would need a huge tank to hold enough hydrogen gas to have a typical mileage range. Clearly, hydrogen must be stored as a liquid or in some other way. Is this feasible? Because of its very low boiling point (20 K), storage of liquid hydrogen requires a superinsulated container that can withstand high pressures. Storage in this manner would be both expensive and hazardous because of the potential for explosion. Thus storage of hydrogen in the individual automobile as a liquid does not seem practical.

Metal hydrides are discussed in Chapter 18.

A much better alternative seems to be the use of metals that absorb hydrogen to form solid metal hydrides:

$$H_2(g) + M(s) \rightarrow MH_2(s)$$

To use this method of storage, hydrogen gas would be pumped into a tank containing the solid metal, where it would be absorbed to form the hydride, whose volume would be little more than that of the metal. This hydrogen would then be available for combustion in the engine by release of $H_2(g)$ from the hydride as needed:

$$MH_2(s) \rightarrow M(s) + H_2(g)$$

Several types of solids that absorb hydrogen to form hydrides are being studied for use in hydrogen-powered vehicles.

Chemical Impact

Anaerobic Engines: Energy Without Oxygen

A fireman frantically tries to rescue a person trapped in a burning building, but his chainsaw dies from a lack of oxygen. This potential tragedy could be averted if the chainsaw were powered by an engine whose fuel did not require oxygen—an anaerobic engine. Such an engine has been suggested by two Canadian scientists* who demonstrated that the compound $(CH_3)_3COOC(CH_3)_3$, called di-tert-butylperoxide or DTBP, will decompose exothermically when compressed in the cylinder of an engine, thus providing energy to run the engine without oxygen. The decomposition reaction is

$$(CH_3)_3COOC(CH_3)_3(g) \rightarrow C_2H_6(g) + 2(CH_3)_2CO(g)$$

Under conditions where oxygen is plentiful (aerobic conditions) the engine runs normally with the DTBP fuel reacting with O_2 to produce CO_2 and H_2O. However, when the oxygen supply is severely limited, as in the smoke from a fire, the engine continues to run (although more slowly) from the energy produced by the decomposition of DTBP described by the above equation. Thus the DTBP fuel would allow a chainsaw to be used under the conditions encountered at the scene of the fire.

The Canadian scientists have also suggested that large anaerobic engines could be developed to power equipment used in mine rescues and other situations where lack of oxygen would prevent an engine with normal fuel from operating.

*H. O. Pritchard and P. Q. E. Clothier, "Anaerobic Operation of an Internal Combustion Engine," *J. Chem. Soc., Chem. Commun.* (1986), p. 1529.

Other Energy Alternatives

Many other energy sources are being considered for future use. The western states, especially Colorado, contain huge deposits of *oil shale*, which consists of a complex carbon-based material called kerogen contained in porous rock formations. These deposits have the potential of being a larger energy source than the vast petroleum deposits of the Middle East. The main problem with oil shale is that the trapped fuel is not fluid and cannot be pumped. To recover the fuel, the rock must be heated to a temperature of 250°C or higher to decompose the kerogen to smaller molecules that produce gaseous and liquid products. This process is expensive and yields large quantities of waste rock, which have a negative environmental impact.

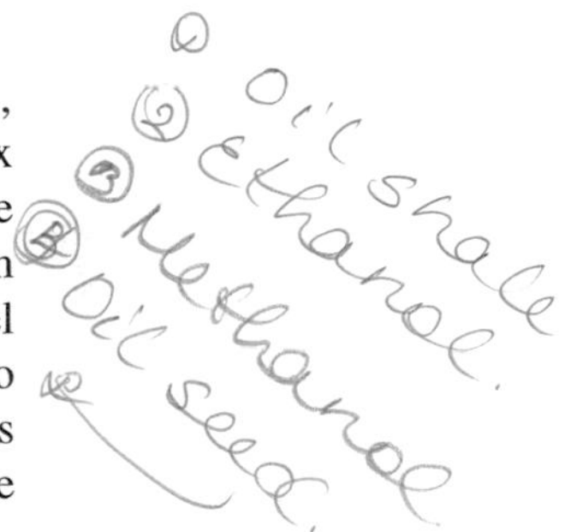

Ethanol (C_2H_5OH) is another fuel with the potential to supplement, if not replace, gasoline. The most common method of producing ethanol is fermentation, a process in which sugar is changed to alcohol by the action of yeast. The sugar can come from virtually any source, including fruits and grains, although fuel-grade ethanol would probably come mostly from corn. Car engines can burn pure alcohol or *gasohol,* an alcohol-gasoline mixture (10% ethanol in gasoline), with little modification. Gasohol is now widely available in the United States. The use of pure alcohol as a motor fuel is not feasible in most of the United States because it does not vaporize easily when temperatures are low. However, pure ethanol could be a very practical fuel in warm climates. For example, in Brazil large quantities of ethanol fuel are being produced for cars.

Methanol (CH_3OH), an alcohol similar to ethanol, which has been used successfully for many years in race cars, is now being evaluated as a motor fuel in California. A major gasoline retailer has agreed to install pumps at 25 locations to dispense a fuel that is 85% methanol and 15% gasoline for use in specially prepared automobiles. The California Energy Commission feels that methanol has great potential for providing a secure, long-term energy supply that would help alleviate air-quality problems. Arizona and Colorado are also considering methanol as a major source of portable energy.

Another potential source of liquid fuels is oil squeezed from seeds (*seed oil*). For example, some farmers in North Dakota, South Africa, and Australia are now using sunflower oil to replace diesel fuel. Oil seeds, found in a wide variety of plants, can be processed to produce an oil composed mainly of carbon and hydrogen, which of course reacts with oxygen to produce carbon dioxide, water, and heat. It is hoped that oil-seed plants can be developed that will thrive under soil and climatic conditions unsuitable for corn and wheat. The main advantage of seed oil as a fuel is that it is renewable. Ideally, fuel would be grown just like food crops.

FOR REVIEW

Summary

Energy can be defined as the capacity to do work or to produce heat. The law of conservation of energy (or the first law of thermodynamics) states that energy can be converted from one form to another, but it cannot be created or destroyed. Heat is a means for the transfer of energy. Chemical reactions can either evolve or absorb heat, depending on whether the potential energy stored in the bonds of the reactant molecules is greater or less than the energy stored in the bonds of the product molecules.

When a reaction evolves heat, it is said to be exothermic; a reaction that absorbs heat from its surroundings is endothermic. Energy is a state function; that is, a change in energy in going from one state to another is independent of the pathway.

Heat flow associated with a chemical reaction is measured by calorimetry, a technique that involves measuring temperature changes when a body absorbs or discharges heat. For a process carried out at constant pressure, the heat flow equals the change in enthalpy (ΔH).

Since enthalpy is a state function, the change in enthalpy in going from a given set of reactants to a given set of products is the same regardless of whether the reaction takes place in one step or in a series of steps (a principle known as Hess's law). We can therefore calculate the enthalpy change for an overall reaction by summing the ΔH values from a series of steps that yield that net reaction. Also, it is often convenient to calculate the ΔH value for a reaction from values for the standard enthalpies of formation (ΔH_f°) of the reactants and products. The standard enthalpy of formation is the enthalpy change that accompanies the formation of 1 mole of a compound from its elements, where all substances are in their standard states. For a given reaction

$$\Delta H_{\text{reaction}} = \Sigma \Delta H_f^\circ(\text{products}) - \Sigma \Delta H_f^\circ(\text{reactants})$$

The practical aspects of energy production from fossil fuels involve problems of supply and environmental impact. The combustion of carbon-based fuels produces carbon dioxide, which is involved, along with atmospheric water, in controlling the surface temperature of the earth. Increased carbon dioxide concentration in the atmosphere may produce gradual global warming due to the greenhouse effect. Alternative fuels, such as syngas from coal, hydrogen from the breakdown of water, and ethanol from fermentation of sugar, are being studied as replacements for petroleum products.

Key Terms

Section 6.1
energy
law of conservation of energy
potential energy
kinetic energy
heat
work
pathway
state function
system
surroundings
exothermic
endothermic
thermodynamics
first law of thermodynamics
internal energy

Section 6.2
enthalpy
calorimeter
calorimetry
heat capacity
specific heat capacity
molar heat capacity
constant pressure calorimetry
constant volume calorimetry
bomb calorimeter

Section 6.3
Hess's law

Section 6.4
standard enthalpy of formation
standard state

Section 6.5
fossil fuels
petroleum
natural gas
coal
greenhouse effect

Section 6.6
syngas

Exercises

A blue exercise number indicates that the answer to that exercise appears at the back of this book and a solution appears in the Solutions Guide.

Potential and Kinetic Energy

1. Consider the diagram at the right. Ball A is allowed to fall and strike ball B. Assume that all of ball A's energy is transferred to ball B, at point I, and that there is no loss of energy to other sources. What will be the kinetic energy and the potential energy of ball B at point II? For a falling object, the potential energy is given by PE = mgz, where m is the mass in kilograms, g is the gravitational constant (9.8 m/s^2), and z is the distance in meters.

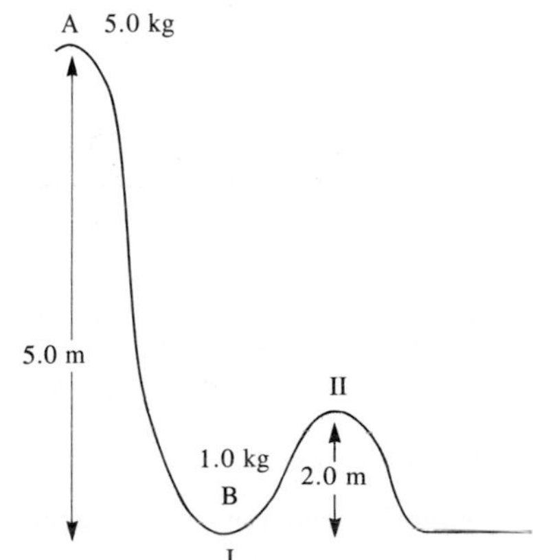

2. Calculate the kinetic energy of a baseball (mass = 5.25 oz) with a velocity of 1.0×10^2 mi/h.

3. Which has the greater kinetic energy, an object with a mass of 2.0 kg and a velocity of 1.0 m/s, or an object with a mass of 1.0 kg and a velocity of 2.0 m/s?

4. Baseball sportscasters often comment that it is harder to hit a slow pitch for a home run than it is to hit a fast ball. In terms of kinetic energy, why is this so?

5. Calculate the kinetic energy of a 1.0×10^{-5}-g object with a velocity of 2.0×10^5 cm/s.

Heat and Work

6. Calculate ΔE for each of the following cases:
   a. $q = +51$ kJ, $w = -15$ kJ
   b. $q = +100.$ kJ, $w = -65$ kJ
   c. $q = -65$ kJ, $w = -20.$ kJ
   d. $q = -47$ kJ, $w = +88$ kJ
   e. $q = +82$ kJ, $w = +47$ kJ

7. In which of the five cases in Exercise 6 does the system do work on the surroundings?

8. The volume of an ideal gas is decreased from 5.0 L to 5.0 mL at a constant pressure of 2.0 atm. Calculate the work.

9. A balloon filled with 39.1 mol of helium has a volume of 875 L at 0.0°C and 1.00 atm pressure. The temperature of the balloon is increased to 38.0°C, and it expands to a volume of 998 L, with the pressure remaining constant. Calculate q, w, and ΔE for the helium in the balloon. (The molar heat capacity for helium gas is 20.8 J/°C mol.)

10. Consider a mixture of air and gasoline vapor in a cylinder with a piston. The original volume is 40. cm^3. If the combustion of this mixture releases 950. J of energy, to what volume will the gases expand against a constant pressure of 650. torr, if all of the energy of combustion is converted into work to push back the piston?

11. A balloon contains 313 g He at a pressure of 1.00 atm. The volume of the balloon is 1910 L. The temperature is decreased by 15°C and the volume decreases to 1840 L. Calculate q, w, and ΔE for the helium in the balloon. (The molar heat capacity of helium gas is 20.8 J/°C mol.)

Properties of Enthalpy

12. Are the following processes exothermic or endothermic?
    a. When solid KBr is dissolved in water, the solution gets colder.
    b. Natural gas (CH_4) is burned in a forced-air furnace.
    c. When concentrated H_2SO_4 is added to water, the solution gets very hot.
    d. Water is boiled in a teakettle.

13. The equation for the fermentation of glucose to alcohol and carbon dioxide is

$$C_6H_{12}O_6 \rightarrow 2C_2H_5OH + 2CO_2$$

The enthalpy change for the reaction is -67 kJ. Is the reaction exothermic or endothermic? Is energy, in the form of heat, absorbed or evolved as the reaction occurs? Calculate the enthalpy change per gram of glucose reacted.

14. The reaction

$$SO_3(g) + H_2O(l) \rightarrow H_2SO_4(aq)$$

is the last step in the commercial production of sulfuric acid. The enthalpy change for this reaction is -266 kJ. In designing a sulfuric acid plant, is it necessary to provide for heating or cooling the reaction mixture? Explain your answer.

15. The enthalpy change for the following reaction is -891 kJ:

$$CH_4(g) + 2O_2(g) \rightarrow CO_2(g) + 2H_2O(l)$$

Calculate the enthalpy change for each of these cases:
   a. 1.00 g of methane is burned in an excess of oxygen.
   b. 1.00×10^3 L of methane gas at 740. torr and 25°C is burned in an excess of oxygen.

16. For the reaction

$$S(s) + O_2(g) \rightarrow SO_2(g) \qquad \Delta H° = -296 \text{ kJ/mol}$$

   a. How much heat is evolved when 275 g of sulfur is burned in excess O_2?
   b. How much heat is evolved when 25 mol of sulfur is burned in excess O_2?
   c. How much heat is evolved when 150. g of sulfur dioxide is produced?

17. For the reaction

$$B_2H_6(g) + 3O_2(g) \rightarrow B_2O_3(s) + 3H_2O(g) \qquad \Delta H° = -2035 \text{ kJ}$$

how much heat is released when each of the following amounts of diborane, B_2H_6, is burned?
   a. 1.0 g
   b. 1.0 kg
   c. 1.0 mol
   d. 1.0×10^2 mol
   e. 2.6×10^{22} molecules
   f. a mixture of 10.0 g B_2H_6 and 10.0 g O_2

Calorimetry and Heat Capacity

18. Why is it that ΔH is obtained directly using a coffee-cup calorimeter and ΔE is obtained using a bomb calorimeter?

19. The specific heat capacity of aluminum is 0.900 J/°C g.
    a. How much energy is needed to raise the temperature of a 8.50×10^2-g block of aluminum from 22.8°C to 94.6°C?
    b. What is the molar heat capacity of aluminum?

20. The specific heat capacity of graphite is 0.71 J/°C g. How much heat is required to
a. raise the temperature of 1.0 mol of graphite by 1.0°C?
b. raise the temperature of 850 g of graphite by 150°C?
c. raise the temperature of 75 kg of graphite from 294 K to 348 K?

21. It takes 78.2 J to raise the temperature of 45.6 g of lead by 13.3°C. What is the specific heat capacity of lead? What is the molar heat capacity of lead?

22. A 28.2-g sample of nickel is heated to 99.8°C and placed in a coffee-cup calorimeter containing 150.0 g of water at a temperature of 23.5°C. After the metal cools, the final temperature of metal and water is 25.0°C. Calculate the specific heat capacity of nickel, assuming that no heat escapes to the surroundings or is transferred to the calorimeter.

23. A coffee-cup calorimeter initially contains 125 g of water, at a temperature of 24.2°C. Potassium bromide (10.5 g), also at 24.2°C, is added to the water, and the final temperature is 21.1°C. What is the heat of solution (the heat accompanying the dissolving of the salt) of potassium bromide in J/g and kJ/mol? Assume the specific heat capacity of the solution is 4.18 J/°C g and that no heat is transferred to the surroundings or to the calorimeter.

24. In a coffee-cup calorimeter, 100.0 mL of 1.0 *M* NaOH and 100.0 mL of 1.0 *M* HCl are mixed. Both solutions were originally at 24.6°C. After the reaction, the temperature is 31.3°C. Assuming all solutions have a density of 1.0 g/cm^3 and a specific heat capacity of 4.18 J/°C g, what is the enthalpy change for the neutralization of HCl by NaOH? Assume that no heat is lost to the surroundings or the calorimeter.

25. In a coffee-cup calorimeter, 50.0 mL of 0.100 *M* $AgNO_3$ and 50.0 mL of 0.100 *M* HCl are mixed. The following reaction occurs:

$$Ag^+(aq) + Cl^-(aq) \rightarrow AgCl(s)$$

If the two solutions were initially at 22.6°C and the final temperature is 23.4°C, calculate ΔH for the above reaction. Assume there is 100.0 g of combined solution with a specific heat capacity of 4.18 J/°C g.

26. Camphor ($C_{10}H_{16}O$) has a heat of combustion of 5903.6 kJ/mol. A sample of camphor with a mass of 0.1204 g is burned in a bomb calorimeter. The temperature increases by 2.28°C. What is the heat capacity of the calorimeter?

27. A 0.1964-g sample of quinone ($C_6H_4O_2$) is burned in a bomb calorimeter that has a heat capacity of 1.56 kJ/°C. The temperature of the calorimeter increases by 3.2°C. Calculate the energy of combustion of quinone per gram and per mole.

28. The combustion of 0.1584 g of benzoic acid increases the temperature of a bomb calorimeter by 2.54°C. Calculate the heat capacity of the calorimeter. (The energy released by combustion of benzoic acid is 26.42 kJ/g.) A 0.2130-g sample of vanillin ($C_8H_8O_3$) is then burned in the same calorimeter. The temperature increases by 3.25°C. What is the energy of combustion per gram of vanillin and per mole of vanillin?

29. A swimming pool, 10.0 m by 4.0 m, is filled to a depth of 3.0 m with water at a temperature of 20.2°C. How much energy is required to raise the temperature of the water to 30.0°C?

30. In a bomb calorimeter, the bomb is surrounded by water that must be added, and the amount of water is not necessarily constant from experiment to experiment. The mass of water must be measured for each experiment and the heat capacity of the calorimeter is broken down into two parts: for the water and for the calorimeter components. If a calorimeter contains 1.00 kg of water and has a total heat capacity of 10.84 kJ/°C, what is the heat capacity of the calorimeter components?

Hess's Law

31. Given the following data:

$$S(s) + \tfrac{3}{2}O_2(g) \rightarrow SO_3(g) \qquad \Delta H^\circ = -395.2 \text{ kJ}$$
$$2SO_2(g) + O_2(g) \rightarrow 2SO_3(g) \qquad \Delta H^\circ = -198.2 \text{ kJ}$$

calculate ΔH° for the reaction

$$S(s) + O_2(g) \rightarrow SO_2(g)$$

32. Given the following data:

$$N_2(g) + 2O_2(g) \rightarrow 2NO_2(g) \qquad \Delta H^\circ = 67.7 \text{ kJ}$$
$$N_2(g) + 2O_2(g) \rightarrow N_2O_4(g) \qquad \Delta H^\circ = 9.7 \text{ kJ}$$

calculate ΔH° for the dimerization of NO_2:

$$2NO_2(g) \rightarrow N_2O_4(g)$$

33. Given the following data:

$$H_2(g) + \tfrac{1}{2}O_2(g) \rightarrow H_2O(l) \qquad \Delta H^\circ = -285.8 \text{ kJ}$$

$$N_2O_5(g) + H_2O(l) \rightarrow 2HNO_3(l) \qquad \Delta H^\circ = -76.6 \text{ kJ}$$

$$\tfrac{1}{2}N_2(g) + \tfrac{3}{2}O_2(g) + \tfrac{1}{2}H_2(g) \rightarrow HNO_3(l) \qquad \Delta H^\circ = -174.1 \text{ kJ}$$

calculate the ΔH° for the reaction

$$2N_2(g) + 5O_2(g) \rightarrow 2N_2O_5(g)$$

34. Given the following data:

$$Fe_2O_3(s) + 3CO(g) \rightarrow 2Fe(s) + 3CO_2(g) \quad \Delta H° = -28 \text{ kJ}$$

$$3Fe_2O_3(s) + CO(g) \rightarrow 2Fe_3O_4(s) + CO_2(g) \quad \Delta H° = -59 \text{ kJ}$$

$$Fe_3O_4(s) + CO(g) \rightarrow 3FeO(s) + CO_2(g) \quad \Delta H° = +38 \text{ kJ}$$

calculate $\Delta H°$ for the reaction

$$FeO(s) + CO(g) \rightarrow Fe(s) + CO_2(g)$$

35. The standard enthalpy of combustion of solid carbon to form carbon dioxide is −393.7 kJ per mol of carbon, and the standard enthalpy of combustion of carbon monoxide to form carbon dioxide is −283.3 kJ per mol of CO. Use these data to calculate $\Delta H°$ for the reaction

$$2C(s) + O_2(g) \rightarrow 2CO(g)$$

36. Given the following data:

$$C_2H_2(g) + \tfrac{5}{2}O_2(g) \rightarrow 2CO_2(g) + H_2O(l) \quad \Delta H° = -1300. \text{ kJ}$$

$$C(s) + O_2(g) \rightarrow CO_2(g) \quad \Delta H° = -394 \text{ kJ}$$

$$H_2(g) + \tfrac{1}{2}O_2(g) \rightarrow H_2O(l) \quad \Delta H° = -286 \text{ kJ}$$

calculate $\Delta H°$ for the reaction

$$2C(s) + H_2(g) \rightarrow C_2H_2(g)$$

37. Given the following data:

$$2O_3(g) \rightarrow 3O_2(g) \quad \Delta H° = -427 \text{ kJ}$$

$$O_2(g) \rightarrow 2O(g) \quad \Delta H° = +495 \text{ kJ}$$

$$NO(g) + O_3(g) \rightarrow NO_2(g) + O_2(g) \quad \Delta H° = -199 \text{ kJ}$$

calculate $\Delta H°$ for the reaction

$$NO(g) + O(g) \rightarrow NO_2(g)$$

38. The bombardier beetle uses an explosive discharge as a defensive measure. The chemical reaction involved is the oxidation of hydroquinone by hydrogen peroxide to produce quinone and water:

$$C_6H_4(OH)_2(aq) + H_2O_2(aq) \rightarrow C_6H_4O_2(aq) + 2H_2O(l)$$

Calculate $\Delta H°$ for this reaction from the following data:

$$C_6H_4(OH)_2(aq) \rightarrow C_6H_4O_2(aq) + H_2(g) \quad \Delta H° = +177.4 \text{ kJ}$$

$$H_2(g) + O_2(g) \rightarrow H_2O_2(aq) \quad \Delta H° = -191.2 \text{ kJ}$$

$$H_2(g) + \tfrac{1}{2}O_2(g) \rightarrow H_2O(g) \quad \Delta H° = -241.8 \text{ kJ}$$

$$H_2O(g) \rightarrow H_2O(l) \quad \Delta H° = -43.8 \text{ kJ}$$

39. Given the following data:

$$O_2(g) + H_2(g) \rightarrow 2OH(g) \quad \Delta H° = +77.9 \text{ kJ}$$

$$O_2(g) \rightarrow 2O(g) \quad \Delta H° = +495 \text{ kJ}$$

$$H_2(g) \rightarrow 2H(g) \quad \Delta H° = +435.9 \text{ kJ}$$

calculate $\Delta H°$ for the reaction

$$O(g) + H(g) \rightarrow OH(g)$$

40. Calculate $\Delta H°$ for the reaction:

$$N_2H_4(l) + O_2(g) \rightarrow N_2(g) + 2H_2O(l)$$

given the following data:

$$2NH_3(g) + 3N_2O(g) \rightarrow 4N_2(g) + 3H_2O(l) \quad \Delta H° = -1010. \text{ kJ}$$

$$N_2O(g) + 3H_2(g) \rightarrow N_2H_4(l) + H_2O(l) \quad \Delta H° = -317 \text{ kJ}$$

$$2NH_3(g) + \tfrac{1}{2}O_2(g) \rightarrow N_2H_4(l) + H_2O(l) \quad \Delta H° = -143 \text{ kJ}$$

$$H_2(g) + \tfrac{1}{2}O_2(g) \rightarrow H_2O(l) \quad \Delta H° = -286 \text{ kJ}$$

Standard Enthalpies of Formation

41. Give the definition of the standard enthalpy of formation for a substance. Write separate reactions for the formation of NaCl, H_2O, $C_6H_{12}O_6$, and $PbSO_4$ that have $\Delta H°$ values equal to $\Delta H_f°$ for each compound.

42. Use the values of $\Delta H_f°$ in Appendix 4 to calculate $\Delta H°$ for the following reactions:
   a. $2NH_3(g) + 3O_2(g) + 2CH_4(g) \rightarrow 2HCN(g) + 6H_2O(g)$
   b. $Ca_3(PO_4)_2(s) + 3H_2SO_4(l) \rightarrow 3CaSO_4(s) + 2H_3PO_4(l)$
   c. $NH_3(g) + HCl(g) \rightarrow NH_4Cl(s)$
   d. $SiCl_4(l) + 2H_2O(l) \rightarrow SiO_2(s) + 4HCl(aq)$
   e. $MgO(s) + H_2O(l) \rightarrow Mg(OH)_2(s)$

43. The Ostwald process for the commercial production of nitric acid from ammonia and oxygen involves the following steps:

$$4NH_3(g) + 5O_2(g) \rightarrow 4NO(g) + 6H_2O(g)$$
$$2NO(g) + O_2(g) \rightarrow 2NO_2(g)$$
$$3NO_2(g) + H_2O(l) \rightarrow 2HNO_3(aq) + NO(g)$$

a. Use the values of ΔH_f° in Appendix 4 to calculate the value of ΔH° for each of the above reactions.
b. Write the overall equation for the production of nitric acid by the Ostwald process by combining the above equations. (Water is also a product.) Is the overall reaction exothermic or endothermic?

44. Calculate ΔH° for each of the following reactions using the data in Appendix 4:

$$4Na(s) + O_2(g) \rightarrow 2Na_2O(s)$$
$$2Na(s) + 2H_2O(l) \rightarrow 2NaOH(aq) + H_2(g)$$
$$2Na(s) + CO_2(g) \rightarrow Na_2O(s) + CO(g)$$

Use these values to show why a water or carbon dioxide fire extinguisher might not be effective in putting out a sodium fire.

45. The reusable booster rockets of the space shuttle use a mixture of aluminum and ammonium perchlorate as fuel. A possible reaction is

$$3Al(s) + 3NH_4ClO_4(s) \rightarrow Al_2O_3(s) + AlCl_3(s) + 3NO(g) + 6H_2O(g)$$

Calculate ΔH° for this reaction.

46. The space shuttle orbiter utilizes the oxidation of methyl hydrazine by dinitrogentetroxide for propulsion. The balanced reaction is

$$5N_2O_4(l) + 4N_2H_3CH_3(l) \rightarrow 12H_2O(g) + 9N_2(g) + 4CO_2(g)$$

Calculate ΔH° for this reaction.

47. Does the reaction in Exercise 45 or that in Exercise 46 produce more energy per kilogram of the reactant mixture (stoichiometric amounts)?

48. Water gas is produced from the reaction of steam with coal. Assuming coal is pure graphite, calculate ΔH° for this reaction:

$$C(s) + H_2O(g) \rightarrow H_2(g) + CO(g)$$

49. Titanium (IV) oxide is one of the most commonly used white paint pigments. It is produced by the reaction

$$TiCl_4(g) + 2H_2O(g) \rightarrow TiO_2(s) + 4HCl(g)$$

Calculate ΔH° for this reaction.

50. Calculate ΔH° for each of the following reactions, which occur in the atmosphere:
a. $C_2H_4(g) + O_3(g) \rightarrow CH_3CHO(g) + O_2(g)$
b. $O_3(g) + NO(g) \rightarrow NO_2(g) + O_2(g)$
c. $SO_3(g) + H_2O(l) \rightarrow H_2SO_4(aq)$
d. $2NO(g) + O_2(g) \rightarrow 2NO_2(g)$

51. For the reaction:

$$2ClF_3(g) + 2NH_3(g) \rightarrow N_2(g) + 6HF(g) + Cl_2(g) \qquad \Delta H^\circ = -1196 \text{ kJ}$$

calculate ΔH_f° for $ClF_3(g)$.

52. The enthalpy of combustion of ethene gas, $C_2H_4(g)$, is -1411.1 kJ/mol at 298 K. Given the following enthalpies of formation, calculate ΔH_f° for $C_2H_4(g)$.

$CO_2(g)$	-393.5 kJ/mol
$H_2O(l)$	-285.9 kJ/mol

Energy Consumption and Sources

53. Assume that the energy in Exercise 29 comes from the combustion of methane (CH_4). What volume of methane, measured at STP, must be burned? ($\Delta H^\circ_{combustion}$ for $CH_4 = -802$ kJ/mol CH_4.)

54. Ethanol (C_2H_5OH) has been proposed as an alternative fuel. Calculate the enthalpy of combustion per gram of ethanol.

55. Syngas can be burned directly or converted to methanol. Calculate ΔH° for the reaction

$$CO(g) + 2H_2(g) \rightarrow CH_3OH(l)$$

56. Photosynthetic plants use the following reaction to produce glucose, cellulose, etc.

$$6CO_2(g) + 6H_2O(l) \xrightarrow{\text{Sunlight}} C_6H_{12}O_6(s) + 6O_2(g)$$

How might extensive destruction of forests exacerbate the greenhouse effect?

57. Some automobiles and buses have been equipped to burn propane (C_3H_8) as a fuel. Compare the amount of energy that can be obtained per gram of $C_3H_8(g)$ to that per gram of gasoline, assuming that gasoline is octane, $C_8H_{18}(l)$. (See Sample Exercise 6.11.) Look up the physical properties of propane. What disadvantages are there to using propane instead of gasoline as a fuel?

58. What safety hazards might be associated with storing hydrogen as a metal hydride?

Additional Exercises

59. Consider a gas at an initial pressure of 5.0 atm and an initial volume of 1.0 L that undergoes a change to 2.0 atm and 4.0 L by two different pathways.

Pathway 1

$$\begin{matrix} P_i = 5.0 \text{ atm} \\ V_i = 1.0 \text{ L} \end{matrix} \rightarrow \begin{matrix} P_i = 5.0 \text{ atm} \\ V_f = 4.0 \text{ L} \end{matrix} \rightarrow \begin{matrix} P_f = 2.0 \text{ atm} \\ V_f = 4.0 \text{ L} \end{matrix}$$

Pathway 2

$$\begin{matrix} P_i = 5.0 \text{ atm} \\ V_i = 1.0 \text{ L} \end{matrix} \rightarrow \begin{matrix} P_f = 2.0 \text{ atm} \\ V_i = 1.0 \text{ L} \end{matrix} \rightarrow \begin{matrix} P_f = 2.0 \text{ atm} \\ V_f = 4.0 \text{ L} \end{matrix}$$

These pathways are summarized on the following graph of P versus V:

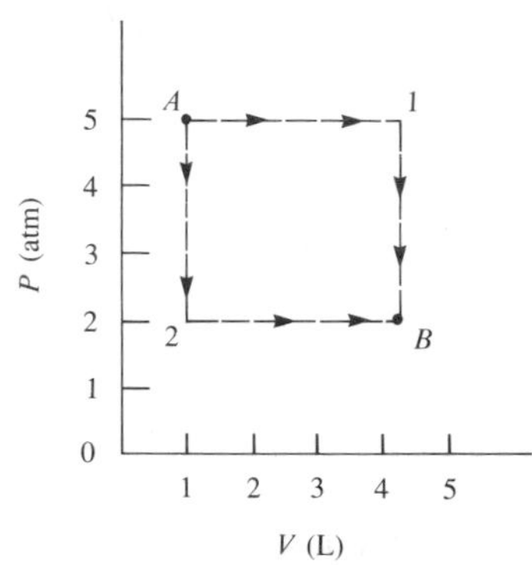

Calculate the work associated with the two pathways. Is work a state function? Explain.

60. A lead bullet with a mass of 7.8 g and a velocity of 5.0×10^4 cm/s strikes a 1.00-kg wooden block and becomes embedded in the wood. If both the bullet and the block are initially at a temperature of 25.0°C, what is the final temperature? Assume no heat loss to the surroundings and that all of the kinetic energy of the bullet is converted to heat. Specific heat capacities: wood, 2.1 J/°C g; lead, 0.13 J/°C g.

61. Consider the reaction

$$2HCl(aq) + Ba(OH)_2(aq) \rightarrow BaCl_2(aq) + 2H_2O(l) \qquad \Delta H = -118 \text{ kJ}$$

How much heat is evolved when 100.0 mL of 0.500 M HCl is reacted with 300.0 mL of 0.500 M $Ba(OH)_2$?

62. On Easter Sunday, April 3, 1983, nitric acid spilled from a tank car near downtown Denver, Colorado. The spill was neutralized with sodium carbonate. The reaction is

$$2HNO_3(aq) + Na_2CO_3(s) \rightarrow 2NaNO_3(aq) + H_2O(l) + CO_2(g)$$

a. Calculate $\Delta H°$ for this reaction. Approximately 2.0×10^4 gal of nitric acid was spilled. Assuming that the acid was 70.0% HNO_3 by mass (the rest is water) with a density of 1.42 g/cm^3, how much sodium carbonate was required for complete neutralization and how much heat was evolved? [$\Delta H_f°(NaNO_3(aq)) = -467$ kJ/mol]

b. According to *The Denver Post* for April 4, 1983, authorities feared a volatile reaction might occur during the neutralization. Considering the magnitude of $\Delta H°$, what was their major concern?

63. A piece of chocolate cake contains about 400 Calories. A nutritional Calorie is equal to one thousand calories (thermochemical calories), or one kilocalorie, which is equal to 4.18 kJ. How many 8-in-high steps must a 180-lb man climb to expend the 400 Cal from the piece of cake? See Exercise 1 for the formula for potential energy.

64. Hess's law is really just another statement of the first law of thermodynamics. Explain.

65. Hydrazine (N_2H_4) is used as a fuel in liquid-fueled rockets and oxygen (assume $O_2(g)$ for this problem) or dinitrogentetroxide ($N_2O_4(l)$) can be used as the oxidizing agent. In both cases the products are nitrogen gas and gaseous water. Write balanced equations for the two reactions and calculate $\Delta H°$ using standard enthalpies of formation. Compare these values to the result in Exercise 46. Which of the three combinations (per kilogram of the stoichiometric reactant mixture) is the most efficient rocket fuel?

66. The combustion reaction in an oxyacetylene torch is

$$C_2H_2(g) + \tfrac{5}{2}O_2(g) \rightarrow 2CO_2(g) + H_2O(g)$$

Calculate $\Delta H°$ for this reaction. Assume that all of the heat generated by this reaction is used in raising the temperature of the carbon dioxide and water produced. What is the maximum temperature of the flame of an oxyacetylene torch? At constant pressure, the molar heat capacities of carbon dioxide gas and water vapor are 37.1 J/°C mol and 33.6 J/°C mol, respectively. Assume an initial temperature of 25.0°C.

67. Consider the following changes:

a. $H_2O(g) \rightarrow H_2O(l)$
b. $H_2(g) + Cl_2(g) \rightarrow 2HCl(g)$
c. $2H_2(g) + O_2(g) \rightarrow 2H_2O(g)$
d. $Xe(g) + F_2(g) \rightarrow XeF_2(s)$
e. $NiCl_2 \cdot 6H_2O(s) \rightarrow NiCl_2(s) + 6H_2O(g)$
f. $CO_2(s) \rightarrow CO_2(g)$

At constant pressure, in which of the changes is work done by the system on the surroundings? by the surroundings on the system? In which of them is no work done?

68. If all of the heat in Sample Exercise 6.3 comes from the combustion of propane (C_3H_8), what mass of propane must be burned?

69. What is a state function? Enthalpy and internal energy are state functions as a direct consequence of the first law of thermodynamics. Why is this so?

70. Write reactions for which the enthalpy change will be:

a. $\Delta H_f°$ for solid aluminum oxide
b. the standard enthalpy of combustion of liquid ethanol, $C_2H_5OH(l)$

c. the standard enthalpy of neutralization of barium hydroxide solution by hydrochloric acid
d. ΔH_f° for gaseous vinyl chloride, $C_2H_3Cl(g)$
e. the enthalpy of combustion of liquid benzene, $C_6H_6(l)$
f. the enthalpy of solution of solid ammonium bromide

71. Why is it useful to define standard enthalpies of formation?

72. Calculate ΔH° for the reaction

$$2K(s) + 2H_2O(l) \rightarrow 2KOH(aq) + H_2(g)$$

A 5.00-g chunk of potassium is dropped into 1.00 L of water at 24.0°C. What is the final temperature of the water after the above reaction occurs? Assume that all of the heat is used in raising the temperature of the water. (Never do this. It is very dangerous!)

73. The bomb of a certain bomb calorimeter has a volume of 50.0 mL. It will be charged with oxygen to a pressure of 125 atm at 22°C. What is the maximum mass of benzoic acid, $C_7H_6O_2$, that can be burned in this calorimeter if you want to have 25 times as much oxygen in the bomb as will be needed?

74. High quality hi-fi power amplifiers generate large amounts of heat. To dissipate the heat and prevent damage to the electronic devices, heat radiating metal fins are used. Would it be better to make these fins out of iron or aluminum? Why? (See Table 6.1 for specific heat capacities.)

CHAPTER 7

Atomic Structure and Periodicity

CONTENTS

In the last 200 years a great deal of experimental evidence has accumulated to support the atomic model. This theory has proved both extremely useful and physically reasonable. When atoms were first suggested by the Greek philosophers Democritus and Leucippus about 400 B.C., and in fact for the following 20 centuries, no convincing experimental evidence was available to support their existence. The concept of atoms was based mostly on intuition. The first real scientific data were gathered by Lavoisier and others from quantitative measurements of chemical reactions. The results of these stoichiometric experiments led John Dalton to propose the first systematic atomic theory. Dalton's theory, although crude, has stood the test of time extremely well. To this day we remain firmly convinced of the value of the atomic model.

Once we came to "believe in" atoms, it was logical to ask: What is the nature of an atom? Does an atom have parts, and, if so, what are they? In Chapter 2 we considered some of the experiments most important in shedding light on the nature of the atom. Now we will see how the atomic theory has evolved to its present state.

One of the most striking things about the chemistry of the elements is the periodic repetition of properties. There are several groups of elements that show great similarities in chemical behavior. As we saw in Chapter 2, these similarities led to the development of the periodic table of the elements. In this chapter we will see that the modern theory of atomic structure accounts for periodicity in terms of the arrangement of electrons. This is probably the most important success of the current atomic model.

However, before we examine atomic structure, we must consider the nature of electromagnetic radiation, which plays a central role in the study of the atom.

< Light diffracted by the closely-spaced grooves on a compact disc.

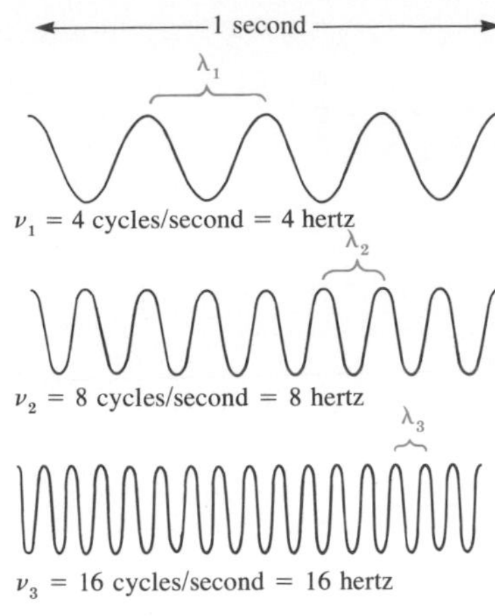

Figure 7.1

The nature of waves. Note that the radiation with the shortest wavelength has the highest frequency.

7.1 Electromagnetic Radiation

Purpose

- To characterize electromagnetic radiation in terms of wavelength, frequency, and speed.

One of the ways that energy travels through space is by **electromagnetic radiation.** The light from the sun, the energy used to cook food in a microwave oven, the X rays used by dentists, and the radiant heat from a fireplace are all examples of electromagnetic radiation. Although these forms of radiant energy seem quite different, they all exhibit the same type of wavelike behavior and travel at the speed of light in a vacuum.

Waves have three primary characteristics: wavelength, frequency, and speed. **Wavelength** (symbolized by the Greek letter lambda, λ) is the *distance between two consecutive peaks or troughs in a wave,* as shown in Fig. 7.1. The **frequency** (symbolized by the Greek letter nu, ν) is defined as the *number of waves (cycles) per second that pass a given point in space.* Since all types of electromagnetic radiation travel at the speed of light, short-wavelength radiation must have a high frequency. You can see this in Fig. 7.1, where three waves are shown traveling between two points at constant speed. Note that the wave with the shortest wavelength (λ_3) has the highest frequency and the wave with the longest wavelength (λ_1) has the lowest frequency. This implies an inverse relationship between wavelength and frequency, that is, $\lambda \propto 1/\nu$, or

Wavelength (λ) and frequency (ν) are inversely related.

$$\lambda\nu = c$$

where λ is the wavelength in meters, ν is the frequency in cycles per second (abbreviated s), and c is the speed of light (2.9979×10^8 m/s). In the SI system cycles is understood and the unit becomes 1/s, or s^{-1}, which is called the *hertz* (abbreviated Hz).

c = speed of light = 2.9979×10^8 m/s

Electromagnetic radiation is classified as shown in Fig. 7.2. Radiation provides an important means of energy transfer. For example, the energy from the sun reaches the earth mainly in the form of visible and ultraviolet radiation, while the glowing coals of a fireplace transmit heat energy by infrared radiation. In a microwave oven the water molecules in food absorb microwave radiation, which in-

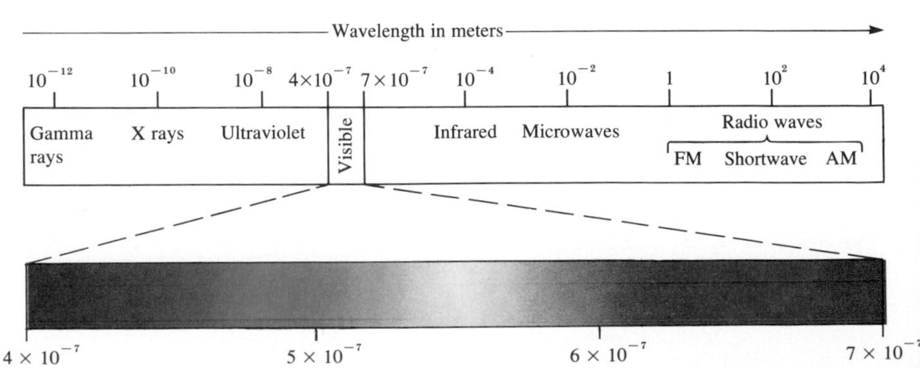

Figure 7.2

Classification of electromagnetic radiation. Spectrum adapted by permission from C. W. Keenan, D. C. Kleinfelter, and J. H. Wood, *General College Chemistry,* 6th edition, Harper & Row, Publishers, Inc., 1980.

Chemical Impact

Solar Polar Bears

The polar bear, a regal beast that has dominated the world's arctic regions for thousands of years, can exist in its incredibly hostile environment partly because its fur is an almost perfect absorber and converter of solar energy.

What color is a polar bear's fur? White, the obvious answer, is incorrect. The hairs on a polar bear are completely colorless and transparent. The bear appears white because of the way the rough inner surfaces of the hollow hairs reflect visible light. The most interesting characteristic of each of these hollow fibers is its ability to act as a solar converter designed to trap ultraviolet light and transmit it to the bear's black skin. In the summer, the sun directly supplies up to 25% of the bear's total energy requirements. This allows the polar bear to be very active in pursuing prey, while still building up the layers of blubber needed to survive the winter. This ingenious system also insures that, although the skin of the polar bear is very warm, the outer layers of the fur remain at about the same temperature as the bear's surroundings. Because of the small temperature difference between its fur and the air, the bear loses very little energy through heat leakage to the environment.

A polar bear on Hudson Bay, Canada.

Human society could benefit greatly from polar bear fur as we continue to search for more efficient energy supplies. Clearly, covering our roofs with polar bear pelts is not a viable alternative, but polar bear hair provides an excellent model for possible synthetic optical fibers to convert the sun's energy to other energy forms.

creases their motions. This energy is then transferred to other types of molecules via collisions, causing an increase in the food's temperature. As we proceed in the study of chemistry, we will consider many of the classes of electromagnetic radiation and the ways in which they affect matter.

When a strontium salt is dissolved in methanol (with a little water) and ignited, it gives a brilliant red flame. The red color is produced by emission of light as electrons, excited by the energy of the burning methanol, fall back to their ground states.

Sample Exercise 7.1

The brilliant red colors seen in fireworks are due to the emission of light with wavelengths around 650 nm when strontium salts such as $Sr(NO_3)_2$ and $SrCO_3$ are heated. (This can be easily demonstrated in the lab by dissolving one of these salts in methanol that contains a little water and igniting the mixture in an evaporating dish.) Calculate the frequency of red light of wavelength 6.50×10^2 nm.

Solution

We can convert wavelength to frequency using the equation

$$\lambda\nu = c \quad \text{or} \quad \nu = \frac{c}{\lambda}$$

Sample Exercise 7.1, continued

where $c = 2.9979 \times 10^8$ m/s. In this case $\lambda = 6.50 \times 10^2$ nm. Since

$$6.50 \times 10^2 \cancel{\text{nm}} \times \frac{1 \text{ m}}{10^9 \cancel{\text{nm}}} = 6.50 \times 10^{-7} \text{ m}$$

$$\nu = \frac{c}{\lambda} = \frac{2.9979 \times 10^8 \cancel{\text{m}}/\text{s}}{6.50 \times 10^{-7} \cancel{\text{m}}} = 4.61 \times 10^{14} \text{ s}^{-1} = 4.61 \times 10^{14} \text{ Hz}$$

7.2 The Nature of Matter

Purpose

- To introduce the concept of quantized energy.
- To show that light has both wave and particulate properties.
- To describe how diffraction experiments were used to demonstrate the dual nature of all matter.

It is probably fair to say that at the end of the nineteenth century physicists were feeling rather smug. Available theories could explain phenomena as diverse as the motions of the planets and the dispersion of visible light by a prism. Rumor has it that students were being discouraged from pursuing physics as a career because it was felt that all the major problems had been solved, or at least described in terms of the current physical theories.

At the end of the nineteenth century, the idea prevailed that matter and energy were distinct. Matter was thought to consist of particles, while energy in the form of light (electromagnetic radiation) was described as a wave. Particles were things that had mass and whose position in space could be specified. Waves were described as massless and delocalized; that is, their position in space could not be specified. It was also assumed that there was no intermingling of matter and light. Everything known before 1900 seemed to fit neatly into this view.

At the beginning of the twentieth century, however, certain experimental results suggested that this picture was incorrect. The first important advance came in 1901 from the German physicist Max Planck. Studying the radiation profiles emitted by solid bodies heated to incandescence, Planck found that the results could not be explained in terms of the physics of his day, which held that matter could absorb or emit any quantity of energy. Planck could only account for these observations by postulating that energy can be gained or lost only in *whole-number multiples* of the quantity $h\nu$, where h is a constant called **Planck's constant,** determined by experiment to have the value 6.626×10^{-34} J s. That is, the change in energy for a system, ΔE, can be represented by the equation

Energy can be gained or lost only in integer multiples of $h\nu$.

$$\Delta E = nh\nu$$

where n is an integer (1, 2, 3, . . .), h is Planck's constant, and ν is the frequency of the electromagnetic radiation absorbed or emitted.

Planck's constant = 6.626×10^{-34} J s.

Planck's result was a real surprise. It had always been assumed that the energy of matter was continuous, which meant that the transfer of any quantity of energy was possible. Now it seemed clear that energy is in fact **quantized** and can only occur in discrete units of size $h\nu$. Each of these small "packets" of energy is called a *quantum*. A system can transfer energy only in whole *quanta*. Thus energy seems to have particulate properties.

Sample Exercise 7.2

The blue color in fireworks is often achieved by heating copper(I) chloride (CuCl) to about 1200°C. Then the compound emits blue light having a wavelength of 450 nm. What is the increment of energy (the quantum) that is emitted at 4.50×10^2 nm by CuCl?

Solution

The quantum of energy can be calculated from the equation

$$\Delta E = h\nu$$

The frequency ν for this case can be calculated as follows:

$$\nu = \frac{c}{\lambda} = \frac{2.9979 \times 10^8 \cancel{\text{m}}/\text{s}}{4.50 \times 10^{-7} \cancel{\text{m}}} = 6.66 \times 10^{14} \text{ s}^{-1}$$

So $\quad \Delta E = h\nu = (6.626 \times 10^{-34} \text{ J} \cancel{\text{s}})(6.66 \times 10^{14} \cancel{\text{s}}^{-1}) = 4.41 \times 10^{-19} \text{ J}$

A sample of CuCl emitting light at 450 nm can only lose energy in increments of 4.41×10^{-19} J, the size of the quantum in this case.

The next important development in the knowledge of atomic structure came when Albert Einstein (see Figure 7.3) proposed that electromagnetic radiation is itself quantized. Einstein suggested that electromagnetic radiation can be viewed as

Figure 7.3

Albert Einstein (1879–1955) was born in Germany. Nothing in his early development suggested genius; even at the age of 9 he did not speak clearly, and his parents feared that he might be handicapped. When asked what profession Einstein should follow, his school principal replied, "It doesn't matter; he'll never make a success of anything." When he was 10, Einstein entered the Luitpold Gymnasium (high school), which was typical of German schools of that time in being harshly disciplinarian. There he developed a deep suspicion of authority and a scepticism that encouraged him to question and doubt—valuable qualities in a scientist. In 1905, while a patent clerk in Switzerland, Einstein published a paper explaining the photoelectric effect via the quantum theory. For this revolutionary thinking he received a Nobel Prize in 1921. Highly regarded by this time, he worked in Germany until 1933, when Hitler's persecution of the Jews forced him to come to the United States. He worked at the Institute for Advanced Studies at Princeton University until his death in 1955.

Einstein was undoubtedly the greatest physicist of our age. Even if someone else had derived the theory of relativity, his other work would have ensured his ranking as the second greatest physicist of his time. Our concepts of space and time were radically changed by ideas he first proposed when he was 26 years old. From then until the end of his life, he attempted unsuccessfully to find a single unifying theory that would explain all physical events.

a stream of ''particles'' called **photons.** The energy of each photon is given by the expression

$$E_{\text{photon}} = h\nu = \frac{hc}{\lambda}$$

where h is Planck's constant, ν is the frequency of the radiation, and λ is the wavelength of the radiation.

In a related development, Einstein derived the famous equation

$$\boldsymbol{E = mc^2}$$

in his *special theory of relativity* published in 1905. The main significance of this equation is that *energy has mass*. This is more apparent if we rearrange the equation to the following form:

$$m = \frac{E}{c^2}$$

E ← Energy; m ← Mass; c ← Speed of light

Using this form of the equation, we can calculate the mass associated with a given quantity of energy. For example, we can calculate the mass of a photon. For electromagnetic radiation of wavelength λ, the energy of each photon is given by the expression

$$E = \frac{hc}{\lambda}$$

Note that the mass of a photon depends on its wavelength. The mass of a photon at rest is thought to be zero, although we never observe it at rest.

Then the mass of a photon of light with wavelength λ is given by

$$m = \frac{E}{c^2} = \frac{hc/\lambda}{c^2} = \frac{h}{\lambda c}$$

Do photons really have mass? The answer appears to be yes. In 1922 American physicist Arthur Compton performed experiments involving collisions of X rays and electrons which showed that photons do exhibit the apparent mass calculated from the above equation.

We can summarize the important conclusions from the work of Planck and Einstein as follows:

Energy is quantized. It can occur only in discrete units called quanta.

Electromagnetic radiation, which was previously thought to exhibit only wave properties, seems to show certain characteristics of particulate matter as well. This phenomenon is sometimes referred to as the **dual nature of light** and is illustrated in Fig. 7.4.

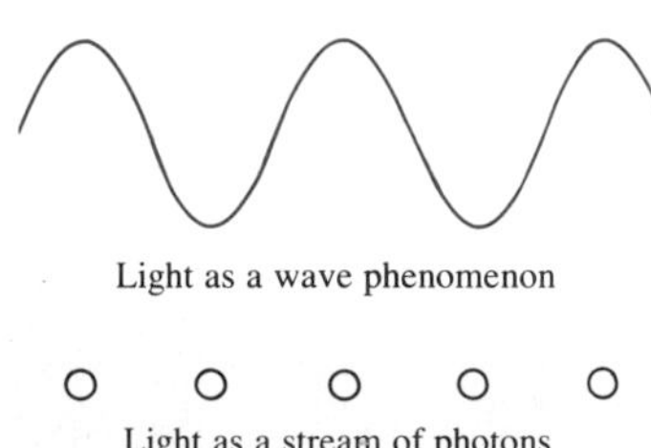

Figure 7.4

Electromagnetic radiation exhibits wave properties and particulate properties. The energy of each photon of the radiation is related to the wavelength and frequency by the equation $E_{\text{photon}} = h\nu = hc/\lambda$.

Thus light, which was previously thought to be purely wavelike, was found to have certain characteristics of particulate matter. But is the opposite also true? That is, does matter that is normally assumed to be particulate exhibit wave properties? This question was raised in 1923 by a young French physicist named Louis de Broglie (1892–1987). To see how de Broglie supplied the answer to this question, recall that the relationship between mass and wavelength for electromagnetic radiation is $m = h/\lambda c$. For a particle with velocity v, the corresponding expression is

$$m = \frac{h}{\lambda v}$$

Rearranging to solve for λ, we have

$$\lambda = \frac{h}{mv}$$

This equation, called de Broglie's equation, allows us to calculate the wavelength for a particle, as shown in Sample Exercise 7.3.

Do not confuse ν (frequency) with v (velocity).

Sample Exercise 7.3

Compare the wavelength for an electron (mass = 9.11×10^{-31} kg) traveling at a speed of 1.0×10^{7} m/s with that for a ball (mass = 0.10 kg) traveling at 35 m/s.

Solution

We use the equation $\lambda = h/mv$ where

$$h = 6.626 \times 10^{-34}\ \text{J s} \quad \text{or} \quad 6.626 \times 10^{-34}\ \text{kg m}^2/\text{s}$$

since

$$1\ \text{J} = 1\ \text{kg m}^2/\text{s}^2$$

For the electron

$$\lambda_e = \frac{6.626 \times 10^{-34}\ \dfrac{\cancel{\text{kg}}\ \cancel{\text{m}} \cdot \text{m}}{\cancel{\text{s}}}}{(9.11 \times 10^{-31}\ \cancel{\text{kg}})(1.0 \times 10^{7}\ \cancel{\text{m}}/\cancel{\text{s}})} = 7.27 \times 10^{-11}\ \text{m}$$

For the ball

$$\lambda_b = \frac{6.626 \times 10^{-34}\ \dfrac{\cancel{\text{kg}}\ \cancel{\text{m}} \cdot \text{m}}{\cancel{\text{s}}}}{(0.10\ \cancel{\text{kg}})(35\ \cancel{\text{m}}/\cancel{\text{s}})} = 1.9 \times 10^{-34}\ \text{m}$$

(top) Light diffracted by the closely-spaced grooves on a compact disc shows the visible spectrum. (bottom) Prismatic diffraction.

Notice from Sample Exercise 7.3 that the wavelength associated with the ball is incredibly short. On the other hand, the wavelength of the electron, although still quite small, happens to be on the same order as the spacing between the atoms in a typical crystal. This is important because, as we will see presently, it provides a means for testing de Broglie's equation.

Diffraction results when light is scattered from a regular array of points or lines. The diffraction of light from the ridges and grooves of a compact disc is shown in the photograph. The colors result because the various wavelengths of visible light are not all scattered in the same way. The colors are "separated," giving the same effect as light passing through a prism. Just as a regular arrangement of ridges and grooves produces diffraction, so does a regular array of atoms or ions in a crystal. For example, when X rays are directed onto a crystal of sodium chloride, with its regular array of Na^+ and Cl^- ions, the scattered radiation produces a **diffraction pattern** of bright spots and dark areas on a photographic plate, as shown in Fig. 7.5(a) on page 270. This occurs because the scattered light can

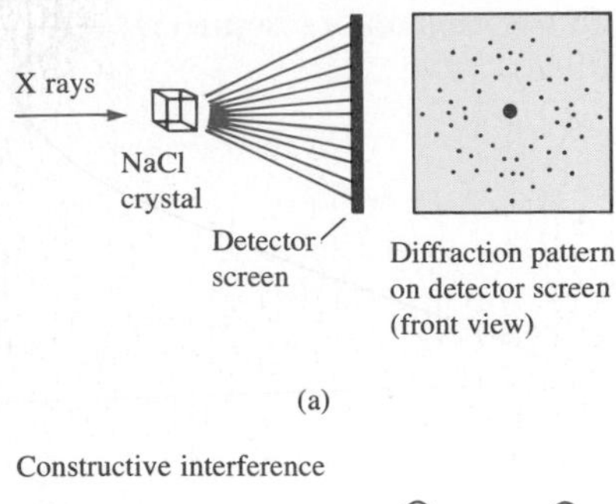

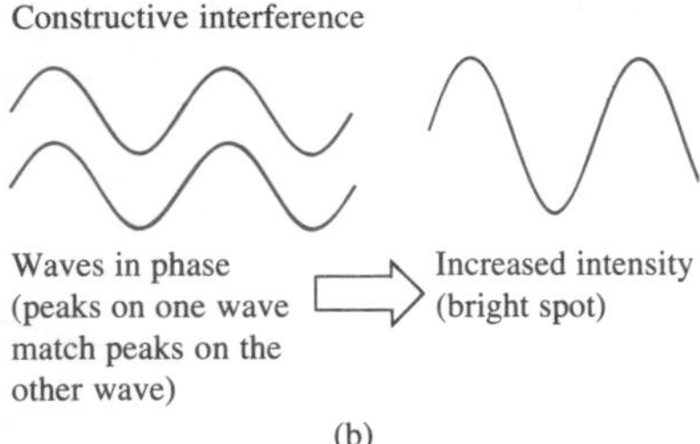

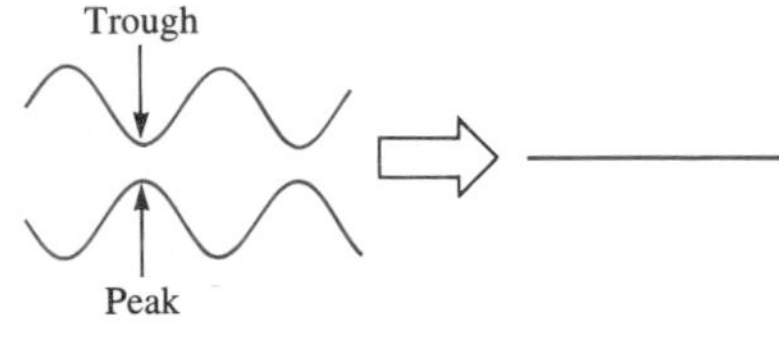

Figure 7.5

(a) Diffraction occurs when electromagnetic radiation is scattered from a regular array of objects, such as the ions in a crystal of sodium chloride. The large spot in the center is from the main incident beam of X rays. (b) Bright spots in the diffraction pattern result from *constructive interference* of waves. The waves are in phase; that is, their peaks match. (c) Dark areas result from *destructive interference* of waves. The waves are out of phase; the peaks of one wave coincide with the troughs of another wave.

interfere constructively (the peaks and troughs of the beams are in phase) to produce a bright spot [Fig. 7.5(b)] or destructively (the peaks and troughs are out of phase) to produce a dark area [Fig. 7.5(c)].

A diffraction pattern can only be explained in terms of waves. Thus this phenomenon provides a test for the postulate that particles such as electrons have wavelengths. As we saw in Sample Exercise 7.3, an electron with a velocity of 10^7 m/s (easily achieved by acceleration of the electron in an electric field) has a wavelength of about 10^{-10} m, which is roughly the distance between the ions in a crystal such as sodium chloride. This is important because diffraction occurs most efficiently when the spacing between the scattering points is about the same as the wavelength. Thus if electrons really do have an associated wavelength, a crystal should diffract electrons. An experiment to test this idea was carried out in 1927 by Davisson and Germer at the Bell Laboratories. When they directed a beam of electrons at a nickel crystal, they observed a diffraction pattern similar to that seen from the diffraction of X rays. This result verified de Broglie's relationship, at least for electrons. Larger chunks of matter, such as balls, have such small wavelengths (Sample Exercise 7.3) that they are impossible to verify experimentally. However, we believe that all matter obeys de Broglie's equation.

Now we have come full circle. Electromagnetic radiation, which at the turn of the twentieth century was thought to be a pure wave form, was found to possess particulate properties. Conversely electrons, which were thought to be particles, were found to have a wavelength associated with them. The significance of these results is that matter and energy are not distinct. Energy is really a form of matter and all matter shows the same types of properties. That is, *all matter exhibits both particulate and wave properties*. Large pieces of matter, like baseballs, exhibit predominantly particulate properties. The associated wavelength is so small that it is not observed. Very small pieces of matter, such as photons, while showing some particulate properties, exhibit predominantly wave properties. Pieces of matter with intermediate mass, such as electrons, show clearly both the particulate and wave properties of matter.

7.3 The Atomic Spectrum of Hydrogen

Purpose

- To show that the line spectrum of hydrogen demonstrates the quantized nature of the energy of its electron.

As we saw in Chapter 2, key information about the atom came from several experiments carried out in the early twentieth century, in particular, Thomson's discovery of the electron and Rutherford's discovery of the nucleus. Another important experiment was the study of the emission of light by excited hydrogen atoms. When a sample of hydrogen gas receives a high-energy spark, the H_2 molecules absorb energy, and some of the H—H bonds are broken. The resulting hydrogen atoms are *excited;* that is, they contain excess energy, which they release by emitting light of various wavelengths to produce what is called the *emission spectrum* of the hydrogen atom.

To understand the significance of the hydrogen emission spectrum, we must first describe the **continuous spectrum** that results when white light is passed through a prism, as shown in Fig. 7.6(a). This spectrum, like the rainbow produced when sunlight is dispersed by raindrops, contains *all* the wavelengths of visible light. In contrast, when the hydrogen emission spectrum in the visible region is passed through a prism, as shown in Fig. 7.6(b), we see only a few lines, each of which corresponds to a discrete wavelength. The hydrogen emission spectrum is called a **line spectrum.**

A rainbow over the Serengeti Plain, Tanzania.

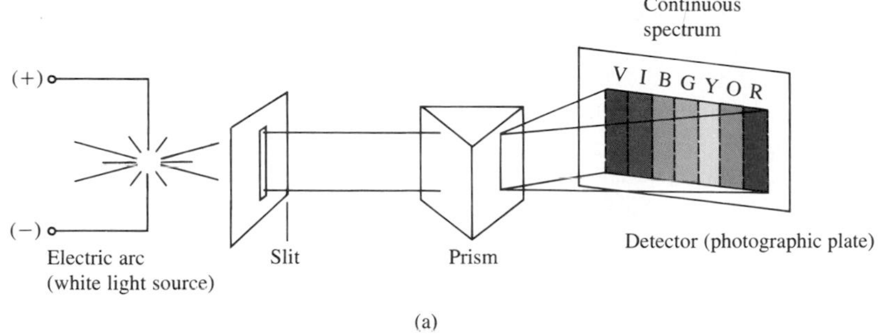

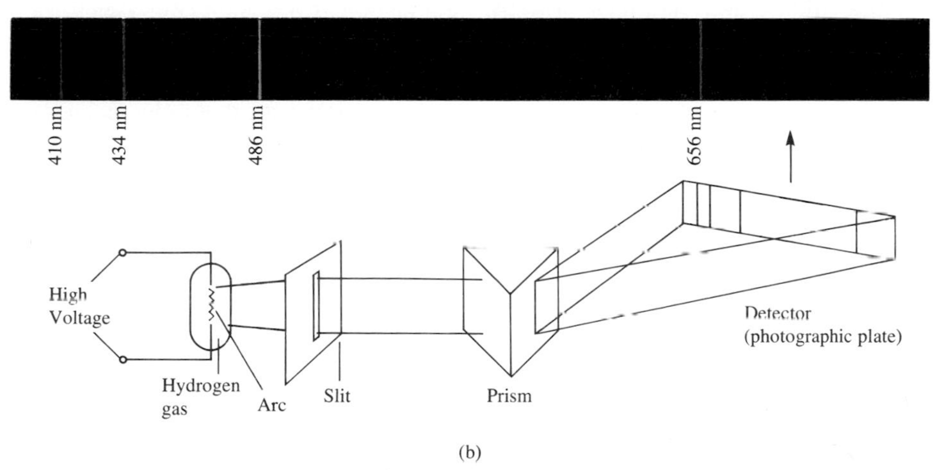

Figure 7.6

(a) A continuous spectrum containing all wavelengths of visible light (indicated by the initial letters of the colors of the rainbow). (b) The hydrogen line spectrum contains only a few discrete wavelengths. Spectrum adapted by permission from C. W. Keenan, D. C. Kleinfelter, and J. H. Wood, *General College Chemistry,* 6th edition, Harper & Row, Publishers, Inc., 1980.

What is the significance of the line spectrum of hydrogen? It indicates that *only certain energies are allowed for the electron in the hydrogen atom*. In other words, the energy of the electron in the hydrogen atom is *quantized*. This observation ties in perfectly with the postulates of Max Planck discussed in Section 7.2. Changes in energy between discrete energy levels in hydrogen will produce only certain wavelengths of emitted light, as shown in Fig. 7.7. For example, a given change in energy from a high to a lower level would give a wavelength of light that can be calculated from Planck's equation:

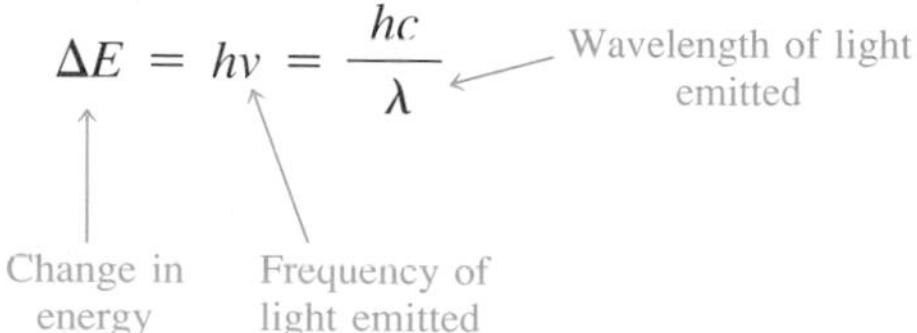

$$\Delta E = h\nu = \frac{hc}{\lambda}$$

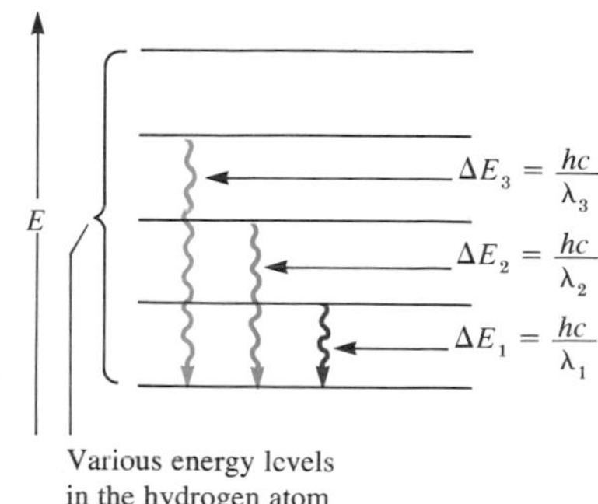

Figure 7.7

A change between two discrete energy levels emits a photon of light.

Chemical Impact

Spectra and Space

Every element has a characteristic signature: its unique line spectrum produced by the movement of electrons between quantized energy levels. (The spectrum of hydrogen is shown in Fig. 7.6.) The characteristic spectrum of an element is a sure way to verify its presence (or absence) in a given sample. Spectra are particularly useful for the analyses of extraterrestrial material, since collecting actual samples is so difficult. Although we have directly analyzed rocks from the moon and Mars (at great expense), most of our information about the materials in space comes from the light emitted or absorbed by these materials. For example, we have obtained a detailed knowledge of the composition of the sun and other stars by looking for the characteristic spectral lines of the elements in the light emitted by these sources.

Molecules also have characteristic spectral lines that can be used to identify them. The first spectroscopic observations of a comet were carried out by Giovanni Donati in 1864 when he identified species such as C_2, $(CN)_2$, C_3, and CH in the Comet Tempel. Since then, using increasingly sensitive instruments, scientists have found more than 60 different molecules in space, including ethanol (C_2H_5OH), carbon monoxide, hydrogen cyanide (HCN), methyl cyanide (CH_3CN), and water.

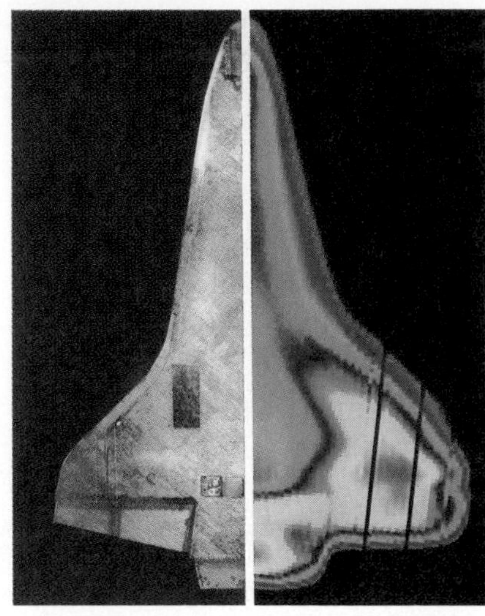

An infrared thermogram of the space shuttle Columbia.

Electromagnetic radiation is also useful to space scientists in other ways. For example, the microwave radiation that is left over from the postulated "big bang" has yielded an age for the universe of between 15 and 25 billion years. Infrared radiation is especially useful for determining the temperature of space objects. The accompanying photo is an infrared picture of the shuttle Orbiter Columbia as it returned to earth in 1982, in which the hottest areas appear red and the coolest areas appear violet.

Suggested Reading

P. B. Kelter, W. E. Snyder, and C. S. Buchar, "Using NASA and the Space Program to Help High School and College Students Learn Chemistry (Part II—The Current State of Chemistry at NASA)" *J. Chem. Ed.* **64** (1987), p. 228.

The energy of the electron in the hydrogen atom is quantized.

The discrete line spectrum of hydrogen shows that only certain energies are possible; that is, the electron energy levels are quantized. In contrast, if any energy level were allowed, the emission spectrum would be continuous.

7.4 The Bohr Model

Purpose

- To describe the development of the Bohr model for the hydrogen atom.

In 1913 a Danish physicist named Niels Bohr, aware of the experimental results we have just discussed, developed a **quantum model** for the hydrogen atom. Bohr proposed that the *electron in a hydrogen atom moves around the nucleus only in certain allowed circular orbits*. He calculated the radii for these allowed orbits by using the theories of classical physics and by making some new assumptions.

From classical physics Bohr knew that a particle in motion tends to move in a straight line and can be made to travel in a circle only by application of a force toward the center of the circle. Thus Bohr reasoned that the tendency of the revolving electron to fly off the atom must be just balanced by its attraction for the positively charged nucleus. But classical physics also decreed that a charged particle under acceleration should radiate energy. Since an electron revolving around the nucleus constantly changes its direction, it is constantly accelerating. Therefore, the electron should emit light and lose energy, and thus be drawn into the nucleus. This, of course, does not correlate with the existence of stable atoms.

Niels Bohr at 37 years of age, photographed in 1922 when he received the Nobel Prize for Physics.

Clearly, an atomic model based solely on the theories of classical physics was untenable. Bohr also knew that the correct model had to account for the experimental spectrum of hydrogen, which showed that only certain electron energies (and thus only certain orbits) were allowed. The experimental data were absolutely clear on this point. Bohr found that his model would fit the experimental results if he assumed that the angular momentum of the electron (angular momentum equals the product of mass, velocity, and orbital radius) could only occur in certain increments. It wasn't clear why this should be true, but with this assumption, Bohr's model gave hydrogen atom energy levels consistent with the hydrogen emission spectrum. The model is represented pictorially in Fig. 7.8 on page 274.

Although we will not show the derivation here, the most important equation to come from Bohr's model is the expression for the *energy levels available to the electron in the hydrogen atom:*

$$E = -2.178 \times 10^{-18}\ \text{J}\left(\frac{Z^2}{n^2}\right) \tag{7.1}$$

The "J" in Equation (7.1) stands for joules.

in which n is an integer (the larger the value of n, the larger the orbit radius) and Z is the nuclear charge. Using Equation (7.1), Bohr was able to calculate hydrogen atom energy levels that exactly matched the values obtained by experiment.

The negative sign in Equation (7.1) simply means that the energy of the electron bound to the nucleus is lower than it would be if the electron were at an infinite distance ($n = \infty$) from the nucleus, where there is no interaction and the energy is zero:

$$E = -2.178 \times 10^{-18}\ \text{J}\left(\frac{Z^2}{\infty}\right) = 0$$

The energy of the electron in any orbit is negative relative to this reference state.

Equation (7.1) can be used to calculate the change in energy of an electron and the wavelength of light emitted or absorbed when the electron changes orbits. For example, suppose an electron in level $n = 6$ of an excited hydrogen atom falls back to level $n = 1$ as the hydrogen atom returns to its lowest possible energy state, its **ground state.** We use Equation (7.1) with $Z = 1$ since the hydrogen nucleus contains a single proton. The energies corresponding to the two states are

The ground state is the lowest possible energy state of an atom.

For n = 6: $$E_6 = -2.178 \times 10^{-18}\ \text{J}\left(\frac{1^2}{6^2}\right) = -6.05 \times 10^{-20}\ \text{J}$$

For n = 1: $$E_1 = -2.178 \times 10^{-18}\ \text{J}\left(\frac{1^2}{1^2}\right) = -2.178 \times 10^{-18}\ \text{J}$$

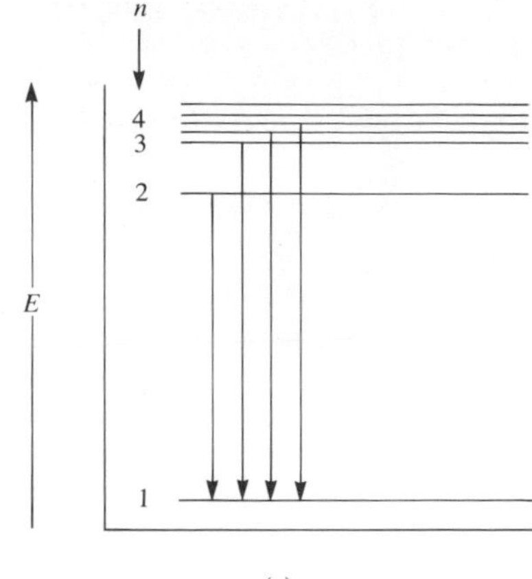

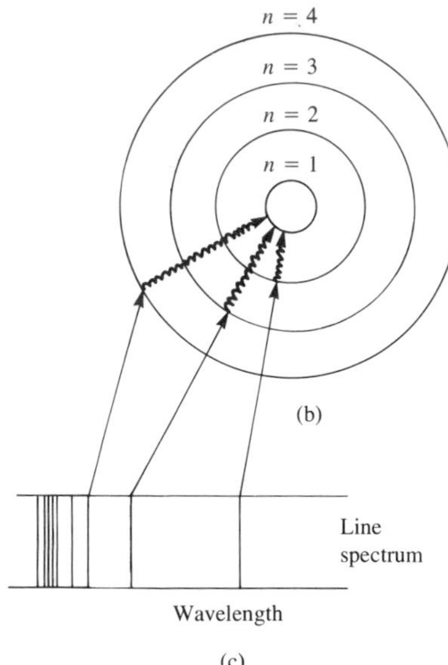

Figure 7.8

Electronic transitions in the Bohr model for the hydrogen atom. (a) An energy-level diagram for electronic transitions. (b) An orbit-transition diagram, which accounts for the experimental spectrum. (c) The resulting line spectrum on a photographic plate.

Note that for $n = 1$ the electron has a more negative energy than it does for $n = 6$, which means that the electron is more tightly bound in the smallest allowed orbit.

The change in energy, ΔE, when the electron falls from $n = 6$ to $n = 1$ is

$$\begin{aligned}\Delta E &= \text{energy of final state} - \text{energy of initial state} \\ &= E_1 - E_6 = (-2.178 \times 10^{-18}\ \text{J}) - (-6.05 \times 10^{-20}\ \text{J}) \\ &= -2.117 \times 10^{-18}\ \text{J}\end{aligned}$$

The negative sign for the *change* in energy indicates that the atom has *lost* energy and is now in a more stable state. The energy is carried away from the atom by the production (emission) of a photon.

The wavelength of the emitted photon can be calculated from the equation

$$\Delta E = h\left(\frac{c}{\lambda}\right) \quad \text{or} \quad \lambda = \frac{hc}{\Delta E}$$

where ΔE represents the change in energy of the atom, which equals the energy of the emitted photon. We have

$$\lambda = \frac{hc}{\Delta E} = \frac{(6.626 \times 10^{-34}\ \cancel{\text{J}}\ \cancel{\text{s}})(2.9979 \times 10^{8}\ \text{m}/\cancel{\text{s}})}{2.117 \times 10^{-18}\ \cancel{\text{J}}} = 9.383 \times 10^{-8}\ \text{m}$$

Note that for this calculation the absolute value of ΔE is used. In this case the direction of energy flow is indicated by saying that a photon of wavelength 9.383×10^{-8} m has been *emitted* from the hydrogen atom. Simply plugging the negative value of ΔE into the equation would produce a negative value for λ, which is physically meaningless.

Sample Exercise 7.4

Calculate the energy required to excite the hydrogen electron from level $n = 1$ to level $n = 2$. Also calculate the wavelength of light that must be absorbed by a hydrogen atom in its ground state to reach this excited state.*

Solution

Using Equation (7.1) with $Z = 1$, we have

$$E_1 = -2.178 \times 10^{-18}\ \text{J}\left(\frac{1^2}{1^2}\right) = -2.178 \times 10^{-18}\ \text{J}$$

$$E_2 = -2.178 \times 10^{-18}\ \text{J}\left(\frac{1^2}{2^2}\right) = -5.445 \times 10^{-19}\ \text{J}$$

$$\Delta E = E_2 - E_1 = (-5.445 \times 10^{-19}\ \text{J}) - (-2.178 \times 10^{-18}\ \text{J}) = 1.633 \times 10^{-18}\ \text{J}$$

The positive value for ΔE indicates that the system has gained energy. The wavelength of light that must be *absorbed* to produce this change is

*After this exercise we will no longer show cancellation marks. However, the same process for canceling units applies throughout this text.

Sample Exercise 7.4, continued

$$\lambda = \frac{hc}{\Delta E} = \frac{(6.626 \times 10^{-34} \text{ J s})(2.9979 \times 10^{8} \text{ m/s})}{1.633 \times 10^{-18} \text{ J}}$$

$$= 1.216 \times 10^{-7} \text{ m}$$

Note from Fig. 7.2 that the light required to produce the transition from the $n = 1$ to $n = 2$ level in hydrogen lies in the ultraviolet region.

At this time we must emphasize two important points about the Bohr model:

1. The model correctly fits the quantized energy levels of the hydrogen atom and postulates only certain allowed circular orbits for the electron.
2. As the electron becomes more tightly bound, its energy becomes more negative relative to the zero-energy reference state (corresponding to the electron being at infinite distance from the nucleus). As the electron is brought closer to the nucleus, energy is released from the system.

Using Equation (7.1), we can derive a general equation for the electron moving from one level ($n_{initial}$) to another level (n_{final}):

$$\Delta E = \text{energy of level } n_{final} - \text{energy of level } n_{initial}$$

$$= E_{final} - E_{initial}$$

$$= (-2.178 \times 10^{-18} \text{ J})\left(\frac{1^2}{n_{final}^2}\right) - (-2.178 \times 10^{-18} \text{ J})\left(\frac{1^2}{n_{initial}^2}\right)$$

$$= -2.178 \times 10^{-18} \text{ J}\left(\frac{1}{n_{final}^2} - \frac{1}{n_{initial}^2}\right) \quad (7.2)$$

Equation (7.2) can be used to calculate the energy change between *any* two energy levels in a hydrogen atom, as shown in Sample Exercise 7.5.

Sample Exercise 7.5

Calculate the energy required to remove the electron from a hydrogen atom in its ground state.

Solution

Removing the electron from a hydrogen atom in its ground state corresponds to taking the electron from $n_{initial} = 1$ to $n_{final} = \infty$. Thus

$$\Delta E = -2.178 \times 10^{-18} \text{ J}\left(\frac{1}{n_{final}^2} - \frac{1}{n_{initial}^2}\right)$$

$$= -2.178 \times 10^{-18} \text{ J}\left(\frac{1}{\infty} - \frac{1}{1^2}\right)$$

$$= -2.178 \times 10^{-18} \text{ J}(0 - 1) = 2.178 \times 10^{-18} \text{ J}$$

The energy required to remove the electron from a hydrogen atom in its ground state is 2.178×10^{-18} J.

Chemical Impact

Fireworks

The art of using mixtures of chemicals to produce explosives is an ancient one. Black powder—a mixture of potassium nitrate, charcoal, and sulfur—was being used in China well before 1000 A.D., and has been subsequently used through the centuries in military explosives, in construction blasting, and for fireworks. The du Pont Company, now a major chemical manufacturer, started out as a manufacturer of black powder. In fact, the founder, Eleuthère du Pont, learned the manufacturing technique from none other than Lavoisier.

Before the nineteenth century fireworks were mainly confined to rockets and loud bangs. Orange and yellow colors came from the presence of charcoal and iron filings. However, with the great advances in chemistry in the nineteenth century, new compounds found their way into fireworks. Salts of copper, strontium, and barium added brilliant colors. Magnesium and aluminum metals gave a dazzling white light. Fireworks, in fact, have changed very little since then.

How do fireworks produce their brilliant colors and loud bangs? Actually, only a handful of different chemicals are responsible for most of the spectacular effects. To produce the noise and flashes, an oxidizer (an oxidizing agent) and a fuel (a reducing agent) are used. A common mixture involves potassium perchlorate ($KClO_4$) as the oxidizer and aluminum and sulfur as the fuel. The perchlorate oxidizes the fuel in a very exothermic reaction, which produces a brilliant flash, due to the aluminum, and a loud report from the rapidly expanding gases produced. For a color effect an element with a colored emission spectrum is included. Recall that the electrons in atoms can be raised to higher-energy orbitals when the atoms absorb energy. The excited atoms can then release this excess

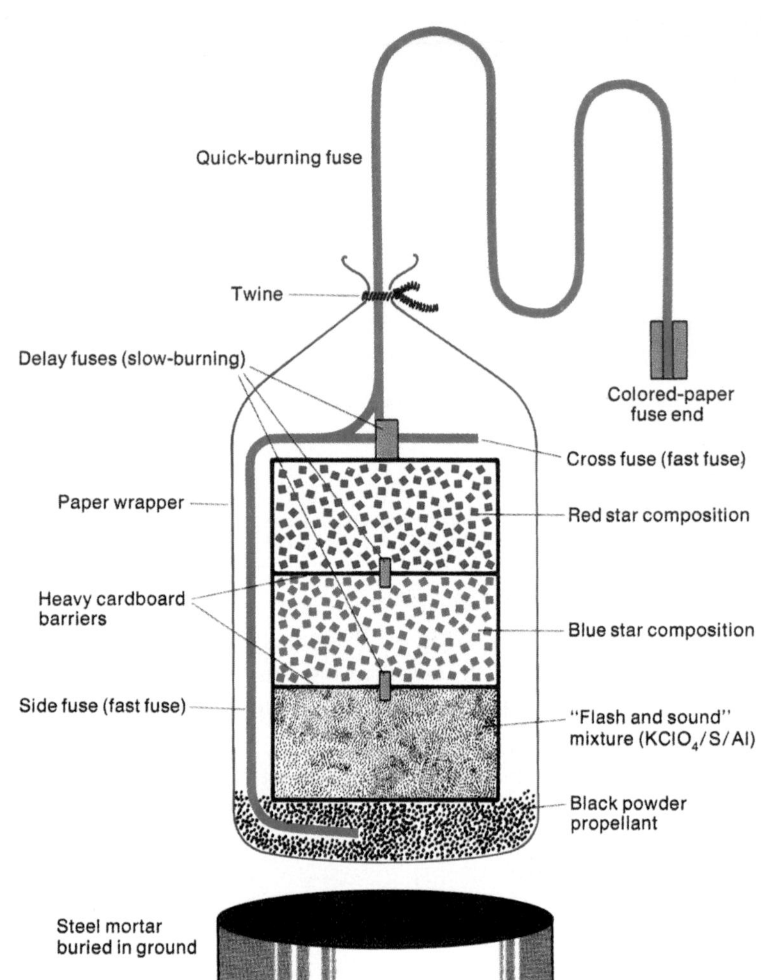

A typical aerial shell used in fireworks displays. Time-delayed fuses cause a shell to explode in stages. In this case a red starburst occurs first, followed by a blue starburst, and finally a flash and loud report. From *Chemical & Engineering News,* June 29, 1981, p. 24. Reprinted with permission. Copyright 1981 American Chemical Society.

Chemical Impact

energy by emitting light of specific wavelengths, often in the visible region. In fireworks the energy to excite the electrons comes from the reaction between the oxidizer and fuel.

Yellow colors in fireworks are due to the 589-nm emission of sodium ions. Red colors come from strontium salts emitting at 606 nm and 636–688 nm. This red color is familiar from highway safety flares. Barium salts give a green color in fireworks, due to a series of emission lines between 505 nm and 535 nm. A really good blue color, however, is hard to obtain. Copper salts give a blue color, emitting in the 420–460 nm region. But difficulties occur because another commonly used oxidizing agent potassium chlorate ($KClO_3$) reacts with copper salts to form copper chlorate, a highly explosive compound which is dangerous to store. Paris green, a copper salt containing arsenic, was once extensively used but is now considered to be too toxic.

A typical aerial shell is shown on the previous page. The shell is launched from a mortar (a steel cylinder) using black powder as the propellant. Time-delayed fuses are used to fire the shell in stages. A list of chemicals commonly used in fireworks is given in the table.

Although you might think that the chemistry of fireworks is simple, the achievement of the vivid white flashes and the brilliant colors requires complex combinations of chemicals. For example, because the white flashes produce high flame temperatures, the colors tend to wash out. Thus oxidizers, such as $KClO_4$, are sometimes used with fuels that produce relatively low flame temperatures. An added difficulty, however, is that perchlorates are very sensitive to accidental ignition and are therefore quite hazardous. Another problem arises from the use of sodium salts. Because sodium produces an extremely bright yellow emission, sodium salts cannot be used when other colors are desired. Carbon-based fuels also give a yellow flame that masks other colors, and this limits the use of organic compounds as fuels. You can see that the manufacture of fireworks that produce the desired effects and are also safe to handle requires careful selection of chemicals. And of course there is still the dream of a deep blue flame.

Chemicals Commonly Used in the Manufacture of Fireworks

Oxidizers	Fuels	Special effects
Potassium nitrate	Aluminum	Red flame: strontium nitrate, strontium carbonate
Potassium chlorate	Magnesium	Green flame: barium nitrate, barium chlorate
Potassium perchlorate	Titanium	Blue flame: copper carbonate, copper sulfate, copper oxide
Ammonium perchlorate	Charcoal	Yellow flame: sodium oxalate, cryolite (Na_3AlF_6)
Barium nitrate	Sulfur	White flame: magnesium, aluminum
Barium chlorate	Antimony sulfide	Gold sparks: iron filings, charcoal
Strontium nitrate	Dextrin	White sparks: aluminum, magnesium, aluminum-magnesium alloy, titanium
	Red gum	Whistle effect: potassium benzoate or sodium salicylate
	Polyvinyl chloride	White smoke: mixture of potassium nitrate and sulfur
		Colored smoke: mixture of potassium chlorate, sulfur, and organic dye

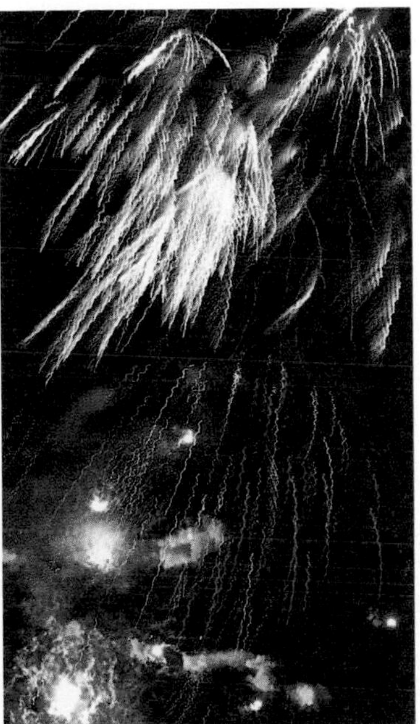

An exploding aerial shell.

At first Bohr's model appeared to be very promising. The energy levels calculated by Bohr closely agreed with the values obtained from the hydrogen emission spectrum. However, when Bohr's model was applied to atoms other than hydrogen, it did not work at all. Although some attempts were made to adapt the model using elliptical orbits, it was concluded that Bohr's model is fundamentally incorrect. The model is, however, very important historically, because it showed that the observed quantization of energy in atoms could be explained by making rather simple assumptions. Bohr's model paved the way for later theories. It is important to realize, however, that the current theory of atomic structure is in no way derived from the Bohr model. Electrons do *not* move around the nucleus in circular orbits, as we shall see later in this chapter.

Although Bohr's model fits the energy levels for hydrogen, it is a fundamentally incorrect model for the hydrogen atom.

7.5 The Wave Mechanical Model of the Atom

Purpose

- To show how standing waves can be used to describe electrons in atoms.
- To describe the Heisenberg uncertainty principle.
- To explain the significance of electron probability distributions.

By the mid-1920s it had become apparent that the Bohr model could not be made to work. A totally new approach was needed. Three physicists were at the forefront of this effort: Werner Heisenberg, Louis de Broglie, and Erwin Schrödinger. As we have already seen, de Broglie originated the idea that the electron, previously considered to be a particle, also shows wave properties. Pursuing this line of reasoning, Schrödinger, an Austrian physicist, decided to attack the problem of atomic structure by giving emphasis to the wave properties of the electron. To Schrödinger and de Broglie, the electron bound to the nucleus seemed similar to a **standing wave,** and they began research on a **wave mechanical model** of the atom.

The most familiar example of standing waves occurs in association with musical instruments such as guitars or violins, where a string attached at both ends vibrates to produce the musical tone. The waves are described as standing, since they are stationary; the waves do not travel along the length of the string. The motions of the string can be explained as a combination of simple waves of the type shown in Fig. 7.9. The dots in this figure indicate the nodes, or points of zero lateral (sideways) displacement for a given wave. Note that there are limitations on the allowed wavelengths of the standing wave. Each end of the string is fixed, so there is always a node at each end. This means that there must be a whole number of *half* wavelengths in any of the allowed motions of the string (see Fig. 7.9).

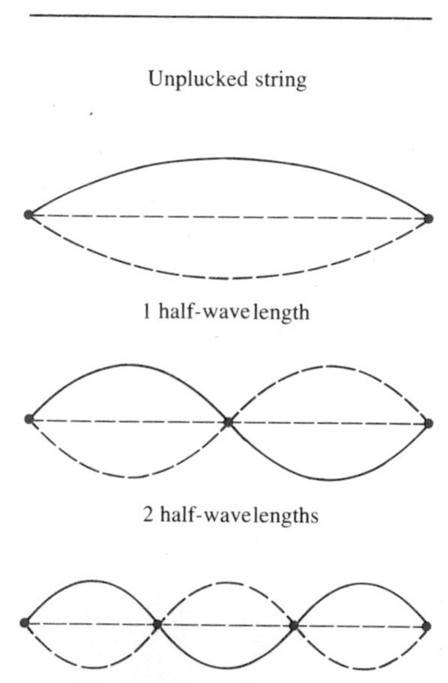

Figure 7.9

The standing waves caused by the vibration of a guitar string fastened at both ends. Each dot represents a node (a point of zero displacement)

A similar situation results when the electron in the hydrogen atom is imagined to be a standing wave. As shown in Fig. 7.10, only certain circular orbits have a circumference into which a whole number of wavelengths of the standing electron wave will "fit." All other orbits would produce destructive interference of the standing electron wave and are not allowed. This seemed like a possible explanation for the observed quantization of the hydrogen atom, so Schrödinger worked out a model for the hydrogen atom in which the electron was assumed to behave as a standing wave.

It is important to recognize that Schrödinger could not be sure that this idea would work. The test had to be whether or not the model would correctly fit the experimental data on hydrogen and other atoms. The physical principles for describing standing waves were well known in 1925 when Schrödinger decided to treat the electron in this way. His mathematical treatment is too complicated to be detailed here. However, the form of Schrödinger's equation is

$$\hat{H}\psi = E\psi$$

where ψ, called the **wave function,** is a function of the coordinates (x, y, and z) of the electron's position in three-dimensional space, and where $\hat{H}$ represents a set of mathematical instructions called an *operator*. In this case, the operator contains mathematical terms that produce the total energy of the atom when they are applied to the wave function. E represents the total energy of the atom (the sum of the potential energy due to the attraction between the proton and electron and the kinetic energy of the moving electron). When this equation is analyzed, many solutions are found. Each solution consists of a wave function, ψ, which is characterized by a particular value of E. A specific wave function is often called an **orbital.**

To illustrate the most important ideas of the wave mechanical model of the atom, we will first concentrate on the wave function corresponding to the lowest energy for the hydrogen atom. This wave function is called the 1s orbital. The first point of interest is the meaning of the word *orbital*. One thing is clear: an orbital is *not* a Bohr orbit. The electron in the hydrogen 1s orbital is not moving around the nucleus in a circular orbit. How, then, is the electron moving? The answer is somewhat surprising: *we do not know*. The wave function gives us no information about the detailed pathway of the electron. This is somewhat disturbing. When we solve problems involving the motions of particles in the macroscopic world, we are able to predict their pathways. For example, when two billiard balls with known velocities collide, we can predict their motions after the collision. However, we cannot predict the electron's motion from the 1s orbital function. Does this mean that the theory is wrong? Not necessarily: we have already learned that an electron does not behave much like a billiard ball, so we must examine the situation closely before we discard the theory.

Werner Heisenberg, who was also involved in the development of the wave mechanical model for the atom, discovered a very important principle that helps us to understand the meaning of orbitals—the **Heisenberg uncertainty principle.** Heisenberg's mathematical analysis led him to a surprising conclusion: *there is a fundamental limitation to just how precisely we can know both the position and momentum of a particle at a given time*. Stated mathematically, the uncertainty principle is

$$\Delta x \cdot \Delta(mv) \geq \frac{h}{4\pi}$$

where Δx is the uncertainty in a particle's position, $\Delta(mv)$ is the uncertainty in a particle's momentum, and h is Planck's constant. Thus, the minimum uncertainty in the product $\Delta x \cdot \Delta(mv)$ is $h/4\pi$. What this equation really says is that the more accurately we know a particle's position, the less accurately we can know its momentum and vice versa. This limitation is so small for large particles such as baseballs or billiard balls that it is unnoticed. However, for a small particle such as the

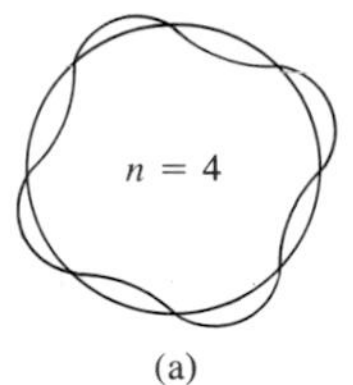

(a)

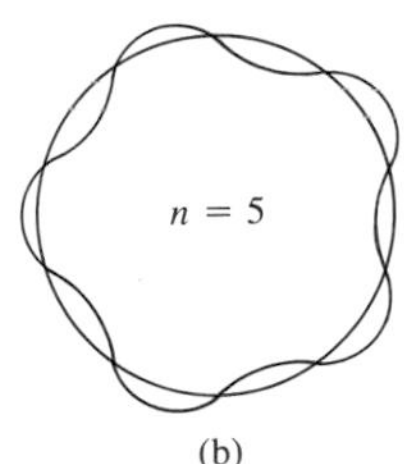

(b)

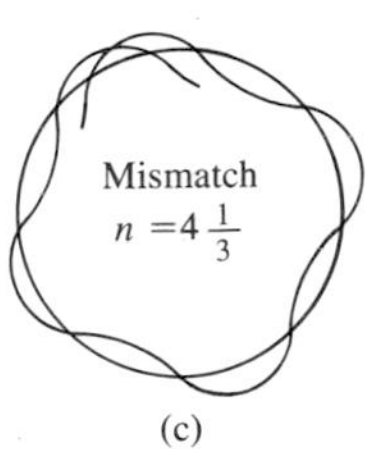

(c)

Figure 7.10

The hydrogen electron visualized as a standing wave around the nucleus. The circumference of a particular circular orbit would have to correspond to a whole number of wavelengths, as shown in (a) and (b), or else destructive interference occurs, as shown in (c). This is consistent with the fact that only certain electron energies are allowed; the atom is quantized. (Although this idea encouraged scientists to use a wave theory, it does not mean that the electron really travels in circular orbits.)

electron, the limitation becomes quite important. Applied to the electron, the uncertainty principle implies that we cannot know the exact motion of the electron as it moves around the nucleus. It is therefore not appropriate to assume that the electron is moving around the nucleus in a well-defined orbit as in the Bohr model.

Sample Exercise 7.6

Although, as we will see later, it is impossible to specify the radius of an atom precisely, the hydrogen atom has a radius on the order of 0.05 nm. Assuming that we know the position of an electron to an accuracy of 1% of the hydrogen radius, calculate the uncertainty in the velocity of the electron using the Heisenberg uncertainty principle. Then compare this value to the uncertainty in the velocity of a ball of mass 0.2 kg and radius 0.05 m whose position is known to an accuracy of 1% of its radius.

Solution

From Heisenberg's uncertainty principle the smallest possible uncertainty in the product $\Delta x \cdot \Delta(mv)$ is $h/4\pi$, that is,

$$\Delta x \cdot \Delta(mv) = \frac{h}{4\pi}$$

We will assume that the masses of the electron and the ball are constant, which means that the uncertainty in mv is due only to an uncertainty in velocity; that is,

$$\Delta(mv) = m(\Delta v)$$

With this assumption the uncertainty principle becomes

$$\Delta x \cdot \Delta(mv) = \Delta x \cdot m \cdot \Delta v = \frac{h}{4\pi}$$

or

$$\Delta x \cdot \Delta v = \frac{h}{m \cdot 4\pi}$$

We will use the equation in this form to solve for the uncertainty in velocity (Δv) for the electron and the ball.

For the electron the uncertainty in position (Δx) is 1% of 0.05 nm, or

$$\Delta x = (0.01)(0.05 \text{ nm}) = 5 \times 10^{-4} \text{ nm}$$

Converting to meters gives

$$5 \times 10^{-4} \text{ nm} \times \frac{10^{-9} \text{ m}}{1 \text{ nm}} = 5 \times 10^{-13} \text{ m}$$

The values of the constants are

$$m = \text{mass of the electron} = 9.11 \times 10^{-31} \text{ kg}$$

$$h = 6.626 \times 10^{-34} \text{ J s} = 6.626 \times 10^{-34} \frac{\text{kg m}^2 \text{ s}}{\text{s}^2}$$

$$\pi = 3.14$$

Sample Exercise 7.6, continued

We solve for Δv:

$$\Delta v = \frac{h}{\Delta x \cdot m \cdot 4\pi} = \frac{6.626 \times 10^{-34} \frac{\text{kg m}^2\ \text{s}}{\text{s}^2}}{(5 \times 10^{-13}\ \text{m})(9.11 \times 10^{-31}\ \text{kg})(4)(3.14)}$$

$$= 1 \times 10^{8}\ \text{m/s}$$

Thus if we know the electron's position with a minimum uncertainty of 5×10^{-13} m, the uncertainty in the electron's velocity is at least 1×10^8 m/s. This is a very large number; in fact, it is the same magnitude as the speed of light ($c = 3 \times 10^8$ m/s). At this level of uncertainty we have virtually no idea of the velocity of the electron.

For the ball the uncertainty in position (Δx) is 1% of 0.05 m, or 5×10^{-4} m. Thus the minimum uncertainty in velocity is

$$\Delta v = \frac{h}{\Delta x \cdot m \cdot 4\pi} = \frac{6.626 \times 10^{-34} \frac{\text{kg m}^2\ \text{s}}{\text{s}^2}}{(5 \times 10^{-4}\ \text{m})(0.2\ \text{kg})(4)(3.14)}$$

$$= 5 \times 10^{-31}\ \text{m/s}$$

This means there is a very small (undetectable) uncertainty in our measurements of the speed of a ball. Note that this uncertainty is not due to the limitations of measuring instruments; Δv is an *inherent* uncertainty.

Thus the uncertainty principle is negligible in the world of macroscopic objects but is very important for objects with small masses, such as the electron.

The Physical Meaning of a Wave Function

Given the limitations indicated by the uncertainty principle, what then is the physical meaning of a wave function for an electron? Although the function itself has no easily visualized meaning, the square of the wave function does have a definite physical significance. *The square of the function indicates the probability of finding an electron at a particular point in space.* For example, suppose we have two positions in space, one defined by the coordinates x_1, y_1, and z_1, and the other by the coordinates x_2, y_2, and z_2. The relative probability of finding the electron at positions 1 and 2 is given by substituting the values of x, y, and z for the two positions into the wave function, squaring the function value, and computing the following ratio:

Probability is the likelihood, or odds, that something will occur.

$$\frac{[\psi(x_1, y_1, z_1)]^2}{[\psi(x_2, y_2, z_2)]^2} = \frac{N_1}{N_2}$$

The quotient N_1/N_2 is the ratio of the probabilities of finding the electron at positions 1 and 2. For example, if the value of the ratio N_1/N_2 is 100, the electron is 100 times more likely to be found at position 1 than at position 2. The model gives no information concerning when the electron will be at either position or how it moves between

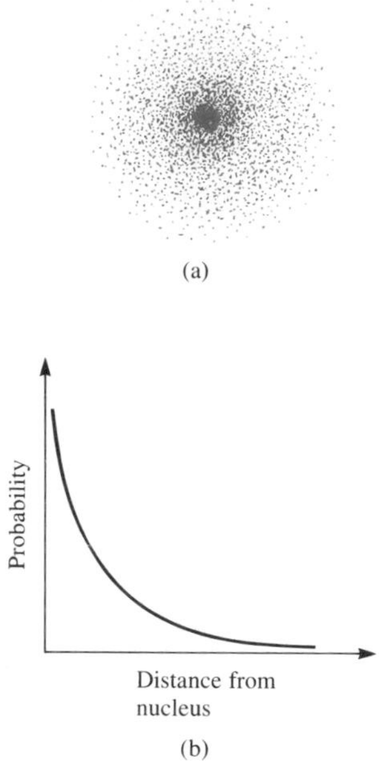

Figure 7.11

(a) The probability distribution for the hydrogen 1*s* orbital in three-dimensional space. (b) The probability of finding the electron at points along a line drawn from the nucleus outward in any direction for the hydrogen 1*s* orbital.

the positions. This vagueness is consistent with the concept of the Heisenberg uncertainty principle.

The square of the wave function is most conveniently represented as a **probability distribution,** in which the intensity of color is used to indicate the probability value at a given point in space. The probability distribution for the hydrogen 1*s* orbital is shown in Fig. 7.11(a). The best way to think about this diagram is as a three-dimensional time exposure, with the electron as a tiny moving light. The more times the electron visits a particular point, the darker the negative becomes. Thus the darkness of a point indicates the probability of finding an electron at that position. This diagram is sometimes known as an *electron density map;* electron density and electron probability mean the same thing.

Another way of representing the electron probability distribution for the 1*s* orbital is to calculate the probability at points along a line drawn outward in any direction from the nucleus. The result is shown in Fig. 7.11(b). Note that the probability of finding the electron at a particular position is greatest close to the nucleus and that it drops off rapidly as the distance from the nucleus increases. We are also interested in knowing the *total* probability of finding the electron in the hydrogen atom at a particular *distance* from the nucleus. Imagine that the space around the hydrogen nucleus is made up of a series of thin spherical shells (rather like layers in an onion), as shown in Fig. 7.12(a). When the total probability of finding the electron in each spherical shell is plotted versus the distance from the nucleus, the plot in Fig. 7.12(b) is obtained. This graph is called the **radial probability distribution.**

The maximum in the curve occurs because of two opposing effects. The probability of finding an electron at a particular position is greatest near the nucleus, but the volume of the spherical shell increases with distance from the nucleus. Therefore, as we move away from the nucleus, the probability of finding the electron at a given position decreases, but we are summing more positions. Thus the total probability increases to a certain radius and then decreases as the electron probability at each position becomes very small. For the hydrogen 1*s* orbital, the maximum radial probability (the distance at which the electron is most likely to be found) occurs at a distance of 5.29×10^{-2} nm or .529 Å from the nucleus. Interestingly, this is exactly the radius of the innermost orbit in the Bohr model. Note that in Bohr's model the electron is assumed to have a circular path and so is *always* found at this distance. In the wave mechanical model, the specific electron motions are unknown, and this is the *most probable* distance at which the electron is found

1 Å = 10^{-10} m; the angstrom is most often used as the unit for atomic radius because of its convenient size. Another convenient unit is the picometer (= 10^{-12} m).

One more characteristic of the hydrogen 1*s* orbital that we must consider is its size. As we can see from Fig. 7.11, the size of this orbital cannot be precisely

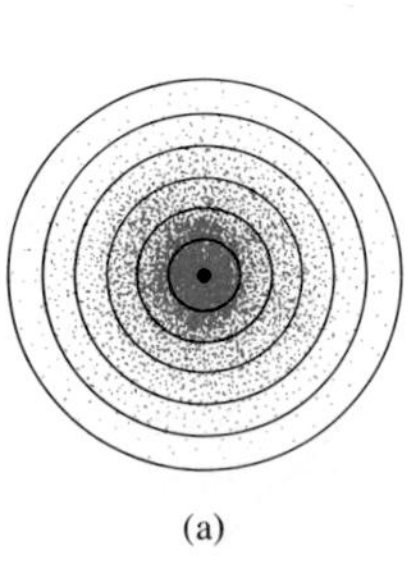

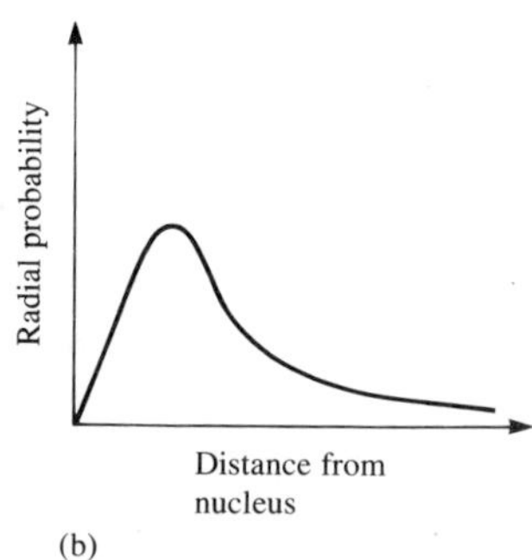

Figure 7.12

(a) Cross section of the hydrogen 1*s* orbital probability distribution divided into successive thin spherical shells. (b) The radial probability distribution. A plot of the total probability of finding the electron in each thin spherical shell as a function of distance from the nucleus.

defined, since the probability never becomes zero (although it drops to an extremely small value at large values of r). So, in fact, the hydrogen 1*s* orbital has no distinct size. However, it is useful to have a definition of relative orbital size. *The normally accepted arbitrary definition of the size of the hydrogen 1s orbital is the radius of the sphere that encloses 90% of the total electron probability*. That is, 90% of the time the electron is inside this sphere.

So far we have described only the lowest-energy wave function in the hydrogen atom, the 1*s* orbital. Hydrogen has many other orbitals, which we will describe in the next section.

7.6 Quantum Numbers

Purpose

- To explain the quantum numbers n, ℓ, and m_ℓ.

When we solve the Schrödinger equation for the hydrogen atom, we find many wave functions (orbitals) that satisfy it. Each of these orbitals is characterized by a series of numbers called **quantum numbers,** which describe various properties of the orbital:

The **principal quantum number** (n) can have integral values: 1, 2, 3, ... The principal quantum number is related to the size and energy of the orbital. As n increases, the orbital becomes larger and the electron spends more time further from the nucleus. An increase in n also means higher energy, because the electron is less tightly bound to the nucleus, and the energy is less negative.

The **azimuthal quantum number** (ℓ) can have integral values from 0 to $n - 1$ for each value of n. This quantum number relates to the shape of atomic orbitals. The value of ℓ for a particular orbital is commonly assigned a letter: $\ell = 0$ is called *s*; $\ell = 1$ is called *p*; $\ell = 2$ is called *d*; and $\ell = 3$ is called *f*. This system arises from early spectral studies and is summarized in Table 7.1.

The **magnetic quantum number** (m_ℓ) can have integral values between ℓ and $-\ell$, including zero. The value of m_ℓ relates to the orientation of the orbital in space relative to the other orbitals in the atom.

$n = 1,2,3, \ldots$
$\ell = 0,1, \ldots (n - 1)$
$m_\ell = -\ell, \ldots 0, \ldots +\ell$

The first four levels of orbitals in the hydrogen atom are listed with their quantum numbers in Table 7.2 on page 284. Note that each set of orbitals with a given value of ℓ (sometimes called a **subshell**) is designated by giving the value of n and the letter for ℓ. Thus an orbital where $n = 2$ and $\ell = 1$ is symbolized as 2*p*. There are three 2*p* orbitals, which have different orientations in space. We will describe these orbitals in the next section.

The Azimuthal Quantum Numbers and Corresponding Atomic Orbitals					
Value of ℓ	0	1	2	3	4
Letter used	*s*	*p*	*d*	*f*	*g*

Table 7.1

Number of orbitals per subshell
$s = 1$
$p = 3$
$d = 5$
$f = 7$
$g = 9$

Table 7.2

Quantum Numbers for the First Four Levels of Orbitals in the Hydrogen Atom

n	ℓ	Orbital designation	m_ℓ	Number of orbitals
1	0	1*s*	0	1
2	0	2*s*	0	1
	1	2*p*	−1, 0, +1	3
3	0	3*s*	0	1
	1	3*p*	−1, 0, 1	3
	2	3*d*	−2, −1, 0, 1, 2	5
4	0	4*s*	0	1
	1	4*p*	−1, 0, 1	3
	2	4*d*	−2, −1, 0, 1, 2	5
	3	4*f*	−3, −2, −1, 0, 1, 2, 3	7

Sample Exercise 7.7

For principal quantum level $n = 5$, determine the number of subshells (different values of ℓ) and give the designation of each.

Solution

For $n = 5$, the allowed values of ℓ run from 0 to 4 ($n - 1 = 5 - 1$). Thus the subshells and their designations are

$\ell = 0$	$\ell = 1$	$\ell = 2$	$\ell = 3$	$\ell = 4$
5*s*	5*p*	5*d*	5*f*	5*g*

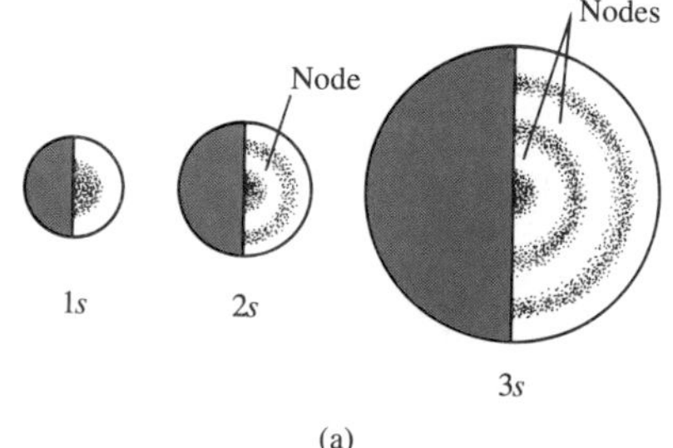

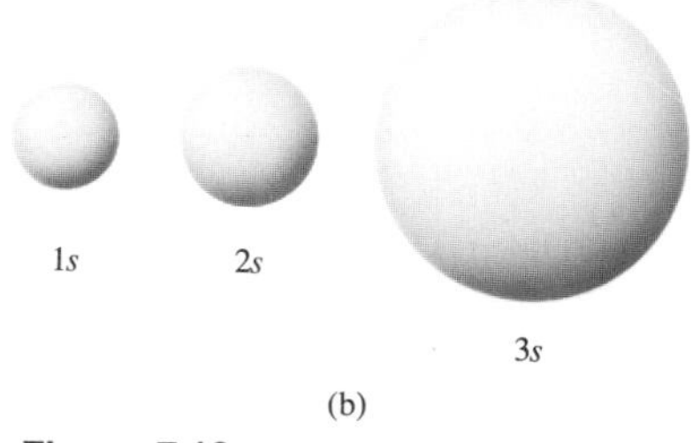

Figure 7.13

Two representations of the hydrogen 1*s*, 2*s*, and 3*s* orbitals. (a) The electron probability distribution. (b) The surface that contains 90% of the total electron probability (the size of the orbital, by definition).

7.7 Orbital Shapes and Energies

Purpose

- To describe the shapes of orbitals designated by *s*, *p*, *d*, and *f*, and to discuss orbital energies.

We have seen that the meaning of an orbital is illustrated most clearly by a probability distribution. Each orbital in the hydrogen atom has a unique probability distribution. We also saw that another means of representing an orbital is by the surface that surrounds 90% of the total electron probability. These two types of representations for the hydrogen 1*s*, 2*s*, and 3*s* orbitals are shown in Fig. 7.13. Note the characteristic spherical shape of each of the *s* orbitals. Note also that the 2*s* and 3*s* orbitals contain areas of high probability separated by areas of zero probability. These latter areas are called **nodal surfaces,** or simply **nodes.** The number of nodes increases as n increases. For *s* orbitals the number of nodes is given by $n - 1$. For our purposes, however, we will think of *s* orbitals only in terms of their overall spherical shape, which becomes larger as the value of n increases.

The two types of representations for the 2*p* orbitals (there are no 1*p* orbitals) are shown in Fig. 7.14. Note that the *p* orbitals are not spherical like *s* orbitals but have

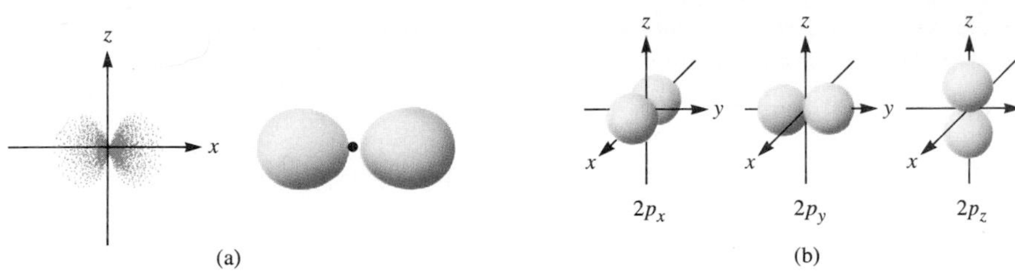

Figure 7.14

Representation of the $2p$ orbitals. (a) The electron probability distribution and the boundary surface for a $2p$ orbital. (b) The boundary surface representations of all three $2p$ orbitals.

two **lobes** separated by a node at the nucleus. The p orbitals are labeled according to the axis of the xyz coordinate system along which the lobes lie. For example, the $2p$ orbital with lobes along the x axis is called the $2p_x$ orbital.

n value
↓
$2p_x$ ← orientation in space
↑
ℓ value

As you might expect from our discussion of the s orbitals, the $3p$ orbitals have a more complex probability distribution than that of the $2p$ orbitals (see Fig. 7.15), but they can still be represented by the same boundary surface shapes. The surfaces just grow larger as the value of n increases.

There are no d orbitals that correspond to principal quantum levels $n = 1$ and $n = 2$. The d orbitals ($\ell = 2$) first occur in level $n = 3$. The five $3d$ orbitals have the shapes shown in Fig. 7.16. The d orbitals have two different fundamental shapes. Four of the orbitals (d_{xz}, d_{yz}, d_{xy}, and $d_{x^2-y^2}$) have four lobes centered in the plane indicated in the orbital label. Note that d_{xy} and $d_{x^2-y^2}$ are both centered in the xy plane; however, the lobes of the $d_{x^2-y^2}$ lie *along* the x and y axes, while the lobes of d_{xy} lie *between* the axes. The fifth orbital, d_{z^2}, has a unique shape with two lobes along the z axis and a "belt" centered in the xy plane. The d orbitals for levels $n > 3$ look like the $3d$ orbitals but have larger lobes.

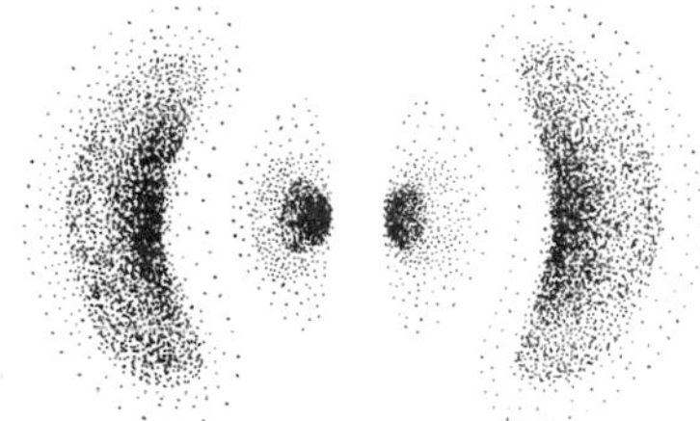

Figure 7.15

A cross section of the electron probability distribution for a $3p$ orbital.

The f orbitals first occur in level $n = 4$, and, as might be expected, they have shapes even more complex than those of the d orbitals. Fig. 7.17 shows representations

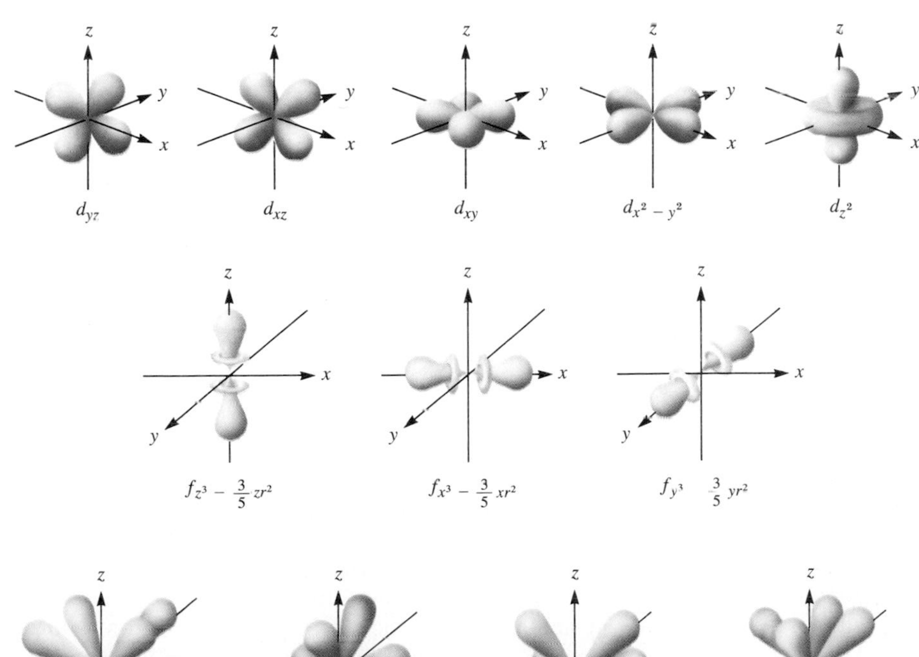

Figure 7.16

Representation of the $3d$ orbitals in terms of their boundary surfaces.

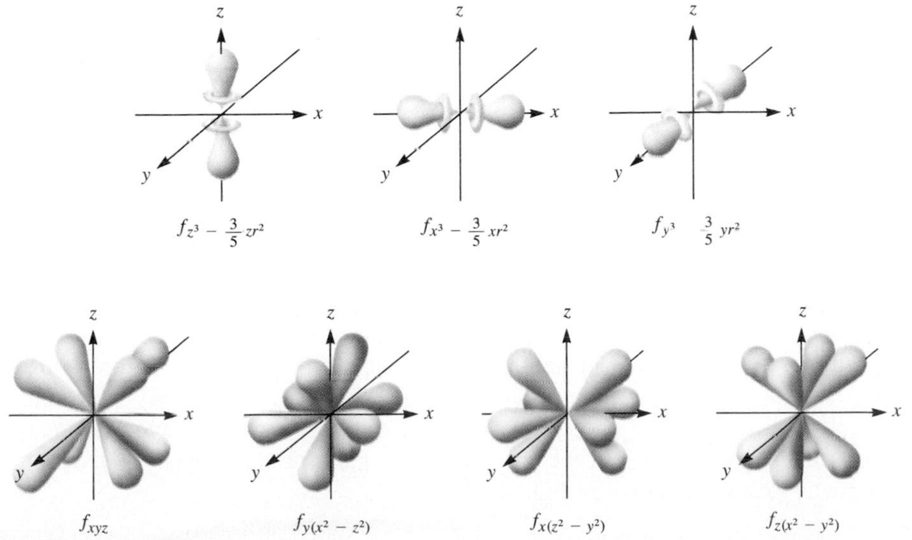

Figure 7.17

Representation of the $4f$ orbitals in terms of their boundary surfaces.

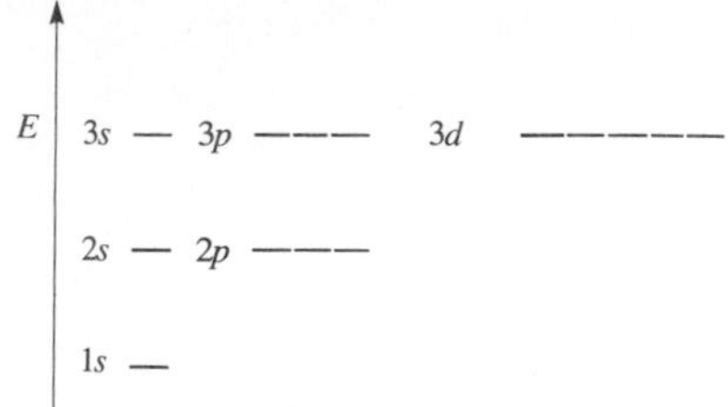

Figure 7.18

Orbital energy levels for the hydrogen atom.

of the $4f$ orbitals ($\ell = 3$) along with their designations. These orbitals are not involved in the bonding in any of the compounds we will consider in this text. Their shapes and labels are simply included for completeness.

So far we have talked about the shapes of the hydrogen atomic orbitals but not about their energies. For the hydrogen atom the energy of a particular orbital is determined by its value of n. Thus *all* orbitals with the same value of n have the *same energy*—they are said to be **degenerate.** This is shown in Fig. 7.18, where the energies for the orbitals in the first three quantum levels for hydrogen are shown.

Hydrogen's single electron can occupy any of its atomic orbitals. However, in the lowest energy state, the ground state, the electron resides in the 1s orbital. If energy is put into the atom, the electron can be transferred to a higher-energy orbital producing an excited state.

A Summary of the Hydrogen Atom

- In the wave mechanical model, the electron is viewed as a standing wave. This representation leads to a series of wave functions (orbitals) that describe the possible energies and spatial distributions available to the electron.
- In agreement with the Heisenberg uncertainty principle, the model cannot specify the detailed electron motions. Instead the square of the wave function represents the probability distribution of the electron in that orbital. This allows us to picture orbitals in terms of probability distributions, or electron density maps.
- The size of an orbital is arbitrarily defined as the surface that contains 90% of the total electron probability.
- The hydrogen atom has many types of orbitals. In the ground state the single electron resides in the 1s orbital. The electron can be excited to higher-energy orbitals if energy is put into the atom.

7.8 Electron Spin and the Pauli Principle

Purpose

- To define electron spin and the electron spin quantum number.
- To explain the Pauli exclusion principle.

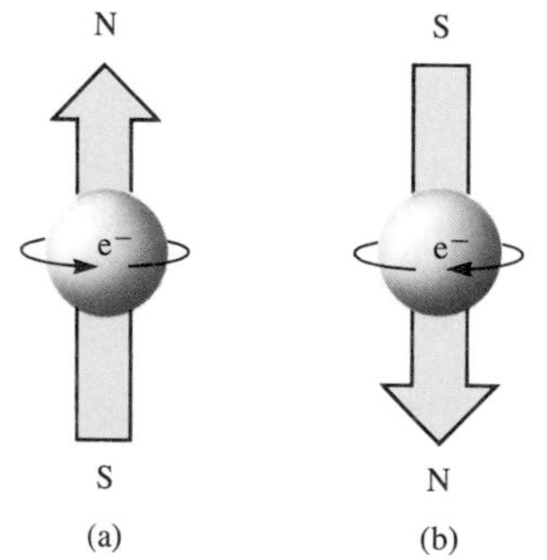

Figure 7.19

A picture of the spinning electron. Spinning in one direction, the electron produces the magnetic field oriented as shown in (a). Spinning in the opposite direction, it gives a magnetic field of the opposite orientation, as shown in (b).

An important property of the electron that we have not yet discussed is **electron spin.** The concept that the electron spins on its axis was developed by Samuel Goudsmit and George Uhlenbeck while they were graduate students at the University of Leyden in the Netherlands. They found that a fourth quantum number (in addition to n, ℓ, and m_ℓ) was necessary to account for the details of the emission spectra of atoms. This new quantum number, the **electron spin quantum number** (m_s), can have only one of two values, $+\frac{1}{2}$ and $-\frac{1}{2}$, which we interpret to mean that the electron can spin in one of two opposite directions. Since a spinning charge produces a magnetic field, the two opposing spins produce oppositely directed magnetic fields, as shown in Fig. 7.19.

For our purposes the main significance of electron spin is connected with the postulate of Austrian physicist Wolfgang Pauli (1900–1958): *in a given atom no two electrons can have the same set of four quantum numbers* (n, ℓ, m_ℓ, and m_s). This is called the **Pauli exclusion principle.** Since electrons in the same orbital have the same values of n, ℓ, and m_ℓ, this postulate says that they must have different values of m_s. Then, since only two values of m_s are allowed, *an orbital can only hold two electrons, and they must have opposite spins.* This principle will have important consequences as we use the atomic model to account for the electron arrangements of the atoms in the periodic table.

$m_s = +\frac{1}{2}$ or $-\frac{1}{2}$

Each orbital can hold a maximum of two electrons.

7.9 Polyelectronic Atoms

Purpose

- To show how the wave mechanical model can be applied to atoms besides hydrogen.

The wave mechanical model gives a description of the hydrogen atom that agrees very well with experimental data. However, the model would not be very useful if it did not account for the properties of all of the other atoms as well.

To see how the model applies to **polyelectronic atoms,** that is, atoms with more than one electron, let's consider helium, which has two protons in its nucleus and two electrons:

2+ e^- e^-

There are three energy contributions that must be considered in the description of the helium atom: (1) the kinetic energy of the electrons as they move around the nucleus, (2) the potential energy of attraction between the nucleus and the electrons, and (3) the potential energy of repulsion between the two electrons.

Although the atom can be readily described in terms of the wave mechanical model, the Schrödinger equation that results cannot be solved exactly. The difficulty arises in dealing with the repulsion between the electrons. Since the electron pathways are unknown, the electron repulsions cannot be calculated exactly. This is called the *electron correlation problem*.

The electron correlation problem occurs with all polyelectronic atoms. To treat these systems using the wave mechanical model, we must make approximations. Most commonly, the approximation used is to treat each electron as if it were moving in a *field of charge that is the net result of the nuclear attraction and the average repulsions of all the other electrons*. To see how this is done, let's consider the neutral helium atom and the He^+ ion:

2+ e^- e^- He   2+ e^- He^+

What energy is required to remove an electron from each of these species? Experiments show that 2372 kJ of energy is required to remove one electron from all of the atoms in a mole of helium. Removing the one electron from each ion in a mole of He^+ ions requires 5248 kJ of energy. Thus it takes more than twice as much energy to remove an electron from an He^+ ion as it does from an He atom.

Why such a large difference? In both cases the nucleus has a 2+ charge. However, in the helium atom there are two electrons that exert electron-electron repulsions, while in the He^+ ion there is only one electron and thus no electron-electron repulsion. That is, the large difference in the energies required to remove one electron must be due to the electron-electron repulsions in the neutral atom. Each electron in the He atom is much less tightly bound to the nucleus than is the electron in the He^+ ion. In other words, the effectiveness of the positively charged nucleus in binding the electrons has been decreased by the repulsions between the electrons. Thus the *effect of the electron repulsions can be thought of as reducing the nuclear charge* to an apparent value of less than 2+ toward a particular electron, as shown below:

Z_{eff} = Z − effect of electron repulsion.

The *apparent* nuclear charge, or the **effective nuclear charge,** is designated Z_{eff}. For a helium atom Z_{eff}, the charge ''felt'' by each electron, is less than 2. In general,

$$\text{Effective nuclear charge} = Z_{eff} = Z_{actual} - (\text{effect of electron repulsions})$$

where $Z_{actual} = Z$, the atomic number (number of protons).

This simplification allows us to treat each electron individually, where each electron is viewed as moving under the influence of a positive nuclear charge of Z_{eff}. This simplified atom has one electron, like hydrogen, but with a positive nuclear charge of Z_{eff} instead of 1. We therefore can find the energy and wave function for each helium electron by substituting Z_{eff} in place of $Z = 1$ in the hydrogen wave mechanical equations. When we do this, we find that both helium electrons reside in a 1*s* orbital (but have opposite spin quantum numbers). The orbital is spherical, like that for the hydrogen atom, but smaller because Z_{eff} is larger than 1. The larger nuclear charge draws each of the electrons closer to the nucleus and binds each more tightly than the electron in hydrogen. The increased nuclear charge of the helium atom is more important than the repulsions between the two electrons, so that each of the electrons in helium is bound more tightly than the electron in the hydrogen atom.

This same procedure can be used for all polyelectronic atoms. For example, consider the sodium atom, which has 11 protons in its nucleus and 11 electrons. To find the energy and orbital characteristics of each of sodium's electrons, we treat each electron individually and assume that it is influenced by a nuclear charge of Z_{eff} as illustrated below:

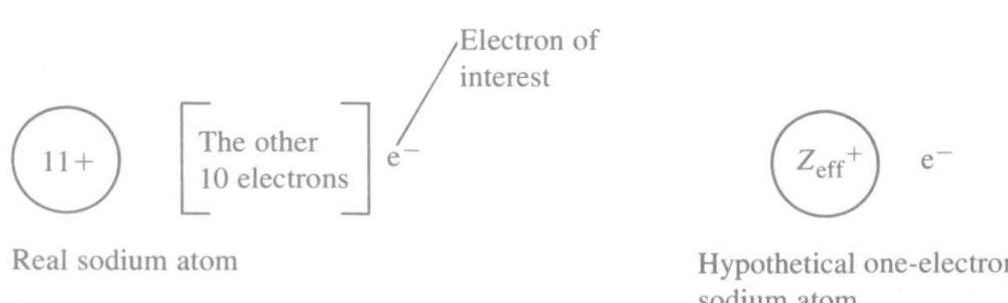

By treating each electron in turn, we can eventually obtain a complete (though approximate) description of the sodium atom.

The value of this approximation is that we can describe any polyelectronic atom one electron at a time using the mathematical treatment developed for hydrogen. The orbitals obtained in this type of treatment are *hydrogenlike orbitals* (the same types of orbitals with the same shapes), but their sizes and energies differ from those of the corresponding hydrogen orbitals. These differences occur because in a polyelectronic atom each electron is attracted by a nuclear charge of Z_{eff} rather than a charge of 1. One especially important difference between polyelectronic atoms and the hydrogen atom is that for hydrogen all orbitals with the same value of n have the same energy (are degenerate). This is not the case for polyelectronic atoms as we will see in Section 7.12.

It is important to understand that each electron in a polyelectronic atom has its own value of Z_{eff}, which can be calculated from the experimental energy required to remove that electron from the atom. We will discuss this procedure in more detail in Section 7.12.

7.10 The History of the Periodic Table

Purpose

- To trace the development of the periodic table.

The modern periodic table contains a tremendous amount of useful information. In this section we will discuss the origin of this valuable tool; later we will see how the wave mechanical model for the atom explains the periodicity of chemical properties. Certainly the greatest triumph of the wave mechanical model is its ability to account for the arrangement of the elements in the periodic table.

The periodic table was originally constructed to represent the patterns observed in the chemical properties of the elements. As chemistry progressed during the eighteenth and nineteenth centuries, it became evident that the earth is composed of a great many elements with very different properties. Things are much more complicated than the simple model of earth, air, fire, and water suggested by the ancients. At first, the array of elements and properties was bewildering. Gradually, however, patterns were noticed.

The first chemist to recognize patterns was Johann Dobereiner, who found several groups of three elements that have similar properties, for example, chlorine, bromine, and iodine. However, as Dobereiner attempted to expand this model of **triads** (as he called them) to the rest of the known elements, it became clear that it was severely limited.

The next notable attempt was made by the English chemist John Newlands, who in 1864 suggested that elements should be arranged in **octaves,** based on the idea that certain properties seemed to repeat for every eighth element in a way similar to the musical scale, which repeats for every eighth tone. But, although this model managed to group several elements with similar properties, it was not generally successful.

The present form of the periodic table was conceived independently by two chemists: the German Julius Lothar Meyer and Dmitri Ivanovich Mendeleev, a

Figure 7.20

Dmitri Ivanovich Mendeleev (1834–1907), born in Siberia as the youngest of 17 children, taught chemistry at the University of St. Petersburg (now Leningrad). In 1860 Mendeleev heard the Italian chemist Cannizzaro lecture on a reliable method for determining the correct atomic masses of the elements. This important development paved the way for Mendeleev's own brilliant contribution to chemistry—the periodic table. In 1861 Mendeleev returned to St. Petersburg, where he wrote a book on organic chemistry. Later Mendeleev also wrote a book on inorganic chemistry, and he was struck by the fact that the systematic approach characterizing organic chemistry was lacking in inorganic chemistry. In attempting to systematize inorganic chemistry, he eventually arranged the elements in the form of the periodic table.

Mendeleev was a versatile genius who was interested in many fields of science. He worked on many problems associated with Russia's natural resources, such as coal, salt, and various metals. Being particularly interested in the petroleum industry, he visited the United States in 1876 to study the Pennsylvania oil fields. His interests also included meteorology and hot-air balloons. In 1887 he made an ascent in a balloon to study a total eclipse of the sun.

Russian (Fig. 7.20). Usually Mendeleev is given most of the credit, because it was he who showed how useful the table could be in predicting the existence and properties of still unknown elements. For example, in 1872 when Mendeleev first published his table (see Fig. 7.21), the elements gallium, scandium, and germanium were unknown. Mendeleev correctly predicted the existence and properties of these elements from gaps in his periodic table. The data for germanium (which Mendeleev called *ekasilicon*) are shown in Table 7.3. Note the excellent agreement between the actual values and Mendeleev's predictions, which were based on the properties of other members in the group of elements similar to germanium.

TABELLE II

REIHEN	GRUPPE I. — R^2O	GRUPPE II. — RO	GRUPPE III. — R^2O^3	GRUPPE IV. RH^4 RO^2	GRUPPE V. RH^3 R^2O^5	GRUPPE VI. RH^2 RO^3	GRUPPE VII. RH R^2O^7	GRUPPE VIII. — RO^4
1	H = 1							
2	Li = 7	Be = 9,4	B = 11	C = 12	N = 14	O = 16	F = 19	
3	Na = 23	Mg = 24	Al = 27,3	Si = 28	P = 31	S = 32	Cl = 35,5	
4	K = 39	Ca = 40	— = 44	Ti = 48	V = 51	Cr = 52	Mn = 55	Fe = 56, Co = 59, Ni = 59, Cu = 63.
5	(Cu = 63)	Zn = 65	— = 68	— = 72	As = 75	Se = 78	Br = 80	
6	Rb = 85	Sr = 87	?Yt = 88	Zr = 90	Nb = 94	Mo = 96	— = 100	Ru = 104, Rh = 104, Pd = 106, Ag = 108.
7	(Ag = 108)	Cd = 112	In = 113	Sn = 118	Sb = 122	Te = 125	J = 127	
8	Cs = 133	Ba = 137	?Di = 138	?Ce = 140	—	—	—	— — — —
9	(—)	—	—	—	—	—	—	
10	—	—	?Er = 178	?La = 180	Ta = 182	W = 184	—	Os = 195, Ir = 197, Pt = 198, Au = 199.
11	(Au = 199)	Hg = 200	Tl = 204	Pb = 207	Bi = 208	—	—	
12	—	—	—	Th = 231	—	U = 240	—	— — — —

Figure 7.21

Mendeleev's early periodic table, published in 1872. Note the spaces left for missing elements with atomic weights 44, 68, 72, and 100. (From *Annalen der Chemie und Pharmacie,* VIII, Supplementary Volume for 1872, page 511.)

Comparison of the Properties of Germanium as Predicted by Mendeleev and as Actually Observed

Properties of germanium	Predicted in 1871	Observed in 1886
Atomic weight	72	72.3
Density	5.5 g/cm^3	5.47 g/cm^3
Specific heat	0.31 J/(°C g)	0.32 J/(°C g)
Melting point	Very high	960°C
Oxide formula	RO_2	GeO_2
Oxide density	4.7 g/cm^3	4.70 g/cm^3
Chloride formula	RCl_4	$GeCl_4$
bp of chloride	100°C	86°C

Table 7.3

Predicted Properties of Elements 113 and 114

Property	Element 113	Element 114
Chemically like	thallium	lead
Atomic weight	297	298
Density	16 g/ml	14 g/ml
Melting point	430°C	70°C
Boiling point	1100°C	150°C

Table 7.4

Using his table, Mendeleev was also able to correct several values for atomic weights. For example, the original atomic weight of 76 for indium was based on the assumption that indium oxide had the formula InO. This atomic weight placed indium, which has metallic properties, among the nonmetals. Mendeleev assumed the atomic weight was probably incorrect and proposed that the formula of indium oxide was really In_2O_3. Based on this correct formula, indium has an atomic weight of approximately 113, placing the element among the metals. Mendeleev also corrected the atomic weights of beryllium and uranium.

Because of its obvious usefulness, Mendeleev's periodic table was almost universally adopted, and it remains one of the most valuable tools at the chemist's disposal. For example, it is still used to predict the properties of elements yet to be discovered, as shown in Table 7.4.

A current version of the periodic table is shown inside the front cover of this book. The only fundamental difference between this table and that of Mendeleev is that it lists the elements in order by atomic number rather than by atomic weight. The reason for this will become clear later in this chapter as we explore the electron arrangements of the atom. Another recent format of the table is discussed in the following section.

7.11 The Aufbau Principle and the Periodic Table

Purpose

- To explain the Aufbau principle.

We can use the wave mechanical model of the atom to show how the electron arrangements in the hydrogenlike atomic orbitals of the various atoms account for the organization of the periodic table. Our main assumption here is that all atoms have the same type of orbitals as have been described for the hydrogen atom. *As protons are added one by one to the nucleus to build up the elements, electrons are similarly added to these hydrogenlike orbitals.* This is called the **Aufbau principle.**

Aufbau is German for "building up."

Hydrogen has one electron, which occupies the 1*s* orbital in its ground state. The configuration for hydrogen is written as $1s^1$, which can be represented by the following *orbital diagram:*

		1*s*	2*s*	2*p*
H:	$1s^1$	↑		☐☐☐

H (Z = 1)
He (Z = 2)
Li (Z = 3)
Be (Z = 4)
B (Z = 5)
etc.

The arrow represents an electron spinning in a particular direction.

The next element, *helium,* has two electrons. Since two electrons with opposite spins can occupy an orbital, according to the Pauli exclusion principle, the electrons for helium are in the 1*s* orbital with opposite spins, producing a $1s^2$ configuration:

		1*s*	2*s*	2*p*
He:	$1s^2$	↑↓		☐☐☐

Lithium has three electrons, two of which can go into the 1*s* orbital before the orbital is filled. Since the 1*s* orbital is the only orbital for $n = 1$, the third electron will occupy the lowest-energy orbital with $n = 2$, or the 2*s* orbital, giving a $1s^22s^1$ configuration:

We will see in Section 7.12 why the 2*s* orbital is lower in energy than the 2*p* orbital.

		1*s*	2*s*	2*p*
Li:	$1s^22s^1$	↑↓	↑	☐☐☐

The next element, *beryllium,* has four electrons, which occupy the 1*s* and 2*s* orbitals:

		1*s*	2*s*	2*p*
Be:	$1s^22s^2$	↑↓	↑↓	☐☐☐

Boron has five electrons, four of which occupy the 1*s* and 2*s* orbitals. The fifth electron goes into the second type of orbital with $n = 2$, the 2*p* orbitals.

		1*s*	2*s*	2*p*
B:	$1s^22s^22p^1$	↑↓	↑↓	↑ ☐☐

Since all of the 2*p* orbitals have the same energy (are degenerate), it does not matter which 2*p* orbital the electron occupies.

Carbon is the next element and has six electrons: two electrons occupy the 1*s* orbital; two occupy the 2*s* orbital; and two occupy 2*p* orbitals. Since there are three 2*p* orbitals with the same energy, the mutually repulsive electrons will occupy *separate* 2*p* orbitals.

It is found that single electrons in degenerate orbitals tend to have the same spins. This behavior is summarized by **Hund's rule** (named for the German physicist F. H. Hund), which states that *the lowest energy configuration for an atom is the one having the maximum number of unpaired electrons allowed by the Pauli principle in a particular set of degenerate orbitals. In addition, all of the unpaired electrons have parallel spins.*

For an atom with unfilled subshells, the lowest energy is achieved by electrons occupying separate orbitals with parallel spins, as far as allowed by the Pauli exclusion principle.

The configuration for carbon could be written $1s^22s^22p^12p^1$ to indicate that the electrons occupy separate 2*p* orbitals. However, the configuration is usually given as $1s^22s^22p^2$, and it is understood that the electrons are in different $2p$ orbitals. The orbital diagram for carbon is

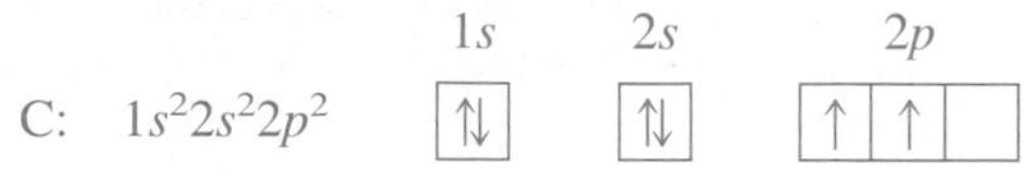

		$1s$	$2s$	$2p$
C:	$1s^22s^22p^2$	⇅	⇅	↑ ↑ □

Note the parallel spins for the unpaired electrons in the $2p$ orbitals, as required by Hund's rule.

The configuration for *nitrogen,* which has seven electrons, is $1s^22s^22p^3$. The three electrons in $2p$ orbitals occupy separate orbitals and have parallel spins:

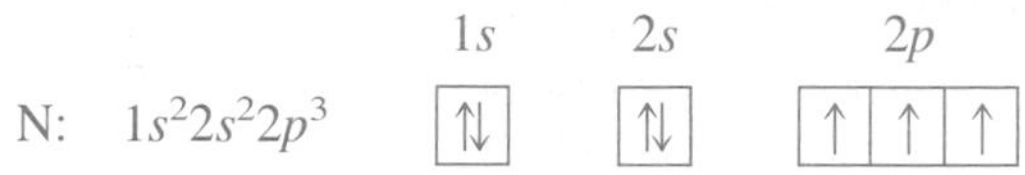

		$1s$	$2s$	$2p$
N:	$1s^22s^22p^3$	⇅	⇅	↑ ↑ ↑

The configuration for *oxygen,* which has eight electrons, is $1s^22s^22p^4$. One of the $2p$ orbitals is now occupied by a pair of electrons with spins paired, as required by the Pauli exclusion principle:

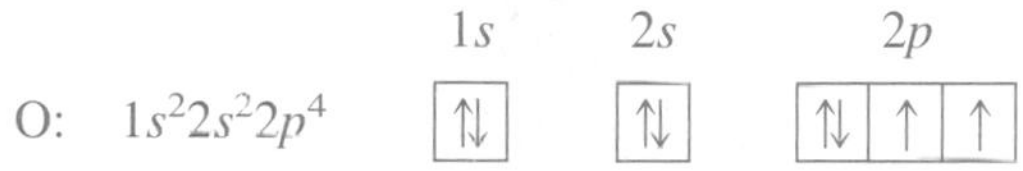

		$1s$	$2s$	$2p$
O:	$1s^22s^22p^4$	⇅	⇅	⇅ ↑ ↑

The orbital diagrams and electron configurations for *fluorine* (nine electrons) and *neon* (ten electrons) are given below:

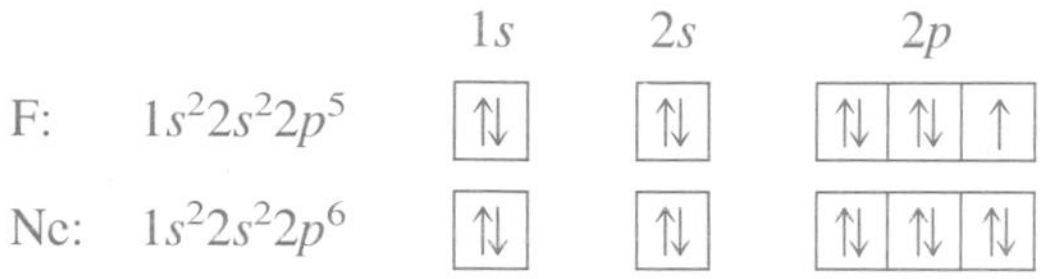

		$1s$	$2s$	$2p$
F:	$1s^22s^22p^5$	⇅	⇅	⇅ ⇅ ↑
Ne:	$1s^22s^22p^6$	⇅	⇅	⇅ ⇅ ⇅

With neon, the orbitals with $n = 1$ and $n = 2$ are now completely filled.

For *sodium,* the first ten electrons occupy the $1s$, $2s$, and $2p$ orbitals, and the eleventh electron must occupy the first orbital with $n = 3$, the $3s$ orbital. The electron configuration for sodium is $1s^22s^22p^63s^1$. To avoid writing the inner-level electrons, this configuration is often abbreviated as [Ne]$3s^1$ where [Ne] represents the electron configuration of neon, $1s^22s^22p^6$.

[Ne] is shorthand for $1s^22s^22p^6$.

The next element, *magnesium,* has the configuration $1s^22s^22p^63s^2$, or [Ne]$3s^2$. Then the next six elements, *aluminum* through *argon,* have configurations obtained by filling the $3p$ orbitals one electron at a time. Figure 7.22 summarizes the electron configurations of the first 18 elements by giving the number of electrons in the type of orbital occupied last.

At this point it is useful to introduce the concept of **valence electrons,** that is, *the electrons in the outermost principal quantum level of an atom.* The valence

H $1s^1$							He $1s^2$
Li $2s^1$	Be $2s^2$	B $2p^1$	C $2p^2$	N $2p^3$	O $2p^4$	F $2p^5$	Ne $2p^6$
Na $3s^1$	Mg $3s^2$	Al $3p^1$	Si $3p^2$	P $3p^3$	S $3p^4$	Cl $3p^5$	Ar $3p^6$

Figure 7.22

The electron configurations in the type of orbital occupied last for the first 18 elements.

Elemental potassium.

[Ar] is shorthand for $1s^22s^22p^63s^23p^6$.

Turnings of calcium metal.

Chromium is often used to plate bumpers and hood ornaments, such as this statue of Mercury found on a 1929 Buick.

electrons of the nitrogen atom, for example, are the $2s$ and $2p$ electrons. For the sodium atom the valence electron is the electron in the $3s$ orbital, and so on. Valence electrons are the most important electrons to chemists, because they are involved in bonding, as we will see in the next two chapters. The inner electrons are known as **core electrons.**

Note in Fig. 7.22 that a very important pattern is developing: *the elements in the same group (vertical column of the periodic table) have the same valence electron configuration*. Remember that Mendeleev originally placed the elements in groups based on similarities in chemical properties. Now we understand the reason behind these groupings. Elements with the same valence electron configuration show similar chemical behavior.

The element after argon is *potassium*. Since the $3p$ orbitals are fully occupied in argon, we might expect the next electron to go into a $3d$ orbital (recall that for $n = 3$ the orbitals are $3s$, $3p$, and $3d$). However, the chemistry of potassium is clearly very similar to that of lithium and sodium, indicating that the last electron in potassium occupies the $4s$ orbital instead of one of the $3d$ orbitals, a conclusion confirmed by many types of experiments. The electron configuration of potassium is

$$\text{K:}\quad 1s^22s^22p^63s^23p^64s^1 \quad \text{or} \quad [\text{Ar}]4s^1$$

The next element is *calcium:*

$$\text{Ca:}\quad [\text{Ar}]4s^2$$

The next element, *scandium,* begins a series of ten elements (scandium through zinc) called the **transition metals,** whose configurations are obtained by adding electrons to the five $3d$ orbitals. The configuration of scandium is

$$\text{Sc:}\quad [\text{Ar}]4s^23d^1$$

That of *titanium* is $\quad$ Ti: $[\text{Ar}]4s^23d^2$

And that of *vanadium* is $\quad$ V: $[\text{Ar}]4s^23d^3$

Chromium is the next element. The expected configuration is $[\text{Ar}]4s^23d^4$. However, the observed configuration is

$$\text{Cr:}\quad [\text{Ar}]4s^13d^5$$

The explanation for this configuration of chromium is beyond the scope of this book. In fact, chemists are still disagreeing over the exact cause of this anomaly. Note, however, that the observed configuration has both the $4s$ and $3d$ orbitals half-filled. This is a good way to remember the correct configuration.

The next four elements, *manganese* through *nickel,* have the expected configurations:

$$\text{Mn:}\quad [\text{Ar}]4s^23d^5 \qquad \text{Co:}\quad [\text{Ar}]4s^23d^7$$
$$\text{Fe:}\quad [\text{Ar}]4s^23d^6 \qquad \text{Ni:}\quad [\text{Ar}]4s^23d^8$$

The configuration for *copper* is expected to be $[\text{Ar}]4s^23d^9$. However, the observed configuration is

$$\text{Cu:}\quad [\text{Ar}]4s^13d^{10}$$

In this case, a half-filled $4s$ orbital and a filled set of $3d$ orbitals characterize the actual configuration.

K	Ca	Sc	Ti	V	Cr	Mn	Fe	Co	Ni	Cu	Zn	Ga	Ge	As	Se	Br	Kr
$4s^1$	$4s^2$	$3d^1$	$3d^2$	$3d^3$	$4s^13d^5$	$3d^5$	$3d^6$	$3d^7$	$3d^8$	$4s^13d^{10}$	$3d^{10}$	$4p^1$	$4p^2$	$4p^3$	$4p^4$	$4p^5$	$4p^6$

Figure 7.23

Electron configurations for potassium through krypton. The transition metals (scandium through zinc) have the general configuration $[Ar]4s^23d^n$, except for chromium and copper.

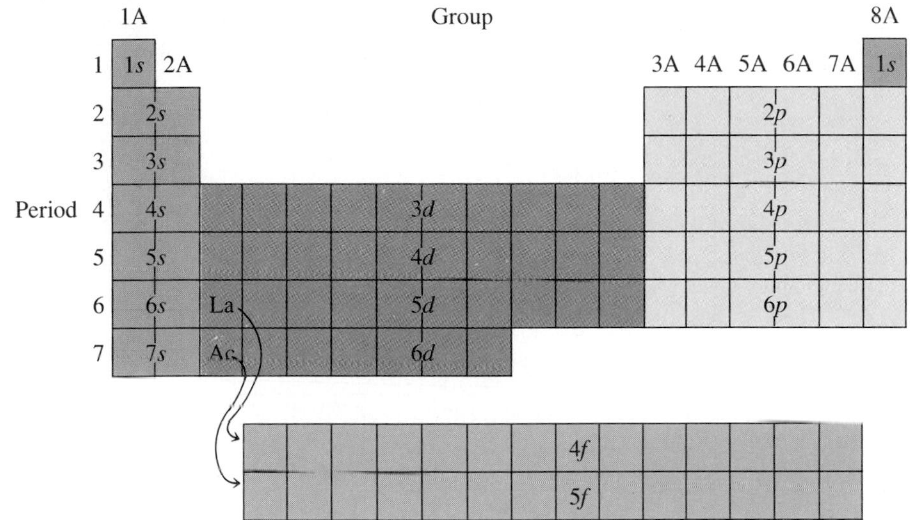

Figure 7.24

The orbitals being filled for elements in various parts of the periodic table. Note that in going along a horizontal row (a period), the $(n + 1)s$ orbital fills before the nd orbital. The group labels indicate the number of valence electrons (ns plus np electrons) for the elements in each group.

Zinc has the expected configuration:

$$\text{Zn:}\quad [\text{Ar}]4s^23d^{10}$$

The configurations of the transition metals are shown in Fig. 7.23. After that, the next six elements, *gallium* through *krypton,* have configurations that correspond to filling the $4p$ orbitals (see Fig. 7.23).

The entire periodic table is represented in Fig. 7.24 in terms of which orbitals are being filled. The valence electron configurations are given in Fig. 7.25 on page 296. From these two figures, note the following additional points:

The $(n + 1)s$ orbital fills before the nd orbitals.

1. The $(n + 1)s$ orbitals always fill before the nd orbitals. For example, the $5s$ orbitals fill in rubidium and strontium before the $4d$ orbitals fill in the second row of transition metals (yttrium through cadmium).

Lanthanides are elements in which the $4f$ orbitals are being filled.

2. After lanthanum, which has the configuration $[Xe]6s^25d^1$, a group of 14 elements called the **lanthanide series,** or the lanthanides, occurs. This series of elements corresponds to the filling of the seven $4f$ orbitals. Note that sometimes one electron occupies a $5d$, instead of a $4f$, orbital. This occurs because the energies of the $4f$ and $5d$ orbitals are very similar.

Representative Elements | *d*-Transition Elements | Representative Elements | Noble gases

Group numbers

Period number, highest occupied electron level

Period	1A ns^1	2A ns^2											3A ns^2np^1	4A ns^2np^2	5A ns^2np^3	6A ns^2np^4	7A ns^2np^5	8A ns^2np^6
1	1 H $1s^1$																	2 He $1s^2$
2	3 Li $2s^1$	4 Be $2s^2$											5 B $2s^22p^1$	6 C $2s^22p^2$	7 N $2s^22p^3$	8 O $2s^22p^4$	9 F $2s^22p^5$	10 Ne $2s^22p^6$
3	11 Na $3s^1$	12 Mg $3s^2$											13 Al $3s^23p^1$	14 Si $3s^23p^2$	15 P $3s^23p^3$	16 S $3s^23p^4$	17 Cl $3s^23p^5$	18 Ar $3s^23p^6$
4	19 K $4s^1$	20 Ca $4s^2$	21 Sc $4s^23d^1$	22 Ti $4s^23d^2$	23 V $4s^23d^3$	24 Cr $4s^13d^5$	25 Mn $4s^23d^5$	26 Fe $4s^23d^6$	27 Co $4s^23d^7$	28 Ni $4s^23d^8$	29 Cu $4s^13d^{10}$	30 Zn $4s^23d^{10}$	31 Ga $4s^24p^1$	32 Ge $4s^24p^2$	33 As $4s^24p^3$	34 Se $4s^24p^4$	35 Br $4s^24p^5$	36 Kr $4s^24p^6$
5	37 Rb $5s^1$	38 Sr $5s^2$	39 Y $5s^24d^1$	40 Zr $5s^24d^2$	41 Nb $5s^14d^4$	42 Mo $5s^14d^5$	43 Tc $5s^14d^6$	44 Ru $5s^14d^7$	45 Rh $5s^14d^8$	46 Pd $4d^{10}$	47 Ag $5s^14d^{10}$	48 Cd $5s^24d^{10}$	49 In $5s^25p^1$	50 Sn $5s^25p^2$	51 Sb $5s^25p^3$	52 Te $5s^25p^4$	53 I $5s^25p^5$	54 Xe $5s^25p^6$
6	55 Cs $6s^1$	56 Ba $6s^2$	57 La* $6s^25d^1$	72 Hf $4f^{14}6s^25d^2$	73 Ta $6s^25d^3$	74 W $6s^25d^4$	75 Re $6s^25d^5$	76 Os $6s^25d^6$	77 Ir $6s^25d^7$	78 Pt $6s^15d^9$	79 Au $6s^15d^{10}$	80 Hg $6s^25d^{10}$	81 Tl $6s^26p^1$	82 Pb $6s^26p^2$	83 Bi $6s^26p^3$	84 Po $6s^26p^4$	85 At $6s^26p^5$	86 Rn $6s^26p^6$
7	87 Fr $7s^1$	88 Ra $7s^2$	89 Ac** $7s^26d^1$	104 Unq $7s^26d^2$	105 Unp $7s^26d^3$	106 Unh $7s^26d^4$	107 Uns $7s^26d^5$	108	109 Une $7s^26d^7$									

f-Transition Elements

*Lanthanides	58 Ce $6s^24f^15d^1$	59 Pr $6s^24f^35d^0$	60 Nd $6s^24f^45d^0$	61 Pm $6s^24f^55d^0$	62 Sm $6s^24f^65d^0$	63 Eu $6s^24f^75d^0$	64 Gd $6s^24f^75d^1$	65 Tb $6s^24f^95d^0$	66 Dy $6s^24f^{10}5d^0$	67 Ho $6s^24f^{11}5d^0$	68 Er $6s^24f^{12}5d^0$	69 Tm $6s^24f^{13}5d^0$	70 Yb $6s^24f^{14}5d^0$	71 Lu $6s^24f^{14}5d^1$
**Actinides	90 Th $7s^25f^0d^2$	91 Pa $7s^25f^26d^1$	92 U $7s^25f^36d^1$	93 Np $7s^25f^46d^1$	94 Pu $7s^25f^66d^0$	95 Am $7s^25f^76d^0$	96 Cm $7s^25f^76d^1$	97 Bk $7s^25f^96d^0$	98 Cf $7s^25f^{10}6d^0$	99 Es $7s^25f^{11}6d^0$	100 Fm $7s^25f^{12}6d^0$	101 Md $7s^25f^{13}6d^0$	102 No $7s^25f^{14}6d^0$	103 Lr $7s^25f^{14}6d^1$

Figure 7.25

The periodic table with atomic symbols, atomic numbers, and partial electron configurations.

Actinides are elements in which the 5*f* orbitals are being filled.

3. After actinium, which has the configuration $[Rn]7s^26d^1$, a group of 14 elements called the **actinide series,** or the actinides, occurs. This series corresponds to the filling of the seven 5*f* orbitals. Note that sometimes one or two electrons occupy the 6*d* orbitals instead of the 5*f* orbitals, because these orbitals have very similar energies.

The group label tells the total number of valence electrons for that group.

4. The group labels for the Groups 1A, 2A, 3A, 4A, 5A, 6A, 7A, and 8A indicate the *total number* of valence electrons for the atoms in these groups. For example, all the elements in Group 5A have the configuration ns^2np^3. (The *d* electrons fill one period late and are usually not counted as valence electrons.) The meaning of the group labels for the transition metals is not as clear as for the A group elements, and these will not be used in this text.

1	2	3	4	5	6	7	8	9	10	11	12	13	14	15	16	17	18
1 H																	2 He
3 Li	4 Be											5 B	6 C	7 N	8 O	9 F	10 Ne
11 Na	12 Mg											13 Al	14 Si	15 P	16 S	17 Cl	18 Ar
19 K	20 Ca	21 Sc	22 Ti	23 V	24 Cr	25 Mn	26 Fe	27 Co	28 Ni	29 Cu	30 Zn	31 Ga	32 Ge	33 As	34 Se	35 Br	36 Kr
37 Rb	38 Sr	39 Y	40 Zr	41 Nb	42 Mo	43 Tc	44 Ru	45 Rh	46 Pd	47 Ag	48 Cd	49 In	50 Sn	51 Sb	52 Te	53 I	54 Xe
55 Cs	56 Ba	57 La	72 Hf	73 Ta	74 W	75 Re	76 Os	77 Ir	78 Pt	79 Au	80 Hg	81 Tl	82 Pb	83 Bi	84 Po	85 At	86 Rn
87 Fr	88 Ra	89 Ac	104 Unq	105 Unp	106 Unh	107 Uns	108	109 Une									

Lanthanide series	58 Ce	59 Pr	60 Nd	61 Pm	62 Sm	63 Eu	64 Gd	65 Tb	66 Dy	67 Ho	68 Er	69 Tm	70 Yb	71 Lu
Actinide series	90 Th	91 Pa	92 U	93 Np	94 Pu	95 Am	96 Cm	97 Bk	98 Cf	99 Es	100 Fm	101 Md	102 No	103 Lr

Figure 7.26

A form of the periodic table recommended by IUPAC.

5. The groups labeled 1A, 2A, 3A, 4A, 5A, 6A, 7A, and 8A are often called the **main-group, or representative, elements.** Remember that every member of these groups has the same valence-electron configuration.

Recently, the International Union of Pure and Applied Chemistry (IUPAC), a body of scientists organized to standardize scientific conventions, has recommended a new form for the periodic table, which the American Chemical Society has adopted (see Fig. 7.26). In this new version the group number indicates the number of *s*, *p*, and *d* electrons added since the last noble gas. We will not use the new format in this book, but you should be aware that the familiar periodic table may be soon replaced by this or a similar format.

The results considered in this section are very important. We have seen that the wave mechanical model can be used to explain the arrangement of elements in the periodic table. This model allows us to understand that the similar chemistry exhibited by the members of a given group arises from the fact that they all have the same valence-electron configuration. Only the principal quantum number of the occupied orbitals changes in going down a particular group.

It is important to be able to give the electron configuration for each of the main-group elements. This is most easily done by using the periodic table. If you understand how the table is organized, it is not necessary to memorize the order in which the orbitals fill. Review Fig. 7.24 and Fig. 7.25 to make sure that you understand the correspondence between the orbitals and the periods and groups.

Predicting the configurations of the transition metals (3*d*, 4*d*, and 5*d* elements), the lanthanides (4*f* elements), and the actinides (5*f* elements) is somewhat more difficult because there are many exceptions of the type encountered in the first-row transition metals (the 3*d* elements). You should memorize the configurations of chromium and copper, the two exceptions in the first-row transition metals, since these elements are often encountered.

When an electron configuration is given in this text, the orbitals are listed in the order in which they fill.

Cr: $[Ar]4s^1 3d^5$

Cu: $[Ar]4s^1 3d^{10}$

Sample Exercise 7.8

Give the electron configurations for sulfur (S), cadmium (Cd), hafnium (Hf), and radium (Ra), using the periodic table inside the front cover of this book.

Solution

Sulfur is element 16 and resides in Period 3, where the 3*p* orbitals are being filled (see Fig. 7.27). Since sulfur is the fourth among the "3*p* elements," it must have four 3*p* electrons. Its configuration is

$$\text{S:} \quad 1s^2 2s^2 2p^6 3s^2 3p^4 \quad \text{or} \quad [\text{Ne}]3s^2 3p^4$$

Cadmium is element 48 and is located in Period 5 at the end of the 4*d* transition metals, as shown in Fig. 7.27. It is the tenth element in the series and thus has ten electrons in the 4*d* orbitals, in addition to the two electrons in the 5*s* orbital. The configuration is

$$\text{Cd:} \quad 1s^2 2s^2 2p^6 3s^2 3p^6 4s^2 3d^{10} 4p^6 5s^2 4d^{10} \quad \text{or} \quad [\text{Kr}]5s^2 4d^{10}$$

Hafnium is element 72 and is found in Period 6, as shown in Fig. 7.27. Note that it occurs just after the lanthanide series. Thus the 4*f* orbitals are already filled. Hafnium is the second member of the 5*d* transition series and has two 5*d* electrons. The configuration is

$$\text{Hf:} \quad 1s^2 2s^2 2p^6 3s^2 3p^6 4s^2 3d^{10} 4p^6 5s^2 4d^{10} 5p^6 6s^2 4f^{14} 5d^2 \quad \text{or} \quad [\text{Xe}]6s^2 4f^{14} 5d^2$$

Radium is element 88 and is in Period 7 (and Group 2A), as shown in Fig. 7.27. Thus radium has two electrons in the 7*s* orbital, and the configuration is

$$\text{Ra:} \quad 1s^2 2s^2 2p^6 3s^2 3p^6 4s^2 3d^{10} 4p^6 5s^2 4d^{10} 5p^6 6s^2 4f^{14} 5d^{10} 6p^6 7s^2 \quad \text{or} \quad [\text{Rn}]7s^2$$

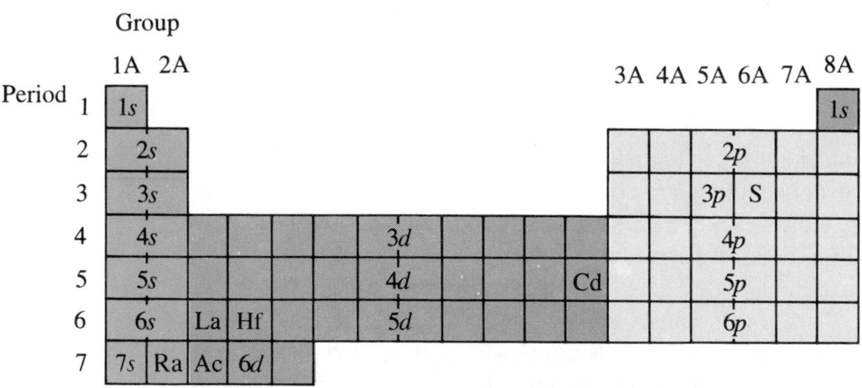

Figure 7.27

The positions of the elements considered in Sample Exercise 7.8.

7.12 Further Development of the Polyelectronic Model

Purpose

- To show how effective nuclear charge varies with position on the periodic table.

The simplest model for polyelectronic atoms views the atom one electron at a time, with each electron moving under the influence of an apparent nuclear charge of Z_{eff}^{+}. This model results in hydrogenlike orbitals occupied in a way that allows us to explain the patterns observed in the properties of elements—patterns summarized by the periodic table. Now we need to consider this model in more detail. By calculating actual values of Z_{eff} and by observing how these values change from element to element, we can "flesh out" and develop the model to explain and predict atomic properties.

In the terms of this polyelectronic model, the energy required to remove an electron from an atom depends mainly on two factors: the effective nuclear charge, and the average distance of the electron from the nucleus.

We will use Equation (7.1) to calculate values for Z_{eff}.

$$E = -2.18 \times 10^{-18}\ \text{J}\left(\frac{Z^2}{n^2}\right)$$

This expression for the hydrogen energy levels can be derived from the wave mechanical model, although it originally was derived from the Bohr model. This equation only applies to one-electron species such as H, He^{+}, Li^{2+}, and so on.

Although this equation originally comes from Bohr's model, which is not physically correct, it also can be derived from the wave mechanical model. Thus this equation accurately describes the energy levels of one-electron species and is often used to treat systems of this type. For example, Equation (7.1) has been shown to agree with the measured energies for the H atom and the He^{+}, Li^{2+}, and Be^{3+} ions, all of which have only one electron.

Remember that the energy of a single electron reflects how tightly bound that electron is to the nucleus; the more negative the energy, the more tightly the electron is bound. Given a positive sign, it represents *the quantity of energy required to remove the electron from the gaseous atom or ion,* which is called the **ionization energy.** We will discuss this concept more fully in the next section. To calculate ionization energies we usually convert Equation (7.1) to express energy per mole of atoms by multiplying by Avogadro's number:

$$E = \left(\frac{6.022 \times 10^{23}}{\text{mol}}\right)(-2.18 \times 10^{-18}\ \text{J})\left(\frac{Z^2}{n^2}\right)$$

$$= -\left(\frac{Z^2}{n^2}\right)(1.31 \times 10^{6})\ \text{J/mol} = -\left(\frac{Z^2}{n^2}\right)(1310)\ \text{kJ/mol}$$

Sample Exercise 7.9

Calculate the ionization energy of the Be^{3+} ion with the electron in its ground state.

Sample Exercise 7.9, continued

Solution

Be is $1s^2\ 2s^2$; Be^{3+} is $1s^1$.

The Be^{3+} ion has one electron in the $1s$ orbital ($n = 1$) in its ground state. Since beryllium has four protons in its nucleus ($Z = 4$), the energy of the electron in the $1s$ orbital of Be^{3+} can be calculated using Equation (7.1) as follows:

$$E = -\left(\frac{Z^2}{n^2}\right)(1310)\ \text{kJ/mol} = -\left(\frac{4^2}{1^2}\right)(1310)\ \text{kJ/mol}$$

$$= 2.10 \times 10^4\ \text{kJ/mol}$$

Recall that this energy (E) is negative since the electron is bound to the nucleus. An equal amount of energy must be applied to remove the electron. Thus, the ionization energy is *positive* and is 2.10×10^4 kJ/mol.

Recall that in the simplified model of the polyelectronic atom introduced in Section 7.9, we consider only one electron at a time. We assume that electron is bound by a positive charge Z_{eff}, where Z_{eff} represents the charge that the electron "feels." Thus Z_{eff} represents the net effect of the attraction of the nuclear charge and the repulsion of the other electrons. In this way we change from a very complicated real atom to a much simpler hypothetical atom:

To calculate the value of Z_{eff} for a particular electron in a polyelectronic atom, we can modify Equation (7.1) by substituting Z_{eff} for Z:

Energy of given electron in polyelectronic atom

$$= -\left(\frac{Z_{eff}^2}{n^2}\right)(1310)\ \text{kJ/mol} \qquad (7.3)$$

Since this quantity of energy must be furnished to remove the electron, the ionization energy is the *positive* value of Equation (7.3). Thus the ionization energy of a given electron in a polyelectronic atom is:

$$\left(\frac{Z_{eff}^2}{n^2}\right)(1310)\ \text{kJ/mol} \qquad (7.4)$$

We can see that, if the ionization energy of an electron is known from experiment, we can use Equation (7.4) to calculate Z_{eff} for that electron. That is, since ionization energy is a measurable quantity, we have a means of finding values for Z_{eff}. To see how this works, we will calculate the Z_{eff} values for the $1s$ electrons and the $3s$ electron in the sodium atom. The ionization energy for a $1s$ electron is known from experiment to be 1.39×10^5 kJ/mol and for a $3s$ electron is 4.95×10^2 kJ/mol.

For the 1s electron (n = *1*)

$$1.39 \times 10^5 \text{ kJ/mol} = \text{ionization energy for a } 1s \text{ electron}$$

$$= \left(\frac{Z_{eff}^2}{n^2}\right)(1310) \text{ kJ/mol}$$

$$1.39 \times 10^5 \text{ kJ/mol} = Z_{eff}^2\left(\frac{1310 \text{ kJ/mol}}{1^2}\right)$$

$$Z_{eff}^2 = \frac{1.39 \times 10^5 \frac{\text{kJ}}{\text{mol}}}{1310 \frac{\text{kJ}}{\text{mol}}} = 1.06 \times 10^2$$

$$Z_{eff} = 10.3$$

For the 3s electron (n = *3*)

$$4.95 \times 10^2 \text{ kJ/mol} = \text{ionization energy for a } 3s \text{ electron}$$

$$= \left(\frac{Z_{eff}^2}{n^2}\right)(1310) \text{ kJ/mol}$$

$$4.95 \times 10^2 \text{ kJ/mol} = Z_{eff}^2\left(\frac{1310 \text{ kJ/mol}}{3^2}\right)$$

$$Z_{eff}^2 = \frac{(4.95 \times 10^2)(9)\frac{\text{kJ}}{\text{mol}}}{1310 \frac{\text{kJ}}{\text{mol}}} = 3.40$$

$$Z_{eff} = 1.84$$

The calculated Z_{eff} value for an electron in a sodium 1*s* orbital is 10.3. Does this value make sense? It does. A 1*s* electron would be significantly affected only by the nucleus and the other 1*s* electron. The outer electrons would not have much effect on it because they are further out from the nucleus. Thus, for a 1*s* electron, the actual charge of 11 on the sodium nucleus is only slightly reduced (from 11 to 10.3) by electron repulsions between the two 1*s* electrons. Note that since both 1*s* electrons are in an identical environment, each one has a Z_{eff} value of 10.3.

The value of Z_{eff} calculated from Equation (7.4) for the sodium 3*s* electron is 1.84. Let's see how we can use this result to help us understand our model better. The sodium 3*s* valence electron "sees" the nucleus through the ten inner electrons, whose electron repulsions toward the 3*s* electron effectively reduce the nuclear "pull" on this electron. We say that the inner electrons *screen,* or *shield,* the 3*s* electron from some of the nuclear charge (see Fig. 7.28). If these ten electrons were completely effective as shielding electrons, the effective nuclear charge would be 11 − 10 = 1. However, Z_{eff} is calculated to be 1.84. The **shielding** is not totally effective. Why not?

We can understand this by looking at the radial probability distribution for an electron in a 3*s* orbital, shown in Fig. 7.29(a). Note that although the 3*s* electron spends most of the time far from the nucleus, it still has a small but significant probability of being close to the nucleus. We can see this is so from noting the two

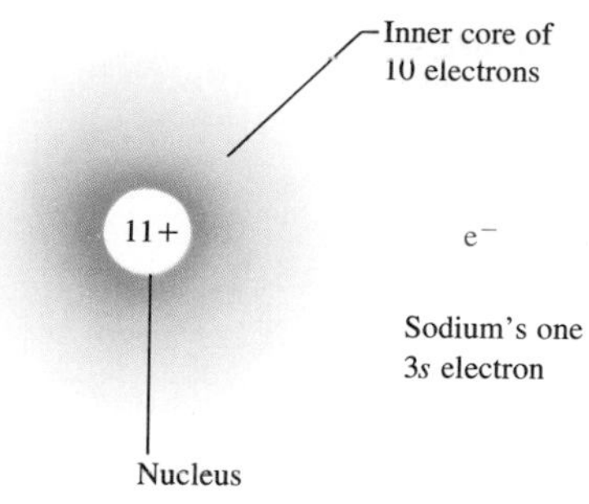

Figure 7.28

The 3*s* electron in the sodium atom "sees" the nucleus through a screen of ten inner electrons. Because of this, the effective nuclear charge is much less than the actual nuclear charge of 11.

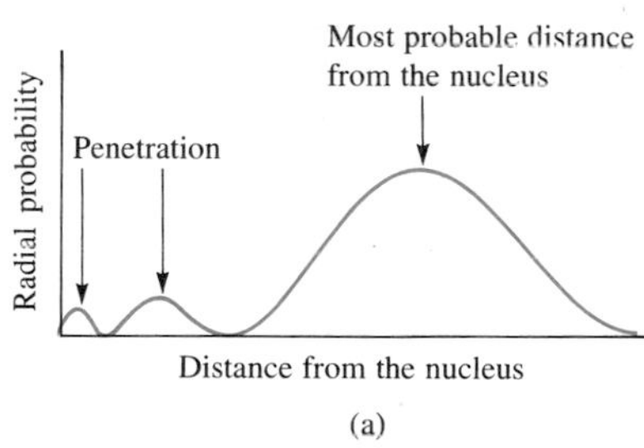

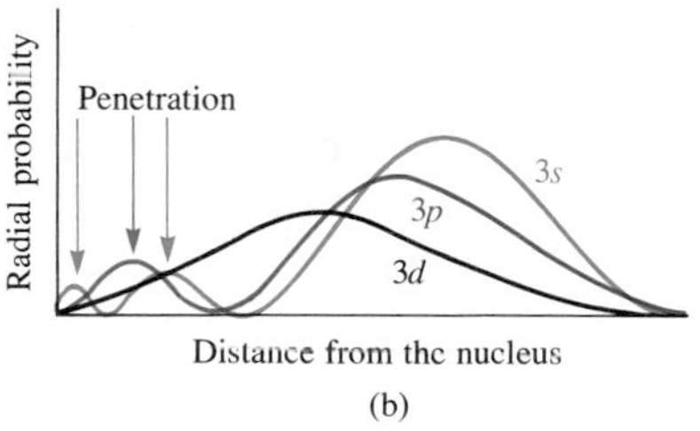

Figure 7.29

(a) The radial probability distribution for an electron in a 3*s* orbital. Although a 3*s* electron is mostly found far from the nucleus, there is a small but significant probability (shown by the arrows) of its being found close to the nucleus. The 3*s* electron penetrates the shield of inner electrons. (b) The radial probability distribution for the 3*s*, 3*p*, and 3*d* orbitals. The arrows indicate that the *s* orbital (blue arrows) allows greater electron penetration than the *p* orbital (red arrow) does; the *d* orbital allows minimal electron penetration.

inner peaks on the distribution. So we can say that the 3*s* electron in sodium *penetrates* the ten core electrons. *This penetration decreases the effectiveness of the shielding and thus increases the value of* Z_{eff} *for the* 3*s electron*. Because of this **penetration effect,** Z_{eff} is 1.84 instead of 1.00, the value it would be if shielding were perfect.

The amount of penetration depends on the orbital type, as we can see from Fig. 7.29(b), which shows the radial probability distributions of the orbitals for $n = 3$. For a given value of n, an electron in the *s* orbital always penetrates to the greatest extent, followed by one in a *p* orbital, then one in a *d* orbital. Although it is not applicable for $n = 3$, an electron in an *f* orbital always penetrates least for a given n. This can be summarized as follows:

$$ns > np > nd > nf$$

↑ Most penetration ↑ Least penetration

Orbital Filling Across a Period

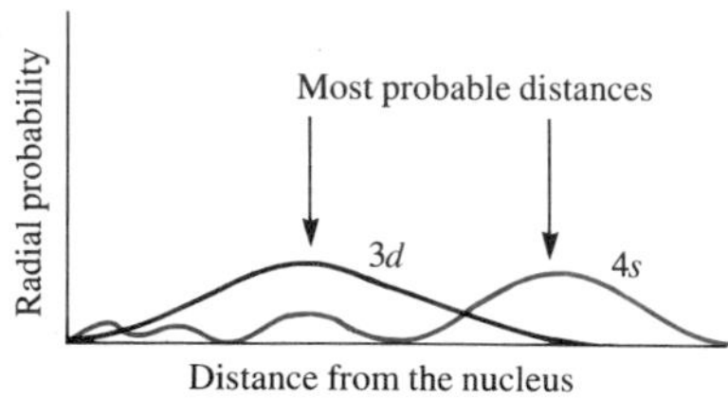

Figure 7.30

Radial probability distributions for the 3*d* and 4*s* orbitals. Note that the most probable distance of the electron from the nucleus for the 3*d* orbital is less than that for the 4*s*. However, the 4*s* orbital allows more electron penetration close to the nucleus and thus is preferred over the 3*d* orbital.

Recall that in building up the periodic table of elements we found that for a given value of n, the order of orbital filling is first the *ns* orbital, then the *np* orbital, then the $(n + 1)s$ orbital, and then the *nd* orbital. We can now explain why this happens. Because of the penetration effect, electrons will prefer the *ns* orbital over the *np* orbital, and the *np* orbital over the *nd* orbital, and so on. In terms of orbital energies,

$$E_{ns} < E_{np} < E_{nd} < E_{nf}$$

Electrons always fill the lowest-energy orbitals first.

The penetration effect also explains why the 4*s* orbital fills before the 3*d* orbital. Recall that potassium has the electron configuration $1s^22s^22p^63s^23p^64s^1$ rather than the expected $1s^22s^22p^63s^23p^63d^1$. We can explain this by observing that an electron in a 4*s* orbital penetrates much more than an electron in a 3*d* orbital does, as shown graphically in Fig. 7.30. Note that although the most probable distance from the nucleus for an electron in a 3*d* orbital is less than that for an electron in a 4*s* orbital, the 4*s* electron has a significant probability of penetrating close to the nucleus. This penetration ability means that the potassium atom in its lowest-energy state has its last electron in the 4*s* orbital rather than in the 3*d* orbital.

In polyelectronic atoms, only orbitals with the same *n* and *ℓ* values are degenerate. In contrast, for the hydrogen atom all orbitals with the same *n* value are degenerate.

Next, we want to consider the nature of the orbitals in a polyelectronic atom. How exactly do the hydrogenlike orbitals in these atoms differ from the orbitals of hydrogen? For example, let's compare the 1*s* orbital in sodium to that in hydrogen. An electron in the 1*s* orbital of sodium ''sees'' a positive nuclear charge of about 10.3. How does this large effective nuclear charge affect the nature of the sodium 1*s* orbitals? Since an electron in a sodium 1*s* orbital is attracted to the nucleus with a charge 10 times that for a hydrogen 1*s* electron, the sodium 1*s* electron will be bound more tightly (it will have lower energy) than the hydrogen 1*s* electron and will have a smaller average distance from the nucleus. Thus the sodium 1*s* orbital is lower in energy and smaller than the hydrogen 1*s* orbital. Both orbitals still have the same spherical shape. The differences are in size and energy.

The nature of electron shielding can best be understood by considering the Z_{eff} values (calculated from Equation 7.4) for the highest-energy electron in each of the elements hydrogen through sodium, as shown in Fig. 7.31. Note that Z_{eff} for a 1*s*

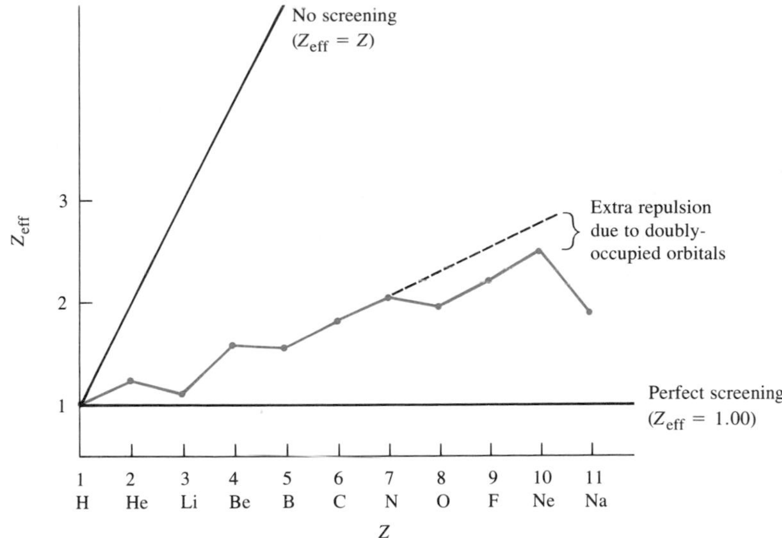

Figure 7.31

Calculated Z_{eff} values for the highest-energy electron in each of the elements hydrogen through sodium. Notice from the graph that Z_{eff} generally increases across a period going from left to right and increases going down a group.

electron increases in going from hydrogen to helium. The two electrons in the 1*s* orbital of helium do not completely shield each other from the nuclear charge. In general, it is true that *electrons in the same type of orbital are not very effective in shielding each other*. On the other hand, the *core electrons closer to the nucleus do provide effective shielding for outer electrons*. For example, note that Z_{eff} decreases in going from helium ($1s^2$) to lithium ($1s^22s^1$), because the 1*s* electrons are effective in shielding the electron in the 2*s* orbital.

Electrons sharing an orbital do not shield each other as well as core electrons shield outer electrons.

The increase in Z_{eff} (shown in Fig. 7.31) in going from lithium ($1s^22s^1$) to beryllium ($1s^22s^2$) is expected since the 2*s* electrons do not shield each other completely. The decrease in Z_{eff} in going from beryllium ($1s^22s^2$) to boron ($1s^22s^22p^1$) suggests that the electrons in 2*s* effectively shield the 2*p* electron. This is sensible in view of the greater penetration of the 2*s* electrons compared to the 2*p* electron. The 2*s* electrons spend time closer to the nucleus than the 2*p* electrons, where they provide effective shielding. The steady increase in Z_{eff} in going from boron ($1s^22s^22p^1$) to carbon ($1s^22s^22p^2$) to nitrogen ($1s^22s^22p^3$) is expected because the 2*p* electrons are not effective in shielding each other from the increasing nuclear charge. The drop in Z_{eff} in going from nitrogen ($1s^22s^22p^3$) to oxygen ($1s^22s^22p^4$) can be explained as follows. In nitrogen each 2*p* electron is in a separate orbital. When the extra electron for oxygen is added, one 2*p* orbital is now doubly occupied. This extra electron repulsion decreases Z_{eff} for the electrons in the doubly occupied orbital, making either of these electrons easier to remove. In going from oxygen ($1s^22s^22p^4$) to fluorine ($1s^22s^22p^5$) to neon ($1s^22s^22p^6$), Z_{eff} increases because in each case the electron being considered is in a doubly occupied orbital. Note that the Z_{eff} values for oxygen, fluorine, and neon are different from those of boron, carbon, and nitrogen by an amount that represents the extra electron repulsion due to the doubly occupied orbital, as shown in Fig. 7.31.

Sample Exercise 7.10

Explain why Z_{eff} for the highest-energy electron on sodium is so much smaller than Z_{eff} for the highest-energy electron on neon.

Sample Exercise 7.10, continued

Solution

Neon has an electron configuration of $1s^22s^22p^6$; all of the $n = 2$ orbitals are filled. Because the $2p$ electrons do not shield each other very effectively, we observe a relatively large value of Z_{eff} for the $2p$ electrons. On the other hand, sodium has an electron configuration of $1s^22s^22p^63s^1$; the highest-energy electron in sodium (the $3s$) is effectively shielded from the nuclear charge by the core electrons. The Z_{eff} value for the $3s$ electron therefore is much smaller than Z_{eff} for the $2p$ electrons on neon. Note, however, that the shielding is not perfect, that is, $Z_{eff} \neq 1$, because the $3s$ electron penetrates the core electrons to some extent.

7.13 Periodic Trends in Atomic Properties

Purpose

- To show general trends in ionization energy, electron affinity, and atomic radius in the periodic table.

We have developed a fairly complete picture of polyelectronic atoms. Although the model is rather crude because the nuclear attractions and electron repulsions are simply lumped together to produce an effective nuclear charge, it is very successful in accounting for the periodic table of elements. Using the concepts of shielding and penetration, we have been able to explain the order of orbital filling and the electron configurations of the atoms. We will next use the model to account for the observed trends in several important atomic properties: ionization energy, electron affinity, and atomic size.

Ionization Energy

As we saw in Section 7.12, ionization energy is the energy required to remove an electron from a gaseous atom or ion

$$X(g) \rightarrow X^+(g) + e^-$$

where the atom or ion is assumed to be in its ground state.

To introduce some of the characteristics of ionization energy, we will consider the energy required to remove several electrons in succession from aluminum in the gaseous state. The ionization energies are

$$\begin{aligned} Al(g) &\rightarrow Al^+(g) + e^- && I_1 = 580 \text{ kJ/mol} \\ Al^+(g) &\rightarrow Al^{2+}(g) + e^- && I_2 = 1815 \text{ kJ/mol} \\ Al^{2+}(g) &\rightarrow Al^{3+}(g) + e^- && I_3 = 2740 \text{ kJ/mol} \\ Al^{3+}(g) &\rightarrow Al^{4+}(g) + e^- && I_4 = 11{,}600 \text{ kJ/mol} \end{aligned}$$

Several important points can be illustrated from these results. In a stepwise ionization process, it is always the highest-energy electron (the one bound least tightly) that is removed first. The energy required to remove the highest-energy electron of an atom is called the **first ionization energy** (I_1). The first electron removed from the aluminum atom comes from the $3p$ orbital (Al has the electron configuration $[Ne]3s^23p^1$). The second electron comes from the $3s$ orbital (since Al^+ has the configuration $[Ne]3s^2$). Note that the value of I_1 is considerably smaller than the value of I_2, the **second ionization energy.**

This makes sense for several reasons. The primary factor is simply charge. Note that the first electron is removed from a neutral atom (Al) whereas the second electron is removed from a 1+ ion (Al^+). The increase in positive charge binds the electrons more firmly and the ionization energy increases. The same trend shows up in the third (I_3) and fourth (I_4) ionization energies where the electron is removed from the Al^{2+} and Al^{3+} ions, respectively.

The increase in successive ionization energies for an atom can also be interpreted using our simple model for polyelectronic atoms. The increase in ionization energy from I_1 to I_2 makes sense because the first electron is removed from a $3p$ orbital that is higher in energy than the $3s$ orbital from which the second electron is removed. The largest jump in ionization energy by far occurs in going from the third ionization energy (I_3) to the fourth (I_4). This is because I_4 corresponds to removing a core electron (Al^{3+} has the configuration $1s^22s^22p^6$), and core electrons are bound much more tightly than valence electrons.

Table 7.5 gives the values of ionization energies for all of the Period 3 elements. Note the large jump in energy in each case in going from removal of valence electrons to removal of core electrons.

The values of the first ionization energy for the elements in the first five periods of the periodic table are graphed in Fig. 7.32 on page 306. Note that in general as we go *across a period from left to right, the first ionization energy increases*. This is consistent with our idea that electrons in the same principal quantum level do not completely shield the increasing nuclear charge from added protons. Thus, there is a general increase in both the Z_{eff} and ionization energy values as electrons are added to a given principal quantum level.

On the other hand, *first ionization energy decreases in going down a group*. This can be most clearly seen by focusing on the Group 1A elements (the alkali

First ionization energy increases across a period and decreases down a group.

General increase →

General decrease ↑

Successive Ionization Energies in Kilojoules per Mole for the Elements in Period 3.

Element	I_1	I_2	I_3	I_4	I_5	I_6	I_7
Na	495	4560					
Mg	735	1445	7730	Core electrons*			
Al	580	1815	2740	11,600			
Si	780	1575	3220	4350	16,100		
P	1060	1890	2905	4950	6270	21,200	
S	1005	2260	3375	4565	6950	8490	27,000
Cl	1255	2295	3850	5160	6560	9360	11,000
Ar	1527	2665	3945	5770	7230	8780	12,000

*Note the large jump in ionization energy in going from removal of valence electrons to removal of core electrons.

Table 7.5

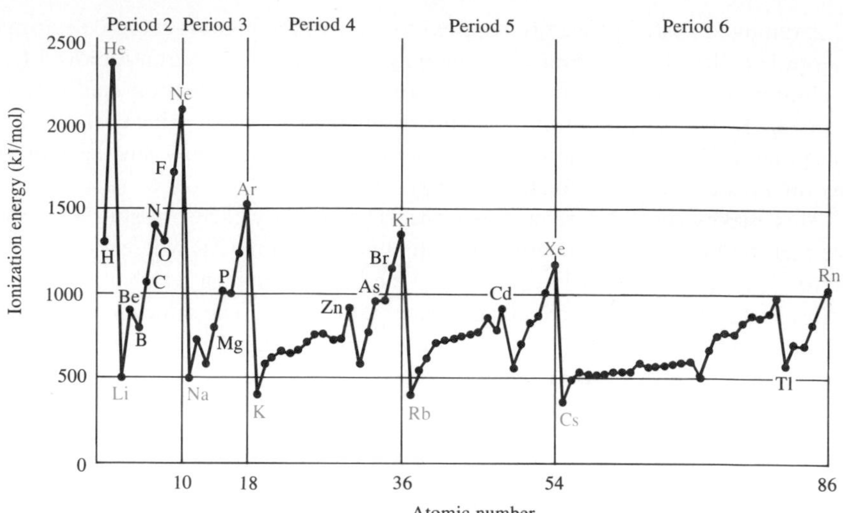

Figure 7.32

The values of first ionization energy for the elements in the first five periods. In general, ionization energy decreases in going down a group. For example, note the decrease in values for Group 1A and Group 8A. In general, ionization energy increases in going left to right across a period. For example, note the sharp increase going across Period 2 from lithium through neon.

metals) and the Group 8A elements (the noble gases), as shown in Table 7.6. The main reason for the decrease in ionization energy in going down a group is that the electrons being removed are, on the average, farther from the nucleus. As n increases, the size of the orbital increases, and the electron is easier to remove.

In Fig. 7.32 we see that there are some discontinuities in ionization energy in going across a period. For example, for Period 2, discontinuities occur in going from beryllium to boron and from nitrogen to oxygen. These exceptions to the normal trend can be explained in terms of how the effective nuclear charge depends on the electron configuration. The decrease in ionization energy in going from beryllium to boron reflects the fact that the electrons in the filled $2s$ orbital provide some shielding for electrons in the $2p$ orbital from the nuclear charge. The decrease in ionization energy in going from nitrogen to oxygen reflects the extra electron repulsions in the doubly occupied oxygen $2p$ orbital.

First Ionization Energies for the Alkali Metals and Noble Gases

Atom	I_1 (kJ/mol)
Group 1A	
Li	520.
Na	495
K	419
Rb	409
Cs	382
Group 8A	
He	2377
Ne	2088
Ar	1527
Kr	1356
Xe	1176
Rn	1042

Table 7.6

Sample Exercise 7.11

The first ionization energy for phosphorus is 1060 kJ/mol, and that for sulfur is 1005 kJ/mol. Why?

Solution

Phosphorus and sulfur are neighboring elements in Period 3 of the periodic table and have the following valence-electron configurations: phosphorus is $3s^23p^3$, and sulfur is $3s^23p^4$.

Ordinarily, the first ionization energy increases as we go across a period, so we might expect sulfur to have a greater ionization energy than phosphorus. However, in this case the fourth p electron in sulfur must be placed in an already occupied orbital. The electron-electron repulsions that result cause this electron to be more easily removed than might be expected.

Sample Exercise 7.12

Consider atoms with the following electron configurations:

$$1s^22s^22p^6$$
$$1s^22s^22p^63s^1$$
$$1s^22s^22p^63s^2$$

Which atom has the largest first ionization energy, and which one has the smallest second ionization energy? Explain your choices.

Solution

The atom with the largest value of I_1 is the one with the configuration $1s^22s^22p^6$ (this is the neon atom) because this element is found at the right end of Period 2. Since the $2p$ electrons do not shield each other very effectively, Z_{eff} will be relatively large, and thus I_1 will also be large. The other configurations given include $3s$ electrons. These electrons are effectively shielded by the core electrons and are farther from the nucleus than the $2p$ electrons in neon. Thus I_1 for these atoms will be smaller than for neon.

The atom with the smallest value of I_2 is the one with the configuration $1s^22s^22p^63s^2$ (the magnesium atom). For magnesium both I_1 and I_2 involve valence electrons. For the atom with the configuration $1s^22s^22p^63s^1$ (sodium), the second electron lost (corresponding to I_2) is a core electron (from a $2p$ orbital).

Electron Affinity

Electron affinity *is the energy change associated with the addition of an electron to a gaseous atom:*

$$X(g) + e^- \rightarrow X^-(g)$$

Because two different conventions have been used, there is a good deal of confusion in the chemical literature about the signs for electron affinity values. Electron affinity has been defined in many textbooks as the energy *released* when an electron is added to a gaseous atom. This convention requires that a positive sign be attached to an exothermic addition of an electron to an atom, which opposes normal thermodynamic conventions. Therefore, in this book we define electron affinity as a *change* in energy, which means that if the addition of the electron is exothermic, the corresponding value for electron affinity will carry a negative sign.

The sign convention for electron affinity values follows the convention for energy changes used in Chapter 6.

Electron affinity values for the first 20 elements in the periodic table are shown in Fig. 7.33 on page 308. Note that the *more negative* the energy, the greater the quantity of energy that is released. Although electron affinities generally become more negative in going across a period, there are several exceptions to this rule in each period. The dependence of electron affinity on atomic number can be explained by considering the changes in effective nuclear charge. For example, the fact that much less energy is released when an electron is added to a nitrogen atom than to a carbon atom reflects the difference in the electron configurations of these atoms. When an electron is added to nitrogen ($1s^22s^22p^3$) to form the N^- ion

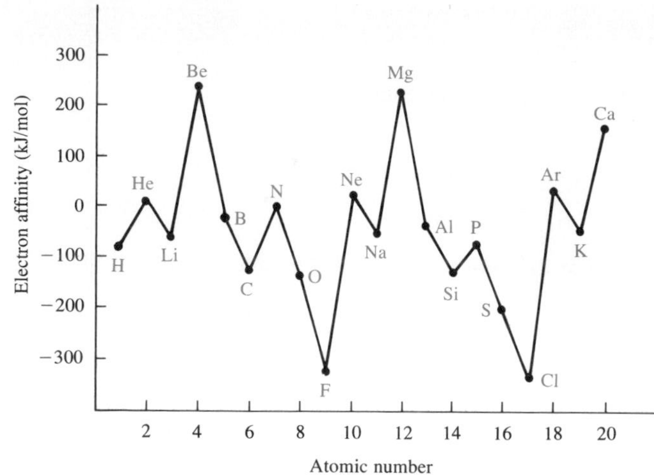

Figure 7.33

The electron affinity values for the first 20 elements. A negative sign for a given value indicates an exothermic process; energy is released when an electron is added.

$(1s^22s^22p^4)$, the extra electron enters a $2p$ orbital that already contains one electron. The extra repulsion between these two electrons in an orbital causes less energy to be released than when an electron is added to carbon $(1s^22s^22p^2)$ to form the C^- ion $(1s^22s^22p^3)$.

In going down a group, electron affinity should become more positive (less energy released) since the electron is added at increasing distances from the nucleus. While this is generally the case, the changes in electron affinity in going down most groups are relatively small and numerous exceptions occur. This behavior is demonstrated by the electron affinities of the Group 7A elements (the halogens) shown in Table 7.7. Note that the range of values is quite small compared to the changes that typically occur across a period. Also note that although chlorine, bromine, and iodine show the expected trend, the energy released when an electron is added to fluorine is smaller than might be expected. This has been attributed to the small size of the $2p$ orbitals. Because the electrons must be very close together in these orbitals, there are unusually large electron-electron repulsions. In the other halogens with their larger orbitals, the repulsions are not as severe.

Electron Affinities of the Halogens

Atom	Electron affinity (kJ/mol)
F	−327.8
Cl	−348.7
Br	−324.5
I	−295.2

Table 7.7

Atomic Radius

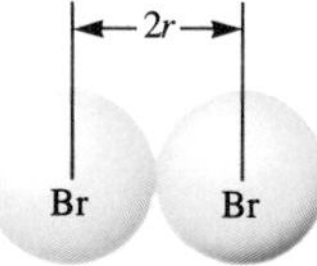

Figure 7.34

The radius of an atom (r) is defined as half the distance between the nuclei in a molecule consisting of identical atoms.

Atomic radius decreases across a period and increases down a group.

Just as the size of an orbital cannot be specified exactly, the size of an atom cannot be precisely defined. **Atomic radii** must be obtained by measuring the distances between atoms in chemical compounds. For example, in the bromine molecule, the distance between the two nuclei is known to be 2.28 Å (228 pm). The bromine atomic radius is assumed to be half this distance, or 1.14 angstroms, as shown in Fig. 7.34. Measurements of this type have led to the values of atomic radii for the elements (see Fig. 7.35). Note that atomic radius decreases in going from left to right across a period. This can be explained in terms of the effective nuclear charge, which increases in going from left to right. This means the valence electrons are drawn closer to the nucleus, decreasing the size of the atom.

Atomic radius increases going down a group. The increasing principal quantum number of the valence orbitals means larger orbitals and an increase in atomic size.

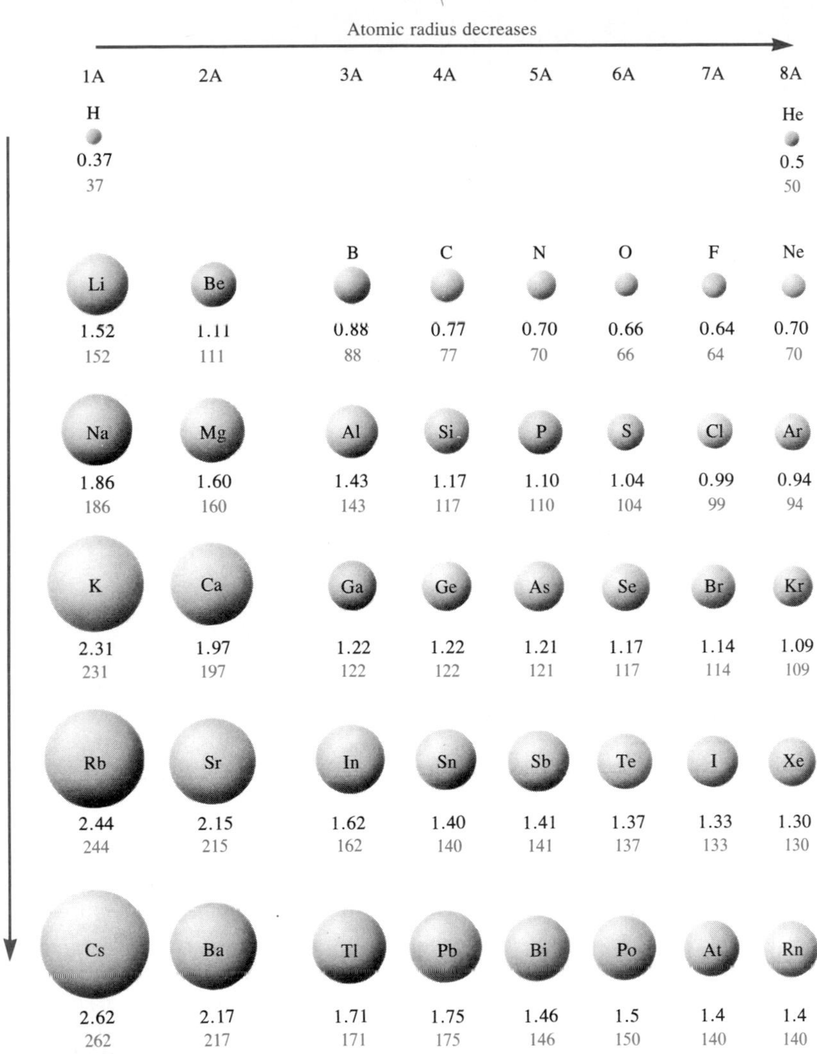

Figure 7.35

Atomic radii in angstroms and picometers (in blue) for selected atoms. Note that atomic radius decreases going across a period and increases going down a group.

Sample Exercise 7.13

Predict the trend in radius of the following ions: Be^{2+}, Mg^{2+}, Ca^{2+}, and Sr^{2+}.

Solution

All these ions are formed by removing two electrons from an atom of a Group 2A element. In going from beryllium to strontium, we are going down the group, so the sizes increase.

$$\underset{\substack{\uparrow \\ \text{Smallest radius}}}{Be^{2+}} < Mg^{2+} < Ca^{2+} < \underset{\substack{\uparrow \\ \text{Largest radius}}}{Sr^{2+}}$$

7.14 The Properties of a Group: The Alkali Metals

Purpose

- To show what types of information can be obtained from the periodic table.

We have seen that the periodic table originated as a way to portray the systematic properties of the elements. Mendeleev was primarily responsible for first showing its usefulness in correlating and predicting the elemental properties. In this section we will summarize much of the information available from the table. We will also illustrate the usefulness of the table by discussing the properties of a representative group, the alkali metals.

Information Contained in the Periodic Table

1. The essence of the periodic table is that the groups of representative elements exhibit similar chemical properties that change in a regular way. The wave mechanical model of the atom has allowed us to understand the basis for the similarity of properties in a group—that each group member has the same valence-electron configuration. *It is the number and type of valence electrons that primarily determine an atom's chemistry.*
2. One of the most valuable types of information available from the periodic table is the electron configuration of any representative element. If you understand the organization of the table, you will not need to memorize electron configurations for these elements. Although the predicted electron configurations for transition metals are sometimes incorrect, this is not a serious problem. You should however memorize the configuration of two exceptions, chromium and copper, since these 3*d* transition elements are found in many important compounds.
3. As we mentioned in Chapter 2, certain groups in the periodic table have special names. These are summarized in Fig. 7.36. Groups are often referred to by these names, so you should learn them.

Metals and nonmetals were first discussed in Chapter 2.

4. The most basic division of the elements in the periodic table is into metals and nonmetals. The most important chemical property of a metal is the tendency to give up electrons to form a positive ion; metals tend to have low ionization energies. The metallic elements are found on the left side of the table, as shown in Fig. 7.36. The most chemically reactive metals are found on the lower left-hand portion of the table where the ionization energies are smallest. The most distinctive chemical property of a nonmetal is the ability to gain electrons to form an anion when reacting with a metal. Thus nonmetals are elements with large ionization energies and the most negative electron affinities. The nonmetals are found on the right side of the table, with the most reactive ones in the upper right-hand corner, except for the noble gas elements, which are quite unreactive. The division into metals and nonmetals shown in Fig. 7.36 is only approximate. Many elements along the division line exhibit both metallic and nonmetallic properties under certain circumstances. These elements are often called **metalloids,** or sometimes *semimetals*.

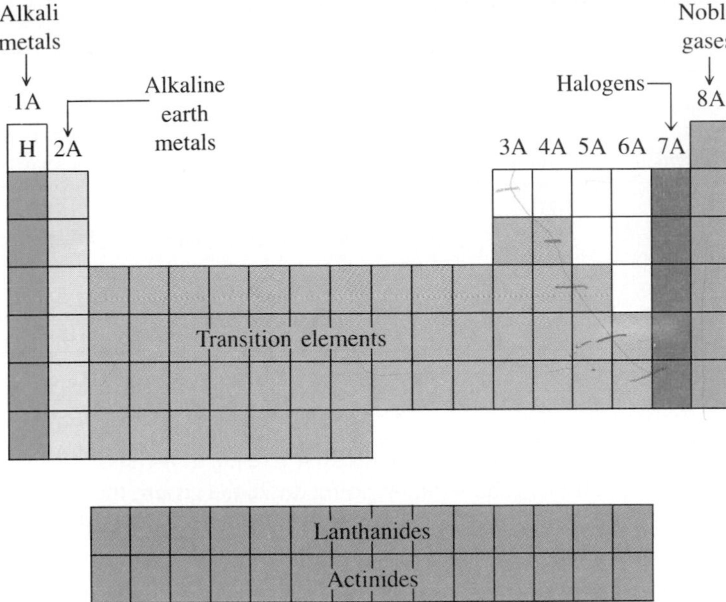

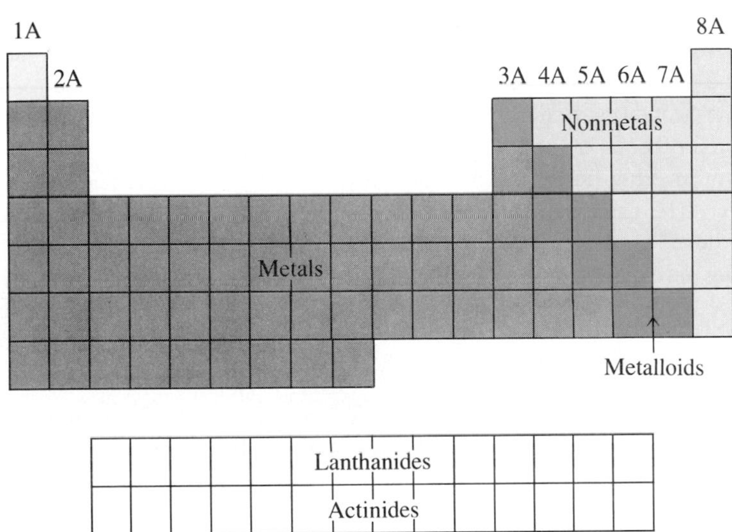

Figure 7.36

Special names for groups in the periodic table.

The Alkali Metals

The metals of Group 1A, the alkali metals, illustrate very well the relationships among the properties of the elements in a group. Lithium, sodium, potassium, rubidium, cesium, and francium are the most chemically reactive of the metals. We will not discuss francium here since it occurs in nature only in very small quantities. Although hydrogen is found in Group 1A on the periodic table, it behaves as a nonmetal, in contrast to the other members of that group. The fundamental reason for hydrogen's nonmetallic character is its very small size (see Fig. 7.35). The electron in the small 1*s* orbital is bound tightly to the nucleus.

Hydrogen will be discussed further in Chapter 18.

Properties of Five Alkali Metals

Element	Valence-electron configuration	Density at 25°C (g/cm^3)	mp (°C)	bp (°C)	First ionization energy (kJ/mol)	Atomic radius (Å)	Ionic (M^+) radius (Å)
Li	$2s^1$	0.53	180	1330	520	1.52	0.60
Na	$3s^1$	0.97	98	892	495	1.86	0.95
K	$4s^1$	0.86	64	760	419	2.31	1.33
Rb	$5s^1$	1.53	39	688	409	2.44	1.48
Cs	$6s^1$	1.87	29	690	382	2.62	1.69

Table 7.8

Some important properties of the first five alkali metals are shown in Table 7.8. The data in Table 7.8 show that in going down the group, the first ionization energy decreases and the atomic radius increases. This agrees with the general trends discussed in Section 7.13.

The overall increase in density in going down Group 1A is typical of all groups. This occurs because atomic mass generally increases more rapidly than atomic size. Thus there is more mass per unit volume for each succeeding element.

Other groups will be discussed in Chapters 18 and 19.

The smooth decrease in melting point and boiling point in going down Group 1A is not typical; in most other groups more complicated behavior occurs. Note that the melting point of cesium is only 29°C. Cesium can be melted readily using only the heat from your hand. This is very unusual—metals typically have rather high melting points. For example, tungsten melts at 3410°C. The only other metals with low melting points are mercury (mp, −38°C) and gallium (mp, 30°C).

Oxidation-reduction reactions were discussed in Chapter 4.

The chemical property most characteristic of a metal is the ability to lose its valence electrons. The Group 1A elements are very reactive. They have low ionization energies and react readily with nonmetals to form ionic solids. A typical example involves the reaction of sodium with chlorine to form sodium chloride:

$$2Na(s) + Cl_2(g) \rightarrow 2NaCl(s)$$

where sodium chloride consists of Na^+ and Cl^- ions. This is an oxidation-reduction reaction in which chlorine oxidizes sodium. In the reactions between metals and nonmetals, it is typical for the nonmetal to behave as the oxidizing agent and the metal to behave as the reducing agent, as shown by the following reactions:

$$2Na(s) + S(s) \rightarrow Na_2S(s)$$
Contains Na^+ and S^{2-} ions

$$6Li(s) + N_2(g) \rightarrow 2Li_3N(s)$$
Contains Li^+ and N^{3-} ions

$$4Na(s) + O_2(g) \rightarrow 2Na_2O(s)$$
Contains Na^+ and O^{2-} ions

Potassium reacts violently with water.

For reactions of the types shown above, the relative reducing powers of the alkali metals can be predicted from the first ionization energies listed in Table 7.8. Since it is much easier to remove an electron from a cesium atom than from a lithium atom, cesium should be the better reducing agent. The expected trend in reducing ability is

$$Cs > Rb > K > Na > Li$$

Chemical Impact

Lithium: Behavior Medicine

More and more people in our society seem to be suffering from the debilitating effects of mania and depression, but the alkali metal lithium can provide help for many. In fact, over three million prescriptions for lithium carbonate are filled annually by retail pharmacies.

Although the details are not well understood, the lithium ion seems to alleviate mood disorders by affecting the way that brain cells respond to neurotransmitters, a class of molecules that facilitate the transmission of nerve impulses.

Specifically, it is thought that the lithium ion may interfere with a complex cycle of reactions that relays and amplifies messages carried to the cells by neurotransmitters and hormones. It is theorized that exaggerated forms of behavior, such as mania or depression, arise from the overactivity of this cycle. The fact that lithium inhibits this cycle may be responsible for its moderating effect on behavior.

There is a growing collection of evidence that violent behavior may result at least partially from improper regulation of neurotransmitters and hormones. For example, a study in Finland showed that violent criminals, especially arsonists, often had low levels of serotonin, a common neurotransmitter. Studies are now underway to determine whether lithium might be effective in altering violent behavior.

This order is observed experimentally for direct reactions between the solid alkali metals and nonmetals. However, this is not the order of reducing ability found when the alkali metals react in aqueous solution. For example, the reduction of water by an alkali metal is very vigorous and exothermic:

$$2M(s) + 2H_2O(l) \rightarrow H_2(g) + 2M^+(aq) + 2OH^-(aq) + \text{energy}$$

The order of reducing abilities observed for this reaction for the first three group members is

$$\text{Li} > \text{K} > \text{Na}$$

In the gas phase potassium loses an electron more easily than sodium, and sodium more easily than lithium. Thus it is surprising that lithium is the best reducing agent toward water.

This reversal occurs because the formation of the M^+ ions in aqueous solution is strongly influenced by the hydration of these ions by the polar water molecules. The hydration energy of an ion represents the change in energy that occurs when water molecules attach to the M^+ ion. The hydration energies for the Li^+, Na^+, and K^+ ions shown in Table 7.9 indicate that the process is exothermic in each case. However, nearly twice as much energy is released by the hydration of the Li^+ ion as for the K^+ ion. This difference is caused by size effects; the Li^+ ion is much smaller than the K^+ ion, and thus its *charge density* (charge per unit volume) is also much greater. This means that the polar water molecules are more strongly attracted to the small Li^+ ion. Because the Li^+ ion is so strongly hydrated, its formation from the lithium atom occurs more readily than the formation of the K^+ ion from the potassium atom. Although a potassium atom in the gas phase loses its valence electron more readily than a lithium atom in the gas phase, the opposite is true in aqueous solution. This anomaly is an example of the importance of the polarity of the water molecule in aqueous reactions.

Hydration Energies for Li^+, Na^+, and K^+ Ions

Ion	Hydration energy (kJ/mol)
Li^+	−510
Na^+	−402
K^+	−314

Table 7.9

There is one more surprise involving the highly exothermic reactions of the alkali metals with water. Experiments show that in water lithium is the best reducing agent, so we might expect that lithium should react the most violently with water. However, this is not true. Sodium and potassium react much more vigorously. Why is this so? The answer lies in the relatively high melting point for lithium. When sodium and potassium react with water, the heat evolved causes them to melt, giving a larger area of contact with water. Lithium, on the other hand, does not melt under these conditions and reacts more slowly. This illustrates the important principle (which we will discuss in detail in Chapter 12) that the energy of a reaction and the rate at which it occurs are not necessarily related.

In this section we have seen that the trends in atomic properties summarized by the periodic table can be a great help in understanding the chemical behavior of the elements. This fact will be emphasized over and over as we proceed in our study of chemistry.

FOR REVIEW

Summary

Energy can travel through space as electromagnetic radiation, which is characterized by wavelength (λ), frequency (ν), and speed ($c = 2.9979 \times 10^8$ m/s). These three properties are related by the equation $\lambda\nu = c$. Originally, energy was assumed to be continuous, but Planck showed that energy could be lost or gained only in quanta, which are whole-number multiples of $h\nu$ (h is *Planck's constant,* 6.626×10^{-34} J s). Einstein postulated that electromagnetic radiation itself can be viewed as a stream of particles called photons, each with energy $h\nu$. Electromagnetic radiation therefore exhibits both wave and particulate properties. The postulate by de Broglie that particles possess wave characteristics was verified by diffraction experiments showing that all matter exhibits the same properties.

A continuous spectrum results from dispersing white light. However, when the emission spectrum produced by excited hydrogen atoms is dispersed, only selected wavelengths are seen, giving a line spectrum. This shows that only certain energies are allowed for the electron in the hydrogen atom.

The Bohr model of the hydrogen atom speculated that the electron in a hydrogen atom moves around the nucleus in certain allowed circular orbits and that the line spectrum was produced when an electron moved from one orbit to another. The Bohr model, however, was shown to be incorrect.

The wave mechanical model for the hydrogen atom employs the concept of standing waves. The electron is described by a wave function, often called an orbital, which does not give the exact position for the electron at a given time. This is consistent with the Heisenberg uncertainty principle, which states that it is impossible to know both the position and the momentum of a particle accurately at a given time. However, we can give a physical meaning to the wave function: the square of the function is related to the probability of finding an electron at a particular point in space. Probability distributions (electron density maps) are used to define orbital shapes. Orbitals are characterized by the quantum numbers n, ℓ, and m_ℓ.

Another property of the electron is its spin. An electron appears to spin in two ways giving rise to two values for the electron spin quantum number (m_s). This concept leads to the Pauli exclusion principle: no two electrons can have the same set of four quantum numbers.

Using this principle along with the wave mechanical model, we can predict the electronic configurations of the atoms in the periodic table. As protons are added one by one to the nucleus to build up the elements, electrons are added to the orbitals. This is called the Aufbau principle.

The greatest triumph of the wave mechanical model is its use to explain the observed periodic properties of the elements as summarized by the periodic table. The chemical properties of an atom are determined by the arrangement of the valence electrons (the electrons in the outermost principal quantum level). Elements in the same group of the periodic table have the same valence-electron configurations, which explains why group members show similar chemical properties.

In polyelectronic atoms the repulsions among electrons can be assumed to reduce the actual nuclear charge to an apparent value called the effective nuclear charge (Z_{eff}), which is always less than Z (the atomic number). The ionization energy, the energy required to remove an electron from a gaseous atom or ion, provides a means for calculating Z_{eff}.

The periodic trends in atomic properties such as ionization energy, electron affinity, and atomic radius can be explained in terms of the concepts of shielding and penetration and their influence on the size of Z_{eff}.

Key Terms

Section 7.1
electromagnetic radiation
wavelength
frequency

Section 7.2
Planck's constant
quantization
photon
$E = mc^2$
dual nature of light
diffraction
diffraction pattern

Section 7.3
continuous spectrum
line spectrum

Section 7.4
quantum model
ground state

Section 7.5
standing wave
wave mechanical model
wave function
orbital
Heisenberg uncertainty principle
probability distribution
radial probability distribution

Section 7.6
quantum numbers
principal quantum number
azimuthal quantum number
magnetic quantum number
subshell

Section 7.7
nodal surface
node
degenerate

Section 7.8
electron spin
electron spin quantum number
Pauli exclusion principle

Section 7.9
polyelectronic atoms
effective nuclear charge

Section 7.11
Aufbau principle
Hund's rule
valence electrons
core electrons
transition metals
lanthanide series
actinide series
main-group elements (representative elements)

Section 7.12
ionization energy
shielding
penetration effect

Section 7.13
first ionization energy
second ionization energy
electron affinity
atomic radii

Section 7.14
metalloids

Exercises

A blue exercise number indicates that the answer to that exercise appears at the back of this book and a solution appears in the Solutions Guide.

Light and Matter

1. What experimental evidence leads to the quantum theory for light?
2. An FM radio station broadcasts at 99.5 MHz. What is the wavelength of the radio waves?
3. The laser in an audio compact disc player uses light with a wavelength of 7.80×10^2 nm. What is the frequency of this light? What is the energy of a single photon of this light?
4. Microwave radiation has a wavelength on the order of 1.0 cm. Calculate the frequency and the energy of a single photon of this radiation. Calculate the energy of an Avogadro's number of photons of this electromagnetic radiation.
5. It takes 7.21×10^{-19} J of energy to remove an electron from an iron atom. What is the maximum wavelength of light that can do this?
6. The work function of an element is the energy required to remove an electron from the surface of the solid. The work function for lithium is 279.7 kJ/mol (that is, it takes 279.7 kJ of energy to remove one mole of electrons from one mole of Li atoms on the surface of Li metal). What is the maximum wavelength of light that can remove an electron from an atom in lithium metal?
7. The primary visible emissions from mercury are at 404.7 nm and 435.8 nm. Calculate the frequencies of this light. Calculate the energy of a single photon and a mole of photons of light with each of these wavelengths.
8. The ionization energy of gold is 890.1 kJ/mol. Is light with a wavelength of 225 nm capable of ionizing a gold atom in the gas phase?
9. It takes 492 kJ to remove one mole of electrons from the surface of solid gold. How much energy does it take to remove a single electron from the surface of gold? What is the maximum wavelength of light capable of doing this?
10. It takes 208.4 kJ of energy to remove one mole of electrons from the atoms on the surface of rubidium metal. If rubidium metal is irradiated with 254-nm light, what is the maximum kinetic energy the photoelectrons can have?

Hydrogen Atom: The Bohr Model

11. What do we mean when we say that something is quantized? In the Bohr model of the hydrogen atom, what is quantized?
12. Calculate the wavelength of light emitted in each of the following spectral transitions in the hydrogen atom:
    a. $n = 3 \rightarrow n = 2$
    b. $n = 4 \rightarrow n = 2$
    c. $n = 2 \rightarrow n = 1$
    d. $n = 4 \rightarrow n = 3$
    e. $n = 5 \rightarrow n = 4$
    f. $n = 5 \rightarrow n = 3$
13. What is the maximum wavelength of light capable of removing an electron from a hydrogen atom for each of the three energy states characterized by $n = 1$, $n = 2$, and $n = 3$?
14. Using vertical lines, indicate the spectral transitions from Exercise 12 on the energy-level diagram for the hydrogen atom (see Fig. 7.8).
15. Using vertical lines, indicate the transitions that correspond to the ionizations in Exercise 13 on the energy-level diagram for the hydrogen atom (see Fig. 7.8).
16. An electron is excited from the $n = 1$ ground state to the $n = 3$ state in a hydrogen atom. Which of the following statements are true? Correct the false statements to make them true.
    a. It takes more energy to ionize the electron from $n = 3$ than from the ground state.
    b. The electron is further from the nucleus in the $n = 3$ state than in the $n = 1$ state.
    c. The wavelength of light emitted if the electron drops from $n = 3$ to $n = 2$ will be shorter than the wavelength of light emitted if the electron falls to $n = 1$ from $n = 3$.
    d. The wavelength of light emitted when the electron returns to the ground state from $n = 3$ will be the same as the wavelength of light absorbed to go from $n = 1$ to $n = 3$.
    e. With $n = 3$, the electron is in the first excited state.
17. Calculate the longest and shortest wavelengths of light emitted by electrons in the hydrogen atom that begin in the $n = 6$ state and then fall to states with smaller values of n.
18. An excited hydrogen atom emits light with a frequency of 1.141×10^{14} Hz to reach the energy level for which $n = 4$. In what principal quantum level did the electron begin?

Wave Mechanics, Quantum Numbers, and Orbitals

19. Calculate the de Broglie wavelength for each of the following:
    a. a proton with a velocity 5.0% of the speed of light
    b. an electron with a velocity 15% of the speed of light
    c. the fastest measured fast ball (a 5.2-oz baseball with a velocity of 100.8 mph)
20. Neutron diffraction is used in determining the structures of molecules.

a. Calculate the de Broglie wavelength of a neutron moving at 1.00% of the speed of light.
b. Calculate the velocity of a neutron with a wavelength of 75 pm. (1 pm = 10^{-12} m)

21. Calculate the velocities of electrons with the de Broglie wavelengths of 1.0×10^2 nm and 1.0 nm.

22. What would the value of Planck's constant have to be in order for the wavelength of the baseball in Exercise 19, part c, to be 5.0 cm? 5.0×10^2 nm?

23. For what sizes of particles and velocities must one consider quantum effects?

24. Summarize some of the evidence supporting the wave properties of matter.

25. The Heisenberg uncertainty principle can also be expressed as:

$$\Delta E \Delta t \geq \frac{h}{4\pi}$$

where E represents energy and t represents time. Show that the units for this form are the same as the units for the form used in this chapter:

$$\Delta x \cdot \Delta(mv) \geq \frac{h}{4\pi}$$

26. Using the Heisenberg uncertainty principle, calculate Δx for each of the following:
a. an electron with $\Delta v = 0.100$ m/s
b. a baseball (mass = 145 g) with $\Delta v = 0.100$ m/s
c. How does the answer in part a compare to the size of a hydrogen atom?
d. How does the answer in part b correspond to the size of a baseball?

27. In the hydrogen atom, what information do we get from the values of the quantum numbers n, ℓ, and m_ℓ?

28. What are the possible values for the quantum numbers n, ℓ, and m_ℓ?

29. Which of the following orbital designations are incorrect: 1s, 1p, 7d, 9s, 3f, 4f, 2d?

30. Which of the following sets of quantum numbers are not allowed in the hydrogen atom? For the sets of quantum numbers that are incorrect, state what is wrong in each set.
a. $n = 2$, $\ell = 1$, $m_\ell = -1$
b. $n = 1$, $\ell = 1$, $m_\ell = 0$
c. $n = 8$, $\ell = 7$, $m_\ell = -6$
d. $n = 1$, $\ell = 0$, $m_\ell = 2$
e. $n = 3$, $\ell = 2$, $m_\ell = 2$
f. $n = 4$, $\ell = 3$, $m_\ell = 4$
g. $n = 0$, $\ell = 0$, $m_\ell = 0$
h. $n = 2$, $\ell = -1$, $m_\ell = 1$

31. Are the possible values of ℓ for an electron with $n = 2$ the same as for an electron with $n = 3$?

32. How many orbitals can have the designation: 5p, 3d_{z^2}, 4d, $n = 5$, $n = 4$?

33. In defining the sizes of orbitals, why must we use an arbitrary value, such as 90% of the probability of finding an electron in that region?

34. From the diagrams of 2p and 3p orbitals in Fig. 7.14 and Fig. 7.15, draw a rough graph of the square of the wave function for these orbitals in the direction of one of the lobes.

35. How do the 2p orbitals differ from each other?

36. How do the 2p and 3p orbitals differ from each other?

37. What is a nodal surface in an atomic orbital?

38. What is the physical significance of the value of ψ^2 at a particular point?

39. The wave function for a 1s orbital is

$$\psi_{1s} = 2\left(\frac{1}{a_0}\right)^{1/2} e^{-r/a_0}$$

where r is the distance of the electron from the nucleus, a_0 is a constant (5.3×10^{-11}), and e is the base of the natural logarithm, that is, 2.71828 . . . . Calculate the values of ψ_{1s} and $\psi_{1s}{}^2$ at $r = 0$, $r = a_0$, and $r = 2a_0$.

Polyelectronic Atoms

40. Why can we not account exactly for the repulsions among electrons in a polyelectronic atom?

41. Calculate the ionization energy (in kJ/mol) for each of the following one-electron species, using the Bohr model for hydrogen:
a. H b. He^+ c. Li^{2+} d. C^{5+} e. Fe^{25+}

42. What would be the relative sizes of the 1s orbitals in the species in Exercise 41?

43. We define a spin quantum number. Do we know for a fact that an electron spins?

44. What is the maximum number of electrons in an atom that can have these quantum numbers:
a. $n = 4$
b. $n = 5$, $m_\ell = +1$
c. $n = 5$, $m_s = +\frac{1}{2}$
d. $n = 3$, $\ell = 2$
e. $n = 2$, $\ell = 1$
f. $n = 0$, $\ell = 0$, $m_\ell = 0$
g. $n = 2$, $\ell = 1$, $m_\ell = -1$, $m_s = -\frac{1}{2}$
h. $n = 3$
i. $n = 2$, $\ell = 2$
j. $n = 1$, $\ell = 0$, $m_\ell = 0$

45. Write the expected electron configurations for the following atoms: Sc, Fe, S, P, Cs, Eu, Pt, Xe, Br, Se.

46. Write the expected electron configurations for the following atoms: K, Rb, Fr, Pu, Sb, Os, Pd, Cd, Pb, I.

47. Using Figure 7.25, list elements (ignore the lanthanides and actinides) that have ground state electron configurations that differ from those we would expect from their positions in the periodic table.

48. The electron configurations in Figure 7.25 are for atoms in the gas phase and are determined experimentally. Would you expect the electron configurations to be the same in the solid and liquid states as in the gas phase?

49. Write an electron configuration for each of the following:
a. the smallest halogen
b. the alkali metal with only $2p$ and $3p$ electrons
c. the three lightest alkaline earth metals
d. the Group 3A element in the same period as Sn
e. the nonmetallic elements in Group 4A
f. the (as yet undiscovered) noble gas after radon

50. What do we mean when we say that a $4s$ electron is more penetrating than a $3d$ electron?

51. Which of the following are possible sets of quantum numbers for an electron? For the sets of quantum numbers that are not possible, state what is wrong with each set.
a. $n = 1,\ \ell = 0,\ m_\ell = 1,\ m_s = +\frac{1}{2}$
b. $n = 9,\ \ell = 7,\ m_\ell = -6,\ m_s = -\frac{1}{2}$
c. $n = 2,\ \ell = 1,\ m_\ell = 0,\ m_s = 0$
d. $n = 1,\ \ell = 1,\ m_\ell = 1,\ m_s = +\frac{1}{2}$
e. $n = 3,\ \ell = 2,\ m_\ell = -3,\ m_s = +\frac{1}{2}$
f. $n = 4,\ \ell = 0,\ m_\ell = 0,\ m_s = -\frac{1}{2}$

52. The successive ionization energies for boron are 0.8006, 2.4270, 3.6598, 25.0257, and 32.8266 MJ/mol. Calculate Z_{eff} for each of these electrons. Assume the boron atom and the successive ions are in the ground state.

53. Give a possible set of values of the four quantum numbers for all of the electrons in a boron atom, a carbon atom, a nitrogen atom, and an oxygen atom, if each is in the ground state.

54. Why do we emphasize the valence electrons in an atom when discussing atomic properties?

55. Give a possible set of values for the four quantum numbers of all of the valence electrons in each of the following atoms if each is in its ground state: Al, As, Sb, S, Se.

56. A certain oxygen atom has the electron configuration $1s^2 2s^2 2p_x^2 2p_y^2$. How many unpaired electrons are present? Is this an excited state of oxygen? In going from this state to the ground state would energy be released or absorbed?

57. How many unpaired electrons are there in the ground state electron configurations of Sc, Ti, Al, Sn, Te, and Br?

58. How many electrons in an atom can have the designation: $1s$, $2p$, $3p_x$, $6f$, $2d_{xy}$?

59. Identify the following elements:
a. an excited state of this element has the electron configuration $1s^2 2s^2 2p^5 3s^1$
b. the ground state electron configuration is $[\text{Ne}]3s^2 3p^4$
c. an excited state of this element has the electron configuration $[\text{Kr}]5s^2 4d^6 5p^2 6s^1$
d. the ground state electron configuration contains three unpaired $6p$ electrons

The Periodic Table and Periodic Properties

60. Arrange the following groups of atoms in order of increasing size.
a. Be, Mg, Ca
b. Te, I, Xe
c. Ga, Ge, In
d. As, N, F
e. S, Cl, F

61. Arrange the atoms in Exercise 60 in order of increasing first ionization energy.

62. Arrange the atoms in Exercise 60 in order of increasing Z_{eff} for the highest-energy electron.

63. From your answers to the previous three exercises, how does the size and ionization energy of atoms in a given period depend on the effective nuclear charge?

64. The first ionization energies of Ge, As, and Se are 0.7622, 0.944, and 0.9409 MJ/mol, respectively. Rationalize these values in terms of electron configurations.

65. Many times the claim is made that subshells half filled with electrons are particularly stable. Can you suggest a physical basis for this claim?

66. Calculate Z_{eff} for the highest-energy electron in Ge, As, and Se, using the data in Exercise 64.

67. We expect the atomic radius to increase going down a group in the periodic table. Can you suggest why the atomic radius of hafnium breaks this rule? (See data below.)

Atomic radii in Å			
Sc	1.57	Ti	1.477
Y	1.693	Zr	1.593
La	1.915	Hf	1.476

68. Predict some of the properties of element 117 (the symbol is Uus following conventions proposed by the International Union of Pure and Applied Chemistry, or IUPAC).
a. What will be its electron configuration?
b. What element will it most resemble chemically?
c. What will be the formulas of the neutral binary compounds it forms with sodium, magnesium, carbon, and oxygen?
d. What oxyanions would you expect Uus to form?

69. Order each of the following sets from the least exothermic electron affinity to the most.
a. O, S
b. F, Cl, Br, I
c. N, O, F

70. The changes in electron affinity as one goes down a group in the periodic table are not nearly as large as the variations in ionization energies. Why?

71. The electron affinity of sodium is −52.9 kJ/mol. In principle, should it be possible to synthesize a compound containing the sodide ion, Na^-?

72. From the information in Table 7.8, estimate values of the ionization energy and melting point for francium.

73. Which would have the more favorable (more negative) electron affinity, the oxygen atom or the O^- ion? Explain your answer.

74. In each of the following sets, which atom or ion has the smallest radius?
a. Li, Na, K
b. P, As
c. O^+, O, O^-
d. S, Cl, Kr
e. Pd, Ni, Cu

75. In each of the following sets, which atom or ion has the smallest ionization energy?
a. Cs, Ba, La
b. Zn, Ga, Ge
c. Tl, In, Sn
d. Tl, Sn, As
e. O, O^-, O^{2-}

76. The electron affinities of the elements from aluminum to chlorine are −44, −120, −74, −200.4, and −348.7 kJ/mol, respectively. Rationalize the trend in these values.

77. Three elements have the electron configurations $1s^22s^22p^63s^23p^6$, $1s^22s^22p^63s^2$, and $1s^22s^22p^63s^23p^64s^1$. The first ionization energies of the three elements (not in the same order) are 0.4189, 0.7377, and 1.5205 MJ/mol. The atomic radii are 1.60, 0.94, and 1.97 Å. Identify the three elements and match the appropriate values of ionization energy and atomic radius to each configuration.

78. Use data in this chapter to determine:
a. the electron affinity of Mg^{2+}
b. the electron affinity of Al^+
c. the ionization energy of Cl^-
d. the ionization energy of Cl
e. the electron affinity of Cl^+

79. The first ionization energies of the so-called coinage metals, Cu, Ag, and Au are 745.5, 731.0, and 890.1 kJ/mol, respectively. Is this the trend you might have expected? In the *CRC Handbook of Chemistry and Physics,* look up the first ionization energies of the following sets of elements: (Sc, Y, La), (Ti, Zr, Hf), (Fe, Ru, Os), (Ga, In, Tl), and (As, Sb, Bi). Which groups follow the same trend as the coinage metals?

80. Calculate Z_{eff} for the highest energy electron in Cu, Ag, and Au using data in Exercise 79. (Assume an *s* electron is being removed.)

81. Why do the successive ionization energies of an atom always increase?

82. Note the successive ionization energies for silicon given in Table 7.5. Would you expect to see any large jumps between successive ionization energies of silicon as you removed all of the electrons, one-by-one, beyond those shown in the table?

83. For each of the following pairs of elements

(Li and K) (S and Sc) (B and N) (F and Cl)

pick the one with
a. more favorable (exothermic) electron affinity
b. higher ionization energy
c. larger size

The Alkaline Metals

84. Many more anhydrous lithium salts are hygroscopic (readily absorb water) than those of the other alkali metals. Explain.

85. An ionic compound of potassium and oxygen has the empirical formula KO. Would you expect this compound to be potassium(II) oxide or potassium peroxide? Explain.

86. Complete and balance the equations for the following reactions:
a. $Li(s) + O_2(g) \rightarrow$
b. $K(s) + S(s) \rightarrow$
c. $Cs(s) + H_2O(l) \rightarrow$
d. $Na(s) + Cl_2(g) \rightarrow$

87. Cesium was discovered in natural mineral waters in 1860 by R. W. Bunsen and G. R. Kirchoff using the spectroscope they invented in 1859. The name came from the Latin (*caesius,* sky blue) for the prominent blue line observed for this element at 455.5 nm. Calculate the frequency and energy of a photon of this light.

88. Small daily doses (1–2 g) of lithium carbonate taken orally are often given to treat manic-depressive psychoses. This dosage maintains the level of lithium ion in the blood at approximately 1×10^{-3} mol/L.
a. What is the formula of lithium carbonate?
b. How many grams of lithium are present per liter of blood in these patients?

89. Give the name and formula of each of the binary compounds formed from the following elements:
a. Li and N
b. Na and Br
c. K and S
d. Li and P
e. Rb and H
f. Na and H

90. Predict the atomic number of the next alkali metal after Francium, and give its ground state electron configuration.

Additional Exercises

91. The bright yellow light emitted by a sodium vapor lamp consists of two emission lines at 589.0 nm and 589.6 nm. What are the frequency and the energy of a photon of light at each of these wavelengths? What are the energies in kJ/mol?

92. Spectroscopists use emission spectra to confirm the presence of an element in materials of unknown composition. How is this possible?

93. Total radial probability distributions for the helium, neon, and argon atoms are shown below. How can one interpret the shapes of these curves in terms of electron configurations, quantum numbers, and effective nuclear charges?

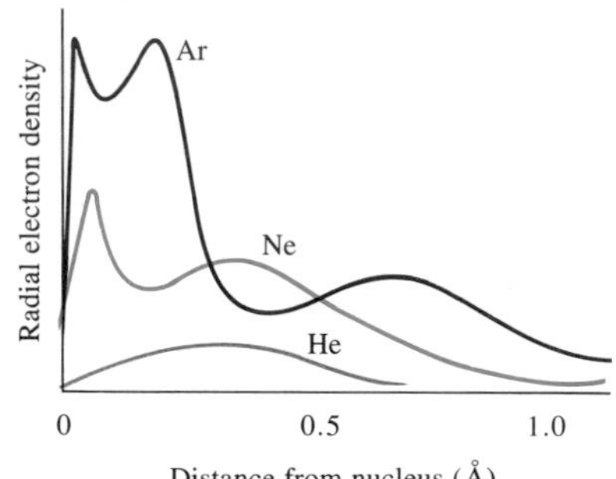

94. How many unpaired electrons are in each of the following in the ground state: O, O^+, O^-, Fe, Mn, S, F, Ar?

95. On which quantum numbers does the energy of an electron depend in each of the following?
a. a one-electron atom or ion
b. an atom or ion with more than one electron

96. The wave function for the $2p_z$ orbital in the hydrogen atom is

$$\psi_{2p_z} = \frac{1}{4\sqrt{2\pi}}\left(\frac{Z}{a_0}\right)^{3/2} \sigma e^{-\sigma/2} \cos\theta$$

where a_0 is the value for the radius of the first Bohr orbit in meters (5.3×10^{-11}), σ is $Z(r/a_0)$, r is the value for the distance from the nucleus in meters, and θ is an angle. Calculate the value of $\psi_{2p_z}{}^2$ at $r = a_0$ for $\theta = 0$ (z axis) and for $\theta = 90°$ (xy plane).

97. Elements with very large ionization energies also tend to have highly exothermic electron affinities. Explain. Which group of elements would you expect to be an exception to this statement?

98. Define each of the following terms.
a. photon of light
b. quantum number
c. ground state
d. excited state

99. Using the element phosphorus as an example, write the equation for a process in which the energy change will be equal to each of the following:
a. the ionization energy
b. the electron affinity

100. The work function is the energy required to remove an electron from an atom on the surface of a metal. How does this definition differ from that for ionization energy?

101. In the hydrogen atom what is the physical significance of the state in which $n = \infty$, and $E = 0$?

102. Which of the following electron configurations correspond to an excited state? Identify the atoms and write the ground-state electron configuration where appropriate.
a. $1s^2 2s^2 3p^1$
b. $1s^2 2s^2 2p^6$
c. $1s^2 2s^2 2p^4 3s^1$
d. $[Ar]4s^2 3d^5 4p^1$

103. Look up the first ionization energies of lithium, sodium, and potassium, and use them to calculate a value for the effective nuclear charge for the highest energy electron in each of these atoms. What conclusion can you draw about the value of Z_{eff} for members of the same group in the periodic table? Test this conclusion for another group.

104. Does the minimization of electron-electron repulsions correlate with Hund's rule?

105. The electron affinity for sulfur is more exothermic than that for oxygen. How do you account for this?

106. The ionization energy of sodium is 495 kJ/mol. Will gaseous sodium atoms be ionized by light with a wavelength of 589 nm?

107. What role does the repulsion between electrons play in the trend in ionization energies from lithium to neon?

108. Give possible values for the quantum numbers of the valence electrons in an atom of titanium (Ti).

109. Photogrey lenses incorporate small amounts of silver chloride in the glass of the lens. When light hits the AgCl particles, the following reaction occurs:

$$AgCl \xrightarrow{h\nu} Ag + Cl$$

The silver metal that is formed causes the lenses to darken. The enthalpy change for this reaction is 3.10×10^2 kJ/mol. Assuming all of this energy must be supplied by light, what is the maximum wavelength of light that can cause this reaction?

110. How many electrons in an atom can have the following sets of quantum numbers?
a. $n = 3$
b. $n = 2, \ell = 0$
c. $n = 2, \ell = 2$
d. $n = 2, \ell = 0\ m_\ell = 0, m_s = +\frac{1}{2}$

111. Although no currently known elements contain electrons in g orbitals in the ground state, it is possible that these elements will be found or that electrons in excited states of known elements could be in g orbitals. For g orbitals, the value of ℓ is 4. What is the lowest value of n for which g orbitals could exist? What are the possible values of m_ℓ? How many electrons could a set of g orbitals hold?

112. What is the most important relationship between elements in the same group in the periodic table?

113. Using the data in this chapter calculate ΔH for each of the following processes:
   a. $2Cu^+(g) \rightarrow Cu(g) + Cu^{2+}(g)$ (For Cu, I_1 = 746 kJ/mol, I_2 = 1958 kJ/mol.)
   b. $Na^-(g) + Na^+(g) \rightarrow 2Na(g)$ (Electron affinity for Na = −52 kJ/mol.)
   c. $Mg^{2+}(g) + K(g) \rightarrow Mg^+(g) + K^+(g)$
   d. $Na(g) + Cl(g) \rightarrow Na^+(g) + Cl^-(g)$
   e. $Mg(g) + F(g) \rightarrow Mg^+(g) + F^-(g)$
   f. $Mg^+(g) + F(g) \rightarrow Mg^{2+}(g) + F^-(g)$
   g. $Mg(g) + 2F(g) \rightarrow Mg^{2+}(g) + 2F^-(g)$

114. Write equations corresponding to the following energy terms:
   a. the fourth ionization energy of Se
   b. the electron affinity of S^-
   c. the electron affinity of Fe^{3+}
   d. the ionization energy of Mg
   e. the work function of Mg (See Exercise 6.)

115. Using only the periodic table inside the front cover of the text write the expected electron configurations for:
   a. the third element in Group 5A
   b. element number 116
   c. an element with three unpaired $5d$ electrons in the ground state
   d. Ti, Ni, and Os

116. In the ground state of antimony, Sb:
   a. How many electrons have ℓ = 1 as one of their quantum numbers?
   b. How many electrons have m_ℓ = 0?
   c. How many electrons have m_ℓ = 1?

117. An ion having a 4+ charge and a mass of 49.9 amu has two electrons with principal quantum number n = 1, eight electrons with n = 2, and ten electrons with n = 3. Supply as many of the properties for the ion as possible from the information given. *Hint:* In forming ions, the $4s$ electrons are lost before the $3d$ electrons.
   a. the atomic number
   b. total number of s electrons
   c. total number of p electrons
   d. total number of d electrons
   e. the number of neutrons in the nucleus
   f. the mass of 3.01×10^{23} atoms
   g. the ground state electron configuration of the neutral atom

118. Consider the following ionization energies for aluminum:

$$Al(g) \rightarrow Al^+(g) + e^- \qquad I_1 = 580 \text{ kJ/mol}$$
$$Al^+(g) \rightarrow Al^{2+}(g) + e^- \qquad I_2 = 1815 \text{ kJ/mol}$$
$$Al^{2+}(g) \rightarrow Al^{3+}(g) + e^- \qquad I_3 = 2740 \text{ kJ/mol}$$
$$Al^{3+}(g) \rightarrow Al^{4+}(g) + e^- \qquad I_4 = 11{,}600 \text{ kJ/mol}$$

   a. Account for the increasing trend in the values of the ionization energies.
   b. Explain the large increase between I_3 and I_4.
   c. Which one of the four ions has the greatest electron affinity? Explain.
   d. List the four aluminum ions given above in order of increasing size and explain your ordering. (*Hint:* Remember most of the size of an atom or ion is due to its electrons.)

119. Answer the following questions assuming that m_s could have three values rather than two and that the rules for n, ℓ, and m_ℓ are the normal ones.
   a. How many electrons would an orbital be able to hold?
   b. How many elements would the first and second periods in the periodic table contain?
   c. How many elements would be contained in the first transition metal series?
   d. How many electrons would the set of $4f$ orbitals be able to hold?

120. Assume we are in another universe with different physical laws. Electrons in this universe are described by four quantum numbers with meanings similar to those we use. We will call these quantum numbers p, q, r, and s. The rules for these quantum numbers are

p = 1, 2, 3, 4, 5, . . .

q takes on odd integer values and $q \leq p$

r takes on all even integer values from $-q$ to $+q$. (Zero is considered an even number.)

$s = +\frac{1}{2}$ or $-\frac{1}{2}$

   a. Sketch what the first three periods of the periodic table might look like in this universe.
   b. How many electrons can have p = 3?
   c. How many electrons can have p = 4, q = 3, r = 2?
   d. How many electrons can have p = 4, q = 3?
   e. How many electrons can have p = 3, q = 0, r = 0?
   f. What are the possible values of q and r for p = 5?
   g. How many electrons can have p = 6?

121. While Mendeleev predicted the existence of several undiscovered elements, he did not predict the existence of the noble gases, the lanthanides, or the actinides. Propose some reasons why Mendeleev was not able to predict the existence of the noble gases.

CHAPTER 8

Bonding: General Concepts

As we examine the world around us, we find it to be composed almost entirely of compounds and mixtures of compounds: rocks, coal, soil, petroleum, trees, and human bodies are all complex mixtures of chemical compounds in which different kinds of atoms are bound together. Substances composed of unbound atoms do exist in nature, but they are very rare. Examples are the argon in the atmosphere and the helium mixed with natural gas reserves.

The manner in which atoms are bound together has a profound effect on chemical and physical properties. For example, graphite is a soft, slippery material used as a lubricant in locks, and diamond is one of the hardest materials known, valuable both as a gemstone and in industrial cutting tools. Why do these materials, both composed solely of carbon atoms, have such different properties? The answer, as we will see, lies in the bonding in these substances.

Silicon and carbon are next to each other in Group 4A on the periodic table. From our knowledge of periodic trends, we might expect SiO_2 and CO_2 to be very similar. But SiO_2 is the empirical formula of silica, which is found in sand and quartz, and carbon dioxide is a gas, a product of respiration. Why are they so different? We will be able to answer this question after we have developed models for bonding.

Molecular bonding and structure play the central role in determining the course of all chemical reactions, many of which are vital to our survival. Later in this book we will demonstrate their importance by showing how enzymes facilitate complex chemical reactions, how genetic characteristics are transferred, and how hemoglobin in the blood carries oxygen throughout the body. All of these fundamental biological reactions hinge on the geometric structures of molecules, sometimes depending on very subtle differences in molecular shape to channel the chemical reaction one way rather than another.

Many of the world's current problems require fundamentally chemical answers: disease and pollution control, the search for new energy sources, the development of new fertilizers to increase crop yields, the improvement of the protein content in various staple grains, and many more. To understand the behavior of natural materials, we must understand the

< A polarized light micrograph (×100) of crystals of thiamine (vitamin B_1). Essential for nutrition, thiamine's bonding is partly covalent and partly ionic.

CONTENTS

(left) Quartz grows in beautiful, regular crystals. (right) Diamond and graphite.

nature of chemical bonding and the factors that control the structures of compounds. In this chapter we will present various classes of compounds that illustrate the different types of bonds and then develop models to describe the structure and bonding that characterize materials found in nature. Later these models will be useful in understanding chemical reactions.

8.1 Types of Chemical Bonds

Purpose

- To explain why an ionic bond is formed.
- To explain why a covalent bond is formed.
- To introduce the polar covalent bond.

What is a chemical bond? There is no simple and yet complete answer to this question. In Chapter 2 we defined bonds as forces that hold groups of atoms together and make them function as a unit.

There are many types of experiments we can perform to determine the fundamental nature of materials. For example, we can study physical properties such as melting point, hardness, and electrical and thermal conductivity. We can also study solubility characteristics and the properties of the resulting solutions. To determine the charge distribution in a molecule, we can study its behavior in an electric field. We can obtain information about the strength of a bonding interaction by measuring the energy required to break the bond, the **bond energy.**

There are several ways in which atoms can interact with one another to form aggregates. We will consider several specific examples to illustrate the various types of chemical bonds.

Earlier, we saw that when solid sodium chloride is dissolved in water, the resulting solution conducts electricity, a fact that convinces us that sodium chloride is composed of Na^+ and Cl^- ions. So when sodium and chlorine react to form sodium chloride, electrons are transferred from the sodium atoms to the chlorine atoms to form Na^+ and Cl^- ions, which then aggregate to form solid sodium chloride. Why does this happen? The best simple answer is that *the system can achieve the lowest possible energy by behaving in this way*. The attraction of a chlorine atom for the extra electron and the very strong mutual attractions of the oppositely charged ions provide the driving forces for the process. The resulting solid sodium chloride is a very sturdy material; it has a melting point of approximately 800°C. The bonding forces that produce this great thermal stability result from the electrostatic attractions of the closely packed, oppositely charged ions. This is an example of **ionic bonding.** Ionic substances are formed when an atom that loses electrons relatively easily reacts with an atom that has a high affinity for electrons. That is, an **ionic compound** results when a metal reacts with a nonmetal.

The energy of interaction between a pair of ions can be calculated using **Coulomb's law:**

$$E = 2.31 \times 10^{-19}\ \text{J nm}\left(\frac{Q_1Q_2}{r}\right)$$

where E has units of joules, r is the distance between the ion centers in nm, and Q_1 and Q_2 are the numerical ion charges.

For example, in solid sodium chloride the distance between the centers of the Na^+ and Cl^- ions is 2.76 Å (0.276 nm), and the ionic energy per pair of ions is

$$E = 2.31 \times 10^{-19}\ \text{J nm}\left[\frac{(+1)(-1)}{0.276\ \text{nm}}\right] = -8.37 \times 10^{-19}\ \text{J}$$

where the negative sign indicates an attractive force. That is, the *ion pair has lower energy than the separated ions*. For a mole of pairs of Na^+ and Cl^- ions, the energy of interaction is

$$E = \left(-8.37 \times 10^{-19}\ \frac{\text{J}}{\text{ion pair}}\right)\left(6.022 \times 10^{23}\ \frac{\text{ion pair}}{\text{mol}}\right)$$
$$= -504\ \frac{\text{kJ}}{\text{mol}}$$

Coulomb's law can also be used to calculate the repulsive energy when two like-charged ions are brought together. In this case the calculated value of the energy will have a positive sign.

We have seen that a bonding force develops when two very different types of atoms react to form oppositely charged ions. But how does a bonding force develop between two identical atoms? Let's explore this situation from a very simple point of view by considering the energy terms that result when two hydrogen atoms are brought close together, as shown in Fig. 8.1(a). When hydrogen atoms are brought

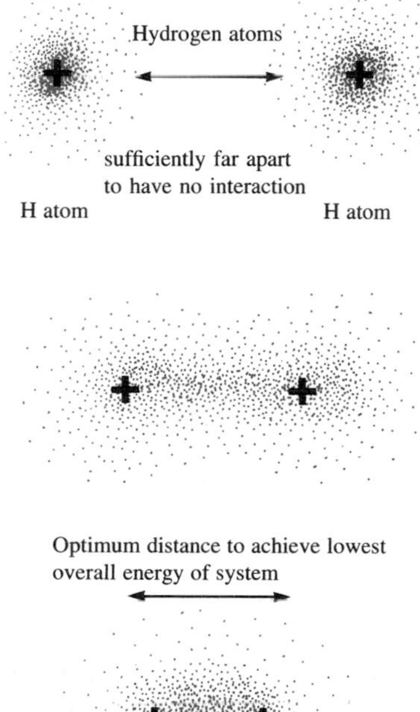

Figure 8.1a

The interaction of two hydrogen atoms.

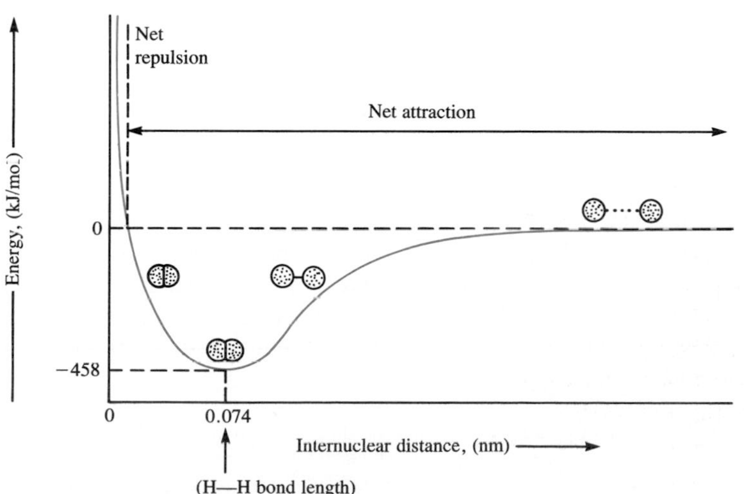

Figure 8.1b

Energy profile as a function of the distance between the nuclei of the hydrogen atoms. As the atoms approach each other, the energy decreases until the distance reaches 0.074 nm (0.74 Å) and then begins to increase again due to repulsions.

close together, there are two unfavorable energy terms, proton-proton repulsion and electron-electron repulsion, and one favorable term, proton-electron attraction. Under what conditions will the H_2 molecule be favored over the separated hydrogen atoms? That is, what conditions will favor bond formation? The answer lies in the strong tendency in nature for any system to achieve the lowest possible energy. A bond will form, that is, the two hydrogen atoms will exist as a molecular unit, if the system can lower its total energy in the process.

A bond will form if the energy of the aggregate is lower than that of the separated atoms.

In this case, then, the hydrogen atoms will position themselves so that the system will achieve the lowest possible energy; the system will act to minimize the sum of the positive (repulsive) energy terms and the negative (attractive) energy term. The distance where the energy is minimum is called the **bond length.** The total energy of this system as a function of distance between the hydrogen nuclei was shown in Fig. 8.1(b). Note several important features of this diagram:

Potential energy was discussed in Chapter 6.

The energy terms involved are the potential energy that results from the attractions and repulsions among the charged particles and the kinetic energy due to the motions of the electrons.

The zero point of energy is defined with the atoms at infinite separation.

At very short distances the energy rises steeply because of the importance of the repulsive forces when the atoms are very close together.

The bond length is the distance at which the system has minimum energy.

In the H_2 molecule the electrons reside primarily in the space between the two nuclei where they are attracted simultaneously by both protons. This positioning is precisely what leads to the stability of the H_2 molecule compared to two separated hydrogen atoms. The potential energy of each electron is lowered because of the increased attractive forces in this area. When we say that a bond is formed between the hydrogen atoms, we mean that the H_2 molecule is more stable than two separated hydrogen atoms by a certain quantity of energy (the bond energy).

We can also think of a bond in terms of forces. The simultaneous attraction of each electron by the two protons generates a force that pulls the protons toward each other and that just balances the proton-proton and electron-electron repulsive forces at the distance corresponding to the bond length.

The type of bonding we encounter in the hydrogen molecule and in many other molecules where *electrons are shared by nuclei* is called **covalent bonding.**

Ionic and covalent bonds are the extreme bond types.

So far we have considered two extreme types of bonding. In ionic bonding the participating atoms are so different that one or more electrons are transferred to form oppositely charged ions. The bonding results from electrostatic interactions. In covalent bonding two identical atoms share electrons equally. The bonding results from the mutual attraction of the two nuclei for the shared electrons. Between these extremes are intermediate cases in which the atoms are not so different that electrons are completely transferred but are different enough so that unequal sharing results, forming what is called a **polar covalent bond.** An example of this type of bond occurs in the hydrogen fluoride (HF) molecule. When a sample of hydrogen fluoride gas is placed in an electric field, the molecules tend to orient themselves as shown in Fig. 8.2, with the fluoride end closest to the positive pole and the hydrogen end closest to the negative pole. This result implies that the HF molecule has the following charge distribution:

$$\underset{\delta+}{\mathrm{H}}\!-\!\underset{\delta-}{\mathrm{F}}$$

where δ (delta) is used to indicate a fractional charge. This same effect was noted in Chapter 4 where many of water's unusual properties were attributed to the polar O—H bonds in the H_2O molecule.

The most logical explanation for the development of the partial positive and negative charges on the atoms (bond polarity) in such molecules as HF and H_2O is that the electrons in the bonds are not shared equally. For example, we can account for the polarity of the HF molecule by assuming that the fluorine atom has a stronger attraction for the shared electrons than the hydrogen atom. Likewise, in the H_2O molecule the oxygen atom appears to attract the shared electrons more strongly than the hydrogen atoms do. Because bond polarity has important chemical implications, we find it useful to quantify the ability of an atom to attract shared electrons. In the next section we show how this is done.

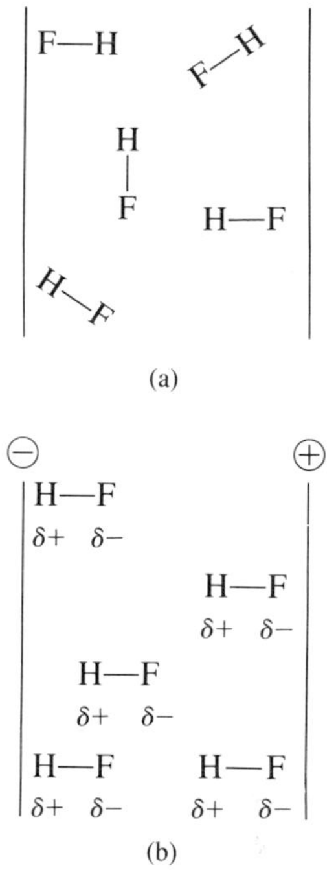

F—H F—H H—F H—F H—F

(a)

⊖ ⊕

H—F δ+ δ− H—F δ+ δ− H—F δ+ δ− H—F δ+ δ− H—F δ+ δ−

(b)

Figure 8.2

The effect of an electric field on hydrogen fluoride molecules. (a) When no electric field is present the molecules are randomly oriented. (b) When the field is turned on the molecules tend to line up with their negative ends toward the positive pole and their positive ends toward the negative pole.

8.2 Electronegativity

Purpose

- To discuss the nature of bonds in terms of electronegativity.

The different affinities of atoms for the electrons in a bond are described by a property called **electronegativity:** *the ability of an atom in a molecule to attract shared electrons to itself.*

The most widely accepted method for determining values of electronegativity is that of Linus Pauling, an American scientist who has won the Nobel Prizes for both chemistry and peace. To understand Pauling's model, consider a hypothetical molecule HX. The relative electronegativities of the H and X atoms are determined by comparing the measured H—X bond energy and the "expected" H—X bond energy, which is an average of the H—H and X—X bond energies:

$$\text{Expected H—X bond energy} = \frac{\text{H—H bond energy} + \text{X—X bond energy}}{2}$$

The difference (Δ) between the actual (measured) and expected bond energies is

$$\Delta = (\text{H—X})_{\text{act}} - (\text{H—X})_{\text{exp}}$$

If H and X have identical electronegativities, $(\text{H—X})_{\text{act}}$ and $(\text{H—X})_{\text{exp}}$ are the same and Δ is 0. On the other hand, if X has a greater electronegativity than H, the shared electron(s) will tend to be closer to the X atom. The molecule will be polar, with the following charge distribution:

$$\underset{\delta+\ \ \delta-}{\text{H—X}}$$

Note that this bond can be viewed as having an ionic, as well as a covalent, component. The electrostatic attraction between the partially charged H and X atoms will lead to a greater bond strength. Thus $(\text{H—X})_{\text{act}}$ will be larger than $(\text{H—X})_{\text{exp}}$. The greater the difference in the electronegativities of the atoms, the greater the ionic component of the bond and the greater the value of Δ. Thus the relative electronegativities of H and X can be assigned from the Δ values.

Increasing electronegativity →

Decreasing electronegativity ↓

						H 2.1										
Li 1.0	Be 1.5											B 2.0	C 2.5	N 3.0	O 3.5	F 4.0
Na 0.9	Mg 1.2											Al 1.5	Si 1.8	P 2.1	S 2.5	Cl 3.0
K 0.8	Ca 1.0	Sc 1.3	Ti 1.5	V 1.6	Cr 1.6	Mn 1.5	Fe 1.8	Co 1.9	Ni 1.9	Cu 1.9	Zn 1.6	Ga 1.6	Ge 1.8	As 2.0	Se 2.4	Br 2.8
Rb 0.8	Sr 1.0	Y 1.2	Zr 1.4	Nb 1.6	Mo 1.8	Tc 1.9	Ru 2.2	Rh 2.2	Pd 2.2	Ag 1.9	Cd 1.7	In 1.7	Sn 1.8	Sb 1.9	Te 2.1	I 2.5
Cs 0.7	Ba 0.9	La–Lu 1.0–1.2	Hf 1.3	Ta 1.5	W 1.7	Re 1.9	Os 2.2	Ir 2.2	Pt 2.2	Au 2.4	Hg 1.9	Tl 1.8	Pb 1.9	Bi 1.9	Po 2.0	At 2.2
Fr 0.7	Ra 0.9	Ac 1.1	Th 1.3	Pa 1.4	U 1.4	Np–No 1.4–1.3										

Figure 8.3

The Pauling electronegativity values. Electronegativity generally increases across a period and decreases down a group.

Electronegativity values have been determined by this process for virtually all of the elements; the results are given in Fig. 8.3. Note that electronegativity generally increases going from left to right across a period and decreases going down a group for the representative elements. The range of electronegativity values is from 4.0 for fluorine to 0.7 for cesium.

The relationship between electronegativity and bond type is shown in Table 8.1. For identical atoms (an electronegativity difference of zero), the electrons in the bond are shared equally and no polarity develops. When two atoms with very different electronegativities interact, electron transfer usually occurs, to produce the ions that make up an ionic substance. Intermediate cases give polar covalent bonds with unequal electron sharing.

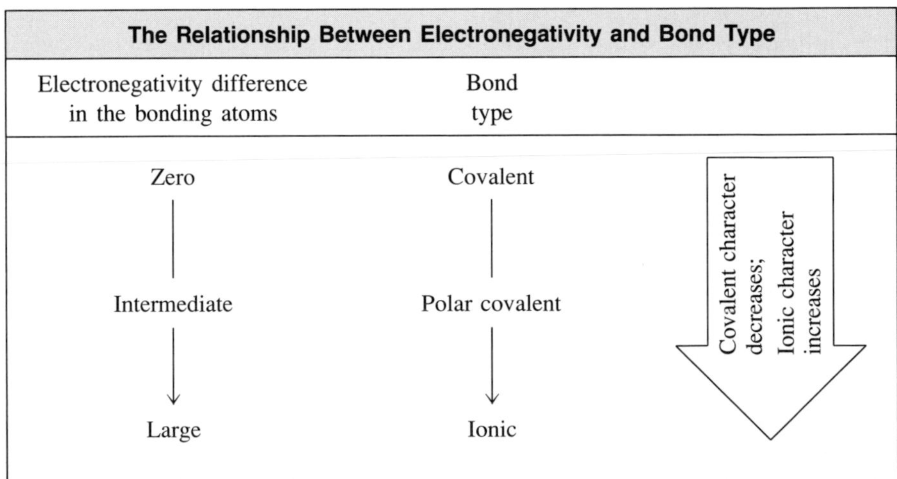

The Relationship Between Electronegativity and Bond Type

Electronegativity difference in the bonding atoms	Bond type
Zero	Covalent
↓	↓
Intermediate	Polar covalent
↓	↓
Large	Ionic

Table 8.1

Sample Exercise 8.1

Order the following bonds according to polarity: H—H, O—H, Cl—H, S—H, and F—H.

Solution

The polarity of the bond increases as the difference in electronegativity increases. From the electronegativity values in Fig. 8.3, the following variation in bond polarity is expected (the electronegativity value appears in parentheses below each element):

	H—H	< S—H	< Cl—H	< O—H	< F—H
	(2.1)(2.1)	(2.5)(2.1)	(3.0)(2.1)	(3.5)(2.1)	(4.0)(2.1)
Electronegativity difference	0	0.4	0.9	1.4	1.9

Covalent bond → Polar covalent bond (Polarity increases)

8.3 Bond Polarity and Dipole Moments

Purpose

- To define the relationship between bond polarity and molecular polarity.

We have seen that when hydrogen fluoride is placed in an electric field, the molecules have a preferential orientation (Fig. 8.2). This follows from the charge distribution in the HF molecule, which has a positive end and a negative end. A molecule like HF that has a center of positive charge and a center of negative charge is said to be *dipolar,* or to have a **dipole moment.** The dipolar character of a molecule is often represented by an arrow pointing to the negative charge center with the tail of the arrow indicating the positive center of charge:

$$\underset{\delta+}{+}\!\!\longrightarrow_{\delta-}$$

Of course, any diatomic (two-atom) molecule that has a polar bond will also show a molecular dipole moment. Polyatomic molecules can also exhibit dipolar behavior. For example, because the oxygen atom in the water molecule has a greater electronegativity than the hydrogen atoms, the molecular charge distribution is that shown in Fig. 8.4(a). Because of this charge distribution, the water molecule behaves in an electric field as if it had two centers of charge—one positive and one negative—as shown in Fig. 8.4(b). The water molecule has a dipole moment. The same type of behavior is observed for the NH_3 molecule (Fig. 8.5 on page 330). Some molecules have polar bonds but do not have a dipole moment. This occurs when the individual bond polarities are arranged in such a way that they cancel each

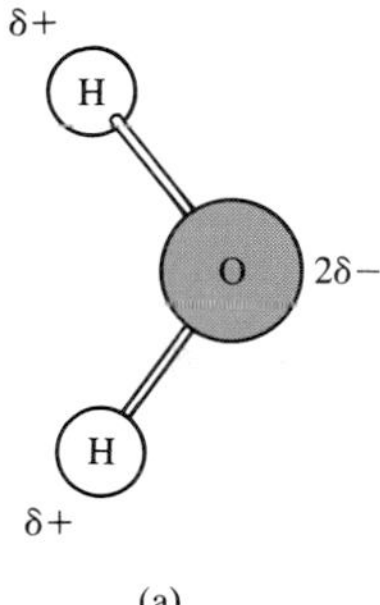

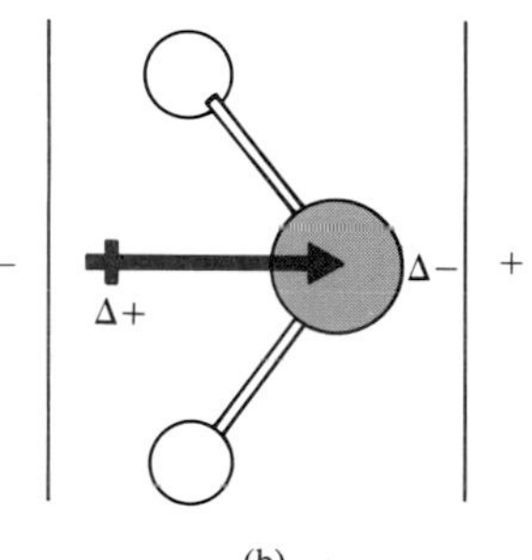

Figure 8.4

(a) The charge distribution in the water molecule. (b) The water molecule in an electric field.

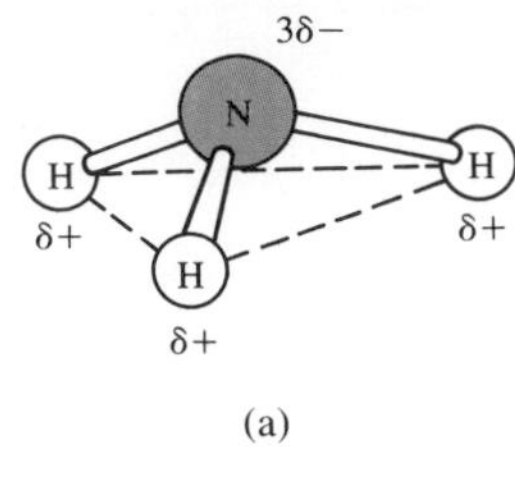

(a)

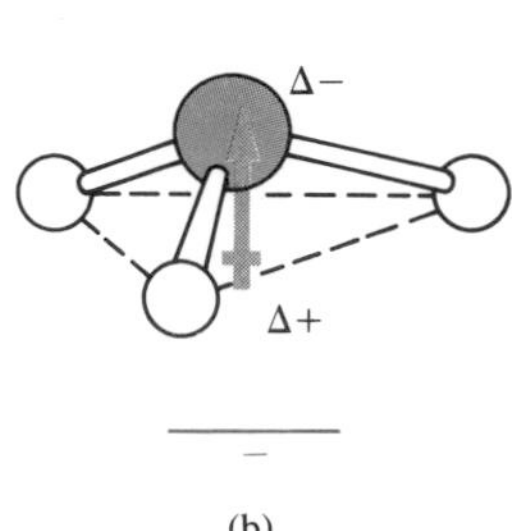

(b)

Figure 8.5

(a) The structure and charge distribution of the ammonia molecule. The polarity of the N—H bonds occurs because nitrogen has a greater electronegativity than hydrogen. (b) The dipole moment of the ammonia molecule oriented in an electric field.

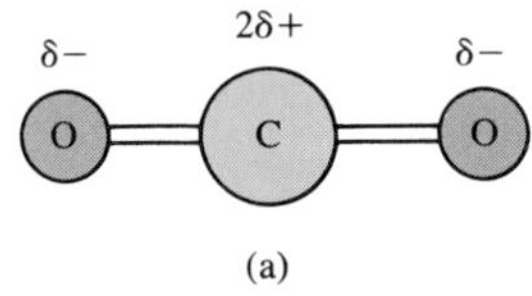

(a)

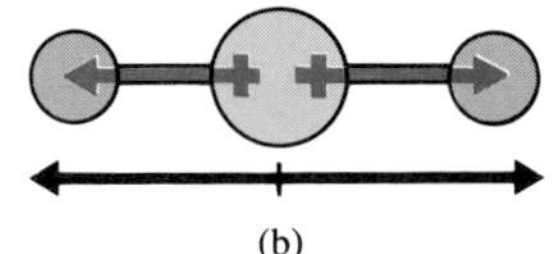

(b)

Figure 8.6

(a) The carbon dioxide molecule. (b) The opposed bond polarities cancel out, and the carbon dioxide molecule has no dipole moment.

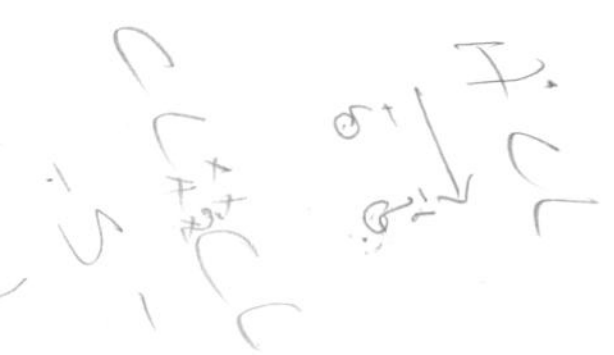

other out. An example is the CO_2 molecule, which is a linear molecule and has the charge distribution shown in Fig. 8.6. In this case, the opposing bond polarities cancel out and the carbon dioxide molecule does not have a dipole moment. There is no preferential way for this molecule to line up in an electric field. (Try to find a preferred orientation to make sure you understand this concept.)

There are many cases besides that of carbon dioxide where the bond polarities oppose and exactly cancel each other. Some common types of molecules with polar bonds but no dipole moment are shown in Table 8.2.

Types of Molecules with Polar Bonds but No Resulting Dipole Moment

Type		Cancellation of polar bonds	Example
Linear molecules with two identical bonds	B—A—B	⟵+ +⟶	CO_2
Planar molecules with three identical bonds 120° apart	B, A, B, B, 120°		SO_3
Tetrahedral molecules with four identical bonds 109.5° apart	B, A, B, B, B		CCl_4

Table 8.2

Sample Exercise 8.2

For each of the following molecules, show the direction of the bond polarities and indicate which ones have a dipole moment: HCl, Cl_2, SO_3 (a planar molecule with the oxygen atoms spaced evenly around the central sulfur atom), CH_4 (tetrahedral (see Table 8.2) with the carbon atom at the center), and H_2S (V-shaped with the sulfur atom at the point).

Solution

The HCl molecule: From Fig. 8.3 we can see that the electronegativity of chlorine (3.0) is greater than that of hydrogen (2.1). Thus the chlorine will be partially negative, and the hydrogen will be partially positive. The HCl molecule has a dipole moment:

$$\underset{\delta+}{H}\text{———}\underset{\delta-}{Cl}$$
$$+\!\!\longrightarrow$$

The Cl_2 molecule: The two chlorine atoms share the electrons equally. No bond polarity occurs, and the Cl_2 molecule has no dipole moment.

Sample Exercise 8.2, continued

The SO_3 molecule: The electronegativity of oxygen (3.5) is greater than that of sulfur (2.5). This means that each oxygen will have a partial negative charge, and the sulfur will have a partial positive charge:

$$\begin{array}{c} O\ \delta- \\ | \\ S\ 3\delta+ \\ \delta-O \quad O\ \delta- \end{array}$$

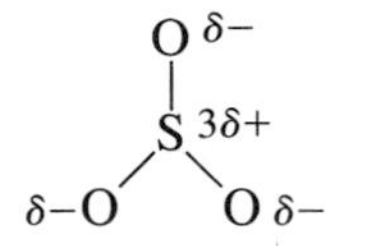

The bond polarities arranged symmetrically as shown cancel, and the molecule has no dipole moment. This molecule is the second type shown in Table 8.2.

The presence of polar bonds does not always yield a polar molecule.

The CH_4 molecule: Carbon has a slightly higher electronegativity (2.5) than does hydrogen (2.1). This leads to small partial positive charges on the hydrogen atoms and a small partial negative charge on the carbon:

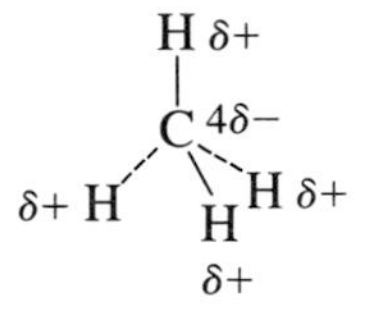

$$\begin{array}{c} H\ \delta+ \\ | \\ C\ 4\delta- \\ \delta+\ H \quad H \quad H\ \delta+ \\ \delta+ \end{array}$$

This case is similar to the third type in Table 8.2, and the bond polarities cancel. The molecule has no dipole moment.

The H_2S molecule: Since the electronegativity of sulfur (2.5) is greater than that of hydrogen (2.1), the sulfur will have a partial negative charge and the hydrogen atoms will have a partial positive charge, which can be represented as follows:

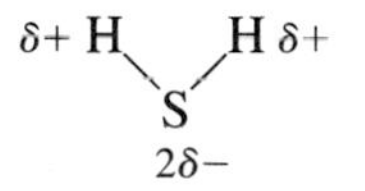

$$\begin{array}{c} \delta+\ H \quad H\ \delta+ \\ S \\ 2\delta- \end{array}$$

This case is analogous to the water molecule, and the polar bonds result in a dipole moment oriented as shown:

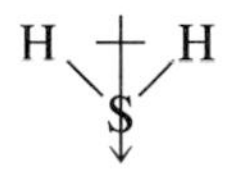

$$\begin{array}{c} H \quad + \quad H \\ S \\ \downarrow \end{array}$$

8.4 Ions: Electron Configurations and Sizes

Purpose

- To show how to predict the formulas of ionic compounds.
- To discuss the factors governing ion size.

The description of the arrangements of electrons in atoms that emerged from the wave mechanical model has helped a great deal in our understanding of what constitutes a stable compound. In virtually every case the atoms in a stable compound have a noble gas arrangement of electrons. Nonmetallic elements achieve a

Atoms in stable compounds usually have a noble gas electron configuration.

noble gas electron configuration either by sharing electrons with other nonmetals to form covalent bonds or by taking electrons from metals to form ions. In the second case the nonmetals form anions and the metals form cations. The generalizations that apply to electron configurations in stable compounds are as follows:

- When *two nonmetals* react to form a covalent bond, they share electrons in a way that completes the valence electron configurations of both atoms. That is, both nonmetals attain noble gas electron configurations.
- When *a nonmetal and a representative group metal* react to form a binary ionic compound, the ions form so that the valence electron configuration of the nonmetal is completed and the valence orbitals of the metal are emptied. In this way both ions achieve noble gas electron configurations.

With a few exceptions these generalizations apply to the vast majority of compounds and are important to remember. We will deal with covalent bonds more thoroughly later. Next we will consider what implications these rules hold for ionic compounds.

Predicting Formulas of Ionic Compounds

To illustrate the principles of electron configurations in stable compounds, we will consider the formation of an ionic compound from calcium and oxygen. We can predict what compound will form by considering the valence electron configurations of the two atoms.

$$\text{Ca:} \quad [\text{Ar}]4s^2$$
$$\text{O:} \quad [\text{He}]2s^2 2p^4$$

A bauxite mine in Northern Queensland, Australia. Bauxite contains Al_2O_3, the main source of aluminum.

From Fig. 8.3 we see that the electronegativity of oxygen (3.5) is much greater than that of calcium (1.0). Because of this large difference, electrons will be transferred from calcium to oxygen to form an oxygen anion and a calcium cation. How many electrons are transferred? We can base our prediction on the observation that noble gas configurations are the most stable. Note that oxygen needs two electrons to fill its $2s$ and $2p$ valence orbitals and achieve the configuration of neon ($1s^2 2s^2 2p^6$). And by losing two electrons, calcium can achieve the configuration of argon. Two electrons will therefore be transferred as shown:

$$\underset{\curvearrowright 2e^-}{\text{Ca} + \text{O}} \rightarrow \text{Ca}^{2+} + \text{O}^{2-}$$

To predict the formula of the ionic compound, we must recognize that chemical compounds are always electrically neutral—they have the same quantities of positive and negative charges. In this case we must have equal numbers of Ca^{2+} and O^{2-} ions, and the empirical formula of the compound is CaO.

The same principles can be applied to many other cases. For example, consider the compound formed from aluminum and oxygen. Aluminum has the configuration $[Ne]3s^2 3p^1$. To achieve the neon configuration, aluminum must lose three electrons to form the Al^{3+} ion. Thus the ions will be Al^{3+} and O^{2-}. Since the compound

Common Ions with Noble Gas Configurations in Ionic Compounds					
Group 1A	Group 2A	Group 3A	Group 6A	Group 7A	Electron configuration
H^-, Li^+	Be^{2+}				[He]
Na^+	Mg^{2+}	Al^{3+}	O^{2-}	F^-	[Ne]
K^+	Ca^{2+}		S^{2-}	Cl^-	[Ar]
Rb^+	Sr^{2+}		Se^{2-}	Br^-	[Kr]
Cs^+	Ba^{2+}		Te^{2-}	I^-	[Xe]

Table 8.3

must be electrically neutral, there must be three O^{2-} ions for every two Al^{3+} ions, and the compound has the empirical formula Al_2O_3.

Table 8.3 shows common elements that form ions with noble gas electron configurations in ionic compounds. In losing electrons to form cations, metals in Group 1A lose one electron, those in Group 2A lose two electrons, and those in Group 3A lose three electrons. In gaining electrons to form anions, nonmetals in Group 7A (the halogens) gain one electron and those in Group 6A gain two electrons. Hydrogen typically behaves as a nonmetal and can gain one electron to form the hydride ion (H^-), which has the electron configuration of helium.

There are some important exceptions to the rules we have been following here. For example, tin forms both Sn^{2+} and Sn^{4+} ions, and lead forms both Pb^{2+} and Pb^{4+} ions. Bismuth forms Bi^{3+} and Bi^{5+} ions, and thallium forms Tl^+ and Tl^{3+} ions. There are no simple explanations for the behavior of these ions. For now, just note them as exceptions to the very useful rule that ions generally adopt noble gas electron configurations in ionic compounds. Our discussion here refers to representative metals. The transition metals exhibit more complicated behavior, forming a variety of ions that will be considered in Chapter 20.

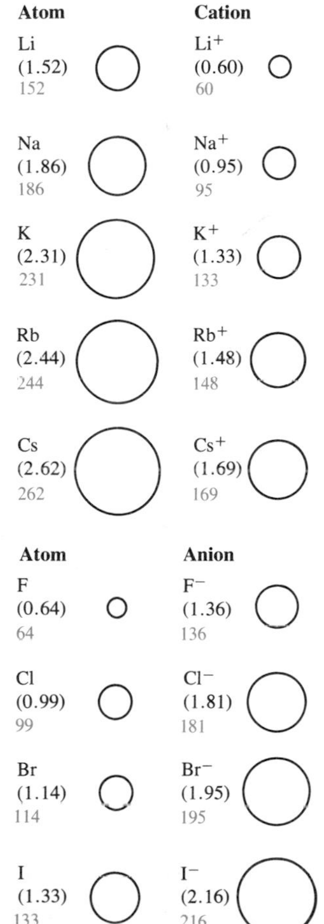

Figure 8.7

Sizes of some ions and their parent atoms. Note that cations are smaller and anions are larger than their parent atoms. The radii are given in angstroms (parentheses) and picometers (blue).

Sizes of Ions

Ion size plays an important role in determining the structure and stability of ionic solids, the properties of ions in aqueous solution, and the biological effects of ions. Various factors influence ionic size. We will first consider the relative sizes of an ion and its parent atom. Since a positive ion is formed by taking electrons from a neutral atom, the resulting cation is smaller than its parent atom. The opposite is true for negative ions; the addition of electrons to a neutral atom produces an anion significantly larger than its parent atom. Selected examples of this behavior are shown in Fig. 8.7.

It is also important to know how the sizes of ions vary depending on the positions of the elements in the periodic table. Figure 8.8 on page 334 shows the sizes of the most important ions (each with a noble gas configuration) related to their position in the periodic table. Note that ion size increases going down a group. The changes that occur horizontally are complicated because there is a change from predominantly metals on the left-hand side of the periodic table to nonmetals on the right-hand side. A given period thus contains both elements that give up valence electrons to form cations and ones that accept electrons to form anions.

One trend worth noting involves the relative sizes of a set of **isoelectronic ions**—*ions containing the same number of electrons*. Consider the ions O^{2-}, F^-,

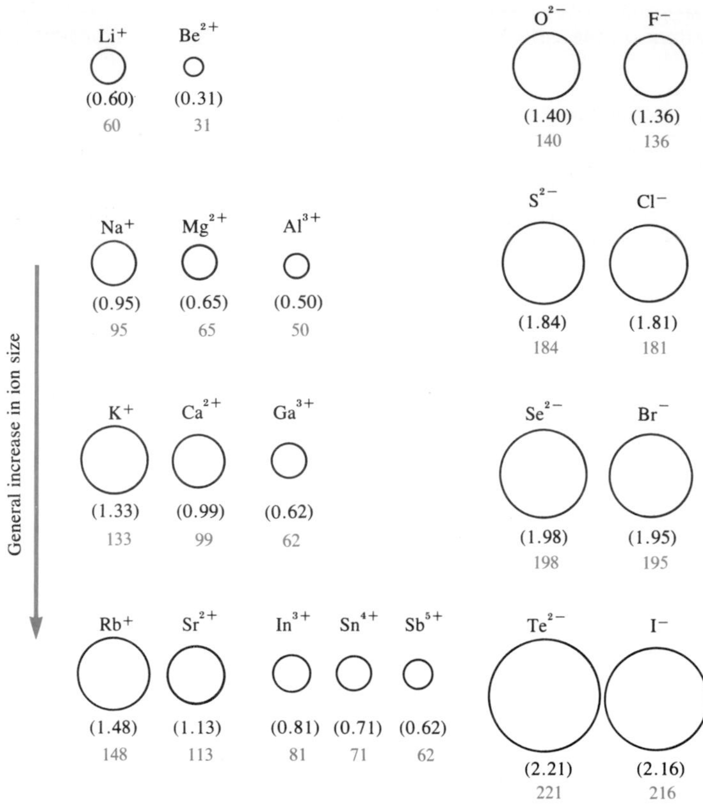

Figure 8.8

Sizes of ions related to positions of elements in the periodic table. Note that size generally increases going down a group. Also note that in a series of isoelectronic ions size decreases with increasing atomic number. The ionic radii are given in units of angstroms (parentheses) and picometers (blue).

Na^+, Mg^{2+}, and Al^{3+}. Each of these ions has the neon electron configuration (confirm this for yourself). How do the sizes of these ions vary? In general, there are two important facts to consider in predicting the relative sizes of ions: the number of electrons and the number of protons. Since the ions being considered here are isoelectronic, the number of electrons is ten in each case. Electron repulsions should therefore be about the same in all cases. However, the number of protons increases from eight to thirteen as we go from the O^{2-} ion to the Al^{3+} ion. Thus, in going from O^{2-} to Al^{3+}, the ten electrons experience increasing attraction by the increasing positive charge of the nucleus, which causes the ions to become smaller. You can confirm this by looking at the sizes of these ions as shown in Fig. 8.8. In general, for a series of isoelectronic ions, the size decreases as the nuclear charge (Z) increases.

For isoelectronic ions, size decreases as Z increases.

Sample Exercise 8.3

Arrange the ions Se^{2-}, Br^-, Rb^+, and Sr^{2+} in order of decreasing size.

Sample Exercise 8.3, continued

Solution

This is an isoelectronic series of ions with the electron configuration of krypton. Since these ions all have the same number of electrons, their sizes will depend on nuclear charge (Z). The Z values are 34 for Se^{2-}, 35 for Br^-, 37 for Rb^+, and 38 for Sr^{2+}. Since the nuclear charge is greatest for the Sr^{2+} ion, it will be the smallest of these ions. The Se^{2-} ion with the smallest value of Z will be the largest:

$$\underset{\text{Largest}}{\underset{\uparrow}{Se^{2-}}} > Br^- > Rb^+ > \underset{\text{Smallest}}{\underset{\uparrow}{Sr^{2+}}}$$

Sample Exercise 8.4

Choose the largest ion in each of the following groups:

a. Li^+, Na^+, K^+, Rb^+, Cs^+ **b.** Ba^{2+}, Cs^+, I^-, Te^{2-}

Solution

a. The ions are all from Group 1A elements. Since size increases going down a group (the ion with the greatest number of electrons is the largest), Cs^+ is the largest ion.

Ion size increases going down a group.

b. This is an isoelectronic series of ions, all of which have the electron configuration of xenon. The ion with the smallest nuclear charge will be the largest ion:

$$\underset{Z=52}{Te^{2-}} > \underset{Z=53}{I^-} > \underset{Z=55}{Cs^+} > \underset{Z=56}{Ba^{2+}}$$

8.5 Formation of Binary Ionic Compounds

Purpose

- To define lattice energy and to show how it can be calculated.

In this section we will introduce the factors influencing the formation and structures of binary ionic compounds. We know that metals and nonmetals react by transferring electrons to form cations and anions that are mutually attractive. The resulting ionic solid forms because the aggregated oppositely charged ions have lower energy than the original elements. Just how strongly the ions attract each other in the solid state is indicated by the **lattice energy**—*the change in energy that takes place when separated gaseous ions are packed together to form an ionic solid:*

$$M^+(g) + X^-(g) \rightarrow MX(s)$$

The structures of ionic solids will be discussed in detail in Chapter 10.

The lattice energy is often defined as the energy *released* when an ionic solid forms from its ions. However, in this book, the sign of an energy term will be determined from the system's point of view: negative if the process is exothermic; positive if endothermic. Thus lattice energy will have a negative sign.

We can illustrate the energy changes involved in the formation of an ionic solid by considering the formation of solid lithium fluoride from its elements:

$$Li(s) + \tfrac{1}{2}F_2(g) \rightarrow LiF(s)$$

To see the energy terms associated with this process, we will take advantage of the fact that energy is a state function and break this reaction down into steps, the sum of which gives the overall reaction:

STEP 1

Sublimation of solid lithium. Sublimation involves taking a substance from the solid state to the gaseous state:

$$Li(s) \rightarrow Li(g)$$

The energy of sublimation for $Li(s)$ is 161 kJ/mol.

Lithium fluoride.

STEP 2

Ionization of lithium atoms to form Li^+ ions in the gas phase:

$$Li(g) \rightarrow Li^+(g) + e^-$$

This process corresponds to the first ionization energy for lithium, which is 520 kJ/mol.

STEP 3

Dissociation of fluorine molecules. We need to form 1 mole of fluorine atoms by breaking the F—F bond in $\frac{1}{2}$ mole of F_2 molecules:

$$\tfrac{1}{2}F_2(g) \rightarrow F(g)$$

The energy required to break this bond is known from experiment to be 154 kJ/mol. In this case we are breaking the bonds in a half mole of fluorine, so the energy required for this step is 154 kJ/2, or 77 kJ.

STEP 4

Formation of F^- ions from fluorine atoms in the gas phase:

$$F(g) + e^- \rightarrow F^-(g)$$

We have defined the energy change for this process as the electron affinity of fluorine, which is -328 kJ/mol.

STEP 5

Formation of solid lithium fluoride from the gaseous Li^+ and F^- ions:

$$Li^+(g) + F^-(g) \rightarrow LiF(s)$$

This corresponds to the definition of the lattice energy for LiF, which is known to be -1047 kJ/mol.

Since the sum of these five processes gives the overall reaction desired, the sum of the individual energy changes gives the overall energy change:

	Process	Energy change (kJ)
	$Li(s) \rightarrow Li(g)$	161
	$Li(g) \rightarrow Li^+(g) + e^-$	520
	$\frac{1}{2}F_2(g) \rightarrow F(g)$	77
	$F(g) + e^- \rightarrow F^-(g)$	−328
	$Li^+(g) + F^-(g) \rightarrow LiF(s)$	−1047
Overall:	$Li(s) + \frac{1}{2}F_2(g) \rightarrow LiF(s)$	−617 kJ (per mole of LiF)

This process is summarized by the energy diagram in Fig. 8.9. Note that the formation of solid lithium fluoride from its elements is highly exothermic, mainly because of the very large negative lattice energy. A great deal of energy is released when the ions combine to form the solid. In fact, note that the energy released when an electron is added to a fluorine atom to form the F^- ion (328 kJ/mol) is not enough to remove an electron from lithium (520 kJ/mol). That is, when a metallic lithium atom reacts with a nonmetallic fluorine atom to form *separated* ions,

$$Li(g) + F(g) \rightarrow Li^+(g) + F^-(g)$$

the process is endothermic and thus unfavorable. Clearly then, the main impetus for the formation of an ionic compound rather than a covalent compound in this case results from the strong mutual attractions of the Li^+ and F^- ions in the solid. The lattice energy is the dominant energy term.

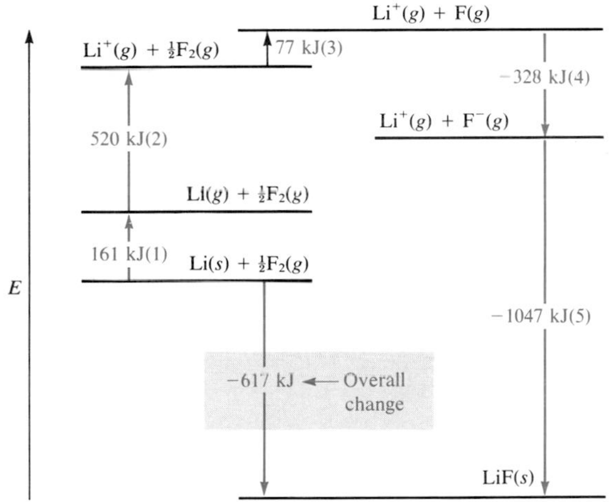

Figure 8.9

The energy changes involved in the formation of solid lithium fluoride from its elements. The numbers in parentheses refer to the reaction steps discussed on page 336.

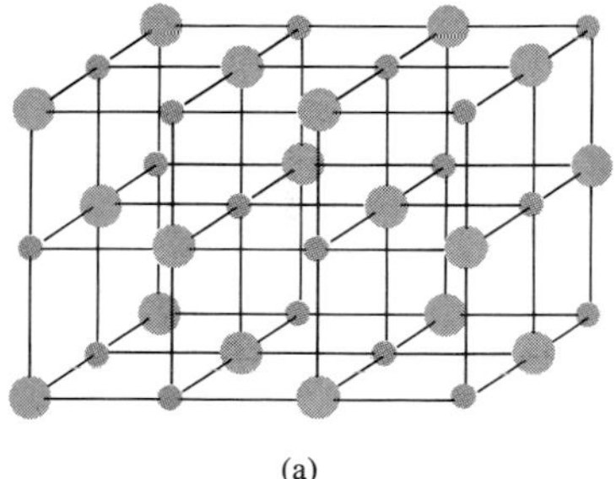

(a)

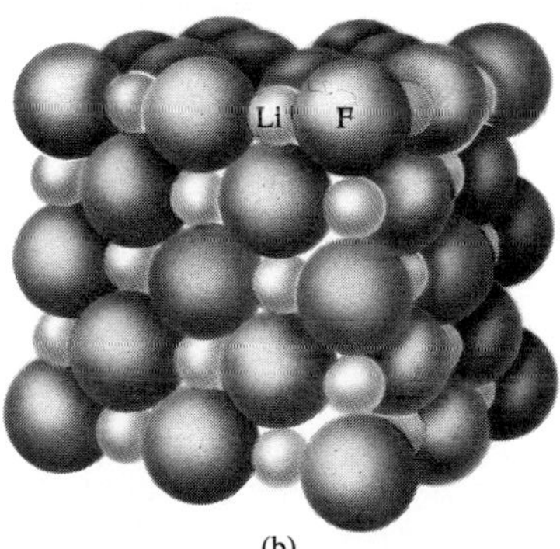

(b)

Figure 8.10

The structure of lithium fluoride. (a) Represented by a ball and stick model. Note that each Li^+ ion is surrounded by six F^- ions, and each F^- ion is surrounded by six Li^+ ions. (b) Represented with the ions shown as spheres. The structure is determined by packing the spherical ions in a way that both maximizes the ionic attractions and minimizes the ionic repulsions.

The structure of the solid lithium fluoride is represented in Fig. 8.10. Note the alternating arrangement of the Li^+ and F^- ions, and that each Li^+ is surrounded by six F^- ions and each F^- ion is surrounded by six Li^+ ions. This structure can be rationalized by assuming that the ions behave as hard spheres that pack in a way that both maximizes the attractions among the oppositely charged ions and minimizes the repulsions among the identically charged ions.

All of the binary ionic compounds formed by an alkali metal and a halogen have the structure shown in Fig. 8.10, except the cesium salts. The arrangement of ions shown in Fig. 8.10 is often called the *sodium chloride structure* after the most common substance that possesses it.

Lattice Energy Calculations

In the discussion of the energetics of the formation of solid lithium fluoride, we emphasized the importance of lattice energy in contributing to the stability of ionic solids. Lattice energy can be represented by Coulomb's law,

$$\text{Lattice energy} = k\left(\frac{Q_1Q_2}{r}\right)$$

where k is a proportionality constant that depends on the structure of the solid and the electron configurations of the ions, Q_1 and Q_2 are the charges on the ions, and r is the shortest distance between the centers of the cations and anions. Note that the lattice energy has a negative sign when Q_1 and Q_2 have opposite signs. This is to be expected, since bringing cations and anions together is an exothermic process. Also note that the process becomes more exothermic as the ionic charges increase and the distances between the ions in the solid decrease.

The importance of charge in ionic solids can be illustrated by comparing the energies involved in the formation of NaF(*s*) and MgO(*s*) where the ions involved, Na^+, F^-, Mg^{2+}, and O^{2-}, are isoelectronic. The energy diagram for the formation of the two ionic solids is given in Fig. 8.11. There are several important features:

The energy released when gaseous Mg^{2+} and O^{2-} ions combine to form MgO(*s*) is much greater (more than four times greater) than that released when gaseous Na^+ and F^- ions combine to form NaF(*s*).

The energy required to remove two electrons from the magnesium atom (735 kJ/mol for the first and 1445 kJ/mol for the second for a total of 2180 kJ/mol) is much greater than the energy required to remove an electron from a sodium atom (495 kJ/mol).

Energy (737 kJ/mol) is required to add two electrons to the oxygen atom in the gas phase. Addition of the first electron is exothermic (−141 kJ/mol), but addition of the second electron is quite endothermic (878 kJ/mol).

Since the equation for lattice energy contains the product Q_1Q_2, the lattice energy for a solid with 2^+ and 2^- ions should be four times that for a solid with 1^+ and 1^- ions. That is

$$\frac{(+2)(-2)}{(+1)(-1)} = 4$$

For MgO and NaF the observed ratio of lattice energies (see Fig. 8.11) is:

$$\frac{-3925 \text{ kJ}}{-923 \text{ kJ}} = 4.25$$

In view of the facts that twice as much energy is required to remove the second electron from magnesium as to remove the first and that addition of an electron to the gaseous O^- ion is quite endothermic, it seems puzzling that magnesium oxide contains Mg^{2+} and O^{2-} ions rather than Mg^+ and O^- ions. The answer to this lies in the lattice energy. Note that the lattice energy for combining gaseous Mg^{2+} and O^{2-} ions to form MgO(*s*) is 3000 kJ/mol more negative than that for combining gaseous Na^+ and F^- ions to form NaF(*s*). Thus the energy released in forming a solid containing Mg^{2+} and O^{2-} ions rather than Mg^+ and O^- ions more than compensates for the energies required for the processes that produce the Mg^{2+} and O^{2-} ions.

If there is so much lattice energy to be gained in going from singly charged to doubly charged ions in the case of magnesium oxide, why then does solid sodium fluoride contain Na^+ and F^- ions rather than Na^{2+} and F^{2-} ions? We can answer this question by recognizing that both Na^+ and F^- ions have the neon electron configuration. Removal of an electron from Na^+ would require an extremely large quantity of energy, since it would be a $2p$ electron. Conversely, the addition of an electron to F^- would require use of the relatively large $3s$ orbital, which is also an unfavorable process. We can say that for sodium fluoride the extra energy required to form the doubly charged ions is greater than the gains in lattice energy that would result.

This comparison of the energies involved in the formation of sodium fluoride and magnesium oxide illustrates that a variety of factors operates to determine the composition and structure of ionic compounds. The most important of these factors involve the balancing of the energies required to form highly charged ions and the energy released when highly charged ions combine to form the solid.

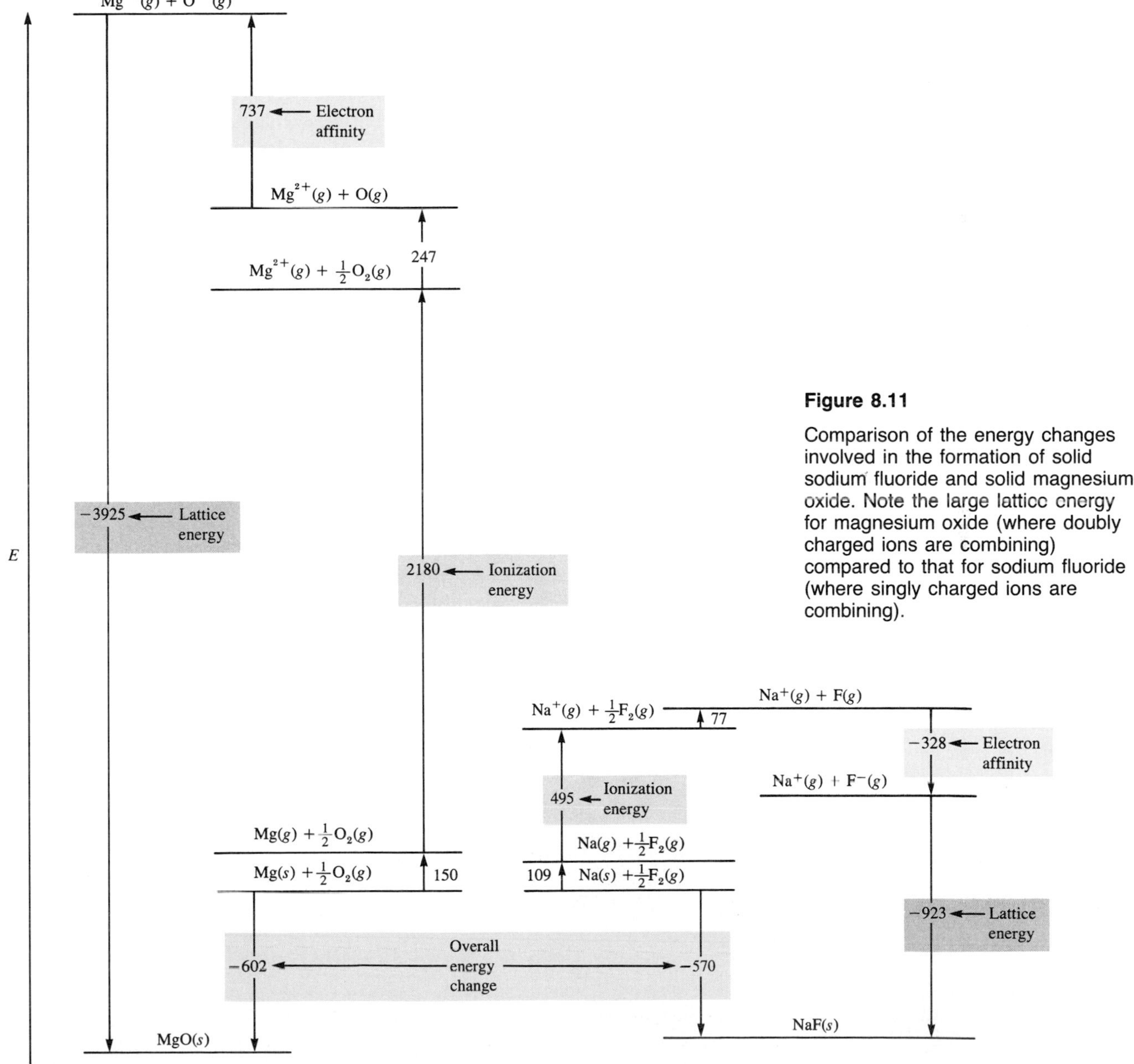

Figure 8.11

Comparison of the energy changes involved in the formation of solid sodium fluoride and solid magnesium oxide. Note the large lattice energy for magnesium oxide (where doubly charged ions are combining) compared to that for sodium fluoride (where singly charged ions are combining).

8.6 Partial Ionic Character of Covalent Bonds

Purpose

- To show the relationship between electronegativity and the ionic character of a bond.

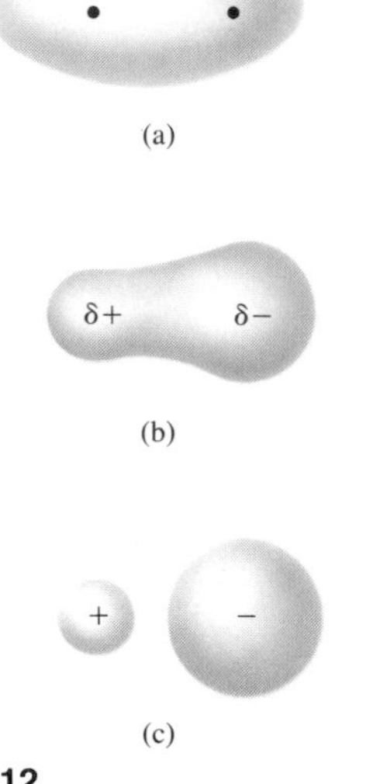

Figure 8.12

The three possible types of bonds: (a) a covalent bond formed between identical atoms; (b) a polar covalent bond, with both ionic and covalent components; and (c) an ionic bond, with no electron sharing.

When atoms with different electronegativities react to form compounds, the electrons are not shared equally. The possible result is a polar covalent bond or, in the case of a large electronegativity difference, a complete transfer of one or more electrons to form ions. The cases are summarized in Fig. 8.12.

How well can we tell the difference between an ionic bond and a polar covalent bond? The only honest answer to this question is that there are probably no totally ionic bonds. The evidence for this statement comes from calculations of the percent ionic character in various binary compounds. These calculations are based on comparisons of the measured dipole moments for molecules of the type X—Y with the calculated dipole moments for the completely ionic case, X^+Y^-. The percent ionic character of a bond is then defined as follows:

$$\text{Percent ionic character of a bond} = \left(\frac{\text{measured dipole moment of X—Y}}{\text{calculated dipole moment of } X^+Y^-}\right) \times 100$$

Application of this definition to various ionic compounds gives the results shown in Fig. 8.13, where percent ionic character is plotted versus the difference in the electronegativity values of X and Y. Note from this plot that ionic character increases with electronegativity difference, as expected. However, none of the compounds reaches 100% ionic character, even though compounds with the maximum possible electronegativity differences are considered. Thus, according to this definition, no compounds are completely ionic. This is in contrast to the usual classification of these compounds. All of the compounds shown in Fig. 8.13 with more than 50% ionic character are normally considered to be ionic.

Another complication in identifying ionic compounds is that many substances contain polyatomic ions. For example, NH_4Cl contains NH_4^+ and Cl^- ions, and Na_2SO_4 contains Na^+ and SO_4^{2-} ions. The ammonium and sulfate ions are held

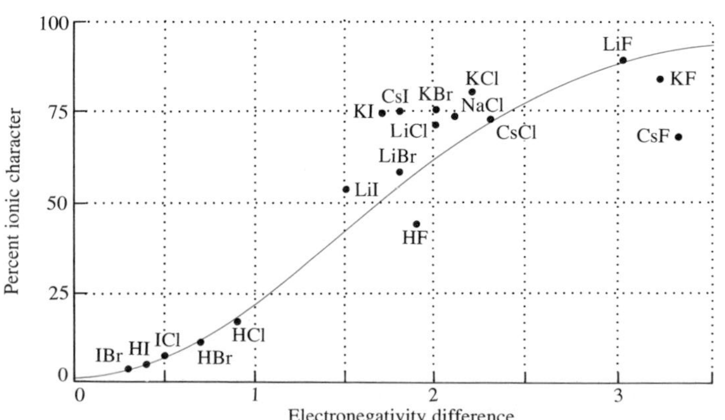

Figure 8.13

The relationship between the ionic character of a covalent bond and the electronegativity difference of the bonded atoms.

together by covalent bonds. Thus calling NH_4Cl and Na_2SO_4 ionic compounds is somewhat ambiguous.

We will avoid these problems by adopting an operational definition of ionic compounds: *any solid that conducts an electric current when melted or dissolved in water will be classified as ionic*. Also, the generic term *salt* will be used interchangeably with *ionic compound* in this book.

Molten NaCl conducts an electric current indicating the presence of mobile Na^+ and Cl^- ions.

8.7 The Covalent Chemical Bond: A Model

Purpose

- To discuss the covalent bonding model.

Before we develop specific models of covalent chemical bonding, it will be helpful if we summarize some of the concepts that have been introduced in this chapter.

What is a chemical bond? A chemical bond can be viewed as a force that causes a group of atoms to behave as a unit.

Why do chemical bonds occur? There is no principle of nature that states that bonds are favored or disfavored. Bonds are neither inherently "good" nor inherently "bad" as far as nature is concerned, but they result from the tendency of a system to seek its lowest possible energy. From a simplistic point of view, bonds occur where collections of atoms are more stable (lower in energy) than are the separate atoms. For example, approximately 1652 kJ of energy is required to break a mole of methane (CH_4) molecules into separate C and H atoms. Or, to take the opposite view, 1652 kJ of energy is released when 1 mole of methane is formed from 1 mole of gaseous C atoms and 4 moles of gaseous H atoms. Thus we can say that 1 mole of CH_4 molecules in the gas phase is 1652 kJ lower in energy than 1 mole of carbon atoms plus 4 moles of hydrogen atoms. Methane is therefore a stable molecule relative to the separated atoms.

We find it useful to interpret molecular stability in terms of a model called a *chemical bond*. To understand why this model has been invented, let's continue with methane, which consists of four hydrogen atoms arranged around a carbon atom at the corners of a tetrahedron:

A tetrahedron has four equal triangular faces.

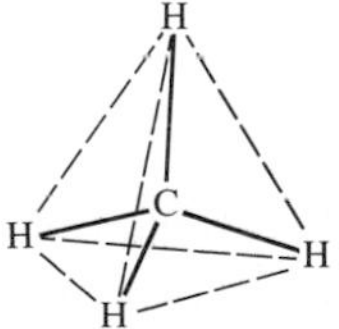

Given this structure, it is natural to envision four individual C—H interactions (we call them bonds). The energy of stabilization of CH_4 is divided equally among them to give an average C—H bond energy per mole of C—H bonds:

$$\frac{1652 \text{ kJ}}{4} = 413 \text{ kJ}$$

Next consider methyl chloride, which consists of CH_3Cl molecules having the structure

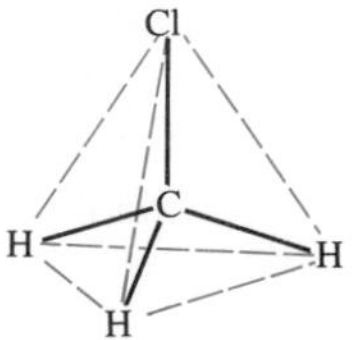

From experiment, it has been determined that approximately 1578 kJ of energy is required to break down 1 mole of gaseous CH_3Cl molecules into gaseous carbon, chlorine, and hydrogen atoms. The reverse process can be represented as:

$$C(g) + Cl(g) + 3H(g) \rightarrow CH_3Cl(g) + 1578 \text{ kJ/mol}$$

A mole of gaseous methyl chloride is lower in energy by 1578 kJ than its separate gaseous atoms. Thus a mole of methyl chloride is held together by 1578 kJ of energy. Again, it is very useful to divide this energy into individual bonds. Methyl chloride can be visualized as containing one C—Cl bond and three C—H bonds. If we assume arbitrarily that a C—H interaction represents the same quantity of energy in any situation (that is, that the strength of a C—H bond is independent of its molecular environment), we can do the following bookkeeping:

$$\begin{aligned} \text{1 mol of C—Cl bonds plus 3 mol of C—H bonds} &= 1578 \text{ kJ} \\ \text{C—Cl bond energy} + 3(\text{average C—H bond energy}) &= 1578 \text{ kJ} \\ \text{C—Cl bond energy} + 3(413 \text{ kJ/mol}) &= 1578 \text{ kJ} \\ \text{C—Cl bond energy} &= 1578 - 1239 = 339 \text{ kJ/mol} \end{aligned}$$

These assumptions allow us to associate given quantities of energy with C—H and C—Cl bonds.

It is important to note that the bond concept is a human invention. Bonds simply provide a method for dividing up the energy evolved when a stable molecule is formed from its component atoms. A *bond thus represents a quantity of energy* obtained from the overall molecular energy in a rather arbitrary way. This is not to say that the bonding concept is a bad idea. In fact, the modern concept of the chemical bond, conceived by the American chemists G. N. Lewis and Linus Pauling, is one of the most useful ideas chemists have ever conceived.

Models: An Overview

The framework of chemistry, like that of any science, consists of models—attempts to explain how nature operates on the microscopic level, based on experiences in the macroscopic world. To understand chemistry it is essential to understand its models and how they are used. We will use the concept of bonding to reemphasize the important characteristics of models including their origin, structure, and uses.

Bonding is a model proposed to explain molecular stability.

Models originate from our observations of the properties of nature. For example, the concept of bonds arose from the observations that most chemical processes involve collections of atoms and that chemical reactions involve rearrangements of the ways the atoms are grouped. So to understand reactions we must understand the forces that bind atoms together.

The reason nature seeks the lowest energy state is discussed in Chapter 16.

In natural processes there is a tendency toward lower energy. Collections of atoms therefore occur because the aggregated state has lower energy than the separated atoms. Why? As we have seen earlier in this chapter, the best explanation for

Chemical Impact

Testing for Toxicity by Computer

When a new compound that will come in contact with humans is synthesized, the compound's safety must be tested. Currently this is done with expensive bacterial or animal tests. However, computers are now being introduced to screen new compounds to detect those that are likely to be toxic in some way.

The fundamental idea is that the toxicity of a compound is due to some part or parts of the molecular structure. That is, certain types of molecular fragments seem to always lead to a particular kind of toxicity, such as carcinogenicity (ability to cause cancer). The new procedure involves entering data on known structure/toxicity relationships into the computer database. When a new compound is prepared, its structure is compared by the computer to this database. If the new molecule contains a fragment that is known to cause a toxic response, this information will alert the researcher to the possible problems. The hope is that this approach will prevent many compounds from being tested only to be rejected after much time and money already has been spent. This should make products both safer and less expensive.

the energy change involves either atoms sharing electrons or atoms transferring electrons to become ions. In the case of electron sharing, we find it convenient to assume that individual bonds occur between pairs of atoms. Let's explore the validity of this assumption and see how it is useful.

In a diatomic molecule such as H_2, it is natural to assume that a bond exists between the atoms that holds them together. It is also useful to assume that individual bonds are present in polyatomic molecules such as CH_4. So instead of thinking of CH_4 as a unit with a stabilization energy of 1652 kJ per mole, we choose to think of CH_4 as containing four C—H bonds, each worth 413 kJ of energy per mole of bonds. Without this concept of individual bonds in molecules, chemistry would be hopelessly complicated. There are millions of different chemical compounds, and if each of these compounds had to be considered as an entirely new entity, the task of understanding chemical behavior would be overwhelming.

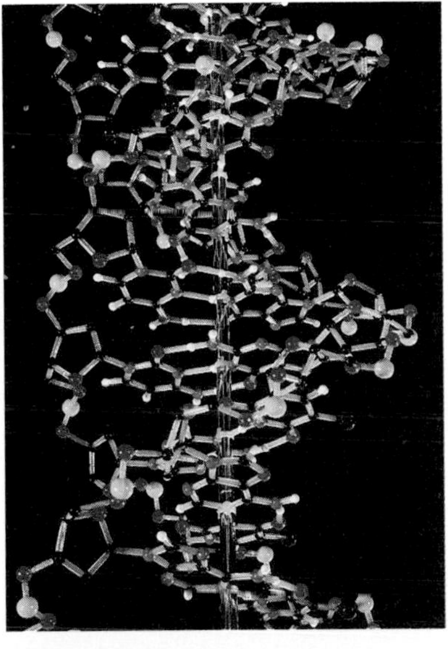

The concept of individual bonds makes it much easier to deal with complex molecules such as DNA. A small segment of a DNA molecule is shown here using ball and stick models.

The bonding model provides a framework to systematize chemical behavior by enabling us to think of molecules as collections of common fundamental components. For example, a typical biomolecule, such as a protein, contains hundreds of atoms and might seem discouragingly complex. However, if we think of a protein as being constructed with individual bonds: C—C, C—H, C—N, C—O, N—H, and so on, it helps tremendously in predicting and understanding the protein's behavior. The essential idea is that we expect a given bond to behave about the same in any molecular environment. Used in this way, the model of the chemical bond has helped chemists to systematize the reactions of the millions of existing compounds.

In addition to being very useful, the bonding model is also physically sensible. It makes sense that atoms can form stable groups by sharing electrons. Shared electrons give a lower energy state because they are simultaneously attracted by two nuclei.

Also, as we will see in the next section, bond energy data support the existence of discrete bonds that are relatively independent of the molecular environment. It is very important to remember, however, that the chemical bond is only a model. While our concept of discrete bonds in molecules agrees with many of our observations, some molecular properties require that we think of a molecule as a whole, with the electrons free to move through the entire molecule. This is called *delocalization* of the electrons, a concept that will be discussed more completely in the next chapter.

Models Have Several Fundamental Properties:

Understand both the power and limitations of chemical models.

- Models are human inventions, always based on an incomplete understanding of how nature works. *A model does not equal reality.*
- Models are often wrong. This property derives from the first property. Models are based on speculation and are always oversimplifications.
- Models tend to become more complicated as they age. As flaws are discovered in our models, we patch them by adding more assumptions.
- It is very important to understand the assumptions inherent in a particular model before you use it to interpret observations or to make predictions. Simple models usually involve very restrictive assumptions and can only be expected to yield qualitative information. Asking for a sophisticated explanation from a simple model is like expecting to get an accurate mass for a diamond using a bathroom scale.

  If a model is to be used effectively, we must understand its strengths and weaknesses and ask only appropriate questions. An illustration of this point is the simple Aufbau principle used to explain the electron configurations of the elements. Although this model correctly predicts the configuration for most atoms, chromium and copper do not agree with the predictions. Detailed studies show that the configurations of chromium and copper result from complex electron interactions that are not taken into account in the model. But this does not mean that we should discard the simple model that is so useful for most atoms. Instead, we should apply it with caution and not expect it to be correct in every case.
- When a model is wrong, we often learn much more than when it is right. If a model makes a wrong prediction, it usually means we do not understand some fundamental characteristics of nature. We often learn by making mistakes (try to remember that when you get back your next chemistry test).

8.8 Covalent Bond Energies and Chemical Reactions

Purpose

- To describe the relationship of bond energy to bond multiplicity.
- To show how bond energies can be used to calculate heats of reaction.

In this section we will consider the energies associated with various types of bonds and see how the bonding concept is useful in dealing with the energies of chemical reactions. One important consideration is to establish the sensitivity of a particular type of bond to its molecular environment. For example, consider the stepwise decomposition of methane shown below:

Process	*Energy required (kJ/mol)*
$CH_4(g) \rightarrow CH_3(g) + H(g)$	435
$CH_3(g) \rightarrow CH_2(g) + H(g)$	453
$CH_2(g) \rightarrow CH(g) + H(g)$	425
$CH(g) \rightarrow C(g) + H(g)$	339
	Total = 1652

$$\text{Average} = \frac{1652}{4} = 413$$

Although a C—H bond is being broken in each case, the energy required varies in a nonsystematic way. This shows that the C—H bond is somewhat sensitive to its environment. We use the *average* of these individual bond dissociation energies even though this quantity only approximates the energy associated with a C—H bond in a particular molecule. Similar measurements on other types of bonds have provided the values for bond energies listed in Table 8.4.

Average Bond Energies (kJ/mol)

Single bonds						Multiple Bonds	
H—H	432	N—H	391	I—I	149	C=C	614
H—F	565	N—N	160.	I—Cl	208	C≡C	839
H—Cl	427	N—F	272	I—Br	175	O=O	495
H—Br	363	N—Cl	200.			C=O	799
H—I	295	N—Br	243	S—H	347	C≡O	1072
		N—O	201	S—F	327	N=O	607
C—H	413	O—H	467	S—Cl	253	N=N	418
C—C	347	O—O	146	S—Br	218	N≡N	941
C—N	305	O—F	190.	S—S	266	C≡N	891
C—O	358	O—Cl	203			C=N	615
C—F	485	O—I	234	Si—Si	226		
C—Cl	339			Si—H	323		
C—Br	276	F—F	154	Si—C	301		
C—I	240.	F—Cl	253	Si—O	368		
C—S	259	F—Br	237				
		Cl—Cl	239				
		Cl—Br	218				
		Br—Br	193				

Table 8.4

So far, we have discussed bonds where one pair of electrons is shared. This type of bond is called a **single bond.** As we will see in more detail later, atoms sometimes share two pairs of electrons, forming a **double bond,** or share three pairs of electrons, forming a **triple bond.** The bond energies for these *multiple bonds* are also given in Table 8.4.

Bond Lengths for Selected Bonds			
Bond	Bond type	Bond length (Å)	Bond energy (kJ/mol)
C—C	single	1.54	347
C=C	double	1.34	614
C≡C	triple	1.20	839
C—O	single	1.43	358
C=O	double	1.23	799
C—N	single	1.43	305
C=N	double	1.38	615
C≡N	triple	1.16	891

Table 8.5

There is also a relationship between the number of shared electron pairs and the bond length. As the number of shared electrons increases, the bond length shortens. This relationship is shown for selected bonds in Table 8.5.

Bond Energy and Enthalpy

Enthalpy changes for chemical reactions were discussed in Chapter 6.

Bond energy values can be used to calculate approximate energies for reactions. To illustrate how this is done, we will calculate the change in energy that accompanies the following reaction:

$$H_2(g) + F_2(g) \rightarrow 2HF(g)$$

This reaction involves breaking one H—H and one F—F bond and forming two H—F bonds. To break bonds energy must be *added* to the system, which is an endothermic process. This means that the energy terms associated with bond breaking will have *positive* signs. The formation of a bond *releases* energy, an exothermic process, and the energy terms associated with bond making will carry a *negative* sign. We can write the enthalpy change for the reaction as follows:

ΔH = sum of the energies required to break old bonds (positive signs) plus the energies released from the formation of new bonds (negative signs)

This leads to the expression

$$\Delta H = \underbrace{\Sigma D \text{ (bonds broken)}}_{\text{Energy required}} - \underbrace{\Sigma D \text{ (bonds formed)}}_{\text{Energy released}}$$

where Σ represents the sum of terms and D represents the bond energy per mole of bonds. (D *always* has a positive sign.)

In the case of the formation of HF:

$$\Delta H = D_{\text{H—H}} + D_{\text{F—F}} - 2D_{\text{H—F}}$$

$$= 1 \text{ mol} \times \frac{432 \text{ kJ}}{\text{mol}} + 1 \text{ mol} \times \frac{154 \text{ kJ}}{\text{mol}} - 2 \text{ mol} \times \frac{565 \text{ kJ}}{\text{mol}} = -544 \text{ kJ}$$

Thus when 1 mole of $H_2(g)$ and 1 mole of $F_2(g)$ react to form 2 moles of $HF(g)$, 544 kJ of energy should be released.

Sample Exercise 8.5

Using the bond energies listed in Table 8.4, calculate ΔH for the reaction of methane with chlorine and fluorine to give freon-12, CF_2Cl_2.

$$CH_4(g) + 2Cl_2(g) + 2F_2(g) \rightarrow CF_2Cl_2(g) + 2HF(g) + 2HCl(g)$$

Solution

The idea here is to break the bonds in the reactants to give individual atoms and then assemble these atoms into the products by forming new bonds:

$$\text{Reactants} \xrightarrow[\text{required}]{\text{Energy}} \text{atoms} \xrightarrow[\text{released}]{\text{Energy}} \text{products}$$

We then combine the energy changes to calculate ΔH:

$$\Delta H = \text{energy required to break bonds} - \text{energy released when bonds form}$$

where the minus sign gives the correct sign to the energy terms for the exothermic processes.

Reactant bonds broken:

CH_4	4 mol C—H	$4\text{ mol} \times \frac{413\text{ kJ}}{\text{mol}} = 1652\text{ kJ}$
$2Cl_2$	2 mol Cl—Cl	$2\text{ mol} \times \frac{239\text{ kJ}}{\text{mol}} = 478\text{ kJ}$
$2F_2$	2 mol F—F	$2\text{ mol} \times \frac{154\text{ kJ}}{\text{mol}} = 308\text{ kJ}$
		Total energy required = 2438 kJ

Product bonds formed:

CF_2Cl_2	2 mol C—F	$2\text{ mol} \times \frac{485\text{ kJ}}{\text{mol}} = 970\text{ kJ}$
	and	
	2 mol C—Cl	$2\text{ mol} \times \frac{339\text{ kJ}}{\text{mol}} = 678\text{ kJ}$
HF	2 mol H—F	$2\text{ mol} \times \frac{565\text{ kJ}}{\text{mol}} = 1130\text{ kJ}$
HCl	2 mol H—Cl	$2\text{ mol} \times \frac{427\text{ kJ}}{\text{mol}} = 854\text{ kJ}$
		Total energy released = 3632 kJ

We now can calculate ΔH:

$$\begin{aligned}\Delta H &= \text{energy required to break bonds} - \text{energy released when bonds form}\\ &= 2438\text{ kJ} - 3632\text{ kJ}\\ &= -1194\text{ kJ}\end{aligned}$$

Since the sign of the value for the enthalpy change is negative, this means that 1194 kJ of energy is released per mole of CF_2Cl_2 formed.

8.9 The Localized Electron Bonding Model

Purpose

- To introduce the localized electron model.

So far we have discussed the general characteristics of the chemical bonding model and have seen that properties such as bond strength and polarity can be assigned to individual bonds. In this section we introduce a specific model used to describe covalent bonds. We need a simple model that can be easily applied even to very complicated molecules and that can be used routinely by chemists to interpret and organize the wide variety of chemical phenomena. The model that serves this purpose is the **localized electron (LE) model,** which assumes that *a molecule is composed of atoms that are bound together by sharing pairs of electrons using the atomic orbitals of the bound atoms*. Electron pairs in the molecule are assumed to be localized on a particular atom or in the space between two atoms. Those pairs of electrons localized on an atom are called **lone pairs,** and those found in the space between the atoms are called **bonding pairs.**

As we will apply it, the LE model has three parts:

1. Description of the valence electron arrangement in the molecule using Lewis structures, discussed in the next section.
2. Prediction of the geometry of the molecule, using the valence shell electron pair repulsion (VSEPR) model, discussed in Section 8.13.
3. Description of the types of atomic orbitals used by the atoms to share electrons or hold lone pairs. We will discuss orbitals in Chapter 9.

8.10 Lewis Structures

Purpose

- To show how to write Lewis structures.

The **Lewis structure** of a molecule shows how the valence electrons are arranged among the atoms in the molecule. These representations are named after G. N. Lewis (Fig. 8.14). The rules for writing Lewis structures are based on observations of thousands of molecules. From experiment, chemists have learned that the *most important requirement for the formation of a stable compound is that the atoms achieve noble gas electron configurations*.

We have already seen that when metals and nonmetals react to form binary ionic compounds, electrons are transferred and the resulting ions typically have noble gas electron configurations. An example is the formation of KBr where the K^+ ion has the [Ar] electron configuration and the Br^- ion has the [Kr] electron configuration. In writing Lewis structures, the rule is that *only the valence electrons are included*. Using dots to represent electrons, the Lewis structure for KBr is

Lewis structures show only valence electrons.

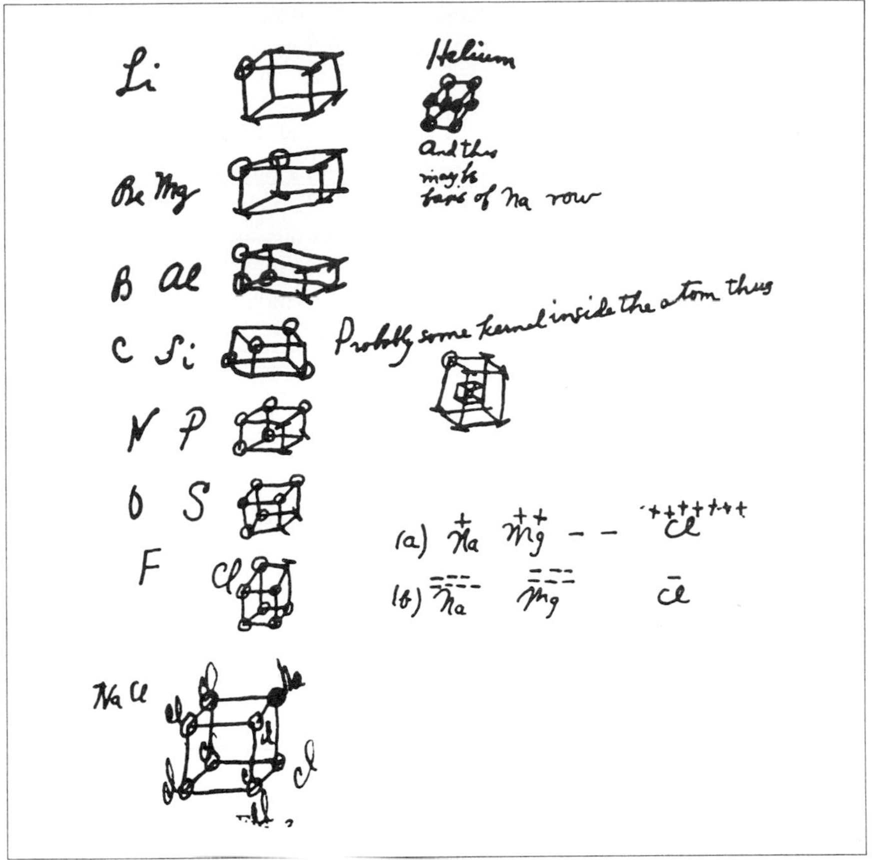

Figure 8.14

G. N. Lewis conceived the octet rule while lecturing to a class of general chemistry students in 1902. This is his original sketch. (G. N. Lewis, *Valence,* Dover Publication, Inc., New York, 1966).

No dots are shown on the K^+ ion since it has no valence electrons. The Br^- ion is shown with eight electrons since it has a filled valence shell.

Next we will consider Lewis structures for molecules with covalent bonds, involving elements in the first and second periods. The principle of achieving a noble gas electron configuration applies to these elements as follows:

Hydrogen forms stable molecules where it shares two electrons. That is, it follows a **duet rule.** For example, when two hydrogen atoms, each with one electron, combine to form the H_2 molecule, we have

By sharing electrons, each hydrogen in H_2, in effect, has two electrons; that is, each hydrogen has a filled valence shell.

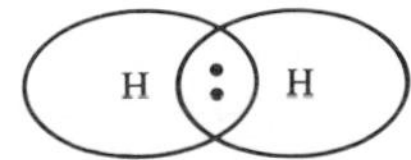

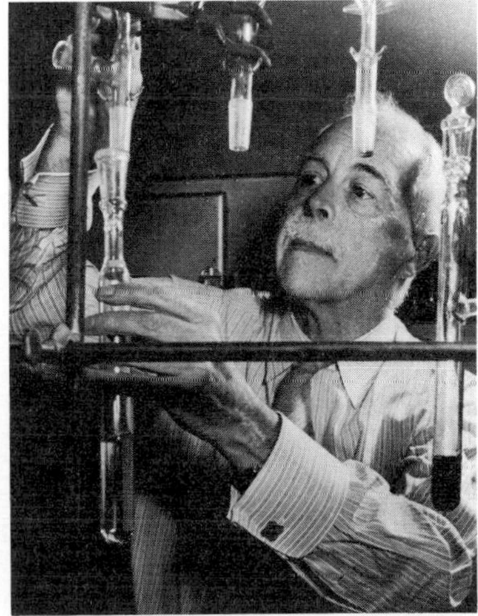

G. N. Lewis.

Helium does not form bonds because its valence orbitals are already filled; it is a noble gas. Helium has the electron configuration $1s^2$ and can be represented by the Lewis structure

$$\text{He}:$$

The second row nonmetals carbon through fluorine form stable molecules when they are surrounded by enough electrons to fill the valence orbitals, that is, the $2s$ and the three $2p$ orbitals. Since eight electrons are required to fill these orbitals, these elements typically obey the **octet rule;** they are surrounded by eight electrons. An example is the F_2 molecule, which has the following Lewis structure:

Carbon, nitrogen, oxygen, and fluorine always obey the octet rule in stable molecules.

$$:\ddot{\underset{\cdot\cdot}{F}}\cdot \longrightarrow :\ddot{\underset{\cdot\cdot}{F}}:\ddot{\underset{\cdot\cdot}{F}}: \longleftarrow \cdot\ddot{\underset{\cdot\cdot}{F}}:$$

F atom with seven valence electrons | F_2 molecule | F atom with seven valence electrons

Note that each fluorine atom in F_2 is, in effect, surrounded by eight electrons, two of which are shared with the other atom. This is a *bonding pair* of electrons, as discussed earlier. Each fluorine atom also has three pairs of electrons not involved in bonding. These are the *lone pairs*.

Neon does not form bonds since it already has an octet of valence electrons (it is a noble gas). The Lewis structure is

$$:\ddot{\underset{\cdot\cdot}{\text{Ne}}}:$$

Note that only the valence electrons of the neon atom ($2s^22p^6$) are represented by the Lewis structure. The $1s^2$ electrons are core electrons and take no part in chemical reactions.

From the discussion above we can formulate the following rules for writing Lewis structures of molecules containing atoms from the first two periods.

Rules for Writing Lewis Structures

- Sum the valence electrons from all the atoms. Do not worry about keeping track of which electrons come from which atoms. It is the *total* number of electrons that is important.
- Use a pair of electrons to form a bond between each pair of bound atoms.
- Arrange the remaining electrons to satisfy the duet rule for hydrogen and the octet rule for the second-row elements.

To see how these rules are applied, we will draw the Lewis structures of a few molecules. We will first consider the water molecule and follow the rules above.

STEP 1

We sum the *valence* electrons for H_2O as shown:

$$\underset{\text{H}}{1} + \underset{\text{H}}{1} + \underset{\text{O}}{6} = 8 \text{ valence electrons}$$

STEP 2

Using a pair of electrons per bond, we draw in the two O—H single bonds:

H—O—H

Note that *a line instead of a pair of dots is used to indicate each pair of bonding electrons*. This is the standard notation.

STEP 3

We distribute the remaining electrons around the atoms to achieve a noble gas electron configuration for each atom. Since four electrons have been used in forming the two bonds, four electrons (8 − 4) remain to be distributed. Hydrogen is satisfied with two electrons (duet rule), but oxygen needs eight electrons to have a noble gas configuration. Thus the remaining four electrons are added to oxygen as two lone pairs. Dots are used to represent the lone pairs:

H—Ö—H Lone pairs (with lone pairs above and below O)

H—O—H represents H : O : H

This is the correct Lewis structure for the water molecule. Each hydrogen has two electrons and the oxygen has eight as shown below.

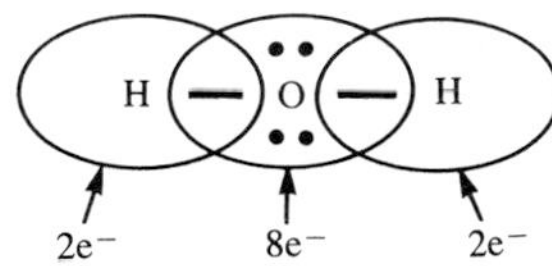

As a second example, let's write the Lewis structure for carbon dioxide. Summing the valence electrons gives

$$\underset{C}{4} + \underset{O}{6} + \underset{O}{6} = 16$$

After forming a bond between the carbon and each oxygen,

O—C—O

the remaining electrons are distributed to achieve noble gas configurations on each atom. In this case we have twelve electrons (16 − 4) remaining after the bonds are drawn. The distribution of these electrons is determined by a trial-and-error process. We have six pairs of electrons to distribute. Suppose we try three pairs on each oxygen to give

:Ö—C—Ö:

Is this correct? To answer this we need to check two things:

1. The total number of electrons. There are sixteen valence electrons in this structure, which is the correct number.
2. The octet rule for each atom. Each oxygen has eight electrons, but the carbon only has four. This cannot be the correct Lewis structure.

How can we arrange the sixteen available electrons to achieve an octet for each atom? Suppose there are two shared pairs between the carbon and each oxygen:

O=C=O represents O::C::O

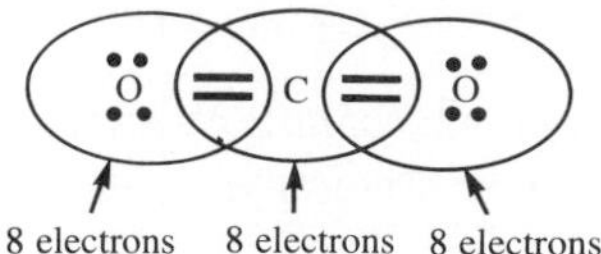

8 electrons 8 electrons 8 electrons

Now each atom is surrounded by eight electrons, and the total number of electrons is sixteen, as required. This is the correct Lewis structure for carbon dioxide, which has two double bonds.

Finally, let's consider the Lewis structure of the CN^- (cyanide) ion. Summing the valence electrons, we have

$$CN^-$$
$$4 + 5 + 1 = 10$$

Note that the negative charge means an extra electron must be added. After drawing a single bond (C—N), we distribute the remaining electrons to achieve a noble gas configuration for each atom. Eight electrons remain to be distributed. We can try various possibilities, for example:

C—N (with lone pairs)

This structure is incorrect, because C and N have only six electrons each instead of eight. The correct arrangement is

$$[:C\equiv N:]^-$$

(Satisfy yourself that both carbon and nitrogen have eight electrons.)

Sample Exercise 8.6

Give the Lewis structure for each of the following:

a. HF **b.** N_2 **c.** NH_3 **d.** CH_4 **e.** CF_4 **f.** NO^+

Solution

In each case we apply the three rules for writing Lewis structures. Recall that lines are used to indicate shared electron pairs and that dots are used to indicate nonbonding pairs (lone pairs). We have the following tabulated results:

	Total valence electrons	Draw single bonds	Calculate number of electrons remaining	Use remaining electrons to achieve noble gas configurations	Check number of electrons
a. HF	$1 + 7 = 8$	H—F	6	H—F: (three lone pairs on F)	H, 2 F, 8
b. N_2	$5 + 5 = 10$	N—N	8	:N≡N:	N, 8

Sample Exercise 8.6, continued

	Total valence electrons	Draw single bonds	Calculate number of electrons remaining	Use remaining electrons to achieve noble gas configurations	Check number of electrons
c. NH_3	5 + 3(1) = 8	H—N—H with H below N	2	H—N̈—H with H below N	H, 2 N, 8
d. CH_4	4 + 4(1) = 8	H—C—H with H above and below C	0	H—C—H with H above and below C	H, 2 C, 8
e. CF_4	4 + 4(7) = 32	F—C—F with F above and below C	24	:F̤̈—C—F̤̈: with :F̈: above and :F̤: below C	F, 8 C, 8
f. NO^+	5 + 6 − 1 = 10	N—O	8	[:N≡O:]$^+$	N, 8 O, 8

When writing Lewis structures, don't worry about which electrons come from which atoms in a molecule. The best way to look at a molecule is to regard it as a new entity that uses all of the available valence electrons of the atoms to achieve the lowest possible energy.* The valence electrons belong to the molecule, rather than to the individual atoms. Simply distribute all valence electrons so that the various rules are satisfied, without regard to the origin of each particular electron.

8.11 Exceptions to the Octet Rule

Purpose

- To show how to write Lewis structures for certain special cases.

The localized electron model is a simple but very successful model, and the rules we have used for Lewis structures apply to most molecules. However, with such a simple model, some exceptions are inevitable. Boron, for example, tends to form compounds where the boron atom has fewer than eight electrons around it—it

*In a sense this approach corrects for the fact that the localized electron model overemphasizes that a molecule is simply a sum of its parts, that is, that the atoms retain their individual identities in the molecule.

does not have a complete octet. Boron trifluoride (BF_3), a gas at normal temperatures and pressures, reacts very energetically with molecules such as water and ammonia that have available electron pairs (lone pairs). The violent reactivity of BF_3 with electron-rich molecules arises because the boron atom is electron-deficient. Boron trifluoride has 24 valence electrons. The Lewis structure that seems most consistent with the properties of BF_3 is

```
    ..
   :F:
    |
    B
   / \
 :F:  :F:
  ..   ..
```

Note that in this structure boron has only six electrons around it. The octet rule for boron can be satisfied by drawing a structure with a double bond, such as

```
   :F:
    ||
    B
   / \
 :F:  :F:
  ..   ..
```

However, since fluorine is much more electronegative than boron, this does not make sense. In fact, experiments indicate that each B—F bond is a single bond in accordance with the first Lewis structure. This structure is also consistent with the reactivity of BF_3 toward electron-rich molecules, for example, toward NH_3 to form H_3NBF_3:

```
                ..                          ..
 H             F:            H             F:
   \          / ..             \          / ..
            ..                                  ..
 H — N: + B — F:   ⟶       H — N — B — F:
            ..                                  ..
   /          \ ..             /          \ ..
 H             F:            H             F:
               ..                          ..
```

In this stable compound boron has an octet of electrons.

It is characteristic of boron to form molecules where the boron atom is electron-deficient. On the other hand, carbon, nitrogen, oxygen, and fluorine can be counted on to obey the octet rule.

Some atoms exceed the octet rule. This behavior is observed only for those elements in Period 3 of the periodic table and beyond. To see how this arises, we will consider the Lewis structure for sulfur hexafluoride (SF_6). The sum of the valence electrons is

$$6 + 6(7) = 48 \text{ electrons}$$

Indicating the single bonds gives the structure on the left below:

```
     F                    ..
  F  |  F             .. :F: ..
   \ | /             :F   |   F:
     S               ..\  |  / ..
   / | \                  S
  F  |  F            .. / | \ ..
     F               :F   |   F:
                      .. :F: ..
                          ..
```

We have used 12 electrons to form the S—F bonds, which leaves 36 electrons. Since fluorine always follows the octet rule, we complete the six fluorine octets to give the structure on the right above. This structure uses all 48 valence electrons for SF_6, but sulfur has 12 electrons around it; that is, sulfur *exceeds* the octet rule. How can this happen?

To answer this question we need to consider the different types of valence orbitals characteristic of second- and third-period elements. The second-row elements have 2*s* and 2*p* valence orbitals, and the third-row elements have 3*s*, 3*p*, and 3*d* orbitals. The 3*s* and 3*p* orbitals fill with electrons in going from sodium to argon, but the 3*d* orbitals remain empty. For example, the valence orbital diagram for a sulfur atom is

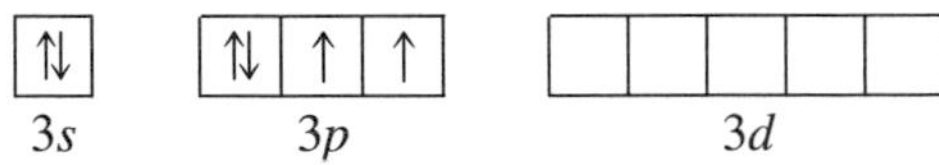

The localized electron model assumes that the empty 3*d* orbitals can be used to accommodate extra electrons. Thus the sulfur atom in SF_6 can have 12 electrons around it by using the 3*s* and 3*p* orbitals to hold 8 electrons with the extra 4 electrons placed in the formerly empty 3*d* orbitals.

Third-row elements can exceed the octet rule.

Lewis Structures: Comments About the Octet Rule

- The second-row elements C, N, O, and F should always be assumed to obey the octet rule.
- The second-row elements B and Be often have fewer than eight electrons around them in their compounds. These electron deficient compounds are very reactive.
- The second-row elements never exceed the octet rule, since their valence orbitals (2*s* and 2*p*) can accommodate only eight electrons.
- Third-row and heavier elements often satisfy the octet rule but can exceed the octet rule by using their empty valence *d* orbitals.
- When writing the Lewis structure for a molecule, satisfy the octet rule for the atoms first. If electrons remain after the octet rule has been satisfied, then place them on the elements having available *d* orbitals (elements in the third period or beyond).

Sample Exercise 8.7

Write the Lewis structure for PCl_5.

Solution

We can follow the same stepwise procedure we used for sulfur hexafluoride on page 354.

STEP 1

Sum the valence electrons.

$$\underset{\text{P}}{\underset{\uparrow}{5}} + 5(\underset{\text{Cl}}{\underset{\uparrow}{7}}) = 40 \text{ electrons}$$

Sample Exercise 8.7, continued

STEP 2

Indicate single bonds between bound atoms.

```
    Cl  Cl
    | /
Cl—P
    | \
    Cl  Cl
```

STEP 3

Distribute the remaining electrons. In this case, 30 electrons (40 − 10) remain. These are used to satisfy the octet rule for each chlorine atom. The final Lewis structure is

```
      ..
     :Cl: ..
       |  /Cl:
 ..    | /  ..
:Cl—P
 ..    | \  ..
       |  \Cl:
     :Cl: ..
      ..
```

Note that phosphorus, which is a third-row element, has exceeded the octet rule by two electrons.

In the PCl_5 and SF_6 molecules, the central atoms (P and S, respectively) must have the extra electrons. However, in molecules having more than one atom that can exceed the octet rule, it is not always clear which atom should have the extra electrons. Consider the Lewis structure for the triiodide ion (I_3^-), which has

$$\underset{\substack{\uparrow \\ \text{I}}}{3(7)} + \underset{\substack{\uparrow \\ 1-\text{ charge}}}{1} = 22 \text{ valence electrons}$$

Indicating the single bonds gives I—I—I. At this point 18 electrons (22 − 4) remain. Trial and error will convince you that one of the iodine atoms must exceed the octet rule, but *which* one?

The rule we will follow is that *when it is necessary to exceed the octet rule for one of several third-row (or higher) elements, assume that the extra electrons should be placed on the central atom.*

Thus for I_3^- the Lewis structure is

```
 ..  ...  ..   -
[:I—I—I:]
 ..  ..   ..
```

where the central iodine exceeds the octet rule. This structure agrees with known properties of I_3^-.

Sample Exercise 8.8

Write the Lewis structure for each molecule or ion:

a. ClF_3 **b.** XeO_3 **c.** $RnCl_2$ **d.** $BeCl_2$ **e.** ICl_4^-

Sample Exercise 8.8, continued

Solution

a. The chlorine atom (third row) accepts the extra electrons.

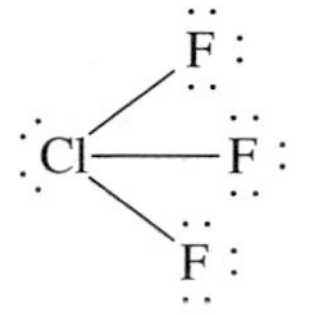

b. All atoms obey the octet rule.

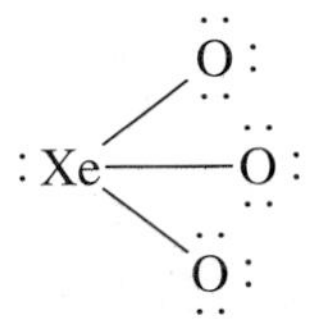

c. Radon, a noble gas in Period 6, accepts the extra electrons.

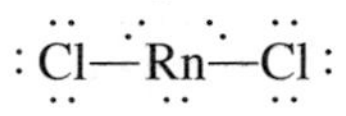

d. Beryllium is electron deficient.

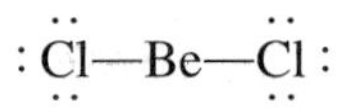

e. Iodine exceeds the octet rule.

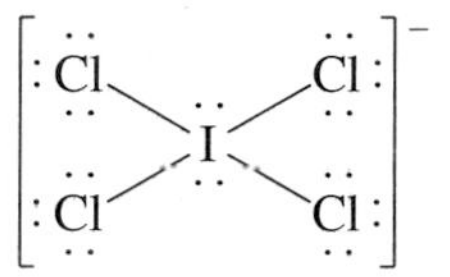

8.12 Resonance

Purpose

- To illustrate the concept of resonance.
- To show how to write resonance structures.

Sometimes more than one valid Lewis structure (one that obeys the rules we have outlined) is possible for a given molecule. Consider the Lewis structure for the nitrate ion (NO_3^-), which has 24 valence electrons. To achieve an octet of electrons around each atom, a structure like this is required:

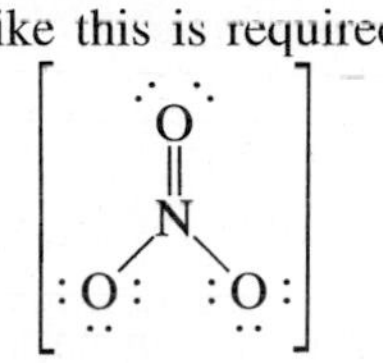

If this structure accurately represents the bonding in NO_3^-, there should be two types of N---O bonds observed in the molecule: one shorter bond (the double bond) and two identical longer ones (the two single bonds). However, experiments clearly show that NO_3^- exhibits only *one* type of N---O bond with a length and strength between those expected for a single bond and a double bond. Thus, although the structure we have shown above is a valid Lewis structure, it does *not* correctly represent the bonding in NO_3^-. This is a serious problem, and it means that the model must be modified.

Look again at the proposed Lewis structure for NO_3^-. There is no reason for choosing a particular oxygen atom to have the double bond. There are really three valid Lewis structures:

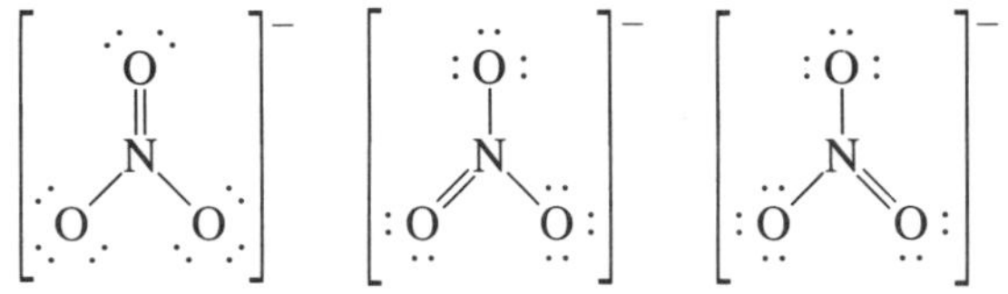

Is any of these structures a correct description of the bonding in NO_3^-? No, because NO_3^- does not have one double and two single bonds—it has three equivalent bonds. We can solve this problem by making the following assumption: the correct description of NO_3^- is *not given by any one* of the three Lewis structures, but is given only by the *superposition of all three.*

The nitrate ion does not exist as any of the three extreme structures but exists as an average of all three. **Resonance** *occurs when more than one valid Lewis structure can be written for a particular molecule.* The resulting electron structure of the molecule is given by the average of these **resonance structures.** This situation is usually represented by double-headed arrows as follows:

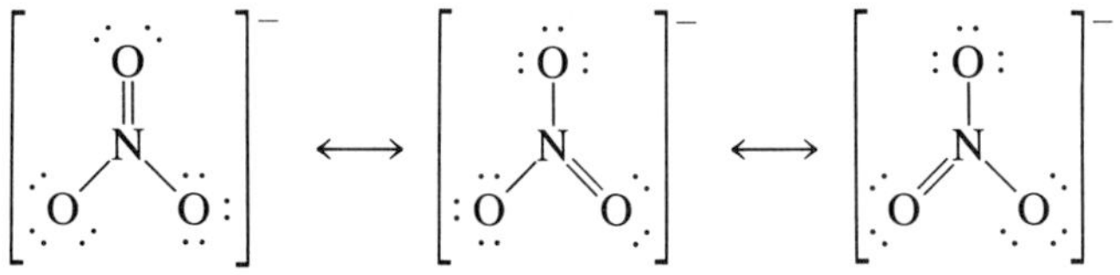

Note that in all of these resonance structures the arrangement of the nuclei is the same. Only the placement of the electrons differs. The arrows do not indicate that the molecule ''flips'' from one resonance structure to another. They simply show that the *actual structure is an average of the three resonance structures.*

The concept of resonance is necessary because the localized electron model postulates that electrons are localized between a given pair of atoms. However, nature doesn't really operate this way. Electrons are really delocalized—they can move around the entire molecule. The valence electrons in the NO_3^- molecule distribute themselves to provide equivalent N---O bonds. Resonance is necessary to compensate for the defective assumption of the localized electron model. However, this model is so useful that we retain the concept of localized electrons and add resonance to allow the model to treat species like NO_3^-.

Sample Exercise 8.9

Describe the electron arrangement in the nitrite anion (NO_2^-) using the localized electron model.

Sample Exercise 8.9, continued

Solution

We will follow the usual procedure for obtaining the Lewis structure for the NO_2^- ion.

In NO_2^- there are 5 + 2(6) + 1 = 18 valence electrons.

Indicating the single bonds gives the structure

$$O—N—O$$

The remaining 14 electrons (18 − 4) can be distributed to produce these structures:

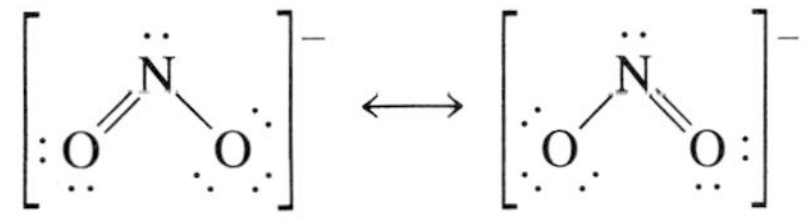

$$\left[:\ddot{O}{=}\ddot{N}{-}\ddot{\underset{..}{O}}:\right]^- \longleftrightarrow \left[:\ddot{\underset{..}{O}}{-}\ddot{N}{=}\ddot{O}:\right]^-$$

This is a resonance situation. Two equivalent Lewis structures can be drawn. *The electronic structure of the molecule is not correctly represented by either resonance structure but by the average of the two.* There are two equivalent N---O bonds, each one intermediate between a single and double bond.

Odd-Electron Molecules

Relatively few molecules formed from nonmetals contain odd numbers of electrons. One common example is nitric oxide (NO) which is formed when nitrogen and oxygen gases react at the high temperatures in automobile engines. Nitric oxide is emitted into the air, where it immediately reacts with oxygen to form gaseous nitrogen dioxide (NO_2), another odd-electron molecule.

Since the localized electron model is based on pairs of electrons, it does not handle odd-electron cases in a natural way. To treat odd-electron molecules, a more sophisticated model is needed.

Formal Charge

Molecules or polyatomic ions containing atoms that can exceed the octet rule often have many nonequivalent Lewis structures, all of which obey the rules for writing Lewis structures. For example, as we will see in detail below, the sulfate ion has a Lewis structure with all single bonds and several Lewis structures that contain double bonds. How do we decide which of the many possible Lewis structures best describes the actual bonding in sulfate? One method is to estimate the charge on each atom in the various possible Lewis structures and use these charges to select the most appropriate structure(s). We will see below how this is done, but first we must decide on a method to evaluate atomic charges in molecules.

In Chapter 4 we discussed one system for obtaining charges, called oxidation states. However, in assigning oxidation states we always count *both* of the shared electrons as belonging to the more electronegative atom in a bond. This practice leads to highly exaggerated estimates of charge. In other words, although oxidation states are useful for bookkeeping electrons in redox reactions, they are not realistic estimates of the actual charges on individual atoms in a molecule, and so they are not suitable for judging the appropriateness of Lewis structures. Another definition

Equivalent Lewis structures contain the same numbers of single and multiple bonds. For example, the resonance structures for O_3

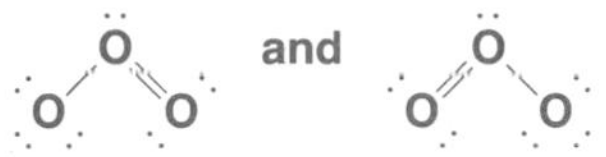

$$:\ddot{\underset{..}{O}}{-}\ddot{O}{=}\ddot{O} \quad \text{and} \quad \ddot{O}{=}\ddot{O}{-}\ddot{\underset{..}{O}}:$$

are equivalent Lewis structures. These are equally important in describing the bonding in O_3. Nonequivalent Lewis structures contain different numbers of single and multiple bonds.

of the charge on an atom in a molecule, the **formal charge,** can, however, be used to evaluate Lewis structures.

The concept of formal charge is based on knowing two things about a given atom:

1. the number of valence electrons on the free neutral atom (which has zero charge because the number of electrons equals the number of protons),
2. the number of valence electrons "belonging" to the atom in a molecule.

We then compare these numbers. If in the molecule the atom has the same number of valence electrons as it does in the free state, the positive and negative charges just balance, and it has a formal charge of zero. If the atom has one more valence electron in a molecule than it has as a free atom, it has a formal charge of -1 and so on. Thus, the formal charge on an atom in a molecule is defined as

$$\begin{aligned}\text{Formal charge} = &\ (\text{number of valence electrons on free atom}) \\ &- (\text{number of valence electrons assigned in the molecule})\end{aligned}$$

To compute the formal charge of an atom in a molecule we assign the valence electrons in the molecule to the various atoms, making the following assumptions:

1. Lone pair electrons belong entirely to the atom in question.
2. Shared electrons are *divided equally* between the two sharing atoms.

Thus the number of valence electrons assigned to a given atom is calculated as follows:

$$\begin{aligned}(\text{Valence electrons})_{\text{assigned}} = &\ (\text{number of lone pair electrons}) \\ &+ \tfrac{1}{2}\,(\text{number of shared electrons})\end{aligned}$$

We will illustrate the procedures for calculating formal charges by considering two of the possible Lewis structures for the sulfate ion, which has 32 valence electrons. For the Lewis structure

$$\left[\begin{array}{c} :\ddot{\text{O}}: \\ | \\ :\ddot{\text{O}}-\text{S}-\ddot{\text{O}}: \\ | \\ :\ddot{\text{O}}: \end{array}\right]^{2-}$$

each oxygen atom has 6 lone pair electrons and shares 2 electrons with the sulfur atom. Thus, using the above assumptions, each oxygen is assigned 7 valence electrons:

$$\text{Valence electrons assigned to each oxygen} = \underset{\substack{\uparrow \\ \text{Lone} \\ \text{pair} \\ \text{electrons}}}{6} \text{ plus } \underset{\substack{\uparrow \\ \text{Shared} \\ \text{electrons}}}{\tfrac{1}{2}(2)} = 7$$

$$\text{Formal charge on oxygen} = \underset{\substack{\uparrow \\ \text{Valence electrons} \\ \text{on a free O atom}}}{6} \text{ minus } \underset{\substack{\uparrow \\ \text{Valence electrons} \\ \text{assigned to each O} \\ \text{in } SO_4^{2-}}}{7} = -1$$

The formal charge on each oxygen is -1.

For the sulfur atom there are no lone pair electrons and eight electrons are shared with the oxygen atoms. Thus for sulfur:

$$\text{Valence electrons assigned to sulfur} = \underset{\substack{\uparrow \\ \text{Lone pair electrons}}}{0} \text{ plus } \underset{\substack{\uparrow \\ \text{Shared electrons}}}{\tfrac{1}{2}(8)} = 4$$

$$\text{Formal charge on sulfur} = \underset{\substack{\uparrow \\ \text{Valence electrons on free S atom}}}{6} \text{ minus } \underset{\substack{\uparrow \\ \text{Valence electrons assigned to S in } SO_4^{2-}}}{4} = 2$$

A second possible Lewis structure is

$$\left[\begin{array}{ccccc} & & :\ddot{O}: & & \\ & & \| & & \\ :\ddot{O}- & & S & & -\ddot{O}: \\ & & \| & & \\ & & :\ddot{O}: & & \end{array}\right]^{2-}$$

In this case the formal charges are:

For oxygen atoms with single bonds:

Valence electrons assigned $= 6 + \frac{1}{2}(2) = 7$

Formal charge $= 6 - 7 = -1$

For oxygen atoms with double bonds:

$$\text{Valence electrons assigned} = 4 + \underset{\substack{\uparrow \\ \text{Each double bond has 4 electrons.}}}{\tfrac{1}{2}(4)} = 6$$

Formal charge $= 6 - 6 = 0$

For the sulfur atom:

Valence electrons assigned $= 0 + \frac{1}{2}(12) = 6$

Formal charge $= 6 - 6 = 0$

We will use two fundamental assumptions about formal charges to evaluate Lewis structures:

1. Atoms in molecules try to achieve formal charges as close to zero as possible.
2. Any negative formal charges are expected to reside on the most electronegative atoms.

We can use these principles to evaluate the two Lewis structures for sulfate given previously. Notice that in the structure with only single bonds each oxygen has a formal charge of -1 while the sulfur has a formal charge of $+2$. In contrast, in the structure with two double bonds and two single bonds, the sulfur and two oxygen atoms have a formal charge of 0 while two oxygens have a formal charge of -1.

Based on the assumptions given above, the structure with two double bonds is preferred—it has lower formal charges and the −1 formal charges are on electronegative oxygen atoms. Thus, for the sulfate ion we might expect resonance structures such as

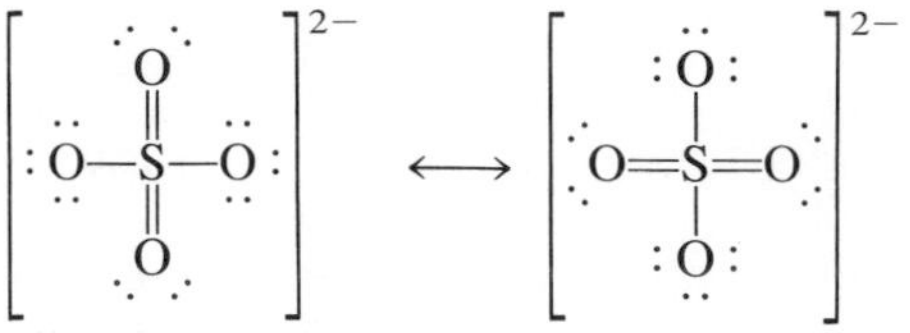

[sulfate resonance structures: $[O{=}S(O^-)_2{=}O]^{2-}$ with S=O double bonds to top and bottom O, single bonds to left and right O ⟷ S=O double bonds to left and right O, single bonds to top and bottom O]$^{2-}$

to more closely describe the bonding than the Lewis structure with only single bonds.

Rules Governing Formal Charge

- To calculate the formal charge on an atom:
  1. Take the sum of the lone pair electrons and one-half of the shared electrons. This is the number of valence electrons assigned to the atom in the molecule.
  2. Subtract the number of assigned electrons from the number of valence electrons on the free, neutral atom to obtain the formal charge.
- The sum of the formal charges of all atoms in a given molecule or ion must equal the overall charge on that species.
- If nonequivalent Lewis structures exist for a species, those with formal charges closest to zero and with any negative formal charges on the most electronegative atoms are considered to best describe the bonding in the molecule or ion.

Sample Exercise 8.10

Give possible Lewis structures for XeO_3, an explosive compound of xenon. Which Lewis structure or structures are most appropriate according to the formal charges?

Solution

For XeO_3 (26 valence electrons) we can draw the following possible Lewis structures (formal charges are indicated in parentheses):

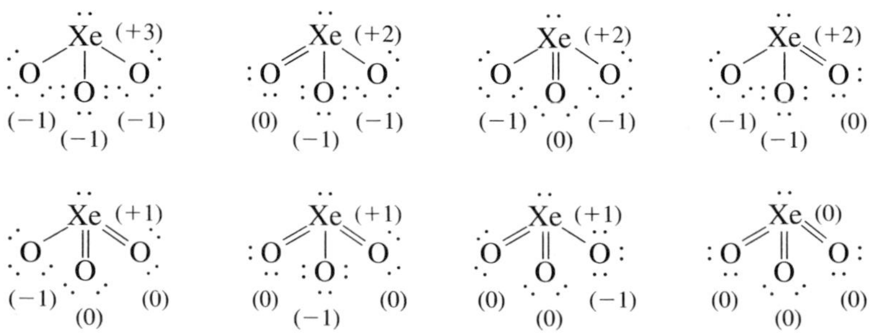

Xe (+3); O (−1), O (−1), O (−1)

Xe (+2); O (0), O (−1), O (−1)

Xe (+2); O (−1), O (0), O (−1)

Xe (+2); O (−1), O (−1), O (0)

Xe (+1); O (−1), O (0), O (0)

Xe (+1); O (0), O (−1), O (0)

Xe (+1); O (0), O (0), O (−1)

Xe (0); O (0), O (0), O (0)

Sample Exercise 8.10, continued

Based on the ideas of formal charge, we would predict that the Lewis structures with the lower values of formal charge would be most appropriate for describing the bonding in XeO_3.

As a final note there are a couple of cautions about formal charge to keep in mind. First, although formal charges are closer to actual atomic charges in molecules than are oxidation states, formal charges are still only estimates of charge—they should not be taken as actual atomic charges. Second, the evaluation of Lewis structures using formal charge ideas can lead to erroneous predictions. Tests based on experiments must be used to make the final decisions on the correct description of bonding in a molecule or polyatomic ion.

8.13 Molecular Structure: The VSEPR Model

Purpose

- To describe how molecular geometry can be predicted from the number of electron pairs.

The structures of molecules play a very important role in determining their chemical properties. As we will see later, this is particularly important for biological molecules; a slight change in the structure of a large biomolecule can completely destroy its usefulness to a cell or may even change the cell from a normal one to a cancerous one.

Many accurate methods now exist for determining **molecular structure,** the three-dimensional arrangement of atoms in a molecule. These methods must be used if precise information about structure is required. However, it is often useful to be able to predict the approximate molecular structure of a molecule. In this section we consider a simple model that allows us to do this. This model, called the **valence shell electron pair repulsion (VSEPR) model,** is useful in predicting the geometries of molecules formed from nonmetals. The main postulate of this model is that *the structure around a given atom is determined principally by minimizing electron-pair repulsions*. The idea here is that the bonding and nonbonding pairs around a given atom will be positioned as far apart as possible. To see how this model works, we will first consider the molecule $BeCl_2$, which has the Lewis structure

$$:\ddot{\underset{..}{Cl}}—Be—\ddot{\underset{..}{Cl}}:$$

$BeCl_2$ has only four electrons around Be and is expected to be very reactive with electron-pair donors.

Note that there are two pairs of electrons around the beryllium atom. What arrangement of these electron pairs allows them to be as far apart as possible to minimize the repulsions? Clearly, the best arrangement places the pairs on opposite sides of the beryllium atom at 180° from each other:

$$—Be—$$
$$180°$$

This is the maximum possible separation for two electron pairs. Once we have determined the optimum arrangement of the electron pairs around the central atom, we can specify the molecular structure of $BeCl_2$, that is, the positions of the atoms. Since each electron pair on beryllium is shared with a chlorine atom, the molecule has a **linear structure** with a 180° bond angle:

Cl—Be—Cl
180°

Next, let's consider BF_3, which has the Lewis structure

:F:
|
:F—B—F:

Here the boron atom is surrounded by three pairs of electrons. What arrangement will minimize the repulsions? The electron pairs are farthest apart at angles of 120°:

B 120°

Since each of the electron pairs is shared with a fluorine atom, the molecular structure will be

F
|
B 120°
F F

This is a planar (flat) and triangular molecule, which is commonly described as a **trigonal planar structure.**

Next, let's consider the methane molecule, which has the Lewis structure

H
|
H—C—H
|
H

There are four pairs of electrons around the central carbon atom. What arrangement of these electron pairs best minimizes the repulsions? First, let's try a square planar arrangement:

—C— 90°

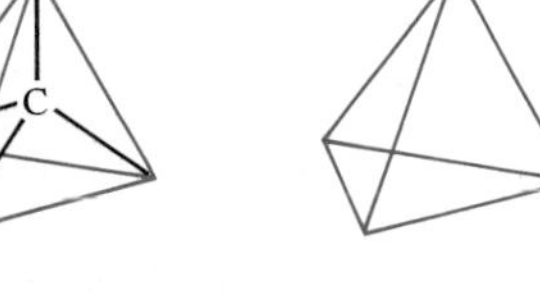

The carbon atom and the electron pairs are centered in the plane of the paper, and the angles between the pairs are all 90°.

Is there another arrangement with angles greater than 90° that would put the electron pairs even further away from each other? The answer is yes. The **tetrahedral arrangement** has angles of approximately 109.5°:

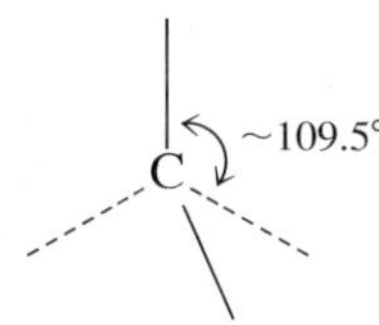

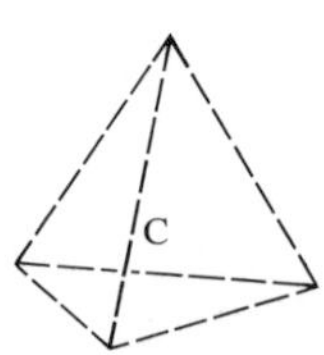

It can be shown that this is the maximum possible separation of four pairs around a given atom. This means that *whenever four pairs of electrons are present around an atom, they should always be arranged tetrahedrally.*

Now that we have the electron-pair arrangement that gives the least repulsion, we can determine the positions of the atoms and thus the molecular structure of CH_4. In methane each of the four electron pairs is shared between the carbon atom and a hydrogen atom. Thus the hydrogen atoms are placed as in Fig. 8.15, and the molecule has a tetrahedral structure with the carbon atom at the center.

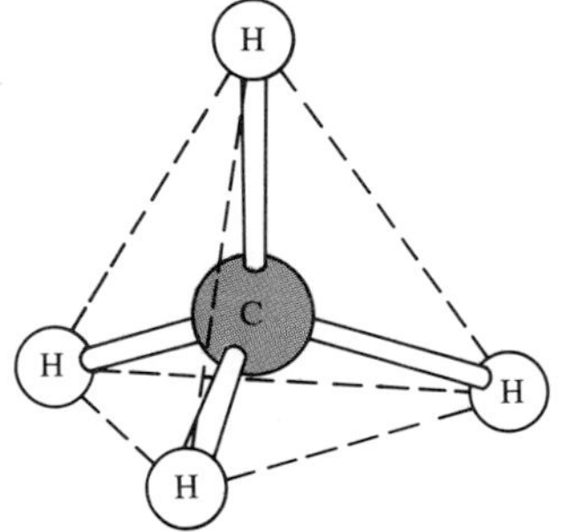

Figure 8.15

The molecular structure of methane. The tetrahedral arrangement of electron pairs produces a tetrahedral arrangement of hydrogen atoms.

Recall that the main idea of the VSEPR model is to find the arrangement of electron pairs around the central atom that minimizes the repulsions. Then we can determine the molecular structure from knowing how the electron pairs are shared with the peripheral atoms. Use the following steps to predict the structure of a molecule using the VSEPR model.

- Draw the Lewis structure for the molecule.
- Count the electron pairs and arrange them in the way that minimizes repulsion (that is, put the pairs as far apart as possible).
- The positions of the atoms are determined from the way the electron pairs are shared.
- The name of the molecular structure is determined from the positions of the atoms.

We will predict the structure of ammonia (NH_3) using this stepwise approach.

STEP 1

Draw the Lewis structure:

```
      ..
    H—N—H
      |
      H
```

STEP 2

Count the pairs of electrons and arrange them to minimize repulsions. The NH_3 molecule has four pairs of electrons: three bonding pairs, and one nonbonding pair. From the discussion of the methane molecule, we know that the best arrangement of four electron pairs is a tetrahedral array as shown in Fig. 8.16(a) on the following page.

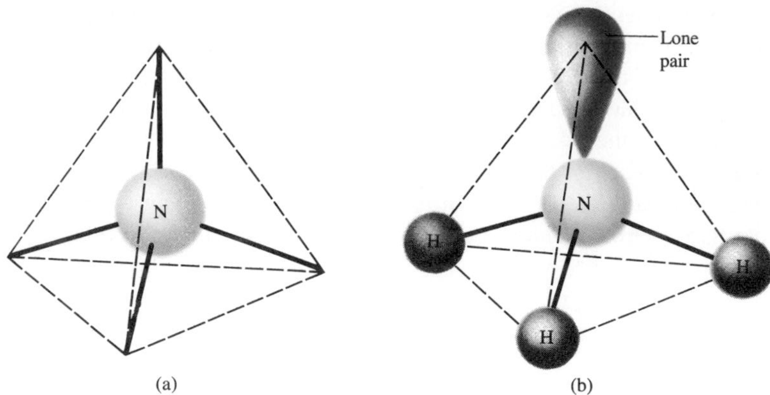

Figure 8.16

(a) The tetrahedral arrangement of electron pairs around the nitrogen atom in the ammonia molecule. (b) Three of the electron pairs around nitrogen are shared with hydrogen atoms as shown and one is a lone pair. Although the arrangement of *electron pairs* is tetrahedral, as in the methane molecule, the hydrogen atoms in the ammonia molecule occupy only three corners of the tetrahedron. A lone pair occupies the fourth corner.

STEP 3

Determine the positions of the atoms. The three H atoms share electron pairs as shown in Fig. 8.16(b).

STEP 4

Name the molecular structure. It is very important to recognize that the *name* of the molecular structure is always based on the *positions of the atoms*. The placement of the electron pairs determines the structure, but the name is based on the positions of the atoms. Thus it is incorrect to say that the NH_3 molecule is tetrahedral. It has a tetrahedral arrangement of electron pairs but not a tetrahedral arrangement of atoms. The molecular structure of ammonia is a **trigonal pyramid** (one side is different from the other three) rather than a tetrahedron.

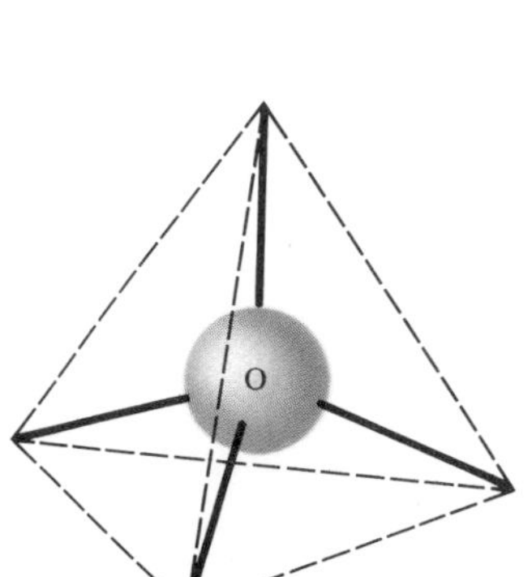

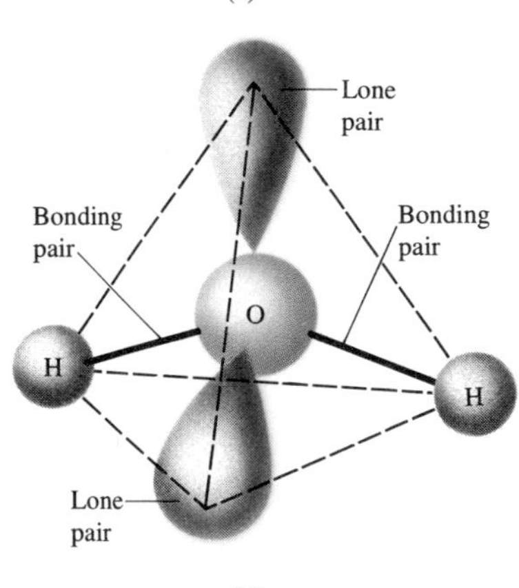

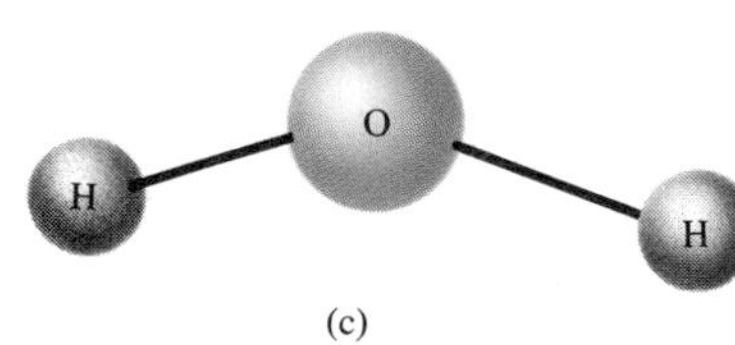

Figure 8.17

(a) The tetrahedral arrangement of the four electron pairs around oxygen in the water molecule. (b) Two of the electron pairs are shared between oxygen and the hydrogen atoms and two are lone pairs. (c) The V-shaped molecular structure of the water molecule.

Sample Exercise 8.11

Describe the molecular structure of the water molecule.

Solution

The Lewis structure for water is

$$\mathrm{H{-}\ddot{\underset{\cdot\cdot}{O}}{-}H}$$

There are four pairs of electrons: two bonding pairs, and two nonbonding pairs. To minimize repulsions, these are best arranged in a tetrahedral array as shown in Fig. 8.17(a). Although H_2O has a tetrahedral arrangement of electron pairs, it is not a tetrahedral molecule. The atoms in the H_2O molecule form a V shape as shown in Fig. 8.17(b) and (c).

From Sample Exercise 8.11 we see that the H_2O molecule is V-shaped, or bent, because of the presence of the lone pairs. If no lone pairs were present, the molecule would be linear, and the polar bonds would cancel and the molecule would have no dipole moment. This would make water very different from the polar substance so familiar to us.

From the previous discussion we would predict that the H—X—H bond angle (where X is the central atom) in CH_4, NH_3, and H_2O should be the tetrahedral angle of 109.5°. Experimental studies, however, show that the actual bond angles are those given in Fig. 8.18. What significance do these results have for the VSEPR model? One possible point of view is that we should be pleased to have the observed angles so close to the tetrahedral angle. The opposite view is that the deviations are significant enough to require modification of the simple model so that it can more accurately handle similar cases. We will take the latter view.

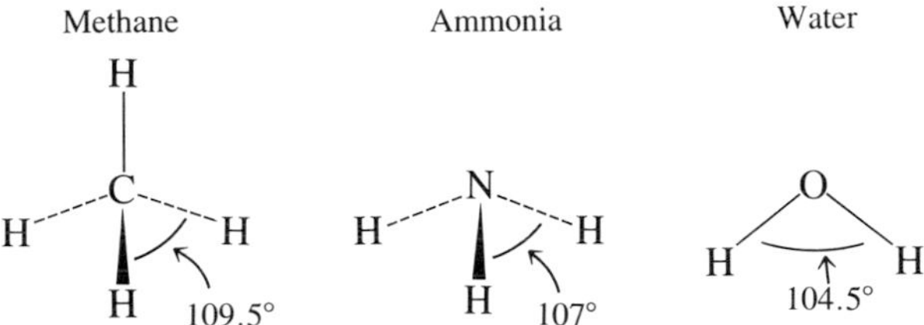

Figure 8.18

The bond angles in the CH_4, NH_3, and H_2O molecules. Note that the bond angle between bonding pairs decreases as the number of lone pairs increases.

Let us examine the following data:

	CH_4	NH_3	H_2O
Number of lone pairs	0	1	2
Bond angle	109.5°	107°	104.5°

One interpretation of the trend observed here is that lone pairs require more space than bonding pairs; in other words, as the number of lone pairs increases, the bonding pairs are increasingly squeezed together.

This interpretation seems to make physical sense if we think in the following terms. A bonding pair is shared between two nuclei and the electrons can be close to either nucleus. They are relatively confined between the two nuclei. A lone pair is localized on only one nucleus and both electrons will be close only to that nucleus, as shown schematically in Fig. 8.19. These pictures help us understand why a lone pair may require more space near an atom than a bonding pair.

As a result of these observations, we make the following addition to the original postulate of the VSEPR model: *lone pairs require more room than bonding pairs and tend to compress the angles between the bonding pairs.*

So far we have considered cases with two, three, and four electron pairs around the central atom. These are summarized in Table 8.6. For five pairs of electrons, there are several possible choices. The one that produces minimum repulsion is a **trigonal bipyramid.** Note from Table 8.6 that this arrangement has two different angles, 90° and 120°. As the name suggests, the structure formed by this arrangement of pairs consists of two trigonal-based pyramids that share a common base. Six pairs of electrons can best be arranged around a given atom with 90° angles to form an **octahedral structure,** as shown in Table 8.6 on the following page.

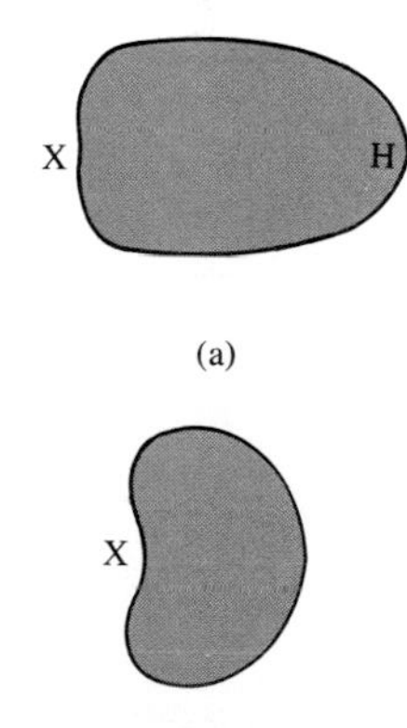

Figure 8.19

(a) In a bonding pair of electrons, the electrons are shared by two nuclei. (b) In a lone pair, both electrons must be close to a single nucleus and tend to take up more of the space around that atom.

Arrangements of Electron Pairs Around an Atom Yielding Minimum Repulsion

Number of electron pairs		Arrangement of electron pairs
2	Linear	:—A—:
3	Trigonal planar	
4	Tetrahedral	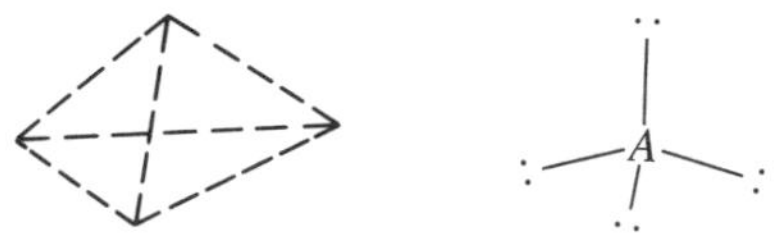
5	Trigonal bipyramidal	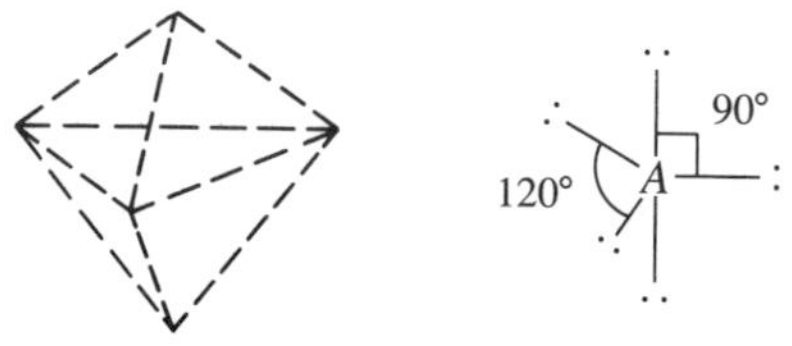 90°, 120°
6	Octahedral	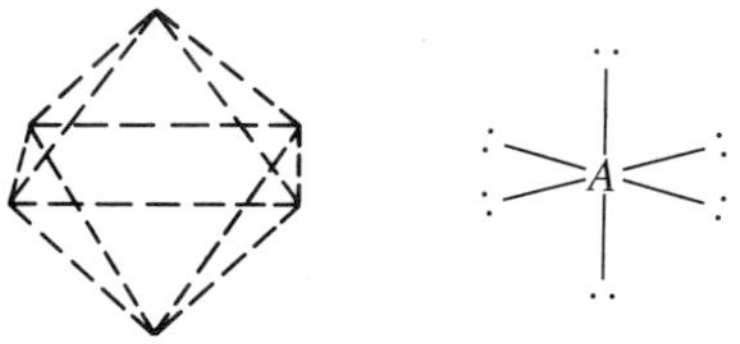

Table 8.6

In order to use the VSEPR model to determine the geometric structures of molecules, you should memorize the relationships between the number of electron pairs and their best arrangement.

Sample Exercise 8.12

When phosphorus reacts with excess chlorine gas, the compound phosphorus pentachloride (PCl_5) is formed. In the gaseous and liquid states, this substance consists of PCl_5 molecules, but in the solid state it consists of a 1:1 mixture of PCl_4^+ and PCl_6^- ions. Predict the geometric structures of PCl_5, PCl_4^+, and PCl_6^-.

Solution

Lewis structure for PCl_5

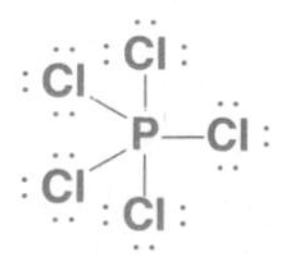

The Lewis structure for PCl_5 is shown in the margin. Five pairs of electrons around the phosphorus atom require a trigonal bipyramidal arrangement (see Table 8.6 above). When the chlorine atoms are included, a trigonal bipyramidal molecule results, as shown below.

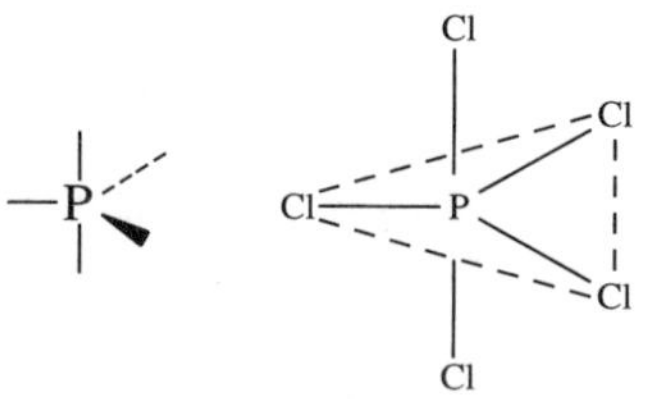

Sample Exercise 8.12, continued

The Lewis structure for the PCl_4^+ ion [$5 + 4(7) - 1 = 32$ valence electrons] is shown in the margin. There are four pairs of electrons surrounding the phosphorus atom in the PCl_4^+ ion, which requires a tetrahedral arrangement of the pairs, as shown in the figure in the margin. Since each pair is shared with a chlorine atom, a tetrahedral PCl_4^+ cation results.

Lewis structure for PCl_4^+

```
       ..
      :Cl:
   ..  |  ..
  :Cl—P—Cl:
   ..  |  ..
      :Cl:
       ..
```

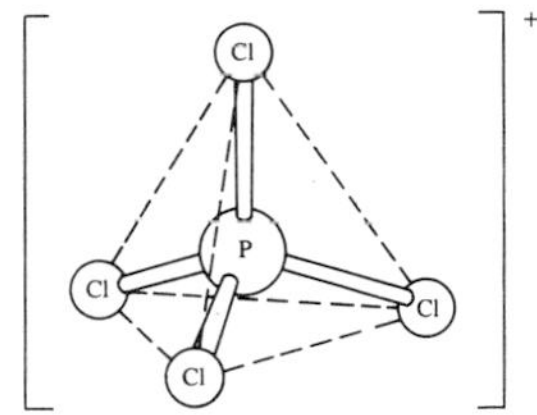

Tetrahedral PCl_4^+ cation.

The Lewis structure for PCl_6^- [$5 + 6(7) + 1 = 48$ valence electrons] is

```
 ⌈ ..  :Cl:  .. ⌉ −
 | :Cl\  |  /Cl: |
 |  ..  \ | /  .. |
 |  ..    P    .. |
 | :Cl/  |  \Cl: |
 ⌊ ..  :Cl:  .. ⌋
        ..
```

Since phosphorus is surrounded by six pairs of electrons, an octahedral arrangement is required to minimize repulsions, as shown below on the left. Since each electron pair is shared with a chlorine atom, an octahedral PCl_6^- anion is predicted.

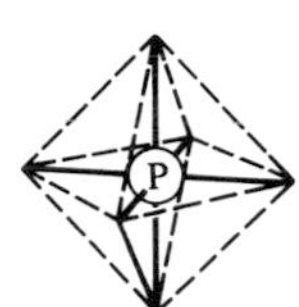

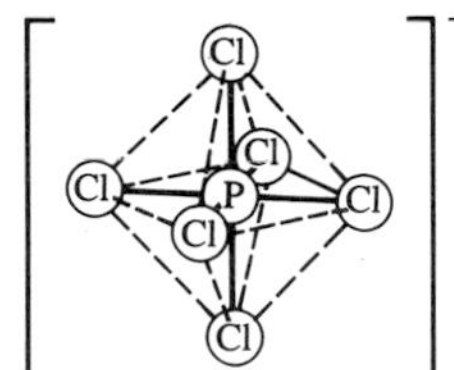

Sample Exercise 8.13

Because the noble gases have filled *s* and *p* valence orbitals, they were not expected to be chemically reactive. In fact, for many years these elements were called *inert gases* because of this supposed inability to form any compounds. However, in the early 1960s several compounds of krypton, xenon, and radon were synthesized. For example, a team at the Argonne National Laboratory produced the stable colorless compound xenon tetrafluoride (XeF_4). Predict its structure and whether it has a dipole moment.

Solution

The Lewis structure for XeF_4 is

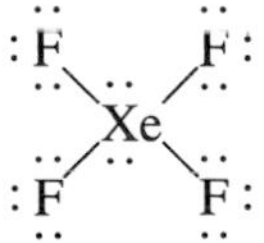

The xenon atom in this molecule is surrounded by six pairs of electrons, which means an octahedral arrangement:

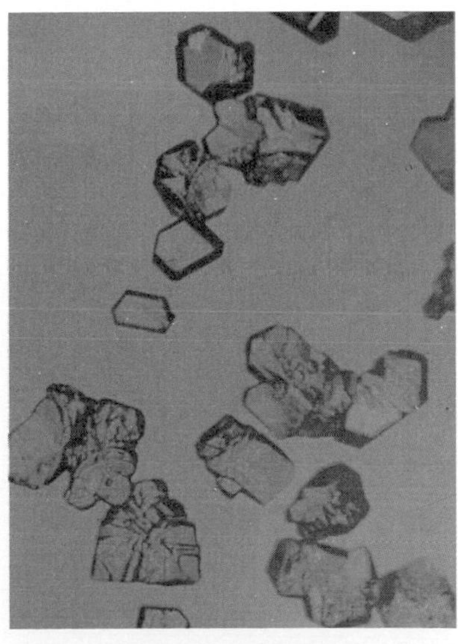

Xenon tetrafluoride crystals.

Sample Exercise 8.13, continued

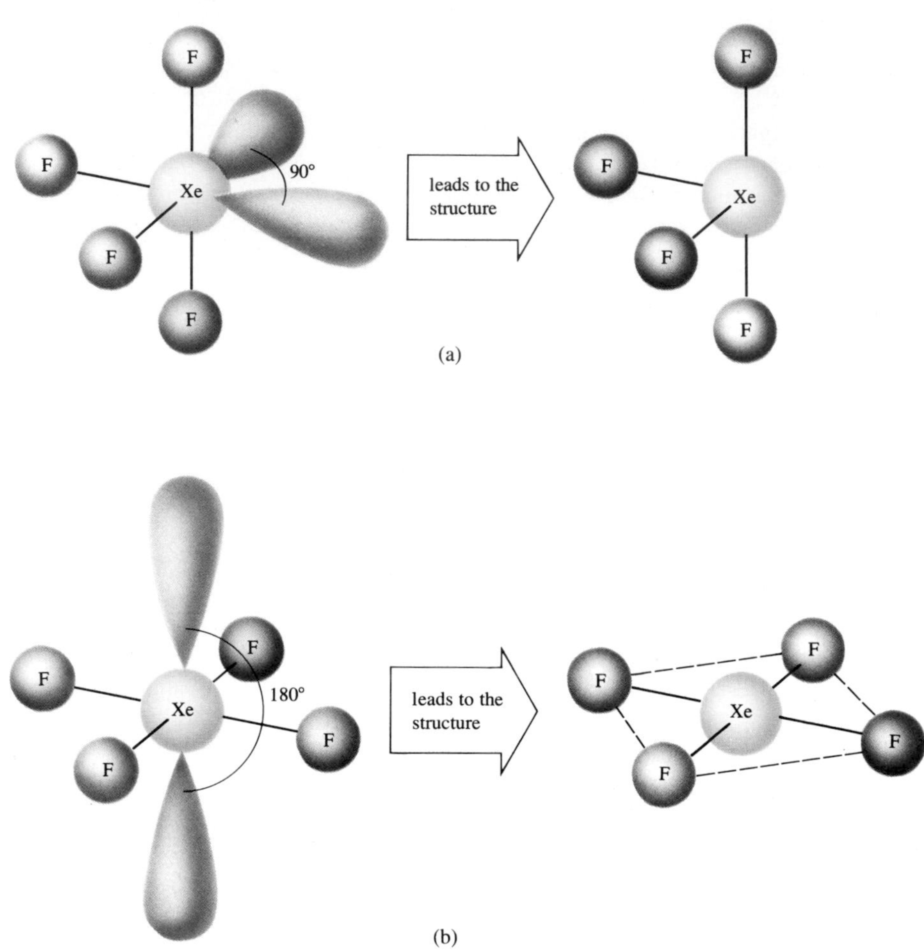

Figure 8.20

Possible electron pair arrangements for XeF_4. Since arrangement (a) has lone pairs at 90° from each other, it is less favorable than arrangement (b), where the lone pairs are at 180°.

The structure predicted for this molecule will depend on how the lone pairs and bonding pairs are arranged. Consider the two possibilities shown in Fig. 8.20. The bonding pairs are indicated by the presence of the fluorine atoms. Since the structure predicted differs in the two cases, we must decide which of these arrangements is preferable. The key is to look at the lone pairs. In the structure in part (a) the lone pair–lone pair angle is 90°; in the structure in part (b) the lone pairs are separated by 180°. Since lone pairs require more room than bonding pairs, a structure with two lone pairs at 90° is unfavorable. Thus the arrangement in Fig. 8.20(b) is preferred, and the molecular structure is predicted to be square planar. Note that this molecule is *not* described as being octahedral. There is an *octahedral arrangement of electron pairs,* but the *atoms* form a **square planar** structure.

Although each Xe—F bond is polar (fluorine has a greater electronegativity than xenon), the square planar arrangement of these bonds causes the polarities to cancel.

Sample Exercise 8.13, continued

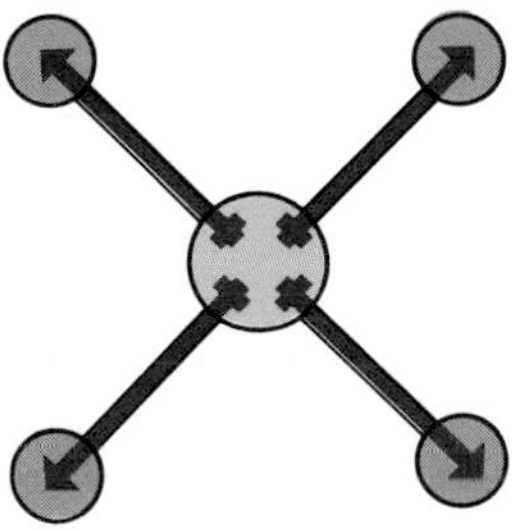

Thus XeF_4 has no dipole moment.

We can further illustrate the use of the VSEPR model for molecules or ions with lone pairs by considering the triiodide ion (I_3^-). The Lewis structure for I_3^- is

$$[:\ddot{\underset{..}{I}}-\underset{..}{\ddot{I}}-\ddot{\underset{..}{I}}:]^-$$

The central iodine atom has five pairs around it, which requires a trigonal bipyramidal arrangement. Several possible arrangements of the lone pairs are shown in Fig. 8.21. Note that structures (a) and (b) have lone pairs at 90°, whereas in (c) all lone pairs are at 120°. Thus structure (c) is preferred. The resulting molecular structure for I_3^- is linear.

$$[I—I—I]^-$$

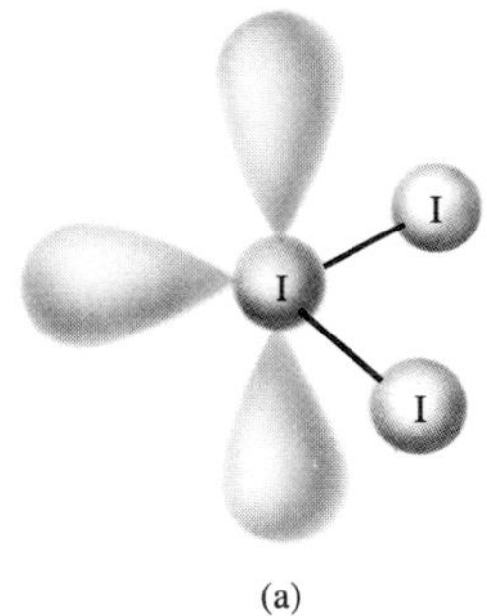

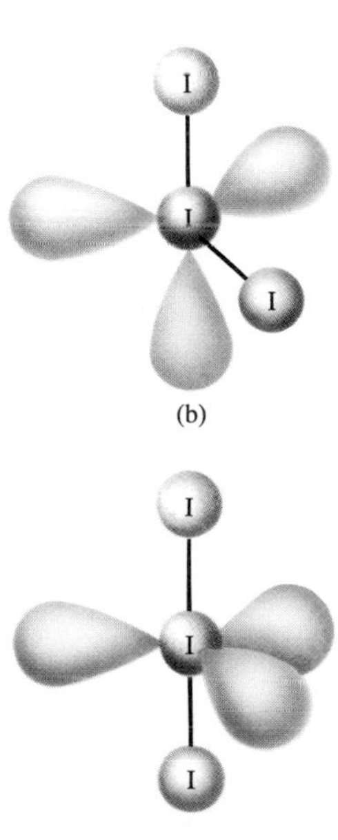

Figure 8.21

Three possible arrangements of the electron pairs in the I_3^- ion. Arrangement (c) is preferred because there are no 90° lone pair–lone pair interactions.

The VSEPR Model and Multiple Bonds

So far in our treatment of the VSEPR model, we have not considered any molecules with multiple bonds. To see how these molecules are handled by this model, let's consider the NO_3^- ion, which requires three resonance structures to describe its electronic structure:

$$\left[\begin{array}{c} \ddot{O} \\ \| \\ N \\ :\ddot{O}:\ \ :\ddot{O}: \end{array}\right]^- \longleftrightarrow \left[\begin{array}{c} :\ddot{O}: \\ | \\ N \\ \ddot{O}\ \ :\ddot{O}: \end{array}\right]^- \longleftrightarrow \left[\begin{array}{c} :\ddot{O}: \\ | \\ N \\ :\ddot{O}:\ \ \ddot{O} \end{array}\right]^-$$

The NO_3^- ion is known to be planar with 120° bond angles.

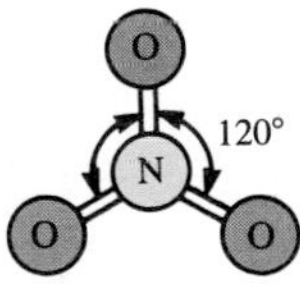

This planar structure is the one expected for three pairs of electrons around a central atom, which means that *a double bond should be counted as one effective pair* in using the VSEPR model. This makes sense because the two pairs of electrons involved in the double bond are *not* independent pairs. Both of the electron pairs must be in the space between the nuclei of the two atoms in order to form the double bond. In other words, the double bond acts as one center of electron density to repel the other pairs of electrons. The same holds true for triple bonds. This leads us to another general rule: *for the VSEPR model multiple bonds count as one effective electron pair.*

The molecular structure of nitrate also shows us one more important point: *when a molecule exhibits resonance, any one of the resonance structures can be used to predict the molecular structure using the VSEPR model.* These rules are illustrated in Sample Exercise 8.14.

Sample Exercise 8.14

Predict the molecular structure of the sulfur dioxide molecule. Is this molecule expected to have a dipole moment?

Solution

First, we must determine the Lewis structure for the SO_2 molecule, which has 18 valence electrons. The expected resonance structures are

$$:\ddot{O}=\ddot{S}-\ddot{O}: \longleftrightarrow :\ddot{O}-\ddot{S}=\ddot{O}:$$

To determine the molecular structure, we must count the electron pairs around the sulfur atom. In each resonance structure the sulfur has one lone pair, one pair in a single bond, and one double bond. Counting the double bond as one pair yields three effective pairs around the sulfur. According to Table 8.6, a trigonal planar arrangement is required, which yields a V-shaped molecule:

120°
O S O

Thus the structure of the SO_2 molecule is expected to be V-shaped, with a 120° bond angle. The molecule has a dipole moment directed as shown.

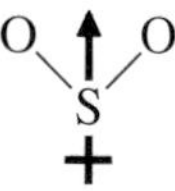

Since the molecule is V-shaped, the polar bonds do not cancel.

It should be noted at this point that lone pairs that are oriented at least 120° from other pairs do not produce significant distortions of bond angles. For example, the angle in the SO_2 molecule is actually quite close to 120°. We will follow the general principle that *a 120° angle provides lone pairs with enough space so that distortions do not occur. Angles less than 120° are distorted when lone pairs are present.*

Molecules Containing No Single Central Atom

So far we have considered molecules consisting of one central atom surrounded by other atoms. The VSEPR model can be readily extended to more complicated molecules, such as methanol (CH_3OH). This molecule is represented by the following Lewis structure:

```
     H
     |   ..
 H—C—O—H
     |   ..
     H
```

The molecular structure can be predicted from the arrangement of pairs around the carbon and oxygen atoms. Note that there are four pairs of electrons around the carbon, which calls for a tetrahedral arrangement as shown in Fig. 8.22(a). The oxygen also has four pairs, which requires a tetrahedral arrangement. However, in this case the tetrahedron will be slightly distorted by the space requirements of the lone pairs [Fig. 8.22(b)]. The overall geometric arrangement for the molecule is shown in Fig. 8.22(c).

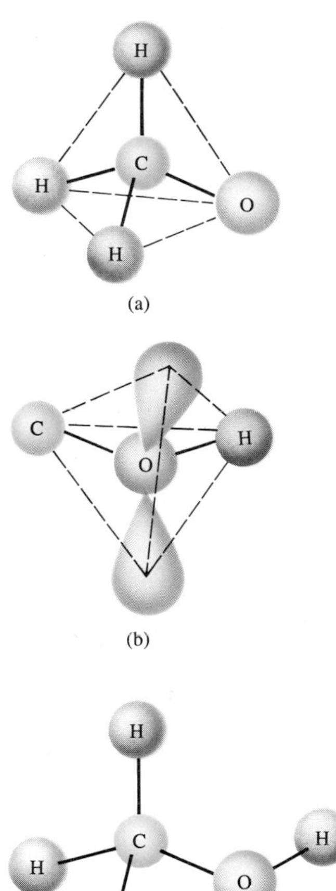

Figure 8.22

The molecular structure of methanol. (a) The arrangement of electron pairs and atoms around the carbon atom. (b) The arrangement of bonding and lone pairs around the oxygen atom. (c) The molecular structure.

Summary of the VSEPR Model

The rules for using the VSEPR model to predict molecular structure are

1. Determine the Lewis structure(s) for the molecule.
2. For molecules with resonance structures, use any of the structures to predict the molecular structure.
3. Sum the electron pairs around the central atom.
4. In counting pairs, count each multiple bond as a single effective pair.
5. The arrangement of the pairs is determined by minimizing electron pair repulsions. These arrangements are shown in Table 8.6.
6. Lone pairs require more space than bonding pairs. Choose an arrangement that gives the lone pairs as much room as possible. Recognize that the lone pairs may produce a slight distortion of the structure at angles less than 120°.

The VSEPR Model—How Well Does It Work?

The VSEPR model is very simple. There are only a few rules to remember, yet the model correctly predicts the molecular structures of most molecules formed from nonmetallic elements. Molecules of any size can be treated by applying the VSEPR model to each appropriate atom (those bonded to at least two other atoms) in the molecule. Thus we can use this model to predict the structures of molecules with hundreds of atoms. It does, however, fail in a few instances. For example, phosphine (PH_3), which has a Lewis structure analogous to that of ammonia,

```
  ..          ..
H—P—H      H—N—H
  |           |
  H           H
```

would be predicted to have a molecular structure similar to that for NH_3 with bond angles of approximately 107°. However, the bond angles of phosphine are actually 94°. There are ways of explaining this structure, but more rules have to be added to the model.

This again illustrates the point that simple models are bound to have exceptions. In introductory chemistry we want to use simple models that fit the majority of cases; we are willing to accept a few failures rather than complicate the model. The amazing thing about the VSEPR model is that such a simple model predicts correctly the structures of so many molecules.

Chemical Impact

Chemical Structure and Communication: Semiochemicals

In this chapter we have stressed the importance of being able to predict the three-dimensional structure of a molecule. Molecular structure is important because of its effect on chemical reactivity. This is especially true in biological systems, where reactions must be efficient and highly specific. Among the hundreds of types of molecules in the fluids of a typical biological system, the appropriate reactants must find and react only with each other—they must be very discriminating. This specificity depends largely on structure. The molecules are constructed so that only the appropriate partners can approach each other in a way that allows reaction.

Another area where molecular structure is central is in the use of molecules as a means of communication. Examples of chemical communication occur in humans in the conduction of nerve impulses across synapses, the control of the manufacture and storage of key chemicals in cells, and the senses of smell and taste. Plants and animals also use chemical communication. For example, ants lay down a chemical trail so that other ants can find a particular food supply. Ants also warn their fellow workers of approaching danger by emitting certain chemicals.

Molecules convey messages by fitting into appropriate receptor sites in a very specific way, which is determined by their structure. When a molecule occupies a receptor site, chemical processes are stimulated that produce the appropriate response. Sometimes receptors can be fooled, as in the use of artificial sweeteners—molecules fit the sites on the taste buds that stimulate a "sweet" response in the brain, but they are not metabolized in the same way as natural sugars. Similar deception is useful in insect control. If an area is sprayed with synthetic female sex attractant molecules, the males of that species become so confused that mating does not occur.

A *semiochemical* is a molecule that delivers a message between members of the same or different species of plant or animal. There are three groups of these chemical messengers: allomones, kairomones, and pheromones. Each is of great ecological importance.

An *allomone* is defined as a chemical that somehow gives adaptive advantage to the producer. For example, leaves of the black walnut tree contain a herbicide, juglone, that appears after the leaves fall to the ground. Juglone is not toxic to grass or certain grains, but it is effective against plants such as apple trees that would compete for the available water and food supplies.

Antibiotics are also allomones since the microorganisms produce them to inhibit other species from growing near them.

Many plants produce bad-tasting chemicals to protect themselves from plant-eating insects and animals. The familiar compound nicotine deters animals from eating the tobacco plant. The millipede sends an unmistakable "back off" message by squirting a predator with benzaldehyde and hydrogen cyanide.

Defense is not the only use of allomones, however. Flowers use scent to attract pollinating insects. Honeybees, for instance, are guided to alfalfa flowers by a series of sweet-scented compounds.

Kairomones are chemical messengers that bring advantageous news to the receiver, and the floral scents are kairomones from the honeybees' viewpoint. Many predators are guided by kairomones emitted by their food. For example, apple skins exude a chemical that attracts the codling moth larva. In some cases kairomones help the underdog. Certain marine mollusks can pick up the "scent" of their predators, the sea stars, and make their escape.

Pheromones are chemicals that affect receptors of the same species as the donor. That is, they are specific within a species. *Releaser pheromones* cause an immediate reaction in the receptor, and *primer pheromones* cause long-term effects. Examples of releaser pheromones are the sex attractants of insects, generated in some species by the males and in others by the females. Sex pheromones have also been found in plants and mammals.

Alarm pheromones are highly volatile compounds (ones easily changed to a gas) released to warn of danger. Honeybees produce isoamyl acetate ($C_7H_{14}O_2$) in their sting

Chemical Impact

glands. Because of its high volatility, this compound does not linger after the state of alert is over. Social behavior in insects is characterized by the use of *trail pheromones,* which are used to indicate a food source. Social insects such as bees, ants, wasps, and termites use these substances. Since trail pheromones are less volatile compounds, the indicators persist for some time.

Primer pheromones, which cause long-term behavioral changes, are harder to isolate and identify. One example, however, is the "queen substance" produced by queen honeybees. All the eggs in a colony are laid by one queen bee. If she is removed from the hive or dies, the worker bees are activated by the absence of the queen substance and begin to feed royal jelly to bee larvae in order to raise a new queen. The queen substance also prevents the development of the workers' ovaries so that only the queen herself can produce eggs.

Many studies of insect pheromones are now underway in the hope that they will provide a method of controlling insects that is more efficient and safer than the current chemical pesticides. Pest control will be discussed in Chapter 24.

A queen honeybee and her attendants.

FOR REVIEW

Summary

Chemical bonds hold groups of atoms together. Bonding occurs when a group of atoms can lower its total energy by aggregating. Bonds can be classified into several types. In an ionic bond there is a transfer of electrons to form ions; in a covalent bond electrons are shared. Between these extremes, in a polar covalent bond electrons are shared unequally. The percent ionic character of the bond in a diatomic molecule XY can be defined as

$$\text{Percent ionic character} = \left(\frac{\text{measured dipole moment of X} - \text{Y}}{\text{calculated dipole moment of X}^+\text{Y}^-}\right) \times 100$$

Electronegativity is defined as the relative ability of an atom in a molecule to attract the electrons shared in a bond. The electronegativity difference of the atoms involved in a bond determines the polarity of that bond. The spatial arrangement of polar bonds determines the overall polarity, or dipole moment, of a molecule.

Stable molecules usually contain atoms that have filled valence orbitals. Nonmetals bonding to each other achieve noble gas configurations in their valence orbitals by covalent bonding. A nonmetal and a representative group metal achieve the same result by transferring electrons to form ions.

Ions have significantly different sizes from their parent atoms. Cations are smaller because the parent atom has lost electrons to become an ion. Anions, because they have more electrons than the parent atom, are larger. In general, ion size increases going down a group. Isoelectronic ions (having the same number of electrons) decrease in size with increasing Z values.

Lattice energy is the change in energy that takes place when separated gaseous ions are packed together to form an ionic solid. Bond energy, the energy necessary to break a covalent bond, varies with the number of shared electron pairs. A single bond involves one electron pair; a double bond, two electron pairs; and a triple bond, three electron pairs. Bond energies can be used to calculate the enthalpy change for a reaction.

The Lewis structure of a molecule shows how the valence electrons are arranged among the atoms. The duet rule for hydrogen and the octet rule for second-row elements reflect the observation that atoms tend to fill their valence orbitals. Elements in the third row and beyond can exceed the octet rule because of the availability of empty *d* orbitals.

Sometimes more than one valid Lewis structure can be drawn for the same molecule, which is accounted for by the concept of resonance. The actual electronic structure is represented by the superposition of the resonance structures in such cases.

When several nonequivalent Lewis structures can be drawn for a molecule, formal charge is often used to choose the most appropriate structure(s). The VSEPR model is very useful in predicting the geometries of molecules formed from nonmetals. The principal postulate of this model is that the structure around a given atom is determined by minimizing electron-pair repulsions.

Key Terms

Section 8.1
bond energy
ionic bonding
ionic compound
Coulomb's law
bond length
covalent bonding
polar covalent bond

Section 8.2
electronegativity

Section 8.3
dipole moment

Section 8.4
isoelectronic ions

Section 8.5
lattice energy

Section 8.8
single bond
double bond
triple bond

Section 8.9
localized electron (LE) model
lone pair
bonding pair

Section 8.10
Lewis structure
duet rule
octet rule

Section 8.12
resonance
resonance structure
formal charge

Section 8.13
molecular structure
valence shell electron pair repulsion (VSEPR) model
linear structure
trigonal planar structure
tetrahedral structure
trigonal pyramid
trigonal bipyramid
octahedral structure
square planar structure

Exercises

A blue exercise number indicates that the answer to that exercise appears at the back of this book and a solution appears in the Solutions Guide.

Chemical Bonds and Electronegativity

1. Tell the difference between the following pairs of terms:
   a. electronegativity and electron affinity
   b. covalent bond and polar covalent bond
   c. polar covalent bond and ionic bond

2. Use Coulomb's law,

$$E = k\left(\frac{Q_1Q_2}{r}\right)$$

to calculate the energy of interaction for the following two arrangements of charges in arbitrary units. Assume a value of $k = 1.0$.

a. (+1) ←1 unit of distance→ (−1) ←∞→ (+1) ←1 unit→ (−1)

b.

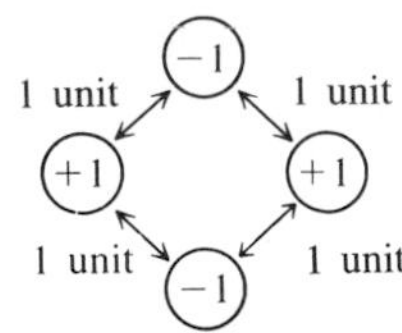

3. Without using Fig. 8.3, predict the order of increasing electronegativity in each of the following groups of elements:
a. C, N, O
b. S, Se, Cl
c. Si, Ge, Sn
d. Tl, S, Ge
e. Na, K, Rb
f. B, O, Ga

4. Without using Fig. 8.3, predict which bond in each of the following groups will be the most polar:
a. C—F, Si—F, Ge—F
b. P—Cl, S—Cl
c. S—F, S—Cl, S—Br
d. Ti—Cl, Si—Cl, Ge—Cl
e. C—H, Si—H, Sn—H
f. Al—Br, Ga—Br, In—Br, Tl—Br

5. Repeat Exercises 3 and 4. This time use the values for the electronegativities of the elements given in Fig. 8.3. Are there any differences in your answers?

6. a. Although not often used, the electronegativities of the noble gases are quite high (Ne = 4.4; Ar = 4.0; Kr = 2.9; Xe = 2.6). Explain these high values relative to those of the other elements using an alternative definition of the electronegativity

$$\text{Electronegativity} \propto (\text{I.E.} - \text{E.A.})$$

where I.E. is the ionization energy and E.A. is the electron affinity using the sign convention of this book.
b. Elemental fluorine (F_2) reacts directly with xenon to produce three different compounds: XeF_2, XeF_4, XeF_6. Would you expect any of the other noble gases to react with fluorine?

Ionic Compounds

7. For each of the following groups, place the atoms and ions in order of decreasing size:
a. Cu, Cu^+, Cu^{2+}
b. Ni^{2+}, Pd^{2+}, Pt^{2+}
c. O^{2-}, S^{2-}, Se^{2-}
d. La^{3+}, Eu^{3+}, Gd^{3+}, Yb^{3+}
e. Te^{2-}, I^-, Xe, Cs^+, Ba^{2+}, La^{3+}

8. Write electron configurations for
a. the cations: Mg^{2+}, Sn^{2+}, K^+, Al^{3+}, Tl^+, As^{3+}
b. the anions: N^{3-}, O^{2-}, F^-, Te^{2-}
c. the most stable ion formed by each of the elements: Be, Rb, Ba, Se, I

9. Which of the following ions have noble gas electron configurations?
a. Fe^{2+}, Fe^{3+}, Sc^{3+}, Co^{3+}
b. Tl^+, Te^{2-}, Cr^{3+}
c. Pu^{4+}, Ce^{4+}, Ti^{4+}
d. Ba^{2+}, Pt^{2+}, Mn^{2+}

10. Define the term *isoelectronic*. When comparing sizes of ions of elements in the same period in the periodic table, why is it advantageous to compare isoelectronic species?

11. Give three ions that are isoelectronic with the krypton atom.

12. Which compound in each of the following pairs of ionic substances has the most exothermic lattice energy? Justify your answers.
a. NaCl, KCl
b. LiF, LiCl
c. $Mg(OH)_2$, MgO
d. $Fe(OH)_2$, $Fe(OH)_3$
e. NaCl, Na_2O
f. MgO, BaS

13. Some of the important properties of ionic compounds are
I. Low electrical conductivity as solids, and high conductivity in solution or when molten
II. Relatively high melting and boiling points
III. Brittleness
IV. Solubility in polar solvents
How does the concept of ionic bonding discussed in this chapter account for these properties?

14. Use the following data to estimate ΔH_f° for sodium chloride.

$$Na(s) + \tfrac{1}{2}Cl_2(g) \rightarrow NaCl(s)$$

Lattice energy	−786 kJ/mol
Ionization energy for Na	495 kJ/mol
Electron affinity of Cl	−348 kJ/mol
Bond energy of Cl_2	239 kJ/mol
Enthalpy of sublimation for Na	108 kJ/mol

15. Use the following data to estimate ΔH_f° for barium chloride:

$$Ba(s) + Cl_2(g) \rightarrow BaCl_2(s)$$

Lattice energy	−2056 kJ/mol
First ionization energy of Ba	503 kJ/mol
Second ionization energy of Ba	965 kJ/mol
Electron affinity of Cl	−348 kJ/mol
Bond energy of Cl_2	239 kJ/mol
Enthalpy of sublimation of Ba	178 kJ/mol

16. Consider the following energy changes:

	ΔH(kJ/mol)
$Mg(g) \rightarrow Mg^+(g) + e^-$	735
$Mg^+(g) \rightarrow Mg^{2+}(g) + e^-$	1445
$O(g) + e^- \rightarrow O^-(g)$	−141
$O^-(g) + e^- \rightarrow O^{2-}(g)$	+878

a. Magnesium oxide exists as $Mg^{2+}O^{2-}$ and not as Mg^+O^-. Explain.
b. What simple experiment could be done to confirm that magnesium oxide does not exist as Mg^+O^-?

17. Use the following data to estimate ΔH for the reaction $S^-(g) + e^- \rightarrow S^{2-}(g)$. Include an estimate of uncertainty.

	ΔH_f^0	Lattice energy	I.E. of M	ΔH_{sub} of M
Li_2S	−500	−2472	520	162
Na_2S	−365	−2203	496	108
K_2S	−381	−2052	419	90
Rb_2S	−361	−1949	409	82
Cs_2S	−360	−1850	382	78

$$S(s) \rightarrow S(g) \qquad \Delta H = 277 \text{ kJ/mol}$$
$$S(g) + e^- \rightarrow S^-(g) \qquad \Delta H = -200 \text{ kJ/mol}$$

Assume all values are known to ±1 kJ.

18. Using data in Exercise 16, calculate ΔH for the reaction $O(g) + 2e^- \rightarrow O^{2-}(g)$.

19. The second electron affinity values for both oxygen and sulfur are unfavorable (endothermic). Explain.

20. Rationalize the following lattice energy values:

Compound	Lattice energy (kJ/mol)
CaSe	−2862
Na_2Se	−2130
CaTe	−2721
Na_2Te	−2095

21. The lattice energies of the oxides and chlorides of iron(II) and iron(III) are: −2631, −3865, −5359, and −14,774 kJ/mol. Match the appropriate formula to each lattice energy. Explain your answer.

22. Predict the empirical formulas of the ionic compounds formed from the following pairs of elements. Name each compound.

a. Li and N
b. Ga and O
c. Rb and Cl
d. Ba and S

Bond Energies

23. Use bond energies to predict ΔH for the isomerization of methyl isocyanide to acetonitrile.

$$CH_3NC(g) \rightarrow CH_3CN(g)$$

24. Use bond energy values in Table 8.4 to estimate ΔH for each of the following reactions in the gas phase:

a. $H_2 + Cl_2 \rightarrow 2HCl$
b. $N_2 + 3H_2 \rightarrow 2NH_3$
c. $H_2C{=}CH_2 + Br_2 \rightarrow CH_2BrCH_2Br$ (ethylene drawn as structural formula, H's on each C)
d. $C_2H_4 + H_2O_2 \rightarrow$ $CH_2(OH){-}CH_2(OH)$ (drawn with OH above each CH_2)

25. Compare your answers from parts a and b of Exercise 24 to ΔH values calculated for each reaction from standard enthalpies of formation in Appendix 4. Do enthalpy changes calculated from bond energies give a reasonable estimate of the actual values?

26. Use bond energies to calculate the value of $\Delta H°$ for the oxidation of methylhydrazine by dinitrogentetroxide, and compare this value to the one you obtained in Exercise 46 of Chapter 6.

27. Calculate $\Delta H°$ for the following reaction using bond energies.

$$C_2H_4(g) + O_3(g) \rightarrow CH_3CHO(g) + O_2(g)$$

Compare this result to that obtained using standard enthalpies of formation in Exercise 50 of Chapter 6.

28. Three processes that have been used for the industrial manufacture of acrylonitrile, an important chemical used in the manufacture of plastics, synthetic rubber, and fibers are shown below. Use bond energy values (Tables 8.4 and 8.5) to estimate ΔH for each of the reactions.

a. $CH_2{-}CH_2$ (ring with O) $+ HCN \rightarrow HOCH_2{-}CH_2{-}C{\equiv}N$ (drawn with H's above and below each C)

$HOCH_2CH_2CN \rightarrow H_2C{=}CH{-}C{\equiv}N + H_2O$ (drawn as structural formula)

b. $4CH_2{=}CHCH_3 + 6NO \xrightarrow[Ag]{700°C} 4CH_2{=}CHCN + 6H_2O + N_2$

The nitrogen-oxygen bond energy in nitric oxide, NO, is 630 kJ/mol.

c. $2CH_2{=}CHCH_3 + 2NH_3 + 3O_2 \xrightarrow[425\text{–}510°C]{\text{Catalyst}} 2CH_2{=}CHCN + 6H_2O$

29. Is the elevated temperature noted in parts b and c of Exercise 28 needed to provide energy to endothermic reactions?

30. Acetic acid is responsible for the sour taste of vinegar. It can be manufactured using the following reaction:

$$CH_3OH + CO \rightarrow CH_3C({=}O){-}OH$$

Use tabulated values of bond energies (Table 8.4) to estimate ΔH for this reaction.

31. Use bond energies (Table 8.4), values of electron affinities (Table 7.7), and the ionization energy of hydrogen (1312 kJ/mol) to estimate ΔH for each of the following reactions:

a. $HF(g) \rightarrow H^+(g) + F^-(g)$
b. $HCl(g) \rightarrow H^+(g) + Cl^-(g)$

c. $HI(g) \rightarrow H^+(g) + I^-(g)$
d. $H_2O(g) \rightarrow H^+(g) + OH^-(g)$
(Electron affinity of $OH(g) = -180$ kJ/mol)

32. The standard enthalpies of formation of $S(g)$, $F(g)$, $SF_4(g)$, and $SF_6(g)$ are +278.8, +79.0, −775, and −1209 kJ/mol, respectively.
a. Use these data to estimate the energy of an S—F bond.
b. Compare your calculated values to the value given in Table 8.4. What conclusions can you draw?
c. Why are the ΔH_f° values for $S(g)$ and $F(g)$ not equal to zero, since sulfur and fluorine are elements?

33. Use the following standard enthalpies of formation to estimate the N—H bond energy in ammonia. Compare this to the value in Table 8.4.

$N(g)$, 472.7 kJ/mol
$H(g)$, 216.0 kJ/mol
$NH_3(g)$, −46.1 kJ/mol

34. The standard enthalpy of formation of $N_2H_4(g)$ is 95.4 kJ/mol. Use this and the data in Exercise 33 to estimate the N—N single bond energy. Compare this to the value in Table 8.4.

35. What is the relationship between ΔH_f° for $H(g)$ given in Exercise 33 and the H—H bond energy in Table 8.4.

Lewis Structures and Resonance

36. Draw Lewis structures that obey the octet rule for each of the following:
a. HCN
b. PH_3
c. $CHCl_3$
d. NH_4^+
e. BF_4^-
f. SeF_2

37. Write a Lewis structure that obeys the octet rule for each of the following molecules and ions. In each case the first atom listed is the central atom.
a. $POCl_3$, SO_4^{2-}, XeO_4, PO_4^{3-}, ClO_4^-
b. NF_3, SO_3^{2-}, PO_3^{3-}, ClO_3^-
c. ClO_2^-, SCl_2, PCl_2^-

38. Considering your answers to Exercise 37, what conclusions can you draw concerning the structures of species containing the same number of atoms and the same number of valence electrons?

39. Write Lewis structures for the following. Show all resonance structures where applicable.
a. NO_2^-, HNO_2, NO_3^-, HNO_3
b. SO_4^{2-}, HSO_4^-, H_2SO_4
c. CN^-, HCN
d. OCN^-, SCN^-, N_3^-
e. C_2N_2 (atomic arrangement is NCCN)

40. Some of the important pollutants in the atmosphere are ozone, sulfur dioxide, and sulfur trioxide. Write Lewis structures for these three molecules.

41. Peroxyacetyl nitrate, or PAN, is a product of photochemical smog. Write Lewis structures, including resonance structures, for PAN. The skeletal arrangement of atoms is

```
   H  O        O
   |  |       /
H—C—C—O—O—N
   |          \
   H           O
```

42. A toxic cloud covered Bhopal, India, in December, 1984, when water got into a tank of methyl isocyanate, and the product escaped into the atmosphere. Methyl isocyanate is used in the production of many pesticides. Draw the Lewis structures of methyl isocyanate, CH_3NCO, including resonance forms.

43. S_2Cl_2 and SCl_2 are important industrial chemicals used primarily in the vapor-phase vulcanization of certain rubbers. Draw Lewis structures for S_2Cl_2 and SCl_2.

44. Benzene (C_6H_6) consists of a six-membered ring of carbon atoms with one hydrogen bonded to each carbon. Write Lewis structures for benzene, including resonance structures.

45. An important observation supporting the need for resonance in the localized electron model was that there are only three different structures of dichlorobenzene ($C_6H_4Cl_2$). How does this fact support the need for the concept of resonance?

46. Borazine ($B_3N_3H_6$), has often been called "inorganic" benzene. Write Lewis structures for borazine. Borazine contains a six-membered ring of alternating boron and nitrogen atoms.

47. Write Lewis structures for the following molecules, which have central atoms that do not obey the octet rule: PF_5, BrF_3, $Be(CH_3)_2$, BCl_3, $XeOF_4$ (Xe is the central atom), XeF_6, SeF_4.

48. ClF_3 and BrF_3 are both used to fluorinate uranium to produce UF_6 in the processing and reprocessing of nuclear fuel. Draw Lewis structures for ClF_3 and BrF_3.

49. Write Lewis structures for CO_3^{2-}, HCO_3^-, and H_2CO_3. When acid is added to an aqueous solution containing carbonate or bicarbonate ions, carbon dioxide gas is formed. We generally say that carbonic acid (H_2CO_3) is unstable. Use bond energies to estimate ΔH for the reaction

$$H_2CO_3 \rightarrow CO_2 + H_2O$$

Specify a possible cause for the instability of carbonic acid.

50. Nitrous oxide (N_2O) has three possible Lewis structures:

$$\ddot{N}{=}N{=}\ddot{O} \longleftrightarrow :N{\equiv}N{-}\ddot{O}: \longleftrightarrow :\ddot{N}{-}N{\equiv}O:$$

Given the following bond lengths,

N—N	1.67 Å	N=O	1.15 Å
N=N	1.20 Å	N—O	1.47 Å
N≡N	1.10 Å		

rationalize the observations that the N—N bond length in N_2O is 1.12 Å and that the N—O bond length is 1.19 Å.

51. Consider the following bond lengths:

C—O 1.43 Å C=O 1.23 Å C≡O 1.09 Å

In the CO_3^{2-} ion, all three C—O bonds have identical bond lengths of 1.36 Å. Why?

52. Order the following species with respect to carbon-oxygen bond length (longest to shortest):

CO, CO_2, CO_3^{2-}, CH_3OH

What is the order from the weakest to the strongest carbon-oxygen bond?

53. Place the species below in order of the shortest to the longest nitrogen-oxygen bond.

H_2NOH, N_2O, NO^+, NO_2^-, NO_3^-

Formal Charge

54. Assign formal charges to the resonance structures for N_2O in Exercise 50. Can you eliminate any of the resonance structures on the basis of formal charge? Is this consistent with observation?

55. Use formal charge to rationalize why BF_3 does not follow the octet rule.

56. Assign formal charges to the atoms in carbon monoxide. Use these to explain why CO has a much smaller dipole moment than would be expected on the basis of electronegativity.

57. Draw Lewis structures that obey the octet rule for the following species. Assign the formal charge for each central atom.
a. $OPCl_3$ c. ClO_4^- e. SO_2Cl_2 g. ClO_3^-
b. SO_4^{2-} d. PO_4^{3-} f. XeO_4 h. NO_4^{3-}

58. Draw Lewis structures for the species in Exercise 57 that involve minimum formal charges.

Molecular Structure and Polarity

59. Predict the molecular structure and bond angles for each molecule or ion in Exercises 36, 37, and 39.

60. Predict the molecular structure and bond angles for each of the following ions:
a. I_3^- b. ClF_3 c. IF_4^+ d. SF_5^+

61. Predict the molecular structure and bond angles for each of the following:
a. SeO_3^{2-} b. SeH_2 c. SeO_4^{2-}

62. Predict the molecular structure and bond angles for each of the following:
a. BrF_5 b. KrF_4 c. IF_6^+

63. The addition of antimony pentafluoride (SbF_5) to liquid hydrogen produces a solution that is a *superacid*. Superacids are capable of acting as acids toward many compounds that we normally expect not to act as bases. For example, in the following reaction HF acts as a base:

$$SbF_5 + 2HF \rightarrow SbF_6^- + H_2F^+$$

Write Lewis structures for and predict the molecular structures of the reactants and products in this reaction.

64. Which of the following molecules have dipole moments? For the molecules that are polar, indicate the polarity of each bond and the direction of the net dipole moment of the molecule.
a. CH_2Cl_2, $CHCl_3$, CCl_4
b. CO_2, N_2O
c. PH_3, NH_3, AsH_3

65. Write Lewis structures and predict the molecular structures of the following:
a. chromate ion
b. dichromate ion
c. thiosulfate ion ($S_2O_3^{2-}$)
d. peroxydisulfate ion ($S_2O_8^{2-}$)

66. What two requirements must be satisfied for a molecule to be polar?

67. Write Lewis structures and predict the molecular structures of the following:
a. OCl_2, Br_3^-, BeH_2, BH_2^-
b. BCl_3, NF_3, IF_3
c. CF_4, SeF_4, XeF_4
d. IF_5, AsF_5
Which of the above compounds have dipole moments?

68. Write a Lewis structure and predict the molecular structure and polarity of each of the following sulfur fluorides: SF_2, SF_4, SF_6, and S_2F_4 (exists as $F_3S—SF$).
Predict the F—S—F bond angles in each molecule.

69. The molecules BF_3, CF_4, CO_2, PF_5, and SF_6 are all nonpolar, even though they all contain polar bonds. Why?

70. Two molecules exist with the formula N_2F_2. The Lewis structures are

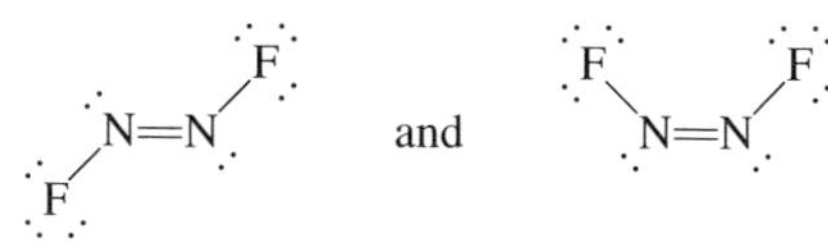

a. What are the N—N—F bond angles in the two molecules?
b. What is the polarity of each molecule?

71. Draw Lewis structures and predict whether each of the following are polar or nonpolar.
a. HOCN
b. BeF_2
c. KrF_4
d. CF_2Cl_2
e. H_2NNH_2 (hydrazine)
f. H_2CO

72. Order each of the following groups of molecules from most polar to least polar.
a. $CHCl_3$, CH_2Cl_2, CH_3Cl, CH_4
b. CF_2Cl_2, CH_2F_2, CHF_3
c. CHF_3, $CHCl_3$, $CHBr_3$, CHI_3

Additional Exercises

73. Although both Br_3^- and I_3^- ions are known, the F_3^- ion does not exist. Explain.

74. There are three possible structures of $PF_3(CH_3)_2$, where P is the central atom. Draw them and describe how measurements of dipole moments might be used to distinguish among them.

75. Look up the energies of the bonds in CO and N_2. Although the bond in CO is stronger, CO is considerably more reactive than N_2. Give a possible explanation.

76. Carbon tetrachloride and silicon tetrachloride both exist as nonpolar liquids. When $CCl_4(l)$ is added to water, distinct layers form. When $SiCl_4(l)$ is added to water, a violent reaction occurs.

$$SiCl_4(l) + 2H_2O(l) \rightarrow SiO_2(s) + 4HCl(aq)$$

Explain why $SiCl_4$ is so much more reactive toward water than CCl_4.

77. Which member of the following pairs would you expect to be more stable? Justify each choice.
a. SO_4 or SO_4^{2-}
b. NF_5 or PF_5
c. OF_6 or SF_6
d. BH_3 or BH_4^-
e. MgF or MgO
f. CsCl or $CsCl_2$
g. KBr or K_2Br

78. Many times extra stability is characteristic of a molecule or ion in which resonance is possible. How could this be used to explain the acidities of the following compounds? (The acidic hydrogen is marked by an asterisk.) Part c shows resonance in the phenol ring C_6H_5.

a. H—C(=O)—OH* b. CH_3—C(=O)—CH=C(OH*)—CH_3 c. C_6H_5—OH* (benzene ring)

79. Would you expect the electronegativity of titanium to be the same in the species Ti, Ti^{2+}, Ti^{3+}, and Ti^{4+}? Explain your answer.

80. Write a Lewis structure and predict the molecular structure of each of the following:
a. SiF_4 b. SeF_4 c. KrF_4
Why do these three molecules have different molecular structures? Which, if any, are polar?

81. Write a Lewis structure and predict the molecular structure of each of the following:
a. BF_3 b. PF_3 c. BrF_3
Why are the molecular structures different? Which, if any, molecules are polar?

82. Give a rationalization for the octet rule in terms of orbitals.

83. Use bond energies (Table 8.4) to estimate ΔH for the following reactions:
a. $C_3H_8(g) + 5O_2(g) \rightarrow 3CO_2(g) + 4H_2O(g)$
b. $C_6H_{12}O_6(s) \rightarrow 2CO_2(g) + 2C_2H_5OH(l)$
The structure of glucose, $C_6H_{12}O_6$, is:

84. Draw a Lewis structure for the N,N-dimethylformamide molecule. The skeletal structure is

H—C(=O)—N(CH_3)—CH_3

Various types of evidence lead to the conclusion that there is some double bond character to the CN bond. Draw one or more resonance structures that support this observation.

85. Disulfurdinitrogen, S_2N_2, exists as a ring of alternating sulfur and nitrogen atoms. S_2N_2 will polymerize to polythiazyl, which acts as a metallic conductor of electricity along the polymer chain. Draw a Lewis structure for S_2N_2.

86. For each of the following, write an equation that corresponds to the energy given:
a. lattice energy of NaCl
b. lattice energy of NH_4Br
c. lattice energy of MgS
d. ΔH_f^0 of $O(g)$
e. O—O double bond energy beginning with $O_2(g)$ as a reactant

87. Aluminum chloride exists as discrete Al_2Cl_6 molecules in the gas phase. The skeletal structure is

(two Al atoms bridged by two Cl atoms, each Al also bonded to two terminal Cl atoms)

a. Complete the Lewis structure.
b. Predict the bond angles around each Al atom.
c. Is the molecule polar or nonpolar?

88. Refer back to Exercises 57 and 58. Would you make the same prediction for the molecular structure for each case using the Lewis structure obtained in Exercise 57 as compared to the one obtained in Exercise 58?

89. The xenon-oxygen bond energy in XeO_3 is 84 kJ/mol. Which of the Lewis structures in Sample Exercise 8.10 is most consistent with this fact?

CHAPTER 9

Covalent Bonding: Orbitals

In Chapter 8 we discussed the fundamental concepts of bonding and introduced the most widely used simple model for covalent bonding: the localized electron model. We saw the usefulness of a bonding model as a means for systematizing chemistry by allowing us to look at molecules in terms of individual bonds. We also saw that molecular structure can be predicted by minimizing electron-pair repulsions. In this chapter we will examine bonding models in more detail, particularly focusing on the role of orbitals.

CONTENTS

9.1 Hybridization and the Localized Electron Model

Purpose

- To show how special atomic orbitals are formed in covalent bonding.

As we saw in Chapter 8, the localized electron model views a molecule as a collection of atoms bound together by sharing electrons using their atomic orbitals. The arrangement of valence electrons is represented by the Lewis structure (or structures, where resonance occurs), and the molecular geometry can be predicted from the VSEPR model. In this section we will describe the atomic orbitals used to share electrons and hence to form the bonds.

< An abstract computer graphic of a molecular structure. Complex structures can be graphed easily by computers, allowing us to see the general molecular shape.

sp^3 Hybridization

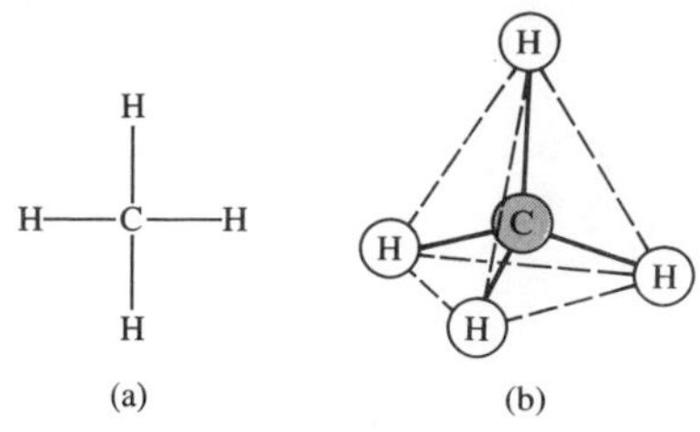

Figure 9.1

(a) The Lewis structure of the methane molecule. (b) The tetrahedral molecular geometry of the methane molecule.

Let us reconsider the bonding in methane, which has the Lewis structure and molecular geometry shown in Fig. 9.1. In general, we assume that bonding involves only the valence orbitals. This means that the hydrogen atoms in methane use 1*s* orbitals. The valence orbitals of a carbon atom are the 2*s* and 2*p* orbitals shown in Fig. 9.2. In thinking about how carbon can use these orbitals to bond to the hydrogen atoms, we can see two related problems:

1. Using the 2*p* and 2*s* atomic orbitals will lead to two different types of C—H bonds: (a) those from the overlap of a 2*p* orbital of carbon and a 1*s* orbital of hydrogen (there will be three of these) and (b) those from the overlap of a 2*s* orbital of carbon and a 1*s* orbital of hydrogen (there will be one of these). This is a problem because methane is known to have four identical C—H bonds.
2. Since the carbon 2*p* orbitals are mutually perpendicular, we might expect the three C—H bonds formed with these orbitals to be oriented at 90° angles:

H 90°
90° C—H
90°
H

However, the methane molecule is known by experiment to be tetrahedral with bond angles of 109.5°.

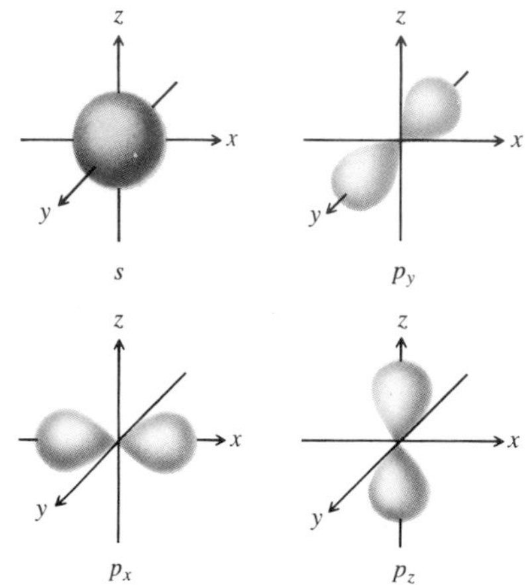

Figure 9.2

The valence orbitals on a free carbon atom: 2*s*, $2p_x$, $2p_y$, and $2p_z$.

This analysis leads to one of two conclusions: either the localized electron model is completely wrong or carbon adopts a set of atomic orbitals other than its "native" 2*s* and 2*p* orbitals to bond to the hydrogen atoms in forming the methane molecule. The second conclusion seems more reasonable. After all, we have already seen how useful the localized electron model is. The 2*s* and 2*p* orbitals present on an *isolated* carbon atom may not be the best set of orbitals for bonding. It seems possible that a new set of atomic orbitals might better serve the carbon atom in forming molecules. To account for the known structure of methane, there must be four equivalent atomic orbitals, arranged tetrahedrally. In fact, such a set of orbitals can be obtained quite readily by combining the carbon 2*s* and 2*p* orbitals, as shown schematically in Fig. 9.3. This mixing of the native atomic orbitals to form special orbitals for bonding is called **hybridization.** The four new orbitals are called sp^3 orbitals since they are formed from one 2*s* and three 2*p* orbitals (s^1p^3). Similarly, we say that the carbon atom undergoes **sp^3 hybridization,** or is sp^3 hybridized. The four sp^3 orbitals are identical in shape, each one having a large lobe and a small lobe. The four orbitals are oriented in space so that the large lobes form a tetrahedral arrangement, as shown in Fig. 9.3.

Hybridization is a modification of the localized electron model to account for the observation that atoms often seem to use special atomic orbitals in forming molecules.

The hybridization of the carbon 2*s* and 2*p* orbitals can also be represented by an orbital energy-level diagram, as shown in Fig. 9.4. Note that electrons have been omitted because we do not need to be concerned with the electron arrangements on the individual atoms—it is the total number of electrons and the arrangement of these electrons in the *molecule* that are important. We are assuming that carbon atomic orbitals are rearranged to accommodate the best electron arrangement for the molecule as a whole. The new sp^3 atomic orbitals on carbon are used to share electron pairs with the 1*s* orbitals from the four hydrogen atoms, as shown in Fig. 9.5.

sp^3 hybridization gives a tetrahedral set of orbitals.

At this point let's summarize the bonding in the methane molecule. The experimentally known structure of this molecule can be explained if we assume that

Figure 9.3

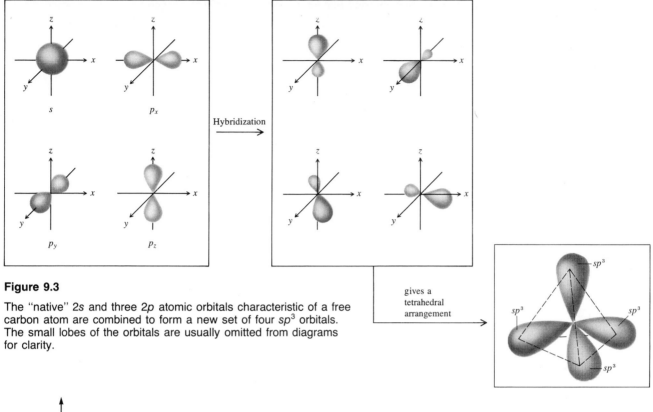

The "native" 2*s* and three 2*p* atomic orbitals characteristic of a free carbon atom are combined to form a new set of four sp^3 orbitals. The small lobes of the orbitals are usually omitted from diagrams for clarity.

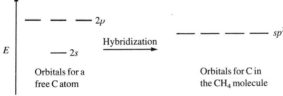

Figure 9.4

An energy-level diagram showing the formation of four sp^3 orbitals.

carbon adopts a special set of atomic orbitals. These new orbitals are obtained by combining the 2*s* and the three 2*p* orbitals of the carbon atom to produce four identically shaped orbitals that are oriented toward the corners of a tetrahedron and are used to bond to the hydrogen atoms. Thus the four sp^3 orbitals on carbon in methane are postulated to account for its known structure.

Remember this principle: *whenever a set of equivalent tetrahedral atomic orbitals is required by an atom, it will adopt a set of sp^3 orbitals;* the atom will become sp^3 hybridized.

It is really not surprising that an atom in a molecule adopts a different set of atomic orbitals, called **hybrid orbitals,** from those it has in the free state. It does not seem unreasonable that, to achieve minimum energy, an atom uses one set of atomic orbitals in the free state and a different set in a molecule. This is consistent with the idea that a molecule is more than simply a sum of its parts. What the atoms

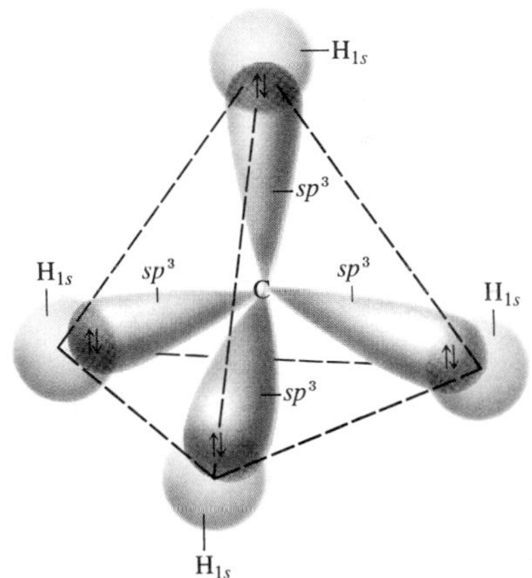

Figure 9.5

The tetrahedral set of four sp^3 orbitals of the carbon atom are used to share electron pairs with the four 1*s* orbitals of the hydrogen atoms to form the four equivalent C—H bonds. This accounts for the known tetrahedral structure of the CH_4 molecule.

in a molecule were like before the molecule was formed is not so important as how the electrons are best arranged in the molecule. The individual atoms are assumed to respond as needed to achieve the minimum energy for the molecule.

Sample Exercise 9.1

Describe the bonding in the ammonia molecule using the localized electron model.

Solution

A complete description of the bonding involves three steps:

1. writing the Lewis structure
2. determining the arrangement of electron pairs using the VSEPR model
3. determining the hybrid atomic orbitals used in bonding in the molecule

The Lewis structure for NH_3 is

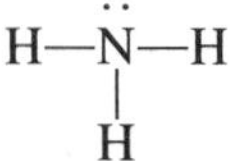

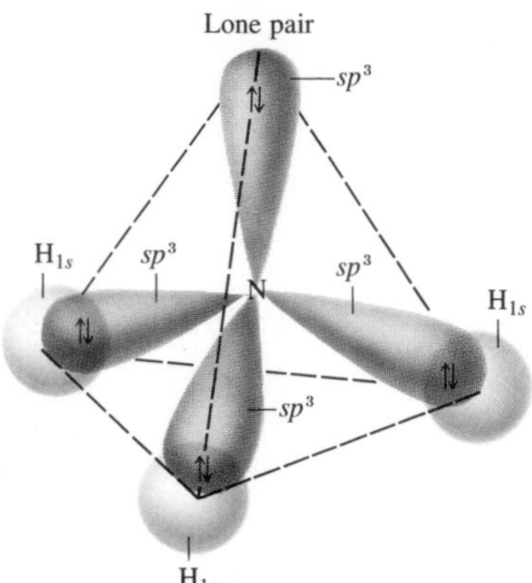

Figure 9.6

The nitrogen atom in ammonia is sp^3 hybridized.

The four electron pairs around the nitrogen atom require a tetrahedral arrangement to minimize repulsions. We have seen that a tetrahedral set of sp^3 hybrid orbitals is obtained by combining the 2*s* and three 2*p* orbitals. In the NH_3 molecule three of the sp^3 orbitals are used to form bonds to the three hydrogen atoms and the fourth sp^3 orbital holds the lone pair, as shown in Fig. 9.6.

sp^2 Hybridization

Ethylene (C_2H_4) is an important starting material in the manufacture of plastics. The C_2H_4 molecule has 12 valence electrons and the following Lewis structure:

```
H       H
 \     /
  C = C
 /     \
H       H
```

A double bond acts as one effective electron pair.

We saw in Chapter 8 that a double bond acts as one effective pair, so in the ethylene molecule each carbon is surrounded by three effective pairs. This requires a trigonal planar arrangement with bond angles of 120°. What orbitals do the carbon atoms in this molecule employ? The molecular geometry requires a set of orbitals in one plane at angles of 120°. Since the 2*s* and 2*p* valence orbitals of carbon do not have the required arrangement, we need a set of hybrid orbitals.

sp^2 hybridization gives a trigonal planar arrangement of atomic orbitals.

The sp^3 orbitals we have just considered will not work, because they are at angles of 109.5° rather than the required 120°. In ethylene the carbon atom must hybridize in a different manner. A set of three orbitals arranged at 120° angles in the same plane can be obtained by combining one *s* orbital and two *p* orbitals, as shown in Fig. 9.7. The orbital energy-level diagram for this arrangement is shown in Fig. 9.8. Since one 2*s* and two 2*p* orbitals are used to form these hybrid orbitals, this is called ***sp^2* hybridization.** Note from Fig. 9.7 that the plane of the sp^2 hybridized

Figure 9.7

The hybridization of the *s*, p_x, and p_y atomic orbitals results in the formation of three sp^2 orbitals centered in the *xy* plane. The large lobes of the orbitals lie in the plane at angles of 120° and point toward the corners of a triangle.

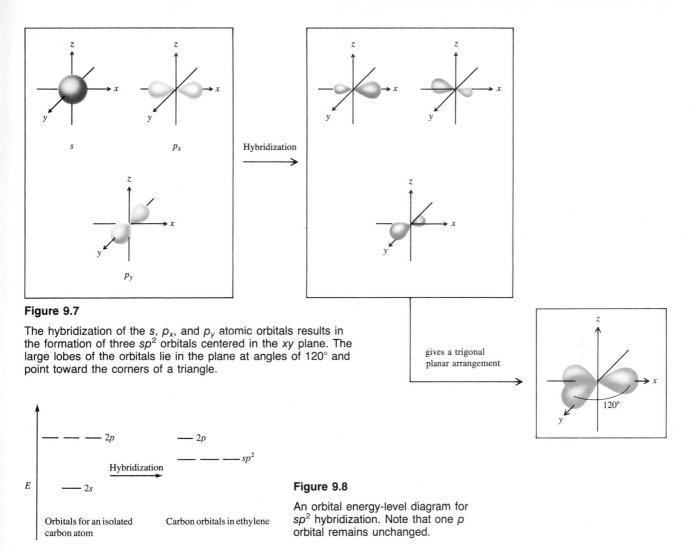

Figure 9.8

An orbital energy-level diagram for sp^2 hybridization. Note that one *p* orbital remains unchanged.

orbitals is determined by which *p* orbitals are used. Since in this case we have arbitrarily decided to use the p_x and p_y orbitals, the hybrid orbitals are centered in the *xy* plane.

In forming the sp^2 orbitals, one 2*p* orbital on carbon has not been used. This remaining *p* orbital (p_z) is oriented perpendicular to the plane of the sp^2 orbitals, as shown in Fig. 9.9.

Now we will see how these orbitals are used to form the bonds in ethylene. The three sp^2 orbitals on each carbon are used to share electrons, as shown in Fig. 9.10. In each of these bonds, the electron pair is shared in an area centered on a line running between the atoms. This type of covalent bond is called a **sigma (σ) bond.** In the ethylene molecule the σ bonds are formed using sp^2 orbitals on each carbon atom and the 1*s* orbital on each hydrogen atom.

How can we explain the double bond between the carbon atoms? In the σ bond the electron pair occupies the space between the carbon atoms. The second bond

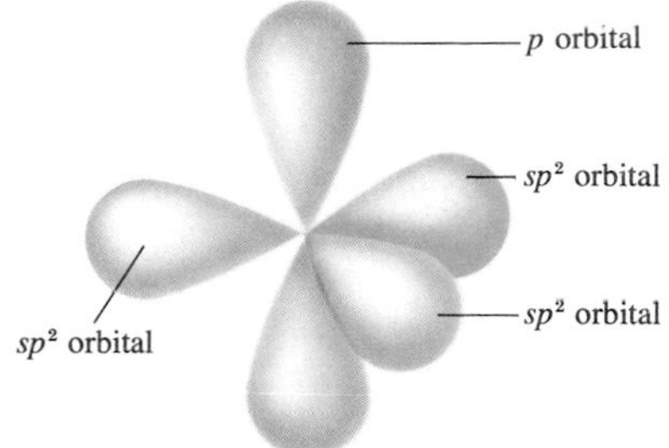

Figure 9.9

When an *s* and two *p* orbitals are mixed to form a set of three sp^2 orbitals, one *p* orbital remains unchanged and is perpendicular to the plane of the hybrid orbitals.

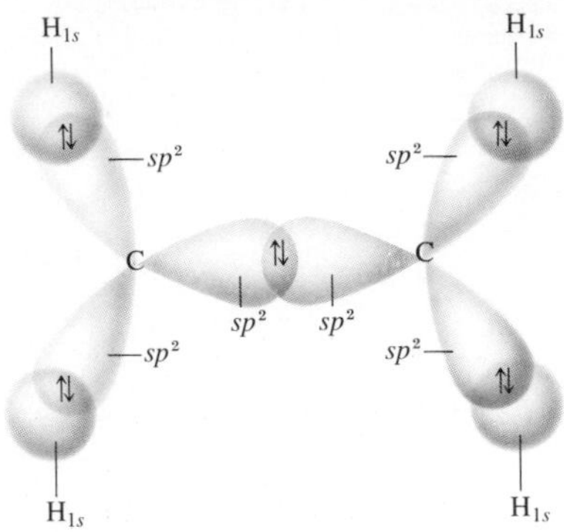

Figure 9.10

The σ bonds in ethylene. Note that for each bond the shared electron pair occupies the region directly between the atoms.

must therefore result from sharing an electron pair in the space *above and below* the σ bond. This type of bond can be formed using the $2p$ orbital perpendicular to the sp^2 hybrid orbitals on each carbon atom (refer to Fig. 9.9). These parallel p orbitals can share an electron pair, which occupies the space above and below a line joining the atoms, to form a **pi (π) bond,** as shown in Fig. 9.11.

Note that σ bonds are formed from orbitals whose lobes point toward each other, but π bonds result from parallel orbitals. A *double bond always consists of one σ bond,* where the electron pair is located directly between the atoms, *and one π bond,* where the shared pair occupies the space above and below the σ bond.

We can now completely specify the orbitals used to form the bonds in the ethylene molecule. As shown in Fig. 9.12, the carbon atoms use sp^2 hybrid orbitals to form the σ bonds to the hydrogen atoms and to each other and use p orbitals to form the π bond with each other. Note that we have accounted fully for the Lewis structure of ethylene with its carbon-carbon double bond and carbon-hydrogen single bonds.

This example illustrates an important general principle of this model: *whenever an atom is surrounded by three effective pairs, a set of sp^2 hybrid orbitals is required.*

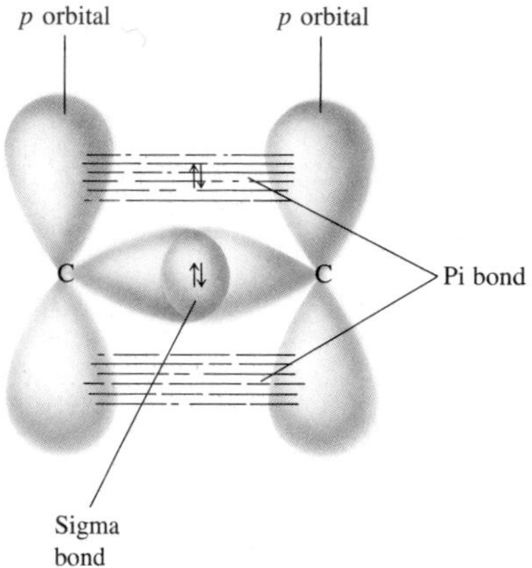

Figure 9.11

A carbon-carbon double bond consists of a σ bond and a π bond. In the σ bond the shared electrons occupy the space directly between the atoms. The π bond is formed from the unhybridized p orbitals on the two carbon atoms. In a π bond the shared electron pair occupies the space above and below a line joining the atoms.

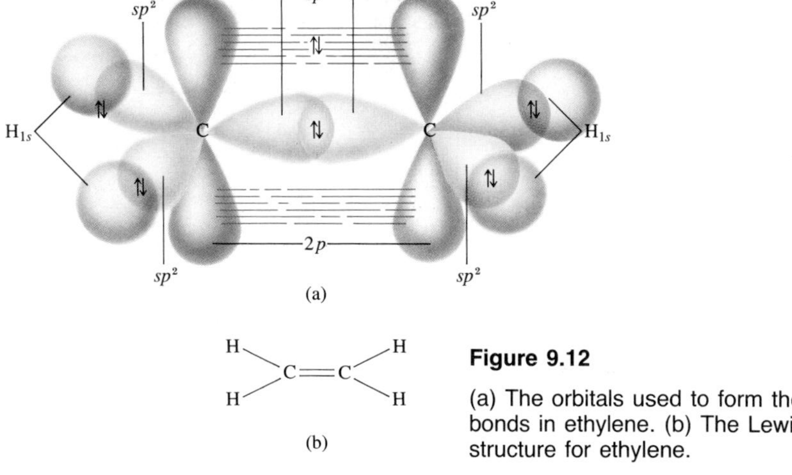

$$\begin{matrix} H & & & & H \\ & C & = & C & \\ H & & & & H \end{matrix}$$

(b)

Figure 9.12

(a) The orbitals used to form the bonds in ethylene. (b) The Lewis structure for ethylene.

sp Hybridization

Another type of hybridization occurs in carbon dioxide, which has the following Lewis structure:

$$:\ddot{O}=C=\ddot{O}:$$

In the CO_2 molecule the carbon atom has two effective pairs that will be arranged at an angle of 180°. We therefore need a pair of atomic orbitals oriented in opposite directions. This requires a new type of hybridization, since neither sp^3 nor sp^2 hybrid orbitals will fit this case. To obtain two hybrid orbitals arranged at 180° requires ***sp* hybridization,** involving one s orbital and one p orbital, as shown in Fig. 9.13.

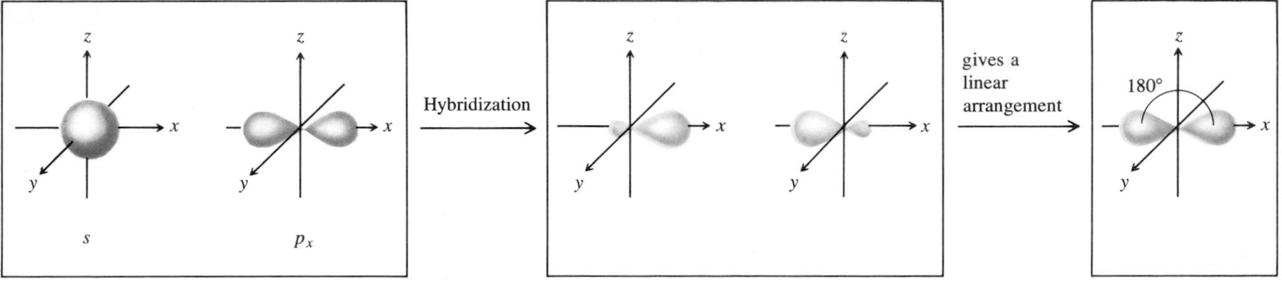

Figure 9.13

When one *s* orbital and one *p* orbital are hybridized, a set of two *sp* orbitals oriented at 180° results.

In terms of this model, *two effective pairs around an atom will always require sp hybridization of that atom.* The *sp* orbitals of carbon in carbon dioxide can be seen in Fig. 9.14, and the corresponding orbital energy-level diagram for their formation is given in Fig. 9.15. These *sp* hybrid orbitals are used to form the σ bonds between carbon and the oxygen atoms. Note that two $2p$ orbitals remain unchanged on the *sp* hybridized carbon. These are used to form the π bonds to the oxygen atoms.

In the CO_2 molecule, each oxygen atom has three effective pairs around it, requiring a trigonal planar arrangement of the pairs. Since a trigonal set of hybrid orbitals requires sp^2 hybridization, each oxygen atom is sp^2 hybridized. One *p* orbital on each oxygen is unchanged and is used for the π bond to the carbon atom.

Now we are ready to describe the bonding in carbon dioxide. The *sp* orbitals on carbon form σ bonds with the sp^2 orbitals on the two oxygen atoms (Fig. 9.14). The remaining sp^2 orbitals on the oxygen atoms hold lone pairs. The π bonds between the carbon atom and each oxygen atom are formed by the overlap of parallel $2p$ orbitals. The *sp* hybridized carbon atom has two unhybridized *p* orbitals, pictured in Fig. 9.16. Each of these *p* orbitals is used to form a π bond with an oxygen atom

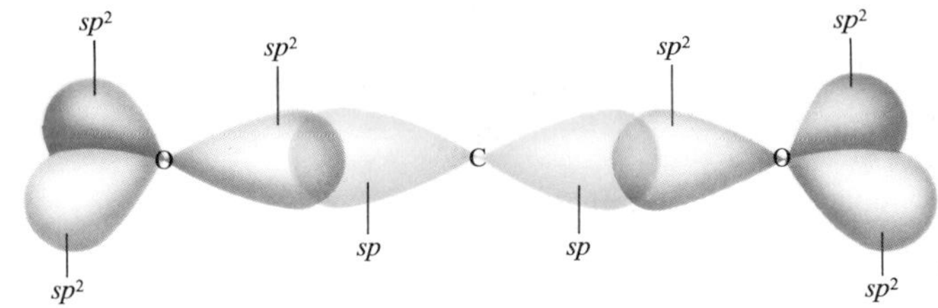

Figure 9.14

The hybrid orbitals in the CO_2 molecule.

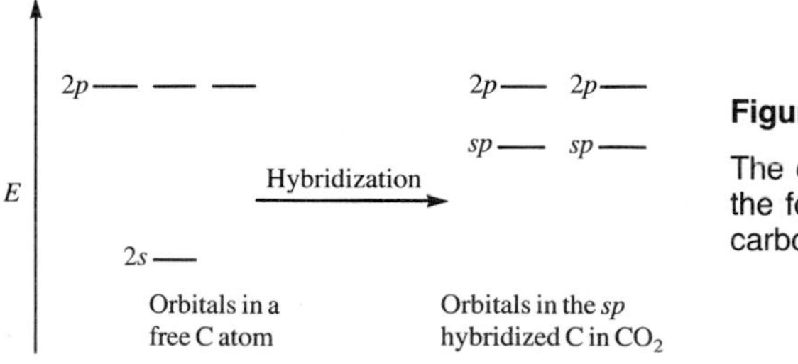

Figure 9.15

The orbital energy-level diagram for the formation of *sp* hybrid orbitals of carbon.

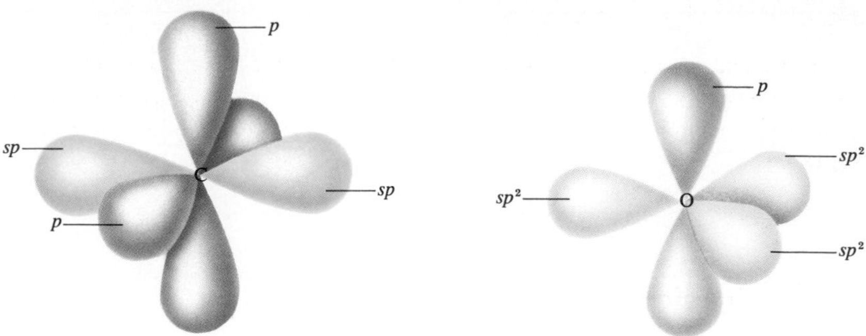

Figure 9.16

The orbitals of an *sp* hybridized carbon atom.

Figure 9.17

The orbital arrangement for an sp^2 hybridized oxygen atom.

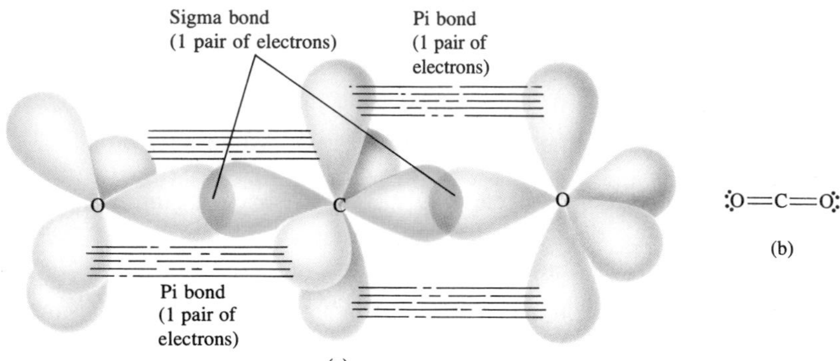

Figure 9.18

(a) The orbitals used to form the bonds in carbon dioxide. Note that the carbon-oxygen double bonds each consist of one σ bond and one π bond. (b) The Lewis structure for carbon dioxide.

(see Fig. 9.17). The total bonding picture for the CO_2 molecule is shown in Fig. 9.18. Note that this picture of the bonding neatly explains the arrangement of electrons predicted by the Lewis structure.

Sample Exercise 9.2

Describe the bonding in the N_2 molecule.

Solution

The Lewis structure for the nitrogen molecule is

$$:N{\equiv}N:$$

where each nitrogen atom is surrounded by two effective pairs. (Remember that multiple bonds count as one effective pair.) This gives a linear arrangement (180°) requiring a pair of oppositely directed orbitals. This situation requires *sp* hybridization. Each nitrogen atom in the nitrogen molecule has two *sp* hybrid orbitals and two unchanged *p* orbitals, as shown in Fig. 9.19(a). The *sp* orbitals are used to form the σ bond between the nitrogen atoms and to hold lone pairs, as shown in Fig. 9.19(b). The *p* orbitals are used to form the two π bonds [see Fig. 9.19(c)]; each pair of overlapping parallel *p* orbitals holds one electron pair. Such bonding ac-

Sample Exercise 9.2, continued

counts for the electron arrangement given by the Lewis structure. The triple bond consists of a σ bond (overlap of two *sp* orbitals) and two π bonds (each one from an overlap of two *p* orbitals). In addition, a lone pair occupies an *sp* orbital on each nitrogen atom.

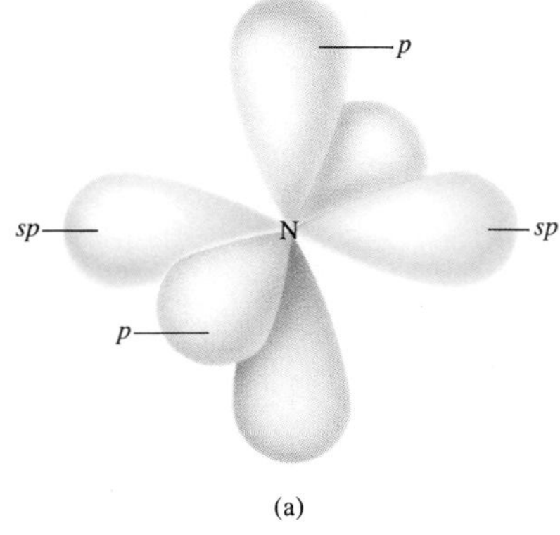

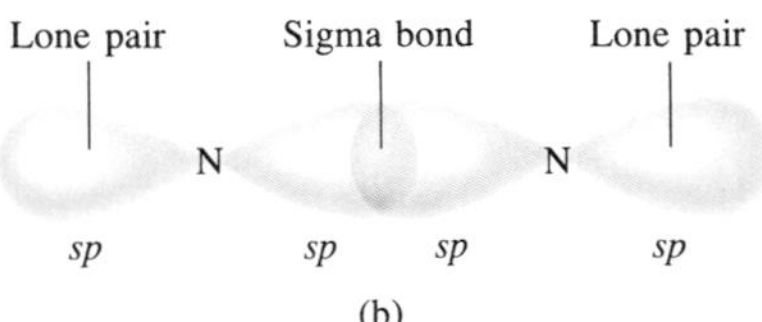

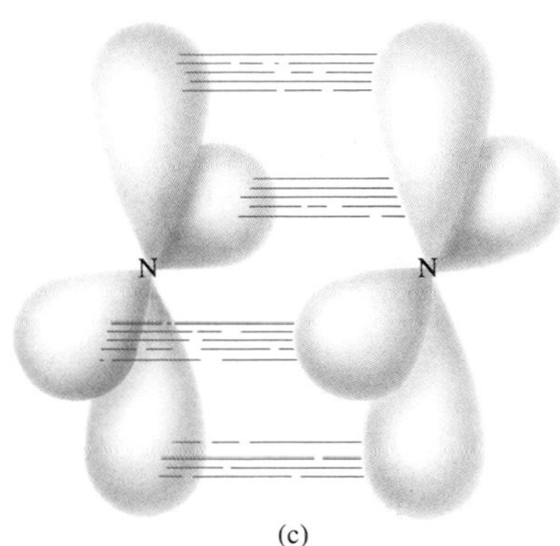

Figure 9.19

An *sp* hybridized nitrogen atom. There are two *sp* hybrid orbitals and two unhybridized *p* orbitals. (b) The sigma bond in the N_2 molecule. (c) The two π bonds in N_2 are formed when electron pairs are shared between two sets of parallel *p* orbitals.

*dsp*³ Hybridization

To illustrate the treatment of a molecule in which the central atom exceeds the octet rule, consider the bonding in the phosphorus pentachloride molecule (PCl_5). The Lewis structure

```
     ..
    :Cl:  ..
     |   Cl:
 ..  | /  ..
:Cl—P
 ..  | \  ..
     |   Cl:
    :Cl:  ..
     ..
```

shows that the phosphorus atom is surrounded by five electron pairs. Since five pairs require a trigonal bipyramidal arrangement, we need a trigonal bipyramidal set of atomic orbitals on phosphorus. Such a set of orbitals is formed by ***dsp*³ hybridization** of one *d* orbital, one *s* orbital, and three *p* orbitals, as shown in Fig. 9.20.

The dsp^3 hybridized phosphorus atom in the PCl_5 molecule uses its five dsp^3 orbitals to share electrons with the five chlorine atoms. *Note that a set of five effective pairs around a given atom always requires a trigonal bipyramidal arrangement, which in turn requires* dsp^3 *hybridization of that atom.*

The Lewis structure for PCl_5 shows that each chlorine atom is surrounded by four electron pairs. This requires a tetrahedral arrangement, which in turn requires a set of four sp^3 orbitals on each chlorine atom.

Now we can describe the bonding in the PCl_5 molecule. The five P—Cl σ bonds are formed by sharing electrons between a dsp^3 orbital on the phosphorus atom and an sp^3 orbital on each chlorine.* The other sp^3 orbitals on each chlorine hold lone pairs. This is shown in Fig. 9.21.

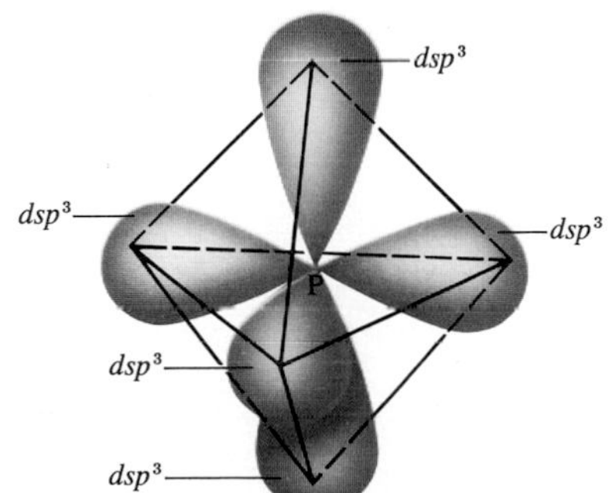

Figure 9.20

A set of dsp^3 hybrid orbitals on a phosphorus atom. Note that the set of five dsp^3 orbitals has a trigonal bipyramidal arrangement. (Each dsp^3 orbital also has a small lobe that is not shown in this diagram.)

*Although we have no way of proving conclusively that each chlorine atom is sp^3 hybridized, we assume that minimizing electron-pair repulsions is as important for peripheral atoms as for the central atom. Thus we will apply the VSEPR model and hybridization to both central and peripheral atoms.

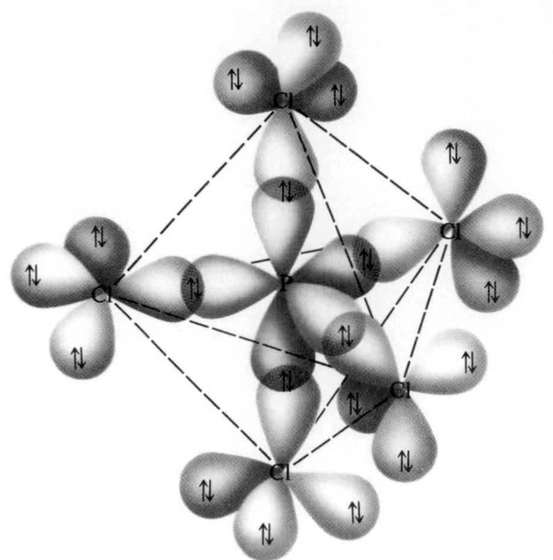

Figure 9.21

The orbitals used to form the bonds in the PCl_5 molecule. The phosphorus uses a set of five dsp^3 orbitals to share electron pairs with sp^3 orbitals on the five chlorine atoms. The other sp^3 orbitals on each chlorine atom hold lone pairs.

Sample Exercise 9.3

Describe the bonding in the triiodide ion (I_3^-).

Solution

The Lewis structure for I_3^-

$$\left[:\ddot{\underset{..}{I}}-\overset{..}{\underset{..}{I}}-\ddot{\underset{..}{I}}:\right]^-$$

shows that the central iodine atom has five pairs of electrons (see Section 8.11). A set of five pairs requires a trigonal bipyramidal arrangement, which in turn requires a set of dsp^3 orbitals. The outer iodine atoms have four pairs of electrons, which calls for a tetrahedral arrangement and sp^3 hybridization.

Thus the central iodine is dsp^3 hybridized. Three of these hybrid orbitals hold lone pairs, and two of them overlap with sp^3 orbitals of the other two iodine atoms to form σ bonds.

d^2sp^3 Hybridization

Some molecules have six pairs of electrons around a central atom; an example is sulfur hexafluoride (SF_6), which has the Lewis structure:

```
        ..
       :F:
   ..   |   ..
  :F\   |   /F:
   ..\  |  /..
        S
   ../  |  \..
  :F/   |   \F:
   ..   |   ..
       :F:
        ..
```

d^2sp^3 hybridization gives six orbitals arranged octahedrally.

This requires an octahedral arrangement of pairs and, in turn, an octahedral set of six hybrid orbitals; or **d^2sp^3 hybridization,** in which two d orbitals, one s orbital, and three p orbitals are combined (see Fig. 9.22). Note that *six electron pairs around an atom are always arranged octahedrally and require d^2sp^3 hybridization of the atom.* Each of the d^2sp^3 orbitals on the sulfur atom is used to bond to a fluorine atom. Since there are four pairs on each fluorine atom, the fluorine atoms are assumed to be sp^3 hybridized.

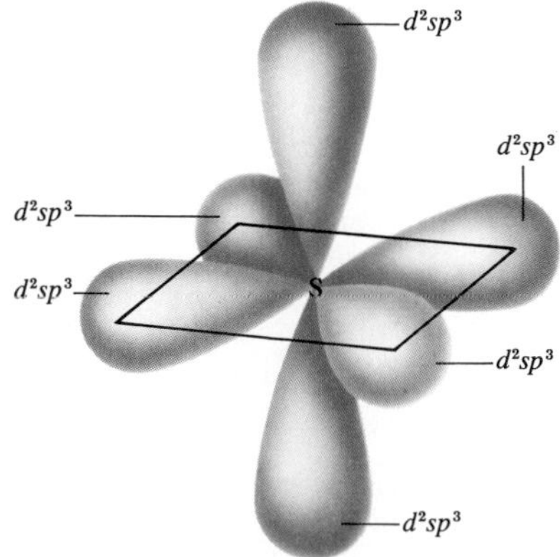

Figure 9.22

An octahedral set of d^2sp^3 orbitals on a sulfur atom. The small lobe of each hybrid orbital has been omitted for clarity.

Sample Exercise 9.4

How is the xenon atom in XeF_4 hybridized?

Solution

As seen in Sample Exercise 8.13, XeF_4 has six pairs of electrons around xenon that are arranged octahedrally to minimize repulsions. An octahedral set of six atomic orbitals is required to hold these electrons, and the xenon atom is d^2sp^3 hybridized.

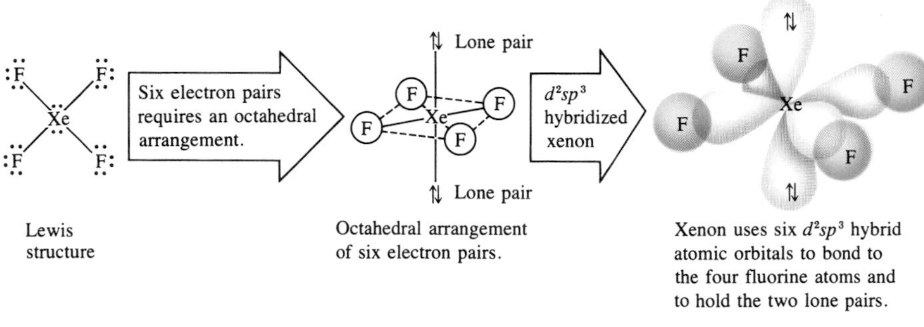
Six electron pairs requires an octahedral arrangement.
Lone pair
Lone pair
d^2sp^3 hybridized xenon
Xe
F
Lewis structure
Octahedral arrangement of six electron pairs.
Xenon uses six d^2sp^3 hybrid atomic orbitals to bond to the four fluorine atoms and to hold the two lone pairs.

The Localized Electron Model: A Summary

The description of a molecule using the localized electron model involves three distinct steps.

- Draw the Lewis structure(s).
- Determine the arrangement of electron pairs using the VSEPR model.
- Specify the hybrid orbitals needed to accommodate the electron pairs.

It is important to do the steps in this order. For a model to be successful, it must follow nature's priorities. In the case of bonding, it seems clear that the tendency for a molecule to minimize its energy is more important than the maintenance of the characteristics of atoms as they exist in the free state. The atoms adjust to meet the "needs" of the molecule. When considering the bonding in a particular molecule, therefore, we always start with the molecule rather than the component atoms. In the molecule the electrons will be arranged to give each atom a noble gas configuration, where possible, and to minimize electron-pair repulsions. We then assume that the atoms adjust their orbitals by hybridization to allow the molecule to adopt the structure that gives the minimum energy.

In applying the localized electron model, we must remember not to overemphasize the characteristics of the separate atoms. It is not where the valence electrons originate that is important; it is where they are needed in the molecule to achieve stability. In the same vein, it is not the orbitals in the isolated atom that matter, but what orbitals the molecule requires for minimum energy.

The requirements for the various types of hybridization are summarized in Fig. 9.23 on the following page.

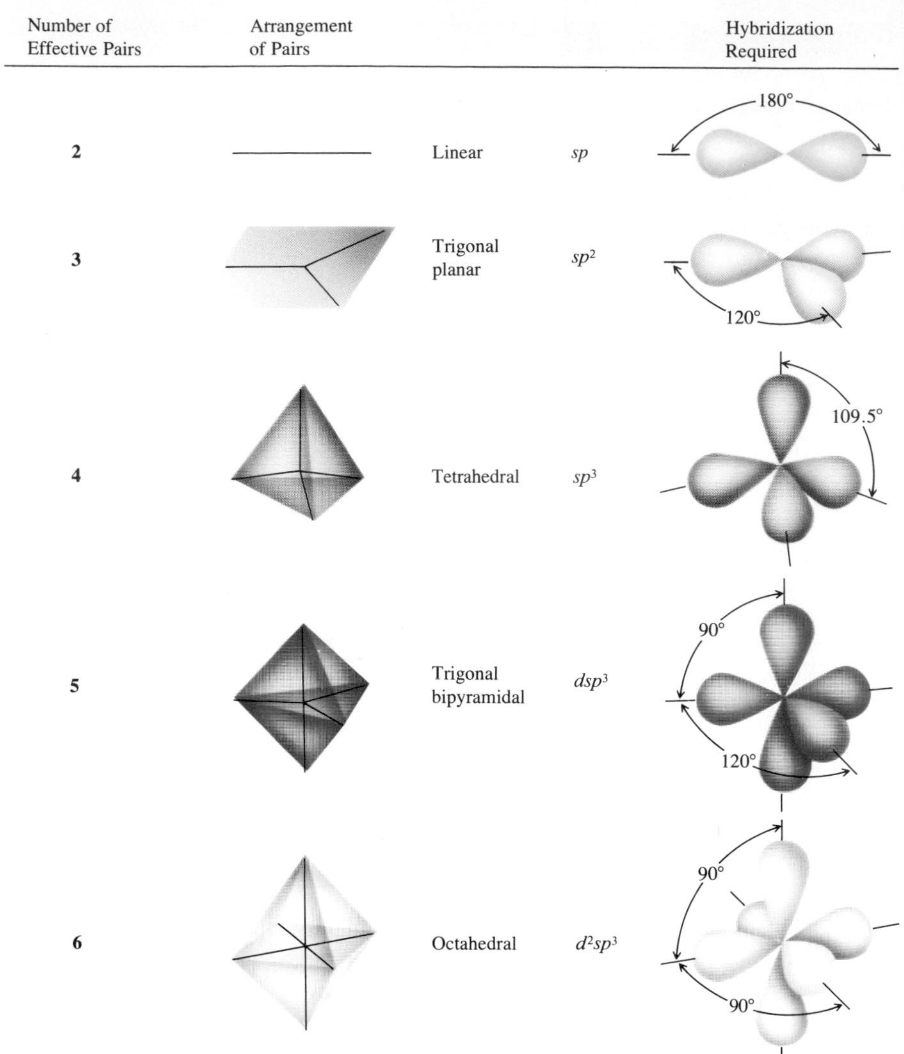

Number of Effective Pairs
Arrangement of Pairs
Hybridization Required
2
Linear
sp
180°
3
Trigonal planar
sp²
120°
4
Tetrahedral
sp³
109.5°
5
Trigonal bipyramidal
dsp³
90°
120°
6
Octahedral
d²sp³
90°
90°

Figure 9.23

The relationship of the number of effective pairs, their spatial arrangement, and the hybrid orbital set required.

Sample Exercise 9.5

For each of the following molecules or ions, predict the hybridization of each atom and describe the molecular structure:

a. CO **b.** BF_4^- **c.** XeF_2

Solution

a. The CO molecule has 10 valence electrons and its Lewis structure is

$$:C{\equiv}O:$$

Each atom has two effective pairs, which means that both are *sp* hybridized. The triple bond consists of a σ bond produced by overlap of an *sp* orbital from each atom and two π bonds produced by overlap of $2p$ orbitals from each atom. The lone pairs are in *sp* orbitals. Since the CO molecule has only two atoms, it must be linear.

Sample Exercise 9.5, continued

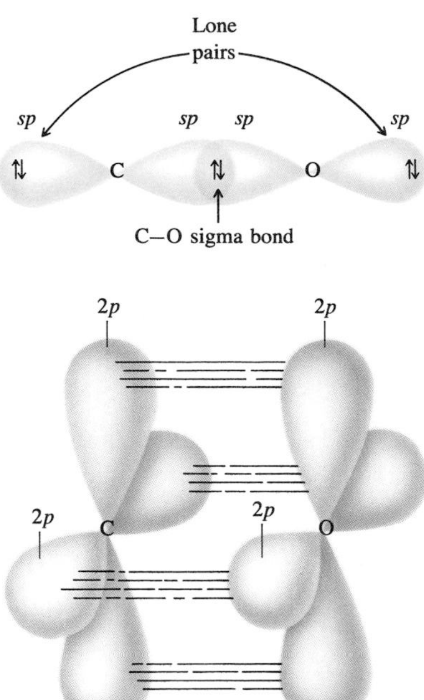

b. The BF_4^- ion has 32 valence electrons. The Lewis structure shows four pairs of electrons around the boron atom, which means a tetrahedral arrangement:

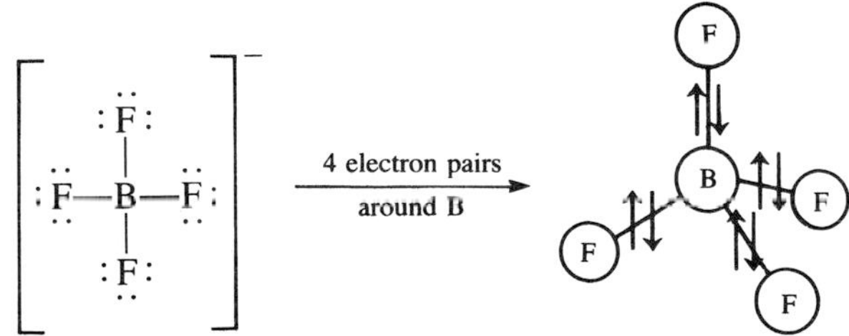

This requires sp^3 hybridization of the boron atom. Each fluorine atom also has four electron pairs and can be assumed to be sp^3 hybridized (only one sp^3 orbital is shown for each fluorine atom). The BF_4^- ion's molecular structure is tetrahedral.

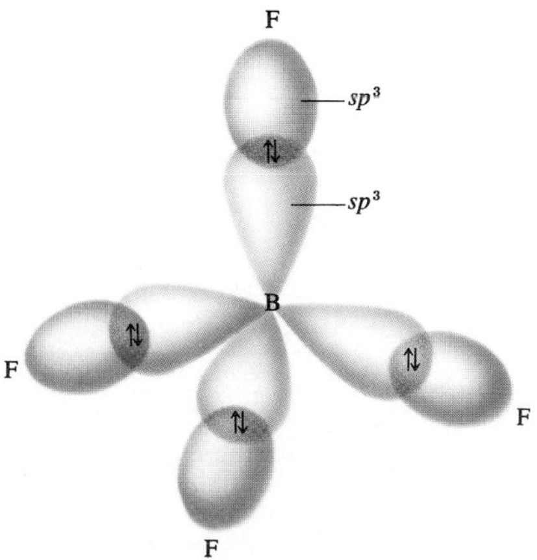

Sample Exercise 9.5, continued

c. The XeF_2 molecule has 22 valence electrons. The Lewis structure shows five electron pairs on the xenon atom, which requires a trigonal bipyramidal arrangement:

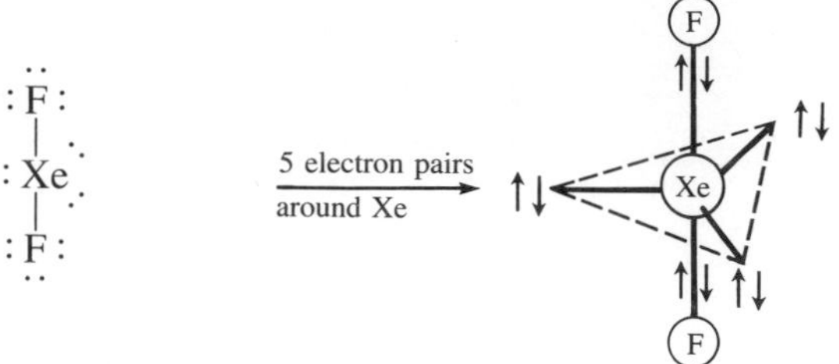

Note that the lone pairs are placed in the plane where they are 120° apart. To accommodate five pairs at the vertices of a trigonal bipyramid requires that the xenon atom adopt a set of five dsp^3 orbitals. Each fluorine atom has four electron pairs and can be assumed to be sp^3 hybridized. The XeF_2 molecule has a linear arrangement of atoms.

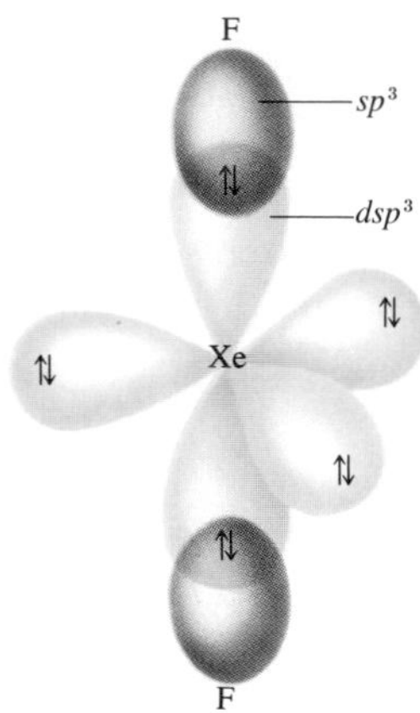

9.2 The Molecular Orbital Model

Purpose

- To show how molecular orbitals are formed in a molecule.
- To define bond order and demonstrate how to calculate it.

We have seen that the localized electron model is of great value in interpreting the structure and bonding of molecules. However, there are some problems with this model. For example, it incorrectly assumes that electrons are localized, and so the concept of resonance must be added. Also, the model does not deal effectively with molecules containing unpaired electrons. And finally, the model gives no direct information about bond energies.

Another model often used to describe bonding is the **molecular orbital model.** To introduce the assumptions, methods, and results of this model, we will consider the simplest of all molecules, H_2, which consists of two protons and two electrons. A very stable molecule, H_2 is lower in energy than the separated hydrogen atoms by 432 kJ/mol.

Molecular orbital theory parallels the atomic theory discussed in Chapter 7.

Since the hydrogen molecule consists of protons and electrons, the same components found in separated hydrogen atoms, it seems reasonable to use a theory similar to the atomic theory discussed in Chapter 7, which assumes that the electrons in an atom exist in orbitals of a given energy. Can we apply this same type of model to the hydrogen molecule? Yes, in fact, describing the H_2 molecule in terms of quantum mechanics is quite straightforward.

However, even though it is formulated rather easily, this problem cannot be solved exactly. The difficulty is the same as that in dealing with polyelectronic atoms—the electron correlation problem. Since we do not know the details of the electron movements, we cannot deal with the electron-electron interactions in a specific way. We need to make approximations that allow solution of the problem but do not destroy the model's physical integrity. The success of these approximations can only be measured by comparing predictions based on theory with experimental observations. In this case we will see that the simplified model works very well.

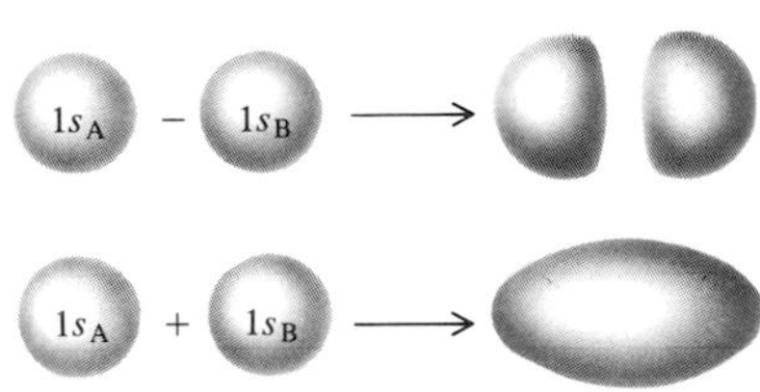

Figure 9.24

The combination of hydrogen 1*s* atomic orbitals to form molecular orbitals.

Just as atomic orbitals are solutions to the quantum mechanical treatment of atoms, **molecular orbitals (MOs)** are solutions to the molecular problem. Molecular orbitals have many of the same characteristics as atomic orbitals. Two of the most important are that they can hold two electrons with opposite spins and that the square of the molecular orbital wave function indicates electron probability.

We will now describe the bonding in the hydrogen molecule using this model. The first step is to obtain the hydrogen molecule's orbitals, a process that is greatly simplified if we assume that the molecular orbitals can be constructed from the hydrogen 1*s* atomic orbitals.

When the quantum mechanical equations for the hydrogen molecule are solved, two molecular orbitals result, which can be represented as

$$MO_1 = 1s_A + 1s_B$$
$$MO_2 = 1s_A - 1s_B$$

where $1s_A$ and $1s_B$ represent the 1*s* orbitals from the two separated hydrogen atoms. This process is shown schematically in Fig. 9.24.

The orbital properties of most interest are size, shape (described by the electron probability distribution), and energy. These properties for the hydrogen molecular orbitals are represented in Fig. 9.25. From Fig. 9.25, we can note several important points:

1. The electron probability of both molecular orbitals is centered along the line passing through the two nuclei. For MO_1 the greatest electron probability is *between* the nuclei, and for MO_2 it is on *either side* of the nuclei. This type of electron distribution is described as *sigma* (σ), as in the localized electron model. Accordingly, we refer to MO_1 and MO_2 as **sigma (σ) molecular orbitals.**
2. In the molecule only the molecular orbitals are available for occupation by electrons. The 1*s* atomic orbitals of the hydrogen atoms no longer exist, because the H_2 molecule—a new entity—has its own set of new orbitals.

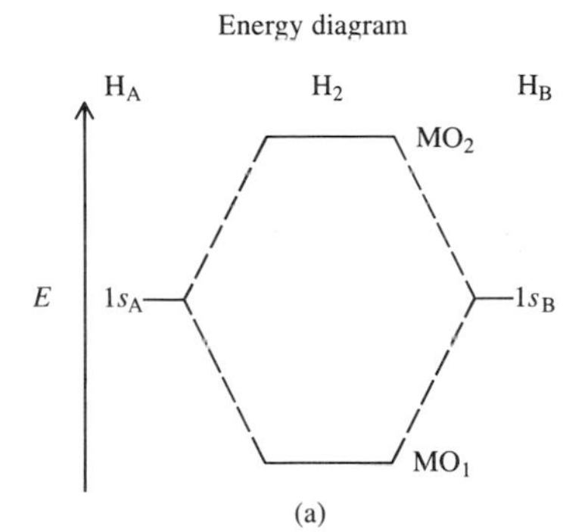

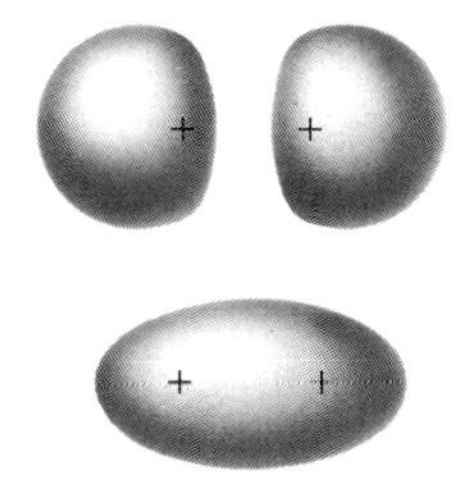

Figure 9.25

(a) The molecular orbital energy-level diagram for the H_2 molecule. (b) The shapes of the molecular orbitals are obtained by squaring the wave functions for MO_1 and MO_2.

3. MO_1 is lower in energy than the $1s$ orbitals of free hydrogen atoms, while MO_2 is higher in energy than the $1s$ orbitals. This fact has very important implications for the stability of the H_2 molecule, since if the two electrons (one from each hydrogen atom) occupy the lower-energy MO_1 they will have lower energy than they do in the two separate hydrogen atoms. This situation favors molecule formation, because nature tends to seek the lowest energy state. That is, the driving force for molecule formation is that the molecular orbital available to the two electrons has lower energy than the atomic orbitals these electrons occupy in the separated atoms. This situation is *pro-bonding*.

   On the other hand, if the two electrons were forced to occupy the higher energy MO_2, they would be definitely *antibonding*. In this case, these electrons would have lower energy in the separated atoms than in the molecule, and the separated state would be favored. Of course, since the lower-energy MO_1 *is* available, the two electrons occupy that MO and the molecule is stable.

   We have seen that the molecular orbitals of the hydrogen molecule fall into two classes: bonding and antibonding. A **bonding molecular orbital** is *lower in energy than the atomic orbitals of which it is composed.* Electrons in this type of orbital will favor the molecule; that is, they will favor bonding. An **antibonding molecular orbital** is *higher in energy than the atomic orbitals of which it is composed.* Electrons in this type of orbital will favor the separated atoms (they are antibonding). Figure 9.26 illustrates these ideas.

Bonding will result if the molecule has lower energy than the separated atoms.

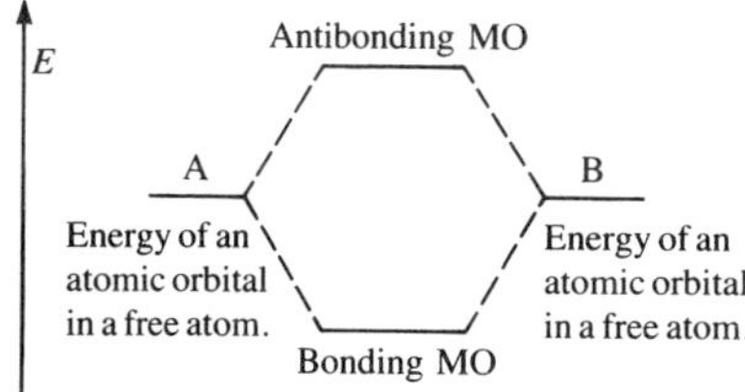

Figure 9.26

Bonding and antibonding molecular orbitals (MOs).

4. Figure 9.25 shows that for the bonding molecular orbital in the H_2 molecule, the electrons have the greatest probability of being between the nuclei. This is exactly what we would expect, since the electrons can lower their energies by being simultaneously attracted by both nuclei. On the other hand, the electron distribution for the antibonding molecular orbital is such that the electrons are mainly outside the space between the nuclei. This type of distribution is not expected to provide any bonding force. In fact, it causes the electrons to be higher in energy than in the separated atoms. Thus the molecular orbital model produces electron distributions and energies that agree with our basic ideas of bonding. This fact reassures us that the model is physically reasonable.
5. The labels on molecular orbitals indicate their symmetry (shape), the parent atomic orbitals, and whether they are bonding or antibonding. Antibonding character is indicated by an asterisk. For the H_2 molecule, both MOs have σ symmetry and both are constructed from hydrogen $1s$ atomic orbitals. The molecular orbitals for H_2 are therefore labeled as follows:

$$MO_1 = \sigma_{1s}$$
$$MO_2 = \sigma_{1s}^*$$

6. Molecular electron configurations can be written in much the same way as atomic configurations. Since the H_2 molecule has two electrons in the σ_{1s} molecular orbital, the electron configuration is σ_{1s}^2.
7. Each molecular orbital can hold two electrons, but the spins must be opposite.
8. Orbitals are conserved. The number of molecular orbitals will always be the same as the number of atomic orbitals used to construct them.

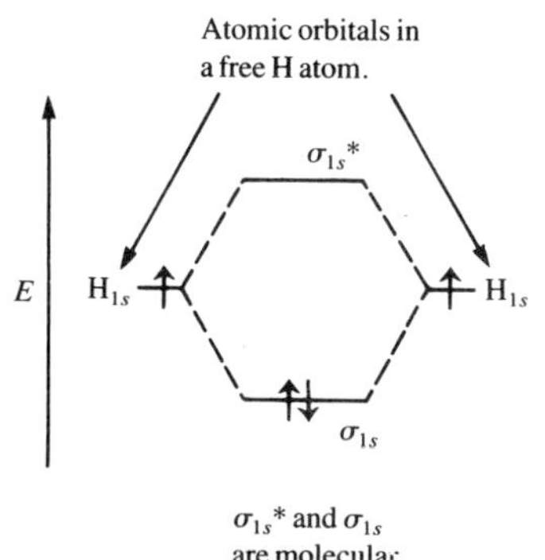

Figure 9.27

A molecular orbital energy-level diagram for the H_2 molecule.

Many of the above points are summarized in Fig. 9.27.

Now suppose we could form the H_2^- ion from a hydride ion (H^-) and a hydrogen atom. Would this species be stable? Since the H^- ion has the configuration $1s^2$ and the H atom has a $1s^1$ configuration, we will use $1s$ atomic orbitals to construct the MO diagram for the H_2^- ion, as shown in Fig. 9.28. The electron configuration for H_2^- is $\sigma_{1s}^2(\sigma_{1s}^*)^1$.

The key idea is that the H_2^- ion will be stable if it has a lower energy than its separated parts. From Fig. 9.28 we see that in going from the separated H^- ion and H atom to the H_2^- ion, two electrons are lowered in energy and one electron is raised in energy. In other words, two electrons are bonding and one electron is antibonding. Since more electrons favor bonding, H_2^- is a stable entity—a bond has formed. But how do the bond strengths in the molecules of H_2 and H_2^- compare?

In the formation of the H_2 molecule, two electrons are lowered in energy, and no electrons are raised in energy compared to the parent atoms. When H_2^- is formed, two electrons are lowered in energy and one is raised, producing *a net lowering of the energy of only one electron*. This means that H_2 is *twice as stable* as H_2^- with respect to their separated components. In other words, the bond in the H_2 molecule is predicted to be about twice as strong as the bond in the H_2^- ion.

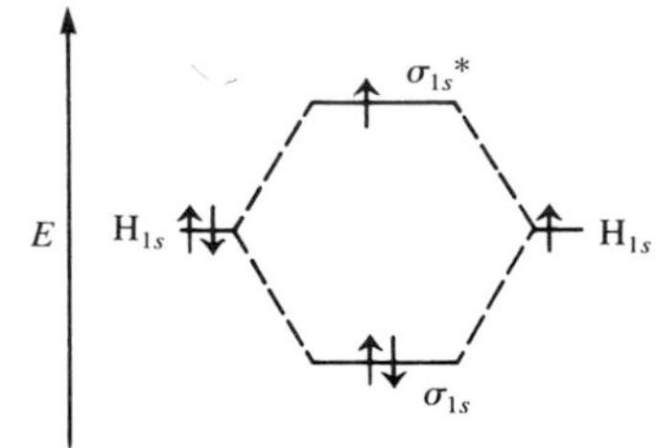

Figure 9.28

The molecular orbital energy-level diagram for the H_2^- ion.

Bond Order

To indicate bond strength we use the concept of bond order. **Bond order** is *the difference between the number of bonding electrons and the number of antibonding electrons, divided by two.*

$$\text{Bond order} = \frac{\text{number of bonding electrons} - \text{number of antibonding electrons}}{2}$$

We divide by 2 because, from the localized electron model, we are used to thinking of bonds in terms of *pairs* of electrons.

Since the H_2 molecule has two bonding electrons and no antibonding electrons, the bond order is

$$\text{Bond order} = \frac{2 - 0}{2} = 1$$

The H_2^- ion has two bonding electrons and one antibonding electron; the bond order is

$$\text{Bond order} = \frac{2 - 1}{2} = \frac{1}{2}$$

Bond order is an indication of bond strength because it reflects the difference between the number of bonding electrons and the number of antibonding electrons. *Larger bond order means greater bond strength.*

We will now apply the molecular orbital model to the helium molecule (He_2). Does this model predict that this molecule will be stable? Since the He atom has a $1s^2$ configuration, $1s$ orbitals are used to construct the molecular orbitals, and the molecule will have four electrons. From the diagram shown in Fig. 9.29, it is apparent that two electrons are raised in energy and two are lowered in energy. Thus the bond order is zero:

$$\frac{2 - 2}{2} = 0$$

This implies that the He_2 molecule is *not* stable with respect to the two free He atoms, which agrees with the observation that helium gas consists of individual He atoms.

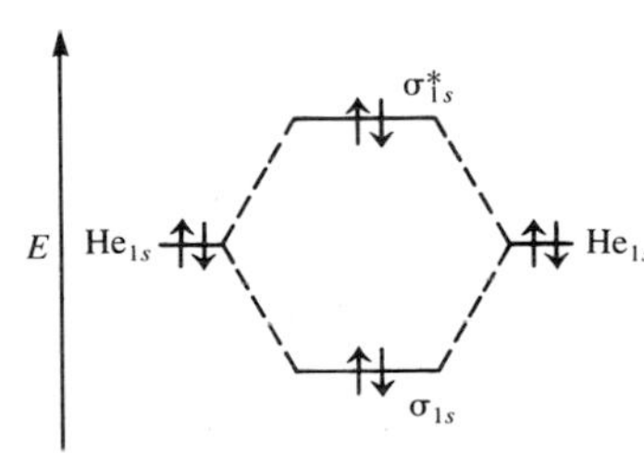

Figure 9.29

The molecular orbital energy-level diagram for the He_2 molecule.

9.3 Bonding in Homonuclear Diatomic Molecules

Purpose

- To discuss the bonding in certain molecules of the general formula X_2.
- To relate paramagnetism to the filling of molecular orbitals.
- To correlate bond order, bond energy, and bond length.

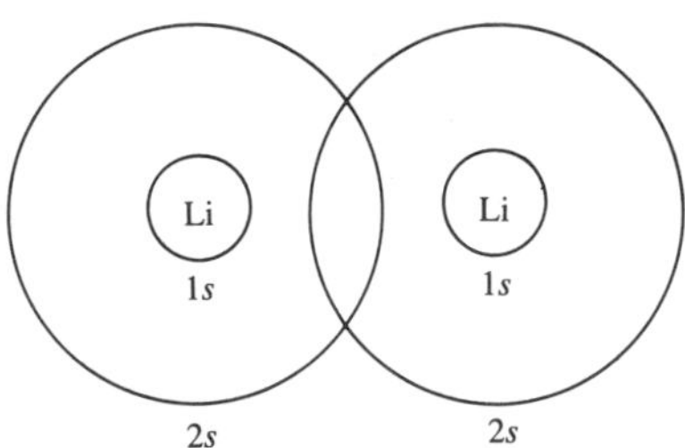

Figure 9.30

The relative sizes of the lithium 1*s* and 2*s* atomic orbitals.

Only valence atomic orbitals contribute to MOs.

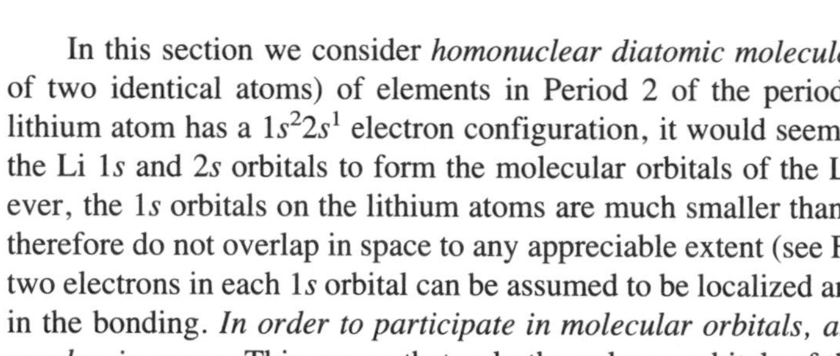

In this section we consider *homonuclear diatomic molecules* (those composed of two identical atoms) of elements in Period 2 of the periodic table. Since the lithium atom has a $1s^22s^1$ electron configuration, it would seem that we should use the Li 1*s* and 2*s* orbitals to form the molecular orbitals of the Li_2 molecule. However, the 1*s* orbitals on the lithium atoms are much smaller than the 2*s* orbitals and therefore do not overlap in space to any appreciable extent (see Fig. 9.30). Thus the two electrons in each 1*s* orbital can be assumed to be localized and not to participate in the bonding. *In order to participate in molecular orbitals, atomic orbitals must overlap in space.* This means that only the valence orbitals of the atoms contribute significantly to the molecular orbitals of a particular molecule.

The molecular orbital diagram of the Li_2 molecule and the shapes of its bonding and antibonding MOs are shown in Fig. 9.31. The electron configuration for Li_2 (valence electrons only) is $\sigma_{2s}{}^2$ and the bond order is:

$$\frac{2-0}{2} = 1$$

Thus Li_2 is a stable molecule (has lower energy than two separated lithium atoms). However, this does not mean that Li_2 is the most stable form of elemental lithium. In fact, at normal temperature and pressure, lithium exists as a solid containing many lithium atoms bound to each other.

For the beryllium molecule (Be_2) the bonding and antibonding orbitals both contain two electrons. In this case the bond order is $(2-2)/2 = 0$, and since Be_2 is not more stable than two separated Be atoms, no molecule forms. However, beryllium metal contains many beryllium atoms bonded to each other and is stable for reasons we will discuss in Chapter 10.

Since the boron atom has a $1s^22s^22p^1$ configuration, we describe the B_2 molecule by considering how *p* atomic orbitals combine to form molecular orbitals.

Beryllium metal.

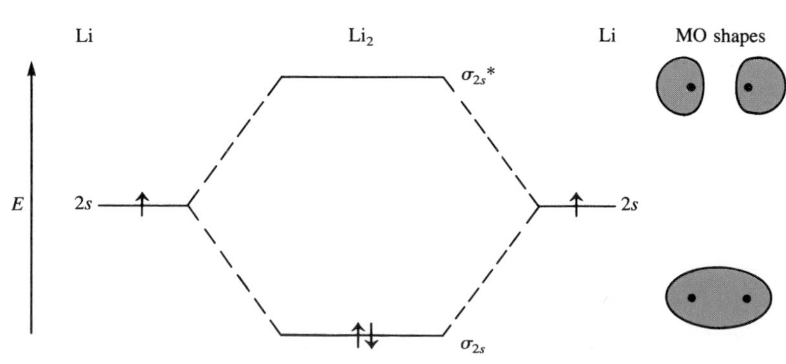

Figure 9.31

The molecular orbital energy-level diagram for the Li_2 molecule.

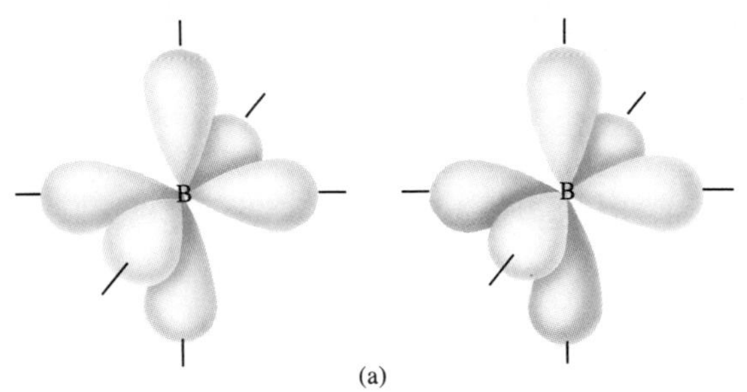

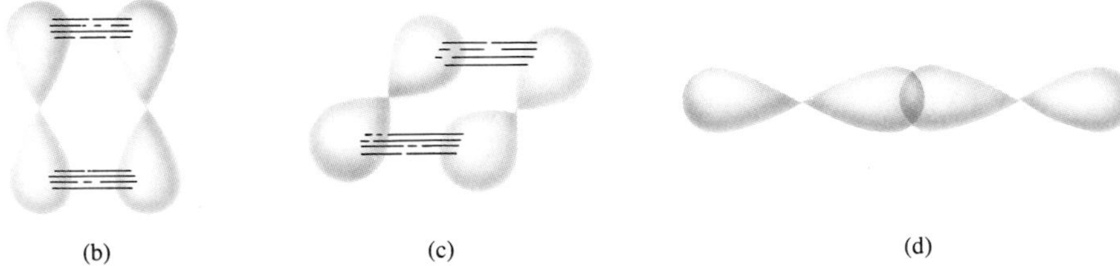

Figure 9.32

(a) The three mutually perpendicular 2*p* orbitals on two adjacent boron atoms. Two pairs of parallel *p* orbitals can overlap as shown in (b) and (c), and the third pair can overlap head-on as shown in (d).

Recall that *p* orbitals have two lobes and that they occur in sets of three mutually perpendicular orbitals [see Fig. 9.32(a)]. When two B atoms approach each other, two pairs of *p* orbitals can overlap in a parallel fashion [Fig. 9.32(b) and (c)] and one pair can overlap head-on [Fig. 9.32(d)].

First, let's consider the molecular orbitals from the head-on overlap, as shown in Fig. 9.33(a). Note that the electrons in the bonding MO are, as expected, concentrated between the nuclei, and the electrons in the antibonding MO are concentrated outside the area between the two nuclei. Also both of these MOs are σ molecular orbitals. The *p* orbitals that overlap in a parallel fashion also produce bonding and antibonding orbitals [Fig. 9.33(b)]. Since the electron probability lies above and

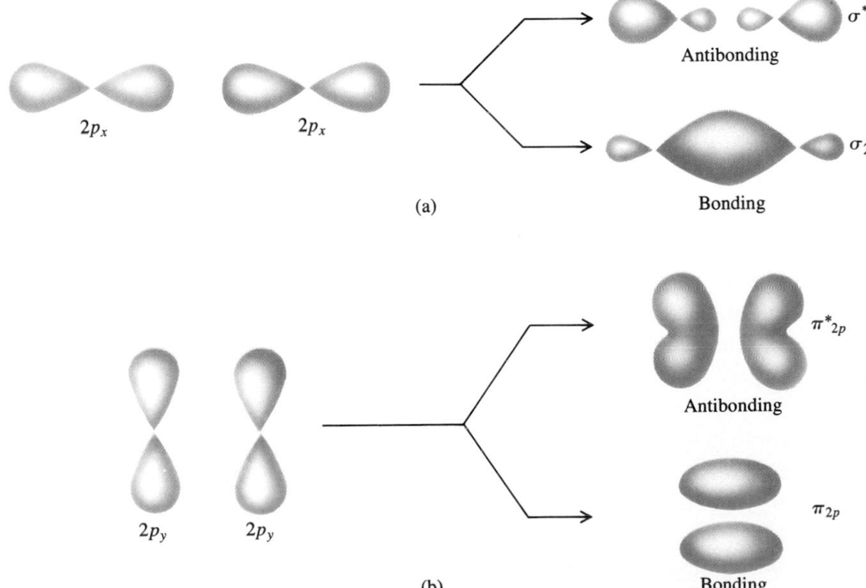

Figure 9.33

(a) The two *p* orbitals on the boron atom that overlap head-on produce two σ molecular orbitals, one bonding and one antibonding.
(b) Two *p* orbitals that lie parallel overlap to produce two π molecular orbitals, one bonding and one antibonding.

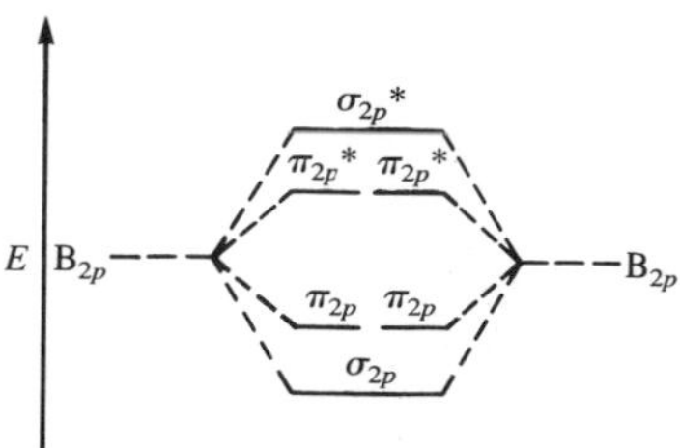

Figure 9.34

The *expected* molecular orbital energy-level diagram for the combination of the $2p$ orbitals on two boron atoms.

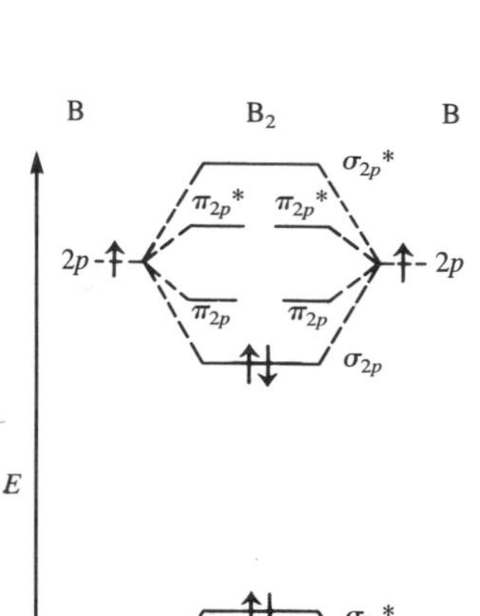

Figure 9.35

The *expected* molecular orbital energy-level diagram for the B_2 molecule.

below the line between the nuclei, both the orbitals are **pi (π) molecular orbitals.** They are designated as π_{2p} for the bonding MO and $\pi_{2p}*$ for the antibonding MO.

Let's try to make an educated guess about the relative energies of the σ and π molecular orbitals formed from the $2p$ atomic orbitals. Would we expect the electrons to prefer the σ bonding orbital (where the electron probability is concentrated in the area between the nuclei) or the π bonding orbital? The σ orbital would seem to have the lower energy since the electrons are closest to the two nuclei. This agrees with the observation that σ interactions are stronger than π interactions.

Figure 9.34 gives the molecular orbital energy-level diagram *expected* when the two sets of $2p$ orbitals on the boron atoms combine to form molecular orbitals. Note that there are two π bonding orbitals at the same energy (degenerate orbitals) formed from the two pairs of parallel p orbitals, and there are two degenerate π antibonding orbitals. The energy of the π_{2p} orbitals is expected to be higher than that of the σ_{2p} orbital because σ interactions are generally stronger than π interactions.

To construct the total molecular orbital diagram for the B_2 molecule, we make the assumption that the $2s$ and $2p$ orbitals combine separately (in other words, there is no $2s$-$2p$ mixing). The resulting diagram is shown in Fig. 9.35. Note that B_2 has six *valence* electrons. (Remember the $1s$ orbitals and electrons are assumed not to participate in the bonding.) This diagram predicts the bond order:

$$\frac{4-2}{2} = 1$$

Therefore, B_2 should be a stable molecule.

Paramagnetism

At this point we need to discuss an additional molecular property—magnetism. Most materials have no magnetism until they are placed in a magnetic field. However, in the presence of such a field, magnetism of two types can be induced. **Paramagnetism** causes the substance to be attracted into the inducing magnetic field. **Diamagnetism** causes the substance to be repelled from the inducing magnetic field. Figure 9.36 illustrates how paramagnetism is measured. The sample is weighed with the electromagnet turned off and then weighed again with the electromagnet turned on. An increase in weight when the field is turned on indicates the sample is paramagnetic. Studies have shown that *paramagnetism is associated with unpaired electrons, and diamagnetism is associated with paired electrons.* Any substance that has both paired and unpaired electrons will exhibit paramagnetism, since the effect of paramagnetism is much stronger than that of diamagnetism.

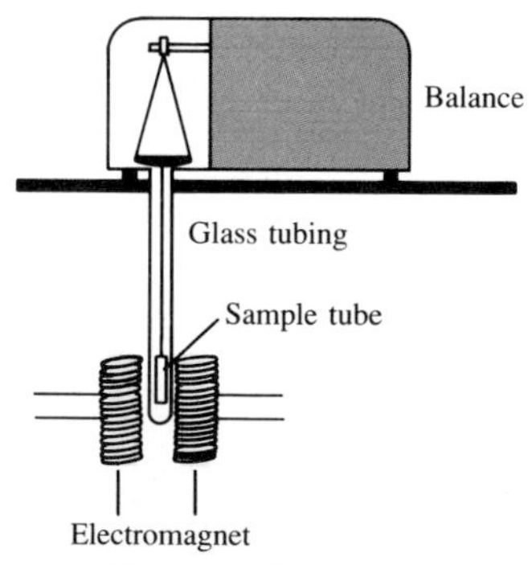

Figure 9.36

An apparatus used to measure the paramagnetism of a sample. A paramagnetic sample will appear heavier when the electromagnet is turned on because the sample is attracted into the inducing magnetic field.

The molecular orbital energy-level diagram represented in Fig. 9.35 predicts that the B_2 molecule will be diamagnetic since the MOs contain only paired electrons. However, experiments show that B_2 is actually paramagnetic with two unpaired electrons. Why does the model yield the wrong prediction? This is yet another illustration of how models are developed and used. In general, we try to use the simplest possible model that accounts for all of the important observations. In this case, although the simplest model successfully describes all of the diatomic molecules up to B_2, it certainly is suspect if it cannot describe the B_2 molecule correctly. This means we must either discard the model or find a way to modify it.

Let's reconsider one assumption that we made. In our treatment of B_2, we have assumed that the *s* and *p* orbitals combine separately to form molecular orbitals. Calculations show that when the *s* and *p* orbitals are allowed to mix in the same molecular orbital, a different energy-level diagram results for B_2 (see Fig. 9.37). Note that even though the *s* and *p* contributions to the MOs are no longer separate, we retain the simple orbital designations. The energies of π_{2p} and σ_{2p} orbitals are reversed by *p-s* mixing, and the σ_{2s} and the σ_{2s}* are no longer equally spaced relative to the energy of the free 2*s* orbital.

When the six valence electrons of the B_2 molecule are placed in the modified energy-level diagram, each of the last two electrons goes into one of the degenerate π_{2p} orbitals. This produces a paramagnetic molecule in agreement with experimental results. Thus when the model is extended to allow *p-s* mixing in molecular orbitals, it predicts the correct magnetism. Note that the bond order is $(4 - 2)/2 = 1$, as before.

The remaining homonuclear diatomic molecules of the Period 2 elements can be described using the same molecular orbitals as for B_2 (see Fig. 9.37) and inserting the correct number of electrons. These are summarized in Fig. 9.38, together

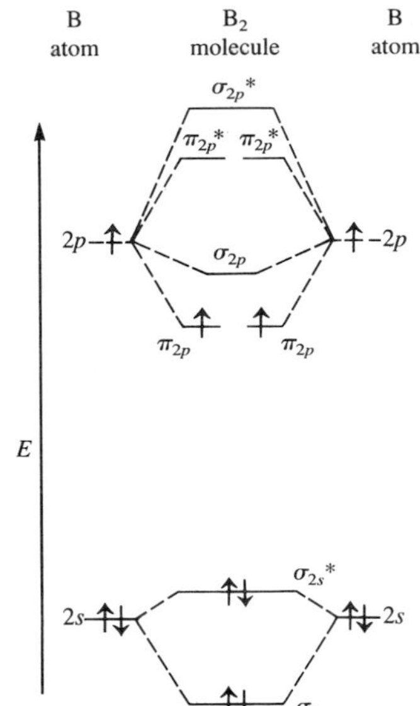

Figure 9.37

The *correct* molecular orbital energy-level diagram for the B_2 molecule. When *p-s* mixing is allowed, the energies of the σ_{2p} and π_{2p} orbitals are reversed. The two electrons from the B 2*p* orbitals now occupy separate, degenerate π_{2p} molecular orbitals and thus have parallel spins. Therefore, this diagram explains the observed paramagnetism of B_2.

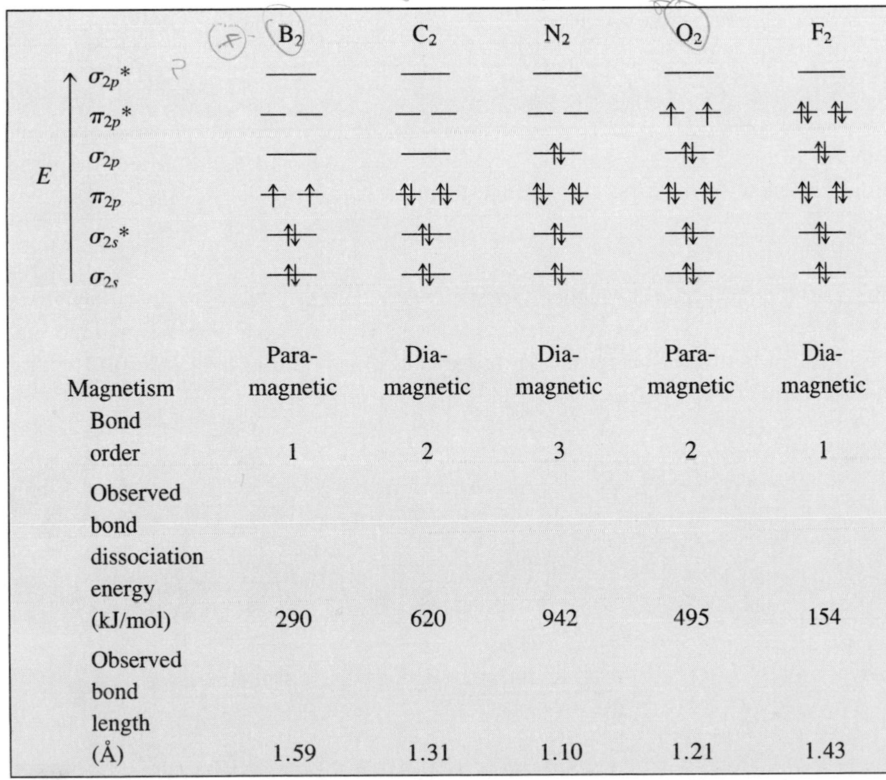

	B_2	C_2	N_2	O_2	F_2
σ_{2p}*	—	—	—	—	—
π_{2p}*	— —	— —	— —	↑ ↑	↑↓ ↑↓
σ_{2p}	—	—	↑↓	↑↓	↑↓
π_{2p}	↑ ↑	↑↓ ↑↓	↑↓ ↑↓	↑↓ ↑↓	↑↓ ↑↓
σ_{2s}*	↑↓	↑↓	↑↓	↑↓	↑↓
σ_{2s}	↑↓	↑↓	↑↓	↑↓	↑↓
Magnetism	Paramagnetic	Diamagnetic	Diamagnetic	Paramagnetic	Diamagnetic
Bond order	1	2	3	2	1
Observed bond dissociation energy (kJ/mol)	290	620	942	495	154
Observed bond length (Å)	1.59	1.31	1.10	1.21	1.43

Bond energy increases with bond order; bond length decreases with increasing bond order.

Figure 9.38

The molecular orbital energy-level diagrams, bond orders, bond energies, and bond lengths for the diatomic molecules B_2 through F_2. There is evidence that for O_2 and F_2 the σ_{2p} orbital is lower in energy than the π_{2p} orbitals. However, since this does not affect the predicted magnetism or bond order for these molecules, we will use the same order of MOs for all of the diatomic molecules of the Period 2 elements.

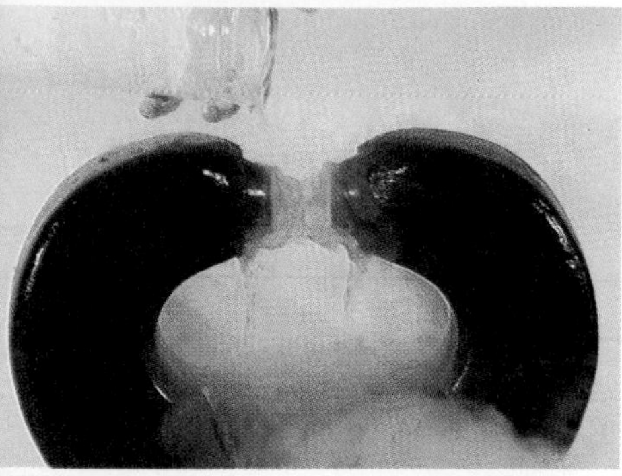

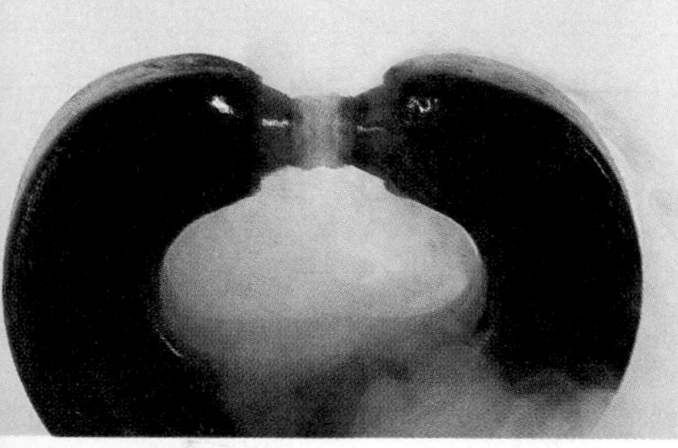

Figure 9.39

When liquid oxygen is poured into the space between the poles of a strong magnet, it remains there until it boils away. This attraction of liquid oxygen for the magnetic field demonstrates the paramagnetism of the O_2 molecule.

with experimentally obtained bond strengths and lengths. Several significant points arise from these results.

1. There are definite correlations between bond order, bond energy, and bond length. As the bond order predicted by the molecular orbital model increases, the bond energy increases and the bond length decreases. This is a clear indication that the bond order predicted by the model accurately reflects bond strength, and it strongly supports the reasonableness of the MO model.
2. Comparison of the bond energies of the B_2 and F_2 molecules indicates that bond order cannot automatically be associated with a particular bond energy. Although both molecules have a bond order of 1, the bond in B_2 appears to be about twice as strong as the bond in F_2. As we will see in our later discussion of the halogens, F_2 has an unusually weak single bond due to larger than usual electron-electron repulsions (there are 14 valence electrons on the small F_2 molecule).
3. Note the very large bond energy associated with the N_2 molecule, which the molecular orbital model predicts will have a bond order of 3, a triple bond. The very strong bond in N_2 is the principal reason that many nitrogen-containing compounds are used as high explosives. The reactions involving these explosives give the very stable N_2 molecule as a product, thus releasing large quantities of energy.
4. The O_2 molecule is known to be paramagnetic. This can be very convincingly demonstrated by pouring liquid oxygen between the poles of a strong magnet, as shown in Fig. 9.39. The oxygen remains there until it evaporates. Significantly, the molecular orbital model correctly predicts oxygen's paramagnetism, while the localized electron model predicts a diamagnetic molecule.

Sample Exercise 9.6

For the species O_2, O_2^+, and O_2^-, give the electron configuration and the bond order for each. Which has the strongest bond?

Solution

The O_2 molecule has 12 valence electrons (6 + 6); O_2^+ has 11 valence electrons (6 + 6 − 1); and O_2^- has 13 valence electrons (6 + 6 + 1). We will assume that the ions can be treated using the same molecular orbital diagram as for the neutral diatomic molecule:

	O_2	O_2^+	O_2^-
σ_{2p}^*	—	—	—
π_{2p}^*	↑ ↑	↑ —	↑↓ ↑
σ_{2p}	↑↓	↑↓	↑↓
π_{2p}	↑↓ ↑↓	↑↓ ↑↓	↑↓ ↑↓
σ_{2s}^*	↑↓	↑↓	↑↓
σ_{2s}	↑↓	↑↓	↑↓

Sample Exercise 9.6, continued

The electron configuration for each species can then be taken from the diagram:

O_2: $(\sigma_{2s})^2(\sigma_{2s}{}^*)^2(\pi_{2p})^4(\sigma_{2p})^2(\pi_{2p}{}^*)^2$

$O_2{}^+$: $(\sigma_{2s})^2(\sigma_{2s}{}^*)^2(\pi_{2p})^4(\sigma_{2p})^2(\pi_{2p}{}^*)^1$

$O_2{}^-$: $(\sigma_{2s})^2(\sigma_{2s}{}^*)^2(\pi_{2p})^4(\sigma_{2p})^2(\pi_{2p}{}^*)^3$

The bond orders are

$$\text{For } O_2\text{:} \quad \frac{8-4}{2} = 2$$

$$\text{For } O_2{}^+\text{:} \quad \frac{8-3}{2} = 2.5$$

$$\text{For } O_2{}^-\text{:} \quad \frac{8-5}{2} = 1.5$$

Thus $O_2{}^+$ is expected to have the strongest bond of the three species.

Sample Exercise 9.7

Use the molecular orbital model to predict the bond order and magnetism of each molecule:

a. Ne_2 **b.** P_2

Solution

a. The valence orbitals for Ne are $2s$ and $2p$. Thus we can use the molecular orbitals we have already constructed for diatomic molecules of Period 2 elements. The Ne_2 molecule has 16 valence electrons (8 from each atom). Placing these electrons in the appropriate molecular orbitals produces the following diagram:

E ↑

$\sigma_{2p}{}^*$ ⇅

$\pi_{2p}{}^*$ ⇅ ⇅

σ_{2p} ⇅

π_{2p} ⇅ ⇅

$\sigma_{2s}{}^*$ ⇅

σ_{2s} ⇅

The bond order is $(8 - 8)/2 = 0$, and Ne_2 does not exist.

b. The P_2 molecule contains phosphorus atoms from the third row of the periodic table. We will assume that diatomic molecules of the Period 3 elements can be treated in a way very similar to that we have used so far. The only change will be that the molecular orbitals will be formed from $3s$ and $3p$ atomic orbitals. The P_2 molecule has 10 valence electrons (5 from each phosphorus atom). The resulting molecular orbital diagram is shown on the following page.

Sample Exercise 9.7, continued

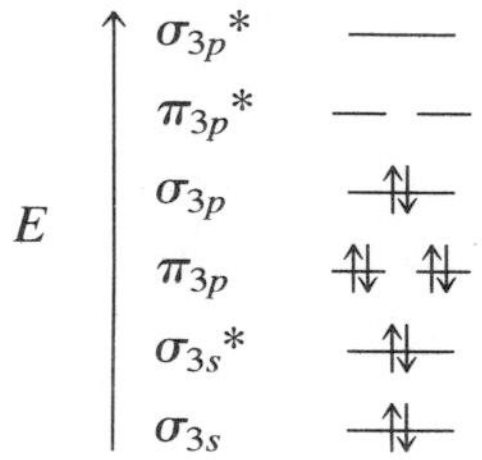

$\sigma_{3p}{}^*$	—
$\pi_{3p}{}^*$	— —
σ_{3p}	↑↓
π_{3p}	↑↓ ↑↓
$\sigma_{3s}{}^*$	↑↓
σ_{3s}	↑↓

The molecule has a bond order of 3 and is expected to be diamagnetic.

9.4 Bonding in Heteronuclear Diatomic Molecules

Purpose

- To use the molecular orbital model to treat bonding between two different atoms.

In this section we will deal with selected examples of **heteronuclear** (different atoms) **diatomic molecules.** A special case involves molecules containing atoms adjacent to each other in the periodic table. Since the atoms involved in such a molecule are so similar, we can use the molecular orbital diagram for homonuclear molecules. For example, we can predict the bond order and magnetism of nitric oxide (NO) by placing its 11 valence electrons (5 from nitrogen and 6 from oxygen) in the molecular orbital energy-level diagram shown in Fig. 9.40. The molecule should be paramagnetic and has a bond order of

$$\frac{8-3}{2} = 2.5$$

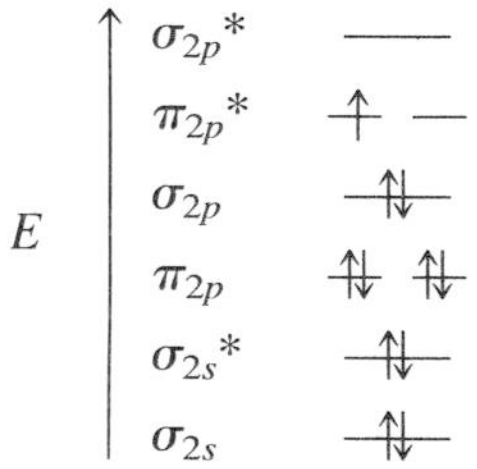

Experimentally, nitric oxide is indeed found to be paramagnetic. Note that this odd electron molecule is described very naturally by the MO model. In contrast, the localized electron model, in the simple form used in this text, cannot be used readily to treat such molecules.

$\sigma_{2p}{}^*$	—
$\pi_{2p}{}^*$	↑ —
σ_{2p}	↑↓
π_{2p}	↑↓ ↑↓
$\sigma_{2s}{}^*$	↑↓
σ_{2s}	↑↓

Figure 9.40

The molecular orbital energy-level diagram for the NO molecule. The bond order is 2.5.

Sample Exercise 9.8

Use the molecular orbital model to predict the magnetism and bond order of the NO^+ and CN^- ions.

Solution

The NO^+ ion has 10 valence electrons $(5 + 6 - 1)$. The CN^- ion also has 10 valence electrons $(4 + 5 + 1)$. Both ions are therefore diamagnetic and have a bond order derived from the equation

Sample Exercise 9.8, continued

$$\frac{8-2}{2} = 3$$

The molecular orbital diagram for these two ions is the same (see Fig. 9.41).

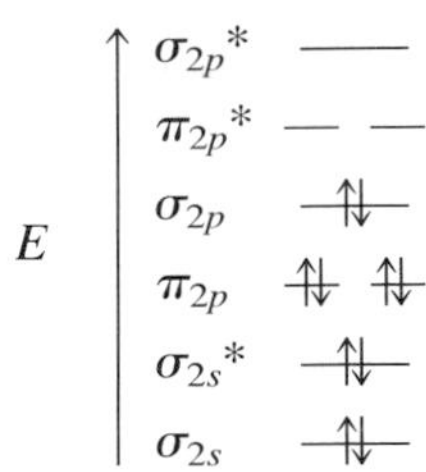

Figure 9.41

The molecular orbital energy-level diagram for both the NO^+ and CN^- ions.

When the two atoms of a diatomic molecule are very different, the energy-level diagram for homonuclear molecules can no longer be used. A new diagram must be devised for each molecule. We will illustrate this case by considering the hydrogen fluoride (HF) molecule. The electron configurations of the hydrogen and fluorine atoms are $1s^1$ and $1s^22s^22p^5$, respectively. To keep things as simple as possible, we will assume that fluorine uses only one of its $2p$ orbitals to bond to hydrogen. Thus the molecular orbitals for HF will be composed of fluorine $2p$ and hydrogen $1s$ orbitals. Figure 9.42 gives the partial molecular orbital energy-level diagram for HF, focusing only on the orbitals involved in the bonding. We are assuming that fluorine's other valence electrons remain localized on the fluorine. The $2p$ orbital of fluorine is shown at a lower energy than the $1s$ orbital of hydrogen on the diagram because fluorine binds its valence electrons more tightly. Thus the $2p$ electron on a free fluorine atom is at lower energy than the $1s$ electron on a free hydrogen atom. The diagram predicts that the HF molecule should be stable since both electrons are lowered in energy relative to their energy in the free hydrogen and fluorine atoms, and this is the driving force for bond formation.

Because the fluorine $2p$ orbital is lower in energy than the hydrogen $1s$ orbital, the electrons prefer to be closer to the fluorine atom. That is, the σ molecular orbital containing the bonding electron pair shows greater electron probability close to the fluorine (see Fig. 9.43). The electron pair is not shared equally. This causes the fluorine atom to have a slight excess of negative charge and leaves the hydrogen atom partially positive. This is *exactly* the bond polarity observed for HF. Thus the molecular orbital model accounts in a straightforward way for the different electronegativities of hydrogen and fluorine and the resulting unequal charge distribution.

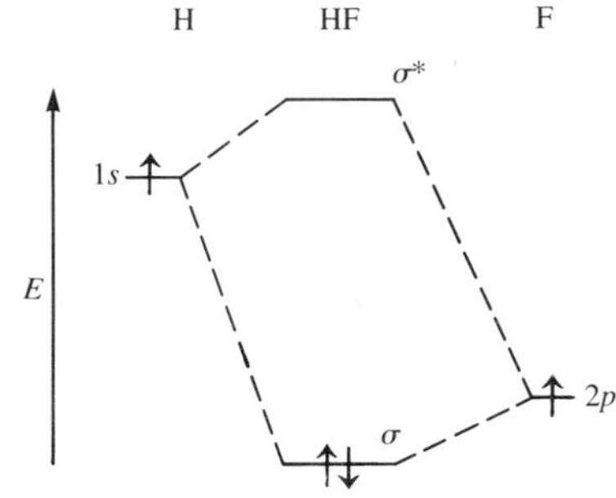

Figure 9.42

A partial molecular orbital energy-level diagram for the HF molecule.

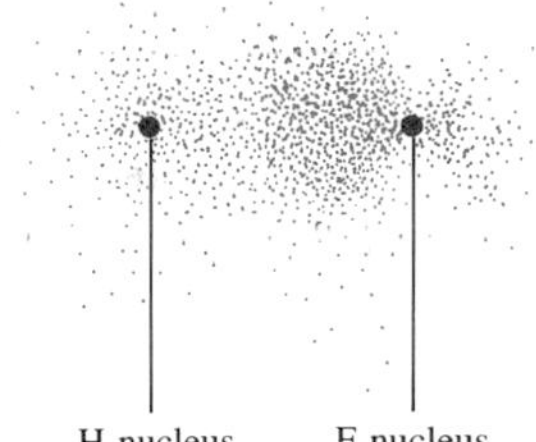

Figure 9.43

The electron probability distribution in the bonding molecular orbital of the HF molecule. Note the greater electron density close to the fluorine atom.

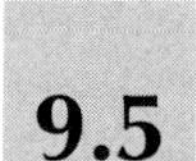

9.5 Combining the Localized Electron and Molecular Orbital Models

Purpose

- To show how the need for resonance is eliminated if the localized electron and molecular orbital models are combined.

One of the main difficulties with the localized electron model is its assumption that electrons are localized. This problem is most apparent with molecules for which several valid Lewis structures can be drawn. It is clear that none of these structures taken alone adequately describes the electronic structure of the molecule. The concept of resonance was invented to solve this problem. However, even with resonance included, the localized electron model does not describe molecules and ions such as O_3 and NO_3^- in a very satisfying way.

It would seem that the ideal bonding model would be one with the simplicity of the localized electron model but with the delocalization characteristic of the molecular orbital model. We can achieve this by combining the two models to describe molecules that require resonance. Note that for species such as O_3 and NO_3^- the double bond changes position in the resonance structures (see Fig. 9.44). Since a double bond involves one σ and one π bond, there is a σ bond between all bound atoms in each resonance structure. It is really the π bond that has different locations in the various resonance structures.

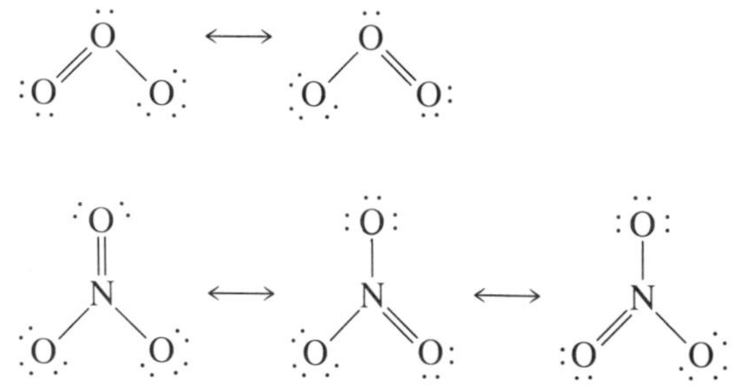

Figure 9.44

The resonance structures for O_3 and NO_3^-. Note that it is the double bond that occupies various positions in the resonance structures.

In molecules that require resonance, it is the π bonding that is most clearly delocalized.

We can conclude that the σ bonds in a molecule can be described as being localized with no apparent problems. It is the π bonding that must be treated as being delocalized. Thus, for molecules that require resonance we will use the localized electron model to describe the σ bonding and the molecular orbital model to describe the π bonding. This allows us to keep the bonding model as simple as possible and yet give a more physically accurate description of such molecules.

We will illustrate the general method by considering the bonding in benzene, an important industrial chemical that must be handled carefully because it is a known carcinogen. The benzene molecule (C_6H_6) consists of a planar hexagon of carbon atoms with one hydrogen atom bound to each carbon atom, as shown in Fig. 9.45(a). In the molecule all six C—C bonds are known to be equivalent. To explain this fact, the localized electron model must invoke resonance [see Fig. 9.45(b)].

A better description of the bonding in benzene results when we use a combination of the models, as described above. In this description it is assumed that the σ bonds to carbon involve sp^2 orbitals, as shown in Fig. 9.46. These σ bonds are all centered in the plane of the molecule.

Since each carbon atom is sp^2 hybridized, a p orbital perpendicular to the plane of the ring remains on each carbon atom. These six p orbitals can be used to form π molecular orbitals, as shown in Fig. 9.47(a). The electrons in the resulting π molecular orbitals are delocalized above and below the plane of the ring, as shown in Fig.

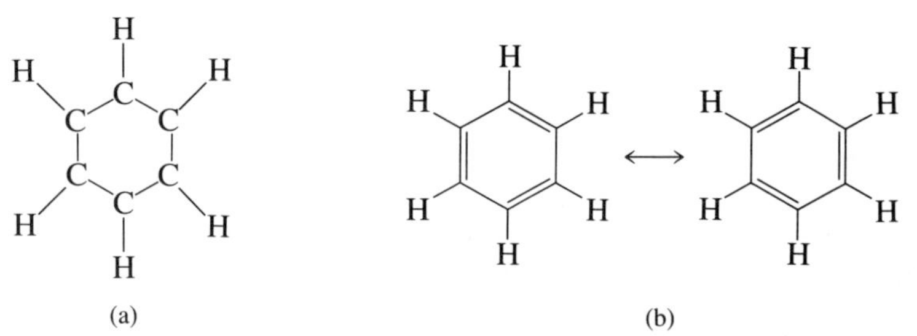

Figure 9.45

(a) The benzene molecule consists of a ring of six carbon atoms with one hydrogen atom bound to each carbon; all atoms are in the same plane. All of the C—C bonds are known to be equivalent. (b) Two of the resonance structures for the benzene molecule. The localized electron model must invoke resonance to account for the six equal C—C bonds.

(a)

(b)

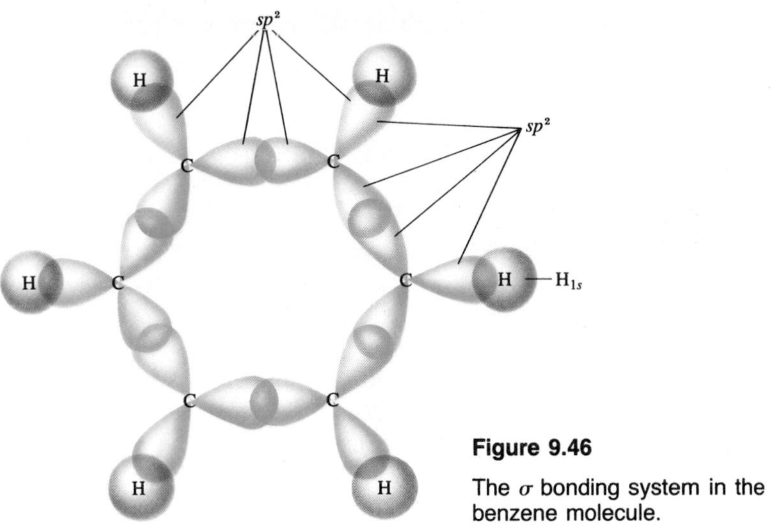

Figure 9.46

The σ bonding system in the benzene molecule.

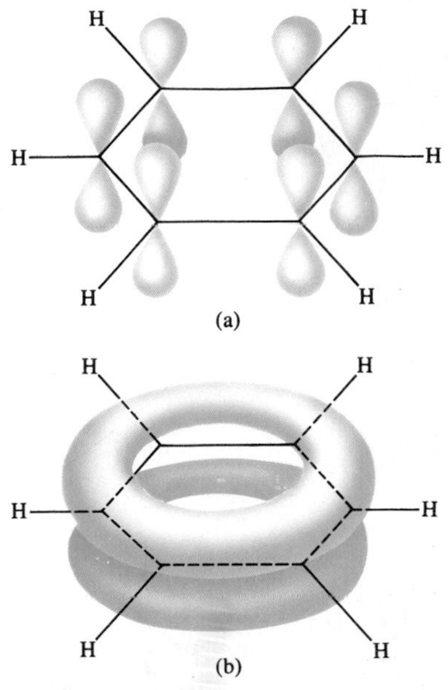

Figure 9.47

(a) The π molecular orbital system in benzene is formed by combining the six p orbitals from the six sp^2 hybridized carbon atoms. (b) The electrons in the resulting π molecular orbitals are delocalized over the entire ring of carbon atoms, giving six equivalent bonds. A composite of these orbitals is represented here.

9.47(b). This gives six equivalent C—C bonds, as required by the known structure of the benzene molecule. The benzene structure is often written as

to indicate the **delocalized π bonding** in the molecule.

Very similar treatments can be applied to other planar molecules for which resonance is required by the localized electron model. For example, the NO_3^- ion can be described using the π molecular orbital system shown in Fig. 9.48. In this molecule each atom is assumed to be sp^2 hybridized, which leaves one p orbital on each atom perpendicular to the plane of the ion. These p orbitals can combine to form the π molecular orbital system.

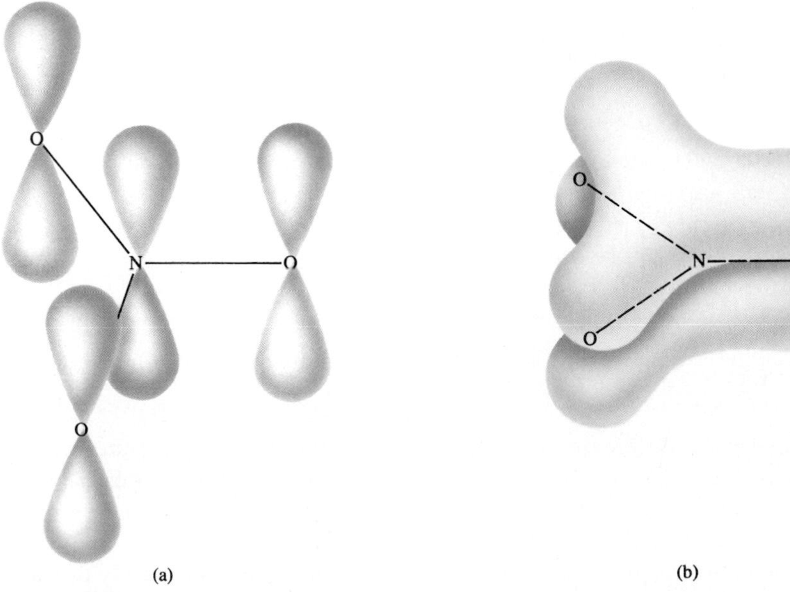

Figure 9.48

(a) The p orbitals used to form the π bonding system in the NO_3^- ion.
(b) A representation of the delocalization of the electrons in the π molecular orbital system of the NO_3^- ion.

FOR REVIEW

Summary

In this chapter we have emphasized the role of orbitals in the covalent bonding in molecules. The two most useful models are the localized electron model and the molecular orbital model.

In the localized electron model, a molecule is pictured as a group of atoms sharing electron pairs using atomic orbitals. Because the requirements of the molecule are different from those of separated atoms, the original atomic orbitals are combined to form new hybrid orbitals, which are used by the atoms in the molecule. The specific set of hybrid orbitals needed is determined by the arrangement of electron pairs that gives the minimum repulsion. For example, an atom surrounded by four electron pairs requires a tetrahedral arrangement and a set of sp^3 hybrid orbitals; a trigonal planar arrangement requires sp^2 hybridization, and a linear arrangement requires *sp* hybridization.

When an electron pair is shared in the area centered on a line joining the atoms, the covalent bond is called a σ bond. A π bond occurs when *p* orbitals overlap in a parallel fashion. The shared electron pair then occupies the space above and below the line joining the bonded atoms.

Multiple bonds involve σ and π bonds. A double bond consists of a σ bond and a π bond; a triple bond involves a σ bond and two π bonds.

In the molecular orbital model for covalent bonding, the molecule is assumed to be a new entity consisting of positively charged nuclei and electrons. We use calculations similar to those used for atoms to determine the allowed orbitals in the molecule. The molecular orbitals are constructed from the valence orbitals of the atoms involved in forming the molecule.

Molecular orbitals are classified in two ways:

1. MOs are classified according to their energy. A bonding MO is an orbital that is lower in energy than the original atomic orbitals. In going from the free atoms to the molecule, electrons occupying bonding MOs are lowered in energy. This favors formation of a stable molecule (bonds are formed). An antibonding MO is an orbital that is higher in energy than the original atomic orbitals. In going from the free atoms to the molecule, electrons in antibonding MOs are raised in energy. These electrons oppose bonding.
2. MOs are classified according to their shape (symmetry). Sigma (σ) molecular orbitals have their electron probability centered about a line passing through the two nuclei. Pi (π) molecular orbitals have their electron probability concentrated above and below the line passing through the nuclei.

Bond order is an index of bond strength:

$$\text{Bond order} = \frac{\text{number of bonding electrons} - \text{number of antibonding electrons}}{2}$$

The greater the difference between the number of electrons that are lowered in energy (bonding electrons) and the number of electrons that are raised in energy (antibonding electrons) when the molecule forms, the greater will be the stability of the molecule (the stronger the bonding).

The main strengths of the molecular orbital model are: it correctly predicts the relative bond strength and magnetism of simple diatomic molecules (the correct prediction of the paramagnetism of the oxygen molecule is an important example); it accounts for bond polarity; and it correctly portrays electrons as being delocalized in polyatomic molecules. The main disadvantage of the molecular orbital model is that it is difficult to apply qualitatively to polyatomic molecules.

The concept of resonance is required for certain molecules because the localized electron model incorrectly assumes that electrons are located between a given pair of atoms in a molecule. A more physically accurate description of these molecules can be given by combining the localized electron and molecular orbital models. The σ bonds in these molecules can be regarded as localized, while the π bonding is best described as delocalized.

Key Terms

Section 9.1
hybridization
sp^3 hybridization
hybrid orbitals
sp^2 hybridization
sigma (σ) bond
pi (π) bond
sp hybridization
dsp^3 hybridization
d^2sp^3 hybridization

Section 9.2
molecular orbital model
molecular orbital (MO)
sigma (σ) molecular orbital
bonding molecular orbital
antibonding molecular orbital
bond order

Section 9.3
pi (π) molecular orbital
paramagnetism
diamagnetism

Section 9.4
heteronuclear diatomic molecule

Section 9.5
delocalized π bonding

Exercises

A blue exercise number indicates that the answer to that exercise appears at the back of this book and a solution appears in the Solutions Guide.

The Localized Electron Model and Hybrid Orbitals

1. For each of the following molecules, write the Lewis structure, predict the molecular structure including bond angles, give the expected hybrid orbitals on the central atom, and predict the overall polarity.

a. CF_4
b. NF_3
c. OF_2
d. BF_3
e. BeH_2
f. TeF_4
g. AsF_5
h. KrF_2
i. KrF_4
j. SeF_6
k. $XeOF_4$
l. $XeOF_2$
m. XeO_4

2. Predict the hybrid orbitals used by the sulfur atom(s) in each of the following:

a. SO_2
b. SO_3
c. $S_2O_3^{2-}$ $[S—S(O)(O)—O]^{2-}$
d. $S_2O_8^{2-}$ $[O—S(O)(O)—O—O—S(O)(O)—O]^{2-}$
e. SO_3^{2-}
f. SO_4^{2-}
g. SF_2
h. SF_4
i. SF_6
j. $F_3S—SF$

3. Phosphorus forms many compounds with three-dimensional cage structures. Some are shown below. Complete the Lewis structures showing all unshared electron pairs. What is the hybridization of phosphorus in each of the molecules? Predict approximate values of the O—P—O and P—O—P bond angles.

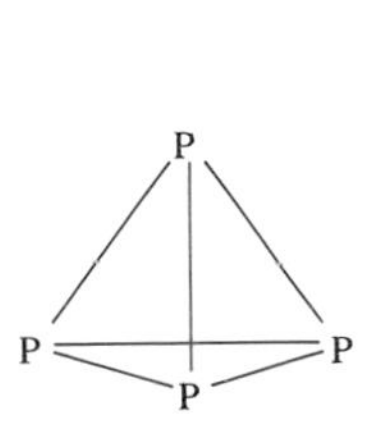

a. P_4

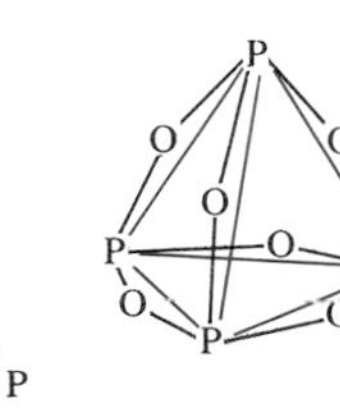

b. P_4O_6

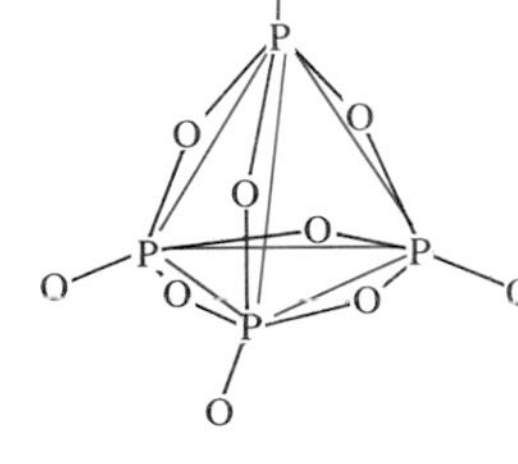

c. P_4O_{10}

4. Why must all six atoms in C_2H_4 be in the same plane?

5. The allene molecule has the following Lewis structure:

$H_2C{=}C{=}CH_2$

Are all four hydrogen atoms in the same plane? If not, what is the spatial relationship among them? Why?

6. Biacetyl and acetoin are added to margarine to make it taste more like butter.

$CH_3—C(=O)—C(=O)—CH_3$ (Biacetyl) $CH_3—CH(OH)—C(=O)—CH_3$ (Acetoin)

Biacetyl Acetoin

Complete the Lewis structures, predict values for all C—C—O bond angles, and give the hybridization of the carbon atoms in these two compounds. Are the four carbons and two oxygens in biacetyl in the same plane?

7. How many σ bonds and how many π bonds are there in biacetyl and acetoin (see Exercise 6)?

8. Many important compounds in the chemical industry are derivatives of ethylene, C_2H_4. Two of them are methyl methacrylate and acrylonitrile.

Methyl methacrylate: $H_2C=C(CH_3)—C(=O)—O—CH_3$ (angles d, e, f) Acrylonitrile: $H_2C=CH—C\equiv N$ (angles a, b, c)

Methyl methacrylate Acrylonitrile

Complete the Lewis structures, showing all lone pairs. Give approximate values for bond angles a through f. Give the hybridization of all carbon atoms. In methylmethacrylate, how many of the atoms in the molecule lie in the same plane?

9. How many σ bonds and how many π bonds are there in methyl methacrylate and acrylonitrile (see Exercise 8)?

10. One of the first drugs to be approved for use in treatment of acquired immune deficiency syndrome (AIDS) is azidothymidine (AZT). Complete the Lewis structure of AZT.

a. How many carbon atoms use sp^3 hybridization?
b. How many carbon atoms use sp^2 hybridization?
c. What atom is sp hybridized?
d. How many σ bonds are in the molecule?
e. How many π bonds are in the molecule?
f. Which ring(s) is/are planar?
g. What is the N—N—N bond angle in the azide group?
h. What is the H—O—C bond angle in the side group attached to the five membered ring?
i. What is the hybridization of the oxygen atom in the —CH_2OH group?

11. Cyanamide (H_2NCN), an important industrial chemical, is produced by the following steps:

$$CaC_2 + N_2 \longrightarrow CaNCN + C$$

$$CaNCN \xrightarrow{\text{Acid}} \underset{\text{Cyanamide}}{H_2NCN}$$

Calcium cyanamide (CaNCN) is used as a direct application fertilizer, weed killer, and cotton defoliant. It is also used to make cyanamide, dicyandiamide, and melamine plastics:

$$H_2NCN \xrightarrow{\text{Acid}} \underset{\text{Dicyandiamide}}{NCNC(NH_2)_2}$$

$$NCNC(NH_2) \xrightarrow[NH_3]{\text{Heat}} \text{Melamine } (C_3N_3(NH_2)_3)$$

Melamine (π bonds not shown)

a. Draw Lewis structures for NCN^{2-}, H_2NCN, dicyandiamide, and melamine, including resonance structures where appropriate.
b. Give the hybridization of the C and N atoms in each species.
c. How many σ bonds and how many π bonds are in each species?
d. Is the ring in melamine planar?
e. There are three different C—N distances in dicyandiamide, NCNC (NH_2), and the molecule is nonlinear. Of all the resonance structures you drew for this molecule, predict which should be the most important.

12. The red color of tomatoes is due to the presence of the compound lycopene.

$$\left[(CH_3)_2C{=}CH—CH_2—CH_2—C(CH_3){=}CH—CH{=}CH—C(CH_3){=}CH—CH{=}CH—C(CH_3){=}CH—CH{=} \right]_2$$

Note: The right half of lycopene is just the reverse of the left half (shown).

a. How many carbon atoms are sp^3 hybridized? sp^2 hybridized? sp hybridized?
b. How many σ and π bonds are there between carbon atoms in the entire molecule?

The Molecular Orbital Model

13. Describe the differences between bonding and antibonding molecular orbitals in terms of electron distribution and energy.

14. Which of the following are predicted by the molecular orbital model to be stable diatomic species?
a. H_2^+, H_2, H_2^-, H_2^{2-}
b. He_2^{2+}, He_2^+, He_2
c. Be_2, B_2, Li_2

15. Using the molecular orbital model to describe the bonding in O_2^+, O_2, O_2^-, and O_2^{2-}, predict the bond orders and the relative bond lengths for these four species. How many unpaired electrons are in each one?

16. Does the molecular orbital model or the localized electron model account better for the bonding in nitric oxide (NO)? Explain.

17. What are the relationships among bond order, bond energy, and bond length? Which of these quantities can be measured?

18. Show how two $2p$ atomic orbitals can combine to form a σ and a π molecular orbital.

19. Show how a H $1s$ atomic orbital and a F $2p$ atomic orbital overlap to form bonding and antibonding molecular orbitals in the hydrogen fluoride molecule. Are these molecular orbitals σ or π molecular orbitals?

20. Using the molecular orbital model, write electron configurations for the following diatomic species and calculate the bond orders. Which ones are paramagnetic?

a. H_2	d. CN^+	g. N_2
b. B_2	e. CN	h. N_2^+
c. F_2	f. CN^-	i. N_2^-

21. Place the species CN^+, CN, and CN^- in order of increasing bond length and increasing bond energy. Use your results from Exercise 20.

22. Place the species N_2, N_2^+, and N_2^- in order of increasing bond length and increasing bond energy. Use your results from Exercise 20.

23. The molecules N_2 and CO are isoelectronic but their properties are quite different. Although as a first approximation we often use the same molecular orbital diagram for both, suggest how the molecular orbitals in N_2 and CO might be different.

24. As compared to CO and O_2, CS and S_2 are very unstable molecules. Give an explanation based on the relative abilities of the sulfur and oxygen atoms to form π bonds.

25. Construct a molecular orbital energy level diagram for the P_2 molecule.

26. Acetylene (C_2H_2) can be produced from the reaction of calcium carbide (CaC_2) with water. Use both the localized electron and molecular orbital models to describe the bonding in the acetylide anion (C_2^{2-}). What neutral, heteronuclear diatomic molecule is isoelectronic with the acetylide anion?

27. Use Figures 9.42 and 9.43 to answer the following questions:
a. Would the bonding molecular orbital in HF place greater electron density near the H or the F atom? Why?
b. Would the bonding molecular orbital have greater fluorine $2p$ character, greater hydrogen $1s$ character, or an equal contribution from both? Why?
c. Answer the previous two questions for the antibonding molecular orbital in HF.

28. Use orbital energy diagrams like that shown for B_2 in Figure 9.37 to answer the following questions:
a. The first ionization energy of N_2 (1501 kJ/mol) is greater than the first ionization energy of atomic nitrogen (1402 kJ/mol). Explain.
b. Would you expect F_2 to have a lower or higher first ionization energy than atomic fluorine? Why?

29. The diatomic molecule OH exists in the gas phase. The bond length and bond energy have been measured to be 97.06 pm and 424.7 kJ/mol, respectively. Assume the OH molecule is analogous to the HF molecule discussed in the chapter and that molecular orbitals result from the overlap of a p_z orbital from oxygen and the $1s$ orbital of hydrogen.
a. Draw pictures of the sigma bonding and antibonding molecular orbitals in OH.
b. Which of the two molecular orbitals will have the greater hydrogen $1s$ character?
c. Can the $2p_x$ orbital of oxygen form molecular orbitals with the $1s$ orbital of hydrogen? Explain.
d. Knowing that only the $2p$ orbitals of oxygen will interact significantly with the $1s$ orbital of hydrogen, complete the molecular orbital energy level diagram for OH. Place the correct number of electrons in the energy levels.
e. Estimate the bond order for OH.
f. Predict whether the bond order of OH^+ will be greater than, less than, or the same as that of OH. Explain.

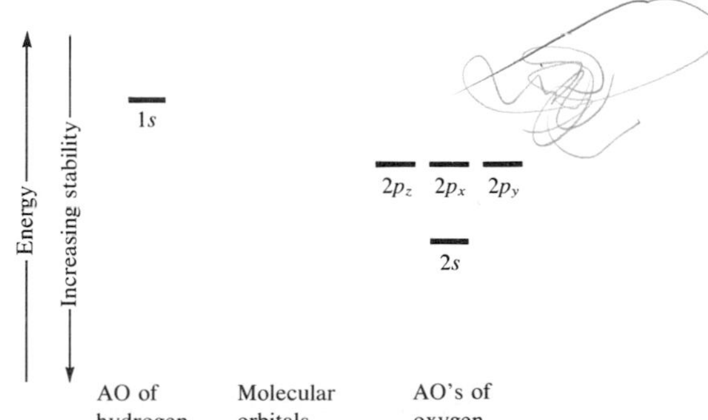

Additional Exercises

30. A variety of chlorine oxide fluorides and related cations and anions are known. They tend to be powerful oxidizing and fluorinating agents. $FClO_3$ is the most stable of this group of compounds and has been studied as an oxidizing component of rocket propellents. Draw Lewis structures, predict molecular structures, and describe the bonding (in terms of hybrid orbitals) for the following:
    a. FClO
    b. $FClO_2$
    c. $FClO_3$
    d. F_3ClO
    e. F_3ClO_2

31. $FClO_2$ and F_3ClO can both gain a fluoride ion to form stable anions. F_3ClO and F_3ClO_2 will both lose a fluoride ion to form stable cations. Draw Lewis structures and describe the hybrid orbitals used by chlorine in these ions.

32. Antimony pentafluoride exists in the liquid phase not as discrete SbF_5 molecules but as a polymer (a large aggregate of small molecules). Describe the molecular structure and the hybrid orbitals in discrete SbF_5 molecules. What hybrid orbitals does antimony employ in the polymer?

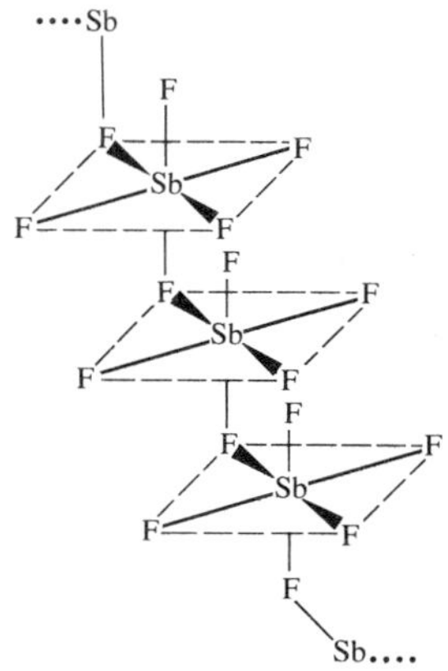

    Experimental observations indicate that the fluorine atoms are in three distinct types of chemical environments in liquid antimony pentafluoride. Indicate the different types of fluorine atoms in the polymer.

33. Indigo is the dye used in coloring blue jeans. The term *navy blue* is derived from the use of indigo to dye British naval uniforms in the eighteenth century. The structure of the indigo molecule is

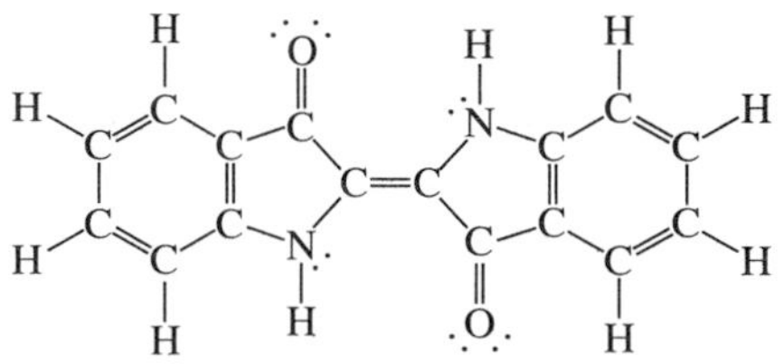

    a. What other resonance structures are possible?
    b. How many σ bonds and π bonds exist in the molecule?
    c. Are the two sets of rings coplanar or at an angle to each other?
    d. What hybrid orbitals are used by the carbon atoms in the indigo molecule?

34. Two structures can be drawn for cyanuric acid.

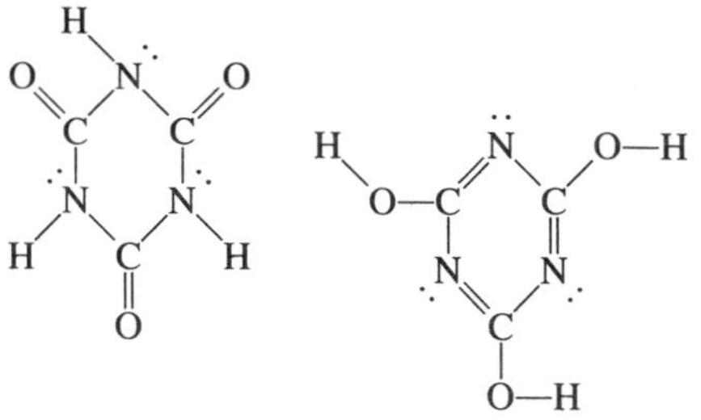

    a. Are these two resonance structures of the same molecule? Why or why not?
    b. Give the hybridization of the carbon and nitrogen atoms in each structure.
    c. Use bond energies (Table 8.4) to predict which form is more stable; i.e., which contains the strongest bonds?

35. Using bond energies from Table 8.4, estimate the barrier to rotation about a carbon-carbon double bond. To do this, consider what must happen to go from

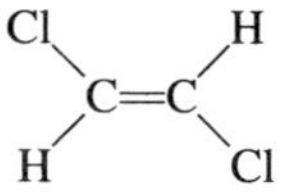

    to

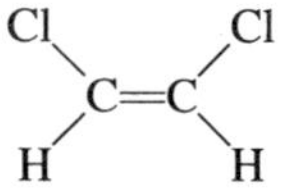

    in terms of making and breaking chemical bonds; i.e., what must happen in terms of the π bond?

36. Describe the bonding in the O_3 molecule and the NO_2^- ion using the localized electron model. How would the molecular orbital model describe the π bonding in these two species?

37. The structure of vitamin C, ascorbic acid, is as follows:

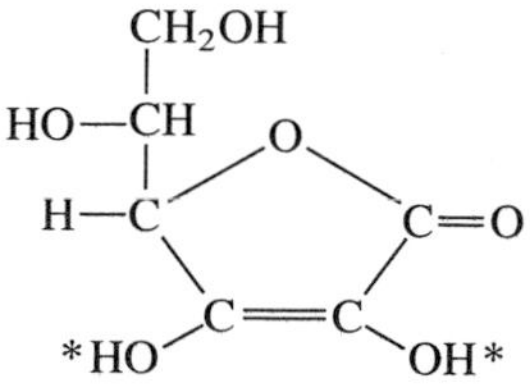

    Indicate which carbon atoms use sp, sp^2, or sp^3 hybrid orbitals. Are all five atoms in the ring in the same plane? Why or why not? Ascorbic acid is a weak acid. The two hydrogens marked with asterisks are the acidic hydrogens. Draw resonance structures for the $C_6H_6O_6^{2-}$ anion.

38. Cholesterol ($C_{27}H_{46}O$) has the following structure:

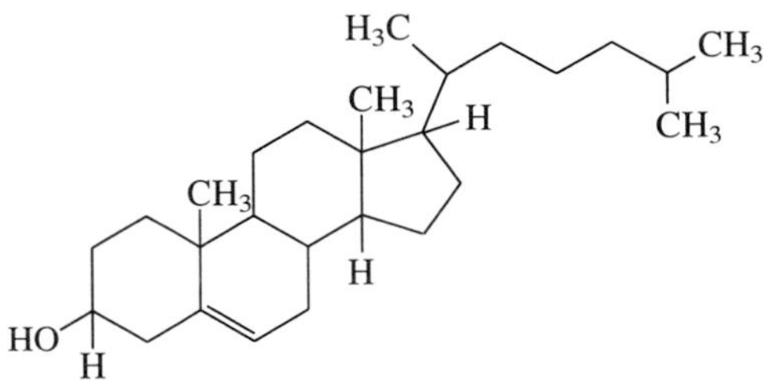

H_3C CH_3 CH_3 H CH_3 CH_3 H HO H

In such shorthand structures, each point where lines meet is a carbon atom. Draw the complete structure showing all carbon and hydrogen atoms. (There will be four bonds to each carbon atom.) Indicate which carbon atoms use sp^2 or sp^3 hybrid orbitals. Are all carbon atoms in the same plane, as implied by the structure?

39. Describe the bonding in NO^+, NO^-, and NO, using both the localized electron and molecular orbital models. From the molecular orbital model, predict the order of bond energies and bond lengths for the nitrogen-oxygen bond in the three species.

40. Describe the bonding in the first excited state of N_2 (the one closest in energy to the ground state) using the molecular orbital model. What differences do you expect in the properties of the molecule in the ground state and the first excited state? (An excited state of a molecule corresponds to an electron arrangement other than that giving the lowest possible energy.)

41. Describe the bonding in the Be_2 molecule using the localized electron model. How does this compare to the description of the bonding in Be_2 using the molecular orbital model?

42. What type of experiment can be done to determine if a material is paramagnetic?

43. A Lewis structure obeying the octet rule can be drawn for O_2 as follows:

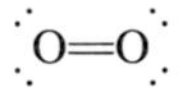

:Ö=Ö:

Use the molecular orbital energy-level diagram for O_2 to show that the above Lewis structure corresponds to an excited state.

44. Complete the Lewis structures of the following molecules. Predict the molecular structure and polarity, bond angles, and hybrid orbitals used by the atoms marked by asterisks for each molecule.

a. $COCl_2$

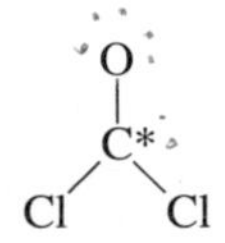

O
|
C*
/ \
Cl Cl

b. N_2F_2

F—N*—N*—F

c. N_2O_3

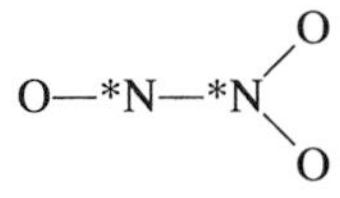

O—*N—*N(—O)(—O)

d. COS

O—C*—S

e. ICl_3

Cl—I*—Cl
|
Cl

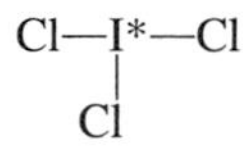

45. Which of the species in Exercise 44 show resonance? Draw all resonance structures for each. Include formal charges in all structures.

46. What is incorrect about the following statement: "The methane molecule (CH_4) is a tetrahedral molecule because the carbon atom is sp^3 hybridized"? What is the correct statement about the relationship between the molecular structure and hybridization in methane?

47. Complete the following resonance structures for $OPCl_3$:

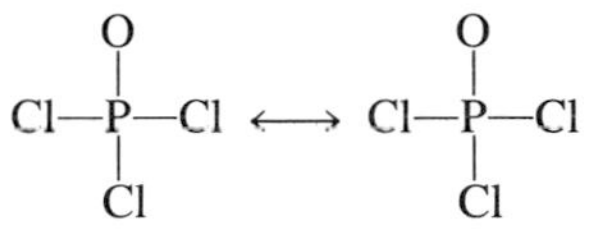

O | Cl—P—Cl | Cl (A) ⟷ O | Cl—P—Cl | Cl (B)

a. Would you predict the same molecular structure from each resonance structure?
b. What is the hybridization of P in each structure?
c. What orbitals can the P-atom use to form the π bond in structure B?
d. Which resonance structure would be favored on the basis of formal charges?

48. The N_2O molecule is linear and polar.
a. On the basis of this experimental evidence, which arrangement, NNO or NON, is correct? Explain your answer.
b. On the basis of your answer in (a) above, write the Lewis structure of N_2O (including resonance forms). Give the formal charge on each atom and the hybridization of the central atom.
c. How would the multiple bonding in :N≡N—Ö: be described in terms of orbitals?

CHAPTER 10

Liquids and Solids

CONTENTS

You have only to think about water to appreciate how different the three states of matter are. Flying, swimming, and ice skating are all done in contact with water in its various forms. Clearly the arrangements of the water molecules must be significantly different in its gas, liquid, and solid forms.

In Chapter 5 we saw that a gas can be pictured as a substance whose component particles are far apart, are in rapid random motion, and exert relatively small forces on each other. The kinetic molecular model was constructed to account for the ideal behavior that most gases approach at high temperatures and low pressures.

Solids are obviously very different from gases. Gases have low density, high compressibility, and completely fill a container. Solids have much greater densities, are compressible only to a very slight extent, and are rigid—a solid maintains its shape irrespective of its container. These properties indicate that the components of a solid are close together and exert large attractive forces on each other.

The properties of liquids lie somewhere between those of solids and of gases, but not midway between, as can be seen from some of the properties of the three states of water. For example, compare the enthalpy change for the melting of ice at 0°C (the heat of fusion) to that for vaporizing liquid water at 100°C (the heat of vaporization):

$$H_2O(s) \rightarrow H_2O(l) \qquad \Delta H^\circ_{fus} = 6.02 \text{ kJ/mol}$$
$$H_2O(l) \rightarrow H_2O(g) \qquad \Delta H^\circ_{vap} = 40.7 \text{ kJ/mol}$$

These values show a much greater change in structure in going from the liquid to the gaseous state than in going from the solid to the liquid. This suggests that there are extensive attractive forces among the molecules in liquid water, similar to but not as strong as those in the solid state.

< Icicles on an iceberg in Antarctica. The open crystal structure of ice which results from maximizing the hydrogen bonding makes ice less dense than water and enables it to float.

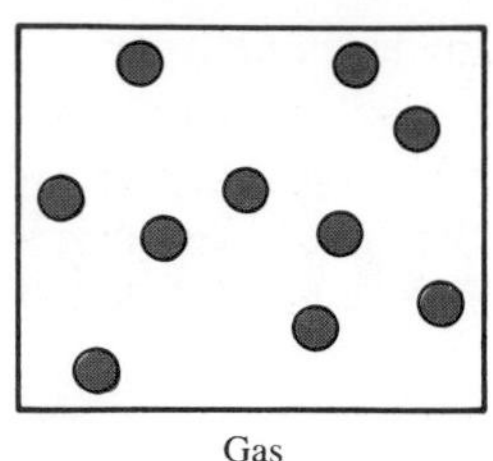

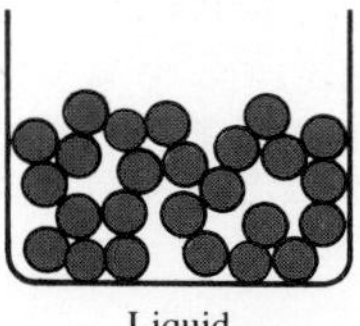

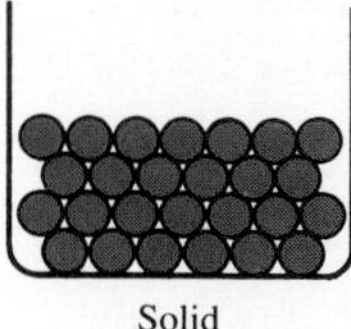

Figure 10.1

The three states of matter.

Densities of the Three States of Water

State	Density (g/cm^3)
Solid (0°C, 1 atm)	0.9168
Liquid (25°C, 1 atm)	0.9971
Gas (400°C, 1 atm)	3.26×10^{-4}

Table 10.1

The relative similarity of the liquid and solid states can also be seen in the densities of the three states of water. As shown in Table 10.1, the densities for liquid and solid water are quite close. Compressibilities can also be used to explore the relationship among water's states. At 25°C, the density of liquid water changes from 0.99707 g/cm^3 at a pressure of 1 atm to 1.046 g/cm^3 at 1065 atm. Given the large change in pressure, this is a very small variation in the density. Ice also shows little variation in density with increased pressure. On the other hand, at 400°C, the density of gaseous water changes from 3.26×10^{-4} g/cm^3 at 1 atm pressure to 0.157 g/cm^3 at 242 atm—a huge variation.

The conclusion is clear. The liquid and solid states show many similarities and are strikingly different from the gaseous state (see Fig. 10.1). We must bear this in mind as we develop models for the structures of solids and liquids.

10.1 Intermolecular Forces

Purpose

- To define dipole-dipole forces, hydrogen bonding, and London dispersion forces.
- To describe the effects these forces have on the properties of liquids and solids.

In Chapters 8 and 9 we saw that atoms can form stable units called molecules by sharing electrons. This is called *intramolecular* (within the molecule) bonding. In this chapter we consider the properties of the **condensed states** of matter (liquids and solids) and the forces that cause the aggregation of the components of a substance to form a liquid or a solid. These forces may involve covalent or ionic bonding, or they may involve weaker interactions usually called **intermolecular forces** (because they occur between, rather than within, molecules).

Intermolecular forces were introduced in Chapter 5 to explain nonideal gas behavior.

It is important to recognize that when a substance like water changes from solid to liquid to gas, *the molecules remain intact*. The changes of state are due to changes in the forces *among* the molecules rather than those *within* the molecules. In ice, as we will see later in this chapter, the molecules are virtually locked in place, although they can vibrate about their positions. If energy is added, the motions of the molecules increase, and they eventually achieve the greater movement and disorder characteristic of liquid water. The ice has melted. As more energy is

Remember that temperature is a measure of the random motions of the particles in a substance.

added, the gaseous state is eventually reached, with the individual molecules far apart and interacting relatively little. However, the gas still consists of water molecules. It would take much more energy to overcome the covalent bonds and decompose the water molecules into their component atoms. This can be seen by comparing the energy needed to vaporize 1 mole of liquid water (40.7 kJ) to that needed to break the O—H bonds in 1 mole of water molecules (934 kJ).

Dipole-Dipole Forces

As we saw in Section 8.3, molecules with polar bonds often behave in an electric field as if they had a center of positive charge and a center of negative charge. That is, they exhibit a dipole moment. Molecules with dipole moments can attract each other electrostatically by lining up so that the positive and negative ends are close to each other, as shown in Fig. 10.2(a). This is called a **dipole-dipole attraction.** In a condensed state such as a liquid, the dipoles find the best compromise between attraction and repulsion, as shown in Fig. 10.2(b).

Dipole-dipole forces are typically only about 1% as strong as covalent or ionic bonds, and they rapidly become weaker as the distance between the dipoles increases. At low pressures in the gas phase, where the molecules are far apart, these forces are relatively unimportant.

Particularly strong dipole-dipole forces, however, are seen among molecules in which hydrogen is bound to a highly electronegative atom, such as nitrogen, oxygen, or fluorine. Two factors account for the strengths of these interactions: the great polarity of the bond and the close approach of the dipoles, allowed by the very small size of the hydrogen atom. Because dipole-dipole attractions of this type are so unusually strong, they are given a special name—**hydrogen bonding.** Figure 10.3 shows hydrogen bonding among water molecules.

Hydrogen bonding has a very important effect on various physical properties. For example, the boiling points for the covalent hydrides of the elements in Groups 4A, 5A, 6A, and 7A are given in Fig. 10.4. Note that the nonpolar tetrahedral hydrides of Group 4A show a steady increase in boiling point with molecular weight (that is, in going down the group), while for the other groups, the lightest member has an unexpectedly high boiling point. Why? The answer lies in the especially large hydrogen-bonding interactions that exist among the smallest molecules with

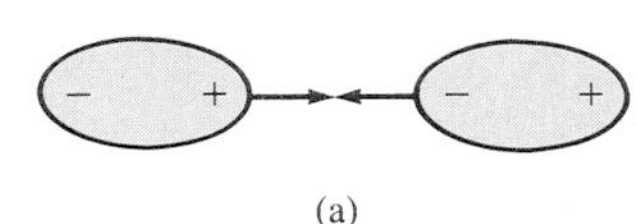

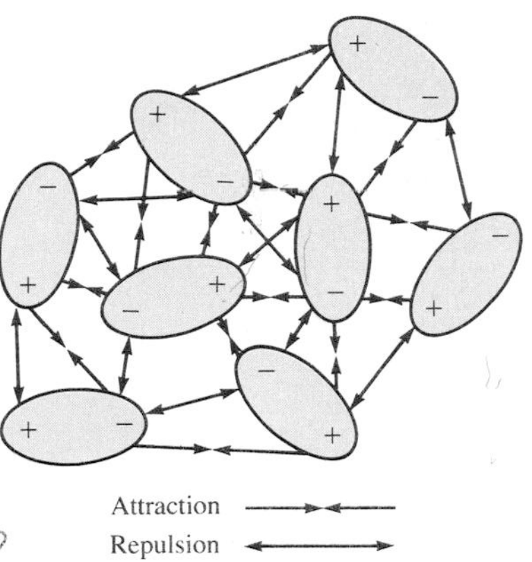

Figure 10.2

(a) The electrostatic interaction of two polar molecules. (b) The interaction of many dipoles in a condensed state.

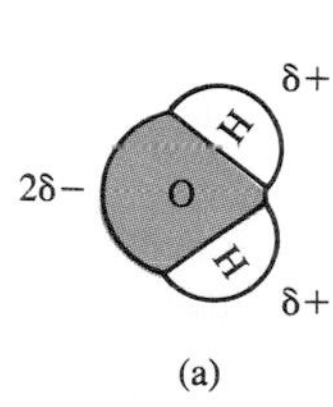

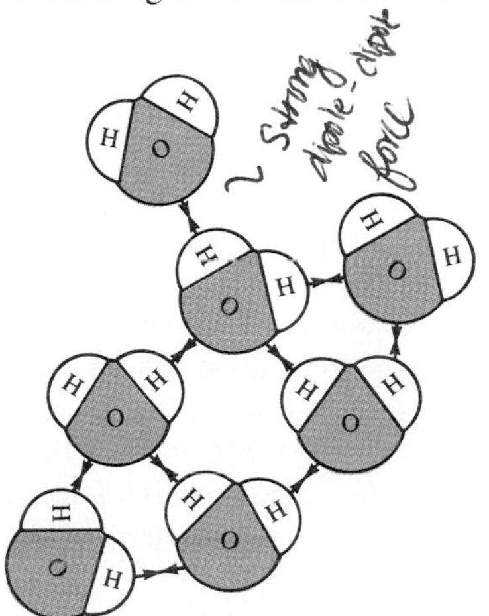

Figure 10.3

(a) The polar water molecule.
(b) Hydrogen bonding among water molecules. Note that the small size of the hydrogen atom allows for close interactions.

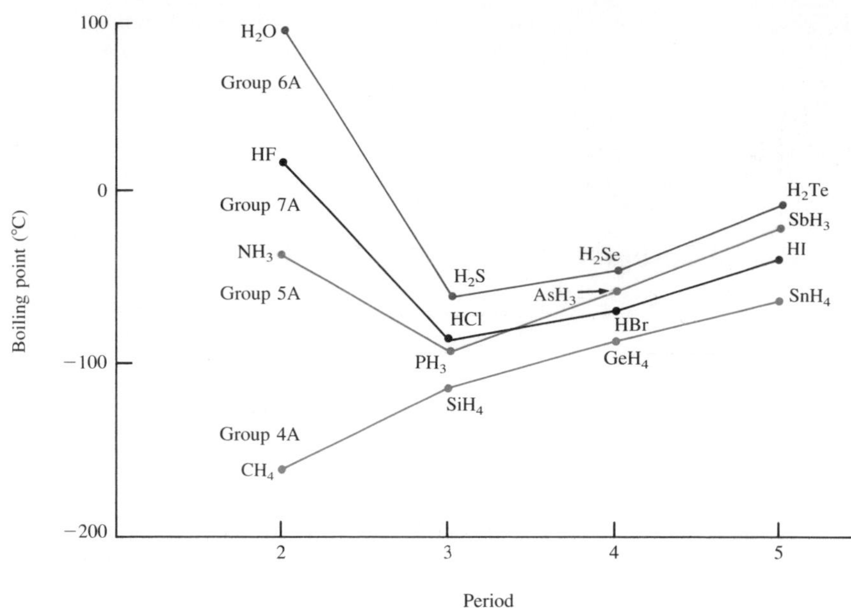

Figure 10.4

The boiling points of the covalent hydrides of elements in Groups 4A, 5A, 6A, and 7A.

the most polar X—H bonds. These unusually strong hydrogen bonding forces are due primarily to two factors. One factor is the relatively large electronegativity values of the lightest elements in each group, which leads to especially polar X—H bonds. The second factor is the small size of the first element of each group, which allows for the close approach of the dipoles further strengthening the intermolecular forces. Because the interactions among the molecules containing the lightest elements in Groups 4A, 5A, and 6A are so strong, an unusually large quantity of energy must be supplied to overcome these interactions and separate the molecules to produce the gaseous state. These molecules will remain together in the liquid state even at high temperatures; hence the very high boiling points.

Boiling point will be defined precisely in Section 10.8.

London Dispersion Forces (Van der Waals)

Even molecules without dipole moments must exert forces on each other. We know this because all substances—even the noble gases—exist in the liquid and solid states under certain conditions. The relatively weak forces that exist among noble gas atoms and nonpolar molecules are called **London dispersion forces.** To understand the origin of these forces, let us consider a pair of noble gas atoms. Although we usually assume that the electrons of an atom are uniformly distributed about the nucleus, this is apparently not true at every instant. Atoms can develop a momentary nonsymmetrical electron distribution that produces a temporary dipolar arrangement of charge. This *instantaneous dipole* can then *induce* a similar dipole in a neighboring atom, as shown in Fig. 10.5(a). This phenomenon leads to an interatomic attraction that is both weak and short-lived, but which can be very significant for large atoms (see below). For these interactions to become strong enough to produce a solid, the motions of the atoms must be greatly slowed down. This explains, for instance, why the noble gas elements have such low freezing points (see Table 10.2).

The Freezing Points of the Group 8A Elements

Element	Freezing point (°C)
Helium	−269.7
Neon	−248.6
Argon	−189.4
Krypton	−157.3
Xenon	−111.9

Table 10.2

Note from Table 10.2 that the freezing point rises going down the group. There are two principal factors that cause this trend. First, as the atomic mass increases, the average velocity of an atom at a given temperature decreases, allowing the atom

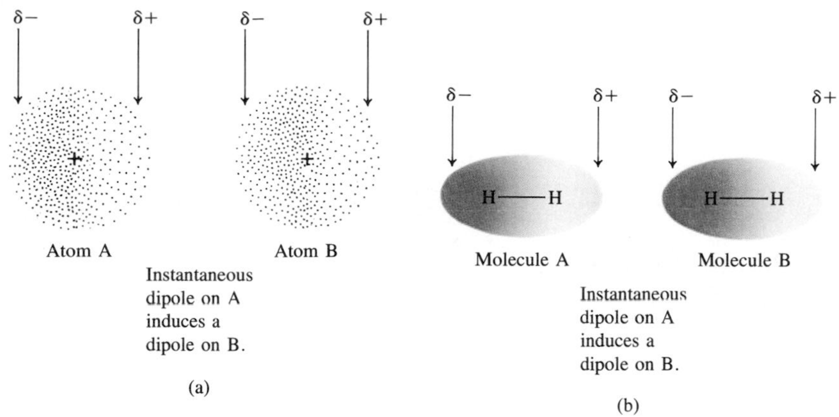

Figure 10.5

(a) An instantaneous polarization can occur on atom A, creating an instantaneous dipole. This dipole creates an induced dipole on neighboring atom B. (b) Nonpolar molecules such as H_2 can develop instantaneous and induced dipoles.

to "lock into" the solid more readily. Second, as the atomic number increases, the number of electrons increases, and there is an increased chance of the occurrence of momentary dipole interactions. We say that large atoms with many electrons exhibit a higher *polarizability* than small atoms. Thus the importance of London dispersion forces greatly increases as the size of the atom increases.

These same ideas also apply to nonpolar molecules such as H_2, CH_4, CCl_4, and CO_2 [see Fig. 10.5(b)]. Since none of these molecules has a permanent dipole moment, their principal means of attracting each other is through London dispersion forces.

10.2 The Liquid State

Purpose

- To describe some properties of liquids: surface tension, capillary action, and viscosity.

Liquids and liquid solutions are vital to our lives. Of course, water is the most important liquid. Besides being essential to life, it provides a medium for food preparation, for transportation, for cooling in many types of machines and industrial processes, for recreation, for cleaning, and for a myriad of other uses.

Liquids exhibit many characteristics that help us understand their nature. We have already mentioned their low compressibility, lack of rigidity, and high density compared to gases. Many of the properties of liquids give us direct information about the forces that exist among the particles. For example, when a liquid is poured onto a solid surface, it tends to bead as droplets, a phenomenon that depends on the intermolecular forces. Although molecules in the interior of the liquid are completely surrounded by other molecules, those at the liquid surface are subject to attractions only from the side and from below (Fig. 10.6). The effect of this uneven pull on the surface molecules draws them into the body of the liquid and causes a droplet of liquid to assume the shape that has the minimum surface area—a sphere.

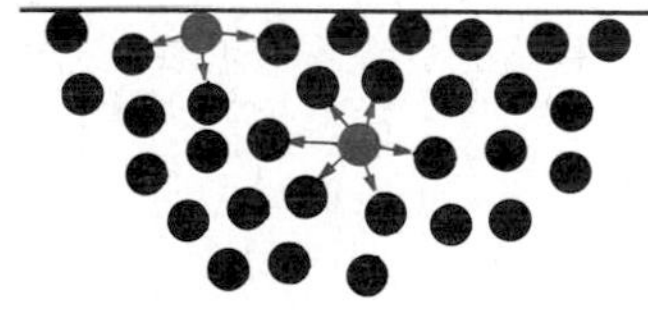

Figure 10.6

A molecule in the interior of a liquid is attracted to the molecules surrounding it, while a molecule at the surface of a liquid is attracted only by molecules below it and on each side.

To increase a liquid's surface area, molecules must move from the interior of the liquid to the surface. This requires energy, since some intermolecular forces must be overcome. The resistance of a liquid to an increase in its surface area is

For a given volume, a sphere has a smaller surface area than any other shape.

called the **surface tension** of the liquid. As we would expect, liquids with relatively large intermolecular forces tend to have relatively high surface tensions. Thus we can readily predict the surface tension if we know the polarity of the particles comprising the liquid, as seen in Sample Exercise 10.1.

Sample Exercise 10.1

Would you expect liquid carbon tetrachloride (CCl_4) or liquid chloroform ($CHCl_3$) to have the higher surface tension?

Solution

The dispersion forces in molecules with large atoms are quite significant and are often actually more important than dipole-dipole forces. In fact the surface tension of $CCl_4(l)$ is only slightly less than that of $CHCl_3(l)$.

Both carbon tetrachloride and chloroform have molecules with a tetrahedral arrangement of atoms around carbon. However, because chloroform has three bonds of one type and one of another, it has a dipole moment. Carbon tetrachloride, with four equal bonds, does not have a dipole moment. Because of the polar character of its molecules, we predict larger intermolecular forces in liquid chloroform and thus a higher surface tension.

The composition of glass is discussed in Section 10.5.

Polar liquids also exhibit **capillary action,** the spontaneous rising of a liquid in a narrow tube, as shown in Fig. 10.7(a). Two different types of forces are responsible for this property: *cohesive forces,* the intermolecular forces among the molecules of the liquid; and *adhesive forces,* the forces between the liquid molecules and their container. We have already seen how cohesive forces operate among polar molecules. Adhesive forces occur when a container is made of a substance that has polar bonds. For example, glass contains many oxygen atoms with partial negative charges that are attractive to the positive end of a polar molecule such as water. This ability of water to "wet" glass makes it creep up the walls of the tube where the water surface touches the glass. This, however, tends to increase the surface area of the water which is opposed by the cohesive forces that try to minimize the surface. Thus, because water has both strong cohesive (intermolecular) forces and strong adhesive forces to glass, it "pulls itself" up a glass capillary tube (a tube with a small diameter) to a height where the weight of the column of water just balances the water's tendency to be attracted to the glass surface. The concave shape of the meniscus [see Fig. 10.7(a)] shows that water's adhesive forces toward the glass are stronger than its cohesive forces. A nonpolar liquid such as mercury [see Fig. 10.7(b)] shows a lower level in a capillary tube and a convex meniscus. This behavior is characteristic of a liquid in which the cohesive forces are stronger than the adhesive forces toward glass.

Beads of water on a waxed car finish.

Another property of liquids strongly dependent on intermolecular forces is **viscosity,** a measure of a liquid's resistance to flow. As might be expected, liquids with large intermolecular forces tend to be highly viscous. For example, glycerol, whose structure is

```
      H
      |
   H—C—O—H
      |
   H—C—O—H
      |
   H—C—O—H
      |
      H
```

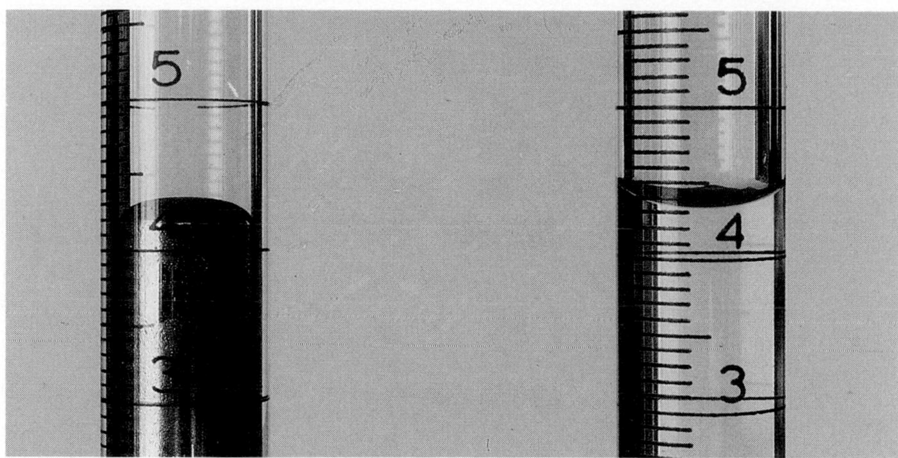

Nonpolar liquid mercury forms a convex meniscus in a glass tube while polar water forms a concave meniscus.

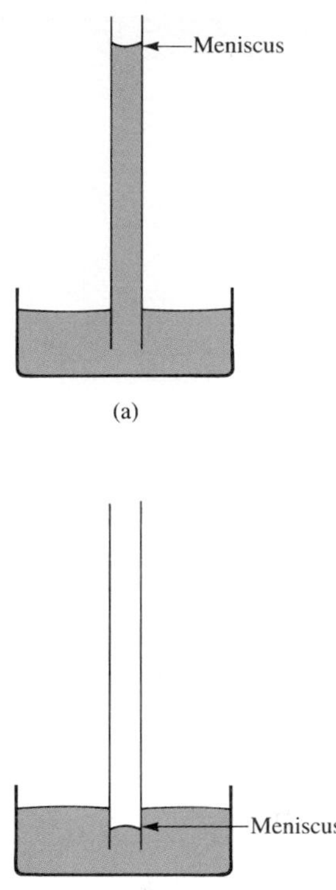

Figure 10.7

(a) A polar liquid such as water rises in a capillary tube. (b) A nonpolar liquid such as mercury shows a depression of level in the tube.

has an unusually high viscosity mainly due to a high capacity to form hydrogen bonds.

Molecular complexity also leads to higher viscosity because very large molecules can become entangled with each other. For example, nonviscous gasoline contains molecules of the type $CH_3—(CH_2)_n—CH_3$, where n varies from about 3 to 8. However, grease, which is very viscous, contains much larger molecules in which n varies from 20 to 25.

In many respects, the development of a structural model for liquids presents greater challenges than developing such a model for the other two states of matter. In the gaseous state, the particles are so far apart and are moving so rapidly that intermolecular forces are negligible under most circumstances. This means we can use a relatively simple model for gases. In the solid state, although the intermolecular forces are large, the molecular motions are minimal and fairly simple models are again possible. The liquid state, however, has both strong intermolecular forces *and* significant molecular motions. Such a situation precludes the use of really simple models for liquids. Recent advances in spectroscopy, the study of the manner in which substances interact with electromagnetic radiation, make it possible to follow the very rapid changes that occur in liquids. As a result, our models of liquids are becoming more accurate. As a starting point, a typical liquid might best be viewed as containing a large number of regions where the arrangements of the components are similar to those found in the solid, but with more disorder, and a smaller number of regions where holes are present. The situation is highly dynamic, with rapid fluctuations occurring in both types of regions.

10.3 An Introduction to Structures and Types of Solids

Purpose

- To contrast crystalline and amorphous solids.
- To introduce X-ray diffraction as a means for structure determination.

Figure 10.8

Several crystalline solids. (clockwise from upper left) Rhodochrosite; calcite with malachite; pyrite; amethyst.

There are many ways to classify solids, but the broadest categories are **crystalline solids,** those with a highly regular arrangement of their components, and **amorphous solids,** those with considerable disorder in their structures.

The regular arrangement of the components of a crystalline solid at the microscopic level produces the beautiful, characteristic shapes of crystals, such as those shown in Fig. 10.8. The positions of the components in a crystalline solid are usually represented by a **lattice,** a three-dimensional system of points designating the centers of the components (atoms, ions, or molecules). The *smallest repeating unit* of the lattice is called the **unit cell.** Thus a particular lattice can be generated by repeating the unit cell in all three dimensions to form the extended structure. Three common unit cells and their lattices are shown in Fig. 10.9.

Although we will concentrate on crystalline solids in this book, there are many important noncrystalline (amorphous) materials. An example is common glass, which is best pictured as a solution in which the components are "frozen in place" before they can achieve an ordered arrangement. Although glass is a solid (it has a rigid shape), a great deal of disorder exists in its structure.

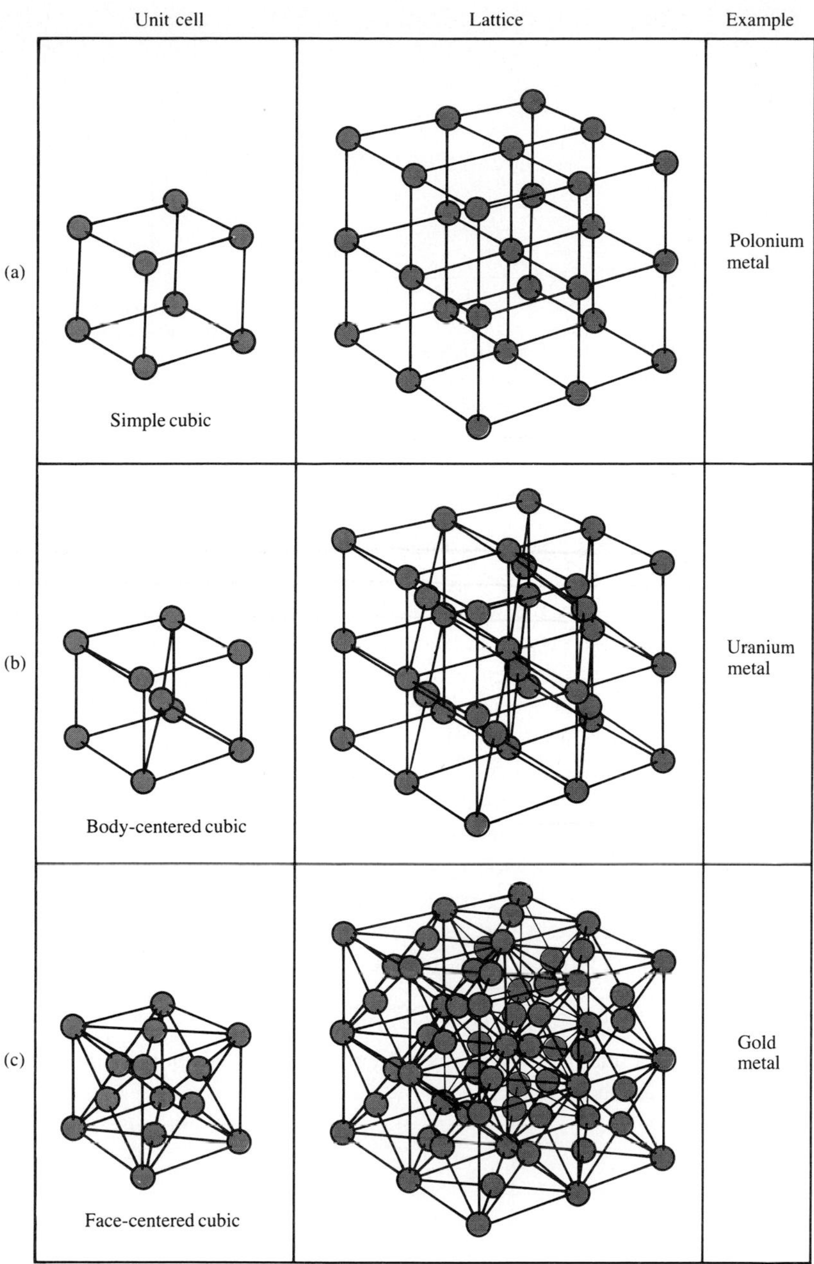

Figure 10.9

Three cubic unit cells and the corresponding lattices.

X-ray Analysis of Solids

The structures of crystalline solids are most commonly determined by **X-ray diffraction.** Diffraction occurs when beams of light are scattered from a regular array of points or lines in which the spacings between the components are comparable to the wavelength of the light. Diffraction is due to constructive interference when the waves of parallel beams are in phase and to destructive interference when the waves are out of phase.

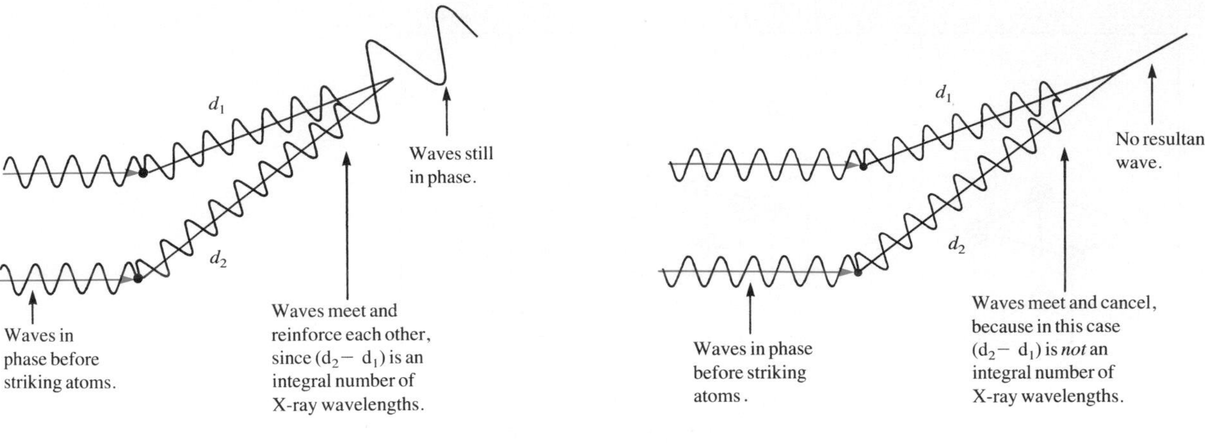

d_1
d_2
Waves still in phase.
Waves in phase before striking atoms.
Waves meet and reinforce each other, since ($d_2 - d_1$) is an integral number of X-ray wavelengths.
(a)
d_1
d_2
No resultant wave.
Waves in phase before striking atoms.
Waves meet and cancel, because in this case ($d_2 - d_1$) is *not* an integral number of X-ray wavelengths.
(b)

Figure 10.10

X rays that are scattered from two different atoms may either (a) reinforce or (b) cancel each other when they meet, depending on whether they are in phase or out of phase.

When X rays of a single wavelength are directed at a crystal, a diffraction pattern is obtained, as we saw in Fig. 7.5. The light and dark areas on the photographic plate occur because the waves scattered from various atoms may reinforce or cancel each other when they meet (see Fig. 10.10). The key to whether the waves reinforce or cancel is the difference in distance traveled by the waves after they strike the atoms. The waves are in phase before they are reflected, so if the difference in distance traveled after reflection is an *integral number of wavelengths,* the waves will still be in phase when they meet again.

Since the distance traveled after reflection depends on the distance between the atoms, the diffraction pattern can be used to determine the interatomic spacings. The exact relationship can be worked out using the diagram in Fig. 10.11, which shows two in-phase waves being reflected by atoms in two different layers in a crystal. The extra distance traveled by the lower wave is the sum of the distances xy and yz, and the waves will be in phase after reflection if

$$xy + yz = n\lambda \tag{10.1}$$

where n is an integer and λ is the wavelength of the X rays. Using trigonometry (see Fig. 10.11), we can show that

$$xy + yz = 2d \sin \theta \tag{10.2}$$

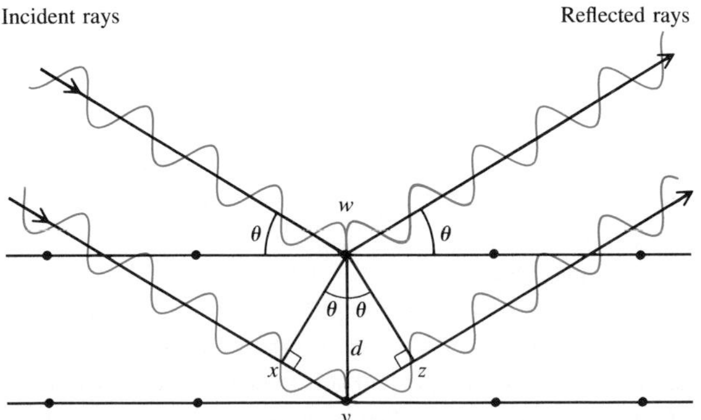

Incident rays
Reflected rays
w
θ
θ
θ
θ
d
x
z
y

Figure 10.11

Reflection of X rays of wavelength λ from a pair of atoms in two different layers of a crystal. The lower wave travels an extra distance equal to the sum of xy and yz. If this distance is an integral number of wavelengths ($n = 1, 2, 3 \ldots$), the waves will reinforce each other when they exit the crystal.

where d is the distance between the atoms and θ is the angle of incidence and reflection. Combining Equation (10.1) and Equation (10.2) gives

$$n\lambda = 2d \sin \theta \tag{10.3}$$

Equation (10.3) is called the **Bragg equation** after William Henry Bragg (1862–1942) and his son William Lawrence Bragg (1890–1972), who shared the Nobel prize in physics in 1915 for their pioneering work in X-ray crystallography.

A diffractometer is a computer-controlled instrument used for carrying out the X-ray analysis of crystals. It rotates the crystal with respect to the X-ray beam and collects the data produced by the scattering of the X rays from the various planes of atoms in the crystal. The results are then analyzed by computer. The techniques for crystal structure analysis have reached a level of sophistication that allows the determination of very complex structures, such as those important in biological systems. Using X-ray diffraction, we can gather data on bond lengths and angles, and in doing so can test the predictions of our models of molecular geometry.

Sample Exercise 10.2

X rays of wavelength 1.54 Å were used to analyze an aluminum crystal. A reflection was produced at $\theta = 19.3°$. Assuming $n = 1$, calculate the distance d between the planes of atoms producing this reflection.

Solution

To determine the distance between the planes, we use Equation (10.3) with $n = 1$, $\lambda = 1.54$ Å, and $\theta = 19.3°$. Since $2d \sin \theta = n\lambda$,

$$d = \frac{n\lambda}{2 \sin \theta} = \frac{(1)(1.54\text{ Å})}{(2)(0.3305)} = 2.33\text{ Å}$$

Types of Crystalline Solids

There are many different types of crystalline solids. For example, although both sugar and salt dissolve readily in water, the properties of the resulting solutions are quite different. The salt solution readily conducts an electric current, while the sugar solution does not. This behavior arises from the nature of the components in these two solids. Common salt (NaCl) is an ionic solid; it contains Na^+ and Cl^- ions. When solid sodium chloride dissolves in the polar water, sodium and chloride ions are distributed throughout the resulting solution and are free to conduct electrical current. Table sugar (sucrose), on the other hand, is composed of neutral molecules that are dispersed throughout the water when the solid dissolves. No ions are present, and the resulting solution does not conduct electricity. These examples illustrate two important types of solids: **ionic solids,** represented by sodium chloride; and **molecular solids,** represented by sucrose.

A third type of solid is represented by elements such as graphite and diamond (both pure carbon), boron, silicon, and all metals. These substances all contain atoms covalently bonded to each other; we will call them **atomic solids.** Examples of the three types of solids are shown in Fig. 10.12.

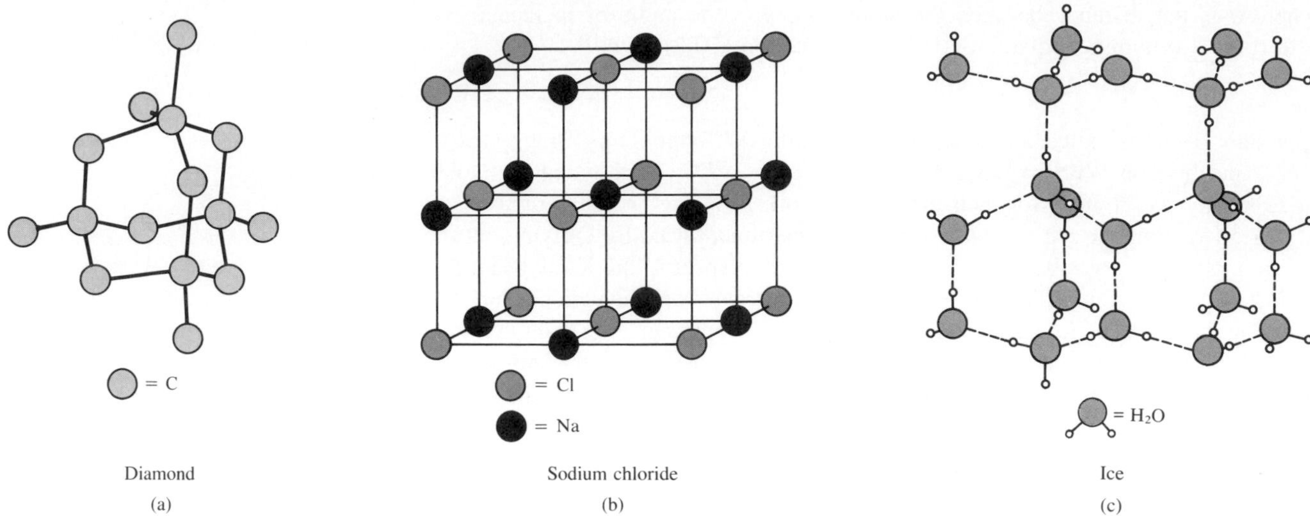

Figure 10.12

Examples of three types of crystalline solids. Only part of the structure is shown in each case. (a) An atomic solid. (b) An ionic solid. (c) A molecular solid. The dashed lines show the hydrogen bonding among the polar water molecules.

The internal forces in a solid determine the properties of the solid.

The properties of a solid are determined primarily by the nature of the forces that hold the solid together. For example, although argon, copper, and diamond all form atomic solids, they have strikingly different properties. Argon has a very low melting point (−189°C), while diamond and copper melt at high temperatures (about 3500°C and 1083°C, respectively). Copper is an excellent conductor of electricity, while argon and diamond are both insulators. Copper can be easily changed in shape; it is both malleable (will form thin sheets) and ductile (can be pulled into a wire). Diamond, on the other hand, is the hardest natural substance known. The marked differences in properties among these three atomic solids are due to bonding differences. We will explore the bonding in solids in the next two sections.

10.4 Structure and Bonding in Metals

Purpose

- To discuss the concept of closest packing of metal atoms.
- To describe two models for bonding in metals.
- To define and classify alloys.

Metals are characterized by high thermal and electrical conductivity, malleability, and ductility. As we will see, these properties can be traced to the nondirectional covalent bonding found in metallic crystals.

A metallic crystal can be pictured as containing spherical atoms packed together and bonded to each other equally in all directions. We can model such a structure by packing uniform, hard spheres in a manner that most efficiently uses

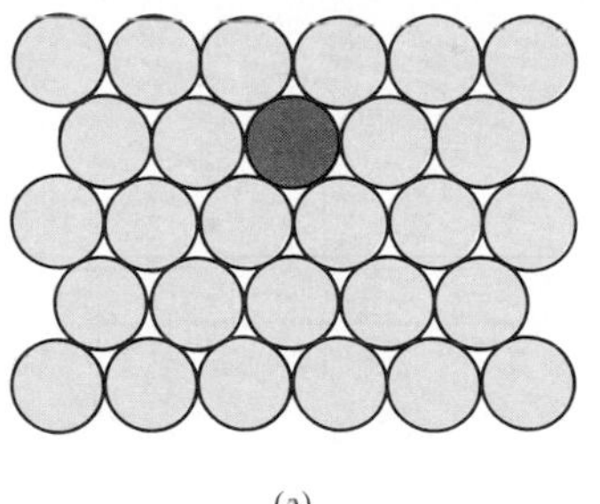
(a)

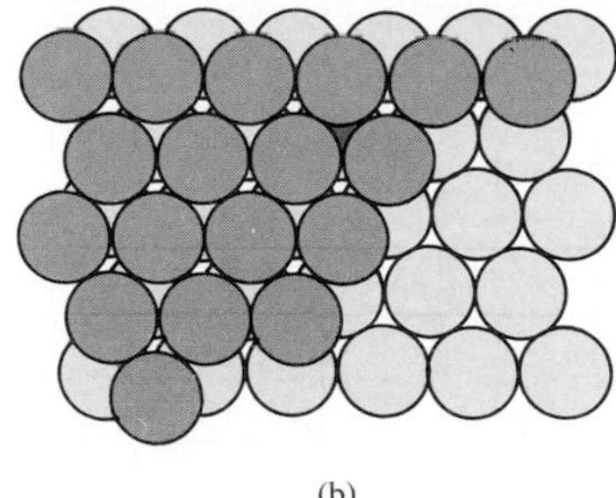
(b)

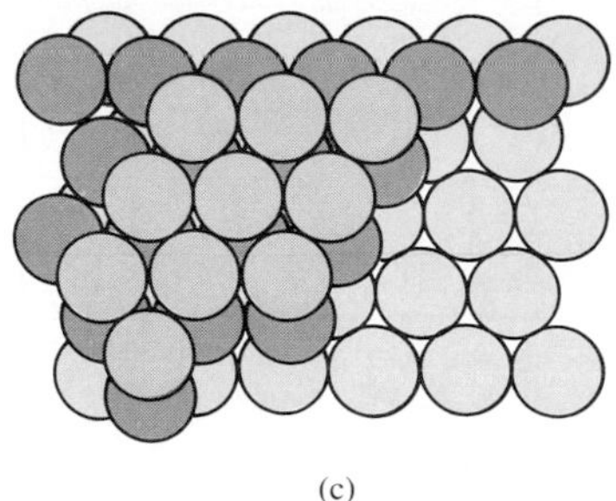
(c)

Figure 10.13

The closest packing arrangement of uniform spheres. (a) A typical layer where each sphere is surrounded by six others. (b) The second layer is like the first, but it is displaced so that each sphere in the second layer occupies a dimple in the first layer. (c) The spheres in the third layer can occupy dimples in the second so that the spheres in the third layer lie directly over those in the first layer (*aba*), or they can occupy dimples in the second layer so that no spheres in the third layer lie above any in the first layer (*abc*).

the available space. Such an arrangement is called **closest packing.** The spheres are packed in layers, as shown in Fig. 10.13(a), where each sphere is surrounded by six others. In the second layer the spheres do not lie directly over those in the first layer. Instead each one occupies an indentation (or dimple) formed by three spheres in the first layer [see Fig. 10.13(b)]. In the third layer the spheres can occupy the dimples of the second layer in two possible ways. They can occupy positions so that each sphere in the third layer lies directly over a sphere in the first layer (the *aba* arrangement), or they can occupy positions so that no sphere in the third layer lies over one in the first layer (the *abc* arrangement).

The closest packing model for metallic crystals assumes that metal atoms are uniform, hard spheres.

The *aba* arrangement has the *hexagonal* unit cell shown in Fig. 10.14, and the resulting structure is called the **hexagonal closest packed (hcp) structure.** The *abc* arrangement has a *face centered cubic* unit cell, as shown in Fig. 10.15, and the resulting structure is called the **cubic closest packed (ccp) structure.** Note that in

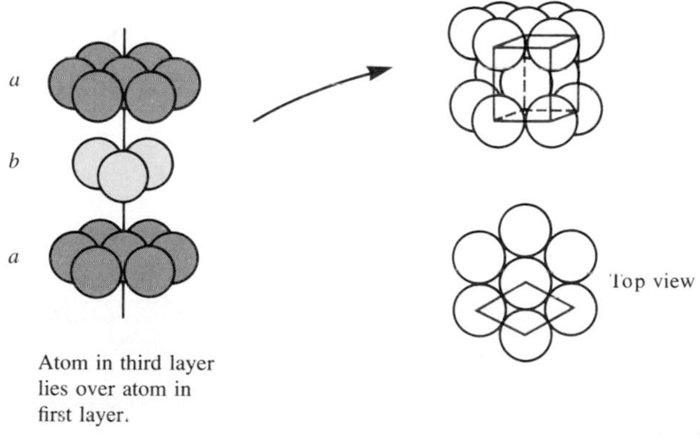
a
b
a
Top view
Atom in third layer lies over atom in first layer.

Figure 10.14

When spheres are closest packed so that the spheres in the third layer are directly over those in the first layer (*aba*), the unit cell is the hexagonal prism illustrated here in red.

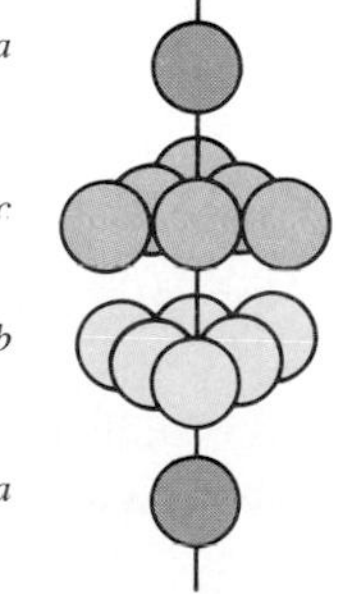
a
c
b
a
An atom in every fourth layer lies over an atom in the first layer.

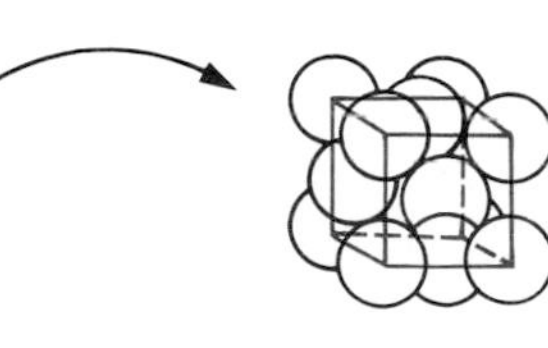

Figure 10.15

When spheres are packed in the *abc* arrangement, the unit cell is face-centered cubic. To make the cubic arrangement easier to see, the vertical axis has been tilted as shown.

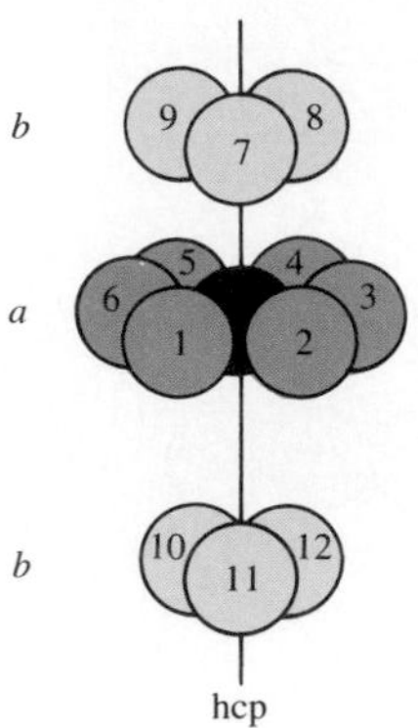

Figure 10.16

The indicated sphere has 12 nearest neighbors.

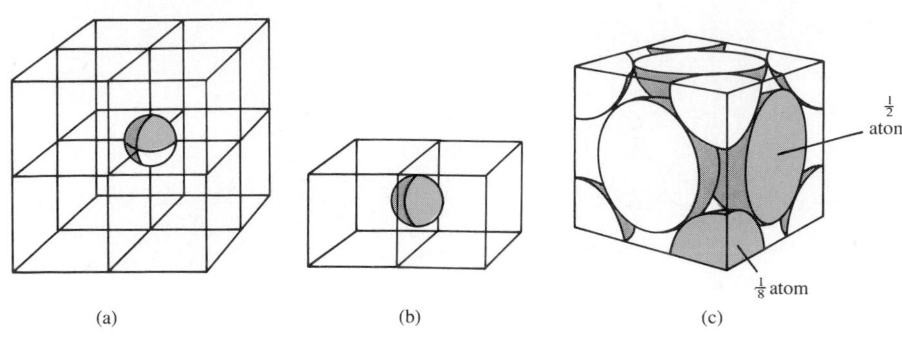

Figure 10.17

The net number of spheres in a face-centered cubic unit cell. (a) Note that the sphere on a corner of the colored cell is shared with 7 other unit cells. Thus $\frac{1}{8}$ of such a sphere lies within a given unit cell. Since there are 8 corners in a cube, there are 8 of these $\frac{1}{8}$ pieces, or 1 net sphere. (b) The sphere on the center of each face is shared by two unit cells, and thus each unit cell has $\frac{1}{2}$ of each of these types of spheres. There are 6 of these $\frac{1}{2}$ spheres to give 3 net spheres. (c) Thus the face-centered cubic unit cell contains 4 net spheres.

the hcp structure the spheres in every other layer occupy the same vertical position (*ababab* . . .), while in the ccp structure the spheres in every fourth layer occupy the same vertical position (*abcabca* . . .). A characteristic of both structures is that each sphere has 12 equivalent nearest neighbors: 6 in the same layer, 3 in the layer above, and 3 in the layer below (that form the dimples). This is illustrated for the hcp structure in Fig. 10.16.

Knowing the *net* number of spheres (atoms) in a particular unit cell is important for many applications involving solids. To illustrate how to find the net number of spheres in a unit cell, we will consider a face-centered cubic unit cell (Fig. 10.17). Note that this unit cell is defined by the *centers* of the spheres on the cube's corners. Thus 8 cubes share a given sphere, so $\frac{1}{8}$ of this sphere lies inside each unit cell. Since a cube has 8 corners, there are $8 \times \frac{1}{8}$ pieces, or enough to put together 1 whole sphere. The spheres at the center of each face are shared by 2 unit cells, so $\frac{1}{2}$ of each lies inside a particular unit cell. Since the cube has 6 faces, we have $6 \times \frac{1}{2}$ pieces, or enough to construct 3 whole spheres. Thus the net number of spheres in a face-centered cubic unit cell is

$$\left(8 \times \frac{1}{8}\right) + \left(6 \times \frac{1}{2}\right) = 4$$

Sample Exercise 10.3

Silver crystallizes in a cubic closest packed structure. The radius of a silver atom is 1.44 Å. Calculate the density of solid silver.

Solution

Density is mass per unit volume. Thus we need to know how many silver atoms occupy a given volume in the crystal. The structure is cubic closest packed, which means the unit cell is face-centered cubic, as shown in the accompanying figure.

Sample Exercise 10.3, continued

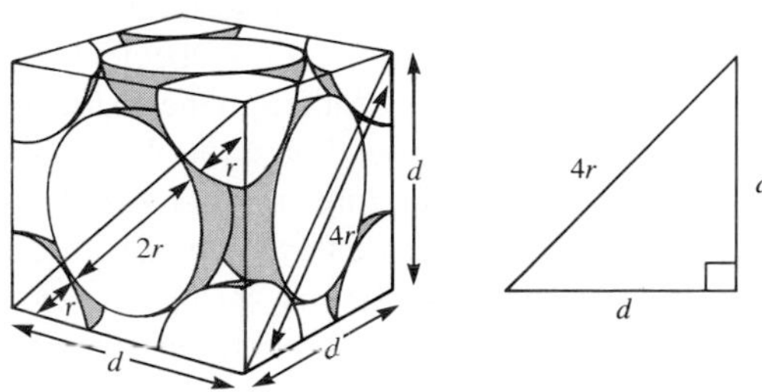

We must find the volume of this unit cell for silver and the net number of atoms it contains. Note that in this structure the atoms touch along the diagonals for each face and not along the edges of the cube. Thus the length of the diagonal is $r + 2r + r$, or $4r$. We use this fact to find the length of the edge of the cube by the Pythagorean theorem:

$$d^2 + d^2 = (4r)^2$$
$$2d^2 = 16r^2$$
$$d^2 = 8r^2$$
$$d = \sqrt{8r^2} = r\sqrt{8}$$

Since $r = 1.44$ Å for a silver atom,

$$d = (1.44\ \text{Å})(\sqrt{8}) = 4.07\ \text{Å}$$

The volume of the unit cell is d^3, which is $(4.07\ \text{Å})^3$, or $67.4\ \text{Å}^3$. We convert this to cubic centimeters as follows:

$$67.4\ \text{Å}^3 \times \left(\frac{1.00 \times 10^{-8}\ \text{cm}}{\text{Å}}\right)^3 = 6.74 \times 10^{-23}\ \text{cm}^3$$

Since we know that the net number of atoms in the face-centered cubic unit cell is 4, we have 4 silver atoms contained in a volume of $6.74 \times 10^{-23}\ \text{cm}^3$. The density is therefore

$$\text{Density} = \frac{\text{mass}}{\text{volume}} = \frac{(4\ \text{atoms})(107.9\ \text{g/mol})(1\ \text{mol}/6.022 \times 10^{23}\ \text{atoms})}{6.74 \times 10^{-23}\ \text{cm}^3}$$
$$= 10.6\ \text{g/cm}^3$$

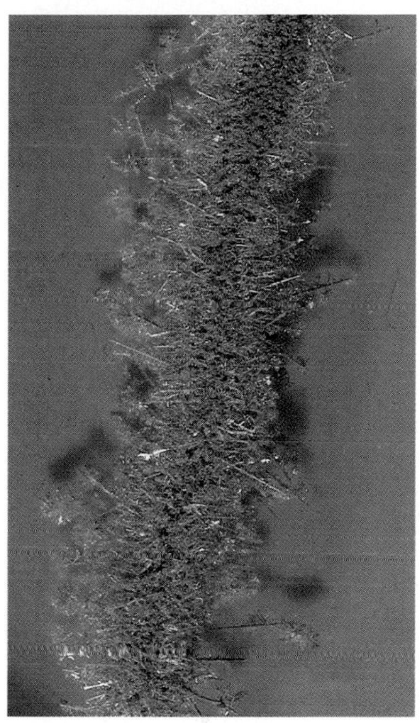

Crystalline silver contains cubic closest packed silver atoms.

Examples of metals that are cubic closest packed are aluminum, iron, copper, cobalt, and nickel. Magnesium and zinc are hexagonal closest packed. Calcium and certain other metals can crystallize in either of these structures. Some metals, however, assume structures that are not closest packed. For example, the alkali metals have structures characterized by a *body-centered cubic (bcc) unit cell* (see Fig. 10.9), where the spheres touch along the body diagonal of the cube. In this structure each sphere has 8 nearest neighbors (count the number of atoms around the atom at the center of the unit cell), as compared to 12 in the closest packed structures. Why a particular metal adopts the structure it does is not well understood.

Bonding in Metals

Any successful bonding model for metals must account for the typical physical properties of metals: malleability, ductility, and the efficient and uniform conduction of heat and electricity in all directions. Although the shapes of most pure metals can be changed relatively easily, most metals are durable and have high melting points. These facts indicate that the bonding in most metals is both *strong* and *nondirectional*. That is, although it is difficult to separate metal atoms, it is relatively easy to move them, provided the atoms stay in contact with each other.

The simplest picture that explains these observations is the **electron sea model,** which envisions a regular array of metal cations in a ''sea'' of valence electrons (see Fig. 10.18). The mobile electrons can conduct heat and electricity, and the cations can be easily moved around as the metal is hammered into a sheet or pulled into a wire.

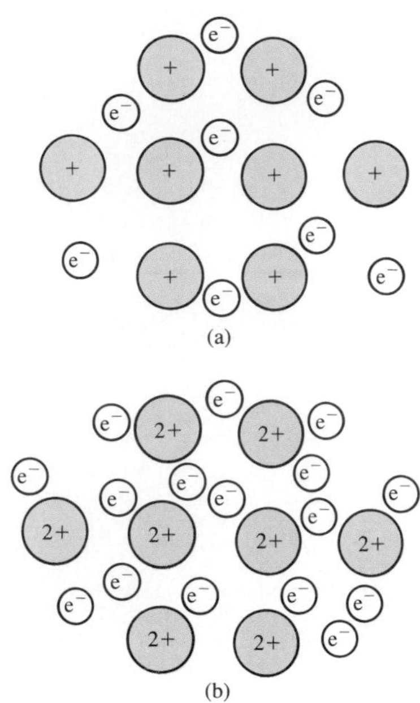

Figure 10.18

The electron sea model for metals postulates a regular array of cations in a "sea" of valence electrons. (a) Representation of an alkali metal (Group 1A) with one valence electron. (b) Representation of an alkaline earth metal (Group 2A) with two valence electrons.

A related model that gives a more detailed view of the electron energies and motions is the **band model,** or molecular orbital (MO) model, for metals. In this model the electrons are assumed to travel around the metal crystal in molecular orbitals formed from the valence atomic orbitals of the metal atoms (Fig. 10.19).

Recall that in the MO model for the gaseous Li_2 molecule (Section 9.3), two widely spaced molecular orbital energy levels (bonding and antibonding) result when two identical atomic orbitals interact. However, when many metal atoms interact, as in a metal crystal, the large number of resulting molecular orbitals become more closely spaced and finally form a virtual continuum of levels, called **bands,** as shown in Fig. 10.19.

As an illustration, picture a magnesium metal crystal, which has an hcp structure. Since each magnesium atom has one 3*s* and three 3*p* valence atomic orbitals, a crystal with *n* magnesium atoms has available *n*(3*s*) and 3*n*(3*p*) orbitals to form the molecular orbitals, as illustrated in Fig. 10.20. Note that the core electrons are localized, as shown by their presence in the energy ''well'' around each magnesium atom. However, the valence electrons occupy closely spaced molecular orbitals, which are only partially filled.

The existence of empty molecular orbitals close in energy to filled molecular orbitals explains the thermal and electrical conductivity of metal crystals. Metals conduct electricity and heat very efficiently because of the availability of highly

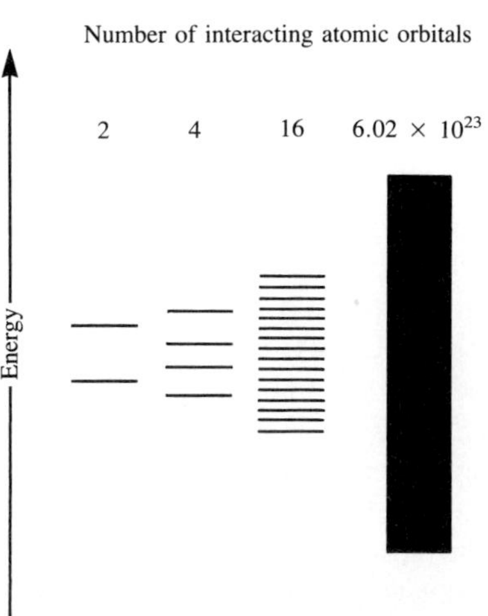

Figure 10.19

The molecular orbital energy levels produced when various numbers of atomic orbitals interact. Note that for two atomic orbitals two rather widely spaced energy levels result. (Recall the description of H_2 in Section 9.2.) As more atomic orbitals are available to form molecular orbitals, the resulting energy levels are more closely spaced, finally producing a band of very closely spaced orbitals.

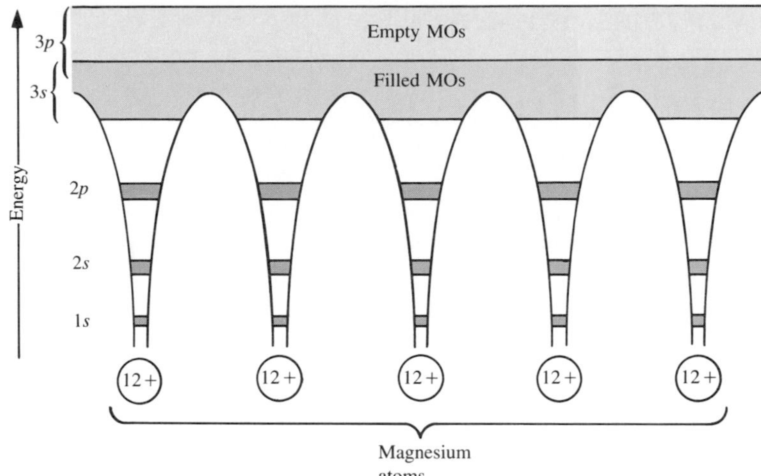

Figure 10.20

A representation of the energy levels (bands) in a magnesium crystal. The electrons in the 1*s*, 2*s*, and 2*p* orbitals are close to the nuclei and thus are localized on each magnesium atom as shown. However, the 3*s* and 3*p* valence orbitals overlap and mix to form molecular orbitals. Electrons in these energy levels can travel throughout the crystal.

mobile electrons. For example, when an electrical potential is placed across a strip of metal, for current to flow electrons must be free to move from the negative to the positive areas of the metal. In the band model for metals, mobile electrons are furnished when electrons in filled molecular orbitals are excited into empty ones. These conduction electrons are free to travel throughout the metal crystal as dictated by the potential imposed on the metal. The molecular orbitals occupied by these conducting electrons are called **conduction bands.** These mobile electrons also account for the efficiency of the conduction of heat through metals. When one end of a metal rod is heated, the mobile electrons can rapidly transmit the thermal energy to the other end.

Metal Alloys

Because of the nature of the structure and bonding of metals, other elements can be introduced into a metallic crystal relatively easily to produce substances called alloys. An **alloy** is best defined as *a substance that contains a mixture of elements and has metallic properties.* Alloys can be conveniently classified into two types.

In a **substitutional alloy** some of the host metal atoms are *replaced* by other metal atoms of similar size. For example, in brass approximately one-third of the atoms in the host copper metal have been replaced by zinc atoms, as shown in Fig. 10.21(a). Sterling silver (93% silver and 7% copper), pewter (85% tin, 7% copper, 6% bismuth and 2% antimony), and plumber's solder (67% lead and 33% tin) are other examples of substitutional alloys.

An **interstitial alloy** is formed when some of the interstices (holes) in the closest packed metal structure are occupied by small atoms, as shown in Fig. 10.21(b). Steel, the best known interstitial alloy, contains carbon atoms in the holes of an iron crystal. The presence of the interstitial atoms changes the properties of the host metal. Pure iron is relatively soft, ductile, and malleable due to the absence of strong directional bonding. The spherical metal atoms can be rather easily moved with respect to each other. However, when carbon, which forms strong directional bonds, is introduced into an iron crystal, the presence of the directional carbon-iron bonds makes the resulting alloy harder, stronger, and less ductile than pure iron. The amount of carbon directly affects the properties of steel. *Mild steels,* containing

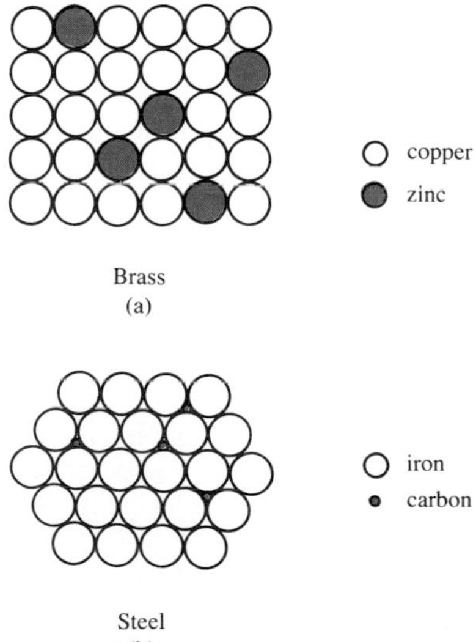

Figure 10.21

Two types of alloys.

Chemical Impact

Superconductivity

Although metals such as copper and aluminum are good conductors of electricity, up to 20% of the total energy is wasted in transmission by resistance heating of the wires. However, this loss of energy may be avoidable. Certain kinds of materials undergo a remarkable transition as they are cooled; their electrical resistance changes virtually to zero. These substances, called **superconductors,** conduct electricity with no wasted heat energy.

There is, of course, a catch. In the past superconductivity has been observed only at very low temperatures and is thus very expensive to maintain. In 1911, mercury was observed to exhibit superconductivity, but only at ~4 K, the boiling point of liquid helium. In the next 75 years several niobium alloys showed superconductivity at temperatures as high as 23 K. However, in 1986 researchers realized that certain metallic oxides are superconductors at temperatures well above 23 K. The plot below shows how rapidly our knowledge of superconductors is increasing.

The current class of high temperature superconductors is called *perovskites,* a well-known class of compounds that contain copper, an alkaline earth metal, a lanthanide metal, and oxygen. The unit cell for one of these compounds with formula $YBa_2Cu_3O_x$, in which $x = 6.527$, is shown below. In this structure the small open circles indicate the expected oxygen positions, but some of these positions are vacant in the known materials—thus the variable oxygen content.

In addition to exhibiting superconductivity at temperatures above the boiling point of liquid nitrogen (77 K), these ceramic materials can sustain very high currents, an important characteristic necessary for any large scale application.

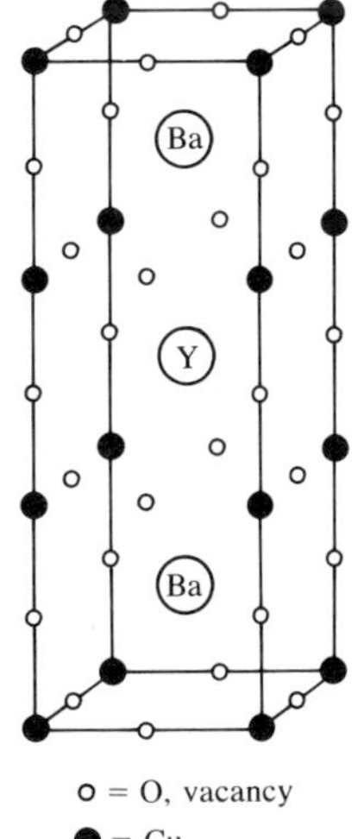

What does all this mean for the future? Superconducting materials could easily cause a revolution similar to that brought about by solid state electronics. For example, one can envision electrical power transmitted with virtually no loss of energy. Also, new generations of superconducting electromagnets could lead to particle accelerators operating at unprecedented energies, the development of nuclear fusion power, the ability to produce incredibly accurate three dimensional images of the human body for diagnosis of disease, and the development of superfast magnetically levitated trains.

There are, of course, many practical problems. For example, metal conductors can be easily bent into coils to allow construction of electromagnets, but anyone who has ever dropped a ceramic plate knows how brittle such materials are. However, there seems little doubt that ingenious solutions will be found to the problems. The answers are already beginning to appear.

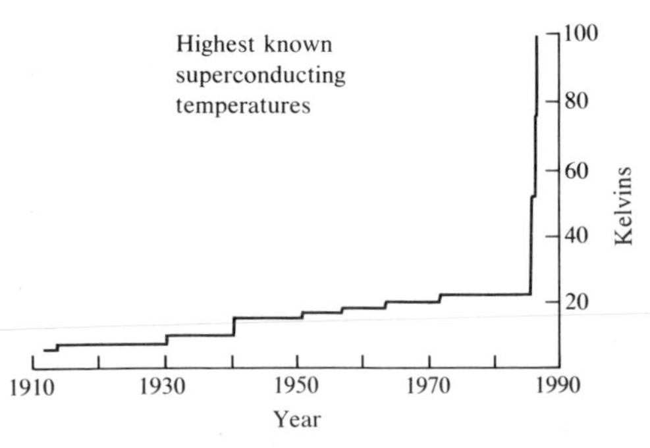

A magnet is levitated over a superconducting ceramic immersed in liquid nitrogen.

The Composition of the Two Brands of Steel Tubing Most Commonly Used to Make Lightweight Racing Bicycles					
Brand of tubing	% C	% Si	% Mn	% Mo	% Cr
Reynolds	0.25	0.25	1.3	0.20	—
Columbus	0.25	0.30	0.65	0.20	1.0

Table 10.3

less than 0.2% carbon, are ductile and malleable and used for nails, cables, and chains. *Medium steels,* containing 0.2–0.6% carbon, are harder than mild steels and are used in rails and structural steel beams. *High-carbon steels,* containing 0.6–1.5% carbon, are tough and hard and are used for springs, tools, and cutlery.

Many types of steel also contain elements in addition to iron and carbon. Such steels are often called *alloy steels* and can be viewed as being mixed interstitial (carbon) and substitutional (other metals) alloys. Bicycle frames, for example, are constructed from a wide variety of alloy steels. The compositions of the two brands of steel tubing most commonly used in expensive racing bicycles are given in Table 10.3.

The manufacture of steel will be discussed in detail in Chapter 24.

10.5 Carbon and Silicon: Network Atomic Solids

Purpose

- To show how the bonding in elemental carbon and silicon accounts for the widely different properties of their compounds.
- To explain how a semiconductor works.

Many atomic solids contain strong directional covalent bonds. We will call these substances **network solids.** In contrast to metals, these materials are typically brittle and do not efficiently conduct heat or electricity. To illustrate network solids, in this section we will discuss two very important elements, carbon and silicon, and some of their compounds.

Carbon, which occurs in the *allotropes* (different forms) diamond and graphite, is a typical network solid. In diamond, the hardest naturally occurring substance, each carbon atom is surrounded by a tetrahedral arrangement of other carbon atoms, as shown in Fig. 10.22(a). This structure is stabilized by covalent bonds, which, in terms of the localized electron model, are formed by the overlap of sp^3 hybridized carbon atomic orbitals.

It is also useful to consider the bonding among the carbon atoms in diamond in terms of the molecular orbital model. Energy-level diagrams for diamond and a typical metal are given in Fig. 10.23 on the following page. Recall that the conductivity of metals can be explained by postulating that electrons are excited from filled levels into the very near empty levels, or conduction bands. However, note that in the energy-level diagram for diamond there is *a large gap between the filled and the empty levels*. This means that electrons cannot be easily transferred to the empty conduction bands. As a result, diamond is not expected to be a good electrical conductor. In fact, this prediction of the model agrees exactly with the observed

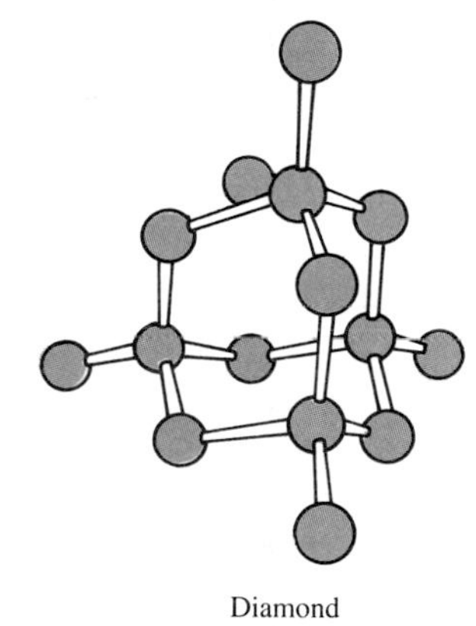

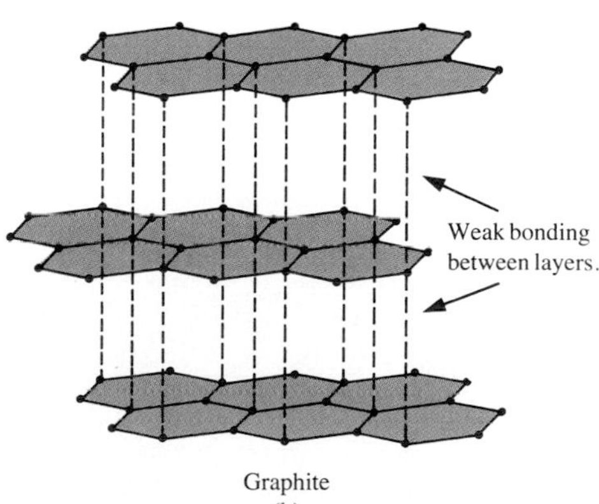

Figure 10.22

The structures of the carbon allotropes.

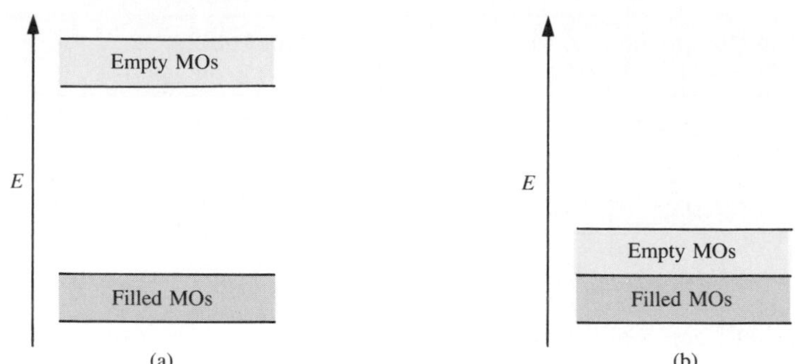

Figure 10.23

Partial representation of the molecular orbital energies in (a) diamond and (b) a typical metal.

behavior of diamond, which is known to be an electrical *insulator*—it does not conduct an electrical current.

Graphite, the other allotrope of elemental carbon, is very different from diamond. While diamond is hard, basically colorless, and an insulator, graphite is slippery, black, and a conductor. These differences, of course, arise from the differences in bonding in the two types of solids. In contrast to the tetrahedral arrangement of carbon atoms in diamond, the structure of graphite is based on layers of carbon atoms arranged in fused six-membered rings, as shown in Fig. 10.22(b). Each carbon atom in a particular layer of graphite is surrounded by three other carbon atoms in a trigonal planar arrangement with 120° bond angles. The localized electron model predicts sp^2 hybridization in this case. The three sp^2 orbitals on each carbon are used to form σ bonds to three other carbon atoms. One $2p$ orbital remains unhybridized on each carbon and is perpendicular to the plane of carbon atoms, as shown in Fig. 10.24. These orbitals combine to form a group of closely spaced π molecular orbitals that are important in two ways. First, they contribute significantly to the stability of the graphite layers because of the π bond formation. Second, the π molecular orbitals with their delocalized electrons account for the electrical conductivity of graphite. These closely spaced orbitals are exactly analogous to the conduction bands found in metal crystals.

Graphite is often used as a lubricant in locks (where oil is undesirable because it collects dirt). The slipperiness that is characteristic of graphite can be explained by noting that graphite has very strong bonding *within* the layers of carbon atoms but

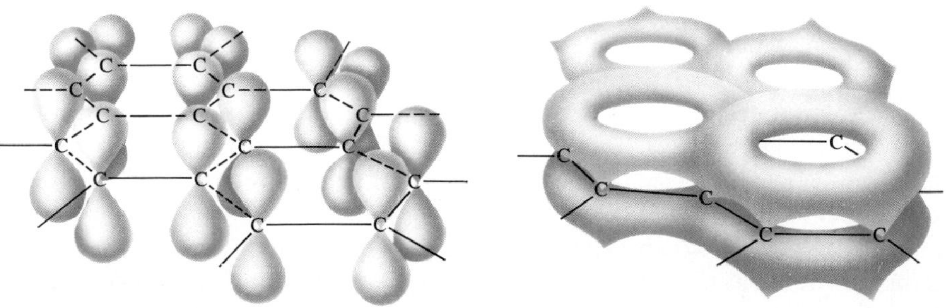

Figure 10.24

The p orbitals perpendicular to the plane of the carbon ring system in graphite can combine to form an extensive π-bonding network.

little bonding *between* the layers (the valence electrons are all used to form σ and π bonds among carbons within the layers). This arrangement allows the layers to slide past one another quite readily. Graphite's layered structure is quite obvious when viewed with a high-magnification electron microscope (Fig. 10.25). This is in contrast to diamond which has uniform bonding in all directions in the crystal.

Figure 10.25

A high resolution image of crystalline graphite (×2,500,000) taken by an electron microscope.

Because of their extreme hardness, diamonds are extensively used in industrial cutting implements. Thus it is desirable to convert cheaper graphite to diamond. As we might expect from the higher density of diamond (3.5 g/cm^3) compared to that of graphite (2.2 g/cm^3), this transformation can be accomplished by applying very high pressures to graphite. The application of 150,000 atm of pressure at 2800°C converts graphite virtually completely to diamond. The high temperature is required to break the strong bonds in graphite so the rearrangement can occur.

Silicon is an important constituent of the compounds that make up the earth's crust. In fact, silicon is to geology as carbon is to biology. Just as carbon compounds are the basis for most biologically significant systems, silicon compounds are fundamental to most of the rocks, sands, and soils found in the earth's crust. However, although carbon and silicon are next to each other in Group 4A of the periodic table, the carbon-based compounds of biology and the silicon-based compounds of geology have markedly different structures. Carbon compounds typically contain long strings of carbon-carbon bonds, while the most stable silicon compounds involve chains with silicon-oxygen bonds. The *most important silicon compounds contain silicon and oxygen.*

The bonding in the CO_2 molecule was described in Section 9.1.

The fundamental silicon-oxygen compound is **silica,** which has the empirical formula SiO_2. Knowing the properties of the similar compound carbon dioxide, one might expect silica to be a gas that contains discrete SiO_2 molecules. In fact, nothing could be further from the truth—quartz and some types of sand are typical of the materials composed of silica. What accounts for this difference? The answer lies in the bonding.

Recall that the Lewis structure for CO_2 is

$$\ddot{\underset{\cdot\cdot}{O}}=C=\ddot{\underset{\cdot\cdot}{O}}$$

and that each C=O bond is a combination of a σ bond, involving a carbon *sp* hybrid orbital, and a π bond, involving a carbon 2*p* orbital. On the contrary, silicon cannot use its valence 3*p* orbitals to form strong π bonds to oxygen, mainly because of the larger size of the silicon atom and its orbitals, which results in less effective overlap with the smaller oxygen orbitals. Therefore, instead of forming π bonds, the silicon atom satisfies the octet rule by forming single bonds to four oxygen atoms, as shown in the representation of the structure of quartz in Fig. 10.26. Note that each silicon atom is at the center of a tetrahedral arrangement of oxygen atoms, which are shared with other silicon atoms. Although the empirical formula for quartz is SiO_2, the structure is based on a *network* of SiO_4 tetrahedra with shared oxygen atoms rather than discrete SiO_2 molecules. It is obvious that the differing abilities of carbon and silicon to form π bonds with oxygen have profound effects on the structures and properties of CO_2 and SiO_2.

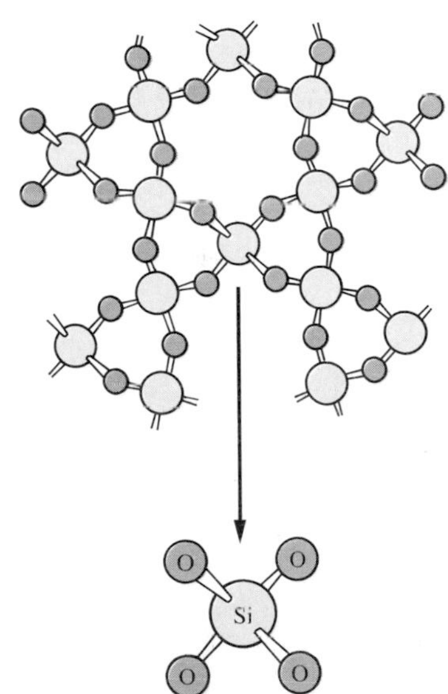

Figure 10.26

The structure of quartz (empirical formula SiO_2). Quartz contains chains of SiO_4 tetrahedra that share oxygen atoms.

Compounds closely related to silica and found in most rocks, soils, and clays are the **silicates.** Like silica, the silicates are based on interconnected SiO_4 tetrahedra. However, in contrast to silica, where the O:Si ratio is 2:1, silicates have O:Si ratios greater than 2:1 and contain silicon-oxygen *anions*. This means that to

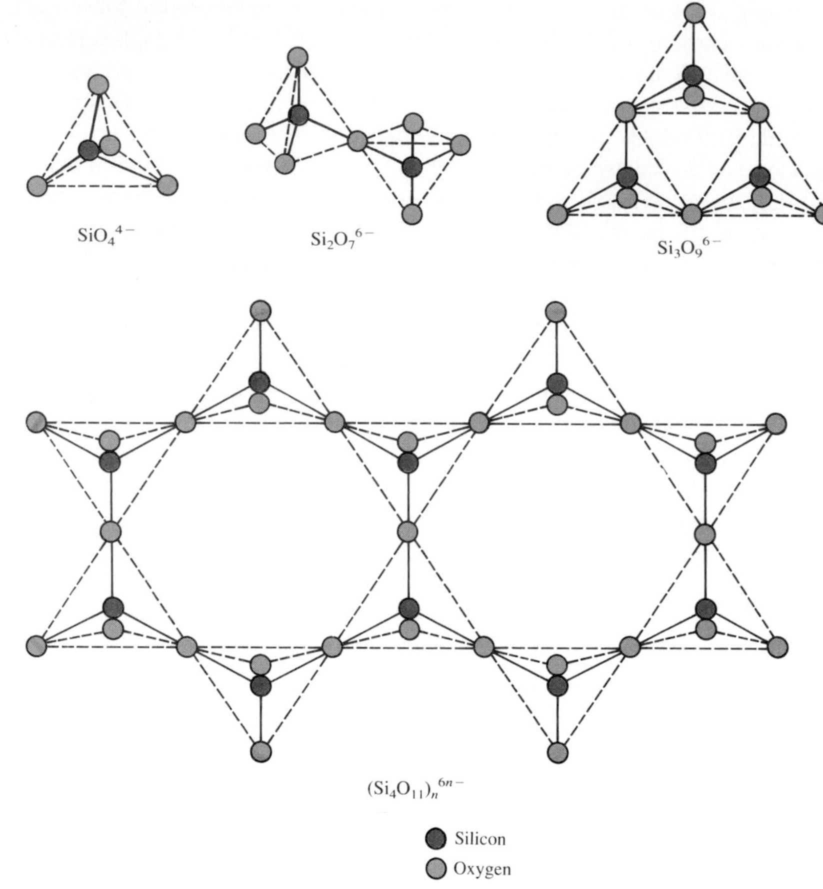

Figure 10.27

Examples of silicate anions, all of which are based on SiO_4^{4-} tetrahedra.

(a)

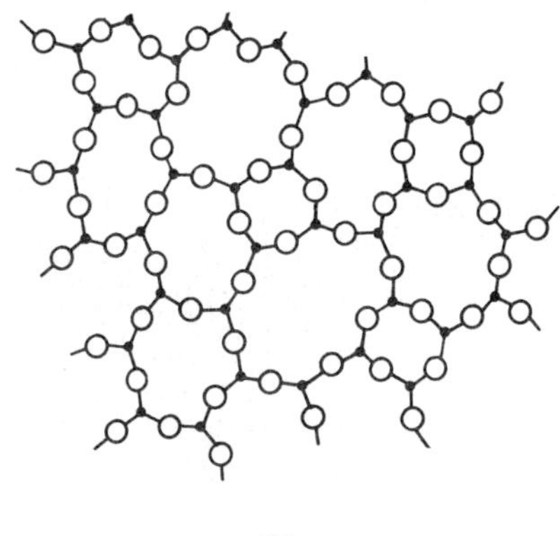

(b)

Figure 10.28

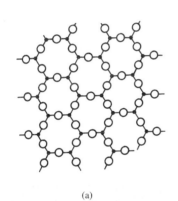

Two-dimensional representations of (a) a quartz crystal and (b) a quartz glass.

form the neutral solid silicates, cations are needed to balance the excess negative charge. In other words, silicates are salts containing metal cations and polyatomic silicon-oxygen anions. Examples of important silicate anions are shown in Fig. 10.27.

When silica is heated above its melting point (about 1600°C) and cooled rapidly, an amorphous solid called a **glass** results (see Fig. 10.28). Note that a glass contains a good deal of disorder, in contrast to the crystalline nature of quartz. Glass more closely resembles a very viscous solution than it does a crystalline solid. Common glass results when substances such as Na_2CO_3 are added to the silica melt, which is then cooled. The properties of glass can be varied greatly by varying the additives. For example, addition of B_2O_3 produces a glass (called borosilicate glass) that expands and contracts little under large temperature changes. Thus it is useful for labware and cooking utensils. The most common brand name for this glass is Pyrex. The addition of K_2O produces an especially hard glass that can be ground to the precise shapes needed for eyeglass and contact lenses. The compositions of several types of glass are shown in Table 10.4.

Compositions of Some Common Types of Glass							
	Percentages of various components						
Type of glass	SiO_2	CaO	Na_2O	B_2O_3	Al_2O_3	K_2O	MgO
Window (soda-lime glass)	72	11	13	—	0.3	3.8	—
Cookware (aluminosilicate glass)	55	15	—	—	20	—	10
Heat-resistant (borosilicate glass)	76	3	5	13	2	0.5	—
Optical	69	12	6	0.3	—	12	—

Table 10.4

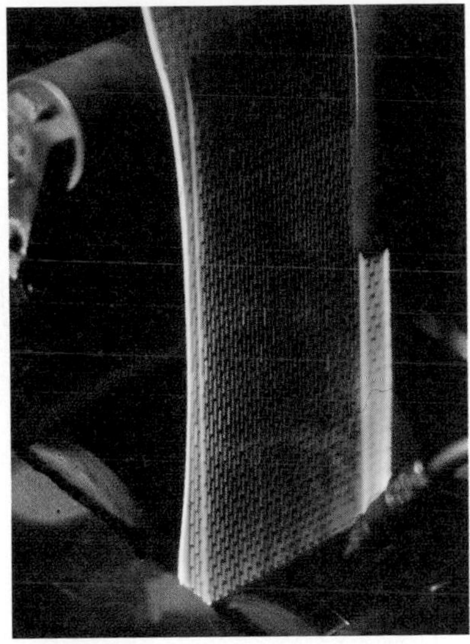

Making sheets of glass at Corning Glassworks.

Ceramics

Ceramics are typically made from clays (which contain silicates) and hardened by firing at high temperatures. Ceramics are a class of nonmetallic materials that are strong, brittle, and resistant to heat and attack by chemicals.

Like glass, ceramics are based on silicates, but with that the resemblance ends. Glass can be melted and remelted as often as desired, but once a ceramic has been hardened, it is resistant to extremely high temperatures. This behavior results from the very different structures of glasses and ceramics. A glass is a *homogeneous*, noncrystalline "frozen solution," and a ceramic is *heterogeneous*. A ceramic contains two phases: minute crystals of silicates are suspended in a glassy cement.

To understand how ceramics harden, it is necessary to know something about the structure of clays. Clays are formed by the weathering action of water and carbon dioxide on the mineral feldspar, which is a mixture of silicates with empirical formulas such as $K_2O \cdot Al_2O_3 \cdot 6SiO_2$ and $Na_2O \cdot Al_2O_3 \cdot 6SiO_2$. Feldspar is really an **aluminosilicate** in which aluminum as well as silicon atoms are part of the oxygen-bridged polyanion. The weathering of feldspar produces kaolinite, consisting of tiny thin platelets with the empirical formula $Al_2Si_2O_5(OH)_4$. When dry, the platelets cling together; when water is present, they can slide over one another, giving clay its plasticity. As clay dries, the platelets begin to interlock again. When the remaining water is driven off during firing, the silicates and cations form a glass that binds the tiny crystals of kaolinite.

Ceramics have a very long history. Rocks, which are natural ceramic materials, served as the earliest tools. Later, clay vessels dried in the sun or baked in fires served as containers for food and water. These early vessels were no doubt crude and quite porous. With the discovery of glazing, which probably occurred about 3000 B.C. in Egypt, pottery became more serviceable as well as more beautiful. Prized porcelain is essentially the same material as crude earthenware, but specially selected clays and glazings are used for porcelain and the clay object is fired at a very high temperature.

Early Bronze Age (~3000 B.C.) teapot from the Middle East.

Semiconductors

Elemental silicon has the same structure as diamond, as might be expected from its position in the periodic table (in Group 4A directly under carbon). Recall that in

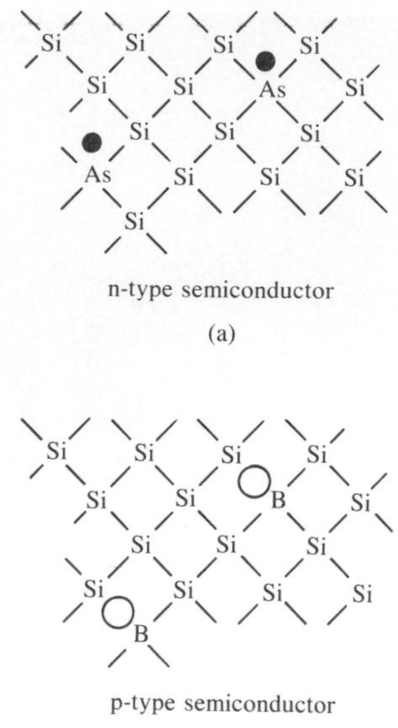

Figure 10.29

(a) A silicon crystal doped with arsenic, which has one extra valence electron. (b) A silicon crystal doped with boron, which has one less electron than silicon.

diamond there is a large energy gap between the filled and empty molecular orbitals (Fig. 10.26). This gap prevents excitation of electrons to the empty molecular orbitals (conduction bands) and makes diamond an insulator. In silicon the situation is similar but the energy gap is smaller. A few electrons can cross the gap at 25°C, making silicon a *semiconducting element,* or **semiconductor.** In addition, at higher temperatures, where more energy is available to excite electrons into the conduction bands, the conductivity of silicon increases. This is typical behavior for a semiconducting element and is in contrast to that of metals, whose conductivity decreases with increasing temperature.

The small conductivity of silicon can be enhanced at normal temperatures if the silicon crystal is *doped* with certain other elements. For example, when a small fraction of silicon atoms is replaced by arsenic atoms, each having *one more* valence electron than silicon, extra electrons become available for conduction, as shown in Fig. 10.29(a). This produces an **n-type semiconductor,** a substance whose conductivity is increased by doping it with atoms having more valence electrons than the atoms in the host crystal. These extra electrons lie close in energy to the conduction bands and can easily be excited into these levels, where they can conduct an electric current [see Fig. 10.30(a)].

We can also enhance the conductivity of silicon by doping the crystal with an element such as boron, which has only three valence electrons, one *less* than silicon. Because boron has one less electron than is required to form the bonds to the surrounding silicon atoms, an electron vacancy, or *hole,* is created, as shown in Fig. 10.29(b). As an electron fills this hole, it leaves a new hole, and this process can be repeated. Thus the hole advances through the crystal in a direction opposite to movement of the electrons jumping to fill the hole. Another way of thinking about this phenomenon is that in pure silicon each atom has four valence electrons, and the low-energy molecular orbitals are exactly filled. Replacing silicon atoms with boron atoms leaves vacancies in these molecular orbitals, as shown in Fig. 10.30(b). This means there is only one electron in some of the molecular orbitals, and these unpaired electrons can function as conducting electrons. Thus the substance becomes a better conductor. When semiconductors are doped with atoms having fewer valence electrons than the atoms of the host crystal, they are called **p-type semiconductors,** so named because the positive holes can be viewed as the charge carriers.

Electrons must be in singly occupied molecular orbitals to conduct a current.

Most important applications of semiconductors involve connection of a p-type and an n-type to form a **p-n junction.** Figure 10.31(a) shows a typical junction; the

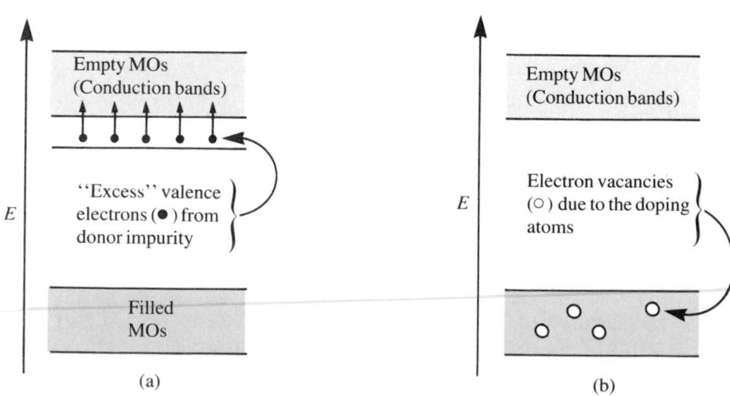

Figure 10.30

Energy-level diagrams for (a) an n-type semiconductor and (b) a p-type semiconductor.

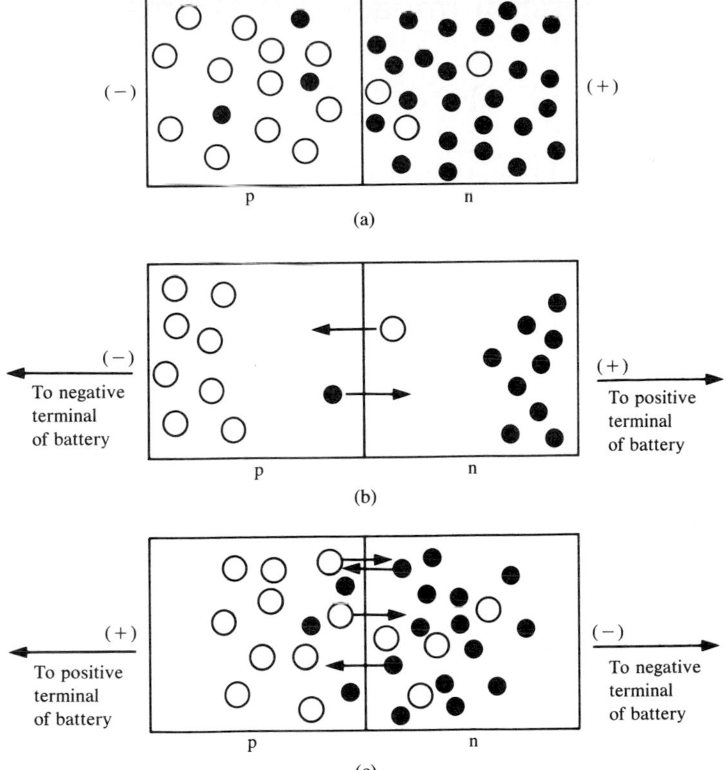

Figure 10.31

The p-n junction involves the contact of a p-type and an n-type semiconductor. (a) The charge carriers of the p-type region are holes (○). In the n-type region the charge carriers are electrons (•). (b) No current flows (reverse bias). (c) Current readily flows (forward bias). Note that each electron that crosses the boundary leaves a hole behind. Thus the electrons and the holes move in opposite directions.

dark circles represent excess electrons in the n-type semiconductor, and the white circles represent holes (electron vacancies) in the p-type semiconductor. At the junction a small number of electrons migrate from the n-type region into the p-type region, where there are vacancies in the low-energy molecular orbitals. The effect of these migrations is to place a negative charge on the p-type region (since it now has a surplus of electrons) and a positive charge on the n-type region (since it has lost electrons, leaving holes in its low-energy molecular orbitals). This charge build-up, called the *contact potential,* or *junction potential,* prevents further migration of electrons.

Now suppose an external electrical potential is applied by connecting the negative terminal of a battery to the p-type region and the positive terminal to the n-type region. The situation represented in Fig. 10.31(b) results. Electrons are drawn toward the positive terminal, and the resulting holes move toward the negative terminal—exactly opposite to the natural flow of electrons at the p-n junction. The junction resists the imposed current flow in this direction and is said to be under *reverse bias*. No current flows through the system.

On the other hand, if the battery is connected so that the negative terminal is connected to the n-type region and the positive terminal is connected to the p-type region [Fig. 10.31(c)], the movement of electrons (and holes) is in the favored direction. The junction has low resistance, and a current flows easily. The junction is said to be under *forward bias*.

A p-n junction makes an excellent *rectifier,* a device that produces direct current (flows in one direction) from an alternating current (flows in both directions

Chemical Impact

Shocked Quartz and Disappearing Dinosaurs

Why did the dinosaurs disappear over such a short span of time about 65 million years ago? Recently scientists have speculated that the collision of a large meteorite or comet with the earth was responsible. Some of the evidence that supports this theory involves unusual iridium and niobium concentrations in the rocks from this period (see "Chemical Impact: Transition Metals in Action" in Chapter 20). New supporting evidence from researchers in the U.S. Geological Survey office in Denver is based on quartz sampled from five sites around the world. This so-called shocked quartz shows multiple lamellae, or parallel fracture planes, caused by shearing forces from high-pressure shock waves. The USGS scientists say that 25% of the quartz in the samples from 65-million-year-old rock show the dislocations characteristically caused by extreme shock. This strongly indicates a very large object (as much as 6 miles wide) hit the earth at very high speed, causing a dust cloud that enveloped the earth, blocking out life-giving sunlight.

Suggested Reading

R. Monastersky, "Extinction upon Impact?" *Science News* **131** (May 16, 1987): 309.

Printed circuits are discussed in a Chemical Impact feature in this chapter.

alternately). When placed in a circuit where the potential is constantly reversing, a p-n junction transmits current only under forward bias, thus converting the alternating current to a direct current. Radios, computers, and other electrical devices formerly used bulky, unreliable vacuum tubes as rectifiers. The p-n junction has revolutionized electronics; modern solid-state components contain p-n junctions in printed circuits.

10.6 Molecular Solids

Purpose

- To describe the bonding in molecular solids.

So far we have considered solids in which atoms occupy the lattice positions, and in most cases such a crystal can be considered to be one giant molecule. However, there are many types of solids that contain discrete molecular units at each lattice position. A common example is ice, where the lattice positions are occupied by water molecules [see Fig. 10.12(c)]. Other examples are dry ice (solid carbon dioxide), some forms of sulfur that contain S_8 molecules [Fig. 10.32(a)], and certain forms of phosphorus that contain P_4 molecules [Fig. 10.32(b)]. These substances are characterized by strong covalent bonding *within* the molecules but relatively weak forces *between* the molecules. For example, it takes only 6 kJ of energy to melt 1 mole of solid water (ice) because only intermolecular (H_2O—H_2O) interactions must be overcome. However, 470 kJ of energy is required to break a mole of covalent O—H bonds. The differences between the covalent bonds within the molecules and the forces between the molecules are apparent from the comparison of the interatomic and intermolecular distances in solids shown in Table 10.5.

A "steaming" piece of dry ice.

Comparison of Atomic Separations Within Molecules (Covalent Bonds) and Between Molecules (Intermolecular Interactions)

Solid	Distance between atoms in molecule*	Closest distance between molecules in the solid
P_4	2.20 Å	3.8 Å
S_8	2.06 Å	3.7 Å
Cl_2	1.99 Å	3.6 Å

*The shorter distances within the molecules indicate stronger bonding.

Table 10.5

The forces that exist among the molecules in a molecular solid depend on the nature of the molecules. Many molecules such as CO_2, I_2, P_4, and S_8 have no dipole moment, and the intermolecular forces are the relatively weak London dispersion forces. Because these forces are usually small, we might expect all of these substances to be gaseous at 25°C, as is the case for carbon dioxide. However, as the size of the molecules increases, the London forces become larger, causing many of these substances to be solids at 25°C.

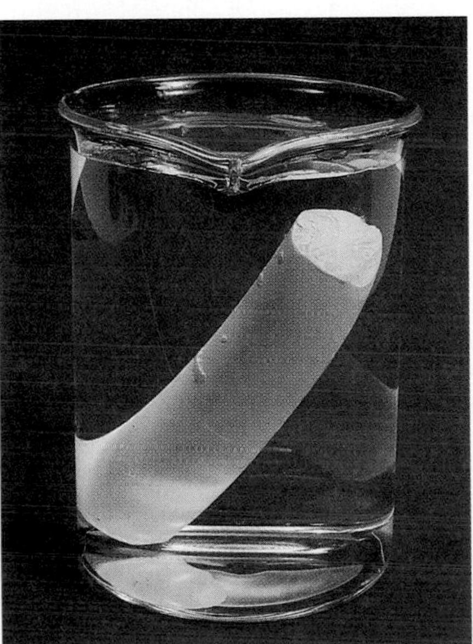

Figure 10.32

(left) Sulfur crystals (yellow) contain S_8 molecules. (right) White phosphorus contains P_4 molecules. It is so reactive with the oxygen in air that it must be stored under water.

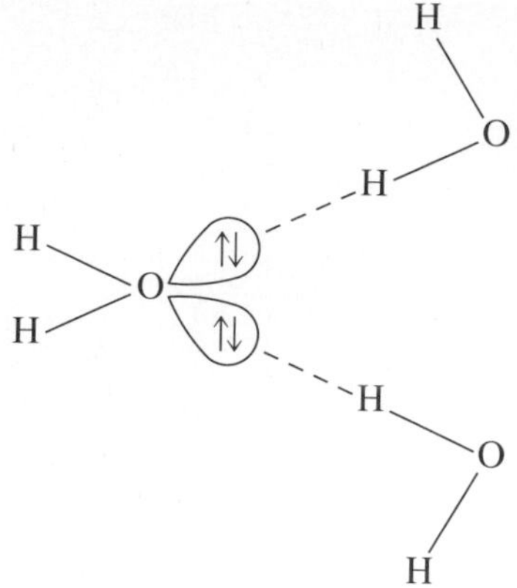

When molecules do have dipole moments, their intermolecular forces are significantly greater, especially when hydrogen bonding is possible. Water molecules are particularly well suited to interact with each other because each molecule has two polar O—H bonds and two lone pairs on the oxygen atom. This can lead to the association of four hydrogen atoms with each oxygen: two by covalent bonds and two by dipole forces, as shown in the figure to the left. Note the two relatively short covalent oxygen-hydrogen bonds and the two longer oxygen-hydrogen dipole interactions that can be seen in the ice structure in Fig. 10.12(c).

10.7 Ionic Solids

Purpose

- To model the structures of ionic solids using the packing of spheres.

Ionic solids are stable, high-melting substances held together by the strong electrostatic forces that exist between oppositely charged ions. The principles governing the structures of ionic solids were introduced in Section 8.5. In this section we will review and extend these principles.

The structures of most binary ionic solids, such as sodium chloride, can be explained by the closest packing of spheres. Typically, the larger ions, usually the anions, are packed in one of the closest packing arrangements (hcp or ccp), and the smaller cations fit into holes among the close packed anions. The packing is done in a way that maximizes the electrostatic attractions among oppositely charged ions and minimizes the repulsions among ions with like charges.

There are three types of holes in closest packed structures:

1. Trigonal holes are formed by three spheres in the same layer [Fig. 10.33(a)].
2. Tetrahedral holes are formed when a sphere sits in the dimple of three spheres in an adjacent layer [Fig. 10.33(b)].
3. Octahedral holes are formed between two sets of three spheres in adjoining layers of the closest packed structures [Fig. 10.33(c)].

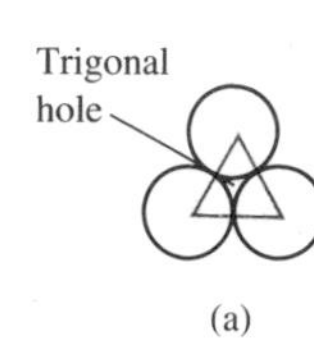

(a)

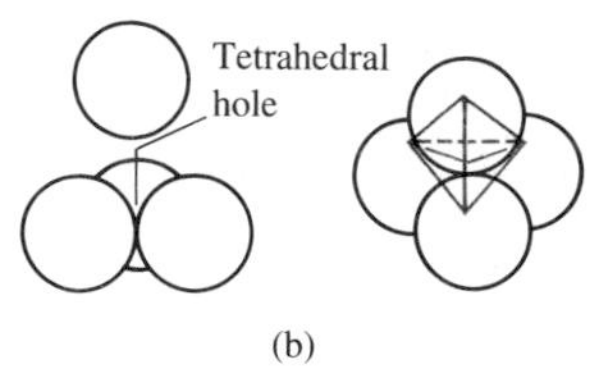

(b)

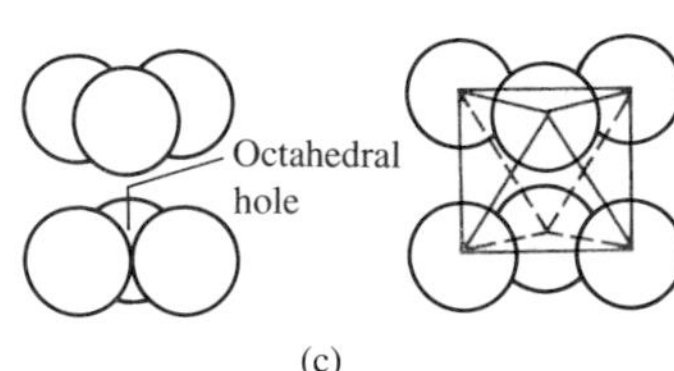

(c)

Figure 10.33

The holes that exist among closest packed uniform spheres. (a) The trigonal hole formed by three spheres in a given plane. (b) The tetrahedral hole formed when a sphere occupies a dimple in an adjacent layer. (c) The octahedral hole formed by six spheres in two adjacent layers.

Closest packed structures contain twice as many tetrahedral holes as packed spheres. Closest packed structures contain the same number of octahedral holes as packed spheres.

For spheres of a given diameter, the holes increase in size in the order:

$$\text{trigonal} < \text{tetrahedral} < \text{octahedral}$$

In fact, trigonal holes are so small that they are never occupied in binary ionic compounds. Whether the tetrahedral or octahedral holes in a given binary ionic solid are occupied depends mainly on the *relative* sizes of the anion and cation. For example, in zinc sulfide the S^{2-} ions (ionic radius = 1.8 Å) are arranged in a cubic closest packed structure with the smaller Zn^{2+} ions (ionic radius = 0.7 Å) in the tetrahedral holes. The location of the tetrahedral holes in the face-centered cubic unit cell of the ccp structure is shown in Fig. 10.34(a). Note from this figure that there are eight tetrahedral holes in the unit cell. Also recall from the discussion in Section 10.4 that there are four net spheres in the face-centered cubic unit cell. Thus there are *twice as many tetrahedral holes as packed anions* in the closest packed structure. Zinc sulfide must have the same number of S^{2-} ions and Zn^{2+} ions to

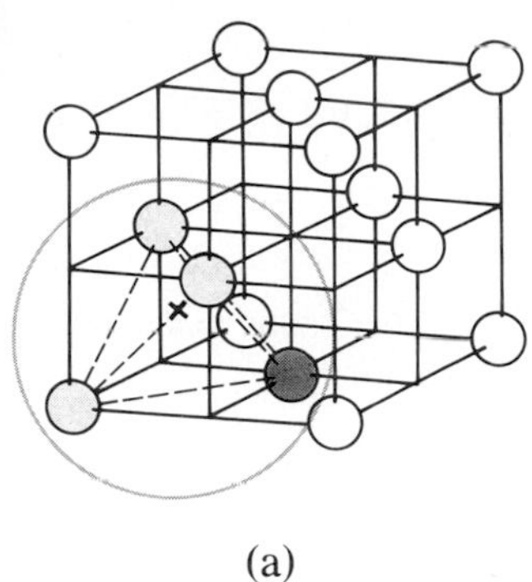

(a)

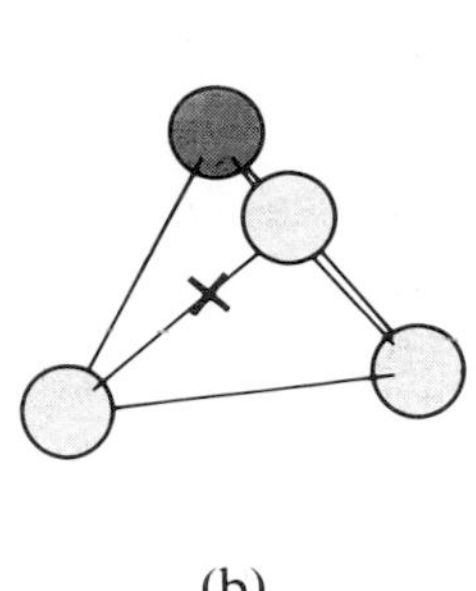

(b)

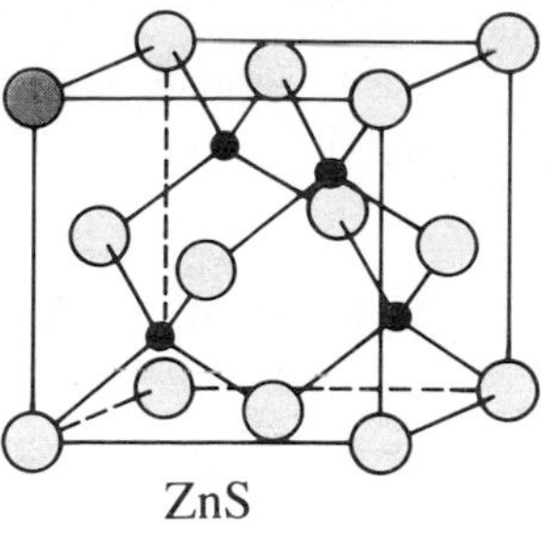

(c)

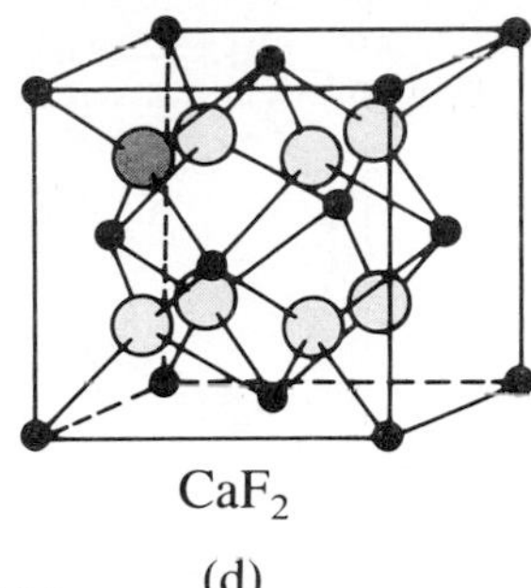

(d)

Figure 10.34

(a) The location (*x*) of a tetrahedral hole in the face-centered cubic unit cell. (b) One of the tetrahedral holes. (c) The unit cell for ZnS where the S^{2-} ions (○) are closest packed with the Zn^{2+} ions (●) in alternate tetrahedral holes. (d) The unit cell for CaF_2 where the Ca^{2+} ions (●) are closest packed with the F^- ions (○) in all of the tetrahedral holes.

achieve electrical neutrality. Thus in the zinc sulfide structure only *half* of the tetrahedral holes contain Zn^{2+} ions, as shown in Fig. 10.34(c).

The structure of calcium fluoride (CaF_2) can be similarly explained. However, in this case the structure is best visualized as a cubic closest packed array of Ca^{2+} ions with F^- ions in *all* of the tetrahedral holes [Fig. 10.34(d)]. This produces the required 2:1 stoichiometry.

The structure of sodium chloride can be described in terms of a cubic closest packed array of Cl^- ions with Na^+ ions in all of the octahedral holes. The location of the octahedral holes in the face-centered cubic unit cell is shown in Fig. 10.35(a). The easiest octahedral hole to find in this structure is the one at the center of the cube. Note that this hole is surrounded by six spheres, as is required to form an octahedron. The remaining octahedral holes are shared with other unit cells and are more difficult to visualize. However, it can be shown that the number of octahedral holes in the ccp structure is the *same* as the number of packed anions. Figure 10.35(b) shows the structure for sodium chloride that results from Na^+ ions filling all of the octahedral holes in a ccp array of Cl^- ions.

A great variety of ionic solids exists. Our purpose in this section is not to give an exhaustive treatment of ionic solids, but to emphasize the fundamental principles governing their structures. As we have seen, the most useful model for explaining the structures of these solids regards the ions as hard spheres that are packed to maximize attractions and minimize repulsions.

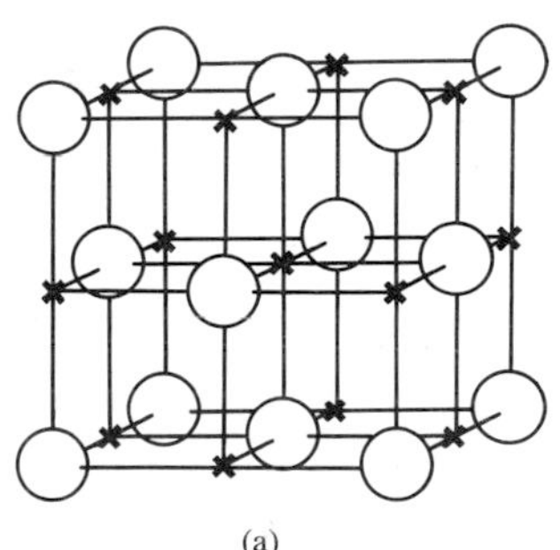

(a)

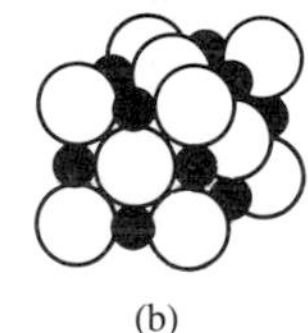

(b)

Figure 10.35

(a) The locations (*x*) of the octahedral holes in the face-centered cubic unit cell. (b) Representation of the unit cell for solid NaCl. The Cl^- ions (white spheres) have a ccp arrangement with Na^+ ions (black spheres) in all of the octahedral holes.

Sample Exercise 10.4

Determine the net number of Na^+ and Cl^- ions in the sodium chloride unit cell.

Solution

Note from Fig. 10.35(b) that the Cl^- ions are cubic closest packed and thus form a face-centered cubic unit cell. There is a Cl^- ion on each corner and one at the center of each face of the cube. Thus the net number of Cl^- ions present in a unit cell is

$$8(\tfrac{1}{8}) + 6(\tfrac{1}{2}) = 4$$

The Na^+ ions occupy the octahedral holes located in the center of the cube and midway along each edge. The Na^+ ion in the center of the cube is contained entirely in the unit cell, while those on the edges are shared by four unit cells (four cubes

Sample Exercise 10.4, continued

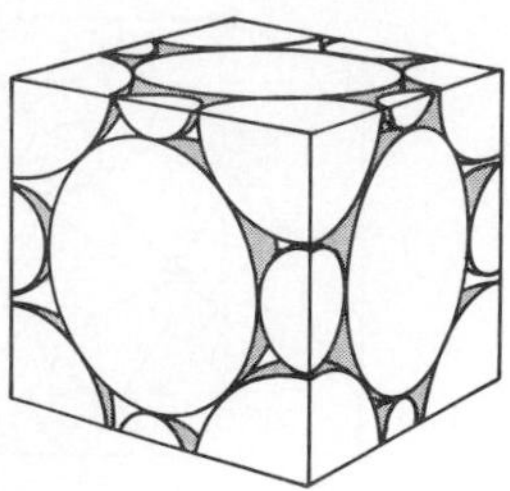

share a common edge). Since the number of edges in a cube is 12, the net number of Na^+ ions present is

$$1(1) + 12(\tfrac{1}{4}) = 4$$

We have shown that the net number of ions in a unit cell is 4 Na^+ ions and 4 Cl^- ions, which agrees with the 1:1 stoichiometry of sodium chloride.

In this chapter we have considered various types of solids. Table 10.6 summarizes these types of solids and some of their properties.

Types and Properties of Solids

Type of solid:	Molecular	Atomic: Network	Atomic: Metallic	Ionic
Structural unit:	Molecule	Atom	Atom	Ion
Type of bonding:	Polar molecules: dipole-dipole interactions Nonpolar molecules: London dispersion forces	Highly directional covalent bonds	Nondirectional covalent bonds involving electrons that are delocalized throughout the crystal	Electrostatic
Typical properties:	Soft	Hard	Wide range of hardness	Hard
	Low melting point	High melting point	Wide range of melting points	High melting point
	Insulator	Insulator	Conductor	Insulator
Examples:	Ice (solid H_2O) Dry ice (solid CO_2)	Diamond	Silver Iron Brass	Sodium chloride Calcium fluoride

Table 10.6

Sample Exercise 10.5

Using Table 10.6, classify each of the following substances according to the type of solid it forms:

a. gold
b. carbon dioxide
c. lithium fluoride
d. krypton

Solution

a. Solid gold is an atomic solid with metallic properties.
b. Solid carbon dioxide contains nonpolar carbon dioxide molecules and is a molecular solid.

Chemical Impact

Miracle Coatings

Imagine a pair of plastic-lensed sunglasses that are unscratchable, even if you drop them on concrete or rub them with sandpaper. Research may make such glasses possible, along with cutting tools that never need sharpening, special glass for windshields and buildings that cannot be scratched by wind-blown sand, and speakers that reproduce sound with a crispness unimagined, at any cost, until now. The secret of all of these marvels is a thin diamond coating. Diamond is so hard that virtually nothing can scratch it. A thin diamond coating on a speaker cone limits resonance and gives a remarkably pure tone.

But how do you coat something with diamond? It is nearly impossible to melt diamond (mp ~ 3500°C), and even if it could be melted, the object being coated would itself melt immediately at this temperature. Surprisingly, a diamond coating can be applied quite easily to something even as fragile as plastic. First, the surface is bathed with a mixture of gaseous methane and hydrogen at low pressures. Then the methane is decomposed by an energy source such as microwaves, depositing a thin diamond coating of carbon atoms from the decomposed methane.

A similar process is being developed by Gregory Girolami, a chemist at the University of Illinois. In this case, the coating is titanium carbide, which melts at 3000°C and has a hardness similar to that of diamond. The coating is produced by tetraneopentyl titanium

$$\left[\begin{array}{ccccc} & & \mathrm{H} & & \\ & & | & & \\ & \mathrm{H} & \mathrm{H{-}C{-}H} & \mathrm{H} & \\ & | & | & | & \\ \mathrm{H{-}} & \mathrm{C} & \mathrm{{-}C{-}} & \mathrm{C} & \\ & | & | & | & \\ & \mathrm{H} & \mathrm{H{-}C{-}H} & \mathrm{H} & \\ & & | & & \\ & & \mathrm{H} & & \end{array}\right]_4 \!\!\mathrm{{-}Ti}$$

which decomposes at only 150°C.

Clearly, the coating of soft, scratchable materials with super-tough substances such as diamond or titanium carbide will have tremendous economic implications for both industry and consumer.

Sample Exercise 10.5, continued

c. Solid lithium fluoride contains Li^+ and F^- ions and is a binary ionic solid.

d. Solid krypton contains krypton atoms that can interact only through London dispersion forces. It is an atomic solid but has properties characteristic of a molecular solid with nonpolar molecules.

10.8 Vapor Pressure and Changes of State

Purpose

- To define the vapor pressure of a liquid.
- To discuss the features of heating curves.

Now that we have considered the general properties of the three states of matter, we can explore the processes by which matter changes state. One very familiar example of a change in state occurs when a liquid evaporates from an open container. This is clear evidence that the molecules of a liquid can escape the liquid's

Vapor is the usual term for the gas phase of a substance that exists as a solid or liquid at 25°C and 1 atm.

surface and form a gas. Called **vaporization,** or *evaporation,* this process is endothermic because energy is required to overcome the relatively strong intermolecular forces in the liquid. The energy required to vaporize 1 mole of a liquid at a pressure of 1 atm is called the **heat of vaporization,** or the **enthalpy of vaporization,** and is usually symbolized as $\mathbf{\Delta H_{vap}}$.

The endothermic nature of vaporization has great practical significance; in fact, one of the most important roles that water plays in our world is to act as a coolant. Because of the strong hydrogen bonding among its molecules in the liquid state, water has an unusually large heat of vaporization (40.7 kJ/mol). A significant portion of the sun's energy that reaches earth is spent evaporating water from the oceans, lakes, and rivers rather than warming the earth. The vaporization of water is also crucial to the body's temperature control system through evaporation of perspiration.

Vapor Pressure

When a liquid is placed in a closed container, the amount of liquid at first decreases but eventually becomes constant. The decrease occurs because there is an initial net transfer of molecules from the liquid to the vapor phase (Fig. 10.36). However, as the number of vapor molecules increases, so does the rate of return of these molecules to the liquid. The process by which vapor molecules reform a liquid is called **condensation.** Eventually enough vapor molecules are present above the liquid so that the rate of condensation equals the rate of evaporation (see Fig. 10.37). *At this point no further net change occurs in the amount of liquid or vapor because the two opposite processes exactly balance each other;* the system is at **equilibrium.** Note that this system is highly *dynamic* on the molecular level—molecules are constantly escaping from and entering the liquid at a high rate. However, there is no *net* change because the two opposite processes just *balance* each other.

A system at equilibrium is dynamic on the molecular level, but shows no macroscopic changes.

The pressure of the vapor present at equilibrium is called the *equilibrium vapor pressure,* or more commonly, the **vapor pressure** of the liquid. A simple barometer can measure the vapor pressure of a liquid, as shown below in Fig. 10.38(a). The liquid is injected at the bottom of the tube of mercury and floats to the surface because the mercury is so dense. The liquid evaporates at the top of the column,

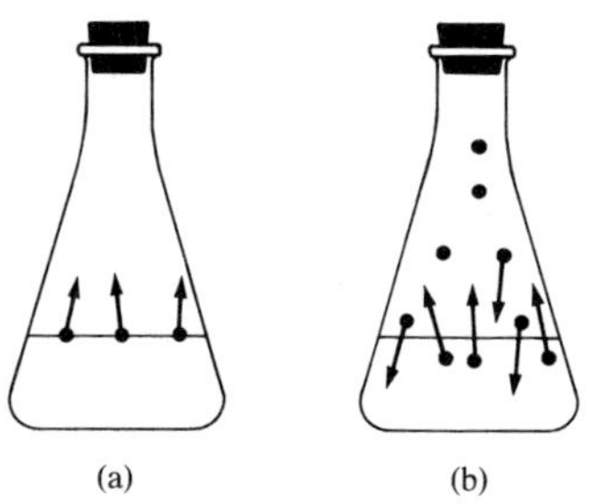

Figure 10.36

Behavior of a liquid in a closed container. (a) Initially net evaporation occurs as molecules are transferred from the liquid to the vapor phase, so the amount of liquid decreases. (b) As the number of vapor molecules increases, the rate of return to the liquid (condensation) increases, until finally the rate of condensation equals the rate of evaporation. The system is at equilibrium, and no further changes occur in the amounts of vapor or liquid.

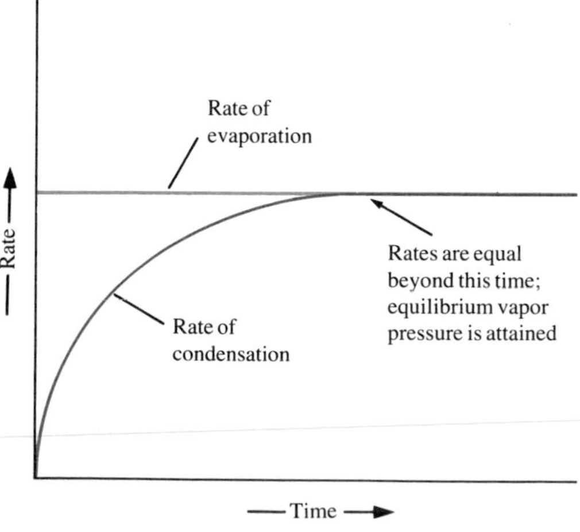

Figure 10.37

The rates of condensation and evaporation over time for a liquid sealed in a closed container. The rate of evaporation remains constant and the rate of condensation increases as the number of molecules in the vapor phase increases, until the two rates become equal. At this point, the equilibrium vapor pressure is attained.

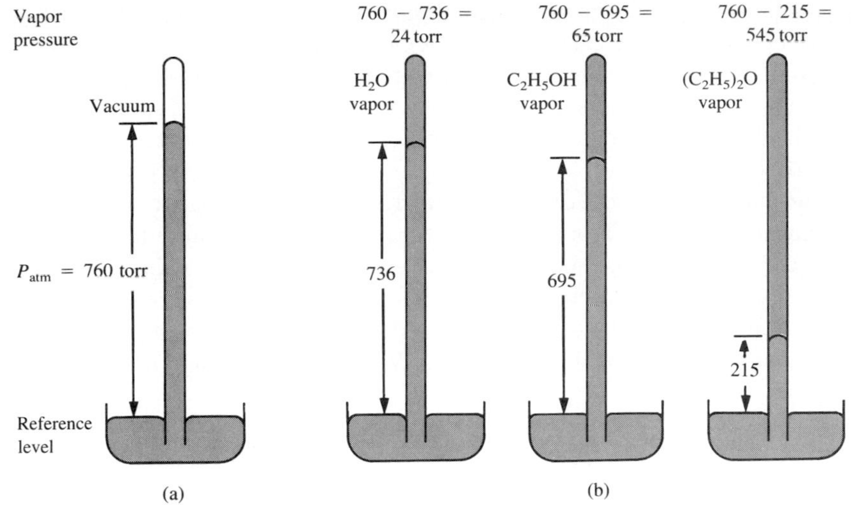

Figure 10.38

(a) The vapor pressure of a liquid can be easily measured using a simple barometer of the type shown here. (b) The three liquids water, ethanol (C_2H_5OH), and diethyl ether [$(C_2H_5)_2O$] have quite different vapor pressures. Ether is by far the most volatile of the three.

producing a vapor whose pressure pushes some mercury out of the tube. When the system reaches equilibrium, the vapor pressure can be determined from the change in the height of the mercury column since

$$P_{\text{atmosphere}} = P_{\text{vapor}} + P_{\text{Hg column}}$$

Thus

$$P_{\text{vapor}} = P_{\text{atmosphere}} - P_{\text{Hg column}}$$

The vapor pressures of liquids vary widely [see Fig. 10.38(b)]. Liquids with high vapor pressures are said to be *volatile*—they evaporate rapidly from an open dish.

The vapor pressure of a liquid is affected by two main factors: *molecular weight* and *intermolecular forces*. Molecular weight is important because at a given temperature heavy molecules have lower velocities than light molecules and thus have a much smaller tendency to escape from the liquid surface. A liquid with a high molecular weight tends to have a small vapor pressure. Liquids in which the intermolecular forces are large also have relatively low vapor pressures because the molecules need high energies to escape to the vapor phase. For example, although water has a much lower molecular weight than that of diethyl ether, the strong hydrogen bonding forces that exist among water molecules in the liquid cause water's vapor pressure to be much lower than that of diethyl ether [see Fig. 10.38(b)].

Measurements of the vapor pressure for a given liquid at several temperatures show that *vapor pressure increases significantly with temperature*. Figure 10.39 illustrates the distribution of molecular velocities present in a liquid at two different

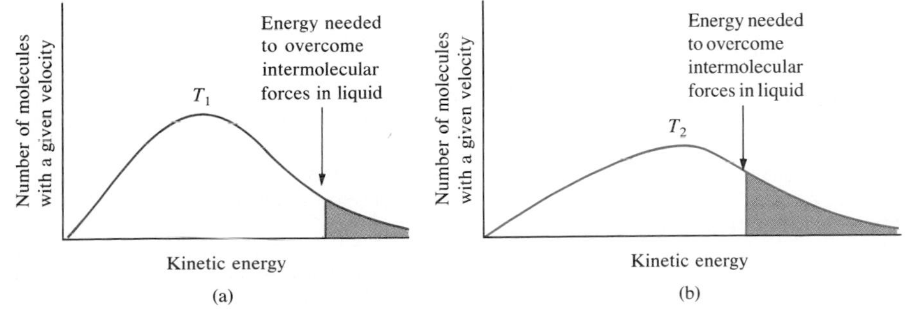

Figure 10.39

The number of molecules in a liquid with a given velocity versus kinetic energy at two temperatures. Part a shows a lower temperature than that in part b. Note that the proportion of molecules with enough energy to escape the liquid to the vapor phase (indicated by shaded areas) increases dramatically with temperature. This causes vapor pressure to increase markedly with temperature.

The Vapor Pressure of Water as a Function of Temperature	
T (°C)	*P* (torr)
0.0	4.579
10.0	9.209
20.0	17.535
25.0	23.756
30.0	31.824
40.0	55.324
60.0	149.4
70.0	233.7
90.0	525.8

Table 10.7

temperatures. To overcome the intermolecular forces in a liquid, a molecule must have a minimum kinetic energy. As the temperature of the liquid is increased, the fraction of molecules having sufficient energy to overcome these forces and escape to the vapor phase increases markedly. Thus the vapor pressure of a liquid increases dramatically with temperature. Values for water at several temperatures are given in Table 10.7.

The quantitative nature of the temperature dependence of vapor pressure can be determined graphically. Plots of vapor pressure versus temperature for water, ethanol, and diethyl ether are shown in Fig. 10.40(a). Note the nonlinear increase in vapor pressure for all the liquids as the temperature is increased. We find that a straight line can be obtained by plotting $\ln(P_{vap})$ versus $1/T$, where T is the Kelvin temperature, as shown in Fig. 10.40(b). We can represent this behavior by the equation

$$\ln(P_{vap}) = -\frac{\Delta H_{vap}}{R}\left(\frac{1}{T}\right) + C \qquad (10.4)$$

Natural logarithms are reviewed in Appendix 1.2.

where ΔH_{vap} is the enthalpy of vaporization, R is the universal gas constant, and C is a constant characteristic of a given liquid. The symbol ln means that the natural logarithm of the vapor pressure should be taken.

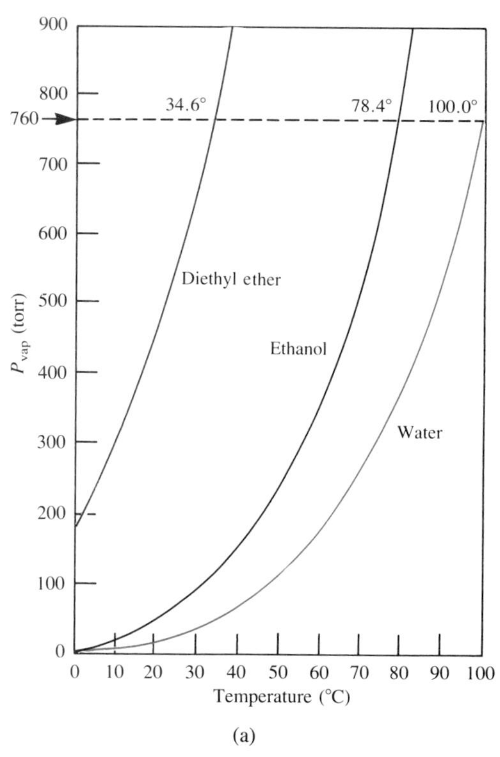

(a)

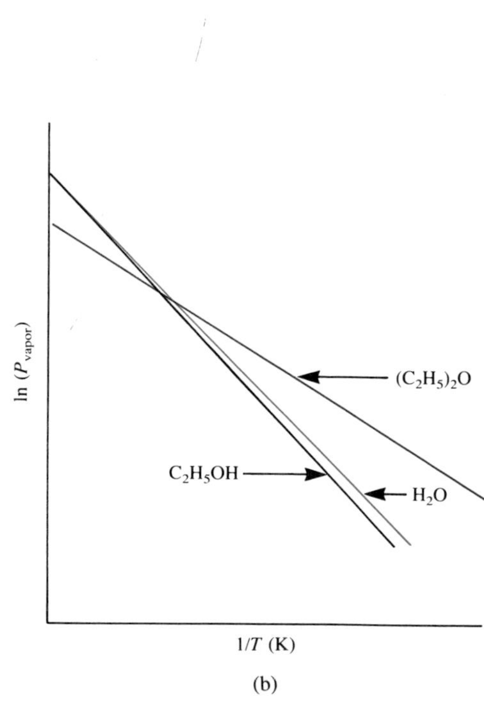

(b)

Figure 10.40

(a) The vapor pressures of water, ethanol, and diethyl ether as a function of temperature. (b) Plots of $\ln(P_{vap})$ versus $1/T$ (Kelvin temperature) for water, ethanol, and diethyl ether.

Equation (10.4) is the equation of a straight line of the form $y = mx + b$, where

$$y = \ln(P_{\text{vap}})$$

$$x = \frac{1}{T}$$

$$m = \text{slope} = -\frac{\Delta H_{\text{vap}}}{R}$$

$$b = \text{intercept} = C$$

Sample Exercise 10.6

Using the plots in Fig. 10.40(b), determine whether water or diethyl ether has the larger enthalpy of vaporization.

Solution

When $\ln(P_{\text{vap}})$ is plotted versus $1/T$, the slope of the resulting straight line is

$$-\frac{\Delta H_{\text{vap}}}{R}$$

Note from Fig. 10.40(b) that the slopes of the lines for water and diethyl ether are both negative, as expected, and that the line for ether has the smaller slope. Thus ether has the smaller value of ΔH_{vap}. This makes sense because the hydrogen bonding in water causes it to have a relatively large enthalpy of vaporization.

Equation (10.4) is important for several reasons. For example, we can determine the heat of vaporization for a liquid by measuring P_{vap} at several temperatures and then evaluating the slope of a plot of $\ln(P_{\text{vap}})$ versus $1/T$. On the other hand, if we know the value of ΔH_{vap} and P_{vap} at one temperature, we can use Equation (10.4) to calculate P_{vap} at another temperature. This can be done by recognizing that the constant C does not depend on temperature. Thus at two temperatures, T_1 and T_2, we can solve Equation (10.4) for C and then write the equality

$$\ln(P_{\text{vap}}^{T_1}) + \frac{\Delta H_{\text{vap}}}{RT_1} = C = \ln(P_{\text{vap}}^{T_2}) + \frac{\Delta H_{\text{vap}}}{RT_2}$$

This can be rearranged to

$$\ln(P_{\text{vap}}^{T_1}) - \ln(P_{\text{vap}}^{T_2}) = \frac{\Delta H_{\text{vap}}}{R}\left(\frac{1}{T_2} - \frac{1}{T_1}\right)$$

or

$$\ln\left(\frac{P_{\text{vap}}^{T_1}}{P_{\text{vap}}^{T_2}}\right) = \frac{\Delta H_{\text{vap}}}{R}\left(\frac{1}{T_2} - \frac{1}{T_1}\right) \qquad (10.5)$$

Equation 10.5 is called the Clausius-Clapeyron equation.

Sample Exercise 10.7

The vapor pressure of water at 25°C is 23.8 torr, and the heat of vaporization of water is 43.9 kJ/mol. Calculate the vapor pressure of water at 50°C.

Solution

We will use Equation (10.5):

In solving this problem, we ignore the fact that ΔH_{vap} is slightly temperature dependent.

$$\ln\left(\frac{P_{vap}^{T_1}}{P_{vap}^{T_2}}\right) = \frac{\Delta H_{vap}}{R}\left(\frac{1}{T_2} - \frac{1}{T_1}\right)$$

For water we have

$$P_{vap}^{T_1} = 23.8 \text{ torr}$$
$$T_1 = 25 + 273 = 298 \text{ K}$$
$$T_2 = 50 + 273 = 323 \text{ K}$$
$$\Delta H_{vap} = 43.9 \text{ kJ/mol} = 43{,}900 \text{ J/mol}$$
$$R = 8.3145 \text{ J/K mol}$$

Thus,

$$\ln\left(\frac{23.8 \text{ torr}}{P_{vap}^{T_2} \text{ (torr)}}\right) = \frac{43{,}900 \text{ J/mol}}{8.3145 \text{ J/K mol}}\left(\frac{1}{323 \text{ K}} - \frac{1}{298 \text{ K}}\right)$$

$$\ln\left(\frac{23.8}{P_{vap}^{T_2}}\right) = -1.37$$

Taking the antilog (see Appendix 1.2) of both sides gives

$$\frac{23.8}{P_{vap}^{T_2}} = 0.254$$

$$P_{vap}^{T_2} = 93.6 \text{ torr}$$

Figure 10.41
Sublimation of iodine.

Like liquids, solids have vapor pressures. Figure 10.41 shows iodine vapor in equilibrium with solid iodine in a closed flask. Under normal conditions iodine *sublimes;* that is, it goes directly from the solid to the gaseous state without passing through the liquid state. **Sublimation** also occurs with dry ice (solid carbon dioxide).

Changes of State

What happens when a solid is heated? Typically, it will melt to form a liquid. If the heating continues, the liquid will at some point boil and form the vapor phase. This process can be represented by a **heating curve:** a plot of temperature versus time for a process where energy is added at a constant rate.

Phase changes of carbon dioxide are discussed in Section 10.9.

The heating curve for water is given in Fig. 10.42. As energy flows into the ice, the random vibrations of the water molecules increase and the temperature rises. Eventually the molecules become so energetic that they break loose from their lattice positions, and the change from solid to liquid occurs. This is indicated by a plateau at 0°C on the heating curve. At this temperature, called the *melting point,* all

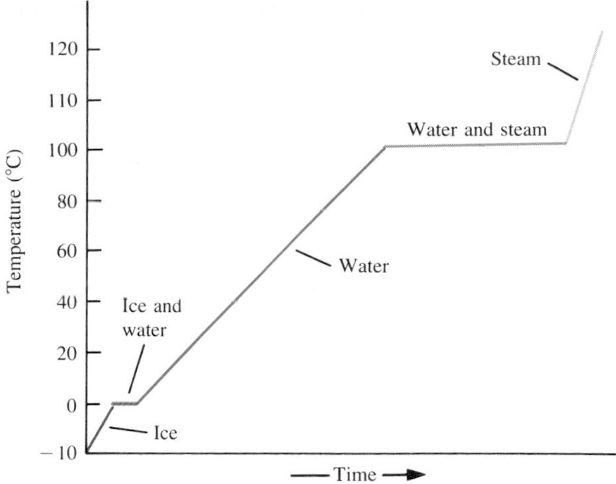

Figure 10.42

The heating curve for a given quantity of water where energy is added at a constant rate. The plateau at the boiling point is longer than the plateau at the melting point because it takes almost seven times more energy (and thus seven times the heating time) to vaporize liquid water than to melt ice. The slopes of the other lines are different because the different states of water have different molar heat capacities (the energy required to raise the temperature of 1 mole of a substance by 1°C).

of the added energy is used to disrupt the ice structure by breaking the hydrogen bonds, thus increasing the potential energy of the water molecules. The enthalpy change that occurs at the melting point when a solid melts is called the **heat of fusion,** (or more accurately, the **enthalpy of fusion,**) ΔH_{fus}. The melting points and enthalpies of fusion for several representative solids are listed in Table 10.8.

Ionic solids such as NaCl and NaF have very high melting points and enthalpies of fusion because of the strong ionic forces in these solids. At the other extreme is $O_2(s)$, a molecular solid containing nonpolar molecules with weak intermolecular forces.

The temperature remains constant until the solid has completely changed to liquid, then it begins to increase again. At 100°C the liquid water reaches its *boiling point,* and the temperature then remains constant as the added energy is used to vaporize the liquid. When the liquid is completely changed to vapor, the temperature again begins to rise. Note that changes of state are physical changes; although intermolecular forces have been overcome, no chemical bonds have been broken. If the water vapor were heated to much higher temperatures, the water molecules would break down into the individual atoms. This would be a chemical change since covalent bonds are broken. We no longer have water after this occurs.

The melting and boiling points will be defined more precisely later in this section.

The melting and boiling points for a substance are determined by the vapor pressures of the solid and liquid states. Figure 10.43 shows the vapor pressures of solid and liquid water as functions of temperature near 0°C. Note that below 0°C the vapor pressure of ice is less than the vapor pressure of liquid water. Also note that

Melting Points and Enthalpies of Fusion for Several Representative Solids

Compound	Melting point (°C)	Enthalpy of fusion (kJ/mol)
O_2	−218	0.45
HCl	−114	1.99
HI	−51	2.87
CCl_4	−23	2.51
$CHCl_3$	−64	9.20
H_2O	0	6.01
NaF	992	29.3
NaCl	801	30.2

Table 10.8

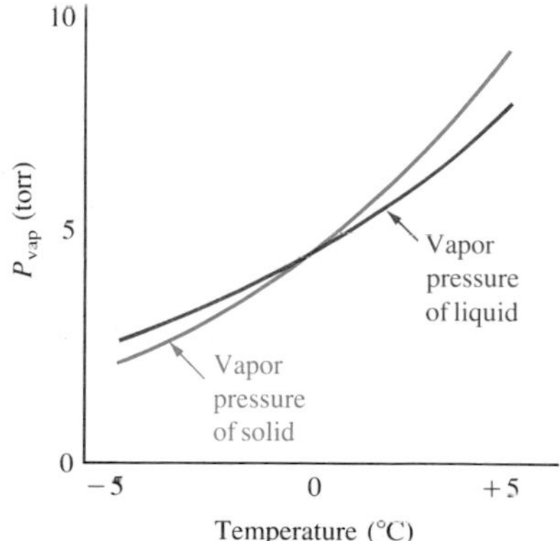

Figure 10.43

The vapor pressures of solid and liquid water as a function of temperature. The data for liquid water below 0°C are obtained from supercooled water. The data for solid water above 0°C are estimated by extrapolation of vapor pressure from below 0°C.

the vapor pressure of ice has a larger temperature dependence than that of the liquid. That is, the vapor pressure of ice increases more rapidly for a given rise in temperature than does the vapor pressure of water. Thus, as the temperature of the solid is increased, a point is eventually reached where the *liquid and solid have identical vapor pressures*. This is the melting point.

These concepts can be demonstrated experimentally using the apparatus illustrated in Fig. 10.44, where ice occupies one compartment and liquid water the other. Consider the following cases.

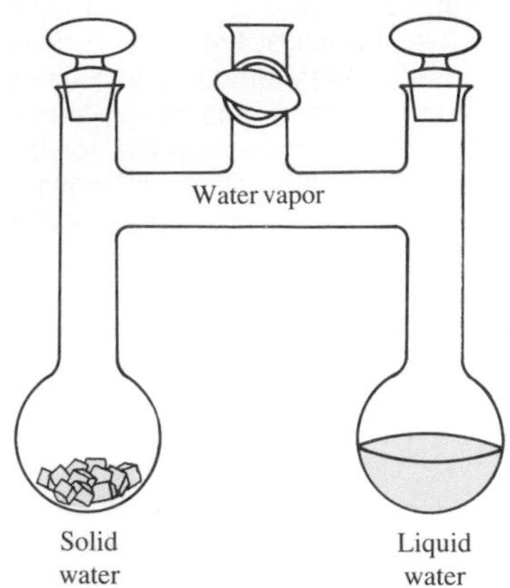

Figure 10.44

An apparatus that allows solid and liquid water to interact only through the vapor state.

Case 1. A temperature at which the vapor pressure of the solid is greater than that of the liquid. At this temperature the solid requires a higher pressure than the liquid does to be in equilibrium with the vapor. Thus, as vapor is released from the solid to try to achieve equilibrium, the liquid will absorb vapor in an attempt to reduce the vapor pressure to its equilibrium value. The net effect is a conversion from solid to liquid through the vapor phase. In fact, no solid can exist under these conditions. The amount of solid will steadily decrease and the volume of liquid will increase. Finally, there will be only liquid in the right compartment, which will come to equilibrium with the water vapor, and no further changes will occur in the system. This temperature must be *above the melting point* of ice, since only the liquid state can exist.

Case 2. A temperature at which the vapor pressure of the solid is less than that of the liquid. This is the opposite of the situation in Case 1. In this case, the liquid requires a higher pressure than the solid does to be in equilibrium with the vapor. So the liquid will gradually disappear, and the amount of ice will increase. Finally, only the solid will remain, which will achieve equilibrium with the vapor. This temperature must be *below the melting point* of ice, since only the solid state can exist.

Case 3. A temperature at which the vapor pressures of the solid and liquid are identical. In this case, the solid and liquid states have the same vapor pressure, so they can coexist in the apparatus at equilibrium simultaneously with the vapor. This temperature represents the *melting point* where both the solid and liquid states can exist.

We can now describe the melting point of a substance more precisely. The **normal melting point** is defined as *the temperature at which the solid and liquid states have the same vapor pressure under conditions where the total pressure is 1 atmosphere*.

Boiling occurs when the vapor pressure of a liquid becomes equal to the pressure of its environment. The **normal boiling point** of a liquid is *the temperature at which the vapor pressure of the liquid is exactly 1 atmosphere*. This concept is illustrated in Fig. 10.45. At temperatures where the vapor pressure of the liquid is less than 1 atmosphere, no bubbles of vapor can form because the pressure on the surface of the liquid is greater than the pressure in any spaces in the liquid where the bubbles are trying to form. Only when the liquid reaches a temperature at which the pressure of vapor in the spaces in the liquid is 1 atmosphere can bubbles form and boiling occur.

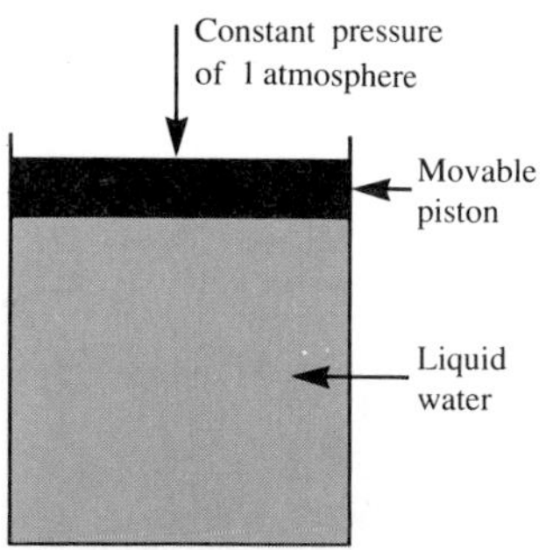

Figure 10.45

Water in a closed system with a pressure of 1 atm exerted on the piston. No bubbles can form within the liquid as long as the vapor pressure is less than 1 atm.

But changes of state do not always occur exactly at the boiling point or melting point. For example, water can be readily **supercooled;** that is, it can be cooled below 0°C at 1 atm pressure and remain in the liquid state. Supercooling occurs because, as it is cooled, the water may not achieve the degree of organization

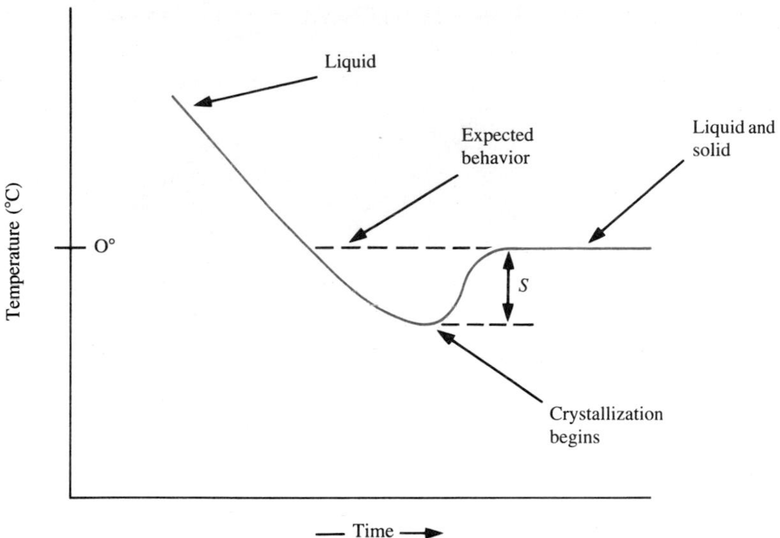

Figure 10.46

The supercooling of water. The extent of supercooling is given by *S*.

necessary to form ice at 0°C, and thus it continues to exist as the liquid. At some point the correct ordering occurs and ice rapidly forms, releasing energy in the exothermic process and bringing the temperature back up to the melting point, where the remainder of the water freezes (see Fig. 10.46).

A liquid can also be **superheated,** or raised to temperatures above its boiling point, especially if it is heated rapidly. Superheating can occur because bubble formation in the interior of the liquid requires that many high-energy molecules gather in the same vicinity, and this may not happen at the boiling point, especially if the liquid is heated rapidly. If the liquid becomes superheated, the vapor pressure in the liquid is greater than the atmospheric pressure. Once a bubble does form, since its internal pressure is greater than that of the atmosphere, it can burst before rising to the surface, blowing the surrounding liquid out of the container. This is called *bumping,* and has ruined many experiments. It can be avoided by adding boiling chips to the flask containing the liquid. Boiling chips are bits of porous ceramic material containing trapped air that escapes on heating, forming tiny bubbles that act as "starters" for vapor bubble formation. This allows a smooth onset of boiling as the boiling point is reached.

10.9 Phase Diagrams

Purpose

- To discuss the features of phase diagrams.

A **phase diagram** is a convenient way of representing the phases of a substance as a function of temperature and pressure. For example, the phase diagram for water (Fig. 10.47 on page 456) shows which state exists at a given temperature and pressure. It is important to recognize that a phase diagram describes conditions and events in a *closed* system of the type represented in Fig. 10.45, where no material can escape into the surroundings and no air is present.

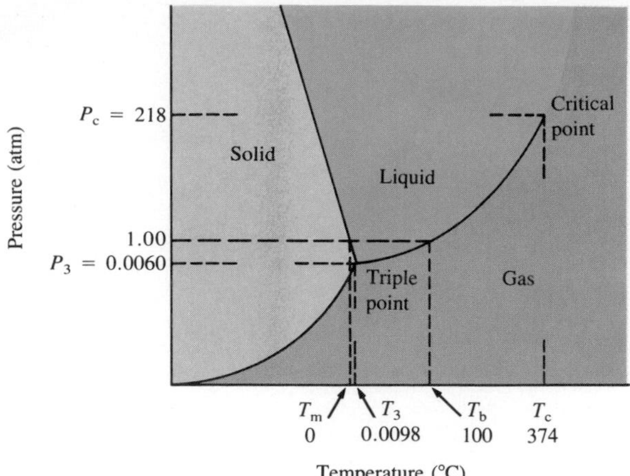

Figure 10.47

The phase diagram for water. T_m represents the normal melting point; T_3 and P_3 denote the triple point; T_b represents the normal boiling point; T_c represents the critical temperature; P_c represents the critical pressure. The negative slope of the solid/liquid line reflects the fact that the density of ice is less than that of liquid water.

To show how to interpret the phase diagram for water, we will consider heating experiments at several pressures, shown by the dashed lines in Fig. 10.48.

Experiment 1. Pressure is 1 atm. This experiment begins with the cylinder shown in Fig. 10.45 completely filled with ice at a temperature of −20°C, and the piston exerting a pressure of 1 atm directly on the ice (there is no air space). Since at temperatures below 0°C the vapor pressure of ice is less than 1 atm—which is the constant external pressure on the piston—no vapor is present in the cylinder. As the cylinder is heated, ice is the only component until the temperature reaches 0°C, where the ice changes to liquid water as energy is added. This is the normal melting point of water. When the solid has completely changed to liquid, the temperature again rises. At this point, the cylinder contains only liquid water. *No vapor is present* because the vapor pressure of liquid water under these conditions is less than 1 atmosphere, the constant external pressure

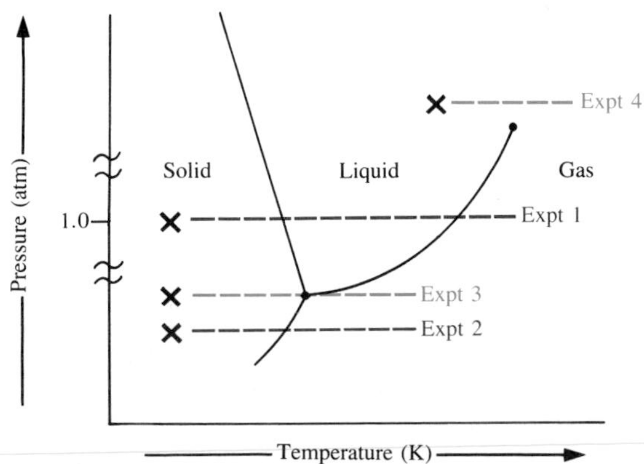

Figure 10.48

Diagrams of various heating experiments on samples of water in a closed system.

on the piston. Heating continues until the temperature of the liquid water reaches 100°C. At this point, the vapor pressure of liquid water is 1 atm, and boiling occurs, with the liquid changing to vapor. This is the normal boiling point of water. After the liquid has been completely converted to steam, the temperature again rises as the heating continues. The cylinder now contains only water vapor.

Experiment 2. Pressure is 2.0 torr. Again we start with ice as the only component in the cylinder at −20°C. The pressure exerted by the piston in this case is only 2.0 torr. As heating proceeds, the temperature rises to −10°C, where the ice changes directly to vapor, a process called sublimation. Sublimation occurs when the vapor pressure of ice is equal to the external pressure, which in this case is only 2.0 torr. No liquid water appears under these conditions because the vapor pressure of liquid water is always greater than 2.0 torr and thus it cannot exist at this pressure. If liquid water were placed in a cylinder under such a low pressure, it would vaporize immediately.

Experiment 3. Pressure is 4.588 torr. Again we start with ice as the only component in the cylinder at −20°C. In this case the pressure exerted on the ice by the piston is 4.588 torr. As the cylinder is heated, no new phase appears until the temperature reaches 0.0098°C. At this point, called the **triple point,** solid and liquid water have identical vapor pressures of 4.588 torr. Thus *at 0.0098°C and 4.588 torr all three states of water are present.* In fact, *only* under these conditions can all three states of water coexist.

Experiment 4. Pressure is 225 atm. In this experiment we start with liquid water in the cylinder at 300°C; the pressure exerted by the piston on the water is 225 atm. Liquid water can be present at this temperature because of the high external pressure. As the temperature increases, something happens that we did not see in the first three experiments: the liquid gradually changes into a vapor but goes through an intermediate "fluid" region, which is neither true liquid nor vapor. This is quite unlike the behavior at lower temperatures and pressures, say at 100°C and 1 atm, where the temperature remains constant while a definite phase change from liquid to vapor occurs. This unusual behavior occurs because the conditions are beyond the critical point for water. The **critical temperature** can be defined as the temperature above which the vapor cannot be liquefied no matter what pressure is applied. The **critical pressure** is the pressure required to produce liquefaction *at* the critical temperature. Together, the critical temperature and the critical pressure define the **critical point.** For water the critical point is 374°C and 218 atm. Note that the liquid/vapor line on the phase diagram for water ends at the critical point. Beyond this point the transition from one state to another involves the intermediate "fluid" region just described.

Applications of the Phase Diagram for Water

There are several additional interesting features of the phase diagram for water. Note that the solid/liquid boundary line has a negative slope. This means that the melting point of water *decreases* as the external pressure *increases*. This behavior, which is opposite to that observed for most substances, occurs because the density of ice is *less* than that of liquid water at the melting point. The maximum density of water occurs at 4°C; when liquid water freezes, its volume increases.

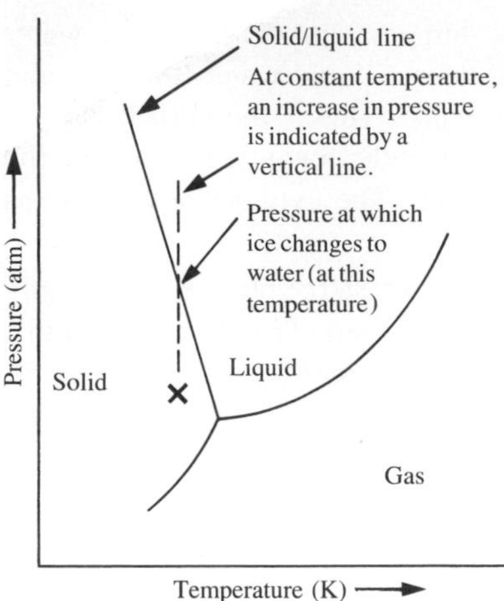

Figure 10.49

The phase diagram for water. At point *x* on the phase diagram, water is a solid. However, as the external pressure is increased while the temperature remains constant (indicated by the vertical dotted line), the solid/liquid line is crossed and the ice melts.

We can account for the effect of pressure on the melting point of water using the following reasoning. At the melting point, liquid and solid water coexist—they are in dynamic equilibrium, since the rate at which ice is melting is just balanced by the rate at which the water is freezing. What happens if we apply pressure to this system? When subjected to increased pressure, matter reduces its volume. This behavior is most dramatic for gases but also occurs for condensed states. Since a given mass of ice at 0°C has a larger volume than the same mass of liquid water, the system can reduce its volume in response to the increased pressure by changing to liquid. Thus, at 0°C and an external pressure greater than 1 atm, water is liquid. In other words, the freezing point of water is less than 0°C when the pressure is greater than 1 atm.

Brian Orser, silver medalist at the 1988 Winter Olympics at Calgary.

Figure 10.49 illustrates the effect of pressure on ice. At the point *X* on the phase diagram, ice is subjected to increased pressure at constant temperature. Note that as the pressure is increased, the solid/liquid line is crossed, indicating that the ice melts. This is exactly what happens in ice skating. The narrow blade of the skate exerts a large pressure, since the skater's weight is supported by the small area of the blade. The ice under the blade melts because of the pressure, providing lubrication. After the blade passes, the liquid refreezes, as normal pressure returns. Without this lubrication effect due to the thawing ice, ice skating would not be the smooth, graceful activity that it can be.

Ice's lower density has other implications. When water freezes in a pipe or an engine block, it will expand and break the container. This is why water pipes are insulated in cold climates and antifreeze is used in water-cooled engines. The lower density of ice also means that ice formed on rivers and lakes will float, providing a layer of insulation that helps prevent bodies of water from freezing solid in the winter. Aquatic life can therefore continue to live through periods of freezing temperatures.

A liquid boils at the temperature where the vapor pressure of the liquid equals the external pressure. Thus the boiling point of a substance, like the melting point, depends on the external pressure. This is why water boils at different temperatures at different elevations (see Table 10.9), and any cooking carried out in boiling water will be affected by this variation. For example, it takes longer to hard-boil an egg in Leadville, Colorado (elevation: 10,150 ft), than in Chicago (sea level), since water boils at a lower temperature in Leadville.

Water boils at 89°C in Leadville, Colorado.

Boiling Point of Water at Various Locations			
Location	Feet above sea level	P_{atm} (torr)	Boiling point (°C)
Top of Mt. Everest, Tibet	29,028	240	70
Top of Mt. McKinley, Alaska	20,320	340	79
Top of Mt. Whitney, Calif.	14,494	430	85
Leadville, Colo.	10,150	510	89
Top of Mt. Washington, N.H.	6,293	590	93
Boulder, Colo.	5,430	610	94
Madison, Wis.	900	730	99
New York City, N.Y.	10	760	100
Death Valley, Calif.	−282	770	100.3

Table 10.9

Ice fishing in Michigan.

As we mentioned earlier, the phase diagram for water describes a closed system. Therefore, we must be very cautious in using the phase diagram to explain the behavior of water in a natural setting, such as on the earth's surface. For example, in dry climates (low humidity), snow and ice seem to sublime—a minimum amount of slush is produced. Wet clothes put on an outside line at temperatures below 0°C freeze, then dry while frozen. However, the phase diagram (Fig. 10.47) shows that ice should *not* be able to sublime at normal atmospheric pressures. What is happening in these cases? Ice in the natural environment is not in a closed system. The pressure is provided by the atmosphere rather than by a solid piston. This means that the vapor produced over the ice can escape from the immediate region as soon as it is formed. The vapor does not come to equilibrium with the solid, and the ice slowly disappears. Sublimation, which seems forbidden by the phase diagram, does in fact occur under these conditions.

The Phase Diagram for Carbon Dioxide

The phase diagram for carbon dioxide (Fig. 10.50) differs from that for water. The solid/liquid line has a positive slope, since solid carbon dioxide is more dense than liquid carbon dioxide. The triple point for carbon dioxide occurs at 5.1 atm and −56.6°C, and the critical point occurs at 72.8 atm and 31°C. At a pressure of 1 atm, solid carbon dioxide sublimes at −78°C, a property that leads to its common name, *dry ice*. No liquid phase occurs under normal atmospheric conditions, making dry ice a convenient refrigerant.

Fighting a gasoline fire with a carbon dioxide fire extinguisher.

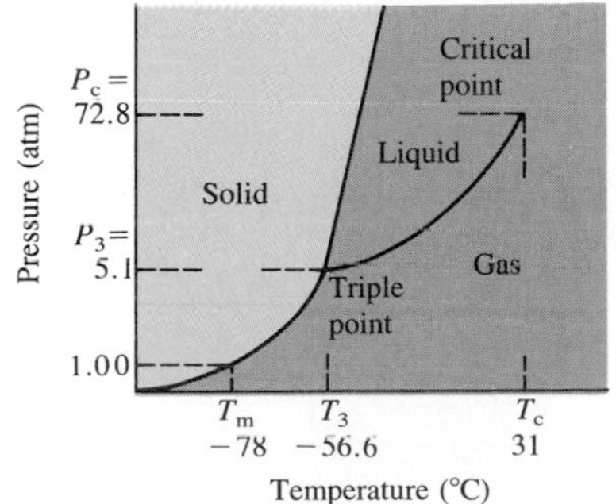

Figure 10.50

The phase diagram for carbon dioxide. The liquid state does not exist at a pressure of 1 atmosphere. The solid/liquid line has a positive slope since the density of solid carbon dioxide is greater than that of liquid carbon dioxide.

Chemical Impact

Transistors and Printed Circuits

Transistors have had an immense impact on the technology of electronic devices for which signal amplification is needed, such as communications equipment and computers. Before the invention of the transistor at Bell Laboratories in 1947, amplification was provided exclusively by vacuum tubes, which were both bulky and unreliable. The first electronic digital computer, ENIAC, built at the University of Pennsylvania, had 19,000 vacuum tubes and consumed 150,000 watts of electricity. Because of the discovery and development of the transistor and the printed circuit, a hand-held calculator run by a small battery has the same computing power as ENIAC.

A *junction transistor* is made by joining n-type and p-type semiconductors so as to form an n-p-n or a p-n-p junction. The former type is shown in Fig. 10.51. In this diagram the input signal (to be amplified) occurs in circuit 1 which has a small resistance and a forward-biased n-p junction (junction 1). As the voltage of the input signal to this circuit varies, the current in the circuit varies, which means there is a change in the number of electrons crossing the n-p junction. Circuit 2 has a relatively large resistance and is under reverse bias. The key to the operation of the transistor is that current only flows in circuit 2 when electrons crossing junction 1 also cross junction 2 and travel to the positive terminal. Since the current in circuit 1 determines the number of electrons crossing junction 1, the number of electrons available to cross junction 2 is also directly proportional to the current in circuit 1. The current in circuit 2 therefore varies depending on the current in circuit 1.

The voltage (V), current (I), and resistance (R) in a circuit are related by the equation

$$V = IR$$

Since circuit 2 has a large resistance, a given current in circuit 2 produces a larger voltage than the same current in circuit 1, which has a small resistance. Thus a signal of variable voltage in circuit 1, such as might be produced by a human voice on a telephone, is reproduced in circuit 2, but with much greater voltage changes. That is, the input signal has been *amplified* by the junction transistor. This device replaces the large vacuum tube and is a tiny component of a printed circuit on a silicon chip.

Silicon chips are really "planar" transistors constructed from thin layers of n-type and p-type regions connected by conductors. A chip less than 1 cm wide can contain hundreds of printed circuits and be used in computers, radios, and televisions.

A printed circuit has many n-p-n junction transistors. Figure 10.52 illustrates the formation of one transistor area. The chip begins as a thin wafer of silicon that has been doped

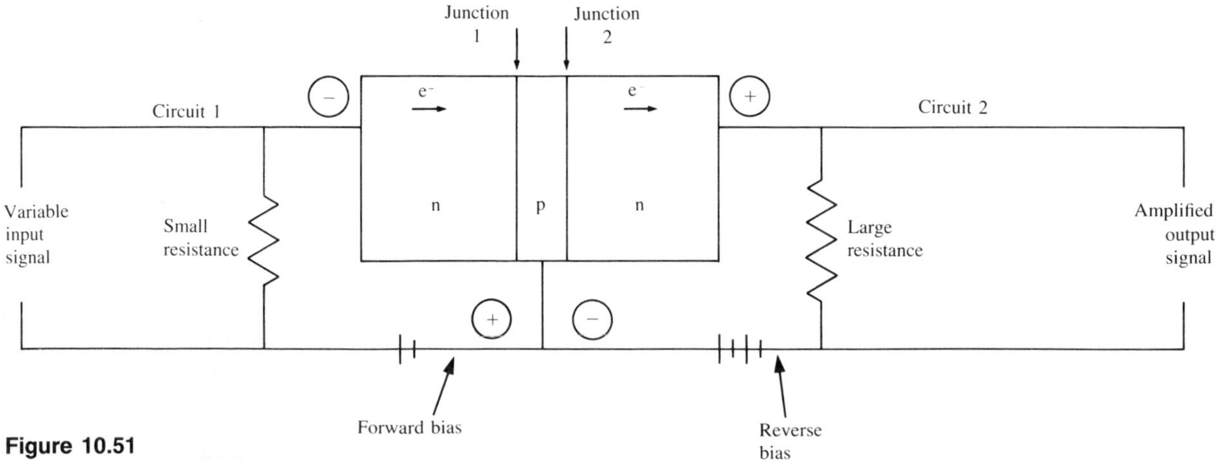

Figure 10.51

A schematic of two circuits connected by a transistor. The signal in circuit 1 is amplified in circuit 2.

Chemical Impact

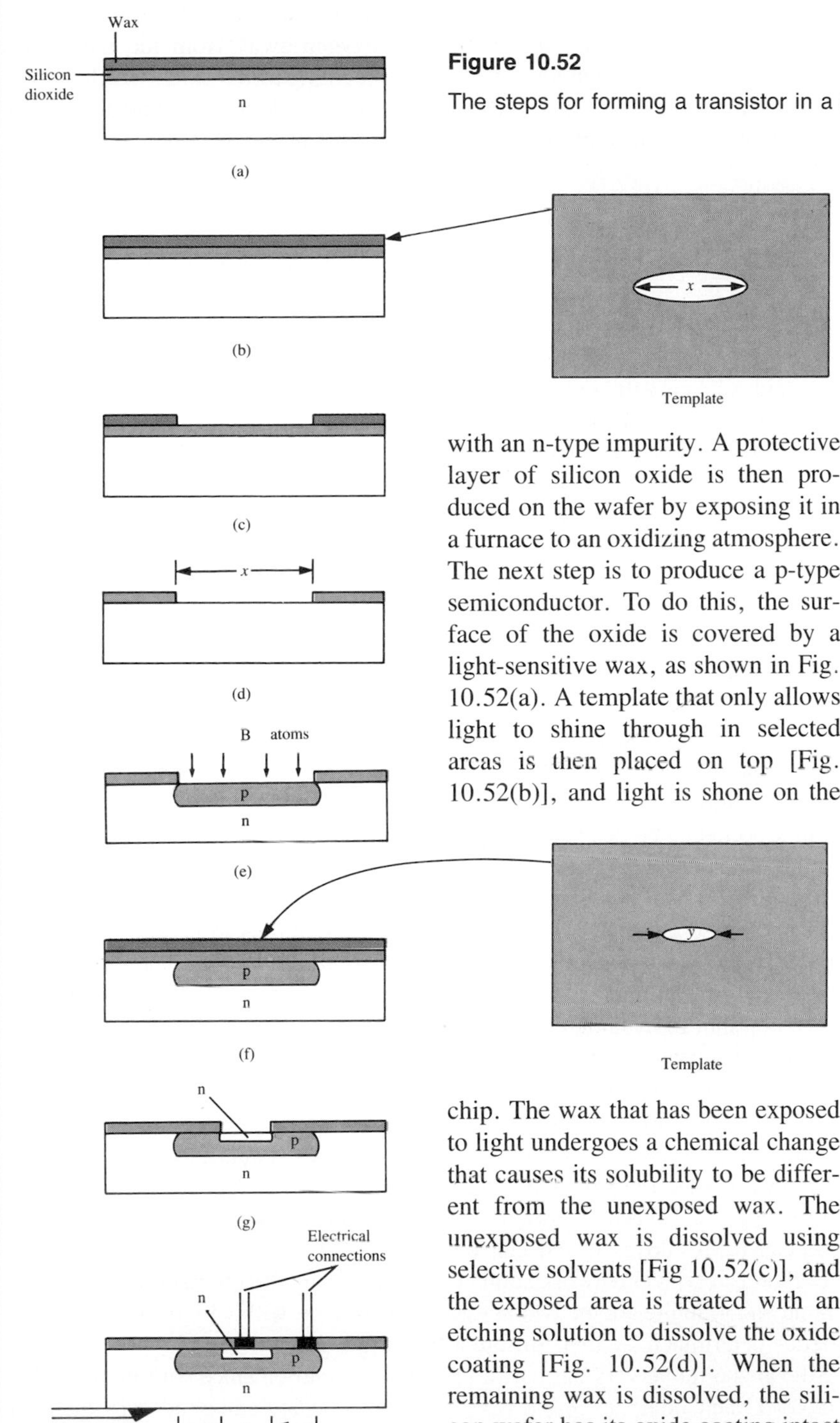

Figure 10.52

The steps for forming a transistor in a crystal of initially pure silicon.

with an n-type impurity. A protective layer of silicon oxide is then produced on the wafer by exposing it in a furnace to an oxidizing atmosphere. The next step is to produce a p-type semiconductor. To do this, the surface of the oxide is covered by a light-sensitive wax, as shown in Fig. 10.52(a). A template that only allows light to shine through in selected areas is then placed on top [Fig. 10.52(b)], and light is shone on the chip. The wax that has been exposed to light undergoes a chemical change that causes its solubility to be different from the unexposed wax. The unexposed wax is dissolved using selective solvents [Fig 10.52(c)], and the exposed area is treated with an etching solution to dissolve the oxide coating [Fig. 10.52(d)]. When the remaining wax is dissolved, the silicon wafer has its oxide coating intact except at the one spot (of diameter x), as shown in Fig. 10.52(d).

Exposing the wafer to a p-type impurity such as boron at about 1000°C causes a p-type semiconductor area to be formed in the exposed spot as the boron atoms diffuse into the silicon crystal [Fig. 10.52(e)]. Next, to form a small n-type area in the center of the p-type region, the wafer is again placed in the oxidizing furnace to be recoated over its entire surface with oxide. Then a new wax covering is applied, which is illuminated through a template with a transparent area indicated by y [Fig. 10.52(f)]. The wax and oxide are then removed from the illuminated area, and the wafer is exposed to an n-type impurity to form a small n-type region as shown in Fig. 10.52(g). Next, conductors are layered onto the chip giving the finished transistor [Fig. 10.52(h)], which has two circuits connected through an n-p-n junction (see Fig. 10.51). This transistor then becomes a part of a larger circuit layered onto the chip and interconnected by conductors.

The method given here for producing a printed circuit does not represent the latest technology in this field. The manufacture of printed circuits is a highly competitive business and changes in methodology occur almost daily.

Carbon dioxide is often used in fire extinguishers, where it exists as a liquid at 25°C under high pressures. Liquid carbon dioxide released from the extinguisher into the environment at 1 atm immediately changes to a vapor. Being heavier than air, this vapor smothers the fire by keeping oxygen away from the flame. The liquid/vapor transition is highly endothermic, so cooling also results, which helps to put out the fire. The "fog" produced by a carbon dioxide extinguisher is not solid carbon dioxide but rather moisture frozen from the air.

FOR REVIEW

Summary

Liquids and solids are the two condensed states of matter; their formation is caused by intermolecular forces among the molecules, atoms, or ions. Dipole-dipole forces are attractions between molecules that have dipole moments. These forces are particularly strong for molecules containing hydrogen bonded to a highly electronegative element such as nitrogen, oxygen, or fluorine. In these cases the resulting dipole forces are called hydrogen bonds and account for the unusually high boiling points of the hydrides of the first elements in Groups 4A, 5A, 6A, and 7A. London dispersion forces caused by instantaneous dipoles are important in substances composed of nonpolar molecules. Liquids exhibit various properties—surface tension, capillary action, and viscosity—that depend on the strengths of the intermolecular forces.

The two broadest categories of solids are crystalline and amorphous solids. Crystalline solids have a regular arrangement, or lattice, of component particles, the smallest repeating unit of which is the unit cell. The arrangement of particles in a crystalline solid can be determined by X-ray diffraction techniques, and the interatomic distances in the crystal can be related to the wavelength and angle of reflection of the X rays by the Bragg equation.

The structure of metals is modeled by assuming atoms to be uniform spheres and packing them as compactly as possible. There are two closest packing arrangements: hexagonal and cubic.

The bonding in metallic crystals can be described in terms of the electron sea model (valence electrons are assumed to move freely about the metal cations) or in terms of the band model (electrons are assumed to travel through the metal crystal in molecular orbitals formed from the valence atomic orbitals of the metal atoms). In the band model the large number of available atomic orbitals form closely spaced energy levels. Electricity can be readily conducted by electrons in conduction bands, which are molecular orbitals containing only a single electron.

Metals form alloys, which can be classified as substitutional or interstitial.

Carbon is a typical network solid containing strong directional covalent bonds. Diamond and graphite are the two allotropes of carbon, with very different physical properties determined by their different bonding.

Silicon should be very similar in properties to carbon since carbon and silicon are next to each other in Group 4A, but in fact their compounds are markedly dissimilar. Silica, the fundamental silicon-oxygen compound, does not contain discrete SiO_2 molecules but rather a network of interconnected SiO_4 tetrahedra. Silicates, salts containing polyatomic anions of silicon and oxygen, are components of rocks, soil, clay, glass, and ceramics.

Semiconductors are formed when pure silicon is doped with other elements. An n-type semiconductor contains atoms with more valence electrons than the silicon atom, and a p-type semiconductor has atoms with fewer valence electrons than silicon. The modern electronics industry is based on devices with p-n junctions.

Molecular solids consist of discrete molecular units held together by relatively weak intermolecular forces. Ice is an example. Ionic solids, on the other hand, are held together by strong electrostatic attractions. The crystal structure of ionic solids can often be described by fitting the smaller ions (usually cations) into holes in the closest packed structures formed by the larger ions (usually anions).

The phase change between liquid and vapor is called vaporization, or evaporation, and the energy required to change 1 mole of liquid to its vapor is the heat of vaporization (ΔH_{vap}). Condensation is the reverse process by which vapor molecules return to the liquid state. When the rates of evaporation and condensation exactly balance in a closed container, the system is at equilibrium, and the resulting pressure of the vapor is called the vapor pressure.

Volatility of liquids depends on both molecular weight and intermolecular forces. Liquids composed of heavier molecules tend to be less volatile, as do liquids with strong intermolecular forces. The normal melting point of a solid is defined as the temperature at which the solid and its liquid have identical vapor pressures. The normal boiling point of a liquid occurs at a temperature where the vapor pressure of the liquid is 1 atmosphere.

A phase diagram for a substance shows which state exists at a given temperature and pressure. The triple point on a phase diagram represents the temperature and pressure where all three states coexist. The critical point is defined by the critical temperature and critical pressure. Critical temperature is that temperature above which the vapor cannot be liquefied no matter what pressure is applied. The critical pressure is the pressure required to produce liquefaction at the critical temperature.

Key Terms

Section 10.1

condensed states
intermolecular forces
dipole-dipole attraction
hydrogen bonding
London dispersion forces

Section 10.2

surface tension
capillary action

Section 10.3

crystalline solid
amorphous solid
lattice
unit cell
X-ray diffraction
ionic solid
molecular solid
atomic solid

Section 10.4

closest packing
hexagonal closest packed (hcp) structure
cubic closest packed (ccp) structure
band model
alloy
substitutional alloy
interstitial alloy

Section 10.5

network solid
silica
silicate
glass
ceramic
semiconductor
n-type semiconductor
p-type semiconductor
p-n junction

Section 10.8

vaporization
heat of vaporization
enthalpy of vaporization (ΔH_{vap})
vapor pressure
condensation
equilibrium
sublimation
heating curve
enthalpy of fusion (ΔH_{fus})
normal melting point
normal boiling point
supercooled
superheated

Section 10.9

phase diagram
triple point
critical temperature
critical pressure
critical point

Exercises

A blue exercise number indicates that the answer to that exercise appears at the back of this book and a solution appears in the Solutions Guide.

Intermolecular Forces and Physical Properties

1. Describe the relationship between the polarity of individual molecules and the nature and strength of intermolecular forces.

2. List the major types of intermolecular forces in order of increasing strength. Is there some overlap? That is, can the strongest London dispersion forces be greater than some dipole-dipole forces? Give examples of such instances.

3. Describe the relationship between molecular size and the strength of London dispersion forces.

4. Is it possible for the dispersion forces in a particular substance to be stronger than the hydrogen bonding forces in another substance? Explain your answer.

5. How do the following physical properties depend on the strength of intermolecular forces?
   a. surface tension
   b. viscosity
   c. melting point
   d. boiling point
   e. vapor pressure

6. Does the nature of intermolecular forces change when a substance goes from a solid to a liquid, or from a liquid to a gas? What causes a substance to undergo a phase change?

7. Identify the most important types of interparticle forces present in the solids of each of the following substances:
   a. $BaSO_4$
   b. H_2S
   c. Xe
   d. C_2H_6
   e. Cs
   f. Hg
   g. P_4
   h. H_2O
   i. NH_4Cl
   j. steel
   k. teflon, $CF_3(CF_2CF_2)_nCF_3$
   l. polyethylene, $CH_3(CH_2CH_2)_nCH_3$
   m. $CHCl_3$
   n. Ge
   o. NO
   p. BF_3

8. Predict which substance in each of the following pairs would have the greater dipole-dipole forces.
   a. CO_2 or OCS
   b. PF_3 or PF_5
   c. SF_2 or SF_6
   d. SO_3 or SO_2

9. How do typical dipole-dipole forces differ from hydrogen bonds? In what ways are they similar?

10. For which molecule in each of the following pairs would you expect the hydrogen bonding to other molecules of the same type to be greater?
    a. $CH_3CH_2CH_2NH_2$ or $H_2NCH_2CH_2NH_2$
    b. $B(OH)_3$ or BH_3
    c. CH_3OH or H_2CO
    d. HF or HI

11. Rationalize the difference in boiling points for each of the following pairs of substances:
    a. *n*-pentane $CH_3CH_2CH_2CH_2CH_3$ 36.2°C
       neopentane $C(CH_3)_4$ (structure: H_3C—C—CH_3 with CH_3 above and below the central C) 9.5°C
    b. dimethyl ether CH_3OCH_3 −25°C
       ethanol CH_3CH_2OH 79°C
    c. HF 20°C
       HCl −85°C
    d. $TiCl_4$ 136°C
       LiCl 1360°C
    e. HCl −85°C
       LiCl 1360°C
    f. LiCl 1360°C
       CsCl 1290°C

12. Rationalize the following differences in physical properties in terms of intermolecular forces. Compare the first three substances to each other, compare the last three to each other, and then compare all six. Can you account for any anomalies?

	bp (°C)	mp (°C)	ΔH_{vap} (kJ/mol)
Benzene, C_6H_6	80	6	33.9
Naphthalene, $C_{10}H_8$	218	80	51.5
Carbon tetrachloride	76	−23	31.8
Acetone, CH_3COCH_3	56	−95	31.8
Acetic acid, CH_3CO_2H	118	17	39.7
Benzoic acid, $C_6H_5CO_2H$	249	122	68.2

13. Consider the following enthalpy changes:

$$F^- + HF \rightarrow FHF^- \qquad \Delta H = -155 \text{ kJ/mol}$$

$$(CH_3)_2C{=}O + HF \rightarrow (CH_3)_2C{=}O\text{---}HF \qquad \Delta H = -46 \text{ kJ/mol}$$

$$H_2O(g) + HOH(g) \rightarrow H_2O\text{---}HOH \text{ (in ice)} \qquad \Delta H = -21 \text{ kJ/mol}$$

How do the strengths of hydrogen bonds vary with the electronegativity of the element to which hydrogen is bonded? Where in the above series would you expect hydrogen bonds of the following type to fall?

$>$N---HO— and $>$N---H—N$<$

14. The strongest known hydrogen bond is found in potassium bifluoride, KHF_2. Potassium bifluoride is produced by reacting KF and HF in water. What structure would you predict for the bifluoride ion (FHF^-)? Why?

15. Why is ΔH_{vap} for water much greater than ΔH_{fus}? What does this say about changes in intermolecular forces in going from liquid to solid to vapor?

16. Using the heats of fusion and vaporization for water, calculate the change in enthalpy for the sublimation of water:

$$H_2O(s) \rightarrow H_2O(g)$$

Using the ΔH value given in Exercise 13 and the number of hydrogen bonds formed to each water molecule, estimate what portion of the intermolecular forces in ice can be accounted for by hydrogen bonding.

17. Oil of wintergreen, or methyl salicylate, has the following structure:

(benzene ring with C(=O)—OCH$_3$ and ortho OH) mp = −8°C

Methyl-4-hydroxybenzoate is another molecule with exactly the same molecular formula; it has the following structure:

(benzene ring with C(=O)—OCH$_3$ and para OH) mp = 127°C

Account for the large difference in the melting points of the two substances.

18. Consider the following melting point data:

Compound	NaCl	$MgCl_2$	$AlCl_3$	$SiCl_4$	PCl_3	SCl_2	Cl_2
mp (°C)	801	708	190	−70	−91	−78	−101
Compound	NaF	MgF_2	AlF_3	SiF_4	PF_5	SF_6	F_2
mp (°C)	997	1396	1040	−90	−94	−56	−220

Account for the trends in melting points in terms of interparticle forces.

19. In each of the following groups of substances, pick the one that has the given property. Justify each answer.
   a. Highest boiling point: Hg, NaCl, or N_2
   b. Smallest surface tension: H_2O, CH_3CN, or CH_3OH
   c. Lowest freezing point: H_2, CH_4, or CO
   d. Smallest vapor pressure at 25°C: SiO_2, CO_2, or H_2O
   e. Greatest viscosity: $CH_3CH_2CH_2CH_3$, CH_3CH_2OH, or $HOCH_2CH_2OH$
   f. Strongest hydrogen bonding: NH_3, PH_3, or SbH_3
   g. Greatest heat of vaporization: HF, HCl, HBr, or HI
   h. Smallest enthalpy of fusion: H_2O, CO_2, MgO, or Li_2O

20. The heats of vaporization of the hydrogen halides are

HF	7.5 kJ/mol
HCl	16.1 kJ/mol
HBr	17.6 kJ/mol
HI	19.7 kJ/mol

The boiling points for the hydrogen halides follow the same trend as those of the hydrides of the elements in Group 5 and Group 6. Can you account for any discrepancies between these data and your predictions in part g of Exercise 19? *Hint:* Think about the effects of aggregates in the vapor phase.

21. How could you tell experimentally if TiO_2 is an ionic solid or a network solid? What would you predict on the basis of electronegativity differences?

22. Distinguish between each of the following:
   a. polarizability and polarity
   b. London dispersion forces and dipole-dipole forces
   c. intermolecular forces and intramolecular forces

23. Titanium(IV) chloride is a liquid that boils at 136°C. What might explain why $TiCl_4$ exists as discrete covalent molecules rather than as an ionic substance?

Properties of Liquids

24. In what ways are liquids similar to solids? In what ways are liquids similar to gases?

25. Define critical temperature and critical pressure. In terms of the kinetic molecular theory, why is it impossible for a substance to exist as a liquid above its critical temperature?

26. What is the relationship between critical temperature and intermolecular forces?

27. Use the kinetic molecular theory to explain why a liquid gets cooler as it evaporates.

28. The shape of the meniscus of water in a glass tube is different from that of mercury in a glass tube. Why?

H_2O in glass

Hg in glass

What would be the shape of the meniscus of water in a polyethylene tube? (Polyethylene can be represented as $CH_3(CH_2)_nCH_3$ where n is a large number on the order of 1000.)

29. Will water rise to a greater height by capillary action in a glass tube or in a polyethylene tube of the same diameter?

30. Some of the physical properties of H_2O and D_2O are as follows:

	H_2O	D_2O
Density at 20°C (g/mL)	0.997	1.108
Boiling point (°C)	100.00	101.41
Melting point (°C)	0.00	3.79
ΔH_{vap} (kJ/mol)	40.7	41.61
ΔH_{fus} (kJ/mol)	6.01	6.3

Account for the differences. (Note: D is a symbol often used for ^2H, the deuterium isotope of hydrogen.)

31. Hydrogen peroxide (H_2O_2) is a syrupy liquid with a relatively low vapor pressure and a normal boiling point of 152.2°C. Rationalize the differences of these physical properties from those of water.

32. Carbon diselenide, CSe_2, is a liquid at room temperature. The normal boiling point is 125°C, and the melting point is −45.5°C. Carbon disulfide, CS_2, is also a liquid at room temperature with normal boiling and melting points of 46.5°C and −111.6°C, respectively. How do the strengths of the interparticle forces vary from CO_2 to CS_2 to CSe_2? Explain your answer.

Structures and Properties of Solids

33. Distinguish between the items in the following pairs:
a. crystalline solid and amorphous solid
b. ionic solid and molecular solid
c. molecular solid and network solid
d. metallic solid and network solid

34. Will a crystalline solid or an amorphous solid give a simpler X-ray diffraction pattern? Why?

35. Is it possible to generalize that amorphous solids always have weaker or stronger interparticle forces than crystalline solids? Explain your answer.

36. What type of solid will each of the following substances form?
a. CO_2
b. SiO_2
c. Si
d. CH_4
e. Ru
f. I_2
g. KBr
h. H_2O
i. NaOH
j. U
k. $CaCO_3$
l. PH_3

37. When a metal was exposed to X rays, it emitted X rays of a different wavelength. The emitted X rays were diffracted by a LiF crystal (d = 201 pm) and first-order diffraction (n = 1 in the Bragg equation) was detected at an angle of 34.68°. Calculate the wavelength of the X ray emitted by the metal.

38. The value of $2d$ for mica, a silicate mineral, is 19.93 Å. What would be the angle for first-order diffraction (n = 1 in the Bragg equation) of X rays from a molybdenum X-ray source (λ = 0.712 Å)?

39. X rays from a copper X-ray tube (λ = 1.54 Å) were diffracted at an angle of 14.22° by a crystal of silicon. Assuming first-order diffraction (n = 1 in the Bragg equation), what is the interplanar spacing in silicon?

40. Consider the following values for lattice energies (Section 8.5) and interionic distances (center to center) for some of the alkali metal halides:

	Lattice energy (kJ/mol)	Interionic distance (Å)
LiF	−1034	2.01
LiCl	−840	2.57
LiBr	−781	2.75
LiI	−718	3.02
NaF	−914	2.31
NaCl	−770	2.81
NaBr	−728	2.98
NaI	−680	3.23
KCl	−701	3.14
RbCl	−682	3.28
CsCl	−629	3.56
CsF	−744	3.00

Plot the lattice energies of these compounds as a function of interionic distance and as a function of the reciprocal of the interionic distance. Which plot is most nearly linear? Why? Are there any discrepancies in the linear plot? Account for them.

41. A metallic solid with atoms in a face-centered cubic unit cell with an edge length of 3.92 Å has a density of 21.45 g/cm^3. Calculate the atomic weight and the atomic radius of the metal. What element might this metal be?

42. The unit cell for a pure xenon fluoride is shown below.
a. What is the formula of the compound?

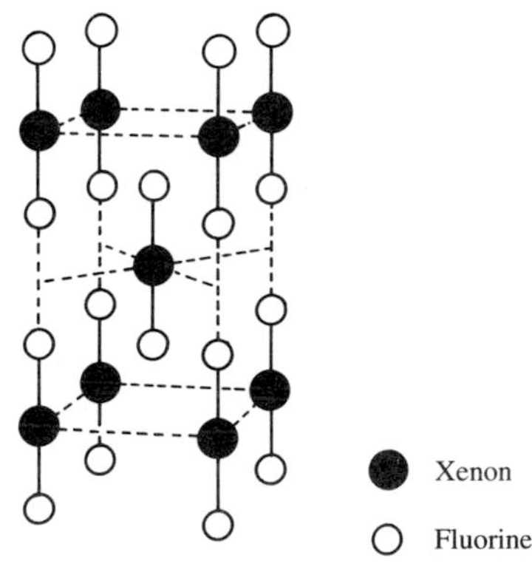

b. The unit cell of XeF_2 has a height of 702 pm, and the edge of the square base is 432 pm. Calculate the density of XeF_2.

43. Superalloys have been made of nickel and aluminum. (See *Scientific American,* October 1986, p. 159.) The alloy owes its strength to the formation of an ordered phase, called the *gamma-prime phase,* in which Al atoms are at the corners of a cubic unit cell and Ni atoms are at the face centers. What is the composition (relative numbers of atoms) for this phase of the nickel-aluminum superalloy?

44. Cobalt exists in two crystalline forms. Below 417°C cobalt is in the α-form, which has a hexagonal closest packed structure and a density of 8.90 g/cm^3. Above 417°C cobalt is in the β-form, a cubic closest packed arrangement. The atomic radius of cobalt is 1.25 Å. Is there a change in the density of cobalt in going from the α-form to the β-form?

45. Iridium (Ir) has a face-centered cubic unit cell with an edge length of 3.833 Å. The density of iridium is 22.61 g/cm^3. Use these data to calculate a value for Avogadro's number.

46. Titanium metal has a body-centered cubic unit cell. The density of titanium is 4.50 g/cm^3. Calculate the edge length of the unit cell and a value for the atomic radius of titanium. (*Hint:* In a body-centered arrangement of spheres, the spheres touch across the body diagonal.)

47. Tungsten metal, which has the highest melting point of all the elements except for carbon, exists in a body-centered cubic structure. The atomic radius of tungsten is 139 pm. Calculate the density of solid tungsten.

48. In solid KCl the smallest distance between a potassium ion and a chloride ion is 314 pm. Calculate the length of the edge of the unit cell and the density of KCl, assuming it has the same structure as sodium chloride.

49. The unit cell for nickel arsenide is shown below. What is the formula of this compound?

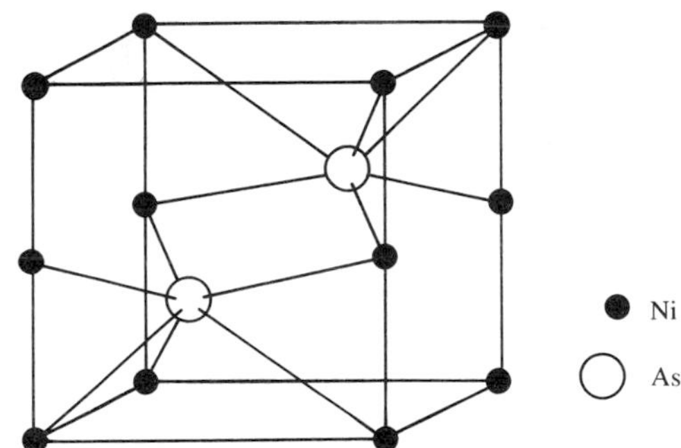

50. Perovskite is a mineral containing calcium, titanium, and oxygen. A diagram of its unit cell is shown below. What is the formula of perovskite?

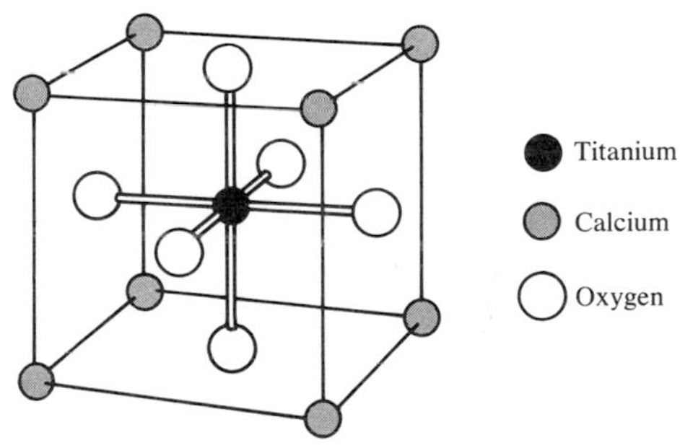

51. Materials, containing the elements Y, Ba, Cu, and O, that are superconductors (electrical resistance equals zero) at temperatures above that of liquid nitrogen were recently discovered. The structures of these materials are based on the perovskite structure. Were they to have the ideal perovskite structure the superconductor would have the structure shown in part (a) of the accompanying figure on page 468.
a. What is the formula of this ideal perovskite material?
b. How is this structure related to the perovskite structure shown in Exercise 50?

These materials, however, do not act as superconductors unless they are deficient in oxygen. The structure of the actual superconducting phase appears to be that shown in part (b) of the figure.
c. What is the formula of this material?

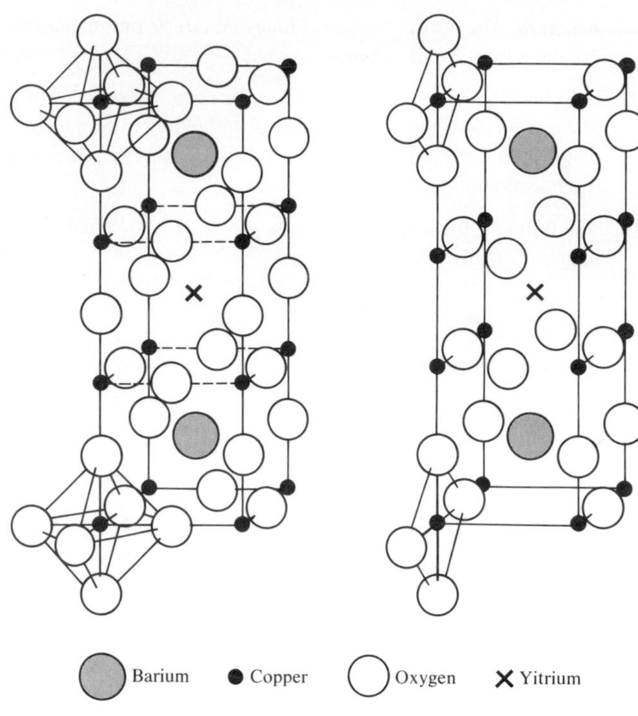

Ideal perovskite structure (a)

Actual structure of superconductor (b)

52. The structures of some common crystalline substances are shown below. Show that the net composition of each unit cell corresponds to the correct formula of each substance.

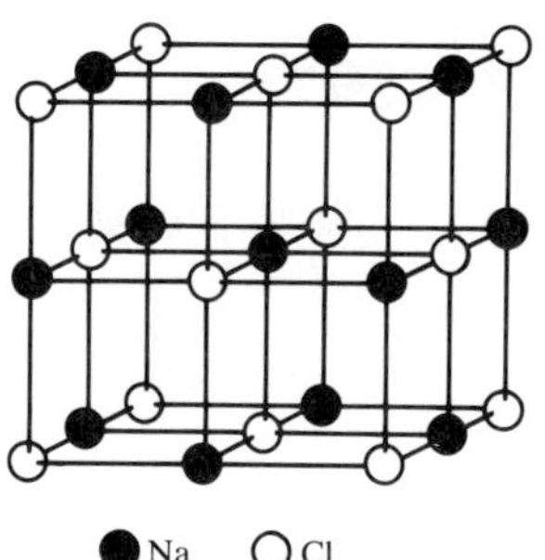

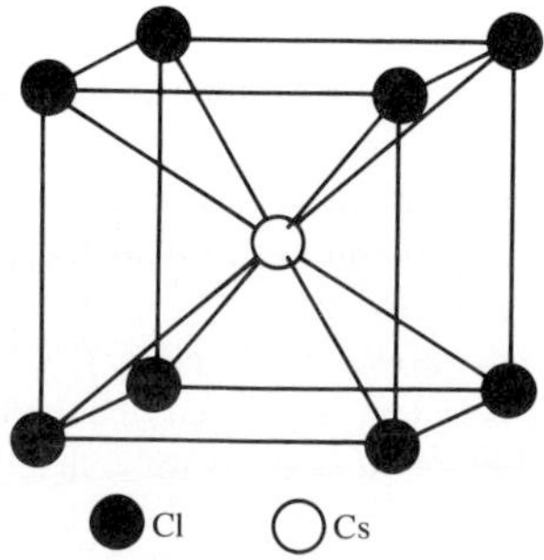

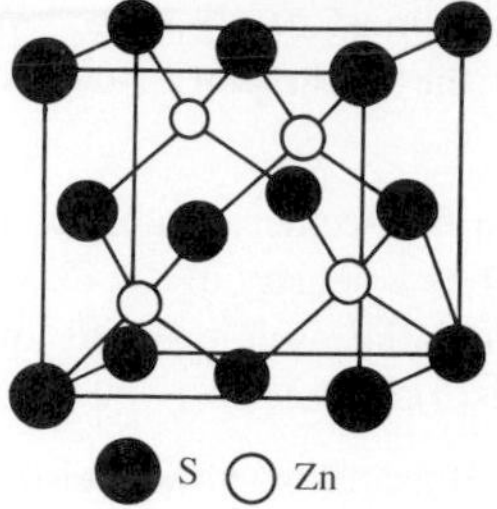

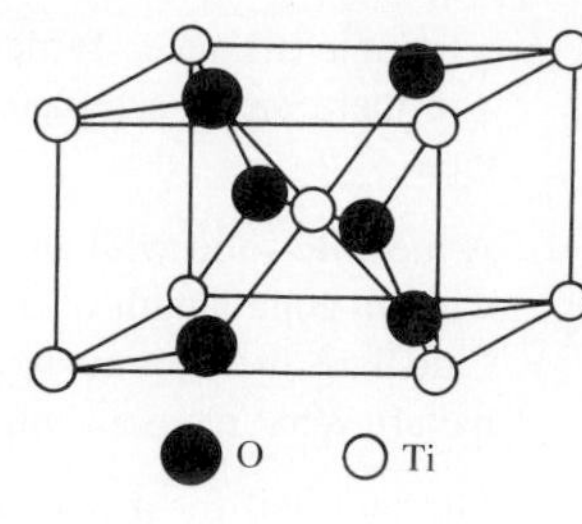

53. Nickel has a face-centered cubic unit cell. The density of nickel is 6.84 g/cm^3. Calculate a value for the atomic radius of nickel.

54. The ionic radius of gold is 144 pm and the density is 19.32 g/cm^3. Does elemental gold have a face-centered cubic structure or body-centered cubic structure?

55. Use the band model to describe differences among insulators, conductors, and semiconductors.

56. Use the band model to explain why each of the following increases the conductivity of a semiconductor:
   a. increasing the temperature
   b. irradiating with light
   c. adding an impurity

57. Selenium is a semiconductor used in photocopying machines. What type of semiconductor would be formed if a small amount of indium impurity is added to pure selenium?

58. The Group 3/Group 5 semiconductors are composed of equal amounts of atoms from Group 3A and Group 5A, for example InP and GaAs. These types of semiconductors are used in light emitting diodes and solid state lasers. What would you add to make a p-type semiconductor from pure GaAs? How would you dope pure GaAs to make an n-type semiconductor?

59. The band gap in aluminum phosphide, AlP, is 2.5 electron volts (1 eV = 1.6×10^{-19} J). What wavelength of light is emitted by an AlP diode?

60. An aluminum antimonide solid state laser emits light with a wavelength of 730 nm. Calculate the band gap in joules.

Phase Changes and Phase Diagrams

61. Define each of the following:
   a. condensation
   b. evaporation
   c. sublimation
   d. supercooled liquid

62. Describe what is meant by a dynamic equilibrium in terms of the vapor pressure of a volatile liquid.

63. The temperature inside a pressure cooker is 115°C. Use Equation 10.5 to calculate the vapor pressure of water inside the

pressure cooker. What would be the temperature inside the pressure cooker if the vapor pressure of water was 3.5 atm?

64. What pressure would have to be applied to steam at 350°C to condense the steam to liquid water?

65. How much energy does it take to convert 0.50 kg of ice at −20°C to steam at 250°C? Specific heat capacities: ice, 2.1 J/g °C; liquid, 4.2 J/g °C; steam, 1.8 J/g °C. $\Delta H_{vap} = 40.7$ kJ/mol, $\Delta H_{fus} = 6.0$ kJ/mol.

66. What is the final temperature when 0.850 kJ of energy is added to 10.0 g ice at 0°C?

67. In regions with dry climates, evaporative coolers are used to cool air. A typical electric air conditioner is rated at 1.00×10^4 Btu/hr (1 Btu, or British thermal unit = amount of energy to raise the temperature of 1 lb of water by 1°F). How much water must be evaporated each hour to dissipate as much heat as a typical electric air conditioner?

68. Plot the following data and determine ΔH_{vap} for magnesium and lithium. In which metal is the bonding stronger?

Vapor pressure (mm Hg)	Temperature (°C) Li	Mg
1.	750.	620.
10.	890.	740.
100.	1080.	900.
400.	1240.	1040.
760.	1310.	1110.

69. The enthalpy of vaporization of acetone is 32.0 kJ/mol. The normal boiling point of acetone is 56.5°C. What is the vapor pressure of acetone at 25.0°C?

70. At an elevation of 5300 feet the atmospheric pressure is about 630. torr. What would be the boiling point of acetone ($\Delta H_{vap} = 32.0$ kJ/mol) at this elevation? What would be the vapor pressure at 25°C at this elevation?

71. The enthalpy of vaporization of mercury is 59.1 kJ/mol. The normal boiling point of mercury is 357°C. What is the vapor pressure of mercury at 25°C?

72. How does each of the following affect the rate of evaporation of a liquid in an open dish?
a. intermolecular forces
b. temperature
c. surface area

73. Some water is placed in a sealed glass container connected to a vacuum pump (a device used to pump gases from a container), and the pump is turned on. The water appears to boil and then freezes. Explain these changes using the phase diagram for water. What would happen to the ice if the vacuum pump was left on indefinitely?

74. Consider the phase diagram given below. What phases are present at points A through H? Identify the triple point, normal boiling point, normal freezing point, and critical point.

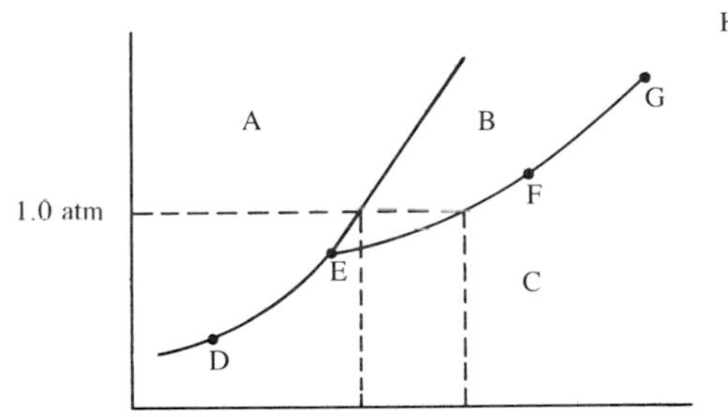

75. Describe how a phase diagram can be constructed from the heating curve for a substance.

76. A substance has the following properties.

		Specific heat capacities	
ΔH_{vap}	20 kJ/mol	$C_{(s)}$	3.0 J/g °C
ΔH_{fus}	5 kJ/mol	$C_{(l)}$	2.5 J/g °C
bp	75°C	$C_{(g)}$	1.0 J/g °C
mp	−15°C		

Sketch a heating curve for the substance starting at −50°C.

77. Why is a burn from steam typically much more severe than a burn from boiling water?

78. A 0.250-g chunk of sodium metal is cautiously dropped into a mixture of 50.0 g of water and 50.0 g of ice at 0°C. The reaction is

$$2Na(s) + 2H_2O(l) \rightarrow 2NaOH(aq) + H_2(g) \qquad \Delta H = -368 \text{ kJ}$$

Will the ice melt? Assuming the final mixture has a specific heat capacity of 4.18 J/g °C calculate the final temperature.

79. Use the accompanying phase diagram for sulfur to answer the questions on the following page:

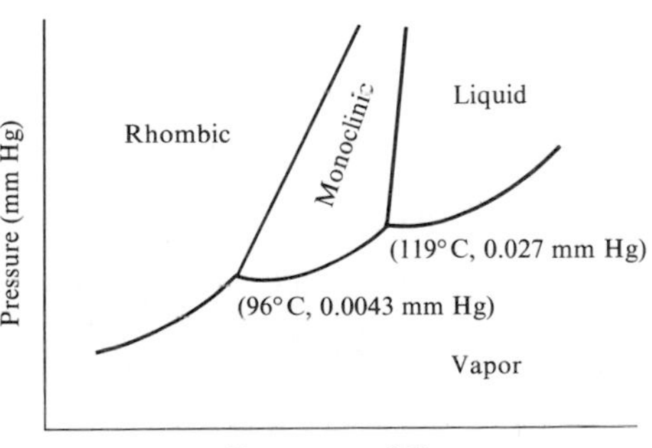

a. How many triple points are in the phase diagram?
b. What phases are in equilibrium at each of the triple points?
c. What phase is stable at room temperature and 1.0 atm pressure?
d. Can monoclinic sulfur exist in equilibrium with sulfur vapor?

Additional Exercises

80. Boron nitride (BN) exists in two forms. The first is a slippery solid formed from the reaction of BCl_3 with NH_3, followed by heating in an ammonia atmosphere at 750°C. Subjecting the first form of BN to a pressure of 85,000 atm at 1800°C produces a second form that is the second hardest substance known. Both forms of BN remain solids to 3000°C. Suggest structures for the two forms of BN.

81. A crystal of hafnium (Hf) was exposed to X rays from a Mo X-ray tube ($\lambda = 71.2$ pm). First-order diffraction ($n = 1$ in the Bragg equation) was observed at an angle of 5.564°. The density of hafnium is 13.28 g/cm^3. Assuming that the distance calculated from the Bragg equation gives the edge length of the cubic unit cell, does hafnium exist in a body-centered or face-centered cubic arrangement? Calculate the atomic radius of hafnium.

82. From the following data for liquid nitric acid, determine its heat of vaporization and normal boiling point.

Temperature (°C)	Vapor pressure (mm Hg)
0	14.4
10	26.6
20	47.9
30	81.3
40	133
50	208
80	670

83. What is the unit cell for each of the following two-dimensional arrays of circles? How many of each type of circle are in each unit cell?

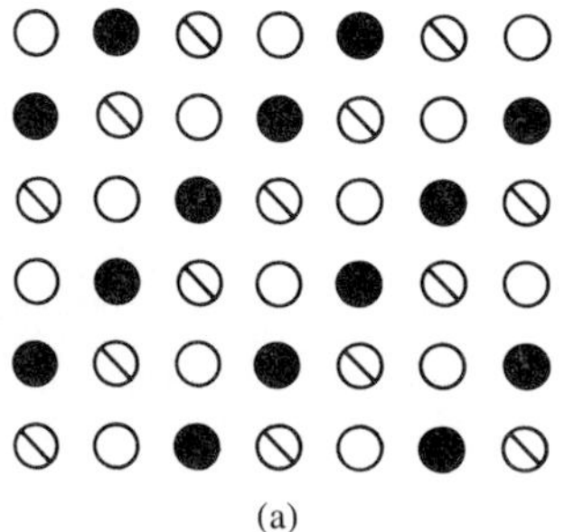
(a)

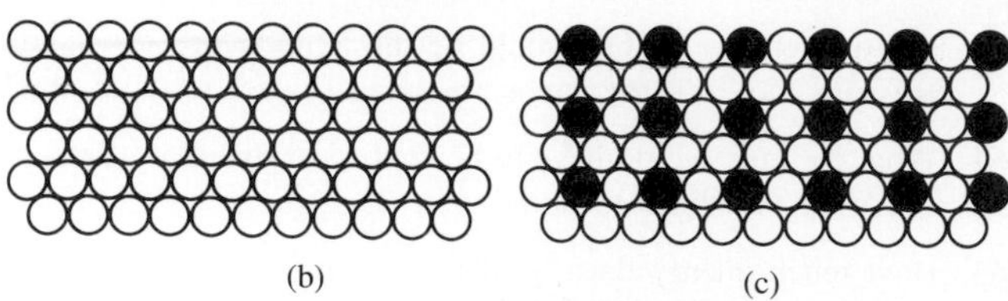
(b) (c)

84. When wet laundry is hung on a clothesline on a cold winter day, it will freeze, but eventually dry. Explain.

85. What fraction of the total volume of a cubic closest packed structure is occupied by atoms?

86. Many organic acids, such as acetic acid (CH_3CO_2H), whose structure is shown in Exercise 87, exist in the gas phase as hydrogen-bonded dimers (two molecule units). Draw a reasonable structure for such dimers.

87. Rationalize the following boiling points:

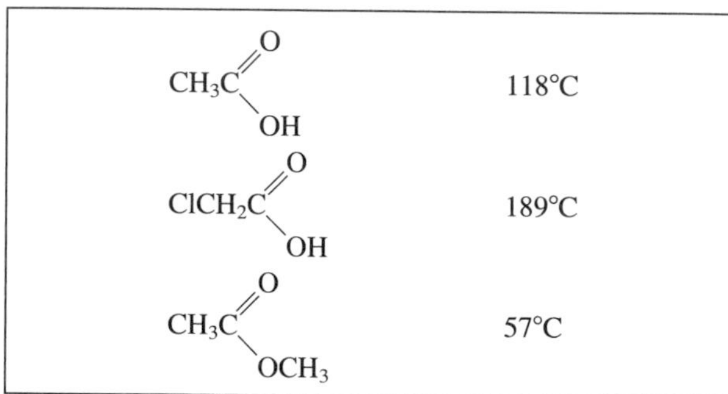

88. Use the diagram of the unit cell for the hexagonal closest packed structure in Fig. 10.14 to calculate the net number of atoms in the unit cell.

89. Some ice cubes with a total mass of 475 g are placed in a microwave oven and subjected to 750. W (750. J/s) of energy for 5.00 min. Does the water boil? If not, what is the final temperature of the water? (Assume all of the energy of the microwaves is absorbed by the water and that no heat is lost from the water.)

90. Calculate the enthalpy of vaporization and the normal boiling point of methanol using the following data:

Temperature (°C)	Vapor pressure (mm Hg)
−6.0°C	20.0
5.0°C	40.0
12.1°C	60.0
21.2°C	100.0
49.9°C	400.0

91. Use the following phase diagram for carbon to answer the following questions:

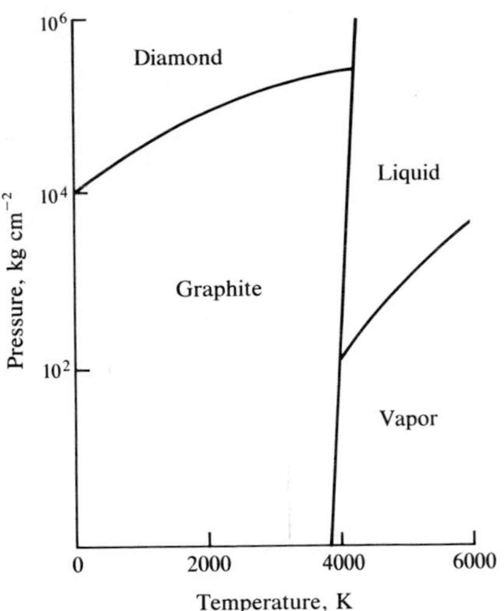

a. How many triple points are in the phase diagram?
b. What phases can coexist at each triple point?
c. What happens if graphite is subjected to very high pressures at room temperature?
d. If we assume that the density increases with an increase in pressure, which is more dense, graphite or diamond?

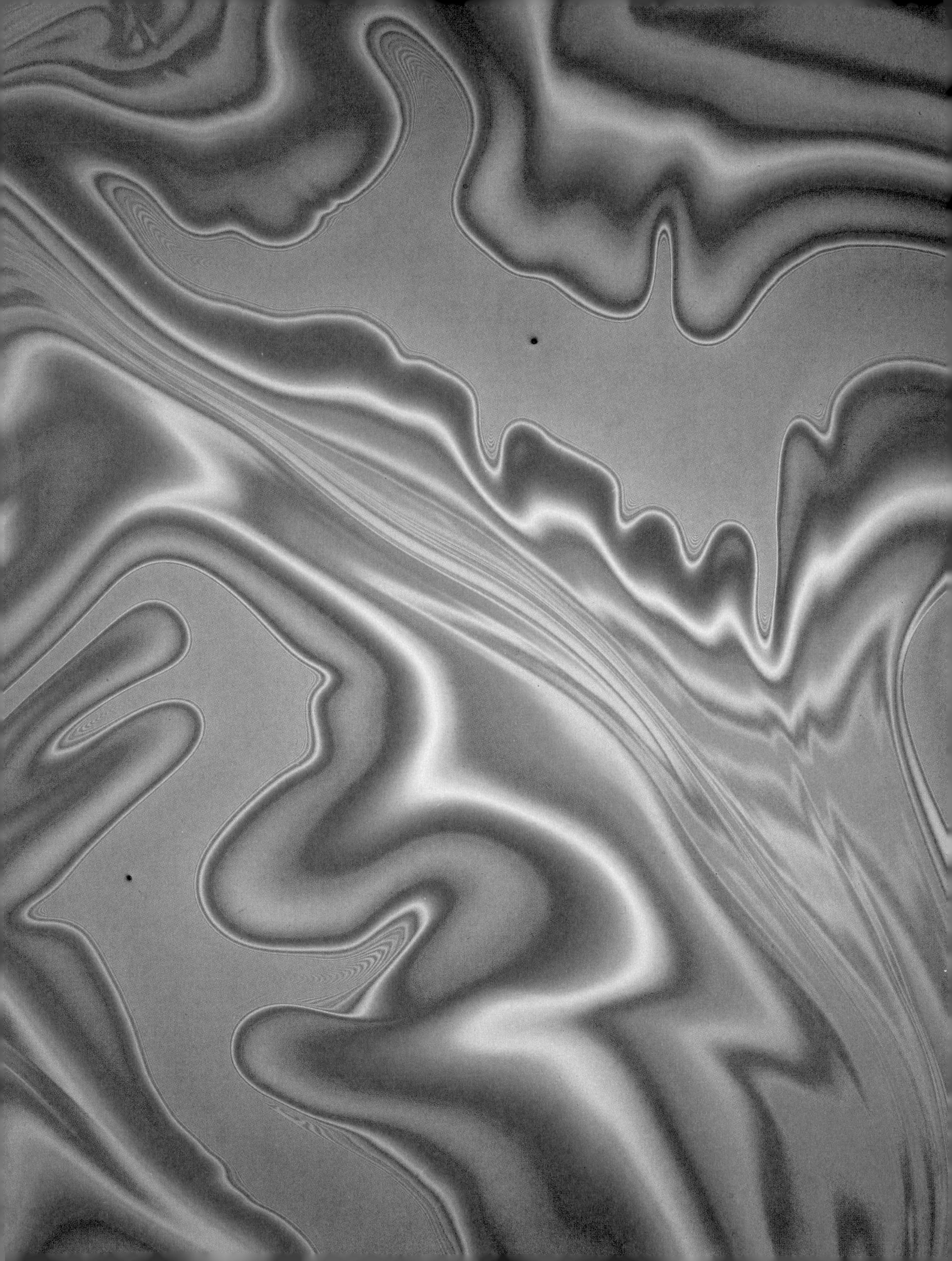

CHAPTER 11

Properties of Solutions

Most of the substances we encounter in daily life are mixtures: wood, milk, gasoline, shampoo, steel, air, and many others. When the components of a mixture are uniformly intermingled, that is, when a mixture is homogeneous, it is called a **solution.** Solutions can be gases, liquids, or solids, as shown in Table 11.1, but we will be concerned in this chapter with the properties of liquid solutions, particularly those containing water. As we saw in Chapter 4, many essential chemical reactions occur in aqueous solutions, since water is capable of dissolving so many substances.

Various Types of Solutions

Example	State of solution	State of solute	State of solvent
Air, natural gas	Gas	Gas	Gas
Vodka in water, antifreeze	Liquid	Liquid	Liquid
Brass, steel	Solid	Solid	Solid
Carbonated water (soda)	Liquid	Gas	Liquid
Sea water, sugar solution	Liquid	Solid	Liquid
Hydrogen in platinum	Solid	Gas	Solid

Table 11.1

CONTENTS

< Light dispersed by a thin film of oil floating on water. The oil does not dissolve in water because it is nonpolar and thus immiscible with polar water.

11.1 Solution Composition

Purpose

- To define various ways of describing solution composition.

A solute is the substance being dissolved. The solvent is the dissolving medium.

$$\text{Molarity} = \frac{\text{moles of solute}}{\text{liters of solution}}$$

Because a mixture, unlike a chemical compound, has a variable composition, the relative amounts of substances in a solution must be specified. The qualitative terms *dilute* (relatively little solute present) and *concentrated* (relatively large amount of solute) are often used to describe solution content, but we need to define solution composition more precisely in order to perform calculations. For example, in dealing with the stoichiometry of solution reactions in Chapter 4, we found it useful to describe solution composition in terms of **molarity,** or the number of moles of solute per liter of solution.

Other ways of describing solution composition are also useful. **Mass percent** (sometimes called *weight percent*) is the percent by mass of the solute in the solution:

$$\text{Mass percent} = \left(\frac{\text{grams of solute}}{\text{grams of solution}}\right) \times 100$$

When liquids are mixed, the liquid present in the largest amount is called the solvent.

Another way of describing solution composition is the **mole fraction** (symbolized by the Greek letter chi, χ), the ratio of the number of moles of a given component to the total number of moles of solution. For a two-component solution, where n_A and n_B represent the number of moles of the two components,

$$\text{Mole fraction of component A} = \chi_A = \frac{n_A}{n_A + n_B}$$

In very dilute aqueous solutions, the molality and the molarity are almost the same.

Still another way of describing solution composition is **molality** (symbolized by m), the number of moles of solute per *kilogram of solvent:*

$$\text{Molality} = \frac{\text{moles of solute}}{\text{kilograms of solvent}}$$

Sample Exercise 11.1

A solution is prepared by mixing 1.00 g of ethanol (C_2H_5OH) with 100.0 g of water to give a final volume of 101 mL. Calculate the molarity, mass percent, mole fraction, and molality of ethanol in this solution.

Solution

Since molarity depends on the volume of the solution, it changes slightly with temperature. Molality is independent of temperature since it depends on mass.

Molarity: The moles of ethanol can be obtained from its molecular weight (46.07):

$$1.00 \text{ g } C_2H_5OH \times \frac{1 \text{ mol } C_2H_5OH}{46.07 \text{ g } C_2H_5OH} = 2.17 \times 10^{-2} \text{ mol } C_2H_5OH$$

$$\text{Volume} = 101 \text{ mL} \times \frac{1 \text{ L}}{1000 \text{ mL}} = 0.101 \text{ L}$$

Sample Exercise 11.1, continued

$$\text{Molarity of } C_2H_5OH = \frac{\text{moles of } C_2H_5OH}{\text{liters of solution}} = \frac{2.17 \times 10^{-2}\text{ mol}}{0.101\text{ L}}$$

$$= 0.215\ M$$

Mass percent:

$$\text{Mass percent } C_2H_5OH = \left(\frac{\text{grams of } C_2H_5OH}{\text{grams of solution}}\right) \times 100$$

$$= \left(\frac{1.00\text{ g } C_2H_5OH}{100.0\text{ g } H_2O + 1.00\text{ g } C_2H_5OH}\right) \times 100$$

$$= 0.990\%\ C_2H_5OH$$

Mole fraction:

$$\text{Mole fraction of } C_2H_5OH = \frac{n_{C_2H_5OH}}{n_{C_2H_5OH} + n_{H_2O}}$$

$$n_{H_2O} = 100.0\text{ g } H_2O \times \frac{1\text{ mol } H_2O}{18.0\text{ g } H_2O} = 5.56\text{ mol}$$

$$\chi_{C_2H_5OH} = \frac{2.17 \times 10^{-2}\text{ mol}}{2.17 \times 10^{-2}\text{ mol} + 5.56\text{ mol}}$$

$$= \frac{2.17 \times 10^{-2}}{5.58} = 0.00389$$

Molality:

$$\text{Molality of } C_2H_5OH = \frac{\text{moles of } C_2H_5OH}{\text{kilograms of } H_2O} = \frac{2.17 \times 10^{-2}\text{ mol}}{100.0\text{ g} \times \frac{1\text{ kg}}{1000\text{ g}}}$$

$$= \frac{2.17 \times 10^{-2}\text{ mol}}{0.1000\text{ kg}}$$

$$= 0.217\ m$$

Another concentration measure sometimes encountered is **normality** (designated by *N*). Normality is defined as the number of *equivalents* per liter of solution, where the definition of an equivalent depends on the reaction taking place in the solution. *For an acid-base reaction,* the equivalent is the mass of acid or base that can furnish or accept exactly 1 mole of protons (H^+ ions). In Table 11.2 note, for example, that the equivalent weight of sulfuric acid is the molecular weight divided by 2, since each mole of H_2SO_4 can furnish 2 moles of protons. The equivalent weight of calcium hydroxide is also half of the molecular weight, since each mole of $Ca(OH)_2$ contains 2 moles of OH^- ions that can react with 2 moles of protons. Note that the equivalent is defined so that 1 equivalent of acid will react with exactly 1 equivalent of base.

The definition of an equivalent depends on the reaction taking place in the solution.

?

The Molecular Weight, Equivalent Weight, and Relationship of Molarity and Normality for Several Acids and Bases

Acid or base	Molecular weight	Equivalent weight	Relationship of molarity and normality
HCl	36.5	36.5	$1\ M = 1\ N$
H_2SO_4	98	$\frac{98}{2} = 49$	$1\ M = 2\ N$
H_3PO_4	98	$\frac{98}{3} = 32.6$	$1\ M = 3\ N$
NaOH	40	40	$1\ M = 1\ N$
$Ca(OH)_2$	74	$\frac{74}{2} = 37$	$1\ M = 2\ N$

Table 11.2

Oxidation-reduction half-reactions were discussed in Section 4.10.

For oxidation-reduction reactions, the equivalent is defined as the quantity of oxidizing or reducing agent that can accept or furnish 1 mole of electrons. Thus 1 equivalent of reducing agent will react with exactly 1 equivalent of oxidizing agent. The equivalent weight of an oxidizing or reducing agent can be calculated from the number of electrons in its half-reaction. For example, MnO_4^- reacting in acidic solution absorbs five electrons to produce Mn^{2+}:

$$MnO_4^- + 5e^- + 8H^+ \rightarrow Mn^{2+} + 4H_2O$$

Since the MnO_4^- ion present in 1 mole of $KMnO_4$ consumes 5 moles of electrons, the equivalent weight is the molecular weight divided by 5:

$$\text{Equivalent weight of } KMnO_4 = \frac{\text{mol. wt.}}{5} = \frac{158}{5} = 31.6$$

Sample Exercise 11.2

A modern 12-volt lead storage battery of the type used in automobiles.

The electrolyte in automobile lead storage batteries is a 3.75 *M* sulfuric acid solution that has a density of 1.230 g/mL. Calculate the mass percent, molality, mole fraction, and normality of the sulfuric acid.

Solution

The density of the solution in grams per liter is:

$$1.230\ \frac{\text{g}}{\text{mL}} \times \frac{1000\ \text{mL}}{1\ \text{L}} = 1.230 \times 10^3\ \text{g/L}$$

Thus 1 liter of this solution contains 1230 g of the mixture of sulfuric acid and water. Since the solution is 3.75 *M*, we know that 3.75 mol of H_2SO_4 are present per liter of solution. The number of grams of H_2SO_4 present is

$$3.75\ \text{mol} \times \frac{98.1\ \text{g}\ H_2SO_4}{1\ \text{mol}} = 368\ \text{g}\ H_2SO_4$$

The amount of water present in 1 liter of solution is obtained from the difference

$$1230\ \text{g solution} - 368\ \text{g}\ H_2SO_4 = 862\ \text{g}\ H_2O$$

Sample Exercise 11.2, continued

$$862 \text{ g } H_2O \times \frac{1 \text{ mol } H_2O}{18.0 \text{ g } H_2O} = 47.9 \text{ mol } H_2O$$

Since we now know the masses of the solute and solvent, we can calculate the mass percent.

$$\text{Mass percent } H_2SO_4 = \frac{\text{grams of } H_2SO_4}{\text{grams of solution}} \times 100 = \frac{368 \text{ g}}{1230 \text{ g}} \times 100$$

$$= 29.9\% \ H_2SO_4$$

From the moles of solute and the mass of solvent we can calculate the molality.

$$\text{Molality of } H_2SO_4 = \frac{\text{moles } H_2SO_4}{\text{kilograms of } H_2O}$$

$$= \frac{3.75 \text{ mol } H_2SO_4}{862 \text{ g } H_2O \times \dfrac{1 \text{ kg } H_2O}{1000 \text{ g } H_2O}} = 4.35 \ m$$

The mole fraction of solute can be obtained from moles of solute and solvent.

$$\text{Mole fraction of } H_2SO_4 = \frac{\text{moles of } H_2SO_4}{\text{moles of } H_2SO_4 + \text{moles of } H_2O}$$

$$= \frac{3.75}{3.75 + 47.9} = 0.0726$$

Since each sulfuric acid molecule can furnish two protons, 1 mol of H_2SO_4 represents 2 equivalents. Thus a solution with 3.75 mol of H_2SO_4 per liter contains $2 \times 3.75 = 7.50$ equivalents per liter, and the normality is 7.50 *N*.

11.2 The Energies of Solution Formation

Purpose

- To define the heat of solution and discuss its various energy components.

Dissolving solutes in liquids is very common. We dissolve salt in the water used to cook vegetables, sugar in iced tea, stains in cleaning fluid, gaseous carbon dioxide in water to make soda water, ethanol in gasoline to make gasohol, and so on.

Solubility is important in other ways. Because the pesticide DDT is fat-soluble, it is retained and concentrated in animal tissues where it causes detrimental effects. This is why DDT, even though it is effective for killing mosquitos, has been banned in the United States. The solubility of various vitamins is important in determining correct dosages. The insolubility of barium sulfate means it can be safely used to improve X rays of the gastrointestinal tract, even though Ba^{2+} ions are quite toxic.

A bald eagle's nest in Everglades National Park, Florida. The presence of DDT in many birds causes their eggs' shells to become too thin.

Polar solvents dissolve polar solutes; nonpolar solvents dissolve nonpolar solutes.

What factors affect solubility? The cardinal rule of solubility is *like dissolves like*. We find that we must use a polar solvent to dissolve a polar or ionic solute and a nonpolar solvent to dissolve a nonpolar solute. Now we will see why this is true. To simplify the discussion, we will assume that the formation of a liquid solution takes place in three distinct steps:

STEP 1

Breaking up the solute into individual components (expanding the solute).

STEP 2

Overcoming intermolecular forces in the solvent to make room for the solute (expanding the solvent.)

STEP 3

Allowing the solute and solvent to interact to form the solution.

These steps are illustrated in Fig. 11.1. Steps 1 and 2 require energy, since forces must be overcome to expand the solute and solvent. Step 3 usually releases energy. In other words, Steps 1 and 2 are endothermic, and Step 3 is exothermic. The enthalpy change associated with the formation of the solution, called the **enthalpy (heat) of solution** (ΔH_{soln}), is the sum of the ΔH values for the steps:

The enthalpy of solution is the sum of energies used in expanding both solvent and solute and energy of solvent-solute interaction.

$$\Delta H_{soln} = \Delta H_1 + \Delta H_2 + \Delta H_3$$

where ΔH_{soln} may have a positive sign (energy absorbed) or a negative sign (energy released), as shown in Fig. 11.2.

To illustrate the importance of the various energy terms in the equation for ΔH_{soln}, we will consider two specific cases. First, we know that oil is not soluble in water. When oil tankers leak, the petroleum forms an oil slick that floats on the water and is eventually carried onto the beaches. We can explain the immiscibility

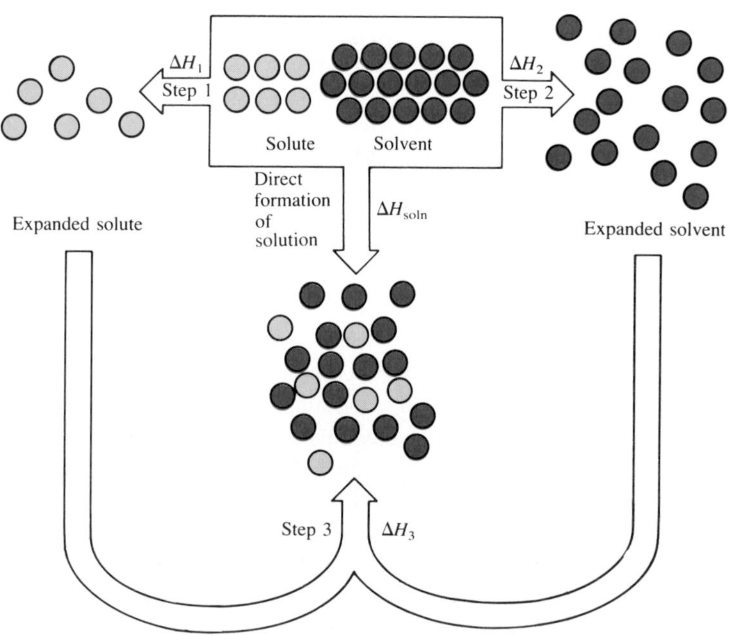

Figure 11.1

The formation of a liquid solution can be divided into three steps: (1) expanding the solute, (2) expanding the solvent, and (3) combining the expanded solute and solvent to form the solution.

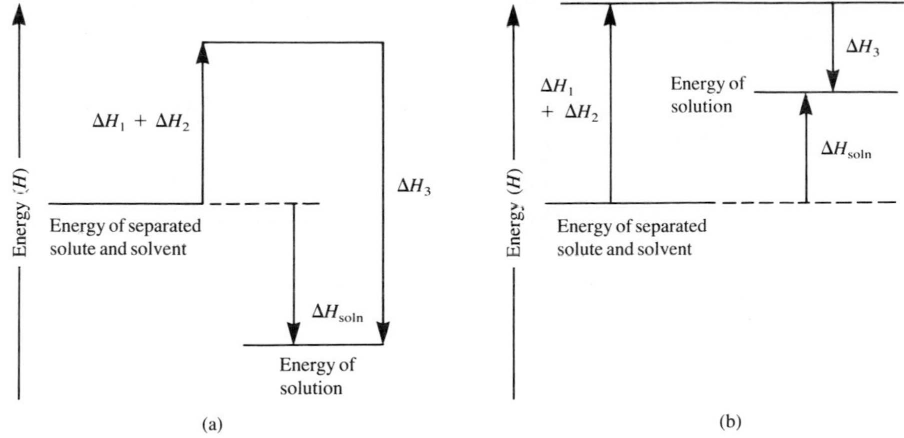

Figure 11.2

The heat of solution (a) ΔH_{soln} has a negative sign (the process is exothermic) if Step 3 releases more energy than that required by Steps 1 and 2. (b) ΔH_{soln} has a positive sign (the process is endothermic) if Steps 1 and 2 require more energy than is released in Step 3. (If the energy changes for Steps 1 and 2 equal that for Step 3, then ΔH_{soln} is zero.)

of oil and water by considering the energy terms involved. Oil is a mixture of nonpolar molecules that interact through London dispersion forces, which depend on molecule size. We expect ΔH_1 to be small for a typical nonpolar solute, but it will be relatively large for the large oil molecules. The term ΔH_3 will be small, since interactions between the nonpolar solute molecules and the polar water molecules will be negligible. However, ΔH_2 will be large and positive because it takes considerable energy to overcome the hydrogen bonding forces among the water molecules to expand the solvent. Thus ΔH_{soln} will be large and positive mainly because of the ΔH_2 term. Since a large amount of energy would have to be expended to form an oil-water solution, this process does not occur to any appreciable extent. These same arguments hold true for any nonpolar solute and polar solvent—the combination of a nonpolar solute and a highly polar solvent is not expected to produce a solution.

ΔH_1 is expected to be small for nonpolar solutes but can be large for large molecules.

As a second case, let's consider the solubility of an ionic solute, such as sodium chloride, in water. Here the term ΔH_1 is large and positive because the strong ionic forces in the crystal must be overcome, and ΔH_2 is large and positive because hydrogen bonds must be broken. Finally, ΔH_3 is large and negative because of the strong interactions between the ions and the water molecules. In fact, the exothermic and endothermic terms essentially cancel, as shown from the known values:

$$\mathrm{NaCl}(s) \rightarrow \mathrm{Na^+}(g) + \mathrm{Cl^-}(g) \qquad \Delta H_1 = 786 \text{ kJ/mol}$$

$$\mathrm{H_2O}(l) + \mathrm{Na^+}(g) + \mathrm{Cl^-}(g) \rightarrow \mathrm{Na^+}(aq) + \mathrm{Cl^-}(aq) \qquad \Delta H_{hyd} = \Delta H_2 + \Delta H_3 = -783 \text{ kJ/mol}$$

Here the **enthalpy (heat) of hydration** (ΔH_{hyd}) combines the terms ΔH_2 (for expanding the solvent) and ΔH_3 (for solvent-solute interactions). The heat of hydration represents the enthalpy change associated with the dispersal of a gaseous solute in water. Thus the heat of solution for dissolving sodium chloride is the sum of ΔH_1 and ΔH_{hyd}:

$$\Delta H_{soln} = 786 \text{ kJ/mol} - 783 \text{ kJ/mol} = 3 \text{ kJ/mol}$$

Note that ΔH_{soln} is small but positive; the dissolving process requires a small amount of energy. Then why is NaCl so soluble in water? The answer lies in nature's tendency toward disorder. That is, processes naturally run in the direction

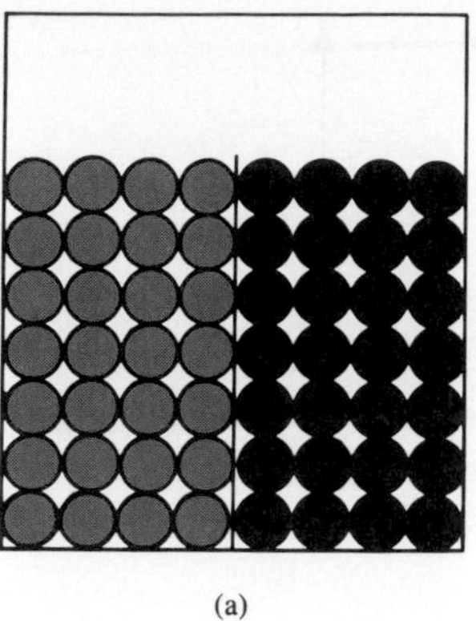

(a)

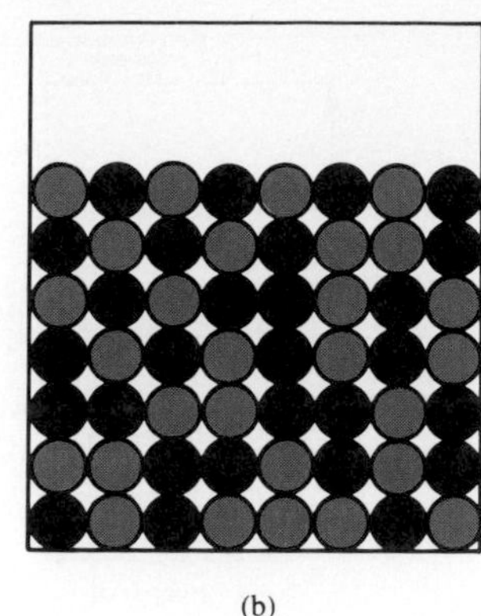

(b)

Figure 11.3

(a) Red and black spheres separated by a partition in a closed container. (b) The spheres after the partition is removed and the container has been shaken for some time.

that leads to greater disorder. For example, imagine equal numbers of red and black spheres separated by a partition, as shown in Fig. 11.3(a). If we remove the partition and shake the container, the spheres will mix [Fig. 11.3(b)], and no amount of shaking will cause them to return to the state of separated red and black. Why? The mixed, or disordered, state is simply much more likely to occur (more probable) than the original separate, or ordered, state because there are many more ways of placing the spheres to give a mixed state than a separated state. This is a general principle. *One factor that favors a process is an increase in disorder.*

The factors that act as driving forces for a process are discussed more fully in Chapter 16.

But energy considerations are also important. *Processes that require large amounts of energy tend not to occur.* Since dissolving 1 mole of solid NaCl requires only a small amount of energy, the solution will form because of the large increase in disorder occurring when the solute and solvent mix.

The various possible cases for solution formation are summarized in Table 11.3. Note that in two cases, polar-polar and nonpolar-nonpolar, the heat of solution is small. In these cases, the solution forms because of the increase in disorder. In the other cases (polar-nonpolar and nonpolar-polar), the heat of solution is large and positive, and the energy effects prevent the solution from forming. These results explain the rule of "like dissolves like."

The Energy Terms for Various Types of Solutes and Solvents

	ΔH_1	ΔH_2	ΔH_3	ΔH_{soln}	Outcome
Polar solvent, polar solute	Large	Large	Large, negative	Small	Solution forms
Polar solvent, nonpolar solute	Small	Large	Small	Large, positive	No solution forms
Nonpolar solvent, nonpolar solute	Small	Small	Small	Small	Solution forms
Nonpolar solvent, polar solute	Large	Small	Small	Large, positive	No solution forms

Table 11.3

Sample Exercise 11.3

Decide whether liquid hexane (C_6H_{14}) or liquid methanol (CH_3OH) is the more appropriate solvent for the substances grease ($C_{20}H_{42}$) and potassium iodide (KI).

Solution

Hexane is a nonpolar solvent because it contains nonpolar C—H bonds. Thus hexane will work best for the nonpolar solute grease. Methanol has an O—H group that makes it significantly polar. Thus it will serve as the better solvent for the ionic solid KI.

11.3 Factors Affecting Solubility

Purpose

- To show how molecular structure, pressure, and temperature affect solubility.

Structure Effects

In the last section we saw that solubility is favored if the solute and solvent have similar polarities. Since it is the molecular structure that determines polarity, there should be a definite connection between structure and solubility. Vitamins provide an excellent example of the relationship of molecular structure, polarity, and solubility.

Recently, there has been considerable publicity about the pros and cons of consuming large quantities of vitamins. For example, large doses of vitamin C have been advocated to combat various illnesses, including the common cold. Vitamin E has been extolled as a youth-preserving elixir and a protector against the carcinogenic (cancer-causing) effects of certain chemicals. However, there are possible detrimental effects from taking large amounts of some vitamins, depending on their solubilities.

Vitamins can be divided into two classes: *fat-soluble* (vitamins A, D, E, and K) and *water-soluble* (vitamins B and C). The reason for the differing solubility characteristics can be seen by comparing the structures of vitamins A and C (Fig. 11.4). Vitamin A, composed mostly of carbon and hydrogen atoms that have similar electronegativities, is virtually nonpolar. This causes it to be soluble in nonpolar materials such as body fat, which is also largely composed of carbon and hydrogen, but

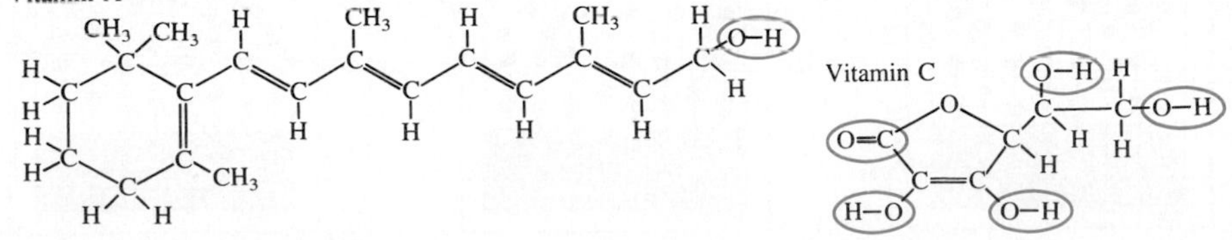

Figure 11.4

The molecular structures of vitamin A (nonpolar, fat-soluble) and vitamin C (polar, water-soluble). The circles indicate polar bonds. Note that vitamin C contains far more polar bonds than vitamin A.

not soluble in polar solvents such as water. On the other hand, vitamin C has many polar O—H and C—O bonds, making the molecule polar and thus water-soluble. We often describe nonpolar materials like vitamin A as *hydrophobic* (water-fearing) and polar substances like vitamin C as *hydrophilic* (water-loving).

Because of their solubility characteristics, the fat-soluble vitamins can build up in the fatty tissues of the body. This has both positive and negative effects. Since these vitamins can be stored, the body can tolerate for a time a diet deficient in vitamins A, D, E, or K. Conversely, if excessive amounts of these vitamins are consumed, their build-up can lead to the illness *hypervitaminosis*.

In contrast, the water-soluble vitamins are excreted by the body and must be consumed regularly. This fact was first recognized when the British navy discovered that scurvy, a disease often suffered by sailors, could be prevented if the sailors regularly ate fresh limes (which are a good source of vitamin C) when aboard ship (hence the name ''limey'' for the British sailor).

Carbonation in a bottle of soda.

Pressure Effects

While pressure has little effect on the solubilities of solids or liquids, it does significantly increase the solubility of a gas. Carbonated beverages, for example, are always bottled at high pressures of carbon dioxide to ensure a high concentration of carbon dioxide in the liquid. The fizzing that occurs when you open a can of soda results from the escape of gaseous carbon dioxide because the atmospheric pressure of CO_2 is much lower than that used in the bottling process.

The increase in gas solubility with pressure can be understood from Fig. 11.5. Figure 11.5(a) shows a gas in equilibrium with a solution; that is, the gas molecules are entering and leaving the solution at the same rate. If the pressure is suddenly increased [Fig. 11.5(b)], the number of gas molecules per unit volume increases, and the gas enters the solution at a higher rate than it leaves. As the concentration of dissolved gas increases, the rate of the escape of the gas also increases until a new equilibrium is reached [Fig. 11.5(c)], where the solution contains more dissolved gas than before.

The relationship between gas pressure and the concentration of dissolved gas is given by **Henry's law:**

$$P = kC$$

where P represents the partial pressure of the gaseous solute above the solution, C represents the concentration of the dissolved gas, and k is a constant characteristic of a particular solution. In words, Henry's law states that *the amount of a gas*

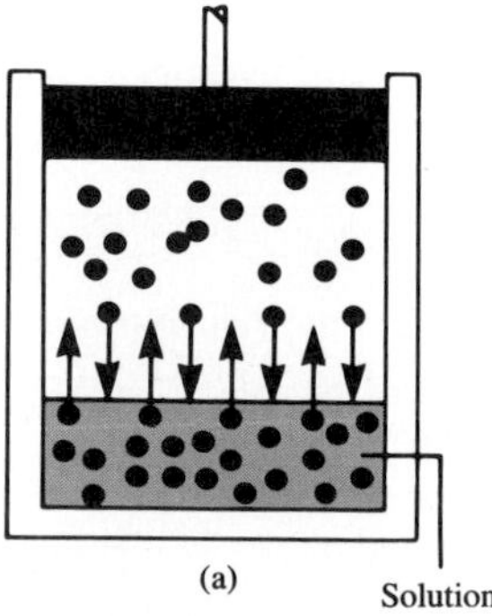

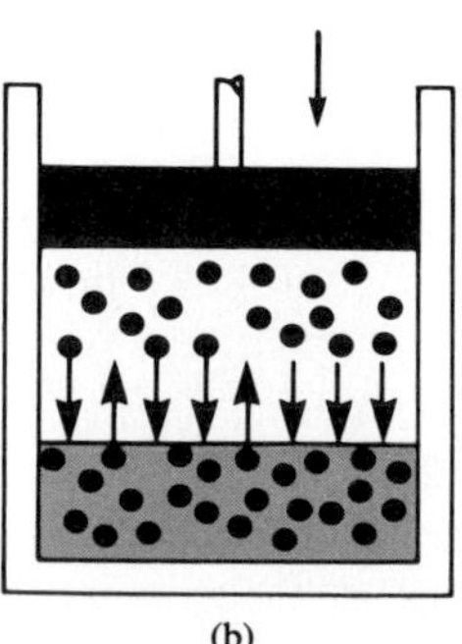

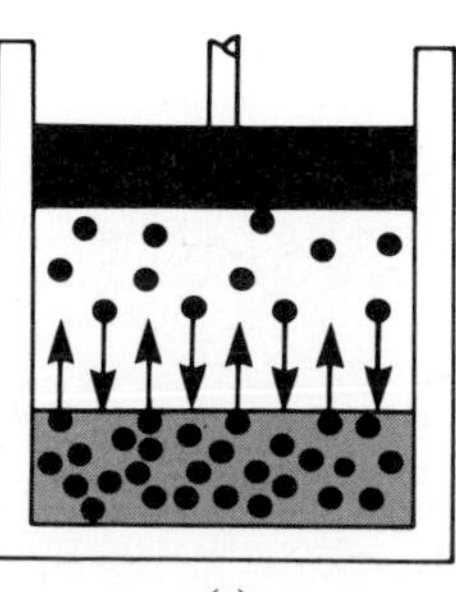

Figure 11.5

(a) A gaseous solute in equilibrium with a solution. (b) The piston is pushed in, which increases the pressure of the gas and the number of gas molecules per unit volume. This causes an increase in the rate at which the gas enters the solution, so the concentration of dissolved gas increases. (c) The greater gas concentration in the solution causes an increase in the rate of escape. A new equilibrium is reached.

Chemical Impact

The Lake Nyos Tragedy

On August 21, 1986, a cloud of gas suddenly boiled from Lake Nyos in Cameroon, killing nearly 2000 people. Although at first it was speculated that the gas was hydrogen sulfide, it now seems clear it was carbon dioxide. What would cause Lake Nyos to emit this huge, suffocating cloud of CO_2? Although the answer may never be known for certain, many scientists believe that the lake suddenly "turned over," bringing to the surface water that contained huge quantities of dissolved carbon dioxide. Lake Nyos is a deep lake that is thermally stratified: layers of warm, less dense water near the surface float on the colder, denser water layers nearer the lake's bottom. Under normal conditions the lake stays this way: there is little mixing among the different layers. Scientists believe that over hundreds or thousands of years, carbon dioxide gas had seeped into the cold water at the lake's bottom and dissolved in great amounts because of the large pressure of CO_2 present (in accordance with Henry's law). For some reason on August 21, 1986, the lake apparently suffered an overturn, possibly due to wind or to unusual cooling of the lake's surface by monsoon clouds. This caused water that was greatly supersaturated with CO_2 to reach the surface and release tremendous quantities of gaseous CO_2 that suffocated humans and animals before they knew what hit them—a tragic, monumental illustration of Henry's law.

Lake Nyos in Cameroon.

dissolved in a solution is directly proportional to the pressure of the gas above the solution.

Henry's law is obeyed most accurately for dilute solutions of gases that do not dissociate in or react with the solvent. For example, Henry's law is obeyed by oxygen gas in water, but it does *not* correctly represent the behavior of gaseous hydrogen chloride in water because of the dissociation reaction

$$HCl(g) \xrightarrow{H_2O} H^+(aq) + Cl^-(aq)$$

William Henry (1774–1836), a close friend of John Dalton, formulated his law in 1801.

Henry's law holds only when there is no chemical reaction between solute and solvent.

Sample Exercise 11.4

A certain soft drink is bottled so that a bottle at 25°C contains CO_2 gas at a pressure of 5.0 atm over the liquid. Assuming the partial pressure of CO_2 in the atmosphere is 4.0×10^{-4} atm, calculate the concentration of CO_2 in the soda before and after the bottle is opened. The Henry's law constant for CO_2 in aqueous solution is 32 L atm/mol at 25°C.

Sample Exercise 11.4, continued

Solution

We can write Henry's law for CO_2 as

$$P_{CO_2} = k_{CO_2}C_{CO_2}$$

where $k_{CO_2} = 32$ L atm/mol. In the *unopened* bottle, $P_{CO_2} = 5.0$ atm and

$$C_{CO_2} = \frac{P_{CO_2}}{k_{CO_2}} = \frac{5.0 \text{ atm}}{32 \text{ L atm/mol}} = 0.16 \text{ mol/L}$$

In the *opened* bottle, the CO_2 in the soda eventually reaches equilibrium with the atmospheric CO_2. So $P_{CO_2} = 4.0 \times 10^{-4}$ atm and

$$C_{CO_2} = \frac{P_{CO_2}}{k_{CO_2}} = \frac{4.0 \times 10^{-4} \text{ atm}}{32 \text{ L atm/mol}} = 1.2 \times 10^{-5} \text{ mol/L}$$

Note the large change in concentration of CO_2. This is why soda goes ''flat'' after being open for a while.

Temperature Effects

ΔH°_{soln} refers to the formation of a 1.0 *M* ideal solution and is not necessarily relevant to the process of dissolving a solid in a saturated solution. Thus ΔH°_{soln} is of limited use in predicting the variation of solubility with temperature.

Everyday experiences of dissolving substances like sugar may lead you to think that solubility always increases with temperature. This is not the case. The dissolving of a solid occurs *more rapidly* at higher temperatures, but the amount of solid that can be dissolved may increase or decrease with increasing temperature. The effect of temperature on the solubility in water of several common solids is shown in Fig. 11.6. Note that, although the solubility of most solids increases with temperature, the solubilities of some substances (such as sodium sulfate and cerium sulfate) decrease with increasing temperature.

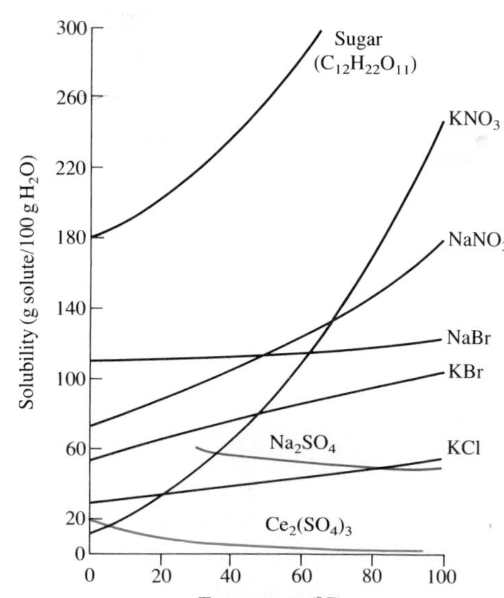

Figure 11.6

The solubilities of several solids as a function of temperature. Note that while most substances become more soluble in water with increasing temperature, sodium sulfate and cerium sulfate become less soluble.

Chemical Impact

Recyclable Heat

The current rise in athletic activities has led to many injuries and has necessitated new products to treat those injuries. Of the many instant hot and cold packs available (also see Chemical Impact: Heat Packs in Chapter 6), one of the most interesting types produces heat by the exothermic precipitation of a solid from a solution. The heat pack looks like a tiny air mattress with two compartments. The larger one contains a viscous supersaturated solution of sodium thiosulfate ($Na_2S_2O_3$) dissolved in water, and the smaller one contains crystals of solid sodium thiosulfate.

Because sodium thiosulfate crystals do not readily precipitate from aqueous solution, water can be greatly supersaturated with the solid. That is, much more solid $Na_2S_2O_3$ can be dissolved in water than should be possible at a given temperature. However, as soon as the supersaturated solution is exposed to a seed crystal of $Na_2S_2O_3$, to provide a pattern for crystal growth, the solid forms, releasing a considerable quantity of energy as heat over several hours. The pack is activated by squeezing out a seed crystal so that it comes into contact with the solution. The resulting formation of the solid gives a pleasantly warm temperature of 48°C (118°F).

This heat pack can be recycled by simply placing it in boiling water to redissolve the sodium thiosulfate. As the pack cools, the solid remains dissolved—it again becomes supersaturated. The pack can be reused until the supply of seed crystals runs out.

Predicting the temperature dependence of solubility is very difficult. For example, although there is some correlation between the sign of ΔH°_{soln} and the variation of solubility with temperature, important exceptions exist.* The only sure way to determine the temperature dependence of a solid's solubility is by experiment.

The behavior of gases appears less complex. The solubility of a gas in water typically decreases with increasing temperature, as is shown for several cases in Fig. 11.7. This temperature effect has important environmental implications because of the widespread use of water from lakes and rivers for industrial cooling. After being used, the water is returned to its natural source at a higher than ambient temperature (**thermal pollution** has occurred). Because it is warmer, this water contains less than the normal concentration of oxygen and is also less dense; it tends to "float" on the colder water below, thus blocking normal oxygen absorption. This effect can be especially important in deep lakes. The warm upper layer can seriously decrease the amount of oxygen available to aquatic life in the deeper layers of the lake.

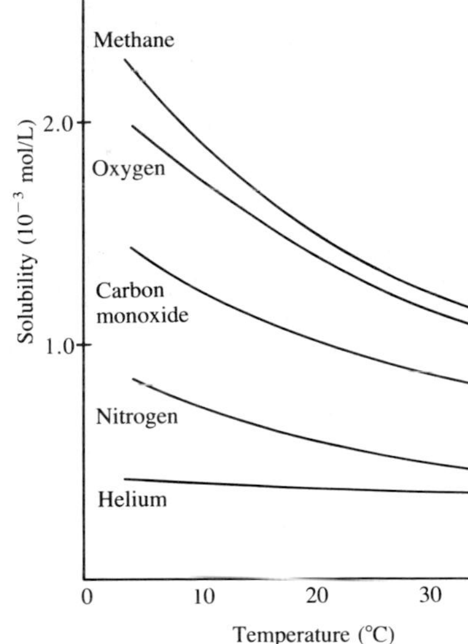

Figure 11.7

The solubilities of several gases as a function of temperature at a constant pressure of 1 atmosphere of gas above the solution.

The decreasing solubility of gases with increased temperature is also responsible for the formation of *boiler scale*. As we will see in more detail in Chapter 14, the bicarbonate ion is formed when carbon dioxide is dissolved in water containing the carbonate ion:

$$CO_3^{2-}(aq) + CO_2(aq) + H_2O(l) \rightarrow 2HCO_3^-(aq)$$

*For more information see R. S. Treptow, "Le Châtelier's Principle Applied to the Temperature Dependence of Solubility," *J. Chem. Ed.* **61** (1984): 499.

Figure 11.8

A hot-water pipe cut in half to show the collected mineral deposits almost clogging the pipe.

When the water also contains Ca^{2+} ions, this reaction is especially important—calcium bicarbonate is soluble in water, but calcium carbonate is insoluble. When the water is heated, the carbon dioxide is driven off. In order for the system to replace the lost carbon dioxide, the reverse reaction occurs:

$$2HCO_3^-(aq) \rightarrow H_2O(l) + CO_2(aq) + CO_3^{2-}(aq)$$

This reaction, however, also increases the concentration of carbonate ions, causing solid calcium carbonate to form. This solid is the boiler scale that coats the walls of containers such as industrial boilers and tea kettles. Boiler scale reduces the efficiency of heat transfer and can lead to blockage of pipes (see Fig. 11.8).

11.4 The Vapor Pressures of Solutions

Purpose

- To show how a solution's vapor pressure is affected by the concentration of solute and the interactions of solute and solvent.

Liquid solutions have physical properties significantly different from those of the pure solvent, a fact that has great practical importance. For example, we add antifreeze to the water in a car's cooling system to prevent freezing in winter and boiling in summer. We also melt ice on sidewalks and streets by spreading salt. These preventatives work because of the solute's effect on the solvent's properties.

A nonvolatile solute has no tendency to escape from solution into the vapor phase.

To explore how a nonvolatile solute affects a solvent, we will consider the experiment represented in Fig. 11.9, in which a sealed container encloses a beaker containing an aqueous sugar solution and a beaker containing pure water. Gradually, the volume of the sugar solution increases and the volume of the pure water decreases. Why? We can explain this observation if the vapor pressure of the pure solvent is greater than that of the solution. Under these conditions, the pressure of vapor necessary to achieve equilibrium with the pure solvent is greater than that required to reach equilibrium with the solution. Thus, as the pure solvent emits vapor to attempt to reach equilibrium, the solution absorbs vapor to try to lower the vapor pressure toward its equilibrium value. This process results in a net transfer of water from the pure liquid through the vapor phase to the solution. The system can only reach an equilibrium vapor pressure when all of the water is transferred to the solution. This experiment is just one of many observations indicating that the presence of a *nonvolatile solute lowers the vapor pressure of a solvent.*

The presence of a nonvolatile solute reduces the tendency of solvent molecules to escape.

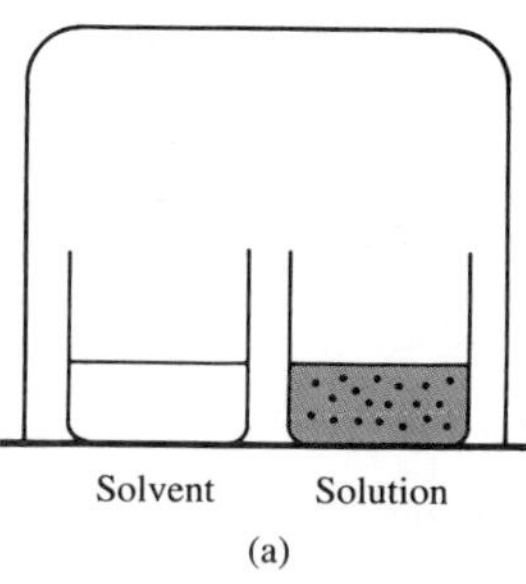

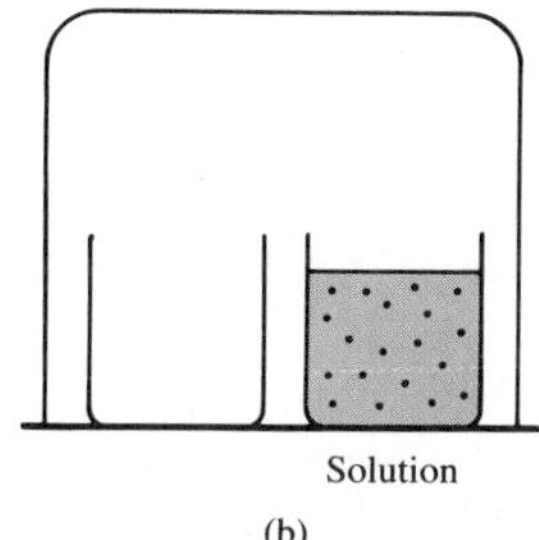

Figure 11.9

A solution and a pure solvent in a closed environment. (a) Initial stage. (b) After a period of time, the solvent is completely transferred to the solution.

We can account for this behavior in terms of the simple model shown in Fig. 11.10. The dissolved nonvolatile solute decreases the number of solvent molecules per unit volume. Thus it lowers the number of solvent molecules at the surface, and it should proportionately lower the escaping tendency of the solvent molecules. For example, in a solution consisting of half nonvolatile solute molecules and half solvent molecules, we might expect the observed vapor pressure to be half that of the pure solvent, since only half as many molecules can escape. In fact, this is what is observed.

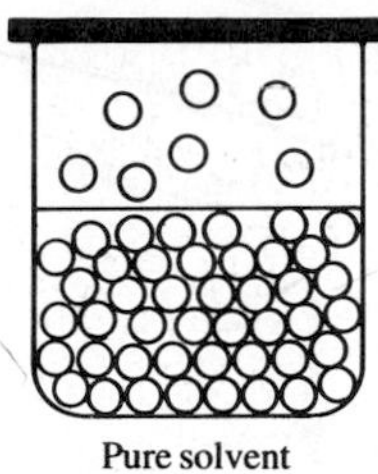

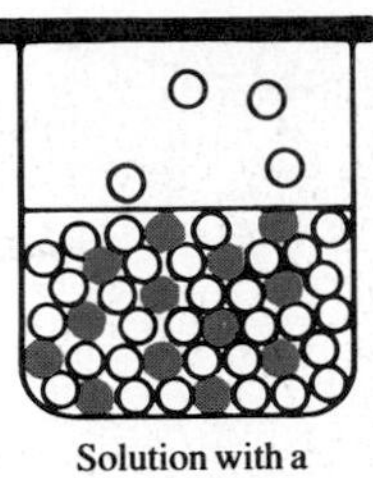

Figure 11.10

The presence of a nonvolatile solute inhibits the escape of solvent molecules from the liquid and so lowers the vapor pressure of the solvent.

Detailed studies of the vapor pressures of solutions containing nonvolatile solutes were carried out by François M. Raoult. His results are described by the equation known as **Raoult's law:**

$$P_{soln} = \chi_{solvent}P^0_{solvent}$$

where P_{soln} is the observed vapor pressure of the solution, $\chi_{solvent}$ is the mole fraction of solvent, and $P^0_{solvent}$ is the vapor pressure of the pure solvent. Note that for a solution of half solute and half solvent molecules, $\chi_{solvent}$ is 0.5, so the vapor pressure of the solution is half that of the pure solvent. On the other hand, for a solution where three-fourths of the solution molecules are solvent, $\chi_{solvent} = \frac{3}{4} = 0.75$, and $P_{soln} = 0.75P^0_{solvent}$. The idea is that the nonvolatile solute simply dilutes the solvent.

Raoult's law is a linear equation of the form $y = mx + b$, where $y = P_{soln}$, $x = \chi_{solvent}$, $m = P^0_{solvent}$, and $b = 0$. Thus a plot of P_{soln} versus $\chi_{solvent}$ gives a straight line with a slope equal to $P^0_{solvent}$, as is shown in Fig. 11.11.

Raoult's law states that the vapor pressure of a solution is directly proportional to the mole fraction of solvent present.

Sample Exercise 11.5

Calculate the expected vapor pressure at 25°C for a solution prepared by dissolving 158.0 g of common table sugar (sucrose, mol. wt. = 342.3) in 643.5 cm^3 of water. At 25°C the density of water is 0.9971 g/cm^3 and the vapor pressure is 23.76 torr.

Solution

We will use Raoult's law in the form

$$P_{soln} = \chi_{H_2O}P^0_{H_2O}$$

To calculate the mole fraction of water in the solution, we must first determine the number of moles of sucrose:

$$\text{mol sucrose} = 158.0 \text{ g sucrose} \times \frac{1 \text{ mol sucrose}}{342.3 \text{ g sucrose}}$$

$$= 0.4616 \text{ mol sucrose}$$

To determine the moles of water present, we first convert volume to mass using the density:

$$643.5 \text{ cm}^3 \text{ H}_2\text{O} \times \frac{0.9971 \text{ g H}_2\text{O}}{\text{cm}^3 \text{ H}_2\text{O}} = 641.6 \text{ g H}_2\text{O}$$

The number of moles of water is therefore

$$641.6 \text{ g H}_2\text{O} \times \frac{1 \text{ mol H}_2\text{O}}{18.01 \text{ g H}_2\text{O}} = 35.63 \text{ mol H}_2\text{O}$$

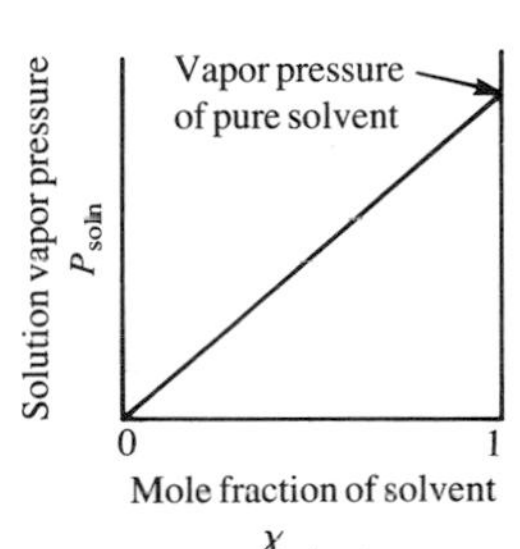

Figure 11.11

For a solution that obeys Raoult's law, a plot of P_{soln} versus $\chi_{solvent}$ gives a straight line.

Sample Exercise 11.5, continued

The mole fraction of water in the solution is

$$\chi_{H_2O} = \frac{\text{mol } H_2O}{\text{mol } H_2O + \text{mol sucrose}} = \frac{35.63 \text{ mol}}{35.63 \text{ mol} + 0.4616 \text{ mol}}$$

$$= \frac{35.63 \text{ mol}}{36.09 \text{ mol}} = 0.9872$$

Then

$$P_{\text{soln}} = \chi_{H_2O} P^0_{H_2O} = (0.9872)(23.76 \text{ torr}) = 23.46 \text{ torr}$$

Thus the vapor pressure of water has been lowered from 23.76 torr in the pure state to 23.46 torr in the solution. The vapor pressure has been lowered by 0.30 torr.

The phenomenon of the lowering of the vapor pressure gives us a convenient way to "count" molecules and thus provides a means for experimentally determining molecular weights. Suppose a certain weight of a compound is dissolved in a solvent and the vapor pressure of the resulting solution is measured. Using Raoult's law, we can determine the number of moles of solute present. Since the mass of this number of moles is known, we can calculate the molecular weight, as shown in Sample Exercise 11.6.

Photomicrograph of urea crystals.

Sample Exercise 11.6

A solution was prepared by adding 20.0 g of urea to 125 g of water at 25°C, a temperature at which pure water has a vapor pressure of 23.76 torr. The observed vapor pressure of the solution was found to be 22.67 torr. Calculate the molecular weight of urea.

Solution

Raoult's law can be rearranged to give

$$\chi_{H_2O} = \frac{P_{\text{soln}}}{P^0_{H_2O}}$$

This form allows us to determine the mole fraction of water in the solution:

$$\chi_{H_2O} = \frac{22.67 \text{ torr}}{23.76 \text{ torr}} = 0.9541$$

However, we are interested in the urea. To find its molecular weight, we must find the number of moles represented by 20.0 g. We can calculate the number of moles of urea from the definition of χ_{H_2O}:

$$\chi_{H_2O} = \frac{\text{mol } H_2O}{\text{mol } H_2O + \text{mol urea}} = \frac{n_{H_2O}}{n_{H_2O} + n_{\text{urea}}}$$

From the mass of water used to prepare the solution, we have

$$n_{H_2O} = 125 \text{ g } H_2O \times \frac{1 \text{ mol } H_2O}{18.0 \text{ g } H_2O} = 6.94 \text{ mol } H_2O$$

Since $\chi_{H_2O} = 0.9541$, we deduce the following:

Sample Exercise 11.6, continued

$$\chi_{H_2O} = 0.9541 = \frac{n_{H_2O}}{n_{H_2O} + n_{urea}} = \frac{6.94}{6.94 + n_{urea}}$$

or

$$0.9541\ (6.94 + n_{urea}) = 6.94$$

Solving for the number of moles of urea gives

$$n_{urea} = \frac{6.94 - 6.62}{0.9541} = 0.335 \text{ mol}$$

Since 20.0 g of urea was originally dissolved, 0.335 mol of urea weighs 20.0 g. Thus

$$\frac{20.0 \text{ g}}{0.335 \text{ mol}} = 59.7 \text{ g/mol}$$

The value for the molecular weight of urea determined in this experiment is thus 59.7. Urea has the formula $(NH_2)_2CO$ and a molecular weight of 60.0, so the result obtained in this experiment agrees well with the known molecular weight.

We can also use vapor pressure measurements to characterize solutions. For example, 1 mole of sodium chloride dissolved in water lowers the vapor pressure approximately twice as much as expected because the solid ionizes completely when it dissolves. Thus vapor pressure measurements can give valuable information about the nature of the solute after it dissolves.

The lowering of vapor pressure depends on the number of solute particles present in the solution.

Sample Exercise 11.7

Predict the vapor pressure of a solution prepared by mixing 35.0 g of solid Na_2SO_4 (mol. wt. = 142) with 175 g of water at 25°C. The vapor pressure of pure water at 25°C is 23.76 torr.

Solution

First, we need to know the mole fraction of H_2O.

$$n_{H_2O} = 175 \text{ g } H_2O \times \frac{1 \text{ mol } H_2O}{18.0 \text{ g } H_2O} = 9.72 \text{ mol } H_2O$$

$$n_{Na_2SO_4} = 35.0 \text{ g } Na_2SO_4 \times \frac{1 \text{ mol } Na_2SO_4}{142 \text{ g } Na_2SO_4} = 0.246 \text{ mol } Na_2SO_4$$

It is essential to recognize that when 1 mol of solid Na_2SO_4 dissolves, it produces 2 mol of Na^+ ions and 1 mol of SO_4^{2-} ions. Thus the number of solute particles present in this solution is three times the number of moles of solute dissolved:

$$n_{solute} = 3(0.246) = 0.738 \text{ mol}$$

$$\chi_{H_2O} = \frac{n_{H_2O}}{n_{solute} + n_{H_2O}} = \frac{9.72 \text{ mol}}{0.738 \text{ mol} + 9.72 \text{ mol}} = \frac{9.72 \text{ mol}}{10.458} = 0.929$$

Now we can use Raoult's law to predict the vapor pressure:

$$P_{soln} = \chi_{H_2O} P^0_{H_2O} = (0.929)(23.76 \text{ torr}) = 22.1 \text{ torr}$$

Nonideal Solutions

Any solution that obeys Raoult's law is called an **ideal solution.** Raoult's law is to solutions what the ideal gas law is to gases. As with gases, ideal behavior for solutions is never perfectly achieved, but is sometimes closely approached. Nearly ideal behavior is often observed when the solute-solute, solvent-solvent, and solute-solvent interactions are very similar. That is, in solutions where the solute and solvent are very much alike, the solute simply acts to dilute the solvent. However, if the solvent has a special affinity for the solute, such as if hydrogen bonding occurs, the tendency of the solvent molecules to escape will be lowered more than expected. The observed vapor pressure will be *lower* than the value predicted by Raoult's law; there will be a *negative deviation from Raoult's law.*

Strong solute-solvent interaction gives a vapor pressure lower than that predicted by Raoult's law.

When a solute and solvent release large quantities of energy in the formation of a solution, that is, when ΔH_{soln} is large and negative, we can assume that strong interactions exist between the solute and solvent. In this case we expect a negative deviation from Raoult's law.

So far we have assumed that the solute is nonvolatile and so does not contribute to the vapor pressure over the solution. However, for liquid-liquid solutions where both components are volatile, a modified form of Raoult's law applies:

$$P_{TOTAL} = P_A + P_B = \chi_A P_A^0 + \chi_B P_B^0$$

where P_{TOTAL} represents the total vapor pressure of a solution containing A and B, χ_A and χ_B are the mole fractions of A and B, P_A^0 and P_B^0 are the vapor pressures of pure A and pure B, and P_A and P_B are the partial pressures resulting from molecules of A and of B in the vapor above the solution.

Liquid-liquid solutions obeying this form of Raoult's law are said to be *ideal.* However, as with solutions containing nonvolatile solutes, deviations from Raoult's law are often observed. These can be positive or negative, as shown in Fig. 11.12.

Again, large negative heats of solution indicate especially strong solute-solvent interactions, and such solutions are expected to show *negative* deviations from

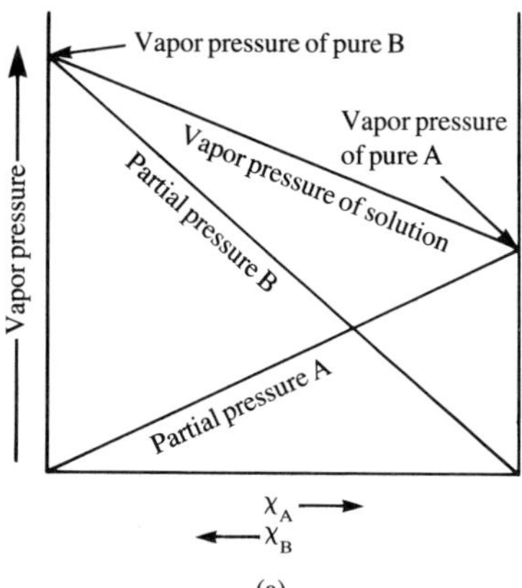

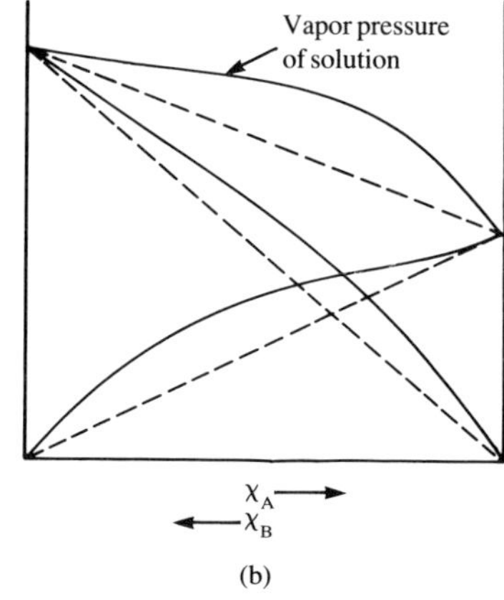

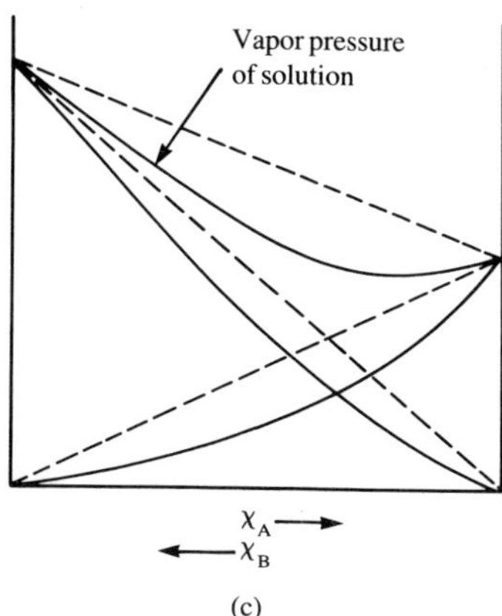

Figure 11.12

Vapor pressure for a solution of two volatile liquids. (a) The behavior predicted for an ideal liquid-liquid solution by Raoult's law. (b) A solution for which P_{TOTAL} is larger than the value calculated from Raoult's law. This solution shows a positive deviation from Raoult's law. (c) A solution for which P_{TOTAL} is smaller than the value calculated from Raoult's law. This solution shows a negative deviation from Raoult's law.

Raoult's law. Both components will have a lower escaping tendency in the solution than in the pure liquids. This behavior is illustrated by an acetone-water solution where the molecules can hydrogen bond effectively.

$$\begin{array}{l} CH_3 \diagdown \qquad\qquad\qquad\quad \diagup H \\ \quad C{=}O{-}{-}{-}H{-}O \\ CH_3 \diagup \quad \delta- \quad \delta+ \end{array}$$

In contrast, if two liquids mix endothermically, this indicates that the solute-solvent interactions are weaker than the interactions among the molecules in the pure liquids. More energy is required to expand the liquids than is released when the liquids are mixed. In this case the molecules in the solution have a higher tendency to escape than expected, and *positive* deviations from Raoult's law are observed. An example of this case is provided by a solution of ethanol and hexane, whose Lewis structures are as follows:

$$H{-}\underset{H}{\overset{H}{C}}{-}\underset{H}{\overset{H}{C}}{-}\ddot{\underset{\cdot\cdot}{O}}{-}H \qquad H{-}\underset{H}{\overset{H}{C}}{-}\underset{H}{\overset{H}{C}}{-}\underset{H}{\overset{H}{C}}{-}\underset{H}{\overset{H}{C}}{-}\underset{H}{\overset{H}{C}}{-}\underset{H}{\overset{H}{C}}{-}H$$

Ethanol Hexane

The polar ethanol and the nonpolar hexane molecules are not able to interact effectively. Thus the enthalpy of solution is positive, as is the deviation from Raoult's law.

Finally, for a solution of very similar liquids, such as benzene and toluene,

Benzene Toluene (CH_3)

the enthalpy of solution is very close to zero, and thus the solution closely obeys Raoult's law (ideal behavior).

A summary of the behavior of various types of solutions is given in Table 11.4.

Summary of the Behavior of Various Types of Solutions

Interactive forces between solute (A) and solvent (B) particles	ΔH_{soln}	ΔT for solution formation	Deviation from Raoult's law	Example
A↔A, B↔B ≡ A↔B	Zero	Zero	None (ideal solution)	Benzene-toluene
A↔A, B↔B < A↔B	Negative (exothermic)	Positive	Negative	Acetone-water
A↔A, B↔B > A↔B	Positive (endothermic)	Negative	Positive	Ethanol-hexane

Table 11.4

Sample Exercise 11.8

A solution is prepared by mixing 5.81 g of acetone (C_3H_6O, mol. wt. = 58.1) and 11.9 g of chloroform ($HCCl_3$, mol. wt. = 119.4). At 35°C this solution has a total vapor pressure of 260 torr. Is this an ideal solution? The vapor pressures of pure acetone and pure chloroform at 35°C are 345 torr and 293 torr, respectively.

Solution

To decide whether or not this solution behaves ideally, we calculate the expected vapor pressure using Raoult's law:

$$P_{\text{TOTAL}} = \chi_A P_A^0 + \chi_C P_C^0$$

where A stands for acetone and C stands for chloroform. The calculated value can then be compared to the observed vapor pressure.

First, we must calculate the number of moles of acetone and chloroform:

$$5.81 \text{ g acetone} \times \frac{1 \text{ mol acetone}}{58.1 \text{ g acetone}} = 0.100 \text{ mol acetone}$$

$$11.9 \text{ g chloroform} \times \frac{1 \text{ mol chloroform}}{119 \text{ g chloroform}} = 0.100 \text{ mol chloroform}$$

Since the solution contains equal numbers of moles of acetone and chloroform:

$$\chi_A = 0.500 \quad \text{and} \quad \chi_C = 0.500$$

The expected vapor pressure is

$$P_{\text{TOTAL}} = (0.500)(345 \text{ torr}) + (0.500)(293 \text{ torr}) = 319 \text{ torr}$$

Comparing this value to the observed pressure of 260 torr shows that the solution does not behave ideally. The observed value is lower than that expected. This negative deviation from Raoult's law can be explained in terms of the following hydrogen bonding interaction:

$(CH_3)_2C{=}O^{\delta-}\cdots H^{\delta+}{-}CCl_3$

Acetone   Chloroform

which lowers the tendency of these molecules to escape from the solution.

11.5 Boiling-Point Elevation and Freezing-Point Depression

Purpose

- To explain the effect of a solute on the boiling and freezing points of a solvent.

In the preceding section we saw how a solute affects the vapor pressure of a liquid solvent. Because changes of state depend on vapor pressure, the presence of a solute also affects the freezing point and boiling point of a solvent. Freezing-point depression, boiling-point elevation, and osmotic pressure (discussed in Section 11.6) are called **colligative properties.** As we will see, they are grouped together because they depend only on the number, and not on the identity, of the solute particles in an ideal solution. Because of their direct relationship to the number of solute particles, the colligative properties are very useful for characterizing the nature of a solute after it is dissolved in a solvent and for determining molecular weights of substances.

The relationships between vapor pressure and changes of state were discussed in Section 10.8.

Boiling-Point Elevation

The normal boiling point of a liquid occurs at the temperature where the vapor pressure is equal to 1 atmosphere. We have seen that a nonvolatile solute lowers the vapor pressure of the solvent. Therefore, such a solution must be heated to a higher temperature than the boiling point of the pure solvent to reach a vapor pressure of 1 atmosphere. This means that *a nonvolatile solute elevates the boiling point of the solvent*. Figure 11.13 shows the phase diagram for an aqueous solution containing a nonvolatile solute. Note that the liquid/vapor line is shifted to higher temperatures than those for pure water.

Normal boiling point was defined in Section 10.8.

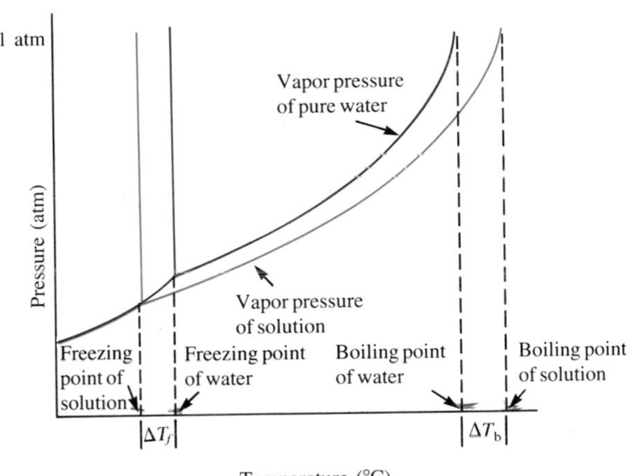

Figure 11.13

Phase diagrams for pure water (red lines) and for an aqueous solution containing a nonvolatile solute (blue lines). Note that the boiling point of the solution is higher than that of pure water. Conversely, the freezing point of the solution is lower than that of pure water. The effect of a nonvolatile solute is to extend the liquid range of a solvent.

As you might expect, the magnitude of the boiling-point elevation depends on the concentration of the solute. The change in boiling point can be represented by the equation

$$\Delta T = K_b m_{\text{solute}}$$

where ΔT is the boiling-point elevation, or the difference between the boiling point of the solution and that of the pure solvent; K_b is a constant that is characteristic of the solvent and is called the **molal boiling-point elevation constant;** and m_{solute} is the *molality* of the solute in the solution.

Values of K_b for some common solvents are given in Table 11.5 on the following page.

The molecular weight of a solute can be determined from the observed boiling-point elevation, as shown in Sample Exercise 11.9.

Solvent	Boiling point (°C)	K_b (°C kg/mol)	Freezing point (°C)	K_f (°C kg/mol)
Water (H_2O)	100.0	0.51	0	1.86
Carbon tetrachloride (CCl_4)	76.5	5.03	−22.99	30.
Chloroform ($CHCl_3$)	61.2	3.63	−63.5	4.70
Benzene (C_6H_6)	80.1	2.53	5.5	5.12
Carbon disulfide (CS_2)	46.2	2.34	−111.5	3.83
Ethyl ether ($C_4H_{10}O$)	34.5	2.02	−116.2	1.79
Camphor ($C_{10}H_{16}O$)	208.0	5.95	179.8	40.

Molal Boiling-Point Elevation Constants (K_b) and Freezing-Point Depression Constants (K_f) for Several Solvents

Table 11.5

Sample Exercise 11.9

A solution was prepared by dissolving 18.00 g of glucose in 150.0 g of water. The resulting solution was found to have a boiling point of 100.34°C. Calculate the molecular weight of glucose. Glucose is a molecular solid that is present as individual molecules in solution.

Solution

We make use of the equation

$$\Delta T = K_b m_{\text{solute}}$$

where

$$\Delta T = 100.34°\text{C} - 100.00°\text{C} = 0.34°\text{C}$$

From Table 11.5, for water, $K_b = 0.51$. The molality of this solution then can be calculated by rearranging the boiling-point elevation equation to give

$$m_{\text{solute}} = \frac{\Delta T}{K_b} = \frac{0.34°\text{C}}{0.51 \dfrac{°\text{C kg}}{\text{mol}}} = 0.67 \text{ mol/kg}$$

The solution was prepared using 0.150 kg of water. Using the definition of molality, we can find the number of moles of glucose in the solution.

$$m_{\text{solute}} = 0.67 \text{ mol/kg} = \frac{\text{mol solute}}{\text{kg solvent}} = \frac{n_{\text{glucose}}}{0.1500 \text{ kg}}$$

$$n_{\text{glucose}} = (0.670 \text{ mol/kg})(0.1500 \text{ kg}) = 0.10 \text{ mol}$$

Thus 0.10 mol of glucose weighs 18.00 g, and 1.0 mol of glucose weighs 180 g (10×18.00 g). The molecular weight of glucose is 180.

Freezing-Point Depression

When a solute is dissolved in a solvent, the freezing point of the solution is lower than that of the pure solvent. Why? Recall that the vapor pressures of ice and liquid water are the same at 0°C. Suppose a solute is dissolved in water. The resulting solution will not freeze at 0°C because *the water in the solution has a lower vapor pressure than that of pure ice*. No ice will form under these conditions. However, the vapor pressure of ice decreases more rapidly than that of liquid water as the temperature decreases. Therefore, as the solution is cooled, the vapor pressure of the ice and that of the liquid water in the solution will eventually become equal. The temperature at which this occurs is the new freezing point of the solution and is below 0°C. The freezing point has been *depressed*.

Melting point and freezing point both refer to the temperature where the solid and liquid coexist.

We can account for this behavior in terms of the simple model shown in Fig. 11.14. The presence of the solute lowers the rate at which molecules in the liquid return to the solid state. Thus, for an aqueous solution, only the liquid state is found at 0°C. As the solution is cooled, the rate at which water molecules leave the solid ice decreases until this rate and the rate of formation of ice become equal and equilibrium is reached. This is the freezing point of the water in the solution.

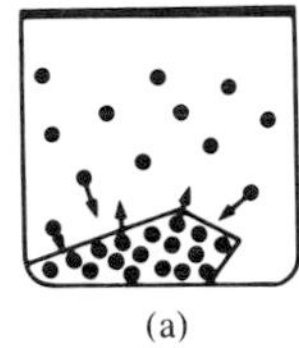
(a)

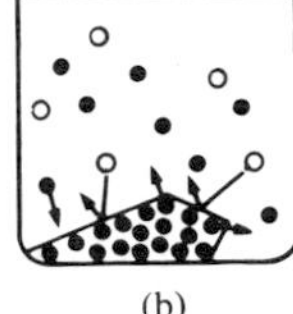
(b)

Figure 11.14

(a) Ice in equilibrium with liquid water. (b) Ice in equilibrium with liquid water containing a dissolved solute (shown in color).

Because a solute lowers the freezing point of water, compounds such as sodium chloride and calcium chloride are often spread on streets and sidewalks to prevent ice from forming in freezing weather. Of course, if the outside temperature is lower than the freezing point of the resulting salt solution, ice forms anyway. So this procedure is not effective at extremely cold temperatures.

The solid/liquid line for an aqueous solution is shown on the phase diagram for water in Fig. 11.13. Since the presence of a solute elevates the boiling point and depresses the freezing point of the solvent, adding a solute has the effect of extending the liquid range.

The equation for freezing-point depression is analogous to that for boiling-point elevation:

$$\Delta T = K_f m_{\text{solute}}$$

where ΔT is the freezing-point depression, or the difference between the freezing point of the pure solvent and that of the solution; and K_f is a constant that is characteristic of a particular solvent and is called the **molal freezing-point depression constant.** Values of K_f for common solvents are listed in Table 11.5.

Like the boiling-point elevation, the observed freezing-point depression can be used to determine molecular weights and to characterize solutions.

Shovelling salt onto an icy sidewalk.

Sample Exercise 11.10

What mass of ethylene glycol ($C_2H_6O_2$, mol. wt. = 62.1), the main component of antifreeze, must be added to 10.0 L of water to produce a solution for use in a car's radiator that freezes at −10.0°F (−23.3°C)? Assume the density of water is exactly 1 g/mL.

Solution

The freezing point must be lowered from 0°C to −23.3°C. To determine the molality of ethylene glycol needed to accomplish this, we can use the equation

$$\Delta T = K_f m_{\text{solute}}$$

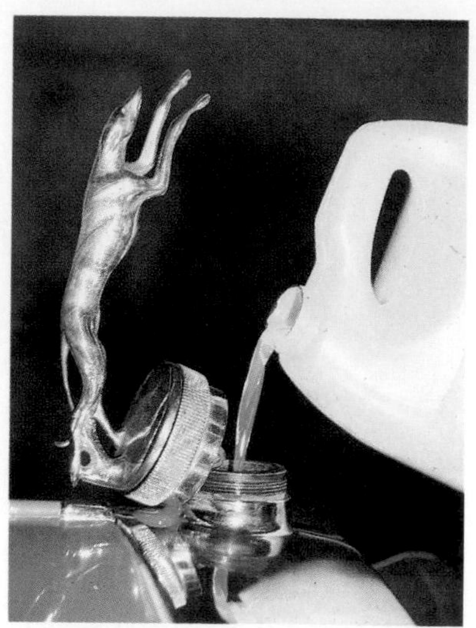

The addition of antifreeze lowers the freezing point of water in a car's radiator in this case, a 1928 Lincoln.

Sample Exercise 11.10, continued

where $\Delta T = 23.3°C$ and $K_f = 1.86$ (from Table 11.5). Solving for the molality gives

$$m_{\text{solute}} = \frac{\Delta T}{K_f} = \frac{23.3°C}{1.86 \dfrac{°C\ \text{kg}}{\text{mol}}} = 12.5\ \text{mol/kg}$$

This means that 12.5 mol of ethylene glycol must be added per kilogram of water. We have 10.0 L, or 10.0 kg, of water. Therefore, the total number of moles of ethylene glycol needed is

$$\frac{12.5\ \text{mol}}{\text{kg}} \times 10.0\ \text{kg} = 1.25 \times 10^2\ \text{mol}$$

The mass of ethylene glycol needed is

$$1.25 \times 10^2\ \text{mol} \times \frac{62.1\ \text{g}}{\text{mol}} = 7.8 \times 10^3\ \text{g (or 7.8 kg)}$$

Sample Exercise 11.11

Thyroxine is a human hormone that controls metabolism. A sample weighing 0.546 g was dissolved in 15.0 g of benzene, and the freezing-point depression was determined to be 0.240°C. Calculate the molecular weight of thyroxine.

Solution

From Table 11.5, K_f for benzene is 5.12°C kg/mol, so the molality of thyroxine is

$$m_{\text{thyroxine}} = \frac{\Delta T}{K_f} = \frac{0.240°C}{5.12°C\ \text{kg/mol}} = 4.69 \times 10^{-2}\ \text{mol/kg}$$

The moles of thyroxine can be obtained from the definition of molality:

$$4.69 \times 10^{-2}\ \text{mol/kg} = m_{\text{solute}} = \frac{\text{mol thyroxine}}{0.0150\ \text{kg benzene}}$$

or

$$\text{mol thyroxine} = \left(4.69 \times 10^{-2}\ \frac{\text{mol}}{\text{kg}}\right)(0.0150\ \text{kg}) = 7.04 \times 10^{-4}\ \text{mol}$$

Since 0.546 g of thyroxine was dissolved, 7.04×10^{-4} mol of thyroxine has a mass of 0.546 g, and

$$\frac{0.546\ \text{g}}{7.04 \times 10^{-4}\ \text{mol}} = \frac{x\ \text{g}}{1.00\ \text{mol}}$$

$$x = 776\ \text{g/mol}$$

Thus the molecular weight of thyroxine is 776.

11.6 Osmotic Pressure

Purpose

- To explain osmosis and describe its applications.

Osmotic pressure, another of the colligative properties, can be understood from Fig. 11.15. A solution and pure solvent are separated by a **semipermeable membrane,** which allows *solvent but not solute* molecules to pass through. As time passes, the volume of the solution increases and that of the solvent decreases. This flow of solvent into the solution through the semipermeable membrane is called **osmosis.** Eventually the liquid levels stop changing, indicating that the system has reached equilibrium. Because the liquid levels are different at this point, there is a greater hydrostatic pressure on the solution than on the pure solvent. This excess pressure is called the **osmotic pressure.**

We can take another view of this phenomenon, as illustrated in Fig. 11.16. Osmosis can be prevented by applying a pressure to the solution. *The pressure that just stops the osmosis is equal to the osmotic pressure of the solution.* A simple model to explain osmotic pressure can be constructed as shown in Fig. 11.17. The membrane allows only solvent molecules to pass through. However, the initial rates of solvent transfer to and from the solution are not the same. The solute particles interfere with the passage of solvent, so the rate of transfer is slower from the solution to the solvent than in the reverse direction. Thus there is a net transfer of solvent molecules into the solution, which causes the solution volume to increase. As the solution level rises in the tube, the resulting pressure exerts an extra "push" on the solvent molecules in the solution, forcing them back through the membrane. Eventually, enough pressure develops so that the solvent transfer becomes equal in both directions. At this point, equilibrium is achieved and the levels stop changing.

Osmotic pressure can be used to characterize solutions and determine molecular weights as can the other colligative properties, but osmotic pressure is particularly useful because a small concentration of solute produces a relatively large osmotic pressure.

Experiments show that the dependence of the osmotic pressure on solution concentration is represented by the equation

$$\pi = MRT$$

where π is the osmotic pressure in atmospheres, M is the molarity of the solute, R is the gas law constant, and T is the Kelvin temperature.

A molecular weight determination using osmotic pressure is illustrated in Sample Exercise 11.12.

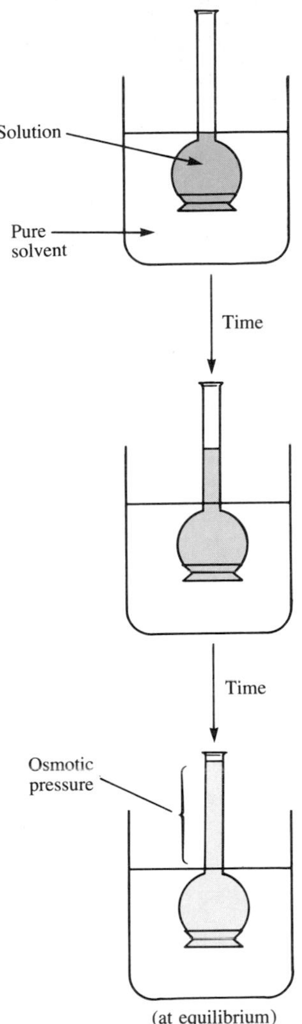

Figure 11.15

A tube with a bulb on the end is covered by a semipermeable membrane. The solution is inside the tube and is bathed in the pure solvent. There is a net transfer of solvent molecules into the solution until the hydrostatic pressure equalizes the solvent flow in both directions. The solvent level shows little change because the narrow stem of the bulb has a very small volume.

Sample Exercise 11.12

To determine the molecular weight of a certain protein, 1.00×10^{-3} g of it was dissolved in enough water to make 1.00 mL of solution. The osmotic pressure of this solution was found to be 1.12 torr at 25.0°C. Calculate the molecular weight of the protein.

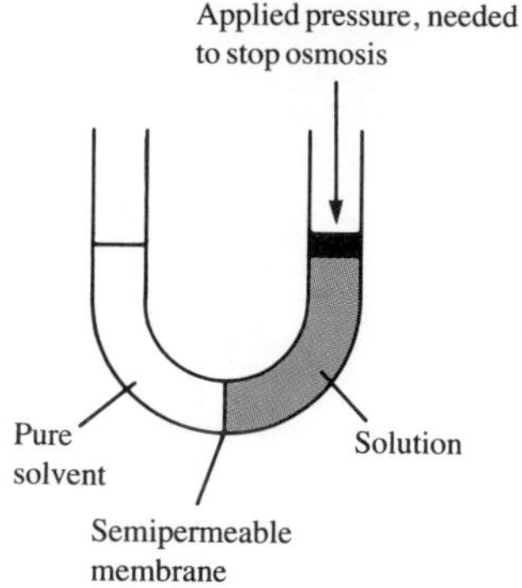

Figure 11.16

The normal flow of solvent into the solution (osmosis) can be prevented by applying an external pressure to the solution. The pressure required to stop the osmosis is equal to the osmotic pressure of the solution.

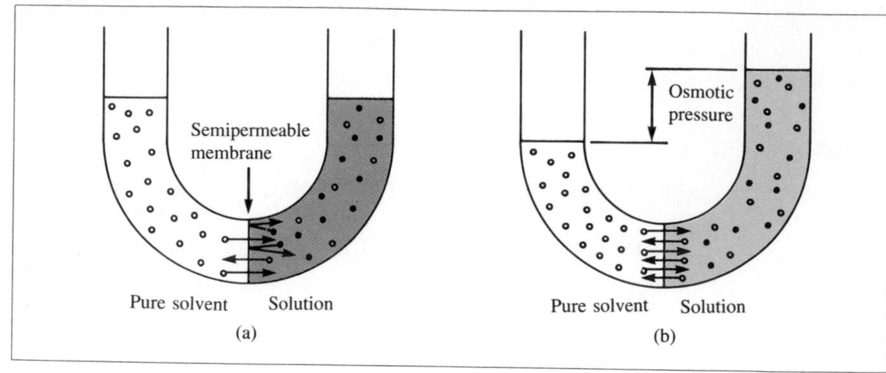

Figure 11.17

(a) A pure solvent and its solution with a nonvolatile solute are separated by a semipermeable membrane through which solvent molecules (○) can pass but solute molecules (●) cannot. The rate of solvent transfer is greater from solvent to solution than from solution to solvent. (b) The system at equilibrium, where the rate of solvent transfer is the same in both directions.

Sample Exercise 11.12, continued

Solution

We use the equation

$$\pi = MRT$$

In this case we have

$$\pi = 1.12 \text{ torr} \times \frac{1 \text{ atm}}{760 \text{ torr}} = 1.47 \times 10^{-3} \text{ atm}$$

$$R = 0.08206 \text{ L atm/K mol}$$

$$T = 25.0 + 273 = 298 \text{ K}$$

Note that the osmotic pressure must be converted to atmospheres because of the units of R.

Solving for M gives

$$M = \frac{1.47 \times 10^{-3} \text{ atm}}{(0.08206 \text{ L atm/K mol})(298 \text{ K})} = 6.01 \times 10^{-5} \text{ mol/L}$$

Since 1.00×10^{-3} g of protein was dissolved in 1 mL of solution, the mass of protein per liter of solution is 1.00 g. The solution's concentration is 6.01×10^{-5} mol/L. This concentration is produced from 1.00×10^{-3} g of protein per milliliter, or 1.00 g/L. Thus 6.01×10^{-5} mol of protein has a mass of 1.00 g and

$$\frac{1.00 \text{ g}}{6.01 \times 10^{-5} \text{ mol}} = \frac{x \text{ g}}{1.00 \text{ mol}}$$

$$x = 1.66 \times 10^4$$

Measurements of osmotic pressure generally give much more accurate molecular weight values than those from freezing-point or boiling-point changes.

The molecular weight of the protein is 1.66×10^4. This molecular weight may seem very large, but it is relatively small for a protein.

In osmosis a semipermeable membrane prevents transfer of *all* solute particles. A similar phenomenon called **dialysis** occurs at the walls of most plant and animal cells. However, in this case the membrane allows transfer of both solvent molecules and *small* solute molecules and ions. One of the most important applications of dialysis is the use of artificial kidney machines to purify the blood. The blood is passed through a cellophane tube, which acts as the semipermeable membrane. The tube is immersed in a dialyzing solution (see Fig. 11.18). This "washing" solution contains the same concentrations of ions and small molecules as blood, but has none of the waste products normally removed by the kidneys. The resulting dialysis of waste products cleanses the blood.

Solutions that have identical osmotic pressures are said to be **isotonic solutions.** Fluids administered intravenously must be isotonic with body fluids. For example, if cells are bathed in a hypertonic solution, which is a solution having an osmotic pressure higher than that of the cell fluids, the cells will shrivel because of a net transfer of water out of the cells. This phenomenon is called *crenation*. The opposite phenomenon, called *lysis,* occurs when cells are bathed in a hypotonic solution, a solution with an osmotic pressure lower than that of the cell fluids. In this case, the cells rupture because of the flow of water into the cells.

The brine used in pickling causes the cucumbers to shrivel.

We can use the phenomenon of crenation to our advantage. Food can be preserved by treating its surface with a solute that gives a solution that is hypertonic to bacteria cells. Bacteria on the food then tend to shrivel and die. This is why salt can be used to protect meat and sugar can be used to protect fruit.

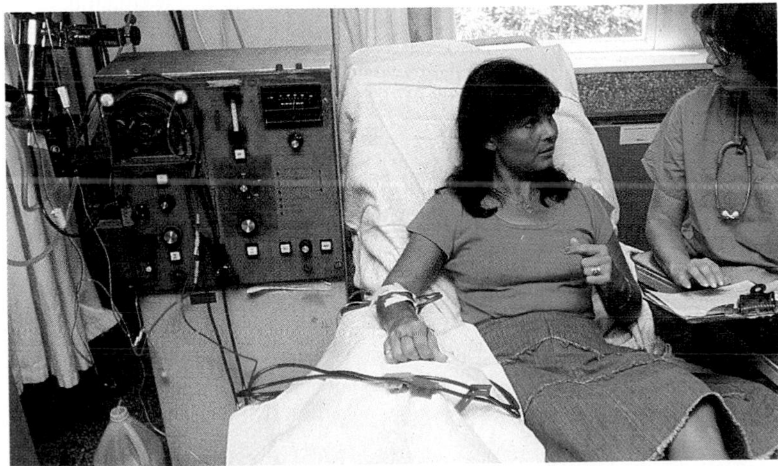

A patient undergoing hemodialysis.

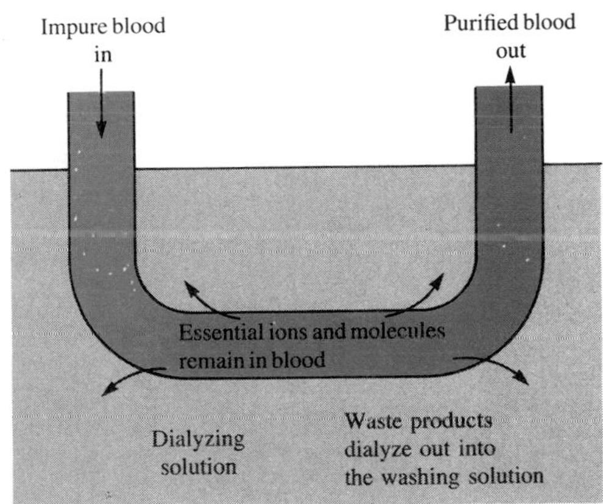

Figure 11.18

Representation of the functioning of an artificial kidney.

Sample Exercise 11.13

What concentration of sodium chloride in water is needed to produce an aqueous solution isotonic with blood ($\pi = 7.70$ atm at 25°C)?

Solution

We can calculate the molarity of the solute from the equation

$$\pi = MRT \qquad \text{or} \qquad M = \frac{\pi}{RT}$$

Sample Exercise 11.13, continued

$$M = \frac{7.70 \text{ atm}}{(0.08206 \text{ L atm/K mol})(298 \text{ K})} = 0.315 \text{ mol/L}$$

This represents the total molarity of solute particles. But NaCl gives two ions per formula unit. Therefore, the concentration of NaCl needed is 0.315/2, or 0.158 *M*. That is,

$$\underset{0.1575\ M}{\text{NaCl}} \rightarrow \underbrace{\underset{0.1575\ M}{\text{Na}^+} + \underset{0.1575\ M}{\text{Cl}^-}}_{0.315\ M}$$

Reverse Osmosis

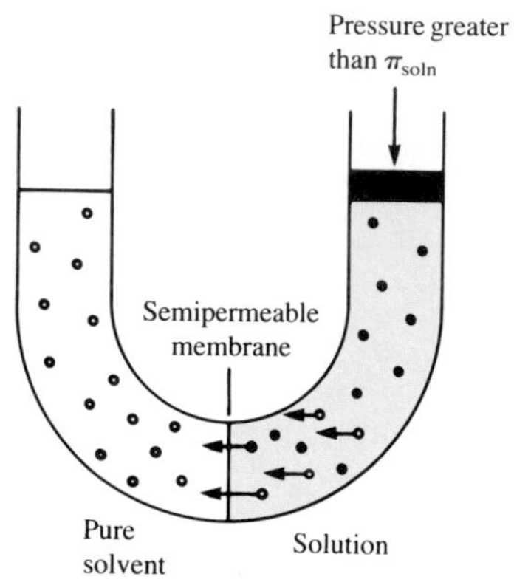

Figure 11.19

Reverse osmosis. A pressure greater than the osmotic pressure of the solution is applied, which causes a net flow of solvent molecules (○) from the solution to the pure solvent. The solute molecules (●) remain behind.

If a solution in contact with pure solvent across a semipermeable membrane is subjected to an external pressure larger than its osmotic pressure, **reverse osmosis** occurs. The pressure will cause a net flow of solvent from the solution to the solvent, as shown in Fig. 11.19. In reverse osmosis the semipermeable membrane acts as a "molecular filter" to remove solute particles. This fact is potentially applicable to the **desalination** (removal of dissolved salts) of sea water (Fig. 11.20), which is highly hypertonic to body fluids and thus is not drinkable.

As the population of the sun belt areas of the United States increases, more demand will be placed on the limited supplies of fresh water there. One obvious source of fresh water is from the desalination of sea water. Various schemes have been suggested, including solar evaporation, reverse osmosis, and even a plan for towing icebergs from Antarctica (remember that pure water freezes out of a solution). The problem, of course, is that all of the available processes are expensive and so far only small-scale operations have been attempted. However, as water shortages increase, desalination might become necessary.

A small-scale, manually operated, reverse osmosis desalinator has been developed by the U.S. Navy to provide fresh water on life rafts (Fig. 11.21). Potable water can be supplied by this desalinator at the rate of 1.25 gallons of water per hour—enough to keep 25 people alive. This compact desalinator weighs only 10 pounds and will replace the bulky cases of fresh water now stored in Navy life rafts.

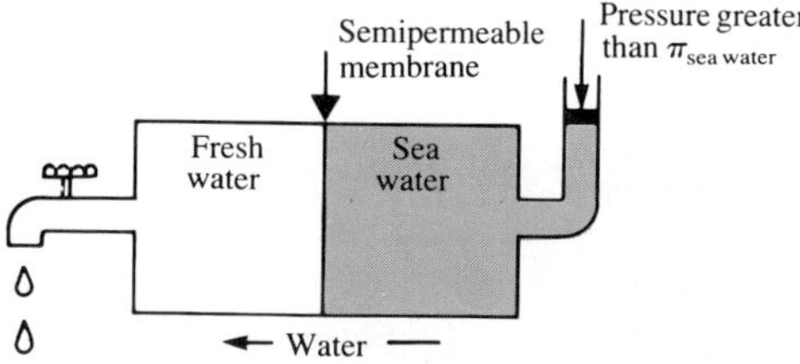

Figure 11.20

Desalination of sea water. When a pressure greater than the osmotic pressure of sea water is applied, water is forced through the membrane, leaving the solutes (dissolved salts such as sodium chloride and magnesium chloride) behind.

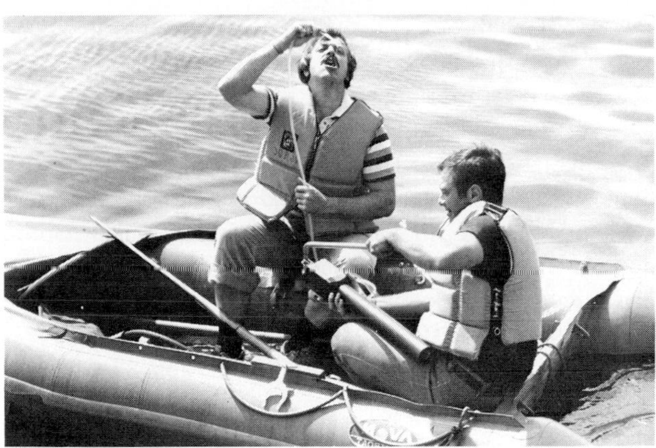

Figure 11.21

The essential part of the small desalinator developed by the Navy for life rafts is a cellophane-like membrane wrapped around a tube with holes in it. Sea water is forced through the tube by a hand-operated pump at pressures of about 70 atmospheres to produce water with a salt content only 40% higher than that from a typical tap.

An iceberg floating in Baffin Bay, in the Arctic. Pure water freezes out of a solution.

11.7 Colligative Properties of Electrolyte Solutions

Purpose

- To show how the colligative properties of electrolyte solutions can be used to characterize the solute.

As we have seen previously, the colligative properties of solutions depend on the total concentration of solute particles. For example, a 0.10 *m* glucose solution shows a freezing-point depression of 0.186°C:

$$\Delta T = K_f m = (1.86°\text{C kg/mol})(0.10 \text{ mol/kg}) = 0.186°\text{C}$$

On the other hand, a 0.10 *m* sodium chloride solution should show a freezing-point depression of 0.37°C since the solution is 0.10 *m* Na^+ ions and 0.10 *m* Cl^- ions. Therefore, the solution is 0.20 *m* in solute particles and $\Delta T = (1.86°\text{C kg/mol})(0.20 \text{ mol/kg}) = 0.37°\text{C}$.

The relationship between the moles of solute dissolved and the moles of particles in solution is usually expressed using the **van't Hoff factor** (*i*):

Dutch chemist J. H. van't Hoff (1852–1911) received the first Nobel prize in chemistry in 1901.

$$i = \frac{\text{moles of particles in solution}}{\text{moles of solute dissolved}}$$

The *expected* value for *i* can be calculated for a salt by noting the number of ions per formula unit. For example, for NaCl, *i* is 2; for K_2SO_4, *i* is 3; and for $Fe_3(PO_4)_2$, *i* is 5. These calculated values assume that when a salt dissolves, it

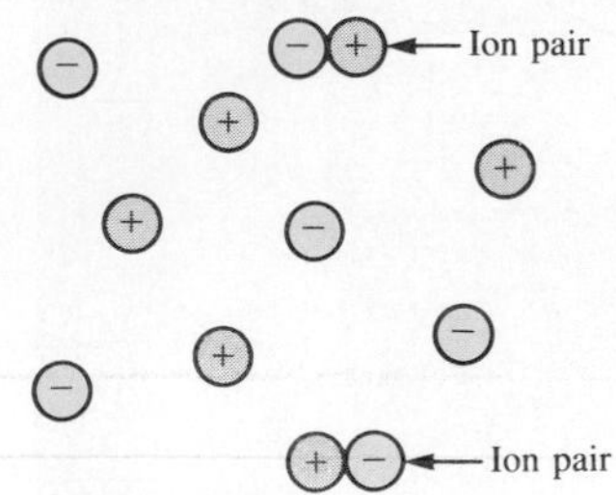

Figure 11.22

In an aqueous solution a few ions aggregate, forming ion pairs that behave as a unit.

completely dissociates into its component ions, which then move around independently. This assumption is not always true. For example, the freezing-point depression observed for 0.10 *m* NaCl is 1.87 times that for 0.10 *m* glucose rather than twice as great. That is, for a 0.10 *m* NaCl solution the observed value for i is 1.87 rather than 2. Why? The best explanation is that **ion pairing** occurs in solution (see Fig. 11.22). At a given instant a small percentage of the sodium and chloride ions are paired and thus count as a single particle. In general, ion pairing is most important in concentrated solutions. As the solution becomes more dilute, the ions are farther apart and less ion pairing occurs. For example, in a 0.0010 *m* NaCl solution, the observed value of i is 1.97, which is very close to the expected value.

Ion pairing occurs to some extent in all electrolyte solutions. Table 11.6 shows expected and observed values of i for a given concentration of various electrolytes. Note that the deviation of i from the expected value tends to be greatest where the ions have multiple charges. This is expected because ion pairing ought to be most important for highly charged ions.

Expected and Observed Values of the van't Hoff Factor for 0.05 *m* Solutions of Several Electrolytes

Electrolyte	i (expected)	i (observed)
NaCl	2.0	1.9
$MgCl_2$	3.0	2.7
$MgSO_4$	2.0	1.3
$FeCl_3$	4.0	3.4
HCl	2.0	1.9
Glucose*	1.0	1.0

*A nonelectrolyte shown for comparison.

Table 11.6

The colligative properties of electrolyte solutions are described by including the van't Hoff factor in the appropriate equation. For example, for changes in freezing and boiling points, the modified equation is

$$\Delta T = imK$$

where K represents the freezing-point depression or boiling-point elevation constant for the solvent.

For the osmotic pressure of electrolyte solutions, the equation is

$$\pi = iMRT$$

Sample Exercise 11.14

The observed osmotic pressure for a 0.10 *M* solution of $Fe(NH_4)_2(SO_4)_2$ at 25°C is 10.8 atm. Compare the expected and experimental values for i.

Solution

The ionic solid $Fe(NH_4)_2(SO_4)_2$ dissociates in water to produce 5 ions:

$$Fe(NH_4)_2(SO_4)_2 \xrightarrow{H_2O} Fe^{2+} + 2NH_4^{+} + 2SO_4^{2-}$$

Sample Exercise 11.14, continued

Thus the expected value for i is 5. We can obtain the experimental value for i by using the equation for osmotic pressure

$$\pi = iMRT \quad \text{or} \quad i = \frac{\pi}{MRT}$$

where $\pi = 10.8$ atm, $M = 0.10$ mol/L, $R = 0.08206$ L atm/K mol, and $T = 25 + 273 = 298$ K. Substituting these values into the equation gives

$$i = \frac{\pi}{MRT} = \frac{10.8 \text{ atm}}{\left(0.10 \frac{\text{mol}}{\text{L}}\right)\left(0.08206 \frac{\text{L atm}}{\text{K mol}}\right)(298 \text{ K})} = 4.42$$

The experimental value for i is less than the expected value, presumably because of ion pairing.

11.8 Colloids

Purpose

- To define a colloid and explain how it is stabilized.

Mud can be suspended in water by vigorous stirring. When the stirring stops, most of the particles rapidly settle out, but even after several days some of the smallest particles remain suspended. Although undetected in normal lighting, their presence can be demonstrated by shining a beam of intense light through the suspension. The beam is visible from the side because the light is scattered by the suspended particles (Fig. 11.23). In a true solution, on the other hand, the beam is

Figure 11.23
The Tyndall effect.

Types of Colloids			
Examples	Dispersing medium	Dispersed substance	Colloid type
Fog, aerosol sprays	Gas	Liquid	Aerosol
Smoke, airborne bacteria	Gas	Solid	Aerosol
Whipped cream, soap suds	Liquid	Gas	Foam
Milk, mayonnaise	Liquid	Liquid	Emulsion
Paint, clays, gelatin	Liquid	Solid	Sol
Marshmallow, polystyrene foam	Solid	Gas	Solid foam
Butter, cheese	Solid	Liquid	Solid emulsion
Ruby glass	Solid	Solid	Solid sol

Table 11.7

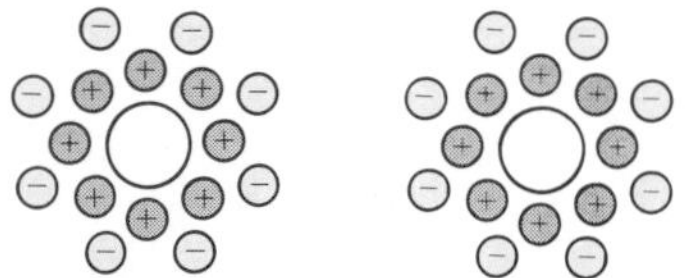

Figure 11.24

A representation of two colloidal particles. In each the center particle is surrounded by a layer of positive ions with negative ions in the outer layer. Thus although the particles are electrically neutral, they still repel each other because of their outer negative layer of ions.

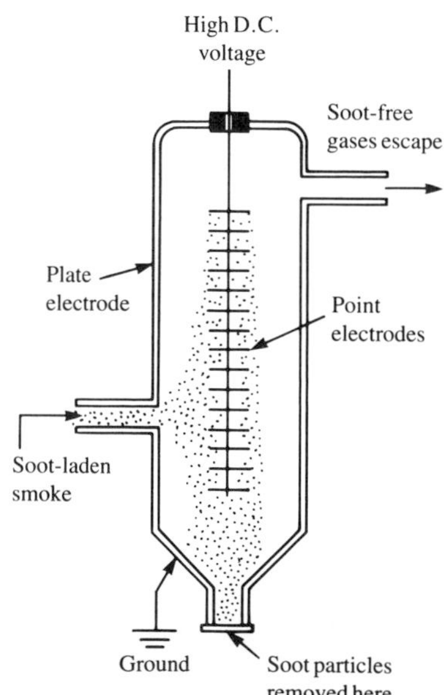

Figure 11.25

The Cottrell precipitator installed in a smokestack. The charged plates attract the colloidal particles because of their ion layers and thus remove them from the smoke.

invisible from the side because the individual ions and molecules dispersed in the solution are too small to scatter visible light.

The scattering of light by particles is called the **Tyndall effect** and is often used to distinguish between a suspension and a true solution.

A suspension of tiny particles in some medium is called a *colloidal dispersion,* or a **colloid.** The suspended particles are single large molecules or aggregates of molecules or ions ranging in size from 1 to 1000 nanometers. Colloids are classified according to the states of the dispersed phase and the dispersing medium. Table 11.7 summarizes various types of colloids.

What stabilizes a colloid? Why do the particles remain suspended rather than forming larger aggregates and precipitating out? The answer is complicated, but the main factor seems to be *electrostatic repulsion*. A colloid, like all other macroscopic substances, is electrically neutral. However, when a colloid is placed in an electric field, the dispersed particles all migrate to the same electrode and thus must all have the same charge. How is this possible? The center of a colloidal particle (a tiny ionic crystal, a group of molecules, or a single large molecule) attracts from the medium a layer of ions, all of the same charge. This group of ions in turn attracts another layer of oppositely charged ions, as shown in Fig. 11.24. Because the colloidal particles all have an outer layer of ions with the same charge, they repel each other and do not easily aggregate to form particles that are large enough to precipitate.

The destruction of a colloid, called **coagulation,** can usually be accomplished either by heating or by adding an electrolyte. Heating increases the velocities of the colloidal particles, causing them to collide with enough energy that the ion barriers are penetrated and the particles can aggregate. Because this is repeated many times, the particle grows to a point where it settles out. Adding an electrolyte neutralizes the adsorbed ion layers. This is why clay suspended in rivers is deposited where the river reaches the ocean, forming the deltas characteristic of large rivers like the Mississippi. The high salt content of the sea water causes the colloidal clay particles to coagulate.

The removal of soot from smoke is another example of the coagulation of a colloid. When smoke is passed through an electrostatic precipitator (Figure 11.25), the suspended solids are removed. The use of precipitators has produced an immense improvement in the air quality of heavily industrialized cities.

Chemical Impact

Organisms and Ice Formation

The ice-cold waters of the polar oceans are teeming with fish that seem immune to freezing. One might think that these fish have some kind of antifreeze in their blood. However, studies show that they are protected from freezing in a very different way from the way antifreeze protects our cars. As we have seen in this chapter, solutes such as sugars, salts, and ethylene glycol lower the temperature at which the solid and liquid phases of water can coexist. However, the fish could not tolerate high concentrations of solutes in their blood because of the osmotic pressure effects. Instead, they are protected by proteins in their blood. These proteins allow the water in the bloodstream to be supercooled—exist below 0°C—without forming ice. They apparently coat the surface of each tiny ice crystal, as soon as it begins to form, preventing it from growing to a size that would cause biological damage.

Although it might at first seem surprising, this research on polar fish has attracted the attention of ice cream manufacturers. Premium quality ice cream is smooth; it does not have large ice crystals in it. The makers of ice cream would like to incorporate these polar fish proteins, or molecules that behave similarly, into ice cream to prevent growth of ice crystals during storage.

Fruit and vegetable growers have a similar interest: they also want to prevent ice formation that damages their crops during an unusual cold wave. However, this is a very different kind of problem than keeping polar fish from freezing. Many types of fruit and vegetables are colonized by bacteria that manufacture a protein that *encourages* freezing by acting as a nucleating agent to start an ice crystal. Chemists have identified the offending protein in the bacteria and the gene that is responsible for making it. They have learned to modify the genetic material of these bacteria in such a way that removes their ability to make the protein that encourages ice crystal formation. If testing shows that these modified bacteria have no harmful effects on the crop or the environment, the original bacteria strain will be replaced with the new form so that ice crystals will not form so readily when a cold snap occurs.

An Antarctic fish, *Chaerophalus aceratus*.

FOR REVIEW

Summary

Solution composition can be described in terms of molarity (the moles per liter of solution), mass percent (the ratio of grams of solute to grams of solution multiplied by 100), mole fraction (the moles of a given component per total number of moles of all components of the solution), or molality (the moles of solute per kilogram of solvent).

Normality is defined as the number of equivalents per liter of solution. The definition of an equivalent depends on the reaction taking place in the solution. For an acid-base reaction, the equivalent is the mass of acid (or base) that can furnish (or accept) exactly 1 mole of protons. For an oxidation-reduction reaction, the equivalent is the quantity of reducing (or oxidizing) agent that can furnish (or accept) 1 mole of electrons.

The enthalpy of solution (ΔH_{soln}) is the enthalpy change accompanying solution formation. It can be broken down into terms expressing energies associated with overcoming interactions between solute particles and interactions between solvent particles (endothermic processes) and a term expressing energies associated with the formation of interactions between solute and solvent particles (an exothermic process).

Various factors affect solubility. Molecular structure determines the polarity of a molecule. Polar molecules tend to be soluble in polar solvents and nonpolar molecules tend to be soluble in nonpolar solvents. Pressure has little effect on the solubilities of liquids or solids, but increased pressure increases the solubility of gases. Henry's law, which applies to gases that do not react chemically with the solvent, states that the concentration of a gaseous solute in a solvent is directly proportional to the partial pressure of the same gas above the solution. An increase in temperature decreases the solubility of a gas, but the effect of temperature on the solubility of a solid varies with the identity of the solid.

The vapor pressure of a solution containing a nonvolatile solute is always less than the vapor pressure of the pure solvent, since the presence of the solute decreases the number of solvent molecules per unit volume and thus proportionately lowers the escaping tendency of the solvent molecules. Raoult's law states that the vapor pressure of a solution is directly proportional to the mole fraction of the solvent. Raoult's law holds for an ideal solution. However, if a solute has a special affinity for the solvent, nonideal behavior is seen in the form of a negative deviation from Raoult's law (a lower vapor pressure than is predicted).

The colligative properties of solutions, which depend on the number of solute particles present, include boiling-point elevation, freezing-point depression, and osmotic pressure. Because a nonvolatile solute lowers the freezing point and raises the boiling point, it extends the liquid state of the solvent.

Osmosis occurs when a pure solvent and its solution are separated by a semipermeable membrane that prevents the passage of solute molecules. Under these circumstances solvent molecules flow from the solvent into the solution. Osmotic pressure, the exact pressure that must be applied to a solution to prevent osmosis, is directly related to the molarity of the solution. Reverse osmosis occurs when an external pressure larger than the solution's osmotic pressure is applied. The effect is to cause a net flow of solvent from the solution into the pure solvent.

To describe the colligative properties of electrolyte solutions, the number of ions must be taken into account by use of the van't Hoff factor (i). Because of ion pairing, the experimental value for i is often significantly less than the expected value.

A colloid is a suspension of tiny particles in solution. Colloids are stabilized because of the electrostatic repulsions between the ion layers surrounding the individual particles. A colloid can be coagulated by heating or by adding an electrolyte.

Key Terms

Section 11.1
molarity
mass percent
mole fraction
molality
normality

Section 11.2
enthalpy (heat) of solution
enthalpy (heat) of hydration

Section 11.3
Henry's law
thermal pollution

Section 11.4
Raoult's law
ideal solution

Section 11.5
colligative properties
molal boiling-point elevation constant
molal freezing-point depression constant

Section 11.6
semipermeable membrane
osmosis
osmotic pressure
dialysis
isotonic solution
reverse osmosis
desalination

Section 11.7
van't Hoff factor
ion pairing

Section 11.8
Tyndall effect
colloid
coagulation

Exercises

A blue exercise number indicates that the answer to that exercise appears at the back of this book and a solution appears in the Solutions Guide.

Solution Review: If you have trouble with these exercises, review Sections 4.1 to 4.3 in Chapter 4.

1. A solution is prepared by dissolving 125 g of sucrose, $C_{12}H_{22}O_{11}$, in enough water to produce 1.00 L of solution. What is the molarity of this solution?

2. What mass of sodium oxalate, $Na_2C_2O_4$, is needed to prepare 0.250 L of a 0.100 *M* solution?

3. A solution is prepared by diluting 25.00 mL of a 0.308 *M* solution of $NiCl_2$ to a final volume of 0.500 L. What is the concentration of nickel chloride in this solution? What are the concentrations of nickel ions and chloride ions in this solution?

4. A 158.5-mg sample of pure Cu metal is dissolved in a small amount of concentrated nitric acid giving Cu^{2+} ions in solution. It is then diluted to a final volume of 1.00 L.
   a. What is the molar concentration of Cu^{2+} ions in the final solution?
   b. Assuming the density of the solution is 1.0 g/cm^3, what is the concentration of Cu^{2+} ions in parts per million? (See Exercise 15 in Chapter 4.)

5. An aqueous solution contains 2.8 mg of Cd^{2+} ions per mL of solution.
   a. What is the concentration of Cd^{2+} ions in ppm and ppb? (See Exercise 15 in Chapter 4.)
   b. What is the concentration of Cd^{2+} ions in mol/L?

6. A solution is 1.06×10^{-3} *M* in $Ca(NO_3)_2$.
   a. What are the molar concentrations of calcium cations and nitrate anions in this solution?
   b. What is the mass (in grams) of the Ca^{2+} ions in one milliliter of this solution?
   c. How many nitrate ions are there in 1.0×10^{-6} L (1 μL) of this solution?

7. What volume of 0.25 *M* HCl solution must be diluted to prepare 1.00 L of 0.040 *M* HCl?

8. Write equations showing the ions present when the following strong electrolytes are dissolved in water:

   a. HNO_3 d. $SrBr_2$ g. NH_4NO_3
   b. Na_2SO_4 e. $KClO_4$ h. $CuSO_4$
   c. $AlCl_3$ f. NH_4Br i. NaOH

Concentration of Solutions

9. A commonly purchased disinfectant is a 3.0% (by mass) solution of hydrogen peroxide (H_2O_2) in water. Assuming the density of the solution is 1.0 g/cm^3, calculate the molarity, molality, and mole fraction of H_2O_2.

10. Common commercial acids and bases are aqueous solutions with the following properties:

	Density (g/cm^3)	Mass percent of solute
Hydrochloric acid	1.19	38
Nitric acid	1.42	70.
Sulfuric acid	1.84	95
Acetic acid	1.05	99
Ammonia	0.90	28

Calculate the molarity, molality, and mole fraction of each of the above reagents.

11. A solution is prepared by mixing 50.0 mL of toluene ($C_6H_5CH_3$, $d = 0.867$ g/cm^3) with 125 mL of benzene (C_6H_6, $d = 0.874$ g/cm^3). Assuming that the volumes add upon mixing, what are the mass percent, mole fraction, molality, and molarity of the toluene?

12. An aqueous antifreeze solution is 40.0% ethylene glycol ($C_2H_6O_2$) by mass. The density of the solution is 1.05 g/cm^3. Calculate the molality, molarity, and mole fraction of the ethylene glycol.

13. A 1.37 *M* solution of citric acid ($H_3C_6H_5O_7$) in water has a density of 1.10 g/cm^3. What are the mass percent, molality, and mole fraction of the citric acid?

14. Is molarity or molality dependent on temperature? Explain your answer. Why is molality, and not molarity, used in the equations describing freezing-point depression and boiling-point elevation?

15. What are the molarity and mole fraction of a 1.0 *m* solution of acetone (CH_3COCH_3) in ethanol (C_2H_5OH)? (Density of acetone = 0.788 g/cm^3; density of ethanol = 0.789 g/cm^3) Assume the volumes of acetone and ethanol add.

16. In a solution of acetone and ethanol, the mole fraction of acetone is 0.40. What is the concentration of acetone as a mass percent?

17. A solution is made by dissolving 25 g of NaCl in enough water to make 1.0 L of solution. Assume the density of the solution is 1.0 g/cm^3. Calculate the mass percent, molarity, molality, and mole fraction of NaCl.

18. A solution is prepared by dissolving 50.0 g of cesium chloride (CsCl) in 50.0 g of water. The density of the solution is 1.58 g/cm^3. What are the mass percent, molarity, molality, and mole fraction of the cesium chloride?

Energetics of Solutions and Solubility

19. What specific feature of chemical substances do we refer to when we say that "like dissolves like"?

20. Rationalize the trend in water solubility for the following simple alcohols:

Alcohol	Solubility (g/100 g H_2O at 20°C)
Methanol, CH_3OH	Soluble in all proportions
Ethanol, CH_3CH_2OH	Soluble in all proportions
Propanol, $CH_3CH_2CH_2OH$	Soluble in all proportions
Butanol, $CH_3(CH_2)_2CH_2OH$	8.14
Pentanol, $CH_3(CH_2)_3CH_2OH$	2.64
Hexanol, $CH_3(CH_2)_4CH_2OH$	0.59
Heptanol, $CH_3(CH_2)_5CH_2OH$	0.09

21. Which ion in each of the following pairs would you expect to be more strongly hydrated? Why?
    a. Na^+ or Mg^{2+}
    b. Mg^{2+} or Be^{2+}
    c. Fe^{2+} or Fe^{3+}
    d. F^- or Br^-
    e. Cl^- or ClO_4^-
    f. ClO_4^- or SO_4^{2-}

22. The lattice energy* of KCl is −715 kJ/mol and the enthalpy of hydration is −684 kJ/mol. Calculate the enthalpy of solution per mole of solid KCl. Describe the process to which this enthalpy change applies.

23. Use the following data to calculate the enthalpy of hydration for cesium iodide and cesium hydroxide.

	Lattice energy*	ΔH_{soln}
CsI(*s*)	−604 kJ/mol	33 kJ/mol
CsOH(*s*)	−724 kJ/mol	−72 kJ/mol

24. Based on your answers to Exercise 23, which ion, OH^- or I^-, is more strongly hydrated?

25. What factors cause one solute to be more strongly hydrated than another? For each of the following pairs predict which substance would be more soluble in water:
    a. CH_3CH_2OH or $CH_3CH_2CH_3$
    b. $CHCl_3$ or CCl_4
    c. CH_3CO_2H or $CH_3(CH_2)_{14}CO_2H$
    d. Na_2S or CuS
    e. $AlCl_3$ or Al_2O_3
    f. CO_2 or SiO_2

*Lattice energy was defined in Chapter 8 as the energy change for the process:

$$M^+(g) + X^-(g) \rightarrow MX(s)$$

26. Which solvent, water or carbon tetrachloride, would you choose to dissolve each of the following:
    a. $Cu(NO_3)_2$
    b. CS_2
    c. CH_3CO_2H
    d. $CH_3(CH_2)_{16}CH_2OH$
    e. HCl
    f. C_6H_6

27. While $Al(OH)_3$ is insoluble in water, NaOH is very soluble. Explain in terms of lattice energies.

28. In flushing and cleaning columns used in liquid chromatography, a series of solvents is used. Hexane, chloroform, methanol, and water are passed through the column in that order. Rationalize the order in terms of intermolecular forces and the mutual solubility (miscibility) of the solvents.

29. A typical detergent is sodium dodecylsulfate, or SDS, $CH_3(CH_2)_{10}CH_2SO_4^-Na^+$. In aqueous solution, small aggregates of detergent anions called *micelles* form. Propose a structure for the micelles.

30. Define hydrophobic and hydrophilic.

31. Is Henry's law valid for the solubility of CO_2 in a basic solution? *Hint:* $CO_2 + OH^- \rightarrow HCO_3^-$.

32. Calculate the solubility of O_2 in water at a partial pressure of O_2 of 120 torr at 25°C. The Henry's law constant for O_2 is 7.8×10^2 atm L/mol.

33. Rationalize the temperature dependence of the solubility of a gas in terms of the kinetic molecular theory.

Vapor Pressures of Solutions

34. A solution is prepared by mixing 50.0 g of glucose with 600.0 g of water. What is the vapor pressure of this solution at 25°C? (At 25°C the vapor pressure of pure water is 23.8 torr.)

35. A solution is made by mixing 50.0 g of acetone and 50.0 g of methanol. What is the vapor pressure of this solution at 25°C? What is the composition of the vapor expressed as a mole fraction? Assume ideal solution and gas behavior. (At 25°C the vapor pressures of pure acetone and pure methanol are 271 torr and 143 torr, respectively.) Is is reasonable to assume that this is an ideal solution? Why or why not?

36. Which of the following will have the lowest total vapor pressure at 25°C?
    a. pure water
    b. a solution of glucose in water with $\chi_{glucose} = 0.01$
    c. a solution of sodium chloride in water with $\chi_{NaCl} = 0.01$
    d. a solution of methanol in water with $\chi_{CH_3OH} = 0.2$ [Consider the vapor pressure of both methanol (see Exercise 35) and water.]

37. In terms of intermolecular forces, what gives rise to positive and negative deviations from Raoult's law?

38. If a solution shows positive deviations from Raoult's law, would you expect it to have a higher or lower boiling point than if it were ideal? Why?

39. For each case choose the pair of substances you would expect to give the most nearly ideal solution. Explain your choices.

a. CF_3CF_3 and $CF_3CF_2CF_3$ or H_2O and $CH_3\overset{\overset{O}{\|}}{C}CH_3$
b. $CH_3(CH_2)_5CH_3$ and $CH_3(CH_2)_4CH_3$ or CHF_3 and $CH_3\text{—}O\text{—}CH_3$
c. H_3PO_4 and H_2O or CCl_4 and CF_4

40. A solution is prepared by mixing 1.0 mol of methanol (CH_3OH) and 1.0 mol of propanol ($CH_3CH_2CH_2OH$). What is the composition of the vapor at 40°C? At 40°C the vapor pressures of pure methanol and pure propanol are 303 torr and 44.6 torr, respectively.

41. A solution is made by dissolving 25.8 g of urea, CH_4N_2O, a nonelectrolyte, in 275 g of water. Calculate the vapor pressures of this solution at 25°C and 45°C. (The vapor pressure of pure water is 23.8 torr at 25°C and 71.9 torr at 45°C.)

42. Pentane, C_5H_{12}, and hexane, C_6H_{14}, form an ideal solution. At 25°C the vapor pressures of pentane and hexane are 511 and 150 torr, respectively. A solution is prepared by mixing 25 mL of pentane (density, 0.63 g/mL) with 45 mL of hexane (density, 0.66 g/mL).
a. What is the vapor pressure of this solution?
b. What is the composition by mole fraction of pentane in the vapor in equilibrium with this solution?

43. What is the composition of a pentane-hexane solution that has a vapor pressure of 350 torr at 25°C? What is the composition of the vapor in equilibrium with this solution? (See Exercise 42 for vapor pressures of pure hexane and pentane.)

Colligative Properties

44. A solution is prepared by dissolving 4.9 g of sucrose ($C_{12}H_{22}O_{11}$) in 175 g of water. Calculate the boiling point, freezing point, and osmotic pressure of this solution at 25°C. Sucrose is a nonelectrolyte. Assume that the molality and molarity of this solution are equal.

45. Calculate the freezing-point depression and osmotic pressure at 25°C of an aqueous solution of 1.0 g/L of a protein (mol. wt. = 9.0×10^4) if the density of the solution is 1.0 g/cm^3.

46. Considering your answer to Exercise 45, which colligative property, freezing-point depression or osmotic pressure, would be better used to determine the molecular weights of large molecules. Explain your answer.

47. A 20.0-mg sample of a protein is dissolved in water to make 25.0 mL of solution. The osmotic pressure of the solution is 0.56 torr at 25°C. What is the molecular weight of the protein?

48. Reserpine is a natural product isolated from the roots of the shrub *Rauwolfia serpentina*. It was first synthesized in 1956 by Nobel prize winner R. B. Woodward. It is used as a tranquilizer and sedative. When 1.00 g of reserpine is dissolved in 25.0 g of camphor, the freezing-point depression is 2.63°C (K_f for camphor is 40.°C kg/mol). Calculate the molality of the solution and the molecular weight of reserpine.

49. Cellophane is a form of cellulose that has undergone the viscose process. The structure of the cellophane polymer is

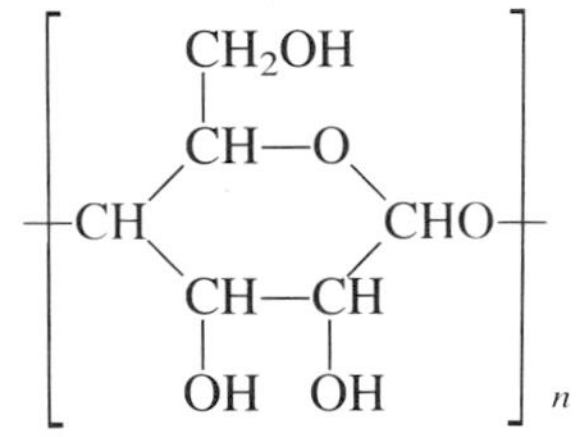

$$\left[\begin{array}{c} CH_2OH \\ | \\ CH\text{—}O \\ \diagup \qquad \diagdown \\ \text{—}CH \qquad\qquad CHO\text{—} \\ \diagdown \qquad \diagup \\ CH\text{—}CH \\ |\qquad | \\ OH\quad OH \end{array}\right]_n$$

What structural features allow cellophane to act as a semipermeable membrane for water? (*Hint:* What types of intermolecular forces are possible between cellophane and water?)

50. Propylene glycol is produced commercially by the reaction of propylene oxide with water at high temperature:

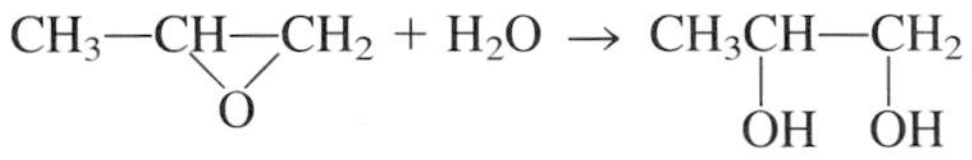

$$CH_3\text{—}\underset{\diagdown\, O\, \diagup}{CH\text{—}CH_2} + H_2O \rightarrow CH_3\underset{OH}{\underset{|}{CH}}\text{—}\underset{OH}{\underset{|}{CH_2}}$$

Calculate the freezing point and boiling point of a solution of 200.0 g of propylene glycol in 400.0 g of water.

51. A 1.60-g sample of a mixture of naphthalene ($C_{10}H_8$) and anthracene ($C_{14}H_{10}$) is dissolved in 20.0 g of benzene (C_6H_6). The freezing point of the solution is 2.8°C. What is the composition as mass percent of the sample mixture? The freezing point of benzene is 5.5°C, and K_f is 5.12°C kg/mol.

52. What mass of urea, $(NH_2)_2CO$, must be dissolved in 150.0 g of water to give a solution with a freezing point of −3.00°C?

53. If the fluid inside a tree is approximately 0.1 *M* more concentrated in solute than the ground water that bathes the roots, how high will a column of fluid rise in the tree? Assume the density of the fluid is 1.0 g/cm^3. (The density of mercury is 13.6 g/cm^3.)

54. Before refrigeration was common, many foods were preserved by salting them heavily, and many fruits were preserved by mixing them with a large amount of sugar (fruit preserves). How do salt and sugar act as preservatives?

Properties of Electrolyte Solutions

55. Sea water is approximately 0.5 *M* NaCl. What is the minimum pressure that must be applied at 25°C to purify sea water by reverse osmosis?

56. Consider the following:

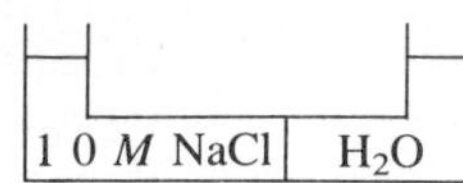

What would happen to the level of liquid in the two arms if the semipermeable membrane was permeable to:
a. H_2O only
b. H_2O, Na^+, and Cl^-

57. Place the following solutions in order by size of freezing-point depression:
a. 1.0 *m* glucose in water
b. 1.0 *m* NaCl in water
c. 1.0 *m* HOCl in water (HOCl is a weak acid)
d. 1.0 *m* $MgCl_2$ in water

58. From the following:

pure water
solution of sucrose ($\chi = 0.01$) in water
solution of NaCl ($\chi = 0.01$) in water
solution of $CaCl_2$ ($\chi = 0.01$) in water

choose the one with the
a. highest freezing point
b. lowest freezing point
c. highest boiling point
d. lowest boiling point
e. highest osmotic pressure

59. Calculate the freezing point of a 0.5 *m* solution of $Ca(NO_3)_2$ in water. (Assume $i = 3.0$.)

60. Would you expect the measured freezing point of 0.5 *m* $Ca(NO_3)_2$ to be higher or lower than the value you calculated in Exercise 59? Explain.

61. A 1.00-g sample of each of the following compounds is dissolved separately in 25.0 g of water. The freezing-point depression for solution is as follows:

Compound	ΔT
$Co(NH_2CH_2CH_2NH_2)_2Cl_3$	0.52°C
$Co(NH_2CH_2CH_2NH_2)_3Cl_3$	0.87°C

What is the experimental value of *i* for each of the compounds?

62. Why is the observed freezing-point depression for electrolyte solutions sometimes less than the calculated value? Is the error greater for concentrated or dilute solutions?

63. Use the following data for three solutions of $CaCl_2$ to calculate the apparent value of the van't Hoff factor.

Molality	Freezing-point depression
0.091	0.440
0.279	1.330
0.475	2.345

64. The freezing-point depression of a 0.091 molal solution of CsCl is 0.302°C. The freezing-point depression of a 0.091 molal solution of $CaCl_2$ is 0.440°C. In which solution does ion association appear to be greater? Explain your answer.

Additional Exercises

65. The solubility of benzoic acid,

$$C_6H_5-\underset{\underset{O}{\|}}{C}-O-H$$

is 0.34 g/100 mL in water at 25°C and is 10.0 g/100 mL in benzene, C_6H_6, at 25°C. Rationalize this solubility behavior.

66. Would benzoic acid be more or less soluble in a basic aqueous solution than it is in water?

67. Specifications for lactated Ringer's solution, which is used for intravenous (IV) injections, are as follows for each 100 mL of solution:

285–315 mg Na^+
14.1–17.3 mg K^+
4.9–6.0 mg Ca^{2+}
368–408 mg Cl^-
231–261 mg lactate, $C_3H_5O_3^-$

a. Specify the amounts of NaCl, KCl, $CaCl_2 \cdot 2H_2O$, and $NaC_3H_5O_3$ needed to prepare 100 mL of lactated Ringer's solution.
b. What is the range of the osmotic pressure of the solution at 37°C, given the above specifications?

68. When pure methanol is mixed with water the solution gets warmer to the touch. Would you expect this solution to be ideal? Explain.

69. The vapor over a pentane-hexane solution at 25°C has a mole fraction of pentane equal to 0.15 at 25°C. What is the mole fraction of pentane in the solution? (See Exercise 42 for the vapor pressures of the pure liquids.)

70. The normal boiling point of methanol is 64.7°C. A solution containing a nonvolatile solute in methanol has a vapor pressure of 710.0 torr at 64.7°C. What is the mole fraction of methanol in this solution?

71. A solid consists of a mixture of $NaNO_3$ and $Mg(NO_3)_2$. When 6.5 g of the solid is dissolved in 50.0 g of water, the freezing

point is lowered by 5.4°C. What is the composition, by mass, of the solid?

72. Anthraquinone contains only carbon, hydrogen, and oxygen. When 4.80 mg of anthraquinone is burned, 14.22 mg of CO_2 and 1.66 mg of H_2O are produced. The freezing point of camphor is lowered by 22.3°C when 1.32 g of anthraquinone is dissolved in 11.4 g of camphor. Calculate the empirical and molecular formulas of anthraquinone.

73. A forensic chemist is given a white solid that is suspected of being pure cocaine ($C_{17}H_{21}NO_4$, mol.wt. = 303.35). She dissolves 1.22 ± 0.01 g of the solid in 15.60 ± 0.01 g of benzene. The freezing point is lowered by 1.32 ± 0.04°C.
   a. What is the molecular weight of the substance? Assuming that the percent uncertainty in the calculated molecular weight is the same as the percent uncertainty in the temperature change, calculate the uncertainty in the molecular weight.
   b. Could the chemist unequivocally state that the substance is cocaine? For example, is the uncertainty small enough to distinguish cocaine from codeine ($C_{18}H_{21}NO_3$, mol.wt. = 299.36)?
   c. Assuming the absolute uncertainties in the measurements of temperature and mass remain unchanged, how could the chemist improve the precision of her results?

74. A 0.15-g sample of a purified protein is dissolved in water to give 2.0 mL of solution. The osmotic pressure is found to be 18.6 torr at 25°C. Calculate the protein's molecular weight.

75. An unknown compound contains only carbon, hydrogen, and oxygen. Combustion analysis of the compound gives mass percents of 31.57% C and 5.30% H. The molecular weight is determined by measuring the freezing-point depression of an aqueous solution. A freezing point of −5.20°C is recorded for a solution made by dissolving 10.56 g of the compound in 25.0 mL of water. Determine the empirical formula, molecular weight, and molecular formula of the compound. Assume the compound is a nonelectrolyte.

76. The most concentrated aqueous solution of NaOH that can be prepared is approximately 50% NaOH by mass. Calculate the mole fraction and molality of NaOH in this solution.

77. A bottle of wine contains 12.5% ethanol by volume. The density of ethanol (C_2H_5OH) is 0.79 g/cm^3. Calculate the concentration of ethanol in wine as mass percent and molality.

78. In lab you need 100 mL of each of the following solutions. Explain how you would proceed using the given information.
   a. 2.0 *m* KCl in water (density unknown to you)
   b. 15% NaOH by mass in water (density unknown to you)
   c. 25% NaOH by mass in CH_3OH (d = 0.79 g/cm^3)
   d. 30% C_2H_5OH by volume in water

79. Calculate the freezing point and boiling point of an antifreeze solution that is 40.0% by mass of ethylene glycol ($HOCH_2CH_2OH$) in water.

80. How would you prepare 1.0 L of an aqueous solution of sucrose ($C_{12}H_{22}O_{11}$) that would have an osmotic pressure of 15 atm at a temperature of 25°C?

81. How would you prepare 2.00 L of an aqueous solution of $CaCl_2$ that would have a freezing point of −10.0°C?

82. Calculate the freezing point of each of the following solutions. (For parts b and c assume complete dissociation.)
   a. 5.0 g of glucose ($C_6H_{12}O_6$) in 25 g of H_2O
   b. 5.0 g of NaCl in 25 g of H_2O
   c. 2.0 g of $Al(NO_3)_3$ in 15 g of H_2O
   d. 1.0 g of benzoic acid ($C_6H_5CO_2H$) in 10.0 g benzene

83. In Exercise 73 in Chapter 5 you calculated the pressure of CO_2 in a bottle of sparkling wine assuming that the CO_2 was insoluble in water. This was a bad assumption. Redo this problem assuming that CO_2 obeys Henry's law. Use the data given in that problem to calculate the partial pressure of CO_2 in the gas phase and the solubility of CO_2 in the wine at 25°C. The Henry's law constant for CO_2 is 32 L atm/mol at 25°C.

CHAPTER 12

Chemical Kinetics

It is important to understand several characteristics of chemical reactions. First, a reaction is defined by its reactants and products, whose identity must be learned by experiment. Once the reactants and products are known, we can write and balance the equation for the reaction and carry out stoichiometric calculations. A second very important characteristic of a reaction is its spontaneity. Spontaneity refers to the *inherent tendency* for the process to occur; however, it implies nothing about speed. *Spontaneous does not mean fast.* There are many spontaneous reactions that are so slow that no apparent reaction occurs over a period of weeks or years at normal temperatures. For example, there is a strong inherent tendency for gaseous hydrogen and oxygen to combine,

$$2H_2(g) + O_2(g) \rightarrow 2H_2O(l)$$

but in fact the two gases can coexist indefinitely at 25°C. Similarly, the gaseous reactions

$$H_2(g) + Cl_2(g) \rightarrow 2HCl(g)$$
$$N_2(g) + 3H_2(g) \rightarrow 2NH_3(g)$$

are both highly likely to occur from a thermodynamic standpoint, but we observe no reactions under normal conditions. In addition, the process of changing diamond to graphite is spontaneous but is so slow that it is not detectable.

To be useful, reactions must occur at a reasonable rate. To produce the 20 million tons of ammonia needed each year for fertilizer, we cannot

CONTENTS

< Burning lycopodium powder. The immense surface area of the tiny particles of the powder causes it to react rapidly with oxygen.

The burning of gasoline, the growth of plants, and the metabolism of food to provide energy for human activities all depend on chemical reactions.

simply mix nitrogen and hydrogen gases at 25°C and wait for them to react. It is not enough to understand the stoichiometry and thermodynamics of a reaction; we must also understand the factors that govern the rate of the reaction. The area of chemistry that concerns reaction rates is called **chemical kinetics.**

One of the main goals of chemical kinetics is to understand the steps by which a reaction takes place. This series of steps is called the *reaction mechanism*. Understanding the mechanism allows us to find ways to facilitate the reaction. For example, the Haber process for the production of ammonia requires high temperatures to achieve commercially feasible reaction rates, but even higher temperatures (and more cost) would be required without the use of iron oxide, which speeds up the reaction.

Chapter 16 deals with the spontaneity of reactions using the principles of thermodynamics.

In this chapter we will consider the main ideas of chemical kinetics. We will explore rate laws, reaction mechanisms, and simple models for chemical reactions.

12.1 Reaction Rates

Purpose

- To define reaction rate and to show how rates can be measured from experimental data.

To introduce the concept of the rate of a reaction, we will consider the decomposition of nitrogen dioxide, a gas that causes air pollution. Nitrogen dioxide decomposes to nitric oxide and oxygen as follows:

The kinetics of air pollution is discussed in Section 12.8.

$$2NO_2(g) \rightarrow 2NO(g) + O_2(g)$$

Suppose in a particular experiment we start with a flask of nitrogen dioxide at 300°C and measure the concentrations of nitrogen dioxide, nitric oxide, and oxygen as the nitrogen dioxide decomposes. The results of this experiment are summarized in Table 12.1, and the data are plotted in Fig. 12.1 on page 516.

Note from these results that the concentration of the reactant (NO_2) decreases with time and the concentrations of the products (NO and O_2) increase with time. Chemical kinetics deals with the speed at which these changes occur. The speed, or *rate,* of a process is defined as the change in a given quantity over a specific period of time. For chemical reactions, the quantity that changes is the amount or concentration of a reactant or product. So the **reaction rate** of a chemical reaction is defined as the *change in concentration of a reactant or product per unit time:*

$$\text{Rate} = \frac{\text{concentration of A at time } t_2 - \text{concentration of A at time } t_1}{t_2 - t_1}$$

$$= \frac{\Delta[A]}{\Delta t}$$

[A] means concentration of A in mol/L.

Concentrations of Reactant and Products as a Function of Time for the Reaction $2NO_2(g) \rightarrow 2NO(g) + O_2(g)$ (at 300°C)

	Concentration (mol/L)		
Time (±1s)	NO_2	NO	O_2
0	0.0100	0	0
50	0.0079	0.0021	0.0011
100	0.0065	0.0035	0.0018
150	0.0055	0.0045	0.0023
200	0.0048	0.0052	0.0026
250	0.0043	0.0057	0.0029
300	0.0038	0.0062	0.0031
350	0.0034	0.0066	0.0033
400	0.0031	0.0069	0.0035

Table 12.1

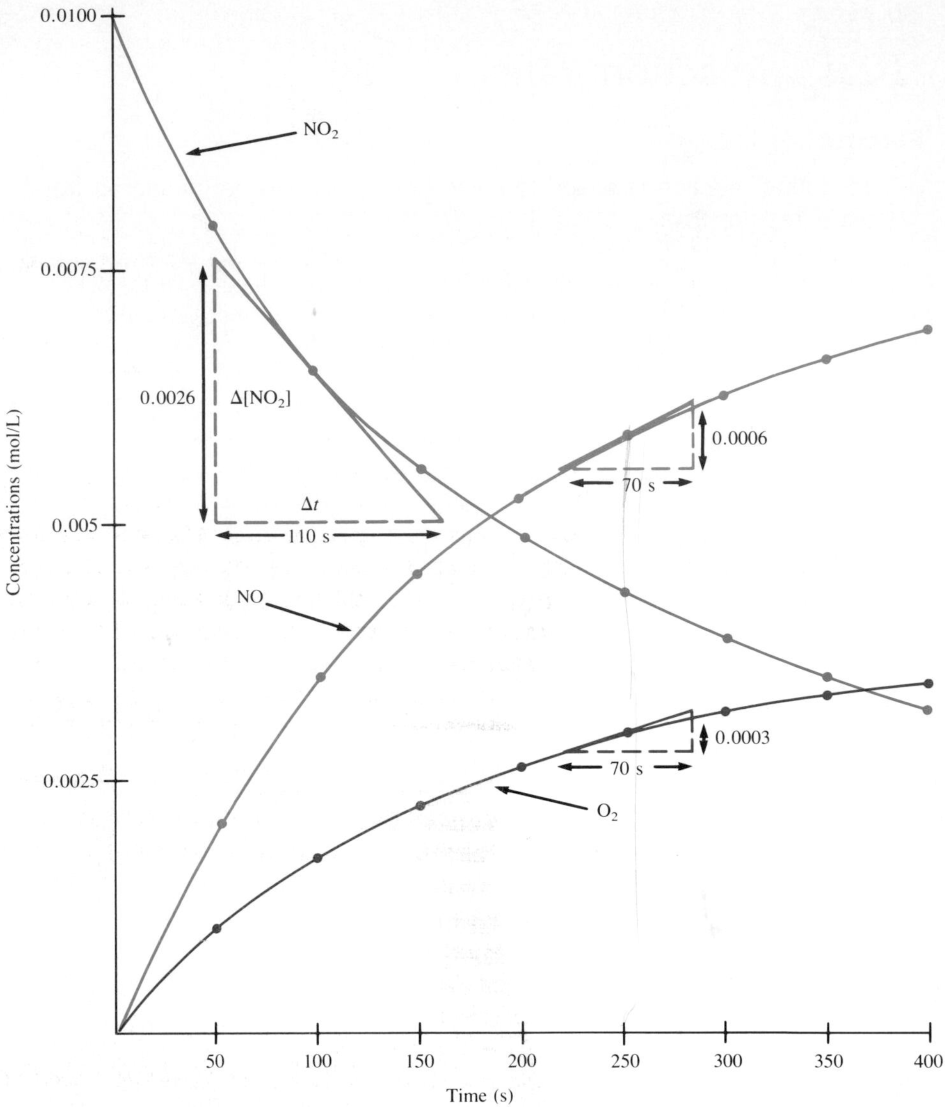

Figure 12.1

Starting with a flask of nitrogen dioxide at 300°C, the concentrations of nitrogen dioxide, nitric oxide, and oxygen are plotted versus time.

where A is the reactant or product being considered and the square brackets indicate concentration in mol/L. As usual, the symbol Δ indicates a *change* in a given quantity.

Now let us calculate the average rate at which the concentration of NO_2 changes over the first 50 seconds of the reaction, using the data given in Table 12.1.

$$\begin{aligned}\text{Rate} &= \frac{\Delta[NO_2]}{\Delta t} \\ &= \frac{[NO_2]_{t=50} - [NO_2]_{t=0}}{50\text{ s} - 0\text{ s}} \\ &= \frac{0.0079\text{ mol/L} - 0.0100\text{ mol/L}}{50\text{ s}} \\ &= -4.2 \times 10^{-5}\text{ mol/L s}\end{aligned}$$

Note that since the concentration of NO_2 decreases with time, $\Delta[NO_2]$ is a negative quantity. Because it is customary to work with *positive* reaction rates, we define the rate of this particular reaction as

$$\text{Rate} = -\frac{\Delta[NO_2]}{\Delta t}$$

Since the concentrations of reactants always decrease with time, any rate expression involving a reactant will include a negative sign. The average rate of this reaction from 0 second to 50 seconds is then

$$\begin{aligned}\text{Rate} &= -\frac{\Delta[NO_2]}{\Delta t}\\ &= -(-4.2 \times 10^{-5}\ \text{mol/L s})\\ &= 4.2 \times 10^{-5}\ \text{mol/L s}\end{aligned}$$

Average Rate (in mol/L s) of Decomposition of Nitrogen Dioxide as a Function of Time*

$-\frac{\Delta[NO_2]}{\Delta t}$	Time period (s)
4.2×10^{-5}	$0 \rightarrow 50$
2.8×10^{-5}	$50 \rightarrow 100$
2.0×10^{-5}	$100 \rightarrow 150$
1.4×10^{-5}	$150 \rightarrow 200$
1.0×10^{-5}	$200 \rightarrow 250$

*Note that the *rate* decreases with time.

Table 12.2

The average rates for this reaction during several other time intervals are given in Table 12.2. Note that the rate is not constant, but decreases with time. The rates given in Table 12.2 are *average* rates over 50-second time intervals. The value of the rate at a particular time (the **instantaneous rate**) can be obtained by computing the slope of a line tangent to the curve at that point. Figure 12.1 shows a tangent drawn at $t = 100$ seconds. The *slope* of this line gives the rate at $t = 100$ seconds as follows:

Appendix 1.3 reviews slopes of straight lines.

$$\begin{aligned}\text{Slope of the tangent line} &= \frac{\text{change in } y}{\text{change in } x}\\ &= \frac{\Delta[NO_2]}{\Delta t}\end{aligned}$$

But

$$\text{Rate} = -\frac{\Delta[NO_2]}{\Delta t}$$

Therefore

$$\begin{aligned}\text{Rate} &= -(\text{slope of the tangent line})\\ &= -\left(\frac{-0.0026\ \text{mol/L}}{110\ \text{s}}\right)\\ &= 2.4 \times 10^{-5}\ \text{mol/L s}\end{aligned}$$

So far, we have discussed the rate of this reaction only in terms of the reactant. The rate can also be defined in terms of the products. However, in doing so we must take into account the coefficients in the balanced equation for the reaction, because the stoichiometry determines the relative rates of consumption of reactants and generation of products. For example, in the reaction we are considering,

$$2NO_2(g) \rightarrow 2NO(g) + O_2(g)$$

both the reactant NO_2 and the product NO have a coefficient of 2, so NO is produced at the same rate as NO_2 is consumed. We can verify this from Fig. 12.1. Note that the curve for NO is the same shape as the curve for NO_2 except that it is inverted, or flipped over. This means that, at any point in time, the slope of the tangent to the curve for NO will be the negative of the slope to the curve for NO_2. (Verify this at the point t = 100 seconds on both curves.) In the balanced equation, the product O_2 has a coefficient of 1, which means it is produced half as fast as NO since NO has a coefficient of 2. That is, the rate of NO production is twice the rate of O_2 production.

We can also verify this fact from Fig. 12.1. For example, at t = 250 seconds,

$$\text{Slope of the tangent to the NO curve} = \frac{6.0 \times 10^{-4}\ \text{mol/L}}{70\ \text{s}}$$

$$= 8.6 \times 10^{-6}\ \text{mol/L s}$$

$$\text{Slope of the tangent to the } O_2 \text{ curve} = \frac{3.0 \times 10^{-4}\ \text{mol/L}}{70\ \text{s}}$$

$$= 4.3 \times 10^{-6}\ \text{mol/L s}$$

The slope at t = 250 seconds on the NO curve is twice the slope of that point on the O_2 curve, showing that the rate of production of NO is twice that of O_2.

The rate information can be summarized as follows:

Rate of consumption of NO_2	=	rate of production of NO	=	2(rate of production of O_2)
$-\dfrac{\Delta[NO_2]}{\Delta t}$	=	$\dfrac{\Delta[NO]}{\Delta t}$	=	$2\left(\dfrac{\Delta[O_2]}{\Delta t}\right)$

We have seen that the rate of a reaction is not constant, but that it changes with time. This is because the concentrations change with time (Fig. 12.1).

Because the reaction rate changes with time, and because the rate is different (by factors that depend on the coefficients in the balanced equation) depending on which reactant or product is being studied, we must be very specific when we describe a rate for a chemical reaction.

12.2 Rate Laws: An Introduction

Purpose

- To describe the two types of rate laws.

Chemical reactions are *reversible*. In our discussion of the decomposition of nitrogen dioxide, we have so far considered only the *forward reaction* as shown here:

$$2NO_2(g) \rightarrow 2NO(g) + O_2(g)$$

However, the *reverse reaction* can also occur. As NO and O_2 accumulate, they can react to reform NO_2:

$$O_2(g) + 2NO(g) \rightarrow 2NO_2(g)$$

When gaseous NO_2 is placed in an otherwise empty container, initially the dominant reaction is

$$2NO_2(g) \rightarrow 2NO(g) + O_2(g)$$

and the change in the concentration of NO_2 ($\Delta[NO_2]$) depends only on the forward reaction. However, after a period of time, enough products accumulate so that the reverse reaction becomes important. Now $\Delta[NO_2]$ depends on the *difference in the rates of the forward and reverse reactions*. This complication can be avoided if we study the rate of a reaction under conditions where the reverse reaction makes only a negligible contribution. Typically this means that we must study a reaction at a point soon after the reactants are mixed, before the products have had time to build up to significant levels. From now on in this book we will deal only with rates studied under such conditions.

When forward and reverse reaction rates are equal, there will be no changes in the concentrations of reactants or products. This is called chemical equilibrium and is discussed fully in Chapter 13.

Because we will always choose conditions where the reverse reaction can be neglected, the *reaction rate will depend only on the concentrations of the reactants*. For the decomposition of nitrogen dioxide, we can write

$$\text{Rate} = k[NO_2]^n \qquad (12.1)$$

Such an expression, which shows how the rate depends on the concentrations of reactants, is called a **rate law.** The proportionality constant k, called the **rate constant,** and n, called the **order** of the reactant, must both be determined by experiment. The order of a reactant can be positive or negative and can be an integer or a fraction. For the relatively simple reactions we will consider in this book, the orders will generally be positive integers.

Note two important points about Equation (12.1).

1. The concentrations of the products do not appear in the rate law because the reaction rate is being studied under conditions where the reverse reaction does not contribute to the overall rate.
2. The value of the exponent n must be determined by experiment; it cannot be written from the balanced equation.

Before we go further we must define exactly what we mean by the term *rate* in Equation (12.1). In Section 12.1 we saw that reaction rate means a change in concentration per unit time. However, which reactant or product concentration do we choose in defining the rate? For example, for the decomposition of NO_2 to produce O_2 and NO, considered in Section 12.1, we could define the rate in terms of any of these three species. However, since O_2 is produced only half as fast as NO_2, we must be careful to specify which species we are talking about in a given case. For instance, we might choose to define the reaction rate in terms of the consumption of NO_2:

$$\text{Rate} = -\frac{\Delta[NO_2]}{\Delta t} = k[NO_2]^n$$

On the other hand, we could define the rate in terms of the production of O_2:

$$\text{Rate}' = \frac{\Delta[O_2]}{\Delta t} = k'[NO_2]^n$$

Note that because $2NO_2$ molecules are consumed for every O_2 molecule produced

$$\text{Rate} = 2 \times \text{Rate}'$$

or

$$k[NO_2]^n = 2k'[NO_2]^n$$

and

$$k = 2 \times k'$$

Thus the value of the rate constant depends on how the rate is defined.

In this text we will always be careful to define exactly what is meant by the rate for a given reaction so there will be no confusion about which specific rate constant is being used.

Types of Rate Laws

Notice that the rate law we have used to this point expresses rate as a function of concentration. For example, for the decomposition of NO_2 we have defined

$$\text{Rate} = -\frac{\Delta[NO_2]}{\Delta t} = k[NO_2]^n$$

which tells us (once we have determined the value of n) exactly how the rate depends on the concentration of the reactant, NO_2. A rate law that expresses how the *rate depends on concentration* is technically called the **differential rate law,** but it is often simply called the **rate law.** Thus when we use the term *the rate law* in this text, we mean the expression that gives the rate as a function of concentration.

The name "differential" rate law comes from a mathematical term. We will regard it simply as a label. The terms differential rate law and rate law will be used interchangeably in this text.

A second kind of rate law, the **integrated rate law,** will also be important in our study of kinetics. The integrated rate law expresses how the *concentrations depend on time.* Although we will not consider the details here, a given differential rate law is always related to a certain type of integrated rate law and vice-versa. That is, if we determine the differential rate law for a given reaction, we automatically know the form of the integrated rate law for the reaction. This means that once we determine experimentally either type of rate law for a reaction we also know the other one.

Which rate law we choose to determine by experiment often depends on what types of data are easiest to collect. If we can conveniently measure how the rate changes as the concentrations are changed, we can readily determine the differential (rate/concentration) rate law. On the other hand, if it is more convenient to measure the concentration as a function of time, we can determine the form of the integrated (concentration/time) rate law. We will discuss how rate laws are actually determined in the next several sections.

A 17-year cicada emerges.

Why are we interested in determining the rate law for a reaction? How does it help us? It helps us because we can work backward from the rate law to find the steps by which the reaction occurs. Most chemical reactions do not take place in a single step, but result from a series of sequential steps. To understand a chemical reaction we must learn what these steps are. For example, a chemist who is designing an insecticide may study the reactions involved in the process of insect growth to see what type of molecule may interrupt this series of reactions. Or an industrial chemist may be trying to make a reaction occur faster, using less expensive conditions. To accomplish this, he or she must know which step is slowest because it is that step that must be speeded up. Thus a chemist is usually not interested in a rate law for its own sake but because of what it tells about the steps by which a reaction occurs. We will develop a process for finding the reaction steps in this chapter.

Rate Laws: A Summary

- There are two types of rate laws.
  1. The differential rate law (often called simply the rate law) shows how the rate of a reaction depends on concentrations.
  2. The integrated rate law shows how the concentrations of species in the reaction depend on time.
- Because we will consider reactions only under conditions where the reverse reaction is unimportant, our rate laws will involve only concentrations of reactants.
- Because the differential and integrated rate laws for a given reaction are related in a well-defined way, the experimental determination of *either* of the rate laws is sufficient.
- Experimental convenience usually dictates which type of rate law is determined experimentally.
- The importance of the rate law for a reaction arises mainly from the fact that we can usually infer the individual steps involved in the reaction from the specific form of the rate law.

12.3 Determining the Form of the Rate Law

Purpose

- To learn methods to determine the rate law for a reaction.

The first step in understanding how a given chemical reaction occurs is to determine the *form* of the rate law. In this section we will explore ways to obtain the differential rate law for a reaction. First, we will consider the decomposition of dinitrogen pentoxide in carbon tetrachloride solution:

$$2N_2O_5(soln) \rightarrow 4NO_2(soln) + O_2(g)$$

Data for this reaction at 45°C are listed in Table 12.3 and plotted in Fig. 12.2 on the next page. In this reaction the oxygen gas escapes from the solution and thus does not react with the nitrogen dioxide, so we do not have to be concerned about the effects of the reverse reaction at any time over the life of the reaction. That is, the reverse reaction is negligible at all times over the course of this reaction.

Evaluation of the reaction rate at concentrations of N_2O_5 of 0.90 *M* and 0.45 *M*, by taking the slopes of the tangents to the curve at these points (see Fig. 12.2), gives

$[N_2O_5]$	Rate (mol/L s)
0.90 *M*	5.4×10^{-4}
0.45 *M*	2.7×10^{-4}

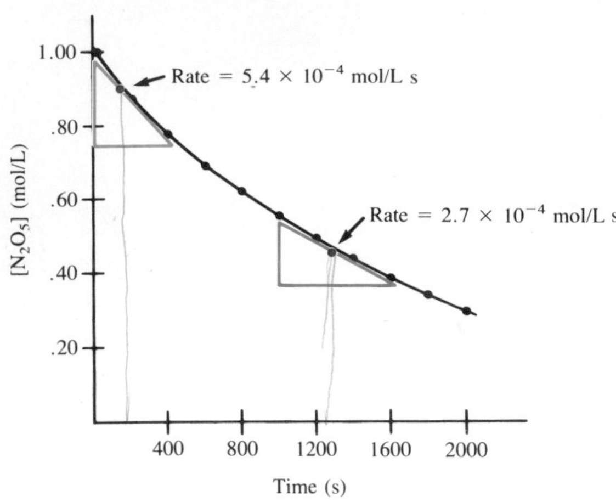

Figure 12.2

A plot of the concentration of N_2O_5 as a function of time for the reaction $2N_2O_5(soln) \rightarrow 4NO_2(soln) + O_2(g)$ (at 45°C). Note that the reaction rate at $[N_2O_5] = 0.90\ M$ is twice that at $[N_2O_5] = 0.45\ M$.

Concentration/Time Data for the Reaction $2N_2O_5(soln) \rightarrow 4NO_2(soln) + O_2(g)$ (at 45°C)	
$[N_2O_5]$ (mol/L)	Time (s)
1.00	0
0.88	200
0.78	400
0.69	600
0.61	800
0.54	1000
0.48	1200
0.43	1400
0.38	1600
0.34	1800
0.30	2000

Table 12.3

Note that when $[N_2O_5]$ is halved, the rate is also halved. This means that the rate of this reaction depends on the concentration of N_2O_5 to the *first power*. In other words, the (differential) rate law for this reaction is

$$\text{Rate} = -\frac{\Delta[N_2O_5]}{\Delta t} = k[N_2O_5]$$

First order: rate = k[A]. Doubling the concentration of A doubles the reaction rate.

Thus the reaction is *first order* in N_2O_5. Note that for this reaction the order is *not* the same as the coefficient of N_2O_5 in the balanced equation for the reaction. This reemphasizes the fact that the order of a particular reactant must be obtained by *observing* how the reaction rate depends on the concentration of that reactant.

We have seen that by determining the instantaneous rate at two different reactant concentrations, the rate law for the decomposition of N_2O_5 is shown to have the form

$$\text{Rate} = -\frac{\Delta[A]}{\Delta t} = k[A]$$

where A represents N_2O_5.

Method of Initial Rates

The value of the initial rate is determined for each experiment at the same value of t as close to $t = 0$ as possible.

The most common method for experimentally determining the form of the rate law for a reaction is the **method of initial rates.** The **initial rate** of a reaction is the instantaneous rate determined just after the reaction begins (just after $t = 0$). The idea is to determine the instantaneous rate before the initial concentrations of reactants have changed significantly. Several experiments are carried out using different initial concentrations, and the initial rate is determined for each run. The results are then compared to see how the initial rate depends on the initial concentrations. This allows the form of the rate law to be determined. We will illustrate the method of initial rates using the following reaction:

$$NH_4^+(aq) + NO_2^-(aq) \rightarrow N_2(g) + 2H_2O(l)$$

Initial Rates from Three Experiments for the Reaction $NH_4^+(aq) + NO_2^-(aq) \rightarrow N_2(g) + 2H_2O(l)$			
Experiment	Initial concentration of NH_4^+	Initial concentration of NO_2^-	Initial rate (mol/L s)
1	0.100 *M*	0.0050 *M*	1.35×10^{-7}
2	0.100 *M*	0.010 *M*	2.70×10^{-7}
3	0.200 *M*	0.010 *M*	5.40×10^{-7}

Table 12.4

Table 12.4 gives initial rates obtained from three experiments involving different initial concentrations of reactants. The general form of the rate law for this reaction is

$$\text{Rate} = -\frac{\Delta[NH_4^+]}{\Delta t} = k[NH_4^+]^n[NO_2^-]^m$$

We can determine the values of n and m by observing how the initial rate depends on the initial concentrations of NH_4^+ and NO_2^-. In Experiments 1 and 2, where the initial concentration of NH_4^+ remains the same but the initial concentration of NO_2^- doubles, the observed initial rate also doubles. Since

$$\text{Rate} = k[NH_4^+]^n[NO_2^-]^m$$

we have for Experiment 1:

$$\text{Rate} = 1.35 \times 10^{-7} \text{ mol/L s} = k(0.100 \text{ mol/L})^n(0.0050 \text{ mol/L})^m$$

and for Experiment 2:

$$\text{Rate} = 2.70 \times 10^{-7} \text{ mol/L s} = k(0.100 \text{ mol/L})^n(0.010 \text{ mol/L})^m$$

The ratio of these rates is

$$\frac{\text{Rate 2}}{\text{Rate 1}} = \underbrace{\frac{2.70 \times 10^{-7} \text{ mol/L s}}{1.35 \times 10^{-7} \text{ mol/L s}}}_{2.00} = \frac{\cancel{k(0.100 \text{ mol/L})^n}(0.010 \text{ mol/L})^m}{\cancel{k(0.100 \text{ mol/L})^n}(0.0050 \text{ mol/L})^m}$$

$$= \underbrace{\frac{(0.010 \text{ mol/L})^m}{(0.0050 \text{ mol/L})^m}}_{(2.0)^m}$$

Rates 1, 2, and 3 were determined at the same value of t (very close to $t = 0$).

Thus
$$\frac{\text{Rate 2}}{\text{Rate 1}} = 2.00 = (2.0)^m$$

which means the value of m is 1. The rate law for this reaction is first order in the reactant NO_2^-.

A similar analysis of the results for Experiments 2 and 3 yields the ratio

$$\frac{\text{Rate 3}}{\text{Rate 2}} = \frac{5.40 \times 10^{-7} \text{ mol/L s}}{2.70 \times 10^{-7} \text{ mol/L s}} = \frac{(0.200 \text{ mol/L})^n}{(0.100 \text{ mol/L})^n}$$

$$= 2.00 = \left(\frac{0.200}{0.100}\right)^n = (2.00)^n$$

The value of n is also 1.

We have shown that the values of n and m are both 1 and the rate law is

$$\text{Rate} = k[NH_4^+][NO_2^-]$$

This rate law is first order in both NO_2^- and NH_4^+. Note that it is merely a coincidence that n and m have the same values as the coefficients of NH_4^+ and NO_2^- in the balanced equation for the reaction.

Overall reaction order is the sum of the orders for each reactant.

The **overall reaction order** is the sum of n and m. For this reaction, $n + m = 2$. The reaction is second order overall.

The value of the rate constant (k) can now be calculated using the results of *any* of the three experiments shown in Table 12.4. From the data for Experiment 1, we know that

$$\text{Rate} = k[NH_4^+][NO_2^-]$$
$$1.35 \times 10^{-7} \text{ mol/L s} = k(0.100 \text{ mol/L})(0.0050 \text{ mol/L})$$

Then

$$k = \frac{1.35 \times 10^{-7} \text{ mol/L s}}{(0.100 \text{ mol/L})(0.0050 \text{ mol/L})} = 2.7 \times 10^{-4} \text{ L/mol s}$$

Sample Exercise 12.1

The reaction between bromate ions and bromide ions in acidic aqueous solution is given by the following equation:

$$BrO_3^-(aq) + 5Br^-(aq) + 6H^+(aq) \rightarrow 3Br_2(l) + 3H_2O(l)$$

Table 12.5 gives the results from four experiments. Using these data, determine the orders for all three reactants, the overall reaction order, and the value of the rate constant.

The Results from Four Experiments to Study the Reaction $BrO_3^-(aq) + 5Br^-(aq) + 6H^+(aq) \rightarrow 3Br_2(l) + 3H_2O(l)$

Experiment	Initial concentration of BrO_3^- (mol/L)	Initial concentration of Br^- (mol/L)	Initial concentration of H^+ (mol/L)	Measured initial rate (mol/L s)
1	0.10	0.10	0.10	8.0×10^{-4}
2	0.20	0.10	0.10	1.6×10^{-3}
3	0.20	0.20	0.10	3.2×10^{-3}
4	0.10	0.10	0.20	3.2×10^{-3}

Table 12.5

Solution

The general form of the rate law for this reaction is

$$\text{Rate} = k[BrO_3^-]^n[Br^-]^m[H^+]^p$$

Sample Exercise 12.1, continued

We can determine the values of n, m, and p by comparing the rates from the various experiments. To determine the value of n, we use the results from Experiments 1 and 2, in which only $[BrO_3^-]$ changes:

$$\frac{\text{Rate 2}}{\text{Rate 1}} = \frac{1.60 \times 10^{-3}\ \text{mol/L s}}{8.00 \times 10^{-4}\ \text{mol/L s}} = \frac{k(0.20\ \text{mol/L})^n(0.10\ \text{mol/L})^m(0.10\ \text{mol/L})^p}{k(0.10\ \text{mol/L})^n(0.10\ \text{mol/L})^m(0.10\ \text{mol/L})^p}$$

$$2.0 = \left(\frac{0.20\ \text{mol/L}}{0.10\ \text{mol/L}}\right)^n = (2.0)^n$$

Thus n is equal to 1.

To determine the value of m, we use the results from Experiments 2 and 3, in which only $[Br^-]$ changes:

$$\frac{\text{Rate 3}}{\text{Rate 2}} = \frac{3.20 \times 10^{-3}\ \text{mol/L s}}{1.60 \times 10^{-3}\ \text{mol/L s}} = \frac{k(0.20\ \text{mol/L})^n(0.20\ \text{mol/L})^m(0.10\ \text{mol/L})^p}{k(0.20\ \text{mol/L})^n(0.10\ \text{mol/L})^m(0.10\ \text{mol/L})^p}$$

$$2.0 = \left(\frac{0.20\ \text{mol/L}}{0.10\ \text{mol/L}}\right)^m = (2.0)^m$$

Thus m is equal to 1.

To determine the value of p, we use the results from Experiments 1 and 4, in which $[BrO_3^-]$ and $[Br^-]$ are constant but $[H^+]$ differs:

$$\frac{\text{Rate 4}}{\text{Rate 1}} = \frac{3.20 \times 10^{-3}\ \text{mol/L s}}{8.00 \times 10^{-4}\ \text{mol/L s}} = \frac{k(0.10\ \text{mol/L})^n(0.10\ \text{mol/L})^m(0.20\ \text{mol/L})^p}{k(0.10\ \text{mol/L})^n(0.10\ \text{mol/L})^m(0.10\ \text{mol/L})^p}$$

$$4.0 = \left(\frac{0.20\ \text{mol/L}}{0.10\ \text{mol/L}}\right)^p$$

$$4.0 = (2.0)^p = (2.0)^2$$

Thus p is equal to 2.

The rate of this reaction is first order in BrO_3^- and Br^- and second order in H^+. The overall reaction order is $n + m + p = 4$.

The rate law can now be written

$$\text{Rate} = k[BrO_3^-][Br^-][H^+]^2$$

The value of the rate constant k can be calculated from the results of any of the four experiments. For Experiment 1, the initial rate is 8.0×10^{-4} mol/L s and $[BrO_3^-] = 0.100\ M$, $[Br^-] = 0.100\ M$, and $[H^+] = 0.100\ M$. Using these values in the rate law gives

$$8.0 \times 10^{-4}\ \text{mol/L s} = k(0.10\ \text{mol/L})(0.10\ \text{mol/L})(0.10\ \text{mol/L})^2$$

$$8.0 \times 10^{-4}\ \text{mol/L s} = k(1.0 \times 10^{-4}\ \text{mol}^4/\text{L}^4)$$

$$k = \frac{8.0 \times 10^{-4}\ \text{mol/L s}}{1.0 \times 10^{-4}\ \text{mol}^4/\text{L}^4} = 8.0\ \text{L}^3/\text{mol}^3\ \text{s}$$

Check: Verify that the same value of k can be obtained from the results of the other experiments.

12.4 The Integrated Rate Law

Purpose

- To develop rate laws relating concentration to reaction time and to show how they can be used to determine reaction order.

The rate laws we have considered so far express the rate as a function of the reactant concentrations. It is also useful to be able to express the reactant concentrations as a function of time, given the (differential) rate law for the reaction. In this section we show how this is done.

We will proceed by first looking at reactions involving a single reactant:

$$aA \rightarrow \text{products}$$

all of which have a rate law of the form

$$\text{Rate} = -\frac{\Delta[A]}{\Delta t} = k[A]^n$$

We will develop the integrated rate laws individually for the cases $n = 1$ (first order), $n = 2$ (second order), and $n = 0$ (zero order).

First-Order Rate Laws

For the reaction

$$2N_2O_5(soln) \rightarrow 4NO_2(soln) + O_2(g)$$

we have found that the rate law is

$$\text{Rate} = -\frac{\Delta[N_2O_5]}{\Delta t} = k[N_2O_5]$$

Since the rate of this reaction depends on the concentration of N_2O_5 to the first power, it is a **first-order reaction.** This means that if the concentration of N_2O_5 in a flask were suddenly doubled, the rate of production of NO_2 and O_2 would also double. This rate law can be put into a different form using a calculus operation known as integration, which yields the expression

$$\ln[N_2O_5] = -kt + \ln[N_2O_5]_0$$

Appendix 1.2 contains a review of logarithms.

where ln indicates the natural logarithm, t is the time, $[N_2O_5]$ is the concentration of N_2O_5 at time t, and $[N_2O_5]_0$ is the initial concentration of N_2O_5 (at $t = 0$, the start of the experiment). Note that such an equation, called the integrated rate law, expresses the *concentration of the reactant as a function of time.*

For a chemical reaction of the form

$$aA \rightarrow \text{products}$$

where the kinetics are first order in [A], the rate law is

$$\text{Rate} = -\frac{\Delta[A]}{\Delta t} = k[A]$$

and the **integrated first-order rate law** is

$$\ln[A] = -kt + \ln[A]_0 \qquad (12.2)$$

There are several important things to note about Equation (12.2):

An integrated rate law relates concentration to reaction time.

1. The equation shows how the concentration of A depends on time. If the initial concentration of A and the rate constant k are known, the concentration of A at any time can be calculated.
2. Equation (12.2) is of the form $y = mx + b$, where a plot of y versus x is a straight line with slope m and intercept b. In Equation (12.2),

$$y = \ln[A] \qquad x = t \qquad m = -k \qquad b = \ln[A]_0$$

Thus, for a first-order reaction, plotting the natural logarithm of concentration versus time always gives a straight line. This fact is often used to test whether a reaction is first order or not. For the reaction

For a first-order reaction, a plot of ln[A] versus t is always a straight line.

$$aA \rightarrow \text{products}$$

the *reaction is first order in A if a plot of ln[A] versus t is a straight line*. Conversely, if the plot is not a straight line, the reaction is not first order in A.

3. This integrated rate law for a first-order reaction can also be expressed in terms of a *ratio* of [A] and $[A]_0$ as follows:

$$\ln\left(\frac{[A]_0}{[A]}\right) = kt$$

Sample Exercise 12.2

The decomposition of N_2O_5 in the gas phase was studied at constant temperature.

$$2N_2O_5(g) \rightarrow 4NO_2(g) + O_2(g)$$

The following results were collected:

$[N_2O_5]$ (mol/L)	Time (s)
0.1000	0
0.0707	50
0.0500	100
0.0250	200
0.0125	300
0.00625	400

Using these data, verify that the rate law is first order in $[N_2O_5]$ and calculate the value of the rate constant, where the rate $= -\Delta[N_2O_5]/\Delta t$.

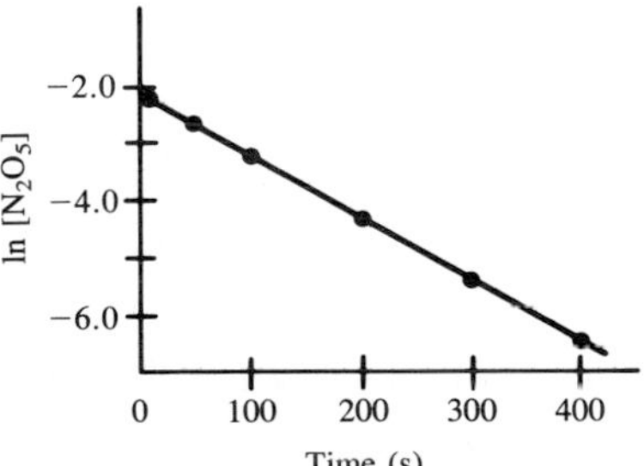

Figure 12.3
A plot of $\ln[N_2O_5]$ versus time.

Solution

We can verify that the rate law is first order in $[N_2O_5]$ by constructing a plot of $\ln[N_2O_5]$ versus time. The values of $\ln[N_2O_5]$ at various times are given below, and the plot of $\ln[N_2O_5]$ versus time is shown in Fig. 12.3.

Sample Exercise 12.2, continued

$\ln[N_2O_5]$	Time (s)
−2.303	0
−2.649	50
−2.996	100
−3.689	200
−4.382	300
−5.075	400

The fact that the plot is a straight line confirms that the reaction is first order in N_2O_5, since it follows the equation $\ln[N_2O_5] = -kt + \ln[N_2O_5]_0$.

Since the reaction is first order, the slope of the line equals $-k$ where

$$\text{Slope} = \frac{\text{change in } y}{\text{change in } x} = \frac{\Delta y}{\Delta x} = \frac{\Delta(\ln[N_2O_5])}{\Delta t}$$

Since the first and last points are exactly on the line, we will use these points to calculate the slope:

$$\text{Slope} = \frac{-5.075 - (-2.303)}{400 \text{ s} - 0 \text{ s}} = \frac{-2.772}{400 \text{ s}} = -6.93 \times 10^{-3} \text{ s}^{-1}$$

$$k = -(\text{slope}) = 6.93 \times 10^{-3} \text{ s}^{-1}$$

Sample Exercise 12.3

Using the data given in Sample Exercise 12.2, calculate $[N_2O_5]$ at 150. s after the start of the reaction.

Solution

We know from Sample Exercise 12.2 that $[N_2O_5] = 0.0500$ mol/L at 100 s and $[N_2O_5] = 0.0250$ mol/L at 200 s. Since 150 s is halfway between 100 s and 200 s, it is tempting to assume that we can simply use an arithmetic average to obtain $[N_2O_5]$ at that time. This is incorrect, because it is $\ln[N_2O_5]$, not $[N_2O_5]$, that depends directly on t. To calculate $[N_2O_5]$ after 150 s, we must use Equation (12.2):

$$\ln[N_2O_5] = -kt + \ln[N_2O_5]_0$$

where $t = 150.$ s, $k = 6.93 \times 10^{-3}$ s^{-1} (as determined in Sample Exercise 12.2), and $[N_2O_5]_0 = 0.100$ mol/L.

$$\ln([N_2O_5])_{t=150} = -(6.93 \times 10^{-3} \text{ s}^{-1})(150. \text{ s}) + \ln(0.100)$$
$$= -1.040 - 2.303 = -3.343$$
$$[N_2O_5]_{t=150} = \text{antilog}(-3.343) = 0.0353 \text{ mol/L}$$

The antilog operation means to exponentiate (see Appendix 1.2).

Note that this value of $[N_2O_5]$ is *not* halfway between 0.0500 mol/L and 0.0250 mol/L.

Half-Life of a First-Order Reaction

The time required for a reactant to reach half of its original concentration is called the **half-life of a reaction** and is designated by the symbol $t_{1/2}$. For example, we can calculate the half-life of the decomposition reaction discussed in Sample Exercise 12.2. The data plotted in Fig. 12.4 show that the half-life for this reaction is 100 seconds. We can see this by considering the following numbers:

$[N_2O_5]$(mol/L)	t (s)		
0.100	0	$\Delta t = 100$ s;	$\dfrac{[N_2O_5]_{t=100}}{[N_2O_5]_{t=0}} = \dfrac{0.050}{0.100} = \dfrac{1}{2}$
0.0500	100	$\Delta t = 100$ s;	$\dfrac{[N_2O_5]_{t=200}}{[N_2O_5]_{t=100}} = \dfrac{0.025}{0.050} = \dfrac{1}{2}$
0.0250	200	$\Delta t = 100$ s;	$\dfrac{[N_2O_5]_{t=300}}{[N_2O_5]_{t=200}} = \dfrac{0.0125}{0.0250} = \dfrac{1}{2}$
0.0125	300		

Note that it *always* takes 100 seconds for $[N_2O_5]$ to be halved in this reaction.

A general formula for the half-life of a first-order reaction can be derived from the integrated rate law for the general reaction,

$$aA \rightarrow \text{products}$$

If the reaction is first order in [A],

$$\ln\left(\frac{[A]_0}{[A]}\right) = kt$$

By definition, when $t = t_{1/2}$,

$$[A] = \frac{[A]_0}{2}$$

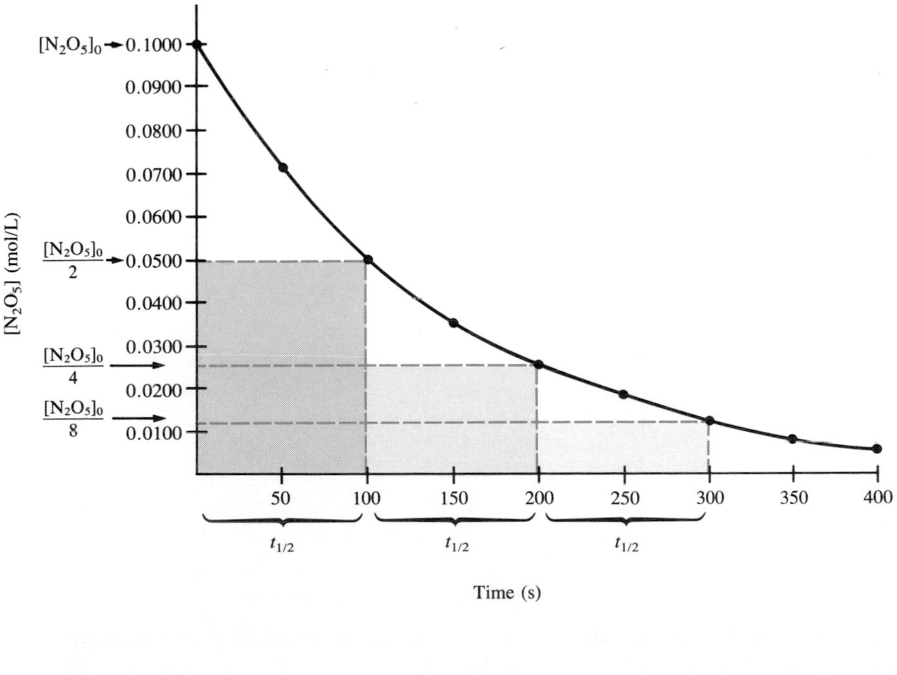

Figure 12.4

A plot of $[N_2O_5]$ versus time for the decomposition reaction of N_2O_5.

Then, for $t = t_{1/2}$, the integrated rate law becomes

$$\ln\left(\frac{[A]_0}{[A]_0/2}\right) = kt_{1/2}$$

or

$$\ln(2) = kt_{1/2}$$

Substituting the value of ln(2) and solving for $t_{1/2}$ gives

$$t_{1/2} = \frac{0.693}{k} \qquad (12.3)$$

For a first-order reaction, $t_{1/2}$ is independent of the initial concentration.

This is the *general equation for the half-life of a first-order reaction*. Equation (12.3) can be used to calculate $t_{1/2}$ if k is known or k if $t_{1/2}$ is known. Note that for a first-order reaction, *the half-life does not depend on concentration*.

Sample Exercise 12.4

A certain first-order reaction has a half-life of 20.0 min.

a. Calculate the rate constant for this reaction.

b. How much time is required for this reaction to be 75% complete?

Solution

a. Solving Equation (12.3) for k gives

$$k = \frac{0.693}{t_{1/2}} = \frac{0.693}{20.0 \text{ min}} = 3.47 \times 10^{-2} \text{ min}^{-1}$$

b. We use the integrated rate law in the form

$$\ln\left(\frac{[A]_0}{[A]}\right) = kt$$

If the reaction is 75% complete, 75% of the reactant has been consumed, leaving 25% in the original form:

$$\frac{[A]}{[A]_0} \times 100 = 25$$

This means that

$$\frac{[A]}{[A]_0} = 0.25 \quad \text{or} \quad \frac{[A]_0}{[A]} = \frac{1}{0.25} = 4.0$$

Then

$$\ln\left(\frac{[A]_0}{[A]}\right) = \ln(4.0) = kt = \left(\frac{3.47 \times 10^{-2}}{\text{min}}\right)t$$

and

$$t = \frac{\ln(4.0)}{\dfrac{3.47 \times 10^{-2}}{\text{min}}} = 40.\ \text{min}$$

Thus it takes 40. min for this particular reaction to reach 75% completion.

Sample Exercise 12.4, continued

Let's consider another way of solving this problem using the definition of half-life. After one half-life the reaction has gone 50% to completion. If the initial concentration were 1.0 mol/L, after one half-life the concentration would be 0.50 mol/L. One more half-life would produce a concentration of 0.25 mol/L. Comparing 0.25 mol/L to the original 1.0 mol/L shows that 25% of the reactant is left after two half-lives. This is a general result. (What percentage of reactant remains after three half-lives?) Two half-lives for this reaction is 2(20.0 min), or 40.0 min, which agrees with the above answer.

Second-Order Rate Laws

For a general reaction involving a single reactant,

$$\mathrm{aA} \rightarrow \text{products}$$

which is second order in A, the rate law is

$$\text{Rate} = -\frac{\Delta[\mathrm{A}]}{\Delta t} = k[\mathrm{A}]^2 \tag{12.4}$$

Second order: rate = $k[\mathrm{A}]^2$. Doubling the concentration of A quadruples the reaction rate; tripling the concentration of A increases the rate by nine times.

The **integrated second-order rate law** has the form

$$\frac{1}{[\mathrm{A}]} = kt + \frac{1}{[\mathrm{A}]_0} \tag{12.5}$$

Note the following characteristics of Equation (12.5):

1. A plot of $1/[\mathrm{A}]$ versus t will produce a straight line with a slope equal to k.
2. Equation (12.5) shows how [A] depends on time and can be used to calculate [A] at any time t, provided k and $[\mathrm{A}]_0$ are known.

For second-order reactions, a plot of 1/[A] versus t will be linear.

When one half-life of the second-order reaction has elapsed ($t = t_{1/2}$), by definition,

$$[\mathrm{A}] = \frac{[\mathrm{A}]_0}{2}$$

Equation (12.5) then becomes

$$\frac{1}{\frac{[\mathrm{A}]_0}{2}} = kt_{1/2} + \frac{1}{[\mathrm{A}]_0}$$

$$\frac{2}{[\mathrm{A}]_0} - \frac{1}{[\mathrm{A}]_0} = kt_{1/2}$$

$$\frac{1}{[\mathrm{A}]_0} = kt_{1/2}$$

Solving for $t_{1/2}$ gives *the expression for the half-life of a second-order reaction:*

$$t_{1/2} = \frac{1}{k[\mathrm{A}]_0} \tag{12.6}$$

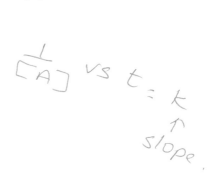

Sample Exercise 12.5

When two identical molecules combine, the resulting molecule is called a dimer.

Butadiene reacts to form its dimer according to the equation

$$2C_4H_6(g) \rightarrow C_8H_{12}(g)$$

The following data were collected for this reaction at a given temperature:

$[C_4H_6]$ (mol/L)	Time (±1s)
0.01000	0
0.00625	1000
0.00476	1800
0.00370	2800
0.00313	3600
0.00270	4400
0.00241	5200
0.00208	6200

a. Is this reaction first order or second order?

b. What is the value of the rate constant for the reaction?

c. What is the half-life for the reaction under the conditions of this experiment?

Solution

a. To decide whether the rate law for this reaction is first order or second order, we must see whether the plot of $\ln[C_4H_6]$ versus time is a straight line (first order) or the plot of $1/[C_4H_6]$ versus time is a straight line (second order). The data necessary to make these plots are as follows:

t (s)	$\frac{1}{[C_4H_6]}$	$\ln[C_4H_6]$
0	100	−4.605
1000	160	−5.075
1800	210	−5.348
2800	270	−5.599
3600	320	−5.767
4400	370	−5.915
5200	415	−6.028
6200	481	−6.175

The resulting plots are shown in Fig. 12.5. Since the $\ln[C_4H_6]$ versus t plot [Fig. 12.5(a)] is not a straight line, the reaction is *not* first order. The reaction is, however, second order, as shown by the linearity of the $1/[C_4H_6]$ versus t plot [Fig. 12.5(b)]. Thus we can now write the rate law for this second-order reaction:

$$\text{Rate} = -\frac{\Delta[C_4H_6]}{\Delta t} = k[C_4H_6]^2$$

Sample Exercise 12.5, continued

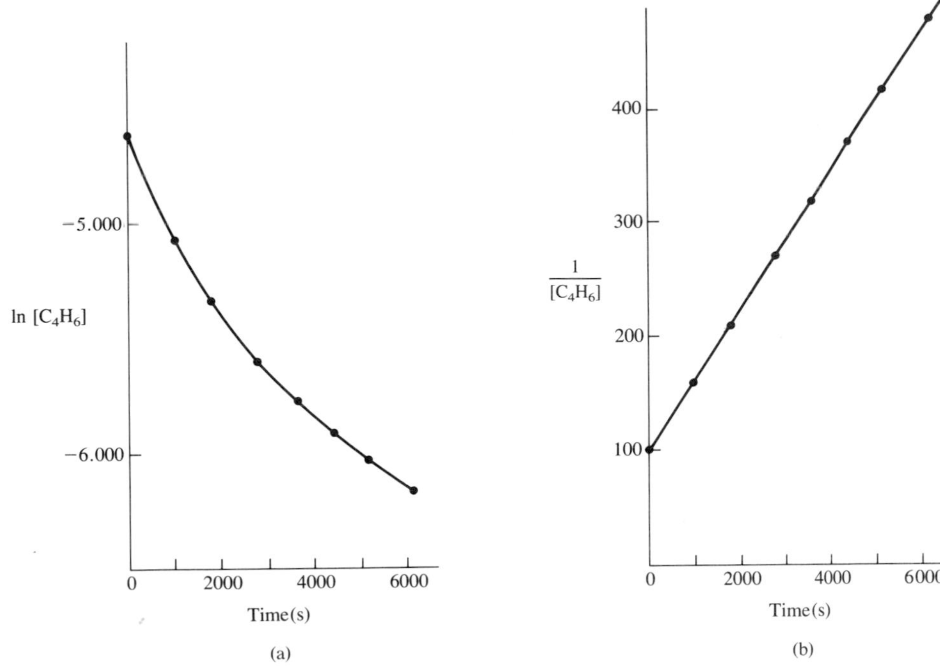

Figure 12.5
(a) A plot of $\ln[C_4H_6]$ versus t. (b) A plot of $1/[C_4H_6]$ versus t.

b. For a second-order reaction, a plot of $1/[C_4H_6]$ versus t produces a straight line of slope k. In terms of the standard equation for a straight line, $y = mx + b$, we have $y = 1/[C_4H_6]$ and $x = t$. Thus the slope of the line can be expressed as follows:

$$\text{Slope} = \frac{\Delta y}{\Delta x} = \frac{\Delta\left(\frac{1}{[C_4H_6]}\right)}{\Delta t}$$

Using the points at $t = 0$ and $t = 6200$, we can find the rate constant for the reaction:

$$k = \text{slope} = \frac{(481 - 100)\ \text{L/mol}}{(6200 - 0)\ \text{s}} = \frac{381}{6200}\ \text{L/mol s} = 6.14 \times 10^{-2}\ \text{L/mol s}$$

c. The expression for the half-life of a second-order reaction is

$$t_{1/2} = \frac{1}{k[A]_0}$$

In this case $k = 6.14 \times 10^{-2}$ L/mol s (from part b) and $[A]_0 = [C_4H_6]_0 = 0.01000\ M$ (the concentration at $t = 0$). Thus

Sample Exercise 12.5, continued

$$t_{1/2} = \frac{1}{(6.14 \times 10^{-2}\ \text{L/mol s})(1.000 \times 10^{-2}\ \text{mol/L})} = 1.63 \times 10^3\ \text{s}$$

The initial concentration of C_4H_6 is halved in 1630 s.

For a second-order reaction, $t_{1/2}$ is dependent on $[A]_0$. For a first-order reaction, $t_{1/2}$ is independent of $[A]_0$.

It is important to recognize the difference between the half-life for a first-order reaction and the half-life for a second-order reaction. For a second-order reaction, $t_{1/2}$ depends on both k and $[A]_0$; for a first-order reaction, $t_{1/2}$ depends only on k. For a first-order reaction, a constant time is required to reduce the concentration of the reactant by half, and then by half again, and so on, as the reaction proceeds. In Sample Exercise 12.5 we saw that this is *not* true for a second-order reaction. For that second-order reaction, we found that the first half-life (the time required to go from $[C_4H_6] = 0.010\ M$ to $[C_4H_6] = 0.0050\ M$) is 1630 seconds. We can estimate the second half-life from the concentration data as a function of time. Note that to reach $0.0024\ M$ C_4H_6 (approximately 0.0050/2) requires 5200 seconds of reaction time. Thus to get from $0.0050\ M$ C_4H_6 to $0.0024\ M$ C_4H_6 takes 3570 seconds (5200 − 1630). The second half-life is much longer than the first. This pattern is characteristic of second-order reactions. In fact, *for a second-order reaction, each successive half-life is double the preceding one* (provided the effects of the reverse reaction can be ignored, as we are assuming here). Prove this to yourself by examining the equation $t_{1/2} = 1/(k[A]_0)$.

For each successive half-life, $[A]_0$ is halved. Since $t_{1/2} = 1/k[A]_0$, $t_{1/2}$ doubles.

Zero-Order Rate Laws

Most reactions involving a single reactant show either first-order or second-order kinetics. However, sometimes such a reaction can be a **zero-order reaction.** The rate law for a zero-order reaction is

$$\text{Rate} = k[A]^0 = k(1) = k$$

A zero-order reaction has a constant rate.

For a zero-order reaction, the rate is constant. It does not change with concentration as it does for first-order or second-order reactions.

The **integrated rate law for a zero-order reaction** is

$$[A] = -kt + [A]_0 \qquad (12.7)$$

In this case a plot of [A] versus t gives a straight line of slope $-k$, as shown in Fig. 12.6.

The expression for the half-life of a zero-order reaction can be obtained from the integrated rate law. By definition, $[A] = [A]_0/2$ when $t = t_{1/2}$, so

$$\frac{[A]_0}{2} = -kt_{1/2} + [A]_0$$

or

$$kt_{1/2} = \frac{[A]_0}{2}$$

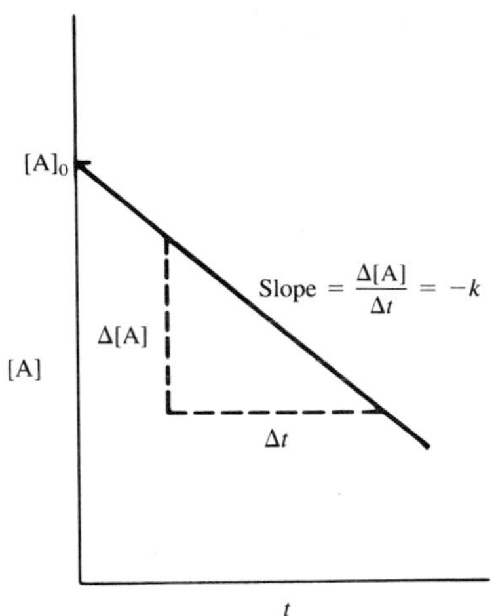

Figure 12.6

A plot of [A] versus t for a zero-order reaction.

Solving for $t_{1/2}$ gives

$$t_{1/2} = \frac{[A]_0}{2k} \tag{12.8}$$

Zero-order reactions are most often encountered when a substance such as a metal surface or an enzyme is required for the reaction to occur. For example, the decomposition reaction

$$2N_2O(g) \longrightarrow 2N_2(g) + O_2(g)$$

occurs on a hot platinum surface. When the platinum surface is completely covered with N_2O molecules, an increase in the concentration of N_2O has no effect on the rate since only those N_2O molecules on the surface can react. Under these conditions *the rate is a constant* because it is controlled by what happens on the platinum surface rather than by the total concentration of N_2O, as illustrated in Fig. 12.7. This reaction can also occur at high temperatures with no platinum surface present, but under these conditions it is not zero order.

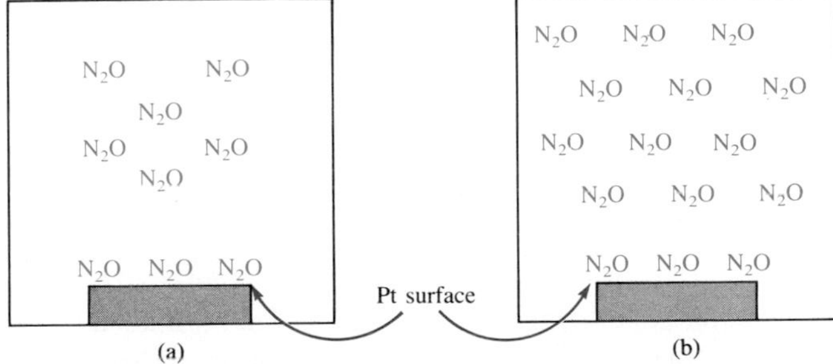

Figure 12.7

The decomposition reaction $2N_2O(g) \rightarrow 2N_2(g) + O_2(g)$ takes place on a platinum surface. Although $[N_2O]$ is twice as great in (b) as (a), the rate of decomposition of N_2O is the same in both cases since the platinum surface can only accommodate a certain number of molecules. As a result, this reaction is zero order.

Integrated Rate Laws for Reactions with More than One Reactant

So far we have considered the integrated rate laws for simple reactions with only one reactant. Special techniques are required to deal with more complicated reactions. Let's consider the reaction

$$BrO_3^-(aq) + 5Br^-(aq) + 6H^+(aq) \rightarrow 3Br_2(l) + 3H_2O(l)$$

From experimental evidence we know that the rate law is

$$\text{Rate} = -\frac{\Delta[BrO_3^-]}{\Delta t} = k[BrO_3^-][Br^-][H^+]^2$$

Suppose we run this reaction under conditions where $[BrO_3^-]_0 = 1.0 \times 10^{-3}\ M$, $[Br^-]_0 = 1.0\ M$, and $[H^+]_0 = 1.0\ M$. As the reaction proceeds, $[BrO_3^-]$ decreases significantly, but because the Br^- ion and H^+ ion concentrations are so large initially, relatively little of these two reactants is consumed. Thus $[Br^-]$ and $[H^+]$ remain *approximately constant*. In other words, under the conditions where the Br^- ion and H^+ ion concentrations are much larger than the BrO_3^- ion concentration, we can assume that throughout the reaction

$$[Br^-] = [Br^-]_0 \quad \text{and} \quad [H^+] = [H^+]_0$$

This means that the rate law can be written

$$\text{Rate} = k[Br^-]_0[H^+]^2_0[BrO_3^-] = k'[BrO_3^-]$$

where, since $[Br^-]_0$ and $[H^+]_0$ are constant,

$$k' = k[Br^-]_0[H^+]^2_0$$

The rate law

$$\text{Rate} = k'[BrO_3^-]$$

is first order. However, since this rate law was obtained by simplifying a more complicated one, it is called a **pseudo-first-order rate law.** Under the conditions of this experiment, a plot of $\ln[BrO_3^-]$ versus t will give a straight line where the slope is equal to $-k'$. Since $[Br^-]_0$ and $[H^+]_0$ are known, the value of k can be calculated from the equation

$$k' = k[Br^-]_0[H^+]^2_0$$

which can be rearranged to give

$$k = \frac{k'}{[Br^-]_0[H^+]^2_0}$$

Note that the kinetics of complicated reactions can be studied by observing the behavior of one reactant at a time. If the concentration of one reactant is much smaller than the concentrations of the others, then the amounts of those reactants present in large concentrations will not change significantly and can be regarded as constant. The change in concentration with time of the reactant present in a relatively small amount can then be used to determine the order of the reaction in that component. This technique allows us to determine rate laws for complex reactions.

12.5 Rate Laws: A Summary

Purpose

- To summarize the two types of rate laws and the methods by which they can be determined.

In the last several sections we have developed the following important points:

1. To simplify the rate laws for reactions we have always assumed that the rate is being studied under conditions where only the forward reaction is important. This produces rate laws that only contain reactant concentrations.
2. There are two types of rate laws.
   a. The differential rate law (often called *the rate law*) shows how the rate depends on the concentrations. The forms of the rate laws for zero-order, first-order, and second-order kinetics of reactions with single reactants are shown in Table 12.6.
   b. The integrated rate law shows how concentration depends on time. The integrated rate laws corresponding to zero-order, first-order, and second-order kinetics of one-reactant reactions are given in Table 12.6.
3. Whether we determine the differential rate law or the integrated rate law depends on the type of data that can be collected conveniently and accurately. Once we have experimentally determined either type of rate law we can write the other for a given reaction.
4. The most common method for experimentally determining the differential rate law is the method of initial rates. Several experiments are run at different initial

Summary of the Kinetics for Reactions of the Type aA → Products That Are Zero, First, or Second Order in [A]

	Order		
	Zero	First	Second
Rate law	Rate = k	Rate = $k[A]$	Rate = $k[A]^2$
Integrated rate law	$[A] = -kt + [A]_0$	$\ln[A] = -kt + \ln[A]_0$	$\frac{1}{[A]} = kt + \frac{1}{[A]_0}$
Plot needed to give a straight line	$[A]$ versus t	$\ln[A]$ versus t	$\frac{1}{[A]}$ versus t
Relationship of rate constant to the slope of straight line	Slope = $-k$	Slope = $-k$	Slope = k
Half-life	$t_{1/2} = \frac{[A]_0}{2k}$	$t_{1/2} = \frac{0.693}{k}$	$t_{1/2} = \frac{1}{k[A]_0}$

Table 12.6

concentrations and the instantaneous rates are determined for each at the same value of t as close to $t = 0$ as possible. The idea is to evaluate the rate before the concentrations change significantly from the initial values. By comparing the initial rates and the initial concentrations the dependence of the rate on the concentrations of various reactants can be obtained—that is, the order in each reactant can be determined.

5. To experimentally determine the integrated rate law for a reaction, concentrations are measured at various values of t as the reaction proceeds. Then the job is to see which integrated rate law correctly fits the data. Typically this is done by ascertaining which type of plot gives a straight line. This is described for one-reactant reactions in Table 12.6. Once the correct straight line plot is found, the correct integrated rate law can be chosen and the value of k obtained from the slope. Also the (differential) rate law for the reaction can then be written.
6. The integrated rate law for a reaction that involves several reactants can be treated by choosing conditions such that the concentration of only one reactant varies in a given experiment. This is done by having the concentration of one reactant small compared to the concentrations of all of the others, causing a rate law such as

$$\text{Rate} = k[\text{A}]^n[\text{B}]^m[\text{C}]^p$$

to reduce to

$$\text{Rate} = k'[\text{A}]^n$$

where $k' = k[\text{B}]_0{}^m[\text{C}]_0{}^p$ and $[\text{B}]_0 \gg [\text{A}]_0$ and $[\text{C}]_0 \gg [\text{A}]_0$. The value of n is obtained by determining whether a plot of $\ln[\text{A}]$ versus t is linear ($n = 1$) or a plot of $1/[\text{A}]$ versus t is linear ($n = 2$). The value of k' is determined from the slope of the appropriate plot. The values of m, p, and k are found by determining the value of k' at several different concentrations of B and C.

12.6 Reaction Mechanisms

Purpose

- To explore the relationship between the reaction pathway and the rate law.

Most chemical reactions occur by a *series of steps* called the **reaction mechanism.** To understand a reaction we must know its mechanism, and one of the main purposes for studying kinetics is to learn as much as possible about the steps involved in a reaction. In this section we explore some of the fundamental characteristics of reaction mechanisms.

Consider the reaction between nitrogen dioxide and carbon monoxide:

$$NO_2(g) + CO(g) \rightarrow NO(g) + CO_2(g)$$

The rate law for this reaction is known from experiment to be

$$\text{Rate} = k[NO_2]^2$$

As we will see below, this reaction is more complicated than it appears from the balanced equation. This is quite typical; the balanced equation for a reaction tells us the reactants, the products, and the stoichiometry but gives no direct information about the reaction mechanism.

A balanced equation does not tell us *how* the reactants become products.

For the reaction between nitrogen dioxide and carbon monoxide, the mechanism is thought to involve the following steps:

$$NO_2(g) + NO_2(g) \xrightarrow{k_1} NO_3(g) + NO(g)$$

$$NO_3(g) + CO(g) \xrightarrow{k_2} NO_2(g) + CO_2(g)$$

where k_1 and k_2 are the rate constants of the individual reactions. In this mechanism, gaseous NO_3 is an **intermediate,** a species that is neither a reactant nor a product but that is formed and consumed in the reaction sequence.

An intermediate is formed in one step and used up in a subsequent step and so is never seen as a product.

Each of these two reactions is called an **elementary step,** *a reaction whose rate law can be written from its molecularity.* **Molecularity** is defined as the number of species that must collide to produce the reaction indicated by that step. A reaction involving one molecule is called a **unimolecular step.** Reactions involving the collision of two and three species are termed **bimolecular** and **termolecular,** respectively. Termolecular steps are quite rare, because the probability of three molecules colliding simultaneously is very small. Examples of these three types of elementary steps and the corresponding rate laws are shown in Table 12.7. Note from Table 12.7 that the rate law for an elementary step follows *directly* from the molecularity of that step. For example, for a bimolecular step the rate law is always second order, either of the form $k[A]^2$ for a step with a single reactant or of the form $k[A][B]$ for a step involving two reactants.

The prefix *uni-* means one, *bi-* means two, and *ter-* means three.

A unimolecular elementary step is always first order, a bimolecular step is always second order, and so on.

Examples of Elementary Steps

Elementary step	Molecularity	Rate law
$A \rightarrow$ products	*Uni*molecular	Rate = $k[A]$
$A + A \rightarrow$ products ($2A \rightarrow$ products)	*Bi*molecular	Rate = $k[A]^2$
$A + B \rightarrow$ products	*Bi*molecular	Rate = $k[A][B]$
$A + A + B \rightarrow$ products ($2A + B \rightarrow$ products)	*Ter*molecular	Rate = $k[A]^2[B]$
$A + B + C \rightarrow$ products	*Ter*molecular	Rate = $k[A][B][C]$

Table 12.7

We can now define a reaction mechanism more precisely. It is a *series of elementary steps that must satisfy two requirements:*

1. The sum of the elementary steps must give the overall balanced equation for the reaction.
2. The mechanism must agree with the experimentally determined rate law.

To see how these requirements are applied, we will consider the mechanism given above for the reaction of nitrogen dioxide and carbon monoxide. First, note that the sum of the two steps gives the overall balanced equation:

$$
\begin{array}{r}
NO_2(g) + NO_2(g) \rightarrow NO_3(g) + NO(g) \\
NO_3(g) + CO(g) \rightarrow NO_2(g) + CO_2(g) \\
\hline
\cancel{NO_2}(g) + NO_2(g) + \cancel{NO_3}(g) + CO(g) \rightarrow \cancel{NO_3}(g) + NO(g) + \cancel{NO_2}(g) + CO_2(g)
\end{array}
$$

Overall reaction: $NO_2(g) + CO(g) \rightarrow NO(g) + CO_2(g)$

A reaction is only as fast as its slowest step.

The first requirement for a correct mechanism is met. To see if the mechanism meets the second requirement, we need to introduce a new idea: the **rate-determining step.** Multistep reactions often have one step that is much slower than all the others. Reactants can become products only as fast as they can get through this slowest step. That is, the overall reaction can be no faster than the slowest or rate-determining step in the sequence. An analogy for this situation is the pouring of water rapidly into a container through a funnel. The water collects in the container at a rate that is essentially determined by the size of the funnel opening and not by the rate of pouring.

Which is the rate-determining step in the reaction of nitrogen dioxide and carbon monoxide? Let's *assume* that the first step is rate-determining and the second step is relatively fast:

$$
\begin{array}{rl}
NO_2(g) + NO_2(g) \rightarrow NO_3(g) + NO(g) & \text{Slow (rate-determining)} \\
NO_3(g) + CO(g) \rightarrow NO_2(g) + CO_2(g) & \text{Fast}
\end{array}
$$

What we have really assumed here is that the formation of NO_3 occurs much more slowly than its reaction with CO. The rate of CO_2 production is then controlled by the rate of formation of NO_3 in the first step. Since this is an elementary step, we can write the rate law from the molecularity. The bimolecular first step has the rate law

$$\text{Rate of formation of } NO_3 = \frac{\Delta[NO_3]}{\Delta t} = k_1[NO_2]^2$$

Since the overall reaction rate can be no faster than the slowest step,

$$\text{Overall rate} = k_1[NO_2]^2$$

Note that this rate law agrees with the experimentally determined rate law given earlier. The mechanism we assumed above satisfies the two requirements stated earlier and *may* be the correct mechanism for the reaction.

How does a chemist deduce the mechanism for a given reaction? The rate law is always determined first. Then, using chemical intuition and following the two rules given above, the chemist constructs a possible mechanism. *A mechanism can never be proved absolutely*. We can only say that a mechanism that satisfies the two requirements is *possibly* correct. Deducing mechanisms for chemical reactions can be difficult and requires skill and experience. We will only touch on this process in this text.

Sample Exercise 12.6

The balanced equation for the reaction of the gases nitrogen dioxide and fluorine is

$$2NO_2(g) + F_2(g) \rightarrow 2NO_2F(g)$$

The experimentally determined rate law is

$$\text{Rate} = k[NO_2][F_2]$$

Sample Exercise 12.6, continued

A suggested mechanism for this reaction is

$$NO_2 + F_2 \xrightarrow{k_1} NO_2F + F \quad \text{Slow}$$

$$F + NO_2 \xrightarrow{k_2} NO_2F \quad \text{Fast}$$

Is this an acceptable mechanism? That is, does it satisfy the two requirements?

Solution

The first requirement for an acceptable mechanism is that the sum of the steps should give the balanced equation:

$$\begin{array}{rl} & NO_2 + F_2 \rightarrow NO_2F + F \\ & F + NO_2 \rightarrow NO_2F \\ \hline & 2NO_2 + F_2 + F \rightarrow 2NO_2F + F \\ \text{Overall reaction:} & 2NO_2 + F_2 \rightarrow 2NO_2F \end{array}$$

The first requirement is met.

The second requirement is that the mechanism must agree with the experimentally determined rate law. Since the proposed mechanism states that the first step is rate-determining, the overall reaction rate must be that of the first step. The first step is bimolecular, so the rate law is

$$\text{Rate} = k_1[NO_2][F_2]$$

This has the same form as the experimentally determined rate law. The proposed mechanism is acceptable because it satisfies both requirements. (Note that we have not proved it is *the correct* mechanism.)

Mechanisms with Fast Forward and Reverse First Steps

A common type of reaction mechanism is one involving a first step in which *both* the forward and reverse reactions are very fast compared to the second step. An example of this type of mechanism is that for the decomposition of ozone to oxygen. The balanced reaction is

$$2O_3(g) \rightarrow 3O_2(g)$$

The observed rate law is

$$\text{Rate} = k\frac{[O_3]^2}{[O_2]}$$

Note that this rate law is unusual in that it contains the concentration of a *product*. The mechanism proposed for this process is

$$O_3 \underset{k_{-1}}{\overset{k_1}{\rightleftharpoons}} O_2 + O$$

$$O + O_3 \xrightarrow{k_2} 2O_2$$

The double arrows in the first step indicate that both the forward and reverse reactions are important. They have the rate constants k_1 and k_{-1}, respectively.

For this mechanism we will assume that *both* the forward and reverse reactions of the first step are very fast compared to the second step. This means that the second step is rate-determining. Therefore, the rate for the overall reaction is equal to the rate of this step:

$$\text{Rate} = k_2[\text{O}][\text{O}_3]$$

This rate law does not have the same form as the experimentally determined rate law. For one thing it contains the concentration of the intermediate, an oxygen atom. We can remove [O] and obtain a rate law that agrees with experiment by making an additional assumption. We will assume that the rates of the forward and reverse reactions in the first step are equal. Since

$$\text{Rate of forward reaction} = k_1[\text{O}_3]$$

and

$$\text{Rate of reverse reaction} = k_{-1}[\text{O}_2][\text{O}]$$

the assumption of equal rates gives

$$k_1[\text{O}_3] = k_{-1}[\text{O}_2][\text{O}]$$

We solve for [O]:

$$[\text{O}] = \frac{k_1[\text{O}_3]}{k_{-1}[\text{O}_2]}$$

Now we substitute for [O] in the rate law for the second step:

$$\text{Rate} = k_2[\text{O}][\text{O}_3] = k_2\left(\frac{k_1[\text{O}_3]}{k_{-1}[\text{O}_2]}\right)[\text{O}_3] = \frac{k_2k_1[\text{O}_3]^2}{k_{-1}[\text{O}_2]}$$

$$= k\frac{[\text{O}_3]^2}{[\text{O}_2]}$$

where k is a composite constant representing k_2k_1/k_{-1}.

This rate law, *derived* by postulating the two elementary steps and making assumptions about the relative rates of these steps, agrees with the experimental rate law. Since this mechanism (the elementary steps *plus* the assumptions) also gives the correct overall stoichiometry, it is an acceptable mechanism for the decomposition of ozone to oxygen.

Sample Exercise 12.7

The gas-phase reaction of chlorine with chloroform is described by the equation

$$Cl_2(g) + CHCl_3(g) \rightarrow HCl(g) + CCl_4(g)$$

The rate law determined from experiment is an example of one with a noninteger order:

$$\text{Rate} = k[Cl_2]^{1/2}[CHCl_3]$$

A proposed mechanism for this reaction is shown in the following reaction sequence:

Sample Exercise 12.7, continued

$$Cl_2(g) \underset{k_{-1}}{\overset{k_1}{\rightleftharpoons}} 2Cl(g) \qquad \text{Both fast with equal rates}$$

$$Cl(g) + CHCl_3(g) \xrightarrow{k_2} HCl(g) + CCl_3(g) \qquad \text{Slow}$$

$$CCl_3(g) + Cl(g) \xrightarrow{k_3} CCl_4(g) \qquad \text{Fast}$$

Is this an acceptable mechanism for the reaction?

Solution

Two questions must be answered. First, does the mechanism give the correct overall stoichiometry? Adding the three steps does give the correct balanced equation:

$$\begin{aligned} Cl_2(g) &\rightleftharpoons 2Cl(g) \\ Cl(g) + CHCl_3(g) &\longrightarrow HCl(g) + CCl_3(g) \\ CCl_3(g) + Cl(g) &\longrightarrow CCl_4(g) \end{aligned}$$

$$Cl_2(g) + Cl(g) + CHCl_3(g) + CCl_3(g) + Cl(g) \longrightarrow 2Cl(g) + HCl(g) + CCl_3(g) + CCl_4(g)$$

$$\text{Overall reaction:} \quad Cl_2(g) + CHCl_3(g) \longrightarrow HCl(g) + CCl_4(g)$$

Second, does the mechanism agree with the observed rate law? Since the overall reaction rate will be determined by the rate of the slowest step,

$$\text{Overall rate} = \text{rate of second step} = k_2[Cl][CHCl_3]$$

Since the chlorine atom is an intermediate, we must find a way to eliminate [Cl] in the rate law. This can be done using the equal rates of the forward and reverse reactions of the first step:

$$k_1[Cl_2] = k_{-1}[Cl]^2$$

Solving for $[Cl]^2$ gives

$$[Cl]^2 = \frac{k_1[Cl_2]}{k_{-1}}$$

Taking the square root of both sides gives

$$[Cl] = \left(\frac{k_1}{k_{-1}}\right)^{1/2} [Cl_2]^{1/2}$$

then

$$\text{Rate} = k_2[Cl][CHCl_3] = k_2\left(\frac{k_1}{k_{-1}}\right)^{1/2} [Cl_2]^{1/2}[CHCl_3] = k[Cl_2]^{1/2}[CHCl_3]$$

where

$$k = k_2\left(\frac{k_1}{k_{-1}}\right)^{1/2}$$

The rate law derived from the mechanism now agrees with the experimentally observed rate law. This mechanism satisfies the two requirements and thus is an acceptable mechanism.

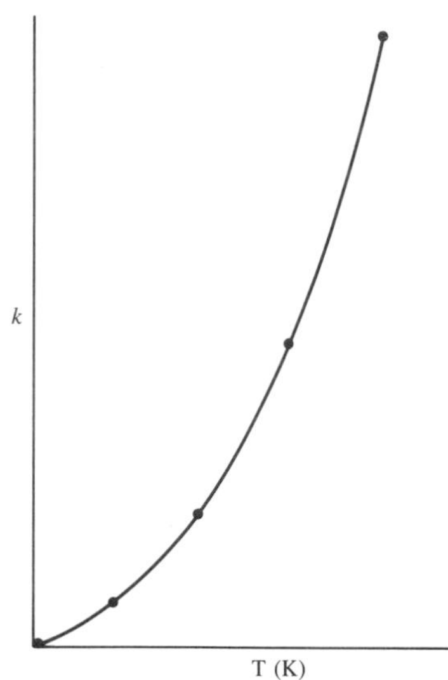

Figure 12.8

A plot showing the exponential dependence of the rate constant on absolute temperature. The exact temperature dependence of k varies with the reaction in question. This plot represents the behavior of a rate constant that doubles for every increase in temperature of 10 K.

12.7 A Model for Chemical Kinetics

Purpose

- To discuss the temperature dependence of reaction rates.
- To describe the collision model.
- To define and show how to calculate activation energy.

How do chemical reactions occur? We already have given some indications. For example, we have seen that the rates of chemical reactions depend on the concentrations of the reacting species. The initial rate for the reaction

$$aA + bB \rightarrow \text{products}$$

can be described by the rate law

$$\text{Rate} = k[A]^n[B]^m$$

where the order of each reactant depends on the detailed reaction mechanism. This explains why reaction rates depend on concentration. But what about some of the other factors affecting reaction rates? For example, how does temperature affect the speed of a reaction?

We can answer this question qualitatively from our experience. We have refrigerators because food spoilage is retarded at low temperatures. The combustion of wood occurs at a measurable rate only at high temperatures. An egg cooks in boiling water much faster at sea level than in Leadville, Colorado (elevation 10,000 feet), where the boiling point of water is approximately 90°C. These observations and others lead us to conclude that *chemical reactions speed up when the temperature is increased*. Experiments have shown that virtually all rate constants show an exponential increase with absolute temperature, as represented in Fig. 12.8.

In this section we discuss a model used to account for the observed characteristics of reaction rates. This model, called the **collision model,** is built around the central idea that *molecules must collide to react*. We have already seen how this assumption explains the concentration dependence of reaction rates. Now we need to consider whether this model can account for the observed temperature dependence of reaction rates.

Food and other supplies on Ross Island left by the 1910 Scott expedition to the South Pole. The low temperatures that prevail in this region greatly slow the reactions that cause spoilage and corrosion.

The kinetic molecular theory of gases predicts that an increase in temperature raises molecular velocities and so increases the frequency of collisions between molecules. This idea agrees with the observation that reaction rates are greater at higher temperatures. Thus there is qualitative agreement between the collision model and experimental observations. However, it is found that the rate of reaction is much smaller than the calculated collision frequency in a collection of gas particles. This must mean that *only a small fraction of the collisions produces a reaction*. Why?

This question was first addressed in the 1880s by Svante Arrhenius. He proposed the existence of a *threshold energy,* called the **activation energy,** that must be overcome to produce a chemical reaction. Such a proposal makes sense, as we can see by considering the decomposition of BrNO in the gas phase:

$$2BrNO(g) \rightarrow 2NO(g) + Br_2(g)$$

In this reaction two Br—N bonds must be broken and one Br—Br bond formed.

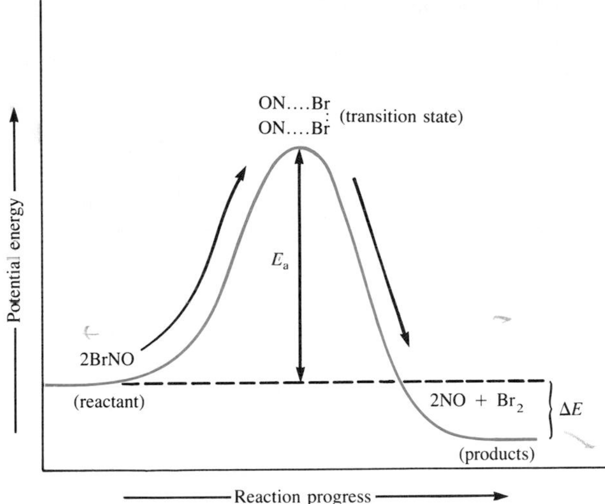

Figure 12.9

The change in potential energy as a function of reaction progress for the reaction $2BrNO \rightarrow 2NO + Br_2$. The activation energy (E_a) represents the energy needed to disrupt the BrNO molecules so that they can form products. The quantity ΔE represents the net change in energy in going from reactant to products.

Breaking a Br—N bond requires considerable energy (243 kJ/mol), which must come from somewhere. The collision model postulates that the energy comes from the kinetic energies possessed by the reacting molecules before the collision. This kinetic energy is changed into potential energy as the molecules are distorted during a collision to break bonds and rearrange the atoms into the product molecules.

We can envision the reaction process as shown in Fig. 12.9. The arrangement of atoms found at the top of the potential energy "hill," or barrier, is called the **activated complex,** or **transition state.** The conversion of BrNO to NO and Br_2 is exothermic, as indicated by the fact that the products have lower potential energy than the reactant. However, ΔE has no effect on the rate of the reaction. Rather the rate depends on the size of the activation energy, E_a.

The higher the activation energy, the slower the reaction at a given temperature.

The main point here is that a certain minimum energy is required for two BrNO molecules to "get over the hill" so that products can form. This energy is furnished by the energy of the collision. A collision between two BrNO molecules with small kinetic energies will not have enough energy to get over the barrier. At a given temperature only a certain fraction of the collisions possesses enough energy to be effective, or to result in product formation.

We can be more precise by recalling from Chapter 5 that a distribution of velocities occurs in a sample of gas molecules. Therefore, a distribution of collision energies also occurs, as shown in Fig. 12.10 for two different temperatures. Figure 12.10 also shows the activation energy for the reaction in question. Only collisions with energy greater than the activation energy are able to react (get over the barrier). At the lower temperature, T_1 [Fig. 12.10(a)], the fraction of effective collisions is quite small. However, as the temperature is increased to T_2 [Fig. 12.10(b)], the fraction of collisions with the required activation energy increases dramatically. When the temperature is doubled, the fraction of effective collisions much more than doubles. In fact, the fraction of effective collisions increases *exponentially* with temperature. This is encouraging for our theory; remember that rates of reactions are observed to increase exponentially with temperature. Arrhenius postulated that the number of collisions having the activation energy is a fraction of the total number of collisions given by the expression:

$$\text{Number of collisions with the activation energy} = (\text{total number of collisions})e^{-E_a/RT}$$

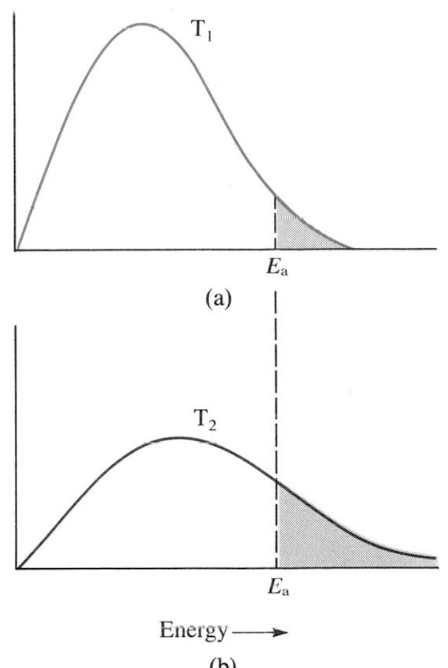

Figure 12.10

Plot showing the number of collisions with a particular energy at (a) T_1 and (b) T_2, where $T_2 > T_1$.

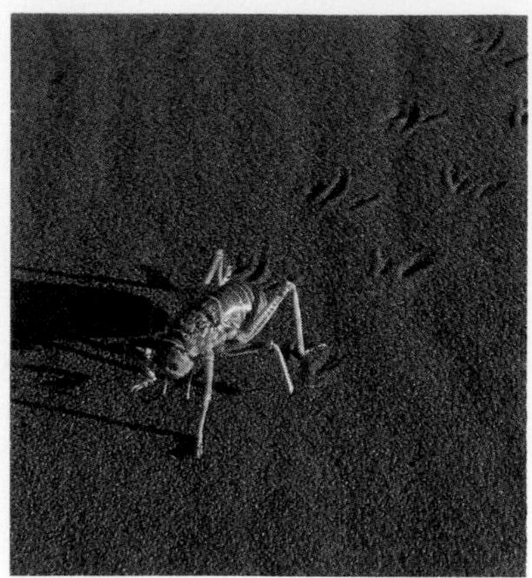

A dune cricket in Namibia, Africa. The frequency of a cricket's chirps depends on the temperature of the cricket.

where E_a is the activation energy, R is the universal gas constant, and T is the Kelvin temperature. The factor $e^{-E_a/RT}$ represents the fraction of collisions with energy E_a or greater at temperature T.

We have seen that not all molecular collisions are effective in producing chemical reactions because a minimum energy is required for the reaction to occur. There is, however, another complication. Experiments show that the *observed reaction rate is considerably smaller than the rate of collisions with enough energy to surmount the barrier*. This means that many collisions, even though they have the required energy, still do not produce a reaction. Why not?

The answer lies in the **molecular orientations** during collisions. We can illustrate this using the reaction between two BrNO molecules, as shown in Fig. 12.11. Some collision orientations can lead to reaction, and others cannot. Therefore, we must include a correction factor to allow for collisions with nonproductive molecular orientations.

To summarize, two requirements must be satisfied for reactants to collide successfully (to rearrange to products):

1. The collision must involve enough energy to produce the reaction; that is, the collision energy must equal or exceed the activation energy.
2. The relative orientation of the reactants must allow formation of any new bonds necessary to produce products.

Taking these factors into account, we can represent the rate constant as

$$k = zpe^{-E_a/RT}$$

where z is the collision frequency, p is called the **steric factor** (always less than one) and reflects the fraction of collisions with effective orientations, and $e^{-E_a/RT}$ represents the fraction of collisions with sufficient energy to produce a reaction. This expression is most often written in form

$$k = Ae^{-E_a/RT} \tag{12.9}$$

which is called the **Arrhenius equation.** In this equation A replaces zp and is called the **frequency factor** for the reaction.

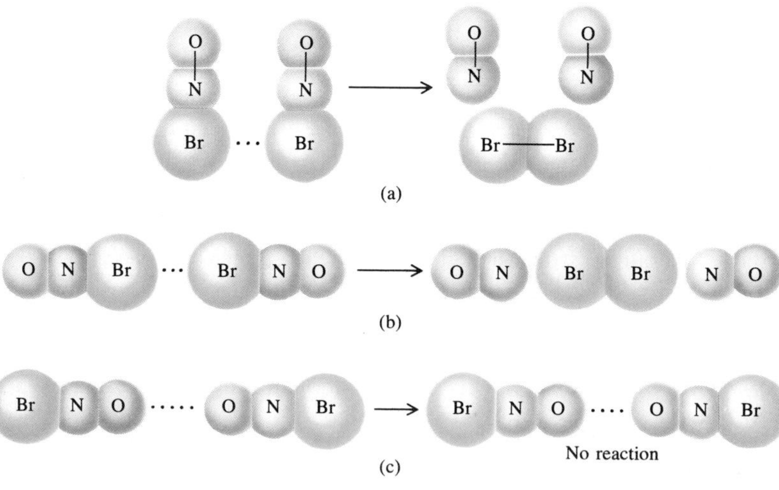

Figure 12.11

Several possible orientations for a collision between two BrNO molecules. Orientations (a) and (b) can lead to a reaction, but orientation (c) cannot.

Taking the natural logarithm of each side of the Arrhenius equation gives

$$\ln(k) = -\frac{E_a}{R}\left(\frac{1}{T}\right) + \ln(A) \qquad (12.10)$$

Equation (12.10) is a linear equation of the type $y = mx + b$, where $y = \ln(k)$, $m = -E_a/R$ = slope, $x = 1/T$, and $b = \ln(A)$ = intercept. Thus, for a reaction where the rate constant obeys the Arrhenius equation, a plot of ln(k) versus $1/T$ gives a straight line. The slope and intercept can be used to determine, respectively, the values of E_a and A characteristic of that reaction. The fact that most rate constants obey the Arrhenius equation to a good approximation indicates that the collision model for chemical reactions is physically reasonable.

Sample Exercise 12.8

The reaction

$$2N_2O_5(g) \rightarrow 4NO_2(g) + O_2(g)$$

was studied at several temperatures and the following values of k were obtained:

k (s^{-1})	T (°C)
2.0×10^{-5}	20
7.3×10^{-5}	30
2.7×10^{-4}	40
9.1×10^{-4}	50
2.9×10^{-3}	60

Calculate the value of E_a for this reaction.

Solution

To obtain the value of E_a, we need to construct a plot of ln(k) versus $1/T$. First we must calculate values of ln(k) and $1/T$ as shown below:

T (°C)	T (K)	$1/T$ (K)	k (s^{-1})	ln(k)
20	293	3.41×10^{-3}	2.0×10^{-5}	−10.82
30	303	3.30×10^{-3}	7.3×10^{-5}	−9.53
40	313	3.19×10^{-3}	2.7×10^{-4}	−8.22
50	323	3.10×10^{-3}	9.1×10^{-4}	−7.00
60	333	3.00×10^{-3}	2.9×10^{-3}	−5.84

The plot of ln(k) versus $1/T$ is shown in Fig. 12.12, where the slope

$$\frac{\Delta \ln(k)}{\Delta\left(\frac{1}{T}\right)}$$

is found to be -1.2×10^4 K. The value of E_a can be determined by solving the following equation.

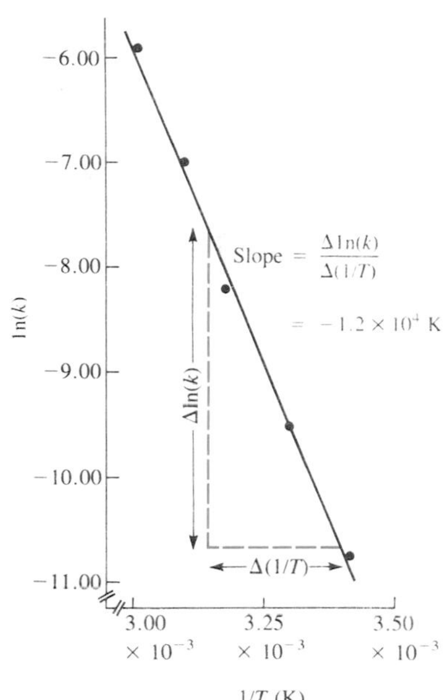

Figure 12.12

Plot of ln(k) versus $1/T$ for the reaction $2N_2O_5(g) \rightarrow 4NO_2(g) + O_2(g)$. The value of the activation energy for this reaction can be obtained from the slope of the line, which equals $-E_a/R$.

Sample Exercise 12.8, continued

$$\text{Slope} = -\frac{E_a}{R}$$

$$E_a = -R(\text{slope}) = -(8.3145 \text{ J/K mol})(-1.2 \times 10^4 \text{ K})$$
$$= 1.0 \times 10^5 \text{ J/mol}$$

Thus the value of the activation energy for this reaction is 1.0×10^5 J/mol.

The most common procedure for finding E_a for a reaction involves measuring the rate constant k at several temperatures and then plotting $\ln(k)$ versus $1/T$, as shown in Sample Exercise 12.8. However, E_a can also be calculated from the values of k at only two temperatures by using a formula that can be derived as follows from Equation (12.10).

At temperature T_1 where the rate constant is k_1,

$$\ln(k_1) = -\frac{E_a}{RT_1} + \ln(A)$$

At temperature T_2 where the rate constant is k_2,

$$\ln(k_2) = -\frac{E_a}{RT_2} + \ln(A)$$

Subtracting the first equation from the second gives

$$\ln(k_2) - \ln(k_1) = \left(-\frac{E_a}{RT_2} + \ln(A)\right) - \left(-\frac{E_a}{RT_1} + \ln(A)\right)$$

$$= -\frac{E_a}{RT_2} + \frac{E_a}{RT_1}$$

$$\ln\left(\frac{k_2}{k_1}\right) = \frac{E_a}{R}\left(\frac{1}{T_1} - \frac{1}{T_2}\right) \qquad (12.11)$$

Thus the values of k_1 and k_2 measured at temperatures T_1 and T_2 can be used to calculate E_a, as shown in Sample Exercise 12.9.

Sample Exercise 12.9

The gas-phase reaction between methane and diatomic sulfur is given by the equation

$$CH_4(g) + 2S_2(g) \rightarrow CS_2(g) + 2H_2S(g)$$

At 550°C the rate constant for this reaction is 1.1 L/mol s, and at 625°C the rate constant is 6.4 L/mol s. Using these values, calculate E_a for this reaction.

Solution

The relevant data are shown in the table on the next page.

Sample Exercise 12.9, continued

k (L/mol s)	T (°C)	T (K)
$1.1 = k_1$	550	$823 = T_1$
$6.4 = k_2$	625	$898 = T_2$

Substituting these values into Equation (12.11) gives

$$\ln\left(\frac{6.4}{1.1}\right) = \frac{E_a}{8.3145\ \text{J/K mol}}\left(\frac{1}{823\ \text{K}} - \frac{1}{898\ \text{K}}\right)$$

Solving for E_a gives

$$E_a = \frac{(8.3145\ \text{J/K mol})\ln\left(\frac{6.4}{1.1}\right)}{\left(\frac{1}{823}\ \text{K} - \frac{1}{898\ \text{K}}\right)}$$

$$= 1.4 \times 10^5\ \text{J/mol}$$

12.8 Catalysis

Purpose

- To explain how a catalyst speeds up a reaction.
- To discuss heterogeneous and homogeneous catalysis.

We have seen that the rate of a reaction increases dramatically with temperature. If a particular reaction does not occur fast enough at normal temperatures, we can speed it up by raising the temperature. However, sometimes this is not feasible. For example, living cells can survive only in a rather narrow temperature range, and the human body is designed to operate at an almost constant temperature of 98.6°F. But many of the complicated biochemical reactions keeping us alive would be much too slow at this temperature without intervention. We survive only because the body contains many substances called **enzymes,** which increase the rates of these reactions. In fact, almost every biologically important reaction is assisted by a specific enzyme.

Although it is possible to use higher temperatures to speed up commercially important reactions, such as the Haber process for synthesizing ammonia, this is very expensive. In a chemical plant an increase in temperature means significantly increased costs for energy. The use of an appropriate catalyst allows a reaction to proceed rapidly at a relatively low temperature, and can therefore hold down production costs.

A **catalyst** is *a substance that speeds up a reaction without being consumed.* Just as virtually all vital biological reactions are assisted by enzymes (biological

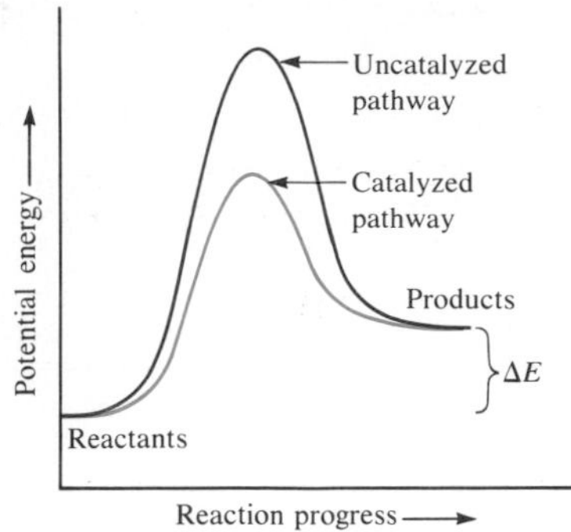

Figure 12.13

Energy plots for a catalyzed and an uncatalyzed pathway for a given reaction.

catalysts), almost all industrial processes also involve the use of catalysts. For example, the production of sulfuric acid uses vanadium(V) oxide, and the Haber process uses a mixture of iron and iron oxide.

How does a catalyst work? Remember that for each reaction a certain energy barrier must be surmounted. How can we make a reaction occur faster without raising the temperature to increase the molecular energies? The solution is to provide a new pathway for the reaction, one with a *lower activation energy*. That is what a catalyst does, as is shown in Fig. 12.13. Because the catalyst allows the reaction to occur with a lower activation energy, a much larger fraction of collisions is effective at a given temperature, and the reaction rate is increased. This effect is illustrated in Fig. 12.14. Note from this diagram that although a catalyst lowers the activation energy (E_a) for a reaction, it does not affect the energy difference (ΔE) between products and reactants.

Catalysts are classified as homogeneous or heterogeneous. A **homogeneous catalyst** is one that is *present in the same phase as the reacting molecules*. A **heterogeneous catalyst** exists *in a different phase,* usually as a solid.

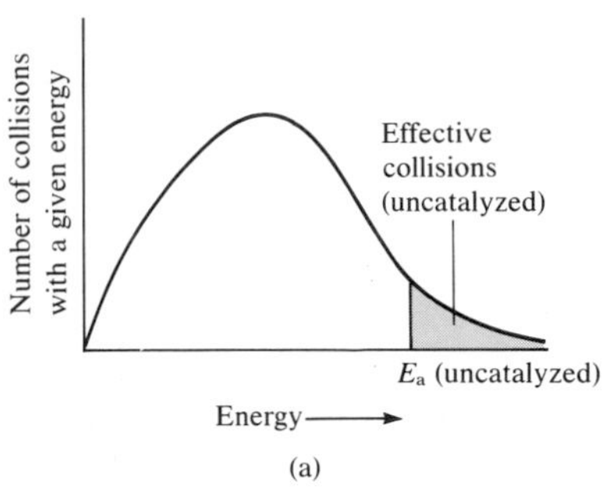

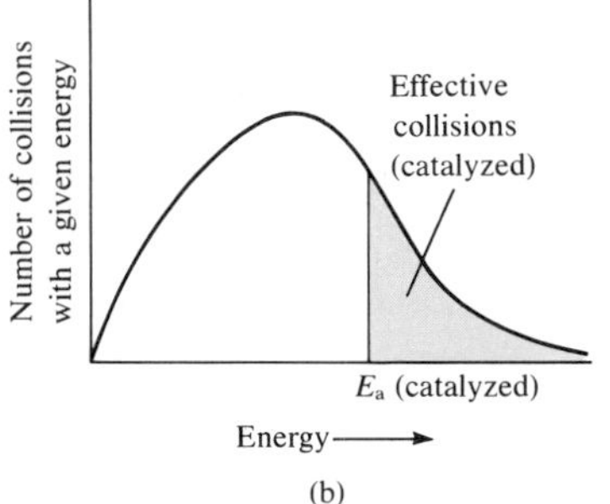

Figure 12.14

Effect of a catalyst on the number of reaction-producing collisions. Because a catalyst provides a reaction pathway with a lower activation energy, a much greater fraction of the collisions is effective for the catalyzed pathway (b) than for the uncatalyzed pathway (a) (at a given temperature). This allows reactants to become products at a much higher rate, even though there is no temperature increase.

Heterogeneous Catalysis

Heterogeneous catalysis most often involves gaseous reactants being adsorbed on the surface of a solid catalyst. **Adsorption** refers to the collection of one substance on the surface of another substance; **absorption** refers to the penetration of one substance into another. Water is *absorbed* by a sponge.

An important example of heterogeneous catalysis occurs in the hydrogenation of unsaturated hydrocarbons, compounds composed mainly of carbon and hydrogen with some carbon-carbon double bonds. Hydrogenation is an important industrial process used to change unsaturated fats, occurring as oils, to saturated fats (solid shortenings such as Crisco) in which the C═C bonds have been converted to C—C bonds, through addition of hydrogen.

A simple example of hydrogenation involves ethylene:

$$\underset{\text{Ethylene}}{H_2C{=}CH_2}(g) + H_2(g) \rightarrow \underset{\text{Ethane}}{H{-}\overset{H}{\underset{H}{C}}{-}\overset{H}{\underset{H}{C}}{-}H}\ (g)$$

This reaction is quite slow at normal temperatures, mainly because the strong bond in the hydrogen molecule results in a large activation energy for the reaction. However, the reaction rate can be greatly increased by using a solid catalyst of platinum, palladium, or nickel. The hydrogen and ethylene adsorb on the catalyst surface, where the reaction occurs. The main function of the catalyst apparently is to allow formation of metal-hydrogen interactions that weaken the H—H bonds and facilitate the reaction. The mechanism is illustrated in Fig. 12.15.

Typically, heterogeneous catalysis involves four steps:

1. adsorption of the reactants,
2. migration of the adsorbed reactants on the surface,
3. reaction of the adsorbed substances, and
4. escape, or *desorption,* of the products.

Heterogeneous catalysis also occurs in the oxidation of gaseous sulfur dioxide to gaseous sulfur trioxide. This process is especially interesting because it illustrates both positive and negative consequences of chemical catalysis. The negative side is the formation of damaging air pollutants.

Recall that sulfur dioxide, a toxic gas with a choking odor, is formed whenever sulfur-containing fuels are burned. However, it is sulfur trioxide that causes most of the environmental damage, mainly through the production of acid rain. When sulfur trioxide combines with a droplet of water, sulfuric acid is formed:

$$H_2O(l) + SO_3(g) \rightarrow H_2SO_4(aq)$$

This sulfuric acid can cause considerable damage to vegetation, buildings and statues, and fish populations.

Sulfur dioxide is *not* rapidly oxidized to sulfur trioxide in clean, dry air. Why, then, is there a problem? The answer is catalysis. Dust particles and water droplets catalyze the reaction between SO_2 and O_2 in the air.

On the positive side, the heterogeneous catalysis of the oxidation of SO_2 is used to advantage in the manufacture of sulfuric acid, where the reaction of O_2 and SO_2 to form SO_3 is catalyzed by a solid mixture of platinum and vanadium (V) oxide.

Heterogeneous catalysis is also utilized in the catalytic converters in automobile exhaust systems. The exhaust gases, containing compounds such as nitric oxide, carbon monoxide, and unburned hydrocarbons, are passed through a converter containing beads of solid catalyst (see Fig. 12.16). The catalyst promotes the conversion of carbon monoxide to carbon dioxide, hydrocarbons to carbon dioxide and water, and nitric oxide to nitrogen gas to lessen the environmental impact of the exhaust gases. However, this beneficial catalysis can, unfortunately, be accompanied by the unwanted catalysis of the oxidation of SO_2 to SO_3, which reacts with the moisture present to form sulfuric acid.

Because of the complex nature of the reactions that take place in the converter, a mixture of catalysts is used. The most effective catalytic materials are transition metal oxides and noble metals such as palladium and platinum.

One consequence of the widespread use of catalytic converters has been the need to remove lead from gasoline. Tetraethyl lead was used for more than 50 years as a very effective "octane-booster" due to its antiknocking characteristics. However, lead quickly destroys much of a converter's catalytic efficiency. This poisoning effect has necessitated the removal of lead from gasoline, which has caused a search for other antiknock additives and the redesigning of engines to run on lower-octane gasoline.

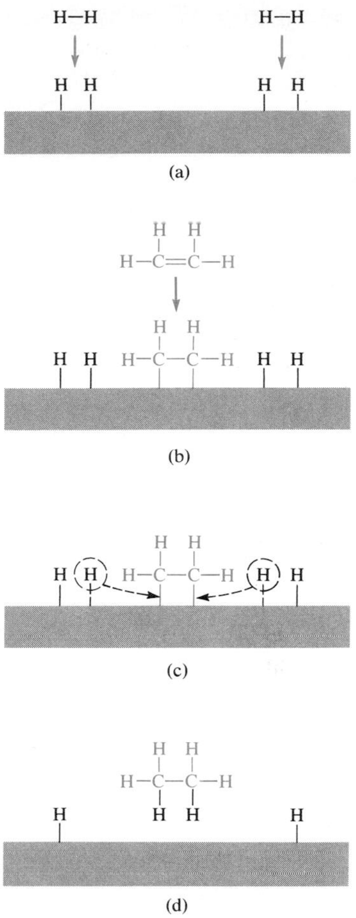

(a)

(b)

(c)

(d)

Figure 12.15

Heterogeneous catalysis of the hydrogenation of ethylene. (a) Hydrogen is adsorbed onto the metal surface, forming metal-hydrogen bonds and breaking the H—H bonds. (b) The C—C π bond in ethylene is broken and metal-carbon bonds are formed during adsorption. (c) The adsorbed molecules and atoms migrate toward each other on the metal surface, forming new C—H bonds. (d) The C atoms in ethane (C_2H_6) have completely saturated bonding capacities and so cannot bind strongly to the metal surfaces. The C_2H_6 molecule thus escapes.

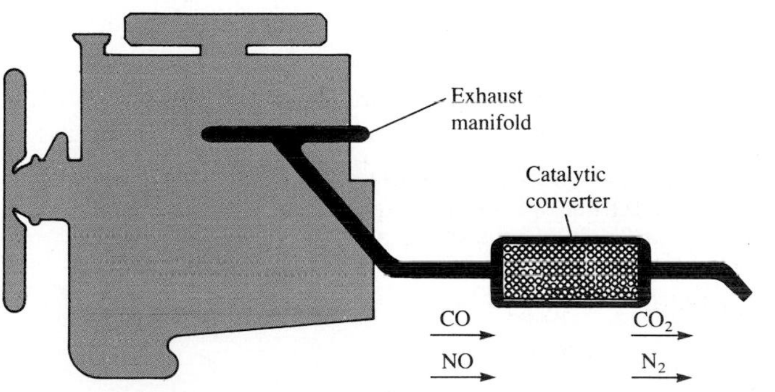

Figure 12.16

The exhaust gases from an automobile engine are passed through a catalytic converter to minimize environmental damage.

Chemical Impact

Enzymes: Nature's Catalysts

The most impressive examples of homogeneous catalysis occur in nature, where the complex reactions necessary for plant and animal life are made possible by enzymes. Enzymes are large molecules specifically tailored to facilitate a given type of reaction. Usually enzymes are proteins, an important class of biomolecules constructed from α-amino acids that have the general structure

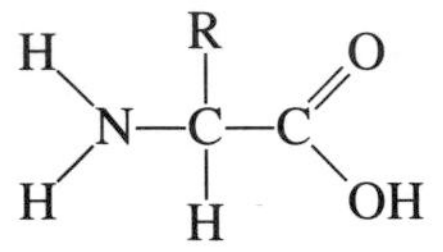

$$\mathrm{H_2N{-}CHR{-}C({=}O){-}OH}$$

where R represents any one of 20 different substituents. These amino acid molecules can be "hooked together" to form a polymer (a word meaning "many parts") called a *protein*. The general structure of a protein can be represented as follows:

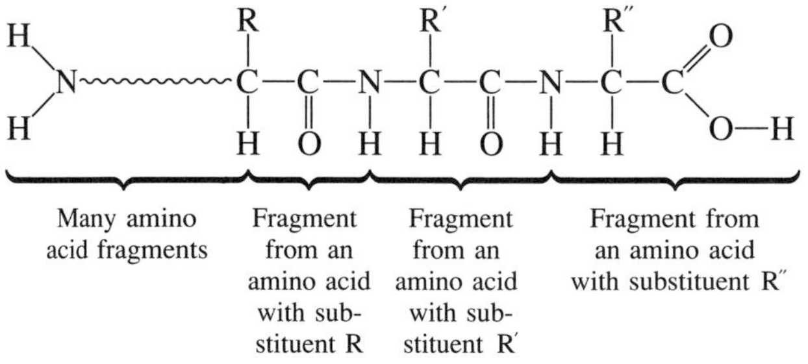

$$\mathrm{H_2N} \sim\sim\sim \mathrm{CHR{-}C({=}O){-}NH{-}CHR'{-}C({=}O){-}NH{-}CHR''{-}C({=}O){-}O{-}H}$$

Many amino acid fragments; Fragment from an amino acid with substituent R; Fragment from an amino acid with substituent R′; Fragment from an amino acid with substituent R″

Since specific proteins are needed by the human body, the proteins in food must be broken into their constituent amino acids, which are then used to construct new proteins in the body's cells. The reaction in which a protein is broken down one amino acid at a time is shown in Fig. 12.17. Note that in this reaction a water molecule reacts with a protein molecule to produce an amino acid and a new protein containing one less amino acid. Without the enzymes found in human cells, this reaction would be much too slow to be useful. One of these enzymes is *carboxypeptidase-A,* a zinc-containing protein (Fig. 12.18).

Carboxypeptidase-A captures the protein to be acted on (called the *substrate*) in a special groove and positions the substrate so that the end is in the *active site,* where the catalysis occurs (Fig. 12.19). Note that the Zn^{2+} ion bonds to the oxygen of the

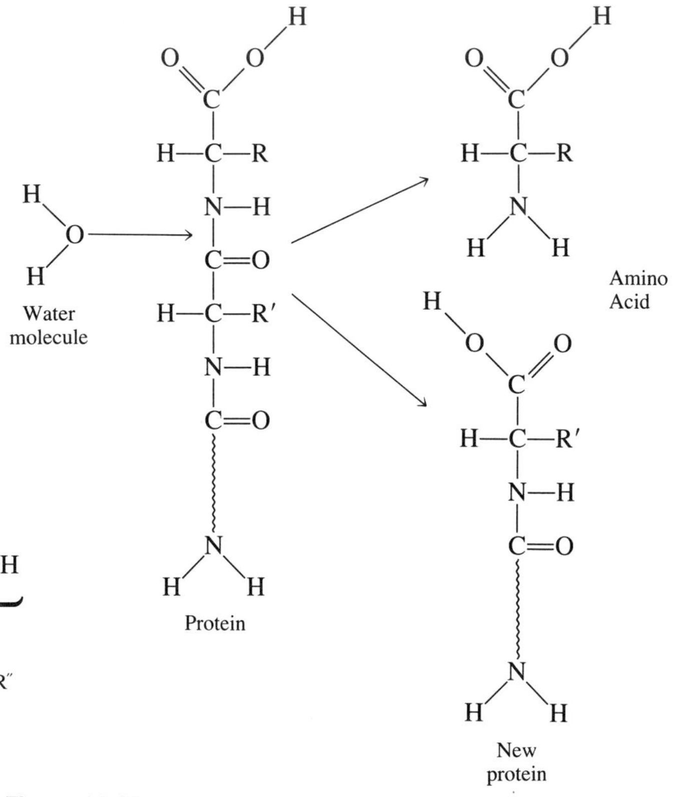

Water molecule: $\mathrm{H{-}O{-}H}$ → Protein: $\mathrm{HO{-}C({=}O){-}CHR{-}NH{-}C({=}O){-}CHR'{-}NH{-}C({=}O)} \sim\sim\sim \mathrm{NH_2}$ → Amino Acid: $\mathrm{HO{-}C({=}O){-}CHR{-}NH_2}$ + New protein: $\mathrm{HO{-}C({=}O){-}CHR'{-}NH{-}C({=}O)} \sim\sim\sim \mathrm{NH_2}$

Figure 12.17

The removal of the end amino acid from a protein by reaction with a molecule of water. The products are an amino acid and a new, smaller protein.

Chemical Impact

C=O (carbonyl) group. This polarizes the electron density in the carbonyl group, allowing the neighboring C—N bond to be broken much more easily. When the reaction is completed, the remaining portion of the substrate protein and the newly formed amino acid are released by the enzyme.

The process just described for carboxypeptidase-A is characteristic of the behavior of other enzymes. Enzyme catalysis can be represented by the series of reactions shown below:

$$E + S \underset{k_{-1}}{\overset{k_1}{\rightleftharpoons}} E \cdot S$$

$$E \cdot S \xrightarrow{k_2} E + P$$

where E represents the enzyme, S represents the substrate, E · S represents the enzyme-substrate complex, and P represents the products. The enzyme and substrate form a complex, where the reaction occurs. The enzyme then releases the products and is ready to repeat the process. The most amazing thing about enzymes is their efficiency. Because an enzyme plays its catalytic role over and over and very rapidly, only a tiny amount of enzyme is required. This makes the isolation of enzymes for study quite difficult.

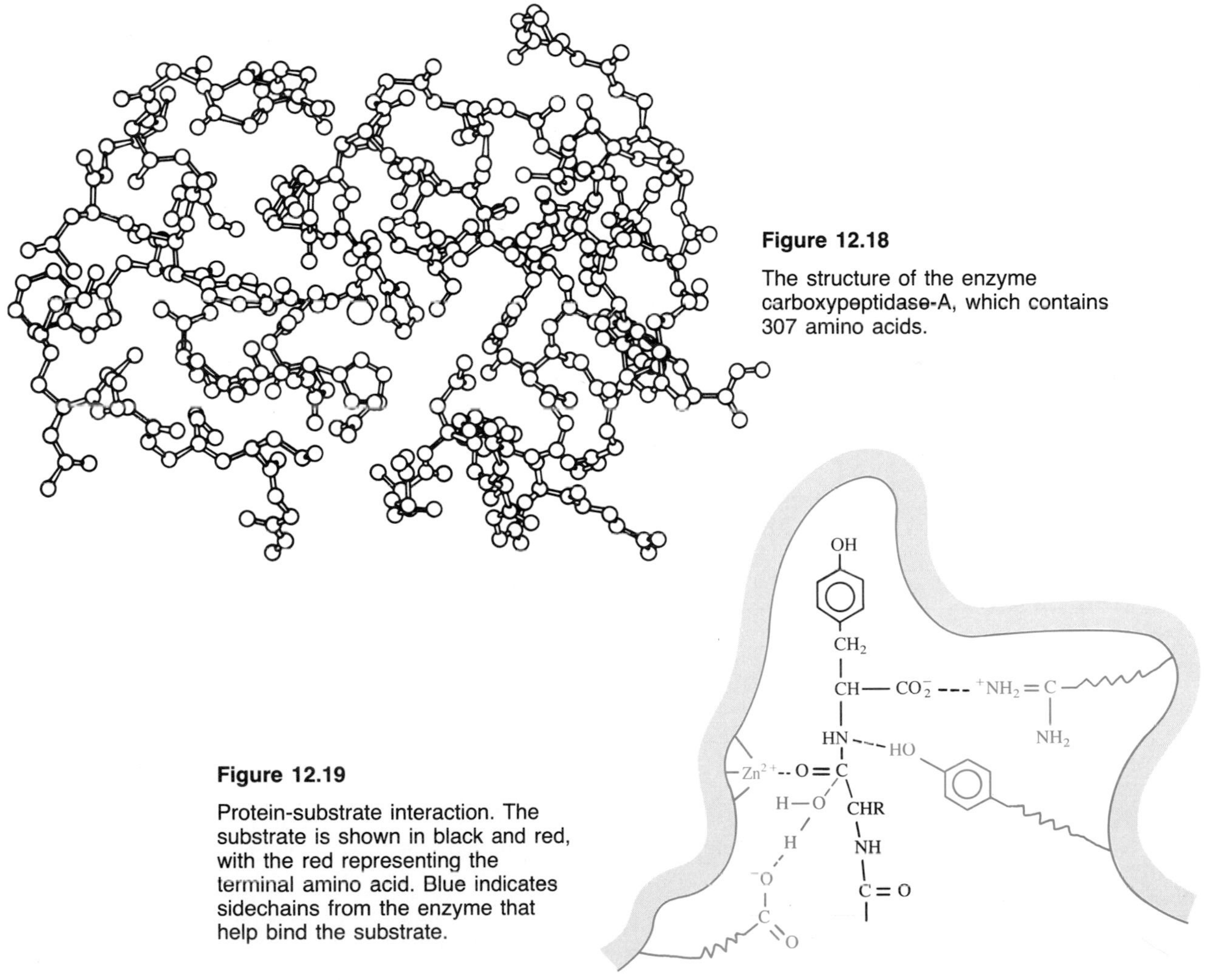

Figure 12.18

The structure of the enzyme carboxypeptidase-A, which contains 307 amino acids.

Figure 12.19

Protein-substrate interaction. The substrate is shown in black and red, with the red representing the terminal amino acid. Blue indicates sidechains from the enzyme that help bind the substrate.

Homogeneous Catalysis

A homogeneous catalyst exists in the same phase as the reacting molecules. There are many examples in both the gas and liquid phases. One example is the unusual catalytic behavior of nitric oxide toward ozone. In the troposphere, that part of the atmosphere closest to earth, nitric oxide catalyzes ozone production, but in the upper atmosphere it catalyzes the decomposition of ozone. Both of these effects are unfortunate environmentally.

In the lower atmosphere NO is produced in any high-temperature combustion process where N_2 is present. The reaction

$$N_2(g) + O_2(g) \rightarrow 2NO(g)$$

is very slow at normal temperatures because of the very strong N≡N and O=O bonds. However, at elevated temperatures, such as those found in the internal combustion engines of automobiles, significant quantities of NO form. Some of this NO is converted back to N_2 in the catalytic converter, but significant amounts escape into the atmosphere to react with oxygen:

$$2NO(g) + O_2(g) \rightarrow 2NO_2(g)$$

In the atmosphere NO_2 can absorb light and decompose as follows:

$$NO_2(g) \xrightarrow{\text{Light}} NO(g) + O(g)$$

The oxygen atom is very reactive and can combine with oxygen molecules to form ozone:

$$O_2(g) + O(g) \rightarrow O_3(g)$$

Although O_2 is represented here as the oxidizing agent for NO, the actual oxidizing agent is probably some type of peroxide compound produced by reaction of oxygen with pollutants. The direct reaction of NO and O_2 is very slow.

Ozone is a powerful oxidizing agent that can react with other air pollutants to form substances irritating to the eyes and lungs, and is itself very toxic.

In this series of reactions, nitric oxide is acting as a true catalyst because it assists the production of ozone without being consumed itself. This can be seen by summing the reactions:

$$\begin{array}{rcl} NO(g) + \frac{1}{2}O_2(g) & \longrightarrow & NO_2(g) \\ NO_2(g) & \xrightarrow{\text{Light}} & NO(g) + O(g) \\ O_2(g) + O(g) & \longrightarrow & O_3(g) \\ \hline \frac{3}{2}O_2(g) & \longrightarrow & O_3(g) \end{array}$$

In the upper atmosphere the presence of nitric oxide has the opposite effect—the depletion of ozone. The series of reactions involved is

$$\begin{array}{rcl} NO(g) + O_3(g) & \longrightarrow & NO_2(g) + O_2(g) \\ O(g) + NO_2(g) & \longrightarrow & NO(g) + O_2(g) \\ \hline O(g) + O_3(g) & \longrightarrow & 2O_2(g) \end{array}$$

Nitric oxide is again catalytic, but here its effect is to change O_3 to O_2. This is a potential problem because O_3, which absorbs ultraviolet light, is necessary to protect us from the harmful effects of this high-energy radiation. That is, we want O_3 in the upper atmosphere to block ultraviolet radiation from the sun, but not in the lower atmosphere where we would have to breathe it and its oxidation products.

The ozone layer is also threatened by *freons,** a group of stable, noncorrosive compounds used as refrigerants and as propellants in aerosol cans. The most commonly used substance of this type is freon-12, CCl_2F_2. The chemical inertness of freons makes them valuable but also creates a problem since they remain a long time in the environment. Eventually, they migrate into the upper atmosphere to be decomposed by high-energy light. Among the decomposition products are chlorine atoms:

$$CCl_2F_2(g) \xrightarrow{\text{Light}} CClF_2(g) + Cl(g)$$

These chlorine atoms can catalyze the decomposition of ozone:

$$\begin{array}{rcl} Cl(g) + O_3(g) & \longrightarrow & ClO(g) + O_2(g) \\ O(g) + ClO(g) & \longrightarrow & Cl(g) + O_2(g) \\ \hline O(g) + O_3(g) & \longrightarrow & 2O_2(g) \end{array}$$

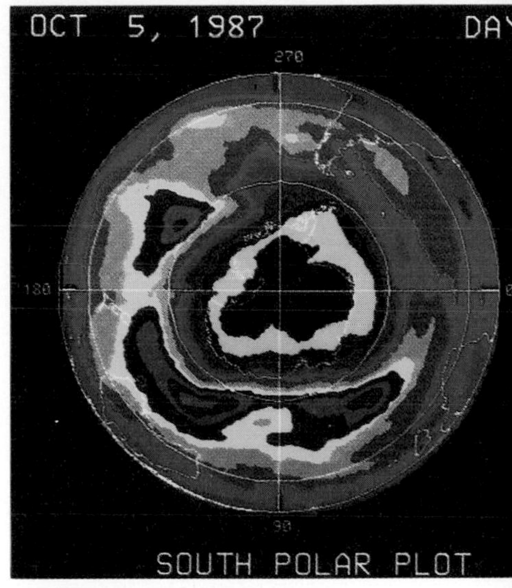

A pictorial representation of the ozone concentrations over Antarctica.

The problem of freons has been brought into strong focus by the recent discovery of a mysterious "hole" in the ozone layer in the stratosphere over Antarctica. Studies performed there to find the reason for the hole have found unusually high levels of chlorinemonoxide (ClO). This strongly implicates the freons in the atmosphere as being at least partially responsible for the ozone destruction in the area.

Because they pose environmental problems, freons have been banned by the U.S. government for use in aerosol cans, but they are still widely used in air conditioners and refrigerators. However, the chemical industry is actively seeking adequate substitutes for these applications also.

FOR REVIEW

Summary

Even though a chemical reaction may have a strong tendency to occur thermodynamically, it is not necessarily fast. Chemical kinetics is the study of the factors that control the rates of chemical reactions. The reaction rate is defined as the change in concentration of a reactant or product per unit of time. Kinetic measurements are usually made under conditions where the reverse reaction is insignificant. Then only the forward reaction is important in determining the rate.

There are two kinds of rate laws: the differential rate law (or just rate law) and the integrated rate law. The rate law for a reaction of the type

$$aA \rightarrow \text{products}$$

describes the dependence of the rate on the concentration of A:

$$\text{Rate} = -\frac{\Delta[A]}{\Delta t} = k[A]^n$$

*For more information see S. Elliott and F. S. Rowland, "Chlorofluorocarbons and Stratospheric Ozone," *J. Chem. Ed.* **64** (1987): 387.

where k is called the rate constant and n is the order of the reaction in A. The rate is defined as the negative of $\Delta[A]/\Delta t$ because [A] is decreasing (as the reactants are used up in the reaction). The value for n (the order) cannot be determined from the balanced equation for the reaction, but must be found experimentally. For a first-order reaction n is 1 and the rate $= k[A]$. For a second-order reaction n is 2 and the rate $= k[A]^2$. For a zero-order reaction n is 0 and the rate $= k$.

A common experimental method for determining the rate is the method of initial rates in which several runs are carried out with different initial concentrations, and the rate is measured for each at a value of t close to $t = 0$.

An integrated rate law expresses reactant concentrations as a function of time. The integrated rate law for a first-order reaction is

$$\ln[A] = -kt + \ln[A]_0$$

where $[A]_0$ is the initial concentration of A. Thus we can calculate [A] at any time, given $[A]_0$ and k. For a reaction that shows first-order kinetics, a plot of ln[A] versus time is a straight line.

The half-life ($t_{1/2}$) of a reaction is the time required for a reactant to reach half of its original concentration. For first-order reactions the half-life is independent of concentration:

$$t_{1/2} = \frac{0.693}{k}$$

The integrated rate law for a second-order reaction is

$$\frac{1}{[A]} = kt + \frac{1}{[A]_0}$$

Consequently, if a plot of 1/[A] versus time for a given reaction is a straight line, then the rate must be second order in A. For second-order kinetics, the half-life is dependent on both k and $[A]_0$:

$$t_{1/2} = \frac{1}{k[A]_0}$$

The integrated rate law for a zero-order reaction is

$$[A] = -kt + [A]_0$$

and the half-life is

$$t_{1/2} = \frac{[A]_0}{2k}$$

For a reaction with several reactants, the overall reaction order is the sum of the individual orders. We can determine the orders for this type of reaction by making the concentration of a particular reactant small in comparison to those of the other reactants.

A reaction mechanism is a series of elementary steps by which a given reaction occurs. The rate law for each step can be written from its molecularity, that is, the number of species that must collide to produce the reaction indicated by that step. A unimolecular step is always first order, and a bimolecular step is always second order.

For a mechanism to be acceptable, the sum of the elementary steps must give the overall balanced equation for the reaction, and the mechanism must give a rate law that agrees with the experimentally determined rate law. Multistep reactions often have one step that is much slower than the others. This is called the rate-determining step, since the overall rate of the reaction is limited by the rate of this step.

Chemical kinetics can be described by the collision model, which assumes that molecules must collide to react. In terms of this model, a certain threshold energy called the activation energy (E_a) must be overcome for a collision to produce a chemical reaction. E_a is the energy needed to rearrange the bonds in the molecules to form products. Both the rate of a reaction and the effect of temperature on the rate depend on the size of E_a.

Molecular orientation during a collision is also important, since not every collision will have the molecules aligned in an effective way. The factors affecting k can be described by the Arrhenius equation

$$k = Ae^{-E_a/RT}$$

where A is a factor reflecting collision frequency and molecular orientation, and $e^{-E_a/RT}$ represents the fraction of collisions with sufficient energy to produce a reaction. The value of E_a can be determined by measuring values of k at two or more temperatures.

A catalyst is a substance that speeds up a reaction without being consumed. A catalyst operates by providing a lower energy pathway for the reaction in question. Enzymes are biological catalysts. A homogeneous catalyst is present in the same phase as the reacting molecules, and a heterogeneous catalyst exists in a different phase.

Key Terms

chemical kinetics

Section 12.1

reaction rate
instantaneous rate

Section 12.2

rate law
rate constant
order
(differential) rate law
integrated rate law

Section 12.3

method of initial rates
initial rate
overall reaction order

Section 12.4

first-order reaction
integrated first-order rate law
half-life of a reaction
integrated second-order rate law
zero-order reaction
integrated zero-order rate law
pseudo-first-order rate law

Section 12.6

reaction mechanism
intermediate
elementary step
molecularity
unimolecular step
bimolecular step
termolecular step
rate-determining step

Section 12.7

collision model
activation energy
activated complex (transition state)
molecular orientations
steric factor
Arrhenius equation
frequency factor

Section 12.8

enzyme
catalyst
homogeneous catalyst
heterogeneous catalyst
adsorption
absorption

Exercises

A blue exercise number indicates that the answer to that exercise appears at the back of this book and a solution appears in the Solutions Guide.

Reaction Rates

1. Thiosulfate ion is oxidized by iodine according to the following reaction:

$$2S_2O_3^{2-}(aq) + I_2(aq) \rightarrow S_4O_6^{2-}(aq) + 2I^-(aq)$$

If 0.0080 mol of $S_2O_3^{2-}$ is consumed in 1.0 L of solution each second, what is the rate of consumption of I_2? At what rates are $S_4O_6^{2-}$ and I^- produced?

2. In the Haber process for the production of ammonia,

$$N_2(g) + 3H_2(g) \rightarrow 2NH_3(g)$$

what is the relationship between the rate of production of ammonia and the rate of consumption of hydrogen?

3. What are the units for each of the following if concentrations are expressed in moles per liter?
   a. rate of a chemical reaction
   b. rate constant for a zero-order rate law*
   c. rate constant for a first-order rate law
   d. rate constant for a second-order rate law
   e. rate constant for a third-order rate law

4. The rate law for the reaction

$$Cl_2(g) + CHCl_3(g) \rightarrow HCl(g) + CCl_4(g)$$

is

$$\text{Rate} = k[Cl_2]^{1/2}[CHCl_3]$$

What are the units for k?

5. Concentrations of trace substances in the atmosphere are often expressed in molecules per cubic centimeter. What are the units for the quantities in Exercise 3 if concentrations are expressed this way?

6. Hydroxyl radical (OH) is an important oxidizing agent in the atmosphere. At 298 K, the rate constant for the reaction of OH with benzene is 1.24×10^{-12} cm³/molecule s. Calculate the value of the rate constant in L/mol s.

7. The decomposition of hydrogen iodide on finely divided gold at 150°C is zero order with respect to HI. The rate defined below is constant at 1.20×10^{-4} mol/L s.

$$2HI(g) \xrightarrow{Au} H_2(g) + I_2(g)$$

$$\text{Rate} = -\frac{\Delta[HI]}{\Delta t} = k = 1.20 \times 10^{-4} \text{ mol/L s}$$

   a. If an experiment has an initial HI concentration of 0.250 mol/L, what is the concentration of HI after 25 minutes?
   b. How long will it take for all of the HI to decompose?
   c. Give the rates of formation for H_2 and I_2.

Rate Laws from Experimental Data: Initial Rates Method

8. The reaction

$$2NO(g) + Cl_2(g) \rightarrow 2NOCl(g)$$

was studied at −10°C. The following results were obtained where

$$\text{Rate} = -\frac{\Delta[Cl_2]}{\Delta t}$$

$[NO]_0$ (mol/L)	$[Cl_2]_0$ (mol/L)	Initial rate (mol/L min)
0.10	0.10	0.18
0.10	0.20	0.35
0.20	0.20	1.45

   a. What is the rate law?
   b. What is the value of the rate constant?

9. The reaction

$$2I^-(aq) + S_2O_8^{2-}(aq) \rightarrow I_2(aq) + 2SO_4^{2-}(aq)$$

was studied at 25°C. The following results were obtained where

$$\text{Rate} = -\frac{\Delta[S_2O_8^{2-}]}{\Delta t}$$

$[I^-]_0$ (mol/L)	$[S_2O_8^{2-}]_0$ (mol/L)	Initial rate (mol/L s)
0.080	0.040	12.50×10^{-6}
0.040	0.040	6.250×10^{-6}
0.080	0.020	5.560×10^{-6}
0.032	0.040	4.350×10^{-6}
0.060	0.030	6.410×10^{-6}

   a. Determine the rate law.
   b. Calculate a value for the rate constant for each experiment and an average value for the rate constant.

*In the Exercises the term "rate law" always means differential rate law.

10. The reaction

$$I^-(aq) + OCl^-(aq) \rightarrow IO^-(aq) + Cl^-(aq)$$

was studied and the following data were obtained:

$[I^-]_0$ (mol/L)	$[OCl^-]_0$ (mol/L)	Initial rate (mol/L s)
0.12	0.18	7.91×10^{-2}
0.060	0.18	3.95×10^{-2}
0.030	0.090	9.88×10^{-3}
0.24	0.090	7.91×10^{-2}

a. What is the rate law?
b. Calculate the rate constant.

11. The decomposition of nitrosyl chloride was studied:

$$2NOCl(g) \rightleftharpoons 2NO(g) + Cl_2(g)$$

The following data were obtained where

$$\text{Rate} = -\frac{\Delta[NOCl]}{\Delta t}$$

$[NOCl]_0$ (molecules/cm^3)	Initial rate (molecules/cm^3 s)
3.0×10^{16}	5.98×10^4
2.0×10^{16}	2.66×10^4
1.0×10^{16}	6.64×10^3
4.0×10^{16}	1.06×10^5

a. What is the rate law?
b. Calculate the rate constant.
c. Calculate the rate constant for the concentrations given in moles per liter.

12. The following data were obtained for the reaction

$$2ClO_2(aq) + 2OH^-(aq) \rightarrow ClO_3^-(aq) + ClO_2^-(aq) + H_2O(l)$$

where

$$\text{Rate} = -\frac{\Delta[ClO_2]}{\Delta t}$$

$[ClO_2]$ (mol/L)	$[OH^-]$ (mol/L)	Initial rate (mol/L s)
0.0500	0.100	5.75×10^{-2}
0.100	0.100	2.30×10^{-1}
0.100	0.050	1.15×10^{-1}

Determine the rate law and the value of the rate constant.

Integrated Rate Laws from Experimental Data

13. The dimerization of butadiene was studied at 500 K.

$$2C_4H_6(g) \rightarrow C_8H_{12}(g)$$

The following data were obtained where

$$\text{Rate} = -\frac{\Delta[C_4H_6]}{\Delta t}$$

Time (s)	$[C_4H_6]$ (mol/L)
195	1.6×10^{-2}
604	1.5×10^{-2}
1246	1.3×10^{-2}
2180	1.1×10^{-2}
6210	0.68×10^{-2}

Determine the form of the integrated rate law and the rate constant for this reaction. Write the rate law for the reaction. (These are actual experimental data, so they may not give an exactly straight line.)

14. Determine the form of the integrated rate law for the decomposition of benzene diazonium chloride,

$$C_6H_5N_2Cl(aq) \rightarrow C_6H_5Cl(l) + N_2(g)$$

from the following data, which were collected at 50°C and 1 atm:

Time (s)	N_2 evolved (mL)
6	19.3
9	26.0
14	36.0
22	45.0
30	50.4
∞	58.3

The total solution volume was 40.0 mL. Also write the rate law.

15. The decomposition of hydrogen peroxide was studied and the data in the table here and on page 560 were obtained at a particular temperature where

$$\text{Rate} = -\frac{\Delta[H_2O_2]}{\Delta t}$$

Time (s)	$[H_2O_2]$ (mol/L)
0	1.0
120 ± 1	0.91
300 ± 1	0.78
600 ± 1	0.59

Time (s)	$[H_2O_2]$ (mol/L)
1200 ± 1	0.37
1800 ± 1	0.22
2400 ± 1	0.13
3000 ± 1	0.082
3600 ± 1	0.050

Determine the integrated rate law and the value of the rate constant. Also write the rate law.

16. The rate of the reaction:

$$NO_2(g) + CO(g) \rightarrow NO(g) + CO_2(g)$$

depends only on the concentration of nitrogen dioxide below 225°C. At a temperature below 225°C the following data were collected:

Time (s)	$[NO_2]$ (mol/L)
0	0.500
1.20×10^3	0.444
3.00×10^3	0.381
4.50×10^3	0.340
9.00×10^3	0.250
1.80×10^4	0.174

Determine the integrated rate law and the value of the rate constant at this temperature. Also write the rate law.

17. The rate of the reaction

$$O(g) + NO_2(g) \rightarrow NO(g) + O_2(g)$$

was studied. This reaction is one step of the nitric oxide–catalyzed destruction of ozone in the upper atmosphere.

a. In the first set of experiments, NO_2 was in large excess, at a concentration of 1.0×10^{13} molecules/cm^3.

Time (s)	[O] (atoms/cm^3)
0	5.0×10^9
1.0×10^{-2}	1.9×10^9
2.0×10^{-2}	6.8×10^8
3.0×10^{-2}	2.5×10^8

What is the order of the reaction with respect to oxygen atoms?

b. The reaction is known to be first order with respect to NO_2. What are the overall rate law and the value of the rate constant?

18. The reaction

$$NO(g) + O_3(g) \rightarrow NO_2(g) + O_2(g)$$

was studied by performing two experiments. In the first experiment the rate of disappearance of NO was followed in a large excess of O_3. The results were as follows ($[O_3]$ remains effectively constant at 1.0×10^{14} molecules/cm^3):

Time (ms)	[NO] (molecules/cm^3)
0	6.0×10^8
100 ± 1	5.0×10^8
500 ± 1	2.4×10^8
700 ± 1	1.7×10^8
1000 ± 1	9.9×10^7

In the second experiment [NO] was held constant at 2.0×10^{14} molecules/cm^3. The data for the disappearance of O_3 are as follows.

Time (ms)	$[O_3]$ (molecules/cm^3)
0	1.0×10^{10}
50 ± 1	8.4×10^9
100 ± 1	7.0×10^9
200 ± 1	4.9×10^9
300 ± 1	3.4×10^9

a. What is the order with respect to each reactant?
b. What is the overall rate law?
c. What is the value of the rate constant from each set of experiments?

$$\text{Rate} = k'[NO]^x \qquad \text{Rate} = k''[O_3]^y$$

d. What is the value of the rate constant for the overall rate law?

$$\text{Rate} = k[NO]^x[O_3]^y$$

Half-Life

19. Radioactive decay is a process that follows a first-order rate law: Rate $= kN$, where N is the number of nuclei and the rate is the nuclear disintegrations per unit time. The half-life for the decay of ^{239}Pu is 24,360 yr. The half-life for the decay of ^{241}Pu is 13 yr.

a. Calculate the rate constants for the decay of ^{239}Pu and ^{241}Pu.
b. Which species decays more rapidly?
c. If one starts with 5.0 g of pure ^{241}Pu, how many disintegrations per second occur initially? What mass of ^{241}Pu is left after 1.0 yr, 10. yr, 100. yr?

20. A certain first-order reaction is 45.0% complete in 65 s. What are the rate constant and the half-life of this reaction?

21. The radioactive isotope ^{32}P decays by first-order kinetics and has a half-life of 14.3 days. How long does it take for 95.0% of a sample of ^{32}P to decay?

22. It took 143 s for 50.0% of a particular substance to decompose. If the initial concentration was 0.060 M and the decomposition reaction follows second-order kinetics, what is the value of the rate constant?

23. A first-order reaction is 38.5% complete in 480 s.
    a. Calculate the rate constant.
    b. What is the value of the half-life?
    c. How long will it take for the reaction to go to 25%, 75%, and 95% completion?

24. Radioactive copper-64 decays by first-order kinetics with a half-life of 12.8 days.
    a. What is the value of k in s^{-1}?
    b. A sample contains 28.0 mg of ^{64}Cu. How many decays will be produced in the first second?
    c. A chemist obtains a fresh sample of ^{64}Cu and measures its radioactivity. She then determines that to do the experiment she has in mind, the radioactivity cannot go below 3% of the initial measured value. How long does she have to do the experiments?

Reaction Mechanisms

25. Define each of the following:
    a. elementary step
    b. reaction mechanism
    c. rate-determining step

26. Write the rate laws for the following elementary reactions:
    a. $CH_3NC(g) \rightarrow CH_3CN(g)$
    b. $O_3(g) + NO(g) \rightarrow O_2(g) + NO_2(g)$
    c. $O_3(g) \rightarrow O_2(g) + O(g)$
    d. $O_3(g) + O(g) \rightarrow 2O_2(g)$
    e. $^{14}_{6}C \rightarrow {}^{14}_{7}N + \beta$ particle (nuclear decay)

27. Is the mechanism

$$NO + Cl_2 \xrightarrow{k_1} NOCl_2$$

$$NOCl_2 + NO \xrightarrow{k_2} 2NOCl$$

consistent with the results you obtained in Exercise 8? If so, which step is the rate-determining step?

28. The mechanism for the decomposition of hydrogen peroxide is:

$$H_2O_2 \rightarrow 2OH$$

$$H_2O_2 + OH \rightarrow H_2O + HO_2$$

$$HO_2 + OH \rightarrow H_2O + O_2$$

Using your results from Exercise 15, specify which step is the rate-determining step.

29. The reaction

$$2NO(g) + O_2(g) \rightarrow 2NO_2(g)$$

exhibits the rate law

$$\text{Rate} = k[NO]^2[O_2]$$

Which of the following mechanisms is consistent with this rate law?

a.	$NO + O_2 \rightarrow NO_2 + O$	Slow
	$O + NO \rightarrow NO_2$	Fast
b.	$NO + O_2 \rightleftharpoons NO_3$	Both fast with equal rates
	$NO_3 + NO \rightarrow 2NO_2$	Slow
c.	$2NO \rightarrow N_2O_2$	Slow
	$N_2O_2 + O_2 \rightarrow N_2O_4$	Fast
	$N_2O_4 \rightarrow 2NO_2$	Fast
d.	$2NO \rightleftharpoons N_2O_2$	Both fast with equal rates
	$N_2O_2 \rightarrow NO_2 + O$	Slow
	$O + NO \rightarrow NO_2$	Fast

30. The reaction

$$2NO_2Cl \rightarrow 2NO_2 + Cl_2$$

follows the rate law

$$\text{Rate} = k[NO_2Cl]$$

The mechanism is

$$NO_2Cl \xrightarrow{k_1} NO_2 + Cl$$

$$NO_2Cl + Cl \xrightarrow{k_2} NO_2 + Cl_2$$

Which is the rate-determining step?

31. The reaction

$$5Br^-(aq) + BrO_3^-(aq) + 6H^+(aq) \rightarrow 3Br_2(aq) + 3H_2O(l)$$

obeys the mechanism:

$BrO_3^-(aq) + H^+(aq) \underset{k_{-1}}{\overset{k_1}{\rightleftharpoons}} HBrO_3(aq)$	Both fast with equal rates
$HBrO_3(aq) + H^+(aq) \underset{k_{-2}}{\overset{k_2}{\rightleftharpoons}} H_2BrO_3^+$	Both fast with equal rates
$Br^-(aq) + H_2BrO_3^+(aq) \xrightarrow{k_3} (Br{-}BrO_2)(aq) + H_2O(l)$	Slow
$(Br{-}BrO_2)(aq) + 4H^+(aq) + 4Br^-(aq) \rightarrow$ Products	Fast

Write the rate law for this reaction.

Temperature Dependence of Rate Constants and the Collision Model

32. Why is the rate of a reaction affected by each of the following?
    a. frequency of collisions
    b. kinetic energy of collisions
    c. orientation of collisions

33. Why are termolecular steps infrequently seen in chemical reactions?

34. Define what is meant by unimolecular and bimolecular steps.

35. Which of the following reactions would you expect to have the larger rate at room temperature? Why? (*Hint:* Think of which would have the lower activation energy.)

$$2Ce^{4+}(aq) + Hg_2^{2+}(aq) \rightarrow 2Ce^{3+}(aq) + 2Hg^{2+}(aq)$$

$$H_3O^+(aq) + OH^-(aq) \rightarrow 2H_2O(l)$$

36. At 25°C the first-order rate constant for a reaction is $2.0 \times 10^3\ s^{-1}$. The activation energy is 15.0 kJ/mol. What is the value of the rate constant at 75°C?

37. A first-order reaction has rate constants of $4.6 \times 10^{-2}\ s^{-1}$ and $8.1 \times 10^{-2}\ s^{-1}$ at 0°C and 20°C, respectively. What is the value of the activation energy?

38. For the reaction:

$$H_2(g) + I_2(g) \rightarrow 2HI(g)$$

where the rate law is

$$-\frac{\Delta[H_2]}{\Delta t} = k[H_2][I_2]$$

the rate constants are 2.45×10^{-4} L/mol s and 0.950 L/mol s at 302°C and 508°C, respectively.
a. Calculate the values of E_a and A for this reaction.
b. Calculate the value of k at 375°C.

39. Experimental values for the temperature dependence of the rate constant for the gas-phase reaction

$$NO + O_3 \rightarrow NO_2 + O_2$$

are as follows:

T (K)	k (L/mol s)
195	1.08×10^9
230	2.95×10^9
260	5.42×10^9
298	12.0×10^9
369	35.5×10^9

Make the appropriate graph using these data and determine the activation energy for this reaction.

40. Chemists commonly use a rule of thumb that an increase of 10 K in temperature doubles the rate of a reaction. What must the activation energy be for this statement to be true for a temperature increase from 25°C to 35°C?

41. The rate constant for the gas-phase decomposition of N_2O_5,

$$N_2O_5 \rightarrow 2NO_2 + \tfrac{1}{2}O_2$$

has the following temperature dependence:

T (K)	k (s^{-1})
338	4.9×10^{-3}
318	5.0×10^{-4}
298	3.5×10^{-5}

Make an appropriate graph using these data and determine the activation energy for this reaction.

42. For each of the following reaction profiles, indicate:
a. the positions of reactants and products
b. the activation energy
c. ΔE for the reaction

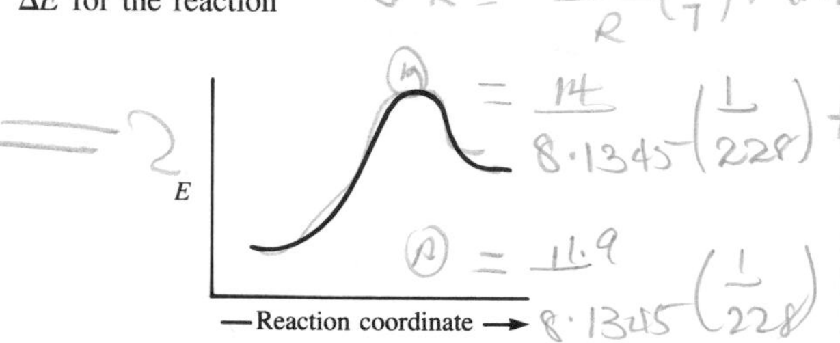

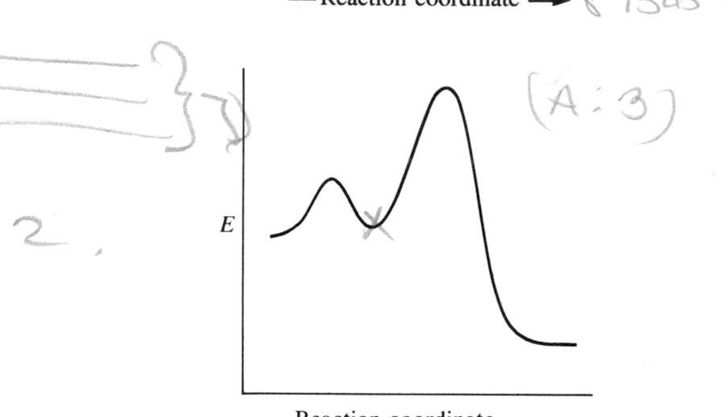

43. Draw a rough sketch of the energy profile for each of the following cases:
a. $\Delta E = +10$ kJ/mol, $E_a = 25$ kJ/mol
b. $\Delta E = -10$ kJ/mol, $E_a = 50$ kJ/mol
c. $\Delta E = -50$ kJ/mol, $E_a = 50$ kJ/mol
Which of the reactions will have the greatest rate at 298 K? Assume the frequency factor A is the same for all three reactions.

44. The activation energy for the reaction

$$H_2(g) + I_2(g) \rightarrow 2HI(g)$$

is 167 kJ/mol and ΔE for the reaction is +28 kJ/mol. What is the activation energy for the decomposition of HI?

Catalysis

45. Define each of the following:
a. homogeneous catalyst
b. heterogeneous catalyst

46. How does a catalyst increase the rate of a chemical reaction?

47. Should the concentration of a homogeneous catalyst appear in the rate law?

48. Would a given reaction have the same rate law for both a catalyzed and an uncatalyzed pathway? Explain.

49. One pathway for the destruction of ozone in the upper atmosphere is

$$O_3(g) + NO(g) \rightarrow NO_2(g) + O_2(g) \quad \text{Slow}$$

$$NO_2(g) + O(g) \rightarrow NO(g) + O_2(g) \quad \text{Fast}$$

Overall reaction: $O_3(g) + O(g) \rightarrow 2O_2(g)$

a. Which species is a catalyst?
b. Which species is an intermediate?
c. E_a for the uncatalyzed reaction

$$O_3(g) + O(g) \rightarrow 2O_2$$

is 14.0 kJ. E_a for the same reaction when catalyzed is 11.9 kJ. What is the ratio of the rate constant for the catalyzed reaction to that for the uncatalyzed reaction at 25°C? Assume the frequency factor A is the same for each reaction.

50. One of the concerns about the use of freons is that they will migrate to the upper atmosphere where chlorine atoms can be generated by the reaction

$$\underset{\text{Freon 12}}{CCl_2F_2} \xrightarrow{h\nu} CF_2Cl + Cl$$

Chlorine atoms can also act as a catalyst for the destruction of ozone. The activation energy for the reaction

$$Cl + O_3 \rightarrow ClO + O_2$$

is 2.1 kJ/mol. Which is the more effective catalyst for the destruction of ozone, Cl or NO? (See Exercise 49.)

51. Assuming the mechanism for the hydrogenation of C_2H_4 given in Section 12.8 is correct, would you predict that the product of the reaction of C_2H_4 with D_2 would be $CH_2D—CH_2D$ or $CHD_2—CH_3$?

52. For enzyme-catalyzed reactions that follow the mechanism

$$E + S \rightleftharpoons E \cdot S$$

$$E \cdot S \rightleftharpoons E + P$$

a graph of the rate as a function of [S], the concentration of the substrate, has the following general appearance:

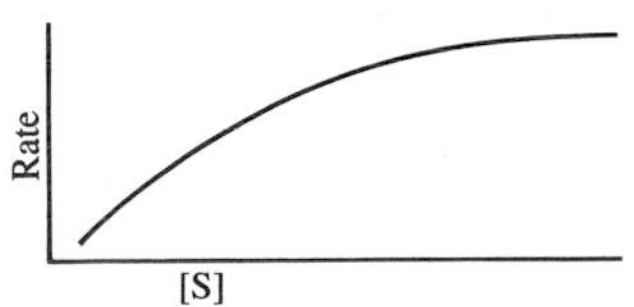

Note that at high substrate concentrations the rate no longer changes with [S]. Suggest a reason for this.

Additional Exercises

53. Distinguish among initial rate, average rate, and instantaneous rate of a chemical reaction.

54. The reaction

$$H_2SeO_3(aq) + 6I^-(aq) + 4H^+(aq) \rightarrow Se(s) + 2I_3^-(aq) + 3H_2O(l)$$

was studied at 0°C. The following data were obtained:

$[H_2SeO_3]_0$	$[H^+]_0$	$[I^-]_0$	Initial rate (mol/L s)
1.0×10^{-4}	2.0×10^{-2}	2.0×10^{-2}	1.66×10^{-7}
2.0×10^{-4}	2.0×10^{-2}	2.0×10^{-2}	3.33×10^{-7}
3.0×10^{-4}	2.0×10^{-2}	2.0×10^{-2}	4.99×10^{-7}
1.0×10^{-4}	4.0×10^{-2}	2.0×10^{-2}	6.66×10^{-7}
1.0×10^{-4}	1.0×10^{-2}	2.0×10^{-2}	0.42×10^{-7}
1.0×10^{-4}	2.0×10^{-2}	4.0×10^{-2}	13.2×10^{-7}
1.0×10^{-4}	1.0×10^{-2}	4.0×10^{-2}	3.36×10^{-7}

These relationships hold only if there is an insignificant amount of I_3^- present. What is the rate law and the value of the rate constant? $\left(\text{Assume Rate} = -\frac{\Delta[H_2SeO_3]}{\Delta t}.\right)$

55. A study was made of the effect of the hydroxide concentration on the rate of the reaction:

$$I^-(aq) + OCl^-(aq) \rightarrow IO^-(aq) + Cl^-(aq)$$

The following data were obtained:

$[I^-]_0$ (mol/L)	$[OCl^-]_0$ (mol/L)	$[OH^-]_0$ (mol/L)	Initial rate (mol/L s)
0.0013	0.012	0.10	9.4×10^{-3}
0.0026	0.012	0.10	18.7×10^{-3}
0.0013	0.006	0.10	4.7×10^{-3}
0.0013	0.018	0.10	14.0×10^{-3}
0.0013	0.012	0.05	18.7×10^{-3}
0.0013	0.012	0.20	4.7×10^{-3}
0.0013	0.018	0.20	7.0×10^{-3}

Determine the rate law and the value of the rate constant for this reaction.

56. The following data were obtained for the gas-phase decomposition of dinitrogenpentoxide,

$$2N_2O_5(g) \rightarrow 4NO_2(g) + O_2(g)$$

$[N_2O_5]_0$ mol/L	Initial rate (mol/L s)
0.075	8.90×10^{-4}
0.190	2.26×10^{-3}
0.275	3.26×10^{-3}
0.410	4.85×10^{-3}

Defining the rate as $-\dfrac{\Delta[N_2O_5]}{\Delta t}$, write the rate law and calculate the value of the rate constant.

57. The thermal degradation of silk was studied by Kuruppillai, Hersh, and Tucker (''Historic Textile and Paper Materials, *ACS Advances in Chemistry Series,* No. 212, 1986) by measuring the tensile strength of silk fibers at various times of exposure to elevated temperature. The loss of tensile strength follows first order kinetics

$$\frac{\Delta s}{\Delta t} = -ks$$

where s is the strength of the fiber retained after heating and k is the first-order rate constant. The effects of adding a deacidifying agent and an antioxidant to the silk were studied and the following data were obtained.

	Strength retained (%)		
Heating time (Days)	Untreated	Deacidifying agent	Antioxidant
0.00	100.0	100.1	114.6
1.00	67.9	60.8	65.2
2.00	38.9	26.8	28.1
3.00	16.1	—	11.3
6.00	6.8	—	—

a. Determine the first-order rate constants for the thermal degradation of silk for each of the three experiments.
b. Does either of the two additives appear to retard the degradation of silk?

58. Calculate the half-life for the thermal degradation of silk for each of the three experiments in Exercise 57.

59. Sulfuryl chloride, SO_2Cl_2, decomposes to sulfur dioxide, SO_2, and chlorine, Cl_2, by a first-order reaction in the gas phase. The following data were obtained when a sample containing 5.00×10^{-2} mol of sulfuryl chloride was heated to 600 K $\pm$ 1 K in a 5.00×10^{-1} L container.

Time (hours)	0.00	1.00	2.00	4.00	8.00	16.00
Pressure (atm)	4.93	5.60	6.34	7.33	8.56	9.52

Defining the rate as $-\dfrac{\Delta[SO_2Cl_2]}{\Delta t}$,

a. Determine the value of the rate constant for the decomposition of sulfuryl chloride at 600 K.
b. What is the half-life of the reaction?
c. What would be the pressure in the vessel after 0.500 h and after 12.0 h?
d. What fraction of the sulfuryl chloride remains after 20.0 h?

60. The decomposition of many substances on the surface of a heterogeneous catalyst shows the following behavior.

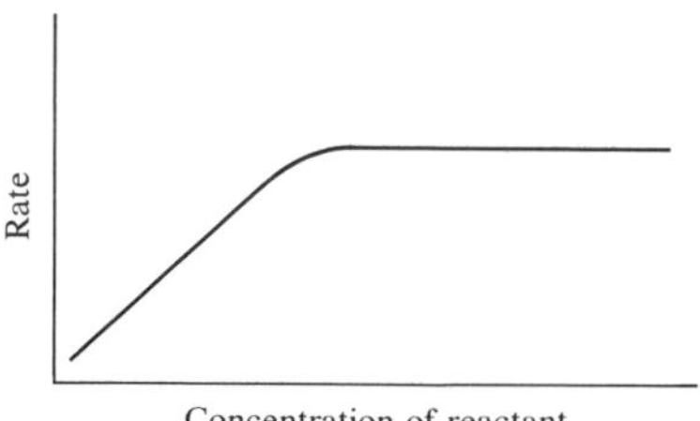

How do you account for the rate law changing from first order to zero order in the concentration of reactant?

61. The decomposition of NH_3 to N_2 and H_2 was studied on two surfaces.

Surface	E_a(kJ/mol)
W	163
Os	197

Without a catalyst the activation energy is 335 kJ/mol.

a. Which surface is the better heterogeneous catalyst for the decomposition of NH_3? Why?
b. How many times faster is the reaction at 298 K on the W surface compared with the reaction with no catalyst present?
c. The decomposition reaction on the two surfaces obeys a rate law of the form:

$$\text{Rate} = k\frac{[NH_3]}{[H_2]}$$

How can you explain the inverse dependence of the rate on the H_2 concentration?

62. The reaction

$$I^-(aq) + OCl^-(aq) \rightarrow IO^-(aq) + Cl^-(aq)$$

is believed to occur by the following mechanism:

$$OCl^- + H_2O \underset{k_{-1}}{\overset{k_1}{\rightleftharpoons}} HOCl + OH^- \quad \text{Both fast with equal rates}$$

$$I^- + HOCl \xrightarrow{k_2} HOI + Cl^- \qquad \text{Slow}$$

$$HOI + OH^- \xrightarrow{k_3} H_2O + IO^- \qquad \text{Fast}$$

Write the rate law for this reaction.
Note: Since the reaction is in aqueous solution, the effective concentration of water is constant. Thus the rate of the forward reaction in the first step can be written:

$$\text{Rate} = k[H_2O][OCl^-] = k_1[OCl^-]$$

63. The rate law for the reaction

$$BrO_3^-(aq) + 3SO_3^{2-}(aq) \rightarrow Br^-(aq) + 3SO_4^{2-}(aq)$$

is

$$\text{Rate} = k[BrO_3^-][SO_3^{2-}][H^+]$$

The first step in the mechanism is

$$SO_3^{2-}(aq) + H^+(aq) \xrightarrow{k_1} HSO_3^-(aq) \qquad \text{Fast}$$

The second step is rate-determining. Write a possible second step for the mechanism.

64. In the gas phase the production of phosgene from chlorine and carbon monoxide proceeds by the following mechanism:

$$Cl_2 \underset{k_{-1}}{\overset{k_1}{\rightleftharpoons}} 2Cl \qquad \text{Both fast with equal rates}$$

$$Cl + CO \underset{k_{-2}}{\overset{k_2}{\rightleftharpoons}} COCl \qquad \text{Both fast with equal rates}$$

$$COCl + Cl_2 \xrightarrow{k_3} COCl_2 + Cl \qquad \text{Slow}$$

Overall reaction: $CO + Cl_2 \longrightarrow COCl_2$

a. Write the rate law for this reaction.
b. Which species are intermediates?

CHAPTER 13

Chemical Equilibrium

CONTENTS

In doing stoichiometry calculations we assumed that reactions proceeded to completion, that is, until one of the reactants ran out. Many reactions *do* proceed essentially to completion. For such reactions it can be assumed that the reactants are quantitatively converted to products and that the amount of limiting reactant that remains is negligible. On the other hand, there are many chemical reactions that stop far short of completion. An example is the dimerization of nitrogen dioxide:

$$NO_2(g) + NO_2(g) \rightarrow N_2O_4(g)$$

The reactant NO_2 is a dark brown gas, and the product N_2O_4 is a colorless gas. When NO_2 is placed in an evacuated, sealed glass vessel at 25°C, the initial dark brown color decreases in intensity as it is converted to colorless N_2O_4. However, even over a long period of time, the contents of the reaction vessel do not become colorless. Instead, the intensity of the brown color eventually becomes constant, which means that the concentration of NO_2 is no longer changing. This observation is a clear indication that the reaction has stopped short of completion. In fact, the system has reached **chemical equilibrium,** *the state where the concentrations of all reactants and products remain constant with time.*

Any chemical reactions carried out in a closed vessel will reach equilibrium. For some reactions the equilibrium position so favors the products that the reaction appears to have gone to completion. We say that the equilibrium position for such reactions lies *far to the right,* in the direction of the products. For example, when gaseous hydrogen and oxygen are mixed in stoichiometric quantities and react to form water vapor, the reaction proceeds essentially to completion. The amounts of the reactants that remain when the system reaches equilibrium are so tiny as to be negligible. By contrast, some reactions occur only to a slight extent. For example,

< The shell of a red abalone, *Hallotis Refuscens.* The shell is made primarily of calcium carbonate deposited in layers.

when solid CaO is placed in a closed vessel at 25°C, the decomposition to solid Ca and gaseous O_2 is virtually undetectable. In cases like this, the equilibrium position is said to lie *far to the left,* in the direction of the reactants.

In this chapter we will discuss how and why a chemical system comes to equilibrium and the characteristics of equilibrium. In particular, we will discuss how to calculate the concentrations of the reactants and products present for a given system at equilibrium.

13.1 The Equilibrium Condition

Purpose

- To discuss how equilibrium is established.

Equilibrium is a dynamic situation.

Since no changes occur in the concentrations of reactants or products in a reaction system at equilibrium, it may appear that everything has stopped. However, this is not the case. On the molecular level, there is frantic activity. Equilibrium is not static, but is a highly *dynamic* situation. The concept of chemical equilibrium is analogous to two cities connected by a bridge. Suppose the traffic flow on the bridge is the same in both directions. It is obvious that there is motion, since one can see the cars traveling across the bridge, but the number of people in each city is not changing because there is an equal flow of people entering and leaving. The result is no *net* change in population.

To see how this concept applies to chemical reactions, consider the reaction between steam and carbon monoxide in a closed vessel at a high temperature where the reaction takes place rapidly:

$$H_2O(g) + CO(g) \rightleftharpoons H_2(g) + CO_2(g)$$

Assume that the same number of moles of gaseous CO and gaseous H_2O are placed in a closed vessel and allowed to react. The plots of the concentrations of reactants and products versus time are shown in Fig. 13.1. Note that since CO and H_2O were originally present in equal molar quantities, and since they react in a 1:1 ratio, the concentrations of the two gases are always equal. Also since H_2 and CO_2 are formed in equal amounts, they are always present at the same concentrations.

Figure 13.1 is a profile of the progress of the reaction. When CO and H_2O are mixed, they immediately begin to react to form H_2 and CO_2. This leads to a decrease in the concentrations of the reactants, but the concentrations of the products,

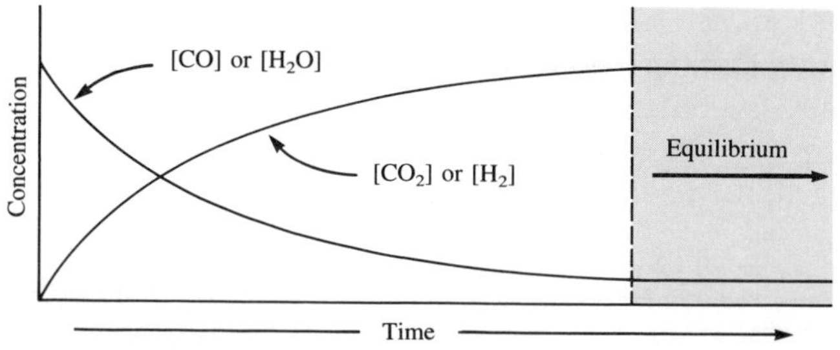

Figure 13.1

The changes in concentrations with time for the reaction $H_2O(g)$ + $CO(g) \rightleftharpoons H_2(g) + CO_2(g)$ when equimolar quantities of $H_2O(g)$ and $CO(g)$ are mixed.

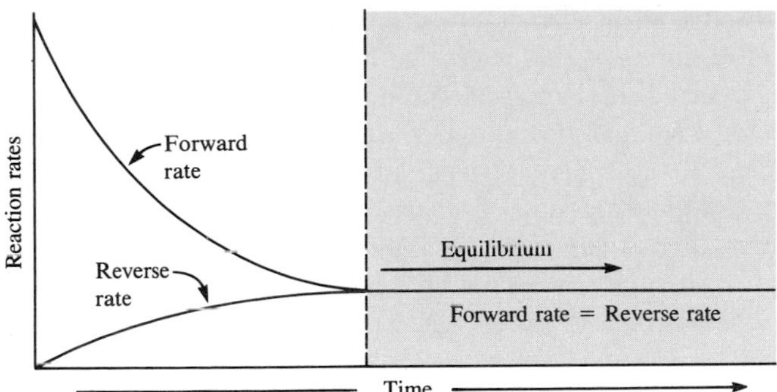

Figure 13.2

The changes with time in the rates of forward and reverse reactions for $H_2O(g) + CO(g) \rightleftharpoons H_2(g) + CO_2(g)$ when equimolar quantities of $H_2O(g)$ and $CO(g)$ are mixed. The rates do not change in the same way with time because the forward reaction has a much larger rate constant than the reverse reaction.

which were initially at zero, are increasing. Beyond a certain time, indicated by the dashed line in Fig. 13.1, the concentrations of reactants and products no longer change—equilibrium has been reached. Unless the system is somehow disturbed, no further changes in concentrations will occur. Note that although the equilibrium position lies far to the right, the concentrations of reactants never go to zero; the reactants will always be present in small but constant concentrations.

Why does equilibrium occur? We have seen in Chapter 12 that molecules react by colliding with one another, and the more collisions, the faster the reaction. This is why reaction rates depend on concentrations. In this case the concentrations of H_2O and CO are lowered by the forward reaction:

$$H_2O + CO \rightarrow H_2 + CO_2$$

As the concentrations of the reactants decrease, the forward reaction slows down (Fig. 13.2). As in the bridge traffic analogy, there is also a reverse direction:

$$H_2 + CO_2 \rightarrow H_2O + CO$$

Initially in this experiment no H_2 and CO_2 were present, and this reverse reaction could not occur. However, as the forward reaction proceeds, the concentrations of H_2 and CO_2 build up and the rate of the reverse reaction increases (Fig. 13.2) as the forward reaction slows down. Eventually the concentrations reach levels where the rate of the forward reaction equals the rate of the reverse reaction. The system has reached equilibrium.

The sublimation of iodine.

A double arrow ($\rightleftharpoons$) is used to show that a reaction can occur in either direction.

The equilibrium position of a reaction—left, right, or somewhere in between—is determined by many factors: the initial concentrations, the relative energies of the reactants and products, and the relative degree of "organization" of the reactants and products. Energy and organization come into play because nature tries to achieve minimum energy and maximum disorder, as we will show in detail in Chapter 16. For now, we will simply view the equilibrium phenomenon in terms of the rates of opposing reactions.

The relationship between equilibrium and thermodynamics is explored in Section 16.8.

The Characteristics of Chemical Equilibrium

To explore the important characteristics of chemical equilibrium, we will consider the synthesis of ammonia from elemental nitrogen and hydrogen:

$$N_2(g) + 3H_2(g) \rightleftharpoons 2NH_3(g)$$

This process is of great commercial value because ammonia is an important fertilizer for the growth of corn and other crops. Ironically, this beneficial process was

The United States produces about 20 million tons of ammonia annually.

discovered in Germany just before World War I in a search for ways to produce nitrogen-based explosives. In the course of this work, German chemist Fritz Haber pioneered the large-scale production of ammonia.

Molecules with strong bonds have large activation energies and tend to react slowly at 25°C.

When gaseous nitrogen, hydrogen, and ammonia are mixed in a closed vessel at 25°C, no apparent change in the concentrations occurs over time, irrespective of the original amounts of the gases. It would seem that equilibrium has been attained. However, this is not necessarily true.

There are two possible reasons why the concentrations of the reactants and products of a given chemical reaction remain unchanged when mixed:

1. The system is at chemical equilibrium.
2. The forward and reverse reactions are so slow that the system moves toward equilibrium at a rate that cannot be detected.

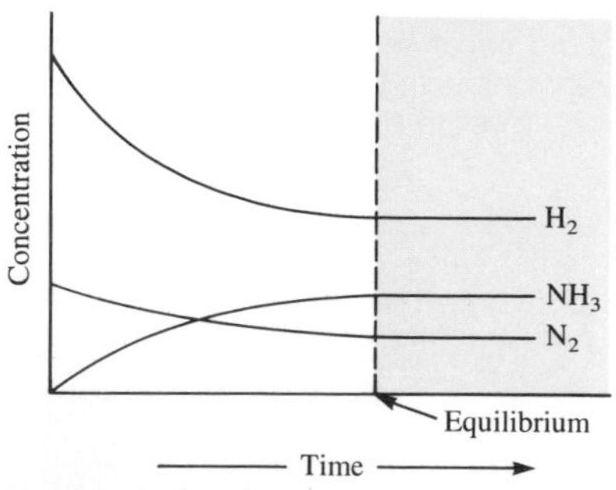

Figure 13.3

A concentration profile for the reaction $N_2(g) + 3H_2(g) \rightleftharpoons 2NH_3(g)$ when only $N_2(g)$ and $H_2(g)$ are mixed initially.

The second reason applies to the nitrogen, hydrogen, and ammonia mixture at 25°C. As we saw in Chapters 8 and 9, the N_2 molecule has a very strong triple bond (941 kJ/mol) and thus is very unreactive. The H_2 molecule also has an unusually strong single bond (432 kJ/mol). So mixtures of N_2, H_2 and NH_3 at 25°C can exist with no apparent change over long periods of time, unless a catalyst is introduced to speed up the forward and reverse reactions. Under appropriate conditions the system does reach equilibrium, as shown in Fig. 13.3. Note that because of the reaction stoichiometry, H_2 disappears three times as fast as N_2 does, and NH_3 forms twice as fast as N_2 disappears.

13.2 The Equilibrium Constant

Purpose

- To introduce the law of mass action and to show how to calculate values for the equilibrium constant.

The law of mass action is based on experimental observation.

Science is fundamentally empirical—it is based on experiment. The development of the equilibrium concept is typical. Based on observations of many chemical reactions, two Norwegian chemists, Cato Maximilian Guldberg (1836–1902) and Peter Waage (1833–1900), proposed in 1864 the **law of mass action** as a general description of the equilibrium condition. Guldberg and Waage postulated that for a reaction of the type

$$jA + kB \rightleftharpoons lC + mD$$

where A, B, C, and D represent chemical species and j, k, l, and m are their coefficients in the balanced equation, the law of mass action is represented by the following **equilibrium expression:**

$$K = \frac{[C]^l[D]^m}{[A]^j[B]^k}$$

The square brackets indicate the concentrations of the chemical species *at equilibrium,* and K is a constant called the **equilibrium constant.**

Sample Exercise 13.1

Write the equilibrium expression for the following reaction:

$$4NH_3(g) + 7O_2(g) \rightleftharpoons 4NO_2(g) + 6H_2O(g)$$

Solution

Applying the law of mass action gives

Coefficient of NO_2 ↘ ↙ Coefficient of H_2O

$$K = \frac{[NO_2]^4[H_2O]^6}{[NH_3]^4[O_2]^7}$$

↑ Coefficient of NH_3 ↖ Coefficient of O_2

The square brackets indicate concentration in units of mol/L.

The value of the equilibrium constant at a given temperature can be calculated if we know the equilibrium concentrations of the reaction components, as illustrated in Sample Exercise 13.2.

Sample Exercise 13.2

The following equilibrium concentrations were observed for the Haber process at 127°C:

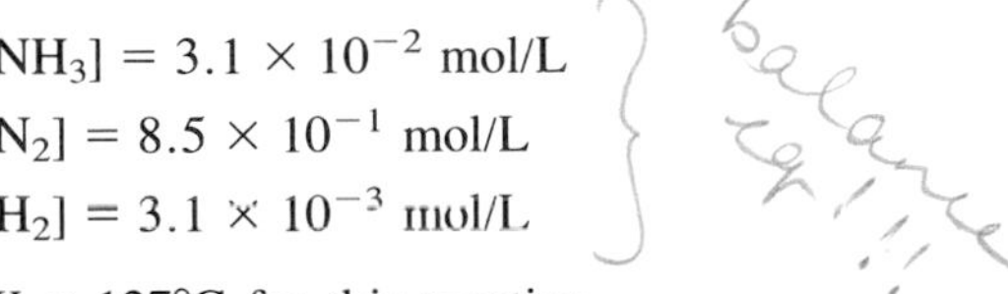

$$[NH_3] = 3.1 \times 10^{-2} \text{ mol/L}$$
$$[N_2] = 8.5 \times 10^{-1} \text{ mol/L}$$
$$[H_2] = 3.1 \times 10^{-3} \text{ mol/L}$$

a. Calculate the value of K at 127°C for this reaction.

b. Calculate the value of the equilibrium constant at 127°C for the reaction

$$2NH_3(g) \rightleftharpoons N_2(g) + 3H_2(g)$$

c. Calculate the value of the equilibrium constant at 127°C for the reaction given by the equation

$$\tfrac{1}{2}N_2(g) + \tfrac{3}{2}H_2(g) \rightleftharpoons NH_3(g)$$

Solution

a. The balanced equation for the Haber process is

$$N_2(g) + 3H_2(g) \rightleftharpoons 2NH_3(g)$$

Thus

$$K = \frac{[NH_3]^2}{[N_2][H_2]^3} = \frac{(3.1 \times 10^{-2} \text{ mol/L})^2}{(8.5 \times 10^{-1} \text{ mol/L})(3.1 \times 10^{-3} \text{ mol/L})^3}$$
$$= 3.8 \times 10^4 \text{ L}^2/\text{mol}^2$$

Sample Exercise 13.2, continued

b. This reaction is written in the reverse order from the equation given in part a. This leads to the equilibrium expression

$$K' = \frac{[N_2][H_2]^3}{[NH_3]^2}$$

which is the reciprocal of the expression used in part a. So

$$K' = \frac{[N_2][H_2]^3}{[NH_3]^2} = \frac{1}{K} = \frac{1}{3.8 \times 10^4 \text{ L}^2/\text{mol}^2} = 2.6 \times 10^{-5} \text{ mol}^2/\text{L}^2$$

c. We use the law of mass action:

$$K'' = \frac{[NH_3]}{[N_2]^{1/2}[H_2]^{3/2}}$$

If we compare this expression to that obtained in part a, we see that since

$$\frac{[NH_3]}{[N_2]^{1/2}[H_2]^{3/2}} = \left(\frac{[NH_3]^2}{[N_2][H_2]^3}\right)^{1/2}$$

$$K'' = K^{1/2}$$

Thus $K'' = K^{1/2} = (3.8 \times 10^4 \text{ L}^2/\text{mol}^2)^{1/2} = 1.9 \times 10^2 \text{ L/mol}$

We can draw some important conclusions from the results of Sample Exercise 13.2. For a reaction of the form

$$jA + kB \rightleftharpoons lC + mD$$

the equilibrium expression is

$$K = \frac{[C]^l[D]^m}{[A]^j[B]^k}$$

If this reaction is reversed, then the new equilibrium expression is

$$K' = \frac{[A]^j[B]^k}{[C]^l[D]^m} = \frac{1}{K}$$

If the original reaction is multiplied by some factor n to give

$$njA + nkB \rightleftharpoons nlC + nmD$$

the equilibrium expression becomes

$$K'' = \frac{[C]^{nl}[D]^{nm}}{[A]^{nj}[B]^{nk}} = K^n$$

We Can Summarize These Conclusions About the Equilibrium Expression as Follows:

- The equilibrium expression for a reaction is the reciprocal of that for the reaction written in reverse.

- When the balanced equation for a reaction is multiplied by a factor n, the equilibrium expression for the new reaction is the original expression raised to the nth power. Thus $K_{new} = (K_{original})^n$.
- The units for K are determined by the powers of the various concentration terms. The units for K therefore depend on the reaction being considered.

The units of equilibrium constants are discussed at the end of Section 13.3.

The law of mass action is widely applicable. It correctly describes the equilibrium behavior of an amazing variety of chemical systems in solution and in the gas phase. Although, as we will see later, corrections must be applied in certain cases, such as for concentrated aqueous solutions and for gases at high pressures, the law of mass action provides a remarkably accurate description of all types of chemical equilibria.

The law of mass action applies to solution and gaseous equilibria.

Consider again the ammonia synthesis reaction. The equilibrium constant, K, always has the same value at a given temperature. At 500°C the value of K is 6.0×10^{-2} L²/mol². Whenever N_2, H_2, and NH_3 are mixed together at this temperature, the system will always come to an equilibrium position such that

$$\frac{[NH_3]^2}{[N_2][H_2]^3} = 6.0 \times 10^{-2} \text{ L}^2/\text{mol}^2$$

This expression has the same value at 500°C, *regardless of the amounts of the gases that are mixed together initially.*

Although the special ratio of products to reactants defined by the equilibrium expression is constant for a given reaction system at a given temperature, the *equilibrium concentrations will not always be the same.* Table 13.1 gives three sets of data for the synthesis of ammonia, showing that even though the individual sets of equilibrium concentrations are quite different for the different situations, the *equilibrium constant, which depends on the ratio of the concentrations, remains the same* (within experimental error). Note that subscript zeros indicate initial concentrations.

Anhydrous ammonia being applied before crops are planted.

Each *set of equilibrium concentrations* is called an **equilibrium position.** It is essential to distinguish between the equilibrium constant and the equilibrium positions for a given reaction system. There is only *one* equilibrium constant for a

Results of Three Experiments for the Reaction $N_2(g) + 3H_2(g) \rightleftharpoons 2NH_3(g)$

Experiment	Initial concentrations	Equilibrium concentrations	$K = \frac{[NH_3]^2}{[N_2][H_2]^3}$
I	$[N_2]_0 = 1.000\ M$ $[H_2]_0 = 1.000\ M$ $[NH_3]_0 = 0$	$[N_2] = 0.921\ M$ $[H_2] = 0.763\ M$ $[NH_3] = 0.157\ M$	$K = 6.02 \times 10^{-2}$ L²/mol²
II	$[N_2]_0 = 0$ $[H_2]_0 = 0$ $[NH_3]_0 = 1.000\ M$	$[N_2] = 0.399\ M$ $[H_2] = 1.197\ M$ $[NH_3] = 0.203\ M$	$K = 6.02 \times 10^{-2}$ L²/mol²
III	$[N_2]_0 = 2.00\ M$ $[H_2]_0 = 1.00\ M$ $[NH_3]_0 = 3.00\ M$	$[N_2] = 2.59\ M$ $[H_2] = 2.77\ M$ $[NH_3] = 1.82\ M$	$K = 6.02 \times 10^{-2}$ L²/mol²

Table 13.1

For a reaction at a given temperature, there are many equilibrium positions but only one value for K.

particular system at a particular temperature, but there are an *infinite* number of equilibrium positions. The specific equilibrium position adopted by a system depends on the initial concentrations, but the equilibrium constant does not.

Sample Exercise 13.3

The following results were collected for two experiments involving the reaction at 600°C between gaseous sulfur dioxide and oxygen to form gaseous sulfur trioxide:

Experiment 1		**Experiment 2**	
Initial	Equilibrium	Initial	Equilibrium
$[SO_2]_0 = 2.00\ M$	$[SO_2] = 1.50\ M$	$[SO_2]_0 = 0.500\ M$	$[SO_2] = 0.590\ M$
$[O_2]_0 = 1.50\ M$	$[O_2] = 1.25\ M$	$[O_2]_0 = 0$	$[O_2] = 0.0450\ M$
$[SO_3]_0 = 3.00\ M$	$[SO_3] = 3.50\ M$	$[SO_3]_0 = 0.350\ M$	$[SO_3] = 0.260\ M$

Show that the equilibrium constant is the same in both cases.

Solution

The balanced equation for the reaction is

$$2SO_2(g) + O_2(g) \rightleftharpoons 2SO_3(g)$$

From the law of mass action

$$K = \frac{[SO_3]^2}{[SO_2]^2[O_2]}$$

For Experiment 1:

$$K_1 = \frac{(3.50\ \text{mol/L})^2}{(1.50\ \text{mol/L})^2(1.25\ \text{mol/L})} = 4.36\ \text{L/mol}$$

For Experiment 2:

$$K_2 = \frac{(0.260\ \text{mol/L})^2}{(0.590\ \text{mol/L})^2(0.0450\ \text{mol/L})} = 4.32\ \text{L/mol}$$

The value of K is constant, within experimental error.

13.3 Equilibrium Expressions Involving Pressures

Purpose

- To show how K and K_p are related.

So far we have been describing equilibria involving gases in terms of concentrations. Equilibria involving gases can also be described in terms of pressures. The relationship between the pressure and the concentration of a gas can be seen from the ideal gas equation:

$$PV = nRT \quad \text{or} \quad P = \left(\frac{n}{V}\right)RT = CRT$$

The ideal gas equation is discussed in Section 5.3.

where C equals n/V, or the number of moles of gas (n) per unit volume (V). Thus C represents the *molar concentration of the gas*.

For the ammonia synthesis reaction, the equilibrium expression can be written in terms of concentrations

$$K = \frac{[NH_3]^2}{[N_2][H_2]^3} = \frac{C_{NH_3}{}^2}{(C_{N_2})(C_{H_2}{}^3)}$$

or in terms of the *equilibrium partial pressures of the gases*

$$K_p = \frac{P_{NH_3}{}^2}{(P_{N_2})(P_{H_2}{}^3)}$$

In this book K is used for an equilibrium constant in terms of concentrations, and K_p represents an equilibrium constant in terms of partial pressures.

K involves concentrations; K_p involves pressures.

Sample Exercise 13.4

The reaction for the formation of nitrosyl chloride

$$2NO(g) + Cl_2(g) \rightleftharpoons 2NOCl(g)$$

was studied at 25°C. The pressures at equilibrium were found to be

$$P_{NOCl} = 1.2 \text{ atm}$$
$$P_{NO} = 5.0 \times 10^{-2} \text{ atm}$$
$$P_{Cl_2} = 3.0 \times 10^{-1} \text{ atm}$$

Calculate the value of K_p for this reaction at 25°C.

Solution

For this reaction

$$K_p = \frac{P_{NOCl}{}^2}{(P_{NO}{}^2)(P_{Cl_2})} = \frac{(1.2 \text{ atm})^2}{(5.0 \times 10^{-2} \text{ atm})^2(3.0 \times 10^{-1} \text{ atm})}$$
$$= 1.9 \times 10^3 \text{ atm}^{-1}$$

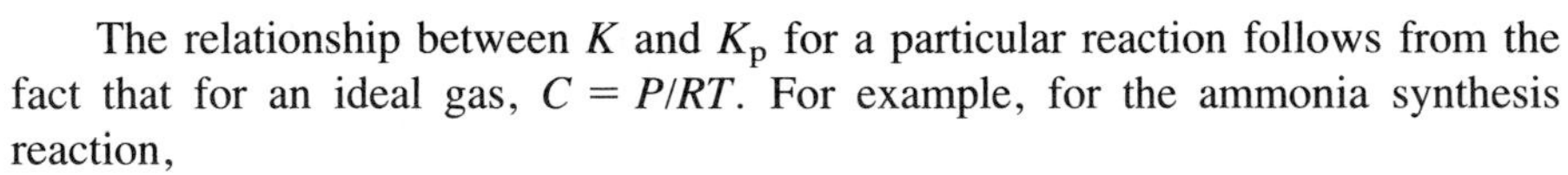

The relationship between K and K_p for a particular reaction follows from the fact that for an ideal gas, $C = P/RT$. For example, for the ammonia synthesis reaction,

$$K = \frac{[NH_3]^2}{[N_2][H_2]^3} = \frac{C_{NH_3}{}^2}{(C_{N_2})(C_{H_2}{}^3)}$$
$$= \frac{\left(\frac{P_{NH_3}}{RT}\right)^2}{\left(\frac{P_{N_2}}{RT}\right)\left(\frac{P_{H_2}}{RT}\right)^3} = \frac{P_{NH_3}{}^2}{(P_{N_2})(P_{H_2}{}^3)} \times \frac{\left(\frac{1}{RT}\right)^2}{\left(\frac{1}{RT}\right)^4}$$

$P = CRT$

or

$C = \dfrac{P}{RT}$

$$= \frac{P_{NH_3}{}^2}{(P_{N_2})(P_{H_2}{}^3)}(RT)^2$$
$$= K_p(RT)^2$$

However, for the synthesis of hydrogen fluoride from its elements,

$$H_2(g) + F_2(g) \rightleftharpoons 2HF(g)$$

the relationship between K and K_p is given by

$$K = \frac{[HF]^2}{[H_2][F_2]} = \frac{C_{HF}{}^2}{(C_{H_2})(C_{F_2})}$$
$$= \frac{\left(\frac{P_{HF}}{RT}\right)^2}{\left(\frac{P_{H_2}}{RT}\right)\left(\frac{P_{F_2}}{RT}\right)} = \frac{P_{HF}{}^2}{(P_{H_2})(P_{F_2})}$$
$$= K_p$$

Thus for this reaction, K is equal to K_p. This equality occurs because the sum of the coefficients on either side of the balanced equation is identical, so the terms in RT cancel out. In the equilibrium expression for the ammonia synthesis reaction, the sum of the powers in the numerator is different from that in the denominator, and K does not equal K_p.

For the general reaction

$$jA + kB \rightleftharpoons lC + mD$$

the relationship between K and K_p is

$$K_p = K(RT)^{\Delta n}$$

where Δn is the sum of the coefficients of the *gaseous* products minus the sum of the coefficients of the *gaseous* reactants. This equation is quite easy to derive from the definitions of K and K_p and the relationship between pressure and concentration. For the above general reaction,

$$K_p = \frac{(P_C{}^l)(P_D{}^m)}{(P_A{}^j)(P_B{}^k)} = \frac{(C_C \times RT)^l(C_D \times RT)^m}{(C_A \times RT)^j(C_B \times RT)^k}$$
$$= \frac{(C_C{}^l)(C_D{}^m)}{(C_A{}^j)(C_B{}^k)} \times \frac{(RT)^{l+m}}{(RT)^{j+k}} = K(RT)^{(l+m)-(j+k)}$$
$$= K(RT)^{\Delta n}$$

Δn always involves products minus reactants.

where $\Delta n = (l + m) - (j + k)$, the difference in the sums of the coefficients for the gaseous products and reactants.

Sample Exercise 13.5

Using the value of K_p obtained in Sample Exercise 13.4, calculate the value of K at 25°C for the reaction

$$2NO(g) + Cl_2(g) \rightleftharpoons 2NOCl(g)$$

Sample Exercise 13.5, continued

Solution

From the value of K_p, we can calculate K using

$$K_p = K(RT)^{\Delta n}$$

where $T = 25 + 273 = 298$ K and

$$\Delta n = 2 - (2 + 1) = -1$$

Sum of product coefficients (2); Sum of reactant coefficients (2 + 1)

Thus

$$K_p = K(RT)^{-1} = \frac{K}{RT}$$

or

$$K = K_p(RT)$$
$$= \left(\frac{1.9 \times 10^3}{\text{atm}}\right)\left(0.08206 \frac{\text{L atm}}{\text{K mol}}\right)(298 \text{ K})$$
$$= 4.6 \times 10^4 \text{ L/mol}$$

We have seen that the units of the equilibrium constant depend on the specific reaction. For example, for the reaction

$$H_2(g) + F_2(g) \rightleftharpoons 2HF(g)$$

the units for K and K_p can be found as follows:

$$K = \frac{C_{HF}^2}{(C_{H_2})(C_{F_2})} = \frac{(\text{mol/L})^2}{(\text{mol/L})(\text{mol/L})} \Rightarrow \text{no units}$$

$$K_p = \frac{P_{HF}^2}{(P_{H_2})(P_{F_2})} = \frac{(\text{atm})^2}{(\text{atm})(\text{atm})} \Rightarrow \text{no units}$$

For this reaction, neither K nor K_p has units, and K is equal to K_p. The identical powers in numerator and denominator cause the units to cancel. For equilibrium expressions in which the powers in the numerator and denominator are not the same, the equilibrium constants will have units and K will not equal K_p. You should note, however, that the units are often omitted when equilibrium constants are listed in tables.

The equilibrium constants used in this chapter involve *observed* concentrations and pressures. These K values have units (unless the units cancel because the powers in the numerator and denominator are the same). However, when corrections are made for nonideal behavior, the resulting equilibrium constants (called thermodynamic equilibrium constants) are unitless. Thus it is quite common to see K values given without units. (*See* Letter to the Editor, F. G. Helfferich, *J. Chem. Ed.*, 64 (1987): 1069.)

13.4 Heterogeneous Equilibria

Purpose

- To show how condensed phases are treated in constructing the equilibrium expression.

So far, we have discussed equilibria only for systems in the gas phase, where all reactants and products are gases. These are **homogeneous equilibria.** However, many equilibria involve more than one phase and are called **heterogeneous equilibria.** For example, the thermal decomposition of calcium carbonate in the commercial preparation of lime occurs by a reaction involving both solid and gas phases:

Lime is among the top five chemicals manufactured in the United States in terms of amount produced.

$$CaCO_3(s) \rightleftharpoons \underset{\substack{\uparrow \\ \text{Lime}}}{CaO(s)} + CO_2(g)$$

Straightforward application of the law of mass action leads to the equilibrium expression

$$K' = \frac{[CO_2][CaO]}{[CaCO_3]}$$

The concentrations of pure liquids and solids are constant.

However, experimental results show that the *position of a heterogeneous equilibrium does not depend on the amounts of pure solids or liquids present.* The fundamental reason for this behavior is that the concentrations of pure solids and liquids cannot change. Thus the equilibrium expression for the decomposition of solid calcium carbonate might be represented as

$$K' = \frac{[CO_2]C_1}{C_2}$$

K doesn't include [ ]'s of solids & liquids because these [ ]'s cannot change.

where C_1 and C_2 are constants representing the concentrations of the solids CaO and $CaCO_3$, respectively. This expression can be rearranged to give

$$\frac{C_2K'}{C_1} = K = [CO_2]$$

We can generalize from this result as follows. If pure solids or pure liquids are involved in a chemical reaction, their concentrations *are not included in the equilibrium expression* for the reaction. This simplification occurs *only* with pure solids or liquids, not with solutions or gases, since in these last two cases the concentrations can vary.

For example, in the decomposition of liquid water to gaseous hydrogen and oxygen

$$2H_2O(l) \rightleftharpoons 2H_2(g) + O_2(g)$$

where

$$K = [H_2]^2[O_2] \quad \text{and} \quad K_p = (P_{H_2}{}^2)(P_{O_2})$$

water is not included in either equilibrium expression because it is a pure liquid. However, if the reaction were carried out under conditions where the water is a gas rather than a liquid:

$$2H_2O(g) \rightleftharpoons 2H_2(g) + O_2(g)$$

then

$$K = \frac{[H_2]^2[O_2]}{[H_2O]^2} \quad \text{and} \quad K_p = \frac{(P_{H_2}{}^2)(P_{O_2})}{P_{H_2O}{}^2}$$

because the concentration or pressure of water vapor can change.

Sample Exercise 13.6

Write the expressions for K and K_p for the following processes:

a. The decomposition of solid phosphorus pentachloride to liquid phosphorus trichloride and chlorine gas.

b. Deep-blue solid copper(II) sulfate pentahydrate is heated to drive off water vapor to form white solid copper(II) sulfate.

Hydrated copper(II) sulfate on the left. Water applied to anhydrous copper(II) sulfate, on the right, forms the hydrated compound.

Solution

a. The reaction is

$$PCl_5(s) \rightleftharpoons PCl_3(l) + Cl_2(g)$$

The equilibrium expressions are

$$K = [Cl_2] \quad \text{and} \quad K_p = P_{Cl_2}$$

In this case neither the pure solid PCl_5 nor the pure liquid PCl_3 is included in the equilibrium expressions.

b. The reaction is

$$CuSO_4 \cdot 5H_2O(s) \rightleftharpoons CuSO_4(s) + 5H_2O(g)$$

The equilibrium expressions are

$$K = [H_2O]^5 \quad \text{and} \quad K_p = P_{H_2O}{}^5$$

The solids are not included.

13.5 Applications of the Equilibrium Constant

Purpose

- To show how the equilibrium constant is used to predict the direction a system will move to reach equilibrium.
- To demonstrate the calculation of equilibrium concentrations given initial concentrations.

Knowing the equilibrium constant for a reaction allows us to predict several important features of the reaction: the tendency of the reaction to occur (but not the speed of the reaction), whether or not a given set of concentrations represents an equilibrium condition, and the equilibrium position that will be achieved from a given set of initial concentrations.

The Extent of a Reaction

The inherent tendency for a reaction to occur is indicated by the magnitude of the equilibrium constant. A value of K much larger than 1 means that at equilibrium the reaction system will consist of mostly products—the equilibrium lies to the right. Another way of saying this is that reactions with very large equilibrium constants go essentially to completion. On the other hand, a very small value of K means that the system at equilibrium will consist of mostly reactants—the equilibrium position is far to the left. The given reaction does not occur to any significant extent.

It is important to understand that *the size of* K *and the time required to reach equilibrium are not directly related.* The time required to achieve equilibrium depends on the reaction rate, which is determined by the size of the activation energy. The size of K is determined by thermodynamic factors such as the difference in energy between products and reactants. This difference is represented in Fig. 13.4 and will be discussed in detail in Chapter 16.

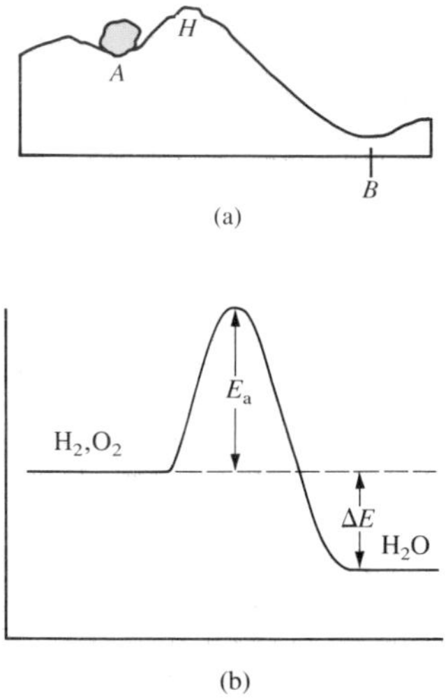

Figure 13.4

(a) A physical analogy illustrating the difference between thermodynamic and kinetic stabilities. The boulder is thermodynamically more stable (lower potential energy) in position B than in position A but cannot get over the hump H. (b) The reactants H_2 and O_2 have a strong tendency to form H_2O. That is, H_2O has lower energy than H_2 and O_2. However, the large activation energy (E_a) prevents the reaction at 25°C. In other words, the magnitude of K for the reaction depends on ΔE, but the reaction rate depends on E_a.

Reaction Quotient

When reactants and products of a given chemical reaction are mixed, it is useful to know whether the mixture is at equilibrium, and if not, in which direction the system will shift to reach equilibrium. If the concentration of one of the reactants or products is zero, the system will shift in the direction that produces the missing component. However, if all the initial concentrations are nonzero, it is more difficult to determine the direction of the move toward equilibrium. To determine the shift in such cases, we use the **reaction quotient (Q).** The reaction quotient is obtained by applying the law of mass action, using *initial concentrations* instead of equilibrium concentrations. For example, for the synthesis of ammonia

$$N_2(g) + 3H_2(g) \rightleftharpoons 2NH_3(g)$$

the expression for the reaction quotient is

$$Q = \frac{[NH_3]_0^2}{[N_2]_0[H_2]_0^3}$$

where the subscript zeros indicate initial concentrations.

To determine in which direction a system will shift to reach equilibrium, we compare the values of Q and K. There are three possible cases:

1. Q *is equal to* K. The system is at equilibrium; no shift will occur.
2. Q *is greater than* K. In this case, the ratio of initial concentrations of products to initial concentrations of reactants is too large. To reach equilibrium, a net change of products to reactants must occur. The system *shifts to the left,* consuming products and forming reactants, until equilibrium is achieved.
3. Q *is less than* K. In this case, the ratio of initial concentrations of products to initial concentrations of reactants is too small. The *system must shift to the right,* consuming reactants and forming products, to attain equilibrium.

Sample Exercise 13.7

For the synthesis of ammonia at 500°C, the equilibrium constant is 6.0×10^{-2} L^2/mol^2. Predict the direction in which the system will shift to reach equilibrium in each of the following cases:

a. $[NH_3]_0 = 1.0 \times 10^{-3}\ M$; $[N_2]_0 = 1.0 \times 10^{-5}\ M$; $[H_2]_0 = 2.0 \times 10^{-3}\ M$
b. $[NH_3]_0 = 2.00 \times 10^{-4}\ M$; $[N_2]_0 = 1.50 \times 10^{-5}\ M$; $[H_2]_0 = 3.54 \times 10^{-1}\ M$
c. $[NH_3]_0 = 1.0 \times 10^{-4}\ M$; $[N_2]_0 = 5.0\ M$; $[H_2]_0 = 1.0 \times 10^{-2}\ M$

Solution

a. First we calculate the value of Q:

$$Q = \frac{[NH_3]_0^2}{[N_2]_0[H_2]_0^3} = \frac{(1.0 \times 10^{-3}\ \text{mol/L})^2}{(1.0 \times 10^{-5}\ \text{mol/L})(2.0 \times 10^{-3}\ \text{mol/L})^3}$$
$$= 1.2 \times 10^7\ L^2/\text{mol}^2$$

Since $K = 6.0 \times 10^{-2}\ L^2/mol^2$, Q is much greater than K. To attain equilibrium, the concentrations of the products must be decreased and the concentrations of the reactants increased. The system will shift to the left:

$$N_2 + 3H_2 \leftarrow 2NH_3$$

b. We calculate the value of Q:

$$Q = \frac{[NH_3]_0^2}{[N_2]_0[H_2]_0^3} = \frac{(2.00 \times 10^{-4}\ \text{mol/L})^2}{(1.50 \times 10^{-5}\ \text{mol/L})(3.54 \times 10^{-1}\ \text{mol/L})^3}$$
$$= 6.01 \times 10^{-2}\ L^2/\text{mol}^2$$

In this case, $Q = K$, so the system is at equilibrium. No shift will occur.

c. The value of Q is

$$Q = \frac{[NH_3]_0^2}{[N_2]_0[H_2]_0^3} = \frac{(1.0 \times 10^{-4}\ \text{mol/L})^2}{(5.0\ \text{mol/L})(1.0 \times 10^{-2}\ \text{mol/L})^3}$$
$$= 2.0 \times 10^{-3}\ L^2/\text{mol}^2$$

Sample Exercise 13.7, continued

Here Q is less than K, so the system will shift to the right to attain equilibrium by increasing the concentration of the product and decreasing the reactant concentrations:

$$N_2 + 3H_2 \rightarrow 2NH_3$$

Calculating Equilibrium Pressures and Concentrations

A typical equilibrium problem involves finding the equilibrium concentrations (or pressures) of reactants and products, given the value of the equilibrium constant and the initial concentrations (or pressures). However, since such problems sometimes become complicated mathematically, we will develop useful strategies for solving them by considering cases for which we know one or more equilibrium concentrations (or pressures).

Sample Exercise 13.8

Dinitrogen tetroxide in its liquid state was used as one of the fuels on the lunar lander for the NASA Apollo missions. In the gas phase it decomposes to gaseous nitrogen dioxide:

$$N_2O_4(g) \rightleftharpoons 2NO_2(g)$$

Consider an experiment in which gaseous N_2O_4 was placed in a flask and allowed to reach equilibrium at a temperature where $K_p = 0.133$ atm. At equilibrium, the pressure of N_2O_4 was found to be 2.71 atm. Calculate the equilibrium pressure of $NO_2(g)$.

Solution

We know that the equilibrium pressures of the gases NO_2 and N_2O_4 must satisfy the relationship

$$K_p = \frac{P_{NO_2}{}^2}{P_{N_2O_4}} = 0.133 \text{ atm}$$

Since we know $P_{N_2O_4}$, we can simply solve for P_{NO_2}:

$$P_{NO_2}{}^2 = K_p(P_{N_2O_4}) = (0.133 \text{ atm})(2.71 \text{ atm}) = 0.360 \text{ atm}^2$$

Therefore $$P_{NO_2} = \sqrt{0.360 \text{ atm}^2} = 0.600 \text{ atm}$$

Sample Exercise 13.9

At a certain temperature a 1.00-L flask initially contained 0.298 mol of $PCl_3(g)$ and 8.70×10^{-3} mol of $PCl_5(g)$. After the system had reached equilibrium, 2.00×10^{-3} mol of $Cl_2(g)$ was found in the flask. Gaseous PCl_5 decomposes according to the reaction

$$PCl_5(g) \rightleftharpoons PCl_3(g) + Cl_2(g)$$

Calculate the equilibrium concentrations of all species and the value of K.

Sample Exercise 13.9, continued

Solution

The equilibrium expression for this reaction is

$$K = \frac{[Cl_2][PCl_3]}{[PCl_5]}$$

To find the value of K, we must calculate the equilibrium concentrations of all species and then substitute these quantities into the equilibrium expression. The best method for finding the equilibrium concentrations is to begin with the initial concentrations, which we will define as the concentrations before any shift toward equilibrium has occurred. We will then modify these initial concentrations appropriately to find the equilibrium concentrations.

The initial concentrations are

$$[Cl_2]_0 = 0$$

$$[PCl_3]_0 = \frac{0.298 \text{ mol}}{1.00 \text{ L}} = 0.298 \ M$$

$$[PCl_5]_0 = \frac{8.70 \times 10^{-3} \text{ mol}}{1.00 \text{ L}} = 8.70 \times 10^{-3} \ M$$

Next we find the change required to reach equilibrium. Since no Cl_2 was initially present but 2.00×10^{-3} M Cl_2 is present at equilibrium, 2.00×10^{-3} mol of PCl_5 must have decomposed to form 2.00×10^{-3} mol of Cl_2 and 2.00×10^{-3} mol of PCl_3. In other words, to reach equilibrium, the reaction shifted to the right:

$$\underset{\substack{\uparrow \\ \text{Net amount of } PCl_5 \\ \text{decomposed}}}{\overset{PCl_5(g)}{2.00 \times 10^{-3} \text{ mol}}} \rightarrow \overset{PCl_3(g)}{2.00 \times 10^{-3} \text{ mol}} + \overset{Cl_2(g)}{2.00 \times 10^{-3} \text{ mol}}$$

(Net amounts of products formed)

Now we apply this change to the initial concentrations to obtain the equilibrium concentrations:

$$[Cl_2] = \underset{[Cl_2]_0}{0} + \frac{2.00 \times 10^{-3} \text{ mol}}{1.00 \text{ L}} = 2.00 \times 10^{-3} \ M$$

$$[PCl_3] = \underset{[PCl_3]_0}{0.298 \ M} + \frac{2.00 \times 10^{-3} \text{ mol}}{1.00 \text{ L}} = 0.300 \ M$$

$$[PCl_5] = \underset{[PCl_5]_0}{8.70 \times 10^{-3} \ M} - \frac{2.00 \times 10^{-3} \text{ mol}}{1.00 \text{ L}} = 6.70 \times 10^{-3} \ M$$

These equilibrium concentrations can now be used to find K:

$$K = \frac{[Cl_2][PCl_3]}{[PCl_5]} = \frac{(2.00 \times 10^{-3} \text{ mol/L})(0.300 \text{ mol/L})}{6.70 \times 10^{-3} \text{ mol/L}}$$

$$= 8.96 \times 10^{-2} \text{ mol/L}$$

Sometimes we are not given any of the equilibrium concentrations (or pressures), only the initial values. Then we must use the stoichiometry of the reaction to express concentrations (or pressures) at equilibrium in terms of the initial values. This is illustrated in Sample Exercise 13.10.

Sample Exercise 13.10

Carbon monoxide reacts with steam to produce carbon dioxide and hydrogen. At 700 K the equilibrium constant is 5.10. Calculate the equilibrium concentrations of all species if 1.000 mol of each component is mixed in a 1.000-L flask.

Solution

The balanced equation for the reaction is

$$CO(g) + H_2O(g) \rightleftharpoons CO_2(g) + H_2(g)$$

and

$$K = \frac{[CO_2][H_2]}{[CO][H_2O]} = 5.10$$

Next, we calculate the initial concentrations:

$$[CO]_0 = [H_2O]_0 = [CO_2]_0 = [H_2]_0 = \frac{1.000 \text{ mol}}{1.000 \text{ L}} = 1.000\ M$$

Is the system at equilibrium, and if not, which way will it shift to reach the equilibrium position? These questions can be answered by calculating Q:

$$Q = \frac{[CO_2]_0[H_2]_0}{[CO]_0[H_2O]_0} = \frac{(1.000 \text{ mol/L})(1.000 \text{ mol/L})}{(1.000 \text{ mol/L})(1.000 \text{ mol/L})} = 1.000$$

Since Q is less than K, the system is not initially at equilibrium but must shift to the right.

What are the equilibrium concentrations? As before, we start with the initial concentrations and modify them to obtain the equilibrium concentrations. We must ask this question: how much will the system shift to the right to attain the equilibrium condition? In Sample Exercise 13.9 the change needed for the system to reach equilibrium was given. However, in this case we do not have this information.

Since the required change in concentrations is unknown at this point, we will define it in terms of x. Let's assume that x mol/L of CO must react for the system to reach equilibrium. This means that the initial concentration of CO will decrease by x mol/L:

$$\underset{\text{Equilibrium}}{\underset{\uparrow}{[CO]}} = \underset{\text{Initial}}{\underset{\uparrow}{[CO]_0}} - \underset{\text{Change}}{\underset{\uparrow}{x}}$$

Since each CO molecule reacts with one H_2O molecule, the concentration of water vapor must also decrease by x mol/L:

$$[H_2O] = [H_2O]_0 - x$$

Sample Exercise 13.10, continued

As the reactant concentrations decrease, the product concentrations increase. Since all of the coefficients are 1 in the balanced reaction, 1 mol of CO reacting with 1 mol of H_2O will produce 1 mol of CO_2 and 1 mol of H_2. Or in the present case, to reach equilibrium, x mol/L of CO will react with x mol/L of H_2O to give an additional x mol/L of CO_2 and x mol/L of H_2:

$$x\text{CO} + x\text{H}_2\text{O} \longrightarrow x\text{CO}_2 + x\text{H}_2$$

Thus the initial concentrations of CO_2 and H_2 will increase by x mol/L:

$$[\text{CO}_2] = [\text{CO}_2]_0 + x$$

$$[\text{H}_2] = [\text{H}_2]_0 + x$$

Now we have all of the equilibrium concentrations defined in terms of the initial concentrations and the change x:

Initial concentration (mol/L)	Change (mol/L)	Equilibrium concentration (mol/L)
$[\text{CO}]_0 = 1.000$	$-x$	$1.000 - x$
$[\text{H}_2\text{O}]_0 = 1.000$	$-x$	$1.000 - x$
$[\text{CO}_2]_0 = 1.000$	$+x$	$1.000 + x$
$[\text{H}_2]_0 = 1.000$	$+x$	$1.000 + x$

Note that the sign of x is determined by the direction of the shift. In this example, the system shifts to the right, so the product concentrations increase and the reactant concentrations decrease. Also note that because the coefficients in the balanced equation are all 1, the magnitude of the change is the same for all species.

Now since we know that the equilibrium concentrations must satisfy the equilibrium expression, we can find the value of x by substituting these concentrations into the expression:

$$K = 5.10 = \frac{[\text{CO}_2][\text{H}_2]}{[\text{CO}][\text{H}_2\text{O}]} = \frac{(1.000 + x)(1.000 + x)}{(1.000 - x)(1.000 - x)} = \frac{(1.000 + x)^2}{(1.000 - x)^2}$$

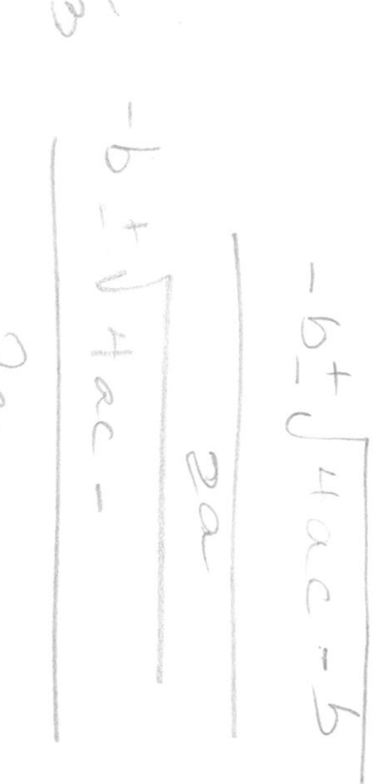

Since the right side of the equation is a perfect square, the solution of the problem can be simplified by taking the square root of both sides:

$$\sqrt{5.10} = 2.26 = \frac{1.000 + x}{1.000 - x}$$

Multiplying and collecting terms gives

$$x = 0.387 \text{ mol/L}$$

Thus the system shifts to the right, consuming 0.387 mol/L of CO and 0.387 mol/L of H_2O and forming 0.387 mol/L of CO_2 and 0.387 mol/L of H_2.

Now the equilibrium concentrations can be calculated:

$$[\text{CO}] = [\text{H}_2\text{O}] = 1.000 - x = 1.000 - 0.387 = 0.613\ M$$

$$[\text{CO}_2] = [\text{H}_2] = 1.000 + x = 1.000 + 0.387 = 1.387\ M$$

Sample Exercise 13.10, continued

Check: These values can be checked by substituting them back into the equilibrium expression to make sure they give the correct value for K:

$$K = \frac{[CO_2][H_2]}{[CO][H_2O]} = \frac{(1.387)^2}{(0.613)^2} = 5.12$$

This result is the same as the given value of K (5.10) within round-off error so the answer must be correct.

Sample Exercise 13.11

The reaction for the formation of gaseous hydrogen fluoride from hydrogen and fluorine has an equilibrium constant of 1.15×10^2 at a certain temperature. In a particular experiment 3.000 mol of each component was added to a 1.500-L flask. Calculate the equilibrium concentrations of all species.

Solution

The balanced equation for the reaction is

$$H_2(g) + F_2(g) \rightleftharpoons 2HF(g)$$

The equilibrium expression is

$$K = 1.15 \times 10^2 = \frac{[HF]^2}{[H_2][F_2]}$$

We first calculate the initial concentrations:

$$[HF]_0 = [H_2]_0 = [F_2]_0 = \frac{3.000 \text{ mol}}{1.500 \text{ L}} = 2.000\ M$$

Then we find the value of Q:

$$Q = \frac{[HF]_0{}^2}{[H_2]_0[F_2]_0} = \frac{(2.000)^2}{(2.000)(2.000)} = 1.000$$

Since Q is much less than K, the system must shift to the right to reach equilibrium.

What change in the concentrations is necessary? Since this is presently unknown, we will define the change needed in terms of x. Let x equal the number of moles per liter of H_2 consumed to reach equilibrium. The stoichiometry of the reaction shows that x mol/L of F_2 will also be consumed and $2x$ mol/L of HF will be formed:

$$\begin{array}{ccccc} H_2(g) & + & F_2(g) & \rightarrow & 2HF(g) \\ x \text{ mol/L} & + & x \text{ mol/L} & \rightarrow & 2x \text{ mol/L} \end{array}$$

Now the equilibrium concentrations can be expressed in terms of x:

Initial concentration (mol/L)	Change (mol/L)	Equilibrium concentration (mol/L)
$[H_2]_0 = 2.000$	$-x$	$[H_2] = 2.000 - x$
$[F_2]_0 = 2.000$	$-x$	$[F_2] = 2.000 - x$
$[HF]_0 = 2.000$	$+2x$	$[HF] = 2.000 + 2x$

Sample Exercise 13.11, continued

To solve for x, we substitute the equilibrium concentrations into the equilibrium expression:

$$K = 1.15 \times 10^2 = \frac{[HF]^2}{[H_2][F_2]} = \frac{(2.000 + 2x)^2}{(2.000 - x)^2}$$

The right side of this equation is a perfect square, so taking the square root of both sides gives

$$\sqrt{1.15 \times 10^2} = \frac{2.000 + 2x}{2.000 - x}$$

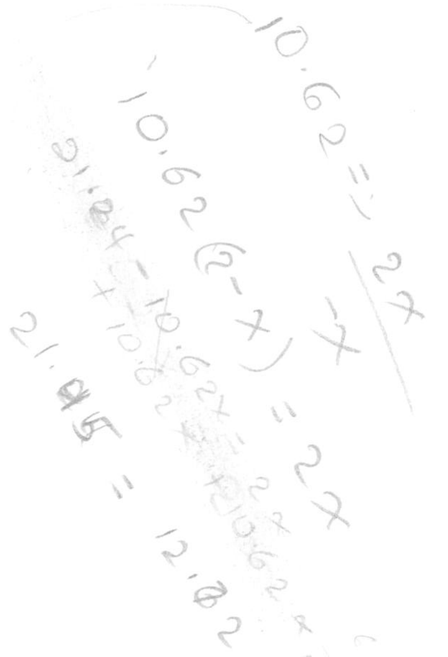

which yields $x = 1.528$. The equilibrium concentrations can now be calculated:

$$[H_2] = [F_2] = 2.000\ M - x = 0.472\ M$$
$$[HF] = 2.000\ M + 2x = 5.056\ M$$

Check: Checking these values by substituting them into the equilibrium expression gives

$$\frac{[HF]^2}{[H_2][F_2]} = \frac{(5.056)^2}{(0.472)^2} = 1.15 \times 10^2$$

which agrees with the given value of K.

13.6 Solving Equilibrium Problems

Purpose

- To generalize the procedure for doing equilibrium calculations.

We have seen most strategies needed to solve equilibrium problems. The typical procedure for analyzing a chemical equilibrium problem can be summarized:

Procedure for Solving Equilibrium Problems

- Write the balanced equation for the reaction.
- Write the equilibrium expression using the law of mass action.
- List the initial concentrations.
- Calculate Q and determine the direction of the shift to equilibrium.
- Define the change needed to reach equilibrium, and define the equilibrium concentrations by applying the change to the initial concentrations.
- Substitute the equilibrium concentrations into the equilibrium expression, and solve for the unknown.
- Check your calculated equilibrium concentrations by making sure they give the correct value of K.

So far, we have been careful to choose systems where we can solve for the unknown by taking the square root of both sides of the equation. However, this type of system is not really very common, and we must now consider a more typical problem. Suppose for a synthesis of hydrogen fluoride from hydrogen and fluorine, 3.000 moles of H_2 and 6.000 moles of F_2 are mixed in a 3.000-liter flask. The equilibrium constant for the synthesis reaction at this temperature is 1.15×10^2. We calculate the equilibrium concentration of each component as follows:

STEP 1

We begin as usual, by writing the balanced equation for the reaction:

$$H_2(g) + F_2(g) \rightleftharpoons 2HF(g)$$

STEP 2

The equilibrium expression is

$$K = 1.15 \times 10^2 = \frac{[HF]^2}{[H_2][F_2]}$$

STEP 3

The initial concentrations are

$$[H_2]_0 = \frac{3.000 \text{ mol}}{3.000 \text{ L}} = 1.000\ M$$

$$[F_2]_0 = \frac{6.000 \text{ mol}}{3.000 \text{ L}} = 2.000\ M$$

$$[HF]_0 = 0$$

STEP 4

There is no need to calculate Q, since no HF is initially present, and we know that the system must shift to the right to reach equilibrium.

STEP 5

If we let x represent the number of moles per liter of H_2 consumed to reach equilibrium, we can represent the equilibrium concentrations as follows:

Initial concentration (mol/L)	Change (mol/L)	Equilibrium concentration (mol/L)
$[H_2]_0 = 1.000$	$-x$	$[H_2] = 1.000 - x$
$[F_2]_0 = 2.000$	$-x$	$[F_2] = 2.000 - x$
$[HF]_0 = 0$	$+2x$	$[HF] = 0 + 2x$

STEP 6

Substituting the equilibrium concentrations into the equilibrium expression gives

$$K = 1.15 \times 10^2 = \frac{[HF]^2}{[H_2][F_2]} = \frac{(2x)^2}{(1.000 - x)(2.000 - x)}$$

Since the right side of this equation is not a perfect square, we cannot take the square root of both sides, but must use some other procedure.

First do the indicated multiplication:

$$(1.000 - x)(2.000 - x)(1.15 \times 10^2) = (2x)^2$$

or

$$(1.15 \times 10^2)x^2 - 3.000(1.15 \times 10^2)x + 2.000(1.15 \times 10^2) = 4x^2$$

and collect terms

$$(1.11 \times 10^2)x^2 - (3.45 \times 10^2)x + 2.30 \times 10^2 = 0$$

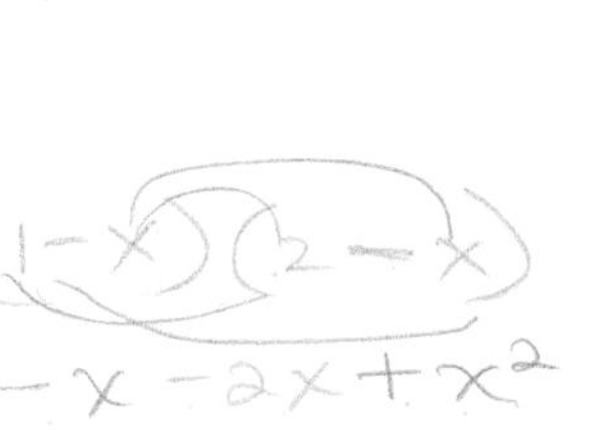

This is a quadratic equation of the general form

$$ax^2 + bx + c = 0$$

where the roots can be obtained from the quadratic formula:

Use of the quadratic formula is explained in Appendix 1.4.

$$x = \frac{-b \pm \sqrt{b^2 - 4ac}}{2a}$$

In this example, $a = 1.11 \times 10^2$, $b = -3.45 \times 10^2$, and $c = 2.30 \times 10^2$. Substituting these values into the quadratic formula gives two values for x:

$$x = 2.14 \text{ mol/L} \quad \text{and} \quad x = 0.968 \text{ mol/L}$$

Both of these results cannot be valid (since a *given* set of initial concentrations leads to only *one* equilibrium position). How can we choose between them? Since the expression for the equilibrium concentration of H_2 is

$$[H_2] = 1.000\ M - x$$

the value of x cannot be 2.14 mol/L (because subtracting 2.14 M from 1.000 M gives a negative concentration of H_2, which is physically impossible). Thus the correct value for x is 0.968 mol/L, and the equilibrium concentrations are as follows:

$$[H_2] = 1.000\ M - 0.968\ M = 3.2 \times 10^{-2}\ M$$
$$[F_2] = 2.000\ M - 0.968\ M = 1.032\ M$$
$$[HF] = 2(0.968\ M) = 1.936\ M$$

STEP 7

We can check these concentrations by substituting them into the equilibrium expression

$$\frac{[HF]^2}{[H_2][F_2]} = \frac{(1.936)^2}{(3.2 \times 10^{-2})(1.032)} = 1.13 \times 10^2$$

This value is in close agreement with the given value for K (1.15×10^2) so the calculated equilibrium concentrations are correct.

This procedure is further illustrated for a problem involving pressures in Sample Exercise 13.12.

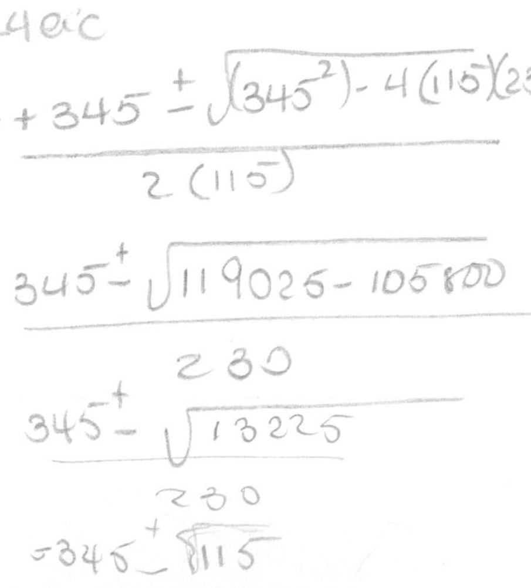

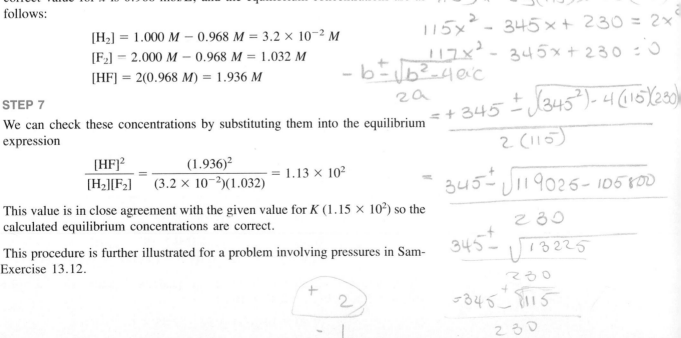

Sample Exercise 13.12

Gaseous hydrogen iodide is synthesized from hydrogen gas and iodine vapor at a temperature where the equilibrium constant is 1.00×10^2. Suppose 5.000×10^{-1} atm of HI, 1.000×10^{-2} atm of H_2, and 5.000×10^{-3} atm of I_2 are mixed in a 5.000-L flask. Calculate the equilibrium pressures of all species.

Solution

The balanced equation for this process is

$$H_2(g) + I_2(g) \rightleftharpoons 2HI(g)$$

and the equilibrium expression in terms of pressure is

$$K_p = \frac{P_{HI}{}^2}{(P_{H_2})(P_{I_2})} = 1.00 \times 10^2$$

The given initial pressures are

$$P_{HI}{}^0 = 5.000 \times 10^{-1} \text{ atm}$$
$$P_{H_2}{}^0 = 1.000 \times 10^{-2} \text{ atm}$$
$$P_{I_2}{}^0 = 5.000 \times 10^{-3} \text{ atm}$$

The value of Q for this system is

$$Q = \frac{(P_{HI}{}^0)^2}{(P_{H_2}{}^0)(P_{I_2}{}^0)} = \frac{(5.000 \times 10^{-1} \text{ atm})^2}{(1.000 \times 10^{-2} \text{ atm})(5.000 \times 10^{-3} \text{ atm})} = 5.000 \times 10^3$$

Since Q is greater than K, the system will shift to the left to reach equilibrium.

So far we have used moles or concentrations in stoichiometric calculations. However, it is equally valid to use pressures for a gas-phase system at constant temperature and volume, because in this case pressure is directly proportional to the number of moles:

$$P = n\left(\frac{RT}{V}\right) \longleftarrow \text{Constant if constant } T \text{ and } V$$

So we can represent the change needed to achieve equilibrium in terms of pressures.

Let x be the change in pressure (in atm) of H_2 as the system shifts left toward equilibrium. This leads to the following equilibrium pressures:

Initial pressure (atm)	Change (atm)	Equilibrium pressure (atm)
$P_{HI}{}^0 = 5.000 \times 10^{-1}$	$-2x$	$P_{HI} = 5.000 \times 10^{-1} - 2x$
$P_{H_2}{}^0 = 1.000 \times 10^{-2}$	$+x$	$P_{H_2} = 1.000 \times 10^{-2} + x$
$P_{I_2}{}^0 = 5.000 \times 10^{-3}$	$+x$	$P_{I_2} = 5.000 \times 10^{-3} + x$

Substitution into the equilibrium expression gives

$$K_p = \frac{(P_{HI})^2}{(P_{H_2})(P_{I_2})} = \frac{(5.000 \times 10^{-1} - 2x)^2}{(1.000 \times 10^{-2} + x)(5.000 \times 10^{-3} + x)}$$

Multiplying and collecting terms yields the quadratic equation where $a = 9.60 \times 10^1$, $b = 3.5$, and $c = -2.45 \times 10^{-1}$:

$$(9.60 \times 10^1)x^2 + 3.5x - (2.45 \times 10^{-1}) = 0$$

Sample Exercise 13.12, continued

From the quadratic formula, the correct value for x is $x = 3.55 \times 10^{-2}$ atm. The equilibrium pressures can now be calculated from the expressions involving x:

$$P_{HI} = 5.000 \times 10^{-1} \text{ atm} - 2(3.55 \times 10^{-2}) \text{ atm} = 4.29 \times 10^{-1} \text{ atm}$$
$$P_{H_2} = 1.000 \times 10^{-2} \text{ atm} + 3.55 \times 10^{-2} \text{ atm} = 4.55 \times 10^{-2} \text{ atm}$$
$$P_{I_2} = 5.000 \times 10^{-3} \text{ atm} + 3.55 \times 10^{-2} \text{ atm} = 4.05 \times 10^{-2} \text{ atm}$$

Check:

$$\frac{P_{HI}{}^2}{P_{H_2} \cdot P_{I_2}} = \frac{(4.29 \times 10^{-1})^2}{(4.55 \times 10^{-2})(4.05 \times 10^{-2})} = 99.9$$

This agrees well with the given value of K (1.00×10^2), so the calculated equilibrium concentrations are correct.

Treating Systems That Have Small Equilibrium Constants

We have seen that fairly complicated calculations are often necessary to solve equilibrium problems. However, under certain conditions, simplifications are possible that greatly reduce the mathematical difficulties. For example, gaseous NOCl decomposes to form the gases NO and Cl_2. At 35°C the equilibrium constant is 1.6×10^{-5} mol/L. In an experiment in which 1.0 mole of NOCl is placed in a 2.0-liter flask, what are the equilibrium concentrations?

The balanced equation is

$$2NOCl(g) \rightleftharpoons 2NO(g) + Cl_2(g)$$

and

$$K = \frac{[NO]^2[Cl_2]}{[NOCl]^2} = 1.6 \times 10^{-5} \text{ mol/L}$$

The initial concentrations are

$$[NOCl]_0 = \frac{1.0 \text{ mol}}{2.0 \text{ L}} = 0.50\ M \qquad [NO]_0 = 0 \qquad [Cl_2]_0 = 0$$

Since there are no products initially, the system will move to the right to reach equilibrium. We will define x as the change in concentration of Cl_2 needed to reach equilibrium. The changes in the concentrations of NOCl and NO can then be obtained from the balanced equation:

$$2NOCl(g) \longrightarrow 2NO(g) + Cl_2(g)$$
$$2x \longrightarrow 2x + x$$

The concentrations can be summarized as follows:

Initial concentration (mol/L)	Change (mol/L)	Equilibrium concentration (mol/L)
$[NOCl]_0 = 0.50$	$-2x$	$[NOCl] = 0.50 - 2x$
$[NO]_0 = 0$	$+2x$	$[NO] = 0 + 2x = 2x$
$[Cl_2]_0 = 0$	$+x$	$[Cl_2] = 0 + x = x$

The equilibrium concentrations must satisfy the equilibrium expression:

$$K = 1.6 \times 10^{-5} = \frac{[NO]^2[Cl_2]}{[NOCl]^2} = \frac{(2x)^2(x)}{(0.50 - 2x)^2}$$

Multiplying and collecting terms will give an equation with terms containing x^3, x^2, and x, which requires complicated methods to solve directly. However, we can avoid this situation by recognizing that since K is so small (1.6×10^{-5} mol/L), the system will not proceed far to the right to reach equilibrium. That is, *x represents a relatively small number*. The consequence of this fact is that the term $(0.50 - 2x)$ can be approximated by 0.50. That is, when x is small,

$$0.50 - 2x \approx 0.50$$

Approximations can simplify complicated math, but their validity should be carefully checked.

Making this approximation allows is to simplify the equilibrium expression:

$$1.6 \times 10^{-5} = \frac{(2x)^2(x)}{(0.50 - 2x)^2} \approx \frac{(2x)^2(x)}{(0.50)^2} = \frac{4x^3}{(0.50)^2}$$

Solving for x^3 gives

$$x^3 = \frac{(1.6 \times 10^{-5})(0.50)^2}{4} = 1.0 \times 10^{-6}$$

and $x = 1.0 \times 10^{-2}$ (mol/L).

How valid is this approximation? If $x = 1.0 \times 10^{-2}$, then

$$0.50 - 2x = 0.50 - 2(1.0 \times 10^{-2}) = 0.48$$

The difference between 0.50 and 0.48 is 0.02, or 4% of the initial concentration of NOCl, a relatively small discrepancy that will have little effect on the outcome. That is, since $2x$ is very small compared to 0.50, the value of x obtained in the approximate solution should be very close to the exact value. We use this approximate value of x to calculate the equilibrium concentrations:

$$[NOCl] = 0.50 - 2x \approx 0.50\ M$$
$$[NO] = 2x = 2(1.0 \times 10^{-2}\ M) = 2.0 \times 10^{-2}\ M$$
$$[Cl_2] = x = 1.0 \times 10^{-2}\ M$$

Check:

$$\frac{[NO]^2[Cl_2]}{[NOCl]^2} = \frac{(2.0 \times 10^{-2})^2(1.0 \times 10^{-2})}{(0.50)^2} = 1.6 \times 10^{-5}$$

Since the given value of K is 1.6×10^{-5}, these calculated concentrations are correct.

This problem was much easier to solve than it appeared at first, because the *small value of* K *and the resulting small shift to the right to reach equilibrium allowed simplification*. A similar problem is handled in Sample Exercise 13.13.

Sample Exercise 13.13

Gaseous phosphorus pentachloride decomposes to gaseous phosphorus trichloride and chlorine at a temperature where $K = 1.00 \times 10^{-3}$ mol/L. Suppose 2.00 mol of

Sample Exercise 13.13, continued

phosphorus pentachloride in a 2.00-L vessel is allowed to come to equilibrium. Calculate the equilibrium concentrations of all species.

Solution

The balanced equation and equilibrium expression are

$$PCl_5(g) \rightleftharpoons PCl_3(g) + Cl_2(g)$$

$$K = 1.00 \times 10^{-3} = \frac{[PCl_3][Cl_2]}{[PCl_5]}$$

The concentrations are

Initial concentration (mol/L)	Change (mol/L)	Equilibrium concentration (mol/L)
$[PCl_5]_0 = 2.000\text{ mol}/2.00\text{ L} = 1.00$	$-x$	$[PCl_5] = 1.00 - x$
$[PCl_3]_0 = 0$	$+x$	$[PCl_3] = 0 + x$
$[Cl_2]_0 = 0$	$+x$	$[Cl_2] = 0 + x$

Substitution in the equilibrium expression gives

$$K = 1.00 \times 10^{-3} = \frac{[PCl_3][Cl_2]}{[PCl_5]} = \frac{(x)(x)}{1.00 - x} \approx \frac{x^2}{1.00}$$

Note that since K is small, we can assume, at least temporarily, that the change in concentration of PCl_5 is negligible. The approximate value of x is then obtained as follows:

$$x^2 \approx (1.00)(1.00 \times 10^{-3})$$
$$x \approx 3.16 \times 10^{-2}$$

Now we must check the validity of this approximation. Since $x = 3.16 \times 10^{-2}$

$$1.00 - x = 1.00 - 0.03 = 0.97$$

Thus $1.00 - x \approx 1.00$ is a good approximation. This is a relatively small difference that will have little effect on the calculated value of x. The equilibrium concentrations therefore are

$$[PCl_5] = 1.00 - x = 0.97\ M \approx 1.00\ M$$
$$[Cl_2] = [PCl_3] = x = 3.16 \times 10^{-2}\ M$$

Check:

$$\frac{[Cl_2][PCl_3]}{[PCl_5]} = \frac{(3.16 \times 10^{-2})(3.16 \times 10^{-2})}{1.00} = 1.00 \times 10^{-3}$$

This agrees with the given value of K.

13.7 Le Châtelier's Principle

Purpose

- To show how to predict the changes that occur when a system at equilibrium is disturbed.

It is important to understand the factors that control the *position* of a chemical equilibrium. For example, when a chemical is manufactured, the chemists and chemical engineers in charge of production want to choose conditions that favor the desired product as much as possible. That is, they want the equilibrium to lie far to the right. When Fritz Haber was developing the process for the synthesis of ammonia, he did extensive studies on how the equilibrium concentration of ammonia depended on the conditions of temperature and pressure. Some of his results are given in Table 13.2. Note that the equilibrium amount of NH_3 increases with an increase in pressure but decreases as the temperature is increased. Thus the amount of NH_3 present at equilibrium is favored by conditions of low temperature and high pressure.

However, this is not the whole story. Carrying out the process at low temperatures is not feasible, because then the reaction is too slow. Even though the equilibrium tends to shift to the right as the temperature is lowered, the attainment of equilibrium would be much too slow at low temperatures to be practical. This emphasizes once again that we must study both the thermodynamics and the kinetics of a reaction before we really understand the factors that control it.

We can qualitatively predict the effects of changes in concentration, pressure, and temperature on a system at equilibrium by using **Le Châtelier's principle,** which states that *if a change is imposed on a system at equilibrium, the position of the equilibrium will shift in a direction that tends to reduce that change*. Although this rule sometimes oversimplifies the situation, it works remarkably well.

The Percent by Mass of NH_3 at Equilibrium in a Mixture of N_2, H_2, and NH_3 as a Function of Temperature and Total Pressure*

	Total pressure		
Temperature (°C)	300 atm	400 atm	500 atm
400	48% NH_3	55% NH_3	61% NH_3
500	26% NH_3	32% NH_3	38% NH_3
600	13% NH_3	17% NH_3	21% NH_3

*Each experiment was begun with a 3:1 mixture of H_2 and N_2.

Table 13.2

The Effect of a Change in Concentration

To see how we can predict the effects of change in concentration on a system at equilibrium, we will consider the ammonia synthesis reaction. Suppose there is an

Blue anhydrous cobalt(II) chloride and pink hydrated cobalt(II) chloride. Since the reaction $CoCl_2(s) + 6H_2O(g) \rightarrow CoCl_2 \cdot 6H_2O(s)$ is shifted to the right by water vapor, $CoCl_2$ is often used in novelty devices to detect humidity.

equilibrium position described by these concentrations:

$$[N_2] = 0.399\ M \qquad [H_2] = 1.197\ M \qquad [NH_3] = 0.202\ M$$

What will happen if 1.000 mol/L of N_2 is suddenly injected into the system? We can answer this question by calculating the value of Q. The concentrations before the system adjusts are

$$[N_2]_0 = 0.399\ M + \underset{\text{Added } N_2}{\underset{\uparrow}{1.000\ M}} = 1.399\ M$$

$$[H_2]_0 = 1.197\ M$$

$$[NH_3]_0 = 0.202\ M$$

Note we are labeling these as "initial concentrations" because the system is no longer at equilibrium. Then

$$Q = \frac{[NH_3]_0^{\,2}}{[N_2]_0[H_2]_0^{\,3}} = \frac{(0.202)^2}{(1.399)(1.197)^3} = 1.70 \times 10^{-2}$$

Since we are not given the value of K, we must calculate it from the first set of equilibrium concentrations:

$$K = \frac{[NH_3]^2}{[N_2][H_2]^3} = \frac{(0.202)^2}{(0.399)(1.197)^3} = 5.96 \times 10^{-2}$$

As we might have expected, Q is less than K because the concentration of N_2 was increased.

The system will shift to the right to come to the new equilibrium position. Rather than do the calculations, we simply summarize the results:

Equilibrium position I		Equilibrium position II
$[N_2] = 0.399\ M$ $[H_2] = 1.197\ M$ $[NH_3] = 0.202\ M$	1.000 mol/L $\Longrightarrow$ of N_2 added	$[N_2] = 1.348\ M$ $[H_2] = 1.044\ M$ $[NH_3] = 0.304\ M$

(top) At a higher temperature, brown $NO_2(g)$ is favored. (bottom) As the temperature decreases, a shift in equilibrium from brown $NO_2(g)$ to colorless $N_2O_4(g)$ occurs.

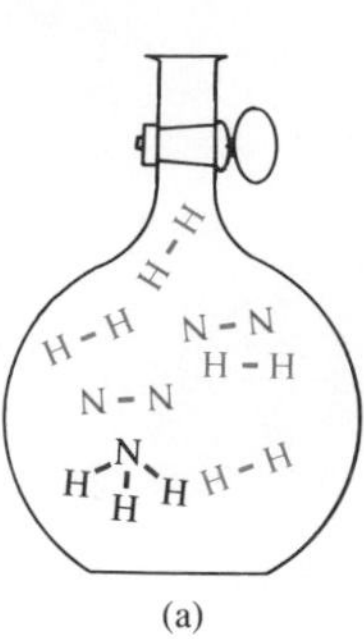

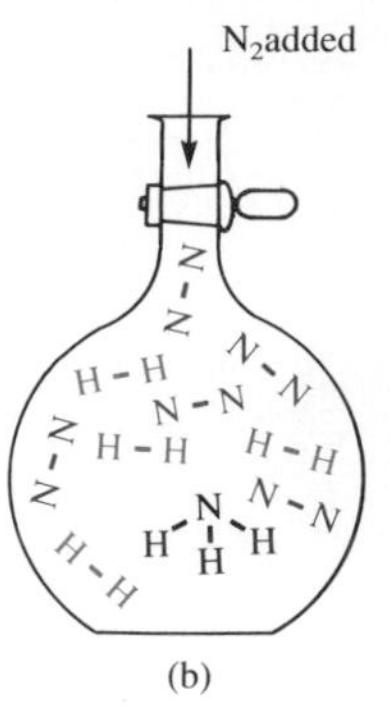

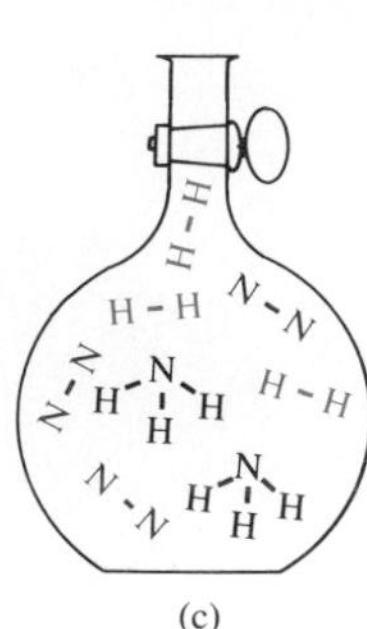

Figure 13.5

(a) The initial equilibrium mixture of N_2, H_2, and NH_3. (b) Addition of N_2. (c) The new equilibrium position for the system containing more N_2 (due to addition of N_2), less H_2, and more NH_3 than in (a).

Note from these data that the equilibrium position does in fact shift to the right: the concentration of H_2 decreases, the concentration of NH_3 increases, and, of course, since nitrogen is added, the concentration of N_2 shows an increase relative to the amount present at the original equilibrium position. (However, the nitrogen showed a decrease relative to the amount present immediately after addition of the 1.000 mol of N_2.)

We can predict this shift qualitatively by using Le Châtelier's principle. Since the change imposed is the addition of nitrogen, Le Châtelier's principle predicts that the system will shift in a direction that consumes nitrogen. This reduces the effect of the addition. Thus Le Châtelier's principle correctly predicts that adding nitrogen will cause the equilibrium to shift to the right (see Fig. 13.5).

The system shifts in the direction that compensates for the imposed change.

If ammonia had been added instead of nitrogen, the system would have shifted to the left to consume ammonia. So another way of stating Le Châtelier's principle is to say that *if a reactant or product is added to a system at equilibrium, the system will shift away from the added component. If a reactant or product is removed, the system will shift toward the removed component.*

Sample Exercise 13.14

Arsenic can be extracted from its ores by first reacting the ore with oxygen (called *roasting*) to form solid As_4O_6, which is then reduced using carbon:

$$As_4O_6(s) + 6C(s) \rightleftharpoons As_4(g) + 6CO(g)$$

Predict the direction of the shift of the equilibrium position in response to each of the following changes in conditions:

a. Addition of carbon monoxide

b. Addition or removal of carbon or arsenic(III) oxide (As_4O_6)

c. Removal of gaseous arsenic (As_4)

Solution

a. Le Châtelier's principle predicts that the shift will be away from the substance whose concentration is increased. The equilibrium position will shift to the left when carbon monoxide is added.

b. Since the amount of a pure solid has no effect on the equilibrium position, changing the amount of carbon or arsenic(III) oxide will have no effect.

Sample Exercise 13.14, continued

c. If gaseous arsenic is removed, the equilibrium position will shift to the right to form more products. In industrial processes, the desired product is often continuously removed from the reaction system to increase the yield.

The Effect of a Change in Pressure

Basically, there are three ways to change the pressure of a reaction system involving gaseous components:

1. Add or remove a gaseous reactant or product.
2. Add an inert gas (one not involved in the reaction).
3. Change the volume of the container.

We have already considered the addition or removal of a reactant or product. When an inert gas is added, there is no effect on the equilibrium position. *The addition of an inert gas increases the total pressure but has no effect on the concentrations or partial pressures of the reactants or products.* Thus the system remains at the original equilibrium position.

When the volume of the container is changed, the concentrations (and thus the partial pressures) of both reactants and products are changed. We could calculate Q and predict the direction of the shift. However, for systems involving gaseous components, there is an easier way: we focus on the volume. The central idea is that *when the volume of the container holding a gaseous system is reduced, the system responds by reducing its own volume. This is done by decreasing the total number of gaseous molecules in the system.*

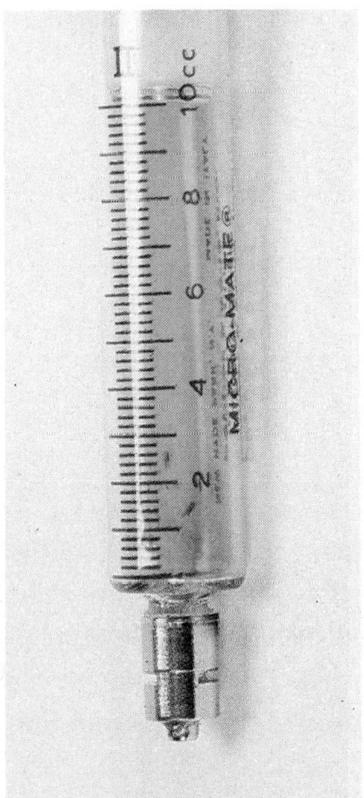

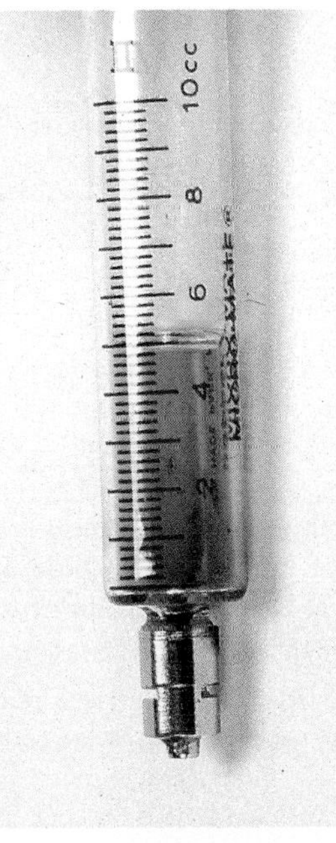

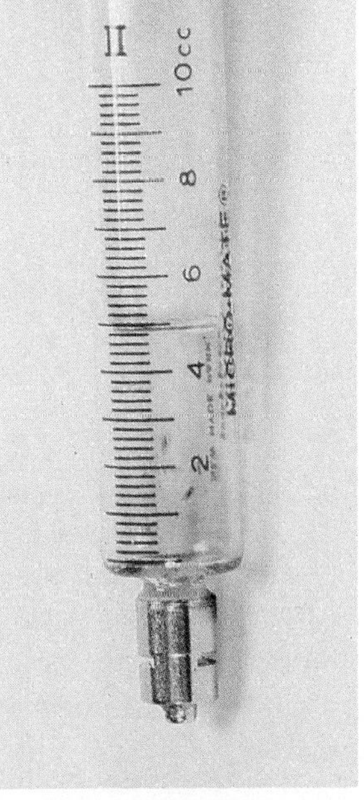

(left) Brown $NO_2(g)$ and colorless $N_2O_4(g)$ in equilibrium in a syringe. (middle) The volume is suddenly decreased, giving a greater concentration of both N_2O_4 and NO_2 (indicated by the darker brown color). (right) A few seconds after the sudden volume decrease, the color is much lighter brown as the equilibrium shifts from brown $NO_2(g)$ to colorless $N_2O_4(g)$ as predicted by Le Châtelier's principle, since in the equilibrium

$$2NO_2(g) \rightleftharpoons N_2O_4(g)$$

the product side has the smaller number of molecules.

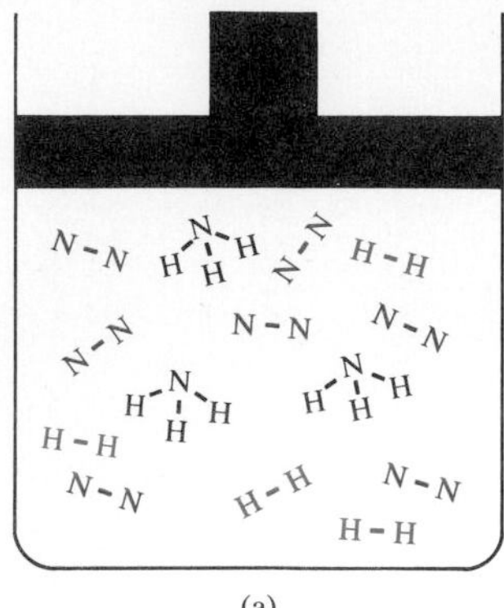

(a)

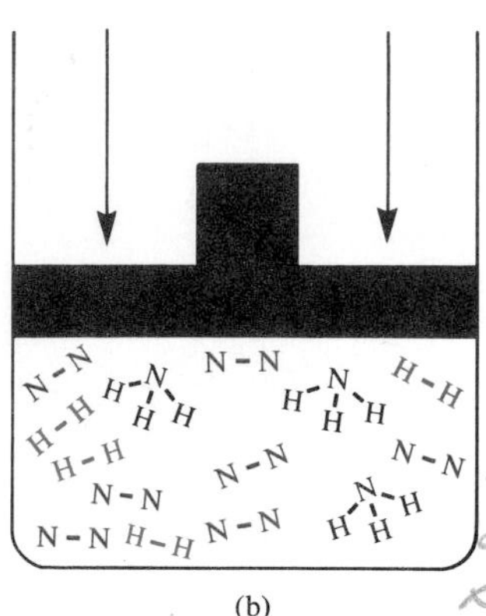

(b)

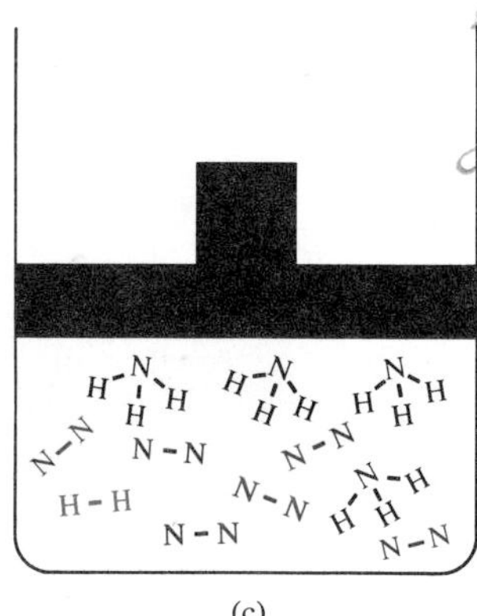

(c)

Figure 13.6

(a) A mixture of $NH_3(g)$, $N_2(g)$, and $H_2(g)$ at equilibrium. (b) The volume is suddenly decreased. (c) The new equilibrium position for the system containing more NH_3 and less N_2 and H_2. The reaction $N_2(g) + 3H_2(g) \rightleftharpoons 2NH_3(g)$ shifts to the right (toward the side with fewer molecules) when the container volume is decreased.

To see that this is true, we can rearrange the ideal gas law to give

$$V = \left(\frac{RT}{P}\right)n$$

or at constant T and P

$$V \propto n$$

That is, at constant temperature and pressure, the volume of a gas is directly proportional to the number of moles of gas present.

Suppose we have a mixture of the gases nitrogen, hydrogen, and ammonia at equilibrium (Fig. 13.6). If we suddenly reduce the volume, what will happen to the equilibrium position? The reaction system can reduce its volume by reducing the number of molecules present. This means that the reaction

$$N_2(g) + 3H_2(g) \rightleftharpoons 2NH_3(g)$$

will shift to the right, since in this direction four molecules (one of nitrogen and three of hydrogen) react to produce two molecules (of ammonia), thus *reducing the total number of gaseous molecules present*. The new equilibrium position will be further to the right than the original one. That is, the equilibrium position will shift toward the side of the reaction involving the smaller number of gaseous molecules in the balanced equation.

The opposite is also true. When the container volume is increased, the system will shift so as to increase its volume. An increase in volume in the ammonia synthesis system will produce a shift to the left to increase the total number of gaseous molecules present.

Sample Exercise 13.15

Predict the shift in equilibrium position that will occur for each of the following processes when the volume is reduced:

a. The preparation of liquid phosphorus trichloride by the reaction

$$P_4(s) + 6Cl_2(g) \rightleftharpoons 4PCl_3(l)$$

b. The preparation of gaseous phosphorus pentachloride according to the equation

$$PCl_3(g) + Cl_2(g) \rightleftharpoons PCl_5(g)$$

c. The reaction of phosphorus trichloride with ammonia:

$$PCl_3(g) + 3NH_3(g) \rightleftharpoons P(NH_2)_3(g) + 3HCl(g)$$

Solution

a. Since P_4 and PCl_3 are a pure solid and a pure liquid, respectively, we need to consider only the effect of the change in volume on Cl_2. The volume is decreased, so the position of the equilibrium will shift to the right, since the reactant side contains six gaseous molecules and the product side has none.

b. Decreasing the volume will shift the given reaction to the right, since the product side contains only one gaseous molecule while the reactant side has two.

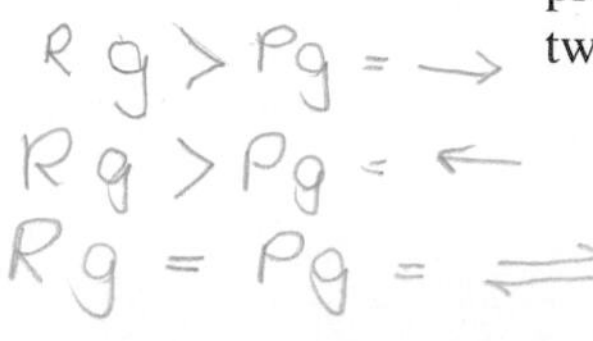

Sample Exercise 13.15, continued

c. Both sides of the balanced reaction equation have four gaseous molecules. A change in volume will have no effect on the equilibrium position. There is no shift in this case.

The Effect of a Change in Temperature

It is important to realize that although the changes we have just discussed may alter the equilibrium *position,* they do not alter the equilibrium *constant.* For example, the addition of a reactant shifts the equilibrium position to the right but has no effect on the value of the equilibrium constant; the new equilibrium concentrations satisfy the original equilibrium constant.

The effect of temperature on equilibrium is different, however, because *the value of K changes with temperature.* We can use Le Châtelier's principle to predict the direction of the change.

The synthesis of ammonia from nitrogen and hydrogen is exothermic. We can represent this by treating energy as a product:

$$N_2(g) + 3H_2(g) \rightleftharpoons 2NH_3(g) + 92\ \text{kJ}$$

If energy is added to this system at equilibrium by heating it, Le Châtelier's principle predicts that the shift will be in the direction that consumes energy, that is, to the left. Note that this shift decreases the concentration of NH_3 and increases the concentrations of N_2 and H_2, thus *decreasing the value of K.* The experimentally observed change in K with temperature for this reaction is indicated in Table 13.3. The value of K decreases with increased temperature, as predicted.

Observed Value of *K* for the Ammonia Synthesis Reaction as a Function of Temperature*

Temperature (K)	K (L^2/mol^2)
500	90
600	3
700	0.3
800	0.04

*For this exothermic reaction, the value of K decreases as the temperature increases, as predicted by Le Châtelier's principle.

Table 13.3

On the other hand, for an endothermic reaction, such as the decomposition of calcium carbonate,

$$556\ \text{kJ} + CaCO_3(s) \rightleftharpoons CaO(s) + CO_2(g)$$

an increase in temperature will cause the equilibrium to shift to the right and the value of K to increase.

In summary, to use Le Châtelier's principle to describe the effect of a temperature change on a system at equilibrium, treat energy as a reactant (in an endothermic process) or as a product (in an exothermic process), and predict the direction of the shift in the same way as when an actual reactant or product is added or removed. Although Le Châtelier's principle cannot predict the size of the change in K, it does correctly predict the direction of the change.

Sample Exercise 13.16

For each of the following reactions, predict how the value of K changes as the temperature is increased.

a. $N_2(g) + O_2(g) \rightleftharpoons 2NO(g) \qquad \Delta H° = 181\ \text{kJ}$

b. $2SO_2(g) + O_2(g) \rightleftharpoons 2SO_3(g) \qquad \Delta H° = -198\ \text{kJ}$

Solution

a. This is an endothermic reaction as indicated by the positive value for $\Delta H°$. Energy can be viewed as a reactant, and K increases (the equilibrium shifts to the right) as the temperature is increased.

b. This is an exothermic reaction (energy can be regarded as a product). As the temperature is increased, the value of K decreases (the equilibrium shifts to the left).

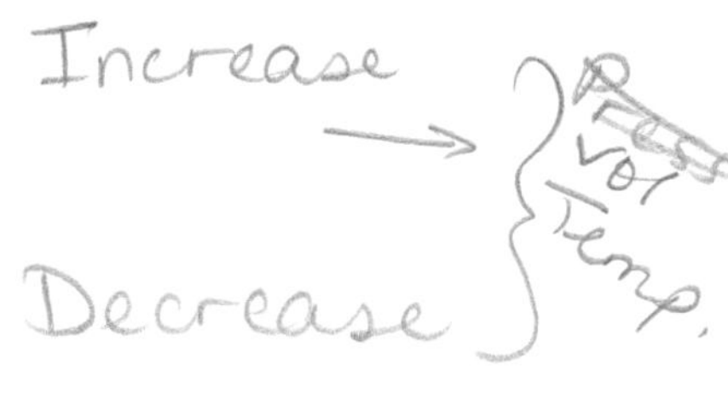

We have seen how Le Châtelier's principle can be used to predict the effect of several types of changes on a system at equilibrium. To summarize these ideas, Table 13.4 shows how various changes affect the equilibrium position of the endothermic reaction

$$N_2O_4(g) \rightleftharpoons 2NO_2(g) \qquad \Delta H° = 58\ \text{kJ}$$

Shifts in the Equilibrium Position for the Reaction 58 kJ + $N_2O_4(g) \rightleftharpoons 2NO_2(g)$

Change	Shift
Addition of $N_2O_4(g)$	Right
Addition of $NO_2(g)$	Left
Removal of $N_2O_4(g)$	Left
Removal of $NO_2(g)$	Right
Addition of $He(g)$	None
Decrease container volume	Left
Increase container volume	Right
Increase temperature	Right
Decrease temperature	Left

Table 13.4

FOR REVIEW

Summary

For stoichiometry calculations we assume that reactions run to completion. However, when a chemical reaction is carried out in a closed vessel, the system achieves chemical equilibrium, the state where the concentrations of both reactants and products remain constant over time. Although a particular equilibrium concentration may be small, it is never zero. Equilibrium is a highly dynamic state; reactants are converted continually to products, and vice versa, as molecules collide with each other. At equilibrium the rates of the forward and reverse reactions are equal.

The law of mass action is a general description of the equilibrium condition. It states that for a reaction of the type

$$jA + kB \rightleftharpoons lC + mD$$

the equilibrium expression is given by

$$K = \frac{[C]^l[D]^m}{[A]^j[B]^k}$$

where K is the equilibrium constant.

For each reaction system at a given temperature, there is only one value for the equilibrium constant but an infinite number of possible equilibrium positions. An equilibrium position is defined by a particular set of concentrations (equilibrium concentrations) that satisfies the equilibrium expression. The specific equilibrium position attained by a system depends on the initial concentrations. An equilibrium position never depends on the amount of a pure liquid or a pure solid, and therefore these substances are not included in the equilibrium expression.

For a gas-phase reaction, equilibrium positions can also be described in terms of partial pressures. K_p, the equilibrium constant in terms of pressures, is related to K as follows:

$$K_p = K(RT)^{\Delta n}$$

where Δn is the sum of the coefficients of the gaseous products minus the sum of the coefficients of the gaseous reactants.

Homogeneous equilibria involve reactions where all reactants and products are in the same phase. However, many systems involve different phases and their equilibria are described as heterogeneous.

The value of K allows us to predict the inherent tendency of a reaction to occur. A small value of K means that the equilibrium position lies far to the left, and therefore the reaction will not occur to any great extent. A large value of K means that the reaction will go essentially to completion—the equilibrium position will lie far to the right. However, the value of K gives no information concerning how rapidly equilibrium will be achieved.

The reaction quotient (Q) applies the law of mass action to initial rather than equilibrium concentrations. We compare Q to K to determine in which direction a chemical system will shift to reach equilibrium. If Q is equal to K, the system is already at equilibrium; if Q is less than K, the system will shift to the right to reach equilibrium; if Q is greater than K the system will shift to the left to reach equilibrium.

To find the concentrations that characterize a given equilibrium position, it is usually best to start with the given initial concentrations (or pressures), define the change needed to reach equilibrium, and apply this change to the initial concentrations (or pressures) to define the equilibrium concentrations (or pressures).

Le Châtelier's principle allows us to predict qualitatively the effects of changes in concentration, pressure, and temperature on a system at equilibrium. The principle states that if a change is imposed, the equilibrium position will shift in a direction that tends to compensate for the imposed change.

Key Terms

chemical equilibrium

Section 13.2
law of mass action
equilibrium expression
equilibrium constant
equilibrium position

Section 13.4
homogeneous equilibria
heterogeneous equilibria

Section 13.5
reaction quotient (Q)

Section 13.7
Le Châtelier's principle

Exercises

A blue exercise number indicates that the answer to that exercise appears at the back of this book and a solution appears in the Solutions Guide.

Characteristics of Chemical Equilibrium

1. Characterize a system at chemical equilibrium with respect to each of the following:
   a. the rates of the forward and reverse reactions
   b. the overall composition of the reaction mixture

2. Is the following statement true or false? "Reactions with large equilibrium constants are very fast." Explain your answer.

3. Consider the following reaction:

$$H_2O(g) + CO(g) \rightleftharpoons H_2(g) + CO_2(g)$$

   Amounts of H_2O, CO, H_2, and CO_2 are put into a flask so that the composition corresponds to an equilibrium position. If the CO placed in the flask is labeled with radioactive ^{14}C, will ^{14}C be found only in CO molecules for an indefinite period of time? Why or why not?

4. Consider the same reaction as in Exercise 3. In one experiment 1.0 mol of $H_2O(g)$ and 1.0 mol of $CO(g)$ are put into a flask and heated to 350°C. In a second experiment 1.0 mol of $H_2(g)$ and 1.0 mol of $CO_2(g)$ are put into another flask with the same volume as the first. This mixture is also heated to 350°C. After equilibrium is reached, will there be any difference in the composition of the mixtures in the two flasks?

The Equilibrium Constant

5. Distinguish between the terms *equilibrium constant* and *equilibrium position*.

6. Distinguish between the terms *equilibrium constant* and *reaction quotient*.

7. Write the equilibrium expression (for K) for each of the following gas-phase reactions, which occur in the atmosphere:
   a. $NO(g) + O_3(g) \rightleftharpoons NO_2(g) + O_2(g)$
   b. $O_3(g) \rightleftharpoons O_2(g) + O(g)$
   c. $Cl(g) + O_3(g) \rightleftharpoons ClO(g) + O_2(g)$
   d. $2O_3(g) \rightleftharpoons 3O_2(g)$

8. Write the equilibrium expression (for K_p) for each reaction in Exercise 7.

9. For which of the reactions in Exercise 7 is K_p equal to K?

10. In atmospheric chemistry the concentrations of trace substances are usually expressed in units of molecules/cm^3. Below are data for some gas-phase reactions at 300 K (from the NASA publication "Chemical Kinetics and Photochemical Data for Use in Atmospheric Modeling, Evaluation Number 5"). K^* indicates the use of molecules/cm^3 for concentration. Calculate K and K_p for each reaction.

a. $HO_2(g) + NO_2(g) \rightleftharpoons HO_2NO_2(g)$

$K^* = 1.26 \times 10^{-11}$

b. $CH_3O_2(g) + NO_2(g) \rightleftharpoons CH_3O_2NO_2(g)$

$K^* = 2.09 \times 10^{-12}$

11. At 127°C, $K = 2.6 \times 10^{-5}$ mol²/L² for the reaction

$$2NH_3(g) \rightleftharpoons N_2(g) + 3H_2(g)$$

Calculate K_p at this temperature.

12. Write expressions for K for the following reactions:

a. $P_4(s) + 5O_2(g) \rightleftharpoons P_4O_{10}(s)$
b. $NH_4NO_3(s) \rightleftharpoons N_2O(g) + 2H_2O(g)$
c. $CO_2(g) + NaOH(s) \rightleftharpoons NaHCO_3(s)$
d. $S_8(s) + 8O_2(g) \rightleftharpoons 8SO_2(g)$

13. Write expressions for K_p for the following reactions:

a. $2Fe(s) + \frac{3}{2}O_2(g) \rightleftharpoons Fe_2O_3(s)$
b. $CO_2(g) + MgO(s) \rightleftharpoons MgCO_3(s)$
c. $C(s) + H_2O(g) \rightleftharpoons CO(g) + H_2(g)$
d. $4KO_2(s) + 2H_2O(g) \rightleftharpoons 4KOH(s) + 3O_2(g)$

14. At a given temperature $K = 278$ for the reaction

$$2SO_2(g) + O_2(g) \rightleftharpoons 2SO_3(g)$$

Calculate values of K for the following reactions at this temperature.

a. $SO_2(g) + \frac{1}{2}O_2(g) \rightleftharpoons SO_3(g)$
b. $2SO_3(g) \rightleftharpoons 2SO_2(g) + O_2(g)$
c. $SO_3(g) \rightleftharpoons SO_2(g) + \frac{1}{2}O_2(g)$
d. $4SO_2(g) + 2O_2(g) \rightleftharpoons 4SO_3(g)$

15. For the reaction

$$H_2(g) + Br_2(g) \rightleftharpoons 2HBr(g)$$

$K_p = 3.5 \times 10^4$ at 1495 K. What is the value of K_p for the following reactions at 1495 K?

a. $HBr(g) \rightleftharpoons \frac{1}{2}H_2(g) + \frac{1}{2}Br_2(g)$
b. $2HBr(g) \rightleftharpoons H_2(g) + Br_2(g)$
c. $\frac{1}{2}H_2(g) + \frac{1}{2}Br_2(g) \rightleftharpoons HBr(g)$

16. Write the expression for K_p for the reaction

$$2O_3(g) \rightleftharpoons 3O_2(g)$$

What is the relationship between K_p for each of the following reactions and K_p for the one above?

a. $O_3(g) \rightleftharpoons \frac{3}{2}O_2(g)$ b. $3O_2(g) \rightleftharpoons 2O_3(g)$

17. At 427°C a 1.0-L flask contains 20.0 mol of H_2, 18.0 mol of CO_2, 12.0 mol of H_2O, and 5.9 mol of CO at equilibrium. Calculate K for the reaction

$$CO_2(g) + H_2(g) \rightleftharpoons CO(g) + H_2O(g)$$

18. At a particular temperature a 3.0-L flask contains 3.5 mol of HI, 4.1 mol of H_2, and 0.3 mol of I_2 in equilibrium. Calculate K at this temperature for the reaction

$$H_2(g) + I_2(g) \rightleftharpoons 2HI(g)$$

19. An equilibrium mixture contains 0.60 g of solid carbon and the gases carbon dioxide and carbon monoxide at partial pressures of 2.9 atm and 2.6 atm, respectively. Calculate K_p for the reaction

$$C(s) + CO_2(g) \rightleftharpoons 2CO(g)$$

20. A sample of solid ammonium chloride was placed in an evacuated container and then heated so that it decomposed to ammonia gas and hydrogen chloride gas. After heating, the total pressure in the container was found to be 4.4 atm. Calculate K_p at this temperature for the decomposition reaction

$$NH_4Cl(s) \rightleftharpoons NH_3(g) + HCl(g)$$

21. A sample of gaseous PCl_5 was introduced into an evacuated flask so that the pressure of pure PCl_5 would be 0.50 atm at 523 K. However, PCl_5 decomposes to gaseous PCl_3 and Cl_2 and the actual pressure in the flask was found to be 0.84 atm. Calculate K_p for the decomposition reaction

$$PCl_5(g) \rightleftharpoons PCl_3(g) + Cl_2(g)$$

at 523 K. Calculate K at this temperature.

22. A flask was filled with 2.0 mol of gaseous SO_2 and 2.0 mol of gaseous NO_2 and heated. After equilibrium was reached, it was found that 1.3 mol of gaseous NO was present. Assume that the reaction

$$SO_2(g) + NO_2(g) \rightleftharpoons SO_3(g) + NO(g)$$

occurs under these conditions. Calculate the value of the equilibrium constant for this reaction.

Equilibrium Calculations

23. The equilibrium constant is 0.0900 at 25°C for the reaction

$$H_2O(g) + Cl_2O(g) \rightleftharpoons 2HOCl(g)$$

For which of the following sets of conditions is the system at equilibrium? For those that are not at equilibrium, in which direction will the system shift?

a. $P_{H_2O} = 200.$ torr, $P_{Cl_2O} = 49.8$ torr, $P_{HOCl} = 21.0$ torr
b. $P_{H_2O} = 296$ torr, $P_{Cl_2O} = 15.0$ torr, $P_{HOCl} = 20.0$ torr
c. A 2.0-L flask contains 0.084 mol of HOCl, 0.080 mol of Cl_2O, and 0.98 mol of H_2O.
d. A 3.0-L flask contains 0.25 mol of HOCl, 0.0010 mol of Cl_2O, and 0.56 mol of H_2O.

24. At a particular temperature $K_p = 0.133$ atm for the reaction

$$N_2O_4(g) \rightleftharpoons 2NO_2(g)$$

Which of the following conditions correspond to equilibrium positions?

a. $P_{NO_2} = 0.144$ atm, $P_{N_2O_4} = 0.156$ atm
b. $P_{NO_2} = 0.175$ atm, $P_{N_2O_4} = 0.102$ atm
c. $P_{NO_2} = 0.056$ atm, $P_{N_2O_4} = 0.048$ atm
d. $P_{NO_2} = 0.064$ atm, $P_{N_2O_4} = 0.0308$ atm

25. Ethyl acetate is synthesized in a nonreacting solvent (not water) according to the following:

$$\underset{\text{Acetic acid}}{CH_3CO_2H} + \underset{\text{Ethanol}}{C_2H_5OH} \rightleftharpoons \underset{\text{Ethyl acetate}}{CH_3CO_2C_2H_5} + H_2O \qquad K = 2.2$$

Which of the following mixtures are at equilibrium?

a. $[CH_3CO_2C_2H_5] = 0.22\ M$, $[H_2O] = 0.10\ M$, $[CH_3CO_2H] = 0.010\ M$, $[C_2H_5OH] = 0.010\ M$
b. $[CH_3CO_2C_2H_5] = 0.22\ M$, $[H_2O] = 0.0020\ M$, $[CH_3CO_2H] = 0.0020\ M$, $[C_2H_5OH] = 0.10\ M$
c. $[CH_3CO_2C_2H_5] = 0.88\ M$, $[H_2O] = 0.12\ M$, $[CH_3CO_2H] = 0.044\ M$, $[C_2H_5OH] = 6.0\ M$
d. $[CH_3CO_2C_2H_5] = 4.4\ M$, $[H_2O] = 4.4\ M$, $[CH_3CO_2H] = 0.88\ M$, $[C_2H_5OH] = 10.0\ M$

26. For the reaction in Exercise 25, what must the concentration of water be for a mixture with $[CH_3CO_2C_2H_5] = 2.0\ M$, $[CH_3CO_2H] = 0.10\ M$, $[C_2H_5OH] = 5.0\ M$ to be at equilibrium? Why is water included in the equilibrium expression for this reaction?

27. At 900°C $K_p = 1.04$ atm for the reaction

$$CaCO_3(s) \rightleftharpoons CaO(s) + CO_2(g)$$

At a low temperature, dry ice (solid CO_2), calcium oxide, and calcium carbonate are introduced into a 50.0-L reaction chamber. The temperature is raised to 900°C. For the following mixtures, will the initial amount of calcium oxide increase, decrease, or remain the same as the system moves toward equilibrium?

a. 655 g of $CaCO_3$, 95.0 g of CaO, 58.4 g of CO_2
b. 780 g of $CaCO_3$, 1.00 g of CaO, 23.76 g of CO_2
c. 0.14 g of $CaCO_3$, 5000 g of CaO, 23.76 g of CO_2
d. 715 g of $CaCO_3$, 813 g of CaO, 4.82 g of CO_2

28. At 25°C, $K = 0.090$ for the reaction

$$H_2O(g) + Cl_2O(g) \rightleftharpoons 2HOCl(g)$$

Calculate the concentrations of all species at equilibrium for each of the following cases:

a. 1.0 g of H_2O and 2.0 g of Cl_2O are mixed in a 1.0-L flask.
b. 1.0 mol of pure HOCl is placed in a 2.0-L flask.

29. Iodine is sparingly soluble in pure water. However, it dissolves in solutions containing excess iodide ion.

$$I^-(aq) + I_2(aq) \rightleftharpoons I_3^-(aq) \qquad K = 710\ \text{L/mol}$$

What is the ratio of $[I_3^-]$ to $[I_2]$ if 0.10 mol of I_2 is added to 1.0 L of KI solution having each of the following concentrations?

a. $[I^-] = 0.20\ M$ b. $[I^-] = 1.5\ M$ c. $[I^-] = 5.0\ M$

30. At 2200°C, $K = 0.050$ for the reaction

$$N_2(g) + O_2(g) \rightleftharpoons 2NO(g)$$

What is the partial pressure of NO in equilibrium with N_2 and O_2 at initial pressures of 0.80 atm and 0.20 atm, respectively?

31. At 35°C, $K = 1.6 \times 10^{-5}$ mol/L for the reaction

$$2NOCl(g) \rightleftharpoons 2NO(g) + Cl_2(g)$$

Calculate the concentrations of all species at equilibrium for each of the following original mixtures:

a. 2.0 mol of pure NOCl in a 2.0-L flask
b. 2.0 mol of NO and 1.0 mol of Cl_2 in a 1.0-L flask
c. 1.0 mol of NOCl and 1.0 mol of NO in a 1.0-L flask
d. 3.0 mol of NO and 1.0 mol of Cl_2 in a 1.0-L flask
e. 2.0 mol of NOCl, 2.0 mol of NO, and 1.0 mol of Cl_2 in a 1.0-L flask
f. 1.0 mol/L concentration of all three gases

32. At 1100 K, $K_p = 0.25\ \text{atm}^{-1}$ for the following reaction:

$$2SO_2(g) + O_2(g) \rightleftharpoons 2SO_3(g)$$

Calculate the equilibrium partial pressures of SO_2, O_2, and SO_3 produced from an initial mixture in which $P_{SO_2} = P_{O_2} = 0.50$ atm and $P_{SO_3} = 0$.

33. At a particular temperature, $K = 1.0 \times 10^2$ for the reaction

$$H_2(g) + F_2(g) \rightleftharpoons 2HF(g)$$

a. In an experiment, 2.0 mol of H_2 and 2.0 mol of F_2 are introduced into a 1.0-L flask. Calculate the concentrations of all species when equilibrium is reached.
b. To the equilibrium mixture in part a, an additional 0.50 mol of H_2 is added. Calculate the new equilibrium concentrations of H_2, F_2, and HF.

34. For the reaction

$$N_2O_4(g) \rightleftharpoons 2NO_2(g)$$

$K_p = 0.25$ atm at a certain temperature. What are the equilibrium partial pressures of NO_2 and N_2O_4 for each of the following initial conditions?

a. pure NO_2 at a pressure of 0.050 atm
b. pure N_2O_4 at a pressure of 0.040 atm
c. a mixture with $P_{N_2O_4} = P_{NO_2} = 1.0$ atm

35. At a certain temperature, $K = 1.1 \times 10^3$ L/mol for the reaction

$$Fe^{3+}(aq) + SCN^-(aq) \rightleftharpoons FeSCN^{2+}(aq)$$

Calculate the concentrations of Fe^{3+}, SCN^-, and $FeSCN^{2+}$ if 0.10 mol of $Fe(NO_3)_3$ is added to 1.0 L of 2.0 *M* KSCN.

Le Châtelier's Principle

36. The fragrances of many naturally occurring substances are due to the presence of organic compounds called *esters*. For example, the ester ethyl butyrate smells like pineapple. Ethyl butyrate can be made by the following reaction:

$$\underset{\text{Butyric acid}}{CH_3CH_2CH_2\overset{\overset{O}{\|}}{C}{-}OH} + \underset{\text{Ethanol}}{CH_3CH_2OH}$$

$$\rightleftharpoons \underset{\text{Ethyl butyrate}}{CH_3CH_2CH_2\overset{\overset{O}{\|}}{C}{-}OCH_2CH_3} + H_2O$$

Butyric acid has an objectionable odor. If you were to choose a solvent for preparing ethyl butyrate, which of the following would be the best choice: water, 95% ethanol (5% water), 100% ethanol, or acetonitrile (CH_3CN, a nonreactive solvent)?

37. Changing the pressure in a reaction vessel by changing the volume may shift the position of a gas phase equilibrium, while changing the pressure by adding an inert gas will not. Why?

38. How will the equilibrium position of a gas phase reaction be affected by changing the volume of the reaction vessel? Are there reactions that will not have their equilibria shifted by a change in volume? Explain.

39. An important reaction in the commercial production of hydrogen is

$$CO(g) + H_2O(g) \rightleftharpoons H_2(g) + CO_2(g)$$

How will this system at equilibrium shift in each of the four following cases?

a. Gaseous carbon dioxide is removed.
b. Water vapor is added.
c. The pressure is increased by adding helium gas.
d. The temperature is increased. (Use values of ΔH_f° in Appendix 4 to calculate ΔH°.)

40. Novelty devices for predicting rain contain cobalt(II) chloride and are based on the following equilibrium:

$$\underset{\text{Purple}}{CoCl_2(s) + 6H_2O(g)} \rightleftharpoons \underset{\text{Pink}}{CoCl_2 \cdot 6H_2O(s)}$$

What color will such an indicator be if rain is imminent?

41. What will happen to the number of moles of SO_3 in equilibrium with SO_2 and O_2 in the reaction

$$2SO_3(g) \rightleftharpoons 2SO_2(g) + O_2(g) \qquad \Delta H^\circ = 197 \text{ kJ}$$

in each of the following cases?

a. Oxygen gas is added.
b. The pressure is increased by decreasing the volume.
c. The pressure is increased by adding argon gas.
d. The temperature is decreased.
e. A catalyst is added. (See Exercise 44.)
f. Gaseous sulfur dioxide is removed.

42. In which direction will the position of the equilibrium

$$H_2(g) + I_2(g) \rightleftharpoons 2HI(g)$$

be shifted for each of the following changes:

a. $H_2(g)$ is added.
b. $I_2(g)$ is removed.
c. $HI(g)$ is removed.
d. Some $Ar(g)$ is added.
e. The volume of the container is doubled.
f. The temperature is increased. (For HI, $\Delta H_f^\circ = 25.9$ kJ/mol)

43. Use bond energies (Table 8.4) to estimate ΔH for the reaction to prepare methanol from hydrogen and carbon monoxide:

$$CO(g) + 2H_2(g) \rightleftharpoons CH_3OH(g)$$

If you wanted to use this reaction for producing methanol commercially, would high or low temperatures favor a maximum yield?

Kinetics and Equilibrium

44. A catalyst cannot affect the position of an equilibrium, since it lowers the activation energy for both the forward and reverse reactions by the same amount and thus changes their rates by the same factor. Why was the discovery of an appropriate catalyst vital for the use of the Haber process for ammonia production?

45. Which of the following will be affected by the addition of a catalyst?

a. The value of the equilibrium constant.
b. The value of k_f, the forward rate constant.
c. The value of k_r, the reverse rate constant.
d. The value of the forward rate.
e. The value of the reverse rate.
f. The value of the ratio k_f/k_r.

46. Why is it impossible for a catalyst to catalyze only the forward reaction of a system?

47. Consider the following diagram for the *uncatalyzed* reaction

$$N_2 + O_2 \rightleftharpoons 2NO$$

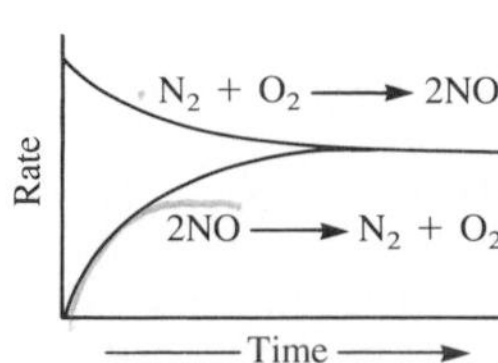

What will a similar diagram for the catalyzed reaction look like?

48. The rate law for the forward reaction of the system

$$BrO_3^-(aq) + 3SO_3^{2-}(aq) \rightleftharpoons Br^-(aq) + 3SO_4^{2-}(aq)$$

in acidic solution is

$$\text{Rate} = k_1[BrO_3^-][SO_3^{2-}][H^+]$$

Write a possible rate law for the reverse reaction. *Hint:* Think of the form of the equilibrium constant expression and the fact that at equilibrium $\text{Rate}_f = \text{Rate}_r$.

49. The rate law for the forward reaction of the system

$$CO(g) + Cl_2(g) \rightleftharpoons COCl_2(g)$$

is

$$\text{Rate} = k_1[CO][Cl_2]^{3/2}$$

Write a possible rate law for the reverse reaction. (See hint in Exercise 48.)

50. Ammonia is produced by the Haber process in which nitrogen and hydrogen are reacted directly using an iron mesh impregnated with oxides as a catalyst. For the reaction

$$N_2(g) + 3H_2(g) \rightleftharpoons 2NH_3(g)$$

equilibrium constants as a function of temperature are

300°C, 4.34×10^{-3} atm^{-2}
500°C, 1.45×10^{-5} atm^{-2}
600°C, 2.25×10^{-6} atm^{-2}

a. Is the reaction exothermic or endothermic?
b. Typical conditions used for producing ammonia are temperatures of 400 to 500°C and total pressures of 200 to 600 atm. Why are these conditions chosen?

Additional Exercises

51. At 25°C, $K_p \approx 1 \times 10^{-31}$ for the reaction

$$N_2(g) + O_2(g) \rightleftharpoons 2NO(g)$$

a. Calculate the concentration of NO, in molecules/cm^3, that can exist in equilibrium in air at 25°C. In air, $P_{N_2} = 0.8$ atm and $P_{O_2} = 0.2$ atm.
b. Typical concentrations of NO in relatively pristine environments range from 10^8 to 10^{10} molecules/cm^3. Why is there a discrepancy between these values and your answer to part a?

52. Given the following equilibrium constants at 427°C,

$$Na_2O(s) \rightleftharpoons 2Na(l) + \tfrac{1}{2}O_2(g) \qquad K_1 = 2 \times 10^{-25}$$
$$NaO(g) \rightleftharpoons Na(l) + \tfrac{1}{2}O_2(g) \qquad K_2 = 2 \times 10^{-5}$$
$$Na_2O_2(s) \rightleftharpoons 2Na(l) + O_2(g) \qquad K_3 = 5 \times 10^{-29}$$
$$NaO_2(s) \rightleftharpoons Na(l) + O_2(g) \qquad K_4 = 3 \times 10^{-14}$$

determine the values for the equilibrium constants for the following:

a. $Na_2O(s) + \tfrac{1}{2}O_2(g) \rightleftharpoons Na_2O_2(s)$
b. $NaO(g) + Na_2O(s) \rightleftharpoons Na_2O_2(s) + Na(l)$
c. $2NaO(g) \rightleftharpoons Na_2O_2(s)$

Hint: When reaction equations are added, the equilibrium expressions are multiplied.

53. Calculate a value for the equilibrium constant for the reaction

$$O_2(g) + O(g) \rightleftharpoons O_3(g)$$

given

$$NO_2(g) \overset{h\nu}{\rightleftharpoons} NO(g) + O(g) \qquad K = 6.8 \times 10^{-49}$$
$$O_3(g) + NO(g) \rightleftharpoons NO_2(g) + O_2(g) \qquad K = 5.8 \times 10^{-34}$$

(See hint in Exercise 52.)

54. At 90°C the equilibrium constant is 6.8×10^{-2} for the reaction

$$H_2(g) + S(s) \rightleftharpoons H_2S(g)$$

If 0.15 mol hydrogen and 1.0 mol sulfur are heated to 90.0°C in a 1.0 L container, what will be the partial pressure of H_2S at equilibrium?

55. At a certain temperature, the equilibrium constant for the gas-phase reaction between carbon monoxide and oxygen to produce carbon dioxide is 5.0×10^3 L/mol. Calculate the concentrations of all species at equilibrium if 2.0 mol each of CO and O_2 are placed in a 5.0 L vessel and allowed to come to equilibrium.

56. A system at equilibrium is described by the equation

$$\text{Energy} + SO_2Cl_2(g) \rightleftharpoons SO_2(g) + Cl_2(g)$$

Why does the temperature of the system increase when SO_2 is added to the system at equilibrium?

57. Nitric oxide and bromine at initial partial pressures of 98.4 and 41.3 torr, respectively, were allowed to react at 300. K. At equilibrium the total pressure was 110.5 torr. The reaction is

$$2NO(g) + Br_2(g) \rightleftharpoons 2NOBr(g)$$

a. Calculate the value of K_p.
b. What would be the partial pressures of all species if NO and Br_2, both at an initial partial pressure of 0.30 atm, were allowed to come to equilibrium at this temperature?

58. Hydrogen for use in ammonia production is produced by the reaction

$$CH_4(g) + H_2O(g) \underset{750°C}{\overset{\text{Ni catalyst}}{\rightleftharpoons}} CO(g) + 3H_2(g)$$

What will happen to a reaction mixture at equilibrium if:
a. $H_2O(g)$ is removed.
b. The temperature is increased. (Use Appendix 4 to calculate $\Delta H°$.)
c. An inert gas is added.
d. $CO(g)$ is removed.
e. The Ni catalyst is removed.

CHAPTER 14

Acids and Bases

In this chapter we reencounter two very important classes of compounds, acids and bases. We will explore their interactions and apply the fundamentals of chemical equilibria discussed in Chapter 13 to systems involving proton-transfer reactions.

Acid-base chemistry is important in a wide variety of everyday applications. There are complex systems in our bodies that carefully control the acidity of our blood, since even small deviations may lead to serious illness and death. The same sensitivity is seen in other life forms. If you have ever had tropical fish or goldfish, you know how important it is to monitor and control the acidity of the water in the aquarium.

Acids and bases are also important industrially. For example, the vast quantity of sulfuric acid manufactured in the United States each year is needed to produce fertilizers, polymers, steel, and many other materials (see the Chemical Impact in Chapter 3).

The influence of acids on living things has assumed special importance in the United States, Canada, and Europe in recent years as a result of the phenomenon of acid rain (see the Chemical Impact in Chapter 5). This problem is complex and has diplomatic and economic overtones that make it all the more difficult to solve.

CONTENTS

< Goldfish in an aquarium where the pH must be carefully controlled to maintain healthy fish.

14.1 The Nature of Acids and Bases

Purpose

- To discuss two models of acids and bases and to relate equilibrium concepts to acid dissociation.

***Don't* taste chemicals!**

Acids were first recognized as a class of substances that taste sour. Vinegar tastes sour because it is a dilute solution of acetic acid; citric acid is responsible for the sour taste of a lemon. Bases, sometimes called *alkalis,* are characterized by their bitter taste and slippery feel. Commercial preparations for unclogging drains are highly basic.

Acids and bases were first discussed in Section 4.2.

The first person to recognize the essential nature of acids and bases was Svante Arrhenius. Based on his experiments with electrolytes, Arrhenius postulated that *acids produce hydrogen ions in aqueous solution, while bases produce hydroxide ions*. At the time, the **Arrhenius concept** of acids and bases was a major step forward in quantifying acid-base chemistry, but this concept is limited because it applies only to aqueous solutions and allows for only one kind of base—the hydroxide ion. A more general definition of acids and bases was suggested by the Danish chemist Johannes Brönsted and the English chemist Thomas Lowry. In terms of the **Brönsted-Lowry model,** *an acid is a proton (H^+) donor, and a base is a proton acceptor*. For example, when gaseous HCl dissolves in water, each HCl molecule donates a proton to a water molecule, and so qualifies as a Brönsted-Lowry acid. The molecule that accepts the proton in this case, water, is a Brönsted-Lowry base.

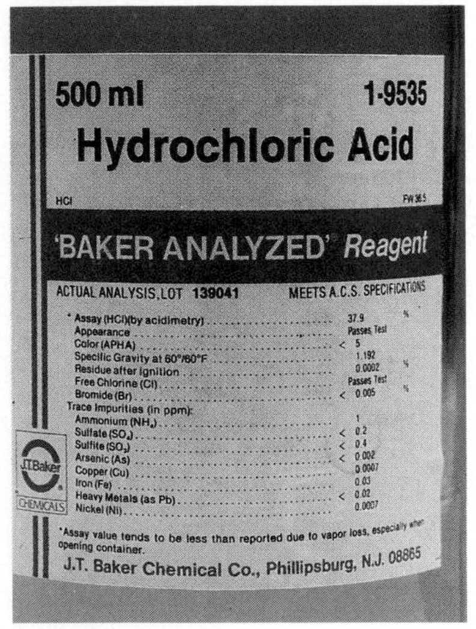

The label on a bottle of concentrated hydrochloric acid.

To understand how water can act as a base, we need to remember that the oxygen of the water molecule has two unshared electron pairs, either of which can form a covalent bond with an H^+ ion. When gaseous HCl dissolves, the following reaction occurs:

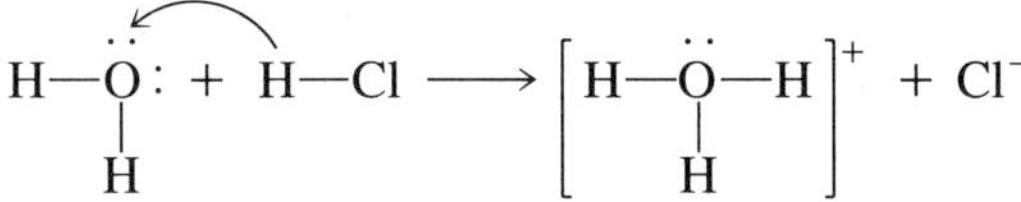

Note that the proton is transferred from the HCl molecule to the water molecule to form H_3O^+, which is called the **hydronium ion.**

The general reaction that occurs when an acid is dissolved in water can best be represented as

Recall that (*aq*) means the substance is hydrated.

$$\underset{\text{Acid}}{HA(aq)} + \underset{\text{Base}}{H_2O(l)} \rightleftharpoons \underset{\text{Conjugate acid}}{H_3O^+(aq)} + \underset{\text{Conjugate base}}{A^-(aq)} \qquad (14.1)$$

This representation emphasizes the significant role of the polar water molecule in pulling the proton from the acid. Note that the **conjugate base** is everything that remains of the acid molecule after a proton is lost. The **conjugate acid** is formed when the proton is transferred to the base. A **conjugate acid-base pair** consists of two substances related to each other by the donating and accepting of a single proton. In Equation (14.1) there are two conjugate acid-base pairs: HA and A^-, and H_2O and H_3O^+.

It is important to note that Equation (14.1) really represents *a competition for the proton between the two bases H_2O and A^-*. If H_2O is a much stronger base than A^-, that is, if H_2O has a much greater affinity for H^+ than does A^-, the equilibrium position will be far to the right; most of the acid dissolved will be in the ionized form. Conversely, if A^- is a much stronger base than H_2O, the equilibrium position will lie far to the left. In this case most of the acid dissolved will be present at equilibrium as HA.

The equilibrium expression for the reaction given in Equation (14.1) is

$$K_a = \frac{[H_3O^+][A^-]}{[HA]} = \frac{[H^+][A^-]}{[HA]} \qquad (14.2)$$

where K_a is called the **acid dissociation constant.** Both $H_3O^+(aq)$ and $H^+(aq)$ are commonly used to represent the hydrated proton. In this book we will often use simply H^+, but you should remember that it is hydrated in aqueous solutions.

In this chapter we will always represent an acid as simply dissociating. This does not mean we are using the Arrhenius model for acids. Since water does not affect the equilibrium position, it is simply easier to leave it out of the acid dissociation reaction.

In Chapter 13 we saw that the concentration of a pure solid or a pure liquid is always omitted from the equilibrium expression. In a dilute solution we can assume that the concentration of liquid water remains essentially constant when an acid is dissolved. Thus the term $[H_2O]$ is not included in Equation (14.2), and the equilibrium expression for K_a has the same form as that for the simple dissociation

$$HA(aq) \rightleftharpoons H^+(aq) + A^-(aq)$$

You should not forget, however, that water plays an important role in causing the acid to dissociate.

Note that K_a is the equilibrium constant for the reaction in which a proton is removed from HA to form the conjugate base A^-. We use K_a to represent *only* this type of reaction. Knowing this, you can write the K_a expression for any acid, even one that is totally unfamiliar to you. As you do Sample Exercise 14.1, focus on the definition of the reaction corresponding to K_a.

Sample Exercise 14.1

Write the simple dissociation reaction (omitting water) for each of the following acids:

a. hydrochloric acid (HCl)
b. acetic acid ($HC_2H_3O_2$)
c. the ammonium ion (NH_4^+)
d. the anilinium ion ($C_6H_5NH_3^+$)
e. the hydrated aluminum(III) ion $[Al(H_2O)_6]^{3+}$

Solution

a. $HCl(aq) \rightleftharpoons H^+(aq) + Cl^-(aq)$
b. $HC_2H_3O_2(aq) \rightleftharpoons H^+(aq) + C_2H_3O_2^-(aq)$
c. $NH_4^+(aq) \rightleftharpoons H^+(aq) + NH_3(aq)$
d. $C_6H_5NH_3^+(aq) \rightleftharpoons H^+(aq) + C_6H_5NH_2(aq)$

Sample Exercise 14.1, continued

e. Although this formula looks complicated, writing the reaction is simple if you concentrate on the meaning of K_a. Removing a proton, which can only come from one of the water molecules, leaves one OH^- and five H_2O molecules attached to the Al^{3+} ion. So the reaction is

$$[Al(H_2O)_6]^{3+}(aq) \rightleftharpoons H^+(aq) + [Al(H_2O)_5OH]^{2+}(aq)$$

The Brönsted-Lowry model is not limited to aqueous solutions; it can be extended to reactions in the gas phase. For example, we discussed the reaction between gaseous hydrogen chloride and ammonia when we studied diffusion (Section 5.7):

$$NH_3(g) + HCl(g) \rightleftharpoons NH_4Cl(s)$$

In this reaction a proton is donated by the hydrogen chloride to the ammonia, as shown by these Lewis structures:

$$\mathrm{H{-}\overset{H}{\underset{H}{N}}{:}\quad H{-}\ddot{\underset{\cdot\cdot}{Cl}}{:} \rightleftharpoons \left[H{-}\overset{H}{\underset{H}{N}}{-}H\right]^{+} + \left[:\ddot{\underset{\cdot\cdot}{Cl}}:\right]^{-}}$$

Note that this is not considered an acid-base reaction according to the Arrhenius concept.

14.2 Acid Strength

Purpose

- To relate acid strength to the position of the dissociation equilibrium.
- To discuss the autoionization of water.

The strength of an acid is defined by the equilibrium position of its dissociation reaction:

$$HA(aq) + H_2O(l) \rightleftharpoons H_3O^+(aq) + A^-(aq)$$

A **strong acid** is one for which *this equilibrium lies far to the right*. This means that almost all the original HA is dissociated at equilibrium [see Fig. 14.1(a)]. There is an important connection between the strength of an acid and that of its conjugate base. *A strong acid yields a weak conjugate base*—one that has a low affinity for a proton. A strong acid can also be described as an acid whose conjugate base is a much weaker base than water (see Fig. 14.2). In this case the water molecules win the competition for the H^+ ions.

Conversely, a **weak acid** is one for which *the equilibrium lies far to the left*. Most of the acid originally placed in the solution is still present as HA at equilibrium. That is, a weak acid dissociates only to a very small extent in aqueous solution [see Fig. 14.1(b)]. In contrast to a strong acid, a weak acid has a conjugate base

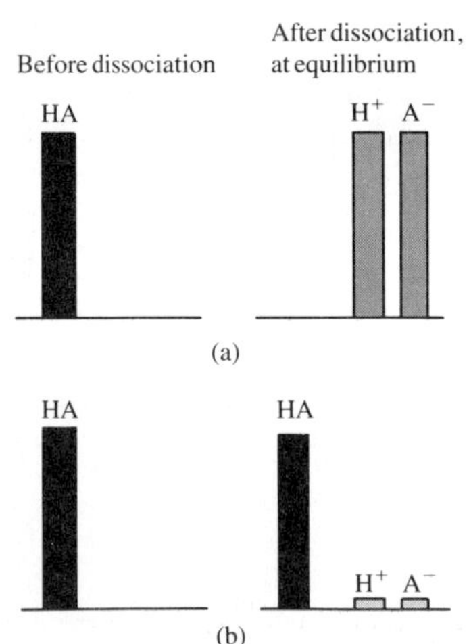

Figure 14.1

Graphical representation of the behavior of acids of different strengths in aqueous solution. (a) A strong acid. (b) A weak acid.

that is a much stronger base than water. In this case a water molecule is not very successful in pulling an H^+ ion from the conjugate base. *A weak acid yields a relatively strong conjugate base.*

A strong acid has a weak conjugate base.

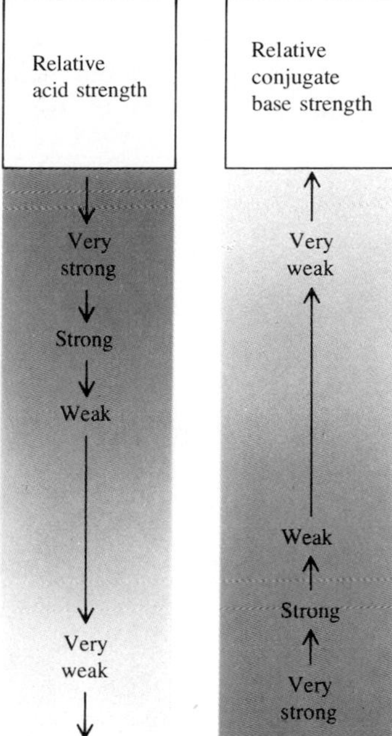

Figure 14.2

The relationship of acid strength and conjugate base strength for the dissociation reaction

$$\underset{\text{Acid}}{HA(aq)} + H_2O(l) \rightleftharpoons H_3O^+(aq) + \underset{\text{Conjugate base}}{A^-(aq)}$$

The various ways of describing the strength of an acid are summarized in Table 14.1.

Various Ways to Describe Acid Strength		
Property	Strong acid	Weak acid
K_a value	K_a is large	K_a is small
Position of the dissociation equilibrium	Far to the right	Far to the left
Equilibrium concentration of H^+ compared to original concentration of HA	$[H^+] \approx [HA]_0$	$[H^+] \ll [HA]_0$
Strength of conjugate base compared to that of water	A^- much weaker base than H_2O	A^- much stronger base than H_2O

Table 14.1

$\ll$ means much less than
$\gg$ means much greater than

The common strong acids are sulfuric acid ($H_2SO_4(aq)$), hydrochloric acid ($HCl(aq)$), nitric acid ($HNO_3(aq)$), and perchloric acid ($HClO_4(aq)$). Sulfuric acid is actually a **diprotic acid,** an acid having two acidic protons. The acid H_2SO_4 is a strong acid, virtually 100% dissociated in water:

Perchloric acid can explode if handled improperly.

$$H_2SO_4(aq) \rightarrow H^+(aq) + HSO_4^-(aq)$$

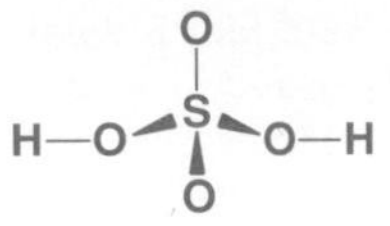

Sulfuric acid

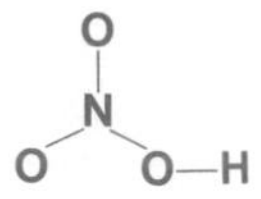

Nitric acid

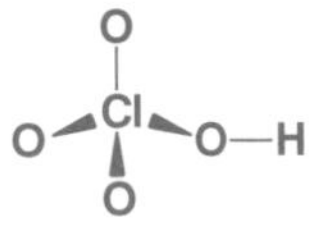

Perchloric acid

Appendix 5.1 contains a table of K_a values.

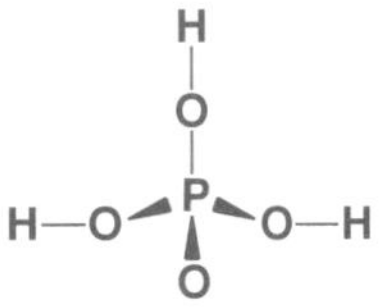

Phosphoric acid

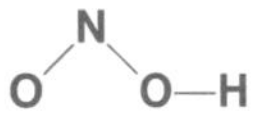

Nitrous acid

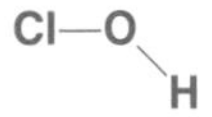

Hypochlorous acid

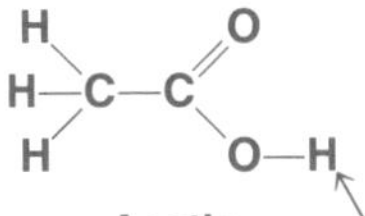

Acetic acid

Acidic H

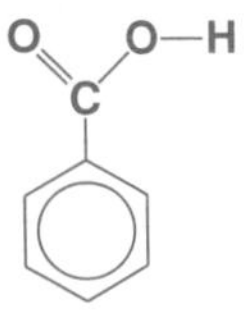

Benzoic acid

but the HSO_4^- ion is a weak acid:

$$HSO_4^-(aq) \rightleftharpoons H^+(aq) + SO_4^{2-}(aq)$$

Most acids are **oxyacids,** in which the acidic proton is attached to an oxygen atom. The strong acids mentioned above, except hydrochloric acid, are typical examples. Many common weak acids, such as phosphoric acid (H_3PO_4), nitrous acid (HNO_2), and hypochlorous acid (HOCl), are also oxyacids. **Organic acids,** those with a carbon-atom backbone, commonly contain the **carboxyl group:**

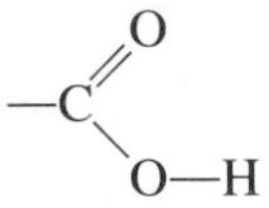

$$-C(=O)-O-H$$

Acids of this type are usually weak. Examples are acetic acid (CH_3COOH), often written $HC_2H_3O_2$, and benzoic acid (C_6H_5COOH).

There are some important acids in which the acidic proton is attached to an atom other than oxygen. The most significant of these are the hydrohalic acids *HX*, where *X* represents a halogen atom.

Table 14.2 contains a list of common **monoprotic acids** (those having *one* acidic proton) and their K_a values. Note that the strong acids are not listed. When a strong acid molecule such as HCl, for example, is placed in water, the position of the dissociation equilibrium

$$HCl(aq) \rightleftharpoons H^+(aq) + Cl^-(aq)$$

lies so far to the right that [HCl] cannot be measured accurately. This prevents an accurate calculation of K_a:

$$K_a = \frac{[H^+][Cl^-]}{[HCl]}$$

↖Very small and highly uncertain

Values of K_a for Some Common Monoprotic Acids

Formula	Name	Value of K_a*
HSO_4^-	Hydrogen sulfate ion	1.2×10^{-2}
$HClO_2$	Chlorous acid	1.2×10^{-2}
$HC_2H_2ClO_2$	Monochloracetic acid	1.35×10^{-3}
HF	Hydrofluoric acid	7.2×10^{-4}
HNO_2	Nitrous acid	4.0×10^{-4}
$HC_2H_3O_2$	Acetic acid	1.8×10^{-5}
$[Al(H_2O)_6]^{3+}$	Hydrated aluminum(III) ion	1.4×10^{-5}
HOCl	Hypochlorous acid	3.5×10^{-8}
HCN	Hydrocyanic acid	6.2×10^{-10}
NH_4^+	Ammonium ion	5.6×10^{-10}
HOC_6H_5	Phenol	1.6×10^{-10}

↑ Increasing acid strength

*The units of K_a are mol/L, but are customarily omitted.

Table 14.2

Sample Exercise 14.2

Using Table 14.2, arrange the following species according to their strength as bases: H_2O, F^-, Cl^-, NO_2^-, CN^-.

Solution

Remember that water is a stronger base than the conjugate base of a strong acid, but a weaker base than the conjugate base of a weak acid. This leads to the following order:

$$Cl^- < H_2O < \text{conjugate bases of weak acids}$$

Weakest bases ⟶ Strongest bases

We can order the remaining conjugate bases by recognizing that the strength of an acid is *inversely related* to the strength of its conjugate base. Since from Table 14.2 we have

$$K_a \text{ for HF} > K_a \text{ for } HNO_2 > K_a \text{ for HCN}$$

the base strengths increase as follows:

$$F^- < NO_2^- < CN^-$$

The combined order of increasing base strength is

$$Cl^- < H_2O < F^- < NO_2^- < CN^-$$

Water as an Acid and a Base

A substance is said to be *amphoteric* if it can behave either as an acid or as a base. Water is the most common **amphoteric substance.** We can see this clearly in the **autoionization** of water, which involves the transfer of a proton from one water molecule to another to produce a hydroxide ion and a hydronium ion:

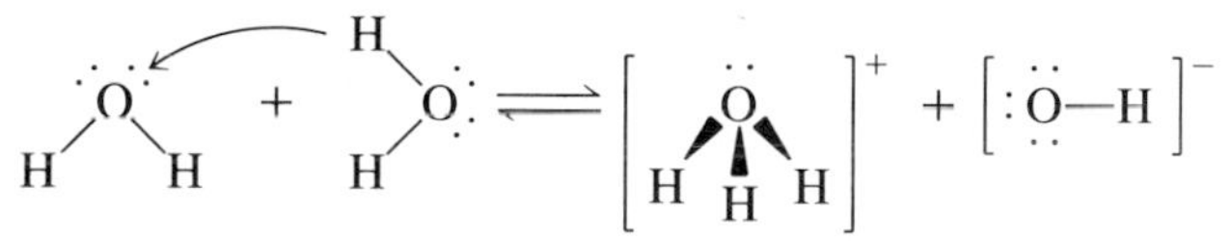

$$H_2\ddot{O}: + H_2\ddot{O}: \rightleftharpoons [H_3\ddot{O}]^+ + [:\ddot{O}\text{—}H]^-$$

In this reaction one water molecule acts as an acid by furnishing a proton, and the other acts as a base by accepting the proton.

Autoionization can occur in other liquids besides water. For example, in liquid ammonia the autoionization reaction is

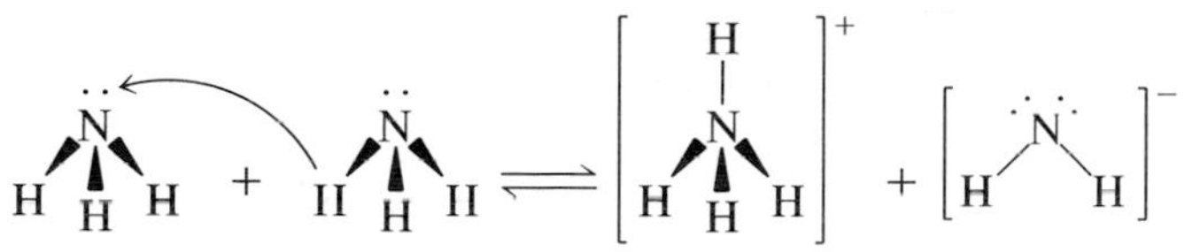

$$H_3\ddot{N} + H_3\ddot{N} \rightleftharpoons [NH_4]^+ + [H_2\ddot{N}:]^-$$

The autoionization reaction for water

$$2H_2O(l) \rightleftharpoons H_3O^+(aq) + OH^-(aq)$$

leads to the equilibrium expression

$$K_w = [H_3O^+][OH^-] = [H^+][OH^-]$$

where K_w, called the **ion-product constant** (or the *dissociation constant*), always refers to the autoionization of water.

Experiment shows that at 25°C

$$[H^+] = [OH^-] = 1.0 \times 10^{-7}\ M$$

which means that at 25°C

$K_w = [H^+][OH^-] = 1.0 \times 10^{-14}$

$$K_w = [H^+][OH^-] = (1.0 \times 10^{-7}\ \text{mol/L})(1.0 \times 10^{-7}\ \text{mol/L})$$
$$= 1.0 \times 10^{-14}\ \text{mol}^2/\text{L}^2$$

The units are customarily omitted.

It is important to recognize the meaning of K_w. In any aqueous solution at 25°C, *no matter what it contains*, the product of $[H^+]$ and $[OH^-]$ must always equal 1.0×10^{-14}. There are three possible situations:

1. A neutral solution, where $[H^+] = [OH^-]$.
2. An acidic solution, where $[H^+] > [OH^-]$.
3. A basic solution, where $[OH^-] > [H^+]$.

In each case, however, at 25°C

$$K_w = [H^+][OH^-] = 1.0 \times 10^{-14}$$

Sample Exercise 14.3

Calculate $[H^+]$ or $[OH^-]$ as required for each of the following solutions at 25°C, and state whether the solution is neutral, acidic, or basic.

a. $1.0 \times 10^{-5}\ M\ OH^-$

b. $1.0 \times 10^{-7}\ M\ OH^-$

c. $10.0\ M\ H^+$

Solution

a. $K_w = [H^+][OH^-] = 1.0 \times 10^{-14}$. Since $[OH^-]$ is $1.0 \times 10^{-5}\ M$, solving for $[H^+]$ gives

$$[H^+] = \frac{1.0 \times 10^{-14}}{[OH^-]} = \frac{1.0 \times 10^{-14}}{1.0 \times 10^{-5}} = 1.0 \times 10^{-9}\ M$$

Since $[OH^-] > [H^+]$, the solution is basic.

b. As in part a, solving for $[H^+]$ gives

$$[H^+] = \frac{1.0 \times 10^{-14}}{[OH^-]} = \frac{1.0 \times 10^{-14}}{1.0 \times 10^{-7}} = 1.0 \times 10^{-7}\ M$$

Since $[H^+] = [OH^-]$, the solution is neutral.

Sample Exercise 14.3, continued

c. Solving for $[OH^-]$ gives

$$[OH^-] = \frac{1.0 \times 10^{-14}}{[H^+]} = \frac{1.0 \times 10^{-14}}{10.0} = 1.0 \times 10^{-15}\ M$$

Since $[H^+] > [OH^-]$, the solution is acidic.

Since K_w is an equilibrium constant, it varies with temperature. The effect of temperature is considered in Sample Exercise 14.4.

Sample Exercise 14.4

At 60°C the value of K_w is 1×10^{-13}.

a. Using Le Châtelier's principle, predict whether the reaction

$$2H_2O(l) \rightleftharpoons H_3O^+(aq) + OH^-(aq)$$

is exothermic or endothermic.

b. Calculate $[H^+]$ and $[OH^-]$ in a neutral solution at 60°C.

Solution

a. K_w *increases* from 1×10^{-14} at 25°C to 1×10^{-13} at 60°C. Le Châtelier's principle states that if a system at equilibrium is heated, it will adjust to consume energy. Since the value of K_w increases with temperature, we must think of energy as a reactant, and so the process must be endothermic.

b. At 60°C

$$[H^+][OH^-] = 1 \times 10^{-13}$$

For a neutral solution

$$[H^+] = [OH^-] = \sqrt{1 \times 10^{-13}} = 3 \times 10^{-7}\ M$$

14.3 The pH Scale

Purpose

- To define pH, pOH, and pK and to introduce general methods for solving acid-base problems.

Because $[H^+]$ in an aqueous solution is typically quite small, the **pH scale** provides a convenient way to represent solution acidity. The pH is a log scale based on 10 where

The pH scale is a compact way to represent solution acidity.

$$pH = -\log[H^+]$$

Appendix 1.2 has a review of logs.

Thus for a solution where

$$[H^+] = 1.0 \times 10^{-7}\ M$$
$$pH = -(-7.00) = 7.00$$

For $\Delta[H^+] = 10$, $\Delta pH = 1$.

At this point we need to discuss significant figures for logarithms. The rule is that *the number of decimal places in the log is equal to the number of significant figures in the original number*. Thus

$$[H^+] = \underset{\text{2 significant figures}}{1.0} \times 10^{-9}\ M$$
$$pH = \underset{\text{2 decimal places}}{9.00}$$

Similar log scales are used for representing other quantities, for example:

$$pOH = -\log[OH^-]$$
$$pK = -\log K$$

The pH decreases as $[H^+]$ increases, and vice versa.

Since pH is a log scale based on 10, *the pH changes by 1 for every power of 10 change in $[H^+]$*. For example, a solution of pH 3 has an H^+ concentration 10 times that of a solution of pH 4 and 100 times that of a solution of pH 5. Also note that because pH is defined as $-\log[H^+]$, *the pH decreases as $[H^+]$ increases*. The pH scale and the pH values for several common substances are shown in Fig. 14.3.

The pH of a solution is usually measured using a pH meter, an electronic device with a probe that can be inserted into a solution of unknown pH. The probe contains an acidic aqueous solution enclosed by a special glass membrane that allows migration of H^+ ions. If the unknown solution has a different pH from the solution in the probe, an electrical potential results, which is registered on the meter (see Fig. 14.4).

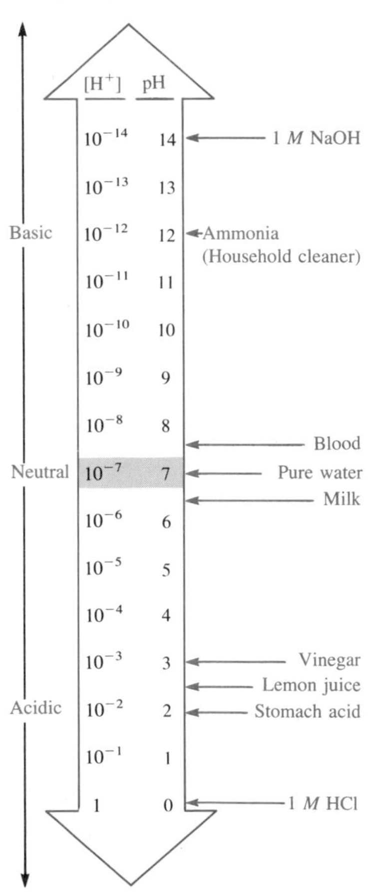

Figure 14.3

The pH scale and pH values of some common substances.

Sample Exercise 14.5

Calculate pH and pOH for each of the following solutions at 25°C:

a. $1.0 \times 10^{-3}\ M\ OH^-$ **b.** $1.0\ M\ H^+$.

Solution

a.

$$[H^+] = \frac{K_w}{[OH^-]} = \frac{1.0 \times 10^{-14}}{1.0 \times 10^{-3}} = 1.0 \times 10^{-11}\ M$$

$$pH = -\log[H^+] = -\log(1.0 \times 10^{-11}) = 11.00$$

$$pOH = -\log[OH^-] = -\log(1.0 \times 10^{-3}) = 3.00$$

b.

$$[OH^-] = \frac{K_w}{[H^+]} = \frac{1.0 \times 10^{-14}}{1.0} = 1.0 \times 10^{-14}\ M$$

$$pH = -\log[H^+] = -\log(1.0) = 0.00$$

$$pOH = -\log[OH^-] = -\log(1.0 \times 10^{-14}) = 14.00$$

It is useful to consider the log form of the expression

$$K_w = [H^+][OH^-]$$

That is,

$$\log K_w = \log[H^+] + \log[OH^-]$$

or

$$-\log K_w = -\log[H^+] - \log[OH^-]$$

Thus

$$pK_w = pH + pOH \tag{14.3}$$

Since $K_w = 1.0 \times 10^{-14}$,

$$pK_w = -\log(1.0 \times 10^{-14}) = 14.00$$

Thus, for *any* aqueous solution at 25°C, pH and pOH add up to 14.00.

$$pH + pOH = 14.00 \tag{14.4}$$

Figure 14.4

A typical pH meter.

The pH meter is discussed more fully in Section 17.4.

Sample Exercise 14.6

The pH of a sample of human blood was measured to be 7.41 at 25°C. Calculate pOH, $[H^+]$, and $[OH^-]$ for the sample.

Solution

Since pH + pOH = 14.00,

$$pOH = 14.00 - pH = 14.00 - 7.41 = 6.59$$

To find $[H^+]$ we must go back to the definition of pH:

$$pH = -\log[H^+]$$

Thus

$$7.41 = -\log[H^+] \quad \text{or} \quad \log[H^+] = -7.41$$

We need to know the *antilog* of −7.41. As shown in Appendix 1.2, taking the antilog is the same as exponentiation, that is,

$$\text{antilog}(n) = 10^n$$

There are different methods for carrying out the antilog operation on various calculators. The most common are the $\boxed{Y^X}$ key and the two-key $\boxed{\text{INV}}$ $\boxed{\text{LOG}}$ sequence. Consult the user's manual for your calculator to find out how to do the antilog operation.

Since $pH = -\log[H^+]$,

$$-pH = \log[H^+]$$

and $[H^+]$ can be calculated by taking the antilog of −pH:

$$[H^+] = \text{antilog}(-pH)$$

In the present case

$$[H^+] = \text{antilog}(-pH) = \text{antilog}(-7.41) = 10^{-7.41} = 3.9 \times 10^{-8}$$

Similarly, $[OH^-] = \text{antilog}(-pOH)$, and

$$[OH^-] = \text{antilog}(-6.59) = 10^{-6.59} = 2.6 \times 10^{-7}\ M$$

Now that we have considered all of the fundamental definitions relevant to acid-base solutions, we can proceed to a quantitative description of the equilibria present in these solutions. The main reason that acid-base problems sometimes seem difficult is that a typical aqueous solution contains many components so the problems tend to be complicated. However, you can deal with these problems successfully if you use the following general strategies:

- *Think chemistry*. Focus on the solution components and their reactions. It will almost always be possible to choose one reaction that is the most important.
- *Be systematic*. Acid-base problems require a step-by-step approach.
- *Be flexible*. Although all acid-base problems are similar in many ways, important differences do occur. Treat each problem as a separate entity. Do not try to force a given problem into matching any you have solved before. Look for both the similarities and the differences.
- *Be patient*. The complete solution to a complicated problem cannot be seen immediately in all its detail. Pick the problem apart into its workable steps.
- *Be confident*. Look within the problem for the solution, and let the problem guide you. Assume that you can think it out. Do not rely on memorizing solutions to problems. In fact, memorizing solutions is usually detrimental because you tend to try to force a new problem to be the same as one you have seen before. *Understand and think; don't just memorize*.

14.4 Calculating the pH of Strong Acid Solutions

Purpose

- To demonstrate the systematic treatment of solutions of strong acids.

When we deal with acid-base equilibria, *it is essential to focus on the solution components and their chemistry*. For example, what species are present in a 1.0 *M* solution of HCl? Since hydrochloric acid is a strong acid, we assume that it is completely dissociated. Thus, although the label on the bottle says 1.0 *M* HCl, the solution contains virtually no HCl molecules. Typically, container labels indicate the substance(s) used to make up the solution, but do not necessarily describe the solution components after dissolution. Thus a 1.0 *M* HCl solution contains H^+ and Cl^- ions rather than HCl molecules.

The next step in dealing with aqueous solutions is to determine which components are significant and which can be ignored. We need to focus on the **major species,** those solution components present in relatively large amounts. In 1.0 *M* HCl, for example, the major species are H^+, Cl^-, and H_2O. Since this is a very acidic solution, OH^- is present only in tiny amounts and is classed as a minor species. In attacking acid-base problems, the importance of *writing the major spe-*

cies in the solution as the first step cannot be overemphasized. *This single step is the key to solving these problems successfully.*

To illustrate the main ideas involved, let us calculate the pH of 1.0 *M* HCl. We first list the major species: H^+, Cl^-, and H_2O. Since we want to calculate the pH, we will focus on those major species that can furnish H^+. Obviously, we must consider H^+ from the dissociation of HCl. However, H_2O also furnishes H^+ by autoionization, which is often represented by the simple dissociation reaction

Always write the major species present in the solution.

$$H_2O(l) \rightleftharpoons H^+(aq) + OH^-(aq)$$

But is autoionization an important source of H^+ ions? In pure water at 25°C, $[H^+]$ is 10^{-7} *M*. In 1.0 *M* HCl solution, the water will produce even less than 10^{-7} *M* H^+, since by Le Châtelier's principle, the H^+ from the dissociated HCl will drive the position of the water equilibrium to the left. Thus the amount of H^+ contributed by water is negligible compared to the 1.0 *M* H^+ from the dissociation of HCl. Therefore, we can say that $[H^+]$ in the solution is 1.0 *M*. The pH is then

The H^+ from the strong acid drives the equilibrium $H_2O \rightleftharpoons H^+ + OH^-$ to the left.

$$pH = -\log[H^+] = -\log(1.0) = 0$$

Sample Exercise 14.7

a. Calculate the pH of 0.10 *M* HNO_3.

b. Calculate the pH of 1.0×10^{-10} *M* HCl.

Solution

a. Since HNO_3 is a strong acid, the major species in solution are:

$$H^+, \quad NO_3^-, \quad \text{and} \quad H_2O$$

The concentration of HNO_3 is virtually zero, since the acid completely dissociates in water. Also, $[OH^-]$ will be very small because the H^+ ions from the acid will drive the equilibrium

$$H_2O(l) \rightleftharpoons H^+(aq) + OH^-(aq)$$

to the left. That is, this is an acidic solution where $[H^+] \gg [OH^-]$ and so $[OH^-] \ll 10^{-7}$ *M*. The sources of H^+ are

1. H^+ from HNO_3 (0.10 *M*).
2. H^+ from H_2O

The number of H^+ ions contributed by the autoionization of water will be very small compared to the 0.10 *M* contributed by the HNO_3 and can be neglected. Since the dissolved HNO_3 is the only important source of H^+ ions in this solution,

In pure water, only 10^{-7} *M* H^+ is produced.

$$[H^+] = 0.10\ M \quad \text{and} \quad pH = -\log(0.10) = 1.00$$

b. Normally, in an aqueous solution of HCl the major species are H^+, Cl^-, and H_2O. However, in this case the amount of HCl in solution is so small that it has no effect; the only major species is H_2O. Thus the pH will be that of pure water, or pH = 7.00.

14.5 Calculating the pH of Weak Acid Solutions

Purpose

- To demonstrate the systematic treatment of solutions of weak acids.
- To show how to calculate percent dissociation.

Since a weak acid dissolved in water can be viewed as a prototype of almost any equilibrium occurring in aqueous solution, we will proceed carefully and systematically. Although some of the procedures we develop here may seem superfluous, they will become essential as the problems become more complicated. We will develop the necessary strategies by calculating the pH of a 1.00 *M* solution of HF ($K_a = 7.2 \times 10^{-4}$).

First, always write the major species present in the solution.

The first step, as always, is to *write the major species in the solution*. From its small K_a value, we know that hydrofluoric acid is a weak acid and will be dissociated only to a slight extent. Thus, when we write the major species, the hydrofluoric acid will be represented in its dominant form, as HF. The major species in solution are: HF and H_2O.

The next step (since this is a pH problem) is to decide which of the major species can furnish H^+ ions. Actually, both major species can do so:

$$\mathrm{HF}(aq) \rightleftharpoons \mathrm{H}^+(aq) + \mathrm{F}^-(aq) \qquad K_a = 7.2 \times 10^{-4}$$

$$\mathrm{H_2O}(l) \rightleftharpoons \mathrm{H}^+(aq) + \mathrm{OH}^-(aq) \qquad K_w = 1.0 \times 10^{-14}$$

But in aqueous solutions typically one source of H^+ can be singled out as dominant. By comparing K_a for HF to K_w for H_2O, we see that hydrofluoric acid, although weak, is still a much stronger acid than water. Thus we will assume that hydrofluoric acid will be the dominant source of H^+. We will ignore the tiny contribution by water.

Therefore, it is the dissociation of HF that will determine the equilibrium concentration of H^+ and hence the pH:

$$\mathrm{HF}(aq) \rightleftharpoons \mathrm{H}^+(aq) + \mathrm{F}^-(aq)$$

The equilibrium expression is

$$K_a = 7.2 \times 10^{-4} = \frac{[\mathrm{H}^+][\mathrm{F}^-]}{[\mathrm{HF}]}$$

To solve the equilibrium problem, we follow the procedures developed in Chapter 13 for gas-phase equilibria. First, we list the initial concentrations, the *concentrations before the reaction of interest has proceeded to equilibrium*. Before any HF dissociates, the concentrations of the species in the equilibrium are:

$$[\mathrm{HF}]_0 = 1.00\ M \qquad [\mathrm{F}^-]_0 = 0 \qquad [\mathrm{H}^+]_0 = 10^{-7}\ M \approx 0$$

(Note that the zero value for $[H^+]_0$ is an approximation, since we are neglecting the H^+ ions from the autoionization of water.)

The next step is to determine the change required to reach equilibrium. Since some HF will dissociate to come to equilibrium (but that amount is presently unknown), we let x be the change in the concentration of HF that is required to achieve equilibrium. That is, we assume that x mol/L of HF will dissociate to produce x mol/L H^+ and x mol/L F^- as the system adjusts to its equilibrium position. Now the equilibrium concentrations can be defined in terms of x:

$$[HF] = [HF]_0 - x = 1.00 - x$$
$$[F^-] = [F^-]_0 + x = 0 + x = x$$
$$[H^+] = [H^+]_0 + x \approx 0 + x = x$$

Substituting these equilibrium concentrations into the equilibrium expression gives

$$K_a = 7.2 \times 10^{-4} = \frac{[H^+][F^-]}{[HF]} = \frac{(x)(x)}{1.00 - x}$$

This expression produces a quadratic equation that can be solved using the quadratic formula, as for the gas-phase systems in Chapter 13. However, since K_a for HF is so small, HF will dissociate only slightly, and x is expected to be small. This will allow us to simplify the calculation. If x is very small compared to 1.00, the term in the denominator can be approximated as follows:

$$1.00 - x \approx 1.00$$

The equilibrium expression then becomes

$$7.2 \times 10^{-4} = \frac{(x)(x)}{1.00 - x} \approx \frac{(x)(x)}{1.00}$$

which yields

$$x^2 \approx (7.2 \times 10^{-4})(1.00) = 7.2 \times 10^{-4}$$
$$x \approx \sqrt{7.2 \times 10^{-4}} = 2.7 \times 10^{-2}$$

The validity of an approximation should always be checked.

How valid is the approximation that $[HF] = 1.00\ M$? Because this question will arise often in connection with acid-base equilibrium calculations, we will consider it carefully. *The validity of the approximation depends on how much accuracy we demand for the calculated value of $[H^+]$.* Typically, the K_a values for acids are known to an accuracy of only about ±5%. It is reasonable therefore to apply this figure when determining the validity of the approximation

$$[HA]_0 - x \approx [HA]_0$$

We will use the following test.

We first calculate the value of x by making the approximation

$$K_a = \frac{x^2}{[HA]_0 - x} \approx \frac{x^2}{[HA]_0}$$

where

$$x^2 \approx K_a[HA]_0 \quad \text{and} \quad x \approx \sqrt{K_a[HA]_0}$$

We then compare the sizes of x and $[HA]_0$. If the expression

$$\frac{x}{[HA]_0} \times 100$$

is less than or equal to 5%, the value of x is so small that the approximation

$$[HA]_0 - x \approx [HA]_0$$

will be considered valid.

In our example

$$x = 2.7 \times 10^{-2} \text{ mol/L}$$
$$[HA]_0 = [HF]_0 = 1.00 \text{ mol/L}$$

and

$$\frac{x}{[HA]_0} \times 100 = \frac{2.7 \times 10^{-2}}{1.00} \times 100 = 2.7\%$$

The approximation we made is considered valid, and the value of x calculated using that approximation is acceptable. Thus

$$x = [H^+] = 2.7 \times 10^{-2}\ M \quad \text{and} \quad pH = -\log(2.7 \times 10^{-2}) = 1.57$$

This problem illustrates all the important steps for solving a typical equilibrium problem involving a weak acid. These steps are summarized as follows:

Solving Weak Acid Equilibrium Problems

- List the major species in the solution.
- Choose the species that can produce H^+, and write balanced equations for the reactions producing H^+.
- Using the values of the equilibrium constants for the reactions you have written, decide which equilibrium will dominate in producing H^+.
- Write the equilibrium expression for the dominant equilibrium.
- List the initial concentrations of the species participating in the dominant equilibrium.
- Define the change needed to achieve equilibrium; that is, define x.
- Write the equilibrium concentrations in terms of x.
- Substitute the equilibrium concentrations into the equilibrium expression.
- Solve for x the ''easy'' way; that is, by assuming that $[HA]_0 - x \approx [HA]_0$.
- Use the 5% rule to verify whether the approximation is valid.
- Calculate $[H^+]$ and pH.

A table of K_a values for various weak acids is given in Appendix 5.1.

We use this systematic approach in Sample Exercise 14.8.

Sample Exercise 14.8

The hypochlorite ion (OCl^-) is a strong oxidizing agent often found in household bleaches and disinfectants. It is also the active ingredient that forms when swimming pool water is treated with chlorine. In addition to its oxidizing abilities, the hypochlorite ion has a relatively high affinity for protons (it is a much stronger base

Sample Exercise 14.8, continued

than Cl^-, for example) and forms the weakly acidic hypochlorous acid (HOCl, $K_a = 3.5 \times 10^{-8}$). Calculate the pH of a 0.100 *M* aqueous solution of hypochlorous acid.

Bacteria in swimming pools are controlled by use of chlorine.

Solution

STEP 1

We list the major species. Since HOCl is a weak acid and remains mostly undissociated, the major species in a 0.100 *M* HOCl solution are

$$\text{HOCl} \quad \text{and} \quad H_2O$$

STEP 2

Both species can produce H^+:

$$HOCl(aq) \rightleftharpoons H^+(aq) + OCl^-(aq) \qquad K_a = 3.5 \times 10^{-8}$$
$$H_2O(l) \rightleftharpoons H^+(aq) + OH^-(aq) \qquad K_w = 1.0 \times 10^{-14}$$

STEP 3

Since HOCl is a significantly stronger acid than H_2O, it will dominate in the production of H^+.

STEP 4

We therefore use the following equilibrium expression:

$$K_a = 3.5 \times 10^{-8} = \frac{[H^+][OCl^-]}{[HOCl]}$$

STEP 5

The initial concentrations appropriate for this equilibrium are

$$[HOCl]_0 = 0.100\ M$$
$$[OCl^-]_0 = 0$$
$$[H^+]_0 \approx 0 \qquad \text{(We neglect the contribution from } H_2O\text{.)}$$

STEP 6

Since the system will reach equilibrium by the dissociation of HOCl, let x be the amount of HOCl (in mol/L) that dissociates in reaching equilibrium.

STEP 7

The equilibrium concentrations in terms of x are

$$[HOCl] = [HOCl]_0 - x = 0.100 - x$$
$$[OCl^-] = [OCl^-]_0 + x = 0 + x = x$$
$$[H^+] = [H^+]_0 + x \approx 0 + x = x$$

STEP 8

Substituting these concentrations into the equilibrium expression gives

$$K_a = 3.5 \times 10^{-8} = \frac{(x)(x)}{0.100 - x}$$

Sample Exercise 14.8, continued

STEP 9

Since K_a is so small, we can expect a small value for x. Thus we make the approximation $[HA]_0 - x \approx [HA]_0$, or $0.100 - x \approx 0.100$, which leads to the expression

$$K_a = 3.5 \times 10^{-8} = \frac{x^2}{0.100 - x} \approx \frac{x^2}{0.100}$$

Solving for x gives

$$x = 5.9 \times 10^{-5}$$

STEP 10

The approximation $0.100 - x \approx 0.100$ must be validated. To do this, we compare x to $[HOCl]_0$:

$$\frac{x}{[HA]_0} \times 100 = \frac{x}{[HOCl]_0} \times 100 = \frac{5.9 \times 10^{-5}}{0.100} \times 100 = 0.059\%$$

Since this value is much less than 5%, the approximation is considered valid.

STEP 11

We calculate $[H^+]$ and pH:

$$[H^+] = x = 5.9 \times 10^{-5}\ M \quad \text{and} \quad pH = 4.23$$

The pH of a Mixture of Weak Acids

The same systematic approach applies to all solution equilibria.

Sometimes a solution contains two weak acids of very different strengths. This case is considered in Sample Exercise 14.9. Note that the steps are again followed (though not labeled).

Sample Exercise 14.9

Calculate the pH of a solution that contains 1.00 M HCN ($K_a = 6.2 \times 10^{-10}$) and 5.00 M HNO_2 ($K_a = 4.0 \times 10^{-4}$). Also calculate the concentration of cyanide ion (CN^-) in this solution at equilibrium.

Solution

Since HCN and HNO_2 are both weak acids and are largely undissociated, the major species in the solution are:

$$HCN, \quad HNO_2, \quad \text{and} \quad H_2O$$

All three of these components produce H^+.

$$HCN(aq) \rightleftharpoons H^+(aq) + CN^-(aq) \qquad K_a = 6.2 \times 10^{-10}$$
$$HNO_2(aq) \rightleftharpoons H^+(aq) + NO_2^-(aq) \qquad K_a = 4.0 \times 10^{-4}$$
$$H_2O(l) \rightleftharpoons H^+(aq) + OH^-(aq) \qquad K_w = 1.0 \times 10^{-14}$$

Sample Exercise 14.9, continued

A mixture of three acids might lead to a very complicated problem. However, the situation is greatly simplified by the fact that, even though HNO_2 is a weak acid, it is much stronger than the other two acids present (as revealed by the K values). Thus HNO_2 can be assumed to be the dominant producer of H^+, and we will focus on the equilibrium expression:

$$K_a = 4.0 \times 10^{-4} = \frac{[H^+][NO_2^-]}{[HNO_2]}$$

The initial concentrations, the definition of x, and the equilibrium concentrations are as follows:

Initial concentration (mol/L)		Equilibrium concentration (mol/L)
$[HNO_2]_0 = 5.00$ $[NO_2^-]_0 = 0$ $[H^+]_0 \approx 0$	x mol/L HNO_2 dissociates $\longrightarrow$	$[HNO_2] = 5.00 - x$ $[NO_2^-] = x$ $[H^+] = x$

Substituting the equilibrium concentrations in the equilibrium expression and making the approximation that $5.00 - x = 5.00$ gives

$$K_a = 4.0 \times 10^{-4} = \frac{(x)(x)}{5.00 - x} \approx \frac{x^2}{5.00}$$

We solve for x:

$$x = 4.5 \times 10^{-2}$$

Using the 5% rule, we show that the approximation is valid:

$$\frac{x}{[HNO_2]_0} \times 100 = \frac{4.5 \times 10^{-2}}{5.00} \times 100 = 0.90\%$$

Therefore

$$[H^+] = x = 4.5 \times 10^{-2}\ M \quad \text{and} \quad pH = 1.35$$

We also want to calculate the equilibrium concentration of cyanide ion in this solution. The CN^- ions in this solution come from the dissociation of HCN:

$$HCN(aq) \rightleftharpoons H^+(aq) + CN^-(aq)$$

Although the position of this equilibrium lies far to the left and does not contribute *significantly* to $[H^+]$, HCN is the *only source* of CN^-. Thus we must consider the extent of the dissociation of HCN to calculate $[CN^-]$. The equilibrium expression for the above reaction is

$$K_a = 6.2 \times 10^{-10} = \frac{[H^+][CN^-]}{[HCN]}$$

We know $[H^+]$ for this solution from the results for the first part of the problem. It is important to understand that *there is only one kind of* H^+ *in this solution*. It does not matter from which acid the H^+ ions originate. The equilibrium value of

Sample Exercise 14.9, continued

$[H^+]$ for the HCN dissociation is 4.5×10^{-2} *M*, even though the H^+ was contributed almost entirely from the dissociation of HNO_2. What is [HCN] at equilibrium? We know $[HCN]_0 = 1.00$ *M*, and since K_a for HCN is so small, a negligible amount of HCN will dissociate. Thus

$$[HCN] = [HCN]_0 - \text{amount of HCN dissociated} \approx [HCN]_0 = 1.00\ M$$

Since $[H^+]$ and [HCN] are known, we can find $[CN^-]$ from the equilibrium expression:

$$K_a = 6.2 \times 10^{-10} = \frac{[H^+][CN^-]}{[HCN]} = \frac{(4.5 \times 10^{-2})[CN^-]}{1.00}$$

$$[CN^-] = \frac{(6.2 \times 10^{-10})(1.00)}{4.5 \times 10^{-2}} = 1.4 \times 10^{-8}\ M$$

Note the significance of this result. Since $[CN^-] = 1.4 \times 10^{-8}$ *M*, and HCN is the only source of CN^-, this means that only 1.4×10^{-8} mol/L of HCN has dissociated. This is a very small amount compared to the initial concentration of HCN, which is exactly what we would expect from its very small K_a value, and [HCN] = 1.00 *M* as assumed.

Percent Dissociation

It is often useful to specify the amount of weak acid that has dissociated in achieving equilibrium in an aqueous solution. The **percent dissociation** is defined as follows:

$$\text{Percent dissociation} = \frac{\text{amount dissociated (mol/L)}}{\text{initial concentration (mol/L)}} \times 100 \qquad (14.5)$$

For example, we found earlier that in a 1.00 *M* solution of HF, $[H^+] = 2.7 \times 10^{-2}$ *M*. To reach equilibrium, 2.7×10^{-2} mol/L of the original 1.00 *M* HF dissociates, so

$$\text{Percent dissociation} = \frac{2.7 \times 10^{-2}\ \text{mol/L}}{1.00\ \text{mol/L}} \times 100 = 2.7\%$$

For a given weak acid, the percent dissociation increases as the acid becomes more dilute. For example, the percent dissociation of acetic acid ($HC_2H_3O_2$, $K_a = 1.8 \times 10^{-5}$) is significantly greater in a 0.10 *M* solution than in a 1.0 *M* solution, as is demonstrated in Sample Exercise 14.10.

Sample Exercise 14.10

Calculate the percent dissociation of acetic acid ($K_a = 1.8 \times 10^{-5}$) in each of the following solutions:

a. 1.00 *M* $HC_2H_3O_2$
b. 0.100 *M* $HC_2H_3O_2$

Sample Exercise 14.10, continued

Solution

a. Since acetic acid is a weak acid, the major species in this solution are $HC_2H_3O_2$ and H_2O. Both species are weak acids, but acetic acid is much stronger than water. Thus the dominant equilibrium will be

$$HC_2H_3O_2(aq) \rightleftharpoons H^+(aq) + C_2H_3O_2^-(aq)$$

and the equilibrium expression is

$$K_a = 1.8 \times 10^{-5} = \frac{[H^+][C_2H_3O_2^-]}{[HC_2H_3O_2]}$$

The initial concentrations, definition of x, and the equilibrium concentrations are as follows:

Initial concentration (mol/L)		Equilibrium concentration (mol/L)
$[HC_2H_3O_2]_0 = 1.00\ M$ $[C_2H_3O_2^-]_0 = 0$ $[H^+]_0 \approx 0$	x mol/L $HC_2H_3O_2$ $\longrightarrow$ dissociates	$[HC_2H_3O_2] = 1.00 - x$ $[C_2H_3O_2^-] = x$ $[H^+] = x$

Inserting the equilibrium concentrations into the equilibrium expression and making the usual approximation that x is small compared to $[HA]_0$ gives

$$K_a = 1.8 \times 10^{-5} = \frac{[H^+][C_2H_3O_2^-]}{[HC_2H_3O_2]} = \frac{(x)(x)}{1.00 - x} \approx \frac{x^2}{1.00}$$

Thus

$$x^2 \approx 1.8 \times 10^{-5} \quad \text{and} \quad x \approx 4.2 \times 10^{-3}$$

The approximation $1.00 - x \approx 1.00$ is valid by the 5% rule so

$$[H^+] = x = 4.2 \times 10^{-3}\ M$$

The percent dissociation is

$$\frac{[H^+]}{[HC_2H_3O_2]_0} \times 100 = \frac{4.2 \times 10^{-3}}{1.00} \times 100 = 0.42\%$$

b. This is a similar problem except that in this case $[HC_2H_3O_2]_0 = 0.100\ M$. Analysis of the problem leads to the expression

$$K_a = 1.8 \times 10^{-5} = \frac{[H^+][C_2H_3O_2^-]}{[HC_2H_3O_2]} = \frac{(x)(x)}{0.100 - x} \approx \frac{x^2}{0.100}$$

Thus

$$x = [H^+] - 1.3 \times 10^{-3}\ M$$

and

$$\text{Percent dissociation} = \frac{1.3 \times 10^{-3}}{0.10} \times 100 = 1.3\%$$

The more dilute the weak acid solution, the greater the percent dissociation.

The results in Sample Exercise 14.10 show two important facts. The concentration of H^+ ion at equilibrium is smaller in the 0.10 *M* acetic acid solution than in the 1.0 *M* acetic acid solution, as we would expect. However, the percent dissociation is significantly greater in the 0.10 *M* solution than in the 1.0 *M* solution. This is a general result. *For solutions of any weak acid HA, $[H^+]$ decreases as $[HA]_0$ decreases, but the percent dissociation increases as $[HA]_0$ decreases.* This phenomenon can be explained as follows.

Consider the weak acid HA with the initial concentration $[HA]_0$, where at equilibrium

$$[HA] = [HA]_0 - x \approx [HA]_0$$
$$[H^+] = [A^-] = x$$

Thus

$$K_a = \frac{[H^+][A^-]}{[HA]} \approx \frac{(x)(x)}{[HA]_0}$$

Now suppose enough water is added suddenly to dilute the solution by a factor of 10. The new concentrations before any adjustment occurs are

$$[A^-]_{new} = [H^+]_{new} = \frac{x}{10}$$

$$[HA]_{new} = \frac{[HA]_0}{10}$$

and Q, the reaction quotient, is

$$Q = \frac{\left(\frac{x}{10}\right)\left(\frac{x}{10}\right)}{\frac{[HA]_0}{10}} = \frac{1(x)(x)}{10[HA]_0} = \frac{1}{10}K_a$$

Since Q is less than K_a, the system must adjust to the right to reach the new equilibrium position. Thus the percent dissociation increases when the acid is diluted. This behavior is summarized in Fig. 14.5. In Sample Exercise 14.11 we see how the percent dissociation can be used to calculate the K_a value for a weak acid.

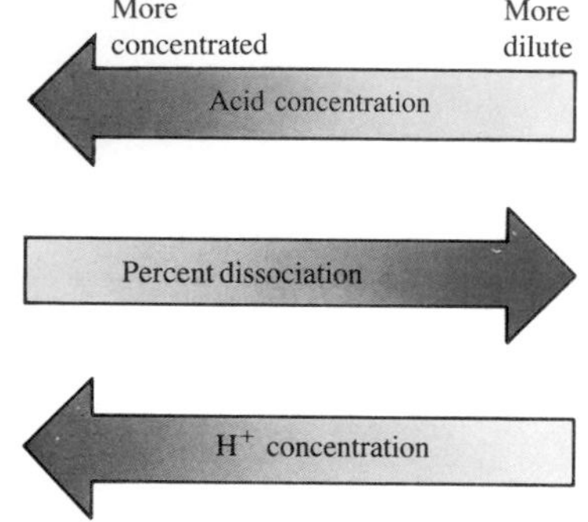

Figure 14.5

The effect of dilution on the percent dissociation and $[H^+]$ of a weak acid solution.

Sample Exercise 14.11

Lactic acid ($HC_3H_5O_3$) is a waste product that accumulates in muscle tissue during exertion, leading to pain and a feeling of fatigue. In a 0.100 *M* aqueous solution, lactic acid is 3.7% dissociated. Calculate the value of K_a for this acid.

Solution

From the small value for the percent dissociation, it is clear that $HC_3H_5O_3$ is a weak acid. Thus the major species in the solution are the undissociated acid and water:

$$HC_3H_5O_3 \text{ and } H_2O$$

But although $HC_3H_5O_3$ is a weak acid, it is much stronger than water and will be the dominant source of H^+ in the solution. The dissociation reaction is

$$HC_3H_5O_3(aq) \rightleftharpoons H^+(aq) + C_3H_5O_3^-(aq)$$

Sample Exercise 14.11, continued

and the equilibrium expression is

$$K_a = \frac{[H^+][C_3H_5O_3^-]}{[HC_3H_5O_3]}$$

The initial and equilibrium concentrations are as follows:

Initial concentration		Equilibrium concentration
$[HC_3H_5O_3]_0 = 0.10\ M$ $[C_3H_5O_3^-]_0 = 0$ $[H^+]_0 \approx 0$	$\xrightarrow[\text{HC}_3\text{H}_5\text{O}_3 \text{ dissociates}]{x \text{ mol/L}}$	$[HC_3H_5O_3] = 0.10 - x$ $[C_3H_5O_3^-] = x$ $[H^+] = x$

The change needed to reach equilibrium can be obtained from the percent dissociation and Equation (14.5). For this acid

$$\text{Percent dissociation} = 3.7\% = \frac{x}{[HC_3H_5O_3]_0} \times 100 = \frac{x}{0.10} \times 100$$

and

$$x = \frac{3.7}{100}(0.10) = 3.7 \times 10^{-3} \text{ mol/L}$$

Now we can calculate the equilibrium concentrations:

$[HC_3H_5O_3] = 0.10 - x = 0.10\ M$ (to the correct number of significant figures)

$[C_3H_5O_3^-] = [H^+] = x = 3.7 \times 10^{-3}\ M$

These concentrations can now be used to calculate the value of K_a for lactic acid:

$$K_a = \frac{[H^+][C_3H_5O_3^-]}{[HC_3H_5O_3]} = \frac{(3.7 \times 10^{-3})(3.7 \times 10^{-3})}{0.10} = 1.4 \times 10^{-4}$$

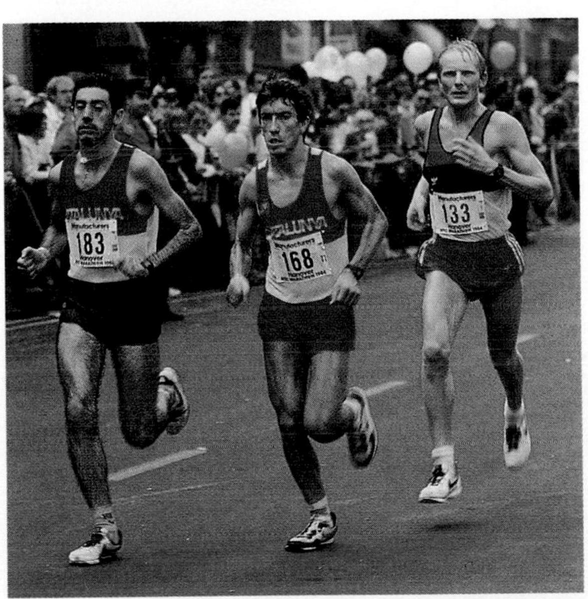

Strenuous exercise, such as in the New York City Marathon, causes a build-up of lactic acid in tissue.

14.6 Bases

Purpose

- To introduce equilibria involving strong and weak bases.
- To show how to calculate pH for basic solutions.

In a basic solution pH > 7.

According to the Arrhenius concept, a base is a substance that produces OH^- ions in aqueous solution. According to the Brönsted-Lowry model, a base is a proton acceptor. The bases sodium hydroxide (NaOH) and potassium hydroxide (KOH) fulfill both criteria. They contain OH^- ions in the solid lattice and, behaving as strong electrolytes, dissociate completely when dissolved in aqueous solution:

$$NaOH(s) \rightarrow Na^+(aq) + OH^-(aq)$$

leaving virtually no undissociated NaOH. Thus a 1.0 *M* NaOH solution really contains 1.0 *M* Na^+ and 1.0 *M* OH^-. Because of their complete dissociation, NaOH and KOH are called **strong bases** in the same sense as we defined strong acids.

All the hydroxides of the Group 1A elements (LiOH, NaOH, KOH, RbOH, and CsOH) are strong bases, but only NaOH and KOH are common laboratory reagents because the lithium, rubidium, and cesium compounds are expensive. The alkaline earth (Group 2A) hydroxides—$Ca(OH)_2$, $Ba(OH)_2$, and $Sr(OH)_2$—are also strong bases. For these compounds, two moles of hydroxide ion are produced for every mole of metal hydroxide dissolved in aqueous solution.

The alkaline earth hydroxides are not very soluble and are used only when the solubility factor is not important. In fact, the low solubility of these bases can sometimes be an advantage. For example, many antacids are suspensions of metal hydroxides such as aluminum hydroxide and magnesium hydroxide. The low solubility of these compounds prevents a large hydroxide ion concentration that would harm the tissues of the mouth, esophagus, and stomach. Yet these suspensions furnish plenty of hydroxide ion to react with the stomach acid, since the salts dissolve as this reaction proceeds.

Calcium carbonate is also used in scrubbing, as discussed in Section 5.9.

Calcium hydroxide, $Ca(OH)_2$, often called **slaked lime,** is widely used in industry because it is inexpensive and plentiful. For example, slaked lime is used in scrubbing stack gases to remove sulfur dioxide from the exhaust of power plants and factories. In the scrubbing process a suspension of slaked lime is sprayed into the stack gases to react with sulfur dioxide gas according to the following steps:

$$SO_2(g) + H_2O(l) \rightleftharpoons H_2SO_3(aq)$$

$$Ca(OH)_2(aq) + H_2SO_3(aq) \rightleftharpoons CaSO_3(s) + 2H_2O(l)$$

Slaked lime is also widely used in water treatment plants for softening hard water, which involves the removal of ions such as Ca^{2+} and Mg^{2+}, which hamper the action of detergents. The softening method most often employed in water treatment plants is the **lime-soda process,** in which *lime* (CaO) and *soda ash* (Na_2CO_3) are added to the water. As we will see in more detail later in this chapter, the CO_3^{2-} ion reacts with water to produce the HCO_3^- ion. When the lime is added to the water, it forms slaked lime

$$CaO(s) + H_2O(l) \rightarrow Ca(OH)_2(aq)$$

which then reacts with the HCO_3^- ion from the added soda ash and a Ca^{2+} ion in the hard water to produce calcium carbonate:

$$Ca(OH)_2(aq) + \underset{\nearrow \atop \text{From hard water}}{Ca^{2+}(aq)} + 2HCO_3^-(aq) \rightarrow 2CaCO_3(s) + 2H_2O(l)$$

Thus for every mole of $Ca(OH)_2$ consumed, 1 mole of Ca^{2+} is removed from the hard water, thereby softening it. Some hard water naturally contains bicarbonate ions. In this case, no soda ash is needed—simply adding the lime produces the softening.

Calculating the pH of a strong base solution is relatively simple, as illustrated in Sample Exercise 14.12.

Sample Exercise 14.12

Calculate the pH of a 5.0×10^{-2} *M* NaOH solution.

Solution

The major species in this solution are

$$\underbrace{Na^+, \quad OH^-}_{\text{From NaOH}} \quad \text{and} \quad H_2O$$

Although autoionization of water also produces OH^- ions, the pH will be dominated by the OH^- ions from the dissolved NaOH. Thus, in the solution

$$[OH^-] = 5.0 \times 10^{-2} \ M$$

and the concentration of H^+ can be calculated from K_w:

$$[H^+] = \frac{K_w}{[OH^-]} = \frac{1.0 \times 10^{-14}}{5.0 \times 10^{-2}} = 2.0 \times 10^{-13} \ M$$

$$pH = 12.70$$

Note this is a basic solution for which

$$[OH^-] > [H^+] \quad \text{and} \quad pH > 7$$

The added OH^- from the salt has shifted the water autoionization equilibrium

$$H_2O(l) \rightleftharpoons H^+(aq) + OH^-(aq)$$

to the left, significantly lowering $[H^+]$ compared to that in pure water.

A base does not have to contain hydroxide ion.

Many types of proton acceptors (bases) do not contain the hydroxide ion. However, when dissolved in water, these substances increase the concentration of hydroxide ion because of their reaction with water. For example, ammonia reacts with water as follows:

$$NH_3(aq) + H_2O(l) \rightleftharpoons NH_4^+(aq) + OH^-(aq)$$

The ammonia molecule accepts a proton and thus functions as a base. Water is the acid in this reaction. Note that even though the base ammonia contains no hydroxide ion, it still increases the concentration of hydroxide ion to yield a basic solution.

Bases like ammonia typically have at least one unshared pair of electrons that is capable of forming a bond with a proton. The reaction of an ammonia molecule with a water molecule can be represented as follows:

$$H_3N: + H-\ddot{O}(H): \rightleftharpoons [H-NH_3]^+ + :\ddot{O}-H^-$$

There are many bases like ammonia that produce hydroxide ion by reaction with water. In most of these bases, the lone pair is located on a nitrogen atom. Some examples are

Methylamine (H_3C, H, H on N) | Dimethylamine (H_3C, H, CH_3 on N) | Trimethylamine (H_3C, CH_3, CH_3 on N) | Ethylamine (H, H, C_2H_5 on N) | Pyridine

Note that the first four bases can be thought of as substituted ammonia molecules with hydrogen atoms replaced by methyl (CH_3) or ethyl (C_2H_5) groups. The pyridine molecule is like benzene

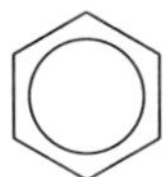

except that a nitrogen atom replaces one of the carbon atoms in the ring. The general reaction between a base (B) and water is given by:

$$\underset{\text{Base}}{B(aq)} + \underset{\text{Acid}}{H_2O(l)} \rightleftharpoons \underset{\text{Conjugate acid}}{BH^+(aq)} + \underset{\text{Conjugate base}}{OH^-(aq)} \quad (14.6)$$

The equilibrium constant for this general solution is

Appendix 5.3 contains a table of K_b values.

$$K_b = \frac{[BH^+][OH^-]}{[B]}$$

where K_b *always refers to the reaction of a base with water to form the conjugate acid and the hydroxide ion.*

Bases of the type represented by B in Equation (14.6) compete with OH^-, a very strong base, for the H^+ ion. Thus their K_b values tend to be small (for example, for ammonia, $K_b = 1.8 \times 10^{-5}$), and they are called **weak bases.** The values of K_b for some common weak bases are listed in Table 14.3.

Table 14.3 **Values of K_b for Some Common Weak Bases**

Name	Formula	Conjugate acid	K_b
Ammonia	NH_3	NH_4^+	1.8×10^{-5}
Methylamine	CH_3NH_2	$CH_3NH_3^+$	4.38×10^{-4}
Ethylamine	$C_2H_5NH_2$	$C_2H_5NH_3^+$	5.6×10^{-4}
Aniline	$C_6H_5NH_2$	$C_6H_5NH_3^+$	3.8×10^{-10}
Pyridine	C_5H_5N	$C_5H_5NH^+$	1.7×10^{-9}

Typically, pH calculations for solutions of weak bases are very similar to those for weak acids, as illustrated by Sample Exercises 14.13 and 14.14.

Refer to the steps for solving weak acid equilibrium problems on page 624. Use the same systematic approach for weak base equilibrium problems.

Sample Exercises 14.13

Calculate the pH for a 15.0 *M* solution of NH_3 ($K_b = 1.8 \times 10^{-5}$).

Solution

Since ammonia is a weak base, as can be seen from its small K_b value, most of the dissolved NH_3 will remain as NH_3. Thus the major species in solution are

$$NH_3 \quad \text{and} \quad H_2O$$

Both of these substances can produce OH^- according to the reactions:

$$NH_3(aq) + H_2O(l) \rightleftharpoons NH_4^+(aq) + OH^-(aq) \qquad K_b = 1.8 \times 10^{-5}$$
$$H_2O(l) \rightleftharpoons H^+(aq) + OH^-(aq) \qquad K_w = 1.0 \times 10^{-14}$$

However, the contribution from water can be neglected, since $K_b \gg K_w$. The equilibrium for NH_3 will dominate, and the equilibrium expression to be used is

$$K_b = 1.8 \times 10^{-5} = \frac{[NH_4^+][OH^-]}{[NH_3]}$$

The appropriate concentrations are

Initial concentration (mol/L)		Equilibrium concentration (mol/L)
$[NH_3]_0 = 15.0$ $[NH_4^+]_0 = 0$ $[OH^-]_0 \approx 0$	x mol/L NH_3 reacts with H_2O to reach equilibrium $\longrightarrow$	$[NH_3] = 15.0 - x$ $[NH_4^+] = x$ $[OH^-] = x$

Substituting the equilibrium concentrations in the equilibrium expression and making the usual approximation gives

$$K_b = 1.8 \times 10^{-5} = \frac{[NH_4^+][OH^-]}{[NH_3]} = \frac{(x)(x)}{15.0 - x} \approx \frac{x^2}{15.0}$$

Thus

$$x \approx 1.6 \times 10^{-2}$$

The 5% rule validates the approximation, and so

$$[OH^-] = 1.6 \times 10^{-2}\ M$$

Since we know that K_w must be satisfied for this solution, we can calculate $[H^+]$ as follows:

$$[H^+] = \frac{K_w}{[OH^-]} = \frac{1.0 \times 10^{-14}}{1.6 \times 10^{-2}} = 6.3 \times 10^{-13}\ M$$

Therefore

$$\text{pH} = -\log(6.3 \times 10^{-13}) = 12.20$$

A table of K_b values for bases is also given in Appendix 5.3.

Sample Exercise 14.13 illustrates how a typical weak base equilibrium problem should be solved. Note two additional important points:

1. We calculated $[H^+]$ from K_w and then calculated the pH, but another method is available. The pOH could have been calculated from $[OH^-]$ and then used in Equation (14.3):

$$pK_w = 14.00 = pH + pOH$$
$$pH = 14.00 - pOH$$

2. In a 15.0 M NH_3 solution, the equilibrium concentrations of NH_4^+ and OH^- are each 1.6×10^{-2} M. Only a small percentage,

$$\frac{1.6 \times 10^{-2}}{15.0} \times 100 = 0.11\%$$

of the ammonia reacts with water. Bottles containing 15.0 M NH_3 solution are often labeled 15.0 M NH_4OH, but as you can see from these results, 15.0 M NH_3 is actually a much more accurate description of the solution contents.

Sample Exercise 14.14

Calculate the pH of a 1.0 M solution of methylamine ($K_b = 4.38 \times 10^{-4}$).

Solution

Since methylamine (CH_3NH_2) is a weak base, the major species in solution are

$$CH_3NH_2 \quad \text{and} \quad H_2O$$

Both are bases; however water can be neglected as a source of OH^-, so the dominant equilibrium is

$$CH_3NH_2(aq) + H_2O(l) \rightleftharpoons CH_3NH_3^+(aq) + OH^-(aq)$$

and

$$K_b = 4.38 \times 10^{-4} = \frac{[CH_3NH_3^+][OH^-]}{[CH_3NH_2]}$$

The concentrations are as follows:

Initial concentration (mol/L)		Equilibrium concentration (mol/L)
$[CH_3NH_2]_0 = 1.0$ $[CH_3NH_3^+]_0 = 0$ $[OH^-]_0 \approx 0$	x mol/L CH_3NH_2 reacts with H_2O → to reach equilibrium	$[CH_3NH_2] = 1.0 - x$ $[CH_3NH_3^+] = x$ $[OH^-] = x$

Substituting the equilibrium concentrations in the equilibrium expression and making the usual approximation gives

$$K_b = 4.38 \times 10^{-4} = \frac{[CH_3NH_3^+][OH^-]}{[CH_3NH_2]} = \frac{(x)(x)}{1.0 - x} \approx \frac{x^2}{1.0}$$

$$x \approx 2.1 \times 10^{-2}$$

Sample Exercise 14.14, continued

The approximation is valid by the 5% rule, so

$$[OH^-] = x = 2.1 \times 10^{-2}\ M$$
$$pOH = 1.68$$
$$pH = 14.00 - 1.68 = 12.32$$

Chemical Impact

Amines

We have seen that many bases have nitrogen atoms with one lone pair and can be viewed as substituted ammonia molecules, with the general formula $R_xNH_{(3-x)}$. Compounds of this type are called **amines.** Amines are widely distributed in animals and plants, and complex amines often serve as messengers or regulators. For example, in the human nervous system, there are two amine stimulants, *norepinephrine* and *adrenaline*.

Norepinephrine

Adrenaline

Ephedrine, widely used as a decongestant, was a known drug in China over 2000 years ago. Indians in Mexico and the Southwest have used the hallucinogen *mescaline,* extracted from peyote cactus, for centuries.

Ephedrine

Mescaline

Many other drugs, such as codeine and quinine, are amines, but they are usually not used in their pure amine forms. Instead, they are treated with an acid to become acid salts. An example of an acid salt is ammonium chloride, obtained by the reaction:

$$NH_3 + HCl \rightarrow NH_4Cl$$

Peyote cactus growing on a rock.

Amines can also be protonated in this way. The resulting acid salt, written as AHCl (where A represents the amine), contains AH^+ and Cl^-. In general, the acid salts are more stable and more soluble in water than the parent amines. For instance, the parent amine of the well-known local anaesthetic *novocaine* is water-insoluble, while the acid salt is much more soluble.

Novocaine hydrochloride

14.7 Polyprotic Acids

Purpose

- To describe the dissociation equilibria of acids with more than one acidic proton.

Some important acids, such as sulfuric acid (H_2SO_4) and phosphoric acid (H_3PO_4), can furnish more than one proton and are called **polyprotic acids.** A polyprotic acid always dissociates in a *stepwise* manner, one proton at a time. For example, the diprotic (two-proton) acid *carbonic acid* (H_2CO_3), which is so important in maintaining a constant pH in human blood, dissociates in the following steps:

$$H_2CO_3(aq) \rightleftharpoons H^+(aq) + HCO_3^-(aq) \qquad K_{a_1} = \frac{[H^+][HCO_3^-]}{[H_2CO_3]} = 4.3 \times 10^{-7}$$

$$HCO_3^-(aq) \rightleftharpoons H^+(aq) + CO_3^{2-}(aq) \qquad K_{a_2} = \frac{[H^+][CO_3^{2-}]}{[HCO_3^-]} = 5.6 \times 10^{-11}$$

The successive K_a values for the dissociation equilibria are designated K_{a_1} and K_{a_2}. Note that the conjugate base HCO_3^- of the first dissociation equilibrium becomes the acid in the second step.

Carbonic acid is formed when carbon dioxide gas is dissolved in water. In fact, the first dissociation step for carbonic acid is best represented by the reaction

$$CO_2(aq) + H_2O(l) \rightleftharpoons H^+(aq) + HCO_3^-(aq)$$

since relatively little H_2CO_3 actually exists in solution. However, it is convenient to consider CO_2 in water as H_2CO_3 so that we can treat such solutions using the familiar dissociation reactions for weak acids.

Phosphoric acid is a **triprotic acid** (three protons) that dissociates in the following steps:

$$H_3PO_4(aq) \rightleftharpoons H^+(aq) + H_2PO_4^-(aq) \qquad K_{a_1} = \frac{[H^+][H_2PO_4^-]}{[H_3PO_4]} = 7.5 \times 10^{-3}$$

$$H_2PO_4^-(aq) \rightleftharpoons H^+(aq) + HPO_4^{2-}(aq) \qquad K_{a_2} = \frac{[H^+][HPO_4^{2-}]}{[H_2PO_4^-]} = 6.2 \times 10^{-8}$$

$$HPO_4^{2-}(aq) \rightleftharpoons H^+(aq) + PO_4^{3-}(aq) \qquad K_{a_3} = \frac{[H^+][PO_4^{3-}]}{[HPO_4^{2-}]} = 4.8 \times 10^{-13}$$

For a typical weak polyprotic acid,

$$K_{a_1} > K_{a_2} > K_{a_3}$$

That is, the acid involved in each step of the dissociation is successively weaker, as shown by the stepwise dissociation constants given in Table 14.4. These values indicate that the loss of a second or third proton occurs less readily than the loss of the first proton. This is not surprising; as the negative charge on the acid increases, it becomes more difficult to remove the positively charged proton.

Stepwise Dissociation Constants for Several Common Polyprotic Acids				
Name	Formula	K_{a_1}	K_{a_2}	K_{a_3}
Phosphoric acid	H_3PO_4	7.5×10^{-3}	6.2×10^{-8}	4.8×10^{-13}
Arsenic acid	H_3AsO_4	5×10^{-3}	8×10^{-8}	6×10^{-10}
Carbonic acid	H_2CO_3	4.3×10^{-7}	5.6×10^{-11}	
Sulfuric acid	H_2SO_4	Large	1.2×10^{-2}	
Sulfurous acid	H_2SO_3	1.5×10^{-2}	1.0×10^{-7}	
Hydrosulfuric acid	H_2S	1.0×10^{-7}	1.3×10^{-13}	
Oxalic acid	$H_2C_2O_4$	6.5×10^{-2}	6.1×10^{-5}	
Ascorbic acid (vitamin C)	$H_2C_6H_6O_6$	7.9×10^{-5}	1.6×10^{-12}	

Table 14.4

A table of K_a values for polyprotic acids is also given in Appendix 5.2.

Although we might expect the pH calculations for solutions of polyprotic acids to be complicated, the most common cases are surprisingly straightforward. To illustrate, we will consider a typical case, phosphoric acid, and a unique case, sulfuric acid.

Phosphoric Acid

Phosphoric acid is typical of most weak polyprotic acids in that the successive K_a values are very different. For example, the ratios of successive K_a values (from Table 14.4) are

$$\frac{K_{a_1}}{K_{a_2}} = \frac{7.5 \times 10^{-3}}{6.2 \times 10^{-8}} = 1.2 \times 10^5$$

$$\frac{K_{a_2}}{K_{a_3}} = \frac{6.2 \times 10^{-8}}{4.8 \times 10^{-13}} = 1.3 \times 10^5$$

Thus the relative acid strengths are

$$H_3PO_4 \gg H_2PO_4^- \gg HPO_4^{2-}$$

This means that in a solution prepared by dissolving H_3PO_4 in water, *only the first dissociation step makes an important contribution to [H⁺]*. This greatly simplifies the pH calculations for phosphoric acid solutions, as is illustrated in Sample Exercise 14.15.

For a typical polyprotic acid in water, only the first dissociation step is important in determining the pH.

Sample Exercise 14.15

Calculate the pH of a 5.0 *M* H_3PO_4 solution and the equilibrium concentrations of the species H_3PO_4, $H_2PO_4^-$, HPO_4^{2-}, and PO_4^{3-}.

Solution

The major species in solution are

$$H_3PO_4 \quad \text{and} \quad H_2O$$

None of the dissociation products of H_3PO_4 is written, since the K_a values are all so

Sample Exercise 14.15, continued

small that they will be minor species. The dominant equilibrium will be the dissociation of H_3PO_4:

$$H_3PO_4(aq) \rightleftharpoons H^+(aq) + H_2PO_4^-(aq)$$

where
$$K_{a_1} = 7.5 \times 10^{-3} = \frac{[H^+][H_2PO_4^-]}{[H_3PO_4]}$$

The concentrations are as follows:

Initial concentration (mol/L)		Equilibrium concentration (mol/L)
$[H_3PO_4]_0 = 5.0$ $[H_2PO_4^-]_0 = 0$ $[H^+]_0 \approx 0$	x mol/L H_3PO_4 dissociates $\longrightarrow$	$[H_3PO_4] = 5.0 - x$ $[H_2PO_4^-] = x$ $[H^+] = x$

Substituting the equilibrium concentrations into the expression for K_{a_1} and making the usual approximation gives

$$K_{a_1} = 7.5 \times 10^{-3} = \frac{[H^+][H_2PO_4^-]}{[H_3PO_4]} = \frac{(x)(x)}{5.0 - x} \approx \frac{x^2}{5.0}$$

Thus

$$x \approx 1.9 \times 10^{-1}$$

Since 1.9×10^{-1} is less than 5% of 5.0, the approximation is acceptable, and

$$[H^+] = x = 0.19\ M$$
$$pH = 0.72$$

So far, we have determined that

$$[H^+] = [H_2PO_4^-] = 0.19\ M$$

and

$$[H_3PO_4] = 5.0 - x = 4.8\ M$$

The concentration of HPO_4^{2-} can be obtained by using the expression for K_{a_2}:

$$K_{a_2} = 6.2 \times 10^{-8} = \frac{[H^+][HPO_4^{2-}]}{[H_2PO_4^-]}$$

where

$$[H^+] = [H_2PO_4^-] = 0.19\ M$$

Thus

$$[HPO_4^{2-}] = K_{a_2} = 6.2 \times 10^{-8}\ M$$

To calculate $[PO_4^{3-}]$, we use the expression for K_{a_3} and the values of $[H^+]$ and

Sample Exercise 14.15, continued

$[HPO_4^{2-}]$ calculated previously:

$$K_{a_3} = \frac{[H^+][PO_4^{3-}]}{[HPO_4^{2-}]} = 4.8 \times 10^{-13} = \frac{0.19[PO_4^{3-}]}{(6.2 \times 10^{-8})}$$

$$[PO_4^{3-}] = \frac{(4.8 \times 10^{-13})(6.2 \times 10^{-8})}{0.19} = 1.6 \times 10^{-19}\ M$$

These results show that the second and third dissociation steps do not make an important contribution to $[H^+]$. This is apparent from the fact that $[HPO_4^{2-}]$ is 6.2×10^{-8} *M*, which means that only 6.2×10^{-8} mol/L of $H_2PO_4^-$ has dissociated. The value of $[PO_4^{3-}]$ shows that the dissociation of HPO_4^{2-} is even smaller. We must, however, use the second and third dissociation steps to calculate $[HPO_4^{2-}]$ and $[PO_4^{3-}]$ since these steps are the only sources of these ions.

Sulfuric Acid

Sulfuric acid is unique among the common acids in that it is *a strong acid in its first dissociation step and a weak acid in its second step:*

$$H_2SO_4(aq) \rightarrow H^+(aq) + HSO_4^-(aq) \qquad K_{a_1} \text{ is very large}$$

$$HSO_4^-(aq) \rightleftharpoons H^+(aq) + SO_4^{2-}(aq) \qquad K_{a_2} = 1.2 \times 10^{-2}$$

Sample Exercise 14.16 illustrates how to calculate the pH for sulfuric acid solutions.

Sample Exercise 14.16

Calculate the pH of a 1.0 *M* H_2SO_4 solution.

Solution

The major species in the solution are

$$H^+, \quad HSO_4^-, \quad \text{and} \quad H_2O$$

where the first two ions are produced by the complete first dissociation step of H_2SO_4. The concentration of H^+ in this solution will be at least 1.0 *M*, since this amount is produced by the first dissociation step of H_2SO_4. We must now answer this question: does the HSO_4^- ion dissociate enough to produce a significant contribution to the concentration of H^+? This question can be answered by calculating the equilibrium concentrations for the dissociation reaction of HSO_4^-:

$$HSO_4^-(aq) \rightleftharpoons H^+(aq) + SO_4^{2-}(aq)$$

where

$$K_{a_2} = 1.2 \times 10^{-2} = \frac{[H^+][SO_4^{2-}]}{[HSO_4^-]}$$

Sample Exercise 14.16, continued

The concentrations are as follows:

Initial concentration (mol/L)		Equilibrium concentration (mol/L)
$[HSO_4^-]_0 = 1.0$ $[SO_4^{2-}]_0 = 0$ $[H^+]_0 = 1.0$	x mol/L HSO_4^- dissociates $\longrightarrow$ to reach equilibrium	$[HSO_4^-] = 1.0 - x$ $[SO_4^{2-}] = x$ $[H^+] = 1.0 + x$

Note that $[H^+]_0$ is not equal to zero, as it usually is for a weak acid, because the first dissociation step has already occurred. Substituting the equilibrium concentrations into the expression for K_{a_2} and making the usual approximation gives

$$K_{a_2} = 1.2 \times 10^{-2} = \frac{[H^+][SO_4^{2-}]}{[HSO_4^-]} = \frac{(1.0 + x)(x)}{1.0 - x} \approx \frac{(1.0)(x)}{(1.0)}$$

Thus

$$x \approx 1.2 \times 10^{-2}$$

Since 1.2×10^{-2} is 1.2% of 1.0, the approximation is valid according to the 5% rule. Note that x is not equal to $[H^+]$ in this case. Instead

$$[H^+] = 1.0\ M + x = 1.0\ M + (1.2 \times 10^{-2})\ M$$
$$= 1.0\ M \quad \text{(to the correct number of significant figures)}$$

Thus the dissociation of HSO_4^- does not make a significant contribution to the concentration of H^+, and

$$[H^+] = 1.0\ M \quad \text{and} \quad \text{pH} = 0.00$$

Only in dilute H_2SO_4 solutions does the second dissociation step contribute significantly to $[H^+]$.

Sample Exercise 14.16 illustrates the most common case for sulfuric acid in which only the first dissociation makes an important contribution to the concentration of H^+. In solutions more dilute than 1.0 M (for example, 0.10 M H_2SO_4), the dissociation of HSO_4^- is important, and solving the problem requires use of the quadratic formula, as shown in Sample Exercise 14.17.

Sample Exercise 14.17

Calculate the pH of a 1.00×10^{-2} M H_2SO_4 solution.

Solution

The major species in solution are

$$H^+, \quad HSO_4^-, \quad \text{and} \quad H_2O$$

Proceeding as in Sample Exercise 14.16, we consider the dissociation of HSO_4^-,

Sample Exercise 14.17, continued

which leads to the following concentrations:

Initial concentration (mol/L)		Equilibrium concentration (mol/L)
$[HSO_4^-]_0 = 0.0100$ $[SO_4^{2-}]_0 = 0$ $[H^+]_0 = 0.0100$ ↑ From dissociation of H_2SO_4	x mol/L HSO_4^- dissociates ⟶ to reach equilibrium	$[HSO_4^-] = 0.0100 - x$ $[SO_4^{2-}] = x$ $[H^+] = 0.0100 + x$

Substituting the equilibrium concentrations into the expression for K_{a_2} gives

$$1.2 \times 10^{-2} = K_{a_2} = \frac{[H^+][SO_4^{2-}]}{[HSO_4^-]} = \frac{(0.0100 + x)(x)}{(0.0100 - x)}$$

If we make the usual approximation, then $0.010 + x \approx 0.010$ and $0.010 - x \approx 0.010$, and we have

$$1.2 \times 10^{-2} = \frac{(0.0100 + x)(x)}{(0.0100 - x)} \approx \frac{(0.0100)x}{(0.0100)}$$

The calculated value of x is

$$x = 1.2 \times 10^{-2} = 0.012$$

This value is larger than 0.010, clearly a ridiculous result. Thus we cannot make the usual approximation and must instead solve the quadratic equation. The expression

$$1.2 \times 10^{-2} = \frac{(0.0100 + x)(x)}{(0.0100 - x)}$$

leads to

$$(1.2 \times 10^{-2})(0.0100 - x) = (0.0100 + x)(x)$$
$$(1.2 \times 10^{-4}) - (1.2 \times 10^{-2})x = (1.0 \times 10^{-2})x + x^2$$
$$x^2 + (2.2 \times 10^{-2})x - (1.2 \times 10^{-4}) = 0$$

This equation can be solved using the quadratic formula

$$x = \frac{-b \pm \sqrt{b^2 - 4ac}}{2a}$$

where $a = 1$, $b = 2.2 \times 10^{-2}$, and $c = -1.2 \times 10^{-4}$. Use of the quadratic formula gives one negative root (which cannot be correct) and one positive root,

$$x = 4.5 \times 10^{-3}$$

Thus

$$[H^+] = 0.0100 + x = 0.0100 + 0.0045 = 0.0145$$

and

$$pH = 1.84$$

Sample Exercise 14.17, continued

Note that in this case the second dissociation step produces about half as many H^+ ions as the initial step does.

This problem can also be solved by successive approximations, a method illustrated in Appendix 1.4.

Characteristics of Weak Polyprotic Acids

- Typically, successive K_a values are so much smaller than the first value that only the first dissociation step makes a significant contribution to the equilibrium concentration of H^+. This means that the calculation of the pH for a solution of a typical weak polyprotic acid is identical to that for a solution of a weak monoprotic acid.
- Sulfuric acid is unique in being a strong acid in its first dissociation step and a weak acid in its second step. For relatively concentrated solutions of sulfuric acid (1.0 *M* or higher), the large concentration of H^+ from the first dissociation step represses the second step, which can be neglected as a contributor of H^+ ions. For dilute solutions of sulfuric acid, the second step does make a significant contribution, and the quadratic equation must be used to obtain the total H^+ concentration.

14.8 Acid-Base Properties of Salts

Purpose

- To explain why certain salts give acidic or basic solutions and to show how to calculate the pH of these solutions.

Salt is simply another name for *ionic compound*. When a salt dissolves in water, we assume that it breaks up into its ions, which move about independently, at least in dilute solutions. Under certain conditions these ions can behave as acids or bases. In this section we explore such reactions.

Salts That Produce Neutral Solutions

Recall that the conjugate base of a strong acid has virtually no affinity for protons as compared to that of the water molecule. This is why strong acids completely dissociate in aqueous solution. Thus, when anions such as Cl^- and NO_3^- are placed in water, they do not combine with H^+ and have no effect on the pH. Cations such as K^+ and Na^+ from strong bases have no affinity for H^+ nor can they produce H^+, and so they too have no effect on the pH of an aqueous solution. *Salts that consist of* *the cations of strong bases and the anions of strong acids have no effect on $[H^+]$ when dissolved in water*. This means that aqueous solutions of salts such as KCl, NaCl, $NaNO_3$, and KNO_3 are neutral (have a pH of 7).

The salt of a strong acid and a strong base gives a neutral solution.

Salts That Produce Basic Solutions

In an aqueous solution of sodium acetate ($NaC_2H_3O_2$), the major species are:

$$Na^+, \quad C_2H_3O_2^-, \quad \text{and} \quad H_2O$$

What are the acid-base properties of each component? The Na^+ ion has neither acid nor base properties. The $C_2H_3O_2^-$ ion is the conjugate base of acetic acid, a weak acid. This means that $C_2H_3O_2^-$ has a significant affinity for a proton and is a base. Finally, water is a weakly amphoteric substance.

The pH of this solution will be determined by the $C_2H_3O_2^-$ ion. Since $C_2H_3O_2^-$ is a base, it will react with the best proton donor available. In this case, water is the *only* source of protons, and the reaction between the acetate ion and water is

$$C_2H_3O_2^-(aq) + H_2O(l) \rightleftharpoons HC_2H_3O_2(aq) + OH^-(aq) \qquad (14.7)$$

Note that this reaction, which yields a basic solution, involves a *base reacting with water to produce hydroxide ion and a conjugate acid*. We have defined K_b as the equilibrium constant for such a reaction. In this case

$$K_b = \frac{[HC_2H_3O_2][OH^-]}{[C_2H_3O_2^-]}$$

The value of K_a for acetic acid is well known (1.8×10^{-5}). But how can we obtain the K_b value for the acetate ion? The answer lies in the relationships among K_a, K_b, and K_w. Note that when the expression for K_a for acetic acid is multiplied by the expression for K_b for the acetate ion, the result is K_w:

$$K_a \times K_b = \frac{[H^+]\cancel{[C_2H_3O_2^-]}}{\cancel{[HC_2H_3O_2]}} \times \frac{\cancel{[HC_2H_3O_2]}[OH^-]}{\cancel{[C_2H_3O_2^-]}} = [H^+][OH^-] = K_w$$

This is a very important result. For any weak acid and its conjugate base,

$$K_a \times K_b = K_w$$

Thus, when either K_a or K_b is known, the other can be calculated. For the acetate ion,

$$K_b = \frac{K_w}{K_a \text{ (for } HC_2H_3O_2)} = \frac{1.0 \times 10^{-14}}{1.8 \times 10^{-5}} = 5.6 \times 10^{-10}$$

This is the K_b value for the reaction described by Equation (14.7). Note that it is obtained from the K_a value of the parent weak acid, in this case acetic acid. The sodium acetate solution is an example of an important general case. *For any salt whose cation has neutral properties (such as Na^+ or K^+) and whose anion is the conjugate base of a weak acid, the aqueous solution will be basic*. The K_b value for the anion can be obtained from the relationship $K_b = K_w/K_a$. Equilibrium calculations of this type are illustrated in Sample Exercise 14.18.

A basic solution is formed if the anion of the salt is the conjugate base of a weak acid.

Sample Exercise 14.18

Calculate the pH of a 0.30 M NaF solution. The K_a value for HF is 7.2×10^{-4}.

Solution

The major species in solution are:

$$Na^+, \quad F^-, \quad \text{and} \quad H_2O$$

Since HF is a weak acid, the F^- ion must have a significant affinity for protons, and the dominant reaction will be

$$F^-(aq) + H_2O(l) \rightleftharpoons HF(aq) + OH^-(aq)$$

which yields the K_b expression

$$K_b = \frac{[HF][OH^-]}{[F^-]}$$

The value of K_b can be calculated from K_w and the K_a value for HF:

$$K_b = \frac{K_w}{K_a \text{ (for HF)}} = \frac{1.0 \times 10^{-14}}{7.2 \times 10^{-4}} = 1.4 \times 10^{-11}$$

The concentrations are as follows:

Initial concentration (mol/L)		Equilibrium concentration (mol/L)
$[F^-]_0 = 0.30$ $[HF]_0 = 0$ $[OH^-]_0 \approx 0$	x mol/L F^- reacts with H_2O to reach equilibrium $\longrightarrow$	$[F^-] = 0.30 - x$ $[HF] = x$ $[OH^-] = x$

Thus $$K_b = 1.4 \times 10^{-11} = \frac{[HF][OH^-]}{[F^-]} = \frac{(x)(x)}{0.30 - x} \approx \frac{x^2}{0.30}$$

$$x \approx 2.0 \times 10^{-6}$$

The approximation is valid by the 5% rule, and so

$$[OH^-] = x = 2.0 \times 10^{-6} \ M$$
$$pOH = 5.69$$
$$pH = 14.00 - 5.69 = 8.31$$

As expected, the solution is basic.

Base Strength in Aqueous Solution

To emphasize the concept of base strength, let us consider the basic properties of the cyanide ion. One relevant reaction is the dissociation of hydrocyanic acid in water:

$$HCN(aq) + H_2O(l) \rightleftharpoons H_3O^+(aq) + CN^-(aq) \qquad K_a = 6.2 \times 10^{-10}$$

Since HCN is such a weak acid, CN^- appears to be a *strong* base, showing a very high affinity for H^+ *compared to H_2O*, with which it is competing. However, we also need to look at the reaction in which cyanide ion reacts with water:

$$CN^-(aq) + H_2O(l) \rightleftharpoons HCN(aq) + OH^-(aq)$$

where

$$K_b = \frac{K_w}{K_a} = \frac{1.0 \times 10^{-14}}{6.2 \times 10^{-10}} = 1.6 \times 10^{-5}$$

In this reaction CN^- appears to be a weak base; the K_b value is only 1.6×10^{-5}. What accounts for this apparent difference in base strength? The key idea is that in the reaction of CN^- with H_2O, *CN^- is competing with OH^- for H^+, instead of competing with H_2O,* as it does in the HCN dissociation reaction. These equilibria show the following relative base strengths:

$$OH^- > CN^- > H_2O$$

Similar arguments can be made for other "weak" bases, such as ammonia, the acetate ion, the fluoride ion, and so on.

Salts That Produce Acidic Solutions

Some salts produce acidic solutions when dissolved in water. For example, when solid NH_4Cl is dissolved in water, NH_4^+ and Cl^- ions are present, with NH_4^+ behaving as a weak acid:

$$NH_4^+(aq) \rightleftharpoons NH_3(aq) + H^+(aq)$$

The Cl^- ion, having virtually no affinity for H^+ in water, does not affect the pH of the solution.

In general, *salts of which the cation is the conjugate acid of a weak base produce acidic solutions.*

Sample Exercise 14.19

Calculate the pH of a 0.10 *M* NH_4Cl solution. The K_b value for NH_3 is 1.8×10^{-5}.

Solution

The major species in solution are:

$$NH_4^+, \quad Cl^-, \quad \text{and} \quad H_2O$$

Note that both NH_4^+ and H_2O can produce H^+. The dissociation reaction for the NH_4^+ ion is

$$NH_4^+(aq) \rightleftharpoons NH_3(aq) + H^+(aq)$$

for which

$$K_a = \frac{[NH_3][H^+]}{[NH_4^+]}$$

Note that although the K_b value for NH_3 is given, the reaction corresponding to K_b

Sample Exercise 14.19, continued

is not appropriate here, since NH_3 is not a major species in the solution. Instead, the given value of K_b is used to calculate K_a for NH_4^+ from the relationship:

$$K_a \times K_b = K_w$$

Thus

$$K_a \text{ (for } NH_4^+) = \frac{K_w}{K_b \text{ (for } NH_3)} = \frac{1.0 \times 10^{-14}}{1.8 \times 10^{-5}} = 5.6 \times 10^{-10}$$

Although NH_4^+ is a very weak acid, as indicated by its K_a value, it is stronger than H_2O and thus will dominate in the production of H^+. Thus we will focus on the dissociation reaction of NH_4^+ to calculate the pH in this solution.

We solve the weak acid problem in the usual way:

Initial concentration (mol/L)		Equilibrium concentration (mol/L)
$[NH_4^+]_0 = 0.10$ $[NH_3]_0 = 0$ $[H^+]_0 \approx 0$	x mol/L NH_4^+ dissociates → to reach equilibrium	$[NH_4^+] = 0.10 - x$ $[NH_3] = x$ $[H^+] \approx x$

Thus

$$5.6 \times 10^{-10} = K_a = \frac{[H^+][NH_3]}{[NH_4^+]} = \frac{(x)(x)}{0.10 - x} = \frac{x^2}{0.10}$$

$$x \approx 7.5 \times 10^{-6}$$

The approximation is valid by the 5% rule, so

$$[H^+] = x = 7.5 \times 10^{-6}\ M \quad \text{and} \quad pH = 5.13$$

A second type of salt that produces an acidic solution is one that contains a *highly charged metal ion*. For example, when solid aluminum chloride ($AlCl_3$) is dissolved in water, the resulting solution is significantly acidic. Although the Al^{3+} ion is not itself a Brönsted-Lowry acid, the hydrated ion $Al(H_2O)_6^{3+}$ formed in water is a weak acid:

$$Al(H_2O)_6^{3+}(aq) \rightleftharpoons Al(OH)(H_2O)_5^{2+}(aq) + H^+(aq)$$

Section 14.9 contains a further discussion of the acidity of hydrated ions.

The high charge on the metal ion polarizes the O—H bonds in the attached water molecules, making the hydrogens in these water molecules more acidic than those in free water molecules. Typically, the higher the charge on the metal ion, the stronger the acidity of the hydrated ion.

Sample Exercise 14.20

Calculate the pH of a 0.010 M $AlCl_3$ solution. The K_a value for $Al(H_2O)_6^{3+}$ is 1.4×10^{-5}.

Sample Exercise 14.20, continued

Solution

The major species in solution are:

$$Al(H_2O)_6^{3+}, \quad Cl^-, \quad \text{and} \quad H_2O$$

Since the $Al(H_2O)_6^{3+}$ ion is a stronger acid than water, the dominant equilibrium is

$$Al(H_2O)_6^{3+}(aq) \rightleftharpoons Al(OH)(H_2O)_5^{2+}(aq) + H^+(aq)$$

and

$$1.4 \times 10^{-5} = K_a = \frac{[Al(OH)(H_2O)_5^{2+}][H^+]}{[Al(H_2O)_6^{3+}]}$$

This is a typical weak acid problem which we can solve with the usual procedures.

Initial concentration (mol/L)		Equilibrium concentration (mol/L)
$[Al(H_2O)_6^{3+}]_0 = 0.010$ $[Al(OH)(H_2O)_5^{2+}]_0 = 0$ $[H^+]_0 \approx 0$	$\xrightarrow{x \text{ mol/L}}$ $Al(OH_2)_6^{3+}$ dissociates to reach equilibrium	$[Al(H_2O)_6^{3+}] = 0.010 - x$ $[Al(OH)(H_2O)_5^{2+}] = x$ $[H^+] = x$

Thus

$$1.4 \times 10^{-5} = K_a = \frac{[Al(OH)(H_2O)_5^{2+}][H^+]}{[Al(H_2O)_6^{3+}]} = \frac{(x)(x)}{0.010 - x} \approx \frac{x^2}{0.010}$$

$$x \approx 3.7 \times 10^{-4}$$

Since the approximation is valid by the 5% rule,

$$[H^+] = x = 3.7 \times 10^{-4}\ M \quad \text{and} \quad pH = 3.43$$

So far we have considered salts in which only one of the ions has acidic or basic properties. For many salts, such as ammonium acetate ($NH_4C_2H_3O_2$), both ions can affect the pH of the aqueous solution. Because the equilibrium calculations for these cases can be quite complicated, we will consider only the qualitative aspects of such problems. We can predict whether the solution will be basic, acidic, or neutral by comparing the K_a value for the acidic ion to the K_b value for the basic ion. If the K_a value for the acidic ion is larger than the K_b value for the basic ion, the solution will be acidic. If the K_b value is larger than the K_a value, the solution will be basic. Equal K_a and K_b values mean a neutral solution. These facts are summarized in Table 14.5.

Qualitative Prediction of pH for Solutions of Salts for Which Both Cation and Anion Have Acidic or Basic Properties

$K_a > K_b$	$pH < 7$ (acidic)
$K_b > K_a$	$pH > 7$ (basic)
$K_a - K_b$	$pH = 7$ (neutral)

Table 14.5

Sample Exercise 14.21

Predict whether an aqueous solution of each of the following salts will be acidic, basic, or neutral:

a. $NH_4C_2H_3O_2$ **b.** NH_4CN **c.** $Al_2(SO_4)_3$

Sample Exercise 14.21, continued

Solution

a. The ions in solution are NH_4^+ and $C_2H_3O_2^-$. As we mentioned previously, K_a for NH_4^+ is 5.6×10^{-10}, and K_b for $C_2H_3O_2^-$ is 5.6×10^{-10}. Thus K_a for NH_4^+ is equal to K_b for $C_2H_3O_2^-$, and the solution will be neutral (pH = 7).

b. The solution will contain NH_4^+ and CN^- ions. The K_a value for NH_4^+ is 5.6×10^{-10} and

$$K_b \text{ (for } CN^-) = \frac{K_w}{K_a \text{ (for HCN)}} = 1.6 \times 10^{-5}$$

Since K_b for CN^- is much larger than K_a for NH_4^+, CN^- is a much stronger base than NH_4^+ is an acid. This solution will be basic.

c. The solution will contain $Al(H_2O)_6^{3+}$ and SO_4^{2-} ions. The K_a value for $Al(H_2O)_6^{3+}$ is 1.4×10^{-5}, as given in Sample Exercise 14.20. We must calculate K_b for SO_4^{2-}. The HSO_4^- ion is the conjugate acid of SO_4^{2-}, and its K_a value is K_{a_2} for sulfuric acid, or 1.2×10^{-2}. Therefore

$$K_b \text{ (for } SO_4^{2-}) = \frac{K_w}{K_{a_2} \text{ (for sulfuric acid)}}$$

$$= \frac{1.0 \times 10^{-14}}{1.2 \times 10^{-2}} = 8.3 \times 10^{-13}$$

This solution will be acidic, since K_a for $Al(H_2O)_6^{3+}$ is much greater than K_b for SO_4^{2-}.

The acid-base properties of aqueous solutions of various salts are summarized in Table 14.6.

Acid-Base Properties of Various Types of Salts

Type of salt	Examples	Comment	pH of solution
Cation is from strong base; anion is from strong acid	KCl, KNO_3, NaCl, $NaNO_3$	Neither acts as an acid or a base	Neutral
Cation is from strong base; anion is from weak acid	$NaC_2H_3O_2$, KCN, NaF	Anion acts as a base; cation has no effect on pH	Basic
Cation is conjugate acid of weak base; anion is from strong acid	NH_4Cl, NH_4NO_3	Cation acts as acid; anion has no effect on pH	Acidic
Cation is conjugate acid of weak base; anion is conjugate base of weak acid	$NH_4C_2H_3O_2$, NH_4CN	Cation acts as an acid; anion acts as a base	Acidic if $K_a > K_b$, basic if $K_b > K_a$, neutral if $K_a = K_b$
Cation is highly charged metal ion; anion is from strong acid	$Al(NO_3)_3$, $FeCl_3$	Hydrated cation acts as an acid; anion has no effect on pH	Acidic

Table 14.6

14.9 The Effect of Structure on Acid-Base Properties

Purpose

- To show how bond strength and polarity affect acid-base properties.

Further aspects of acid strengths are discussed in Section 19.7.

We have seen that when a substance is dissolved in water, it produces an acidic solution if it can donate protons and produces a basic solution if it can accept protons. What structural properties of a molecule cause it to behave as an acid or as a base?

Any molecule containing a hydrogen atom is potentially an acid. However, many such molecules show no acidic properties. For example, molecules containing C—H bonds, such as chloroform ($CHCl_3$) and nitromethane (CH_3NO_2), do not produce acidic aqueous solutions because a C—H bond is both strong and nonpolar and thus there is no tendency to donate protons. On the other hand, although the H—Cl bond in gaseous hydrogen chloride is slightly stronger than a C—H bond, it is much more polar, and this molecule readily dissociates when dissolved in water.

Thus there are two main factors that determine whether a molecule containing an X—H bond will behave as a Brönsted-Lowry acid: the strength of the bond and the polarity of the bond.

The importance of these factors is clearly demonstrated by the relative acid strengths of the hydrogen halides. The bond polarities vary as shown

$$\underset{\text{Most polar}}{\underset{\uparrow}{\text{H—F}}} > \text{H—Cl} > \text{H—Br} > \underset{\text{Least polar}}{\underset{\uparrow}{\text{H—I}}}$$

because electronegativity decreases going down the group. Based on the high polarity of the H—F bond, we might expect hydrogen fluoride to be a very strong acid. In fact, among HX molecules, HF is the only weak acid ($K_a = 7.2 \times 10^{-4}$) when dissolved in water. The H—F bond is unusually strong, as shown in Table 14.7, and thus is difficult to break. This contributes significantly to the reluctance of the HF molecules to dissociate in water.

Bond Strengths and Acid Strengths for Hydrogen Halides

H—X bond	Bond strength (kJ/mol)	Acid strength in water
H—F	565	Weak
H—Cl	427	Strong
H—Br	363	Strong
H—I	295	Strong

Table 14.7

Another important class of acids are the oxyacids, which as we saw in Section 14.2 characteristically contain the grouping H—O—X. Several series of oxyacids are listed with their K_a values in Table 14.8 on the following page. Note from these data that for a given series the acid strength increases with an increase in the number

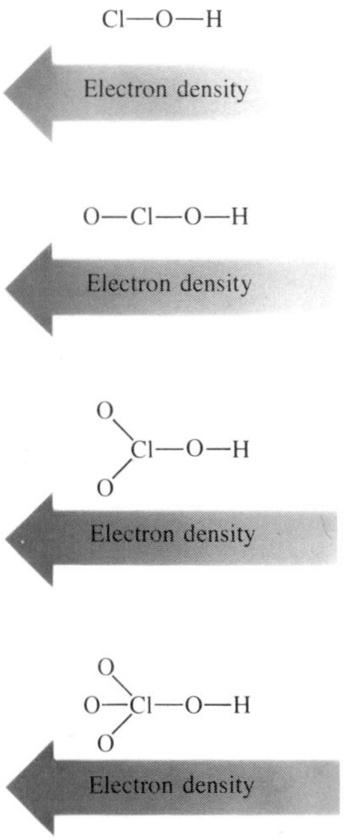

Several Series of Oxyacids and Their K_a Values		
Oxyacid	Structure	K_a value
$HClO_4$	H—O—Cl(—O)(=O)(—O)	Large (~10^7)
$HClO_3$	H—O—Cl(—O)(—O)	~1
$HClO_2$	H—O—Cl—O	1.2×10^{-2}
$HClO$	H—O—Cl	3.5×10^{-8}
H_2SO_4	H—O—S(—O—H)(—O)(—O)	Large
H_2SO_3	H—O—S(—O—H)(—O)	1.5×10^{-2}
HNO_3	H—O—N(—O)(—O)	Large
HNO_2	H—O—N—O	4.0×10^{-4}

Table 14.8

Figure 14.6

The effect of the number of attached oxygens on the H—O bond in a series of chlorine oxyacids. As the number of oxygen atoms attached to the chlorine atom increases, they become more effective at withdrawing electron density from the H—O bond, thereby weakening and polarizing it. This increases the tendency for the molecule to produce a proton, and so its acid strength increases.

of oxygen atoms attached to the central atom. For example, in the series containing chlorine and a varying number of oxygen atoms, HOCl is a weak acid, but the acid strength is successively greater as the number of oxygen atoms increases. This happens because the very electronegative oxygen atoms are able to draw electrons away from the chlorine atom and the O—H bond, as shown in Fig. 14.6. The net effect is to both polarize and weaken the O—H bond; this effect is more important as the number of attached oxygen atoms increases. This means that a proton is most readily produced by the molecule with the largest number of attached oxygen atoms ($HClO_4$).

This type of behavior is also observed for hydrated metal ions. Earlier in this chapter we saw that highly charged metal ions such as Al^{3+} produce acidic solutions. The acidity of the water molecules attached to the metal ion is increased by the attraction of electrons to the positive metal ion:

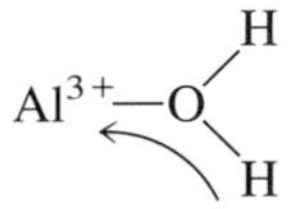

$$Al^{3+}\text{—}O(H)(H)$$

The greater the charge on the metal ion, the more acidic the hydrated ion becomes.

For acids containing the H—O—X grouping, the greater the ability of X to draw electrons toward itself, the greater is the acidity of the molecule. Since the electronegativity of X reflects its ability to attract the electrons involved in bonding, we might expect acid strength to depend on the electronegativity of X. In fact, there is an excellent correlation between the electronegativity of X and the acid strength for oxyacids, as shown in Table 14.9.

Comparison of Electronegativity of X and K_a Value for a Series of Oxyacids

Acid	X	Electronegativity of X	K_a for acid
HOCl	Cl	3.0	4×10^{-8}
HOBr	Br	2.8	2×10^{-9}
HOI	I	2.5	2×10^{-11}
$HOCH_3$	CH_3	2.3 (for carbon in CH_3)	~0

Table 14.9

14.10 Acid-Base Properties of Oxides

Purpose

- To show how to predict whether an oxide will produce an acidic or basic solution.

We have just seen that molecules containing the grouping H—O—X can behave as acids and that the acid strength depends on the electron-withdrawing ability of X. But substances with this grouping can also behave as bases, if hydroxide ion instead of a proton is produced. What determines which behavior will occur? The answer lies mainly in the nature of the O—X bond. If X has a relatively high electronegativity, the O—X bond will be covalent and strong. When the compound containing the H—O—X grouping is dissolved in water, the O—X bond will remain intact. It will be the polar and relatively weak H—O bond that will tend to break, releasing a proton. On the other hand, if X has a very low electronegativity, the O—X bond will be ionic and subject to being broken in polar water. Examples are the ionic substances NaOH and KOH that dissolve in water to give the metal cation and the hydroxide ion.

A compound containing the H—O—X group will produce an acidic solution in water if the O—X bond is strong and covalent. If the O—X bond is ionic, the compound will produce a basic solution in water.

We can use these principles to explain the acid-base behavior of oxides when they are dissolved in water. For example, when a covalent oxide such as sulfur trioxide is dissolved in water, an acidic solution results because sulfuric acid is formed:

$$SO_3(g) + H_2O(l) \rightarrow H_2SO_4(aq)$$

The structure of H_2SO_4 is shown in the margin. In this case, the strong, covalent O—S bonds remain intact and the H—O bonds break to produce protons. Other common covalent oxides that react with water to form acidic solutions are sulfur dioxide, carbon dioxide, and nitrogen dioxide, as shown by the following reactions:

$$SO_2(g) + H_2O(l) \rightarrow H_2SO_3(aq)$$

$$CO_2(g) + H_2O(l) \rightarrow H_2CO_3(aq)$$

$$2NO_2(g) + H_2O(l) \rightarrow HNO_3(aq) + HNO_2(aq)$$

Thus, when a covalent oxide dissolves in water, an acidic solution forms. These oxides are called **acidic oxides.**

On the other hand, when an ionic oxide dissolves in water, a basic solution results, as shown by the following reactions:

$$CaO(s) + H_2O(l) \rightarrow Ca(OH)_2(aq)$$

$$K_2O(s) + H_2O(l) \rightarrow 2KOH(aq)$$

These reactions can be explained by recognizing that the oxide ion has a high affinity for protons and reacts with water to produce hydroxide ions:

$$O^{2-}(aq) + H_2O(l) \rightarrow 2OH^-(aq)$$

Thus the most ionic oxides, such as those of the Group 1A and 2A metals, produce basic solutions when they are dissolved in water. As a result, these oxides are called **basic oxides.**

14.11 The Lewis Acid-Base Model

Purpose

- To define acids and bases in terms of electron pairs.

We have seen that the first successful conceptualization of acid-base behavior was proposed by Arrhenius. This useful but limited model was replaced by the more general Brönsted-Lowry model. An even more general model for acid-base behavior was suggested by G. N. Lewis in the early 1920s. A **Lewis acid** is an *electron-pair acceptor,* and a **Lewis base** is an *electron-pair donor*. The three models are summarized in Table 14.10.

Three Models for Acids and Bases

Model	Definition of acid	Definition of base
Arrhenius	H^+ producer	OH^- producer
Brönsted-Lowry	H^+ donor	H^+ acceptor
Lewis	Electron-pair acceptor	Electron-pair donor

Table 14.10

Note that Brönsted-Lowry acid-base reactions (proton donor–proton acceptor reactions) are encompassed by the Lewis model. For example, the reaction between

a proton and an ammonia molecule

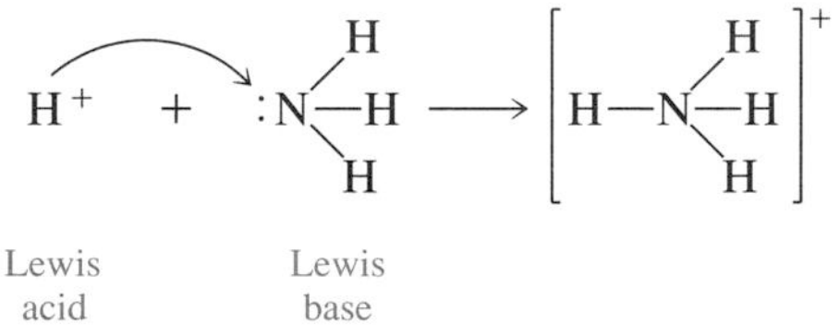

$$H^+ + :NH_3 \longrightarrow [H{-}NH_3]^+$$

Lewis acid — Lewis base

can be represented as a reaction between an electron-pair acceptor (H^+) and an electron-pair donor (NH_3). The same holds true for a reaction between a proton and a hydroxide ion:

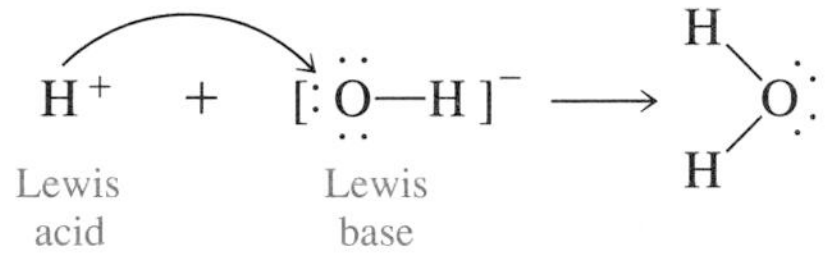

$$H^+ + [:\ddot{O}{-}H]^- \longrightarrow H_2\ddot{O}:$$

Lewis acid — Lewis base

The real value of the Lewis model for acids and bases is that it covers many reactions that do not involve Brönsted-Lowry acids. For example, consider the gas-phase reaction between boron trifluoride and ammonia:

The Lewis model encompasses the Brönsted-Lowry model, but the reverse is not true.

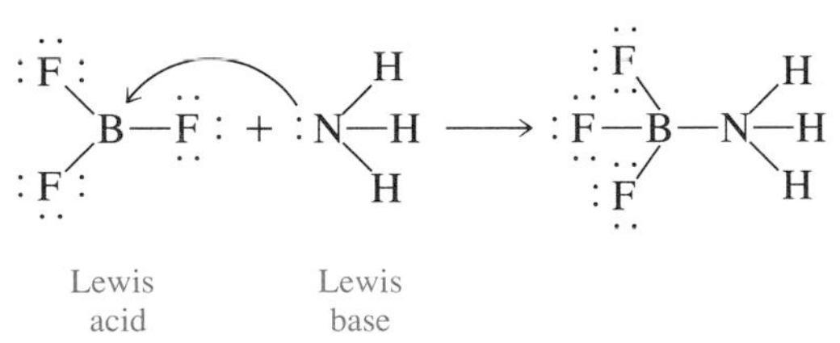

$$BF_3 + :NH_3 \longrightarrow F_3B{-}NH_3$$

Lewis acid — Lewis base

Here the electron-deficient BF_3 molecule (there are only six electrons around the boron) completes its octet by reacting with NH_3, which has a lone pair of electrons. In fact, as mentioned in Chapter 8, the electron deficiency of boron trifluoride makes it very reactive toward any electron-pair donor. That is, it is a strong Lewis acid.

The hydration of a metal ion, such as Al^{3+}, can be also viewed as a Lewis acid-base reaction:

$$Al^{3+} + 6\,(:\ddot{O}H_2) \longrightarrow [Al(\ddot{O}H_2)_6]^{3+}$$

Lewis acid — Lewis base

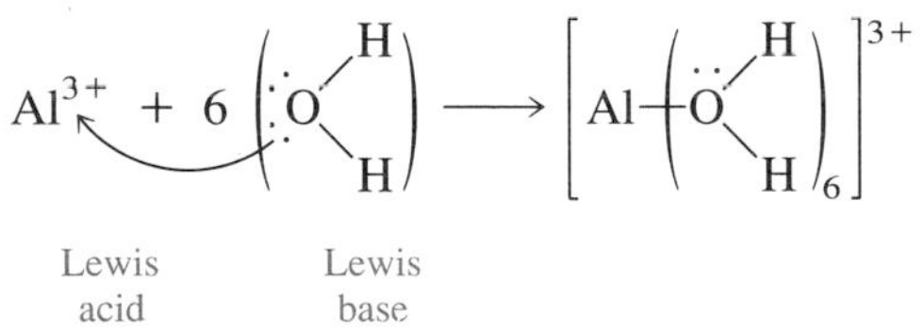

Here the Al^{3+} ion accepts one electron pair from each of six water molecules.

In addition, the reaction between a covalent oxide and water to form a Brönsted-Lowry acid can be defined as a Lewis acid-base reaction. An example is the reaction between sulfur trioxide and water:

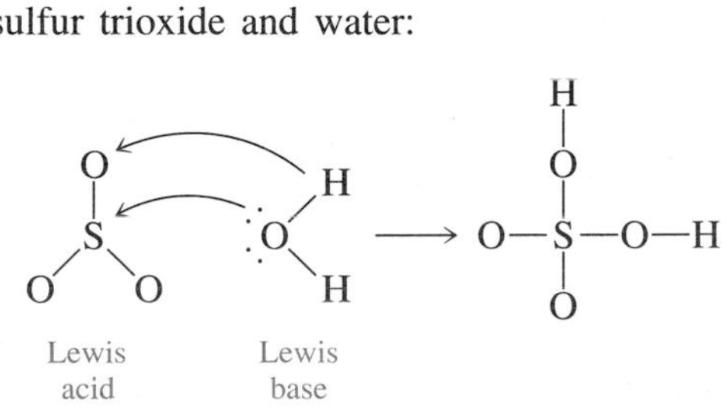

$$SO_3 + :\ddot{O}H_2 \longrightarrow O{-}S(O)(OH){-}O{-}H$$

Lewis acid — Lewis base

Note that as the water molecule attaches to sulfur trioxide, a proton shift occurs to form sulfuric acid.

Sample Exercise 14.22

For each reaction, identify the Lewis acid and base.

a. $Ni^{2+}(aq) + 6NH_3(aq) \rightarrow Ni(NH_3)_6{}^{2+}(aq)$

b. $H^+(aq) + H_2O(l) \rightleftharpoons H_3O^+(aq)$

Solution

a. Each NH_3 molecule donates an electron pair to the Ni^{2+} ion:

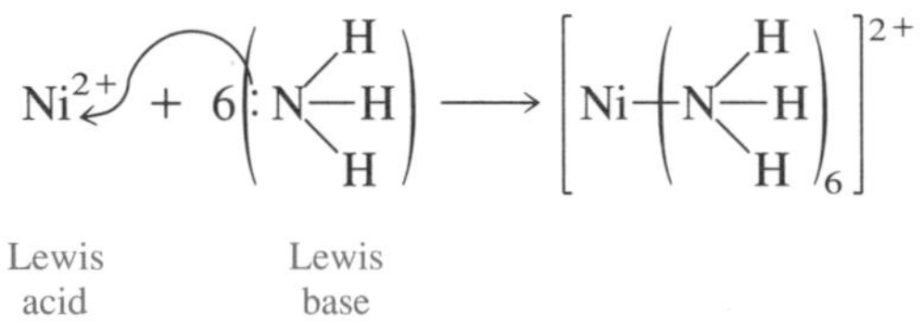

$$Ni^{2+} + 6\left(:NH_3\right) \longrightarrow \left[Ni\left(NH_3\right)_6\right]^{2+}$$

Lewis acid Lewis base

The nickel(II) ion is the Lewis acid, and ammonia is the Lewis base.

b. The proton is the Lewis acid and the water molecule is the Lewis base:

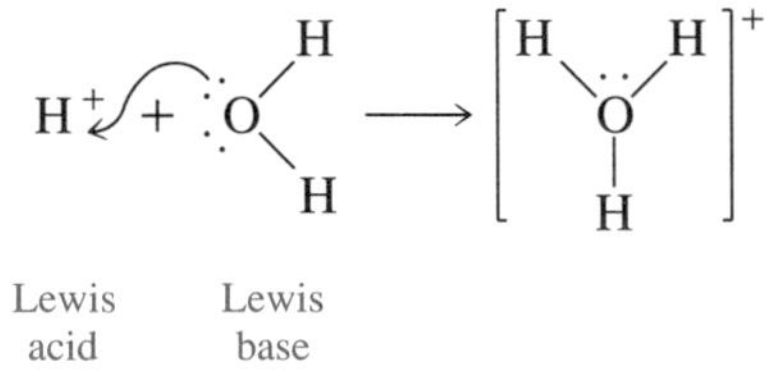

$$H^+ + :\ddot{O}H_2 \longrightarrow \left[H_3O\right]^+$$

Lewis acid Lewis base

14.12 Strategy for Solving Acid-Base Problems: A Summary

In this chapter we have encountered many different situations involving aqueous solutions of acids and bases, and in the next chapter we will encounter still more. In solving for the equilibrium concentrations in these aqueous solutions, it is tempting to create a pigeonhole for each possible situation and to memorize the procedures necessary to deal with that particular case. This approach is just not practical, and usually leads to frustration: too many pigeonholes are required and there seems to be an infinite number of cases. But you can handle any case successfully by taking a systematic, patient, and thoughtful approach. When analyzing an acid-base equilibrium problem, do *not* ask yourself how a memorized solution can be used to solve the problem. Instead, ask this question: *what are the major species in the solution and what is their chemical behavior?*

Chemical Impact

Self-Destructing Paper

The New York City Public Library has 88 miles of bookshelves, and on 36 miles of these shelves the books are quietly disintegrating between their covers. In fact, it is estimated that 40% of the books in the major research collections in the United States will soon be too fragile to handle.

This problem results from the acidic paper widely used in printing books in the last century. Ironically, books have survived from the eighteenth, seventeenth, sixteenth, and even fifteenth century. Gutenberg Bibles contain paper that is in remarkably good condition. In those days, paper was made by hand from linen or rags, but in the nineteenth century demand for cheap paper skyrocketed. Paper manufacturers found that paper could be made economically, by machine, using wood pulp. To size the paper (that is, fill in microscopic holes to lower absorption of moisture and prevent seeping or spreading of inks), alum ($Al_2(SO_4)_3$) was added in large amounts. Because the hydrated aluminum ion, $Al(OH_2)_6{}^{3+}$, is an acid ($K_a \approx 10^{-5}$), paper manufactured using alum is quite acidic. Over time this acidity causes the paper fibers to disintegrate; the pages of books fall apart when they are used.

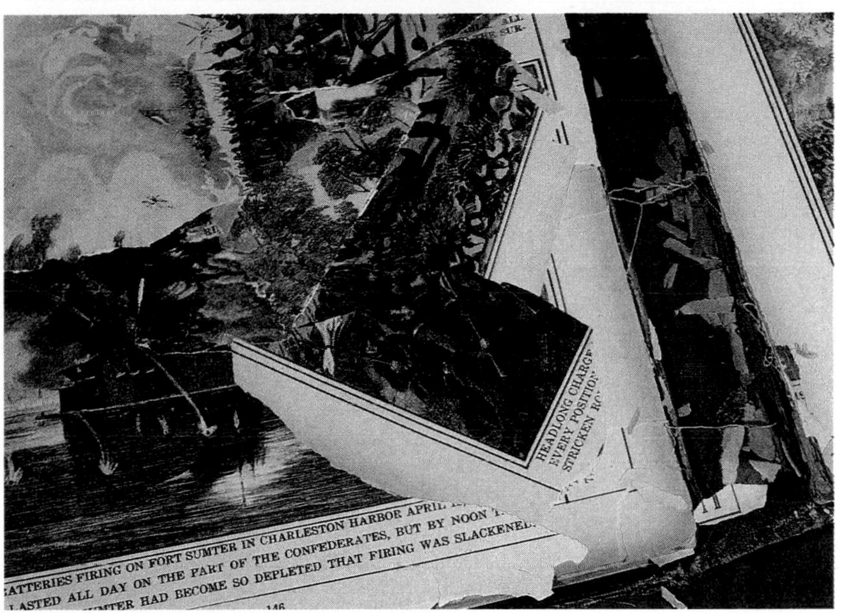

A book ravaged by the decomposition of acidic paper.

One could transfer the contents of the threatened books to microfilm, but that is a very slow and expensive process. Can the books be chemically treated to neutralize the acid and stop the deterioration? Yes. In fact, you know enough chemistry at this point to design the treatment patented in 1936 by Otto Schierholz. He dipped individual pages in solutions of alkaline earth bicarbonate salts ($MgHCO_3$, $CaHCO_3$, etc.). The $HCO_3{}^-$ ions present in these solutions react with the H^+ in the paper to give CO_2 and H_2O. This treatment works well and is used today to preserve especially important works, but it is slow and labor intensive.

It would be much more economical if large numbers of books could be treated at one time without disturbing the bindings. However, soaking entire books in an aqueous solution is out of the question. A logical question then is: Are there gaseous bases that could be used to neutralize the acid? Certainly; the organic amines (general formula, RNH_2) are bases, and those with low molecular weights are gases under normal conditions. Experiments in which books were treated using ammonia, butylamine ($CH_3CH_2CH_2CH_2NH_2$), and other amines have shown that the method works, but only for a short time. The amines do enter the paper and neutralize the acid, but being volatile they gradually evaporate leaving the paper in its original acidic condition.

A different but especially promising treatment involves diethylzinc, $(CH_3CH_2)_2Zn$, which boils at 117°C and 1 atm. Diethylzinc (DEZ) reacts

Chemical Impact

with oxygen or water to produce ZnO as follows:

$$(CH_3CH_2)_2Zn(g) + 7O_2(g) \rightarrow ZnO(s) + 4CO_2(g) + 5H_2O(g)$$

$$(CH_3CH_2)_2Zn(g) + H_2O(g) \rightarrow ZnO(s) + 2CH_3CH_3(g)$$

The solid zinc oxide produced in these reactions is deposited among the paper fibers, and, being a basic oxide, it neutralizes the acid present as shown in the equation

$$ZnO + 2H^+ \rightarrow Zn^{2+} + H_2O$$

One major problem is that DEZ ignites spontaneously on contact with air. Therefore this treatment must be carried out in a chamber, filled mainly with $N_2(g)$, where the amount of O_2 present can be rigorously controlled. The pressure in the chamber must be maintained well below one atmosphere both to lower the boiling point of DEZ and to remove excess moisture from the book's pages. Several major DEZ fires have slowed its implementation as a book preservative. However, the Library of Congress has designed a new DEZ treatment plant, which should be in full operation by 1990, that includes a chamber large enough for 7500 to 9000 books to be treated at one time.

The most important part of doing a complicated acid-base equilibrium problem is the analysis you do at the beginning of a problem:

What major species are present?

Does a reaction occur that can be assumed to go to completion?

What equilibrium dominates the solution?

Let the problem guide you. Be patient.

The following steps outline a general strategy for solving problems involving acid-base equilibria.

Solving Acid-Base Problems

- STEP 1
  List the major species in solution.
- STEP 2
  Look for reactions that can be assumed to go to completion, for example, a strong acid dissociating or H^+ reacting with OH^-.
- STEP 3
  For a reaction that can be assumed to go to completion:
  a. Determine the concentrations of the products.
  b. Write down the major species in solution after the reaction.
- STEP 4
  Look at each major component of the solution and decide if it is an acid or a base.
- STEP 5
  Pick the equilibrium that will control the pH. Use known values of the dissociation constants for the various species to help decide on the dominant equilibrium.
  a. Write the equation for the reaction and the equilibrium expression.

b. Compute the initial concentrations (assuming the dominant equilibrium has not yet occurred, that is, no acid dissociation, etc.).
c. Define x.
d. Compute the equilibrium concentrations in terms of x.
e. Substitute the concentrations into the equilibrium expression and solve for x.
f. Check the validity of the approximation.
g. Calculate the pH and other concentrations as required.

Although these steps may seem somewhat cumbersome, especially for simpler problems, they will become increasingly helpful as the aqueous solutions become more complicated. If you develop the habit of approaching acid-base problems systematically, the more complex cases will be much easier to manage.

FOR REVIEW

Summary

In this chapter we have developed several models of acid-base behavior and have applied fundamental equilibrium principles to calculate the pH values for solutions of acids and bases.

Arrhenius postulated that acids produce H^+ ions in solutions and bases produce OH^- ions. The Brönsted-Lowry model is more general: an acid is a proton donor, and a base is a proton acceptor. Water acts as a Brönsted-Lowry base when it accepts a proton from an acid to form a hydronium ion:

$$\underset{\text{Acid}}{HA(aq)} + \underset{\text{Base}}{H_2O(l)} \rightleftharpoons \underset{\text{Conjugate acid}}{H_3O^+(aq)} + \underset{\text{Conjugate base}}{A^-(aq)}$$

A conjugate base is everything that remains of the acid molecule after the proton is lost. A conjugate acid is formed when a proton is transferred to the base. Two substances related in this way are called a conjugate acid-base pair.

The equilibrium expression for the dissociation of an acid in water is

$$K_a = \frac{[H^+][A^-]}{[HA]}$$

where H_3O^+ is simplified to H^+, $[H_2O]$ is not included because it is assumed to be constant, and K_a is called the acid dissociation constant. The strength of an acid is defined by the position of the dissociation equilibrium. A small value of K_a denotes a weak acid, one that does not dissociate to any great extent in aqueous solution. A strong acid is one for which the dissociation equilibrium lies far to the right—the K_a value is very large.

Strong acids such as nitric, hydrochloric, sulfuric, and perchloric acids have weak conjugate bases, which have a low affinity for a proton. The strength of an acid is inversely related to the strength of its conjugate base.

Water is an amphoteric substance because it can behave either as an acid or as a base. The autoionization of water reveals this property, since one water molecule transfers a proton to another water molecule to produce a hydronium ion and a hydroxide ion:

$$2H_2O(l) \rightleftharpoons H_3O^+(aq) + OH^-(aq)$$

This leads to the equilibrium expression

$$K_w = [H^+][OH^-]$$

where K_w is called the ion-product constant. It has been experimentally shown that at 25°C in pure water $[H^+] = [OH^-] = 1.0 \times 10^{-7}\ M$. Thus at 25°C

$$K_w = 1.0 \times 10^{-14}$$

In an acidic solution, $[H^+]$ is greater than $[OH^-]$. In a basic solution, $[OH^-]$ is greater than $[H^+]$. In a neutral solution, $[H^+]$ is equal to $[OH^-]$.

To describe $[H^+]$ in aqueous solutions, we often use the pH scale, where

$$pH = -\log[H^+]$$

Since pH is a log scale based on 10, the pH changes by 1 for every change by a power of 10 in $[H^+]$. Because pH is defined as $-\log[H^+]$, the pH decreases as $[H^+]$ increases.

The percent dissociation of a weak acid is defined as follows:

$$\text{Percent dissociation} = \frac{\text{amount dissociated (mol/L)}}{\text{initial concentration (mol/L)}} \times 100$$

This is another measure of the strength of an acid: the larger the percent dissociation, the stronger the acid. As $[HA]_0$ decreases, $[H^+]$ decreases, but the percent dissociation increases. That is, dilution causes an increase in the percent dissociation.

Strong bases are hydroxide salts, such as NaOH or KOH, that dissociate completely in water. Bases do not have to contain the hydroxide ion; they can be species that remove a proton from water, producing the hydroxide ion. For example, ammonia is a base because it can accept a proton from water

$$NH_3(aq) + H_2O(l) \rightleftharpoons NH_4^+(aq) + OH^-(aq)$$

The equilibrium expression is

$$K_b = \frac{[NH_4^+][OH^-]}{[NH_3]}$$

where K_b always refers to a reaction in which a base reacts with water to produce the conjugate acid and hydroxide ion. Because bases like ammonia must compete with the hydroxide ion for the proton, values of K_b for these bases are typically much less than 1, and these substances are called weak bases.

A polyprotic acid is an acid with more than one acidic proton, which dissociates in a stepwise fashion with a K_a value for each step. Typically, for a weak polyprotic acid

$$K_{a_1} > K_{a_2} > K_{a_3}$$

Sulfuric acid is unique in that it is a strong acid in the first dissociation step and a weak acid in the second step.

Salts can exhibit neutral, acidic, or basic properties when dissolved in water. Salts that contain the cations of strong bases and the anions of strong acids produce neutral aqueous solutions. A basic solution is produced when the dissolved salt has a neutral cation and an anion that is the conjugate base of a weak acid. An acidic solution is produced when the dissolved salt has a cation that is the conjugate acid of a weak base and a neutral anion. Acidic solutions are also produced by salts containing a highly charged metal cation. For example, the hydrated ion $Al(H_2O)_6^{3+}$ is a weak acid.

Most substances that function as acids or bases contain the H—O—X grouping. Molecules where the O—X bond is strong tend to behave as acids, especially when X is highly electronegative, because the X—O bond is polarized and weakened. When X has low electronegativity (if X is a metal, for example), the O—X bond tends to be ionic and hydroxide ion forms when these substances are dissolved in water.

The Lewis model for acids and bases generalizes the concept of acid-base behavior in terms of electron pairs. A Lewis acid is an electron-pair acceptor, and a Lewis base is an electron-pair donor. The value of the Lewis model is that it covers many reactions that do not involve Brönsted-Lowry acids and bases.

Key Terms

Section 14.1
Arrhenius concept
Brönsted-Lowry model
hydronium ion
conjugate base
conjugate acid
conjugate acid-base pair
acid dissociation constant

Section 14.2
strong acid
weak acid
diprotic acid
oxyacids
organic acids
carboxyl group
monoprotic acid
amphoteric substance
autoionization
ion-product constant

Section 14.3
pH scale

Section 14.4
major species

Section 14.5
percent dissociation

Section 14.6
strong bases
slaked lime
lime-soda process
weak base
amine

Section 14.7
polyprotic acid
triprotic acid

Section 14.8
salt

Section 14.10
acidic oxide
basic oxide

Section 14.11
Lewis acid
Lewis base

Exercises

A blue exercise number indicates that the answer to that exercise appears at the back of this book and a solution appears in the Solutions Guide.

Nature of Acids and Bases

1. Classify each of the following as a strong acid, weak acid, strong base, or weak base in aqueous solution.
a. HNO_2
b. H_3PO_4
c. CH_3NH_2
d. NaOH
e. NH_3
f. HF
g. $HC(=O)—OH$
h. $H_2NCH_2CH_2NH_2$
i. H_2SO_4

2. Use Table 14.2 to order the following from the strongest to the weakest acid:

$$HNO_3,\quad H_2O,\quad HOCl,\quad NH_4^+$$

3. Use Table 14.2 to order the following from the strongest to the weakest base:

$$NO_3^-,\quad H_2O,\quad OCl^-,\ NH_3$$

4. You may need Table 14.2 to answer the following questions:
a. Which is the stronger acid, HI or H_2O?
b. Which is the stronger acid, H_2O or $HClO_2$?
c. Which is the stronger acid, HF or HCN?

5. You may need Table 14.2 to answer the following questions:
a. Which is the stronger base, I^- or H_2O?
b. Which is the stronger base, H_2O or ClO_2^-?
c. Which is the stronger base, F^- or CN^-?

6. For each of the following reactions, identify the acid, the base, the conjugate base, and the conjugate acid:
   a. $H_2O + H_2O \rightleftharpoons H_3O^+ + OH^-$
   b. $CH_3O^- + CH_3\overset{\overset{\displaystyle O}{\|}}{C}CH_3 \rightleftharpoons CH_3OH + CH_3\overset{\overset{\displaystyle O^-}{|}}{C}{=}CH_2$
   c. $H_2S + NH_3 \rightleftharpoons HS^- + NH_4^+$
   d. $H_2SO_4 + H_2O \rightleftharpoons H_3O^+ + HSO_4^-$
   e. $H^+ + OH^- \rightleftharpoons H_2O$
   f. $H_2PO_4^- + H_2O \rightleftharpoons H_3PO_4 + OH^-$
   g. $H_2PO_4^- + H_2O \rightleftharpoons HPO_4^{2-} + H_3O^+$
   h. $H_2PO_4^- + H_2PO_4^- \rightleftharpoons H_3PO_4 + HPO_4^{2-}$
   i. $Fe(H_2O)_6^{3+} + H_2O \rightleftharpoons Fe(H_2O)_5(OH)^{2+} + H_3O^+$
   j. $HCN + CO_3^{2-} \rightleftharpoons CN^- + HCO_3^-$
   k. $CO_2 + 2H_2O \rightleftharpoons HCO_3^- + H_3O^+$

7. Write the dissociation reaction and the corresponding equilibrium expression for each of the following acids in water.
   a. H_3PO_4
   b. $H_2PO_4^-$
   c. HPO_4^{2-}
   d. HNO_2
   e. $Ti(H_2O)_6^{4+}$
   f. HCN
   g. acetic acid, CH_3CO_2H ($HC_2H_3O_2$)
   h. phenol, C_6H_5OH
   i. benzoic acid, $C_6H_5CO_2H$
   j. glycine, $H_2NCH_2CO_2H$

8. Write the reaction and the corresponding K_b equilibrium expression for each of the following substances acting as bases in water:
   a. PO_4^{3-}
   b. HPO_4^{2-}
   c. $H_2PO_4^-$
   d. NH_3
   e. CN^-
   f. pyridine, C_5H_5N
   g. glycine, $NH_2CH_2CO_2H$
   h. ethylamine, $CH_3CH_2NH_2$
   i. aniline, $C_6H_5NH_2$
   j. dimethylamine, $(CH_3)_2NH$

9. Define each of the following using the Arrhenius model.
   a. strong acid
   b. strong base
   c. weak acid
   d. weak base

10. Hydride ion (H^-) and methoxide ion (CH_3O^-) have much greater affinities for H^+ than the OH^- ion does. Write equations for the reactions that occur when NaH and $NaOCH_3$ are dissolved in water.

11. Why is H_3O^+ the strongest acid and OH^- the strongest base that can exist in aqueous solutions?

12. Write balanced equations that describe the following reactions.
   a. The dissociation of perchloric acid in water.
   b. The dissociation of propanoic acid, $CH_3CH_2CO_2H$, in water.
   c. The reaction between acetic acid and sodium hydroxide.
   d. The dissociation of ammonium ion in water.

Autoionization of Water and the pH Scale

13. Give the conditions for a neutral solution at 25°C, in terms of $[H^+]$, pH, and the relationship between $[H^+]$ and $[OH^-]$.

14. Values of K_w as a function of temperature are as follows:

Temp (°C)	K_w
0	1.14×10^{-15}
25	1.00×10^{-14}
35	2.09×10^{-14}
40	2.92×10^{-14}
50	5.47×10^{-14}

   a. Is the autoionization of water exothermic or endothermic?
   b. What is the pH of pure water at 50°C?
   c. Restate your answers to Exercise 13 for water at 50°C. Which of the three criteria for neutrality is most general?
   d. From a plot of $\ln(K_w)$ versus $1/T$ (using the Kelvin scale) estimate K_w at 37°C, normal physiological temperature.
   e. What is the pH of a neutral solution at 37°C?

15. Calculate the pH of each solution:
   a. $[H^+] = 1.4 \times 10^{-3}$ *M*
   b. $[H^+] = 2.5 \times 10^{-10}$ *M*
   c. $[H^+] = 6.1$ *M*
   d. $[OH^-] = 3.5 \times 10^{-2}$ *M*
   e. $[OH^-] = 8 \times 10^{-11}$ *M*
   f. $[OH^-] = 5.0$ *M*
   g. pOH = 10.5
   h. pOH = 2.3

16. Calculate $[H^+]$ and $[OH^-]$ for each solution:
   a. pH = 7.41 (the normal pH of blood)
   b. pH = 15.3
   c. pH = −1.0
   d. pH = 3.2
   e. pOH = 5.0
   f. pOH = 9.6

17. How many significant figures are there in the numbers: 10.78, 6.78, 0.78? If these were pH values, to how many significant figures can you express the $[H^+]$? Explain any discrepancies between your answers to the two questions.

Solutions of Acids

18. What are the major species present in 0.250 *M* solutions of each of the following acids?
   a. HCl
   b. HBr
   c. $HClO_4$
   d. HNO_3
   e. HNO_2
   f. CH_3CO_2H ($HC_2H_3O_2$)
   g. NH_4Cl
   h. HCN

19. Calculate the pH of each of the following solutions of a strong acid in water.
   a. 0.1 *M* HCl
   b. 0.1 *M* HNO_3
   c. 0.1 *M* $HClO_4$
   d. 3.0×10^{-5} *M* HCl
   e. 2.0×10^{-2} *M* HNO_3
   f. 4.0 *M* HNO_3

20. Calculate the pH of a 1.0×10^{-12} M solution of HCl in water. Before checking your answer, decide whether or not it makes sense.

21. A solution is prepared by adding 50.0 mL of concentrated hydrochloric acid and 20.0 mL of concentrated nitric acid to 300 mL of water. Water is added until the final volume is 500.0 mL. Calculate $[H^+]$, $[OH^-]$, and pH for this solution. (See Exercise 10 in Chapter 11 for the composition of the concentrated reagents.)

22. Using the K_a values in Table 14.2, calculate the pH values of the solutions in Exercise 18.

23. Using the K_a values given in Table 14.2, calculate the concentrations of all species present and the pH for each of the following:
    a. 0.20 M $HC_2H_3O_2$
    b. 1.5 M HNO_2
    c. 0.020 M HF
    d. 0.83 M lactic acid

$$\left(CH_3\overset{\overset{\displaystyle OH}{|}}{C}HCO_2H,\ K_a = 1.38 \times 10^{-4} \right)$$

24. Formic acid (HCO_2H) is secreted by ants. Calculate $[H^+]$ and the pH of a 0.025 M solution of formic acid ($K_a = 1.8 \times 10^{-4}$).

25. Calculate the pH of a 0.50 M solution of iodic acid (HIO_3, $K_a = 0.17$).

26. Boric acid (H_3BO_3) is commonly used in eyewash solutions in chemistry laboratories to neutralize bases splashed in the eye. It acts as a monoprotic acid, but the dissociation reaction is slightly different from that of other acids:

$$B(OH)_3 + H_2O \rightleftharpoons B(OH)_4^- + H^+ \qquad K_a = 5.8 \times 10^{-10}$$

Calculate the pH of a 0.50 M solution of boric acid.

27. A solution is prepared by dissolving 0.56 g of benzoic acid ($C_6H_5CO_2H$, $K_a = 6.4 \times 10^{-5}$) in enough water to make 1.0 L of solution. Calculate $[C_6H_5CO_2H]$, $[C_6H_5CO_2^-]$, $[H^+]$, $[OH^-]$, and the pH in this solution.

28. At 25°C a saturated solution of benzoic acid (see Exercise 27) has a pH of 2.8. Calculate the water-solubility of benzoic acid in moles per liter and grams per 100 milliliters.

29. A solution with a volume of 250.0 mL is prepared by diluting 20.0 mL of glacial acetic acid with water. Calculate $[H^+]$ and the pH of this solution. Assume glacial acetic acid is pure liquid acetic acid, with a density of 1.05 g/cm^3.

30. Calculate the pH of each of the following:
    a. a solution containing 0.10 M HCl and 0.10 M HOCl
    b. a solution containing 0.050 M HNO_3 and 0.50 M $HC_2H_3O_2$

31. A solution is prepared by adding 50.0 mL of 0.050 M HCl to 150.0 mL of 0.10 M HNO_3. Calculate the concentrations of all species in this solution.

32. Calculate the pH of a 0.0010 M solution of H_2SO_4.

33. What types of measurements, other than pH measurements, can be made to determine the extent of dissociation of an acid in water?

34. In a 0.100 M solution of HF, the percent dissociation is 8.1%. Calculate K_a.

35. Calculate the percent dissociation of the acid in each of the following solutions:
    a. 0.50 M acetic acid
    b. 0.050 M acetic acid
    c. 0.0050 M acetic acid

36. Calculate the percent dissociation of a 0.22 M solution of chlorous acid ($HClO_2$, $K_a = 1.2 \times 10^{-2}$).

37. Using the K_a values in Table 14.2, calculate the percent dissociation in a 0.100 M solution of each of the following acids:
    a. hypochlorous acid (HOCl)
    b. hydrocyanic acid (HCN)
    c. hydrochloric acid (HCl)

38. The pH of a 0.063 M solution of hypobromous acid (HOBr but usually written HBrO) is 4.95. Calculate K_a.

39. Trichloroacetic acid (CCl_3CO_2H) is a corrosive acid that is used to precipitate proteins. The pH of a 0.050 M solution of trichloroacetic acid is 1.4. Calculate K_a.

40. Using the K_a values in Table 14.4, and only the first dissociation step, calculate the pH of 0.10 M solutions of each of the following polyprotic acids:
    a. H_3PO_4
    b. H_3AsO_4
    c. H_2CO_3

41. Calculate $[H^+]$, $[OH^-]$, $[H_3PO_4]$, $[H_2PO_4^-]$, $[HPO_4^{2-}]$, and $[PO_4^{3-}]$ for a 0.100 M solution of H_3PO_4.

42. Calculate $[CO_3^{2-}]$ in a 0.010 M solution of CO_2 in water (H_2CO_3). If all the CO_3^{2-} in this solution comes from the reaction

$$HCO_3^- \rightleftharpoons H^+ + CO_3^{2-}$$

what percent of the H^+ ions in the solution are a result of the dissociation of HCO_3^-? When acid is added to a solution of sodium hydrogen carbonate ($NaHCO_3$), vigorous bubbling occurs. How is this reaction related to the existence of carbonic acid (H_2CO_3) molecules in aqueous solution?

Solutions of Bases

43. Use Table 14.3 to help to order the following bases from strongest to weakest:

$$NO_3^-, \quad H_2O, \quad NH_3, \quad CH_3NH_2$$

44. Use Table 14.3 to help to order the following acids from strongest to weakest:

$$HNO_3, \quad H_2O, \quad NH_4^+, \quad CH_3NH_3^+$$

45. Use Table 14.3 to help to answer the following questions:
    a. Which is the stronger base: NO_3^- or NH_3?
    b. Which is the stronger base: H_2O or NH_3?
    c. Which is the stronger base: OH^- or NH_3?
    d. Which is the stronger base: NH_3 or CH_3NH_2?

46. Use Table 14.3 to help to answer the following questions:
    a. Which is the stronger acid: HNO_3 or NH_4^+?
    b. Which is the stronger acid: H_2O or NH_4^+?
    c. Which is the stronger acid: NH_4^+ or $CH_3NH_3^+$?

47. Thallium(I) hydroxide is a strong base used in the synthesis of some organic compounds. Calculate the pH of a solution containing 2.48 g of TlOH per liter.

48. Calculate $[OH^-]$, pOH, and pH for each of the following:
    a. 0.25 *M* NaOH
    b. 0.00040 *M* $Ba(OH)_2$
    c. a solution containing 25 g of KOH per liter
    d. a solution containing 150.0 g of NaOH per liter

49. The presence of what element most commonly results in basic properties for an organic compound?

50. What are the major species present in 0.150 *M* solutions of each of the following bases?
    a. KOH
    b. CsOH
    c. NH_3
    d. pyridine
    e. methylamine

51. Using data from Table 14.3, calculate the pH of each of the solutions in Exercise 50.

52. Calculate $[OH^-]$, $[H^+]$, and pH of 0.200 *M* solutions of each of the following amines (the missing K_b values are found in Table 14.3):
    a. ethylamine
    b. diethylamine, $(C_2H_5)_2NH$, $K_b = 1.3 \times 10^{-3}$
    c. triethylamine, $(C_2H_5)_3N$, $K_b = 4.0 \times 10^{-4}$
    d. aniline
    e. pyridine
    f. hydroxylamine, $HONH_2$, $K_b = 1.1 \times 10^{-8}$

53. Draw the structures of the conjugate acids of ephedrine and mescaline. See page 637 for structures of the amines.

54. Codeine is a derivative of morphine and is used as an analgesic, narcotic, or antitussive. It was once commonly used in cough syrups, but is now available only by prescription because of its addictive properties. The formula of codeine is $C_{18}H_{21}NO_3$ and the pK_b is 6.05. Calculate the pH of a 10.0-mL solution containing 5.0 mg of codeine.

55. What is the percent ionization in each of the following solutions?
    a. 0.10 *M* NH_3
    b. 0.010 *M* NH_3
    c. 0.10 *M* CH_3NH_2

56. For the reaction of hydrazine (N_2H_4) in water,

$$H_2NNH_2 + H_2O \rightleftharpoons H_2NNH_3^+ + OH^-$$

K_b is 3.0×10^{-6}. Calculate the pH of a 2.0 *M* solution of hydrazine in water.

57. Quinine ($C_{20}H_{24}N_2O_2$) is the most important alkaloid derived from cinchona bark. It is used as an antimalarial drug.

H₂C=CH, H, H, N, HO—CH, H₃CO, N

Quinine

For quinine, $pK_{b_1} = 5.1$ and $pK_{b_2} = 9.7$. (Recall that $pK_b = -\log K_b$.) A gram of quinine will dissolve in 1900.0 mL of water. Calculate the pH of a saturated aqueous solution of quinine. Consider only the reaction $Q + H_2O \rightleftharpoons QH^+ + OH^-$ described by pK_{b_1}.

58. The pH of a 0.016 *M* solution of *p*-toluidine ($CH_3C_6H_4NH_2$) in water is 8.60. Calculate K_b.

59. The pH of a 1.00×10^{-3} *M* solution of pyrrolidine is 10.82. Calculate K_b.

CH_2—CH_2, CH_2, CH_2, NH

Pyrrolidine

Acid-Base Properties of Salts

60. Derive an expression for the relationship between pK_a and pK_b for a conjugate acid-base pair. (Recall that $pK = -\log K$.)

61. Are solutions of the following salts acidic, basic, or neutral? For those that are not neutral, write balanced chemical equations for the reactions causing the solution to be acidic or basic. The relevant K_a and K_b values are found in Tables 14.2 and 14.4.
a. KCl
b. $NaNO_3$
c. $NaNO_2$
d. NH_4NO_3
e. NH_4NO_2
f. $NaHCO_3$
g. $NH_4C_2H_3O_2$
h. NaF

62. Calculate the pH of each of the following solutions:
a. 0.10 *M* CH_3NH_3Cl
b. 0.050 *M* NaCN
c. 0.20 *M* Na_2CO_3 (consider only the reaction $CO_3^{2-} + H_2O \rightleftharpoons HCO_3^- + OH^-$)
d. 0.12 *M* $NaNO_2$
e. 0.45 *M* NaOCl

63. Arrange the following 0.10 *M* solutions in order from most acidic to most basic:

KOH, KBr, KCN, NH_4Br, NH_4CN, HCN

64. Arrange the following 0.10 *M* solutions in order from most acidic to most basic:

H_2O, KNO_2, HNO_3, HNO_2, NH_4NO_3, NH_4NO_2

65. Is an aqueous solution of $NaHSO_4$ acidic, basic, or neutral? What reaction with water occurs? If solid Na_2CO_3 is added to a solution of $NaHSO_4$, what reaction can occur between the CO_3^{2-} and HSO_4^- ions?

66. Sodium azide (NaN_3) is sometimes added to water to kill bacteria. Calculate the concentration of all species in a 0.010 *M* solution of NaN_3. The K_a value for hydrazoic acid (HN_3) is 1.9×10^{-5}.

67. Given that the K_a value for acetic acid is 1.8×10^{-5} and the K_a value for hypochlorous acid is 3×10^{-8}, which is the stronger base, OCl^- or $C_2H_3O_2^-$?

Relationships Between Structure and Strengths of Acids and Bases

68. For oxyacids, how does acid strength depend on
a. strength of the bond to the acidic hydrogen atom?
b. electronegativity of the element bonded to the acidic hydrogen?
c. the number of oxygen atoms?

69. How does the strength of a conjugate base depend on the factors listed in Exercise 68?

70. Place the species in each of the following groups in order of increasing acid strength. Give your reason for the particular order you chose in each group.
a. HBrO, $HBrO_2$, $HBrO_3$
b. H_3AsO_4, $H_2AsO_4^-$, $HAsO_4^{2-}$

71. Order the following from the strongest to weakest acid:
a. $HClO_2$, $HBrO_2$, HIO_2
b. H_3PO_4, H_3AsO_4

72. Place the species in each of the following groups in order of increasing base strength. Give your reasoning in each case.
a. BrO^-, BrO_2^-, BrO_3^-
b. $H_2PO_4^-$, HPO_4^{2-}, PO_4^{3-}

73. Will the following oxides give acidic, basic, or neutral solutions when dissolved in water? Write reactions to justify your answers.
a. CaO
b. Na_2O
c. SO_2
d. Cl_2O
e. P_4O_{10}
f. NO_2

Lewis Acids and Bases

74. Define each of the following:
a. Arrhenius acid
b. Brönsted-Lowry acid
c. Lewis acid

75. Which of the definitions in Exercise 74 is most general? Write reactions to justify your answer.

76. Identify the Lewis acid and the Lewis base in each of the following reactions:
a. $B(OH)_3 + H_2O \rightleftharpoons B(OH)_4^- + H^+$
b. $Ag^+ + 2NH_3 \rightleftharpoons Ag(NH_3)_2^+$
c. $BF_3 + NH_3 \rightleftharpoons F_3BNH_3$
d. $I_2 + I^- \rightleftharpoons I_3^-$
e. $Zn(OH)_2 + 2OH^- \rightleftharpoons Zn(OH)_4^{2-}$

77. Aluminum hydroxide is an amphoteric substance. It can act as either a Brönsted-Lowry base or a Lewis acid. Write a reaction showing $Al(OH)_3$ acting as a base toward H^+ and as an acid toward OH^-.

78. Zinc hydroxide is an amphoteric substance. Write equations that describe $Zn(OH)_2$ acting as a Brönsted-Lowry base toward H^+ and as a Lewis acid toward OH^-.

79. In terms of orbitals and electron arrangements, what must be present for a molecule or ion to act as a Lewis acid? What must be present for a molecule or ion to act as a Lewis base?

80. Would you expect Fe^{3+} or Fe^{2+} to be the stronger Lewis acid? Explain your answer.

81. $TiCl_4$ is a liquid at room temperature. It exists as discrete $TiCl_4$ molecules rather than as an ionic solid composed of Ti^{4+} and Cl^- ions. How can the Lewis acid-base model be used to rationalize these facts?

Additional Exercises

82. Liquid ammonia is sometimes used as a solvent for chemical reactions. It undergoes the autoionization reaction:

$$2NH_3 \rightleftharpoons NH_4^+ + NH_2^-$$

a. What species correspond to H^+ and OH^- in liquid ammonia?
b. What is the condition for a neutral solution in liquid ammonia?
c. Sodium metal reacts with liquid ammonia in a manner analogous to the reaction of sodium metal with water. Write a balanced chemical equation describing this reaction.
d. What advantages might there be to using liquid ammonia as a solvent?

83. Alka-Seltzer employs the reaction between citric acid and sodium bicarbonate to generate its fizz ($CO_2(g)$). The structure of citric acid and values of the acid dissociation constants are

$$\begin{array}{ll} \quad CH_2\text{—}CO_2H & K_{a_1} = 8.4 \times 10^{-4} \\ HO\text{—}C\text{—}CO_2H & K_{a_2} = 1.8 \times 10^{-5} \\ \quad CH_2\text{—}CO_2H & K_{a_3} = 4.0 \times 10^{-6} \end{array}$$

Calculate the value of the equilibrium constant for each of the following reactions (we abbreviate citric acid as H_3Cit), using the K_a values for citric acid and for carbonic acid:

$$H_3Cit + HCO_3^- \rightleftharpoons H_2Cit^- + H_2O + CO_2$$
$$H_3Cit + 3HCO_3^- \rightleftharpoons Cit^{3-} + 3H_2O + 3CO_2$$

Hint: When reactions are added, the corresponding equilibrium constants are multiplied.

84. The osmotic pressure of a 1.00×10^{-2} *M* solution of cyanic acid (HOCN) is 217.2 torr at 25°C. Calculate K_a for HOCN from this result.

85. a. The principal equilibrium in a solution of $NaHCO_3$ is

$$HCO_3^- + HCO_3^- \rightleftharpoons H_2CO_3 + CO_3^{2-}$$

Calculate the value of the equilibrium constant for this reaction.
b. At equilibrium, what is the relationship between $[H_2CO_3]$ and $[CO_3^{2-}]$?
c. Using the equilibrium

$$H_2CO_3 \rightleftharpoons 2H^+ + CO_3^{2-}$$

derive an expression for the pH of the solution in terms of K_{a_1} and K_{a_2} using the result from part b.
d. What is the pH of a solution of $NaHCO_3$?

86. Calculate the value for the equilibrium constant for each of the following reactions:
a. $NH_3 + H_3O^+ \rightleftharpoons NH_4^+ + H_2O$
b. $NO_2^- + H_3O^+ \rightleftharpoons HNO_2 + H_2O$
c. $NH_4^+ + CH_3CO_2^- \rightleftharpoons NH_3 + CH_3CO_2H$
d. $H_3O^+ + OH^- \rightleftharpoons 2H_2O$
e. $NH_4^+ + OH^- \rightleftharpoons NH_3 + H_2O$
f. $HNO_2 + OH^- \rightleftharpoons H_2O + NO_2^-$

87. Hemoglobin (abbreviated Hb) is a protein that is responsible for the transport of oxygen in the blood of mammals. Each hemoglobin molecule contains four iron atoms that are the binding sites for O_2 molecules. The oxygen binding is pH dependent. The relevant equilibrium reaction is as follows.

$$HbH_4^{4+} + 4O_2 \rightleftharpoons Hb(O_2)_4 + 4H^+$$

Use Le Châtelier's principle to answer the following.
a. What form of hemoglobin, HbH_4^{4+} or $Hb(O_2)_4$, is favored in the lungs? What form is favored in the cells?
b. When a person hyperventilates, the concentration of CO_2 in the blood is decreased. How does this affect the oxygen-binding equilibrium? How does breathing into a paper bag help to counteract this effect?
c. When a person has suffered a cardiac arrest, injection of a sodium bicarbonate solution is given. Why is this necessary?

88. Construct a thermochemical cycle for the following reaction in terms of bond energies, ionization energies, and electron affinities given in Chapters 7 and 8:

$$HX(g) \rightleftharpoons H^+(g) + X^-(g)$$

a. Calculate ΔH for the above reaction when X = F, Cl, Br, and I.
b. How are the results of this thermochemical cycle consistent with statements in Section 14.9?

89. Use the results of Exercise 88 to explain two things that will increase the strength of an acid.

90. Using your conclusions from Exercise 89, place the species in each of the following groups in order of increasing acid strength:
a. H_2O, H_2S, H_2Se (bond energies: H—O, 463 kJ/mol; H—S, 363 kJ/mol; H—Se, 276 kJ/mol)
b. CH_3CO_2H, FCH_2CO_2H, F_2CHCO_2H, F_3CCO_2H
c. NH_4^+, $CH_3NH_3^+$, $HONH_3^+$

Give reasons for the orders you chose.

91. Using your conclusions from Exercise 90, place the species in each of the following groups in order of increasing base strength:
a. OH^-, SH^-, SeH^-

b. NH_3, PH_3 (bond energies: N—H, 386 kJ/mol; P—H, 322 kJ/mol)
c. NH_3, $HONH_2$

92. Construct a thermochemical cycle for the reaction

$$HX(aq) \rightleftharpoons H^+(aq) + X^-(aq)$$

by adding hydration energies to the factors considered in Exercise 88. In what way can hydration energies affect the strength of an acid?

93. Calculate the pH of a 1.0×10^{-7} *M* solution of NaOH in water. Consider the amount of hydroxide ion from the autoionization of water and the effect of the added hydroxide ion on the position of the autoionization equilibrium.

94. Making use of the assumptions we ordinarily make in calculating the pH of an aqueous solution of a weak acid, calculate the pH of a 1.0×10^{-6} *M* solution of hypobromous acid (HBrO, $K_a = 2 \times 10^{-9}$). What is wrong with your answer? Why is it wrong? Without trying to solve the problem, tell what has to be included to solve the problem correctly.

Applications of Aqueous Equilibria

Much important chemistry, including almost all of the chemistry of the natural world, occurs in aqueous solution. We have already introduced one very significant class of aqueous equilibria, acid-base reactions. In this chapter we consider more applications of acid-base chemistry and introduce two additional types of aqueous equilibria, those involving the solubility of salts and the formation of complex ions.

The interplay of acid-base, solubility, and complex ion equilibria is often important in natural processes, such as the weathering of minerals, the uptake of nutrients by plants, and tooth decay. For example, limestone ($CaCO_3$) will dissolve in water made acidic by dissolved carbon dioxide:

$$CO_2(aq) + H_2O(l) \rightleftharpoons H^+(aq) + HCO_3^-(aq)$$
$$H^+(aq) + CaCO_3(s) \rightleftharpoons Ca^{2+}(aq) + HCO_3^-(aq)$$

This process and its reverse account for the formation of limestone caves and the stalactites and stalagmites found therein. The acidic water (containing carbon dioxide) dissolves the underground limestone deposits, thereby forming a cavern. As the water drips from the ceiling of the cave, the carbon dioxide is lost to the air and solid calcium carbonate forms by the reverse of the above process to produce stalactites on the ceiling and stalagmites where the drops hit the cave floor.

Before we consider the other types of aqueous equilibria, we will deal with acid-base equilibria in more detail.

< Font's Point, Borrego Badlands and Vallecito Mountains in the Anza-Borrego Desert State Park, California.

CONTENTS

15.1 Solutions of Acids or Bases Containing a Common Ion

Purpose

- To study the effect of a common ion on acid dissociation equilibria.

In Chapter 14 we were concerned with calculating the equilibrium concentrations of species (particularly H^+ ions) in solutions containing an acid or a base. In this section we discuss solutions that contain not only the weak acid HA, but also its salt NaA. Although this appears to be a new type of problem, we will see that this case can be handled rather easily using the procedures developed in Chapter 14.

Suppose we have a solution containing the weak acid hydrofluoric acid (HF, $K_a = 7.2 \times 10^{-4}$) and its salt sodium fluoride (NaF). Recall that when a salt dissolves in water, it breaks up completely into its ions—it is a strong electrolyte:

$$NaF(s) \xrightarrow{H_2O(l)} Na^+(aq) + F^-(aq)$$

Since hydrofluoric acid is a weak acid and only slightly dissociated, the major species in the solution are HF, Na^+, F^-, and H_2O. The **common ion** in this solution is F^-, since it is produced by both hydrofluoric acid and sodium fluoride. What effect does the presence of the dissolved sodium fluoride have on the dissociation equilibrium of hydrofluoric acid?

To answer this question we compare the extent of dissociation of hydrofluoric acid in two different solutions, the first containing 1.0 *M* HF and the second containing 1.0 *M* HF and 1.0 *M* NaF. By Le Châtelier's principle, we would expect the dissociation equilibrium for HF

$$HF(aq) \rightleftharpoons H^+(aq) + F^-(aq)$$

in the second solution to be *driven to the left by the presence of F^- ions from the NaF*. The extent of dissociation of HF will be *less* in the presence of dissolved NaF:

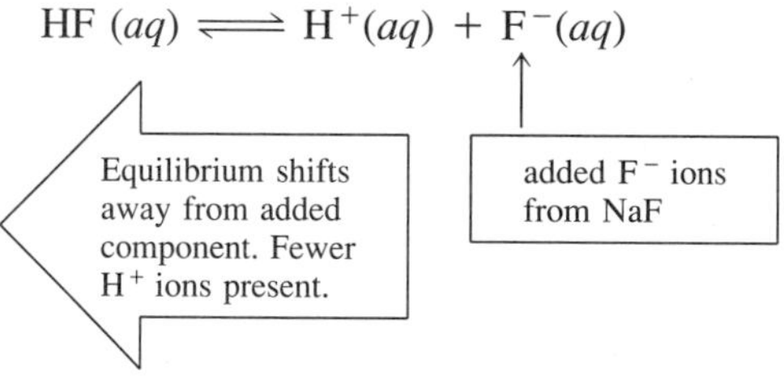

The shift in equilibrium position that occurs because of the addition of an ion already involved in the equilibrium reaction is called the **common ion effect.** This effect makes a solution of NaF and HF less acidic than a solution of HF alone.

The common ion effect is an application of Le Châtelier's principle.

The common ion effect is quite general. For example, when solid NH_4Cl is added to a 1.0 *M* NH_3 solution, the ammonium ions produced

$$NH_4Cl(s) \xrightarrow{H_2O} NH_4^+(aq) + Cl^-(aq)$$

cause the position of the ammonia-water equilibrium to shift to the left. This reduces the equilibrium concentration of OH^- ions.

$$NH_3(aq) + H_2O(l) \rightleftharpoons NH_4^+(aq) + OH^-(aq)$$

The common ion effect is also important in solutions of polyprotic acids. The production of protons by the first dissociation step greatly inhibits the succeeding dissociation steps, which of course also produce protons, the common ion in this case. We will see later in this chapter that the common ion effect is also important in dealing with the solubility of salts.

Equilibrium Calculations

The procedures for finding the pH of a solution containing a weak acid or base plus a common ion are very similar to the procedures, which we have covered in Chapter 14, for solutions containing the acids or bases alone. For example, in the case of a weak acid the only important difference is that the initial concentration of the anion A^- is not zero in a solution that also contains the salt NaA. Sample Exercise 15.1 illustrates a typical example using the same general approach we developed in Chapter 14.

Sample Exercise 15.1

In Section 14.4 we found that the equilibrium concentration of H^+ in a 1.0 M HF solution is 2.7×10^{-2} M and the percent dissociation of HF is 2.7%. Calculate $[H^+]$ and the percent dissociation of HF in a solution containing 1.0 M HF ($K_a = 7.2 \times 10^{-4}$) and 1.0 M NaF.

Solution

As the aqueous solutions we consider become more complex, it is more important than ever to be systematic and to *focus on the chemistry* occurring in the solution before thinking about mathematical procedures. The way to do this is *always* to write the major species first and consider the chemical properties of each one.

In a solution containing 1.0 M HF and 1.0 M NaF, the major species are:

$$HF, F^-, Na^+, \text{ and } H_2O$$

Since Na^+ ions have neither acidic nor basic properties, and since water is such a weak acid or base, the important species are HF and F^-; they participate in the acid dissociation equilibrium that controls $[H^+]$ in this solution. That is, the position of the equilibrium

$$HF(aq) \rightleftharpoons H^+(aq) + F^-(aq)$$

will determine $[H^+]$ in the solution. The equilibrium expression is

$$K_a = \frac{[H^+][F^-]}{[HF]} = 7.2 \times 10^{-4}$$

The important concentrations are shown in the table on page 672.

Sample Exercise 15.1, continued

Initial concentration (mol/L)		Equilibrium concentration (mol/L)
$[HF]_0 = 1.0$ (from dissolved HF)		$[HA] = 1.0 - x$
$[F^-]_0 = 1.0$ (from dissolved NaF)	$\xrightarrow{x \text{ mol/L HF dissociates}}$	$[F^-] = 1.0 + x$
$[H^+]_0 = 0$ (neglect contribution from H_2O)		$[H^+] = x$

Note that $[F^-]_0 = 1.0\ M$ due to the dissolved sodium fluoride, and that this concentration increases when the acid dissociates, producing F^- as well as H^+. Then

$$K_a = 7.2 \times 10^{-4} = \frac{[H^+][F^-]}{[HF]} = \frac{(x)(1.0 + x)}{1.0 - x} \approx \frac{(x)(1.0)}{1.0}$$

(since x is expected to be small).

Now solving for x gives

$$x = \frac{1.0}{1.0}(7.2 \times 10^{-4}) = 7.2 \times 10^{-4}$$

Noting that x is small compared to 1.0, the approximate solution is acceptable. Thus

$$[H^+] = x = 7.2 \times 10^{-4}\ M \qquad \text{(The pH is 3.14.)}$$

The percent dissociation of HF in this solution is

$$\frac{[H^+]}{[HF]_0} \times 100 = \frac{7.2 \times 10^{-4}\ M}{1.0\ M} \times 100 = 0.072\%$$

Compare these values for $[H^+]$ and percent dissociation of HF to those for a 1.0 M HF solution, where $[H^+] = 2.7 \times 10^{-2}\ M$ and the percent dissociation is 2.7%. The large difference shows clearly that the presence of the F^- ions from the dissolved NaF greatly inhibits the dissociation of HF. The position of the acid dissociation equilibrium has been shifted to the left by the presence of F^- ions from NaF.

15.2 Buffered Solutions

Purpose

- To explain the characteristics of buffered solutions.
- To show how to calculate a buffer pH given the concentrations of the buffering materials.

The most important application of acid-base solutions containing a common ion is for buffering. A **buffered solution** is one that *resists a change in its pH* when either hydroxide ions or protons are added. The most important practical example of a buffered solution is our blood, which can absorb the acids and bases produced in biological reactions without changing its pH. A constant pH for blood is vital because cells can survive only in a very narrow pH range.

The most important buffering system in the blood involves HCO_3^- and H_2CO_3.

A buffered solution may contain a *weak* acid and its salt (for example, HF and NaF) or a *weak* base and its salt (for example, NH_3 and NH_4Cl). By choosing the appropriate components, a solution can be buffered at virtually any pH.

In treating buffered solutions in this chapter, we will start by considering the equilibrium calculations. We will then use these results to show how buffering works. That is, we will answer the question: How does a buffered solution resist changes in pH when an acid or base is added?

As you do the calculations associated with buffered solutions, it is important to keep in mind that these are just solutions containing weak acids or bases, and the procedures required are the same ones we have already developed. Be sure to use the systematic approach introduced in Chapter 14.

The systematic approach developed in Chapter 14 for weak acids and bases applies to buffered solutions.

Sample Exercise 15.2

A buffered solution contains 0.50 *M* acetic acid ($HC_2H_3O_2$, $K_a = 1.8 \times 10^{-5}$) and 0.50 *M* sodium acetate ($NaC_2H_3O_2$).

Notice as you do this problem that it is exactly like examples you have seen in Chapter 14.

a. Calculate the pH of this solution.

Solution

a. The major species in the solution are:

$HC_2H_3O_2$,	Na^+,	$C_2H_3O_2^-$	and	H_2O
↑	↑	↑		↑
Weak acid	Neither acid nor base	Base (conjugate base of $HC_2H_3O_2$)		Very weak acid or base

Examination of the solution components leads to the conclusion that the acetic acid dissociation equilibrium, which involves both $HC_2H_3O_2$ and $C_2H_3O_2^-$, will control the pH of the solution:

$$HC_2H_3O_2(aq) \rightleftharpoons H^+(aq) + C_2H_3O_2^-(aq)$$

$$K_a = 1.8 \times 10^{-5} = \frac{[H^+][C_2H_3O_2^-]}{[HC_2H_3O_2]}$$

The concentrations are as follows:

Initial concentration (mol/L)		Equilibrium concentration (mol/L)
$[HC_2H_3O_2]_0 = 0.50$ $[C_2H_3O_2^-]_0 = 0.50$ $[H^+]_0 \approx 0$	x mol/L of $HC_2H_3O_2$ dissociates to reach equilibrium →	$[HC_2H_3O_2] = 0.50 - x$ $[C_2H_3O_2^-] = 0.50 + x$ $[H^+] \approx x$

Then

$$K_a = 1.8 \times 10^{-5} = \frac{[H^+][C_2H_3O_2^-]}{[HC_2H_3O_2]} = \frac{(x)(0.50 + x)}{0.50 - x} \approx \frac{(x)(0.50)}{0.50}$$

and

$$x = 1.8 \times 10^{-5}$$

Sample Exercise 15.2, continued

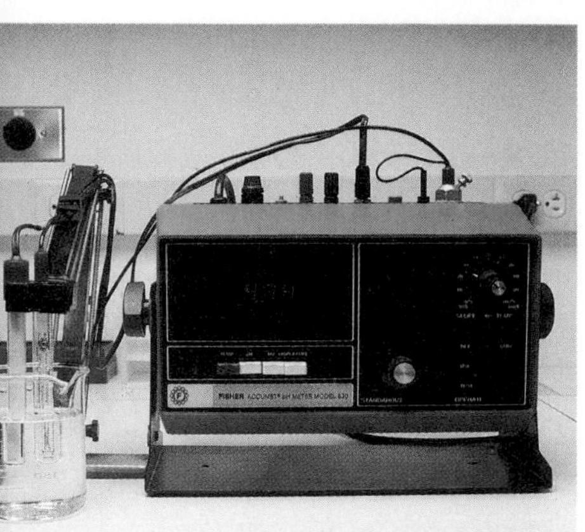

A digital pH meter shows the pH of the buffered solution to be 4.74.

Since the approximation is valid by the 5% rule,

$$[H^+] = x = 1.8 \times 10^{-5}\ M \quad \text{and} \quad \text{pH} = 4.74$$

b. Calculate the change in pH that occurs when 0.010 mol of solid NaOH is added to 1.0 L of the buffered solution. Compare this pH change to that which occurs when 0.010 mol of solid NaOH is added to 1.0 L of water.

Solution

b. Since the added solid NaOH will completely dissociate, the major species in solution *before any reaction occurs* are $HC_2H_3O_2$, Na^+, $C_2H_3O_2^-$, OH^-, and H_2O. Note that the solution contains a relatively large amount of the very strong base hydroxide ion, which has a great affinity for protons. The best source of protons is the acetic acid, and the reaction that will occur is

$$OH^- + HC_2H_3O_2 \rightarrow H_2O + C_2H_3O_2^-$$

Although acetic acid is a weak acid, hydroxide ion is such a strong base that the reaction above will *proceed essentially to completion* (until the OH^- ions are consumed).

The best approach to this problem involves two distinct steps: (1) assume the reaction goes to completion and carry out the stoichiometric calculations; then (2) carry out the equilibrium calculations.

(1) *The stoichiometry problem.* The reaction is

	$HC_2H_3O_2$	+	OH^-	→	$C_2H_3O_2^-$	+	H_2O
Before reaction:	$1.0\ L \times 0.50\ M$ = 0.50 mol		0.010 mol		$1.0\ L \times 0.50\ M$ = 0.50 mol		
After reaction:	0.50 − 0.01 = 0.49 mol		.010 − .010 = 0 mol		0.50 + 0.01 = 0.51 mol		

Note that 0.01 mol of $HC_2H_3O_2$ has been converted to 0.01 mol of $C_2H_3O_2^-$ by the added OH^-.

(2) *The equilibrium problem.* After the reaction between OH^- and $HC_2H_3O_2$ is complete, the major species in solution are:

$$HC_2H_3O_2,\quad Na^+,\quad C_2H_3O_2^-,\quad \text{and}\quad H_2O$$

The dominant equilibrium involves dissociation of acetic acid.

This problem is then very similar to that in part a. The only difference is that the addition of 0.01 mol of OH^- has consumed some $HC_2H_3O_2$ and produced some $C_2H_3O_2^-$ to give the following concentrations:

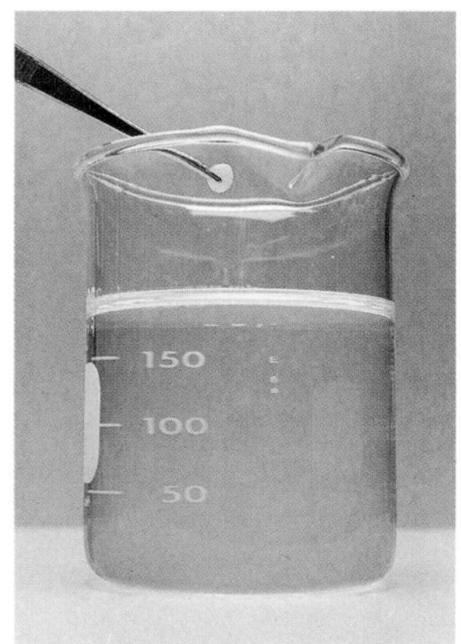

A pellet of sodium hydroxide being added to water.

Initial concentration (mol/L)		Equilibrium concentration (mol/L)
$[HC_2H_3O_2]_0 = 0.49$ $[C_2H_3O_2^-]_0 = 0.51$ $[H^+]_0 \approx 0$	$\xrightarrow[\text{dissociates}]{x \text{ mol/L } HC_2H_3O_2}$	$[HC_2H_3O_2] = 0.49 - x$ $[C_2H_3O_2^-] = 0.51 + x$ $[H^+] \approx x$

Sample Exercise 15.2, continued

Note that the initial concentrations are defined after the reaction with OH^- is complete but before the system adjusts to equilibrium. Following the usual procedures gives

$$K_a = 1.8 \times 10^{-5} = \frac{[H^+][C_2H_3O_2^-]}{[HC_2H_3O_2]} = \frac{(x)(0.51 + x)}{0.49 - x} \approx \frac{(x)(0.51)}{0.49}$$

$$x \approx 1.7 \times 10^{-5}$$

The approximations are valid by the 5% rule, so

$$[H^+] = x = 1.7 \times 10^{-5} \quad \text{and} \quad \text{pH} = 4.76$$

The change in pH produced by addition of 0.01 mol of OH^- to this buffered solution is then

$$\underset{\substack{\uparrow \\ \text{New solution}}}{4.76} - \underset{\substack{\uparrow \\ \text{Original solution}}}{4.74} = +0.02$$

The pH increased by 0.02 pH units. Now compare this to what happens when 0.01 mol of solid NaOH is added to 1.0 L of water to give 0.01 *M* NaOH, where $[OH^-] = 0.01\ M$ and

$$[H^+] = \frac{K_w}{[OH^-]} = \frac{1.0 \times 10^{-14}}{1.0 \times 10^{-2}} = 1.0 \times 10^{-12}$$

$$\text{pH} = 12.00$$

Thus the change in pH is

$$\underset{\substack{\uparrow \\ \text{New solution}}}{12.00} - \underset{\substack{\uparrow \\ \text{Pure water}}}{7.00} = +5.00$$

The increase is 5.00 pH units. Note how well the buffered solution resists a change in pH, as compared to pure water. The pH of pure water can be changed rather easily by the addition of a suitable substance.

(top) Pure water at pH 7.00. (bottom) When 0.01 mol NaOH is added to 1.0 L of pure water, the pH jumps to 12.00.

Sample Exercise 15.2 is a typical buffer problem. It contains all of the concepts that you need to know to handle buffered solutions containing weak acids. Pay special attention to the following points:

1. Buffered solutions are simply solutions of weak acids or bases containing a common ion. The pH calculations on buffered solutions require exactly the same procedures previously introduced in Chapter 14. *This is not a new type of problem.*

2. When a strong acid or base is added to a buffered solution, it is best to deal with the stoichiometry of the resulting reaction first. After the stoichiometric calculations are completed, then consider the equilibrium calculations. This procedure can be represented as follows:

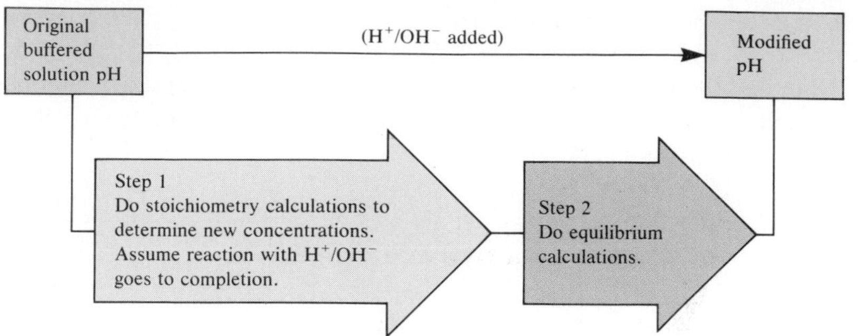

Buffering: How Does It Work?

Sample Exercise 15.2 demonstrates the ability of a buffered solution to absorb hydroxide ions without a significant change in pH. *But how does a buffer work?* Suppose a buffered solution contains relatively large quantities of a weak acid HA and its conjugate base A^-. When hydroxide ions are added to the solution, since the weak acid represents the best source of protons, the following reaction occurs:

$$OH^- + HA \rightarrow A^- + H_2O$$

The net result is that OH^- ions are not allowed to accumulate, but are replaced by A^- ions.

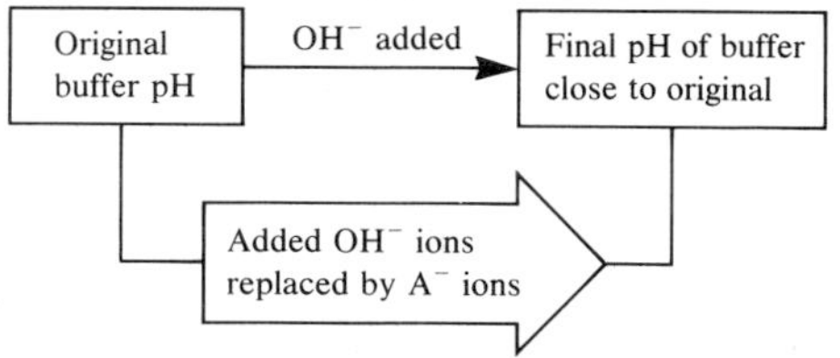

The stability of the pH under these conditions can be understood by examining the equilibrium expression for the dissociation of HA:

$$K_a = \frac{[H^+][A^-]}{[HA]} \quad \text{or, rearranging,} \quad [H^+] = K_a\frac{[HA]}{[A^-]}$$

In a buffered solution the pH is governed by the ratio $[HA]/[A^-]$.

In other words, the *equilibrium concentration of H^+, and thus the pH, is determined by the ratio $[HA]/[A^-]$.* When OH^- ions are added, HA is converted to A^-, and the ratio $[HA]/[A^-]$ decreases. However, *if the amounts of HA and A^- originally present are very large compared to the amount of OH^-* added, the change in the $[HA]/[A^-]$ ratio will be small.

In Sample Exercise 15.2

$$\frac{[HA]}{[A^-]} = \frac{0.50}{0.50} = 1.0 \qquad \text{Initially}$$

$$\frac{[HA]}{[A^-]} = \frac{0.49}{0.51} = 0.96 \qquad \text{After adding 0.01 mol/L of } OH^-$$

The change in the ratio $[HA]/[A^-]$ is very small. Thus the $[H^+]$ and the pH remain essentially constant.

The essence of buffering, then, is that [HA] and [A^-] are large compared to the amount of OH^- added. Thus when the OH^- is added, the concentrations of HA and A^- change but only by small amounts. Under these conditions, the [HA]/[A^-] ratio and thus the [H^+] are virtually constant.

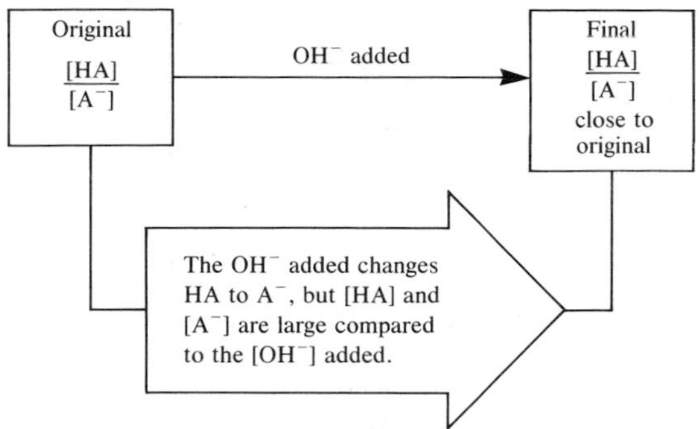

Similar reasoning applies when protons are added to a buffered solution of a weak acid and a salt of its conjugate base. Because the A^- ion has a high affinity for H^+, the added H^+ ions react with A^- to form the weak acid:

$$H^+ + A^- \rightarrow HA$$

and free H^+ ions do not accumulate. In this case there will be a net change of A^- to HA. However, if [A^-] and [HA] are large compared to the [H^+] added, little change in the pH will occur.

The form of the acid dissociation equilibrium expression

$$[H^+] = K_a\frac{[HA]}{[A^-]} \tag{15.1}$$

is often useful for calculating [H^+] in buffered solution, since [HA] and [A^-] are known. For example, to calculate [H^+] in a buffered solution containing 0.10 M HF ($K_a = 7.2 \times 10^{-4}$) and 0.30 M NaF, we simply substitute into Equation (15.1):

$$[H^+] = \underset{K_a\nearrow}{(7.2 \times 10^{-4})}\frac{\overset{[HF]\searrow}{0.10}}{\underset{\nwarrow[F^-]}{0.30}} = 2.4 \times 10^{-4}\ M$$

Another useful form of Equation (15.1) can be obtained by taking the negative log of both sides:

$$-\log[H^+] = -\log(K_a) - \log\left(\frac{[HA]}{[A^-]}\right)$$

That is,

$$pH = pK_a - \log\left(\frac{[HA]}{[A^-]}\right)$$

or, where inverting the log term reverses the sign:

$$pH = pK_a + \log\left(\frac{[A^-]}{[HA]}\right) = pK_a + \log\left(\frac{[base]}{[acid]}\right) \tag{15.2}$$

This log form of the expression for K_a is called the **Henderson-Hasselbalch equation** and is useful for calculating the pH of solutions when the ratio $[HA]/[A^-]$ is known.

For a particular buffering system (acid/conjugate base pair), all solutions that have the same ratio $[A^-]/[HA]$ will have the same pH. For example, a buffered solution containing 5.0 *M* $HC_2H_3O_2$ and 3.0 *M* $NaC_2H_3O_2$ will have the same pH as one containing 0.050 *M* $HC_2H_3O_2$ and 0.030 *M* $NaC_2H_3O_2$. This can be shown as follows:

System	$[A^-]/[HA]$
5.0 *M* $HC_2H_3O_2$ and 3.0 *M* $NaC_2H_3O_2$	$\frac{3.0\ M}{5.0\ M} = 0.60$
0.050 *M* $HC_2H_3O_2$ and 0.030 *M* $NaC_2H_3O_2$	$\frac{0.030\ M}{0.050\ M} = 0.60$

and

$$pH = pK_a + \log\left(\frac{[C_2H_3O_2^-]}{[HC_2H_3O_2]}\right) = 4.74 + \log(0.60) = 4.74 - 0.22 = 4.52$$

Note that in using this equation we have assumed that the equilibrium concentrations of A^- and HA are equal to the initial concentrations. That is, we are assuming the validity of the approximations

$$[A^-] = [A^-]_0 + x \approx [A^-]_0 \quad \text{and} \quad [HA] = [HA]_0 - x \approx [HA]_0$$

where x is the amount of acid that dissociates. Since the initial concentrations of HA and A^- are relatively large in a buffered solution, this assumption is generally acceptable.

Sample Exercise 15.3

Calculate the pH of a solution containing 0.75 *M* lactic acid ($K_a = 1.4 \times 10^{-4}$) and 0.25 *M* sodium lactate. Lactic acid ($HC_3H_5O_3$) is a common constituent of biological systems. For example, it is found in milk and is present in human muscle tissue during exertion.

Solution

The major species in solution are

$$HC_3H_5O_3, \quad Na^+, \quad C_3H_5O_3^-, \quad \text{and} \quad H_2O$$

Since Na^+ has no acid-base properties and H_2O is a weak acid or base, the pH will be controlled by the lactic acid dissociation equilibrium:

$$HC_3H_5O_3(aq) \rightleftharpoons H^+(aq) + C_3H_5O_3^-(aq)$$

$$K_a = \frac{[H^+][C_3H_5O_3^-]}{[HC_3H_5O_3]} = 1.4 \times 10^{-4}$$

Since $[HC_3H_5O_3]_0$ and $[C_3H_5O_3^-]_0$ are relatively large,

Sample Exercise 15.3, continued

$$[HC_3H_5O_3] \approx [HC_3H_5O_3]_0 = 0.75\ M$$

and

$$[C_3H_5O_3^-] \approx [C_3H_5O_3^-]_0 = 0.25\ M$$

Thus using the rearranged K_a expression

$$[H^+] = K_a \frac{[HC_3H_5O_3]}{[C_3H_5O_3^-]} = (1.4 \times 10^{-4}) \frac{(0.75\ M)}{(0.25\ M)} = 4.2 \times 10^{-4}\ M$$

and

$$pH = -\log(4.2 \times 10^{-4}) = 3.38$$

Alternatively, we could use the Henderson-Hasselbalch equation

$$pH = pK_a + \log\left(\frac{[C_3H_5O_3^-]}{[HC_3H_5O_3]}\right) = 3.85 + \log\left(\frac{0.25\ M}{0.75\ M}\right) = 3.38$$

Buffered solutions can also be formed from a weak base and the corresponding conjugate acid. In these solutions the weak base B reacts with any H^+ added

$$B + H^+ \rightarrow BH^+$$

and the conjugate acid BH^+ reacts with any added OH^-

$$BH^+ + OH^- \rightarrow B + H_2O$$

The approach needed to perform pH calculations for these systems is virtually identical to that used above. This makes sense because as is true of all buffered solutions, a weak acid (BH^+) and a weak base (B) are present. A typical case is illustrated in Sample Exercise 15.4.

Sample Exercise 15.4

A buffered solution contains 0.25 M NH_3 ($K_b = 1.8 \times 10^{-5}$) and 0.40 M NH_4Cl.

a. Calculate the pH of this solution.

Solution

a. The major species in solution are:

$$NH_3,\ \underbrace{NH_4^+,\ Cl^-}_{\text{From the dissolved } NH_4Cl}\ \text{and}\ H_2O$$

Since Cl^- is such a weak base and water is a weak acid or base, the important equilibrium is

$$NH_3(aq) + H_2O(l) \rightleftharpoons NH_4^+(aq) + OH^-(aq)$$

and

$$K_b = 1.8 \times 10^{-5} = \frac{[NH_4^+][OH^-]}{[NH_3]}$$

Sample Exercise 15.4, continued

The appropriate concentrations are as follows:

Initial concentration (mol/L)		Equilibrium concentration (mol/L)
$[NH_3]_0 = 0.25$ $[NH_4^+]_0 = 0.40$ $[OH^-] \approx 0$	$\xrightarrow[\text{reacts with } H_2O]{x \text{ mol/L } NH_3}$	$[NH_3] = 0.25 - x$ $[NH_4^+] = 0.40 + x$ $[OH^-] \approx x$

Then

$$K_b = 1.8 \times 10^{-5} = \frac{[NH_4^+][OH^-]}{[NH_3]} = \frac{(0.40 + x)(x)}{0.25 - x} \approx \frac{(0.40)(x)}{0.25}$$

$$x = 1.1 \times 10^{-5}$$

Since the approximations are valid by the 5% rule,

$$[OH^-] = x = 1.1 \times 10^{-5}$$
$$pOH = 4.95$$
$$pH = 14.00 - 4.95 = 9.05$$

This case is typical of a buffered solution in that the initial and equilibrium concentrations of buffering materials are essentially the same.

Alternative Solution There is another way of looking at this problem. Since the solution contains relatively large quantities of *both* NH_4^+ and NH_3, we can use the equilibrium

$$NH_3(aq) + H_2O(l) \rightleftharpoons NH_4^+(aq) + OH^-(aq)$$

to calculate $[OH^-]$ and then calculate $[H^+]$ from K_w as we have just done. Or we can use the dissociation equilibrium for NH_4^+,

$$NH_4^+(aq) \rightleftharpoons NH_3(aq) + H^+(aq)$$

to calculate $[H^+]$ directly. *Either choice will give the same answer,* since the same equilibrium concentrations of NH_3 and NH_4^+ must satisfy both equilibria.

We can obtain the K_a value for NH_4^+ from the given K_b value for NH_3, since $K_a \times K_b = K_w$:

$$K_a = \frac{K_w}{K_b} = \frac{1.0 \times 10^{-14}}{1.8 \times 10^{-5}} = 5.6 \times 10^{-10}$$

Then using the Henderson-Hasselbalch equation, we have

$$pH = pK_a + \log\left(\frac{[\text{base}]}{[\text{acid}]}\right)$$

$$= 9.25 + \log\left(\frac{0.25\ M}{0.40\ M}\right) = 9.25 - 0.20 = 9.05$$

b. Calculate the pH of the solution that results when 0.10 mol of gaseous HCl is added to 1.0 L of the buffered solution from part a.

Sample Exercise 15.4, continued

Solution

b. *Before any reaction occurs,* the solution contains the following major species:

$$NH_3, \quad NH_4^+, \quad Cl^-, \quad H^+, \quad \text{and} \quad H_2O$$

What reaction can occur? We know that H^+ will not react with Cl^- to form HCl. In contrast to Cl^-, the NH_3 molecule has a great affinity for protons (this is demonstrated by the fact that NH_4^+ is such a weak acid ($K_a = 5.6 \times 10^{-10}$). Thus NH_3 will react with H^+ to form NH_4^+:

Remember: Think about the chemistry first. Ask yourself if a reaction will occur among the major species.

$$NH_3(aq) + H^+(aq) \rightarrow NH_4^+(aq)$$

Since this reaction can be assumed to go essentially to completion to form the very weak acid NH_4^+, we will do the stoichiometry calculations before we consider the equilibrium calculations. That is, we will let the reaction run to completion and then consider the equilibrium.

The stoichiometry calculations for this process are

	NH_3	+	H^+	→	NH_4^+
Before reaction:	(1.0 L)(0.25 *M*) = 0.25 mol		0.10 mol ↑ limiting reactant		(1.0 L)(0.40 *M*) = 0.40 mol
After reaction:	0.25 − 0.10 = 0.15 mol		0		0.40 + 0.10 = 0.50 mol

After the reaction goes to completion, the solution contains the major species

$$NH_3, \quad NH_4^+, \quad Cl^-, \quad \text{and} \quad H_2O$$

and

$$[NH_3]_0 = \frac{0.15 \text{ mol}}{1.0 \text{ L}} = 0.15\ M$$

$$[NH_4^+]_0 = \frac{0.50 \text{ mol}}{1.0 \text{ L}} = 0.50\ M$$

We can use the Henderson-Hasselbalch equation where

$$[\text{base}] = [NH_3] \approx [NH_3]_0 = 0.15\ M$$
$$[\text{acid}] = [NH_4^+] \approx [NH_4^+]_0 = 0.50\ M$$

Then

$$\text{pH} = \text{p}K_a + \log\left(\frac{[NH_3]}{[NH_4^+]}\right)$$

$$= 9.25 + \log\left(\frac{0.15\ M}{0.50\ M}\right) = 9.25 - 0.52 = 8.73$$

Note that the addition of HCl only slightly decreases the pH, as we would expect in a buffered solution.

We can now summarize the most important characteristics of buffered solutions.

- Buffered solutions contain relatively large concentrations of a weak acid and the corresponding weak base. They can involve a weak acid HA and the conjugate base A^- or a weak base B and the conjugate acid BH^+.
- When H^+ is added to a buffered solution, it reacts essentially to completion with the weak base present

$$H^+ + A^- \rightarrow HA$$

or

$$H^+ + B \rightarrow BH^+$$

- When OH^- is added to a buffered solution it reacts essentially to completion with the weak acid present

$$OH^- + HA \rightarrow A^- + H_2O$$

or

$$OH^- + BH^+ \rightarrow B + H_2O$$

- The pH in the buffered solution is determined by the ratio of the concentrations of the weak acid and weak base. As long as this ratio remains virtually constant, the pH will remain virtually constant. This will be the case as long as the concentrations of the buffering materials (HA and A^- or B and BH^+) are large compared to the amounts of H^+ or OH^- added.

15.3 Buffer Capacity

Purpose

- To describe the meaning of buffer capacity.

A buffer with a large capacity contains large concentrations of the buffering components.

The **buffering capacity** of a buffered solution is defined in terms of the amount of protons or hydroxide ions it can absorb without a significant change in pH. A buffer with a large capacity contains large concentrations of buffering components and so can absorb a relatively large amount of protons or hydroxide ions and show little pH change. *The pH of a buffered solution is determined by the ratio* $[A^-]/[HA]$. *The capacity of a buffered solution is determined by the magnitudes of* $[HA]$ *and* $[A^-]$.

Sample Exercise 15.5

Calculate the change in pH that occurs when 0.010 mol of gaseous HCl is added to 1.0 L of each of the following solutions:

Sample Exercise 15.5, continued

Solution A: 5.00 *M* $HC_2H_3O_2$ and 5.00 *M* $NaC_2H_3O_2$
Solution B: 0.050 *M* $HC_2H_3O_2$ and 0.050 M $NaC_2H_3O_2$

For acetic acid, $K_a = 1.8 \times 10^{-5}$.

Solution

For both solutions the initial pH can be determined from the Henderson-Hasselbalch equation:

$$pH = pK_a + \log\left(\frac{[C_2H_3O_2^-]}{[HC_2H_3O_2]}\right)$$

In each case $[C_2H_3O_2^-] = [HC_2H_3O_2]$ so that for both A and B initially

$$pH = pK_a + \log(1) = pK_a = -\log(1.8 \times 10^{-5}) = 4.74$$

After the addition of HCl to each of these solutions, the major species *before any reaction occurs* are:

$$HC_2H_3O_2,\quad Na^+,\quad C_2H_3O_2^-,\quad \underbrace{H^+,\quad Cl^-,}_{\text{From the added HCl}}\quad \text{and}\quad H_2O$$

Will any reactions occur among these species? Note that we have a relatively large quantity of H^+, which will readily react with any effective base. We know that Cl^- will not react with H^+ to form HCl in water. However, $C_2H_3O_2^-$ will react with H^+ to form the weak acid $HC_2H_3O_2$:

$$H^+(aq) + C_2H_3O_2^-(aq) \rightarrow HC_2H_3O_2(aq)$$

Because $HC_2H_3O_2$ is a weak acid, we assume this reaction runs to completion; the 0.010 mol of added H^+ will convert 0.010 mol of $C_2H_3O_2^-$ to 0.010 mol of $HC_2H_3O_2$.

For solution A (since the solution volume is 1.0 L, the number of moles equals the molarity):

	H^+	+	$C_2H_3O_2^-$	$\rightarrow$	$HC_2H_3O_2$
Before reaction:	0.01 *M*		5.00 *M*		5.00 *M*
After reaction:	0		4.99 *M*		5.01 *M*

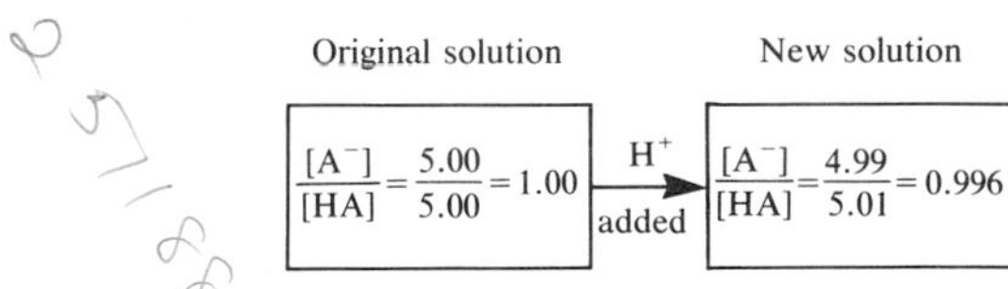

The new pH can be obtained by substituting the new concentrations into the Henderson-Hasselbalch equation:

$$pH = pK_a + \log\left(\frac{[C_2H_3O_2^-]}{[HC_2H_3O_2]}\right)$$

$$= 4.74 + \log\left(\frac{4.99}{5.01}\right) = 4.74 + 0.0017 = 4.74$$

There is virtually no change in pH for solution A when 0.010 mol of gaseous HCl is added.

Sample Exercise 15.5, continued

For solution B:

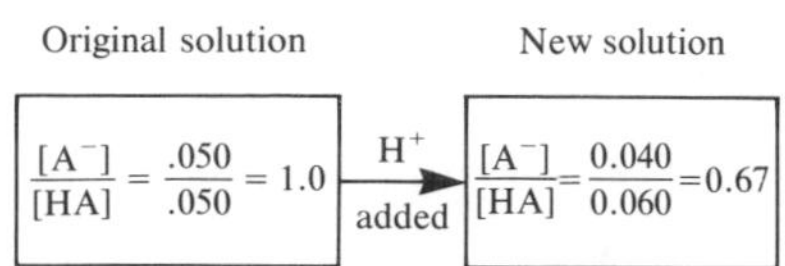

	H^+	+	$C_2H_3O_2^-$	→	$HC_2H_3O_2$
Before reaction:	0.01 *M*		0.050 *M*		0.050 *M*
After reaction:	0		0.040 *M*		0.060 *M*

The new pH is

$$\text{pH} = 4.74 + \log\left(\frac{0.040}{0.060}\right)$$

$$= 4.74 - 0.18 = 4.56$$

Although the pH change for solution B is small, a change did occur, which is in contrast to solution A.

These results show that solution A, which contains much larger quantities of buffering components, has a much higher buffering capacity than solution B.

We have seen that the pH of a buffered solution depends on the ratio of the concentrations of buffering components. When this ratio is least affected by added protons or hydroxide ions, the solution is the most resistant to change in pH. To find the ratio that gives optimum buffering, let's suppose we have a buffered solution containing a large concentration of acetate ion and only a small concentration of acetic acid. Addition of protons to form acetic acid will produce a relatively large *percent* change in the concentration of acetic acid and so will produce a relatively large change in the ratio $[C_2H_3O_2^-]/[HC_2H_3O_2]$ (see Table 15.1). Similarly, if hydroxide ions are added to remove some acetic acid, the percent change in the concentration of acetic acid is again large. The same effects are seen if the initial concentration of acetic acid is large and that of acetate ion is small.

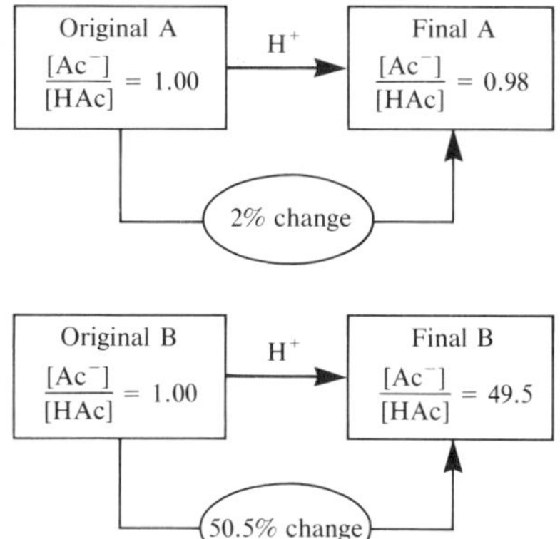

Because large changes in the ratio $[A^-]/[HA]$ will produce large changes in pH, we want to avoid this situation for the most effective buffering. This type of reasoning leads us to the general conclusion that optimum buffering will occur when

Change in $[C_2H_3O_2^-]/[HC_2H_3O_2]$ for Two Solutions When 0.01 mol of H^+ Is Added to 1.0 L of Each

Solution	$\left(\frac{[C_2H_3O_2^-]}{[HC_2H_3O_2]}\right)_{orig}$	$\left(\frac{[C_2H_3O_2^-]}{[HC_2H_3O_2]}\right)_{new}$	Change	Percent change
A	$\frac{1.00\ M}{1.00\ M} = 1.00$	$\frac{0.99\ M}{1.01\ M} = 0.98$	1.00 → 0.98	2.00%
B	$\frac{1.00\ M}{0.01\ M} = 100$	$\frac{0.99\ M}{0.02\ M} = 49.5$	100 → 49.5	50.5%

Table 15.1

[HA] is equal to $[A^-]$. It is under this condition that the ratio $[A^-]/[HA]$ is most resistant to change when H^+ or OH^- is added. This means that when choosing the buffering components for a specific application, we want $[A^-]/[HA]$ to equal 1, and since

$$\text{pH} = \text{p}K_a + \log\left(\frac{[A^-]}{[HA]}\right) = \text{p}K_a + \log(1) = \text{p}K_a$$

the pK_a *of the weak acid to be used in the buffer should be as close as possible to the desired pH*. For example, suppose we need a buffered solution with a pH of 4.00. The most effective buffering will occur when [HA] is equal to $[A^-]$. From the Henderson-Hasselbalch equation,

$$\underset{\substack{\uparrow \\ \text{4.00 is wanted}}}{\text{pH}} = \text{p}K_a + \log\left(\frac{[A^-]}{[HA]}\right)$$

(4.00 is wanted; Ratio = 1 for most effective buffer)

That is, $4.00 = \text{p}K_a + \log(1) = \text{p}K_a + 0$ and $\text{p}K_a = 4.00$

Thus the best choice of a weak acid is one that has pK_a = 4.00, or $K_a = 1.0 \times 10^{-4}$.

Sample Exercise 15.6

A chemist needs a solution buffered at pH 4.30 and can choose from the following acids (and their sodium salts):

a. chloroacetic acid ($K_a = 1.35 \times 10^{-3}$)
b. propanoic acid ($K_a = 1.3 \times 10^{-5}$)
c. benzoic acid ($K_a = 6.4 \times 10^{-5}$)
d. hypochlorous acid ($K_a = 3.5 \times 10^{-8}$)

Calculate the ratio $[HA]/[A^-]$ required for each system to yield a pH of 4.30. Which system will work best?

Solution

A pH of 4.30 corresponds to

$$[H^+] = 10^{-4.30} = \text{antilog}(-4.30) = 5.0 \times 10^{-5}\ M$$

Since K_a values rather than pK_a values are given for the various acids, we use Equation (15.1)

$$[H^+] = K_a\frac{[HA]}{[A^-]}$$

rather than the Henderson-Hasselbalch equation. We substitute the required $[H^+]$ and K_a for each acid into Equation (15.1) to calculate the ratio $[HA]/[A^-]$ needed in each case.

Sample Exercise 15.6, continued

Acid	$[H^+] = K_a \frac{[HA]}{[A^-]}$	$\frac{[HA]}{[A^-]}$
a. chloroacetic	$5.0 \times 10^{-5} = 1.35 \times 10^{-3}\left(\frac{[HA]}{[A^-]}\right)$	3.7×10^{-2}
b. propanoic	$5.0 \times 10^{-5} = 1.3 \times 10^{-5}\left(\frac{[HA]}{[A^-]}\right)$	3.8
c. benzoic	$5.0 \times 10^{-5} = 6.4 \times 10^{-5}\left(\frac{[HA]}{[A^-]}\right)$	0.78
d. hypochlorous	$5.0 \times 10^{-5} = 3.5 \times 10^{-8}\left(\frac{[HA]}{[A^-]}\right)$	1.4×10^3

Since $[HA]/[A^-]$ for benzoic acid is closest to 1, for buffering a solution at pH 4.3, the system of benzoic acid and its sodium salt will be the best choice among those given. This demonstrates the principle that the optimum buffering system has a pK_a value close to the desired pH. The pK_a for benzoic acid is 4.19.

15.4 Titrations and pH Curves

Purpose

- To demonstrate how to calculate pH at any point in an acid-base titration.

As we saw in Chapter 4, a titration is commonly used to analyze for the amount of acid or base in a solution. This process involves a solution of known concentration (the titrant) delivered from a buret into the unknown solution until the substance being analyzed is just consumed (the stoichiometric or equivalence point, which is usually signaled by the color change of an indicator). In this section we will discuss the pH changes that occur during an acid-base titration. We will later use this information to show how an appropriate indicator can be chosen for a particular titration.

The progress of an acid-base titration is often monitored by plotting the pH of the solution being analyzed as a function of the amount of titrant added. Such a plot is called a **pH curve,** or **titration curve.**

Strong Acid–Strong Base Titrations

The reaction for a strong acid–strong base titration is

$$H^+(aq) + OH^-(aq) \rightarrow H_2O(l)$$

To compute $[H^+]$ at a given point in the titration, we must determine the amount of H^+ that remains at that point and divide by the total volume of the solution. Before we proceed, we need to define a new unit, which is especially convenient when the volume of a solution is given in milliliters. For example, since titrations usually involve small quantities (burets are typically graduated in milliliters), the mole is inconveniently large. Therefore, we will use the **millimole** (abbreviated **mmol**), which as the prefix indicates is a thousandth of a mole:

1 millimole = 1×10^{-3} mol

1 mL = 1×10^{-3} L

$$\frac{\text{mmol}}{\text{mL}} = \frac{\text{mol}}{\text{L}} = M$$

$$1 \text{ mmol} = \frac{1 \text{ mol}}{1000} = 10^{-3} \text{ mol}$$

So far we have defined molarity only in terms of moles per liter. We can now define it in terms of millimoles per milliliter, as shown below:

$$\text{Molarity} = \frac{\text{mol of solute}}{\text{L of solution}} = \frac{\dfrac{\text{mol of solute}}{1000}}{\dfrac{\text{L of solution}}{1000}} = \frac{\text{mmol of solute}}{\text{mL of solution}}$$

A 1.0 M solution thus contains 1.0 mole of solute per liter of solution or, *equivalently,* 1.0 millimole of solute per milliliter of solution. Just as we obtain the number of moles of solute from the product of the volume in liters and the molarity, we obtain the number of millimoles of solute from the product of the volume in milliliters and the molarity:

$$\text{Number of mmol} = \text{volume (in mL)} \times \text{molarity}$$

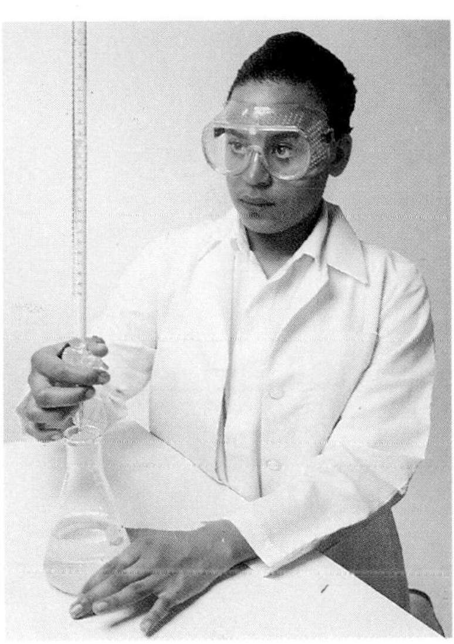

A student performs a titration.

We will illustrate the calculations involved in a strong acid–strong base titration by considering the titration of 50.0 mL of 0.200 M HNO_3 with 0.100 M NaOH. We will calculate the pH of the solution at selected points during the course of the titration, where specific volumes of 0.100 M NaOH have been added.

A. No NaOH has been added.

Since HNO_3 is a strong acid and is completely dissociated, the solution contains the major species

$$H^+, \quad NO_3^-, \quad \text{and} \quad H_2O,$$

and the pH is determined by the H^+ from the nitric acid. Since 0.20 M HNO_3 contains 0.20 M H^+,

$$[H^+] = 0.20\ M \quad \text{and} \quad \text{pH} - 0.70$$

B. 10.0 mL of 0.100 M NaOH has been added.

In the mixed solution before any reaction occurs, the major species are:

$$H^+, \quad NO_3^-, \quad Na^+, \quad OH^-, \quad \text{and} \quad H_2O$$

Note that large quantities of both H^+ and OH^- are present. On mixing, the 1.0 mmol (10.0 mL × 0.10 M) of added OH^- will react with 1.0 mmol of H^+ to form water:

	H^+	+	OH^-	→	H_2O
Before reaction:	50.0 mL × 0.200 M = 10.0 mmol		10.0 mL × 0.10 M = 1.00 mmol		
After reaction:	10.0 − 1.0 = 9.0 mmol		1.0 − 1.0 = 0		

After the reaction the solution contains:

H^+, NO_3^-, Na^+, and H_2O (the OH^- ions have been consumed)

and the pH will be determined by the H^+ remaining:

The final solution volume is the sum of the original volume of HNO_3 and the volume of added NaOH.

$$[H^+] = \frac{\text{mmol } H^+ \text{ left}}{\text{volume of solution (mL)}} = \frac{9.0 \text{ mmol}}{(50.0 + 10.0) \text{ mL}} = 0.15\ M$$

(50.0: Original volume of HNO_3 solution; 10.0: Volume of NaOH added)

$$\text{pH} = -\log(0.15) = 0.82$$

C. 20.0 mL (total) of 0.100 *M* NaOH has been added.

We consider this point from the perspective that a total of 20.0 mL of NaOH has been added to the *original* solution, rather than that 10.0 mL has been added to the solution from point B. It is best to go back to the original solution each time, so that a mistake made at an earlier point does not show up in each succeeding calculation. As before, the added OH^- will react with H^+ to form water:

	H^+	+	OH^-	$\rightarrow$	H_2O
Before reaction:	50.0 mL × 0.200 *M* = 10.0 mmol		20.0 mL × 0.10 *M* = 2.0 mmol		
After reaction:	10.0 − 2.0 = 8.0 mmol		2.0 − 2.0 = 0 mmol		

After the reaction

$$[H^+] = \frac{8.0 \text{ mmol}}{(50.0 + 20.0) \text{ mL}} = 0.11\ M$$

(8.0 mmol: H^+ remaining)

$$\text{pH} = 0.94$$

D. 50.0 mL (total) of 0.100 *M* NaOH has been added.

Proceeding exactly as for points B and C, the pH is found to be 1.30.

E. 100.0 mL (total) of 0.100 *M* NaOH has been added.

At this point the amount of NaOH that has been added is

$$100.0 \text{ mL} \times 0.100\ M = 10.0 \text{ mmol}$$

The original amount of nitric acid was

$$50.0 \text{ mL} \times 0.200\ M = 10.0 \text{ mmol}$$

Enough OH^- has been added to react exactly with the H^+ from the nitric acid. This is the *stoichiometric point,* or **equivalence point,** of the titration. At this point the major species in solution are

Na^+, NO_3^-, and H_2O

Since Na^+ has no acid or base properties and NO_3^- is the anion of the strong acid HNO_3 and is therefore a very weak base, neither NO_3^- nor Na^+ affects the pH, and the solution is neutral (the pH is 7.0).

F. 150.0 mL (total) of 0.100 *M* NaOH has been added.

The titration reaction is

	H^+	+	OH^-	$\rightarrow$	H_2O
Before reaction:	50.0 mL × 0.200 *M* = 10.0 mmol		150.00 mL × 0.100 *M* = 15.0 mmol		
After reaction:	10.0 − 10.0 = 0 mmol		15.0 − 10.0 = 5.0 mmol ↑ Excess OH^- added		

Now OH^- is *in excess* and will determine the pH.

$$[OH^-] = \frac{\text{mmol OH}^- \text{ in excess}}{\text{volume (mL)}} = \frac{5.0 \text{ mmol}}{(50.0 + 150.0) \text{ mL}} = \frac{5.0 \text{ mmol}}{200.0 \text{ mL}} = 0.025\ M$$

Since $[H^+][OH^-] = 1.0 \times 10^{-14}$,

$$[H^+] = \frac{1.0 \times 10^{-14}}{2.5 \times 10^{-2}} = 4.0 \times 10^{-13}\ M \quad \text{and} \quad \text{pH} = 12.40$$

G. 200.0 mL (total) 0.100 *M* NaOH has been added.

Proceeding as for point F, the pH is found to be 12.60.

The results of these calculations are summarized by the pH curve shown in Fig. 15.1. Note that the pH changes very gradually until the titration is close to the equivalence point, where a dramatic change occurs. This behavior is due to the fact that early in the titration there is a relatively large amount of H^+ in the solution, and the addition of a given amount of OH^- thus produces a small change in pH. However, near the equivalence point $[H^+]$ is relatively small, and the addition of a small amount of OH^- produces a large change.

The pH curve in Fig. 15.1, typical of the titration of a strong acid with a strong base, has the following characteristics:

Before the equivalence point $[H^+]$ (and hence the pH) can be calculated by dividing the number of millimoles of H^+ left by the total volume of the solution in milliliters.

At the equivalence point, the pH is 7.0.

After the equivalence point $[OH^-]$ can be calculated by dividing the number of millimoles of excess OH^- by the total volume of the solution. Then $[H^+]$ is obtained from K_w.

The titration of a strong base with a strong acid requires reasoning very similar to that used above, except, of course, that OH^- is in excess before the equivalence point and H^+ is in excess after the equivalence point. The pH curve for the titration of 100.0 mL of 0.50 *M* NaOH with 1.0 *M* HCl is shown in Fig. 15.2.

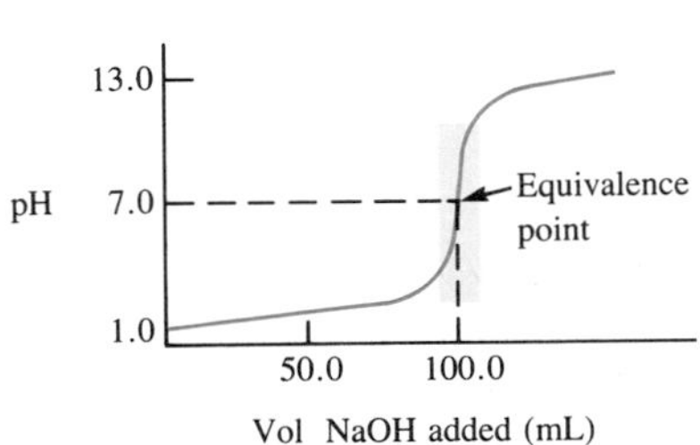

Figure 15.1

The pH curve for the titration of 50.0 mL of 0.200 *M* HNO_3 with 0.100 *M* NaOH. Note that the equivalence point occurs at 100.0 mL of NaOH added, the point where exactly enough OH^- has been added to react with all of the H^+ originally present. The pH of 7 at the equivalence point is characteristic of a strong acid–strong base titration.

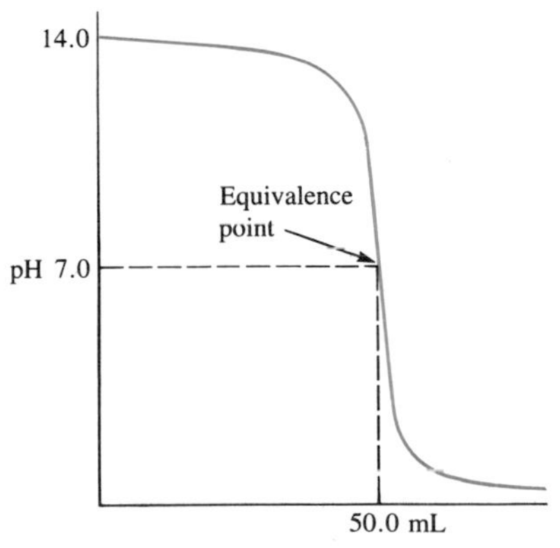

Figure 15.2

The pH curve for the titration of 100.0 mL of 0.50 *M* NaOH with 1.0 *M* HCl. The equivalence point occurs at 50.00 mL of HCl added, since at this point 5.0 mmol of H^+ ions has been added to react with the original 5.0 mmol of OH^- ions.

Titrations of Weak Acids with Strong Bases

We have seen that since strong acids and strong bases are completely dissociated, the calculations to obtain the pH curves for titrations involving the two are quite straightforward. When the acid being titrated is a weak acid, there is one main difference: to calculate $[H^+]$ after a certain amount of strong base has been added, we must deal with the weak acid dissociation equilibrium. In fact we have dealt with this same situation earlier in this chapter when we treated buffered solutions. Calculation of the pH curve for a titration of a weak acid with a strong base really amounts to a series of buffer problems. In performing these calculations it is very important to remember that even though the acid is weak, it *reacts essentially to completion* with hydroxide ion, a very strong base.

Calculating the pH curve for a weak acid–strong base titration involves a two-step procedure:

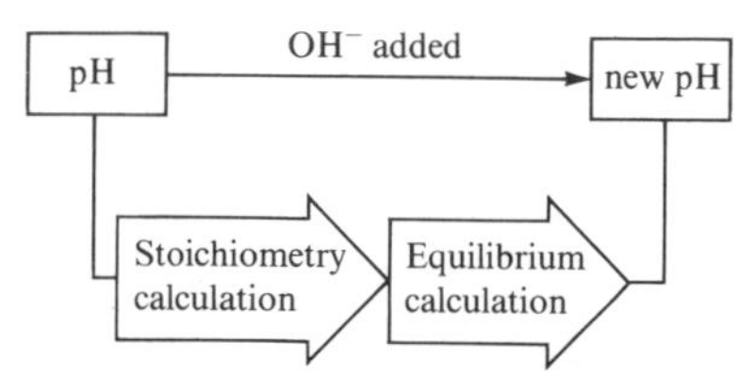

STEP 1

A stoichiometry problem. The reaction of hydroxide ion with the weak acid is assumed to run to completion, and the concentrations of the acid *remaining* and the conjugate base *formed* are determined.

STEP 2

An equilibrium problem. The position of the weak acid equilibrium is determined, and the pH is calculated.

Treat the stoichiometry and equilibrium problems separately.

It is *essential* to do these steps *separately*. Note that the procedures necessary to do these problems have all been used before.

As an illustration, we will consider the titration of 50.0 mL of 0.10 *M* acetic acid ($HC_2H_3O_2$, $K_a = 1.8 \times 10^{-5}$) with 0.10 *M* NaOH. As before, we will calculate the pH at various points representing volumes of added NaOH.

A. No NaOH has been added.

This is a typical weak acid calculation of the type introduced in Chapter 14. The pH is 2.87. (Check this yourself.)

B. 10.0 mL of 0.10 *M* NaOH has been added.

The major species in the mixed solution *before any reaction takes place* are:

$$HC_2H_3O_2, \quad OH^-, \quad Na^+, \quad \text{and} \quad H_2O$$

The strong base OH^- will react with the strongest proton donor, which in this case is $HC_2H_3O_2$.

The Stoichiometry Problem.

You are again doing exactly the same type of calculation already seen in Chapter 14.

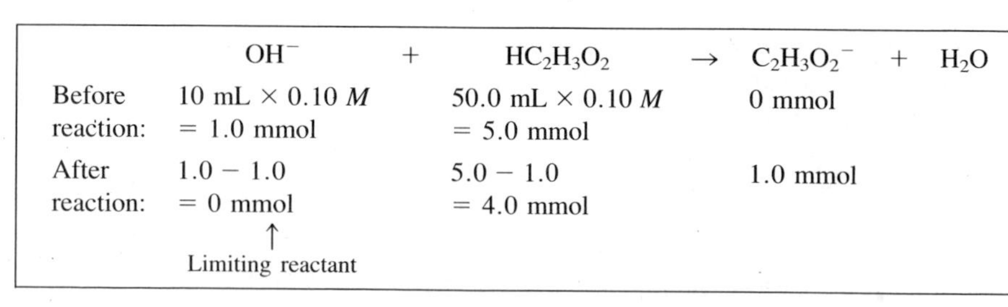

	OH^-	+	$HC_2H_3O_2$	→	$C_2H_3O_2^-$	+	H_2O
Before reaction:	10 mL × 0.10 *M* = 1.0 mmol		50.0 mL × 0.10 *M* = 5.0 mmol		0 mmol		
After reaction:	1.0 − 1.0 = 0 mmol ↑ Limiting reactant		5.0 − 1.0 = 4.0 mmol		1.0 mmol		

The Equilibrium Problem. We examine the major components left in the solution *after the reaction takes place* to decide on the dominant equilibrium. The major species are:

$$HC_2H_3O_2, \quad C_2H_3O_2^-, \quad Na^+, \quad \text{and} \quad H_2O$$

Since $HC_2H_3O_2$ is a much stronger acid than H_2O, and since $C_2H_3O_2^-$ is the conjugate base of $HC_2H_3O_2$, the pH will be determined by the position of the acetic acid dissociation equilibrium:

$$HC_2H_3O_2(aq) \rightleftharpoons H^+(aq) + C_2H_3O_2^-(aq)$$

and

$$K_a = \frac{[H^+][C_2H_3O_2^-]}{[HC_2H_3O_2]}$$

We follow the usual steps to complete the equilibrium calculations:

Initial concentration		Equilibrium concentration
$[HC_2H_3O_2]_0 = \frac{4.0 \text{ mmol}}{(50.0 + 10.0) \text{ mL}} = \frac{4.0}{60.0}$		$[HC_2H_3O_2] = \frac{4.0}{60.0} - x$
$[C_2H_3O_2^-]_0 = \frac{1.0 \text{ mmol}}{(50.0 + 10.0) \text{ mL}} = \frac{1.0}{60.0}$	$\xrightarrow[\text{dissociates}]{x \text{ mmol/mL } HC_2H_3O_2}$	$[C_2H_3O_2^-] = \frac{1.0}{60.0} + x$
$[H^+]_0 \approx 0$		$[H^+] \approx x$

The initial concentrations are defined after the reaction with OH^- has gone to completion but before any dissociation of $HC_2H_3O_2$ occurs.

$$1.8 \times 10^{-5} = K_a = \frac{[H^+][C_2H_3O_2^-]}{[HC_2H_3O_2]} = \frac{x\left(\frac{1.0}{60.0} + x\right)}{\frac{4.0}{60.0} - x} \approx \frac{x\left(\frac{1.0}{60.0}\right)}{\frac{4.0}{60.0}} = \left(\frac{1.0}{4.0}\right)x$$

$$x = \left(\frac{4.0}{1.0}\right)(1.8 \times 10^{-5}) = 7.2 \times 10^{-5} = [H^+]$$

$$\text{pH} = 4.14$$

Note that the approximations made are well within the 5% rule.

C. 25.0 mL (total) of 0.10 *M* NaOH has been added.

The procedure here is very similar to that used at point B and will only be summarized briefly. The stoichiometry problem is

	OH^-	+	$HC_2H_3O_2$	→	$C_2H_3O_2^-$	+	H_2O
Before reaction:	25.0 mL × 0.10 *M* = 2.5 mmol		5.0 mL × 0.10 *M* = 5.0 mmol		0 mmol		
After reaction:	2.5 − 2.5 = 0		5.0 − 2.5 = 2.5 mmol		2.5 mmol		

After the reaction, the major species in solution are

$$HC_2H_3O_2, \quad C_2H_3O_2^-, \quad Na^+, \quad \text{and} \quad H_2O$$

The equilibrium that will control the pH is

$$HC_2H_3O_2(aq) \rightleftharpoons H^+(aq) + C_2H_3O_2^-(aq)$$

Initial concentration		Equilibrium concentration
$[HC_2H_3O_2]_0 = \dfrac{2.5 \text{ mmol}}{(50.0 + 25.0) \text{ mL}}$		$[HC_2H_3O_2] = \dfrac{2.5}{75.0} - x$
$[C_2H_3O_2^-]_0 = \dfrac{2.5 \text{ mmol}}{(50.0 + 25.0) \text{ mL}}$	$\xrightarrow[\text{dissociates}]{x \text{ mmol/mL } HC_2H_3O_2}$	$[C_2H_3O_2^-] = \dfrac{2.5}{75.0} + x$
$[H^+]_0 \approx 0$		$[H^+] \approx x$

$$1.8 \times 10^{-5} = K_a = \frac{[H^+][C_2H_3O_2^-]}{[HC_2H_3O_2]} = \frac{x\left(\frac{2.5}{75.0} + x\right)}{\frac{2.5}{75.0} - x} \approx \frac{x\left(\frac{2.5}{75.0}\right)}{\frac{2.5}{75.0}}$$

$x = 1.8 \times 10^{-5} = [H^+]$

$pH = 4.74$

This is a special point in the titration because it is *halfway to the equivalence point*. The original solution, 50.0 milliliters of 0.10 M $HC_2H_3O_2$, contained 5.0 mmol of $HC_2H_3O_2$. Thus 5.0 mmol of OH^- is required to reach the equivalence point, that is, 50 mL of NaOH, since

$$(50.0 \text{ mL})(0.100\ M) = 5.00 \text{ mmol}$$

After 25.0 mL of NaOH has been added, half of the original $HC_2H_3O_2$ has been converted to $C_2H_3O_2^-$. At this point in the titration $[HC_2H_3O_2]_0$ is equal to $[C_2H_3O_2^-]_0$. We can neglect the effect of dissociation; that is,

At this point, half of the acid has been used up, so

$[HC_2H_3O_2] = [C_2H_3O_2^-]$

$$[HC_2H_3O_2] = [HC_2H_3O_2]_0 - x \approx [HC_2H_3O_2]_0$$
$$[C_2H_3O_2^-] = [C_2H_3O_2^-]_0 + x \approx [C_2H_3O_2^-]_0$$

The expression for K_a at the halfway point is

$$K_a = \frac{[H^+][C_2H_3O_2^-]}{[HC_2H_3O_2]} = \frac{[H^+][C_2H_3O_2^-]_0}{[HC_2H_3O_2]_0} = [H^+]$$

Equal at the halfway point

Then *at the halfway point* in the titration,

$$[H^+] = K_a \quad \text{and} \quad pH = pK_a$$

D. 40.0 mL (total) of 0.10 M NaOH has been added.

The procedures required here are the same as those used for points B and C. The pH is 5.35. (Check this yourself.)

E. 50.0 mL (total) of 0.10 M NaOH has been added.

This is the equivalence point of the titration; 5.0 mmol of OH^- has been added, which will just react with the 5.0 mmol of $HC_2H_3O_2$ originally present. At this point, the solution contains the major species:

$$Na^+, \quad C_2H_3O_2^-, \quad \text{and} \quad H_2O$$

Note that the solution contains $C_2H_3O_2^-$, which is a base. Remember that a base will combine with a proton, and the only source of protons in the solution is water. Thus the reaction will be

$$C_2H_3O_2^-(aq) + H_2O(l) \rightleftharpoons HC_2H_3O_2(aq) + OH^-(aq)$$

This is a *weak base* reaction characterized by K_b:

$$K_b = \frac{[HC_2H_3O_2][OH^-]}{[C_2H_3O_2^-]} = \frac{K_w}{K_a} = \frac{1.0 \times 10^{-14}}{1.8 \times 10^{-5}} = 5.6 \times 10^{-10}$$

The relevant concentrations are as follows:

Initial concentration (before any $C_2H_3O_2^-$ reacts with H_2O)		Equilibrium concentration
$[C_2H_3O_2^-]_0 = \frac{5.0 \text{ mmol}}{(50.0 + 50.0) \text{ mL}} = 0.050\ M$	$\xrightarrow[\text{with } H_2O]{x \text{ mmol/mL } C_2H_3O_2^- \text{ reacts}}$	$[C_2H_3O_2^-] = 0.050 - x$
$[OH^-]_0 \approx 0$		$[OH^-] \approx x$
$[HC_2H_3O_2]_0 = 0$		$[HC_2H_3O_2] = x$

$$5.6 \times 10^{-10} = K_b = \frac{[HC_2H_3O_2][OH^-]}{[C_2H_3O_2^-]} = \frac{(x)(x)}{0.050 - x} \approx \frac{x^2}{0.050}$$

$$x = 5.3 \times 10^{-6}$$

Since the approximation is valid by the 5% rule,

$$[OH^-] = 5.3 \times 10^{-6}\ M$$

and

$$[H^+][OH^-] = K_w = 1.0 \times 10^{-14}$$

$$[H^+] = 1.9 \times 10^{-9}\ M$$

$$pH = 8.72$$

This is another important result: *the pH at the equivalence point of a titration of a weak acid with a strong base is always greater than 7*. This is because the anion of the acid, which remains in solution at the equivalence point, is a base. In contrast, for the titration of a strong acid with a strong base, the pH at the equivalence point is 7.0, because the anion remaining in this case is *not* an effective base.

The pH at the equivalence point of a titration of a weak acid with a strong base is always greater than 7.

F. 60.0 mL (total) of 0.10 *M* NaOH has been added.

At this point, excess OH^- has been added:

	OH^-	+	$HC_2H_3O_2$	$\rightarrow$	$C_2H_3O_2^-$	+ H_2O
Before reaction:	$60.0 \text{ mL} \times 0.10\ M = 6.0 \text{ mmol}$		$50.0 \text{ mL} \times 0.10\ M = 5.0 \text{ mmol}$		0 mmol	
After reaction:	$6.0 - 5.0 = 1.0 \text{ mmol in excess}$		$5.0 - 5.0 = 0$		5.0 mmol	

After the reaction, the solution contains the major species:

$$Na^+, \quad C_2H_3O_2^-, \quad OH^-, \quad \text{and} \quad H_2O$$

There are two bases in this solution, OH^- and $C_2H_3O_2^-$. However, since $C_2H_3O_2^-$ is a weak base compared to OH^-, the amount of OH^- produced by reaction of $C_2H_3O_2^-$ with H_2O will be small compared to the excess OH^- already in solution. You can see this by looking at point E, where only $5.3 \times 10^{-6}\ M\ OH^-$ was produced by $C_2H_3O_2^-$. The amount in this case will be even smaller, since the excess OH^- will push the K_b equilibrium to the left.

Thus the pH will be determined by the excess OH^-.

$$[OH^-] = \frac{\text{mmol of } OH^- \text{ in excess}}{\text{volume (in mL)}} = \frac{1.0 \text{ mmol}}{(50.0 + 60.0) \text{ mL}}$$

$$= 9.1 \times 10^{-3}\ M$$

and

$$[H^+] = \frac{1.0 \times 10^{-14}}{9.1 \times 10^{-3}} = 1.1 \times 10^{-12}\ M$$

$$\text{pH} = 11.96$$

G. 75.0 mL (total) of 0.10 *M* NaOH has been added.

The procedure is very similar to that for point F. The pH is 12.30. (Check this.)

The pH curve for this titration is shown in Fig. 15.3. It is important to note the differences between this curve and that in Fig. 15.1. For example, the shapes of the plots are quite different before the equivalence point, although they are very similar after that point. (The shapes of the strong and weak acid curves are the same after the equivalence points, because excess OH^- controls the pH in this region in both cases.) Near the beginning of the titration of the weak acid, the pH increases more rapidly than it does in the strong acid case. It levels off near the halfway point and

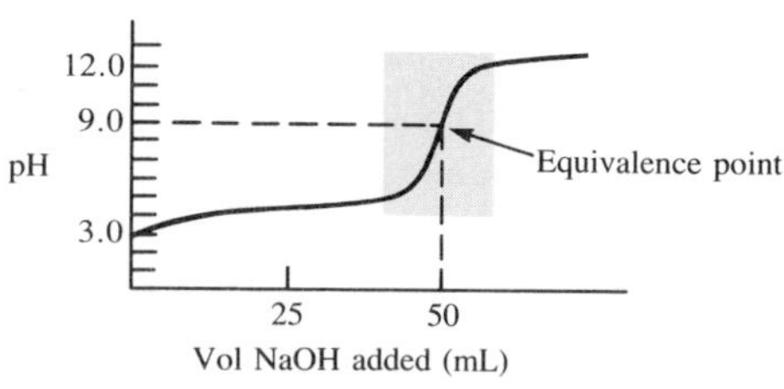

Figure 15.3

The pH curve for the titration of 50.0 mL of 0.100 *M* $HC_2H_3O_2$ with 0.100 *M* NaOH. Note that the equivalence point occurs at 50.0 mL of NaOH added, where the amount of added OH^- ions exactly equals the original amount of acid. The pH at the equivalence point is greater than 7.0 because the $C_2H_3O_2^-$ ion present at this point is a base and reacts with water to produce OH^-.

then increases rapidly again. The leveling off near the halfway point is caused by buffering effects. Earlier in this chapter we saw that optimum buffering occurs when [HA] is equal to $[A^-]$. This is exactly the case at the halfway point of the titration. As we can see from the curve, the pH changes least as hydroxide ion is added in this region.

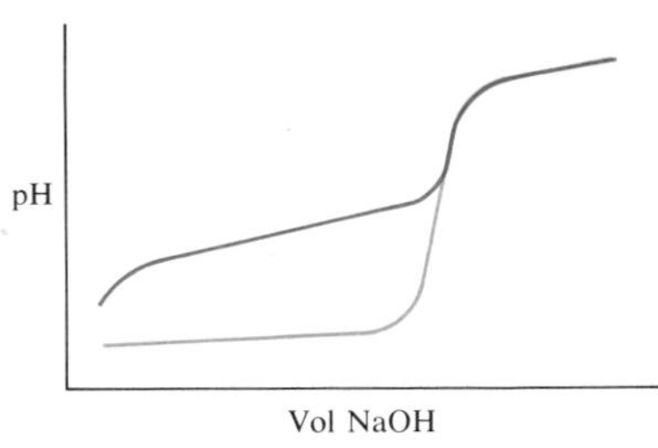

The other notable difference between the curves for strong and weak acids is the value of the pH at the equivalence point. For the titration of a strong acid, the equivalence point occurs at pH 7. For the titration of a weak acid, the pH at the equivalence point is greater than 7 because of the basicity of the conjugate base of the weak acid.

It is important to understand that the equivalence point in an acid-base titration is *defined by the stoichiometry, not by the pH*. The equivalence point occurs when enough titrant has been added to react exactly with all of the acid or base being titrated.

The equivalence point is defined by the stoichiometry, not by the pH.

Sample Exercise 15.7

Hydrogen cyanide gas (HCN) is a powerful respiratory inhibitor that is highly toxic. It is a very weak acid ($K_a = 6.2 \times 10^{-10}$) when dissolved in water. If a 50.0 mL sample of 0.100 *M* HCN is titrated with 0.100 *M* NaOH, calculate the pH of the solution

a. after 8.00 mL of 0.100 *M* NaOH has been added

Solution

a. *The stoichiometry problem*. After 8.00 mL of 0.100 *M* NaOH has been added, the titration reaction is

	HCN	+	OH^-	→	CN^-	+ H_2O
Before reaction:	50.0 mL × 0.100 *M* = 5.00 mmol		8.00 mL × 0.100 *M* = 0.800 mmol		0 mmol	
After reaction:	5.00 − 0.80 = 4.20 mmol		0		0.800 mmol	

The equilibrium problem. Since the solution contains the major species

$$HCN, \quad CN^-, \quad Na^+, \quad \text{and} \quad H_2O$$

the position of the acid dissociation equilibrium

$$HCN(aq) \rightleftharpoons H^+(aq) + CN^-(aq)$$

will determine the pH.

Initial concentration		Equilibrium concentration
$[HCN]_0 = \dfrac{4.2 \text{ mmol}}{(50.0 + 8.0) \text{ mL}}$		$[HCN] = \dfrac{4.2}{58.0} - x$
$[CN^-]_0 = \dfrac{0.800 \text{ mmol}}{(50.0 + 8.0) \text{ mL}}$	$\xrightarrow[\text{dissociates}]{x \text{ mmol/mL HCN}}$	$[CN^-] = \dfrac{0.80}{58.0} + x$
$[H^+]_0 \approx 0$		$[H^+] \approx x$

Sample Exercise 15.7, continued

Substituting into the expression for K_a gives

The approximations made here are well within the 5% rule.

$$6.2 \times 10^{-10} = K_a = \frac{[H^+][CN^-]}{[HCN]} = \frac{x\left(\frac{0.80}{58.0} + x\right)}{\frac{4.2}{58.0} - x} \approx \frac{x\left(\frac{0.80}{58.0}\right)}{\left(\frac{4.2}{58.0}\right)} = x\left(\frac{0.80}{4.2}\right)$$

$x = 3.3 \times 10^{-9}\ M = [H^+]$

pH = 8.49

b. at the halfway point in the titration

Solution

b. The amount of HCN originally present can be obtained from the original volume and molarity:

$$50.0 \text{ mL} \times 0.100\ M = 5.00 \text{ mmol}$$

Thus the halfway point will occur when 2.50 mmol of OH^- has been added:

$$\text{Volume of NaOH (in mL)} \times 0.100\ M = 2.50 \text{ mmol } OH^-$$

or

$$\text{Volume of NaOH} = 25.0 \text{ mL}$$

As was pointed out previously, at the halfway point [HCN] is equal to $[CN^-]$ and pH is equal to pK_a. Thus after 25.0 mL of 0.100 M NaOH has been added

$$pH = pK_a = -\log(6.2 \times 10^{-10}) = 9.21$$

c. at the equivalence point

Solution

c. The equivalence point will occur when a total of 5.00 mmol of OH^- has been added. Since the NaOH solution is 0.100 M, the equivalence point occurs when 50.0 mL of NaOH has been added. This amount will form 5.00 mmol of CN^-. The major species in solution at the equivalence point are:

$$CN^-, \quad Na^+, \quad \text{and} \quad H_2O$$

Thus the reaction that will control the pH involves the basic cyanide ion extracting a proton from water:

$$CN^-(aq) + H_2O(l) \rightleftharpoons HCN(aq) + OH^-(aq)$$

and

$$K_b = \frac{K_w}{K_a} = \frac{1.0 \times 10^{-14}}{6.2 \times 10^{-10}} = 1.6 \times 10^{-5} = \frac{[HCN][OH^-]}{[CN^-]}$$

Sample Exercise 15.7, continued

Initial concentration		Equilibrium concentration
$[CN^-]_0 = \dfrac{5.00 \text{ mmol}}{(50.0 + 50.0) \text{ mL}}$ $= 5.00 \times 10^{-2} M$		$[CN^-] = (5.00 \times 10^{-2}) - x$
$[HCN]_0 = 0$	$\xrightarrow[\text{with } H_2O]{x \text{ mmol/mL of } CN^- \text{ reacts}}$	$[HCN] = x$
$[OH^-]_0 \approx 0$		$[OH^-] \approx x$

Substituting the equilibrium concentrations into the expression for K_b and solving in the usual way gives

$$[OH^-] = x = 8.9 \times 10^{-4}$$

Then from K_w we have

$$[H^+] = 1.1 \times 10^{-11} \quad \text{and} \quad pH = 10.96$$

Two important conclusions can be drawn from a comparison of the titration of 50.0 mL of 0.1 M acetic acid covered earlier in this section and that of 50.0 mL of 0.1 M hydrocyanic acid analyzed in Sample Exercise 15.7. First, the same amount of 0.1 M NaOH is required to reach the equivalence point in both cases. The fact that HCN is a much weaker acid than $HC_2H_3O_2$ has no bearing on the amount of base required. It is the *amount* of acid, not its strength, that determines the equivalence point. Second, the *pH value* at the equivalence point *is* affected by the acid strength. For the titration of acetic acid, the pH at the equivalence point is 8.72; for the titration of hydrocyanic acid, the pH at the equivalence point is 10.96. This difference occurs because the CN^- ion is a much stronger base than the $C_2H_3O_2^-$ ion. Also, the pH at the halfway point of the titration is much higher for HCN than for $HC_2H_3O_2$, again because of the greater base strength of the CN^- ion (or equivalently, the smaller acid strength of HCN).

The amount of acid present, not its strength, determines the equivalence point.

The strength of a weak acid has a significant effect on the shape of its pH curve. Figure 15.4 shows pH curves for 50-mL samples of 0.10 M solutions of various acids titrated with 0.10 M NaOH. Note that the equivalence point occurs in each case when the same volume of 0.10 M NaOH has been added, but that the shapes of the curves are dramatically different. The weaker the acid, the greater the pH value at the equivalence point. In particular, note that the vertical region that surrounds the equivalence point becomes shorter as the acid being titrated becomes weaker. We will see in the next section that the choice of an indicator for such a titration is more limited.

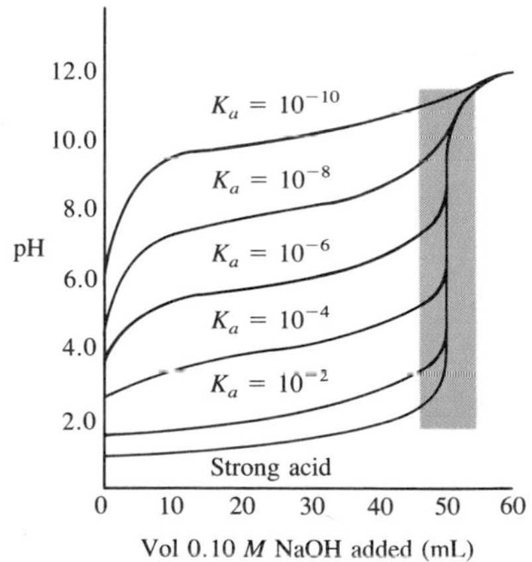

Figure 15.4

The pH curves for the titrations of 50.0-mL samples of 0.10 M acids with various K_a values with 0.10 M NaOH.

Besides being used to analyze for the amount of acid or base in a solution, titrations can also be used to determine the values of equilibrium constants, as shown in Sample Exercise 15.8.

Calculation of K_a

Sample Exercise 15.8

A chemist has synthesized a monoprotic weak acid and wants to determine its K_a value. To do so, the chemist dissolves 2.00 mmol of the solid acid in 100.0 mL of water and titrates the resulting solution with 0.0500 *M* NaOH. After 20.0 mL of NaOH has been added, the pH is 6.00. What is the K_a value for the acid?

Solution

The stoichiometry problem. We represent the monoprotic acid as HA. The titration reaction is

2.00 mmol HA

↓ add OH^-

1.00 mmol HA
1.00 mmol A^-

	HA	+	OH^-	→	A^-	+	H_2O
Before reaction:	2.00 mmol		20.0 mL × 0.0500 *M* = 1.00 mmol		0 mmol		
After reaction:	2.00 − 1.00 = 1.00 mmol		1.0 − 1.0 = 0		1.00 mmol		

The equilibrium problem. After the reaction the solution contains the major species

$$HA, \quad A^-, \quad Na^+, \quad \text{and} \quad H_2O$$

The pH will be determined by the equilibrium

$$HA(aq) \rightleftharpoons H^+(aq) + A^-(aq)$$

$$K_a = \frac{[H^+][A^-]}{[HA]}$$

Initial concentration		Equilibrium concentration
$[HA]_0 = \frac{1.00 \text{ mmol}}{(100.0 + 20.0) \text{ mL}} = 8.33 \times 10^{-3}\ M$		$[HA] = 8.33 \times 10^{-3} - x$
$[A^-]_0 = \frac{1.00 \text{ mmol}}{(100.0 + 20.0) \text{ mL}} = 8.33 \times 10^{-3}\ M$	x mmol/mL HA dissociates →	$[A^-] = 8.33 \times 10^{-3} + x$
$[H^+]_0 \approx 0$		$[H^+] \approx x$

Note that x is known here because the pH at this point is known to be 6.00. Thus

$$x = [H^+] = \text{antilog}(-\text{pH}) = 1.0 \times 10^{-6}\ M$$

Substituting the equilibrium concentrations into the expression for K_a allows calculation of the value of K_a:

$$K_a = \frac{[H^+][A^-]}{[HA]} = \frac{x(8.33 \times 10^{-3} + x)}{(8.33 \times 10^{-3}) - x}$$

Sample Exercise 15.8, continued

$$= \frac{(1.0 \times 10^{-6})(8.33 \times 10^{-3} + 1.0 \times 10^{-6})}{(8.33 \times 10^{-3}) - (1.0 \times 10^{-6})}$$

$$\approx \frac{(1.0 \times 10^{-6})(8.33 \times 10^{-3})}{8.33 \times 10^{-3}} = 1.0 \times 10^{-6}$$

There is an easier way to think about this problem: the original solution contained 2.00 mmol of HA; and since 20.0 mL of added 0.0500 *M* NaOH contains 1.00 mmol of OH^-, this is the halfway point in the titration, where [HA] is equal to $[A^-]$.

Thus $$[H^+] = K_a = 1.0 \times 10^{-6}$$

Titrations of Weak Bases with Strong Acids

Titrations of weak bases with strong acids can be treated using the procedures we've introduced previously. As always, you should *think first about the major species in solution* and decide whether a reaction occurs that runs essentially to completion. If such a reaction does occur, let it run to completion and do the stoichiometric calculations. Then choose the dominant equilibrium and calculate the pH.

The calculations involved for the titration of a weak base with a strong acid are shown by the titration of 100.0 mL of 0.050 *M* NH_3 with 0.10 *M* HCl.

Before the addition of any HCl.

1. Major species: NH_3 and H_2O. NH_3 is a base and will seek a source of protons. In this case, H_2O is the only available source.
2. No reactions occur that go to completion, since NH_3 cannot readily take a proton from H_2O. That is, the K_b value for NH_3 is small.
3. The equilibrium that controls the pH involves the reaction of ammonia with water:

$$NH_3(aq) + H_2O(l) \rightleftharpoons NH_4^+(aq) + OH^-(aq)$$

Use K_b to calculate $[OH^-]$. Although NH_3 is a weak base (compared to OH^-), it produces much more OH^- in this reaction than is produced from the autoionization of H_2O.

Before the equivalence point.

1. Major species (before any reaction):

$$NH_3, \underbrace{H^+, Cl^-,}_{\text{From added HCl}} \text{ and } H_2O$$

2. The NH_3 will react with H^+ from the added HCl:

$$NH_3(aq) + H^+(aq) \rightleftharpoons NH_4^+(aq)$$

This reaction proceeds essentially to completion. The NH_3 readily reacts with a free proton. This is much different from the previous case where H_2O was the only source of protons. Using the known volume of the 0.10 *M* HCl added, do the stoichiometric calculations.

3. After the reaction of NH_3 with H^+ is run to completion, the solution contains the following major species:

$$NH_3, \quad \underset{\substack{\uparrow \\ \text{Formed in} \\ \text{titration reaction}}}{NH_4^+}, \quad Cl^-, \quad \text{and} \quad H_2O$$

Thus the solution contains NH_3 and NH_4^+, and the equilibria involving these species will determine $[H^+]$. You can use either the dissociation reaction of NH_4^+:

$$NH_4^+(aq) \rightleftharpoons NH_3(aq) + H^+(aq)$$

or the reaction of NH_3 with H_2O:

$$NH_3(aq) + H_2O(l) \rightleftharpoons NH_4^+(aq) + OH^-(aq)$$

At the equivalence point.

1. By definition, the equivalence point occurs when all of the original NH_3 is converted to NH_4^+. Thus the major species in solution are: NH_4^+, Cl^-, and H_2O.
2. No reactions occur that go to completion.
3. The dominant equilibrium will be the dissociation of the weak acid NH_4^+, for which

$$K_a = \frac{K_w}{K_b \text{ (for } NH_3)}$$

Beyond the equivalence point.

1. Excess HCl has been added and the major species are H^+, NH_4^+, Cl^-, and H_2O.
2. No reaction occurs that goes to completion.
3. Although NH_4^+ will dissociate, it is such a weak acid that $[H^+]$ will be determined simply by the excess H^+:

$$[H^+] = \frac{\text{mmol of } H^+ \text{ in excess}}{\text{mL of solution}}$$

The results of these calculations are shown in Table 15.2. The pH curve is shown in Fig. 15.5.

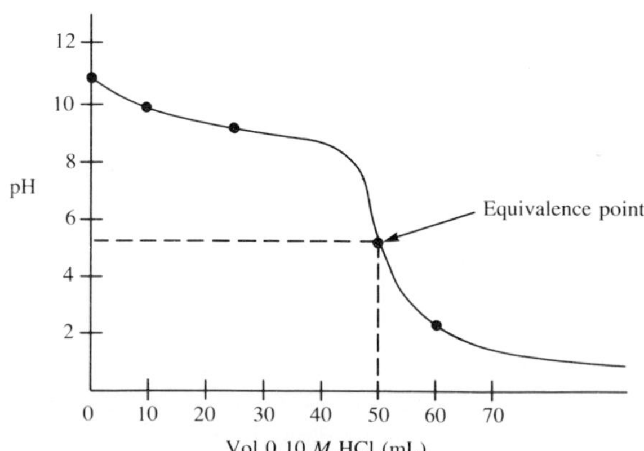

Figure 15.5

The pH curve for the titration of 100.0 mL of 0.050 *M* NH_3 with 0.10 *M* HCl. Note the pH at the equivalence point is less than 7, since the solution contains the weak acid NH_4^+.

Summary of Results for the Titration of 100.0 mL 0.050 M NH_3 with 0.10 M HCl				
Volume of 0.10 M HCl added (mL)	$[NH_3]_0$	$[NH_4^+]_0$	$[H^+]$	pH
0	0.05 M	0	1.1×10^{-11} M	10.96
10.0	$\frac{4.0 \text{ mmol}}{(100 + 10) \text{ mL}}$	$\frac{1.0 \text{ mmol}}{(100 + 10) \text{ mL}}$	1.4×10^{-10} M	9.85
25.0*	$\frac{2.5 \text{ mmol}}{(100 + 25) \text{ mL}}$	$\frac{2.5 \text{ mmol}}{(100 + 25) \text{ mL}}$	5.6×10^{-10} M	9.25
50.0†	0	$\frac{5.0 \text{ mmol}}{(100 + 50) \text{ mL}}$	4.3×10^{-6} M	5.36
60.0‡	0	$\frac{5.0 \text{ mmol}}{(100 + 60) \text{ mL}}$	$\frac{1.0 \text{ mmol}}{160 \text{ mL}} = 6.2 \times 10^{-3}$ M	2.21

*Halfway point
†Equivalence point
‡$[H^+]$ determined by the 1.0 mmol of excess H^+

Table 15.2

15.5 Acid-Base Indicators

Purpose

- To explain how acid-base indicators work.

There are two common methods for determining the equivalence point of an acid-base titration:

1. Plot a titration curve using a pH meter (see Fig. 14.4) to monitor pH. The center of the vertical region of the pH curve indicates the equivalence point (for example, see Figs. 15.1 through 15.5).
2. Use an **acid-base indicator,** which marks the end point of a titration by changing color. Although the *equivalence point of a titration, defined by the stoichiometry, is not necessarily the same as the end point* (where the indicator changes color), we can choose an indicator so that the two points are close enough that the error is negligible.

The most common acid-base indicators are complex molecules that are themselves weak acids and are represented by HIn. They exhibit one color when the proton is attached to the molecule and a different color when the proton is absent. For example, **phenolphthalein,** a commonly used indicator, is colorless in its HIn form and pink in its In^-, or basic, form. The actual structures of phenolphthalein are shown in Fig. 15.6 on the following page.

Methyl orange indicator is yellow in basic solution and red in acidic solution.

(Colorless acid form, HIn)

(Pink base form, In^-)

Figure 15.6

The acid and base forms of the indicator phenolphthalein. In the acid form (HIn), the molecule is colorless. When a proton (plus H_2O) is removed to give the base form (In^-), the color changes to pink.

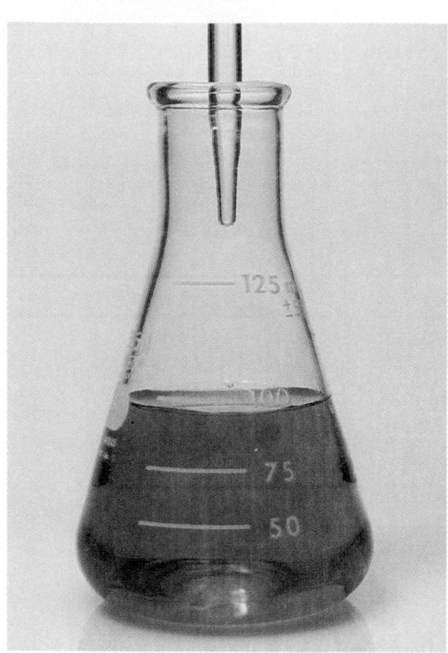

The indicator phenolphthalein is pink in basic solution and colorless in acidic solution.

To see how molecules such as phenolphthalein function as indicators, consider the following equilibrium for some hypothetical indicator HIn, a weak acid with $K_a = 1.0 \times 10^{-8}$.

$$\underset{\text{Red}}{\mathrm{HIn}(aq)} \rightleftharpoons \mathrm{H}^+(aq) + \underset{\text{Blue}}{\mathrm{In}^-(aq)}$$

$$K_a = \frac{[\mathrm{H}^+][\mathrm{In}^-]}{[\mathrm{HIn}]}$$

By rearranging, we get

$$\frac{K_a}{[\mathrm{H}^+]} = \frac{[\mathrm{In}^-]}{[\mathrm{HIn}]}$$

Suppose we add a few drops of this indicator to an acidic solution whose pH is 1.0 ($[\mathrm{H}^+] = 1.0 \times 10^{-1}$). Then

$$\frac{K_a}{[\mathrm{H}^+]} = \frac{1. \times 10^{-8}}{1.0 \times 10^{-1}} = 10^{-7} = \frac{[\mathrm{In}^-]}{[\mathrm{HIn}]} = \frac{1}{10{,}000{,}000}$$

The *end point* is defined by the change in color of the indicator. The *equivalence point* is defined by the reaction stoichiometry.

This ratio shows that the predominant form of the indicator in this solution is HIn, and the color of the solution will be red. If OH^- is added to this solution, $[H^+]$ decreases and the equilibrium, by Le Châtelier's principle, shifts to the right, changing HIn to In^-. At some point, enough of the In^- form will be present in the solution so that a bluish tint will be noticeable. That is, a color change from red to reddish purple will occur.

How much In^- must be present for the human eye to detect that the color is different from the original one? For most indicators, about a tenth of the initial form must be converted to the other form before a new color is apparent. We will assume, then, that in the titration of an acid with a base, the color change will occur at a pH where

$$\frac{[\mathrm{In}^-]}{[\mathrm{HIn}]} = \frac{1}{10}$$

Sample Exercise 15.9

Bromthymol blue, an indicator with a K_a value of approximately 1.0×10^{-7}, is yellow in its HIn form (in acids) and blue in its In^- form (in bases). Suppose we put a few drops of this indicator in a strongly acidic solution. If the solution is then titrated with NaOH, at what pH will the indicator color change first be visible?

Solution

For bromthymol blue,

$$K_a = 1.0 \times 10^{-7} = \frac{[\mathrm{H}^+][\mathrm{In}^-]}{[\mathrm{HIn}]}$$

We assume the color change is visible when

$$\frac{[\mathrm{In}^-]}{[\mathrm{HIn}]} = \frac{1}{10}$$

That is, we assume we can see the first hint of a greenish tint (yellow plus a little blue) when the solution contains one part blue and ten parts yellow (see Fig. 15.7). Thus

$$K_a = 1.0 \times 10^{-7} = \frac{[\mathrm{H}^+](1)}{10}$$

$$[\mathrm{H}^+] = 1.0 \times 10^{-6} \quad \text{or} \quad \mathrm{pH} = 6.0$$

The color change is first visible at pH 6.0.

The Henderson-Hasselbalch equation is very useful in determining the pH at which an indicator changes color. For example, Equation (15.2) applied to the general indicator HIn is

$$\mathrm{pH} = \mathrm{p}K_a + \log\left(\frac{[\mathrm{In}^-]}{[\mathrm{HIn}]}\right)$$

where K_a is the dissociation constant for the acid form of the indicator (HIn). Since we assume that the color change is visible when

$$\frac{[\mathrm{In}^-]}{[\mathrm{HIn}]} = \frac{1}{10}$$

we have the following equation for determining the pH at which the color change occurs:

$$\mathrm{pH} = \mathrm{p}K_a + \log(\tfrac{1}{10}) = \mathrm{p}K_a - 1$$

(a)

(b)

(c)

Figure 15.7

(a) Yellow acid form of bromthymol blue; (b) a greenish tint is seen when the solution contains 1 part blue and 10 parts yellow; (c) blue basic form.

For bromthymol blue ($K_a = 1 \times 10^{-7}$, or $pK_a = 7$), the pH at the color change is

$$pH = 7 - 1 = 6$$

as we calculated in Sample Exercise 15.9.

When a basic solution is titrated, the indicator HIn will initially exist as In^- in solution, but as acid is added, HIn will form. In this case, the color will change when there is a mixture of 10 parts In^- and 1 part HIn. That is, a color change from blue to a blue-green color will occur (See Fig. 15.7) due to the presence of some of the yellow HIn molecules. This color change will be first visible when

$$\frac{[In^-]}{[HIn]} = \frac{10}{1}$$

Note that this is the reciprocal of the ratio for the titration of an acid. Substituting this ratio into the Henderson-Hasselbalch equation gives

$$pH = pK_a + \log(\tfrac{10}{1}) = pK_a + 1$$

For bromthymol blue ($pK_a = 7$), we have a color change at

$$pH = 7 + 1 = 8$$

In summary, when we use bromthymol blue for the titration of an acid, the starting form will be HIn (yellow), and the color change will occur at a pH of about 6. When bromthymol blue is used for the titration of a base, the starting form is In^- (blue), and the color change occurs at a pH of about 8. Thus the useful pH range for bromthymol blue is

$$pK_a \text{ (bromthymol blue)} \pm 1 = 7 \pm 1$$

or from 6 to 8. The useful pH ranges for several common indicators are shown in Fig. 15.8.

When we choose an indicator for a titration, we want the indicator end point (where the color changes) and the titration equivalence point to be as close as possible. Choosing an indicator is easier if there is a large change in pH near the

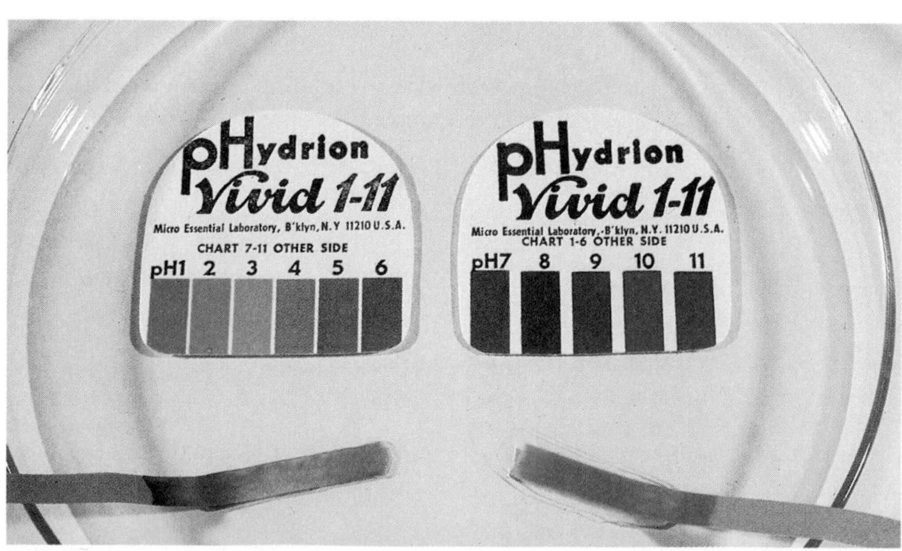

Universal indicator paper can be used to estimate the pH of a solution.

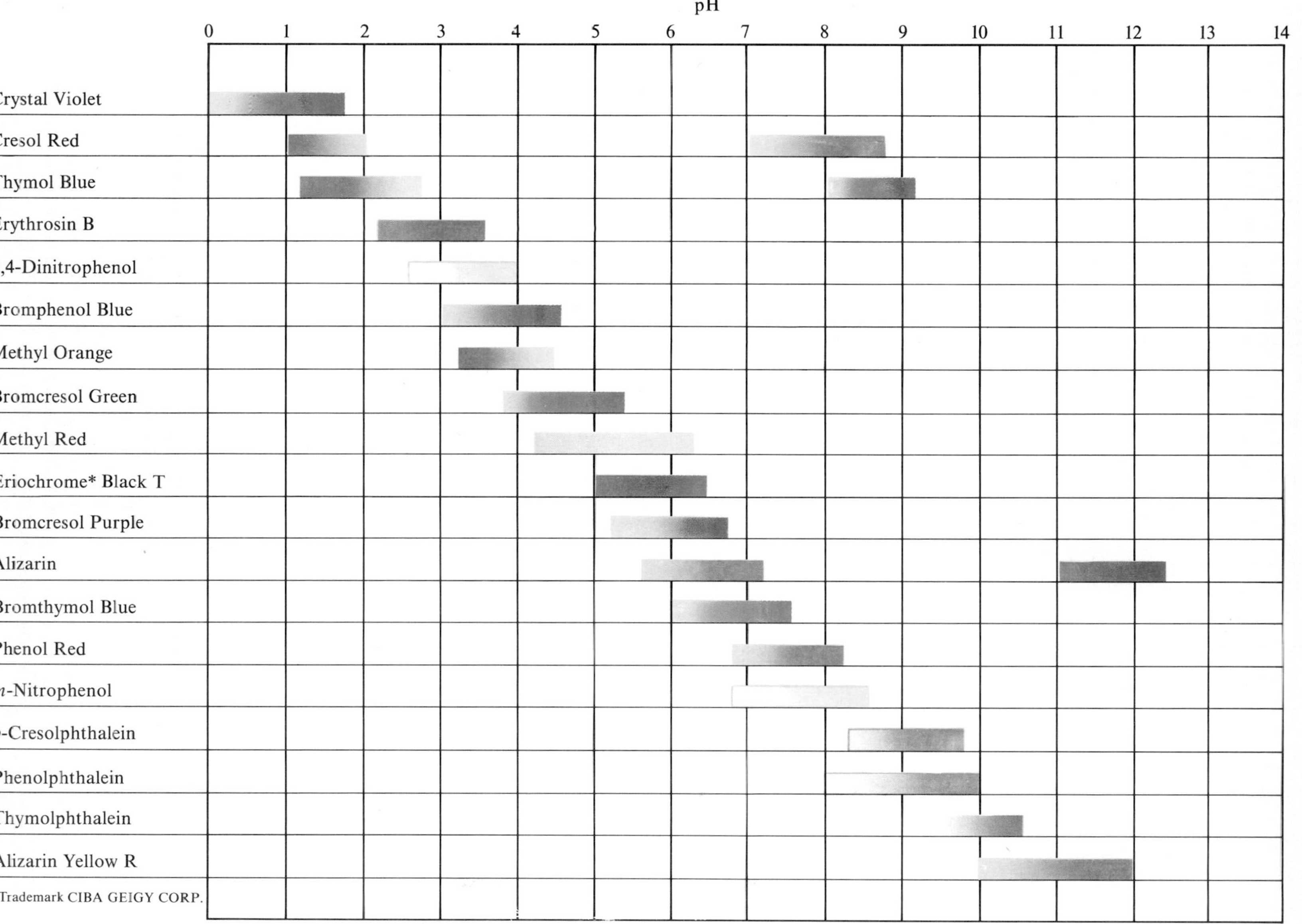

Figure 15.8

The useful pH ranges for several common indicators. Note that most indicators have a useful range of about two pH units, as predicted by the expression $pK_a \pm 1$.

equivalence point of the titration. The dramatic change in pH near the equivalence point in a strong acid–strong base titration (Figs. 15.1 and 15.2) produces a sharp end point; that is, the complete color change (from the acid-to-base or base-to-acid colors) usually occurs over one drop of added titrant.

What indicator should we use for the titration of 100.00 mL of 0.100 *M* HCl with 0.100 *M* NaOH? We know that the equivalence point occurs at pH 7.0. In the initially acidic solution, the indicator will be predominantly in the HIn form. As OH^- ions are added, the pH will increase rather slowly at first (Fig. 15.1) and then will rise rapidly at the equivalence point. This sharp change causes the indicator dissociation equilibrium

$$\text{HIn} \rightleftharpoons \text{H}^+ + \text{In}^-$$

to shift suddenly to the right, producing enough In^- ions to give a color change. Since we are titrating an acid and the indicator is predominantly in the acid form initially, the first observable color change will occur at a pH where

$$\frac{[\text{In}^-]}{[\text{HIn}]} = \frac{1}{10}$$

Thus

$$\text{pH} = \text{p}K_a + \log(\tfrac{1}{10}) = \text{p}K_a - 1$$

If we want an indicator that will change color at pH 7, we can use this relationship to find the pK_a value for a suitable indicator:

$$\text{pH} = 7 = \text{p}K_a - 1 \quad \text{or} \quad \text{p}K_a = 7 + 1 = 8$$

Thus an indicator with a pK_a value of 8 ($K_a = 1 \times 10^{-8}$) will change color at about pH 7 and will be ideal to mark the end point for a strong acid–strong base titration.

How crucial is it for a strong acid–strong base titration that the indicator change color exactly at pH 7? We can answer this question by examining the pH change near the equivalence point of the titration of 100 mL of 0.10 *M* HCl with 0.10 *M* NaOH. The data for a few points at or near the equivalence point are shown in Table 15.3. Note that in going from 99.99 mL to 100.01 mL of added NaOH solution (about half of a drop), the pH changes from 5.3 to 8.7—a very dramatic change. This behavior leads to the following general conclusions about indicators for a strong acid–strong base titration:

Selected pH Values near the Equivalence Point in the Titration of 100.0 mL of 0.10 *M* HCl with 0.10 *M* NaOH

NaOH added (mL)	pH
99.99	5.3
100.00	7.0
100.01	8.7

Table 15.3

Indicator color changes will be sharp, occurring with the addition of a single drop of titrant.

There is a wide choice of suitable indicators. The results will agree within one drop of titrant, using indicators with end points as far apart as pH 5 and pH 9 (see Fig. 15.9).

The titration of weak acids is somewhat different. Figure 15.4 shows that the weaker the acid being titrated, the smaller the vertical area around the equivalence point, and there is much less flexibility in choosing the indicator. We must choose an indicator whose useful pH range has a midpoint as close as possible to the pH at the equivalence point. For example, we saw earlier that in the titration of 0.1 *M* $HC_2H_3O_2$ with 0.1 *M* NaOH, the pH at the equivalence point is 8.7 (see Fig. 15.3). A good indicator choice would be phenolphthalein, since its useful pH range is 8–10. Thymol blue would also be acceptable, but methyl red would not. The choice of an indicator is illustrated graphically in Fig. 15.10.

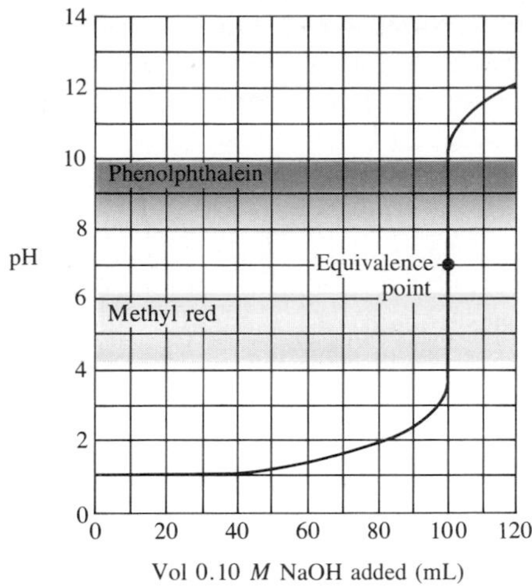

Figure 15.9

The pH curve for the titration of 100.0 mL of 0.10 M HCl with 0.10 M NaOH. Note that phenolphthalein and methyl red have end points at virtually the same amounts of added NaOH.

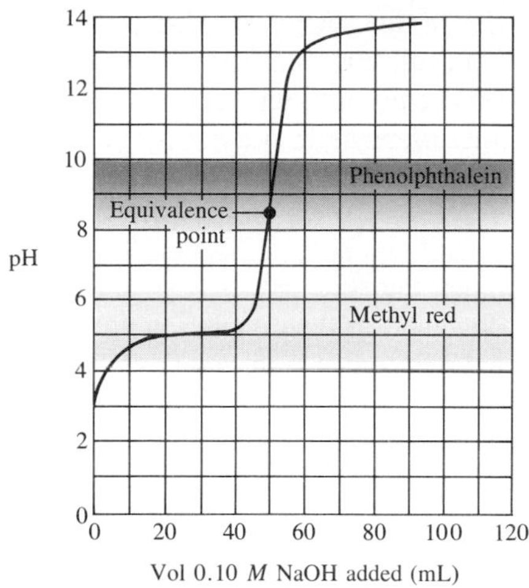

Figure 15.10

The pH curve for the titration of 50 mL of 0.1 M $HC_2H_3O_2$ with 0.1 M NaOH. Phenolphthalein will give an end point very close to the equivalence point of the titration. Methyl red would change color well before the equivalence point (so the end point would be very different from the equivalence point) and would not be a suitable indicator for this titration.

15.6 Solubility Equilibria and the Solubility Product

Purpose

- To show how to calculate the solubility product of a salt given its solubility, and vice versa.
- To demonstrate the prediction of relative solubilities from K_{sp} values.
- To explain the effect of pH and a common ion on the solubility of a salt.

Adding F^- to drinking water is controversial. See "Fluoridation of Water" by Bette Hileman, *Chem. and Eng. News,* Aug. 1, 1988, p. 26.

Solubility is a very important phenomenon. The fact that substances such as sugar and table salt dissolve in water allows us to flavor foods easily. The fact that calcium sulfate is less soluble in hot water than in cold water causes it to coat tubes in boilers, reducing thermal efficiency. Tooth decay involves solubility: when food lodges between the teeth, acids form that dissolve tooth enamel, which contains a mineral called hydroxyapatite, $Ca_5(PO_4)_3OH$. Tooth decay can be reduced by treating teeth with fluoride (Chemical Impact, p. 729). Fluoride replaces the hydroxide in hydroxyapatite to produce the corresponding fluorapatite, $Ca_5(PO_4)_3F$, and calcium fluoride, CaF_2, both of which are less soluble in acids than the original enamel. Another important consequence of solubility involves the use of a suspension of barium sulfate to improve the clarity of X rays of the gastrointestinal tract. The very low solubility of barium sulfate, which contains the toxic ion Ba^{2+}, makes ingestion of the compound safe.

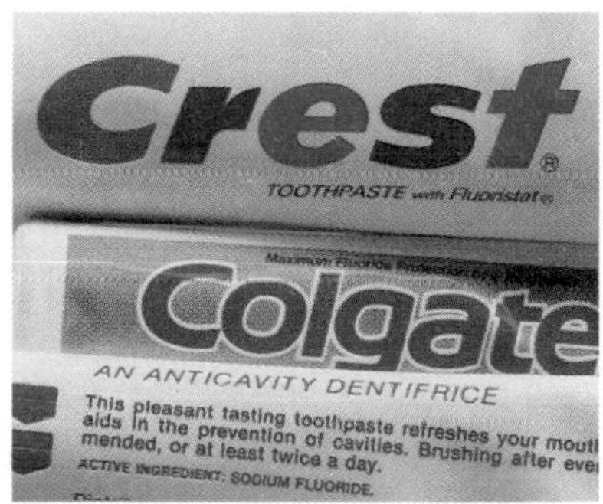

Most popular toothpastes have fluoride added.

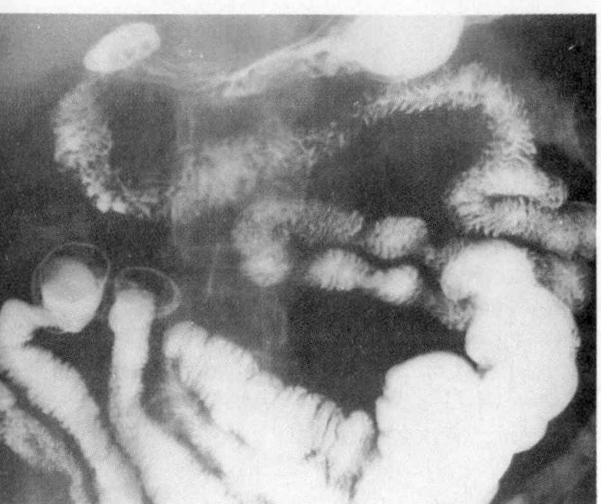

An X ray of the upper gastrointestinal tract using barium sulfate.

In this section we consider the equilibria associated with solids dissolving to form aqueous solutions. When a typical ionic solid dissolves in water, it dissociates completely into separate hydrated cations and anions. For example, calcium fluoride dissolves in water as follows:

$$CaF_2(s) \xrightarrow{H_2O} Ca^{2+}(aq) + 2F^-(aq)$$

When the solid salt is first added to the water, no Ca^{2+} and F^- ions are present. However, as the dissolution proceeds, the concentrations of Ca^{2+} and F^- increase, and it becomes more and more likely that these ions will collide and reform the solid phase. Thus two competing processes are occurring: the above reaction and the reverse reaction:

$$Ca^{2+}(aq) + 2F^-(aq) \rightarrow CaF_2(s)$$

Ultimately, dynamic equilibrium is reached:

$$CaF_2(s) \rightleftharpoons Ca^{2+}(aq) + 2F^-(aq)$$

Then no more solid dissolves (the solution is said to be *saturated*).

We can write an equilibrium expression for this process according to the law of mass action:

$$K_{sp} = [Ca^{2+}][F^-]^2$$

where $[Ca^{2+}]$ and $[F^-]$ are expressed in mol/L. The constant K_{sp} is called the **solubility product constant,** or simply the **solubility product** for the equilibrium expression.

Pure liquids and pure solids are never included in an equilibrium expression (Section 13.4).

Since CaF_2 is a pure solid, it is not included in the equilibrium expression. The fact that the amount of excess solid present does not affect the position of the solubility equilibrium might seem strange at first; more solid means more surface area exposed to the solvent, which would seem to result in greater solubility. This is not the case, however. When the ions in solution reform the solid, they do so on the surface of the solid. Thus, doubling the surface area of the solid not only doubles the rate of dissolving, it also doubles the rate of reformation of the solid. The amount of excess solid present therefore has no effect on the equilibrium position. Similarly, although either increasing the surface area by grinding up the solid or stirring the solution speeds up the attainment of equilibrium, neither procedure changes the amount of solid dissolved at equilibrium. Neither the amount of excess solid nor the size of the particles present will shift the *position* of the solubility equilibrium.

K_{sp} is an equilibrium constant; solubility is an equilibrium position.

It is very important to distinguish between the *solubility* of a given solid and its *solubility product*. The solubility product is an *equilibrium constant* and has only *one* value for a given solid at a given temperature. Solubility, on the other hand, is an *equilibrium position* and has an *infinite number* of possible values at a given temperature, depending on the other conditions (such as the presence of a common ion). The K_{sp} values at 25°C for many common ionic solids are listed in Table 15.4. The units are customarily omitted.

Solving solubility equilibria problems requires many of the same procedures we have used to deal with acid-base equilibria, as illustrated in Sample Exercises 15.10 and 15.11.

K_{sp} Values at 25°C for Common Ionic Solids

Ionic solid	K_{sp} (at 25°C)	Ionic solid	K_{sp} (at 25°C)	Ionic solid	K_{sp} (at 25°C)
Fluorides		Hg_2CrO_4*	2×10^{-9}	$Co(OH)_2$	2.5×10^{-16}
BaF_2	2.4×10^{-5}	$BaCrO_4$	8.5×10^{-11}	$Ni(OH)_2$	1.6×10^{-16}
MgF_2	6.4×10^{-9}	Ag_2CrO_4	9.0×10^{-12}	$Zn(OH)_2$	4.5×10^{-17}
PbF_2	4×10^{-8}	$PbCrO_4$	2×10^{-16}	$Cu(OH)_2$	1.6×10^{-19}
SrF_2	7.9×10^{-10}	Carbonates		$Hg(OH)_2$	3×10^{-26}
CaF_2	4.0×10^{-11}	$NiCO_3$	1.4×10^{-7}	$Sn(OH)_2$	3×10^{-27}
Chlorides		$CaCO_3$	8.7×10^{-9}	$Cr(OH)_3$	6.7×10^{-31}
$PbCl_2$	1.6×10^{-5}	$BaCO_3$	1.6×10^{-9}	$Al(OH)_3$	2×10^{-32}
$AgCl$	1.6×10^{-10}	$SrCO_3$	7×10^{-10}	$Fe(OH)_3$	4×10^{-38}
Hg_2Cl_2*	1.1×10^{-18}	$CuCO_3$	2.5×10^{-10}	$Co(OH)_3$	2.5×10^{-43}
Bromides		$ZnCO_3$	2×10^{-10}	Sulfides	
$PbBr_2$	4.6×10^{-6}	$MnCO_3$	8.8×10^{-11}	MnS	2.3×10^{-13}
$AgBr$	5.0×10^{-13}	$FeCO_3$	2.1×10^{-11}	FeS	3.7×10^{-19}
Hg_2Br_2*	1.3×10^{-22}	Ag_2CO_3	8.1×10^{-12}	NiS	3×10^{-21}
Iodides		$CdCO_3$	5.2×10^{-12}	CoS	5×10^{-22}
PbI_2	1.4×10^{-8}	$PbCO_3$	1.5×10^{-15}	ZnS	2.5×10^{-22}
AgI	1.5×10^{-16}	$MgCO_3$	1×10^{-15}	SnS	1×10^{-26}
Hg_2I_2*	4.5×10^{-29}	Hg_2CO_3*	9.0×10^{-15}	CdS	1.0×10^{-28}
Sulfates		Hydroxides		PbS	7×10^{-29}
$CaSO_4$	6.1×10^{-5}	$Ba(OH)_2$	5.0×10^{-3}	CuS	8.5×10^{-45}
Ag_2SO_4	1.2×10^{-5}	$Sr(OH)_2$	3.2×10^{-4}	Ag_2S	1.6×10^{-49}
$SrSO_4$	3.2×10^{-7}	$Ca(OH)_2$	1.3×10^{-6}	HgS	1.6×10^{-54}
$PbSO_4$	1.3×10^{-8}	$AgOH$	2.0×10^{-8}	Phosphates	
$BaSO_4$	1.5×10^{-9}	$Mg(OH)_2$	8.9×10^{-12}	Ag_3PO_4	1.8×10^{-18}
Chromates		$Mn(OH)_2$	2×10^{-13}	$Sr_3(PO_4)_2$	1×10^{-31}
$SrCrO_4$	3.6×10^{-5}	$Cd(OH)_2$	5.9×10^{-15}	$Ca_3(PO_4)_2$	1.3×10^{-32}
		$Pb(OH)_2$	1.2×10^{-15}	$Ba_3(PO_4)_2$	6×10^{-39}
		$Fe(OH)_2$	1.8×10^{-15}	$Pb_3(PO_4)_2$	1×10^{-54}

*Contains Hg_2^{2+} ions. $K_{sp} = [Hg_2^{2+}][X^-]^2$ for Hg_2X_2 salts.

Table 15.4

Sample Exercise 15.10

Copper(I) bromide has a measured solubility of 2.0×10^{-4} mol/L at 25°C. Calculate its K_{sp} value.

Solution

In this experiment the solid was placed in contact with water. Thus before any reaction occurred, the system contained solid CuBr and H_2O. The process that occurs is the dissolving of CuBr to form the separated Cu^+ and Br^- ions:

$$CuBr(s) \rightleftharpoons Cu^+(aq) + Br^-(aq)$$

where

$$K_{sp} = [Cu^+][Br^-]$$

Sample Exercise 15.10, continued

Initially, the solution contains no Cu^+ or Br^-, so the initial concentrations are

$$[Cu^+]_0 = [Br^-]_0 = 0$$

The equilibrium concentrations can be obtained from the measured solubility of CuBr, which is 2.0×10^{-4} mol/L. This means that 2.0×10^{-4} mol of solid CuBr dissolves per 1.0 L of solution to come to equilibrium with the excess solid. The reaction is

$$\text{CuBr}(s) \rightarrow \text{Cu}^+(aq) + \text{Br}^-(aq)$$

Thus

$$2.0 \times 10^{-4} \text{ mol/L CuBr}(s) \rightarrow 2.0 \times 10^{-4} \text{ mol/L Cu}^+(aq) + 2.0 \times 10^{-4} \text{ mol/L Br}^-(aq)$$

We can now write the equilibrium concentrations:

$$\begin{aligned}[Cu^+] &= [Cu^+]_0 + \text{change to reach equilibrium} \\ &= 0 + 2.0 \times 10^{-4} \text{ mol/L}\end{aligned}$$

and

$$\begin{aligned}[Br^-] &= [Br^-]_0 + \text{change to reach equilibrium} \\ &= 0 + 2.0 \times 10^{-4} \text{ mol/L}\end{aligned}$$

These equilibrium concentrations allow us to calculate the value of K_{sp} for CuBr:

$$\begin{aligned}K_{sp} &= [Cu^+][Br^-] = (2.0 \times 10^{-4} \text{ mol/L})(2.0 \times 10^{-4} \text{ mol/L}) \\ &= 4.0 \times 10^{-8} \text{ mol}^2/\text{L}^2 = 4.0 \times 10^{-8}\end{aligned}$$

The units for K_{sp} values are usually omitted.

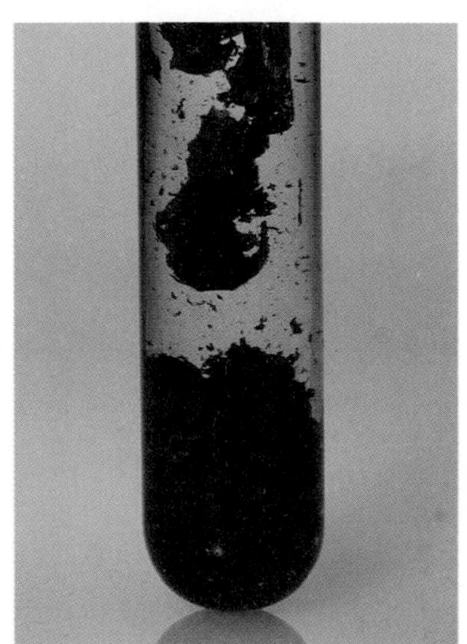

Precipitation of bismuth sulfide.

Sample Exercise 15.11

Calculate the K_{sp} value for bismuth sulfide (Bi_2S_3), which has a solubility of 1.0×10^{-15} mol/L at 25°C.

Solution

The system initially contains H_2O and solid Bi_2S_3, which dissolves as follows:

$$\text{Bi}_2\text{S}_3(s) \rightleftharpoons 2\text{Bi}^{3+}(aq) + 3\text{S}^{2-}(aq)$$

Therefore $$K_{sp} = [Bi^{3+}]^2[S^{2-}]^3$$

Since no Bi^{3+} and S^{2-} ions were present in solution before the Bi_2S_3 dissolved,

$$[Bi^{3+}]_0 = [S^{2-}]_0 = 0$$

Thus the equilibrium concentrations of these ions will be determined by the amount of salt that dissolves to reach equilibrium, which in this case is 1.0×10^{-15} mol/L. Since each Bi_2S_3 unit contains $2Bi^{3+}$ and $3S^{2-}$ ions:

$$\text{Bi}_2\text{S}_3(s) \rightarrow 2\text{Bi}^{3+}(aq) + 3\text{S}^{2-}(aq)$$

Sample Exercise 15.11, continued

Thus

$$1.0 \times 10^{-15}\ \text{mol/L}\ Bi_2S_3(s) \rightarrow 2(1.0 \times 10^{-15}\ \text{mol/L})\ Bi^{3+}(aq) + 3(1.0 \times 10^{-15}\ \text{mol/L})\ S^{2-}(aq)$$

Sulfide is a very basic anion and really exists in water as HS^-. We will not consider this complication.

The equilibrium concentrations are

$$[Bi^{3+}] = [Bi^{3+}]_0 + \text{change} = 0 + 2.0 \times 10^{-15}\ \text{mol/L}$$

$$[S^{2-}] = [S^{2-}]_0 + \text{change} = 0 + 3.0 \times 10^{-15}\ \text{mol/L}$$

Then $K_{sp} = [Bi^{3+}]^2[S^{2-}]^3 = (2.0 \times 10^{-15})^2(3.0 \times 10^{-15})^3 = 1.1 \times 10^{-73}$

Solubilities must be expressed in mol/L in K_{sp} calculations.

We have seen that the experimentally determined solubility of an ionic solid can be used to calculate its K_{sp} value.* The reverse is also possible: the solubility of an ionic solid can be calculated if its K_{sp} value is known.

Sample Exercise 15.12

The K_{sp} value for copper(II) iodate, $Cu(IO_3)_2$, is 1.4×10^{-7} at 25°C. Calculate its solubility at 25°C.

Solution

The system initially contains H_2O and solid $Cu(IO_3)_2$, which dissolves according to the following equilibrium:

$$Cu(IO_3)_2(s) \rightleftharpoons Cu^{2+}(aq) + 2IO_3^-(aq)$$

Therefore $$K_{sp} = [Cu^{2+}][IO_3^-]^2$$

To find the solubility of $Cu(IO_3)_2$, we must find the equilibrium concentrations of the Cu^{2+} and IO_3^- ions. We do this in the usual way by specifying the initial concentrations (before any solid has dissolved) and then defining the change required to reach equilibrium. Since in this case we do not know the solubility, we will assume that x mol/L of the solid dissolves to reach equilibrium. The 1:2 stoichiometry of the salt means that

$$x\ \text{mol/L}\ Cu(IO_3)_2(s) \rightarrow x\ \text{mol/L}\ Cu^{2+}(aq) + 2x\ \text{mol/L}\ IO_3^-(aq)$$

The concentrations are as follows:

Initial concentration (mol/L) (before any $Cu(IO_3)_2$ dissolves)		Equilibrium concentration (mol/L)
$[Cu^{2+}]_0 = 0$ $[IO_3^-]_0 = 0$	x mol/L dissolves $\longrightarrow$ to reach equilibrium	$[Cu^{2+}] = x$ $[IO_3^-] = 2x$

*This calculation assumes that all of the dissolved solid is present as separated ions. In some cases, such as $CaSO_4$, large numbers of ion pairs exist in solution, so this method yields an incorrect value for K_{sp}.

Sample Exercise 15.12, continued

Substituting the equilibrium concentrations into the expression for K_{sp} gives

$$1.4 \times 10^{-7} = K_{sp} = [Cu^{2+}][IO_3^{-}]^2 = (x)(2x)^2 = 4x^3$$

Then $$x = \sqrt[3]{3.5 \times 10^{-8}} = 3.3 \times 10^{-3} \text{ mol/L}$$

Thus the solubility of solid $Cu(IO_3)_2$ is 3.3×10^{-3} mol/L.

Relative Solubilities

A salt's K_{sp} value gives us information about its solubility. However, we must be careful in using K_{sp} values to predict the *relative* solubilities of a group of salts. There are two possible cases:

1. The salts being compared produce the same number of ions. For example, consider

$$\begin{array}{ll} AgI(s) & K_{sp} = 1.5 \times 10^{-16} \\ CuI(s) & K_{sp} = 5.0 \times 10^{-12} \\ CaSO_4(s) & K_{sp} = 6.1 \times 10^{-5} \end{array}$$

Each of these solids dissolves to produce two ions:

$$\text{Salt} \rightleftharpoons \text{cation} + \text{anion}$$
$$K_{sp} = [\text{cation}][\text{anion}]$$

If x is the solubility in mol/L, then at equilibrium

$$[\text{cation}] = x$$
$$[\text{anion}] = x$$
$$K_{sp} = [\text{cation}][\text{anion}] = x^2$$
$$x = \sqrt{K_{sp}} = \text{solubility}$$

In this case we can compare the solubilities for these solids by comparing the K_{sp} values:

$$CaSO_4(s) > CuI(s) > AgI(s)$$

Most soluble; largest K_{sp} — Least soluble; smallest K_{sp}

2. The salts being compared produce different numbers of ions. For example, consider

$$\begin{array}{ll} CuS(s) & K_{sp} = 8.5 \times 10^{-45} \\ Ag_2S(s) & K_{sp} = 1.6 \times 10^{-49} \\ Bi_2S_3(s) & K_{sp} = 1.1 \times 10^{-73} \end{array}$$

These salts produce different numbers of ions when they dissolve. The K_{sp} values cannot be compared *directly* to determine relative solubilities. In fact, if we calculate the solubilities (using the procedure in Sample Exercise 15.12), we obtain the results summarized in Table 15.5. The order of solubilities is

Calculated Solubilities for CuS, Ag_2S, and Bi_2S_3 at 25°C		
Salt	K_{sp}	Calculated solubility (mol/L)
CuS	8.5×10^{-45}	9.2×10^{-23}
Ag_2S	1.6×10^{-49}	3.4×10^{-17}
Bi_2S_3	1.1×10^{-73}	1.0×10^{-15}

Table 15.5

$$\underset{\text{Most soluble}}{Bi_2S_3(s)} > Ag_2S(s) > \underset{\text{Least soluble}}{CuS(s)}$$

which is opposite to the order of the K_{sp} values.

Remember that relative solubilities can be predicted by comparing K_{sp} values *only* for salts that produce the same total number of ions.

Common Ion Effect

So far we have considered ionic solids dissolved in pure water. We will now see what happens when the water contains an ion in common with the dissolving salt. For example, consider the solubility of solid silver chromate (Ag_2CrO_4, $K_{sp} = 9.0 \times 10^{-12}$) in a 0.100 M solution of $AgNO_3$. Before any Ag_2CrO_4 dissolves, the solution contains the major species Ag^+, NO_3^-, and H_2O, with solid Ag_2CrO_4 on the bottom of the container. Since NO_3^- is not found in Ag_2CrO_4, we can ignore it. The relevant initial concentrations (before any Ag_2CrO_4 dissolves) are

$$[Ag^+]_0 = 0.100\ M \text{ (from the dissolved } AgNO_3)$$
$$[CrO_4^{2-}]_0 = 0$$

The system comes to equilibrium as Ag_2CrO_4 dissolves according to the reaction

$$Ag_2CrO_4(s) \rightleftharpoons 2Ag^+(aq) + CrO_4^{2-}(aq)$$

for which

$$K_{sp} = [Ag^+]^2[CrO_4^{2-}] = 9.0 \times 10^{-12}$$

We assume that x mol/L of Ag_2CrO_4 dissolves to reach equilibrium, which means that

$$x \text{ mol/L } Ag_2CrO_4(s) \rightarrow 2x \text{ mol/L } Ag^+(aq) + x \text{ mol/L } CrO_4^{2-}$$

Now we can specify the equilibrium concentrations in terms of x:

$$[Ag^+] = [Ag^+]_0 + \text{change} = 0.100 + 2x$$
$$[CrO_4^{2-}] = [CrO_4^{2-}]_0 + \text{change} = 0 + x = x$$

Substituting these concentrations into the expression for K_{sp} gives

$$9.0 \times 10^{-12} = [Ag^+]^2[CrO_4^{2-}] = (0.100 + 2x)^2(x)$$

The mathematics required here can be complicated, since the multiplication of terms on the right-hand side produces an expression that contains an x^3 term. However, as is usually the case, we can make simplifying assumptions. Since the K_{sp}

Silver chromate is a brown solid.

value for Ag_2CrO_4 is small (the position of the equilibrium lies far to the left), x is expected to be small compared to 0.100 M. Therefore, $0.100 + 2x \approx 0.100$, which allows simplification of the expression:

$$9.0 \times 10^{-12} = (0.100 + 2x)^2(x) \approx (0.100)^2(x)$$

Then

$$x \approx \frac{9.0 \times 10^{-12}}{(0.100)^2} = 9.0 \times 10^{-10} \text{ mol/L}$$

Since x is, in fact, much less than 0.100 M, the assumption is valid by the 5% rule. Thus

$$\text{Solubility of } Ag_2CrO_4 \text{ in } 0.100\ M\ AgNO_3 = x = 9.0 \times 10^{-10} \text{ mol/L}$$

and the equilibrium concentrations are

$$[Ag^+] = 0.100 + 2x = 0.100 + 2(9.0 \times 10^{-10}) = 0.100\ M$$
$$[CrO_4^{2-}] = x = 9.0 \times 10^{-10}\ M$$

Now we compare the solubilities of Ag_2CrO_4 in pure water and in 0.100 M $AgNO_3$:

$$\text{Solubility of } Ag_2CrO_4 \text{ in pure water} = 1.3 \times 10^{-4} \text{ mol/L}$$
$$\text{Solubility of } Ag_2CrO_4 \text{ in } 0.100\ M\ AgNO_3 = 9.0 \times 10^{-10} \text{ mol/L}$$

Note that the solubility of Ag_2CrO_4 is much less in the presence of Ag^+ ions from $AgNO_3$. This is another example of the common ion effect. The solubility of a solid is lowered if the solution already contains ions common to the solid.

Sample Exercise 15.13

Calculate the solubility of solid CaF_2 ($K_{sp} = 4.0 \times 10^{-11}$) in a 0.025 M NaF solution.

Solution

Before any CaF_2 dissolves, the solution contains the major species Na^+, F^-, and H_2O. The solubility equilibrium for CaF_2 is

$$CaF_2(s) \rightleftharpoons Ca^{2+}(aq) + 2F^-(aq)$$

and

$$K_{sp} = 4.0 \times 10^{-11} = [Ca^{2+}][F^-]^2$$

Initial concentration (mol/L) (before any CaF_2 dissolves)		Equilibrium concentration (mol/L)
$[Ca^{2+}]_0 = 0$ $[F^-]_0 = 0.025\ M$ From 0.025 M NaF	x mol/L CaF_2 dissolves → to reach equilibrium	$[Ca^{2+}] = x$ $[F^-] = 0.025 + 2x$ From NaF From CaF_2

Substituting the equilibrium concentrations into the expression for K_{sp} gives

$$K_{sp} = 4.0 \times 10^{-11} = [Ca^{2+}][F^-]^2 = (x)(0.025 + 2x)^2$$

Sample Exercise 15.13, continued

Assuming that $2x$ is negligible compared to 0.025 (since K_{sp} is small) gives

$$4.0 \times 10^{-11} \approx (x)(0.025)^2$$
$$x \approx 6.4 \times 10^{-8}$$

The assumption is valid by the 5% rule and

$$\text{Solubility} = x = 6.4 \times 10^{-8} \text{ mol/L}$$

Thus 6.4×10^{-8} mol of solid CaF_2 dissolves per liter of the 0.025 *M* NaF solution.

pH and Solubility

The pH of a solution can also affect a salt's solubility. For example, magnesium hydroxide dissolves according to the equilibrium

$$Mg(OH)_2(s) \rightleftharpoons Mg^{2+}(aq) + 2OH^-(aq)$$

Addition of OH^- ions (an increase in pH) will, by the common ion effect, force the equilibrium to the left, decreasing the solubility of $Mg(OH)_2$. On the other hand, an addition of H^+ ions (a decrease in pH) increases the solubility, because OH^- ions are removed from solution by reacting with the added H^+ ions. In response to the lower concentration of OH^-, the equilibrium position moves to the right. This is why a suspension of solid $Mg(OH)_2$, known as milk of magnesia, dissolves as required in the stomach to combat excess acidity.

This idea also applies to salts with other types of anions. For example, the solubility of silver phosphate, Ag_3PO_4, is greater in acid than in pure water, because the PO_4^{3-} ion is a strong base that reacts with H^+ to form the HPO_4^{2-} ion. The reaction

$$H^+ + PO_4^{3-} \rightarrow HPO_4^{2-}$$

lowers the concentration of PO_4^{3-} and shifts the solubility equilibrium

$$Ag_3PO_4(s) \rightleftharpoons 3Ag^+(aq) + PO_4^{3-}(aq)$$

to the right, increasing the solubility of silver phosphate.

Silver chloride (AgCl), however, has the same solubility in acid as in pure water. Why? The Cl^- ion is a very weak base; that is, HCl is a very strong acid, and no HCl molecules are formed. Thus addition of H^+ to a solution containing Cl^- does not affect $[Cl^-]$ and has no effect on the solubility of a chloride salt.

The general rule is that if the anion X^- is an effective base, that is, if HX is a weak acid, the salt MX will show increased solubility in an acidic solution. Examples of common anions that are effective bases are OH^-, S^{2-}, CO_3^{2-}, $C_2O_4^{2-}$, and CrO_4^{2-}. Salts containing these anions are much more soluble in acid than in pure water.

As mentioned at the beginning of this chapter, one practical result of the increased solubility of carbonates in acid is the formation of huge limestone caves such as Mammoth Cave in Kentucky or Carlsbad Caverns in New Mexico. Carbon dioxide dissolved in ground water makes it acidic, increasing the solubility of calcium carbonate and eventually producing huge caverns. As the carbon dioxide escapes to the air, the pH of the dripping water goes up and the calcium carbonate precipitates, forming stalactites and stalagmites.

Limestone formations in Bridal Cave, Missouri.

15.7 Precipitation and Qualitative Analysis

Purpose

- To show how to predict if precipitation will occur when solutions are mixed.
- To describe the use of selective precipitation to separate a mixture of ions in solution.

So far we have considered solids dissolving in solutions. Now we will consider the reverse process—the formation of a solid from solution. When solutions are mixed, various reactions can occur. We have already considered acid-base reactions in some detail. In this section we show how to predict whether a precipitate will form when two solutions are mixed. We will use the **ion product,** which is defined just like the expression for K_{sp} for a given solid except that *initial concentrations are used* instead of equilibrium concentrations. For solid CaF_2, the expression for the ion product (Q) is written

$$Q = [Ca^{2+}]_0[F^-]_0^2$$

If we add a solution containing Ca^{2+} ions to a solution containing F^- ions, a precipitate may or may not form, depending on the concentrations of these ions in the resulting mixed solution. To predict whether precipitation will occur, we consider the relationship between Q and K_{sp}:

Q is used here in a very similar way to the use of the reaction quotient in Chapter 13.

If Q is greater than K_{sp}, precipitation occurs and will continue until the concentrations satisfy K_{sp}.

If Q is less than K_{sp}, no precipitation occurs.

Sample Exercise 15.14

A solution is prepared by adding 750.0 mL of 4.0×10^{-3} M $Ce(NO_3)_3$ to 300.0 mL of 2.0×10^{-2} M KIO_3. Will $Ce(IO_3)_3$ ($K_{sp} = 1.9 \times 10^{-10}$) precipitate from this solution?

Solution

First, we calculate $[Ce^{3+}]_0$ and $[IO_3^-]_0$ in the mixed solution before any reaction occurs:

$$[Ce^{3+}]_0 = \frac{(750.0 \text{ mL})(4.0 \times 10^{-3} \text{ mmol/mL})}{(750.0 + 300.0) \text{ mL}} = 2.86 \times 10^{-3}\ M$$

$$[IO_3^-]_0 = \frac{(300.0 \text{ mL})(2.0 \times 10^{-2} \text{ mmol/mL})}{(750.0 + 300.0) \text{ mL}} = 5.71 \times 10^{-3}\ M$$

For $Ce(IO_3)_3(s)$, $K_{sp} = [Ce^{3+}][IO_3^-]^3$.

$$Q = [Ce^{3+}]_0[IO_3^-]_0^3 = (2.86 \times 10^{-3})(5.71 \times 10^{-3})^3 = 5.32 \times 10^{-10}$$

Since Q is greater than K_{sp}, $Ce(IO_3)_3$ will precipitate from the mixed solution.

Sometimes we will want to do more than simply predict whether precipitation will occur; we will want to calculate the equilibrium concentrations in the solution after precipitation occurs. For example, let us calculate the equilibrium concentrations of Pb^{2+} and I^- ions in a solution formed by mixing 100.0 mL of 0.050 *M* $Pb(NO_3)_2$ and 200.0 mL of 0.100 *M* NaI. First, we must determine whether solid PbI_2 ($K_{sp} = 1.4 \times 10^{-8}$) forms when the solutions are mixed. To do so, we need to calculate $[Pb^{2+}]_0$ and $[I^-]_0$ before any reaction occurs:

$$[Pb^{2+}]_0 = \frac{\text{mmol of } Pb^{2+}}{\text{mL of solution}} = \frac{(100.0 \text{ mL})(0.050 \text{ mmol/mL})}{300.0 \text{ mL}} = 1.67 \times 10^{-2}\ M$$

$$[I^-]_0 = \frac{\text{mmol of } I^-}{\text{mL of solution}} = \frac{(200.0 \text{ mL})(0.100 \text{ mmol/mL})}{300.0 \text{ mL}} = 6.67 \times 10^{-2}\ M$$

The ion product for PbI_2 is

$$Q = [Pb^{2+}]_0[I^-]_0^2 = (1.67 \times 10^{-2})(6.67 \times 10^{-2})^2 = 7.43 \times 10^{-5}$$

Since Q is greater than K_{sp}, a precipitate of PbI_2 will form.

Note that since the K_{sp} for PbI_2 is quite small (1.4×10^{-8}), only very small quantities of Pb^{2+} and I^- can coexist in aqueous solution. In other words, when Pb^{2+} and I^- are mixed, most of these ions will precipitate out as PbI_2. That is,

The equilibrium constant for formation of solid PbI_2 is $1/K_{sp}$, or 7×10^7, so this equilibrium lies far to the right.

$$Pb^{2+}(aq) + 2I^-(aq) \rightarrow PbI_2(s)$$

which is the reverse of the dissolution reaction, goes essentially to completion.

Doing acid-base problems, we have learned that it is best to run "complete" reactions and do the stoichiometry calculations before considering the equilibrium. So we let the system go completely in the direction toward which it tends and then adjust it back to equilibrium. If we let Pb^{2+} and I^- react to completion, we have

	Pb^{2+}	+	$2I^-$	$\rightarrow$	PbI_2
Before reaction:	(100.0 mL)(0.0500 *M*) = 5.00 mmol		(200.0 mL)(0.100 *M*) = 20.0 mmol		The amount of PbI_2 formed does not influence the equilibrium
After reaction:	0 mmol		20.0 − 2(5.00) = 10.0 mmol		

In this reaction 10 mmol of I^- is in excess.

At equilibrium $[Pb^{2+}]$ is not actually zero because the reaction does not quite go to completion. The best way to think about this is that once the PbI_2 is formed, a very small amount redissolves to come to equilibrium. Since I^- is in excess, the PbI_2 is dissolving into a solution that contains 10.0 mmol of I^- per 300.0 ml of solution, or $3.33 \times 10^{-2}\ M\ I^-$.

We could state this problem as follows: what is the solubility of solid PbI_2 in a 3.33×10^{-2} *M* NaI solution? The lead iodide dissolves according to the equation:

$$PbI_2(s) \rightleftharpoons Pb^{2+}(aq) + 2I^-(aq)$$

The concentrations are as follows:

Initial concentration (mol/L)		Equilibrium concentration (mol/L)
$[Pb^{2+}]_0 = 0$ $[I^-]_0 = 3.33 \times 10^{-2}$	$\xrightarrow[\text{dissolves}]{x \text{ mol/L } PbI_2(s)}$	$[Pb^{2+}] = x$ $[I^-] = 3.33 \times 10^{-2} + 2x$

Substituting into the expression for K_{sp} gives

$$K_{sp} = 1.4 \times 10^{-8} = [Pb^{2+}][I^-]^2 = (x)(3.33 \times 10^{-2} + 2x)^2 \approx (x)(3.33 \times 10^{-2})^2$$

Then $$[Pb^{2+}] = x = 1.3 \times 10^{-5}\ M$$

$$[I^-] = 3.33 \times 10^{-2}\ M$$

Note that $3.33 \times 10^{-2} \gg 2x$, so the approximation is valid.

Sample Exercise 15.15

A solution is prepared by mixing 150.0 mL of $1.00 \times 10^{-2}\ M$ $Mg(NO_3)_2$ and 250.0 mL of $1.00 \times 10^{-1}\ M$ NaF. Calculate the concentrations of Mg^{2+} and F^- at equilibrium with solid MgF_2 ($K_{sp} = 6.4 \times 10^{-9}$).

Solution

The first step is to determine whether solid MgF_2 forms. To do this, we need to calculate the concentrations of Mg^{2+} and F^- in the mixed solution and find Q:

$$[Mg^{2+}]_0 = \frac{\text{mmol of } Mg^{2+}}{\text{mL of solution}} = \frac{(150.0\ \text{mL})(1.00 \times 10^{-2}\ M)}{400.0\ \text{mL}} = 3.75 \times 10^{-3}\ M$$

$$[F^-]_0 = \frac{\text{mmol of } F^-}{\text{mL of solution}} = \frac{(250.0\ \text{mL})(1.00 \times 10^{-1}\ M)}{400.0\ \text{mL}} = 6.25 \times 10^{-2}\ M$$

For $MgF_2(s)$, $K_{sp} = [Mg^{2+}][F^-]^2$.

$$Q = [Mg^{2+}]_0[F^-]_0^2 = (3.75 \times 10^{-3})(6.25 \times 10^{-2})^2 = 1.46 \times 10^{-5}$$

Since Q is greater than K_{sp}, solid MgF_2 will form.

The next step is to run the precipitation reaction to completion:

	Mg^{2+}	+	$2F^-$	$\rightarrow$	$MgF_2(s)$
Before reaction:	$(150.0)(1.00 \times 10^{-2})$ $= 1.5$ mmol		$(250.0)(1.00 \times 10^{-1})$ $= 25.0$ mmol		
After reaction:	0 mmol		$25.0 - 2(1.50)$ $= 22.0$ mmol		

Note that excess F^- remains after the precipitation reaction goes to completion. The concentration is

$$[F^-]_{excess} = \frac{22.0\ \text{mmol}}{400.0\ \text{mL}} = 5.50 \times 10^{-2}\ M$$

Although we have assumed that the Mg^{2+} is completely consumed, we know $[Mg^{2+}]$ will not be zero at equilibrium. We can compute the equilibrium $[Mg^{2+}]$ by letting MgF_2 redissolve to satisfy the expression for K_{sp}. How much MgF_2 will dissolve in a $5.50 \times 10^{-2}\ M$ NaF solution? We proceed as usual:

$$MgF_2(s) \rightleftharpoons Mg^{2+}(aq) + 2F^-(aq)$$

$$K_{sp} = [Mg^{2+}][F^-]^2 = 6.4 \times 10^{-9}$$

Sample Exercise 15.15, continued

Initial concentration (mol/L)		Equilibrium concentration (mol/L)
$[Mg^{2+}]_0 = 0$ $[F^-]_0 = 5.50 \times 10^{-2}$	$\xrightarrow[\text{dissolves}]{x \text{ mol/L } MgF_2(s)}$	$[Mg^{2+}] = x$ $[F^-] = 5.50 \times 10^{-2} + 2x$

$$K_{sp} = 6.4 \times 10^{-9} = [Mg^{2+}][F^-]^2 = (x)(5.50 \times 10^{-2} + 2x)^2 \approx (x)(5.50 \times 10^{-2})^2$$

The approximations made here fall within the 5% rule.

$$[Mg^{2+}] = x = 2.1 \times 10^{-6}\ M$$

$$[F^-] = 5.50 \times 10^{-2}\ M$$

Selective Precipitation

Mixtures of metal ions in aqueous solution are often separated by **selective precipitation,** that is, by using a reagent whose anion forms a precipitate with only one or a few of the metal ions in the mixture. For example, suppose we have a solution containing both Ba^{2+} and Ag^+ ions. If NaCl is added to the solution, AgCl precipitates as a white solid; but since $BaCl_2$ is soluble, the Ba^{2+} ions remain in solution.

Sample Exercise 15.16

A solution contains $1.0 \times 10^{-4}\ M\ Cu^+$ and $2.0 \times 10^{-3}\ M\ Pb^{2+}$. If a source of I^- is added to this solution, will PbI_2 ($K_{sp} = 1.4 \times 10^{-8}$) or CuI ($K_{sp} = 5.3 \times 10^{-12}$) precipitate first? Specify the concentration of I^- necessary to begin precipitation of each salt.

Solution

For PbI_2, the K_{sp} expression is

$$1.4 \times 10^{-8} = K_{sp} = [Pb^{2+}][I^-]^2$$

Since $[Pb^{2+}]$ in this solution is known to be $2.0 \times 10^{-3}\ M$, the greatest concentration of I^- that can be present without causing precipitation of PbI_2 can be calculated from the K_{sp} expression:

$$1.4 \times 10^{-8} = [Pb^{2+}][I^-]^2 = (2.0 \times 10^{-3})[I^-]^2$$

$$[I^-] = 2.6 \times 10^{-3}\ M$$

Any I^- in excess of this concentration will cause solid PbI_2 to form.

Similarly, for CuI, the K_{sp} expression is

$$5.3 \times 10^{-12} = K_{sp} = [Cu^+][I^-] = (1.0 \times 10^{-4})[I^-]$$

and

$$[I^-] = 5.3 \times 10^{-8}\ M$$

A concentration of I^- in excess of $5.3 \times 10^{-8}\ M$ will cause formation of solid CuI.

Sample Exercise 15.16, continued

As I^- is added to the mixed solution, CuI will precipitate first since the $[I^-]$ required is less. Therefore, Cu^+ could be separated from Pb^{2+} using this reagent.

We can compare K_{sp} values to find relative solubilities because FeS and MnS produce the same number of ions in solution.

Since metal sulfide salts differ dramatically in their solubilities, the sulfide ion is often used to separate metal ions by selective precipitation. For example, consider a solution containing a mixture of 10^{-3} *M* Fe^{2+} and 10^{-3} *M* Mn^{2+}. Since FeS ($K_{sp} = 3.7 \times 10^{-19}$) is much less soluble than MnS ($K_{sp} = 2.3 \times 10^{-13}$), careful addition of S^{2-} to the mixture will precipitate Fe^{2+} as FeS, leaving Mn^{2+} in solution.

One real advantage of the sulfide ion as a precipitating reagent is that because it is basic, its concentration can be controlled by regulating the pH of the solution. H_2S is a diprotic acid that dissociates in two steps:

$$H_2S \rightleftharpoons H^+ + HS^- \qquad K_{a_1} = 1.0 \times 10^{-7}$$

$$HS^- \rightleftharpoons H^+ + S^{2-} \qquad K_{a_2} = 1.3 \times 10^{-13}$$

Note from the small K_{a_2} value that S^{2-} ions have a high affinity for protons. In an acidic solution (large $[H^+]$), $[S^{2-}]$ will be relatively small, since under these conditions the dissociation equilibria will lie far to the left. On the other hand, in basic solutions $[S^{2-}]$ will be relatively large, since the very small value of $[H^+]$ will pull both equilibria to the right, producing S^{2-}.

This means that the most insoluble sulfide salts, such as CuS ($K_{sp} = 8.5 \times 10^{-45}$) and HgS ($K_{sp} = 1.6 \times 10^{-54}$), can be precipitated from an acidic solution, leaving the more soluble ones, such as MnS ($K_{sp} = 2.3 \times 10^{-13}$) and NiS ($K_{sp} = 3 \times 10^{-21}$), still dissolved. The manganese and nickel sulfides can then be precipitated by making the solution slightly basic. This procedure is diagrammed in Fig. 15.11.

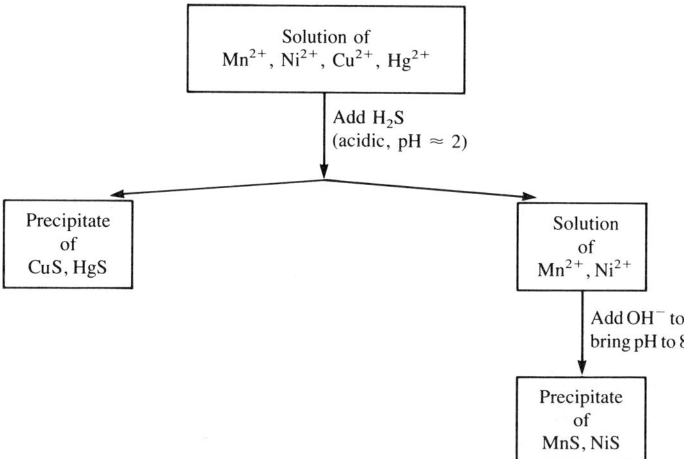

Figure 15.11

The separation of Cu^{2+} and Hg^{2+} from Ni^{2+} and Mn^{2+} using H_2S. At a low pH, $[S^{2-}]$ is relatively low and only the very insoluble HgS and CuS precipitate. When OH^- is added to lower $[H^+]$, the value of $[S^{2-}]$ increases, and MnS and NiS precipitate.

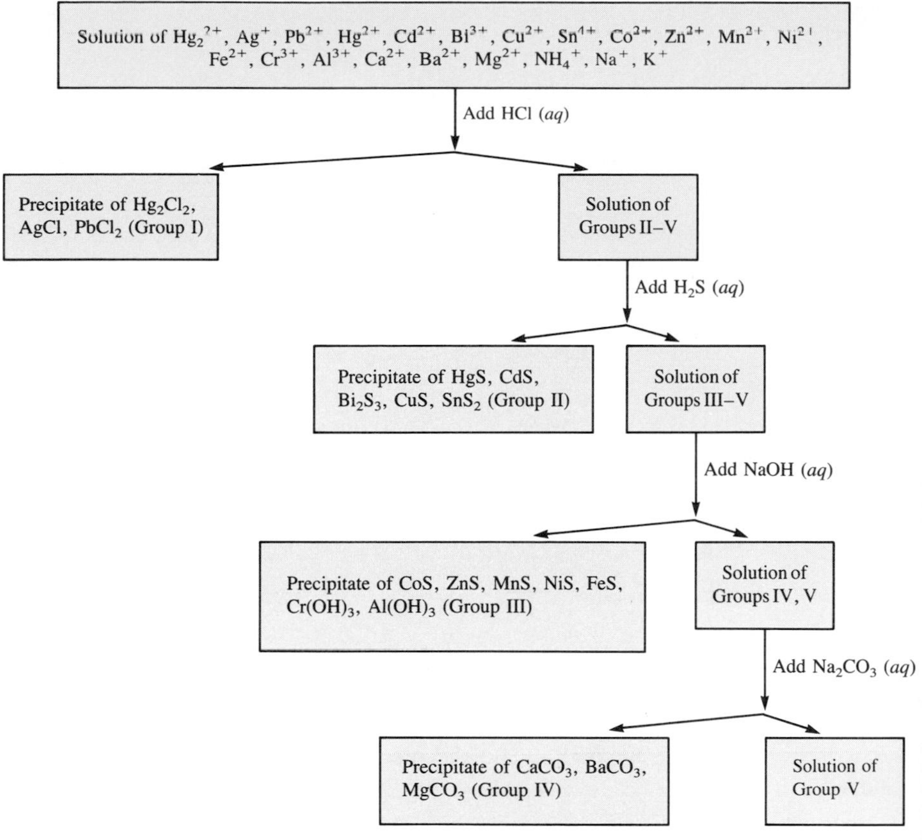

Figure 15.12

A schematic diagram of the classic method for separating the common cations by selective precipitation.

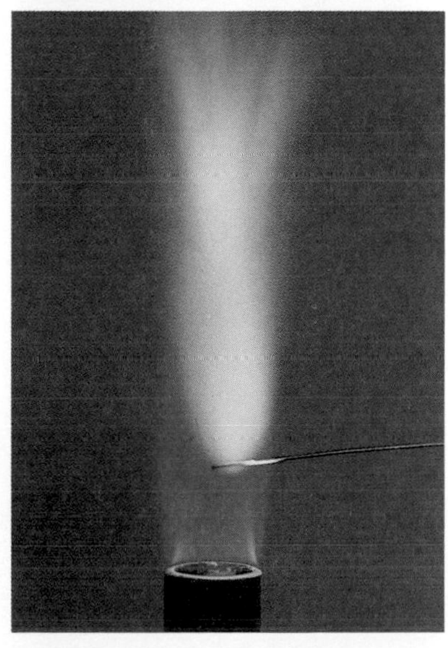

(top) Flame test for potassium.
(bottom) Flame test for sodium.

Qualitative Analysis

The classic scheme for **qualitative analysis** of a mixture containing all of the common cations (listed in Fig. 15.12) involves first separating them into five major groups based on solubilities. (These groups are not directly related to the groups of the periodic table.) Each group is then treated further to separate and identify the individual ions. We will be concerned here only with separation of the major groups.

Group I—insoluble chlorides.

When dilute aqueous HCl is added to a solution containing a mixture of the common cations, only Ag^+, Pb^{2+}, and Hg_2^{2+} will precipitate out as insoluble chlorides. All other chlorides are soluble and remain in solution. The Group I precipitate is removed, leaving the other ions in solution for treatment with sulfide ion.

Group II—sulfides insoluble in acid solution.

After the insoluble chlorides are removed, the solution is still acidic, since HCl was added. If H_2S is added to this solution, only the most insoluble sulfides (those

From left to right, cadmium sulfide, chromium(III) hydroxide, aluminum hydroxide, and nickel(II) hydroxide.

of Hg^{2+}, Cd^{2+}, Bi^{3+}, Cu^{2+}, and Sn^{4+}) will precipitate, since $[S^{2-}]$ is relatively low because of the high concentration of H^+. The more soluble sulfides will remain dissolved under these conditions, and the precipitate of the insoluble salts is removed.

Group III—sulfides insoluble in basic solution.

The solution is made basic at this stage and more H_2S is added. As we saw earlier, a basic solution produces a higher $[S^{2-}]$, which leads to precipitation of the more soluble sulfides. The cations precipitated as sulfides at this stage are Co^{2+}, Zn^{2+}, Mn^{2+}, Ni^{2+}, and Fe^{2+}. If any Cr^{3+} and Al^{3+} ions are present, they will also precipitate, but as insoluble hydroxides (remember the solution is now basic). The precipitate is separated from the solution containing the rest of the ions.

Group IV—insoluble carbonates.

At this point, all of the cations have been precipitated except those from Groups 1A and 2A of the periodic table. The Group 2A cations form insoluble carbonates and can be precipitated by the addition of CO_3^{2-}. For example, Ba^{2+}, Ca^{2+}, and Mg^{2+} form solid carbonates and can be removed from the solution.

Group V—alkali metal and ammonium ions.

The only ions remaining in solution at this point are the Group 1A cations and the NH_4^+ ion, all of which form soluble salts with the common anions. The Group 1A cations are usually identified by the characteristic colors they produce when heated in a flame. These colors are due to the emission spectra of these ions.

The qualitative analysis scheme for cations based on the selective precipitation procedure described above is summarized in Fig. 15.12.

15.8 Complex Ion Equilibria

Purpose

- To apply the principles of equilibrium to the formation of complex ions.
- To show how complex ion formation can increase the solubility of a salt.

A **complex ion** is a charged species consisting of a metal ion surrounded by *ligands*. A ligand is simply a Lewis base—a molecule or ion having a lone pair that can be donated to an empty orbital on the metal ion to form a covalent bond. Some common ligands are H_2O, NH_3, Cl^-, and CN^-. The number of ligands attached to a metal ion is called the *coordination number*. The most common coordination numbers are 6, for example, in $Co(H_2O)_6{}^{2+}$ and $Ni(NH_3)_6{}^{2+}$; 4, for example, in $CoCl_4{}^{2-}$ and $Cu(NH_3)_4{}^{2+}$; and 2, for example, in $Ag(NH_3)_2{}^+$; but others are known.

The properties of complex ions will be discussed in more detail in Chapter 20. For now, we will just look at the equilibria involving these species. Metal ions add ligands one at a time in steps characterized by equilibrium constants called **formation constants,** or **stability constants.** For example, when solutions containing Ag^+ ions and NH_3 molecules are mixed, the following reactions take place:

$$Ag^+ + NH_3 \rightleftharpoons Ag(NH_3)^+ \qquad K_1 = 2.1 \times 10^3$$

$$Ag(NH_3)^+ + NH_3 \rightleftharpoons Ag(NH_3)_2{}^+ \qquad K_2 = 8.2 \times 10^3$$

where K_1 and K_2 are the formation constants for the two steps. In a solution containing Ag^+ and NH_3, all of the species NH_3, Ag^+, $Ag(NH_3)^+$, and $Ag(NH_3)_2{}^+$ exist at equilibrium. Calculating the concentrations of all these components can be complicated. However, usually the total concentration of ligand is much larger than the total concentration of metal ion, and approximations can greatly simplify the problems.

For example, consider a solution prepared by mixing 100.0 mL of 2.0 *M* NH_3 with 100.0 mL of 1.0×10^{-3} *M* $AgNO_3$. *Before any reaction occurs,* the mixed solution contains the major species Ag^+, $NO_3{}^-$, NH_3, and H_2O. What reaction or reactions will occur in this solution? From our discussions of acid-base chemistry, we know that one reaction is

$$NH_3(aq) + H_2O(l) \rightleftharpoons NH_4{}^+(aq) + OH^-(aq)$$

However, we are interested in the reaction between NH_3 and Ag^+ to form complex ions, and since the position of the above equilibrium lies far to the left (K_b for NH_3 is 1.8×10^{-5}), we can neglect the amount of NH_3 used up in the reaction with water. So, before any complex ion formation, the concentrations in the mixed solution are

$$[Ag^+]_0 = \frac{(100.0 \text{ mL})(1.0 \times 10^{-3}\ M)}{(200.0 \text{ mL})} = 5.0 \times 10^{-4}\ M$$

↖Total volume

$$[NH_3]_0 = \frac{(100.0 \text{ mL})(2.0\ M)}{(200.0 \text{ mL})} = 1.0\ M$$

As mentioned already, the Ag^+ ion reacts with NH_3 in a stepwise fashion to form $AgNH_3^+$ and then $Ag(NH_3)_2^+$:

$$Ag^+ + NH_3 \rightleftharpoons Ag(NH_3)^+ \qquad K_1 = 2.1 \times 10^3$$

$$AgNH_3^+ + NH_3 \rightleftharpoons Ag(NH_3)_2^+ \qquad K_2 = 8.2 \times 10^3$$

Since both K_1 and K_2 are large, and since there is a large excess of NH_3, *both reactions can be assumed to go essentially to completion*. This is equivalent to writing the net reaction in the solution as follows:

	Ag^+	+	$2NH_3$	→	$Ag(NH_3)_2^+$
Before reaction:	$5.0 \times 10^{-4}\ M$		$1.0\ M$		0
After reaction:	0		$1.0 - 2(5.0 \times 10^{-4}) \approx 1.0\ M$ ↗ Twice as much NH_3 as Ag^+ is required		$5.0 \times 10^{-4}\ M$

Note that in this case we have used molarities to do the stoichiometry calculations and we have assumed this reaction to be complete, using all of the original Ag^+ to form $Ag(NH_3)_2^+$. In reality, a *very small* amount of the $Ag(NH_3)_2^+$ formed will dissociate to produce small amounts of $Ag(NH_3)^+$ and Ag^+. However, since the amount of $Ag(NH_3)_2^+$ dissociating will be so small, we can safely assume that $[Ag(NH_3)_2^+]$ is $5.0 \times 10^{-4}\ M$ at equilibrium. Also, we know that since so little NH_3 has been consumed, $[NH_3]$ is $1.0\ M$ at equilibrium. We can use these concentrations to calculate $[Ag^+]$ and $[Ag(NH_3)^+]$ using the K_1 and K_2 expressions.

To calculate the equilibrium concentration of $Ag(NH_3)^+$, we use

$$K_2 = 8.2 \times 10^3 = \frac{[Ag(NH_3)_2^+]}{[Ag(NH_3)^+][NH_3]}$$

since $[Ag(NH_3)_2^+]$ and $[NH_3]$ are known. Rearranging and solving for $[Ag(NH_3)^+]$ gives

$$[Ag(NH_3)^+] = \frac{[Ag(NH_3)_2^+]}{K_2[NH_3]} = \frac{5.0 \times 10^{-4}}{(8.2 \times 10^3)(1.0)} = 6.1 \times 10^{-8}\ M$$

Now the equilibrium concentration of Ag^+ can be calculated using K_1:

$$K_1 = 2.1 \times 10^3 = \frac{[Ag(NH_3)^+]}{[Ag^+][NH_3]} = \frac{6.1 \times 10^{-8}}{[Ag^+](1.0)}$$

$$[Ag^+] = \frac{6.1 \times 10^{-8}}{(2.1 \times 10^3)(1.0)} = 2.9 \times 10^{-11}\ M$$

So far, we have assumed that $Ag(NH_3)_2^+$ is the dominant silver-containing species in solution. Is this a valid assumption? The calculated concentrations are

$$[Ag(NH_3)_2^+] = 5.0 \times 10^{-4}\ M$$

$$[AgNH_3^+] = 6.1 \times 10^{-8}\ M$$

$$[Ag^+] = 2.9 \times 10^{-11}\ M$$

Essentially all of the Ag^+ ions originally present end up in $Ag(NH_3)_2^+$.

These values clearly support the conclusion that

$$[Ag(NH_3)_2^+] \gg [AgNH_3^+] \gg [Ag^+]$$

Thus the assumption that $[Ag(NH_3)_2^+]$ is dominant is valid, and the calculated concentrations are correct.

This analysis shows that, although complex ion equilibria have many species present and look complicated, the calculations are actually quite straightforward if the ligand is in large excess.

Sample Exercise 15.17

Calculate the concentrations of Ag^+, $Ag(S_2O_3)^-$, and $Ag(S_2O_3)_2^{3-}$ in a solution prepared by mixing 150.0 mL of 1.0×10^{-3} M $AgNO_3$ with 200.0 mL of 5.00 M $Na_2S_2O_3$. The stepwise formation equilibria are

$$Ag^+ + S_2O_3^{2-} \rightleftharpoons Ag(S_2O_3)^- \qquad K_1 = 7.4 \times 10^8$$

$$Ag(S_2O_3)^- + S_2O_3^{2-} \rightleftharpoons Ag(S_2O_3)_2^{3-} \qquad K_2 = 3.9 \times 10^4$$

Solution

The concentrations of ligand and metal ion in the mixed solution *before any reaction occurs* are

$$[Ag^+]_0 = \frac{(150.0 \text{ mL})(1.0 \times 10^{-3}\ M)}{(150.0 \text{ mL} + 200.0 \text{ mL})} = 4.29 \times 10^{-4}\ M$$

$$[S_2O_3^{2-}]_0 = \frac{(200.0 \text{ mL})(5.0\ M)}{(150.0 \text{ mL} + 200.0 \text{ mL})} = 2.86\ M$$

Since $[S_2O_3^{2-}]_0 \gg [Ag^+]_0$, and since K_1 and K_2 are large, both formation reactions can be assumed to go to completion, and the net reaction in the solution is

	Ag^+	+	$2S_2O_3^{2-}$	→	$Ag(S_2O_3)_2^{3-}$
Before reaction:	$4.29 \times 10^{-4}\ M$		$2.86\ M$		0
After reaction:	~0		$2.86 - 2(4.29 \times 10^{-4}) \approx 2.86\ M$		$4.29 \times 10^{-4}\ M$

Note that Ag^+ is limiting, and that the amount of $S_2O_3^{2-}$ consumed is negligible. Also note that since all of these species are in the same solution, the molarities can be used to do the stoichiometry problem.

Of course, the concentration of Ag^+ is not zero at equilibrium, and there is some $Ag(S_2O_3)^-$ in the solution. To calculate the concentrations of these species, we must use the K_1 and K_2 expressions. We can calculate the concentration of $Ag(S_2O_3)^-$ from K_2:

$$3.9 \times 10^4 = K_2 = \frac{[Ag(S_2O_3)_2^{3-}]}{[Ag(S_2O_3)^-][S_2O_3^{2-}]} = \frac{4.29 \times 10^{-4}}{[Ag(S_2O_3)^-](2.86)}$$

$$[Ag(S_2O_3)^-] = 3.8 \times 10^{-9}\ M$$

We can calculate $[Ag^+]$ from K_1:

$$7.4 \times 10^8 = K_1 = \frac{[Ag(S_2O_3)^-]}{[Ag^+][S_2O_3^{2-}]} = \frac{3.8 \times 10^{-9}}{[Ag^+](2.86)}$$

$$[Ag^+] = 1.8 \times 10^{-18}\ M$$

Sample Exercise 15.17, continued

These results show that

$$[Ag(S_2O_3)_2^{3-}] \gg [Ag(S_2O_3)^-] \gg [Ag^+]$$

Thus the assumption that essentially all of the original Ag^+ is converted to $Ag(S_2O_3)_2^{3-}$ is valid, and the calculations are correct.

Complex Ions and Solubility

Often ionic solids that are very nearly water-insoluble must be dissolved somehow in aqueous solutions. For example, when the various qualitative analysis groups are precipitated out, the precipitates must be redissolved to separate the ions within each group. Consider a solution of cations that contains Ag^+, Pb^{2+}, and Hg_2^{2+}, among others. When dilute aqueous HCl is added to this solution, the Group I ions will form the insoluble chlorides AgCl, $PbCl_2$, and Hg_2Cl_2. Once this mixed precipitate is separated from the solution, it must be redissolved to identify the cations individually. How can this be done? We know that some solids are more soluble in acidic than in neutral solutions. What about chloride salts? For example, can AgCl be dissolved by using a strong acid? The answer is no, because Cl^- ions have virtually no affinity for H^+ ions in aqueous solution. The position of the dissolution equilibrium

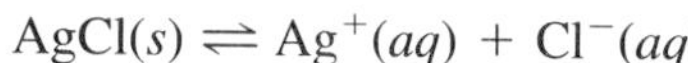

$$AgCl(s) \rightleftharpoons Ag^+(aq) + Cl^-(aq)$$

is not affected by the presence of H^+.

How can we pull the dissolution equilibrium to the right, even though Cl^- is an extremely weak base? The key is to lower the concentration of Ag^+ in solution by forming complex ions. For example, Ag^+ reacts with excess NH_3 to form the stable complex ion $Ag(NH_3)_2^+$. As a result, AgCl is quite soluble in concentrated ammonia solutions. The relevant reactions are

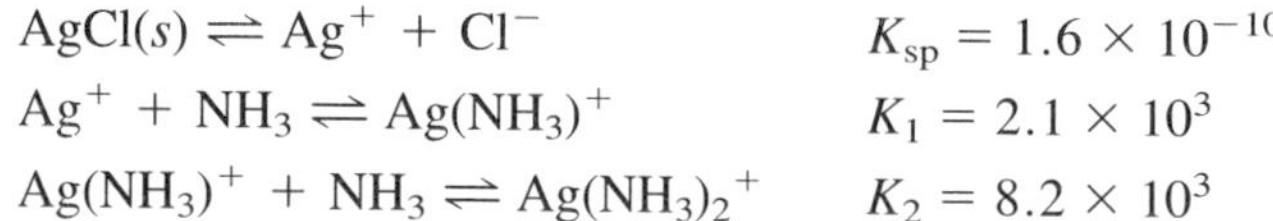

$$AgCl(s) \rightleftharpoons Ag^+ + Cl^- \qquad K_{sp} = 1.6 \times 10^{-10}$$
$$Ag^+ + NH_3 \rightleftharpoons Ag(NH_3)^+ \qquad K_1 = 2.1 \times 10^3$$
$$Ag(NH_3)^+ + NH_3 \rightleftharpoons Ag(NH_3)_2^+ \qquad K_2 = 8.2 \times 10^3$$

The Ag^+ ion produced by dissolving solid AgCl combines with NH_3 to form $Ag(NH_3)_2^+$, which causes more AgCl to dissolve, until the point at which

$$[Ag^+][Cl^-] = K_{sp} = 1.6 \times 10^{-10}$$

Here $[Ag^+]$ refers only to the Ag^+ ion that is present as a separate species in solution. It is *not* the total silver content of the solution, which is

$$[Ag]_{\text{total dissolved}} = [Ag^+] + [AgNH_3^+] + [Ag(NH_3)_2^+]$$

As we saw in the previous section, virtually all of the Ag^+ from the dissolved AgCl ends up in the complex $Ag(NH_3)_2^+$, so we can represent the dissolving of solid AgCl in excess NH_3 by the equation

$$AgCl(s) + 2NH_3(aq) \rightleftharpoons Ag(NH_3)_2^+(aq) + Cl^-(aq)$$

Since this equation is the *sum of the three stepwise reactions* given above, the equilibrium constant for the reaction is the product of the constants for the three

(top) Aqueous ammonia is added to silver chloride (white). (bottom) The silver chloride, insoluble in water, dissolves to form $Ag(NH_3)_2^+(aq)$ and $Cl^-(aq)$.

reactions. (Demonstrate this to yourself by multiplying together the three expressions for K_{sp}, K_1 and K_2.) The equilibrium expression is

When reactions are added, the equilibrium constant for the overall process is the product of the constants for the individual reactions.

$$K = \frac{[Ag(NH_3)_2^+][Cl^-]}{[NH_3]^2}$$

$$= K_{sp} \times K_1 \times K_2 = (1.6 \times 10^{-10})(2.1 \times 10^3)(8.2 \times 10^3) = 2.8 \times 10^{-3}$$

Using this expression, we will now calculate the solubility of solid AgCl in a 10.0 M NH_3 solution. If we let x be the solubility (in mol/L) of AgCl in this solution, we can then write the following expressions for the equilibrium concentrations of the pertinent species:

$[Cl^-] = x$ ← x mol/L of AgCl dissolves to produce x mol/L of Cl^- and x mol/L of $Ag(NH_3)_2^+$

$[Ag(NH_3)_2^+] = x$ ←

$[NH_3] = 10.0 - 2x$ ← Formation of x mol/L of $Ag(NH_3)_2^+$ requires $2x$ mol/L of NH_3, since each complex ion contains two NH_3 ligands

Substituting these concentrations into the equilibrium expression gives

$$K = 2.8 \times 10^{-3} = \frac{[Ag(NH_3)_2^+][Cl^-]}{[NH_3]^2} = \frac{(x)(x)}{(10.0 - 2x)^2} = \frac{x^2}{(10.0 - 2x)^2}$$

No approximations are necessary here. Taking the square root of both sides of the equation gives

$$\sqrt{2.8 \times 10^{-3}} = \frac{x}{10.0 - 2x}$$

$$x = 0.48 \text{ mol/L} = \text{solubility of AgCl}(s) \text{ in } 10.0\ M\ NH_3$$

Thus the solubility of AgCl in 10.0 M NH_3 is much greater than its solubility in pure water, which is

$$\sqrt{K_{sp}} = 1.3 \times 10^{-5} \text{ mol/L}$$

We have considered two strategies for dissolving a water-insoluble ionic solid. If the *anion* of the solid is a good base, the solubility is greatly increased by acidifying the solution. In cases where the anion is not sufficiently basic, the ionic solid can often be dissolved in a solution containing a ligand that forms stable complex ions with its *cation*.

Sometimes solids are so insoluble that combinations of reactions are needed to dissolve them. For example, to dissolve the extremely insoluble HgS ($K_{sp} = 10^{-54}$), it is necessary to use a mixture of concentrated HCl and concentrated HNO_3, called *aqua regia*. The H^+ ions in the aqua regia react with the S^{2-} ions to form H_2S, and Cl^- reacts with Hg^{2+} to form various complex ions including $HgCl_4^{2-}$. In addition, NO_3^- oxidizes S^{2-} to elemental sulfur. These processes lower the concentrations of Hg^{2+} and S^{2-} and thus promote the solubility of HgS.

Since the solubility of many salts increases with temperature, simple heating is sometimes enough to make a salt sufficiently soluble. For example, earlier in this section we considered the mixed chloride precipitates of the Group I ions—$PbCl_2$, AgCl, and Hg_2Cl_2. The effect of temperature on the solubility of $PbCl_2$ is such that we can precipitate $PbCl_2$ with cold aqueous HCl and then redissolve it by heating the solution to near boiling. The silver and mercury(I) chlorides remain precipitated, since they are not significantly soluble in hot water. However, solid AgCl can be dissolved using aqueous ammonia. The solid Hg_2Cl_2 reacts with NH_3 to form a mixture of elemental mercury and $HgNH_2Cl$:

$$Hg_2Cl_2(s) + 2NH_3(aq) \rightarrow \underset{\text{White}}{HgNH_2Cl(s)} + \underset{\text{Black}}{Hg(l)} + NH_4^+(aq) + Cl^-(aq)$$

The mixed precipitate appears gray. This is an oxidation-reduction reaction, in which one mercury(I) ion in Hg_2Cl_2 is oxidized to Hg^{2+} in $HgNH_2Cl$ and the other mercury(I) ion is reduced to Hg, or elemental mercury.

The treatment of the Group I ions is summarized in Fig. 15.13. Note that the presence of Pb^{2+} is confirmed by adding CrO_4^{2-}, which forms bright yellow lead(II) chromate ($PbCrO_4$). Also note that H^+ added to a solution containing $Ag(NH_3)_2^+$ and Cl^- reacts with the NH_3 to form NH_4^+, destroying the $Ag(NH_3)_2^+$ complex. Silver chloride can then reform:

$$2H^+(aq) + Ag(NH_3)_2^+(aq) + Cl^-(aq) \rightarrow 2NH_4^+(aq) + AgCl(s)$$

Note that the qualitative analysis of cations by selective precipitation involves all of the types of reactions we have discussed and represents an excellent application of the principles of chemical equilibrium.

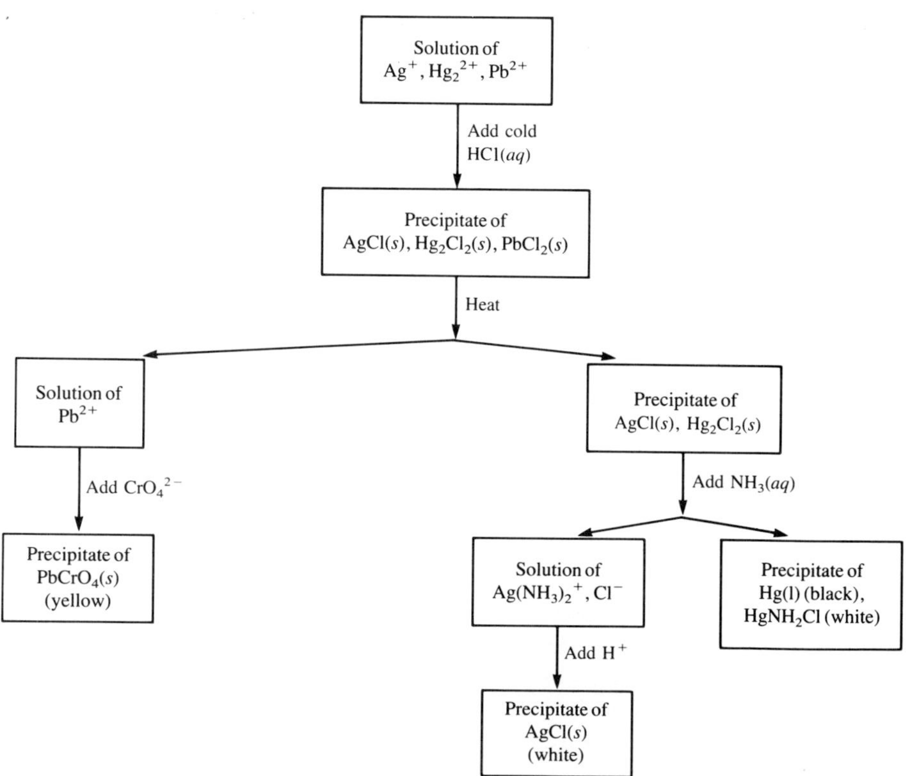

Figure 15.13

The separation of the Group I ions in the classic scheme of qualitative analysis.

Chemical Impact

The Chemistry of Teeth

If dental chemistry continues to progress at the present rate, tooth decay may soon be a thing of the past. Cavities are holes that develop in tooth enamel, which is composed of the mineral hydroxyapatite, $Ca_5(PO_4)_3OH$. Recent research has shown that there is constant dissolving and reforming of the tooth mineral in the saliva at the tooth's surface. Demineralization (dissolving of tooth enamel) is mainly caused by weak acids in the saliva created by bacteria as they metabolize carbohydrates in food. (The solubility of $Ca_5(PO_4)_3OH$ in acidic saliva should come as no surprise to you if you understand how pH affects the solubility of a salt with basic anions.)

In the first stages of tooth decay parts of the tooth surface become porous and spongy and develop swiss-cheese-like holes that, if untreated, eventually turn into cavities (see photo). However, recent results indicate that if the affected tooth is bathed in a solution containing appropriate amounts of Ca^{2+}, PO_4^{3-}, and F^-, it remineralizes. Because the F^- replaces OH^- in the tooth mineral [$Ca_5(PO_4)_3OH$ is changed to $Ca_5(PO_4)_3F$], the remineralized area is more resistant to future decay since fluoride is a weaker base than hydroxide ion. In addition it has been shown that the presence of Sr^{2+} in the remineralizing fluid significantly increases resistance to decay.

If these results hold up under further study, the work of dentists will change dramatically. Dentists will be much more involved in preventing damage to teeth than in repairing damage that has already occurred. One can picture the routine use of a remineralization rinse that will repair problem areas before they become cavities. Dental drills could join leeches as a medical anachronism.

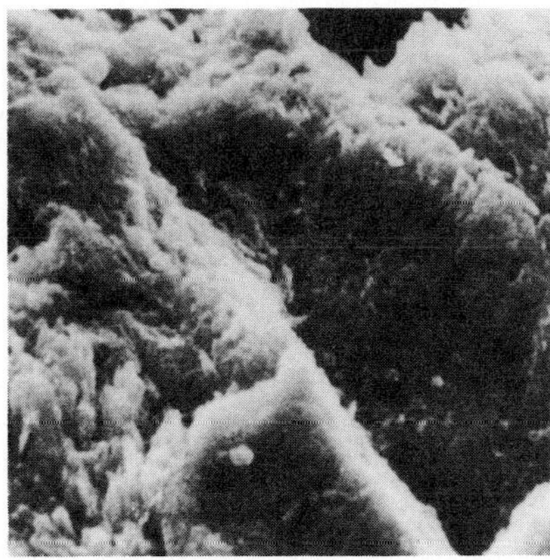

Tooth decay lesion.

FOR REVIEW

Summary

When a weak acid HA and its salt NaA are dissolved in the same aqueous solution, a buffering effect is observed. A buffered solution is one that resists a change in pH when either hydroxide ions or protons are added. A buffer may contain a weak acid and its salt or a weak base and its salt. In a buffered solution the pH is governed by the ratio $[A^-]/[HA]$. The Henderson-Hasselbalch equation gives the relationship between pH, pK_a, and the ratio $[A^-]/[HA]$:

$$pH = pK_a + \log\left(\frac{[A^-]}{[HA]}\right)$$

The capacity of a buffer depends on the amounts of HA and A^- present. Buffering results because the relatively large amounts of HA and A^- cause the ratio $[A^-]/[HA]$ to be almost constant when H^+ or OH^- is added. Optimum buffering results when [HA] is equal to $[A^-]$.

The progress of an acid-base titration is often monitored by plotting the pH of the solution versus the volume of added titrant to give a pH curve, or titration curve. Strong acid–strong base titrations have a very sharp change in pH near the stoichiometric, or equivalence, point, the point at which the original acid (or base) in the solution has been exactly consumed by the titrant base (or acid). When a weak acid is titrated with a strong base, the shape of the titration curve is quite different. Initially, the curve rises sharply, then levels off because of the buffering effects of the weak acid and salt mixture. At the equivalence point of a titration of a weak acid with a strong base, the pH is greater than 7 because of the basicity of the conjugate base of the weak acid. For a titration of a weak base with a strong acid, the pH at the equivalence point is always less than 7 because of the acidity of the conjugate acid of the weak base. For a strong acid–strong base titration, the pH at the equivalence point equals 7. The equivalence point of a titration is determined by stoichiometry, not by pH.

Another way of following an acid-base titration is by the use of an indicator that marks the end point by changing color. Most commonly used indicators are themselves weak acids and exist in two forms, HIn and In^-, that exhibit different colors. For most indicators, the ratio of the minor form to the major form must be $\frac{1}{10}$ before a color change is apparent. The goal in choosing an indicator is to have the equivalence point and the end point coincide.

The principles of equilibrium can also be applied when a solid dissolves in water. The solubility product (K_{sp}) is an equilibrium constant defined by the law of mass action. Solubility is an equilibrium position. The K_{sp} value for a salt can be determined by measuring its solubility. The solubility of a salt can be calculated if its K_{sp} value is known.

Addition of an ion common to a salt will push the dissolution equilibrium to the left, reducing the salt's solubility. This is known as the common ion effect. The solubility of a salt MX that is only slightly soluble in pure water can be increased if the solubility equilibrium

$$MX(s) \rightleftharpoons M^+(aq) + X^-(aq)$$

can be shifted to the right by lowering $[M^+]$ or $[X^-]$. If X^- is a good base, addition of H^+ will form HX and lower $[X^-]$, thus increasing the salt's solubility. The equilibrium concentration of M^+ can be lowered by adding a ligand to form complex ions with it.

We can predict whether a precipitate will form when two solutions are mixed by using the ion product (Q), which is defined just like K_{sp} except that initial concentrations of the ions are used. If Q is greater than K_{sp}, the salt will precipitate.

Mixtures of ions can be separated by selective precipitation, a process in which reagents are added to precipitate single ions or small groups of ions.

Key Terms

Section 15.1
common ion
common ion effect

Section 15.2
buffered solution
Henderson-Hasselbalch equation

Section 15.3
buffering capacity

Section 15.4
pH curve (titration curve)
millimole (mmol)
equivalence point (stoichiometric point)

Section 15.5
acid-base indicator
phenolphthalein

Section 15.6
solubility product constant (solubility product)

Section 15.7
ion product
selective precipitation
qualitative analysis

Section 15.8
complex ion
formation constants (stability constants)

Exercises

A blue exercise number indicates that the answer to that exercise appears at the back of this book and a solution appears in the Solutions Guide.

Buffers

1. What is meant by a common ion? How does the presence of a common ion affect an equilibrium such as

$$HNO_2(aq) \rightleftharpoons H^+(aq) + NO_2^-(aq)$$

2. What components must be present in order to have a buffered solution? For what purposes are buffers used?

3. Write reactions to show how a solution containing acetic acid and sodium acetate acts as a buffer when a strong acid or base is added.

4. A buffer is made by dissolving $NaHCO_3$ and Na_2CO_3 in water. Write equations to show how this buffer neutralizes added H^+ and OH^-.

5. What is meant by the capacity of a buffer? How do the following buffers differ in capacity? How do they differ in pH?

0.01 *M* acetic acid and 0.01 *M* sodium acetate

0.1 *M* acetic acid and 0.1 *M* sodium acetate

1.0 *M* acetic acid and 1.0 *M* sodium acetate

6. Is it necessary that the concentration of the weak acid and weak base in a buffered solution be equal? Why or why not?

7. Derive an equation analogous to the Henderson-Hasselbalch equation but relating pOH and pK_b of a buffered solution composed of a weak base and its conjugate acid, such as NH_3 and NH_4^+.

8. Calculate the pH of each of the following solutions:
   a. 0.10 *M* propanoic acid ($HC_3H_5O_2$, $K_a = 1.3 \times 10^{-5}$)
   b. 0.10 *M* sodium propanoate ($NaC_3H_5O_2$)
   c. pure H_2O
   d. 0.10 *M* $HC_3H_5O_2$ and 0.10 *M* $NaC_3H_5O_2$

9. Calculate the pH after 0.020 mol of HCl is added to 1.0 L of each of the four solutions in Exercise 8.

10. Calculate the pH after 0.020 mol of NaOH is added to 1.0 L of each of the four solutions in Exercise 8.

11. Using tabulated K_a and K_b values, calculate the pH of each of the following solutions:
    a. a solution containing 0.10 *M* HNO_2 and 0.15 *M* $NaNO_2$
    b. 25.0 g of pure (glacial) acetic acid and 40.0 g of sodium acetate diluted to 500. mL with water
    c. 50.0 mL of 1.0 *M* HOCl and 30.0 mL of 0.80 *M* NaOH diluted to 250 mL with water
    d. 100.0 g of NH_4Cl and 65.0 g of NaOH diluted to 1.0 L with water
    e. 26.4 g of sodium acetate and 50.0 mL of 6.00 *M* hydrochloric acid diluted to 5.00×10^2 mL with water

    Hint: In parts c, d, and e, reactions occur that go to completion.

12. Calculate the pH of each of the following buffered solutions:
    a. 0.10 *M* acetic acid and 0.25 *M* sodium acetate
    b. 0.25 *M* acetic acid and 0.10 *M* sodium acetate
    c. 0.080 *M* acetic acid and 0.20 *M* sodium acetate
    d. 0.20 *M* acetic acid and 0.080 *M* sodium acetate

13. Calculate the ratio $[NH_3]/[NH_4^+]$ for each of the following ammonia/ammonium chloride buffered solutions:
    a. pH = 9.00
    b. pH = 8.80
    c. pH = 10.00
    d. pH = 9.60

14. What mass of solid NaOH must be added to 1.0 L of 2.0 *M* $HC_2H_3O_2$ to produce a solution buffered at each pH?
    a. pH = pK_a
    b. pH = 4.00
    c. pH = 5.00

15. A buffered solution is made by adding 75.0 g of sodium acetate to 500.0 mL of a 0.64 *M* solution of acetic acid. What is the pH of the final solution? (Assume no volume change.)

16. Calculate the pH after 0.010 mol of gaseous HCl is added to 250.0 mL of each of the following buffered solutions:
    a. 0.050 *M* NH_3 and 0.15 *M* NH_4Cl
    b. 0.50 *M* NH_3 and 1.5 *M* NH_4Cl

17. Do the two initial buffered solutions in Exercise 16 differ in their pH or their capacity? What advantage is there in having a buffer with a greater capacity?

18. In blood the normal pH is 7.41. What is the ratio of CO_2 (usually written H_2CO_3) to HCO_3^- in blood?

$$H_2CO_3 \rightleftharpoons HCO_3^- + H^+ \qquad K_a = 4.3 \times 10^{-7}$$

19. Which of the following result in buffered solutions when equal volumes of the two solutions are mixed?
    a. 0.1 *M* HCl and 0.1 *M* NH_4Cl
    b. 0.1 *M* HCl and 0.1 *M* NH_3
    c. 0.2 *M* HCl and 0.1 *M* NH_3
    d. 0.1 *M* HCl and 0.2 *M* NH_3

20. Which of the following are buffers?
    a. A solution containing 0.1 *M* KNO_3 and 0.1 *M* HNO_3
    b. A solution containing 0.1 *M* $NaNO_2$ and 0.15 *M* HNO_2
    c. 250 mL of a solution that was originally 0.10 *M* acetic acid and has had 0.5 g KOH added
    d. A solution containing 0.10 *M* Na_2CO_3 and 0.05 *M* Na_3PO_4

Acid-Base Titrations

21. Calculate the value of the equilibrium constant for each of the following reactions in aqueous solution:
    a. $HC_2H_3O_2 + OH^- \rightleftharpoons C_2H_3O_2^- + H_2O$

b. $C_2H_3O_2^- + H^+ \rightleftharpoons HC_2H_3O_2$
c. $HC_2H_3O_2 + NH_3 \rightleftharpoons C_2H_3O_2^- + NH_4^+$
d. $HCl + NaOH \rightleftharpoons NaCl + H_2O$

Which of these reactions are useful for chemical analysis by titration?

22. Consider the titration of a generic weak acid HA with a strong base that gives the following titration curve:

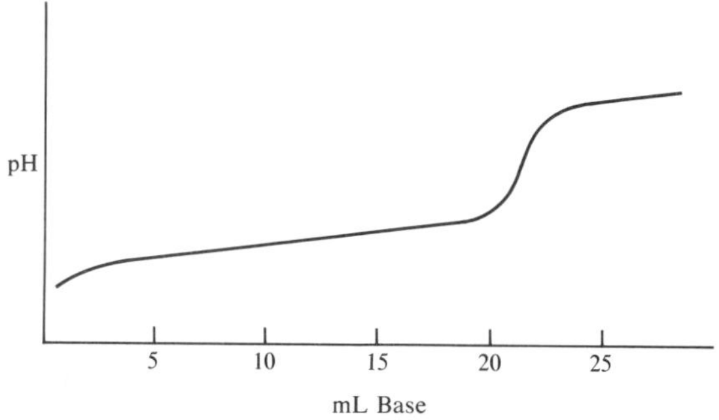

On the curve indicate the points that correspond to the following:

a. the equivalence point
b. the buffer region
c. pH = pK_a
d. pH depends only on $[HA]_0$
e. pH depends only on $[A^-]_0$
f. pH depends only on the amount of excess strong base added

23. A 25.0-mL sample of 0.100 *M* lactic acid ($HC_3H_5O_3$, pK_a = 3.86) is titrated with 0.100 *M* NaOH solution. Calculate the pH after the addition of 0.0 mL, 4.0 mL, 8.0 mL, 12.5 mL, 20.0 mL, 24.0 mL, 24.5 mL, 24.9 mL, 25.0 mL, 25.1 mL, 26 mL, 28 mL, and 30. mL of the NaOH. Plot the results of your calculations as pH versus milliliters of NaOH added.

24. Repeat the procedure in Exercise 23, but for the titration of 25.0 mL of 0.100 *M* NH_3 ($K_b = 1.8 \times 10^{-5}$) with 0.100 *M* HCl.

25. Repeat the procedure in Exercise 23, but for the titration of 25.0 mL of 0.100 *M* pyridine with 0.100 *M* hydrochloric acid (K_b for pyridine is 1.7×10^{-9}). Do not do the points at 24.9 and 25.1 mL.

26. Repeat the procedure in Exercise 23, but for the titration of 25.0 mL of 0.100 *M* NaOH with 0.100 *M* HNO_3.

27. Show that the pH at the halfway point of the titration of a weak acid with a strong base is equal to pK_a.

28. Is it possible to determine the K_a of a weak acid at points other than the halfway point of a titration?

29. Calculate the pH at the equivalence point for each of the following titrations:

a. 0.104 g of sodium acetate ($K_b = 5.6 \times 10^{-10}$) is dissolved in 25 mL of water and titrated with 0.0996 *M* HCl
b. 50.00 mL of 0.0426 *M* HOCl ($K_a = 3.5 \times 10^{-8}$) is titrated with 0.1028 *M* NaOH
c. 50.0 mL of 0.205 *M* HBr (a strong acid) is titrated with 0.356 *M* KOH

30. Estimate the pH of a solution in which crystal violet is yellow and methyl orange is red. (See Fig. 15.8.)

31. Estimate the pH of a solution in which bromcresol green is blue and thymol blue is yellow. (See Fig. 15.8.)

32. A solution has a pH of 9.0. What would be the color of the solution if each of the following indicators were added? (See Fig. 15.8.)

a. methyl orange
b. alizarin
c. bromcresol green
d. thymol blue

33. A solution has a pH of 4.5. What would be the color of the solution if each of the following indicators were added? (See Fig. 15.8.)

a. methyl orange
b. alizarin
c. bromcresol green
d. phenolphthalein

34. Is it possible for thymol blue to contain only a single —CO_2H group and no other acidic or basic functional group? Why or why not?

35. Define the equivalence point and the end point of a titration. Why does one choose an indicator so that the two points coincide? Is it necessary that the pH of the two points be within ±0.01 pH unit of each other? Why or why not?

36. Why does an indicator change from its acid to its base color over a range of pH values?

37. Which of the indicators in Fig. 15.8 should be used for doing the titrations in Exercises 23 and 25?

38. Which of the indicators in Fig. 15.8 should be used for doing the titrations in Exercises 24 and 26?

39. Methyl red has the following structure:

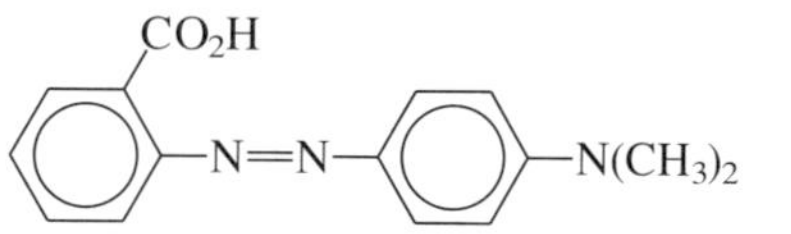

CO_2H

—N=N— —$N(CH_3)_2$ $K_a = 5.0 \times 10^{-6}$

It undergoes a color change from red to yellow as a solution gets more basic. Calculate an approximate pH range for which methyl red is useful. What are the color change and the pH at the color change when a weak acid is titrated with a strong base using methyl red as an indicator? What are the color change and the pH at the color change when a weak base is titrated with a strong acid using methyl red as an indicator? For which of the titrations in Exercises 23–26 would methyl red be a suitable indicator?

Solubility Equilibria

40. Write balanced dissolution reactions and expressions for the solubility products of the following:

a. $AgC_2H_3O_2$ (silver acetate)

b. MnS
c. $Al(OH)_3$
d. $Ca_3(PO_4)_2$
e. $Ca_5(PO_4)_3(OH)$ (hydroxyapatite, the major component of tooth enamel)
f. PbI_2
g. Cu_2S

41. Calculate values of K_{sp} for the following salts using the information given:
a. A sample of 4.8×10^{-5} mol of calcium oxalate (CaC_2O_4) dissolves in 1.0 L of water to produce a saturated solution.

$$CaC_2O_4(s) \rightleftharpoons Ca^{2+}(aq) + C_2O_4^{2-}(aq)$$

b. The concentration of Pb^{2+} in a solution saturated with $PbBr_2$ is 2.14×10^{-2} *M*.
c. The molar solubility of BiI_3 is 1.32×10^{-5} mol/L.
d. The solubility of iron(II) oxalate (FeC_2O_4) is 65.9 mg/L at 25°C.
e. A 0.100 L sample of a saturated solution of copper(II) periodate ($Cu(IO_4)_2$) contains 0.146 g of the dissolved salt.
f. The solubility of lithium carbonate (Li_2CO_3) is 5.48 g/L.

42. Calculate the solubility of each of the following compounds in moles per liter and grams per liter:
a. $Al(OH)_3$, $K_{sp} = 2 \times 10^{-32}$
b. $Be(IO_4)_2$, $K_{sp} = 1.57 \times 10^{-9}$
c. $CaSO_4$, $K_{sp} = 6.1 \times 10^{-5}$
d. $CaCO_3$, $K_{sp} = 8.7 \times 10^{-9}$
e. $MgNH_4PO_4$, $K_{sp} = 3 \times 10^{-13}$
f. Hg_2Cl_2, $K_{sp} = 1.1 \times 10^{-18}$ (Hg_2^{2+} is the cation in solution)
g. $SrSO_4$, $K_{sp} = 3.2 \times 10^{-7}$
h. Ag_2CO_3, $K_{sp} = 8.1 \times 10^{-12}$
i. $Ag_2Cl_2O_7$, $K_{sp} = 2 \times 10^{-7}$ ($Cl_2O_7^{2-}$ is the anion in solution)
j. Cu_2S, $K_{sp} = 2 \times 10^{-47}$

43. Under what circumstances can the relative solubilities of two salts be compared by directly comparing values of the solubility products?

44. Calculate the solubility (in moles per liter) of $Fe(OH)_3$ ($K_{sp} \cong 4 \times 10^{-38}$) in each of the following:
a. water (assume pH is 7.0 and constant)
b. a solution buffered at pH = 5.0
c. a solution buffered at pH = 11.0

45. Of the substances in Exercise 42, which will show increased solubility as the pH of the solution becomes more acidic? Write equations for the reactions that occur to increase the solubility.

46. The K_{sp} of hydroxyapatite, $Ca_5(PO_4)_3OH$, is 6.8×10^{-37}. Calculate the solubility of hydroxyapatite in pure water in moles per liter. How is the solubility of hydroxyapatite affected by adding acid? When hydroxyapatite is treated with fluoride, the mineral fluorapatite, $Ca_5(PO_4)_3F$, forms. The K_{sp} of this substance is 1×10^{-60}. Calculate the solubility of fluorapatite in water. How do these calculations provide a rationale for the fluoridation of drinking water?

47. The K_{sp} for lead iodide (PbI_2) is 1.4×10^{-8}. Calculate the solubility of lead iodide in each of the following:
a. water
b. 0.10 *M* $Pb(NO_3)_2$
c. 0.010 *M* NaI

48. Which of the following metal ions will be precipitated by a 0.1 *M* H_2S solution as sulfides at pH = 0.5? Which will be precipitated as sulfides at pH = 10.0? Assume the initial metal concentration is 0.010 *M*.

$$Cu^{2+},\ Ni^{2+},\ Mn^{2+},\ Hg^{2+}$$

49. Describe how you could separate the ions in each of the following groups by selective precipitation:
a. Ag^+, Mg^{2+}, Cu^{2+}
b. Pb^{2+}, Ca^{2+}, Fe^{2+}
c. Cl^-, Br^-, I^-
d. Pb^{2+}, Bi^{3+}

50. Silica (SiO_2) can undergo the following reactions.

$$SiO_2(s) + 2H_2O(l) \rightleftharpoons H_4SiO_4(aq) \qquad K = 2 \times 10^{-3}$$

Silicic acid

$$H_4SiO_4(aq) \rightleftharpoons H_3SiO_4^-(aq) + H^+(aq) \quad pK_a = 9.46$$

Will silica be more soluble in an acidic or basic solution? Why?

51. What are the concentrations of all of the ions in solution after 100.0 mL of 0.020 *M* $Pb(NO_3)_2$ and 100.0 mL of 0.020 *M* NaCl are mixed?

52. A solution is prepared by mixing 75.0 mL of 0.020 *M* $BaCl_2$ and 125 mL of 0.040 *M* H_2SO_4. What are the concentrations of barium and sulfate ions in this solution? Assume only SO_4^{2-} ions (no HSO_4^-) are present.

Complex Ion Equilibria

53. Write equations for the stepwise and overall formation of each of the following complex ions:
a. $Co(NH_3)_6^{2+}$
b. $Fe(SCN)_6^{3-}$
c. $Ag(NH_3)_2^+$
d. $Ni(CN)_4^{2-}$
e. $FeCl_4^-$

54. Given the following

$$Mn^{2+}(aq) + C_2O_4^{2-}(aq) \rightleftharpoons MnC_2O_4(aq) \qquad K_1 = 7.9 \times 10^3$$

$$MnC_2O_4(aq) + C_2O_4^{2-}(aq) \rightleftharpoons Mn(C_2O_4)_2^{2-}(aq) \qquad K_2 = 7.9 \times 10^1$$

calculate the value for the overall formation constant for $Mn(C_2O_4)_2^{2-}$:

$$K = \frac{[Mn(C_2O_4)_2^{2-}]}{[Mn^{2+}][C_2O_4^{2-}]^2}$$

55. Concentrated ammonia is added to an aqueous solution of copper(II) sulfate. Initially, a white precipitate forms. As more ammonia is added, the precipitate dissolves and the solution becomes a deep bluish-purple. Write equations describing the reactions that are occurring. (*Hint*: Cu^{2+} reacts with NH_3 to form $Cu(NH_3)_4^{2+}$.)

56. The overall formation constant for HgI_4^{2-} is 1.0×10^{30}. That is,

$$1.0 \times 10^{30} = \frac{[HgI_4^{2-}]}{[Hg^{2+}][I^-]^4}$$

What is the concentration of Hg^{2+} in 500.0 mL of a solution that was originally 0.010 *M* Hg^{2+} and had 65 g of KI added to it? The reaction is:

$$Hg^{2+}(aq) + 4I^-(aq) \rightleftharpoons HgI_4^{2-}(aq)$$

57. A solution is prepared by adding 0.10 mol of $Ni(NH_3)_6Cl_2$ to 0.50 L of 3.0 *M* NH_3. Calculate $[Ni(NH_3)_6^{2+}]$ and $[Ni^{2+}]$ in this solution (K (overall) for $Ni(NH_3)_6^{2+}$ is 5.5×10^8).

58. As sodium chloride solution is added to a solution of silver nitrate, a white precipitate forms. Ammonia is added to the mixture and the precipitate dissolves. When potassium bromide solution is then added, a pale yellow precipitate appears. When a solution of sodium thiosulfate is added, the yellow precipitate dissolves. Finally, potassium iodide is added to the solution and a yellow precipitate forms. Write reactions for all of the changes mentioned above. What conclusions can you draw concerning the sizes of the K_{sp} values for AgCl, AgBr, and AgI? What can you say about the relative values of the formation constants of $Ag(NH_3)_2^+$ and $Ag(S_2O_3)_2^{3-}$?

Additional Exercises

59. Will a precipitate of $Cd(OH)_2$ form if 1.0 mL of 1.0 *M* $Cd(NO_3)_2$ is added to 1.0 L of 5.0 *M* NH_3?

$$Cd^{2+} + 4NH_3 \rightleftharpoons Cd(NH_3)_4^{2+} \qquad K = 1.0 \times 10^7$$

$$Cd(OH)_2 \rightleftharpoons Cd^{2+} + 2OH^- \qquad K_{sp} = 5.9 \times 10^{-15}$$

60. Borax ($Na_2B_4O_7 \cdot 10H_2O$) dissolves in water according to the reaction

$$Na_2B_4O_7 \cdot 10H_2O(s) \rightarrow 2Na^+ + 3H_2O + 2B(OH)_3 + 2B(OH)_4^-$$

Boric acid reacts with water according to the following reaction:

$$B(OH)_3 + H_2O \rightleftharpoons B(OH)_4^- + H^+ \qquad K_a = 5.8 \times 10^{-10}$$

a. Calculate the pH of the resulting solution when 28.6 g of borax is dissolved to make 1.0 L of solution.
b. If 100. mL of 0.10 *M* NaOH is added to the solution in part a, what is the new pH?

61. Tris(hydroxymethyl)aminomethane, commonly called TRIS or Trizma, is often used as a buffer in biochemical studies. Its buffering range is pH 7–9, and K_b is 1.19×10^{-6} for the reaction:

$$\underset{\text{TRIS}}{(HOCH_2)_3CNH_2} + H_2O \rightleftharpoons \underset{\text{TRISH}^+}{(HOCH_2)_3CNH_3^+} + OH^-$$

a. What is the optimum pH for TRIS buffers?
b. Calculate the ratio TRIS/TRISH$^+$ at pH = 7.0 and at pH = 9.0.
c. A buffer is prepared by diluting 50.0 g of TRIS base and 65.0 g of TRIS hydrochloride (written as TRISHCl) to a total volume of 2.0 L. What is the pH of this buffer? What is the pH after 0.50 mL of 12 *M* HCl is added to a 200.0-mL portion of the buffer?

62. Cacodylic acid, $(CH_3)_2AsO_2H$, a toxic compound that behaves as a weak acid, is used in preparing buffered solutions. For the following reaction $pK_a = 6.19$:

$$(CH_3)_2AsO_2H + H_2O \rightleftharpoons H_3O^+ + (CH_3)_2AsO_2^-$$

Calculate the masses of cacodylic acid and sodium cacodylate that should be used to prepare 500.0 mL of a buffer at pH = 6.60 that has a total concentration of all arsenic-containing species equal to 0.25 *M*; that is,

$$[(CH_3)_2AsO_2H] + [(CH_3)_2AsO_2^-] = 0.25\ M$$

63. You have the following reagents on hand.

Solids (pK_a of acid form is given)	
$(CH_3)_2AsO_2Na$ (6.19)	Sodium acetate (4.74)
TRISHCl (8.08)	Potassium fluoride (3.14)
Benzoic acid (4.19)	Ammonium chloride (9.26)
Solutions	
5.0 *M* HCl	2.6 *M* NaOH
Glacial acetic acid	

What combinations of reagents would you use to prepare buffers at the following pH values?

a. 3.0
b. 4.0
c. 5.0
d. 7.0
e. 9.0

64. a. Calculate the pH of a buffered solution that is 0.10 *M* in $C_6H_5CO_2H$ (benzoic acid, $K_a = 6.4 \times 10^{-5}$) and 0.10 *M* in $C_6H_5CO_2Na$.
b. Calculate the pH after 20.0% (by moles) of the benzoic acid is converted to benzoate anion by addition of base. Use the dissociation equilibrium:

$$C_6H_5CO_2H \rightleftharpoons C_6H_5CO_2^- + H^+$$

c. Do the same as in part b, but use the following equilibrium to calculate the pH:

$$C_6H_5CO_2^- + H_2O \rightleftharpoons C_6H_5CO_2H + OH^-$$

d. Do your answers in parts b and c agree? Why or why not?

65. One method for determining the purity of aspirin is to hydrolyze it with NaOH solution and then titrate the remaining NaOH. The reaction of aspirin with NaOH is as follows.

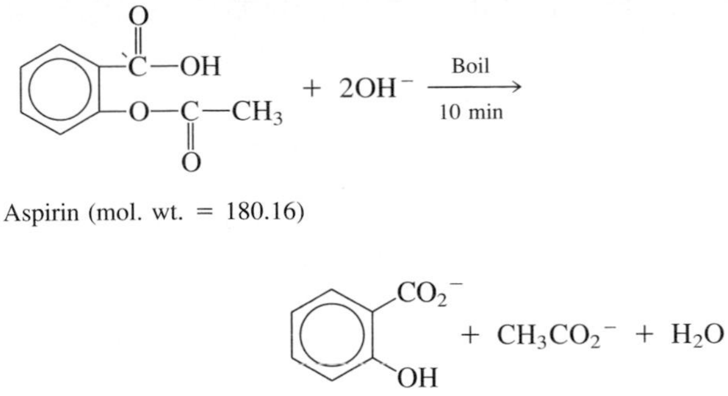

$$\text{(acetylsalicylic acid: } C_6H_4(COOH)(O{-}C(=O){-}CH_3)) + 2OH^- \xrightarrow[\text{10 min}]{\text{Boil}}$$

Aspirin (mol. wt. = 180.16)

$$C_6H_4(CO_2^-)(OH) + CH_3CO_2^- + H_2O$$

Salicylate ion   Acetate ion

A sample of aspirin with a mass of 1.427 g was boiled in 50.00 mL of 0.500 M NaOH. After the solution was cooled, it took 31.92 mL of 0.289 M HCl to titrate the excess NaOH. Calculate the purity of the aspirin. What indicator should be used for this titration? Why?

66. Another way to treat data from a pH titration is to graph the absolute value of the change in pH per milliliter added versus milliliters added (ΔpH/mL versus mL added). Make such graphs for the calculations you did in Exercise 23. What advantage might this method have over the traditional method for treating titration data?

67. Potassium hydrogen phthalate, known as KHP (mol. wt. = 204.22 g/mol), can be obtained in high purity and is used to determine the concentration of solutions of strong bases by the reaction:

$$HP^- + OH^- \rightarrow H_2O + P^{2-}$$

If a typical titration experiment begins with about 0.5 g of KHP and has a final volume of about 100 mL, what would be an appropriate indicator to use? The pK_a for HP^- is 5.51.

68. The equilibrium constant for the following reaction is 1.0×10^{23}.

$$Cr^{3+}(aq) + H_2EDTA^{2-}(aq) \rightleftharpoons CrEDTA^- + 2H^+$$

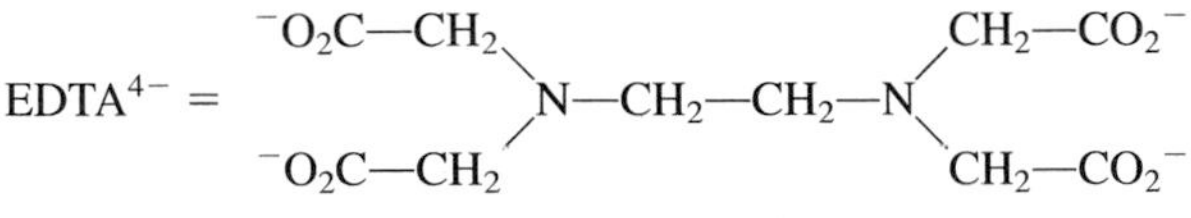

$$EDTA^{4-} = \begin{matrix} {}^-O_2C{-}CH_2 \\ {}^-O_2C{-}CH_2 \end{matrix} \!\!> N{-}CH_2{-}CH_2{-}N <\!\! \begin{matrix} CH_2{-}CO_2^- \\ CH_2{-}CO_2^- \end{matrix}$$

Ethylenediaminetetraacetate

EDTA is used as a complexing agent in chemical analysis. Solutions of EDTA, usually as the disodium salt Na_2H_2EDTA, are used to treat heavy metal poisoning. Calculate $[Cr^{3+}]$ at equilibrium in a solution originally 0.0010 M in Cr^{3+} and 0.050 M in H_2EDTA^{2-} and buffered at pH = 6.0.

69. Calculate the concentration of Pb^{2+} in each of the following:
a. saturated solution of $Pb(OH)_2$; $K_{sp} = 1.2 \times 10^{-15}$
b. saturated solution of $Pb(OH)_2$ buffered at pH = 13.0
c. 0.010 mol of $Pb(NO_3)_2$ added to 1.0 L of aqueous solution, buffered at pH = 13.0 and containing 0.050 M Na_4EDTA. Does $Pb(OH)_2$ precipitate from this solution? For the reaction:

$$Pb^{2+} + EDTA^{4-} \rightleftharpoons PbEDTA^{2-} \qquad K = 1.1 \times 10^{18}$$

70. Calculate the concentration of Fe^{3+} in blood (pH = 7.41) using the solubility equilibrium of $Fe(OH)_3$ ($K_{sp} = 4 \times 10^{-38}$). Iron levels in blood serum range from 60 to 150 μg/100 mL. How can you account for the discrepancy between the two values?

71. Using the K_{sp} for $Cu(OH)_2$ (1.6×10^{-19}) and the overall formation constant for $Cu(NH_3)_4^{2+}$ (1.0×10^{13}) calculate a value for the equilibrium constant for the reaction:

$$Cu(OH)_2(s) + 4NH_3(aq) \rightleftharpoons Cu(NH_3)_4^{2+}(aq) + 2OH^-(aq)$$

72. Use the value of the equilibrium constant you calculated in Exercise 71 to calculate the solubility (in mol/L) of $Cu(OH)_2$ in 5.0 M NH_3. In 5.0 M NH_3 the concentration of OH^- is about 0.010 M.

73. For which salt in each of the following groups will the solubility depend on pH?
a. AgF, AgCl, AgBr
b. PbO, $PbCl_2$
c. $Sr(NO_3)_2$, $Sr(NO_2)_2$
d. $Ni(NO_3)_2$, $Ni(CN)_2$

74. Use the equilibrium constants in Exercise 50 to calculate the molar solubility (in mol/L) of silica at a pH of 7.0 and a pH of 10.0.

75. In the chapter discussion of precipitate formation, we ran the precipitation reaction to completion and then let some of the precipitate redissolve to get back to equilibrium. To see why, redo Sample Exercise 15.15 where

Initial concentration (mol/L)		Equilibrium concentration (mol/L)
$[Mg^{2+}]_0 = 3.75 \times 10^{-3}$ $[F^-]_0 = 6.25 \times 10^{-2}$	$\xrightarrow[\text{reacts to form } MgF_2]{y \text{ mol/L } Mg^{2+}}$	$[Mg^{2+}] = 3.75 \times 10^{-3} - y$ $[F^-] = 6.25 \times 10^{-2} - 2y$

76. Calculate the final concentrations of $K^+(aq)$, $C_2O_4^{2-}(aq)$, $Ba^{2+}(aq)$, and $Br^-(aq)$ in a solution prepared by adding 0.100 L of 0.20 M $K_2C_2O_4$ to 0.150 L of 0.25 M $BaBr_2$. (For BaC_2O_4, $K_{sp} = 2.3 \times 10^{-8}$.)

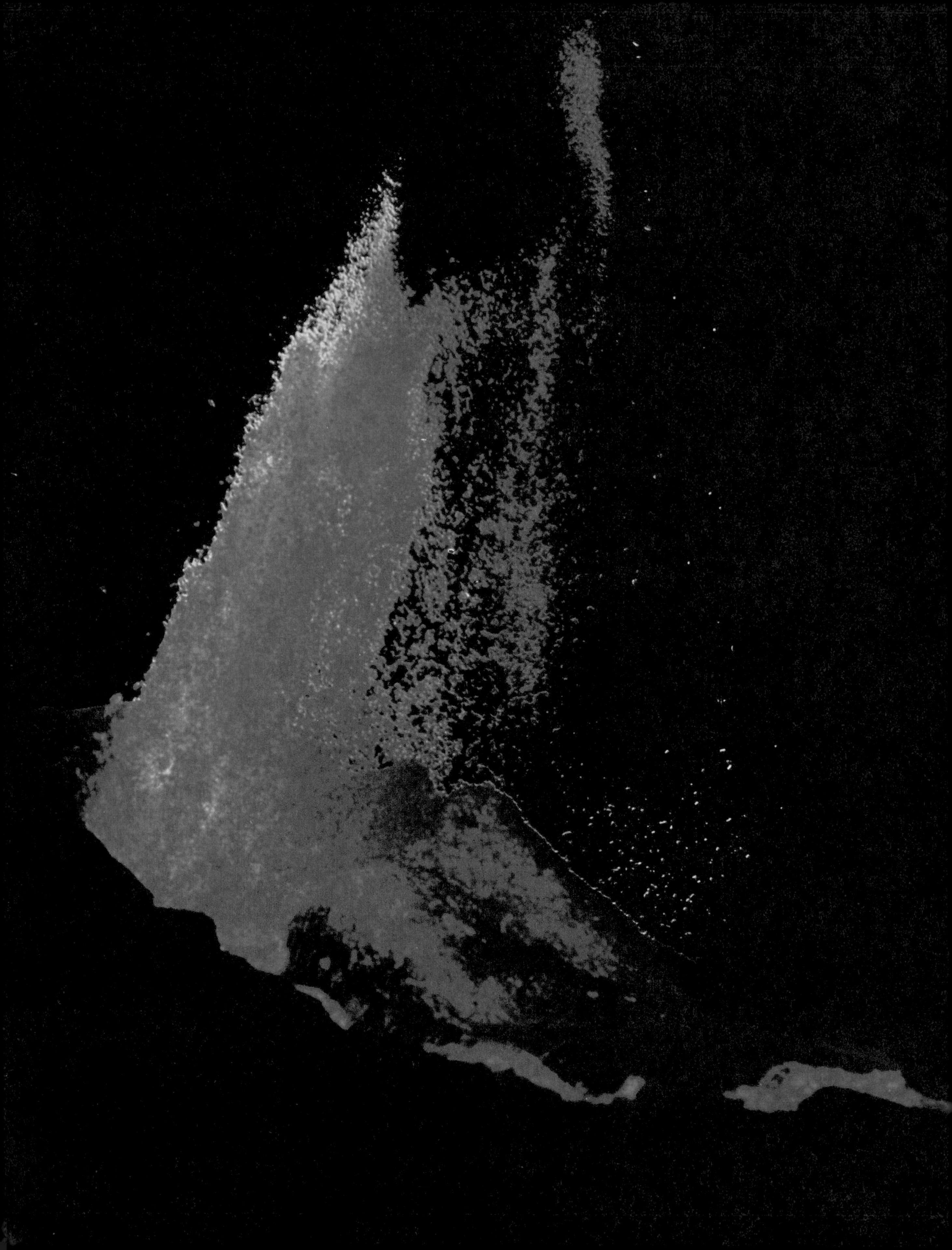

CHAPTER 16

Spontaneity, Entropy, and Free Energy

The *first law of thermodynamics* is a statement of the law of conservation of energy: energy can be neither created nor destroyed. In other words, *the energy of the universe is constant.* Although the total energy is constant, the various forms of energy can be interchanged in physical and chemical processes. For example, if you drop a book, some of the initial potential energy of the book is changed to kinetic energy, which is transferred to the atoms in the air and the floor as random motion. The net effect of this process is to change a given quantity of potential energy to exactly the same quantity of thermal energy. Energy has been converted from one form to another, but the same quantity of energy exists before and after the process.

Now let's consider a chemical example. When methane is burned in excess oxygen, the major reaction is

$$CH_4(g) + 2O_2(g) \rightarrow CO_2(g) + 2H_2O(g) + \text{energy}$$

This reaction produces a quantity of energy, which is released as heat. This energy flow results from a lowering of the potential energy stored in the bonds of CH_4 and O_2 as they react to form CO_2 and H_2O. This is illustrated in Fig. 16.1. Potential energy has been converted to thermal energy, but the energy content of the universe has remained constant.

The first law of thermodynamics is used mainly for energy bookkeeping, that is, to answer questions such as:

How much energy is involved in the change?
Does energy flow into or out of the system?
What form does the energy finally assume?

CONTENTS

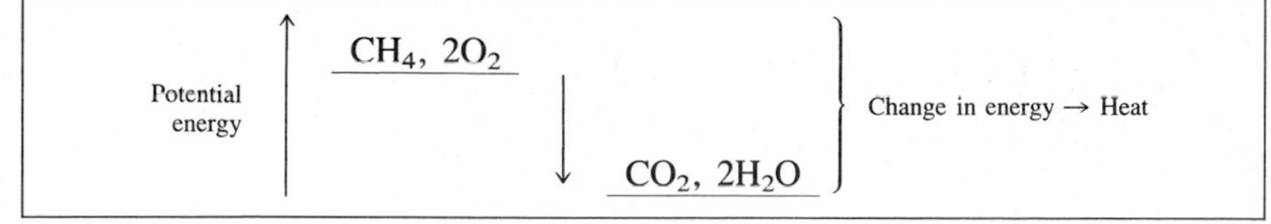

Figure 16.1

When methane and oxygen react to form carbon dioxide and water, the products have lower potential energy than the reactants. This change in potential energy results in energy flow (heat) to the surroundings.

< An eruption of the Kilauea volcano. Vast quantities of energy are transferred to the environment as heat by the volcano.

The first law of thermodynamics: the energy of the universe is constant.

Although the first law of thermodynamics provides the means for accounting for energy, it gives no hint as to *why* a particular process occurs in a given direction. This is the main question to be considered in this chapter.

16.1 Spontaneous Processes and Entropy

Purpose

- To define a spontaneous process.
- To define entropy in terms of positional probability.

Spontaneous does not mean fast.

A process is said to be *spontaneous* if it *occurs without outside intervention.* **Spontaneous processes** may be fast or slow. As we will see in this chapter, thermodynamics can tell us the *direction* in which a process will occur but can say nothing about the *speed* of the process. As we saw in Chapter 12, the rate of a reaction depends on many factors, such as activation energy, temperature, concentration, and catalysts, and we were able to explain these effects using a simple collision model. In describing a chemical reaction, the discipline of chemical kinetics focuses on the pathway between reactants and products; thermodynamics only considers the initial and final states and does not require knowledge of the pathway between reactants and products (see Fig. 16.2).

We will see in this chapter that thermodynamics lets us predict whether a process will occur but gives no information about the amount of time required for the process. For example, according to the principles of thermodynamics, a diamond should change spontaneously to graphite. The fact that we do not observe this

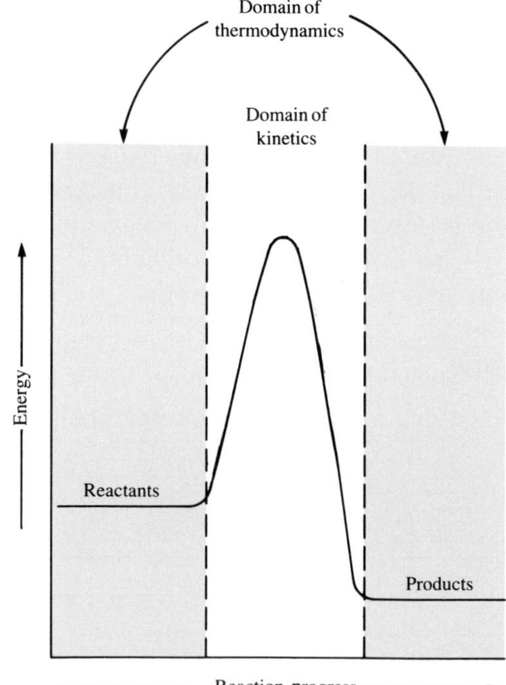

Figure 16.2

The rate of a reaction depends on the pathway from reactants to products; this is the domain of kinetics. Thermodynamics tells us whether or not a reaction is spontaneous based only on the properties of the reactants and products. The predictions of thermodynamics do not require knowledge of the pathway between reactants and products.

process does not mean the prediction is wrong; it simply means the process is very slow. Thus we need both thermodynamics and kinetics to describe reactions fully.

To explore the idea of spontaneity, consider the following physical and chemical processes:

A ball rolls down a hill but never spontaneously rolls back up the hill.

If exposed to air and moisture, steel rusts spontaneously. However, the iron oxide in rust does not spontaneously change back to iron metal and oxygen gas.

A gas fills its container uniformly. It never spontaneously collects at one end of the container.

Heat flow always occurs from a hot object to a cooler one. The reverse process never occurs spontaneously.

Wood burns spontaneously in an exothermic reaction to form carbon dioxide and water, but wood is not formed when carbon dioxide and water are heated together.

At temperatures below 0°C, water spontaneously freezes, and at temperatures above 0°C, ice spontaneously melts.

What thermodynamic principle will provide an explanation of why, under a given set of conditions, each of these diverse processes occurs in one direction and never in the reverse? In searching for an answer, we could explain the behavior of a ball on a hill in terms of gravity. But what does gravity have to do with the rusting of a nail or the freezing of water? In the early development of thermodynamics, it was thought that exothermicity might be the key, that a process would be spontaneous if it was exothermic. Although this factor does appear to be important, since many spontaneous processes are exothermic, it is not the total answer. For example, the melting of ice, which occurs spontaneously at temperatures greater than 0°C, is an endothermic process.

What common characteristic causes the processes listed to be spontaneous in one direction only? After many years of observation, scientists have concluded that the characteristic common to all spontaneous processes is an increase in a property called **entropy** (S). *The driving force for a spontaneous process is an increase in the entropy of the universe.*

Plant materials burn to form carbon dioxide and water.

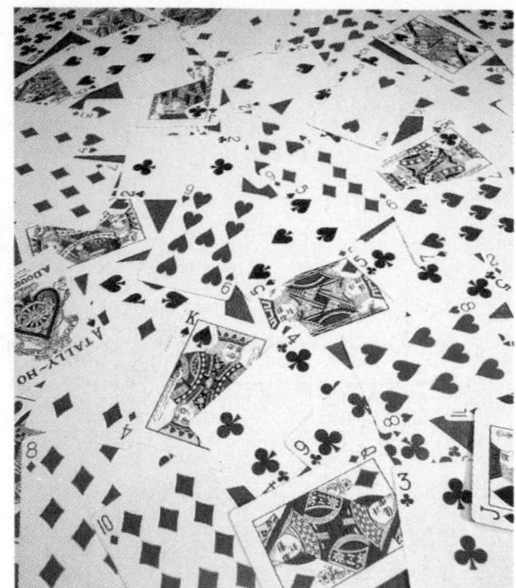

A disordered pile of playing cards.

What is entropy? Although there is no simple definition that is completely accurate, *entropy can be viewed as a measure of randomness or disorder*. The natural progression of things is from order to disorder, from lower entropy to higher entropy. You only have to think about the condition of your room to be convinced of this. Your room naturally tends to get messy (disordered), because an ordered room requires everything to be in its place. There are simply many more ways for things to be out of place than for them to be in their places.

As another example, suppose you have a deck of playing cards ordered in some particular way. You throw these cards into the air and pick them all up at random. Looking at the new sequence of the cards, you would be very surprised to find that it matched the original order. Such an event would be possible, but *very improbable*. There are billions of ways for the deck to be disordered, but only one way to be ordered according to your definition. Thus the chances of picking the cards up out of order are much greater than of picking them up in order. It is natural for disorder to increase.

Probability refers to likelihood.

Entropy is a thermodynamic function that describes the *number of arrangements* (positions and/or energy levels) that are *available to a system* existing in a given state. Entropy is closely associated with probability. The key concept is that the more ways a particular state can be achieved, the greater is the likelihood (probability) of finding that state. In other words, *nature spontaneously proceeds toward the states that have the highest probabilities of existing*. This is not a surprising conclusion at all. The difficulty comes in connecting this concept to real-life processes. For example, what does the spontaneous rusting of steel have to do with probability?

Understanding the connection between entropy and spontaneity will allow us to answer such questions. We will begin to explore this connection by considering a very simple process, the expansion of an ideal gas into a vacuum, as represented in Fig. 16.3. Why is this process spontaneous? The driving force is probability. Because there are more ways of having the gas evenly spread throughout the container than there are ways for it to be in any other possible state, the gas spontaneously attains the uniform distribution.

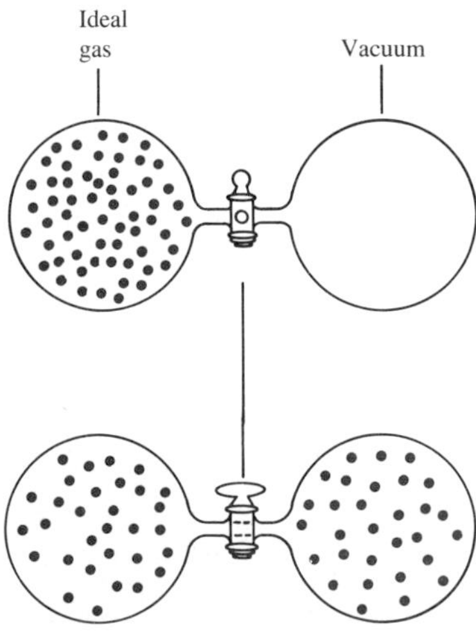

Figure 16.3

The expansion of an ideal gas into an evacuated bulb.

To understand this conclusion, we will greatly simplify the system and consider some of the possible arrangements of only four gas molecules in the two-bulbed container (Fig. 16.4). How many ways can each arrangement (state) be achieved? Arrangement I can be achieved in only one way—all the molecules must be in one end. Arrangement II can be achieved in four ways, as shown in Table 16.1. Each configuration that gives a particular arrangement is called a *microstate*. Arrangement I has one microstate, and arrangement II has four microstates. Arrangement III can be achieved in six ways (six microstates), as shown in Table 16.1. *Which arrangement is most likely to occur?* The one that can be achieved in the greatest number of ways. Thus arrangement III is most probable; the relative probabilities of arrangements III, II, and I are 6:4:1. We have discovered an important principle: the probability of occurrence of a particular arrangement (state) depends on the number of ways (microstates) in which that arrangement can be achieved.

The consequences of this principle are dramatic for large numbers of molecules. One gas molecule in the flask in Fig. 16.4 has one chance in two of being in the left bulb. We say that the probability of finding the molecule in the left bulb is $\frac{1}{2}$. For two molecules in the flask, there is one chance in two of finding each molecule in the left bulb so there is one chance in four ($\frac{1}{2} \times \frac{1}{2} = \frac{1}{4}$) that *both* mole-

The Microstates That Give a Particular Arrangement (State)

Arrangement	Microstates	
I		(ABCD)—( )
II	(BCD)—(A)	(ABD)—(C)
	(ACD)—(B)	(ABC)—(D)
III	(AB)—(CD)	(BC)—(AD)
	(AC)—(BD)	(BD)—(AC)
	(AD)—(BC)	(CD)—(AB)

Table 16.1

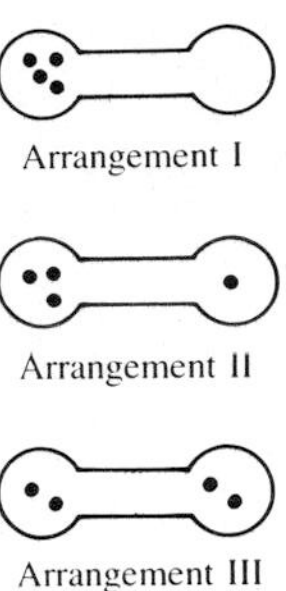

Figure 16.4

Three possible arrangements (states) of four molecules in a two-bulbed flask.

cules will be in the left bulb. As the number of molecules increases, the relative probability of finding all of them in the left bulb decreases, as shown in Table 16.2. For 1 mole of gas, the probability of finding all the molecules in the left bulb is so small that this arrangement would "never" occur.

For two molecules in the flask, there are four possible microstates:

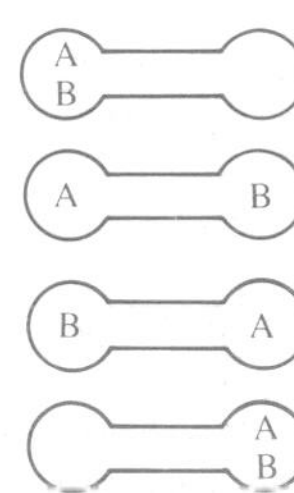

Thus there is one chance in four of finding

Probability of Finding All the Molecules in the Left Bulb as a Function of the Total Number of Molecules

Number of molecules	Relative probability of finding all molecules in the left bulb
1	$\frac{1}{2}$
2	$\frac{1}{2} \times \frac{1}{2} = \frac{1}{2^2} = \frac{1}{4}$
3	$\frac{1}{2} \times \frac{1}{2} \times \frac{1}{2} = \frac{1}{2^3} = \frac{1}{8}$
5	$\frac{1}{2} \times \frac{1}{2} \times \frac{1}{2} \times \frac{1}{2} \times \frac{1}{2} = \frac{1}{2^5} = \frac{1}{32}$
10	$\frac{1}{2^{10}} = \frac{1}{1024}$
n	$\frac{1}{2^n} = \left(\frac{1}{2}\right)^n$
6×10^{23} (1 mole)	$\left(\frac{1}{2}\right)^{6\times10^{23}} = 10^{-(2\times10^{23})}$

Table 16.2

Thus a gas placed in one end of a container will spontaneously expand to fill the entire vessel evenly because, for a large number of gas molecules, there is a huge number of microstates in which equal numbers of molecules are in both ends. On the other hand, the opposite process,

although not impossible, is *highly* improbable since only one microstate leads to this arrangement. Therefore, the process does not occur spontaneously.

The type of probability we have been considering in this example is called **positional probability** because it depends on the number of configurations in space (positional microstates) that yield a particular state. A gas expands into a vacuum to give a uniform distribution because the expanded state has the highest positional probability, that is, the largest entropy, of the states available to the system.

Solid, liquid, and gaseous states were compared in Chapter 10.

Solids are more ordered than liquids or gases and thus have lower entropy.

Positional probability is also illustrated by changes of state. In general, positional entropy, or disorder, increases in going from solid to liquid to gas. A mole of a substance has a much smaller volume in the solid state than it does in the gaseous state. In the solid state, the molecules are close together, with relatively few positions available to them; in the gaseous state, the molecules are far apart, with many more positions available to them. The liquid state is somewhat closer to the solid state than it is to the gaseous state. We can summarize these comparisons as follows:

$$S_{solid} < S_{liquid} \ll S_{gas}$$

The tendency to mix is actually due to the increased volume available to the particles of each component of the mixture. For example, when two liquids are mixed, the molecules of each liquid have more available space and thus more available positions.

Positional entropy is also very important in the formation of solutions. In Chapter 11 we saw that solution formation is favored by the natural tendency for substances to mix. We can now be more precise. The entropy change associated with the mixing of two pure substances is expected to be positive. There is an increase in disorder because there are many more microstates for the mixed condition than for the separated condition. Thus the increase in positional entropy associated with mixing favors the formation of solutions.

Sample Exercise 16.1

For each of the following pairs, choose the substance with the higher positional entropy (per mole) at a given temperature.

a. solid CO_2 and gaseous CO_2

b. N_2 gas at 1 atm and N_2 gas at 1.0×10^{-2} atm

Solution

a. Since a mole of gaseous CO_2 has the greater volume by far, the molecules have many more available positions than in a mole of solid CO_2. Thus gaseous CO_2 has the higher positional entropy.

b. A mole of N_2 gas at 1×10^{-2} atm has 100 times the volume (at a given temperature) of a mole of N_2 gas at 1 atm. Thus N_2 gas at 1×10^{-2} atm has the higher positional entropy.

A "steaming" piece of dry ice (solid carbon dioxide).

Sample Exercise 16.2

Predict the sign of the entropy change for each of the following processes:

a. Sugar is added to water to form a solution.
b. Iodine vapor condenses on a cold surface to form crystals.

Solution

a. The sugar molecules become randomly dispersed in the water when the solution forms, and their positional disorder is increased. There will be an increase in entropy, or ΔS is positive.
b. Gaseous iodine is forming a solid. This process involves a change from a relatively large volume to a much smaller volume, which results in lower positional disorder. For this process ΔS is negative (the entropy decreases).

16.2 Entropy and the Second Law of Thermodynamics

Purpose

- To state the second law of thermodynamics in terms of entropy.

We have seen that processes are spontaneous when they result in an increase in disorder. Nature always moves toward the most probable state available to it. We can state this principle in terms of entropy: *in any spontaneous process there is always an increase in the entropy of the universe*. This is the **second law of thermodynamics.** Contrast this with the first law of thermodynamics, which tells us that the energy of the universe is constant. Energy is conserved in the universe, but entropy is not. In fact, the second law can be paraphrased as follows: *the entropy of the universe is increasing*.

The total energy of the universe is constant, but the entropy is increasing.

As in Chapter 6, we find it convenient to divide the universe into a system and the surroundings. Thus we can represent the change in the entropy of the universe as

$$\Delta S_{univ} = \Delta S_{sys} + \Delta S_{surr}$$

where ΔS_{sys} and ΔS_{surr} represent the changes in entropy that occur in the system and surroundings, respectively.

To predict whether a given process will be spontaneous, we must know the sign of ΔS_{univ}. If ΔS_{univ} is positive, the entropy of the universe increases, and the process is spontaneous in the direction written. If ΔS_{univ} is negative, the process is spontaneous in the *opposite* direction. If ΔS_{univ} is zero, the process has no tendency to occur and the system is at equilibrium. To predict whether a process is spontaneous, we must consider the entropy changes that occur in the system and in the surroundings.

Sample Exercise 16.3

In a living cell, large molecules are assembled from simple ones. Is this process consistent with the second law of thermodynamics?

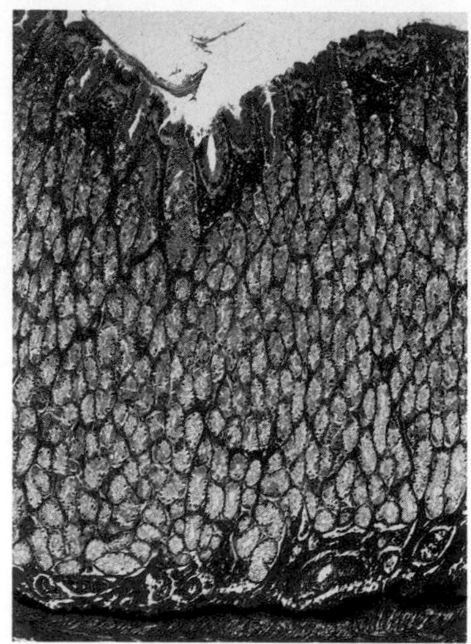

Mucosa in a human stomach.

Sample Exercise 16.3, continued

Solution

To reconcile the operation of an order-producing cell with the second law of thermodynamics, we must remember that ΔS_{univ}, not ΔS_{sys}, must be positive for a process to be spontaneous. A process for which ΔS_{sys} is negative can be spontaneous if ΔS_{surr} is large and positive. The operation of a cell is such a process.

16.3 The Effect of Temperature on Spontaneity

Purpose

- To discuss the important characteristics of entropy changes in the surroundings.
- To apply the relationship between ΔS_{surr}, ΔH, and T (K).

To explore the interplay of ΔS_{sys} and ΔS_{surr} in determining the sign of ΔS_{univ}, we will first discuss the change of state for one mole of water from liquid to gas,

$$H_2O(l) \rightarrow H_2O(g)$$

considering the water to be the system and everything else the surroundings.

What happens to the entropy of water in this process? A mole of liquid water (18 grams) has a volume of approximately 18 mL. A mole of gaseous water at 1 atmosphere and 100°C occupies a volume of approximately 31 liters. Clearly there are many more positions available to the water molecules in a volume of 31 L than in 18 mL and the vaporization of water is favored by this increase in positional probability. That is, for this process the entropy of the system increases; ΔS_{sys} has a positive sign.

What about the entropy change in the surroundings? Although we will not prove it here, entropy changes in the surroundings are determined primarily by the flow of energy into or out of the system as heat. To understand this, suppose an exothermic process transfers 50 J of energy as heat to the surroundings, where it becomes thermal energy, that is, kinetic energy associated with the random motions of atoms. Thus this flow of energy into the surroundings increases the random motions of atoms there and thereby increases the entropy of the surroundings. The sign of ΔS_{surr} is positive. When an endothermic process occurs in the system, it produces the opposite effect. Heat flows from the surroundings to the system, and the random motions of the atoms in the surroundings decrease, decreasing the entropy of the surroundings. The vaporization of water is an endothermic process. Thus for this change of state, ΔS_{surr} is negative.

In an endothermic process, heat flows from the surroundings. In an exothermic process, heat flows to the surroundings.

Remember it is the sign of ΔS_{univ} that tells us whether the vaporization of water is spontaneous or not. We have seen that ΔS_{sys} is positive and favors the process and that ΔS_{surr} is negative and unfavorable. Thus the components of ΔS_{univ} are in opposition. Which one controls the situation? The answer *depends on the temperature*. We know that at a pressure of 1 atmosphere, water changes spontaneously from liquid to gas at all temperatures above 100°C. Below 100°C, the opposite process (condensation) is spontaneous.

Since ΔS_{sys} and ΔS_{surr} are in opposition to one another for the vaporization of water, the temperature must have an effect on the relative importance of these two terms. To understand why this is so, we must discuss in more detail the factors that control the entropy changes in the surroundings. The central idea is that *the entropy changes in the surroundings are primarily determined by heat flow.* An exothermic process in the system increases the entropy of the surroundings, because the energy flow increases the random motions in the surroundings. This means that exothermicity is an important driving force for spontaneity. In earlier chapters we have seen that a system tends to undergo changes that lower its energy. We now understand the reason for that tendency. When a system at constant temperature moves to a lower energy, the energy it gives up is transferred to the surroundings, leading to an increase in entropy there.

The significance of exothermicity as a driving force *depends on the temperature at which the process occurs*. That is, the magnitude of ΔS_{surr} depends on the temperature at which the heat is transferred. We will not attempt to prove this fact here. Instead, we offer an analogy. Suppose that you have $50 to give away. Giving it to a millionaire would not create much of an impression—a millionaire has money to spare. However, to a poor college student, $50 would represent a significant sum and would be received with considerable joy. The same principle can be applied to energy transfer via the flow of heat. If 50 J of energy is transferred to the surroundings, the impact of that event depends greatly on the temperature. If the temperature of the surroundings is very high, the atoms there are in rapid motion. The 50 J of energy will not make a large percent change in these motions. On the other hand, if 50 J of energy is transferred to the surroundings at a very low temperature, where atomic motion is slow, the energy will cause a large percent change in the motions. The impact of the transfer of a given quantity of energy as heat to or from the surroundings will be greatest at low temperatures.

For our purposes there are two important characteristics of the entropy changes that occur in the surroundings:

1. *The sign of ΔS_{surr} depends on the direction of the heat flow.* At constant temperature an exothermic process in the system causes heat to flow into the surroundings, increasing the random motions of the atoms and thereby increasing the entropy of the surroundings. For this case, ΔS_{surr} is positive. The opposite is true for an endothermic process in a system at constant temperature. This principle is often stated in terms of energy. An important driving force in nature results from the tendency of a system to achieve the lowest possible energy.
2. *The magnitude of ΔS_{surr} depends on the temperature.* The transfer of a given quantity of energy as heat produces a much greater percent change in the randomness of the surroundings at a low temperature than it does at a high temperature. Thus ΔS_{surr} depends directly on the quantity of heat transferred and inversely on temperature. In other words, the tendency for the system to lower its energy becomes a more important driving force at lower temperatures.

In a process occurring at constant temperature, the tendency for the system to lower its energy is due to the resulting positive ΔS_{surr}.

$$\begin{array}{c}\text{Driving force}\\ \text{provided by}\\ \text{the energy flow}\\ \text{(heat)}\end{array} = \begin{array}{c}\text{magnitude of the}\\ \text{entropy change of}\\ \text{the surroundings}\end{array} = \frac{\text{quantity of heat (J)}}{\text{temperature (K)}}$$

These ideas are summarized as follows:

Exothermic process: ΔS_{surr} = **positive**
Endothermic process: ΔS_{surr} = **negative**

$$\text{Exothermic process:} \quad \Delta S_{surr} = +\frac{\text{quantity of heat (J)}}{\text{temperature (K)}}$$

$$\text{Endothermic process:} \quad \Delta S_{surr} = -\frac{\text{quantity of heat (J)}}{\text{temperature (K)}}$$

We can express ΔS_{surr} in terms of the change in enthalpy (ΔH) for a process occurring at constant pressure, since

$$\text{Heat flow (constant } P) = \text{change in enthalpy} = \Delta H$$

When no subscript is present the quantity (for example, ΔH) refers to the system.

Recall that ΔH consists of two parts: a sign and a number. The *sign* indicates the direction of flow, where a plus sign means into the system (endothermic) and a minus sign means out of the system (exothermic). The *number* indicates the quantity of energy.

Combining all these concepts produces the following definition of ΔS_{surr} for a reaction that takes place under conditions of constant temperature (Kelvin) and pressure:

$$\Delta S_{surr} = -\frac{\Delta H}{T}$$

The minus sign is necessary because the sign of ΔH is determined with respect to the reaction system, and this equation expresses a property of the surroundings. This means that if the reaction is exothermic, ΔH has a negative sign, but since heat flows into the surroundings, ΔS_{surr} is positive.

Sample Exercise 16.4

Stibnite contains Sb_2S_3.

In the metallurgy of antimony, the pure metal is recovered via different reactions, depending on the composition of the ore. For example, iron is used to reduce antimony in sulfide ores:

$$Sb_2S_3(s) + 3Fe(s) \rightarrow 2Sb(s) + 3FeS(s) \qquad \Delta H = -125 \text{ kJ}$$

and carbon is used to reduce it in oxide ores:

$$Sb_4O_6(s) + 6C(s) \rightarrow 4Sb(s) + 6CO(g) \qquad \Delta H = 778 \text{ kJ}$$

Calculate ΔS_{surr} for each of these reactions at 25°C and 1 atm.

Solution

We use

$$\Delta S_{surr} = -\frac{\Delta H}{T}$$

where

$$T = 25 + 273 = 298 \text{ K}$$

For the sulfide ore reaction,

$$\Delta S_{surr} = -\frac{-125 \text{ kJ}}{298 \text{ K}} = 0.419 \text{ kJ/K} = 419 \text{ J/K}$$

Sample Exercise 16.4, continued

Note that ΔS_{surr} is positive, as it should be since this reaction is exothermic and heat flow occurs to the surroundings, increasing the randomness of the surroundings.

For the oxide ore reaction,

$$\Delta S_{surr} = -\frac{778 \text{ kJ}}{298} = -2.61 \text{ kJ/K} = -2.61 \times 10^3 \text{ J/K}$$

In this case, ΔS_{surr} is negative because heat flow occurs from the surroundings to the system.

We have seen that the spontaneity of a process is determined by the entropy change it produces in the universe. We have also seen that ΔS_{univ} has two components, ΔS_{sys} and ΔS_{surr}. If for some process both ΔS_{sys} and ΔS_{surr} are positive, then ΔS_{univ} is positive, and the process is spontaneous. If, on the other hand, both ΔS_{sys} and ΔS_{surr} are negative, the process will not occur in the direction indicated but will be spontaneous in the opposite direction. Finally, if ΔS_{sys} and ΔS_{surr} have opposite signs, the spontaneity of the process depends on the sizes of the opposing terms. These cases are summarized in Table 16.3.

Interplay of ΔS_{sys} and ΔS_{surr} in Determining the Sign of ΔS_{univ}

Signs of entropy changes			
ΔS_{sys}	ΔS_{surr}	ΔS_{univ}	Process spontaneous?
+	+	+	Yes
−	−	−	No (reaction will occur in opposite direction)
+	−	?	Yes, if ΔS_{sys} has a larger magnitude than ΔS_{surr}
−	+	?	Yes, if ΔS_{surr} has a larger magnitude than ΔS_{sys}

Table 16.3

The relative importance of ΔS_{sys} and ΔS_{surr} in determining ΔS_{univ} can be clearly seen in the thermodynamics of solution formation. The entropy of mixing, represented by ΔS_{sys}, is positive and thus favorable. However, ΔS_{surr}, which equals $-\Delta H/T$, varies depending on the heat of solution. As we discussed in Section 11.3, ΔH tends to be large and positive when dissimilar materials are mixed. This means ΔS_{surr} will be large and negative, a situation that opposes solution formation. Conversely, when similar materials are mixed, ΔH tends to be small, leading to a small value for ΔS_{surr}. Under these circumstances the large, positive value of ΔS_{sys} is dominant, and the solution forms. These relationships are summarized in Table 16.4 on the following page for several types of solution. (We have just explained the like-dissolves-like rule on the basis of entropy.)

As always, ΔH refers to ΔH_{sys}.

The rule that like dissolves like makes sense in terms of entropy.

Relative Contributions of ΔS_{sys} and ΔS_{surr} to the Thermodynamics of Solution Formation

Type of solution	Signs of thermodynamic quantities				Is solution favored?
	ΔH	ΔS_{surr}	ΔS_{sys}	ΔS_{univ}	
Polar–polar	+ or − (small)	− or + (small)	+	+	Yes
Nonpolar–nonpolar	+ or − (small)	− or + (small)	+	+	Yes
Polar–nonpolar	+ (large)	− (large)	+	−	No

Table 16.4

We can now understand why spontaneity is often dependent on temperature and thus why water spontaneously freezes below 0°C and melts above 0°C. The term ΔS_{surr} is temperature-dependent. Since

$$\Delta S_{surr} = -\frac{\Delta H}{T}$$

at constant pressure, the value of ΔS_{surr} changes markedly with temperature. The magnitude of ΔS_{surr} will be very small at high temperatures and will increase as the temperature decreases. That is, exothermicity is most important as a driving force at low temperatures.

16.4 Free Energy

Purpose

- To define free energy and relate it to spontaneity.

So far we have used ΔS_{univ} to predict the spontaneity of a process. However, another thermodynamic function is also related to spontaneity and is especially useful in dealing with the temperature dependence of spontaneity. This function is called the **free energy** (G) and is defined by the relationship

The symbol G for free energy honors Josiah Willard Gibbs (1839–1903) who was Professor of Mathematical Physics at Yale University from 1871 to 1903 and laid the foundations of many areas of thermodynamics.

$$G = H - TS$$

where H is the enthalpy, T is the Kelvin temperature, and S is the entropy.

For a process that occurs at constant temperature, the change in free energy (ΔG) is given by the equation

$$\Delta G = \Delta H - T\Delta S$$

Note that all quantities here refer to the system. From this point on, we will follow the usual convention that when no subscript is included, the quantity refers to the system.

To see how this equation relates to spontaneity, we divide both sides of the equation by $-T$ to produce

$$-\frac{\Delta G}{T} = -\frac{\Delta H}{T} + \Delta S$$

Remember that at constant temperature and pressure

$$\Delta S_{\text{surr}} = -\frac{\Delta H}{T}$$

So we can write

$$-\frac{\Delta G}{T} = -\frac{\Delta H}{T} + \Delta S = \Delta S_{\text{surr}} + \Delta S = \Delta S_{\text{univ}}$$

We have shown that

$$\Delta S_{\text{univ}} = -\frac{\Delta G}{T} \quad \text{at constant } T \text{ and } P$$

This is a very important result. It means that a process carried out at constant temperature and pressure will be spontaneous only if ΔG is negative. That is, *a process (at constant* T *and* P*) is spontaneous in the direction in which the free energy decreases.*

Now we have two functions that can be used to predict spontaneity: the entropy of the universe, which applies to all processes; and free energy, which can be used for processes carried out at constant temperature and pressure. Since so many chemical reactions occur under the latter conditions, free energy is the more useful to chemists.

Let's use the free energy equation to predict the spontaneity of the melting of ice:

$$H_2O(s) \rightarrow H_2O(l)$$

for which $\Delta H^\circ = 6.03 \times 10^3$ J/mol and $\Delta S^\circ = 22.1$ J/K mol

Results of the calculations of ΔS_{univ} and ΔG° at -10°C, 0°C, and 10°C are shown in Table 16.5. These data predict that the process is spontaneous at 10°C; that is, ice melts at this temperature since ΔS_{univ} is positive and ΔG° is negative. The opposite is true at -10°C, where water freezes spontaneously.

The superscript degree symbol (°) indicates all substances are in their standard states.

Results of the Calculation of ΔS_{univ} and ΔG° for the Process $H_2O(s) \rightarrow H_2O(l)$ at −10°C, 0°C, and 10°C*

T (°C)	T (K)	ΔH° (J/mol)	ΔS° (J/K mol)	$\Delta S_{\text{surr}} = -\frac{\Delta H^\circ}{T}$ (J/K mol)	$\Delta S_{\text{univ}} = \Delta S^\circ + \Delta S_{\text{surr}}$ (J/K mol)	$T\Delta S^\circ$ (J/mol)	$\Delta G^\circ = \Delta H^\circ - T\Delta S^\circ$ (J/mol)
−10	263	6.03×10^3	22.1	−22.9	−0.8	5.81×10^3	$+2.2 \times 10^2$
0	273	6.03×10^3	22.1	−22.1	0	6.03×10^3	0
10	283	6.03×10^3	22.1	−21.3	+0.8	6.25×10^3	-2.2×10^2

*Note that at 10°C, ΔS° (ΔS_{sys}) controls, and the process occurs even though it is endothermic. At −10°C, the magnitude of ΔS_{surr} is larger than that of ΔS°, so the process is spontaneous in the opposite (exothermic) direction.

Table 16.5

Case	Result
ΔS positive, ΔH negative	Spontaneous at all temperatures
ΔS positive, ΔH positive	Spontaneous at high temperatures (where exothermicity is relatively unimportant)
ΔS negative, ΔH negative	Spontaneous at low temperatures (where exothermicity is dominant)
ΔS negative, ΔH positive	Process not spontaneous at *any* temperature (reverse process is spontaneous at *all* temperatures)

Various Possible Combinations of ΔH and ΔS for a Process and the Resulting Dependence of Spontaneity on Temperature

Table 16.6

Why is this so? The answer lies in the fact that ΔS_{sys} (ΔS°) and ΔS_{surr} oppose each other. The term ΔS° favors the melting of ice because of the increase in positional entropy, and ΔS_{surr} favors the freezing of water because it is an exothermic process. At temperatures below 0°C, the change of state occurs in the exothermic direction because ΔS_{surr} is larger in magnitude than ΔS_{sys}. But, above 0°C, the change occurs in the direction in which ΔS_{sys} is favorable, since in this case ΔS_{sys} is larger in magnitude than ΔS_{surr}. At 0°C, the *opposing tendencies just balance,* and the two states coexist; there is no driving force in either direction. An equilibrium exists between the two states of water. Note that ΔS_{univ} is equal to 0 at 0°C.

We can reach the same conclusions by examining ΔG°. At -10°C, ΔG° is positive because the ΔH° term is larger than the $T\Delta S^\circ$ term. The opposite is true at 10°C. At 0°C, ΔH° is equal to $T\Delta S^\circ$ and ΔG° is equal to 0. This means that solid H_2O and liquid H_2O have the same free energy at 0°C ($\Delta G^\circ = G_{(l)} - G_{(s)}$), and the system is at equilibrium.

We can understand the temperature dependence of spontaneity by examining the behavior of ΔG. For a process occurring at constant temperature and pressure,

$$\Delta G = \Delta H - T\Delta S$$

If ΔH and ΔS favor opposite processes, spontaneity will depend on temperature in such a way that the exothermic direction will be favored at low temperatures. For example, for the process

$$H_2O(s) \rightarrow H_2O(l)$$

ΔH is positive and ΔS is positive. The natural tendency for this system to lower its energy is in opposition to its natural tendency to increase its positional randomness. At low temperatures, ΔH dominates, and at high temperatures, ΔS dominates. The various possible cases are summarized in Table 16.6.

Sample Exercise 16.5

At what temperatures is the following process spontaneous at 1 atm?

$$Br_2(l) \rightarrow Br_2(g)$$

$$\Delta H^\circ = 31.0 \text{ kJ/mol} \quad \text{and} \quad \Delta S^\circ = 93.0 \text{ J/K mol}$$

What is the normal boiling point of liquid Br_2?

Sample Exercise 16.5, continued

Solution

The vaporization process will be spontaneous at all temperatures where $\Delta G°$ is negative. Note that $\Delta S°$ favors the vaporization process because of the increase in positional entropy, and $\Delta H°$ favors the *opposite* process, which is exothermic. These opposite tendencies will exactly balance at the boiling point of liquid Br_2, since at this temperature liquid and gaseous Br_2 are in equilibrium ($\Delta G° = 0$). We can find this temperature by setting $\Delta G° = 0$ in the equation

$$\Delta G° = \Delta H° - T\Delta S°$$

$$0 = \Delta H° - T\Delta S°$$

$$\Delta H° = T\Delta S°$$

Then $$T = \frac{\Delta H°}{\Delta S°} = \frac{3.10 \times 10^4 \text{ J/mol}}{93.0 \text{ J/K mol}} = 333 \text{ K}$$

At temperatures above 333 K, $T\Delta S°$ will have a larger magnitude than $\Delta H°$, and $\Delta G°$ (or $\Delta H° - T\Delta S°$) will be negative. Above 333 K, the vaporization process will be spontaneous; the opposite process will occur spontaneously below this temperature. At 333 K, liquid and gaseous Br_2 coexist in equilibrium. These observations can be summarized as follows (the pressure is 1 atm in each case):

1. $T > 333$ K. The term $\Delta S°$ controls. The increase in entropy occurring when liquid Br_2 is vaporized is dominant.
2. $T < 333$ K. The process is spontaneous in the direction in which it is exothermic. The term $\Delta H°$ controls.
3. $T = 333$ K. The opposing driving forces are just balanced ($\Delta G° = 0$), and the liquid and gaseous phases of bromine coexist. This is the normal boiling point.

16.5 Entropy Changes in Chemical Reactions

Purpose

- To apply positional probability to chemical reactions.
- To relate molecular complexity to entropy.

The second law of thermodynamics tells us that a process will be spontaneous if the entropy of the universe increases when the process occurs. We saw in Section 16.4 that for a process at constant temperature and pressure, we can use the change in free energy of the system to predict the sign of ΔS_{univ} and thus the direction in which it is spontaneous. So far we have applied these ideas only to physical processes, such as changes of state and the formation of solutions. However, the main business of chemistry is studying chemical reactions, and therefore we want to apply the second law to reactions.

The first step is to consider the entropy changes accompanying chemical reactions that occur under conditions of constant temperature and pressure. As for the other types of processes we have considered, the entropy changes in the *surroundings* are determined by the heat flow that occurs as the reaction takes place. However, the entropy changes in the *system* (the reactants and products of the reaction) are determined by positional probability.

For example, in the ammonia synthesis reaction

$$N_2(g) + 3H_2(g) \rightarrow 2NH_3(g)$$

four reactant molecules become two product molecules, lowering the number of independent units in the system, which leads to less positional disorder. *Fewer molecules mean fewer possible configurations*. To help clarify this, consider a special container with a million compartments, each large enough to hold a hydrogen molecule. Thus there are a million ways one H_2 molecule can be placed in this container. But suppose we break the H—H bond and place the two independent H atoms in the same container. A little thought will convince you that there are *many* more than a million ways to place the two separate atoms. The number of arrangements possible for the two independent atoms is much greater than the number for the molecule in which the atoms are bound together. Thus for the process

$$H_2 \rightarrow 2H$$

positional entropy increases.

Does positional entropy increase or decrease when the following reaction takes place?

$$4NH_3(g) + 5O_2(g) \rightarrow 4NO(g) + 6H_2O(g)$$

In this case 9 gaseous molecules are changed to 10 gaseous molecules, and the positional entropy increases. There are more independent units as products than as reactants. In general, when a reaction involves gaseous molecules, *the change in positional entropy is dominated by the relative numbers of molecules of gaseous reactants and products*. If the number of molecules of the gaseous products is greater than the number of molecules of the gaseous reactants, positional entropy typically increases, and ΔS will be positive for the reaction.

Sample Exercise 16.6

Predict the sign of $\Delta S°$ for each of the following reactions:

a. The thermal decomposition of solid calcium carbonate

$$CaCO_3(s) \rightarrow CaO(s) + CO_2(g)$$

b. The oxidation of SO_2 in air

$$2SO_2(g) + O_2(g) \rightarrow 2SO_3(g)$$

Solution

a. Since in this reaction a gas is produced from a solid reactant, the positional entropy will increase, and $\Delta S°$ will be positive.

b. Here three molecules of gaseous reactants become two molecules of gaseous products. Since the number of gas molecules decreases, positional entropy decreases, and $\Delta S°$ will be negative.

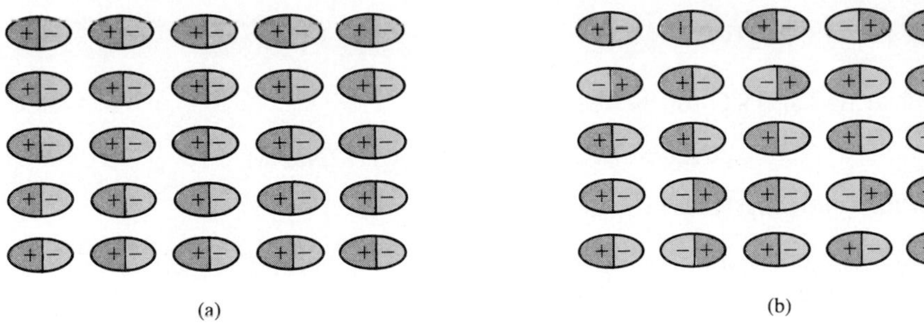

Figure 16.5

(a) A perfect crystal of hydrogen chloride at 0 K; the dipolar HCl molecules are represented by (+|−). The entropy is zero ($S = 0$) for this perfect crystal at 0 K. (b) As the temperature rises above 0 K, lattice vibrations allow some dipoles to change their orientations, producing some disorder and an increase in entropy ($S > 0$).

In thermodynamics it is the *change* in a certain function that is usually important. The change in enthalpy determines if a reaction is exothermic or endothermic at constant pressure. The change in free energy determines if a process is spontaneous at constant temperature and pressure. It is fortunate that changes in thermodynamic functions are sufficient for most purposes, since absolute values for many thermodynamic characteristics of a system, such as enthalpy or free energy, cannot be determined.

However, we can assign absolute entropy values. Consider a solid at 0 K, where molecular motion virtually ceases. If the substance is a perfect crystal, its internal arrangement is absolutely regular [see Fig. 16.5(a)]. There is only *one way* to achieve this perfect order; every particle must be in its place. For example, with N coins there is only one way to achieve the state of all heads. Thus a perfect crystal represents the lowest possible entropy; that is, *the entropy of a perfect crystal at 0 K is zero*. This is a statement of the **third law of thermodynamics.**

A perfect crystal at 0 K is an unattainable ideal, taken as a standard but never actually observed.

As the temperature of a perfect crystal is increased, the random vibrational motions increase, and disorder increases within the crystal [see Fig. 16.5(b)]. Thus the entropy of a substance increases with temperature. Since S is zero for a perfect crystal at 0 K, the entropy value for a substance at a particular temperature can be calculated by knowing the temperature dependence of entropy. (We will not show such calculations here.)

The *standard entropy values* ($S°$) of many common substances at 298 K and 1 atmosphere are listed in Appendix 4. From these values you will see that the entropy of a substance does indeed increase in going from solid to liquid to gas. One especially interesting feature of this table is the very low $S°$ value for diamond. The structure of diamond is highly ordered, with each carbon strongly bound to a tetrahedral arrangement of four other carbon atoms (see Section 10.5, Fig. 10.22). This type of structure allows very little disorder and has a very low entropy, even at 298 K. Graphite has a slightly higher entropy because its layered structure allows for a little more disorder.

The standard entropy values represent the increase in entropy that occurs when a substance is heated from 0 K to 298 K at 1 atmosphere pressure.

Because *entropy is a state function of the system* (it is not pathway-dependent), the entropy change for a given chemical reaction can be calculated by taking the difference between the standard entropy values of products and those of the reactants:

$$\Delta S°_{\text{reaction}} = \Sigma S°_{\text{products}} - \Sigma S°_{\text{reactants}}$$

where, as usual, Σ represents the sum of the terms. It is important to note that entropy is an extensive property (it depends on the amount of substance present). This means that *the number of moles of a given reactant or product must be taken into account*.

Sample Exercise 16.7

Calculate $\Delta S°$ at 25°C for the reaction

$$2NiS(s) + 3O_2(g) \rightarrow 2SO_2(g) + 2NiO(s)$$

given the following standard entropy values:

Substance	$S°$ (J/K mol)
$SO_2(g)$	248
$NiO(s)$	38
$O_2(g)$	205
$NiS(s)$	53

Solution

Since

$$\begin{aligned}\Delta S° &= \Sigma S°_{\text{products}} - \Sigma S°_{\text{reactants}} \\ &= 2S°_{SO_2(g)} + 2S°_{NiO(s)} - 2S°_{NiS(s)} - 3S°_{O_2(s)} \\ &= 2\text{ mol}\left(248\frac{\text{J}}{\text{K mol}}\right) + 2\text{ mol}\left(38\frac{\text{J}}{\text{K mol}}\right) \\ &\quad -2\text{ mol}\left(53\frac{\text{J}}{\text{K mol}}\right) - 3\text{ mol}\left(205\frac{\text{J}}{\text{K mol}}\right) \\ &= 496\text{ J/K} + 76\text{ J/K} - 106\text{ J/K} - 615\text{ J/K} \\ &= -149\text{ J/K}\end{aligned}$$

We would expect $\Delta S°$ to be negative since the number of gaseous molecules decreases in this reaction.

Sample Exercise 16.8

Calculate $\Delta S°$ for the reduction of aluminum oxide by hydrogen gas:

$$Al_2O_3(s) + 3H_2(g) \rightarrow 2Al(s) + 3H_2O(g)$$

Use the following standard entropy values:

Substance	$S°$ (J/K mol)
$Al_2O_3(s)$	51
$H_2(g)$	131
$Al(s)$	28
$H_2O(g)$	189

Solution

$$\begin{aligned}\Delta S° &= \Sigma S°_{\text{products}} - \Sigma S°_{\text{reactants}} \\ &= 2S°_{Al(s)} + 3S°_{H_2O(g)} - 3S°_{H_2(g)} - S°_{Al_2O_3(s)}\end{aligned}$$

Sample Exercise 16.8, continued

$$= 2\text{ mol}\left(28\frac{\text{J}}{\text{K mol}}\right) + 3\text{ mol}\left(189\frac{\text{J}}{\text{K mol}}\right)$$

$$-3\text{ mol}\left(131\frac{\text{J}}{\text{K mol}}\right) - 1\text{ mol}\left(51\frac{\text{J}}{\text{K mol}}\right)$$

$$= 56\text{ J/K} + 567\text{ J/K} - 393\text{ J/K} - 51\text{ J/K}$$

$$= 179\text{ J/K}$$

The reaction considered in Sample Exercise 16.8 involves 3 moles of hydrogen gas on the reactant side and 3 moles of water vapor on the product side. Would you expect ΔS to be large or small for such a case? We have assumed that ΔS depends on the relative numbers of molecules of gaseous reactants and products. Based on that assumption, ΔS should be near zero for the present reaction. However, ΔS is large and positive. Why is this so? The large value for ΔS results from the difference in the entropy values for hydrogen gas and water vapor. The reason for this difference can be traced to the difference in molecular structure. Because it is a nonlinear, triatomic molecule, H_2O has more rotational and vibrational motions (see Fig. 16.6) than does the diatomic H_2 molecule. Thus the standard entropy value for $H_2O(g)$ is greater than that for $H_2(g)$. Generally, *the more complex the molecule, the higher the standard entropy value.*

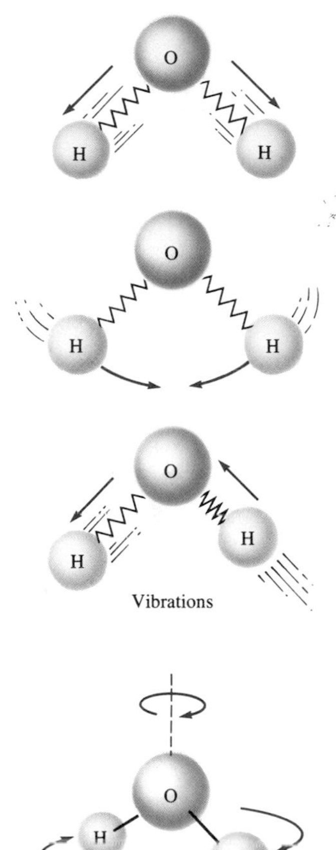

Figure 16.6

The H_2O molecule can vibrate and rotate in several ways, some of which are shown here. This freedom of motion leads to a higher entropy for water than for a substance like hydrogen, which has a simple diatomic molecule with fewer possible motions.

16.6 Free Energy and Chemical Reactions

Purpose

- To show how to calculate the standard free energy change in a chemical reaction.
- To define the standard free energy of formation and show how to use it to predict spontaneity.

For chemical reactions we are often interested in the **standard free energy change** ($\Delta G°$), *the change in free energy that will occur if the reactants in their standard states are converted to products in their standard states.* For example, for the ammonia synthesis reaction at 25°C

$$N_2(g) + 3H_2(g) \rightleftharpoons 2NH_3(g) \quad \Delta G° = -33.3\text{ kJ} \tag{16.1}$$

This $\Delta G°$ value represents the change in free energy when 1 mole of nitrogen gas at 1 atm reacts with 3 moles of hydrogen gas at 1 atm to produce 2 moles of gaseous NH_3 at 1 atm.

It is important to recognize that the standard free energy for a reaction is not measured directly. For example, we can measure heat flow in a calorimeter to determine $\Delta H°$, but we cannot measure $\Delta G°$ this way. The value of $\Delta G°$ for the ammonia synthesis in Equation (16.1) was *not* obtained by mixing 1 mole of N_2 and 3 moles of H_2 in a flask and measuring the change in free energy as 2 moles of NH_3

formed. For one thing if we mixed 1 mole of N_2 and 3 moles of H_2 in a flask, the system would go to equilibrium rather than to completion. Also, we have no instrument that measures free energy. But, while we cannot directly measure ΔG° for a reaction, we can calculate it from other measured quantities, as we will see later in this section.

The value of ΔG° tells us nothing about the rate of a reaction, only its eventual equilibrium position.

Why is it useful to know ΔG° for a reaction? As we will see in more detail later in this chapter, knowing the ΔG° values for several reactions allows us to compare the relative tendency of these reactions to occur. The more negative the value of ΔG°, the further a reaction will go to the right to reach equilibrium. We must use standard state free energies to make this comparison because free energy changes with pressure or concentration. Thus to get an accurate comparison of reaction tendencies we must compare all reactions under the same pressure or concentration conditions. We will have more to say about the significance of ΔG° later.

There are several ways to calculate ΔG°. One common method uses the equation

$$\Delta G^\circ = \Delta H^\circ - T\Delta S^\circ$$

which applies to a reaction carried out at constant temperature. For example, for the reaction

$$C(s) + O_2(g) \rightarrow CO_2(g)$$

the values of ΔH° and ΔS° are known to be -393.5 kJ and 3.05 J/K, respectively, and ΔG° can be calculated at 298 K as follows:

$$\begin{aligned}\Delta G^\circ &= \Delta H^\circ - T\Delta S^\circ \\ &= -3.935 \times 10^5 \text{ J} - (298 \text{ K})(3.05 \text{ J/K}) \\ &= -3.944 \times 10^5 \text{ J} \\ &= -394.4 \text{ kJ (per mole of } CO_2)\end{aligned}$$

Sample Exercise 16.9

Consider the reaction

$$2SO_2(g) + O_2(g) \rightarrow 2SO_3(g)$$

carried out at 25°C and 1 atm. Calculate ΔH°, ΔS°, and ΔG°, using the following data:

Substance	ΔH_f° (kJ/mol)	S° (J/K mol)
$SO_2(g)$	−297	248
$SO_3(g)$	−396	257
$O_2(g)$	0	205

Solution

The value of ΔH° can be calculated from the enthalpies of formation using the equation we discussed in Section 6.4:

$$\Delta H^\circ = \Sigma \Delta H_{f\,(\text{products})}^\circ - \Sigma \Delta H_{f\,(\text{reactants})}^\circ$$

Then

$$\begin{aligned}\Delta H^\circ &= 2\Delta H_{f\,(SO_3(g))}^\circ - 2\Delta H_{f\,(SO_2(g))}^\circ - \Delta H_{f\,(O_2(g))}^\circ \\ &= 2 \text{ mol}(-396 \text{ kJ/mol}) - 2 \text{ mol}(-297 \text{ kJ/mol}) - 0\end{aligned}$$

Sample Exercise 16.9, continued

$$= -792 \text{ kJ} + 594 \text{ kJ}$$
$$= -198 \text{ kJ}$$

The value of $\Delta S°$ can be calculated using the standard entropy values and the equation discussed in Section 16.5:

$$\Delta S° = \Sigma S°_{\text{products}} - \Sigma S°_{\text{reactants}}$$

Thus

$$\Delta S° = 2S°_{SO_3(g)} - 2S°_{SO_2(g)} - S°_{O_2(g)}$$
$$= 2 \text{ mol}(257 \text{ J/K mol}) - 2 \text{ mol}(248 \text{ J/K mol}) - 1 \text{ mol}(205 \text{ J/K mol})$$
$$= 514 \text{ J/K} - 496 \text{ J/K} - 205 \text{ J/K}$$
$$= -187 \text{ J/K}$$

We would expect $\Delta S°$ to be negative since three molecules of gaseous reactants give two molecules of gaseous products.

The value of $\Delta G°$ can now be calculated from the equation:

$$\Delta G° = \Delta H° - T\Delta S°$$
$$= -198 \text{ kJ} - (298 \text{ K})\left(-187 \frac{\text{J}}{\text{K}}\right)\left(\frac{1 \text{ kJ}}{1000 \text{ J}}\right)$$
$$= -198 \text{ kJ} + 55.7 \text{ kJ} = -142 \text{ kJ}$$

Like enthalpy, *free energy is a state function,* and we can use procedures for finding ΔG similar to those for finding ΔH using Hess's law.

To illustrate this second method for calculating the free energy change, we will obtain $\Delta G°$ for the reaction

$$2CO(g) + O_2(g) \rightarrow 2CO_2(g) \tag{16.2}$$

from the following data:

$$2CH_4(g) + 3O_2(g) \rightarrow 2CO(g) + 4H_2O(g) \quad \Delta G° = -1088 \text{ kJ} \tag{16.3}$$
$$CH_4(g) + 2O_2(g) \rightarrow CO_2(g) + 2H_2O(g) \quad \Delta G° = -801 \text{ kJ} \tag{16.4}$$

Note that $CO(g)$ is a reactant in Equation (16.2), so Equation (16.3) must be reversed, since $CO(g)$ is a product in that reaction as written. When a reaction is reversed, the sign of $\Delta G°$ is also reversed. In Equation (16.4) $CO_2(g)$ is a product, as it is in Equation (16.2), but only one molecule of CO_2 is formed. Thus Equation (16.4) must be multiplied by 2, which means the $\Delta G°$ value for Equation (16.4) must also be multiplied by 2. Free energy is an extensive property since it is defined by two extensive properties, H and S.

Reversed Equation (16.3)

$$2CO(g) + 4H_2O(g) \rightarrow 2CH_4(g) + 3O_2(g) \qquad \Delta G° = -(-1088 \text{ kJ})$$

2 × Equation (16.4)

$$2CH_4(g) + 4O_2(g) \rightarrow 2CO_2(g) + 4H_2O(g) \qquad \Delta G° = 2(-801 \text{ kJ})$$

$$2CO(g) + O_2(g) \rightarrow 2CO_2(g) \qquad \Delta G° = -(-1088 \text{ kJ}) + 2(-801 \text{ kJ})$$
$$= -514 \text{ kJ}$$

This example shows that ΔG values for reactions are manipulated in exactly the same way as ΔH values.

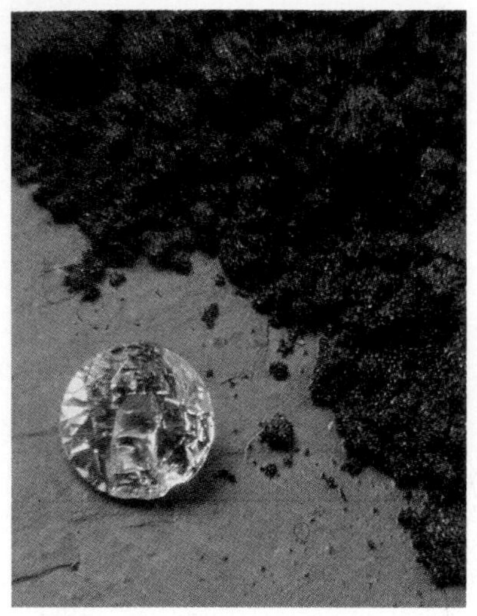

Diamond and graphite.

Sample Exercise 16.10

Using the following data (at 25°C)

$$C_{(s)}^{\text{diamond}} + O_2(g) \rightarrow CO_2(g) \qquad \Delta G^\circ = -397 \text{ kJ} \qquad (16.5)$$

$$C_{(s)}^{\text{graphite}} + O_2(g) \rightarrow CO_2(g) \qquad \Delta G^\circ = -394 \text{ kJ} \qquad (16.6)$$

calculate ΔG° for the reaction

$$C_{(s)}^{\text{diamond}} \rightarrow C_{(s)}^{\text{graphite}}$$

Solution

We reverse Equation (16.6) to make graphite a product, as required, and then add the new equation to Equation (16.5):

$$C_{(s)}^{\text{diamond}} + O_2(g) \rightarrow CO_2(g) \qquad \Delta G^\circ = -397 \text{ kJ}$$

Reversed Equation (16.6)

$$CO_2(g) \rightarrow C_{(s)}^{\text{graphite}} + O_2(g) \qquad \Delta G^\circ = -(-394 \text{ kJ})$$

$$C_{(s)}^{\text{diamond}} \rightarrow C_{(s)}^{\text{graphite}} \qquad \Delta G^\circ = -397 \text{ kJ} + 394 \text{ kJ} = -3 \text{ kJ}$$

Since ΔG° is negative for this process, diamond should spontaneously change to graphite at 25°C and 1 atm. However, the reaction is so slow under these conditions that we do not observe the process. This is another example of kinetic rather than thermodynamic control of a reaction. We can say that diamond is kinetically stable with respect to graphite even though it is thermodynamically unstable.

In Sample Exercise 16.10, we saw that the process

$$C_{(s)}^{\text{diamond}} \rightarrow C_{(s)}^{\text{graphite}}$$

is spontaneous but very slow at 25°C and 1 atm. The reverse process can be made to occur at high temperatures and pressures. Diamond has a more compact structure and thus a higher density than graphite, so the exertion of very high pressure causes it to become thermodynamically favored. If high temperatures are also used to make the process fast enough to be feasible, diamonds can be made from graphite. The conditions usually used involve temperatures greater than 1000°C and pressures of about 10^5 atm. About half of all industrial diamonds are made this way.

The standard state of an element is its most stable state at 25°C and 1 atm.

A third method for calculating the free energy change for a reaction uses standard free energies of formation. The **standard free energy of formation** (ΔG_f°) of a substance is defined as the *change in free energy that accompanies the formation of 1 mole of that substance from its constituent elements with all reactants and products in their standard states*. For the formation of glucose ($C_6H_{12}O_6$) the appropriate reaction is

$$6C(s) + 6H_2(g) + 3O_2(g) \rightarrow C_6H_{12}O_6(s)$$

The standard free energy associated with this process is called the free energy of formation of glucose. Values of the standard free energy of formation are useful in calculating ΔG° for specific chemical reactions using the equation

$$\Delta G^\circ = \Sigma\Delta G^\circ_{f\ (products)} - \Sigma\Delta G^\circ_{f\ (reactants)}$$

Calculating ΔG° is very similar to calculating ΔH°, as shown in Section 6.4.

Values of ΔG°_f for many common substances are listed in Appendix 4. Note that, analogous to the enthalpy of formation, *the standard free energy of formation of an element in its standard state is zero.*

Sample Exercise 16.11

Methanol is a high-octane fuel used in high-performance racing engines. Calculate ΔG° for the reaction

$$2CH_3OH(g) + 3O_2(g) \rightarrow 2CO_2(g) + 4H_2O(g)$$

given the following free energies of formation:

Substance	ΔG°_f (kJ/mol)
$CH_3OH(g)$	−163
$O_2(g)$	0
$CO_2(g)$	−394
$H_2O(g)$	−229

Oxygenated fuels are being considered for use in passenger cars to reduce carbon monoxide pollution problems.

Solution

We use the equation

$$\begin{aligned}\Delta G^\circ_f &= \Sigma\Delta G^\circ_{f\ (products)} - \Sigma\Delta G^\circ_{f\ (reactants)}\\ &= 2\Delta G^\circ_{f\ (CO_2(g))} + 4\Delta G^\circ_{f\ (H_2O(g))} - 3\Delta G^\circ_{f\ (O_2(g))} - 2\Delta G^\circ_{f\ (CH_3OH(g))}\\ &= 2\text{ mol}(-394\text{ kJ/mol}) + 4\text{ mol}(-229\text{ kJ/mol}) - 3(0) - 2\text{ mol}(-163\text{ kJ/mol})\\ &= -1378\text{ kJ}\end{aligned}$$

The large magnitude and the negative sign of ΔG° indicate that this reaction is very favorable thermodynamically.

Sample Exercise 16.12

A chemical engineer wants to determine the feasibility of making ethanol (C_2H_5OH) by reacting water with ethylene (C_2H_4) according to the equation

$$C_2H_4(g) + H_2O(l) \rightarrow C_2H_5OH(l)$$

Is this reaction spontaneous under standard conditions?

Solution

To determine the spontaneity of this reaction under standard conditions, we must determine ΔG° for the reaction. We can do this using standard free energies of formation at 25°C from Appendix 4:

$$\begin{aligned}\Delta G^\circ_{f\ (C_2H_5OH(l))} &= -175\text{ kJ/mol}\\ \Delta G^\circ_{f\ (H_2O(l))} &= -237\text{ kJ/mol}\\ \Delta G^\circ_{f\ (C_2H_4(g))} &= 68\text{ kJ/mol}\end{aligned}$$

Sample Exercise 16.12, continued

Then
$$\begin{aligned}\Delta G^\circ &= \Delta G^\circ_{f\ (C_2H_5OH(l))} - \Delta G^\circ_{f\ (H_2O(l))} - \Delta G^\circ_{f\ (C_2H_4(g))} \\ &= -175\text{ kJ} - (-237\text{ kJ}) - 68\text{ kJ} \\ &= -6\text{ kJ}\end{aligned}$$

Thus the process is spontaneous under standard conditions at 25°C.

Although the reaction considered in Sample Exercise 16.12 is spontaneous, other features of the reaction must be studied to see if the process is feasible. For example, the chemical engineer will need to study the kinetics of the reaction to determine whether it is fast enough to be useful and, if it is not, whether a catalyst can be found to enhance the rate. In doing these studies the engineer must remember that ΔG° depends on temperature:

$$\Delta G^\circ = \Delta H^\circ - T\Delta S^\circ$$

Thus if the process must be carried out at high temperatures to be fast enough to be feasible, ΔG° must be recalculated at that temperature from the ΔH° and ΔS° values for the reaction.

16.7 The Dependence of Free Energy on Pressure

Purpose

- To relate free energy to pressure.

In this chapter we have seen that a system at constant temperature and pressure will proceed spontaneously in the direction that lowers its free energy. This is why reactions proceed until they reach equilibrium. As we will see later in this section, the equilibrium position represents the lowest free energy value available to a particular reaction system. The free energy of a reaction system changes as the reaction proceeds, because free energy is dependent on the pressure of a gas or on the concentration of species in solution. We will deal only with the pressure dependence of the free energy of an ideal gas. The dependence of free energy on concentration can be developed using similar reasoning.

To understand the pressure dependence of free energy, we need to know how pressure affects the thermodynamic functions that comprise free energy, that is, enthalpy and entropy (recall that $G = H - TS$). For an ideal gas, enthalpy is not pressure-dependent. However, entropy *does* depend on pressure because of its dependence on volume. Consider 1 mole of an ideal gas at a given temperature. At a volume of 10.0 L, the gas has many more positions available for the molecules than if its volume is 1.0 L. The positional entropy is greater in the larger volume. In summary, at a given temperature for 1 mole of ideal gas

$$S_{\text{large volume}} > S_{\text{small volume}}$$

or, since pressure and volume are inversely related,

$$S_{\text{low pressure}} > S_{\text{high pressure}}$$

We have shown qualitatively that the entropy and therefore the free energy of an ideal gas depend on its pressure. Using a more detailed argument, which we will not consider here, it can be shown that

$$G = G^\circ + RT\ln(P)$$

where G° is the free energy of the gas at a pressure of 1 atmosphere, G is the free energy of the gas at a pressure of P atmospheres, R is the universal gas constant, and T is the Kelvin temperature.

To see how the change in free energy for a reaction depends on pressure, we will consider the ammonia synthesis reaction

$$N_2(g) + 3H_2(g) \longrightarrow 2NH_3(g)$$

In general, $\Delta G = \Sigma G_{\text{products}} - \Sigma G_{\text{reactants}}$

For this reaction $\Delta G = 2G_{NH_3} - G_{N_2} - 3G_{H_2}$

where

$$\begin{aligned} G_{NH_3} &= G^\circ_{NH_3} + RT\ln(P_{NH_3}) \\ G_{N_2} &= G^\circ_{N_2} + RT\ln(P_{N_2}) \\ G_{H_2} &= G^\circ_{H_2} + RT\ln(P_{H_2}) \end{aligned}$$

See Appendix 1.2 if you need a review of logarithms.

Substituting these values into the equation gives

$$\begin{aligned} \Delta G &= 2[G^\circ_{NH_3} + RT\ln(P_{NH_3})] - [G^\circ_{N_2} + RT\ln(P_{N_2})] - 3[G^\circ_{H_2} + RT\ln(P_{H_2})] \\ &= 2G^\circ_{NH_3} - G^\circ_{N_2} - 3G^\circ_{H_2} + 2RT\ln(P_{NH_3}) - RT\ln(P_{N_2}) - 3RT\ln(P_{H_2}) \\ &= \underbrace{(2G^\circ_{NH_3} - G^\circ_{N_2} - 3G^\circ_{H_2})}_{\Delta G^\circ_{\text{reaction}}} + RT[2\ln(P_{NH_3}) - \ln(P_{N_2}) - 3\ln(P_{H_2})] \end{aligned}$$

The first term in parentheses is ΔG° for the reaction. Thus we have

$$\Delta G = \Delta G^\circ_{\text{reaction}} + RT[2\ln(P_{NH_3}) - \ln(P_{N_2}) - 3\ln(P_{H_2})]$$

and since

$$\begin{aligned} 2\ln(P_{NH_3}) &= \ln(P_{NH_3}{}^2) \\ -\ln(P_{N_2}) &= \ln\left(\frac{1}{P_{N_2}}\right) \\ -3\ln(P_{H_2}) &= \ln\left(\frac{1}{P_{H_2}{}^3}\right) \end{aligned}$$

the equation becomes

$$\Delta G = \Delta G^\circ + RT\ln\left(\frac{P_{NH_3}{}^2}{(P_{N_2})(P_{H_2}{}^3)}\right)$$

But the term

$$\frac{P_{NH_3}{}^2}{(P_{N_2})(P_{H_2}{}^3)}$$

is the reaction quotient (Q) discussed in Section 13.5. Therefore we have

$$\Delta G = \Delta G^\circ + RT\ln(Q)$$

where Q is the reaction quotient (from the law of mass action), T is the temperature (K), R is the gas law constant and is equal to 8.3145 J/K mol, ΔG° is the free energy change for the reaction with all reactants and products at a pressure of 1 atmosphere, and ΔG is the free energy change for the reaction for the specified pressures of reactants and products.

Sample Exercise 16.13

One method for synthesizing methanol (CH_3OH) involves reacting carbon monoxide and hydrogen gases:

$$CO(g) + 2H_2(g) \rightarrow CH_3OH(l)$$

Calculate ΔG at 25°C for this reaction where carbon monoxide gas at 5.0 atm and hydrogen gas at 3.0 atm are converted to liquid methanol.

Solution

To calculate ΔG for this process, we use the equation

$$\Delta G = \Delta G^\circ + RT\ \ln(Q)$$

We must first compute ΔG° from standard free energies of formation (see Appendix 4). Since

$$\Delta G^\circ_{f\ (CH_3OH(l))} = -166\ \text{kJ}$$
$$\Delta G^\circ_{f\ (H_2(g))} = 0$$
$$\Delta G^\circ_{f\ (CO(g))} = -137\ \text{kJ}$$
$$\Delta G^\circ = -166\ \text{kJ} - (-137\ \text{kJ}) - 0 = -29\ \text{kJ} = -2.9 \times 10^4\ \text{J}$$

Note that this is the value of ΔG° for the reaction of 1 mol CO with 2 mol H_2 to produce 1 mol CH_3OH. We might call this the value of ΔG° for one "round" of the reaction or for one mole of the reaction. Thus the ΔG° value might better be written as -2.9×10^4 J/mol of reaction or -2.9×10^4 J/mol rxn.

Note in this case ΔG is defined for one mole of the reaction, that is, for 1 mol $CO(g)$ reacting with 2 mol $H_2(g)$ to form 1 mol $CH_3OH(l)$. Thus ΔG, ΔG°, and $RT\ \ln(Q)$ all have units of J/mol of reaction. In this case the units of R are actually J/K mol of reaction, although they are usually not written this way.

We can now calculate ΔG using

$$\Delta G^\circ = -2.9 \times 10^4\ \text{J/mol rxn}$$
$$R = 8.3145\ \text{J/K mol}$$
$$T = 273 + 25 = 298\ \text{K}$$
$$Q = \frac{1}{(P_{CO})(P_{H_2}{}^2)} = \frac{1}{(5.0)(3.0)^2} = 2.2 \times 10^{-2}$$

Note that the pure liquid methanol is not included in the calculation of Q. Then

$$\begin{aligned}\Delta G &= \Delta G^\circ + RT\ \ln(Q)\\ &= (-2.9 \times 10^4\ \text{J/mol rxn}) + (8.3145\ \text{J/K mol rxn})(298\ \text{K})\ \ln(2.2 \times 10^{-2})\\ &= (-2.9 \times 10^4\ \text{J/mol rxn}) - (9.4 \times 10^3\ \text{J/mol rxn}) = -3.8 \times 10^4\ \text{J/mol rxn}\\ &= -38\ \text{kJ/mol rxn}\end{aligned}$$

Note that ΔG is significantly more negative than ΔG°, implying that the reaction is more spontaneous at reactant pressures greater than 1 atm. We might expect this result from Le Châtelier's principle.

16.8 Free Energy and Equilibrium

Purpose

- To define equilibrium in terms of minimum free energy.
- To show how the value of K is related to $\Delta G°$.

When the components of a given chemical reaction are mixed, they will proceed, rapidly or slowly depending on the kinetics of the process, to the equilibrium position. In Chapter 13 we defined the equilibrium position as the point at which the forward and reverse reaction rates are equal. In this chapter we look at equilibrium from a thermodynamic point of view, and we find that *the equilibrium point occurs at the lowest value of free energy available to the reaction system.* As it turns out, the two definitions give the same equilibrium state, which must be the case for both the kinetic and thermodynamic models to be valid.

To understand the relationship of free energy to equilibrium, let's consider the following simple hypothetical reaction:

$$A(g) \rightleftharpoons B(g)$$

where 1.0 mole of gaseous A is initially placed in a reaction vessel at a pressure of 2.0 atm. The free energies for A and B are diagrammed as shown in Fig. 16.7(a). As A reacts to form B, the total free energy of the system changes, yielding the following results:

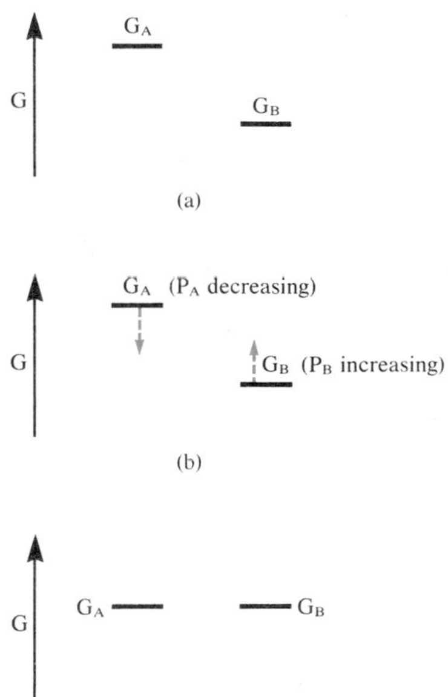

Figure 16.7

(a) The initial free energies of A and B. (b) As $A(g)$ changes to $B(g)$, the free energy of A decreases and that of B increases. (c) Eventually, pressures of A and B are achieved such that $G_A = G_B$, the equilibrium position.

$$\text{Free energy of A} = G_A = G_A^\circ + RT\ln(P_A)$$

$$\text{Free energy of B} = G_B = G_B^\circ + RT\ln(P_B)$$

$$\text{Total free energy of system} = G = G_A + G_B$$

As A changes to B, G_A will decrease because P_A is decreasing [Fig. 16.7(b)]. In contrast, G_B will increase since P_B is increasing. The reaction will proceed to the right as long as the total free energy of the system decreases (as long as G_B is less than G_A). At some point the pressures of A and B reach the values P_A^e and P_B^e that make G_A equal to G_B. *The system has reached equilibrium* [Fig. 16.7(c)]. Since A at pressure P_A^e and B at pressure P_B^e have the same free energy (G_A equals G_B), then ΔG is zero for A at pressure P_A^e changing to B at pressure P_B^e. *The system has reached minimum free energy*. There is no longer any driving force to change A to B or B to A, so the system remains at this position (the pressures of A and B remain constant).

Suppose that for the experiment described above, the plot of free energy versus the mole fraction of A reacted is defined as shown in Fig. 16.8(a) on the following page. In this experiment, minimum free energy is reached when 75% of A has been changed to B. At this point, the pressure of A is 0.25 times the original pressure, or

$$(0.25)(2.0 \text{ atm}) = 0.50 \text{ atm}$$

The pressure of B is

$$(0.75)(2.0 \text{ atm}) = 1.5 \text{ atm}$$

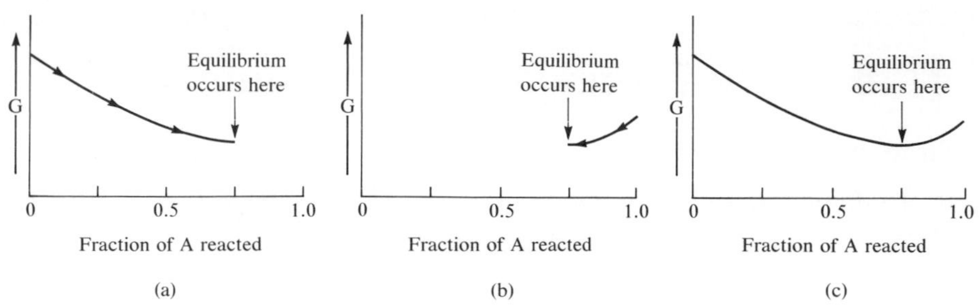

Figure 16.8

(a) The change in free energy to reach equilibrium beginning with 1.0 mole of A(g) at $P_A = 2.0$ atm. (b) The change in free energy to reach equilibrium, beginning with 1.0 mole of B(g) at $P_B = 2.0$ atm. (c) The free energy profile for $A(g) \rightleftharpoons B(g)$ in a system containing 1.0 mole (A plus B) at $P_{TOTAL} = 2.0$ atm. Each point on the curve corresponds to the total free energy of the system for a given combination of A and B.

Since this is the equilibrium position, we can use the equilibrium pressures to calculate a value for K for the reaction in which A is converted to B at this temperature:

$$K = \frac{P_B^e}{P_A^e} = \frac{1.5 \text{ atm}}{0.50 \text{ atm}} = 3.0$$

For the reaction $A(g) \rightleftharpoons B(g)$ the pressure is always constant during the reaction, since the same number of gas molecules is always present.

Exactly the same equilibrium point would be achieved if we placed 1.0 mole of pure B(g) in the flask at a pressure of 2.0 atm. In this case, B would change to A until equilibrium ($G_B = G_A$) is reached. This is shown in Fig. 16.8(b).

The overall free energy curve for this system is shown in Fig. 16.8(c). Note that any mixture of A(g) and B(g) containing 1.0 mole of A plus B at a total pressure of 2.0 atm will react until it reaches the minimum in the curve.

In summary, when substances undergo a chemical reaction, the reaction proceeds to the minimum free energy (equilibrium), which corresponds to the point where $G_{products} = G_{reactants}$ or

$$\Delta G = G_{products} - G_{reactants} = 0$$

We can now establish a quantitative relationship between free energy and the value of the equilibrium constant. We have seen that

$$\Delta G = \Delta G^\circ + RT \ln(Q)$$

and at equilibrium ΔG equals 0 and Q equals K.

So
$$\Delta G = 0 = \Delta G^\circ + RT \ln(K)$$

or
$$\Delta G^\circ = -RT \ln(K)$$

We must note the following characteristics of this very important equation.

Case 1: $\Delta G^\circ = 0$. When ΔG° equals zero for a particular reaction, the free energies of the reactants and products are equal when all components are in the standard states (1 atm for gases). The system is at equilibrium when the pressures of all reactants and products are 1 atm, which means that K equals 1.

Case 2: $\Delta G^\circ < 0$. In this case ΔG° ($G^\circ_{products} - G^\circ_{reactants}$) is negative, which means that

$$G^\circ_{products} < G^\circ_{reactants}$$

If a flask contains the reactants and products, all at 1 atm, the system will *not* be at equilibrium. Since $G^\circ_{products}$ is less than $G^\circ_{reactants}$, the system will adjust to the right to reach equilibrium. In this case, K will be *greater than 1*, since the pressures of the products at equilibrium will be greater than 1 atm and the pressures of the reactants at equilibrium will be less than 1 atm.

Case 3: $\Delta G^\circ > 0$. Since ΔG° ($G^\circ_{products} - G^\circ_{reactants}$) is positive

$$G^\circ_{reactants} < G^\circ_{products}$$

If a flask contains the reactants and products, all at 1 atmosphere, the system will *not* be at equilibrium. In this case, the system will adjust to the left (toward the reactants, which have a lower free energy) to reach equilibrium. The value of K will be *less than 1*, since at equilibrium the pressures of the reactants will be greater than 1 atm and the pressures of the products will be less than 1 atm.

These results are summarized in Table 16.7. The value of K for a specific reaction can be calculated from the equation

$$\Delta G^\circ = -RT \ln(K)$$

as is shown in Sample Exercises 16.14 and 16.15.

Qualitative Relationship between the Change in Standard Free Energy and the Equilibrium Constant for a Given Reaction

ΔG°	K
$\Delta G^\circ = 0$	$K = 1$
$\Delta G^\circ < 0$	$K > 1$
$\Delta G^\circ > 0$	$K < 1$

Table 16.7

Sample Exercise 16.14

Consider the ammonia synthesis reaction

$$N_2(g) + 3H_2(g) \rightleftharpoons 2NH_3(g)$$

where $\Delta G^\circ = -33.3$ kJ per mole of N_2 consumed at 25°C. For each of the following mixtures of reactants and products at 25°C, predict the direction in which the system will shift to reach equilibrium:

a. $P_{NH_3} = 1.00$ atm, $P_{N_2} = 1.47$ atm, $P_{H_2} = 1.00 \times 10^{-2}$ atm
b. $P_{NH_3} = 1.00$ atm, $P_{N_2} = 1.00$ atm, $P_{H_2} = 1.00$ atm

Solution

a. We can predict the direction of reaction to equilibrium by calculating the value of ΔG using the equation

$$\Delta G = \Delta G^\circ + RT \ln(Q)$$

Sample Exercise 16.14, continued

The units of ΔG, $\Delta G°$, and $RT \ln(Q)$ are all per "mole of reaction," although only the "per mole" is indicated for R (as is customary).

where

$$Q = \frac{P_{NH_3}{}^2}{(P_{N_2})(P_{H_2}{}^3)} = \frac{(1.00)^2}{(1.47)[(1.00 \times 10^{-2})^3]} = 6.80 \times 10^5$$

$$T = 25 + 273 = 298 \text{ K}$$

$$R = 8.3145 \text{ J/K mol}$$

and $\Delta G° = -33.3 \text{ kJ/mol} = -3.33 \times 10^4 \text{ J/mol}$

Then
$$\Delta G = (-3.33 \times 10^4 \text{ J}) + (8.3145 \text{ J/K mol})(298 \text{ K}) \ln(6.8 \times 10^5)$$
$$= (-3.33 \times 10^4 \text{ J/mol}) + (3.33 \times 10^4 \text{ J/mol}) = 0$$

Since $\Delta G = 0$, the reactants and products have the same free energies at these partial pressures. The system is already at equilibrium, and no shift will occur.

b. The partial pressures given here are all 1.00 atm, which means that the system is in the standard state. That is,

$$\Delta G = \Delta G° + RT \ln(Q) = \Delta G° + RT \ln \frac{(1.00)^2}{(1.00)(1.00)^3}$$
$$= \Delta G° + RT \ln(1.00) = \Delta G° + 0 = \Delta G°$$

For this reaction at 25°C

$$\Delta G° = -33.3 \text{ kJ/mol}$$

The negative value for $\Delta G°$ means that in their standard states the products have a lower free energy than the reactants. Thus the system will move to the right to reach equilibrium. That is, K is greater than 1.

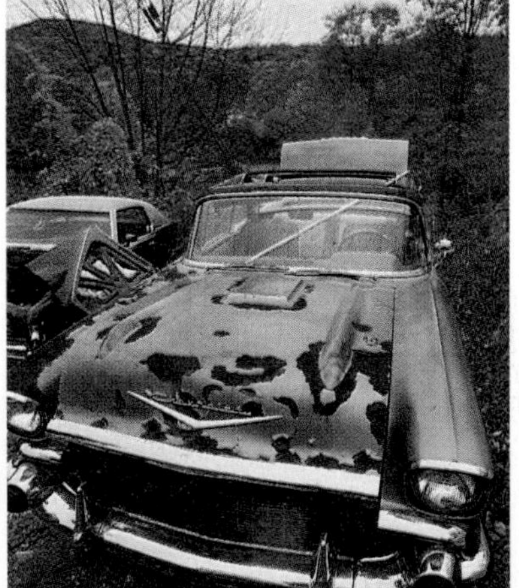

Formation of rust on bare steel is a spontaneous process.

Sample Exercise 16.15

The overall reaction for the corrosion of iron by oxygen (rusting) is

$$4Fe(s) + 3O_2(g) \rightleftharpoons 2Fe_2O_3(s)$$

Using the following data, calculate the equilibrium constant for this reaction at 25°C.

Substance	$\Delta H_f°$ (kJ/mol)	$S°$ (J/K mol)
$Fe_2O_3(s)$	−826	90
$Fe(s)$	0	27
$O_2(g)$	0	205

Solution

To calculate K for this reaction we will use the equation

$$\Delta G° = -RT \ln(K)$$

Sample Exercise 16.15, continued

We must first calculate $\Delta G°$ from

$$\Delta G° = \Delta H° - T\Delta S°$$

where

$$\Delta H° = 2\Delta H_{f\ (Fe_2O_3(s))}^{\circ} - 3\Delta H_{f\ (O_2(g))}^{\circ} - 4\Delta H_{f\ (Fe(s))}^{\circ}$$
$$= 2\text{ mol}(-826\text{ kJ/mol}) - 0 - 0$$
$$= -1652\text{ kJ} = -1.652 \times 10^6\text{ J}$$

$$\Delta S° = 2S_{Fe_2O_3}^{\circ} - 3S_{O_2}^{\circ} - 4S_{Fe}^{\circ}$$
$$= 2\text{ mol}(90\text{ J/K mol}) - 3\text{ mol}(205\text{ J/K mol}) - 4\text{ mol}(27\text{ J/K mol})$$
$$= -543\text{ J/K}$$

and

$$T = 273 + 25 = 298\text{ K}$$

Then

$$\Delta G° = \Delta H° - T\Delta S° = (-1.652 \times 10^6\text{ J}) - (298\text{ K})(-543\text{ J/K})$$
$$= -1.490 \times 10^6\text{ J}$$

and

$$\Delta G° = -RT\ln(K) = -1.490 \times 10^6\text{ J} = -(8.3145\text{ J/K mol})(298\text{ K})\ln(K)$$

Thus

$$\ln(K) = \frac{1.490 \times 10^6}{2.48 \times 10^3} = 601$$

and

$$K = e^{601}$$

In terms of base 10:

$$K = 10^{261}$$

This is a very large equilibrium constant. The rusting of iron is clearly very favorable from a thermodynamic point of view.

The units of ΔG, $\Delta G°$, and $RT\ln(Q)$ are all per "mole of reaction," although only the "per mole" is indicated for R (as is customary).

16.9 Free Energy and Work

Purpose

- To relate work done to the change in free energy.

One of the main reasons we are interested in physical and chemical processes is that we want to use them to do work for us, and we want this work done as efficiently and economically as possible. We have already seen that at constant temperature and pressure, the sign of the change in free energy tells us whether a given process is spontaneous or not. This is very useful information because it prevents us from wasting effort on a process that has no inherent tendency to occur. Although a thermodynamically favorable chemical reaction may not occur to any appreciable extent because it is too slow, it makes sense in this case to try to find a catalyst to speed up the reaction. On the other hand, if the reaction is prevented from occurring by its thermodynamic characteristics, we would be wasting our time looking for a catalyst.

In addition to its qualitative usefulness (telling us whether a process is spontaneous), the change in free energy is important quantitatively because it can tell us how much work can be done with a given process. In fact the *maximum possible useful work obtainable from a process at constant temperature and pressure is equal to the change in free energy:*

$$w_{max} = \Delta G$$

This relationship explains why this function is called the *free* energy. Under certain conditions, ΔG for a spontaneous process represents the energy that is *free to do useful work*. On the other hand, for a process that is not spontaneous, the value of ΔG tells us the minimum amount of work that must be *expended* to make the process occur.

Knowing the value of ΔG for a process thus gives us valuable information about how close the process is to 100% efficiency. For example, when gasoline is burned in a car's engine, the work produced is about 20% of the maximum work available.

For reasons we will only briefly introduce in this book, the amount of work we actually obtain from a spontaneous process is *always* less than the maximum possible amount.

To explore this idea more fully, let's consider an electrical current flowing through the starter motor of a car. The current is generated from a chemical change in a battery, and we can calculate ΔG for the battery reaction and so determine the energy available to do work. Can we use all of that energy to do work? No, because a current flowing through a wire causes frictional heating, and the more current that flows, the more heat is produced. This heat represents wasted energy—it is not useful for running the starter motor. We can minimize this energy waste by running very low currents through the motor circuit. However, no current flowing would be necessary to eliminate frictional heating entirely, and we cannot derive any work from the motor if no current flows. This represents the difficulty in which nature places us. Using a process to do work requires that some of the energy be wasted, and usually the faster we run the process, the more energy we waste.

Achieving the maximum work available from a spontaneous process can only occur via a hypothetical pathway. Any real pathway wastes energy. If we could discharge the battery infinitely slowly by an infinitesimal current flow, we would achieve the maximum useful work. Also, if we could then recharge the battery using an infinitesimally small current, exactly the same amount of energy would be used to return the battery to its original state. After we cycle the battery in this way, the universe (the system and surroundings) is exactly the same as it was before the cyclic process. This is a **reversible process** (see Fig. 16.9).

However, if the battery is discharged to run the starter motor and then recharged using a *finite* current flow as is the case in reality, *more* work will always be required to recharge the battery than the battery produces as it discharges. This means that even though the battery (the system) has returned to its original state, the surroundings have not, because the surroundings had to furnish a net amount of work as the battery was cycled. The *universe is different* after this cyclic process is performed, and this function is called an **irreversible process.** *All real processes are irreversible.*

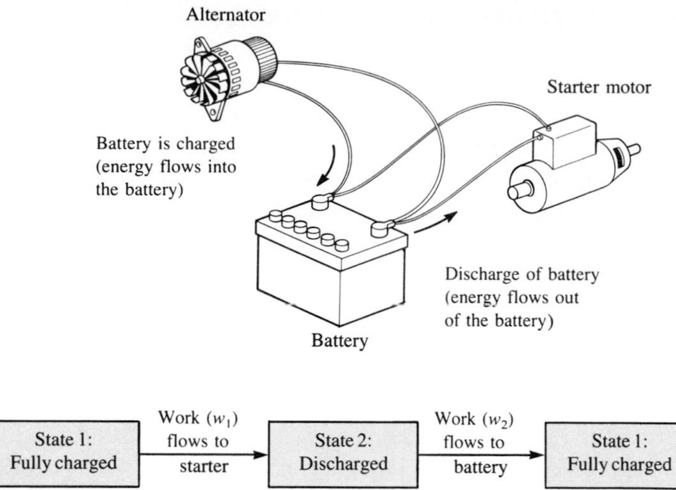

Figure 16.9

A battery can do work by sending current to a starter motor. The battery can then be recharged by forcing current through it. If the current flow in both processes is infinitesimally small, $w_1 = w_2$. This is a *reversible process.* But if the current flow is finite, as it would be in any real case, $w_2 > w_1$. This is an *irreversible process* (the *universe is different* after the cyclic process occurs). All real processes are irreversible.

In general, after any real cyclic process is carried out in a system, the surroundings have less ability to do work and contain more thermal energy. In other words, *in any real cyclic process, work is changed to heat in the surroundings, and the entropy of the universe increases.* This is another way of stating the second law of thermodynamics.

Thus thermodynamics tells us the work potential of a process and then tells us that we can never achieve this potential. In this spirit, thermodynamicist Henry Bent has paraphrased the first two laws of thermodynamics as follows:

First law: You can't win, you can only break even.

Second law: You can't break even.

When energy is used to do work, it becomes less organized and less concentrated, and thus less useful.

The ideas we have discussed in this section are applicable to the energy crisis that will probably increase in severity over the next 25 years. The crisis is obviously not one of supply; the first law tells us that the universe contains a constant supply of energy. The problem is the availability of *useful* energy. *As we use energy, we degrade its usefulness.* For example, when gasoline reacts with oxygen in the combustion reaction, the change in potential energy results in heat flow. Thus the energy concentrated in the bonds of the gasoline and oxygen molecules ends up *spread* over the surroundings as thermal energy, where it is much more difficult to harness for useful work. This is a way in which the entropy of the universe increases: concentrated energy becomes spread out—more disordered and less useful. Thus the crux of the energy problem is that we are rapidly consuming the concentrated energy found in fossil fuels. It took millions of years to concentrate the sun's energy in these fuels, and we will consume those same fuels in a few hundred years. Thus, we must consider carefully our use of these energy sources and use them as wisely as possible so as not to degrade their usefulness.

FOR REVIEW

Summary

The first law of thermodynamics, which states that the energy of the universe is constant, gives us no clue as to why a particular process occurs in a given direction. A process is said to be spontaneous if it occurs without outside intervention. The driving force for a spontaneous process is described in terms of entropy (S). Entropy is a thermodynamic function that can be viewed as a measure of randomness, or disorder. It describes the number of arrangements (positions and/or energy levels) available to a system existing in a given state. Nature spontaneously proceeds toward states that have the highest probabilities of occurring. The second law of thermodynamics states that for any spontaneous process there is always an increase in the entropy of the universe. Using entropy, thermodynamics predicts the direction in which a process will occur. However, it cannot predict the rate at which the process will occur; that must be done using the principles of kinetics.

We can predict whether a process is spontaneous by considering the entropy changes that occur in the system and the surroundings:

$$\Delta S_{univ} = \Delta S_{sys} + \Delta S_{surr}$$

For a process at constant temperature and pressure, ΔS_{sys} is dominated by changes in positional entropy, and ΔS_{surr} is determined by heat flow:

$$\Delta S_{surr} = -\frac{\Delta H}{T}$$

The sign of ΔS_{surr} depends on the direction of the heat flow: ΔS_{surr} is positive for an exothermic process and negative for an endothermic process. The magnitude of ΔS_{surr}, however, depends on both the quantity of energy that flows as heat and the temperature at which the energy is transferred. Thus the significance of exothermicity as a driving force for a process depends on the temperature at which the process occurs.

For a chemical reaction, entropy changes in the system are dominated by the change in the number of gaseous molecules. Fewer gaseous molecules on the product side means a decrease in entropy, and vice versa. Molecular structure also plays a role. In general, a complex molecule has a higher standard entropy than a simpler one.

The third law of thermodynamics states that the entropy of a perfect crystal at 0 K is zero. Above this temperature, the molecules will begin to vibrate and disorder will occur.

Free energy (G) is a thermodynamic function defined by the relationship

$$G = H - TS$$

A process occurring at constant temperature and pressure is spontaneous in the direction in which the free energy decreases.

The standard free energy change ($\Delta G°$) is the change in free energy that will occur if the reactants in their standard states are converted to the products in their standard states. Since free energy is a state function, we can use procedures similar to those used with Hess's law to calculate $\Delta G°$, for example,

$$\Delta G^\circ = \Sigma \Delta G^\circ_{f\ (products)} - \Sigma \Delta G^\circ_{f\ (reactants)}$$

where ΔG°_f, the standard free energy of formation of a substance, is the change in free energy accompanying the formation of 1 mole of that substance from its constituent elements with all reactants and products in their standard states. The standard free energy of formation of an element in its standard state is zero.

Free energy depends on pressure according to the equation

$$G = G^\circ + RT \ln(P)$$

The ΔG value for a reaction at conditions other than standard conditions can be calculated from

$$\Delta G = \Delta G^\circ + RT \ln(Q)$$

where Q is the reaction quotient.

The equilibrium position for a chemical reaction occurs where the free energy is the minimum value available to the system. The relationship between ΔG° and the equilibrium constant K is

$$\Delta G^\circ = -RT \ln(K)$$

When ΔG° is equal to zero, the system is at equilibrium with the products and reactants in their standard states ($K = 1$). When ΔG° is less than zero, then $G^\circ_{products}$ is less than $G^\circ_{reactants}$, and the system will adjust to the right from standard conditions to attain equilibrium (K is greater than 1). When ΔG° is greater than zero, $G^\circ_{reactants}$ is less than $G^\circ_{products}$, and the system will shift to the left from standard conditions to attain equilibrium (K is less than 1).

The maximum possible useful work obtainable from a process at constant temperature and pressure is equal to the change in free energy:

$$w_{max} = \Delta G$$

In any real process the actual work obtained is less than w_{max}. When energy is used to do work, the total energy of the universe remains constant, but it is rendered less useful. After being used to do work, concentrated energy is spread out in the surroundings as thermal energy, and this is the crux of our energy problem.

Key Terms

Section 16.1
spontaneous process
entropy
positional probability

Section 16.2
second law of thermodynamics

Section 16.4
free energy

Section 16.5
third law of thermodynamics

Section 16.6
standard free energy change
standard free energy of formation

Section 16.8
equilibrium (thermodynamic definition)

Section 16.9
reversible process
irreversible process

Exercises

A blue exercise number indicates that the answer to that exercise appears at the back of this book and a solution appears in the Solutions Guide.

Spontaneity and Entropy

1. Define what is meant by a spontaneous process.

2. Which of the following processes require energy as they occur?
   a. Salt dissolves in H_2O.
   b. A clear solution becomes a uniform color after a few drops of dye are added.
   c. A cell produces proteins from amino acids.
   d. Iron rusts.
   e. A house is built.
   f. A satellite is launched into orbit.
   g. A satellite falls back to earth.

3. Consider the following energy levels, each capable of holding two objects:

$$E = 2\text{ kJ} \quad \_\_\_\_\_$$
$$E = 1\text{ kJ} \quad \_\_\_\_\_$$
$$E = 0 \quad \underline{\text{XX}}$$

   Draw all of the possible arrangements of the two identical particles (represented by X) in the three energy levels. What total energy is most likely, that is, occurs the greatest number of times? Assume the particles are indistinguishable from each other.

4. Do Exercise 3 with two particles, A and B, which can be distinguished from each other.

5. Do Exercise 3 assuming each level can hold three objects and there are three indistinguishable particles, X.

6. Do Exercise 3 assuming each level can hold three objects and there are three particles (A, B, and C), which can be distinguished from each other.

7. a. Define entropy.
   b. Is entropy a form of energy?

8. Which of the following involve an increase in the entropy of the system under consideration?
   a. melting of a solid
   b. evaporation of a liquid
   c. sublimation
   d. freezing
   e. mixing
   f. separation
   g. diffusion

9. In rolling two dice, what *total* number is the most likely to occur? Is there an energy reason why this number is favored? Would energy have to be spent to increase the probability of getting a particular number (that is, to cheat)?

10. Entropy can be calculated by a relationship proposed by Ludwig Boltzmann:

$$S = k \ln(W)$$

   where $k = 1.38 \times 10^{-23}$ J/K and W is the number of ways a particular state can be obtained. (This equation is engraved on Boltzmann's tombstone.) Calculate S for the three arrangements of particles in Table 16.1.

11. Describe how the following changes affect the positional entropy of a substance:
   a. increase in volume of a gas at constant T
   b. increase in temperature of a gas at constant V
   c. increase in pressure of a gas at constant T

12. Choose the compound with the greatest positional entropy in each case:
   a. 1 mol of H_2 at STP or 1 mol of N_2O at STP
   b. 1 mol of H_2 at STP or 1 mol of H_2 at 100°C and 0.5 atm
   c. 1 mol of N_2 at STP or 1 mol of N_2 at 100 K and 2.0 atm
   d. 1 mol of $H_2O(s)$ at 0°C or 1 mol of $H_2O(l)$ at 20°C

Entropy and the Second Law of Thermodynamics: Free Energy

13. Define each of the following:
   a. system
   b. surroundings

14. Using the second law of thermodynamics and the following relationships

$$\Delta S_{univ} = \Delta S_{sys} + \Delta S_{surr}$$

$$\Delta S_{surr} = -\frac{\Delta H}{T}$$

$$\Delta G = \Delta H - T\Delta S$$

   show that $\Delta G < 0$ for a spontaneous process at constant temperature and pressure. (The unlabeled quantities refer to the system as is the usual convention.)

15. The synthesis of glucose directly from CO_2 and H_2O and the synthesis of proteins directly from amino acids are both nonspontaneous processes under standard conditions. Yet it is necessary for these to occur in order for life to exist. In light of the second law of thermodynamics, how can life exist?

16. For each of the following pairs of substances, which substance has the greater value of $S°$?
   a. glucose ($C_6H_{12}O_6$) or sucrose ($C_{12}H_{22}O_{11}$)
   b. H_2O (at 0 K) or H_2O (at 0°C)
   c. $H_2O(l)$ (at 25°C) or $H_2S(g)$ (at 25°C)
   d. N_2O (at 0 K) or He (at 10 K)

e. $N_2O(g)$ (1 atm, at 25°C) or $He(g)$ (1 atm, at 25°C)
f. $HF(g)$ (1 atm, at 25°C) or $HCl(g)$ (1 atm, at 25°C)

17. Predict the sign of $\Delta S°$ for each of the following changes:
a.

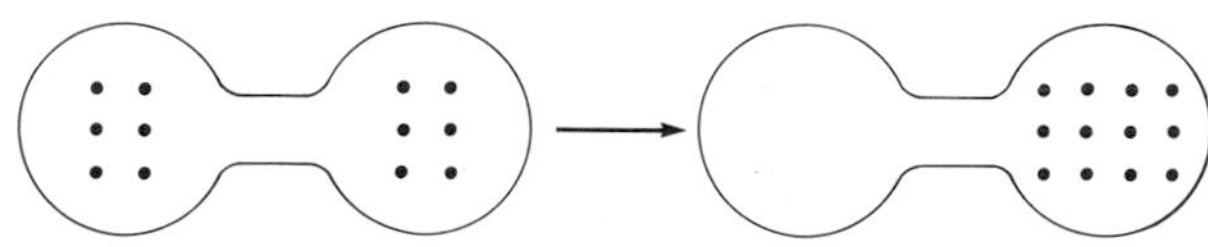

b. $AgCl(s) \rightarrow Ag^+(aq) + Cl^-(aq)$
c. $2H_2(g) + O_2(g) \rightarrow 2H_2O(l)$
d. $Na(s) + \frac{1}{2}Cl_2(g) \rightarrow NaCl(s)$
e. $HCl(g) \rightarrow H^+(aq) + Cl^-(aq)$
f. $NaCl(s) \rightarrow Na^+(aq) + Cl^-(aq)$

18. Calculate $\Delta S°$ for each of the following reactions:
a. $H_2(g) + \frac{1}{2}O_2(g) \rightarrow H_2O(g)$
b. $3O_2(g) \rightarrow 2O_3(g)$
c. $N_2(g) + O_2(g) \rightarrow 2NO(g)$

19. For the reaction

$$CS_2(g) + 3O_2(g) \rightarrow CO_2(g) + 2SO_2(g)$$

$\Delta S°$ is equal to -143 J/K. Use this value and data from Appendix 4 to calculate the value of $S°$ for $CS_2(g)$.

20. For the reaction:

$$2Al(s) + 3Br_2(l) \rightarrow 2AlBr_3(s)$$

$\Delta S°$ is equal to -144 J/K. Use this value and data from Appendix 4 to calculate the value of $S°$ for solid aluminum bromide.

21. Given the values of ΔH and ΔS, which of the following changes will be spontaneous at constant T and P?
a. $\Delta H = +25$ kJ, $\Delta S = +5$ J/K, $T = 300$ K
b. $\Delta H = +25$ kJ, $\Delta S = +100$ J/K, $T = 300$ K
c. $\Delta H = -10$ kJ, $\Delta S = +5$ J/K, $T = 298$ K
d. $\Delta H = -10$ kJ, $\Delta S = -40$ J/K, $T = 200$ K
e. $\Delta H = -10$ kJ, $\Delta S = -40$ J/K, $T = 300$ K
f. $\Delta H = -10$ kJ, $\Delta S = -40$ J/K, $T = 500$ K
g. $\Delta H = +5$ kJ, $\Delta S = -5$ J/K, $T = 10$ K
h. $\Delta H = +5$ kJ, $\Delta S = -5$ J/K, $T = 1000$ K

22. The boiling point of chloroform ($CHCl_3$) is 61.7°C. The enthalpy of vaporization is 31.4 kJ/mol. Calculate the entropy of vaporization.

23. For mercury, the enthalpy of vaporization is 58.51 kJ/mol and the entropy of vaporization is 92.92 J/K mol. What is the normal boiling point of mercury?

24. The melting point of tungsten is the second highest among the elements (only that of carbon is higher). The melting point of tungsten is 3650 K. The enthalpy of fusion is 35.23 kJ/mol. What is the entropy of fusion?

25. Two crystalline forms of white phosphorus are known. Both forms contain P_4 molecules, but the molecules are packed together in different ways. The α form is always obtained when the liquid freezes. However, below -76.9°C, the α form spontaneously converts to the β form:

$$P_4(s, \alpha) \rightarrow P_4(s, \beta)$$

a. Predict the signs of ΔH and ΔS for this process.
b. Predict which form of phosphorus has the more ordered crystalline structure.

Free Energy and Chemical Reactions

26. From data in Appendix 4, calculate $\Delta H°$, $\Delta S°$, and $\Delta G°$ for each of the following reactions:
a. $CH_4(g) + 2O_2(g) \rightarrow CO_2(g) + 2H_2O(g)$
b. $6CO_2(g) + 6H_2O(l) \rightarrow C_6H_{12}O_6(s) + 6O_2(g)$ (Glucose)
c. $P_4O_{10}(s) + 6H_2O(l) \rightarrow 4H_3PO_4(aq)$
d. $HCl(g) + NH_3(g) \rightarrow NH_4Cl(s)$

27. For the reaction:

$$SF_4(g) + F_2(g) \rightarrow SF_6(g)$$

the value of $\Delta G°$ is -374 kJ/mol. Use this value and data from Appendix 4 to calculate the value of $\Delta G_f°$ for $SF_4(g)$.

28. The value of $\Delta G°$ for the reaction

$$2Al(OH)_3(s) \rightarrow Al_2O_3(s) + 3H_2O(g)$$

is +7 kJ. Use this value and data from Appendix 4 to calculate the value of the standard free energy of formation of aluminum hydroxide.

29. Acrylonitrile is the starting material used in the manufacture of acrylic fibers (U.S. production capacity is more than 2 million pounds). Three industrial processes for the production of acrylonitrile are given below. Using data from Appendix 4, calculate $\Delta S°$, $\Delta H°$, and $\Delta G°$ for each process. For part a, assume that $T = 25$°C; for part b, $T = 70$°C; and for part c, $T = 700$°C.

a. $CH_2{-}CH_2(g)$ (bridged by O; Ethylene oxide) $+ HCN(g) \rightarrow CH_2{=}CHCN(l) + H_2O(l)$ (Acrylonitrile)

b. $HC{\equiv}CH(g) + HCN(g) \xrightarrow[70\text{–}90°C]{CaCl_2 \cdot HCl} CH_2{=}CHCN(l)$

c. $4CH_2{=}CHCH_3(g) + 6NO(g) \xrightarrow[Ag]{700°C} 4CH_2{=}CHCN(g) + 6H_2O(g) + N_2(g)$

30. Consider two reactions for the production of ethanol:

$$C_2H_4(g) + H_2O(g) \rightarrow CH_3CH_2OH(l)$$
$$C_2H_6(g) + H_2O(g) \rightarrow CH_3CH_2OH(l) + H_2(g)$$

Which would be the more thermodynamically feasible? Why?

31. When most biological enzymes are heated they lose their catalytic activity. The change

original enzyme $\rightarrow$ new form

that occurs upon heating is endothermic and spontaneous. Is the structure of the original enzyme or its new form more ordered? Explain your answer.

32. For the reaction

$$H_2(g) \rightarrow 2H(g)$$

a. Predict the signs of ΔH and ΔS.
b. Would the reaction be more spontaneous at high or low temperatures?

33. Hydrogen cyanide is produced industrially by the following reaction:

$$2NH_3(g) + 3O_2(g) + 2CH_4(g) \xrightarrow[\text{Pt-Rh}]{1000^\circ C} 2HCN(g) + 6H_2O(g)$$

Is the high temperature needed for thermodynamic, or kinetic, reasons?

34. The autoionization of water at 25°C,

$$H_2O(l) \rightleftharpoons H^+(aq) + OH^-(aq)$$

has the equilibrium constant 1.00×10^{-14}. Calculate ΔG° for this process at 25°C.

35. The Ostwald process for the commercial production of nitric acid involves three steps:

$$4NH_3(g) + 5O_2(g) \xrightarrow[825^\circ C]{\text{Pt}} 4NO(g) + 6H_2O(g)$$
$$2NO(g) + O_2(g) \rightarrow 2NO_2(g)$$
$$3NO_2(g) + H_2O(l) \rightarrow 2HNO_3(l) + NO(g)$$

a. Calculate ΔH°, ΔS°, ΔG°, and K (at 298 K) for each of the three steps in the Ostwald process (see Appendix 4).
b. Calculate the equilibrium constant for the first step at 825°C.
c. What is the reason for the high temperature in the first step?

36. One of the reactions that destroys ozone in the upper atmosphere is

$$NO(g) + O_3(g) \rightleftharpoons NO_2(g) + O_2(g)$$

Using data from Appendix 4, calculate ΔG° and K (at 298 K) for this reaction.

37. Hydrogen sulfide can be removed from natural gas by the reaction

$$2H_2S(g) + SO_2(g) \rightleftharpoons 3S(s) + 2H_2O(g)$$

Calculate ΔG° and K (at 298 K) for this reaction. Would this reaction be favored at a high or low temperature?

38. Cells use the hydrolysis of adenosine triphosphate, abbreviated as ATP, as a source of energy.

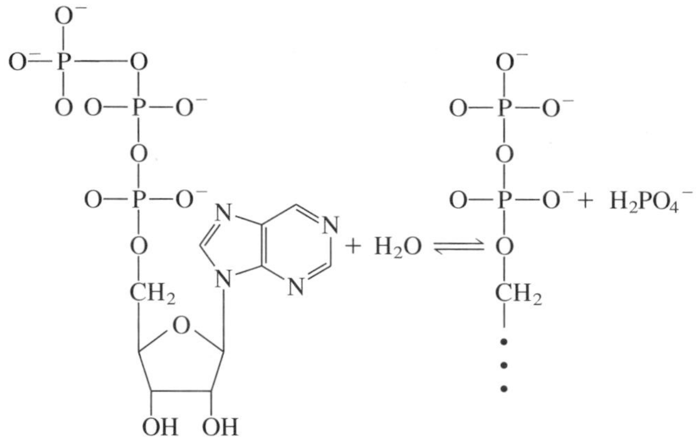

ATP $+ H_2O \rightleftharpoons$ ADP $+ H_2PO_4^-$ (P_i)

For this reaction, $\Delta G^\circ = -30.5$ kJ/mol.
a. Calculate K at 25°C.
b. If all of the free energy from the metabolism of glucose

$$C_6H_{12}O_6(s) + 6O_2(g) \rightarrow 6CO_2(g) + 6H_2O(l)$$

goes into ATP, how many ATP molecules can be produced for every molecule of glucose?

39. Carbon monoxide is toxic because it binds much more strongly to the iron in hemoglobin than does O_2. The equilibrium constant for the binding of CO is about two hundred times that for the binding of O_2. That is, for the reactions

$$\text{(heme)Fe} + O_2 \xrightleftharpoons{K_{O_2}} \text{(heme)Fe} - O_2$$
$$\text{(heme)Fe} + CO \xrightleftharpoons{K_{CO}} \text{(heme)Fe} - CO$$

$K_{CO}/K_{O_2} = 2.1 \times 10^2$. Calculate the difference in ΔG° for the binding of CO and O_2 to hemoglobin at 25°C.

Free Energy and Pressure

40. Calculate ΔG for the reaction

$$NO(g) + O_3(g) \rightarrow NO_2(g) + O_2(g)$$

for these conditions:

$T = 298$ K

$P_{NO} = 1.00 \times 10^{-6}$ atm, $P_{O_3} = 2.00 \times 10^{-6}$ atm

$P_{NO_2} = 1.00 \times 10^{-7}$ atm, $P_{O_2} = 1.00 \times 10^{-3}$ atm

For ΔG° use your answer from Exercise 36.

41. Calculate ΔG for the reaction

$$2H_2S(g) + SO_2(g) \rightleftharpoons 3S(s) + 2H_2O(g)$$

for these conditions at 25°C:

$$P_{H_2S} = 1.0 \times 10^{-4} \text{ atm}$$
$$P_{SO_2} = 1.0 \times 10^{-2} \text{ atm}$$
$$P_{H_2O} = 3.0 \times 10^{-2} \text{ atm}$$

See Exercise 37 for $\Delta G°$.

42. Using data from Appendix 4, calculate $\Delta H°$, $\Delta S°$, and K (at 298 K) for the synthesis of ammonia by the Haber process:

$$N_2(g) + 3H_2(g) \rightleftharpoons 2NH_3(g)$$

Calculate ΔG for this reaction under the following conditions (assume an uncertainty of ± 1 in all quantities):

a. $T = 298$ K, $P_{N_2} = P_{H_2} = 200$ atm, $P_{NH_3} = 50$ atm
b. $T = 298$ K, $P_{N_2} = 200$ atm, $P_{H_2} = 600$ atm, $P_{NH_3} = 200$ atm
c. $T = 100$ K, $P_{N_2} = 50$ atm, $P_{H_2} = 200$ atm, $P_{NH_3} = 10$ atm
d. $T = 700$ K, $P_{N_2} = 50$ atm, $P_{H_2} = 200$ atm, $P_{NH_3} = 10$ atm

43. Consider the following reaction:

$$H_2O(g) + Cl_2O(g) \rightleftharpoons 2HOCl(g) \qquad K_{298} = 0.090$$

For $Cl_2O(g)$:

$$\Delta G_f° = 97.9 \text{ kJ/mol}$$
$$\Delta H_f° = 80.3 \text{ kJ/mol}$$
$$S° = 266.1 \text{ J/K mol}$$

a. Calculate $\Delta G°$ for the reaction using the equation $\Delta G° = -RT \ln(K)$.
b. Use bond energy values (Table 8.4) to estimate $\Delta H°$ for the reaction.
c. Use the results from parts a and b to estimate $\Delta S°$ for the reaction.
d. Estimate $\Delta H_f°$ and $S°$ for $HOCl(g)$. (Assume these values are unavailable from tabulated data.)
e. Estimate the value of K at 500 K.
f. Estimate ΔG at 25°C when $P_{H_2O} = 18$ torr, $P_{Cl_2O} = 2.0$ torr, and $P_{HOCl} = 0.10$ torr.

44. A rough rule of thumb is that for a chemical reaction

$$\Delta S° \approx 100(\Delta n)$$

where Δn is the change in the number of moles of gaseous components and $\Delta S°$ has units of J/K.

a. Use this approximation and the values of bond energies (Table 8.4) to estimate $\Delta S°$, $\Delta H°$, and $\Delta G°$ for the reaction

$$CO(g) + 2H_2(g) \rightleftharpoons CH_3OH(g)$$

b. Compare the above values to those calculated from thermodynamic data in Appendix 4.
c. If you were to propose using this reaction on an industrial scale, would you choose a high or low temperature?

45. Acrylic acid ($CH_2{=}CHCO_2H$) is an important monomer in the chemical industry. One method for the production of acrylic acid is the reaction of carbon monoxide with ethylene oxide:

$$\underset{\text{O}}{CH_2{-}CH_2}(g) + C{\equiv}O(g) \xrightarrow{Ni(CO)_4} \underset{H}{\overset{H}{\ }}C{=}C\underset{C(=O){-}OH}{\overset{H}{\ }} \quad (l)$$

a. Use the bond energies (Table 8.4) and the approximation given in Exercise 44 to estimate $\Delta H°$, $\Delta S°$, and $\Delta G°$ for this reaction.
b. Under what temperature conditions should this reaction be run?

Additional Exercises

46. If you calculate a value for $\Delta G°$ for a reaction using the values of $\Delta G_f°$ in Appendix 4 and get a negative number, is it correct to say that the reaction is always spontaneous?

47. When the environment is contaminated by a toxic or potentially toxic substance, for example, from a chemical spill or the use of insecticides, the substance tends to disperse. How is this consistent with the second law of thermodynamics? In terms of the second law, which requires the least work: cleaning the environment after it has been contaminated, or trying to prevent the contamination before it occurs? Explain your answer.

48. A green plant synthesizes glucose by photosynthesis as shown in the reaction

$$6CO_2(g) + 6H_2O(l) \rightarrow C_6H_{12}O_6(s) + 6O_2(g)$$

Animals use glucose as a source of energy:

$$C_6H_{12}O_6(s) + 6O_2(g) \rightarrow 6CO_2(g) + 6H_2O(l)$$

If we were to assume that both of these processes occur to the same extent in a cyclic process, what thermodynamic property must have a nonzero value?

49. Using the relationships $\Delta G° = \Delta H° - T\Delta S° = -RT \ln(K)$ derive the equation

$$\ln(K) = -\frac{\Delta H°}{RT} + \frac{\Delta S°}{R}$$

Show that, for a system at equilibrium, the equilibrium will shift to the right for an endothermic process when the temperature is increased.

50. Use the equation you derived in Exercise 49 to determine $\Delta H°$ and $\Delta S°$ for the autoionization of water:

$$H_2O(l) \rightleftharpoons H^+(aq) + OH^-(aq)$$

T (°C)	K
0	1.14×10^{-15}
25	1.00×10^{-14}
35	2.09×10^{-14}
40	2.92×10^{-14}
50	5.47×10^{-14}

51. Using the free energy profile for a simple one-step reaction, show that at equilibrium $K = k_f/k_r$, where k_f and k_r are the rate constants for the forward and reverse reactions. *Hint:* Use the relationship $\Delta G° = -RT\ln(K)$ and represent k_f and k_r using the Arrhenius equation ($k = Ae^{-E_a/RT}$).

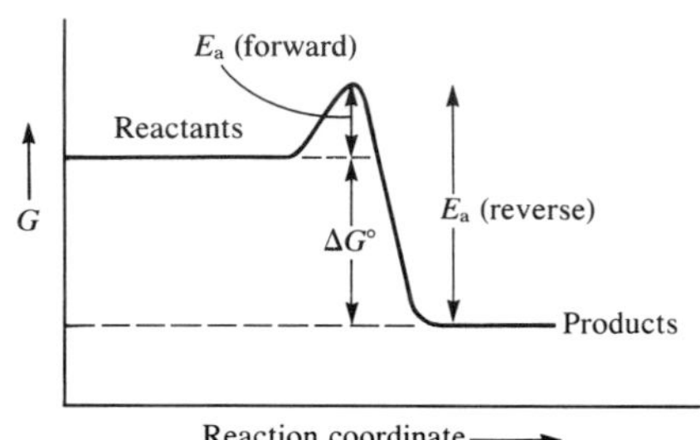

52. Why is the following statement false? "A catalyst can increase the rate of a forward reaction but not the rate of the reverse reaction."

53. Elemental sulfur can exist in two crystalline forms, rhombic and monoclinic. Calculate the temperature for the conversion of monoclinic sulfur to rhombic sulfur given the following data:

	$\Delta H_f°$ (kJ/mol)	$S°$ (J/K mol)
S (rhombic)	0	31.88
S (monoclinic)	0.30	32.55

54. Using data from Appendix 4, calculate $\Delta H°$, $\Delta S°$, and $\Delta G°$ for the following reactions that produce acetic acid:

$$CH_4(g) + CO_2(g) \rightarrow CH_3\overset{\overset{\large O}{\|}}{C}\text{—}OH(g)$$

$$CH_3OH(g) + CO(g) \rightarrow CH_3\overset{\overset{\large O}{\|}}{C}\text{—}OH(g)$$

Which reaction would you choose as a commercial method for producing acetic acid, CH_3CO_2H? What temperature conditions would you choose for the reaction?

55. In the text we derived the equation

$$\Delta G = \Delta G° + RT\ln(Q)$$

for gaseous reactions where the quantities in Q were expressed in units of pressure. We can also use units of mol/L for the quantities in Q. With this in mind, calculate ΔG for the reaction

$$H_2O(l) \rightleftharpoons H^+(aq) + OH^-(aq)$$

under the following conditions, at 25°C:

a. $[H^+] = [OH^-] = 1.00 \times 10^{-7}\ M$
b. $[H^+] = 1.00 \times 10^{-5}\ M$, $[OH^-] = 1.00 \times 10^{-9}\ M$
c. $[H^+] = [OH^-] = 1.00 \times 10^{-10}\ M$
d. $[H^+] = 10.0\ M$, $[OH^-] = 1.00 \times 10^{-7}\ M$
e. $[H^+] = 1.00\ M$, $[OH^-] = 1.00\ M$

See Exercise 34 for the $\Delta G°$ value. Based on the calculated ΔG values, in what direction will the system shift to equilibrium for each of the five sets of conditions? Are these results consistent with Le Châtelier's principle?

56. Many biochemical reactions that occur in cells require relatively high concentrations of potassium ion (K^+). The concentration of K^+ in muscle cells is about 0.15 M. The concentration of K^+ in blood plasma is about 0.0050 M. The high internal concentration in cells is maintained by pumping K^+ from the plasma. How much work must be done to transport 1.0 mol of K^+ from the blood to the inside of a muscle cell at 37°C, normal body temperature? When 1.0 mol of K^+ is transferred from blood to the cells, do any other ions have to be transported? Why or why not? Much of the ATP (see Exercise 38) formed from metabolic processes is used in providing energy for transport of cellular components. How much ATP must be hydrolyzed to provide the energy for the transport of 1.0 mol of K^+?

57. Comment on how this refrain from a popular song might be taken as a statement of the second law of thermodynamics.

> And the seasons they go 'round and 'round,
> And the painted ponies go up and down.
> We're captive on the carousel of time.
> We can't return; we can only look behind
> from where we came,
> And go 'round and 'round and 'round in the
> circle game.

58. Entropy has been described as "time's arrow." Interpret this view of entropy.

59. The lines from T. S. Eliot's *The Hollow Men (V)*

> . . . this is the way the world ends–
> not with a bang, but a whimper

have been said by some to describe the eventual heat death of the universe. Comment on this view.
(Excerpt from *The Hollow Men* in *Collected Poems 1909–1962* by T. S. Eliot, copyright 1936 by Harcourt Brace Jovanovich, Inc.; copyright © 1963, 1964 by T. S. Eliot. Reprinted by permission of the publisher.)

60. Human DNA contains almost twice as much information as is needed to code for all of the substances produced in the body. Likewise, the digital data sent from Voyager II contained one redundant bit out of every two bits of information. The proposed space telescope will transmit three redundant bits for every bit of information. How is entropy related to the transmission of information? What do you think is accomplished by having so many redundant bits of information in both DNA and the space probes?

Electrochemistry

Electrochemistry is an important component of a general chemistry course since it is one of the most important interfaces between chemistry and everyday life. Every time you start your car, turn on your calculator, look at your digital watch, or listen to a radio at the beach, you are depending on electrochemical reactions. Our society sometimes seems to run almost entirely on batteries. Certainly the advent of small, dependable batteries along with silicon-chip technology has made possible the tiny calculators, tape recorders, and clocks that we take for granted.

Electrochemistry is important in other less obvious ways. For example, the corrosion of iron, which has tremendous economic implications, is an electrochemical process. In addition, many important industrial materials such as aluminum, chlorine, and sodium hydroxide are prepared by electrolytic processes. In analytical chemistry, electrochemical techniques employ electrodes that are specific for a given molecule or ion, such as H^+ (pH meters), F^-, Cl^-, and many others. These increasingly important methods are used to analyze for trace pollutants in natural waters or for the tiny quantities of chemicals in human blood that may signal the development of a specific disease.

Electrochemistry is best defined as *the study of the interchange of chemical and electrical energy.* It is primarily concerned with two processes that involve oxidation-reduction reactions: the generation of an electric current from a chemical reaction, and the opposite process, the use of a current to produce chemical change.

< The patina on a bronze statue in Union Square, New York City. Under pristine atmospheric conditions, copper forms an external layer of greenish copper carbonate, but the presence of SO_2 in polluted air produces copper sulfate instead.

CONTENTS

17.1 Galvanic Cells

Purpose

- To review oxidation and reduction.
- To define the components of an electrochemical cell.
- To distinguish between a galvanic and an electrolytic cell.
- To define cell potentials.

As we discussed in detail in Section 4.10, an **oxidation-reduction (redox) reaction** involves a transfer of electrons from the **reducing agent** to the **oxidizing agent.** Recall that **oxidation** involves a *loss of electrons* (an increase in oxidation number) and that **reduction** involves a *gain of electrons* (a decrease in oxidation number).

To understand how a redox reaction can be used to generate a current, let's consider the reaction between MnO_4^- and Fe^{2+}:

$$8H^+(aq) + MnO_4^-(aq) + 5Fe^{2+}(aq) \rightarrow Mn^{2+}(aq) + 5Fe^{3+}(aq) + 4H_2O(l)$$

In this reaction, Fe^{2+} is oxidized and MnO_4^- is reduced; electrons are transferred from Fe^{2+} (the reducing agent) to MnO_4^- (the oxidizing agent).

Balancing half-reactions is discussed in Section 4.11.

It is useful to break a redox reaction into **half-reactions,** one involving oxidation and one involving reduction. For the reaction above, the half-reactions are

$$8H^+ + MnO_4^- + 5e^- \rightarrow Mn^{2+} + 4H_2O$$

$$5(Fe^{2+} \rightarrow Fe^{3+} + e^-)$$

The multiplication of the second half-reaction by 5 indicates that this reaction must occur five times for each time the first reaction occurs. The balanced overall reaction is the sum of the half-reactions.

When the reaction between MnO_4^- and Fe^{2+} occurs in solution, the electrons are transferred directly when the reactants collide. Thus no useful work is obtained from the chemical energy involved in the reaction, which instead is released as heat. How can we harness this energy? The key is to separate physically the oxidizing agent and the reducing agent, thus requiring the electron transfer to occur through a wire. The current produced in the wire by the electron flow can then be directed through a device, such as an electric motor, to produce useful work.

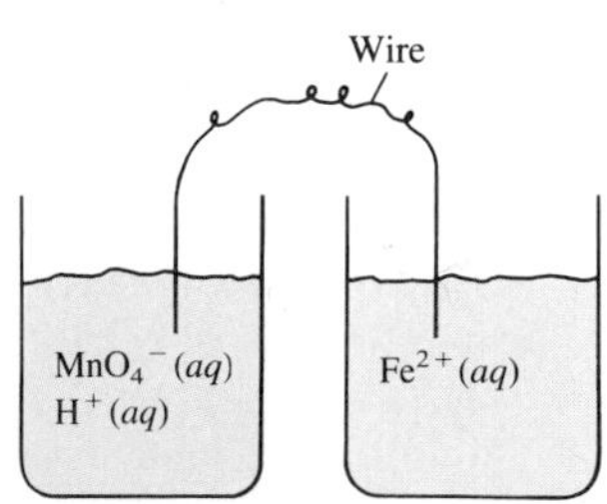

Figure 17.1

Schematic of a method to separate the oxidizing and reducing agents in a redox reaction. (The solutions also contain counter ions to balance the charge.)

For example, consider the system illustrated in Fig. 17.1. If our reasoning has been correct, electrons should flow through the wire from Fe^{2+} to MnO_4^-. However, when we construct the apparatus as shown, no flow of electrons is apparent. Why? Careful observation would show that when we connected the wires from the two compartments, current would flow for an instant and then cease. The current stops flowing because of charge build-ups in the two compartments. If electrons flowed from the right to the left compartment in the apparatus as shown, the left compartment (receiving electrons) would become negatively charged, and the right compartment (losing electrons) would become positively charged. Creating a charge separation of this type would require a large amount of energy. Thus sustained electron flow cannot occur under these conditions.

We can, however, solve this problem very simply. The solutions must be connected (without allowing the two solutions to mix) so that ions can also flow to keep

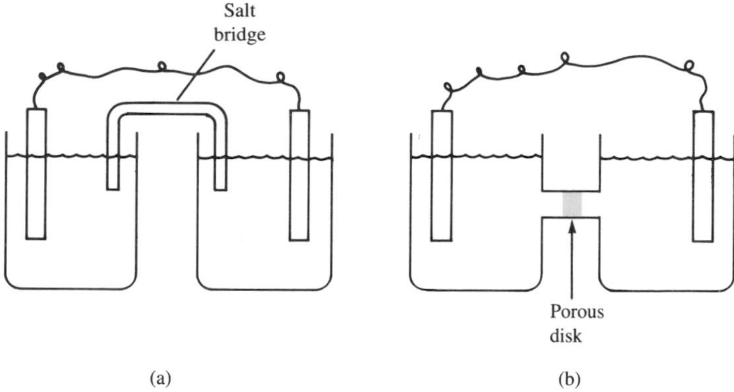

Figure 17.2

Galvanic cells can contain a salt bridge as in (a) or a porous-disk connection as in (b). A salt bridge contains a strong electrolyte held in a Jello-like matrix. A porous disk contains tiny passages that allow hindered flow of ions.

the net charge in each compartment zero. This can be accomplished by using a **salt bridge** (a U-tube filled with an electrolyte) or a **porous disk** in a tube connecting the two solutions (see Fig. 17.2). Either of these devices allows ion flow without extensive mixing of the solutions. When we make the provision for ion flow, the circuit is complete. Electrons flow through the wire from reducing agent to oxidizing agent, and ions flow from one compartment to the other to keep the net charge zero.

We have now covered all of the essential characteristics of a **galvanic cell,** *a device in which chemical energy is changed to electrical energy*. (The opposite process is called *electrolysis* and will be considered in Section 17.7.)

A galvanic cell uses a spontaneous redox reaction to produce a current that can be used to do work.

The reaction in an electrochemical cell occurs at the interface between the electrode and the solution where the electron transfer occurs. The electrode at which *oxidation* occurs is called the **anode;** the electrode at which *reduction* occurs is called the **cathode** (see Fig. 17.3).

Oxidation occurs at the anode. Reduction occurs at the cathode.

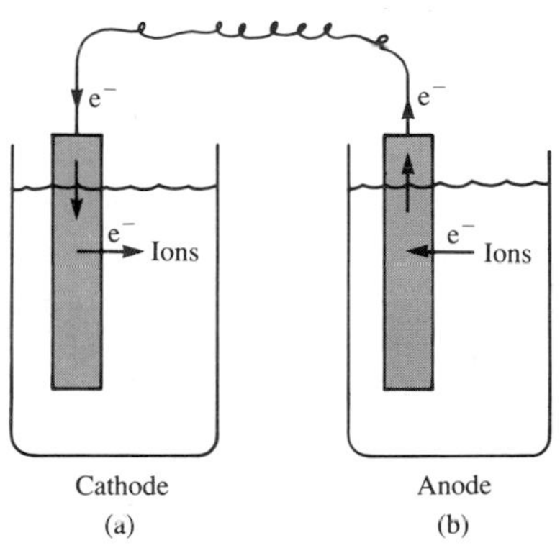

Figure 17.3

An electrochemical process involves electron transfer at the interface between the electrode and the solution. (a) The species in solution are receiving electrons from the cathode. (b) The species in solution are losing electrons to the anode.

Cell Potential

A galvanic cell consists of an oxidizing agent in one compartment with the power to pull electrons through a wire from a reducing agent in the other compartment. The "pull," or driving force, on the electrons is called the **cell potential** ($\mathscr{E}_{cell}$), or the **electromotive force** (emf) of the cell. The unit of electrical potential is the **volt,** abbreviated V and defined as 1 joule of work per coulomb of charge transferred.

A volt is 1 joule of work per coulomb of charge transferred: 1 V = 1 J/C.

How can we measure the cell potential? One possible instrument is a crude **voltmeter,** which works by drawing current through a known resistance. However,

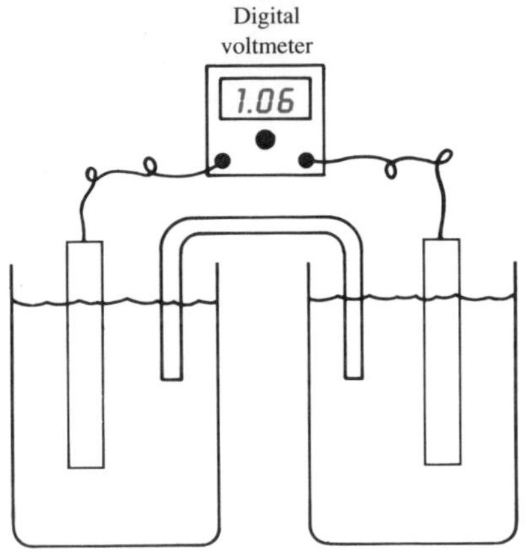

Figure 17.4

Digital voltmeters draw only a negligible current and are convenient to use.

when current flows through a wire, the frictional heating that occurs wastes some of the potentially useful energy of the cell. A traditional voltmeter will therefore measure a potential that is less than the maximum cell potential. The key to determining the maximum potential is to do the measurement under conditions of zero current so no energy is wasted. Traditionally this has been accomplished by inserting a variable voltage device (powered from an external source) in *opposition* to the cell potential. The voltage on this instrument, called a **potentiometer,** is adjusted until no current flows in the cell circuit. Under such conditions, the cell potential is equal in magnitude and opposite in sign to the voltage setting of the potentiometer and is the *maximum* cell potential, since no energy is wasted heating the wire. More recently, advances in electronic technology have allowed the design of *digital voltmeters* that draw only a negligible amount of current (see Fig. 17.4). Since these instruments are more convenient to use, they have replaced potentiometers in the modern laboratory.

17.2 Standard Reduction Potentials

Purpose

- To describe how standard reduction potentials are assigned in terms of the standard hydrogen electrode.
- To demonstrate the combination of half-reactions to form the cell reaction.
- To characterize a galvanic cell.

The name *galvanic cell* honors Luigi Galvani (1737–1798), an Italian scientist generally credited with the discovery of electricity. These cells are sometimes called *voltaic cells* after Alessandro Volta (1745–1827), another Italian, who first constructed cells of this type around 1800.

The reaction in a galvanic cell is always an oxidation-reduction reaction that can be broken down into two half-reactions. It would be convenient to assign a potential to *each* half-reaction so that when we construct a cell from a given pair of half-reactions, we can obtain the cell potential by summing the half-cell potentials. For example, the observed potential for the cell shown in Fig. 17.5(a) is 0.76 volt and the cell reaction* is

$$2H^+(aq) + Zn(s) \rightarrow Zn^{2+}(aq) + H_2(g)$$

For this cell the anode compartment contains a zinc metal electrode with Zn^{2+} and SO_4^{2-} ions in aqueous solution. The anode reaction is the oxidation half-reaction:

$$Zn \rightarrow Zn^{2+} + 2e^-$$

The zinc metal, in producing Zn^{2+} ions that go into solution, is giving up electrons, which flow through the wire. For now, we will assume that all cell components are in their standard states, so in this case the solution in the anode compartment will contain 1 M Zn^{2+}. The cathode reaction of this cell is

$$2H^+ + 2e^- \rightarrow H_2$$

*In this text we will follow the convention of indicating the physical states of the reactants and products only in the overall redox reaction. For simplicity, half-reactions will *not* include the physical states.

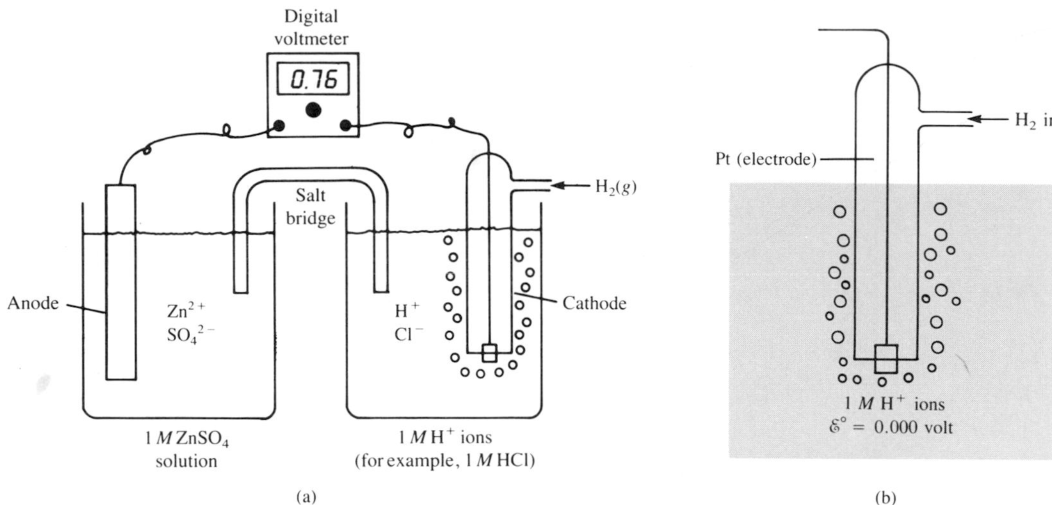

Figure 17.5

(a) A galvanic cell involving the reactions $Zn \rightarrow Zn^{2+} + 2e^-$ (at the anode) and $2H^+ + 2e^- \rightarrow H_2$ (at the cathode) has a potential of 0.76 V. (b) The standard hydrogen electrode where $H_2(g)$ at 1 atm is passed over a platinum electrode in contact with 1 M H^+ ions. This electrode process (assuming ideal behavior) is arbitrarily assigned a value of exactly zero volts.

The cathode consists of a platinum electrode (used because it is a chemically inert conductor) in contact with 1 M H^+ ions and bathed by hydrogen gas at 1 atm. Such an electrode, called the **standard hydrogen electrode,** is shown in Fig. 17.5(b).

Although we can measure the *total* potential of this cell (0.76 V), there is no way to measure the potentials of the individual electrode processes. Thus, if we want potentials for the half-reactions (half-cells), we must arbitrarily divide the total cell potential. For example, if we assign the reaction

$$2H^+ + 2e^- \rightarrow H_2$$

where

$$[H^+] = 1\ M \quad \text{and} \quad P_{H_2} = 1\ \text{atm}$$

a potential of exactly zero volts, then the reaction

$$Zn \rightarrow Zn^{2+} + 2e^-$$

will have a potential of 0.76 volt since

$$\underset{0.76\ V}{\underset{\uparrow}{\mathscr{E}°_{cell}}} = \underset{0\ V}{\underset{\uparrow}{\mathscr{E}°_{H^+ \rightarrow H_2}}} + \underset{0.76\ V}{\underset{\uparrow}{\mathscr{E}°_{Zn \rightarrow Zn^{2+}}}}$$

where the superscript ° indicates *standard states* are employed. In fact, by setting the standard potential for the half-reaction $2H^+ + 2e^- \rightarrow H_2$ equal to zero, we can assign values to all other half-reactions.

Standard states were discussed in Section 6.4.

For example, the measured potential for the cell shown in Fig. 17.6 is 1.10 V. The cell reaction is

$$Zn(s) + Cu^{2+}(aq) \rightarrow Zn^{2+}(aq) + Cu(s)$$

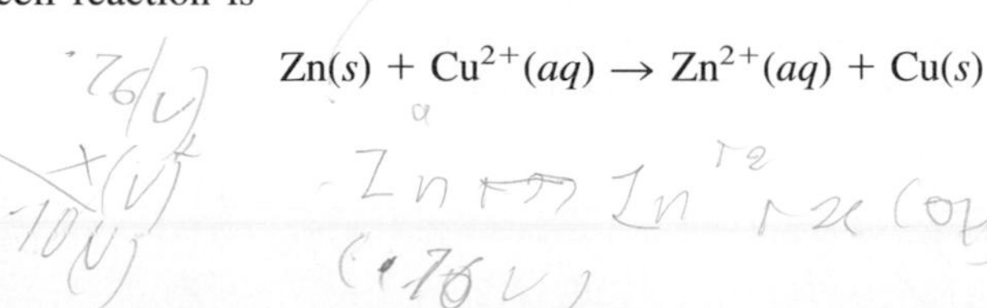

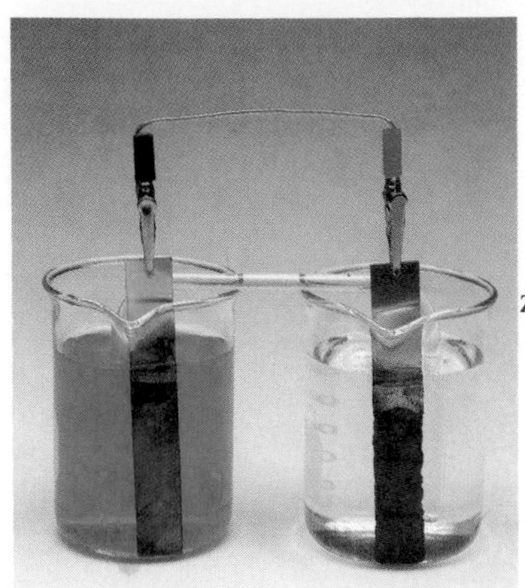

Copper being plated onto the copper metal cathode on the left, and zinc dissolving from the zinc metal anode on the right. Note the salt bridge connecting the two solutions which allows the ion flow to balance the electron flow through the wire.

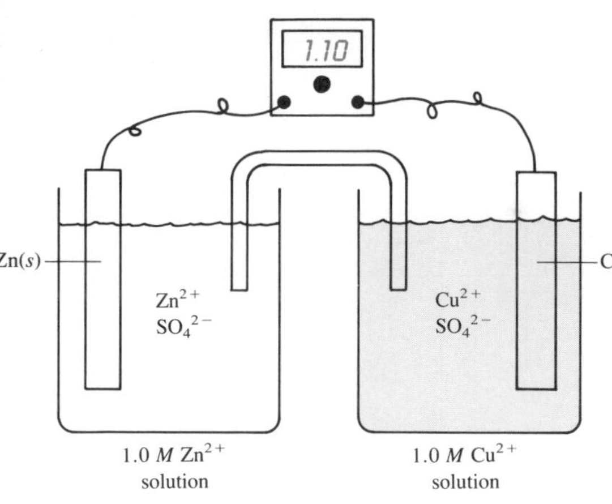

Figure 17.6

A galvanic cell involving the half-reactions $Zn \rightarrow Zn^{2+} + 2e^-$ (anode) and $Cu^{2+} + 2e^- \rightarrow Cu$ (cathode), with $\mathscr{E}^\circ_{cell} = 1.10$ V.

which can be divided into the half-reactions

$$\text{Anode:} \quad Zn \rightarrow Zn^{2+} + 2e^-$$

$$\text{Cathode:} \quad Cu^{2+} + 2e^- \rightarrow Cu$$

Then

$$\mathscr{E}^\circ_{cell} = \mathscr{E}^\circ_{Zn \rightarrow Zn^{2+}} + \mathscr{E}^\circ_{Cu^{2+} \rightarrow Cu}$$

Since $\mathscr{E}^\circ_{Zn \rightarrow Zn^{2+}}$ was earlier assigned a value of 0.76 V, the value of $\mathscr{E}^\circ_{Cu^{2+} \rightarrow Cu}$ must be 0.34 V because

$$1.10 \text{ V} = 0.76 \text{ V} + 0.34 \text{ V}$$

The standard hydrogen potential is the reference potential against which all half-reaction potentials are assigned.

The scientific community has universally accepted the half-reaction potentials based on the assignment of zero volts to the process $2H^+ + 2e^- \rightarrow H_2$ at standard conditions where ideal behavior is assumed. However, before we can use these values to calculate cell potentials, we need to understand several essential characteristics of half-cell potentials.

The currently accepted convention is to give the potentials of half-reactions as *reduction* processes, for example:

$$2H^+ + 2e^- \rightarrow H_2$$

$$Cu^{2+} + 2e^- \rightarrow Cu$$

$$Zn^{2+} + 2e^- \rightarrow Zn$$

All half-reactions are given as reduction processes in standard tables.

The $\mathscr{E}^\circ$ values corresponding to these half-reactions are called **standard reduction potentials.** Standard reduction potentials for the most common half-reactions are given in Table 17.1 and Appendix 5.5.

Combining two half-reactions to obtain a balanced oxidation-reduction reaction often requires two manipulations:

When a half-reaction is reversed, the sign of $\mathscr{E}^\circ$ is reversed.

When a half-reaction is multiplied by an integer, $\mathscr{E}^\circ$ remains the same.

1. One of the reduction half-reactions must be reversed (since redox reactions must involve a substance being oxidized and a substance being reduced), which means that the *sign* of the potential for this half-reaction must also be *reversed.*
2. Since the number of electrons lost must equal the number gained, the half-reactions must be multiplied by integers as necessary to achieve the balanced equation. However, the *value of $\mathscr{E}^\circ$ is not changed* when a half-reaction is multiplied by an integer. Since a standard reduction potential is an *intensive property* (it does not depend on how many times the reaction occurs), the potential is *not* multiplied by the integer required to balance the cell reaction.

Half-reaction	$\mathscr{E}°$ (V)	Half-reaction	$\mathscr{E}°$ (V)
Standard Reduction Potentials at 25°C (298 K) for Many Common Half-reactions			
$F_2 + 2e^- \rightarrow 2F^-$	2.87	$O_2 + 2H_2O + 4e^- \rightarrow 4OH^-$	0.40
$Ag^{2+} + e^- \rightarrow Ag^+$	1.99	$Cu^{2+} + 2e^- \rightarrow Cu$	0.34
$Co^{3+} + e^- \rightarrow Co^{2+}$	1.95	$Hg_2Cl_2 + 2e^- \rightarrow 2Hg + 2Cl^-$	0.34
$H_2O_2 + 2H^+ + 2e^- \rightarrow 2H_2O$	1.78	$AgCl + e^- \rightarrow Ag + Cl^-$	0.22
$Ce^{4+} + e^- \rightarrow Ce^{3+}$	1.70	$SO_4^{2-} + 4H^+ + 2e^- \rightarrow H_2SO_3 + H_2O$	0.20
$PbO_2 + 4H^+ + SO_4^{2-} + 2e^- \rightarrow PbSO_4 + 2H_2O$	1.69	$Cu^{2+} + e^- \rightarrow Cu^+$	0.16
$MnO_4^- + 4H^+ + 3e^- \rightarrow MnO_2 + 2H_2O$	1.68	$2H^+ + 2e^- \rightarrow H_2$	0.00
$2e^- + 2H^+ + IO_4^- \rightarrow IO_3^- + H_2O$	1.60	$Fe^{3+} + 3e^- \rightarrow Fe$	−0.036
$MnO_4^- + 8H^+ + 5e^- \rightarrow Mn^{2+} + 4H_2O$	1.51	$Pb^{2+} + 2e^- \rightarrow Pb$	−0.13
$Au^{3+} + 3e^- \rightarrow Au$	1.50	$Sn^{2+} + 2e^- \rightarrow Sn$	−0.14
$PbO_2 + 4H^+ + 2e^- \rightarrow Pb^{2+} + 2H_2O$	1.46	$Ni^{2+} + 2e^- \rightarrow Ni$	−0.23
$Cl_2 + 2e^- \rightarrow 2Cl^-$	1.36	$PbSO_4 + 2e^- \rightarrow Pb + SO_4^{2-}$	−0.35
$Cr_2O_7^{2-} + 14H^+ + 6e^- \rightarrow 2Cr^{3+} + 7H_2O$	1.33	$Cd^{2+} + 2e^- \rightarrow Cd$	−0.40
$O_2 + 4H^+ + 4e^- \rightarrow 2H_2O$	1.23	$Fe^{2+} + 2e^- \rightarrow Fe$	−0.44
$MnO_2 + 4H^+ + 2e^- \rightarrow Mn^{2+} + 2H_2O$	1.21	$Cr^{3+} + e^- \rightarrow Cr^{2+}$	−0.50
$IO_3^- + 6H^+ + 5e^- \rightarrow \frac{1}{2}I_2 + 3H_2O$	1.20	$Cr^{3+} + 3e^- \rightarrow Cr$	−0.73
$Br_2 + 2e^- \rightarrow 2Br^-$	1.09	$Zn^{2+} + 2e^- \rightarrow Zn$	−0.76
$VO_2^+ + 2H^+ + e^- \rightarrow VO^{2+} + H_2O$	1.00	$2H_2O + 2e^- \rightarrow H_2 + 2OH^-$	−0.83
$AuCl_4^- + 3e^- \rightarrow Au + 4Cl^-$	0.99	$Mn^{2+} + 2e^- \rightarrow Mn$	−1.18
$NO_3^- + 4H^+ + 3e^- \rightarrow NO + 2H_2O$	0.96	$Al^{3+} + 3e^- \rightarrow Al$	−1.66
$ClO_2 + e^- \rightarrow ClO_2^-$	0.954	$H_2 + 2e^- \rightarrow 2H^-$	−2.23
$2Hg^{2+} + 2e^- \rightarrow Hg_2^{2+}$	0.91	$Mg^{2+} + 2e^- \rightarrow Mg$	−2.37
$Ag^+ + e^- \rightarrow Ag$	0.80	$La^{3+} + 3e^- \rightarrow La$	−2.37
$Hg_2^{2+} + 2e^- \rightarrow 2Hg$	0.80	$Na^+ + e^- \rightarrow Na$	−2.71
$Fe^{3+} + e^- \rightarrow Fe^{2+}$	0.77	$Ca^{2+} + 2e^- \rightarrow Ca$	−2.76
$O_2 + 2H^+ + 2e^- \rightarrow H_2O_2$	0.68	$Ba^{2+} + 2e^- \rightarrow Ba$	−2.90
$MnO_4^- + e^- \rightarrow MnO_4^{2-}$	0.56	$K^+ + e^- \rightarrow K$	−2.92
$I_2 + 2e^- \rightarrow 2I^-$	0.54	$Li^+ + e^- \rightarrow Li$	−3.05
$Cu^+ + e^- \rightarrow Cu$	0.52		

Table 17.1

Consider a galvanic cell based on the redox reaction

$$Fe^{3+}(aq) + Cu(s) \rightarrow Cu^{2+}(aq) + Fe^{2+}(aq)$$

The pertinent half-reactions are

$$Fe^{3+} + e^- \rightarrow Fe^{2+} \qquad \mathscr{E}° = 0.77\ \text{V} \qquad (1)$$

$$Cu^{2+} + 2e^- \rightarrow Cu \qquad \mathscr{E}° = 0.34\ \text{V} \qquad (2)$$

To balance the cell reaction and calculate the standard cell potential, reaction (2) must be reversed:

$$Cu \rightarrow Cu^{2+} + 2e^- \qquad -\mathscr{E}° = -0.34\ \text{V}$$

Note the change in sign for the $\mathscr{E}°$ value. Now, since each Cu atom produces two electrons but each Fe^{3+} ion accepts only one electron, reaction (1) must be multiplied by 2:

$$2Fe^{3+} + 2e^- \rightarrow 2Fe^{2+} \qquad \mathscr{E}° = 0.77\ \text{V}$$

Note that $\mathscr{E}°$ is not changed in this case.

Now we can obtain the balanced cell reaction by summing the appropriately modified half-reactions:

$$Cu \rightarrow Cu^{2+} + 2e^- \qquad -\mathcal{E}° = -0.34\ V$$
$$2Fe^{3+} + 2e^- \rightarrow 2Fe^{2+} \qquad \mathcal{E}° = 0.77\ V$$

Cell reaction: $$Cu(s) + 2Fe^{3+}(aq) \rightarrow Cu^{2+}(aq) + 2Fe^{2+}(aq) \qquad \mathcal{E}°_{cell} = -0.34\ V + 0.77\ V = 0.43\ V$$

Sample Exercise 17.1

a. Consider a galvanic cell based on the reaction

$$Al^{3+}(aq) + Mg(s) \rightarrow Al(s) + Mg^{2+}(aq)$$

The half-reactions are

$$Al^{3+} + 3e^- \rightarrow Al \qquad \mathcal{E}° = -1.66\ V \qquad (1)$$
$$Mg^{2+} + 2e^- \rightarrow Mg \qquad \mathcal{E}° = -2.37\ V \qquad (2)$$

Give the balanced cell reaction, and calculate $\mathcal{E}°$ for the cell.

b. A galvanic cell is based on the reaction

$$MnO_4^-(aq) + H^+(aq) + ClO_3^-(aq) \rightarrow ClO_4^-(aq) + Mn^{2+}(aq) + H_2O(l)$$

The half-reactions are

$$MnO_4^- + 5e^- + 8H^+ \rightarrow Mn^{2+} + 4H_2O \qquad \mathcal{E}° = 1.51\ V \qquad (1)$$
$$ClO_4^- + 2H^+ + 2e^- \rightarrow ClO_3^- + H_2O \qquad \mathcal{E}° = 1.19\ V \qquad (2)$$

Give the balanced cell reaction, and calculate $\mathcal{E}°$ for the cell.

Solution

a. The half-reaction involving magnesium must be reversed:

$$Mg \rightarrow Mg^{2+} + 2e^- \qquad -\mathcal{E}° = -(-2.37\ V) = 2.37\ V$$

Also, since the two half-reactions involve different numbers of electrons, they must be multiplied by integers as follows:

$$2(Al^{3+} + 3e^- \rightarrow Al) \qquad \mathcal{E}° = -1.66\ V$$
$$3(Mg \rightarrow Mg^{2+} + 2e^-) \qquad -\mathcal{E}° = 2.37\ V$$
$$2Al^{3+}(aq) + 3Mg(s) \rightarrow 2Al(s) + 3Mg^{2+}(aq) \qquad \mathcal{E}°_{cell} = -1.66\ V + 2.37\ V = 0.71\ V$$

b. Half-reaction (2) must be reversed, and both half-reactions must be multiplied by integers to make the number of electrons equal:

$$2(MnO_4^- + 5e^- + 8H^+ \rightarrow Mn^{2+} + 4H_2O) \qquad \mathcal{E}° = 1.51\ V$$
$$5(ClO_3^- + H_2O \rightarrow ClO_4^- + 2H^+ + 2e^-) \qquad -\mathcal{E}° = -1.19\ V$$
$$2MnO_4^-(aq) + 6H^+(aq) + 5ClO_3^-(aq) \rightarrow 2Mn^{2+}(aq) + 3H_2O(l) + 5ClO_4^-(aq) \qquad \mathcal{E}°_{cell} = 1.51\ V - 1.19\ V = 0.32\ V$$

Next we want to consider how to describe a galvanic cell fully, given just its half-reactions. This description will include the cell reaction, the cell potential, and the physical set-up of the cell. Let's consider a galvanic cell based on the following half-reactions:

$$Fe^{2+} + 2e^- \rightarrow Fe \qquad \mathscr{E}^\circ = -0.44 \text{ V}$$
$$MnO_4^- + 5e^- + 8H^+ \rightarrow Mn^{2+} + 4H_2O \qquad \mathscr{E}^\circ = 1.51 \text{ V}$$

In a working galvanic cell, one of these reactions must run in reverse. Which one?

We can answer this question by considering the sign of the potential of a working cell: *a cell will always run spontaneously in the direction that produces a positive cell potential*. Thus, in the present case, it is clear that the half-reaction involving iron must be reversed, since this choice leads to a positive cell potential:

A galvanic cell runs spontaneously in the direction that gives a positive value for $\mathscr{E}_{cell}$.

$$Fe \rightarrow Fe^{2+} + 2e^- \qquad -\mathscr{E}^\circ = 0.44 \text{ V}$$
$$MnO_4^- + 5e^- + 8H^+ \rightarrow Mn^{2+} + 4H_2O \qquad \mathscr{E}^\circ = 1.51 \text{ V}$$

where

$$\mathscr{E}^\circ_{cell} = 0.44 \text{ V} + 1.51 \text{ V} = 1.95 \text{ V}$$

The balanced cell reaction is obtained as follows:

$$5(Fe \rightarrow Fe^{2+} + 2e^-)$$
$$2(MnO_4^- + 5e^- + 8H^+ \rightarrow Mn^{2+} + 4H_2O)$$
$$\overline{2MnO_4^-(aq) + 5Fe(s) + 16H^+(aq) \rightarrow 5Fe^{2+}(aq) + 2Mn^{2+}(aq) + 8H_2O(l)}$$

Now consider the physical set-up of the cell, shown schematically in Fig. 17.7. In the left compartment the active components in their standard states are pure metallic iron (Fe) and 1.0 *M* Fe^{2+}. The anion present (probably NO_3^- or SO_4^{2-}) depends on the iron salt used. In this compartment the anion does not participate in the reaction, but simply balances the charge. The half-reaction that takes place at this electrode is

$$Fe \rightarrow Fe^{2+} + 2e^-$$

which is an oxidation reaction, so this is the anode compartment. The electrode will consist of pure iron metal.

In the right compartment the active components in their standard states are 1.0 *M* MnO_4^-, 1.0 *M* H^+, and 1.0 *M* Mn^{2+}, with appropriate unreacting ions (often called *counter ions*) to balance the charge. The half-reaction in this compartment is

$$MnO_4^- + 5e^- + 8H^+ \rightarrow Mn^{2+} + 4H_2O$$

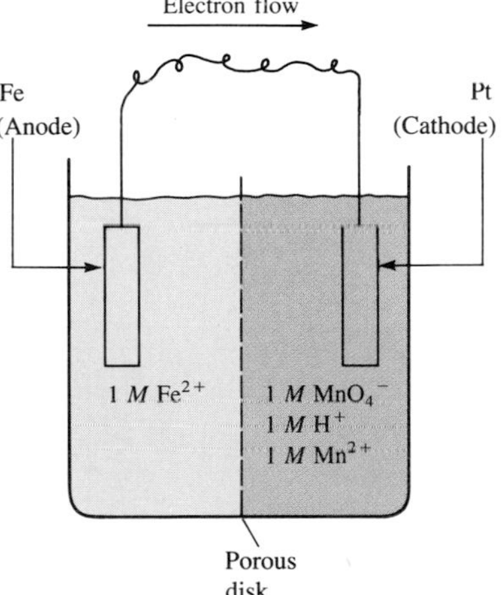

Figure 17.7

The schematic of a galvanic cell based on the half-reactions:
$Fe \rightarrow Fe^{2+} + 2e^-$
$MnO_4^- + 5e^- + 8H^+ \rightarrow Mn^{2+} + 4H_2O$.

which is a reduction reaction, so this is the cathode compartment. Since neither MnO_4^- nor Mn^{2+} ion can serve as the electrode, a nonreacting conductor must be used. The usual choice is platinum.

The next step is to determine the direction of electron flow. In the left compartment the half-reaction is the oxidation of iron:

$$Fe \rightarrow Fe^{2+} + 2e^-$$

In the right compartment the half-reaction is the reduction of MnO_4^-:

$$MnO_4^- + 5e^- + 8H^+ \rightarrow Mn^{2+} + 4H_2O$$

Thus the electrons flow from Fe to MnO_4^- in this cell, or from the anode to the cathode, as is always the case.

A complete description of a galvanic cell usually includes four items:

- The cell potential (always positive for a galvanic cell) and the balanced cell reaction.
- The direction of electron flow, obtained by inspecting the half-reactions and using the directions that give a positive $\mathscr{E}_{cell}$.
- Designation of the anode and cathode.
- The nature of each electrode and the ions present in each compartment. A chemically inert conductor is required if none of the substances participating in the half-reaction is a conducting solid.

Sample Exercise 17.2

Describe completely the galvanic cell based on the following half-reactions under standard conditions:

$$Ag^+ + e^- \rightarrow Ag \qquad \mathscr{E}° = 0.80 \text{ V} \qquad (1)$$

$$Fe^{3+} + e^- \rightarrow Fe^{2+} \qquad \mathscr{E}° = 0.77 \text{ V} \qquad (2)$$

Solution

ITEM 1

Since a positive $\mathscr{E}°_{cell}$ value is required, reaction (2) must run in reverse:

$$Ag^+ + e^- \rightarrow Ag \qquad \mathscr{E}° = 0.80 \text{ V}$$

$$Fe^{2+} \rightarrow Fe^{3+} + e^- \qquad -\mathscr{E}° = -0.77 \text{ V}$$

$$\text{Cell reaction:} \quad Ag^+(aq) + Fe^{2+}(aq) \rightarrow Fe^{3+}(aq) + Ag(s) \qquad \mathscr{E}°_{cell} = 0.03 \text{ V}$$

ITEM 2

Since Ag^+ receives electrons and Fe^{2+} loses electrons in the cell reaction, the electrons will flow from the compartment containing Fe^{2+} to the compartment containing Ag^+.

ITEM 3

Oxidation occurs in the compartment containing Fe^{2+} (electrons flow from Fe^{2+} to Ag^+). Hence this compartment contains the anode. Reduction occurs in the compartment containing Ag^+, so this compartment contains the cathode.

ITEM 4

The electrode in the Ag/Ag^+ compartment will be silver metal, and an inert conductor, such as platinum, must be used in the Fe^{2+}/Fe^{3+} compartment. Appropriate counter ions are assumed to be present. The diagram for this cell is shown in Fig. 17.8.

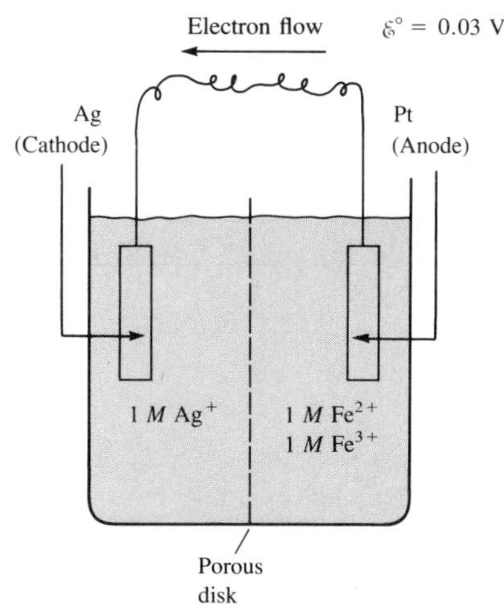

Figure 17.8

Schematic diagram for the galvanic cell based on the half-reactions:

$$Ag^+ + e^- \rightarrow Ag$$
$$Fe^{2+} \rightarrow Fe^{3+} + e^-.$$

17.3 Cell Potential, Electrical Work, and Free Energy

Purpose

- To relate the maximum cell potential to the free energy difference between cell reactants and products.

So far we have considered electrochemical cells in a very practical fashion without much theoretical background. Our next step will be to explore the relationship between thermodynamics and electrochemistry.

The work that can be accomplished when electrons are transferred through a wire depends on the "push" (the thermodynamic driving force) behind the electrons. This driving force is defined in terms of a *potential* (in volts) between two points in the circuit, where a volt represents a joule of work per coulomb of charge transferred:

$$\text{Potential (V)} = \frac{\text{work (J)}}{\text{charge (C)}}$$

Thus 1 joule of work is produced or required (depending on the direction) when 1 coulomb of charge is transferred between two points in the circuit that differ by a potential of 1 volt.

In this book, *work is viewed from the point of view of the system.* Thus work flowing out of the system is indicated by a minus sign. When a cell produces a current, the cell potential is positive, and the current can be used to do work—to run a motor, for instance. Thus the cell potential ($\mathscr{E}$) and the work (w) have opposite signs:

$$\mathscr{E} = \frac{-w \leftarrow \text{Work}}{q \leftarrow \text{Charge}}$$

Then

$$-w = q\mathscr{E}$$

From this equation it can be seen that the maximum work in a cell would be obtained at the maximum cell potential:

$$-w_{\text{max}} = q\mathscr{E}_{\text{max}} \quad \text{or} \quad w_{\text{max}} = -q\mathscr{E}_{\text{max}}$$

However, there is a problem. To obtain electrical work, current must flow. When current flows, some energy is inevitably wasted through frictional heating, and the maximum work is not obtained. This reflects the important general principle introduced in Section 16.9: *in any real, spontaneous process some energy is always wasted—the actual work realized is always less than the calculated maximum.* This is a consequence of the fact that the entropy of the universe must increase in any spontaneous process. Recall from Section 16.9 that the only process from which maximum work could be realized is the hypothetical reversible process, which for a galvanic cell would involve an infinitesimally small current flow and thus an infinite amount of time to do the work. Even though we can never achieve the maximum work through the actual discharge of a galvanic cell, it is possible to measure the

Work is never the maximum possible if any current is flowing.

maximum potential. There is negligible current flow when a cell potential is measured with a potentiometer or an efficient digital voltmeter. No current flow implies no waste of energy, so that the potential measured is the maximum.

Although we can never actually realize the maximum work from a cell reaction, the value for it is still useful in evaluating the efficiency of a real process based on the cell reaction. For example, suppose a certain galvanic cell has a maximum (at zero current) potential of 2.50 V. In a particular experiment 1.33 moles of electrons were passed through this cell at an average actual potential of 2.10 V. The actual work done is

$$w = -q\mathscr{E}$$

where $\mathscr{E}$ represents the actual potential difference at which the current flowed (2.10 V or 2.10 J/C) and q is the quantity of charge in coulombs transferred. The charge on 1 mole of electrons is a constant called the **faraday** (abbreviated F), which has the value *96,485 coulombs of charge per mole of electrons*. Thus q equals the number of moles of electrons times the charge per mole of electrons:

$$q = nF = 1.33 \text{ mol e}^- \times 96{,}485 \text{ C/mol e}^-$$

The faraday was named in honor of Michael Faraday (1791–1867), an Englishman who may have been the greatest experimental scientist of the nineteenth century. Among his many achievements were the invention of the electric motor and generator and the development of the principles of electrolysis.

Then for the experiment above, the actual work is

$$\begin{aligned} w = -q\mathscr{E} &= -(1.33 \text{ mol e}^- \times 96{,}485 \text{ C/mol e}^-) \times (2.10 \text{ J/C}) \\ &= -2.69 \times 10^5 \text{ J} \end{aligned}$$

The calculation of maximum possible work is similar except that the maximum potential is used:

$$\begin{aligned} w_{\text{max}} &= -q\mathscr{E}_{\text{max}} \\ &= -\left(1.33 \text{ mol e}^- \times 96{,}485 \frac{\text{C}}{\text{mol e}^-}\right)\left(2.50 \frac{\text{J}}{\text{C}}\right) \\ &= -3.21 \times 10^5 \text{ J} \end{aligned}$$

Thus in its actual operation the efficiency of this cell is

$$\frac{w}{w_{\text{max}}} \times 100 = \frac{-2.69 \times 10^5 \text{ J}}{-3.21 \times 10^5 \text{ J}} \times 100 = 83.8\%$$

In Section 16.9 we saw that for a process carried out at constant temperature and pressure, the change in free energy equals the maximum useful work obtainable from that process

$$w_{\text{max}} = \Delta G$$

For a galvanic cell

$$w_{\text{max}} = -q\mathscr{E}_{\text{max}} = \Delta G$$

Since

$$q = nF$$

we have

$$\Delta G = -q\mathscr{E}_{\text{max}} = -nF\mathscr{E}_{\text{max}}$$

From now on the subscript on $\mathscr{E}_{\text{max}}$ will be deleted, with the understanding that any potential given in this book is the maximum potential unless specified otherwise.

Thus $$\Delta G = -nF\mathcal{E}$$

For standard conditions,

$$\Delta G^\circ = -nF\mathcal{E}^\circ$$

This equation states that *the maximum cell potential is directly related to the free energy difference between the reactants and the products in the cell*. This relationship is important because it provides an experimental means to obtain ΔG for a reaction. It also confirms that a galvanic cell will run in the direction that gives a positive value for $\mathcal{E}_{cell}$; a positive $\mathcal{E}_{cell}$ value corresponds to a negative ΔG value, which is the condition for spontaneity.

Sample Exercise 17.3

Using the data in Table 17.1, calculate ΔG° for the reaction

$$Cu^{2+}(aq) + Fe(s) \rightarrow Cu(s) + Fe^{2+}(aq)$$

Is this reaction spontaneous?

Solution

The half-reactions are

$Cu^{2+} + 2e^- \rightarrow Cu$	$\mathcal{E}^\circ = 0.34$ V
$Fe \rightarrow Fe^{2+} + 2e^-$	$-\mathcal{E}^\circ = 0.44$ V
$Cu^{2+} + Fe \rightarrow Fe^{2+} + Cu$	$\mathcal{E}^\circ_{cell} = 0.78$ V

We can calculate ΔG° from the equation

$$\Delta G^\circ = -nF\mathcal{E}^\circ$$

Since two electrons are transferred per atom in the reaction, 2 moles of electrons are required per mole of reactants and products. Thus $n = 2$ mol e^-, $F = 96{,}485$ C/mol e^-, and $\mathcal{E}^\circ = 0.78$ V $= 0.78$ J/C. Therefore,

$$\Delta G^\circ = -(2 \text{ mol } e^-)\left(96{,}485 \frac{\text{C}}{\text{mol } e^-}\right)\left(0.78 \frac{\text{J}}{\text{C}}\right)$$

$$= -1.5 \times 10^5 \text{ J}$$

The process is spontaneous, as indicated both by the negative sign of ΔG° and the positive sign of $\mathcal{E}^\circ_{cell}$.

This reaction is used industrially to deposit copper metal from solutions resulting from the dissolving of copper ores.

Sample Exercise 17.4

Using the data from Table 17.1, predict whether 1 *M* HNO_3 will dissolve gold metal to form a 1 *M* Au^{3+} solution.

Solution

The half-reaction for HNO_3 acting as an oxidizing agent is

$$NO_3^- + 4H^+ + 3e^- \rightarrow NO + 2H_2O \qquad \mathcal{E}^\circ = 0.96 \text{ V}$$

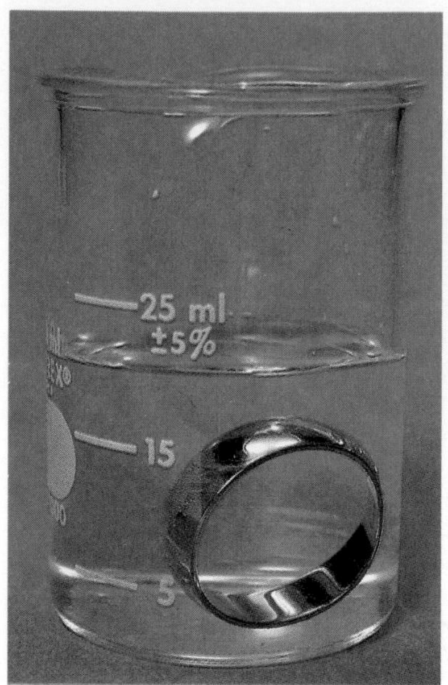

A gold ring does not dissolve in nitric acid.

Sample Exercise 17.4, continued

The reaction for the oxidation of solid gold to Au^{3+} ions is

$$Au \rightarrow Au^{3+} + 3e^- \qquad -\mathscr{E}° = -1.50 \text{ V}$$

The sum of these half-reactions gives the required reaction:

$$Au(s) + NO_3^-(aq) + 4H^+(aq) \rightarrow Au^{3+}(aq) + NO(g) + 2H_2O(l)$$

and

$$\mathscr{E}°_{cell} = 0.96 \text{ V} - 1.50 \text{ V} = -0.54 \text{ V}$$

Since the $\mathscr{E}°$ value is negative, the process will *not* occur under standard conditions. That is, gold will not dissolve in 1 *M* HNO_3 to give 1 *M* Au^{3+}. In fact, a mixture (1:3 by volume) of concentrated nitric and hydrochloric acids, called *aqua regia,* is required to dissolve gold.

17.4 Dependence of Cell Potential on Concentration

Purpose

- To discuss the driving force in concentration cells.
- To quantify the relationship between cell potential and cell concentration.
- To show how to calculate equilibrium constants from cell potentials.

So far we have described cells under standard conditions. In this section we consider the dependence of the cell potential on concentration. Under standard conditions (all concentrations 1 *M*) the cell with the reaction:

$$Cu(s) + 2Ce^{4+}(aq) \rightarrow Cu^{2+}(aq) + 2Ce^{3+}(aq)$$

has a potential of 1.36 V. What will the cell potential be if $[Ce^{4+}]$ is greater than 1.0 *M*? This question can be answered qualitatively in terms of Le Châtelier's principle. An increase in the concentration of Ce^{4+} will favor the forward reaction, and thus increase the driving force on the electrons. The cell potential will increase. On the other hand, an increase in the concentration of a product (Cu^{2+} or Ce^{3+}) will oppose the forward reaction and decrease the cell potential.

These ideas are illustrated in Sample Exercise 17.5.

Sample Exercise 17.5

For the cell reaction

$$2Al(s) + 3Mn^{2+}(aq) \rightarrow 2Al^{3+}(aq) + 3Mn(s) \qquad \mathscr{E}°_{cell} = 0.48 \text{ V}$$

predict whether $\mathscr{E}_{cell}$ is larger or smaller than $\mathscr{E}°_{cell}$ for the following cases:

a. $[Al^{3+}] = 2.0$ *M*, $[Mn^{2+}] = 1.0$ *M*

b. $[Al^{3+}] = 1.0$ *M*, $[Mn^{2+}] = 3.0$ *M*

Sample Exercise 17.5, continued

Solution

a. A product concentration has been raised above 1.0 *M*. This will oppose the cell reaction and will cause $\mathscr{E}_{cell}$ to be less than $\mathscr{E}^\circ_{cell}$ ($\mathscr{E}_{cell} < 0.48$ V).

b. A reactant concentration has been increased above 1.0 *M* and $\mathscr{E}_{cell}$ will be greater than $\mathscr{E}^\circ_{cell}$ ($\mathscr{E}_{cell} > 0.48$ V).

Concentration Cells

Because cell potentials depend on concentration, we can construct galvanic cells where both compartments contain the same components but at different concentrations. For example, in the cell in Fig. 17.9, both compartments contain aqueous $AgNO_3$, but with different molarities. Let's consider the potential of this cell and the direction of electron flow. The half-reaction relevant to both compartments of this cell is

$$Ag^+ + e^- \rightarrow Ag \qquad \mathscr{E}^\circ = 0.80 \text{ V}$$

If the cell had 1 *M* Ag^+ in both compartments,

$$\mathscr{E}^\circ_{cell} = 0.80 \text{ V} - 0.80 \text{ V} = 0 \text{ V}$$

However, in the cell described here, the concentrations of Ag^+ in the two compartments are 1 *M* and 0.1 *M*. Because the concentrations of Ag^+ are unequal, the half-cell potentials will not be identical, and the cell will exhibit a positive voltage. In which direction will the electrons flow in this cell? The best way to think about this question is to recognize that nature will try to equalize the concentrations of Ag^+ in the two compartments. This can be done by transferring electrons from the compartment containing 0.1 *M* Ag^+ to the one containing 1 *M* Ag^+ (right to left in Fig. 17.9). This electron transfer will produce Ag^+ in the right compartment and consume Ag^+ (to form Ag) in the left compartment.

A cell in which both compartments have the same components, but at different concentrations, is called a **concentration cell.** The difference in concentration is the only factor that produces a cell potential in this case, and the voltages are typically small.

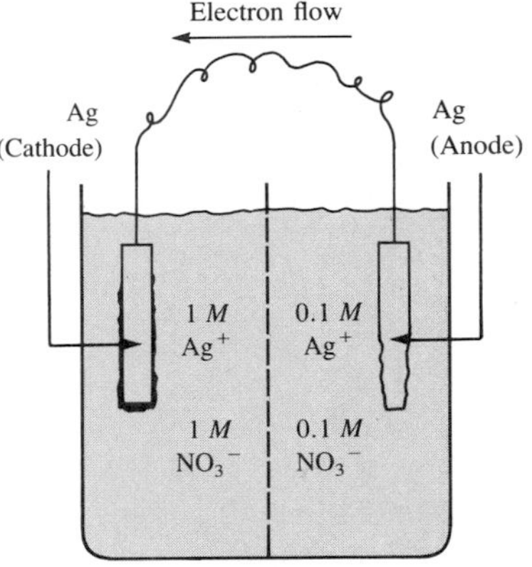

Figure 17.9

A concentration cell that contains a silver electrode and aqueous silver nitrate in both compartments. Because the left compartment contains 1 *M* Ag^+ and the right compartment contains 0.1 *M* Ag^+, there will be a driving force to transfer electrons from right to left. Silver will be deposited on the left electrode, thus lowering the concentration of Ag^+ in the left compartment. In the right compartment the silver electrode dissolves to raise the concentration of Ag^+ in solution.

Sample Exercise 17.6

Determine the direction of electron flow and designate the anode and cathode for the cell in Fig. 17.10.

Solution

The concentrations of Fe^{2+} ion in the compartments can be equalized by transferring electrons from the left compartment to the right. This will cause Fe^{2+} to be formed in the left compartment, and iron metal will be deposited on the right electrode. Since electron flow is from left to right, oxidation occurs in the left compartment (the anode) and reduction occurs in the right (the cathode).

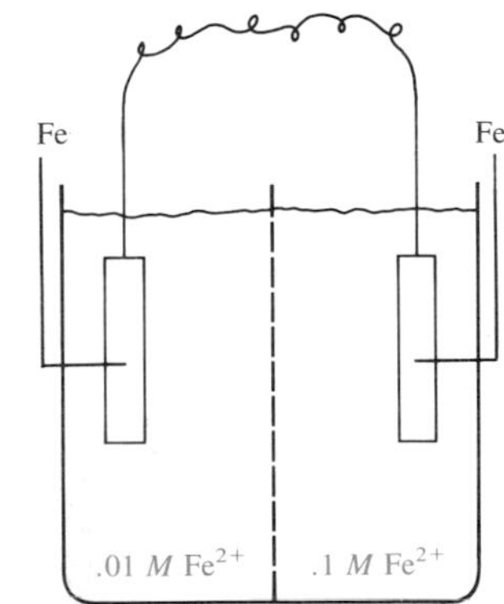

Figure 17.10

A concentration cell containing iron electrodes and different concentrations of Fe^{2+} ion in the two compartments.

The Nernst Equation

The dependence of the cell potential on concentration results directly from the dependence of free energy on concentration. Recall from Chapter 16 that the equation

$$\Delta G = \Delta G^\circ + RT\ln(Q)$$

where Q is the reaction quotient, was used to calculate the effect of concentration on ΔG. Since $\Delta G = -nF\mathscr{E}$ and $\Delta G^\circ = -nF\mathscr{E}^\circ$, the equation becomes

$$-nF\mathscr{E} = -nF\mathscr{E}^\circ + RT\ln(Q)$$

Dividing each side of the equation by $-nF$ gives

$$\mathscr{E} = \mathscr{E}^\circ - \frac{RT}{nF}\ln(Q) \qquad (17.1)$$

Nernst was one of the pioneers in the development of electrochemical theory and is generally given credit for first stating the third law of thermodynamics. He won the Nobel prize in chemistry in 1920.

Equation (17.1), which gives the relationship between the cell potential and the concentrations of the cell components, is commonly called the **Nernst equation,** after the German chemist Hermann Nernst (1864–1941).

The Nernst equation is often given in a form that is valid at 25°C:

$$\mathscr{E} = \mathscr{E}^\circ - \frac{0.0592}{n}\log(Q)$$

Using this relationship we can calculate the potential of a cell in which some or all of the components are not in their standard states.

For example, $\mathscr{E}^\circ_{\text{cell}}$ is 0.48 volt for the galvanic cell based on the reaction

$$2\text{Al}(s) + 3\text{Mn}^{2+}(aq) \longrightarrow 2\text{Al}^{3+}(aq) + 3\text{Mn}(s)$$

However, if

$$[\text{Mn}^{2+}] = 0.50\ M \quad \text{and} \quad [\text{Al}^{3+}] = 1.50\ M$$

the cell potential at 25°C for these concentrations can be calculated using the Nernst equation:

$$\mathscr{E}_{\text{cell}} = \mathscr{E}^\circ_{\text{cell}} - \frac{0.0592}{n}\log(Q)$$

We know that

$$\mathscr{E}^\circ_{\text{cell}} = 0.48\text{ V}$$

and

$$Q = \frac{[\text{Al}^{3+}]^2}{[\text{Mn}^{2+}]^3} = \frac{(1.50)^2}{(0.50)^3} = 18$$

Since the half-reactions are

$$2\text{Al} \longrightarrow 2\text{Al}^{3+} + 6\text{e}^-$$

and

$$3\text{Mn}^{2+} + 6\text{e}^- \longrightarrow 3\text{Mn}$$

we know that

$$n = 6$$

Thus

$$\mathscr{E}_{\text{cell}} = 0.48 - \frac{0.0592}{6}\log(18)$$

$$= 0.48 - \frac{0.0592}{6}(1.26) = 0.48 - 0.01 = 0.47 \text{ V}$$

Note that the cell voltage decreases slightly because of the nonstandard concentrations. This change is consistent with the predictions of Le Châtelier's principle (see Sample Exercise 17.5). In this case, since the reactant concentration is lower than 1.0 M and the product concentration is higher than 1.0 M, $\mathscr{E}_{cell}$ is less than $\mathscr{E}^{\circ}_{cell}$.

The potential calculated from the Nernst equation is the maximum potential before any current flow has occurred. As the cell discharges and current flows from anode to cathode, the concentrations will change, and as a result, $\mathscr{E}_{cell}$ will change. In fact, *the cell will spontaneously discharge until it reaches equilibrium,* at which point

$$Q = K \text{ (the equilibrium constant)} \quad \text{and} \quad \mathscr{E}_{cell} = 0$$

A "dead" battery is one in which the cell reaction has reached equilibrium, and there is no longer any chemical driving force to push electrons through the wire. In other words, *at equilibrium the components in the two cell compartments have the same free energy,* and $\Delta G = 0$ for the cell reaction at the equilibrium concentrations. The cell no longer has the ability to do work.

Sample Exercise 17.7

Describe the cell based on the following half-reactions:

$$VO_2^+ + 2H^+ + e^- \rightarrow VO^{2+} + H_2O \qquad \mathscr{E}^{\circ} = 1.00 \text{ V} \qquad (1)$$

$$Zn^{2+} + 2e^- \rightarrow Zn \qquad \mathscr{E}^{\circ} = -0.76 \text{ V} \qquad (2)$$

where

$T = 25°C$ $\qquad [VO^{2+}] = 1.0 \times 10^{-2}\ M$

$[VO_2^+] = 2.0\ M$ $\qquad [Zn^{2+}] = 1.0 \times 10^{-1}\ M$

$[H^+] = 0.50\ M$

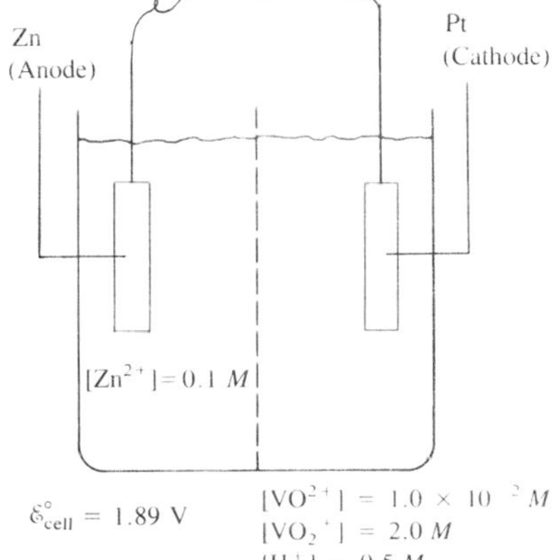

Figure 17.11

Schematic diagram of the cell described in Sample Exercise 17.7.

Solution

The balanced cell reaction is obtained by reversing reaction (2) and multiplying reaction (1) by 2:

2 × reaction (1)	$2VO_2^+ + 4H^+ + 2e^- \rightarrow 2VO^{2+} + 2H_2O$	$\mathscr{E}^{\circ} = 1.00$ V
Reaction (2) reversed	$Zn \rightarrow Zn^{2+} + 2e^-$	$-\mathscr{E}^{\circ} = 0.76$ V
Cell reaction:	$2VO_2^+(aq) + 4H^+(aq) + Zn(s) \rightarrow 2VO^{2+}(aq) + 2H_2O(l) + Zn^{2+}(aq)$	$\mathscr{E}^{\circ}_{cell} = 1.76$ V

Since the cell contains components at concentrations other than 1 M, we must use the Nernst equation, where $n = 2$ (since two electrons are transferred), to calculate the cell potential. At 25°C we can use the equation:

$$\mathscr{E} = \mathscr{E}^{\circ}_{cell} - \frac{0.0592}{n}\log(Q)$$

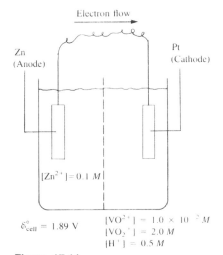

$$= 1.76 - \frac{0.0592}{2}\log\left(\frac{[Zn^{2+}][VO^{2+}]^2}{[VO_2^+]^2[H^+]^4}\right)$$

Sample Exercise 17.7, continued

$$= 1.76 - \frac{0.0592}{2}\log\left(\frac{(1.0 \times 10^{-1})(1.0 \times 10^{-2})^2}{(2.0)^2(0.50)^4}\right)$$

$$= 1.76 - \frac{0.0592}{2}\log(4 \times 10^{-5}) = 1.76 + 0.13 = 1.89 \text{ V}$$

The cell diagram is given in Fig. 17.11.

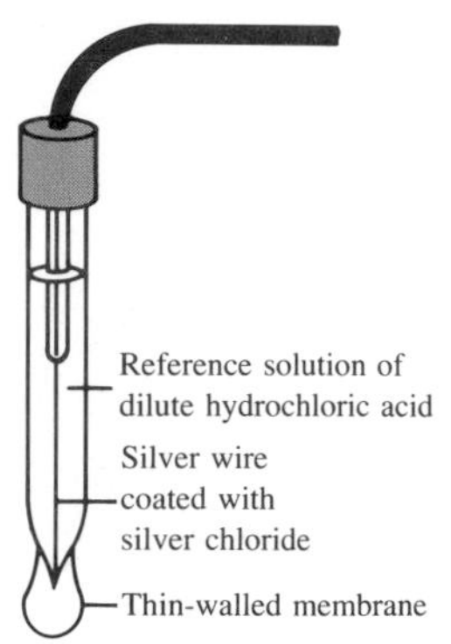

Figure 17.12

A glass electrode contains a reference solution of dilute hydrochloric acid in contact with a thin glass membrane in which a silver wire coated with silver chloride has been embedded. When the electrode is dipped into a solution containing H^+ ions, the electrode potential is determined by the difference in $[H^+]$ between the two solutions.

Ion-Selective Electrodes

Because the cell potential is sensitive to the concentrations of the reactants and products involved in the cell reaction, measured potentials can be used to determine the concentration of an ion. A pH meter (see Fig. 14.4) is a familiar example of an instrument that measures concentration using an observed potential. The pH meter has three main components: a standard electrode of known potential, a special **glass electrode** that changes potential depending on the concentration of H^+ ion in the solution into which it is dipped, and a potentiometer that measures the potential between the two electrodes. The potentiometer reading is automatically converted electronically to a direct reading of the pH of the solution being tested.

The glass electrode (see Fig. 17.12) contains a reference solution of dilute hydrochloric acid in contact with a thin glass membrane. The electrical potential of the glass electrode depends on the difference in $[H^+]$ between the reference solution and the solution into which the electrode is dipped. Thus the electrical potential varies with the pH of the solution being tested.

Electrodes that are sensitive to the concentration of a particular ion are called **ion-selective electrodes,** of which the glass electrode for pH measurement is just one example. Glass electrodes can be made sensitive to ions such as Na^+, K^+, or NH_4^+ by changing the composition of the glass. Other ions can be detected if an appropriate crystal replaces the glass membrane. For example, a crystal of lanthanum(III) fluoride (LaF_3) can be used in an electrode to measure $[F^-]$. Solid silver sulfide (Ag_2S) can be used to measure $[Ag^+]$ and $[S^{2-}]$. Some of the ions that can be detected by ion-selective electrodes are listed in Table 17.2.

Some Ions Whose Concentrations Can Be Detected by Ion-Selective Electrodes

Cations	Anions
H^+	Br^-
Cd^{2+}	Cl^-
Ca^{2+}	CN^-
Cu^{2+}	F^-
K^+	NO_3^-
Ag^+	S^{2-}
Na^+	

Table 17.2

Calculation of Equilibrium Constants for Redox Reactions

The quantitative relationship between $\mathscr{E}°$ and $\Delta G°$ allows calculation of equilibrium constants for redox reactions. For a cell at equilibrium,

$$\mathscr{E}_{\text{cell}} = 0 \quad \text{and} \quad Q = K$$

Applying these conditions to the Nernst equation valid at 25°C,

$$\mathscr{E} = \mathscr{E}° - \frac{0.0592}{n}\log(Q)$$

gives

$$0 = \mathscr{E}° - \frac{0.0592}{n}\log(K)$$

or

$$\log(K) = \frac{n\mathscr{E}°}{0.0592} \quad \text{at 25°C}$$

Sample Exercise 17.8

For the oxidation-reduction reaction

$$S_4O_6^{2-}(aq) + Cr^{2+}(aq) \rightarrow Cr^{3+}(aq) + S_2O_3^{2-}(aq)$$

the appropriate half-reactions are

$$S_4O_6^{2-} + 2e^- \rightarrow 2S_2O_3^{2-} \qquad \mathscr{E}^\circ = 0.17 \text{ V} \qquad (1)$$

$$Cr^{3+} + e^- \rightarrow Cr^{2+} \qquad \mathscr{E}^\circ = -0.50 \text{ V} \qquad (2)$$

Balance the redox reaction and calculate $\mathscr{E}^\circ$ and K (at 25°C).

Solution

To obtain the balanced reaction, reaction (2) must be reversed and multiplied by 2:

2 × reaction (2) reversed

$$2(Cr^{2+} \rightarrow Cr^{3+} + e^-) \qquad -\mathscr{E}^\circ = -(-0.50) \text{ V}$$

Reaction (1)

$$S_4O_6^{2-} + 2e^- \rightarrow 2S_2O_3^{2-} \qquad \mathscr{E}^\circ = 0.17 \text{ V}$$

Cell reaction:

$$2Cr^{2+}(aq) + S_4O_6^{2-}(aq) \rightarrow 2Cr^{3+}(aq) + 2S_2O_3^{2-}(aq) \qquad \mathscr{E}^\circ = 0.67 \text{ V}$$

In this reaction 2 mol of electrons are transferred for every unit of reaction, that is, for every 2 mol of Cr^{2+} reacting with 1 mol of $S_4O_6^{2-}$ to form 2 mol of Cr^{3+} and 2 mol of $S_2O_3^{2-}$. Thus $n = 2$. Then

$$\log(K) = \frac{n\mathscr{E}^\circ}{0.0592} = \frac{2(0.67)}{0.0592} = 22.6$$

The value of K is found by taking the antilog of 22.6:

$$K = 10^{22.6} = 4 \times 10^{22}$$

This very large equilibrium constant is not unusual for a redox reaction.

The blue solution on the left contains Cr^{2+} ions and the green solution contains Cr^{3+} ions.

17.5 Batteries

Purpose

- To discuss the composition and operation of commonly used batteries.

A **battery** is a galvanic cell or, more commonly, *a group of galvanic cells connected in series,* where the potentials of the individual cells add to give the total battery potential. Batteries are a source of direct current and have become an essential source of portable power in our society. In this section we examine the most common types of batteries. Some new batteries currently being developed are described at the end of the chapter.

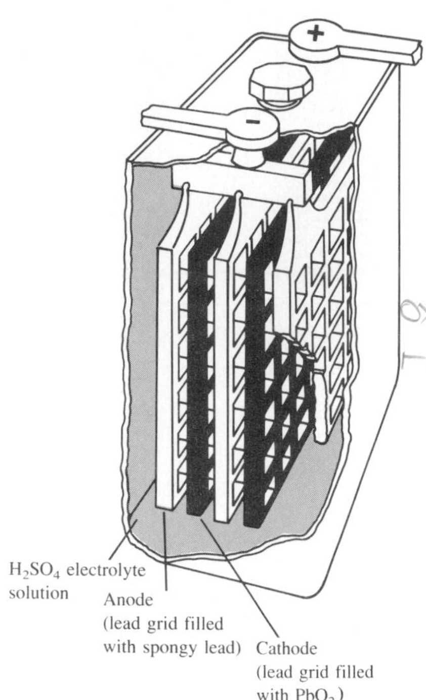

Figure 17.13

One of the six cells in a lead storage battery. The anode consists of a lead grid filled with spongy lead, and the cathode is a lead grid filled with lead dioxide. The cell also contains 38% (by mass) sulfuric acid.

Lead Storage Battery

Since about 1915 when self-starters were first used in automobiles, the **lead storage battery** has been a major factor in making the automobile a practical means of transportation. This type of battery can function for several years under temperature extremes from −30°F to 100°F, and under incessant punishment from rough roads.

In this battery, lead serves as the anode and lead coated with lead dioxide serves as the cathode. The electrodes dip into an electrolyte solution of sulfuric acid. The electrode reactions are

Anode reaction: $$Pb + HSO_4^- \rightarrow PbSO_4 + H^+ + 2e^-$$

Cathode reaction: $$PbO_2 + HSO_4^- + 3H^+ + 2e^- \rightarrow PbSO_4 + 2H_2O$$

Cell reaction: $$Pb(s) + PbO_2(s) + 2H^+(aq) + 2HSO_4^-(aq) \rightarrow 2PbSO_4(s) + 2H_2O(l)$$

The automobile lead storage battery has six cells connected in series. Each cell contains multiple electrodes in the form of grids (Fig. 17.13) and produces approximately 2 volts, to give a total battery potential of about 12 volts. Note from the cell reaction that sulfuric acid is consumed as the battery discharges. This lowers the density of the electrolyte solution from its initial value of about 1.28 g/cm^3 in the fully charged battery. As a result, the condition of the battery can be monitored by using a hydrometer (Fig. 1.11) to measure the density of the sulfuric acid solution. The solid lead sulfate formed in the cell reaction during discharge adheres to the grid surfaces of the electrodes. The battery is recharged by forcing current through the battery in the opposite direction to reverse the cell reaction. A car's battery is continuously charged by an alternator driven by the automobile engine.

An automobile with a dead battery can be ''jump-started'' by connecting its battery to the battery in a running automobile. This process can be dangerous, however, because the resulting flow of current causes electrolysis of water in the dead battery, producing hydrogen and oxygen gases (see Section 17.7 for details). Disconnecting the jumper cables after the disabled car starts causes an arc that can ignite the gaseous mixture. If this happens, the battery may explode, ejecting corrosive sulfuric acid. This problem can be avoided by connecting the ground jumper cable to a part of the engine remote from the battery. Any arc produced when this cable is disconnected will then be harmless.

Using jumper cables to start a car with a dead battery. Notice the ground cable on the assisting car is attached at a point away from the battery to avoid a possible explosion.

Traditional types of storage batteries require periodic ''topping off,'' because the water in the electrolyte solution is depleted by the electrolysis that accompanies the charging process. Recent types of batteries have electrodes made of an alloy of calcium and lead that inhibits the electrolysis of water. These batteries can be sealed since they require no addition of water.

It is rather amazing that in the 75 years lead storage batteries have been used, no better system has been found. Although a lead storage battery does provide excellent service, it has a useful lifetime of 3 to 5 years in an automobile. While it might seem that the battery could undergo an indefinite number of discharge/charge cycles, physical damage from road shock and chemical side-reactions eventually cause it to fail.

Dry Cell Batteries

The calculators, electronic watches, portable radios, and tape players that are so familiar to us are all powered by small, efficient, **dry cell batteries.** The common dry cell battery was invented more than 100 years ago by George Leclanché (1839–

1882), a French chemist. In its *acid version,* the dry cell battery contains a zinc inner case that acts as the anode and a carbon rod in contact with a moist paste of solid MnO_2, solid NH_4Cl, and carbon that acts as the cathode (Fig. 17.14). The half-cell reactions are complex but can be approximated as follows:

Anode reaction: $Zn \rightarrow Zn^{2+} + 2e^-$

Cathode reaction: $2NH_4^+ + 2MnO_2 + 2e^- \rightarrow Mn_2O_3 + 2NH_3 + H_2O$

This cell produces a potential of about 1.5 volts.

In the *alkaline version* of the dry cell battery, the solid NH_4Cl is replaced with KOH or NaOH. In this case the half-reactions can be approximated as follows:

Anode reaction: $Zn + 2OH^- \rightarrow ZnO + H_2O + 2e^-$

Cathode reaction: $2MnO_2 + H_2O + 2e^- \rightarrow Mn_2O_3 + 2OH^-$

The alkaline dry cell lasts longer mainly because the zinc anode corrodes less rapidly under basic conditions than under acidic conditions.

Other types of dry cell batteries include the *silver cell,* which has a Zn anode and a cathode that employs Ag_2O as the oxidizing agent in a basic environment. *Mercury cells,* often used in calculators, have a Zn anode and a cathode involving HgO as the oxidizing agent in a basic medium (see Fig. 17.15).

An especially important type of dry cell is the *nickel-cadmium battery* in which the electrode reactions are

Anode reaction: $Cd + 2OH^- \rightarrow Cd(OH)_2 + 2e^-$

Cathode reaction: $NiO_2 + 2H_2O + 2e^- \rightarrow Ni(OH)_2 + 2OH^-$

As in the lead storage battery, the products adhere to the electrodes. Therefore, a nickel-cadmium battery can be recharged an indefinite number of times.

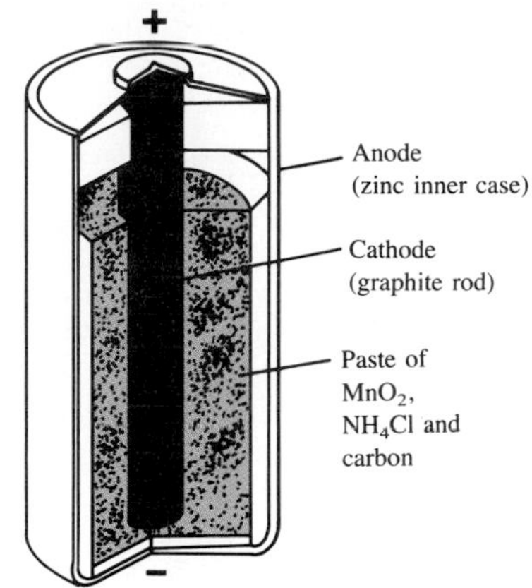

Figure 17.14

A common dry cell battery.

Batteries for electronic watches are, by necessity, very tiny.

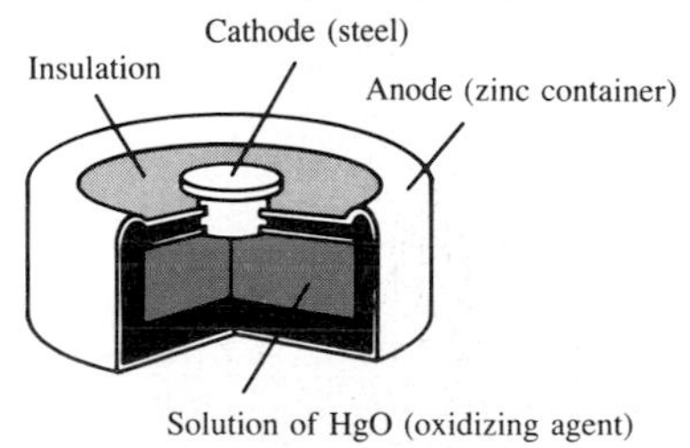

Figure 17.15

A mercury battery of the type used in small calculators.

Fuel Cells

A **fuel cell** is *a galvanic cell for which the reactants are continuously supplied.* To illustrate the principles of fuel cells, let's consider the exothermic redox reaction of methane with oxygen:

$$CH_4(g) + 2O_2(g) \rightarrow CO_2(g) + 2H_2O(g) + \text{energy}$$

Usually the energy from this reaction is released as heat to warm homes and to run machines. In a fuel cell designed to use this reaction, the energy is used to produce an electric current. The electrons will flow from the reducing agent (CH_4) to the oxidizing agent (O_2) through a conductor.

The U.S. space program has supported extensive research to develop fuel cells. The Apollo missions used a fuel cell based on the reaction of hydrogen and oxygen to form water:

$$2H_2(g) + O_2(g) \rightarrow 2H_2O(l)$$

A schematic of a fuel cell that employs this reaction is shown in Fig. 17.16 on the following page. The half-cell reactions are

Anode reaction: $2H_2 + 4OH^- \rightarrow 4H_2O + 4e^-$

Cathode reaction: $4e^- + O_2 + 2H_2O \rightarrow 4OH^-$

A cell of this type weighing about 500 pounds has been designed for space vehicles, but neither this nor any other type of fuel cell is practical enough for general use as

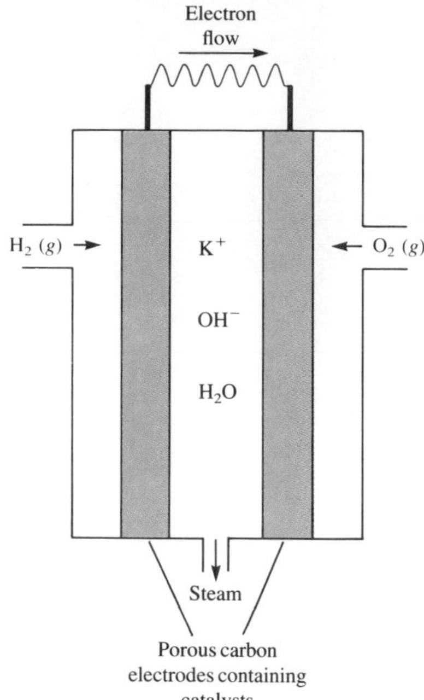

Figure 17.16

Schematic of the hydrogen-oxygen fuel cell.

a source of portable power. Current research on portable electrochemical power sources seems mainly focused on rechargeable storage batteries with high power-to-weight ratios.

Fuel cells are finding some use, however, as permanent power sources. For example, a power plant built in New York City contains stacks of hydrogen-oxygen fuel cells, which can be rapidly put on-line in response to fluctuating power demands. The hydrogen gas is obtained by decomposing the methane in natural gas. A plant of this type is also currently operating in Tokyo.

17.6 Corrosion

Purpose

- To explain the electrochemical nature of corrosion and describe some means for preventing it.

Corrosion can be viewed as the process of returning metals to their natural state—the ores from which they were originally obtained. Corrosion involves oxidation of the metal. Since corroded metal often loses its structural integrity and attractiveness, this spontaneous process has great economic impact. For example, approximately one-fifth of the iron and steel produced annually is used to replace rusted metal.

Some metals, such as copper, gold, silver and platinum, are relatively difficult to oxidize. These are often called noble metals.

Metals corrode because they oxidize easily. Table 17.1 shows that, with the exception of gold, those metals commonly used for structural and decorative purposes all have standard reduction potentials less positive than that of oxygen gas. When any of these half-reactions is reversed (to show oxidation of the metal) and combined with the reduction half-reaction for oxygen, the result is a positive $\mathscr{E}°$ value. Thus the oxidation of most metals by oxygen is spontaneous (although we cannot tell from the potential how fast it will occur).

In view of the large difference in reduction potential between oxygen and most metals, it is surprising that the problem of corrosion does not virtually prevent the use of metals in air. However, most metals develop a thin oxide coating, which tends to protect their internal atoms against further oxidation. The metal that best demonstrates this phenomenon is aluminum. With a reduction potential of -1.7 volts, aluminum should be easily oxidized by O_2. According to the apparent thermodynamics of the reaction, an aluminum airplane could dissolve in a rainstorm. The fact that this very active metal can be used as a structural material is due to the formation of a thin, adherent layer of aluminum oxide, Al_2O_3, more properly represented as $Al_2(OH)_6$, which greatly inhibits further corrosion. The potential of the "passive," oxide-coated aluminum is -0.6 volt, a value that causes it to behave much like a noble metal.

Iron can also form a protective oxide coating. This is not an infallible shield against corrosion, however; when steel is exposed to oxygen in moist air, the oxide that forms tends to scale off and expose new metal surfaces to corrosion.

The corrosion products of noble metals such as copper and silver are complex and affect the use of these metals as decorative materials. Under normal atmospheric conditions, copper forms an external layer of greenish copper carbonate called *patina*. *Silver tarnish* is silver sulfide (Ag_2S), which in thin layers gives the silver surface a richer appearance. Gold, with a positive standard reduction potential of 1.50 volts, significantly larger than that for oxygen (1.23 volts), shows no appreciable corrosion in air.

Corrosion of Iron

Since steel is the main structural material for bridges, buildings, and automobiles, controlling its corrosion is extremely important. To do this we must understand the corrosion mechanism. Instead of being a direct oxidation process as we might expect, the corrosion of iron is an electrochemical reaction, as shown in Fig. 17.17.

Steel has a nonuniform surface because the chemical composition is not completely homogeneous. Also, physical strains leave stress points in the metal. These nonuniformities cause areas where the iron is more easily oxidized *(anodic regions)* than it is at others *(cathodic regions)*. In the anodic regions each iron atom gives up two electrons to form the Fe^{2+} ion:

$$Fe \rightarrow Fe^{2+} + 2e^-$$

The electrons that are released flow through the steel, as they do through the wire of a galvanic cell, to a cathodic region where they react with oxygen:

$$O_2 + 2H_2O + 4e^- \rightarrow 4OH^-$$

The Fe^{2+} ions formed in the anodic regions travel to the cathodic regions through the moisture on the surface of the steel, just as ions travel through a salt bridge in a galvanic cell. In the cathodic regions Fe^{2+} ions react with oxygen to form rust, which is hydrated iron(III) oxide of variable composition:

$$4Fe^{2+}(aq) + O_2(g) + (4 + 2n)H_2O(l) \rightarrow \underbrace{2Fe_2O_3 \cdot nH_2O(s)}_{\text{Rust}} + 8H^+(aq)$$

A rusty shipwreck.

Because of the migration of ions and electrons, rust often forms at sites that are remote from those where the iron dissolved to form pits in the steel. The degree of hydration of the iron oxide affects the color of the rust, which may vary from black to yellow to the familiar reddish-brown.

The electrochemical nature of the rusting of iron explains the importance of moisture in the corrosion process. Moisture must be present to act as a kind of salt bridge between anodic and cathodic regions. Steel does not rust in dry air, a fact that explains why cars last much longer in the arid Southwest than in the relatively humid Midwest. Salt also accelerates rusting, a fact all too easily recognized by car owners in the colder parts of the United States, where salt is used on roads to melt snow and ice. The severity of rusting is greatly increased because the dissolved salt

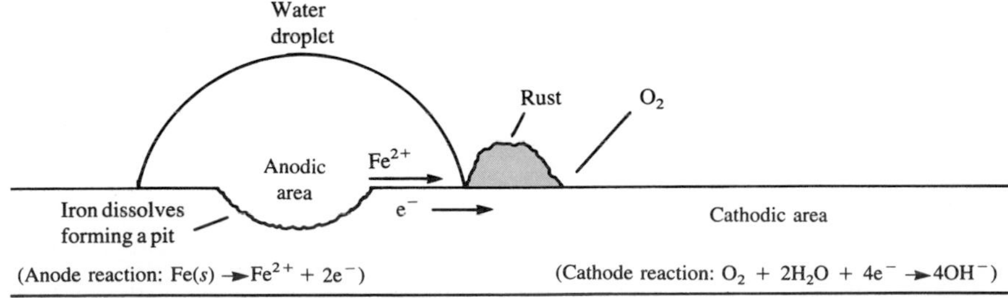

Figure 17.17

The electrochemical corrosion of iron.

Chemical Impact

Refurbishing the Lady

The restoration of the Statue of Liberty in New York harbor represents a fascinating blend of science, technology, and art. The statue consists of copper sheets attached to a framework of iron, which had become so weakened by corrosion during its 100 years of exposure to the elements that it was in danger of collapsing.

Gustave Eiffel, who designed the ingenious support structure, knew from experience that if the copper touched the iron framework, the more active iron would corrode very rapidly. Why does this happen? It is apparent from the reduction potentials

$$Cu^{2+} + 2e^- \rightarrow Cu \qquad \mathscr{E}° = 0.34\ V$$

$$Fe^{2+} + 2e^- \rightarrow Fe \qquad \mathscr{E}° = -0.44\ V$$

that Cu^{2+} will spontaneously oxidize iron. However, as with many electrochemical processes, the situation is more complex than it first appears. Since we are talking about two metal strips touching each other, the question is: Where do the Cu^{2+} ions (the oxidizing agents) come from? In fact, research on this process suggests that the copper simply acts as a conductor for electrons, and that the oxidizing agent is not Cu^{2+} at all but probably O_2 or oxides of N or S.

Whatever the mechanism, Eiffel attempted to combat the corrosion problem by inserting asbestos pads between the copper sheets and the frame. However, this idea did not work, probably because copper is such a good conductor that *any* contact between the two metals anywhere on the statue totally thwarted the effect of the insulation. In fact, the iron framework was found to be so corroded it had to be completely replaced with stainless steel, which is much more corrosion-resistant.

Stainless steel has its own problems, however. In being bent to achieve the intricate shapes needed, the iron bars often became brittle. Flexibility was restored by heating each bar to a very high temperature using a current of 30,000 amperes and then cooling the bar suddenly. (See Section 24.4, for more on the heat treatment of steel.) Unfortunately, this process also destroyed the resistance to corrosion of the stainless steel. That resistance had to be restored by soaking the bars in nitric acid, an oxidizing acid that reforms the protective oxide coating removed by heating.

Another problem faced by the restorers was the removal of layers of coal tar and paint, which had been applied in vain attempts to protect the statue's interior. Although the iron could be cleaned by blasting with aluminum oxide powder, the more fragile copper sheets required a gentler treatment. The restorers discovered that liquid nitrogen (77 K) cracked the paint and caused it to peel away. The coal-tar layer under the paint was removed by blasting with baking soda, $NaHCO_3$, a substance sometimes used by museum curators for polishing dinosaur bones. While this worked very well, it created another chemical problem: where the baking soda seeped between the seams in the copper plates onto the exterior, the statue turned from green to light blue when it rained. After this phenomenon was observed for the first time, workers stationed on the exterior scaffolding immediately cleaned off any leaking $NaHCO_3$.

The Statue of Liberty in New York harbor.

One of the most interesting aspects of the chemistry of the Statue of Liberty is the green patina on its surface. Copper metal exposed to the atmosphere changes from its bright reddish-brown metallic luster, first to an almost black color, then to the familiar green patina. The initial blackening is mainly due to formation of copper oxide and copper sul-

Chemical Impact

fide. The green patina that then forms consists of thin layers of two types of basic copper sulfates: brochantite, $CuSO_4 \cdot 3Cu(OH)_2$, and antlerite, $CuSO_4 \cdot 2Cu(OH)_2$. Crystals of these compounds seem to be cemented onto the copper surface by organic molecules from the air.

In recent years, large areas of the statue's left side have been observed to darken as if the patina were being removed. Scientists speculated that this may be the result of acid rain, that converts brochantite to the more soluble antlerite, which is then washed off by rainwater.

To make any new copper sheets look like the old ones, the restorers transplanted the patina from a weathered piece of copper to the new surface. Applying acetone, an organic solvent, to the weathered surface and scraping with an abrasive cloth caused tiny flakes of patina to fall off. These flakes, applied to the new copper, attached themselves permanently in one to three weeks of exposure to the atmosphere.

The restoration of the Statue of Liberty made use of the latest advances in chemistry as well as facts known to most general chemistry students. It's an example of the fascinating and varied problems faced by chemists as they pursue their profession.

Suggested Reading

Ivars Peterson, "Lessons Learned from a Lady," *Science News* **130** (1986): 392.

on the moist steel surface increases the conductivity of the aqueous solution formed there, and thus accelerates the electrochemical corrosion process. Chloride ions also form very stable complex ions with Fe^{3+}, and this factor tends to encourage the dissolving of the iron, again accelerating the corrosion.

Prevention of Corrosion

Prevention of corrosion is an important way of conserving our natural resources of energy and metals. The primary means of protection is the application of a coating, most commonly paint or metal plating, to protect the metal from oxygen and moisture. Chromium and tin are often used to plate steel (Section 17.8) because they oxidize to form a durable, effective oxide coating. Zinc, also used to coat steel in a process called **galvanizing,** does not form an oxide coating. However, since it is a more active metal than iron, as the potentials for the oxidation half-reactions show,

$$Fe \rightarrow Fe^{2+} + 2e^- \qquad -\mathscr{E}° = 0.44 \text{ V}$$

$$Zn \rightarrow Zn^{2+} + 2e^- \qquad -\mathscr{E}° = 0.76 \text{ V}$$

any oxidation that occurs dissolves zinc rather than iron. Recall that the reaction with the most positive standard potential has the greatest thermodynamic tendency to occur. Thus zinc acts as a "sacrificial" coating on steel.

Alloying is also used to prevent corrosion. *Stainless steel* contains chromium and nickel, both of which form oxide coatings that change steel's reduction potential to one characteristic of the noble metals.

Cathodic protection is a method most often employed to protect steel in buried fuel tanks and pipelines. An active metal, such as magnesium, is connected by a wire to the pipeline or tank to be protected (Fig. 17.18). Because the magnesium is a better reducing agent than iron, electrons are furnished by the magnesium rather than by the iron, keeping the iron from being oxidized. As oxidation occurs, the magnesium anode dissolves, and so it must be replaced periodically.

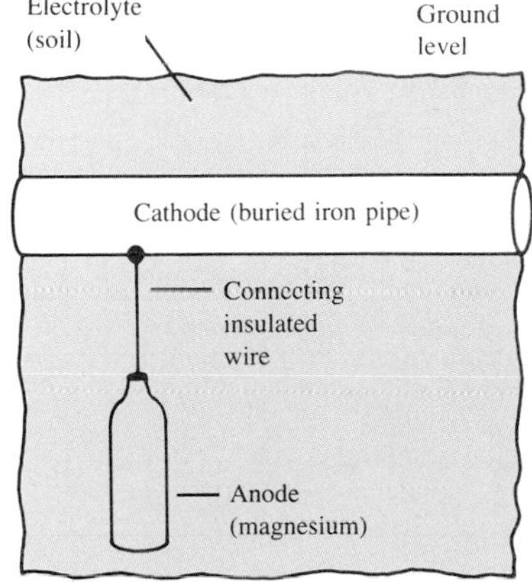

Figure 17.18
Cathodic protection of an underground pipe.

17.7 Electrolysis

Purpose

- To describe the stoichiometry of electrolysis reactions.
- To show how to predict the order of electrolysis of the components of a mixture.

An electrolytic cell uses electrical energy to produce a chemical change that would otherwise not occur spontaneously.

A galvanic cell produces current when an oxidation-reduction reaction proceeds spontaneously. A similar apparatus, an **electrolytic cell,** uses electrical energy to produce chemical change. The process of **electrolysis** involves *forcing a current through a cell to produce a chemical change for which the cell potential is negative;* that is, electrical work causes an otherwise nonspontaneous chemical reaction to occur. Electrolysis has great practical importance; for example, charging a battery, producing aluminum metal, and chrome plating an object are all done electrolytically.

To illustrate the difference between a galvanic cell and an electrolytic cell, consider the cell shown in Fig. 17.19(a) as it runs spontaneously to produce 1.10 V. In this *galvanic* cell the reaction at the anode is

$$Zn \rightarrow Zn^{2+} + 2e^-$$

while at the cathode the reaction is

$$Cu^{2+} + 2e^- \rightarrow Cu$$

Figure 17.19(b) shows an external power source forcing electrons through the cell in the *opposite* direction to that in (a). This requires an external potential greater than 1.10 V, which must be applied in opposition to the natural cell potential. This is an *electrolytic cell.* Notice that since electron flow is opposite in the two cases, the anode and cathode are reversed between (a) and (b). Also ion flow through the salt bridge is opposite in the two cells.

Figure 17.19

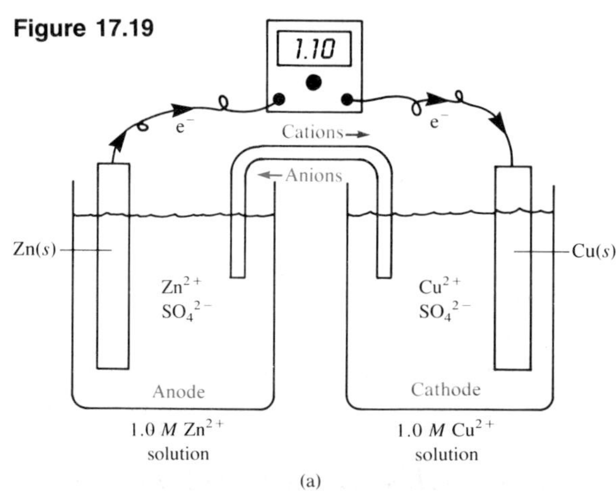

(a) A standard galvanic cell based on the spontaneous reaction

$$Zn + Cu^{2+} \rightarrow Zn^{2+} + Cu$$

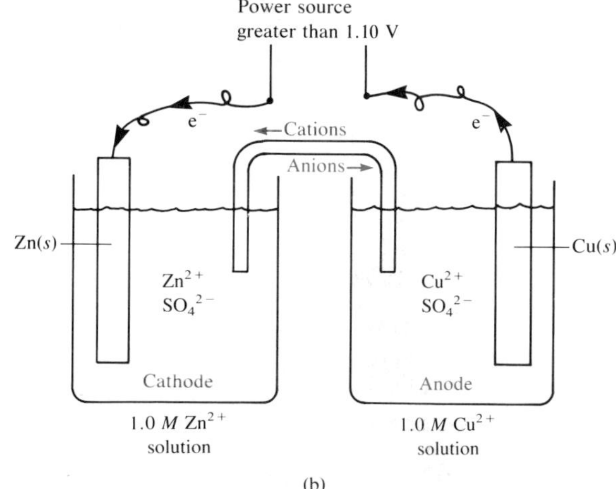

(b) A standard electrolytic cell. A power source forces the reaction $Cu + Zn^{2+} \rightarrow Cu^{2+} + Zn$ to occur.

Now we will consider the stoichiometry of electrolytic processes, that is, *how much chemical change occurs with the flow of a given current for a specified time*. Suppose we wish to determine the mass of copper that is plated out when a current of 10.0 amps (an **ampere** (amp), abbreviated A, is *1 coulomb of charge per second*) is passed for 30.0 minutes through a solution containing Cu^{2+}. *Plating* means depositing the neutral metal on the electrode by reducing the metal ions in solution. In this case each Cu^{2+} ion requires two electrons to become an atom of copper metal:

1 A = 1 C/s

$$Cu^{2+}(aq) + 2e^- \rightarrow Cu(s)$$

This reduction process will occur at the cathode of the electrolytic cell.

To solve this stoichiometry problem, we need the following steps:

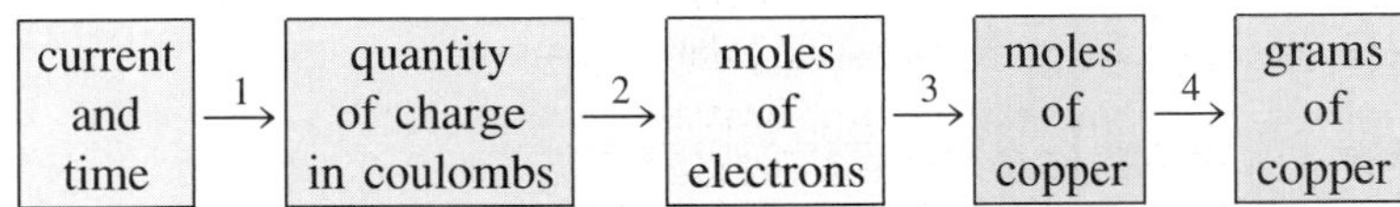

STEP 1

Since an amp is a coulomb of charge per second, we multiply the current by the time in seconds to obtain the total coulombs of charge passed into the Cu^{2+} solution at the cathode:

$$\text{Coulombs of charge} = \text{amps} \times \text{seconds} = \frac{C}{s} \times s$$

$$= 10.0\frac{C}{s} \times 30.0 \text{ min} \times 60.0\frac{s}{\text{min}}$$

$$= 1.80 \times 10^4 \text{ C}$$

STEP 2

Since 1 mole of electrons carries a charge of 1 faraday, or 96,485 coulombs, we can calculate the number of moles of electrons required to carry 1.80×10^4 coulombs of charge:

$$1.80 \times 10^4 \text{ C} \times \frac{1 \text{ mol e}^-}{96{,}485 \text{ C}} = 1.87 \times 10^{-1} \text{ mol e}^-$$

This means that 0.187 mole of electrons flowed into the Cu^{2+} solution.

STEP 3

Each Cu^{2+} ion requires two electrons to become a copper atom. Thus each mole of electrons produces $\frac{1}{2}$ mole of copper metal:

$$1.87 \times 10^{-1} \text{ mol e}^- \times \frac{1 \text{ mol Cu}}{2 \text{ mol e}^-} = 9.35 \times 10^{-2} \text{ mol Cu}$$

STEP 4

We now know the moles of copper metal plated onto the cathode, and we can calculate the mass of copper formed:

$$9.35 \times 10^{-2} \text{ mol Cu} \times \frac{63.546 \text{ g}}{\text{mol Cu}} = 5.94 \text{ g Cu}$$

Sample Exercise 17.9

Sample Exercise 17.9 describes only the half cell of interest. There must also be an anode at which oxidation is occurring.

How long must a current of 5.00 A be applied to a solution of Ag^+ to produce 10.5 g of silver metal?

Solution

In this case, we must use the steps given earlier in reverse:

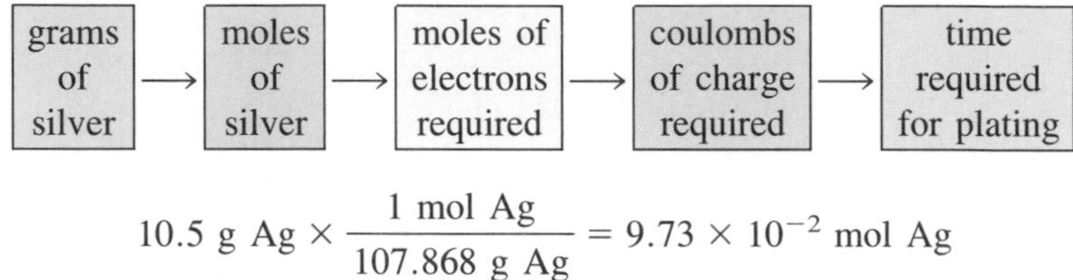

$$10.5 \text{ g Ag} \times \frac{1 \text{ mol Ag}}{107.868 \text{ g Ag}} = 9.73 \times 10^{-2} \text{ mol Ag}$$

Each Ag^+ ion requires one electron to become a silver atom:

$$Ag^+ + e^- \rightarrow Ag$$

Thus 9.73×10^{-2} mol of electrons is required, and we can calculate the quantity of charge carried by these electrons:

$$9.73 \times 10^{-2} \text{ mol e}^- \times \frac{96{,}485 \text{ C}}{\text{mol e}^-} = 9.39 \times 10^3 \text{ C}$$

The 5.00 A (5.00 C/s) current must produce 9.39×10^3 C of charge. Thus

$$\left(5.00\frac{\text{C}}{\text{s}}\right) \times (\text{time, in s}) = 9.39 \times 10^3 \text{ C}$$

$$\text{Time} = \frac{9.39 \times 10^3}{5.00}\text{s} = 1.88 \times 10^3 \text{ s} = 31.3 \text{ min}$$

Electrolysis of Water

We have seen that hydrogen and oxygen combine spontaneously to form water and that the accompanying decrease in free energy can be used to run a fuel cell to produce electricity. The reverse process, which is of course nonspontaneous, can be forced by electrolysis:

Anode reaction:	$2H_2O \rightarrow O_2 + 4H^+ + 4e^-$	$-\mathscr{E}° = -1.23$ V
Cathode reaction:	$4H_2O + 4e^- \rightarrow 2H_2 + 4OH^-$	$\mathscr{E}° = -0.83$ V
Net reaction:	$6H_2O \rightarrow 2H_2 + O_2 + \underbrace{4(H^+ + OH^-)}_{4H_2O}$	$\mathscr{E}° = -2.06$ V
or	$2H_2O \rightarrow 2H_2 + O_2$	

Note that these potentials assume an anode chamber with 1 *M* H^+ and a cathode chamber with 1 *M* OH^-. In pure water, where $[H^+] = [OH^-] = 10^{-7}$ *M*, the potential for the overall process is -1.23 V.

In practice, however, if platinum electrodes connected to a 6-volt battery are dipped into pure water, no reaction is observed because pure water contains so few ions that only a negligible current can flow. However, addition of even a small amount of a soluble salt causes an immediate evolution of bubbles of hydrogen and oxygen, as illustrated in Figure 17.20. A schematic of a complete electrolysis apparatus is given in Fig. 1.12.

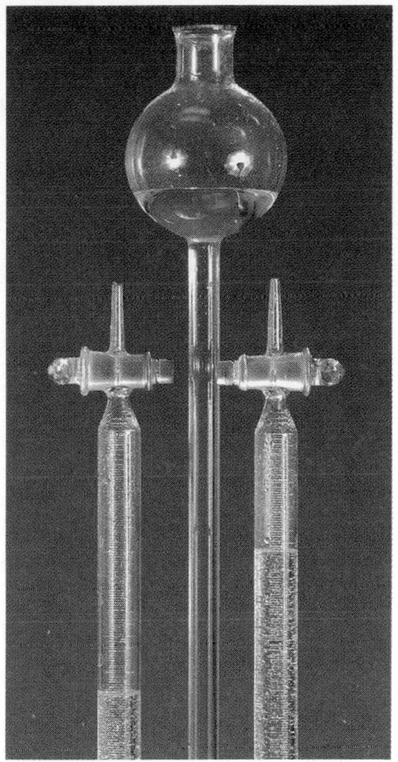

Figure 17.20

The electrolysis of water produces hydrogen gas at the cathode (on the left) and oxygen gas at the anode (on the right). Note that some provision (a salt bridge or porous disk) must be made to allow ion flow.

Electrolysis of Mixtures of Ions

Suppose a solution in an electrolytic cell contains the ions Cu^{2+}, Ag^+, and Zn^{2+}. If the voltage is initially very low and is gradually turned up, in which order will the metals be plated out onto the cathode? This question can be answered by looking at the standard reduction potentials of these ions:

$$Ag^+ + e^- \rightarrow Ag \qquad \mathscr{E}^\circ = 0.80\ V$$
$$Cu^{2+} + 2e^- \rightarrow Cu \qquad \mathscr{E}^\circ = 0.34\ V$$
$$Zn^{2+} + 2e^- \rightarrow Zn \qquad \mathscr{E}^\circ = -0.76\ V$$

Remember that the more *positive* the $\mathscr{E}^\circ$ value, the more the reaction has a tendency to proceed in the direction indicated. Of the three reactions listed, the reduction of Ag^+ occurs most easily, and the order of oxidizing ability is

$$Ag^+ > Cu^{2+} > Zn^{2+}$$

This means that silver will plate out first as the potential is increased, followed by copper, and finally zinc.

Sample Exercise 17.10

An acidic solution contains the ions Ce^{4+}, VO_2^+, and Fe^{3+}. Using the $\mathscr{E}^\circ$ values listed in Table 17.1, give the order of oxidizing ability of these species and predict which one will be reduced at the cathode of an electrolytic cell at the lowest voltage.

Solution

The half-reactions and $\mathscr{E}^\circ$ values are

$$Ce^{4+} + e^- \rightarrow Ce^{3+} \qquad \mathscr{E}^\circ = 1.70\ V$$
$$VO_2^+ + 2H^+ + e^- \rightarrow VO^{2+} + H_2O \qquad \mathscr{E}^\circ = 1.00\ V$$
$$Fe^{3+} + e^- \rightarrow Fe^{2+} \qquad \mathscr{E}^\circ = 0.77\ V$$

The order of oxidizing ability is therefore

$$Ce^{4+} > VO_2^+ > Fe^{3+}$$

The Ce^{4+} ion will be reduced at the lowest voltage in an electrolytic cell.

The principle described in this section is very useful, but it must be applied with some caution. For example, in the electrolysis of an aqueous solution of sodium chloride, we should be able to use $\mathscr{E}^\circ$ values to predict the products. Of the major species in the solution (Na^+, Cl^-, and H_2O) only Cl^- and H_2O can be readily oxidized. The half-reactions (written as oxidization processes) are:

$$2Cl^- \rightarrow Cl_2 + 2e^- \qquad -\mathscr{E}^\circ = -1.36\ V$$
$$2H_2O \rightarrow O_2 + 4H^+ + 4e^- \qquad -\mathscr{E}^\circ = -1.23\ V$$

Chemical Impact

The Chemistry of Sunken Treasure

When the galleon *Atocha* was destroyed on a reef by a hurricane in 1622, it was bound for Spain carrying approximately 47 tons of copper, gold, and silver from the New World. The bulk of the treasure was silver bars and coins packed in wooden chests. When treasure hunter Mel Fisher salvaged the silver in 1985, corrosion and marine growth had transformed the shiny metal into something that looked like coral. Restoring the silver to its original condition required an understanding of the chemical changes that had occurred in 350 years of being submerged in the ocean. Much of this chemistry we have already considered at various places in this text.

As the wooden chests containing the silver decayed, the oxygen supply was depleted, favoring the growth of certain bacteria that use the sulfate ion rather than oxygen as an oxidizing agent to generate energy. As these bacteria consume sulfate ions, they release hydrogen sulfide gas that reacts with silver to form black silver sulfide:

$$2Ag(s) + H_2S(aq) \rightarrow Ag_2S(s) + H_2(g)$$

Thus, over the years, the surface of the silver became covered with a tightly adhering layer of corrosion, which fortunately protected the silver underneath and thus prevented total conversion of the silver to silver sulfide.

Another change that took place as the wood decomposed was the formation of carbon dioxide. This shifted the equilibrium that is present in the ocean,

$$CO_2(aq) + H_2O(l) \rightleftharpoons HCO_3^-(aq) + H^+$$

to the right, producing higher concentrations of HCO_3^-. In turn, the HCO_3^- reacted with Ca^{2+} ions present in the seawater to form calcium carbonate:

$$Ca^{2+}(aq) + HCO_3^-(aq) \rightleftharpoons CaCO_3(s) + H^+(aq)$$

Calcium carbonate is the main component of limestone. Thus, over time, the corroded silver coins and bars became encased in limestone.

Both the limestone formation and the corrosion had to be dealt with. Since $CaCO_3$ contains the basic anion, CO_3^{2-}, acid dissolves limestone

$$2H^+(aq) + CaCO_3(s) \rightarrow Ca^{2+}(aq) + CO_2(g) + H_2O(l)$$

By soaking the mass of coins in a buffered acidic bath for several hours, the individual pieces were separated and the black Ag_2S on the surfaces was revealed. An abrasive could not be used to remove this corrosion; it would have destroyed the details of the engraving—a very valuable feature of the coins to a historian or a collector—and it would have washed away some of the silver. Instead the corrosion reaction

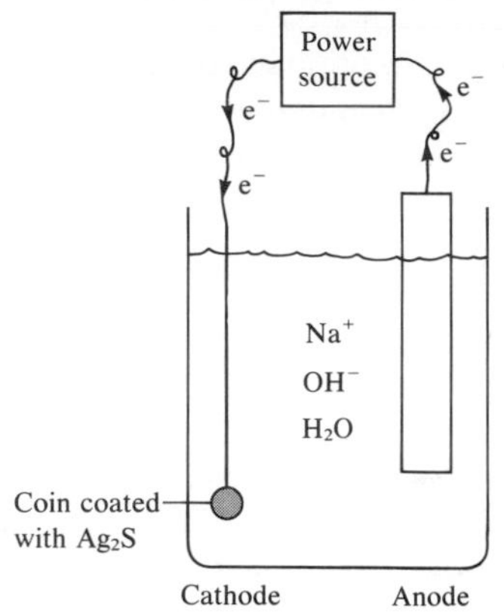

was reversed through electrolytic reduction. The coins were connected to the cathode of an electrolytic cell in a dilute sodium hydroxide solution as represented in the figure.

As electrons flow, the Ag^+ ions in the silver sulfide are reduced to silver metal

$$Ag_2S + 2e^- \rightarrow Ag + S^{2-}$$

As a byproduct, bubbles of hydrogen gas from the reduction of water

$$2H_2O + 2e^- \rightarrow H_2(g) + 2OH^-$$

form on the surface of the coins.

The agitation caused by the bubbles loosens the flakes of metal sulfide and helps clean the coins.

Using these procedures, it has been possible to restore the treasure to very nearly its condition when the *Atocha* sailed many years ago.

Since water has the more positive potential, we would expect to see O_2 produced at the anode because it is easier (thermodynamically) to oxidize H_2O than Cl^-. Actually, this does not happen. As the voltage is increased in the cell, the Cl^- ion is the first to be oxidized. A much higher potential than expected is required to oxidize water. The voltage required in excess of the expected value (called the *overvoltage*) is much greater for the production of O_2 than for Cl_2, which explains why chlorine is produced first.

The causes of overvoltage are very complex. Basically the phenomenon is caused by difficulties in transferring electrons from the species in the solution to the atoms on the electrode across the electrode-solution interface. Because of this situation, $\mathscr{E}°$ values must be used cautiously in predicting the actual order of oxidation or reduction of species in an electrolytic cell.

17.8 Commercial Electrolytic Processes

Purpose

- To discuss the manufacture of aluminum, the chlor-alkali process, and other industrial applications of electrolysis.

The chemistry of metals is characterized by their ability to donate electrons to form ions. Because metals are typically such good reducing agents, most are found in nature in *ores,* mixtures of ionic compounds often containing oxide, sulfide, and silicate anions. The noble metals, such as gold, silver, and platinum, are more difficult to oxidize and are often found as pure metals.

Production of Aluminum

Aluminum is one of the most abundant elements on earth, ranking third behind oxygen and silicon. Since aluminum is a very active metal, it is found in nature as its oxide in an ore called *bauxite* (named after Les Baux, France, where it was discovered in 1821). Production of aluminum metal from its ore proved to be more difficult than production of most other metals. In 1782 Lavoisier recognized aluminum to be a metal "whose affinity for oxygen is so strong that it cannot be overcome by any known reducing agent." As a result, pure aluminum metal remained unknown. Finally, in 1854 a process was found for producing metallic aluminum using sodium, but aluminum remained a very expensive rarity. In fact, it is said that Napoleon III served his most honored guests with aluminum forks and spoons, while the others had to settle for gold and silver utensils.

The breakthrough came in 1886 when two men, Charles M. Hall in the United States and Paul Heroult in France, almost simultaneously discovered a practical electrolytic process for producing aluminum (see Fig. 17.21). The key factor in the *Hall-Heroult process* is the use of molten cryolite (Na_3AlF_6) as the solvent for the aluminum oxide.

Electrolysis is possible only if ions can move to the electrodes. A common method for producing ion mobility is dissolving the substance to be electrolyzed in

Figure 17.21

Charles Martin Hall (1863–1914) was a student at Oberlin College in Ohio when he first became interested in aluminum. One of his professors commented that anyone who could manufacture aluminum cheaply would make a fortune, and Hall decided to give it a try. The 21-year-old Hall worked in a wooden shed near his house with an iron frying pan as a container, a blacksmith's forge as a heat source, and galvanic cells constructed from fruit jars. Using these crude galvanic cells, Hall found that he could produce aluminum by passing a current through a molten Al_2O_3/Na_3AlF_6 mixture. By a strange coincidence, Paul Heroult, a Frenchman who was born and died in the same years as Hall, made the same discovery at about the same time.

water. This is not possible in the case of aluminum because water is more easily reduced than Al^{3+}, as the following standard reduction potentials show:

$$Al^{3+} + 3e^- \rightarrow Al \qquad \mathscr{E}^\circ = -1.66\ V$$

$$2H_2O + 2e^- \rightarrow H_2 + 2OH^- \qquad \mathscr{E}^\circ = -0.83\ V$$

Thus aluminum metal cannot be plated out of an aqueous solution of Al^{3+}.

Ion mobility can also be produced by melting the salt. But the melting point of solid Al_2O_3 is much too high (2050°C) to allow practical electrolysis of the molten oxide. A mixture of Al_2O_3 and Na_3AlF_6, however, has a melting point of 1000°C, and the resulting molten mixture can be used to obtain aluminum metal electrolytically. Because of this discovery by Hall and Heroult, the price of aluminum plunged (see Table 17.3), and its use became economically feasible.

The Price of Aluminum over the Last Century

Date	Price of aluminum ($/lb)*
1855	100,000
1885	100
1890	2
1895	0.50
1970	0.30
1980	0.80

*Note the precipitous drop in price after the discovery of the Hall-Heroult process.

Table 17.3

Bauxite is not pure aluminum oxide (called *alumina*) but also contains the oxides of iron, silicon, and titanium, and various silicate materials. To obtain the pure hydrated alumina ($Al_2O_3 \cdot nH_2O$), the crude bauxite is treated with aqueous sodium hydroxide. Being amphoteric, alumina dissolves in the basic solution:

$$Al_2O_3(s) + 2OH^-(aq) \rightarrow 2AlO_2^-(aq) + H_2O(l)$$

The other metal oxides, which are basic, remain as solids. The solution containing the aluminate ion (AlO_2^-) is separated from the sludge of other oxides and is acidified with carbon dioxide gas, causing the hydrated alumina to reprecipitate:

$$2CO_2(g) + 2AlO_2^-(aq) + (n + 1)H_2O(l) \rightarrow 2HCO_3^-(aq) + Al_2O_3 \cdot nH_2O(s)$$

The purified alumina is then mixed with cryolite and melted, and the aluminum ion is reduced to aluminum metal in an electrolytic cell of the type shown in Fig. 17.22. Because the electrolyte solution contains a large number of aluminum-containing ions, the chemistry is not completely understood, but the alumina probably reacts with the cryolite anion as follows:

$$Al_2O_3 + 4AlF_6^{3-} \rightarrow 3Al_2OF_6^{2-} + 6F^-$$

The electrode reactions are thought to be

Cathode reaction: $AlF_6^{3-} + 3e^- \rightarrow Al + 6F^-$

Anode reaction: $2Al_2OF_6^{2-} + 12F^- + C \rightarrow 4AlF_6^{3-} + CO_2 + 4e^-$

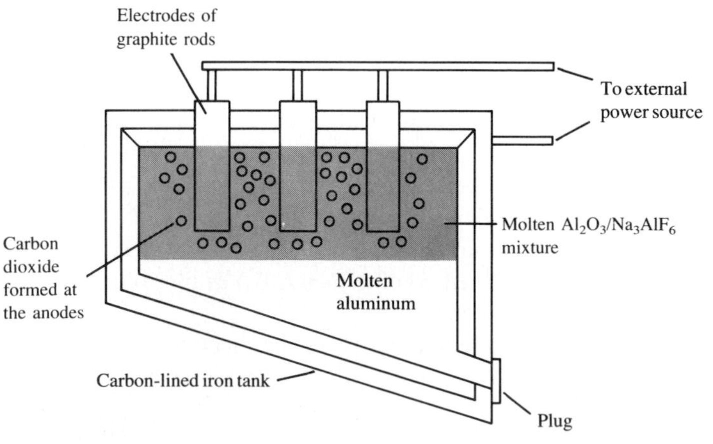

Electrodes of graphite rods

To external power source

Molten Al_2O_3/Na_3AlF_6 mixture

Carbon dioxide formed at the anodes

Molten aluminum

Carbon-lined iron tank

Plug

Figure 17.22

A schematic diagram of an electrolytic cell for producing aluminum by the Hall-Heroult process. Because molten aluminum is more dense than the mixture of molten cryolite and alumina, it settles to the bottom of the cell and is drawn off periodically. The graphite electrodes are gradually eaten away and must be replaced from time to time. The cell operates at a current flow of up to 250,000 A.

The overall cell reaction can be written as

$$2Al_2O_3 + 3C \longrightarrow 4Al + 3CO_2$$

The aluminum produced in this electrolytic process is 99.5% pure. To be useful as a structural material, aluminum is alloyed with metals such as zinc (used for trailer and aircraft construction) and manganese (used for cooking utensils, storage tanks, and highway signs). The production of aluminum consumes about 4.5% of all electricity used in the United States.

Electrorefining of Metals

Purification of metals is another important application of electrolysis. For example, impure copper from the chemical reduction of copper ore is cast into large slabs that serve as the anodes for electrolytic cells. Aqueous copper sulfate is the electrolyte, and thin sheets of ultrapure copper function as the cathodes (see Fig. 17.23).

The main reaction at the anode is

$$Cu \longrightarrow Cu^{2+} + 2e^-$$

Other metals such as iron and zinc are also oxidized from the impure anode:

$$Zn \longrightarrow Zn^{2+} + 2e^-$$
$$Fe \longrightarrow Fe^{2+} + 2e^-$$

Noble-metal impurities in the anode are not oxidized at the voltage used; they fall to the bottom of the cell to form a sludge, which is processed to remove the valuable silver, gold, and platinum.

The Cu^{2+} ions from the solution are deposited onto the cathode

$$Cu^{2+} + 2e^- \longrightarrow Cu$$

producing copper that is 99.95% pure.

Figure 17.23

Ultrapure copper sheets that serve as the cathodes are lowered between slabs of impure copper that serve as the anodes into a tank containing an aqueous solution of copper sulfate ($CuSO_4$). It takes about four weeks for the anodes to dissolve and for the copper to be deposited on the cathodes.

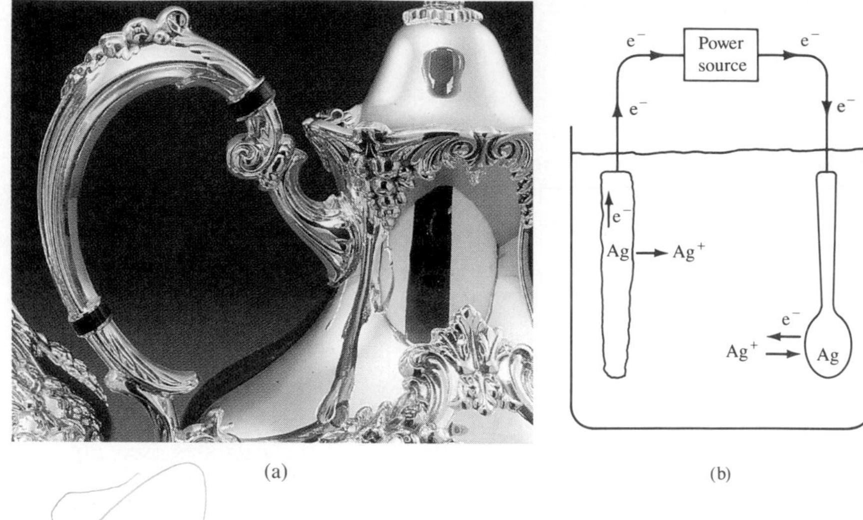

Figure 17.24

(a) A silver-plated teapot. Silver plating is often used to beautify and protect cutlery and items of table service. (b) Schematic of the electroplating of a spoon. The item to be plated is the cathode, and the anode is a silver bar. Silver is plated out at the cathode:

$$Ag^+ + e^- \rightarrow Ag$$

Note that a salt bridge is not needed here since Ag^+ ions are involved at both electrodes.

Metal Plating

Metals that readily corrode can often be protected by the application of a thin coating of a metal that resists corrosion. Examples are "tin" cans, which are actually steel cans with a thin coating of tin, and chrome-plated steel bumpers for automobiles.

An object can be plated by making it the cathode in a tank containing ions of the plating metal. The silver plating of a spoon is shown schematically in Fig. 17.24. In an actual plating process, the solution also contains ligands that form complexes with the silver ion. By lowering the concentration of Ag^+ in this way, a smooth, even coating of silver is obtained.

Electrolysis of Sodium Chloride

Addition of a nonvolatile solute lowers the melting point of the solvent, molten NaCl in this case.

Sodium metal is mainly produced by the electrolysis of molten sodium chloride. Because solid NaCl has a rather high melting point (800°C), it is usually mixed with solid $CaCl_2$ to lower the melting point to about 600°C. The mixture is then electrolyzed in a **Downs cell,** as illustrated in Fig. 17.25, where the reactions are

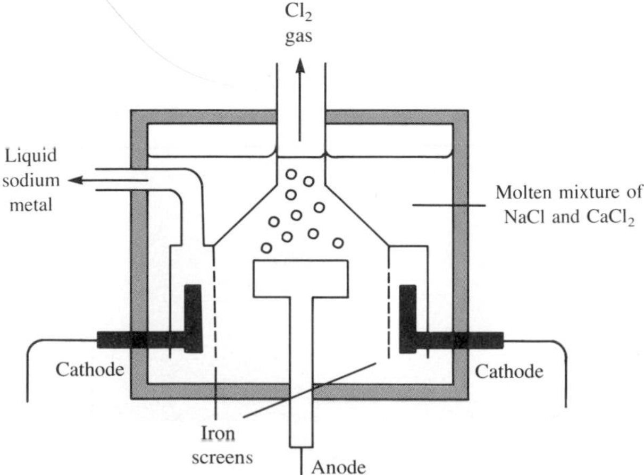

Figure 17.25

The Downs cell for the electrolysis of molten sodium chloride. The cell is designed so that the sodium and chlorine produced cannot come into contact with each other to reform NaCl.

$$\text{Anode reaction:} \quad 2Cl^- \rightarrow Cl_2 + 2e^-$$
$$\text{Cathode reaction:} \quad Na^+ + e^- \rightarrow Na$$

At the temperatures in the Downs cell, the sodium is liquid and is drained off, then cooled, and cast into blocks. Because it is so reactive, sodium must be stored in an inert solvent, such as mineral oil, to prevent its oxidation.

Electrolysis of aqueous sodium chloride (brine) is an important industrial process for the production of chlorine and sodium hydroxide. In fact, this process is the second largest consumer of electricity in the United States, after the production of aluminum. Sodium is not produced in this process under normal circumstances because H_2O is more easily reduced than Na^+, as the standard reduction potentials show:

$$Na^+ + e^- \rightarrow Na \qquad \mathscr{E}° = -2.71 \text{ V}$$
$$2H_2O + 2e^- \rightarrow H_2 + 2OH^- \qquad \mathscr{E}° = -0.83 \text{ V}$$

Hydrogen, not sodium, is produced at the cathode.

For the reasons we discussed in Section 17.7, chlorine gas is produced at the anode. Thus the electrolysis of brine produces hydrogen and chlorine,

$$\text{Anode reaction:} \quad 2Cl^- \rightarrow Cl_2 + 2e^-$$
$$\text{Cathode reaction:} \quad 2H_2O + 2e^- \rightarrow H_2 + 2OH^-$$

leaving a solution containing dissolved NaOH and NaCl.

The contamination of the sodium hydroxide by NaCl can be virtually eliminated using a special **mercury cell** for electrolyzing brine (see Fig. 17.26). In this cell mercury is the conductor at the cathode, and because hydrogen gas has an extremely high overvoltage with a mercury electrode, Na^+ is reduced instead of H_2O. The resulting sodium metal dissolves in the mercury, forming a liquid alloy, which is then pumped to a chamber where the dissolved sodium is reacted with water to produce hydrogen:

$$2Na(s) + 2H_2O(l) \rightarrow 2Na^+(aq) + 2OH^-(aq) + H_2(g)$$

Relatively pure solid NaOH can be recovered from the aqueous solution, and the regenerated mercury is then pumped back to the electrolysis cell. This process, called the **chlor-alkali process,** has often resulted in significant mercury contamination of the environment, but the waste solutions are now carefully treated to remove mercury.

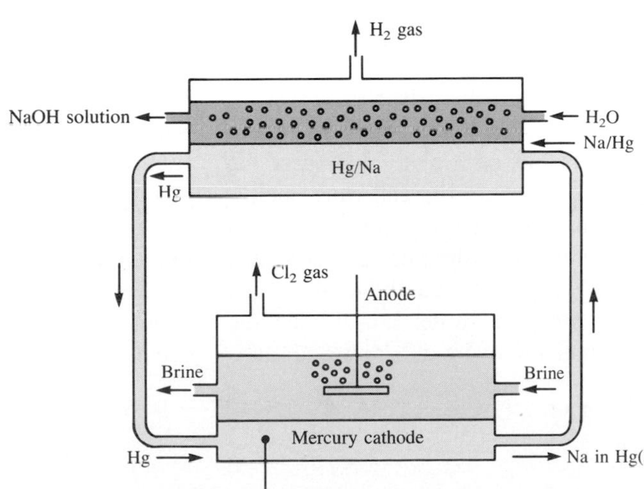

Figure 17.26

The mercury cell for production of chlorine and sodium hydroxide. The large overvoltage required to produce hydrogen at a mercury electrode means that Na^+ ions are reduced rather than water. The sodium formed dissolves in the liquid mercury and is pumped to a chamber where it reacts with water.

Chemical Impact

Batteries of the Future

A possible alternative to the internal combustion engine used in present-day automobiles is an electric motor powered by batteries. This and many other applications have stimulated research into new types of efficient batteries. In fact, some types being developed are capable of producing five to ten times as much power per unit of weight as the lead storage battery.

Ion mobility is needed in a battery to prevent charge polarization. Usually this is accomplished with an aqueous electrolyte. But there is now considerable interest in solid ceramic materials that transport ions very efficiently. For example, in the *beta,* or *sodium-sulfur, battery,* a solid electrolyte, called beta alumina, composed of a mixture of sodium, aluminum, lithium, and magnesium oxides, separates the electrode compartments. Liquid sodium is the reducing agent (the anode), and liquid sulfur is the oxidizing agent (the cathode). Although the beta battery must be operated at 350°C to keep the reactants liquid, the system shows promise because it can store approximately four times as much energy per unit of weight and can undergo approximately three times as many discharge/charge cycles as the lead storage battery.

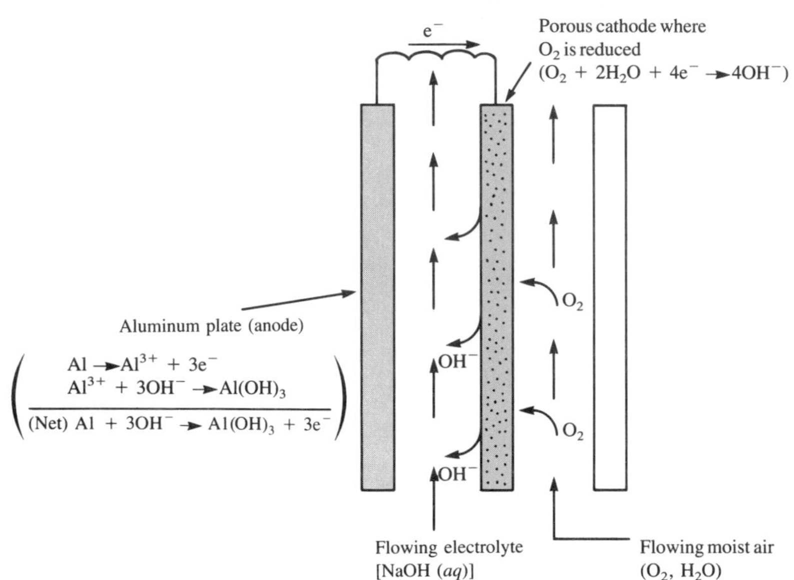

Figure 17.27

The aluminum-air battery. Aluminum is the reducing agent and oxygen gas from the air is the oxidizing agent. As the aluminum loses electrons and dissolves, it forms aluminum hydroxide in the flowing aqueous sodium hydroxide solution.

Another promising power source is the *aluminum-air battery,* which is under development by chemist John Cooper at Lawrence Livermore National Laboratory in California. This is really a combination battery and fuel cell, since it uses a stored quantity of aluminum metal and requires a continuous supply of humid air from the atmosphere. A diagram of the cell is shown in Fig. 17.27. The aluminum-air battery is a flow system, since air and the aqueous sodium hydroxide electrolyte flow past the electrodes. As the cell runs, aluminum and water are consumed, and aluminum hydroxide is produced. Periodically the aluminum plate must be replaced, water added, and the aluminum hydroxide removed. It is estimated that a small car with an aluminum-air battery weighing approximately 500 pounds could travel about 3000 miles before the aluminum needed to be replaced. Every 250 miles during the trip, 6 gallons of water would be needed and the solid aluminum hydroxide would have to be removed.

One new type of battery being studied has no metal electrodes or metal ions [Fig. 17.28(a)]. The electrodes are thin films of giant *polyacetylene* molecules (polymers)

Chemical Impact

of the form shown below:

```
              H   H   H   H   H   H   H   H   H   H   H   H
              |   |   |   |   |   |   |   |   |   |   |   |
~~~~~~~~~~~~~ C = C — C = C — C = C — C = C — C = C — C = C ~~~~~~~~~~~~~
```

The battery is charged by attaching it to a power source, which forces electrons from one polyacetylene electrode to the other. As one electrode gains electrons, positive ions from the electrolyte diffuse into the film to balance the charge. The other electrode, which has given up electrons, absorbs anions to balance the charge [Fig. 17.28(b)]. The battery discharges, generating a current to run an electric motor, as the excess electrons flow from the electron-rich anode to the electron-deficient cathode [Fig. 17.27(c)].

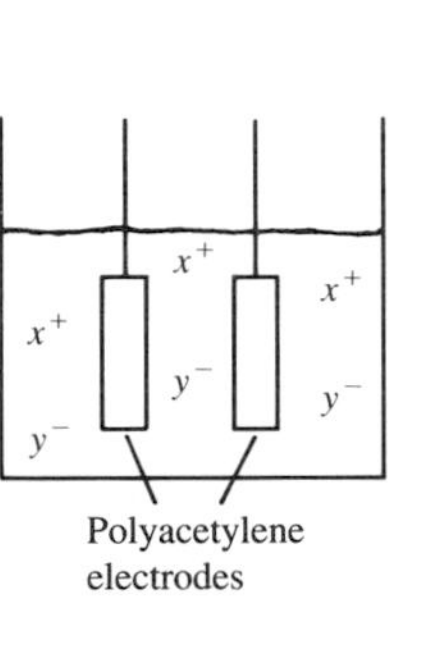

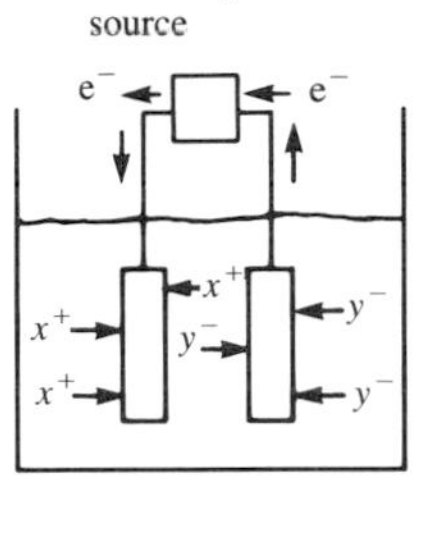

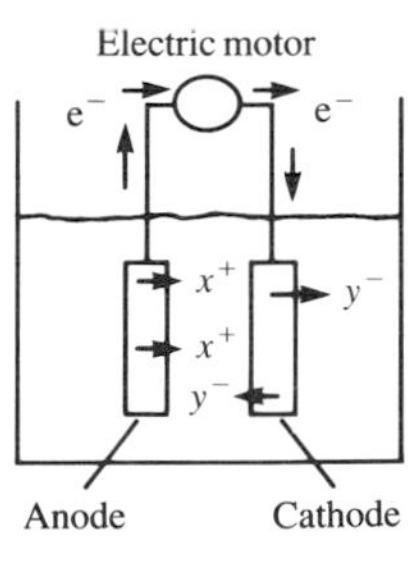

Figure 17.28

A battery with polyacetylene electrodes. (a) The uncharged battery. (b) The charging process using an external power source. (c) The discharging battery running an electric motor.

Suggested Reading

1. Richard C. Alkire, "Electrochemical Engineering," *Journal of Chemical Education* **60** (1983): 274.
2. Jacob Jorné, "Flow Batteries," *American Scientist* **71** (1983): 507.
3. Anthony F. Sammells, "Fuel Cells and Electrochemical Energy Storage," *Journal of Chemical Education* **60** (1983): 320.

FOR REVIEW

Summary

Electrochemistry is the study of the interchange of chemical and electrical energy that occurs through oxidation-reduction (redox) reactions, in which electrons are transferred from a reducing agent to an oxidizing agent. Oxidation is the loss of electrons, and reduction is the gain of electrons. It is useful to break a redox reaction into half-reactions, one involving oxidation and one involving reduction.

When a redox reaction occurs in solution, the electrons are transferred directly, and no useful work can be obtained from the chemical energy involved in the reaction. However, if the oxidizing agent is physically separated from the reducing agent, so that the electrons must flow through a wire from one to the other, the chemical energy can produce useful work. A galvanic cell is a device in which chemical energy is transformed to electrical energy. The opposite process, where electrical energy is used to produce chemical change, is called electrolysis and is accomplished in an electrolytic cell by forcing current through the cell.

Oxidation occurs at the anode of a cell; reduction occurs at the cathode. The driving force on the electrons is called the cell potential ($\mathscr{E}_{cell}$). A galvanic cell always runs in the direction that gives a positive value for $\mathscr{E}_{cell}$. Standard reduction potentials are assigned to half-reactions written as reduction processes by arbitrarily assigning a potential of zero volts to the half-reaction $2H^+ + 2e^- \rightarrow H_2$ under standard conditions.

The unit of electrical potential is the volt (V), defined as a joule of work per coulomb of charge:

$$\mathscr{E}\ (\text{V}) = \frac{-\text{work (J)}}{\text{charge (C)}} = \frac{-w}{q}$$

where the sign for work is determined from the point of view of the system (cell). The maximum work can be calculated from the equation

$$-w_{max} = q\mathscr{E}_{max}$$

where $\mathscr{E}_{max}$ represents the cell potential when no current flows. Actual work is always less than the maximum possible work since energy is wasted through frictional heating when a current flows through a wire.

For a process carried out at constant temperature and pressure, the change in free energy equals the maximum useful work obtainable from that process

$$\Delta G = w_{max} = -q\mathscr{E}_{max} = -nF\mathscr{E}$$

where F (the faraday) is the charge of 1 mole of electrons (96,485 coulombs) and n is the number of moles of electrons transferred in the reaction.

A concentration cell is a galvanic cell in which both compartments have the same components, but at different concentrations. The electrons flow in the direction that tends to equalize the concentrations.

The Nernst equation gives the relationship between the cell potential and the concentrations of the cell components:

$$\mathscr{E} = \mathscr{E}^\circ - \frac{RT}{nF}\ \ln(Q)$$

When the cell is at 25°C, the Nernst equation can be written

$$\mathscr{E} = \mathscr{E}^\circ - \frac{0.0592}{n}\ \log(Q)$$

When the cell is at equilibrium $Q = K$, and $\mathscr{E}_{cell} = 0$. Thus the Nernst equation can be used to calculate the values of equilibrium constants for oxidation-reduction reactions.

A battery is a galvanic cell, or group of cells connected in series, that serves as a source of direct current. The lead storage battery has a lead anode, a cathode of lead coated with PbO_2 and an electrolyte of sulfuric acid. Dry cell batteries do not have liquid electrolytes but a moist paste of either solid manganese dioxide, carbon, and ammonium chloride (the acid version) or potassium hydroxide or sodium hydroxide (the alkaline version). A case of zinc acts as the anode, and a carbon rod acts as the electrode for the cathode.

Fuel cells are galvanic cells for which the reactants are continuously supplied. For example, a hydrogen-oxygen fuel cell is based on the reaction of hydrogen and oxygen gases to form water.

Corrosion involves the oxidation of metals to form mainly oxides and sulfides. Some metals, such as aluminum, form a thin protective coating of oxide that inhibits its further corrosion. The corrosion of iron is an electrochemical process; different parts of the iron surface form anodic and cathodic regions. The Fe^{2+} ions formed in the anodic regions migrate through the moisture on the iron surface to the cathodic regions, and the electrons flow through the metal to the cathodic regions, where they react with oxygen in the air. Corrosion can be prevented by coating the iron with paint or a layer of metal such as chromium, tin, or zinc; by alloying; and by cathodic protection.

Aluminum is commercially produced by the electrolysis of a molten alumina-cryolite mixture in the Hall-Heroult process. Metals such as copper can be refined electrolytically. The chlor-alkali process is used to manufacture chlorine and sodium hydroxide by electrolysis.

Key Terms

electrochemistry

Section 17.1

oxidation-reduction (redox) reaction
reducing agent
oxidizing agent
oxidation
reduction
half-reactions
salt bridge
porous disk
galvanic cell
anode
cathode
cell potential (electromotive force)
volt
voltmeter
potentiometer

Section 17.2

standard hydrogen electrode
standard reduction potential

Section 17.3

faraday

Section 17.4

concentration cell
Nernst equation
glass electrode
ion-selective electrode

Section 17.5

battery
lead storage battery
dry cell battery
fuel cell

Section 17.6

corrosion
galvanizing
cathodic protection

Section 17.7

electrolytic cell
electrolysis
ampere

Section 17.8

Downs cell
mercury cell
chlor-alkali process

Exercises

A blue exercise number indicates that the answer to that exercise appears at the back of this book and a solution appears in the Solutions Guide.

Review of Oxidation-Reduction Reactions

1. Define oxidation and reduction in terms of change in oxidation number and electron loss or gain.

2. Assign oxidation numbers to all of the atoms in each of the following:

a. HNO_3	e. $MgSO_4$	i. $Na_2C_2O_4$
b. $CuCl_2$	f. Ag	j. CO_2
c. O_2	g. $PbSO_4$	k. $(NH_4)_2Ce(SO_4)_3$
d. H_2O_2	h. PbO_2	l. Cr_2O_3

3. Which of the following equations represent oxidation-reduction reactions? For those that do, indicate the oxidizing agent,

the reducing agent, the species being oxidized, and the species being reduced.
a. $CH_4(g) + H_2O(g) \rightarrow CO(g) + 3H_2(g)$
b. $2AgNO_3(aq) + Cu(s) \rightarrow Cu(NO_3)_2(aq) + 2Ag(s)$
c. $AgCl(s) + 2NH_3(aq) \rightarrow Ag(NH_3)_2^+(aq) + Cl^-(aq)$
d. $Zn(s) + 2HCl(aq) \rightarrow ZnCl_2(aq) + H_2(g)$
e. $2H^+(aq) + 2CrO_4^{2-}(aq) \rightarrow Cr_2O_7^{2-}(aq) + H_2O(l)$
f. $2Fe(s) + 8HCl(aq) \rightarrow 2HFeCl_4(aq) + 3H_2(g)$

4. Balance each of the following equations by the half-reaction method. These reactions occur in acidic solution.
a. $Cr(s) + NO_3^-(aq) \rightarrow Cr^{3+}(aq) + NO(g)$
b. $Al(s) + MnO_4^-(aq) \rightarrow Al^{3+}(aq) + Mn^{2+}(aq)$
c. $CH_3OH(aq) + Ce^{4+}(aq) \rightarrow CO_2(aq) + Ce^{3+}(aq)$
d. $SO_3^{2-}(aq) + MnO_4^-(aq) \rightarrow SO_4^{2-}(aq) + Mn^{2+}(aq)$

5. Balance each of the following equations by the half-reaction method. These reactions occur in basic solution.
a. $PO_3^{3-}(aq) + MnO_4^-(aq) \rightarrow PO_4^{3-}(aq) + MnO_2(s)$
b. $Mg(s) + OCl^-(aq) \rightarrow Mg(OH)_2(s) + Cl^-(aq)$
c. $Ni(CN)_4^{2-}(aq) + Ce^{4+}(aq)$
$\rightarrow Ce(OH)_3(s) + Ni(OH)_2(s) + CO_3^{2-}(aq) + NO_3^-(aq)$
d. $H_2CO(aq) + Ag(NH_3)_2^+(aq)$
$\rightarrow HCO_2^-(aq) + Ag(s) + NH_3(aq)$

Galvanic Cells

6. What is the difference between a galvanic and an electrolytic cell?

7. Why is it necessary to use a salt bridge in a galvanic cell?

8. Define the following:
a. cathode
b. anode
c. oxidation half-reaction
d. reduction half-reaction

9. Sketch the galvanic cells based on the following overall reactions. Show direction of electron flow, direction of migration of ions from the salt bridge, and identify the cathode and anode. Give the overall balanced reaction. Assume all concentrations are 1.0 *M* and all partial pressures are 1.0 atm.
a. $Cr^{3+}(aq) + Cl_2(g) \rightleftharpoons Cr_2O_7^{2-}(aq) + Cl^-(aq)$
b. $Cu^{2+}(aq) + Mg(s) \rightleftharpoons Mg^{2+}(aq) + Cu(s)$
c. $IO_3^-(aq) + Fe^{2+}(aq) \rightleftharpoons Fe^{3+}(aq) + I_2(s)$
d. $Zn(s) + Ag^+(aq) \rightleftharpoons Zn^{2+}(aq) + Ag(s)$

10. Draw diagrams for the galvanic cells utilizing the following half-cells. Write balanced equations for the overall cell reactions.

Cathode	Anode
a. Cl_2/Cl^-	Br^-/Br_2
b. IO_4^-/IO_3^-	Mn^{2+}/MnO_4^-
c. Ni^{2+}/Ni	Al/Al^{3+}
d. Co^{2+}/Co^{3+}	Fe^{3+}/Fe^{2+}

Cell Potentials, Standard Reduction Potentials, and Free Energy

11. Calculate $\mathscr{E}°$ values for the cells in Exercises 9 and 10. Standard reduction potentials are found in Table 17.1.

12. The saturated calomel electrode, abbreviated SCE, is often used as a reference electrode in making electrochemical measurements. The SCE is composed of mercury in contact with a saturated solution of calomel (Hg_2Cl_2). The electrolyte solution is saturated KCl. $\mathscr{E}_{SCE}$ is +0.242 V with respect to the standard hydrogen electrode. What will be the measured potential for each of the following cells, containing an SCE and the given half-cell components? In each case, indicate whether the SCE is the cathode or the anode. Standard reduction potentials are found in Table 17.1.
a. Cu^{2+}/Cu ($Cu^{2+} + 2e^- \rightarrow Cu$)
b. Fe^{3+}/Fe^{2+} ($Fe^{3+} + e^- \rightarrow Fe^{2+}$)
c. $Ag/AgCl$ ($AgCl + e^- \rightarrow Ag + Cl^-$)
d. Al/Al^{3+} ($Al^{3+} + 3e^- \rightarrow Al$)
e. Ni/Ni^{2+} ($Ni^{2+} + 2e^- \rightarrow Ni$)

13. Calculate $\mathscr{E}°$ values for the following cells. Which reactions are spontaneous as written under standard conditions? Balance the reactions that are not already balanced. Standard reduction potentials are found in Table 17.1.
a. $2Ag^+(aq) + Cu(s) \rightleftharpoons Cu^{2+}(aq) + 2Ag(s)$
b. $Zn^{2+}(aq) + Ni(s) \rightleftharpoons Ni^{2+}(aq) + Zn(s)$
c. $MnO_4^-(aq) + I^-(aq) \rightleftharpoons I_2(aq) + Mn^{2+}(aq)$
d. $MnO_4^-(aq) + F^-(aq) \rightleftharpoons F_2(g) + Mn^{2+}(aq)$

14. Answer the following questions using data from Table 17.1 (all under standard conditions):
a. Is $H^+(aq)$ capable of oxidizing $Cu(s)$ to $Cu^{2+}(aq)$?
b. Is $H^+(aq)$ capable of oxidizing $Mg(s)$?
c. Is $Fe^{3+}(aq)$ capable of oxidizing $I^-(aq)$?
d. Is $Fe^{3+}(aq)$ capable of oxidizing $Br^-(aq)$?

15. Answer the following questions using data from Table 17.1 (all under standard conditions):
a. Is $H_2(g)$ capable of reducing $Ag^+(aq)$?
b. Is $H_2(g)$ capable of reducing $Ni^{2+}(aq)$?
c. Is $Fe^{2+}(aq)$ capable of reducing VO_2^+?
d. Is $Fe^{2+}(aq)$ capable of reducing $Cr^{3+}(aq)$ to $Cr^{2+}(aq)$?
e. Is $Fe^{2+}(aq)$ capable of reducing $Cr^{3+}(aq)$ to $Cr(s)$?
f. Is $Fe^{2+}(aq)$ capable of reducing $Sn^{2+}(aq)$ to $Sn(s)$?

16. Using data from Table 17.1 place the following in order of increasing strength as oxidizing agents (all under standard conditions):

MnO_4^-, Cl_2, $Cr_2O_7^{2-}$, Mg^{2+}, Fe^{2+}, Fe^{3+}

17. Using data from Table 17.1 place the following in order of increasing strength as reducing agents (all under standard conditions):

Cr^{3+}, H_2, Zn, Li, F^-, Fe^{2+}

18. Consider only the species (standard conditions)

$$Br^-,\ Br_2,\ H^+,\ H_2,\ La^{3+},\ Ca,\ Cd$$

in answering the following questions. Give reasons for your answers.

a. Which is the strongest oxidizing agent?
b. Which is the strongest reducing agent?
c. Which species can be oxidized by MnO_4^- in acid?
d. Which species can be reduced by $Zn(s)$?

19. Consider only the species (standard conditions)

$$Ce^{4+},\ Ce^{3+},\ Fe^{2+},\ Fe^{3+},\ Fe,\ Mg^{2+},\ Mg,\ Ni^{2+},\ Sn$$

in answering the following questions. Give reasons for your answers. (Use data from Table 17.1.)

a. Which is the strongest oxidizing agent?
b. Which is the strongest reducing agent?
c. Will iron dissolve in a 1.0 *M* solution of Ce^{4+}? If so, will there be Fe^{2+} or Fe^{3+} ions in the solution?
d. Which of the species can be oxidized by $H^+(aq)$?
e. Which of the species can be reduced by $H_2(g)$?
f. Would you use Mg or Sn to reduce Fe^{3+} to Fe^{2+}?

20. Use the table of standard reduction potentials (Table 17.1) to pick a reagent that is capable of each of the following oxidations (under standard conditions in acid solution):

a. oxidize Hg to Hg_2^{2+} but not oxidize Hg_2^{2+} to Hg^{2+}
b. oxidize Br^- to Br_2 but not oxidize Cl^- to Cl_2
c. oxidize Mn to Mn^{2+} but not oxidize Ni to Ni^{2+}

21. Use the table of standard reduction potentials (Table 17.1) to pick a reagent that is capable of each of the following reductions (under standard conditions in acidic solution):

a. reduce Fe^{3+} to Fe^{2+} but not reduce Fe^{2+} to Fe
b. reduce Ag^+ to Ag but not reduce H_2O_2 to O_2

22. Is it possible for a reagent to reduce I_2 to I^- but not to reduce Cu^{2+} to Cu? What range of standard reduction potentials could such a reagent have?

23. The standard reduction potential for the half-reaction

$$Sn^{4+} + 2e^- \rightarrow Sn^{2+}$$

is +0.15 V. Use data from Table 17.1 to answer the following questions.

a. Is it possible to find a reducing agent that will reduce Sn^{4+} to Sn^{2+} and not reduce Sn^{2+} to Sn? If so, name a reducing agent that can accomplish this.
b. Is it possible to find an oxidizing agent that will oxidize Sn to Sn^{2+} but not oxidize Sn^{2+} to Sn^{4+}? If so, name an oxidizing agent that can do this.

24. Hydrogen peroxide can function either as an oxidizing agent or as a reducing agent. Referring to Table 17.1, write half-reactions for H_2O_2 acting in these roles. Calculate $\mathcal{E}^\circ$ for the reaction

$$2H_2O_2 \rightarrow 2H_2O + O_2$$

25. A 1.0 *M* solution of Cu^+ ion is not stable even if there is no exposure to oxygen in the air. Use data from Table 17.1 to predict what reaction occurs. Calculate $\mathcal{E}^\circ$, ΔG°, and *K* for this reaction at 25°C.

26. Chlorine dioxide (ClO_2), which is produced by the reaction

$$2NaClO_2(aq) + Cl_2(g) \rightarrow 2ClO_2(g) + 2NaCl(aq)$$

has been tested as a disinfectant for municipal water treatment.

a. Using data from Table 17.1 calculate $\mathcal{E}^\circ$, ΔG°, and *K* at 25°C for the production of ClO_2.
b. One of the concerns in using ClO_2 as a disinfectant is that the carcinogenic chlorate ion (ClO_3^-) might be a by-product. It can be formed from the reaction

$$ClO_2(g) \rightleftharpoons ClO_3^-(aq) + Cl^-(aq)$$

Balance the equation for the decomposition of ClO_2.

27. A patent attorney has asked for your advice concerning the merits of a patent application claiming the invention of an aqueous single galvanic cell capable of producing a 12-V potential. Comment.

28. Under standard conditions, what reaction occurs, if any, when each of the following operations is performed?

a. Crystals of I_2 are added to a solution of NaCl.
b. Cl_2 gas is bubbled into a solution of NaI.
c. A silver wire is placed in a solution of $CuCl_2$.
d. A lead wire is placed in a solution of $CuCl_2$.
e. A solution of $FeSO_4$ is allowed to sit exposed to air.

29. For reactions that occur in Exercise 28, write a balanced equation and calculate $\mathcal{E}^\circ$, ΔG°, and *K* at 25°C.

30. Calculate ΔG° and *K* at 25°C for the reactions in Exercises 9 and 10.

31. The amount of manganese in steel is determined by changing it to permanganate ion. The steel is first dissolved in nitric acid, producing Mn^{2+} ions. These ions are then oxidized to the deeply colored MnO_4^- ions by periodate ion (IO_4^-) in acid solution.

a. Complete and balance an equation describing each of the above reactions.
b. Calculate $\mathcal{E}^\circ$, ΔG°, and *K* at 25°C for each reaction.

32. Combine the equations

$$\Delta G^\circ = -nF\mathcal{E}^\circ \quad \text{and} \quad \Delta G^\circ = \Delta H^\circ - T\Delta S^\circ$$

to derive an expression for $\mathcal{E}^\circ$ as a function of temperature. Describe how one can graphically determine ΔH° and ΔS° from measurements of $\mathcal{E}^\circ$ at different temperatures.

33. a. Calculate $\mathcal{E}^\circ$ and ΔG° for the reaction at 298 K:

$$2H_2O(l) \rightleftharpoons 2H_2(g) + O_2(g)$$

b. Calculate $\Delta H°$ and $\Delta S°$ for the reaction from thermodynamic data in Appendix 4.
c. Calculate the values of $\mathscr{E}°$ and $\Delta G°$ at temperatures of 90°C and 0°C.

34. What property would you look for in designing a reference half-cell that would produce a potential relatively stable with respect to temperature? (See Exercise 32.)

35. The equation $\Delta G° = -nF\mathscr{E}°$ can also be applied to half-reactions. For the half-reaction:

$$Ag^+ + e^- \rightarrow Ag \qquad \mathscr{E}° = 0.80 \text{ V}$$

Calculate the value of $\Delta G_f°$ for $Ag^+(aq)$.

36. Use standard reduction potentials to estimate $\Delta G_f°$ for $Fe^{2+}(aq)$ and $Fe^{3+}(aq)$.

37. Estimate $\mathscr{E}°$ for the half-reaction:

$$2H_2O + 2e^- \rightarrow H_2 + 2OH^-$$

given the following values of $\Delta G_f°$:

$H_2O(l) = -237$ kJ/mol
$H_2(g) = 0.0$
$OH^-(aq) = -157$ kJ/mol

Compare this value of $\mathscr{E}°$ to the value of $\mathscr{E}°$ given in Table 17.1.

38. Calculate $\mathscr{E}°$ for the reaction:

$$CH_3OH(l) + \tfrac{3}{2}O_2(g) \rightarrow CO_2(g) + 2H_2O(l)$$

using values of $\Delta G_f°$ in Appendix 4. Will $\mathscr{E}°$ increase or decrease with an increase in temperature? (See Exercise 32 for how $\mathscr{E}°$ varies with temperature.)

39. A disproportionation reaction involves a substance that acts as both an oxidizing and reducing agent, producing higher and lower oxidation states of the same element in the products. Which of the following disproportionation reactions are spontaneous under standard conditions?
a. $2Cu^+(aq) \rightarrow Cu^{2+}(aq) + Cu(s)$
b. $2Sn^{2+}(aq) \rightarrow Sn^{4+}(aq) + Sn(aq)$
(See Exercise 23 for the Sn^{4+} reduction potential.)
c. $3Fe^{2+}(aq) \rightarrow 2Fe^{3+}(aq) + Fe(s)$
d. $HClO_2(aq) \rightarrow ClO_3^-(aq) + HClO(aq)$ (Unbalanced)

Use the half-reactions:

$ClO_3^- + 3H^+ + 2e^- \rightarrow HClO_2 + H_2O \qquad \mathscr{E}° = +1.21$ V
$HClO_2 + 2H^+ + 2e^- \rightarrow HClO + H_2O \qquad \mathscr{E}° = +1.65$ V

40. Calculate $\Delta G°$ and K at 25°C for those reactions in Exercise 39 that are spontaneous under standard conditions.

41. For the following half-reaction, $\mathscr{E}° = +0.017$ V:

$$Ag(S_2O_3)_2^{3-} + e^- \rightarrow Ag + 2S_2O_3^{2-}$$

Calculate the value of the equilibrium constant at 25°C for the reaction

$$Ag^+(aq) + 2S_2O_3^{2-}(aq) \rightleftharpoons Ag(S_2O_3)_2^{3-}$$

(*Hint:* Find a half-reaction that can be added to the one above to give the required reaction.)

42. Use the following half-reactions:

$PbSO_4 + 2e^- \rightarrow Pb + SO_4^{2-} \qquad \mathscr{E}° = -0.35$ V
$Pb^{2+} + 2e^- \rightarrow Pb \qquad \mathscr{E}° = -0.13$ V

to calculate the value of the solubility product for $PbSO_4(s)$.

43. The solubility product for $CuI(s)$ is 1.1×10^{-12}. Calculate the value of $\mathscr{E}°$ for the half-reaction:

$$CuI + e^- \rightarrow Cu + I^-$$

The Nernst Equation

44. The Nernst equation can also be applied to half-reactions. Calculate the reduction potential at 25°C of each of the following half-cells:
a. Cu/Cu^{2+} (0.10 *M*)
(The half-reaction is $Cu^{2+} + 2e^- \rightarrow Cu$)
b. Cu/Cu^{2+} (2.0 *M*)
c. Cu/Cu^{2+} (1.0×10^{-4} *M*)
d. MnO_4^- (0.10 *M*)/Mn^{2+} (0.010 *M*) at pH = 3.00 (The half-reaction is: $MnO_4^- + 8H^+ + 5e^- \rightarrow Mn^{2+} + 4H_2O$)
e. MnO_4^- (0.10 *M*)/Mn^{2+} (0.010 *M*) at pH = 1.00

45. The overall reaction in the lead storage battery is

$$Pb(s) + PbO_2(s) + 2H^+(aq) + 2HSO_4^-(aq) \rightarrow 2PbSO_4(s) + 2H_2O(l)$$

a. Calculate $\mathscr{E}$ at 25°C for this battery when $[H_2SO_4] = 4.5$ *M*, that is, $[H^+] = [HSO_4^-] = 4.5$ *M*. At 25°C, $\mathscr{E}° = 2.04$ V for the lead storage battery.
b. For the cell reaction, $\Delta H° = -315.9$ kJ and $\Delta S° = 263.5$ J/K. Calculate $\mathscr{E}°$ at −20°C. (See Exercise 32.)
c. Calculate $\mathscr{E}$ at −20°C when $[H_2SO_4] = 4.5$ *M*.

46. Consider the concentration cell shown below. Calculate the cell potential at 25°C when the concentration of Ag^+ in the beaker on the right is the following:

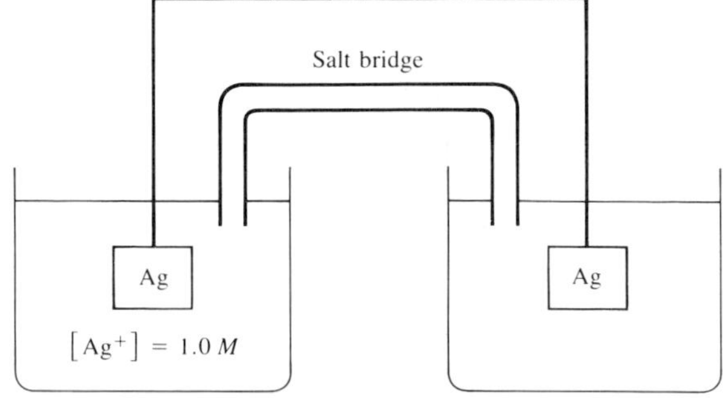

a. 1.0 *M*
b. 2.0 *M*
c. 0.10 *M*
d. 4.0×10^{-5} *M*
e. Calculate the potential when both solutions are 0.10 *M* in Ag^+.

For each case, also identify the cathode, the anode, and the direction in which electrons flow.

47. Can permanganate ion oxidize Fe^{2+} to Fe^{3+} at 25°C under the following conditions (pH = 4.0):

$$[Mn^{2+}] = 1 \times 10^{-6}\ M$$
$$[MnO_4^-] = 0.01\ M$$
$$[Fe^{2+}] = 1 \times 10^{-3}\ M$$
$$[Fe^{3+}] = 1 \times 10^{-6}\ M$$

48. An electrochemical cell consists of a standard hydrogen electrode and a copper metal electrode.
 a. What is the potential of the cell at 25°C if the copper electrode is placed in a solution in which $[Cu^{2+}] = 2.5 \times 10^{-4}$ *M*?
 b. If the copper electrode is placed in a solution of 0.10 *M* NaOH that is saturated with $Cu(OH)_2$, what is the cell potential at 25°C? For $Cu(OH)_2$, $K_{sp} = 1.6 \times 10^{-19}$.
 c. The copper electrode is placed in a solution of unknown $[Cu^{2+}]$. The measured potential at 25°C is 0.195 V. What is $[Cu^{2+}]$?
 d. If you wished to construct a calibration curve for how the cell potential varies with $[Cu^{2+}]$, what would you plot in order to have a straight line? What would be the slope of this line?

49. Consider the following half-reactions:

$$Pt^{2+} + 2e^- \rightarrow Pt \qquad \mathscr{E}° = 1.188\ V$$
$$PtCl_4^{2-} + 2e^- \rightarrow Pt + 4Cl^- \qquad \mathscr{E}° = 0.755\ V$$
$$NO_3^- + 4H^+ + 3e^- \rightarrow NO + 2H_2O \qquad \mathscr{E}° = 0.96\ V$$

Explain why platinum metal will dissolve in aqua regia (a mixture of hydrochloric and nitric acids) but not in either concentrated nitric or concentrated hydrochloric acid individually.

50. The physiological properties of chromium depend on its oxidation state. Chromium(III) ion is essential for an enzyme cofactor called the glucose tolerance factor. Chromium deficiency gives symptoms similar to diabetes. The chromium(VI) ion, on the other hand, is toxic and a potent carcinogen. This is mainly a result of the strength of chromium(VI) as an oxidizing agent.
 a. Consider the following half-reaction:

$$CrO_4^{2-} + 4H_2O + 3e^- \rightarrow Cr(OH)_3 + 5OH^- \qquad \mathscr{E}° = -0.13\ V$$

Note: $Cr(OH)_3$ is a solid.

Using the Nernst equation calculate the potential for this half-reaction at 25°C, where pH = 7.40 and

$$[CrO_4^{2-}] = 1.0 \times 10^{-6}\ M$$

 b. In acidic solution chromium(VI) exists as $Cr_2O_7^{2-}$. Calculate the potential at 25°C for the half-reaction

$$Cr_2O_7^{2-} + 14H^+ + 6e^- \rightarrow 2Cr^{3+} + 7H_2O \qquad \mathscr{E}° = 1.33\ V$$

under conditions where pH = 2.00 and

$$[Cr_2O_7^{2-}] = [Cr^{3+}] = 1.0 \times 10^{-6}\ M$$

51. A chemist wishes to determine the concentration of CrO_4^{2-} electrochemically. A cell is constructed consisting of a saturated calomel electrode (SCE, see Exercise 12) and a silver wire coated with Ag_2CrO_4. The $\mathscr{E}°$ value for the following half-reaction is +0.446 V with respect to the standard hydrogen electrode:

$$Ag_2CrO_4 + 2e^- \rightarrow 2Ag + CrO_4^{2-}$$

 a. Calculate $\mathscr{E}_{cell}$ and ΔG at 25°C for the cell reaction when $[CrO_4^{2-}] = 1.0$ mol/L.
 b. Write the Nernst equation for the cell.
 c. If the coated silver wire is placed in a solution at 25°C in which $[CrO_4^{2-}] = 1.0 \times 10^{-5}$ *M*, what is the expected cell potential?
 d. The measured cell potential at 25°C is 0.504 V when the coated wire is dipped into a solution of unknown $[CrO_4^{2-}]$. What is the $[CrO_4^{2-}]$ for this solution?

52. Write the Nernst equation for the corrosion of iron by oxygen. Will corrosion be a greater problem in an acidic or basic solution?

53. A cleaning solution for glassware is prepared by dissolving potassium dichromate in concentrated sulfuric acid, although this solution is becoming much less commonly used because of the carcinogenic properties of chromium(VI) compounds. An unthinking chemist tried to prepare some cleaning solution from some waste potassium dichromate that had been contaminated by sodium chloride. On adding the solid to some sulfuric acid, he was driven from the lab by some pungent fumes. What happened?

Electrolysis

54. How long will it take to plate out each of the following with a current of 100.0 A?
 a. 1.0 kg of Al from aqueous Al^{3+}
 b. 1.0 g of Ni from aqueous Ni^{2+}
 c. 5.0 mol of Ag from aqueous Ag^+

55. What mass of each of the following substances can be produced in 1.0 h with a current of 15 A?

a. Co from aqueous Co^{2+}
b. Hf from aqueous Hf^{4+}
c. I_2 from aqueous KI
d. Cr from molten CrO_3

56. It took 2.30 min using a current of 2.00 A to plate out all of the silver from 0.250 L of a solution containing Ag^+. What was the original concentration of Ag^+ in the solution?

57. What reaction will take place at the cathode and the anode when each of the following is electrolyzed?
a. molten KF
b. 0.10 *M* KF solution
c. 1.0 *M* H_2O_2 solution containing 1.0 *M* H_2SO_4
d. molten $MgCl_2$
e. 1.0 *M* $AgNO_3$

58. A solution at 25°C is 1.0 *M* in Fe^{2+} and 0.010 *M* in Ag^+. Which metal will be plated on the cathode first when this solution is electrolyzed? (*Hint:* Use the Nernst equation to calculate $\mathscr{E}$ for each half-reaction.)

59. Gold is produced electrochemically from a basic solution of $Au(CN)_4^-$. Gold metal and oxygen gas are produced at the electrodes. Write the overall cell reaction. What volume of pure oxygen gas is produced at 25°C and 740 torr for every kilogram of gold produced?

60. Is it feasible to separate the ions Pt^{2+}, Pd^{2+}, and Ni^{2+} from each other by plating each one in succession if each is present at a concentration of 1.0 *M* in the mixture at 25°C?

$$Ni^{2+} + 2e^- \rightarrow Ni \qquad \mathscr{E}° = -0.23\text{ V}$$
$$Pt^{2+} + 2e^- \rightarrow Pt \qquad \mathscr{E}° = 1.2\text{ V}$$
$$Pd^{2+} + 2e^- \rightarrow Pd \qquad \mathscr{E}° = 0.99\text{ V}$$

61. A solution containing a 3+ metal ion is electrolyzed by a current of 5.0 A for 10.0 min. What is the identity of the metal if 1.18 g of metal was plated out in this electrolysis?

62. It took 74.6 s for a 2.50 A current to plate 0.1086 g of a metal from a solution containing M^{2+} ions. What is the metal?

63. One of the few industrial-scale processes that produce organic compounds electrochemically is used by the Monsanto Company to produce 1,4-dicyanobutane. The reduction reaction is

$$2CH_2{=}CHCN + 2H^+ + 2e^- \rightarrow NC{-}(CH_2)_4{-}CN$$

The NC—$(CH_2)_4$—CN is then chemically reduced by hydrogen to H_2N—$(CH_2)_6$—NH_2, which is used in the production of nylon. What current must be used to produce 1.0×10^3 kg of NC—$(CH_2)_4$—CN per hour?

64. What volume of F_2 gas, at 25°C and 1.00 atm, is produced when molten KF is electrolyzed by a current of 10.0 A for 2.00 h? What mass of potassium metal is produced? At which electrode does each reaction occur?

65. It takes 15 kWh (kilowatt-hours) of electrical energy to produce 1.0 kg of aluminum metal from aluminum oxide by the Hall-Heroult process. Compare this to the amount of energy necessary to melt 1.0 kg of aluminum metal. Why is it economically feasible to recycle aluminum cans? (The enthalpy of fusion for aluminum metal is 10.7 kJ/mol (1 watt = 1 J/s).)

66. In the electrolysis of a sodium chloride solution, what volume of Cl_2 is produced in the same time it takes to produce 6.00 L of $H_2(g)$, both volumes measured at 0°C and 1 atm?

67. What volumes of $H_2(g)$ and $O_2(g)$ at STP are produced from the electrolysis of water by a current of 2.50 A in 15.0 min?

Additional Exercises

68. Describe each of the following:
a. purification of a metal by electrorefining
b. cathodic protection

69. Give three ways in which metals are protected from corrosion.

70. Calculate $\mathscr{E}°$, $\Delta G°$, and K at 25°C for each of the following reactions:
a. $2Cu^+(aq) \rightarrow Cu(s) + Cu^{2+}(aq)$
b. $2Na(s) + 2H_2O(l) \rightleftharpoons 2NaOH(aq) + H_2(g)$
c. $Al(s) + 3H^+(aq) \rightleftharpoons Al^{3+}(aq) + \frac{3}{2}H_2(g)$
d. $Br_2(l) + 2I^-(aq) \rightleftharpoons I_2(s) + 2Br^-(aq)$

71. In 1973 the wreckage of the Civil War ironclad USS *Monitor* was discovered near Cape Hatteras, North Carolina. (The *Monitor* and the CSS *Virginia* [formerly the USS *Merrimack*] fought the first battle between iron-armored ships.) In 1987 investigations were begun to see if the ship could be salvaged. It was reported in *Time* (June 22, 1987) that scientists were considering adding sacrificial anodes of zinc to the rapidly corroding metal hull of the *Monitor*. Describe how attaching zinc to the hull would protect the *Monitor* from further corrosion.

72. What are the advantages and disadvantages in using fuel cells rather than the corresponding combustion reactions to produce electricity?

73. For the following half-reaction, $\mathscr{E}° = -2.07$ V:

$$AlF_6^{3-} + 3e^- \rightarrow Al + 6F^-$$

Calculate the equilibrium constant at 25°C for the reaction:

$$Al^{3+}(aq) + 6F^-(aq) \rightleftharpoons AlF_6^{3-}(aq)$$

74. What is the maximum work that can be obtained from a hydrogen-oxygen fuel cell that produces 1.00 kg of water at 25°C? Why do we say that this is the maximum work that can be obtained?

75. What happens to $\mathscr{E}_{cell}$ as a battery discharges? Is a battery a system at equilibrium? Explain. What is $\mathscr{E}_{cell}$ when a battery reaches equilibrium?

76. Can cobalt metal be oxidized to Co^{2+} by air at 25°C under each of the following conditions?

a. under standard conditions in 1.0 M OH^-
b. under standard conditions in 1.0 M H^+
c. at pH = 7.00

$$Co^{2+} + 2e^- \rightarrow Co \qquad \mathscr{E}° = -0.277\ \text{V}$$

77. The standard reduction potential for the reaction

$$2D^+ + 2e^- \rightarrow D_2$$

where D is deuterium, is −0.0034 V at 25°C. Calculate $\mathscr{E}°$, $\Delta G°$, and K at 25°C for the reaction

$$2H^+(aq) + D_2(g) \rightleftharpoons 2D^+(aq) + H_2(g)$$

78. A crude pH meter is constructed at 25°C using a saturated calomel electrode (SCE, see Exercise 12) and a standard hydrogen electrode (SHE).

a. What is $\mathscr{E}°_{cell}$ when the half cells are the SCE and SHE?
b. Is the hydrogen electrode the cathode or the anode under these conditions?
c. Write the Nernst equation showing how this cell is dependent on $[H^+]$.
d. What is the measured cell potential when the 1.0 M H^+ solution in the hydrogen electrode is replaced by each of the following solutions?

   i. $[H^+] = 1.0 \times 10^{-3}\ M$
   ii. $[H^+] = 2.5\ M$
   iii. $[H^+] = 1.0 \times 10^{-9}\ M$

e. When an unknown acid solution is measured, the observed potential is 0.285 V. What is the pH of the solution?
f. Why is the glass electrode more commonly used than the standard hydrogen electrode to measure pH?

79. The measurement of pH using a glass electrode obeys the Nernst equation. The typical response of a pH meter at 25°C is given by the equation:

$$\mathscr{E}_{meas} = \mathscr{E}_{ref} + 0.05916\ \text{pH}$$

where $\mathscr{E}_{ref}$ contains the potential of the reference electrode and all other potentials that arise in the cell that are not related to the hydrogen ion concentration. Assume that $\mathscr{E}_{ref} = 0.250$ V and that $\mathscr{E}_{meas} = 0.480$ V.

a. What is the uncertainty in the values of pH and $[H^+]$ if the uncertainty in the measured potential is ±1 mV (±0.001 V)?
b. To what accuracy must the potential be measured for the uncertainty in pH to be ±0.02 pH unit?

80. Can NO be produced electrochemically at 25°C by the electrolysis of a 5.0 M aqueous solution of $NaNO_3$?

81. A solution at 25°C contains the following metal ions:

Au^{3+}, $1.0 \times 10^{-6}\ M$
Fe^{3+}, 0.50 M
Ni^{2+}, $1.0 \times 10^{-3}\ M$

What metal will plate out first if this solution is electrolyzed?

82. Beginning with the half-reaction

$$O_2 + 4H^+ + 4e^- \rightarrow 2H_2O \qquad \mathscr{E}° = 1.23\ \text{V}$$

show that $\mathscr{E}° = 0.40$ V at 25°C for the half-reaction

$$O_2 + 2H_2O + 4e^- \rightarrow 4OH^-$$

(*Hint:* Write the Nernst equation for the first half-reaction and use the expression for K_w to get the Nernst equation for the second half-reaction.)

83. The measurement of F^- ion concentration by ion selective electrodes at 25°C obeys the equation:

$$\mathscr{E}_{meas} = \mathscr{E}_{ref} - 0.05916 \log[F^-]$$

a. For a given solution, $\mathscr{E}_{meas}$ is 0.4462 V. If $\mathscr{E}_{ref}$ is 0.2420 V, what is the concentration of F^- in the solution?
b. Hydroxide ion interferes with the measurement of F^-. Taking this into account the response of a fluoride electrode is:

$$\mathscr{E}_{meas} = \mathscr{E}_{ref} - 0.05916 \log([F^-] + k[OH^-])$$

where $k = 1.00 \times 10^1$ and is called the selectivity factor for the electrode response. Calculate $[F^-]$ for the data in part a if the pH is 9.00. What is the percentage error introduced in the $[F^-]$ if the hydroxide interference is ignored?
c. For the $[F^-]$ in part b, what is the maximum pH such that $[F^-]/k[OH^-] = 50.$?
d. At low pH, F^- is mostly converted to HF. The fluoride electrode does not respond to HF. What is the minimum pH at which 99% of the fluoride is present as F^- and only 1% is present as HF?
e. Buffering agents are added to solutions containing fluoride before making measurements with a fluoride selective electrode. Why?

The Representative Elements: Groups 1A Through 4A

So far in this book we have covered the major principles and explored the most important models of chemistry. In particular, we have seen that the chemical properties of the elements can be explained very successfully by the wave mechanical model of the atom. In fact, the most convincing evidence of that model's validity is its ability to relate the observed periodic properties of the elements to the number of valence electrons in their atoms.

We have learned many properties of the elements and their compounds, but we have not discussed extensively the relationship between the chemical properties of a specific element and its position on the periodic table. In this chapter and the next, we will explore the chemical similarities and differences among the elements in the several groups of the periodic table and will try to interpret these data using the wave mechanical model of the atom. In the process we will illustrate a great variety of chemical properties and further demonstrate the practical importance of chemistry.

CONTENTS

18.1 A Survey of the Representative Elements

Purpose

- To give an overview of the representative elements.
- To discuss the relationship of the first member to the rest of the group.
- To give a general introduction to the commercial preparation of some elements.

< Fireworks exploding in the sky over Washington, D.C. The vivid colors displayed are due to the emission spectra of excited elements.

1A	2A											3A	4A	5A	6A	7A	8A
H																	He
Li	Be											B	C	N	O	F	Ne
Na	Mg											Al	Si	P	S	Cl	Ar
K	Ca	Sc	Ti	V	Cr	Mn	Fe	Co	Ni	Cu	Zn	Ga	Ge	As	Se	Br	Kr
Rb	Sr	Y	Zr	Nb	Mo	Tc	Ru	Rh	Pd	Ag	Cd	In	Sn	Sb	Te	I	Xe
Cs	Ba	La	Hf	Ta	W	Re	Os	Ir	Pt	Au	Hg	Tl	Pb	Bi	Po	At	Rn
Fr	Ra	Ac	Rf	Ha	Unh	Uns		Une									

Lathanides	Ce	Pr	Nd	Pm	Sm	Eu	Gd	Tb	Dy	Ho	Er	Tm	Yb	Lu
Actinides	Th	Pa	U	Np	Pu	Am	Cm	Bk	Cf	Es	Fm	Md	No	Lr

Figure 18.1

The periodic table. The elements in the A groups are the representative elements. The elements shown in red are called transition metals. The dark line approximately divides the nonmetals from the metals. The elements that have both metallic and nonmetallic properties (semimetals) are shaded in blue.

The traditional form of the periodic table is shown in Fig. 18.1. Recall that the **representative elements,** whose chemical properties are determined by the valence-level *s* and *p* electrons, are designated Groups 1A through 8A. The **transition metals,** in the center of the table, result from the filling of *d* orbitals. The elements that correspond to the filling of the 4*f* and 5*f* orbitals are listed separately as the **lanthanides** and **actinides,** respectively.

The heavy black line in Fig. 18.1 divides the metals from the nonmetals, except for hydrogen, which appears on the metal side but is a nonmetal. Some elements just on either side of this line, such as silicon and germanium, exhibit both metallic and nonmetallic properties and are often called **metalloids,** or **semimetals.** The fundamental chemical difference between a metal and a nonmetal is that metals tend to lose their valence electrons to form *cations,* usually with the valence-electron configuration of the noble gas from the preceding period, and nonmetals tend to gain electrons to form *anions,* with the electron configuration of the noble gas in the same period. Metallic character is observed to increase going down a given group, which is consistent with the trends in ionization energy, electron affinity, and electronegativity discussed earlier (see Sections 7.13 and 8.2).

Metallic character increases going down a group in the periodic table.

Atomic Size and Group Anomalies

Although the chemical properties of the members of a group have many similarities, there are also important differences. The most dramatic of these differences are usually between the first and second member. For example, hydrogen in Group 1A is a nonmetal, and lithium is a very active metal. This extreme difference results primarily from the very large difference in the atomic radii of hydrogen and lithium, as shown in Fig. 18.2. The small hydrogen atom has a much greater attraction for electrons than do the larger members of Group 1A and forms covalent bonds with nonmetals; the other members of Group 1A lose their valence electrons to nonmetals to form 1+ cations in ionic compounds.

This effect of size is also evident in other groups. For example, the oxides of the metals in Group 2A are all quite basic except for the first member of the series;

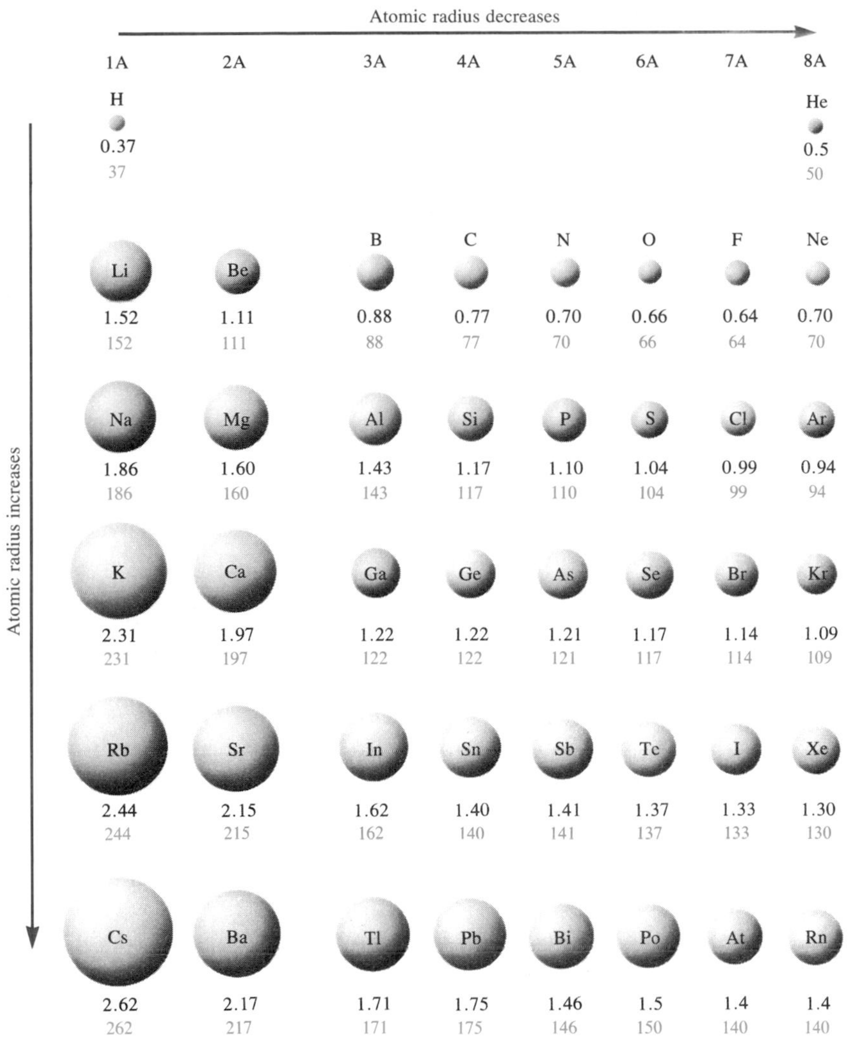

Figure 18.2

The atomic radii of some atoms in angstroms and picometers (in blue).

beryllium oxide (BeO) is amphoteric. Recall from Section 14.10 that the basicity of an oxide depends on its ionic character. Ionic oxides contain the O^{2-} ion, which reacts with water to form two OH^- ions. All the oxides of the Group 2A metals are highly ionic except for beryllium oxide, which has considerable covalent character. The small Be^{2+} ion can effectively polarize the electron "cloud" of the O^{2-} ion, producing significant electron sharing. We see the same pattern in Group 3A, where only the small boron atom behaves as a nonmetal, or sometimes as a semimetal, and aluminum and the other members are active metals.

In Group 4A the effect of size is reflected in the dramatic differences between the chemistry of carbon and that of silicon. The chemistry of carbon is dominated by molecules containing chains of C—C bonds, but silicon compounds mainly contain Si—O bonds rather than Si—Si bonds. Carbon forms such strong C—C single bonds because the relatively small carbon atoms share electrons that are close to the nuclei. The much larger silicon atoms do not form very strong bonds to each other, so compounds containing long chains of silicon atoms are unknown.

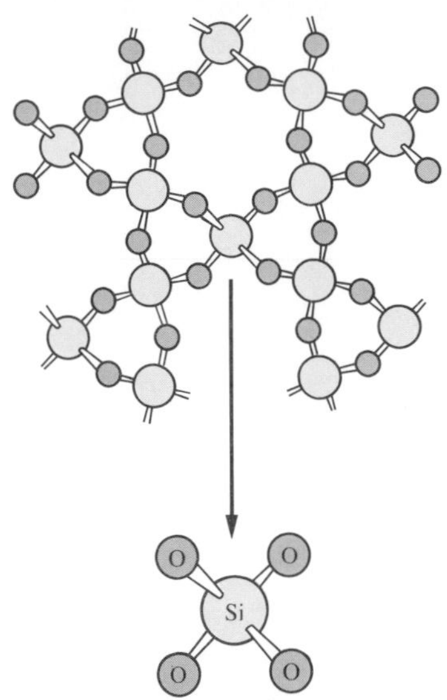

Figure 18.3

The structure of quartz, which has the empirical formula SiO_2. Note that the structure is based on interlocking SiO_4 tetrahedra, where each oxygen atom is shared by two silicon atoms.

Carbon and silicon also differ markedly in their abilities to form π bonds. As we discussed in Section 9.1, carbon dioxide is composed of discrete CO_2 molecules with the Lewis structure $\ddot{O}{=}C{=}\ddot{O}$ where the carbon and oxygen atoms achieve the [Ne] configuration by forming π bonds. In contrast, the structure of silica (empirical formula of SiO_2) is based on SiO_4 tetrahedra with Si—O—Si bridges, as shown in Fig. 18.3. The silicon $3p$ valence orbitals do not overlap very effectively with the smaller oxygen $2p$ orbitals to form π bonds; therefore discrete SiO_2 molecules with the Lewis structure $\ddot{O}{=}Si{=}\ddot{O}$ are not stable. Instead, the silicon atoms achieve a noble gas configuration by forming several Si—O single bonds.

The importance of π bonding for the relatively small elements of the second period also explains the different elemental forms of the members of Groups 5A and 6A. For example, elemental nitrogen exists as very stable N_2 molecules with the Lewis structure $:N{\equiv}N:$. Elemental phosphorus forms larger aggregates of atoms, the simplest being the tetrahedral P_4 molecules found in white phosphorus (Fig. 19.12). Like silicon atoms, the relatively large phosphorus atoms do not form strong π bonds and prefer to achieve a noble gas configuration by forming single bonds to several other phosphorus atoms. In contrast, its very strong π bonds make the N_2 molecule the most stable form of elemental nitrogen. Similarly, in Group 6A the most stable form of elemental oxygen is the O_2 molecule with a double bond, but the larger sulfur atom forms bigger aggregates, such as the cyclic S_8 molecule (see Fig. 19.17), which contains only single bonds.

The relatively large change in size in going from the first to second member of a group also has important consequences for the Group 7A elements. For example, fluorine has a smaller electron affinity than chlorine. This violation of the expected trend can be attributed to the small fluorine $2p$ orbitals, which results in unusually large electron repulsions. The relative weakness of the bond in the F_2 molecule can be explained in terms of the repulsions among the lone pairs, shown in the Lewis structure:

$$:\ddot{\underset{\cdot\cdot}{F}}{-}\ddot{\underset{\cdot\cdot}{F}}:$$

The small size of the fluorine atoms allows close approach of the lone pairs, which leads to much greater repulsions than are found in the Cl_2 molecule with its much larger atoms.

Thus the relatively large increase in atomic radius in going from the first to the second member of a group causes the first element to show properties quite different from the others.

Sand, such as that found at the Great Sand Dunes National Monument in Colorado, is composed of silicon and oxygen.

Abundance and Preparation

Table 18.1 shows the distribution of elements in the earth's crust, oceans, and atmosphere. The major element is, of course, oxygen, which is found in the atmosphere as O_2, in the oceans in H_2O, and in the earth's crust primarily in silicate and carbonate minerals. The second most abundant element, silicon, is found throughout the earth's crust in the silica and silicate minerals that form the basis of most sand, rocks, and soil. The most abundant metals, aluminum and iron, are found in ores in which they are combined with nonmetals, most commonly oxygen. One notable fact revealed by Table 18.1 is the small incidence of most transition metals. Since many of these relatively rare elements are assuming increasing importance in our high-technology society, it is possible that the control of transition metal ores

Distribution (Mass Percent) of the 18 Most Abundant Elements in the Earth's Crust, Oceans, and Atmosphere			
Element	Mass percent	Element	Mass percent
Oxygen	49.2	Titanium	0.58
Silicon	25.7	Chlorine	0.19
Aluminum	7.50	Phosphorus	0.11
Iron	4.71	Manganese	0.09
Calcium	3.39	Carbon	0.08
Sodium	2.63	Sulfur	0.06
Potassium	2.40	Barium	0.04
Magnesium	1.93	Nitrogen	0.03
Hydrogen	0.87	Fluorine	0.03
		All others	0.49

Table 18.1

may ultimately have more significance for world politics than control of petroleum supplies will.

The distribution of elements in living materials is very different from that found in the earth's crust. Table 18.2 shows the distribution of elements in the human body. Oxygen, carbon, hydrogen, and nitrogen form the basis for all biologically important molecules. The other elements, even though they are found in relatively small amounts, are often crucial for life. For example, zinc is found in over 150 different biomolecules in the human body.

Only about one-fourth of the elements occur naturally in the free state. Most are found in a combined state. The *process of obtaining a metal from its ore* is called **metallurgy.** Since the metals in ores are found in the form of cations, the chemistry of *metallurgy always involves reduction of the ions to the elemental metal (with an oxidation state of zero)*. A variety of reducing agents can be used, but carbon is the usual choice because of its wide availability and relatively low cost. As we will see

Metallurgy is discussed in more detail in Chapter 24.

Carbon is the cheapest and most readily available industrial reducing agent for metallic ions.

Abundance of Elements in the Human Body		
Major elements	Mass percent	Trace elements (in alphabetical order)
Oxygen	65.0	Arsenic
Carbon	18.0	Chromium
Hydrogen	10.0	Cobalt
Nitrogen	3.0	Copper
Calcium	1.4	Fluorine
Phosphorus	1.0	Iodine
Magnesium	0.50	Manganese
Potassium	0.34	Molybdenum
Sulfur	0.26	Nickel
Sodium	0.14	Selenium
Chlorine	0.14	Silicon
Iron	0.004	Vanadium
Zinc	0.003	

Table 18.2

in Chapter 24, carbon is the primary reducing agent in the production of steel. Carbon can also be used to produce tin and lead from their oxides:

$$2SnO(s) + C(s) \xrightarrow{\text{Heat}} 2Sn(s) + CO_2(g)$$

$$2PbO(s) + C(s) \xrightarrow{\text{Heat}} 2Pb(s) + CO_2(g)$$

Hydrogen gas can also be used as a reducing agent for metal oxides, as in the production of tin:

$$SnO(s) + H_2(g) \xrightarrow{\text{Heat}} Sn(s) + H_2O(g)$$

Electrolysis is often used to reduce the most active metals. In Chapter 17 we considered the electrolytic production of aluminum metal. The alkali metals are also produced by electrolysis, usually of their molten halide salts.

The preparation of sulfur and the halogens is discussed in Chapter 19.

The preparation of nonmetals varies widely. Elemental nitrogen and oxygen are usually obtained from the **liquefaction** of air, which is based on the principle that a gas cools when it expands. After each expansion, part of the now cooler gas is compressed, and the rest is used to carry away the heat of the compression. The compressed gas is then allowed to expand again. This cycle is repeated many times. Eventually, the remaining gas becomes cold enough to form the liquid state. Because liquid nitrogen and liquid oxygen have different boiling points, they can be separated by the distillation of liquid air. Both substances are important industrial chemicals, nitrogen ranking second in amount manufactured in the United States (~50 billion pounds per year) and oxygen ranking fourth (over 30 billion pounds per year). Hydrogen can be obtained from the electrolysis of water, but more commonly it is obtained from the decomposition of the methane in natural gas. Sulfur is found underground in its elemental form and is recovered using the Frasch process (see Section 19.6). The halogens are obtained by oxidation of the anions from halide salts (see Section 19.7).

18.2 The Group 1A Elements

Purpose

- To consider the thermodynamics and rates of the reactions of alkali metals with water.
- To discuss the oxides of alkali metals.

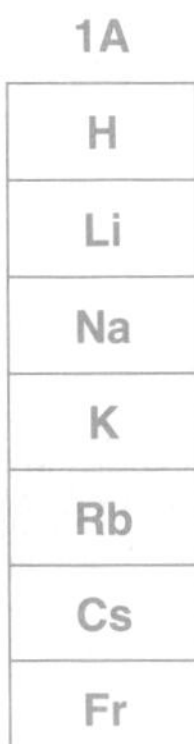

Several properties of the alkali metals are given in Table 7.8.

The Group 1A elements with their ns^1 valence-electron configurations are all very active metals (they lose their valence electrons very readily), except for hydrogen, which behaves as a nonmetal. We will discuss the chemistry of hydrogen in the next section. Many of the properties of the **alkali metals** have been given previously (Section 7.14). The sources and methods of preparation of pure alkali metals are given in Table 18.3. The ionization energies, standard reduction potentials, ionic radii, and melting points for the alkali metals are listed in Table 18.4.

In Section 7.14 we saw that the alkali metals all react vigorously with water to

Sources and Methods of Preparation of the Pure Alkali Metals		
Element	Source	Method of preparation
Lithium	Silicate minerals such as spodumene, $LiAl(Si_2O_6)$	Electrolysis of molten LiCl
Sodium	NaCl	Electrolysis of molten NaCl
Potassium	KCl	Electrolysis of molten KCl
Rubidium	Impurity in lepidolite, $Li_2(F,OH)_2Al_2(SiO_3)_3$	Reduction of RbOH with Mg and H_2
Cesium	Pollucite ($Cs_4Al_4Si_9O_{26} \cdot H_2O$) and an impurity in lepidolite (Fig. 18.4)	Reduction of CsOH with Mg and H_2

Table 18.3

Figure 18.4

Lepidolite is mainly composed of lithium, aluminum, silicon, and oxygen, but it also contains significant amounts of rubidium and cesium.

Selected Physical Properties of the Alkali Metals				
Element	Ionization energy (kJ/mol)	Standard reduction potential (V) for $M^+ + e^- \rightarrow M$	Radius of M^+ (Å)	Melting point (°C)
Lithium	520	−3.05	0.60	180
Sodium	495	−2.71	0.95	98
Potassium	419	−2.92	1.33	63
Rubidium	409	−2.99	1.48	39
Cesium	382	−3.02	1.69	29

Table 18.4

release hydrogen gas:

$$2M(s) + 2H_2O(l) \rightarrow 2M^+(aq) + 2OH^-(aq) + H_2(g)$$

We will reconsider this process briefly because it illustrates several important concepts. Based on the ionization energies, we might expect lithium to be the weakest of the alkali metals as a reducing agent in water. However, the standard reduction potentials indicate that it is the strongest. This reversal results mainly from the very large energy of hydration of the small Li^+ ion. Because of its relatively high charge density, the Li^+ ion very effectively attracts water molecules. A large quantity of energy is released in the process, favoring the formation of the Li^+ ion and making lithium a strong reducing agent in aqueous solution.

We also saw in Section 7.14 that lithium, although it is the strongest reducing agent, reacts more slowly with water than does sodium or potassium. From the discussions in Chapters 12 and 16, we know that the *equilibrium position* for a reaction (in this case indicated by the $\mathscr{E}°$ values) is controlled by thermodynamic factors but that the *rate* of a reaction is controlled by kinetic factors. There is *no* direct connection between these factors. Lithium reacts more slowly with water than does sodium or potassium because as a solid it has a higher melting point than either of them. It does not become molten from the heat of reaction with water as sodium and potassium do, and thus it has a smaller area of contact with the water.

Sodium reacts violently with water.

The relative ease with which the alkali metals lose electrons to form M^+ cations means that they react with nonmetals to form ionic compounds. Although we might expect the alkali metals to react with oxygen to form regular oxides of the general formula M_2O, lithium is the only one that does so in the presence of excess oxygen gas:

$$4Li(s) + O_2(g) \rightarrow 2Li_2O(s)$$

Sodium forms solid Na_2O if the oxygen supply is limited, but in excess oxygen it forms *sodium peroxide:*

$$2Na(s) + O_2(g) \rightarrow Na_2O_2(s)$$

Hydrogen peroxide has the structure

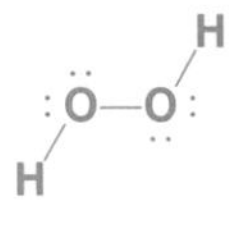

Sodium peroxide contains the basic O_2^{2-} anion and reacts with water to form **hydrogen peroxide** and hydroxide ions:

$$Na_2O_2(s) + 2H_2O(l) \rightarrow 2Na^+(aq) + H_2O_2(aq) + 2OH^-(aq)$$

Hydrogen peroxide is a strong oxidizing agent often used as a bleach for hair and as a disinfectant.

Potassium, rubidium, and cesium react with oxygen to produce **superoxides** of the general formula MO_2, which contain the O_2^- anion. For example, potassium reacts with oxygen as follows:

$$K(s) + O_2(g) \rightarrow KO_2(s)$$

The superoxides release oxygen gas in reactions with water or carbon dioxide:

$$2MO_2(s) + 2H_2O(l) \rightarrow 2M^+(aq) + 2OH^-(aq) + O_2(g) + H_2O_2(aq)$$
$$4MO_2(s) + 2CO_2(g) \rightarrow 2M_2CO_3(s) + 3O_2(g)$$

This chemistry makes superoxides very useful in the self-contained breathing apparatus used by firefighters or as emergency equipment in labs or production facilities in case toxic fumes are released.

The types of compounds formed by the alkali metals with oxygen are summarized in Table 18.5.

Types of Compounds Formed by the Alkali Metals with Oxygen

General formula	Name	Examples
M_2O	Oxide	Li_2O, Na_2O
M_2O_2	Peroxide	Na_2O_2
MO_2	Superoxide	KO_2, RbO_2, CsO_2

Table 18.5

Lithium is the only alkali metal that reacts with gaseous nitrogen to form a **nitride salt** containing the N^{3-} anion:

$$6Li(s) + N_2(g) \rightarrow 2Li_3N(s)$$

Table 18.6 summarizes some important reactions of the alkali metals.

The alkali metal ions are very important for the proper functioning of biological systems, such as nerves and muscles, and Na^+ and K^+ ions are present in all body cells and fluids. In human blood plasma, the concentrations are

$$[Na^+] \approx 0.15\ M \quad \text{and} \quad [K^+] \approx 0.005\ M$$

In the fluids *inside* the cells, the concentrations are reversed:

$$[Na^+] \approx 0.005\ M \quad \text{and} \quad [K^+] \approx 0.16\ M$$

Selected Reactions of the Alkali Metals	
Reaction	Comment
$2M + X_2 \rightarrow 2MX$	X_2 = any halogen molecule
$4Li + O_2 \rightarrow 2Li_2O$	Excess oxygen
$2Na + O_2 \rightarrow NaO_2$	
$M + O_2 \rightarrow MO_2$	M = K, Rb, or Cs
$2M + S \rightarrow M_2S$	
$6Li + N_2 \rightarrow 2Li_3N$	Li only
$12M + P_4 \rightarrow 4M_3P$	
$2M + H_2 \rightarrow 2MH$	
$2M + 2H_2O \rightarrow 2MOH + H_2$	
$2M + 2H^+ \rightarrow 2M^+ + H_2$	Violent reaction!

Table 18.6

Since the concentrations are so different inside and outside the cells, an elaborate mechanism involving selective ligands is needed to transport Na^+ and K^+ ions through the cell membranes.

Recently, studies have been done on the role of the Li^+ ion in the human brain and lithium carbonate has been used extensively in the treatment of manic-depressive patients. The Li^+ ion apparently affects the levels of neurotransmitters, molecules that assist the transmission of messages along the nerve networks. Incorrect concentrations of these molecules can lead to depression or mania.

The effect of lithium on behavior is discussed in a Chemical Impact in Chapter 7.

Sample Exercise 18.1

Predict the products from the following reactants:

a. $Li_3N(s)$ and $H_2O(l)$

b. $KO_2(s)$ and $H_2O(l)$

Solution

a. Solid Li_3N contains the N^{3-} anion, which has a strong attraction for three H^+ ions to form NH_3. Thus the reaction is

$$Li_3N(s) + 3H_2O(l) \rightarrow NH_3(g) + 3Li^+(aq) + 3OH^-(aq)$$

b. Solid KO_2 is a superoxide that characteristically reacts with water to produce O_2, H_2O_2, and OH^-:

$$2KO_2(s) + 2H_2O(l) \rightarrow 2K^+(aq) + 2OH^-(aq) + O_2(g) + H_2O_2(aq)$$

18.3 Hydrogen

Purpose

- To describe the production and uses of hydrogen.
- To discuss types of hydrides.

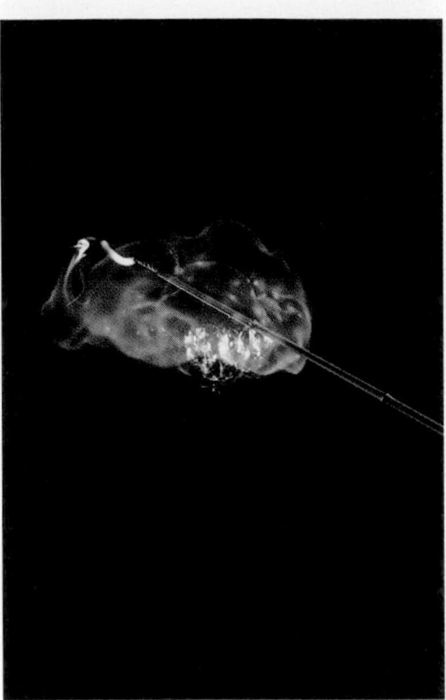

(left) Hydrogen gas being used to blow soap bubbles. (right) As the bubbles float upward, they are lighted using a candle on a long pole. The orange flame is due to the reaction of hydrogen with the oxygen in the air.

Under ordinary conditions of temperature and pressure, hydrogen is a colorless, odorless gas composed of H_2 molecules. Because of its low molecular weight and nonpolarity, hydrogen has a very low boiling point (−253°C) and melting point (−260°C). Hydrogen gas is highly flammable, and mixtures of air containing from 18% to 60% hydrogen by volume are explosive. In a common lecture demonstration hydrogen and oxygen gases are bubbled into soapy water. The resulting bubbles can be ignited with a candle on a long stick, giving a loud explosion.

The major industrial source of hydrogen gas is the reaction of methane and water at high temperatures (800–1000°C) and high pressures (10–50 atmospheres) with a metallic catalyst, often nickel:

$$CH_4(g) + H_2O(g) \xrightarrow[\text{Catalyst}]{\text{Heat, pressure}} CO(g) + 3H_2(g)$$

Large quantities of hydrogen are also formed as a by-product of gasoline production, when hydrocarbons with high molecular weights are broken down (or *cracked*) to produce smaller molecules more suitable for use as a motor fuel (see Section 24.1).

Very pure hydrogen can be produced by electrolysis of water (see Section 17.7), but this method is currently not economically feasible for large-scale production because of the relatively high cost of electricity.

The major industrial use of hydrogen is in the production of ammonia by the Haber process. Large quantities of hydrogen are also used for hydrogenating unsaturated vegetable oils (those containing carbon-carbon double bonds) to produce solid shortenings that are saturated (containing carbon-carbon single bonds):

$$\sim\sim\underset{\text{H}}{\overset{}{}}\overset{\text{H}}{\underset{}{\text{C}}}=\overset{\text{H}}{\text{C}}\sim\sim \xrightarrow{H_2} \sim\sim\overset{\text{H}}{\underset{\text{H}}{\text{C}}}-\overset{\text{H}}{\underset{\text{H}}{\text{C}}}\sim\sim$$

The catalysis of this process was discussed in Section 12.8.

Chemically, hydrogen behaves as a typical nonmetal, forming covalent compounds with other nonmetals and forming salts with very active metals. Binary compounds containing hydrogen are called **hydrides,** of which there are three classes. The **ionic** (or saltlike) **hydrides** are formed when hydrogen combines with the most active metals, those from Groups 1A and 2A. Examples are LiH and CaH_2, which can best be characterized as containing hydride ions (H^-) and metal cations. Because the presence of two electrons in the small 1*s* orbital produces large electron-electron repulsions and the nucleus only has a 1^+ charge, the hydride ion is a strong reducing agent. For example, when ionic hydrides are placed in water, a violent reaction takes place. This reaction results in the formation of hydrogen gas, as seen in the equation

$$LiH(s) + H_2O(l) \rightarrow H_2(g) + Li^+(aq) + OH^-(aq)$$

Covalent hydrides are formed when hydrogen combines with other nonmetals. We have encountered many of these compounds already: HCl, CH_4, NH_3, H_2O, and so on. The most important covalent hydride is water. The polarity of the H_2O molecule leads to many of water's unusual properties. Water has a much higher boiling point than is expected from its molecular weight. It has a large heat of vaporization and a large heat capacity, both of which make it a very useful coolant. Water has a higher density as a liquid than as a solid. This is due to the open structure of ice, which results from maximizing the hydrogen bonding (see Fig. 18.5). Because water is an excellent solvent for ionic and polar substances, it provides an effective medium for life processes. In fact, water is one of the few covalent hydrides that is nontoxic to organisms.

Boiling points of covalent hydrides are discussed in Section 10.1.

The third class of hydrides is the **metallic,** or **interstitial, hydrides,** which are formed when transition metal crystals are treated with hydrogen gas. The hydrogen molecules dissociate at the metal's surface, and the small hydrogen atoms migrate into the crystal structure to occupy holes, or *interstices*. These metal-hydrogen mixtures are more like solid solutions than true compounds. Palladium can absorb about *900 times* its own volume of hydrogen gas. In fact, hydrogen can be purified by placing it under slight pressure in a vessel containing a thin wall of palladium. The hydrogen diffuses into and through the metal wall, leaving the impurities behind.

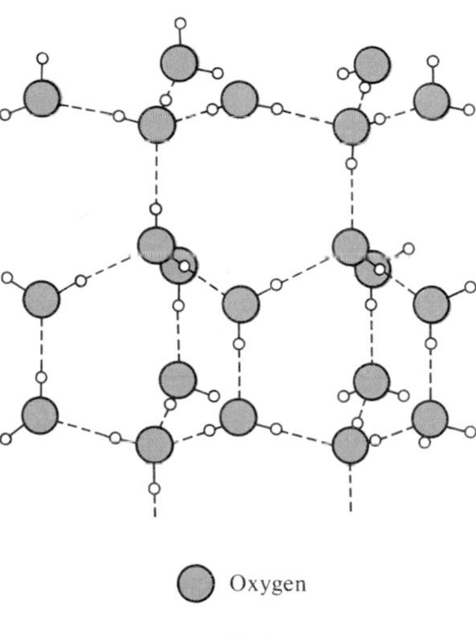

Figure 18.5

The structure of ice, showing the hydrogen bonding.

Although hydrogen can react with transition metals to form compounds such as UH_3 and FeH_6, most of the interstitial hydrides have variable compositions (often called *nonstoichiometric* compositions) with formulas such as $LaH_{2.76}$ and $VH_{0.56}$. The compositions of the nonstoichiometric hydrides vary with the length of exposure of the metal to hydrogen gas and other factors.

When interstitial hydrides are heated, much of the absorbed hydrogen is lost as hydrogen gas. Because of this behavior, these materials offer possibilities for storing hydrogen for use as a portable fuel. The internal combustion engines in current automobiles can burn hydrogen gas with little modification, but storage of enough hydrogen to provide an acceptable mileage range remains a problem. One possible solution might be to use a fuel tank containing a porous solid that includes a transition metal into which the hydrogen gas could be pumped to form the interstitial hydride. The hydrogen gas could then be released when the engine requires additional energy.

See Section 6.6 for a discussion of the feasibility of using hydrogen gas as a fuel.

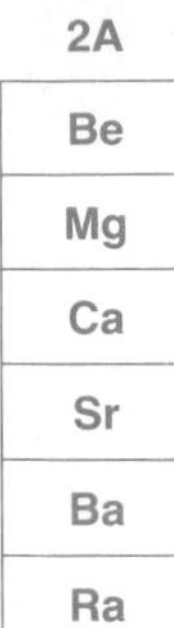

18.4 The Group 2A Elements

Purpose

- To describe the general chemical properties of the alkaline earth metals.
- To discuss three-center bonding in beryllium compounds.

The Group 2A elements (with the valence-electron configuration ns^2) are very reactive, losing their two valence electrons to form ionic compounds containing M^{2+} cations. These elements are commonly called the **alkaline earth metals** because of the basicity of their oxides:

$$MO(s) + H_2O(l) \rightarrow M^{2+}(aq) + 2OH^-(aq)$$

An amphoteric oxide displays both acidic and basic properties.

Only the amphoteric beryllium oxide (BeO) also shows some acidic properties, such as dissolving in aqueous solutions containing hydroxide ions:

$$BeO(s) + 2OH^-(aq) + H_2O(l) \rightarrow Be(OH)_4{}^{2-}(aq)$$

Calcium metal reacting with water to form bubbles of hydrogen gas.

The more active alkaline earth metals react with water as the alkali metals do, producing hydrogen gas:

$$M(s) + 2H_2O(l) \rightarrow M^{2+}(aq) + 2OH^-(aq) + H_2(g)$$

Calcium, strontium, and barium react vigorously at 25°C. The less easily oxidized beryllium and magnesium show no observable reaction with water at 25°C, although magnesium reacts with boiling water. Table 18.7 summarizes various properties, sources, and preparations of the alkaline earth metals.

The heavier alkaline earth metals react with nitrogen or hydrogen at high temperatures to produce ionic nitride or hydride salts, for example:

$$3Ca(s) + N_2(g) \rightarrow Ca_3N_2(s)$$

$$Ca(s) + H_2(g) \rightarrow CaH_2(s)$$

Magnesium, strontium, and barium form similar compounds. Beryllium hydride cannot be formed by direct combination of the elements but can be prepared by the following reaction:

$$BeCl_2 + 2LiH \rightarrow BeH_2 + 2LiCl$$

Hydrogen bridges between the beryllium atoms produce a polymeric structure for BeH_2, as shown in Fig. 18.6. The localized electron model describes this bonding by assuming only one electron pair is available to bind each Be—H—Be cluster. This is called a *three-center bond,* since one electron pair is shared among three atoms. Three-center bonds have also been postulated to explain the bonding in other electron-deficient compounds (compounds where there are fewer electron pairs than bonds) such as the boron hydrides (see Section 18.5).

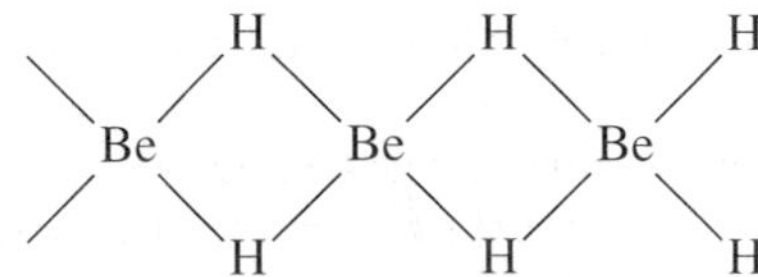

Figure 18.6

The structure of solid BeH_2.

As we saw in Section 18.1, the small size and relatively high electronegativity of the beryllium atom causes its bonds to be more covalent than is usual for a metal. For example, beryllium chloride with the Lewis structure

$$:\ddot{\underset{..}{Cl}}—Be—\ddot{\underset{..}{Cl}}:$$

exists as a linear molecule, as predicted by the VSEPR model. The Be—Cl bonds are covalent, and beryllium is best described as being *sp* hybridized. Note that since

Selected Physical Properties, Sources, and Methods of Preparation for the Group 2A Elements

Element	Radius of M^{2+} (Å)	Ionization energy (kJ/mol) First	Ionization energy (kJ/mol) Second	$\mathscr{E}°$ (V) for $M^{2+} + 2e^- \rightarrow M$	Source	Method of preparation
Beryllium	~0.3	900	1760	−1.70	Beryl ($Be_3Al_2Si_6O_{18}$)	Electrolysis of molten $BeCl_2$
Magnesium	0.65	738	1450	−2.37	Magnesite ($MgCO_3$), dolomite ($MgCO_3 \cdot CaCO_3$), carnallite ($MgCl_2 \cdot KCl \cdot 6H_2O$)	Electrolysis of molten $MgCl_2$
Calcium	0.99	590	1146	−2.76	Various minerals containing $CaCO_3$	Electrolysis of molten $CaCl_2$
Strontium	1.13	549	1064	−2.89	Celestite ($SrSO_4$), strontianite ($SrCO_3$)	Electrolysis of molten $SrCl_2$
Barium	1.35	503	965	−2.90	Baryte ($BaSO_4$), witherite ($BaCO_3$)	Electrolysis of molten $BaCl_2$
Radium	1.40	509	979	−2.92	Pitchblende (1 g of Ra/7 tons of ore)	Electrolysis of molten $RaCl_2$

Table 18.7

the beryllium atom in $BeCl_2$ is very electron-deficient (only four valence electrons surround it), we are not surprised that it is very reactive toward electron-pair donors (Lewis bases) such as ammonia:

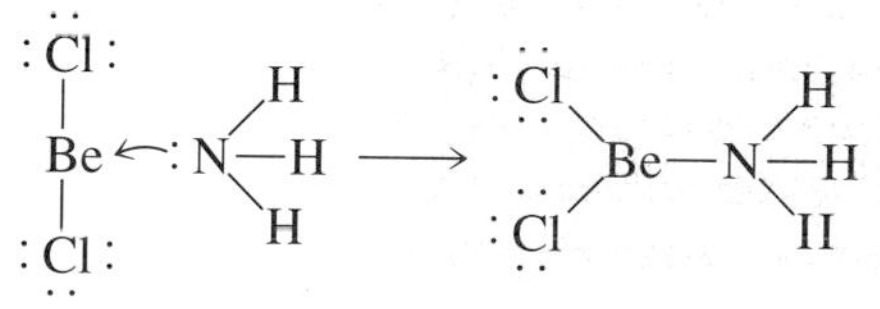

$$Cl-Be(-Cl) \leftarrow :NH_3 \longrightarrow Cl_2Be-NH_3$$

As a solid, $BeCl_2$ achieves an octet of electrons around the beryllium atom by forming an extended structure containing Be in a tetrahedral environment, as shown in Fig. 18.7. The lone pairs on the chlorine atoms are used to form Be—Cl bonds.

The alkaline earth metals have great practical importance. Calcium and magnesium ions are essential for human life. Calcium is found primarily in the structural minerals comprising bones and teeth, and magnesium (as the Mg^{2+} ion) plays a vital role in metabolism and muscle functions. Also, magnesium is commonly used to produce the bright light from photographic flash bulbs from its reaction with oxygen:

$$2Mg(s) + O_2(g) \rightarrow 2MgO(s) + \text{light}$$

Because magnesium metal has a relatively low density and moderate strength, it is a useful structural material, especially if alloyed with aluminum.

Table 18.8 (on the following page) summarizes some important reactions of the alkaline earth metals.

Relatively large concentrations of Ca^{2+} and Mg^{2+} ions are often found in natural water supplies. These ions in this **hard water** interfere with the action of detergents and form precipitates with soap. In Section 14.6 we saw that Ca^{2+} is often removed by precipitation as $CaCO_3$ in large-scale water softening. In individual homes Ca^{2+}, Mg^{2+}, and other cations are removed by **ion exchange.** An **ion-**

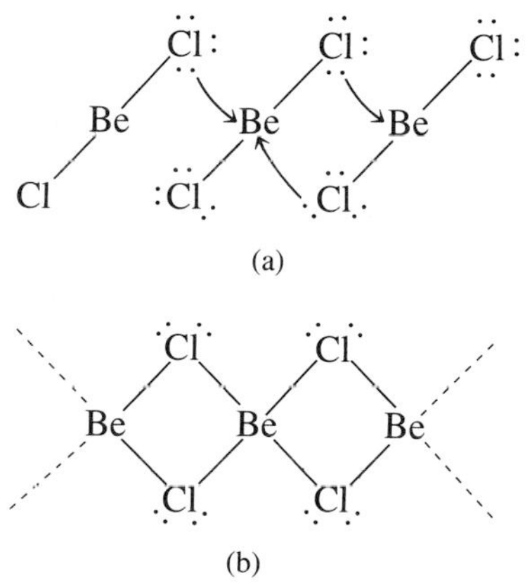

(a)

(b)

Figure 18.7

(a) Solid $BeCl_2$ can be visualized as being formed from many $BeCl_2$ molecules, where lone pairs on the chlorine atoms are used to bond to the beryllium atoms in adjacent $BeCl_2$ molecules. (b) The extended structure of solid $BeCl_2$.

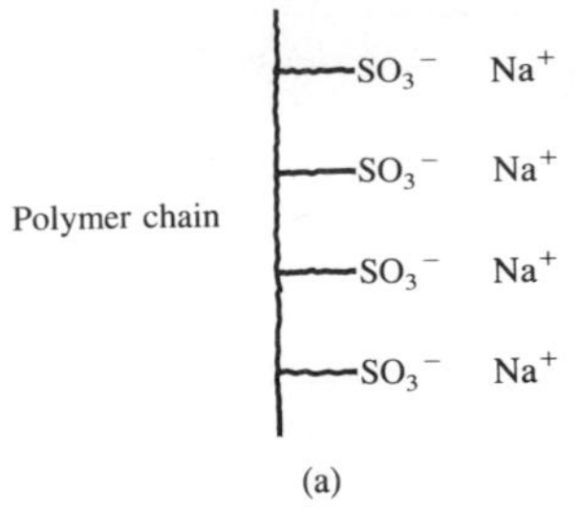

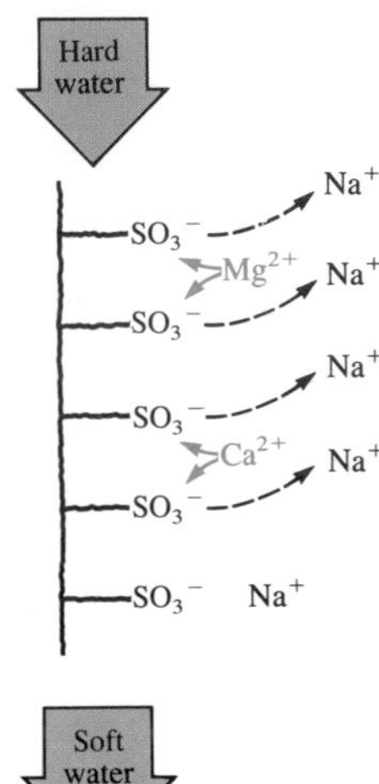

Figure 18.8

(a) A schematic representation of a typical cation-exchange resin.
(b) When hard water is passed over the cation-exchange resin, the Ca^{2+} and Mg^{2+} bind to the resin.

Selected Reactions of the Group 2A Elements

Reaction	Comment
$M + X_2 \rightarrow MX_2$	X_2 = any halogen molecule
$2M + O_2 \rightarrow 2MO$	Ba gives BaO_2 as well
$M + S \rightarrow MS$	
$3M + N_2 \rightarrow M_3N_2$	High temperatures
$6M + P_4 \rightarrow 2M_3P_2$	High temperatures
$M + H_2 \rightarrow MH_2$	M = Ca, Sr or Ba; high temperatures; Mg at high pressure
$M + 2H_2O \rightarrow M(OH)_2 + H_2$	M = Ca, Sr, or Ba
$M + 2H^+ \rightarrow M^{2+} + H_2$	
$Be + 2OH^- + 2H_2O \rightarrow Be(OH)_4^{2-} + H_2$	

Table 18.8

exchange resin consists of large molecules (polymers) that have many ionic sites. A cation-exchange resin is represented schematically in Fig. 18.8(a), showing Na^+ ions bound ionically to the SO_3^- groups that are covalently attached to the resin polymer. When hard water is passed over the resin, Ca^{2+} and Mg^{2+} bind to the resin in place of Na^+, which is released into the solution [Fig. 18.8(b)]. Replacing Mg^{2+} and Ca^{2+} by Na^+ ''softens'' the water because the sodium salts of soap are soluble.

Sample Exercise 18.2

Calculate the amount of time required to produce 1.00×10^3 kg of magnesium metal by the electrolysis of molten $MgCl_2$ using a current of 1.00×10^2 A.

Solution

The reaction for plating magnesium is

$$Mg^{2+} + 2e^- \rightarrow Mg$$

which means that 2 moles of electrons are required for each mole of Mg produced. The number of moles of magnesium in 1.00×10^3 kg is

$$1.00 \times 10^3 \text{ kg} \times \frac{1000 \text{ g}}{\text{kg}} \times \frac{1 \text{ mol Mg}}{24.31 \text{ g}} = 4.11 \times 10^4 \text{ mol Mg}$$

Thus

$$\frac{2 \text{ mol e}^-}{1 \text{ mol Mg}} \times 4.11 \times 10^4 \text{ mol Mg} = 8.22 \times 10^4 \text{ mol e}^-$$

Using the faraday (96,485 C/mol e^-), we can calculate the coulombs of charge:

$$8.22 \times 10^4 \text{ mol e}^- \times \frac{96{,}485 \text{ C}}{\text{mol e}^-} = 7.93 \times 10^9 \text{ C}$$

Since an ampere is a coulomb of charge per second, we can now calculate the time required:

$$\frac{7.93 \times 10^9 \text{ C}}{1.00 \times 10^2 \text{ C/s}} = 7.93 \times 10^7 \text{ s} \quad \text{or} \quad 918 \text{ days}$$

18.5 The Group 3A Elements

Purpose

- To show the general trend from nonmetallic to metallic behavior in Group 3A.

3A
B
Al
Ga
In
Tl

The Group 3A elements (valence-electron configuration ns^2np^1) generally show the increase in metallic character in going down the group that is characteristic of the representative elements. Some physical properties, sources, and methods of preparation for the Group 3A elements are summarized in Table 18.9.

Selected Physical Properties, Sources, and Methods of Preparation for the Group 3A Elements

Element	Radius of M^{3+} (Å)	Ionization energy (kJ/mol)	$\mathscr{E}°$ (V) for $M^{3+} + 3e^- \rightarrow M$	Sources	Method of preparation
Boron	0.2	798	—	Kernite, a form of borax ($Na_2B_4O_7 \cdot 4H_2O$)	Reduction by Mg or H_2
Aluminum	0.51	581	−1.71	Bauxite (Al_2O_3)	Electrolysis of Al_2O_3 in molten Na_3AlF_6
Gallium	0.62	577	−0.53	Traces in various minerals	Reduction with H_2 or electrolysis
Indium	0.81	556	−0.34	Traces in various minerals	Reduction with H_2 or electrolysis
Thallium	0.95	589	0.72	Traces in various minerals	Electrolysis

Table 18.9

Boron is a nonmetal and most of its compounds are covalent. The most interesting compounds of boron are the covalent hydrides called **boranes.** We might expect BH_3 to be the simplest hydride, since boron has three valence electrons to share with three hydrogen atoms. However, this compound is unstable, and the simplest known member of the series is diborane (B_2H_6), with the structure shown in Fig. 18.9(a). In this molecule the terminal B—H bonds are normal covalent bonds, each involving one electron pair. The bridging bonds are three-center bonds similar to those in solid BeH_2. Another interesting borane contains the square pyramidal B_5H_9 molecule [Fig. 18.9(b)], which has four three-center bonds around the base of the

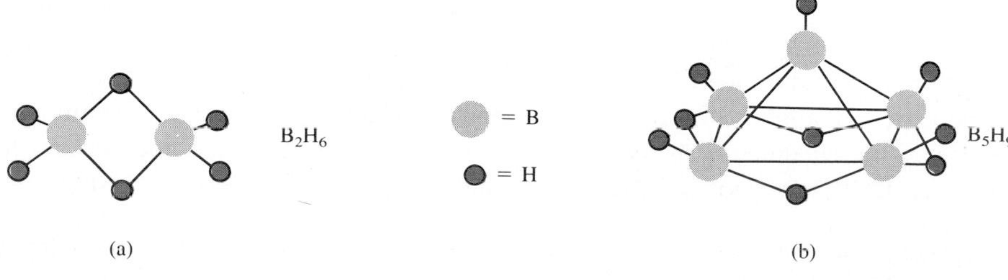

Figure 18.9

(a) The structure of B_2H_6 with its two three-center B—H—B bridging bonds and four "normal" B—H bonds.
(b) The structure of B_5H_9. There are five "normal" B—H bonds to terminal hydrogens and four three-center bridging bonds around the base.

pyramid. Because the boranes are extremely electron-deficient, they are highly reactive. The boranes react very exothermically with oxygen and were once evaluated as potential fuels for rockets in the U.S. space program.

Aluminum, the most abundant metal on earth, has metallic physical properties, such as high thermal and electrical conductivities and a lustrous appearance, but its bonds to nonmetals are significantly covalent. This covalency is responsible for the amphoteric nature of Al_2O_3, which will dissolve in acidic or basic solution, and for the acidity of $Al(H_2O)_6{}^{3+}$ (see Section 14.8):

$$Al(H_2O)_6{}^{3+}(aq) \rightleftharpoons Al(OH)(H_2O)_5{}^{2+}(aq) + H^+(aq)$$

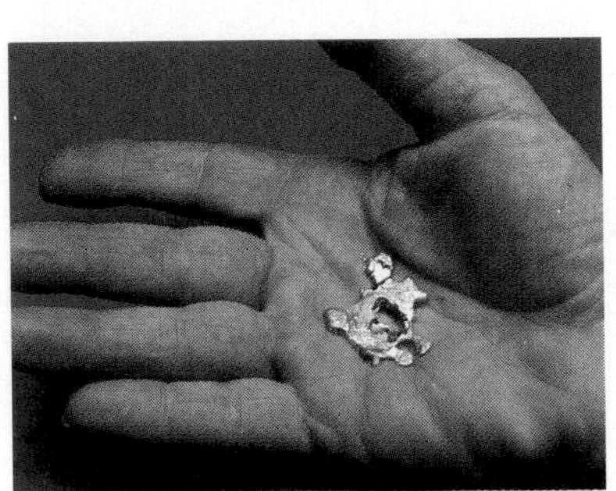

Gallium metal melts in the hand.

One especially interesting property of *gallium* is its unusually low melting point of 29.8°C, which is in contrast to the 660°C melting point of aluminum. Also, gallium's boiling point of ~2400°C means it has the largest liquid range of any metal, and this makes it useful for thermometers, especially to measure high temperatures. Gallium, like water, expands when it freezes. The chemistry of gallium is quite similar to that of aluminum. For example, Ga_2O_3 is amphoteric.

The chemistry of *indium* is similar to that of aluminum and gallium except that compounds containing the 1+ ion are known, such as InCl and In_2O, in addition to those with the more common 3+ ion.

The explanation for the inert pair effect is very complex, involving a relativistic treatment of the atom, and will not be considered here.

The chemistry of *thallium* is completely metallic. For example, Tl_2O_3 is a basic oxide. Both the +1 and +3 oxidation states are quite common for thallium; Tl_2O_3 and Tl_2O, and $TlCl_3$ and TlCl are all well-known compounds. The tendency for the heavier members of Group 3A to exhibit the +1 as well as the expected +3 oxidation state is often called the **inert pair effect.** This effect is also found in Group 4A, where lead and tin exhibit both +4 and +2 oxidation states.

Table 18.10 summarizes some important reactions of the Group 3A elements.

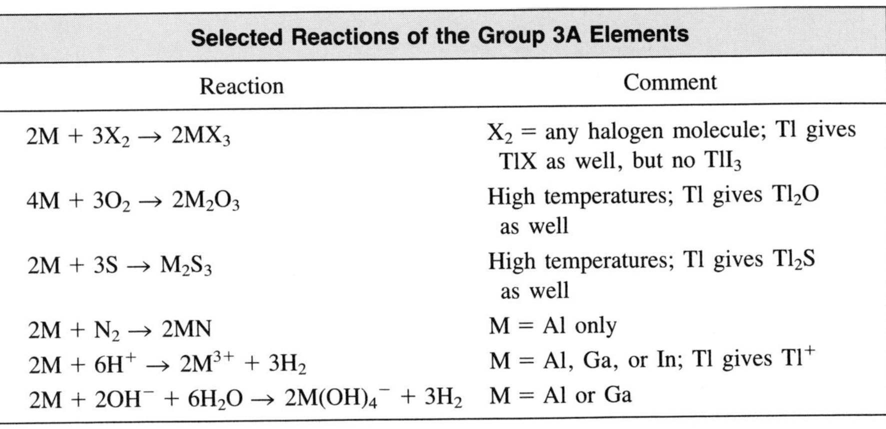

Selected Reactions of the Group 3A Elements

Reaction	Comment
$2M + 3X_2 \rightarrow 2MX_3$	X_2 = any halogen molecule; Tl gives TlX as well, but no TlI_3
$4M + 3O_2 \rightarrow 2M_2O_3$	High temperatures; Tl gives Tl_2O as well
$2M + 3S \rightarrow M_2S_3$	High temperatures; Tl gives Tl_2S as well
$2M + N_2 \rightarrow 2MN$	M = Al only
$2M + 6H^+ \rightarrow 2M^{3+} + 3H_2$	M = Al, Ga, or In; Tl gives Tl^+
$2M + 2OH^- + 6H_2O \rightarrow 2M(OH)_4{}^- + 3H_2$	M = Al or Ga

Table 18.10

The practical importance of the Group 3A elements mostly centers on aluminum. Since the discovery of the electrolytic production process by Hall and Heroult (Section 17.8), aluminum has become a highly important structural material in a wide variety of applications from aircraft bodies to bicycle components. Aluminum is especially valuable because it has a high strength-to-weight ratio and because it protects itself from corrosion by developing a tough, adherent oxide coating.

18.6 The Group 4A Elements

Purpose

- To contrast the chemistry of carbon with that of silicon and the other members of Group 4A.
- To describe carbon oxides.
- To discuss the characteristics of the +2 and +4 oxidation states of tin and lead.

4A
C
Si
Ge
Sn
Pb

Group 4A (with the valence-electron configuration ns^2np^2) contains two of the most important elements on earth: carbon, the fundamental constituent of the molecules necessary for life; and silicon, which forms the basis of the geological world. The change from nonmetallic to metallic properties seen in Group 3A is also apparent in going down Group 4A from carbon, a typical nonmetal, to silicon and germanium, usually considered semimetals, to the metals tin and lead. Table 18.11 summarizes some physical properties, sources, and methods of preparation for the elements in this group.

Selected Physical Properties, Sources, and Methods of Preparation for the Group 4A Elements

Element	Electronegativity	Melting point (°C)	Boiling point (°C)	Sources	Method of preparation
Carbon	2.5	3727 (sublimes)	—	Graphite, diamond, petroleum, coal	—
Silicon	1.8	1410	2355	Silicate minerals, silica	Reduction of K_2SiF_6 with Al, or reduction of SiO_2 with Mg
Germanium	1.8	937	2830	Germanite (mixture of copper, iron, and germanium sulfides)	Reduction of GeO_2 with H_2 or C
Tin	1.8	232	2270	Cassiterite (SnO_2)	Reduction of SnO_2 with C
Lead	1.9	327	1740	Galena (PbS)	Roasting of PbS with O_2 to form PbO_2 and then reduction with C

Table 18.11

All of the Group 4A elements can form four covalent bonds to nonmetals, for example: CH_4, SiF_4, $GeBr_4$, $SnCl_4$, and $PbCl_4$. In each of these tetrahedral molecules, the central atom is described as sp^3 hybridized by the localized electron model. All of these compounds, except those of carbon, can react with Lewis bases to form two additional covalent bonds. For example, $SnCl_4$, which is a fuming liquid (bp = 114°C), can add two chloride ions:

$$SnCl_4 + 2Cl^- \rightarrow SnCl_6^{2-}$$

Carbon compounds cannot react in this way because of the small atomic size of

Table 18.12

Strengths of C—C, Si—Si, and Si—O Bonds	
Bond	Bond energy (kJ/mol)
C—C	347
Si—Si	226
Si—O	368

carbon and because there are no *d* orbitals on carbon to accommodate the extra electrons, as there are on the other elements in the group.

We have seen that carbon also differs markedly from the other members of Group 4A in its ability to form π bonds. This accounts for the completely different structures and properties of CO_2 and SiO_2. From Table 18.12 it is apparent that C—C bonds and Si—O bonds are much stronger than Si—Si bonds. This explains why the chemistry of carbon is dominated by C—C bonds, while that of silicon is dominated by Si—O bonds.

The organic chemistry of carbon is discussed in Chapter 22.

Although graphite is thermodynamically more stable than diamond, the transformation of diamond to graphite is not observed under normal conditions.

Carbon occurs in two allotropic forms—graphite and diamond—whose structures were given in Section 10.5.

Carbon monoxide (CO), one of three oxides of carbon, is an odorless, colorless, and toxic gas at 25°C and 1 atmosphere. It is a by-product of the combustion of carbon-containing compounds when there is a limited oxygen supply. Incidents of carbon monoxide poisoning are especially common in the winter in cold areas of the world when blocked furnace vents limit the availability of oxygen. The bonding in carbon monoxide, which has the Lewis structure :C≡O: is described in terms of *sp* hybridized carbon and oxygen atoms that interact to form one σ and two π bonds.

Carbon dioxide, a linear molecule with the Lewis structure $\ddot{\text{O}}$=C=$\ddot{\text{O}}$ and an *sp* hybridized carbon atom, is a product of human and animal respiration and of the combustion of fossil fuels. It is also produced by fermentation, a process by which the sugar in fruits and grains is changed to ethanol (C_2H_5OH) and carbon dioxide (see Section 24.5):

$$\underset{\text{Glucose}}{C_6H_{12}O_6(aq)} \xrightarrow{\text{Enzymes}} 2C_2H_5OH(aq) + 2CO_2(g)$$

Carbon dioxide dissolves in water to produce an acidic solution:

$$CO_2(aq) + H_2O(l) \rightleftharpoons H^+(aq) + HCO_3^-(aq)$$

Carbon suboxide, the third carbon oxide, is a linear molecule with the Lewis structure $\ddot{\text{O}}$=C=C=C=$\ddot{\text{O}}$ which contains *sp* hybridized carbon atoms.

Silicon, the second most abundant element in the earth's crust, is a semimetal found widely distributed in silica and silicates (see Section 10.5). Approximately 85% of the earth's crust is composed of these substances. Although silicon is found in some steel and aluminum alloys, its major use is in semiconductors for electronic devices (see the Chemical Impact feature at the end of Chapter 10).

Germanium, a relatively rare element, is a semimetal used mainly in the manufacture of semiconductors for transistors and similar electronic devices.

A Persian bronze bottle used to hold eye make-up, circa the first century B.C.

Tin is a soft silvery metal that can be rolled into thin sheets (tin foil) and has been used for centuries in various alloys such as bronze (20% Sn and 80% Cu), solder (33% Sn and 67% Pb), and pewter (85% Sn, 7% Cu, 6% Bi, and 2% Sb). Tin exists as three allotropes: *white tin,* stable at normal temperatures; *gray tin,* stable at temperatures below 13.2°C; and *brittle tin,* found at temperatures above 161°C. When tin is exposed to low temperatures, it gradually changes to the powdery gray tin and crumbles away; this process is known as *tin disease*.

The major current use for tin is as a protective coating for steel, especially for cans used as food containers. The thin layer of tin, applied electrolytically, forms a protective oxide coating that prevents further corrosion.

Tin forms compounds in the +2 and +4 oxidation states. For example, all of the possible tin(II) and tin(IV) halides are known. The tin(IV) halides, except for SnF_4, are all relatively volatile (see Table 18.13) and generally behave much more like molecular than ionic compounds. This is no doubt due to the high charge-to-radius ratio that would be characteristic of Sn^{4+} if it existed. Thus, instead of containing Sn^{4+} and X^- ions, the tin(IV) halides contain covalent Sn—X bonds. The Sn^{4+} ion almost certainly does not exist in these or any other known compounds. The high electronegativity of fluorine makes the Sn—F bonds quite polar, producing large intermolecular forces among SnF_4 molecules. This causes SnF_4 to be much less volatile than the other tin halides.

Boiling Points of the Tin(IV) Halides

Compound	Boiling point (°C)
SnF_4	705 (sublimes)
$SnCl_4$	114
$SnBr_4$	202
SnI_4	364

Table 18.13

Galena.

The tin(II) halides are much less volatile than the corresponding tin(IV) compounds; in fact, they are probably ionic, containing Sn^{2+} and X^- ions. Tin(II) chloride in aqueous solution is commonly used as a reducing agent. Tin(II) fluoride (stannous fluoride) is often added to toothpaste to help prevent tooth decay.

Lead is easily obtained from its ore, galena (PbS). Because lead melts at such a low temperature, it may have been the first pure metal obtained from its ore. We know that lead was used as early as 3000 B.C. by the Egyptians and was later used by the Romans to make eating utensils, glazes on pottery, and even intricate plumbing systems. Lead is, however, very toxic. In fact, the Romans had so much contact with lead that it may have contributed to the demise of their civilization. Analysis of bones from that era shows significant levels of lead.

Although lead poisoning has been known since at least the second century B.C., the incidences of this problem have been relatively isolated. However, the widespread use of tetraethyl lead, $(C_2H_5)_4Pb$, as an antiknock agent in gasoline has increased the lead levels in our environment in this century. Concern about the effects of this lead pollution has caused the U.S. government to require the gradual replacement of the lead in gasoline with other antiknock agents. The largest commercial use of lead (about 1.3 million tons annually) is for electrodes in the lead storage batteries used in automobiles (Section 17.5).

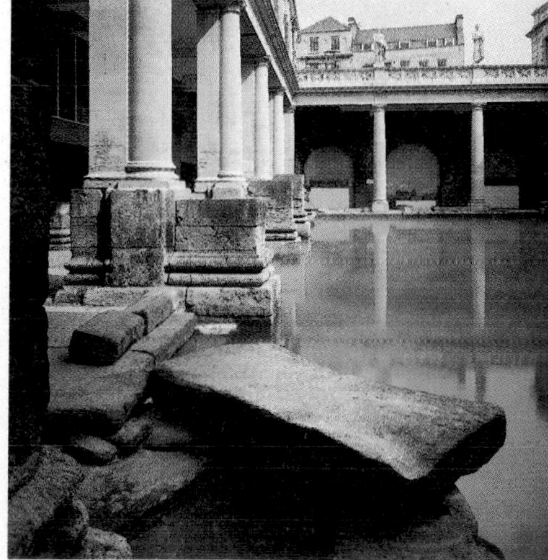
Roman baths such as these in Bath, England, used lead pipes for water.

Lead forms compounds in the +2 and +4 oxidation states. The lead(II) halides, all of which are known, exhibit ionic properties and thus can be assumed to

Lead(II) oxide, known as litharge.

contain Pb^{2+} ions. Only PbF_4 and $PbCl_4$ are known among the possible lead(IV) halides, presumably because lead(IV) oxidizes bromide and iodide ions to give the lead(II) halide and the free halogen:

$$PbX_4 \rightarrow PbX_2 + X_2$$

Lead(IV) chloride decomposes in this fashion at temperatures above 100°C.

Yellow lead(II) oxide, known as *litharge,* is widely used to glaze ceramic ware. Lead(IV) oxide does not exist in nature, but a substance with the formula $PbO_{1.9}$ can be produced in the laboratory by oxidation of lead(II) compounds in basic solution. The nonstoichiometric nature of this compound is caused by defects in the crystal structure. The crystal has some vacancies in positions where there should be oxide ions. These imperfections in the crystal (called *lattice defects*) make lead(IV) oxide an electrical conductor, since the oxide ions jump from hole to hole. This makes possible the use of lead(IV) oxide as an electrode (the anode) in the lead storage battery.

Table 18.14 summarizes some important reactions of the Group 4A elements.

Selected Reactions of the Group 4A Elements

Reaction	Comment
$M + 2X_2 \rightarrow MX_4$	X_2 = any halogen molecule; M = Ge or Sn; Pb gives PbX_2
$M + O_2 \rightarrow MO_2$	M = Ge or Sn; high temperatures; Pb gives PbO or Pb_3O_4
$M + 2H^+ \rightarrow M^{2+} + H_2$	M = Sn or Pb

Table 18.14

FOR REVIEW

Summary

The representative elements have chemical properties determined by their valence *s* and *p* electrons. Although the chemical properties of the members of a group are similar in many ways, there is usually a dramatic difference between the first member and the rest of the group due to size differences. For example, hydrogen in Group 1A is much smaller than lithium and forms covalent bonds because it has a much greater attraction for electrons than do the larger members of Group 1A. The first member of a group forms the strongest π bonds, causing elemental nitrogen and oxygen to exist as N_2 and O_2 molecules, while phosphorus and sulfur exist as P_4 and S_8 molecules.

The most abundant element is oxygen, followed by silicon. The most abundant metals are aluminum and iron, which are found as ores in which they are combined with nonmetals, most commonly oxygen.

The Group 1A elements have the valence-electron configuration ns^1 and lose the one electron readily to form M^+ ions, except for hydrogen, which is a nonmetal.

The alkali metals all react vigorously with water to form M^+ and OH^- ions and hydrogen gas. With oxygen they form a series of oxides. The regular oxide is formed by lithium (Li_2O). Sodium forms the peroxide Na_2O_2 in excess oxygen, and potassium, rubidium, and cesium form the superoxides of general formula MO_2.

Hydrogen can form covalent compounds with other nonmetals and can form salts with very active metals. Hydrogen's binary compounds are called hydrides. Ionic hydrides are formed when hydrogen combines with the Group 1A and Group 2A metals. The hydride ion (H^-) is a strong reducing agent. Covalent hydrides are formed when hydrogen combines with other nonmetals. Metallic hydrides are formed when hydrogen atoms migrate into transition metal crystals.

The Group 2A elements (with the valence-electron configuration ns^2) are called the alkaline earth metals because of the basicity of their oxides. Only the smallest (first) member of the group, beryllium, has an amphoteric oxide. The alkaline earth metals react less vigorously with water than do the alkali metals, beryllium and magnesium showing no reaction at all at 25°C. The heavier alkaline earth metals form ionic nitrides and hydrides. Solid BeH_2 is a polymeric compound that contains three-center bonds in which one electron pair is shared among three atoms.

Hard water contains Ca^{2+} and Mg^{2+} ions, which can be removed by precipitation of $CaCO_3$ or by ion-exchange resins. These polymeric molecules have many ionic sites, usually with Na^+ ions bound to them. When hard water is passed over the resin, the Ca^{2+} and Mg^{2+} ions replace Na^+ on the resin, and are removed from the water.

The Group 3A elements (with the valence-electron configuration ns^2np^1) show increasing metallic character going down the group. Boron is a nonmetal that forms covalent hydrides called *boranes;* these are highly electron-deficient with three-center bonds and are therefore very reactive. Aluminum, a metal, has some covalent characteristics, as do gallium and indium. The chemistry of thallium is completely metallic.

The Group 4A elements (with the valence-electron configuration ns^2np^2) also show a change from nonmetallic to metallic properties in going down the group. However, all of these elements can form covalent bonds to nonmetals. The MX_4 compounds, except those of carbon, can react with Lewis bases to form two additional covalent bonds; CX_4 compounds cannot react in this way because carbon has no $2d$ orbitals and the $3d$ orbitals are too high in energy to be used.

Key Terms

Section 18.1
representative elements
transition metals
lanthanides
actinides
metalloids (semimetals)
metallurgy
liquefaction

Section 18.2
alkali metals
hydrogen peroxide
superoxide
nitride salt

Section 18.3
hydride
ionic hydride
covalent hydride
metallic (interstitial) hydride

Section 18.4
alkaline earth metals
hard water
ion exchange
ion-exchange resin

Section 18.5
boranes
inert pair effect

Exercises

A blue exercise number indicates that the answer to that exercise appears at the back of this book and a solution appears in the Solutions Guide.

Group 1A Elements

1. Although the earth was formed from the same interstellar material as the sun, there is little hydrogen in the earth's atmosphere. How can you explain this?

2. Hydrogen is produced commercially by the reaction of methane with steam:

$$CH_4(g) + H_2O(g) \rightleftharpoons CO(g) + 3H_2(g)$$

   a. Calculate ΔH° and ΔS° for this reaction (use the data in Appendix 4).
   b. What conditions of temperature and pressure would favor the products of this reaction at equilibrium?

3. Hydrogen is also produced commercially by the following reactions:

$$3Fe(s) + 4H_2O(g) \rightleftharpoons Fe_3O_4(s) + 4H_2(g)$$
$$C(s) + H_2O(g) \rightleftharpoons CO(g) + H_2(g)$$

   a. Using the data in Appendix 4, calculate ΔH° and ΔS° for the above reactions.
   b. Under what temperature conditions are each of the above reactions spontaneous at standard conditions?

4. List two major industrial uses of hydrogen.

5. What are the three types of hydrides? How do they differ?

6. Many lithium salts are hygroscopic (absorb water) while the corresponding salts of the other alkali metals are not. Why are lithium salts different from the others?

7. Complete and balance the following reactions:
   a. $Li_3N(s) + HCl(aq) \rightarrow$
   b. $Rb_2O(s) + H_2O(l) \rightarrow$
   c. $Cs_2O_2(s) + H_2O(l) \rightarrow$
   d. $NaH(s) + H_2O(l) \rightarrow$

8. Use the Nernst equation (Section 17.4) to calculate the amount of work that must be done to transport 1.0 mol of K^+ ions from the outside of a cell to the inside. Inside muscle cells $[K^+]$ is about 0.15 mol/L and in blood plasma $[K^+]$ is about 5.0×10^{-3} mol/L.

9. Graph the melting points (Table 18.4) of the alkali metals versus atomic number. Predict the melting point for the element francium. Would you predict francium to be a solid or a liquid at room temperature?

10. Lithium reacts with acetylene in liquid ammonia to produce lithium acetylide ($LiC{\equiv}CH$) and hydrogen gas. Write a balanced equation for this reaction. What type of reaction is this?

11. What evidence supports putting hydrogen in Group 1A of the periodic table? In some periodic tables hydrogen is listed separately from any of the groups. In what ways is hydrogen unlike a Group 1A element?

12. Show how the reaction of NaH with water

$$NaH(s) + H_2O(l) \rightarrow Na^+(aq) + OH^-(aq) + H_2(g)$$

can be described as both an oxidation-reduction and an acid-base reaction.

13. Write balanced equations describing the reaction of potassium metal with each of the following: O_2, S_8, P_4, H_2, H_2O.

14. Write formulas for each of the following: sodium oxide, sodium superoxide, and sodium peroxide.

15. Describe how potassium superoxide can be used in a self-contained breathing apparatus.

Group 2A Elements

16. All of the Group 1A and 2A metals are produced by electrolysis of molten salts. Why?

17. Predict the geometry about beryllium in the compound Cl_2BeNH_3. What hybrid orbitals are used by beryllium and nitrogen in this compound? What type of acid is $BeCl_2$?

18. How does the acidity of the aqueous solutions of the alkaline earth metal ions (M^{2+}) change in going down the group?

19. Would you expect $BeCl_2NH_3$ or another compound to form from the reaction of $BeCl_2$ with an excess of ammonia? Draw Lewis structures of other products that might be produced.

20. Predict a structure of BeF_2 in the gas phase. What structure would you predict for $BeF_2(s)$?

21. Write balanced equations describing the reactions of Ca with each of the following: O_2, S_8, N_2, P_4, H_2, H_2O.

22. Write formulas for barium oxide and barium peroxide.

23. Magnesium nitride and magnesium phosphide are expected to react with water in a fashion similar to lithium nitride. Write equations describing the reactions of magnesium nitride and magnesium phosphide with water.

24. What current is needed to produce 1.00×10^3 kg of Ca metal in 8.00 h from the electrolysis of molten $CaCl_2$? What mass of Cl_2 is produced?

25. Calculate the solubility of $Mg(OH)_2$ in an aqueous solution buffered at pH = 8.0. (See Table 15.4.)

Group 3A Elements

26. Boron hydrides were once evaluated for possible use as rocket fuels. Complete and balance the following reaction:

$$B_2H_6 + O_2 \rightarrow B(OH)_3$$

27. Name each of the following:
   a. TlOH
   b. In_2S_3
   c. Ga_2O_3

28. The compound $AlCl_3$ is quite volatile and appears to exist as a dimer in the gas phase. Propose a structure for this dimer.

29. Assume that element 113 is produced. Predict the radius of its 3+ ion and the first ionization energy given the information in Table 18.9.

30. Lithium aluminum hydride ($LiAlH_4$) is a powerful reducing agent used in the synthesis of organic compounds. Assign oxidation states to all atoms in $LiAlH_4$.

31. Ga_2O_3 is an amphoteric oxide and In_2O_3 is a basic oxide. Write equations describing reactions that illustrate these properties.

32. What type of semiconductor is formed when a Group 3A element is added as an impurity to Si or Ge?

33. Tricalcium aluminate is an important component of Portland cement. It is 44.4% calcium and 20.0% aluminum by weight. The remainder is oxygen.
   a. Calculate the empirical formula of tricalcium aluminate.
   b. The structure of tricalcium aluminate was not determined until 1975. The $Al_6O_{18}{}^{18-}$ anion has the following structure:

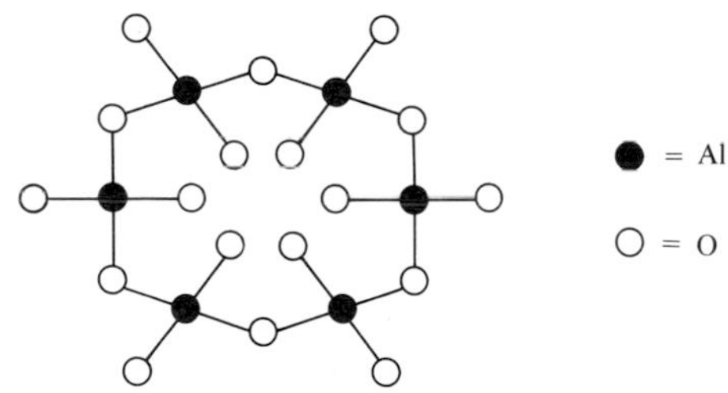

   What is the molecular formula of tricalcium aluminate?
   c. How would you describe the bonding in the $Al_6O_{18}{}^{18-}$ anion?

34. Write balanced equations describing the reactions of In with each of the following: F_2, Cl_2, O_2, and HCl.

35. Write a balanced equation describing the reaction of aluminum metal with concentrated aqueous sodium hydroxide.

36. Is Al_2O_3 an acidic, basic, or amphoteric oxide? Write balanced chemical equations to support your answer.

Group 4A Elements

37. Discuss the importance of the C—C and Si—Si bond strengths and π bonding to the properties of carbon and silicon.

38. Which bond would you expect to be more reactive, a Si—H bond or a C—H bond? Why?

39. Carbon suboxide (C_3O_2) has the following Lewis structure:

$$:\ddot{O}=C=C=C=\ddot{O}:$$

   a. What is the molecular geometry of C_3O_2?
   b. What hybrid orbitals are used by the carbon atoms in C_3O_2?

40. Use bond energies (see Table 8.4) to estimate ΔH for the following reaction.

$$(HO)_2C{=}O \longrightarrow CO_2 + H_2O$$

   Why is it better to write carbonic acid as $CO_2(aq)$ rather than H_2CO_3?

41. Carbon and sulfur form compounds with the formulas CS_2 and C_3S_2. Draw Lewis structures and predict the shapes of these two compounds.

42. Why are the tin(IV) halides more volatile than the tin(II) halides?

43. From the information on the temperature stability of white and gray tin given in this chapter, which form would you expect to have the more ordered structure?

44. The compounds CCl_4 and H_2O do not react with each other. On the other hand, silicon tetrachloride reacts with water according to the equation:

$$SiCl_4(l) + 2H_2O(l) \rightarrow SiO_2(s) + 4HCl(aq)$$

   Discuss the importance of thermodynamics and kinetics in the reactivity of water with $SiCl_4$ as compared to its lack of reactivity with CCl_4.

45. Stannous fluoride (SnF_2) is one of the compounds used to introduce F^- ions into toothpaste. Give another name for stannous fluoride.

46. Silicon carbide (SiC) is an extremely hard substance. Propose a structure for SiC.

47. Hydrofluoric acid (HF) cannot be stored in glass because it reacts with silica (SiO_2) to form the volatile SiF_4. Write a balanced equation for the reaction of HF with glass.

48. What is the proportion of lead(II) to lead(IV) in red lead (Pb_3O_4)?

49. Calculate $\mathscr{E}°$ for the reaction

$$2H_2O(l) + 2Pb(s) + O_2(g) \rightarrow 2Pb(OH)_2(s)$$

which is analogous to the corrosion of iron by O_2 (see Chapter 17). Are lead pipes more easily corroded by oxygen than iron pipes? For $Pb(OH)_2 + 2e^- \rightarrow Pb + 2OH^-$, $\mathscr{E}° = -0.57$ V.

50. Calculate the solubility of $Pb(OH)_2$ in water (in mol/L), given that $K_{sp} = 1.2 \times 10^{-15}$.

51. Which Group 4A elements are capable of reducing H^+ to H_2? Write balanced equations describing the reactions.

Additional Exercises

52. Compounds containing the sodide ion (Na^-) have recently been made.

$$2Na + Crypt \rightarrow [Na(Crypt)]^+Na^-$$

where Crypt represents a cryptand $N[(C_2H_4O)_2C_2H_4]_3N$.

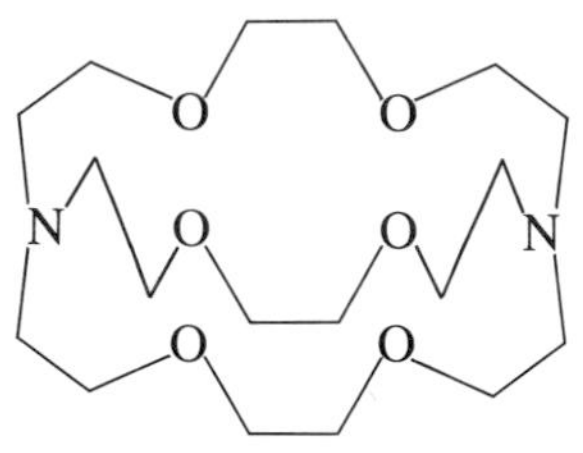

The cryptand encapsulates the Na^+ ion. Why is it necessary to encapsulate Na^+? (See Exercise 71 in Chapter 7.)

53. The Group 1A metals are extremely soluble in liquid ammonia, releasing hydrogen gas on dissolution. As the metal begins to dissolve, a deep blue solution forms. The color is attributed to the presence of solvated electrons, or $e(NH_3)_x^-$. The metal amide (MNH_2) can be isolated by allowing the ammonia to evaporate.
   a. Write a balanced equation for the reaction for the formation of $NaNH_2$.
   b. The solubility of sodium in liquid ammonia at -33.5°C is 251.4 g/kg. Calculate the mass percent, mole fraction, and molality of sodium in this solution.

54. Ionic compounds, such as $KMnO_4$, can be dissolved in nonpolar solvents by adding *crown ethers*. The structure of a typical crown ether is as follows:

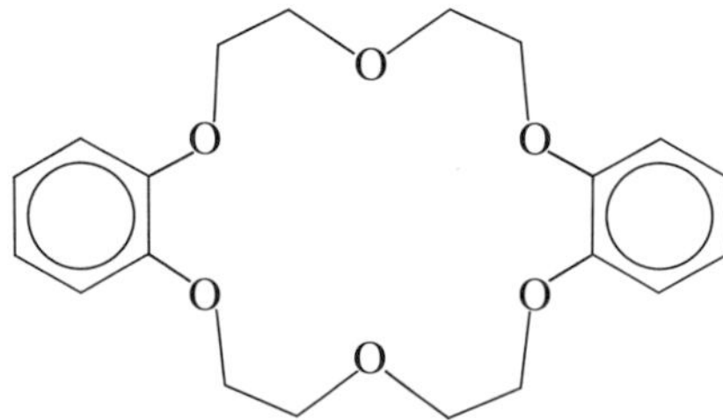

Suggest how crown ethers make $KMnO_4$ soluble in nonpolar solvents.

55. Beryllium is amphoteric, in contrast to its fellow Group 2A metals. Beryllium metal reacts with aqueous NaOH to produce hydrogen gas and $Be(OH)_4^{2-}$. Write a balanced equation for this reaction. What is the oxidizing agent? What is the reducing agent?

56. Give a formula for each binary compound:
   a. beryllium nitride
   b. strontium peroxide

57. The compound $BeSO_4 \cdot 4H_2O$ cannot be easily dehydrated by heating. It dissolves in water to give an acidic solution. How do you account for these observations?

58. Elemental boron is produced by reduction of boron oxide with magnesium to give boron and magnesium oxide. Write a balanced equation for this reaction.

59. The compound with the formula TlI_3 is a black solid. Given the following standard reduction potentials:

$$Tl^{3+} + 2e^- \rightarrow Tl^+ \qquad \mathscr{E}° = +1.25 \text{ V}$$

$$I_3^- + 2e^- \rightarrow 3I^- \qquad \mathscr{E}° = +0.55 \text{ V}$$

would you formulate this compound as thallium(III) iodide or thallium(I) triodide?

60. The compound Ga_2Cl_4 is brown. How could you determine experimentally whether this compound contains two gallium(II) ions or one gallium(I) and one gallium(III) ion? (*Hint:* Consider the electron configurations of the three possible ions.)

61. The resistivity (a measure of electrical resistance) of graphite is (0.4 to 5.0) $\times 10^{-4}$ ohm-cm in the basal plane. (The basal plane is the plane of the six-membered rings of carbon atoms.) The resistivity is 0.2 to 1.0 ohm-cm along the axis perpendicular to the plane. The resistivity of diamond is 10^{14} to 10^{16} ohm-cm and is independent of direction. How can you account for this behavior in terms of the structures of graphite and diamond?

62. Why is graphite a good lubricant? What advantages does it have over grease- or oil-based lubricants?

63. Elemental silicon was not isolated until 1823 when J. J. Berzelius succeeded in reducing K_2SiF_6 with molten potassium.
   a. Write a balanced equation for this reaction.
   b. Describe the bonding in K_2SiF_6.

64. Dioctyltin compounds are used as stabilizers for polyvinyl chloride polymers. (See Chapter 24.) They can be produced by this reaction:

$$2CH_3CH_2CH_2CH_2CH_2CH_2CH_2CH_2X + Sn \rightarrow (C_8H_{17})_2SnX_2$$

where X is a halogen. Predict the structure of $(C_8H_{17})_2SnCl_2$.

65. One reason advanced for the instability of long chains of silicon atoms is that the decomposition involves the transition state shown below:

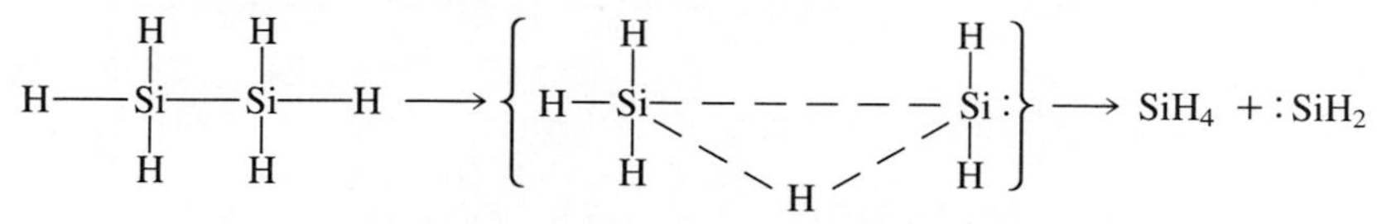

$$\mathrm{H_3Si{-}SiH_3} \longrightarrow \left\{ \mathrm{H{-}SiH_2} \cdots \mathrm{H} \cdots \mathrm{\ddot{S}iH_2} \right\} \longrightarrow \mathrm{SiH_4} + :\mathrm{SiH_2}$$

The activation energy for such a process is 210 kJ/mol, which is less than either the Si—Si or Si—H bond energy. Why would a similar mechanism not be expected to be very important in the decomposition of long carbon chains?

66. What are some of the structural differences between quartz and amorphous SiO_2?

67. Diagonal relationships in the periodic table exist as well as the vertical relationships. For example, Be and Al are similar in some of their properties as are B and Si. Rationalize why these diagonal relationships hold for properties such as size, ionization energy, and electron affinity.

68. What is the inert pair effect? How is it important in the properties of thallium and lead?

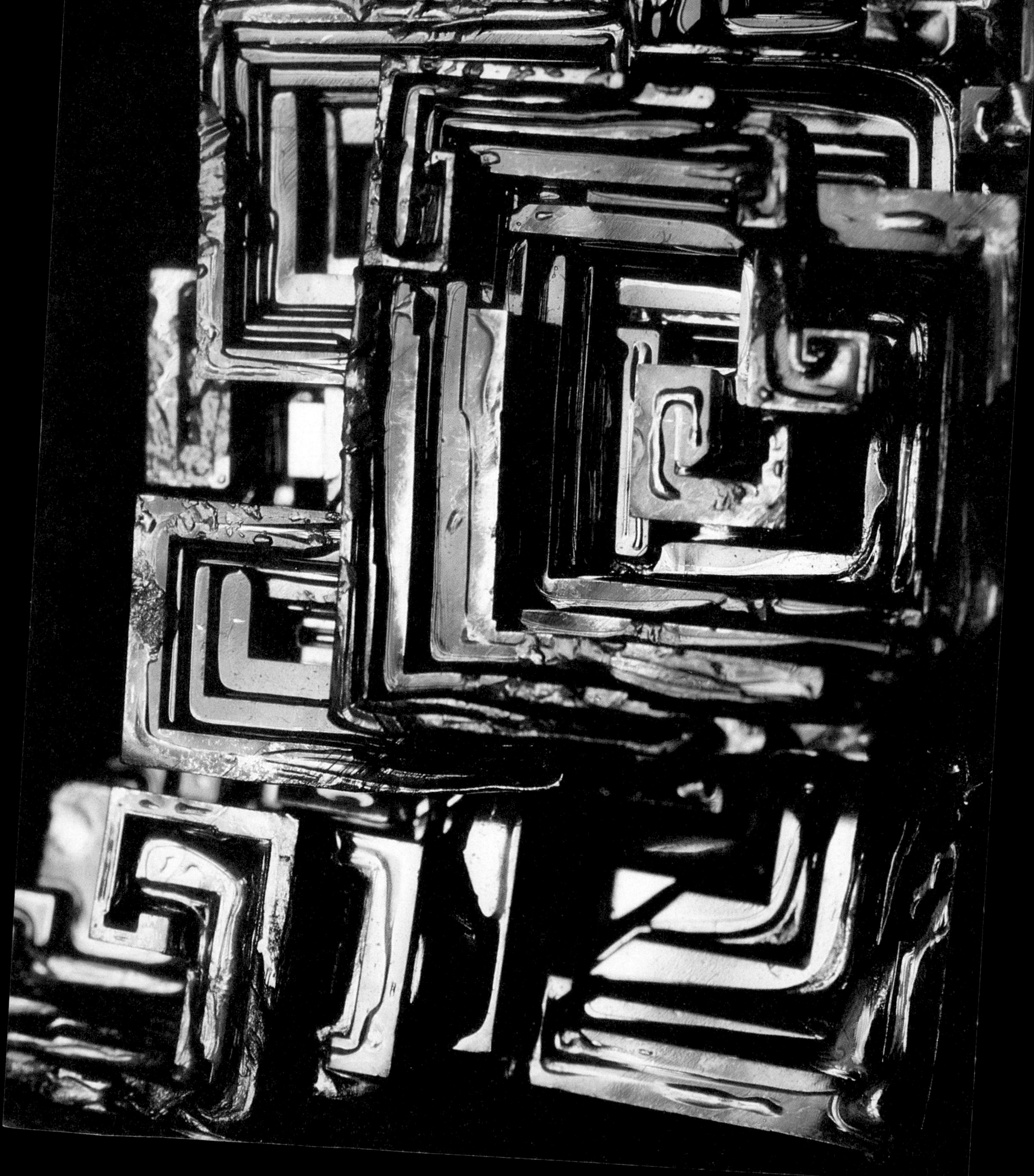

Representative Elements: Groups 5A Through 8A

In Chapter 18 we saw that vertical groups of elements tend to show similar chemical characteristics because they have identical valence-electron configurations. Generally, metallic character increases going down a group, as the electrons are further from the nucleus. The most dramatic change occurs after the first group member and reflects the large increase in atomic size.

As we proceed from Group 1A to Group 7A, the elements change from active metals (electron donors) to strong nonmetals (electron acceptors). Thus it is not surprising that the middle groups show the most varied chemistry: some group members behave principally as metals, others behave mainly as nonmetals, and some show both tendencies. The elements in Groups 5A and 6A show great chemical variety and form many compounds of considerable practical value. The halogens (Group 7A) are nonmetals that are also found in many everyday substances. The elements in Group 8A (the noble gases) are most useful in their elemental forms, but their ability to form compounds, discovered only within the past 30 years, has provided important tests for the theories of chemical bonding.

In this chapter we give an overview of the elements in Groups 5A through 8A, concentrating on the chemistry of the most important elements in these groups: nitrogen, phosphorus, oxygen, sulfur, and the halogens.

CONTENTS

< Bismuth crystals. An element in Group 5A, bismuth exhibits mostly metallic properties.

5A
N
P
As
Sb
Bi

19.1 The Group 5A Elements

Purpose

- To give an overview of the characteristics of the Group 5A elements, with emphasis on bonding.

The Group 5A elements (with the valence-electron configuration ns^2np^3) are prepared as shown in Table 19.1, and they show remarkably varied chemical properties. As usual, metallic character increases going down the group, as is apparent from the electronegativity values (Table 19.1). Nitrogen and phosphorus are nonmetals that can gain three electrons to form 3− anions in salts with active metals, for example, magnesium nitride (Mg_3N_2) and beryllium phosphide (Be_3P_2). The chemistry of these two important elements is discussed in the next two sections.

Selected Physical Properties, Sources, and Methods of Preparation for the Group 5A Elements

Element	Electro-negativity	Sources	Method of preparation
Nitrogen	3.0	Air	Liquefaction of air
Phosphorus	2.1	Phosphate rock ($Ca_3(PO_4)_2$), fluorapatite ($Ca_5(PO_4)_3F$)	$2Ca_3(PO_4)_2 + 6SiO_2 \rightarrow 6CaSiO_3 + P_4O_{10}$ $P_4O_{10} + 10C \rightarrow 4P + 10CO$
Arsenic	2.0	Arsenopyrite (Fe_3As_2, FeS)	Heating arsenopyrite in the absence of air
Antimony	1.9	Stibnite (Sb_2S_3)	Roasting Sb_2S_3 in air to form Sb_2O_3 and then reduction with carbon
Bismuth	1.9	Bismite (Bi_2O_3), bismuth glance (Bi_2S_3)	Roasting Bi_2S_3 in air to form Bi_2O_3 and then reduction with carbon

Table 19.1

Bismuth and *antimony* tend to be metallic, readily losing electrons to form cations. Although these elements have five valence electrons, so much energy is required to remove all five that no ionic compounds containing Bi^{5+} or Sb^{5+} ions are known. Three pentahalides (BiF_5, $SbCl_5$, and SbF_5) are known, but these are molecular rather than ionic compounds. The fact that no other pentahalides of these elements are known no doubt results from the strong oxidizing ability of bismuth and antimony in the +5 oxidation state. In fact, BiF_5 is an excellent fluorinating agent, readily decomposing to fluorine and bismuth trifluoride:

$$BiF_5 \rightarrow BiF_3 + F_2$$

Salts containing Bi^{3+} or Sb^{3+} ions, such as $Sb_2(SO_4)_3$ and $Bi(NO_3)_3$, are quite common. When these salts are dissolved in water, the resulting hydrated cations are very acidic. For example, the reaction of the Bi^{3+} ion with water can be represented as

$$Bi^{3+}(aq) + H_2O(l) \rightarrow BiO^+(aq) + 2H^+(aq)$$

where BiO^+ is called the *bismuthyl ion*. If chloride ion is added to this solution, a white salt called bismuthyl chloride (BiOCl) precipitates. Antimony exhibits similar chemistry in aqueous solution.

The Group 5A elements can form molecules or ions that involve three, five, or six covalent bonds to the Group 5A atom. Examples involving three single bonds are NH_3, PH_3, NF_3, and $AsCl_3$. Each of these molecules has a lone pair of electrons (and thus can behave as a Lewis base) and a pyramidal shape as predicted by the VSEPR model (see Fig. 19.1).

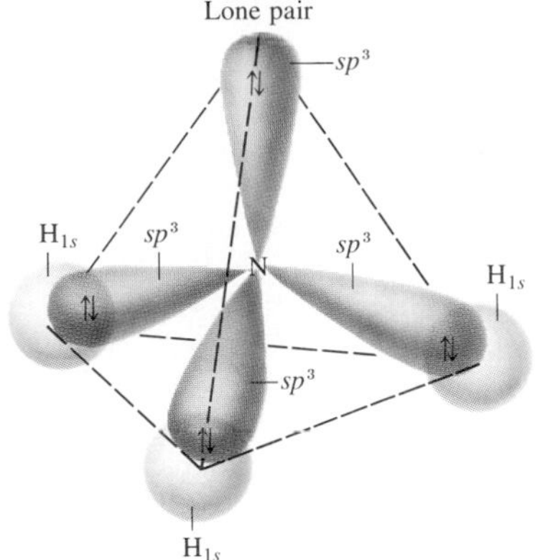

Figure 19.1

The pyramidal shape of the Group 5A MX_3 molecules.

All of the Group 5A elements except nitrogen can form molecules with five covalent bonds (of general formula MX_5). Nitrogen cannot form such molecules because of its small size and lack of available *d* orbitals. The MX_5 molecules have a trigonal bipyramidal shape (see Fig. 19.2) as predicted by the VSEPR model, and the central atom is described as dsp^3 hybridized. The MX_5 molecules can accept an additional electron pair to form ionic species containing six covalent bonds, for example,

$$PF_5 + F^- \rightarrow PF_6^-$$

where the PF_6^- anion has an octahedral shape (see Fig. 19.3) and the phosphorus atom is described as d^2sp^3 hybridized.

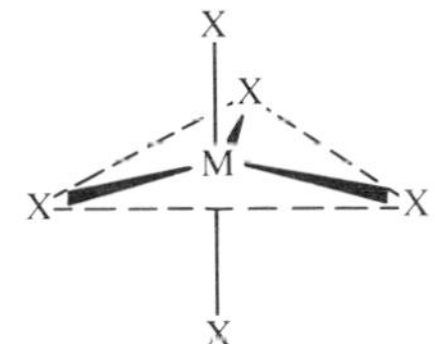

Figure 19.2

The trigonal bipyramidal shape of the MX_5 molecules.

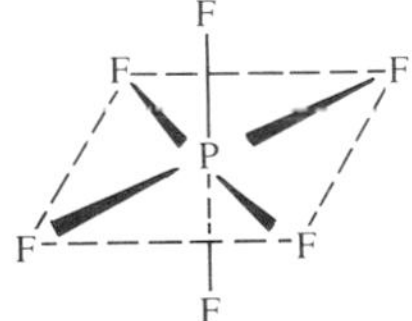

Figure 19.3

The octahedral PF_6^- ion.

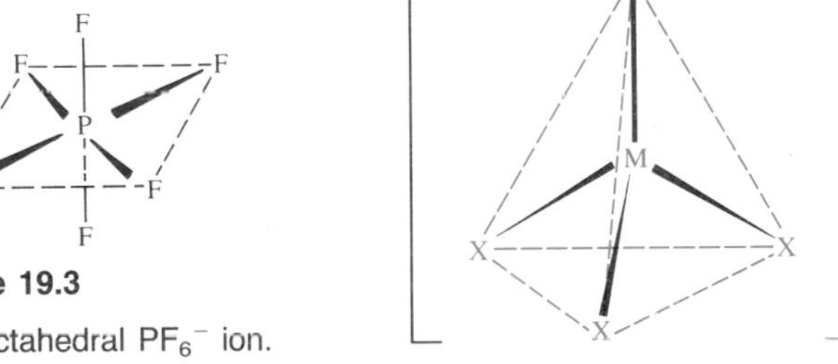

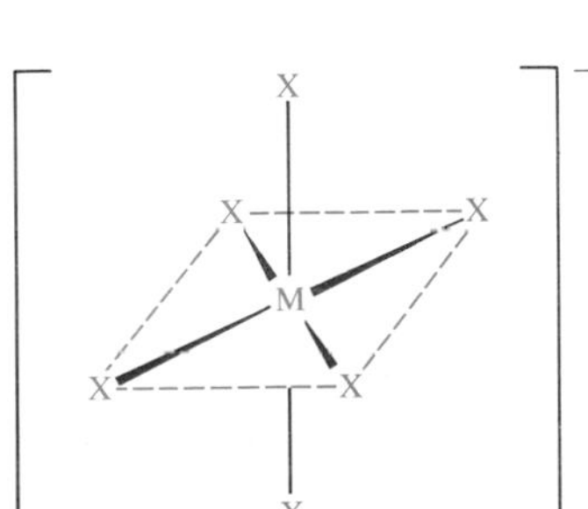

Figure 19.4

The structures of the tetrahedral MX_4^+ and octahedral MX_6^- ions.

Although the MX_5 molecules have a trigonal bipyramidal structure in the gas phase, the solids of many of these compounds contain the ions MX_4^+ and MX_6^- (Fig. 19.4) where the MX_4^+ cation is tetrahedral (the atom represented by M is sp^3 hybridized) and the MX_6^- anion is octahedral (the atom represented by M is d^2sp^3 hybridized). Examples are PCl_5, which in the solid state contains PCl_4^+ and PCl_6^-; and AsF_3Cl_2, which in the solid state contains $AsCl_4^+$ and AsF_6 .

As discussed in Section 18.1, the ability of the Group 5A elements to form π bonds decreases dramatically after nitrogen. This explains why elemental nitrogen exists as N_2 molecules, whereas the other elements in the group exist as larger aggregates containing single bonds. For example, in the gas phase, the elements phosphorus, arsenic, and antimony consist of P_4, As_4, and Sb_4 molecules, respectively.

19.2 The Chemistry of Nitrogen

Purpose

- To discuss the kinetic and thermodynamic importance of the strength of the triple bond in the nitrogen molecule.
- To introduce the nitrogen cycle.
- To describe the structures and uses of major nitrogen-containing compounds.

At the earth's surface, virtually all elemental nitrogen exists as the N_2 molecule with its very strong triple bond (941 kJ/mol). Because of this high bond strength, the N_2 molecule is so unreactive that it can coexist with most other elements under normal conditions without undergoing any appreciable reaction. This property makes nitrogen gas very useful as a medium for experiments involving substances that react with oxygen or water. Such experiments can be done using an inert atmosphere box of the type shown in Fig. 19.5.

The strength of the triple bond in the N_2 molecule is important both thermodynamically and kinetically. Thermodynamically, the great stability of the $N{\equiv}N$ bond means that most binary compounds containing nitrogen decompose exothermically to the elements, for example:

$$N_2O(g) \rightarrow N_2(g) + \tfrac{1}{2}O_2(g) \qquad \Delta H^\circ = -82 \text{ kJ}$$
$$NO(g) \rightarrow \tfrac{1}{2}N_2(g) + \tfrac{1}{2}O_2(g) \qquad \Delta H^\circ = -90 \text{ kJ}$$
$$NO_2(g) \rightarrow \tfrac{1}{2}N_2(g) + O_2(g) \qquad \Delta H^\circ = -34 \text{ kJ}$$
$$N_2H_4(g) \rightarrow N_2(g) + 2H_2(g) \qquad \Delta H^\circ = -95 \text{ kJ}$$
$$NH_3(g) \rightarrow \tfrac{1}{2}N_2(g) + \tfrac{3}{2}H_2(g) \qquad \Delta H^\circ = +46 \text{ kJ}$$

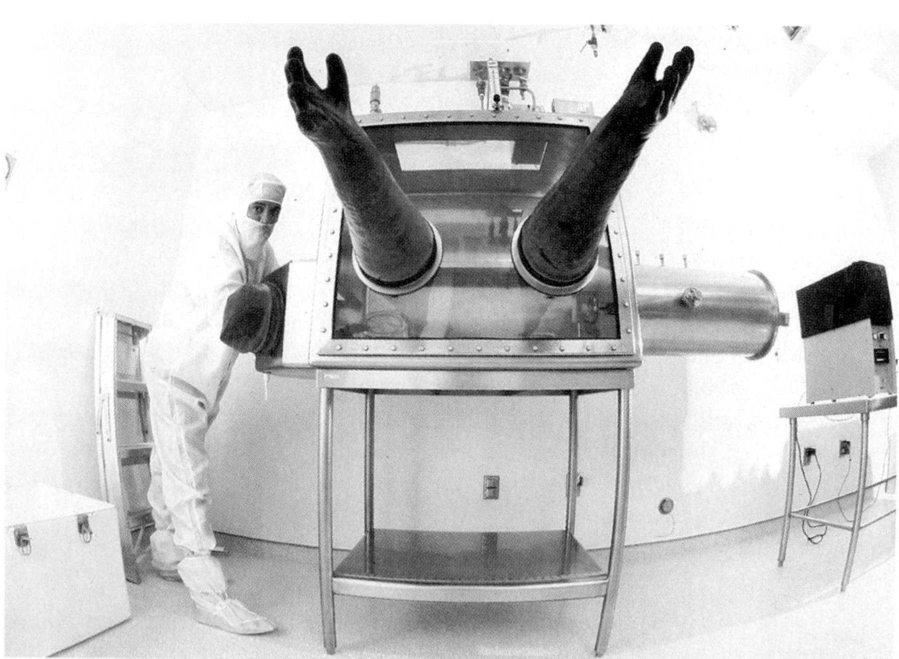

Figure 19.5

An inert atmosphere box used when working with oxygen- or water-sensitive materials. The box is filled with an inert gas such as nitrogen, and work is done through the ports fitted with large rubber gloves.

Of these compounds only ammonia is thermodynamically more stable than its component elements. That is, only for ammonia is energy required (endothermic process, positive value of ΔH°) to decompose the molecule to its elements. For the remaining molecules, energy is released when decomposition to the elements occurs, as a result of the great stability of N_2.

The importance of the thermodynamic stability of N_2 can be clearly seen in the power of nitrogen-based explosives, such as nitroglycerin ($C_3H_5N_3O_9$), which has the structure

```
      H         H         H
      |         |         |
   H—C—————————C—————————C—H
      |         |         |
      O         O         O
      |         |         |
      N         N         N
     / \       / \       / \
    O   O     O   O     O   O
```

Chemical explosives are used to demolish a parking garage in Boston.

When ignited or subjected to sudden impact, nitroglycerin decomposes very rapidly and exothermically:

$$4C_3H_5N_3O_9(l) \rightarrow 6N_2(g) + 12CO_2(g) + 10H_2O(g) + O_2(g) + \text{energy}$$

An explosion occurs; that is, large volumes of gas are produced in a fast, highly exothermic reaction. Note that 4 moles of liquid nitroglycerin produce 29 (6 + 12 + 10 + 1) moles of gaseous products. This alone produces a large increase in volume. However, also note that the products, which include N_2, are very stable molecules with strong bonds. Their formation is therefore accompanied by the release of large quantities of energy as heat, which increases the gaseous volume. The hot, rapidly expanding gases produce a pressure surge and damaging shock wave.

Most high explosives are organic compounds that, like nitroglycerin, contain nitro ($—NO_2$) groups and produce nitrogen and other gases as products. Another example is *trinitrotoluene,* or TNT, a solid at normal temperatures, which decomposes as follows:

$$2C_7H_5N_3O_6(s) \rightarrow 12CO(g) + 5H_2(g) + 3N_2(g) + 2C(s) + \text{energy}$$

CH_3 / NO_2 / NO_2 / NO_2

TNT

Note that 2 moles of solid TNT produce 20 moles of gaseous products plus energy.

The effect of bond strength on the kinetics of reactions involving the N_2 molecule is illustrated by the synthesis of ammonia from nitrogen and hydrogen, a reaction we have discussed many times before. Because a large quantity of energy is required to disrupt the $N \equiv N$ bond, the ammonia synthesis reaction has a negligible rate at room temperature, even though the equilibrium constant is very large ($K \approx 10^6$) at 25°C. Of course, the most direct way to increase the rate is to raise the temperature, but since the reaction is very exothermic,

$$N_2(g) + 3H_2(g) \rightarrow 2NH_3(g) \qquad \Delta H^\circ = -92 \text{ kJ}$$

the value of K decreases significantly with a temperature increase (in fact, at 500°C, $K \approx 10^{-2}$).

Obviously, the kinetics and the thermodynamics of this reaction are in opposition. A compromise must be reached, involving high pressure to force the equilibrium to the right and high temperature to produce a reasonable rate. The **Haber process** for manufacturing ammonia represents such a compromise (see Fig. 19.6).

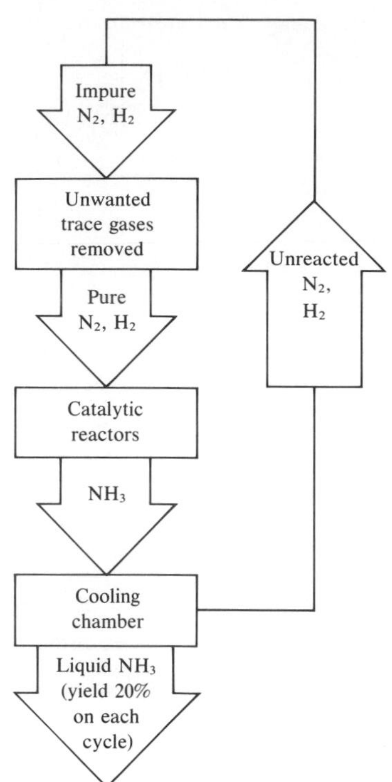

Figure 19.6

A schematic diagram of the Haber process for the manufacture of ammonia.

The process is carried out at a pressure of about 250 atmospheres and a temperature of approximately 400°C. Even higher temperatures would be required except that a catalyst, consisting of a solid iron oxide mixed with small amounts of potassium oxide and aluminum oxide, greatly facilitates the reaction.

Nitrogen is essential to living systems. The problem with nitrogen is not one of supply—we are surrounded by it—but of changing it from the inert N_2 molecule to a form usable by plants and animals. The process of transforming N_2 to other nitrogen-containing compounds is called **nitrogen fixation.** The Haber process is one example of nitrogen fixation. The ammonia produced can be applied to the soil as a fertilizer, since plants can readily employ the nitrogen in ammonia to make the nitrogen-containing biomolecules essential for their growth.

Nitrogen fixation also results from the high-temperature combustion process in automobile engines. The nitrogen in the air drawn into the engine reacts at a significant rate with oxygen to form nitric oxide (NO), which further reacts with oxygen from the air to form nitrogen dioxide (NO_2). This nitrogen dioxide, which is an important contributor to photochemical smog in many urban areas (see Section 12.8), eventually reacts with moisture in the air and reaches the soil to form nitrate salts, which are plant nutrients.

There are also natural nitrogen fixation processes. For example, lightning provides the energy to disrupt N_2 and O_2 molecules in the air, producing highly reactive nitrogen and oxygen atoms that attack other N_2 and O_2 molecules to form nitrogen oxides that eventually become nitrates. Approximately 30 million tons of nitrates are produced this way annually. Another natural nitrogen fixation process is provided by bacteria that reside in the root nodules of plants such as beans, peas, and alfalfa. These **nitrogen-fixing bacteria** readily allow the conversion of nitrogen to ammonia and other nitrogen-containing compounds useful to plants. The efficiency of these bacteria is intriguing: they produce ammonia at soil temperatures and 1 atmosphere pressure, while the Haber process requires severe conditions of 400°C and 250 atmospheres. For obvious reasons, researchers are studying these bacteria intensively.

Nodules on the roots of peas that contain nitrogen-fixing bacteria.

When plants and animals die, they decompose, and the elements they consist of are returned to the environment. In the case of nitrogen, the return of the element to the atmosphere as nitrogen gas, called **denitrification,** is carried out by bacteria that change nitrates to nitrogen. The complex **nitrogen cycle** is summarized in Fig. 19.7. It has been estimated that as much as 10 million tons per year more nitrogen is currently being fixed by natural and human processes than is being returned to the atmosphere. This fixed nitrogen is accumulating in the soil, lakes, rivers, and oceans, where it can promote the growth of algae and other undesirable organisms.

Sample Exercise 19.1

When ammonium nitrite is heated, it decomposes to nitrogen gas and water. Calculate the volume of N_2 gas produced from 1.00 g of solid NH_4NO_2 at 250°C and 1.00 atm.

Solution

The decomposition reaction is

$$NH_4NO_2(s) \xrightarrow{\text{Heat}} N_2(g) + 2H_2O(g)$$

Sample Exercise 19.1, continued

Using the molecular weight of NH_4NO_2 (64.05), we first calculate the moles of NH_4NO_2:

$$1.00 \text{ g } NH_4NO_2 \times \frac{1 \text{ mol } NH_4NO_2}{64.05 \text{ g } NH_4NO_2} = 1.56 \times 10^{-2} \text{ mol } NH_4NO_2$$

Since 1 mol of N_2 is produced for each mole of NH_4NO_2, 1.56×10^{-2} mol of N_2 will be produced in the given experiment. We can calculate the volume of N_2 from thc ideal gas law:

$$PV = nRT$$

In this case we have

$$P = 1.00 \text{ atm}$$
$$n = 1.56 \times 10^{-2} \text{ mol}$$
$$R = 0.08206 \text{ L atm/K mol}$$
$$T = 250 + 273 = 523 \text{ K}$$

and the volume of N_2 is

$$V = \frac{nRT}{P} = \frac{(1.56 \times 10^{-2} \text{ mol})\left(0.08206 \frac{\text{L atm}}{\text{K mol}}\right)(523 \text{ K})}{1.00 \text{ atm}}$$
$$= 0.670 \text{ L}$$

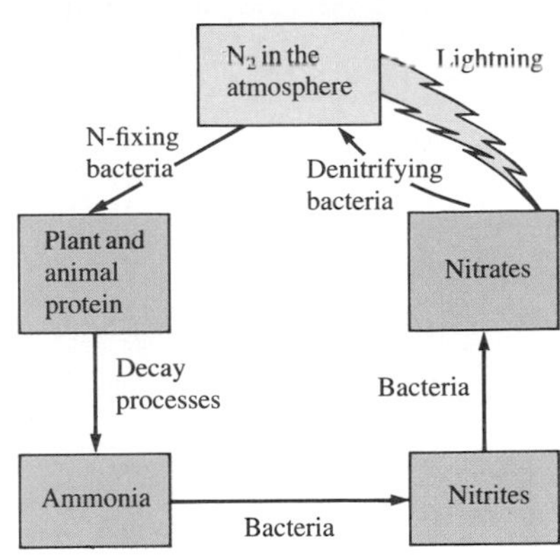

Figure 19.7

The nitrogen cycle. To be used by plants and animals, nitrogen must be converted from N_2 to nitrogen-containing compounds, such as nitrates, ammonia, and proteins. The nitrogen is returned to the atmosphere by natural decay processes.

Nitrogen Hydrides

By far the most important hydride of nitrogen is **ammonia.** A toxic, colorless gas with a pungent odor, ammonia is manufactured in huge quantities (~30 billion pounds per year), mainly for use in fertilizers.

The pyramidal ammonia molecule has a lone pair of electrons on the nitrogen atom (see Fig. 19.1) and polar N—H bonds. This structure leads to a high degree of intermolecular interaction by hydrogen bonding in the liquid state and produces an unusually high boiling point (−33.4°C) for a substance of such low molecular weight. Note, however, that the hydrogen bonding in liquid ammonia is clearly not as important as that in liquid water, which has about the same molecular weight but a much higher boiling point. The water molecule has two polar bonds involving hydrogen and two lone pairs—the right combination for optimum hydrogen bonding—in contrast to the one lone pair and three polar bonds of the ammonia molecule.

Anhydrous ammonia being used as fertilizer.

As we saw in Chapter 14, ammonia behaves as a base and reacts with acids to produce ammonium salts, for example,

$$NH_3(g) + HCl(g) \rightarrow NH_4Cl(s)$$

A second nitrogen hydride of major importance is **hydrazine** (N_2H_4). The Lewis structure of hydrazine (shown on the following page)

$$\begin{array}{ccc} H & & H \\ & \ddot{N}-\ddot{N} & \\ H & & H \end{array}$$

indicates that each nitrogen atom should be sp^3 hybridized with bond angles close to 109.5° (the tetrahedral angle) since the nitrogen atom is surrounded by four electron pairs. The observed structure with bond angles of 112° (see Fig. 19.8) agrees with these predictions. Hydrazine, a colorless liquid with an ammoniacal odor, freezes at 2°C and boils at 113.5°C. This boiling point is quite high for a compound with a molecular weight of 32; this suggests that considerable hydrogen bonding must occur among the polar hydrazine molecules.

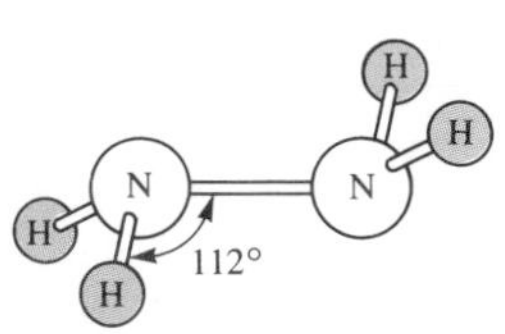

Figure 19.8

The molecular structure of hydrazine (N_2H_4). This arrangement minimizes the repulsion between the lone pairs on the nitrogen atoms by placing them on opposite sides.

Hydrazine is a powerful reducing agent and has been widely used as a rocket propellant. For example, its reaction with oxygen is highly exothermic:

$$N_2H_4(l) + O_2(g) \longrightarrow N_2(g) + 2H_2O(g) \qquad \Delta H^\circ = -622 \text{ kJ}$$

Since hydrazine also reacts vigorously with the halogens, fluorine is often used instead of oxygen as the oxidizer in rocket engines. Substituted hydrazines, where one or more of the hydrogen atoms are replaced by other groups, are also useful rocket fuels. For example, monomethylhydrazine,

$$\begin{array}{ccc} CH_3 & & H \\ & N-N & \\ H & & H \end{array}$$

is used with the oxidizer dinitrogen tetroxide (N_2O_4) to power the U.S. space shuttle orbiter. The reaction is

$$5N_2O_4(l) + 4N_2H_3(CH_3)(l) \longrightarrow 12H_2O(g) + 9N_2(g) + 4CO_2(g)$$

Because of the large number of gaseous molecules produced and the exothermic nature of this reaction, a very high thrust per weight of fuel is achieved. The reaction is also self-starting—it begins immediately when the fuels are mixed—which is a useful property for rocket engines that must be started and stopped frequently.

The space shuttle orbiter uses monomethylhydrazine mixed with an oxidizer as fuel.

Sample Exercise 19.2

Using the bond energies in Table 8.4 (Chapter 8), calculate the approximate value of ΔH for the reaction between gaseous monomethylhydrazine and dinitrogen tetroxide:

$$5N_2O_4(g) + 4N_2H_3(CH_3)(g) \longrightarrow 12H_2O(g) + 9N_2(g) + 4CO_2(g)$$

The bonding in N_2O_4 is described by resonance structures that predict the N—O bonds are intermediate in strength between single and double bonds (assume an average N—O bond energy of 440 kJ/mol).

Solution

To calculate ΔH for this reaction, we must compare the energy necessary to break the bonds of the reactants and the energy released by formation of the bonds in the products

Sample Exercise 19.2, continued

$$5\ \mathrm{O_2N{-}NO_2} + 4\ \mathrm{(H)_2N{-}N(H)(CH_3)} \longrightarrow 12\ \mathrm{H{-}O{-}H} + 9\mathrm{N{\equiv}N} + 4\mathrm{O{=}C{=}O}$$

Breaking bonds requires energy (positive sign), and forming bonds releases energy (negative sign), as summarized in the table below.

$$\Delta H = (19.9 \times 10^3\ \text{kJ}) - (26.1 \times 10^3\ \text{kJ}) = -6.2 \times 10^3\ \text{kJ}$$

The reaction is highly exothermic.

Bonds broken	Energy required (kJ/mol)	Bonds formed	Energy released (kJ/mol)
$5 \times 4 = 20$ N=O	$20 \times 440 = 8.8 \times 10^3$	$12 \times 2 = 24$ O—H	$24 \times 467 = 1.12 \times 10^4$
$5 + 4 = 9$ N—N	$9 \times 160 = 1.4 \times 10^3$	9 N≡N	$9 \times 941 = 8.5 \times 10^3$
$4 \times 3 = 12$ N—H	$12 \times 391 = 4.7 \times 10^3$	$4 \times 2 = 8$ C=O	$8 \times 799 = 6.4 \times 10^3$
$4 \times 3 = 12$ C—H	$12 \times 413 = 5.0 \times 10^3$		
Total	19.9×10^3	Total	26.1×10^3

The use of hydrazine as a rocket propellant is a rather specialized application. The main industrial use of hydrazine is as a "blowing" agent in the manufacture of plastics. Hydrazine decomposes to form nitrogen gas, which causes foaming in the liquid plastic, which results in a porous texture. Another major use of hydrazine is in the production of agricultural pesticides. Of the many hundreds of hydrazine derivatives (substituted hydrazines) that have been tested, 40 are used as fungicides, herbicides, insecticides, and plant growth regulators.

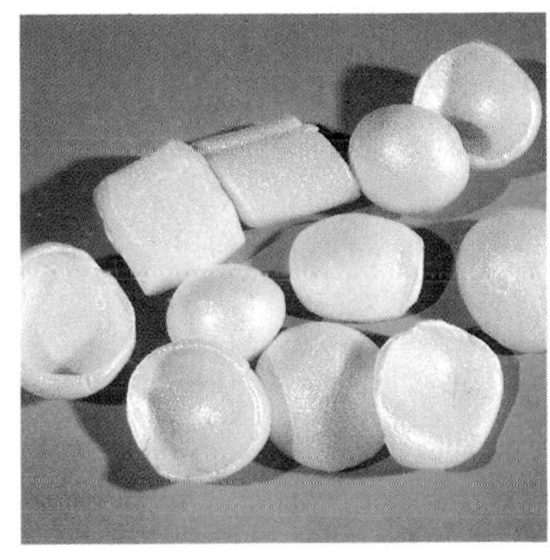

Blowing agents, such as hydrazine which forms nitrogen gas on decomposition, are used to produce porous plastics like these styrofoam chips.

The manufacture of hydrazine involves the oxidation of ammonia by the hypochlorite ion in basic solution:

$$2\mathrm{NH_3}(aq) + \mathrm{OCl^-}(aq) \longrightarrow \mathrm{N_2H_4}(aq) + \mathrm{Cl^-}(aq) + \mathrm{H_2O}(l)$$

Although this reaction looks straightforward, the actual process involves many steps and requires high pressure, high temperature, and catalysis to optimize the yield of hydrazine in the face of many competing reactions.

Nitrogen Oxides

Nitrogen forms a series of oxides in which it has an oxidation state from +1 to +5, as shown in Table 19.2 on the following page.

Dinitrogen monoxide (N_2O), more commonly called *nitrous oxide* or laughing gas, has an inebriating effect and is often used as a mild anesthetic by dentists. Because of its high solubility in fats, nitrous oxide is widely used as a propellant in aerosol cans of whipped cream; it is dissolved in the liquid in the can at high pressure and forms bubbles that produce foaming as the liquid is released from the can. A significant amount of N_2O exists in the atmosphere, mostly produced by soil

Some Common Nitrogen Compounds

Oxidation state of nitrogen	Compound	Formula	Lewis structure*
−3	Ammonia	NH_3	H—N(:)(—H)—H
−2	Hydrazine	N_2H_4	H—N(:)(H)—N(:)(H)—H
−1	Hydroxylamine	NH_2OH	H—N(:)(H)—O(::)—H
0	Nitrogen	N_2	:N≡N:
+1	Dinitrogen monoxide (nitrous oxide)	N_2O	:N̈=N=Ö:
+2	Nitrogen monoxide (nitric oxide)	NO	:Ṅ=Ö:
+3	Dinitrogen trioxide	N_2O_3	(:Ö:)—N(=Ö:)—N̈=Ö:
+4	Nitrogen dioxide	NO_2	:Ö—Ṅ=Ö
+5	Nitric acid	HNO_3	:Ö—N(=Ö:)—Ö—H

*In some cases, additional resonance structures are needed to fully describe the electron distribution.

Table 19.2

microorganisms, and its concentration appears to be gradually increasing. Because it can strongly absorb infrared radiation, nitrous oxide plays a small but probably significant role in controlling the earth's temperature in the same way that atmospheric carbon dioxide and water vapor do (see the discussion of the greenhouse effect in Section 6.5). Some scientists fear that the rapid decrease of tropical rain forests resulting from the development of countries such as Brazil will significantly affect the rate of production of N_2O by soil organisms, and thus will have important effects on the earth's temperature.

In the laboratory, nitrous oxide is prepared by the thermal decomposition of ammonium nitrate:

$$NH_4NO_3(s) \xrightarrow{\text{Heat}} N_2O(g) + 2H_2O(g)$$

Do not **attempt this experiment unless you have the proper safety equipment.**

This experiment must be done carefully because ammonium nitrate can explode. In fact one of the greatest industrial disasters in U.S. history occurred in 1947 in

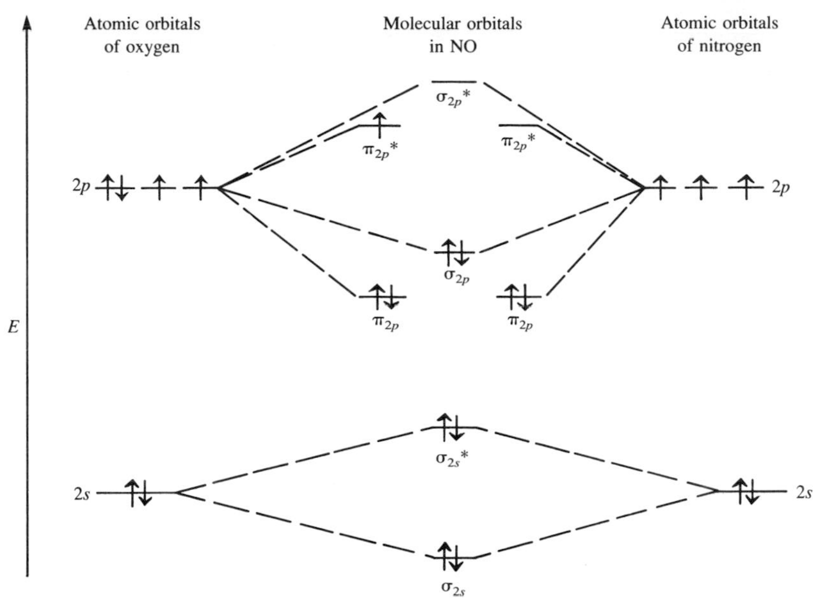

Figure 19.9

The molecular orbital energy-level diagram for nitric oxide (NO). The bond order is 2.5, or (8 − 3)/2.

Texas, when a ship loaded with ammonium nitrate for use as fertilizer exploded and killed nearly 600 people.

Nitrogen monoxide (NO), commonly called nitric oxide, is a colorless gas under normal conditions and can be produced in the laboratory by reacting 6 *M* nitric acid with copper metal:

$$8H^+(aq) + 2NO_3^-(aq) + 3Cu(s) \rightarrow 3Cu^{2+}(aq) + 4H_2O(l) + 2NO(g)$$

When this reaction is run in the air, the nitric oxide is immediately oxidized to brown nitrogen dioxide (NO_2).

Since the NO molecule has an odd number of electrons, it is most conveniently described in terms of the molecular orbital model. The molecular orbital energy-level diagram is shown in Fig. 19.9. Note that the NO molecule should be paramagnetic and have a bond order of 2.5, predictions that are supported by experimental observations. Since the NO molecule has one high-energy electron, it is not surprising that it can be rather easily oxidized to form NO^+, the *nitrosyl ion*. Because an antibonding electron is removed in going from NO to NO^+, the ion should have a stronger bond (the predicted bond order is 3) than the molecule. This is borne out by experiment. The bond lengths and bond energies for nitric oxide and the nitrosyl ion are shown in Table 19.3 on page 862. The nitrosyl ion is formed when nitric oxide and nitrogen dioxide are dissolved in concentrated sulfuric acid:

$$NO(g) + NO_2(g) + 3H_2SO_4(aq) \rightarrow 2NO^+(aq) + 3HSO_4^-(aq) + H_3O^+(aq)$$

The ionic compound $NO^+HSO_4^-$ can be isolated from this solution.

Nitric oxide is thermodynamically unstable and decomposes to nitrous oxide and nitrogen dioxide:

$$3NO(g) \rightarrow N_2O(g) + NO_2(g)$$

Nitrogen dioxide (NO_2) is also an odd-electron molecule and has a V-shaped structure. The brown, paramagnetic NO_2 molecule readily dimerizes to form dinitrogen tetroxide,

A copper penny reacts with nitric acid to produce NO gas, which is immediately oxidized in air to give brown NO_2.

Comparison of the Bond Lengths and Bond Energies for Nitric Oxide and the Nitrosyl Ion		
	NO	NO^+
Bond length (Å)	1.15	1.09
Bond energy (kJ/mol)	630	1020
Bond order (predicted by MO model)	2.5	3

Table 19.3

Production of NO_2 by power plants and automobiles leads to smog (Section 12.8).

$$2NO_2(g) \rightleftharpoons N_2O_4(g)$$

which is diamagnetic and colorless. The value of the equilibrium constant is ~1 for this process at 55°C and, since the dimerization is exothermic, K decreases as the temperature increases.

Sample Exercise 19.3

Use the molecular orbital model to predict the bond order and magnetism of the NO^- ion.

Solution

Using the energy-level diagram for the NO molecule in Fig. 19.9, we can see that NO^- has one more antibonding electron than NO. Thus there will be unpaired electrons in the two π_{2p}^* orbitals, and NO^- will be paramagnetic with a bond order of (8 − 4)/2, or 2. Note that the bond in the NO^- ion is weaker than that in the NO molecule.

The least common of the nitrogen oxides are *dinitrogen trioxide* (N_2O_3), a blue liquid that readily dissociates into gaseous nitric oxide and nitrogen dioxide, and *dinitrogen pentoxide* (N_2O_5), which under normal conditions is a solid that is best viewed as a mixture of NO_2^+ and NO_3^- ions. Although N_2O_5 molecules do exist in the gas phase, they readily dissociate to nitrogen dioxide and oxygen:

$$2N_2O_5(g) \rightleftharpoons 4NO_2(g) + O_2(g)$$

This reaction follows first-order kinetics, as was discussed in Section 12.4.

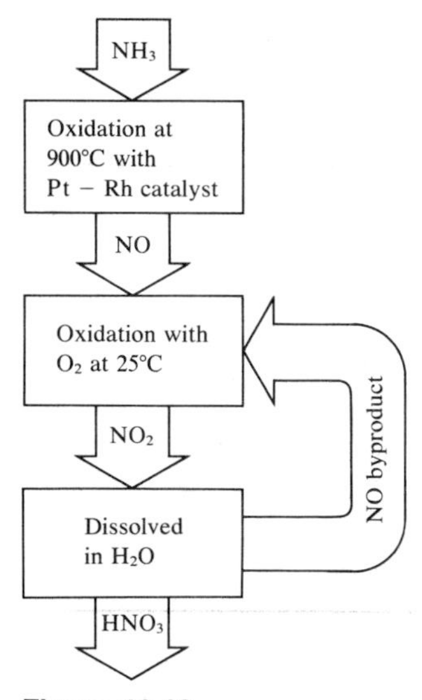

Figure 19.10

The Ostwald process.

Oxyacids of Nitrogen

Nitric acid is an important industrial chemical (~8 million tons are produced annually) used in the manufacture of many products, such as nitrogen-based explosives and ammonium nitrate for use as fertilizer.

Nitric acid is produced commercially by the oxidation of ammonia in the **Ostwald process** (see Fig. 19.10). In the first step of this process, ammonia is oxidized to nitric oxide:

$$4NH_3(g) + 5O_2(g) \rightarrow 4NO(g) + 6H_2O(g) \qquad \Delta H° = -905 \text{ kJ}$$

Although this reaction is highly exothermic, it is very slow at 25°C. There is also a side reaction between nitric oxide and ammonia:

$$4NH_3(g) + 6NO(g) \rightarrow 5N_2(g) + 6H_2O(g)$$

which is particularly undesirable because it traps the nitrogen as very unreactive N_2 molecules. To speed up the desired reaction and minimize the effects of the competing reaction, the ammonia oxidation is carried out using a catalyst of platinum-rhodium alloy heated to 900°C. Under these conditions there is a 97% conversion of the ammonia to nitric oxide.

In the second step nitric oxide is reacted with oxygen to produce nitrogen dioxide:

$$2NO(g) + O_2(g) \rightarrow 2NO_2(g) \qquad \Delta H° = -113 \text{ kJ}$$

This oxidation reaction has a rate constant that *decreases* with increasing temperature. Because of this very unusual behavior, the reaction is carried out at ~25°C and is kept at this temperature by cooling with water.

The third step in the Ostwald process is the absorption of nitrogen dioxide by water:

$$3NO_2(g) + H_2O(l) \rightarrow 2HNO_3(aq) + NO(g) \qquad \Delta H° = -139 \text{ kJ}$$

Preparation of a platinum-rhodium gauze for use in the manufacture of nitric acid.

The gaseous NO produced in the reaction is recycled to be oxidized to NO_2. The aqueous nitric acid from this process is about 50% HNO_3 by mass, which can be increased to 68% by distillation to remove some of the water. The maximum concentration attainable this way is 68% because nitric acid and water form an *azeotrope* at this concentration. The solution can be further concentrated to 95% HNO_3 by treatment with concentrated sulfuric acid, which strongly absorbs water; H_2SO_4 is often used as a *dehydrating (water-removing) agent*.

An *azeotrope* is a solution that, like a pure liquid, distills at a constant temperature without a change in composition.

Nitric acid is a colorless, fuming liquid (bp = 83°C) with a pungent odor; it decomposes in sunlight via the following reaction:

$$4HNO_3(l) \xrightarrow{h\nu} 4NO_2(g) + 2H_2O(l) + O_2(g)$$

As a result, nitric acid turns yellow as it ages because of the dissolved nitrogen dioxide. The common laboratory reagent called concentrated nitric acid is 15.9 *M* HNO_3 (70.4% HNO_3 by mass) and is a very strong oxidizing agent. The resonance structures and molecular structure of HNO_3 are shown in Fig. 19.11. Note that the hydrogen is bound to an oxygen atom, rather than to nitrogen as the formula suggests.

Nitric acid reacts with metal oxides, hydroxides, and carbonates and with other ionic compounds containing basic anions to form nitrate salts; for example,

$$Ca(OH)_2(s) + 2HNO_3(aq) \rightarrow Ca(NO_3)_2(aq) + 2H_2O(l)$$

H, O, N, O, O; ~105°; 120°

(a)

H—O—N(=O)—O ⟷ H—O—N(—O)=O

(b)

Figure 19.11

(a) The molecular structure of HNO_3. (b) The resonance structures of HNO_3.

Nitrate salts are generally very soluble in water.

Nitrous acid (HNO_2) is a weak acid

$$HNO_2(aq) \rightleftharpoons H^+(aq) + NO_2^-(aq) \qquad K_a = 4.0 \times 10^{-4}$$

that forms pale yellow nitrite (NO_2^-) salts. In contrast to nitrates, which are often used as explosives, nitrites are quite stable even at high temperatures. Nitrites are usually prepared by bubbling equal numbers of moles of nitric oxide and nitrogen dioxide into the appropriate aqueous solution of a metal hydroxide, for example,

$$NO(g) + NO_2(g) + 2NaOH(aq) \rightarrow 2NaNO_2(aq) + H_2O(l)$$

19.3 The Chemistry of Phosphorus

Purpose

- To characterize the forms of elemental phosphorus.
- To describe the preparation and structure of the major compounds of phosphorus.

Although phosphorus lies directly below nitrogen in Group 5A of the periodic table, its chemical properties are significantly different from those of nitrogen. The differences arise mainly from four factors: nitrogen's ability to form much stronger π bonds, the greater electronegativity of nitrogen, the larger size of the phosphorus atom, and the availability of empty valence d orbitals on phosphorus.

The chemical differences are apparent in the elemental forms of nitrogen and phosphorus. In contrast to the diatomic form of elemental nitrogen, which is stabilized by strong π bonds, there are several solid forms of phosphorus that all contain aggregates of atoms. *White phosphorus,* which contains discrete tetrahedral P_4 molecules [see Fig. 19.12(a)], is very reactive and bursts into flames on contact with air (it is said to be *pyrophoric*). To prevent this, white phosphorus is commonly stored under water. White phosphorus is quite toxic and the P_4 molecules are very damaging to tissue, particularly the cartilage and bones of the nose and jaw. The much less reactive forms called *black phosphorus* and *red phosphorus* are network solids (see Section 10.5). Black phosphorus has a regular crystalline structure [Fig. 19.12(b)], but red phosphorus is amorphous and is thought to consist of chains of P_4 units [Fig. 19.12(c)]. Red phosphorus can be obtained by heating white phosphorus in the absence of air at 1 atmosphere. Black phosphorus is obtained from either white or red phosphorus by heating at high pressures.

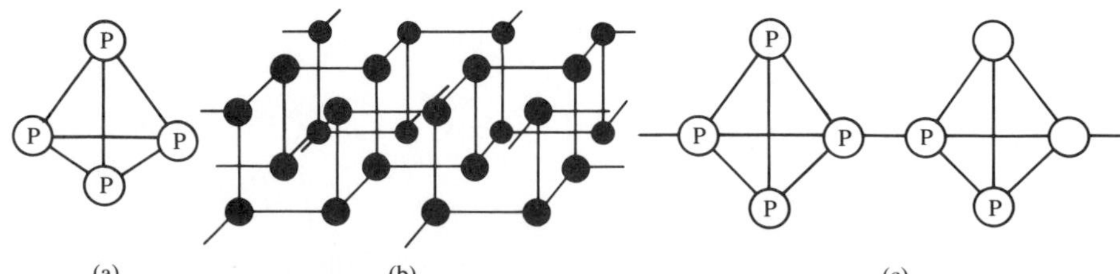

Figure 19.12

(a) The P_4 molecule found in white phosphorus. (b) The crystalline network structure of black phosphorus. (c) The chain structure of red phosphorus.

Even though phosphorus has a lower electronegativity than nitrogen, it will form phosphides (ionic substances containing the P^{3-} anion) such as Na_3P and Ca_3P_2.

Phosphide salts react vigorously with water to produce *phosphine* (PH_3), a toxic, colorless gas:

$$2Na_3P(s) + 6H_2O(l) \rightarrow 2PH_3(g) + 6Na^+(aq) + 6OH^-(aq)$$

Phosphine is analogous to ammonia, although it is a much weaker base ($K_b \approx 10^{-26}$) and much less soluble in water. Because phosphine has a relatively small affinity for protons, phosphonium (PH_4^+) salts are very uncommon and not very stable—only PH_4I, PH_4Cl, and PH_4Br are known.

Phosphine has the Lewis structure

$$\left[\begin{array}{c} \text{H—}\ddot{\text{P}}\text{—H} \\ | \\ \text{H} \end{array}\right]$$

and a pyramidal molecular structure as we would predict from the VSEPR model. However, it has bond angles of 94°, rather than 107° as found in the ammonia molecule. The reasons for this are complex, and we will simply regard phosphine as an exception to the simple version of the VSEPR model that we use.

Phosphorus Oxides and Oxyacids

Phosphorus reacts with oxygen to form oxides in which it has oxidation states of +5 and +3. The oxide P_4O_6 is formed when elemental phosphorus is burned in a limited supply of oxygen, and P_4O_{10} is produced when the oxygen is in excess. These oxides, as shown in Fig. 19.13, can be pictured as being constructed by adding oxygen atoms to the fundamental P_4 structure. The intermediate states, P_4O_7, P_4O_8, and P_4O_9, which contain one, two, and three terminal oxygen atoms, respectively, are also known.

The terminal oxygens are the nonbridging oxygen atoms.

Tetraphosphorus decoxide (P_4O_{10}), which was formerly represented as P_2O_5 and called phosphorus pentoxide, has a great affinity for water and thus is a

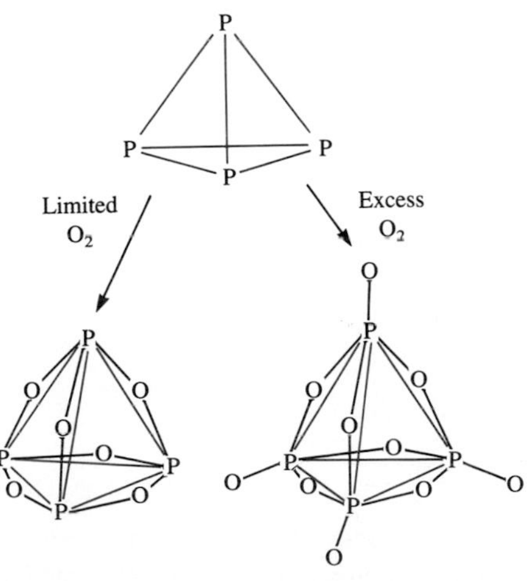

Figure 19.13
The structures of P_4O_6 and P_4O_{10}.

powerful dehydrating agent. For example, it can be used to convert HNO_3 and H_2SO_4 to their parent oxides, N_2O_5 and SO_3, respectively.

When tetraphosphorus decoxide dissolves in water, **phosphoric acid** (H_3PO_4), also called orthophosphoric acid, is produced:

$$P_4O_{10}(s) + 6H_2O(l) \rightarrow 4H_3PO_4(aq)$$

The mineral hydroxyapatite, $Ca_5(PO_4)_3OH$, the principal component of tooth enamel, can be converted to fluorapatite by reaction with fluoride. Fluoride ions added to drinking water and toothpaste prevent tooth decay because fluorapatite is less soluble in the acids of the mouth than hydroxyapatite.

Pure phosphoric acid is a white solid that melts at 42°C. Aqueous phosphoric acid is a much weaker acid ($K_{a_1} \approx 10^{-2}$) than nitric acid or sulfuric acid and is a poor oxidizing agent.

Phosphate minerals are the main source of phosphoric acid. Unlike nitrogen, phosphorus is found in nature exclusively in a combined state, principally as PO_4^{3-} ion in phosphate rock, which is mainly calcium phosphate, $Ca_3(PO_4)_2$, and fluorapatite, $Ca_5(PO_4)_3F$. Fluorapatite can be converted to phosphoric acid by grinding up the phosphate rock and forming a slurry with sulfuric acid:

$$Ca_5(PO_4)_3F(s) + 5H_2SO_4(aq) + 10H_2O(l) \rightarrow HF(aq) + 5CaSO_4 \cdot 2H_2O(s) + 3H_3PO_4(aq)$$

(A similar reaction can be written for the conversion of calcium phosphate.) The solid product $CaSO_4 \cdot 2H_2O$, called *gypsum,* is used to manufacture wallboard for construction.

The process just described, called the *wet process,* produces only impure phosphoric acid. In another procedure, phosphate rock, sand (SiO_2), and coke are heated in an electric furnace to form white phosphorus:

$$12Ca_5(PO_4)_3F + 43SiO_2 + 90C \rightarrow 9P_4 + 90CO + 20(3CaO \cdot 2SiO_2) + 3SiF_4$$

The white phosphorus obtained is burned in air to form phosphorus(V) oxide, which is then combined with water to give phosphoric acid.

Phosphoric acid easily undergoes **condensation reactions,** where a molecule of water is eliminated in the joining of two molecules of acid:

$$\mathrm{H{-}O{-}\overset{\overset{\displaystyle O}{|}}{\underset{\underset{\displaystyle H}{|}}{\underset{|}{\underset{\displaystyle O}{P}}}}{-}O{-}\boxed{H \quad H{-}O}{-}\overset{\overset{\displaystyle O}{|}}{\underset{\underset{\displaystyle H}{|}}{\underset{|}{\underset{\displaystyle O}{P}}}}{-}O{-}H \longrightarrow H{-}O{-}\overset{\overset{\displaystyle O}{|}}{\underset{\underset{\displaystyle H}{|}}{\underset{|}{\underset{\displaystyle O}{P}}}}{-}O{-}\overset{\overset{\displaystyle O}{|}}{\underset{\underset{\displaystyle H}{|}}{\underset{|}{\underset{\displaystyle O}{P}}}}{-}O{-}H + H_2O}$$

The product ($H_4P_2O_7$) is called *pyrophosphoric acid.* Further heating produces polymers, such as *tripolyphosphoric acid* ($H_5P_3O_{10}$), which has the structure

$$\mathrm{H{-}O{-}\overset{\overset{\displaystyle O}{|}}{\underset{\underset{\displaystyle H}{|}}{\underset{|}{\underset{\displaystyle O}{P}}}}{-}O{-}\overset{\overset{\displaystyle O}{|}}{\underset{\underset{\displaystyle H}{|}}{\underset{|}{\underset{\displaystyle O}{P}}}}{-}O{-}\overset{\overset{\displaystyle O}{|}}{\underset{\underset{\displaystyle H}{|}}{\underset{|}{\underset{\displaystyle O}{P}}}}{-}O{-}H}$$

The sodium salt of tripolyphosphoric acid is widely used in detergents because the $P_3O_{10}^{5-}$ anion can form complexes with metal ions such as Mg^{2+} and Ca^{2+}, which would otherwise interfere with detergent action.

Sample Exercise 19.4

What are the molecular structure and the hybridization of the central atom of the phosphoric acid molecule?

Solution

In the phosphoric acid molecule, the hydrogen atoms are attached to oxygens, and the Lewis structure is as shown below. Thus the phosphorus atom is surrounded by four effective pairs, which are arranged tetrahedrally. The atom is sp^3 hybridized.

```
        :O:
   ..    ||   ..
H—O—P—O—H
   ..    |    ..
        :O:
         |
         H
```

Lewis structure

Molecular structure

When the oxide P_4O_6 is placed in water, **phosphorous acid** (H_3PO_3) is formed [Fig. 19.14(a)]. Although the formula suggests a triprotic acid, phosphorous acid is a *diprotic* acid. The hydrogen atom bonded directly to the phosphorus atom is not acidic in aqueous solution; only those hydrogen atoms bonded to the oxygen atoms in H_3PO_3 can be released as protons.

A third oxyacid of phosphorus is *hypophosphorous acid* (H_3PO_2) [Fig. 19.14(b)], which is a monoprotic acid.

```
       H              H
       |              |
H—O—P—O        H—P—O
       |              |
       O              O
       |              |
       H              H
      (a)            (b)
```

Figure 19.14

(a) The structure of phosphorous acid (H_3PO_3). (b) The structure of hypophosphorous acid (H_3PO_2).

Phosphorus in Fertilizers

Phosphorus is essential for plant growth. Although most soil contains large amounts of phosphorus, it is often present as insoluble minerals, which makes it inaccessible to the plants. Soluble phosphate fertilizers are manufactured by treating phosphate rock with sulfuric acid to make **superphosphate of lime,** a mixture of $CaSO_4 \cdot 2H_2O$ and $Ca(H_2PO_4)_2 \cdot H_2O$. If phosphate rock is treated with phosphoric acid, $Ca(H_2PO_4)_2$, or *triple phosphate,* is produced. The reaction of ammonia and phosphoric acid gives *ammonium dihydrogenphosphate* ($NH_4H_2PO_4$), a very efficient fertilizer since it furnishes both phosphorus and nitrogen.

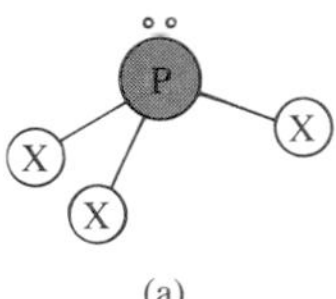

(a)

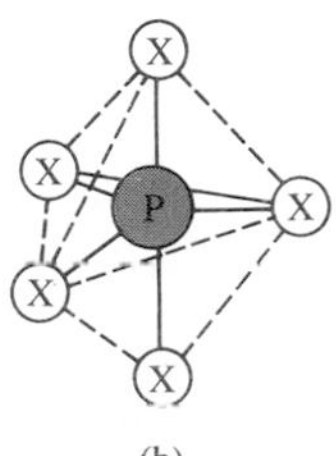

(b)

Figure 19.15

Structures of the phosphorus halides. (a) The PX_3 compounds have pyramidal molecules. (b) The gaseous and liquid phases of the PX_5 compounds are composed of trigonal bipyramidal molecules.

Phosphorus Halides

Phosphorus forms all possible halides of the general formulas PX_3 and PX_5, with the exception of PI_5. The PX_3 molecule has the expected pyramidal structure [Fig. 19.15(a)]. Under normal conditions of temperature and pressure, PF_3 is a colorless gas, PCl_3 is a liquid (bp = 74°C), PBr_3 is a liquid (bp = 175°C), and PI_3 is an unstable red solid (mp = 61°C). All of the PX_3 compounds react with water to produce phosphorous acid:

$$PX_3 + 3H_2O(l) \rightarrow H_3PO_3(aq) + 3HX(aq)$$

In the gaseous and liquid states, the PX_5 compounds have molecules with a trigonal bipyramidal structure [Fig. 19.15(b)]. However, PCl_5 and PBr_5 form ionic solids: solid PCl_5 contains a mixture of octahedral PCl_6^- ions and tetrahedral PCl_4^+ ions, and solid PBr_5 appears to consist of PBr_4^+ and Br^- ions.

The PX_5 compounds react with water to form phosphoric acid:

$$PX_5 + 4H_2O(l) \rightarrow H_3PO_4(aq) + 5HX(aq)$$

6A
O
S
Se
Te
Po

19.4 The Group 6A Elements

Purpose

- To discuss trends in the chemistry of the Group 6A elements.

Although in Group 6A (Table 19.4) there is the usual tendency for metallic properties to increase going down the group, none of the Group 6A elements (the valence-electron configuration is n^2np^4) behaves as a typical metal. The most common chemical behavior of a Group 6A element is to achieve a noble gas electron configuration by adding two electrons to become a 2− anion in ionic compounds with metals. In fact, for most metals, the oxides and sulfides constitute the most common minerals.

The Group 6A elements can form covalent bonds with other nonmetals. For example, they combine with hydrogen to form a series of covalent hydrides of the general formula H_2X. Those members of the group that have valence *d* orbitals available (all except oxygen) commonly form molecules in which they are surrounded by more than eight electrons. Examples are SF_4, SF_6, TeI_4, and $SeBr_4$.

The two heaviest members of Group 6A can lose electrons to form cations. Although they do not lose all six valence electrons because of the high energies that would be required, tellurium and polonium appear to exhibit some chemistry in-

Selected Physical Properties, Sources, and Methods of Preparation for the Group 6A Elements

Element	Electro-negativity	Radius of X^{2-} (Å)	Source	Method of preparation
Oxygen	3.5	1.40	Air	Distillation from liquid air
Sulfur	2.5	1.84	Sulfur deposits	Melted with hot water and pumped to the surface
Selenium	2.4	1.98	Impurity in sulfide ores	Reduction of H_2SeO_4 with SO_2
Tellurium	2.1	2.21	Nagyagite (mixed sulfide and telluride)	Reduction of ore with SO_2
Polonium	2.0	2.30	Pitchblende	

Table 19.4

volving their 4+ cations. However, the chemistry of these Group 6A cations is much more limited than that of the Group 5A elements bismuth and antimony.

In recent years there has been a growing interest in the chemistry of selenium, an element found throughout the environment in trace amounts. Selenium's toxicity has long been known, but recent medical studies have shown an *inverse* relationship between the incidence of cancer and the selenium levels in soil. It has been suggested that the resulting greater dietary intake of selenium by people living in areas of relatively high levels of selenium somehow furnishes protection from cancer. These studies are only preliminary, but selenium is known to be physiologically important (it is involved in the activity of vitamin E and certain enzymes). Also of importance is the fact that selenium (and tellurium) are semiconductors and therefore find some application in the electronics industry.

Polonium was discovered in 1898 by Marie and Pierre Curie in their search for the sources of radioactivity in pitchblende. Polonium has 27 isotopes and is highly toxic and very radioactive. It has been suggested that the isotope ^{210}Po, a natural contaminant of tobacco and an α-particle emitter (see Section 21.1), might be at least partly responsible for the incidence of cancer in smokers.

19.5 The Chemistry of Oxygen

Purpose

- To characterize the forms of elemental oxygen.

It is hard to overstate the importance of oxygen, the most abundant element in and near the earth's crust. Oxygen is present in the atmosphere as oxygen gas and ozone; in soil and rocks in oxide, silicate, and carbonate minerals; in the oceans in water; and in our bodies in water and in a myriad of molecules. In addition, most of the energy we need to live and run our civilization comes from the exothermic reactions of oxygen with carbon-containing molecules.

The most common elemental form of oxygen (O_2) comprises 21% of the volume of the earth's atmosphere. Since nitrogen has a lower boiling point than oxygen, nitrogen can be boiled away from liquid air, leaving oxygen and small amounts of argon, another component of air. Liquid oxygen is a pale blue liquid that freezes at −219°C and boils at −183°C. The paramagnetism of the O_2 molecule can be demonstrated by pouring liquid oxygen between the poles of a strong magnet where it "sticks" as it boils away (see Fig. 9.39). The paramagnetism of the O_2 molecule can be accounted for by the molecular orbital model (Fig. 9.38), which also explains the bond strength.

The other form of elemental oxygen is **ozone** (O_3), a molecule that can be represented by the resonance structures

$$\mathrm{O{=}O{-}O} \longleftrightarrow \mathrm{O{-}O{=}O}$$

The bond angle in the O_3 molecule is 117°, in close agreement with the prediction of the VSEPR model (three effective pairs require a trigonal planar arrangement). That the bond angle is slightly less than 120° can be explained by concluding that more space is required for the lone pair than for the bonding pairs.

Ozone can be prepared by passing an electrical discharge through pure oxygen gas. The electrical energy disrupts the bonds in some O_2 molecules to give oxygen atoms, which react with other O_2 molecules to form O_3. Ozone is much less stable than oxygen at 25°C and 1 atmosphere. For example, $K \approx 10^{-57}$ for the equilibrium

$$3O_2(g) \rightleftharpoons 2O_3(g)$$

A pale blue, highly toxic gas, ozone is a much more powerful oxidizing agent than oxygen. Because of its oxidizing ability, ozone is being considered as a replacement for chlorine in municipal water purification. Chlorine leaves residues of chloro compounds, such as chloroform ($CHCl_3$), which may cause cancer after long-term exposure. One problem with **ozonolysis** is that the water supply is not protected against recontamination since virtually no ozone remains after the initial treatment. With chlorination, significant residual chlorine remains after treatment.

The oxidizing ability of ozone can be highly detrimental, especially when it is formed in the pollution from automobile exhausts (see Section 5.9).

Ozone exists naturally in the upper atmosphere of the earth. The *ozone layer* is especially important because it absorbs ultraviolet light and thus acts as a screen to prevent this radiation, which can cause skin cancer, from penetrating to the earth's surface. When an ozone molecule absorbs this energy, it splits into an oxygen molecule and an oxygen atom.

$$O_3 \xrightarrow{h\nu} O_2 + O$$

If the oxygen molecule and atom collide, they will not stay together as ozone unless a "third body," such as a nitrogen molecule, is present to help absorb the energy of bond formation. The third body absorbs the energy as kinetic energy, which means that the energy originally absorbed as ultraviolet radiation is changed to thermal energy. Thus the ozone prevents the harmful high-energy ultraviolet light from reaching the earth.

Recently, scientists have become concerned that Freons and nitrogen dioxide are promoting the destruction of the ozone layer (Section 12.8).

19.6 The Chemistry of Sulfur

Purpose

- To characterize the elemental forms of sulfur.
- To describe the bonding and structure of the oxycompounds of sulfur.

Sulfur is found in nature both in large deposits of the free element and in widely distributed ores, such as galena (PbS), cinnabar (HgS), pyrite (FeS_2), gypsum ($CaSO_4 \cdot 2H_2O$), epsomite ($MgSO_4 \cdot 7H_2O$), and glauberite ($Na_2SO_4 \cdot CaSO_4$).

About 60% of the sulfur produced in the United States comes from the underground deposits of elemental sulfur found in Texas and Louisiana. This sulfur is recovered using the **Frasch process** developed by Herman Frasch in the 1890s. Superheated water is pumped into the deposit to melt the sulfur (mp = 113°C), which is then forced to the surface by air pressure (see Fig. 19.16). The remaining 40% of sulfur produced in the United States is either a by-product of the purification of fossil fuels before combustion to prevent pollution or from the sulfur dioxide (SO_2) scrubbed from the exhaust gases when sulfur-containing fuels are burned.

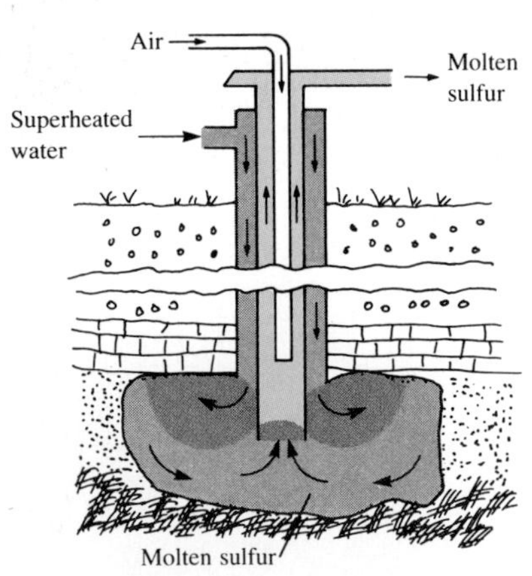

Figure 19.16

The Frasch method for recovering sulfur from underground deposits.

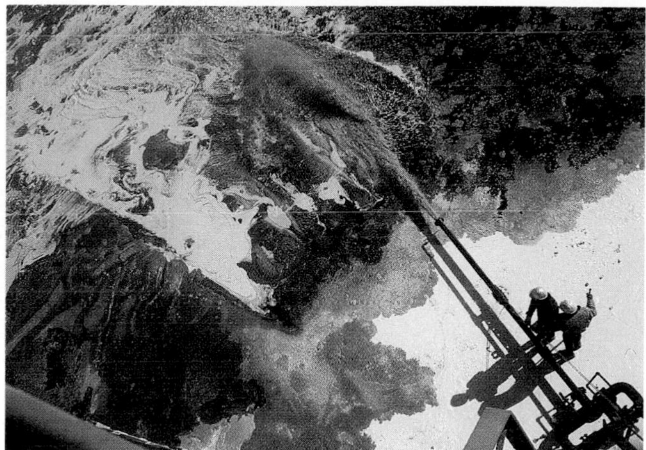

(left) A sulfur deposit. (right) Melted sulfur obtained from underground deposits by the Frasch process.

In contrast to oxygen, elemental sulfur exists as S_2 molecules only in the gas phase at high temperatures. Because sulfur atoms form much stronger σ bonds than π bonds, S_2 is more unstable at 25°C than larger aggregates such as S_6 and S_8 rings and S_n chains (Fig. 19.17). The most stable form of sulfur at 25°C and 1 atmosphere is called *rhombic sulfur* [see Fig. 19.18(a) on the following page], which contains stacked S_8 rings. If rhombic sulfur is melted and heated to 120°C, it forms *monoclinic sulfur* as it cools slowly [Fig. 19.18(b)]. This form also contains S_8 rings, but the rings are stacked differently than they are in rhombic sulfur.

The scrubbing of sulfur dioxide from exhaust gases was discussed in Section 5.9.

As sulfur is heated beyond its melting point, a relatively nonviscous liquid containing S_8 rings forms initially. With continued heating, the liquid becomes highly viscous as the rings first break and then link up to form long chains. Further heating lowers the viscosity because the long chains are broken down as the energetic sulfur atoms break loose. If the liquid is suddenly cooled, a substance called

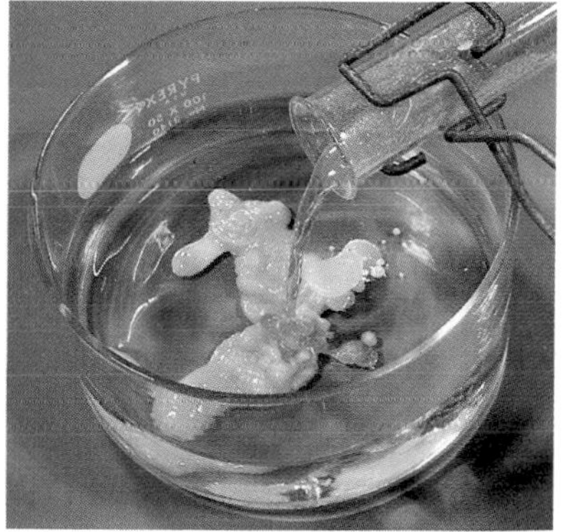

Pouring liquid sulfur in water to produce plastic sulfur.

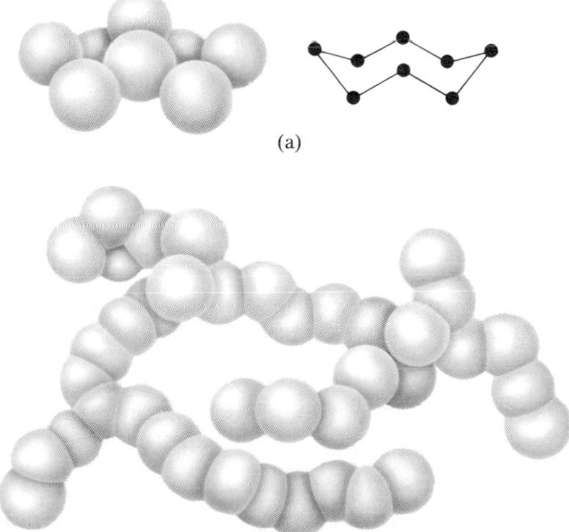

Figure 19.17

(a) The S_8 molecule. (b) Chains of sulfur atoms in viscous liquid sulfur. The chains may contain as many as 10,000 atoms.

Figure 19.18

(left) Crystals of rhombic sulfur.
(right) Crystals of monoclinic sulfur.

plastic sulfur, which contains S_n chains and has rubberlike qualities, is formed. Eventually, this form reverts back to the more stable S_8 rings.

Sulfur Oxides

From its position below oxygen in the periodic table, we might expect the simplest stable oxide of sulfur to have the formula SO. However, *sulfur monoxide,* which can be produced in small amounts when gaseous sulfur dioxide (SO_2) is subjected to an electrical discharge, is very unstable. The difference in the stabilities of the O_2 and SO molecules probably reflects the stronger π bonding between oxygen atoms than between sulfur and oxygen atoms.

Sulfur burns in air with a bright blue flame to give *sulfur dioxide* (SO_2), a colorless gas with a pungent odor, which condenses to a liquid at $-10°C$ and 1 atmosphere. Sulfur dioxide is a very effective antibacterial agent and is often used to preserve stored fruit. Its structure is given in Fig. 19.19.

Sulfur dioxide reacts with oxygen to produce *sulfur trioxide* (SO_3):

$$2SO_2(g) + O_2(g) \rightarrow 2SO_3(g)$$

However, this reaction is very slow in the absence of a catalyst. One of the mysteries during early research on air pollution was how the sulfur dioxide produced from the combustion of sulfur-containing fuels is so rapidly converted to sulfur trioxide. It is now known that dust and other particles can act as heterogeneous catalysts for this process (see Section 12.8). In the preparation of sulfur trioxide for the manufacture of sulfuric acid, a platinum or vanadium(V) oxide (V_2O_5) catalyst is used, and the reaction is carried out at ~500°C, even though this temperature decreases the value of the equilibrium constant for this exothermic reaction.

The preparation of sulfur trioxide provides an example of the compromise that often must be made between thermodynamics and kinetics.

The bonding in the SO_3 molecule is usually described in terms of the resonance structures shown in Fig. 19.20. The molecule is trigonal planar, as predicted by the VSEPR model. Sulfur trioxide is a corrosive gas with a choking odor and forms white fumes of sulfuric acid when it reacts with moisture in air. Thus sulfur trioxide and nitrogen dioxide, which reacts with water to form a mixture of nitrous and nitric acids, are the major culprits in the formation of acid rain.

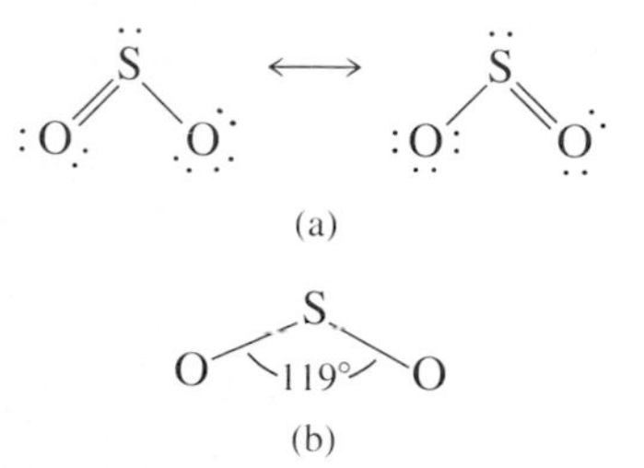

Figure 19.19

(a) Two of the resonance structures for SO_2. (b) SO_2 is a bent molecule with a 119° bond angle, as predicted by the VSEPR model.

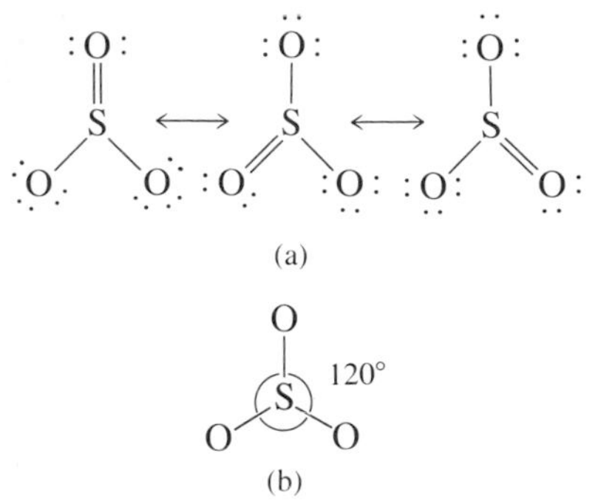

Figure 19.20

(a) Three of the resonance structures of SO_3. (b) SO_3 is a planar molecule with 120° bond angles.

Sulfur trioxide condenses to a colorless liquid at 44.5°C and freezes at 16.8°C to give three solid forms, one containing S_3O_9 rings and the other two containing $(SO_3)_x$ chains (Fig. 19.21).

Oxyacids of Sulfur

Sulfur dioxide dissolves in water to form an acidic solution. The reaction is often represented as

$$SO_2(g) + H_2O(l) \rightarrow H_2SO_3(aq)$$

where H_2SO_3 is called *sulfurous acid*. However, very little H_2SO_3 actually exists in the solution. The major form of sulfur dioxide in water is SO_2, and the acid dissociation equilibria are best represented as

$$SO_2(aq) + H_2O(l) \rightleftharpoons H^+(aq) + HSO_3^-(aq) \qquad K_{a_1} = 1.5 \times 10^{-2}$$

$$HSO_3^-(aq) \rightleftharpoons H^+(aq) + SO_3^{2-}(aq) \qquad K_{a_2} = 1.0 \times 10^{-7}$$

This situation is analogous to the behavior of carbon dioxide in water (Section 14.7). Although H_2SO_3 cannot be isolated, salts of SO_3^{2-} (*sulfites*) and HSO_3^- (*hydrogen sulfites*) are well known.

Sulfur trioxide reacts violently with water to produce the diprotic acid **sulfuric acid:**

$$SO_3(g) + H_2O(l) \rightarrow H_2SO_4(aq)$$

Manufactured in greater amounts than any other chemical, sulfuric acid is usually produced by the *contact process,* which is described at the end of Chapter 3. About 60% of the sulfuric acid manufactured in the United States is used to produce fertilizers from phosphate rock (see Section 19.3). The other 40% is used in lead storage batteries and in petroleum refining, steel manufacturing, and for various purposes in most of the chemical industries.

Because sulfuric acid has a high affinity for water, it is often used as a dehydrating agent. Gases that do not react with sulfuric acid, such as oxygen, nitrogen, and carbon dioxide, are often dried by bubbling them through concentrated solutions of the acid. Sulfuric acid is such a powerful dehydrating agent that it will

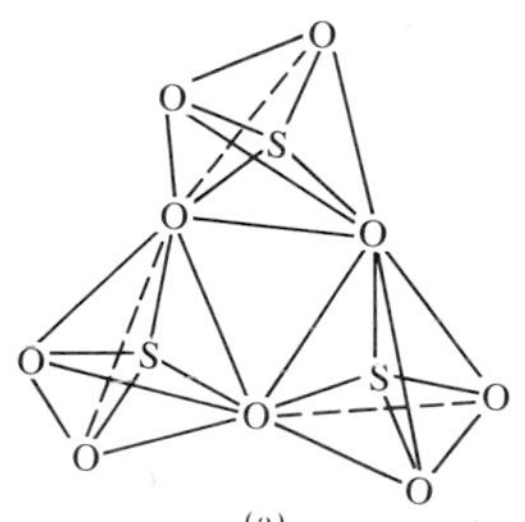

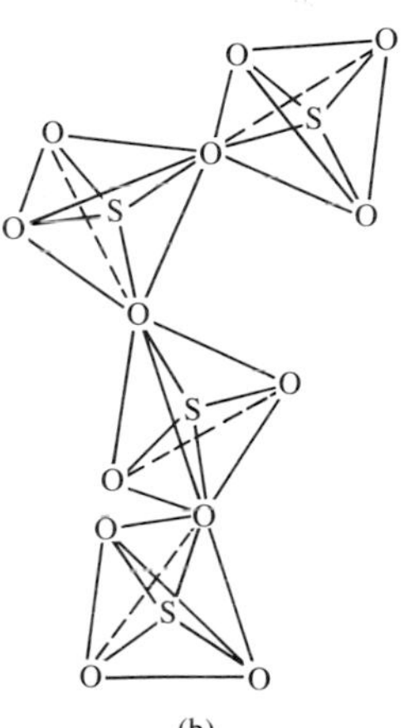

Figure 19.21

Different structures for solid SO_3. (a) S_3O_9 rings. (b) $(SO_3)_x$ chains. In both cases the sulfur atoms are surrounded by a tetrahedral arrangement of oxygen atoms.

Figure 19.22

The reaction of H_2SO_4 with sucrose to produce a blackened column of carbon.

remove hydrogen and oxygen from a substance in a 2:1 ratio even when the substance contains no molecular water. For example, concentrated sulfuric acid reacts vigorously with common table sugar (sucrose), leaving a charred mass of carbon (see Fig. 19.22):

$$\underset{\text{Sucrose}}{C_{12}H_{22}O_{11}(s)} + 11H_2SO_4(conc) \rightarrow 12C(s) + 11H_2SO_4 \cdot H_2O(l)$$

Sulfuric acid is a moderately strong oxidizing agent, especially at high temperatures. Hot concentrated sulfuric acid oxidizes bromide or iodide ions to elemental bromine or iodine, for example,

$$2I^-(aq) + 3H_2SO_4(aq) \rightarrow I_2(aq) + SO_2(aq) + 2H_2O(l) + 2HSO_4^-(aq)$$

Hot sulfuric acid attacks copper metal:

The acidic properties of sulfuric acid solutions are discussed in Section 14.7.

$$Cu(s) + 2H_2SO_4(aq) \rightarrow CuSO_4(aq) + 2H_2O(l) + SO_2(aq)$$

Cold acid does not react.

Other Compounds of Sulfur

Sulfur reacts with both metals and nonmetals to form a wide variety of compounds in which it has a +6, +4, +2, 0, or −2 oxidation state (see Table 19.5). The −2

Common Compounds of Sulfur with Various Oxidation States

Oxidation state of sulfur	Compounds
+6	SO_3 H_2SO_4, SO_4^{2-}, SF_6
+4	SO_2, HSO_3^-, SO_3^{2-}, SF_4
+2	SCl_2
0	S_8 and all other forms of elemental sulfur
−2	H_2S, S^{2-}

Table 19.5

oxidation state occurs in the metal sulfides and in *hydrogen sulfide* (H_2S), a toxic, foul-smelling gas that acts as a diprotic acid when dissolved in water. Hydrogen sulfide is a strong reducing agent in aqueous solution, producing a milky-looking suspension of finely divided sulfur as one of the products. For example, hydrogen sulfide reacts with chlorine in aqueous solution as follows:

$$H_2S(g) + Cl_2(aq) \rightarrow 2H^+(aq) + 2Cl^-(aq) + \underset{\text{Milky suspension of sulfur}}{\underset{\nearrow}{S(s)}}$$

Sulfur also forms the **thiosulfate ion** ($S_2O_3^{2-}$), which has the Lewis structure

The prefix *thio* means sulfur.

$$\left[\begin{array}{c} :\ddot{O}: \\ | \\ :\ddot{O}-S-\ddot{S}: \\ | \\ :\ddot{O}: \end{array}\right]^{2-}$$

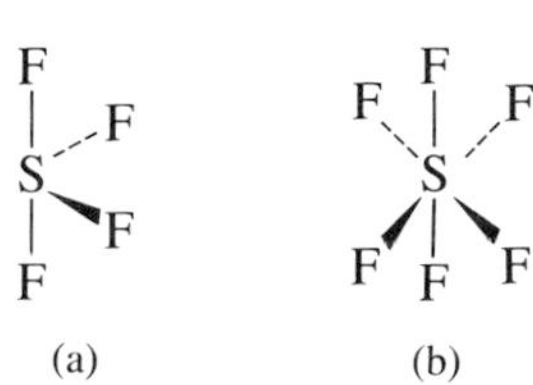

Note that this anion can be viewed as a sulfate ion in which one of the oxygen atoms has been replaced by sulfur, which is reflected in the name *thio*sulfate. The thiosulfate ion can be formed by heating sulfur with a sulfite salt in aqueous solution:

$$S(s) + SO_3^{2-}(aq) \rightarrow S_2O_3^{2-}(aq)$$

One important use of thiosulfate ion is in photography where $S_2O_3^{2-}$ dissolves solid silver bromide by forming a complex with the Ag^+ ion (see the Chemical Impact feature in Section 19.7). Thiosulfate ion is also a good reducing agent and is often used to analyze for iodine:

$$S_2O_3^{2-}(aq) + I_2(aq) \rightarrow S_4O_6^{2-}(aq) + 2I^-(aq)$$

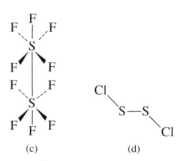

where $S_4O_6^{2-}$ is called the *tetrathionate ion*.

Sulfur reacts with the halogens to form a variety of compounds, such as S_2Cl_2, SF_4, SF_6, and S_2F_{10}. The structures of these molecules are shown in Fig. 19.23.

Figure 19.23

The structures of (a) SF_4, (b) SF_6, (c) S_2F_{10}, and (d) S_2Cl_2.

19.7 The Group 7A Elements

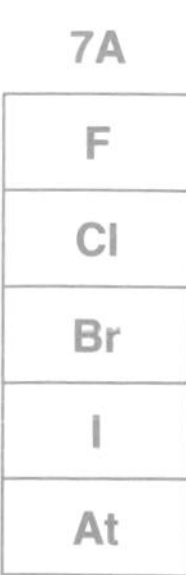

Purpose

- To describe the bonding in compounds of halogens with other nonmetals.
- To explain the acid strengths of the hydrogen halides.
- To discuss the chemistry of oxyhalogen compounds.
- To give the structures of interhalogen compounds.

In our coverage of the representative elements we have progressed from groups of metallic elements (Groups 1A and 2A), through groups in which the lighter members are nonmetals and the heavier members are metals (Groups 3A, 4A, and 5A), to a group of all nonmetals (Group 6A—although some might prefer to call polonium a metal). The Group 7A elements, the **halogens** (with the valence-electron configuration ns^2np^5), are also all nonmetals whose properties generally vary

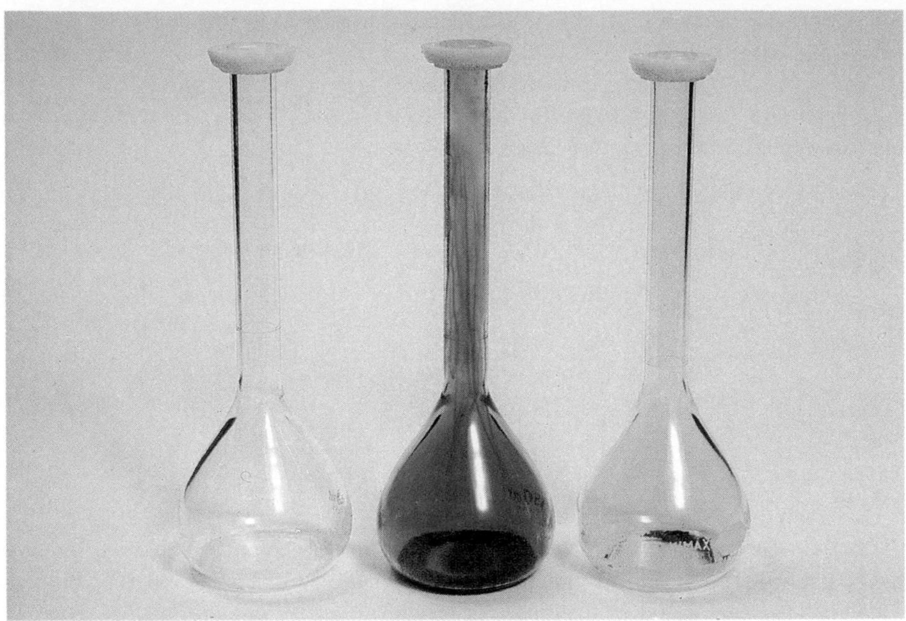

Chlorine, bromine, and iodine.

smoothly going down the group. The only notable exceptions to this are the unexpectedly low values for the electron affinity of fluorine and the bond energy of the F_2 molecule (see Section 18.1). Table 19.6 summarizes the trends in some physical properties of the halogens.

Because of their high reactivities, the halogens are not found as the free elements in nature. Instead they are found as halide ions (X^-) in various minerals and in sea water (see Table 19.7).

Although astatine is a member of Group 7A, its chemistry is of no practical importance because all of its known isotopes are radioactive. The longest-lived isotope, ^{210}At, has a half-life of only 8.3 hours.

The halogens, particularly fluorine, have very high electronegativity values (Table 19.6). They tend to form polar covalent bonds with other nonmetals and ionic bonds with metals in their lower oxidation states. When a metal ion is in a higher oxidation state, such as +3 or +4, the metal-halogen bonds are polar covalent ones. For example, $TiCl_4$ and $SnCl_4$ are both covalent compounds that are liquids under normal conditions.

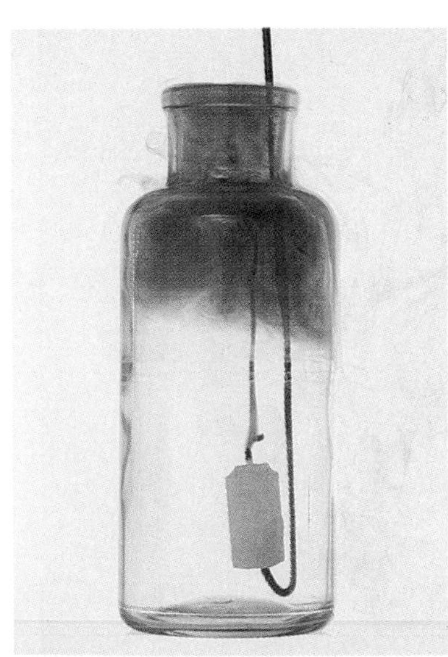

A candle burning in an atmosphere of $Cl_2(g)$. The exothermic reaction, which involves breaking C—C and C—H bonds in the wax and forming C—Cl bonds in their places, [illegible]roduces enough heat to make the [illegible] the region incandescent (a [illegible]).

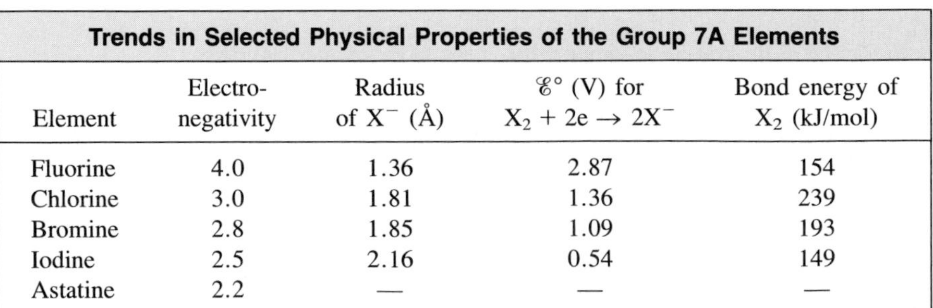

Trends in Selected Physical Properties of the Group 7A Elements

Element	Electro-negativity	Radius of X^- (Å)	$\mathscr{E}°$ (V) for $X_2 + 2e \rightarrow 2X^-$	Bond energy of X_2 (kJ/mol)
Fluorine	4.0	1.36	2.87	154
Chlorine	3.0	1.81	1.36	239
Bromine	2.8	1.85	1.09	193
Iodine	2.5	2.16	0.54	149
Astatine	2.2	—	—	—

Table 19.6

Element	Color and state	Percentage of earth's crust	Melting point (°C)	Boiling point (°C)	Sources	Method of preparation
Fluorine	Pale yellow gas	0.07	−220	−188	Fluorospar (CaF_2), cryolite (Na_3AlF_6), fluorapatite ($Ca_5(PO_4)_3F$)	Electrolysis of molten KHF_2
Chlorine	Yellow-green gas	0.14	−101	−34	Rock salt (NaCl), halite (NaCl), sylvite (KCl)	Electrolysis of aqueous NaCl
Bromine	Red-brown liquid	2.5×10^{-4}	−73	59	Sea water, brine wells	Oxidation of Br^- by Cl_2
Iodine	Violet-black solid	3×10^{-5}	113	184	Seaweed, brine wells	Oxidation of I^- by electrolysis or MnO_2

Table 19.7 Some Physical Properties, Sources, and Methods of Preparation for the Group 7A Elements

Chemical Impact

Photography

In black-and-white photography, light from an object is focused onto a special paper containing an emulsion of solid silver bromide. Silver salts turn dark when exposed to light because the radiant energy stimulates the transfer of an electron to the Ag^+ ion, forming an atom of elemental silver. When photographic paper (*film*) is exposed to light, the areas exposed to the brightest light form the most silver atoms. The next step is the application of a chemical reducing agent to the film, a process called *developing*. The real advantage of silver-based films is that the silver atoms already present from exposure to light catalyze the reduction of millions of Ag^+ ions in the immediate vicinity in the developing process. Thus the effect of exposure to light is greatly intensified in this chemical reduction process. Once the image has been developed, the unchanged solid silver bromide must be removed so that the film is no longer light-sensitive and the image is fixed. A solution of sodium thiosulfate (called *hypo*) is used in this *fixing process:*

$$AgBr(s) + 2S_2O_3^{2-}(aq) \longrightarrow Ag(S_2O_3)_2^{3-}(aq) + Br^-(aq)$$

After the excess silver bromide is dissolved and washed away, the fixed image (the *negative*) is ready to produce the positive print. By shining light through the negative onto a fresh sheet of film and repeating the developing and fixing processes, a black-and-white photograph can be produced.

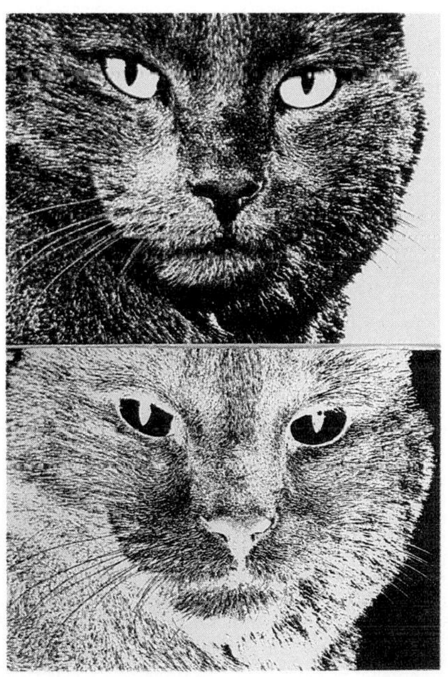

Positive and negative images.

Hydrogen Halides

The hydrogen halides can be prepared by a reaction of the elements:

$$H_2(g) + X_2(g) \rightarrow 2HX(g)$$

This reaction occurs with explosive vigor when fluorine and hydrogen are mixed. On the other hand, hydrogen and chlorine can coexist with little apparent reaction for relatively long periods in the dark. However, ultraviolet light causes an explosively fast reaction, and this is the basis of a popular lecture demonstration, the "hydrogen-chlorine cannon." Bromine and iodine also react with hydrogen, but more slowly.

The hydrogen halides can also be prepared by treating halide salts with acid. For example, hydrogen fluoride and hydrogen chloride can be prepared as follows:

$$CaF_2(s) + H_2SO_4(aq) \rightarrow CaSO_4(s) + 2HF(g)$$

$$2NaCl(s) + H_2SO_4(aq) \rightarrow Na_2SO_4(s) + 2HCl(g)$$

Sulfuric acid is capable of oxidizing Br^- to Br_2 and I^- to I_2 and so cannot be used to prepare hydrogen bromide and hydrogen iodide. However, phosphoric acid, a non-oxidizing acid, can be used to treat bromides and iodides to form the corresponding hydrogen halides.

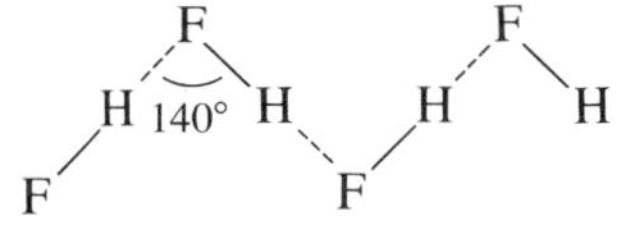

Figure 19.24

The hydrogen bonding among HF molecules in liquid hydrogen fluoride.

Some physical properties of the hydrogen halides are listed in Table 19.8. Note the very high boiling point for hydrogen fluoride, which results from extensive hydrogen bonding among the very polar HF molecules (Fig. 19.24). Fluoride ion has such a high affinity for protons that in concentrated aqueous solutions of hydrogen fluoride, the ion $[F{-}{-}{-}H{-}{-}{-}F]^-$ exists, in which an H^+ ion is centered between two F^- ions.

Some Physical Properties of the Hydrogen Halides

HX	Melting point (°C)	Boiling point (°C)	H—X bond energy (kJ/mol)
HF	−83	20	565
HCl	−114	−85	427
HBr	−87	−67	363
HI	−51	−35	295

Table 19.8

When dissolved in water, the hydrogen halides behave as acids, and all except hydrogen fluoride are completely dissociated. Because water is a much stronger base than Cl^-, Br^-, or I^- ion, the acid strengths of HCl, HBr, and HI cannot be differentiated in water. However, in a less basic solvent, such as glacial (pure) acetic acid, the acids show different strengths of the order

$$\underset{\text{Strongest acid}}{\text{H—I}} > \text{H—Br} > \text{H—Cl} \gg \underset{\text{Weakest acid}}{\text{H—F}}$$

To see why hydrogen fluoride is the only weak acid in water among the HX molecules, let's consider the dissociation equilibrium

$$HX(aq) \rightleftharpoons H^+(aq) + X^-(aq) \qquad \text{where} \qquad K_a = \frac{[H^+][X^-]}{[HX]}$$

from a thermodynamic point of view. Recall that acid strength is reflected by the magnitude of K_a—a small K_a value means a weak acid. Also recall that the value of an equilibrium constant depends on the standard free energy change for the reaction as follows:

$$\Delta G^\circ = -RT \ln(K)$$

As ΔG° becomes more negative, K becomes larger; a *decrease* in free energy favors a given reaction. As we saw in Chapter 16, free energy depends on enthalpy, entropy, and temperature. For a process at constant temperature,

$$\Delta G^\circ = \Delta H^\circ - T\Delta S^\circ$$

Thus, to explain the various acid strengths of the hydrogen halides, we must focus on the factors that determine ΔH° and ΔS° for the acid dissociation reaction.

What energy terms are important in determining ΔH° for the dissociation of HX in water? (Keep in mind that large positive contributions to the value of ΔH° will tend to make ΔG° more highly positive, K_a smaller, and the acid weaker.) One important factor is certainly the H—X bond strength. Note from Table 19.8 that the H—F bond is much stronger than the other H—X bonds. This factor tends to make HF a weaker acid than the others.

Another important contribution to ΔH° is the enthalpy of hydration (see Section 11.2) of X^- (see Table 19.9). As we would expect, the smallest of the halide ions, F^-, has the most negative value—its hydration is the most exothermic. This term favors the dissociation of HF into its ions, more so than it does for the other HX molecules.

So far we have two conflicting factors: the large HF bond energy tends to make HF a weaker acid than the other hydrogen halides, but the enthalpy of hydration favors the dissociation of HF more than that of the others. In comparing data for HF and HCl, the difference in bond energy (138 kJ/mol) is slightly smaller than the difference in the enthalpies of hydration for the anions (144 kJ/mol). If these were the *only* important factors, HF should be a stronger acid than HCl because the large enthalpy of hydration of F^- more than makes up for the large HF bond strength.

The Enthalpies and Entropies of Hydration for the Halide Ions

$$X^-(g) \xrightarrow{H_2O} X^-(aq)$$

X^-	ΔH° (kJ/mol)	ΔS° (J/K mol)
F^-	−510	−159
Cl^-	−366	−96
Br^-	−334	−81
I^-	−291	−64

Table 19.9

Hydration becomes more exothermic as the charge density of an ion increases. Thus, for ions of a given charge, the smallest is most strongly hydrated.

When H_2O molecules cluster around an ion, an ordering effect occurs, and ΔS°_{hyd} is negative.

As it turns out, the *deciding factor is entropy*. Note from Table 19.9 that the entropy of hydration for F^- is much more negative than for the other halides because of the high degree of ordering that occurs as the water molecules associate with the small F^- ion. Remember that a negative change in entropy is unfavorable. Thus although the enthalpy of hydration favors dissociation of HF, the *entropy* of hydration strongly opposes it.

When all of these factors are taken into account, ΔG° for the dissociation of HF in water is positive; that is, K_a is small. In contrast, ΔG° for dissociation of the other HX molecules in water is negative (K_a is large). This example illustrates the complexity of the processes that occur in aqueous solutions and the importance of entropy effects in that medium.

Stomach acid is 0.1 *M* HCl.

In practical terms, **hydrochloric acid** is the most important of the **hydrohalic acids,** the aqueous solutions of the hydrogen halides. About 3 million tons of hydrochloric acid are produced annually for use in cleaning steel before galvanizing and in the manufacture of many other chemicals.

Hydrofluoric acid is used to etch glass by reacting with the silica in glass to form the volatile gas SiF_4:

$$SiO_2(s) + 4HF(aq) \rightarrow SiF_4(g) + 2H_2O(l)$$

Oxyacids and Oxyanions

All the halogens except fluorine combine with various numbers of oxygen atoms to form a series of oxyacids, as shown in Table 19.10. The strengths of these acids vary in direct proportion to the number of oxygen atoms attached to the halogen, as we discussed in Section 14.9.

The only member of the chlorine series that has been obtained in the pure state is *perchloric acid* ($HOClO_3$), a strong acid and powerful oxidizing agent. Because perchloric acid reacts explosively with many organic materials, it must be handled with great caution. The other oxyacids of chlorine are known only in solution, although salts containing their anions are well-known (Fig. 19.25).

The Known Oxyacids of the Halogens*

Oxidation state of halogen	Fluorine	Chlorine	Bromine	Iodine*	General name of acids	General name of salts
+1	HOF	HOCl	HOBr	HOI	Hypohalous acid	Hypohalites, MOX
+3	**	HOClO	**	**	Halous acid	Halites, MXO_2
+5	**	$HOClO_2$	$HOBrO_2$	$HOIO_2$	Halic acid	Halates, MXO_3
+7	**	$HOClO_3$	$HOBrO_3$	$HOIO_3$	Perhalic acid	Perhalates, MXO_4

*Iodine also forms $H_4I_2O_9$ (mesodiperiodic acid) and H_5IO_6 (paraperiodic acid).
**Compound is unknown.

Table 19.10

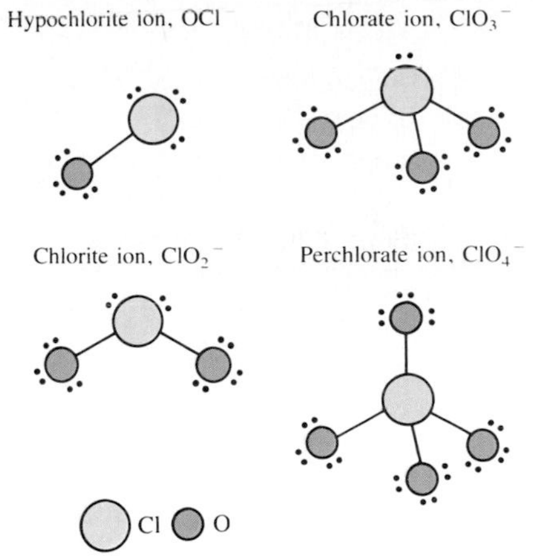

Figure 19.25

The structures of the oxychloro anions.

Hypochlorous acid (HOCl) is formed when chlorine gas is dissolved in cold water:

$$Cl_2(aq) + H_2O(l) \rightleftharpoons HOCl(aq) + H^+(aq) + Cl^-(aq)$$

Note that in this reaction chlorine is both oxidized (from 0 in Cl_2 to $+1$ in HOCl) and reduced (from 0 in Cl_2 to -1 in Cl^-). Such a reaction, *where a given element is both oxidized and reduced,* is called a **disproportionation reaction.** Hypochlorous acid and its salts are strong oxidizing agents, and solutions of them are widely used as household bleaches and disinfectants.

Chlorate salts, such as $KClO_3$, are also strong oxidizing agents and are used as weed killers and as oxidizers in fireworks (see Chapter 7) and explosives.

Fluorine only forms one oxyacid, hypofluorous acid (HOF), but at least two oxides. When fluorine gas is bubbled into a dilute solution of sodium hydroxide, the compound *oxygen difluoride* (OF_2) is formed:

The name for OF_2 is oxygen difluoride rather than difluorine oxide because fluorine has a higher electronegativity than oxygen and thus is named as the anion.

$$4F_2(g) + 3H_2O(l) \rightarrow 6HF(aq) + OF_2(g) + O_2(g)$$

Oxygen difluoride is a pale yellow gas (bp $= -145°C$), which is a strong oxidizing agent. The oxide *dioxygen difluoride* (O_2F_2) is an orange solid that can be prepared by passing an electrical discharge through an equimolar mixture of fluorine and oxygen gases:

$$F_2(g) + O_2(g) \xrightarrow[\text{discharge}]{\text{Electric}} O_2F_2(s)$$

Sample Exercise 19.5

Give the Lewis structure, molecular structure, and hybridization of the oxygen atom for OF_2.

Solution

The OF_2 molecule has 20 valence electrons and the Lewis structure is

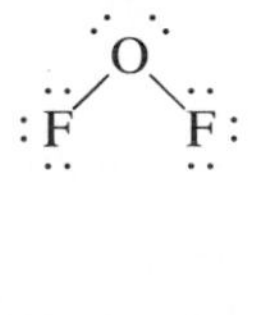

$$:\ddot{\underset{..}{F}}-\ddot{\underset{..}{O}}-\ddot{\underset{..}{F}}:$$

Sample Exercise 19.5, continued

The four effective pairs around the oxygen are arranged tetrahedrally. Therefore the oxygen atom is sp^3 hybridized. The molecule is bent (V-shaped), and the bond angle is predicted to be smaller than 109.5° because of the lone-pair repulsions.

Sample Exercise 19.6

Using the half-reactions

$$I_2 + 2e^- \rightarrow 2I^- \qquad \mathscr{E}° = 0.54 \text{ V}$$

$$2H^+ + 2HOI + 2e^- \rightarrow I_2 + 2H_2O \qquad \mathscr{E}° = 1.45 \text{ V}$$

predict whether I_2 will disproportionate to HOI and I^- in acidic solution.

Solution

We need to determine whether the disproportionation reaction

$$I_2 \rightleftharpoons HOI + I^-$$

is spontaneous. The balanced equation for this process can be obtained by reversing the second half-reaction and adding the two:

$$I_2 + 2e^- \rightarrow 2I^- \qquad \mathscr{E}° = 0.54 \text{ V}$$

$$I_2 + 2H_2O \rightarrow 2H^+ + 2HOI + 2e^- \qquad -\mathscr{E}° = -1.45 \text{ V}$$

$$2I_2 + 2H_2O \rightarrow 2I^- + 2H^+ + 2HOI \qquad \mathscr{E}° = -0.91 \text{ V}$$

or

$$I_2 + H_2O \rightarrow I^- + H^+ + HOI \qquad \mathscr{E}° = -0.91 \text{ V}$$

Since the $\mathscr{E}°$ value for this reaction is negative, the reaction is not spontaneous under standard conditions.

Other Halogen Compounds

The halogens react readily with most nonmetals to form a variety of compounds, some of which are shown in Table 19.11.

Halogens react with each other to form **interhalogen compounds** with the general formula AB_n, where n is typically 1, 3, 5, or 7 and A is the larger of the two halogens. The structures of these compounds (see Fig. 19.26) are predicted accu-

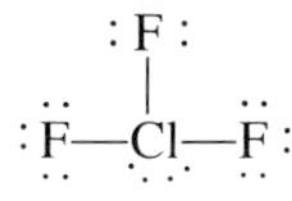

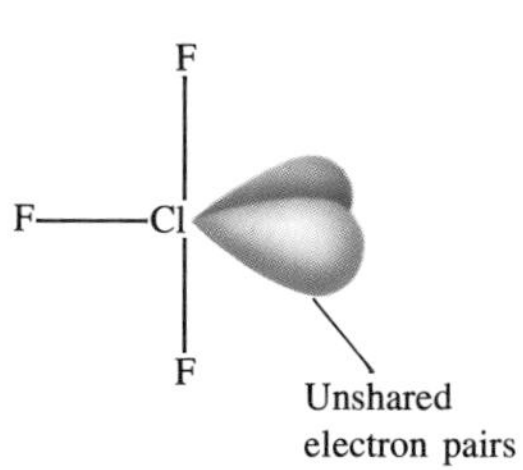

(a) ClF_3 is "T-shaped"

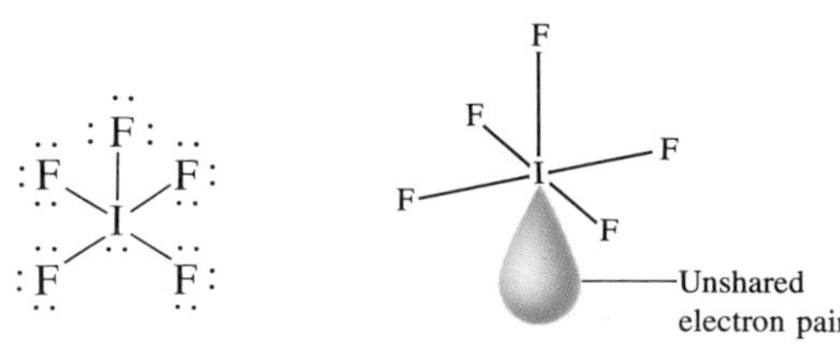

(b) IF_5 is square pyramidal

Figure 19.26

The idealized structures of the interhalogens ClF_3 and IF_5. In reality, the lone pairs cause the bond angles ... slightly less than 90°.

Some Compounds of the Halogens with Nonmetals

Compounds with Group 3A nonmetals	Compounds with Group 4A nonmetals	Compounds with Group 5A nonmetals	Compounds with Group 6A nonmetals	Compounds with Group 7A nonmetals
BX_3 (X = F,Cl,Br,I)	CX_4 (X = F,Cl,Br,I)	NX_3 (X = F,Cl,Br,I)	OF_2	ICl
BF_4^-		N_2F_4	O_2F_2	IBr
	SiF_4		OCl_2	BrF
	SiF_6^{2-}	PX_3 (X = F,Cl,Br,I)	OBr_2	BrCl
	$SiCl_4$	PF_5		ClF
		PCl_5	SF_2	
	GeF_4	PBr_5	SCl_2	ClF_3
	GeF_6^{2-}		S_2F_2	BrF_3
	$GeCl_4$	AsF_3	S_2Cl_2	ICl_3
		AsF_5	SF_4	IF_3
			SCl_4	
		SbF_3	SF_6	ClF_5
		SbF_5		BrF_5
			SeF_4	IF_5
			SeF_6	
			$SeCl_2$	IF_7
			$SeCl_4$	
			$SeBr_4$	
			TeF_4	
			TeF_6	
			$TeCl_4$	
			$TeBr_4$	
			TeI_4	

Table 19.11

rately by the VSEPR model. The interhalogens are volatile, highly reactive compounds that act as strong oxidizing agents. They react readily with water, forming the halide ion of the more electronegative halogen and the hypohalous acid of the less electronegative halogen, for example,

$$ICl(s) + H_2O(l) \rightarrow H^+(aq) + Cl^-(aq) + HOI(aq)$$

The halogens react with carbon to form many commercially important compounds. Especially significant are a group of polymers, or long-chain compounds (see Section 22.5), made from ethylene molecules:

$$\begin{array}{ccc} H & & H \\ & C{=}C & \\ H & & H \end{array}$$

For example, the compound perfluoroethylene

$$\begin{array}{ccc} F & & F \\ & C{=}C & \\ F & & F \end{array}$$

can react with itself to give the polymer

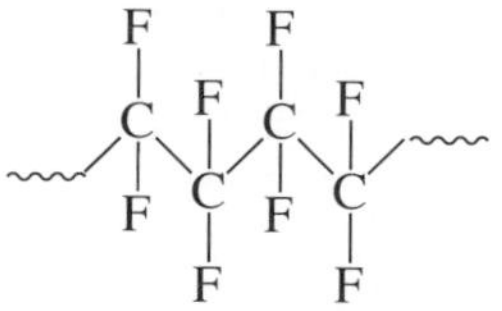

or *Teflon,* a substance with high thermal stability and resistance to chemical attack, used as a coating in many different applications.

The **Freons,** compounds of the general formula CCl_xF_{4-x}, are very stable compounds that are used as coolant fluids in refrigerators and air conditioners.

Chlorine is found in many other useful carbon compounds. For example, the molecule

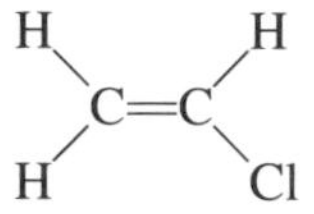

called vinyl chloride, can be polymerized to form *polyvinylchloride,* or PVC:

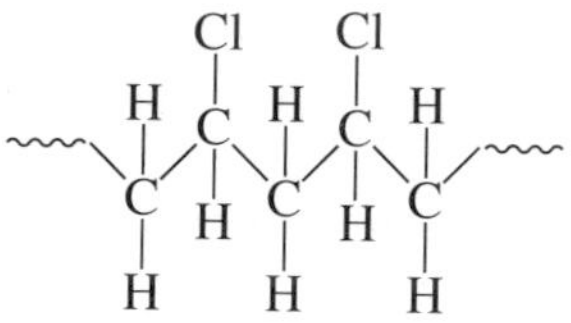

Chlorine is also found in many pesticides.

Bromine is found in dibromoethane ($C_2H_4Br_2$), a compound added to leaded gasoline to form volatile $PbBr_2$, which is carried away in the exhaust, thus preventing lead build-up in the engine. Bromine-containing compounds are also used as flame retardants for cloth (see Section 22.5).

8A
He
Ne
Ar
Kr
Xe
Rn

19.8 The Group 8A Elements

Purpose

- To describe the uses of the noble gases.
- To discuss the structures of xenon compounds.

The Group 8A elements, the **noble gases,** are characterized by filled s and p valence orbitals (electron configuration of $2s^2$ for helium and ns^2np^6 for the others). Because of this, these elements are very unreactive. In fact, no noble gas compounds were known 30 years ago. Selected properties of the Group 8A elements are summarized in Table 19.12.

Helium was identified by its characteristic emission spectrum as a component of the sun before it was found on earth. The major sources of helium on earth are natural gas deposits, where helium was formed from α-particle decay of radioactive elements. The α particle is a helium nucleus that can easily pick up electrons from

Selected Properties of Group 8A Elements

Element	Melting point (°C)	Boiling point (°C)	Atmospheric abundance (% by volume)	Examples of compounds
Helium	−272	−269	5×10^{-4}	None
Neon	−249	−246	1×10^{-3}	None
Argon	−189	−186	9×10^{-1}	None
Krypton	−157	−153	1×10^{-4}	KrF_2
Xenon	−112	−107	9×10^{-6}	XeF_4, XeO_3, XeF_6

Table 19.12

the environment to form a helium atom. Although helium forms no compounds, it is an important substance that is used as a coolant, as a pressurizing gas for rocket fuels, as a diluent in the gases used for deep-sea diving and spaceship atmospheres, and as the gas in lighter-than-air airships (blimps).

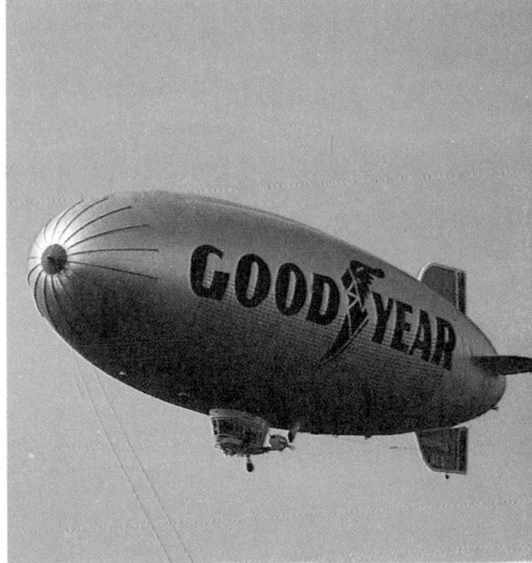

The Goodyear blimp.

Like helium, *neon* and *argon* form no compounds but are used extensively. For example, neon is employed in luminescent lighting (neon signs), and argon is used to provide the noncorrosive atmosphere in incandescent light bulbs, which prolongs the life of the tungsten filament.

Of the Group 8A elements, only *krypton* and *xenon* have been observed to form chemical compounds. The first of these was prepared in 1962 by Neil Bartlett (b. 1932), an English chemist who made the ionic compound $XePtF_6$, which contains the ions Xe^+ and PtF_6^-. The reaction is

$$\underset{\text{Red}}{Xe(g) + PtF_6(g)} \rightarrow \underset{\text{Yellow-orange}}{XePtF_6(s)}$$

Less than a year after Bartlett's report of $XePtF_6$, a group at Argonne National Laboratory near Chicago prepared xenon tetrafluoride by reaction of xenon and fluorine gases in a nickel reaction vessel at 400°C and 6 atmospheres:

$$Xe(g) + 2F_2(g) \rightarrow XeF_4(s)$$

Xenon tetrafluoride forms stable colorless crystals. Two other xenon fluorides, XeF_2 and XeF_6, were synthesized by the group at Argonne, and a highly explosive xenon oxide (XeO_3) was also found. The xenon fluorides react with water to form hydrogen fluoride and oxycompounds, for example:

$$XeF_6(s) + 3H_2O(l) \rightarrow XeO_3(aq) + 6HF(aq)$$
$$XeF_6(s) + H_2O(l) \rightarrow XeOF_4(aq) + 2HF(g)$$

In the past 20 years other xenon compounds have been prepared, for example, XeO_4 (explosive), $XeOF_4$, $XeOF_2$, and XeO_3F_2. These compounds contain discrete molecules with covalent bonds between the xenon and the other atoms. A few compounds of krypton, such as KrF_2 and KrF_4, have also been observed. There is evidence that radon also reacts with fluorine, but the radioactivity of radon makes its chemistry very difficult to study.

Sample Exercise 19.7

Use the VSEPR model to predict whether XeF_6 has an octahedral structure.

Solution

The XeF_6 molecule contains 50 [8 + 6(7)] valence electrons and the Lewis structure is

```
          ..
   ..    :F:    ..
  :F_     |    _F:
   ..  \  |  /  ..
         Xe
   ..  / ·· \   ..
  :F      |     F:
   ..    :F:    ..
          ..
```

The xenon atom has seven effective pairs of electrons around it (one lone pair and six bonding pairs), one more pair than can be accommodated in an octahedral arrangement. Thus XeF_6 will not have an octahedral structure, but should be distorted from this geometry by the extra electron pair. There is experimental evidence that the structure of XeF_6 is not octahedral.

FOR REVIEW

Summary

The Group 5A elements show remarkably varied chemical properties. Nitrogen and phosphorus are nonmetals and can form 3− anions in salts with active metals. Antimony and bismuth tend to be metallic, although no ionic compounds with their 5+ ions are known. The compounds containing bismuth(V) or antimony(V) are molecular rather than ionic. All the Group 5A members except nitrogen form molecules with five covalent bonds. The ability of these elements to form π bonds decreases dramatically after nitrogen. Thus, phosphorus, arsenic, and antimony exist in the gas phase as aggregates of more than two atoms.

The strength of the triple bond in the N_2 molecule is of great importance both thermodynamically and kinetically. Most binary nitrogen-containing compounds decompose exothermically to the elements, a fact that explains the power of nitrogen-based explosives.

The nitrogen cycle is composed of a series of steps describing how nitrogen is cycled in the natural environment. One step involves the transformation of nitrogen gas in the atmosphere into nitrogen-containing compounds useful to plants, a process called nitrogen fixation. The Haber process is one example of nitrogen fixation; another is provided by nitrogen-fixing bacteria in the root nodules of certain plants. Denitrification is the reverse process—the release of nitrogen gas back into the atmosphere.

Ammonia, the most important nitrogen hydride, has pyramidal molecules with polar bonds. Hydrazine (N_2H_4) is a powerful reducing agent. Nitrogen forms a series of oxides in which it has an oxidation state ranging from 1 to 5; the most important are nitric oxide and nitrogen dioxide. Nitric acid (HNO_3) is an important strong acid made by the Ostwald process, in which ammonia is first oxidized to nitric oxide, which is then oxidized to nitrogen dioxide, which is dissolved in water to form nitric acid. Nitrous acid (HNO_2) is a weak acid.

Phosphorus exists in three elemental forms: white, black, and red. Although it has a lower electronegativity than nitrogen, phosphorus forms the P^{3-} anion in phosphides. Phosphine (PH_3) has a structure analogous to that of ammonia but with bond angles close to 90°. Phosphorus reacts with oxygen to form P_4O_6 (oxidation state of +3) and P_4O_{10} (oxidation state of +5). When P_4O_{10} dissolves in water, it forms phosphoric acid (H_3PO_4), a weak acid. Phosphoric acid readily undergoes condensation reactions, giving up water molecules to form polymeric acids.

The Group 6A elements show the usual tendency of increasing metallic properties going down the group. However, none behave as typical metals. Most commonly, these elements achieve noble gas configurations by adding two electrons to form 2− anions. Group 6A elements can also form covalent bonds with other nonmetals.

Oxygen exists in two elemental forms, O_2 and ozone (O_3), a V-shaped molecule. Ozone is a powerful oxidizing agent with important environmental implications.

Sulfur also exists in two elemental forms, rhombic and monoclinic, both of which contain stacks of S_8 rings. Sulfur occurs in nature both as the free element and in ores as sulfides and sulfates. Sulfur forms three oxides. Sulfur monoxide (SO) is unstable. When sulfur dioxide (SO_2) is dissolved in water, sulfurous acid (H_2SO_3) forms, which is analogous in many respects to carbonic acid as a weak acid. Sulfur trioxide (SO_3) dissolves in water to form sulfuric acid (H_2SO_4), an extremely important industrial chemical and a strong acid that is a powerful dehydrating agent. Sulfur forms a variety of compounds in which it has a +6, +4, +2, 0, or −2 oxidation state.

The Group 7A elements, the halogens, are all nonmetals. The halogens form hydrogen halides (HX) that behave as strong acids in water, except for hydrogen fluoride. Hydrofluoric acid is a weak acid because of the strong H—F bond and because the entropy of hydration for the F^- ion is much more negative than it is for the other halide ions. The oxyacids of the halogens become stronger as the number of oxygen atoms attached to the halogen increases. Interhalogens are compounds of two different halogens. Halogen-carbon compounds are important industrially; examples are Teflon, PVC, and the Freons.

The noble gases of Group 8A are characterized by filled *s* and *p* valence orbitals and are generally unreactive. Krypton, xenon, and radon do, however, form compounds with the highly electronegative atoms fluorine and oxygen.

Key Terms

Section 19.2
Haber process
nitrogen fixation
nitrogen-fixing bacteria
denitrification
nitrogen cycle
ammonia
hydrazine
nitric acid
Ostwald process

Section 19.3
phosphoric acid
condensation reactions
phosphorous acid
superphosphate of lime

Section 19.5
ozone
ozonolysis

Section 19.6
Frasch process
sulfuric acid
thiosulfate ion

Section 19.7
halogens
hydrochloric acid
hydrohalic acids
disproportionation reaction
interhalogen compounds
Freons

Section 19.8
noble gases

Exercises

A blue exercise number indicates that the answer to that exercise appears at the back of this book and a solution appears in the Solutions Guide.

Group 5A Elements

1. The oxyanion of nitrogen in which it has the highest oxidation state is the nitrate ion (NO_3^-). The corresponding oxyanion of phosphorus is PO_4^{3-}. The NO_4^{3-} ion is known but not very stable. The PO_3^- ion is not known. Account for these differences in terms of the bonding in the four anions.

2. Using data from Appendix 4 calculate $\Delta H°$, $\Delta S°$, and $\Delta G°$ for the reaction

$$N_2(g) + O_2(g) \rightarrow 2NO(g)$$

Why does NO form in an automobile engine, but then does not readily decompose back to N_2 and O_2 in the atmosphere?

3. What is the hybridization of the nitrogen atoms in the compounds listed in Table 19.2?

4. Which of the compounds in Table 19.2 are capable of resonance? Draw the corresponding resonance structures for those that are.

5. Use bond energies (Table 8.4) to show that the preferred products for the decomposition of N_2O_3 are NO_2 and NO rather than O_2 and N_2O. (The N—O single bond energy is 201 kJ/mol.) *Hint:* Consider the reaction kinetics.

6. Given the information about N_2O_5 in Section 19.2, draw a possible structure for this molecule in the gas phase.

7. What trade-offs must be made between kinetics and thermodynamics in the Haber process for the production of ammonia? How did the discovery of an appropriate catalyst make the process feasible?

8. Oxidation of the cyanide ion produces the stable cynate ion, OCN^-. The fulminate ion, CNO^-, on the other hand, is very unstable. Fulminate salts explode upon being struck; $Hg(CNO)_2$ is used in blasting caps. Draw Lewis structures and assign formal charges for cynate and fulminate ions. Why is the fulminate ion so unstable?

9. Write an equation for the reactions of hydrazine with fluorine gas to produce nitrogen gas and hydrogen fluoride gas. Estimate $\Delta H°$ for this reaction using bond energies from Table 8.4.

10. Write balanced equations that describe how the following can be produced:
    a. $NO(g)$
    b. $N_2O(g)$
    c. $KNO_2(s)$

11. In each of the following pairs of substances, one is stable and known, the other is unstable. For each pair, choose the stable substance and explain why the other is unstable.
    a. NF_5 or PF_5
    b. AsF_5 or AsI_5
    c. NF_3 or NBr_3

12. Write balanced equations for the reactions described in Table 19.1 for the production of Bi and Sb.

13. Predict the relative acid strengths of the following:
    a. H_3PO_4 and H_3PO_3
    b. H_3PO_4, $H_2PO_4^-$, HPO_4^{2-}

14. Isohypophosphonic acid, $H_4P_2O_6$, and diphosphonic acid, $H_4P_2O_5$, are tri- and diprotic acids, respectively. Draw Lewis structures for these acids that are consistent with these facts.

15. Account for the reactivity with oxygen of white phosphorus, red phosphorus, and black phosphorus in terms of their structures.

16. What experiments can be done to verify that bismuthyl chloride (BiOCl) is composed of $BiO^+$and Cl^- ions and not Bi^+ and OCl^- ions?

17. Arsenic reacts with oxygen to form oxides analogous to the phosphorus oxides. These arsenic oxides react with water in a manner analogous to that of phosphorus oxides. Write balanced chemical equations describing the reaction of arsenic with oxygen and the reaction of the oxide with water.

18. Compare the Lewis structures to the molecular orbital view of the bonding in NO, NO^+, and NO^-. Account for any discrepancies between the two models.

19. Sodium tripolyphosphate ($Na_5P_3O_{10}$) is used in many synthetic detergents. Its major effect is to soften the water by complexing Mg^{2+} and Ca^{2+} ions. It also increases the efficiency of surfactants or wetting agents that lower a liquid's surface tension. The p*K* value for the formation of $MgP_3O_{10}^{3-}$ is −8.6. The reaction is $Mg^{2+} + P_3O_{10}^{5-} \rightleftharpoons MgP_3O_{10}^{3-}$. Calculate the concentration of Mg^{2+}in a solution that was originally 50 ppm of Mg^{2+} (50 mg/L of solution) after 40 g of $Na_5P_3O_{10}$ is added to 1.0 L of the solution.

20. Sodium bismuthate ($NaBiO_3$) is used to test for the presence of Mn^{2+} in solution by the reaction:

$$Mn^{2+}(aq) + NaBiO_3(s) \rightarrow MnO_4^-(aq) + BiO_3^{3-}(aq)$$

    a. Balance the above equation.
    b. Given that bismuth does not form double bonds with oxygen in BiO_3^- and that $NaBiO_3$ is relatively insoluble in water, what type of structure must $NaBiO_3$ have that accounts for this behavior?

21. One of the most strongly acidic solutions known is a mixture of antimony pentafluoride (SbF_5) and fluorosulfonic acid (HSO_3F). The important equilibria are

$$SbF_5 + HSO_3F \rightleftharpoons F_5SbOSO_2FH$$

$$F_5SbOSO_2FH + HSO_3F \rightleftharpoons H_2SO_3F^+ + F_5SbOSO_2F^-$$

a. Draw Lewis structures for all of the species shown in the preceding reactions.
b. Such *superacid* solutions are capable of protonating (adding H^+ to) virtually every known organic compound. What is the active protonating agent in these solutions?

22. Draw Lewis structures for the $AsCl_4^+$ and $AsCl_6^-$ ions. What type of reaction (acid-base, oxidation-reduction, etc.) is the following?

$$2AsCl_5(g) \rightarrow AsCl_4AsCl_6(s)$$

23. Complete and balance each of the following:
a. $P_4O_6(s) + O_2(g) \rightarrow$
b. $P_4O_{10}(s) + H_2O(l) \rightarrow$
c. $PCl_5(l) + H_2O(l) \rightarrow$

Group 6A Elements

24. Use bond energies to estimate the maximum wavelength of light that will cause the reaction

$$O_3 \xrightarrow{h\nu} O_2 + O$$

25. Ozone is desirable in the upper atmosphere and undesirable in the lower atmosphere. A dictionary states that ozone has the scent of a fresh spring day. How can these seemingly conflicting statements be reconciled in terms of the chemical properties of ozone?

26. The teflate anion ($OTeF_5^-$) often acts like a halide ion. For example, $P(OTeF_5)_3$ has a structure similar to that of PF_3. Draw Lewis structures for $OTeF_5^-$ and $P(OTeF_5)_3$.

27. The first acid dissociation constants of H_2S, H_2Se, and H_2Te are 1.03×10^{-7}, 1.3×10^{-4}, and 2.3×10^{-3}, respectively.
a. To what do you attribute this trend? (See the data in the table below.)
b. To the nearest power of 10, predict K_a for H_2Po.

Bond energy (kJ/mol)		Electron affinity (kJ/mol)	
H—S	363	S	−200.4
H—Se	276	Se	−195
H—Te	238	Te	−190

28. The structure of TeF_5^- is

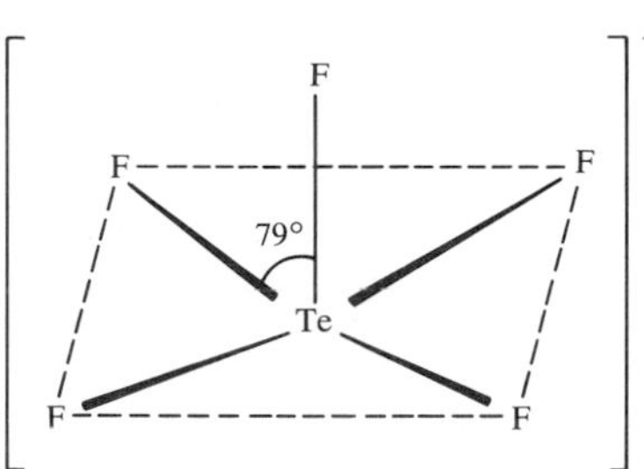

Draw a complete Lewis structure for TeF_5^-. How do you account for the distortion from the ideal square pyramidal geometry?

29. Sulfur will dissolve in an aqueous solution that contains some sulfide ion. Sulfur will then precipitate from this solution if nitric acid is added. Explain this behavior.

30. When nitric acid is added to a solution of sodium sulfide, elemental sulfur forms.
a. What type of reaction is this?
b. Use the table of standard reduction potentials (Table 17.1) to predict the product in the reaction that comes from HNO_3.
c. Write a balanced equation for this process.

31. Write a balanced equation describing the reduction of H_2SeO_4 by SO_2 to produce selenium.

32. The xerographic (dry writing) process was invented in 1938 by C. Carlson. In xerography an image is produced on a photoconductor by exposing it to light. Selenium is commonly used, since its conductivity increases three orders of magnitude upon exposure to light in the range from 400 to 500 nm. What color light should be used to cause selenium to become conductive? (See Figure 20.24.)

33. Cadmium sulfide, which has recently been tested as a photoconductor for xerography, is a semiconducting material with a band gap of 2.42 eV. (1 eV, or electron volt, equals 96.5 kJ/mol.) What is the minimum wavelength of light needed to promote electrons from the valence band to the conduction band in CdS? What color of light is this? (See Figure 20.24.)

34. The SF_5^- ion can be formed by the reaction:

$$CsF + SF_4 \rightarrow Cs^+ + SF_5^-$$

Predict the molecular structure of the SF_5^- ion.

Group 7A Elements

35. Draw Lewis structures and predict geometries for each of the following compounds:
a. ClF_5
b. IF_3

c. Cl_2O_7

$$\text{O}_3\text{Cl—O—ClO}_3$$

d. $FBrO_2$

36. Hypofluorous acid is the most recently prepared of the halogen oxyacids. Weighable amounts were first obtained in 1971 by M. H. Studies and E. N. Appelman, using the fluorination of ice. Hypofluorous acid is exceedingly unstable, decomposing spontaneously, with a half-life of 30 min, to HF and O_2 in a Teflon container at room temperature. It reacts rapidly with water to produce HF, H_2O_2, and O_2. In dilute acid, H_2O_2 is the major product; in dilute base, O_2 is the major product.
 a. Write balanced chemical equations for the reactions described above.
 b. Assign oxidation states to the elements in hypofluorous acid. Does this suggest why hypofluorous acid is so unstable? Explain.

37. Draw the Lewis structure of O_2F_2. Assign oxidation numbers and formal charges to the atoms in O_2F_2. The compound O_2F_2 is a vigorous and potent oxidizing and fluorinating agent. Are oxidation numbers or formal charges more useful in accounting for these properties of O_2F_2?

38. Give two reasons why F_2 is the most reactive of the halogens.

39. Chlorine is used as a disinfectant in the treatment of water supplies.
 a. Calculate $\mathscr{E}°$ for the reaction

$$Cl_2(aq) \rightarrow OCl^-(aq) + Cl^-(aq) \qquad \text{(not balanced)}$$

 b. Is the reaction spontaneous under standard conditions?
 c. What will be the effect of pH on $\mathscr{E}$ for this reaction? *Hint:* HOCl is a weak acid.

Note: For the half-reaction $2OCl^- + 4H^+ + 2e^- \rightarrow 2H_2O + Cl_2$, $\mathscr{E} = 1.63$ V.

40. What is a disproportionation reaction? Use the following reduction potentials

$$ClO_3^- + 3H^+ + 2e^- \rightarrow HClO_2 + H_2O \qquad \mathscr{E}° = 1.21 \text{ V}$$

$$HClO_2 + 2H^+ + 2e^- \rightarrow HClO + H_2O \qquad \mathscr{E}° = 1.65 \text{ V}$$

to predict if $HClO_2$ will disproportionate.

41. Sodium perchlorate is produced by the electrolytic oxidation of aqueous $NaClO_3$:

$$ClO_3^- + H_2O \rightarrow ClO_4^- + H_2$$

More than half of the annual U.S. production is converted to NH_4ClO_4 for use as a rocket fuel. The two booster rockets on the space shuttle use NH_4ClO_4 (70% by mass) as an oxidizer and powdered aluminum (30% by mass) as fuel. About 7×10^5 kg of NH_4ClO_4 is required for each launch.
 a. What is the minimum electrical potential that must be applied to produce $NaClO_4$? For the half-reaction $ClO_4^- + 2H^+ + 2e^- \rightarrow ClO_3^- + H_2O$, $\mathscr{E}° = +1.19$ V.
 b. A possible reaction in the booster rocket is

$$3Al(s) + 3NH_4ClO_4(s) \rightarrow Al_2O_3(s) + AlCl_3(s) + 3NO(g) + 6H_2O(g)$$

Estimate the total heat generated by the booster rocket during a space shuttle launch.

42. Photogray lenses contain small embedded crystals of solid silver chloride. Silver chloride is light-sensitive because of the reaction:

$$AgCl(s) \xrightarrow{h\nu} Ag(s) + Cl$$

Small particles of metallic silver cause the lenses to darken. In the lenses this process is reversible. When the light is removed, the reverse reaction occurs. However, when pure white silver chloride is exposed to sunlight it darkens; the reverse reaction does not occur in the dark.
 a. How do you explain this difference?
 b. Photogray lenses do become permanently dark in time. How do you account for this?

Group 8A Elements

43. The noble gases were among the latest elements discovered; their existence was not predicted by Mendeleev when he published his first periodic table. Why do you think this was so?

44. Although He is the second most abundant element in the universe, it is very rare on earth. Why?

45. In chemistry textbooks written before 1962 the noble gases were referred to as the inert gases. Why do we no longer use the term inert gases?

46. Using the data in Table 19.12, calculate the mass of xenon at 25°C and 1.0 atm in a room 10.0 m × 5.0 m × 3.0 m. How many Xe atoms are in this room? How many Xe atoms do you inhale in one breath (approximately 2 L) of air?

47. Draw Lewis structures and predict the shapes for XeO_3, XeO_4, $XeOF_4$, $XeOF_2$, and XeO_3F_2.

48. Xenon difluoride has proven to be a versatile fluorinating agent. For example, in the reaction

$$C_6H_6(l) + XeF_2(g) \rightarrow C_6H_5F(l) + Xe(g) + HF(g)$$

the byproducts Xe and HF are easily removed leaving pure C_6H_5F. Xenon difluoride is stored in an inert atmosphere free from oxygen and water. Why is this necessary?

Additional Exercises

49. The compound NF_3 is quite stable, but NCl_3 is very unstable (NCl_3 was first synthesized in 1811 by P. L. Dulong, who lost

three fingers and an eye studying its properties). The compounds NBr_3 and NI_3 are unknown, although the explosive $NI_3 \cdot NH_3$ is known. How do you account for the instability of these halides of nitrogen?

50. Many structures of phosphorus- and sulfur-containing compounds are drawn with some P=O and P=S bonds. These bonds are not the typical π bonds we've considered, involving the overlap of two p orbitals. They result instead from the overlap of a d orbital on the phosphorus or sulfur atom and a p orbital on oxygen. This is sometimes used as an explanation for why H_3PO_3 has the first structure below rather than the second:

```
     O           OH
     ‖           |
  H—P—OH     HO—P:
     |           |
     OH          OH
```

Draw a picture showing how a d orbital and a p orbital overlap to form a π bond.

51. The xenon halides, oxyhalides, and oxides are isoelectronic with many other compounds and ions containing halogens. Give a compound or ion, in which iodine is the central atom, that is isoelectronic with each of the following:

a. XeO_4
b. XeO_3
c. XeF_2
d. XeF_4
e. XeF_6
f. $XeOF_3$

52. The energy to break a particular bond is not always constant. It takes about 200 kJ/mol less energy to break the N—Cl bond in NOCl than the one in NCl_3.

$$NOCl \rightarrow NO + Cl \qquad \Delta H° = 158 \text{ kJ/mol}$$
$$NCl_3 \rightarrow NCl_2 + Cl \qquad \Delta H° \approx 375 \text{ kJ/mol}$$

Why is there such a great discrepancy in the apparent N—Cl bond energies? *Hint:* Consider what happens to the nitrogen-oxygen bond in the first reaction.

53. Use data in Appendix 4 to calculate $\Delta H°$, $\Delta S°$, and $\Delta G°$ for the reaction

$$3NO(g) \rightleftharpoons N_2O(g) + NO_2(g)$$

Under what temperature conditions is this reaction spontaneous?

54. Describe the structural changes that occur when plastic sulfur becomes brittle.

55. In many natural waters, nitrogen and phosphorus are the least abundant nutrients available for plant life. Some waters that become polluted from agricultural runoff or municipal sewage become infested with algae. The algae consume most of the dissolved oxygen in the water and fish life dies off as a result. Describe how these events are chemically related.

56. Although the perchlorate and periodate ions have been known for a long time, the perbromate ion proved to be quite elusive and was not synthesized until 1965. Perbromate was synthesized by oxidizing bromate ion with aqueous xenon difluoride, producing xenon, hydrofluoric acid, and perbromate ion. Write a balanced chemical equation for this reaction.

CHAPTER 20

Transition Metals and Coordination Chemistry

Transition metals have many uses in our society. Iron is used for steel, copper for electrical wiring and water pipes, titanium for paint, silver for photographic paper, manganese, chromium, vanadium, and cobalt as additives to steel, platinum for industrial and automotive catalysts, and so on.

One indication of the importance of transition metals is the great concern shown by the U.S. government for continuing the supply of these elements. In 1983 the United States was a net importer of 64 "strategic and critical" minerals, including cobalt, manganese, platinum, palladium, and chromium. All of these metals play a vital role in the U.S. economy and defense, and more than 85% of the required amounts must be imported, mostly from areas that are politically unstable (see Table 20.1).

In addition to being important in industry, transition metal ions also play a vital role in living organisms. For example, complexes of iron provide for the transport and storage of oxygen, molybdenum and iron compounds are catalysts in nitrogen fixation, zinc is found in more than 150 biomolecules in humans, copper and iron play a crucial role in the respiratory cycle, and cobalt is found in essential biomolecules such as vitamin B_{12}.

In this chapter we explore the general properties of transition metals, paying particular attention to the bonding, structure, and properties of the complex ions of these metals.

< Naturally crystallized gold on its quartz matrix. This 3.3-inches-high specimen is from the Colorado Quartz Mine, Mariposa County, California.

CONTENTS

Uses and Sources of Transition Metals Important to the U.S. Economy and Defense

Metal	Uses	Percentage imported	Main sources
Chromium	Stainless steel (especially for parts exposed to corrosive gases and high temperatures)	91%	Republic of South Africa, U.S.S.R., Yugoslavia, Philippines, Turkey
Cobalt	High-temperature alloys in jet engines, magnets, catalysts, drill bits	93%	Zaïre, Belgium, Luxembourg, Zambia, Finland
Manganese	Steelmaking	97%	Gabon, Brazil, Republic of South Africa, France
Platinum and palladium	Catalysts	87%	Republic of South Africa, U.S.S.R., U.K.

Source: Adapted from *Chemical & Engineering News,* May 11, 1981, **59** (19), pp. 20–25. Published 1981 American Chemical Society.

Table 20.1

20.1 The Transition Metals: A Survey

Purpose

- To discuss electron configurations and general trends in the properties of the first-row transition metals.
- To introduce the 4*d* and 5*d* transition series.

General Properties

One of the striking things about the representative elements is that their chemistry changes markedly across a given period as the number of valence electrons changes. The chemical similarities occur mainly within the vertical groups. In contrast, *the transition metals show great similarities within a given period as well as within a given vertical group*. This difference occurs because the last electrons added for transition metals are inner electrons: *d* electrons for the *d*-block transition metals and *f* electrons for the lanthanides and actinides. These inner *d* and *f* electrons cannot as easily participate in bonding as can the valence *s* and *p* electrons. Thus the chemistry of transition elements is not as greatly affected by the gradual change in the number of electrons as is the chemistry of the representative elements.

Group designations are traditionally given on the periodic table for the *d*-block transition metals (Fig. 20.1). However, these designations do not relate as directly

(clockwise from upper left) Calcite stalactites colored by traces of iron. Quartz is often colored by the presence of transition metals such as Mn, Fe, and Ni. Wulfenite contains $PbMoO_4$. Rhodochrosite is a mineral containing $MnCO_3$.

to the chemical behavior of these elements as they do for the representative elements (the A groups), so we will not use them.

As a class, the transition metals behave as typical metals, possessing metallic luster and relatively high electrical and thermal conductivities. Silver is the best conductor of heat and electrical current. However, copper is a close second, which explains copper's wide use in the electrical systems of homes and factories.

In spite of their many similarities, the properties of the transition metals do vary considerably. For example, tungsten has a melting point of 3400°C and is used for

d-block transition elements

Sc	Ti	V	Cr	Mn	Fe	Co	Ni	Cu	Zn
Y	Zr	Nb	Mo	Tc	Ru	Rh	Pd	Ag	Cd
La	Hf	Ta	W	Re	Os	Ir	Pt	Au	Hg
Ac	Unq	Unp	Unh						

f-block transition elements

Lanthanides	Ce	Pr	Nd	Pm	Sm	Eu	Gd	Tb	Dy	Ho	Er	Tm	Yb	Lu
Actinides	Th	Pa	U	Np	Pu	Am	Cm	Bk	Cf	Es	Fm	Md	No	Lr

Figure 20.1

The position of the transition elements on the periodic table. The *d*-block elements correspond to filling the 3*d*, 4*d*, 5*d*, or 6*d* orbitals. The inner transition metals correspond to filling the 4*f* (lanthanides) or 5*f* (actinides) orbitals.

filaments in light bulbs; mercury is a liquid at 25°C. Some transition metals such as iron and titanium are hard and strong and make very useful structural materials; others such as copper, gold, and silver are relatively soft. The chemical reactivity of the transition metals also varies significantly. Some react readily with oxygen to form oxides. Of these metals, some, such as chromium, nickel, and cobalt, form oxides that adhere tightly to the metallic surface, protecting the metal from further oxidation. Others, such as iron, form oxides that scale off, constantly exposing new metal to the corrosion process. On the other hand, the noble metals—primarily gold, silver, platinum, and palladium—do not readily form oxides.

In forming ionic compounds with nonmetals, the transition metals exhibit several typical characteristics:

More than one oxidation state is often found. For example, iron combines with chlorine to form $FeCl_2$ and $FeCl_3$.

The cations are often **complex ions,** species where *the transition metal ion is surrounded by a certain number of ligands* (molecules or ions that behave as Lewis bases). For example, the compound $[Co(NH_3)_6]Cl_3$ contains $Co(NH_3)_6^{3+}$ cations and Cl^- anions.

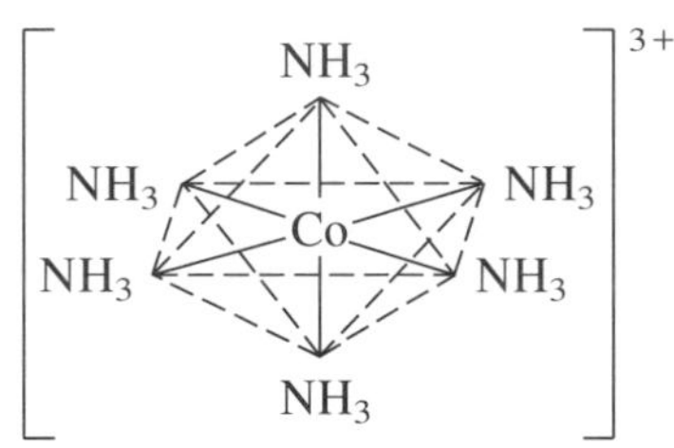

The $Co(NH_3)_6^{3+}$ ion

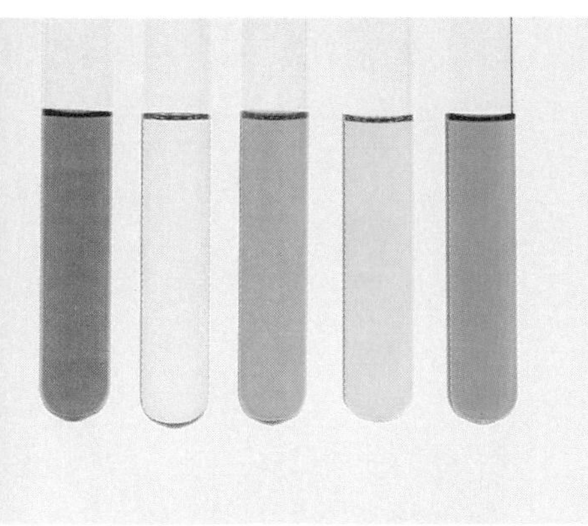

(from left to right) Aqueous solutions containing the metal ions Co^{2+}, Mn^{2+}, Cr^{3+}, Fe^{3+}, and Ni^{2+}.

Most compounds are colored, because the transition metal ion in the complex ion can absorb visible light of specific wavelengths.

Many compounds are paramagnetic (they contain unpaired electrons).

In this chapter we will concentrate on the **first-row transition metals** (scandium through zinc) because they are representative of the other transition series and because they have great practical significance. Some important properties of these elements are summarized in Table 20.2 and are discussed below.

Electron Configurations

The electron configurations of the first-row transition metals were discussed in Section 7.11. The 3*d* orbitals begin to fill after the 4*s* orbital is complete, that is, after calcium ($[Ar]4s^2$). The first transition metal, *scandium,* has one electron in the 3*d* orbitals; the second, *titanium,* has two; and the third, *vanadium,* has three. We would expect *chromium,* the fourth transition metal, to have the electron configuration $[Ar]4s^23d^4$. However, the actual configuration is $[Ar]4s^13d^5$, which shows a half-filled 4*s* orbital and a half-filled set of 3*d* orbitals (one electron in each of the five 3*d* orbitals). It is tempting to say that the configuration results because half-filled "shells" are especially stable. Although there are some reasons to think that this explanation might be true, it is an oversimplification. For instance, tungsten, which is in the same vertical group as chromium, has the configuration

Selected Properties of the First-Row Transition Metals

	Scandium	Titanium	Vanadium	Chromium	Manganese	Iron	Cobalt	Nickel	Copper	Zinc
Atomic number	21	22	23	24	25	26	27	28	29	30
Electron configuration*	$4s^23d^1$	$4s^23d^2$	$4s^23d^3$	$4s^13d^5$	$4s^23d^5$	$4s^23d^6$	$4s^23d^7$	$4s^23d^8$	$4s^13d^{10}$	$4s^23d^{10}$
Atomic radius (Å)	1.61	1.45	1.32	1.27	1.24	1.24	1.25	1.25	1.28	1.33
Ionization energies (eV/atom)										
First	6.54	6.82	6.74	6.77	7.44	7.87	7.86	7.64	7.73	9.39
Second	12.80	13.58	14.65	16.50	15.64	16.18	17.06	18.17	20.29	17.96
Third	24.76	27.49	29.31	30.96	33.67	30.65	33.50	35.17	36.83	39.72
Reduction potential† (V)	−2.08	−1.63	−1.2	−0.91	−1.18	−0.44	−0.28	−0.23	+0.34	−0.76
Common oxidation states	+3	+2,+3, +4	+2,+3, +4,+5	+2,+3, +6	+2,+3, +4,+7	+2,+3	+2,+3	+2	+1,+2	+2
Melting point (°C)	1397	1672	1710	1900	1244	1530	1495	1455	1083	419
Density (g/cm^3)	2.99	4.49	5.96	7.20	7.43	7.86	8.9	8.90	8.92	7.14
Electrical conductivity‡	—	2	3	10	2	17	24	24	97	27

*Each atom has an argon inner-core configuration.
†For the reduction process $M^{2+} + 2e^- \rightarrow M$ (except for scandium, where the ion is Sc^{3+}).
‡Compared to an arbitrarily assigned value of 100 for silver.

Table 20.2

$[Xe]6s^24f^{14}5d^4$, where half-filled *s* and *d* shells are not found. There are several similar cases.

Basically, the chromium configuration occurs because the energies of the 3*d* and 4*s* orbitals are very similar for the first-row transition elements. We saw in Section 7.11 that when electrons are placed in a set of degenerate orbitals, they first occupy each orbital singly to minimize electron repulsions. Since the 4*s* and 3*d* orbitals are virtually degenerate in the chromium atom, we would expect the configuration

Chromium has the electron configuration $[Ar]4s^13d^5$.

A set of orbitals with the same energy is said to be degenerate.

$$4s\ \underline{\uparrow} \qquad 3d\ \underline{\uparrow}\ \underline{\uparrow}\ \underline{\uparrow}\ \underline{\uparrow}\ \underline{\uparrow}$$

rather than

$$4s\ \underline{\uparrow\downarrow} \qquad 3d\ \underline{\uparrow}\ \underline{\uparrow}\ \underline{\uparrow}\ \underline{\uparrow}\ \underline{\ \ }$$

since the second arrangement has greater electron-electron repulsions and thus a higher energy.

Copper has the electron configuration [Ar]$4s^1 3d^{10}$.

The only other unexpected configuration among the first-row transition metals is that of copper, which is [Ar]$4s^1 3d^{10}$ rather than the expected [Ar]$4s^2 3d^9$.

In contrast to the neutral transition metals, where the $3d$ and $4s$ orbitals have very similar energies, the *energy of the* 3d *orbitals in transition metal ions is significantly less than that of the* 4s *orbital*. This means that the electrons remaining after the ion is formed occupy the $3d$ orbitals, since they are lower in energy. *First-row transition metal ions do not have* 4s *electrons*. For example, manganese has the configuration [Ar]$4s^2 3d^5$, while that of Mn^{2+} is [Ar]$3d^5$. The neutral titanium atom has the configuration [Ar]$4s^2 3d^2$, while that of Ti^{3+} is [Ar]$3d^1$.

Oxidation States and Ionization Energies

The transition metals can form a variety of ions by losing one or more electrons. The common oxidation states of these elements are shown in Table 20.2. Note that for the first five metals the maximum possible oxidation state corresponds to the loss of all the $4s$ and $3d$ electrons. For example, the maximum oxidation state of chromium ([Ar]$4s^1 3d^5$) is +6. Toward the right end of the period, the maximum oxidation states are not observed; in fact, the 2+ ions are the most common. The higher oxidation states are not seen for these metals because the $3d$ orbitals become lower in energy as the nuclear charge increases, and the electrons become increasingly difficult to remove. From Table 20.2 we see that ionization energy increases gradually going from left to right across the period. However, the third ionization energy (when an electron is removed from a $3d$ orbital) increases faster than the first ionization energy, clear evidence of the significant decrease in the energy of the $3d$ orbital going across the period (see Fig. 20.2).

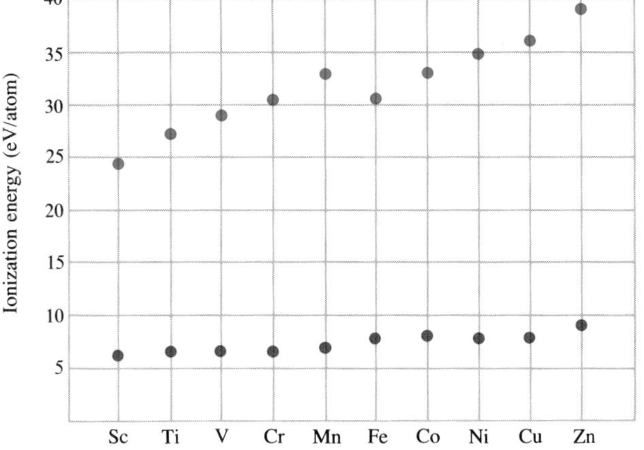

Figure 20.2

Plots of the first (red dots) and third (blue dots) ionization energies for the first-row transition metals.

Standard Reduction Potentials

When a metal acts as a *reducing agent,* the half-reaction is

$$M \rightarrow M^{n+} + ne^-$$

This is the reverse of the conventional listing for half-reactions in tables. Thus, to rank the transition metals in order of reducing ability, it is most convenient to reverse the reactions and the signs given in Table 20.2. The metal with the most positive potential is then the best reducing agent. The transition metals are listed in order of reducing ability in Table 20.3.

Relative Reducing Abilities of the First-Row Transition Metals in Aqueous Solution

Reaction	Potential (V)
$Sc \rightarrow Sc^{3+} + 3e^-$	2.08
$Ti \rightarrow Ti^{2+} + 2e^-$	1.63
$V \rightarrow V^{2+} + 2e^-$	1.2
$Mn \rightarrow Mn^{2+} + 2e^-$	1.18
$Cr \rightarrow Cr^{2+} + 2e^-$	0.91
$Zn \rightarrow Zn^{2+} + 2e^-$	0.76
$Fe \rightarrow Fe^{2+} + 2e^-$	0.44
$Co \rightarrow Co^{2+} + 2e^-$	0.28
$Ni \rightarrow Ni^{2+} + 2e^-$	0.23
$Cu \rightarrow Cu^{2+} + 2e^-$	−0.34

Reducing ability ↑

Table 20.3

Since $\mathscr{E}°$ is zero for the process

$$2H^+ + 2e^- \rightarrow H_2$$

all of the metals except copper can reduce H^+ ions to hydrogen gas in 1 *M* aqueous solutions of strong acid:

$$M(s) + 2H^+(aq) \rightarrow H_2(g) + M^{2+}(aq)$$

As Table 20.3 shows, the reducing abilities of the first-row transition metals generally decrease going from left to right across the period. Only chromium and zinc do not follow this trend.

The 4*d* and 5*d* Transition Series

In comparing the 3*d*, 4*d*, and 5*d* transition series, it is instructive to consider the atomic radii of these elements (Fig. 20.3). Note that there is a general, although not regular, decrease in size going from left to right for each of the series. Also note that, although there is a significant increase in radius in going from the 3*d* to the 4*d* metals, the 4*d* and 5*d* metals are remarkably similar in size. This latter phenomenon is the result of the **lanthanide contraction.** In the **lanthanide series,** consisting of the elements between lanthanum and hafnium (see Fig. 20.1), electrons are filling the 4*f* orbitals. Since the 4*f* orbitals are buried in the interior of these atoms, the

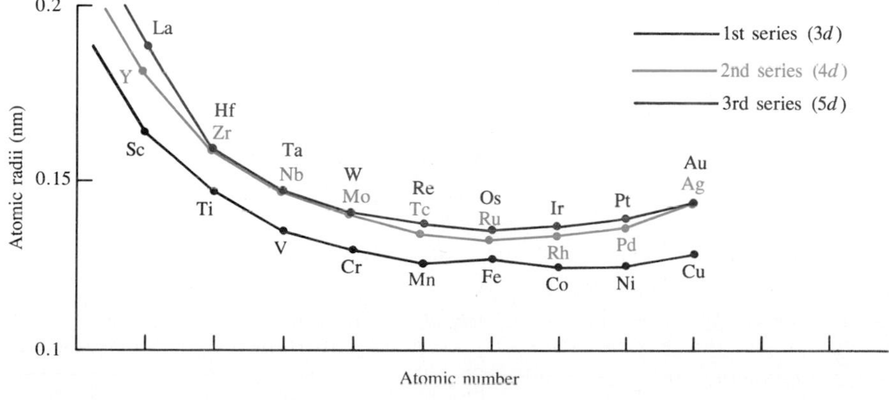

Figure 20.3

Atomic radii of the 3*d*, 4*d*, and 5*d* transition series.

Chemical Impact

Transition Metals in Action

Transition metals are important in many aspects. The following three applications show the varied nature of the contributions transition metals make to everyday life and research. (See also the Chemical Impact feature on cancer therapy in Section 20.4.)

Nitinol is a nickel-titanium alloy that "remembers" a preset shape after it has been deformed; it can be made to return to its original shape by heating. This property of nitinol is being exploited to repair arteries that are partially blocked by plaque or that have weaknesses in their walls.

A nitinol wire is first coiled around a pipe whose outside diameter is identical to the inside diameter of the artery to be repaired. The coil of wire is then heated until its original straw color turns to a deep blue. This sets the "memory" of the alloy for the coiled structure. The wire is subsequently cooled and straightened so that it can be fed into the artery through a catheter (a thin tube) containing a cool saline solution.

When the wire is in place, the body heat causes it to return to its original coiled shape, pushing apart the restricted area in a partially blocked artery or acting as reinforcement for an artery with a weak, bulging wall. Within a few weeks the wire coils are completely incorporated into the arterial walls.

Although overshadowed by gold for making jewelry, platinum has great industrial value. In fact, it is used in the production of perhaps one out of five products in modern society (see Fig. 20.4). Approximately 75 tons of platinum are produced annually, the major portion from the Republic of South Africa. Platinum catalysts are used in the manufacture of nitric acid (see Section 19.2), but the largest single use of the metal is for the platinum-coated ceramic honeycomb catalyst used in automobile exhaust systems to help convert unburned fuel and carbon monoxide to carbon dioxide and water. Also, many airliners use platinum catalysts to convert the harmful ozone found at high altitudes to oxygen.

Because of its durability, malleability, resistance to corrosion, and stability at high temperatures (mp = 1769°C), platinum has many other uses, including jewelry, dies for the manufacture of glass strands for fiberglass, thin protective coatings (100 atoms in thickness) for razor blades, and electrodes in body implant devices such as heart pacemakers. Also, the kilogram is defined by a cylinder of platinum-iridium alloy stored in a special vault in Paris.

Why did the dinosaurs suddenly disappear from the earth about 65 million years ago? Although many hypotheses have been suggested, none has seemed very convincing until recently. Geologists examining core samples containing material that was on the earth's surface when the dinosaurs lived have found an unusually high concentration of iridium. Since this metal is often found in relatively high concentrations in extraterrestrial bodies such as meteorites and comets, it has been suggested that one of these objects may have collided with the earth about 65 million years ago, upsetting the earth's atmosphere and leading to the extinction of many species.

Subsequently, the ratios of niobium isotopes in the same core samples were investigated and found to be much more characteristic of extraterrestrial bodies than of the normal distribution on earth. These niobium studies thus support the iridium data and lend further credence to this hypothesis for the extinction of the vast number of dinosaurs. (For further information see *Science News,* November 19, 1983, p. 329.)

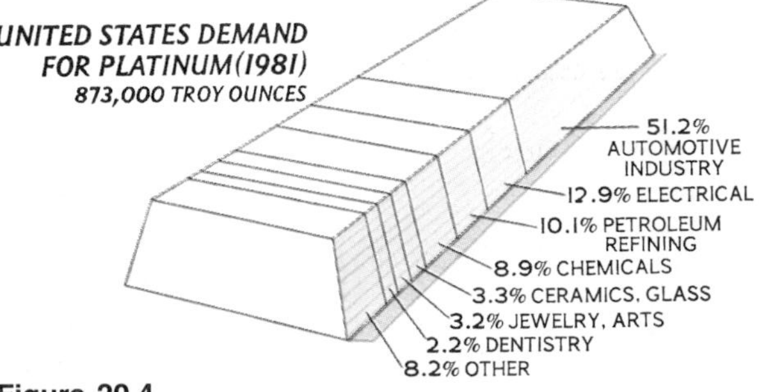

Figure 20.4

Uses of platinum in the U.S. (From *National Geographic,* Vol. 164, No. 5, p. 686, November 1983. Reprinted by courtesy of National Geographic Magazine.)

additional electrons do not add to the atomic size. In fact, the increasing nuclear charge (remember that a proton is added to the nucleus for each electron) causes the radii of the lanthanide elements to decrease significantly going from left to right. This lanthanide contraction just offsets the normal increase in size due to going from one principal quantum level to another. Thus the 5*d* elements, instead of being significantly larger than the 4*d* elements, are almost identical to them in size. This leads to a great similarity in the chemistry of the 4*d* and 5*d* elements in a given vertical group. For example, the chemical properties of hafnium and zirconium are remarkably similar, and they always occur together in nature. Their separation, which is probably more difficult than the separation of any other pair of elements, often requires fractional distillation of their compounds.

In general, the differences between the 4*d* and 5*d* elements in a group increase gradually going from left to right. For example, niobium and tantalum are also quite similar, but less so than zirconium and hafnium.

Niobium was originally called columbium and is still occasionally referred to by that name.

Although generally less well-known than the 3*d* elements, the 4*d* and 5*d* transition metals have certain very useful properties. For example, zirconium and zirconium oxide (ZrO_2) have great resistance to high temperatures and are used, along with niobium and molybdenum alloys, for space vehicle parts that are exposed to high temperatures during reentry into the earth's atmosphere. Niobium and molybdenum are also important alloying materials for certain types of steel. Tantalum, which has a high resistance to attack by body fluids, is often used for replacement of bones. The *platinum group metals*—ruthenium, osmium, rhodium, iridium, palladium, and platinum—are all quite similar and are widely used as catalysts for many types of industrial processes.

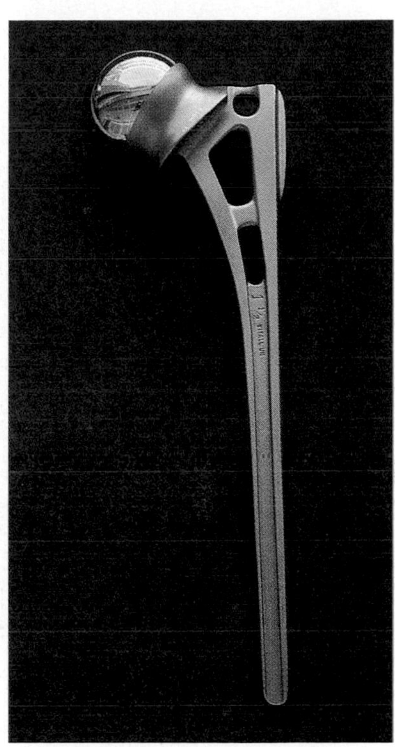

A hip joint constructed from tantalum metal.

20.2 The First-Row Transition Metals

Purpose

- To discuss some chemistry of the 3*d* transition metals.

We have seen that the transition metals are similar in many ways, but also show important differences. We will now explore some of the specific properties of each of the 3*d* transition metals.

Scandium is a rare element that exists in compounds mainly in the +3 oxidation state, for example, in $ScCl_3$, Sc_2O_3, and $Sc_2(SO_4)_3$. The chemistry of scandium strongly resembles that of the lanthanides, with most of its compounds being colorless and diamagnetic. This is not surprising; as we will see in Section 20.6, the color and magnetism of transition metal compounds usually arise from the *d* electrons on the metal ion, and Sc^{3+} has no *d* electrons. Scandium metal, which can be prepared by electrolysis of molten $ScCl_3$, is not widely used because of its rarity, but it is found in some electronic devices, such as high-intensity lamps.

Titanium is widely distributed in the earth's crust (0.6% by mass). Because of its relatively low density and high strength, titanium is an excellent structural material, especially in jet engines where light weight and stability at high temperatures are required. Nearly 5000 kg of titanium alloys are used in each engine of a Boeing 747 jetliner. In addition, the resistance of titanium to chemical attack makes it a useful material for pipes, pumps, and reaction vessels in the chemical industry.

$Ti(H_2O)_6^{3+}$ is purple in solution.

The most familiar compound of titanium is no doubt responsible for the white color of this paper. Titanium dioxide, or more correctly, *titanium(IV) oxide* (TiO_2), is a highly opaque substance used as the white pigment in paper, paint, linoleum, plastics, synthetic fibers, whitewall tires, and cosmetics. Approximately 700,000 tons are used annually in these and other products. Titanium(IV) oxide is widely dispersed in nature, but the main ores are rutile (impure TiO_2) and ilmenite ($FeTiO_3$). Rutile is processed by treatment with chlorine to form volatile $TiCl_4$, which is separated from the impurities and burned to form TiO_2:

$$TiCl_4(g) + O_2(g) \rightarrow TiO_2(s) + 2Cl_2(g)$$

Ilmenite is treated with sulfuric acid to form a soluble sulfate

$$FeTiO_3(s) + 2H_2SO_4(aq) \rightarrow Fe^{2+}(aq) + TiO^{2+}(aq) + 2SO_4^{2-}(aq) + 2H_2O(l)$$

When this aqueous mixture is allowed to stand, under vacuum, solid $FeSO_4 \cdot 7H_2O$ forms first and is removed. The mixture is then heated, and the insoluble titanium(IV) oxide hydrate ($TiO_2 \cdot H_2O$) forms. The water of hydration is driven off by heating to form pure TiO_2:

$$TiO_2 \cdot H_2O(s) \xrightarrow{\text{Heat}} TiO_2(s) + H_2O(g)$$

In its compounds, titanium is most often found in the +4 oxidation state. Examples are TiO_2 and $TiCl_4$, the latter a colorless liquid (bp = 137°C) that fumes in moist air to produce TiO_2:

$$TiCl_4(l) + 2H_2O(l) \rightarrow TiO_2(s) + 4HCl(g)$$

Titanium(III) compounds can be produced by reduction of the +4 state. In aqueous solution, Ti^{3+} exists as the purple $Ti(H_2O)_6^{3+}$ ion, which is slowly oxidized to titanium(IV) by air. Titanium(II) is not stable in aqueous solution but does exist in the solid state in compounds such as TiO and the dihalides of general formula TiX_2.

The manufacture of sulfuric acid was discussed at the end of Chapter 3.

The most common oxidation state for vanadium is +5.

Vanadium is widely spread throughout the earth's crust (0.02% by mass). It is used mostly in alloys with other metals such as iron (80% of vanadium is used in steel) and titanium. The vanadium(V) oxide (V_2O_5) is used as an industrial catalyst in the production of materials such as sulfuric acid.

Pure vanadium can be obtained from the electrolytic reduction of fused salts, such as VCl_2, to produce a metal similar to titanium that is steel-gray, hard, and corrosion resistant. Often the pure element is not required for alloying. For example, *ferrovanadium,* produced by the reduction of a mixture of V_2O_5 and Fe_2O_3 with aluminum, is added to iron to form *vanadium steel,* a hard steel used for engine parts and axles.

Oxidation States and Species for Vanadium in Aqueous Solution

Oxidation state of vanadium	Species in aqueous solution
+5	VO_2^+ (yellow)
+4	VO^{2+} (blue)
+3	$V^{3+}(aq)$ (blue-green)
+2	$V^{2+}(aq)$ (violet)

Table 20.4

The principal oxidation state of vanadium is +5, found in compounds such as the orange V_2O_5 (mp = 650°C) and the colorless VF_5 (mp = 19.5°C). The oxidation states from +5 to +2 all exist in aqueous solution (see Table 20.4). The higher oxidation states, +5 and +4, do not exist as hydrated ions of the type $V^{n+}(aq)$ because the highly charged ion causes the attached water molecules to be very acidic. The H^+ ions are lost to give the oxycations VO_2^+ and VO^{2+}. The hydrated V^{3+} and V^{2+} ions are easily oxidized, and thus can function as reducing agents in aqueous solution.

Although *chromium* is relatively rare, it is a very important industrial material. The chief ore of chromium is chromite ($FeCr_2O_4$), which can be reduced by carbon to give *ferrochrome,*

$$FeCr_2O_4(s) + 4C(s) \rightarrow \underbrace{Fe(s) + 2Cr(s)}_{\text{Ferrochrome}} + 4CO(g)$$

which can be added directly to iron in the steelmaking process. Chromium metal, which is often used to plate steel, is hard and brittle and maintains a bright surface by developing a tough invisible oxide coating.

Chromium commonly forms compounds in which it has the oxidation state +2, +3, or +6, as shown in Table 20.5. The Cr^{2+} (chromous) ion is a powerful reducing agent in aqueous solution. In fact, traces of O_2 in other gases can be removed by bubbling through a Cr^{2+} solution:

$$4Cr^{2+}(aq) + O_2(g) + 4H^+(aq) \rightarrow 4Cr^{3+}(aq) + 2H_2O(l)$$

The chromium(VI) species are excellent oxidizing agents, especially in acidic solution where chromium(VI) as the dichromate ion ($Cr_2O_7^{2-}$) is reduced to the Cr^{3+} ion:

$$Cr_2O_7^{2-}(aq) + 14H^+(aq) + 6e^- \rightarrow 2Cr^{3+}(aq) + 7H_2O(l) \qquad \mathscr{E}° = 1.33 \text{ V}$$

The oxidizing ability of the dichromate ion is strongly pH-dependent, increasing as $[H^+]$ increases, as predicted by Le Châtelier's principle. In basic solution, chromium(VI) exists as the chromate ion, a much less powerful oxidizing agent:

$$CrO_4^{2-}(aq) + 4H_2O(l) + 3e^- \rightarrow Cr(OH)_3(s) + 5OH^-(aq) \qquad \mathscr{E}° = -0.13 \text{ V}$$

The structures of the $Cr_2O_7^{2-}$ and CrO_4^{2-} ions are shown in Fig. 20.5.

Red chromium(VI) oxide (CrO_3) dissolves in water to give a strongly acidic, red-orange solution:

$$2CrO_3(s) + H_2O(l) \rightarrow 2H^+(aq) + Cr_2O_7^{2-}(aq)$$

It is possible to precipitate bright orange dichromate salts, such as $K_2Cr_2O_7$, from these solutions. When made basic, the solution turns yellow and chromate salts such as Na_2CrO_4 can be obtained. A mixture of chromium(VI) oxide and concentrated sulfuric acid, commonly called *cleaning solution,* is a powerful oxidizing medium that can remove organic materials from analytical glassware, yielding a very clean surface.

Manganese is relatively abundant (0.1% of the earth's crust), although no significant sources are found in the United States. The most common use of manganese is in the production of an especially hard steel used for rock crushers, bank vaults, and armor plate. One interesting source of manganese is from *manganese nodules* found on the ocean floor. These roughly spherical "rocks" contain mixtures of manganese and iron oxides as well as smaller amounts of other metals such as cobalt, nickel, and copper. Apparently the nodules were formed at least partly by the action of marine organisms. Because of the abundance of these nodules, there is much interest in developing economical methods for their recovery and processing.

Typical Chromium Compounds

Oxidation state of chromium	Examples of compounds (X = halogen)
+2	CrX_2
+3	CrX_3 Cr_2O_3 (green) $Cr(OH)_3$ (blue-green)
+6	$K_2Cr_2O_7$ (orange) Na_2CrO_4 (yellow) CrO_3 (red)

Table 20.5

(a)

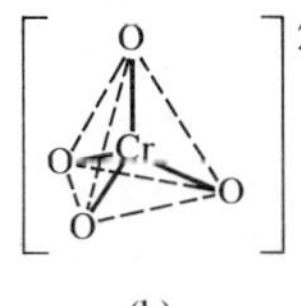

(b)

Figure 20.5

The structures of the chromium(VI) anions: (a) $Cr_2O_7^{2-}$, which exists in acidic solution; and (b) CrO_4^{2-}, which exists in basic solution.

Some Compounds of Manganese in Its Most Common Oxidation States	
Oxidation state of manganese	Examples of compounds
+2	$Mn(OH)_2$ (pink) MnS (salmon) $MnSO_4$ (reddish) $MnCl_2$ (pink)
+4	MnO_2 (dark brown)
+7	$KMnO_4$ (purple)

Table 20.6

Manganese can exist in all oxidation states from +2 to +7, although +2 and +7 are the most common. Manganese(II) forms an extensive series of salts with all of the common anions. In aqueous solution Mn^{2+} forms $Mn(H_2O)_6^{2+}$, which has a light pink color. Manganese(VII) is found in the intensely purple permanganate ion (MnO_4^-). Widely used as an analytical reagent in acidic solution, the MnO_4^- ion behaves as a strong oxidizing agent and the manganese becomes Mn^{2+}:

$$MnO_4^-(aq) + 8H^+(aq) + 5e^- \rightarrow Mn^{2+}(aq) + 4H_2O(l) \qquad \mathscr{E}° = 1.51 \text{ V}$$

Several typical compounds of manganese are shown in Table 20.6.

Iron is the most abundant heavy metal (4.7% of the earth's crust) and the most important to our civilization. It is a white, lustrous, not particularly hard metal that is very reactive toward oxidizing agents. For example, in moist air it is rapidly oxidized by oxygen to form rust, a mixture of iron oxides.

The chemistry of iron mainly involves its +2 and +3 oxidation states. Typical compounds are shown in Table 20.7. In aqueous solutions, iron(II) salts are generally light green because of the presence of $Fe(H_2O)_6^{2+}$. Although the $Fe(H_2O)_6^{3+}$ ion is colorless, aqueous solutions of iron(III) salts are usually yellow to brown in color due to the presence of $Fe(OH)(H_2O)_5^{2+}$, which results from the acidity of $Fe(H_2O)_6^{3+}$ ($K_a = 6 \times 10^{-3}$):

$$Fe(H_2O)_6^{3+}(aq) \rightleftharpoons Fe(OH)(H_2O)_5^{2+}(aq) + H^+(aq)$$

Typical Compounds of Iron	
Oxidation state of iron	Examples of compounds
+2	FeO (black) FeS (brownish-black) $FeSO_4 \cdot 7H_2O$ (green) $K_4Fe(CN)_6$ (yellow)
+3	$FeCl_3$ (brownish-black) Fe_2O_3 (reddish-brown) $K_3Fe(CN)_6$ (red) $Fe(SCN)_3$ (red)
+2, +3 (mixture)	Fe_3O_4 (black) $KFe[Fe(CN)_6]$ (deep blue, "Prussian blue")

Table 20.7

Although *cobalt* is relatively rare, it is found in ores such as smaltite ($CoAs_2$) and cobaltite (CoAsS) in large enough concentrations to make its production economically feasible. Cobalt is a hard, bluish-white metal mainly used in alloys such as stainless steel and stellite, an alloy of iron, copper, and tungsten that is used in surgical instruments.

The chemistry of cobalt involves mainly its +2 and +3 oxidation states, although compounds containing cobalt in the 0, +1, or +4 oxidation state are known. Aqueous solutions of cobalt(II) salts contain the $Co(H_2O)_6^{2+}$ ion, which has a characteristic rose color. Cobalt forms a wide variety of coordination compounds, many of which will be discussed in later sections of this chapter. Some typical cobalt compounds are shown in Table 20.8.

Nickel, which ranks twenty-fourth in elemental abundance in the earth's crust, is found in ores, where it is combined mainly with arsenic, antimony, and sulfur. Nickel metal, a silver-white substance with high electrical and thermal conductivities, is quite resistant to corrosion and is often used for plating more active metals. Nickel is also widely used in the production of alloys such as steel.

Nickel in compounds is almost exclusively in the +2 oxidation state. Aqueous solutions of nickel(II) salts contain the $Ni(H_2O)_6^{2+}$ ion, which has a characteristic emerald green color. Coordination compounds of nickel(II) will be discussed later in this chapter. Some typical nickel compounds are shown in Table 20.9.

Copper roofs and bronze statues, such as the Statue of Liberty (see Chemical Impact feature in Section 17.6), turn green in air because $Cu_3(OH)_4SO_4$ and $Cu_4(OH)_6SO_4$ form.

Copper, widely distributed in nature in ores containing sulfides, arsenides, chlorides, and carbonates, is valued for its high electrical conductivity and its resistance to corrosion. It is widely used for plumbing, and 50% of all copper produced annually is used for electrical applications. Copper is a major constituent in several well-known alloys (see Table 20.10).

Although copper is not highly reactive (it will not reduce H^+ to H_2, for example), the reddish-colored metal does slowly corrode in air, producing the characteristic green *patina* consisting of basic copper sulfate

Typical Compounds of Cobalt

Oxidation state	Examples
+2	$CoSO_4$ (dark blue)
	$[Co(H_2O)_6]Cl_2$ (pink)
	$[Co(H_2O)_6](NO_3)_2$ (red)
	CoS (black)
	CoO (greenish-brown)
+3	CoF_3 (brown)
	Co_2O_3 (charcoal)
	$K_3[Co(CN)_6]$ (yellow)
	$[Co(NH_3)_6]Cl_3$ (yellow)

Table 20.8

Typical Compounds of Nickel

Oxidation state of nickel	Examples of compounds
+2	$NiCl_2$ (yellow)
	$[Ni(H_2O)_6]Cl_2$ (green)
	NiO (greenish-black)
	NiS (black)
	$[Ni(H_2O)_6]SO_4$ (green)
	$[Ni(NH_3)_6](NO_3)_2$ (blue)

Table 20.9

Alloys Containing Copper

Alloy	Composition (% by mass in parentheses)
Brass	Cu (20–97), Zn (2–80), Sn (0–14), Pb (0–12), Mn (0–25)
Bronze	Cu (50–98), Sn (0–35), Zn (0–29), Pb (0–50), P (0–3)
Sterling silver	Cu (7.5), Ag (92.5)
Gold (18-karat)	Cu (5–14), Au (75), Ag (10–20)
Gold (14-karat)	Cu (12–28), Au (58), Ag (4–30)

Table 20.10

Typical Compounds of Copper

Oxidation state of copper	Examples of compounds
+1	Cu_2O (red)
	Cu_2S (black)
	CuCl (white)
+2	CuO (black)
	$CuSO_4 \cdot 5H_2O$ (blue)
	$CuCl_2 \cdot 2H_2O$ (green)
	$[Cu(H_2O)_6](NO_3)_2$ (blue)

Table 20.11

$$3Cu(s) + 2H_2O(l) + SO_2(g) + 2O_2(g) \longrightarrow \underset{\text{Basic copper sulfate}}{Cu_3(OH)_4SO_4}$$

and other similar compounds.

The chemistry of copper principally involves the +2 oxidation state, but many compounds containing copper(I) are also known. Aqueous solutions of copper(II) salts are a characteristic bright blue color due to the presence of the $Cu(H_2O)_6{}^{2+}$ ion. Table 20.11 lists some typical copper compounds.

Although trace amounts of copper are essential for life, copper in large amounts is quite toxic; copper salts are used to kill bacteria, fungi, and algae. For example, paints containing copper are used on ship hulls to prevent fouling by marine organisms.

Widely dispersed in the earth's crust, *zinc* is mainly refined from sphalerite (ZnFe)S, which often occurs with galena (PbS). Zinc is a white, lustrous, very active metal that behaves as an excellent reducing agent and tarnishes rapidly. About 90% of the zinc produced is used for galvanizing steel. Zinc forms colorless salts in which it is in the +2 oxidation state.

20.3 Coordination Compounds

Purpose

- To define terms used in coordination chemistry.
- To discuss some common chelates.
- To give rules for naming coordination compounds.

Transition metal ions characteristically form **coordination compounds,** which are usually colored and often paramagnetic. A coordination compound typically consists of a *complex ion,* a transition metal ion with its attached ligands (see Section 15.8), and **counter ions,** anions or cations as needed to produce a compound with no net charge. The substance $[Co(NH_3)_5Cl]Cl_2$ is a typical coordination compound. The square brackets indicate the composition of the complex ion, in this case $Co(NH_3)_5Cl^{2+}$, and the two Cl^- counter ions are shown outside the brackets. Note that in this compound one Cl^- acts as a ligand along with the five NH_3 molecules. In the solid state this compound consists of the large $Co(NH_3)_5Cl^{2+}$ cations and twice as many Cl^- anions, all packed together as efficiently as possible. When dissolved in water the solid behaves like any ionic solid; the cations and anions are assumed to separate and move about independently:

$$[Co(NH_3)_5Cl]Cl_2(s) \xrightarrow{H_2O} Co(NH_3)_5Cl^{2+}(aq) + 2Cl^-(aq)$$

Coordination compounds have been known since about 1700, but their true nature was not understood until the 1890s when a young Swiss chemist named Alfred Werner proposed that transition metal ions have two types of valence (combining ability). One type of valence, which Werner called the secondary valence, refers to the ability of a metal ion to bind to Lewis bases (ligands) to form complex ions. The other type, the primary valence, refers to the ability of the metal ion to form ionic bonds with oppositely charged ions. Thus Werner explained that the compound, originally written as $CoCl_3 \cdot 5NH_3$, was really $[Co(NH_3)_5Cl]Cl_2$, where the Co^{3+} ion has a primary valence of 3, satisfied by the three Cl^- ions, and a secondary valence of 6, satisfied by the six ligands (five NH_3 and one Cl^-). We now call the primary valence the **oxidation state** and the secondary valence the **coordination number,** which reflects the number of bonds formed between the metal ion and the ligands in the complex ion.

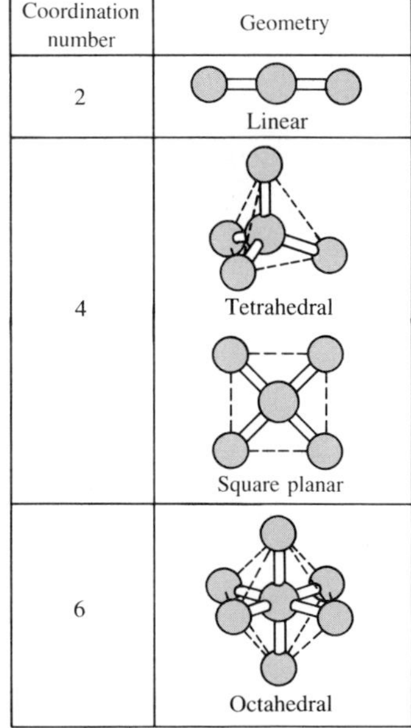

Coordination number
Geometry
2
Linear
4
Tetrahedral
Square planar
6
Octahedral

Figure 20.6

The ligand arrangements for coordination numbers 2, 4, and 6.

Coordination Number

The number of bonds formed by metal ions to ligands in complex ions varies from two to eight depending on the size, charge, and electron configuration of the transition metal ion. As shown in Table 20.12, 6 is the most common coordination number, followed closely by 4, with a few metal ions showing a coordination number of 2. Many metal ions show more than one coordination number, and there is really no simple way to predict what the coordination number will be in a particular case. The typical geometries for the various common coordination numbers are shown in Fig. 20.6. Note that six ligands produce an octahedral arrangement around the metal ion. Four ligands can form either a tetrahedral or a square planar arrangement, and two ligands give a linear structure.

Typical Coordination Numbers for Some Common Metal Ions

M^+	Coordination numbers	M^{2+}	Coordination numbers	M^{3+}	Coordination numbers
Cu^+	2, 4	Mn^{2+}	4, 6	Sc^{3+}	6
Ag^+	2	Fe^{2+}	6	Cr^{3+}	6
Au^+	2, 4	Co^{2+}	4, 6	Co^{3+}	6
		Ni^{2+}	4, 6		
		Cu^{2+}	4, 6	Au^{3+}	4
		Zn^{2+}	4, 6		

Table 20.12

Chemical Impact

Alfred Werner: Coordination Chemist

During the early and middle parts of the nineteenth century, chemists prepared a large number of colored compounds containing transition metals and other substances such as ammonia, chloride ion, cyanide ion, and water. These compounds were very interesting to chemists who were trying to understand the nature of bonding (Dalton's atomic theory of 1808 was very new at this time), and many theories were suggested to explain these substances. The most widely accepted early theory was the *chain theory,* championed by Sophus Mads Jorgensen (1837–1914), professor of chemistry at the University of Copenhagen. The chain theory got its name from the postulate that metal ammine* complexes contain chains of NH_3 molecules. For example, Jorgensen proposed the structure

$$\mathrm{Co}\begin{cases}\mathrm{NH_3{-}Cl} \\ \mathrm{NH_3{-}NH_3{-}NH_3{-}NH_3{-}Cl} \\ \mathrm{NH_3{-}Cl}\end{cases}$$

for the compound $Co(NH_3)_6Cl_3$. In the late nineteenth century this theory was used in classrooms around the world to explain the nature of metal-ammine compounds.

However, in 1890 a young Swiss chemist named Alfred Werner, who had just obtained a Ph.D. in the field of organic chemistry, became so interested in these compounds that he apparently even dreamed about them. In the middle of one night Werner awoke realizing that he had the correct explanation for the constitution of these compounds. Writing furiously the rest of that night and into the late afternoon of the following day, he constructed a scientific paper containing his now famous *coordination theory*. This model postulates an octahedral arrangement of ligands around the Co^{3+} ion, producing the $Co(NH_3)_6{}^{3+}$ complex ion with three Cl^- ions as counter ions. Thus Werner's picture of $Co(NH_3)_6Cl_3$ varied greatly from the chain theory.

In his paper on the coordination theory, Werner explained not only the metal-ammine compounds but most of the other known transition metal compounds, and the importance of his contribution was recognized immediately. He was appointed professor at the University of Zurich where he spent the rest of his life studying coordination compounds and refining his theory. Alfred Werner was a confident, impulsive man of seemingly boundless energy, who was known for his inspiring lectures, his intolerance of incompetence (he once threw a chair at a student who performed poorly on an oral exam), and his intuitive scientific brilliance. For example, he was the first to show that stereochemistry is a general phenomenon, not one exhibited only by carbon, as was previously thought. He also recognized and named many types of isomerism.

For his work on coordination chemistry and stereochemistry, Werner became the fourteenth Nobel Prize winner in chemistry and the first Swiss chemist to be so honored. Werner's work is even more remarkable when one realizes that his ideas preceded by many years any real understanding of the nature of covalent bonds.

*Ammine is the name for NH_3 as a ligand.

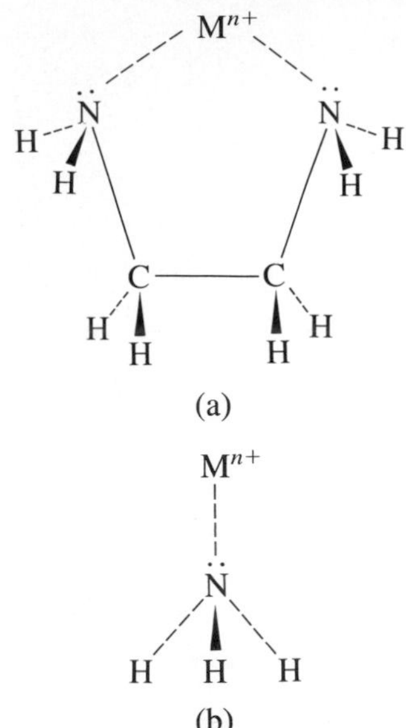

Figure 20.7

(a) The bidentate ligand ethylenediamine can bond to the metal ion through the lone pair on each nitrogen atom, thus forming two coordinate covalent bonds.
(b) Ammonia is a monodentate ligand.

Ligands

A **ligand** is a *neutral molecule or ion having a lone pair that can be used to form a bond to a metal ion*. The formation of a metal-ligand bond therefore can be described as the interaction between a Lewis base (the ligand) and a Lewis acid (the metal ion). This is often called a **coordinate covalent bond.**

A *ligand that can form one bond to a metal ion* is called a **monodentate ligand,** or a **unidentate ligand** (from root words meaning ''one tooth''). Examples of unidentate ligands are shown in Table 20.13.

Some ligands have more than one atom with a lone pair that can be used to bond to a metal ion. Such ligands are said to be **chelating ligands,** or **chelates** (from the Greek word *chela* for claw). A ligand that can form two bonds to a metal ion is called a **bidentate ligand.** A very common bidentate ligand is ethylenediamine (abbreviated en), which is shown coordinating to a metal ion in Fig. 20.7(a). Note the relationship between this ligand and the unidentate ligand ammonia [Fig. 20.7(b)]. Oxalate, another typical bidentate ligand, is shown in Table 20.13.

Ligands that can form more than two bonds to a metal ion are called *polydentate ligands*. Some ligands can form as many as six bonds to a metal ion. The best known example is ethylenediaminetetraacetate (abbreviated EDTA), which is shown in Table 20.13. This ligand virtually surrounds the metal ion (Fig. 20.8),

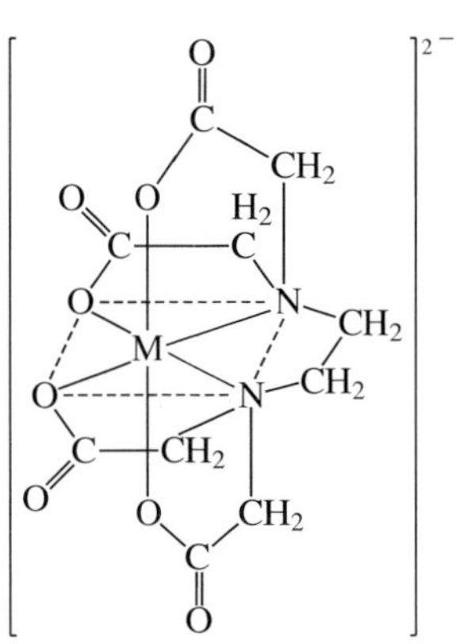

Figure 20.8

The coordination of EDTA with a 2+ metal ion.

Some Common Ligands

Type	Examples
Unidentate	H_2O NH_3 CN^- NO_2^- (nitrite) SCN^- (thiocyanate) OH^- X^- (halides)
Bidentate	Oxalate: $(-)\,:O-C(=O)-C(=O)-O:\,(-)$, both O coordinating to M. Ethylenediamine (en): $H_2N-CH_2-CH_2-NH_2$ ($H_2C—CH_2$), both N coordinating to M
Polydentate	Diethylenetriamine (dien) $H_2N-(CH_2)_2-NH-(CH_2)_2-NH_2$ three coordinating atoms Ethylenediaminetetraacetate (EDTA) $[(-):O-C(=O)-H_2C]_2N-(CH_2)_2-N[CH_2-C(=O)-O:(-)]_2$ Six coordinating atoms

Table 20.13

coordinating through six atoms (a *hexadentate ligand*). As might be expected from the large number of coordination sites, EDTA forms very stable complex ions with most metal ions and is used as a "scavenger" to remove toxic heavy metals such as lead from the human body. It is also used as a reagent to analyze solutions for the metal ion content. EDTA is found in countless consumer products, such as soda, beer, salad dressings, bar soaps, and most cleaners. In these products EDTA ties up trace metal ions that would otherwise catalyze decomposition and produce unwanted precipitates.

Some even more complicated ligands are found in biological systems, where metal ions play crucial roles in catalyzing reactions, transferring electrons, and transporting and storing oxygen. A discussion of these complex ligands will follow in Section 20.8.

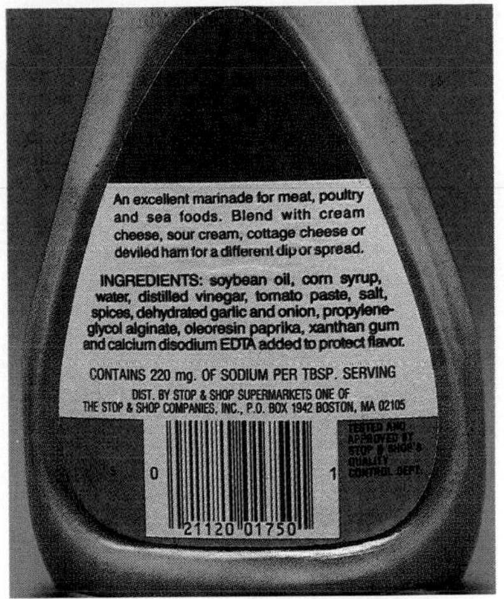

Salad dressing that contains EDTA.

Nomenclature

In Werner's lifetime, no system was used to name coordination compounds. Names of the compounds were commonly based on colors and names of discoverers. As the field expanded and more coordination compounds were identified, an orderly system of nomenclature became necessary. A simplified version of this system is summarized by the following rules.

Rules for Naming Coordination Compounds

- As with any ionic compound, *the cation is named before the anion.*
- In naming a complex ion, *the ligands are named before the metal ion.*
- In naming ligands, *an* o *is added to the root name of an anion.* For example, the halides as ligands are called fluoro, chloro, bromo, and iodo, hydroxide is hydroxo, cyanide is cyano, etc. *For a neutral ligand the name of the molecule is used,* with the exception of H_2O, NH_3, CO, and NO, as illustrated in Table 20.14.
- *The prefixes mono-, di-, tri-, tetra-, penta-, and hexa- are used to denote the number of simple ligands.* The prefixes bis-, tris-, tetrakis-, etc., are also used, especially for more complicated ligands or ones that already contain di-, tri-, etc.
- *The oxidation state of the central metal ion is designated by a Roman numeral in parentheses.*
- *When more than one type of ligand is present, they are named alphabetically.** Prefixes do not affect the order.
- *If the complex ion has a negative charge, the suffix* -ate *is added to the name of the metal.* Sometimes the Latin name is used to identify the metal (see Table 20.15).

The application of these rules is shown in Sample Exercise 20.1.

Names of Some Common Unidentate Ligands	
Neutral molecules	
Aqua	H_2O
Ammine	NH_3
Methylamine	CH_3NH_2
Carbonyl	CO
Nitrosyl	NO
Anions	
Fluoro	F^-
Chloro	Cl^-
Bromo	Br^-
Iodo	I^-
Hydroxo	OH^-
Cyano	CN^-

Table 20.14

Latin Names Used for Some Metal Ions in Anionic Complex Ions	
Metal	Anionic complex base name
Iron	Ferrate
Copper	Cuprate
Lead	Plumbate
Silver	Argentate
Gold	Aurate
Tin	Stannate

Table 20.15

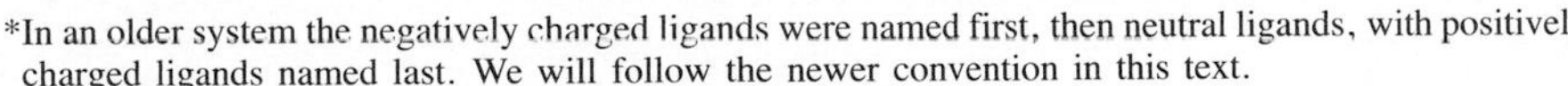

*In an older system the negatively charged ligands were named first, then neutral ligands, with positively charged ligands named last. We will follow the newer convention in this text.

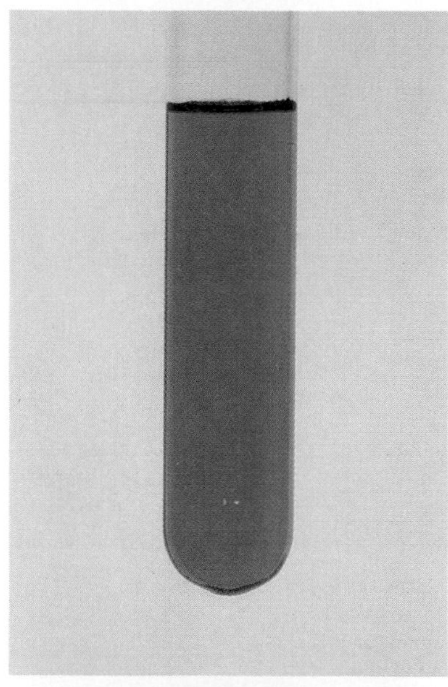

(top) An aqueous solution of $[Co(NH_3)_5Cl]Cl_2$. (bottom) Solid $K_3Fe(CN)_6$.

Sample Exercise 20.1

Give the systematic name for each of the following coordination compounds.

a. $[Co(NH_3)_5Cl]Cl_2$ **b.** $K_3Fe(CN)_6$ **c.** $[Fe(en)_2(NO_2)_2]_2SO_4$

Solution

a. To determine the oxidation state of the metal ion, we examine the charges of all ligands and counter ions. The ammonia molecules are neutral and the chloride ions each have a 1− charge, so the cobalt ion must have a 3+ charge to produce a neutral compound. Thus cobalt has the oxidation state +3, and we use cobalt(III) in the name.

The ligands include one Cl^- ion and five NH_3 molecules. The chloride ion is designated as *chloro,* and each ammonia molecule is designated *ammine.* The prefix *penta-* indicates that there are five NH_3 ligands present. The name of the complex cation is therefore pentaamminechlorocobalt(III). Note that the ligands are named alphabetically, disregarding the prefix. Since the counter ions are chloride ions, the compound is named as a chloride salt:

$$\underbrace{\text{pentaamminechlorocobalt(III)}}_{\text{Cation}}\ \underbrace{\text{chloride}}_{\text{Anion}}$$

b. First, we determine the oxidation state of the iron by considering the other charged species. The compound contains three K^+ ions and six CN^- ions. Therefore the iron must carry a charge of 3+, giving a total of six positive charges to balance the six negative charges. The complex ion present is thus $Fe(CN)_6^{3-}$. The cyanide ligands are each designated *cyano,* and the prefix *hexa-* indicates that six are present. Since the complex ion is an anion, we use the Latin name *ferrate*. The oxidation state is indicated by ferrate(III) in the name. The anion name is therefore hexacyanoferrate(III). The cations are K^+ ions, which are simply named potassium. Putting this together gives the name

$$\underbrace{\text{potassium}}_{\text{Cation}}\ \underbrace{\text{hexacyanoferrate(III)}}_{\text{Anion}}$$

(The common name of this compound is potassium ferricyanide.)

c. We first determine the oxidation state of the iron by looking at the other charged species: four NO_2^- ions and one SO_4^{2-} ion. The ethylenediamine is neutral. Thus the two iron ions must carry a total positive charge of six to balance the six negative charges. This means that each iron has a +3 oxidation state and is designated as iron(III).

Since the name ethylenediamine already contains *di,* we use *bis-* instead of *di-* to indicate the two en ligands. The name for NO_2^- as a ligand is *nitro,* and the prefix *di-* indicates the presence of two NO_2^- ligands. Since the anion is sulfate, the compound's name is

$$\underbrace{\text{bis(ethylenediamine)dinitroiron(III)}}_{\text{Cation}}\ \underbrace{\text{sulfate}}_{\text{Anion}}$$

Since the complex ion is the cation here, the Latin name for iron is not used.

Sample Exercise 20.2

Given the following systematic names, give the formula of each coordination compound.

a. Triamminebromoplatinum(II) chloride

b. Potassium hexafluorocobaltate(III)

Solution

a. *Triammine* signifies three ammonia ligands, and *bromo* indicates one bromide ion as a ligand. The oxidation state of platinum is +2, as indicated by the Roman numeral II. Thus the complex ion is $[Pt(NH_3)_3Br]^+$. One chloride ion is needed to balance the 1+ charge of this cation. The formula of the compound is $[Pt(NH_3)_3Br]Cl$. Note that square brackets enclose the complex ion.

b. The complex ion contains six fluoride ligands attached to a Co^{3+} ion to give CoF_6^{3-}. Note that the *-ate* ending indicates that the complex ion is an anion. The cations are K^+ ions, and three are required to balance the 3− charge on the complex ion. Thus the formula is $K_3[CoF_6]$.

20.4 Isomerism

Purpose

- To introduce and illustrate types of structural isomerism and stereoisomerism.
- To relate molecular structure to optical activity.

When two or more species have the same formula but different properties, they are said to be **isomers.** Although isomers contain exactly the same types and numbers of atoms, the arrangements of the atoms differ, and this leads to different properties. We will consider two main types of isomerism: **structural isomerism,** where the isomers contain the same atoms but one or more bonds differ; and **stereoisomerism,** where all of the bonds in the isomers are the same but the spatial arrangements of the atoms are different. Each of these classes also has subclasses (see Fig. 20.9 on the following page), which we will now consider.

Structural Isomerism

Of the types of structural isomerism, we will first consider **coordination isomerism,** in which the composition of the complex ion varies. For example, $[Cr(NH_3)_5SO_4]Br$ and $[Cr(NH_3)_5Br]SO_4$ are coordination isomers. In the first case, SO_4^{2-} is coordinated to Cr^{3+}, and Br^- is the counter ion; in the second case, the roles of these ions are reversed.

Another example of coordination isomerism is the $[Co(en)_3][Cr(ox)_3]$ and $[Cr(en)_3][Co(ox)_3]$ pair, where ox represents the oxalate ion, a bidentate ligand shown in Table 20.13.

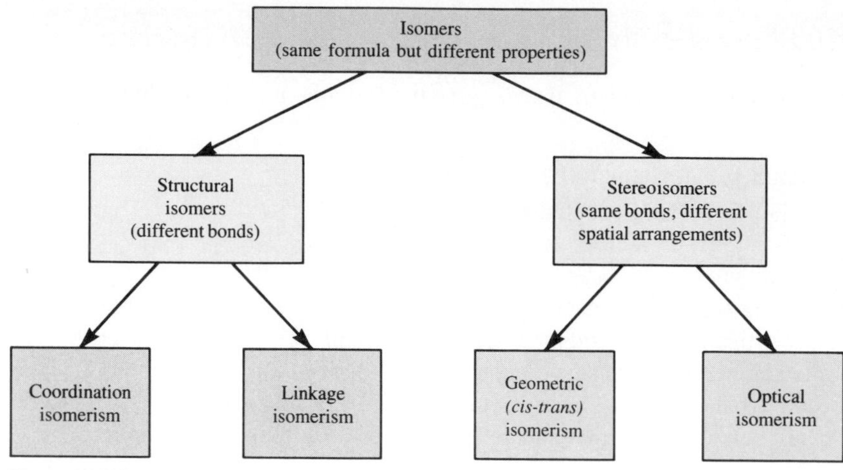

Figure 20.9

Some classes of isomers.

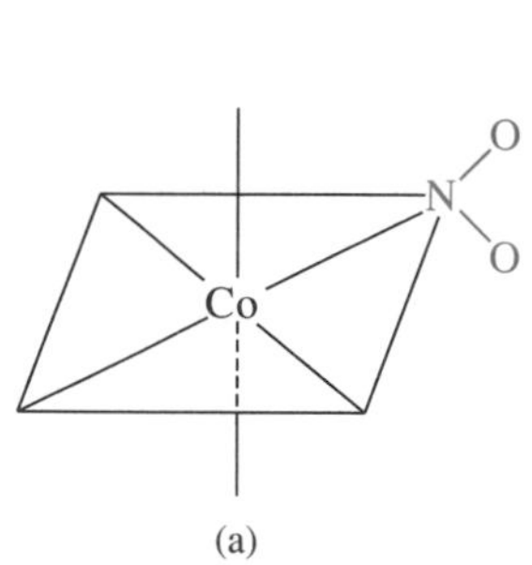

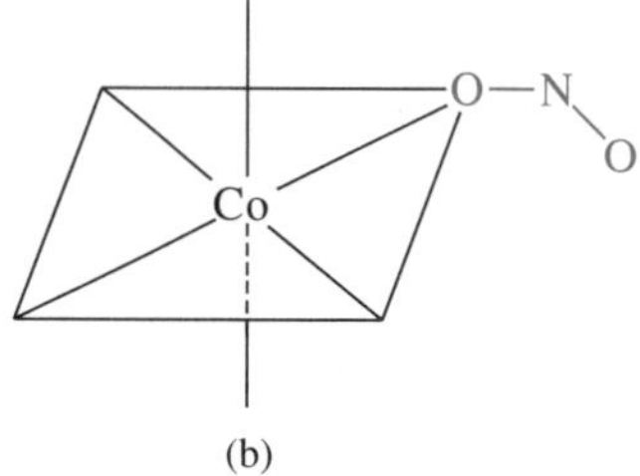

Figure 20.10

As a ligand, NO_2^- can bond to a metal ion (a) through a lone pair on the nitrogen atom or (b) through a lone pair on one of the oxygen atoms.

In a second type of structural isomerism, **linkage isomerism,** the composition of the complex ion is the same, but the point of attachment of at least one of the ligands differs. Two ligands that can attach to metal ions in different ways are thiocyanate (SCN^-), which can bond through lone pairs on the nitrogen or the sulfur atom, and the nitrite ion (NO_2^-), which can bond through lone pairs on the nitrogen or the oxygen atom. For example, the following two compounds are linkage isomers:

$$[Co(NH_3)_4(NO_2)Cl]Cl$$

Tetraamminechloronitrocobalt(III) chloride
(yellow)

$$[Co(NH_3)_4(ONO)Cl]Cl$$

Tetraamminechloronitritocobalt(III) chloride
(red)

In the first case, the NO_2^- ligand is called *nitro* and is attached to Co^{3+} through the nitrogen atom; in the second case, the NO_2^- ligand is called *nitrito* and is attached to Co^{3+} through an oxygen atom (see Fig. 20.10).

Stereoisomerism

Stereoisomers have the same bonds but different spatial arrangements of the atoms. One type, **geometrical isomerism,** or ***cis-trans*** **isomerism,** occurs when atoms or groups of atoms can assume different positions around a rigid ring or bond. An important example is the compound $Pt(NH_3)_2Cl_2$, which has a square planar structure. The two possible arrangements of the ligands are shown in Fig. 20.11. In the ***trans*** **isomer,** the ammonia molecules are across *(trans)* from each other. In the ***cis*** **isomer,** the ammonia molecules are next *(cis)* to each other.

Geometrical isomerism also occurs in octahedral complex ions. For example, the compound $[Co(NH_3)_4Cl_2]Cl$ has *cis* and *trans* isomers (Fig. 20.12).

A second type of stereoisomerism is called **optical isomerism** because the isomers have opposite effects on plane-polarized light. When light is emitted from a

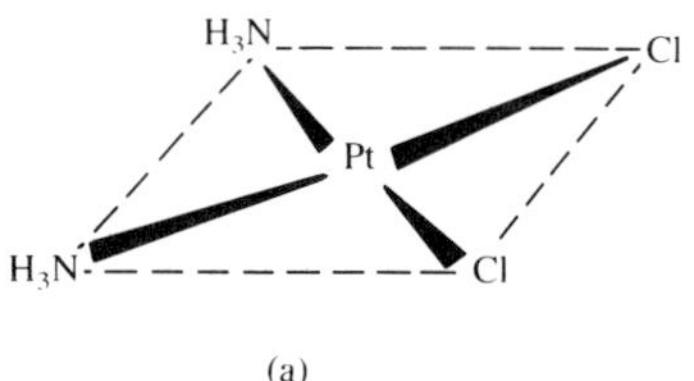

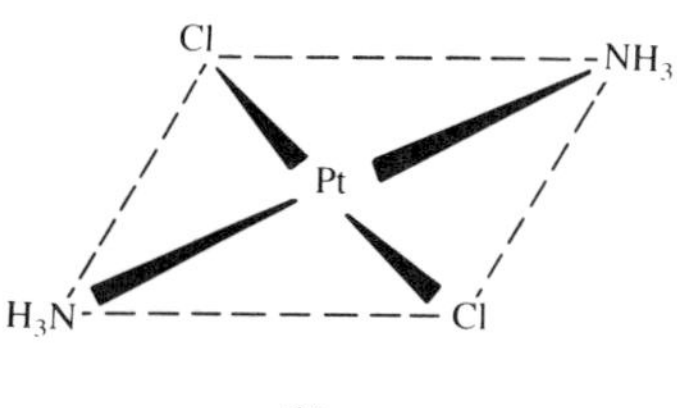

Figure 20.11

(a) The *cis* isomer of $Pt(NH_3)_2Cl_2$ (yellow). (b) The *trans* isomer of $Pt(NH_3)_2Cl_2$ (pale yellow).

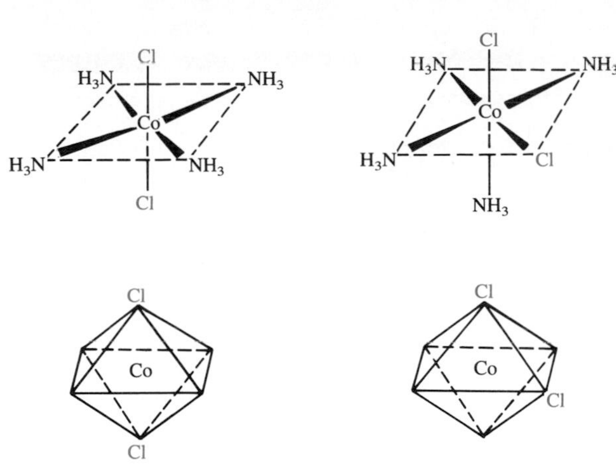

Figure 20.12

(a) The *trans* isomer of $[Co(NH_3)_4Cl_2]^+$. The chloride ligands are directly across from each other. (b) The *cis* isomer of $[Co(NH_3)_4Cl_2]^+$. The chloride ligands in this case share an edge of the octahedron. Because of their different structures, the *trans* isomer of $[Co(NH_3)_4Cl_2]Cl$ is green, and the *cis* isomer is violet.

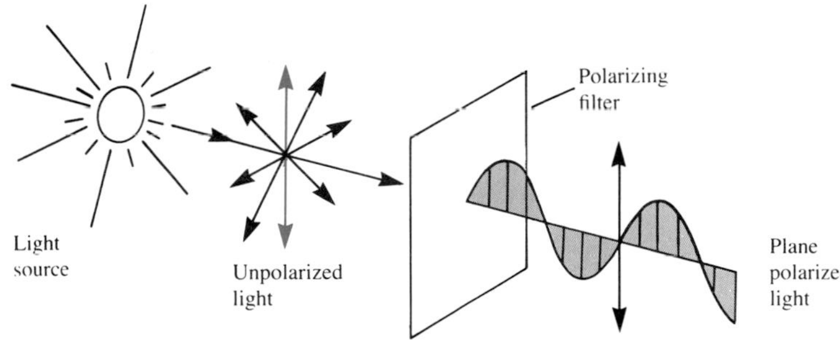

Figure 20.13

Unpolarized light consists of waves vibrating in many different planes (indicated by the arrows). The polarizing filter blocks all waves except those vibrating in a given plane.

source such as a glowing filament, the oscillating electric fields of the photons in the beam are oriented randomly, as shown in Fig. 20.13. If this light is passed through a polarizer, only the photons with electric fields oscillating in a single plane remain, constituting *plane-polarized light*.

In 1815 a French physicist, Jean Biot, showed that certain crystals could rotate the plane of polarization of light. Later it was found that solutions of certain compounds could do the same thing (see Fig. 20.14). Louis Pasteur was the first to

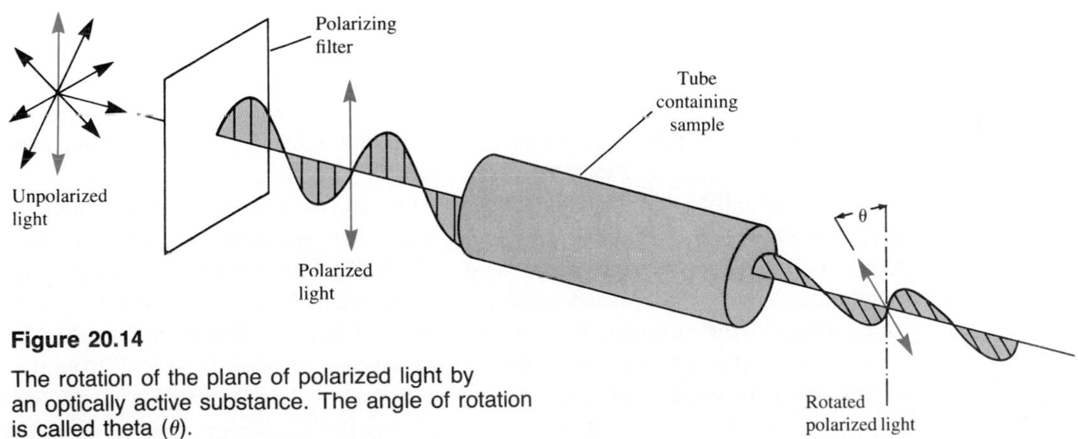

Figure 20.14

The rotation of the plane of polarized light by an optically active substance. The angle of rotation is called theta (θ).

Chemical Impact

The Importance of Being *cis*

Some of the most important advancements of science are the results of accidental discoveries, for example, penicillin, Teflon, and the sugar substitutes cyclamate and aspartame. Another important chance discovery occurred in 1964, when a group of scientists using platinum electrodes to apply an electric field to a colony of *E. coli* bacteria noticed that the bacteria failed to divide but continued to grow, forming long fibrous cells. Further study revealed that cell division was inhibited by small concentrations of *cis*-$Pt(NH_3)_2Cl_2$ and *cis*-$Pt(NH_3)_2Cl_4$ formed electrolytically in the solution.

Cancerous cells multiply very rapidly because cell division is uncontrolled. Thus these and similar platinum complexes were evaluated as antitumor agents, which inhibit the division of cancer cells. The results showed that *cis*-$Pt(NH_3)_2Cl_2$ was active against a wide variety of tumors, including testicular and ovarian tumors, which are very resistant to treatment by more traditional methods. However, although the *cis* complex showed significant antitumor activity, the corresponding *trans* complex had no effect on tumors. This shows the importance of isomerism in biological systems. When drugs are synthesized, great care must be taken to obtain the correct isomer.

Unfortunately, although *cis*-$Pt(NH_3)_2Cl_2$ has proved to be a valuable drug, it has some troublesome side effects, the most serious being kidney damage. As a result, the search continues for even more effective antitumor agents. Promising candidates are shown in Fig. 20.15. Note they are all *cis* complexes.

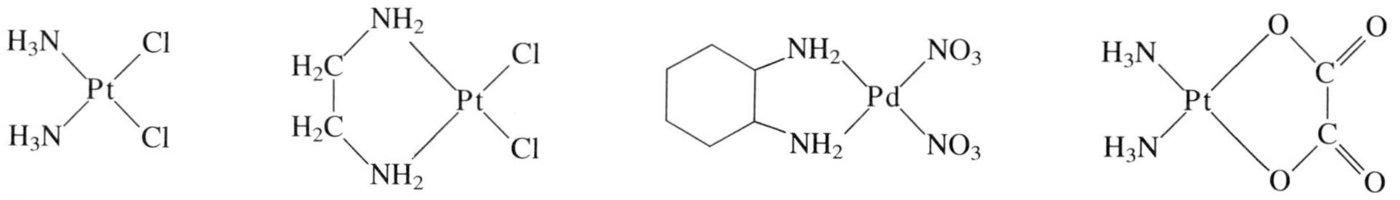

Figure 20.15

Some *cis* complexes of platinum and palladium that show significant antitumor activity. It is thought that the *cis* complexes work by losing two adjacent ligands and forming coordinate covalent bonds to adjacent bases on a DNA molecule.

understand this behavior. In 1848 he noted that solid sodium ammonium tartrate ($NaNH_4C_4H_4O_4$) existed as a mixture of two types of crystals, which he painstakingly separated with tweezers. Separate solutions of these two types of crystals rotated plane-polarized light in exactly opposite directions. This led to a connection between optical activity and molecular structure.

We now realize that optical activity is exhibited by molecules that have *nonsuperimposable mirror images*. Your hands are nonsuperimposable mirror images (Fig. 20.16). The two hands are related like an object and its mirror image; one hand cannot be turned to make it identical to the other. Many molecules show this same feature, for example, the complex ion $[Co(en)_3]^{3+}$ shown in Fig. 20.17. Objects that have nonsuperimposable mirror images are said to be **chiral** (from the Greek word for hand, *cheir*).

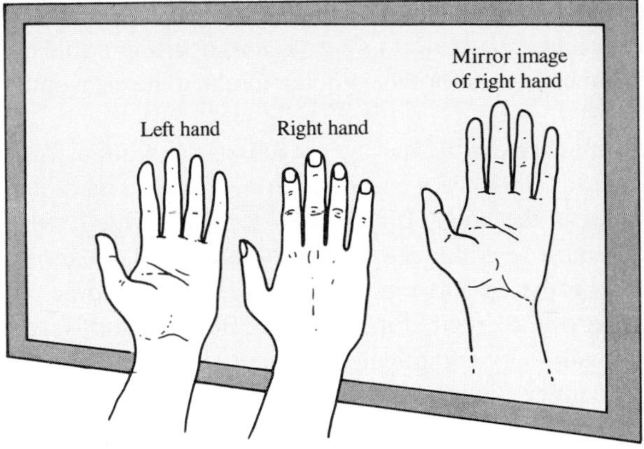
Left hand
Right hand
Mirror image of right hand

Figure 20.16

A human hand has a nonsuperimposable mirror image. Note that the mirror image of the right hand (while identical to the left hand) cannot be turned in any way to make it identical to (superimposable on) the actual right hand.

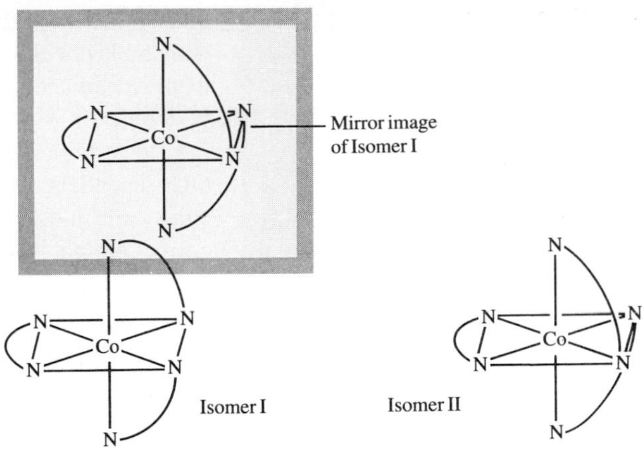
Mirror image of Isomer I
Isomer I
Isomer II

Figure 20.17

Isomers I and II of $Coen_3^{3+}$ are mirror images (the image of I is identical to II) that cannot be superimposed. That is, there is no way that I can be turned in space so that it is the same as II.

The isomers of $[Co(en)_3]^{3+}$ (Fig. 20.17) are nonsuperimposable mirror images called **enantiomers,** which rotate plane-polarized light in opposite directions and are thus optical isomers. The isomer that rotates the plane of light to the right (when viewed down the beam of oncoming light) is said to be *dextrorotatory,* designated by *d*. The isomer that rotates the plane of light to the left is *levorotatory* (*l*). An equal mixture of the *d* and *l* forms in solution, called a *racemic mixture,* does not rotate the plane of the polarized light at all because the two opposite effects cancel.

Geometrical isomers are not necessarily optical isomers. For instance, the *trans* isomer of $[Co(en)_2Cl_2]^+$ shown in Fig. 20.18 is identical to its mirror image. Since

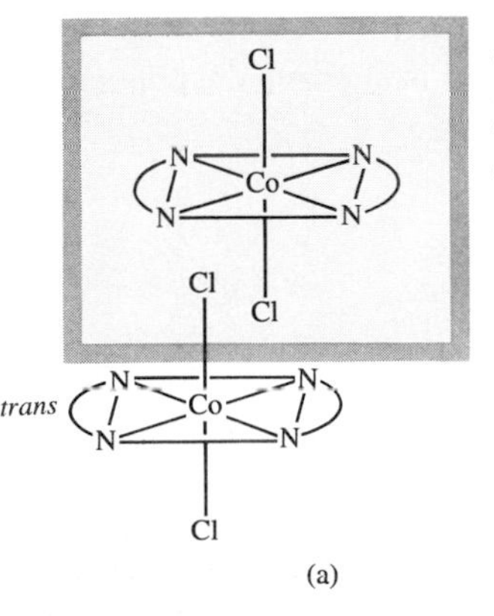
trans
The *trans* isomer and its mirror image are identical. They are not isomers of each other.
(a)

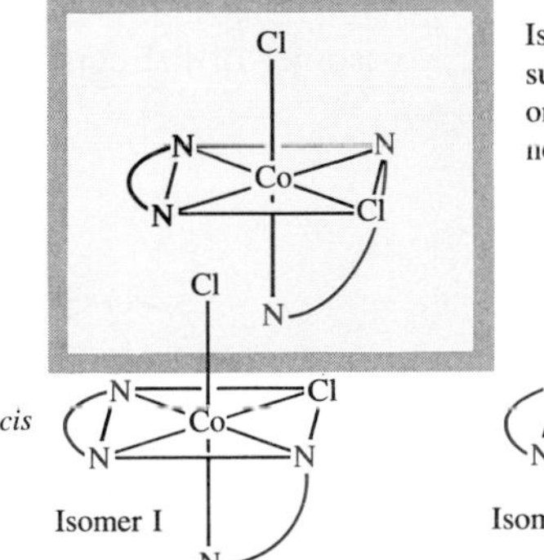
cis
Isomer I
Isomer II
Isomer II cannot be superimposed exactly on isomer I. They are not identical structures.
Isomer II has the same structure as the mirror image of isomer I
(b)

Figure 20.18

(a) The *trans* isomer of $Co(en)_2Cl_2^+$ and its mirror image are identical (superimposable). (b) The *cis* isomer of $Co(en)_2Cl_2^+$ and its mirror image are not superimposable and are thus a pair of optical isomers.

this isomer is superimposable on its mirror image, it does not exhibit optical isomerism and is not chiral. On the other hand, *cis*-$[Co(en)_2Cl_2]^+$ is *not* superimposable on its mirror image; a pair of enantiomers exists for the complex ion, and the *cis* isomer is chiral.

Most importantly, biomolecules are chiral, and their reactions are highly structure dependent. For example, a drug can have a particular effect because its molecules can bind to chiral molecules in the body. But in order to bind correctly, the correct optical isomer of the drug must be administered. Just as the right hand of one person requires the right hand of another to perform a handshake, a given isomer in the body requires a specific isomer of the drug to bind together. Because of this, the syntheses of drugs, which are usually very complicated molecules, must be carried out in a way that produces the correct ''handedness,'' a requirement that greatly adds to the difficulties.

Sample Exercise 20.3

Does the complex ion $[Co(NH_3)Br(en)_2]$ exhibit geometrical isomerism? Does it exhibit optical isomerism?

Solution

The complex ion exhibits geometrical isomerism since the ethylenediamine ligands can be across from or next to each other:

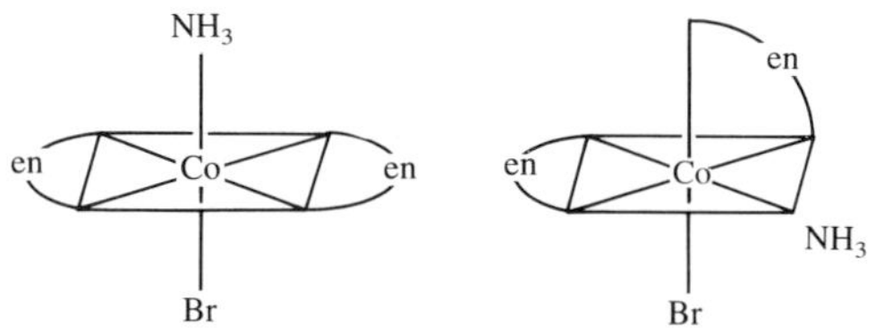

NH$_3$ en Co en Br   en en Co NH$_3$ Br

The *cis* isomer of the complex ion also exhibits optical isomerism since its mirror images

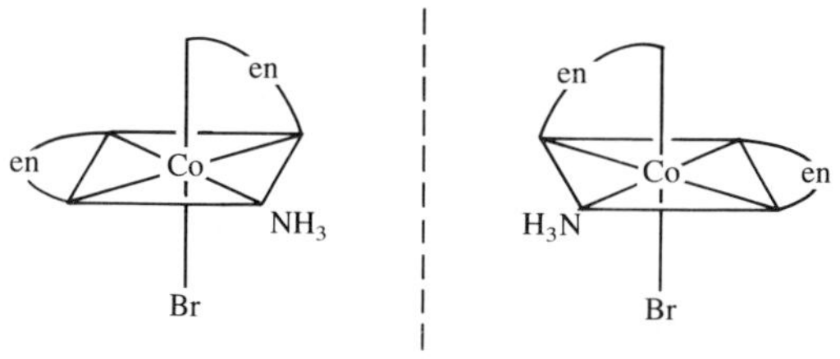

en en Co NH$_3$ Br   en en Co H$_3$N Br

cannot be turned in any way to make them superimposable. Thus these two *cis* isomers are shown to be enantiomers that will rotate plane-polarized light in opposite directions.

20.5 Bonding in Complex Ions: The Localized Electron Model

Purpose

- To use the localized electron model to explain the interactions between a metal ion and ligands.

In Chapters 8 and 9 we considered the localized electron model, a very useful model for describing the bonding in molecules. Recall that a central feature of the model is the formation of hybrid atomic orbitals that are used to share electron pairs to form σ bonds between atoms. This same model can be used to account for the bonding in complex ions, but there are two important points to keep in mind.

1. The VSEPR model for predicting structure *does not work for complex ions.* However, we can safely assume that a complex ion with a coordination number of 6 will have an octahedral arrangement of ligands, and complexes with two ligands will be linear. On the other hand, complex ions with a coordination number of 4 can be either tetrahedral or square planar, and there is no completely reliable way to predict which will occur in a particular case.
2. The interaction between a metal ion and a ligand can be viewed as a Lewis acid-base reaction with the ligand donating a lone pair of electrons to an *empty* orbital of the metal ion to form a coordinate covalent bond:

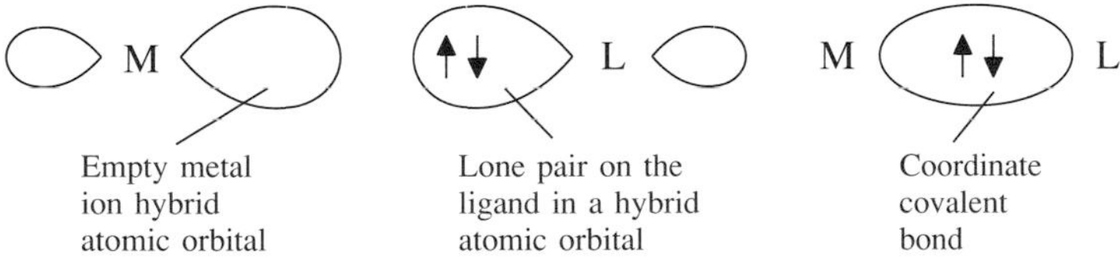

The hybrid orbitals used by the metal ion depend on the number and arrangement of the ligands. For example, to accommodate the lone pair from each ammonia molecule in the octahedral $Co(NH_3)_6^{3+}$ ion requires a set of six empty hybrid atomic orbitals in an octahedral arrangement. As we discussed in Section 9.1, an octahedral set of orbitals is formed by the hybridization of two *d*, one *s*, and three *p* orbitals to give six d^2sp^3 orbitals (see Fig. 20.19 on the following page).

The hybrid orbitals required on the metal ion for four-coordinate complexes depend on whether the structure is tetrahedral or square planar. For a tetrahedral arrangement of ligands, an sp^3 hybrid set is required (see Fig. 20.20 on the following page). For example, in the tetrahedral $CoCl_4^{2-}$ ion, the Co^{2+} can be described as sp^3 hybridized. A square planar arrangement of ligands requires a dsp^2 hybrid orbital set on the metal ion (Fig. 20.20). For example, in the square planar $Ni(CN)_4^{2-}$, the Ni^{2+} is described as dsp^2 hybridized.

A linear complex requires two hybrid orbitals 180° from each other. This arrangement is given by an *sp* hybrid set (Fig. 20.20). Thus in the linear $Ag(NH_3)_2^+$ ion, the Ag^+ can be described as *sp* hybridized.

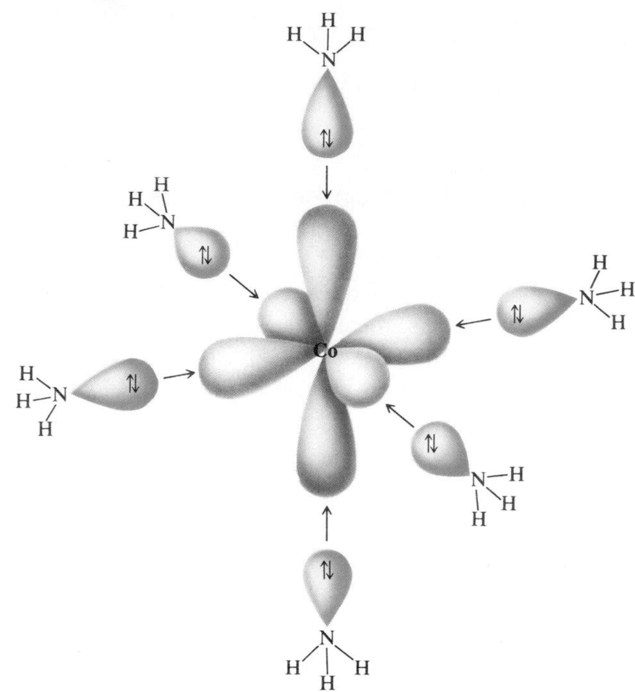

Figure 20.19

A set of six d^2sp^3 hybrid orbitals on Co^{3+} can accept an electron pair from each of six NH_3 ligands to form the $Co(NH_3)_6^{3+}$ ion.

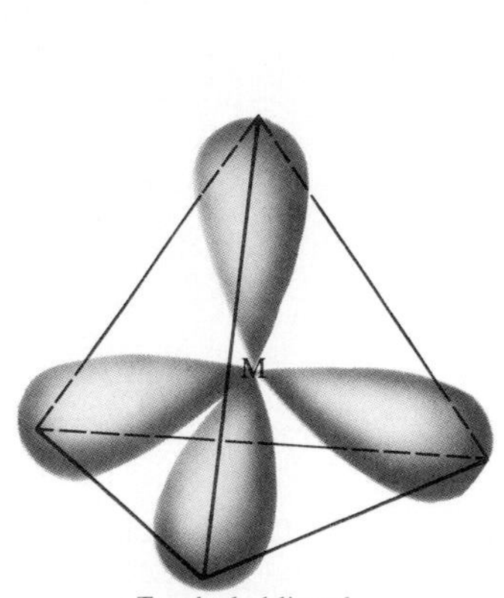

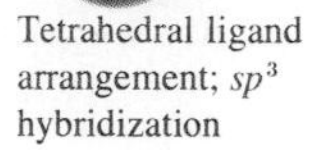

Tetrahedral ligand arrangement; sp^3 hybridization

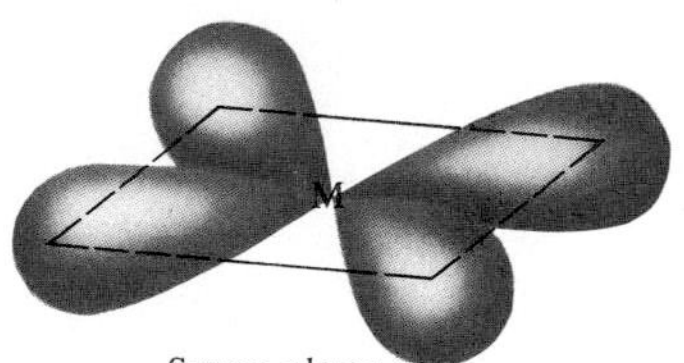

Square planar ligand arrangement; dsp^2 hybridization

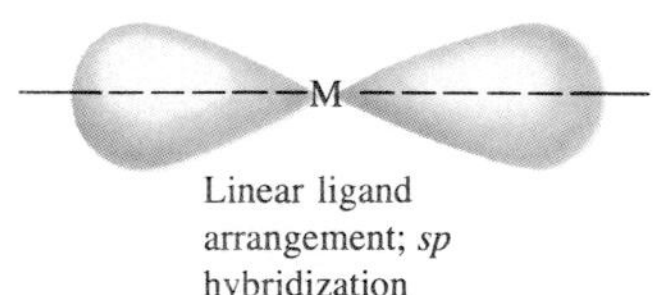

Linear ligand arrangement; sp hybridization

Figure 20.20

The hybrid orbitals required for tetrahedral, square planar, and linear complex ions. The metal ion hybrid orbitals are empty, and the metal ion bonds to the ligands by accepting lone pairs.

Although the localized electron model can account in a general way for metal-ligand bonds, it is rarely used today because it cannot predict important properties of complex ions, such as magnetism and color. Thus we will not pursue the model any further.

20.6 The Crystal Field Model

Purpose

- To describe the crystal field model and use it to explain the magnetism and colors of coordination complexes.

The main reason the localized electron model cannot fully account for the properties of complex ions is that it gives no information about how the energies of the *d* orbitals are affected by complex ion formation. This is critical because, as we will see, the color and magnetism of complex ions result from changes in the energies of the *d* orbitals on the metal ion caused by the metal-ligand interactions.

The **crystal field model** focuses on the energies of the *d* orbitals. In fact, this model is not so much a bonding model as it is an attempt to account for the colors

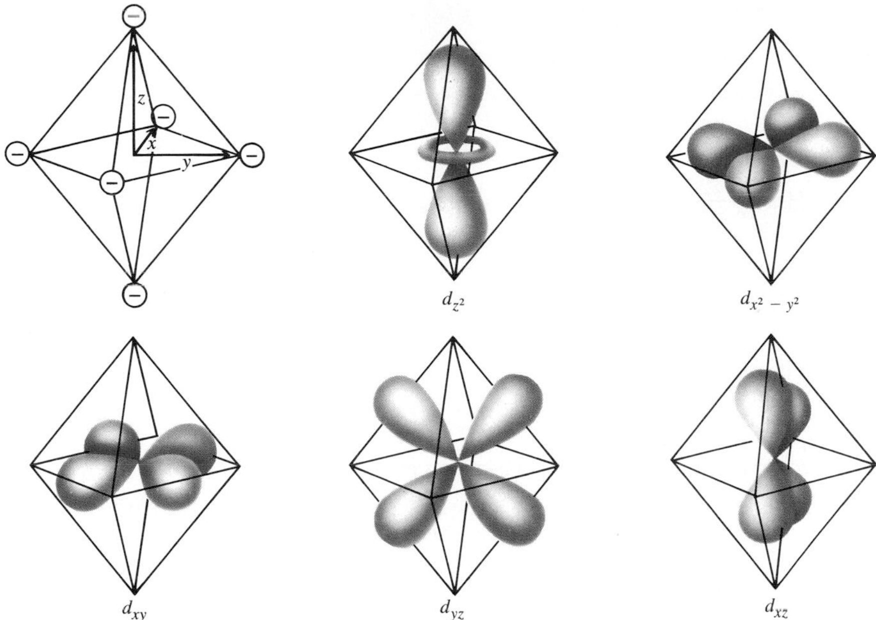

Figure 20.21

An octahedral arrangement of point-charge ligands and the orientation of the 3*d* orbitals.

and magnetic properties of complex ions. In its simplest form, the crystal field model assumes that the ligands can be approximated by *negative point charges* and that metal-ligand bonding is *entirely ionic*.

Octahedral Complexes

We will illustrate the fundamental principles of the crystal field model by applying it to an octahedral complex. Figure 20.21 shows the orientation of the 3*d* orbitals relative to an octahedral arrangement of point-charge ligands. The important thing to note is that two of the orbitals, d_{z^2} and $d_{x^2-y^2}$, point their lobes *directly at* the point-charge ligands and three of the orbitals, d_{xz}, d_{yz}, and d_{xy}, point their lobes *between* the point charges.

To understand the effect of this difference, we need to consider which type of orbital is lower in energy. Because the negative point-charge ligands repel negatively charged electrons, the electrons will first fill the *d* orbitals farthest from the ligands to minimize repulsions. In other words the d_{xz}, d_{yz}, and d_{xy} orbitals (called the t_{2g} set) are at a *lower energy* in the octahedral complex than are the d_{z^2} and $d_{x^2-y^2}$ orbitals (the e_g set). This is shown in Fig. 20.22. The negative point-charge ligands increase the energies of all the *d* orbitals. However, the orbitals that point at the ligands are raised in energy more than those that point between the ligands.

It is this **splitting of the 3*d* orbital energies** (symbolized by Δ) that explains the color and magnetism of complex ions of the first-row transition metal ions. For example, in an octahedral complex of Co^{3+} (a metal ion with six 3*d* electrons), there are two possible ways to place the electrons in the split 3*d* orbitals (Fig. 20.23). If the splitting produced by the ligands is very large, a situation called the **strong-field case,** the electrons will pair in the lower-energy t_{2g} orbitals. This gives a *diamagnetic* complex in which all electrons are paired. On the other hand, if the

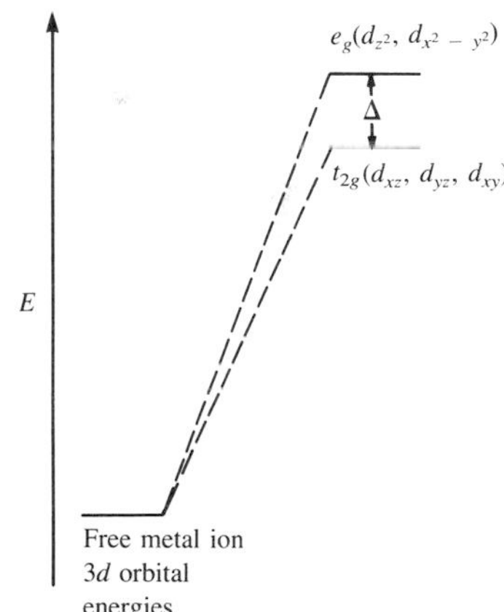

Figure 20.22

The energies of the 3*d* orbitals for a metal ion in an octahedral complex. The 3*d* orbitals are degenerate (all have the same energy) in the free metal ion. In the octahedral complex the orbitals are split into two sets as shown. The difference in energy between the two sets is designated as Δ (delta).

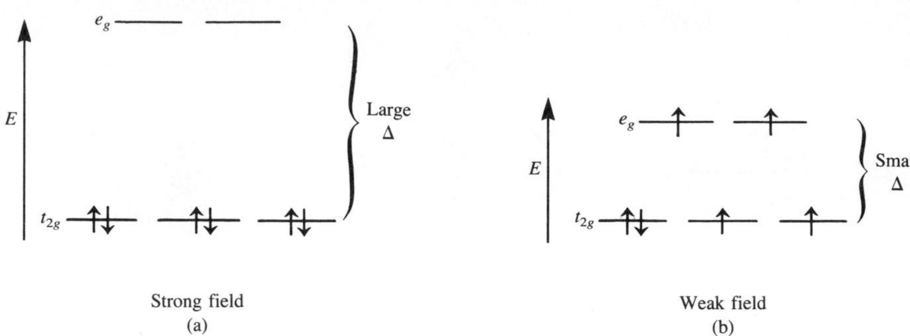

Figure 20.23

Possible electron arrangements in the split $3d$ orbitals in an octahedral complex of Co^{3+} (electron configuration $3d^6$). (a) In a strong field (large Δ value), the electrons fill the t_{2g} set first, giving a diamagnetic complex. (b) In a weak field (small Δ value), the electrons occupy all five orbitals before any pairing occurs.

splitting is small (the **weak-field case**), the electrons will occupy all five orbitals before pairing occurs. In this case, the complex has four unpaired electrons and is *paramagnetic*.

The crystal field model allows us to account for the differences in the magnetic properties of $Co(NH_3)_6^{3+}$ and CoF_6^{3-}. The $Co(NH_3)_6^{3+}$ ion is known to be diamagnetic and thus corresponds to the strong-field case, also called the **low-spin case,** since it yields the *minimum* number of unpaired electrons. In contrast, the CoF_6^{3-} ion, which is known to have four unpaired electrons, corresponds to the weak-field case, also known as the **high-spin case,** since it gives the *maximum* number of unpaired electrons.

Sample Exercise 20.4

The $Fe(CN)_6^{3-}$ ion is known to have one unpaired electron. Does the CN^- ligand produce a strong or weak field?

Solution

Since the ligand is CN^- and the overall complex ion charge is 3−, the metal ion must be Fe^{3+}, which has a $3d^5$ electron configuration. The two possible arrangements of the five electrons in the *d* orbitals split by the octahedrally arranged ligands are

The strong-field case gives one unpaired electron, which agrees with the experimental observation. The CN^- ion is a strong-field ligand toward the Fe^{3+} ion.

From studies of many octahedral complexes, we can arrange ligands in order of their ability to produce *d*-orbital splitting. A partial listing of ligands in this so-called **spectrochemical series** is

$$CN^- > NO_2^- > en > NH_3 > H_2O > OH^- > F^- > Cl^- > Br^- > I^-$$

Strong-field ligands (large Δ) — Weak-field ligands (small Δ)

The ligands are arranged in order of decreasing Δ values toward a given metal ion.

It has also been observed that *the magnitude of Δ for a given ligand increases as the charge on the metal ion increases*. For example, NH_3 is a weak-field ligand toward Co^{2+} but acts as a strong-field ligand toward Co^{3+}. This makes sense; as the metal ion charge increases, the ligands will be drawn closer to the metal ion because of the increased charge density. As the ligands move closer, they cause greater splitting of the *d* orbitals and produce a larger Δ value.

Sample Exercise 20.5

Predict the number of unpaired electrons in the complex ion $[Cr(CN)_6]^{4-}$.

Solution

The net charge of 4− means that the metal ion present must be Cr^{2+} $(-6 + 2 = -4)$, which has a $3d^4$ electron configuration. Since CN^- is a strong-field ligand (see the spectrochemical series), the correct crystal field diagram for $[Cr(CN)_6]^{4-}$ is

e_g — —

E Large Δ

t_{2g} ↑↓ ↑ ↑

The complex ion will have two unpaired electrons. Note that the CN^- ligand produces such a large splitting that two of the electrons will be paired in the same orbital rather than one raised through the energy gap Δ.

We have seen how the crystal field model can account for the magnetic properties of octahedral complexes. The same model can also explain the colors of these complex ions. For example, $Ti(H_2O)_6{}^{3+}$, an octahedral complex of Ti^{3+}, which has a $3d^1$ electron configuration, is violet because it absorbs light in the middle of the visible region of the spectrum (see Fig. 20.24). When a substance absorbs certain wavelengths of light in the visible region, the color of the substance is determined by the wavelengths of visible light that remain. We say that the substance exhibits the color *complementary* to those absorbed. The $Ti(H_2O)_6{}^{3+}$ ion is violet because it absorbs light in the yellow-green region, thus letting red light and blue light pass, which gives the observed violet color. This is shown schematically in Fig. 20.25. Table 20.16 shows the general relationship between the wavelengths of visible light absorbed and the approximate color observed.

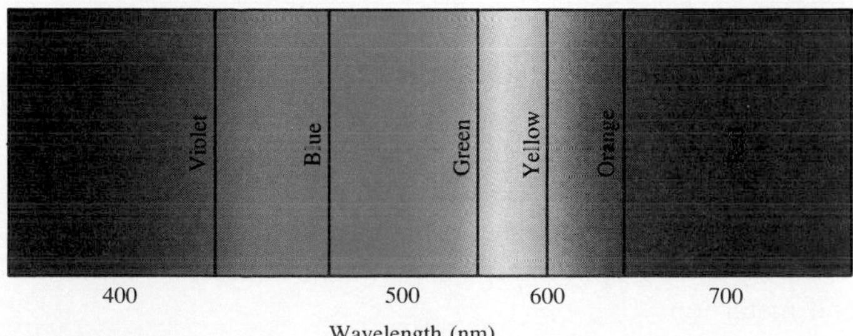

Figure 20.24

The visible spectrum.

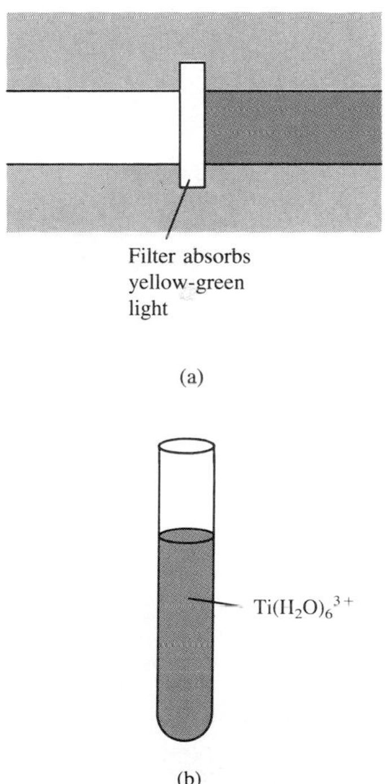

Figure 20.25

(a) When white light shines on a filter that absorbs in the yellow-green region, the emerging light is violet. (b) Because the complex ion $Ti(H_2O)_6{}^{3+}$ absorbs yellow-green light, a solution of it is violet.

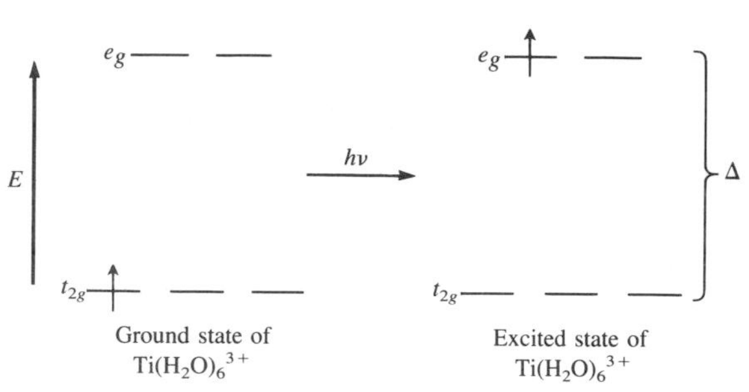

Figure 20.26

The complex ion $Ti(H_2O)_6^{3+}$ can absorb visible light in the yellow-green region to transfer the lone d electron from the t_{2g} to the e_g set.

Approximate Relationship of Wavelength of Visible Light Absorbed to Color Observed

Absorbed wavelength in nm (color)	Observed color
400 (violet)	Greenish-yellow
450 (blue)	Yellow
490 (blue-green)	Red
570 (yellow-green)	Violet
580 (yellow)	Dark blue
600 (orange)	Blue
650 (red)	Green

Table 20.16

The reason that the $Ti(H_2O)_6^{3+}$ ion absorbs specific wavelengths of visible light can be traced to the transfer of the lone d electron between the split d orbitals, as shown in Fig. 20.26. A given photon of light can be absorbed by a molecule only if the wavelength of the light provides exactly the energy needed by the molecule. Or, in other words, the wavelength absorbed is determined by the relationship described by

$$\Delta E = \frac{hc}{\lambda}$$

where ΔE represents the energy spacing in the molecule (we have used simply Δ in this chapter) and λ represents the wavelength of light needed. Because the d-orbital splitting in most octahedral complexes corresponds to the energies of photons in the visible region, octahedral complex ions are usually colored.

Since the ligands coordinated to a given metal ion determine the size of the d-orbital splitting, the color changes as the ligands are changed. This occurs because a change in Δ means a change in the wavelength of light needed to transfer electrons between the t_{2g} and e_g orbitals. Several octahedral complexes of Cr^{3+} and their colors are listed in Table 20.17.

Solutions of $[Cr(NH_3)_6]Cl_3$ (yellow) and $[Cr(NH_3)_5Cl]Cl_2$ (purple).

Several Octahedral Complexes of Cr^{3+} and Their Colors

Isomer	Color
$[Cr(H_2O)_6]Cl_3$	Violet
$[Cr(H_2O)_5Cl]Cl_2$	Blue-green
$[Cr(H_2O)_4Cl_2]Cl$	Green
$[Cr(NH_3)_6]Cl_3$	Yellow
$[Cr(NH_3)_5Cl]Cl_2$	Purple
$[Cr(NH_3)_4Cl_2]Cl$	Violet

Table 20.17

Other Coordination Geometries

Using the same principles developed for octahedral complexes, we will now consider complexes with other geometries. For example, Fig. 20.27 shows a tetrahedral arrangement of point charges in relation to the 3d orbitals of a metal ion. There are two important facts to note.

1. None of the 3d orbitals "point at the ligands" in the tetrahedral arrangement, as the $d_{x^2-y^2}$ and d_{z^2} orbitals do in the octahedral case. Thus the tetrahedrally arranged ligands do not differentiate the d orbitals as much in the tetrahedral as in the octahedral case. That is, the difference in energy between the split d orbitals will be significantly less in tetrahedral complexes. Although we will not derive it here, the tetrahedral splitting is $\frac{4}{9}$ that of the octahedral splitting for a given ligand and metal ion:

$$\Delta_{\text{tet}} = \tfrac{4}{9}\Delta_{\text{oct}}$$

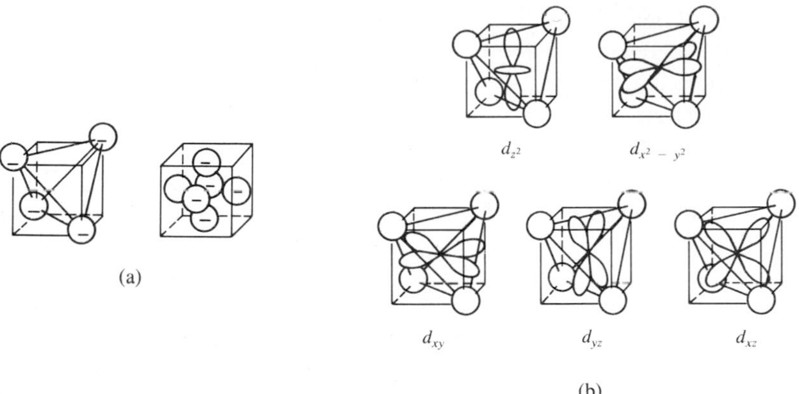

Figure 20.27

(a) Tetrahedral and octahedral arrangements of ligands shown inscribed in cubes. Note that in the two types of arrangements, the point charges occupy opposite parts of the cube: the octahedral point charges are at the centers of the cube faces; the tetrahedral point charges occupy opposite corners of the cube.
(b) The orientations of the 3*d* orbitals relative to the tetrahedral set of point charges.

2. Although not exactly pointing at the ligands, the d_{xy}, d_{xz}, and d_{yz} orbitals are closer to the point charges than are the d_{z^2} and $d_{x^2-y^2}$ orbitals. This means that the *d*-orbital splitting will be opposite to that for the octahedral arrangement. The two arrangements are contrasted in Fig. 20.28. Because the *d*-orbital splitting is relatively small for the tetrahedral case the weak-field case (high-spin case) *always* applies. There are no known ligands powerful enough to produce the strong-field case in a tetrahedral complex.

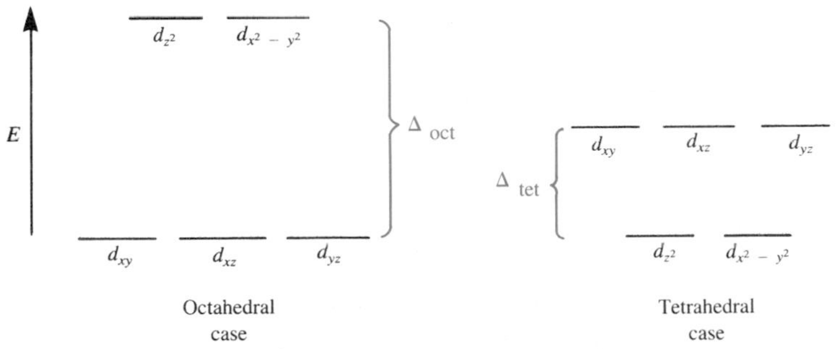

Figure 20.28

The crystal field diagrams for octahedral and tetrahedral complexes. The relative energies of the sets of *d* orbitals are reversed. For a given type of ligand, the splitting is much larger for the octahedral complex ($\Delta_{oct} > \Delta_{tet}$) because in this arrangement the d_{z^2} and $d_{x^2-y^2}$ point their lobes directly at the point charges and are thus relatively high in energy.

Sample Exercise 20.6

Give the crystal field diagram for the tetrahedral complex ion $CoCl_4^{2-}$.

Solution

The complex ion contains Co^{2+}, which has a $3d^7$ electron configuration. The splitting of the *d* orbitals will be small since this is a tetrahedral complex, giving the high-spin case with three unpaired electrons.

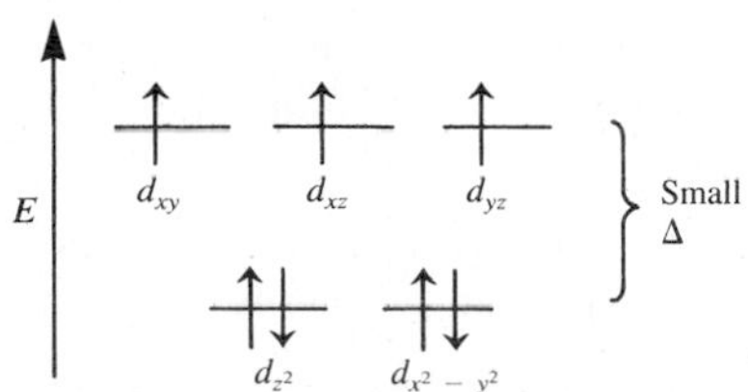

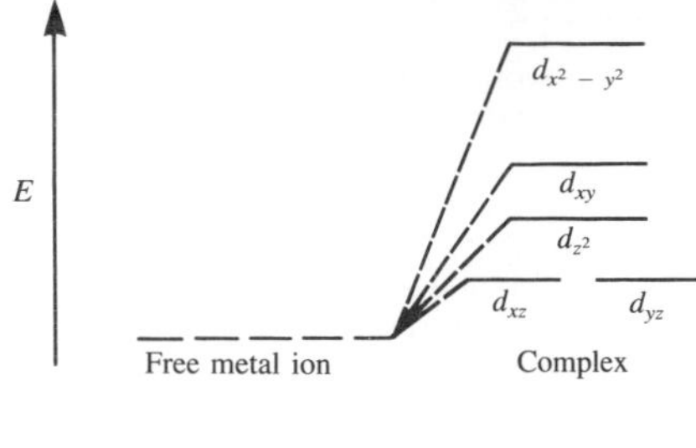

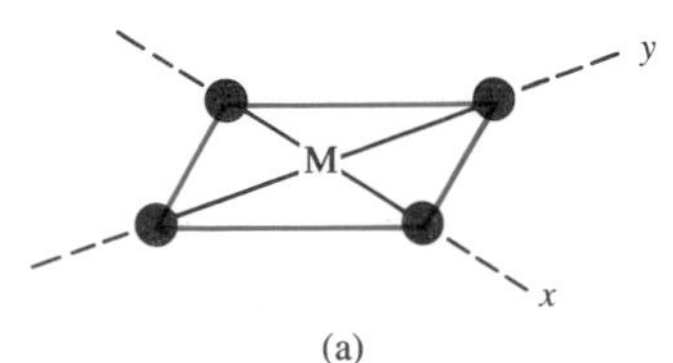

(a)

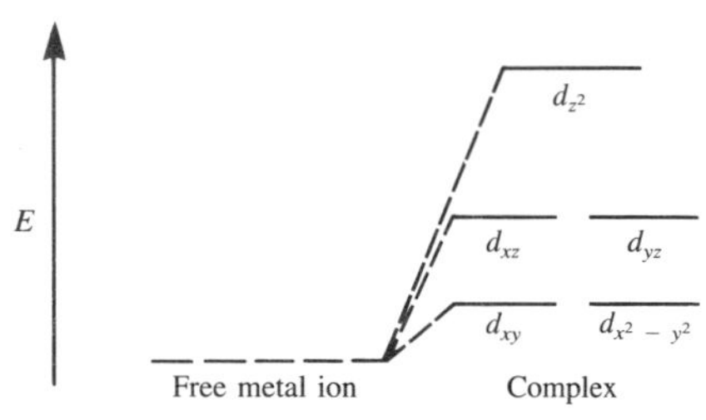

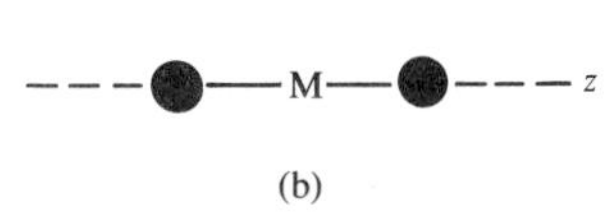

(b)

Figure 20.29

(a) The crystal field diagram for a square planar complex oriented in the *xy* plane with ligands along the *x* and *y* axes. The position of the d_{z^2} orbital is higher than those of the d_{xz} and d_{yz} orbitals because of the "doughnut" of electron density in the *xy* plane. The actual position of d_{z^2} is somewhat uncertain and varies in different square planar complexes. (b) The crystal field diagram for a linear complex where the ligands lie along the *z* axis.

The crystal field model also applies to square planar and linear complexes. The crystal field diagrams for these cases are shown in Fig. 20.29. The ranking of orbitals in these diagrams can be explained by considering the relative orientations of the point charges and the orbitals. The diagram in Fig. 20.28 for the octahedral arrangement can be used to obtain these orientations. We can obtain the square planar complex by removing the two point charges along the *z* axis. This will greatly lower the energy of d_{z^2}, leaving only $d_{x^2-y^2}$, which points at the four remaining ligands as the highest-energy orbital. We can obtain the linear complex from the octahedral arrangement by leaving the two ligands along the *z* axis and removing the four in the *xy* plane. This means only the d_{z^2} points at the ligands and is highest in energy.

20.7 The Molecular Orbital Model

Purpose

- To apply the molecular orbital model to octahedral complexes.

Although quite successful in accounting for the magnetic and spectral (light-absorption) properties of complex ions, the crystal field model has limited use other than explaining those properties dependent on the *d*-orbital splitting. For example, the model gives a very crude and misleading view of the nature of the metal-ligand bonding.

The molecular orbital model gives a much more realistic view of complex ions, especially of the bonding. As we saw in Section 9.2, this model postulates that a new set of orbitals characteristic of the molecule is formed from the atomic orbitals of the component atoms. To illustrate how this model can be applied to complex ions, we will describe the molecular orbitals in an octahedral species of the general formula ML_6^{n+}. To keep things as simple as possible, we will focus only on those ligand orbitals that have lone pairs that interact with the metal ion valence orbitals (3*d*, 4*s*, and 4*p*). There are two important considerations in predicting how atomic orbitals will interact to form molecular orbitals:

1. *Extent of orbital overlap.* Atomic orbitals must have a net overlap in space to form molecular orbitals. Figure 20.30 shows an octahedral arrangement of ligands with lone-pair orbitals. Recall from our previous discussion that two of the metal ion's 3*d* orbitals (d_{z^2} and $d_{x^2-y^2}$) point at the ligands and thus will form molecular orbitals with the ligand lone-pair orbitals. On the other hand, the d_{xy}, d_{xz}, and d_{yz} orbitals point *between* the ligands and will not be involved in the sigma bonding with the ligands. The spherical 4*s* orbital of the metal ion overlaps with all of the ligand lone-pair orbitals, and each of the 4*p* orbitals overlaps with pairs of ligand lone-pair orbitals on the three coordinate axes. Thus both the 4*s* and 4*p* orbitals will be involved in the molecular orbitals in the complex ion.
2. *Relative orbital energies.* Atomic orbitals that are close in energy will interact more than those widely separated in energy.

The molecular orbital model was introduced in Section 9.2.

When we apply these principles to the general complex ML_6^{n+}, we obtain the energy-level diagram shown in Fig. 20.31. Note that the σ_s, σ_p, and σ_d molecular orbitals are bonding MOs; they are *lower* in energy than the ligand orbitals and metal ion orbitals that mix to form them. Electrons in these molecular orbitals will have lower energies than they do in either the metal ion or the ligands; they are therefore the major contributors to the stability of the complex ion.

Because the d_{xz}, d_{yz}, and d_{xy} orbitals (the t_{2g} set) of the metal ion do not overlap with the ligand orbitals, they remain unchanged. Thus they have the same energy in the complex ion as they had in the free metal ion and make no contribution to the stability of the complex. They are called *nonbonding orbitals*.

The e_g* molecular orbitals are antibonding orbitals, since they are higher in energy than the atomic orbitals that mix to form them. Electrons in these orbitals have higher energies than they do in the free metal ion and destabilize the complex ion relative to the separated metal ion and ligands. However, *the most important characteristic of the e_g* orbitals is that they are primarily composed of d_{z^2} and $d_{x^2-y^2}$ atomic orbitals,* with relatively little contribution from ligand orbitals. This lack of mixing is due to the large energy difference between the ligand orbitals and the metal ion 3*d* orbitals. Thus we can see that the molecular orbital model predicts the same type of *d*-orbital splitting as the crystal field model, with the added advantage that it gives a much more realistic picture of the metal-ligand interaction and the origin of the splitting. Thus we can use the molecular orbital model to explain why different ligands produce different magnitudes of splitting. In particular, a ligand with a very electronegative donor atom will have lone-pair orbitals of very low energy (the electrons are very firmly bound to the ligand); these orbitals will not

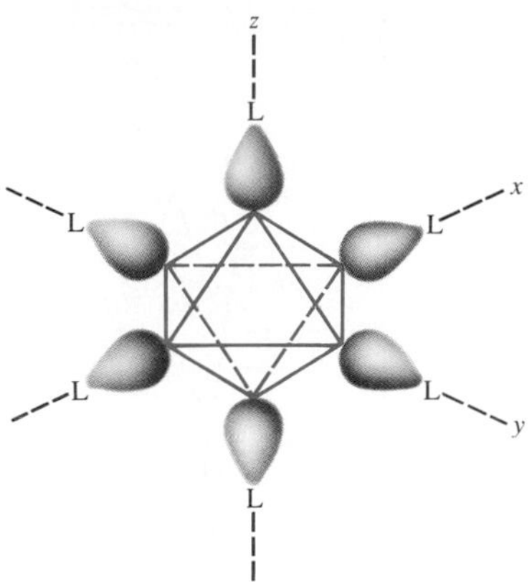

Figure 20.30

An octahedral arrangement of ligands showing their lone-pair orbitals.

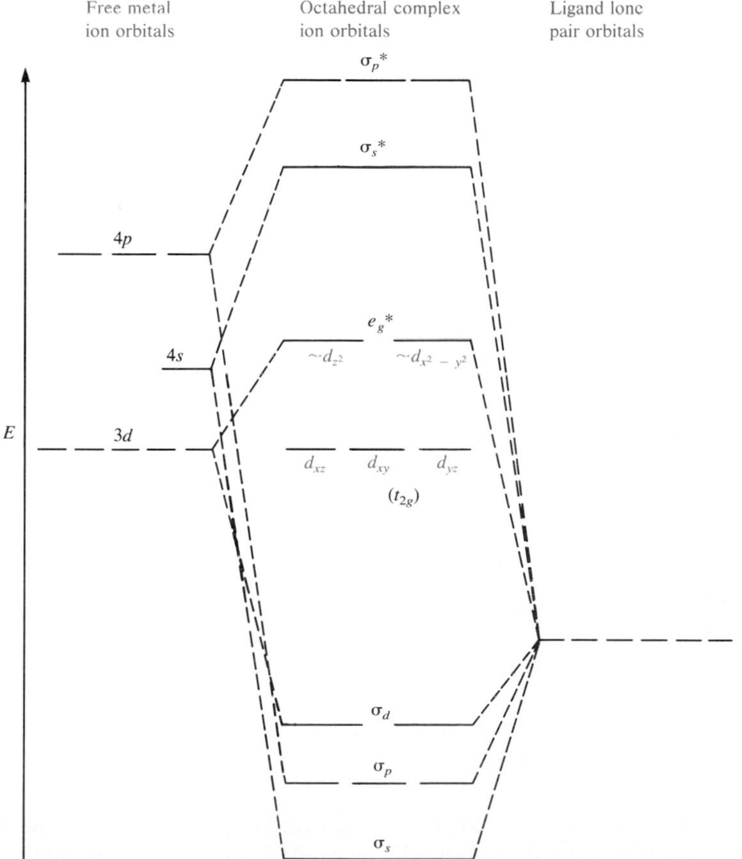

Figure 20.31

The molecular orbital energy-level diagram for an octahedral complex ion ML_6^{n+}. The e_g* molecular orbitals are essentially pure d_{z^2} and $d_{x^2-y^2}$ orbitals of the metal ion. Little mixing with the ligand orbitals occurs because of the large energy difference between the 3*d* and ligand orbitals.

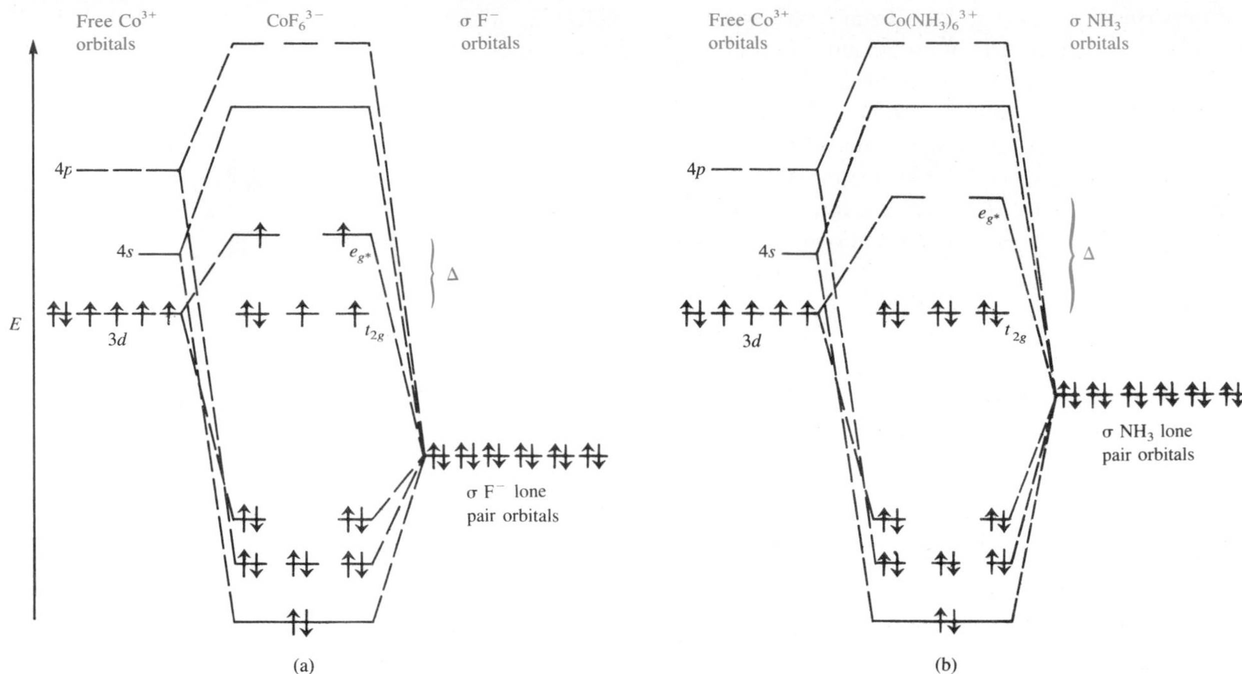

Figure 20.32

(a) The molecular orbital energy-level diagram for CoF_6^{3-}. This yields the high-spin case. (b) The molecular orbital energy-level diagram for $Co(NH_3)_6^{3+}$. This results in the low-spin case.

mix very thoroughly with the metal ion orbitals. This will result in a small difference between the t_{2g} (nonbonding) and e_g* (antibonding) orbitals. For example, in CoF_6^{3-} [see Fig. 20.32(a)] the low-energy orbitals of the very electronegative fluoride ion ligands mix only to a small extent with the cobalt ion orbitals. A small amount of *d*-orbital splitting results, giving the high-spin case.

On the other hand, in $Co(NH_3)_6^{3+}$ the ammonia lone-pair orbitals are closer in energy to the metal ion orbitals, a situation that produces a larger degree of mixing between them. This in turn gives a relatively large amount of *d*-orbital splitting, and the low-spin case results [see Fig. 20.32(b)].

We have seen that the molecular orbital model correctly predicts the *d*-orbital splitting in octahedral complexes, accounting for the magnetic and spectral properties of these species. Moreover, it has a major advantage in that it accounts in a realistic way for metal-ligand bonding. However, it suffers from the disadvantage of being much more complicated to apply than the crystal field model. The molecular orbital model can be applied to all types of complex ions, although we will not pursue the model further in this text.

20.8 The Biological Importance of Coordination Complexes

Purpose

- To describe the function and structure of some biologically important coordination complexes.

The ability of metal ions to coordinate with and release ligands and to easily undergo oxidation and reduction makes them ideal for the transport and storage of oxygen, as electron-transfer agents, as catalysts, and as drugs. Most of the first-row

The First-Row Transition Metals and Their Biological Significance

First-row transition metal	Biological function(s)
Scandium	None known.
Titanium	None known.
Vanadium	None known in humans.
Chromium	Assists insulin in the control of blood sugar; may also be involved in the control of cholesterol.
Manganese	Necessary for a number of enzymatic reactions.
Iron	Component of hemoglobin and myoglobin; involved in the electron-transport chain.
Cobalt	Component of vitamin B_{12}, which is essential for the metabolism of carbohydrates, fats, and proteins.
Nickel	Component of the enzymes urease and hydrogenase.
Copper	Component of several enzymes; assists in iron storage; involved in the production of color pigments of hair, skin, and eyes.
Zinc	Component of insulin and many enzymes.

Table 20.18

transition metals are essential for human health, as is summarized in Table 20.18. We will concentrate on iron's role in biological systems since several of its coordination complexes have been studied extensively.

Iron plays a central role in almost all living cells. In mammals, the principal source of energy comes from the oxidation of carbohydrates, proteins, and fats. Although oxygen is the oxidizing agent for these processes, it does not react directly with these molecules. Instead, the electrons from the breakdown of these nutrients are passed along a complex chain of molecules, called the *respiratory chain,* eventually reaching the O_2 molecule. The principal electron-transfer molecules in the respiratory chain are iron-containing species called **cytochromes,** consisting of two main parts: an iron complex called **heme** and a protein. The structure of the heme complex is shown in Fig. 20.33. Note that it contains an iron ion (it can be either Fe^{2+} or Fe^{3+}) coordinated to a rather complicated planar ligand called a **porphyrin.** As a class, porphyrins all contain the same central ring structure but have different

A protein is a large molecule assembled from amino acids, which have the general structure

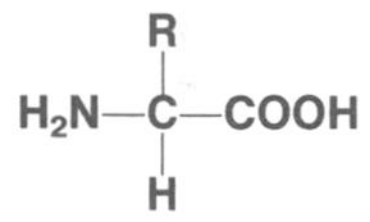

$$H_2N-\overset{\displaystyle R}{\underset{\displaystyle H}{C}}-COOH$$

where R varies.

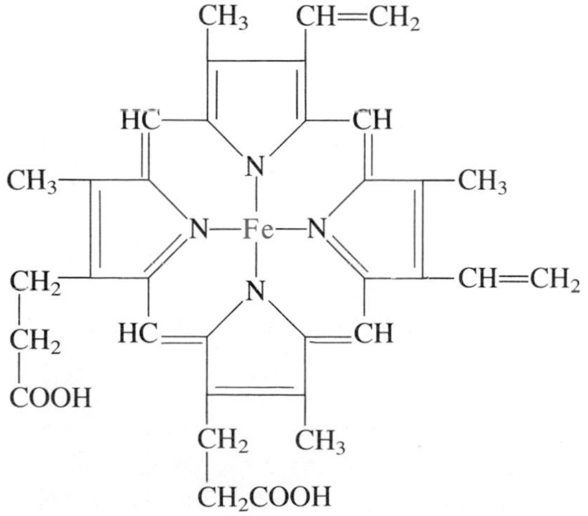

Figure 20.33

The heme complex, in which an Fe^{2+} ion is coordinated to four nitrogen atoms of a planar porphyrin ligand.

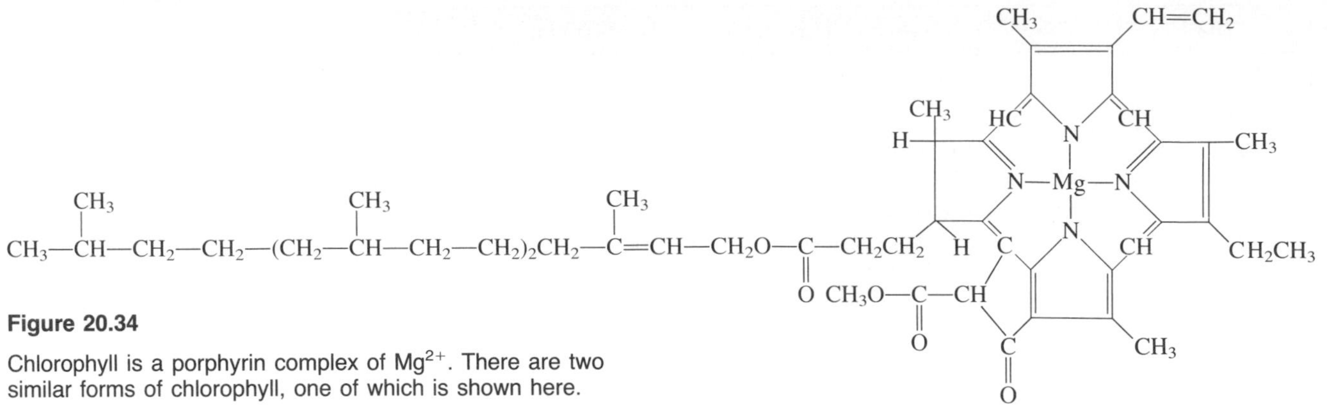

CH_3 $CH=CH_2$
CH_3 HC CH CH_3
H N
N—Mg—N
CH_3 CH_3 CH_3
CH_3—CH—CH_2—CH_2—(CH_2—CH—CH_2—CH_2)$_2$$CH_2$—C=CH—$CH_2$O—C—$CH_2$$CH_2$ H C N CH CH_2CH_3
O CH_3O—C—CH
O C CH_3
O

Figure 20.34

Chlorophyll is a porphyrin complex of Mg^{2+}. There are two similar forms of chlorophyll, one of which is shown here.

substituent groups at the edges of the rings. The various porphyrin molecules act as tetradentate ligands for many metal ions, including iron, cobalt, and magnesium. In fact, *chlorophyll,* a substance essential to the process of photosynthesis, is a magnesium-porphyrin complex of the type shown in Fig. 20.34.

In addition to participating in the transfer of electrons from nutrients to oxygen, iron also plays a principal role in the transport and storage of oxygen in mammalian blood and tissues. Oxygen is stored in a molecule called **myoglobin,** which consists of a heme complex and a protein in a structure very similar to that of the cytochromes. In myoglobin, the Fe^{2+} ion is coordinated to four nitrogen atoms of the porphyrin ring and to a nitrogen atom of the protein chain, as shown in Fig. 20.35. Since Fe^{2+} is normally six-coordinate, this leaves one position open for attachment of an O_2 molecule.

One especially interesting feature of myoglobin is that it involves an O_2 molecule attaching directly to Fe^{2+}. However, if gaseous O_2 is bubbled into an aqueous solution containing heme, the Fe^{2+} is immediately oxidized to Fe^{3+}. This oxidation

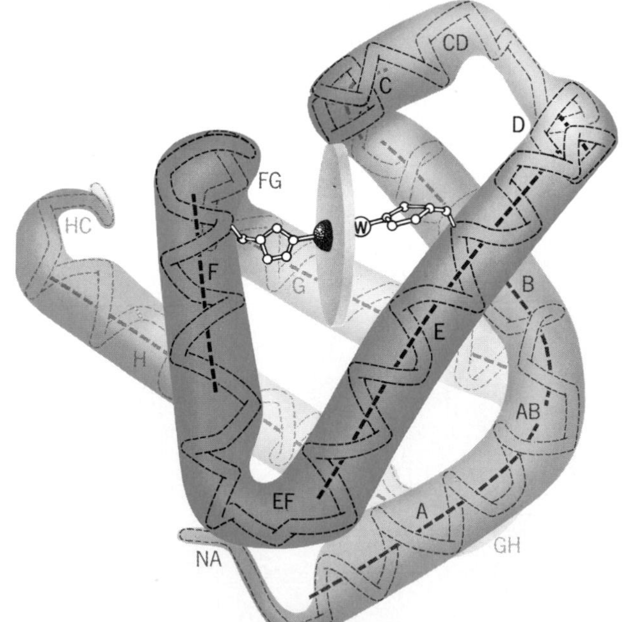

Figure 20.35

A representation of the myoglobin molecule. The Fe^{2+} ion is coordinated to four nitrogen atoms in the porphyrin of the heme (represented by the disk in the figure) and one nitrogen in the protein chain. This leaves a sixth coordination position (indicated by the W) available for an oxygen molecule. Reprinted from R. E. Dickerson and I. Geis from *The Structure and Action of Proteins,* Benjamin/Cummings Publisher, Menlo Park, CA. Illustration © 1969 by Dickerson and Geis.

of the Fe^{2+} in heme does not happen in myoglobin. This fact is of crucial importance because Fe^{3+} does not form a coordinate covalent bond with O_2, and myoglobin would not function if the bound Fe^{2+} could be oxidized. Since Fe^{2+} in the "bare" heme complex can be oxidized, it must be the protein that somehow prevents the oxidation. How? Based on much research, the answer seems to be that the oxidation of Fe^{2+} to Fe^{3+} involves an oxygen bridge between two iron ions (the circles indicate the ligands):

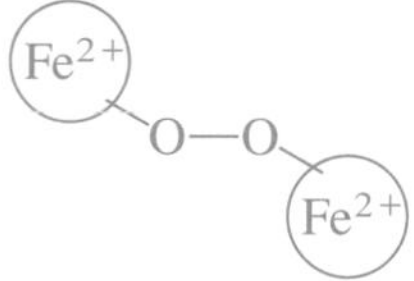

The bulky protein around the heme group in myoglobin prevents two molecules from getting close enough to form the oxygen bridge, and so oxidation of the Fe^{2+} is prevented.

The transport of O_2 in the blood is carried out by **hemoglobin,** a molecule consisting of four myoglobin-like units, as shown in Fig. 20.36. Each hemoglobin can therefore bind four O_2 molecules to form a bright red diamagnetic complex. The diamagnetism occurs because oxygen is a strong-field ligand toward Fe^{2+}, which has a $3d^6$ electron configuration. When the oxygen molecule is released, water molecules occupy the sixth coordination position around each Fe^{2+}, giving a bluish-colored paramagnetic complex (H_2O is a weak-field ligand toward Fe^{2+}) that gives venous blood its characteristic bluish tint.

Hemoglobin dramatically demonstrates how sensitive the function of a biomolecule is to its structure. In certain people, in the synthesis of the proteins needed for hemoglobin, an improper amino acid is inserted into the protein in two places. This

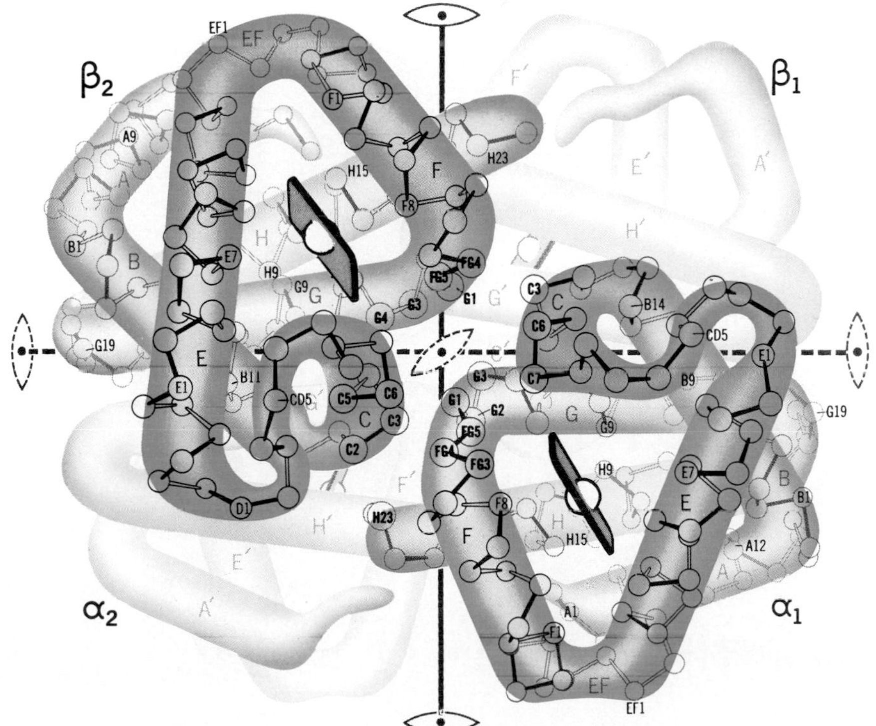

Figure 20.36

A representation of the hemoglobin structure. There are two slightly different types of protein chains (α and β). Each hemoglobin has two α chains and two β chains, each with a heme complex near the center. Thus each hemoglobin molecule can complex with four O_2 molecules. Reprinted from R. E. Dickerson and I. Geis from *The Structure and Action of Proteins,* Benjamin/Cummings Publisher, Menlo Park, CA. Illustration © 1969 by Dickerson and Geis.

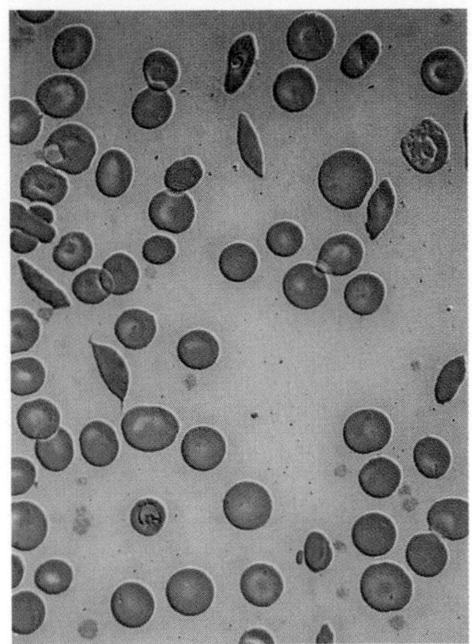

Figure 20.37

Normal red blood cells (round) and sickle-shaped red blood cells.

may not seem very serious, since there are several hundred amino acids present. However, because the incorrect amino acid has a nonpolar substituent instead of the polar one found on the proper amino acid, the hemoglobin drastically changes its shape. The red blood cells are then sickle-shaped rather than disk-shaped, as shown in Fig. 20.37. The misshapen cells can aggregate, causing clogging of tiny capillaries. This condition, known as *sickle cell anemia,* is the subject of intense research.

Our knowledge of the workings of hemoglobin allows us to understand the effects of high altitudes on humans. The reaction between hemoglobin and oxygen can be represented by the following equilibrium:

$$\underset{\text{Hemoglobin}}{Hb(aq)} + 4O_2(g) \rightleftharpoons \underset{\text{Oxyhemoglobin}}{Hb(O_2)_4(aq)}$$

At high altitudes, where the oxygen content of the air is lower, the position of this equilibrium will shift to the left, according to Le Châtelier's principle. Because less oxyhemoglobin is formed, fatigue, dizziness, and even a serious illness called *high-altitude sickness* can result. One way to combat this problem is to use supplemental oxygen, as most high-altitude mountain climbers do. However, this is impractical for people who live at high elevations. In fact, the human body adapts to the lower oxygen concentrations by making more hemoglobin, causing the equilibrium to shift back to the right. Someone moving from Chicago to Boulder, Colorado (5300 feet) would notice the effects of the new altitude for a couple of weeks, but as the hemoglobin level increased the effects would disappear. This change is called *high-altitude acclimatization,* which explains why athletes who want to compete at high elevations should practice there for several weeks prior to the event.

Our understanding of the biological role of iron also allows us to explain the toxicities of substances such as carbon monoxide and the cyanide ion. Both CO and CN^- are very good ligands toward iron and so can interfere with the normal workings of the iron complexes in the body. For example, carbon monoxide has about

Sherpa porters in Nepal are acclimatized to high elevations.

200 times the affinity for the Fe^{2+} in hemoglobin as oxygen does. The resulting stable complex, **carboxyhemoglobin,** prevents the normal uptake of O_2, thus depriving the body of needed oxygen. Asphyxiation can result if enough carbon monoxide is present in the air. The mechanism for the toxicity of the cyanide ion is somewhat different. Cyanide coordinates strongly to cytochrome oxidase, an iron-containing cytochrome enzyme that catalyzes the oxidation-reduction reactions of certain cytochromes. The coordinated cyanide thus prevents the electron-transfer process and rapid death results. Because of its behavior, cyanide is called a *respiratory inhibitor*.

FOR REVIEW

Summary

For the first-row transition metals (scandium through zinc), the last electrons occupy the 3*d* orbitals. Because electrons in these inner orbitals cannot as easily participate in bonding as can electrons in valence *s* and *p* orbitals, the chemistry of the transition elements is not greatly affected by the gradual change in the number of electrons. As a class, these elements have metallic physical and chemical properties. In forming ionic compounds with nonmetals, the transition metals exhibit several typical characteristics: more than one oxidation state is often found, the cations often are complex ions, and most compounds are colored and many are paramagnetic.

First-row transition metal ions can form a variety of ions by losing one or more of their electrons. The maximum possible oxidation state for a given element corresponds to the loss of all the 4*s* and 3*d* electrons, but toward the right end of the period, the maximum possible oxidation state is not seen. Rather, 2+ ions become the most common because the 3*d* electrons become increasingly difficult to remove.

Transition metals typically form coordination compounds, which consist of a complex ion (a transition metal ion with attached ligands) and counter ions (anions or cations to produce a solid with no net charge). A ligand (a Lewis base) is a neutral molecule or ion having a lone pair of electrons that can be shared with an empty orbital on a metal ion (a Lewis acid) to form a coordinate covalent bond. The coordination number is the number of bonds formed by the metal ion to its ligands and can range from 2 to 8 depending on the size, charge, and electron configuration of the metal ion. A coordination number of 6 means an octahedral arrangement of ligands about the central metal ion; 2 means a linear arrangement, and 4 means either a planar or a tetrahedral arrangement.

Ligands can form one bond (unidentate), two bonds (bidentate), or as many as six bonds to the metal ion. Ligands that can bond to a metal ion through more than one atom are called chelates. For example, ethylenediamine (en) is bidentate, and EDTA is hexadentate (forms six bonds).

When two or more species have the same formula but different properties, they are called isomers. Structural isomers contain the same atoms but one or more different bonds. Coordination isomerism is a type of structural isomerism in which the composition of the coordination sphere around the metal varies. In linkage isomerism, the composition of the coordination spheres is the same, but the point of attachment of at least one of the ligands differs.

In stereoisomerism, the isomers have the same bonds but different spatial arrangements. For example, in geometric isomerism, atoms or groups can assume different relative positions in the coordination sphere around the metal. When two identical ligands lie across from each other, the isomer is *trans;* two ligands located next to each other produce the *cis* isomer.

A second type of stereoisomerism is optical isomerism, which is exhibited by molecules that have nonsuperimposable mirror images, that is, by chiral molecules. The isomer pairs, called enantiomers, rotate plane-polarized light in equal but opposite directions.

Bonding in complex ions can be described in terms of the localized electron model. However, even though this model can account for the bonding in complex ions by use of hybrid orbitals, it does not predict important properties such as magnetism and color. The crystal field model is much more successful at predicting these properties by focusing on the energies of the d orbitals of the metal ion. To predict the effects ligands will have on the energies of the d orbitals, the ligands are approximated as negative point charges and the metal-ligand bonding is assumed to be completely ionic. The effect of the point charges in an octahedral complex ion is to split the $3d$ orbitals into two groups with different energies. If the splitting is large (the strong-field case), the electrons pair in the lower-energy orbitals before any jump the energy gap. If the splitting is small (the weak-field case), the electrons occupy all five d orbitals singly before pairing occurs. Ligands can be arranged according to their ability to produce splitting, the spectrochemical series.

The crystal field model also accounts for the colors of complex ions. When a substance absorbs certain wavelengths of light in the visible region, its color is determined by the wavelengths of visible light that remain unabsorbed. Complex ions can absorb light as they transfer electrons between the split d orbitals. The relationship between the wavelength of light absorbed (λ) and the magnitude of the splitting (Δ) is given by

$$\Delta = \frac{hc}{\lambda}$$

The molecular orbital model gives the best description of the actual bonding in complex ions. In terms of this model, the d-orbital splitting occurs because the t_{2g} orbitals are nonbonding, while the e_g orbitals are antibonding molecular orbitals and are raised in energy.

Coordination complexes are vitally important in the chemistry of biological systems, with iron having special importance in electron-transfer processes and the transport and storage of oxygen.

Key Terms

Section 20.1
complex ion
first-row transition metals
lanthanide contraction
lanthanide series

Section 20.3
coordination compound
counter ion
oxidation state
coordination number
ligand
coordinate covalent bond
monodentate (unidentate) ligand
chelating ligand (chelate)
bidentate ligand

Section 20.4
isomers
structural isomerism
stereoisomerism
coordination isomerism
linkage isomerism
geometrical (*cis-trans*) isomerism
trans isomer
cis isomer
optical isomerism
chiral
enantiomers

Section 20.6
crystal field model
d-orbital splitting
strong-field (low-spin) case
weak-field (high-spin) case
spectrochemical series

Section 20.8
cytochromes
heme
porphyrin
myoglobin
hemoglobin
carboxyhemoglobin

Exercises

A blue exercise number indicates that the answer to that exercise appears at the back of this book and a solution appears in the Solutions Guide.

Transition Metals

1. Write electron configurations for the following metals:
   a. Ni c. Zr e. Mo
   b. Cd d. Nd f. Ru

2. Write electron configurations for the following ions:
   a. Ni^{2+} c. Zr^{3+} and Zr^{4+} e. Mo^{4+}
   b. Cd^{2+} d. Nd^{3+} f. Ru^{2+} and Ru^{3+}

3. Write electron configurations for each of the following:
   a. Co, Co^{2+}, Co^{3+}
   b. Pt, Pt^{2+}, Pt^{4+}
   c. Fe, Fe^{2+}, Fe^{3+}
   d. Au, Au^{+}, Au^{3+}
   e. Cu, Cu^{+}, Cu^{2+}

4. What is the maximum mass of titanium that can be produced from 1.00×10^3 kg of ilmenite ($FeTiO_3$)?

5. Use the following ionization energy values to decide whether ilmenite ($FeTiO_3$) is composed of Fe^{2+} and Ti^{4+} ions or Fe^{3+} and Ti^{3+} ions.

Ionization energies (kJ/mol)		
	Fe	Ti
First	759	658
Second	1,561	1,310
Third	2,957	2,652
Fourth	—	4,175

Use the following reduction potentials to answer the same question.

$$TiO^{2+} + 2H^+ + e^- \rightarrow Ti^{3+} + H_2O \qquad \mathscr{E}° = +0.099 \text{ V}$$
$$Fe^{3+} + e^- \rightarrow Fe^{2+} \qquad \mathscr{E}° = +0.77 \text{ V}$$

Which answer is consistent with the information presented in this chapter? Comment on any discrepancies.

6. Molybdenum is obtained as a by-product of copper mining or is mined directly (primary deposits are in the Rocky Mountains in Colorado). In both cases, it is obtained as MoS_2, which is then converted to MoO_3. The MoO_3 can be used directly in the production of stainless steel for high-speed tools, which accounts for about 85% of the molybdenum used. Molybdenum can be purified by dissolving MoO_3 in aqueous ammonia and crystallizing ammonium molybdate. Depending on conditions, either $(NH_4)_2Mo_2O_7$ or $(NH_4)_6Mo_7O_{24} \cdot 4H_2O$ is obtained.
   a. Give names for MoS_2 and MoO_3.
   b. What is the oxidation state of Mo in each of the compounds mentioned above?
   c. Complete and balance the following equations:

$$MoS_2(s) + O_2(g) \longrightarrow MoO_3(s)$$
$$NH_3(aq) + MoO_3(s) \longrightarrow (NH_4)_2Mo_2O_7(s)$$
$$NH_3(aq) + MoO_3(s) \longrightarrow (NH_4)_6Mo_7O_{24} \cdot 4H_2O(s)$$

7. Titanium dioxide, the most widely used white pigment, occurs naturally but is often colored by the presence of impurities. The chloride process is often used in purifying rutile, a mineral form of titanium dioxide.
   a. Show that the unit cell for rutile, shown below, conforms to the formula TiO_2. (*Hint:* Recall the discussion in Section 10.4.)

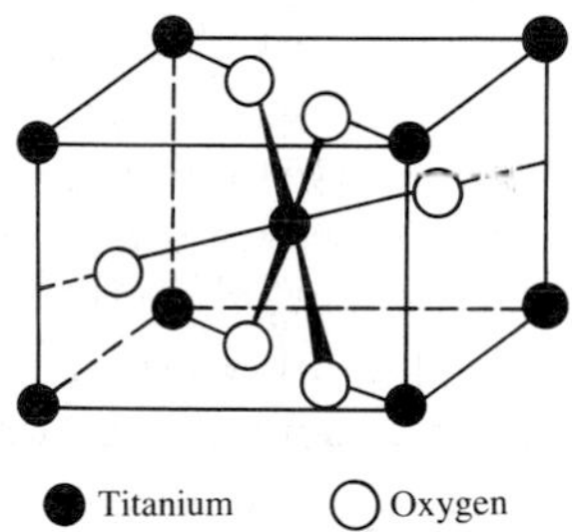

   b. The reactions for the chloride process are

$$2TiO_2(s) + 3C(s) + 4Cl_2(g) \xrightarrow{950°C} 2TiCl_4(g) + CO_2(g) + 2CO(g)$$

$$TiCl_4(g) + O_2(g) \xrightarrow{1000–1400°C} TiO_2(s) + 2Cl_2(g)$$

   Assign oxidation states to the elements in both reactions. Which elements are being reduced, and which are being oxidized? Identify the oxidizing agent and the reducing agent in each reaction.

8. The melting and boiling points of the titanium tetrahalides are given below.

	mp (°C)	bp (°C)
TiF_4	284	—
$TiCl_4$	−24	136.5
$TiBr_4$	38	233.5
TiI_4	155	377

Rationalize these data in terms of the bonding and intermolecular forces for these compounds.

9. Manganese is found in nature in silicate, oxide, and carbonate minerals. Commercially, the most important ores are pyrolu-

site (MnO_2) and rhodochrosite ($MnCO_3$). Give systematic names for each of these compounds.

10. Ores of cobalt have been used to impart a blue color to glass since 2600 B.C. Impure cobalt metal was first isolated in 1735 by G. Brandt, a Swedish chemist. The name is probably derived from *Kobold,* the German word for "goblin" or "evil spirit." The miners of northern Europe thought the spitefulness of such spirits was responsible for ores, which on smelting not only failed to yield the cobalt metal but produced highly toxic fumes (As_4O_6). The principal ores of cobalt are smaltite ($CoAs_2$), cobaltite ($CoAsS$), and linnaeite (Co_3S_4).
   a. Write a reaction for the roasting of smaltite by oxygen, producing cobalt(II) oxide and As_4O_6. Name As_4O_6. Propose a structure for As_4O_6 based on what was discussed about phosphorus chemistry in Chapter 19.
   b. Cobalt is usually obtained as a by-product of the production of copper, nickel, or lead. The ore containing the mixture of metals is first treated with oxygen to form the metal oxides. These oxides are then treated with sulfuric acid, which dissolves the oxides of iron, nickel, and cobalt, but not that of copper, which is left behind. After iron is precipitated as its hydroxide, the basic solution is treated with hypochlorite ion, which oxidizes the Co^{2+} ion to Co^{3+} and leads to precipitation of cobalt(III) hydroxide:

$$Co^{2+}(aq) + OCl^-(aq) + OH^-(aq) \rightarrow Co(OH)_3(s) + Cl^-(aq)$$

   Balance this equation.
   c. Calculate the concentration of Co^{3+} present in a saturated aqueous solution of $Co(OH)_3$ ($K_{sp} = 2.5 \times 10^{-43}$).
   d. Calculate the concentration of Co^{3+} present in a saturated solution of $Co(OH)_3$ buffered at pH = 10.00.

11. Zinc metal tarnishes quickly in moist air. Write a reaction for this process. Using data from Chapter 17 and Appendix 4, calculate $\mathscr{E}°$, $\Delta G°$, and K (at 298 K) for this reaction.

Coordination Compounds

12. Define each of the following:
   a. ligand
   b. chelate
   c. bidentate
   d. complex ion

13. What must a ligand have in order to bond to a metal?

14. What do we mean when we say that a bond is a coordinate covalent bond?

15. Name the following coordination compounds:
   a. $[Co(NH_3)_6]Cl_2$
   b. $[Co(H_2O)_6]I_3$
   c. $K_2[PtCl_4]$
   d. $K_4[PtCl_6]$

16. Name the following complex ions:
   a. $Ru(NH_3)_5Cl^{2+}$
   b. $Fe(CN)_6^{4-}$
   c. $Mn(NH_2CH_2CH_2NH_2)_3^{2+}$
   d. $Co(NH_3)_5NO_2^{2+}$

17. Name the following compounds containing complex ions:
   a. $Na_4[Ni(C_2O_4)_3]$
   b. $K_2[CoCl_4]$
   c. $[Cu(NH_3)_4]SO_4$
   d. $[Co(en)_2(SCN)Cl]Cl$

18. Give formulas for the following:
   a. hexakispyridinecobalt(III) chloride
   b. pentaammineiodochromium(III) iodide
   c. trisethylenediamminenickel(II) bromide
   d. potassium tetracyanonickelate(II)
   e. tetraamminedichloroplatinum(IV)tetrachloroplatinate(II)

19. Give formulas for the following complex ions:
   a. tetrachloroferrate(III) ion
   b. pentaammineaquaruthenium(III) ion
   c. iodopentakispyridineplatinum(IV) ion
   d. amminetrichloroplatinate(II) ion

20. How many bonds could each of the following chelates form with a metal ion?
   a. acetylacetone(acacH)

$$CH_3-\overset{\overset{O}{\|}}{C}-CH_2-\overset{\overset{O}{\|}}{C}-CH_3$$

   b. diethylenetriamine

$$NH_2-CH_2-CH_2-NH-CH_2-CH_2-NH_2$$

   c. cyclam

NH HN
NH HN

   d. porphine

NH N
N HN

21. Define each of the following and give examples:
   a. isomers
   b. structural isomers
   c. stereoisomers

d. coordination isomers
e. linkage isomers
f. geometric isomers
g. optical isomers

22. Draw geometrical isomers of each of the following complex ions:
a. $[Co(C_2O_4)_2(H_2O)_2]^-$
b. $[Pt(NH_3)_4I_2]^{2+}$
c. $[Ir(NH_3)_3Cl_3]$
d. $[Cr(en)(NH_3)_2I_2]^+$

23. Which of the following ligands are capable of linkage isomerism? Explain your answer.

$$SCN^-, N_3^-, NO_2^-, NH_2CH_2CH_2NH_2, OCN^-, I^-$$

24. Draw all geometrical and linkage isomers of square planar $[Pt(NH_3)_2(SCN)_2]$.

25. BAL is a chelating agent used in treating heavy-metal poisoning. It acts as a bidentate ligand. What types of linkage isomers are possible when BAL coordinates to a metal ion?

```
CH2—SH
|
CH—SH
|
CH2—OH
 BAL
```

26. Draw structures of each of the following:
a. *cis*-dichloroethylenediamineplatinum(II)
b. *trans*-dichlorobisethylenediaminecobalt(II)
c. *cis*-tetraamminechloronitrocobalt(III) ion
d. *trans*-tetraamminechloronitritocobalt(III) ion
e. *trans*-diaquabisethylenediaminecopper(II) ion

27. Amino acids can act as ligands toward transition metal ions. The simplest amino acid is glycine, $NH_2CH_2CO_2H$. Draw a structure of the glycinate anion, $NH_2CH_2CO_2^-$, acting as a bidentate ligand. Draw the structural isomers of the square planar complex $Cu(NH_2CH_2CO_2)_2$.

28. A coordination compound of cobalt(III) contains four ammonia molecules, one sulfate ion, and one chloride ion. Addition of aqueous $BaCl_2$ solution to an aqueous solution of the compound gives no precipitate. Addition of aqueous $AgNO_3$ to an aqueous solution of the compound produces a white precipitate. Propose a structure for this coordination compound.

29. The carbonate ion, CO_3^{2-}, can act either as a monodentate or bidentate ligand. Draw a picture of CO_3^{2-} coordinating to a metal ion as a bidentate and as a monodentate ligand. The carbonate ion can also act as a bridge between two metal ions. Draw a picture of a CO_3^{2-} ion bridging between two metal ions.

30. Draw all the geometrical isomers of $[Cr(en)(NH_3)_2BrCl]^+$. Which of these isomers also has an optical isomer? Draw them.

Bonding, Color, and Magnetism in Coordination Compounds

31. Define each of the following:
a. weak-field ligand
b. strong-field ligand
c. low-spin complex
d. high-spin complex

32. Compounds of copper(II) are generally colored, but compounds of copper(I) are not. Why?

33. Would you expect $Cd(NH_3)_4Cl_2$ to be colored? Why or why not?

34. Henry Taube, 1983 Nobel Prize winner in chemistry, has studied the mechanisms of oxidation-reduction reactions of transition metal complexes. In one experiment he and his students studied the following reaction:

$$Cr(H_2O)_6^{2+}(aq) + Co(NH_3)_5Cl^{2+}(aq) \rightarrow Cr(III)\text{ complexes} + Co(II)\text{ complexes}$$

Chromium(III) and cobalt(III) complexes are substitutionally inert (no exchange of ligands) under conditions of the experiment. Chromium(II) and cobalt(II) complexes can exchange ligands very rapidly. One of the products of the reaction is $Cr(H_2O)_5Cl^{2+}$. Is this consistent with the reaction proceeding through formation of $(H_2O)_5Cr—Cl—Co(NH_3)_5$ as an intermediate? Explain your answer.

35. How many unpaired electrons are in the following complex ions?
a. $Ru(NH_3)_6^{2+}$ (low-spin case)
b. $Fe(CN)_6^{3-}$ (low-spin case)
c. $Ni(H_2O)_6^{2+}$
d. $V(en)_3^{3+}$
e. $CoCl_4^{2-}$

36. The complex ion $Ru(phen)_3^{2+}$ has been used as a probe for the structure of DNA.
a. What type of isomerism is found in $Ru(phen)_3^{2+}$?
b. $Ru(phen)_3^{2+}$ is diamagnetic (as are all complex ions of Ru^{2+}). Draw the crystal field diagram for the *d*-orbitals in this complex ion.

(phen = 1,10-phenanthroline = [structure with N N])

37. The complex ion $NiCl_4^{2-}$ contains two unpaired electrons while $Ni(CN)_4^{2-}$ is diamagnetic. Propose structures for these two complex ions.

38. Tetrahedral complexes of Co^{2+} are quite common. Use a *d*-orbital splitting diagram to rationalize the stability of Co^{2+} tetrahedral complex ions.

39. Draw the d-orbital splitting diagrams for the octahedral complex ions of each of the following:

a. Fe^{2+} (high and low spin)
b. Fe^{3+} (high spin)
c. Ni^{2+}
d. Zn^{2+}
e. Co^{2+} (high and low spin)

40. The compound $Ni(H_2O)_6Cl_2$ is green, while $Ni(NH_3)_6Cl_2$ is purple. Predict the predominant color of light absorbed by each compound. Which compound absorbs light with the shorter wavelength? Predict in which compound Δ is greater and whether H_2O or NH_3 is a stronger field ligand. Do your conclusions agree with the spectrochemical series?

41. The complex ion $Fe(CN)_6^{3-}$ is paramagnetic with one unpaired electron. The complex ion $Fe(SCN)_6^{3-}$ has five unpaired electrons. Where does SCN^- lie in the spectrochemical series with respect to CN^-?

42. Would it be better to use octahedral Ni^{2+} complexes or octahedral Cr^{2+} complexes to determine whether a ligand is a high-field or low-field ligand by measuring the number of unpaired electrons? How else could the relative ligand field strengths be determined?

Additional Exercises

43. Nickel can be purified by producing the volatile compound nickel tetracarbonyl, $Ni(CO)_4$. Nickel is the only metal that reacts directly with CO at room temperature. What is the oxidation state of nickel in $Ni(CO)_4$?

44. How would transition metal ions be classified using the Lewis definition of acids and bases?

45. When an aqueous solution of KCN is added to a solution containing Ni^{2+} ion, a precipitate forms, which redissolves on addition of more KCN solution. No precipitate forms when H_2S is bubbled into this solution. Write reactions describing what happens in this solution. [*Hint:* CN^- is a Brönsted-Lowry base ($K_b \approx 10^{-5}$) and a Lewis base.]

46. Enterobactin (shown below), a molecule excreted by certain bacteria, is a chelating agent specific for Fe^{3+}. It is used in the transfer of Fe^{3+} from water-insoluble minerals into the bacterial cells. Coordination to the iron ion is through the oxygen atoms on the catechol rings:

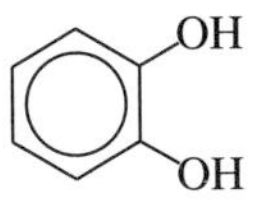

Catechol

a. Through how many atoms does enterobactin coordinate to Fe^{3+}?
b. The protons on the catechol —OH groups are lost in complex formation. Draw an approximate picture of the complex ion $Fe(enterobactin)^{3-}$

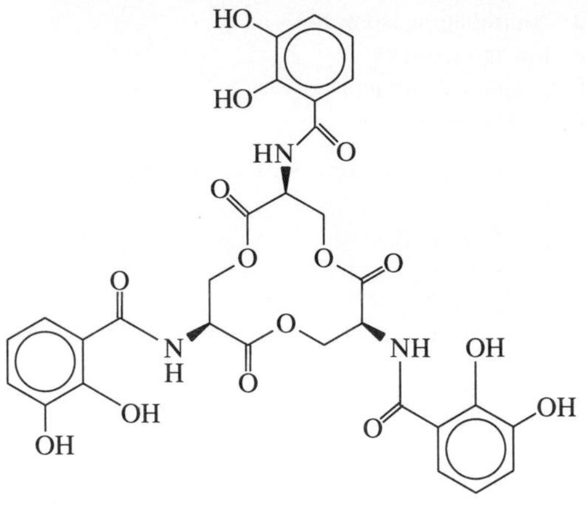

Enterobactin

47. Until the discoveries of Werner, it was thought that carbon had to be present in a compound for it to be optically active. Werner prepared the following compound containing OH^- ions as bridging groups and separated the optical isomers.

a. Draw structures of the two optically active isomers of this compound.
b. What are the oxidation states of the cobalt ions?
c. How many unpaired electrons are present if the complex is the low-spin case?

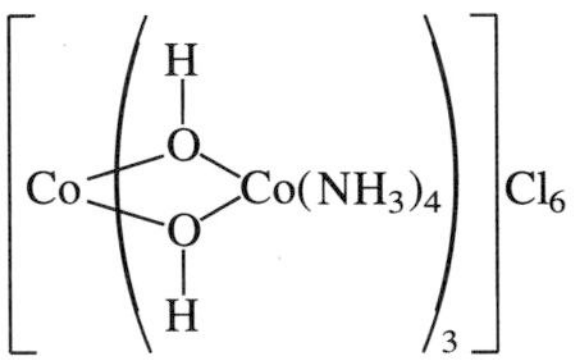

$$\left[\mathrm{Co}\left(\begin{array}{c}\mathrm{H}\\ \mathrm{O}\\ \mathrm{O}\\ \mathrm{H}\end{array}\mathrm{Co(NH_3)_4}\right)_3\right]\mathrm{Cl_6}$$

48. Ammonia and potassium iodide solutions are added to an aqueous solution of $Cr(NO_3)_3$. A solid is isolated (compound A) and the following data are collected:

i. When 0.105 g of compound A was strongly heated in excess O_2, 0.0203 g of CrO_3 was formed.
ii. In a second experiment it took 32.93 mL of 0.100 M HCl to titrate completely the NH_3 present in 0.341 g of compound A.
iii. Compound A was found to contain 73.53% iodine by mass.
iv. The freezing point of water was lowered by 0.62°C when 0.601 g of compound A was dissolved in 10.00 g of H_2O ($K_f = 1.86$).

What is the formula of the compound? What is the structure of the complex ion present? (*Hints:* Cr^{3+} is expected to be six-coordinate with NH_3 and possibly I^- as ligands. The I^- ions will be the counter ions if needed.)

49. Why are CN^- and CO toxic to humans?

50. Using these equilibrium constants:

$$[Cu^{2+}][OH^-]^2 = 1.6 \times 10^{-19}$$

$$\frac{[Cu(NH_3)_4^{2+}]}{[Cu^{2+}][NH_3]^4} = 1.0 \times 10^{13}$$

$$\frac{[Cd(NH_3)_4^{2+}]}{[Cd^{2+}][NH_3]^4} = 1.3 \times 10^{7}$$

$$\frac{[Cd(CN)_4^{2-}]}{[Cd^{2+}][CN^-]^4} = 5.8 \times 10^{18}$$

calculate an equilibrium constant for each of the following:

a. $Cu(OH)_2(s) + 4NH_3 \rightleftharpoons Cu(NH_3)_4^{2+} + 2OH^-$

b. $Cd(NH_3)_4^{2+} + 4CN^- \rightleftharpoons Cd(CN)_4^{2-} + 4NH_3$

(*Hint:* Find reactions that add to give the desired reaction.)

51. Oxalic acid is often used to remove rust stains. Explain what properties of oxalic acid allow it to do this.

52. Why do transition metal ions often have several oxidation states, while other metals generally have one?

53. Compounds of Sc^{3+} are not colored, but those of Ti^{3+} and V^{3+} are. Why?

54. Chelating ligands often form more stable complex ions than the corresponding monodentate ligands with the same donor atoms. For example,

$$Ni^{2+}(aq) + 6NH_3(aq) \rightleftharpoons Ni(NH_3)_6^{2+}(aq) \qquad K_f = 3.2 \times 10^8$$

$$Ni^{2+}(aq) + 3en(aq) \rightleftharpoons Ni(en)_3^{2+}(aq) \qquad K_f = 1.6 \times 10^{18}$$

$$Ni^{2+}(aq) + penten(aq) \rightleftharpoons Ni(penten)^{2+}(aq) \qquad K_f = 2.0 \times 10^{19}$$

where penten is

```
NH2CH2CH2                CH2CH2NH2
         \              /
          N—CH2—CH2—N
         /              \
NH2CH2CH2                CH2CH2NH2
```

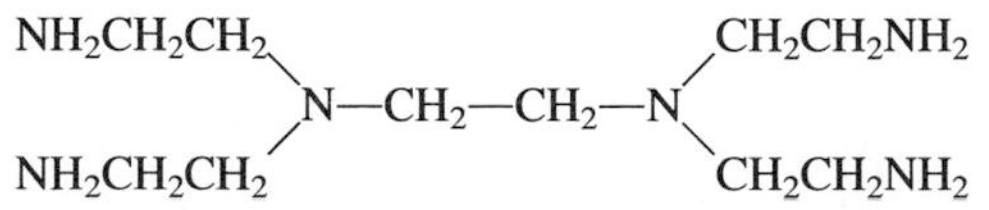

This increased stability is called the chelate effect. Based on bond energies would you expect the enthalpy changes for the above reactions to be very different? What is the order (from least favorable to most favorable) of the entropy changes of the above reactions? How do the values of the formation constants correlate with $\Delta S°$? How can this be used to explain the chelate effect?

55. Acetylacetone (see Exercise 20, part a), abbreviated acacH, is a bidentate ligand. It loses a proton and coordinates as $acac^-$ as shown below:

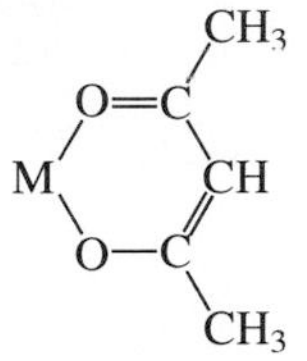

```
          CH3
         /
     O=C
    /     \
  M        CH
    \     //
     O—C
         \
          CH3
```

a. Draw all of the resonance structures for $acac^-$ coordinating to a metal ion.

b. Which of the following complexes are optically active: *cis* $Cr(acac)_2(H_2O)_2$, *trans* $Cr(acac)_2(H_2O)_2$, and $Cr(acac)_3$?

c. Acetylacetone reacts with an ethanol solution of $Eu(NO_3)_3$ to give a compound that is 40.1% C and 4.68% H by mass. Combustion of 0.286 g of the compound gives 0.112 g of Eu_2O_3. What is the molecular formula of the compound formed from the reaction of acetylacetone and europium(III) nitrate? (Assume the compound contains one europium(III) ion.)

56. Draw the structure of the nickel(II) complex of EDTA (see Table 20.13). Is this complex ion optically active?

57. Qualitatively draw the crystal field splitting of the *d*-orbitals in a trigonal planar complex ion. (Let the *z*-axis be perpendicular to the plane of the complex.)

58. Qualitatively draw the crystal field splitting for a trigonal bipyramidal complex ion. (Let the *z*-axis be perpendicular to the trigonal plane.)

59. What is the lanthanide contraction? How does the lanthanide contraction affect the properties of the 4*d* and 5*d* transition metals?

The Nucleus: A Chemist's View

Since the chemistry of an atom is determined by the number and arrangement of its electrons, the properties of the nucleus are not of primary importance to chemists. In the simplest view, the nucleus provides the positive charge to bind the electrons in atoms and molecules. However, a quick reading of any daily newspaper will show you that the nucleus and its properties have an important impact on our society. This chapter considers those aspects of the nucleus about which everyone should have some knowledge.

Several aspects of the nucleus are immediately impressive: its very small size, its very large density, and the magnitude of the energy that holds it together. The radius of a typical nucleus appears to be about 10^{-13} cm. This can be compared to the radius of a typical atom, which is on the order of 10^{-8} cm. A visualization will help you appreciate the small size of the nucleus: if the nucleus of the hydrogen atom were the size of a ping-pong ball, the electron in the 1*s* orbital would be, on the average, 0.5 kilometer (0.3 mile) away. The density of the nucleus is equally impressive—approximately 1.6×10^{14} g/cm^3. A sphere of nuclear material the size of a ping-pong ball would have a mass of 2.5 *billion tons!* In addition, the energies involved in nuclear processes are typically millions of times larger than those associated with normal chemical reactions. This fact makes nuclear processes very attractive for feeding the voracious energy appetite of our civilization.

Atomos, the Greek root of the word atom, means indivisible. It was originally believed that the atom was the ultimate indivisible particle of which all matter was composed. However, as we discussed in Chapter 2, Lord Rutherford showed in 1911 that the atom is not homogeneous, but has a dense, positively charged center surrounded by electrons. Subsequently, scientists have learned that the nucleus of the atom can be subdivided into

CONTENTS

◁ A computer simulation of the head-on collision of two protons, each possessing energy of 20 trillion electron volts.

particles called **neutrons** and **protons.** In fact, in the last two decades it has become apparent that even the protons and neutrons are composed of smaller particles called *quarks*.

For most purposes, the nucleus can be regarded as a collection of **nucleons** (neutrons and protons), and the internal structures of these particles can be ignored. As we discussed in Chapter 2, the number of protons in a particular nucleus is called the **atomic number** (Z) and the sum of the neutrons and protons is the **mass number** (A). Atoms that have identical atomic numbers but different mass number values are called **isotopes.** The general term **nuclide** is applied to each unique atom and is represented using the following symbolism

The atomic number (Z) is the number of protons in a nucleus; the mass number (A) is the sum of protons and neutrons in a nucleus.

$$^{A}_{Z}\mathrm{X}$$

where X represents the symbol for a particular element. For example, the following nuclides constitute the isotopes of carbon: carbon-12 ($^{12}_{6}\mathrm{C}$), carbon-13 ($^{13}_{6}\mathrm{C}$), and carbon-14 ($^{14}_{6}\mathrm{C}$).

21.1 Nuclear Stability and Radioactive Decay

Purpose

- To relate the stability of a nucleus to the number of protons and neutrons.
- To classify the types of radioactive decay.

Nuclear stability is the central topic of this chapter and forms the basis for all the important applications related to nuclear processes. Nuclear stability can be considered from both a kinetic and a thermodynamic point of view. **Thermodynamic stability** refers to the potential energy of a particular nucleus as compared to the sum of the potential energies of its component protons and neutrons. **Kinetic stability** refers to the probability that a nucleus will undergo decomposition to form a different nucleus—a process called **radioactive decay.** We will consider radioactivity in this section.

Many nuclei are radioactive; that is, they decompose forming another nucleus and producing one or more particles. An example is carbon-14, which decays as follows:

$$^{14}_{6}\mathrm{C} \rightarrow {}^{14}_{7}\mathrm{N} + {}^{0}_{-1}\mathrm{e}$$

where ${}^{0}_{-1}\mathrm{e}$ represents an electron, which is called a **beta particle,** or **β particle,** in nuclear terminology. This equation is typical of those representing radioactive decay in that both A and Z must be conserved. That is, the Z values must give the same sum on both sides of the equation ($6 = 7 - 1$), as must the A values ($14 = 14 + 0$).

Of the approximately 2000 known nuclides, only 279 are stable with respect to radioactive decay. Tin has the largest number of stable isotopes—10.

It is instructive to examine how the numbers of neutrons and protons in a nucleus are related to its stability with respect to radioactive decay. Figure 21.1 shows a plot of positions of the stable nuclei as a function of the number of protons

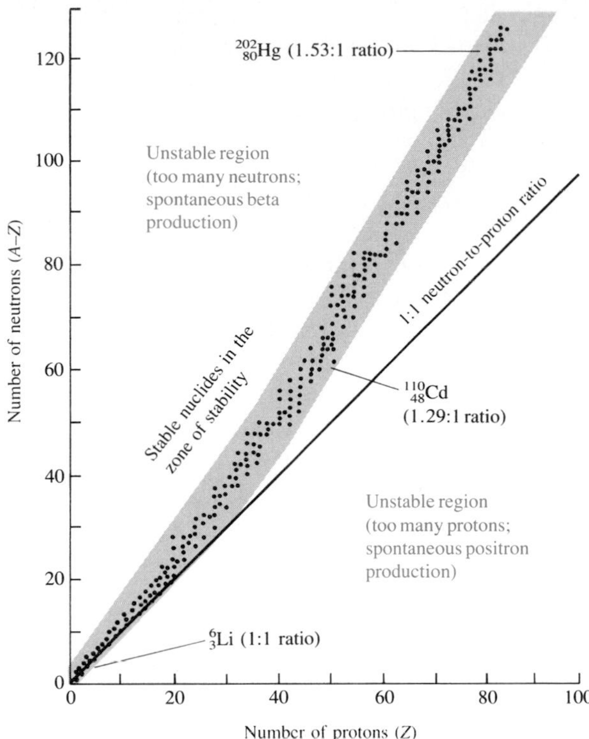

Figure 21.1

The zone of stability. The dots indicate the nuclides that *do not* undergo radioactive decay. Note that as the number of protons in a nuclide increases, the neutron/proton ratio required for stability also increases.

(Z) and the number of neutrons ($A - Z$). The stable nuclides are said to reside in the **zone of stability.**

The following are some important observations concerning radioactive decay:

All nuclides with 84 or more protons are unstable with respect to radioactive decay.

Light nuclides are stable when Z equals $A - Z$, that is, when the neutron/proton ratio is 1. However, for heavier elements the neutron/proton ratio required for stability is greater than 1 and increases with Z.

Certain combinations of protons and neutrons seem to confer special stability. For example, nuclides with even numbers of protons and neutrons are more often stable than those with odd numbers, as shown by the data in Table 21.1.

Number of Stable Nuclides Related to Numbers of Protons and Neutrons

Number of protons	Number of neutrons	Number of stable nuclides	Examples
Even	Even	168	$^{12}_{6}C$, $^{16}_{8}O$
Even	Odd	57	$^{13}_{6}C$, $^{47}_{22}Ti$
Odd	Even	50	$^{19}_{9}F$, $^{23}_{11}Na$
Odd	Odd	4	$^{2}_{1}H$, $^{6}_{3}Li$

NOTE: Even numbers of protons and neutrons seem to favor stability.

Table 21.1

There are also certain specific numbers of protons or neutrons that produce especially stable nuclides. These so-called **magic numbers** are 2, 8, 20, 28, 50, 82, and 126. This behavior parallels that for atoms where certain numbers of electrons (2, 10, 18, 36, 54, and 86) produce special chemical stability (the noble gases).

Types of Radioactive Decay

Radioactive nuclei can undergo decomposition in various ways. These decay processes fall into two categories: those that involve a change in the mass number of the decaying nucleus, and those that do not. We will consider the former type of process first.

α-particle production involves a change in A for the decaying nucleus; β-particle production has no effect on A.

An **alpha particle,** or **α particle,** is a helium nucleus (^4_2He). **Alpha-particle production** is a very common mode of decay for heavy radioactive nuclides. For example, $^{238}_{92}U$, the predominant (99.3%) isotope of natural uranium, decays by α-particle production:

$$^{238}_{92}\text{U} \rightarrow {}^{4}_{2}\text{He} + {}^{234}_{90}\text{Th}$$

Another α-particle producer is $^{230}_{90}Th$:

$$^{230}_{90}\text{Th} \rightarrow {}^{4}_{2}\text{He} + {}^{226}_{88}\text{Ra}$$

Another decay process where the mass number of the decaying nucleus changes is **spontaneous fission,** the splitting of a heavy nuclide into two lighter nuclides with similar mass numbers. Although this process occurs at an extremely slow rate for most nuclides, it is important in some cases, such as for $^{254}_{98}Cf$, where spontaneous fission is the predominant mode of decay.

The most common decay process where the mass number of the decaying nucleus remains constant is **β-particle production.** For example, the thorium-234 nuclide produces a β-particle and protactinium-234:

$$^{234}_{90}\text{Th} \rightarrow {}^{234}_{91}\text{Pa} + {}^{0}_{-1}\text{e}$$

Iodine-131 is also a β-particle producer:

$$^{131}_{53}\text{I} \rightarrow {}^{0}_{-1}\text{e} + {}^{131}_{54}\text{Xe}$$

The β particle is assigned the mass number 0, since its mass is tiny compared to that of a proton or neutron. Since the value of Z is -1 for the β particle, the atomic number for the new nuclide is greater by 1 than for the original nuclide. Thus *the net effect of β-particle production is to change a neutron to a proton.* We therefore expect nuclides that lie above the zone of stability (those nuclides whose neutron/proton ratios are too high) to be β-particle producers.

It should be pointed out that although the β particle is an electron, the emitting nucleus does not contain electrons. As we shall see later in this chapter, a given quantity of energy (which is best regarded as a form of matter) can become a particle (another form of matter) under certain circumstances. The unstable nuclide creates an electron as it releases energy in the decay process. The electron thus results from the decay process rather than being present before the decay occurs. Think of this as somewhat like talking: words are not stored inside us but are formed as we speak. Later in this chapter we will discuss in more detail this very interesting phenomenon where matter in the form of particles and matter in the form of energy can interchange.

A **gamma ray,** or **γ ray,** is a name for a high energy photon. Frequently, γ-ray production accompanies nuclear decays and particle reactions, such as the α-particle decay of $^{238}_{92}U$:

$$^{238}_{92}U \rightarrow {}^{4}_{2}He + {}^{234}_{90}Th + 2\,{}^{0}_{0}\gamma$$

where two γ rays of different energies are produced in addition to the α particle. The emission of γ rays is one way a nucleus with excess energy (in an excited nuclear state) can relax to its ground state.

Positron production occurs for nuclides that are below the zone of stability (those nuclides whose neutron/proton ratios are too small). The positron is a particle with the same mass as the electron but opposite charge. An example of a nuclide that decays by positron production is sodium-22:

$$^{22}_{11}Na \rightarrow {}^{0}_{1}e + {}^{22}_{10}Ne$$

Note that *the net effect is to change a proton to a neutron,* causing the product nuclide to have a higher neutron/proton ratio than the original nuclide did.

Besides being oppositely charged, the positron shows an even more fundamental difference from the electron: it is the *antiparticle* of the electron. When a positron collides with an electron, the particulate matter is changed to electromagnetic radiation in the form of high-energy photons:

$$^{0}_{-1}e + {}^{0}_{1}e \rightarrow {}^{0}_{0}\gamma$$

This process, which is characteristic of matter-antimatter collisions, is called *annihilation* and is another example of the interchange of the forms of matter.

Electron capture is a process where one of the inner-orbital electrons is captured by the nucleus, as illustrated by the process

$$^{201}_{80}Hg + \underset{\text{Inner-orbital electron}}{^{0}_{-1}e} \rightarrow {}^{201}_{79}Au + {}^{0}_{0}\gamma$$

This reaction would have been of great interest to the alchemists, but unfortunately it does not occur at a rate that would make it a practical means for changing mercury to gold. Gamma rays are always produced along with electron capture to release excess energy. The various types of radioactive decay are summarized in Table 21.2.

The Alchemists, an oil on slate painting by Jan van der Straet (1570).

Various Types of Radioactive Processes Showing the Changes That Take Place in the Nuclides

Process	Change in A	Change in Z	Change in neutron/proton ratio	Example
β-particle (electron) production	0	+1	Decrease	$^{227}_{89}Ac \rightarrow {}^{227}_{90}Th + {}^{0}_{-1}e$
Positron production	0	−1	Increase	$^{13}_{7}N \rightarrow {}^{13}_{6}C + {}^{0}_{1}e$
Electron capture	0	−1	Increase	$^{73}_{33}As + {}^{0}_{-1}e \rightarrow {}^{73}_{32}Ge$
α-particle production	−4	−2	Increase	$^{210}_{84}Po \rightarrow {}^{206}_{82}Pb + {}^{4}_{2}He$
γ-ray production	0	0	—	Excited nucleus → ground state nucleus + $^{0}_{0}\gamma$
Spontaneous fission	—	—	—	$^{254}_{98}Cf$ → lighter nuclides + neutrons

Table 21.2

Sample Exercise 21.1

Write balanced equations for each of the following processes:

a. $^{11}_{6}C$ produces a positron.

b. $^{214}_{83}Bi$ produces a β particle.

c. $^{237}_{93}Np$ produces an α particle.

Solution

a. We must find the product nuclide represented by $^{A}_{Z}X$ in the following equation:

$$^{11}_{6}C \rightarrow \underset{\substack{\uparrow \\ \text{Positron}}}{^{0}_{1}e} + ^{A}_{Z}X$$

We can find the identity of $^{A}_{Z}X$ by recognizing that the total of the Z and A values must be the same on both sides of the equation. Thus for X, Z must be $6 - 1 = 5$ and A must be $11 - 0 = 11$. Therefore, $^{A}_{Z}X$ is $^{11}_{5}B$. (The fact that Z is 5 tells us that the nuclide is boron.) Thus the balanced equation is

$$^{11}_{6}C \rightarrow ^{0}_{1}e + ^{11}_{5}B$$

b. Knowing that a β particle is represented by $^{0}_{-1}e$ and that Z and A are conserved, we can write

$$^{214}_{83}Bi \rightarrow ^{0}_{-1}e + ^{214}_{84}X$$

so $^{A}_{Z}X$ must be $^{214}_{84}Po$.

c. Since an α particle is represented by $^{4}_{2}He$, the balanced equation must be

$$^{237}_{93}Np \rightarrow ^{4}_{2}He + ^{233}_{91}Pa$$

Sample Exercise 21.2

In each of the following nuclear reactions, supply the missing particle:

a. $^{195}_{79}Au + ? \rightarrow ^{195}_{78}Pt$

b. $^{38}_{19}K \rightarrow ^{38}_{18}Ar + ?$

Solution

a. Since A does not change and Z decreases by 1, the missing particle must be an electron:

$$^{195}_{79}Au + ^{0}_{-1}e \rightarrow ^{195}_{78}Pt$$

This is an example of electron capture.

b. To conserve Z and A, the missing particle must be a positron:

$$^{38}_{19}K \rightarrow ^{38}_{18}Ar + ^{0}_{1}e$$

Thus potassium-38 decays by positron production.

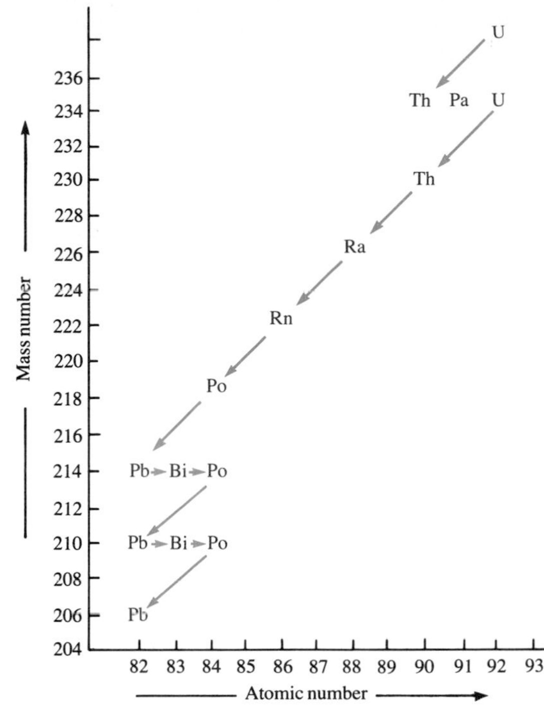

Figure 21.2

The decay series from $^{238}_{92}U$ to $^{206}_{82}Pb$. Each nuclide in the series except $^{206}_{82}Pb$ is unstable, and the successive transformations (shown by the arrows) continue until $^{206}_{82}Pb$ is finally formed. Note that horizontal arrows indicate processes where *A* is unchanged, while diagonal arrows signify that both *A* and *Z* change.

Often a radioactive nucleus cannot reach a stable state through a single decay process. In such a case, a **decay series** occurs until a stable nuclide is formed. A well-known example is the decay series that starts with $^{238}_{92}U$ and ends with $^{206}_{82}Pb$, as shown in Fig. 21.2. Similar series exist for $^{235}_{92}U$:

$$^{235}_{92}U \xrightarrow[\text{decays}]{\text{Series of}} {}^{207}_{82}Pb$$

and for $^{232}_{90}Th$:

$$^{232}_{90}Th \xrightarrow[\text{decays}]{\text{Series of}} {}^{208}_{82}Pb$$

21.2 The Kinetics of Radioactive Decay

Purpose

- To define and show how to calculate the half-life of a radioactive nuclide.

In a sample containing radioactive nuclides of a given type, each nuclide has a certain probability of undergoing decay. Suppose that a sample of 1000 atoms of a certain nuclide produces 10 decay events per hour. This means that over the span of an hour, 1 out of every 100 nuclides will decay. Given that this probability of decay is characteristic for this type of nuclide, we could predict that a 2000-atom sample would give 20 decay events per hour. Thus, for radioactive nuclides the **rate of decay,** which is the negative of the change in the number of nuclides per unit time

Rates of reaction are discussed in Chapter 12.

$$\left(-\frac{\Delta N}{\Delta t}\right)$$

is directly proportional to the number of nuclides (N) in a given sample:

$$\text{Rate} = -\frac{\Delta N}{\Delta t} \propto N$$

The negative sign is included because the number of nuclides is decreasing. We now insert a proportionality constant k to give

$$\text{Rate} = -\frac{\Delta N}{\Delta t} = kN$$

This is the rate law for a first-order process, as we saw in Chapter 12. As shown in Section 12.4, the integrated first-order rate law is

$$\ln\left(\frac{N}{N_0}\right) = -kt$$

where N_0 represents the original number of nuclides (at $t = 0$) and N represents the number *remaining* at time t.

Half-Life

The **half-life** ($t_{1/2}$) of a radioactive sample is defined as the time required for the number of nuclides to reach half of the original value ($N_0/2$). We can use this definition in connection with the integrated first-order rate law (as we did in Section 12.4) to produce the following expression for $t_{1/2}$:

$$t_{1/2} = \frac{\ln(2)}{k} = \frac{0.693}{k}$$

Thus, if the half-life of a radioactive nuclide is known, the rate constant can be easily calculated, and vice versa.

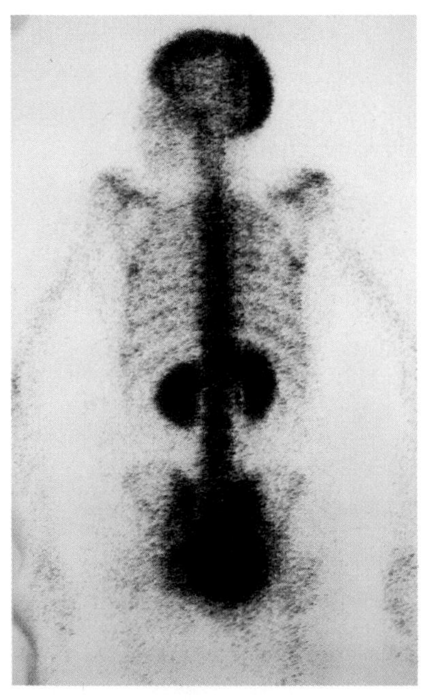

Technetium-99 was used to create this bone scan. The subject has a mild case of arthritis.

Sample Exercise 21.3

Technetium-99 is used to form pictures of internal organs in the body, and is often used to assess heart damage. The rate constant for decay of ${}^{99}_{43}Tc$ is known to be 1.16×10^{-1}/h. What is the half-life of this nuclide?

Solution

The half-life can be calculated from the expression

$$t_{1/2} = \frac{0.693}{k} = \frac{0.693}{1.16 \times 10^{-1}/\text{h}}$$

$$= 6.0 \text{ h}$$

Thus it will take 6.0 h for a given sample of technetium-99 to decrease to half of the original number of nuclides.

Chemical Impact

Stellar Nucleosynthesis

How did all the matter around us originate? The scientific answer to this question is a theory called *stellar nucleosynthesis,* literally, the formation of nuclei in stars.

Many scientists believe that our universe originated as a cloud of neutrons that became unstable and produced an immense explosion, giving this model its name—*the big bang theory*. The model postulates that, following the initial explosion, neutrons decomposed into protons and electrons,

$$^{1}_{0}n \rightarrow {}^{1}_{1}H + {}^{0}_{-1}e$$

which eventually recombined to form clouds of hydrogen. Over the eons, gravitational forces caused many of these hydrogen clouds to contract and heat up sufficiently to reach temperatures where proton fusion was possible, which released large quantities of energy. When the tendency to expand due to the heat from fusion and the tendency to contract due to the forces of gravity are balanced, a stable young star such as our sun can be formed.

Eventually, when the supply of hydrogen is exhausted, the core of the star will again contract with further heating until temperatures are reached where fusion of helium nuclei can occur, leading to the formation of $^{12}_{6}C$ and $^{16}_{8}O$ nuclei. In turn, when the supply of helium nuclei runs out, further contraction and heating will occur, until fusion of heavier nuclei takes place. This process occurs repeatedly, forming heavier and heavier nuclei until iron nuclei are formed. Because the iron nucleus is the most stable of all, energy is required to fuse iron nuclei. This endothermic fusion process cannot furnish energy to sustain the star, and therefore it cools to a small, dense *white dwarf*.

The evolution just described is characteristic of small and medium-sized stars. Much larger stars, however, become unstable at some time during their evolution and undergo a *supernova explosion*. In this explosion, some medium-mass nuclei are fused to form heavy elements. Also, some light nuclei capture neutrons. These neutron-rich nuclei then produce β particles, increasing their atomic number with each event. This eventually leads to heavy nuclei. In fact, almost all nuclei heavier than iron are thought to originate from supernova explosions. The debris of a supernova explosion thus contains a large variety of elements and might eventually form a solar system such as our own.

Although other theories for the origin of matter have been suggested, there is much evidence to support the big bang theory, and it continues to be widely accepted.

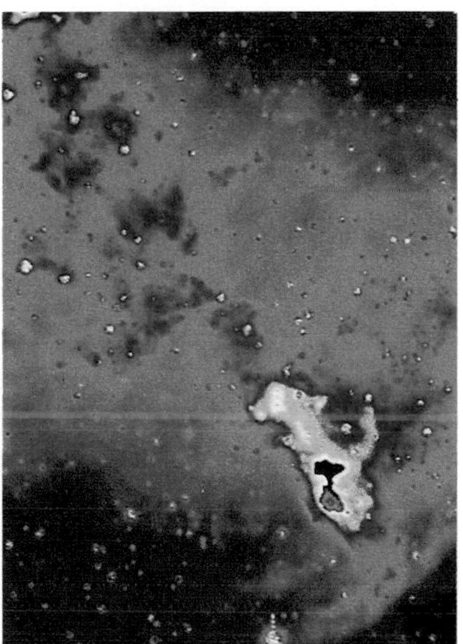

The Dumbbell Nebula, a planetary nebula 1200 light years from earth.

As we saw in Section 12.4, the half-life for a first-order process is constant. This is shown for the β-particle decay of strontium-90 in Fig. 21.3; it takes 28.8 years for each halving of the amount of $^{90}_{38}Sr$. Contamination of the environment with $^{90}_{38}Sr$ poses serious health hazards because of the similar chemistry of strontium and calcium (both are in Group 2A). Strontium-90 in grass and hay is incorporated into cow's milk along with calcium and is then passed on to humans, where it lodges in the bones. Because of its relatively long half-life, it persists for years in humans, causing radiation damage that may lead to cancer.

The harmful effects of radiation will be discussed in Section 21.7.

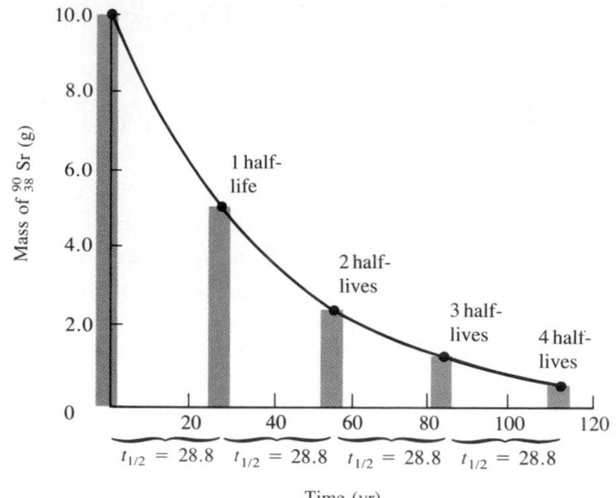

Figure 21.3

The decay of a 10.0-g sample of strontium-90 over time. Note that the half-life is a constant 28.8 years.

Sample Exercise 21.4

The half-life of molybdenum-99 is 67.0 h. How much of a 1.000-mg sample of $^{99}_{42}Mo$ is left after 335 h?

Solution

The easiest way to solve this problem is to recognize that 335 h represents five half-lives for $^{99}_{42}Mo$:

$$335 = 5 \times 67.0$$

We can sketch the change that occurs, as is shown in Fig. 21.4. Thus after 335 h, 0.031 mg of $^{99}_{42}Mo$ remains.

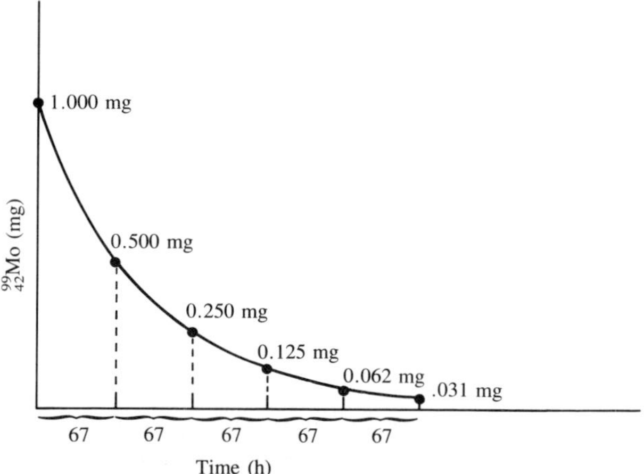

Figure 21.4

The change in the amount of $^{99}_{42}Mo$ with time ($t_{1/2} = 67$ h).

The half-lives of radioactive nuclides vary over a tremendous range. For example, $^{144}_{60}Nd$ has a half-life of 5×10^{15} years, while $^{214}_{84}Po$ has a half-life of 2×10^{-4} second. To give you some perspective on this, the half-lives of the nuclides in the $^{238}_{92}U$ decay series are given in Table 21.3

The Half-Lives of Nuclides in the $^{238}_{92}U$ Decay Series

Nuclide	Particle produced	Half-life
Uranium-238 ($^{238}_{92}U$) ↓	α	4.51×10^9 years
Thorium-234 ($^{234}_{90}Th$) ↓	β	24.1 days
Protactinium-234 ($^{234}_{91}Pa$) ↓	β	6.75 hours
Uranium-234 ($^{234}_{92}U$) ↓	α	2.48×10^5 years
Thorium-230 ($^{230}_{90}Th$) ↓	α	8.0×10^4 years
Radium-226 ($^{226}_{88}Ra$) ↓	α	1.62×10^3 years
Radon-222 ($^{222}_{86}Rn$) ↓	α	3.82 days
Polonium-218 ($^{218}_{84}Po$) ↓	α	3.1 minutes
Lead-214 ($^{214}_{82}Pb$) ↓	β	26.8 minutes
Bismuth-214 ($^{214}_{83}Bi$) ↓	β	19.7 minutes
Polonium-214 ($^{214}_{84}Po$) ↓	α	1.6×10^{-4} second
Lead-210 ($^{210}_{82}Pb$) ↓	β	20.4 years
Bismuth-210 ($^{210}_{83}Bi$) ↓	β	5.0 days
Polonium-210 ($^{210}_{84}Po$) ↓	α	138.4 days
Lead-206 ($^{206}_{82}Pb$)	—	Stable

Table 21.3

21.3 Nuclear Transformations

Purpose

- To show how one element may be changed into another using particle bombardment.

In 1919 Lord Rutherford observed the first **nuclear transformation,** *the change of one element into another*. He found that by bombarding $^{14}_{7}N$ with α particles, the nuclide $^{17}_{8}O$ could be produced:

$$^{14}_{7}N + ^{4}_{2}He \rightarrow ^{17}_{8}O + ^{1}_{1}H$$

Fourteen years later, Irene Curie and her husband Frederick Joliot observed a similar transformation from aluminum to phosphorus,

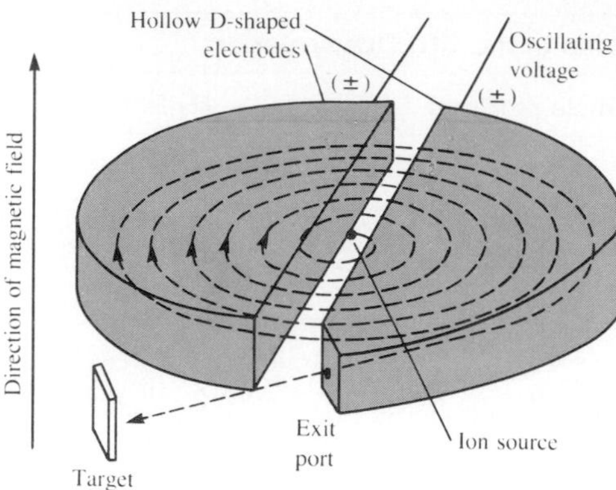

Figure 21.5

A schematic diagram of a cyclotron. The ion is introduced in the center and is pulled back and forth between the hollow D-shaped electrodes by constant reversals of the electric field. Magnets above and below these electrodes produce a spiral path that expands as the particle velocity increases. When the particle has sufficient speed, it exits the accelerator and is directed at the target nucleus.

$$^{27}_{13}\text{Al} + ^{4}_{2}\text{He} \rightarrow ^{30}_{15}\text{P} + ^{1}_{0}\text{n}$$

where $^{1}_{0}\text{n}$ represents a neutron.

Over the years, many other nuclear transformations have been achieved, mostly using **particle accelerators,** which, as the name reveals, are devices used to give particles very high velocities. Because of the electrostatic repulsion between the target nucleus and a positive ion, accelerators are needed when positive ions are used as bombarding particles. The particle, accelerated to a very high velocity, can overcome the repulsion and penetrate the target nucleus, thus effecting the transformation. A schematic diagram of one type of particle accelerator, the **cyclotron,** is shown in Fig. 21.5. The ion is introduced at the center of the cyclotron and is accelerated in an expanding spiral path by use of alternating electrical fields in the presence of a magnetic field. The **linear accelerator** illustrated in Fig. 21.6 employs changing electrical fields to achieve high velocities on a linear pathway.

In addition to positive ions, neutrons are often employed as bombarding particles to effect nuclear transformations. Because neutrons are uncharged and thus not repelled electrostatically by a target nucleus, they are readily absorbed by many nuclei, leading to new nuclides. The most common source of neutrons for this purpose is a fission reactor (see Section 21.6).

By using neutron and positive-ion bombardment, scientists have been able to extend the periodic table. Prior to 1940, the heaviest known element was uranium ($Z = 92$), but in 1940, neptunium ($Z = 93$) was produced by neutron bombardment of $^{238}_{92}\text{U}$. The process initially gives $^{239}_{92}\text{U}$, which decays to $^{239}_{93}\text{Np}$ by β-particle production:

$$^{238}_{92}\text{U} + ^{1}_{0}\text{n} \rightarrow ^{239}_{92}\text{U} \xrightarrow[t_{1/2} = 23 \text{ min}]{} ^{239}_{93}\text{Np} + ^{0}_{-1}\text{e}$$

In the years since 1940, the elements with atomic numbers 93 through 106, called the **transuranium elements,** have been synthesized. Many of these elements

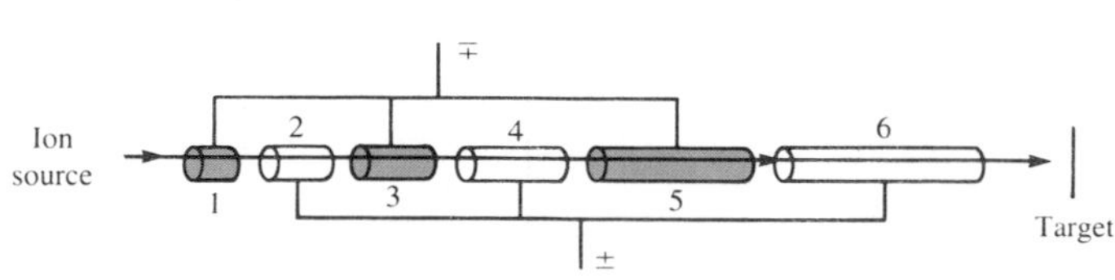

Figure 21.6

Schematic diagram of a linear accelerator, which uses a changing electrical field to accelerate a positive ion along a linear path. As the ion leaves the source, the odd-numbered tubes are negatively charged, and the even-numbered tubes are positively charged. The positive ion is thus attracted into tube 1. As the ion leaves tube 1, the tube polarities are reversed. Now tube 1 is positive, repelling the positive ion; and tube 2 is negative, attracting the positive ion. This process continues, eventually producing high particle velocity.

Syntheses of Some of the Transuranium Elements		
Element	Neutron bombardment	Half-life
Neptunium (Z = 93)	$^{238}_{92}U + ^{1}_{0}n \rightarrow ^{239}_{92}U \rightarrow ^{239}_{93}Np + ^{0}_{-1}e$	2.35 days ($^{239}_{93}Np$)
Plutonium (Z = 94)	$^{239}_{93}Np \rightarrow ^{239}_{94}Pu + ^{0}_{-1}e$	24,400 years ($^{239}_{94}Pu$)
Americium (Z = 95)	$^{239}_{94}Pu + 2\ ^{1}_{0}n \rightarrow ^{241}_{94}Pu \rightarrow ^{241}_{95}Am + ^{0}_{-1}e$	458 years ($^{241}_{95}Am$)
Element	Positive-ion bombardment	Half-life
Curium (Z = 96)	$^{239}_{94}Pu + ^{4}_{2}He \rightarrow ^{242}_{96}Cm + ^{1}_{0}n$	163 days ($^{242}_{96}Cm$)
Californium (Z = 98)	$^{242}_{96}Cm + ^{4}_{2}He \rightarrow ^{245}_{98}Cf + ^{1}_{0}n$ or $^{238}_{92}U + ^{12}_{6}C \rightarrow ^{246}_{98}Cf + 4\ ^{1}_{0}n$	44 minutes ($^{245}_{98}Cf$)
Element 104	$^{249}_{98}Cf + ^{12}_{6}C \rightarrow ^{257}_{104}X + 4\ ^{1}_{0}n$	
Element 105	$^{249}_{98}Cf + ^{15}_{7}N \rightarrow ^{260}_{105}X + 4\ ^{1}_{0}n$	
Element 106	$^{249}_{98}Cf + ^{18}_{8}O \rightarrow ^{263}_{106}X + 4\ ^{1}_{0}n$	

Table 21.4

have very short half-lives, as shown in Table 21.4. As a result, only a few atoms of some have ever been formed. This of course makes the chemical characterization of these elements extremely difficult.

21.4 Detection and Uses of Radioactivity

Purpose

- To discuss radioactivity detection devices.
- To show how objects can be dated using radioactive decay.

Although various instruments measure radioactivity levels the most familiar of them is the **Geiger-Müller counter,** or **Geiger counter** (see Fig. 21.7). This instrument takes advantage of the fact that the high-energy particles from radioactive decay processes produce ions when they travel through matter. The probe of the Geiger counter is filled with argon gas, which can be ionized by a rapidly moving particle as illustrated below. This reaction is demonstrated by the equation:

Geiger counters are often called survey meters in industry.

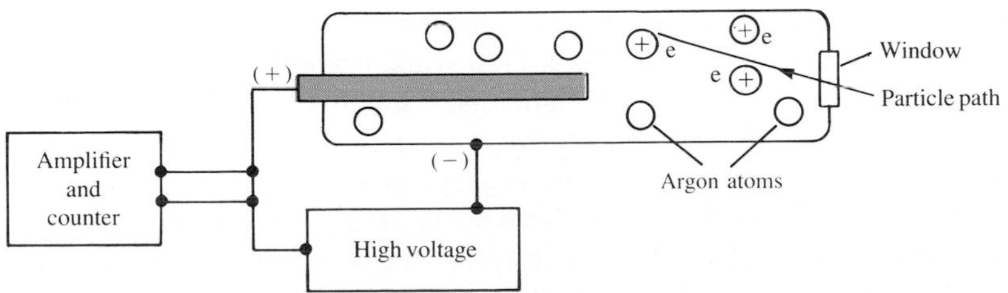

Figure 21.7

A schematic representation of a Geiger-Müller counter. The high-energy radioactive particle enters the window and ionizes argon atoms along its path. The resulting ions and electrons produce a momentary current pulse, which is amplified and counted.

$$\text{Ar}(g) \xrightarrow[\text{particle}]{\text{High-energy}} \text{Ar}^+(g) + \text{e}^-$$

Normally, a sample of argon gas will not conduct a current when an electrical potential is applied. However, the formation of ions and electrons produced by the passage of the high-energy particle allows a momentary current to flow. Electronic devices detect this current flow, and the number of these events can be counted. Thus the decay rate of the radioactive sample can be determined.

Another instrument often used to detect levels of radioactivity is a **scintillation counter,** which takes advantage of the fact that certain substances, such as zinc sulfide, give off light when they are struck by high-energy radiation. A photocell senses the flashes of light that occur as the radiation strikes and thus measures the number of decay events per unit of time.

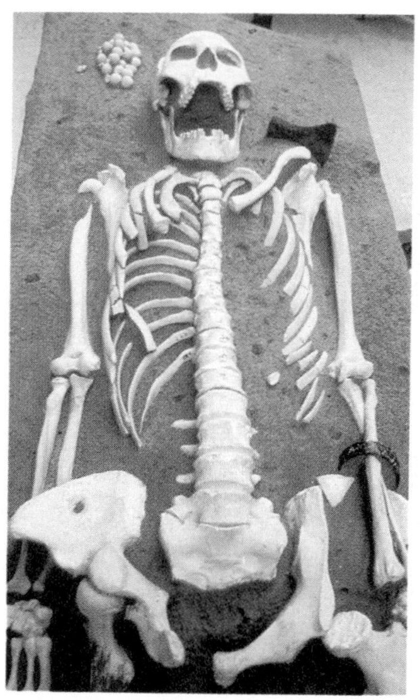

Carbon-14 radioactivity is often used to date human skeletons found at archeological sites.

Dating by Radioactivity

Archaeologists, geologists, and others involved in reconstructing the ancient history of the earth rely heavily on radioactivity to provide accurate dates for artifacts and rocks. A method that has been very important for dating ancient articles made from wood or cloth is **radiocarbon dating,** or **carbon-14 dating,** a technique originated in the 1940s by Willard Libby, an American chemist who received a Nobel Prize for his efforts in this field.

Radiocarbon dating is based on the radioactivity of the nuclide ${}^{14}_{6}\text{C}$, which decays via β-particle production:

$${}^{14}_{6}\text{C} \rightarrow {}^{0}_{-1}\text{e} + {}^{14}_{7}\text{N}$$

Carbon-14 is continuously produced in the atmosphere when high-energy neutrons from space collide with nitrogen-14:

$${}^{14}_{7}\text{N} + {}^{1}_{0}\text{n} \rightarrow {}^{14}_{6}\text{C} + {}^{1}_{1}\text{H}$$

Thus, carbon-14 is continuously produced by this process, and it continuously decomposes through β-particle production. Over the years, the rates for these two processes have become equal, and like a participant in a chemical reaction at equilibrium, the amount of ${}^{14}_{6}\text{C}$ that is present in the atmosphere remains approximately constant.

Carbon-14 can be used to date wood and cloth artifacts because the ${}^{14}_{6}\text{C}$, along with the other carbon isotopes in the atmosphere, reacts with oxygen to form carbon dioxide. A living plant consumes carbon dioxide in the photosynthesis process, and incorporates the carbon, including ${}^{14}_{6}\text{C}$, into its molecules. As long as the plant lives, the ${}^{14}_{6}\text{C}/{}^{12}_{6}\text{C}$ ratio in its molecules remains the same as in the atmosphere because of the continuous uptake of carbon. However, as soon as a tree is cut to make a wooden bowl or a flax plant is harvested to make linen, the ${}^{14}_{6}\text{C}/{}^{12}_{6}\text{C}$ ratio begins to decrease because of the radioactive decay of ${}^{14}_{6}\text{C}$ (the ${}^{12}_{6}\text{C}$ nuclide is stable). Since the half-life of ${}^{14}_{6}\text{C}$ is 5730 years, a wooden bowl found in an archaeological dig that shows a ${}^{14}_{6}\text{C}/{}^{12}_{6}\text{C}$ ratio that is half that found in currently living trees is approximately 5730 years old. This reasoning assumes that the current ${}^{14}_{6}\text{C}/{}^{12}_{6}\text{C}$ ratio is the same as that found in ancient times.

The ${}^{14}_{6}\text{C}/{}^{12}_{6}\text{C}$ ratio is the basis for carbon-14 dating.

Dendrochronologists, scientists who date trees from annual growth rings, have used data collected from long-lived species of trees, such as bristlecone pines and sequoias, to show that the ${}^{14}_{6}\text{C}$ content of the atmosphere has changed significantly over the ages. These data have been used to derive correction factors that allow very accurate dates to be determined from the observed ${}^{14}_{6}C/{}^{12}_{6}C$ ratio in an artifact.

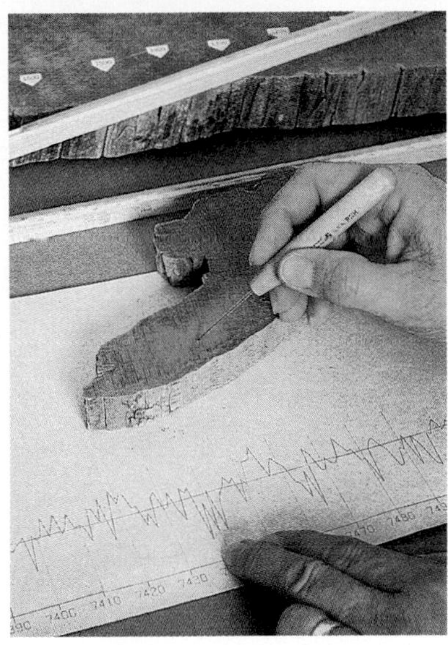

A dendochronologist examines the tree rings of a bristlecone pine tree.

Sample Exercise 21.5

The remnants of an ancient fire in a cave in Africa showed a ${}^{14}_{6}C$ decay rate of 3.1 counts per minute per gram of carbon. Assuming that the decay rate of ${}^{14}_{6}C$ in freshly cut wood (corrected for changes in the ${}^{14}_{6}C$ content of the atmosphere) is 13.6 counts per minute per gram of carbon, calculate the age of the remnants. The half-life of ${}^{14}_{6}C$ is 5730 years.

Solution

The key to solving this problem is to realize that the decay rates given are directly proportional to the number of ${}^{14}C$ nuclides present. Radioactive decay follows first-order kinetics:

$$\text{Rate} = kN$$

Thus

$$\frac{3.1 \text{ counts/min g}}{13.6 \text{ counts/min g}} = \frac{\text{rate at time } t}{\text{rate at time } 0} = \frac{kN}{kN_0}$$

(N: Number of nuclides present at time t; N_0: Number of nuclides present at time 0)

$$= \frac{N}{N_0} = 0.23$$

We can now use the integrated first-order rate law:

$$\ln\left(\frac{N}{N_0}\right) = -kt$$

where

$$k = \frac{0.693}{t_{1/2}} = \frac{0.693}{5730 \text{ yr}}$$

to solve for t, the time elapsed since the campfire.

$$\ln\left(\frac{N}{N_0}\right) = \ln(0.23) = -\left(\frac{0.693}{5730 \text{ yr}}\right)t$$

Solving this equation gives $t = 12{,}000$ yr; the campfire in the cave occurred about 12,000 yr ago.

One drawback of radiocarbon dating is that a fairly large piece of the object (from a half to several grams) must be burned to form carbon dioxide, which is then analyzed for radioactivity. Another method for counting ${}^{14}_{6}C$ nuclides avoids destruction of a significant portion of a valuable artifact. This technique, requiring only about 10^{-3} gram, uses a mass spectrometer (Chapter 3), in which the carbon

Because the half-life of $^{238}_{92}U$ is very long compared to those of the other members of the decay series (Table 21.3) to reach $^{206}_{82}Pb$, the number of nuclides in intermediate stages of decay is negligible. That is, once a $^{238}_{92}U$ nuclide starts to decay, it reaches $^{206}_{82}Pb$ relatively fast.

atoms are ionized and accelerated through a magnetic field which deflects their path. Because of their different masses, the various ions are deflected by different amounts and can be counted separately. This allows a very accurate determination of the $^{14}_{6}C/^{12}_{6}C$ ratio in the sample.

In their attempts to establish the geological history of the earth, geologists have made extensive use of radioactivity. For example, since $^{238}_{92}U$ decays to the stable isotope $^{206}_{82}Pb$, the ratio of $^{206}_{82}Pb$ to $^{238}_{92}U$ in a rock can, under favorable circumstances, be used to estimate the age of the rock.

Sample Exercise 21.6

A rock containing $^{238}_{92}U$ and $^{206}_{82}Pb$ was examined to determine its approximate age. Analysis showed the ratio of $^{206}_{82}Pb$ atoms to $^{238}_{92}U$ atoms to be 0.115. Assuming that no lead was originally present, that all the $^{206}_{82}Pb$ formed over the years has remained in the rock, and that the number of nuclides in intermediate stages of decay between $^{238}_{92}U$ and $^{206}_{82}Pb$ is negligible, calculate the age of the rock. The half-life of $^{238}_{92}U$ is 4.5×10^9 yr.

Solution

This problem can be solved using the integrated first-order rate law:

$$\ln\left(\frac{N}{N_0}\right) = -kt = -\left(\frac{0.693}{4.5 \times 10^9 \text{ yr}}\right)t$$

where N/N_0 represents the ratio of $^{238}_{92}U$ atoms now found in the rock to the number present when the rock was formed. We are assuming that each $^{206}_{82}Pb$ nuclide present must have come from decay of a $^{238}_{92}U$:

$$^{238}_{92}\text{U} \rightarrow {}^{206}_{82}\text{Pb}$$

Thus

$$\boxed{\begin{array}{c}\text{Number of } ^{238}_{92}\text{U atoms}\\ \text{originally present}\end{array}} = \boxed{\begin{array}{c}\text{number of } ^{206}_{82}\text{Pb atoms}\\ \text{now present}\end{array}} + \boxed{\begin{array}{c}\text{number of } ^{238}_{92}\text{U atoms}\\ \text{now present}\end{array}}$$

$$\frac{\text{Atoms of } ^{206}_{82}\text{Pb now present}}{\text{Atoms of } ^{238}_{92}\text{U now present}} = 0.115 = \frac{0.115}{1.000} = \frac{115}{1000}$$

Think carefully about what this means. For every 1115 $^{238}_{92}U$ atoms originally present in the rock, 115 have been changed to $^{206}_{82}Pb$ and 1000 remain as $^{238}_{92}U$. Thus

$$\frac{N}{N_0} = \frac{\overset{\text{Now present}}{^{238}_{92}\text{U}}}{\underbrace{^{206}_{82}\text{Pb} + {}^{238}_{92}\text{U}}_{^{238}_{92}\text{U originally present}}} = \frac{1000}{1115} = 0.8969$$

$$\ln\left(\frac{N}{N_0}\right) = \ln(0.8969) = -\left(\frac{0.693}{4.5 \times 10^9 \text{ yr}}\right)t$$

$$t = 7.1 \times 10^8 \text{ yr}$$

This is the approximate age of the rock. It was formed sometime in the Cambrian period.

Medical Applications of Radioactivity

Although the rapid advances of the medical sciences in recent decades are due to many causes, one of the most important has been the discovery and use of **radiotracers,** radioactive nuclides that can be introduced into organisms in food or drugs and whose pathways can be *traced* by monitoring their radioactivity. For example, the incorporation of nuclides such as $^{14}_{6}C$ and $^{32}_{15}P$ into nutrients has produced important information about metabolic pathways.

Iodine-131 has proved very useful in the diagnosis and treatment of illnesses of the thyroid gland. Patients drink a solution containing small amounts of $Na^{131}I$ and the uptake of the iodine by the thyroid gland is monitored with a scanner (see Fig. 21.8).

Thallium-201 can be used to assess the damage to the heart muscle in a person who has suffered a heart attack, because thallium is concentrated in healthy muscle tissue. Technetium-99 is also taken up by normal heart tissue and is used for damage assessment in a similar way.

Radiotracers provide sensitive and noninvasive methods for learning about biological systems, for detection of disease, for monitoring the action and effectiveness of drugs, and for early detection of pregnancy; and their usefulness should continue to grow. Some useful radiotracers are listed in Table 21.5.

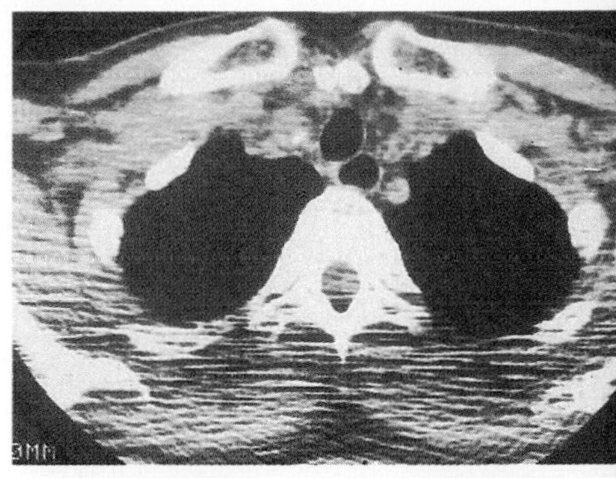

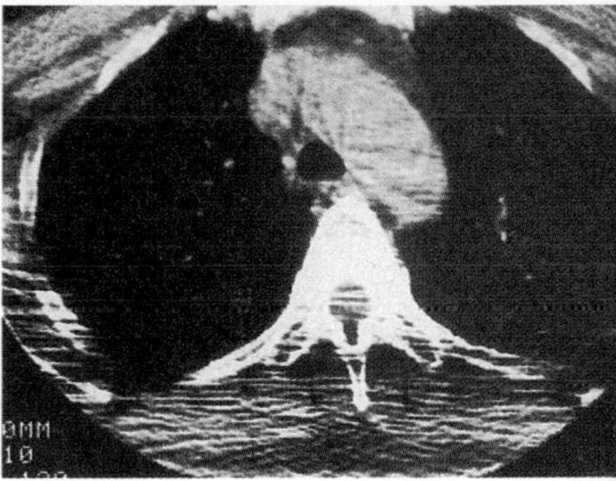

Figure 21.8

After consumption of $Na^{131}I$, the patient's thyroid is scanned for radioactivity levels to determine the efficiency of iodine absorption. (top) A normal thyroid. (bottom) An enlarged thyroid.

Some Radioactive Nuclides, with Half-Lives and Medical Applications as Radiotracers

Nuclide	Half-life	Area of the body studied
^{131}I	8.1 days	Thyroid
^{59}Fe	45.1 days	Red blood cells
^{99}Mo	67 hours	Metabolism
^{32}P	14.3 days	Eyes, liver, tumors
^{51}Cr	27.8 days	Red blood cells
^{87}Sr	2.8 hours	Bones
^{99}Tc	6.0 hours	Heart, bones, liver, and lungs
^{133}Xe	5.3 days	Lungs
^{24}Na	14.8 hours	Circulatory system

Table 21.5

21.5 Thermodynamic Stability of the Nucleus

Purpose

- To discuss the thermodynamic stability of the nucleus.
- To show how to calculate nuclear binding energies.

We can determine the thermodynamic stability of a nucleus by calculating the change in potential energy that would occur if that nucleus were formed from its constituent protons and neutrons. For example, let's consider the hypothetical process of forming a $^{16}_{8}O$ nucleus from eight neutrons and eight protons:

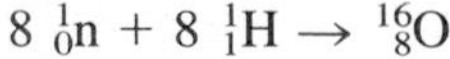

$$8\ ^{1}_{0}n + 8\ ^{1}_{1}H \rightarrow\ ^{16}_{8}O$$

The energy change associated with this process can be calculated by comparing the sum of the masses of eight protons and eight neutrons to that of the oxygen nucleus:

$$\text{Mass of } (8\ {}^{1}_{0}\text{n} + 8\ {}^{1}_{1}\text{H}) = 8(\underset{\text{Mass of } {}^{1}_{0}\text{n}}{1.67493 \times 10^{-24}\text{ g}}) + 8(\underset{\text{Mass of } {}^{1}_{1}\text{H}}{1.67262 \times 10^{-24}\text{ g}})$$

$$= 2.67804 \times 10^{-23}\text{ g}$$

$$\text{Mass of } {}^{16}_{8}\text{O nucleus} = 2.65535 \times 10^{-23}\text{ g}$$

The difference in mass for one nucleus is

$$\text{Mass of } {}^{16}_{8}\text{O} - \text{mass of } (8\ {}^{1}_{0}\text{n} + 8\ {}^{1}_{1}\text{H}) = -2.269 \times 10^{-25}\text{ g}$$

The difference in mass for formation of 1 mole of ${}^{16}_{8}$O nuclei is therefore

$$(-2.269 \times 10^{-25}\text{ g/nucleus})(6.022 \times 10^{23}\text{ nuclei/mol}) = -0.1366\text{ g/mol}$$

Thus 0.1366 g of mass would be lost when 1 mole of oxygen-16 was formed from protons and neutrons. Why does this happen and how can this information be used to calculate the energy change that accompanies this process?

Energy is a form of matter.

The answers to these questions can be found in the work of Albert Einstein. As we discussed in Section 7.2, Einstein's theory of relativity showed that energy should be considered a form of matter. His famous equation,

$$E = mc^2$$

where c is the speed of light, gives the relationship between a quantity of energy and its mass. When a system gains or loses energy, it also gains or loses a quantity of mass, given by E/c^2. Thus the mass of a nucleus is less than that of its component nucleons because the process is so exothermic.

The energy changes associated with normal chemical reactions are small enough that the corresponding mass changes are not detectable.

Einstein's equation in the form

$$\text{Energy change} = \Delta E = \Delta mc^2$$

where Δm is the change in mass, or the **mass defect,** can be used to calculate ΔE for the formation of a nucleus from its component nucleons.

Sample Exercise 21.7

Calculate the change in energy when 1 mol of ${}^{16}_{8}$O nuclei is formed from neutrons and protons.

Solution

We have already calculated that 0.1366 g of mass would be lost in the hypothetical process of assembling 1 mol of ${}^{16}_{8}$O nuclei from the component nucleons. We can calculate the change in energy for this process from

$$\Delta E = \Delta mc^2$$

where

$$c = 3.00 \times 10^8\text{ m/s} \quad \text{and} \quad \Delta m = 0.1366\text{ g/mol} = 1.366 \times 10^{-4}\text{ kg/mol}$$

Thus

$$\Delta E = (-1.366 \times 10^{-4}\text{ kg/mol})(3.00 \times 10^8\text{ m/s})^2 = -1.23 \times 10^{13}\text{ J/mol}$$

Sample Exercise 21.7, continued

The negative sign for the ΔE value indicates that the process is exothermic. Energy, and thus mass, is lost from the system.

The energy changes observed for nuclear processes are extremely large compared to those observed for chemical and physical changes. Thus nuclear processes constitute a potentially valuable energy resource.

The thermodynamic stability of a particular nucleus is normally represented as energy released per nucleon. To illustrate how this quantity is obtained, we will continue to consider ${}^{16}_{8}O$. First, we calculate ΔE per nucleus by dividing the molar value from Sample Exercise 21.7 by Avogadro's number:

$$\Delta E \text{ per } {}^{16}_{8}\text{O nucleus} = \frac{-1.23 \times 10^{13} \text{ J/mol}}{6.022 \times 10^{23} \text{ nuclei/mol}} = -2.04 \times 10^{-11} \text{ J/nucleus}$$

In terms of a more convenient energy unit, a million electron volts (MeV), where

$$1 \text{ MeV} = 1.60 \times 10^{-13} \text{ J}$$

$$\Delta E \text{ per } {}^{16}_{8}\text{O nucleus} = (-2.04 \times 10^{-11} \text{ J/nucleus})\left(\frac{1 \text{ MeV}}{1.60 \times 10^{-13} \text{ J}}\right)$$

$$= -1.28 \times 10^{2} \text{ MeV/nucleus}$$

Next, we can calculate the value of ΔE per nucleon by dividing by A, the sum of neutrons and protons:

$$\Delta E \text{ per nucleon for } {}^{16}_{8}\text{O} = \frac{-1.28 \times 10^{2} \text{ MeV/nucleus}}{16 \text{ nucleons/nucleus}}$$

$$= -7.98 \text{ MeV/nucleon}$$

This means that 7.98 MeV of energy per nucleon would be *released* if ${}^{16}_{8}O$ were formed from neutrons and protons. The energy required to *decompose* this nucleus into its components has the same numeric value but with a positive sign (since energy is required). This is called the **binding energy** per nucleon for ${}^{16}_{8}O$.

The values of the binding energy per nucleon for the various nuclides are shown in Fig. 21.9 on the following page. Note that the most stable nuclei (those requiring the largest energy per nucleon to decompose the nucleus) occur at the top of the curve. The most stable nucleus known is ${}^{56}_{26}Fe$, which has a binding energy per nucleon of 8.79 MeV.

Sample Exercise 21.8

Calculate the binding energy per nucleon for the ${}^{4}_{2}He$ nucleus (atomic masses: ${}^{4}_{2}He = 4.0026$ amu; ${}^{1}_{1}H = 1.0078$ amu).

Solution

First, we must calculate the mass defect (Δm) for ${}^{4}_{2}He$. Since atomic masses (which include the electrons) are given, we must decide how to account for the electron mass:

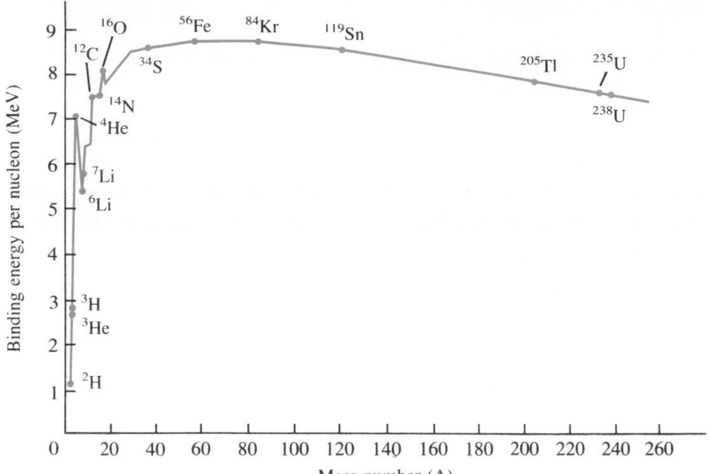

Figure 21.9

The binding energy per nucleon as a function of mass number. The most stable nuclei are at the top of the curve. The most stable nucleus is $^{56}_{26}Fe$.

Sample Exercise 21.8, continued

$$4.0026 = \text{mass of } ^4_2\text{He atom} = \text{mass of } ^4_2\text{He nucleus} + 2m_e$$

Electron mass

$$1.0078 = \text{mass of } ^1_1\text{H atom} = \text{mass of } ^1_1\text{H nucleus} + m_e$$

Thus, since a $^4_2$He nucleus is ''synthesized'' from two protons and two neutrons, we see that

$$\Delta m = \underbrace{(4.0026 - 2m_e)}_{\text{Mass of } ^4_2\text{He nucleus}} - [2\underbrace{(1.0078 - m_e)}_{\substack{\text{Mass of } ^1_1\text{H nucleus} \\ \text{(proton)}}} + 2\underbrace{(1.0087)}_{\text{Mass of neutron}}]$$

$$= 4.0026 - 2m_e - 2(1.0078) + 2m_e - 2(1.0087)$$

$$= 4.0026 - 2(1.0078) - 2(1.0087)$$

$$= -0.0304 \text{ amu}$$

Note that in this case the electron mass cancels out in taking the difference. This will always happen in this type of calculation if the atomic masses are used both for the nuclide of interest and for $^1_1$H. Thus 0.0304 amu of mass is *lost* per $^4_2$He nucleus formed.

The corresponding energy change can be calculated from

$$\Delta E = \Delta mc^2$$

where

$$\Delta m = -0.0304 \frac{\text{amu}}{\text{nucleus}} = \left(-0.0304 \frac{\text{amu}}{\text{nucleus}}\right)\left(1.66 \times 10^{-27} \frac{\text{kg}}{\text{amu}}\right)$$

$$= -5.04 \times 10^{-29} \frac{\text{kg}}{\text{nucleus}}$$

and

$$c = 3.00 \times 10^8 \text{ m/s}$$

Sample Exercise 21.8, continued

Thus $$\Delta E = \left(-5.04 \times 10^{-29} \frac{\text{kg}}{\text{nucleus}}\right)\left(3.00 \times 10^{8} \frac{\text{m}}{\text{s}}\right)^2$$

$$= -4.54 \times 10^{-12} \text{ J/nucleus}$$

This means that 4.54×10^{-12} J of energy is *released* per nucleus formed and that 4.54×10^{-12} J would be required to decompose the nucleus into the constituent neutrons and protons. Thus the binding energy (BE) per nucleon is

$$\text{BE per nucleon} = \frac{4.54 \times 10^{-12} \text{ J/nucleus}}{4 \text{ nucleons/nucleus}}$$

$$= 1.14 \times 10^{-12} \text{ J/nucleon}$$

$$= \left(1.14 \times 10^{-12} \frac{\text{J}}{\text{nucleon}}\right)\left(\frac{1 \text{ MeV}}{1.60 \times 10^{-13} \text{ J}}\right)$$

$$= 7.12 \text{ MeV/nucleon}$$

21.6 Nuclear Fission and Nuclear Fusion

Purpose

- To explore the energetics of nuclear fission and nuclear fusion.
- To describe how a nuclear reactor works.

The graph shown in Fig. 21.9 has very important implications for the use of nuclear processes as sources of energy. Recall that energy is released, that is, ΔE is negative, when a process goes from a less stable to a more stable state. The higher a nuclide is on the curve, the more stable it is. This means that two types of nuclear processes will be exothermic (see Fig. 21.10):

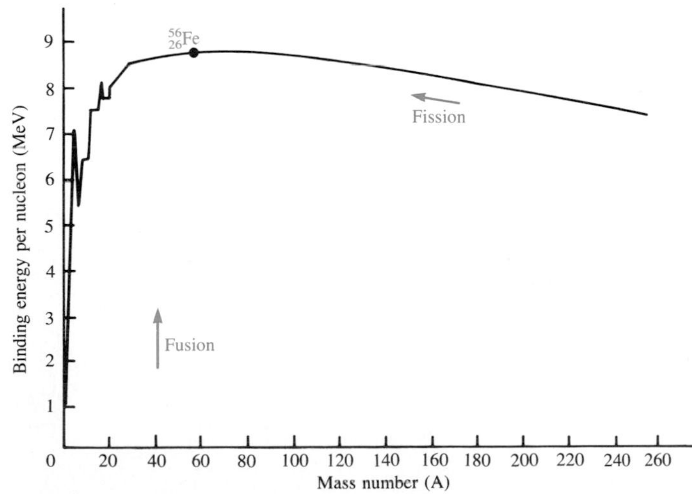

Figure 21.10

Both fission and fusion produce more stable nuclides and are thus exothermic.

1. Combining two light nuclei to form a heavier, more stable nucleus. This process is called **fusion.**
2. Splitting a heavy nucleus into two nuclei with smaller mass numbers. This process is called **fission.**

Because of the large binding energies involved in holding the nucleus together, both of these processes involve energy changes more than a million times larger than those associated with chemical reactions.

Nuclear Fission

Nuclear fission was discovered in the late 1930s when $^{235}_{92}U$ nuclides bombarded with neutrons were observed to split into two lighter elements:

$$^{1}_{0}n + {}^{235}_{92}U \rightarrow {}^{142}_{56}Ba + {}^{91}_{36}Kr + 3\,{}^{1}_{0}n$$

This process, shown schematically in Fig. 21.11, releases 3.5×10^{-11} J of energy per event, which translates to 2.1×10^{13} J per mole of $^{235}_{92}U$. Compare this figure to that for the combustion of methane, which releases only 8.0×10^{5} J of energy per mole. The fission of $^{235}_{92}U$ produces about 26 million times as much energy as the combustion of methane.

The process shown above is only one of the many fission reactions that $^{235}_{92}U$ can undergo. Another is

$$^{1}_{0}n + {}^{235}_{92}U \rightarrow {}^{137}_{52}Te + {}^{97}_{40}Zn + 2\,{}^{1}_{0}n$$

In fact, over 200 different isotopes of 35 different elements have been observed among the fission products of $^{235}_{92}U$.

In addition to the product nuclides, neutrons are also produced in the fission reactions of $^{235}_{92}U$. This means it is possible to produce a self-sustaining fission process—a **chain reaction** (see Fig. 21.12). In order for the fission process to be self-sustaining, at least one neutron from each fission event must go on to split another nucleus. If, on the average, *less than one* neutron causes another fission event, the process dies out and the reaction is said to be **subcritical.** If *exactly one* neutron from each fission event causes another fission event, the process sustains itself at the same level and is said to be **critical.** If *more than one* neutron from each fission event causes another fission event, the process rapidly escalates and the heat

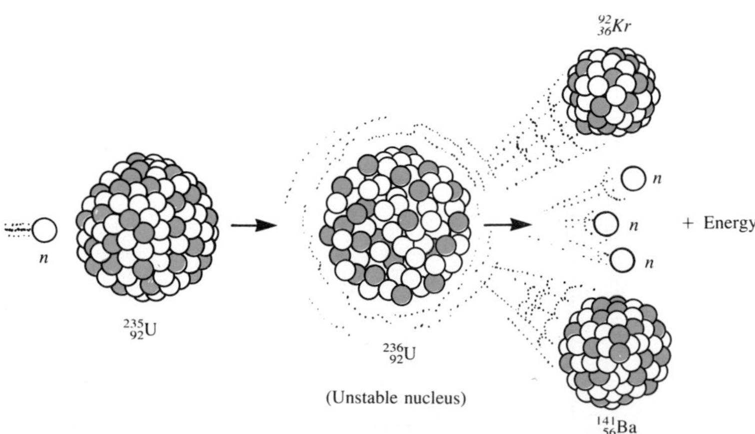

Figure 21.11

On capturing a neutron, the $^{235}_{92}U$ nucleus undergoes fission to produce two lighter nuclides, free neutrons (three typically), and a large amount of energy.

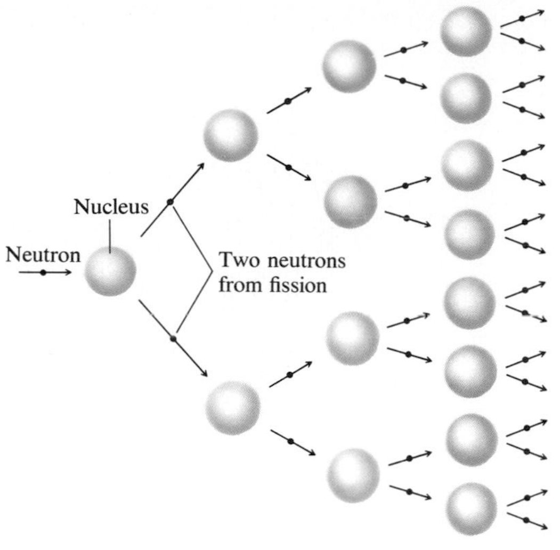

Figure 21.12

Representation of a fission process in which each event produces two neutrons, which can go on to split other nuclei, leading to a self-sustaining chain reaction.

build-up causes a violent explosion. This situation is described as **supercritical.**

To achieve the critical state, a certain mass of fissionable material, called the **critical mass,** is needed. If the sample is too small, too many neutrons escape before they have a chance to cause a fission event, and the process stops. This is illustrated in Fig. 21.13.

During World War II, an intense research effort called the Manhattan Project was carried out by the United States to build a bomb based on the principles of nuclear fission. This program produced the fission bomb, which was used with devastating effects on the cities of Hiroshima and Nagasaki in 1945. Basically, a fission bomb operates by suddenly combining two subcritical masses of fissionable material to form a supercritical mass, thereby producing an explosion of incredible intensity.

Nuclear Reactors

Because of the tremendous energies involved, it seemed desirable to develop the fission process as an energy source to produce electricity. To accomplish this, reactors were designed where controlled fission can occur. The resulting energy is

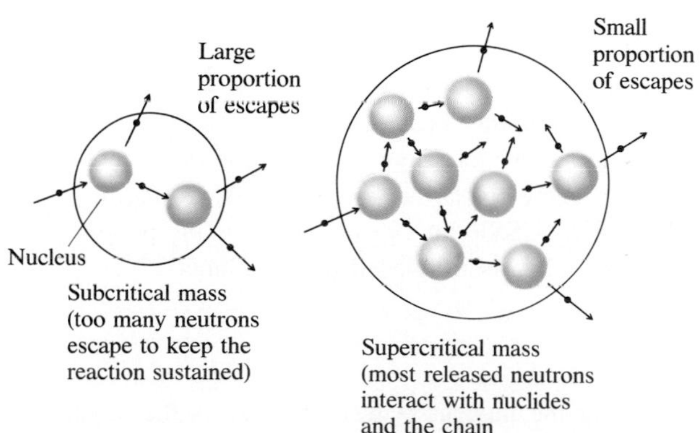

Figure 21.13

If the mass of fissionable material is too small, most of the neutrons escape before causing another fission event, and the process dies out.

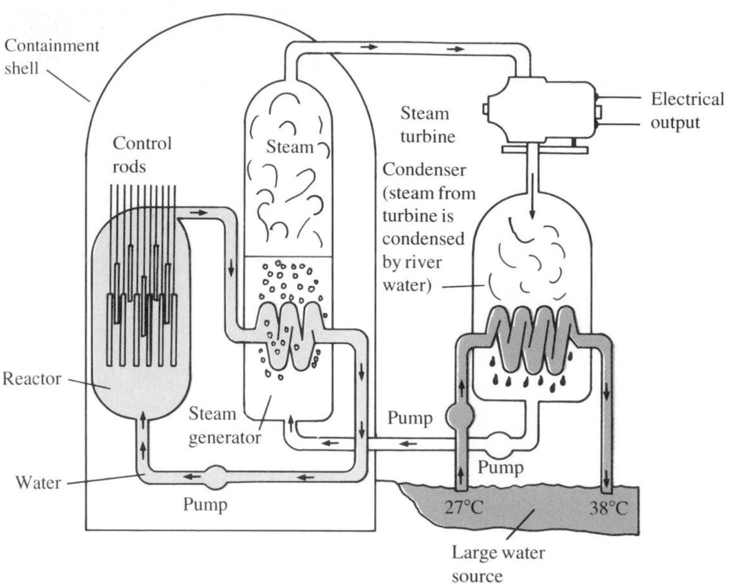

Figure 21.14

A schematic diagram of a nuclear power plant.

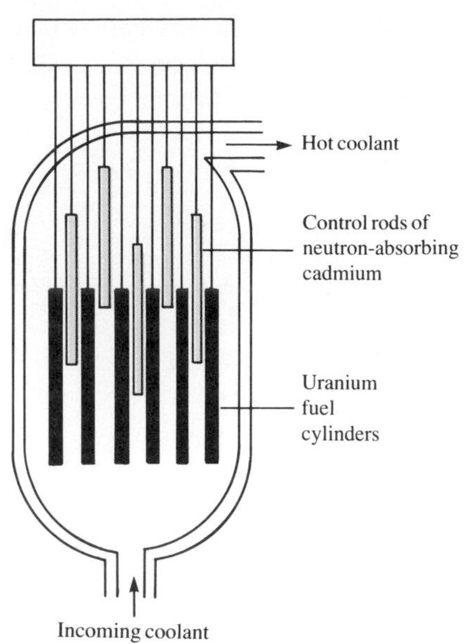

Figure 21.15

A schematic of the reactor core. The position of the control rods determines the level of energy production by regulating the amount of fission taking place.

used to heat water to produce steam to run turbine generators, in much the same way that a coal-burning power plant generates energy. A schematic diagram of a nuclear power plant is shown in Fig. 21.14.

Uranium oxide (refined uranium).

In the **reactor core,** shown in Fig. 21.15, uranium that has been enriched to approximately 3% $^{235}_{92}U$ (natural uranium contains only 0.7% $^{235}_{92}U$) is housed in metal cylinders. A **moderator** surrounds the cylinders to slow down the neutrons so that the uranium fuel can capture them more efficiently. **Control rods,** composed of substances that absorb neutrons, are used to regulate the power level of the reactor. The reactor is designed so that should a malfunction occur, the control rods are automatically inserted into the core to stop the reaction. A liquid (usually water) is circulated through the core to extract the heat generated by the energy of fission; the energy can then be passed on via a heat exchanger to water in the turbine system.

Although the concentration of $^{235}_{92}U$ in the fuel elements is not great enough to allow a supercritical mass to develop in the core, a failure of the cooling system can lead to temperatures high enough to melt the core. As a result, the building housing the core must be designed to contain the core even if melt-down occurs. A great deal of controversy now exists about the efficiency of the safety systems in nuclear power plants. Accidents such as the one at the Three Mile Island facility in Pennsylvania in 1979 and in Chernobyl, USSR, in 1986 have led many people to question the wisdom of continuing to build fission-based power plants.

Breeder Reactors

One potential problem facing the nuclear power industry is the supply of $^{235}_{92}U$. Some scientists have suggested that we have nearly depleted those uranium deposits rich enough in $^{235}_{92}U$ to make production of fissionable fuel economically feasible.

Because of this possibility, **breeder reactors** have been developed, in which fissionable fuel is actually produced while the reactor runs. In the breeder reactors now being studied, the major component of natural uranium, nonfissionable $^{238}_{92}U$, is changed to fissionable $^{239}_{94}Pu$. The reaction involves absorption of a neutron, followed by production of two β particles:

$$^{1}_{0}n + ^{238}_{92}U \rightarrow ^{239}_{92}U$$
$$^{239}_{92}U \rightarrow ^{239}_{93}Np + ^{0}_{-1}e$$
$$^{239}_{93}Np \rightarrow ^{239}_{94}Pu + ^{0}_{-1}e$$

As the reactor runs and $^{235}_{92}U$ is split, some of the excess neutrons are absorbed by $^{238}_{92}U$ to produce $^{239}_{94}Pu$. The $^{239}_{94}Pu$ is then separated out and used to fuel another reactor. Such a reactor thus "breeds" nuclear fuel as it operates.

Although breeder reactors are now used in France, the United States is proceeding slowly with their development because of their controversial nature. One problem involves the hazards in handling plutonium, which flames on contact with air and is very toxic.

Skylab ultraviolet picture of the sun.

Fusion

Large quantities of energy are also produced by the fusion of two light nuclei. In fact, stars produce their energy through nuclear fusion. Our sun, which presently consists of 73% hydrogen, 26% helium, and 1% other elements, gives off vast quantities of energy from fusion of protons to form helium:

$$^{1}_{1}H + ^{1}_{1}H \rightarrow ^{2}_{1}H + ^{0}_{1}e$$
$$^{1}_{1}H + ^{2}_{1}H \rightarrow ^{3}_{2}He$$
$$^{3}_{2}He + ^{3}_{2}He \rightarrow ^{4}_{2}He + 2\ ^{1}_{1}H$$
$$^{3}_{2}He + ^{1}_{1}H \rightarrow ^{4}_{2}He + ^{0}_{1}e$$

Intense research is underway to develop a feasible fusion process because of the ready availability of many light nuclides (deuterium, $^{2}_{1}H$, in sea water, for example) that can serve as fuel in fusion reactors. The major stumbling block is that high temperatures are required to initiate fusion. The forces that bind nucleons together to form a nucleus are effective only at *very small* distances ($\sim 10^{-13}$ cm). Thus, for two protons to bind together and thereby release energy, they must get very close together. But protons, because they are identically charged, repel each other electrostatically. This means that in order to get two protons (or two deuterons) close enough to bind together (the nuclear binding force is *not* electrostatic), they must be "shot" at each other at speeds high enough to overcome the electrostatic repulsion.

The electrostatic repulsion forces between two $^{2}_{1}H$ nuclei are so great that a temperature of 4×10^{7} K is required to give them velocities large enough to cause them to collide with sufficient energy that the nuclear forces can bind the particles together and thus release the binding energy. This situation is represented in Fig. 21.16.

Currently, scientists are studying two types of systems to produce the extremely high temperatures required: high-powered lasers, and heating by electric currents. At present many technical problems remain to be solved, and it is not clear which method will prove more useful or when fusion might become a practical energy source. However, there is still hope that fusion will be a major energy source after the year 2000.

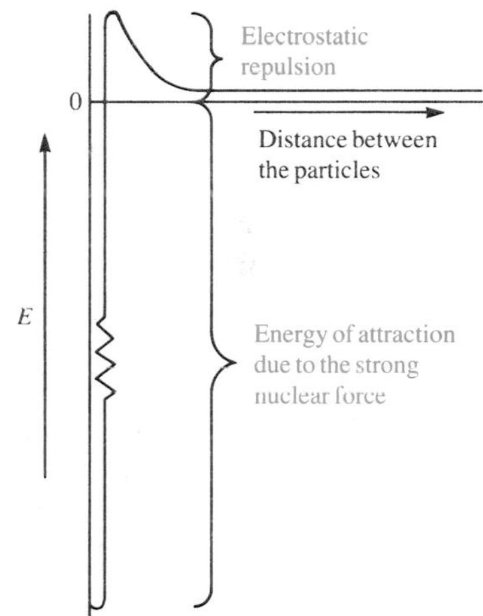

Figure 21.16

A plot of energy versus the separation distance for two $^{2}_{1}H$ nuclei. The nuclei must have sufficient velocities to get over the electrostatic repulsion "hill" and get close enough for the nuclear binding forces to become effective, thus "fusing" the particles into a new nucleus and releasing large quantities of energy. The binding force is at least 100 times the electrostatic repulsion.

Chemical Impact

Nuclear Waste Disposal

Our society has not had a very impressive record for safe disposal of industrial wastes. We have polluted our water and air, and some land areas have become virtually uninhabitable because of the improper burial of chemical wastes. As a result, many people are wary about the radioactive wastes from nuclear reactors. The potential threats of cancer and genetic mutations make these materials especially frightening.

Because of its controversial nature, no nuclear waste generated over the last 40 years has been permanently disposed of. However, in 1982 the U.S. Congress passed the Nuclear Waste Policy Act, which established a timetable for choosing and preparing sites for deep underground disposal of radioactive materials. The program is being funded by a tax of 0.1% per kilowatt hour on electricity generated by nuclear power.

The tentative disposal plan calls for the incorporation of the spent nuclear fuel into blocks of borosilicate glass that will be packed in corrosion-resistant metal containers and then buried in a deep, stable rock formation (see figure).

There are indications that this method will isolate the waste until the radioactivity decays to safe levels. One reassuring indication comes from the natural fission "reactor" at Oklo in Gabon, Africa. Initiated about 2 billion years ago when uranium in ore deposits there formed a critical mass, the "reactor" produced fission and fusion products for several thousand years. Although some of these products have migrated away from the site in the intervening 2 billion years, most have stayed in place. Another indication of the possible success of underground isolation is the disposal program carried out at Oak Ridge National Laboratory over the last 20 years. In this program the waste has been ground up, mixed with cement, fly ash, and

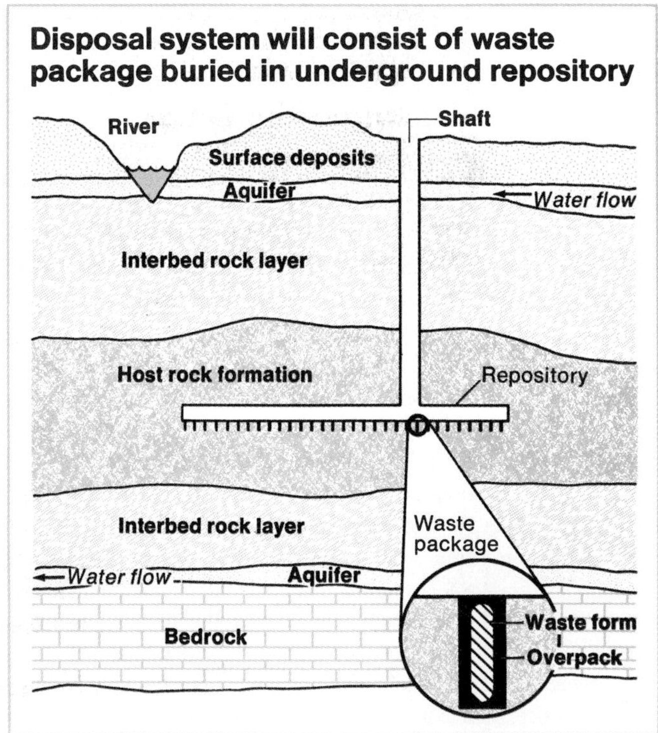

A schematic diagram of the tentative plan for deep underground isolation of nuclear waste. (From *Chemical & Engineering News,* July 18, 1983, pp. 20–38. Reprinted with permission. Copyright 1983 American Chemical Society.)

Chemical Impact

clay, and injected at high pressures into a shale bed 1000 feet underground. This method of injection causes fractures in the rock into which the liquid waste seeps and then solidifies. Monitoring of the wastes has shown little movement of radioactivity. Although this procedure has worked well, the materials involved are not nearly as radioactive as those from power plants, for which this method is not thought suitable. However, this experiment has demonstrated that such wastes can be immobilized.

Choosing the waste repository sites is an especially sensitive issue. Many states have resisted the plan, but Congress has the power to override a state's disapproval. In fact, Congress amended the Nuclear Waste Policy Act in 1987 to make Yucca Mountain in Nevada the primary potential site. Studies are now being carried out to evaluate the feasibility of this site as a safe repository for nuclear waste.

An alternative to permanent isolation of the spent nuclear fuel is monitored, retrievable storage. This method has an advantage since, if reprocessing to remove potentially useful materials such as plutonium becomes feasible, the wastes are still accessible. The argument for this plan is that the technologies and alternatives available to society a few hundred years from now may be so different from those of the present that these wastes could become valuable resources. At the present time, isolation of the wastes seems the most likely event; even so, these underground deposits may become plutonium mines in the future.

21.7 Effects of Radiation

Purpose

- To show how radiation damages human tissue.

Everyone knows that being hit by a train is very serious. The problem is the energy transfer involved. In fact, any source of energy is potentially harmful to organisms. Energy transferred to cells can break chemical bonds and cause malfunctioning of the cell systems. This fact is behind the concern about the ozone layer in the earth's upper atmosphere, which screens out high-energy ultraviolet radiation from the sun. Radioactive elements, which are sources of high-energy particles, are also potentially hazardous, although the effects are usually quite subtle. The reason for the subtlety of radiation damage is that, even though high-energy particles are involved, the quantity of energy actually deposited in tissues *per event* is quite small. However, the resulting damage is no less real, although the effects may not be apparent for years.

The ozone layer is discussed in Section 19.5.

Radiation damage to organisms can be classified as somatic or genetic damage. **Somatic damage** is damage to the organism itself, resulting in sickness or death. The effects may appear almost immediately if a massive dose of radiation is received; for smaller doses, damage may appear years later, usually in the form of cancer. **Genetic damage** is damage to the genetic machinery, which produces malfunctions in the offspring of the organism.

The biological effects of a particular source of radiation depend on several factors:

1. *The energy of the radiation*. The higher the energy content of the radiation, the more damage it can cause. Radiation doses are measured in **rads** (which is

short for *r*adiation *a*bsorbed *d*ose), where 1 rad corresponds to 10^{-2} J of energy deposited per kilogram of tissue.

2. *The penetrating ability of the radiation.* The particles and rays produced in radioactive processes vary in their abilities to penetrate human tissue: γ rays are highly penetrating; β particles can penetrate approximately 1 cm; and α particles are stopped by the skin.
3. *The ionizing ability of the radiation.* Extraction of electrons from biomolecules to form ions is particularly detrimental to their functions. The ionizing ability of radiation varies dramatically. For example, γ rays penetrate very deeply, but cause only occasional ionization. On the other hand, α particles, although not very penetrating, are very effective at causing ionization and produce a dense trail of damage. Thus ingestion of an α-particle producer, such as plutonium, is particularly damaging.
4. *The chemical properties of the radiation source.* When a radioactive nuclide is ingested into the body, its effectiveness in causing damage depends on its residence time. For example, $^{85}_{36}Kr$ and $^{90}_{38}Sr$ are both β-particle producers. However, since krypton is chemically inert, it passes through the body quickly and does not have much time to do damage. Strontium, being chemically similar to calcium, can collect in bones, where it may cause leukemia and bone cancer.

Because of the differences in the behavior of the particles and rays produced by radioactive decay, both the energy dose of the radiation and its effectiveness in causing biological damage must be taken into account. The **rem** (which is short for *r*oentgen *e*quivalent for *m*an) is defined as follows:

$$\text{Number of rems} = (\text{number of rads}) \times \text{RBE}$$

where RBE represents the relative effectiveness of the radiation in causing biological damage.

Table 21.6 shows the physical effects of short-term exposure to various doses of radiation, and Table 21.7 gives the sources and amounts of radiation exposure for a typical person in the United States. Note that natural sources contribute about twice as much as human activities to the total exposure. However, although the nuclear industry contributes only a small percentage of the total exposure, the major controversy associated with nuclear power plants is the *potential* for radiation hazards. These arise mainly from two sources: accidents allowing the release of radioactive materials, and improper disposal of the radioactive products in spent fuel elements. The radioactive products of the fission of $^{235}_{92}U$, although only a small

Effects of Short-Term Exposures to Radiation

Dose (rem)	Clinical effect
0–25	Nondetectable
25–50	Temporary decrease in white blood cell counts
100–200	Strong decrease in white blood cell counts
500	Death of half the exposed population within 30 days after exposure

Table 21.6

Typical Radiation Exposures for a Person Living in the United States (1 millirem = 10^{-3} rem)	
	Exposure (millirems/year)
Cosmic radiation	50
From the earth	47
From building materials	3
In human tissues	21
Inhalation of air	5
Total from natural sources	126
X-ray diagnosis	50
Radiotherapy	10
Internal diagnosis/therapy	1
Nuclear power industry	0.2
TV tubes, industrial wastes, etc.	2
Radioactive fallout	4
Total from human activities	67
Total	193

Table 21.7

percentage of the total products, have half-lives of several hundred years and remain dangerous for a long time. Various schemes have been advanced for the disposal of these wastes. The one that seems to hold the most promise is the incorporation of the wastes into ceramic blocks and the burial of these blocks in geologically stable formations. At present, however, no disposal method has been accepted, and nuclear wastes continue to accumulate in temporary storage facilities.

Disposal of nuclear wastes is discussed in the Chemical Impact feature at the end of Section 21.6.

Even if a satisfactory method for permanent disposal of nuclear wastes is found, there will continue to be concern about the effects of exposure to low levels of radiation. Exposure is inevitable from natural sources such as cosmic rays and radioactive minerals, and many people are also exposed to low levels of radiation from reactors, radioactive tracers, or diagnostic X rays. Currently, we have little reliable information on the long-term effects of low-level exposure to radiation.

Two models of radiation damage, illustrated in Fig. 21.17, have been proposed: the *linear model* and the *threshold model*. The linear model postulates that damage from radiation is proportional to the dose, even at low levels of exposure. Thus any

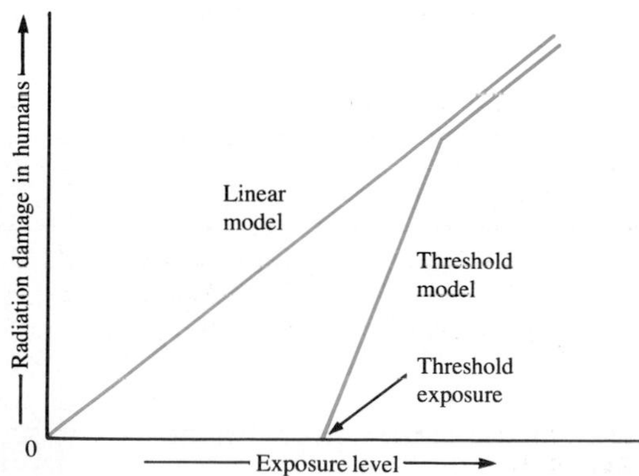

Figure 21.17

The two models for radiation damage. In the linear model, even a small dosage causes a proportional risk. In the threshold model, risk begins only after a certain dosage.

Chemical Impact

Nuclear Physics: An Introduction

Nuclear physics is concerned with the fundamental nature of matter. The central focuses of this area of study are the relationship between a quantity of energy and its mass, given by $E = mc^2$, and the fact that matter can be converted from one form (energy) to another (particulate) in particle accelerators. Collisions between high-speed particles have produced a dazzling array of new particles—hundreds of them. These events can best be interpreted as conversions of kinetic energy into particles. For example, a collision of sufficient energy between a proton and a neutron can produce four particles: two protons, one antiproton, and a neutron:

$$^{1}_{1}\text{H} + ^{1}_{0}\text{n} \rightarrow 2\ ^{1}_{1}\text{H} + ^{1}_{-1}\text{H} + ^{1}_{0}\text{n}$$

where $^{1}_{-1}\text{H}$ is the symbol for an *antiproton,* which has the same mass as a proton but the opposite charge. This process is a little like throwing one baseball at a very high speed into another and having the collision produce four baseballs.

The results of such accelerator experiments have led scientists to postulate the existence of three types of forces important in the nucleus: the *strong force*, the *weak force,* and the *electromagnetic force*. Along with the *gravitational force,* these forces are thought to account for all types of interactions found in matter. These forces are believed to be generated by the exchange of particles between the interacting pieces of matter. For example, gravitational force is thought to be carried by particles called *gravitons*. The electromagnetic force (the classical electrostatic force between charged particles) is assumed to be exerted through the exchange of *photons*. The strong force, not charge-related and only effective at very short distances ($\sim 10^{-13}$ cm), is postulated to involve the exchange of particles called *gluons*. The weak force is 100 times weaker than the strong force and seems to be exerted through the exchange of two types of large particles, the W (has a mass 70 times the proton mass) and the Z (has a mass 90 times the proton mass).

The particles discovered have been classified into several categories. Three of the most important classes are as follows:

1. *Hadrons* are particles that respond to the strong force and have internal structure.
2. *Leptons* are particles that do not respond to the strong force and have no internal structure.

(left) An aerial view of Fermilab, a high energy particle accelerator in Batavia, Illinois. (right) The accelerator tunnel at Fermilab.

Chemical Impact

3. *Quarks* are particles with no internal structure that are thought to be the fundamental constituents of hadrons. Neutrons and protons are hadrons that are thought to be composed of three quarks each.

The world of particle physics appears mysterious and complicated. For example, particle physicists have discovered new properties of matter they call "color," "charm," and "strangeness" and have postulated conservation laws involving these properties. This area of science is extremely important because it should help us to understand the interactions of matter in a more elegant and unified way. For example, the classification of force into four categories is probably necessary only because we do not understand the true nature of forces. All forces may be special cases of a single, all-pervading force field that governs all of nature. In fact, Einstein spent the last 30 years of his life looking for a way to unify the gravitational and electromagnetic forces—without success. Physicists may now be on the verge of accomplishing what Einstein failed to do.

Although the practical aspects of work in nuclear physics are not yet totally apparent, a more fundamental understanding of the way nature operates could lead to presently undreamed-of devices for energy production, communication, and so on, which could revolutionize our lives.

exposure is dangerous. The threshold model, on the other hand, assumes that no significant damage occurs below a certain exposure, called the *threshold exposure*. Note that if the linear model is correct, radiation exposure should be limited to a bare minimum (ideally at the natural levels). If the threshold model is correct, a certain level of radiation exposure beyond natural levels can be tolerated. Most scientists seem to feel that, since there is little evidence available to evaluate these models, it is safest to assume that the linear hypothesis is correct and to minimize radiation exposure.

FOR REVIEW

Summary

The stability of the nucleus, which is composed of neutrons and protons, can be considered from kinetic and thermodynamic points of view. Nuclei that are kinetically unstable decompose into more stable nuclei by radioactive decay, which can involve production of an α particle ($^{4}_{2}He$), which changes both A and Z, production of a β particle ($^{0}_{-1}e$), which increases Z, or production of a positron ($^{0}_{1}e$), which decreases Z. High-energy photons called γ rays can also be produced.

A radioactive nuclide may undergo a number of decay events, called a decay series, until a stable nuclide is formed. Radioactive decay follows first-order kinetics. The half-life of a radioactive sample is the time required for the amount of the nuclide to reach half the original amount. The transuranium elements, those coming after uranium in the periodic table, can be synthesized by particle bombardment of uranium or heavier elements.

Radiocarbon dating uses the ratio $^{14}_{6}C/^{12}_{6}C$ to date any object containing carbon from living sources.

The thermodynamic stability of the nucleus involves a comparison of the energy of the nucleus to that of its component nucleons. When a system gains or loses energy, it also gains or loses a quantity of mass, given by the relationship $E = mc^2$. The change in mass, called the mass defect, can be obtained by comparing the nuclear mass to the sum of the masses of the component nucleons and can be used to calculate the energy of formation of a nucleus from its nucleons. The binding energy is the energy required to decompose the nucleus into protons and neutrons.

Nuclear fusion is the process of combining two light nuclei to form a heavier, more stable nucleus. Fission is the splitting of a heavy nucleus into two lighter nuclei. Nuclear reactors now in use employ controlled fission.

Radiation damage can cause either direct damage to living organisms or genetic damage to the organism's offspring. The biological effects of radiation depend on the energy, the penetrating ability, and the ionizing ability of the radiation, and on the chemical properties of the radiation source.

Key Terms

neutron
proton
nucleon
atomic number
mass number
isotopes
nuclide

Section 21.1
thermodynamic stability
kinetic stability
radioactive decay
beta (β) particle
zone of stability
alpha (α) particle
α-particle production
spontaneous fission
β-particle production
gamma (γ) ray
positron production
electron capture
decay series

Section 21.2
rate of decay
half-life

Section 21.3
nuclear transformation
particle accelerator
cyclotron
linear accelerator
transuranium elements

Section 21.4
Geiger-Müller counter (Geiger counter)
scintillation counter
radiocarbon dating (carbon-14 dating)
radiotracers

Section 21.5
mass defect
binding energy

Section 21.6
fusion
fission
chain reaction
subcritical reaction
critical reaction
supercritical reaction
critical mass
reactor core
moderator
control rods
breeder reactor

Section 21.7
somatic damage
genetic damage
rad
rem

Exercises

A blue exercise number indicates that the answer to that exercise appears at the back of this book and a solution appears in the Solutions Guide.

Radioactive Decay and Nuclear Transformations

1. Write an equation describing the radioactive decay of each of the following nuclides. (The particle produced is shown in parentheses.)
   a. ${}^{3}_{1}H$ (β)
   b. ${}^{8}_{3}Li$ (β followed by α)
   c. ${}^{7}_{4}Be$ (electron capture)
   d. ${}^{8}_{5}B$ (positron)
   e. ${}^{32}_{15}P$ (β)

2. Write an equation describing the radioactive decay of each of the following nuclides:
   a. ${}^{60}Co$ (β)
   b. ${}^{97}Tc$ (electron capture)
   c. ${}^{99}Tc$ (β)
   d. ${}^{239}Pu$ (α)

3. The stable isotopes of boron are boron-10 and boron-11. Four radioactive isotopes with mass numbers 8, 9, 12, and 13 are also known. Predict possible modes of radioactive decay for these four nuclides.

4. Complete the following equations, which describe nuclear reactions that occur in the sun:

a. ${}^{1}_{1}H + {}^{14}_{7}N \rightarrow$ ______ $+ {}^{4}_{2}He$
b. Two helium-3 nuclei undergo fusion, forming helium-4 and two protons.
c. ${}^{1}_{1}H + {}^{1}_{1}H \rightarrow {}^{2}_{1}H +$ ______
d. ${}^{1}_{1}H + {}^{12}_{6}C \rightarrow$ ______

5. Many elements have been synthesized by bombarding relatively heavy atoms with high-energy particles in particle accelerators. Complete the following nuclear reactions, which have been used to synthesize elements:
a. ______ $+ {}^{4}_{2}He \rightarrow {}^{243}_{97}Bk + {}^{1}_{0}n$
b. ${}^{238}_{92}U + {}^{12}_{6}C \rightarrow$ ______ $+ 6\ {}^{1}_{0}n$
c. ${}^{249}_{98}Cf +$ ______ $\rightarrow {}^{263}_{106}Unh + 4\ {}^{1}_{0}n$
d. ${}^{249}_{98}Cf + {}^{10}_{5}B \rightarrow {}^{257}_{103}Lr +$ ______

6. Define fission and fusion. For what types of nuclei are you likely to see each process?

7. When nuclei undergo nuclear transformations, γ rays of characteristic frequencies are observed. How does this fact, along with other information in the chapter on nuclear stability, suggest that a quantum mechanical–like model may apply to the nucleus?

8. There are four stable isotopes of iron with mass numbers 54, 56, 57, and 58. There are also two radioactive isotopes: iron-53 and iron-59. Predict modes of decay for these two isotopes and write a nuclear reaction for each.

9. Uranium-235 undergoes a decay series in which the following particles are produced in succession: α, β, α, β, α, α, α, α, β, α, β.
a. What is the final product of the decay of uranium-235?
b. What are the ten intermediate nuclides?

10. One type of commercial smoke detector contains a minute amount of radioactive americium-241 (${}^{241}Am$), which decays by α-particle production. The α particles ionize molecules in the air allowing it to conduct an electric current. When smoke particles enter, the conductivity of the air is changed and the alarm buzzes.
a. Write the equation for the decay of ${}^{241}_{95}Am$ by α-particle production.
b. The complete decay of ${}^{241}Am$ involves successively α, α, β, α, α, β, α, α, α, β, α, and β production. What is the final stable nucleus produced in this decay series?
c. Identify the eleven intermediate nuclides.

Kinetics of Radioactive Decay

11. How many disintegrations occur in the first second from 1.00 mol of a radioactive nuclide with each of the following half-lives: 12,000 yr? 12 h? 12 min? 12 s?

12. What fraction of the original carbon-14 remains in a piece of wood that was cut from a tree 2200 years ago?

13. What assumptions must be made and what problems arise in using carbon-14 dating?

14. A chemist wishes to do an experiment using ${}^{47}Ca$ (half-life = 4.5 days). He needs 5.0 μg of the nuclide. What mass of ${}^{47}CaCO_3$ must he order if it takes 48 h for delivery from the supplier?

15. Cobalt-60 is commonly used as a source of β particles. How long does it take for 10.0% of a sample of cobalt-60 to decay (the half-life is 5.26 yr)?

16. A living plant contains approximately the same fraction of carbon-14 as in atmospheric carbon dioxide. The observed rate of decay of carbon-14 from a living plant is 15.3 disintegrations per minute per gram of carbon. How many disintegrations per minute per gram of carbon will be measured from a 15,000-yr-old sample? Will radiocarbon dating work well for small samples of 10 mg or less?

17. What is the ratio ${}^{206}Pb/{}^{238}U$ by mass in a rock that is 4.5×10^9 yr old? (For ${}^{238}U$, $t_{1/2} = 4.5 \times 10^9$ yr.)

18. The mass ratios of ${}^{40}Ar$ to ${}^{40}K$ can also be used to date geological materials. Potassium-40 decays by two processes:

$${}^{40}_{19}K \rightarrow {}^{40}_{18}Ar \text{ (electron capture, 10.7\%)} \qquad t_{1/2} = 1.27 \times 10^9 \text{ yr}$$

$${}^{40}_{19}K \rightarrow {}^{40}_{20}Ca + {}^{0}_{-1}e \text{ (89.3\%)}$$

a. Why are ${}^{40}Ar/{}^{40}K$ ratios used to date materials rather than ${}^{40}Ca/{}^{40}K$ ratios?
b. What assumptions must be made using this technique?
c. A sedimentary rock has a ${}^{40}Ar/{}^{40}K$ ratio of 0.95. Calculate the age of the rock.
d. How will the measured age of a rock compare to the actual age if some ${}^{40}Ar$ escaped from the sample?

19. Phosphorus-32 is a commonly used radioactive nuclide in biochemical research, particularly in studies of nucleic acids. The half-life of phosphorus-32 is 14.3 days. What mass of phosphorus-32 is left of an original sample of 175 mg of $Na_3{}^{32}PO_4$ after 35.0 days?

20. The *curie* (Ci) is a commonly used unit for measuring nuclear radioactivity: 1 curie of radiation is equal to 3.7×10^{10} disintegrations per second. This is the number of disintegrations from 1 g of radium in 1 s.
a. What is the activity in mCi (millicuries) of 175 mg of $Na_3{}^{32}PO_4$?
b. What is the activity in mCi of 1.0 mol of plutonium-239 ($t_{1/2}$ = 24,000 yr)?

21. The half-life of sulfur-38 is 2.87 h.
a. What mass of $Na_2{}^{38}SO_4$ has an activity of 10.0 mCi?
b. How long does it take for 99.99% of a sample of sulfur-38 to decay?

22. The first atomic explosion was detonated in the desert north of Alamogordo, New Mexico, on July 16, 1945. What fraction of the strontium-90 ($t_{1/2}$ = 28.8 yr) originally produced by that explosion still remains?

Energy Changes in Nuclear Reactions

23. The sun radiates 3.9×10^{23} J of energy into space every second. What is the rate at which mass is lost from the sun?

24. The earth receives 1.8×10^{14} kJ/s of solar energy. What mass of solar material is converted to energy over a 24-h period to provide the daily amount of solar energy to the earth? What mass of coal would have to be burned to provide the same amount of energy? Coal releases 32 kJ of energy per gram when burned.

25. Calculate the binding energy per nucleon for ^{24}Mg and ^{27}Mg. The atomic masses are ^{24}Mg, 23.9850; and ^{27}Mg, 26.9843. (Since these are atomic masses, they include the mass of magnesium's 12 electrons. See Sample Exercise 21.8.)

26. The atomic mass of lithium-6 is 6.015126 amu. What is the mass of a lithium-6 nucleus? Calculate the total binding energy per mole of lithium-6.

27. Calculate the binding energy per nucleon for $^{2}_{1}H$ and $^{3}_{1}H$. The atomic masses are $^{2}_{1}H$, 2.01410; and $^{3}_{1}H$, 3.01605.

28. Consider the reaction discussed in the Chemical Impact feature on nuclear physics in this chapter

$$^{1}_{1}H + ^{1}_{0}n \rightarrow 2\ ^{1}_{1}H + ^{1}_{0}n + {}_{-1}^{1}H$$

Calculate the energy change for this reaction. Is energy released or absorbed? What is a possible source for this energy?

29. A positron and an electron annihilate each other upon colliding, producing energy:

$$^{0}_{-1}e + ^{0}_{+1}e \rightarrow 2\ ^{0}_{0}\gamma$$

Assuming both γ rays have the same energy, calculate the wavelength of the electromagnetic radiation produced.

30. A small atomic bomb releases energy equivalent to the detonation of 20,000 tons of TNT; a ton of TNT releases 4×10^{9} J of energy when exploded. Using 2×10^{13} J/mol as the energy released by fission of ^{235}U, estimate a value for the critical mass of ^{235}U. What assumptions must be made in the calculation?

31. Using the kinetic molecular theory (Section 5.6), calculate the average kinetic energy of a $^{2}_{1}H$ nucleus at a temperature of 4×10^{7} K. (See Exercise 32 for the appropriate mass value.)

32. Calculate the amount of energy released per gram of hydrogen for the following reaction. The atomic masses are $^{1}_{1}H$, 1.00782; and $^{2}_{1}H$, 2.01410. (*Hint:* Think carefully about how to account for the electron mass.)

$$^{1}_{1}H + ^{1}_{1}H \rightarrow ^{2}_{1}H + ^{0}_{+1}e$$

33. The easiest fusion reaction to initiate is

$$^{2}_{1}H + ^{3}_{1}H \rightarrow ^{4}_{2}He + ^{1}_{0}n$$

Calculate the energy released per nucleus of $^{4}_{2}He$ produced and per mole of $^{4}_{2}He$ produced. The atomic masses are $^{2}_{1}H$, 2.01410; $^{3}_{1}H$, 3.01605; and $^{4}_{2}He$, 4.00260. The masses of the electron and neutron are 5.4858×10^{-4} amu and 1.00866 amu, respectively.

Detection, Uses, and Health Effects of Radiation

34. Why are the chemical properties of a radioactive substance important in assessing its potential health hazards?

35. When using a Geiger-Müller counter to measure radioactivity, it is necessary to maintain the same geometrical orientation between the sample and the Geiger-Müller tube in order to compare different measurements. Why?

36. The typical response of a Geiger-Müller tube is shown below Explain the shape of this curve.

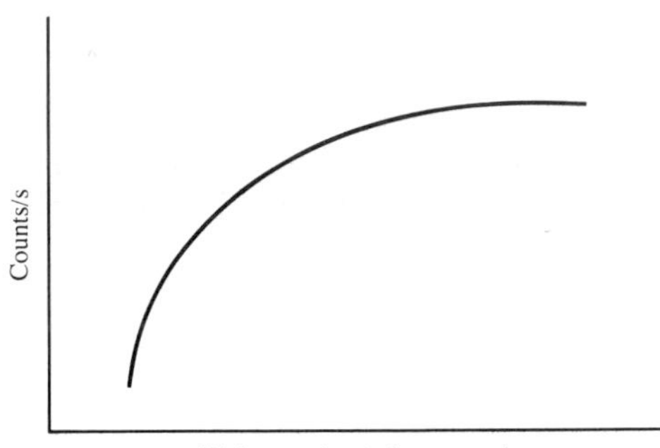

37. Photosynthesis in plants can be represented by the following overall reaction:

$$6CO_2(g) + 6H_2O(l) \xrightarrow{\text{Light}} C_6H_{12}O_6(s) + 6O_2(g)$$

Algae grown in water containing some ^{18}O (in $H_2{}^{18}O$) evolve oxygen with the same isotopic composition as the oxygen in the water. When algae growing in water containing only ^{16}O were furnished carbon dioxide containing ^{18}O, no ^{18}O was found to be evolved. What conclusions about photosynthesis can be drawn from these experiments?

38. How could a radioactive nuclide be used to demonstrate that chemical equilibrium is a dynamic process?

39. There is a trend in the United States toward using coal-fired power plants to generate electricity rather than building new nuclear fission power plants. Is the use of coal-fired power plants without risk? Make a list of the risks to society from the use of each type of power plant.

40. A 0.10-cm^3 sample of a solution containing a radioactive nuclide (5.0×10^{3} disintegrations per minute per milliliter) is injected into a rat. Several minutes later 1.0 cm^3 of blood is removed. The blood shows 48 disintegrations per minute of radioactivity. What is the volume of blood in the rat? What assumptions must be made in performing this calculation?

Additional Exercises

41. Recent attempts to provide a unified field theory for matter have proposed that the fundamental particle is the superstring. A superstring is $\approx 10^{-35}$ m long (10^{-20} times the diameter of a proton). Use the Heisenberg Uncertainty Principle (Chapter 7)

$$\Delta x \cdot \Delta(mv) \geq \frac{h}{4\pi}$$

to calculate the uncertainty in the mass of a superstring. (Assume that $\Delta x = 1 \times 10^{-35}$ m and that the velocity is 10% the speed of light.) Compare this uncertainty in mass to the electron and proton masses.

42. Consider the following information:

    i. The layer of dead skin on our bodies is sufficient to protect us from most α-particle radiation.

    ii. Plutonium is an α-particle producer.

    iii. The chemistry of Pu^{4+} is similar to that of Fe^{3+}.

    iv. $Pu^{4+} + 4e^- \rightarrow Pu \qquad \mathscr{E}° = -1.28$ V

    Why is plutonium one of the most toxic substances known?

43. Fusion processes are more likely to occur for lighter elements, while fission processes are more likely to occur for heavier elements. Explain.

44. Why are elevated temperatures necessary to initiate fusion reactions but not fission reactions?

45. Much of the research on controlled fusion focuses on the problem of how to contain the reacting material. Magnetic fields appear to be the most promising mode of containment. Why is containment such a problem? Why must one resort to magnetic fields for containment?

46. How might the discovery of high temperature superconducting materials affect the feasibility of using fusion as a practical power source?

47. What are the purposes of the moderator and control rods in a fission reactor?

48. Many metals become brittle when subjected to intense radiation from radioactive materials. Explain this observation. How might this fact affect the design of nuclear reactors?

49. Zirconium is one of the few metals that retain their structural integrity upon exposure to radiation. The fuel rods in most nuclear reactors are, therefore, made of zirconium. Answer the following questions about the redox properties of zirconium based on the half-reaction:

$$ZrO_2 \cdot H_2O + H_2O + 4e^- \rightarrow Zr + 4OH^- \qquad \mathscr{E}° = -2.36 \text{ V}$$

a. Is zirconium metal capable of reducing water to form hydrogen gas?
b. Write a balanced equation for the reduction of water by zirconium.
c. Calculate $\mathscr{E}°$, $\Delta G°$, and K for the reduction of water by zirconium metal.
d. The reduction of water by zirconium occurred during the accidents at both Three Mile Island in 1979 and Chernobyl in 1986. At Three Mile Island the hydrogen produced was successfully vented and no chemical explosion occurred. At Chernobyl the hydrogen exploded destroying the reactor. If 1.00×10^3 kg of Zr reacted, what mass of H_2 would be produced? What volume of H_2 at 1 atm and 1000°C would be produced?
e. Considering all of the consequences, do you think it was a correct decision to vent the hydrogen along with other radioactive gases into the atmosphere at Three Mile Island? Explain.

50. In addition to the process described in the text, a second process called the carbon-nitrogen cycle occurs in the sun:

$$^{1}_{1}H + ^{12}_{6}C \rightarrow ^{13}_{7}N + ^{0}_{0}\gamma$$
$$^{13}_{7}N \rightarrow ^{13}_{6}C + ^{0}_{+1}e$$
$$^{1}_{1}H + ^{13}_{6}C \rightarrow ^{14}_{7}N + ^{0}_{0}\gamma$$
$$^{1}_{1}H + ^{14}_{7}N \rightarrow ^{15}_{8}O + ^{0}_{0}\gamma$$
$$^{15}_{8}O \rightarrow ^{15}_{7}N + ^{0}_{+1}e$$
$$^{1}_{1}H + ^{15}_{7}N \rightarrow ^{12}_{6}C + ^{4}_{2}He + ^{0}_{0}\gamma$$

Overall reaction: $\quad 4\,^{1}_{1}H \rightarrow ^{4}_{2}He + 2\,^{0}_{+1}e$

a. What is the catalyst in the above scheme?
b. What nucleons are intermediates?
c. How much energy is released per mole of hydrogen atoms for the overall reaction? (See Exercises 32 and 33 for the appropriate mass values.)

51. Which do you think would be the greater health hazard: the release of a radioactive nuclide of Sr or a radioactive nuclide of Xe into the environment? Assume the amount of radioactivity is the same in each case. Explain your answer on the basis of the chemical properties of Sr and Xe.

CHAPTER 22

Organic Chemistry

CONTENTS

Two Group 4A elements, carbon and silicon, form the basis of most natural substances. Silicon, with its great affinity for oxygen, forms chains and rings containing Si—O—Si bridges to produce the silica and silicates that form the basis for most rocks, sands, and soils. What silicon is to the geological world, carbon is to the biological world. Carbon has the unusual ability of bonding strongly to itself to form long chains or rings of carbon atoms. In addition, carbon forms strong bonds to other nonmetals such as hydrogen, nitrogen, oxygen, sulfur, and the halogens. Because of these bonding properties, there are a myriad of carbon compounds; several million are now known, and the number continues to grow rapidly. Among these many compounds are the **biomolecules,** those responsible for maintaining and reproducing life.

The study of carbon-containing compounds and their properties is called **organic chemistry.** Although a few compounds of carbon, such as its oxides and carbonates, are considered to be inorganic substances, the vast majority are organic compounds that typically contain chains or rings of carbon atoms.

Originally, the distinction between inorganic and organic substances was based on whether or not they were produced by living systems. For example, until the early nineteenth century, it was believed that organic compounds had some sort of "life force" and could only be synthesized by living organisms. This view was dispelled in 1828 when German chemist Friedrich Wöhler (1800–1882) prepared urea from the inorganic salt ammonium cyanate by simple heating:

$$\underset{\text{Ammonium cyanate}}{NH_4OCN} \xrightarrow{\text{Heat}} \underset{\text{Urea}}{H_2N{-}\overset{\displaystyle O}{\overset{\|}{C}}{-}NH_2}$$

< A photomicrograph (×35) of an organic flame retardant additive used to make polymers safer for clothing and building materials.

Since urea is a component of urine, it is clearly an organic material, yet here was clear evidence that it could be produced in the laboratory as well as by living things.

Organic chemistry plays a vital role in our quest to understand living systems. Beyond that, the synthetic fibers, plastics, artificial sweeteners, and drugs that are such an accepted part of modern life are products of industrial organic chemistry. In addition, the energy on which we rely so heavily to power our civilization is based mostly on the organic materials found in coal and petroleum.

Because organic chemistry is such a vast subject, we can give only a brief introduction to it in this book. We will begin with the simplest class of organic compounds, the hydrocarbons, and then show how most other organic compounds can be considered to be derivatives of hydrocarbons.

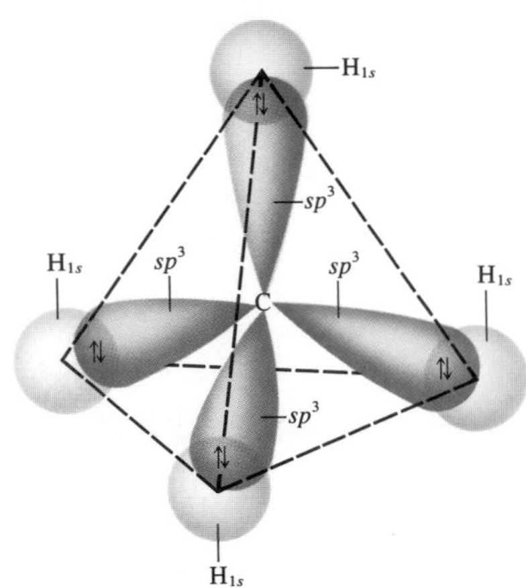

Figure 22.1

The C—H bonds in methane.

22.1 Alkanes: Saturated Hydrocarbons

Purpose

- To describe the nomenclature system of organic chemistry.
- To discuss isomerism in organic molecules.
- To give the rules for naming alkanes.

As the name indicates, **hydrocarbons** are compounds composed of carbon and hydrogen. Those whose carbon-carbon bonds are all single bonds are said to be *saturated,* because each carbon is bound to four atoms, the maximum number. Hydrocarbons containing carbon-carbon multiple bonds are described as being *unsaturated,* since the carbon atoms involved in a multiple bond can react with additional atoms, as shown by the *addition* of hydrogen to ethylene:

$$\underset{\text{Unsaturated}}{\mathrm{H_2C{=}CH_2}} + \mathrm{H_2} \longrightarrow \underset{\text{Saturated}}{\mathrm{H{-}CH_2{-}CH_2{-}H}}$$

Note that each carbon in ethylene is bonded to three atoms (one carbon and two hydrogens) but can bond to one additional atom if one bond of the carbon-carbon double bond is broken.

The simplest member of the saturated hydrocarbons, which are also called the **alkanes,** is *methane* (CH_4). As discussed in Section 9.1, methane has a tetrahedral structure and can be described in terms of a carbon atom using an sp^3 hybrid set of orbitals to bond to the four hydrogen atoms (see Fig. 22.1). The next alkane, the one containing two carbon atoms, is *ethane* (C_2H_6), as shown in Fig. 22.2. Each carbon in ethane is surrounded by four atoms and thus adopts a tetrahedral arrangement and sp^3 hybridization, as expected from the localized electron model.

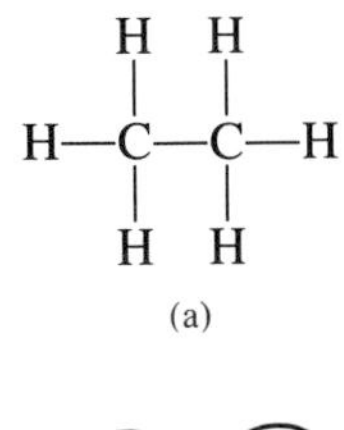

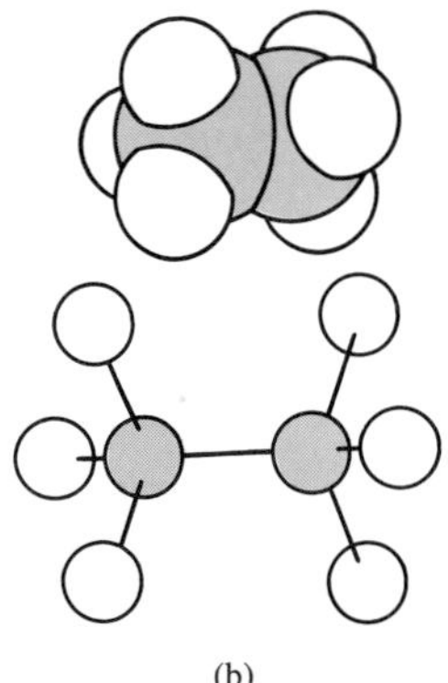

Figure 22.2

(a) The Lewis structure of ethane (C_2H_6). (b) The molecular structure of ethane represented by ball-and-stick and space-filling models.

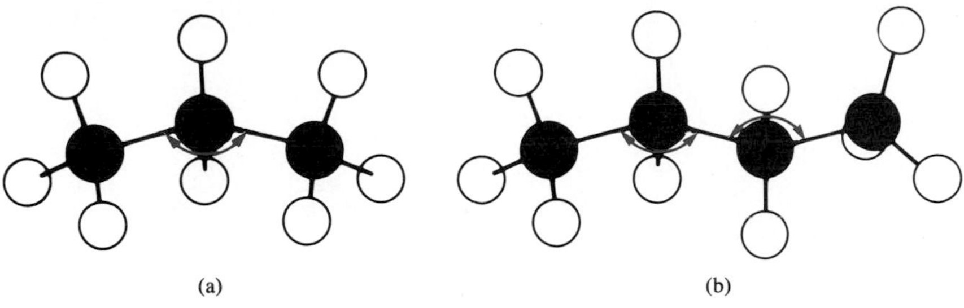

Figure 22.3

The structures of (a) propane ($CH_3CH_2CH_3$) and (b) butane ($CH_3CH_2CH_2CH_3$). Each angle shown in red is 109.5°.

The next two members of the series are *propane* (C_3H_8) and *butane* (C_4H_{10}), shown in Fig. 22.3. Again, each carbon is bonded to four atoms and is described as sp^3 hybridized.

Alkanes in which the carbon atoms form long "strings" or chains are called **normal, straight-chain,** or **unbranched hydrocarbons.** As can be seen from Fig. 22.3, the chains in normal alkanes are not really straight but zig-zag, since the tetrahedral C—C—C angle is 109.5°. The normal alkanes can be represented by the structure

$$\begin{array}{c}H\\ \diagdown\end{array}\!\!\!\!\!\!\begin{array}{c}\\ H-C\\ \\ \end{array}\!\!\!\!\!\!\begin{array}{c}\\ \\ \diagup\\ H\end{array}\left(\begin{array}{c}H\\ |\\ C\\ |\\ H\end{array}\right)_{n}\begin{array}{c}H\\ \diagup\\ C-H\\ \diagdown\\ H\end{array}$$

where n is an integer. Note that each member is obtained from the previous one by inserting a *methylene* (CH_2) group. We can condense the structural formulas by omitting some of the C—H bonds. For example, the general formula for normal alkanes shown above can be condensed to

$$CH_3-(CH_2)_n-CH_3$$

The first ten normal alkanes and some of their properties are listed in Table 22.1. Note that all alkanes can be represented by the general formula C_nH_{2n+2}. For

Selected Properties of the First Ten Normal Alkanes

Name	Formula	Molecular weight	Melting point (°C)	Boiling point (°C)	Number of structural isomers
Methane	CH_4	16	−183	−162	1
Ethane	C_2H_6	30	−172	−89	1
Propane	C_3H_8	44	−187	−42	1
Butane	C_4H_{10}	58	−138	0	2
Pentane	C_5H_{12}	72	−130	36	3
Hexane	C_6H_{14}	86	−95	68	5
Heptane	C_7H_{16}	100	−91	98	9
Octane	C_8H_{18}	114	−57	126	18
Nonane	C_9H_{20}	128	−54	151	35
Decane	$C_{10}H_{22}$	142	−30	174	75

Table 22.1

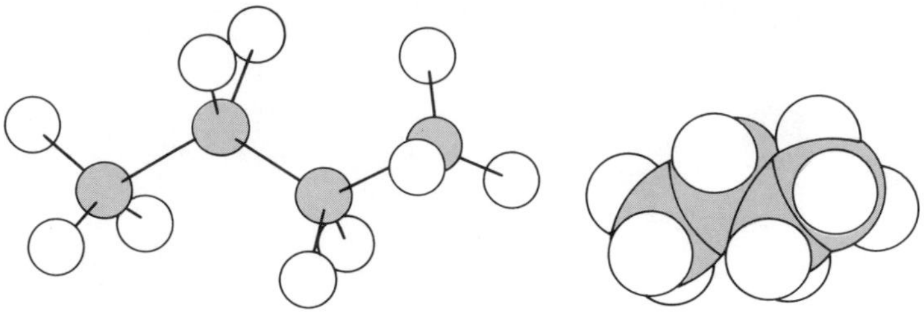

(a)

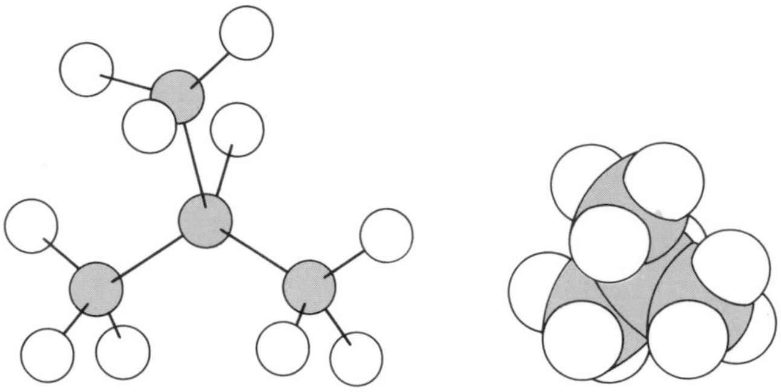

(b)

Figure 22.4

(a) Normal butane (abbreviated *n*-butane). (b) The branched isomer of butane (called isobutane).

example, nonane, which has nine carbon atoms, is represented by $C_9H_{(2\times9)+2}$, or C_9H_{20}. Also note from Table 22.1 that the melting points and boiling points increase as the molecular weights increase, as we would expect.

Isomerism in Alkanes

Butane and all succeeding members of the alkanes exhibit **structural isomerism.** Recall from Section 20.4 that structural isomerism occurs when two molecules have the same atoms but different bonds. For example, butane can exist as a straight-chain molecule (normal butane, or *n*-butane) or with a branched-chain structure (called isobutane), as shown in Fig. 22.4. Because of their different structures, these molecules show different properties. For example, the boiling point of *n*-butane is −0.5°C and that of isobutane is −12°C.

Sample Exercise 22.1

Draw the isomers of pentane.

Solution

Pentane (C_5H_{12}) has the following isomeric structures:

1.

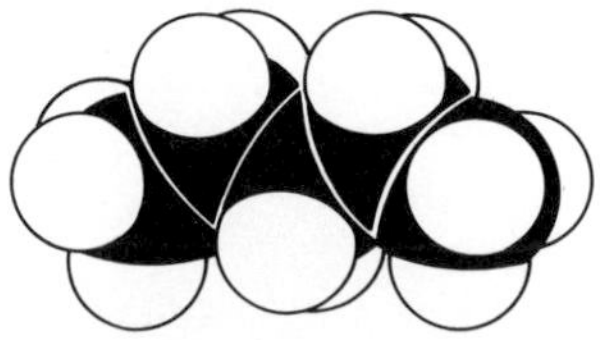

$CH_3—CH_2—CH_2—CH_2—CH_3$

n-Pentane

Chemical Impact

Methane from Termites

Along with carbon dioxide, water vapor, and others, methane is one of the components of the atmosphere that help control the earth's temperature (see the description of the greenhouse effect in Section 6.5). Scientists are therefore very interested in the fact that the methane content of the atmosphere seems to be increasing by about 2% annually.

One source of this increase appears to be termites. When termites consume wood, cellulose (see Section 23.2) is broken down into methane, carbon dioxide, and other compounds by microorganisms that inhabit the termites' digestive system. In fact, it has been estimated that termites produce 165 million tons of methane and 55 million tons of carbon dioxide annually.

Clearly, the amounts of methane and carbon dioxide produced by termites are significant and may be increasing due to changes in the environment that favor a larger termite population. The potential effects of these changes on the earth's atmosphere are being monitored closely by researchers. Too much methane in the atmosphere could lead to significant warming of the earth, resulting in floods from melted ice.

Subterranean termites.

Sample Exercise 22.1, continued

2.

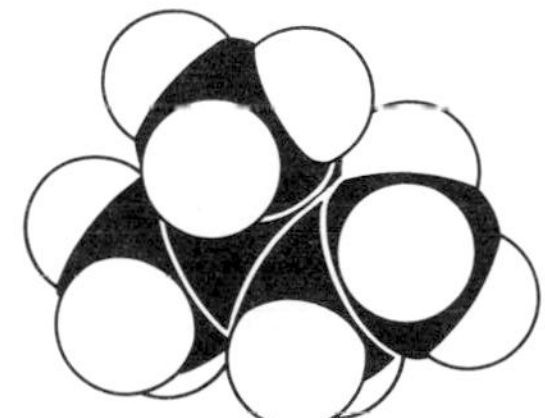

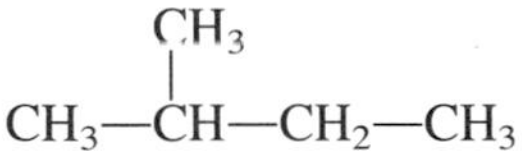

$$\begin{array}{c} \quad CH_3 \\ \quad | \\ CH_3—CH—CH_2—CH_3 \end{array}$$

Isopentane

3.

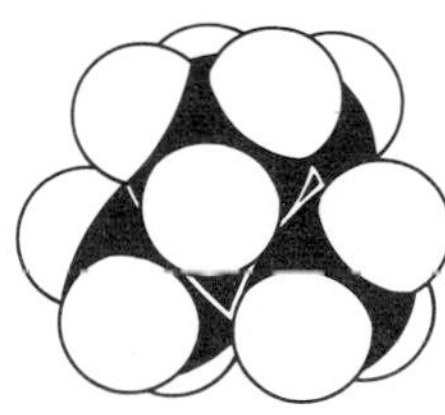

$$\begin{array}{c} CH_3 \\ | \\ CH_3—C—CH_3 \\ | \\ CH_3 \end{array}$$

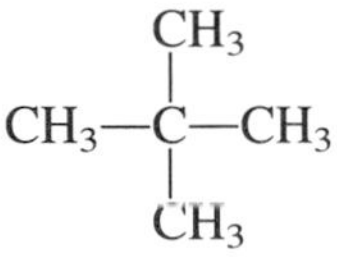

Neopentane

Note that the structures

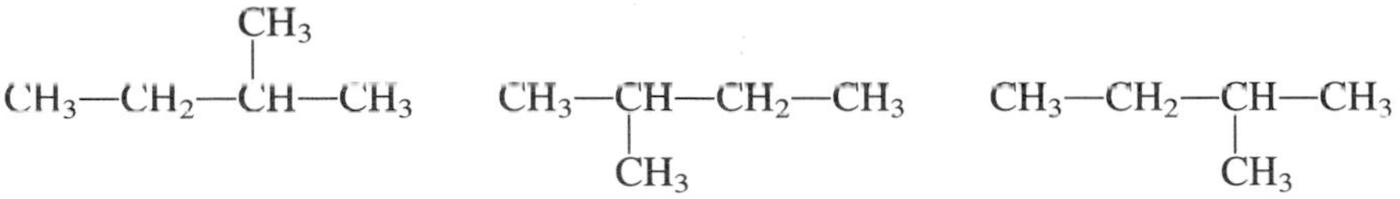

$$\begin{array}{c} \qquad\quad CH_3 \\ \qquad\quad | \\ CH_3—CH_2—CH—CH_3 \end{array} \qquad \begin{array}{c} CH_3—CH—CH_2—CH_3 \\ | \qquad\quad \\ CH_3 \qquad\quad \end{array} \qquad \begin{array}{c} CH_3—CH_2—CH—CH_3 \\ \qquad\quad | \\ \qquad\quad CH_3 \end{array}$$

which might appear to be other isomers, are actually identical to structure 2.

Nomenclature

Because there are literally millions of organic compounds, it would be impossible to remember common names for all of them. We must have a systematic method for naming them. The following rules are used in naming alkanes.

Rules for Naming Alkanes

1. The names of the alkanes beyond butane are obtained by adding the suffix *-ane* to the Greek root for the number of carbon atoms (*pent-* for five, *hex-* for six, etc.). For a branched hydrocarbon, the longest continuous chain of carbon atoms gives the root name for the hydrocarbon. For example, in the alkane

```
                CH3
                 |
                CH2      Six carbons
                 |
                CH2
                 |
CH3—CH2—CH—CH2—CH3
```

   the longest chain contains six carbon atoms, and this compound is named as a hexane.

2. When alkane groups appear as substituents, they are named by dropping the *-ane* and adding *-yl*. For example, —CH_3 is obtained by removing a hydrogen from methane and is called *methyl,* —C_2H_5 is called *ethyl,* —C_3H_7 is called *propyl,* and so on. The compound above is therefore an ethylhexane. (See Table 22.2.)

3. The positions of substituent groups are specified by numbering the longest chain of carbon atoms sequentially, starting at the end closest to the branching. For example, the compound

```
           CH3
            |
CH3—CH2—CH—CH2—CH2—CH3

 1    2    3   4    5    6     Correct numbering
 6    5    4   3    2    1     Incorrect numbering
```

   is called 3-methylhexane. Note that the top set of numbers is correct since the left end of the molecule is closest to the branching, and this gives the smallest number for the position of the substituent. Also, note that a hyphen is written between the number and the substituent name.

4. The location and name of each substituent are followed by the root alkane name. The substituents are listed in alphabetical order, and the prefixes, *di-*, *tri-*, etc., are used to indicate multiple identical substituents.

Sample Exercise 22.2

Draw the structural isomers for the alkane C_6H_{14} and give the systematic name for each one.

Solution

We will proceed systematically, starting with the longest chain and then rearranging the carbons to form the shorter, branched chains.

The Most Common Alkyl Substituents and Their Names

Structure*	Name**
$-CH_3$	Methyl
$-CH_2CH_3$	Ethyl
$-CH_2CH_2CH_3$	Propyl
$CH_3\overset{\vert}{C}HCH_3$	Isopropyl
$-CH_2CH_2CH_2CH_3$	Butyl
$CH_3\overset{\vert}{C}HCH_2CH_3$	*Sec*-butyl
$-CH_2-\underset{\underset{CH_3}{\vert}}{\overset{\overset{H}{\vert}}{C}}-CH_3$	Isobutyl
$-\underset{\underset{CH_3}{\vert}}{\overset{\overset{CH_3}{\vert}}{C}}-CH_3$	*Tert*-butyl

*The bond with one end open shows the point of attachment of the substituent to the carbon chain.
**For the butyl groups, *sec*- indicates attachment to the chain through a secondary carbon, a carbon atom attached to *two* other carbon atoms. The designation *tert*- signifies attachment through a tertiary carbon, a carbon attached to *three* other carbon atoms.

Table 22.2

Sample Exercise 22.2, continued

1. $CH_3CH_2CH_2CH_2CH_2CH_3$ Hexane
   Note that a structure such as

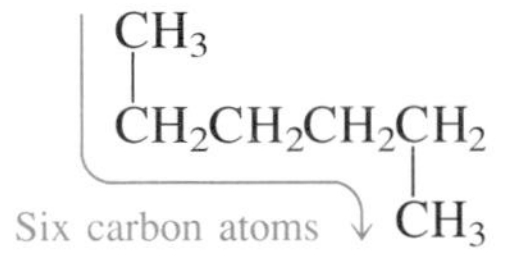

$$\begin{array}{l} CH_3 \\ \vert \\ CH_2CH_2CH_2CH_2 \\ \qquad\qquad\quad\ \vert \\ \qquad\qquad\quad CH_3 \end{array}$$

Six carbon atoms

   may look different but is still hexane since the longest carbon chain has six atoms.

2. We now take one carbon out of the chain and make it a methyl substituent.

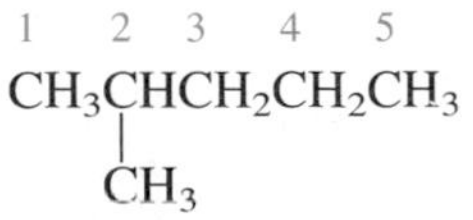

$$\overset{1}{C}H_3\overset{2}{C}H\overset{3}{C}H_2\overset{4}{C}H_2\overset{5}{C}H_3 \quad \text{(with } CH_3 \text{ on carbon 2)}$$

2-Methylpentane

   Since the longest chain has five carbons, this is a substituted pentane: 2-methylpentane. The 2 indicates the position of the methyl group on the chain. Note that if we numbered the chain from the right end, the methyl group would be on carbon 4. We want the smallest possible number, so the numbering shown is correct.

Sample Exercise 22.2, continued

3. The methyl substituent can also be on carbon 3 to give

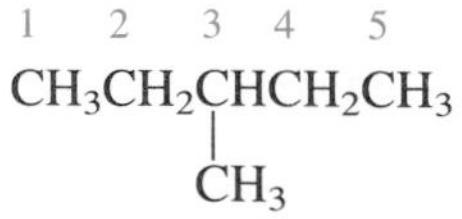

$$\begin{array}{l} \text{1 2 3 4 5} \\ CH_3CH_2\underset{\displaystyle CH_3}{\underset{|}{C}}HCH_2CH_3 \end{array} \qquad \text{3-Methylpentane}$$

Note that we have now exhausted all possibilities for placing a single methyl group on pentane.

4. Next we can take two carbons out of the original six-member chain:

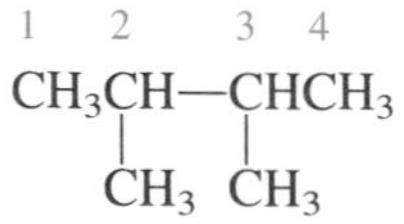

$$CH_3\underset{\displaystyle CH_3}{\underset{|}{C}}H—\underset{\displaystyle CH_3}{\underset{|}{C}}HCH_3 \qquad \text{2,3-Dimethylbutane}$$

(carbons numbered 1 2 3 4)

Since the longest chain now has four carbons, the root name is butane. Since there are two methyl groups, we use the prefix *di-*. The numbers denote that the two methyl groups are positioned on the second and third carbons in the butane chain. Note that when two or more numbers are used, they are separated by a comma.

5. The two methyl groups can also be attached to the same carbon atom as shown here:

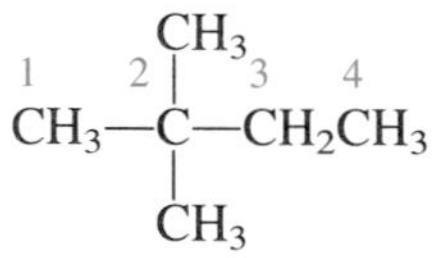

$$CH_3—\overset{\displaystyle CH_3}{\overset{|}{\underset{\displaystyle CH_3}{\underset{|}{C}}}}—CH_2CH_3 \qquad \text{2,2-Dimethylbutane}$$

(carbons numbered 1 2 3 4)

We might also try ethyl-substituted butanes, such as

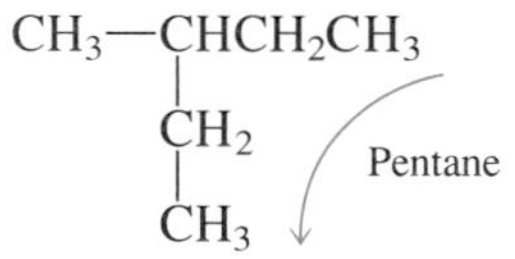

$$CH_3—\underset{\displaystyle \underset{\displaystyle \underset{|}{}CH_3}{CH_2}}{\underset{|}{C}}HCH_2CH_3$$

Pentane

However, note that this is in fact a pentane (3-methylpentane), since the longest chain has five carbon atoms. Thus this is not a new isomer. Trying to reduce the chain to three atoms provides no further isomers either. For example, the structure

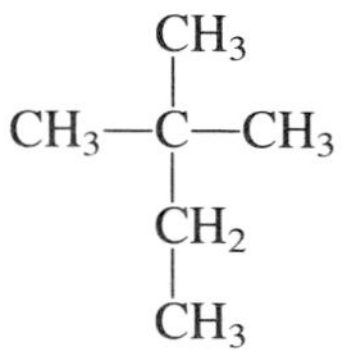

$$CH_3—\overset{\displaystyle CH_3}{\overset{|}{\underset{\displaystyle \underset{\displaystyle \underset{|}{}CH_3}{CH_2}}{\underset{|}{C}}}}—CH_3$$

is really 2,2-dimethylbutane.

Thus there are only five distinct structural isomers of C_6H_{14}: hexane, 2-methylpentane, 3-methylpentane, 2,3-dimethylbutane, and 2,2-dimethylbutane.

Sample Exercise 22.3

Write the structure for each of the following compounds:

a. 4-ethyl-3,5-dimethylnonane **b.** 4-*tert*-butylheptane

Solution

a. The root name nonane signifies a nine-carbon chain. Thus we have

$$\begin{array}{l}
\quad 1 \quad\; 2 \quad\;\; 3 \qquad\quad 4 \qquad\quad 5 \quad 6 \quad\; 7 \quad\; 8 \quad\; 9 \\
CH_3CH_2\underset{\displaystyle CH_3}{\underset{|}{CH}}-\underset{\displaystyle \underset{\displaystyle CH_3}{\underset{|}{CH_2}}}{\underset{|}{CH}}-\underset{\displaystyle CH_3}{\underset{|}{C}}HCH_2CH_2CH_2CH_3
\end{array}$$

b. Heptane signifies a seven-carbon chain, and the *tert*-butyl group is

$$H_3C-\underset{\displaystyle CH_3}{\underset{|}{\overset{|}{C}}}-CH_3$$

Thus we have

$$\begin{array}{l}
\quad 1 \quad\;\; 2 \quad\;\;\; 3 \quad\;\; 4 \;\; 5 \quad\;\; 6 \quad\;\; 7 \\
CH_3CH_2CH_2\underset{\displaystyle H_3C-\underset{\displaystyle CH_3}{\underset{|}{C}}-CH_3}{\underset{|}{C}}HCH_2CH_2CH_3
\end{array}$$

Reactions of Alkanes

Because they are saturated compounds and because the C—C and C—H bonds are relatively strong, the alkanes are fairly unreactive. For example, at 25°C they do not react with acids, bases, or strong oxidizing agents. This chemical inertness makes them valuable as lubricating materials and as the backbone for structural materials such as plastics.

At a sufficiently high temperature, alkanes do react vigorously and exothermically with oxygen, and these **combustion reactions** are the basis for their widespread use as fuels. For example, the reaction of butane with oxygen is

$$2C_4H_{10}(g) + 13O_2(g) \rightarrow 8CO_2(g) + 10H_2O(g)$$

The alkanes can also undergo **substitution reactions,** primarily where halogen atoms replace hydrogen atoms. For example, methane can be successively chlorinated as follows:

$$CH_4 + Cl_2 \xrightarrow{h\nu} \underset{\text{Chloromethane}}{CH_3Cl} + HCl$$

$$CH_3Cl + Cl_2 \xrightarrow{h\nu} \underset{\text{Dichloromethane}}{CH_2Cl_2} + HCl$$

Pouring motor oil, a mixture containing many different alkanes.

$$CH_2Cl_2 + Cl_2 \xrightarrow{h\nu} \underset{\text{Trichloromethane (chloroform)}}{CHCl_3} + HCl$$

$$CHCl_3 + Cl_2 \xrightarrow{h\nu} \underset{\text{Tetrachloromethane (carbon tetrachloride)}}{CCl_4} + HCl$$

Note that the products of the last two reactions have two names; the systematic one is given first, followed by the common name in parentheses. (This format will be used throughout this chapter for compounds that have common names.) Also, the $h\nu$ above each arrow signifies that ultraviolet light is used to furnish the energy to break the Cl—Cl bond to produce chlorine atoms:

$$Cl_2 \rightarrow Cl\cdot + Cl\cdot$$

A chlorine atom has an unpaired electron, indicated by the dot, which makes it very reactive and able to attack the C—H bond.

As we mentioned before, substituted methanes that contain both chlorine and fluorine as substituents, with the general formula CF_xCl_{4-x}, are called *Freons*. These substances are very unreactive and are extensively used as coolant fluids in refrigerators and air conditioners. Unfortunately, their chemical inertness allows them to remain in the atmosphere so long that they eventually reach altitudes where they may be a threat to the protective ozone layer (see Section 12.8).

Alkanes can also undergo **dehydrogenation reactions** in which hydrogen atoms are removed and the product is an unsaturated hydrocarbon. For example, in the presence of chromium(III) oxide at high temperatures, ethane can be dehydrogenated, yielding ethylene:

$$CH_3CH_3 \xrightarrow[500°C]{Cr_2O_3} \underset{\text{Ethylene}}{CH_2{=}CH_2} + H_2$$

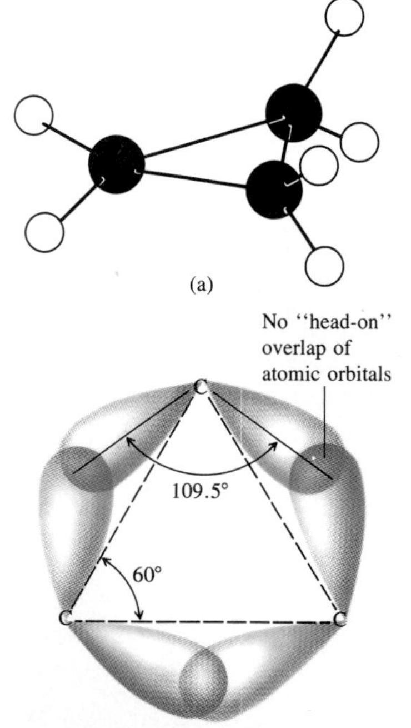

Figure 22.5

(a) The molecular structure of cyclopropane (C_3H_6). (b) The overlap of the sp^3 orbitals that form the C—C bonds in cyclopropane.

Cyclic Alkanes

Besides forming chains, carbon atoms can also form rings. The simplest of the **cyclic alkanes** (general formula C_nH_{2n}) is cyclopropane (C_3H_6), as shown in Fig. 22.5(a). Since the carbon atoms in cyclopropane form an equilateral triangle with 60° bond angles, their sp^3 hybrid orbitals do not overlap head-on as they do in normal alkanes [Fig. 22.5(b)]. The result is unusually weak, or *strained,* C—C bonds, and the cyclopropane molecule is much more reactive than straight-chain propane. The carbon atoms in cyclobutane (C_4H_8) form a square with 90° bond angles, and cyclobutane is also quite reactive.

The next two members of the series, cyclopentane (C_5H_{10}) and cyclohexane (C_6H_{12}), are quite stable because their rings have bond angles very close to the tetrahedral angles, which allows the sp^3 hybrid orbitals on adjacent carbon atoms to overlap "head-on" and form normal C—C bonds, which are quite strong. To attain the tetrahedral angles, the cyclohexane ring must "pucker," that is, become nonplanar. Cyclohexane can exist in two forms, the *chair* and the *boat* forms, as shown in Fig. 22.6. The two hydrogen atoms above the ring in the boat form are quite close to each other, and the resulting repulsion between them causes the chair form to be preferred. At 25°C more than 99% of cyclohexane is in the chair form.

For simplicity, the cyclic alkanes are often represented by the following structures:

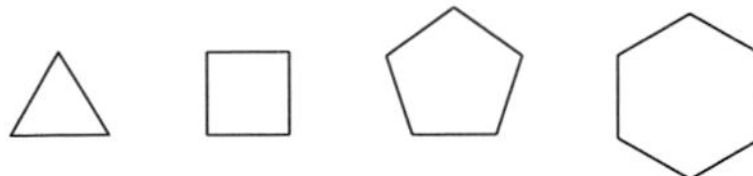

Thus the structure

CH_3

represents methylcyclopropane.

The nomenclature for cycloalkanes follows the same rules as for the other alkanes except that the root name is preceded by the prefix *cyclo-*. The ring is numbered so as to give the smallest substituent numbers possible.

Chair

(a)

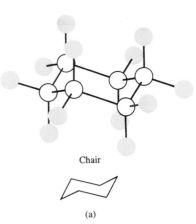

These two H atoms repel each other

Boat

(b)

Figure 22.6

The (a) chair and (b) boat forms of cyclohexane.

Sample Exercise 22.4

Name each of the following cyclic alkanes:

a. $CH_3—CH—CH_3$; CH_3

b.

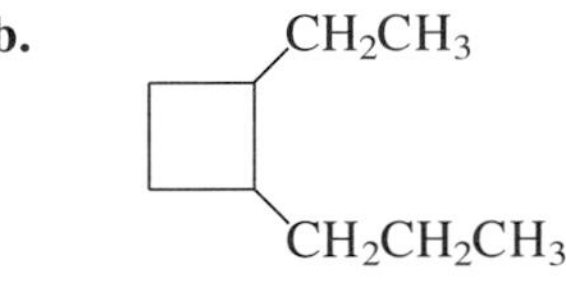

Solution

a. The six-carbon cyclohexane ring is numbered as follows:

$CH_3—CH—CH_3$; 1, 2, 3, 4, 5, 6 ; CH_3

There is an isopropyl group at carbon 1 and a methyl group at carbon 3. The name is 1-isopropyl-3-methylcyclohexane, since the alkyl groups are named in alphabetical order.

b. This is a cyclobutane ring, which is numbered as follows:

CH_2CH_3 ; 4, 1, 3, 2 ; $CH_2CH_2CH_3$

The name is 1-ethyl-2-propylcyclobutane.

22.2 Alkenes and Alkynes

Purpose

- To introduce unsaturated hydrocarbons.
- To discuss isomerism in alkenes.
- To describe common reactions involving multiple carbon-carbon bonds.

Multiple carbon-carbon bonds result when hydrogen atoms are removed from alkanes. Hydrocarbons that contain a carbon-carbon double bond are called **alkenes,** and have the general formula C_nH_{2n}. The simplest alkene (C_2H_4), commonly known as *ethylene,* has the Lewis structure

$$\begin{matrix} H & & H \\ & C{=}C & \\ H & & H \end{matrix}$$

As we discussed in Section 9.1, each carbon in ethylene can be described as sp^2 hybridized. The C—C σ bond is formed by sharing an electron pair between sp^2 orbitals, and the π bond is formed by sharing a pair of electrons between p orbitals (Fig. 22.7).

The systematic nomenclature for alkenes is quite similar to that for alkanes.

1. The root hydrocarbon name ends in *-ene* rather than *-ane*. Thus the systematic name for C_2H_4 is *ethene* and that for C_3H_6 is propene.
2. In alkenes with more than three carbon atoms, the location of the double bond is indicated by the lowest numbered carbon atom involved in the bond. Thus $CH_2{=}CHCH_2CH_3$ is called 1-butene, and $CH_3CH{=}CHCH_3$ is called 2-butene.

Note from Fig. 22.7 that the p orbitals on the two carbon atoms in ethylene must be lined up (parallel) to allow formation of the π bond. This prevents rotation of the two CH_2 groups relative to each other at ordinary temperatures, and is in contrast to alkanes where free rotation is possible (see Fig. 22.8). The restricted rotation about doubly bonded carbon atoms means that alkenes exhibit ***cis-trans* isomerism.** For example, there are two stereoisomers of 2-butene (Fig. 22.9). Identical substituents on the same side of the double bond are designated *cis* and on opposite sides are labeled *trans*.

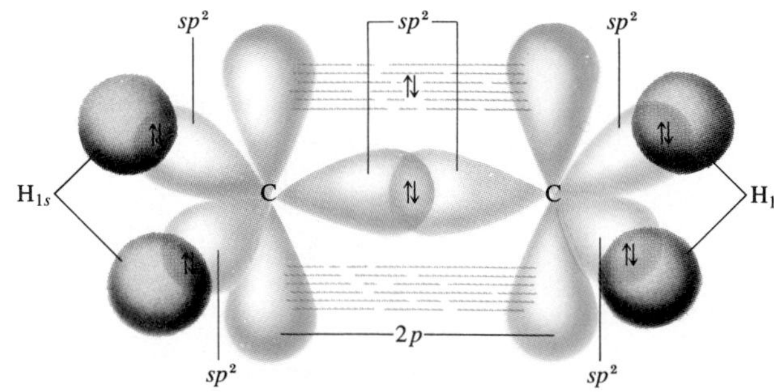

Figure 22.7
The bonding in ethylene.

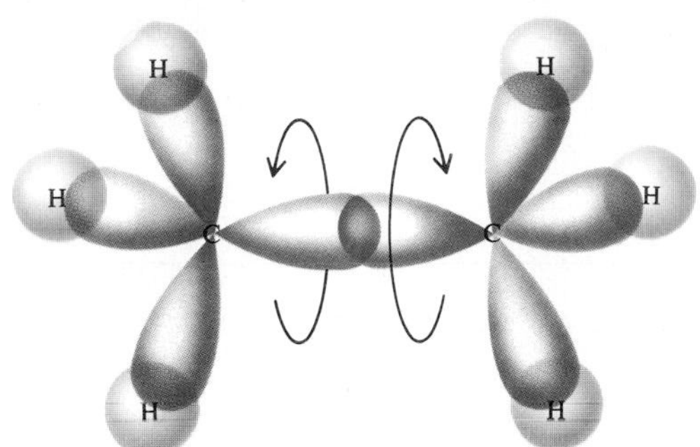

Figure 22.8

The bonding in ethane.

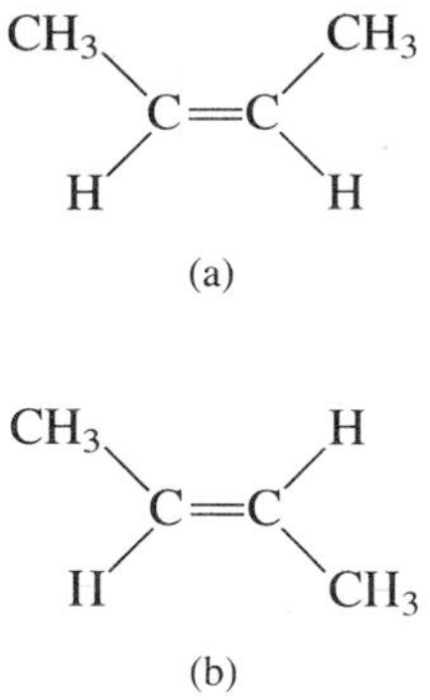

CH_3 CH_3
C=C
H H

(a)

CH_3 H
C=C
H CH_3

(b)

Figure 22.9

The two stereoisomers of 2-butene: (a) *cis*-2-butene and (b) *trans*-2-butene.

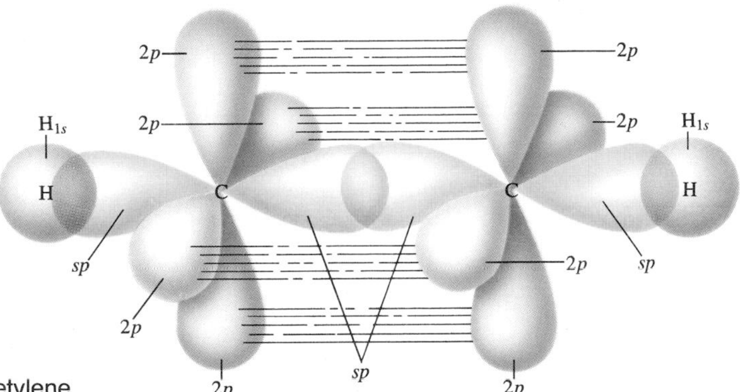

Figure 22.10

The bonding in acetylene.

Alkynes are unsaturated hydrocarbons containing a triple carbon-carbon bond. The simplest alkyne is C_2H_2, commonly called *acetylene,* which has the systematic name *ethyne*. As discussed in Section 9.1, the triple bond in acetylene can best be described as one σ bond between two *sp* hybrid orbitals on the two carbon atoms and two π bonds involving two $2p$ orbitals on each carbon atom (Fig. 22.10).

The nomenclature for alkynes involves use of *-yne* as a suffix to replace the *-ane* of the parent alkane. Thus the molecule $CH_3CH_2C{\equiv}CCH_3$ has the name 2-pentyne.

Like alkanes, unsaturated hydrocarbons can exist as ringed structures, for example:

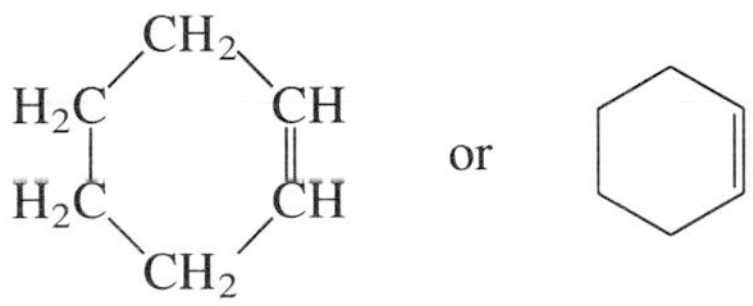

CH_2
H_2C CH
H_2C CH
CH_2

or

Cyclohexene

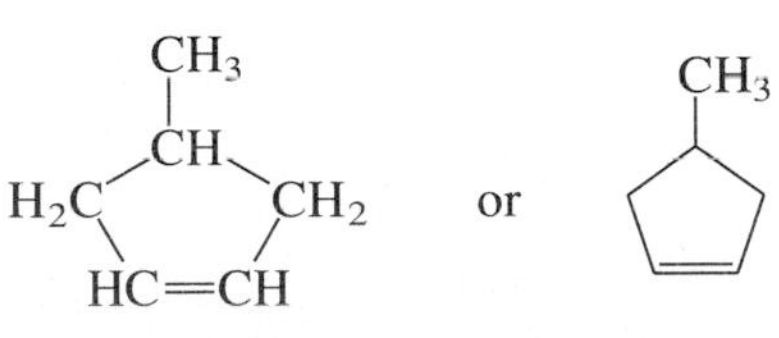

CH_3
CH
H_2C CH_2
HC=CH

or

CH_3

4-Methyl-cyclopentene

For cyclic alkenes, number through the double bond toward the substituent.

Sample Exercise 22.5

Name each of the following molecules:

a.

$$\begin{array}{l} \text{H} \quad\quad\quad \text{CH}_3 \\ \quad \diagdown \quad \diagup \\ \quad\; \text{C}{=}\text{C} \\ \quad \diagup \quad \diagdown \\ \text{CH}_3\text{CH}_2\underset{\displaystyle \text{CH}_3}{\underset{|}{\text{CH}}} \quad \text{H} \end{array}$$

b.

$$\text{CH}_3\text{CH}_2\text{C}{\equiv}\text{C}\underset{\underset{\displaystyle \text{CH}_3}{|}}{\underset{\displaystyle \text{CH}_2}{\underset{|}{\text{C}}}}\text{HCH}_2\text{CH}_3$$

Solution

a. The longest chain contains six carbon atoms and the chain is numbered as follows:

$$\begin{array}{l} \quad\quad\quad\quad \text{H}\;{}^{3} \quad {}^{2}\;\overset{1}{\text{CH}_3} \\ {}^{6}\quad{}^{5}\quad{}^{4} \quad \text{C}{=}\text{C} \\ \text{CH}_3\text{CH}_2\underset{\displaystyle \text{CH}_3}{\underset{|}{\text{CH}}} \quad\quad \text{H} \end{array}$$

Thus the hydrocarbon is a 2-hexene. Since the hydrogen atoms are on opposite sides of the double bond, the molecule is the *trans* isomer. The name is 4-methyl-*trans*-2-hexene.

b. The longest chain of carbon atoms is seven, and the chain is numbered as shown (giving the triple bond the lowest possible number):

$$\overset{1}{\text{CH}_3}\overset{2}{\text{CH}_2}\overset{3}{\text{C}}{\equiv}\overset{4}{\text{C}}\underset{\underset{\displaystyle \text{CH}_3}{|}}{\underset{\displaystyle \text{CH}_2}{\underset{|}{\overset{5}{\text{C}}}}}\text{H}\overset{6}{\text{CH}_2}\overset{7}{\text{CH}_3}$$

The hydrocarbon is a 3-heptyne. The full name is 5-ethyl-3-heptyne, where the position of the triple bond is indicated by the lower-numbered carbon atom involved in this bond.

Reactions of Alkenes and Alkynes

Because alkenes and alkynes are unsaturated, their most important reactions are **addition reactions,** in which π bonds, which are weaker than the C—C σ bonds, are broken, and new σ bonds are formed to the atoms being added. For example, **hydrogenation reactions** involve the addition of hydrogen atoms:

$$\underset{\text{1-Propene}}{\text{CH}_2{=}\text{CHCH}_3} + \text{H}_2 \xrightarrow{\text{Catalyst}} \underset{\text{Propane}}{\text{CH}_3\text{CH}_2\text{CH}_3}$$

In order for this reaction to proceed rapidly at normal temperatures, a catalyst of platinum, palladium, or nickel is employed. The catalyst serves to help break the relatively strong H—H bond, as was discussed in Section 12.8. Hydrogenation of alkenes is an important industrial process, particularly in the manufacture of solid shortenings where unsaturated fats (fats containing double bonds), which are generally liquid, are converted to solid saturated fats.

Halogenation of unsaturated hydrocarbons involves addition of halogen atoms, for example:

$$CH_2{=}CHCH_2CH_2CH_3 + Br_2 \longrightarrow CH_2BrCHBrCH_2CH_2CH_3$$

1-Pentene 1,2-Dibromopentane

Another important reaction of certain unsaturated hydrocarbons is **polymerization,** a process in which many small molecules are joined together to form a large molecule. This will be discussed in Section 22.5.

The reaction of solid calcium carbide with water produces acetylene, C_2H_2, a flammable gas.

22.3 Aromatic Hydrocarbons

Purpose

- To discuss the structures and reactivities of aromatic hydrocarbons.

A special class of cyclic unsaturated hydrocarbons is known as **aromatic hydrocarbons.** The simplest of these is benzene (C_6H_6) with a planar ring structure as shown in Fig. 22.11(a). In the localized electron model of the bonding in benzene, resonance structures of the type shown in Fig. 22.11(b) are used to account for the known equivalence of all of the carbon-carbon bonds. But as we discussed in Section 9.5, the best description of the benzene molecule assumes that sp^2 hybrid orbitals on each carbon are used to form the C—C and C—H σ bonds, with the remaining $2p$ orbital on each carbon used to form π molecular orbitals. The delocalization of these π electrons is usually shown by a circle inside the ring [Fig. 22.11(c)].

The delocalization of the π electrons makes the benzene ring behave quite differently from a typical unsaturated hydrocarbon. As we have seen previously, unsaturated hydrocarbons typically undergo rapid addition reactions. Benzene does not. Rather it undergoes substitution reactions where *hydrogen atoms are replaced by other atoms,* for example:

$$C_6H_6 + Cl_2 \xrightarrow{FeCl_3} C_6H_5Cl + HCl$$

Chlorobenzene

$$C_6H_6 + HNO_3 \xrightarrow{H_2SO_4} C_6H_5NO_2 + H_2O$$

Nitrobenzene

$$C_6H_6 + CH_3Cl \xrightarrow{AlCl_3} C_6H_5CH_3 + HCl$$

Toluene

H, H, H, H, H, H; C, C, C, C, C, C; 120°, 120°, 120°

(a)

(b)

(c)

Figure 22.11

(a) The structure of benzene, a planar ring system in which all bond angles are 120°. (b) Two of the resonance structures of benzene. (c) The usual representation of benzene. The circle represents the electrons in the delocalized π system. All C—C bonds in benzene are equivalent.

In each case the substance shown over the arrow is needed to catalyze these substitution reactions.

Substitution reactions are characteristic of saturated hydrocarbons, and addition reactions are characteristic of unsaturated ones. The fact that benzene reacts more like a saturated hydrocarbon indicates the great stability of the delocalized π electron system.

The nomenclature of benzene derivatives is similar to the nomenclature for saturated ring systems. If there is more than one substituent, numbers are used to indicate substituent positions. For example, the compound

is named 1,2-dichlorobenzene. Another nomenclature system uses the prefix *ortho-* (*o-*) for two adjacent substituents, *meta-* (*m-*) for two substituents with one carbon between them, and *para-* (*p-*) for two substituents opposite each other. When benzene is used as a substituent, it is called the **phenyl group.** Examples of some aromatic compounds are shown in Fig. 22.12.

Benzene is the simplest aromatic molecule. More complex aromatic systems can be viewed as consisting of a number of "fused" benzene rings. Some examples are given in Table 22.3.

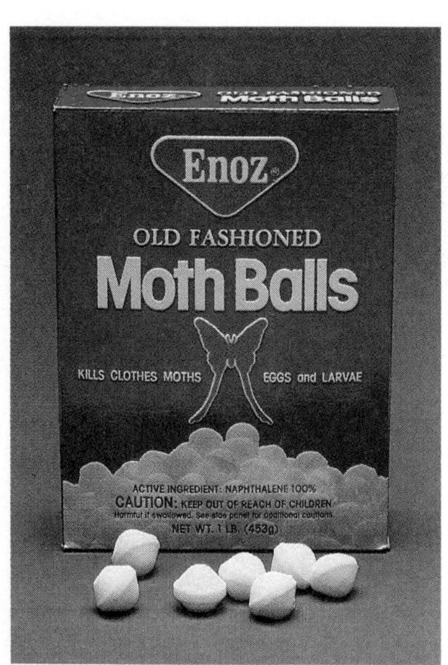

Mothballs contain napthalene.

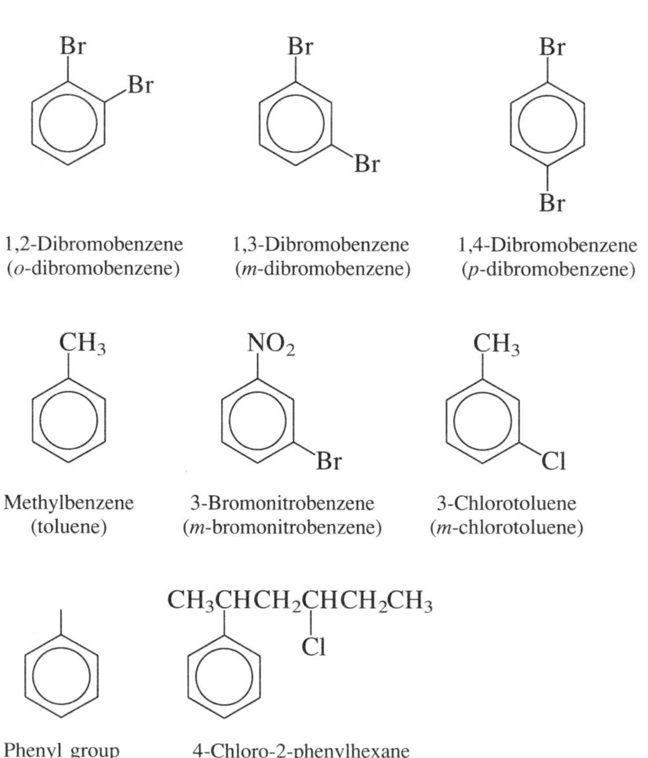

Figure 22.12

Some selected substituted benzenes and their names. Common names are given in parentheses.

More Complex Aromatic Systems

Structural formula	Name	Use or effect
	Naphthalene	Mothballs
	Anthracene	Dyes
	Phenanthrene	Dyes, explosives, and synthesis of drugs
	3,4-Benzpyrene	Active carcinogen found in smoke and smog

Table 22.3

Dyed nylon rope used to climb mountains in Nepal.

22.4 Hydrocarbon Derivatives

Purpose

- To introduce the basic functional groups and give a few characteristics of each one.

The vast majority of organic molecules contain elements in addition to carbon and hydrogen. However, most of these substances can be viewed as **hydrocarbon derivatives,** molecules that are fundamentally hydrocarbons but that have additional atoms or groups of atoms called **functional groups.** The common functional groups are listed in Table 22.4. Because each functional group exhibits characteristic chemistry, we will consider them separately.

Alcohols

Alcohols are characterized by the presence of the hydroxyl group (—OH). Some common alcohols are shown in Table 22.5. The systematic name for an alcohol is obtained by replacing the final *-e* of the parent hydrocarbon by *-ol*. The position of the —OH group is specified by a number (where necessary) chosen so that it is the smallest of the substituent numbers. Alcohols are classified according to the number of hydrocarbon fragments bonded to the carbon where the —OH group is attached,

The Common Functional Groups

Class	Functional group	General formula*	Example
Halohydrocarbons	—X (F,Cl,Br,I)	R—X	CH_3I Iodomethane (methyl iodide)
Alcohols	—OH	R—OH	CH_3OH Methanol (methyl alcohol)
Ethers	—O—	R—O—R′	CH_3OCH_3 Dimethyl ether
Aldehydes	$-\overset{\overset{\displaystyle O}{\|}}{C}-H$	$R-\overset{\overset{\displaystyle O}{\|}}{C}-H$	CH_2O Methanal (formaldehyde)
Ketones	$-\overset{\overset{\displaystyle O}{\|}}{C}-$	$R-\overset{\overset{\displaystyle O}{\|}}{C}-R'$	CH_3COCH_3 Propanone (dimethyl ketone or acetone)
Carboxylic acids	$-\overset{\overset{\displaystyle O}{\|}}{C}-OH$	$R-\overset{\overset{\displaystyle O}{\|}}{C}-OH$	CH_3COOH Ethanoic acid (acetic acid)
Esters	$-\overset{\overset{\displaystyle O}{\|}}{C}-O-$	$R-\overset{\overset{\displaystyle O}{\|}}{C}-O-R'$	$CH_3COOCH_2CH_3$ Ethyl acetate
Amines	$-NH_2$	$R-NH_2$	CH_3NH_2 Aminomethane (methylamine)

*R and R′ represent hydrocarbon fragments.

Table 22.4

Some Common Alcohols

Formula	Systematic name	Common name
CH_3OH	Methanol	Methyl alcohol
CH_3CH_2OH	Ethanol	Ethyl alcohol
$CH_3CH_2CH_2OH$	1-Propanol	*n*-Propyl alcohol
$CH_3\underset{\underset{\displaystyle OH}{\|}}{C}HCH_3$	2-Propanol	Isopropyl alcohol

Table 22.5

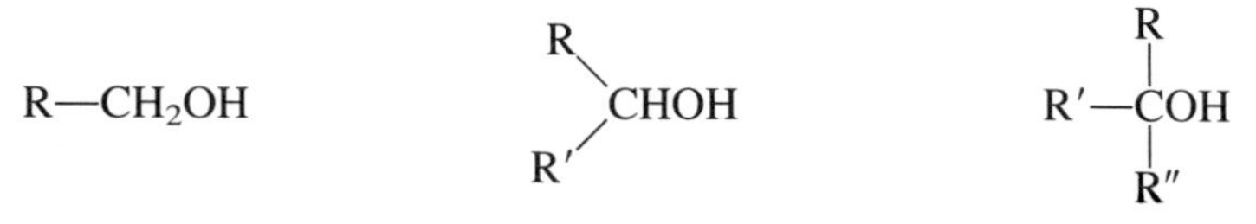

$R-CH_2OH$

Primary alcohol (one R group)

$\begin{matrix} R \\ & \searrow CHOH \\ R' \nearrow \end{matrix}$

Secondary alcohol (two R groups)

$R'-\overset{\overset{\displaystyle R}{|}}{\underset{\underset{\displaystyle R''}{|}}{C}}OH$

Tertiary alcohol (three R groups)

where R, R′, and R″ (which may be the same or different) represent hydrocarbon fragments.

Sample Exercise 22.6

For each of the following alcohols, give the systematic name and specify whether the alcohol is primary, secondary, or tertiary.

a. $CH_3\underset{\displaystyle OH}{\underset{|}{C}}HCH_2CH_3$

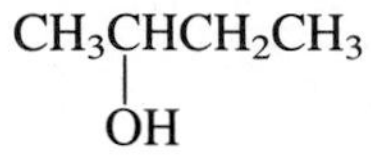

b. $ClCH_2CH_2CH_2OH$

c. $CH_3\overset{\displaystyle CH_3}{\overset{|}{\underset{\displaystyle OH}{\underset{|}{C}}}}CH_2CH_2CH_2CH_2Br$

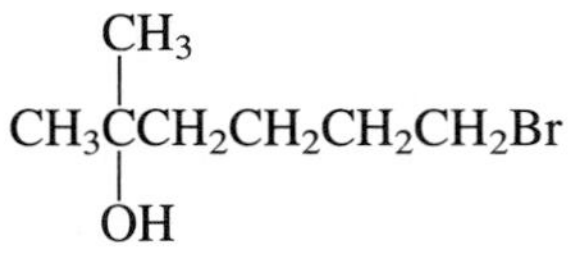

Solution

a. The chain is numbered as follows:

$$\overset{1}{CH_3}\overset{2}{\underset{\displaystyle OH}{\underset{|}{C}}H}\overset{3}{CH_2}\overset{4}{CH_3}$$

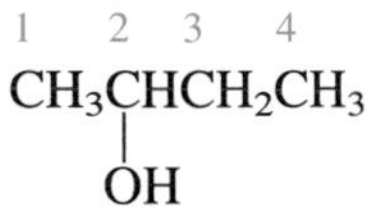

The compound is called 2-butanol, since the —OH group is located at the number 2 position of a four-carbon chain. Note that the carbon to which the —OH is attached also has $—CH_3$ and $—CH_2CH_3$ groups attached:

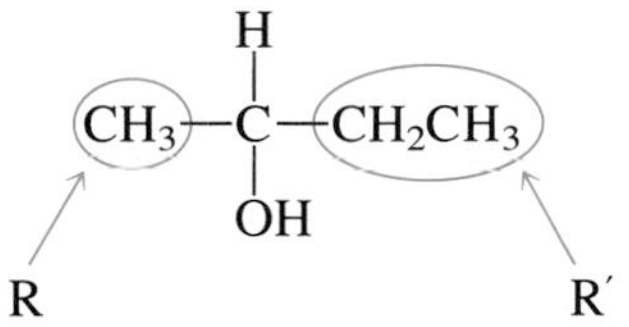

$$\underset{R}{CH_3}—\overset{\displaystyle H}{\overset{|}{\underset{\displaystyle OH}{\underset{|}{C}}}}—\underset{R'}{CH_2CH_3}$$

Thus this is a *secondary* alcohol.

b. The chain is numbered as follows:

$$Cl—\overset{3}{CH_2}—\overset{2}{CH_2}—\overset{1}{CH_2}—OH$$

The name is 3-chloro-1-propanol. This is a *primary* alcohol:

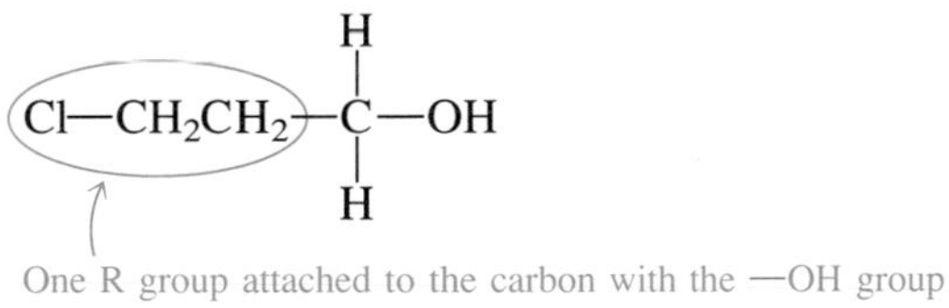

$$Cl—CH_2CH_2—\overset{\displaystyle H}{\overset{|}{\underset{\displaystyle H}{\underset{|}{C}}}}—OH$$

One R group attached to the carbon with the —OH group

c. The chain is numbered as follows:

$$\overset{1}{CH_3}—\overset{\displaystyle CH_3}{\overset{|}{\underset{\displaystyle OH}{\underset{|}{\overset{2}{C}}}}}—\overset{3}{CH_2}—\overset{4}{CH_2}—\overset{5}{CH_2}—\overset{6}{CH_2Br}$$

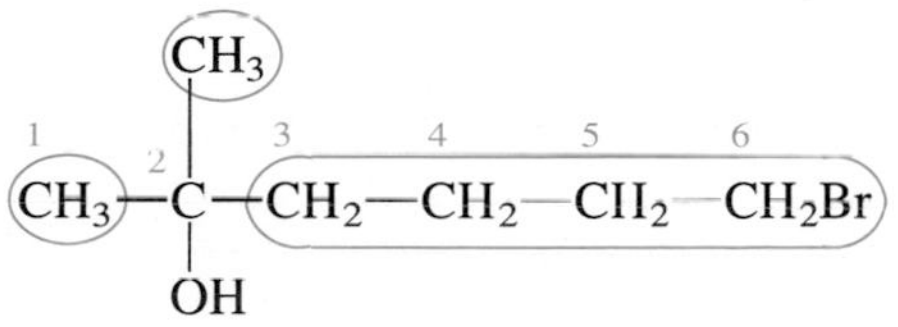

The name is 6-bromo-2-methyl-2-hexanol. This is a *tertiary* alcohol since the carbon where the —OH is attached also has three other R groups attached.

Alcohols usually have much higher boiling points than might be expected from their molecular weights. For example, both methanol and ethane have a molecular weight of 30, but the boiling point for methanol is 65°C while that for ethane is −89°C. This difference can be understood if we consider the types of intermolecular attractions that occur in these liquids. Ethane molecules are nonpolar and exhibit only weak London dispersion interactions. However, the polar —OH group of methanol produces extensive hydrogen bonding similar to that found in water (see Section 10.1), which results in the relatively high boiling point.

Although there are many important alcohols, the simplest ones, methanol and ethanol, have the greatest commercial value. Methanol, also known as *wood alcohol* because it was formerly obtained by heating wood in the absence of air, is prepared industrially (~4 million tons annually in the United States) by the hydrogenation of carbon monoxide:

$$CO + 2H_2 \xrightarrow[ZnO/Cr_2O_3]{400°C} CH_3OH$$

Methanol is used as a starting material for the synthesis of acetic acid and many types of adhesives, fibers, and plastics. It is also used (and such use may increase) as a motor fuel. Methanol is highly toxic to humans and can lead to blindness and death if ingested.

Fermentation and the wine industry are discussed in Section 24.5.

Ethanol is the alcohol found in beverages such as beer, wine, and whiskey; it is produced by the fermentation of glucose in corn, barley, grapes, etc.

$$\underset{\text{Glucose}}{C_6H_{12}O_6} \xrightarrow{\text{Yeast}} \underset{\text{Ethanol}}{2CH_3CH_2OH} + 2CO_2$$

Wine fermentation tanks at Half Moon Bay, California.

The reaction is catalyzed by the enzymes found in yeast, and can proceed only until the alcohol content reaches approximately 13% (that found in most wines), at which point the yeast can no longer survive. Beverages with higher alcohol content are made by distilling the fermentation mixture.

Ethanol, like methanol, can be burned in the internal combustion engines of automobiles and is now commonly added to gasoline to form gasohol (see Section 6.6). It is also used in industry as a solvent and for the preparation of acetic acid. The commercial production of ethanol (a half million tons per year in the United States) is carried out by reaction of water with ethylene:

$$CH_2{=}CH_2 + H_2O \xrightarrow[\text{Catalyst}]{\text{Acid}} CH_3CH_2OH$$

Many polyhydroxyl (more than one —OH group) alcohols are known, the most important being *1,2-ethanediol* (ethylene glycol),

$$\begin{array}{l} H_2C{-}OH \\ \;\;| \\ H_2C{-}OH \end{array}$$

a toxic substance that is the major constituent of most automobile antifreeze solutions.

The simplest aromatic alcohol is

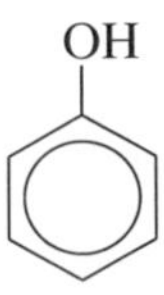

commonly called **phenol.** Most of the 1 million tons of phenol produced annually in the United States is used to produce polymers for adhesives and plastics.

Aldehydes and Ketones

Aldehydes and ketones contain the **carbonyl group**

$$>C=O$$

The aldehyde vanillin gives the pleasant odor to vanilla beans.

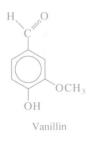

In **ketones** this group is bonded to two carbon atoms as in acetone,

$$CH_3-\underset{\displaystyle O}{\underset{\|}{C}}-CH_3$$

In **aldehydes** the carbonyl group is bonded to at least one hydrogen atom, as in formaldehyde,

$$H-\underset{\displaystyle O}{\underset{\|}{C}}-H$$

or acetaldehyde,

$$CH_3-\underset{\displaystyle O}{\underset{\|}{C}}-H$$

The systematic name for an aldehyde is obtained from the parent alkane by removing the final *-e* and adding *-al*. For ketones the final *-e* is replaced by *-one*, and a number indicates the position of the carbonyl group where necessary. Examples of common aldehydes and ketones are shown in Fig. 22.13.

Ketones often have useful solvent properties (acetone is found in nail polish remover, for example) and are frequently used in industry for this purpose. Aldehydes typically have strong odors. Vanillin is responsible for the pleasant odor in

H, O, C, OCH_3, OH

Vanillin

$H-\overset{\displaystyle O}{\overset{\|}{C}}-H$ Methanal (formaldehyde)

$CH_3-\overset{\displaystyle O}{\overset{\|}{C}}-H$ Ethanal (acetaldehyde)

$CH_3-\overset{\displaystyle O}{\overset{\|}{C}}-CH_3$ 2-Propanone (acetone)

$CH_3-\overset{\displaystyle O}{\overset{\|}{C}}-CH_2CH_2CH_3$ 2-Pentanone

$CH_3\underset{\displaystyle Cl}{\underset{|}{C}}HCH_2\overset{\displaystyle O}{\overset{\|}{C}}-H$ 3-Chlorobutanal

H, O, C Benzaldehyde

$CH_3\underset{\displaystyle O}{\underset{\|}{C}}CH_2CH_3$ 2-Butanone (methyl ethyl ketone, or MEK)

O, CH_3, C Methyl phenyl ketone

Figure 22.13

Some common ketones and aldehydes.

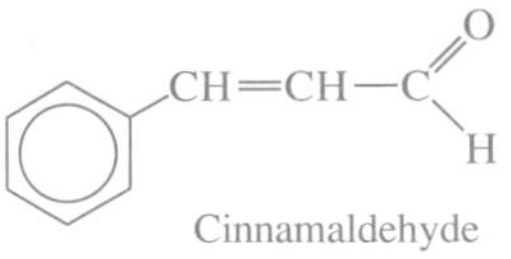

$C_6H_5CH{=}CH{-}CHO$

Cinnamaldehyde

$CH_3CH_2CH_2CHO$

Butyraldehyde

vanilla beans; cinnamaldehyde produces the characteristic odor of cinnamon. On the other hand, the unpleasant odor in rancid butter arises from the presence of butyraldehyde.

Aldehydes and ketones are most often produced commercially by the oxidation of alcohols. For example, oxidation of a *primary* alcohol gives the corresponding aldehyde:

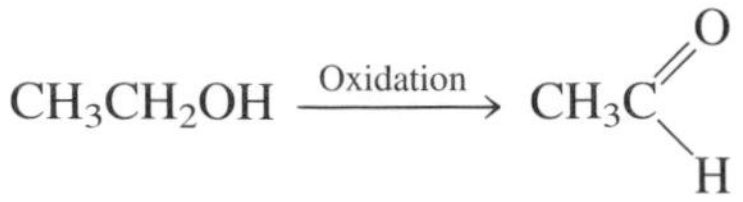

$$CH_3CH_2OH \xrightarrow{\text{Oxidation}} CH_3CHO$$

Oxidation of a *secondary* alcohol results in a ketone:

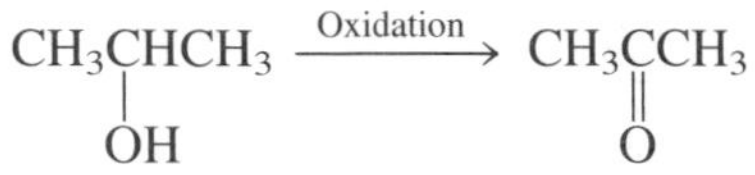

$$CH_3CH(OH)CH_3 \xrightarrow{\text{Oxidation}} CH_3C(=O)CH_3$$

Carboxylic Acids and Esters

Carboxylic acids are characterized by the presence of the **carboxyl group**

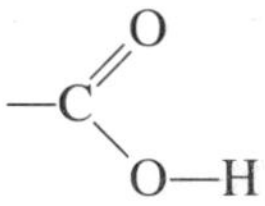

$$-C(=O)-O-H$$

to give an acid of the general formula RCOOH. Typically, these molecules are weak acids in aqueous solution (see Section 14.5). Organic acids are named from the parent alkane by dropping the final *-e* and adding *-oic*. Thus CH_3COOH, commonly called acetic acid, has the systematic name ethanoic acid since the parent alkane is ethane. Other examples of carboxylic acids are shown in Fig. 22.14.

Many carboxylic acids are synthesized by oxidizing primary alcohols with a strong oxidizing agent. For example, ethanol can be oxidized to acetic acid using potassium permanganate:

$$CH_3CH_2OH \xrightarrow{KMnO_4(aq)} CH_3COOH$$

A carboxylic acid reacts with an alcohol to form an **ester** and a water molecule. For example, the reaction of acetic acid and ethanol produces ethyl acetate and water:

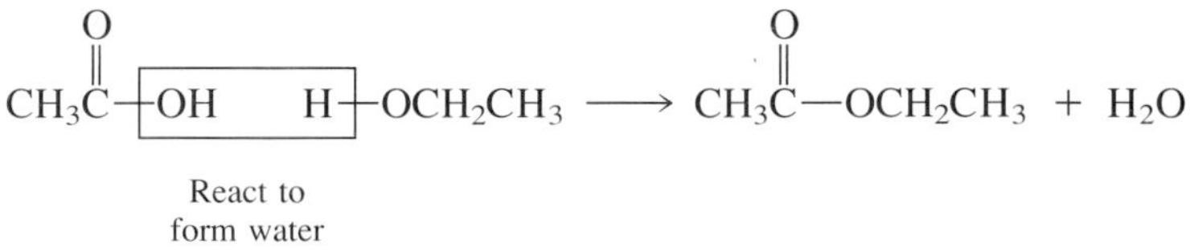

$$CH_3C(=O)\boxed{-OH \quad H-}OCH_2CH_3 \longrightarrow CH_3C(=O)-OCH_2CH_3 + H_2O$$

React to form water

Esters often have a sweet, fruity odor that is in contrast to the often pungent odors of the parent carboxylic acids. For example, the odor of bananas is from *n*-amyl acetate,

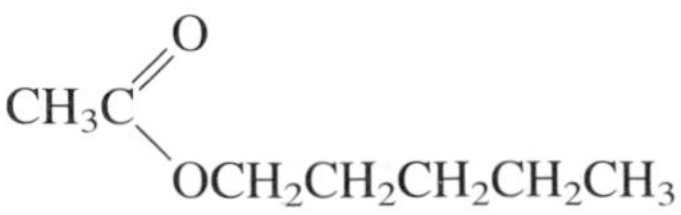

$$CH_3C(=O)OCH_2CH_2CH_2CH_2CH_3$$

$CH_3CH_2CH_2COOH$

Butanoic acid

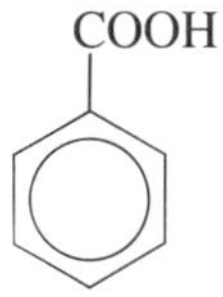

C_6H_5COOH

Benzoic acid

$CH_3CH(Br)CH_2CH_2COOH$

4-Bromopentanoic acid

$Cl_3C{-}COOH$

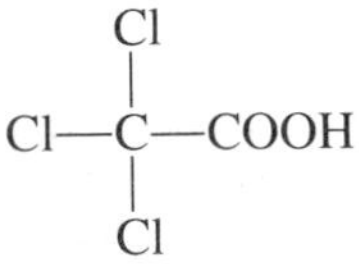

Trichloroethanoic acid (trichloroacetic acid)

Figure 22.14

Some carboxylic acids.

and that of oranges is from *n*-octyl acetate,

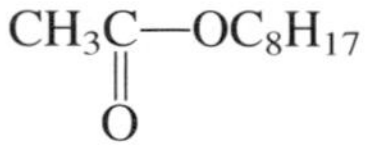

$$CH_3\underset{\underset{O}{\|}}{C}—OC_8H_{17}$$

A very important ester is formed from the reaction of salicylic acid and acetic acid:

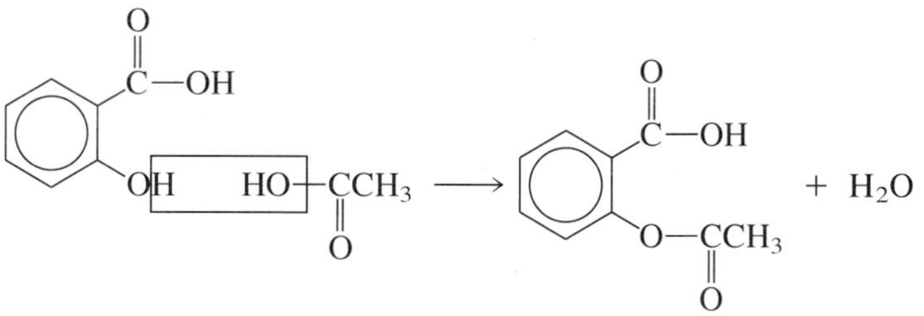

$$C_6H_4(OH)(COOH) + HO—\underset{\underset{O}{\|}}{C}CH_3 \longrightarrow C_6H_4(COOH)(O—\underset{\underset{O}{\|}}{C}CH_3) + H_2O$$

Salicylic acid   Acetic acid   Acetylsalicylic acid

The product is acetylsalicylic acid, commonly known as *aspirin*, which is used in huge quantities as an analgesic (pain-killer).

Amines

Amines are probably best viewed as derivatives of ammonia in which one or more N—H bonds are replaced by N—C bonds. The resulting amines are classified as *primary* if one N—C bond is present, *secondary* if they contain two N—C bonds, and *tertiary* if all three N—H bonds in NH_3 have been replaced by N—C bonds (Fig. 22.15). Examples of some common amines are given in Table 22.6.

Common names are often used for simple amines; the systematic nomenclature for more complex molecules uses the name *amino-* for the $—NH_2$ functional group.

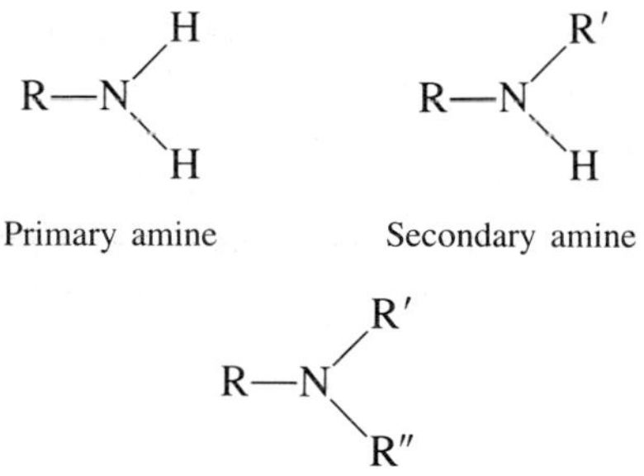

Primary amine: R—N(H)(H)   Secondary amine: R—N(R′)(H)   Tertiary amine: R—N(R′)(R″)

Figure 22.15

The general formulas for primary, secondary, and tertiary amines. R, R′, and R″ represent carbon-containing substituents.

Some Common Amines

Formula	Common name	Type
CH_3NH_2	Methylamine	Primary
$CH_3CH_2NH_2$	Ethylamine	Primary
$(CH_3)_2NH$	Dimethylamine	Secondary
$(CH_3)_3N$	Trimethylamine	Tertiary
$C_6H_5NH_2$	Aniline	Primary
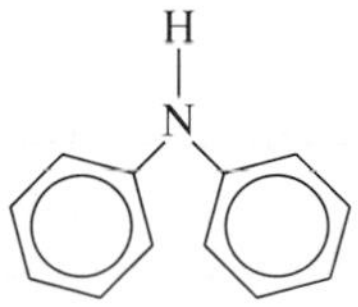 $(C_6H_5)_2NH$	Diphenylamine	Secondary

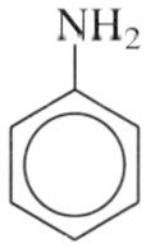

Table 22.6

For example, the molecule

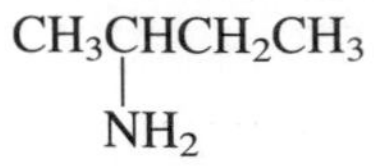

$$\underset{\displaystyle NH_2}{CH_3\underset{|}{C}HCH_2CH_3}$$

is named 2-aminobutane.

Many amines have unpleasant "fishlike" odors. For example, the odors associated with decaying animal and human tissues are due to amines such as putrescine ($H_2NCH_2CH_2CH_2NH_2$) and cadaverine ($H_2NCH_2CH_2CH_2CH_2CH_2NH_2$).

Aromatic amines are primarily used to make dyes. Since many of them are carcinogenic, they must be handled with great care.

22.5 Polymers

Purpose

- To show how polymers are formed from monomers via two types of polymerization reactions.

Polymers are large, usually chainlike molecules, which are built from small molecules called *monomers*. Polymers form the basis for synthetic fibers, rubbers, and plastics and have played a leading role in the revolution brought about in daily life by chemistry during the past 50 years.

The simplest and one of the best known polymers is *polyethylene,* which is constructed from ethylene monomers:

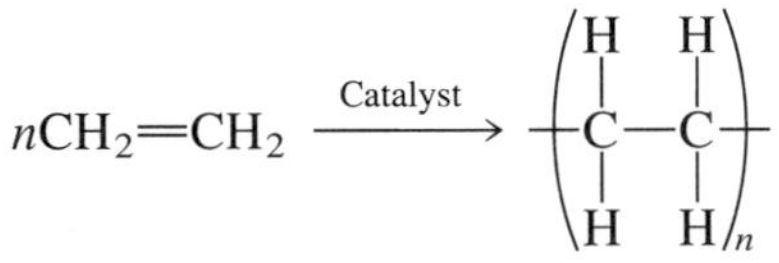

$$nCH_2{=}CH_2 \xrightarrow{\text{Catalyst}} \left(\begin{array}{cc} H & H \\ | & | \\ -C- & C- \\ | & | \\ H & H \end{array} \right)_n$$

where n represents a large number (usually several thousand). Polyethylene is a tough, flexible plastic used for piping, bottles, electrical insulation, packaging films, garbage bags, and many other purposes. Its properties can be varied by using substituted ethylene monomers. For example, when tetrafluoroethylene is the monomer, the polymer Teflon is obtained:

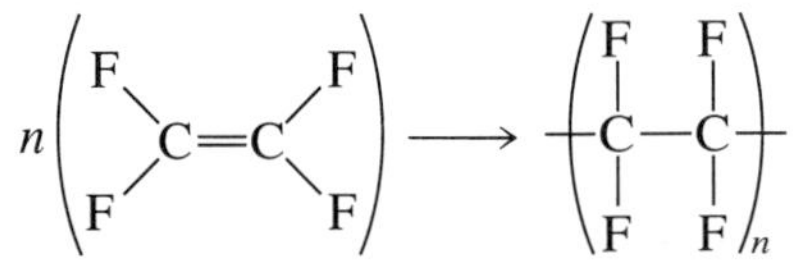

$$n\left(\begin{array}{c} F \qquad\quad F \\ \diagdown\quad\;\diagup \\ C{=}C \\ \diagup\quad\;\diagdown \\ F \qquad\quad F \end{array} \right) \longrightarrow \left(\begin{array}{cc} F & F \\ | & | \\ -C- & C- \\ | & | \\ F & F \end{array} \right)_n$$

Tetrafluorethylene Teflon

Because of the resistance of the strong C—F bonds to chemical attack, Teflon is an inert, tough, and nonflammable material widely used for electrical insulation, nonstick coatings for cooking utensils, and bearings for low-temperature applications.

Other polyethylene-type polymers are made from monomers containing chloro, methyl, cyano, and phenyl substituents, as summarized in Table 22.7. In each case, the double carbon-carbon bond in the substituted ethylene monomer becomes a

Some Common Synthetic Polymers, Their Monomers, and Applications

Monomer: Name and formula		Polymer: Name and formula	Uses
Ethylene	$H_2C{=}CH_2$	Polyethylene $-(CH_2-CH_2)_n-$	Plastic piping, bottles, electrical insulation, toys
Propylene	$H_2C{=}C(H)(CH_3)$	Polypropylene $-(CH(CH_3)-CH_2-CH(CH_3)-CH_2)_n-$	Film for packaging, carpets, lab wares, toys
Vinyl chloride	$H_2C{=}C(H)(Cl)$	Polyvinyl chloride (PVC) $-(CH_2-CH(Cl))_n-$	Piping, siding, floor tile, clothing, toys
Acrylonitrile	$H_2C{=}C(H)(CN)$	Polyacrylonitrile (PAN) $-(CH_2-CH(CN))_n-$	Carpets, fabrics
Tetrafluoroethylene	$F_2C{=}CF_2$	Teflon $-(CF_2-CF_2)_n-$	Cooking utensils, electrical insulation, bearings
Styrene	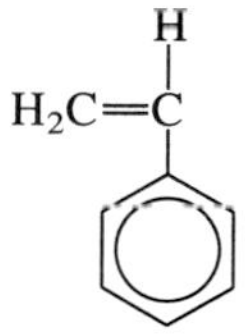 $H_2C{=}C(H)(C_6H_5)$	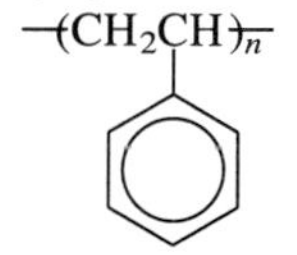 Polystyrene $-(CH_2CH(C_6H_5))_n-$	Containers, thermal insulation, toys
Butadiene	$H_2C{=}C(H)-C(H){=}CH_2$	Polybutadiene $-(CH_2CH{=}CHCH_2)_n-$	Tire tread, coating resin
Butadiene and styrene	(See above.)	Styrene-butadiene rubber 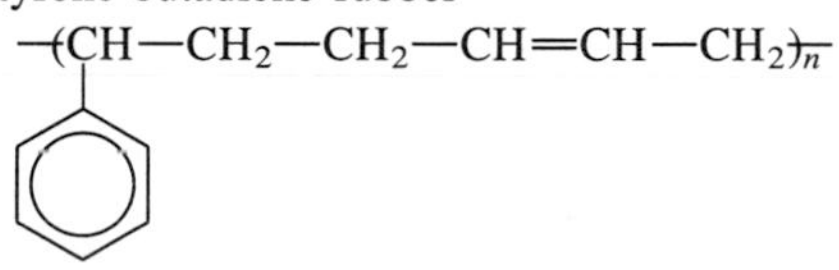 $-(CH(C_6H_5)-CH_2-CH_2-CH{=}CH-CH_2)_n-$	Synthetic rubber

Table 22.7

single bond in the polymer. The different substituents lead to a wide variety of properties.

The polyethylene polymers illustrate one of the major types of polymerization reactions, called **addition polymerization,** in which the monomers simply ''add together'' to form the polymer, with no other products. The polymerization process is initiated with a **free radical** (a species with an unpaired electron) such as the hydroxyl radical (HO·). The free radical attacks and breaks the π bond of an ethylene molecule to form a new free radical,

$$\begin{matrix} H & & H \\ & C{=}C & \\ H & \uparrow & H \\ & \dot{O}H & \end{matrix} \longrightarrow \begin{matrix} & H & H \\ & | & | \\ H- & C- & C\cdot \\ & | & | \\ & OH & H \end{matrix}$$

which can then attack another ethylene molecule:

$$\begin{matrix} H & & H \\ & C{=}C & \\ H & & H \end{matrix} \longrightarrow \begin{matrix} & H & H & H & H \\ & | & | & | & | \\ H- & C- & C- & C- & C\cdot \\ & | & | & | & | \\ & OH & H & H & H \end{matrix}$$

$$\begin{matrix} & H & H \\ & | & | \\ H- & C- & C\cdot \\ & | & | \\ & HO & H \end{matrix}$$

Repetition of this process thousands of times creates a long-chain polymer. Termination of the growth of the chain occurs when *two radicals* react to form a bond, a process that consumes two radicals without producing any others.

Another common type of polymerization is **condensation polymerization,** in which a small molecule, such as water, is formed for each extension of the polymer chain. The most familiar polymer produced by condensation is *nylon*. Nylon is a **copolymer,** since two different types of monomers combine to form the chain; a **homopolymer** is the result of polymerizing a single type of monomer. One common form of nylon is produced when hexamethylenediamine and adipic acid react by splitting out a water molecule to form a C—N bond:

$$\underset{\text{Hexamethylenediamine}}{\begin{matrix} H & & H \\ & N{-}(CH_2)_6{-}N & \\ H & & H \end{matrix}} \quad \underset{\text{Adipic acid}}{\begin{matrix} O & & O \\ & C{-}(CH_2)_4{-}C & \\ H{-}O & & O{-}H \end{matrix}} \longrightarrow$$

$$\begin{matrix} H & & & H & O & & & O \\ & & & | & \| & & & \\ & N{-}(CH_2)_6{-} & & N- & C{-}(CH_2)_4{-} & & C & \\ H & & & & & & & O{-}H \end{matrix} + H_2O$$

The molecule formed, called a **dimer** (two monomers joined), can undergo further condensation reactions since it has an amino group at one end and a carboxyl group at the other. Thus both ends are free to react with another monomer. Repetition of this process leads to a long chain of the type

$$\left(\begin{matrix} H & & H & O & & O \\ | & & | & \| & & \| \\ N- & (CH_2)_6- & N- & C- & (CH_2)_4- & C \end{matrix} \right)_n$$

which is the basic structure of nylon. The reaction to form nylon occurs quite readily and is often used as a lecture demonstration (see Fig. 22.16). The properties

Figure 22.16

The reaction to form nylon can be carried out at the interface of two immiscible liquid layers in a beaker. The bottom layer contains adipoyl chloride,

$$\mathrm{Cl{-}\underset{\underset{O}{\|}}{C}{-}\!(CH_2)_4\!{-}\underset{\underset{O}{\|}}{C}{-}Cl}$$

in CCl_4, and the top layer contains hexamethylenediamine,

$$\mathrm{H_2N{-}\!(CH_2)_6\!{-}NH_2}$$

dissolved in water. A molecule of HCl is formed as each C—N bond forms.

of nylon can be varied by changing the number of carbon atoms in the chain of the acid or amine monomer.

More than 1 million tons of nylon are produced annually in the United States for use in clothing, carpets, rope, and so on. Many other types of condensation polymers are also produced. For example, Dacron is a copolymer formed from the condensation reaction of ethylene glycol (a dialcohol) and *p*-terephthalic acid (a dicarboxylic acid):

$$\mathrm{HOCH_2CH_2O\boxed{H\quad HO}{-}\overset{\overset{O}{\|}}{C}{-}C_6H_4{-}\overset{\overset{O}{\|}}{C}{-}O{-}H}$$

Ethylene glycol ↓ H_2O *p*-Terephthalic acid

The repeating unit of Dacron is

$$\left(\mathrm{OCH_2CH_2{-}O{-}\overset{\overset{O}{\|}}{C}{-}C_6H_4{-}\overset{\overset{O}{\|}}{C}}\right)_n$$

Note that this polymerization involves a carboxylic acid and an alcohol to form an ester group:

$$\mathrm{R{-}O{-}\overset{\overset{O}{\|}}{C}{-}R_1}$$

Thus Dacron is called a **polyester.** By itself or blended with cotton, Dacron is widely used in fibers for the manufacture of clothing.

More details concerning the manufacturing processes for producing polymers are given in Section 24.2.

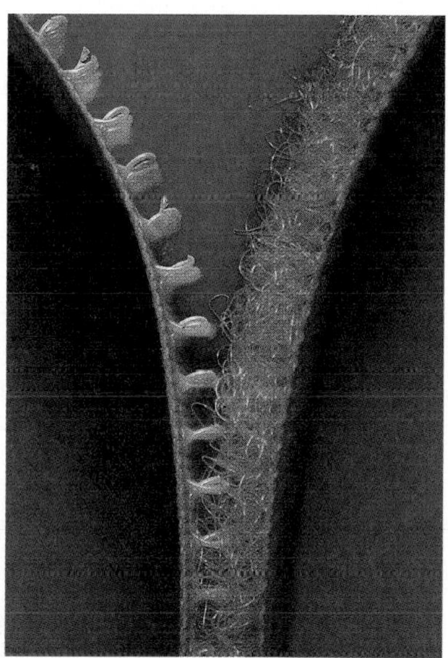

A nylon Velcro fastener.

FOR REVIEW

Summary

The study of carbon-containing compounds and their properties is called organic chemistry. Most organic compounds contain chains or rings of carbon atoms. The organic molecules responsible for maintaining and reproducing life are called biomolecules.

Hydrocarbons are organic compounds composed of carbon and hydrogen. Those that contain only C—C single bonds are saturated and are called alkanes, and those with carbon-carbon multiple bonds are unsaturated. Unsaturated hydrocarbons can become saturated by the addition of hydrogen, halogens, and so on. All alkanes can be represented by the general formula C_nH_{2n+2}.

Methane (CH_4) is the simplest alkane, and the next three in the series are ethane (C_2H_6), propane (C_3H_8), and butane (C_4H_{10}). In a saturated hydrocarbon, each carbon atom is bonded to four other atoms and is described as being sp^3 hybridized. Alkanes containing long chains of carbon atoms are called normal, or straight-chain, hydrocarbons.

Structural isomerism in alkanes involves formation of branched structures. Specific rules for systematically naming alkanes indicate the point of attachment of any substituent group, the length of the root chain, and so on.

Alkanes can undergo combustion reactions to form carbon dioxide and water or substitution reactions in which hydrogen atoms are replaced by other atoms. Alkanes can also undergo dehydrogenation to form unsaturated hydrocarbons. Cyclic alkanes are saturated hydrocarbons with ring structures.

Hydrocarbons with carbon-carbon double bonds are called alkenes. The simplest alkene is ethylene (C_2H_4), which contains sp^2 hybridized carbon atoms. Because the p orbitals on the carbon atoms in ethylene form a π bond, free rotation of the two CH_2 groups relative to each other does not occur (as it does in alkanes). This restriction means that alkenes exhibit *cis-trans* isomerism.

Alkynes are unsaturated hydrocarbons with a triple carbon-carbon bond. The simplest in the series is acetylene (C_2H_2), containing sp hybridized carbon atoms.

Unsaturated hydrocarbons undergo addition reactions such as hydrogenation (addition of hydrogen) and halogenation (addition of halogen atoms). Ethylene and substituted ethylene molecules can undergo polymerization, a process by which many molecules (monomers) are joined together to form a large chainlike molecule.

Aromatic hydrocarbons, such as benzene, are stable ring compounds with delocalized π electron systems. They undergo substitution reactions rather than the addition reactions characteristic of typical alkenes.

Organic molecules that contain elements in addition to carbon and hydrogen can be viewed as hydrocarbon derivatives: hydrocarbons with functional groups. Each functional group exhibits characteristic chemical properties.

Alcohols contain the —OH group and tend to have strong hydrogen bonding in the liquid state. Aldehydes and ketones contain the carbonyl group:

$$\gt C{=}O$$

In aldehydes this group is bonded to at least one hydrogen atom. Carboxylic acids are characterized by the carboxyl group

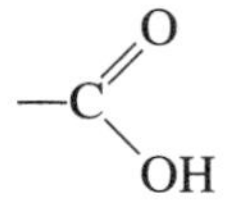

$$-C\begin{matrix}{=}O\\ \diagdown OH\end{matrix}$$

They can react with alcohols to form esters. Amines can be thought of as derivatives of ammonia in which one or more N—H bonds have been replaced by an N—C bond.

Polymers can be formed by addition polymerization in which monomers add together via a free radical mechanism. Polymers can also be formed by condensation polymerization, which involves the splitting out of a small molecule (often water) between two monomers to form a dimer, which then undergoes further condensation.

Key Terms

biomolecule
organic chemistry

Section 22.1
hydrocarbons
alkanes
normal (straight-chain or unbranched) hydrocarbons
structural isomerism
combustion reaction
substitution reaction
dehydrogenation reaction
cyclic alkanes

Section 22.2
alkenes
cis-trans isomerism
alkynes
addition reaction
hydrogenation reaction
halogenation
polymerization

Section 22.3
aromatic hydrocarbons
phenyl group

Section 22.4
hydrocarbon derivatives
functional group
alcohols
phenol
carbonyl group
ketones
aldehydes
carboxylic acids
carboxyl group
esters
amines

Section 22.5
polymers
addition polymerization
free radical
condensation polymerization
copolymer
homopolymer
dimer
polyester

Exercises

A blue exercise number indicates that the answer to that exercise appears at the back of this book and a solution appears in the Solutions Guide.

Hydrocarbons

1. Draw the five structural isomers of hexane (C_6H_{14}). Give systematic names for each. Which isomer would you expect to have the highest boiling point? Why?

2. Name each of the following:

a.
$$\begin{matrix} CH_3 & CH_3 & \\ | & | & \\ CH- & C- & CH_2CH_2CH_3 \\ | & | & \\ CH_3 & CH_3 & \end{matrix}$$

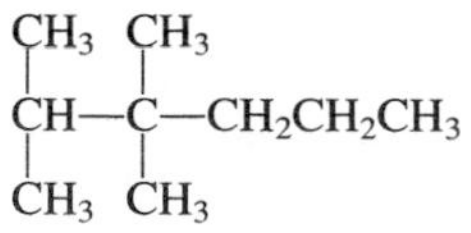

b.
$$\begin{matrix} & CH_2CH_2CH(CH_3)_2 & & & & & \\ & | & & & & & \\ CH_3- & C- & CH_2- & CH_2- & CH- & CH_2- & CH_3 \\ & | & & & | & & \\ & CH_3 & & & CH_2CH_3 & & \end{matrix}$$

c.
$$\begin{matrix} CH_3- & CH_2- & CH_2- & CH- & CH_3 \\ & & & | & \\ & & & CH_2CH_3 & \end{matrix}$$

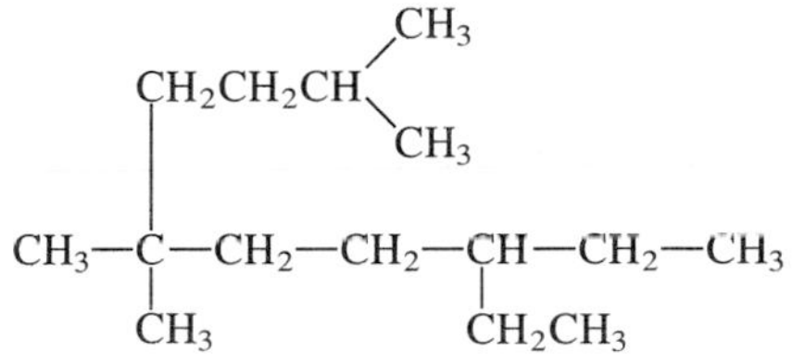

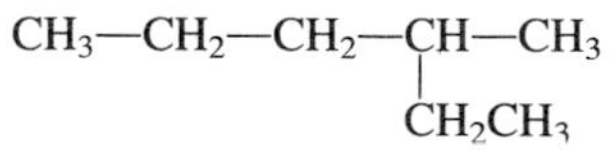

3. The normal (unbranched) hydrocarbons are often referred to as the straight-chain hydrocarbons. What does this name refer to? Does this mean that all carbon atoms in a straight-chain hydrocarbon really have a linear arrangement? Explain your answer.

4. Draw the structural formula for each of the following:
   a. 2-methylpentane
   b. 2,2,4-trimethylpentane (Also called isooctane, this substance is the reference, or 100 level, for octane ratings.)
   c. 2-*tert*-butylpentane
   d. The name given in part c is incorrect. Give the correct name for this hydrocarbon.

5. Name each of the following alkenes:

   a. $CH_2{=}CH{-}CH_2{-}CH_3$

   b. $(CH_3)_2C{=}CH{-}CH_3$

   c. $CH_3CH_2CH(CH_3){-}CH{=}CH{-}CH(CH_3)_2$

6. Give the structure of each of the following:
   a. 3-hexene
   b. 2,4-heptadiene
   c. 2-methyl-3-octene

7. Give the structure of each of the following aromatic hydrocarbons:
   a. *o*-xylene
   b. *p*-di-*tert*-butylbenzene
   c. *m*-diethylbenzene

8. Name each of the following:

   a. $Cl{-}CH_2{-}CH_2{-}CH(Cl){-}CH_3$

   b. $CH_3CH_2CH_2CCl_3$

   c. $(CH_3)_2CCl{-}CH(Cl){-}CH(CH_3)(CH_2CH_3)$

   d. CH_2FCH_2F

   e. Cl on a benzene ring

   f. Cl on a cyclohexane ring

   g. Cl on a cyclohexene ring

9. Name the following compounds:

   a. cyclopropane ring with $-CH_3$

   b. cyclohexane ring with $-C(CH_3)_3$

   c. cyclopentene ring with $-CH_3$ and CH_3 substituents

   d. $ClCH{=}CH_2$

   e. cyclopentene ring with $-CH_3$ and CH_3 substituents

   f. cyclohexane ring with two Cl on the same carbon

10. Cumene is the starting material for the industrial production of acetone and phenol. The structure of cumene is

    benzene ring with $-CH(CH_3)_2$

    Give the systematic name for cumene.

11. Draw the structure for the hydrocarbon 2-ethyl-3-methyl-5-isopropylhexane. The name given here is incorrect. Give the correct systematic name.

12. Give structures for each of the following:
    a. difluoromethane
    b. 1-bromo-1,2-dichloropropane
    c. *m*-difluorobenzene

Isomerism

13. Distinguish between structural and geometrical isomerism.

14. Distinguish between isomerism and resonance.

15. *Cis-trans* isomerism is also possible in molecules with rings. Draw the *cis* and *trans* isomers of 1,2-dimethylcyclohexane.

16. Draw all the structural and geometric isomers of $C_3H_4Cl_2$.

17. Draw all the isomers of dimethylnaphthalene.

18. Draw all of the structural and geometrical isomers of bromochloropropene.

19. Draw all of the isomers of difluoroethene. Which are polar?

20. There is only one compound that is named 1,2-dichloroethane, but there are two distinct compounds that can be named 1,2-dichloroethene. Why?

21. Tautomerism is a property exhibited by molecules that differ in the position of a hydrogen atom. Use bond energies (Table 8.4) to predict which tautomer in each of the following pairs is more stable:

    a. $CH_3{-}C(OH){=}CH_2$ or $CH_3{-}C({=}O){-}CH_3$

b. C_6H_5—C(=O)—N(H)—CH_3 or C_6H_5—C(OH)=N—CH_3

c. H_3C—C(=O)—CH_2—C(=O)—CH_3 or H_3C—C(=O)—CH=C(OH)—CH_3

22. Polychlorinated dibenzo-*p*-dioxins, or PCDDs, are highly toxic substances that are present in trace amounts as by-products of some chemical manufacturing processes. They have been implicated in a number of environmental incidents, for example, Love Canal and herbicide spraying in Vietnam. The structure of dibenzo-*p*-dioxin, with the numbering convention, is

(dibenzo-*p*-dioxin, positions 1, 2, 3, 4, 6, 7, 8, 9)

The most toxic PCDD is 2,3,7,8-tetrachloro-CDD. Draw the structure of this compound and the structures of two other isomers containing four chlorine atoms.

Functional Groups

23. Identify the functional groups present in the following drugs.

a. aspirin

(C_6H_4 with CO_2H and OCCH$_3$ (C=O))

b. morphine

(structure with HO, O, HO, N—CH_3)

c. naloxone (a narcotic antagonist)

(structure with HO, O, OH, O, N—CH_2—CH=CH_2)

24. Mycomycin is a naturally occurring antibiotic produced by the fungus *Nocardia acidophilus*. The molecular formula is $C_{13}H_{10}O_2$, and the systematic name of the substance is 3,5,7,8-tridecatetraene-10,12-diynoic acid. Draw the structure of mycomycin.

25. Menthol has the systematic name 2-isopropyl-5-methylcyclohexanol. Draw the structure of menthol.

26. Mimosine is a natural product found in large quantities in the seeds and foliage of some legume plants and has been shown to cause inhibition of hair growth and hair loss in mice.

(ring structure with H, H, O=, OH, H, N—CH_2—CH(NH_2)—CO_2H)

Mimosine, $C_8H_{10}N_2O_4$

a. What functional groups are present in mimosine?
b. Give the hybridization of the eight carbon atoms in mimosine.
c. How many sigma and pi bonds are found in mimosine?

27. Minoxidil, $C_9H_{15}N_5O$, is a compound, produced by the Upjohn Company, that has been approved as a treatment of some types of male pattern baldness.

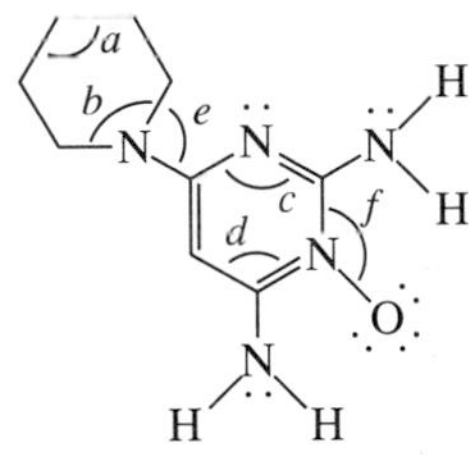

a. Would minoxidil be more soluble in acidic or basic aqueous solution? Explain.
b. Give the hybridization of the five nitrogen atoms in minoxidil.
c. Give the hybridization of each of the nine carbon atoms in minoxidil.
d. Give approximate values of the bond angles marked *a*, *b*, *c*, *d*, *e*, and *f*.
e. Including all of the hydrogen atoms, how many sigma bonds exist in minoxidil?
f. How many pi bonds exist in minoxidil?

28. Many drugs such as morphine (see Exercise 23) are treated with strong inorganic acids. The most commonly used form of morphine is morphine hydrochloride. Draw the structure of morphine hydrochloride. Why are many drugs treated in this way?

29. Ethyl caprate is an ester used in the manufacture of wine bouquets. It is sometimes called "cognac essence." Combustion

analysis of a sample of ethyl caprate shows it to be 71.89% carbon, 12.13% hydrogen, and 15.98% oxygen by mass. Hydrolysis (reaction with water) of the ester yields ethanol and an acid. The molecular weight of the acid is 172. What is the molecular formula of ethyl caprate?

30. Draw an isomer specified by each of the following:
a. an aldehyde that is an isomer of acetone
b. an ether that is an isomer of 2-propanol
c. a geometrical isomer of *cis*-2-butene
d. a primary amine that is an isomer of trimethylamine
e. a secondary amine that is an isomer of trimethylamine
f. a primary alcohol that is an isomer of 2-propanol

31. Identify each of the following compounds as a carboxylic acid, ester, ketone, aldehyde, or amine.
a. anthraquinone, an important starting material in the manufacture of dyes

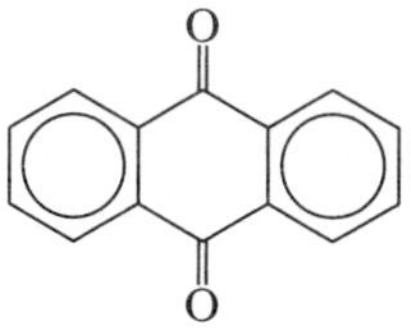

b. 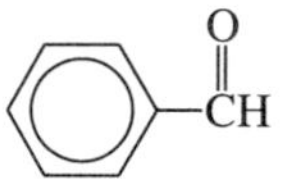 $C_6H_5-\overset{O}{\overset{\|}{C}}H$

c. $CH_3-\overset{O}{\overset{\|}{C}}-CH_2\underset{CH_3}{\underset{|}{C}}HCH_3$

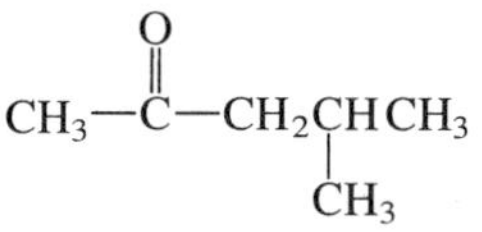

d. $C_6H_5-NH-C_6H_5$

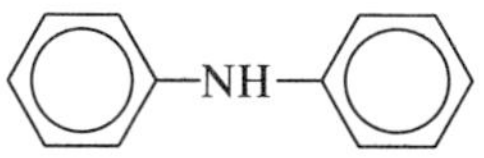

32. The structure of ephedrine (adrenaline) is:

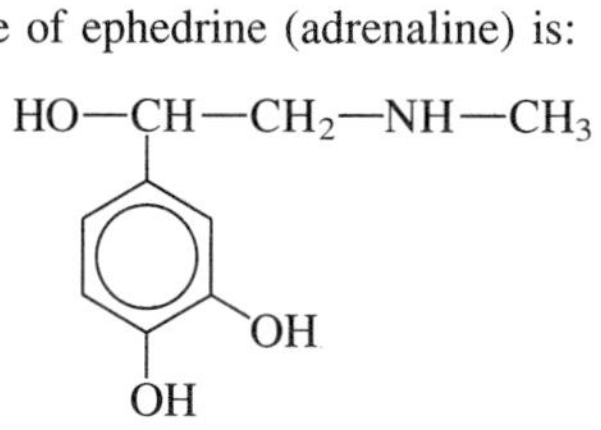

HO—CH—CH₂—NH—CH₃ (the CH bonded to a benzene ring bearing two OH groups)

What functional groups are present?

Reactions of Organic Compounds

33. Distinguish between substitution and addition reactions. Give an example of each type.

34. Consider the reaction of propane with chlorine.
a. How many monochloro products are formed? Draw their structures.
b. How many dichloro products can be formed? Draw their structures.

35. Complete the following reactions:
a. $CH_2CH_2 + Br_2 \rightarrow$
b.  $C_6H_6 + Br_2 \xrightarrow{Fe}$
c. $CH_3CO_2H + CH_3OH \rightarrow$

36. Reagents such as HCl, HBr, and HOH can add across carbon-carbon double bonds. The addition occurs so that the hydrogen atom in the reagent attaches to the carbon atom in the double bond that already has the greater number of hydrogen atoms bonded to it. With this rule in mind, draw the structure of the major product in each of the following reactions:
a. $CH_3CH{=}CH_2 + HCl \rightarrow$
b. $CH_3CH{=}CH_2 + H_2O \rightarrow$
c. 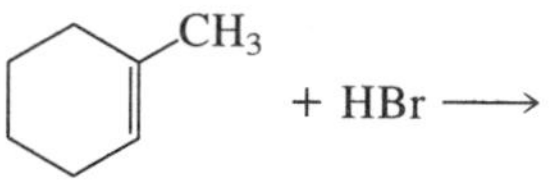 1-methylcyclohexene (ring with CH_3 on the double-bond carbon) $+ HBr \longrightarrow$
d. 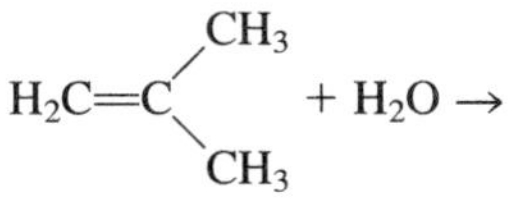 $H_2C{=}C(CH_3)_2 + H_2O \rightarrow$
e. 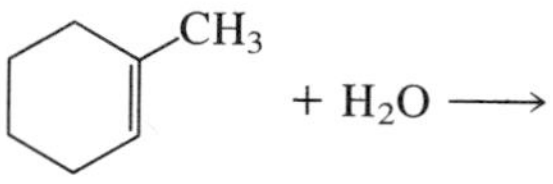 1-methylcyclohexene (ring with CH_3 on the double-bond carbon) $+ H_2O \longrightarrow$
f. $CH_3-C{\equiv}CH + 2HCl \rightarrow$

37. Give the structure of the product resulting from the oxidation of each of the following alcohols:
a. CH_3CH_2OH
b. 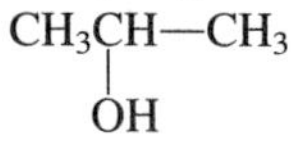 $CH_3\underset{OH}{\underset{|}{C}}H-CH_3$
c. cyclopentyl—OH
d. 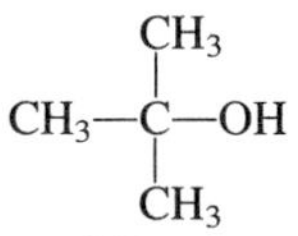 $CH_3-\overset{CH_3}{\overset{|}{\underset{CH_3}{\underset{|}{C}}}}-OH$
e. 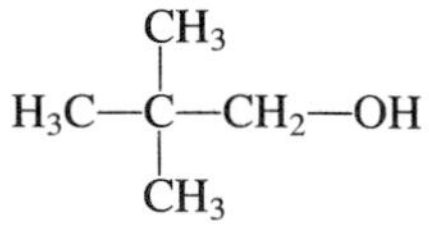 $H_3C-\overset{CH_3}{\overset{|}{\underset{CH_3}{\underset{|}{C}}}}-CH_2-OH$

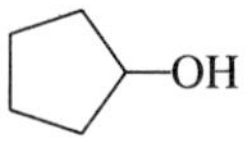

38. Oxidation of an aldehyde yields a carboxylic acid.

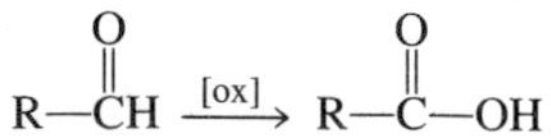

$$R-\overset{O}{\overset{\|}{C}}H \xrightarrow{[ox]} R-\overset{O}{\overset{\|}{C}}-OH$$

Thus, it is very difficult to get only the pure aldehyde from the oxidation of a primary alcohol.

a. Draw the structures for the products of the following oxidations.

i. 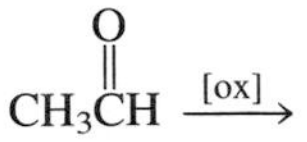 $CH_3\overset{\displaystyle O}{\overset{\|}{C}}H \xrightarrow{[ox]}$

ii. 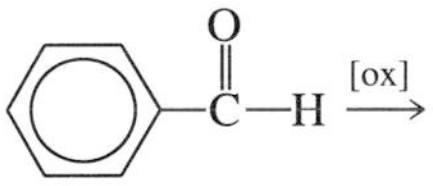 $C_6H_5-\overset{\displaystyle O}{\overset{\|}{C}}-H \xrightarrow{[ox]}$

iii. 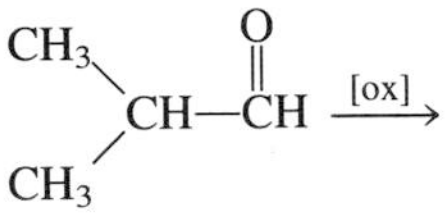 $(CH_3)_2CH-\overset{\displaystyle O}{\overset{\|}{C}}H \xrightarrow{[ox]}$

b. Which of the reactions in Exercise 37 would result in a mixture of an aldehyde and an acid as products? Draw the structures of the acids formed.

39. How would you make each of the following:

a. 1,2-dibromopropane from propene
b. 1,2-dibromopropane from propyne
c. butyl acetate
d. ethyl butyrate
e. acetone from an alcohol

Polymers

40. Define and give an example of each of the following:

a. addition polymer
b. condensation polymer
c. copolymer

41. Kevlar, used in bulletproof vests, is made by the condensation polymerization of the monomer

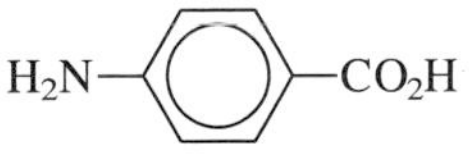

$$H_2N-C_6H_4-CO_2H$$

Draw the structure of a portion of the Kevlar chain.

42. The polyester formed from lactic acid

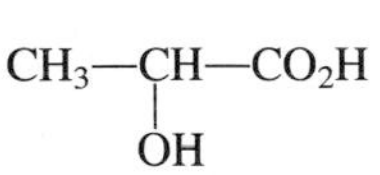

$$CH_3-\underset{\displaystyle OH}{\underset{|}{C}H}-CO_2H$$

is used for tissue implants and surgical sutures that will dissolve in the body. Draw the structure of a portion of this polymer.

43. Polyimides are polymers that are tough and stable at temperatures of up to 400°C. They are used as a protective coating on the quartz fibers used in fiber optics. What monomers were used to make the following polyimide:

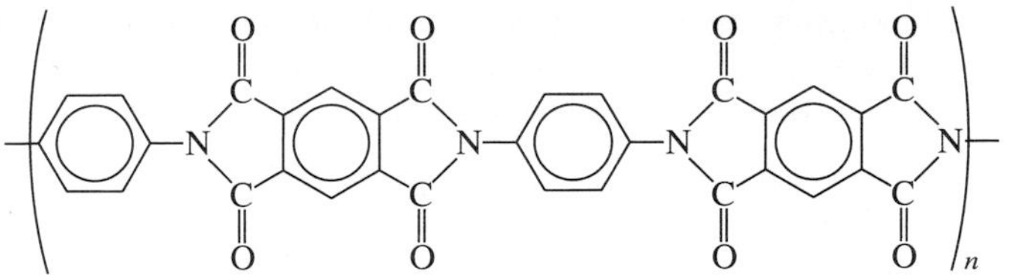

$$\left(\!-C_6H_4-N\!\left<\!\!\begin{smallmatrix}C(=O)\\C(=O)\end{smallmatrix}\!\!\right>\!C_6H_2\!\left<\!\!\begin{smallmatrix}C(=O)\\C(=O)\end{smallmatrix}\!\!\right>\!N-C_6H_4-N\!\left<\!\!\begin{smallmatrix}C(=O)\\C(=O)\end{smallmatrix}\!\!\right>\!C_6H_2\!\left<\!\!\begin{smallmatrix}C(=O)\\C(=O)\end{smallmatrix}\!\!\right>\!N-\right)_n$$

44. Ethylene oxide

$$\underset{\displaystyle O}{CH_2-CH_2}$$

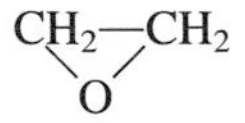

is an important industrial chemical. While most ethers are unreactive, ethylene oxide is quite reactive. It resembles C_2H_4 in its reactions in that many addition reactions occur across the C—O bond.

a. Why is ethylene oxide so reactive? (*Hint:* Consider the bond angle in ethylene oxide as compared to those predicted by the VSEPR model.)
b. Ethylene oxide undergoes addition polymerization forming a polymer used in many applications requiring a non-ionic surfactant. Draw the structure of this polymer.

45. Polycarbonates are a class of thermoplastic polymers that are used in the plastic lenses of eyeglasses and in the shells of bicycle helmets. A polycarbonate is made from the reaction of bisphenol A (BPA) with phosgene ($COCl_2$):

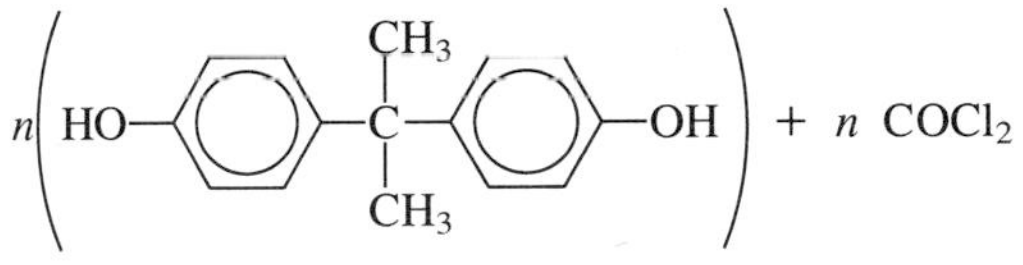

$$n\left(HO-C_6H_4-\underset{\displaystyle CH_3}{\overset{\displaystyle CH_3}{\overset{|}{\underset{|}{C}}}}-C_6H_4-OH\right) + n\ COCl_2$$

BPA

$$\xrightarrow[\text{catalyst}]{2n\ NaOH} \text{Polycarbonate} + 2n\ NaCl + 2n\ H_2O$$

Phenol, C_6H_5OH, is used to terminate the polymer (stop its growth).

a. Draw the structure of the polycarbonate chain formed from the above reaction.
b. Is this reaction a condensation or addition polymerization?

46. Polystyrene can be made more rigid by copolymerizing styrene with divinylbenzene,

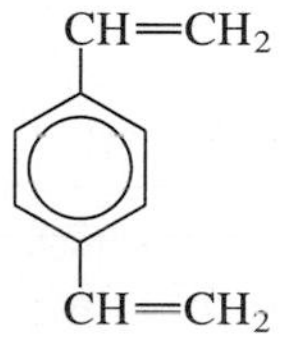

$$CH_2=CH-C_6H_4-CH=CH_2$$

What does the divinylbenzene do? Why is the copolymer more rigid?

47. In which polymer, polyethylene or polyvinylchloride, would you expect to find the stronger intermolecular forces, assuming the average chain lengths are equal?

48. What monomer(s) must be used to produce the following polymers?

a. 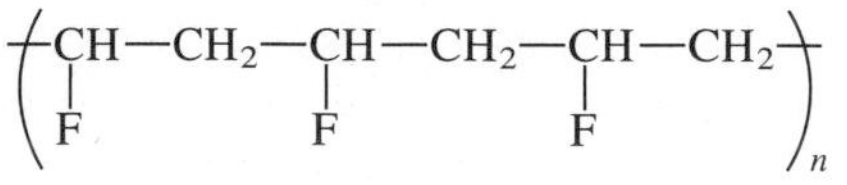 $\left(\underset{\mid\atop F}{CH}-CH_2-\underset{\mid\atop F}{CH}-CH_2-\underset{\mid\atop F}{CH}-CH_2\right)_n$

b. 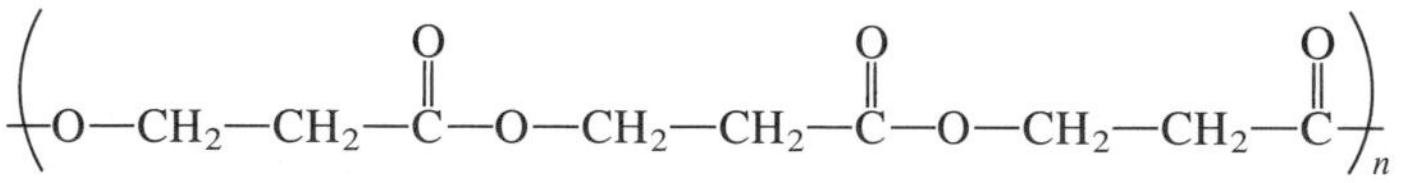 $\left(O-CH_2-CH_2-\overset{O\atop \|}{C}-O-CH_2-CH_2-\overset{O\atop \|}{C}-O-CH_2-CH_2-\overset{O\atop \|}{C}\right)_n$

c. $\left(O-CH_2-CH_2-O-\overset{O\atop \|}{C}-CH_2-CH_2-\overset{O\atop \|}{C}-O-CH_2-CH_2-O-\overset{O\atop \|}{C}-CH_2-CH_2-\overset{O\atop \|}{C}\right)_n$

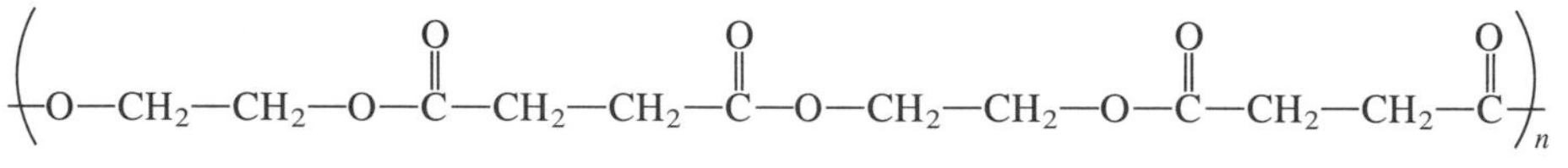

d. 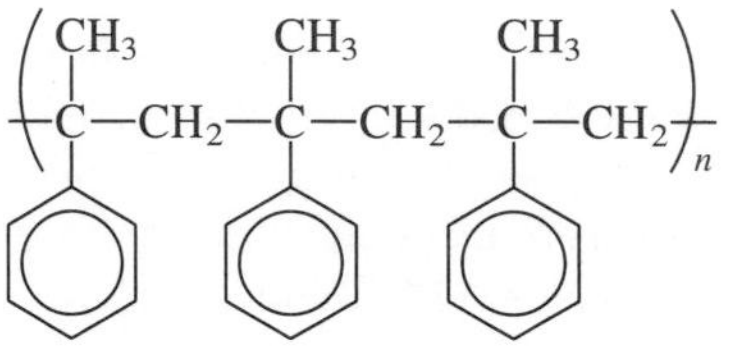 $\left(\overset{CH_3\atop \mid}{\underset{\mid\atop C_6H_5}{C}}-CH_2-\overset{CH_3\atop \mid}{\underset{\mid\atop C_6H_5}{C}}-CH_2-\overset{CH_3\atop \mid}{\underset{\mid\atop C_6H_5}{C}}-CH_2\right)_n$ (each C_6H_5 drawn as a benzene ring)

e. 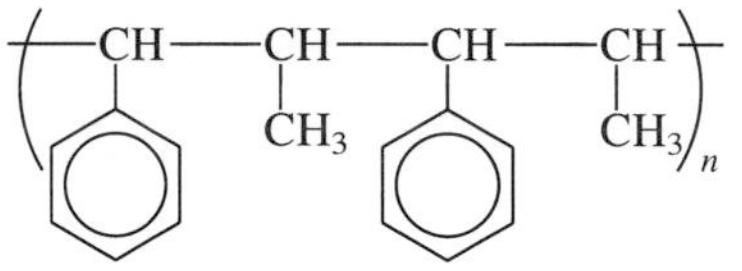 $\left(\underset{\mid\atop C_6H_5}{CH}-\underset{\mid\atop CH_3}{CH}-\underset{\mid\atop C_6H_5}{CH}-\underset{\mid\atop CH_3}{CH}\right)_n$ (each C_6H_5 drawn as a benzene ring)

49. Classify the polymers in Exercise 48 as condensation or addition polymers. Which are copolymers?

Additional Exercises

50. Isoprene is the repeating unit in natural rubber. The structure of isoprene is

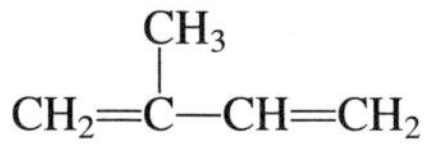

$$CH_2{=}\overset{CH_3\atop \mid}{C}-CH{=}CH_2$$

Give a systematic name for isoprene.

51. The structure of Dewar's benzene is

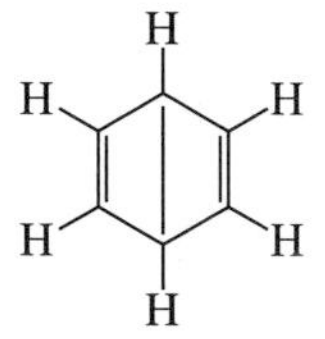

Is this another resonance structure for benzene, or would you expect Dewar's benzene to be a different substance?

52. Estimate ΔH for the following reactions, using bond energies (Table 8.4):

$$3CH_2{=}CH_2(g) + 3H_2(g) \rightarrow 3CH_3{-}CH_3(g)$$

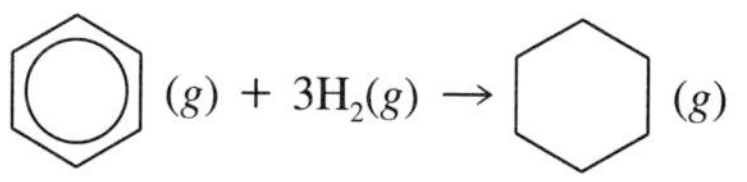

$$\text{benzene}(g) + 3H_2(g) \rightarrow \text{cyclohexane}(g)$$

The enthalpies of formation of $C_6H_6(g)$ and $C_6H_{12}(g)$ are 82.9 and -90.3 kJ/mol respectively. Calculate $\Delta H°$ for the two reactions using standard enthalpies of formation from Appendix 4. How do you account for any differences between results obtained by the two methods?

53. The Amoco Chemical Company has successfully raced a car with a plastic engine. Many of the engine parts, including piston skirts, connecting rods, and valve-train components, were made of a polymer called *Torlon:*

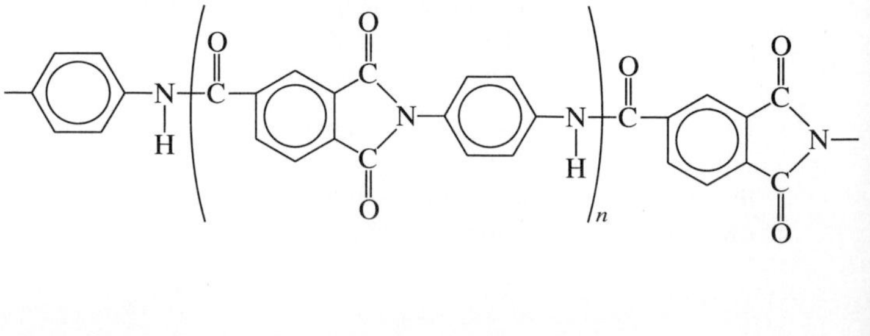

What monomers are used to make this polymer?

54. A urethane linkage occurs when an alcohol adds across the carbon-nitrogen double bond in an isocyanate.

$$\underset{\text{Alcohol}}{R{-}O{-}H} + \underset{\text{Isocyanate}}{O{=}C{=}N{-}R'} \longrightarrow \underset{\text{A urethane}}{RO{-}\overset{\overset{\displaystyle O}{\|}}{C}{-}\underset{\underset{\displaystyle H}{|}}{N}{-}R'}$$

Polyurethanes are formed from the copolymerization of a diol with a diisocyanate. Polyurethanes are used in foamed insulation and a variety of other construction materials. What is the structure of the polyurethane formed by the reaction:

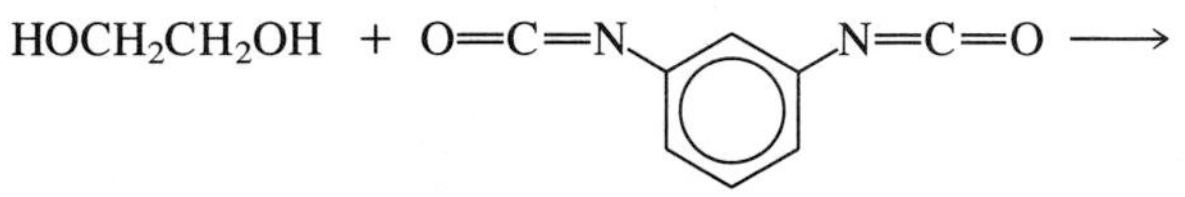

$$HOCH_2CH_2OH \;+\; O{=}C{=}N{-}C_6H_4{-}N{=}C{=}O \longrightarrow$$

55. Look up the structures of the following drugs in *The Merck Index*. What functional groups are present? Give a use for each drug.
    a. dopa
    b. amphetamine
    c. ephedrine
    d. acetaminophen
    e. phenacetin

56. Many drugs are sold using trade names or common names. The structures of such substances can be found using the Cross Index of Names at the back of *The Merck Index*. Use *The Merck Index* to find the structure of each of the following and its primary use.
    a. Ecotrin
    b. Tylenol
    c. Lasix
    d. Keflex
    e. Nalfon
    f. Voltaren

57. Read the label on an over-the-counter medication (Tylenol, cough syrup, etc.) or cosmetic. Look up the structures of the organic compounds listed in a reference such as *The Merck Index*. From the information given, speculate on the function of each of the compounds you looked up.

CHAPTER 23

Biochemistry

CONTENTS

Biochemistry, which is the study of the chemistry of living systems, is a vast and exciting field in which important discoveries about how life is maintained and how diseases occur are being made every day. In particular, there has been a rapid growth in the understanding of how living cells manufacture the molecules necessary for life.

We cannot hope to cover even the majority of the important aspects of this field in this chapter; we will concentrate on the major types of biomolecules that support living systems. First, however, we will survey the elements found in living systems and briefly describe the constitution of a cell.

At present 30 elements are known to be essential or are strongly suspected to be essential to human life. These **essential elements** are shown in Fig. 23.1. The most abundant elements are hydrogen, carbon, nitrogen, and oxygen; but sodium, magnesium, potassium, calcium, phosphorus, sulfur, and chlorine are also present in relatively large amounts. Although present only in trace amounts, the first-row transition metals are essential for the action of many enzymes. For example, zinc, one of the **trace elements,** is part of over 160 biologically important molecules. The functions of the essential elements are summarized in Table 23.1. As more studies are performed, other elements will probably be found to be essential.

Life is organized around the functions of the **cell,** the smallest unit in living things that has the properties normally associated with life, such as reproduction, metabolism, mutation, and sensitivity to external stimuli. There are two fundamental types of cells: *procaryotes,* those without nuclei; and *eucaryotes,* those that have nuclei. Bacteria are examples of procaryotic organisms, and plant and animal cells are examples of eucaryotic cells.

As the fundamental building blocks of all living systems, cells aggregate to form tissues, which in turn are assembled into the organs that make up complex living systems. Thus, to understand how life is maintained and

< Foliose lichens with fruiting bodies. While living organisms contain some macromolecules, the same bonding considerations apply as with simpler molecules.

1	2											3	4	5	6	7	8
1 H																	2 He
3 Li	4 Be											5 B	6 C	7 N	8 O	9 F	10 Ne
11 Na	12 Mg											13 Al	14 Si	15 P	16 S	17 Cl	18 Ar
19 K	20 Ca	21 Sc	22 Ti	23 V	24 Cr	25 Mn	26 Fe	27 Co	28 Ni	29 Cu	30 Zn	31 Ga	32 Ge	33 As	34 Se	35 Br	36 Kr
37 Rb	38 Sr	39 Y	40 Zr	41 Nb	42 Mo	43 Tc	44 Ru	45 Rh	46 Pd	47 Ag	48 Cd	49 In	50 Sn	51 Sb	52 Te	53 I	54 Xe
55 Cs	56 Ba	57 La	72 Hf	73 Ta	74 W	75 Re	76 Os	77 Ir	78 Pt	79 Au	80 Hg	81 Tl	82 Pb	83 Bi	84 Po	85 At	86 Rn
87 Fr	88 Ra	89 Ac	104 Unq	105 Unp	106 Unh	107 Uns	108 Uno	109 Une									
				58 Ce	59 Pr	60 Nd	61 Pm	62 Sm	63 Eu	64 Gd	65 Tb	66 Dy	67 Ho	68 Er	69 Tm	70 Yb	71 Lu
				90 Th	91 Pa	92 U	93 Np	94 Pu	95 Am	96 Cm	97 Bk	98 Cf	99 Es	100 Fm	101 Md	102 No	103 Lw

Figure 23.1

The chemical elements essential for life. Those most abundant in living systems are shown as purple. Nineteen elements, called the trace elements, are shown as green.

reproduced we must learn how cells operate on the molecular level. This is the main thrust of biochemistry.

The composition and organization of a typical animal cell is shown in Fig. 23.2. The *nucleus,* which contains the *chromosomes,* is separated from the cell fluid (the *cytoplasm*) by a membrane. The chromosomes store both the hereditary information necessary for cell reproduction and the codes for the manufacture of essential biomolecules. The cytoplasm contains a variety of subcellular structures that carry out various cell functions. For example, the *mitochondria* process nutrients and produce energy to be used as the cell requires it. The *lysosomes* contain enzymes for "digesting" nutrients such as proteins. The *ribosomes* are the sites for protein synthesis. The cell membrane encloses the cytoplasm, protects the cell components, and allows the passage of nutrients, ions, and wastes.

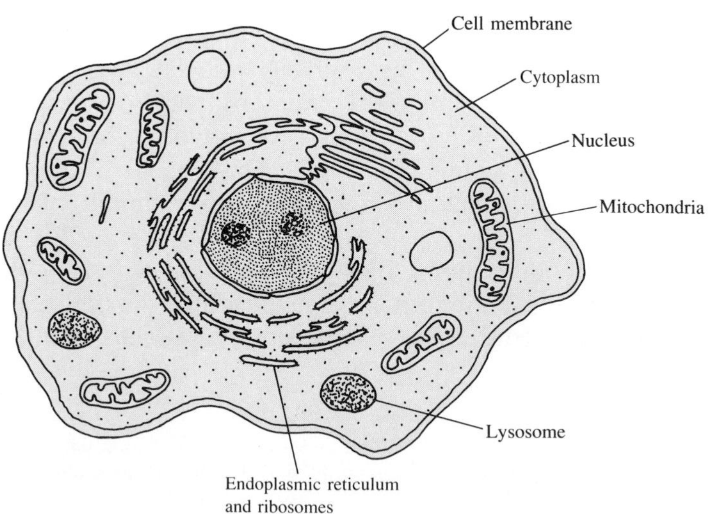

Figure 23.2

A typical animal cell (eucaryotic).

The Essential Elements and Some of Their Functions		
Element	Percent by mass in the human body	Function
Oxygen	65	Component of water and many organic compounds
Carbon	18	Component of all organic compounds
Hydrogen	10	Component of water and many inorganic and organic compounds
Nitrogen	3	Component of both inorganic and organic compounds
Calcium	1.5	Major component of bone; essential to some enzymes and to muscle action
Phosphorus	1.2	Essential in cellular synthesis and energy transfer
Potassium	0.2	Cation in intracellular fluid
Chlorine	0.2	Anion inside and outside the cells
Sulfur	0.2	Component of proteins and some other organic compounds
Sodium	0.1	Cation in extracellular fluid
Magnesium	0.05	Essential to some enzymes
Iron	<0.05	In hemoglobin, myoglobin, and other proteins (Section 20.8)
Zinc	<0.05	Essential to many enzymes
Cobalt	<0.05	Found in vitamin B_{12}
Copper	<0.05	Essential to several enzymes
Iodine	<0.05	Essential to thyroid hormones
Selenium	<0.01	Essential to some enzymes
Fluorine	<0.01	In teeth and bones
Nickel	<0.01	Essential to some enzymes
Molybdenum	<0.01	Essential to some enzymes
Silicon	<0.01	Found in connective tissue
Chromium	<0.01	Essential in carbohydrate metabolism
Vanadium, tin, manganese, lithium, boron, arsenic, lead, cadmium	<0.01	Exact function not known

Table 23.1

This peacock feather and spider web are constructed of fibrous proteins.

23.1 Proteins

Purpose

- To describe the levels of structure and the functions of proteins.

We have seen that many useful synthetic materials are polymers. Thus it should not be surprising that a great many natural materials are also polymers: starch, hair, silicate chains in soil and rocks, silk and cotton fibers, and the cellulose in woody plants, to name only a few.

In this section we consider a class of natural polymers, the **proteins,** which make up about 15% of our bodies, and have molecular weights that range from approximately 6000 to over 1,000,000. Proteins have many functions in the human body. **Fibrous proteins** provide structural integrity and strength for many types of tissue and are the main components of muscle, hair, and cartilage. Other proteins, usually called **globular proteins** because of their roughly spherical shape, are the "worker" molecules of the body. These proteins transport and store oxygen and nutrients, act as catalysts for the thousands of reactions that make life possible, fight invasion by foreign objects, participate in the body's many regulatory systems, and transport electrons in the complex process of metabolizing nutrients.

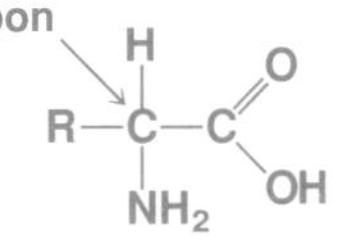

α-Carbon

$$\mathrm{R{-}\underset{NH_2}{\overset{H}{C}}{-}C({=}O)OH}$$

The building blocks of all proteins are the ***α*-amino acids,** where R may represent H, CH_3, or more complex substituents. These molecules are called α-amino acids because the amino group ($-NH_2$) is always attached to the α-carbon, the one next to the carboxyl group ($-CO_2H$). The 20 amino acids most commonly found in proteins are shown in Fig. 23.3.

Note from Fig. 23.3 that the amino acids are grouped into polar and nonpolar classes, determined by the R groups, or **side chains.** Nonpolar side chains contain mostly carbon and hydrogen atoms, while polar side chains contain nitrogen and oxygen atoms. This difference is important because polar side chains are *hydrophilic* (water-loving), while nonpolar side chains are *hydrophobic* (water-fearing), and this greatly affects the three-dimensional structure of the resulting protein.

The protein polymer is built by condensation reactions between amino acids, for example,

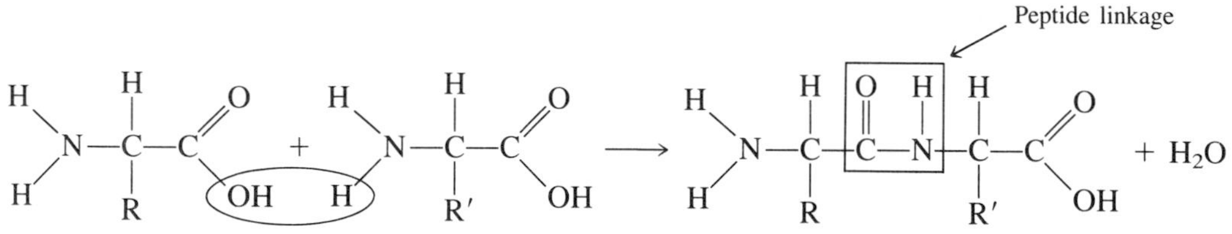

$$\mathrm{H_2N{-}\underset{R}{\overset{H}{C}}{-}C({=}O)OH + H{-}N(H){-}\underset{R'}{\overset{H}{C}}{-}C({=}O)OH \longrightarrow H_2N{-}\underset{R}{\overset{H}{C}}{-}\boxed{\overset{O}{\overset{\|}{C}}{-}\overset{H}{\overset{|}{N}}}{-}\underset{R'}{\overset{H}{C}}{-}C({=}O)OH + H_2O}$$

Peptide linkage

The product shown above is called a **dipeptide.** This name is used because the structure

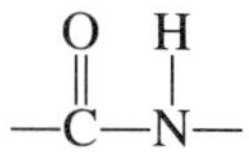

$$-\overset{O}{\overset{\|}{C}}-\overset{H}{\overset{|}{N}}-$$

is called a **peptide linkage** by biochemists. (The same grouping is called an amide by organic chemists.) Additional condensation reactions lengthen the chain to produce a **polypeptide** and eventually a protein.

The peptide linkage is also found in nylon (see Section 22.5).

You can imagine that with 20 amino acids, which can be assembled in any order, there is essentially infinite variety possible in the construction of proteins. This flexibility allows an organism to tailor proteins for the many types of functions that must be carried out.

The order or sequence of amino acids in the protein chain is called the **primary structure,** conveniently indicated by using three-letter codes for the amino acids

Figure 23.3

The 20 α-amino acids found in most proteins are shown at right. The R group is shown in color.

Nonpolar R groups

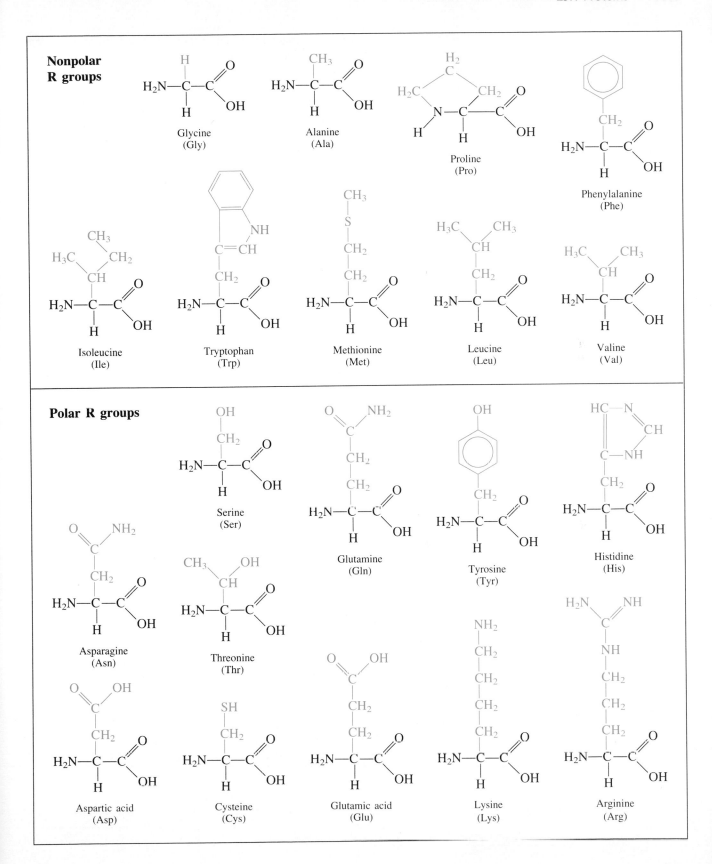

Glycine (Gly)

Alanine (Ala)

Proline (Pro)

Phenylalanine (Phe)

Isoleucine (Ile)

Tryptophan (Trp)

Methionine (Met)

Leucine (Leu)

Valine (Val)

Polar R groups

Serine (Ser)

Glutamine (Gln)

Tyrosine (Tyr)

Histidine (His)

Asparagine (Asn)

Threonine (Thr)

Aspartic acid (Asp)

Cysteine (Cys)

Glutamic acid (Glu)

Lysine (Lys)

Arginine (Arg)

(see Fig. 23.3), where it is understood that the terminal carboxyl group is on the right and the terminal amino group is on the left. For example, one possible sequence for a tripeptide containing the amino acids lysine, alanine, and leucine is

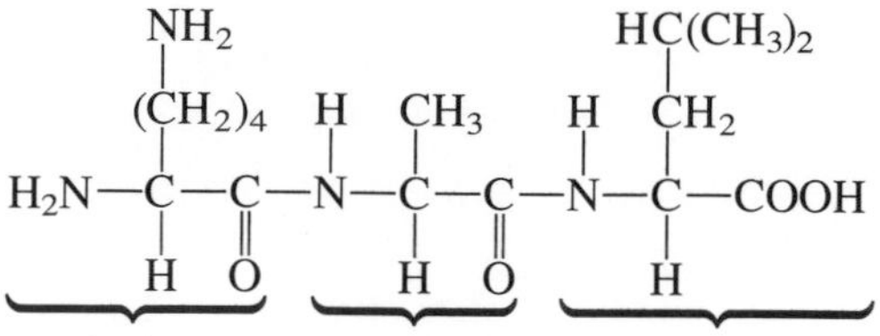

```
             NH2                          HC(CH3)2
              |                              |
            (CH2)4  H   CH3       H      CH2
              |     |    |        |       |
      H2N—C—C—N—C—C—N—C—COOH
              |   ||       |   ||          |
              H   O        H   O           H
         \_______/     \_______/      \_________/
           Lysine        Alanine        Leucine
```

which is represented in the shorthand notation as

lys-ala-leu

Sample Exercise 23.1

Write the sequences of all possible tripeptides composed of the amino acids tyrosine, histidine, and cysteine.

Solution

There are six possible sequences:

tyr-his-cys	his-tyr-cys	cys-tyr-his
tyr-cys-his	his-cys-tyr	cys-his-tyr

Note from Sample Exercise 23.1 that there are six sequences possible for a polypeptide with three given amino acids. There are three possibilities for the first amino acid (any one of the three given amino acids); there are two possibilities for the second amino acid (one has already been accounted for); there is only one possibility left for the third amino acid. Thus the number of sequences is $3 \times 2 \times 1 = 6$. The product $3 \times 2 \times 1$ is often written 3! (and is called 3 factorial). Similar reasoning shows that for a polypeptide with four amino acids, there are 4!, or $4 \times 3 \times 2 \times 1 = 24$, possible sequences.*

Sample Exercise 23.2

What number of possible sequences exists for a polypeptide composed of 20 different amino acids?

Solution

The answer is 20!, or

$$20 \times 19 \times 18 \times 17 \times 16 \times \cdots \times 5 \times 4 \times 3 \times 2 \times 1 = 2.43 \times 10^{18}$$

*Most calculators have a key for factorials.

A striking example of the importance of the primary structure of polypeptides can be seen in the differences between *oxytocin* and *vasopressin*. Both of these molecules are nine-unit polypeptides, and they differ by only two amino acids (Fig. 23.4); yet they have completely different functions in the human body. Oxytocin is a hormone that triggers contraction of the uterus and milk secretion. Vasopressin raises blood pressure levels and regulates kidney function.

cys-tyr-ile-gln-asn-cys-pro-leu-gly
(a)

cys-tyr-phe-gln-asn-cys-pro-arg-gly
(b)

Figure 23.4

The amino acid sequences in (a) oxytocin and (b) vasopressin. The differing amino acids are boxed.

A second level of structure in proteins, beyond the sequence of amino acids, is the arrangement of the chain of the long molecule. The **secondary structure** is determined to a large extent by hydrogen bonding between lone pairs on an oxygen in the carbonyl group of an amino acid with a hydrogen atom attached to a nitrogen of another amino acid:

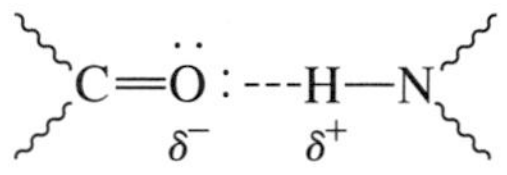

$$>C{=}\underset{\delta^-}{\ddot{O}}\colon{-}{-}{-}\underset{\delta^+}{H}{-}N<$$

Such interactions can occur *within* the chain coils to form a spiral structure called an **α-helix,** as shown in Fig. 23.5 and Fig. 23.6. This type of secondary

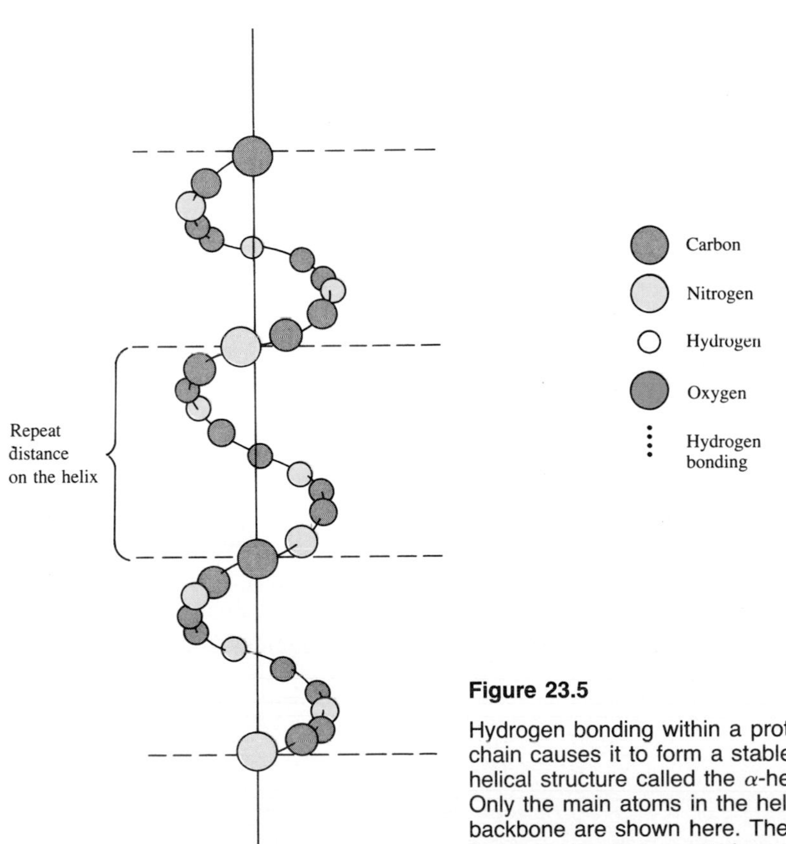

Figure 23.5

Hydrogen bonding within a protein chain causes it to form a stable helical structure called the α-helix. Only the main atoms in the helical backbone are shown here. The hydrogen bonds are not shown.

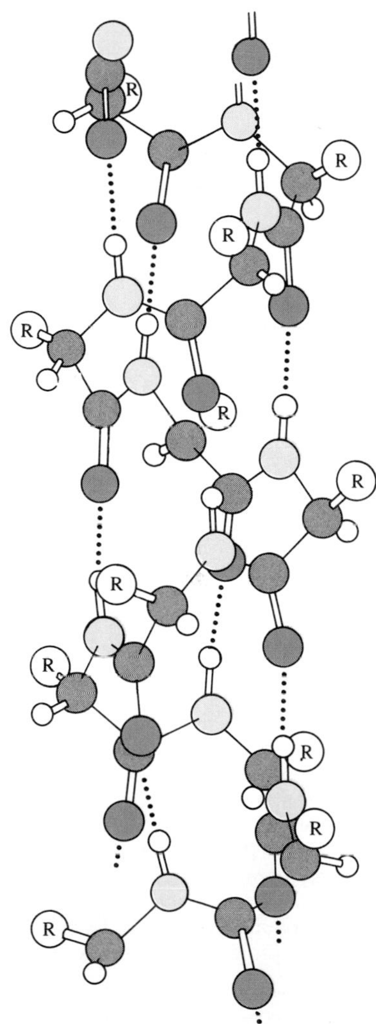

Figure 23.6

Ball-and-stick model of a portion of a protein chain in the α-helical arrangement showing the hydrogen bonding interactions.

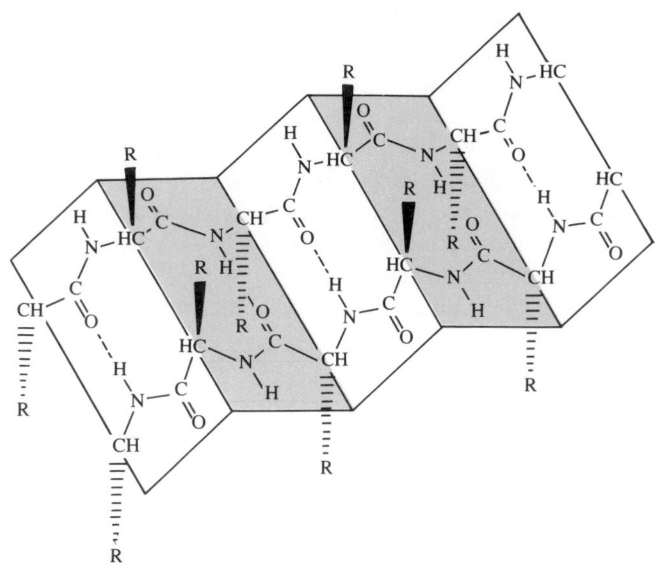

Figure 23.7

When hydrogen bonding occurs between protein chains rather than within them, a stable structure (the pleated sheet) results. This structure contains many protein chains and is found in natural fibers, such as silk, and also in muscles.

structure gives the protein elasticity (springiness) and is found in the fibrous proteins in wool, hair, and tendons. Hydrogen bonding can also occur *between different* protein chains, joining them together in an arrangement called a **pleated sheet,** as shown in Fig. 23.7. Silk contains this arrangement of proteins, making its fibers flexible but very strong and resistant to stretching. The pleated sheet is also found in muscle fibers. The hydrogen bonds in the α-helical protein are called *intrachain* (within a given protein chain), and those in the pleated sheet are said to be *interchain* (between protein chains).

As you might imagine, a molecule as large as a protein has a great deal of flexibility and can assume a variety of overall shapes. The specific shape that a protein assumes depends on its function. For long, thin structures, such as hair, wool and silk fibers, and tendons, an elongated shape is required. This may involve an α-helical secondary structure, as found in the protein α-keratin in hair and wool or in the collagen found in tendons [Fig. 23.8(a)], or it may involve a pleated-sheet secondary structure, as found in silk [Fig. 23.8(b)]. Many of the proteins in the body that have nonstructural functions are globular, such as myoglobin (Fig. 23.9). Note that the secondary structure of myoglobin is basically α-helical. However, in the areas where the chain bends to give the protein its compact globular structure, the α-helix breaks down to give a secondary configuration known as the **random-coil arrangement.**

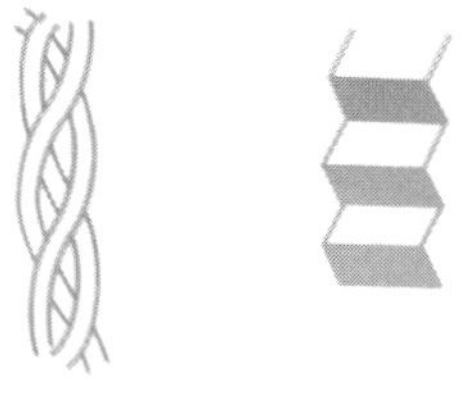

(a) (b)

Figure 23.8

(a) Collagen, a protein found in tendons, consists of three protein chains (each with an α-helical structure) twisted together to form a super helix. The result is a long, relatively narrow protein. (b) The pleated-sheet arrangement of many proteins bound together to form the elongated protein found in silk fibers.

The overall shape of the protein, long and narrow or globular, is called its **tertiary structure** and is maintained by several different types of interactions: hydrogen bonding, dipole-dipole interactions, ionic bonds, covalent bonds, and London dispersion forces between nonpolar groups. These bonds, which represent all of the bonding types discussed in this text, are summarized in Fig. 23.10.

The amino acid *cysteine*

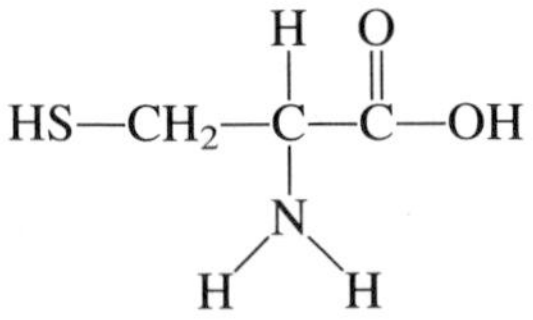

$$HS{-}CH_2{-}\overset{\displaystyle H}{\overset{|}{\underset{\underset{\displaystyle N(H)_2}{|}}{C}}}{-}\overset{\displaystyle O}{\overset{\|}{C}}{-}OH$$

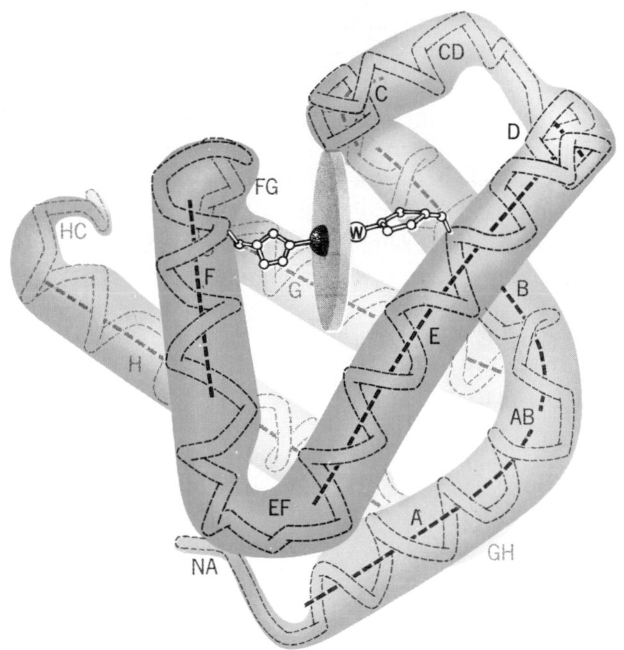

Figure 23.9

The protein myoglobin. Reprinted from R. E. Dickerson and I. Geis from *The Structure and Action of Proteins,* Benjamin/Cummings Publisher, Menlo Park, CA. Illustration © 1969 by Dickerson and Geis.

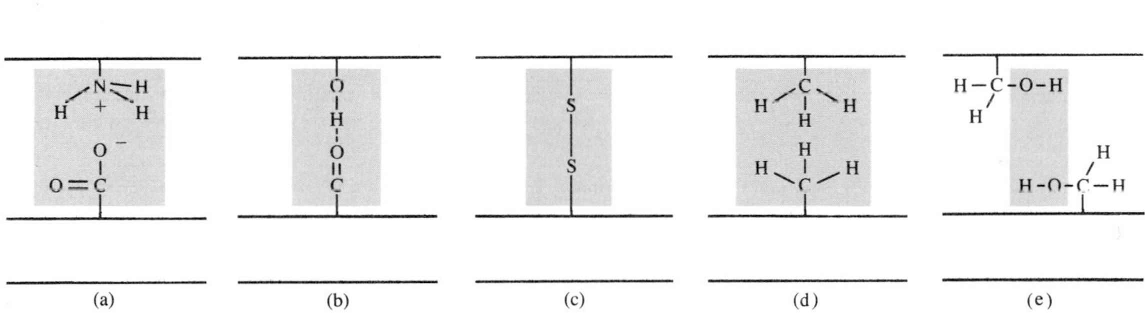

Figure 23.10

Summary of the various types of interactions that stabilize the tertiary structure of a protein: (a) ionic, (b) hydrogen bonding, (c) covalent, (d) London dispersion, and (e) dipole-dipole.

plays a special role in stabilizing the tertiary structure of many proteins because the —SH groups on two cysteines can react in the presence of an oxidizing agent to form a S—S bond called a **disulfide linkage:**

$$\text{C—CH}_2\text{—S—H} + \text{H—S—CH}_2\text{—C} \longrightarrow \text{C—CH}_2\text{—}\boxed{\text{S—S}}\text{—CH}_2\text{—C}$$

A practical application of the chemistry of disulfide bonds is permanent waving of hair, as summarized in Fig. 23.11 on the following page. The S—S linkages in the protein of hair are broken by treatment with a reducing agent. The hair is then set in

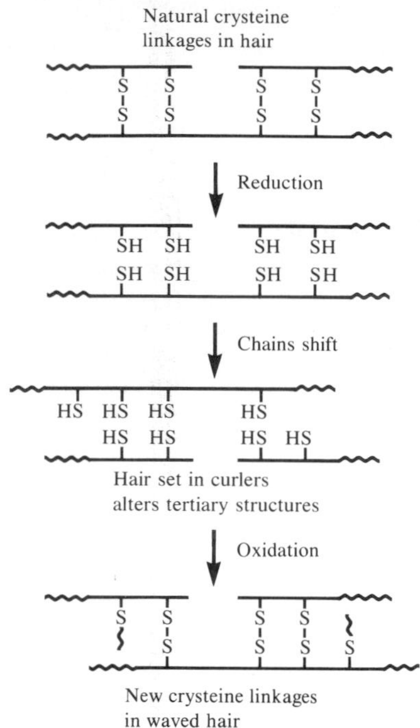

Figure 23.11

The permanent waving of hair.

curlers to change the tertiary protein structure to the desired shape. Then treatment with an oxidizing agent causes new S—S bonds to form, which make the hair protein retain the new structure.

The three-dimensional structure of a protein is crucial to its function. The process of breaking down this structure is called **denaturation** (Fig. 23.12). For example, the denaturation of egg proteins occurs when an egg is cooked. Any source of energy can cause denaturation of proteins and is thus potentially dangerous to living organisms. For example, ultraviolet or X-ray radiation or nuclear radioactivity can disrupt protein structure, which may lead to cancer or genetic damage. Protein damage is also caused by chemicals like benzene, trichloroethane, and 1,2-dibromoethane (called EDB). The metals lead and mercury, which have a very high affinity for sulfur, cause protein denaturation by disrupting disulfide bonds between protein chains.

The tremendous flexibility in the various levels of protein structure allows the tailoring of proteins for a wide range of specific functions, some of which are given in Table 23.2.

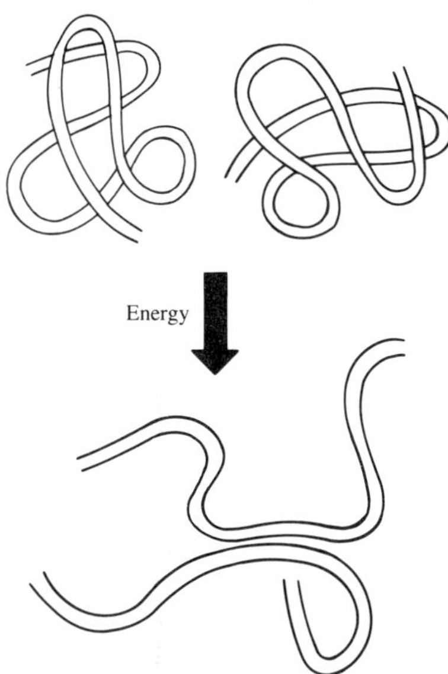

Figure 23.12

A schematic representation of the thermal denaturation of a protein.

Table 23.2

Some Functions of Proteins

1. *Structure*. Proteins provide the high tensile strength of tendons, bones, and skin. Cartilage, hair, wool, fingernails, and claws are mainly protein. Viruses have an outer layer of protein around a nucleic acid core. Proteins called *histones* are bound tightly to DNA in the cells of higher organisms, helping to fold and wrap the DNA in an orderly fashion in chromosomes.
2. *Movement*. Proteins are the major components of muscles and are directly responsible for the ability of muscles to contract. The swimming of sperm is the result of contraction of protein filaments in their tails. The same is true of movements of chromosomes during cell division.
3. *Catalysis*. Nearly all chemical reactions in living organisms are catalyzed by enzymes, which are almost always proteins.
4. *Transport*. Oxygen is carried from the lungs to tissues by the protein hemoglobin in red blood cells. The protein transferrin transports iron in blood plasma from the intestines where the iron is absorbed to the spleen where it is stored and to the liver and bone marrow where it is used for synthesis. Proteins in the membranes of cells allow the passage of various molecules and ions.
5. *Storage*. The protein ferritin stores iron in the liver, spleen, and bone marrow.
6. *Energy transformation and storage*. Myosin, a protein in muscle, transforms chemical energy into useful work. Rhodopsin, a protein in the retina of the eye, traps light energy and, working with other membrane components, converts it into the electrical energy of a nerve impulse. Receptor protein molecules that combine with specific small molecules are responsible for the transmission of nerve impulses.
7. *Protection*. *Antibodies* are special proteins that are synthesized in response to foreign substances and cells, such as bacterial cells. They then bind to those substances or cells, which are called *antigens,* and provide us with immunity to various diseases. We acquire antibodies either from having had the disease or from receiving inactivated viruses in vaccines. Hay fever and food allergies are also caused by the interaction of antibodies with antigens. Interferon, a small protein made and released by cells when they are exposed to a virus, protects other cells against viral infection. Blood-clotting proteins protect against bleeding (hemorrhage). Some antibiotics are polypeptides, which are smaller than proteins.

8. *Control*. *Hormones* are chemical substances produced in the body that have specific effects on the activity of certain organs. Some hormones are proteins. For example, human growth hormone is a protein. Some hormones such as insulin and glucagon, which are made in the pancreas and control carbohydrate metabolism, are polypeptides. Expression of genetic information is also under the control of proteins.
9. *Buffering*. Because proteins contain both acidic and basic groups on their side chains, they can neutralize both acids and bases. Therefore, proteins provide some buffering for blood and tissues.

Table 23.2, *continued*

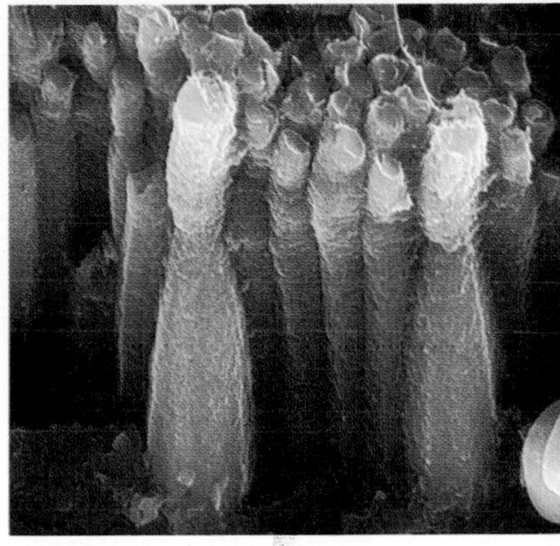

Rods and cones in the human eye (×2500).

Enzymes

Enzymes are proteins tailored to catalyze specific biological reactions. Without the several hundred enzymes now known, life would be impossible. Enzymes are impressive for their tremendous efficiency (typically 1 to 10 million times as efficient as inorganic catalysts) and their incredible selectivity—they ignore the thousands of molecules in body fluids for which they were not designed.

Although the mechanisms of catalytic activity are complex and not fully understood in most cases, a theory called the **lock-and-key model** (see Fig. 23.13) seems to fit many enzymes. This model postulates that the shapes of the reacting molecule, the **substrate,** and the enzyme are such that they fit together much like a key fits a specific lock. The substrate and enzyme attach to each other through hydrogen bonding, ionic bonding, metal ion–ligand bonding, or some combination of these, in such a way that the part of the substrate where the reaction is to occur occupies the active site of the enzyme. This process is summarized schematically in Fig. 23.14 on page 1022 for the enzyme carboxypeptidase binding a protein. After the reaction occurs, the products are liberated, and the enzyme is ready for a new substrate. Because the enzyme cycles ("turns over") so rapidly, only a tiny amount of enzyme is required. This has made the isolation and study of enzymes quite difficult.

The enzyme carboxypeptidase assists in the digestion of proteins by catalyzing the breaking of the peptide linkage that attaches the end amino acid to the protein:

$$\sim\!\!\sim\text{C}(=\text{O})-\text{N(H)}-\text{C(H)(R)}-\text{C}(=\text{O})\text{OH} + \text{H}_2\text{O} \xrightarrow{\text{Carboxypeptidase}} \sim\!\!\sim\text{C}(=\text{O})\text{OH} + \text{H}_2\text{N}-\text{C(H)(R)}-\text{C}(=\text{O})\text{OH}$$

The active site contains Zn^{2+}, which coordinates to the carbonyl group of the end amino acid. There is also a hydrophobic pocket nearby to accommodate a large hydrophobic R group on the terminal amino acid. There are other interactions between groups on the enzymes and substrate, which help hold the substrate in place.

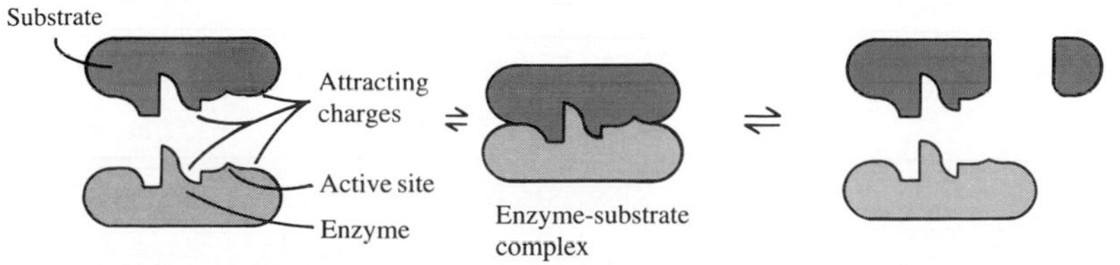

Figure 23.13

Schematic diagram of the lock-and-key model.

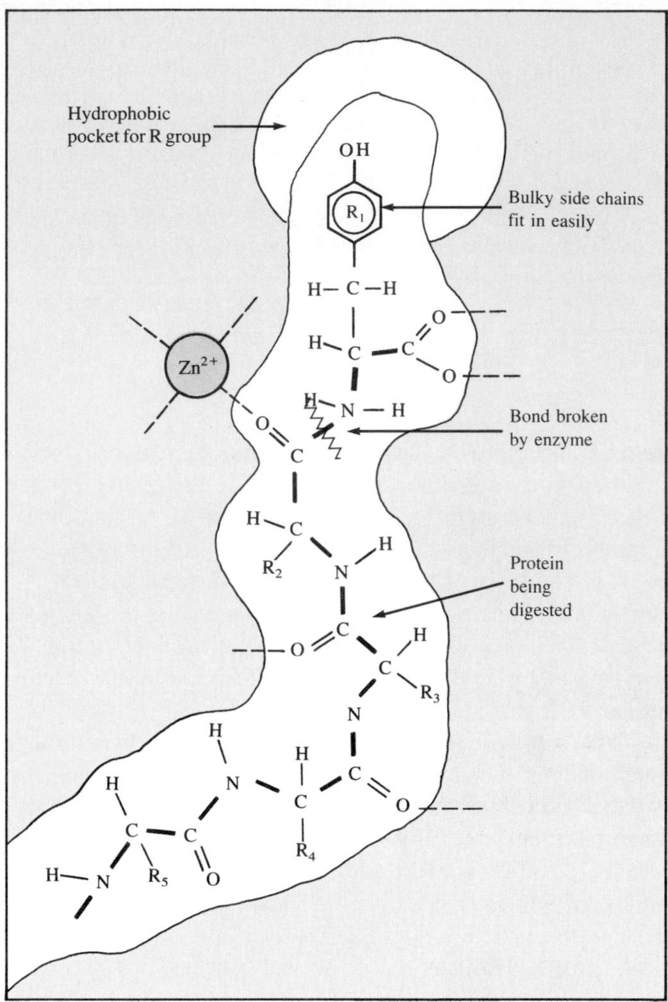

Figure 23.14

Schematic of the binding of a protein by carboxypeptidase.

When a substance other than the substrate is bound to the enzyme's active site, the enzyme is said to be *inhibited*. If the inhibition is permanent, the enzyme is said to be inactivated. Some of the most powerful toxins act by inhibiting or inactivating key enzymes. "Nerve gases," chemicals used as weapons, act in this way. Nerves transfer messages using small molecules called *neurotransmitters*. An example of a neurotransmitter is acetylcholine.

$$CH_3-\overset{\overset{O}{\|}}{C}-O-CH_2-CH_2-\overset{+}{N}(CH_3)_3$$

Acetylcholine

Once the nerve impulse has been transmitted, for example, to a muscle, the acetylcholine molecules must be destroyed because their continued presence would cause overstimulation and lead to convulsions, paralysis, and possibly death. The acetylcholine molecules are removed by reaction with water.

$$[CH_3COOCH_2CH_2N(CH_3)_3]^+ + H_2O \rightarrow \underset{\text{Acetic acid}}{CH_3COOH} + \underset{\text{Choline}}{[HOCH_2CH_2N(CH_3)_3]^+}$$

This reaction is catalyzed by the enzyme acetylcholinesterase. The nerve gas diisopropyl phosphorofluoridate (shortened to DIPF) attaches itself permanently to the active site of this enzyme, thus blocking the enzyme's ability to catalyze the removal of acetylcholine. The result is usually death.

$$
\begin{array}{ccccccccc}
 & CH_3 & & & F & & & CH_3 & \\
 & | & & & | & & & | & \\
H- & C & -O- & & P & & -O- & C & -H \\
 & | & & & | & & & | & \\
 & CH_3 & & & O & & & CH_3 &
\end{array}
$$

DIPF

There are many other ways that enzymes can be deactivated, in addition to blockage of the active site by foreign molecules. For example, in **metalloenzymes,** which contain a metal ion at the active site, substitution of a different metal ion for the original one usually causes the enzyme to malfunction. And, of course, anything that denatures the enzyme protein destroys its activity.

Because enzymes are so crucial for healthy life and because we hope to learn how to mimic their efficiency in our industrial catalysts, the study of enzymes occupies a prominent role in chemical research.

23.2 Carbohydrates

Purpose

- To describe the structures and isomerism of simple carbohydrates.
- To describe the structures and functions of polymers of glucose.

Another class of biologically important molecules is the **carbohydrates,** which serve as a food source for most organisms and as a structural material for plants. Because many carbohydrates have the empirical formula CH_2O, it was originally believed that these substances were hydrates of carbon, and that accounts for the name.

Most important carbohydrates, such as starch and cellulose, are polymers composed of monomers called **monosaccharides,** or **simple sugars.** The monosaccharides are polyhydroxy ketones and aldehydes. The most important contain five carbon atoms (**pentoses**) or six carbon atoms (**hexoses**). One important hexose is *fructose,* a sugar found in honey and fruit, whose structure is

General name of sugar	Number of carbon atoms
Triose	3
Tetrose	4
Pentose	5
Hexose	6
Heptose	7
Octose	8
Nonose	9

$$
\begin{array}{c}
CH_2OH \\
| \\
C{=}O \\
| \\
HO-{}^*C-H \\
| \\
H-{}^*C-OH \\
| \\
H-{}^*C-OH \\
| \\
CH_2OH
\end{array}
$$

Fructose

where the asterisks indicate chiral carbon atoms. In Section 20.4 we saw that molecules with nonsuperimposable mirror images exhibit optical isomerism. A carbon atom with four *different* groups bonded to it in a tetrahedral arrangement *always* has

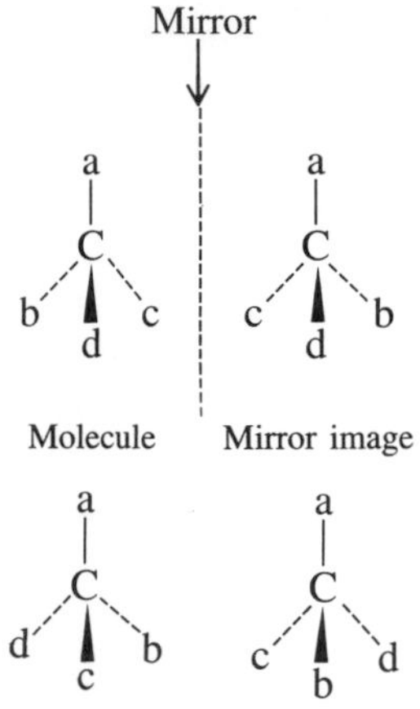

Mirror

Molecule Mirror image

Figure 23.15

When a tetrahedral carbon atom has four different substituents, there is no way that its mirror image can be superimposed. The lower two forms show possible orientations of the molecule. Compare these to the mirror image and note they cannot be superimposed.

a nonsuperimposable mirror image (see Fig. 23.15), which gives rise to a pair of optical isomers. For example, the simplest sugar, glyceraldehyde,

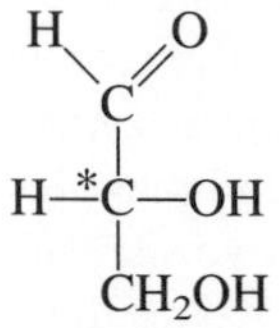

H—C=O
H—*C—OH
CH_2OH

which has one chiral carbon, has two optical isomers, as shown in Fig. 23.16.

In fructose each of the three chiral carbon atoms satisfies the requirement of being surrounded by four different groups, which leads to a total of 2^3 or eight isomers that differ in their ability to rotate polarized light. The particular isomer whose structure is given above is called D-fructose. Generally, monosaccharides have one isomer that is more common in nature than the others. The most important pentoses and hexoses are shown in Table 23.3.

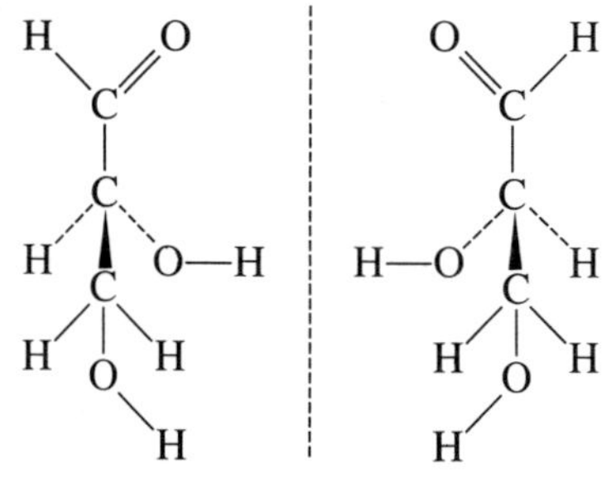

Figure 23.16

The mirror-image optical isomers of glyceraldehyde. Note that these mirror images cannot be superimposed.

Sample Exercise 23.3

Determine the number of chiral carbon atoms in the following pentose:

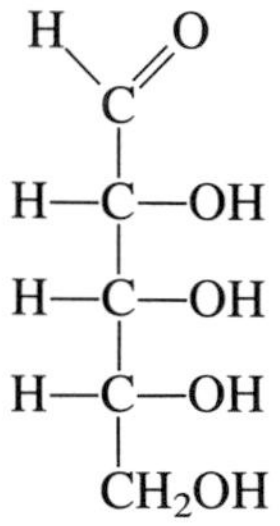

H—C=O
H—C—OH
H—C—OH
H—C—OH
CH_2OH

Solution

We must look for carbon atoms that have four different substituents. The top carbon has only three substituents and thus cannot be chiral. The three carbon atoms shown in red each have four different groups attached to them as indicated below.

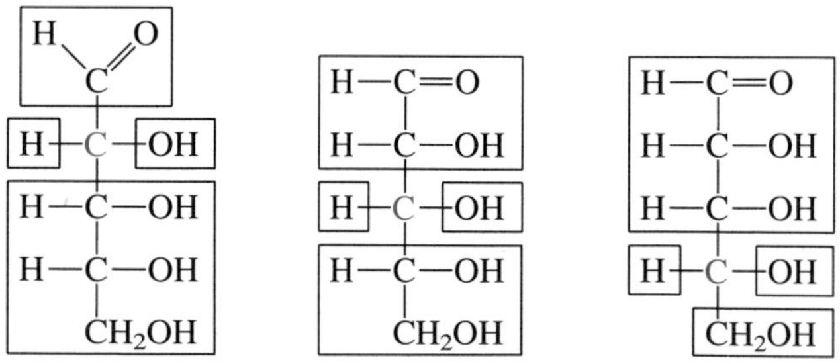

H—C=O | H—C=O | H—C=O
H—C—OH | H—C—OH | H—C—OH
H—C—OH | H—C—OH | H—C—OH
H—C—OH | H—C—OH | H—C—OH
CH_2OH | CH_2OH | CH_2OH

The fifth carbon atom has only three types of substituents (it has two hydrogen atoms) and is not chiral.

Thus the three chiral carbon atoms in this pentose are the ones shown in red.

Some Important Monosaccharides

Pentoses			
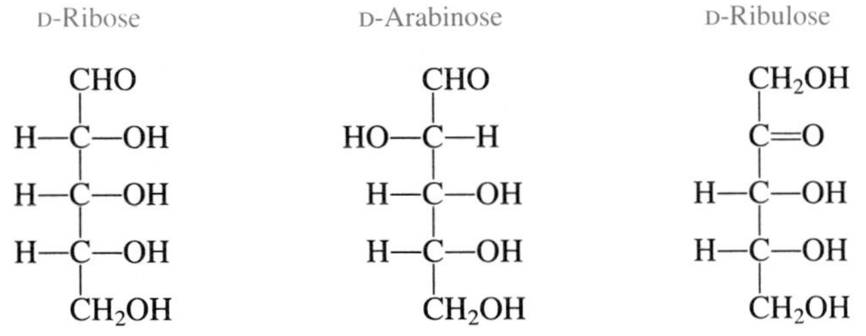 D-Ribose	D-Arabinose	D-Ribulose	
CHO H—C—OH H—C—OH H—C—OH CH_2OH	CHO HO—C—H H—C—OH H—C—OH CH_2OH	CH_2OH C=O H—C—OH H—C—OH CH_2OH	
Hexoses			
D-Glucose	D-Mannose	D-Galactose	D-Fructose
CHO H—C—OH HO—C—H H—C—OH H—C—OH CH_2OH	CHO HO—C—H HO—C—H H—C—OH H—C—OH CH_2OH	CHO H—C—OH HO—C—H HO—C—H H—C—OH CH_2OH	CH_2OH C=O HO—C—H H—C—OH H—C—OH CH_2OH

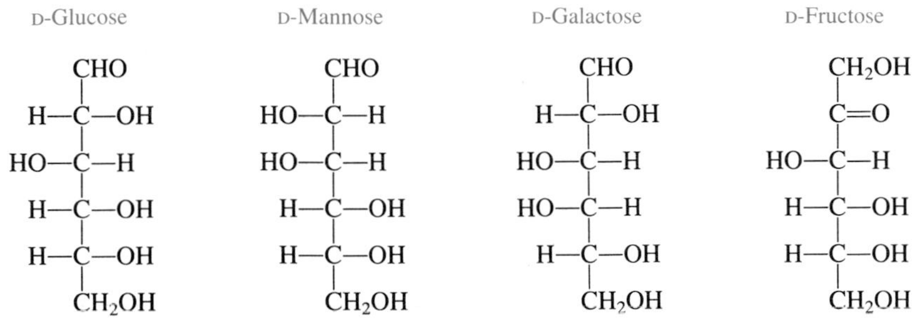

Table 23.3

Sample Exercise 23.3, continued

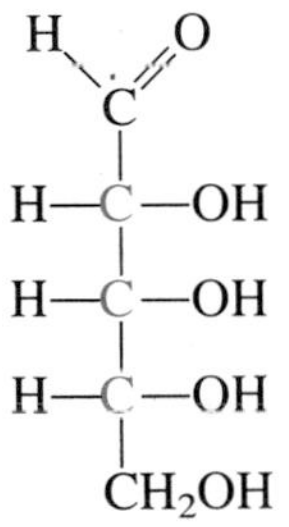

```
H   O
 \ //
  C
  |
H—C—OH
  |
H—C—OH
  |
H—C—OH
  |
  CH2OH
```

Note that D-ribose and D-arabinose, shown in Table 23.3, are two of the eight isomers of this pentose.

Although we have so far represented the monosaccharides as straight-chain molecules, they usually cyclize, or form a ring structure, in aqueous solution. Figure 23.17 shows this reaction for fructose. Note that a new bond is formed between the oxygen of the terminal hydroxyl group and the carbon of the ketone group. In the cyclic form fructose is a five-membered ring containing a C—O—C bond. The same type of reaction can occur between a hydroxyl group and an aldehyde group, as shown for D-glucose in Fig. 23.18. In this case a six-membered ring is formed.

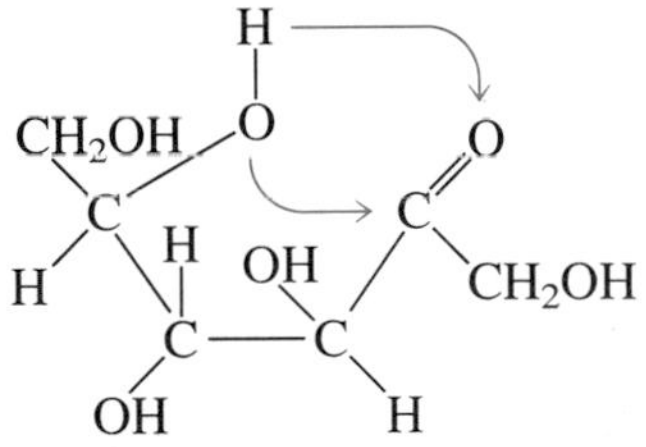

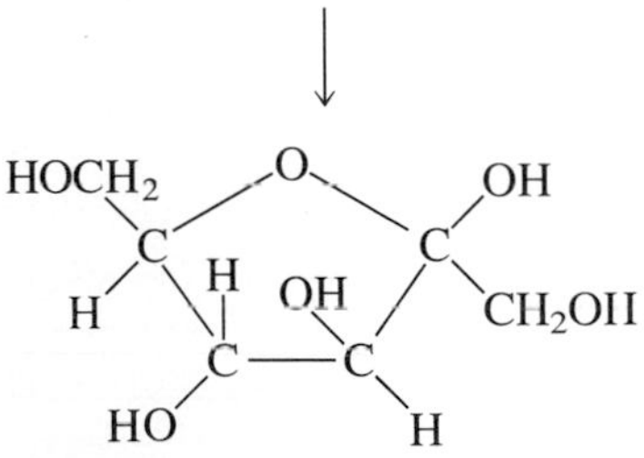

Figure 23.17

The cyclization of D-fructose.

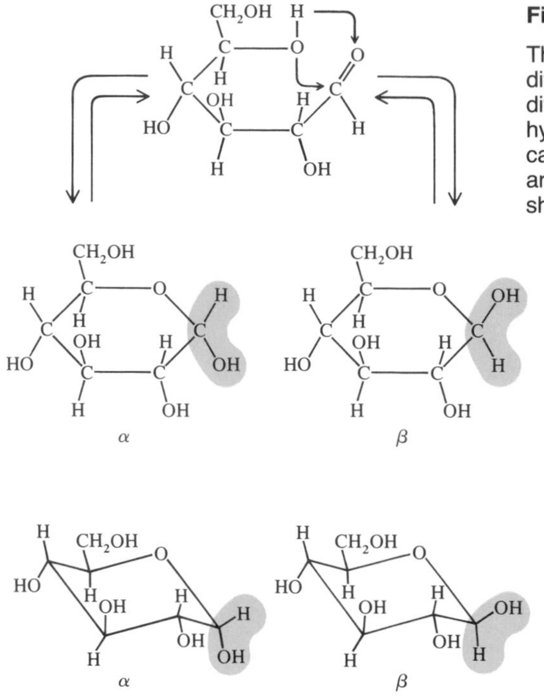

Figure 23.18

The cyclization of glucose. Two different rings are possible; they differ in the orientation of the hydroxy group and hydrogen on one carbon, as indicated. The two forms are designated α and β and are shown here in two representations.

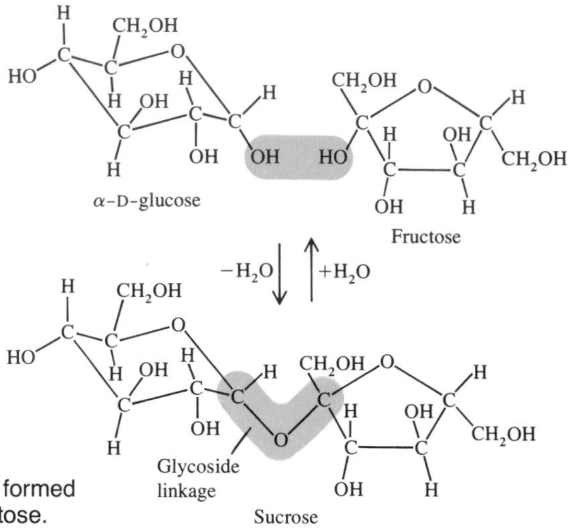

Figure 23.19

Sucrose is a disaccharide formed from α-D-glucose and fructose.

These jungle plants store energy in the form of starch.

More complex carbohydrates are formed by combining monosaccharides. For example, **sucrose,** common table sugar, is a **disaccharide** formed from glucose and fructose by elimination of water to form a C—O—C bond between the rings, which is called a **glycoside linkage** (Fig. 23.19). When sucrose is consumed in food, the above reaction is reversed. An enzyme in saliva (α-amylase) catalyzes the breakdown of this disaccharide.

Large polymers of many monosaccharide units are called polysaccharides and can form when each ring forms two glycoside linkages, as shown in Fig. 23.19. Three of the most important of these polymers are starch, cellulose, and glycogen. All of these substances are polymers of glucose, differing from each other in the nature of the glycoside linkage, the amount of branching, and molecular weight.

Starch is a polymer of α-D-glucose and consists of two parts: *amylose,* a straight-chain polymer of α-glucose [see Fig. 23.20(a)]; and *amylopectin,* a highly branched polymer of α-glucose with a molecular weight that is 10–20 times that of amylose. Branching occurs when a third glycoside linkage attaches a branch to the main polymer chain.

Starch, the carbohydrate reservoir in plants, is the form in which glucose is stored by the plant for later use as cellular fuel. Glucose is stored in this high-molecular-weight form because this results in less stress on the plant's internal structure by osmotic pressure. Remember from Section 11.6 that it is the concentration of solute molecules (or ions) that determines the osmotic pressure. Thus com-

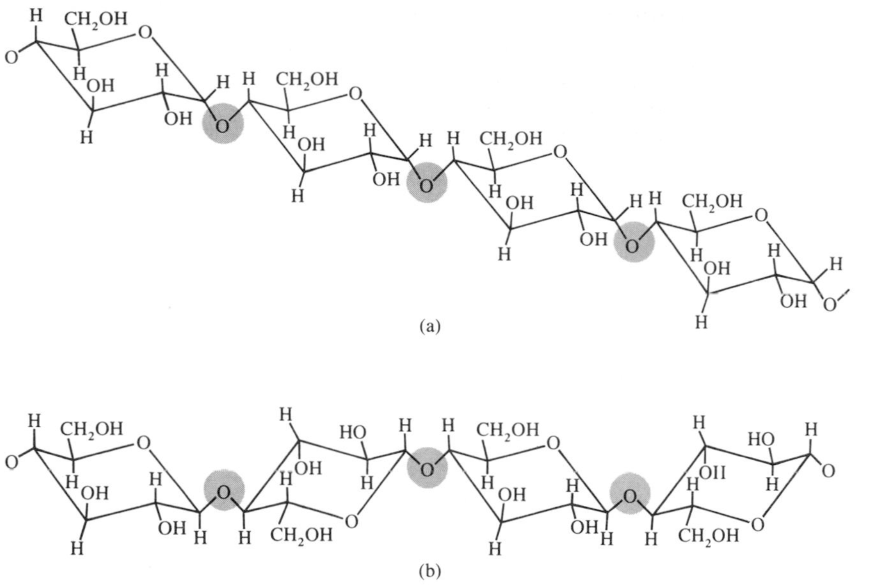

(a)

(b)

Figure 23.20

(a) The polymer amylose is a major component of starch and is made up of α-D-glucose monomers. (b) The polymer cellulose, which consists of β-D-glucose monomers.

bining the individual glucose molecules into one large chain keeps the concentration of solute molecules relatively low and minimizes the osmotic pressure.

Cellulose, the major structural component of woody plants and natural fibers such as cotton, is a polymer of β-D-glucose and has the structure shown in Fig. 23.20(b). Note that the β-glycoside linkages in cellulose give the glucose rings a different relative orientation than is found in starch. Although this difference may seem minor, it has very important consequences. The human digestive system contains α-glycosidases, enzymes that can catalyze breakage of the α-glycoside bonds in starch. These enzymes are not effective on the β-glycoside bonds of cellulose, presumably because the different structure causes a poor fit between the enzyme's active site and the carbohydrate. The enzymes necessary to cleave β-glycoside linkages, the β-glycosidases, are found in bacteria that exist in the digestive tracts of termites, cows, deer, and many other animals. Thus, unlike humans, these animals can derive nutrition from cellulose.

Glycogen, the carbohydrate reservoir in animals, has a structure like that of amylopectin but with more branching. It is this branching that is thought to facilitate the rapid breakdown of glycogen into glucose when energy is required.

23.3 Nucleic Acids

Purpose

- To describe the structures and functions of nucleic acids.

Life is possible only because each cell, when it divides, can transmit the vital information about how it works to the next generation. It has been known for a long time that this process involves the chromosomes in the nucleus of the cell. Only since 1953, however, have scientists understood the molecular basis of this intriguing cellular "talent."

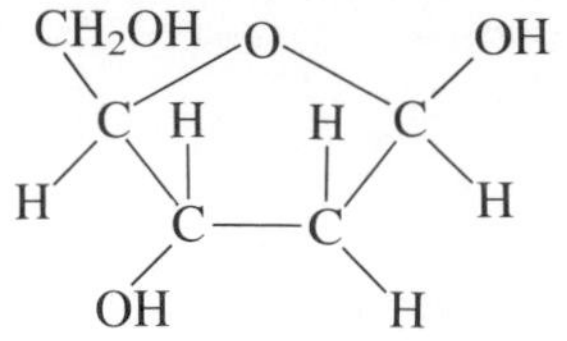

(a)

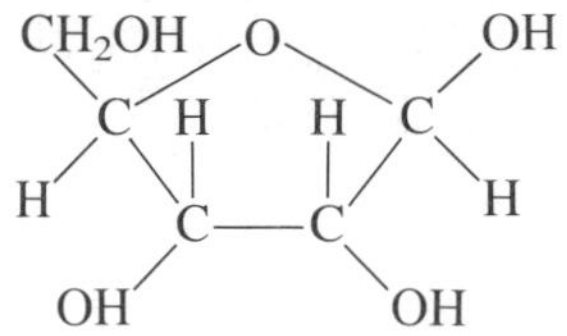

(b)

Figure 23.21

The structure of the pentoses (a) deoxyribose and (b) ribose. Deoxyribose is the sugar molecule present in DNA; ribose is found in RNA.

The substance that stores and transmits the genetic information is a polymer called **deoxyribonucleic acid (DNA),** a huge molecule with a molecular weight as high as several billion. Together with other similar nucleic acids called the **ribonucleic acids (RNA),** DNA is also responsible for the synthesis of the various proteins needed by the cell to carry out its life functions. The RNA molecules, which are found in the cytoplasm outside the nucleus, are much smaller than DNA polymers, with molecular weights of only 20,000 to 40,000.

The monomers of the nucleic acids, called **nucleotides,** are composed of three distinct parts:

1. A *five-carbon sugar,* deoxyribose in DNA and ribose in RNA (Fig. 23.21).
2. A *nitrogen-containing organic base* of the type shown in Fig. 23.22.
3. A *phosphoric acid molecule* (H_3PO_4).

The base and the sugar combine as shown in Fig. 23.23(a) to form a unit that in turn reacts with phosphoric acid to create the nucleotide, which is an ester [see Fig. 23.23(b)]. The nucleotides are connected by undergoing condensation reactions that eliminate water to give a polymer of the type represented in Fig. 23.24; such a polymer can contain a *billion* units.

The key to DNA's functioning is its *double helical structure with complementary bases on the two strands*. The bases form hydrogen bonds to each other, as shown in Fig. 23.25. Note that the structures of cytosine and guanine make them perfect partners for hydrogen bonding, and they are *always* found as pairs on the two strands of DNA. Thymine and adenine form similar hydrogen-bonding pairs.

There is much evidence to suggest that the two strands of DNA unwind during cell division and that new complementary strands are constructed on the unraveled strands (Fig. 23.26). Because the bases on the strands always pair in the same way—cytosine with guanine and thymine with adenine—each unraveled strand serves as a template for attaching the complementary bases (along with the rest of the nucleotide). This process results in two double-helix DNA structures that are identical to the original one. Each new double strand contains one strand from the

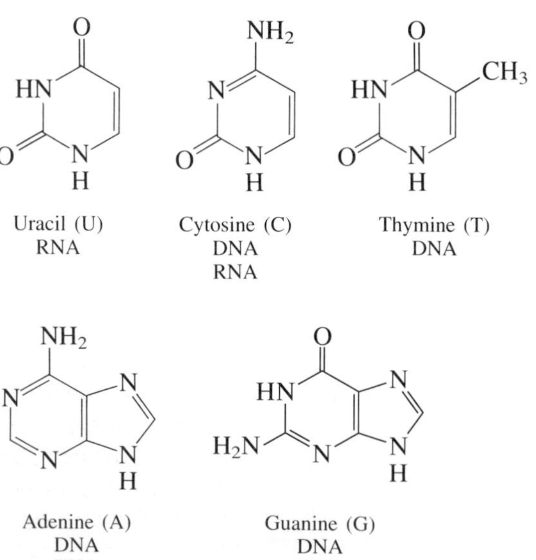

Figure 23.22

The organic bases found in DNA and RNA.

Figure 23.23

(a) Adenosine is formed by reaction of adenine and ribose. (b) The reaction of phosphoric acid with adenosine to form the ester adenosine 5-phosphoric acid, a nucleotide.

Adenine + Ribose $\xrightarrow{-H_2O}$ Adenosine

(a)

Phosphoric acid + Adenosine $\xrightarrow{-H_2O}$ Adenosine 5-phosphoric acid

(b)

original DNA double helix and one newly synthesized strand. This replication of DNA allows for the transmission of genetic information when cells divide.

The other major function of DNA is **protein synthesis.** A given segment of the DNA, called a **gene,** contains the code for a specific protein. These codes transmit the primary structure of the protein (the sequence of amino acids) to the construction "machinery" of the cell. There is a specific code for each amino acid in the protein, which ensures that the correct amino acid will be inserted as the protein chain grows. A code consists of a set of three bases called a **codon.**

DNA stores the genetic information, and RNA molecules are responsible for transmitting this information to the ribosomes, where protein synthesis actually occurs. This complex process involves, first, the construction of a special RNA molecule called **messenger RNA (mRNA).** The mRNA is built in the cell nucleus on the appropriate section of DNA (the gene); the double helix is "unzipped," and the complementarity of the bases is utilized in a process similar to that for DNA replication. The mRNA then migrates into the cytoplasm of the cell where, with the assistance of the ribosomes, the protein is synthesized.

Small RNA fragments, called **transfer RNA (tRNA),** are tailored to find specific amino acids and attach them to the growing protein chain as dictated by the codons in the mRNA. Transfer RNA has a lower molecular weight than messenger RNA and consists of a chain of 75 to 80 nucleotides, including the bases adenine, cytosine, guanine, and uracil, among others. The chain folds back onto itself in various places as the complementary bases along the chain form hydrogen bonds. The tRNA decodes the genetic message from the mRNA using a complementary triplet of bases called an **anticodon.** It is the nature of the anticodon that governs which amino acid will be brought to the protein under construction.

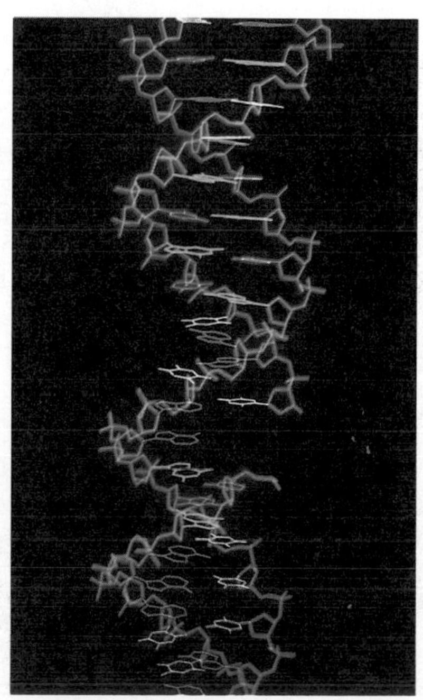

Computer graphic image of the base pairs of DNA.

The protein is built as follows: a tRNA molecule brings an amino acid to the mRNA; the anticodon of the tRNA must complement the codon of the mRNA (see

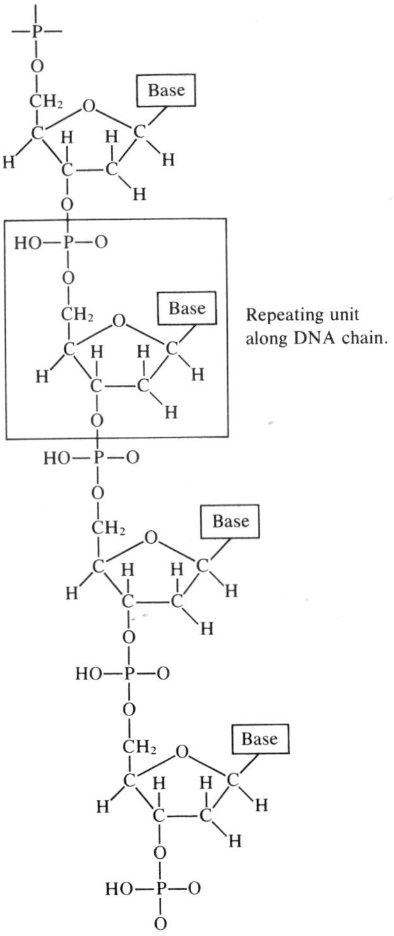

Base

Base

Repeating unit along DNA chain.

Base

Base

Figure 23.24

A portion of a typical nucleic acid chain. Note that the backbone consists of sugar-phosphate esters.

A=T
G≡C
A=T
C≡G
T=A
A=T
C≡G
G≡C
T=A
T=A
T=A

(a)

Thymine
Adenine
Deoxyribose
Deoxyribose

(b)

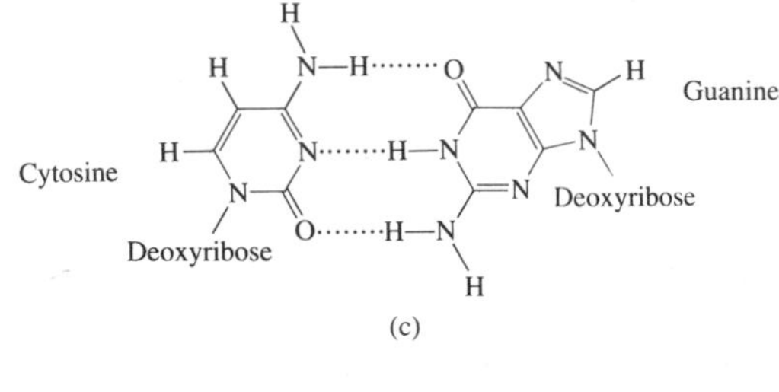

Figure 23.25

(a) The DNA double helix contains two sugar-phosphate backbones with the bases from the two strands hydrogen bonded to each other. The complementarity of the (b) thymine-adenine and (c) cytosine-guanine pairs.

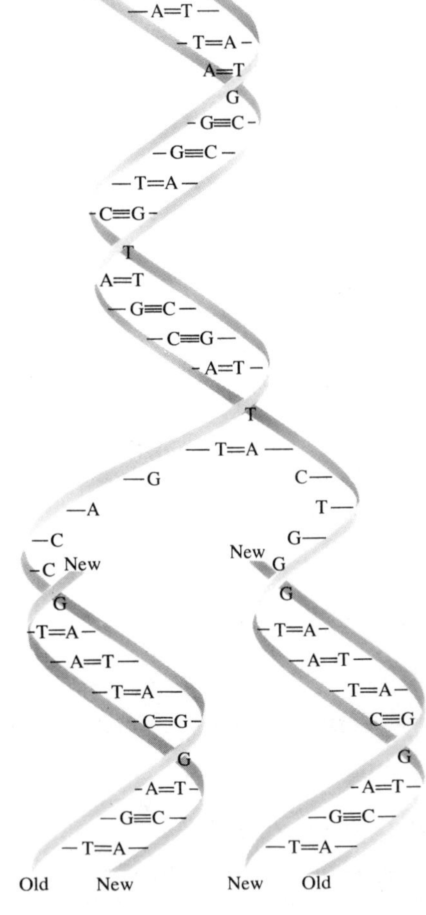

Figure 23.26

During cell division, the original DNA double helix unwinds and new complementary strands are constructed on each original strand.

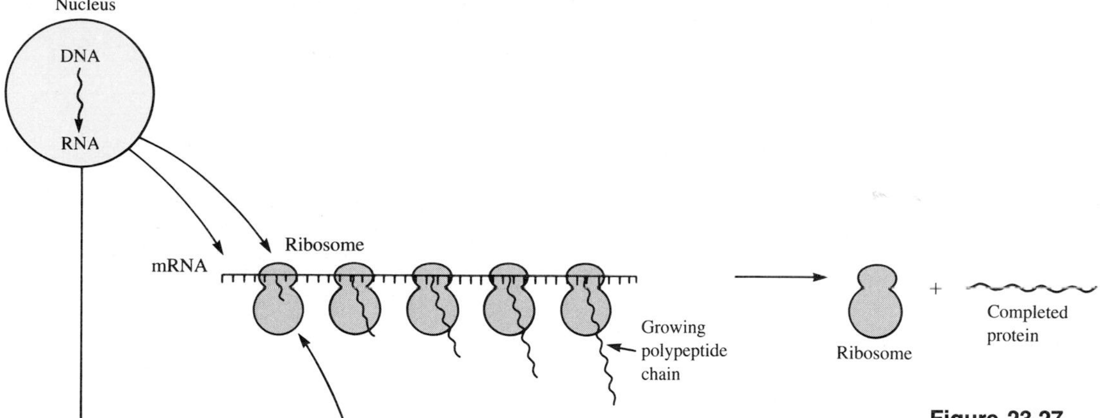

Figure 23.27

The mRNA molecule, constructed from a specific gene on the DNA, is used as the pattern to construct a given protein with the assistance of ribosomes. The tRNA molecules attach to specific amino acids and put them in place as called for by the codons on the mRNA.

Fig. 23.27). Once this amino acid is in place, another tRNA moves to the second codon site of the mRNA with its specific amino acid. The two amino acids link via a peptide bond, and the tRNA on the first codon breaks away. The process is repeated down the chain, always matching the tRNA anticodon to the mRNA codon.

23.4 Lipids

Purpose

- To describe the classes of lipids.

The **lipids** are a group of substances classified according to a solubility characteristic. They are defined as water-insoluble substances that can be extracted from cells by nonpolar organic solvents such as ether and benzene. The lipids found in the human body can be divided into four classes according to molecular structure: fats, phospholipids, waxes, and steroids.

The most common **fats** are esters composed of the polyhydroxy alcohol glycerol and long-chain carboxylic acids called **fatty acids** (see Table 23.4 on page 1032). *Tristearin,* the most common animal fat, is typical of these substances.

$$\begin{array}{l} CH_2{-}OH + H{-}O{-}\overset{\displaystyle O}{\overset{\|}{C}}{-}(CH_2)_{16}{-}CH_3 \\ | \\ CH{-}OH + H{-}O{-}\overset{\displaystyle O}{\overset{\|}{C}}{-}(CH_2)_{16}{-}CH_3 \\ | \\ CH_2{-}OH + H{-}O{-}\overset{\displaystyle O}{\overset{\|}{C}}{-}(CH_2)_{16}{-}CH_3 \end{array} \longrightarrow \begin{array}{l} CH_2{-}O{-}\overset{\displaystyle O}{\overset{\|}{C}}{-}(CH_2)_{16}{-}CH_3 \\ | \\ CH{-}O{-}\overset{\displaystyle O}{\overset{\|}{C}}{-}(CH_2)_{16}{-}CH_3 \\ | \\ CH_2{-}O{-}\overset{\displaystyle O}{\overset{\|}{C}}{-}(CH_2)_{16}{-}CH_3 \end{array} + 3H_2O$$

Glycerol   Three stearic acid molecules   Tristearin

Some Common Fatty Acids and Their Sources

Name	Formula	Source
Saturated		
Arachidic acid	$CH_3(CH_2)_{18}—COOH$	Peanut oil
Butyric acid	$CH_3(CH_2)_2—COOH$	Butter
Caproic acid	$CH_3(CH_2)_4—COOH$	Butter
Lauric acid	$CH_3(CH_2)_{10}—COOH$	Coconut oil
Stearic acid	$CH_3(CH_2)_{16}—COOH$	Animal and vegetable fats
Unsaturated		
Oleic acid	$CH_3(CH_2)_7CH{=}CH(CH_2)_7—COOH$	Corn oil
Linoleic acid	$CH_3(CH_2)_4CH{=}CH—CH_2—CH{=}CH(CH_2)_7—COOH$	Linseed oil
Linolenic acid	$CH_3CH_2CH{=}CH—CH_2CH{=}CH—CH_2—CH{=}CH—(CH_2)_7COOH$	Linseed oil

Table 23.4

Fats that are esters of glycerol are called **triglycerides** and have the general structure

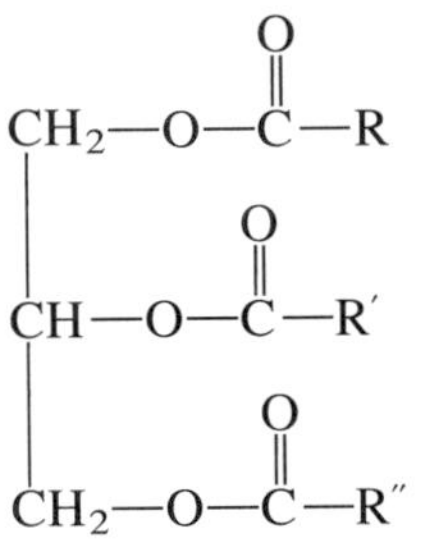

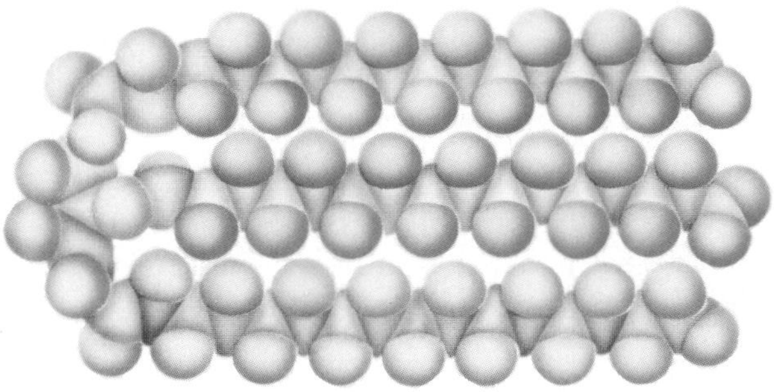

$$\begin{array}{l} CH_2—O—\overset{\overset{\displaystyle O}{\|}}{C}—R \\ | \\ CH—O—\overset{\overset{\displaystyle O}{\|}}{C}—R' \\ | \\ CH_2—O—\overset{\overset{\displaystyle O}{\|}}{C}—R'' \end{array}$$

Unsaturated compounds contain one or more C=C bonds.

where the three R groups may be the same or different and may be saturated or unsaturated. Vegetable fats tend to be unsaturated and usually occur as oily liquids; animal fats are saturated and occur as solids.

Triglycerides can be decomposed by treatment with aqueous sodium hydroxide. The products are glycerol and the fatty acid salts; the latter are known as soaps. This process is called **saponification.**

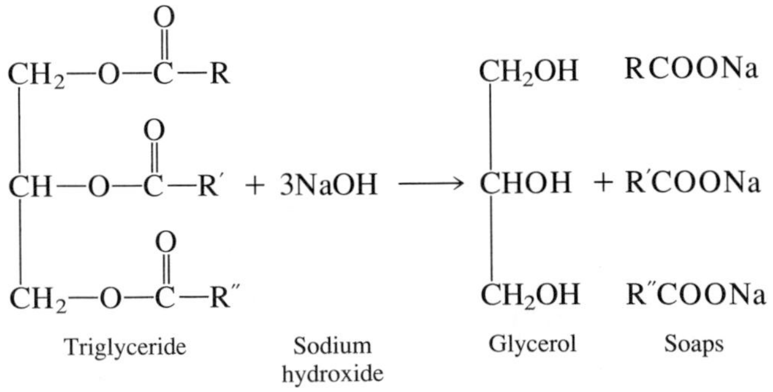

$$\begin{array}{l} CH_2—O—\overset{\overset{\displaystyle O}{\|}}{C}—R \\ | \\ CH—O—\overset{\overset{\displaystyle O}{\|}}{C}—R' \\ | \\ CH_2—O—\overset{\overset{\displaystyle O}{\|}}{C}—R'' \end{array} + 3NaOH \longrightarrow \begin{array}{l} CH_2OH \\ | \\ CHOH \\ | \\ CH_2OH \end{array} + \begin{array}{l} RCOONa \\ \\ R'COONa \\ \\ R''COONa \end{array}$$

Triglyceride — Sodium hydroxide — Glycerol — Soaps

Much of what we call dirt is nonpolar. Grease, for example, consists of long-chain hydrocarbons. However, water, the solvent most commonly available to us, is very polar and will not dissolve "greasy dirt." We need to add something to the water that is somehow compatible with both the polar water and the nonpolar grease. Fatty acid anions are perfect for this role since they have a long nonpolar tail and a polar head. For example, the stearate anion can be represented as

Like dissolves like.

$$CH_3-CH_2-CH_2-CH_2-CH_2-CH_2-CH_2-CH_2-CH_2-CH_2-CH_2-CH_2-CH_2-CH_2-CH_2-CH_2-CH_2-C(=O)-O^-$$

Such ions can be dispersed in water because they form **micelles** (Fig. 23.28). These aggregates of fatty acid anions have the water-incompatible tails in the interior while the anionic parts (the polar heads) point outward to interact with the polar water molecules. A soap "solution" is not a true solution; it does not contain *individual* fatty-acid anions dispersed in the water but rather groups of ions (micelles). Thus a soap-water mixture is really a suspension of micelles in water. Because the relatively large micelles scatter light, soapy water looks cloudy.

Soap dissolves grease by taking the grease molecules into the nonpolar interior of the micelle (Fig. 23.29) where they can be carried away by the water. Soap can be viewed as an emulsifying agent, since it acts to suspend the normally incompatible grease in the water. Because of this ability to assist water in "wetting" and suspending nonpolar materials, soap is also called a *wetting agent,* or **surfactant.**

A major disadvantage is that soap anions form precipitates with the cations in hard water, principally Ca^{2+} and Mg^{2+}. This "soap scum" dulls clothes and drastically reduces soap's cleaning efficiency. Water can be softened with slaked lime (Section 14.6) or by ion exchange (Section 18.4). In addition, a huge industry has developed to produce artificial soaps, called detergents. Any molecule that has nonpolar and polar areas similar to those in the fatty acid anions should act to emulsify grease in water. The most widely used class of detergents is the alkylbenzene sulfonates, for example,

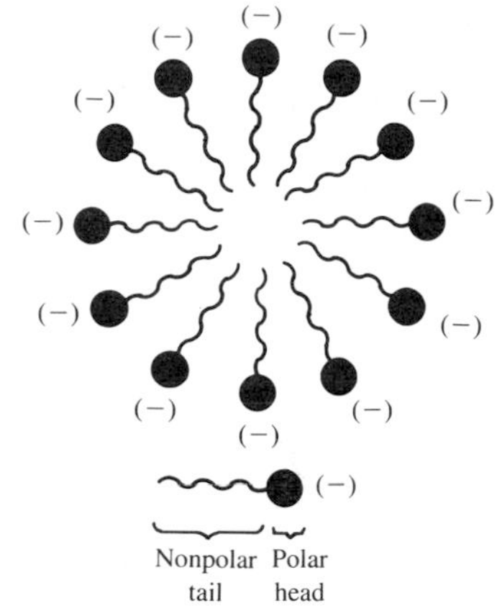

Figure 23.28

The structure of a micelle of fatty acid anions. The cations of the salt also surround the negatively charged micelle.

$$CH_3-CH_2-CH_2-CH_2-CH_2-CH_2-CH_2-CH_2-CH_2-CH_2-CH_2-CH_2-CH_2-C_6H_4-\underbrace{S(=O)_2-O^-}_{\text{Sulfonate group}}\ Na^+$$

which can be synthesized from petroleum-based raw materials. Detergent anions have the advantage of not forming insoluble solids with Ca^{2+} and Mg^{2+} ions.

Phospholipids are similar in structure to fats in that they are esters of glycerol. However, unlike fats, they contain only two fatty acids. The third ester linkage involves a phosphate group, which gives phospholipids two distinct parts: the long nonpolar "tail," and the polar substituted phosphate "head" (Fig. 23.30). Because of this dual nature, phospholipids tend to form **bilayers** in aqueous solution with the tails in the interior and the polar heads interfacing with the polar water molecules, as shown in Fig. 23.31(a). Note that this behavior is very similar to that of fatty-acid anions. The bilayers of larger phospholipids can close to form *vesicles* [Fig. 23.31(b)].

Phospholipids form a significant portion of cell membranes. Figure 23.32 shows a cell membrane in the form of a phospholipid bilayer with proteins distributed in it. The cell membrane must first protect the workings of the cell from the extracellular fluids that bathe it. Its second function is to allow nutrients and other necessary chemicals to enter the cell while allowing waste products to leave the cell.

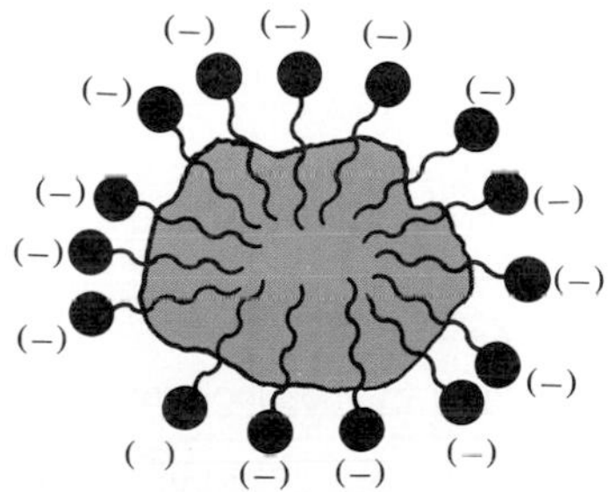

Figure 23.29

Soap micelles absorb grease molecules into their interiors so that the molecules are suspended (emulsified) in the water and can be washed away.

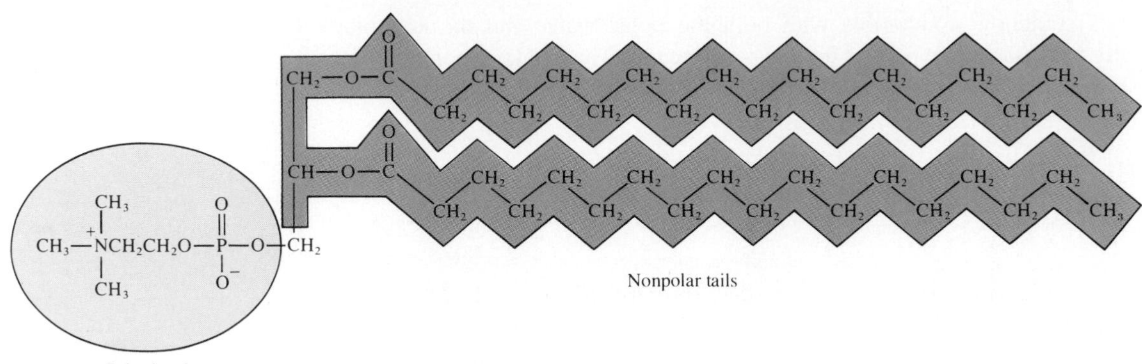

Figure 23.30

Lecithin, a phospholipid, with its long nonpolar tails and polar substituted phosphate head.

The detailed operation of the membrane is very complex and not well understood; however, according to the most widely accepted model, called the *fluid mosaic model,* some small uncharged molecules such as water, oxygen, and carbon dioxide diffuse freely through the lipid bilayer, while other substances pass through "gates and passages" provided by specific proteins imbedded in the membrane. The proteins in the membrane also provide the means for communication between cells, via hormones and other messenger molecules.

Like fats and phospholipids, **waxes** are esters. Unlike the former classes, they involve monohydroxy alcohols instead of glycerol. For example, *beeswax,* a substance secreted by the wax glands of bees, is mainly myricyl palmitate,

$$CH_3(CH_2)_{14}—\overset{\overset{\displaystyle O}{\|}}{C}—O—(CH_2)_{29}—CH_3$$

formed from palmitic acid,

$$CH_3(CH_2)_{14}—C\begin{matrix} \diagup\!\!\!\!= O \\ \diagdown OH \end{matrix}$$

and the alcohol

$$CH_3(CH_2)_{29}—OH$$

Waxes are low-melting solids that furnish waterproof coatings on leaves and fruit and on the skins and feathers of animals. Waxes are also important commer-

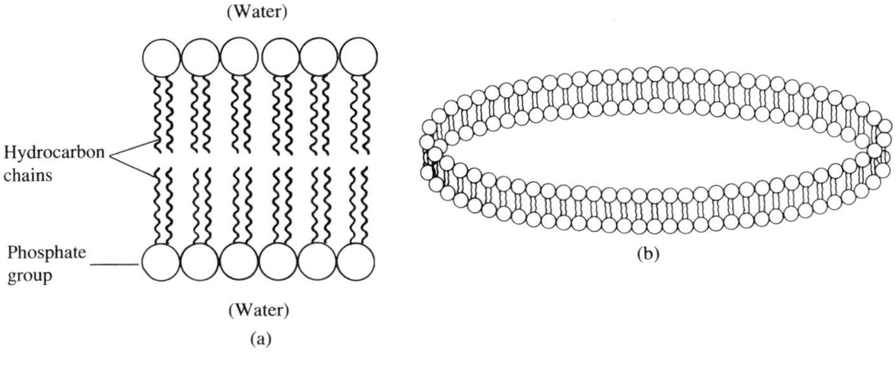

(Water)

Hydrocarbon chains

Phosphate group

(Water)

(a)

(b)

Figure 23.31

(a) A bilayer of phospholipids in aqueous solution. The nonpolar tails remain in the interior of the bilayer with the polar heads interacting with water molecules. (b) The bilayers can close to form a spherical vesicle shown here in cross section.

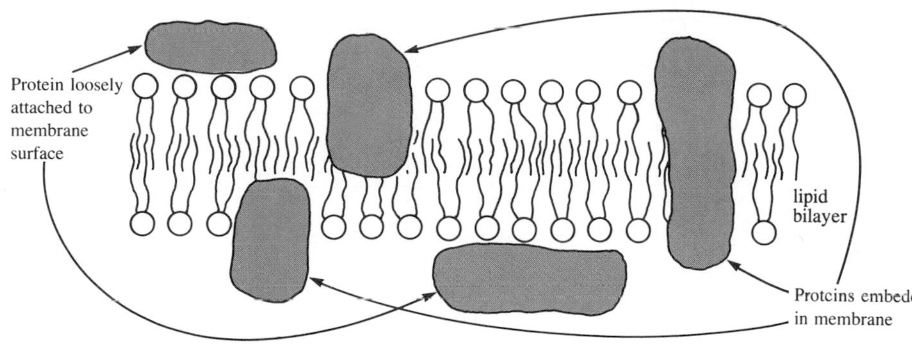

Figure 23.32

A representation of a cell membrane consisting of proteins imbedded in a bilayer of phospholipids that are oriented with their polar heads outward.

cially. For example, whale oil is largely composed of the wax cetyl palmitate. It has been used in so many products, including cosmetics and candles, that the blue whale has been hunted almost to extinction.

Steroids are a class of lipids that have a characteristic fused carbon-ring structure of the type

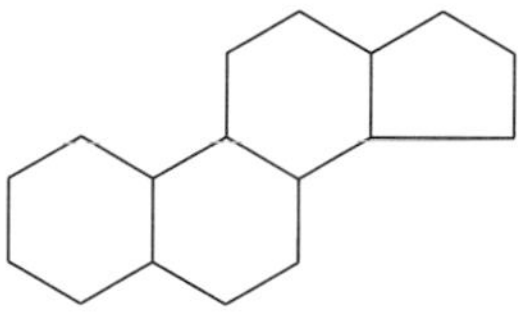

Steroids include four groups: cholesterol, adrenocorticoid hormones, sex hormones, and bile acids.

Olives have a shiny wax coating.

Cholesterol [Fig. 23.33(a)] is found in virtually all organisms and is the starting material for the formation of the bile acids, steroid hormones, and vitamin D [see Fig. 23.33(b)]. Although cholesterol is essential for human life, it has been implicated in the formation of plaque on the walls of arteries (a process called arteriosclerosis, or hardening of the arteries), which leads to eventual clogging. The phenomenon seems especially important in the arteries that supply blood to the heart. Blockage of these arteries leads to heart damage that often results in death from a heart attack.

The **adrenocorticoid hormones,** synthesized in the adrenal gland, are involved in the regulation of water and the electrolyte balance and in the metabolism of proteins and carbohydrates. For example, *cortisol* [Fig. 23.33(c)] slows the construction of proteins so that the amino acids normally used for this purpose can be used by the liver to synthesize extra glucose.

Of the **sex hormones,** the most important male hormone is *testosterone* [Fig. 23.33(d)], a hormone that controls the growth of the reproductive organs and hair and the development of the muscle structure and deep voice characteristic of males. There are two types of female sex hormones of particular significance: *progesterone* [Fig. 23.33(e)], and a group of estrogens, one of which is *estradiol* [Figure 23.33(f)]. These hormones cause the periodic changes in the ovaries and the uterus responsible for the menstrual cycle. During pregnancy, a high level of progesterone is maintained, which prevents ovulation. This effect has led to the use of progesterone-type compounds as birth-control drugs. One of the most common of these is ethynodiol diacetate [Figure 23.33(g)].

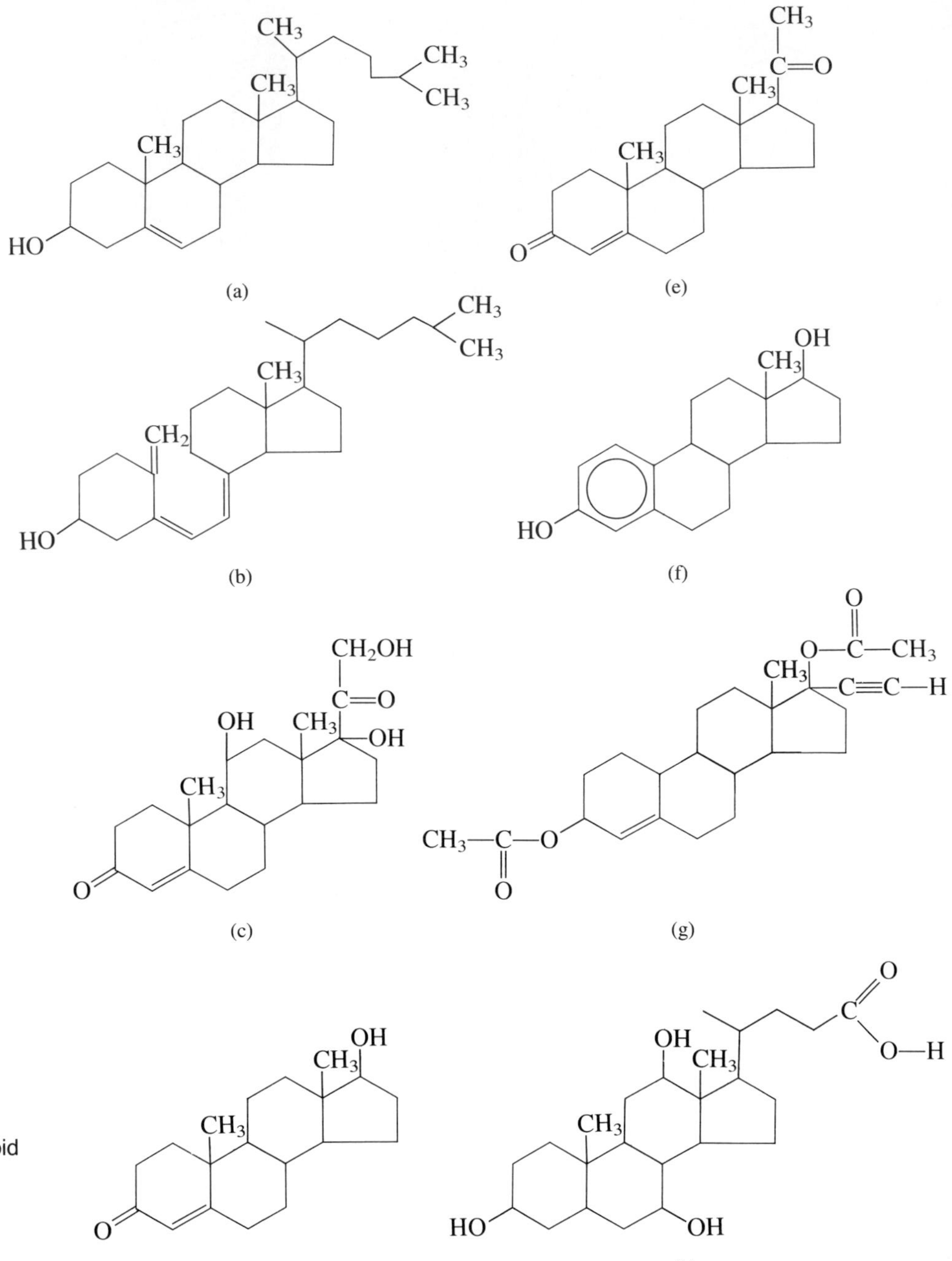

Figure 23.33

Several common steroids and steroid derivatives. (a) cholesterol, (b) vitamin D_3, (c) cortisol, (d) testosterone, (e) progesterone, (f) estradiol, (g) ethynodiol diacetate, and (h) cholic acid.

The **bile acids** are produced from cholesterol found in the liver and are stored in the gall bladder. The primary human bile acid is *cholic acid* [shown in Fig. 23.33(h) above], a substance that aids in the digestion of fats by emulsifying them in the intestine. Bile acids can also dissolve cholesterol ingested in food. They are therefore important in limiting cholesterol in the body, since too much cholesterol can be detrimental to human health.

Chemical Impact

The Chemistry of Vision

Vision, the most remarkable of our senses, is a complex chemical phenomenon. The human eye is roughly spherical with an opening in the front to admit light. The light falls on a rear surface lined with cone-shaped and rod-shaped cells. Each eye contains 7 million cones for detecting color and 120 million rods to detect white light and to provide sharpness of visual images. The molecules responsible for vision are attached to the tops of the rods and cones. We will consider here the function of one of them, rhodopsin.

Rhodopsin has two parts: a protein portion called *opsin,* and a small aldehyde portion called *retinal*. The structure of retinal is shown in its all *trans* form in Fig. 23.34(a), where all substituents around the double bonds are *trans*. It is interesting to note that vitamin A, a substance known to aid vision, especially night vision, has the same structure as retinal except that the terminal aldehyde functional group is replaced by a —CH_2OH group to give an alcohol. Retinal can occur in other isomeric forms, one of which is called the 11-*cis* form because of the *cis* arrangement at carbon 11 [see Fig. 23.34(b)]. It is in this *cis* form that retinal is bound to opsin. When rhodopsin absorbs light, the retinal is isomerized to the all-*trans* form, which separates from the opsin.

When the two portions separate, the natural reddish-purple color of rhodopsin is lost. In addition, the cell to which rhodopsin is attached becomes excited. This receptor cell then excites other cells and sends a message to the brain. Normally, five closely spaced receptor cells must be excited to produce the sensation of vision. This means that only five photons of light are necessary to stimulate the eye, representing a total energy of only 10^{-18} J.

After the activation of rhodopsin and the separation of the *trans* form of retinal, the retinal returns to the 11-*cis* form and reconnects to opsin. This process is relatively slow, which is one reason our eyes need time to adapt to low-light conditions. Also, intensely bright light causes saturation of the light receptors, which causes temporary blindness since there is no attached retinal to absorb additional photons of light.

Color discrimination is possible because cone cells occur in three groups: those receptive to blue light, those receptive to green light, and those receptive to yellow-red light. Each type can absorb light in a range around its primary color. Thus light in the blue-green region excites both the blue and green receptors.

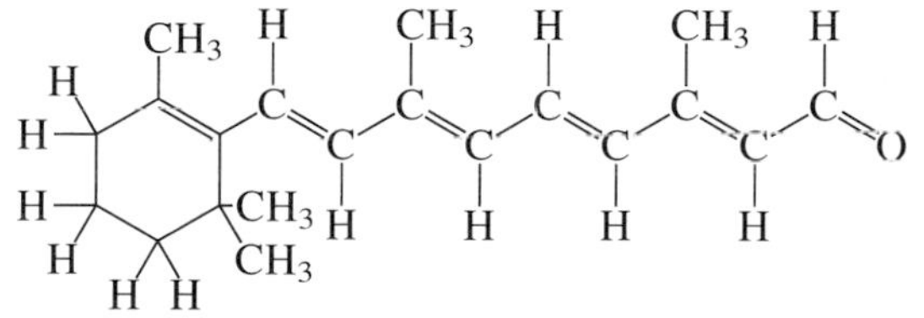

(a)

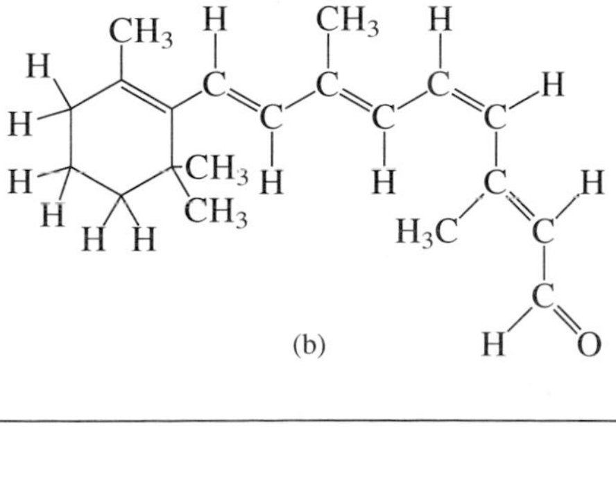

(b)

Figure 23.34

The retinal portion of rhodopsin. (a) The *trans* form. (b) The 11-*cis* form. Carbon 11 is shown in red.

FOR REVIEW

Summary

Thirty elements are currently known to be essential or are suspected to be essential for life. The most abundant in the human body are hydrogen, carbon, nitrogen, and oxygen; but sodium, magnesium, potassium, sulfur, and chlorine are also present in large amounts. Other elements found only in trace amounts, such as zinc, are essential for the action of many enzymes.

Proteins are a class of natural polymers with molecular weights ranging from 6000 to 1,000,000. Fibrous proteins are employed in the human body for structural purposes in muscle, hair, and cartilage. Globular proteins are molecules that transport and store oxygen and nutrients, act as catalysts, help regulate the body's systems, fight foreign objects, and so on.

The building blocks of proteins are the α-amino acids, which can be divided into polar and nonpolar classes, depending on whether the side chain (the R group) attached to the α-carbon is hydrophilic or hydrophobic.

A protein polymer is built by successive condensation reactions that produce peptide linkages:

$$H_2N-CH(R)-C(=O)-OH + H-NH-CH(R')-C(=O)-OH \longrightarrow H_2N-CH(R)-\boxed{C(=O)-NH}-CH(R')-C(=O)-OH + H_2O$$

Peptide linkage

The order or sequence of amino acids in the protein chain is called the primary structure. Differences in primary structure are what tailor proteins for specific functions.

The secondary structure refers to the arrangement of the protein chain, which is determined to a large extent by hydrogen bonding between atoms on different amino acids. When such interactions occur within the chain, a spiral structure called an α-helix results. When the hydrogen bonds are between different chains, a pleated sheet results. The overall shape of the protein is called its tertiary structure. The breakdown of tertiary protein structure, called denaturation, can be caused by energy sources or a variety of chemicals.

Enzymes are proteins that act as catalysts in biological reactions. The specificity and efficiency of an enzyme result from its structure, which is tailored to a specific type of reacting molecule (substrate).

Carbohydrates serve as food sources for most organisms and as structural materials for plants. Simple carbohydrates, called monosaccharides, are most commonly five-carbon and six-carbon polyhydroxy ketones and aldehydes that contain chiral carbon atoms. Monosaccharides combine to form more complex carbohydrates. For example, sucrose is a disaccharide, and starch and cellulose are polymers of D-glucose.

When a cell divides, the genetic information is transmitted via deoxyribonucleic acid (DNA), which has a double helical structure whose two strands are held together by hydrogen bonding between pairs of organic bases—one from each strand. The bases occur only in specific pairs. During cell division, the double helix unravels, and a new polymer forms along each strand of the original DNA to form two double helical DNA molecules. The DNA contains segments called genes, which store the structural information for specific proteins. Various types of ribonucleic acid (RNA) molecules assist in protein synthesis.

Lipids are water-insoluble substances found in cells and can be divided into four classes: fats, phospholipids, waxes, and steroids. Fats are esters composed of glycerol (a polyhydroxy alcohol) and fatty acids (carboxylic acids that contain long-chain hydrocarbons). Soaps are sodium salts of fatty acids. Soaps act as surfactants by forming micelles, with the nonpolar tails of the molecules in the interior and the polar heads toward the exterior where they can interact with the polar water molecules.

Steroids have the characteristic fused-ring structure

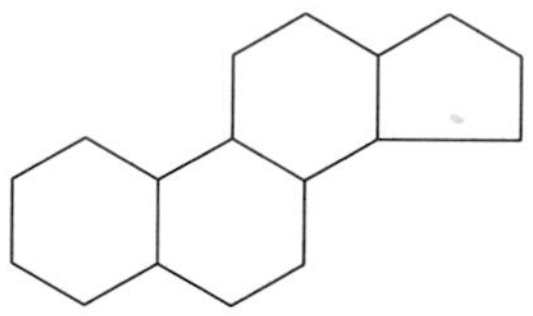

and include cholesterol, the adrenocorticoid hormones, the sex hormones, and the bile acids.

Key Terms

biochemistry
essential elements
trace elements
cell

Section 23.1

proteins
α-amino acids
side chain
dipeptide
peptide linkage
polypeptide
primary structure
secondary structure
α-helix
pleated sheet
random-coil arrangement
tertiary structure
disulfide linkage
denaturation
enzymes
lock-and-key model
metalloenzyme

Section 23.2

carbohydrates
monosaccharides (simple sugars)
sucrose
disaccharide
glycoside linkage
starch
cellulose
glycogen

Section 23.3

deoxyribonucleic acid (DNA)
ribonucleic acid (RNA)
nucleotides
protein synthesis
gene
codon
messenger RNA (mRNA)
transfer RNA (tRNA)
anticodon

Section 23.4

lipids
fats
fatty acids
micelle
surfactant
phospholipids
bilayer
waxes
steroids

Exercises

A blue exercise number indicates that the answer to that exercise appears at the back of this book and a solution appears in the Solutions Guide.

Proteins and Amino Acids

1. Glycine can exist in water in two forms as shown below; K_a for the carboxylic acid group is 4.3×10^{-3}; K_b for the amino group is 6.0×10^{-3}.

$$H_2N{-}CH_2\overset{\overset{\displaystyle O}{\|}}{C}OH \quad \text{or} \quad {}^+H_3N{-}CH_2{-}\overset{\overset{\displaystyle O}{\|}}{C}{-}O^-$$

   a. Would you expect the position of the following equilibrium to lie significantly to the right or the left?

$$H_2NCH_2CO_2H \rightleftharpoons {}^+H_3NCH_2CO_2{}^-$$

   b. Two ions of glycine are possible. Which of these would be predominant in a solution with $[H^+] = 1.0\ M$, or with $[OH^-] = 1.0\ M$?

2. Aspartame, the artificial sweetener marketed under the name Nutra-Sweet, is a methyl ester of a dipeptide. The structure of aspartame is

$$H_2N{-}\underset{\underset{\displaystyle CH_2CO_2H}{|}}{CH}{-}\overset{\overset{\displaystyle O}{\|}}{C}{-}NH{-}\overset{\overset{\displaystyle CO_2CH_3}{|}}{CH}{-}CH_2{-}C_6H_5$$

   a. What two amino acids are used to prepare aspartame?
   b. There is concern that methanol may be produced by the decomposition of aspartame. From what portion of the molecule can methanol be produced? Write an equation for this reaction.

3. When pure crystalline amino acids are heated, decomposition generally occurs before melting. Account for this observation. (*Hint:* See Exercise 1.)

4. Two new amino acids, amino malonic acid and β-carboxy aspartic acid, have recently been isolated from the bacteria *E. coli*. Amino malonic acid has also been found in mammals and seems to be associated with the formation of arteriosclerotic plaque. The structures of these amino acids are

$$\underset{\text{Amino malonic acid}}{HO_2C{-}\underset{\underset{\displaystyle NH_2}{|}}{CH}{-}CO_2H} \qquad \underset{\beta\text{-Carboxy aspartic acid}}{H_2N{-}\underset{\underset{\displaystyle HO_2C{-}CH{-}CO_2H}{|}}{CH}{-}CO_2H}$$

Would you classify these as hydrophobic or hydrophilic amino acids? Why?

5. Monosodium glutamate (MSG) is commonly used as a flavoring in foods. Draw the structure of MSG.

6. Distinguish between the primary, secondary, and tertiary structures of a protein. Give examples of the types of forces that maintain each type of structure.

7. Sickle-cell anemia is a disease resulting from abnormal hemoglobin molecules. A single glutamic acid in the hemoglobin of persons suffering from this disease is replaced by valine. How might this substitution affect the structure of hemoglobin?

8. Give an example of amino acids that could give rise to the interactions pictured in Fig. 23.10 that maintain the tertiary structures of proteins.

9. What type of interactions can occur between the side chains of phenylalanine and isoleucine? Aspartic acid and lysine?

10. Draw the structures of the two dipeptides that can be formed from serine and alanine.

11. How many tripeptides can be formed from the amino acids ala, ala, and gly?

12. How many polypeptides that contain 25 amino acids can be made from a mixture of 5 different amino acids?

13. Describe how denaturation and inhibition affect the catalytic activity of an enzyme.

14. The rate of enzyme-catalyzed reactions often depends on temperature, as shown in the graph. How do you account for the shape of this curve?

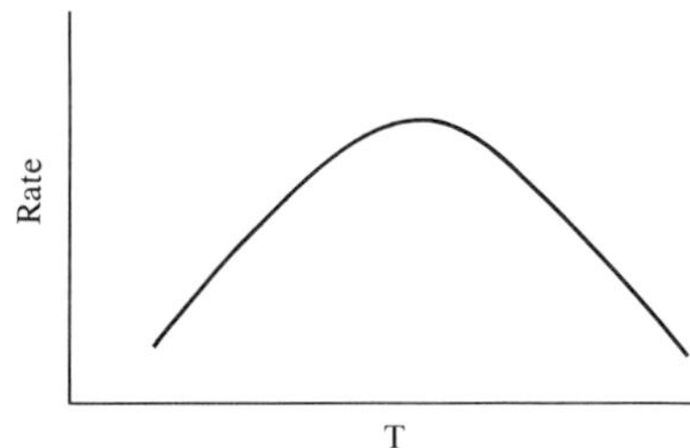

15. Aqueous solutions of amino acids are buffered solutions. Why?

16. Is the primary, secondary, or tertiary structure of an enzyme changed by denaturation?

Carbohydrates

17. Draw cyclic structures for D-ribose and D-mannose.

18. Indicate the chiral carbon atoms found in the carbohydrates D-ribose and D-mannose.

19. Sucrose polyester (SPE) is a material being developed by the Procter and Gamble Company as a fat substitute in cooking oil. SPE, which is indigestible and thus adds no calories to food cooked in it, is the polyester formed from sucrose (which has eight hydroxyl groups) and a given fatty acid. The polyesters containing six, seven, or eight fatty acids attached to the hydroxyl groups as esters are not digestible, while those with five or less are digestible.
   a. Draw the structure of the SPE made from sucrose completely esterified with oleic acid.
   b. How many isomers exist when only seven of the hydroxyl groups are esterified (one hydroxyl group remains)?
   c. How many isomers exist when only six of the hydroxyl groups are esterified?
   d. Based on the discussion of enzymes in Chapters 12 and 23, speculate why the SPE molecules with six, seven, or eight ester functional groups are indigestible while the ones containing fewer fatty acids are digestible.

20. What forces are responsible for the solubility of starch in water?

Optical Isomerism and Chiral Carbon Atoms

21. Distinguish among the following terms: structural isomerism, geometrical isomerism, and optical isomerism.

22. Which of the amino acids in Table 23.3 contain more than one chiral carbon atom? Draw the structures of these amino acids and indicate all chiral carbon atoms.

23. Why is glycine not optically active?

24. How many chiral carbon atoms are there in cholesterol? (See Fig. 23.33(a).)

25. Draw the structure of menthol (2-isopropyl-5-methylcyclohexanol), and indicate the chiral carbon atoms.

26. Which of the isomers of bromochloropropene are optically active? (See Exercise 18 in Chapter 22.)

27. The scent of pine trees mainly results from the presence of α-pinene. Oil of turpentine contains roughly 58–65% α-pinene. The structure of α-pinene is

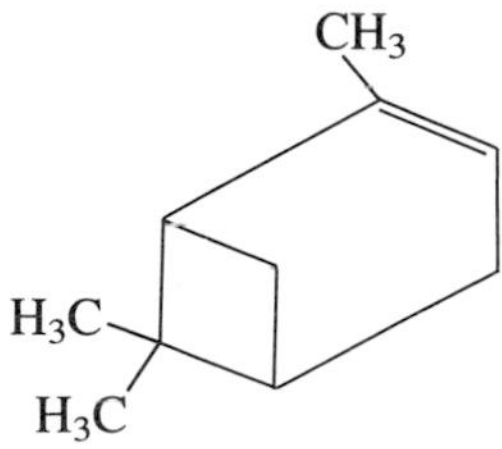

Is α-pinene optically active? If it is, identify the chiral carbon atoms.

Nucleic Acids

28. The compounds adenine, guanine, cytosine, and thymine are called the nucleic acid bases. What structural features in these compounds make them bases?

29. Each human DNA molecule contains roughly 5×10^9 base pairs. The spacing between base pairs along a given chain is about 340 pm. If a single human DNA molecule was stretched to its full length, how long would it be?

30. Describe the structural differences between DNA and RNA.

31. Part of a DNA sequence is A-T-G-C-G-G-C-A-T. What is the complementary sequence?

32. The codons (words) in DNA that identify which amino acid should be in a protein are three bases long. How many such three-letter words can be made from the four bases adenine, cytosine, guanine, and thymine?

33. Which base will hydrogen bond with uracil within an RNA molecule? Draw the structure of this base pair.

34. The base sequences in mRNA that code for some of the amino acids are given below:

   Glu: GAA, GAG
   Val: GUU, GUC, GUA, GUG
   Met: AUG
   Trp: UGG
   Phe: UUU, UUC
   Asp: GAU, GAC

   These sequences are complementary to the sequences in DNA.
   a. Give the corresponding sequences in DNA for the amino acids listed above.
   b. Give a DNA sequence that would code for the peptide: met-met-phe-asp-trp.
   c. How many different DNA sequences can code for the pentapeptide in part b?
   d. What is the peptide that is produced from the DNA sequence: C-T-T-A-C-C-A-A-A?
   e. What other DNA sequences would yield the same tripeptide as in part d?

35. The change of a single base in the DNA sequence for normal hemoglobin can encode for the abnormal hemoglobin giving rise to sickle-cell anemia. Which base in the codon for glu in DNA is replaced to give the codon(s) for val? (See Exercises 7 and 34.)

36. The deletion of a single base from a DNA molecule can be a fatal mutation. Substitution of one base for another is often not as serious a mutation. Why?

37. The compound cisplatin (see Chapter 20) appears to kill cancer cells by inhibiting DNA synthesis. Given the following structural information about cisplatin, the information in Exercise 29, and the fact that the chloride ion is easily displaced

by other donor molecules in cisplatin, speculate on how cisplatin may interact with DNA.

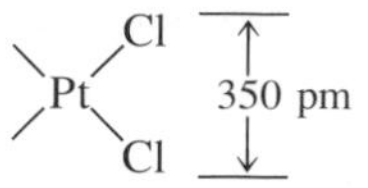

[Structure: Pt with two Cl ligands, Cl–Cl distance 350 pm]

38. A tautomeric form (see Exercise 21 in Chapter 22 for a discussion of tautomerism) of thymine has the structure:

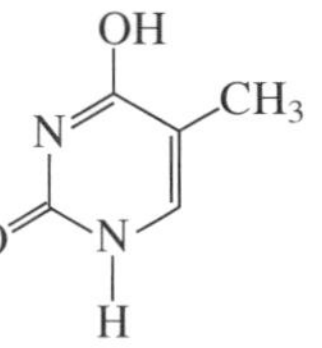

[Structure: pyrimidine ring with OH, CH_3, N, N–H, and O substituents]

a. Use bond energies (Table 8.4) to predict whether this tautomer or the one given in Figure 23.22 is more stable.
b. If the tautomer above, rather than the stable form of thymine, were present in a strand of DNA during replication, what would be the result?

Lipids and Steroids

39. Why can lipids be extracted into organic solvents while carbohydrates cannot?

40. Draw the structure of the triglyceride formed from linoleic acid and glycerol.

41. How many moles of H_2 will it take to completely hydrogenate one mole of the triglyceride in Exercise 40? How many products are possible if only four moles of H_2 react with this triglyceride? Draw the structures of these products.

42. What is a polyunsaturated fat?

43. The oil of deep water fish is rich in omega-3 fatty acids. The presence of these fatty acids in the diet appears to lower blood cholesterol levels. The term *omega-3* applies to the location of a carbon-carbon double bond. Starting with the terminal methyl group (the omega carbon) the double bond is located after the third carbon atom. Draw the structures of the omega-3 fatty acids that contain a total of 16 carbon atoms and 18 carbon atoms, respectively.

44. When one spills a solution of sodium hydroxide on one's hands, the skin feels slippery. Why?

45. Lecithin and sphingomyelin are two phospholipids found in amniotic fluid. During gestation, sphingomyelin levels are relatively constant but lecithin levels increase. Both substances act as surfactants in the lung, and the ratio of lecithin to sphingomyelin in amniotic fluid is used by physicians to assess whether an infant will be able to breathe unassisted upon birth.
a. Look up the structures of these two substances in a reference such as *The Merck Index*.
b. Indicate the hydrophobic and hydrophilic portions of each molecule.

46. Cetyl palmitate is commonly used as a surfactant in many shampoos and related cosmetic products. Look up the structure of cetyl palmitate in an appropriate reference work.

47. Identify the functional group present in the steroids shown in Fig. 23.33.

Additional Exercises

48. How many tetrapeptides can be made that contain two phenylalanines (phe) and two glycines (gly)?

49. How many tripeptides can be formed containing one phenylalanine (phe), one glycine (gly), and one alanine (ala)?

50. Use the values of equilibrium constants given in Exercise 1 to calculate equilibrium constants for the following:
a. $^{+}H_3NCH_2CO_2H + H_2O \rightleftharpoons H_2NCH_2CO_2H + H_3O^+$
b. $H_2NCH_2CO_2^- + H_2O \rightleftharpoons H_2NCH_2CO_2H + OH^-$
c. $^{+}H_3NCH_2CO_2H \rightleftharpoons 2H^+ + H_2NCH_2CO_2^-$

51. The isoelectric point of an amino acid is the pH at which the molecule has no net charge. For glycine that would be the pH at which virtually all glycine molecules are in the form $^{+}H_3NCH_2CO_2^-$. If we assume that the principal equilibrium is

$$2\,{}^{+}H_3NCH_2CO_2^- \rightleftharpoons H_2NCH_2CO_2^- + {}^{+}H_3NCH_2CO_2H \qquad \text{(i)}$$

then at equilibrium

$$[H_2NCH_2CO_2^-] = [{}^{+}H_3NCH_2CO_2H] \qquad \text{(ii)}$$

Use this result and your answer to part c of Exercise 50 to calculate the pH at which Equation (ii) is true. This will be the isoelectric point of glycine.

52. Look up the structures of the following vitamins in *The Merck Index*. Classify each as water-soluble or fat-soluble.
a. vitamin A
b. vitamin E
c. vitamin K_5
d. vitamin K_6

53. a. Use bond energies (Table 8.4) to estimate ΔH for the reaction of two molecules of glycine to form a peptide linkage.
b. Would you predict ΔS to favor the formation of peptide linkages between two molecules of glycine?
c. Would you predict the formation of proteins to be a spontaneous process?

54. The reaction to form a phosphate ester linkage between two nucleotides can be approximated as follows:

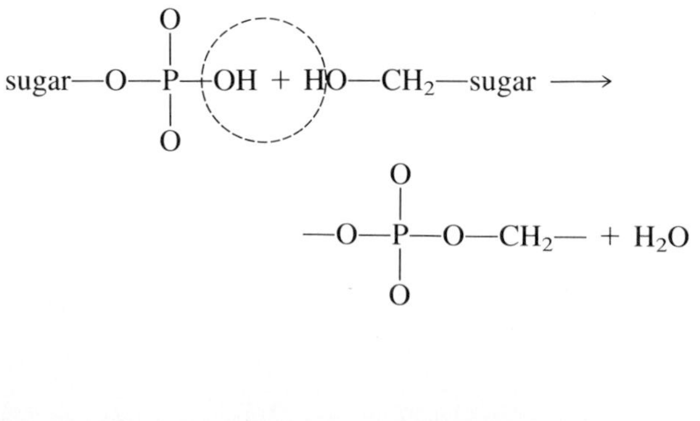

$$\text{sugar—O—}\overset{\overset{\text{O}}{|}}{\underset{\underset{\text{O}}{|}}{\text{P}}}\text{—OH} + \text{HO—CH}_2\text{—sugar} \longrightarrow \text{—O—}\overset{\overset{\text{O}}{|}}{\underset{\underset{\text{O}}{|}}{\text{P}}}\text{—O—CH}_2\text{—} + H_2O$$

Would you predict the formation of a dinucleotide from two nucleotides to be a spontaneous process?

55. Considering your answers to Exercises 53 and 54, how can you justify the existence of proteins and nucleic acids in light of the second law of thermodynamics?

56. The structure of tartaric acid is

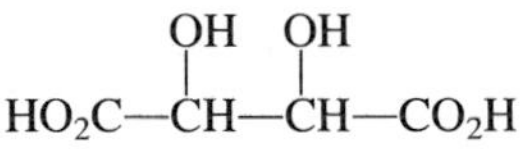

a. Is the form of tartaric acid pictured below optically active? Why or why not?

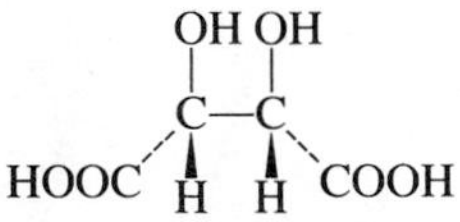

Note: The dashed lines show groups behind the plane of the page. The wedges show groups in front of the plane.

b. Draw the optically active forms of tartaric acid.

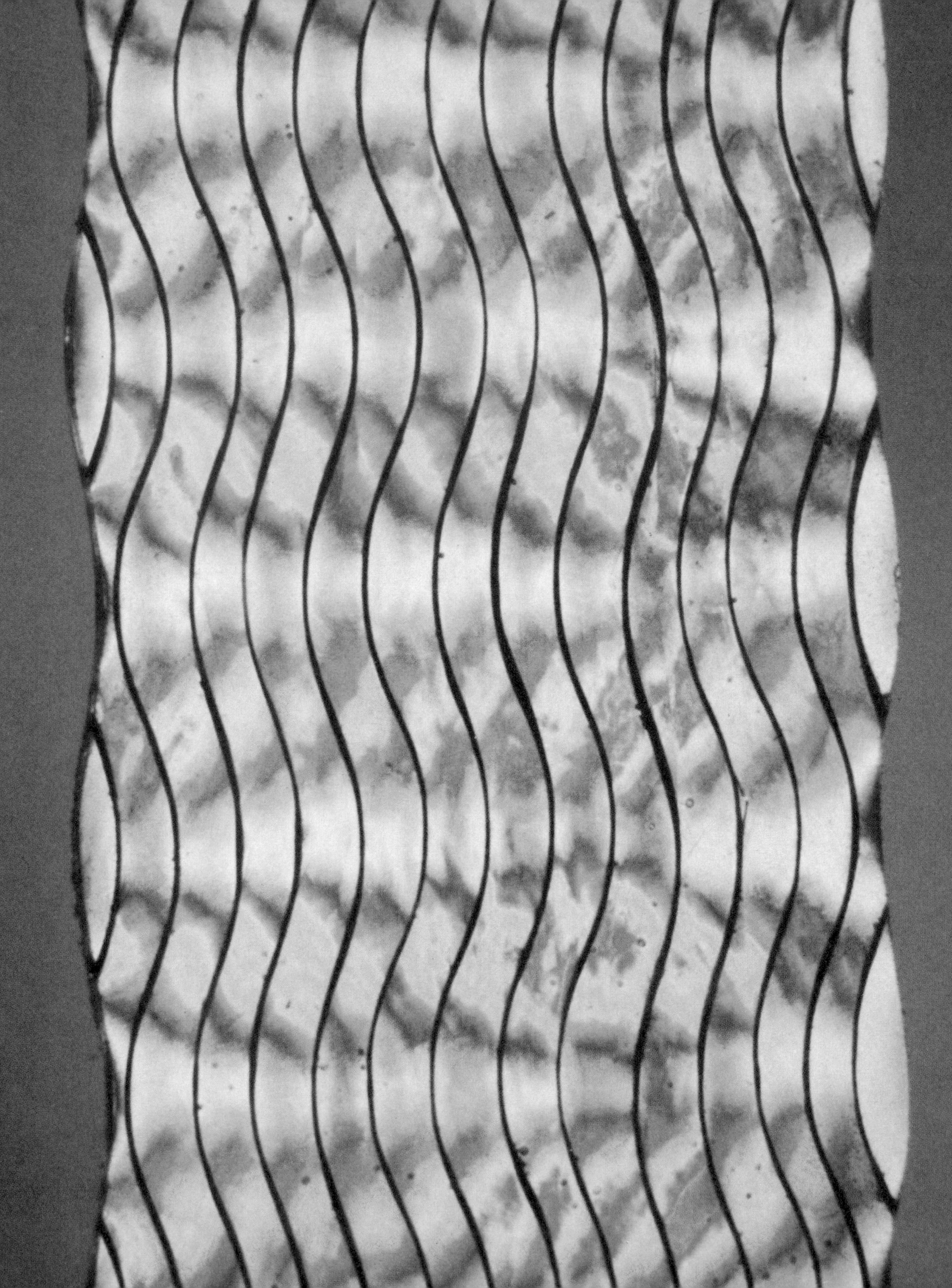

Industrial Chemistry

CONTENTS

The impact of chemistry on our lives is due in no small measure to the many industries that process and manufacture chemicals to provide the fuels, fabrics, fertilizers, food preservatives, detergents, and many other products that affect us daily. Although we have referred quite often to "chemistry in the real world," the main emphasis so far in this text has been to present concepts of chemistry. In this chapter we turn to the practical world of chemistry and consider some of the major industries.

The chemical industry can be subdivided in terms of three basic types of activities:

1. the isolation of naturally occurring substances for use as raw materials,
2. the processing of raw materials by chemical reactions to manufacture commercial products, and
3. the use of chemicals to provide a service.

A given industry may participate in one, two, or all of these activities.

We have already seen many examples of industrial chemicals. For example, in Chapter 17 we saw that chlorine and sodium hydroxide (caustic soda) are produced by electrolysis of brine. The brine comes directly from brine wells or is prepared from mined sodium chloride. Also, we have referred many times to the manufacture of ammonia by the Haber process, using nitrogen obtained from the air and hydrogen obtained from the decomposition of methane in natural gas. The most important inorganic acid is sulfuric acid (see the Chemical Impact feature in Chapter 3) obtained by stepwise oxidation of sulfur that is mined by the Frasch process. The oxygen comes from the air.

The main product of the phosphorus industry is phosphoric acid, which is prepared by treatment of phosphate rock with sulfuric acid (see Section 19.3). Lime (CaO), produced by heating limestone ($CaCO_3$), is

< Stress in a flexible composite made up of graphite fibers embedded in an epoxy matrix shows a tapestry of color. The specimen was placed under tension and photographed through crossed polarizing prisms.

often converted to slaked lime $(Ca(OH)_2)_x$ (see Section 14.6). Petroleum serves as the source for a large number of raw materials. For example, ethylene, the starting material for many polymers, is obtained by breaking down (or *cracking*) the larger molecules in petroleum.

Producing chemicals on a large industrial scale is very different from an academic laboratory experiment. Some of the important differences are:

In the academic laboratory, convenience is typically the most important consideration. Because the amounts are small, hazardous materials can be handled using fume hoods, safety shields, and so on; and expense, although always a consideration, is not a primary factor. However, for any industrial process, economy and safety are critical.

In industry, containers and pipes are metal rather than glass, and corrosion is a constant problem. In addition, since the progress of reactions cannot be monitored visually, gauges must be used.

In the laboratory, any by-products of a reaction are simply disposed of, but in industry, they are usually recycled or sold. If no current market exists for a given by-product, the manufacturer tries to develop a market.

Industrial processes often run at very high temperatures and pressures and ideally are *continuous flow,* meaning that reactants are added and products are extracted continuously. In the laboratory, reactions are run in batches and typically at much lower temperatures and pressures.

The many criteria that must be satisfied to make a process feasible on the industrial scale mean that great care must be taken in the development of each process to ensure safe and economical operation. The development of an industrial chemical process typically involves the following steps:

STEP 1

A need for a particular product is identified.

STEP 2

The relevant chemistry is studied on a small scale in a laboratory. Various ways of producing the desired material are evaluated in terms of costs and potential hazards.

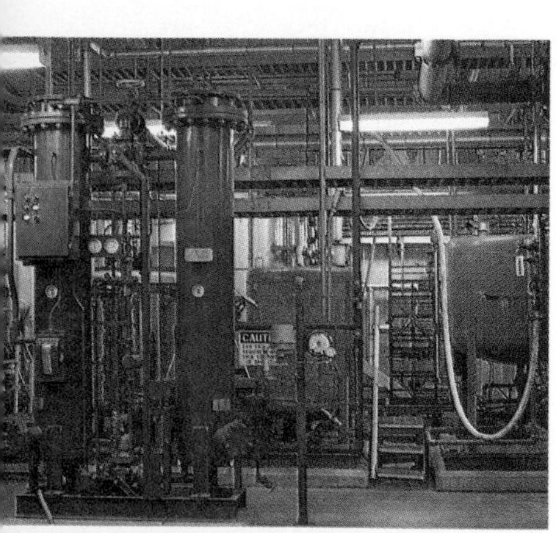

A pilot plant is used to scale up processes from the lab to the manufacturing plant.

STEP 3

The data are evaluated by chemists, chemical engineers, business managers, safety engineers, and others to determine which one of the possibilities is most feasible.

STEP 4

A *pilot plant test* of the process is carried out. The scale of the pilot plant is between that of the laboratory and that of a manufacturing plant. The purpose of this test is to make sure the reaction is efficient at a larger scale, to test reactor (reaction container) designs, to determine the costs of the process, to evaluate the hazards, and to gather information on environmental impact.

STEP 5

A manufacturing facility to carry out the process is designed and constructed.

In this chapter we consider several industries, including the production of polymers, steel, and wine, to illustrate the wide range of chemical reactions and techniques used in the commercial world.

24.1 The Petrochemical Industry

Purpose

- To discuss the major aspects of the production of petrochemicals.

As we saw in Section 6.5, petroleum represents an extremely valuable natural resource. The viscous, dark brown liquid pumped from the world's oil fields is a complex mixture of alkanes, alkenes, cycloalkanes, aromatic compounds, and various inorganic components. It contains literally thousands of compounds.

Many of the top twenty-five chemicals produced in the United States (shown in Table 24.1 on pages 1048–1049) are made from the cracking of petroleum. Note that most of the chemicals listed in the table are simple substances that are used in the manufacture of other materials.

More than 90% of the petroleum produced is used for fuels for heating and transportation. However, petroleum is also an extremely valuable source of **petrochemicals,** which serve as raw materials for the chemical industry. Petrochemicals are produced in two major ways: as by-products of refinery operations, and from steam cracking of petroleum fractions.

Chemicals from Refineries

Even though petrochemicals represent only a small part (~5%) of the total output of petroleum refineries, which are designed mainly to produce fuels, refineries are an important source of these chemicals. Petroleum is first washed to remove acids and salts, then transferred to a distillation tower where various fractions are separated by fractional distillation. The approximate compositions of the fractions are shown in Table 24.2 on page 1049.

Distillation is a physical separation of petroleum components, but most refinery operations also involve chemical reactions to maximize the output of the most useful product. **Cracking** is a process whereby large molecules are broken down to smaller ones by breaking carbon-carbon bonds. This can be done at high temperatures in a process called **pyrolysis,** or **thermal cracking,** or it can be done at lower temperatures using a catalyst in a process called **catalytic cracking.** Most cracking in refineries is now carried out catalytically because of lower energy costs and better control. The catalyst most commonly employed is a silica-alumina mixture (SiO_2-Al_2O_3), and temperatures are typically about 500°C. The very finely divided catalyst is usually "fluidized" by the moving stream of gaseous reactants, and a hydrogen atmosphere is often used to promote formation of alkanes (**hydrocracking**) and to remove nitrogen and sulfur impurities in the form of ammonia and hydrogen sulfide, respectively.

The 25 Chemicals Produced in the Largest Quantities in the United States in 1986				
Substance	Formula	Production in 1986 (billions of pounds)	Method of preparation	Uses
1. Sulfuric acid	H_2SO_4	73.6	Oxidation of S to SO_3, which is reacted with H_2O (Chapter 3)	Manufacture of fertilizers, detergents, explosives, petroleum, plastics, pesticides, pharmaceuticals, dyes, storage batteries, and metals
2. Nitrogen	N_2	48.6	Distillation of liquid air	Manufacture of ammonia, nitric acid and nitrates; for purging and protecting; as a low-temperature coolant
3. Oxygen	O_2	33.0	Distillation of liquid air	Manufacture of steel; rocket fuel; medical uses (respiration)
4. Ethylene	C_2H_4	32.8	Cracking of petroleum (see Section 24.1)	Manufacture of ethylene glycol, ethanol, vinyl chloride, ethylene oxide, polyethylene, and other polymers; as a ripening agent for fruit
5. Lime	CaO	30.3	Thermal decomposition of limestone ($CaCO_3$)	Manufacture of glass, cement, insecticides, and petroleum
6. Ammonia	NH_3	28.0	Haber process (see Section 19.2)	As a fertilizer; in the manufacture of nitric acid, explosives, and fertilizers
7. Sodium hydroxide (caustic soda)	$NaOH$	22.0	Electrolysis of brine (see Section 17.8)	Manufacture of rayon, cellophane, petroleum, pulp and paper, detergents, and aluminum
8. Chlorine	Cl_2	21.0	Electrolysis of brine (see Section 17.8)	Manufacture of pesticides and bleach; for water purification
9. Phosphoric acid	H_3PO_4	18.4	Treatment of phosphate rock with H_2SO_4 (see Section 19.3)	Manufacture of fertilizers, detergents, and pharmaceuticals; pickling and rust-proofing steel; water treatment; flavoring in beverages
10. Propylene	C_3H_6	17.3	Cracking of petroleum	Manufacture of polypropylene, propylene oxide, isopropyl alcohol, and acetone
11. Sodium carbonate (soda ash)	Na_2CO_3	17.2	Reaction of CaO, NH_3 and $NaCl$	Manufacture of glass, pulp and paper, soaps and detergents, petroleum, and aluminum
12. Ethylene dichloride (dichloroethane)	$C_2H_4Cl_2$	14.5	Reaction of C_2H_4 and Cl_2	Manufacture of vinyl chloride, paint removers, solvents, and rubber
13. Nitric acid	HNO_3	13.1	Reaction of NH_3 and O_2 (see Section 19.2)	Manufacture of explosives, fertilizers, dyes, and pharmaceuticals
14. Urea	$(NH_2)_2CO$	12.1	Reaction of NH_3 and CO_2	Manufacture of fertilizers, plastics, adhesives, animal feeds, and flameproofing agents

(continued)

Table 24.1

Substance	Formula	Production in 1986 (billions of pounds)	Method of preparation	Uses
15. Ammonium nitrate	NH_4NO_3	11.1	Reaction of NH_3 and HNO_3	Manufacture of fertilizers, explosives, nitrous oxide, and insecticides
16. Benzene	C_6H_6	10.2	Cracking and reforming petroleum	Manufacture of styrene, phenol, detergents, cyclohexane, dyes, paint removers, rubber cement, and gasoline; as a solvent
17. Ethylbenzene	$C_2H_5C_6H_5$	8.9	From petroleum refining and reacting C_6H_6 and C_2H_4	Manufacture of styrene; as a solvent
18. Carbon dioxide	CO_2	8.5	Recovered as a by-product in various industrial processes	Manufacture of various chemicals; for beverage carbonation and fire extinguishing
19. Vinyl chloride	CH_2CHCl	8.4	Thermal decomposition of ethylene dichloride	Manufacture of polyvinylchloride and other polymers
20. Styrene	$C_6H_5CHCH_2$	7.8	Dehydrogenation of ethylbenzene	Manufacture of polystyrene plastics and other polymers
21. Terephthalic acid	$C_6H_4(COOH)_2$	7.7	Oxidation of *p*-xylene	Manufacture of polyester resins and other polymers
22. Methanol	CH_3OH	7.3	Reaction of CO and H_2	Manufacture of acetic acid, formaldehyde, aviation fuels, and antifreeze; as a solvent
23. Hydrochloric acid	HCl	6.0	Reaction of NaCl and H_2SO_4	Manufacture of pharmaceuticals, PVC, alkyl chlorides, and rubber; metal cleanser before galvanizing
24. Ethylene oxide	C_2H_4O	5.9	Oxidation of ethylene	Manufacture of ethylene glycol for antifreeze
25. Formaldehyde	H_2CO	5.9	Oxidation of methanol	Manufacture of polymer resins

Table 24.1 (*continued*)

The Fractions Obtained from the Distillation of Petroleum

Percentage of total volume	Boiling point (°C)	Approximate number of carbon atoms	Common names of products
1–2%	< 30	C_1–C_4	Light hydrocarbons (methane, ethane, propane, butane)
15–30%	30–200	C_4–C_{12}	Naphtha or straight-run gasoline
5–20%	200–300	C_{12}–C_{15}	Kerosene
10–40%	300–400	C_{15}–C_{25}	Gas oil
Undistilled	> 400	$> C_{25}$	Residual oil, paraffin, asphalt

Table 24.2

A refinery in Carson, California.

Typical reactions in the cracking operation are

$$n\text{-}C_{30}H_{62} \rightarrow CH_3(CH_2)_5CH{=}CH_2 + CH_3(CH_2)_{10}CH_3 + CH_3(CH_2)_7CH{=}CH_2$$

$$\text{(cyclohexane with } CH_2CH_2R\text{)} \longrightarrow \text{(benzene)} + RCH{=}CH_2 + 3H_2$$

In the continuous flow process for the production of gasoline, molecules with about eight carbon atoms are removed as product, and heavier molecules are recycled to be broken down further.

Another refinery process is **catalytic reforming,** in which alkanes and cycloalkanes are converted to aromatic compounds to improve the octane rating of gasoline. For example, toluene can be obtained from *n*-heptane:

$$\underset{n\text{-Heptane}}{CH_3(CH_2)_5CH_3} \xrightarrow[\text{Catalyst}]{500^\circ} \underset{\text{Toluene}}{C_6H_5CH_3} + 4H_2$$

The catalyst for this process usually contains platinum and one or more other metals such as rhenium, which specifically promotes conversion of cyclic alkanes to benzenes. Catalytic reforming has replaced coal tar as the major source of BTX (benzene, toluene, and xylenes) for the chemical industry.

Catalytic reforming is also a major source of hydrogen gas. In fact, the rapid increase in the use of ammonia as a fertilizer resulted from the need to use this by-product hydrogen. Petroleum companies constructed many plants in the 1950s to manufacture ammonia by the Haber process,

$$N_2(g) + 3H_2(g) \xrightarrow[\text{High temperature and pressure}]{\text{Catalyst}} 2NH_3(g)$$

They convinced farmers to use the ammonia directly on their fields before planting. The use of ammonia escalated so rapidly that hydrogen production from reforming could not keep up with the demand. Today, much of the hydrogen used to make ammonia comes from steam reforming of methane:

$$CH_4(g) + 2H_2O(g) \xrightarrow[\text{High temperature}]{\text{Catalyst}} CO_2(g) + 4H_2(g)$$

Other refining processes, which are mainly used to increase gasoline octane ratings, are alkylation, polymerization, and isomerization. In the first two processes, small gaseous molecules are changed to the larger ones useful in gasoline. **Alkylation** involves reactions such as that of 2-methylpropane (isobutane) with a smaller alkene (a propene, butene, or pentene), for example,

$$CH_2{=}CH{-}CH_3 + CH_3\underset{\displaystyle CH_3}{\underset{|}{C}}HCH_3 \longrightarrow CH_3\underset{\displaystyle CH_3}{\underset{|}{C}}H{-}\underset{\displaystyle CH_3}{\underset{|}{C}}HCH_2CH_3$$

or

$$\underset{\displaystyle CH_3}{CH_3\underset{|}{C}}HCH_2\underset{\displaystyle CH_3}{\underset{|}{C}}HCH_3$$

The process of combining two, three, or four alkenes to form a larger molecule is called **polymerization** by the oil industry, but **oligomerization** is a more appropriate name, since the products are dimers, trimers, and tetramers rather than long-chain polymers. An example of oligomerization is

Oligomer is a term used to describe a molecule assembled by reaction of only a few monomers, in contrast to the term *polymer* for very long-chain molecules.

$$3CH_3{-}CH{=}CH_2 \xrightarrow{\text{Catalyst}} CH_3{-}\underset{\displaystyle CH_3}{\underset{|}{CH}}{-}CH_2{-}\underset{\displaystyle CH_3}{\underset{|}{CH}}{-}CH{=}CH{-}CH_3$$

In **isomerization** straight-chain alkanes are converted to branched-chain alkanes, for example,

$$CH_3CH_2CH_2CH_3 \xrightarrow{\text{Catalyst}} CH_3\underset{\displaystyle CH_3}{\underset{|}{C}}HCH_3$$

Isomerization is typically not a separate refinery process, but takes place in connection with cracking and reforming operations.

Refineries are always run continuously, with petroleum fed in and products removed on a 24-hour basis. Although all refineries are similar, no two are the same. Operations vary depending on the composition of the crude oil being processed and the seasonality of the products; gasoline is usually a major product in the spring and summer, when driving increases, and the production of heating oil is emphasized in the fall and winter. On the other hand, a refinery with a relatively large petrochemical market will maximize its production of ethylene, propylene, BTX, and other commonly used chemicals.

Chemicals from Petrochemical Plants

Although refineries are an important source of chemicals, the great majority of chemicals from petroleum are produced in *petrochemical plants* designed to process petroleum fractions to produce various chemicals. The petrochemical industry consumes approximately 7% of all the petroleum used in the United States.

The portion of the petroleum used to make petrochemicals is called the **feedstock** and can vary from *gas oil,* which contains the heavy hydrocarbons, to the *light hydrocarbons* (ethane, propane, and the butanes). What feedstock is used at a given time depends mostly on the price. Because such a large percentage of petroleum goes to produce fuels, the fuel prices determine the price of various fractions. In the late summer the price of propane typically increases because home heating is about to increase. Then, in winter, butane prices often rise since butane is added to gasoline to improve its volatility in cold weather. Such price fluctuations affect the cost of feedstocks. Because the prices of the chemicals produced are determined mainly by the price of the feedstock (50–90% of the total cost), petrochemical plants must be able to switch quickly to the cheapest feedstock. Thus *the key to petrochemical plant design is flexibility*. Present economic conditions favor the light hydrocarbon feedstocks (65–75% of the total used).

Natural gas burning.

Another consideration in the design of petrochemical plants is that the natural gas now being recovered is "drier" (contains less ethane, propane, and heavier

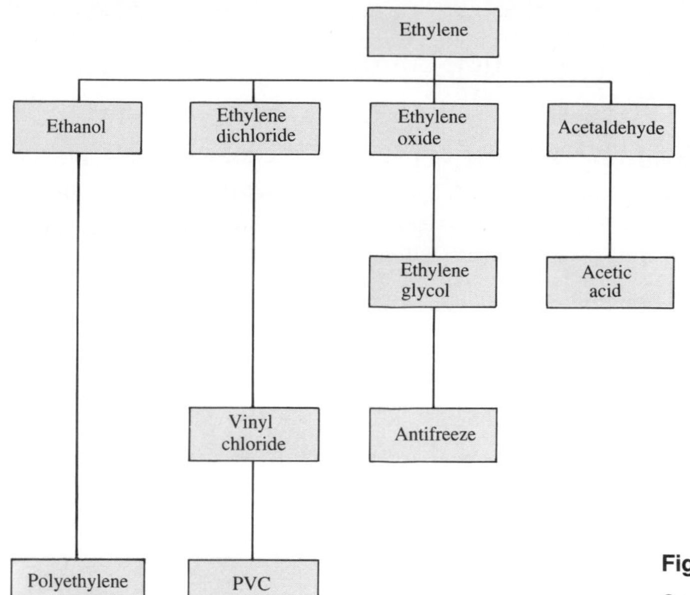

Figure 24.1

Some uses of ethylene.

hydrocarbons that are more easily liquefied than methane). This may lead to a heavier dependence on gas oil feedstocks in the future unless wells with ''wet'' gas are discovered.

The fundamental operation in a petrochemical plant is **steam cracking,** a process whereby hydrocarbon molecules are broken into small fragments by steam at very high temperatures (~1000°C). The major product of steam cracking is ethylene; the current U.S. *capacity* is ~38 billion pounds per year. *Actual production* of ethylene is about 35 billion pounds per year; some of its uses are shown in Fig. 24.1. Ethylene is an important starting material; of the estimated 250 billion pounds of organic chemicals produced annually in the United States, approximately 100 billion pounds (40%) are derived from ethylene. Many other chemicals are obtained from steam cracking, with the mix depending on the feedstock used.

Capacity refers to the production capability when all plants run at maximum output.

The petrochemical industry is an important and highly competitive business in which flexibility and innovative solutions to chemical and engineering problems are essential if an operation is to remain profitable.

24.2 Polymers

Purpose

- To discuss the types of polymers and their properties.

It has been estimated that approximately 50% of the industrial chemists in the United States work in some area of polymer chemistry, a fact which illustrates just how important polymers are to our economy and standard of living. In Section 22.5, we described the major types of polymers and their uses. In this section we focus on the development of the polymer industry and on the processes used to manufacture several important polymers. These polymers are essential to the production of goods ranging from toys to roofing materials.

The Development and Properties of Polymers

The development of the polymer industry is a striking example of the importance of serendipity in the progress of science. Many discoveries in polymer chemistry arose from accidental observations on which scientists followed up.

The age of plastics might be traced to a day in 1846 when Christian Schoenbein, a chemistry professor at the University of Basel in Switzerland, spilled a flask containing nitric and sulfuric acids. In his hurry to clean up the spill, he grabbed his wife's cotton apron, which he then rinsed out and hung up in front of a hot stove to dry. Instead of drying, the apron flared and burned.

Very interested in this event, Schoenbein repeated the reaction under more controlled conditions and found that the new material, which he correctly concluded to be nitrated cellulose, had some surprising properties. As he had experienced, the nitrated cellulose was extremely flammable and, under certain circumstances, highly explosive. In addition, he found that it could be molded at moderate temperatures to give objects that were, upon cooling, tough but elastic. Predictably, the explosive nature of the substance was initially of more interest than its other properties, and cellulose nitrate rapidly became the basis for smokeless gun powder. Although Schoenbein's discovery cannot be described as a truly synthetic polymer because he simply found a way to modify the natural polymer cellulose (see Section 23.2), it formed the basis for a large number of industries that grew up to produce photographic films, artificial fibers, and molded objects of all types.

Casting polyurethane plastic.

The first synthetic polymers were produced as by-products of various organic reactions and were regarded as unwanted contaminants. Thus the first preparations of many of the polymers now regarded as essential to our modern lifestyle were thrown away in disgust. One chemist who refused to be defeated by the "tarry" products obtained when he reacted phenol and formaldehyde was the Belgian-American chemist Leo H. Baekeland (1863–1944). Baekeland's work resulted in the first completely synthetic plastic, called Bakelite, a substance that when molded to a certain shape under high pressure and temperature cannot be softened again or dissolved. Bakelite is a **thermoset polymer.** In contrast, cellulose nitrate is a **thermoplastic polymer;** that is, it can be remelted after it has been molded.

The discovery of Bakelite in 1907 spawned a large plastics industry, producing telephones, billiard balls, and insulators for electrical devices. During the early days of polymer chemistry, there was a great deal of controversy about the nature of these materials. Although the German chemist Hermann Staudinger speculated in 1920 that polymers were very large molecules held together by strong chemical bonds, most chemists of the time assumed that these materials were much like colloids, where small molecules are aggregated into large units by forces weaker than chemical bonds.

One chemist who contributed greatly to the understanding of polymers as giant molecules was Wallace H. Carothers of the DuPont Chemical Company. Among his accomplishments was the preparation of nylon (see Section 22.5). The nylon story illustrates further the importance of serendipity in scientific research. When nylon is made, by the reaction of a dicarboxylic acid and a diamine,

$$n\left\{\begin{matrix}H \\ H\end{matrix}\!\!>\!N\!-\!(CH_2)_6\!-\!N\!<\!\!\begin{matrix}H \\ H\end{matrix} \quad \begin{matrix}O \\ HO\end{matrix}\!\!>\!C\!-\!(CH_2)_4\!-\!C\!<\!\!\begin{matrix}O \\ OH\end{matrix}\right\}$$

Hexamethylenediamine $\quad H_2O \quad$ Adipic acid

$$\downarrow$$

$$\left(\overset{H}{\overset{|}{N}}-(CH_2)_6-\overset{H}{\overset{|}{N}}-\overset{O}{\overset{\|}{C}}-(CH_2)_4-\overset{O}{\overset{\|}{C}}\right)_n$$

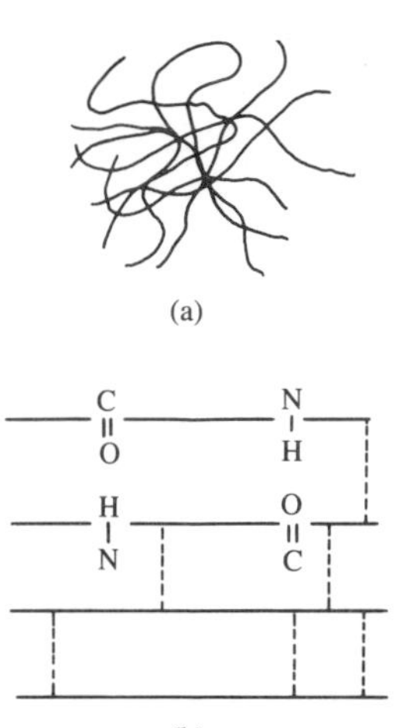

Figure 24.2

(a) Raw (amorphous) nylon with random chain orientations. (b) Nylon fibers after drawing.

the resulting product is a sticky material with little structural integrity. Because of this, it was initially put aside as having no apparently useful characteristics. However, Julian Hill, a chemist in the Carothers research group, one day put a small ball of this nylon on the end of a stirring rod and drew it away from the remaining sticky mass, forming a string. He noticed the silky appearance and strength of this thread and realized that nylon could be drawn into useful fibers.

The reason for this behavior of nylon is now understood. When nylon is first formed, the individual polymer chains are oriented randomly, like cooked spaghetti, and the substance is highly amorphous. However, when drawn out into a thread, the chains tend to line up (the nylon becomes more crystalline), which leads to increased hydrogen bonding between N—H and C=O groups on adjacent chains, as shown in Fig. 24.2. This increase in crystallinity and the resulting increase in hydrogen-bonding interactions lead to strong fibers and a highly useful material. Commercially, nylon is produced by forcing the raw material through a *spinneret,* a plate containing small holes, which forces the polymer chains to line up.

Another property that adds strength to polymers is **crosslinking,** the existence of covalent bonds between adjacent chains. The structure of Bakelite is highly crosslinked, which accounts for the strength and toughness of this polymer. Another example of crosslinking occurs in the manufacture of rubber. Raw natural rubber consists of chains of the type

$$\sim\sim CH_2-CH_2-CH=\underset{CH_3}{\underset{|}{C}}-CH_2-CH_2-CH=\underset{CH_3}{\underset{|}{C}}-CH_2\sim\sim$$

Charles Goodyear tried for many years to change natural rubber into a useful product. In 1839 he accidentally dropped some rubber containing sulfur on a hot stove. Noting that the rubber did not melt as expected, Goodyear pursued this lead and developed vulcanization.

and is a soft, sticky material unsuitable for tires. However, in 1839 Charles Goodyear (1800–1860), an American chemist, accidentally found that if sulfur is added to rubber and the mixture is heated, a process called **vulcanization,** the resulting rubber is still elastic (reversibly stretchable) but much stronger. This change in character occurs because of the formation of covalent bonds to sulfur atoms that link the chains.

Polyethylene and substituted polyethylenes (see Section 22.5) are mainstays of the polymer industry. The discovery of *Teflon,* a very important substituted polyethylene, is another illustration of the role of chance in chemical research. In 1938 a DuPont chemist named Roy Plunkett was studying the chemistry of gaseous perfluoroethylene:

$$\begin{matrix}F \\ F\end{matrix}\!\!>\!C\!=\!C\!<\!\!\begin{matrix}F \\ F\end{matrix}$$

Chemical Impact

Wallace Hume Carothers

Wallace H. Carothers, a brilliant organic chemist who was principally responsible for the development of nylon and the first synthetic rubber (Neoprene), was born in 1896 in Burlington, Iowa. As a youth, Carothers was fascinated by tools and mechanical devices and spent many hours experimenting. In 1915 he entered Tarkio College in Missouri and so excelled in chemistry that even before his graduation, he was made a chemistry instructor.

Carothers eventually moved to the University of Illinois at Urbana-Champaign, where he was appointed to the faculty when he completed his Ph.D. in organic chemistry in 1924. He moved to Harvard University in 1926 and then to DuPont in 1928, to participate in a new program in fundamental research. At DuPont, Carothers headed the organic chemistry division and during his ten years there played a prominent role in laying the foundations of polymer chemistry.

By the age of 33, Carothers had become a world-famous chemist whose advice was sought by almost everyone working in polymers. He was the first industrial chemist to be elected to the prestigious National Academy of Sciences.

Carothers was an avid reader of poetry and a lover of classical music. Unfortunately, he also suffered from severe bouts of depression that finally led to his suicide in 1937 in a Philadelphia hotel room, where he drank a cyanide solution. He was 41 years old. Despite the brevity of his career, Carothers was truly one of the finest American chemists of all time. His great intellect, his love of chemistry, and his insistence on perfection produced his special genius.

He synthesized about 100 pounds of the chemical and stored it in steel cylinders. When one of the cylinders produced no perfluoroethylene gas when the valve was opened, the cylinder was cut open to reveal a white powder. This powder turned out to be a polymer of perfluoroethylene, which was eventually developed into Teflon.

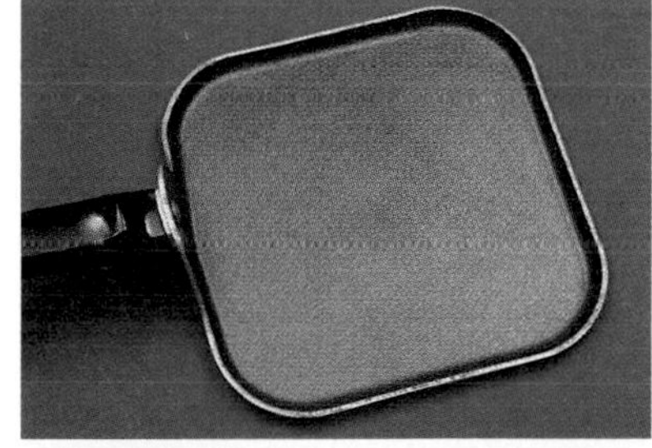

Teflon-coated frying pan. Teflon is a polymer of C_2F_4.

Polymers Based on Ethylene

A large section of the polymer industry involves the production of macromolecules made from ethylene or substituted ethylenes. As we saw in Section 22.5, ethylene molecules polymerize by addition, after the double bond has been broken by some initiator:

$$\begin{array}{ccccccccccccccccccc}
 & & H & & H & & H & & H & & & & H & & H & & H & & H \\
 & & | & & | & & | & & | & & & & | & & | & & | & & | \\
X & - & C & - & C\cdot & \curvearrowright & C & = & C & \longrightarrow & X & - & C & - & C & - & C & - & C\cdot \\
 & & | & & | & & | & & | & & & & | & & | & & | & & | \\
 & & H & & H & & H & & H & & & & H & & H & & H & & H
\end{array}$$

This process continues by adding new ethylene molecules to eventually give polyethylene, a thermoplastic material.

There are two forms of polyethylene: low-density polyethylene (LDPE) and high-density polyethylene (HDPE). The chains in LDPE contain many branches and thus do not pack as tightly as those in HDPE, which consist of mostly straight chain molecules.

Traditionally, LDPE has been manufactured at high pressure ($\sim$20,000 psi) and high temperature (500°C). These severe reaction conditions require specially designed equipment, and for safety reasons the reaction is usually run behind a

psi is the abbreviation for pounds per square inch; 15 psi $\approx$ 1 atm.

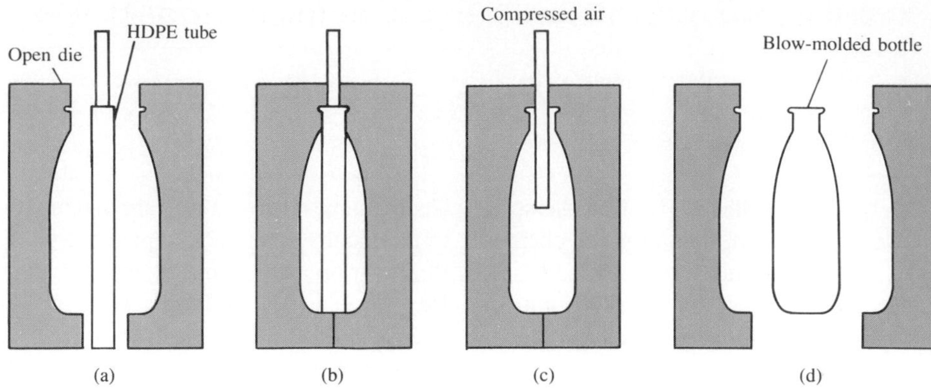

Figure 24.3

A major use of HDPE is for blow-molded objects such as bottles for soft drinks, shampoos, bleaches, and so on. (a) A tube composed of HDPE is inserted into the mold (die). (b) The die closes, sealing the bottom of the tube. (c) Compressed air is forced into the warm HDPE tube, and it expands to take the shape of the die. (d) The molded bottle is removed from the die.

reinforced concrete barrier. More recently, lower reaction pressures and temperatures have become possible through the use of catalysts. One catalytic system using triethylaluminum, $Al(C_2H_5)_3$, and titanium(IV) chloride was developed by Karl Ziegler in Germany and Giulio Natta in Italy. Although this is a very efficient catalyst, it catches fire on contact with air and must be handled very carefully. A safer catalytic system was developed at Phillips Petroleum Company. It uses a chromium(III) oxide (Cr_2O_3) and aluminosilicate catalyst and has largely taken over in the United States. The product of the catalyzed reaction is highly linear (unbranched) and is often called *linear low-density polyethylene*. It is very similar to HDPE.

The major use of LDPE is in manufacturing the tough transparent film that is used in packaging so many consumer goods. Two-thirds of the approximately 10 billion pounds of low-density polyethylene produced annually in the United States is used for this purpose. The major use of HDPE is for blow-molded products such as bottles for consumer products (see Fig. 24.3).

The useful properties of polyethylene are due primarily to its high molecular weight. Although the strengths of the interactions between specific points on the nonpolar chains are quite small, the chains are so long that these small attractions accumulate to a very significant value, so that the chains stick together very tenaciously. There is also a great deal of physical tangling of the lengthy chains. The combination of these interactions gives the polymer strength and toughness. However, a material like polyethylene can be melted and formed into a new shape (thermoplastic behavior) because in the melted state the molecules can readily flow past one another.

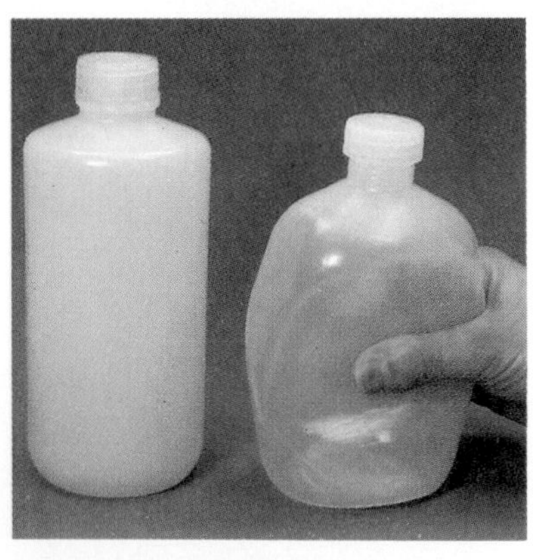

Two polyethylene bottles. The one on the left is made with high-density polyethylene while the one on the right is made with low-density polyethylene.

Since a high molecular weight gives a polymer useful properties, one might think that the goal would be to produce polymers with chains as long as possible. However, this is not the case—polymers become much harder to process as the molecular weights increase. Most industrial operations require that the polymer flow through pipes as it is processed. But as the chain lengths increase, viscosity also increases. In practice, the upper limit of a polymer's molecular weight is set by the flow requirements of the manufacturing process. Thus the final product often reflects a compromise between the optimum properties for applications and for ease of processing.

Although many polymer properties are greatly influenced by molecular weight, other important properties are not. For example, chain length does not affect a polymer's resistance to chemical attack. Physical properties such as color, refractive index, hardness, density, and electrical conductivity are also not greatly influenced by molecular weight.

We have seen that one way to change the strength of a polymeric material is to vary the chain length. Another method for modifying polymer behavior is to vary the substituents. For example, if we use a monomer of the type

```
H       H
 \     /
  C=C
 /     \
H       X
```

the properties of the resulting polymer depend on the identity of X. The simplest example is polypropylene, whose monomer is

```
H       H
 \     /
  C=C
 /     \
H       CH3
```

and that has the form

```
H  CH3   H  CH3   H  CH3   H  CH3   H  CH3
 \ /      \ /      \ /      \ /      \ /
~C----C----C----C----C----C----C----C----C----C~
     / \      / \      / \      / \      / \
    H   H    H   H    H   H    H   H    H   H
```

The CH_3 groups can be arranged on the same side of the chain (called an **isotactic chain**) as shown above, can alternate (called a **syndiotactic chain**) as follows:

```
H  CH3  CH3  H   H  CH3  CH3  H
 \ /      \ /     \ /      \ /
~C----C----C----C----C----C----C----C~
     / \      / \      / \      / \
    H   H    H   H    H   H    H   H
```

or can be randomly distributed (called an **atactic chain**).

The chain arrangement has a significant effect on the polymer properties. Most polypropylene is made by using the Ziegler-Natta catalyst, $Al(C_2H_5)_3 \cdot TiCl_4$, which produces highly isotactic chains that pack together quite closely. As a result, polypropylene is more crystalline, and therefore stronger and harder, than polyethylene. The major uses of polypropylene are for molded parts (40%), fibers (35%), and packaging films (10%). Polypropylene fibers are especially useful for athletic wear because they do not absorb water from perspiration as cotton does. Rather the moisture is wicked away from the skin to the surface of the polypropylene garment where it can evaporate. The annual U.S. production of polypropylene is about 7 billion pounds.

Another related polymer, **polystyrene,** is constructed from the monomer styrene,

```
H       H
 \     /
  C=C
 /     \
H       (benzene ring)
```

Plastic films made from ethylene and its derivatives are used to package most consumer products. This photo shows molten plastic being formed into a bag.

Pure polystyrene is too brittle for many uses, so most polystyrene-based polymers are really *copolymers* of styrene and butadiene,

$$\mathrm{H_2C{=}CH{-}CH{=}CH_2}$$

thus incorporating bits of butadiene rubber in the polystyrene matrix. The resulting polymer is very tough and is often used as a substitute for wood in furniture.

Another polystyrene-based product is acrylonitrile-butadiene-styrene (ABS), a tough, hard, and chemically resistant plastic used for pipes and items such as radio housings, telephone cases, and golf club heads for which shock resistance is an essential property. Originally, ABS was produced by copolymerization of the three monomers:

$$\underset{\text{Acrylonitrile}}{\mathrm{H_2C{=}CH{-}CN}} + \underset{\text{Styrene}}{\mathrm{H_2C{=}CH{-}C_6H_5}} + \underset{\text{Butadiene}}{\mathrm{H_2C{=}CH{-}CH{=}CH_2}} \longrightarrow \left(\mathrm{CH_2{-}\underset{CN}{CH}{-}CH_2{-}\underset{C_6H_5}{CH}{-}CH_2{-}\underset{CH_2=CH}{CH}}\right)_n$$

It is now prepared by a special process called *grafting,* in which butadiene is polymerized first, and then the cyanide and phenyl substituents are added chemically.

PVC: Additives Make the Difference

Another high-volume polymer, **polyvinyl chloride (PVC),** is constructed from the monomer vinyl chloride,

$$\mathrm{H_2C{=}CHCl}$$

The production of this polymer is quite representative of the processes used in the polymer industry and will be discussed in detail.

In pure form, PVC is a hard, brittle substance that decomposes easily at the temperatures necessary to process it, and is thus almost useless. That it has become a high-volume plastic ($\sim$8 billion pounds per year produced in the United States by 1987) is a tribute to chemical innovation. By use of the proper additives, PVC can be made rigid or highly flexible, and it can be tailored for use in inexpensive plastic novelty items or for precision engineering applications (see Table 24.3).

A *plasticizer* is a compound added to a polymer to soften it. The type and amount of plasticizer determine the pliability of the final product. One theory suggests that plasticizers cause softening by insertion between polymer chains, where they are held in place only by London dispersion forces. This lessens polymer-polymer interactions and softens the material.

The development of PVC illustrates the interplay of logic and serendipity and the optimization of properties for both processing and applications. PVC production has been beset with difficulties from the beginning, but solutions have been found for each problem by a combination of chemical deduction and trial and error. For example, many additives have been found that provide temperature stability so that PVC can be processed as a melt and so that PVC products can be used at high temperatures. However, there is still controversy among chemists about exactly how PVC decomposes thermally, and thus the reason that these stabilizers work is

Consumption of PVC in the United States in 1984

Consumption	(millions of pounds)	Consumption	(millions of pounds)
Construction	3775	*Transportation*	235
Flooring	350	Upholstery, trim, and tops	135
Panels and siding	305	Floor mats	25
Pool and pond liners	40	Bumper parts	25
Roofing	30	Other	50
Pipe and pipe fittings*	2805		
Windows and other profiles	175	*Recreation*	195
		Records	105
Weatherstripping	45	Toys	45
Other	25	Other	45
Packaging	650	*Apparel*	185
Sheet	125	Footwear	95
Film	310	Handbags	30
Bottles	175	Luggage	35
Closures and other uses	40	Other	25
Electrical	475	*Medical*	150
Wire and cable	400	Tubing	50
Connectors and plugs	60	Blood and intravenous bags	50
Other	15	Other	50
Furnishings	365	*Miscellaneous*	290
Upholstery	105	Credit cards	25
Wall covering	50	Garden hose	50
Appliances	105	Traffic cones	20
Window shades, blinds, and awnings	35	Tool handles	15
		Adhesives and sealants	90
Tablecloths and mats	50	Other	90
Other	20		
		Total	6320

*Note that 45% of the total is used for pipe.
SOURCE: Reprinted with permission from *Chemical & Engineering News*, June 28, 1984, p. 30. Copyright 1984 American Chemical Society.

Table 24.3

not well understood. Also, there are approximately a hundred different plasticizers available for PVC, but the theory of its plasticization is too primitive to predict accurately what compounds might work even better.

PVC was discovered by a German chemical company in 1912, but its brittleness and thermal instability proved so problematical that in 1926 the company stopped paying the fees that maintained its patents. That same year Waldo Semon, a chemist at B. F. Goodrich, found that PVC could be made flexible by the addition of phosphate and phthalate esters. Semon also found that white lead, $Pb_3(OH)_2(CO_3)_2$, provided thermal stability. These advances led to the beginning of significant U.S. industrial production of PVC (~4 million pounds per year by 1936). In an attempt to further improve PVC, T. L. Gresham (also a chemist at B. F. Goodrich) tried approximately a thousand compounds, searching for a better plasticizer, and found that di-2-ethylhexylphthalate,

$$C_6H_4\left[-C(=O)-O-CH_2-CH(CH_2CH_3)-CH_2-CH_2-CH_2-CH_3\right]_2$$

gave the best results. This compound remains the most common plasticizer for PVC. Table 24.4 lists the types of additives commonly used in the production of PVC.

Recently, there has been concern about the carcinogenicity of di-2-ethylhexyl phthalate, and other phthalate esters are being studied as possible replacements.

Although the exact mechanism of the thermal decomposition of PVC remains unknown, most chemists agree that the chlorine atoms play an important role. Compounds such as lead phosphite, lead sulfate, or lead phthalate are added to provide anions less reactive than chloride and to provide lead ions to combine with the released chloride ions. As a beneficial side effect, the lead chloride formed gives the PVC added electrical resistance, making lead stabilizers particularly useful in producing PVC for electrical wire insulation.

One major use of PVC is for pipes in plumbing systems. Here, even though the cheap lead stabilizers would be preferred from an economic standpoint, the possibility that the toxic lead will be leached from the pipes into the drinking water necessitates the use of more expensive tin and antimony compounds as thermal stabilizers. Because 45% of the annual production of PVC goes for pipe, this is a huge market for companies that manufacture additives, and the competition is very intense. The

Types of Additives Commonly Used in the Production of PVC

Type of additive	Effect	Examples
Plasticizer	Softens the material	Di-2-ethylhexyl phthalate
Heat stabilizer	Increases resistance to thermal decomposition	Lead salts, tin compounds, antimony compounds
Ultraviolet absorber	Prevents damage by sunlight	Titanium(IV) oxide (TiO_2), substituted benzophenones $R-C_6H_4-C(=O)-C_6H_4-R$
Flame retardant	Lowers flammability	Antimony(III) oxide (Sb_2O_3)
Biocide	Prevents bacterial or fungal attack	Arsenic and tin compounds

Table 24.4

most recently developed low-cost PVC thermal stabilizer is a mixture of antimony and calcium salts, which was produced to replace the tin compounds made increasingly costly as the price of tin rose from \$1.67 per pound in 1971 to \$6.02 in 1983. Although, in contrast, antimony cost only \$0.90 per pound in 1983, it has not yet found widespread use because its presence makes processing more difficult.

Outdoor applications of PVC often require that it contain ultraviolet light absorbers to protect against damage from sunlight. For pigmented applications such as vinyl siding, window frames, and building panels, titanium(IV) oxide is usually used. For applications where the PVC must be transparent, various aromatic organic compounds are added.

The additives used in PVC in the largest amounts are the plasticizers, but one detrimental effect of these additives is an increase in flammability. Rigid PVC, which contains little plasticizer, is quite flame-resistant because of the high average chloride content. However, as more plasticizer is added for flexibility, the flammability increases to the point where fire retardants must be added, the most common being antimony(III) oxide (Sb_2O_3). As the PVC is heated, this oxide forms antimony(III) chloride ($SbCl_3$), which migrates into the flame where it inhibits free radicals (see the discussion of flame retardancy later in this section). Because antimony(III) oxide is a white salt, it cannot be used for transparent or darkly colored PVC. In these cases sodium antimonate (Na_3SbO_4), a transparent salt, is used.

Once the additives have been chosen for a particular PVC, the materials must be blended. This is often done in a dry-blending process, giving a powdered form that is then used for fabrication of the final product. The powdered mixture also can be melted and formed into pellets.

The production of PVC provides a good case study of an industrial process. It illustrates many of the factors that must be taken into account when any product is manufactured: effectiveness of the product, cost, ease of production, safety, and environmental impact.

Elastomers: Stretchable Polymers

Elastomers are materials that recover their shape after a deforming force. The fundamental elastomer is natural rubber, obtained as a latex from a tree native to tropical America but now grown primarily in the tropical Orient. When the Japanese conquered the major rubber-growing areas at the start of World War II, a crash program was undertaken in the United States to develop a suitable substitute. The results of this program were phenomenal; the U.S. production of synthetic rubber increased from none in 1940 to ~800,000 tons in 1944.

The term *rubber* was apparently coined by Joseph Priestley, best known for the discovery of oxygen, because of its ability to rub out pencil marks.

Natural rubber is primarily *cis*-polyisoprene,

$$\left(\!-CH_2-C(CH_3){=}CH-CH_2CH_2-C(CH_3){=}CH-CH_2-\!\right)_n$$

which, after crosslinking by vulcanization, is especially useful for tires because of its unusual resistance to heat.

Synthetic rubber, which currently accounts for approximately 50% of the total market, is primarily made from copolymerization of styrene (one part) and butadiene (three parts).

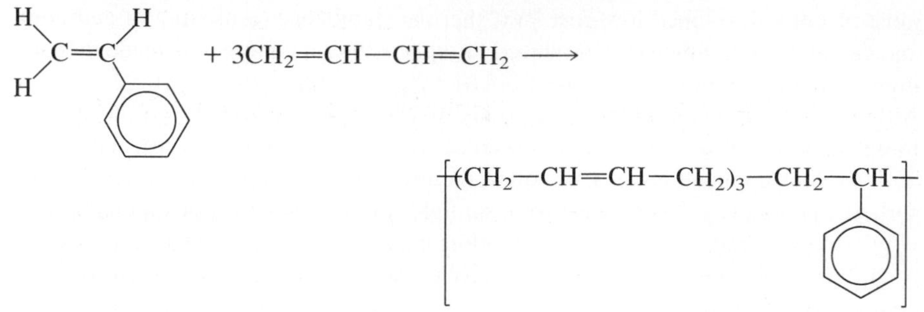

$$\text{H}_2\text{C}{=}\text{CH}{-}\text{C}_6\text{H}_5 + 3\text{CH}_2{=}\text{CH}{-}\text{CH}{=}\text{CH}_2 \longrightarrow \left[\!\!-(\text{CH}_2{-}\text{CH}{=}\text{CH}{-}\text{CH}_2)_3{-}\text{CH}_2{-}\underset{\text{C}_6\text{H}_5}{\text{CH}}-\!\!\right]$$

This polymer, called SBR rubber, was developed during World War II and is used principally (75%) for the manufacture of tires. Other types of synthetic elastomers are shown in Table 24.5.

$\Delta G = \Delta H - T\Delta S$

Elastomers are characterized by a highly random (noncrystalline) arrangement of the chains in the relaxed state. When the material is stretched, the chains become more ordered. The tendency to snap back to the original state can be described in terms of entropy. Because the energy of interaction between the nonpolar chains is relatively small, the change in free energy is dominated by the positive ΔS term when the elastomer goes from the stretched, relatively ordered state to the relaxed random state. The positive ΔS term results in a negative value for ΔG, making the relaxing process spontaneous.

Common Elastomers

Type	Monomer(s)	Uses
Polychloroprene (Neoprene)	$CH_2{=}C(Cl){-}CH{=}CH_2$	Wire covering, automotive drive belts
Polyisoprene	$CH_2{=}C(CH_3){-}CH{=}CH_2$	Tires
Nitrile	$CH_2{=}CH(CN)$ + $CH_2{=}CH{-}CH{=}CH_2$	Automotive hoses and gaskets
Polybutadiene	$CH_2{=}CH{-}CH{=}CH_2$	Tires

Table 24.5

Flammability of Polymers

One important property of polymers, especially those used for fabrics and construction materials, is flammability. The United States has a serious problem with domestic fires, which cause thousands of deaths and billions of dollars in property damage each year. One way to combat this problem is to make synthetic materials flame-retardant. To see how this might be done, we must first understand what happens when something burns. A relatively simple combustion process, that between methane and oxygen, is illustrated in the following Chemical Impact feature. Note that the process takes place via a free-radical chain reaction.

Polymer combustion, although a more complicated process, still involves free radicals. The burning of a polymer (Fig. 24.4) is a cyclic process in which thermal

Chemical Impact

The Mechanism of Methane Combustion

The reaction between methane and oxygen,

$$CH_4 + 2O_2 \rightarrow CO_2 + 2H_2O$$

occurs by formation of free radicals (fragments with an unpaired electron) that undergo a series of reactions, producing new radicals in a chain reaction. The radicals are initially formed when energy from an ignition source is fed into the system. The reactions in the chain can be classified into four types: *initiation,* when radicals are formed; *propagation,* when one or more radicals react to form a new radical; *branching,* when one radical reacts to form two radicals; and *termination,* when two radicals react to form a nonradical.

Examples of the various types of steps are as follows, where R · represents a carbon-containing radical and M is a nonreacting substance that absorbs energy to give M* as the R—H bond forms:

Initiation

$$CH_4 \rightarrow CH_3\cdot + H\cdot$$
$$O_2 \rightarrow 2O\cdot$$

Propagation

$$CH_4 + H\cdot \rightarrow CH_3\cdot + H_2$$
$$CH_4 + HO\cdot \rightarrow CH_3\cdot + H_2O$$
$$CH_3\cdot + O\cdot \rightarrow CH_2O + H\cdot$$
$$CH_2O + HO\cdot \rightarrow CHO\cdot + H_2O$$
$$CH_2O + H\cdot \rightarrow CHO\cdot + H_2$$
$$CHO\cdot \rightarrow CO + H\cdot$$
$$CO + HO\cdot \rightarrow CO_2 + H\cdot$$

Branching

$$H\cdot + O_2 \rightarrow HO\cdot + O\cdot$$

Termination

$$H\cdot + R\cdot + M \rightarrow RH + M^*$$

energy is used to break bonds to form polymer fragments. These fragments then react exothermically with oxygen by a free-radical chain mechanism and thus furnish energy to sustain combustion. The polymer can be made flame retardant by providing some means of interrupting this cycle. This is done by solid-phase inhibition or by vapor-phase inhibition. In **solid-phase inhibition** the polymer surface is changed to help prevent fragmentation. If the polymer is extensively crosslinked at the surface, a carbonaceous char forms that insulates the underlying polymer, depriving the combustion process of its fuel. As a result, the flame cools, and the fire goes out. **Vapor-phase inhibition** involves changing the chemistry of the combustion process. Substances called free-radical inhibitors are built into the polymer. They migrate into the flame to produce additional radical-termination steps or to replace highly reactive radicals with less active radicals. Their effect is to greatly decrease the combustion rate either by radical scavenging:

$$H\cdot + HBr \longrightarrow H_2 + Br\cdot$$
$$HO\cdot + HBr \longrightarrow H_2O + Br\cdot$$

or by radical recombination:

$$HO\cdot + Na\cdot \longrightarrow NaOH$$
$$NaOH + H\cdot \longrightarrow H_2O + Na\cdot$$
$$HO\cdot + H\cdot \xrightarrow{\text{Catalyst}} H_2O$$

Many polymers have natural flame retardancy. For example, a polymer containing halogen atoms or aromatic groups burns less readily because large quantities of energy are required to break the bonds in these substances and because these substances tend to char naturally.

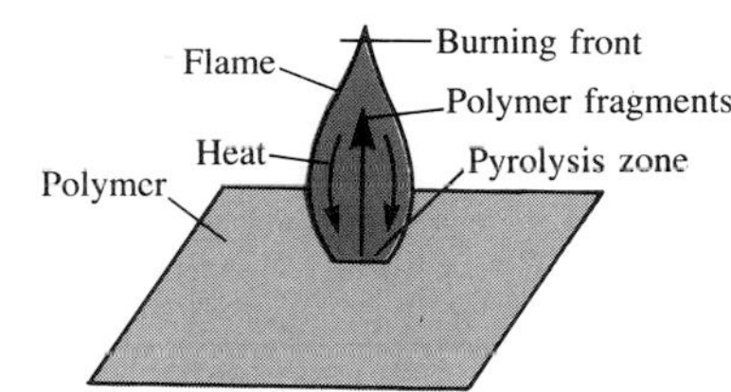

Figure 24.4

The combustion of a polymer. In the pyrolysis zone, the polymer is broken into small fragments as bonds are disrupted by the thermal energy. These fragments then migrate into the higher-temperature flame zone, where they react with oxygen by free-radical chain reactions. The heat from these exothermic reactions is fed down to the polymer surface to provide energy for more pyrolysis.

Chemical Impact

Supernylon

For the first time since the early work by Carothers and his co-workers at DuPont, a new type of nylon appears ready to assume a dominant role in the marketplace. Tradenamed Stanyl and developed in the Netherlands, the new nylon polymer reportedly has a higher melting point and greater toughness and temperature stability than traditional nylon polymers—properties that should make it ideal for the manufacture of molded parts and fibers.

For the last 50 years the nylon market has been dominated by two polymers. One is made from hexamethylenediamine (1,6-diaminohexane) and adipic acid:

$$H_2N{-}(CH_2)_6{-}NH_2 + HO{-}\overset{\displaystyle O}{\overset{\|}{C}}{-}(CH_2)_4{-}\overset{\displaystyle O}{\overset{\|}{C}}{-}OH \longrightarrow \left(\overset{\displaystyle H}{\overset{|}{N}}{-}(CH_2)_6{-}\overset{\displaystyle H}{\overset{|}{N}}{-}\underset{\displaystyle O}{\underset{\|}{C}}{-}(CH_2)_4{-}\underset{\displaystyle O}{\underset{\|}{C}} \right)_n$$

Nylon-66

This polymer is called *nylon-66* because the two starting materials have six carbon atoms each. The other, *nylon-6,* comes from a cyclic amide:

$$\text{cyclic amide } (CH_2)_5C(=O)NH \xrightarrow[250°C]{H_2O,\ \text{Acetic acid}} \left(\overset{\displaystyle H}{\overset{|}{N}}{-}(CH_2)_5{-}\overset{\displaystyle O}{\overset{\|}{C}} \right)_n$$

Nylon-6

The new nylon polymer is formed by reaction of 1,4-diaminobutane (DAB) and adipic acid:

$$H_2N{-}(CH_2)_4{-}NH_2 + HO{-}\overset{\displaystyle O}{\overset{\|}{C}}{-}(CH_2)_4{-}\overset{\displaystyle O}{\overset{\|}{C}}{-}OH \longrightarrow \left(\overset{\displaystyle H}{\overset{|}{N}}{-}(CH_2)_4{-}\overset{\displaystyle H}{\overset{|}{N}}{-}\overset{\displaystyle O}{\overset{\|}{C}}{-}(CH_2)_4{-}\overset{\displaystyle O}{\overset{\|}{C}} \right)_n$$

It is called *nylon-46*.

Initial laboratory tests indicate that nylon-46 has very useful physical properties. Its tensile strength, stiffness, resistance to abrasion, and ability to withstand impact are greater than those of nylon-66 or nylon-6.

Nylon-46 was originally prepared by Wallace Carothers. Since the dark brown polymer of relatively low molecular weight did not appear to have promising properties, it was not investigated further at that time.

The Dutch company made a salt from DAB and adipic acid, which is converted to a low-molecular-weight "prepolymer" in water and then collected as a solid. When heated to 250°C in an atmosphere of steam and nitrogen, the solid polymerizes further to an average molecular weight of approximately 30,000.

Although Stanyl is in the relatively early stages of development, it may well assume a major role in the nylon industry.

Polymers that are naturally flammable have flame retardants either built into the polymer itself or added as separate molecules. The most common additives are phosphorus compounds, which retard flames by solid-phase inhibition as a result of the formation of acids that promote char formation.

Halogens are the substances most often used to modify the polymer itself to decrease flammability. The carbon-halogen bonds break in the flame, releasing the halogen atoms which can then form hydrogen halides that behave as free-radical inhibitors. For example, styrene becomes less flammable if halogens are substituted for hydrogen atoms on the phenyl ring.

24.3 Pesticides

Purpose

- To describe herbicides and insecticides.

Of the problems we confront in growing food, one of the most serious is that of pests: weeds that compete for soil nutrients and water, fungi that infect plants, worms that eat roots and foliage, and rodents that consume huge quantities of stored grain. During this century we have made great strides in effectively fighting these pests by breeding hardier plant species and developing chemical pesticides.

Herbicides

A pesticide that kills weeds is called a **herbicide.** Weeds are formidable opponents of food production, causing an estimated 10% loss in agriculture production, which costs approximately $12 billion annually. In addition, more than $6 billion is spent each year to control weeds. Chemical weedkillers (~500 million pounds per year) account for about 60% of all pesticides used in the United States. Currently about 180 different herbicides are used against the estimated 1500 species of weeds that interfere with food crops.

Chemical weed control is not new. For example, farmers in ancient times used salt to combat weeds. The modern history of herbicides began around 1850 with the use of mixtures of sodium chloride and lime and the use of other inorganic compounds such as iron sulfate, sulfuric acid, copper nitrate, and various arsenic compounds. However, these compounds suffer from the disadvantages of being nonselective (they kill part of the crop they are intended to protect) and requiring large amounts.

A major breakthrough in weed control occurred with the discovery in the early 1940s of 2,4-dichlorophenoxyacetic acid, or 2,4-D,

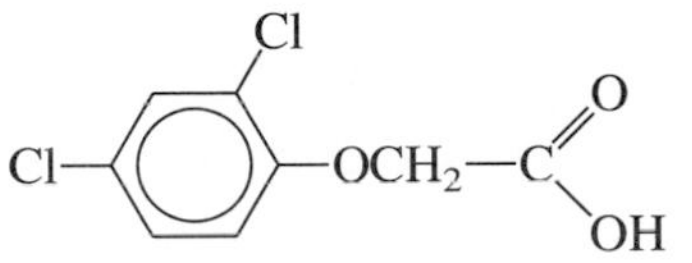

$$\text{Cl}-\text{C}_6\text{H}_3(\text{Cl})-\text{OCH}_2-\text{C}(=\text{O})\text{OH}$$

2,4-D

which kills broadleaf weeds while having no detrimental effects on narrowleaf plants. In addition to being much more selective, 2,4-D is applied in amounts 100 to 1000 times *lower* than the inorganic herbicides previously in use. A postemergence herbicide (one applied after the plants are growing), 2,4-D has been widely used to control weeds in major crops such as wheat, barley, rice, corn, and sugarcane, as well as in lawns.

A closely related compound, 2,4,5-trichlorophenoxyacetic acid (2,4,5-T), is shown below.

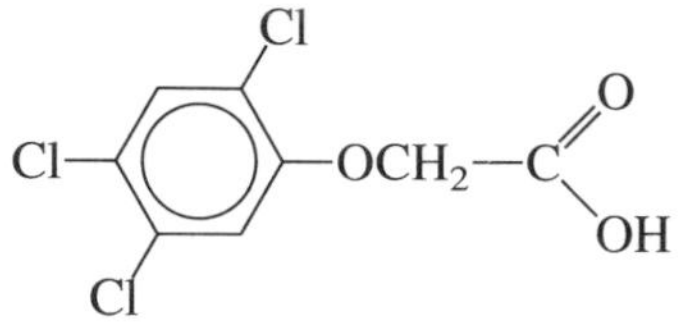

2,4,5-T

It was first marketed in 1948 and has found wide use in controlling brush along highways, railways, and cattle rangeland.

In the past several years, 2,4,5-T has been the focus of controversy because of its use in Vietnam to defoliate trees so as to destroy the cover of enemy soldiers. The substance used, called Agent Orange after the color of its storage drums, was actually a 50:50 mixture of the butyl esters of 2,4,5-T and 2,4-D. Over 12 million gallons were sprayed on the jungles of Vietnam. It has been claimed that exposure to this defoliant resulted in birth defects, spontaneous abortions, skin irritations, and cancer. Although much controversy remains, it is now thought by many scientists that any damage to humans by the herbicide was probably due to the impurity 2,3,7,8-tetrachlorodibenzo-*p*-dioxin (often called simply dioxin):

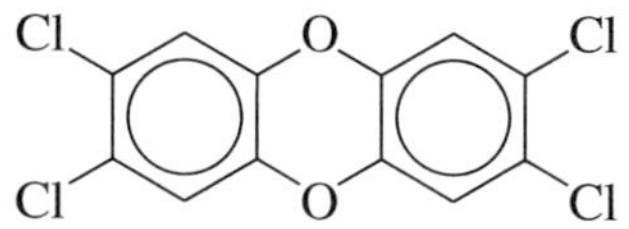

2,3,7,8-Tetrachlorodibenzo-*p*-dioxin

However, concerns about the safety of 2,4,5-T remain, and severe restrictions have been placed on its use.

Many other herbicides have entered the marketplace since the development of the phenoxy-based compounds. For example, the main herbicides used on corn are atrazine and alachlor. Another very popular herbicide for the control of broadleaf weeds is trifluralin, which is used on more than 40 different crops.

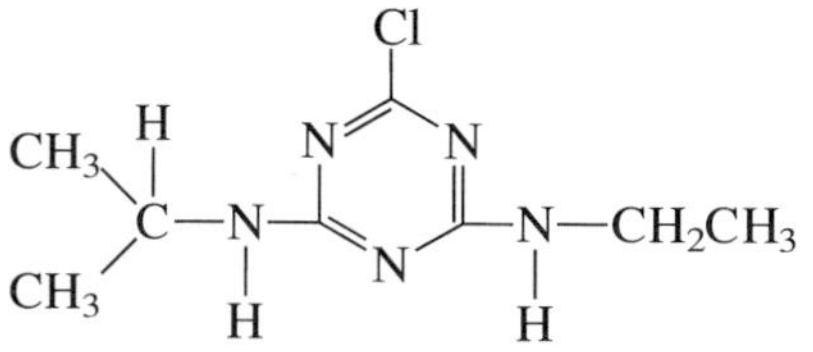

Atrazine

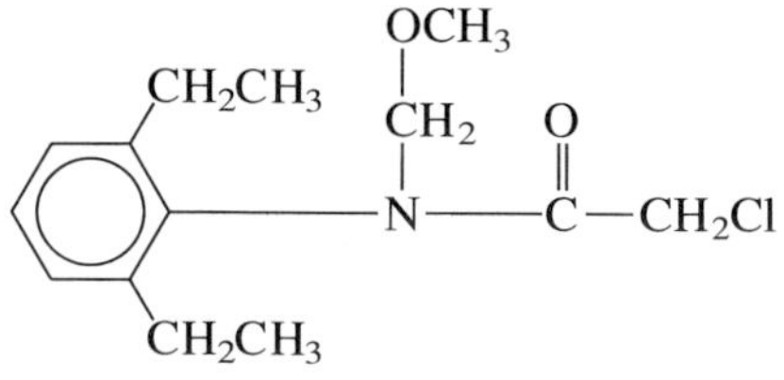

Alachlor

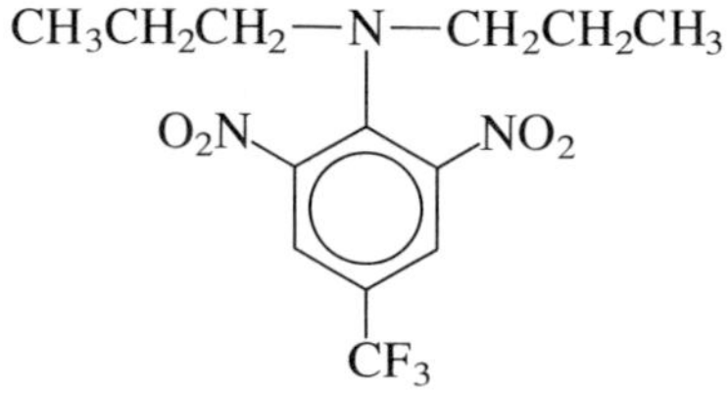

Trifluralin

Chemical Impact

Allelopathic Chemicals: Nature's Herbicides

In recent decades, great improvements have been made in the production of food crops, mainly through the use of fertilizers, pesticides, and herbicides. However, the use of chemicals has also had detrimental effects.

One means of weed control currently under study is the use of chemicals that the plants themselves or associated organisms produce to inhibit the growth of competing plants. Such compounds, called **allelopathic chemicals,** are attractive as potential herbicides because they are natural products. While this, of course, does not automatically make them safe—many natural chemicals are quite toxic to humans—it does make sense to take advantage of mechanisms already in place in nature. Natural control systems tend to be both specific and efficient.

It has long been known that certain plant species tend to become predominant in ecosystems. While this may be partly due to their efficient use of water and nutrients, it is also often due to their release of allelopathic agents that adversely affect neighboring plants. For example, sunflowers produce substances that interfere with the growth of weeds such as velvetleaf, pigweed, jimsonweed, and wild mustard. Black walnut trees inhibit the growth of certain plants near them, and sorghum seems to release chemicals that control weed growth in crops planted later in the same soil. Most current research in allelopathy involves attempts to isolate and identify specific active chemicals. Once these substances are identified, the next decision to be made will be whether plants can be used as natural "factories" for these chemicals or whether (probably more realistically) they would be more efficiently produced in chemical manufacturing facilities.

Scientists have done tremendous amounts of research to determine the mode of action of herbicides. For example, 2,4-D is believed to kill plants by imitating a natural growth regulator. This causes the plant to grow too fast and thus interferes with respiration and the transport of nutrients. Atrazine disrupts photosynthesis and deprives the plant of photosynthetic products necessary for survival. Alachlor inhibits the synthesis of proteins, and trifluralin interferes with cell division.

Food plants that are not harmed by herbicides usually have one or more mechanisms for detoxifying them. For instance, corn has three ways of breaking down atrazine into chemicals harmless to the corn plant.

Herbicides are usually not directly harmful to humans because they act by mechanisms important only to plant survival. However, the level of herbicide use is a concern, because some herbicides are potential carcinogens. As a result, methods of integrated weed control, involving mechanical cultivation and crop rotation as well as herbicide application, are being used more frequently. Some natural control programs have also been carried out in which insects are released that feed on a particular type of weed, but these programs have been small and isolated. One way to minimize the dangers of herbicides is to find new compounds that are so effective that application rates can be kept very low. Many such high-potency herbicides are under development by major chemical companies. There is also much interest in allelopathic substances (see the Chemical Impact feature above), which some plants produce to protect themselves from competitive plants.

Insecticides

Insects represent nearly 80% of the total animal mass in the world and pose formidable competition for the food supply. It has been estimated by the World Health Organization that as much as one-third of the agricultural produce grown in the world is consumed or destroyed by insects.

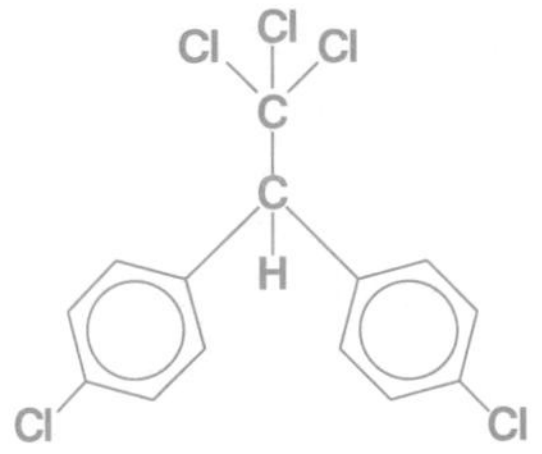

As with herbicides, the first widely used insecticides were inorganic compounds. Examples include sulfur and compounds of fluorine, copper, boron, zinc, or arsenic. One inorganic compound still used extensively against the potato beetle is lead hydrogen arsenate ($PbHAsO_4$). Although quite toxic to humans, $PbHAsO_4$ does not present a residue problem in the potato because the crop grows underground where it is not affected by the insecticide applied to the foliage.

DDT

The initials DDT come from a common name for this chemical: dichlorodiphenyl trichloroethane.

Most modern insecticides are organic compounds. The first highly successful organic insecticide was **DDT,** or 1,1-bis(4-chlorophenyl)-2,2,2-trichloroethane, a compound that revolutionized insect control. DDT is rapidly absorbed by insects and is stable in the environment so that its effects last a long time. The latter property has in fact turned out to be a liability. Because of its stability and because it is nonpolar and dissolves in fatty tissues, DDT is incorporated into the food chain and has caused damage to certain animals. As a result, DDT has been banned in the United States and in many other countries.

Methoxychlor

After the effectiveness of DDT became apparent and before its detrimental qualities were known, several other chlorinated hydrocarbons were developed as insecticides; they suffer from many of the same problems. Recently, attempts have been made to design chlorinated hydrocarbon molecules that give polar products when metabolized by animals, thus making them water-soluble and excretable. An example is methoxychlor, a biodegradable analogue of DDT. Its structure is shown on the left.

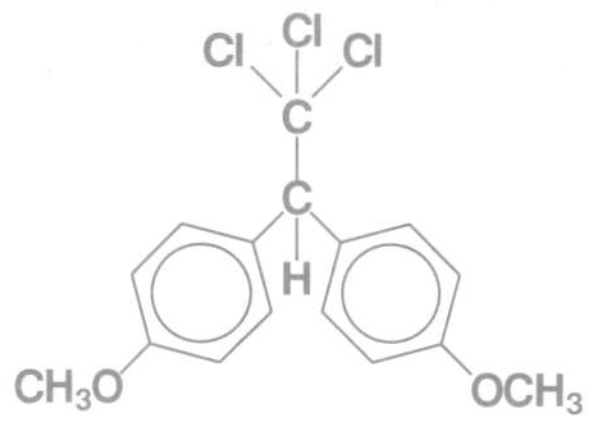

Because of the environmental problems and the carcinogenicity of many chlorinated hydrocarbon insecticides, they are gradually being replaced by other classes of compounds, such as the organic phosphates. Most of these molecules are closely related to the nerve gases developed for human warfare and so are quite toxic. In the manufacturing process, great care is taken to prevent worker exposure. However, because they decompose in moist air within a couple of weeks, they do not offer severe residue problems.

One of the first organophosphates marketed as an insecticide is *parathion,*

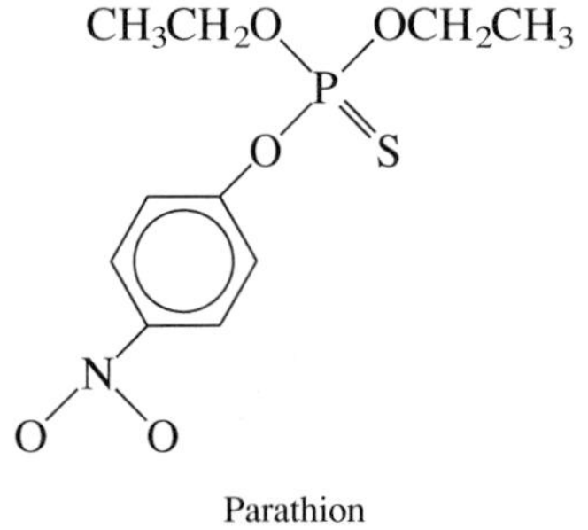

Parathion

which is highly toxic to humans. Another commonly used organophosphate insecticide is *malathion,* which is much less toxic than parathion:

$$CH_3{-}O{-}P(=S)(OCH_3){-}S{-}CH(C(=O){-}O{-}CH_2{-}CH_3){-}CH_2{-}C(=O){-}O{-}CH_2{-}CH_3$$

Malathion

A crop duster spraying pesticides on corn.

Over the years more than 100,000 organophosphorus compounds have been screened for use as insecticides, and several are currently used. Although these materials suffer from the disadvantage of being toxic to humans, they do biodegrade into harmless products and thus do not persist in the environment or get into the food chain. However, because they decompose easily, they must be sprayed on the plants more often, increasing the cost of treatment. This latter problem is avoided in some cases because certain of the organophosphorus compounds act as systemic insecticides. That is, they are absorbed into the plants' circulation systems. This means that respraying is unnecessary even as new growth occurs.

Another class of insecticides are the **carbamates** of which *Sevin,*

$$C_{10}H_7{-}O{-}C(=O){-}N(H){-}CH_3$$

Sevin

used to control insects on cotton, vegetables, and fruit, is an example. The carbamates can be viewed as derivatives of carbamic acid

$$H{-}O{-}C(=O){-}N(H){-}H$$

where the hydrogen atoms are replaced by various hydrocarbons. Many carbamate insecticides are now in use; they are generally safer to handle than the organophosphates but are more expensive to produce.

There has been much recent interest in attempting to control insects by manipulating the chemicals they themselves produce as sex attractants and growth regulators. A **pheromone** is a chemical produced by an insect, which, when released, evokes a response in another insect of the same species. There has been special interest in sex pheromones, which, among other functions, allow males to find females. For example, one of the sex pheromones of the European corn borer is the compound 11-tetradecenyl acetate:

$$\begin{array}{ccc} CH_3CH_2 & & (CH_2)_9CH_2O\overset{\overset{\displaystyle O}{\|}}{C}CH_3 \\ & C{=}C & \\ H & & H \end{array}$$

11-Tetradecenyl acetate

Scientists have found that males of the Iowa strain of this borer show maximum response to a mixture of 96% of *cis* isomer and 4% of the *trans* isomer of this compound, while males of the New York strain respond best to the opposite ratio. This again demonstrates the importance of molecular structure in biological systems as well as the specificity of the insect pheromone systems. It is hoped that insect sex pheromones can be used to combat insect infestations by serving as bait in traps for males or by being sprayed in relatively large quantities on fields to confuse the males so that they cannot find females with whom to mate.

Agricultural scientists are also studying chemicals that control growth in insects in the hope of discovering highly specific insecticides that prevent the development of particular insect species but are harmless to humans and useful insects.

24.4 Metallurgy and Iron and Steel Production

Purpose

- To describe processes for concentrating minerals in ores.
- To discuss the major processes in hydrometallurgy.
- To describe the blast furnace, open hearth, and basic oxygen processes for steelmaking.
- To discuss the thermodynamics of steelmaking.

Pyrite crystals.

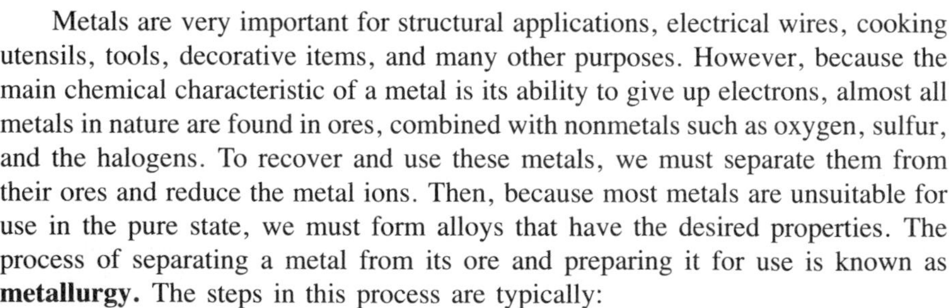

Metals are very important for structural applications, electrical wires, cooking utensils, tools, decorative items, and many other purposes. However, because the main chemical characteristic of a metal is its ability to give up electrons, almost all metals in nature are found in ores, combined with nonmetals such as oxygen, sulfur, and the halogens. To recover and use these metals, we must separate them from their ores and reduce the metal ions. Then, because most metals are unsuitable for use in the pure state, we must form alloys that have the desired properties. The process of separating a metal from its ore and preparing it for use is known as **metallurgy.** The steps in this process are typically:

1. mining
2. pretreatment of the ore
3. reduction to the free metal
4. purification of the metal (refining)
5. alloying

Common Minerals Found in Ores	
Anion	Examples
None (free metal)	Au, Ag, Pt, Pd, Rh, Ir, Ru, As, Sb, Bi
Oxide	Fe_2O_3 (hematite) Fe_3O_4 (magnetite) Al_2O_3 (bauxite) SnO_2 (cassiterite)
Sulfide	PbS (galena) ZnS (sphalerite) FeS_2 (pyrites) HgS (cinnabar) Cu_2S (chalcocite)
Chloride	NaCl (rock salt) KCl (sylvite) $KCl \cdot MgCl_2$ (carnalite)
Carbonate	$FeCO_3$ (siderite) $CaCO_3$ (limestone) $MgCO_3$ (magnesite) $MgCO_3 \cdot CaCO_3$ (dolomite)
Sulfate	$CaSO_4 \cdot 2H_2O$ (gypsum) $BaSO_4$ (barite)
Silicate	$Be_3Al_2Si_6O_{18}$ (beryl) $Al_2(Si_2O_8)(OH)_4$ (kaolinite) $LiAl(SiO_3)_2$ (spodumene)

Table 24.6

An ore can be viewed as a mixture containing **minerals** (relatively pure metal compounds) and **gangue** (sand, clay, and rock). Some typical minerals are listed in Table 24.6. Although silicate minerals are the most common in the earth's crust, they are typically very hard and difficult to process, making metal extraction relatively expensive. Therefore, other ores are used when available.

After mining, an ore must be treated to remove the gangue and to concentrate the mineral. The ore is first pulverized and then processed in a variety of devices, including cyclone separators (see Fig. 24.5), inclined vibrating tables, and flotation tanks.

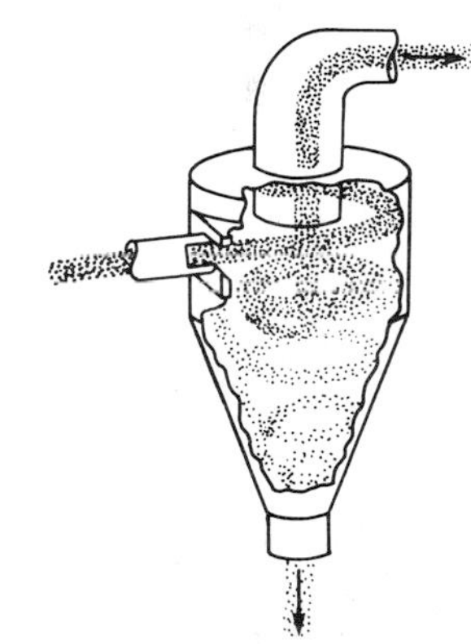

Figure 24.5

A schematic diagram of a cyclone separator. The ore is pulverized and blown into the separator. The more dense mineral particles are thrown toward the walls by centrifugal force and fall down the funnel. The lighter particles (gangue) tend to stay closer to the center and are drawn out through the top by the stream of air.

In the **flotation process,** the crushed ore is fed into a tank containing a water-oil-detergent mixture. Because of the difference in the surface characteristics of the mineral particles and the silicate rock particles, the oil wets the mineral particles. A stream of air blown through the mixture causes tiny bubbles to form on the oil-covered pieces, which then float to the surface where they can be skimmed off.

After the mineral has been concentrated, it is often chemically altered in preparation for the reduction step. For example, nonoxide minerals are often converted to oxides before reduction. Carbonates and hydroxides can be converted by simple heating:

$$CaCO_3(s) \xrightarrow{\text{Heat}} CaO(s) + CO_2(g)$$

$$Mg(OH)_2(s) \xrightarrow{\text{Heat}} MgO(s) + H_2O(g)$$

Sulfide minerals can be converted to oxides by heating in air at temperatures below their melting points, a process called **roasting:**

$$2ZnS(s) + 3O_2(g) \xrightarrow{\text{Heat}} 2ZnO(s) + 2SO_2(g)$$

As we have seen earlier, sulfur dioxide causes severe problems if released into the atmosphere, and modern roasting operations collect this gas and use it in the manufacture of sulfuric acid.

The method chosen to reduce the metal ion to the free metal, a process called **smelting,** depends on the affinity of the metal ion for electrons. Some metals are good enough oxidizing agents that the free metal is produced in the roasting process. For example, the roasting reaction for cinnabar is

$$HgS(s) + O_2(g) \xrightarrow{\text{Heat}} Hg(l) + SO_2(g)$$

where the Hg^{2+} is reduced by electrons donated by the S^{2-} ion, which is then further oxidized by O_2 to form SO_2.

The roasting of a more active metal produces the metal oxide, which must be reduced to obtain the free metal. The most common reducing agents are coke (impure carbon), carbon monoxide, and hydrogen. The following are some common examples of the reduction process:

$$Fe_2O_3(s) + 3CO(g) \xrightarrow{\text{Heat}} 2Fe(l) + 3CO_2(g)$$

$$WO_3(s) + 3H_2(g) \xrightarrow{\text{Heat}} W(l) + 3H_2O(g)$$

$$ZnO(s) + C(s) \xrightarrow{\text{Heat}} Zn(l) + CO(g)$$

The most active metals, such as aluminum and the alkali metals, must be reduced electrolytically, usually from molten salts (see Section 17.8).

The metal obtained in the reduction step is invariably impure and must be refined. The methods of refining include electrolytic refining (see Section 17.8), oxidation of impurities (as for iron, see below), and distillation of low-boiling metals such as mercury and zinc. One process used when highly pure metals are needed is **zone refining.** In this process a bar of the impure metal travels through a heater (see Fig. 24.6), which causes melting and recrystallizing of the metal as the bar cools. Purification of the metal occurs because as the crystal reforms, the metal ions are likely to fit much better in the crystal lattice than are the atoms of impurities. Thus the impurities tend to be excluded and carried to the end of the bar. Several repetitions of this process give a very pure metal bar.

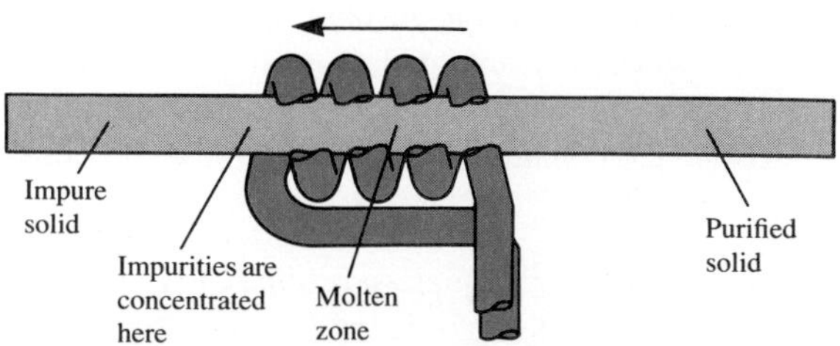

Figure 24.6

A schematic representation of zone refining.

Hydrometallurgy

The metallurgical processes we have considered so far are usually called **pyrometallurgy** (*pyro* means at high temperatures). These traditional methods require large quantities of energy and have two other serious problems: atmospheric pollution (mainly by sulfur dioxide) and relatively high costs that make treatment of low-grade ores economically unfeasible.

In the last hundred years, a different process, **hydrometallurgy** (*hydro* means water) has been employed to extract metals from ores by use of aqueous chemical solutions, a process called **leaching.** The first two uses of hydrometallurgy were for the extraction of gold from low-grade ores and for the production of aluminum oxide, or alumina, from bauxite, an aluminum-bearing ore.

Gold is sometimes found in ores in the elemental state, but it usually occurs in relatively small concentrations. A process called **cyanidation** treats the crushed ore with an aqueous cyanide solution in the presence of air to dissolve the gold by forming the complex ion $Au(CN)_2^-$:

$$4Au(s) + 8CN^-(aq) + O_2(g) + 2H_2O(l) \rightarrow 4Au(CN)_2^-(aq) + 4OH^-(aq)$$

Pure gold is then recovered by reacting the solution of $Au(CN)_2^-$ with zinc powder to reduce Au^+ to Au:

$$2Au(CN)_2^-(aq) + Zn(s) \rightarrow 2Au(s) + Zn(CN)_4^{2-}(aq)$$

The extraction of alumina from bauxite (the Bayer process) leaches the ore with sodium hydroxide at high temperatures and pressures to dissolve the amphoteric aluminum oxide:

$$Al_2O_3(s) + 2OH^-(aq) \rightarrow 2AlO_2^-(aq) + H_2O(l)$$

This process leaves behind solid impurities such as SiO_2, Fe_2O_3, and TiO_2, which are not appreciably soluble in basic solution. After the solid impurities are removed, the pH of the solution is lowered by addition of carbon dioxide. The pure aluminum oxide reforms and is then electrolyzed to produce aluminum metal (see Section 17.8).

As illustrated by these processes, hydrometallurgy involves two distinct steps: *selective leaching* of a given metal ion from the ore, and recovery of the metal ion from the solution by *selective precipitation* as an ionic compound.

The leaching agent can simply be water if the metal-containing compound is a water-soluble chloride or sulfate. However, most commonly, the metal is present in a water-insoluble substance that must somehow be dissolved. The leaching agents used in such cases are usually aqueous solutions containing acids, bases, oxidizing agents, salts, or some combination of these. Often the dissolving process involves the formation of complex ions. For example, when an ore containing water-insoluble lead sulfate is treated with an aqueous sodium chloride solution, the soluble complex ion $PbCl_4^{2-}$ is formed:

$$PbSO_4(s) + 4Na^+(aq) + 4Cl^-(aq) \rightarrow 4Na^+(aq) + PbCl_4^{2-}(aq) + SO_4^{2-}(aq)$$

Formation of a complex ion also occurs in the cyanidation process for the recovery of gold. However, since the gold is present in the ore as particles of metal, it must first be oxidized by oxygen to produce Au^+, which then reacts with CN^- to

Examples of Methods for Recovery of Metal Ions from Leaching Solutions		
Method		Examples
Precipitation of a salt		$Cu^{2+}(aq) + S^{2-}(aq) \rightarrow CuS(s)$
		$Cu^{+}(aq) + HCN(aq) \rightarrow CuCN(s) + H^{+}(aq)$
Reduction	Chemical	$Au^{+}(aq) + Fe^{2+}(aq) \rightarrow Au(s) + Fe^{3+}(aq)$
		$Cu^{2+}(aq) + Fe(s) \rightarrow Cu(s) + Fe^{2+}(aq)$
		$Ni^{2+}(aq) + H_2(g) \rightarrow Ni(s) + 2H^{+}(aq)$
	Electrolytic	$Cu^{2+}(aq) + 2e^{-} \rightarrow Cu(s)$
		$Al^{3+}(aq) + 3e^{-} \rightarrow Al(s)$
Reduction plus precipitation		$2Cu^{2+}(aq) + 2Cl^{-}(aq) + H_2SO_3(aq) + H_2O(l) \rightarrow 2CuCl(s) + 3H^{+}(aq) + HSO_4^{-}(aq)$

Table 24.7

form the soluble $Au(CN)_2^-$ species. Thus, in this case, the leaching process involves a combination of oxidation and complexation.

Sometimes just oxidation is used. For example, insoluble zinc sulfide can be converted to soluble zinc sulfate by pulverizing the ore and suspending it in water to form a slurry through which oxygen is bubbled:

$$ZnS(s) + 2O_2(aq) \rightarrow Zn^{2+}(aq) + SO_4^{2-}(aq)$$

One advantage of hydrometallurgy over the traditional processes is that sometimes the leaching agent can be pumped directly into the ore deposits in the earth. For example, aqueous sodium carbonate (Na_2CO_3) can be injected into uranium-bearing ores to form water-soluble complex carbonate ions.

Precipitation reactions are discussed in Section 15.7.

Recovering the metal ions from the leaching solution involves forming an insoluble solid containing the metal ion to be recovered. This step may involve addition of an anion to form an insoluble salt, reduction to the solid metal, or a combination of reduction and precipitation of a salt. Examples of these processes are shown in Table 24.7. Because of its suitability for treating low-grade ores economically and without significant pollution, hydrometallurgy is becoming more popular for recovering many important metals such as copper, nickel, zinc, and uranium.

The Metallurgy of Iron

Iron is present in the earth's crust in many types of minerals. *Iron pyrite* (FeS_2) is widely distributed but is not suitable for production of metallic iron and steel because it is almost impossible to remove the last traces of sulfur. The presence of sulfur makes the resulting steel too brittle to be useful. *Siderite* ($FeCO_3$) is a valuable iron mineral that can be converted to iron oxide by heating. The iron oxide minerals are *hematite* (Fe_2O_3), the most abundant, and *magnetite* (Fe_3O_4, really $FeO \cdot Fe_2O_3$). Taconite ores contain iron oxides mixed with silicates and are more difficult to process than the others. However, taconite ores are being increasingly used as the more desirable ores are consumed.

To concentrate the iron in iron ores, advantage is taken of the natural magnetism of Fe_3O_4 (hence its name, magnetite). The Fe_3O_4 particles can be separated from the gangue by magnets. The ores that are not magnetic are often converted to

Fe_3O_4; hematite is partially reduced to magnetite, while siderite is first converted to FeO thermally, then oxidized to Fe_2O_3, and then reduced to Fe_3O_4:

$$FeCO_3(s) \xrightarrow{\text{Heat}} FeO(s) + CO_2(g)$$
$$4FeO(s) + O_2(g) \longrightarrow 2Fe_2O_3(s)$$
$$3Fe_2O_3(s) + C(s) \longrightarrow 2Fe_3O_4(s) + CO(g)$$

Sometimes the nonmagnetic ores are concentrated by flotation processes.

The most commonly used reduction process for iron takes place in the **blast furnace** (Fig. 24.7). The raw materials required are concentrated iron ore, coke, and limestone (which serves as a *flux* to trap impurities). The furnace, which is approximately 25 feet in diameter, is charged from the top with a mixture of iron ore, coke, and limestone. A very strong blast (~350 miles per hour) of hot air is injected at the bottom where the oxygen reacts with the carbon in the coke to form carbon monoxide, the reducing agent for the iron. The temperature of the charge increases as it travels down the furnace, with reduction of the iron to iron metal occurring in steps.

$$3Fe_2O_3 + CO \rightarrow 2Fe_3O_4 + CO_2$$
$$Fe_3O_4 + CO \rightarrow 3FeO + CO_2$$
$$FeO + CO \rightarrow Fe + CO_2$$

Iron can reduce carbon dioxide,

$$Fe + CO_2 \rightarrow FeO + CO$$

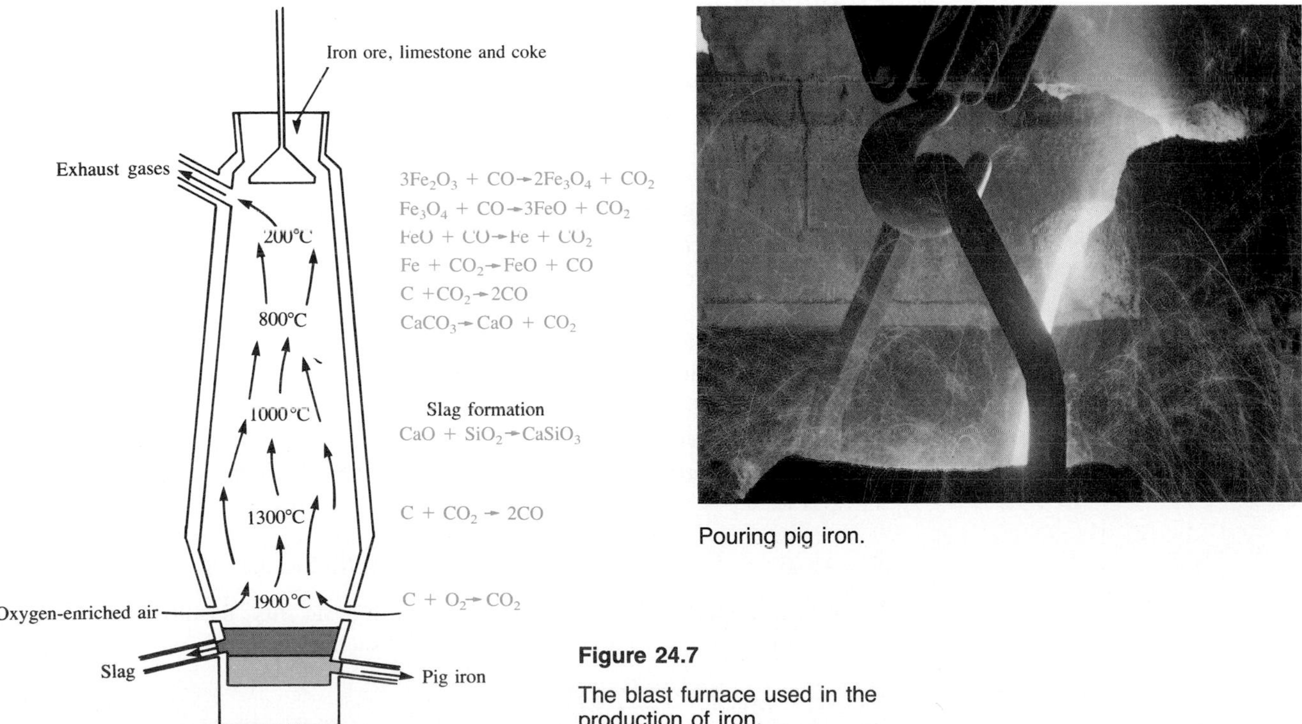

Pouring pig iron.

Figure 24.7

The blast furnace used in the production of iron.

so complete reduction of the iron occurs only if the carbon dioxide is destroyed by adding excess coke:

$$CO_2 + C \rightarrow 2CO$$

The limestone ($CaCO_3$) in the charge loses carbon dioxide, or *calcines,* in the hot furnace and combines with silica and other impurities to form **slag,** which is mostly molten calcium silicate, $CaSiO_3$,

$$CaO + SiO_2 \rightarrow CaSiO_3$$

and alumina (Al_2O_3). The slag floats on the molten iron and is skimmed off. The gas that escapes from the top of the furnace contains carbon monoxide, which is combined with air to form carbon dioxide. The energy released in this exothermic reaction is collected in a heat exchanger and used in heating the furnace.

The iron collected from the blast furnace, called **pig iron,** is quite impure. It contains ~90% iron, ~5% carbon, ~2% manganese, ~1% silicon, ~0.3% phosphorus, and ~0.04% sulfur (from impurities in the coke). The production of 1 ton of pig iron requires approximately 1.7 tons of iron ore, 0.5 ton of coke, 0.25 ton of limestone, and 2 tons of air.

Iron oxide can also be reduced in a **direct reduction furnace,** which operates at much lower temperatures (1300–2000°F) than a blast furnace and produces a solid "sponge iron" rather than molten iron. Because of the milder reaction conditions, the direct reduction furnace requires a higher grade of iron ore (with fewer impurities) than that used in a blast furnace. The iron from the direct reduction furnace is called DRI (directly reduced iron) and contains ~95% iron, with the balance mainly silica and alumina.

Production of Steel

Steel is an alloy and can be classified as **carbon steel,** which contains up to about 1.5% carbon, or **alloy steel,** which contains carbon plus other metals such as Cr, Co, Mn, and Mo. The wide range of mechanical properties associated with steel is determined by its chemical composition and by the heat treatment of the final product.

The production of iron from its ore is fundamentally a reduction process, but the conversion of iron to steel is basically an oxidation process in which unwanted impurities are eliminated. Oxidation is carried out in various ways, but the two most common are the open hearth process and the basic oxygen process.

In the oxidation reactions of steelmaking, the manganese, phosphorus, and silicon in the impure iron react with oxygen to form oxides, which in turn react with appropriate fluxes to form slag. Sulfur enters the slag primarily as sulfides, and excess carbon forms carbon monoxide or carbon dioxide. The flux chosen depends on the major impurities present. If manganese is the chief impurity, an acidic flux of silica is used:

$$MnO(s) + SiO_2(s) \xrightarrow{\text{Heat}} MnSiO_3(l)$$

If the main impurity is silicon or phosphorus, a basic flux, usually lime (CaO) or magnesia (MgO), is needed to give reactions such as

$$SiO_2(s) + MgO(s) \xrightarrow{\text{Heat}} MgSiO_3(l)$$

$$P_4O_{10}(s) + 6CaO(s) \xrightarrow{\text{Heat}} 2Ca_3(PO_4)_2(l)$$

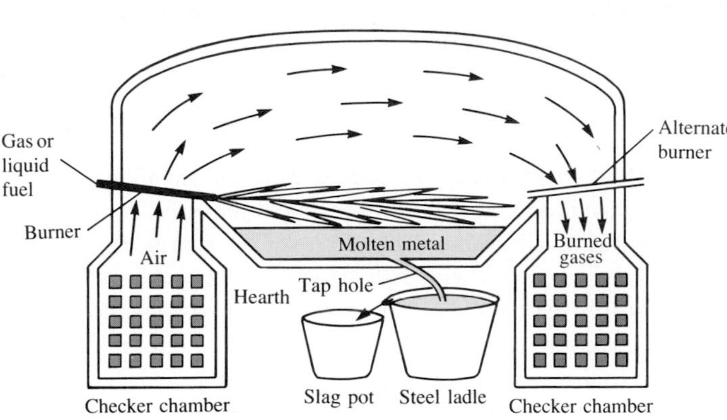

Figure 24.8

A schematic diagram of the open hearth process for steelmaking. The checker chambers contain bricks that absorb heat from gases passing over the molten charge. The flow of air and gases is reversed periodically.

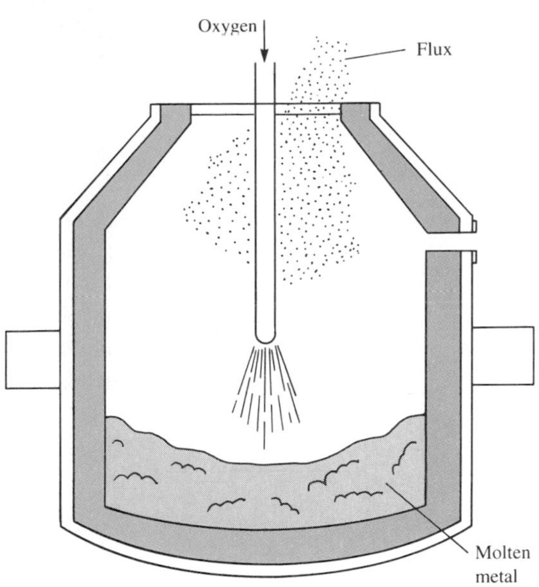

Figure 24.9

The basic oxygen process for steelmaking.

Whether an acidic or a basic slag will be needed is a factor that must be considered when a furnace is constructed so that its refractory linings will be compatible with the slag. Silica bricks would deteriorate quickly in the presence of basic slag, and magnesia or lime bricks would dissolve in acid slag.

The **open hearth process** (Fig. 24.8) uses a dishlike container that holds 100–200 tons of molten iron. An external heat source is required to keep the iron molten, and a concave roof over the container reflects heat back toward the iron surface. A blast of air or oxygen is passed over the surface of the iron to react with impurities. Silicon and manganese are oxidized first and enter the slag, followed by oxidation of carbon to carbon monoxide, which causes agitation and foaming of the molten bath. The exothermic oxidation of carbon raises the temperature of the bath, causing the limestone flux to calcine:

$$CaCO_3 \xrightarrow{\text{Heat}} CaO + CO_2$$

The resulting lime floats to the top of the molten mixture (an event called the *lime boil*), where it combines with phosphates, sulfates, and silicates. Next comes the refining process, which involves continued oxidation of carbon and other impurities. Because the melting point increases as the carbon content decreases, the bath temperatures must be increased during this phase of the operation. If the carbon content falls below that desired in the final product, coke or pig iron may be added.

The final composition of the steel is "fine-tuned" after the charge is poured. For example, aluminum is sometimes added at this stage to remove trace amounts of oxygen via the reaction

$$4Al + 3O_2 \rightarrow 2Al_2O_3$$

Alloying metals, such as vanadium, chromium, titanium, manganese, and nickel, are also added to give the steel the properties needed for a specific application.

The processing of a batch of steel by the open hearth process is quite slow, taking up to eight hours. The **basic oxygen process** is much faster. Molten pig iron and scrap iron are placed in a barrel-shaped container (Fig. 24.9) that can hold as much as 300 tons of material. A high-pressure blast of oxygen is directed at the surface of the molten iron, oxidizing impurities in a manner very similar to that used in the open hearth process. Fluxes are added after the oxygen blow begins. One advantage of this process is that the exothermic oxidation reactions proceed so rapidly that they produce enough heat to raise the temperature nearly to the boiling point of iron without an external heat source. Also, at these high temperatures only about an hour is needed to complete the oxidation processes.

The **electric arc method,** which was once used only for small batches of specialty steels, is being utilized more and more in the steel industry. In this process, an electric arc between carbon electrodes is used to melt the charge. This means that no impurities are added to the steel with fuel, since no fuel is needed. Also, higher temperatures are possible than in the open hearth or basic oxygen processes, and this leads to more effective removal of sulfur and phosphorus impurities. Oxygen is added in this process so that the oxide impurities in the steel can be controlled effectively.

Heat Treatment of Steel

Refer to Section 10.3 for a review of packing and crystal lattices.

One way of producing the desired physical properties in steel is by controlling the chemical composition (see Table 24.8). Another method for tailoring the properties of steel involves heat treatment. Pure iron exists in two different crystalline forms, depending on the temperature. At any temperature below 912°C, iron has a body-centered cubic structure and is called α-iron. Between 912°C and 1394°C, iron has a face-centered cubic structure called austentite, or γ-iron. At 1394°C, iron changes to δ-iron, a body-centered cubic structure identical to α-iron.

When iron is alloyed with carbon, which fits into holes among the iron atoms to form the interstitial alloy carbon steel, the situation becomes even more complex. For example, the temperature at which α-iron changes to austentite is lowered by

Percent Composition and Uses of Various Types of Steel

Type of steel	% Carbon	% Manganese	% Phosphorus	% Sulfur	% Silicon	% Nickel	% Chromium	% Other	Uses
Plain carbon	≤1.35	≤1.65	≤0.04	≤0.05	≤0.60	—	—	—	Sheet steel, tools
High-strength (low-alloy)	≤0.25	≤1.65	≤0.04	≤0.05	0.15–0.9	0.4–1	0.3–1.3	Cu (0.2–0.6) Sb (0.01–0.08) V (0.01–0.08)	Transportation equipment, structural beams
Alloy	≤1.00	≤3.50	≤0.04	≤0.05	0.15–2.0	0.25–10.0	0.25–4.0	Mo (0.08–4.0) V (0–0.2) W (0–18) Co (0–5)	Automobile and aircraft engine parts
Stainless	0.03–1.2	1.0–10	0.04–0.06	≤0.03	1–3	1–22	4.0–27	—	Engine parts, steam turbine parts, kitchen utensils
Silicon	—	—	—	—	0.5–5.0	—	—	—	Electric motors and transformers

Table 24.8

Chemical Impact

Steel Bicycle Frames

One interesting use of steel is for the construction of bicycle frames (see Fig. 24.10). A bicycle frame is built from several steel tubes, whose composition varies greatly depending on the intended market for the bicycle. Inexpensive bicycles are made from plain low-carbon steel, which starts as a flat strip and through a series of rolling operations is formed into a tube and welded (see Fig. 24.11). Because the steel is not very strong, this seamed tubing must have thick walls, and the resulting frame is heavy and rather unresponsive.

Expensive, lightweight bicycles are constructed from seamless tubing, which starts out as a solid bar (called a *billet*) of alloy steel that is pierced and rolled while hot (see Fig. 24.12) to approximately the desired dimensions and then is cold-drawn to its final dimensions. This tubing typically has very thin walls in the center of the tube and thicker walls toward the ends for strength, as shown in Fig. 24.13. A tube of this type is said to be double-butted. The wall thicknesses are typically 0.5 millimeter at the thinnest points and 0.8 millimeter at the thickest points. Frames constructed from this light, thin-walled tubing are strong, light, and responsive.

Seamless tubing is made from alloy steel with 0.4% or less carbon content. Although the strength increases with carbon content, a steel with a carbon content greater than 0.4% is not ductile enough to be cold-drawn. The other elements typically found in the alloy steels used in high-quality bicycle frames are manganese, chromium, molybdenum, niobium, nickel, vanadium, and silicon. (The compositions of two of the best-known brands of bicycle tubing are shown in Table 10.3.)

The bicycle frame is a testament to the strength and utility of steel. Although other materials such as titanium, aluminum alloys and composites are being used with more frequency, steel remains the choice of most frame manufacturers because of its strength, economy, and convenience.

Figure 24.10

A bicycle with a high-quality frame.

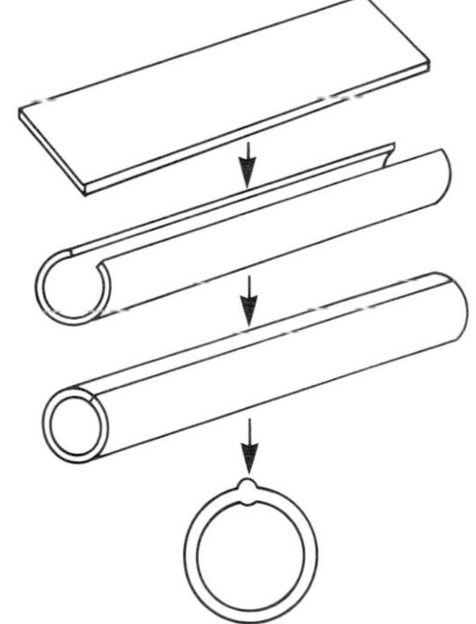

Figure 24.11

Plain low-carbon steel, which starts as a sheet and is rolled and welded, is used for low-cost bicycle frames.

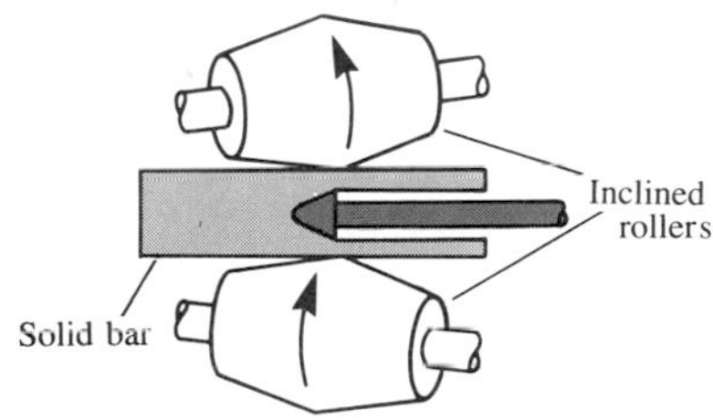

Figure 24.12

Manufacture of a tube for an expensive lightweight bicycle frame involves piercing a bar of solid steel to form a thin-walled tube.

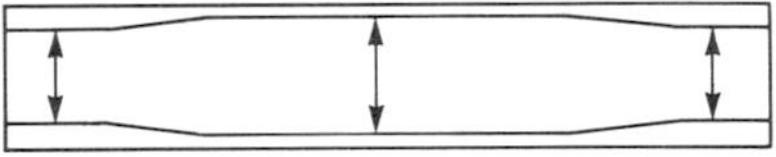

Figure 24.13

The tubing in a racing bicycle needs to be strong at positions of stress but still be as light as possible. This is accomplished by double-butting, a process by which the tube ends are made thicker than the center.

about 200°C. Also, at high temperatures iron and carbon react by an endothermic reversible reaction to form an iron carbide called *cementite:*

$$3Fe + C + \underset{(Heat)}{energy} \rightleftharpoons \underset{Cementite}{Fe_3C}$$

By Le Châtelier's principle, we can predict that cementite will become more stable relative to iron and carbon as the temperature is increased. This is the observed result.

Thus steel is really a mixture of iron metal in one of its crystal forms, carbon, and cementite. The proportions of these components are very important in determining the physical properties of steel.

When steel is heated to temperatures in the region of 1000°C, much of the carbon is converted to cementite. If the steel is then allowed to cool slowly, the equilibrium shown above shifts to the left, and small crystals of carbon precipitate, giving a steel that is relatively ductile. If the cooling is very rapid, the equilibrium does not have time to adjust. The cementite is trapped, and the steel has a high cementite content, making it quite brittle. The proportions of carbon crystals and cementite can be ''fine-tuned'' to give the desired properties by heating to intermediate temperatures followed by rapid cooling, a process called **tempering.** The rate of heating and cooling determines not only the amounts of cementite present but also the size of its crystals and the crystal form of the iron.

24.5 The Wine Industry

Purpose

- To describe the basic steps of winemaking and the chemical analysis of wine.

Although winemaking has a long history and, until recently, has involved more art than science, it has become a large, highly technical industry in the United States, particularly in California. The winemaking process is especially interesting because the character of the product is determined by a raw material (grapes) that varies greatly, depending on soil, weather conditions, and the extent of maturation when harvested. Also, wine is an unusual product because its quality is evaluated so subjectively. From a chemist's viewpoint, there are many interesting questions to consider. What chemical constituents are most responsible for the taste and aroma of wine? Can efficient modern processes replace the relatively expensive traditional methods without adversely affecting the product? How do fertilizers and pesticides affect the chemical constitution of the grapes?

The production of wine is based on the **fermentation** of grape juice. This process, which was first described chemically by Gay-Lussac in 1810, involves the transformation of sugar to ethanol. For example, the fermentation of glucose ($C_6H_{12}O_6$) can be written as

$$C_6H_{12}O_6 \rightarrow 2C_2H_5OH + 2CO_2$$

The science of winemaking, or **enology,** began about 1860, when Louis Pasteur recognized the importance of yeasts in the fermentation process and showed that substances such as glycerol, acetic acid, and acetaldehyde were also products of fermentation, in addition to ethanol and carbon dioxide. It is now known that there are at least 12 steps in the overall fermentation process, each step being catalyzed by one or more enzymes. Twenty-two enzymes, six coenzymes, and several metal ions have been identified as necessary ingredients in the reaction sequence.

Although the fermentation process is at the heart of winemaking, there is much more to wine production than that, as is shown in Fig. 24.14. After the grapes are harvested, they are crushed to form a *must*. For white wines, the grape skins are removed after crushing; for red wines, the skins are left in.

At this point, the must is analyzed for total acid and sugar content and adjustments are made as required. Sulfur dioxide is then added to kill any bacteria and "wild" yeasts prior to the addition of a pure yeast culture. For a white wine, fermentation proceeds until the wine contains ~12% alcohol by volume. The fermentation of a red wine is interrupted when the appropriate amount of colored materials have been leached from the skins, which are then filtered off (called *pressing*). The fermentation is then allowed to run to completion.

After fermentation, the yeast is removed, and the wine is stored in containers for aging. During the aging process, the sulfur dioxide and acid levels are monitored to ensure that oxidation or bacterial spoilage does not occur. Before bottling, the wine is filtered, and a substance such as gelatin is added to cause aggregation and settling of colloidal suspensions of gums, pectins, and tannins that cause the wine to look hazy. This process is called *fining* the wine.

Grape variety
↙↗ ↘↖
Climate ↔ Soil
↓
Harvested grapes
↓ Crushing operation
Grape must
↓ Fermentation (yeast)
Young wine
↓ Aging
Wine

Figure 24.14

A schematic diagram of the wine production process.

We will now examine some of the chemistry involved in winemaking, considering first the effects of the type of grape and the growing conditions. Grapes are a complex mixture of water, simple sugars, starches, organic acids, amino acids, aromatic alcohols (phenolic compounds), metal ions, vitamins, enzymes, volatile esters, and many other trace substances. The relative amounts of these materials depend on the grape variety, among other things. For example, native American grapes, such as the Concord variety, have low sugar and high acid levels compared to European varieties. Growing conditions are also very important in determining the chemical composition of the grapes. Studies have shown, for example, that the use of nitrogen-containing fertilizers significantly increases the amino acid content of the grapes.

The two major sugars in grapes are *glucose* and *fructose,* which occur in about equal amounts. A significant amount of *sucrose* is also present in many types of grapes. Sugar content is very important because it determines the most general characteristic of a wine—whether it is sweet or dry (lack of sweetness). As the fermentation of sugar proceeds, the alcohol content increases to a point (~12% by volume) where the yeast can no longer survive. If the sugar has been virtually consumed when the yeast dies and fermentation ceases, the resulting wine will be very dry. On the other hand, if sugar remains when the alcohol content reaches 12%, the wine will have a sweetness that depends on the amount of sugar left.

The most important organic acids present in wine are *tartaric acid* and *malic acid*. Although organic acids are not present in high percentages in grapes, they are very important in determining the pH of the wine. Besides affecting the taste, the pH of wine has important implications for its clarity and stability. For example, a pH between 3 and 4 allows yeast to grow, but prevents the growth of bacteria that cause spoilage.

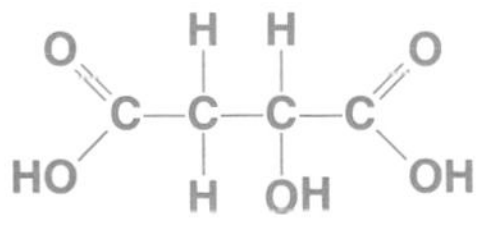

$HOOC{-}CH_2{-}CH(OH){-}COOH$

Malic acid

$HOOC{-}CH(OH){-}CH(OH){-}COOH$

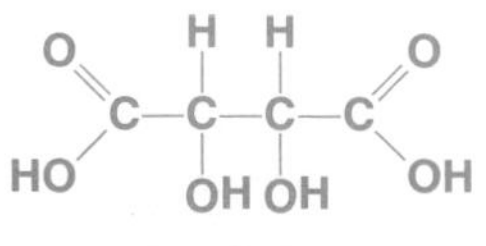

Tartaric acid

The color of the grape skins is due primarily to a group of compounds called *anthocyanins,* which have the following general structure where R represents —OH or —OCH_3 and R′ represents —OH, —OCH_3, or —H in some cases.

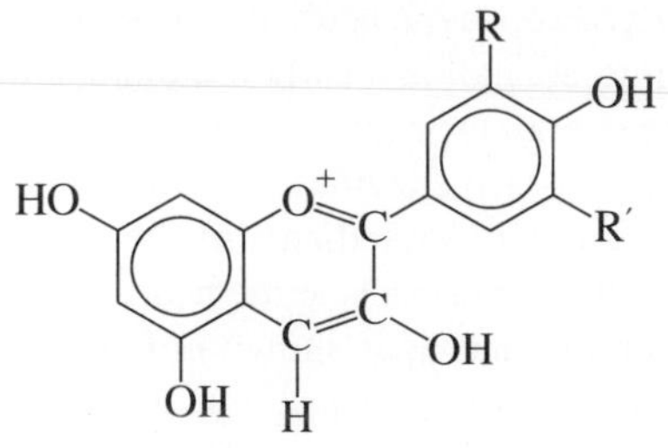

These cations are especially interesting because they behave much like the pH indicators discussed in Chapter 15. They react with water as follows:

$$A^+ + H_2O \rightleftharpoons AOH + H^+$$

where the A^+ form is red and the AOH form is colorless. Thus, as the pH of the wine decreases ($[H^+]$ increases), the equilibrium shifts to the left and the wine is more highly colored.

Another interesting phenomenon is that a red wine tends to lose its color when treated with sulfur dioxide because SO_2 reacts with water to form HSO_3^-:

$$SO_2(g) + H_2O(l) \rightleftharpoons H^+(aq) + HSO_3^-(aq)$$

The HSO_3^- reacts with the anthocyanins to form colorless compounds.

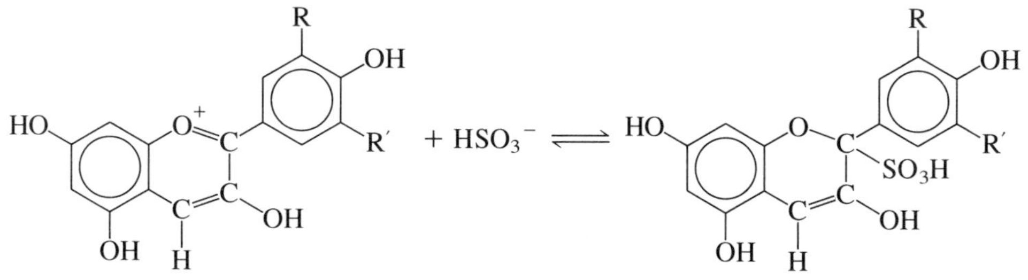

As the sulfur dioxide is gradually lost from the wine, the color returns because this equilibrium shifts to the left, reforming the red anthocyanin.

Sample Exercise 24.1

If the pK_a of a certain anthocyanin A^+ is 3.0, at what pH of the wine will $[A^+]$ and [AOH] be equal?

Solution

The relevant reaction is

$$A^+ + H_2O \rightleftharpoons AOH + H^+$$

where $pK_a = 3.0$ and thus $K_a = 1 \times 10^{-3}$. The equilibrium expression is

$$K_a = \frac{[AOH][H^+]}{[A^+]} = 1 \times 10^{-3} \quad \text{and} \quad \frac{[AOH]}{[A^+]} = \frac{1 \times 10^{-3}}{[H^+]}$$

For $[AOH]/[A^+] = 1$, $[H^+] = 1 \times 10^{-3}$. Thus at pH 3, the concentrations of A^+ and AOH will be equal.

Sample Exercise 24.2

Will the decolorization of red wine by the addition of sulfur dioxide be most noticeable at a high or a low pH?

Solution

At a low pH (high $[H^+]$), the equilibrium

$$SO_2(g) + H_2O(l) \rightleftharpoons H^+(aq) + HSO_3^-(aq)$$

will be shifted to the left, allowing only relatively small amounts of HSO_3^- to be present in solution. Thus the decolorization equilibrium will also lie to the left. This means that decolorization by sulfur dioxide will be less noticeable at a low pH.

As we mentioned before, the chemistry of fermentation involves at least 12 enzyme-catalyzed steps. The process is carried out between 65°F and 75°F. Although higher temperatures produce more rapid fermentation, the quality of the wine suffers because of the formation of unwanted by-products. Sulfur dioxide is added to the fermentation mixture to inhibit bacteria and to help prevent the formation of acetic acid.

From an esthetic point of view, the most important part of the winemaking process involves the refining that takes place after fermentation. High-molecular-weight substances such as proteins can cause cloudiness; salts of tartaric acid can precipitate, producing sediment; and bacteria can cause "browning" of the color and unpleasant tastes and odors. To avoid these problems, various steps are taken to refine and stabilize the wine after fermentation is complete. First, the wine is separated from the fermentation solids (the *lees*) as soon as possible to avoid contamination of the wine with hydrogen sulfide and other sulfide-containing compounds from the decomposition of dead yeast cells. The filtered wine is then stored to allow suspended particles to settle, but this is insufficient to produce the crystal clear wine desired by consumers. Fining agents are added to break up colloidal suspensions

A California vineyard.

and to absorb contaminants. The most commonly used fining agent is *bentonite,* a complex aluminum silicate clay that swells greatly in solution. The surface area of the bentonite platelets is about 750 square meters per gram, and these platelets are highly efficient in absorbing proteins from the wine. Activated carbon also has a very high surface area and is often used as an absorbent to clarify wine. In addition, enzymes are used to break down colloids of pectins, which are polysaccharides found in various fruits. All of these operations are carried out in the absence of oxygen, since oxygen can react with many of the trace components to produce aldehydes, acids, and other unwanted products that cause deterioration of color and taste.

Louis Pasteur in his laboratory, 1889.

During fermentation most wine becomes supersaturated with potassium hydrogen tartrate ($KC_4H_5O_4$), which will eventually precipitate out after the wine is bottled. Although this solid is not harmful, it is esthetically unacceptable to the consumer. This problem can be solved during winemaking by cooling the wine to temperatures just above its freezing point ($\sim$20°F) and storing at this temperature for several weeks. Because the solubility of $KC_4H_5O_4$ decreases with temperature, it precipitates out and can be removed by filtering at low temperature.

The area where modern chemistry is most helpful in ensuring the quality of wine is the analysis of the components at various stages of the winemaking process. Chemical analysis of wine was pioneered by Pasteur, who designed several simple tests that allowed him to elucidate the fermentation process and to understand wine spoilage. The most common analysis performed on wine is the determination of its alcohol content both for quality control and tax purposes. The most accurate analytical method for determination of ethanol is titration with a potassium dichromate solution (see Section 4.11). Other methods, including density determinations and chromatography, are also used, depending on speed and accuracy requirements.

The accurate analysis of sulfur dioxide content is important because control of its concentration during winemaking is essential and for reasons of consumer health. The legal limit for SO_2 in wine is approximately 300 milligrams per liter. Analysis for SO_2 usually involves distillation to expel the gas from the wine, but may also involve titration with an iodate solution.

Various techniques are used to analyze for trace components such as fluoride, lead, arsenic, and boron. The most common method is atomic absorption (AA) spectroscopy, a technique that involves exciting the electrons of the element in question and measuring the emitted radiation as the atom relaxes to its ground state.

The U.S. limit for carbon dioxide content in nonsparkling wines is 2.8 grams per liter. Wines with CO_2 content above this limit are subject to much higher taxes. Analysis for CO_2 is carried out by using an enzyme (carbonic anhydrase) to aid in the conversion of CO_2 to HCO_3^-, which is then titrated with a standard acid solution.

Chromatography was introduced in Section 1.9.

The total acidity of wine is determined by simple titration with a standard solution of sodium hydroxide. The determination of individual acids is usually carried out using chromatography.

Accurate analysis for the components in wine presents a great challenge to the analytical chemist because of the extraordinary complexity of wine. However steady progress is being made in the development of accurate and automated tests that will allow increasingly effective quality control. Winemaking involves a varied and complicated chemistry, and the wine industry is rapidly expanding its scientific approach and becoming a chemical industry.

FOR REVIEW

Summary

Most of the differences between industrial processes and laboratory experiments arise from the large difference in scale. Because of the large quantities of chemicals involved and the high temperatures and pressures usually used, economy and safety are of paramount importance in industry. Also, continuous flow rather than batch processing is generally preferred in industry, and by-products must be minimized unless there is a market for them.

Chemicals made from petroleum are called petrochemicals and are produced as by-products of refinery operations or from steam cracking of petroleum feedstocks. Petroleum fractions are separated by distillation and then undergo cracking, a process that breaks large molecules into smaller ones. Thermal cracking (pyrolysis) is done at high temperatures; catalytic cracking uses a catalyst at lower temperatures; hydrocracking employs a hydrogen atmosphere. In catalytic reforming, alkanes and cycloalkanes are converted to aromatic compounds.

The manufacture of polymers is a major chemical industry. Polymer strength can often be increased by drawing, as in the case of nylon, to line up the polymer chains and thus increase the hydrogen bonding between them. Crosslinking, where covalent bonds exist between adjacent chains, also increases the polymer strength. For example, Bakelite is heavily crosslinked, as is vulcanized rubber. Thermoplastic polymers can be remelted after the initial molding has taken place, but thermoset polymers cannot.

A major section of the polymer industry produces macromolecules from ethylene. There are two forms of polyethylene, low-density (LDPE) and high-density (HDPE), with different properties arising mainly from the degree of chain branching. Polyethylene properties can also be modified by replacing one or more hydrogen atoms on the ethylene monomer with other groups.

Copolymerization involves forming a polymer chain using two or more different monomers. Polymeric properties are often modified by additives that can provide temperature stability, elasticity (plasticity), protection against ultraviolet radiation, and flame retardancy.

A polymer can be made flame retardant by solid-phase or vapor-phase inhibition. If a polymer is extensively crosslinked at the surface, a carbonaceous char forms that protects the underlying polymer. In vapor-phase inhibition, free-radical inhibitors built into the polymer migrate into the flame to cause free-radical termination.

Pesticides include herbicides that kill weeds and insecticides that kill insects. Some important herbicides are 2,4-D and 2,4,5-T; DDT, parathion, and Sevin are insecticides.

Pheromones (chemicals produced by an insect, which, when released, evoke a response in another insect of the same species) can also be used to control insect populations.

The process of separating a metal from its ore and preparing it for use is known as metallurgy. An ore is a mixture of minerals (relatively pure metal compounds) and gangue (sand, clay, and rock). Minerals can be concentrated by various methods before being processed. Minerals are often converted to the metal oxides (roasted) before the metal ion is reduced in a process called smelting.

Hydrometallurgy extracts metals from ores using aqueous chemical solutions in a process called leaching. An example of this is cyanidation, in which gold is leached with an aqueous cyanide solution.

Iron is often concentrated from iron ore by magnetic methods. The most commonly used reduction process for iron takes place in the blast furnace. The raw materials required are concentrated iron ore, coke, and limestone (which serves as a flux to trap impurities). The iron oxide (Fe_2O_3) is reduced by carbon monoxide to iron metal. The impure product, pig iron, is about 90% iron. Another process using the direct reduction furnace, which yields sponge iron, is coming into increased use.

Steel is a carbon-containing alloy of iron that is manufactured by oxidizing the impurities in pig iron or sponge iron in either the open hearth process or the basic oxygen process. Other metals are often added to form alloy steels.

Winemaking (enology) is a growing chemical industry in the United States. Chemical technology is being increasingly applied to improve the quality and efficiency of the production of wine.

Key Terms

Section 24.1
petrochemicals
cracking
polymerization
feedstock
steam cracking

Section 24.2
thermoset polymer
thermoplastic polymer
crosslinking
vulcanization
polyethylene
isotactic chain
syndiotactic chain
atactic chain
polystyrene
polyvinyl chloride (PVC)
elastomers
inhibition

Section 24.3
herbicide
DDT
pheromone

Section 24.4
metallurgy
minerals
gangue
flotation process
roasting
smelting
zone refining
pyrometallurgy
hydrometallurgy
leaching
cyanidation
blast furnace
slag
pig iron
direct reduction furnace
carbon steel
alloy steel
open hearth process
basic oxygen process
electric arc method
tempering

Section 24.5
fermentation
enology

Exercises

A blue exercise number indicates that the answer to that exercise appears at the back of this book and a solution appears in the Solutions Guide.

Commercial Chemistry

1. List some differences between chemistry on an industrial scale and chemistry on the laboratory scale.

2. Chlorinated hydrocarbons (DDT, PCBs, dioxins, etc.) are widely used and have recently been the cause of much environmental concern. From the point of view of the chemical industry, why do you think so many chlorinated materials were produced in the first place? (*Hint:* Consider the major source for the production of Cl_2.)

3. It is very common to run industrial processes at high temperatures and pressures. Why? What trade-offs are involved in running reactions at elevated temperatures and pressures?

4. Catalytic reactions tend to be more common in industrial plants than in academic laboratories. Account for this.

5. Anthraquinone is an important intermediate in the production of dyes. Traditionally, it was prepared by the reaction of phthalic anhydride with benzene:

(i) Phthalic anhydride + benzene $\xrightarrow{AlCl_3}$ [CO_2H] $\xrightarrow[180°C]{H_2SO_4}$ Anthraquinone (81–90% yield)

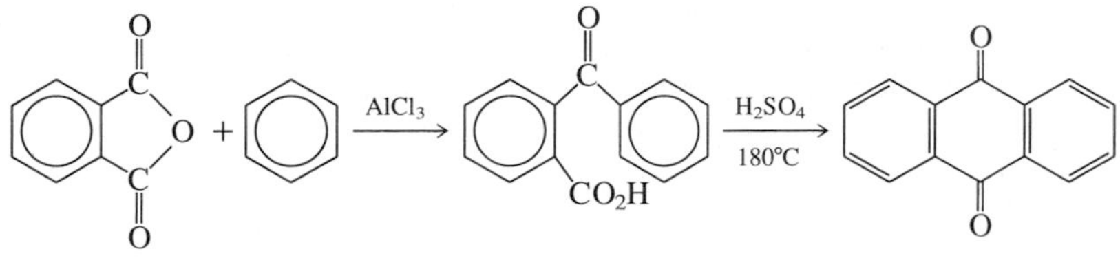

In 1974, the company Badische Anilin Soda Fabrik developed an alternative route:

(ii) 2 Styrene ($CH{=}CH_2$) $\xrightarrow{H^+}$ [CH_3] (83% yield) $\xrightarrow[\text{Vapor phase}]{O_3/V_2O_5}$ Anthraquinone (52% yield)

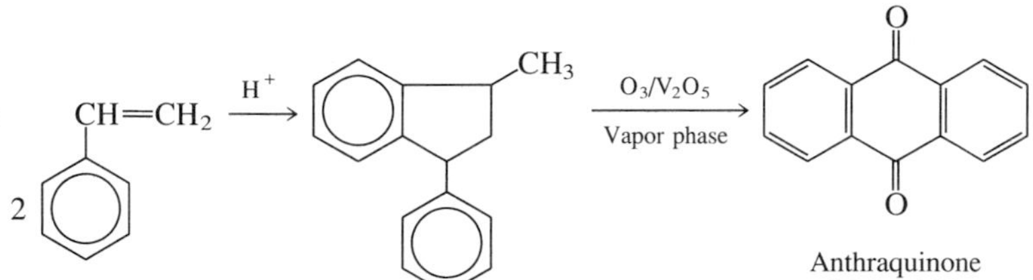

Process (iii) was announced in 1977 and process (iv) was introduced in 1978 by Bayer-Ciba-Geigy:

(iii) $2C_6H_6$ Benzene or Benzophenone + CO $\xrightarrow[\text{FeCl}_3,\ 4\text{ hr}]{200°C,\ 700\text{ psi}}$ (80% yield)

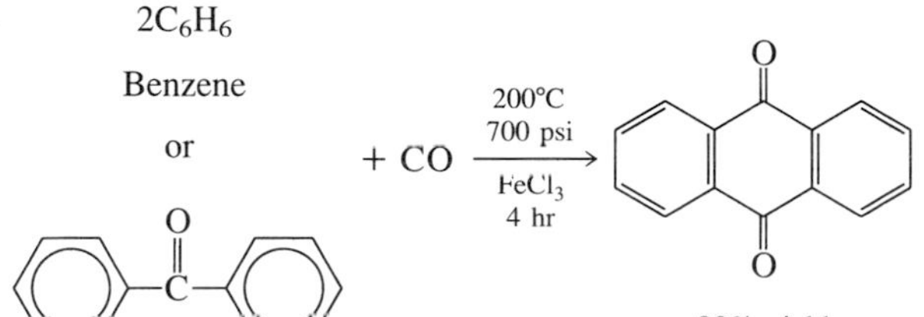

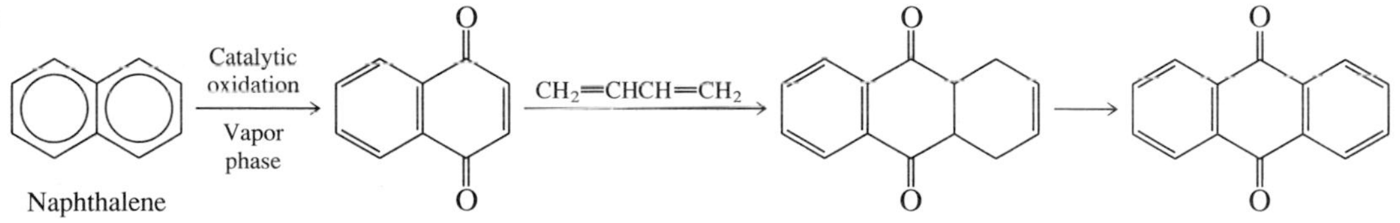

(iv) Naphthalene $\xrightarrow[\text{Vapor phase}]{\text{Catalytic oxidation}}$ $\xrightarrow{CH_2{=}CHCH{=}CH_2}$ $\longrightarrow$ (90% yield)

a. What is the overall yield of the process outlined in (ii)?

b. What is the theoretical yield of anthraquinone (mol. wt. 208) from 1000. kg of styrene? What will be the actual yield?

c. Given the prices in the table at right, calculate the cost of raw materials per pound of anthraquinone produced by both processes (i) and (ii).

d. Would either process (i) or (ii) have an advantage over the other from the standpoint of waste disposal?

Compound	Mol. wt.	Cost/ pound	Pounds used
Phthalic anhydride	148	$0.32	148
Anhydrous $AlCl_3$	133	$0.57	293
Styrene	104	$0.42	208
Benzene	78	$0.13	78

e. Given the prices in the table at right, calculate the raw material cost per pound of anthraquinone produced by processes (iii) and (iv).
f. Are there any economic advantages to process (iii) or (iv)?

	Mol. wt.	Cost/pound
Benzene	78	$0.13
Benzophenone	182	$2.40
Naphthalene	128	$0.22
1,3-Butadiene	54	$0.29

6. Hexamethylenediamine (1,6-diaminohexane) is one of the raw materials used in the production of nylon. Several processes are available for the production of hexamethylenediamine. Two of them are as follows:

(i) Furfural $\xrightarrow[\text{H}_2\text{O, 400°C}]{\text{Zn/Cr/Mo}}$ furan $\xrightarrow[\text{Raney Ni, 100–150 atm}]{\text{H}_2\text{, 120°C}}$ tetrahydrofuran

$$\text{tetrahydrofuran} \xrightarrow{\text{HCl}} Cl{-}(CH_2)_4{-}Cl \xrightarrow{\text{NaCN}} NC{-}(CH_2)_4{-}CN \xrightarrow[\text{Catalyst}]{H_2} H_2N{-}(CH_2)_6{-}NH_2$$

(ii) $$CH_2{=}CH{-}CH{=}CH_2 + 2HCN \xrightarrow[\text{Various catalysts}]{\text{Several steps}} NC{-}(CH_2)_4{-}CN \xrightarrow[\text{5000 psi, 25–250°C}]{H_2\text{, Co catalyst}} H_2N{-}(CH_2)_6{-}NH_2$$

The following are the relevant prices:

	Cost/pound
Furfural	$0.75
Butadiene	$0.29

a. At the quoted prices, assuming all things such as other costs and yields are comparable, which process has the economic advantage?
b. Furfural is derived from corncobs. Butadiene is derived from petroleum. Assuming the raw petroleum represents most of the cost of butadiene, what percentage increase in crude oil prices will make the two processes equivalent from the cost standpoint?

Petroleum Chemistry

7. Distinguish among the following: steam cracking, catalytic cracking, thermal cracking, and hydrocracking. What is the end result of all these cracking processes?

8. Define and give an example of each of the following:
a. catalytic reforming
b. alkylation
c. polymerization
d. isomerization

9. Which of the processes in Exercises 7 and 8 lead to longer hydrocarbon chains? Shorter hydrocarbon chains?

10. Ethylene oxide is an important intermediate in the chemical industry. Its major use is in the production of ethylene glycol for antifreeze.

$$\underset{\text{O}}{CH_2{-}CH_2} \xrightarrow[H_2SO_4,\ 50\text{–}70°C]{H_2O} HOCH_2CH_2OH$$

This reaction is similar to the addition reactions of alkenes, for example:

$$CH_2{=}CH_2 \xrightarrow[H_2SO_4]{H_2O} CH_3CH_2OH$$

Use bond energies (Table 8.4) to estimate ΔH for the above two reactions.

11. The oxo reaction (reaction with carbon monoxide) is useful for increasing chain length and producing alcohols and aldehydes. Some examples of oxo reactions are given below. Why are two products obtained in the first reaction as compared to only one product in the other two?

$$CH_3CH{=}CH_2 + CO + H_2 \xrightarrow{[Co(CO)_4]_2}$$

$$CH_3CH_2CH_2\overset{O}{\overset{\|}{C}}H + (CH_3)_2CH{-}\overset{O}{\overset{\|}{C}}H$$

$$CH_2{=}CH_2 + CO + 2H_2 \xrightarrow{Fe(CO)_4} CH_3CH_2CH_2OH$$

$$CH_2{=}CH_2 + CO + H_2 \xrightarrow{[Co(CO)_4]_2} CH_3CH_2\overset{\overset{\displaystyle O}{\|}}{C}H$$

Polymer Chemistry

12. Define and give an example of each of the following:
 a. free-radical polymerization
 b. addition polymerization
 c. copolymer
 d. condensation polymerization

13. Distinguish between a thermoset polymer and a thermoplastic polymer.

14. Kel-F is a polymer with the structure

$$\left(\begin{array}{cccccc} F & F & F & F & F & F \\ | & | & | & | & | & | \\ -C- & C- & C- & C- & C- & C- \\ | & | & | & | & | & | \\ Cl & F & Cl & F & Cl & F \end{array}\right)_n$$

 a. What is the monomer for Kel-F?
 b. Would you expect Kel-F to be more or less reactive than Teflon?

15. Explain how plasticizers and crosslinking agents are used to change the physical properties of polymers.

16. Isotactic polypropylene is harder than atactic polypropylene. Explain.

17. Explain why each of the following changes in a polymer will lead to decreased flammability.
 a. replacing hydrogen atoms with halogen atoms
 b. replacing carbon atoms singly bonded to each other in the polymer chain with aromatic rings
 c. highly crosslinking a polymer

18. Which of the following polymers would be stronger or more rigid? Explain your choices.
 a. the copolymer of ethylene glycol and terephthalic acid or the copolymer of 1,2-diaminoethane and terephthalic acid
 b. the polymer of $HO{-}(CH_2)_6{-}CO_2H$ or that of

$$HO-C_6H_4-CO_2H$$

 c. polyacetylene or polyethylene

19. How does crosslinking affect the elasticity of an elastomer?

20. Draw the structures of the polymers formed from the monomers listed in Table 24.5.

21. Stretch a rubber band while holding it gently to your lips. Then slowly let it go back to the unstretched state while still in contact with your lips.
 a. What happens to the temperature of the rubber band on stretching?
 b. Is the stretching an exothermic or endothermic process?
 c. Explain the above result in terms of interparticle forces.
 d. What is the sign of ΔS and ΔG for stretching the rubber band?
 e. Give the molecular explanation for the sign of ΔS for stretching.

22. All of the following observations about the aging of an automobile are related. Explain the relationship among these three observations.
 i. On hot days a waxy coating may appear on the inside of the glass of a car.
 ii. The "new-car" smell gradually goes away.
 iii. Vinyl seat covers become brittle and crack.

23. Why can polypropylene be machined into parts while polyethylene cannot? Why is this property dependent on the type of polypropylene polymer being considered?

24. When manufacturers list the properties of polymers, an average molecular weight is given. For example, Carbowax 20 M is a polymer of ethylene glycol, $HOCH_2CH_2OH$, with an average molecular weight of 20,000.
 a. Draw a structure of Carbowax 20 M.
 b. On the average, how many monomer units are in a molecule of Carbowax 20 M?
 c. Does this mean that every molecule in a sample of Carbowax 20 M is the same length? Why or why not?

25. When acrylic polymers are burned, toxic fumes are produced. For example, in many airplane fires, more passenger deaths have been caused by breathing toxic fumes than by the fire itself. Using polyacrylonitrile as an example, what would you expect to be one of the most toxic, gaseous combustion products created in the reaction?

26. How do the physical properties of polymers depend on chain length and extent of chain branching?

27. Speculate as to why nylon-46 is stronger than nylon-6. (*Hint:* Consider the strengths of interchain force.)

Pesticides

28. Sarin, tabun, and VX are nerve agents used in chemical warfare. Look up the structures of these substances in *The Merck Index* and compare them to the structures of malathion and parathion.

Toxicities are listed as LD-50 doses. The LD-50 dose is defined as the amount of material that will cause the deaths of 50% of the subjects who are given that dose. Compare the toxicities of the nerve agents to the toxicities of malathion and parathion.

29. Chlorinated phenols are the precursors to the herbicides 2,4-D and 2,4,5-T

Phenol + Cl_2 → 2,4-Dichlorophenol $\xrightarrow[NaOH]{ClCH_2CO_2H}$ (2,4-Dichlorophenoxyacetic acid)

1,2,4,5-Tetrachlorobenzene $\xrightarrow[OH^-]{H_2O}$ 2,4,5-Trichlorophenol $\xrightarrow[NaOH]{ClCH_2CO_2H}$ 2,4,5-Trichlorophenoxyacetic acid

Polychlorinated dibenzo-*p*-dioxins (PCDDs) are toxic impurities found in trace amounts in 2,4-D and 2,4,5-T. (See Exercise 22 in Chapter 22.) Two molecules of a chlorinated phenol react with each other to produce HCl and a PCDD. Draw the structures of the PCDDs produced from the following reactions:

a. 2,4,6-trichlorophenol + 2,4,5-trichlorophenol ⟶ PCDD + 2HCl

b. 2 (2,4,6-trichlorophenol) ⟶ PCDD + 2HCl

c. 2,4,5-trichlorophenol + 2,4-dichlorophenol → PCDD + 2HCl

Do you think it would be feasible to produce dioxin-free 2,4-D and 2,4,5-T? Levels of dioxin in the parts-per-billion range are significant from a toxicological standpoint.

30. Considering your answers to Exercises 28 and 29, why is there substantial impetus for producing allelopathic chemicals and insect pheromones for use as pesticides?

31. Use the Cross Index of Names at the back of *The Merck Index* to find the structures and uses of the following agricultural chemicals:
a. Lasso
b. Roundup
c. Aldicarb

Metallurgy

32. Define and give an example of each of the following:
a. roasting
b. smelting
c. flotation
d. leaching
e. gangue

33. What are the advantages and disadvantages of hydrometallurgy?

34. Use standard reduction potentials to calculate $\mathscr{E}°$, $\Delta G°$, and K (at 298 K) for the reaction that is used in production of gold:

$$2Au(CN)_2^-(aq) + Zn(s) \rightarrow 2Au(s) + Zn(CN)_4^{2-}(aq)$$

The relevant half-reactions are:

$$Au(CN)_2^- + e^- \rightarrow Au + 2CN^- \quad \mathscr{E}° = -0.60\ V$$
$$Zn(CN)_4^{2-} + 2e^- \rightarrow Zn + 4CN^- \quad \mathscr{E}° = -1.26\ V$$

35. What is the role of kinetics and thermodynamics in the effect of the following reaction on the properties of steel?

$$3Fe + C \rightleftharpoons Fe_3C$$

36. Use the data in Appendix 4 for the following:
a. Calculate $\Delta H°$ and $\Delta S°$ for the reaction $3Fe_2O_3 + CO \rightarrow 2Fe_3O_4 + CO_2$ that occurs in a blast furnace as shown in Fig. 24.7.
b. Assume $\Delta H°$ and $\Delta S°$ are independent of temperature. Calculate $\Delta G°$ at 800°C for this reaction.

37. Describe the process by which a metal is purified by zone refining.

Winemaking

38. Using the data in Appendix 4, calculate $\Delta H°$, $\Delta S°$, $\Delta G°$, and K (at 298 K) for the fermentation of glucose to ethanol and carbon dioxide.

39. A wine is 13% ethanol by volume. The density of ethanol is 0.79 g/cm^3, and the density of water is 1.0 g/cm^3. Assuming

the wine is only ethanol and water and that there is no volume change in mixing ethanol and water, calculate the concentration of ethanol in the wine in terms of each of the following:

a. mass percent
b. molality
c. molarity

40. Sparkling wines are fermented in the bottle in which they are sold (the Methode Champenoise). What are the bubbles in a sparkling wine? The bottom of a sparkling wine bottle is shaped differently from a still wine bottle (see figure at right) and the glass of a sparkling wine bottle is thicker. Explain.

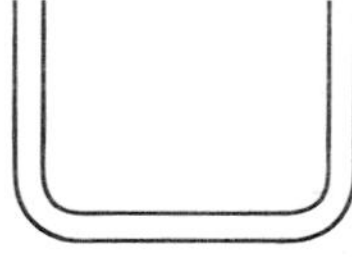
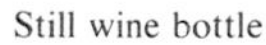
Still wine bottle

Sparkling wine bottle

41. The pH of a wine sample is 3.0. What is the predominant form of malic acid ($K_{a_1} = 4.0 \times 10^{-4}$, $K_{a_2} = 8.9 \times 10^{-6}$) and tartaric acid ($K_{a_1} = 9.2 \times 10^{-4}$, $K_{a_2} = 4.3 \times 10^{-5}$) in this wine?

42. Give a chemical reason for allowing some time for a wine to "breathe" between opening the bottle and drinking the wine.

43. Describe how the color of wine can be affected by its pH.

APPENDIX ONE Mathematical Procedures

A1.1 Exponential Notation

The numbers characteristic of scientific measurements are often very large or very small; thus it is convenient to express them using powers of 10. For example, the number 1,300,000 can be expressed as 1.3×10^6, which means multiply 1.3 by 10 six times, or

$$1.3 \times 10^6 = 1.3 \times \underbrace{10 \times 10 \times 10 \times 10 \times 10 \times 10}_{10^6\ =\ 1\ \text{million}}$$

Note that each multiplication by 10 moves the decimal point one place to the right:

$$\begin{aligned} 1.3 \times 10 &= 13. \\ 13 \times 10 &= 130. \\ 130 \times 10 &= 1300. \\ &\vdots \end{aligned}$$

Thus the easiest way to interpret the notation 1.3×10^6 is that it means move the decimal point in 1.3 to the right six times:

$$1.3 \times 10^6 = 1\underbrace{3\,0\,0\,0\,0\,0}_{1\,2\,3\,4\,5\,6} = 1{,}300{,}000$$

Using this notation, the number 1985 can be expressed as 1.985×10^3. Note that the usual convention is to write the number that appears before the power of 10 as a number between 1 and 10. To end up with the number 1.985, which is between 1 and 10, we had to move the decimal point three places to the left. To compensate for that, we must multiply by 10^3, which says that to get the intended number we start with 1.985 and move the decimal point three places to the right; that is:

$$1.985 \times 10^3 = 1\underbrace{9\,8\,5.}_{1\,2\,3}$$

Some other examples are given below.

Number	Exponential notation
5.6	5.6×10^0 or 5.6×1
39	3.9×10^1
943	9.43×10^2
1126	1.126×10^3

So far, we have considered numbers greater than 1. How do we represent a number such as 0.0034 in exponential notation? We start with a number between 1 and 10 and *divide* by the appropriate power of 10:

$$0.0034 = \frac{3.4}{10 \times 10 \times 10} = \frac{3.4}{10^3} = 3.4 \times 10^{-3}$$

Division by 10 moves the decimal point one place to the *left*. Thus the number

$$0.\underbrace{0\,0\,0\,0\,0\,0\,1}_{7\;6\;5\;4\;3\;2\;1}4$$

can be written as 1.4×10^{-7}.

To summarize, we can write any number in the form

$$N \times 10^{\pm n}$$

where N is between 1 and 10 and the exponent n is an integer. If the sign preceding n is positive, it means the decimal point in N should be moved n places to the right. If a negative sign precedes n, the decimal point in N should be moved n places to the left.

Multiplication and Division

When two numbers expressed in exponential notation are multiplied, the initial numbers are multiplied and the exponents of 10 are added:

$$(M \times 10^m)(N \times 10^n) = (MN) \times 10^{m+n}$$

For example (to two significant figures, as required),

$$(3.2 \times 10^4)(2.8 \times 10^3) = 9.0 \times 10^7$$

When the numbers are multiplied, if a result greater than 10 is obtained for the initial number, the decimal point is moved one place to the left and the exponent of 10 is increased by 1:

$$\begin{aligned}(5.8 \times 10^2)(4.3 \times 10^8) &= 24.9 \times 10^{10} \\ &= 2.49 \times 10^{11} \\ &= 2.5 \times 10^{11} \quad \text{(two significant figures)}\end{aligned}$$

Division of two numbers expressed in exponential notation involves normal division of the initial numbers and *subtraction* of the exponent of the divisor from that of the dividend. For example,

$$\frac{4.8 \times 10^8}{\underbrace{2.1 \times 10^3}_{\text{Divisor}}} = \frac{4.8}{2.1} \times 10^{(8-3)} = 2.3 \times 10^5$$

If the initial number resulting from the division is less than 1, the decimal point is moved one place to the right and the exponent of 10 is decreased by 1. For example,

$$\begin{aligned}\frac{6.4 \times 10^3}{8.3 \times 10^5} = \frac{6.4}{8.3} \times 10^{(3-5)} &= 0.77 \times 10^{-2} \\ &= 7.7 \times 10^{-3}\end{aligned}$$

Addition and Subtraction

To add or subtract numbers expressed in exponential notation, *the exponents of the numbers must be the same*. For example, to add 1.31×10^5 and 4.2×10^4, we must rewrite one number so that the exponents of both are the same. The number 1.31×10^5 can be written 13.1×10^4, since moving the decimal point one place to the right can be compensated for by decreasing the exponent by 1. Now we can add the numbers:

$$\begin{array}{r} 13.1 \times 10^4 \\ +\ 4.2 \times 10^4 \\ \hline 17.3 \times 10^4 \end{array}$$

In correct exponential notation the result is expressed as 1.73×10^5.

To perform addition or subtraction with numbers expressed in exponential notation, only the initial numbers are added or subtracted. The exponent of the result is the same as those of the numbers being added or subtracted. To subtract 1.8×10^2 from 8.99×10^3, we write

$$\begin{array}{r} 8.99 \times 10^3 \\ -0.18 \times 10^3 \\ \hline 8.81 \times 10^3 \end{array}$$

Powers and Roots

When a number expressed in exponential notation is taken to some power, the initial number is taken to the appropriate power and the exponent of 10 is multiplied by that power:

$$(N \times 10^n)^m = N^m \times 10^{m \cdot n}$$

For example,*

$$\begin{aligned} (7.5 \times 10^2)^3 &= 7.5^3 \times 10^{3 \cdot 2} \\ &= 422 \times 10^6 \\ &= 4.22 \times 10^8 \\ &= 4.2 \times 10^8 \qquad \text{(two significant figures)} \end{aligned}$$

When a root it taken of a number expressed in exponential notation, the root of the initial number is taken and the exponent of 10 is divided by the number representing the root:

$$\sqrt{N \times 10^n} = (N \times 10^n)^{1/2} = \sqrt{N} \times 10^{n/2}$$

For example,

$$\begin{aligned} (2.9 \times 10^6)^{1/2} &= \sqrt{2.9} \times 10^{6/2} \\ &= 1.7 \times 10^3 \end{aligned}$$

Because the exponent of the result must be an integer, we may sometimes have to change the form of the number so that the power divided by the root equals an

*Refer to the instruction booklet for your calculator for directions concerning how to take roots and powers of numbers.

integer, for example:

$$\sqrt{1.9 \times 10^3} = (1.9 \times 10^3)^{1/2} = (0.19 \times 10^4)^{1/2}$$
$$= \sqrt{0.19} \times 10^2$$
$$= 0.44 \times 10^2$$
$$= 4.4 \times 10^1$$

In this case, we moved the decimal point one place to the left and increased the exponent from 3 to 4 to make $n/2$ an integer.

The same procedure is followed for roots other than square roots, for example:

$$\sqrt[3]{6.9 \times 10^5} = (6.9 \times 10^5)^{1/3} = (0.69 \times 10^6)^{1/3}$$
$$= \sqrt[3]{0.69} \times 10^2$$
$$= 0.88 \times 10^2$$
$$= 8.8 \times 10^1$$

and

$$\sqrt[3]{4.6 \times 10^{10}} = (4.6 \times 10^{10})^{1/3} = (46 \times 10^9)^{1/3}$$
$$= \sqrt[3]{46} \times 10^3$$
$$= 3.6 \times 10^3$$

A1.2 Logarithms

A logarithm is an exponent. Any number N can be expressed as follows:

$$N = 10^x$$

For example,

$$1000 = 10^3$$
$$100 = 10^2$$
$$10 = 10^1$$
$$1 = 10^0$$

The common, or base 10, logarithm of a number is the power to which 10 must be taken to yield that number. Thus, since $1000 = 10^3$,

$$\log 1000 = 3$$

Similarly,

$$\log 100 = 2$$
$$\log 10 = 1$$
$$\log 1 = 0$$

For a number between 10 and 100, the required exponent of 10 will be between 1 and 2. For example, $65 = 10^{1.8129}$; that is, $\log 65 = 1.8129$. For a number between 100 and 1000, the exponent of 10 will be between 2 and 3. For example, $650 = 10^{2.8129}$ and $\log 650 = 2.8129$.

A number N greater than 0 and less than 1 can be expressed as follows:

$$N = 10^{-x} = \frac{1}{10^x}$$

For example,

$$0.001 = \frac{1}{1000} = \frac{1}{10^3} = 10^{-3}$$

$$0.01 = \frac{1}{100} = \frac{1}{10^2} = 10^{-2}$$

$$0.1 = \frac{1}{10} = \frac{1}{10^1} = 10^{-1}$$

Thus

$$\log 0.001 = -3$$
$$\log 0.01 = -2$$
$$\log 0.1 = -1$$

Although common logs are often tabulated, the most convenient method for obtaining such logs is to use an electronic calculator. On most calculators the number is first entered and then the log key is punched. The log of the number then appears in the display.* Some examples are given below. You should reproduce these results on your calculator in order to be sure that you can find common logs correctly.

Number	Common log
36	1.56
1849	3.2669
0.156	−0.807
1.68×10^{-5}	−4.775

Note that the number of digits after the decimal point in a common log is equal to the number of significant figures in the original number.

Since logs are simply exponents, they are manipulated according to the rules for exponents. For example, if $A = 10^x$ and $B = 10^y$, then their product is

$$A \cdot B = 10^x \cdot 10^y = 10^{x+y}$$

and

$$\log AB = x + y = \log A + \log B$$

For division, we have

$$\frac{A}{B} = \frac{10^x}{10^y} = 10^{x-y}$$

and

$$\log \frac{A}{B} = x - y = \log A - \log B$$

*Refer to the instruction booklet for your calculator for the exact sequence to obtain logarithms.

For a number raised to a power, we have

$$A^n = (10^x)^n = 10^{nx}$$

and

$$\log A^n = nx = n \log A$$

It follows that

$$\log \frac{1}{A^n} = \log A^{-n} = -n \log A$$

or, for $n = 1$,

$$\log \frac{1}{A} = -\log A$$

When a common log is given, to find the number it represents, we must carry out the process of exponentiation. For example, if the log is 2.673, then $N = 10^{2.673}$. The process of exponentiation is also called taking the antilog, or the inverse logarithm. This operation is usually carried out on calculators in one of two ways. The majority of calculators require that the log be entered first and then the keys [INV] and [LOG] pressed in succession. For example to find $N = 10^{2.673}$, we enter 2.673 and then press [INV] and [LOG]. The number 471 will be displayed; that is, $N = 471$. Some calculators have a [10^x] key. In that case, the log is entered first and then the [10^x] key is pressed. Again, the number 471 will be displayed.

Natural logarithms, another type of logarithm, are based on the number 2.7183, which is referred to as e. In this case, a number is represented as $N = e^x = 2.7183^x$. For example:

$$N = 7.15 = e^x$$

$$\ln 7.15 = x = 1.967$$

To find the natural log of a number using a calculator, the number is entered and then the [ln] key is pressed. Use the following examples to check your technique for finding natural logs with your calculator:

Number (e^x)	Natural log (x)
784	6.664
1.61×10^3	7.384
1.00×10^{-7}	-16.118
1.00	0

If a natural logarithm is given, to find the number it represents, exponentiation to the base e (2.7183) must be carried out. With many calculators this is done using a key marked [e^x] (the natural log is entered, with the correct sign, and then the [e^x] key is pressed). The other common method for exponentiation to base e is to enter the natural log and then press the [INV] and [ln] keys in succession. The following examples will help you check your technique.

ln N (x)	N (e^x)
3.256	25.9
−5.169	5.69×10^{-3}
13.112	4.95×10^5

Since natural logarithms are simply exponents, they are also manipulated according to the mathematical rules for exponents given above for common logs.

A1.3 Graphing Functions

In interpreting the results of a scientific experiment, it is often useful to make a graph. If possible, the function to be graphed should be in a form that gives a straight line. The equation for a straight line (a *linear equation*) can be represented by the general form

$$y = mx + b$$

where y is the *dependent variable*, x is the *independent variable*, m is the *slope*, and b is the *intercept* with the y axis.

To illustrate the characteristics of a linear equation, the function $y = 3x + 4$ is plotted in Fig. A.1. For this equation $m = 3$ and $b = 4$. Note that the y intercept occurs when $x = 0$. In this case the intercept is 4, as can be seen from the equation ($b = 4$).

The slope of a straight line is defined as the ratio of the rate of change in y to that in x:

$$m = \text{slope} = \frac{\Delta y}{\Delta x}$$

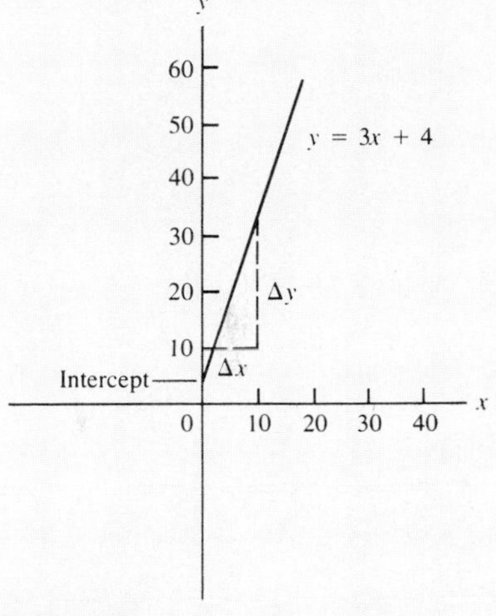

Figure A.1

Graph of the linear equation $y = 3x + 4$.

For the equation $y = 3x + 4$, y changes three times as fast as x (since x has a coefficient of 3). Thus the slope in this case is 3. This can be verified from the graph. For the triangle shown in Fig. A.1,

$$\Delta y = 34 - 10 = 24 \quad \text{and} \quad \Delta x = 10 - 2 = 8$$

Thus

$$\text{Slope} = \frac{\Delta y}{\Delta x} = \frac{24}{8} = 3$$

The above example illustrates a general method for obtaining the slope of a line from the graph of that line. Simply draw a triangle with one side parallel to the y axis and the other parallel to the x axis as shown in Fig. A.1. Then determine the lengths of the sides to give Δy and Δx, respectively, and compute the ratio $\Delta y/\Delta x$.

Sometimes an equation that is not in standard form can be changed to the form $y = mx + b$ by rearrangement or mathematical manipulation. An example is the equation $k = Ae^{-E_a/RT}$ described in Section 12.7, where A, E_a and R are constants, k is the dependent variable, and $1/T$ is the independent variable. This equation can be changed to standard form by taking the natural logarithm of both sides,

$$\ln k = \ln Ae^{-E_a/Rt} = \ln A + \ln e^{-E_a/RT} = \ln A - \frac{E_a}{RT}$$

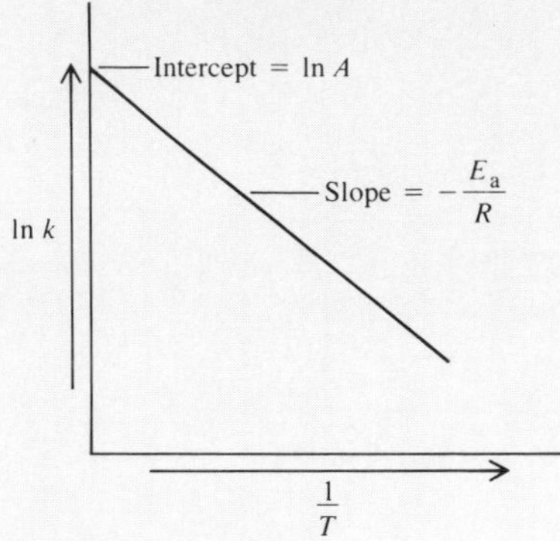

Figure A.2

Graph of ln k versus $1/T$.

noting that the log of a product is equal to the sum of the logs of the individual terms and that the natural log of $e^{-E_a/RT}$ is simply the exponent $-E_a/RT$. Thus, in standard form, the equation $k = Ae^{-E_a/RT}$ is written

$$\underbrace{\ln k}_{\substack{\uparrow \\ y}} = \underbrace{-\frac{E_a}{R}}_{\substack{\uparrow \\ m}} \underbrace{\left(\frac{1}{T}\right)}_{\substack{\uparrow \\ x}} + \underbrace{\ln A}_{\substack{\uparrow \\ b}}$$

A plot of ln k versus $1/T$ (see Fig. A.2) gives a straight line with slope $-E_a/R$ and intercept ln A.

Other linear equations that are useful in the study of chemistry are listed in standard form in Table A.1.

Some Useful Linear Equations in Standard Form

Equation ($y = mx + b$)	What is Plotted (y vs. x)	Slope (m)	Intercept (b)	Section in Text
$[A] = -kt + [A]_0$	$[A]$ vs. t	$-k$	$[A]_0$	12.4
$\ln [A) = -kt + \ln [A]_0$	$\ln [A]$ vs. t	$-k$	$\ln [A]_0$	12.4
$\frac{1}{[A]} = kt + \frac{1}{[A]_0}$	$\frac{1}{[A]}$ vs. t	k	$\frac{1}{[A]_0}$	12.4
$\ln P_{vap} = -\frac{\Delta H_{vap}}{R}\left(\frac{1}{T}\right) + C$	$\ln P_{vap}$ vs. $\frac{1}{T}$	$\frac{-\Delta H_{vap}}{R}$	C	10.8

Table A.1

A1.4 Solving Quadratic Equations

A *quadratic equation,* a polynomial in which the highest power of x is 2, can be written as

$$ax^2 + bx + c = 0$$

One method for finding the two values of x that satisfy a quadratic equation is to use the *quadratic formula:*

$$x = \frac{-b \pm \sqrt{b^2 - 4ac}}{2a}$$

where a, b, and c are the coefficients of x^2 and x and the constant, respectively. For example, in determining $[H^+]$ in a solution of 1.0×10^{-4} M acetic acid the following expression arises:

$$1.8 \times 10^{-5} = \frac{x^2}{1.0 \times 10^{-4} - x}$$

which yields

$$x^2 + (1.8 \times 10^{-5})x - 1.8 \times 10^{-9} = 0$$

where $a = 1$, $b = 1.8 \times 10^{-5}$, and $c = -1.8 \times 10^{-9}$. Using the quadratic formula, we have

$$x = \frac{-b \pm \sqrt{b^2 - 4ac}}{2a}$$

$$= \frac{-1.8 \times 10^{-5} \pm \sqrt{3.24 \times 10^{-10} - (4)(1)(-1.8 \times 10^{-9})}}{2(1)}$$

$$= \frac{-1.8 \times 10^{-5} \pm \sqrt{3.24 \times 10^{-10} + 7.2 \times 10^{-9}}}{2}$$

$$= \frac{-1.8 \times 10^{-5} \pm \sqrt{7.5 \times 10^{-9}}}{2}$$

$$= \frac{-1.8 \times 10^{-5} \pm 8.7 \times 10^{-5}}{2}$$

Thus

$$x = \frac{6.9 \times 10^{-5}}{2} = 3.5 \times 10^{-5}$$

and

$$x = \frac{-10.5 \times 10^{-5}}{2} = -5.2 \times 10^{-5}$$

Note that there are two roots as there always will be for a polynomial in x^2. In this case x represents a concentration of H^+ (see Section 14.3). Thus the positive root is the one that solves the problem, since a concentration cannot be a negative number.

A second method for solving quadratic equations is by *successive approximations,* a systematic method of trial and error. A value of x is guessed and substituted into the equation everywhere x (or x^2) appears, except for one place. For example, for the equation

$$x^2 + (1.8 \times 10^{-5})x - 1.8 \times 10^{-9} = 0$$

we might guess $x = 2 \times 10^{-5}$. Substituting that value into the equation gives

$$x^2 + (1.8 \times 10^{-5})(2 \times 10^{-5}) - 1.8 \times 10^{-9} = 0$$

or $\quad x^2 = 1.8 \times 10^{-9} - 3.6 \times 10^{-10} = 1.4 \times 10^{-9}$

Thus $\quad x = 3.7 \times 10^{-5}$

Note that the guessed value of x (2×10^{-5}) is not the same as the value of x that is calculated (3.7×10^{-5}) after inserting the estimated value. This means that $x = 2 \times 10^{-5}$ is not the correct solution, and we must try another guess. We take the calculated value (3.7×10^{-5}) as our next guess:

$$x^2 + (1.8 \times 10^{-5})(3.7 \times 10^{-5}) - 1.8 \times 10^{-9} = 0$$

$$x^2 = 1.8 \times 10^{-9} - 6.7 \times 10^{-10} = 1.1 \times 10^{-9}$$

Thus $\quad x = 3.3 \times 10^{-5}$

Now we compare the two values of x again:

Guessed: $x = 3.7 \times 10^{-5}$

Calculated: $x = 3.3 \times 10^{-5}$

These values are closer but not close enough. Next we try 3.3×10^{-5} as our guess:

$$x^2 + (1.8 \times 10^{-5})(3.3 \times 10^{-5}) - 1.8 \times 10^{-9} = 0$$

$$x^2 = 1.8 \times 10^{-9} - 5.9 \times 10^{-10} = 1.2 \times 10^{-9}$$

Thus $$x = 3.5 \times 10^{-5}$$

Again we compare:

$$\text{Guessed: } x = 3.3 \times 10^{-5}$$

$$\text{Calculated: } x = 3.5 \times 10^{-5}$$

Next we guess $x = 3.5 \times 10^{-5}$ to give

$$x^2 + (1.8 \times 10^{-5})(3.5 \times 10^{-5}) - 1.8 \times 10^{-9} = 0$$

$$x^2 = 1.8 \times 10^{-9} - 6.3 \times 10^{-10} = 1.2 \times 10^{-9}$$

Thus $$x = 3.5 \times 10^{-5}$$

Now the guessed value and the calculated value are the same; we have found the correct solution. Note that this agrees with one of the roots found with the quadratic formula in the first method above.

To further illustrate the method of successive approximations, we will solve Sample Exercise 14.17 using this procedure. In solving for $[H^+]$ for 0.010 *M* H_2SO_4, we obtain the following expression:

$$1.2 \times 10^{-2} = \frac{x(0.010 + x)}{0.010 - x}$$

which can be rearranged to give

$$x = (1.2 \times 10^{-2})\left(\frac{0.010 - x}{0.010 + x}\right)$$

We will guess a value for x, substitute it into the right side of the equation, and then calculate a value for x. In guessing a value for x, we know it must be less than 0.010, since a larger value will make the calculated value for x negative and the guessed and calculated values will never match. We start by guessing $x = 0.005$.

The results of the successive approximations are shown in the following table:

Trial	Guessed value for x	Calculated value for x
1	0.0050	0.0040
2	0.0040	0.0051
3	0.00450	0.00455
4	0.00452	0.00453

Note that the first guess was close to the actual value and that there was oscillation between 0.004 and 0.005 for the guessed and calculated values. For trial 3, an average of these values was used as the guess, and this led rapidly to the correct value (0.0045 to the correct number of significant figures). Also, note that it is useful to carry extra digits until the correct value is obtained. That value can then be rounded off to the correct number of significant figures.

The method of successive approximations is especially useful for solving polynomials containing x to a power of 3 or higher. The procedure is the same as for quadratic equations: substitute a guessed value for x into the equation for every x term but one, and then solve for x. Continue this process until the guessed and calculated values agree.

A1.5 Uncertainties in Measurements

Like all the physical sciences, chemistry is based on the results of measurements. Every measurement has an inherent uncertainty; so if we are to use the results of measurements to reach conclusions, we must be able to estimate the sizes of these uncertainties.

For example, the specification for a commercial 500-mg acetaminophen (the active painkiller in Tylenol) tablet is that each batch of tablets must contain 450–550 mg of acetaminophen per tablet. Suppose that chemical analysis gave the following results for a batch of acetaminophen tablets: 428 mg, 479 mg, 442 mg, and 435 mg. How can we use these results to decide if the batch of tablets meets the specification? Although the details of how to draw such conclusions from measured data are beyond the scope of this text, we will consider some aspects of how this is done. We will focus here on the types of experimental uncertainty, the expression of experimental results, and a simplified method for estimating experimental uncertainty when several types of measurement contribute to the final result.

Types of Experimental Error

There are two types of experimental uncertainty (error). A variety of names is applied to these types of errors:

$$\text{Precision} \leftrightarrow \text{Random error} \equiv \text{Indeterminate error}$$
$$\text{Accuracy} \leftrightarrow \text{Systematic error} \equiv \text{Determinate error}$$

The difference between the two types of error is well illustrated by the attempts to hit a target shown in Fig. 1.8 in Chapter 1.

Random error is associated with every measurement. To obtain the last significant figure for any measurement, we must always make an estimate. For example, we interpolate between the marks on a meter stick, a buret, or a balance. The precision of replicate measurements (repeated measurements of the same type) reflects the size of the random errors. Precision refers to the reproducibility of replicate measurements.

The accuracy of a measurement refers to how close it is to the true value. An inaccurate result occurs as a result of some flaw (systematic error) in the measurement: the presence of an interfering substance, incorrect calibration of an instrument, operator error, etc. The goal of chemical analysis is to eliminate systematic error, but random errors can only be minimized. In practice, an experiment is almost always done in order to find an unknown value (the true value is not known—someone is trying to obtain that value by doing the experiment). In this case the precision of several replicate determinations is used to assess the accuracy of the result. The results of the replicate experiments are expressed as an average (which we assume is close to the true value) with an error limit that gives some indication of how close the average value may be to the true value. The error limit represents the uncertainty of the experimental result.

Expression of Experimental Results

If we perform several measurements, such as for the analysis for acetaminophen in painkiller tablets, the results should express two things: the average of the measurements and the size of the uncertainty.

There are two common ways of expressing an average: the mean and the median. The mean ($\bar{x}$) is the arithmetic average of the results, or

$$\text{Mean} = \bar{x} = \sum_{i=1}^{n} \frac{x_i}{n} = \frac{x_1 + x_2 \cdots x_n}{n}$$

where Σ means take the sum of the values. The mean is equal to the sum of all the measurements divided by the number of measurements. For the acetaminophen results given previously, the mean is

$$\bar{x} = \frac{428 + 479 + 442 + 435}{4} = 446 \text{ mg}$$

The median is the value that lies in the middle among the results. Half of the measurements are above the median and half are below the median. For results of 465 mg, 485 mg, and 492 mg, the median is 485 mg. When there is an even number of results, the median is the average of the two middle results. For the acetaminophen results, the median is

$$\frac{442 + 435}{2} = 438 \text{ mg}$$

There are several advantages to using the median. If a small number of measurements is made, one value can greatly affect the mean. Consider the results for the analysis of acetaminophen: 428 mg, 479 mg, 442 mg, and 435 mg. The mean is 446 mg, which is larger than three of the four weights. The median is 438 mg, which lies near the three values that are relatively close to one another.

In addition to expressing an average value for a series of results, we must also express the uncertainty. This usually means expressing either the precision of the measurements or the observed range of the measurements. The range of a series of measurements is defined by the smallest value and the largest value. For the analytical results on the acetaminophen tablets, the range is from 428 mg to 479 mg. Using this range, we can express the results by saying that the true value lies between 428 mg and 479 mg. That is, we can express the amount of acetaminophen in a typical tablet as 446 ± 33 mg, where the error limit is chosen to give the observed range (approximately).

The most common way to specify precision is by the standard deviation, s, which for a small number of measurements is given by the formula

$$s = \left[\frac{\sum_{i=1}^{n} (x_i - \bar{x})^2}{n - 1} \right]^{1/2}$$

where x_i is an individual result, $\bar{x}$ is the average (either mean or median), and n is the total number of measurements. For the acetaminophen example, we have

$$s = \left[\frac{(428 - 446)^2 + (479 - 446)^2 + (442 - 446)^2 + (435 - 446)^2}{4 - 1} \right]^{1/2} = 23$$

Thus we can say the amount of acetaminophen in the typical tablet in the batch of tablets is 446 mg with a sample standard deviation of 23 mg. Statistically this means that any additional measurement has a 68% probability (68 chances out of 100) of being between 423 mg (446 − 23) and 469 mg (446 + 23). Thus the standard deviation is a measure of the precision of a given type of determination.

The standard deviation gives us a means of describing the precision of a given type of determination using a series of replicate results. However, it is also useful to be able to estimate the precision of a procedure that involves several measurements by combining the precisions of the individual steps. That is, we want to answer the following question: How do the uncertainties propagate when we combine the results of several different types of measurements? There are many ways to deal with the propagation of uncertainty. We will discuss only one simple method here.

A Simplified Method for Estimating Experimental Uncertainty

To illustrate this method, we will consider the determination of the density of an irregularly shaped solid. In this determination we make three measurements. First, we measure the mass of the object on a balance. Next, we must obtain the volume of the solid. The easiest method for doing this is to partially fill a graduated cylinder with a liquid and record the volume. Then we add the solid and record the volume again. The difference in the measured volumes is the volume of the solid. We can then calculate the density of the solid from the equation

$$D = \frac{M}{V_2 - V_1}$$

where M is the mass of the solid, V_1 is the initial volume of liquid in the graduated cylinder, and V_2 is the volume of liquid plus solid. Suppose we get the following results:

$$M = 23.06 \text{ g}$$
$$V_1 = 10.4 \text{ mL}$$
$$V_2 = 13.5 \text{ mL}$$

The calculated density is

$$\frac{23.06 \text{ g}}{13.5 \text{ mL} - 10.4 \text{ mL}} = 7.44 \text{ g/mL}$$

Now suppose that the precision of the balance used is ±0.02 g and that the volume measurements are precise to ±0.05 mL. How do we estimate the uncertainty of the density? We can do this by assuming a worst case. That is, we assume the largest uncertainties in all measurements, and see what combinations of measurements will give the largest and smallest possible results (the greatest range). Since the density is the mass divided by the volume, the largest value of the density will be that obtained using the largest possible mass and the smallest possible volume:

Largest possible mass = 23.06 + .02

$$D_{\text{max}} = \frac{23.08}{13.45 - 10.45} = 7.69 \text{ g/mL}$$

Smallest possible V_2 ↗ ↖ Largest possible V_1

The smallest value of the density is

$$D_{\text{min}} = \frac{\overset{\text{Smallest possible mass}}{23.04}}{\underset{\text{Largest possible } V_2}{13.55} - \underset{\text{Smallest possible } V_1}{10.35}} = 7.20 \text{ g/mL}$$

Thus, the calculated range is from 7.20 to 7.69 and the average of these values is 7.44. The error limit is the number that gives the high and low range values when added and subtracted from the average. Therefore we can express the density as 7.44 ± 0.25 g/mL, which is the average value plus or minus the quantity that gives the range calculated by assuming the largest uncertainties.

Analysis of the propagation of uncertainties is useful in drawing qualitative conclusions from the analysis of measurements. For example, suppose that we obtained the above results for the density of an unknown alloy and we want to know if it is one of the following alloys:

Alloy A: $D = 7.58$ g/mL

Alloy B: $D = 7.42$ g/mL

Alloy C: $D = 8.56$ g/mL

We can safely conclude that the alloy is not C. But the values of the densities for alloys A and B are both within the inherent uncertainty of our method. To distinguish between A and B, we need to improve the precision of our determination: the obvious choice is to improve the precision of the volume measurement.

The worst case method is very useful in estimating uncertainties when the results of several measurements are combined to calculate a result. We assume the maximum uncertainty in each measurement and calculate the minimum and maximum possible result. These extreme values describe the range and thus the error limit.

APPENDIX TWO The Quantitative Kinetic Molecular Model

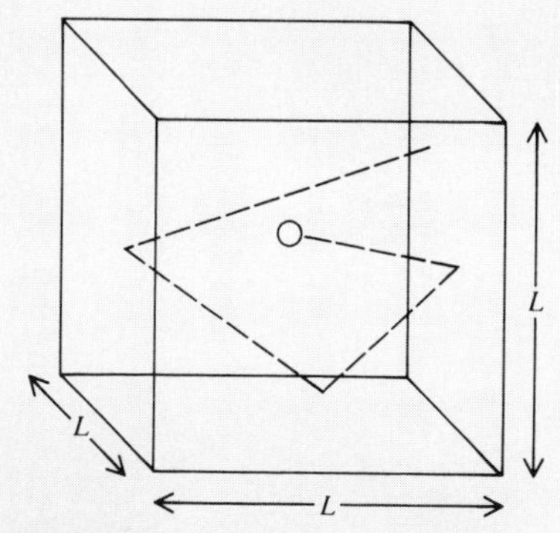

Figure A.3

An ideal gas particle in a cube whose sides are of length L. The particle collides elastically with the walls in a random, straight-line motion.

We have seen that the kinetic molecular model successfully accounts for the properties of an ideal gas. This appendix will show in some detail how the postulates of the kinetic molecular model lead to an equation corresponding to the experimentally obtained ideal gas equation.

Recall that the particles of an ideal gas are assumed to be volumeless, to have no attraction for each other, and to produce pressure on their container by colliding with the container walls.

Suppose there are n moles of an ideal gas in a cubical container with sides each of length L. Assume each gas particle has a mass m and that it is in rapid, random, straight-line motion colliding with the walls, as shown in Fig. A.3. The collisions will be assumed to be *elastic*—no loss of kinetic energy occurs. We want to compute the force on the walls from the colliding gas particles and then, since pressure is force per unit area, to obtain an expression for the pressure of the gas.

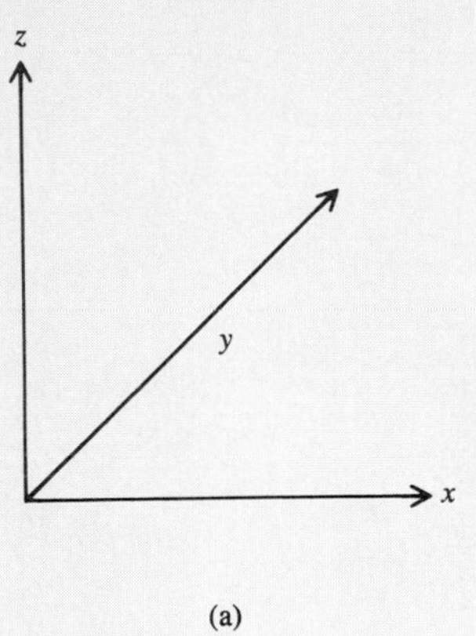

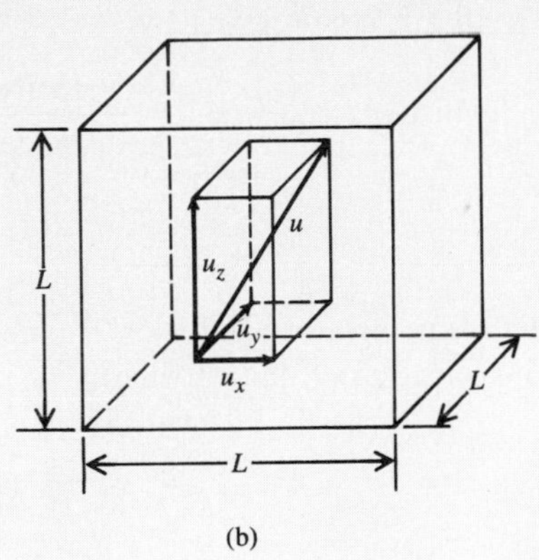

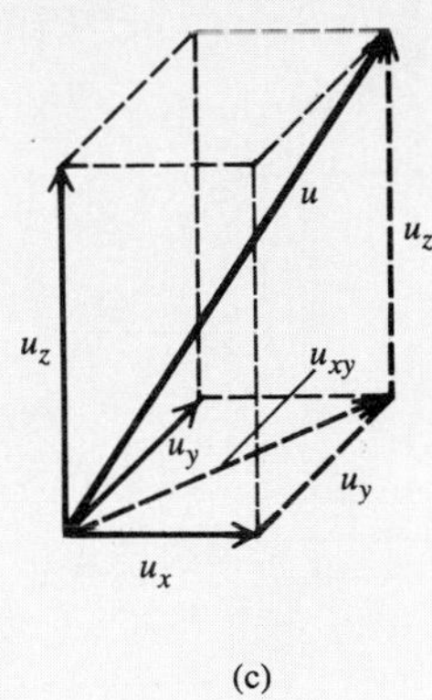

Figure A.4

(a) The Cartesian coordinate axes.

(b) The velocity u of any gas particle can be broken down into three mutually perpendicular components, u_x, u_y, and u_z. This can be represented as a rectangular solid with sides u_x, u_y, and u_z and body diagonal u.

(c) In the xy plane,

$$u_x^2 + u_y^2 = u_{xy}^2$$

by the Pythagorean theorem. Since u_{xy} and u_z are also perpendicular,

$$u^2 = u_{xy}^2 + u_z^2 = u_x^2 + u_y^2 + u_z^2$$

Before we can derive the expression for the pressure of a gas, we must first discuss some characteristics of velocity. Each particle in the gas has a particular velocity, u, which can be broken into components u_x, u_y, and u_z, as shown in Fig. A.4. First, using u_x and u_y and the Pythagorean theorem, we can obtain u_{xy} as shown in Fig. A.4(c):

$$u_{xy}^2 = u_x^2 + u_y^2$$

Hypotenuse of right triangle ↗ (u_{xy}^2) — Sides of right triangle ↖ (u_x^2, u_y^2)

Then, constructing another triangle as shown in Fig. A.4(c), we find

$$u^2 = u_{xy}^2 + u_z^2$$

or

$$u^2 = \underbrace{u_x^2 + u_y^2}_{u_{xy}^2} + u_z^2$$

Now let's consider how an "average" gas particle moves. For example, how often does this particle strike the two walls of the box that are perpendicular to the x axis? It is important to realize that only the x component of the velocity affects the particle's impacts on these two walls, as shown in Fig. A.5(a) on the next page. The larger the x component of the velocity, the faster the particle travels between these two walls, and the more impacts per unit of time it will make on these walls. Remember, the pressure of the gas is due to these collisions with the walls.

The collision frequency (collisions per unit of time) with the two walls that are

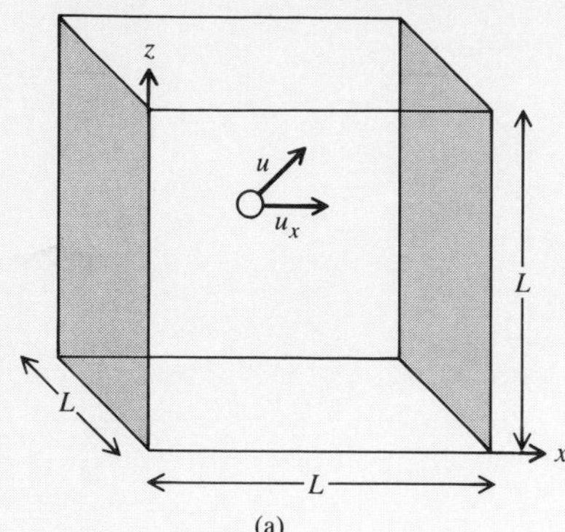

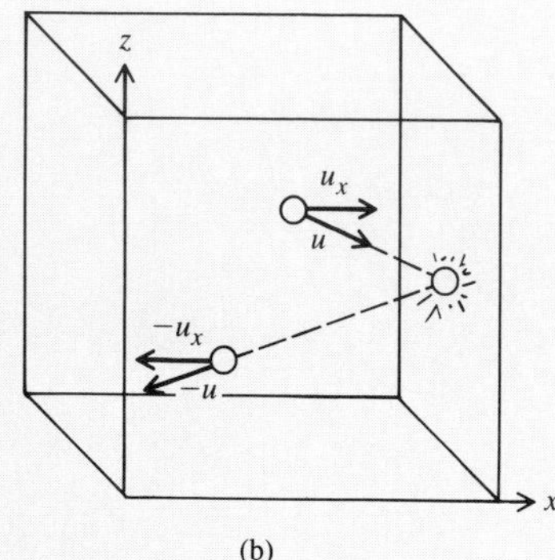

Figure A.5

(a) Only the x component of the gas particle's velocity affects the frequency of impacts on the shaded walls, the walls that are perpendicular to the x axis. (b) For an elastic collision, there is an exact reversal of the x component of the velocity and of the total velocity. The change in momentum (final − initial) is then

$$-mu_x - mu_x = -2mu_x$$

perpendicular to the x axis is given by

$$(\text{collision frequency})_x = \frac{\text{velocity in the } x \text{ direction}}{\text{distance between the walls}}$$

$$= \frac{u_x}{L}$$

Next, what is the force of a collision? Force is defined as mass times acceleration (change in velocity per unit of time):

$$F = ma = m\left(\frac{\Delta u}{\Delta t}\right)$$

where F represents force, a represents acceleration, Δu represents a change in velocity, and Δt represents a given length of time.

Since we assume that the particle has constant mass, we can write

$$F = \frac{m\Delta u}{\Delta t} = \frac{\Delta(mu)}{\Delta t}$$

The quantity mu is the momentum of the particle (momentum is the product of mass and velocity), and the expression $F = \Delta(mu)/\Delta t$ implies that force is the change in momentum per unit of time. When a particle hits a wall perpendicular to the x axis, as shown in Fig. A.5(b), an elastic collision results in an *exact reversal* of the x component of velocity. That is, the *sign,* or direction, of u_x reverses when the particle collides with one of the walls perpendicular to the x axis. Thus, the final momentum is the *negative,* or opposite, of the initial momentum. Remember that an elastic collision means that there is no change in the *magnitude* of the velocity. The change in momentum in the x direction is then

$$\text{Change in momentum} = \Delta(mu_x) = \text{final momentum} - \text{initial momentum}$$

$$= -mu_x - mu_x$$

(↗ Final momentum in x direction; ↖ Initial momentum in x direction)

$$= -2mu_x$$

But we are interested in the force the gas particle exerts on the walls of the box. Since we know that every action produces an equal but opposite reaction, the change in momentum with respect to the wall on impact is $-(-2mu_x)$, or $2mu_x$.

Recall that since force is the change in momentum per unit of time,

$$\text{Force}_x = \frac{\Delta(mu_x)}{\Delta t}$$

for the walls perpendicular to the x axis.

This expression can be obtained by multiplying the change in momentum per impact by the number of impacts per unit of time:

$$\text{Force}_x = (2mu_x)\left(\frac{u_x}{L}\right) = \text{Change in momentum per unit of time}$$

(↗ Change in momentum per impact; ↖ Impacts per unit of time)

That is,

$$\text{Force}_x = \frac{2mu_x^2}{L}$$

So far, we have considered only the two walls of the box perpendicular to the x axis. We can assume that the force on the two walls perpendicular to the y axis is given by

$$\text{Force}_y = \frac{2mu_y^2}{L}$$

and that on the two walls perpendicular to the z axis by

$$\text{Force}_z = \frac{2mu_z^2}{L}$$

Since we have shown that

$$u^2 = u_x^2 + u_y^2 + u_z^2$$

the total force on the box is

$$\begin{aligned}\text{Force}_{\text{TOTAL}} &= \text{Force}_x + \text{Force}_y + \text{Force}_z \\ &= \frac{2mu_x^2}{L} + \frac{2mu_y^2}{L} + \frac{2mu_z^2}{L} \\ &= \frac{2m}{L}(u_x^2 + u_y^2 + u_z^2) = \frac{2m}{L}(u^2)\end{aligned}$$

Now since we want the average force, we use the average of the square of the velocity $(\overline{u^2})$ to obtain

$$\overline{\text{Force}_{\text{TOTAL}}} = \frac{2m}{L}(\overline{u^2})$$

Next, we need to compute the pressure (force per unit of area)

$$\text{Pressure due to ``average'' particle} = \frac{\overline{\text{Force}_{\text{TOTAL}}}}{\text{Area}_{\text{TOTAL}}}$$

$$= \frac{\frac{2m\overline{u^2}}{L}}{6L^2} = \frac{m\overline{u^2}}{3L^3}$$

(6 ↗ The 6 sides of the cube; L^2 ↖ Area of each side)

Since the volume V of the cube is equal to L^3, we can write

$$\text{Pressure} = P = \frac{m\overline{u^2}}{3V}$$

So far we have considered the pressure on the walls due to a single, ''average'' particle. Of course, we want the pressure due to the entire gas sample. The number of particles in a given gas sample can be expressed as follows:

$$\text{Number of gas particles} = nN_{\text{A}}$$

where n is the number of moles and N_{A} is Avogadro's number.

The total pressure on the box due to n moles of a gas is therefore

$$P = nN_A \frac{m\overline{u^2}}{3V}$$

Next we want to express the pressure in terms of the kinetic energy of the gas molecules. Kinetic energy (the energy due to motion) is given by $\frac{1}{2}mu^2$ where m is the mass and u the velocity. Since we are using the average of the velocity squared ($\overline{u^2}$), and since $m\overline{u^2} = 2(\frac{1}{2}m\overline{u^2})$, we have

$$P = \left(\frac{2}{3}\right)\frac{nN_A(\frac{1}{2}m\overline{u^2})}{V}$$

or

$$\frac{PV}{n} = \left(\frac{2}{3}\right)N_A(\tfrac{1}{2}m\overline{u^2})$$

Thus, based on the postulates of the kinetic molecular model, we have been able to derive an equation that has the same form as the ideal gas equation,

$$\frac{PV}{n} = RT$$

This agreement between experiment and theory supports the validity of the assumptions made in the kinetic molecular model about the behavior of gas particles, at least for the limiting case of an ideal gas.

APPENDIX THREE Spectral Analysis

Although volumetric and gravimetric analyses are still very commonly used, spectroscopy is the technique most often used for modern chemical analysis. *Spectroscopy* is the study of electromagnetic radiation emitted or absorbed by a given chemical species. Since the quantity of radiation absorbed or emitted can be related to the quantity of the absorbing or emitting species present, this technique can be used for quantitative analysis. There are many spectroscopic techniques, since electromagnetic radiation spans a wide range of energies to include X rays, ultraviolet, infrared, and visible light, and microwaves, to name a few of its familiar forms. However, we will consider here only one procedure, which is based on the absorption of visible light.

If a liquid is colored, it is because some component of the liquid absorbs visible light. In a solution the greater the concentration of the light-absorbing substance, the more light is absorbed, and the more intense the color of the solution.

The quantity of light absorbed by a substance can be measured by a *spectrophotometer,* shown schematically in Fig. A.6. This instrument consists of a source that emits all wavelengths of light in the visible region (wavelengths of ~400–700 nm); a monochromator, which selects a given wavelength of light; a sample holder for the solution being measured; and a detector, which compares the intensity of incident light, I_0, to the intensity of light after it has passed through the sample, I. The ratio I/I_0, called the *transmittance,* is a measure of the fraction of light that passes

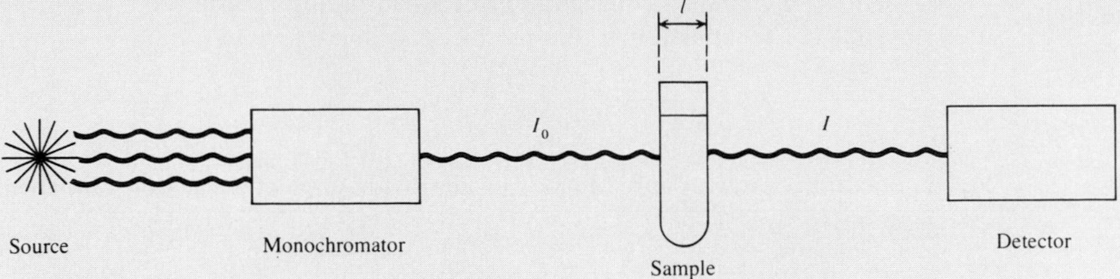

Figure A.6

A schematic diagram of a simple spectrophotometer. The source emits all wavelengths of visible light, which are dispersed using a prism or grating and then focused, one wavelength at a time, onto the sample. The detector compares the intensity of the incident light (I_0) to the intensity of the light after it has passed through the sample (I).

through the sample. The amount of light absorbed is given by the *absorbance, A,* where

$$A = -\log\frac{I}{I_0}$$

The absorbance can be expressed by the *Beer-Lambert law:*

$$A = \epsilon l c$$

where ϵ is the molar absorptivity or the molar extinction coefficient (in L/mol cm), l is the distance the light travels through the solution (in cm), and c is the concentration of the absorbing species (in mol/L). The Beer-Lambert law is the basis for using spectroscopy in quantitative analysis. If ϵ and l are known, measuring A for a solution allows us to calculate the concentration of the absorbing species in the solution.

Suppose we have a pink solution containing an unknown concentration of $Co^{2+}(aq)$ ions. A sample of this solution is placed in a spectrophotometer, and the absorbance is measured at a wavelength where ϵ for $Co^{2+}(aq)$ is known to be 12 L/mol cm. The absorbance A is found to be 0.60. The width of the sample tube is 1.0 cm. We want to determine the concentration of $Co^{2+}(aq)$ in the solution. This problem can be solved by a straightforward application of the Beer-Lambert law,

$$A = \epsilon l c$$

where

$$A = 0.60$$

$$\epsilon = \frac{12\text{ L}}{\text{mol cm}}$$

$$l = \text{light path} = 1.0\text{ cm}$$

Solving for the concentration gives

$$c = \frac{A}{\epsilon l} = \frac{0.60}{\left(12\,\frac{\text{L}}{\text{mol cm}}\right)(1.0\text{ cm})} = 5.0 \times 10^{-2}\text{ mol}/L$$

To obtain the unknown concentration of an absorbing species from the measured absorbance, we must know the product ϵl, since

$$c = \frac{A}{\epsilon l}$$

We can obtain the product ϵl by measuring the absorbance of a solution of *known* concentration, since

$$\epsilon l = \frac{A}{c}$$

↙ Measured using a spectrophotometer (A)

↖ Known from making up the solution (c)

However, a more accurate value of the product ϵl can be obtained by plotting A versus c for a series of solutions. Note that the equation $A = \epsilon lc$ gives a straight line with slope ϵl when A is plotted against c.

For example, consider the following typical spectroscopic analysis. A sample of steel from a bicycle frame is to be analyzed to determine its manganese content. The procedure involves weighing out a sample of the steel, dissolving it in strong acid, treating the resulting solution with a very strong oxidizing agent to convert all the manganese to permanganate ion (MnO_4^-), and then using spectroscopy to determine the concentration of the intensely purple MnO_4^- ions in the solution. To do this, however, the value of ϵl for MnO_4^- must be determined at an appropriate wavelength. The absorbance values for four solutions with known MnO_4^- concentrations were measured to give the following data:

Solution	Concentration of MnO_4^- (mol/L)	Absorbance
1	7.00×10^{-5}	0.175
2	1.00×10^{-4}	0.250
3	2.00×10^{-4}	0.500
4	3.50×10^{-4}	0.875

A plot of absorbance versus concentration for the solutions of known concentration is shown in Fig. A.7. The slope of this line (change in A/change in c) is 2.48×10^3 L/mol. This quantity represents the product ϵl.

A sample of the steel weighing 0.1523 g was dissolved and the unknown amount of manganese was converted to MnO_4^- ions. Water was then added to give a solution with a final volume of 100.0 mL. A portion of this solution was placed in a spectrophotometer, and its absorbance was found to be 0.780. Using these data, we want to calculate the percent manganese in the steel. The MnO_4^- ions from the manganese in the dissolved steel sample show an absorbance of 0.780. Using the Beer-Lambert law, we calculate the concentration of MnO_4^- in this solution:

$$c = \frac{A}{\epsilon l} = \frac{0.780}{2.48 \times 10^3 \text{ L/mol}} = 3.15 \times 10^{-4} \text{ mol/L}$$

There is a more direct way for finding c. Using a graph such as that in Fig. A.7

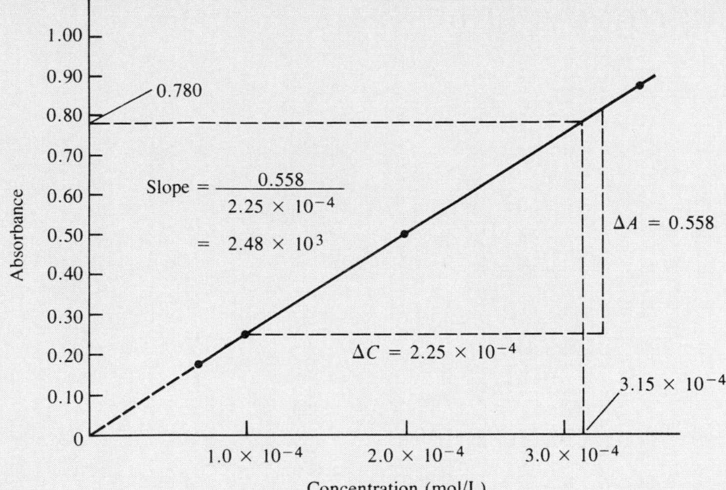

Figure A.7

A plot of absorbance versus concentration of MnO_4^- in a series of solutions of known concentration.

(often called a Beer's law plot), we can read the concentration that corresponds to $A = 0.780$. This interpolation is shown by dashed lines on the graph. By this method, $c = 3.15 \times 10^{-4}$ mol/L, which agrees with the value obtained above.

Recall that the original 0.1523-g steel sample was dissolved, the manganese was converted to permanganate, and the volume was adjusted to 100.0 mL. We now know that the $[MnO_4^-]$ in that solution is 3.15×10^{-4} *M*. Using this concentration, we can calculate the total number of moles of MnO_4^- in that solution:

$$\text{mol of } MnO_4^- = 100.0 \text{ mL} \times \frac{1 \text{ L}}{1000 \text{ mL}} \times 3.15 \times 10^{-4} \frac{\text{mol}}{\text{L}}$$

$$= 3.15 \times 10^{-5} \text{ mol}$$

Since each mole of manganese in the original steel sample yields a mole of MnO_4^-, that is,

$$1 \text{ mol of Mn} \xrightarrow{\text{Oxidation}} 1 \text{ mol of } MnO_4^-$$

the original steel sample must have contained 3.15×10^{-5} mol of manganese. The mass of manganese present in the sample is

$$3.15 \times 10^{-5} \text{ mol of Mn} \times \frac{54.938 \text{ g of Mn}}{1 \text{ mol of Mn}} = 1.73 \times 10^{-3} \text{ g of Mn}$$

Since the steel sample weighed 0.1523 g, the percent manganese in the steel is

$$\frac{1.73 \times 10^{-3} \text{ g of Mn}}{1.523 \times 10^{-1} \text{ g of sample}} \times 100 = 1.14\%$$

This example illustrates a typical use of spectroscopy in quantitative analysis. The steps commonly involved are as follows:

1. Preparation of a calibration plot (a Beer's law plot) by measuring the absorbance values of a series of solutions with known concentrations.
2. Measurement of the absorbance of the solution of unknown concentration.
3. Use of the calibration plot to determine the unknown concentration.

APPENDIX FOUR Selected Thermodynamic Data

Note: All values are assumed precise to at least ±1.

Substance and state	ΔH_f° (kJ/mol)	ΔG_f° (kJ/mol)	S° (J/K mol)
Aluminum			
Al(*s*)	0	0	28
Al_2O_3(*s*)	−1676	−1582	51
$Al(OH)_3$(*s*)	−1277		
$AlCl_3$(*s*)	−704	−629	111
Barium			
Ba(*s*)	0	0	67
$BaCO_3$(*s*)	−1219	−1139	112
BaO(*s*)	−582	−552	70
$Ba(OH)_2$(*s*)	−946		
$BaSO_4$(*s*)	−1465	−1353	132
Beryllium			
Be(*s*)	0	0	10
BeO(*s*)	−599	−569	14
$Be(OH)_2$(*s*)	−904	−815	47
Bromine			
Br_2(*l*)	0	0	152
Br_2(*g*)	31	3	245
Br_2(*aq*)	−3	4	130
Br^-(*aq*)	−121	−104	82
HBr(*g*)	−36	−53	199
Cadmium			
Cd(*s*)	0	0	52
CdO(*s*)	−258	−228	55
$Cd(OH)_2$(*s*)	−561	−474	96
CdS(*s*)	−162	−156	65
$CdSO_4$(*s*)	−935	−823	123
Calcium			
Ca(*s*)	0	0	41
CaC_2(*s*)	−63	−68	70
$CaCO_3$(*s*)	−1207	−1129	93
CaO(*s*)	−635	−604	40
$Ca(OH)_2$(*s*)	−987	−899	83
$Ca_3(PO_4)_2$(*s*)	−4126	−3890	241
$CaSO_4$(*s*)	−1433	−1320	107
$CaSiO_3$(*s*)	−1630	−1550	84
Carbon			
C(*s*) (graphite)	0	0	6
C(*s*) (diamond)	2	3	2
CO(*g*)	−110.5	−137	198
CO_2(*g*)	−393.5	−394	214
CH_4(*g*)	−75	−51	186
CH_3OH(*g*)	−201	−163	240
CH_3OH(*l*)	−239	−166	127
H_2CO(*g*)	−116	−110	219
HCOOH(*g*)	−363	−351	249
Carbon, *continued*			
HCN(*g*)	135.1	125	202
C_2H_2(*g*)	227	209	201
C_2H_4(*g*)	52	68	219
CH_3CHO(*g*)	−166	−129	250
C_2H_5OH(*l*)	−278	−175	161
C_2H_6(*g*)	−84.7	−32.9	229.5
C_3H_6(*g*)	20.9	62.7	266.9
C_3H_8(*g*)	−104	−24	270
C_2H_4O(*g*) (ethylene oxide)	−53	−13	242
$CH_2{=}CHCN$(*l*)	152	190	274
CH_3COOH(*l*)	−484	−389	160
$C_6H_{12}O_6$(*s*)	−1275	−911	212
CCl_4	−135	−65	216
Chlorine			
Cl_2(*g*)	0	0	223
Cl_2(*aq*)	−23	7	121
Cl^-(*aq*)	−167	−131	57
HCl(*g*)	−92	−95	187
Chromium			
Cr(*s*)	0	0	24
Cr_2O_3(*s*)	−1128	−1047	81
CrO_3(*s*)	−579	−502	72
Copper			
Cu(*s*)	0	0	33
$CuCO_3$(*s*)	−595	−518	88
Cu_2O(*s*)	−170	−148	93
CuO(*s*)	−156	−128	43
$Cu(OH)_2$(*s*)	−450	−372	108
CuS(*s*)	−49	−49	67
Fluorine			
F_2(*g*)	0	0	203
F^-(*aq*)	−333	−279	−14
HF(*g*)	−271	−273	174
Hydrogen			
H_2(*g*)	0	0	131
H(*g*)	217	203	115
H^+(*aq*)	0	0	0
OH^-(*aq*)	−230	−157	−11
H_2O(*l*)	−286	−237	70
H_2O(*g*)	−242	−229	189
Iodine			
I_2(*s*)	0	0	116
I_2(*g*)	62	19	261
I_2(*aq*)	23	16	137
I^-(*aq*)	−55	−52	106

Appendix Four (continued)

Substance and state	ΔH_f° (kJ/mol)	ΔG_f° (kJ/mol)	S° (J/K mol)
Iron			
Fe(s)	0	0	27
Fe_3C(s)	21	15	108
$Fe_{0.95}O$(s) (wustite)	−264	−240	59
FeO	−272	−255	61
Fe_3O_4(s) (magnetite)	−1117	−1013	146
Fe_2O_3(s) (hematite)	−826	−740	90
FeS(s)	−95	−97	67
FeS_2(s)	−178	−166	53
$FeSO_4$(s)	−929	−825	121
Lead			
Pb(s)	0	0	65
PbO_2(s)	−277	−217	69
PbS(s)	−100	−99	91
$PbSO_4$(s)	−920	−813	149
Magnesium			
Mg(s)	0	0	33
$MgCO_3$(s)	−1113	−1029	66
MgO(s)	−602	−569	27
$Mg(OH)_2$(s)	−925	−834	64
Manganese			
Mn(s)	0	0	32
MnO(s)	−385	−363	60
Mn_3O_4(s)	−1387	−1280	149
Mn_2O_3(s)	−971	−893	110
MnO_2(s)	−521	−466	53
MnO_4^-(aq)	−543	−449	190
Mercury			
Hg(l)	0	0	76
Hg_2Cl_2(s)	−265	−211	196
$HgCl_2$(s)	−230	−184	144
HgO(s)	−90	−59	70
HgS(s)	−58	−49	78
Nickel			
Ni(s)	0	0	30
$NiCl_2$(s)	−316	−272	107
NiO(s)	−241	−213	38
$Ni(OH)_2$(s)	−538	−453	79
NiS(s)	−93	−90	53
Nitrogen			
N_2(g)	0	0	192
NH_3(g)	−46	−17	193
NH_3(aq)	−80	−27	111
NH_4^+(aq)	−132	−79	113
NO(g)	90	87	211
NO_2(g)	34	52	240
Nitrogen, *continued*			
N_2O(g)	82	104	220
N_2O_4(g)	10	98	304
N_2O_4(l)	−20	97	209
N_2O_5(s)	−42	134	178
N_2H_4(l)	51	149	121
$N_2H_3CH_3$(l)	54	180	166
HNO_3(aq)	−207	−111	146
HNO_3(l)	−174	−81	156
NH_4ClO_4(s)	−295	−89	186
NH_4Cl(s)	−314	−203	96
Oxygen			
O_2(g)	0	0	205
O(g)	249	232	161
O_3(g)	143	163	239
Phosphorus			
P(s) (white)	0	0	41
P(s) (red)	−18	−12	23
P(s) (black)	−39	−33	23
P_4(g)	59	24	280
PF_5(g)	−1578	−1509	296
PH_3(g)	5	13	210
H_3PO_4(s)	−1279	−1119	110
H_3PO_4(l)	−1267	—	—
H_3PO_4(aq)	−1288	−1143	158
P_4O_{10}(s)	−3110	−2698	229
Potassium			
K(s)	0	0	64
KCl(s)	−436	−408	83
$KClO_3$(s)	−391	−290	143
$KClO_4$(s)	−433	−304	151
K_2O(s)	−361	−322	98
K_2O_2(s)	−496	−430	113
KO_2(s)	−283	−238	117
KOH(s)	−425	−379	79
KOH(aq)	−481	−440	9.20
Silicon			
SiO_2(s) (quartz)	−911	−856	42
$SiCl_4$(l)	−687	−620	240
Silver			
Ag(s)	0	0	43
Ag^+(aq)	105	77	73
AgBr(s)	−100	−97	107
AgCN(s)	146	164	84
AgCl(s)	−127	−110	96
Ag_2CrO_4(s)	−712	−622	217
AgI(s)	−62	−66	115
Ag_2O(s)	−31	−11	122
Ag_2S(s)	−32	−40	146

Appendix Four (continued)

Substance and state	ΔH_f° (kJ/mol)	ΔG_f° (kJ/mol)	S° (J/K mol)
Sodium			
$Na(s)$	0	0	51
$Na^+(aq)$	−240	−262	59
$NaBr(s)$	−360	−347	84
$Na_2CO_3(s)$	−1131	−1048	136
$NaHCO_3(s)$	−948	−852	102
$NaCl(s)$	−411	−384	72
$NaH(s)$	−56	−33	40
$NaI(s)$	−288	−282	91
$NaNO_2(s)$	−359		
$NaNO_3(s)$	−467	−366	116
$Na_2O(s)$	−416	−377	73
$Na_2O_2(s)$	−515	−451	95
$NaOH(s)$	−427	−381	64
$NaOH(aq)$	−470	−419	50
Sulfur			
$S(s)$ (rhombic)	0	0	32
$S(s)$ (monoclinic)	0.3	0.1	33
$S^{2-}(aq)$	33	86	−15
$S_8(g)$	102	50	431
$SF_6(g)$	−1209	−1105	292
$H_2S(g)$	−21	−34	206
$SO_2(g)$	−297	−300	248
$SO_3(g)$	−396	−371	257
$SO_4^{2-}(aq)$	−909	−745	20
$H_2SO_4(l)$	−814	−690	157
$H_2SO_4(aq)$	−909	−745	20
Tin			
$Sn(s)$ (white)	0	0	52
$Sn(s)$ (gray)	−2	0.1	44
Tin, *continued*			
$SnO(s)$	−285	−257	56
$SnO_2(s)$	−581	−520	52
$Sn(OH)_2(s)$	−561	−492	155
Titanium			
$TiCl_4(g)$	−763	−727	355
$TiO_2(s)$	−945	−890	50
Uranium			
$U(s)$	0	0	50
$UF_6(s)$	−2137	−2008	228
$UF_6(g)$	−2113	−2029	380
$UO_2(s)$	−1084	−1029	78
$U_3O_8(s)$	−3575	−3393	282
$UO_3(s)$	−1230	−1150	99
Xenon			
$Xe(g)$	0	0	170
$XeF_2(g)$	−108	−48	254
$XeF_4(s)$	−251	−121	146
$XeF_6(g)$	−294		
$XeO_3(s)$	402		
Zinc			
$Zn(s)$	0	0	42
$ZnO(s)$	−348	−318	44
$Zn(OH)_2(s)$	−642		
$ZnS(s)$ (wurtzite)	−193		
$ZnS(s)$ (zinc blende)	−206	−201	58
$ZnSO_4(s)$	−983	−874	120

APPENDIX FIVE Equilibrium Constants and Reduction Potentials

A5.1 Values of K_a for Some Common Monoprotic Acids

Name	Formula	Value of K_a
Hydrogen sulfate ion	HSO_4^-	1.2×10^{-2}
Chlorous acid	$HClO_2$	1.2×10^{-2}
Monochloracetic acid	$HC_2H_2ClO_2$	1.35×10^{-3}
Hydrofluoric acid	HF	7.2×10^{-4}
Nitrous acid	HNO_2	4.0×10^{-4}
Formic acid	HCO_2H	1.8×10^{-4}
Lactic acid	$HC_3H_5O_3$	1.38×10^{-4}
Benzoic acid	$HC_7H_5O_2$	6.4×10^{-5}
Acetic acid	$HC_2H_3O_2$	1.8×10^{-5}
Hydrated aluminum(III) ion	$[Al(H_2O)_6]^{3+}$	1.4×10^{-5}
Propanoic acid	$HC_3H_5O_2$	1.3×10^{-5}
Hypochlorous acid	$HOCl$	3.5×10^{-8}
Hypobromous acid	$HOBr$	2×10^{-9}
Hydrocyanic acid	HCN	6.2×10^{-10}
Boric acid	H_3BO_3	5.8×10^{-10}
Ammonium ion	NH_4^+	5.6×10^{-10}
Phenol	HOC_6H_5	1.6×10^{-10}
Hypoiodous acid	HOI	2×10^{-11}

A5.2 Stepwise Dissociation Constants for Several Common Polyprotic Acids

Name	Formula	K_{a_1}	K_{a_2}	K_{a_3}
Phosphoric acid	H_3PO_4	7.5×10^{-3}	6.2×10^{-8}	4.8×10^{-13}
Arsenic acid	H_3AsO_4	5×10^{-3}	8×10^{-8}	6×10^{-10}
Carbonic acid	H_2CO_3	4.3×10^{-7}	5.6×10^{-11}	
Sulfuric acid	H_2SO_4	Large	1.2×10^{-2}	
Sulfurous acid	H_2SO_3	1.5×10^{-2}	1.0×10^{-7}	
Hydrosulfuric acid	H_2S	1.0×10^{-7}	1.3×10^{-13}	
Oxalic acid	$H_2C_2O_4$	6.5×10^{-2}	6.1×10^{-5}	
Ascorbic acid (Vitamin C)	$H_2C_6H_6O_6$	7.9×10^{-5}	1.6×10^{-12}	
Citric acid	$H_3C_6H_5O_7$	8.4×10^{-4}	1.8×10^{-5}	4.0×10^{-6}

A5.3 Values of K_b for Some Common Weak Bases

Name	Formula	Conjugate acid	K_b
Ammonia	NH_3	NH_4^+	1.8×10^{-5}
Methylamine	CH_3NH_2	$CH_3NH_3^+$	4.38×10^{-4}
Ethylamine	$C_2H_5NH_2$	$C_2H_5NH_3^+$	5.6×10^{-4}
Diethylamine	$(C_2H_5)_2NH$	$(C_2H_5)_2NH_2^+$	1.3×10^{-3}
Triethylamine	$(C_2H_5)_3N$	$(C_2H_5)_3NH^+$	4.0×10^{-4}
Hydroxylamine	$HONH_2$	$HONH_3^+$	1.1×10^{-8}
Hydrazine	H_2NNH_2	$H_2NNH_3^+$	3.0×10^{-6}
Aniline	$C_6H_5NH_2$	$C_6H_5NH_3^+$	3.8×10^{-10}
Pyridine	C_5H_5N	$C_5H_5NH^+$	1.7×10^{-9}

A5.4 K_{sp} Values at 25°C for Common Ionic Solids

Ionic solid	K_{sp} (at 25°C)
Fluorides	
BaF_2	2.4×10^{-5}
MgF_2	6.4×10^{-9}
PbF_2	4×10^{-8}
SrF_2	7.9×10^{-10}
CaF_2	4.0×10^{-11}
Chlorides	
$PbCl_2$	1.6×10^{-5}
$AgCl$	1.6×10^{-10}
Hg_2Cl_2*	1.1×10^{-18}
Bromides	
$PbBr_2$	4.6×10^{-6}
$AgBr$	5.0×10^{-13}
Hg_2Br_2*	1.3×10^{-22}
Iodides	
PbI_2	1.4×10^{-8}
AgI	1.5×10^{-16}
Hg_2I_2*	4.5×10^{-29}
Sulfates	
$CaSO_4$	6.1×10^{-5}
Ag_2SO_4	1.2×10^{-5}
$SrSO_4$	3.2×10^{-7}
$PbSO_4$	1.3×10^{-8}
$BaSO_4$	1.5×10^{-9}
Chromates	
$SrCrO_4$	3.6×10^{-5}

Ionic solid	K_{sp} (at 25°C)
Hg_2CrO_4*	2×10^{-9}
$BaCrO_4$	8.5×10^{-11}
Ag_2CrO_4	9.0×10^{-12}
$PbCrO_4$	2×10^{-16}
Carbonates	
$NiCO_3$	1.4×10^{-7}
$CaCO_3$	8.7×10^{-9}
$BaCO_3$	1.6×10^{-9}
$SrCO_3$	7×10^{-10}
$CuCO_3$	2.5×10^{-10}
$ZnCO_3$	2×10^{-10}
$MnCO_3$	8.8×10^{-11}
$FeCO_3$	2.1×10^{-11}
Ag_2CO_3	8.1×10^{-12}
$CdCO_3$	5.2×10^{-12}
$PbCO_3$	1.5×10^{-15}
$MgCO_3$	1×10^{-15}
Hg_2CO_3*	9.0×10^{-15}
Hydroxides	
$Ba(OH)_2$	5.0×10^{-3}
$Sr(OH)_2$	3.2×10^{-4}
$Ca(OH)_2$	1.3×10^{-6}
$AgOH$	2.0×10^{-8}
$Mg(OH)_2$	8.9×10^{-12}
$Mn(OH)_2$	2×10^{-13}
$Cd(OH)_2$	5.9×10^{-15}
$Pb(OH)_2$	1.2×10^{-15}
$Fe(OH)_2$	1.8×10^{-15}

Ionic solid	K_{sp} (at 25°C)
$Co(OH)_2$	2.5×10^{-16}
$Ni(OH)_2$	1.6×10^{-16}
$Zn(OH)_2$	4.5×10^{-17}
$Cu(OH)_2$	1.6×10^{-19}
$Hg(OH)_2$	3×10^{-26}
$Sn(OH)_2$	3×10^{-27}
$Cr(OH)_3$	6.7×10^{-31}
$Al(OH)_3$	2×10^{-32}
$Fe(OH)_3$	4×10^{-38}
$Co(OH)_3$	2.5×10^{-43}
Sulfides	
MnS	2.3×10^{-13}
FeS	3.7×10^{-19}
NiS	3×10^{-21}
CoS	5×10^{-22}
ZnS	2.5×10^{-22}
SnS	1×10^{-26}
CdS	1.0×10^{-28}
PbS	7×10^{-29}
CuS	8.5×10^{-45}
Ag_2S	1.6×10^{-49}
HgS	1.6×10^{-54}
Phosphates	
Ag_3PO_4	1.8×10^{-18}
$Sr_3(PO_4)_2$	1×10^{-31}
$Ca_3(PO_4)_2$	1.3×10^{-32}
$Ba_3(PO_4)_2$	6×10^{-39}
$Pb_3(PO_4)_2$	1×10^{-54}

*Contains Hg_2^{2+} ions. $K_{sp} = [Hg_2^{2+}][X^-]^2$ for Hg_2X_2 salts.

A5.5 Standard Reduction Potentials at 25°C (298 K) for Many Common Half-reactions

Half-reaction	$\mathscr{E}°$ (V)	Half-reaction	$\mathscr{E}°$ (V)
$F_2 + 2e^- \rightarrow 2F^-$	2.87	$O_2 + 2H_2O + 4e^- \rightarrow 4OH^-$	0.40
$Ag^{2+} + e^- \rightarrow Ag^+$	1.99	$Cu^{2+} + 2e^- \rightarrow Cu$	0.34
$Co^{3+} + e^- \rightarrow Co^{2+}$	1.95	$Hg_2Cl_2 + 2e^- \rightarrow 2Hg + 2Cl^-$	0.34
$H_2O_2 + 2H^+ + 2e^- \rightarrow 2H_2O$	1.78	$AgCl + e^- \rightarrow Ag + Cl^-$	0.22
$Ce^{4+} + e^- \rightarrow Ce^{3+}$	1.70	$SO_4{}^{2-} + 4H^+ + 2e^- \rightarrow H_2SO_3 + H_2O$	0.20
$PbO_2 + 4H^+ + SO_4{}^{2-} + 2e^- \rightarrow PbSO_4 + 2H_2O$	1.69	$Cu^{2+} + e^- \rightarrow Cu^+$	0.16
$MnO_4{}^- + 4H^+ + 3e^- \rightarrow MnO_2 + 2H_2O$	1.68	$2H^+ + 2e^- \rightarrow H_2$	0.00
$2e^- + 2H^+ + IO_4{}^- \rightarrow IO_3{}^- + H_2O$	1.60	$Fe^{3+} + 3e^- \rightarrow Fe$	−0.036
$MnO_4{}^- + 8H^+ + 5e^- \rightarrow Mn^{2+} + 4H_2O$	1.51	$Pb^{2+} + 2e^- \rightarrow Pb$	−0.13
$Au^{3+} + 3e^- \rightarrow Au$	1.50	$Sn^{2+} + 2e^- \rightarrow Sn$	−0.14
$PbO_2 + 4H^+ + 2e^- \rightarrow Pb^{2+} + 2H_2O$	1.46	$Ni^{2+} + 2e^- \rightarrow Ni$	−0.23
$Cl_2 + 2e^- \rightarrow 2Cl^-$	1.36	$PbSO_4 + 2e^- \rightarrow Pb + SO_4{}^{2-}$	−0.35
$Cr_2O_7{}^{2-} + 14H^+ + 6e^- \rightarrow 2Cr^{3+} + 7H_2O$	1.33	$Cd^{2+} + 2e^- \rightarrow Cd$	−0.40
$O_2 + 4H^+ + 4e^- \rightarrow 2H_2O$	1.23	$Fe^{2+} + 2e^- \rightarrow Fe$	−0.44
$MnO_2 + 4H^+ + 2e^- \rightarrow Mn^{2+} + 2H_2O$	1.21	$Cr^{3+} + e^- \rightarrow Cr^{2+}$	−0.50
$IO_3{}^- + 6H^+ + 5e^- \rightarrow \frac{1}{2}I_2 + 3H_2O$	1.20	$Cr^{3+} + 3e^- \rightarrow Cr$	−0.73
$Br_2 + 2e^- \rightarrow 2Br^-$	1.09	$Zn^{2+} + 2e^- \rightarrow Zn$	−0.76
$VO_2{}^+ + 2H^+ + e^- \rightarrow VO^{2+} + H_2O$	1.00	$2H_2O + 2e^- \rightarrow H_2 + 2OH^-$	−0.83
$AuCl_4{}^- + 3e^- \rightarrow Au + 4Cl^-$	0.99	$Mn^{2+} + 2e^- \rightarrow Mn$	−1.18
$NO_3{}^- + 4H^+ + 3e^- \rightarrow NO + 2H_2O$	0.96	$Al^{3+} + 3e^- \rightarrow Al$	−1.66
$ClO_2 + e^- \rightarrow ClO_2{}^-$	0.954	$H_2 + 2e^- \rightarrow 2H^-$	−2.23
$2Hg^{2+} + 2e^- \rightarrow Hg_2{}^{2+}$	0.91	$Mg^{2+} + 2e^- \rightarrow Mg$	−2.37
$Ag^+ + e^- \rightarrow Ag$	0.80	$La^{3+} + 3e^- \rightarrow La$	−2.37
$Hg_2{}^{2+} + 2e^- \rightarrow 2Hg$	0.80	$Na^+ + e^- \rightarrow Na$	−2.71
$Fe^{3+} + e^- \rightarrow Fe^{2+}$	0.77	$Ca^{2+} + 2e^- \rightarrow Ca$	−2.76
$O_2 + 2H^+ + 2e^- \rightarrow H_2O_2$	0.68	$Ba^{2+} + 2e^- \rightarrow Ba$	−2.90
$MnO_4{}^- + e^- \rightarrow MnO_4{}^{2-}$	0.56	$K^+ + e^- \rightarrow K$	−2.92
$I_2 + 2e^- \rightarrow 2I^-$	0.54	$Li^+ + e^- \rightarrow Li$	−3.05
$Cu^+ + e^- \rightarrow Cu$	0.52		

GLOSSARY

Accuracy the agreement of a particular value with the true value. (1.3)

Acid a substance that produces hydrogen ions in solution; a proton donor. (4.2)

Acid-base indicator a substance that marks the end point of an acid-base titration by changing color. (15.5)

Acid dissociation constant (K_a) the equilibrium constant for a reaction in which a proton is removed from an acid by H_2O to form the conjugate base and H_3O^+. (14.1)

Acid rain a result of air pollution by sulfur dioxide. (5.9)

Acidic oxide a covalent oxide that dissolves in water to give an acidic solution. (14.10)

Actinide series a group of fourteen elements following actinium in the periodic table, in which the 5*f* orbitals are being filled. (7.11; 18.1)

Activated complex (transition state) the arrangement of atoms found at the top of the potential energy barrier as a reaction proceeds from reactants to products. (12.7)

Activation energy the threshold energy that must be overcome to produce a chemical reaction. (12.7)

Addition polymerization a type of polymerization in which the monomers simply add together to form the polymer, with no other products. (22.5)

Addition reaction a reaction in which atoms add to a carbon-carbon multiple bond. (22.2)

Adsorption the collection of one substance on the surface of another. (12.8)

Air pollution contamination of the atmosphere, mainly by the gaseous products of transportation and production of electricity. (5.9)

Alcohol an organic compound in which the hydroxyl group is a substituent on a hydrocarbon. (22.4)

Aldehyde an organic compound containing the carbonyl group bonded to at least one hydrogen atom. (22.4)

Alkali metal a Group 1A metal. (2.7; 18.2)

Alkaline earth metal a Group 2A metal. (2.7; 18.4)

Alkane a saturated hydrocarbon with the general formula C_nH_{2n+2}. (22.1)

Alkene an unsaturated hydrocarbon containing a carbon-carbon double bond. The general formula is C_nH_{2n}. (22.2)

Alkyne an unsaturated hydrocarbon containing a triple carbon-carbon bond. The general formula is C_nH_{2n-2}. (22.2)

Alloy a substance that contains a mixture of elements and has metallic properties. (10.4)

Alloy steel a form of steel containing carbon plus other metals such as chromium, cobalt, manganese, and molybdenum. (24.4)

Alpha (α) particle a helium nucleus. (21.1)

Alpha particle production a common mode of decay for radioactive nuclides in which the mass number changes. (21.1)

Amine an organic base derived from ammonia in which one or more of the hydrogen atoms are replaced by organic groups. (14.6; 22.4)

α-Amino acid an organic acid in which an amino group and an R group are attached to the carbon atom next to the carboxyl group. (23.1)

Amorphous solid a solid with considerable disorder in its structure. (10.3)

Ampere the unit of electrical current equal to one coulomb of charge per second. (17.7)

Amphoteric substance a substance that can behave either as a acid or as a base. (14.2)

Anion a negative ion. (2.6)

Anode the electrode in a galvanic cell at which oxidation occurs. (17.1)

Antibonding molecular orbital an orbital higher in energy than the atomic orbitals of which it is composed. (9.2)

Aqueous solution a solution in which water is the dissolving medium or solvent. (4.0)

Aromatic hydrocarbon one of a special class of cyclic unsaturated hydrocarbons, the simplest of which is benzene. (22.3)

Arrhenius concept a concept postulating that acids produce hydrogen ions in aqueous solution, while bases produce hydroxide ions. (14.1)

Arrhenius equation the equation representing the rate constant as $k = Ae^{-E_a/RT}$ where A represents the product of the collision frequency and the steric factor, and $e^{-E_a/RT}$ is the fraction of collisions with sufficient energy to produce a reaction. (12.7)

Atactic chain a polymer chain in which the substituent groups such as CH_3 are randomly distributed along the chain. (24.2)

Atmosphere the mixture of gases that surrounds the earth's surface. (5.9)

Atomic number the number of protons in the nucleus of an atom. (2.5)

Atomic radius half the distance between the nuclei in a molecule consisting of identical atoms. (7.13)

Atomic solid a solid that contains atoms at the lattice points. (10.3)

Atomic weight the weighted average mass of the atoms in a naturally occurring element. (2.3)

Aufbau principle the principle stating that as protons are added one by one to the nucleus to build up the elements, electrons are similarly added to hydrogenlike orbitals. (7.11)

Autoionization the transfer of a proton from one molecule to another of the same substance. (14.2)

Avogadro's law equal volumes of gases at the same temperature and pressure contain the same number of particles. (5.2)

Avogadro's number the number of atoms in exactly 12 grams of pure ^{12}C, equal to 6.022×10^{23}. (3.2)

Azimuthal quantum number (ℓ) the quantum number relating to the shape of an atomic orbital, which can assume any integral value from 0 to $n - 1$ for each value of n. (7.6)

Ball-and-stick model a molecular model that distorts the sizes of atoms, but shows bond relationships clearly. (2.6)

Band model a molecular model for metals in which the electrons are assumed to travel around the metal crystal in molecular orbitals formed from the valence atomic orbitals of the metal atoms. (10.4)

Barometer a device for measuring atmospheric pressure. (5.1)

Base a substance that produces hydroxide ions in aqueous solution, a proton acceptor. (4.2)

Base dissociation constant (K_b) the equilibrium constant for the reaction of a base with water to produce the conjugate acid and hydroxide ion. (14.6)

Basic oxide an ionic oxide that dissolves in water to produce a basic solution. (14.10)

Basic oxygen process a process for producing steel by oxidizing and removing the impurities in iron using a high-pressure blast of oxygen. (24.4)

Battery a group of galvanic cells connected in series. (17.5)

Beta (β) particle an electron produced in radioactive decay. (21.1)

Beta particle production a decay process for radioactive nuclides in which the mass number remains constant and the atomic number changes. The net effect is to change a neutron to a proton. (21.1)

Bidentate ligand a ligand that can form two bonds to a metal ion. (20.3)

Bilayer a portion of the cell membrane consisting of phospholipids with their nonpolar tails in the interior and their polar heads interfacing with the polar water molecules. (23.4)

Bimolecular step a reaction involving the collision of two molecules. (12.6)

Binary compound a two-element compound. (2.8)

Binding energy (nuclear) the energy required to decompose a nucleus into its component nucleons. (21.5)

Biochemistry the study of the chemistry of living systems. (23)

Biomolecule a molecule responsible for maintaining and/or reproducing life. (22)

Blast furnace a furnace in which iron oxide is reduced to iron metal by using a very strong blast of hot air to produce carbon monoxide from coke, and then using this gas as a reducing agent for the iron. (24.4)

Bond energy the energy required to break a given chemical bond. (8.1)

Bond length the distance between the nuclei of the two atoms connected by a bond; the distance where the total energy of a diatomic molecule is minimal. (8.1)

Bond order the difference between the number of bonding electrons and the number of antibonding electrons, divided by two. It is an index of bond strength. (9.2)

Bonding molecular orbital an orbital lower in energy than the atomic orbitals of which it is composed. (9.2)

Bonding pair an electron pair found in the space between two atoms. (8.9)

Borane a covalent hydride of boron. (18.5)

Boyle's law the volume of a given sample of gas at constant temperature varies inversely with the pressure. (5.2)

Breeder reactor a nuclear reactor in which fissionable fuel is produced while the reactor runs. (21.6)

Brönsted-Lowry model a model proposing that an acid is a proton donor, and a base is a proton acceptor. (14.1)

Buffer capacity the ability of a buffered solution to absorb protons or hydroxide ions without a significant change in pH; determined by the magnitudes of [HA] and$[A^-]$ in the solution. (15.3)

Buffered solution a solution that resists a change in its pH when either hydroxide ions or protons are added. (15.2)

Calorimetry the science of measuring heat flow. (6.2)

Capillary action the spontaneous rising of a liquid in a narrow tube. (10.2)

Carbohydrate a polyhydroxyl ketone or polyhydroxyl aldehyde or a polymer composed of these. (23.2)

Carbon steel an alloy of iron containing up to about 1.5% carbon. (24.4)

Carboxyhemoglobin a stable complex of hemoglobin and carbon monoxide that prevents normal oxygen uptake in the blood. (20.8)

Carboxyl group the —COOH group in an organic acid. (14.2)

Carboxylic acid an organic compound containing the carboxyl group; an acid with the general formula RCOOH. (22.4)

Catalyst a substance that speeds up a reaction without being consumed. (12.8)

Cathode the electrode in a galvanic cell at which reduction occurs. (17.1)

Cathode rays the "rays" emanating from the negative electrode (cathode) in a partially evacuated tube; a stream of electrons. (2.4)

Cathodic protection a method in which an active metal, such as magnesium, is connected to steel in order to protect it from corrosion. (17.7)

Cation a positive ion. (2.6)

Cell potential (electromotive force) the driving force in a galvanic cell that pulls electrons from the reducing agent in one compartment to the oxidizing agent in the other. (17.1)

Ceramic a nonmetallic material made from clay and hardened by firing at high temperature; it contains minute silicate crystals suspended in a glassy cement. (10.5)

Chain reaction (nuclear) a self-sustaining fission process caused by the production of neutrons which proceed to split other nuclei. (21.6)

Charles's law the volume of a given sample of gas at constant pressure is directly proportional to the temperature in kelvins. (5.2)

Chelating ligand (chelate) a ligand having more than one atom with a lone pair that can be used to bond to a metal ion. (20.3)

Chemical bond the force or, more accurately, the energy, that holds two atoms together in a compound. (2.6)

Chemical change the change of substances into other substances through a reorganization of the atoms; a chemical reaction. (1.8)

Chemical equation a representation of a chemical reaction showing the relative numbers of reactant and product molecules. (3.6)

Chemical equilibrium a dynamic reaction system in which the concentrations of all reactants and products remain constant as a function of time. (13)

Chemical formula the representation of a molecule in which the symbols for the elements are used to indicate the types of atoms present and subscripts are used to show the relative numbers of atoms. (2.6)

Chemical kinetics the area of chemistry that concerns reaction rates. (12.1)

Chemical stoichiometry the calculation of the quantities of material consumed and produced in chemical reactions. (3)

Chirality the quality of having nonsuperimposable mirror images. (20.4)

Chlor-alkali process the process for producing chlorine and sodium hydroxide by electrolyzing brine in a mercury cell. (17.8)

Chromatography the general name for a series of methods for separating mixtures by employing a system with a mobile phase and a stationary phase. (1.9)

Coagulation the destruction of a colloid by causing particles to aggregate and settle out. (11.8)

Codons organic bases in sets of three that form the genetic code. (23.3)

Colligative properties properties of a solution that depend only on the number, and not on the identity, of the solute particles. (11.5)

Collision model a model based on the idea that molecules must collide to react; used to account for the observed characteristics of reaction rates. (12.7)

Colloid a suspension of particles in a dispersing medium. (11.8)

Combustion reaction the vigorous and exothermic reaction that takes place between certain substances, particularly organic compounds, and oxygen. (22.1)

Common ion effect the shift in an equilibrium position caused by the addition or presence of an ion involved in the equilibrium reaction. (15.1)

Complete ionic reaction an equation that shows all substances that are strong electrolytes as ions. (4.6)

Complex ion a charged species consisting of a metal ion surrounded by ligands. (15.8; 20.1)

Compound a substance with constant composition that can be broken down into elements by chemical processes. (1.8)

Concentration cell a galvanic cell in which both compartments contain the same components, but at different concentrations. (17.4)

Condensation the process by which vapor molecules reform a liquid. (10.8)

Condensation polymerization a type of polymerization in which the formation of a small molecule, such as water, accompanies the extension of the polymer chain. (22.5)

Condensation reaction a reaction in which two molecules are joined, accompanied by the elimination of a water molecule. (19.3)

Condensed states of matter liquids and solids. (10.1)

Conduction bands the molecular orbitals that can be occupied by mobile electrons, which are free to travel throughout a metal crystal to conduct electricity or heat. (10.4)

Conjugate acid the species formed when a proton is added to a base. (14.1)

Conjugate acid-base pair two species related to each other by the donating and accepting of a single proton. (14.1)

Conjugate base what remains of an acid molecule after a proton is lost. (14.1)

Continuous spectrum a spectrum that exhibits all the wavelengths of visible light. (7.3)

Control rods rods in a nuclear reactor composed of substances that absorb neutrons. These rods regulate the power level of the reactor. (21.6)

Coordinate covalent bond a metal-ligand bond resulting from the interaction of a Lewis base (the ligand) and a Lewis acid (the metal ion). (20.3)

Coordination compound a compound composed of a complex ion and counter ions sufficient to give no net charge. (20.3)

Coordination isomerism isomerism in a coordination compound in which the composition of the coordination sphere of the metal ion varies. (20.4)

Coordination number the number of bonds formed between the metal ion and the ligands in a complex ion. (20.3)

Copolymer a polymer formed from the polymerization of more than one type of monomer. (22.5)

Core electron an inner electron in an atom; one not in the outermost (valence) principal quantum level. (7.11)

Corrosion the process by which metals are oxidized in the atmosphere. (17.6)

Coulomb's law $E = 2.31 \times 10^{-19}\left(\frac{Q_1Q_2}{r}\right)$, where E is the energy of interaction between a pair of ions, expressed in joules; r is the distance between the ion centers in nm; and Q_1 and Q_2 are the numerical ion charges. (8.1)

Counter ions anions or cations that balance the charge on the complex ion in a coordination compound. (20.3)

Covalent bonding a type of bonding in which electrons are shared by atoms. (8.1)

Cracking a process whereby large molecules of petroleum components are broken down to smaller ones by breaking carbon-carbon bonds. (24.1)

Critical mass the mass of fissionable material required to produce a self-sustaining chain reaction. (21.6)

Critical point the point on a phase diagram at which the temperature and pressure have their critical values; the end point of the liquid-vapor line. (10.9)

Critical pressure the minimum pressure required to produce liquefaction of a substance at the critical temperature. (10.9)

Critical reaction (nuclear) a reaction in which exactly one neutron from each fission event causes another fission event, thus sustaining the chain reaction. (21.6)

Critical temperature the temperature above which vapor cannot be liquefied no matter what pressure is applied. (10.9)

Crosslinking the existence of bonds between adjacent chains in a polymer, thus adding strength to the material. (24.2)

Crystal field model a model used to explain the magnetism and colors of coordination complexes through the splitting of the *d* orbital energies. (20.6)

Crystalline solid a solid with a regular arrangement of its components. (10.3)

Cubic closest packed structure a solid modeled by the closest packing of spheres with an *abcabc* arrangement of layers; the unit cell is face-centered cubic. (10.4)

Cyanidation a process in which crushed gold ore is treated with an aqueous cyanide solution in the presence of air to dissolve the gold. Pure gold is recovered by reduction of the ion to the metal. (24.4)

Cyclotron a type of particle accelerator in which an ion introduced at the center is accelerated in an expanding spiral path by use of alternating electrical fields in the presence of a magnetic field. (21.3)

Cytochromes a series of iron-containing species composed of heme and a protein. Cytochromes are the principal electron-transfer molecules in the respiratory chain. (20.8)

Dalton's law of partial pressures for a mixture of gases in a container, the total pressure exerted is the sum of the pressures that each gas would exert if it were alone. (5.5)

DDT 1,1-bis(4-chlorophenyl)-2,2,2-trichloroethane, the first highly successful organic insecticide, now banned in many countries because of damage to animals. (24.3)

Degenerate orbitals a group of orbitals with the same energy. (7.7)

Dehydrogenation reaction a reaction in which two hydrogen atoms are removed from adjacent carbons of a saturated hydrocarbon, giving an unsaturated hydrocarbon. (22.1)

Delocalization the condition where the electrons in a molecule are not localized between a pair of atoms but can move throughout the molecule. (8.9)

Denaturation the breaking down of the three-dimensional structure of a protein resulting in the loss of its function. (23.1)

Denitrification the return of nitrogen from decomposed matter to the atmosphere by bacteria that change nitrates to nitrogen gas. (19.2)

Density a property of matter representing the mass per unit volume. (1.7)

Deoxyribonucleic acid (DNA) a huge nucleotide polymer having a double helical structure with complementary bases on the two strands. Its major functions are protein synthesis and the storage and transport of genetic information. (23.3)

Desalination the removal of dissolved salts from an aqueous solution. (11.6)

Dialysis a phenomenon in which a semipermeable membrane allows transfer of both solvent molecules and small solute molecules and ions. (11.6)

Diamagnetism a type of magnetism, associated with paired electrons, that causes a substance to be repelled from the inducing magnetic field. (9.3)

Differential rate law an expression that gives the rate of a reaction as a function of concentrations; often called the rate law. (12.2)

Diffraction the scattering of light from a regular array of points or lines, producing constructive and destructive interference. (7.2)

Diffusion the mixing of gases. (5.7)

Dilution the process of adding solvent to lower the concentration of solute in a solution. (4.3)

Dimer a molecule formed by the joining of two identical monomers. (22.5)

Dipole-dipole attraction the attractive force resulting when polar molecules line up so that the positive and negative ends are close to each other. (10.1)

Dipole moment a property of a molecule whose charge distribution can be represented by a center of positive charge and a center of negative charge. (8.3)

Direct reduction furnace a furnace in which iron oxide is reduced to iron metal using milder reaction conditions than in a blast furnace. (24.4)

Disaccharide a sugar formed from two monosaccharides joined by a glycoside linkage. (23.2)

Disproportionation reaction a reaction in which a given element is both oxidized and reduced. (19.7)

Distillation a method for separating the components of a liquid mixture that depends on differences in the ease of vaporization of the components. (1.9)

Disulfide linkage an S—S bond that stabilizes the tertiary structure of many proteins. (23.1)

Double bond a bond in which two pairs of electrons are shared by two atoms. (8.8)

Downs cell a cell used for electrolyzing molten sodium chloride. (17.8)

Dry cell battery a common battery used in calculators, watches, radios, and tape players. (17.5)

Dual nature of light the statement that light exhibits both wave and particulate properties. (7.2)

Effective nuclear charge the apparent nuclear charge exerted on a particular electron, equal to the actual nuclear charge minus the effect of electron repulsions. (7.9)

Effusion the passage of a gas through a tiny orifice into an evacuated chamber. (5.7)

Elastomer a material that recovers its shape after a deforming force is removed. (24.2)

Electrical conductivity the ability to conduct an electric current. (4.2)

Electrochemistry the study of the interchange of chemical and electrical energy. (17)

Electrolysis a process that involves forcing a current through a cell to cause a nonspontaneous chemical reaction to occur. (17.7)

Electrolyte a material that dissolves in water to give a solution that conducts an electric current. (4.2)

Electrolytic cell a cell that uses electrical energy to produce a chemical change that would otherwise not occur spontaneously. (17.7)

Electromagnetic radiation radiant energy that exhibits wavelike behavior and travels through space at the speed of light in a vacuum. (7.1)

Electron a negatively charged particle that moves around the nucleus of an atom. (2.4)

Electron affinity the energy change associated with the addition of an electron to a gaseous atom. (7.13)

Electron capture a process in which one of the inner-orbital electrons in an atom is captured by the nucleus. (21.1)

Electron sea model a model for metals postulating a regular array of cations in a "sea" of electrons. (10.4)

Electron spin quantum number a quantum number representing one of the two possible values for the electron spin; either $+\frac{1}{2}$ or $-\frac{1}{2}$. (7.8)

Electronegativity the tendency of an atom in a molecule to attract shared electrons to itself. (8.2)

Element a substance that cannot be decomposed into simpler substances by chemical or physical means. (1.8)

Elementary step a reaction whose rate law can be written from its molecularity. (12.6)

$E = mc^2$ Einstein's equation proposing that energy has mass; E is energy, m is mass, and c is the speed of light. (7.2)

Empirical formula the simplest whole number ratio of atoms in a compound. (3.5)

Enantiomers isomers that are nonsuperimposable mirror images of each other. (20.4)

End point the point in a titration at which the indicator changes color. (4.9)

Endothermic refers to a reaction where energy (as heat) flows into the system. (6.1)

Energy the capacity to do work or to cause heat flow. (6.1)

Enology the science of winemaking. (24.5)

Enthalpy a property of a system equal to $E + PV$, where E is the internal energy of the system, P is the pressure of the system, and V is the volume of the system. At constant pressure the change in enthalpy equals the energy flow as heat. (6.2)

Enthalpy of fusion the enthalpy change that occurs to melt a solid at its melting point. (10.8)

Entropy a thermodynamic function that measures randomness or disorder. (16.1)

Enzyme a large molecule, usually a protein, that catalyzes biological reactions. (12.8; 23.1)

Equilibrium (thermodynamic definition) the position where the free energy of a reaction system has its lowest possible value. (16.8)

Equilibrium constant the value obtained when equilibrium concentrations of the chemical species are substituted in the equilibrium expression. (13.2)

Equilibrium expression the expression (from the law of mass action) obtained by multiplying the product concentrations and dividing by the multiplied reactant concentrations, with each concentration raised to a power represented by the coefficient in the balanced equation. (13.2)

Equilibrium position a particular set of equilibrium concentrations. (13.2)

Equivalence point (stoichiometric point) the point in a titration when enough titrant has been added to react exactly with the substance in solution being titrated. (4.9; 15.4)

Essential elements the elements definitely known to be essential to human life. (23)

Ester an organic compound produced by the reaction between a carboxylic acid and an alcohol. (22.4)

Exothermic refers to a reaction where energy (as heat) flows out of the system. (6.1)

Exponential notation expresses a number as $N \times 10^M$, a convenient method for representing a very large or very small number and for easily indicating the number of significant figures. (1.4)

Faraday a constant representing the charge on one mole of electrons; 96,485 coulombs. (17.3)

Fat (glyceride) an ester composed of glycerol and fatty acids. (23.4)

Fatty acid a long-chain carboxylic acid. (23.4)

Feedstock the portion of petroleum used as the raw material in a petrochemical plant; it varies from heavy hydrocarbons to light ones. (24.1)

Fermentation (as in winemaking) the conversion of sugar to ethanol using yeast. (24.5)

Filtration a method for separating the components of a mixture containing a solid and a liquid. (1.9)

First law of thermodynamics the energy of the universe is constant; same as the law of conservation of energy. (6.1)

Fission the process of using a neutron to split a heavy nucleus into two nuclei with smaller mass numbers. (21.6)

Flotation process a method of separating the mineral particles in an ore from the gangue that depends on the greater wettability of the mineral pieces. (24.4)

Formal charge the charge assigned to an atom in a molecule or polyatomic ion derived from a specific set of rules. (8.12)

Formation constant (stability constant) the equilibrium constant for each step of the formation of a complex ion by the addition of an individual ligand to a metal ion or complex ion in aqueous solution. (15.8)

Fossil fuel coal, petroleum, or natural gas; consists of carbon-based molecules derived from decomposition of once-living organisms. (6.5)

Frasch process the recovery of sulfur from underground deposits by melting it with hot water and forcing it to the surface by air pressure. (19.6)

Free energy a thermodynamic function equal to the enthalpy (H) minus the product of the entropy (S) and the Kelvin temperature (T); $G = H - TS$. Under certain conditions the change in free energy for a process is equal to the maximum useful work. (16.4)

Free radical a species with an unpaired electron. (22.5)

Frequency the number of waves (cycles) per second that pass a given point in space. (7.1)

Fuel cell a galvanic cell for which the reactants are continuously supplied. (17.5)

Functional group an atom or group of atoms in hydrocarbon derivatives that contains elements in addition to carbon and hydrogen. (22.4)

Fusion the process of combining two light nuclei to form a heavier, more stable nucleus. (21.6)

Galvanic cell a device in which chemical energy from a spontaneous redox reaction is changed to electrical energy that can be used to do work. (17.1)

Galvanizing a process in which steel is coated with zinc to prevent corrosion. (17.6)

Gamma (γ) ray a high-energy photon. (21.1)

Gangue the impurities (such as clay or sand) in an ore. (24.4)

Geiger-Müller counter (Geiger counter) an instrument that measures the rate of radioactive decay based on the ions and electrons produced as a radioactive particle passes through a gas-filled chamber. (21.4)

Gene a given segment of the DNA molecule that contains the code for a specific protein. (23.3)

Geometrical *(cis-trans)* isomerism isomerism in which atoms or groups of atoms can assume different positions around a rigid ring or bond. (20.4; 22.2)

Glass an amorphous solid obtained when silica is mixed with other compounds, heated above its melting point, and then cooled rapidly. (10.5)

Glass electrode an electrode for measuring pH from the potential difference that develops when it is dipped into an aqueous solution containing H^+ ions. (17.4)

Glycosidic linkage a C—O—C bond formed between the rings of two cyclic monosaccharides by the elimination of water. (23.2)

Graham's law of effusion the rate of effusion of a gas is inversely proportional to the square root of the mass of its particles. (5.7)

Gravimetric analysis a method for determining the amount of a given substance in a solution by precipitation, filtration, drying, and weighing. (4.8)

Greenhouse effect a warming effect exerted by the earth's atmosphere (particularly CO_2 and H_2O) due to thermal energy retained by absorption of infrared radiation. (6.5)

Ground state the lowest possible energy state of an atom or molecule. (7.4)

Group (of the periodic table) a vertical column of elements having the same valence electron configuration and showing similar properties. (2.7)

Haber process the manufacture of ammonia from nitrogen and hydrogen, carried out at high pressure and high temperature with the aid of a catalyst. (3.9; 19.2)

Half-life (of a radioactive sample) the time required for the number of nuclides in a radioactive sample to reach half of the original value. (21.2)

Half-life (of a reaction) the time required for a reactant to reach half of its original concentration. (12.4)

Half-reactions the two parts of an oxidation-reduction reaction, one representing oxidation, the other reduction. (4.11; 17.1)

Halogen a Group 7A element. (2.7; 19.7)

Halogenation the addition of halogen atoms to unsaturated hydrocarbons. (22.2)

Hard water water from natural sources that contains relatively large concentrations of calcium and magnesium ions. (18.4)

Heat energy transferred between two objects due to a temperature difference between them. (6.1)

Heat capacity the amount of energy required to raise the temperature of an object by one degree Celsius. (6.2)

Heat of fusion the enthalpy change that occurs to melt a solid at its melting point. (10.8)

Heat of hydration the enthalpy change associated with placing gaseous molecules or ions in water; the sum of the energy needed to expand the solvent and the energy released from the solvent-solute interactions. (11.2)

Heat of solution the enthalpy change associated with dissolving a solute in a solvent; the sum of the energies needed to expand both solvent and solute in a solution and the energy released from the solvent-solute interactions. (11.2)

Heat of vaporization the energy required to vaporize one mole of a liquid at a pressure of one atmosphere. (10.8)

Heating curve a plot of temperature versus time for a substance where energy is added at a constant rate. (10.8)

Heisenberg uncertainty principle a principle stating that there is a fundamental limitation to how precisely both the position and momentum of a particle can be known at a given time. (7.5)

Heme an iron complex. (20.8)

Hemoglobin a biomolecule composed of four myoglobinlike units (proteins plus heme) that can bind and transport four oxygen molecules in the blood. (20.8)

Henderson-Hasselbalch equation an equation giving the relationship between the pH of an acid-base system and the concentrations of base and acid: $\mathrm{pH} = \mathrm{p}K_a + \log\left(\frac{[\text{base}]}{[\text{acid}]}\right)$. (15.2)

Henry's law the amount of a gas dissolved in a solution is directly proportional to the pressure of the gas above the solution. (11.3)

Herbicide a pesticide that kills weeds. (24.3)

Hess's law in going from a particular set of reactants to a particular set of products, the enthalpy change is the same whether the reaction takes place in one step or in a series of steps; in summary, enthalpy is a state function. (6.3)

Heterogeneous equilibrium an equilibrium involving reactants and/or products in more than one phase. (13.4)

Hexagonal closest packed structure a structure composed of closest packed spheres with an *ababab* arrangement of layers; the unit cell is hexagonal. (10.4)

Homogeneous equilibrium an equilibrium system where all reactants and products are in the same phase. (13.4)

Homopolymer a polymer formed from the polymerization of only one type of monomer. (22.5)

Hund's rule the lowest energy configuration for an atom is the one having the maximum number of unpaired electrons allowed by the Pauli exclusion principle in a particular set of degenerate orbitals, with all unpaired electrons having parallel spins. (7.11)

Hybrid orbitals a set of atomic orbitals adopted by an atom in a molecule different from those of the atom in the free state. (9.1)

Hybridization a mixing of the native orbitals on a given atom to form special atomic orbitals for bonding. (9.1)

Hydration the interaction between solute particles and water molecules. (4.1)

Hydride a binary compound containing hydrogen. The hydride ion, H^-, exists in ionic hydrides. The three classes of hydrides are covalent, interstitial, and ionic. (18.3)

Hydrocarbon a compound composed of carbon and hydrogen. (22.1)

Hydrocarbon derivative an organic molecule that contains one or more elements in addition to carbon and hydrogen. (22.4)

Hydrogen bonding unusually strong dipole-dipole attractions that occur among molecules in which hydrogen is bonded to a highly electronegative atom. (10.1)

Hydrogenation reaction a reaction in which hydrogen is added, with a catalyst present, to a carbon-carbon multiple bond. (22.2)

Hydrohalic acid an aqueous solution of a hydrogen halide. (19.7)

Hydrometallurgy a process for extracting metals from ores by use of aqueous chemical solutions. Two steps are involved: selective leaching and selective precipitation. (24.4)

Hydronium ion the H_3O^+ ion; a hydrated proton. (14.1)

Hypothesis one or more assumptions put forth to explain the observed behavior of nature. (1.1)

Ideal gas law an equation of state for a gas, where the state of the gas is its condition at a given time; expressed by $PV = nRT$, where P = pressure, V = volume, n = moles of the gas, R = the universal gas constant, and T = absolute temperature. This equation expresses behavior approached by real gases at high T and low P. (5.3)

Ideal solution a solution whose vapor pressure is directly proportional to the mole fraction of solvent present. (11.4)

Indicator a chemical that changes color and is used to mark the end point of a titration. (4.9; 15.5)

Inert pair effect the tendency for the heavier Group 3A elements to exhibit the +1 as well as the expected +3 oxidation states, and Group 4A elements to exhibit the +2 as well as the +4 oxidation states. (18.5)

Inhibition (of combustion) the process whereby a polymer is made flame-retardant by interrupting the cycle of combustion. (24.2)

Integrated rate law an expression that shows the concentration of a reactant as a function of time. (12.4)

Interhalogen compound a compound formed by the reaction of one halogen with another. (19.7)

Intermediate a species that is neither a reactant nor a product but that is formed and consumed in the reaction sequence. (12.6)

Intermolecular forces relatively weak interactions that occur between molecules. (10.1)

Internal energy a property of a system that can be changed by a flow of work, heat or both; $\Delta E = q + w$, where ΔE is the change in the internal energy of the system, q is heat, and w is work. (6.1)

Ion an atom or a group of atoms that has a net positive or negative charge. (2.6)

Ion exchange (water softening) the process in which an ion-exchange resin removes unwanted ions (for example, Ca^{2+} and Mg^{2+}) and replaces them with Na^+ ions, which do not interfere with soap and detergent action. (18.4)

Ion pairing a phenomenon occurring in solution when oppositely charged ions aggregate and behave as a single particle. (11.7)

Ion-product constant (K_w) the equilibrium constant for the autoionization of water; $K_w = [H^+][OH^-]$. At 25°C, K_w equals 1.0×10^{-14}. (14.2)

Ion-selective electrode an electrode sensitive to the concentration of a particular ion in solution. (17.4)

Ionic bonding the electrostatic attraction between oppositely charged ions. (8.1)

Ionic compound (binary) a compound that results when a metal reacts with a nonmetal to form a cation and an anion. (8.1)

Ionic solid a solid containing cations and anions that dissolves in water to give a solution containing the separated ions which are mobile and thus free to conduct electrical current. (10.3)

Ionization energy the quantity of energy required to remove an electron from a gaseous atom or ion. (7.12)

Irreversible process any real process. When a system undergoes the changes State 1 → State 2 → State 1 by any real pathway, the universe is different than before the cyclic process took place in the system. (16.9)

Isoelectronic ions ions containing the same number of electrons. (8.4)

Isomers species with the same formula but different properties. (20.4)

Isotactic chain a polymer chain in which the substituent groups such as CH_3 are all arranged on the same side of the chain. (24.2)

Isotonic solutions solutions having identical osmotic pressures. (11.6)

Isotopes atoms of the same element (the same number of protons) with different numbers of neutrons. They have identical atomic numbers but different mass numbers. (2.5; 21)

Ketone an organic compound containing the carbonyl group

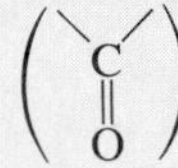

bonded to two carbon atoms. (22.4)

Kinetic energy ($\frac{1}{2}mv^2$) energy due to the motion of an object; dependent on the mass of the object and the square of its velocity. (6.1)

Kinetic molecular theory a model that assumes that an ideal gas is composed of tiny particles (molecules) in constant motion. (5.6)

Lanthanide contraction the decrease in the atomic radii of the lanthanide series elements, going from left to right in the periodic table. (20.1)

Lanthanide series a group of fourteen elements following lanthanum in the periodic table, in which the $4f$ orbitals are being filled. (7.11; 18,1; 20.1)

Lattice a three-dimensional system of points designating the positions of the centers of the components of a solid (atoms, ions, or molecules). (10.3)

Lattice energy the energy change occurring when separated gaseous ions are packed together to form an ionic solid. (8.5)

Law of conservation of energy energy can be converted from one form to another but can be neither created nor destroyed. (6.1)

Law of conservation of mass mass is neither created nor destroyed. (2.2)

Law of definite proportions a given compound always contains exactly the same proportion of elements by mass. (2.2)

Law of mass action a general description of the equilibrium condition; it defines the equilibrium constant expression. (13.2)

Law of multiple proportions a law stating that when two elements form a series of compounds, the ratios of the masses of the second element that combine with one gram of the first element can always be reduced to small whole numbers. (2.2)

Leaching the extraction of metals from ores using aqueous chemical solutions. (24.4)

Lead storage battery a battery (used in cars) in which the anode is lead, the cathode is lead coated with lead dioxide, and the electrolyte is a sulfuric acid solution. (17.5)

Le Châtelier's principle if a change is imposed on a system at equilibrium, the position of the equilibrium will shift in a direction that tends to reduce the effect of that change. (13.7)

Lewis acid an electron-pair acceptor. (14.11)

Lewis base an electron-pair donor. (14.11)

Lewis structure a diagram of a molecule showing how the valence electrons are arranged among the atoms in the molecule. (8.10)

Ligand a neutral molecule or ion having a lone pair of electrons that can be used to form a bond to a metal ion; a Lewis base. (20.1)

Lime-soda process a water-softening method in which lime and soda ash are added to water to remove calcium and magnesium ions by precipitation. (14.6)

Limiting reactant (limiting reagent) the reactant that is completely consumed when a reaction is run to completion. (3.9)

Line spectrum a spectrum showing only certain discrete wavelengths. (7.3)

Linear accelerator a type of particle accelerator in which a changing electrical field is used to accelerate a positive ion along a linear path. (21.3)

Linkage isomerism isomerism involving a complex ion where the ligands are all the same but the point of attachment of at least one of the ligands differs. (20.4)

Lipids water-insoluble substances that can be extracted from cells by nonpolar organic solvents. (23.4)

Liquefaction the transformation of a gas into a liquid. (18.1)

Localized electron (LE) model a model which assumes that a molecule is composed of atoms that are bound together by sharing pairs of electrons using the atomic orbitals of the bound atoms. (8.9)

Lock-and-key model a model for the mechanism of enzyme activity postulating that the shapes of the substrate and the enzyme are such that they fit together as a key fits a specific lock. (23.1)

London dispersion forces the forces, existing among noble gas atoms and nonpolar molecules, that involve an accidental dipole that induces a momentary dipole in a neighbor. (10.1)

Lone pair an electron pair that is localized on a given atom; an electron pair not involved in bonding. (8.9)

Magnetic quantum number m_ℓ, the quantum number relating to the orientation of an orbital in space relative to the other orbitals with the same ℓ quantum number. It can have integral values between ℓ and $-\ell$, including zero. (7.6)

Main-group (representative) elements elements in the groups labeled 1A, 2A, 3A, 4A, 5A, 6A, 7A, and 8A in the periodic table. The group number gives the sum of valence *s* and *p* electrons. (7.11; 18.1)

Major species the components present in relatively large amounts in a solution. (14.3)

Manometer a device for measuring the pressure of a gas in a container. (5.1)

Mass the quantity of matter in an object. (1.2)

Mass defect the change in mass occurring when a nucleus is formed from its component nucleons. (21.5)

Mass number the total number of protons and neutrons in the atomic nucleus of an atom. (2.5; 21)

Mass percent the percent by mass of a component of a mixture (11.1) or of a given element in a compound. (3.4)

Mass spectrometer an instrument used to determine the relative masses of atoms by the deflection of their ions in a magnetic field. (3.1)

Matter the material of the universe. (1.8)

Messenger RNA (mRNA) a special RNA molecule built in the cell nucleus that migrates into the cytoplasm and participates in protein synthesis. (23.3)

Metal an element that gives up electrons relatively easily and is lustrous, malleable, and a good conductor of heat and electricity. (2.7)

Metalloenzyme an enzyme containing a metal ion at its active site. (23.1)

Metalloids (semimetals) elements along the division line in the periodic table between metals and nonmetals. These elements exhibit both metallic and nonmetallic properties. (7.14; 18.1)

Metallurgy the process of separating a metal from its ore and preparing it for use. (18.1; 24.4)

Micelles aggregates of fatty acid anions having their hydrophobic tails in the interior and their polar heads pointing outward to interact with the polar water molecules. (23.4)

Millimeters of mercury (mm Hg) a unit of pressure, also called a torr; 760 mm Hg = 760 torr = 101,325 Pa = 1 standard atmosphere. (5.1)

Mineral a relatively pure compound as found in nature. (24.4)

Mixture a material of variable composition that contains two or more substances. (1.8)

Model (theory) a set of assumptions put forth to explain the observed behavior of matter. The models of chemistry usually involve assumptions about the behavior of individual atoms or molecules. (1.1)

Moderator a substance used in a nuclear reactor to slow down the neutrons. (21.6)

Molal boiling-point elevation constant a constant characteristic of a particular solvent that gives the change in boiling point as a function of solution molality; used in molecular weight determinations. (11.5)

Molal freezing-point depression constant a constant characteristic of a particular solvent that gives the change in freezing point as a function of the solution molality; used in molecular weight determinations. (11.5)

Molality the number of moles of solute per kilogram of solvent in a solution. (11.1)

Molar heat capacity the energy required to raise the temperature of one mole of a substance by one degree Celsius. (6.2)

Molar mass the mass in grams of one mole of molecules or formula units of a substance; also called molecular weight. (3.3)

Molar volume the volume of one mole of an ideal gas; equal to 22.42 liters at STP. (5.4)

Molarity moles of solute per volume of solution in liters. (4.3; 11.1)

Mole (mol) the number equal to the number of carbon atoms in exactly 12 grams of pure ^{12}C: Avogadro's number. One mole represents 6.022×10^{23} units. (3.2)

Mole fraction the ratio of the number of moles of a given component in a mixture to the total number of moles in the mixture. (5.5; 11.1)

Mole ratio (stoichiometry) the ratio of moles of one substance to moles of another substance in a balanced chemical equation. (3.8)

Molecular equation an equation representing a reaction in solution showing the reactants and products in undissociated form, whether they are strong or weak electrolytes. (4.6)

Molecular formula the exact formula of a molecule, giving the types of atoms and the number of each type. (3.5)

Molecular orbital (MO) model a model that regards a molecule as a collection of nuclei and electrons, where the electrons are assumed to occupy orbitals much as they do in atoms, but having the orbitals extend over the entire molecule. In this model the electrons are assumed to be delocalized rather than always located between a given pair of atoms. (9.2)

Molecular orientations (kinetics) orientations of molecules during collisions, some of which can lead to reaction while others cannot. (12.7)

Molecular solid a solid composed of neutral molecules at the lattice points. (10.3)

Molecular structure the three-dimensional arrangement of atoms in a molecule. (8.13)

Molecular weight the mass in grams of one mole of molecules or formula units of a substance; the same as molar mass. (3.3)

Molecularity the number of species that must collide to produce the reaction represented by an elementary step in a reaction mechanism. (12.6)

Molecule a bonded collection of two or more atoms of the same or different elements. (2.6)

Monodentate (unidentate) ligand a ligand that can form one bond to a metal ion. (20.3)

Monoprotic acid an acid with one acidic proton. (14.2)

Monosaccharide (simple sugar) a polyhydroxy ketone or aldehyde containing from three to nine carbon atoms. (23.2)

Myoglobin an oxygen-storing biomolecule consisting of a heme complex and a protein. (20.8)

Natural law a statement that expresses generally observed behavior. (1.1)

Nernst equation an equation relating the potential of an electrochemical cell to the concentrations of the cell components;

$$\mathscr{E} = \mathscr{E}° - \frac{0.0592}{n}\log(Q) \text{ at } 25°\text{C}. \ (17.4)$$

Net ionic equation an equation for a reaction in solution, where strong electrolytes are written as ions, showing only those components that are directly involved in the chemical change. (4.6)

Network solid an atomic solid containing strong directional covalent bonds. (10.5)

Neutralization reaction an acid-base reaction. (4.9)

Neutron a particle in the atomic nucleus with mass virtually equal to the proton's but with no charge. (2.5; 21)

Nitride salt a compound containing the N^{3-} anion. (18.2)

Nitrogen cycle the conversion of N_2 to nitrogen-containing compounds, followed by the return of nitrogen gas to the atmosphere by natural decay processes. (19.2)

Nitrogen fixation the process of transforming N_2 to nitrogen-containing compounds useful to plants. (19.2)

Nitrogen-fixing bacteria bacteria in the root nodules of plants that can convert atmospheric nitrogen to ammonia and other nitrogen-containing compounds useful to plants. (19.2)

Noble gas a Group 8A element. (2.7; 19.8)

Node an area of an orbital having zero electron probability. (7.7)

Nonelectrolyte a substance which, when dissolved in water, gives a nonconducting solution. (4.2)

Nonmetal an element not exhibiting metallic characteristics. Chemically, a typical nonmetal accepts electrons from a metal. (2.7)

Normal boiling point the temperature at which the vapor pressure of a liquid is exactly one atmosphere. (10.8)

Normal melting point the temperature at which the solid and liquid states have the same vapor pressure under conditions where the total pressure on the system is one atmosphere. (10.8)

Normality the number of equivalents of a substance dissolved in a liter of solution. (11.1)

Nuclear atom an atom having a dense center of positive charge (the nucleus) with electrons moving around the outside. (2.4)

Nuclear transformation the change of one element into another. (21.3)

Nucleon a particle in an atomic nucleus, either a neutron or a proton. (21)

Nucleotide a monomer of the nucleic acids composed of a five-carbon sugar, a nitrogen-containing base, and phosphoric acid. (23.3)

Nucleus the small, dense center of positive charge in an atom. (2.4)

Nuclide the general term applied to each unique atom; represented by $^{A}_{Z}X$, where X is the symbol for a particular element. (21)

Octet rule the observation that atoms of nonmetals tend to form the most stable molecules when they are surrounded by eight electrons (to fill their valence orbitals). (8.10)

Open hearth process a process for producing steel by oxidizing and removing the impurities in molten iron using external heat and a blast of air or oxygen. (24.4)

Optical isomerism isomerism in which the isomers have opposite effects on plane-polarized light. (20.4)

Orbital a specific wave function for an electron in an atom. The square of this function gives the probability distribution for the electron. (7.5)

***d*-Orbital splitting** a splitting of the *d*-orbitals of the metal ion in a complex such that the orbitals pointing at the ligands have higher energies than those pointing between the ligands. (20.6)

Order (of reactant) the positive or negative exponent, determined by experiment, of the reactant concentration in a rate law. (12.2)

Organic acid an acid with a carbon-atom backbone; often contains the carboxyl group. (14.2)

Organic chemistry the study of carbon-containing compounds (typically chains of carbon atoms) and their properties. (22)

Osmosis the flow of solvent into a solution through a semipermeable membrane. (11.6)

Osmotic pressure (π) the pressure that must be applied to a solution to stop osmosis; $\pi = MRT$. (11.6)

Ostwald process a commercial process for producing nitric acid by the oxidation of ammonia. (19.2)

Oxidation an increase in oxidation state (a loss of electrons). (4.10; 17.1)

Oxidation-reduction (redox) reaction a reaction in which one or more electrons are transferred. (4.10; 17.1)

Oxidation states a concept that provides a way to keep track of electrons in oxidation-reduction reactions according to certain rules. (4.10)

Oxidizing agent (electron acceptor) a reactant that accepts electrons from another reactant. (4.10; 17.1)

Oxyacid an acid in which the acidic proton is attached to an oxygen atom. (14.2)

Ozone O_3, the form of elemental oxygen in addition to the much more common O_2. (19.5)

Paramagnetism a type of induced magnetism, associated with unpaired electrons, that causes a substance to be attracted into the inducing magnetic field. (9.3)

Partial pressures the independent pressures exerted by different gases in a mixture. (5.5)

Particle accelerator a device used to accelerate nuclear particles to very high speeds. (21.3)

Pascal the SI unit of pressure; equal to newtons per meter squared. (5.1)

Pauli exclusion principle in a given atom no two electrons can have the same set of four quantum numbers. (7.8)

Penetration effect the effect whereby a valence electron penetrates the core electrons, thus reducing the shielding effect and increasing the effective nuclear charge. (7.12)

Peptide linkage the bond resulting from the condensation reaction between amino acids; represented by

$$\overset{\displaystyle O}{\overset{\|}{-C}}-\overset{\displaystyle H}{\overset{|}{N}}- \qquad (23.1)$$

Percent dissociation the ratio of the amount of a substance that is dissociated at equilibrium to the initial concentration of the substance in a solution, multiplied by 100. (14.5)

Percent yield the actual yield of a product as a percentage of the theoretical yield. (3.9)

Periodic table a chart showing all the elements arranged in columns with similar chemical properties. (2.7)

Petrochemicals chemicals obtained from petroleum either by steam cracking or as by-products of refinery operations. They serve as raw materials for the chemical industry. (24.1)

pH curve (titration curve) a plot showing the pH of a solution being analyzed as a function of the amount of titrant added. (15.4)

pH scale a log scale based on 10 and equal to $-\log[H^+]$; a convenient way to represent solution acidity. (14.3)

Phase diagram a convenient way of representing the phases of a substance in a closed system as a function of temperature and pressure. (10.9)

Phenyl group the benzene molecule minus one hydrogen atom. (22.3)

Pheromone a chemical produced by an insect which, when released, evokes a response in another insect of the same species. (24.3)

Phospholipids esters of glycerol containing two fatty acids and a phosphate group. Having nonpolar tails and polar heads, they tend to form bilayers in aqueous solution. (23.4)

Photochemical smog air pollution produced by the action of light on oxygen, nitrogen oxides, and unburned fuel from auto exhaust to form ozone and other pollutants. (5.9)

Photon a quantum of electromagnetic radiation. (7.2)

Physical change a change in the form of a substance, but not in its chemical composition; chemical bonds are not broken in a physical change. (1.9)

Pi (π) bond a covalent bond in which parallel *p* orbitals share an electron pair occupying the space above and below the line joining the atoms. (9.1)

Planck's constant the constant relating the change in energy for a system to the frequency of the electromagnetic radiation absorbed or emitted; equal to 6.626×10^{-34} J s. (7.2)

Polar covalent bond a covalent bond in which the electrons are not shared equally because one atom attracts them more strongly than the other. (8.1)

Polar molecule a molecule that has a permanent dipole moment. (4.1)

Polyatomic ion an ion containing a number of atoms. (2.6)

Polyelectronic atom an atom with more than one electron. (7.9)

Polymer a large, usually chainlike molecule built from many small molecules (monomers). (22.5)

Polymerization a process in which many small molecules (monomers) are joined together to form a large molecule. (22.2; 24.2)

Polypeptide a polymer formed from amino acids joined together by peptide linkages. (23.1)

Polyprotic acid an acid with more than one acidic proton. It dissociates in a stepwise manner, one proton at a time. (14.7)

Porous disk a disk in a tube connecting two different solutions in a galvanic cell that allows ion flow without extensive mixing of the solutions. (17.1)

Porphyrin a planar ligand with a central ring structure and various substituent groups at the edges of the ring. (20.8)

Positional probability a type of probability that depends on the number of arrangements in space that yield a particular state. (16.1)

Positron production a mode of nuclear decay in which a particle is formed having the same mass as an electron but opposite charge. The net effect is to change a proton to a neutron. (21.1)

Potential energy energy due to position or composition. (6.1)

Precipitation reaction a reaction in which an insoluble substance forms and separates from the solution. (4.5)

Precision the degree of agreement among several measurements of the same quantity; the reproducibility of a measurement. (1.3)

Primary structure (of a protein) the order (sequence) of amino acids in the protein chain. (23.1)

Principal quantum number the quantum number relating to the size and energy of an orbital; it can have any positive integer value. (7.6)

Probability distribution the square of the wave function indicating the probability of finding an electron at a particular point in space. (7.5)

Product a substance resulting from a chemical reaction. It is shown to the right of the arrow in a chemical equation. (3.6)

Protein a natural high-molecular-weight polymer formed by condensation reactions between amino acids. (23.1)

Proton a positively charged particle in an atomic nucleus. (2.5; 21)

Pure substance a substance with constant composition. (1.8)

Pyrometallurgy recovery of a metal from its ore by treatment at high temperatures. (24.4)

Qualitative analysis the separation and identification of individual ions from a mixture. (4.7)

Quantization the concept that energy can occur only in discrete units called quanta. (7.2)

Rad a unit of radiation dosage corresponding to 10^{-2} J of energy deposited per kilogram of tissue (from *r*adiation *a*bsorbed *d*ose). (21.7)

Radioactive decay (radioactivity) the spontaneous decomposition of a nucleus to form a different nucleus. (21.1)

Radiocarbon dating (carbon-14 dating) a method for dating ancient wood or cloth based on the rate of radioactive decay of the nuclide $^{14}_{6}C$. (21.4)

Radiotracer a radioactive nuclide, introduced into an organism for diagnostic purposes, whose pathway can be traced by monitoring its radioactivity. (21.4)

Random error an error that has an equal probability of being high or low. (1.3)

Raoult's law the vapor pressure of a solution is directly proportional to the mole fraction of solvent present. (11.4)

Rate constant the proportionality constant in the relationship between reaction rate and reactant concentrations. (12.2)

Rate of decay the change in the number of radioactive nuclides in a sample per unit time. (21.2)

Rate-determining step the slowest step in a reaction mechanism, the one determining the overall rate. (12.6)

Rate law (differential rate law) an expression that shows how the rate of reaction depends on the concentration of reactants. (12.2)

Reactant a starting substance in a chemical reaction. It appears to the left of the arrow in a chemical equation. (3.6)

Reaction mechanism the series of elementary steps involved in a chemical reaction. (12.6)

Reaction quotient a quotient obtained by applying the law of mass action to initial concentrations rather than to equilibrium concentrations. (13.5)

Reaction rate the change in concentration of a reactant or product per unit time. (12.1)

Reactor core the part of a nuclear reactor where the fission reaction takes place. (21.6)

Reducing agent (electron donor) a reactant that donates electrons to another substance to reduce the oxidation state of one of its atoms. (4.10; 17.1)

Reduction a decrease in oxidation state (a gain of electrons). (4.10; 17.1)

Rem a unit of radiation dosage that accounts for both the energy of the dose and its effectiveness in causing biological damage (from *r*oentgen *e*quivalent for *m*an). (21.7)

Resonance a condition occurring when more than one valid Lewis structure can be written for a particular molecule. The actual electronic structure is not represented by any one of the Lewis structures but by the average of all of them. (8.12)

Reverse osmosis the process occurring when the external pressure on a solution causes a net flow of solvent through a semipermeable membrane from the solution to the solvent. (11.6)

Reversible process a cyclic process carried out by a hypothetical pathway, which leaves the universe exactly the same as it was before the process. No real process is reversible. (16.9)

Ribonucleic acid (RNA) a nucleotide polymer that transmits the genetic information stored in DNA to the ribosomes for protein synthesis. (23.3)

Roasting a process of converting sulfide minerals to oxides by heating in air at temperatures below their melting points. (24.4)

Root mean square velocity the square root of the average of the squares of the individual velocities of gas particles. (5.6)

Salt an ionic compound. (14.8)

Salt bridge a U-tube containing an electrolyte that connects the two compartments of a galvanic cell, allowing ion flow without extensive mixing of the different solutions. (17.1)

Scientific method the process of studying natural phenomena, involving observations, forming laws and theories, and testing of theories by experimentation. (1.1)

Scintillation counter an instrument that measures radioactive decay by sensing the flashes of light produced in a substance by the radiation. (21.4)

Second law of thermodynamics in any spontaneous process, there is always an increase in the entropy of the universe. (16.2)

Secondary structure (of a protein) the three-dimensional structure of the protein chain (for example, α-helix, random coil, or pleated sheet). (23.1)

Selective precipitation a method of separating metal ions from an aqueous mixture by using a reagent whose anion forms a precipitate with only one or a few of the ions in the mixture. (4.7; 15.7)

Semiconductor a substance conducting only a slight electrical current at room temperature, but showing increased conductivity at higher temperatures. (10.5)

Semipermeable membrane a membrane that allows solvent but not solute molecules to pass through. (11.6)

Shielding the effect by which the other electrons screen, or shield, a given electron from some of the nuclear charge. (7.12)

SI units International System of units based on the metric system and units derived from the metric system. (1.2)

Side chain (of amino acid) the hydrocarbon group on an amino acid represented by H, CH_3, or a more complex substituent. (23.1)

Sigma (σ) bond a covalent bond in which the electron pair is shared in an area centered on a line running between the atoms. (9.1)

Significant figures the certain digits and the first uncertain digit of a measurement. (1.3)

Silica the fundamental silicon-oxygen compound, which has the empirical formula SiO_2, and forms the basis of quartz and certain types of sand. (10.5)

Silicates salts that contain metal cations and polyatomic silicon-oxygen anions that are usually polymeric. (10.5)

Single bond a bond in which one pair of electrons is shared by two atoms. (8.8)

Smelting a metallurgical process that involves reducing metal ions to the free metal. (24.4)

Solubility the amount of a substance that dissolves in a given volume of solvent at a given temperature. (4.2)

Solubility product constant the constant for the equilibrium expression representing the dissolving of an ionic solid in water. (15.6)

Solute a substance dissolved in a liquid to form a solution. (4.2; 11.1)

Solution a homogeneous mixture. (1.8)

Solvent the dissolving medium in a solution. (4.2)

Somatic damage radioactive damage to an organism resulting in its sickness or death. (21.7)

Space-filling model a model of a molecule showing the relative sizes of the atoms and their relative orientations. (2.6)

Specific heat capacity the energy required to raise the temperature of one gram of a substance by one degree Celsius. (6.2)

Spectator ions ions present in solution that do not participate directly in a reaction. (4.6)

Spectrochemical series a listing of ligands in order based on their ability to produce *d*-orbital splitting. (20.6)

Spontaneous fission the spontaneous splitting of a heavy nuclide into two lighter nuclides. (21.1)

Spontaneous process a process that occurs without outside intervention. (16.1)

Standard atmosphere a unit of pressure equal to 760 mm Hg. (5.1)

Standard enthalpy of formation the enthalpy change that accompanies the formation of one mole of a compound at 25°C from its elements, with all substances in their standard states at that temperature. (6.4)

Standard free energy change the change in free energy that will occur for one unit of reaction if the reactants in their standard states are converted to products in their standard states. (16.6)

Standard free energy of formation the change in free energy that accompanies the formation of one mole of a substance from its constituent elements with all reactants and products in their standard states. (16.6)

Standard hydrogen electrode a platinum conductor in contact with 1 M H^+ ions and bathed by hydrogen gas at one atmosphere. (17.2)

Standard reduction potential the potential of a half-reaction under standard state conditions, as measured against the potential of the standard hydrogen electrode. (17.2)

Standard solution a solution whose concentration is accurately known. (4.3)

Standard state a reference state for a specific substance defined according to a set of conventional definitions. (6.4)

Standard temperature and pressure (STP) the condition 0°C and 1 atmosphere of pressure. (5.4)
Standing wave a stationary wave as on a string of a musical instrument; in the wave mechanical model, the electron in the hydrogen atom is considered to be a standing wave. (7.5)
State function a property that is independent of the pathway. (6.1)
States of matter the three different forms in which matter can exist: solid, liquid, and gas. (1.8)
Steam cracking a process whereby hydrocarbon molecules are broken into small fragments by steam at very high temperatures. (24.1)
Stereoisomerism isomerism in which all the bonds in the isomers are the same but the spatial arrangements of the atoms are different. (20.4)
Steric factor the factor (always less than one) that reflects the fraction of collisions with orientations that can produce a chemical reaction. (12.7)
Steroid one of a class of lipids with a characteristic fused carbon-ring structure that includes cholesterol, hormones, and bile acids. (23.4)
Stoichiometric quantities quantities of reactants mixed in exactly the correct amounts so that all are used up at the same time. (3.9)
Strong acid an acid that completely dissociates to produce an H^+ ion and the conjugate base. (4.2; 14.2)
Strong base a metal hydroxide salt that completely dissociates into its ions in water. (4.2; 14.6)
Strong electrolyte a material which, when dissolved in water, gives a solution that conducts an electric current very efficiently. (4.2)
Structural formula the representation of a molecule in which the relative positions of the atoms are shown and the bonds are indicated by lines. (2.6)
Structural isomerism isomerism in which the isomers contain the same atoms but one or more bonds differ. (20.4; 22.1)
Subcritical reaction (nuclear) a reaction in which less than one neutron causes another fission event and the process dies out. (21.6)
Sublimation the process by which a substance goes directly from the solid to the gaseous state without passing through the liquid state. (10.8)
Subshell a set of orbitals with a given azimuthal quantum number. (7.6)
Substitution reaction (hydrocarbons) a reaction in which an atom, usually a halogen, replaces a hydrogen atom in a hydrocarbon. (22.1)
Supercooling the process of cooling a liquid below its freezing point without its changing to a solid. (10.8)
Supercritical reaction (nuclear) a reaction in which more than one neutron from each fission event causes another fission event. The process rapidly escalates to a violent explosion. (21.6)
Superheating the process of heating a liquid above its boiling point without its boiling. (10.8)
Superoxide a compound containing the O_2^- anion. (18.2)
Surface tension the resistance of a liquid to an increase in its surface area. (10.2)
Surfactant a wetting agent, such as soap, that assists water in wetting and suspending nonpolar materials. (23.4)
Surroundings everything in the universe surrounding a thermodynamic system. (6.1)
Syndiotactic chain a polymer chain in which the substituent groups such as CH_3 are arranged on alternate sides of the chain. (24.2)
Syngas synthetic gas, a mixture of carbon monoxide and hydrogen, obtained by coal gasification. (6.6)
System (thermodynamic) that part of the universe on which attention is to be focused. (6.1)
Systematic error an error that always occurs in the same direction. (1.3)

Tempering a process in steel production that fine-tunes the proportions of carbon crystals and cementite by heating to intermediate temperatures followed by rapid cooling. (24.4)
Termolecular step a reaction involving the simultaneous collision of three molecules. (12.6)
Tertiary structure (of a protein) the overall shape of a protein, long and narrow or globular, maintained by different types of intramolecular interactions. (23.1)
Theoretical yield the maximum amount of a given product that can be formed when the limiting reactant is completely consumed. (3.9)
Theory a set of assumptions put forth to explain some aspect of the observed behavior of matter. (1.1)
Thermal pollution the oxygen-depleting effect on lakes and rivers of using water for industrial cooling and returning it to its natural source at a higher temperature. (11.3)
Thermodynamics the study of energy and its interconversions. (6.1)
Thermodynamic stability (nuclear) the potential energy of a particular nucleus as compared to the sum of the potential energies of its component protons and neutrons. (21.1)
Thermoplastic polymer a substance that when molded to a certain shape under appropriate conditions can later be remelted. (24.2)
Thermoset polymer a substance that when molded to a certain shape under pressure and high temperatures cannot be softened again or dissolved. (24.2)
Third law of thermodynamics the entropy of a perfect crystal at 0 K is zero. (16.5)
Titration a technique in which one solution is used to analyze another. (4.9)
Torr another name for millimeter of mercury (mm Hg). (5.1)
Trace elements metals present only in trace amounts in the human body but essential for the action of many enzymes. (23)

Transfer RNA (tRNA) a small RNA fragment that finds specific amino acids and attaches them to the protein chain as dictated by the codons in mRNA. (23.3)

Transition metals several series of elements in which inner orbitals (*d* or *f* orbitals) are being filled. (7.11; 18.1)

Transuranium elements the elements beyond uranium that are made artificially by particle bombardment. (21.3)

Triple bond a bond in which three pairs of electrons are shared by two atoms. (8.8)

Triple point the point on a phase diagram at which all three states of a substance are present. (10.9)

Tyndall effect the scattering of light by particles in a suspension. (11.8)

Uncertainty (in measurement) the characteristic that any measurement involves estimates and cannot be exactly reproduced. (1.3)

Unimolecular step a reaction step involving only one molecule. (12.6)

Unit cell the smallest repeating unit of a lattice. (10.3)

Unit factor an equivalence statement between units used for converting from one unit to another. (1.5)

Universal gas constant the combined proportionality constant in the ideal gas law; 0.08206 L atm/K mol or 8.3144 J/K mol. (5.3)

Valence electrons the electrons in the outermost principal quantum level of an atom. (7.11)

Valence shell electron pair repulsion (VSEPR) model a model whose main postulate is that the structure around a given atom in a molecule is determined principally by minimizing electron-pair repulsions. (8.13)

van der Waals's equation a mathematical expression for describing the behavior of real gases. (5.8)

van't Hoff factor the ratio of moles of particles in solution to moles of solute dissolved. (11.7)

Vapor pressure the pressure of the vapor over a liquid at equilibrium. (10.8)

Vaporization the change in state that occurs when a liquid evaporates to form a gas. (10.8)

Viscosity the resistance of a liquid to flow. (10.2)

Volt the unit of electrical potential defined as one joule of work per coulomb of charge transferred. (17.1)

Voltmeter an instrument that measures cell potential by drawing electric current through a known resistance. (17.1)

Volumetric analysis a process involving titration of one solution with another. (4.9)

Vulcanization a process in which sulfur is added to rubber and the mixture is heated, causing crosslinking of the polymer chains and thus adding strength to the rubber. (24.2)

Wave function a function of the coordinates of an electron's position in three-dimensional space that describes the properties of the electron. (7.5)

Wave mechanical model a model for the hydrogen atom in which the electron is assumed to behave as a standing wave. (7.5)

Wavelength the distance between two consecutive peaks or troughs in a wave. (7.1)

Wax an ester similar to a fat or a phospholipid but containing a monohydroxy alcohol instead of glycerol. (23.4)

Weak acid an acid that dissociates only slightly in aqueous solution. (4.2; 14.2)

Weak base a base that reacts with water to produce hydroxide ions to only a slight extent in aqueous solution. (4.2; 14.6)

Weak electrolyte a material which, when dissolved in water, gives a solution that conducts only a small electric current. (4.2)

Weight the force exerted on an object by gravity. (1.2)

Work force acting over a distance. (6.1)

X-ray diffraction a technique for establishing the structures of crystalline solids by directing X rays of a single wavelength at a crystal and obtaining a diffraction pattern from which interatomic spaces can be determined. (10.3)

Zone of nuclear stability the area encompassing the stable nuclides on a plot of their positions as a function of the number of protons and the number of neutrons in the nucleus. (21.1)

Zone refining a metallurgical process for obtaining a highly pure metal that depends on continuously melting the impure material and recrystallizing the pure metal. (24.4)

ANSWERS TO SELECTED EXERCISES

Chapter 1

1. (a) A *law* is a concise statement or equation that summarizes a great variety of observations. A *theory* is a hypothesis that has been tested over a length of time. A law is less likely to be challenged or modified than a theory. (b) A *theory* is our explanation of why things behave the way they do, while *experiment* is the process of observing that behavior. Theories attempt to explain the results of experiments and are, in turn, tested by further experiment. (c) A *qualitative* measurement only measures a quality while a *quantitative* measurement attaches a number to the observation. Qualitative observations: The water was hot to the touch. Mercury was found in the drinking water. Quantitative observations: The temperature of the water was 62°C. The concentration of mercury in the drinking water was 1.5 ppm. (d) Both are explanations of experimental observation. A *theory* is a *hypothesis* that has been tested over time and found to still be valid, with, perhaps, some modifications. **2.** No, it is useful whenever a systematic approach of observation and hypothesis testing can be used. **4.** (a) No; (b) Yes; (c) Yes **6.** (a) Inexact; (c) Exact, $\frac{36\text{ in}}{\text{yd}} \times \frac{2.54\text{ cm}}{\text{m}} \times \frac{1\text{ m}}{100\text{ cm}} = \frac{0.9144\text{ yd}}{\text{m}}$ (All conversion factors used are exact); (e) Exact; (f) Inexact **7.** Accuracy is a measure of how close a measurement is to the true value, while precision is a measure of the scatter in a series of identical measurements. **10.** (a) 2; (b) 3; (c) 4; (d) 3; (e) 6; (f) 5 **11.** (a) 1.2×10^{-3}; (b) 4.37×10^{5}; (c) 9.000×10^{2}; (d) 1.06×10^{2}; (e) 1.25904×10^{8}; (f) 1.0012×10^{0} **12.** (a) 6×10^{8}; (b) 5.8×10^{8}; (c) 5.82×10^{8} **13.** (a) 8.41 (2.16 has only three significant figures); (c) 52.5; (e) 0.009 **14.** (a) 467; (b) 0.24; (c) 33.04; (d) 7.5×10^{1} **15.** (a) 6.32×10^{25}; (b) 7.82×10^{-17}; (c) 2.020×10^{-1}; (d) 1.161×10^{-2} **16.** (a) Maximum = 1.04; minimum = 1.00; (b) Maximum = 1.04; minimum = 1.00; (c) Maximum = 1.00; minimum = 0.96. Considering the error limits, (a) and (b) should be expressed to three significant figures and (c) to two significant figures. The division rule differs in (b). The rule says (b) should be expressed to two significant figures. If we do this for (b), we imply that the answer is 1.0 or between 0.995 and 1.05. The actual range is less than this, so we should use the more precise way of expressing uncertainty. **17.** (a) 2%; (b) 2.2% **18.** In a subtraction, the result gets smaller, but the uncertainties are added. If the two numbers are very close, the uncertainty may be larger than the result. **19.** 1 km = 10^3 m = 10^6 mm = 10^{15} pm; (b) 1 g = 10^{-3} kg = 10^3 mg = 10^9 ng; (c) 1 mL = 10^{-3} L = 10^{-3} dm^3 = 1 cm^3; (d) 1 mg = 10^{-6} kg = 10^{-3} g = 10^3 μg = 10^6 ng = 10^9 pg = 10^{12} fg; (e) 1 s = 10^3 ms = 10^9 ns **21.** 1×10^{-1} nm; 1×10^{2} pm **22.** (a) 229 cm, 2.29 m; 157 cm, 1.57 m; (b) 4.0×10^4 km; 4.0×10^7 m **23.** (a) 45.05 m/s; (b) 0.409 s **24.** 475 L **25.** 3×10^9 knots **26.** (a) 68 L = 68 dm^3; (b) 6.55×10^3 cm^3 or 6.55 L; (c) 0.64 cm; (d) 4301 m **27.** (a) 3.0×10^2 mg/mL; (b) 3.0×10^2 kg/m^3; (c) 3.0×10^5 μg/mL; (d) 3.0×10^8 ng/mL; (e) 3.0×10^2 μg/μL **28.** (a) 272 leagues; 1.49×10^3 km; (b) 0.18 km; 1.8×10^4 cm; (c) 96.0 m; 8.2 m; 5.25×10^{-1} cable length; 4.5 fathoms **31.** (a) 0.373 kg; 0.822 lb; (b) 31.1 g; 156 carats; (c) 19.3 cm^3 **32.** (a) Yes, 1 gr ap = 1 gr tr; (b) 1 oz ap = 1.00 oz tr; (c) 0.386 scruple; 7.72 grains ap; (d) 1.296 g **34.** 0.300 m **35.** (a) 4.25 royals; 8.50×10^3 weights; 11 horses; 3.25 royals left; (b) $\frac{52\text{ weights}}{\text{hargus hide}}$; 187 royals **36.** (a) $°C = \frac{8}{5}°A - 45$; (b) $°F = \frac{72}{25}°A - 49$; (c) °C = 75 = °A; (d) 93°C; 199°F; (e) 56°A **37.** 39.2°C; 312.4 K **38.** 77°F **40.** 69.1°F ± 0.2°F **42.** 2.70×10^3 kg/m^3; 168 lb/ft^3 **43.** 3.2×10^{15} g/cm^3 **45.** D_{old} = 8.87 g/cm^3; D_{new} = 7.18 g/cm^3; The difference in mass is accounted for by the difference in the alloy used. **47.** (a) Both are the same mass; (b) 1.0 mL of mercury **48.** (a) 1.0 kg feathers; (b) 100 g water **49.** 1.06 g/cm^3 ± 0.02 g/cm^3 **50.** 8.5 g/cm^3 ± 0.5 g/cm^3 **52.** Solid: own volume, own shape, does not flow. Liquid: own volume, takes shape of container, flows. Gas: takes volume and shape of container, flows. **53.** Homogeneous: only one phase present; Heterogeneous: more than one phase present. (a) Heterogeneous; (b) Heterogeneous; (c) Heterogeneous; (d) Homogeneous; (e) Homogeneous; (f) Homogeneous **54.** (a) Pure; (b) Mixture; (c) Mixture; (d) Pure; (e) Mixture **55.** Iron and uranium are elements. Water and table salt are compounds. **56.** Chemical changes involve the making and breaking of chemical forces (bonds); physical changes do not. The identity of a substance changes after a chemical change, but not after a physical change. **58.** (a) 1×10^{18} kg/km^3; (b) 1×10^9 kg/m^3 **59.** (a) No, because if the volumes were the same the gold idol would have a much greater mass. (b) Mass = 19 kg. It wouldn't be easy to play catch with the idol. **61.** 0.40 ± 0.03 oz/in^3 **62.** Experimental results are the facts with which we deal. Theories are our attempts to rationalize those facts. If the experiment is done properly and the theory can't account for the facts, then the theory is wrong. **63.** 28.9 ¢/L **64.** If the measurement is not precise, then it may be close to a particular "true" value only by chance. The next measurement, or even the next several, may be far from the true value. **65.** 33 ft/s; 22 mi/hr; 1.0×10^1 m/s; 36 km/hr; 1.0×10^1 s

Chapter 2

1. The masses of fluorine are in simple ratios of whole numbers to each other. **2.** ClF_3 **3.** (a) The composition of a substance depends on the numbers of atoms of each element making up the compound (i.e., the formula of the compound) and not on the composition of the mixture from which it was formed. (b) H_2 + $Cl_2 \rightarrow 2HCl$. That is, the volume of HCl produced is twice the volume of H_2 (or Cl_2) used. **4.** H, 1.00; Na, 22.8; Mg, 11.9. The atomic mass of Mg is incorrect. The atomic masses of H and Na are close. There must be an error in the assumed formulas of the compounds. The correct formulas are H_2O, Na_2O, and MgO. **5.** There should be no difference. It does not matter how a substance is produced; the substance remains intact. **6.** (a) Atoms have mass and are neither destroyed nor created by chemical reactions. Therefore, mass is neither created nor destroyed by chemical reactions; mass is conserved. (b) The composition of a substance depends on the number and kinds of atoms that form it. (c) Compounds of the same elements differ only in numbers of atoms of the elements forming them, e.g., NO, N_2O, NO_2. **7.** Some elements exist as molecular substances. For example, hydrogen normally exists as H_2 molecules, not as single hydrogen atoms. **8.** Yes, many questions can be raised from Dalton's theory. For example: What are the masses of atoms? Are the atoms really structureless? What forces hold atoms together in compounds? **10.** Deflection of cathode rays by magnetic and electric fields led to the conclusion that they were negatively charged. The ray was produced at the negative electrode and repelled by the negative pole of the applied electric field. **11.** No, both are electrons. **12.** 3×10^{15} g/cm^3; 0.4 g/cm^3 **13.** 1.28×10^{-12} zirkombs **15.** The atomic number of an element is equal to the number of protons in the nucleus of an atom of that element. The mass number is the sum of the number of protons plus neutrons in the nucleus. The atomic weight is the actual mass of a particular isotope (including electrons). The average mass of an atom is taken from a measurement made from a large number of atoms, and it is this average value we see on the periodic table. **16.** A family is a set of elements in the same vertical column. A family is also called a group. A period is a set of elements in the same horizontal row. **17.** Gold—Au; silver—Ag; mercury—Hg; potassium—K; iron—Fe; antimony—Sb; tungsten—W **18.** Fluorine—F; chlorine—Cl; bromine—Br; sulfur—S; oxygen—O; phosphorus—P **20.** Sn—tin; Pt—platinum; Co—cobalt; Ni—nickel; Mg—magnesium; Ba—barium; K—potassium **22.** The noble gases are He, Ne, Ar, Kr, Xe, and Rn (helium, neon, argon, krypton, xenon, and radon). All isotopes of radon are radioactive. **24.** (a) 8; (b) 8; (c) 18; (d) 3; (e) 5; (f) 3 **26.** (a) ${}^{238}_{94}Pu$; 94 protons, 144 neutrons; (b) ${}^{65}_{29}Cu$; 29 protons, 36 neutrons; (c) ${}^{52}_{24}Cr$; 24 protons, 28 neutrons **27.** (a) ${}^{15}N$; 7 protons, 8 neutrons; (b) ${}^{3}H$; 1 proton, 2 neutrons; (c) ${}^{207}Pb$; 82 protons, 125 neutrons; (d) ${}^{151}Eu$; 63 protons, 88 neutrons **28.** (a) P; (b) I **29.** (a) ${}^{24}_{12}Mg$; 12 protons, 12 neutrons, 12 electrons; (b) ${}^{24}_{12}Mg^{2+}$; 12 protons, 12 neutrons, 10 electrons; (c) ${}^{59}Co^{2+}$; 27 protons, 32 neutrons, 25 electrons; (d) ${}^{59}Co^{3+}$; 27 protons, 32 neutrons, 24 electrons; (e) ${}^{59}Co$; 27 protons, 32 neutrons, 27 electrons; (f) ${}^{79}_{34}Se$; 34 protons, 45 neutrons, 34 electrons **30.** ${}^{118}_{50}Sn^{2+}$

34.

Symbol	Number of protons in nucleus	Number of neutrons in nucleus	Number of electrons	Net charge
${}^{75}_{33}As^{3+}$	33	42	30	3+
${}^{128}_{52}Te^{2-}$	52	76	54	2−
${}^{32}_{16}S$	16	16	16	0
${}^{204}_{81}Tl^{1+}$	81	123	80	1+
${}^{195}_{78}Pt$	78	117	78	0

36. Metals: Mg, Ti, Au, Bi, Ge, Eu, Am; Nonmetals: Si, B, At, Rn, Br **37.** Si, Ge, B, At **38.** a, d, f, h **39.** (a) Na^+, lose one electron; (b) Sr^{2+}, lose two electrons; (c) Ba^{2+}, lose two electrons; (d) I^-, gain one electron; (e) Al^{3+}, lose three electrons; (f) S^{2-}, gain two electrons **40.** Carbon is a nonmetal. Silicon and germanium are metalloids. Tin and lead are metals. Thus, metallic character increases as one goes down a family in the periodic table. **42.** (a) chromium(VI) oxide; (b) chromium(III) oxide; (c) aluminum oxide; (d) selenium dioxide; (e) selenium trioxide **43.** (a) nickel(II) oxide; (b) iron(III) oxide; (c) cerium(IV) oxide; (d) cerium(III) oxide; (e) silver(I) sulfide **44.** (a) sodium chloride; (b) magnesium chloride; (c) rubidium bromide; (d) cesium fluoride; (e) aluminum iodide; (f) hydrogen iodide; (g) nitrogen monoxide or nitric oxide (common name); (h) nitrogen trifluoride **45.** (a) potassium perchlorate; (b) calcium phosphate; (c) aluminum sulfate; (d) lead(II) nitrate; (e) barium sulfite; (f) sodium nitrite; (g) potassium permanganate; (h) potassium dichromate **46.** (a) nitric acid; (b) nitrous acid; (c) phosphoric acid; (d) phosphorous acid; (e) sodium hydrogen sulfate or sodium bisulfate (common); (f) calcium hydrogen sulfite or calcium bisulfite (common); (g) sodium bromate; (h) iron(III) periodate **47.** (a) CsBr; (b) $BaSO_4$; (c) NH_4Cl; (d) ClO **48.** (a) $(NH_4)_2HPO_4$; (b) Hg_2S; (c) SiO_2; (d) Na_2SO_3; (e) $Al(HSO_4)_3$; (f) NCl_3; (g) HBr; (h) $HBrO_2$ **49.** (a) SF_2; (b) SF_6; (c) NaH_2PO_4; (d) Li_3N; (e) $Cr_2(CO_3)_3$; (f) SnF_2; (g) $(NH_4)(C_2H_3O_2)$; (h) $(NH_4)(HSO_4)$ **50.** (a) NaOH; (b) $Al(OH)_3$; (c) HCN; (d) Na_2O_2; (e) $Cu(C_2H_3O_2)_2$; (f) CF_4; (g) PbO; (h) PbO_2 **52.** There should be no difference. The composition of insulin from both sources will be the same and therefore, it will have the same activity regardless of the source. As a practical note, trace contaminants in the two types of insulin may be different. These trace components may be important. **53.** ${}^{98}Tc$; 43 protons and 55 neutrons; ${}^{99}Tc$; 43 protons and 56 neutrons; NH_4TcO_4 **54.** The solid residue must have come from the flask. **55.** $SeO_4{}^{2-}$, selenate; $SeO_3{}^{2-}$, selenite; $TeO_4{}^{2-}$, tellurate; $TeO_3{}^{2-}$, tellurite **58.** (a) BaO, barium oxide; (b) LiH, lithium hydride; (c) InF_3, indium trifluoride or indium(III) fluoride; (d) As_2S_3, diarsenic trisulfide or arsenic(III) sulfide **60.** (a) Xe; (b) Se; (c) Ca; (d) Mo; (e) Pu

Chapter 3

1. The two major isotopes of boron are ${}^{10}B$ and ${}^{11}B$. The listed mass of 10.81 is the average mass of a very large number of boron atoms. **2.** 24.31 amu **3.** 48% ${}^{151}Eu$ and 52% ${}^{153}Eu$ **7.** neon; 20.18 **8.** There are three peaks in the mass spectrum, each 2 mass units apart. This is consistent with two isotopes, differing in mass by two mass units. **10.** (a) 1.0 mol N; 3.0 mol H; (b) 2.0 mol N; 4.0 mol H; (c) 2.0 mol N; 8.0 mol H; 2.0 mol Cr; 7.0 mol O; (d) 1.0 mol Co; 2.0 mol Cl; 12 mol H; 6.0 mol O **11.** (a) 6.0×10^{23} atoms N; 1.8×10^{24} atoms H; (b) 1.2×10^{24} atoms N; 2.4×10^{24} atoms H; (c) 1.2×10^{24} atoms N; 4.8×10^{24} atoms H; 1.2×10^{24} atoms Cr; 4.2×10^{24} atoms O; (d) 6.0×10^{23} atoms Co; 1.2×10^{24} atoms Cl; 7.2×10^{24} atoms H; 3.6×10^{24} atoms O **12.** (a) 14 g N; 3.0 g H; (b) 28 g N; 4.0 g H; (c) 28 g N; 8.0 g H; 1.0×10^2 g Cr; 1.1×10^2 g O; (d) 59 g Co; 71 g Cl; 12 g H; 96 g O **13.** (a) 5.9×10^{-2} mol N; 1.8×10^{-1} mol H; (b) 6.3×10^{-2} mol N; 1.3×10^{-1} mol H; (c) 7.9×10^{-3} mol N; 3.2×10^{-2} mol H; 7.9×10^{-3} mol Cr; 2.8×10^{-2} mol O; (d) 4.2×10^{-3} mol Co; 8.4×10^{-3} mol Cl; 5.0×10^{-2} mol H; $2.5 \times$

10^{-2} mol O **14.** (a) 3.5×10^{22} atoms N; 1.1×10^{23} atoms H; (b) 3.8×10^{22} atoms N; 7.5×10^{22} atoms H; (c) 4.8×10^{21} atoms N; 1.9×10^{22} atoms H; 4.8×10^{21} atoms Cr; 1.7×10^{22} atoms O; (d) 2.5×10^{21} atoms Co; 5.1×10^{21} atoms Cl; 3.0×10^{22} atoms H; 1.5×10^{22} atoms O **15.** (a) 1.40×10^{-2} g N_2; (b) 8.40×10^{-2} g N_2; (c) 4.2×10^{3} g N_2; (d) 4.65×10^{-23} g N_2 **16.** Al_2O_3, 102.0 g/mol; Na_3AlF_6, 210.0 g/mol **17.** 2.841×10^{-3} mol; 1.711×10^{21} molecules **18.** (a) 1.66×10^{-22} mol (100 is exact); (b) 5.549 mol; (c) 8.30×10^{-22} mol (500 is exact); (d) 8.953 mol; (e) 2.49×10^{-22} mol (150 is exact); (f) 0.9393 mol **19.** (a) 294 g/mol; (b) 3.40×10^{-2} mol; (c) 459 g; (d) 1.0×10^{19} molecules; (e) 4.9×10^{21} atoms of nitrogen; (f) 4.9×10^{-13} g or 490 fg; (g) 4.88×10^{-22} g/molecule **21.** (a) 180.0 g/mol; (b) 3×10^{-3} mol; 2×10^{21} molecules **22.** (a) 165.4 g/mol; (b) 3.023 mol; (c) 3.3 g; (d) 5.5×10^{22} atoms of chlorine; (e) 1.6 g chloral hydrate; (f) 1.373×10^{-19} g (500 is exact) **23.** (a) 242.5 g/mol; (b) 150.9 g/mol; (c) 5.0×10^{4} g/mol **24.** (a) 1.9×10^{22} atoms of C; (b) 2.0×10^{22} atoms; (c) 5.7×10^{21} atoms **25.** 4.4×10^{16} molecules of EDB **26.** From Exercise 2.1: (i) 54.30% F; 45.70% S; (ii) 70.37% F; 29.63% S; (iii) 78.08% F; 21.92% S. From Exercise 2.59: Hydrazine, 12.6% H; 87.4% N; Ammonia, 17.8% H; 82.2% N; Hydrogen azide, 2.34% H; 97.7% N. **27.** 13.35% Y; 41.22% Ba; 28.62% Cu; 16.81% O **29.** (a) 46.7% N; (b) 30.4% N; (c) 30.4% N; (d) 63.6% N **31.** 1360 g/mol **33.** XeF_2: %F = 22.45%; XeF_4: %F = 36.66%; XeF_6: %F = 46.47% **34.** (a) 50.00% C; 5.59% H; 44.4% O; (b) 55.80% C; 7.03% H; 37.17% O; (c) 67.90% C; 5.69% H; 26.40% N **36.** The molecular formula gives the actual number of atoms of each element in a molecule (or formula unit) of a compound. The empirical formula gives the simplest whole number ratio of atoms of each element in a molecule. The molecular formula is a whole number multiple of the empirical formula. If that multiplier is one, the molecular and empirical formulas are the same. **37.** (a) $S_4N_4H_4$; (b) $N_3P_3Cl_6$ **38.** (a) 40.0% C; 6.7% H; 53.3% O; (b) 40.0% C; 6.7% H; 53.3% O; (c) 40.0% C; 6.7% H; 53.3% O **39.** All three compounds in Exercise 18 have the same empirical formula, CH_2O, and different molecular formulas. The composition of all three in mass percent is also the same. Therefore, elemental analysis will give us only the empirical formula. **40.** (a) $C_3H_4O_3$; (b) CH; (c) CH; (d) P_2O_5; (e) CH_2O; (f) CH_2O **41.** $C_3H_6O_2$ **42.** $C_7H_5N_3O_6$ **43.** The elemental composition of the confiscated substance: 80.48% C; 10.13% H; 9.39% N. Composition of cocaine: 67.28% C; 6.98% H; 4.62% N; 21.11% O. The composition by mass is not the same. The chemist can conclude the compound is not cocaine, assuming he analyzed a pure substance. **44.** (i) SF_2; (ii) SF_4; (iii) SF_6; Hydrazine: NH_2; Ammonia: NH_3; Hydrogen azide: N_3H **45.** $Na_2S_2O_3$ **48.** TiO_2 **50.** Empirical formula is SN. Molecular formula is S_4N_4 **52.** (a) $SiO_2(s) + 2C(s) \rightarrow Si(s) + 2CO(g)$; (b) $SiCl_4 + 2Mg \rightarrow Si + 2MgCl_2$; (c) $Na_2SiF_6(s) + 4Na(s) \rightarrow Si(s) + 6NaF(s)$ **54.** (a) $12HNO_3(l) + 3P_4O_{10}(s) \rightarrow 4(HPO_3)_3(l) + 6N_2O_5(g)$; (c) $Fe_2S_3(s) + 6HCl(g) \rightarrow 2FeCl_3(s) + 3H_2S(g)$ **56.** (a) $3Ca(OH)_2(aq) + 2H_3PO_4(aq) \rightarrow 6H_2O(l) + Ca_3(PO_4)_2(s)$; b. $Al(OH)_3(s) + 3HCl(g) \rightarrow AlCl_3(s) + 3H_2O(l)$; (c) $2AgNO_3(aq) + H_2SO_4(aq) \rightarrow Ag_2SO_4(s) + 2HNO_3(aq)$ **57.** $CaF_2 \cdot 3Ca_3(PO_4)_2(s) + 10H_2SO_4(aq) + 20H_2O(l) \rightarrow 6H_3PO_4(l) + 2HF(g) + 10CaSO_4 \cdot 2H_2O(s)$ **59.** (a) $4In(s) + 3O_2(g) \rightarrow 2In_2O_3(s)$; (b) $C_6H_{12}O_6(aq) \rightarrow 2C_2H_5OH(aq) + 2CO_2(g)$ **60.** (a) $Cu(s) + 2AgNO_3(aq) \rightarrow 2Ag(s) + Cu(NO_3)_2(aq)$; (b) $Zn(s) + 2HCl(aq) \rightarrow ZnCl_2(aq) + H_2(g)$ **61.** $Pb(NO_3)_2(aq) + H_3AsO_4(aq) \rightarrow PbHAsO_4(s) + 2HNO_3(aq)$ **62.** $2NaCl(aq) + 2H_2O(l) \rightarrow Cl_2(g) + H_2(g) + 2NaOH(aq)$ **63.** 6.51 g Cr_2O_3; 1.20 g N_2; 3.09 g H_2O **65.** 4.4 kg **66.** 4.1×10^{5} kg **68.** (a) $Ba(OH)_2 \cdot 8H_2O(s) + 2NH_4SCN(s) \rightarrow Ba(SCN)_2(s) + 10H_2O(l) + 2NH_3(g)$; (b) 3.1 g NH_4SCN; (c) 5.2 g $Ba(SCN)_2$; 3.7 g H_2O; 0.70 g NH_3 **69.** (a) 76 mg $C_6H_8O_7$; (b) 52 mg CO_2 **70.** (a) 37.0 g $C_4H_6O_3$; (b) 130. g aspirin **71.** $C_8H_{18}(l) + \frac{25}{2}O_2(g) \rightarrow 8CO_2(g) + 9H_2O(g)$ or $2C_8H_{18}(l) + 25O_2(g) \rightarrow 16CO_2(g) + 18H_2O(g)$; 9.6×10^{13} g CO_2 **72.** (a) Stoichiometric mixture. Neither is limiting. (b) I_2 is limiting. (c) Mg is limiting. (d) Mg is limiting. (e) Stoichiometric mixture. Neither is limiting. (f) I_2 is limiting. (g) Stoichiometric mixture. Neither is limiting. (h) I_2 is limiting. (i) Mg is limiting. **73.** (a) Stoichiometric mixture. Neither is limiting. (b) O_2 is limiting. (c) H_2 is limiting. (d) H_2 is limiting. (e) H_2 is limiting. (f) Stoichiometric mixture. Neither is limiting. (g) H_2 is limiting. **75.** 1.88 g Cu_2S; % yield: 93.6% **77.** Ag is limiting; 2.3 g Ag_2S; 1.7 g S_8 remains **78.** 1210 g NH_3; 290. g H_2 is unreacted **79.** 2.81×10^{6} g HCN; 5.62×10^{6} g H_2O **82.** The theoretical yield is: 1.96 g aspirin; % yield: 76.5% **83.** (a) 795 g HMD; (b) % yield: 96.2% **84.** The atomic mass is 54.95; the element is manganese (Mn). **85.** About 700 monomer units **86.** (a) 65.02% Pt; 9.338% N; 2.024% H; 23.63% Cl; (b) 47 g cisplatin; 23 g KCl **87.** Assume In_2O_3 is the formula. The atomic mass is 114.8 g/mol. If the formula is InO: 76.51 g/mol. Mendeleev was correct. **88.** Empirical formula Sb_2O_3; Molecular formula Sb_4O_6. **90.** 9.34% Zn; 90.66% Cu; The Cu remains unreacted. After filtering, washing, and drying, the mass of the unreacted copper could be measured. **93.** 440 g Cl_2 **94.** 1.12×10^{6} g CaO

Chapter 4

1. (a) $NaBr \rightarrow Na^+ + Br^-$; (b) $MgCl_2 \rightarrow Mg^{2+} + 2Cl^-$; (c) $Al(NO_3)_3 \rightarrow Al^{3+} + 3NO_3^-$; (d) $(NH_4)_2SO_4 \rightarrow 2NH_4^+ + SO_4^{2-}$; (e) $HI \rightarrow H^+ + I^-$ **2.** "Slightly soluble" refers to substances that dissolve only to a small extent. A slightly soluble salt may still dissociate completely to ions and, hence, be a strong electrolyte. An example of such a substance is $Mg(OH)_2$. It is a strong electrolyte, but not very soluble. A weak electrolyte is a substance that doesn't dissociate completely to produce ions. A weak electrolyte may be very soluble in water, or it may not be very soluble. Acetic acid is an example of a weak electrolyte that is very soluble in water. **3.** (a), (b), and (c) are strong bases; (d) is a weak base. **4.** Measure the electrical conductivity of a solution and compare it to the conductivity of a solution of equal concentration of a strong electrolyte. **7.** (a) 5.8 g NaCl; Place 5.8 g NaCl in a 1 L volumetric flask; add water to dissolve the NaCl and fill to the mark. (b) Add 40. mL of 2.5 *M* solution to a 1 L volumetric flask; fill to the mark with water. (c) As in (a), using 40. g $NaIO_3$. (d) As in (b), using 50. mL of the 0.20 *M* sodium iodate stock solution. **8.** (a) Dilute 28 mL of concentrated H_2SO_4 to a total volume of 1.00 L with water. (b) Dilute 42 mL of concentrated HCl to a final volume of 1.00 L. (c) Dissolve 120 g $NiCl_2 \cdot 6H_2O$ in water, and add water until the total volume of the solution is 1.00 L. **9.** (a) 0.2677 *M*; (b) 1.255×10^{-3} *M* **10.** (a) $M_{Ca^{2+}} = 0.15$ *M*; $M_{Cl^-} = 0.30$ *M*; (b) $M_{Al^{3+}} = 0.26$ *M*; $M_{NO_3^-} = 0.78$ *M* **11.** (a) $M_{Mg^{2+}} = 1.05 \times 10^{-2}$ *M*; $M_{Cl^-} = 2.10 \times 10^{-2}$ *M*; (b) $M_{NH_4^+} = M_{Br^-} = 1.13 \times 10^{-3}$ *M*; (c) $M_{Na^+} = 1.40 \times 10^{-1}$ *M*; $M_{S^{2-}} = 7.01 \times 10^{-2}$ *M* **14.** 0.69 mol/L **15.** (a) 2.5×10^{-8} *M*; (b) 8.4×10^{-9} *M*; (c) 1.34×10^{-4} *M*; (d) 2.8×10^{-7} *M* **16.** (a) 65 ppm; (b) 5.0×10^{3} ppb **17.** 20.0 ppb; 5.95×10^{-8} *M* **20.** $2.595 \times$

10^{-3} *M* **21.** 100. ppm standard: dilute 10.0 mL of the 1000.0 ppm stock to 100. mL; 75.0 ppm standard: dilute 7.50 mL of the 1000.0 ppm stock to 100. mL; 50.0 ppm standard: dilute 5.00 mL of the 1000.0 ppm stock to 100. mL; 25.0 ppm standard: dilute 2.50 mL of the 1000.0 ppm stock to 100. mL; 10.0 ppm standard: dilute 1.00 mL of the 1000.0 ppm stock to 100. mL.
23. The first procedure would give the most precise concentration.
24. (a) $BaCl_2(aq) + Na_2SO_4(aq) \rightarrow BaSO_4(s) + 2NaCl(aq)$; $Ba^{2+}(aq) + 2Cl^-(aq) + 2Na^+(aq) + SO_4^{2-}(aq) \rightarrow BaSO_4(s) + 2Na^+(aq) + 2Cl^-(aq)$; $Ba^{2+}(aq) + SO_4^{2-}(aq) \rightarrow BaSO_4(s)$; (b) $Pb(NO_3)_2(aq) + 2HCl(aq) \rightarrow PbCl_2(s) + 2HNO_3(aq)$; $Pb^{2+}(aq) + 2NO_3^-(aq) + 2H^+(aq) + 2Cl^-(aq) \rightarrow PbCl_2(s) + 2H^+(aq) + 2NO_3^-(aq)$; $Pb^{2+}(aq) + 2Cl^-(aq) \rightarrow PbCl_2(s)$; (c) $3AgNO_3(aq) + Na_3PO_4(aq) \rightarrow Ag_3PO_4(s) + 3NaNO_3(aq)$; $3Ag^+(aq) + 3NO_3^-(aq) + 3Na^+(aq) + PO_4^{3-}(aq) \rightarrow Ag_3PO_4(s) + 3Na^+(aq) + 3NO_3^-(aq)$; $3Ag^+(aq) + PO_4^{3-}(aq) \rightarrow Ag_3PO_4(s)$; (d) $3NaOH(aq) + Fe(NO_3)_3(aq) \rightarrow Fe(OH)_3(s) + 3NaNO_3(aq)$; $3Na^+(aq) + 3OH^-(aq) + Fe^{3+}(aq) + 3NO_3^-(aq) \rightarrow FeOH_3(s) + 3Na^+(aq) + 3NO_3^-(aq)$; $Fe^{3+}(aq) + 3OH^-(aq) \rightarrow Fe(OH)_3(s)$
25. (a) $Ag^+(aq) + I^-(aq) \rightarrow AgI(s)$; (b) $Cu^{2+}(aq) + S^{2-}(aq) \rightarrow CuS(s)$ **26.** (a) $Ba^{2+}(aq) + SO_4^{2-}(aq) \rightarrow BaSO_4(s)$; (b) $Pb^{2+}(aq) + 2Cl^-(aq) \rightarrow PbCl_2(s)$; (c) No reaction occurs.

27.

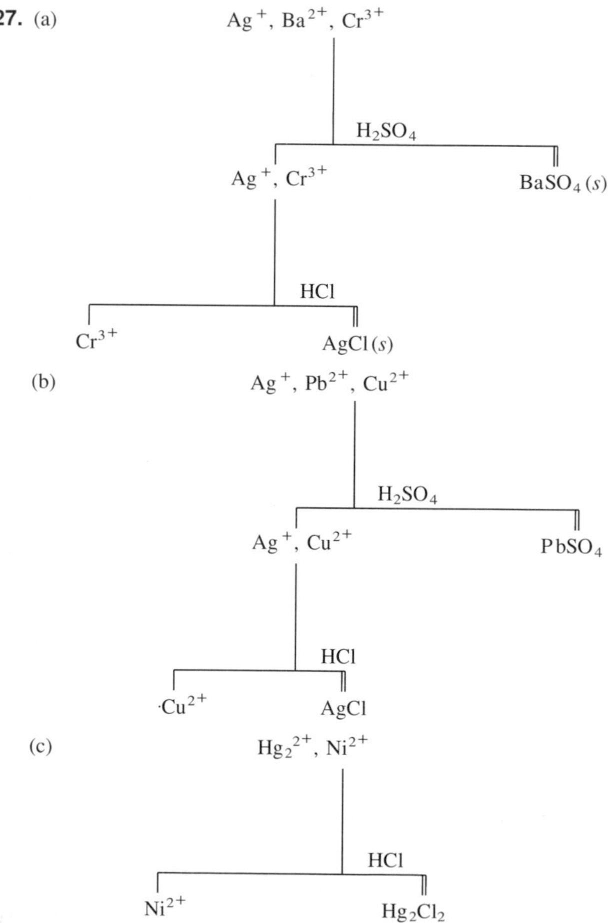

28. Three possibilities are: addition of H_2SO_4 solution to give a white ppt. of $PbSO_4$; addition of HCl solution to give a white ppt. of $PbCl_2$; addition of K_2CrO_4 solution to give a bright yellow ppt. of $PbCrO_4$. **30.** 2.82 g AgBr **31.** 0.750 L or 750 mL **32.** 0.3947% Al **33.** 2.33 g $BaSO_4$ **35.** (a) $2KOH(aq) + Mg(NO_3)_2(aq) \rightarrow Mg(OH)_2(s) + 2KNO_3(aq)$; (b) The precipitate is magnesium hydroxide. (c) 0.583 g $Mg(OH)_2$ **36.** 1.465% Tl_2SO_4 **37.** 67.01% saccharin by mass **39.** (a) $H_3PO_4(aq) + 3NaOH(aq) \rightarrow Na_3PO_4(aq) + 3H_2O(l)$; (b) $2HClO_4(aq) + Mg(OH)_2(s) \rightarrow Mg(ClO_4)_2(aq) + 2H_2O(l)$; (c) $2Al(OH)_3(s) + 3H_2SO_4(aq) \rightarrow Al_2(SO_4)_3(aq) + 6H_2O(l)$; (d) $HCN(aq) + NaOH(aq) \rightarrow NaCN(aq) + H_2O(l)$ **40.** (a) $NH_3(aq) + HNO_3(aq) \rightarrow NH_4NO_3(aq)$; $NH_3(aq) + H^+(aq) + NO_3^-(aq) \rightarrow NH_4^+(aq) + NO_3^-(aq)$; $NH_3(aq) + H^+(aq) \rightarrow NH_4^+(aq)$; (b) $Ba(OH)_2(aq) + 2HCl(aq) \rightarrow 2H_2O(l) + BaCl_2(aq)$; $Ba^{2+}(aq) + 2OH^-(aq) + 2H^+(aq) + 2Cl^-(aq) \rightarrow Ba^{2+}(aq) + 2Cl^-(aq) + 2H_2O(l)$; $2OH^-(aq) + 2H^+(aq) \rightarrow 2H_2O(l)$ **41.** 0.344 *M* **43.** $C_{22}H_{20}O_{13}$
44. (a) 100.0 mL; (b) 50.0 mL; (c) 16.7 mL **45.** 2.590×10^{-2} *M* **46.** 4.6×10^{-2} *M* **47.** 0.942 *M* **49.** (a) 0.8393 *M*; (b) 5.010% **50.** 0.102 mol/L **52.** 0.0777 mol/L **54.** (a) K, +1; O, −2; Mn, +7; (b) Ni, +4; O, −2; (c) K, +1; C, +2; N, −3; Fe, +2; (d) H, +1; O, −2; N, −3; P, +5; (e) P, +3; O, −2; (f) O, −2; Fe, +8/3; (g) O, −2; F, −1; Xe, +6 **55.** +1; +3; +5; +7 **56.** (a) −3; (b) −3; (c) −2; (d) +2; (e) +1
57. (a) O, −2; U, +6; (b) O, −2; As, +3; (c) Na, +1; O, −2; Bi, +5; (d) As, 0; (e) H, +1; O, −2; As, +3; (f) Mg, +2; O, −2; P, +5

58.

Redox?	Oxidizing agent	Reducing agent	Substance oxidized	Substance reduced
(a) Yes	O_2	CH_4	CH_4	O_2
(b) Yes	HCl	Zn	Zn	HCl
(c) No	—	—	—	—
(d) Yes	O_3	NO	NO	O_3
(e) Yes	H_2O_2	H_2O_2	H_2O_2	H_2O_2
(f) Yes	CuCl	CuCl	CuCl	CuCl

59. (a) $C_2H_6(g) + 7/2O_2(g) \rightarrow 2CO_2(g) + 3H_2O(g)$; (b) $Mg(s) + 2HCl(aq) \rightarrow Mg^{2+}(aq) + 2Cl^-(aq) + H_2(g)$; (c) $Cu(s) + 2Ag^+(aq) \rightarrow Cu^{2+}(aq) + 2Ag(s)$ **60.** (a) $3Cu(s) + 6H^+(aq) + 2HNO_3(aq) \rightarrow 3Cu^{2+}(aq) + 2NO(g) + 4H_2O(l)$; (b) $14H^+(aq) + Cr_2O_7^{2-}(aq) + 6Cl^-(aq) \rightarrow 3Cl_2(g) + 2Cr^{3+}(aq) + 7H_2O(l)$; (c) $Pb(s) + 2H_2SO_4(aq) + PbO_2(s) \rightarrow 2PbSO_4(s) + 2H_2O(l)$; (d) $14H^+(aq) + 2Mn^{2+}(aq) + 5NaBiO_3(s) \rightarrow 2MnO_4^-(aq) + 5Bi^{3+}(aq) + 5Na^+(aq) + 7H_2O(l)$ **61.** (a) $2H_2O(l) + Al(s) + MnO_4^-(aq) \rightarrow Al(OH)_4^-(aq) + MnO_2(s)$; (b) $2OH^-(aq) + Cl_2(g) \rightarrow Cl^-(aq) + ClO^-(aq) + H_2O(l)$; (c) $OH^-(aq) + H_2O(l) + NO_2^-(aq) + 2Al(s) \rightarrow NH_3(g) + 2AlO_2^-(aq)$ **62.** (a) $8HCl(aq) + 2Fe(s) \rightarrow 2HFeCl_4(aq) + 3H_2(g)$; (b) $6H^+(aq) + 8I^-(aq) + IO_3^-(aq) \rightarrow 3I_3^-(aq) + 3H_2O(l)$; (d) $64OH^-(aq) + 2CrI_3(s) + 27Cl_2(g) \rightarrow 54Cl^-(aq) + 2CrO_4^{2-}(aq) + 6IO_4^-(aq) + 32H_2O(l)$; (e) $258OH^-(aq) + Fe(CN)_6^{4-}(aq) + 61Ce^{4+}(aq) \rightarrow Fe(OH)_3(s) + 61Ce(OH)_3(s) + 6CO_3^{2-}(aq) + 6NO_3^-(aq) + 36H_2O(l)$ **63.** (a) All three reactions are oxidation reduction reactions. (b) $4NH_3 + 5O_2 \rightarrow 4NO + 6H_2O$; O_2 is the oxidizing agent; NH_3 is the reducing agent. $2NO + O_2 \rightarrow 2NO_2$; O_2 is the oxidizing agent; NO is the reducing agent. $3NO_2 + H_2O \rightarrow 2HNO_3 + NO$; NO_2 is both the oxidizing and reducing agent. (c) 8.8×10^6 g NO **65.** $2H^+(aq) + Mn(s) + 2HNO_3(aq) \rightarrow Mn^{2+}(aq) + 2NO_2(g) + 2H_2O(l)$; $3H_2O(l) + 2Mn^{2+}(aq) + 5IO_4^-(aq) \rightarrow 2MnO_4^-(aq) + 5IO_3^-(aq) + 6H^+(aq)$

67. 1.622×10^{-2} *M* **68.** (a) 4.45×10^{-2} *M*; (b) 24.7 mL **70.** (a) $\%C_2H_5OH = 0.08467\%$; The person is not legally intoxicated. (b) $\%C_2H_5OH = 0.1172\%$; The person is legally intoxicated. **71.** 45.3% Sb_2S_3 **72.** 0.220 g CO_2 **74.** (a) $CoCl_2 \cdot xH_2O(aq) + 2AgNO_3(aq) \rightarrow 2AgCl(s) + Co(NO_3)_2(aq) + xH_2O(l)$; $CoCl_2 \cdot xH_2O(aq) + 2NaOH(aq) \rightarrow Co(OH)_2(s) + 2NaCl(aq) + xH_2O(l)$; $4Co(OH)_2 + O_2 \rightarrow 2Co_2O_3 + 4H_2O$; (b) 29.8% Cl; 24.8% Co; 45.4% is water; (c) $CoCl_2 \cdot 6H_2O$ **75.** $CaCO_3(s) + H_2SO_4(aq) \rightarrow CaSO_4(aq) + H_2O(l) + CO_2(g)$ **78.** 10.34 g $PbCrO_4$ **79.** (a) No, no element shows a change in oxidation number. (b) 3.7 g Fe_2O_3 **80.** (a) 0.0559 g Fe; (b) 0.242 g $Fe(NO_3)_3$; (c) 53.1% **82.** 0.668 g or 6.68×10^2 mg **84.** 39.2% Na_2SO_4 and 60.8% K_2SO_4 **85.** 0.0257 *M* ± 0.0007 *M* **86.** 0.0785 *M* ± 0.0002 *M*

Chapter 5

1. 29.92 in of Hg **2.** (a) 150 atm; (b) 15 M Pa; (c) 1.1×10^5 torr **3.** 65 torr; 8.6×10^{-2} atm; 8.6×10^3 Pa **5.** (a) 620. torr; 0.816 atm; 8.24×10^4 Pa; (b) 935 torr; 1.23 atm; 1.24×10^5 Pa; (c) 810. torr; 495 torr **6.** 1.013×10^5 N/m² or 1.013×10^5 Pa; To exert the same pressure, a column of water will have to be 10.3 m high. **9.** (a) $8.313 \times 10^3 \dfrac{\text{L Pa}}{\text{mol K}}$; (b) $8.313 \dfrac{\text{m}^3\text{ Pa}}{\text{mol K}}$; (c) $8.31 \dfrac{\text{J}}{\text{mol K}}$

10.

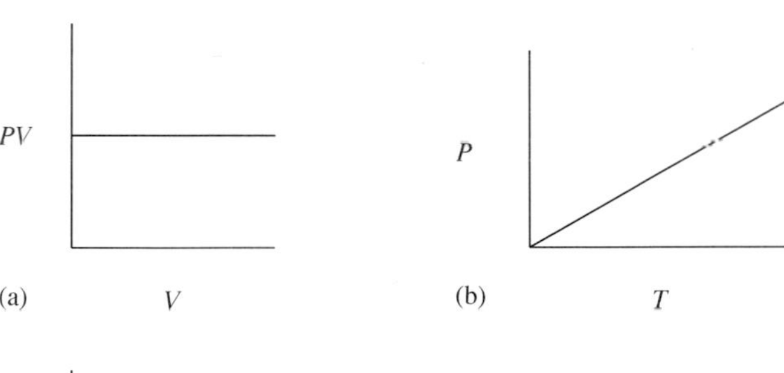

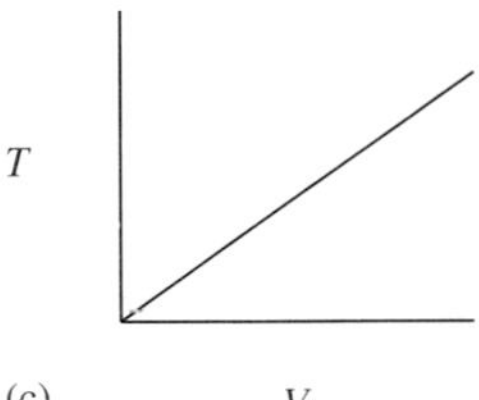

11. 430 torr **12.** 1600 g **13.** 0.09 atm **14.** The volume has increased by 15%. **15.** (a) 1060 torr; (b) 190. torr **16.** 1.1 atm; 2.1 atm **18.** 5.1×10^4 torr **20.** (a) 14.1 L; (b) 4.71×10^{-2} mol; (c) 677 K; (d) $P = 133$ atm **22.** (a) 46.6 atm; (b) 1.02×10^3 K; (c) 171 K **24.** 22.4 L **26.** 660 kPa **28.** 0.531 g O_2 **29.** $P_{total} = 1.4$ atm; $P_{N_2} = 1.1$ atm; $P_{O_2} = .10$ atm; $P_{NO} = 0.18$ atm **31.** 0.284 g N_2 **33.** $P_{H_2} = 0.319$ atm; $P_{He} = 0.161$ atm **35.** 5.77 g/L for $SiCl_4$; 4.59 g/L for $SiHCl_3$ **37.** N_2H_4 **39.** If $Be(C_5H_7O_2)_3$, M = 311 g/mol; If $Be(C_5H_7O_2)_2$, M = 207 g/mol. Data Set I: M = 209 g/mol; Data Set II: M = 202 g/mol. These results are close to the expected value of 207 g/mol for $Be(C_5H_7O_2)_2$. We conclude that beryllium is a divalent element with an atomic weight of 9.0 g/mol. **41.** 1.2 L = 1200 mL **43.** 56 g **44.** 307 L **47.** XeF_4 **48.** 13.3% N **50.** (a) 10.0 L of NO; (b) 1.65×10^4 L; (c) 1.05×10^3 g NO **52.** At 273 K, $(KE)_{avg} = 3.40 \times 10^3$ J/mol; at 546 K, $(KE)_{avg} = 6.81 \times 10^3$ J/mol **54.** No, this is the average energy. There is a distribution of energies. **55.** (a) They will all have the same average kinetic energy since they are all at the same temperature. (b) Flask C; (c) Flask A

57.

	(a)	(b)	(c)	(d)
Average *KE*	Inc	Dec	Same $KE \propto T$	Same
Average velocity	Inc	Dec	Same $1/2mv^2 \propto T$	Same
Gas collisions frequency	Inc	Dec	Inc	Inc
Wall collisions frequency	Inc	Dec	Inc	Inc
Impact *E*	Inc	Dec	Same impact $E \propto KE \propto T$	Same

58. NO **59.** 63.5 g/mol **61.** (a) $P = 12.24$ atm; (b) $P = 12.14$ atm; (c) The ideal gas law is high by 0.10 atm or $0.10/12.14 \times 100 = .82\%$. **63.** For the first experiment in Sample Exercise 5.3, $P = 0.1303 - 0.0001 = 0.1302 \cong 0.13$, which is the same as the measured value. **65.** 5×10^{-7} atm; 1×10^{13} molecules/cm³ **66.** (a) 0.19 torr; (b) 6.6×10^{21} molecules in the 1 m³ of air, (c) 6.6×10^{15} molecules/cm³ **69.** 8×10^7 g NO_2 **70.** For benzene: Mixing ratio = 16 ppbv; 3.9×10^{11} molecules benzene/cm³. For toluene: Mixing ratio = 37 ppbv; 9.0×10^{11} molecules toluene/cm³. **73.** $P = 490$ atm **75.** (a) 1.0×10^4 g; (b) 6.6×10^4 g; (c) 8×10^3 g **76.** 1.6×10^3 g **78.** 0.87 L **80.** 2.26 L of wet acetylene **81.** 45 torr He; 84 torr Ne; 94 torr Ar; $P_{total} = 223$ torr **82.** 3.27×10^{13} molecules; 3.27×10^{10} molecules/cm³ **84.** CH_6N_2 **87.** C_3H_8 should have the largest value of *b*. **88.** The attractions are greatest for CO_2.

Chapter 6

1. $PE = 20.$ J; $KE = 225$ J **2.** $KE = 150$ J **3.** The 1.0 kg object with a velocity of 2.0 m/s has the greater kinetic energy. **5.** $KE = 2.0 \times 10^{-2}$ J **6.** (a) 36 kJ; (b) 35 kJ; (c) −85 kJ **8.** 10. L atm; 1.0×10^3 J **9.** 30.9 kJ; −12.4 kJ; 18.5 kJ **12.** (a) endothermic; (b) exothermic; (c) exothermic; (d) endothermic **13.** The reaction is exothermic. Heat is evolved by the system to the surroundings. −0.37 kJ/g **15.** (a) −55.7 kJ/g; (b) -3.55×10^4 kJ **16.** (a) 2.54×10^3 kJ; (b) 7.4×10^3 kJ; (c) 693 kJ **19.** (a) 54.9 kJ; (b) 24.3 J/mol °C **20.** (a) 8.5 J; (b) 91 kJ; (c) 2.9×10^3 kJ **23.** 2.0×10^1 kJ/mol **24.** −56 kJ/mol **26.** C = 2.05 kJ/°C **27.** −25 kJ/g; −2700 kJ/mol **29.** 4.9×10^6 kJ **30.** 6.66 kJ/°C **31.** −296.1 kJ **33.** 28.4 kJ **34.** −17 kJ **35.** −220.8 kJ **37.** −233 kJ **38.** −202.6 kJ **41.** The enthalpy change for the formation of one mole of a compound at 25°C from its elements, with all substances in their standard states. $Na(s) + \frac{1}{2}Cl_2(g) \rightarrow NaCl(s)$; $H_2(g) + \frac{1}{2}O_2(g) \rightarrow H_2O(l)$; $6C\text{ (graphite)} + 6H_2(g) + 3O_2(g) \rightarrow C_6H_{12}O_6(s)$; $Pb(s) + S(s) + 2O_2(g) \rightarrow PbSO_4(s)$ **42.** (a) −940. kJ; (b) −265 kJ; (c) −176 kJ **43.** (a) $4NH_3(g) + 5O_2(g) \rightarrow 4NO(g) + 6H_2O(g)$; $\Delta H° = -908$ kJ. $2NO(g) + O_2(g) \rightarrow 2NO_2(g)$; $\Delta H° = -112$ kJ. $3NO_2(g) + H_2O(l) \rightarrow 2NHO_3(aq) + NO(g)$; $\Delta H° = -140.$ kJ. (b) $12NH_3(g) + 21O_2(g) \rightarrow 8HNO_3(aq) + 4NO(g) + 14H_2O(g)$. The overall reaction is exothermic, since each step is exothermic. **45.** −2677 kJ **46.** −4594 kJ **47.** We get more energy from the $N_2O_4/N_2H_3CH_3$ mixture. **50.** (a) −361 kJ; (b) −199 kJ **51.** −169 kJ/mol **53.** 1.4×10^5 L **55.** −128 kJ **57.** −46.5 kJ/g for C_3H_8 vs. −43.4 kJ/g for octane. The fuel values of the fuels are very close. An advantage of propane is that it burns more cleanly. The boiling point of propane is −42°C. Thus, it is more difficult to store propane and there are extra safety hazards associated in using the necessary high pressure compressed gas tanks. **59.** Pathway 1: −15 L atm; Pathway 2: −6.0 L atm. No, since w depends on pathway, it cannot be a state function. **60.** $T_f = 25.5°C$ **61.** 2.95 kJ of heat is evolved. **62.** (a) $\Delta H° = -69$ kJ; 6.3×10^7 g Na_2CO_3; 4.1×10^7 kJ heat evolved; (b) The heat might boil the nitric acid resulting in the spreading of acid fumes throughout the city. **64.** If Hess's law were not true, it would be possible to create energy by reversing a reaction using a different series of steps. This violates the law of conservation of energy (First Law). Thus Hess's Law is another statement of the Law of Conservation of Energy. **65.** $N_2H_4(l) + O_2(g) \rightarrow N_2(g) + 2H_2O(g)$; $\Delta H° = -535$ kJ. $2N_2H_4(l) + N_2O_4(l) \rightarrow 3N_2(g) + 4H_2O(g)$; $\Delta H° = -1050$ kJ; $\Delta H = -8.4 \times 10^3$ kJ/kg for $N_2H_4(l) + O_2(g)$; $\Delta H = -6.7 \times 10^3$ kJ/kg for $2N_2H_4(l) + N_2O_4(l)$; $\Delta H = -7.1 \times 10^3$ kJ/kg for $4N_2H_3CH_3(l) + 5N_2O_4(l)$. The hydrazine plus oxygen is the most efficient fuel. **67.** Work done by system on surroundings: (e) and (f); Work done by surroundings on system: (a), (c), and (d); No work: (b). **68.** 2800 g **69.** A state function is a function whose change depends only on the initial and final states and not on how one got from the initial to the final state. If H and E were not state functions, the Law of Conservation of Energy (First Law) would not be true. **70.** (a) $2Al(s) + \frac{3}{2}O_2(g) \rightarrow Al_2O_3(s)$; (b) $C_2H_5OH(l) + 3O_2(g) \rightarrow 2CO_2(g) + 3H_2O(l)$; (c) $Ba(OH)_2(aq) + 2HCl(aq) \rightarrow 2H_2O(l) + BaCl_2(aq)$; (d) $2C(graphite) + \frac{3}{2}H_2(g) + \frac{1}{2}Cl_2(g) \rightarrow C_2H_3Cl(g)$ **72.** 30°C **74.** Al would be the better choice. It has a higher heat capacity and a lower density than Fe. Using Al, the same amount of heat could be dissipated by a smaller mass, keeping the mass of the amplifier down.

Chapter 7

2. 3.02 m **3.** 3.85×10^{14} s^{-1}; 2.55×10^{-19} J **4.** 3.0×10^{10} s^{-1}; 2.0×10^{-23} J; 12 J **5.** 276 nm **6.** 427.7 nm **8.** No, it will take light with a wavelength of 134.4 nm or less to ionize gold. **11.** When something is quantized, it can only have certain discrete values. In the Bohr model of the H-atom, the energy of the electron is quantized. **12.** (a) 656.7 nm; (b) 486.4 nm; (c) 121.6 nm; (d) 1876 nm **13.** 91.2 nm; 364.8 nm; 820.9 nm

14.

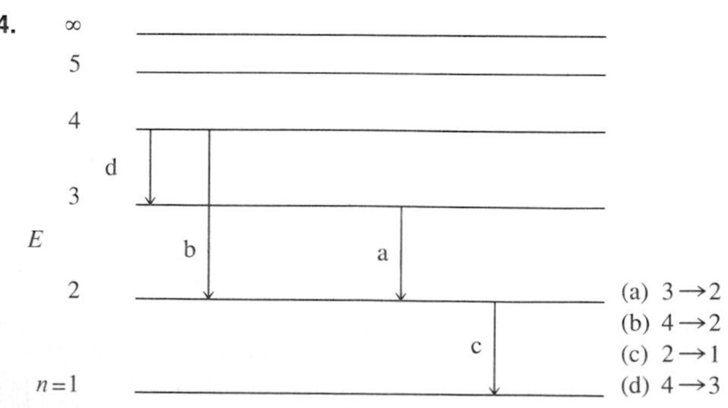

15.

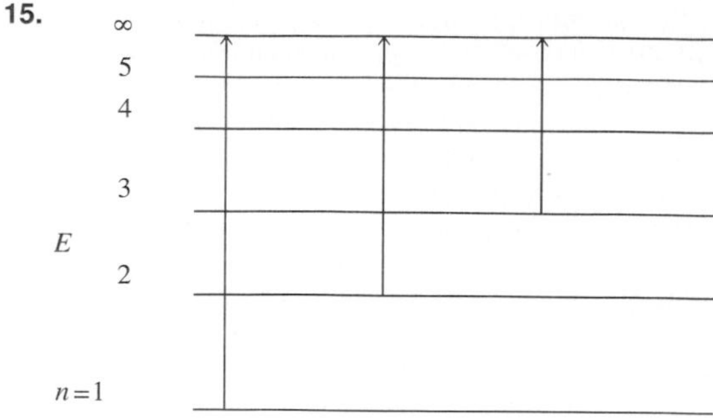

16. (a) False. It takes less energy to ionize an electron from $n = 3$ than the ground state. (b) True. (c) False. The energy difference from $n = 3 \rightarrow n = 2$ is less than the energy difference from $n = 3 \rightarrow n = 1$; thus, the wavelength is larger for $n = 3 \rightarrow n = 2$ than for $n = 3 \rightarrow n = 1$. (d) True. (e) False. $n = 2$ is the first excited state; $n = 3$ is the second excited state. **19.** (a) 2.6×10^{-5} nm; (b) 1.6×10^{-2} nm; (c) 1.0×10^{-25} nm **21.** 7.3×10^3 m/s; 7.3×10^5 m/s **22.** 3.3×10^{-1} J s; 3.3×10^{-6} J s **25.** Units of $\Delta E \times \Delta t$ = J × s, the same as the units of Planck's constant.

$$\text{Units of } \Delta p \Delta x = \text{kg} \times \frac{\text{m}}{\text{s}} \times \text{m} = \frac{\text{kg m}^2}{\text{s}} = \frac{\text{kg m}^2}{\text{s}^2} \times \text{s} = \text{J} \times \text{s}$$

26. (a) 5.79×10^5 nm; (b) 3.64×10^{-33} m; (c) Diameter of H atom is roughly 1.0×10^{-8} cm. The uncertainty in position is much larger than the size of the atom. (d) The uncertainty is insignificant compared to the size of a baseball. **27.** n gives the energy (it completely specifies the energy only for the H-atom or ions with one electron) and the relative size of orbitals. ℓ gives the type (shape) of orbital. m_ℓ gives information about the direction in which the orbital is pointing. **28.** $n = 1, 2, 3, \ldots$; $\ell = 0, 1, 2, \ldots (n-1)$; $m_\ell = -\ell, \ldots -2, -1, 0, 1, 2, \ldots +\ell$ **30.** (b) ℓ must be smaller than n; (d) For $\ell = 0$, $m_\ell = 0$ only allowed value; (f) For $\ell = 3$, m_ℓ can range from −3 to +3; (g) n cannot equal zero; (h) ℓ cannot be a negative number **31.** No, for $n = 2$, the allowed values of ℓ are 0 and 1; for $n = 3$, the allowed values of ℓ are 0, 1, and 2.

34.

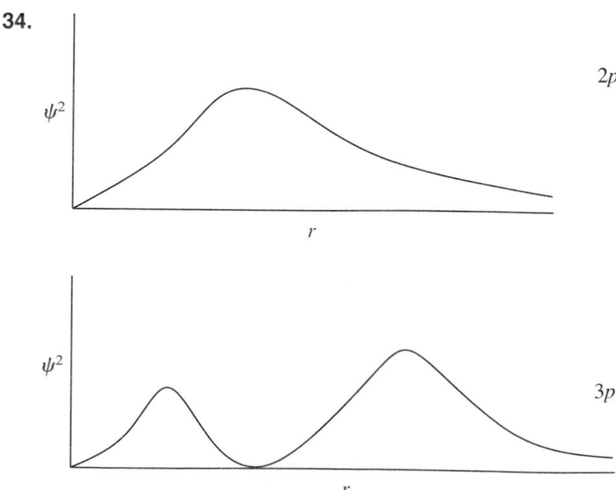

38. ψ^2 gives the probability of finding the electron at that point. **39.** $r = 0$, $\psi_{1s}^2 = 7.6 \times 10^{10}$; $r = a_0$, $\psi_{1s}^2 = 1.0 \times 10^{10}$; $r = 2a_0$, $\psi_{1s}^2 = 1.4 \times 10^9$ **41.** (a) H; 1310 kJ/mol; (b) He^+; 5250 kJ/mol; (c) Li^{2+}; 1.181×10^4 kJ/mol **42.** $H:He^+:Li^{2+}:C^{5+}:Fe^{25+}$; $1:\frac{1}{2}:\frac{1}{3}:\frac{1}{6}:\frac{1}{26}$; 1:0.50:0.33:0.17:0.038 **43.** No, the spin is a convenient model. Since we cannot locate or ''see'' the electron, we cannot see if it is spinning. **44.** (a) 32 electrons; (b) 8 electrons; (c) 25 electrons; (d) 10 electrons; (e) 6 electrons; (f) It is impossible for $n = 0$. No electrons. **45.** Sc: $1s^22s^22p^63s^23p^64s^23d^1$; Fe: $1s^22s^22p^63s^23p^64s^23d^6$; S: $1s^22s^22p^63s^23p^4$; P: $1s^22s^22p^63s^23p^3$; Cs: $1s^22s^22p^63s^23p^64s^23d^{10}4p^65s^24d^{10}5p^66s^1$; Eu: $1s^22s^22p^63s^23p^64s^23d^{10}4p^65s^24d^{10}5p^66s^25d^14f^6$*; Pt: $1s^22s^22p^63s^23p^64s^23d^{10}4p^65s^24d^{10}5p^66s^24f^{14}5d^8$*; Xe: $1s^22s^22p^63s^23p^64s^23d^{10}4p^65s^24d^{10}5p^6$; Br: $1s^22s^22p^63s^23p^64s^23d^{10}4p^5$; Se: $1s^22s^22p^63s^23p^64s^23d^{10}4p^4$

Note:* These electron configurations were written down using only the periodic table. Actual electron configurations are Eu: $[Xe]6s^24f^7$ and Pt: $[Xe]6s^14f^{14}5d^9$. **47. Exceptions: Cr, Cu, Nb, Mo, Tc, Ru, Rh, Pd, Ag, Pt, Au **49.** (a) The smallest halogen is fluorine: $1s^22s^22p^5$; (b) K: $1s^22s^22p^63s^23p^64s^1$; (c) Be: $1s^22s^2$; Mg: $1s^22s^22p^63s^2$; Ca: $1s^22s^22p^63s^23p^64s^2$; (d) In: $[Kr]5s^24d^{10}5p^1$ **51.** (b) and (f) are the only possible sets of quantum numbers. (a) For $\ell = 0$, m_ℓ can only be 0. (c) m_s can only be +1/2 or −1/2. (d) For $n = 1$, ℓ can only be 0. (e) For $\ell = 2$, m_ℓ cannot be −3 **52.** 1.56 (2*p* electron in B); 2.72 (2*s* electron in B^+); 3.34 (2*s* electron in B^{2+}); 4.37 (1*s* electron in B^{3+}); 5.00 (1*s* electron in B^{4+})

53.

B: $1s^22s^22p^1$

	n	ℓ	m_ℓ	m_s
1s	1	0	0	+1/2
1s	1	0	0	−1/2
2s	2	0	0	+1/2
2s	2	0	0	−1/2
2p	2	1	−1	+1/2*

C: $1s^22s^22p^2$

	n	ℓ	m_ℓ	m_s
1s	1	0	0	+1/2
1s	1	0	0	−1/2
2s	2	0	0	+1/2
2s	2	0	0	−1/2
2p	2	1	−1	+1/2†
2p	2	1	0	+1/2†

N: $1s^22s^22p^3$

	n	ℓ	m_ℓ	m_s
1s	1	0	0	+1/2
1s	1	0	0	−1/2
2s	2	0	0	+1/2
2s	2	0	0	−1/2
2p	2	1	−1	+1/2†
2p	2	1	0	+1/2†
2p	2	1	+1	+1/2†

O: $1s^22s^22p^4$

	n	ℓ	m_ℓ	m_s
1s	1	0	0	+1/2
1s	1	0	0	−1/2
2s	2	0	0	+1/2
2s	2	0	0	−1/2
2p	2	1	−1	+1/2
2p	2	1	−1	−1/2
2p	2	1	0	+1/2†
2p	2	1	+1	+1/2†

*This is only one of several possibilities. For example, the 2*p* electron in B could have $m_\ell = -1$, 0, or +1 and $m_s = +1/2$ or −1/2, for a total of six possibilities.

†could have $m_s = -1/2$

54. When atoms interact with each other, it will be the outermost electrons that are involved in those interactions. **56.** There are no unpaired electrons in this oxygen atom. This configuration would be an excited state and in going to the ground state, energy would be released. **57.** Sc: one unpaired electron; Ti: two unpaired electrons; Al: one unpaired electron; Sn: two unpaired electrons; Te: two unpaired electrons; Br: one unpaired electron **60.** (a) Be < Mg < Ca; (b) Xe < I < Te; (c) Ge < Ga < In **61.** (a) Ca < Mg < Be; (b) Te < I < Xe; (c) In < Ga < Ge; (d) As < N < F; (e) S < Cl < F **62.** (a) Be < Mg < Ca; (b) Te < I < Xe; (c) Ga < Ge < In; (d) N < F < As; (e) F < S < Cl **63.** Ionization energy is proportional to Z_{eff}. Size is inversely proportional to Z_{eff}. These conclusions are approximately true for elements in the same period. **65.** If one more electron is added to a half-filled subshell, electron-electron repulsions will increase. **67.** Size also decreases going across a period. Sc & Ti and Y & Zr are adjacent elements. There are 14 elements (the Lanthanides) between La and Hf, making Hf considerably smaller. **68.** (a) $[Rn]7s^25f^{14}6d^{10}7p^5$; (b) It will be in the halogen family and most similar to astatine, At. (c) NaUus, $Mg(Uus)_2$, $C(Uus)_4$, $O(Uus)_2$; (d) $OUus^-$, $UusO_2^-$, $UusO_3^-$, $UusO_4^-$ **69.** (a) O < S; S more exothermic; (b) I < Br < F < Cl; Cl most exothermic; **73.** O because electron-electron repulsions will be much more severe for $O^- + e^- \rightarrow O^{2-}$ than for $O + e^- \rightarrow O^-$. **77.** $1s^22s^22p^63s^23p^6$: Ar: 1.52 MJ/mol: 0.94 Å; $1s^22s^22p^63s^2$: Mg: 0.74 MJ/mol: 1.60 Å; $1s^22s^22p^63s^23p^64s^1$: K: 0.42 MJ/mol: 1.97 Å **78.** (a) −1445 kJ/mol; (b) −580 kJ/mol; (c) +348.7 kJ/mol **79.** No, our generalization was that the first ionization energy decreases as we go down a group. For the coinage metals the first ionization energy increases as we go down the groups. (Fe, Ru, Os) and (Ga, In, Tl) follow the same trend as the coinage metals. Only (As, Sb, Bi) follow the trend we would predict. **80.** $Z_{eff} = 3.02$; $Z_{eff} = 3.73$; $Z_{eff} = 4.94$ **81.** As we remove succeeding electrons, it takes more energy because each succeeding electron is being removed from an increasingly positively charged species. **83.** (a) more favorable E.A.: Li, S, B, Cl; (b) higher I.E.: Li, S, N, F; (c) larger radius: K, Sc, B, Cl **84.** Li^+ ions will be the smallest of the alkali metal cations and will be most strongly attracted to the water molecules. **85.** potassium peroxide **86.** (a) $4Li(s) + O_2(g) \rightarrow 2Li_2O(s)$; (b) $2K(s) + S(s) \rightarrow K_2S(s)$; (c) $2Cs(s) + 2H_2O(l) \rightarrow 2CsOH(aq) + H_2(g)$, (d) $2Na(s) + Cl_2(g) \rightarrow 2NaCl(s)$ **89.** (a) Li_3N; lithium nitride; (b) NaBr; sodium bromide; (c) K_2S; potassium sulfide; (d) Li_3P; lithium phosphide **90.** It should be element 119 with ground state electron configuration: [Rn] $7s^25f^{14}6d^{10}7p^68s^1$. **92.** Each element has a characteristic spectrum. Thus, the presence of the characteristic spectral lines of an element confirms its presence in any particular sample. **94.** O: 2; O^+: 3; O^-: 1; Fe: 4; Mn: 5; S: 2; F: 1; Ar: 0 **97.** A large I.E. can be the result of large Z_{eff}. A large Z_{eff} will also result in a large E.A. The noble gases would be an exception, since Z_{eff} is very large for the element, but is small for the anion since the electron starts a new shell. Thus, the noble gases have a large I.E. but not an exothermic E.A. **98.** (a) A discrete bundle of light energy. (b) A number describing a discrete state of an electron. (c) The lowest energy state in an electron or ion. (d) An allowed energy state that is higher in energy than the ground state. **99.** (a) $P(g) \rightarrow P^+(g) + e^-$; (b) $P(g) + e^- \rightarrow P^-(g)$ **100.** The I.E. is for removal of the electron from the atom in the gas phase. The work function is for the removal of an electron from the solid. **101.** The electron is no longer part of that atom. The proton and the electron are completely separated. **102.** (a) excited state of boron; ground state: $1s^22s^22p^1$; (b) ground state of neon; (c) excited state of fluorine; ground state: $1s^22s^22p^5$ **105.** Electron-electron repulsions are much greater in O^- than S^- because the electron goes into a smaller $2p$ orbital vs. the larger $3p$ orbital on sulfur, resulting in a more favorable (more exothermic) E.A. for sulfur.

108.

Ti: $[Ar]4s^2 3d^2$

	n	ℓ	m_ℓ	m_s
$4s$	4	0	0	$+1/2$
$4s$	4	0	0	$-1/2$
$3d$	3	2	-2	$+1/2$*
$3d$	3	2	-1	$+1/2$

*Only one of 10 possible combinations of ℓ and m_ℓ. For ground state the second d electron should be in a different orbital.

110. (a) 18 electrons; (b) 2 electrons **111.** $n = 5$; $m_\ell = +4$ to -4; 18 electrons **112.** Elements in the same family have similar chemical properties. **113.** (a) 1212 kJ; (b) -443 kJ; (c) -1026 kJ; (d) 146 kJ **114.** (a) $Se^{3+}(g) \rightarrow Se^{4+}(g) + e^-$; (b) $S^-(g) + e^- \rightarrow S^{2-}(g)$; (c) $Fe^{3+}(g) + e^- \rightarrow Fe^{2+}(g)$ **115.** (a) As: $1s^2 2s^2 2p^6 3s^2 3p^6 4s^2 3d^{10} 4p^3$; (b) [Rn] $7s^2 5f^{14} 6d^{10} 7p^4$; (c) Ta: [Xe] $6s^2 4f^{14} 5d^3$ or Ir: [Xe] $6s^2 4f^{14} 5d^7$; (d) Ti: [Ar] $4s^2 3d^2$; Ni: [Ar] $4s^2 3d^8$; Os: [Xe] $6s^2 4f^{14} 5d^6$ **116.** (a) 21; (b) 21; (c) 11 **117.** (a) The atomic number is 24. (b) There are 6 electrons in s orbitals. (c) 12; (d) 2; (e) This is an isotope of $^{50}_{24}Cr$. There are 26 neutrons in the nucleus. (f) $\cong$ 25.0 g; (g) $1s^2 2s^2 2p^6 3s^2 3p^6 4s^1 3d^5$ **119.** (a) Each orbital could hold 3 electrons. (b) First period: 3; second period: 12. (c) 15; (d) 21

Chapter 8

1. (a) Electronegativity: The ability of an atom in a molecule to attract electrons to itself. Electron affinity: The energy change for the reaction:

$$M(g) + e^- \rightarrow M^-(g)$$

The E.A. deals with isolated atoms in the gas phase.

(b) Covalent bond: Sharing of an electron pair. Polar covalent bond: Unequal sharing of an electron pair. (c) Ionic bond: Electrons are no longer shared. Completely unequal sharing (i.e., transfer) of electrons from one atom to another. **2.** (a) -2.0 units; (b) -2.6 units **3.** (a) $C < N < O$; (b) $Se < S < Cl$; (c) $Sn < Ge < Si$; (d) $Tl < Ge < S$ **4.** (a) Ge—F; (b) P—Cl; (c) S—F; (d) Ti—Cl **5.** Order of E.N. from Fig. 8.3: (a) C (2.5) < N (3.0) < O (3.5), same; (b) Se (2.4) < S (2.5) < Cl (3.0), same; (c) Si = Ge = Sn (1.8), different; (d) Tl (1.8) = Ge (1.8) < S (2.5), different. Most polar bonds using actual E.N. values: (a) Si—F and Ge—F equal polarity (Ge—F predicted); (b) P—Cl (same as predicted); (c) S—F (same as predicted); (d) Si—Cl and Ge—Cl equal polarity (TiCl correctly predicted). **6.** E.N. is proportional to (I.E. − E.A.). The ionization energies of the noble gases are very high. This causes the E.N. to also be very high. We also expect F_2 to react with Kr and Rn. **7.** (a) $Cu > Cu^+ > Cu^{2+}$; (b) $Pt^{2+} > Pd^{2+} > Ni^{2+}$; (c) $Se^{2-} > S^{2-} > O^{2-}$ **8.** (a) Mg^{2+}: $1s^2 2s^2 2p^6$; Sn^{2+}: $[Kr]5s^2 4d^{10}$; K^+: $1s^2 2s^2 2p^6 3s^2 3p^6$; Al^{3+}: $1s^2 2s^2 2p^6$; Tl^+: $[Xe]6s^2 4f^{14} 5d^{10}$; As^{3+}: $[Ar]4s^2 3d^{10}$; (b) N^{3-}, O^{2-} and F^-: $1s^2 2s^2 2p^6$; Te^{2-}: $[Kr]5s^2 4d^{10} 5p^6$ **9.** (a) Sc^{3+}; (b) Te^{2-} **10.** Isoelectronic: Same number of electrons. There are two variables, number of protons and number of electrons, that will determine the size of an ion. Keeping the number of electrons constant, we only have to consider the number of protons to predict trends in size. **12.** (a) NaCl, Na^+ smaller than K^+; (b) LiF, F^- smaller than Cl^-; (c) MgO, O^{2-} greater charge than OH^-; (d) $Fe(OH)_3$, Fe^{3+} greater charge than Fe^{2+} **14.** $\Delta H_f^\circ = -412$ kJ/mol **16.** (a) The lattice energy for $Mg^{2+}O^{2-}$ will be much larger than for Mg^+O^-. (b) Mg^+ and O^- both have unpaired electrons. In Mg^{2+} and O^{2-} there are no unpaired electrons. Hence, Mg^+O^- would be paramagnetic; $Mg^{2+}O^{2-}$ would be diamagnetic. Paramagnetism can be detected by measuring the mass of a sample in the presence and absence of a magnetic field. **18.** $\Delta H = +737$ kJ/mol **19.** The extra electron-electron repulsions are much greater than the attraction of the electron for the nucleus.

21.

Compound	Q_1Q_2	Lattice energy
$FeCl_2$	$(+2)(-1)$	-2631 kJ/mol
FeO	$(+2)(-2)$	-5359 kJ/mol
$FeCl_3$	$(+3)(-1)$	-3865 kJ/mol
Fe_2O_3	$(+3)(-2)$	$-14{,}744$ kJ/mol

22. (a) Li_3N (lithium nitride); (b) Ga_2O_3 (gallium oxide); (c) RbCl (rubidium chloride); (d) BaS (barium sulfide) **23.** -42 kJ **24.** (a) -183 kJ; (b) -109 kJ **25.** (a) -184 kJ/mol (-183 from bond energies); (b) -92 kJ/mol (-109 from bond energies). Bond energies seem to give a reasonably good estimate of the enthalpy change for a reaction. **26.** -5681 kJ **28.** (a) -6 kJ; (b) -1373 kJ; (c) -1077 kJ **29.** Since both reactions are highly exothermic, the high temperature is not needed to provide energy. **31.** (a) $\Delta H = 1549$; (b) $\Delta H = 1390$ **32.** (a) Using SF_4 data: $D_{SF} = 342$ kJ/mol; Using SF_6 data: $D_{SF} = 327$ kJ/mol. (b) The S—F bond energy in the table is 327 kJ/mol. The value in the table was based on the S—F bond in SF_6. (c) $S(g)$ and $F(g)$ are not the most stable form of the element at 25°C. The most stable forms are $S_8(s)$ and $F_2(g)$; $\Delta H_f^\circ = 0$ for these two species. **33.** $D_{calc} =$ 389 kJ/mol as compared to 391 kJ/mol in the table.

36. (a) H—C≡N:

(b) H—P̈—H with H bonded below P

(c) H on top; :C̈l—C—C̈l: ; :C̈l: below C

(d) [H—N—H with H above and below N]$^+$

(e) [:F̈: above B; :F̈—B—F̈: ; :F̈: below B]$^-$

(f) :F̈—S̈e—F̈:

37. (a) $POCl_3$:

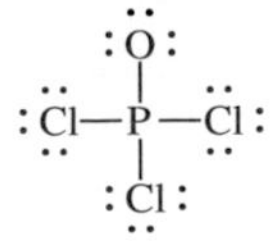

:Ö:
|
:C̈l—P—C̈l:
|
:C̈l:

SO_4^{2-}:

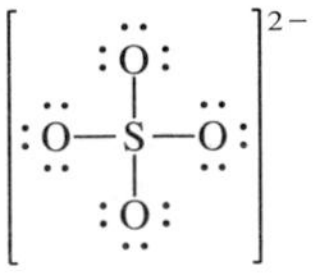

[:Ö: | :Ö—S—Ö: | :Ö:]$^{2-}$

XeO$_4$:

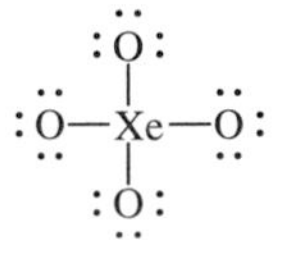

:Ö: | :Ö—Xe—Ö: | :Ö:

PO_4^{3-}:

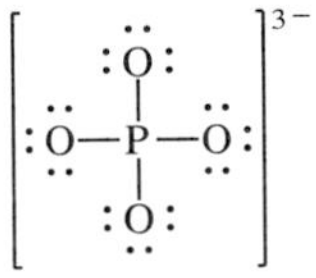

[:Ö: | :Ö—P—Ö: | :Ö:]$^{3-}$

ClO_4^-:

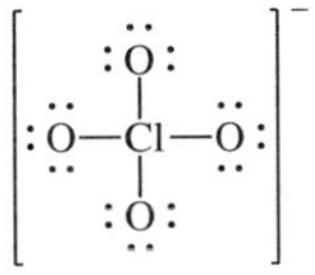

[:Ö: | :Ö—Cl—Ö: | :Ö:]$^-$

(b) NF_3:

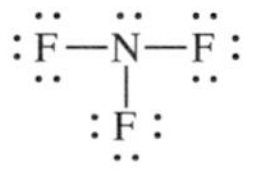

:F̈—N̈—F̈: | :F̈:

SO_3^{2-}:

[:Ö—S̈—Ö: | :Ö:]$^{2-}$

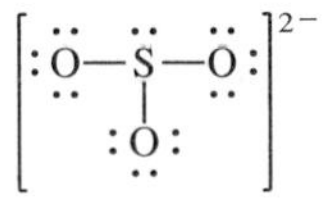

PO_3^{3-}:

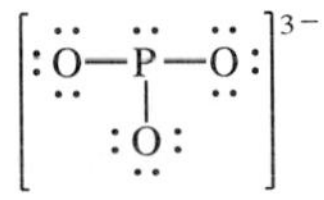

[:Ö—P̈—Ö: | :Ö:]$^{3-}$

ClO_3^-:

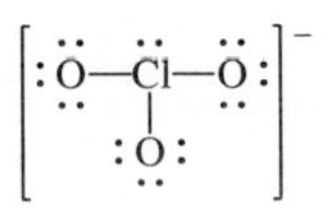

[:Ö—C̈l—Ö: | :Ö:]$^-$

(c) ClO_2^-:

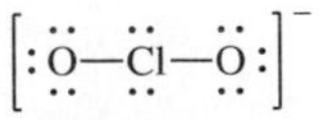

[:Ö—C̈l—Ö:]$^-$

SCl_2:

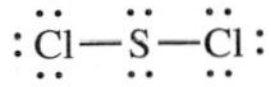

:C̈l—S̈—C̈l:

PCl_2^-:

[:C̈l—P̈—C̈l:]$^-$

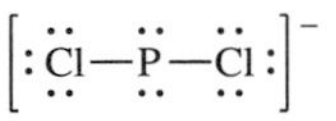

38. Molecules/ions that have the same number of valence electrons and the same number of atoms will have similar Lewis structures.

39. (a) NO_2^-

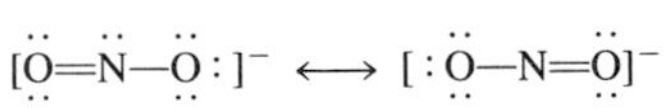

[Ö=N̈—Ö:]$^-$ ⟷ [:Ö—N̈=Ö]$^-$

HNO_2:

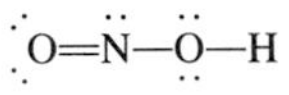

Ö=N̈—Ö—H

NO_3^-:

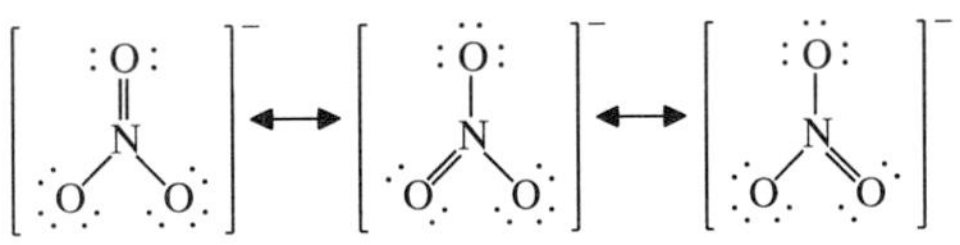

[:O: ‖ N / O O]$^-$ ⟷ [:O: | N // O O]$^-$ ⟷ [:O: | N \\ O O]$^-$

HNO_3:

:O: ‖ N / O O—H ⟷ :O: | N // O O—H

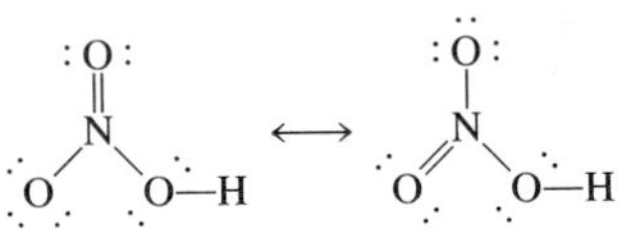

(b) SO_4^{2-} [see Exercise 37(a)]

HSO_4^-:

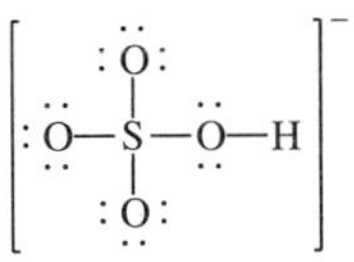

[:Ö: | :Ö—S—Ö—H | :Ö:]$^-$

H_2SO_4:

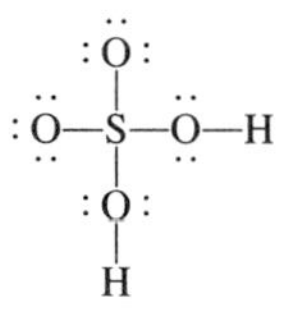

:Ö: | :Ö—S—Ö—H | :Ö: | H

(c) CN^-:

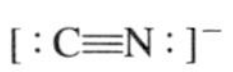

[:C≡N:]$^-$

HCN:

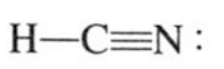

H—C≡N:

40.

Ozone Ö / O \\ O: ⟷ :O / Ö \\\\ O:

Sulfur dioxide S̈ // O \\ O: ⟷ :O / S̈ \\\\ O

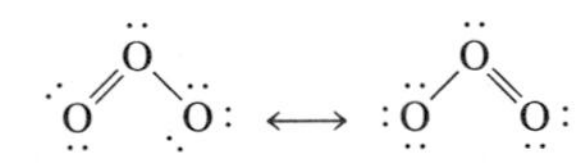

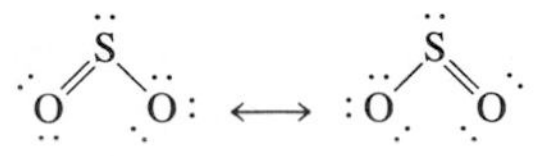

Sulfur trioxide :O: | S // O \\ O: ⟷ :O: ‖ S / O \\ O: ⟷ :O: | S / O \\\\ O

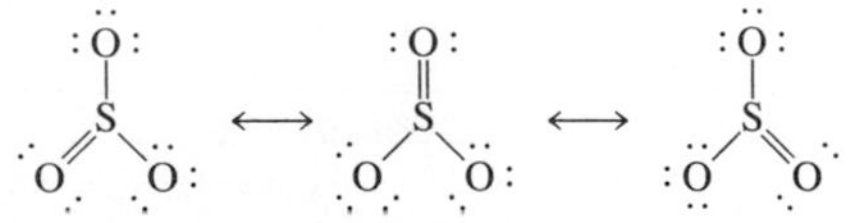

41. PAN $H_3C_2NO_5$:

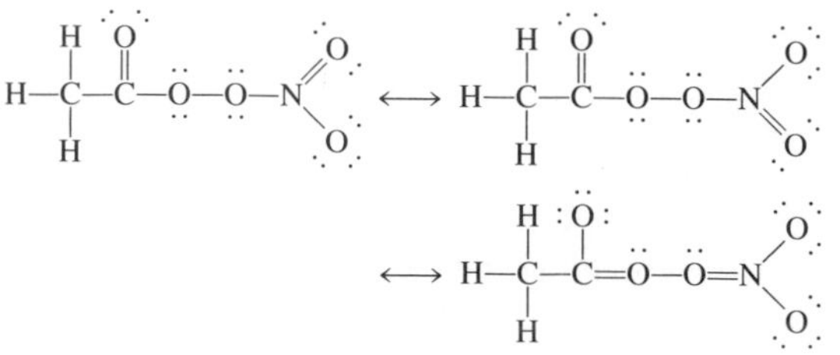

44.

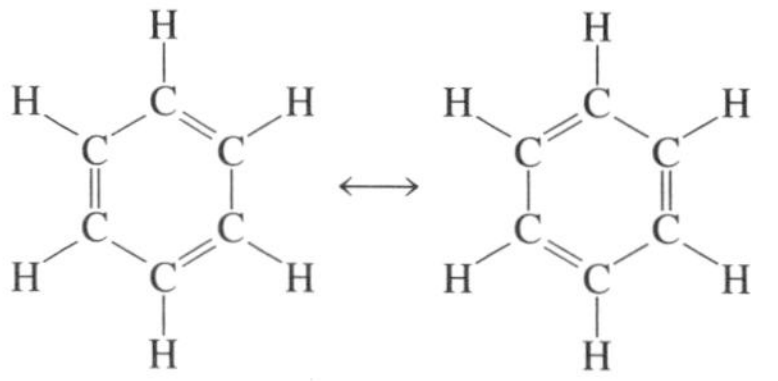

45. With resonance all carbon atoms are equal (we indicate this with a circle in the ring) giving three different structures:

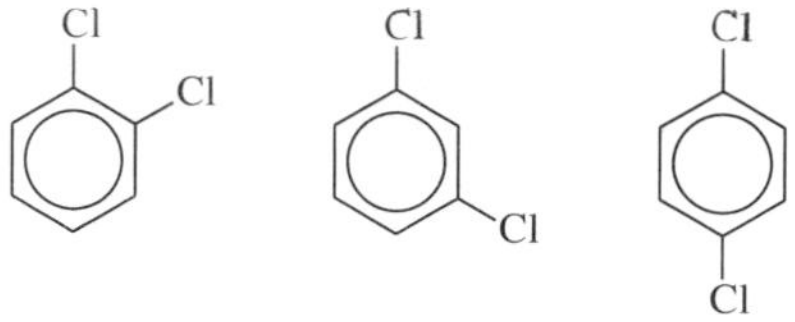

Localized double bonds give four unique structures.

46. Borazine, $B_3N_3H_6$:

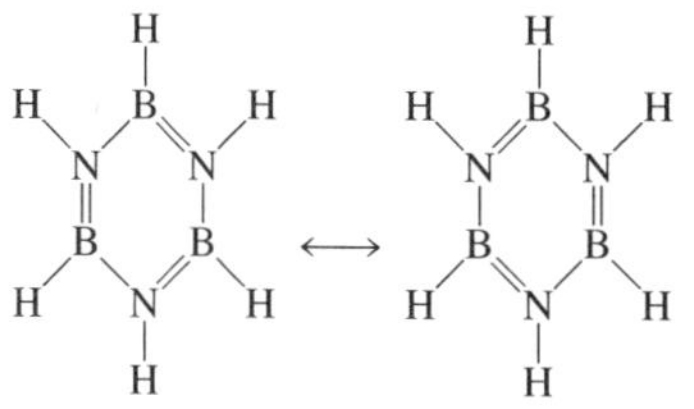

47. PF_5:

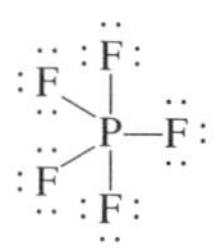

BrF_3:

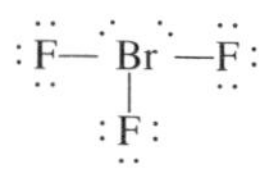

$Be(CH_3)_2$:

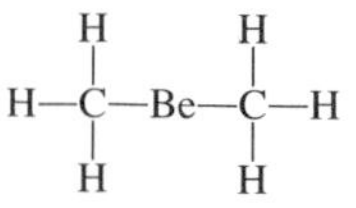

BCl_3:

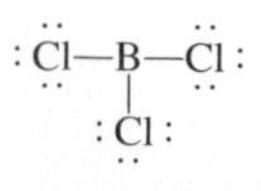

$XeOF_4$:

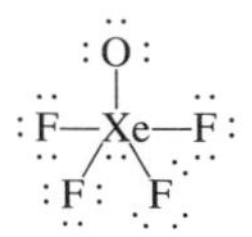

XeF_6:

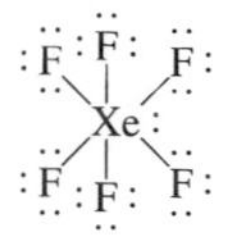

SeF_4:

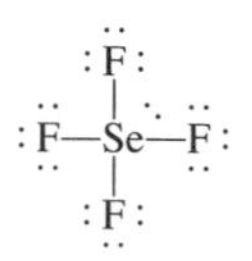

50. The N—N bond is between a double and triple bond; the N—O bond is between a single and double bond; the last resonance structure doesn't appear to be as important as the other two.

$$N{=}N{=}O \longleftrightarrow N{\equiv}N{-}O$$

52. $CH_3OH > CO_3^{2-} > CO_2 > CO$; $CH_3OH < CO_3^{2-} < CO_2 < CO$

54. $N{=}N{=}O$ (−1) (+1) (0) $\longleftrightarrow$ $N{\equiv}N{-}O$ (0) (+1) (−1) $\longleftrightarrow$ $N{-}N{\equiv}O$ (−2) (+1) (+1)

We should eliminate $N{-}N{\equiv}O$ since it has a formal charge of +1 on the most electronegative element. This is consistent with the observation that the N—N bond is between a double and a triple bond and that the N—O bond is between a single and a double bond.

55. The structure

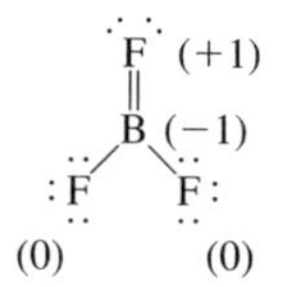

F (+1), B (−1), F (0), F (0)

obeys the octet rule but has a +1 formal charge on the most electronegative element (fluorine) and a negative formal charge on a much less electronegative element (boron). This is just the opposite of what we expect.

56.

$C{\equiv}O$ (−1)(+1)

Electronegativity predicts the opposite polarization. The two opposing effects seem to cancel to give a much less polar molecule than expected.

57. (a) P, FC = +1; (b) S, FC = +2; (c) Cl, FC = +3; (d) P, FC = +1

58.

(a)

(b) $[O=S(=O)(-O)-O]^{2-}$ (each O with lone pairs)

(c) $[O=Cl(=O)(=O)-O]^{-}$ (each O with lone pairs)

(d) $[O=P(-O)(-O)-O]^{3-}$ (each O with lone pairs)

59. [36] (a) Linear, 180°; (b) Trigonal pyramid, <109.5°; (c) Tetrahedral, ≈109.5°; (d) Tetrahedral, 109.5°; (e) Tetrahedral, 109.5°; (f) Bent, <109.5°; [37] (a) All are tetrahedral, ≈109.5°; (b) All are trigonal pyramidal, <109.5°; (c) All are bent (nonlinear), <109.5°; [39] (a) NO_2^- bent, 120°; HNO_2 bent about O and N, HON angle < 109.5°, ONO angle ≈120°; NO_3^- trigonal planar, 120°; HNO_3 trigonal planar about N, 120°, bent about N—O—H, angle < 109.5°; (b) Tetrahedral about S in all three, angle = 109.5°, bent about S—O—H, angle < 109.5°; (c) Both are linear, 180°; (d) All are linear, 180°; (e) Linear, 180° **60.** (a) Linear; 180°; (b) T-shaped; FClF angles are ≈90°. Since the lone pair will take up more space, the FClF bond angles will probably be slightly less than 90°. (c) 120°; 90°; see-saw; (d) 120°; 90°; trigonal bipyramid; **61.** (a) Trigonal pyramid, angles ≤ 109.5°; (b) Bent, ≤109.5°; (c) Tetrahedral, angles = 109.5° **62.** (a) Square pyramid, 90° bond angles; (b) Square planar, 90° bond angles; (c) Octahedral, 90° bond angles

64. (a) CH_2Cl_2 and $CHCl_3$ have dipole moments.
(b) N_2O is polar.

$\delta+\ \delta-$
:N=N=O:

65. (a) CrO_4^{2-} is

$[O-Cr(-O)(-O)-O]^{2-}$ (each O with lone pairs) tetrahedral;

(b) dichromate ion: $Cr_2O_7^{2-}$ has a roughly tetrahedral arrangement of O atoms about each Cr and the Cr—O—Cr bond is bent.

$[O_3Cr-O-CrO_3]^{2-}$ (each O with lone pairs)

66. The two requirements for a polar molecule are: 1. polar bonds; 2. a structure such that the polar bonds do not cancel

67. (a) OCl_2 is

:Cl—O—Cl: bent, polar;

Br_3^- is

:Br—Br—Br: linear, nonpolar;

BeH_2 is

H—Be—H linear, nonpolar;

BH_2^- is

H—B—H bent, polar.

(b) BCl_3 is

BCl_3 (B bonded to three :Cl:) trigonal planar, nonpolar;

NF_3 is

NF_3 (N with lone pair bonded to three F) trigonal pyramid, polar;

IF_3 is

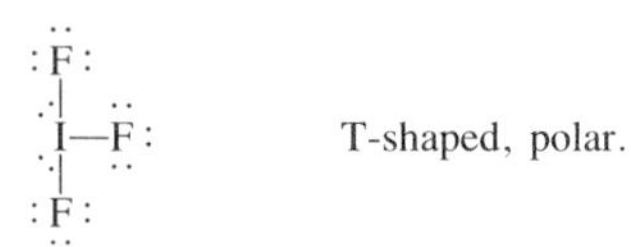

T-shaped, polar.

(c) CF_4 is

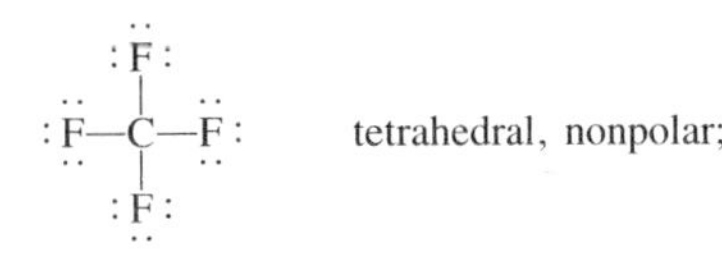

tetrahedral, nonpolar;

SeF_4 is

SeF_4 (Se with lone pair bonded to four F) see-saw, polar;

XeF_4 is

XeF_4 (Xe with two lone pairs bonded to four F) square planar, nonpolar;

(d) IF_5 is

IF_5 (I with lone pair bonded to five F) square pyramid, polar;

AsF_5 is

trigonal bipyramid, nonpolar

68.

:F—S—F: bent, polar;

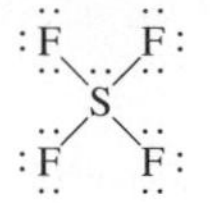 see-saw, polar;

 octahedral, nonpolar;

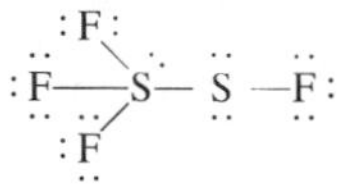 nonlinear, polar

71. (a) 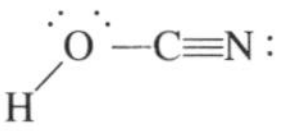 polar;

(b) 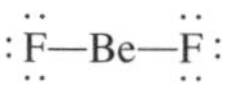 nonpolar;

(c) 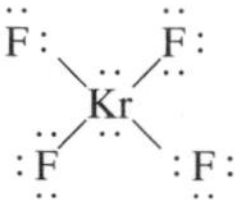 nonpolar;

(d) 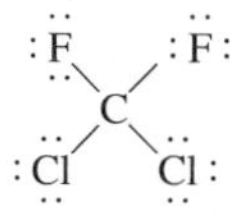 polar

72. (a) $CHCl_3 > CH_2Cl_2 > CH_3Cl > CH_4$; (b) $CHF_3 > CH_2F_2 > CCl_2F_2$

73. The general structure of a trihalide ion is:

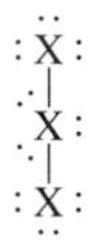

Bromine and iodine are large and have empty *d* orbitals to accommodate the expanded octet. Fluorine is small and its valence shell contains only 2*s* and 2*p* orbitals (4 orbitals) and cannot expand its octet.

75. C≡O 1072 kJ/mol, N≡N 941 kJ/mol. CO is polar while N_2 is nonpolar. This may lead to a greater reactivity. **76.** The Si—Cl bonds are more polar than the C—Cl bonds. **77.** (a) SO_4^{2-}; (b) PF_5; (c) SF_6; (d) BH_4^- **79.** No, we would expect the more highly charged ions to have a greater attraction for an electron. Thus, the electronegativity of an element does depend on its oxidation state.

81. (a) BF_3 is

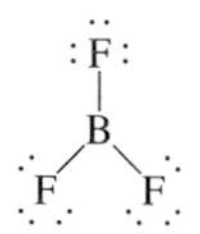 trigonal planar, nonpolar;

(b) PF_3 is

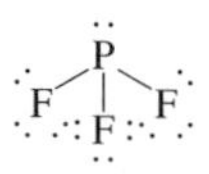 trigonal pyramid, polar;

(c) BrF_3 is

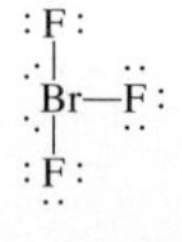 T-shaped, polar

83. (a) −2057 kJ; (b) −203 kJ

84.

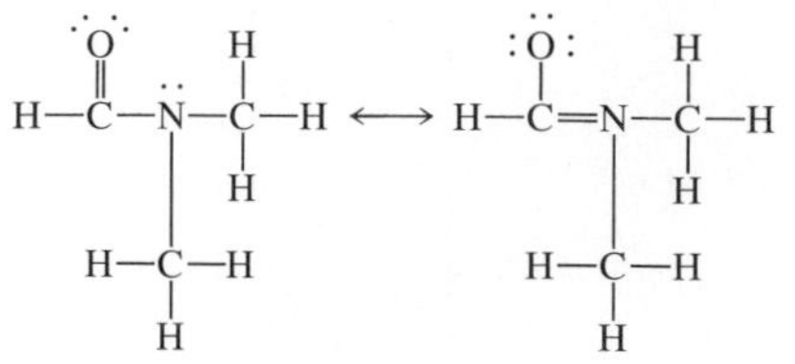

86. (a) $Na^+(g) + Cl^-(g) \rightarrow NaCl(s)$; (b) $NH_4^+(g) + Br^-(g) \rightarrow NH_4Br(s)$; (c) $Mg^{2+}(g) + S^{2-}(g) \rightarrow MgS(s)$; (d) $\frac{1}{2}O_2(g) \rightarrow O(g)$; (e) $O_2(g) \rightarrow 2O(g)$

Chapter 9

1. (a) tetrahedral, 109.5°, sp^3, nonpolar

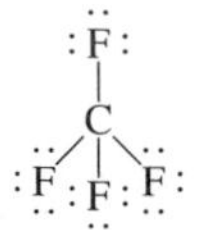

(b) trigonal pyramid, <109.5°, sp^3, polar

(c) bent, <109.5°, sp^3, polar

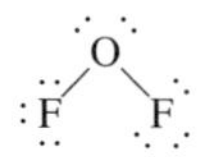

(d) trigonal planar, 120°, sp^2, nonpolar

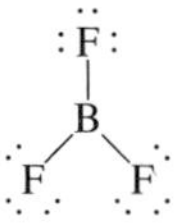

(e) linear, 180°, *sp*, nonpolar

H—Be—H

(f) 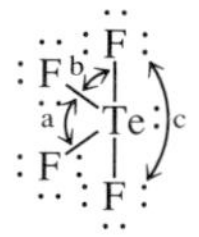

(g) 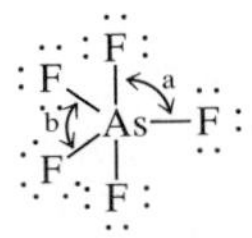

(h) 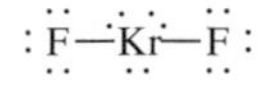

(i) 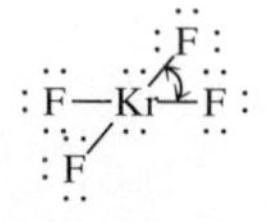

2. (a) bent, sp^2

(only one resonance form shown);

(b) trigonal planar, sp^2

(only one resonance form shown)

(c) tetrahedral, sp^3

(d) tetrahedral geometry about each S, sp^3

(e) trigonal pyramid, sp^3

(f) tetrahedral, sp^3

3. (a) sp^3, P—P—P bond angle 60°

(b) P_4O_6, All P—sp^3, POP angle < 109.5°; OPO angle < 109.5°

5. No, the CH_2 planes are mutually perpendicular.

6. Biacetyl:

All CCO angles are 120°; not in same plane

7. In biacetyl there are 11 σ and 2 π bonds. In acetoin there are 13 σ and 1 π bonds. **10.** (a) 6; (b) 4; (c) The center N in —N═N═N group; (d) 33 σ; (e) 5 π; (f) The six-membered ring is planar; (g) 180°; (h) <109°; (i) sp^3 **14.** (a) H_2^+: stable; H_2: stable; H_2^-: stable; H_2^{2-}: not stable; (b) He_2^{2+}: stable; He_2^+: stable; He_2: not stable; (c) Be_2: not stable; B_2: stable; Li_2: stable **16.** MO model does a better job. It predicts a bond order of 2.5. **17.** Bond energy is proportional to bond order. Bond length is inversely proportional to bond order. Bond energy and length can be measured. **18.** See Fig. 9.33. **19.** See Fig. 9.42. **20.** (a) $(\sigma_{1s})^2$; B.O. = 1; diamagnetic; (b) $(\sigma_{2s})^2(\sigma_{2s}^*)^2(\pi_{2p})^2$; B.O. = 1; paramagnetic; (c) $(\sigma_{2s})^2(\sigma_{2s}^*)^2(\pi_{2p})^4(\sigma_{2p})^2(\pi_{2p}^*)^4$; B.O. = 1; diamagnetic; (d) $(\sigma_{2s})^2(\sigma_{2s}^*)^2(\pi_{2p})^4$; B.O. = 2; diamagnetic; (e) $(\sigma_{2s})^2(\sigma_{2s}^*)^2(\pi_{2p})^4(\sigma_{2p})^1$; B.O. = 2.5; paramagnetic; (f) $(\sigma_{2s})^2(\sigma_{2s}^*)^2(\pi_{2p})^4(\sigma_{2p})^2$; B.O. = 3; diamagnetic **21.** $CN^- < CN < CN^+$ (bond length); $CN^+ < CN < CN^-$ (bond energy) **25.** See Sample Exercise 9.7. **26.** C_2^{2-} has 10 valence electrons. $[:C\equiv C:]^{2-}$; sp hybrid orbitals used; $(\sigma_{2s})^2(\sigma_{2s}^*)^2(\pi_{2p})^4(\sigma_{2p})^2$; B.O. = 3. Both give the same picture, a triple bond composed of one σ and two π-bonds. Both predict the ion to be diamagnetic. Lewis structures deal well with diamagnetic (all electrons paired) species. The Lewis model cannot really predict magnetic properties. C_2^{2-} is isoelectronic with the CO molecule. **27.** (a) The electrons would be closer to F. The F atom is more electronegative than the H atom. The $2p$ orbital of F is lower in energy than the $1s$ orbital of H. (b) The bonding MO would have more fluorine $2p$ character. (c) The anti-bonding MO would place more electron density closer to H and would have a greater contribution from the hydrogen $1s$. **28.** (a) The electron removed from N_2 is in the σ_{2p} molecular orbital which is lower in energy than $2p$ from which the electron in atomic nitrogen is removed. Since the electron removed from N_2 is lower in energy than the electron in N, the ionization energy of N_2 is greater than for N. (b) F_2 should have a lower ionization energy than F. The electron removed from F_2 is in a π_{2p}^* anti-bonding molecular orbital which is higher in energy than the $2p$ atomic orbitals.

30. (a) bent, sp^3 hybrid;

(b) trigonal pyramid, sp^3;

(c) tetrahedral, sp^3;

(d) see-saw, dsp^3;

(e) trigonal bipyramid, dsp^3

31. $F_2ClO_2^-$: see-saw, dsp^3;

F_4ClO^-:

$[F_4ClO]^-$ (Lewis structure: O above Cl, four F around Cl) square pyramid, d^2sp^3;

F_2ClO^+:

$[:\ddot{F}—\ddot{C}l—\ddot{O}:$ with $:\ddot{F}:$ below Cl$]^+$ trigonal pyramid, sp^3;

$F_2ClO_2^+$:

$[$ $:\ddot{O}:$ above Cl; $:\ddot{F}—Cl—\ddot{O}:$; $:\ddot{F}:$ below Cl $]^+$ tetrahedral, sp^3

35. In order to rotate about the double bond, the π bond must be broken, while the sigma bond remains intact. Bond energies are 347 kJ/mol for a C—C and 614 kJ/mol for a C═C. If we take the single bond as the strength of the σ bond, then the strength of the π bond is (614 − 347) = 267 kJ/mol. Thus, 267 kJ/mol must be supplied to rotate about a carbon-carbon double bond.

37.

OH
CH₂
sp³ HO—C—H
O
C C═O
H sp²
C═C
HO OH
sp²

All 5 atoms of the ring must be coplanar.

All eight atoms (the four carbons and one oxygen in the ring and the three oxygens bonded to the ring) lie in the same plane.

—C— —C— H O O O O ⟷ —C— —C— H O O O O

39. Both give the same picture, a triple bond composed of one σ and two π bonds. Both predict the ion to be diamagnetic. Not all carbons are in the same plane. **40.** N_2: $(\sigma_{2s})^2(\sigma_{2s}{}^*)^2(\pi_{2p})^4(\sigma_{2p})^2$ in ground state, B.O. = 3, diamagnetic; 1st excited state: $(\sigma_{2s})^2(\sigma_{2s}{}^*)^2(\pi_{2p})^4(\sigma_{2p})^1(\pi_{2p}{}^*)^1$, B.O. = 2, paramagnetic (2 unpaired electrons) **41.** We can draw a Lewis structure for Be_2 although it does not obey the octet rule. The MO model [See Exercise 14(c)] predicts a bond order of zero; the MO model predicts that Be_2 shouldn't exist. Be_2 has not yet been observed. **42.** Paramagnetic: Unpaired electrons are present. Measure the mass of a substance in the presence and absence of a magnetic field.

43.

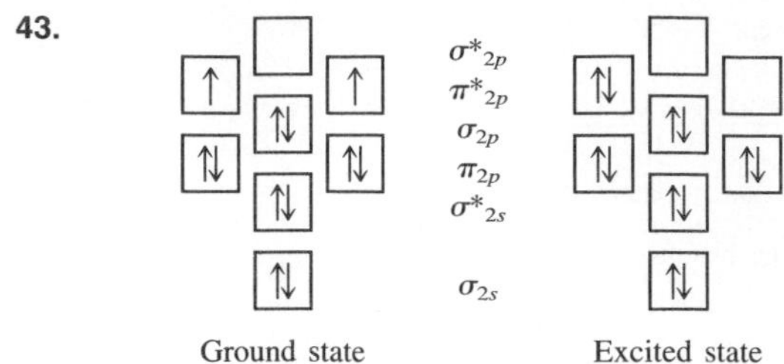

It takes energy to pair electrons in the same orbital. Thus, the structure with no unpaired electrons is at a higher energy; it is the excited state.

Chapter 10

1. There is an electrostatic attraction between the permanent dipoles of the polar molecules. The greater the polarity, the greater the attraction among molecules. **2.** London dispersion (LDF) < dipole-dipole < H-bonding < metallic bonding, covalent network, ionic. Yes, there is considerable overlap. Consider some of the examples in Exercise 12. Benzene (only LDF) has a higher boiling point than acetone (dipole-dipole). Also, there is even more overlap of the stronger forces (metallic, covalent, and ionic). **3.** As the size of the molecule increases, the strength of the London dispersion forces also increases. **4.** Yes **5.** As the strengths of interparticle forces increase; surface tension, viscosity, melting point, and boiling point increase, while the vapor pressure decreases. **6.** The nature of the forces stays the same. As the temperature increases and the phase changes (solid → liquid → gas) occur, a greater fraction of the forces are overcome by the increased thermal (kinetic) energy of the particles. **7.** (a) ionic; (b) LDF, dipole; (c) LDF only; (d) LDF; (e) metallic; (f) metallic; (g) LDF only; (h) H-bonding; (i) ionic; (j) metallic; (k) LDF mostly (C—F bonds are polar, but polymers like teflon are so large the LDF are the predominant interparticle forces.) **8.** (a) OCS: polar; CO_2: nonpolar; (b) PF_3: polar; PF_5: nonpolar **9.** Dipole forces are generally weaker than hydrogen bonds. They are similar in that they arise from an unequal sharing of electrons. We can look at a hydrogen bond as a particularly strong dipole force. **10.** (a) $H_2NCH_2CH_2NH_2$; (b) $B(OH)_3$ **11.** (a) Neopentane is more compact than *n*-pentane. Weaker interparticle forces and a lower boiling point. (b) Ethanol is capable of H-bonding; dimethyl ether is not. (c) HF has stronger H-bonding. (d) LiCl ionic; $TiCl_4$ is a non-polar molecular substance. **12.** Benzene, LDF; Naphthalene, LDF; Carbon tetrachloride, LDF; Acetone, LDF, dipole; Acetic acid, LDF, dipole, H-bonding; Benzoic acid, LDF, dipole, H-bonding; In terms of size: $CCl_4 < C_6H_6 < C_{10}H_8$. The strengths of LDF should be in this order. The physical properties are consistent with this order. We would predict the strength of interparticle forces of the last three molecules to be:

acetone < acetic acid < benzoic acid
↑ polar ↑ H-bond ↑ H-bond, but greater LDF because of greater size

This is consistent with the values given for bp, mp, and H_{vap}. The order of the strengths of interparticle forces based on physical properties is: acetone < CCl_4 < C_6H_6 < acetic acid < naphthalene < benzoic acid. Acetone is lowest because, although it is polar, the LDF in acetone must be very small. Naphthalene has very strong LDF.

14. FHF^- can be looked upon as forming from an HF molecule and a F^- ion

$$\text{F—H} + \text{F}^- \rightarrow \underset{\text{linear}}{[\text{F—H—F}]^-}$$

to give equal F—H distances.

15. Only a fraction of the hydrogen bonds are broken in going from solid to liquid. Most of the hydrogen bonds are still present in the liquid and must be broken during the liquid-to-gas phase change.

17. Both molecules are capable of H-bonding. However, in oil of wintergreen the hydrogen bonding is *intra*molecular.

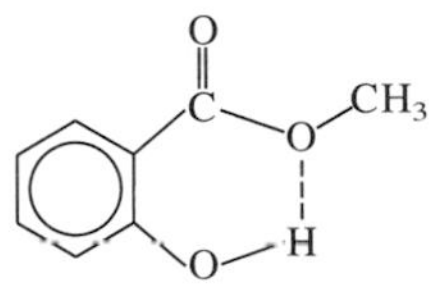

In methyl 4-hydroxybenzoate, the H-bonding is *inter*molecular, resulting in greater intermolecular forces and a higher melting point.

19. (a) NaCl: strong ionic bonding in lattice; (b) CH_3CN: only molecule not capable of H-bonding; (c) H_2: non-polar like CH_4, smaller than CH_4, weaker LDF; (d) SiO_2: covalent network solid vs. a gas and a liquid; (e) $HOCH_2CH_2OH$: (ethylene glycol) greatest amount of H-bonding since two OH groups are present. **20.** The boiling point of HF is higher than might be expected because of especially strong hydrogen bonding in HF. ΔH_{vap} should be high also, but it is lower than expected because many of the H-bonds reform in the gas phase: H—F···H—F. **22.** (a) Polarizability of an atom refers to the ease of distorting the electron cloud. It can also refer to distorting the electron clouds in molecules or ions. Polarity refers to the presence of a permanent dipole moment in a molecule. (b) London dispersion forces are present in all substances. LDF can be described as accidental dipole-induced dipole forces. Dipole-dipole forces involve the attraction of molecules with permanent dipoles for each other. (c) Inter: between; intra: within. For example, in H_2 the covalent bond is an intramolecular force, holding the two H-atoms together in the molecule. The much weaker London dispersion forces are intermolecular forces of attraction. **24.** Liquids and solids both have characteristic volume and are not very compressible. Liquids and gases flow and assume the shape of their container. **25.** *Critical temperature:* The temperature above which a liquid cannot exist, i.e., the vapor cannot be liquified by increased pressure. *Critical pressure:* The pressure that must be applied to a substance at its critical temperature to produce a liquid. At the critical temperature, all molecules have kinetic energies greater than the interparticle forces, and a liquid will not form. **26.** As the interparticle forces increase, the critical temperature increases. **28.** The attraction of H_2O for glass is stronger than H_2O–H_2O attraction. The meniscus is concave to increase the area of contact between glass and H_2O. The Hg–Hg attraction is greater than the Hg–glass attraction. The meniscus is convex to minimize the Hg–glass contact. Polyethylene is a non-polar substance. The H_2O–H_2O attraction is stronger than the H_2O–polyethylene attraction; thus, the meniscus will have a convex shape. **31.** The structure of H_2O_2 (H—O—O—H) produces greater hydrogen bonding than water. Long chains of hydrogen-bonded H_2O_2 molecules then become entangled. **33.** (a) Crystalline solid: Regular, repeating structure. Amorphous solid: Irregular arrangement of atoms or molecules. (b) Ionic solids: Made up of ions held together by ionic bonding. Molecular solid: Made up of discrete covalently bonded molecules held together in the solid by weaker forces (LDF, dipole, or hydrogen bonds). (c) Molecular solid: Discrete molecules. Covalent network solid: No discrete molecules. A covalent network is like a large polymer. The interparticle forces are the covalent bonds between atoms. **34.** Crystalline solid **35.** No; an example is common glass which is primarily amorphous SiO_2 (a covalent network solid) as compared to ice (a crystalline solid held together by weaker H-bonds). **36.** (a) molecular; (b) covalent network; (c) atomic, covalent network; (d) molecular; (e) atomic, metallic; (f) molecular; (g) ionic; (h) molecular **37.** 0.229 nm **39.** 313 pm = 3.13 Å **41.** 195 g/mol; 139 pm = 1.39 Å; platinum **42.** 4.29 g/cm³ **44.** d = 8.85 g/cm³ for β form **45.** 6.038×10^{23} **46.** 3.28×10^{-8} cm = 328 pm; $r = 1.42 \times 10^{-8}$ cm = 142 pm = 1.42 Å **50.** $CaTiO_3$ **52.** (a) NaCl; (b) CsCl; (c) ZnS; (d) TiO_2 **54.** If fcc, d = 19.4 gm/cm³. If bcc, d = 17.8 g/cm³. The measured density is consistent with a face-centered cubic unit cell. **55.** Conductor: partially filled valence band; electrons can move through valence band. Insulator: filled valence band, large band gap; electrons cannot move in valence band and cannot jump to next band. Semiconductor: filled valence band, small band gap; electrons cannot move in valence band but can jump to next band. **56.** (a) As the temperature is increased, more electrons in the valence band have sufficient KE to jump from the valence band to the conduction band. (b) A photon of light is absorbed by an electron which then goes from the valence band to the conduction band. (c) An impurity either adds electrons at an energy near the conduction band (n-type) or adds holes at an energy near the valence band (p-type). **58.** To make a p-type semiconductor we need to dope the material with atoms with fewer valence electrons. The average number of valence electrons is four. We could dope with more of the Group 3A elements or with atoms of Zn or Cd (most commonly used to produce p-type GaAs). To make n-type GaAs we could use an excess of As or dope with a Group 6 element such as S. **59.** 5.0×10^2 nm **61.** (a) condensation: vapor → liquid; (b) evaporation: liquid → vapor, are occurring at equal rates. **63.** 1.7 atm; 410 K or 140°C **65.** 1700 J **67.** 4.64 kg/hr **68.** For Li: 140 kJ/mol. For **62.** Equilibrium: There is no change in composition; the vapor pressure is constant. Dynamic: Two processes, vapor → liquid and liquid → vapor, are occurring at equal rates. **63.** 1.7 atm; 410 K or 140°C **65.** 1700 J **67.** 4.64 kg/hr **68.** For Li: 140 kJ/mol. For Mg: 120 kJ/mol; the bonding is stronger in Li. **69.** 221 torr **72.** (a) As intermolecular forces increase, the rate of evaporation decreases. (b) Increase T: increase rate. (c) Increase surface area: increase rate. **73.** As P is lowered, the water boils. The evaporation of the water is endothermic and the water is cooled to give some ice. If the pump remained on, all the ice would sublime. **74.** A: solid; B: liquid; C: vapor; D: solid + vapor; E: solid + liquid + vapor; F: liquid + vapor; G: liquid + vapor; H: vapor; triple point: E; critical point: G **78.** Heat released is 2.00 kJ but to melt 50 g of ice requires 16.7 kJ. Not all of the ice will melt (remains at 0°C). **79.** (a) 2; (b) Triple point at 96°C: rhombic, monoclinic, vapor; triple point at 119°C: monoclinic, liquid, vapor. (c) rhombic; (d) yes **80.** The two forms of BN have structures similar to graphite and diamond. **82.** 30.6 kJ/mol; BP = 356 K **83.** (a) 3 of each; (b) 2; (c) 1 colored; 3 open **85.** 0.7406 or 74.06%

86.

```
         O---H—O
        //        \
H3C—C            C—CH3
        \        //
         O—H---O
```

88. There are atoms at each of the 8 corners (shared by 8 unit cells) plus one atom inside the unit cell ($\frac{2}{3}$ of one atom and $\frac{1}{6}$ of two others). Thus, there are two atoms in the unit cell.

Chapter 11

1. 0.365 mol/L **2.** 3.35 g $Na_2C_2O_4$ **3.** 1.54×10^{-2} mol/L; $M_{Ni^{2+}} = 1.54 \times 10^{-2}$ mol/L; $M_{Cl^-} = 3.08 \times 10^{-2}$ mol/L
4. (a) $M_{Cu^{2+}} = 2.49 \times 10^{-3}$ mol/L; (b) 160 ppm
5. (a) 2800 ppm; 2.8×10^6 ppb; (b) 2.5×10^{-2} mol/L
6. (a) $M_{Ca^{2+}} = 1.06 \times 10^{-3}$ mol/L; $M_{NO_3^-} = 2.12 \times 10^{-3}$ mol/L; (b) 4.25×10^{-5} g Ca^{2+} ions; (c) 1.28×10^{15} NO_3^- ions
7. 0.16 L = 160 mL **8.** (a) $HNO_3(l) \rightarrow H^+(aq) + NO_3^-(aq)$;
(b) $Na_2SO_4(s) \rightarrow 2Na^+(aq) + SO_4^{2-}(aq)$;
(c) $AlCl_3(s) \rightarrow Al^{3+}(aq) + 3Cl^-(aq)$;
(d) $SrBr_2(s) \rightarrow Sr^{2+}(aq) + 2Br^-(aq)$;
(e) $KClO_4(s) \rightarrow K^+(aq) + ClO_4^-(aq)$;
(f) $NH_4Br(s) \rightarrow NH_4^+(aq) + Br^-(aq)$;
(g) $NH_4NO_3(s) \rightarrow NH_4^+(aq) + NO_3^-(aq)$;
(h) $CuSO_4(s) \rightarrow Cu^{2+}(aq) + SO_4^{2-}(aq)$;
(i) $NaOH(s) \rightarrow Na^+(aq) + OH^-(aq)$
9. 0.88 mol/L; 0.91 mol/kg; $\chi = 1.6 \times 10^{-2}$ **10.** Hydrochloric acid: 12 mol/L; 17 mol/kg; $\chi = 0.23$. Nitric acid: 16 mol/L; 37 mol/kg; $\chi = 0.40$. Sulfuric acid: 18 mol/L; 200 mol/kg; $\chi = 0.78$. Acetic acid: 17 mol/L; 2000 mol/kg; $\chi = 0.97$. Ammonia: 15 mol/L; 23 mol/kg; $\chi = 0.29$. **11.** % toluene = 28.4%; 2.69 mol/L; 4.32 mol/kg; $0.252 = \chi$ **14.** Since volume is temperature dependent and mass isn't, molarity is temperature dependent. In describing T_f and T_b, we are interested in how a temperature depends on composition. Thus, we don't want our expression of composition to be dependent on temperature also. **15.** $\chi_{acetone} = 0.044$; 0.75 *M*
17. 2.5%; 0.43 *M*; 0.44 molal, $\chi = 7.8 \times 10^{-3}$ **19.** The nature of the interparticle forces. **20.** As the hydrocarbon chain gets longer, the solubility decreases. **21.** (a) Mg^{2+}; (b) Be^{2+}; (d) F^-; (e) Cl^- **22.** $KCl(s) \rightarrow K^+(aq) + Cl^-(aq)$ $\Delta H = 31$ kJ/mol
23. CsI: $\Delta H_{hyd} = -571$ kJ/mol; CsOH: $\Delta H_{hyd} = -796$ kJ/mol
25. (a) CH_3CH_2OH; (b) $CHCl_3$; (c) CH_3CO_2H; (d) Na_2S
26. (a) H_2O; (b) CCl_4; (c) H_2O; (d) CCl_4 **27.** Both $Al(OH)_3$ and NaOH are ionic. Since the lattice energy is proportional to charge, the lattice energy of aluminum hydroxide is greater than that of sodium hydroxide. The attraction of water molecules for Al^{3+} and OH^- cannot overcome the extra lattice energy, and $Al(OH)_3$ is insoluble. **30.** Hydrophobic: water hating; hydrophilic: water loving
31. $CO_2 + OH^- \rightarrow HCO_3^-$; No, the reaction of CO_2 with OH^- greatly increases the solubility of CO_2 by forming the soluble bicarbonate ion. **32.** 2.0×10^{-4} mol/L **34.** 23.6 torr **35.** 189 torr. In the vapor, $\chi_{acetone} = 0.512$ and $\chi_{methanol} = 0.488$. It is probably not a good assumption that this solution is ideal because of the possibility of methanol-acetone hydrogen bonding. **36.** The vapor pressure of (c) will be the lowest. **37.** If solute-solvent attraction > solvent-solvent, there is a negative deviation from Raoult's Law. If solute-solvent < solvent-solvent attraction, then there is a positive deviation from Raoult's Law. **38.** The boiling point is lower than if the solution were ideal. **39.** (a) CF_3CF_3 and $CF_3CF_2CF_3$; (b) C_7H_{16} and C_6H_{14} **41.** 23.2 torr at 25°C; 70.0 torr at 45°C
42. (a) 290 torr; (b) In the vapor phase, $\chi_{pen} = 0.68$ **44.** $T_b = 100.042$°C; $T_f = -0.15$°C; $\pi = 2.0$ atm **45.** $\pi = 0.21$ mm Hg; $\Delta T_b = 2.1 \times 10^{-5}$ °C **46.** Osmotic pressure **50.** $T_f = -12.2$°C; $T_b = 103.4$°C, assuming 1.0 atm barometric pressure **51.** 44% naphthalene; 56% anthracene **52.** 14.5 g **55.** >24 atm
56. (a) Water would go from right to left. Level in right arm would go down and the level in left arm would go up. (b) The levels would be equal. The concentration of NaCl would be equal in both chambers. **57.** $MgCl_2 > NaCl > HOCl >$ glucose **59.** $T_f = -3$°C
60. The measured freezing point should be higher because of ion association. **62.** Ion pairing can occur, resulting in fewer particles than expected. Ion pairing will increase as the concentration of electrolyte increases. **65.** Benzoic acid is capable of hydrogen bonding. However, it is more soluble in nonpolar benzene than in water. In benzene, a nonpolar hydrogen-bonded dimer forms. The dimer is nonpolar and thus more soluble in benzene than in water. **66.** Benzoic acid would be more soluble in a basic solution because of the reaction $C_6H_5CO_2H + OH^- \rightarrow C_6H_5CO_2^- + H_2O$. **67.** (a) An analytical balance can weigh to the nearest 0.1 mg. We would use 309.5 mg sodium lactate, 20.0 mg $CaCl_2 \cdot 2H_2O$, 29.9 mg KCl and 601.2 mg NaCl. (b) Osmotic pressure ranges from 6.57 atm to 7.30 atm. **68.** No, ΔH_{soln} for methanol/water is not zero. **72.** Empirical formula: C_7H_4O; Molecular formula: $C_{14}H_8O_2$ **75.** Empirical formula: $C_2H_4O_3$; Molecular formula: $C_4H_8O_6$; Molecular weight = 152 g/mol **77.** 10.%; 2.5 molal **80.** Dissolve 209 g sucrose in water and dilute to 1.0 L in a volumetric flask. **81.** Dissolve 397 g $CaCl_2$ in 2.00 L (2.00 kg) H_2O. This will make slightly more than 2 L. **82.** (a) $T_f = -2.1$°C; (b) $T_f = -13$°C

Chapter 12

1. $-\Delta[I_2]/\Delta t = 0.0040$ mol/L s; $\Delta[S_4O_6^{2-}]/\Delta t = 0.0040$ mol/L s; $\Delta[I^-]/\Delta t = 0.0080$ mol/L s **3.** (a) mol/L s; (b) Rate = *k*; *k* has units of mol/L s; (c) Rate = *k*[A]; *k* must have units of s^{-1}; (d) Rate = $k[A]^2$; *k* must have units L/mol s; (e) L^2/mol^2 s
5. (a) molecules/cm^3 s; (b) molecules/cm^3 s; (c) s^{-1}; (d) cm^3/molecules s; (e) cm^6/molecules^2 s **6.** $k = 7.46 \times 10^8$ L/mol s
7. (a) 0.07 mol/L; (b) $t = 2.08 \times 10^3$ s; (c) Both H_2 and I_2 are formed at the rate of 6.00×10^{-5} mol/L s. **8.** (a) Rate = $k[NO]^2[Cl_2]$; (b) $k = 1.8 \times 10^2$ L^2/mol^2 min **9.** (a) Rate = $k[I^-][S_2O_8^{2-}]$; (b) *k*(L/mol s): 3.9×10^{-3}; 3.9×10^{-3}; 3.5×10^{-3}; 3.4×10^{-3}; 3.6×10^{-3}; $k_{mean} = 3.7 \times 10^{-3}$ L/mol s ± 0.3 L/mol s **11.** (a) Rate = $k[NOCl]^2$; (b) $k = 6.6 \times 10^{-29}$ cm^3/molecules s; The mean value is 6.6×10^{-29} cm^3/molecules s. (c) 4.0×10^{-8} L/mol s **13.** Integrated rate law: $\frac{1}{[C_4H_6]} = kt + \frac{1}{[C_4H_6]_0}$; differential rate law: Rate = $k[C_4H_6]^2$; $k = 1.4 \times 10^{-2}$ L/mol s **14.** Integrated rate law: $\ln[C_6H_5N_2Cl] = -kt + \ln[C_6H_5N_2Cl]_0$; differential rate law: Rate = $k[C_6H_5N_2Cl]$
15. $\ln[H_2O_2] = -kt + \ln[H_2O_2]_0$; Rate = $k[H_2O_2]$; $k = 8.3 \times 10^{-4}$ s^{-1} **17.** (a) First order with respect to O; (b) Rate = $k[NO_2][O]$; $k = 1.0 \times 10^{-11}$ cm^3/molecules s **19.** (a) ^{239}Pu, $k = 2.845 \times 10^{-5}$ yr^{-1}; 9.02×10^{-13} s^{-1}; ^{241}Pu, $k = 0.053$ yr^{-1} = 1.7×10^{-9} s^{-1}; (b) ^{241}Pu; (c) 2.1×10^{13} disintegrations/s; 4.7 g left after 1 yr; 2.9 g ^{241}Pu after 10 years; 0.024 g left after 100 years **20.** $k = 9.2 \times 10^{-3}$ s^{-1}; $t_{1/2} = 75$ s **21.** 62 days **22.** $k = 0.12$ L/mol s **25.** (a) An elementary reaction is one in which the

rate law can be written from the molecularity. (b) The mechanism of a reaction is the series of elementary reactions that occur to give the overall reaction. The sum of all the steps in the mechanism gives the balanced equation. (c) The rate determining step is the slowest elementary reaction in any given mechanism. **26.** (a) Rate = $k[CH_3NC]$ (b) Rate = $k[O_3][NO]$ (c) Rate = $k[O_3]$ **27.** This is a possible mechanism with the second step being rate determining. **29.** b **30.** The first step is rate determining. **32.** (a) The greater the frequency of collisions, the greater the opportunities for molecules to react, and hence, the greater the rate. (b) Chemical reactions involve the making and breaking of chemical bonds. The kinetic energy of the collision can be used to break bonds. (c) For a reaction to occur, it is the reactive portion of each molecule that must be involved in a collision. Thus, only some collisions have the right orientation. **33.** The probability of the simultaneous collision of three molecules with the correct energy and orientation is exceedingly small. **34.** In a unimolecular reaction a single reactant molecule decomposes to products. In a bimolecular reaction, two molecules collide to give products. **35.** $H_3O^+(aq)$ + $OH^-(aq) \rightarrow 2H_2O(l)$ should have the faster rate. H_3O^+ and OH^- will be electrostatically attracted to each other; Ce^{4+} and Hg_2^{2+} will repel each other. **36.** $4.8 \times 10^3\ s^{-1}$ **39.** $E_a = 11.9$ kJ/mol **40.** 52.9 kJ/mol

42.

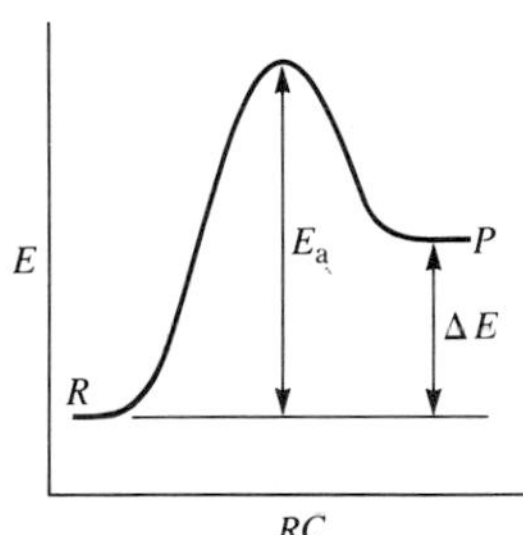

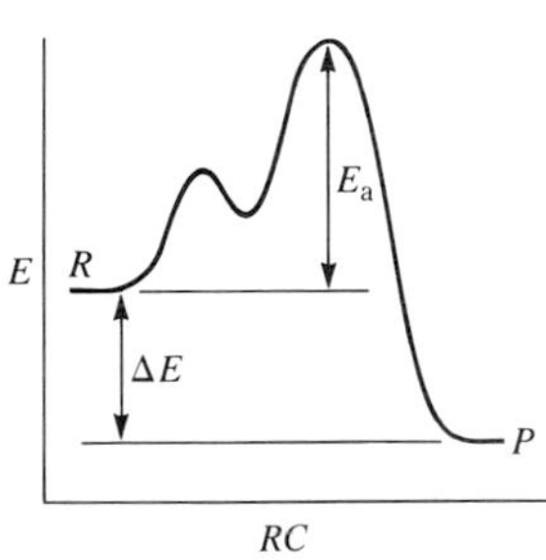

45. (a) A homogeneous catalyst is in the same phase as the reactants. (b) A heterogeneous catalyst is in a different phase than the reactants. **46.** A catalyst increases the rate of a reaction by providing an alternate pathway (mechanism) with energy. **47.** Yes, the catalyst takes part in the mechanism and will appear in the rate law. **48.** No, the catalyzed reaction has a different mechanism and hence, a different rate law. **49.** (a) NO is the catalyst. (b) NO_2 is an intermediate. (c) The catalyzed reaction is 2.33 times faster than the uncatalyzed reaction at 25°C. **52.** At high [S], the enzyme is completely saturated with substrate. **53.** An instantaneous rate is the slope of the tangent line to the graph of [A] vs. *t*. The initial rate is the instantaneous rate very near $t = 0$. The average rate is measured over a period of time and is the slope of the line connecting two points on the graph of [A] vs *t*. **54.** Rate = $k[H_2SeO_3][H^+]^2[I^-]^3$; $k_{ave} = 5.2 \times 10^5\ L^5/mol^5\ s$ **57.** (a) Untreated: $k = 0.465\ day^{-1}$; deacidifying agent: $k = 0.659\ day^{-1}$; antioxidant: $k = 0.779\ day^{-1}$; (b) No, the silk degrades more rapidly with the additives. **58.** Untreated: $t_{1/2} = 1.49$ day; deacidifying agent: $t_{1/2} = 1.05$ day; antioxidant: $t_{1/2} = 0.890$ day **61.** (a) W; lower E_a than Os; (b) 1.5×10^{30} times faster; (c) H_2 decreases the rate of the reaction. For the decomposition to occur NH_3 molecules must be adsorbed on the surface of the catalyst. If H_2 that is produced from the reaction is also adsorbed on the catalyst, then there are fewer sites for NH_3 molecules to be adsorbed, and the rate is decreased. **62.** Rate = $\dfrac{k_2k_1[I^-][OCl^-]}{k_{-1}[OH^-]}$ **63.** $BrO_3^- + HSO_3^- \rightarrow SO_4^{2-} + HBrO_2$ (slow)

Chapter 13

1. (a) The rates of the forward and reverse reactions are equal. (b) There is no net change in the composition. **2.** False **3.** No, equilibrium is a dynamic process. Both reactions, H_2O + $CO \rightarrow H_2 + CO_2$ and $H_2 + CO_2 \rightarrow H_2O + CO$, are occurring. Thus, ^{14}C atoms will be distributed between CO and CO_2. **4.** Both mixtures will give the same equilibrium position. **5.** The equilibrium constant is a number that tells us the relative concentrations (pressures) of reactants and products at equilibrium. An equilibrium position is a set of concentrations that satisfy the equilibrium constant expression. More than one equilibrium position is possible. **6.** The equilibrium constant expression and reaction quotient have the same form. For K_{eq} we use equilibrium concentrations, whereas any set of concentrations can be plugged into the reaction quotient expression. **7.** (a) $K = \dfrac{[NO_2][O_2]}{[NO][O_3]}$; (b) $K = \dfrac{[O_2][O]}{[O_3]}$; (c) $K = \dfrac{[ClO][O_2]}{[Cl][O_3]}$; (d) $K = \dfrac{[O_2]^3}{[O_3]^2}$ **8.** (a) $K_P = \dfrac{P_{NO_2}P_{O_2}}{P_{NO}P_{O_3}}$; (b) $K = \dfrac{P_{O_2}P_O}{P_{O_3}}$; (c) $K_P = \dfrac{P_{ClO}P_{O_2}}{P_{Cl}P_{O_3}}$; (d) $K = \dfrac{P_{O_2}{}^3}{P_{O_3}{}^2}$ **9.** a and c **10.** (a) $K = 7.59 \times 10^9$ L/mol; $K_P = 3.08 \times 10^8\ atm^{-1}$; (b) $K = 1.26 \times 10^9$ L/mol; $K_P = 5.11 \times 10^7\ atm^{-1}$ **14.** (a) 16.7; (b) 3.60×10^{-3}; (c) 6.00×10^{-2}; (d) 7.73×10^4 **16.** $2O_3(g) \rightleftharpoons 3O_2(g)$; $K_P = \dfrac{P_{O_2}{}^3}{P_{O_3}{}^2}$; (a) $K_P' = (K_P)^{1/2}$; (b) $K_P'' = 1/K_P$ **17.** $K = 0.20$ **19.** $K_P =$ 2.3 atm **20.** $K_P = 4.8\ atm^2$ **22.** $K = 3$ **23.** (a) to the right; (b) at equilibrium; (c) at equilibrium; (d) to the left **24.** (a) at equilibrium; (b) $Q = .300$; (c) $Q = 0.065$; (d) at equilibrium **27.** At this temperature all CO_2 will be in the gas phase. (a) Reaction must go to left; the mass of CaO will decrease. (b) At equilibrium; mass of CaO will not change. (c) At equilibrium; mass of CaO will not change. (d) Reaction will go to the right; the mass of CaO will increase. **28.** (a) $[HOCl] = 9.2 \times 10^{-3}$ *M*; $[H_2O] = 5.1 \times 10^{-2}$ *M*; $[Cl_2O] = 1.8 \times 10^{-2}$ *M*; (b) $[H_2O] = [Cl_2O] = 0.22$ *M*; $[HOCl] = 0.065$ *M* **29.** (a) $[I_3^-]/[I_2] = 71$; (b) $[I_3^-]/[I_2] = 1.0 \times 10^3$; (c) $[I_3^-]/[I_2] = 3.5 \times 10^3$ **31.** (a) $[NO] = 0.032$ *M*; $[Cl_2] = 0.016$ *M*; $[NOCl] = 0.97$ *M*; (b) $[NOCl] = 2.0$ *M*; $[NO] = 0.050$ *M*; $[Cl_2] = 0.025$ *M*; (c) $[Cl_2] = 1.6 \times 10^{-5}$ *M*; $[NOCl] = [NO] = 1.0$ *M* **32.** $P_{SO_3} = 0.12$ atm; $P_{SO_2} = 0.38$ atm; $P_{O_2} = 0.44$ atm **34.** (a) $[N_2O_4] = 5.9 \times 10^{-3}$ atm; $[NO_2] = 3.8 \times 10^{-2}$ atm; (b) $[NO_2] = 0.056$ atm; $[N_2O_4] = 0.012$ atm; (c) $[N_2O_4] = 1.2$ atm; $[NO_2] = 0.56$ atm **35.** $[Fe^{3+}] = 4.8 \times 10^{-5}$ *M*; $[FeSCN^{2+}] = 0.10$ *M*; $[SCN^-] = 1.9$ *M* **36.** We can do two things in the choice of solvent. First, we must avoid water. Any extra water we add from the solvent tends to push the equilibrium to the left. This eliminates water and 95% ethanol. Of the remaining two solvents, acetonitrile will not take part in the reaction, whereas ethanol is a reactant. If we use ethanol as the solvent, it will drive the equilibrium to the right, thereby reducing the concentrations of the objectionable butyric acid to a minimum. Thus, the best solvent is 100% ethanol. **37.** When we change the pressure by adding an unreactive gas, we do not change the partial pressure of any of the substances in equilibrium with each other. The equilibrium will not shift in this case. If we change the pressure by changing the volume, we will change the partial pressures of all the substances in equilibrium by the same factor. If there are unequal numbers of gaseous particles on both sides of the equation, the equilibrium will shift. **39.** (a) right; (b) right; (c) no effect; (d) left **41.** (a) increase; (b) increase; (c) no change; (d) increase (e) no change; (f) decrease **43.** Low **44.** Uncatalyzed reaction rate is too slow to be practical. **46.** The catalyst reduces the activation energies for both forward and reverse reactions. **47.** The rates would become equal in less time at a higher value for the catalyzed process; however, the ratio k_f/k_r would be unchanged. **48.** Rate$_r$ = $k_r[Br^-][SO_4^{2-}]^3[H^+]/[SO_3^{2-}]^2$ **51.** (a) 3×10^3 molecules/cm^3; (b) There is more NO in the atmosphere than we would expect from the value of K. The answer must lie in the rates of the reaction. At 25°C the rates of both reactions, $N_2 + O_2 \rightarrow 2NO$ and $2NO \rightarrow N_2 + O_2$, are essentially zero. Very strong bonds must be broken; the activation energy is very high. Nitric oxide is produced in high energy or high temperature environments. Naturally some NO is produced by lightning. The primary manmade source is automobiles. The production of NO is endothermic ($\Delta H_f^\circ = +90$ kJ/mol). At the high temperatures in car engines, K will increase and the rate of the reaction increases, so NO is produced. Once the NO enters the atmosphere, it doesn't decompose to N_2 and O_2 because of the slow rate. **52.** (a) $K = 4 \times 10^3$; (b) $K = 8 \times 10^{-2}$ **54.** P_{H_2S} = 0.28 atm **57.** (a) $K_P = 134$ atm^{-1}; (b) $P_{NO} = 0.052$ atm; P_{Br_2} = 0.18 atm; $P_{NOBr} = 0.25$ atm

Chapter 14

1. (a) weak acid; (b) weak acid; (c) weak base; (d) strong base; (e) weak base; (f) weak acid **2.** $HNO_3 > HOCl > NH_4^+ > H_2O$ **4.** (a) HI; (b) $HClO_2$; (c) HF

6.

Acid	Base	Conjugate base of acid	Conjugate acid of base
(a) H_2O	H_2O	OH^-	H_3O^+
(b) CH_3COCH_3	CH_3O^-	$CH_3COCH_2^-$	CH_3OH
(c) H_2S	NH_3	HS^-	NH_4^+
(d) H_2SO_4	H_2O	HSO_4^-	H_3O^+
(e) H^+	OH^-	H_2O	H_2O
(f) H_2O	$H_2PO_4^-$	OH^-	H_3PO_4
(g) $H_2PO_4^-$	H_2O	HPO_4^{2-}	H_3O^+

7. (a) $H_3PO_4 + H_2O \rightleftharpoons H_3O^+ + H_2PO_4^-$, K_{a_1} = $[H_3O^+][H_2PO_4^-]/[H_3PO_4]$; (b) $H_2PO_4^- + H_2O \rightleftharpoons H_3O^+ + HPO_4^{2-}$, $K_{a_2} = [H_3O^+][HPO_4^{2-}]/[H_2PO_4^-]$; (c) HPO_4^{2-} + $H_2O \rightleftharpoons H_3O^+ + PO_4^{3-}$, $K_{a_3} = [H_3O^+][PO_4^{3-}]/[HPO_4^{2-}]$; (d) $HNO_2 + H_2O \rightleftharpoons H_3O^+ + NO_2^-$, $K_a = [H_3O^+][NO_2^-]/[HNO_2]$; (e) $Ti(H_2O)_6^{4+} + H_2O \rightleftharpoons Ti(H_2O)_5(OH)^{3+} + H_3O^+$, K_a = $[H_3O^+][Ti(H_2O)_5(OH)^{3+}]/[Ti(H_2O)_6^{4+}]$; (f) HCN + $H_2O \rightleftharpoons H_3O^+ + CN^-$, $K_a = [H_3O^+][CN^-]/[HCN]$ **8.** (a) PO_4^{3-} + $H_2O \rightleftharpoons HPO_4^{2-} + OH^-$, $K_b = [OH^-][HPO_4^{2-}]/[PO_4^{3-}]$; (b) $HPO_4^{2-} + H_2O \rightleftharpoons H_2PO_4^- + OH^-$, K_b = $[OH^-][H_2PO_4^-]/[HPO_4^{2-}]$; (c) $H_2PO_4^- + H_2O \rightleftharpoons H_3PO_4 + OH^-$, $K_b = [OH^-][H_3PO_4]/[H_2PO_4^-]$; (d) $NH_3 + H_2O \rightleftharpoons NH_4^+ + OH^-$, $K_b = [NH_4^+][OH^-]/[NH_3]$; (e) $CN^- + H_2O \rightleftharpoons OH^- + HCN$, $K_b = [OH^-][HCN]/[CN^-]$; (f) $C_5H_5N + H_2O \rightleftharpoons C_5H_5NH^+ + OH^-$, $K_b = [OH^-][C_5H_5NH^+]/[C_5H_5N]$ **9.** (a) A strong acid is 100% dissociated in water. (b) A strong base is 100% dissociated in water. (c) A weak acid is much less than 100% dissociated in water. (d) A weak base is one where only a small percent of the molecules react with water to produce OH^-.
11. If we added an acid HX that was a stronger acid than H_3O^+, then the reaction

$$HX + H_2O \rightarrow H_3O^+ + X^-$$

would go to 100% completion because H_2O is a much stronger base than X^-. Thus, strong acids are "leveled" in water. They all generate H_3O^+. If a base stronger than OH^- were added to water, the reaction

$$\text{Base} + H_2O \rightarrow (\text{H base})^+ + OH^-$$

would go to 100% completion because H_2O would be a much stronger acid than (H base)$^+$.
12. (a) $HClO_4(aq) \rightarrow H^+(aq) + ClO_4^-(aq)$; (b) $CH_3CH_2CO_2H(aq) \rightarrow H^+(aq) + CH_3CH_2CO_2^-(aq)$ **13.** $[H^+]$ = $[OH^-]$; $[H^+] = 1 \times 10^{-7}$ M; pH = 7.0 **14.** (a) Since the value of the equilibrium constant increases as the temperature increases, the reaction is endothermic. (b) pH = 6.631; (c) A neutral solution of water at 50°C has $[H^+] = [OH^-]$ (most general); $[H^+] = 2.34 \times 10^{-7}$ M; pH = 6.631; (d) $K_w = 2.35 \times 10^{-14}$; (e) pH = 6.814 **15.** (a) 2.85; (b) 9.60; (c) −0.79; (d) pH = 12.54; (e) 3.9 **16.** (a) $[H^+] = 3.9 \times 10^{-8}$ M; $[OH^-] = 2.6 \times 10^{-7}$ M; (b) $[H^+] = 5 \times 10^{-16}$ M; $[OH^-] = 20$ M; (c) $[H^+] = 10$ M; $[OH^-] = 1 \times 10^{-15}$ M; (d) $[H^+] = 6 \times 10^{-4}$ M; $[OH^-] = 2 \times 10^{-11}$ M **18.** (a) H^+, Cl^-, and H_2O (HCl is a strong acid); (b) H^+, Br^-, and H_2O (HBr is a strong acid); (e) HNO_2 and H_2O (HNO_2 is a weak acid); (f) CH_3CO_2H and H_2O (CH_3CO_2H is a weak acid) **19.** (a) 1.0; (b) 1.0; (c) 1.0; (d) 4.52 **20.** pH = 7.00 **21.** $[H^+] = 1.9$ M; $[OH^-] = 5.3 \times 10^{-15}$ M; pH = −0.27 **22.** (a) pH = 0.602; (b) pH = 0.602; (e) pH = 2.00; (f) pH = 2.67 **23.** pH = 2.72; (b) pH = 1.61; (c) pH = 2.47 **26.** pH = 4.77 **27.** $[H^+] = [Bz^-] = 5.1 \times 10^{-4}$ M; $[HBz] = 4.6 \times 10^{-3} - x = 4.1 \times 10^{-3}$ M; pH = 3.29; $[OH^-] = 2.0 \times 10^{-11}$ M **29.** $[H^+] = 5.02 \times 10^{-3}$ M; pH = 2.299 **30.** (a) pH = 1.00; (b) pH = 1.30 **32.** pH = 2.73 **34.** 7.1×10^{-4} **35.** (a) 0.60%; (b) 1.9%; (c) 5.8% **36.** 21% **39.** 2×10^{-1} **40.** (a) pH = 1.62; (b) pH = 1.7 **41.** $[H_3PO_4] = 7.6 \times 10^{-2}$ M; $[H^+]$ = $[H_2PO_4^-] = 2.4 \times 10^{-2}$ M; $[HPO_4^{2-}] = 6.2 \times 10^{-8}$ M; $[PO_4^{3-}]$ = 1.2×10^{-18} M; $[OH^-] = 4.2 \times 10^{-13}$ M **42.** $[CO_3^{2-}] = 5.6 \times 10^{-11}$ M; 8.5×10^{-5}%; $HCO_3^- + H^+ \rightarrow (H_2CO_3) \rightarrow H_2O + CO_2$. The bubbles are CO_2, and this implies that discrete H_2CO_3 molecules are not stable. Rather we should write $H_2O + CO_2(aq)$ or $CO_2(aq)$ instead of $H_2CO_3(aq)$. **43.** $CH_3NH_2 > NH_3 > H_2O > NO_3^-$ **45.** (a) NH_3; (b) NH_3; (c) OH^-; (d) CH_3NH_2 **47.** pH = 12.049 **48.** (a) $[OH^-] = 0.25$ M; pOH = 0.60; pH = 13.40; (b) $[OH^-] = 8.0 \times 10^{-4}$ M; pOH = 3.10; pH = 10.90; (c) $[OH^-]$ = 0.45 M; pOH = 0.35; pH = 13.65 **50.** (a) K^+, OH^-, H_2O; (c) NH_3, H_2O; (d) pyridine (C_5H_5N), H_2O **51.** (a),(b) 13.176; (c) 11.22; (d) 9.20 **52.** (a) $[OH^-] = 1.0 \times 10^{-2}$ M; $H^+ = 9.7 \times 10^{-13}$ M; pH = 12.01; (b) $[OH^-] = 1.6 \times 10^{-2}$ M; $[H^+] = 6.4 \times 10^{-13}$ M; pH = 12.19; (c) $[OH^-] = 8.9 \times 10^{-3}$ M; $[H^+] = 1.1 \times 10^{-12}$ M; pH = 11.95; (d) $[OH^-] = 8.7 \times 10^{-6}$ M; $[H^+] = 1.1 \times 10^{-9}$ M; pH = 8.94 **55.** (a) 1.3%; (b) 4.2% **56.** pH = 11.39 **58.** 9.9×10^{-10} **60.** $pK_a + pK_b = pK_w$ **61.** (a) neutral; (b) neutral; (c) basic; (d) acidic; (e) acidic **62.** (a) pH = 5.82; (b) pH = 10.95; (c) pH = 11.78 **63.** acidic → basic; HCN, NH_4Br, KBr, NH_4CN, KCN, KOH **65.** The solution is acidic. The

reaction with water is $HSO_4^- + H_2O \rightleftharpoons H_3O^+ + SO_4^{2-}$. The reaction $CO_3^{2-} + HSO_4^- \rightleftharpoons HCO_3^- + SO_4^{2-}$ can occur. **67.** OCl^- is a stronger base than $C_2H_3O_2^-$. **68.** (a) weaker bond = stronger acid; (b) greater electronegativity = stronger acid; (c) more oxygen atoms = stronger acid **69.** (a) weaker bond = weaker base; (b) greater electronegativity = weaker base; (c) more oxygen atoms = weaker base **70.** (a) $HBrO_3 > HBrO_2 > HBrO$; (b) $H_3AsO_4 > H_2AsO_4^- > HAsO_4^{2-}$ **72.** (a) $BrO_3^- < BrO_2^- < BrO^-$; (b) $H_2PO_4^- < HPO_4^{2-} < PO_4^{3-}$ **73.** (a) basic; (b) basic; (c) acidic; (d) acidic **74.** (a) Arrhenius acid: generates H^+ in water; (b) Brönsted-Lowry acid: proton donor; (c) Lewis acid: electron pair acceptor **75.** Lewis definition **76.** (a) $B(OH)_3$ acid; H_2O, base; (b) Ag^+ acid; NH_3, base; (c) BF_3 acid; NH_3, base; (d) I_2 acid; I^- base; (e) $Zn(OH)_2$ acid; OH^- base **77.** $Al(OH)_3(s) + 3H^+(aq) \rightarrow Al^{3+}(aq) + 3H_2O(l)$; $Al(OH)_3(s) + OH^-(aq) \rightarrow Al(OH)_4^-(aq)$ **79.** A Lewis acid must have an empty orbital to accept an electron pair. A Lewis base must have an unshared pair of electrons. **82.** (a) An acid will generate NH_4^+ in liquid ammonia and a base will generate NH_2^-. (b) $[NH_4^+] = [NH_2^-]$; (c) $2Na(s) + 2NH_3(l) \rightarrow 2Na^+ + 2NH_2^- + H_2(g)$; (d) Ammonia is more basic than water. The acidity of substances is enhanced in liquid ammonia. For example, acetic acid is a strong acid in ammonia. **83.** 2.0×10^3; 7.6×10^5 **84.** 3.5×10^{-4} **85.** (a) 1.3×10^{-4}; (b) $[H_2CO_3] = [CO_3^{2-}]$; (c) $pH = \dfrac{pK_{a_1} + pK_{a_2}}{2}$; (d) pH = 8.31 **86.** (a) 1.8×10^9; (b) 2.5×10^3; (c) 3.1×10^{-5}; (d) 1.0×10^{14} **87.** (a) lungs; $Hb(O_2)_4$; cells; $(HbH_4)^{4+}$; (b) $CO_2(H_2O)$ is a weak acid. Removing CO_2 decreases H^+. $Hb(O_2)_4$ is favored and O_2 is not released by hemoglobin in the cells. Breathing into a paper bag increases CO_2 in the blood. (c) CO_2 builds up in the blood and it becomes too acidic, driving the equilibrium to the left. Hemoglobin can't bind O_2 as strongly in the lungs. Bicarbonate ion acts as a base in water and neutralizes the excess acidity. **88.** (a) ΔH = 1549 kJ (HF); 1390. kJ (HCl); 1350. kJ (HBr); 1312 kJ (HI) (b) The ionization of H is common to all acids. An acid will be a stronger acid if the bond to the hydrogen atom is weak or if the electron affinity of X is more favorable. **89.** Weak bonds to H and the presence of X atoms with favorable electron affinities. **90.** (a) $H_2O < H_2S < H_2Se$; (b) $CH_3CO_2H < FCH_2CO_2H < F_2CHCO_2H < F_3CCO_2H$; (c) $CH_3NH_3^+ < NH_4^+ < HONH_3^+$ **94.** pH = 7.35. This answer is impossible. We can't add a small amount of a very weak acid to a neutral solution and get a basic solution. In a correct solution, we would have to take into account the autoionization of water. This is beyond the scope of this textbook.

Chapter 15

1. A common ion is an ion that appears in an equilibrium reaction but came from a source other than that reaction. Addition of a common ion (H^+ or NO_2^-) to the reaction $HNO_2 \rightarrow H^+ + NO_2^-$ will drive the equilibrium to the left. **2.** A buffered solution must contain both a weak acid and a weak base. Buffered solutions are useful for controlling the pH of a solution. **3.** $H^+ + C_2H_3O_2^- \rightarrow HC_2H_3O_2$; $OH^- + HC_2H_3O_2 \rightarrow C_2H_3O_2^- + H_2O$ **4.** $CO_3^{2-} + H^+ \rightarrow HCO_3^-$; $HCO_3^- + OH^- \rightarrow H_2O + CO_3^{2-}$ **5.** The capacity of a buffer is a measure of how much strong acid or base the buffer can neutralize. All the buffers listed have the same pH. The 1.0 M buffer has the greatest capacity; the 0.01 M buffer the least capacity. **8.** (a) 2.94; (b) 8.94; (c) 7.00; (d) 4.89 **9.** (a) 1.70; (b) 5.5; (c) 1.70; (d) 4.7 **10.** (a) 4.3; (b) 12.30; (c) 12.30; (d) 5.1 **11.** (a) 3.58; (b) 4.81; (c) 7.43 **12.** (a) 5.14; (b) 4.34; (c) 5.14; (d) 4.34 **13.** (a) 0.56; (b) 0.35; (c) 5.6 **14.** (a) 40. g NaOH; (b) 12.4 g NaOH (12 g; correct sig. fig.) **16.** (a) 7.97; (b) 8.74 **17.** Because the $[NH_3]/[NH_4^+]$ ratio is the same for each buffer, the pH of each is the same. The two solutions differ in capacity. The buffer with the larger capacity shows a smaller change in pH for the same amount of added hydrogen ion. **19.** Only the combination in (d) results in a buffer. **21.** (a) 1.8×10^9; (b) 5.6×10^4; (c) 3.2×10^4; (d) 1.0×10^{14}. Although all four reactions have large equilibrium constants, only (a), (b), and (d) can be useful in a titration. The solution in (c) results in a buffer; it still contains both a weak acid and a weak base. For a titration to be useful there must be a large change in pH for a small amount of added titrant near the equivalence point. This condition is precisely what a buffer is used to prevent.
22.

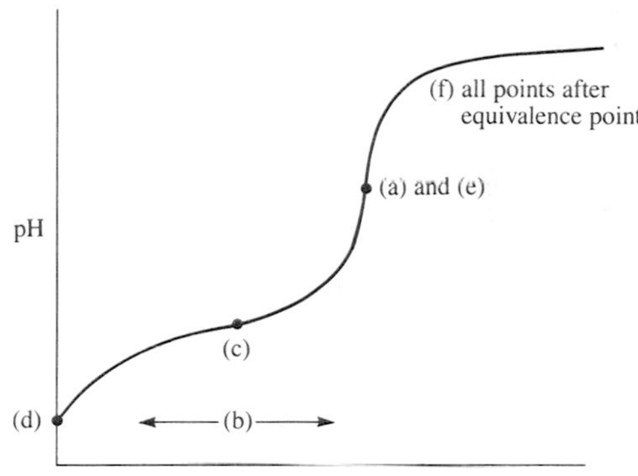

23, 25

mL titrant added	pH Exercise 23	pH Exercise 25
0	2.43	9.12
4	3.14	5.95
8	3.53	5.56
12.5	3.86	5.23
20.	4.46	4.63
24	5.24	3.85
24.5	5.55	3.54
24.9	6.26	—
25.0	8.27	3.27
25.1	10.3	—
26	11.3	2.7
28	11.8	2.2
30.	12.0	2.0

27. Using the equation

$$pH = pK_a + \log\frac{[Base]}{[Acid]},$$

halfway to the equivalence point,

$$[Base] = [Acid]$$

so $pH = pK_a + \log 1 = pK_a$

29. (a) 3.11; (b) 9.97 **30.** pH is less than about .5 **32.** (a) yellow (b) orange (c) blue (d) bluish-green **35.** Equivalence point: moles H^+ = moles base. End point: indicator changes color. They don't have to be as close as 0.01 pH units, since at the equivalence point the pH is changing very rapidly with added titrant. The range over which an indicator changes color only needs to bracket the pH of the equivalence point.
36. The two forms of an indicator are different colors. To see only one color, one form must be in excess over the other. To go from

$$\frac{[HIn]}{[In^-]} = 10 \text{ to } \frac{[HIn]}{[In^-]} = 0.1 \text{ requires a change of 2 pH units.}$$

37. Ex. 23, pH = 8.25, phenolphthalein; Ex. 25, pH = 3.27, 2,4-dinitrophenol **40.** (a) $AgC_2H_3O_2(s) \rightleftharpoons Ag^+(aq) + C_2H_3O_2^-(aq)$, $K_{sp} = [Ag^+][C_2H_3O_2^-]$; (b) $MnS(s) \rightleftharpoons Mn^{2+}(aq) + S^{2-}(aq)$, $K_{sp} = [Mn^{2+}][S^{2-}]$; (c) $Al(OH)_3(s) \rightleftharpoons Al^{3+}(aq) + 3OH^-(aq)$, $K_{sp} = [Al^{3+}][OH^-]^3$; (d) $Ca_3(PO_4)_2(s) \rightleftharpoons 3Ca^{2+}(aq) + 2PO_4^{3-}(aq)$, $K_{sp} = [Ca^{2+}]^3[PO_4^{3-}]^2$; (e) $Ca_5(PO_4)_3(OH)(s) \rightleftharpoons 5Ca^{2+}(aq) + 3PO_4^{3-}(aq) + OH^-(aq)$, $K_{sp} = [Ca^{2+}]^5[PO_4^{3-}]^3[OH^-]$
41. (a) 2.3×10^{-9}; (b) 3.92×10^{-5}; (c) 8.19×10^{-19}; (d) 2.10×10^{-7} **42.** (a) 2×10^{-11} mol/L ($[OH^-] = 1.0 \times 10^{-7}$ *M*); 2×10^{-9} g/L; (b) 7.32×10^{-4} mol/L; 0.286 g/L; (c) 7.8×10^{-3} mol/L; 1.1 g/L; (d) 9.3×10^{-5} mol/L; 9.3×10^{-3} g/L; (e) 7×10^{-5} mol/L; 9×10^{-3} g/L; (f) 6.6×10^{-7} mol/L; 3.1×10^{-4} g/L **43.** If the number of ions in two salts is the same, the K_{sp}'s can be compared. **44.** (a) 4×10^{-17} mol/L; (b) 4×10^{-11} mol/L; (c) 4×10^{-29} mol/L **45.** $Al(OH)_3$, $CaCO_3$, $Mg(NH_4)PO_4$, Ag_2CO_3, and Cu_2S. $CaSO_4$ and $SrSO_4$ will be slightly more soluble in acid. However, K_b for SO_4^{2-} is only about 10^{-12}, and the effect of acid on these salts is not as great as on the others.
46. 2.7×10^{-5} mol/L. The solubility of hydroxyapatite will increase as a solution gets more acidic, since both phosphate and hydroxide can react with H^+. 6×10^{-8} mol/L. The hydroxyapatite in the tooth enamel is converted to the less soluble fluorapatite by fluoride-treated water. The less soluble fluorapatite will then be more difficult to remove, making teeth less susceptible to decay. **48.** NiS, MnS, CuS, and HgS will precipitate at pH = 10; CuS and HgS will precipitate at pH = 0.5. **49.** (a) $NaCl(aq)$, $NH_3(aq)$, $H_2S(aq)$; (b) $NaCl(aq)$, $H_2SO_4(aq)$, $H_2S(aq)$ (basic) **50.** More soluble in base **51.** $[Pb^{2+}] = [Cl^-] = 0.010$ *M*; $[NO_3^-] = 0.020$ *M*; $[Na^+] = 0.010$ *M*

53. (a)

$$\begin{aligned}
Co^{2+} + NH_3 &\rightleftharpoons CoNH_3^{2+} & K_1\\
CoNH_3^{2+} + NH_3 &\rightleftharpoons Co(NH_3)_2^{2+} & K_2\\
Co(NH_3)_2^{2+} + NH_3 &\rightleftharpoons Co(NH_3)_3^{2+} & K_3\\
Co(NH_3)_3^{2+} + NH_3 &\rightleftharpoons Co(NH_3)_4^{2+} & K_4\\
Co(NH_3)_4^{2+} + NH_3 &\rightleftharpoons Co(NH_3)_5^{2+} & K_5\\
Co(NH_3)_5^{2+} + NH_3 &\rightleftharpoons Co(NH_3)_6^{2+} & K_6\\
\hline
Co^{2+} + 6NH_3 &\rightleftharpoons Co(NH_3)_6^{2+} & K_f = K_1K_2K_3K_4K_5K_6
\end{aligned}$$

(b)

$$\begin{aligned}
Fe^{3+} + SCN^- &\rightleftharpoons FeSCN^{2+} & K_1\\
FeSCN^{2+} + SCN^- &\rightleftharpoons Fe(SCN)_2^+ & K_2\\
Fe(SCN)_2^+ + SCN^- &\rightleftharpoons Fe(SCN)_3 & K_3\\
Fe(SCN)_3 + SCN^- &\rightleftharpoons Fe(SCN)_4^- & K_4\\
Fe(SCN)_4^- + SCN^- &\rightleftharpoons Fe(SCN)_5^{2-} & K_5\\
Fe(SCN)_5^{2-} + SCN^- &\rightleftharpoons Fe(SCN)_6^{3-} & K_6\\
\hline
Fe^{3+} + 6SCN^- &\rightleftharpoons Fe(SCN)_6^{3-} & K_f = K_1K_2K_3K_4K_5K_6
\end{aligned}$$

(c)

$$\begin{aligned}
Ag^+ + NH_3 &\rightleftharpoons AgNH_3^+ & K_1\\
AgNH_3^+ + NH_3 &\rightleftharpoons Ag(NH_3)_2^+ & K_2\\
\hline
Ag^+ + 2NH_3 &\rightleftharpoons Ag(NH_3)_2^+ & K_f = K_1K_2
\end{aligned}$$

54. 6.2×10^5

55. An ammonia solution is basic. Initially the reaction that occurs is

$$Cu^{2+}(aq) + 2OH^-(aq) \rightarrow Cu(OH)_2(s).$$

As the concentration of NH_3 increases, the complex ion, $Cu(NH_3)_4^{2+}$, forms $Cu(OH)_2(s) + 4NH_3(aq) \rightleftharpoons Cu(NH_3)_4^{2+}(aq) + 2OH^-(aq)$.

56. $[Hg^{2+}] = 3.3 \times 10^{-32}$ mol/L **59.** No precipitate forms **60.** (a) pH = 9.24; (b) 9.30 **62.** 4.8 g cacodylic acid; 14.4 g sodium cacodylate (14 g, correct sig. fig.) **63.** (a) potassium fluoride + HCl; (b) benzoic acid + NaOH; (c) sodium acetate + acetic acid; (d) $(CH_3)_2AsO_2Na$ + HCl; (e) ammonium chloride + NaOH **65.** 99.7%. The equivalence point is at pH = 7. Bromothymol blue would be the best indicator. **66.** The advantage of this graph is that the end point is much easier to determine accurately. **68.** $[Cr^{3+}] = 2 \times 10^{-37}$ *M* **70.** $[Fe^{3+}] = 2 \times 10^{-18}$ *M*; the lowest level of iron in serum is 1×10^{-5} *M*; the actual concentration of iron is much greater than expected when considering only the solubility of $Fe(OH)_3$. There must be complexing agents present to increase the solubility. **71.** 1.6×10^{-6} **75.** $K_{sp} = 6.4 \times 10^{-9} = [Mg^{2+}][F^-]^2 = (0.00375 - y)(0.0625 - 2y)^2$. This is a cubic equation. No simplifying assumptions can be made since *y* is relatively large.

Chapter 16

1. A spontaneous process is one that occurs without any outside intervention. **2.** (a), (b), (d), and (g) are spontaneous. (c), (e), and (f) require an external source of energy.

3.

2kJ			x		x	xx
1kJ		x		xx	x	
0kJ	xx	x	x			

The most likely total energy is 2kJ.

4.

2kJ			AB			B	A	B	A
1kJ		AB		B	A			A	B
0kJ	AB			A	B	A	B		

The most likely total energy is 2kJ.

7. (a) Entropy is a measure of disorder. (b) The quantity $T\Delta S$ has units of energy. **8.** (a), (b), (c), (e), and (g) involve an increase in entropy. **9.** There are more ways to roll a seven. The seven is not favored by energy; rather it is favored by probability. To change the probability we would have to expend energy (do work). **10.** Arrangement I: $S = 0$; Arrangement II: $S = 1.91 \times 10^{-23}$ J/K; Arrangement III: $S = k = 2.47 \times 10^{-23}$ J/K **11.** (a) increases; (b) no change, but *S* increases due to energy effects; (c) decreases
12. (a) N_2O; (b) H_2 at 100°C **13.** (a) The system is the portion of the universe in which we are interested. (b) The surroundings are everything else in the universe. **14.** $\Delta S_{univ} = \Delta S_{sys} + \Delta S_{surr}$; $\Delta S_{surr} = \frac{-\Delta H_{sys}}{T}$; $\Delta S_{univ} = \Delta S_{sys} - \frac{\Delta H_{sys}}{T}$. For a spontaneous process $\Delta S_{univ} > 0$, so $\Delta S_{sys} - \frac{\Delta H_{sys}}{T} > 0$ or $\Delta H_{sys} - T\Delta S_{sys} < 0$ for a spontaneous process. Since $\Delta G_{sys} = \Delta H_{sys} - T\Delta S_{sys}$, then $\Delta G_{sys} < 0$ for a spontaneous process. **15.** A living cell is not a closed system,

i.e., a system that has no external energy source. **16.** (a) $C_{12}H_{22}O_{11}$; (b) H_2O (0°C); (c) $H_2S(g)$; (d) He (10K) **17.** (a) −; (b) +; (c) −; (d) − **18.** (a) $\Delta S° = -45$ J/K; (b) $\Delta S° = -137$ J/K **19.** $S° = 238$ J/K mol **21.** (a) not spontaneous; (b) spontaneous; (c) spontaneous at all temperatures; (d) spontaneous; (e) not spontaneous **22.** $\Delta S = 93.8$ J/K mol **24.** $\Delta S = 9.65$ J/K mol **25.** (a) $\Delta H°$ (−) and $\Delta S°$ (−); (b) The β form has the more ordered structure. **26.** (a) $\Delta H° = -803$ kJ, $\Delta S° = -4$ J/K, $\Delta G° = -801$ kJ; (b) $\Delta H° = +2802$ kJ, $\Delta S° = -262$ J/K, $\Delta G° = 2880$ kJ **27.** $\Delta G_f° = -731$ kJ/mol **29.** (a) $\Delta H° = -216$ kJ, $\Delta S° = -100.$ J/K (298 K), $\Delta G° = -186$ kJ; (b) $\Delta H° = -210.$ kJ, $\Delta S° = -129$ J/K (343 K), $\Delta G° = -166$ kJ; (c) $\Delta H° = -1468$ kJ, $\Delta S° = 88$ J/K (973 K), $\Delta G° = -1554$ kJ **30.** $C_2H_4(g) + H_2O(g) \rightarrow C_2H_5OH(l)$ would be preferred. **31.** The original enzyme has the more ordered structure. **34.** $\Delta G° = 79.8$ kJ **36.** $\Delta G° = -198$ kJ; $K = 5.30 \times 10^{34}$ **38.** (a) $K = 2.23 \times 10^5$; (b) 94 molecules ATP/molecule glucose **39.** $\Delta G_f°(CO) - \Delta G_f°(O_2) = -13$ kJ/mol **40.** $\Delta G = -188$ kJ **42.** $\Delta H° = -92$ kJ; $\Delta G° = -34$ kJ; $\Delta S° = -190$ J/K; $K = 9.2 \times 10^5$. (a) $\Delta G = -67$ kJ; (b) $\Delta G = -68$ kJ; (c) $\Delta G = -86$ kJ; (d) $\Delta G = -47$ kJ **44.** (a) $\Delta S° = -200$ J/K; $\Delta H° \cong -128$ kJ; $\Delta G° = -68$ kJ; (b) $\Delta H° = -90$ kJ; $\Delta S° = -220$ J/K; $\Delta G° = -24$ kJ; (c) Spontaneous at $T < 410$ K. **46.** No **49.** $\ln(K) = -\Delta H°/RT + \Delta S°/R$. If we graph ln (K) vs 1/T we get a straight line. For an endothermic process the slope is negative. As we increase the temperature, ln (K), and hence K, increase. **50.** $\Delta H° = 57.4$ kJ; $\Delta S° = -75.6$ J/K **52.** The catalyst will change the activation energy of both the forward and reverse reactions but not $\Delta G°$. A state function, $\Delta G°$ depends only on the initial and final conditions, not the path. **53.** 450 K **55.** (a) $\Delta G \cong 0$, at equilibrium; (b) $\Delta G \cong 0$, at equilibrium; (c) $\Delta G = -34.1$ kJ, shift to right; (d) $\Delta G = 45.7$ kJ, shift to left; (e) $\Delta G = 79.9$ kJ, shift to left Le Châtelier's Principle predicts the same results. **56.** $\Delta G = 8.8$ kJ/mol = 0.29 mol ATP. Other ions will have to be transported in order to maintain electroneutrality. **57.** If we go through a cycle, it may look like nothing has changed when we get back to the starting point. This is not true; there has been a change. The change that occurred is that the entropy of the universe has increased. **58.** As any process occurs ΔS_{UNIV} will increase; ΔS_{UNIV} cannot decrease. Time also goes in one direction, just as ΔS_{UNIV} goes in one direction.

Chapter 17

1. Oxidation: increase in oxidation number; loss of electrons. Reduction: decrease in oxidation number; gain of electrons. **2.** (a) H (+1), O (−2), N (+5); (b) Cl (−1), Cu (+2); (c) zero; (d) H (+1), O (−1); (e) Mg (+2), O (−2), S (+6); (f) zero; (g) Pb (+2), O (−2), S (+6); (h) O (−2), Pb (+4); (i) Na (+1), O (−2), C (+3); (j) O (−2), C (+4); (k) H (+1), N (−3), Ce (+4), S (+6), O (−2); (l) O (−2), Cr (+3)

3.

	Redox?	Ox. agent	Red. agent	Substance oxidized	Substance reduced
(a)	yes	H_2O	CH_4	CH_4	H_2O
(b)	yes	$AgNO_3$	Cu	Cu	$AgNO_3$
(c)	no	—	—	—	—
(d)	yes	HCl	Zn	Zn	HCl
(e)	no	—	—	—	—
(f)	yes	HCl	Fe	Fe	HCl

4. (a) $4H^+(aq) + NO_3^-(aq) + Cr(s) \rightarrow Cr^{3+}(aq) + NO(g) + 2H_2O(l)$ (b) $24H^+(aq) + 3MnO_4^-(aq) + 5Al(s) \rightarrow 5Al^{3+}(aq) + 3Mn^{2+}(aq) + 12H_2O(l)$ (c) $H_2O(l) + CH_3OH(aq) + 6Ce^{4+}(aq) \rightarrow 6Ce^{3+}(aq) + CO_2(g) + 6H^+(aq)$ (d) $6H^+(aq) + 5SO_3^{2-}(aq) + 2MnO_4^-(aq) \rightarrow 5SO_4^{2-}(aq) + 2Mn^{2+}(aq) + 3H_2O(l)$ **5.** (a) $H_2O(l) + 3PO_3^{3-}(aq) + 2MnO_4^-(aq) \rightarrow 3PO_4^{3-}(aq) + 2MnO_2(s) + 2OH^-(aq)$; (b) $H_2O(l) + Mg(s) + OCl^-(aq) \rightarrow Mg(OH)_2(s) + Cl^-(aq)$; (c) $170\ OH^-(aq) + 40Ce^{4+}(aq) + Ni(CN)_4^{2-}(aq) \rightarrow Ni(OH)_2(s) + 40Ce(OH)_3(s) + 4CO_3^{2-}(aq) + 4NO_3^-(aq) + 24H_2O(l)$; (d) $5OH^-(aq) + H_2CO(aq) + 4Ag(NH_3)_2^+(aq) \rightarrow 4Ag(s) + 8NH_3(aq) + HCO_3^-(aq) + 3H_2O(l)$ **6.** In a galvanic cell a spontaneous reaction occurs, producing an electric current. In an electrolytic cell electricity is used to force a nonspontaneous reaction to occur. **7.** The salt bridge completes the electrical circuit and allows counter ions to flow into the two cell compartments to maintain electrical neutrality. **8.** (a) cathode: electrode at which reduction occurs. (b) anode: electrode at which oxidation occurs. (c) oxidation half-reaction: half-reaction in which e^- are products. (d) reduction half-reaction: half-reaction in which e^- are reactants. **9.** (a) $7H_2O(l) + 2Cr^{3+}(aq) + 3Cl_2(g) \rightarrow Cr_2O_7^{2-}(aq) + 6Cl^-(aq) + 14H^+(aq)$; Cathode: Pt, Cl_2 bubbled into solution, Cl^- in solution; Anode: Pt, Cr^{3+} and $Cr_2O_7^{2-}$ present. (b) $Cu^{2+}(aq) + Mg(s) \rightarrow Cu(s) + Mg^{2+}(aq)$; Cathode: Cu electrode, Cu^{2+} in solution; Anode: Mg electrode, Mg^{2+} in solution. **10.** $Cl_2(g) + 2Br^-(aq) \rightarrow Br_2(aq) + 2Cl^-(aq)$; Cathode: Pt, with Cl_2 bubbled in, Cl^- in solution; Anode: Pt, with Br_2 and Br^- in solution. (b) $10H^+(aq) + 5IO_4^-(aq) + 8H_2O(l) + 2Mn^{2+}(aq) \rightarrow 5IO_3^-(aq) + 5H_2O(l) + 2MnO_4^-(aq) + 16H^+(aq)$; Cathode: Pt, with IO_4^-, IO_3^-, and H_2SO_4 in solution; Anode: Pt, with Mn^{2+}, MnO_4^-, and H_2SO_4 in solution; (c) $3Ni^{2+}(aq) + 2Al(s) \rightarrow 2Al^{3+}(aq) + 3Ni(s)$; Cathode: Ni, Ni^{2+} in solution; Anode: Al, Al^{3+} in solution. **11.** (9a) 0.03 V; (9b) 2.71 V; (9c) 0.43 V; (9d) 1.56 V; (10a) 0.27 V; (10b) 0.09 V; (10c) 1.43 V; (10b) 1.18 V **12.** (a) 0.10 V, SCE is anode: (b) 0.53 V, SCE is anode; (c) 0.02 V, SCE is cathode **13.** (a) 0.46 V, spontaneous; (b) −0.53 V, not spontaneous; (c) $16H^+ + 2MnO_4^- + 10I^- \rightarrow 5I_2 + 2Mn^{2+} + 8H_2O$; 0.97 V, spontaneous **14.** (a) no; (b) yes; (c) yes **15.** (a) yes; (b) no; (c) yes; (d) no **16.** $Mg^{2+} < Fe^{2+} < Fe^{3+} < Cr_2O_7^{2-} < Cl_2 < MnO_4^-$ **17.** $F^- < Cr^{3+} < Fe^{2+} < H_2 < Zn < Li$ **20.** (a) No half-reaction in this table fits this requirement. However, by changing concentrations, Ag^+ may be able to work. (b) $Cr_2O_7^{2-}$, O_2, MnO_2, and IO_3^- are all possible. **21.** (a) H_2O_2, MnO_4^{2-}, I^-, Cu, OH^-, Hg (in 1 *M* Cl^-), Ag (in 1 *M* Cl^-), H_2SO_3, H_2, Pb, Cd. **22.** Yes, $0.34\text{ V} < \mathscr{E}° < 0.54\text{ V}$ **25.** $2Cu^+ \rightarrow Cu + Cu^{2+}$; $\mathscr{E}°_{cell} = 0.36$ V; $\Delta G° = -35$ kJ; $K = 1.2 \times 10^6$ **26.** (a) $\mathscr{E}°_{cell} = 0.41$ V; $\Delta G° = -79$ kJ; $K = 7.1 \times 10^{13}$; (b) $3H_2O(l) + 6ClO_2(aq) \rightarrow 5ClO_3^-(aq) + Cl^-(aq) + 6H^+(aq)$ **28.** (a) no reaction occurs; (b) $Cl_2(g) + 2I^-(aq) \rightarrow I_2(aq) + 2Cl^-(aq)$; (c) no reaction occurs **29.** (b) $\mathscr{E}° = +0.82$ V; $\Delta G° = -160$ kJ; $K = 5.0 \times 10^{27}$; (d) $\mathscr{E}° = 0.47$ V; $\Delta G° = -91$ kJ; $K = 7.6 \times 10^{15}$; (e) $\mathscr{E}°_{cell} = +0.46$ V; $\Delta G° = -180$ kJ; $K = 1.2 \times 10^{31}$ **30.** (9a) $\Delta G° = -17$ kJ (−20; to correct sig fig); $K = 1 \times 10^3$; (9b) $\Delta G° = -523$ kJ; $K = 3.55 \times 10^{91}$; (9c) $\Delta G° = -210$ kJ; $K = 2.1 \times 10^{36}$; (9d) $\Delta G° = -301$ kJ; $K = 5.01 \times 10^{52}$; (10a) $\Delta G° = -52$ kJ; $K = 1.3 \times 10^9$; (10b) $\Delta G° = -90$ kJ; $K = 2 \times 10^{15}$; (10c) $\Delta G° = -828$ kJ; $K = 1 \times 10^{145}$; (10d) $\Delta G° = -114$ kJ; $K = 8.51 \times 10^{19}$ **32.** $\mathscr{E}° = \dfrac{T\Delta S°}{nF} - \dfrac{\Delta H°}{nF}$; If we graph $\mathscr{E}°$ vs. *T* we should get a straight line. The slope of the line is equal to $\Delta S°/nF$ and the intercept is equal to

$-\Delta H°/nF$. **33.** (a) $\mathcal{E}°_{cell} = -1.23$ V; $\Delta G° = 475$ kJ; (b) $\Delta H° =$ 572 kJ; $\Delta S° = 327$ J/K; (c) at 90°C, $\Delta G° = 453$ kJ; $\mathcal{E}° = -1.17$ V; at 0°C, $\Delta G° = 483$ kJ; $\mathcal{E}° = -1.25$ V **35.** $\Delta G_f° = 77$ kJ/mol **36.** $\Delta G_f°$ [$Fe^{2+}(aq)$] $= -85$ kJ; $\Delta G_f°$ [$Fe^{3+}(aq)$] $= -10$ kJ/mol **39.** (a) and (d) **40.** $2Cu^+(aq) \rightarrow Cu^{2+}(aq) + Cu(s)$; $\Delta G° =$ -35 kJ; $K = 1.2 \times 10^6$; $2HClO_2(aq) \rightarrow ClO_3^-(aq) + H^+(aq) +$ $HClO(aq)$; $\Delta G° = -85$ kJ; $K = 7.3 \times 10^{14}$ **44.** (a) 0.31 V; (b) 0.33 V; (c) 0.22 V; (d) 1.24 V; (e) 1.43 V **45.** (a) 2.12 V; (b) 1.98 V; (c) 2.05 V **47.** yes **48.** (a) $\mathcal{E} =$ $+0.23$ V; (b) $\mathcal{E} = -0.16$ V; (c) $[Cu^{2+}] = 1.2 \times 10^{-5}$ M; (d) Graph $\mathcal{E}$ vs. log $[Cu^{2+}]$. The slope will be 0.0295 V. **50.** (a) $\mathcal{E} = +0.40$ V; (b) $\mathcal{E} = 1.11$ V **52.** $\mathcal{E} = 0.81 \text{ V} - \frac{0.059}{4}\log\left[\frac{[OH^-]^4[Fe^{2+}]^2}{P_{O_2}}\right]$; more acidic; corrosion worse **53.** The pungent fumes were Cl_2. **54.** (a) 30. hours; (b) 33 s **55.** (a) 17 g; (b) 25 g **56.** 1.14×10^{-2} M **57.** (a) Cathode: $K^+ + e^- \rightarrow K$; Anode: $2F^- \rightarrow F_2 + 2e^-$; (b) Cathode: $2H_2O +$ $2e^- \rightarrow H_2 + 2OH^-$; Anode: $2H_2O \rightarrow 4H^+ + O_2 + 4e^-$; (c) Cathode: $H_2O_2 + 2H^+ + 2e^- \rightarrow 2H_2O$; Anode: $H_2O_2 \rightarrow O_2 +$ $2H^+ + 2e^-$ **58.** Ag **61.** Indium, In **65.** The Hall process requires 15 kWh of energy, or 5.4×10^4 kJ. 396 kJ is required to melt Al. It is feasible to recycle Al because it takes less than 1% of the energy required to produce the same amount of Al by the Hall process. **66.** 6.00 L of Cl_2 **70.** (a) $\mathcal{E}°_{cell} = 0.36$ V; $\Delta G° = -35$ kJ; $K = 1.2 \times 10^6$; (b) $\mathcal{E}°_{cell} = 1.88$ V; $\Delta G° = -363$ kJ; $K = 3.24 \times$ 10^{63} **72.** Fuel cells are more efficient in converting chemical energy to electrical energy; they are also less massive. **77.** $\mathcal{E}°_{cell} =$ 0.0034 V; $\Delta G° = -660$ J; $K = 1.3$ **78.** (a) $\mathcal{E}°_{cell} = 0.242$ V; (b) The hydrogen electrode is the anode. (c) $\mathcal{E} = 0.242$ V + 0.059 pH; (d) (i) $\mathcal{E} = 0.419$ V; (ii) $\mathcal{E} = 0.219$ V; (iii) $\mathcal{E} =$ 0.773 V; (e) pH = 0.73; (f) Primarily the reason is convenience. It is inconvenient to deal with the H_2 gas and particularly to keep P_{H_2} constant. H_2 presents a fire hazard. **79.** (a) The uncertainty in pH is ± 0.02 pH units. The uncertainty in $[H^+]$ is $\pm 0.04 \times 10^{-4}$. (b) Measure the potential to the nearest ± 0.001 V (1 mV). **81.** Au

Chapter 18

2. (a) $\Delta H° = +206$ kJ; $\Delta S° = +216$ J/K; (b) Reaction is spontaneous at $T > 950$ K. Pressure won't affect equilibrium. **3.** For $3Fe(s) + 4H_2O(g) \rightarrow Fe_3O_4(s) + 4H_2(g)$, (a) $\Delta H° = -149$ kJ; $\Delta S° = -167$ J/K mol; (b) spontaneous at $T < 892$ K. For $C(s) +$ $H_2O(g) \rightarrow CO(g) + H_2(g)$, (a) $\Delta H° = +132$ kJ; $\Delta S° = +134$ J/K; (b) spontaneous at $T > 985$ K. **4.** (1) ammonia production; (2) hydrogenation of vegetable oils **7.** (a) $Li_3N(s) +$ $3HCl(aq) \rightarrow 3LiCl(aq) + NH_3(aq)$; (b) $Rb_2O(s) +$ $H_2O(l) \rightarrow 2RbOH(aq)$; (c) $Cs_2O_2(s) + 2H_2O(l) \rightarrow 2CsOH(aq) +$ $H_2O_2(aq)$; (d) $NaH(s) + H_2O(l) \rightarrow NaOH(aq) + H_2(g)$ **8.** 8.4 kJ **10.** $2\text{ Li} + 2C_2H_2 \rightarrow 2LiC_2H + H_2$; oxidation reduction **13.** $K(s) + O_2(g) \rightarrow KO_2(s)$; $16K(s) + S_8(s) \rightarrow 8K_2S(s)$; $12K(s) +$ $P_4(s) \rightarrow 4K_3P(s)$; $2K(s) + H_2(g) \rightarrow 2KH(s)$; $2K(s) +$ $2H_2O(l) \rightarrow H_2(g) + 2KOH(aq)$ **14.** sodium oxide: Na_2O; sodium superoxide: NaO_2; sodium peroxide: Na_2O_2; **17.** Trigonal planar; Be uses sp^2 hybrid orbitals; N uses sp^3 hybrid orbitals; $BeCl_2$ is a Lewis acid. **19.** $BeCl_2(NH_3)_2$ would form in excess ammonia. **23.** $Mg_3N_2(s) + 6H_2O(l) \rightarrow 2NH_3(g) + 3Mg^{2+}(aq) + 6OH^-(aq)$; $Mg_3P_2(s) + 6H_2O(l) \rightarrow 2PH_3(g) + 3Mg^{2+}(aq) + 6OH^-(aq)$ **24.** 1.67×10^5 A; 1.77×10^3 kg of Cl_2; **26.** $B_2H_6 +$ $3O_2 \rightarrow 2B(OH)_3$ **27.** (a) thallium(I) hydroxide; (b) indium(III) sulfide; (c) gallium(III) oxide **29.** The radius of the 3^+ ion is about 1.0 Å (100 pm). The ionization energy is close to that of Tl. **31.** $In_2O_3(s) + 6H^+(aq) \rightarrow 2In^{3+}(aq) + 3H_2O(l)$; $In_2O_3(s) +$ $OH^-(aq) \rightarrow$ no reaction; $Ga_2O_3(s) + 2OH^-(aq) +$ $3H_2O(l) \rightarrow 2Ga(OH)_4^-(aq)$; $Ga_2O_3(s) + 6H^+(aq) \rightarrow 2Ga^{3+}(aq) +$ $3H_2O(l)$ **32.** p-type **33.** (a) $Ca_3Al_2O_6$; (b) $Ca_9Al_6O_{18}$; (c) There are covalent bonds between Al and O atoms in the $[Al_6O_{18}]^{18-}$ anion. sp^3 hybrid orbitals on Al overlap with sp^3 on O to form the sigma bonds. **34.** $2In(s) + 3F_2(g) \rightarrow 2InF_3(s)$; $2In(s) +$ $3Cl_2(g) \rightarrow 2InCl_3(s)$; $4In(s) + 3O_2(g) \rightarrow 2In_2O_3(s)$; $2In(s) +$ $6HCl(aq) \rightarrow 3H_2(g) + 2InCl_3(aq)$; **39.** (a) linear; (b) sp **40.** ΔH and ΔS are favorable for the decomposition of H_2CO_3 to CO_2 and H_2O. Hence, H_2CO_3 will spontaneously decompose to CO_2 and H_2O. **41.** :S=C=S:, linear; :S=C=C=C=S:, linear **43.** Gray tin has the more ordered structure. **44.** Thermodynamics predicts both should react with water. The answer must lie in kinetics. $SiCl_4$ reacts because an activated complex can easily form by a water molecule attaching to a silicon atom. Carbon cannot expand its coordination number like silicon and a C—Cl bond must be broken to form an activated complex. **46.** SiC would have a covalent network structure similar to diamond. **47.** $SiO_2(s) + 4HF(aq) \rightarrow SiF_4(g) +$ $2H_2O(l)$ **49.** Fe pipes corrode more easily than Pb pipes. **50.** 6.7×10^{-6} mol/L **51.** $Sn(s) + 2H^+(aq) \rightarrow Sn^{2+}(aq) + H_2(g)$; $Pb(s) + 2H^+(aq) \rightarrow Pb^{2+}(aq) + H_2(g)$; **52.** The purpose of the cryptand is to encapsulate the Na^+ ion so that it does not come in contact with Na^- ion and oxidize it to sodium metal. **55.** $Be(s) +$ $2H_2O(l) + 2OH^-(aq) \rightarrow Be(OH)_4^{2-}(aq) + H_2(g)$. Be is the reducing agent. H_2O is the oxidizing agent. **57.** Be^{2+} ion is a Lewis acid. The ion in solution is $Be(H_2O)_4^{2+}$. The acidic solution results from the reaction: $Be(H_2O)_4^{2+}(aq) \rightleftharpoons Be(H_2O)_3(OH)^+(aq) + H^+(aq)$ **59.** In solution, Tl^{3+} can oxidize I^- to I_3^-. Thus, we expect TlI_3 to be thallium(I) triiodide. **60.** If the compound contained Ga(II), it would be paramagnetic. This can easily be determined by measuring the mass of a sample in the presence and absence of a magnetic field. Paramagnetic compounds will have an apparent greater mass in a magnetic field. **63.** (a) $K_2SiF_6(s) + 4K(l) \rightarrow 6KF(s) + Si(s)$; (b) K_2SiF_6 is an ionic compound, composed of K^+ cations and SiF_6^{2-} anions. The SiF_6^{2-} anion is held together by covalent bonds. **64.** Tetrahedral arrangement around Sn **65.** Carbon is much smaller than Si and cannot form a fifth bond in the transition state. **67.** Size decreases from left to right and increases going down. So going one element right and one element down would result in a similar size for the two elements diagonal to each other. The ionization energies will also be similar for the diagonal elements. Electron affinities are harder to predict, but the similar size and ionization energies would lead to similar properties.

Chapter 19

2. $\Delta H° = 180$ kJ; $\Delta G° = 174$ kJ; $\Delta S° = +25$ J/K. At high temperature the reaction $N_2 + O_2 \rightarrow 2NO$ becomes spontaneous. In the atmosphere, even though $2NO \rightarrow N_2 + O_2$ is spontaneous, it doesn't occur because the rate is too slow. **3.** NH_3: sp^3; N_2H_4: sp^3; NH_2OH: sp^3; N_2: sp; N_2O, central N: sp; NO: sp^2; N_2O_3: both are

sp^2; NO_2: sp^2; HNO_3: sp^2 **4.** Resonance is possible for N_2O, NO, N_2O_3, NO_2, and HNO_3.

5. For

$$O_2N—N{=}O \rightarrow NO_2 + NO$$

the activation energy must in some way involve the breaking of a nitrogen-nitrogen bond. For the reaction

$$O_2N—N{=}O \rightarrow O_2 + N_2O$$

at some point nitrogen-oxygen bonds must be broken. N—N single bonds (160 kJ/mol) are weaker than N—O single bonds (201 kJ/mol). Resonance structures indicate that there is more double bond character to the N—O bonds than the N—N bond. Thus, NO_2 and NO are preferred by kinetics because of the lower activation energy.

8. OCN^-: $[O{=}C{=}N]^- \leftrightarrow [:O—C{\equiv}N:]^- \leftrightarrow [:O{\equiv}C—N:]^-$;

Formal charge 0 0 −1 −1 0 0 +1 0 −2

CNO^-: $[C{=}N{=}O]^- \leftrightarrow [:C{\equiv}N—O:]^- \leftrightarrow [:C—N{\equiv}O:]^-$.

Formal charge −2 +1 0 −1 +1 −1 −3 +1 +1

All of the resonance structures for fulminate involve greater formal charges than in cyanate, making fulminate more reactive (less stable). **10.** (a) $8H^+(aq) + 2NO_3^-(aq) + 3Cu(s) \rightarrow 3Cu^{2+}(aq) + 4H_2O(l) + 2NO(g)$; (b) $NH_4NO_3(s) \xrightarrow{\text{heat}} N_2O(g) + 2H_2O(g)$; (c) $NO(g) + NO_2(g) + 2KOH(aq) \rightarrow 2KNO_2(aq) + H_2O(l)$ **12.** $2Sb_2S_3(s) + 9O_2(g) \rightarrow 2Sb_2O_3(s) + 6SO_2(g)$; $2Sb_2O_3(s) + 3C(s) \rightarrow 4Sb(s) + 3CO_2(g)$. $2Bi_2S_3(s) + 9O_2(g) \rightarrow 2Bi_2O_3(s) + 6SO_2(g)$; $2Bi_2O_3(s) + 3C(s) \rightarrow 4Bi(s) + 3CO_2(g)$ **13.** (a) $H_3PO_4 > H_3PO_3$; (b) $H_3PO_4 > H_2PO_4^- > HPO_4^{2-}$

14. $H_4P_2O_6$:

```
   :O:   :O:
    |     |
H—P—O—P—O—H
    |     |
   :O:   :O:
    |     |
    H     H
```

$H_4P_2O_5$:

```
     :O:   :O:
      |     |
H—O—P—O—P—O—H
      |     |
      H     H
```

18.

	Bond order	No. of unpaired e^-
M.O. NO	2.5	1
NO^+	3	0
NO^-	2	2
Lewis NO $\dot{N}{=}O \leftrightarrow$	$N{=}\dot{O}$	
NO^+	$[:N{\equiv}O:]^+$	
NO^- $N{=}O$		

Lewis structure not adequate for NO and NO^-. M.O. model gives correct results for all three species.

19. 5×10^{-11} *M* **20.** (a) $8H_2O(l) + 2Mn^{2+}(aq) + 5NaBiO_3(s) \rightarrow 2MnO_4^-(aq) + 16H^+(aq) + 5BiO_3^{3-}(aq) + 5Na^+(aq)$; (b) Bismuthate exists as a covalent network: $(BiO_3^-)_x$ **23.** (a) $P_4O_6(s) + 2O_2(g) \rightarrow P_4O_{10}(s)$; (b) $P_4O_{10}(s) + 6H_2O(l) \rightarrow 4H_3PO_4(aq)$; (c) $PCl_5(l) + 4H_2O(l) \rightarrow H_3PO_4(aq) + 5HCl(aq)$

26.

```
      :O: F:
       | /
:F — Te — F: ;    F5TeO—P—OTeF5
     / |                |
   :F: F:             OTeF5
```

27. (a) As we go down the family, K_a increases. This is consistent with the bond to hydrogen getting weaker. (b) Po is below Te, so K_a should be larger. The K_a for H_2Po should be on the order of 10^{-2} or 10^{-1}. **28.** The lone pair takes up more room than the bonding pairs, pushing the four equatorial F's away from the lone pair. **29.** Sulfur forms polysulfide ions, S_n^{2-}, which are soluble in water, *i.e.*, $S_8 + S^{2-} \leftrightarrow S_9^{2-}$. Nitric acid oxidizes S^{2-} to S. **32.** Light from violet to green will work. **34.** Square pyramid **35.** (a) square pyramid; (b) T-shaped; (c) The four O atoms are tetrahedrally arranged about each Cl. The Cl—O—Cl bond angle is close to the tetrahedral angle. **36.** (a) $F_2 + H_2O \rightarrow HOF + HF$; $2HOF \rightarrow 2HF + O_2$; $HOF + H_2O \rightarrow HF + H_2O_2$ (acid); $2OF^- \rightarrow 2F^- + O_2$ (base); (b) HOF: assign +1 to H, −1 to F; Oxidation number of O = 0. Oxygen is very electronegative. A +1 oxidation state would not be very stable.

37.

	:F—	O—	O—	F:
Formal charge	0	0	0	0
Oxidation number	−1	+1	+1	−1

Oxidation numbers are more useful. We are forced to assign +1 as the oxidation number to oxygen. Oxygen is very electronegative and +1 is not a stable oxidation state for this element.

41. (a) A potential greater than 1.19 V must be applied. (b) 7×10^{13} kJ

42. (a) $AgCl(s) \xrightarrow{h\nu} Ag(s) + Cl$;
The reactive chlorine atom is trapped in the crystal. When the light is removed, it reacts with silver atoms to reform AgCl.
(b) Over time chlorine is lost and the dark silver metal is permanent. **44.** Helium is unreactive and doesn't combine with any other elements. It is a very light gas and would easily escape the earth's gravitational pull as the planet was formed. **45.** The heavier members are not really inert. Xe and Kr have been shown to react and form compounds with other elements. **46.** 3×10^{20} atoms; A 2 L breath contains 4×10^{15} atoms of Xe.

47.

```
          :O:
           |
XeO3  :O—X—O:      trigonal pyramid;

          :O:
           |
XeO4  :O—Xe—O:     tetrahedral;
           |
          :O:
```

$XeOF_4$ (Lewis structure: Xe bonded to O and four F) square pyramid;

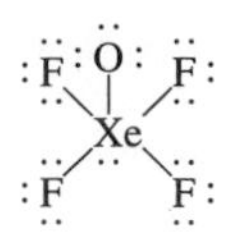

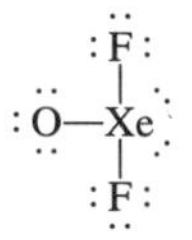

$XeOF_2$ (Lewis structure: O—Xe with two F) T-shaped;

XeO_3F_2 (Lewis structure: Xe—O with two O and two F) trigonal bipyramid

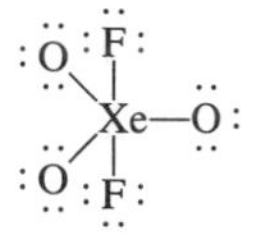

49. As the halogen atoms get larger, it becomes more difficult to fit three halogen atoms around the small N, and the NX_3 molecule becomes less stable. **51.** (a) IO_4^-; (b) IO_3^-; (c) IF_2^-; (d) IF_4^-; (e) IF_6^-; (f) IOF_3^- **52.** For $NCl_3 \rightarrow NCl_2 + Cl$, only the N—Cl bond is being broken. For O—N—Cl → NO + Cl, the NO bond gets stronger (bond order increases from 2.0 to 2.5) making ΔH for the reaction smaller than the energy necessary to break the N—Cl bond. **54.** Plastic sulfur consists of long chains of sulfur atoms. It becomes brittle as the long chains break down into S_8 rings. **55.** The pollution provides nitrogen and phosphorous nutrients so that algae can grow. The algae consume oxygen, causing fish to die. **56.** $H_2O(l) + BrO_3^-(aq) + XeF_2(aq) \rightarrow BrO_4^-(aq) + Xe(g) + 2HF(aq)$

Chapter 20

1. (a) $[Ar]4s^23d^8$; (b) $[Kr]5s^24d^{10}$; (c) $[Kr]5s^24d^2$; (d) $[Xe]6s^25d^14f^3$ or $[Xe]6s^24f^4$ **2.** (a) $[Ar]3d^8$; (b) $[Kr]4d^{10}$; (c) Zr^{3+}: $[Kr]4d^1$; Zr^{4+}: [Kr]; (d) $[Xe]4f^3$ **3.** (a) Co: $[Ar]4s^23d^7$; Co^{2+}: $[Ar]3d^7$; Co^{3+}: $[Ar]3d^6$; (b) Pt: $[Xe]6s^14f^{14}5d^9$; Pt^{2+}: $[Xe]4f^{14}5d^8$; Pt^{4+}: $[Xe]4f^{14}5d^6$; (c) Fe: $[Ar]4s^23d^6$; Fe^{2+}: $[Ar]3d^6$; Fe^{3+}: $[Ar]3d^5$ **4.** 316 kg **5.** From $\mathscr{E}°$ values, we would predict Fe(II) and Ti(IV) to be more stable. The electrochemical data are consistent with the information in the text. These data are for solutions while ionization energies are for gas phase reactions. Solution data are more representative of the conditions from which ilmenite formed. **7.** (a) 4O on faces × ½O/face = 2O atoms; 2O atoms inside body; Total: 4O atoms; 8 corners × ⅛Ti/corner + 1Ti/body center = 2Ti; Formula of unit cell Ti_2O_4 to give the empirical formula TiO_2.

(b)

$$\overset{+4\,-2}{2TiO_2} + \overset{0}{3C} + \overset{0}{4Cl_2} \rightarrow \overset{+4\,-1}{2TiCl_4} + \overset{+4\,-2}{CO_2} + \overset{+2\,-2}{CO}.$$

C is being oxidized. C is the reducing agent. Cl is being reduced. Cl_2 is the oxidizing agent.

$$\overset{+4\,-1}{TiCl_4} + \overset{0}{O_2} \rightarrow \overset{+4\,-2}{TiO_2} + \overset{0}{2Cl_2}.$$

Cl is being oxidized. $TiCl_4$ is the reducing agent. O is being reduced. O_2 is the oxidizing agent.

8. TiF_4: ionic substance containing Ti^{4+} ions and F^- ions. $TiCl_4$, $TiBr_4$, and TiI_4: covalent substances containing discrete, tetrahedral TiX_4 molecules. As the molecule gets larger, the bp and mp increase. **9.** Pyrolusite, manganese(IV) oxide; rhodochrosite, manganese(II) carbonate **10.** (a) $2CoAs_2(s) + 4O_2(g) \rightarrow 2CoO(s) + As_4O_6(s)$; As_4O_6: arsenic(III) oxide; As_4O_6 has a cage structure similar to P_4O_6. (b) $2Co^{2+}(aq) + 4OH^-(aq) + H_2O(l) + OCl^-(aq) \rightarrow 2Co(OH)_3(s) + Cl^-(aq)$; (c) 2.5×10^{-22} *M* ($[OH^-] = 10^{-7}$ *M*); (d) 2.5×10^{-31} *M* **12.** (a) ligand: a Lewis base; (b) chelate: ligand that can form more than one bond; (c) bidentate: ligand that can form two bonds; (d) complex ion: metal ion plus ligands **13.** at least one unshared pair of electrons **14.** Both electrons in the bond originally came from the same atom. **15.** (a) hexaamminecobalt(II) chloride; (b) hexaaquacobalt(III) iodide **16.** (a) pentaaminechlororuthenium(III) ion; (b) hexacyanoferrate(II) ion **17.** (a) sodium *tris*(oxalato) nickelate(II); (b) potassium tetrachlorocobaltate(II) **18.** (a) $[Co(C_5H_5N)_6]Cl_3$; (b) $[Cr(NH_3)_5I]I_2$; (c) $[Ni(NH_2CH_2CH_2NH_2)_3]Br_2$ **19.** (a) $FeCl_4^-$; (b) $[Ru(NH_3)_5H_2O]^{3+}$ **20.** (a) 2; (b) 3; (c) 4; (d) 4 **22.** (a) Water molecules can be *cis* or *trans*. (b) Iodide ions can be *cis* or *trans*.

23.

SCN^- {M—SCN; M—NCS};

NO_2^- {M—NO_2; M—ONO};

OCN^- {M—OCN; M—NCO}

24. SCN^- can be *cis* or *trans* and can bond through S or N.

26. (a) 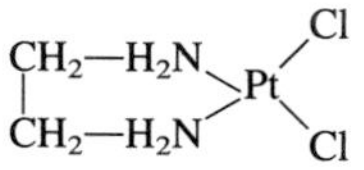 (structure: CH_2—H_2N / CH_2—H_2N chelated to Pt with two Cl);

(b) 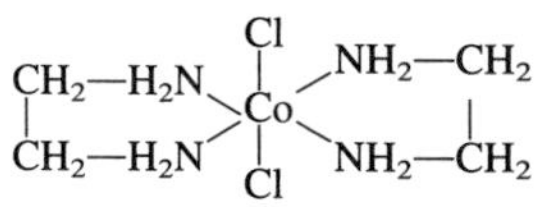 (structure: two CH_2—H_2N / NH_2—CH_2 chelates on Co with two trans Cl);

(c) 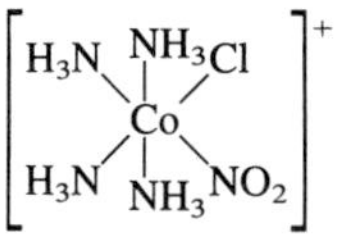 (structure: $[Co(NH_3)_4Cl(NO_2)]^+$ with H_3N, NH_3, Cl, H_3N, NH_3, NO_2 around Co)

31. (a) Ligand that will give complex ions with the maximum number of unpaired electrons. (b) Ligand that will give complex ions with the minimum number of unpaired electrons. **32.** Cu(II) is d^9; Cu(I) is d^{10}. Color is a result of the electron transfer between split *d* orbitals. This cannot occur for the filled *d* orbitals of Cu(I). **34.** Yes. After the oxidation, the ligands on Cr(III) won't exchange. Since Cl^- is in the coordination sphere, it must have formed a bond to Cr(II) before the electron transfer occurred. **37.** $NiCl_4^{2-}$ is tetrahedral, and $Ni(CN)_4^{2-}$ is square planar.

39. (a) Fe^{2+}

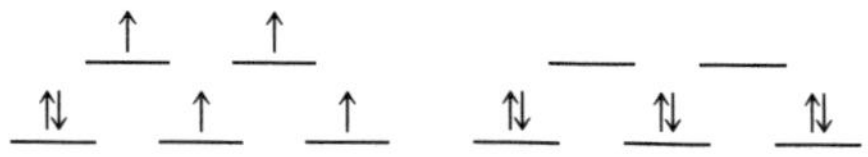

High spin: ↑ ↑ (upper); ⇅ ↑ ↑ (lower)
Low spin: ___ ___ (upper); ⇅ ⇅ ⇅ (lower)

(b) Fe^{3+}

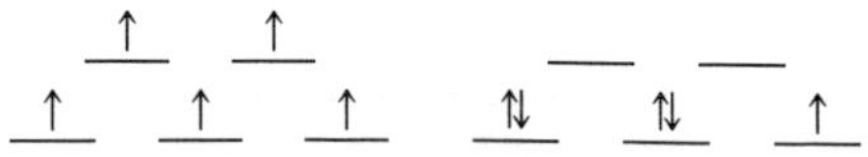

High spin: ↑ ↑ (upper); ↑ ↑ ↑ (lower)
Low spin: ___ ___ (upper); ⇅ ⇅ ↑ (lower)

(c) Ni^{2+}

```
     ↑    ↑
    ___  ___
 ⇅    ⇅    ⇅
___  ___  ___
```

40. $Ni(H_2O)_6Cl_2$ absorbs red light; $Ni(NH_3)_6Cl_2$ absorbs yellow-green light. $Ni(NH_3)_6Cl_2$ absorbs the shorter wavelength light; Δ is larger for $Ni(NH_3)_6Cl_2$. NH_3 is a stronger field ligand than H_2O, consistent with the spectrochemical series. **41.** $Fe(CN)_6^{3-}$: Low spin; $Fe(SCN)_6^{3-}$: High spin; CN^- is a stronger field ligand than SCN^-. **43.** 0 **44.** Lewis acids **48.** $[Cr(NH_3)_5I]I_2$; $[Cr(NH_3)_5I]^{2+}$ **50.** (a) 1.6×10^{-6}; (b) 4.5×10^{11} **51.** $Fe_2O_3(s) + 6H_2C_2O_4(aq) \rightarrow 2Fe(C_2O_4)_3^{3-}(aq) + 3H_2O(l) + 6H^+(aq)$; a soluble complex ion forms. **54.** No, since in all three cases, six bonds are formed between Ni^{2+} and nitrogen. $\Delta S°$ for formation of the complex ion is most negative for $6NH_3$ molecules reacting with a metal ion (seven independent species become one). For penten reacting with a metal ion, two independent species form one, so $\Delta S°$ is less negative. Thus, the chelate effect occurs because the more bonds a chelating agent can form to the metal, the more favorable ΔS is for the formation of the complex ion. **56.** (See Fig. 20.8.) It is optically active.

57.

```
___   ___    d_{x²−y²}, d_{xy}

   ___       d_{z²}

___   ___    d_{xz}, d_{yz}
```

Chapter 21

1. (a) ${}^3_1H \rightarrow {}^{0}_{-1}e + {}^3_2He$; (b) ${}^8_3Li \rightarrow {}^8_4Be + {}^{0}_{-1}e$; ${}^8_4Be \rightarrow 2\,{}^4_2He$; (c) ${}^7_4Be + {}^{0}_{-1}e \rightarrow {}^7_3Li$ **2.** (a) ${}^{60}_{27}Co \rightarrow {}^{60}_{28}Ni + {}^{0}_{-1}e$; (b) ${}^{97}_{43}Tc + {}^{0}_{-1}e \rightarrow {}^{97}_{42}Mo$ **3.** ${}^8B$ and ${}^9B$ might decay by either positron emission or electron capture. ${}^{12}B$ and ${}^{13}B$ contain too many neutrons or too few protons, so ${}^{12}B$ and ${}^{13}B$ are expected to be β-emitters.
4. (a) ${}^1_1H + {}^{14}_7N \rightarrow {}^{11}_6C + {}^4_2He$; (b) $2\,{}^3_2He \rightarrow {}^4_2He + 2\,{}^1_1H$
5. (a) ${}^{240}_{95}Am + {}^4_2He \rightarrow {}^{243}_{97}Bk + {}^1_0n$; (b) ${}^{238}_{92}U + {}^{12}_6C \rightarrow {}^{244}_{98}Cf + 6\,{}^1_0n$ **9.** (a) ${}^{207}_{82}Pb$; (b) ${}^{231}_{90}Th$; ${}^{231}_{91}Pa$; ${}^{227}_{89}Ac$; ${}^{227}_{90}Th$; ${}^{223}_{88}Ra$; ${}^{219}_{86}Rn$; ${}^{215}_{84}Po$; ${}^{221}_{82}Pb$; ${}^{211}_{83}Bi$; ${}^{207}_{81}Tl$ **11.** $t_{1/2} = 12{,}000$ yr, rate $= 1.1 \times 10^{12}$ disintegrations/s; $t_{1/2} = 12$ hr, rate $= 9.6 \times 10^{18}$ disintegrations/s; $t_{1/2} = 12$ min, rate $= 5.8 \times 10^{20}$ disintegrations/s; $t_{1/2} = 12$ s, rate $= 3.5 \times 10^{22}$ disintegrations/s **12.** 0.77 **16.** 2.5 disintegrations per minute per g of C; for 10 mg C: 0.025 disintegrations/min; it would take roughly 40 min to see a single disintegration. The background radiation would probably be much greater than the ${}^{14}C$ activity. Thus, ${}^{14}C$ dating is not practical for small samples. **17.** 0.87
18. (a) The decay of ${}^{40}K$ is not the sole source of ${}^{40}Ca$. (b) Decay of ${}^{40}K$ is the sole source of ${}^{40}Ar$ and no ${}^{40}Ar$ is lost over the years. (c) 4.2×10^9 years old; (d) Measured age would be less than actual age. **20.** (a) 9.7×10^6 mCi; (b) 1.5×10^4 mCi **23.** 4.3×10^6 kg of mass/s **24.** 1.7×10^5 kg of solar material provides 1 day of solar energy to the earth; 5.0×10^{14} kg of coal is needed to provide the same amount of energy. **25.** ${}^{24}_{12}Mg$; 1.324×10^{-12} J/nucleon; for ${}^{27}Mg$, 1.324×10^{-12} J/nucleon **29.** 2.426×10^{-3} nm **32.** 2.0×10^{10} J/g **33.** 1.694×10^{12} J/mol; 2.813×10^{-12} J/atom **34.** The chemical properties may determine where a radioactive material may be concentrated in the body or how easily it may be excreted. **35.** Not all of the emitted radiation enters the Geiger-Müller tube. The fraction of radiation entering the tube must be constant. **36.** The Geiger-Müller tube has a certain response time. After the gas in the tube ionizes to produce a "count," some time must elapse for the gas to return to an electrically neutral state. The response of the tube levels off because at high activities, radioactive particles enter the tube faster than the tube can respond.
37. All evolved O_2 comes from water. **38.** A nonradioactive substance can be put in equilibrium with a radioactive substance. The two materials can then be checked to see if all the radioactivity remains in the original material or if it has been scrambled by the equilibrium. **40.** Assuming that (1) the radionuclide is long lived enough that no significant decay occurs during the time of the experiment, and (2) the total activity is uniformly distributed only in the rat's blood; $V = 10.4$ mL. **41.** 2×10^{-7} kg; this is 2×10^{23} times the electron mass and 1×10^{26} times the proton mass. **42.** (i) and (ii) mean that Pu is not a significant threat outside the body. If Pu gets inside the body, it is easily oxidized to Pu^{4+} (iv) which is chemically similar to Fe^{3+} (iii). Thus, Pu^{4+} will concentrate in tissues where Fe^{3+} is found, such as the bone marrow where red blood cells are produced. Here α particles cause considerable damage.
44. For fusion, a collision of sufficient energy must occur between two positively charged particles to initiate the reaction. This requires high temperatures. In fission, an electrically neutral neutron collides with the positively charged nucleus. This has a much lower activation energy. **45.** The temperatures of fusion reactions are so high that all physical containers would be destroyed. At these high temperatures most of the electrons are stripped from the atoms. A plasma of gaseous ions is formed, which can be controlled by magnetic fields.
47. Moderator: slows the neutrons; Control rods: absorb neutrons to slow or halt fission. **50.** (a) ${}^{12}_6C$; (b) ${}^{13}N$, ${}^{13}C$, ${}^{14}N$, ${}^{15}O$, and ${}^{15}N$ are intermediates. (c) 5.950×10^{11} J/mol H

Chapter 22

1. $CH_3—CH_2—CH_2—CH_2—CH_2—CH_3$, hexane or *n*-hexane (highest b.p., least branched);

```
 CH3
 |
CH3—CH—CH2—CH2—CH3,      2-methylpentane;

         CH3
         |
CH3—CH2—CH—CH2—CH3,      3-methylpentane;

     CH3
     |
CH3—C—CH2—CH3,           2,2-dimethylbutane;
     |
     CH3

    CH3 CH3
     |   |
CH3—CH—CH—CH3,           2,3-dimethylbutane
```

2. (a) 2,3,3-trimethylhexane; (b) 8-ethyl-2,5,5-trimethyldecane; (c) 3-methylhexane **3.** There is only one consecutive chain of C atoms. They are not all in a straight line since the bond angle at each carbon is a tetrahedral angle of 109.5°.

4. (a)

```
CH3 CH—CH2—CH2CH3;
     |
     CH3
```

(b)

```
     CH3
     |
CH3—C—CH2—CH—CH3;
     |      |
     CH3    CH3
```

(c) $CH_3-CH(-C(CH_3)_2-CH_3)-CH_2CH_2CH_3$;

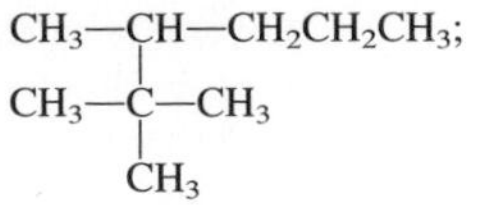

(d) 2,2,3-trimethylhexane

5. (a) 1-butene; (b) 2-methyl-2-butene; (c) 2,5-dimethyl-3-heptene

6. (a) $CH_3-CH_2-CH=CH-CH_2-CH_3$;
(b) $CH_3CH=CHCH=CHCH_2CH_3$;

(c) $CH_3-CH(CH_3)-CH=CHCH_2CH_2CH_2CH_3$

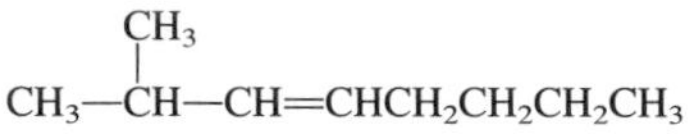

7. (a) (1,2-dimethylbenzene: ring with CH_3, CH_3)

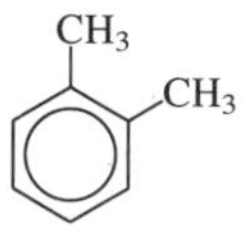

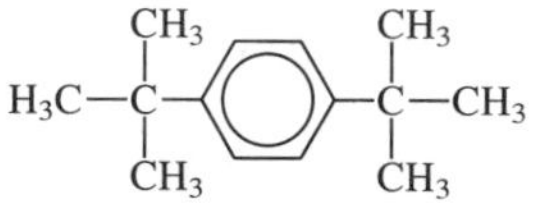

(b) $H_3C-C(CH_3)_2-C_6H_4-C(CH_3)_2-CH_3$

(c) CH_2CH_3 CH_2CH_3

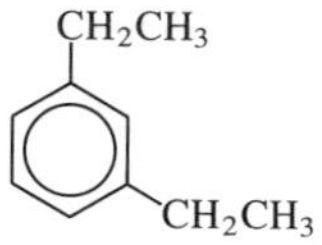

8. (a) 1,3-dichlorobutane; (b) 1,1,1-trichlorobutane; (e) chlorobenzene; (f) chlorocyclohexane **9.** (a) methylcyclopropane; (b) *t*-butylcyclohexane; (c) 3,4-dimethylcyclopentene **10.** isopropylbenzene or 2-phenylpropane

12. (a) $H-CF_2-H$ (H—C(F)(H)—F)

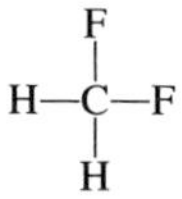

(b) $H-C(Br)(Cl)-C(H)(Cl)-C(H)(H)-H$

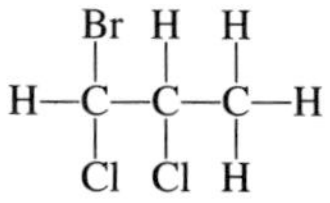

13. Structural isomers: difference in types of bonds, either the kinds of bond present or the way in which bonds connect atoms to each other. Geometrical isomers: same bonds, differ in arrangement in space about rigid bond or ring. **14.** Resonance: all atoms are in the same position. Only the position of π electrons is different. Isomerism: Atoms are in different locations in space.

15.

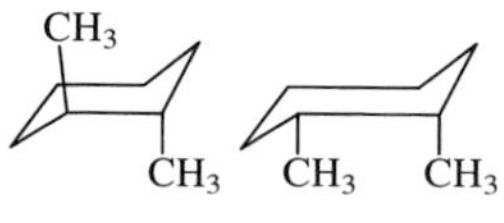

CH_3 / CH_3 — *trans*; CH_3 CH_3 — *cis*

16. $Cl_2C=CH-CH_3$ $CH_2=CCl-CH_2Cl$ $CH_2=CH-CHCl_2$

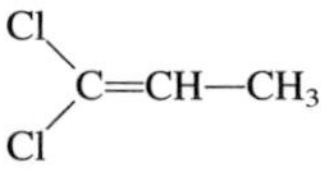

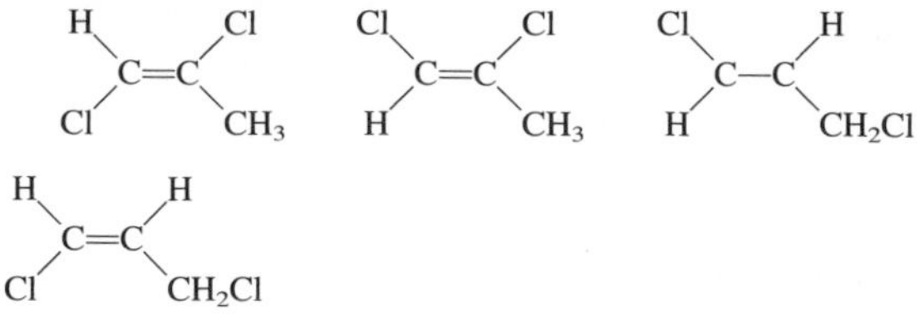

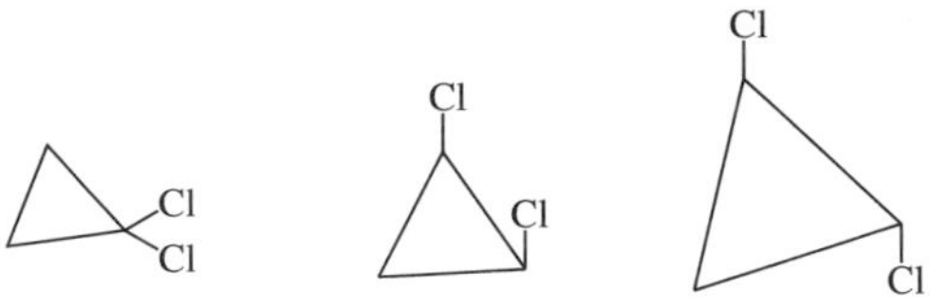

(H)(Cl)C=C(Cl)(CH_3) (Cl)(H)C=C(Cl)(CH_3) (Cl)(H)C=C(H)(CH_2Cl) (H)(Cl)C=C(H)(CH_2Cl)

(1,1-dichloro-2-methylcyclopropane, Cl, Cl; Cl, Cl; Cl, Cl cyclopropanes)

19. $F_2C=CH_2$ polar; $FHC=CHF$ (cis) polar; $FHC=CHF$ (trans) non-polar

21. (a) $CH_3\overset{O}{\overset{\|}{C}}CH_3$ is more stable; it contains the stronger bonds.

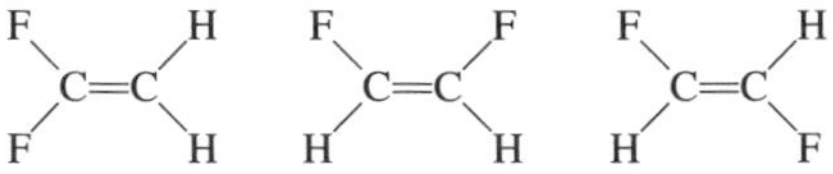

(b) $C_6H_5-\overset{O}{\overset{\|}{C}}-N(H)-CH_3$ is more stable; it contains the stronger bonds.

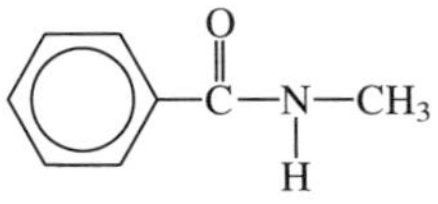

23. (a) CO_2H — carboxylic acid; $O-\overset{O}{\overset{\|}{C}}-CH_3$ — ester; aromatic ring

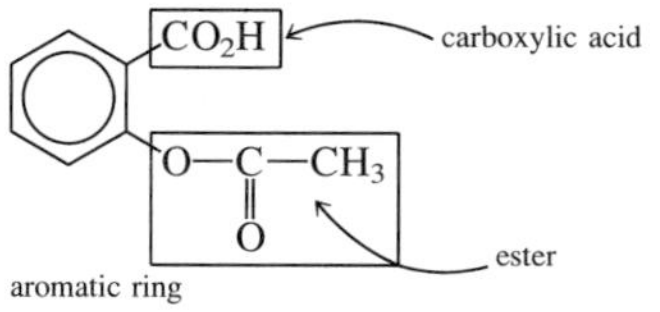

(b) alcohol (phenol) HO; aromatic ring; ether O; amine $N-CH_3$; HO alcohol; alkene

(c) alcohol HO; ether O; OH; amine $N-CH_2CH=CH_2$ alkene; O ketone

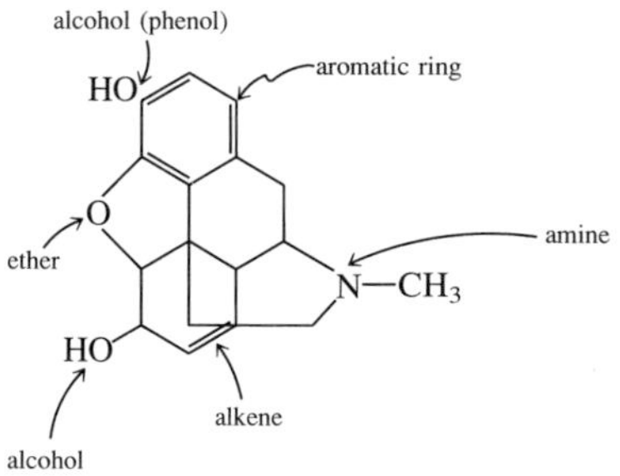

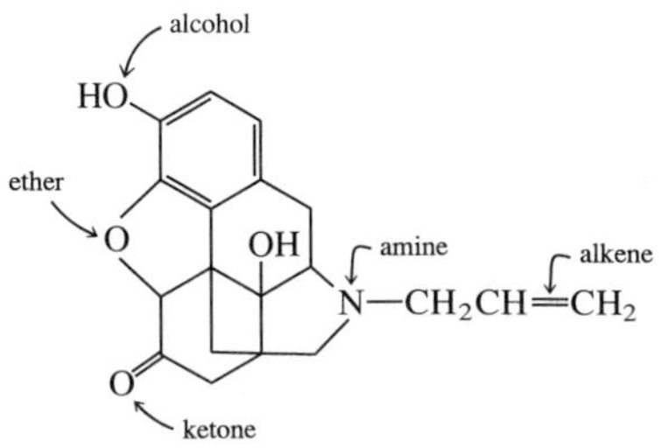

24.

$$HC\equiv C-C\equiv C-HC=C=CH-HC=CH-HC=CH-CH_2-\overset{O}{\overset{\|}{C}}-OH$$

13 12 11 10 9 8 7 6 5 4 3 2 1

26. (a)

$O{=}C_5H_4N{-}CH_2{-}CH(NH_2){-}CO_2H$

NH_2 ← primary amine; ketone; tertiary amine; carboxylic acid

(b) five carbons in rings and in $-CO_2H$: sp^2; $-CH_2-$ and $-CH(NH_2)-$: sp^3;

(c) four pi bonds, 23 sigma bonds

29. $C_{12}H_{24}O_2$

30. (a) $CH_3{-}CH_2{-}C({=}O){-}H$;

(b) $CH_3{-}O{-}CH_2{-}CH_3$;

(c) $CH_3(H)C{=}C(H)CH_3$;

(d) $CH_3CH_2CH_2NH_2$

31. (a) ketone; (b) aldehyde; (c) ketone; (d) amine

33. Substitution: an atom or group is replaced by another atom or group. H on benzene is replaced by Cl.

$C_6H_6 + Cl_2 \longrightarrow C_6H_5{-}Cl + HCl$

Addition: atoms or groups are added to a molecule. Cl_2 adds to ethylene.

$$CH_2{=}CH_2 + Cl_2 \rightarrow CH_2Cl{-}CH_2Cl$$

34. (a) Two monochloro products: $CH_2ClCH_2CH_3$ and $CH_3CHClCH_3$. (b) Four dichloro products: $CHCl_2CH_2CH_3$, $CH_2ClCHClCH_3$, $CH_2ClCH_2CH_2Cl$, and $CH_3CCl_2CH_3$.

35. (a) $CH_2{=}CH_2 + Br_2 \rightarrow CH_2Br{-}CH_2Br$; (b) $C_6H_6 + Br_2 \rightarrow C_6H_5Br + HBr$; (c) $CH_3CO_2H + CH_3OH \rightarrow CH_3CO_2CH_3 + H_2O$

36. (a) $CH_3CH{=}CH_2 + HCl \rightarrow CH_3CHClCH_3$;
(b) $CH_3CH{=}CH_2 + H_2O \rightarrow CH_3CH(OH)CH_3$;

(c) 1-methylcyclohexene $+ HBr \longrightarrow$ 1-bromo-1-methylcyclohexane (CH_3, Br)

37. (a) $CH_3{-}C({=}O){-}H$;

(b) $CH_3{-}C({=}O){-}CH_3$;

(c) cyclopentanone ($C_5H_8{=}O$)

38. (a) i. $CH_3{-}C({=}O){-}OH$

ii. $C_5H_9{-}C({=}O){-}OH$

iii. $(CH_3)_2CHC({=}O){-}OH$

(b)

$$CH_3CH_2OH \xrightarrow{[ox]} CH_3C({=}O)H + CH_3C({=}O)OH$$

$$(CH_3)_3C{-}CH_2OH \xrightarrow{[ox]} (CH_3)_3C{-}C({=}O)H + (CH_3)_3C{-}C({=}O){-}OH$$

39. (a) $CH_3CH{=}CH_2 + Br_2 \rightarrow CH_3CHBrCH_2Br$;
(b) $CH_3C{\equiv}CH + H_2 \rightarrow CH_3CH{=}CH_2 + Br_2 \rightarrow CH_3CHBrCH_2Br$;

(c)

$$CH_3CO_2H + CH_3CH_2CH_2CH_2OH \rightarrow CH_3C({=}O){-}O{-}CH_2CH_2CH_2CH_3 + H_2O$$

40. (a) Addition polymer: polymer formed by adding monomer units to a double bond; Teflon. (b) Condensation polymer: polymer that forms when two monomers combine, eliminating a small molecule; nylon and dacron. (c) Copolymer: polymer formed from more than one monomer; nylon and dacron.

41.

$$\left[NH{-}C_6H_4{-}C({=}O)NH{-}C_6H_4{-}C({=}O)NH{-}C_6H_4{-}C({=}O)\right]_n$$

45. (a)

$$\left[O{-}C_6H_4{-}C(CH_3)_2{-}C_6H_4{-}O{-}C({=}O){-}O{-}C_6H_4{-}C(CH_3)_2{-}C_6H_4{-}O{-}C({=}O)\right]_n$$

(b) condensation

46. Divinyl benzene crosslinks different chains to each other. The chains cannot move past each other because of the crosslinks; thus, the polymer is more rigid. **47.** The stronger interparticle forces would be found in polyvinyl chloride, since there are also dipole-dipole forces in PVC that are not present in polyethylene.

48. (a) $CHF{=}CH_2$; (b) $HO{-}CH_2CH_2{-}CO_2H$; (c) copolymer of $HOCH_2CH_2OH$ and $HO_2CCH_2CH_2CO_2H$
50. 2-methyl-1,3-butadiene **51.** Different substance; bond angles are not the same as in benzene. **52.** For the reaction $3CH_2{=}CH_2 + 3H{-}H \rightarrow 3CH_3{-}CH_3$, $\Delta H = -381$ kJ using bond energies. From enthalpies of formation, $\Delta H° = -410$ kJ/mol. For $C_6H_6(g) + 3H_2(g) \rightarrow C_6H_{12}(g)$ we would get the same ΔH from bond energies, -381 kJ. From enthalpies of formation, $\Delta H° = -173.2$ kJ. Benzene is more stable by 208 kJ/mol (lower in energy) than we expect. This extra stability is evidence for stabilization of the ring by the delocalized π electrons.

54. $\left(\!-O{-}CH_2CH_2O{-}\overset{O}{\overset{\|}{C}}{-}\underset{H}{N}{-}C_6H_4{-}\underset{H}{N}{-}\overset{O}{\overset{\|}{C}}{-}O{-}CH_2CH_2{-}O{-}\overset{O}{\overset{\|}{C}}\underset{H}{N}{-}C_6H_4{-}O{-}\overset{O}{\overset{\|}{C}}{-}\underset{H}{N}\!-\right)_n$

56. (a) Ecotrin: aspirin, analgesic, anti-inflammatory

$C_6H_4(CO_2H)(O\overset{O}{\overset{\|}{C}}CH_3)$

(b) Tylenol: analgesic

$CH_3\overset{O}{\overset{\|}{C}}NH{-}C_6H_4{-}OH$

(c) Lasix: diuretic, antihypertensive

$H_2N{-}SO_2{-}C_6H_2(CO_2H)(Cl){-}NH{-}CH_2{-}C_4H_3O$

Chapter 23

1. (a) $K = 2.6 \times 10^9$; equilibrium lies far to the right;
(b) $^+H_3NCH_2CO_2H$ (1.0 M H^+); $H_2NCH_2CO_2^-$ (1.0 M base)
2. (a) aspartic acid and phenylalanine; (b) Aspartame contains the methyl ester of phenylalanine. This ester can hydrolyze to form methanol, $R{-}CO_2CH_3 + H_2O \rightleftharpoons RCO_2H + CH_3OH$.

3. Crystalline amino acids exist as zwitterions, $^+H_3N{-}\overset{R}{\overset{|}{C}}H{-}CO_2^-$.

The interparticle forces are strong. Before the temperature gets high enough to break the ionic bonds, the amino acid decomposes.

4. They are both hydrophilic amino acids because both contain highly polar R groups.
10.

$H_2N\underset{CH_2OH}{\underset{|}{C}}H{-}\overset{O}{\overset{\|}{C}}NH\underset{CH_3}{\underset{|}{C}}HCO_2H$ ser-ala

$H_2N\underset{CH_3}{\underset{|}{C}}H\overset{O}{\overset{\|}{C}}{-}NH\underset{CH_2OH}{\underset{|}{C}}HCO_2H$ ala-ser

11. Three **13.** Both denaturation and inhibition reduce the catalytic activity of an enzyme. Denaturation changes the structure of an enzyme; inhibition involves the attachment of an incorrect molecule at the active site. **14.** The initial increase in rate is a result of the effect of temperature on the rate constant. At higher temperatures the enzyme begins to denature, losing its activity, and the rate decreases. **15.** An amino acid contains both acidic and basic functional groups. Thus, an amino acid can act as both a weak acid and a weak base; this is the requirement for a buffer.

17.

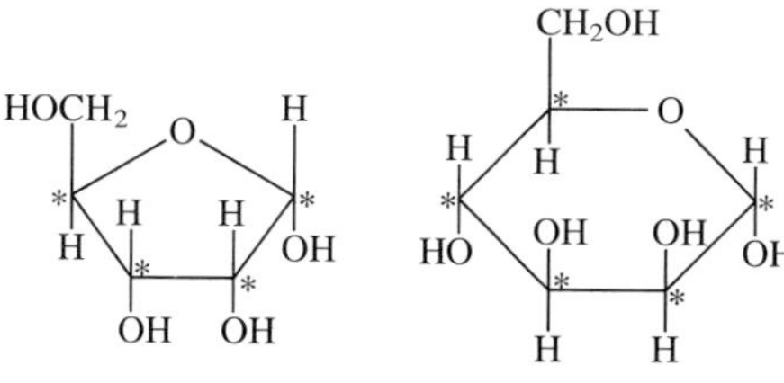

Ribose

18. Chiral carbons are marked with asterisks in 23.17. **21.** Structural isomers: same formula, different functional groups or chain lengths. Geometrical isomers: same functional groups, but different arrangement of some groups in space. Optical isomers: compounds that are nonsuperimposable mirror images of each other.

22.

$CH_3{-}\overset{*}{C}H(CH_2CH_3){-}$ on $H_2N{-}C^*(H){-}CO_2H$

Isoleucine

$CH_3{-}{*}CH{-}OH$ on $H_2N{-}C^*(H){-}CO_2H$

Threonine

24.

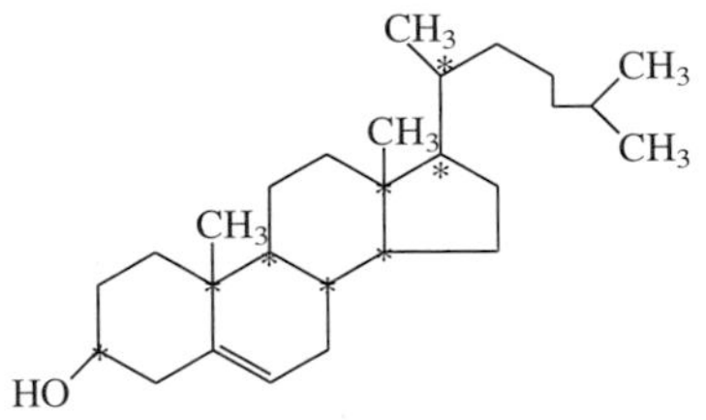

CH_3 CH_3 CH_3 CH_3 CH_3 HO

Eight chiral carbon atoms (marked with *).

27.

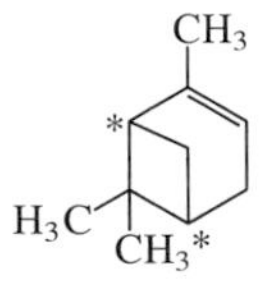

CH_3 H_3C CH_3*

Two chiral carbon atoms are marked with an *.

28. Nitrogen atoms with lone pairs. **29.** 1.7 m **31.** T-A-C-G-C-C-G-T-A **32.** 64

33. Uracil will H-bond to adenine.

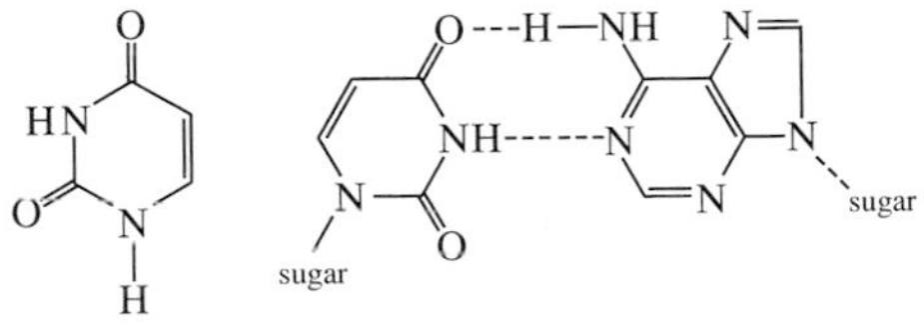

O---H—NH HN NH------N N sugar sugar H

34. (a) glu: CTT, CTC; val: CAA, CAG, CAT, CAC; met: TAC; trp: ACC; phe: AAA, AAG; asp: CTA, CTG;
(b) TAC-TAC-AAA-CTA-ACC;
or or
AAG CTG
(c) four; (d) glu-trp-phe; (e) C-T-C-A-C-C-A-A-A; C-T-T-A-C-C-A-A-G; C-T-C-A-C-C-A-A-G **38.** (a) The structure with two C=O bonds is more stable. (b) It could hydrogen bond to guanine, forming G-T base pair instead of A-T. **39.** Lipids are non-polar and soluble in organic solvents. Carbohydrates contain several —OH groups capable of hydrogen bonding and are soluble in water.

40.

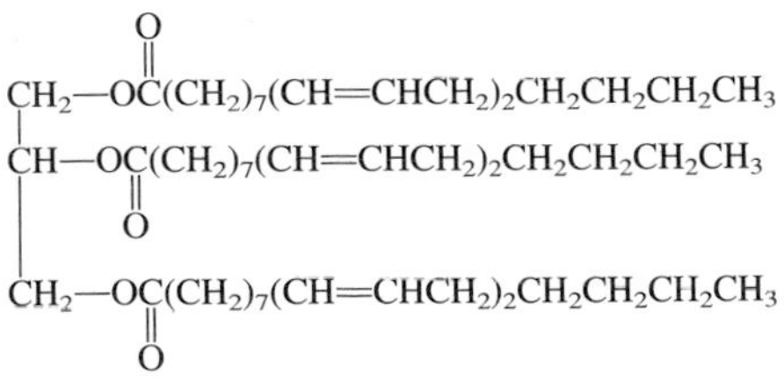

CH_2—$OC(=O)(CH_2)_7(CH{=}CHCH_2)_2CH_2CH_2CH_2CH_3$

CH—$OC(=O)(CH_2)_7(CH{=}CHCH_2)_2CH_2CH_2CH_2CH_3$

CH_2—$OC(=O)(CH_2)_7(CH{=}CHCH_2)_2CH_2CH_2CH_2CH_3$

41. It will take 6 mol of H_2 to hydrogenate this triglyceride completely. Nine products with two double bonds are possible.

45.

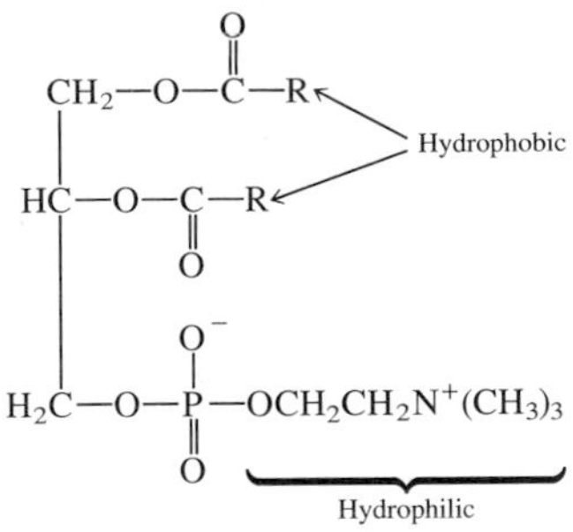

CH_2—O—C(=O)—R ← Hydrophobic

HC—O—C(=O)—R

H_2C—O—P(=O)(O^-)—$OCH_2CH_2N^+(CH_3)_3$

Hydrophilic

Leithin: Hydrophilic

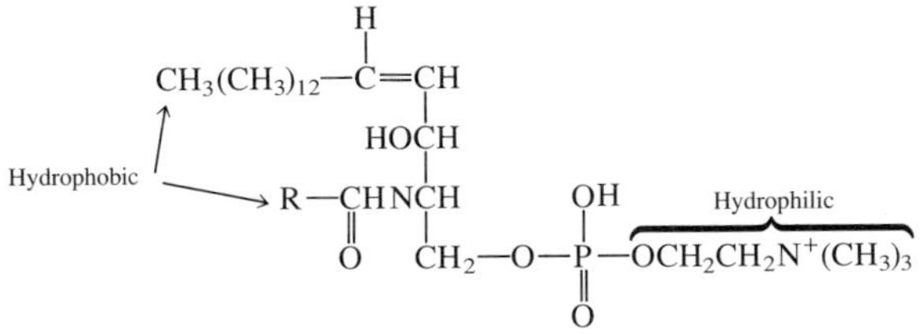

$CH_3(CH_3)_{12}$—C(H)=CH

HOCH

Hydrophobic → R—C(=O)HNCH

CH_2—O—P(=O)(OH)—$OCH_2CH_2N^+(CH_3)_3$ Hydrophilic

46. See *The Merck Index,* 10th edition, #1986, p. 282. **50.** (a) 1.7×10^{-12}; (b) 2.3×10^{-12}; (c) 7.3×10^{-15} **51.** pH = 7.07 **53.** $\Delta H = -23$ kJ. For gly + gly $\rightleftharpoons$ gly-gly, ΔS is negative (unfavorable). ΔG is probably positive because of the negative entropy change. **54.** In the reaction P—O and O—H bonds are broken and P—O and O—H bonds are formed. Thus $\Delta H \cong$ O. $\Delta S <$ O, since two molecules are going to one molecule. Thus, $\Delta G >$ O, not spontaneous.

55. A cell is not a closed system. There is an external source of energy to drive the reaction. A photosynthetic plant uses sunlight and animals use the carbohydrates produced by plants as sources of energy. For a cell $\Delta S_{sys} <$ O, but $\Delta S_{surr} >$ O and

$$|\Delta S_{surr}| > |\Delta S_{sys}|.$$

Therefore ΔS_{univ} increases.

Chapter 24

1. The emphasis in industry is on: economy and safety; metal containers and indirect monitoring of reactions; recycling or selling by-products; and high temperatures, high pressures, and continuous flow systems. **2.** The large amount of Cl_2 produced in the chlor-alkali industry (Cl_2 is produced with NaOH) must be used. Cl_2 reacts readily with many hydrocarbons. These chlorinated hydrocarbons have useful properties as well as undesirable effects (toxicity). **5.** (a) 43%; (b) 1000 kg; 430 kg; (c) i. \$1.20/lb; ii. \$.98/lb; (d) The disposal of $AlCl_3$ and H_2SO_4 in (i) would present a major problem. (e) iii. \$.12/lb; iv. \$.23/lb; (f) Both process (iii), using benzene, and process (iv) are more economical than processes (i) and (ii). Process (iii) using benzene would be best of all in terms of raw material cost. **10.** (a) $\Delta H \cong 0$; $\Delta H = -37$ kJ **11.** In this reaction the CO molecules insert at a double bond. $CH_2{=}CH_2$ is symmetrical, $CH_3CH{=}CH_2$ is not. **14.** (a) $CFCl{=}CF_2$; (b) More reactive,

C—Cl bonds are not as strong as C—F bonds. **15.** Plasticizers make a polymer more flexible. Cross-linking makes a polymer more rigid. **16.** The regular arrangement of the methyl groups in the chain align the chains in a way that leads to stronger forces between chains. **17.** (a) Replacement of hydrogen with halogen results in fewer H· and ·OH radicals in the flame, reducing flame temperature. (b) With aromatic groups present, it is more difficult to "chip off" pieces of the polymer in the pyrolysis zone.

18. (a) The polyamide from 1,2-diaminoethane and terephthalic acid is stronger because of the possibility of hydrogen bonding between chains.

(b) The polymer of HO—C_6H_4—CO_2H

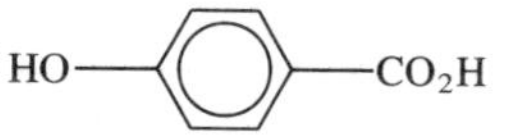

because the chains are stiffer.

19. Cross-linking decreases the elasticity of a polymer.

20. Polychloroprene: $-(CH_2-C(Cl)=CH-CH_2)_n-$

Polyisoprene: $-(CH_2-C(CH_3)=CH-CH_2)_n-$

Polynitrile: $-(CH_2-CH(C{\equiv}N)-CH_2-CH=CH-CH_2)_n-$

Polybutadiene: $-(CH_2-CH=CH-CH_2)_n-$

22. As the car ages the plasticizers escape from the seat covers. (i) The waxy coating is the escaped plasticizer. (ii) The new car smell is the smell of the plasticizers (di-octylphthlate). (iii) Loss of plasticizer causes the vinyl to become brittle.
24. (a) $HO-(CH_2CH_2O)_n-H$; (b) 450 monomer units; (c) No, this is an average molecular weight. **25.** A likely product is the toxic gas HCN. **27.** For a given chain length, there are more hydrogen bonding sites in nylon-46 than nylon-6.

29.

(a) 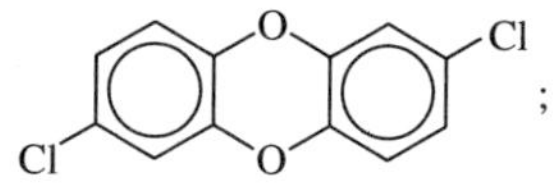 [dibenzo-p-dioxin structure with Cl on each ring (2,7-dichloro)];

(b) 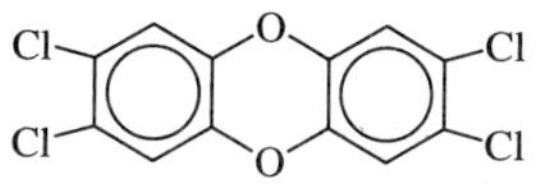 [dibenzo-p-dioxin structure with Cl, Cl on each ring (2,3,7,8-tetrachloro)]

This is the most toxic of the PCDDs.

31. (a) Lasso: herbicide

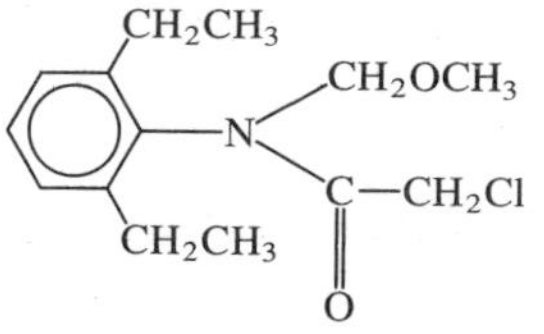

[2,6-diethylphenyl—N(CH₂OCH₃)—C(=O)—CH₂Cl: benzene ring with two CH_2CH_3 groups, attached to N bearing CH_2OCH_3 and $C(=O)-CH_2Cl$]

(b) Roundup: herbicide

$$\left[HOC(=O)-CH_2-NH-CH_2-P(=O)(OH)-O:\right]^- \left[(CH_3)_2CHNH_3\right]^+$$

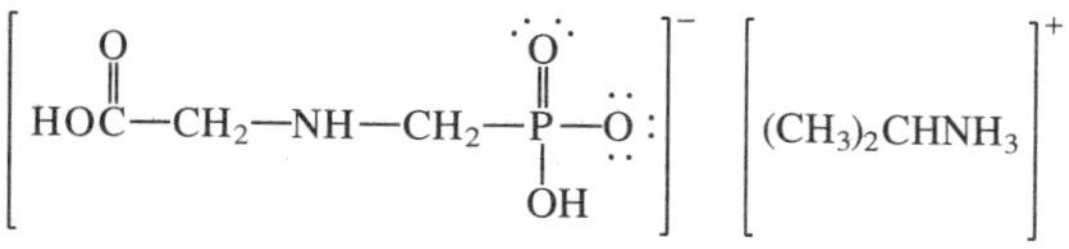

(c) Aldicarb: insecticide, acaricide, nematocide

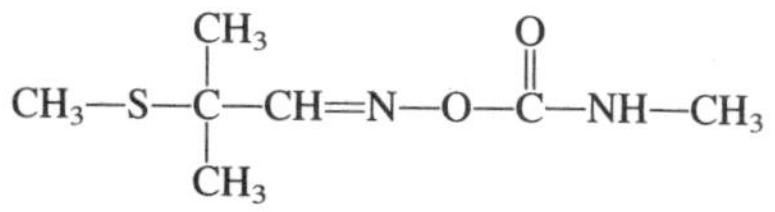

$$CH_3-S-C(CH_3)_2-CH=N-O-C(=O)-NH-CH_3$$

32. (a) Roasting: converting sulfide minerals to oxides by heating in air below their melting points. (b) Smelting: reducing metal ions to the free metal. (c) Flotation: separation of mineral particles in an ore from the gangue that depends on the greater wetability of the mineral particles. (d) Leaching: the extraction of metals from ores using aqueous solutions. (e) Gangue: the impurities (such as clay or sand) in an ore. **34.** $\mathscr{E}^\circ_{cell} = 0.66$ V.; $\Delta G^\circ = -130$ kJ; $K = 2.0 \times 10^{22}$ **36.** (a) $\Delta H^\circ = -39$ kJ; $\Delta S^\circ = 38$ J/K; (b) $\Delta G^\circ = -80$ kJ **39.** (a) 11%; (b) 2.6 molal; (c) 2.2 M **40.** The bubbles are CO_2. The bump increases the surface area of the bottle. This and the extra thickness help withstand the pressure of CO_2 inside the bottle.
41. For malic acid, the major species are H_2mal and Hmal$^-$. For tartaric acid, the major species are H_2tar and Htar$^-$. **42.** Allowing the wine to breathe allows time for SO_2 (used as a preservative) to escape.

PHOTO CREDITS

Chapter 1 p. xviii, D. C. Neckers/Journal of Chemical Education; p. 2, Fred Ward/Black Star; p. 3 (Fig. 1.1, top left), Guy Gillette/Science Source/Photo Researchers, Inc.; p. 3 (Fig. 1.1, top right), Giani Tortoli/Photo Researchers, Inc.; p. 3 (Fig. 1.1, bottom right), Hank Morgan/VHSID Lab/ECE Department University of Massachusetts/Science Source/Photo Researchers, Inc.; p. 3 (Fig. 1.1, bottom left), Hank Morgan/Science Source/Photo Researchers, Inc.; p. 4, The Bettmann Archive; p. 8 [Fig. 1.6 (a)], Tom Pantages; p. 13, Tom Pantages; p. 21, Pat Lanza Field/Bruce Coleman, Inc.; p. 23, G. Anderson/Taurus Photos; p. 24, E. R. Degginger; p. 27, E. R. Degginger; p. 29 (Fig. 1.20), E. R. Degginger. **Chapter 2** p. 36, Manfred Kage/Peter Arnold, Inc.; p. 38, The Granger Collection; p. 39 (Fig. 2.2), The Granger Collection; p. 40 (Fig. 2.3), The Granger Collection; p. 42, The Granger Collection; p. 44 (Fig. 2.6), The Granger Collection; p. 46 (Fig. 2.22), The Bettmann Archive; p. 47 [Fig. 2.23 (b)], David Parker/Science Photo Library/Photo Researchers, Inc; p. 48 (Fig. 2.11), The Bettmann Archive; p. 52 (top left), E. R. Degginger; p. 52 (top right), Ken O'Donoghue; p. 52 (bottom left), Ken O'Donoghue; p. 52 (bottom right), E. R. Degginger; p. 53 [Fig. 2.19 (b)], Dr. Jeremy Burgess/Science Photo Library/Photo Researchers, Inc.; p. 54, Harold & Erica Van Pelt, Los Angeles; p. 56, E. R. Degginger; p. 58 (top), E. R. Degginger; p. 58 (bottom), Tom Pantages; p. 60, E. R. Degginger. **Chapter 3** p. 72, Fundamental Photographs; p. 75, Tom Pantages; p. 77 (Fig. 3.4), Tom Pantages; p. 79, Leslie Zumdahl; p. 82, James H. Carmichael, Jr./Bruce Coleman, Inc.; p. 85, Eli Lilly & Company; p. 87, Manfred Kage/Peter Arnold, Inc.; p. 92, Yoav/Phototake; p. 94, Ken O'Donoghue; p. 96 (both), Fundamental Photographs; p. 97, John Kaprielian/Photo Researchers, Inc.; p. 98 (Fig. 3.10), E. R. Degginger; p. 101, NASA; p. 102, L. S. Stepanowicz/Panographics; p. 106, Norman Owen Tomalin/Bruce Coleman, Inc.; p. 107, Paul Silverman/Fundamental Photographs. **Chapter 4** p. 118, E. R. Degginger; p. 123 (Fig. 4.13), The Bettmann Archive; p. 125 (both), Fundamental Photographs; p. 127, Tom Pantages; p. 129, Lawrence Migdale/Photo Researchers, Inc.; p. 133 (Fig. 4.8), E. R. Degginger; p. 135 (Fig. 4.9), Ken O'Donoghue; p. 136, Ken O'Donoghue; p. 137, E. R. Degginger; p. 150, Ken O'Donoghue; p. 152, Paul Silverman/Fundamental Photographs; p. 155, L. S. Stepanowicz/Panographics; p. 164, E. R. Degginger; p. 165 (Fig. 4.12), E. R. Degginger. **Chapter 5** p. 174, NASA; p. 184, Ken O'Donoghue; p. 187, Tom Pantages; p. 189, M. Lindsay Lambert; p. 193, Timothy O'Keefe/Bruce Coleman, Inc.; p. 194 (both), Ken O'Donoghue; p. 199 (all), Ken O'Donoghue; p. 202 (Fig. 5.18), Photo by Frank Hoffman/Department of Energy/Oak Ridge, Tennessee; p. 203, Ken O'Donoghue; p. 207, Malcolm Lockwood Photography; p. 209 (Fig. 5.27), Peter Menzel/Stock, Boston; p. 210 (Fig. 5.28), John Shaw/Bruce Coleman, Inc.; p. 211, Gary Milburn/Tom Stack and Associates. **Chapter 6** p. 220, Peter Menzel/Stock, Boston; p. 223, E. R. Degginger; p. 226, Bernard Asset/Agence Vandystadt/Photo Researchers, Inc.; p. 230, Daedalus Enterprises, Inc.; p. 233, Malcolm Kirk/Peter Arnold, Inc.; p. 235, Argon National Laboratory; p. 238, E. R. Degginger; p. 242, E. R. Degginger; p. 245, Edward L. Miller/Stock, Boston; p. 246, NASA/Goddard Space Center; p. 247, Paul E. Johnson; p. 249, Mark Heifner/Panographics; p. 252, Ken O'Donoghue. **Chapter 7** p. 262, Robert D. Carr/Bruce Coleman, Inc.; p. 265 (top), Dan Guravich/Photo Researchers, Inc.; p. 265 (bottom), Ken O'Donoghue; p. 267 (Fig. 7.3), The Granger Collection; p. 269 (top), Ken O'Donoghue; p. 269 (bottom), Robert D. Carr/Bruce Coleman, Inc.; p. 271, E. R. Degginger; p. 272, NASA; p. 273, The Granger Collection; p. 277, Charles Ohrstedt/Panographics; p. 290 (Fig. 7.20), The Granger Collection; p. 294 (top), E. R. Degginger; p. 294 (middle), L. S. Stepanowicz/Panographics; p. 294 (bottom), Leslie Zumdahl; p. 312, E. R. Degginger. **Chapter 8** p. 322, Science Photo Library/Science Source/Photo Researchers, Inc.; p. 324 (left), Manfred Kage/Peter Arnold, Inc.; p. 324 (right), Paul Silverman/Fundamental Photographs; p. 332, C. B. & D. W. Frith/Bruce Coleman, Inc.; p. 336, E. R. Degginger; p. 341, Ken O'Donoghue; p. 343, Phil Degginger; p. 349, The Bancroft Library/University of California; p. 370, Argon National Laboratory; p. 375, C. Allan Morgan/Peter Arnold, Inc. **Chapter 9** p. 382, Character House/Science Source/Photo Researchers, Inc.; p. 400, E. R. Degginger; p. 404 (Fig. 9.39, both), Donald Clegg. **Chapter 10** p. 416, Dotte Larsen/Bruce Coleman, Inc.; p. 422, Richard Wood/Taurus Photos; p. 423, Yoav/Phototake; p. 424 (Fig. 10.8, top left), Brian Parker/Tom Stack & Associates; p. 424 (Fig. 10.8, top right), Gary Milburn/Tom Stack & Associates; p. 424 (Fig. 10.8, bottom left), Brian Parker/Tom Stack & Associates; p. 424 (Fig. 10.8, bottom right), Brian Parker/Tom Stack & Associates; p. 431, E. R. Degginger; p. 434, Yoav/Phototake; p. 437 (Fig. 10.25), Professor Williams/Lehigh University, Department of Materials Science & Engineering; p. 439 (top), Corning Glass Works; p. 439 (bottom), E. R. Degginger; p. 442, Bruce Bohor/U.S. Geologic Survey/Photo: G. Irzett; p. 443 (top right), Yoav/Phototake; p. 443 (Fig. 10.32, both), E. R. Degginger; p. 452 (Fig. 10.41), E. R. Degginger; p. 458, Win McNamee/Duomo; p. 459 (top right), John Gerlach/Tom Stack & Associates; p. 459 (bottom right), E. R. Degginger. **Chapter 11** p. 472, Peter Aprahamian/Science Photo Library/Photo Researchers, Inc.; p. 476, Leonard Lessin/Peter Arnold, Inc.; p. 477, Sandy Sprunt, Jr./National Audubon Society/Photo Researchers, Inc.; p. 482, Paul Silverman/Fundamental Photographs; p. 483, T. Orban/Sygma; p. 486 (Fig. 11.8), E. R. Degginger; p. 488, Alfred Pasieka/Taurus Photos; p. 495, Tom Pantages; p. 496, Leslie Zumdahl; p. 499, Grace Moore/Taurus Photos; p. 501 (Fig. 11.21), Robert Sherbow/People Weekly © 1980 Time Inc.; p. 501 (right), E. R. Degginger; p. 503 (Fig. 11.23), Kip Petikolas/Fundamental Photographs; p. 505, Doug Allan/Animals, Animals. **Chapter 12** p. 512, Ken O'Donoghue; p. 514 (top left), Phil Degginger; p. 514 (top right), Norman Owen Tomalin/Bruce Coleman, Inc.; p. 514 (bottom left), Focus On Sports; p. 514 (bottom right), Wendell Metzen/Bruce Coleman, Inc.; p. 520, E. R. Degginger; p. 544, Kevin Schafer/Tom Stack &

Associates; p. 546, David Hughes/Bruce Coleman, Inc.; p. 555, NASA. **Chapter 13** p. 566, D. Cavagnaro/Peter Arnold, Inc.; p. 569, E. R. Degginger; p. 573, Grant Heilman; p. 579, E. R. Degginger; p. 595 (left), Ken O'Donoghue; p. 595 (right, both), Ken O'Donoghue; p. 597 (all), Ken O'Donoghue. **Chapter 14** p. 608, Hans Reinhard/Bruce Coleman, Inc.; p. 610, E. R. Degginger; p. 619 (Fig. 14.4), E. R. Degginger; p. 625, Dan McCoy/Rainbow; p. 631, John Curtis/Taurus Photos; p. 637, Jane Schreibman/Photo Researchers, Inc.; p. 657, E. R. Degginger. **Chapter 15** p. 668, Larry Ulrich Photography; p. 674 (both), Fundamental Photographs; p. 675 (both), Fundamental Photographs; p. 687, Kip Petikolas/Fundamental Photographs; p. 701 (both), E. R. Degginger; p. 702 (both), E. R. Degginger; p. 703 (Fig. 15.7, all), Ken O'Donoghue; p. 704, E. R. Degginger; p. 707, L. S. Stepanowicz/Panographics; p. 708, E. R. Degginger; p. 710, E. R. Degginger; p. 713, E. R. Degginger; p. 715, E. R. Degginger; p. 721 (both), E. R. Degginger; p. 722, E. R. Degginger; p. 726 (both), E. R. Degginger; p. 729, Dr. Featherstone/Eastman Dental Center. **Chapter 16** p. 736, E. R. Degginger/Bruce Coleman, Inc.; p. 739, E. R. Degginger; p. 740, Kristen Brochmann/Fundamental Photographs; p. 742, Yoav/Phototake; p. 744, Manfred Kage/Peter Arnold, Inc.; p. 746, Brian Parker/Tom Stack & Associates; p. 758, Paul Silverman/Fundamental Photographs; p. 759, Peter Menzel/Stock, Boston; p. 766, Hank Morgan/Rainbow. **Chapter 17** p. 778, Diane Schiumo/Fundamental Photographs; p. 784, Ken O'Donoghue; p. 790, The Granger Collection; p. 791, E. R. Degginger; p. 797, E. R. Degginger; p. 798, Fundamental Photographs; p. 799, Paul Silverman/Fundamental Photographs; p. 801, Brian Parker/Tom Stack & Associates; p. 802, Wendell Metzen/Bruce Coleman, Inc.; p. 805, Runk/Schoenberger/Grant Heilman; p. 807 (Fig. 17.20), Yoav/Phototake; p. 809 (Fig. 17.21), Oberlin College Archives; p. 811 (Fig. 17.23), Kennecott; p. 812 [Fig. 17.24(a)], Bill Gallery/Stock, Boston. **Chapter 18** p. 824, Marvin Ickow/Bruce Coleman, Inc.; p. 828, Mark Heifner/Panographics; p. 831, Yoav/Phototake; p. 832 (Fig. 18.4), E. R. Degginger; p. 834 (both), Donald Clegg; p. 836, Tom Pantages; p. 840, E. R. Degginger; p. 842, E. R. Degginger; p. 843 (top), E. R. Degginger; p. 843 (bottom), Michael Holford; p. 844, E. R. Degginger. **Chapter 19** p. 850, Dan McCoy/Rainbow; p. 854 (Fig. 19.5), Roger Rossmeyer/Starlight; p. 855, Miro Vintoniv/Stock, Boston; p. 856, Hugh Spencer/Photo Researchers, Inc.; p. 857, Grant Heilman; p. 858, NASA; p. 859, E. R. Degginger; p. 861, Richard Megna/Fundamental Photographs; p. 863 (Fig. 19.10), Englehard Corporation; p. 871 (top left), E. R. Degginger; p. 871 (top right), Farrell Grehan/Photo Researchers, Inc.; p. 871 (bottom), E. R. Degginger; p. 872 (Fig. 19.18, both), E. R. Degginger; p. 874 (Fig. 19.22), L. S. Stepanowicz/Panographics; p. 876 (top), E. R. Degginger; p. 876 (bottom), Yoav/Phototake; p. 877 (both), E. R. Degginger; p. 885, E. R. Degginger. **Chapter 20** p. 892, Harold & Erica Van Pelt, Los Angeles, CA; p. 895 (all), Paul Silverman/Fundamental Photographs; p. 896, Ken O'Donoghue; p. 901, Tom Pantages; p. 902, E. R. Degginger; p. 909, Tom Pantages; p. 910 (both), Ken O'Donoghue; p. 922, Ken O'Donoghue; p. 930 (Fig. 20.37), David M. Phillips/Taurus Photos; p. 930 (bottom), Galen Rowell/Peter Arnold, Inc. **Chapter 21** p. 938, SSC Central Design Group/Fermi National Accelerator Laboratory; p. 942, The Granger Collection; p. 946, Przekop/Medichrome/The Stock Shop, Inc.; p. 947, Royal Greenwich Observatory/Science Photo Library/Photo Researchers, Inc.; p. 952, University of Pennsylvania; p. 953, Dan McCoy/Rainbow; p. 955 (Fig. 21.8, both), Siu/Peter Arnold, Inc.; p. 962, M. Lazarus/Photo Researchers, Inc.; p. 963, NASA; p. 968 (left), Peter Arnold, Inc.; p. 968 (right), Dan McCoy/Rainbow. **Chapter 22** p. 974, David Gnizal/Phototake; p. 979, Robert C. Simpson/Tom Stack & Associates; p. 983, E. R. Degginger; p. 989, E. R. Degginger; p. 990, Tom Pantages; p. 991, Pam Taylor/Bruce Coleman, Inc.; p. 994, Kevin Schafer/Tom Stack & Associates; p. 995, Kristen Brochmann/Fundamental Photographs; p. 1001 (Fig. 22.16), Donald Clegg; p. 1001 (bottom), Tom Pantages. **Chapter 23** p. 1010, Doug Sokell/Tom Stack & Associates; p. 1013 (top), Pat & Tom Leeson/Photo Researchers, Inc.; p. 1013 (bottom), E. R. Degginger; p. 1021, Ralph Eagle, Jr., M.D./Science Source/Photo Researchers, Inc.; p. 1026, Brian Parker/Tom Stack & Associates; p. 1029, Dan McCoy/Rainbow; p. 1035, Wardene Weisser/Bruce Coleman, Inc. **Chapter 24** p. 1044, Tsu-Wei Chou/University of Delaware; p. 1046, E. R. Degginger; p. 1050, Ellis Herwig/Stock, Boston; p. 1051, John Kaprielian/Photo Researchers, Inc.; p. 1053, E. R. Degginger; p. 1055, E. R. Degginger; p. 1056, E. R. Degginger; p. 1057, Bob Masini/Phototake; p. 1069, Eric Kroll/Taurus; p. 1070, Brian Parker/Tom Stack & Associates; p. 1075, E. R. Degginger; p. 1079 (Fig. 24.10), Tom Pantages; p. 1083, Mickey Gibson/Tom Stack & Associates; p. 1084, The Granger Collection.

INDEX

Page Numbers of Some Important Tables

Physical Constants

Constant	Symbol	Value
Atomic mass unit	amu	1.66054×10^{-27} kg
Avogadro's number	N	6.02214×10^{23} mol^{-1}
Bohr radius	a_0	5.292×10^{-11} m
Boltzmann constant	k	1.38066×10^{-23} J K^{-1}
Charge of an electron	e	1.60218×10^{-19} C
Faraday constant	F	96,485 C mol^{-1}
Gas constant	R	8.31451 J K^{-1} mol^{-1}
		0.08206 L atm K^{-1} mol^{-1}
Mass of an electron	m_e	9.10939×10^{-31} kg
		5.48580×10^{-4} amu
Mass of a neutron	m_n	1.67493×10^{-27} kg
		1.00866 amu
Mass of a proton	m_p	1.67262×10^{-27} kg
		1.00728 amu
Planck's constant	h	6.62608×10^{-34} J s
Speed of light	c	2.997925×10^{8} m s^{-1}